Barcode overleaf

Name	Symbol	Atomic number	Molar mass ($g\ mol^{-1}$)[a]
actinium	Ac	89	227.0278[b]
aluminium	Al	13	26.981 538 5(7)
americium	Am	95	243.0614[b]
antimony	Sb	51	121.760(1)
argon	Ar	18	39.948(1)
arsenic	As	33	74.921 595(6)
astatine	At	85	209.9871[b]
barium	Ba	56	137.327(7)
berkelium	Bk	97	247.0703[b]
beryllium	Be	4	9.012 1831(5)
bismuth	Bi	83	208.980 40(1)
bohrium	Bh	107	270.13[b]
boron	B	5	[10.806, 10.821][d]
bromine	Br	35	[79.901, 79.907][d]
cadmium	Cd	48	112.414(4)
caesium	Cs	55	132.905 451 96(6)
calcium	Ca	20	40.078(4)
californium	Cf	98	251.0796[b]
carbon	C	6	[12.0096, 12.0116][d]
cerium	Ce	58	140.116(1)
chlorine	Cl	17	[35.446, 35.457][d]
chromium	Cr	24	51.9961(6)
cobalt	Co	27	58.933 194(4)
copernicium	Cn	112	285.18[b]
copper	Cu	29	63.546(3)
curium	Cm	96	247.0704[b]
darmstadtium	Ds	110	281.17[b]
dubnium	Db	105	268.13[b]
dysprosium	Dy	66	162.500(1)
einsteinium	Es	99	252.083[b]
erbium	Er	68	167.259(3)
europium	Eu	63	151.964(1)
fermium	Fm	100	257.0951[b]
flerovium	Fl	114	289.19[b]
fluorine	F	9	18.998 403 163(6)
francium	Fr	87	223.0197[b]
gadolinium	Gd	64	157.25(3)
gallium	Ga	31	69.723(1)
germanium	Ge	32	72.630(8)
gold	Au	79	196.966 569(5)
hafnium	Hf	72	178.49(2)
hassium	Hs	108	277.15[b]
helium	He	2	4.002 602(2)
holmium	Ho	67	164.930 33(2)
hydrogen	H	1	[1.007 84, 1.008 11][d]
indium	In	49	114.818(1)
iodine	I	53	126.904 47(3)
iridium	Ir	77	192.217(3)
iron	Fe	26	55.845(2)
krypton	Kr	36	83.798(2)
lanthanum	La	57	138.905 47(7)
lawrencium	Lr	103	262.11[b]
lead	Pb	82	207.2(1)
lithium	Li	3	[6.938, 6.997][d]
livermorium	Lv	116	293.2[b]
lutetium	Lu	71	174.9668(1)
magnesium	Mg	12	[24.304, 24.307][d]
manganese	Mn	25	54.938 044(3)
meitnerium	Mt	109	276.15[b]
mendelevium	[illegible]	[illegible]	[illegible]
mercury	[illegible]	[illegible]	[illegible]
molybdenum	[illegible]	[illegible]	[illegible]
neodymium	[illegible]	[illegible]	[illegible]
neon	[illegible]	[illegible]	[illegible]
neptunium	Np	93	237.0482[b]
nickel	Ni	28	58.6934(4)
niobium	Nb	41	92.906 37(2)
nitrogen	N	7	[14.006 43, 14.007 28][d]
nobelium	No	102	259.1010[b]
osmium	Os	76	190.23(3)
oxygen	O	8	[15.999 03, 15.999 77][d]
palladium	Pd	46	106.42(1)
phosphorus	P	15	30.973 761 998(5)
platinum	Pt	78	195.084(9)
plutonium	Pu	94	244.0642[b]
polonium	Po	84	208.9824[b]
potassium	K	19	39.0983(1)
praseodymium	Pr	59	140.907 66(2)
promethium	Pm	61	144.9127[b]
protactinium	Pa	91	231.035 88(2)[b]
radium	Ra	88	226.0254[b]
radon	Rn	86	222.0176[b]
rhenium	Re	75	186.207(1)
rhodium	Rh	45	102.905 50(2)
roentgenium	Rg	111	280.164[b]
rubidium	Rb	37	85.4678(3)
ruthenium	Ru	44	101.07(2)
rutherfordium	Rf	104	265.1167[b]
samarium	Sm	62	150.36(2)
scandium	Sc	21	44.955 908(5)
seaborgium	Sg	106	271.133[b]
selenium	Se	34	78.971(8)
silicon	Si	14	[28.084, 28.086][d]
silver	Ag	47	107.8682(2)
sodium	Na	11	22.989 769 28(2)
strontium	Sr	38	87.62(1)
sulfur	S	16	[32.059, 32.076][d]
tantalum	Ta	73	180.947 88(2)
technetium	Tc	43	97.9072[b]
tellurium	Te	52	127.60(3)
terbium	Tb	65	158.925 35(2)
thallium	Tl	81	[204.382, 204.385][d]
thorium	Th	90	232.0377(4)[b]
thulium	Tm	69	168.934 22(2)
tin	Sn	50	118.710(7)
titanium	Ti	22	47.867(1)
tungsten	W	74	183.84(1)
ununoctium[c]	Uuo	118	294.21[b]
ununpentium[c]	Uup	115	288.192[b]
ununseptium[c]	Uus	117	294.21[b]
ununtrium[c]	Uut	113	284.178[b]
uranium	U	92	238.028 91(3)[b]
vanadium	V	23	50.9415(1)
xenon	Xe	54	131.293(6)
ytterbium	Yb	70	173.054(5)
yttrium	Y	39	88.905 84(2)
zinc	Zn	30	65.38(2)
zirconium	Zr	40	91.224(2)

(a) All known significant figures are given. Parentheses indicate that the figure is uncertain.
(b) Element has no stable nuclides. The molar mass of the longest lived isotope is given.
(c) Unnamed at November 2014.
(d) The molar mass interval reflects the set of molar mass values in normal materials.
Source: Wieser, ME and Coplen, TB, 'Atomic weights of the elements 2009' *Pure Appl. Chem.*, vol. 83, no. 2, pp. 359–96, © IUPAC 2010; and Commission on Isotopic Abundances and Atomic Weights, www.ciaww.org.

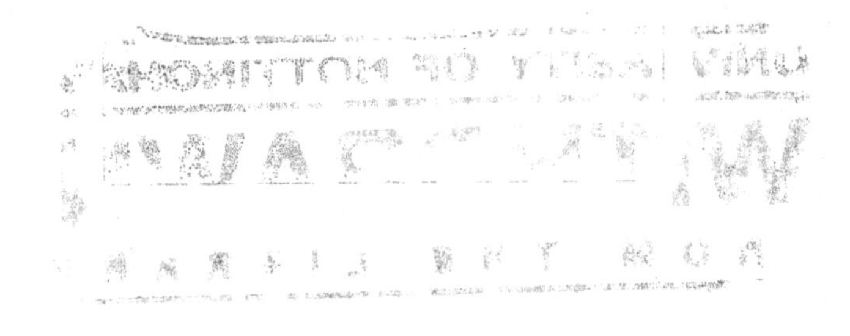

E-CHAPTER

Additional e-chapters are available for this text at **bookshelf.vitalsource.com**.

The code below provides access for one user.

1. To access the chapters, go to **bookshelf.vitalsource.com**, download the Bookshelf® that is right for your computer and enter your unique registration code.
2. Only one account is required to register all your Wiley products.

Already have Bookshelf® installed? Simply redeem your code in Bookshelf® on your desktop or online.

For help, visit support.vitalsource.com.

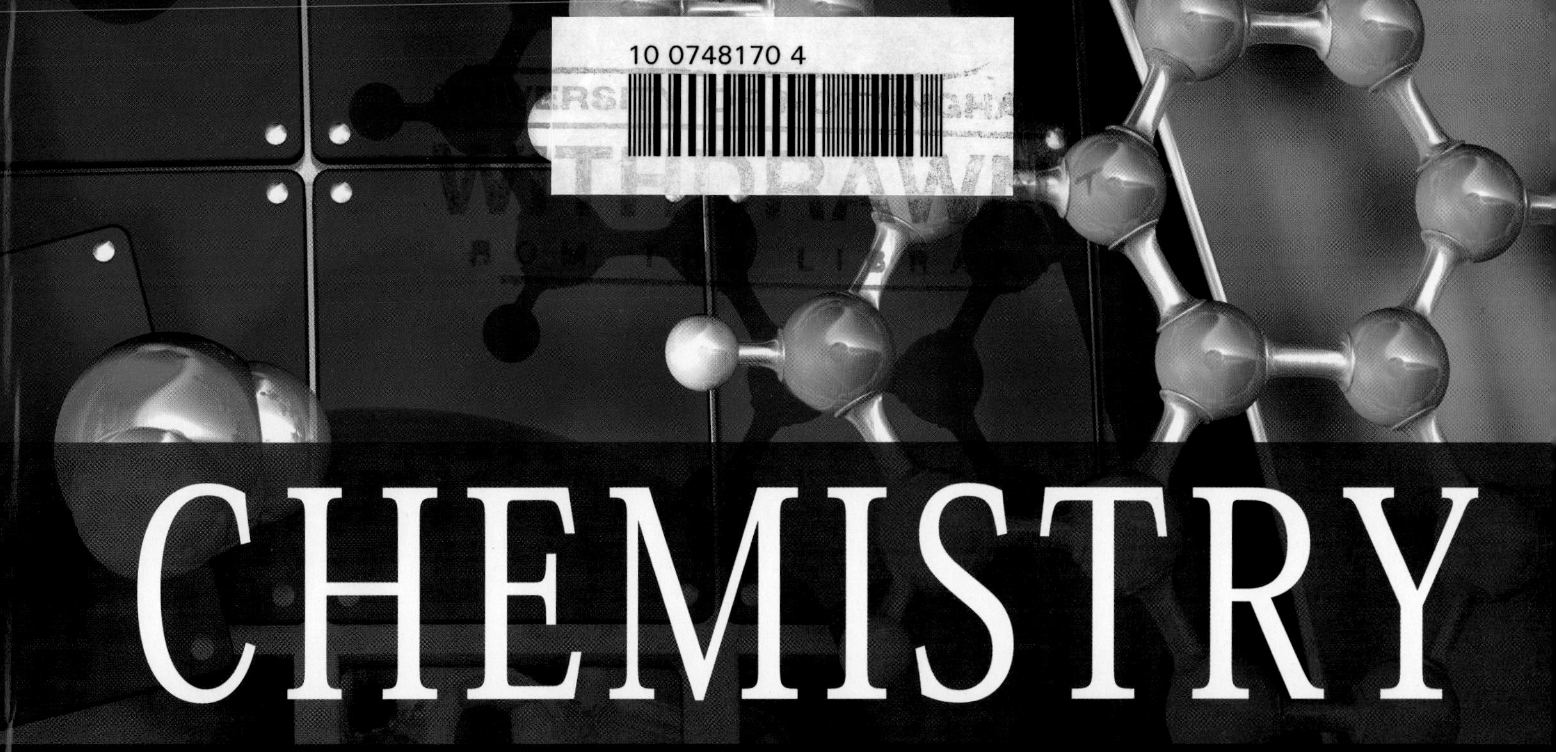
10 0748170 4
CHEMISTRY
3RD EDITION

CHEMISTRY

3RD EDITION

ALLAN BLACKMAN
Auckland University of Technology

STEVEN E BOTTLE
Queensland University of Technology

SIEGBERT SCHMID
University of Sydney

MAURO MOCERINO
Curtin University

UTA WILLE
University of Melbourne

JAMES E BRADY
St John's University

FRED SENESE
Frostburg State University

WILLIAM H BROWN
Beloit College

THOMAS POON
Claremont McKenna College
Scripps College
Pitzer College

JOHN OLMSTEAD III
California State University

GREGORY M WILLIAMS
University of Oregon

WILEY

Published in 2016 by John Wiley & Sons Australia, Ltd
42 McDougall Street, Milton Qld 4064

First edition published in 2008

Typeset in 10/11 Times LT Std Roman

National Library of Australia
Cataloguing-in-Publication data

Author:	Blackman, Allan G., author.
Title:	Chemistry/Allan Blackman, Steve Bottle, Siegbert Schmid, Mauro Mocerino, Uta Wille.
Edition:	Third edition.
ISBN:	9780730311058 (paperback)
Notes:	Includes index.
Subjects:	Chemistry — Textbooks. Chemistry — Study and teaching (Higher)
Other Authors/ Contributors:	Bottle, Steve, author. Schmid, Siegbert, author. Mocerino, Mauro, author. Wille, Uta, author.
Dewey Number:	540

Cover art by Michael Crawford.

Typeset in India by Aptara

Printed in China by
Printplus Limited

10 9 8 7 6 5 4 3 2 1

BRIEF CONTENTS

CONTENTS

Allan Blackman

Allan Blackman is a Professor at the Auckland University of Technology in Auckland, New Zealand. He obtained his BSc(Hons) and PhD degrees from the University of Otago, New Zealand. He has taught all levels of undergraduate chemistry, in the areas of inorganic and physical chemistry, for over 24 years. Allan's research interests lie mainly in the field of coordination chemistry, where he studies the synthesis, structure and reactivity of coordination complexes. He has spent research periods in the US (Indiana University, the University of Minnesota), Australia (the University of Queensland) and France (Universite Joseph Fourier, Grenoble), and has also given numerous undergraduate lectures at the National University of Defense Technology, Changsha, China, where he has been appointed a Guest Professor. Allan regularly appears on TV as a science commentator, and published a monthly newspaper column concerning all things chemical. Outside science, his interests include music and sport.

Steven E Bottle

Professor Steven Bottle is a graduate of the University of Queensland where he completed Honours in Organic Chemistry. After working in various jobs in the pharmaceutical and mining industries, he subsequently undertook a PhD at Griffith University in collaboration with the CSIRO. On completion of his PhD he was awarded an Alexander von Humboldt Fellowship before taking up an academic position at QUT, where he has risen to the rank of full professor and is currently the Science and Engineering Faculty's Director of Research (Quality). Steven is a teaching and research academic with an established reputation for excellence in both pure and applied research, matched with demonstrated teaching capabilities and professional expertise. He has a reputation for inventive and pioneering research and has achieved international recognition for his expertise in the chemistry and applications of free radicals, especially in the context of antioxidant drugs and novel materials. Steven's particular interests include the use of stable nitroxide free radicals in synthesis, polymers and other materials as analytical tools and antioxidant drugs. Stable nitroxide free radicals play critical roles as additives (protecting coatings and plastics), as tools to make new materials and even as new antioxidant medicines. Steven has led much of the modern research on discovering new forms of nitroxide free radicals and applying them in a range of contexts, including as medicinally active compounds, detectors of free radical damage in materials and monitors of particulate pollution that impacts on human health.

Siegbert Schmid

Associate Professor Siegbert Schmid obtained his PhD and Habilitation from the University of Tübingen, Germany, and subsequently accepted a position at the School of Chemistry at the University of Sydney. His research interests lie in the synthesis and structural characterisation of aperiodic and other materials with potential technological applications (e.g. insertion materials for rechargeable batteries). In addition, he is active in chemistry education research and has supervised a number of PhD and Honours students in this area. Siegbert's education research aims to improve current teaching practices and learning outcomes for tertiary level students. He is a Past Chair of the RACI Division of Chemistry Education. His teaching has been recognised with a number of teaching awards, including the Vice Chancellor's Award for Outstanding Teaching (the University of Sydney 2012) and an Office of Learning and Teaching Citation Award (2012) for Excellence in Teaching.

Mauro Mocerino

Associate Professor Mauro Mocerino has enjoyed teaching chemistry at Curtin University for over two decades. During this time he has sought to better understand how students learn chemistry and what can be done to improve their learning. This has developed into a significant component of his research efforts. He also has a strong interest in enhancing the learning in laboratory classes and led the development of a professional development program for those who teach in laboratories. Mauro's other research interests are in the design and synthesis of molecules for specific intermolecular interactions including drug–protein interactions, host–guest interactions, crystal growth modification and corrosion inhibition. He has received numerous awards for his contributions to learning and teaching, including the inaugural Premier's Prize for Excellence in Science Teaching: Tertiary (2003), the Royal Australian Chemical Institute, Division of Chemical Education Medal (2012) and an Office of Learning and Teaching Australian Award for Programs that Enhance Student Learning (2013).

Uta Wille

Uta Wille is a member of the School of Chemistry at the University of Melbourne. She has studied chemistry at the University of Kiel in Germany, where she graduated with a PhD in physical chemistry and afterwards completed a Habilitation in organic chemistry. Uta moved to Australia in 2003 to take up an academic position in the School of Chemistry at Melbourne University. Her research interests lie in the area of physical organic chemistry, environmental free radicals and reaction mechanisms, and she is particularly interested in how environmental radical and non-radical oxidants damage biological and manufactured materials exposed to the atmosphere. Uta teaches chemistry at both undergraduate and graduate levels and enjoys sharing her fascination and passion for chemistry with university students. She is currently Chair of the Academic Programs Committee of the School of Chemistry, Melbourne University.

PREFACE

The third edition of *Chemistry* continues the substantial commitment of Wiley to chemistry education in Australia and New Zealand. *Chemistry, 3rd edition* provides the appropriate mix of organic, inorganic and physical chemistry for the typical local chemistry course.

In this edition, all chapters have been revised, with selected chapter openers and examples being updated to reflect the latest developments in the field of chemistry. New worked examples, practice exercises and end-of-chapter questions have been added throughout, and many of the existing questions have been reworked with new values.

Popular features such as the 'Chemical Connections' have been retained. This hallmark feature highlights the connections between the chemical concepts within the chapter and the local applications of that chemistry in the world around us.

Once again the text is accompanied by a WileyPLUS course featuring algorithmic versions of selected end-of-chapter questions, high-quality molecular animations and interactive revision materials designed to facilitate a rich study experience.

The authors would like to thank the many professional colleagues who have made significant contributions to this project through their feedback and suggestions for the third edition. Once again, the goal was to create a text that is rich with local contexts and examples, and that matches the needs and expectations of chemistry educators in Australia and New Zealand, and this would not have been possible without this crucial input.

The authors and publisher would like to thank the following copyright holders, organisations and individuals for their permission to reproduce copyright material in this book.

Images

• © Felix R. Fischer: **2** (left), **2** (right) Figure 1. Direct imaging of the covalent bond structure in single-molecule chemical reactions using nc-AFM techniques, found at http://www.cchem.berkeley.edu/frfgrp/index.html. Reproduced with permission from Prof. Felix R. Fischer • © Nanomechanics Group: **3** © www.nims.go.jp/nanomechanics • © Alamy Australia Pty Ltd: **7** (right) GL Archive; **48, 1080** (bottom), **1081** © Science Photo Library; **170** Ben Lewis; **194, 218** Bill Bachman; **322** Jim Wileman; **392** Phil Degginger; **432** World History Archive; **534** sciencephotos; **538** Arco Images GmbH; **697** Robert Harding Picture Library Ltd; **792** David J. Green; **798** (bottom) The Science Picture Company; **820** John Boud • © Corbis Australia: **7** (left) Bettmann; **174** Roger Ressmeyer; **298** (top) Philip Evans/Visuals Unlimited; **430** George Steinmetz; **364** John Carnemolla • © IBM Research: **8** (top), **133** (right) Image originally created by IBM Corporation • © Nanomechanics Group: **8** (bottom) From 'Complex patterning by vertical interchange atom manipulation using atomic force microscopy' by Sugimoto et al., *SCIENCE*, vol. 322, Dec. 2008, pp. 413–17 © Nano Characterization Unit, National Institute for Materials Science NIMS. • © ANSTO: **14** (top) Reproduced with permission from ANSTO Australian Nuclear Science and Technology Organisation • © Getty Images Australia: **14** (bottom) CNRI/SPL; **15** Sovfoto; **26** Detlev Van Ravenswaay; **35** CKN; **49, 241** (bottom), **260** (bottom), **265** (top), **265** (bottom), **289** (bottom), **370, 443** (bottom right), **667** (bottom), **667** (top), **681** (top), **681** (middle), **776, 1071** Science Photo Library; **80** BSIP/Contributor; **83, 296** Photo Researchers; **102** (right), **241** (top), **241** (middle), **298** (bottom), **350, 403, 441** (top), **615, 681** (bottom) Charles D Winters; **119** GIPhotoStock; **176** Keystone/Staff; **297, 312** Archive Holdings Inc; **311** Philippe Plailly; **391** Geoff Tompkinson/SPL; **418** Bloomberg; **419** (middle), **419** (left), **419** (right) David M Phillips; **439** Peter Harrison; **510** Andrew Lambert Photography; **533** Charles Falco; **588** Phil Degginger; **600** DR P MARAZZI; **658** Pasieka Pasieka; **759** SSPL; **998** (top) Martin Dohrn/SPL; **998** (bottom) Mark Sykes/SPL; **1048** Jayme Thornton; **1075** James King-Holmes/SPL • © Michael Crawford Infiny: **24** • © Shutterstock: **25** Shots Studio; **39** (bottom) Minerva Studio; **81** AJP; **89** Pakhnyushchy; **103** Petr Student; **105** Art Phaneuf Photography; **118** Mopic; **126** Gary James Calder; **137** (right) Z-art; **137** (left) Africa Studio; **137** (middle) Kordik; **171** Marxon; **175** TravnikovStudio; **219** Laura Stone; **231** JonMilnes; **236** Ditty_about_summer; **252** Pete Niesen; **258** Denis Burden; **259, 299** (right) sheff; **266** CamPot; **275** Yarygin; **286** (middle left) Elena Dijour; **286** (middle right) David Herraez Calzada; **286** (top) Mara008; **287** Byjeng; **289** (top) jps; **290** (bottom), **372** Hung Chung Chih; **299** (left) sezer66; **299** (middle) Ron Frank; **330** Singkham; **331** bogdan ionescu; **332** (left) science photo; **332** (right) Sivolob Igor; **332** (middle) photowind; **348** Vlad61; **349** Dropu; **382** InfinityPhoto; **390** thailoei92; **395** Serdar Tibet; **402** Gyvafoto; **409** Atelier_A; **431** Photodiem; **443** (top right) Denise Kappa; **443** (top left) aguilarphoto; **452** grafvision; **472** Decha Thapanya; **494** Paul Looyen; **495** brackish_nz; **496** Taras Vyshnya; **498** Vereshchagin Dmitry; **546** Prasert Wongchindawest; **547** (right) Byjeng; **547** (left) elen_studio; **567** Julia Reschke; **586** dmitriyd; **587** supapornss; **589** Jens Mayer; **590** Robert Red; **601** Tyler Olson; **604** Razvy; **605** cozyta; **607** (left) mkm3; **607** (right) yuri4u80; **611** Carlos Yudica; **616** Karin Hildebrand Lau; **621** (top) Madlen; **621** (bottom) Albert Lozano; **640** Midkhat Izmaylov; **641** Chaikom; **642** (left) James Steidl; **642** (right) Tyler Boyes; **696** Albo003; **703** Nicolesa; **712** ian woolcock; **758** takayuki; **761** (right) er ryan; **779** Natursports; **783** Mircea BEZERGHEANU; **790** Federico Rostagno; **791** Darla Hallmark; **793** Guryanov Andrey; **798** (top) JPC-PROD; **821** Image Point Fr; **823** Ensuper; **825** Johan Larson; **837** (left) Goncharuk Maksim; **837** (right) Olga Miltsova; **845** Timmary; **850** forbis; **874** EPSTOCK; **875** SpeedKingz; **917** bikeriderlondon; **942** alexpro9500; **943** BW Folsom; **984** TongChuwit; **985** Syda Productions; **1000** Africa Studio; **1003** (top) Moises Fernandez Acosta; **1006** Paul Broadbent; **1010** monticello; **1011** Valentyn Volkov; **1014** Kletr; **1029** Dusan Jankovic; **1046** AN NGUYEN; **1060** eldar nurkovic; **1061** BLACKDAY; **1066** ChinellatoPhoto; **1084** (bottom) © Damian Herde; **1084** (top) Bildagentur Zoonar GmbH • © BIPM — Bureau international: **39** (top) Photograph courtesy of the BIPM • © The Royal Society of Chemistry: **51**(top), **51** (bottom) • © Allan Blackman: **85** (top), **100** (right), **100** (left), **264, 267, 367, 520, 570, 591** (top), **609, 777** • © John Wiley & Sons, Inc.: **85** (bottom), **101, 105, 371, 442, 505** (bottom), **505** (top), **506, 519,** Michael Watson; **104** OPC, Inc.; **108, 260** (top), **281** (bottom), **281** (top), **397, 436, 480, 518, 551,** Andy Washnik; **228, 267** Patrick Watson; **466, 760** Taken by Kari-Ann Tapp; **501** Peter Lerman; **701; 878** *Wiley registry of mass spectral data*, 8th edn, April 2006. Reprinted with permission of John Wiley & Sons, Inc.; **938** From *Organic structures from spectra*, 3rd edn, by Field, Sternhell & Kalman, 2003, John Wiley & Sons Ltd, West Sussex, England; **939** From *Organic structures from spectra*, 3rd edn, by Field, Sternhell & Kalman, 2002, John Wiley & Sons Ltd, West Sussex, England; **997** Photo by Renee Bryon; **1003** (right) Adapted from McCann, M.C., K. Roberts 1991. Architecture of the primary cell wall. In Clive W. Lloyd ed., *The cytoskeletal basis of plant growth and form*, Academic Press; and Hopkins and Huner 2009, *Introduction to plant physiology*, 4th edn, fig. 17.1 and 17.3, John Wiley & Sons, Inc. • © Digital Vision: **102** (left) • © Siegbert Schmid: **124, 133** (middle), **285** • © Australian Synchrotron: **125** Reproduced with permission from the Australian Synchrotron • © Fundamental Photographs: **133** (left), **209, 240, 646** (top), **646** (bottom) Richard Megna; **443** (bottom left) © Larry Stepanowicz • © University Science Books: **146** Figure 1-25 from DeKock R.L., 'Chemical structure and bonding', p. 50 © University Science Books, 1989 • © Armfield Limited: **239** Image courtesy Armfield Ltd. FT67 Hydrogenation Unit • © RMIT: **274** Copyright © 2005 RMIT University, Educational Technology Advancement Group. Photographer Margund Sallowsky • © M.C. Esher Company: **278** M.C. Eschers 'Symmetry Drawing E18' © 2015 The M.C. Escher Company-The Netherlands. All rights reserved. www.mcescher.com • © Jeff Talon: **290** (top) • © iStockphoto: **441** (bottom) alina_hart; **527** JONATHANEASTLAND; **591** (bottom) MOF; **592** Mark Swallow; **608** leezsnow; **716** jgroup • © Stan Sherer: **532** • © Lawrence Berkeley National Laboratory: **617** (left), **617** (right) • © NASA: **623** JPL-Caltech/SETI Institute; **678** (top right), **678** (top left), **678** (middle right), **678** (middle left), **678** (bottom) NASA Ozone Watch • © Lars-Oliver Essen: **682** From Mees et al., 'Crystal structure of a photolyase bound to a CPD-like DNA lesion after in situ repair', *SCIENCE*, vol. 306,

Dec. 2004, p. 1791. Reproduced with permission from Lars-Oliver Essen author • © Timeframe Photography, Inc.: **730, 828, 967** Photography by Charles D. Winters • © Steve Bottle: **761** (left), **882** • © Newspix: **837** (top) News Ltd; **1080** (top) Marc McCormack • © Mauro Mocerino: **841** (left), **841** (middle), **841** (right), **860** • © Simon Lewis: **885** • © The Geis Archives: **1078, 1083** (top left), **1083** (bottom), **1083** (middle) Illustration, Irving Geis. Image from Irving Geis Collection/ Howard Hughes Medical Institute. Rights owned by HHMI. Not to be reproduced without permission. • © Cengage Learning: **1083** (top right) From BETTELHEIM/BROWN/CAMPBELL/ FARR. *Introduction to Organic and Biochemistry with Printed Access Card ThomsonNOWT*, 6E. © 2007 Brooks/Cole, a part of Cengage Learning, Inc. Reproduced by permission. www.cengage.com/permissions • © AAP Newswire: **1085** (bottom) AP/AAP Images • © Portland Press Ltd: **1085** (top) From A.M. Torres et al. 1999, 'Solution structure of a defensin-like peptide from platypus venom', *Biochemical Journal*, vol. 341, pp. 785–94 © Biomechanical Society. Published by Portland Press.

Text

Cengage Learning: **664** From ZUMDAHL/ZUMDAHL. *Chemistry*, 6E. © 2003 Brooks/Cole, a part of Cengage Learning, Inc. Reproduced by permission. www.cengage.com/permissions.

E-chapters

Images

• © Shutterstock: **e-1** wavebreakmedia; **e-2** (left) Australis Photography; **e-2** (right) © Mopic; **e-28** Dmitry Kalinovsky; **e-29** bikeriderlondon; **e-38** Graham Prentice; **e-39** Picsfive; **e-41** Jeremy Smith; **e-43** (bottom) kanusommer; **e-66** Paul Fleet; **e-67** hxdyl; **e-82** Diego Barbieri • © Getty Images Australia: **e-6** Science & Society Picture Library/Contributor; **e-19** BSIP/ Contributor; **e-31** Keystone/Stringer; **e-35** Print Collector/ Contributor; **e-40** BSIP/Contributor; **e-79** (top) Johannes Simon • © Alamy Australia Pty Ltd: **e-7** Heritage Image Partnership Ltd; **e-44** Richard Heyes; **e-76** (top) Tribune Content Agency LLC; **e-87** PHOTOTAKE Inc. • © John Wiley & Sons, Inc.: **e-11** (top left) *Fundamentals of Biochemistry*, 2nd edn, Voet, Voet & Pratt, Fig. 23.46, p. 857. Copyright © 2006 John Wiley & Sons, Inc. Reprinted with permission of John Wiley & Sons, Inc.; **e-11** (top right) *Fundamentals of Biochemistry*, 2nd edn, Voet, Voet & Pratt, Fig. 23.47, p. 857. Copyright © 2006 John Wiley & Sons, Inc. Reprinted with permission of John Wiley & Sons, Inc.; **e-11** (middle right) *Fundamentals of Biochemistry*, 2nd edn, Voet, Voet & Pratt, Fig. 23.50, p. 859. Copyright © 2006 John Wiley & Sons, Inc. Reprinted with permission of John Wiley & Sons, Inc.; **e-14** Pratt and Cornely *Essential Biochemistry*, Fig. 20.2, p. 615. Copyright © 2003 John Wiley & Sons, Inc.; **e-70** D Halliday and R Resnick, *Fundamentals of Physics*, 2nd edn, Revised, John Wiley & Sons, Inc., Copyright 1986. • © Lawrence Kobilinsky, Dr: **e-22** Professor Lawrence Kobilinsky • © Allan Blackman: **e-32** (left), **e-32** (middle), **e-32** (right), **e-59** (bottom) • © Newspix: **e-43** (top) Phil Hillyard; **e-57** (bottom) Todd Martyn Jones • © AAP Newswire: **e-56, e-75** (bottom) AFP/AAP Images • © Image 100: **e-57** (top) • © Reserve Bank of Australia: **e-59** (top) • © Corbis Australia: **e-75** (top) Bettmann • © Darrin Gray: **e-76** (bottom) • © U.S. Department of Energy: **e-85** • © National Institute of Drug Abuse: **e-86** E.D. London, National Institute of Drug Abuse.

Every effort has been made to trace the ownership of copyright material. Information that will enable the publisher to rectify any error or omission in subsequent editions will be welcome. In such cases, please contact the Permissions Section of John Wiley & Sons Australia, Ltd who will arrange for the payment of the usual fee.

John Wiley & Sons, Australia: Terry Burkitt (Publishing Manager), Mark Levings (Executive Publisher), Kylie Challenor (Managing Content Editor), Emma Knight (Project Editor), Emily Nuhn (Publishing Coordinator), Delia Sala (Graphic Designer), Jo Hawthorne (Senior Production Controller), Rebecca Cam (Digital Content Editor), Siale Gilmour (Art Coordinator).

1 The atom

What is the universe made of? This question has occupied human thinking for thousands of years. Some ancient civilisations thought that the universe comprised only four elements — earth, air, fire and water — and everything was made up of a combination of these. Over the past 400 years, the advent of the science called chemistry has allowed us to show that this is not the case. We now know that matter — everything you can see, smell, touch or taste — is made up of atoms, the fundamental building blocks of the universe.

Atoms are incredibly small — far too small to be seen using conventional microscopes. While many experiments over many years have produced results consistent with the existence of atoms, only recently have we been able to 'see' individual atoms and the collections of atoms we call molecules. We now have the technology to observe individual molecules undergoing a chemical reaction, a process in which one chemical substance is converted into another. The images on this page show a molecule called an alkyne, which contains three rings, reacting when heated to over 90 °C to give a molecule containing seven rings. The images were obtained using an atomic force microscope, and the scale on the images (3 Å = 0.000 000 000 3 m) gives an idea of just how tiny atoms truly are.

Our current knowledge of the structure of the atom, and the way in which atoms pack together in three-dimensional space, owes much to experiments carried out by two Australasian-born scientists, Ernest Rutherford (1871–1937, Nobel Prize in chemistry, 1908) and William Lawrence Bragg (1890–1971, Nobel Prize in physics, 1915), both of whom would doubtless be astonished by the images on this page. The New Zealand-born Rutherford was the first to show that the atom consists of a positively charged nucleus surrounded by tiny negatively charged electrons. Bragg (born in Australia),

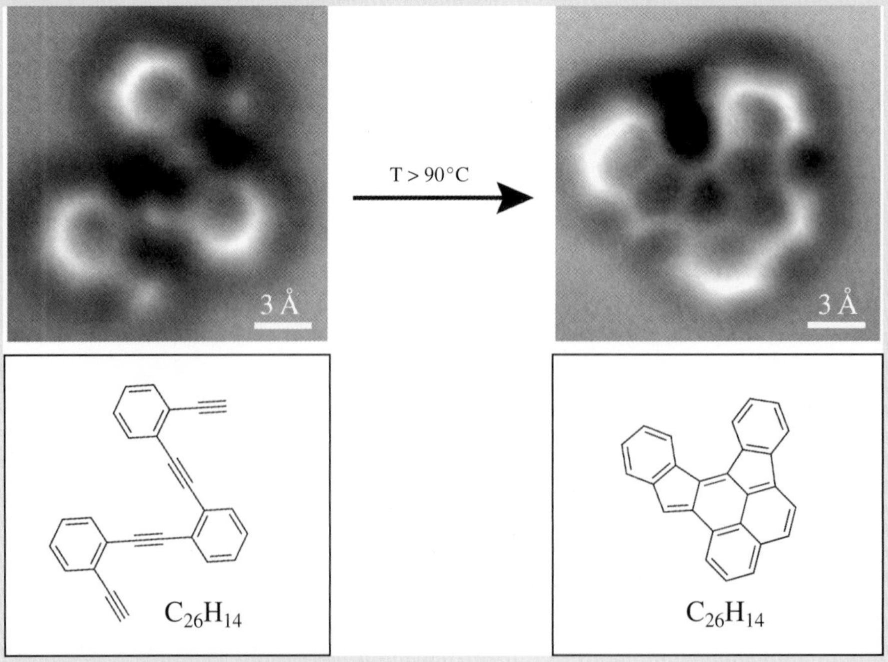

3 Å

T > 90 °C

3 Å

$C_{26}H_{14}$

$C_{26}H_{14}$

together with his British-born father William Henry Bragg, developed the technique of X-ray crystallography, in which X-rays are used to determine the three-dimensional structure of solid matter on the atomic scale. The contribution of the Braggs will be outlined further in chapter 7. This chapter is primarily concerned with the atom. It will examine the contribution of Rutherford and others to the determination of the structure of the atom, and will show how a particular structural feature of the atom forms the basis of the periodic table of the elements.

LEARNING OBJECTIVES

After studying this chapter, you should be able to:

1.1 define atoms, molecules, ions, elements and compounds

1.2 explain how the concept of atoms developed

1.3 describe the structure of the atom

1.4 explain the basis of the periodic table of the elements

1.5 detail the role of electrons in atoms.

1.1 Atoms, molecules, ions, elements and compounds

Chemistry is the study of **matter**. Chemists view matter as being composed of various chemical entities. Before we embark on our study of chemistry, we will briefly define some terms used to describe these entities. This will aid our understanding of the material in this chapter; we will look at these concepts again, in greater detail, in subsequent chapters. **Atoms** are discrete chemical species comprising a central positively charged nucleus surrounded by one or more negatively charged **electrons**. Atoms are always electrically neutral, meaning that the number of electrons is equal to the number of protons in the nucleus. Chemists regard the atom as the fundamental building block of all matter, so it may surprise you to learn that individual atoms are rarely of chemical interest; free atoms (with the exception of the elements helium, neon, argon, krypton, xenon and radon) are usually unstable. Of much greater interest to chemists are **molecules**, which are collections of atoms with a definite structure held together by chemical bonds. The smallest molecules contain just two atoms, while the largest can consist of literally millions. Most gases and liquids consist of molecules, and most solids based on carbon (organic solids) are also molecular. Like atoms, molecules are electrically neutral and are, therefore, uncharged. Molecules are held together by **covalent bonds**, which involve the sharing of electrons between neighbouring atoms.

Ions are chemical species that have either a positive or negative electric charge. Those with a positive charge are called **cations**; those with a negative charge are called **anions** (respectively designated by a $^+$ or $^-$). Ions can be formally derived from either atoms or molecules by the addition or removal of one or more electrons. For example, removing an electron (e^-) from a sodium, Na, atom gives the Na^+ cation.

$$Na \rightarrow Na^+ + e^-$$

Adding an electron to an oxygen molecule, which consists of two oxygen atoms bonded together and is designated O_2, gives the superoxide anion.

$$O_2 + e^- \rightarrow O_2^-$$

Elements are collections of one type of atom only, and the 118 elements known at the time of writing are listed in the periodic table on the inside front cover. **Compounds** are simply substances containing two or more elements in a definite and unchanging proportion. Compounds may be composed of molecules, ions or covalently bonded networks of atoms. Note that we do not have individual 'molecules' of an ionic compound such as sodium chloride. The chemical formula of sodium chloride, NaCl, simply represents the smallest repeating unit in an enormous three-dimensional array of Na^+ ions and Cl^- ions. The same applies to certain covalently bonded structures. For example, quartz, which is composed of an 'infinite' three-dimensional network of covalently bound Si and O atoms, has the chemical formula SiO_2, which refers not to individual SiO_2 'molecules' but to the smallest repeating unit in the network.

All of the above chemical entities (atoms, molecules, ions, elements and compounds) may be involved as **reactants** in **chemical reactions**, processes in which they undergo transformations generally involving the making and/or breaking of chemical bonds, and which usually result in the formation of different chemical species called **products**.

1.2 The atomic theory

Today, we take the existence of atoms for granted. We can explain many aspects of the structure of the atom and, in fact, current technology allows us to 'see' and even manipulate individual atoms. However, scientific evidence for the existence of atoms is relatively recent, and chemistry did not progress very far until that evidence was found.

The concept of atoms began nearly 2500 years ago when the Greek philosopher Leucippus and his student Democritus expressed the belief that matter is ultimately composed of tiny indivisible particles; the word 'atom' is derived from the Greek word *atomos*, meaning 'not cut'. The philosophers' conclusions, however, were not supported by any scientific evidence; they were derived simply from philosophical reasoning. The concept of atoms remained a philosophical belief, having limited scientific usefulness, until the discovery of two laws of chemical combination in the late eighteenth century — the law of conservation of mass and the law of definite proportions. These may be stated as follows.

- The **law of conservation of mass:** No detectable gain or loss of mass occurs in chemical reactions. Mass is conserved.
- The **law of definite proportions:** In a given chemical compound, the elements are always combined in the same proportions by mass.

The French chemist Antoine Lavoisier (1743–1794) proposed the law of conservation of mass as a result of his experiments involving the individual reactions of the elements phosphorus, sulfur, tin and lead with oxygen. He used a large lens to focus the Sun's rays on a sample of each element contained in a closed jar, and the heat caused a chemical reaction to take place. He weighed the closed jar and its contents before and after the chemical reaction and found no difference in mass, leading him to propose the law. (Lavoisier was beheaded following the French Revolution, the judge at his trial reputedly saying 'the Republic has no need of scientists'.) The law of conservation of mass can be alternatively stated as 'mass is neither created nor destroyed in chemical reactions'.

Another French chemist, Joseph Louis Proust (1754–1826), was responsible for the law of definite proportions, following experiments that showed that copper carbonate prepared in the laboratory was identical in composition to copper carbonate that occurs in nature as the mineral malachite. He also showed that the two oxides of tin, SnO and SnO_2, and the two sulfides of iron, FeS and FeS_2, always contain fixed relative masses of their constituent elements. The law states that chemical elements always combine in a definite fixed proportion by mass to form chemical compounds. Thus, if we analyse *any* sample of water (a compound), we *always* find that the ratio of oxygen to hydrogen (the elements that make up water) is 8 to 1 by mass. Similarly, if we form water from oxygen and hydrogen, the mass of oxygen consumed will always be 8 times the mass of hydrogen that reacts. This is true even if there is a large excess of one of them. For instance, if 100 g of oxygen is mixed with 1 g of hydrogen and the reaction to form water is initiated, all the hydrogen would react but only 8 g of oxygen would be consumed; there would be 92 g of oxygen left over. No matter how we try, we cannot alter the chemical composition of the water formed in the reaction.

WORKED EXAMPLE 1.1

Applying the law of definite proportions

The element molybdenum, Mo, combines with sulfur, S, to form a compound commonly called molybdenum disulfide that is useful as a dry lubricant, similar to graphite. It is also used in specialised lithium batteries. A sample of this compound contains 1.50 g of Mo for each 1.00 g of S. If a different sample of the compound contains 2.50 g of S, what mass of Mo does it contain?

Analysis

The law of definite proportions states that the proportions of Mo and S by mass must be the same in both samples. To solve the problem, we will set up the mass ratios for the two samples. In the ratio for the second sample the mass of molybdenum will be an unknown quantity. We will equate the two ratios and solve for the unknown quantity.

Solution

The first sample has a Mo to S mass ratio of:

$$\frac{1.50\text{ g Mo}}{1.00\text{ g S}}$$

We know the mass of S in the second sample, but the mass of Mo is unknown. We do know, however, that the Mo to S mass ratio is the same as that in the first sample. We set up the ratio for the second sample using x for the unknown mass of Mo. Therefore, from the law of definite proportions, we can write:

$$\frac{1.50\text{ g Mo}}{1.00\text{ g S}} = \frac{x}{2.50\text{ g S}}$$

We can solve for x by multiplying both sides of the equation by 2.50 g S, to give:

$$x = \frac{2.50\text{ g S} \times 1.50\text{ g Mo}}{1.00\text{ g S}}$$
$$= 3.75\text{ g Mo}$$

Is our answer reasonable?

To avoid errors, it is always wise to do a rough check of the answer. Usually, some simple reasoning is all we need to see if the answer makes sense. This is how we might do such a check here: Notice that the mass of sulfur in the second sample is more than twice the mass in the first sample. Therefore, we should expect the mass of Mo in the second sample to be somewhat more than twice what it is in the first. The answer we obtained, 3.75 g Mo, is more than twice 1.50 g Mo, so our answer seems to be reasonable.

PRACTICE EXERCISE 1.1 Titanium dioxide, TiO_2, is a compound that is used as a brilliant white pigment in artists' oil colours, as well as in coatings and plastics. A sample of this compound was found to be composed of 1.00 g of titanium and 0.668 g of oxygen. If a second sample of the same compound contains 2.50 g of oxygen, what mass of titanium does it contain?

The laws of conservation of mass and definite proportions served as the experimental foundation for the atomic theory. They prompted the question, 'What must be true about the nature of matter, given the truth of these laws?' In other words, 'What is matter made of?' At the beginning of the nineteenth century, John Dalton (1766–1844), an English scientist, used the Greek concept of atoms to make sense of the laws of conservation of mass and definite proportions. Dalton reasoned that, if atoms really exist, they must have certain properties to account for these laws. He described such properties, and the following list constitutes what we now call **Dalton's atomic theory**.

1. Matter consists of tiny particles called atoms.
2. Atoms are indestructible. In chemical reactions, the atoms rearrange but they do not themselves break apart.
3. In any sample of a pure element, all the atoms are identical in mass and other properties.
4. The atoms of different elements differ in mass and other properties.
5. When atoms of different elements combine to form a given compound, the constituent atoms in the compound are always present in the same fixed numerical ratio.

Dalton's theory easily explained the law of conservation of mass. According to the theory, a chemical reaction is simply a reordering of atoms from one combination to another. If no atoms are gained or lost, and if the masses of the atoms can't change, the mass after the reaction must be the same as the mass before. This explanation of the law of conservation of mass allows us to use a notation system of **chemical equations** to describe chemical reactions. A chemical equation contains the reactants on the left-hand side and the products on the right-hand side, separated by a forward arrow, as demonstrated in the following chemical equation for the formation of liquid water from its gaseous elements.

$$2H_2(g) + O_2(g) \rightarrow 2H_2O(l)$$

The law of conservation of mass requires us to have the same number of each type of atom on each side of the arrow; this being the case, the chemical equation above is described as balanced. We will discuss this concept in detail in chapter 3. Note that this chemical equation also specifies the physical states of the reactants and product. Gases, liquids and solids are abbreviated as (g), (l) and (s), respectively, after each reactant and product.

The law of definite proportions can also be explained by Dalton's theory. According to the theory, a given compound is always composed of atoms of the same elements in the same numerical ratio. Suppose, for example, that elements X and Y combine to form a compound in which the number of atoms of X equals the number of atoms of Y (i.e. the atom ratio is 1 to 1). If the mass of a Y atom is twice that of an X atom, then every time we encounter a sample of this compound, the mass ratio (X to Y) would be 1 to 2. This same mass ratio would exist regardless of the size of the sample so, in samples of this compound, elements X and Y are always present in the same proportion by mass.

Strong support for Dalton's theory came when Dalton and other scientists studied elements that can combine to give at least two compounds. For example, sulfur and oxygen can combine to form both sulfur dioxide, SO_2, and sulfur trioxide, SO_3. The former contains one atom of sulfur and two atoms of oxygen, while the latter contains one atom of sulfur and three atoms of oxygen. Although they have similar chemical formulae, they are different chemically; for example, at room temperature, SO_2 is a colourless gas while SO_3, which melts at 16.8 °C, is a solid or liquid, depending on the temperature of the room. If we analyse samples of SO_2 and SO_3 in which the masses of sulfur are the same, we obtain the results shown in table 1.1.

TABLE 1.1 Mass composition of sulfur dioxide and sulfur trioxide.

Compound	Mass of sulfur	Mass of oxygen
SO_2	1.00 g	1.00 g
SO_3	1.00 g	1.50 g

Note that the ratio of the masses of oxygen in the two samples is one of small whole numbers.

$$\frac{\text{mass of oxygen in sulfur trioxide}}{\text{mass of oxygen in sulfur dioxide}} = \frac{1.50\text{ g}}{1.00\text{ g}} = \frac{3}{2}$$

Similar observations are made when we study other elements that form more than one compound with each other. These observations form the basis of the **law of multiple proportions**, which states that: Whenever two elements form more than one compound, the different masses of one element that combine with the same mass of the other element are in the ratio of small whole numbers.

Dalton's theory explains the law of multiple proportions in a very simple way. Suppose a molecule of sulfur trioxide contains 1 sulfur and 3 oxygen atoms, and a molecule of sulfur dioxide contains 1 sulfur and 2 oxygen atoms (figure 1.1). If we had just one molecule of each, our samples would each have 1 sulfur atom and, therefore, the same mass of sulfur. Then, comparing the oxygen atoms, we find they are in a numerical ratio of 3 to 2. But because oxygen atoms all have the same mass, the mass ratio must also be 3 to 2. The law of multiple proportions was not known before Dalton presented his theory, and its discovery demonstrates science in action. Experimental data suggested to Dalton the existence of atoms, and the atomic theory suggested the relationships that we now call the law of multiple proportions. When found by experiment, the existence of the law of multiple proportions added great support to the atomic theory. In fact, for many years, it was one of the strongest arguments in favour of the existence of atoms.

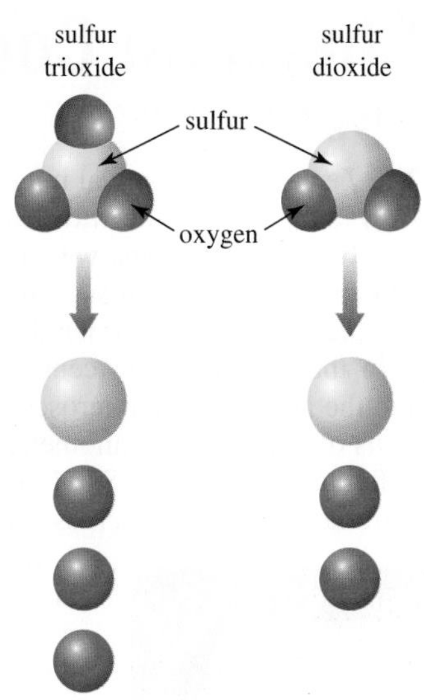

FIGURE 1.1 Compounds containing oxygen and sulfur demonstrate the law of multiple proportions. Represented here are molecules of sulfur trioxide, SO_3, and sulfur dioxide, SO_2. Each has one sulfur atom, and therefore the same mass of sulfur. The oxygen ratio is 3 to 2, both by atoms and by mass.

1.3 The structure of the atom

Even though absolute proof of the existence of atoms was not available around the turn of the twentieth century, scientists were interested in the structure of the atom. While Dalton's theory said that atoms were indestructible and could not be broken apart, experiments around this time showed this was not necessarily true. In particular, the discovery of radiation in the form of X-rays by Wilhelm Röntgen (1845–1923) in 1895 (see figure 1.2) and radioactivity by Antoine Henri Becquerel (1852–1908, pictured in figure 1.3) in 1896 led scientists to believe that the atom was composed of discrete particles, as both forms of radiation involve the release of particles from atoms, thought at that time to be indivisible.

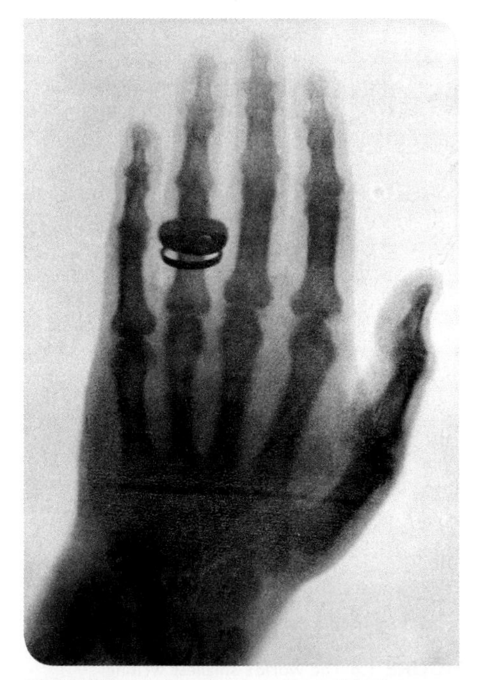

FIGURE 1.2 A reproduction of one of the first ever X-ray images, taken by Wilhelm Röntgen (Nobel Prize in physics, 1901) on 22 December 1895. The hand is that of his wife.

FIGURE 1.3 Antoine Henri Becquerel (Nobel Prize in physics, 1903) discovered radioactivity.

Chemical Connections

Imaging atoms

We cannot use optical microscopes to see atoms. This is because the dimensions of atoms are smaller than the wavelength of visible light. If we use shorter wavelength radiation, such as a beam of electrons, we can obtain images like those shown on page 2. However, the apparatus required to obtain such images is expensive and the samples require a significant degree of preparation and careful handling.

In the late twentieth century, two inventions — the scanning tunnelling microscope (STM) and the atomic force microscope (AFM) — revolutionised the imaging of objects having dimensions of the order of nanometres, and have allowed us to 'see' and, more remarkably, manipulate individual atoms. The STM and AFM operate using the same principle — moving the tip of an extremely fine stylus across a surface at a distance of atomic dimensions.

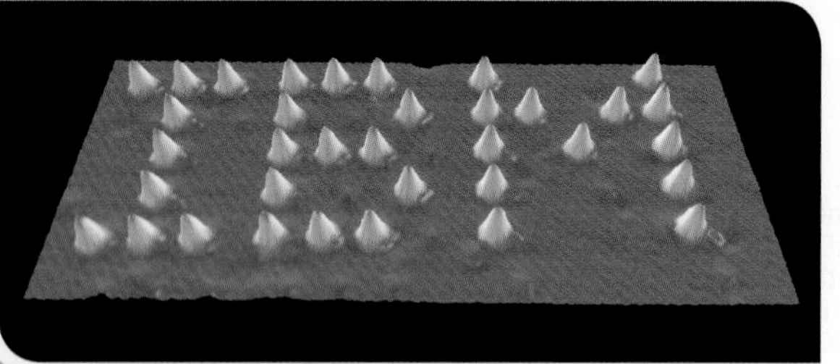

FIGURE 1.4 Individual Xe atoms (blue dots) on a nickel surface manipulated by an STM tip.

In the case of the STM, the surface must be electrically conducting, and this causes a current to flow between the surface and the tip. The magnitude of this current depends on the distance between the tip and the surface, so as the tip is moved across the surface, computer control of the current at a constant value will cause the tip to move up and down, thereby giving a map of the surface. Because of its tiny size, the tip can also be used to move individual atoms. This was first demonstrated in 1989 when Don Eigler, a scientist at IBM, manipulated 35 atoms of xenon on a nickel surface using an STM to spell the name of his employer (figure 1.4).

An AFM (illustrated in figure 1.5) is used to study nonconducting samples. The stylus is moved across the surface of the sample under study. Forces between the tip of the probe and the surface cause the probe to flex as it follows the ups and downs of the bumps that are the individual molecules and atoms. A mirrored surface attached to the probe reflects a laser beam at angles proportional to the amount of deflection of the probe. A sensor picks up the signal from the laser and translates it into data that can be analysed by a computer to give three-dimensional images of the sample's surface.

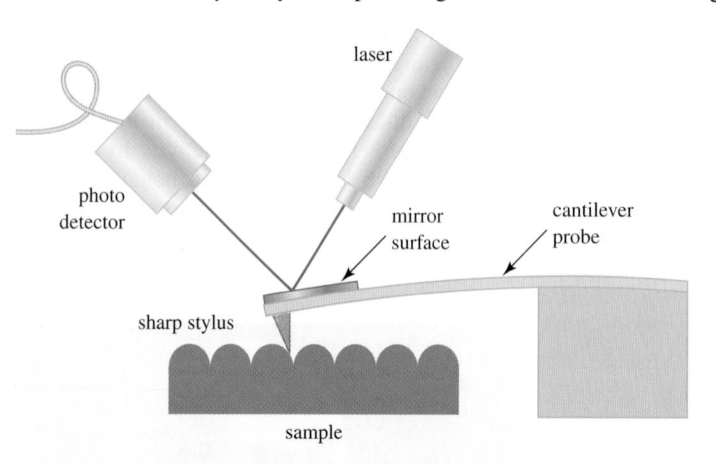

FIGURE 1.5 In an AFM, a sharp stylus attached to the end of a cantilever probe rides up and down over the surface features of the sample. A laser beam, reflected off a mirrored surface at the end of the probe, changes angle as the probe moves up and down. A photodetector reads these changes and sends the information to a computer, which translates the data into an image.

A typical AFM image is shown in figure 1.6. It involves manipulating single atoms to give what is probably the smallest known writing.

FIGURE 1.6 The world's smallest writing? The element symbol for silicon, Si, is spelled out with individual silicon atoms (dark) among tin atoms (light). The silicon atoms were manipulated with the tip of an AFM.

Further evidence for the presence of discrete particles in atoms came from experiments with gas discharge tubes, such as that shown in figure 1.7. When the tube was filled with a low pressure gas and a high voltage was applied between the electrodes, negatively charged particles flowed from the negative electrode (cathode) to the positive electrode (anode).

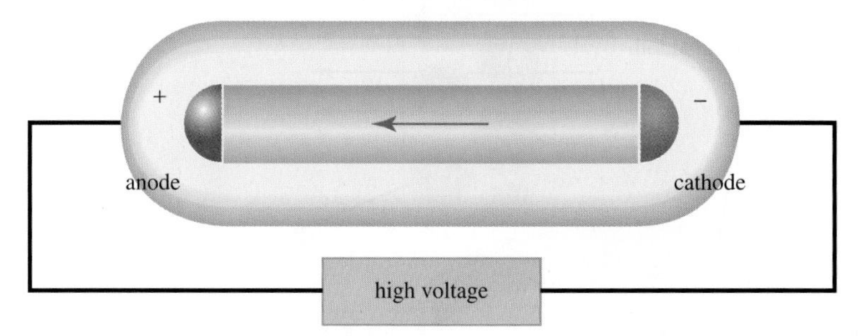

FIGURE 1.7 Diagram of a gas discharge tube.

Because they emanated from the cathode, the particles were called cathode rays. In 1897, the British physicist JJ Thomson passed cathode rays through a magnetic field using the modified discharge tube shown in figure 1.8. The magnetic field caused the path of the cathode rays to bend. Analysis of this effect allowed Thomson to determine the charge to mass ratio of the components of cathode rays, what we now know as electrons.

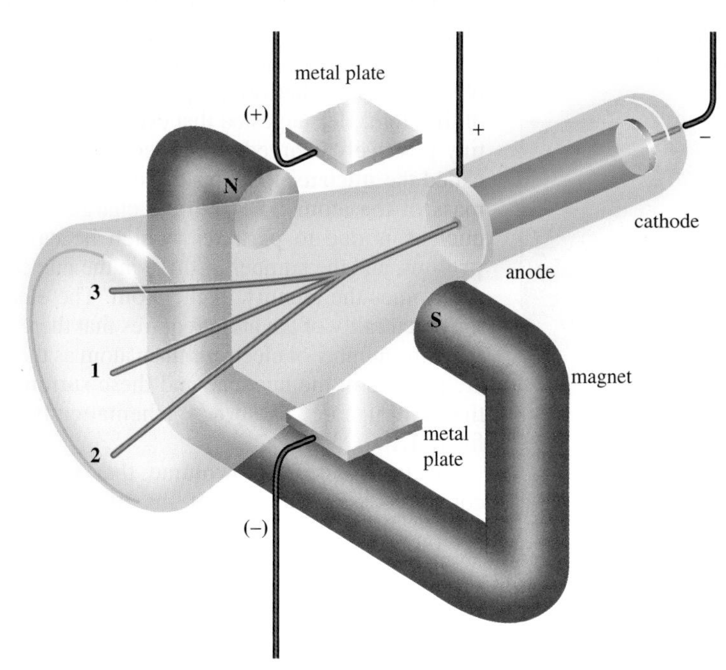

FIGURE 1.8 Diagram of the apparatus used by JJ Thomson to determine the charge to mass ratio of the electron. The cathode ray beam passes between the poles of a magnet and between a pair of metal electrodes that can be given electric charges. The magnetic field tends to bend the beam in one direction (to point 2) while the charged electrodes bend the beam in the opposite direction (to point 3). By adjusting the charge on the electrodes, the two effects can be made to cancel each other (point 1). The amount of charge on the electrodes required to balance the effect of the magnetic field can be used to calculate the charge to mass ratio.

In 1909, the American chemist Robert Millikan determined the charge on an individual electron by measuring the rates at which charged oil drops fell between electrically charged plates. This, combined with Thomson's results, allowed calculation of the mass of an electron as 9.09×10^{-31} kg. The knowledge that atoms were electrically neutral meant that the electron must have a positively charged counterpart, but its exact nature was not known by the early years of the twentieth century. It was the work of the New Zealand-born scientist Ernest Rutherford (1871–1937) that shed light not only on the positively charged component of the atom, but also on the structure of the atom itself. Around 1909, Rutherford, who had already won the Nobel Prize in chemistry in 1908 for his work on the theory of radioactivity, devised his famous gold foil experiment depicted in figure 1.9 (overleaf). Rutherford took an incredibly thin sheet of gold (only a few atoms thick) and bombarded it with a stream of positively charged particles called **alpha particles**.

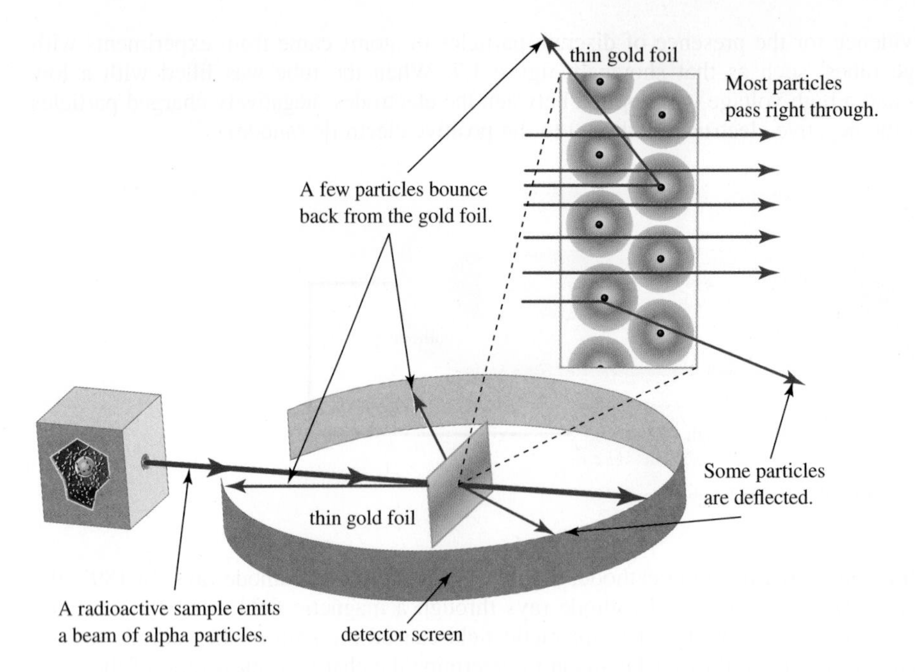
thin gold foil
Most particles
pass right through.
A few particles bounce
back from the gold foil.
Some particles
are deflected.
thin gold foil
A radioactive sample emits
a beam of alpha particles.
detector screen

FIGURE 1.9 Schematic view of Rutherford's gold foil experiment. When a beam of positively charged alpha particles was 'shot' at a thin gold foil, most of them passed straight through the foil. Some, however, were deflected straight back towards the source.

Most of the particles went straight through the foil essentially undisturbed, some were deflected through various angles, and about 1 in 8000 was deflected almost straight back towards the source. Of this amazing observation, Rutherford said, 'It was almost as incredible as if you had fired a fifteen-inch [38.1-centimetre] shell at a piece of tissue paper and it came back and hit you'. To explain his observations, Rutherford proposed a new model of the atom. He suggested that every atom has a tiny positively charged central core, which he called the **nucleus**, that constitutes most of the mass of the atom. The positive charge in the nucleus is due to particles, which he called **protons**, and the number of these in the nucleus determines the identity of the atom. The electrical neutrality of the atom requires that there is the same number of electrons in an atom as there are protons in the nucleus, and these surround the central core, as shown schematically in figure 1.10.

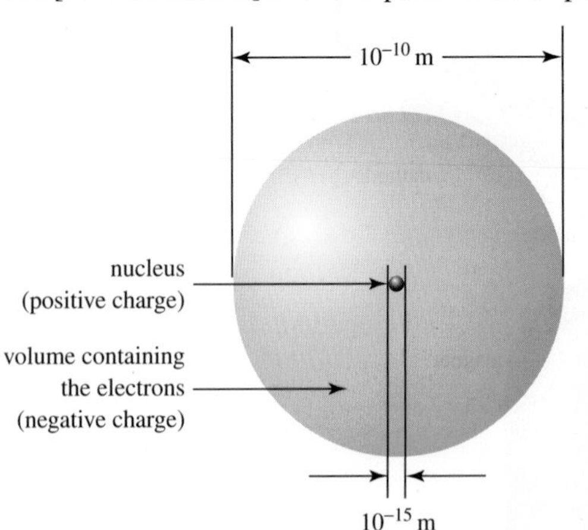
10^{-10} m
nucleus
(positive charge)
volume containing
the electrons
(negative charge)
10^{-15} m

FIGURE 1.10 Diagram (not to scale) showing the nucleus and the volume occupied by the surrounding electrons. If this were drawn to scale, the nucleus would be invisible.

Electrons occupy a volume that is huge compared with the size of the nucleus, but each electron has such a small mass that alpha particles are not deflected by the electrons. Consequently, an alpha particle is deflected only when it passes very near a nucleus, and it bounces straight back only when it collides head-on with a nucleus. Because most of the volume of an atom is essentially empty space, most alpha particles pass through the foil without being affected. From the number of particles deflected and the pattern of deflection, Rutherford calculated that the positive nucleus occupies less than 0.1% of the total atomic volume. To give you an idea of what that means, an atom the size of a rugby stadium would have a nucleus the size of a pea. When Rutherford calculated the nuclear mass based on the number of protons in the nucleus, the value always fell short of the actual mass. In fact, Rutherford found that only about half of the nuclear mass could be accounted for by protons. This led him to suggest that there was another particle in the nucleus that had a mass close to or equal to that of a proton, but with no electric charge. This suggestion initiated a search that finally ended in 1932 with the discovery of the **neutron** by James Chadwick (1891–1974), a British physicist. Because they are found in the nucleus, protons and neutrons are sometimes called **nucleons**. Table 1.2 summarises the **subatomic particles** present in this model of the atom.

TABLE 1.2 Physical data for the electron, proton and neutron.

Particle	Symbol	Charge (C)	Mass (kg)	Mass (u)
electron	e^-	-1.6022×10^{-19}	9.1094×10^{-31}	5.4858×10^{-4}
proton	p	$+1.6022 \times 10^{-19}$	1.6726×10^{-27}	1.0073
neutron	n	0	1.6749×10^{-27}	1.0087

Note: The charge is measured in coulombs (C). The final column gives the mass in atomic mass units (u); $1\ u = 1.66054 \times 10^{-27}$ kg ($\frac{1}{12}$ the mass of the ^{12}C atom) (p. 12).

Over the intervening years, it has been shown that protons and neutrons are themselves composed of still smaller particles called quarks. The existence of quarks has helped us understand how the atomic nucleus can stay together despite the presence of positively charged protons in close proximity. However, quarks are very unstable outside the confines of the atomic nucleus and are of more interest to physicists than chemists.

To examine how atoms are constructed, we will consider the simplest possible atom, hydrogen, with the chemical symbol H. A hydrogen atom consists of a single proton in the nucleus, and a single electron. We designate this as ${}^{1}_{1}H$. We use this terminology for any chemical element X as follows:

$${}^{A}_{Z}X$$

where A is the mass number, Z is the atomic number and X is the **chemical symbol** of the chemical element. The **atomic number (Z)** is the number of protons in the nucleus. The **mass number (A)** is the number of protons in the nucleus plus the number of neutrons (N) in the nucleus.

Note that the atomic number is also equal to the number of electrons in a *neutral* atom (i.e. one in which the number of protons and the number of electrons is the same). A chemical element is defined by its atomic number; all atoms having the same atomic number are atoms of the same element. Therefore, the symbol ${}^{1}_{1}H$ tells us that an atom of hydrogen contains 1 proton ($Z = 1$), 1 electron and 0 neutrons ($A = 1$).

If we were to analyse a sample of hydrogen atoms, we would find that roughly 1 atom in every 6600 would have approximately twice the mass of a ${}^{1}_{1}H$ atom. These heavier atoms belong to an isotope of hydrogen called deuterium. **Isotopes** are atoms of an element with the same number of protons (i.e. the same value of Z) but different numbers of neutrons (i.e. different values of A). Deuterium atoms are symbolised as ${}^{2}_{1}H$, meaning that there is 1 proton ($Z = 1$) and 1 neutron ($A = 2$) in the nucleus. The ${}^{1}_{1}H$ atom is sometimes called protium to distinguish it from deuterium. In chemical terms, deuterium atoms behave essentially identically to hydrogen atoms, but there are some important differences in reactivity when they are bonded to other atoms. In addition, hydrogen also has a third isotope called tritium, ${}^{3}_{1}H$, which has 1 proton and 2 neutrons in the nucleus. It is the least abundant isotope of hydrogen, with only 1 to 10 atoms of tritium in every 10^{18} atoms of hydrogen. Tritium is **radioactive**, meaning that the nucleus is unstable and undergoes spontaneous decay to give an atom of helium, He, a process we will look at in greater detail in chapter 27. Helium atoms are characterised by having 2 protons in the nucleus ($Z = 2$). Helium has two stable isotopes, ${}^{3}_{2}He$ and ${}^{4}_{2}He$, with 1 and 2 neutrons, respectively, in the nucleus. The element with 3 protons in the nucleus, lithium ($Z = 3$), has the stable isotopes ${}^{6}_{3}Li$ and ${}^{7}_{3}Li$ with 3 and 4 neutrons, respectively. Any atom of a specified A and Z is called a **nuclide**. A **radionuclide** is a radioactive nuclide.

WORKED EXAMPLE 1.2

The composition of atoms

The following radioactive isotopes have medical applications. Determine the number of protons, neutrons and electrons in each isotope.

(a) ${}^{165}_{66}Dy$ (used in the treatment of arthritis)

(b) ${}^{131}_{53}I$ (used in the treatment of thyroid cancer)

(c) ${}^{59}_{26}Fe$ (used in studies of iron metabolism)

Analysis

The number of protons is equal to the atomic number (Z), the number of neutrons is found from Z and the mass number (A), and the number of electrons in a neutral atom must equal the number of protons.

Solution

(a) Dy is the chemical symbol for dysprosium. The subscript 66 is Z, which is the number of protons in the nucleus. The superscript 165 is A. We find the number of neutrons by subtracting Z from A: $A - Z = 165 - 66 = 99$ neutrons. Because this is a neutral atom, the number of electrons must equal the number of protons. ${}^{165}_{66}Dy$ has 66 protons, 99 neutrons and 66 electrons.

(b) I is the symbol for iodine. $Z = 53$ and $A = 131$. Z tells us that the nucleus contains 53 protons. Subtracting Z from A, we find that there are 78 neutrons in this isotope. Finally, the atom is neutral, so there are 53 electrons.

(c) Iron, Fe, has $Z = 26$. A neutral atom of ${}^{59}_{26}Fe$ has 26 protons, 26 electrons and $59 - 26 =$ 33 neutrons.

Is our answer reasonable?

In all cases, the number of protons is equal to the number of electrons, as required for neutral atoms. We have followed the rules for calculating the number of neutrons and have carried out the calculations properly. Our answers should therefore be correct.

PRACTICE EXERCISE 1.2 The following radioactive isotopes have medical applications. Determine the number of protons and neutrons in each.

(a) $^{177}_{71}Lu$ (used as an imaging and therapeutic agent)

(b) $^{133}_{54}Xe$ (used in studies of the lungs)

(c) $^{192}_{77}Ir$ (used in the form of an internal wire for cancer treatment)

Inclusion of the atomic number in this terminology is almost redundant when the chemical symbol is included, so it is common to see a shorthand version that excludes this. Thus, we often write $^{1}_{1}H$ as simply ^{1}H, as we know that all atoms of hydrogen have $Z = 1$. Using the same shorthand version, deuterium would be written as ^{2}H and tritium as ^{3}H.

Atomic mass

We saw in table 1.2 that the ^{12}C isotope is used as the basis by which **atomic mass** is measured. The **atomic mass unit (u)** is the mass ($1.660\,54 \times 10^{-27}$ kg) equal to $\frac{1}{12}$ the mass of one atom of ^{12}C, and the masses of all atoms are measured relative to this. The atomic mass unit is also known as the Dalton (Da), particularly in biochemistry. Using this scale, we find that the mass of a single ^{19}F atom is 18.998 403 2 u and that of a single ^{31}P atom is 30.973 762 u. Because both fluorine and phosphorus have only one naturally occurring isotope, we can be sure that any fluorine atom we chose from a macroscopic sample of fluorine would have a mass of 18.998 403 2 u, while any phosphorus atom chosen from a macroscopic sample of phosphorus would have a mass of 30.973 762 u. We therefore say that the atomic mass of fluorine is 18.998 403 2 u and the atomic mass of phosphorus is 30.973 762 u. However, the majority of elements in the periodic table comprise two or more isotopes, and the mass of a single atom chosen at random from a macroscopic sample of these elements would not be constant — it would depend on which isotope was chosen. We therefore define the atomic mass of these elements as the average mass per atom of a naturally occurring sample of atoms of the element. Consider, for example, the element gallium, Ga. There are two naturally occurring isotopes of Ga, namely $^{69}_{31}Ga$ and $^{71}_{31}Ga$, each of which contains 31 protons in the nucleus. Nuclei of the former contain 38 neutrons while those of the latter contain 40. Any naturally occurring sample of gallium will be composed of 60.11% of the $^{69}_{31}Ga$ isotope and 39.89% of the $^{71}_{31}Ga$ isotope. Given the atomic masses of these isotopes ($^{69}_{31}Ga$ = 68.9256 u, $^{71}_{31}Ga$ = 70.9247 u) we can calculate the average atomic mass of Ga by taking the sum of the atomic mass of each isotope multiplied by its abundance as follows.

$$\text{average atomic mass of Ga} = (0.6011 \times 68.9256\,\text{u}) + (0.3989 \times 70.9247\,\text{u}) = 69.72\,\text{u}$$

The average atomic mass of Ga, 69.72 u, is just less than the average of the masses of the two isotopes (69.9252 u) because the lighter $^{69}_{31}Ga$ isotope is more abundant than the heavier $^{71}_{31}Ga$ isotope. Figure 1.11 illustrates the range of isotopic compositions found in four elements, one of which is tin, the element with the largest number of stable isotopes.

While the distribution of isotopes in samples of most elements is essentially constant, there are 12 elements (H, Li, B, C, N, O, Mg, Si, S, Cl, Br and Tl) which show substantial variation in their isotopic compositions, depending on the source of the element. Consider, for example, hydrogen. As we have seen, this element has three isotopes, ^{1}H, ^{2}H and ^{3}H, the last of which we will neglect in this discussion because of its negligible abundance. If we analysed samples of atmospheric methane and natural gas, we would find that the proportion of the ^{2}H isotope in the hydrogen atoms of the former would be greater than that in the latter, and hence the average atomic mass of H in the two samples would be different. Therefore, instead of quoting a single value for the average atomic mass of hydrogen, a range of values of atomic mass is given [1.007 84 u, 1.008 11 u], which corresponds to the lowest and highest values measured in natural samples. Table 1.3 gives the range of atomic mass values for these 12 elements, together with the conventional atomic masses which can be used when the source of the sample is unknown.

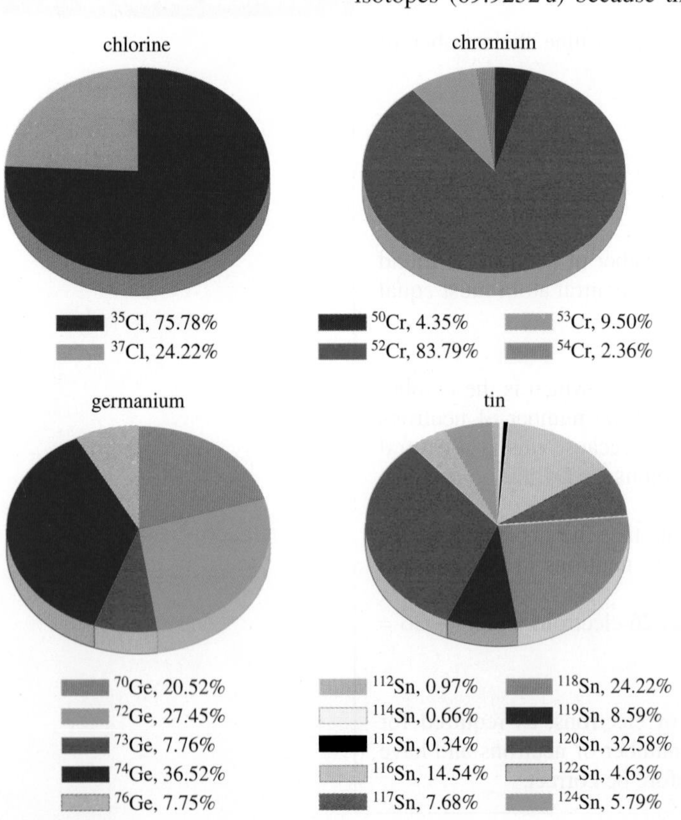

FIGURE 1.11 The natural abundances of the isotopes of chlorine, Cl, chromium, Cr, germanium, Ge, and tin, Sn, illustrate the diversity of isotopic distributions. The mass number and relative abundance of each isotope are indicated.

TABLE 1.3 Atomic mass ranges and conventional atomic masses for H, Li, B, C, N, O, Mg, Si, S, Cl, Br and Tl.

Element name	Symbol	Atomic number	Atomic mass range (u)	Conventional atomic mass (u)
hydrogen	H	1	[1.007 84, 1.008 11]	1.008
lithium	Li	3	[6.938, 6.997]	6.94
boron	B	5	[10.806, 10.821]	10.81
carbon	C	6	[12.0096, 12.0116]	12.011
nitrogen	N	7	[14.006 43, 14.007 28]	14.007
oxygen	O	8	[15.999 03, 15.999 77]	15.999
magnesium	Mg	12	[24.304, 24.307]	24.305
silicon	Si	14	[28.084, 28.086]	28.085
sulfur	S	16	[32.059, 32.076]	32.06
chlorine	Cl	17	[35.446, 35.457]	35.45
bromine	Br	35	[79.901, 79.907]	79.904
thallium	Tl	81	[204.382, 204.385]	204.38

In this book, we will use the conventional atomic masses listed in table 1.3 in any calculations involving these 12 elements.

WORKED EXAMPLE 1.3

Calculating average atomic masses from isotopic abundances

Naturally occurring titanium, Ti, is a mixture of five isotopes and has the following isotopic composition.

$^{46}_{22}$Ti(8.25%), $^{47}_{22}$Ti(7.44%), $^{48}_{22}$Ti(73.72%), $^{49}_{22}$Ti(5.41%), $^{50}_{22}$Ti(5.18%)

The atomic masses of the isotopes are as follows.

$^{46}_{22}$Ti(45.952 6316 u), $^{47}_{22}$Ti(46.951 7631 u), $^{48}_{22}$Ti(47.947 9463 u), $^{49}_{22}$Ti(48.947 8700 u), $^{50}_{22}$Ti(49.944 7912 u)

Use this information to calculate the average atomic mass of titanium.

Analysis

In a sample containing many atoms of titanium, 8.25% of the total mass is contributed by atoms of ^{46}Ti, 7.44% by atoms of ^{47}Ti, 73.72% by atoms of ^{48}Ti, 5.41% by atoms of ^{49}Ti and 5.18% by atoms of ^{50}Ti. This means that, when we calculate the mass of the hypothetical 'average atom' of Ti, we have to weight it according to both the masses of the isotopes and their relative abundance. (Keep in mind, of course, that such an atom does not really exist. This is just a simple way to see how we can calculate the average atomic mass of this element.)

Solution

We will calculate 8.25% of the mass of an atom of ^{46}Ti, 7.44% of an atom of ^{47}Ti, 73.72% of an atom of ^{48}Ti, 5.41% of an atom of ^{49}Ti and 5.18% of an atom of ^{50}Ti. Adding these contributions gives the total mass of the 'average atom'. Therefore:

$$\begin{aligned}\text{average atomic mass of Ti} &= (0.0825 \times 45.952\,631\,6\text{ u}) + (0.0744 \times 46.951\,763\,1\text{ u}) \\ &+ (0.7372 \times 47.947\,946\,3\text{ u}) + (0.0541 \times 48.947\,870\,0\text{ u}) \\ &+ (0.0518 \times 49.944\,791\,2\text{ u}) \\ &= 47.867\text{ u}\end{aligned}$$

Is our answer reasonable?

By far the most abundant isotope is ^{48}Ti, so we would expect the average atomic mass to be close to the mass of this isotope. Our calculated average atomic mass, 47.867 u, is indeed just less than the mass of the ^{48}Ti isotope (47.947 946 3 u), because the lighter ^{46}Ti and ^{47}Ti isotopes are slightly more abundant than the heavier ^{49}Ti and ^{50}Ti isotopes. Hence, we can feel confident our answer is correct.

PRACTICE EXERCISE 1.3

Neon has three naturally occurring isotopes. ^{20}Ne has a mass of 19.9924 u and is 90.48% abundant, ^{21}Ne has a mass of 20.9938 u and is 0.27% abundant, and ^{22}Ne has a mass of 21.9914 u and is 9.25% abundant. Using these data, calculate the average atomic mass of neon.

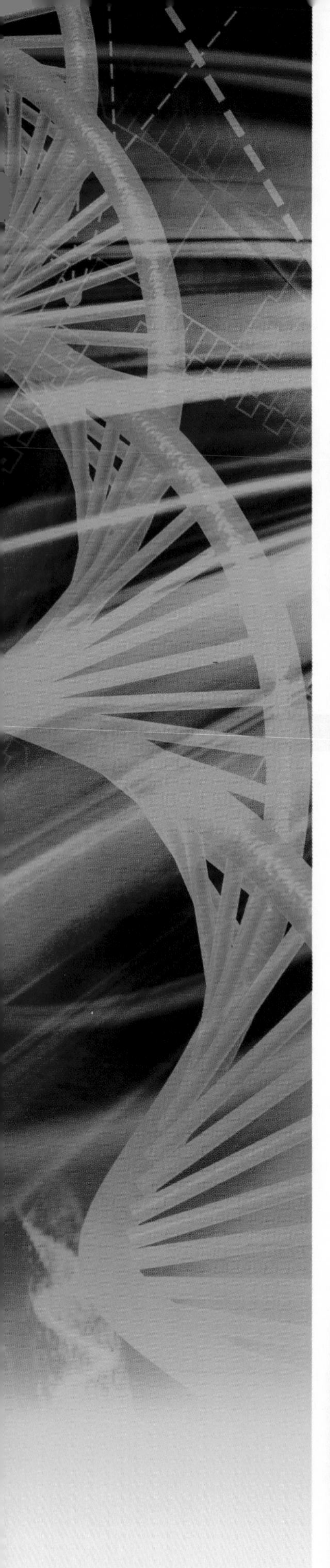

Chemical Connections

Saving lives with isotopes

In this chapter, we have learned that the atom comprises three constituents: protons, neutrons and electrons. The number of protons determines the identity of the atom, while electrons, as we shall see, hold atoms together in molecules. But what is the importance of the neutron in chemistry? The research of Australasian scientists working at the Open Pool Australian Light Water Reactor (OPAL) in Sydney provides an answer to this question. The OPAL reactor, pictured in figure 1.12, is the only nuclear reactor in Australasia and is used solely for scientific applications. It uses uranium enriched in the isotope ${}^{235}_{92}U$ as its nuclear fuel, and consumes about 30 kg of uranium per year. When a ${}^{235}_{92}U$ nucleus undergoes radioactive decay, free neutrons are produced and these are harnessed at OPAL for use in a variety of research studies.

FIGURE 1.12 A view of the OPAL reactor. The reactor core lies at the bottom of the circular section.

Neutrons interact with matter in specific ways, and analysis of the results of these interactions can provide extensive information about the matter being studied. Beams of neutrons from the OPAL reactor are used as the radiation source in a number of instruments designed to investigate a variety of properties of matter. These instruments include diffractometers designed to investigate the three-dimensional arrangement of atoms (particularly hydrogen atoms) in solids, measure stress in materials, study the magnetic properties of substances, and record the motions of atoms in chemical reactions on the millisecond timescale.

Neutrons can also be used to prepare macroscopic amounts of specific isotopes. As we have seen in worked example 1.2, some radioactive isotopes are useful in medical diagnosis and treatment. However, these often do not occur naturally in usable quantities, owing to their short lifetimes. For example, the ${}^{99m}_{43}Tc$ (Tc = technetium, m = metastable) isotope, which is used in over 80% of nuclear medicine imaging procedures, must be prepared via the decay of the ${}^{99}_{42}Mo$ (Mo = molybdenum) isotope, which is itself radioactive and has a half-life of 66 hours. The latter is prepared in the OPAL reactor by the irradiation of a uranium target with neutrons and, following purification, is then shipped to hospitals and other medical facilities throughout Australasia. Other short-lived medical isotopes produced by the OPAL reactor are ${}^{153}_{62}Sm$ (used to relieve pain caused by secondary cancers lodged in bone), ${}^{90}_{39}Y$ (used in cancer brachytherapy), ${}^{51}_{24}Cr$ (used to label red blood cells) and ${}^{131}_{53}I$ (used in the treatment of thyroid cancer and in diagnosis of abnormal liver function).

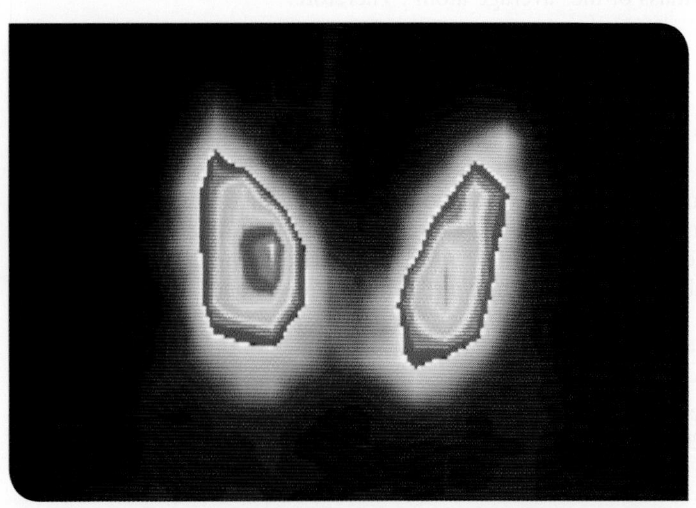

FIGURE 1.13 A scan of a healthy human thyroid gland. This scan is performed by injecting ${}^{99m}_{43}Tc$ into the bloodstream. The ${}^{99m}_{43}Tc$ is taken up by certain organs and the gamma rays emitted are used to produce an image.

1.4 The periodic table of the elements

We have already seen that the elements H, He and Li can be ordered on the basis of increasing atomic number ($Z = 1$, 2 and 3, respectively). If we continue such an ordering, we obtain the **periodic table of the elements**. The periodic table we use today is based primarily on the efforts of a Russian chemist, Dmitri Ivanovich Mendeleev (1834–1907, pictured in figure 1.14), and a German physicist, Julius Lothar Meyer (1830–1895). Working independently, these scientists developed similar periodic tables only a few months apart in 1869. Mendeleev is usually given the credit, however, because he published his version first.

FIGURE 1.14 Dmitri Ivanovich Mendeleev developed the periodic table.

The extraordinary thing about the work of Mendeleev and Meyer was that they knew nothing of the structure of the atom, so were unaware of the concept of atomic number, which is the basis of the modern periodic table. What they did know, however, were the atomic masses of many of the elements. Bear in mind also that not all of the elements had been discovered at this time. Mendeleev was preparing a chemistry textbook for his students at the University of St Petersburg and, looking for some pattern among the properties of the elements, he found that, when he arranged them in order of increasing atomic mass, similar chemical properties were repeated over and over at regular intervals. For instance, the elements lithium, Li, sodium, Na, potassium, K, rubidium, Rb, and caesium, Cs, are soft metals that react vigorously with water. Similarly, the elements that immediately follow each of these also constitute a set with similar chemical properties. Thus, beryllium, Be, follows lithium; magnesium, Mg, follows sodium; calcium, Ca, follows potassium; strontium, Sr, follows rubidium; and barium, Ba, follows caesium. All of these elements form compounds with oxygen having a 1 : 1 metal to oxygen ratio. Mendeleev used such observations to construct his periodic table, which is illustrated in figure 1.15 (overleaf).

At first glance, Mendeleev's original table looks little like the 'modern' table given in figure 1.16 (overleaf). However, a closer look reveals that the rows and columns have been interchanged. The elements in Mendeleev's table are arranged in order of increasing atomic mass. When the sequence is broken at the right places and stacked, the elements fall naturally into columns.

Mendeleev placed elements with similar properties in the same row even when this left occasional gaps in the table. For example, he placed arsenic, As, in the same row as phosphorus because they had similar chemical properties, even though this left gaps in other rows. In a stroke of genius, Mendeleev reasoned, correctly, that the elements that belonged in these gaps had simply not yet been discovered. In fact, on the basis of the location of these gaps, Mendeleev could predict, with astonishing accuracy, the properties of the yet-to-be-found elements, and his predictions helped serve as a guide in the search for them.

The elements tellurium, Te, and iodine, I (note that the German for 'iodine' is *jod*, which has the abbreviation J in Mendeleev's original table), caused Mendeleev some problems. According to the best estimates at that time, the atomic mass of tellurium was greater than that of iodine. Yet, if these elements were placed in the table according to their atomic masses, they would not fall into the proper rows required by their properties. Therefore, Mendeleev switched their order, believing that the atomic mass of tellurium had been incorrectly measured (it had not), and in so doing violated his ordering sequence based on atomic mass.

The table that Mendeleev developed is the basis of the one we use today, but one of the main differences is that Mendeleev's table lacks the elements helium, He, neon, Ne, argon, Ar, krypton, Kr, xenon, Xe, and radon, Rn. In Mendeleev's time, none of these elements had yet been discovered because they are relatively rare and because they have virtually no tendency to undergo chemical reactions. When these elements were finally discovered, beginning in 1894, another problem arose. Two more elements, argon, Ar, and potassium, K, did not fall into the rows required by their properties if they were placed in the table in the order required by their atomic masses. Another switch was necessary and another exception had been found. It became apparent that atomic mass was not the true basis for the periodic repetition of the properties of the elements. With Rutherford's discovery of the structure of the atom, it became apparent that the elements in the periodic table were arranged in order of increasing atomic *number*, not atomic *mass*, and when this was realised it became obvious that Te and I, and Ar and K, were in fact in the correct positions.

Ueber die Beziehungen der Eigenschaften zu den Atomgewichten der Elemente. Von D. Mendelejeff. – Ordnet man Elemente nach zunehmenden Atomgewichten in verticale Reihen so, dass die Horizontalreihen analoge Elemente enthalten, wieder nach zunehmendem Atomgewicht geordnet, so erhält man folgende Zusammenstellung, aus der sich einige allgemeinere Folgerungen ableiten lassen.

			Ti=50	Zr=90	?=180
			V=51	Nb=94	Ta=182
			Cr=52	Mo=96	W=186
			Mn=55	Rh=104,4	Pt=197,4
			Fe=56	Ru=104,4	Ir=198
			Ni=Co=59	Pd=106,6	Os=199
H=1			Cu=63,4	Ag=108	Hg=200
	Be=9,4	Mg=24	Zn=65,2	Cd=112	
	B=11	Al=27,4	?=68	Ur=116	Au=197?
	C=12	Si=28	?=70	Sn=118	
	N=14	P=31	As=75	Sb=122	Bi=210?
	O=16	S=32	Se=79,4	Te=128?	
	F=19	Cl=35,5	Br=80	J=127	
Li=7	Na=23	K=39	Rb=85,4	Cs=133	Tl=204
		Ca=40	Sr=87,6	Ba=137	Pb=207
		?=45	Ce=92		
		?Er=56	La=94		
		?Yt=60	Di=95		
		?In=75,6	Th=118?		

1. Die nach der Grösse des Atomgewichts geordneten Elemente zeigen eine stufenweise Abänderung in den Eigenschaften.
2. Chemisch-analoge Elemente haben entweder übereinstimmende Atomgewichte (Pt, Ir, Os), oder letztere nehmen gleichviel zu (K, Rb, Cs).
3. Das Anordnen nach den Atomgewichten entspricht der *Werthigkeit* der Elemente und bis zu einem gewissen Grade der Verschiedenheit im chemischen Verhalten, z. B. Li, Be, B, C, N, O, F.
4. Die in der Natur verbreitetsten Elemente haben *kleine* Atomgewichte

FIGURE 1.15 Mendeleev's original periodic table, taken from the German chemistry journal *Zeitschrift für Chemie*, 1869, 12, 405–6.

The modern periodic table

The periodic table in use today is shown in figure 1.16, and also on the inside front cover of this book. The horizontal rows are called **periods** and are numbered 1 to 7, while the vertical columns are called **groups** and are numbered 1 to 18. The elements are arranged in order of increasing atomic number across each period, and a new period begins after each group 18 element. The atomic masses are given (generally to four significant figures) below each chemical symbol. While the atomic mass usually increases with atomic number, you can see the exceptions we mentioned previously (Te and I, and Ar and K) as well as Co and Ni. While the isotopic composition and, therefore, the atomic masses of most elements are well established, there are some unstable elements that undergo spontaneous radioactive decay. Given that the isotopic composition of such elements cannot be known, it is usual to simply quote the mass number of the longest lived isotope of the element, and these are given in parentheses in the periodic table. Note that there are discontinuities in the periodic table between elements 56 and 72, and between elements 88 and 104, and these two sets of elements are given below the table itself. The elements from 57 to 71 are called the **lanthanoids** (or, less commonly, the **rare earth elements**). Elements 89 to 103 are called the **actinoids**. The lanthanoids and actinoids are generally situated below the rest of the periodic table, simply to save space and to make the table easier to read; note that the lanthanoid and actinoid elements are chemically distinct from the rest of the elements in the periodic table, and do not belong to any of the groups 1 to 18. The lanthanoids and actinoids are sometimes called the ***f*-block elements**, and a similar terminology is also used elsewhere in the table; elements in groups 1 and 2 are called the ***s*-block elements**, elements in groups 3 to 12 are called the ***d*-block elements**, and elements in groups 13 to 18 are called the ***p*-block elements**. As we will see, *s*, *p*, *d* and *f* refer to orbitals, particular regions in space in the atom where electrons have a high probability of being found. The *d*-block elements are also called **transition metals**.

metals | nonmetals | metalloids

Key: atomic number / Symbol / standard atomic mass

	1 alkali metals	2 alkaline earth metals	3	4	5	6	7	8	9	10	11	12	13	14	15 pnictogens	16 chalcogens	17 halogens	18 noble gases
1	1 H [1.008]																	2 He 4.003
2	3 Li [6.94]	4 Be 9.012											5 B [10.81]	6 C [12.01]	7 N [14.01]	8 O [16.00]	9 F 19.00	10 Ne 20.18
3	11 Na 22.99	12 Mg [24.31]											13 Al 26.98	14 Si [28.09]	15 P 30.97	16 S [32.06]	17 Cl [35.45]	18 Ar 39.95
4	19 K 39.10	20 Ca 40.08	21 Sc 44.96	22 Ti 47.87	23 V 50.94	24 Cr 52.00	25 Mn 54.94	26 Fe 55.85	27 Co 58.93	28 Ni 58.69	29 Cu 63.55	30 Zn 65.38(2)	31 Ga 69.72	32 Ge 72.63	33 As 74.92	34 Se 78.97	35 Br [79.90]	36 Kr 83.80
5	37 Rb 85.47	38 Sr 87.62	39 Y 88.91	40 Zr 91.22	41 Nb 92.91	42 Mo 95.95	43 Tc 97.91	44 Ru 101.1	45 Rh 102.9	46 Pd 106.4	47 Ag 107.9	48 Cd 112.4	49 In 114.8	50 Sn 118.7	51 Sb 121.8	52 Te 127.6	53 I 126.9	54 Xe 131.3
6	55 Cs 132.9	56 Ba 137.3	57–71	72 Hf 178.5	73 Ta 180.9	74 W 183.8	75 Re 186.2	76 Os 190.2	77 Ir 192.2	78 Pt 195.1	79 Au 197.0	80 Hg 200.6	81 Tl [204.4]	82 Pb 207.2	83 Bi 209.0	84 Po (209.0)	85 At (210.0)	86 Rn (222.0)
7	87 Fr (223.0)	88 Ra (226.0)	89–103	104 Rf (265.1)	105 Db (268.1)	106 Sg (271.1)	107 Bh (270.1)	108 Hs (277.2)	109 Mt (276.2)	110 Ds (281.2)	111 Rg (280.2)	112 Cn (285.2)	113 Uut (284.2)	114 Fl (289.2)	115 Uup (288.2)	116 Lv (293.2)	117 Uus (294.2)	118 Uuo (294.2)

lanthanoid series	57 La 138.9	58 Ce 140.1	59 Pr 140.9	60 Nd 144.2	61 Pm 144.9	62 Sm 150.4	63 Eu 152.0	64 Gd 157.3	65 Tb 158.9	66 Dy 162.5	67 Ho 164.9	68 Er 167.3	69 Tm 168.9	70 Yb 173.1	71 Lu 175.0
actinoid series	89 Ac 227.0	90 Th 232.0	91 Pa 231.0	92 U 238.0	93 Np 237.0	94 Pu 244.1	95 Am 243.1	96 Cm 247.1	97 Bk 247.1	98 Cf 251.1	99 Es 252.1	100 Fm 257.1	101 Md 258.1	102 No 259.1	103 Lr 262.1

FIGURE 1.16 The periodic table of the elements. At the time of writing, 118 elements were known. Atomic masses in parentheses refer to the longest-lived isotope of the element.

Individual groups within the periodic table are also known by particular names, although this practice is less prevalent than in the past. Group 1 elements are called **alkali metals**, group 2 elements are called **alkaline earth metals**, group 15 elements are called **pnictogens**, group 16 elements are called **chalcogens**, group 17 elements are called **halogens** and group 18 elements are called **noble gases**. Of these, only the terms halogens and noble gases are in common usage.

All elements on the periodic table belong to one of three categories — metals, nonmetals and metalloids — and the groupings are shown by the different shadings on the periodic table in figure 1.16. **Metals** are generally good conductors of heat and electricity, are malleable (can be beaten into a thin sheet) and ductile (can be drawn out into a wire), and have the usual metallic lustre. Elements that do not have these characteristics are called **nonmetals**, and the majority of these are gases at room temperature and pressure. The properties of **metalloids** lie somewhere between the metals and nonmetals. The most notable property of these elements is the fact that they tend to be semiconductors, and metalloids such as silicon, Si, and germanium, Ge, have therefore found wide use in silicon chips and transistors. Note that the classification of the recently prepared elements 116, 117 and 118 is somewhat arbitrary, as weighable quantities of these have not yet been obtained.

Naming the elements

All of the elements in the periodic table have one-, two- or three-letter abbreviations of their names. The abbreviations of many elements are simply the first one or two letters of their names (e.g. carbon, C, oxygen, O, lithium, Li) but there are quite a number of elements for which the derivation of the abbreviation is not quite so obvious: for example, potassium, K, tin, Sn, lead, Pb, and iron, Fe. Such apparent anomalies occur because of the way that the elements were originally named. When a new element is discovered, the discoverer usually gets to suggest a name for the element, which is then ratified by IUPAC, the International Union of Pure and Applied Chemistry.

Of all the elements on the periodic table, C, S, Fe, Cu, As, Ag, Sn, Sb, Au, Hg, Pb and Bi were known to ancient civilisations so the date of their 'discovery' is not known. Of these, Fe, Cu, Ag, Sn, Sb, Au, Hg and Pb are abbreviations for the Latin names *ferrum*, *cuprum*, *argentum*, *stannum*, *stibium*, *aurum*, *hydrargyrum* and *plumbum*. The earliest known discovery of an element was that of phosphorus, P. It was isolated in 1669 by Hennig Brand from the distillation of urine (he was apparently trying to make silver or gold — unsuccessfully, of course!) and was named after the

Greek word *phosphoros*, meaning 'bringer of light', as the element glows in the dark. Elements have been named after countries (germanium, francium, americium, polonium) and even after the places they were first discovered; the Swedish town of Ytterby has the distinction of having four elements (erbium, Er, ytterbium, Yb, yttrium, Y, and terbium, Tb) named after it, as these were first found in deposits close to the town. Surprisingly few elements have been named after people; at present, only 16 people have been immortalised on the periodic table, and they are listed in table 1.4. It is interesting to note that none of the people listed discovered the element named after them.

TABLE 1.4 People after whom elements have been named.

Name	Brief biography	Element named
Vasilii Yefrafovich von Samarski-Bykhovets (1803–1870)	Chief of staff of the Russian Corps of Mining Engineers	samarium, Sm (element 62)
Johan Gadolin (1760–1852)	Finnish chemist; first person to isolate a lanthanoid element	gadolinium, Gd (element 64)
Pierre (1859–1906) and Marie (1867–1934) Curie	Husband and wife scientific team; Pierre French and Marie Polish by birth; jointly won the Nobel Prize in physics in 1903	curium, Cm (element 96)
Albert Einstein (1879–1955)	Most famous scientist of the twentieth century, if not all time; German by birth; won the Nobel Prize in physics in 1921	einsteinium, Es (element 99)
Enrico Fermi (1901–1954)	Italian physicist; made great advances in the study of nuclear reactions; won the Nobel Prize in physics in 1938	fermium, Fm (element 100)
Dmitri Mendeleev (1834–1907)	Russian chemist; renowned for the development of the periodic table	mendelevium, Md (element 101)
Alfred Nobel (1833–1896)	Swedish inventor of dynamite and patron of the Nobel Prizes	nobelium, No (element 102)
Ernest Lawrence (1901–1958)	American inventor of the cyclotron; won the Nobel Prize in physics in 1939	lawrencium, Lr (element 103)
Ernest Rutherford (1871–1937)	New Zealand physicist/chemist; made seminal contributions to understanding the structure of the atom; won the Nobel Prize in chemistry in 1908	rutherfordium, Rf (element 104)
Glenn Seaborg (1912–1999)	American chemist; first prepared many of the elements beyond uranium in the periodic table; won the Nobel Prize in chemistry in 1951	seaborgium, Sg (element 106)
Niels Bohr (1885–1962)	Danish physicist; studied electronic energy levels within atoms, which aided our understanding of the atom; won the Nobel Prize in physics in 1922	bohrium, Bh (element 107)
Lise Meitner (1878–1968)	Austrian physicist; made fundamental discoveries concerning nuclear fission; controversially never awarded a Nobel Prize	meitnerium, Mt (element 109)
Wilhelm Röntgen (1845–1923)	German physicist; discoverer of X-rays; winner of the inaugural Nobel Prize in physics in 1901	röntgenium, Rg (element 111)
Nicolaus Copernicus (1473–1543)	Polish astronomer; proposed that the Sun, rather than the Earth, was the centre of the solar system	copernicium, Cn (element 112)
Georgii Flerov (1913–1990)	Russian physicist; made significant discoveries in the syntheses of transuranium elements	flerovium (element 114)

1.5 Electrons in atoms

While we have touched briefly on the concept of electrons, we have to this point concentrated primarily on the nucleus of the atom and the way in which the number of protons in the nucleus determines the chemical identity of the atom. However, many of the chemical properties of an atom and, most importantly, its chemical reactivity are determined not by the nucleus but by the electrons.

One of the most interesting things about electrons is that we cannot really say *exactly* where they are at any particular time, so we usually talk about their *most probable* locations. Electrons occupy regions of space called **orbitals** in atoms. Each orbital has a characteristic electron distribution and energy. For example, the lowest energy situation for a hydrogen atom, the **ground state**, occurs when the single electron occupies an orbital in which its most probable distance from the nucleus is 5.29×10^{-11} m. This orbital has a spherical electron distribution. If the ground state hydrogen atom absorbs a specific amount of energy, the electron can be promoted to a higher energy orbital to form an **excited state** in which the electron lies, on average, further from the nucleus. Such a process is called an **electronic transition**, and the electron distribution in the higher energy orbital is dumbbell shaped. Similarly, the electron in an excited-state hydrogen atom can move to a lower energy orbital through the emission of energy, often in the form of light. Indeed, as we will see in chapter 4, such processes are the basis behind both neon and sodium vapour lights. Orbitals have definite energies, so the energy of any electron is dictated by the energy of the orbital it occupies; therefore, an electron in an atom can have only certain well-defined energies. This is a fundamental principle of the science of quantum mechanics called **quantisation**, a phenomenon first proposed by the German physicist Max Planck (1858–1947, Nobel Prize in physics, 1918) in 1900. We will learn more about the quantisation of energy in chapter 4.

Electrons have a single negative charge, and the overall charge on any chemical species is determined by the number of electrons relative to the number of protons; for example, the peroxide ion, O_2^{2-}, has a 2– charge because there are two more electrons than protons in the ion. Similarly, the Li^+ ion contains three protons and two electrons, so it has a single positive charge. In addition to their negative charge, all electrons have an intrinsic property called **spin**. This can have one of two values, which are commonly called 'spin up' and 'spin down' and are often depicted as follows.

↑ (spin up)
↓ (spin down)

Each orbital within an atom can contain a maximum of two electrons, one of which must be spin up and the other spin down.

Chemists are interested in electrons because they constitute the chemical bonds that hold atoms together in molecules. Covalent chemical bonds usually consist of one, two or three pairs of electrons shared between atoms, each pair containing electrons of opposite spin. For a molecule to undergo a chemical reaction, these bonds must usually be broken and new ones made; this requires a reorganisation of the electron pairs between the reactant and product molecules, and the ease with which this can be done determines how fast the reaction occurs. Reactions in which one or more electrons are transferred between chemical species are also known; such reactions, known as **redox reactions**, are important in a huge number of chemical and biochemical processes; in fact, as you are reading this, iron ions and oxygen molecules are busy exchanging electrons in your blood to transport oxygen around your body.

Because of their importance in both chemical structure and chemical reactivity, electrons occupy a central place in chemistry. In the remaining chapters of this book, we will learn more of the properties of atoms and molecules that are predominantly dictated by electrons.

We have learned much about the atom in the years since Rutherford's seminal experiment. Indeed, we have detailed only the very basics of atomic structure in the preceding pages; later chapters will outline some of the amazing complexity of the atom. For the moment, it is sufficient for you to appreciate that the atom is composed of a positively charged central nucleus containing protons and neutrons, which is surrounded by negatively charged electrons that can undergo transitions only between well-defined energy levels. And with only 118 different types of these building blocks, we can construct the universe.

SUMMARY

LO 1 Atoms, molecules, ions, elements and compounds

Atoms are a fundamental building block of all matter. Uncharged collections of atoms bonded together in a definite structure are called molecules. These are held together by covalent bonds that share electrons between adjacent atoms. Ions are charged chemical species that may be derived from both atoms and molecules. Cations are positively charged, while anions are negatively charged. Elements comprise only a single type of atom, while compounds are made up of two or more chemical elements. All of these different chemical entities can be involved as reactants in chemical reactions, in which they are transformed to products.

LO 2 The atomic theory

All matter is composed of atoms. The existence of atoms was proposed on the basis of:

- the law of conservation of mass — mass is conserved in chemical reactions.
- the law of definite proportions — elements are combined in the same proportions by mass in any particular compound.
- the law of multiple proportions — when two elements form more than one compound, the different masses of one element that combine with the same mass of the other element are in the ratio of small whole numbers.

Dalton's atomic theory was the first to propose the existence of atoms on the basis of scientific observations. The basic tenets of his theory are as follows.

1. Matter consists of tiny particles called atoms.
2. Atoms are indestructible. In chemical reactions, the atoms rearrange but they do not themselves break apart.
3. In any sample of a pure element, all the atoms are identical in mass and other properties.
4. The atoms of different elements differ in mass and other properties.
5. When atoms of different elements combine to form a given compound, the constituent atoms in the compound are always present in the same fixed numerical ratio.

Dalton's theory allows us to use chemical equations, in which reactants and products are separated by an arrow, to describe chemical reactions. Such equations are balanced when they contain the same number of each type of atom on each side of the arrow.

Modern apparatus enables us to 'see' individual atoms, and atomic theory is now atomic fact.

LO 3 The structure of the atom

Although Dalton proposed the atom to be indivisible, experiments in the late nineteenth century showed this was not the case. The negatively charged electron was the first subatomic particle to be discovered, while Rutherford's gold foil experiment, in which a thin gold sheet was bombarded with alpha particles, gave evidence for a small, positively charged nucleus. The positive charge is due to subatomic particles called protons, and the number of these in the nucleus determines the identity of the atom in question. The third component of the atom, the neutron, was predicted by Rutherford and found by Chadwick. The atom thus comprises three subatomic particles, the electron, proton and neutron, the latter two collectively being called nucleons. Each type of atom is designated by a chemical symbol, which is determined by its atomic number (Z), the number of protons in the nucleus. The mass number (A) is equal to the number of protons plus the number of neutrons in the nucleus. The terminology used to depict an atom of any element X is ${}^{A}_{Z}X$. All atoms with the same Z are of the same element; however, atoms of the same element can differ in the number of neutrons in the nucleus, and this gives rise to isotopes. Isotopes can be either radioactive (i.e. they decay spontaneously) or stable. A radioactive nucleus is called a radionuclide, while a nuclide is the name given to any atomic nucleus. We can measure atomic mass in atomic mass units (u), where $1\ u = 1.660\,54 \times 10^{-27}$ kg, and is equal to $\frac{1}{12}$ of the mass of one atom of ^{12}C. The atomic mass of any sample of atoms is the weighted average of the masses of the isotopes present in the sample.

LO 4 The periodic table of the elements

The periodic table of the elements contains the 118 known elements arranged in order of increasing atomic number, and was developed by both Mendeleev and Meyer. The horizontal rows are called *periods* and the vertical columns *groups*. Elements in the same group tend to have similar chemical properties. The periodic table is divided into sections according to the electron configuration of the elements, namely the *s*-block elements, the *p*-block elements, the *d*-block elements and the *f*-block elements. The *f*-block elements are divided into the lanthanoids (also sometimes called the rare earth elements) and the actinoids, while the *d*-block elements are also called the transition metals. The elements in certain groups in the periodic table have special names: group 1 elements are called alkali metals, group 2 elements are called alkaline earth metals, group 15 elements are called pnictogens, group 16 elements are called chalcogens, group 17 elements are called halogens and group 18 elements are called noble gases. The elements of the periodic table can be classified as metals, nonmetals, or metalloids.

LO 5 Electrons in atoms

Electrons occupy regions of space called orbitals. The lowest energy arrangement of electrons in the orbitals of an atom is called the ground state. Electrons can be promoted to higher energy orbitals by absorption of energy to give excited states; conversely, electrons in higher energy orbitals can move to lower energy orbitals with the emission of energy, often as light. Such processes are called electronic transitions. The energies of electrons in atoms are determined by the energies of the orbitals within the atom, so electrons in atoms can have only certain well-defined energies. This is called quantisation, a fundamental principle of quantum mechanics. Electrons have a single negative charge, and one of two possible spins. An orbital in an atom can hold a maximum of two electrons, which must be of opposite spin. Covalent bonds comprise one, two or three pairs of electrons. Chemical reactions often involve reorganising these electrons in bond-making and bond-breaking processes. Redox reactions involve the transfer of one or more electrons between chemical species.

KEY CONCEPTS AND EQUATIONS

The law of conservation of mass
(section 1.2)

The total mass of reactants present before a reaction starts equals the total mass of products after the reaction is finished. We can use this law to check whether we have accounted for all the substances formed in a reaction.

The law of definite proportions
(section 1.2)

If we know the mass ratio of the elements in one sample of a compound, the ratio will be the same in a different sample of the same compound.

The law of multiple proportions
(section 1.2)

In different compounds containing the same two elements, the different masses of one element that combine with the same mass of the other element are in a ratio of small whole numbers.

Atomic mass
(section 1.3)

This is used to determine the mass of any atom relative to $\frac{1}{12}$ that of the ^{12}C isotope.

Periodic table of the elements
(section 1.4)

This is a table of the chemical elements arranged in order of increasing atomic number. We can use the periodic table to figure out whether a particular element is a metal, nonmetal or metalloid, predict its chemical reactivity, calculate its number of protons and electrons, obtain its atomic mass and so on. In fact, all of chemistry is contained within the periodic table.

KEY TERMS

REVIEW QUESTIONS

LO 1 Atoms, molecules, ions, elements and compounds

1.1 Define the following terms: atom, covalent bond, ion, cation, anion, element, compound, reactant, chemical reaction, product.

LO 2 The atomic theory

1.2 Name and state the three laws of chemical combination discussed in this chapter.

1.3 Balanced chemical equations have the same number of atoms of each type on either side of the arrow. Which of the three laws discussed in this chapter require this to be the case, and why?

1.4 Which of the laws of chemical combination is used to define the term 'compound'?

1.5 Why does the law of multiple proportions support the existence of atoms?

LO 3 The structure of the atom

1.6 Write the symbol for the isotope that forms the basis of the atomic mass scale. What is the mass of this atom expressed in atomic mass units?

1.7 Why did most of the alpha particles in Rutherford's gold foil experiment pass straight through the foil undeflected? What was the force which resulted in the deflection of some of the alpha particles?

1.8 Where in an atom is nearly all of its mass concentrated? Explain your answer in terms of the particles that contribute to this mass.

1.9 When we calculate the mass of an atom, we generally neglect any contribution to this from electrons in the atom. Why is this?

1.10 Define the terms 'atomic number' and 'mass number'.

1.11 What is an isotope? Why do isotopes of an element exhibit similar chemical behaviour?

1.12 Consider the symbol ${}^{A}_{Z}X$, where X stands for the chemical symbol for an element. What information is given by (a) A and (b) Z?

1.13 Write the symbols (mass number, atomic number and chemical symbol) of the following isotopes. (Use the table of atomic numbers printed inside the front cover, as needed.)
(a) an isotope of silver which contains 60 neutrons
(b) an isotope of tantalum which contains 108 neutrons
(c) an isotope of dysprosium which contains 96 neutrons
(d) an isotope of oxygen which contains 6 neutrons

LO 4 The periodic table of the elements

1.14 What is the chemical symbol for each of the following elements? (a) chlorine (b) sulfur (c) iron (d) silver (e) sodium (f) phosphorus (g) iodine (h) copper (i) mercury (j) calcium

1.15 What is the name of each of the following elements? (a) Li (b) Au (c) U (d) As (e) Co (f) Br (g) Pt (h) B (i) Ne (j) Be

1.16 On what basis did Mendeleev construct his periodic table? On what basis are the elements arranged in the modern periodic table?

1.17 True or false? The elements in the same period of the periodic table exhibit similar chemical behaviour.

1.18 Why did Mendeleev leave gaps in his periodic table?

1.19 Why does the atomic number of an element allow better prediction of its chemical properties than does its mass number?

1.20 On the basis of their positions in the periodic table, why is it not surprising that ${}^{90}Sr$, a dangerous radioactive isotope of strontium, replaces calcium in newly formed bones?

1.21 When nickel-containing ores are refined, commercial amounts of palladium and platinum are also often obtained. Why is this not unexpected?

1.22 Why would you reasonably expect cadmium to be a contaminant in zinc but not in silver?

1.23 Scientists can produce new heavy elements, with atomic numbers greater than 92. Explain why it is very unlikely that a completely new element with an atomic number of less than 92 will ever be discovered.

1.24 In each of the following sets of elements, state which fits the description in parentheses.
(a) Ce, Hg, Si, O, I (halogen)
(b) Pb, W, Ca, Cs, P (transition metal)
(c) Xe, Se, H, Sr, Zr (noble gas)
(d) Th, Sm, Ba, F, Sb (lanthanoid element)
(e) Ho, Mn, Pu, At, Na (actinoid element)

1.25 Calculations show that a rod of platinum 10 cm long and 1 cm in diameter can theoretically be drawn out into a wire nearly 28 000 km long. What is this property of metals called?

1.26 Gold can be hammered into sheets so thin that some light can pass through them. Which property of gold allows such thin sheets to be made?

1.27 Name the elements that exist as diatomic gases (gases that exist as molecules containing two atoms) at 25 °C (room temperature) and 1.013×10^5 Pa (atmospheric pressure).

1.28 Which two elements exist as liquids at room temperature and atmospheric pressure?

1.29 Weighable amounts of the very heavy elements, with atomic numbers greater than 112, have not yet been prepared, and so their bulk physical properties are as yet unknown. Which element, Fl (element 114) or Lv (element 116), would be more likely to exhibit properties of a metalloid?

1.30 Sketch the shape of the periodic table and mark off those areas where we find (a) metals, (b) nonmetals and (c) metalloids.

1.31 What is the name given to the most probable region of space in which an electron might be found?

LO 5 Electrons in atoms

1.32 When electrons of opposite spins occupy an orbital, we say that their spins are paired. Molecules with odd numbers of electrons, therefore, cannot have all of the electron spins paired, and we say that they have unpaired spins. Which of the following molecules *must* have unpaired spins: N_2, F_2, CO, NO, NO_2?

1.33 An atom in an excited state has a higher energy than the same atom in its ground state. Given that neon lights involve neon atoms in excited states, suggest a method by which the excited state atoms might lose the excess energy they have.

1.34 Quantisation is very important on the atomic scale but, in the large scale of our everyday lives, we barely notice it. Why do you think this might be so?

REVIEW PROBLEMS

1.35 Methane is the simplest of a series of compounds collectively called the alkanes, which consist of only carbon and hydrogen and have the general chemical formula C_nH_{2n+2}. For every 1.000 g of C in a sample of methane there is 0.336 g of hydrogen. Which of the following compositions corresponds to that of methane? **LO 2**

(a) 7.317 g carbon, 8.295 g hydrogen
(b) 2.618 g carbon, 5.228 g hydrogen
(c) 3.884 g carbon, 1.305 g hydrogen
(d) 6.911 g carbon, 4.003 g hydrogen
(e) 9.352 g carbon, 7.417 g hydrogen

1.36 One of the substances used to melt ice on footpaths and roads in cold climates is calcium chloride. In this compound, calcium and chlorine are combined in a ratio of 1.00 g of calcium to 1.77 g of chlorine. Which of the following calcium–chlorine mixtures will produce calcium chloride with no calcium or chlorine left over after the reaction is complete? **LO 2**

(a) 3.65 g calcium, 4.13 g chlorine
(b) 0.856 g calcium, 1.56 g chlorine
(c) 2.45 g calcium, 4.57 g chlorine
(d) 1.35 g calcium, 2.39 g chlorine
(e) 5.64 g calcium, 9.12 g chlorine

1.37 The very dense liquid, carbon tetrachloride, was once widely used in dry-cleaning, until it was discovered that it was involved in the destruction of the ozone layer. Any sample of carbon tetrachloride is composed of carbon and chlorine in the mass ratio of 1.00 : 11.8. If a sample of carbon tetrachloride contains 8.35 g of carbon, how much chlorine does it contain? **LO 2**

1.38 A compound of phosphorus and chlorine used in the manufacture of a flame retardant treatment for fabrics contains 1.20 g of phosphorus for every 4.12 g of chlorine. Suppose a sample of this compound contains 6.22 g of chlorine. What mass of phosphorus does it contain? **LO 2**

1.39 With reference to question 1.37, if 5.00 g of carbon combined completely with chlorine to form carbon tetrachloride, what mass of carbon tetrachloride would be formed? **LO 2**

1.40 Refer to the data about the phosphorus–chlorine compound in question 1.38. If 12.5 g of phosphorus combined completely with chlorine to form this compound, what mass of the compound would be formed? **LO 2**

1.41 Combustion of any carbon compound in air forms two major compounds containing only carbon and oxygen. Molecules of one of these compounds contain one atom each of C and O, with the mass ratio of C to O being 1 : 1.332. Molecules of the second compound of carbon and oxygen contain one atom of C and two atoms of O. What mass of oxygen would be combined with each 1.000 g of carbon in this compound? **LO 2**

1.42 Tin forms two compounds with chlorine. In one of them (compound 1), there are two Cl atoms for each Sn atom; in the other (compound 2), there are four Cl atoms for each Sn atom. When combined with the same mass of tin, what would be the ratio of the masses of chlorine in the two compounds? In compound 1, 0.597 g of chlorine is combined with each 1.000 g of tin. What mass of chlorine would be combined with 1.000 g of tin in compound 2? **LO 2**

1.43 The atomic mass unit is defined in terms of the mass of the ^{12}C atom. Given that 1 atomic mass unit corresponds to $1.660\,54 \times 10^{-24}$ g, calculate the mass of 1 atom of ^{12}C in grams. **LO 3**

1.44 Use the mass corresponding to the atomic mass unit given in question 1.43 to calculate the mass of 1 atom of sodium. **LO 3**

1.45 One of the earliest anaesthetics — its first recorded use was in 1844 — was a compound called nitrous oxide, or, more commonly, laughing gas. Molecules of nitrous oxide are composed of two atoms of nitrogen and one atom of oxygen. In this compound, 1.7513 g of nitrogen is combined with 1.0000 g of O. If the atomic mass of O is 16.00 u, use the above information to calculate the atomic mass of the nitrogen. **LO 3**

1.46 Element *X* forms a compound with oxygen in which there are two atoms of *X* for every three atoms of O. In this compound, 1.125 g of *X* is combined with 1.000 g of oxygen. Use the average atomic mass of oxygen to calculate the average atomic mass of *X*. Use your calculated atomic mass to identify element *X*. **LO 3**

1.47 If an atom of ^{12}C had been assigned a relative mass of 24.0000 u, determine the average atomic mass of hydrogen relative to this mass. **LO 3**

1.48 An atom of ^{109}Ag has a mass that is 9.0754 times that of a ^{12}C atom. What is the atomic mass of this isotope of silver expressed in atomic mass units? **LO 3**

1.49 Antimony (Sb) has two stable isotopes. ^{121}Sb has a mass of 120.9038 u and an abundance of 57.36%, while ^{123}Sb has a mass of 122.9042 u and an abundance of 42.64%. Use these data to calculate the average atomic mass of antimony. **LO 3**

1.50 Give the numbers of neutrons, protons and electrons in the atoms of each of the following isotopes. (Use the periodic table printed inside the front cover, as needed.) **LO 3,4**

(a) ^{226}Ra (b) ^{14}C (c) $^{206}_{82}Pb$ (d) $^{23}_{11}Na$

1.51 Give the numbers of protons, electrons and neutrons present in the atoms of each of the following isotopes. (Consult the periodic table printed inside the front cover as necessary.) **LO 3,4**

(a) ^{90}Sr (b) ^{60}Co (c) ^{30}P (d) ^{30}Si

1.52 From the elements Ne, Cs, Sr, Br, Co, Pu, In and O, choose one that fits each of the following descriptions. **LO 4**
(a) a group 2 metal
(b) an element with properties similar to those of aluminium
(c) a transition metal
(d) a noble gas
(e) an actinoid

1.53 Write the names and chemical symbols for three examples of each of the following. (a) nonmetals (b) alkaline earth metals (c) lanthanoids (d) chalcogens **LO 4**

ADDITIONAL EXERCISES

1.54 An atom of an element has 25 protons in its nucleus. **LO 2,3,4**
(a) Is the element a metal, a nonmetal or a metalloid?
(b) On the basis of the average atomic mass, write the symbol for the element's most abundant isotope.
(c) How many neutrons are in the isotope you described in (b)?
(d) How many electrons are in an atom of this element?
(e) How many times heavier than ^{12}C is the average atom of this element?

1.55 The elements *X* and *Y* form a compound having the formula XY_4. When these elements react, it is found that 1.00 g of *X* combines with 0.684 g of *Y*. When 1.00 g of *X* combines with 0.154 g of O, it forms a compound containing two atoms of O for each atom of *X*. Use these data to calculate the atomic masses of *X* and *Y*, and therefore determine the identity of the compound XY_4. **LO 2,3**

1.56 An iron nail is composed of four isotopes with the percentage abundances and atomic masses given in the following table. Calculate the average atomic mass of iron. **LO 3**

Isotope	Percentage abundance	Atomic mass (u)
^{54}Fe	5.80	53.9396
^{56}Fe	91.72	55.9349
^{57}Fe	2.20	56.9354
^{58}Fe	0.28	57.9333

1.57 Arsenic forms two compounds with oxygen. One of these contains three oxygen atoms for each two arsenic atoms, and has a mass ratio of arsenic to oxygen of 3.122 : 1.000. Another compound of arsenic and oxygen contains these elements in the ratio of 1.873 : 1.000. What is the probable formula of the second arsenic–oxygen compound? **LO 2**

1.58 One atomic mass unit corresponds to a mass of $1.660\,54 \times 10^{-24}$ g. Calculate the mass, in grams, of one atom of magnesium. What is the mass of one atom of iron, expressed in grams? Use these two answers to determine how many atoms of Mg are in 24.305 g of magnesium and how many atoms of Fe are in 55.847 g of iron. Compare your answers. What conclusions can you draw from the results of these calculations? Without actually performing any calculations, how many atoms do you think would be in 40.078 g of calcium? **LO 3**

1.59 The diameter of a typical atom is of the order of 10^{-10} m, while the diameter of its nucleus is approximately 10^{-15} m. Use these data to calculate the volume of both an atom and its nucleus. Hence, determine what fraction of the volume of a typical atom is occupied by its nucleus. The volume of a sphere is given by the formula $V = \frac{4}{3}\pi r^3$. **LO 3**

1.60 There are 11 elements in the periodic table that are known to exist as gases at room temperature: hydrogen, helium, nitrogen, oxygen, fluorine, neon, chlorine, argon, krypton, xenon and radon. Which of these gases exist as discrete atoms (*X*), and which exist as molecules (X_2)? **LO 1,4**

1.61 Given below are formulae of some ions. How many electrons are present in each of these ions? **LO 1,3,4,5**
(a) F^-
(b) O_2^-
(c) CO_3^{2-}
(d) Na^+
(e) PO_4^{3-}
(f) ClO_4^-

1.62 We will see in later chapters that the elements in groups 16 and 17 of the periodic table often form negatively charged ions in which they have the same number of electrons as the closest group 18 element. Conversely, the elements in groups 1 and 2 of the periodic table often form positively charged ions in which they have the same number of electrons as the closest group 18 element. With this in mind, predict the formulae of the simplest ions formed by the following elements. **LO 1,4,5**
(a) O
(b) F
(c) Li
(d) Be
(e) K
(f) Br
(g) I
(h) Sr

2 The language of chemistry

Chemistry, like any specialist discipline, has a language all its own. The language of chemistry is not merely about words; it is also about symbols, diagrams, formulae, abbreviations, equations and pictures. A working knowledge of this language is needed to be able to describe and depict chemical elements, compounds and processes, and to communicate and appreciate many of the important concepts of this science. In this chapter we will concentrate on three important areas within the broad field of the language of chemistry:

- measurement and units
- representations of molecules and reactions
- nomenclature, the way in which we name chemical species.

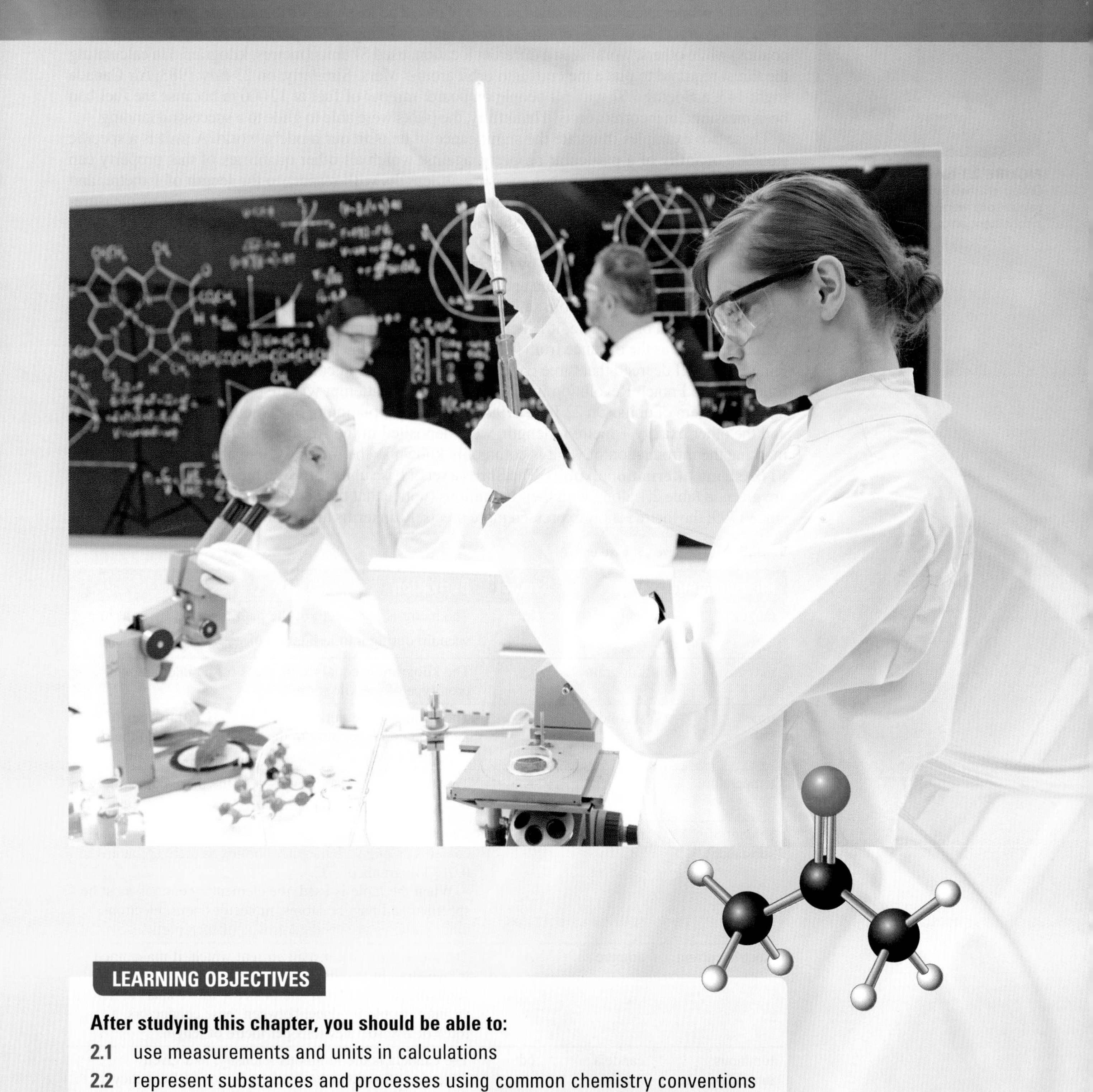

LEARNING OBJECTIVES

After studying this chapter, you should be able to:

2.1 use measurements and units in calculations

2.2 represent substances and processes using common chemistry conventions

2.3 apply basic chemical nomenclature.

2.1 Measurement

Chemistry is a science of measurement. Whenever we weigh a certain mass of a particular substance, evaluate the rate of a chemical reaction or determine the cell potential of an electrochemical cell, we are making measurements. Measurements always have a unit and always have an associated uncertainty. A measurement that is missing either of these is meaningless.

SI units

FIGURE 2.1 NASA's *Mars Climate Orbiter* crashed into Mars after calculations were made using inconsistent units.

On 23 September 1999, NASA, the National Aeronautics and Space Administration of the USA, lost contact with its *Mars Climate Orbiter* spacecraft (figure 2.1) as it was about to enter orbit around Mars. The orbiter eventually crashed into the planet. The loss was attributed to an error that resulted from an incorrect use of units; one team of scientists working on the project used imperial units (feet, pounds) while others, working in a different location, used SI units (metres, kilograms) in calculating the thrust required to place the craft into orbit around Mars. Similarly, on 23 July 1983, Air Canada flight 143, a Boeing 767 with 69 people on board, ran out of fuel at 12 000 m because the fuel had been measured in incorrect units. Thankfully, the pilots were able to glide to a successful landing.

These two examples illustrate the significance of units in our modern world. A **unit** is a specific standard quantity of a particular property, against which all other quantities of that property can be measured. For example, we can measure all lengths with respect to the length of 1 metre, and metre is therefore a unit of length. Humans have recognised the importance of units of measurement for thousands of years. One of the earliest documented units of measurement was the *cubit*, a unit of length used in ancient Egypt equal to the distance between the tip of the middle finger and the elbow. In fact, many units used by the ancient Egyptians were based on parts of the human body, such as the digit, the palm, the hand and the span. Obviously people come in all different shapes and sizes, so this system was far from satisfactory. Over time, a jumble of measuring systems and standards developed. For example, it is believed that in around the year 1100, England's King Henry I defined the *yard* as the distance from the end of his nose to the tip of his thumb, while 200 years later King Edward II decreed that three barleycorns laid end to end constituted an *inch*. It was in France, following the French Revolution in 1789, that a serious attempt was first made to develop an international system of units. On 22 June 1799, two pieces of platinum, one weighing exactly 1 kilogram and the other exactly 1 metre in length, were deposited in the Archives de la République in Paris, marking the introduction of what is commonly known as the metric system and what scientists call **SI (Système International) units**. The SI has a set of base units for seven measured quantities. These are given in table 2.1 along with their definitions (notice that the definition of the metre has changed since 1799; the metre is now expressed more precisely, in terms unimaginable 200 years ago).

TABLE 2.1 The seven SI base units.

Measurement	Unit	Symbol	Derivation
length	metre	m	The metre is the length of the path travelled by light in a vacuum during a time interval of $\frac{1}{299\,792\,458}$ of a second.
mass	kilogram	kg	The kilogram is equal to the mass of the international prototype of the kilogram.
time	second	s	The second is the duration of 9 192 631 770 periods of the radiation corresponding to the transition between the two hyperfine levels of the ground state of the ^{133}Cs atom.
temperature	kelvin	K	The kelvin is the fraction $\frac{1}{273.16}$ of the thermodynamic temperature of the triple point of water.
amount of substance	mole	mol	The mole is the amount of substance of a system which contains as many elementary entities as there are atoms in 0.012 kilograms of ^{12}C. When the mole is used, the elementary entities must be specified and may be atoms, molecules, ions, electrons, other particles or specified groups of such particles.
electric current	ampere	A	The ampere is the constant current which, if maintained in two straight parallel conductors of infinite length, of negligible cross-section and placed 1 metre apart in a vacuum, would produce, between these conductors, a force of 2×10^{-7} newtons per metre of length.
luminous intensity	candela	cd	The candela is the luminous intensity, in a given direction, of a monochromatic radiation of frequency 540×10^{12} hertz and that has a radiant intensity in that direction of $\frac{1}{683}$ watt per steradian.

A working knowledge of the units and symbols given in table 2.1 is essential in chemistry. However, the precise definitions are generally complicated, and it is not necessary to memorise their derivations, which are given for interest only. It should be noted that, at the time of writing, proposals have been made to redefine the mole, the ampere, the kelvin and the kilogram. The international prototype of the kilogram (which is kept in a vault at the International Bureau of Weights and Measures near Paris) has been found to have lost about 50 μg over time, and a definition which does not involve a physical object that is subject to change is thought to be preferable. Similarly, definitions of the mole, the ampere and the kelvin that give exact values to particular physical constants have been proposed. Such redefinitions will make no difference to the use of SI units in everyday life, and will have only a miniscule impact on all but the most precise scientific measurements.

We are all familiar with the SI units for length (**metre**, m), mass (**kilogram**, kg) and time (**second**, s). The SI unit of temperature (**kelvin**, K) has absolute zero (–273.15 °C), the coldest possible temperature, as its zero point. Temperature is more commonly measured in degrees Celsius — a temperature in Celsius may be converted to kelvin by adding 273.15. This means that the freezing point of water, 0 °C, is equal to 273.15 K, while the boiling point of water, 100 °C, is 373.15 K. Note that a temperature *difference* of 1 kelvin (we do not say 1 degree kelvin) is the same as a temperature difference of 1 degree Celsius.

The SI unit of amount of substance, a concept that will be discussed in chapter 3, is the **mole** (mol), while the unit of electrical current, the **ampere**, will be discussed when we study electrochemistry in chapter 12. The SI unit of luminous intensity (the brightness of the radiation emitted by phosphorescent and fluorescent compounds) is the **candela**.

The SI units for *any* physical quantity can be derived from these seven base units. For example, there is no SI base unit for area, but we know that to calculate the area of a rectangular room we multiply its length by its width (i.e. its length in one dimension by its length in the other). Therefore, the SI unit for area is derived by multiplying the unit for length (m) by the unit for length (m) to give m^2 (metre squared, or square metre). Table 2.2 lists several derived SI units that are important in chemistry.

TABLE 2.2 SI derived units commonly used in chemistry.

Measurement	Expression in terms of simpler quantities	Unit	Expression in terms of SI base units
area	length × width[(a)]	square metre	m^2
volume	length × width × height[(a)]	cubic metre	m^3
speed, velocity	distance[(b)]/time	metre per second	$m\ s^{-1}$
acceleration	velocity/time	metre per second squared	$m\ s^{-2}$
density	mass/volume	kilogram per cubic metre	$kg\ m^{-3}$
specific volume	volume/mass	cubic metre per kilogram	$m^3\ kg^{-1}$
force	mass × acceleration	newton, N	$1\ N = 1\ kg\ m\ s^{-2}$
pressure	force/area	pascal, Pa	$1\ Pa = 1\ N\ m^{-2} = 1\ kg\ m^{-1}\ s^{-2}$
energy	force × distance[(b)]	joule, J	$1\ J = 1\ N\ m = 1\ kg\ m^2\ s^{-2}$
power	energy/time	watt, W	$1\ W = 1\ J\ s^{-1} = 1\ kg\ m^2\ s^{-3}$
electric charge	electric current × time	coulomb, C	$1\ C = 1\ A\ s$
electric potential	energy/electric charge	volt, V	$1\ V = 1\ J\ C^{-1} = 1\ kg\ m^2\ s^{-3}\ A^{-1}$

(a) Width and height are simply length in different directions.
(b) Distance is another name for length.

The units for area and volume, square metre and cubic metre, respectively, are probably familiar to you, as are their symbols, m^2 and m^3. However, while you are also probably familiar with the unit of velocity, metre per second, you may perhaps wonder how the symbol $m\ s^{-1}$ is obtained. To obtain a velocity, we divide length by time, and we obtain the unit for velocity by dividing the unit of length (m) by the unit of time (s). We can write the symbol for the resulting unit in two ways: as m/s, or, as we shall do throughout this book, as m s^{-1}. Recall that x^{-1} is simply another way of writing $\frac{1}{x}$, and so, when we write m s^{-1}, this is the same as $m \times \frac{1}{s} = m/s$. Similarly, we obtain an acceleration by dividing velocity by time, and the unit is then derived from the unit of velocity (m s^{-1}) divided by the unit of time (s). The unit is therefore $\frac{m\,s^{-1}}{s} = m\,s^{-2}$ (metre per second squared), where s^{-2} is the same as $\frac{1}{s^2}$. This illustrates a very important concept that is used throughout this book when we perform calculations: *units undergo the same kinds of mathematical operations as the numbers to which they are attached.* We demonstrate this in worked example 2.1.

WORKED EXAMPLE 2.1

Deriving SI units

Heat capacity, which we will meet in chapter 8, is a measure of the heat required to raise the temperature of a particular substance by 1 K. We can obtain heat capacity values by dividing the heat provided by the temperature change obtained. What is the derived SI unit of heat capacity?

Analysis

As we obtain heat capacity values by dividing heat by temperature change, we will obtain the unit for heat capacity by dividing the units of heat by the units of temperature change.

Solution

The SI unit of heat is the joule (J) and the SI unit of temperature, and therefore temperature change, is the kelvin (K). Therefore, the unit of heat capacity is:

$$\frac{J}{K} = J\,K^{-1} \text{ (read as joules per kelvin)}$$

PRACTICE EXERCISE 2.1

The momentum *p* of an object of mass *m* and velocity *u* is given by the equation $p = mu$. What is the derived SI unit of momentum?

Sometimes it is inconvenient to use SI base or derived units. For example, we would not want to use metres to measure the tiny length of an atomic radius or the enormous distance to the nearest star. The SI system addresses this problem by the use of prefixes that divide or multiply the unit by a particular power of ten. For example, the prefix *centi* means the following unit is divided by 100, while the prefix *kilo* signifies that the following unit is multiplied by 1000. Table 2.3 lists the most common prefixes.

TABLE 2.3 Common prefixes for SI units.

Prefix	Symbol	Factor
tera	T	10^{12}
giga	G	10^{9}
mega	M	10^{6}
kilo	k	10^{3}
hecto	h	10^{2}
deca	da	10^{1}
deci	d	10^{-1}
centi	c	10^{-2}
milli	m	10^{-3}
micro	μ	10^{-6}
nano	n	10^{-9}
pico	p	10^{-12}
femto	f	10^{-15}

Each prefix has an associated symbol, which is placed immediately before the base unit. Therefore, km refers to kilometre, a unit of 10^3 m, while cm (centimetre) is a unit of 10^{-2} m.

WORKED EXAMPLE 2.2

Using SI prefixes

Objects that are smaller than about 2×10^{-7} m cannot be seen under an optical microscope. Could we use an optical microscope to observe a virus that is 20 nm long?

Analysis

We need to convert 20 nm to metres, and compare it with 2×10^{-7} m. The SI symbol n stands for *nano* and means 10^{-9}, so we will use this conversion factor.

Solution

We know from table 2.3 that 1 nm = 1×10^{-9} m, and therefore 20 nm = 20×10^{-9} m = 2×10^{-8} m. This is less than 2×10^{-7} m and hence we will not be able to see this virus using an optical microscope.

Is our answer reasonable?

We can check our result by converting the size of the smallest visible object (2×10^{-7} m) to nanometres so that it can be compared directly with the size of the virus; 2×10^{-7} m is equal to 200×10^{-9} m, or 200 nm. The virus is 20 nm long; the smallest object that can be seen under the microscope is 200 nm, so again we conclude that the virus will not be visible.

PRACTICE EXERCISE 2.2

1. Give the abbreviation for: (a) microgram (b) micrometre (c) nanosecond.
2. How many metres are in: (a) 1 nm (b) 1 cm (c) 1 pm?
3. What symbol is used to represent: (a) 10^{-2} g (b) 10^{6} m (c) 10^{-6} s?

Non-SI units

Although SI units have been adopted in nearly all countries (as of 2014, the USA is the only exception), there are many non-SI units in common usage throughout the world. For example, we saw on page 27 that temperatures are much more likely to be measured in degrees Celsius (or degrees Fahrenheit in the USA) than in the SI unit kelvin. Even chemists are not immune to this; we often use the litre, a unit of volume equal to 1000 cm^3, rather than the more correct (but inconveniently large) m^3. The millilitre (1 cm^3) is also often used. Various non-SI units of pressure (atmosphere, millimetre of mercury, torr, bar), rather than the SI unit pascal, are still regularly seen in the chemical literature. As long as use of non-SI units persists, it will occasionally be necessary for you to be able to convert between non-SI and SI units. To do this, we need to know the conversion factor between the units. Then we can obtain an equation to convert between the two. This is illustrated in worked example 2.3.

WORKED EXAMPLE 2.3

Unit conversions

The laws of cricket state that a cricket pitch is 22 yards in length. How long is a cricket pitch in metres given that 1 metre = 1.0936 yards?

Analysis

We are told that 1 metre corresponds to 1.0936 yards; in other words, there are 1.0936 yards in every metre or, to put it yet another way, 1.0936 yards per metre. Therefore, the conversion factor, in terms of yards per metre, is $\frac{1.0936 \text{ yard}}{1 \text{ metre}} = 1.0936 \text{ yard metre}^{-1}$.

However, we are asked to convert from yard to metre, and therefore need the conversion factor in terms of metres per yard (metre yard^{-1}). We obtain this knowing that:

$$\frac{1 \text{ metre}}{1.0936 \text{ yard}} = 0.9144 \text{ metre yard}^{-1}$$

We can now solve the problem by multiplying this conversion factor by 22 yards.

Solution

22 yard × 0.9144 metre yard^{-1} = 20.12 metre

Is our answer reasonable?

We know that 1 metre is slightly longer than 1 yard, so we would expect our numerical answer to be slightly less than 22, which it is. Notice also, as we will see on pages 30–1, that the units in our solution are consistent; yard × metre yard^{-1} = metre.

PRACTICE EXERCISE 2.3

Carry out the following conversions involving pressure units.

1. Convert 0.1 bar to Pa (1 Pa = 1×10^{-5} bar).
2. Convert 642 millimetres of mercury (mmHg) to Pa (1 Pa = 7.50×10^{-3} mmHg).
3. Convert 1.45×10^{4} Pa to atmosphere (atm) (1 Pa = 9.87×10^{-6} atm).

The use of non-SI units in measuring pressure and volume can cause particular problems in chemistry. One of the most important equations in chemistry, which we will meet in chapter 6, is the ideal gas equation, $pV = nRT$ where p = pressure, V = volume, n = amount of substance and T = temperature. R is called the gas constant and, providing that all the other variables in the equation are measured in SI or derived SI units (Pa, m^3, mol and K), it has a value of 8.314 J mol^{-1} K^{-1}. However, as stated above, pressures are often measured in atmosphere, millimetres of mercury, torr, or bar, and volumes are more usually measured in litres, rather than m^3. This means that the value of R depends on the units used, as shown in worked example 2.4.

WORKED EXAMPLE 2.4

The gas constant

Calculate the value and units of the gas constant R if 1.000 mol of a gas occupies a volume of 22.414 L at 1.000 atm pressure and 273.15 K.

Analysis

We are given the values of all the variables in the ideal gas equation except the gas constant. We must then rearrange the ideal gas equation to make R the subject, plug in all the values and solve. We obtain the units of the final answer from this equation also.

Solution

We rearrange the ideal gas equation $pV = nRT$, to make R the subject, by dividing both sides of the equation by nT. This gives:

$$R = \frac{pV}{nT}$$

We now plug in the given values to obtain the value and units of R.

$$\begin{aligned} R &= \frac{pV}{nT} \\ &= \frac{1.000 \text{ atm} \times 22.414 \text{ L}}{1.000 \text{ mol} \times 273.15 \text{ K}} \\ &= 8.206 \times 10^{-2} \text{ atm L K}^{-1} \text{ mol}^{-1} \end{aligned}$$

Is our answer reasonable?

You can see immediately that both the pressure and amount of substance have numerical values of 1.000, which can be ignored in the calculation. At a very rough approximation, we can see that the answer will lie somewhere in the region of $\frac{20}{200}$ (0.1) to $\frac{20}{300}$ (0.066), which it does.

PRACTICE EXERCISE 2.4

Determine the value and units of the gas constant R for the following systems of units.

(a) 1.000 mol of a gas occupies a volume of 22.414×10^{-3} m^3 at 1.000 atm pressure and 273.15 K.

(b) 1.000 mol of a gas occupies a volume of 22.414×10^{-3} m^3 at 1.013 25 bar pressure and 273.15 K.

(c) 1.000 mol of a gas occupies a volume of 22.414×10^{-3} m^3 at 760 mmHg pressure and 273.15 K.

Notice that the method of unit conversion described on p. 29 can also be used for conversion between SI units. For example, suppose we wanted to express 418 cm^3 in m^3; we would need the conversion factor in terms of m^3 cm^{-3}. From the unit prefix 'c', we know that:

$$1 \text{ cm} = 1 \times 10^{-2} \text{ m}$$

$$\text{Therefore, } 1 \text{ cm}^2 = (1 \times 10^{-2} \text{ m})^2 = 1 \times 10^{-4} \text{ m}^2$$
$$\text{and } 1 \text{ cm}^3 = (1 \times 10^{-2} \text{ m})^3 = 1 \times 10^{-6} \text{ m}^3$$

The conversion factor is therefore $\frac{1 \times 10^{-6} \text{ m}^3}{\text{cm}^3} = 1 \times 10^{-6} \text{ m}^3 \text{ cm}^{-3}$, and we carry out the conversion as follows.

$$418 \text{ cm}^3 \times 1 \times 10^{-6} \text{ m}^3 \text{ cm}^{-3} = 418 \times 10^{-6} \text{ m}^3$$

Dimensional analysis

One of the questions most commonly asked by students is 'What equations do I need to learn for the final exam?' If you read this book from cover to cover (or indeed, if you just go to the glossary of equations at the end of the book), you will find a large number of equations, all of which are important to the subject of chemistry in one sense or another, and you might think that it is necessary for you to learn them all by heart. You'll be happy to know that this is not the case.

Dimensional analysis involves using the units of a physical quantity to derive the equation used to determine its value. We first encountered this on page 28 when we stated that '*units undergo the same kinds of mathematical operations as the numbers to which they are attached*'. We then went on to illustrate this in worked example 2.1 (p. 28) when we determined the units of heat capacity (J K^{-1}) through knowledge of the fact that heat capacity is equal to heat (J) divided by temperature change (K). We will now look at this principle more closely, using the stoichiometric equations $M = \frac{m}{n}$ and $c = \frac{n}{V}$, which will be introduced in chapter 3. These are two of the most used equations in chemistry and, without doubt, these equations are the two most often written incorrectly. Yet, all you need to know to determine the correct forms of these equations are the units of molar mass (M), g mol^{-1}, and concentration (c), mol L^{-1}.

The unit of molar mass, g mol^{-1}, is obtained by dividing a physical quantity having the unit gram by a physical quantity having the unit mole. These physical quantities are mass (m) and amount (n), respectively. Knowing this allows us to write the equation relating mass, amount, and molar mass as follows.

$$\text{molar mass unit (g mol}^{-1}\text{)} = \frac{\text{gram}}{\text{mole}}$$

$$\text{Therefore, molar mass} = \frac{\text{physical quantity measured in gram}}{\text{physical quantity measured in mole}} = \frac{\text{mass}}{\text{amount}} = \frac{m}{n}$$

And there you have it — the equation is given by the units. The correct units of molar mass (g mol^{-1}) are obtained *only* if we divide mass (g) by amount (mol). *No other combination of mass and amount will give the correct units of molar mass.*

A similar situation occurs in the case of concentration. The unit of concentration is mol L^{-1}, and is therefore obtained by dividing a physical quantity having the unit mole by a physical quantity having the unit litre. We know that mole is the unit of amount (n) and litre is the unit of volume (V). Therefore, we can write:

$$\text{concentration units (mol L}^{-1}\text{)} = \frac{\text{mole}}{\text{litre}}$$

$$\text{Therefore, concentration} = \frac{\text{physical quantity measured in mole}}{\text{physical quantity measured in litre}} = \frac{\text{amount}}{\text{volume}} = \frac{n}{V}$$

Again, no other combination of amount and volume will give the correct units of concentration.

Dimensional analysis assists you not only in remembering the correct form of an equation, but also in ensuring you have rearranged an equation correctly. Consider, for example, rearranging the above equation for concentration, to make n the subject. The correct rearrangement, obtained by multiplying both sides of the equation by V, gives:

$$n = c \times V$$

and we can easily check this because we know that the units on both sides of the equation must be the same. On the left-hand side, we have units of mol and on the right-hand side we have units of mol $L^{-1} \times L$ = mol. Therefore, we can be confident that we have rearranged the equation correctly.

If we had arranged the variables incorrectly, for example, $n = \frac{c}{V}$ or $n = \frac{V}{c}$, we could quickly see from the units that we had made a mistake; $\frac{c}{V}$ gives units of (mol L^{-1})/L = mol L^{-2}, while $\frac{V}{c}$ gives units of L/(mol L^{-1}) = L^2 mol^{-1}. Neither can be correct as they do not give the units of mol.

This is always a good check whenever you rearrange an equation; *if the units are not the same on both sides of the rearranged equation, you have made a mistake.*

There are situations where dimensional analysis might, at first glance, appear not to work. Consider, for example, the ideal gas equation, $pV = nRT$, which we encountered earlier in this chapter. Given that pressure (p) is measured in Pa, volume (V) in m^3, amount (n) in mol and temperature (T) in K, and the gas constant (R) has units of J mol^{-1} K^{-1}, we can work out the units on both sides of the equation.

The units of pV (left-hand side) are Pa × m^3 = Pa m^3.

The units of nRT (right-hand side) are mol × J mol^{-1} K^{-1} × K = J.

This does not appear to be correct, as it looks as though we have different units on each side of the equation. However, remember that both Pa and J are derived units; from table 2.2 (p. 27) we can see that 1 J = 1 kg m^2 s^{-2} and 1 Pa = 1 kg m^{-1} s^{-2}. Therefore:

$$\text{Pa m}^3 = \text{kg m}^{-1}\,\text{s}^{-2} \times \text{m}^3 = \text{kg m}^2\,\text{s}^{-2} = \text{J}$$

and we can see that the units on both sides of the equation are, in fact, the same.

Dimensional analysis is not limited to these relatively simple equations. It works for *any* equation. This is because the numbers and the units on both sides of an equation must be the same. As worked example 2.5 shows, dimensional analysis can also be used to determine the correct form of an equation if you know the units of all the constituent components.

WORKED EXAMPLE 2.5

Dimensional analysis

An aqueous sugar solution will boil at a temperature slightly greater than 100 °C, depending on how much sugar the solution contains. The magnitude of this boiling point elevation (ΔT) can be calculated from some combination of the molal boiling point elevation constant (K_b) and the molality (b) of the solution. Given that the units of b are mol kg^{-1} and those of K_b are K mol^{-1} kg, what equation should be used to calculate ΔT when this is measured in K?

Analysis

We need to combine the units of b and K_b to give units of K in our final answer. There are three possibilities:

$$\Delta T = \frac{K_b}{b}$$

$$\Delta T = \frac{b}{K_b}$$

$$\Delta T = K_b b$$

We need to determine the final units for ΔT that result from each of these three possibilities. The correct equation will give final units of K.

Solution

We insert the appropriate units into each of the three expressions above.

$$\frac{K_b}{b} \text{ has units of } \frac{\text{K mol}^{-1}\text{ kg}}{\text{mol kg}^{-1}} = \text{K mol}^{-2}\text{ kg}^{2}$$

$$\frac{b}{K_b} \text{ has units of } \frac{\text{mol kg}^{-1}}{\text{K mol}^{-1}\text{ kg}} = \text{K}^{-1}\text{ mol}^{2}\text{ kg}^{-2}$$

$$K_b b \text{ has units of K mol}^{-1}\text{ kg} \times \text{mol kg}^{-1} = \text{K}$$

Therefore, the correct equation to calculate the boiling point elevation is $\Delta T = K_b b$.

Is our answer reasonable?

Given that there is only one way of combining the units of K_b and b to give units of K, we can be confident our answer is correct.

We will learn more about boiling point elevation in chapter 10.

PRACTICE EXERCISE 2.5

Deduce the correct form of each of the following equations by considering the units of all the components of the equation.

(a) Obtain an equation for mass (m, g) in terms of molar mass (M, g mol^{-1}) and amount (n, mol).

(b) Obtain an equation for heat (q, J) in terms of specific heat capacity (c, J g^{-1} K^{-1}), mass (m, g) and temperature change (ΔT, K).

(c) Obtain an equation for the speed of light (c, m s^{-1}) in terms of energy (E, J), Planck's constant (h, J s) and wavelength (λ, m).

Precision and accuracy

Whenever we make a measurement, we want it to be both accurate and precise. These words are often used interchangeably in English, but in fact they have rather different meanings. A measurement's **accuracy** refers to how close the value is to the correct value, while **precision** signifies how reproducible a particular measurement is when made a number of times. As we will see in the next section, the precision of a measurement determines the number of significant figures to which the measurement may be quoted. The concepts of accuracy and precision are illustrated in figure 2.2.

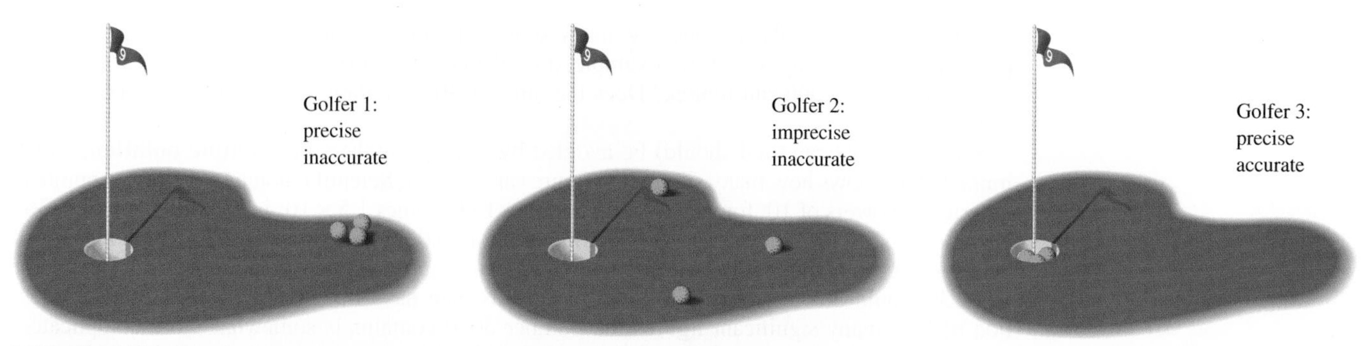

FIGURE 2.2 The difference between precision and accuracy in the game of golf. Golfer 1 hits shots that are precise (because they are tightly grouped), but the accuracy is poor as the balls are not near the target (the 'true' value). Golfer 2 needs help. The shots are neither precise nor accurate. Golfer 3 wins with shots that are precise (tightly grouped) and accurate (in the hole).

PRACTICE EXERCISE 2.6

Four workers weigh a 5.000 g mass three times on several different sets of kitchen scales and obtain the following results.

Worker A: 5.022 g, 4.976 g, 5.008 g
Worker B: 4.836 g, 5.033 g, 4.723 g
Worker C: 5.230 g, 5.231 g, 5.232 g
Worker D: 4.632 g, 4.835 g, 4.926 g

Which set of data has the best precision? Which has the best accuracy?

Uncertainties and significant figures

Every measurement has an associated uncertainty, which results from the limitations of the methods we use to make the measurement. Consider the two thermometers in figure 2.3, which display the same temperature. The divisions on the left thermometer are 1 °C apart and it is therefore certain, barring any imperfections in the manufacture of the thermometer, that the temperature is greater than 24 °C and less than 25 °C. It can be estimated that the top of the fluid column falls about 0.3 of the way between the divisions for 24 °C and 25 °C, so the measurement can be recorded as 24.3 °C. However, we cannot say that the temperature is exactly 24.3 °C, as the last figure is only an estimate, and the left thermometer might be read as 24.2 °C by one observer and 24.4 °C by another. The thermometer on the right is graduated in divisions of 0.1 °C and may therefore be read to a greater precision than the other thermometer. Reading the thermometer on the right, it is now certain, again assuming no flaws in the thermometer, that the temperature lies between 24.3 °C and 24.4 °C, and another figure in the measurement can be estimated by noting that the top of the fluid lies about 0.2 of the way between the divisions. Therefore, the temperature according to the thermometer on the right would be stated as 24.32 °C. By convention in science, all figures in a measurement up to and including the first estimated figure are recorded. The figures recorded according to this convention are called **significant figures**. The thermometer on the right can be read to a greater precision, and the measurement made contains a correspondingly larger number (4) of significant figures. The number of significant figures in a measurement is equal to the number of digits known for certain, plus one that is estimated. Therefore, *there will always be uncertainty in the last significant figure of any measurement*. In the above example, the final figures of both 24.3 °C (the measurement on the left thermometer) and 24.32 °C (the measurement on the right thermometer) are uncertain.

It is usually easy to determine the number of significant figures in a number — simply count the figures. However,

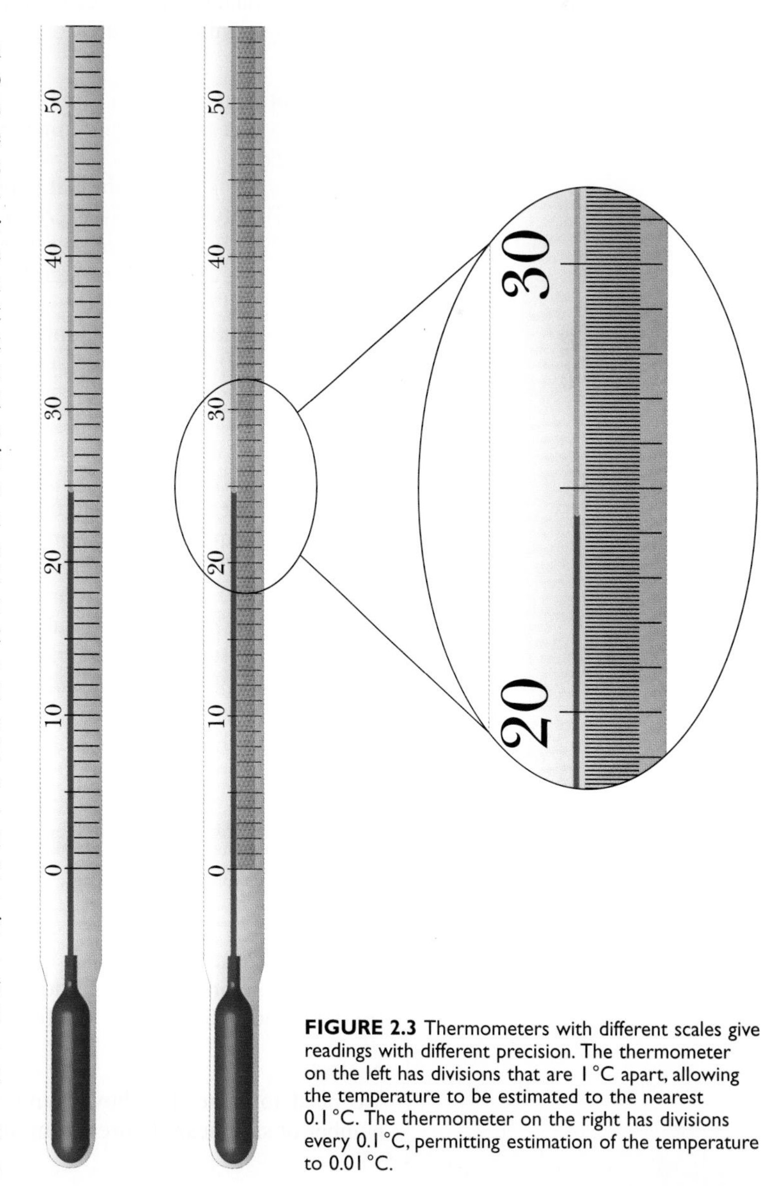

FIGURE 2.3 Thermometers with different scales give readings with different precision. The thermometer on the left has divisions that are 1 °C apart, allowing the temperature to be estimated to the nearest 0.1 °C. The thermometer on the right has divisions every 0.1 °C, permitting estimation of the temperature to 0.01 °C.

it is often not necessarily obvious how many significant figures a number contains when zeros are present at its beginning or end; for example, does the number 0.0023 contain two (23), four (0023) or five (0.0023) significant figures? Does the number 3000 contain one (3) or four (3000) significant figures?

Such problems can (and should) be avoided by writing numbers in **scientific notation**, which immediately shows how many significant figures are present. Scientific notation expresses numbers in terms of powers of 10; for example, the number 15 becomes 1.5×10^1 in scientific notation, 362 becomes 3.62×10^2, and so on. In scientific notation, 0.0023 is written as 2.3×10^{-3}, and it can be seen that the number contains only two significant figures. Similarly, the more precise number 0.002 30 would be written as 2.30×10^{-3}, and this then has three significant figures. The question of how many significant figures the number 3000 contains is somewhat more complicated. In the absence of any context we only know that the 3 is significant for certain, as we have no information to be able to make a judgement on the zeroes, and we would therefore say that 3000 contains 1 significant figure. However, if 3000 was part of a series of measurements that included the numbers 2998, 3001 and 3002, we would say that, in this context, 3000 contains 4 significant figures. With respect to this, any ambiguity can be avoided simply by expressing the number in scientific notation; 3×10^3 contains 1 significant figure, and 3.000×10^3 contains 4 significant figures.

Note that the number of significant figures in a measurement is the same regardless of the units. For example, a length of 2.3×10^{-3} m can be expressed in millimetres (2.3 mm) or nanometres (2.3×10^6 nm), but in each case the values must have the same number of significant figures, as the precision of the measurement does not depend on the units in which it is expressed.

As we shall see below, the number of significant figures to which the answer to any calculation can be expressed depends on the number of significant figures in the data used. It is therefore strongly recommended that, whenever you are recording or manipulating numerical scientific data, you express all numbers in scientific notation. This will help avoid any possible ambiguities and allow you (and others) to see the precision of your data at a glance. Figure 2.4 summarises the rules for using significant figures.

1. All nonzero digits in a number are significant.
2. A zero in a number can either be significant or not, depending on its position in the number.
   (a) A zero is significant if it is positioned:
      (i) between two nonzero digits
         704 has three significant figures
         5.02 has three significant figures
         173.05 has five significant figures
      (ii) at the end of a number that includes a decimal point
         0.500 has three significant figures (0.**500**)
         25.160 has five significant figures (**25.160**)
         3.00 has three significant figures (**3.00**)
   (b) A zero is not significant when it is positioned:
      (i) before the first nonzero digit in a number. In these cases, the zero before the decimal point can be omitted without any loss of information, while any subsequent zeros before the first nonzero digit are merely placeholders
         0.00921 has three significant figures (0.00**921**)
         0.025 01 has four significant figures (0.0**2501**)
      (ii) at the end of a number without a decimal point
         1000 has one significant figure (**1**000)
         590 his two significant figures (**59**0)

Note that some numbers are *exact*, and by definition have an infinite number of significant figures. Examples of exact numbers include those used in counting; if you count 10 objects, you have exactly 10 objects. Numbers that are part of the definitions of units are also considered to be exact. For example, there are exactly 1000 millimetres in 1 metre, and 60 seconds in 1 hour. Exact numbers have no uncertainty.

FIGURE 2.4 Rules for significant figures.

WORKED EXAMPLE 2.6

Significant figures

Determine the number of significant figures in the following numbers.

(a) 0.004 136 (b) 0.1060 (c) 10.01 (d) 200

Analysis

We need to write the above numbers in scientific notation. We can then determine the number of significant figures from the number of figures in the scientific notation.

Solution

The numbers written in scientific notation are as follows.

(a) 4.136×10^{-3} (b) 1.060×10^{-1} (c) 1.001×10^{1} (d) 2×10^{2}

Therefore, a, b and c have four significant figures and d has one.

Is our answer reasonable?

The ambiguity in significant figures comes with zeros. Are they significant, or are they placeholders? In the number 0.004 136, all the zeros are placeholders, and writing this in scientific notation (4.136×10^{-3}) makes this clear — none of the zeros appear in the scientific notation of this number, so they cannot be significant. The numbers in (b) and (c) contain zeros that *are* significant, because they appear in the scientific notation. In (d), by convention only the digit 2 is significant. Taking the rules in figure 2.4 into account, our answer is reasonable.

PRACTICE EXERCISE 2.7

How many significant figures do the following numbers contain?

(a) 3.001 010 (b) 0.004 000 (c) 7 000 010.0 (d) 50

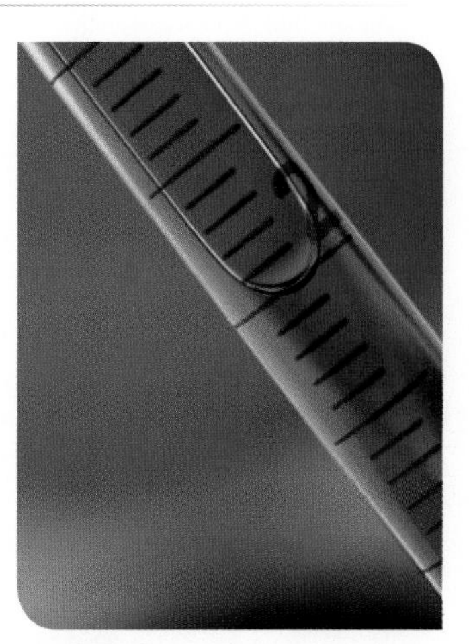

FIGURE 2.5 The 0.1 mL gradations on this pipette imply an uncertainty of ±0.05 mL.

As we saw with the example of the thermometers on page 33, whenever we make a measurement using any type of scientific apparatus, the uncertainty in that measurement will depend on the precision of the apparatus used; the more precise the apparatus, the more significant figures a measurement made using it will contain. For example, consider two 20 mL pipettes, one of which has an uncertainty of ±0.1 mL and the other an uncertainty of ±0.05 mL (± means 'plus or minus'). We would therefore quote the former pipette as measuring (20.0 ± 0.1) mL and the latter as measuring (20.00 ± 0.05) mL (see figure 2.5). This means that the latter pipette is more precise, and delivers between 19.95 mL and 20.05 mL whenever it is used. This type of uncertainty is called an **absolute uncertainty**; in other words, the uncertainty has the same units as the quantity being measured. Because an absolute uncertainty must always be compared with the actual measurement made in order for us to be able to gauge its relative magnitude, it is often more informative to quote the **percentage uncertainty** of a piece of apparatus. In this case, we would calculate the percentage uncertainty of the more precise of the two pipettes as follows.

$$\text{percentage uncertainty} = \frac{\text{absolute uncertainty}}{\text{measured quantity}} \times 100\%$$

$$= \frac{0.05\ \text{mL}}{20.00\ \text{mL}} \times 100\%$$

$$= 0.25\%$$

As we shall see below, percentage uncertainties are important when we are doing calculations involving either multiplication or division.

It should be noted that there are two types of uncertainties in any measurement. Random uncertainties refer to the reproducibility of a measurement, and are generally taken account of in the quoted uncertainty of the apparatus used to make the measurement. These are what give rise to the total uncertainty that we report at the end of a calculation. Systematic uncertainties usually result from deficiencies in the measuring equipment or from human error; for example, when using a pipette which delivers only 19.00 mL rather than the stated 20.00 mL, or using a pH meter which has been incorrectly calibrated. These uncertainties can be eliminated through careful experimental technique.

Uncertainties and significant figures in calculations

We have shown above how to determine the number of significant figures in an individual number. But when we use numbers in a calculation, how many significant figures does the answer contain? And how do uncertainties in numerical measurements translate to an uncertainty in the final answer when these measurements are used in calculations? More often than not, you will be relying on your calculator to carry out calculations; this will generally give an answer containing many more significant figures than are justified and will tell you nothing about the uncertainty in that answer. You therefore need to make use of some rules to appreciate what sort of precision you should give in your final answer.

To determine the appropriate number of significant figures in the answer of a calculation, the following rules apply.

- When a numerical measurement is multiplied by a constant, the number of significant figures in the answer will be the same as that in the numerical measurement.
- When two or more numerical measurements are multiplied or divided, the number of significant figures in the answer should not be greater than the number of significant figures in the least precise measurement.

- When two or more numerical measurements are added or subtracted, the answer should have the same number of decimal places as the measurement with the fewest number of decimal places.

Note that in worked examples and the end-of-chapter exercises we often encounter multi-step calculations. These would often be carried out all in one step on a calculator, and the answer would then be rounded according to the rules given above. However, for illustrative purposes in these examples, we generally carry out each step individually, round each intermediate answer, and then use this rounded answer in the next step of the calculation. This can then result in a discrepancy between the final answers obtained from the one-step and multi-step calculations. One way to minimise the magnitude of this discrepancy is to carry an extra significant figure through each calculation step, and then round the final answer. Worked example 2.7 illustrates these three approaches and shows that, in all cases, any discrepancy is small. Your university may have a preferred way for you to work, and of course you should follow their direction.

WORKED EXAMPLE 2.7

Rounding in multi-step calculations

Analysis of a sample of gas showed that it contained only the elements O and N, in a mass ratio of 40.00 g : 14.01 g, respectively. A separate sample of this gas contained 2.791 g of N. Calculate the mass of O that this sample contained by:

(a) carrying out the calculation in a single step and rounding the final answer

(b) carrying out the calculation as two separate steps and rounding both the intermediate and final answers

(c) carrying out the calculation as two separate steps as in (b) and carrying an extra significant figure through from the intermediate answer.

Analysis

We know the mass ratio of O to N in the first sample of gas, and we know this ratio must be the same in the second sample. Given the mass of N in the second sample, and knowing the O to N ratio, we can calculate the mass of O.

Solution

In all cases, the equation that we will use is:

mass of O in the second sample = the O to N ratio in the first sample × mass of N in the second sample

(a) Carrying out the calculation in a single step and rounding the answer

$$\text{mass of O} = \frac{40.00\text{ g}}{14.01\text{ g}} \times 2.791\text{ g} = 7.969\text{ g}$$

(b) Carrying out the calculation as two separate steps and rounding both the intermediate and final answers

Step 1: Calculate the O to N ratio.

$$\text{O to N ratio} = \frac{40.00\text{ g}}{14.01\text{ g}} = 2.855$$

Step 2: Multiply the ratio by the mass of N.

$$\text{mass of O} = 2.855 \times 2.791\text{ g} = 7.968\text{ g}$$

(c) Carrying out the calculation as two separate steps as in (b) and carrying an extra significant figure through from the intermediate answer

Step 1: Calculate the O to N ratio.

$$\text{O to N ratio} = \frac{40.00\text{ g}}{14.01\text{ g}} = 2.8551$$

Step 2: Multiply the ratio by the mass of N.

$$\text{mass of O} = 2.8551 \times 2.791\text{ g} = 7.969\text{ g}$$

Is our answer reasonable?

The three answers are essentially the same, differing by only 1 in the final figure. This would be expected for a calculation only involving two steps; obviously the more steps in a calculation, the greater the possible discrepancy between the answers. However, you will generally find that any discrepancy occurs in only the final figure of the answer.

We can further illustrate the effect of rounding by considering the following examples involving multiplication, division and subtraction. If we calculate:

$$\frac{3.14\text{ kg} \times 2.751\text{ m}}{0.64\text{ s}}$$

on a ten-digit calculator, the answer is 13.497 093 75, a number with ten figures. But how many of these figures are significant? The numbers in the numerator (top line) contain three and four significant figures, respectively, while the denominator (bottom line) contains two significant figures, and so it makes no sense to have an answer that is many orders of magnitude more precise than any of the numbers from which it is derived. The rule above for multiplication and division states that we can have only as many significant figures in our final answer as there are in the least precise measurement. Therefore, in this example, we can express our answer to only two significant figures, so our final answer becomes:

$$\frac{3.14 \text{ kg} \times 2.751 \text{ m}}{0.64 \text{ s}} = 13 \text{ kg m s}^{-1}$$

Here, we have carried out the calculation as a single step and rounded at the end. Note that, if we carried out the multiplication first as a separate step, rounded, and then carried out the division, our answer would be:

Step 1: $3.14 \text{ kg} \times 2.751 \text{ m} = 8.64 \text{ kg m}$

Step 2: $\frac{8.64 \text{ kg m}}{0.64 \text{ s}} = 14 \text{ kg m s}^{-1}$

If we calculate the sum of the following lengths:

$$\begin{array}{r} 3.247 \text{ m} \\ 41.36 \text{ m} \\ +125.2 \text{ m} \\ \hline 169.8 \text{ m} \end{array}$$

a calculator will give the numerical answer as 169.807. However, using the rule for addition and subtraction, we must give the answer to only a single decimal place, as 125.2 m has the fewest number of decimal places. Therefore, our answer is 169.8 m.

Different rules apply when working with logarithms, as we will do extensively in chapter 11 when considering pH.

- The logarithm of a number can have the same number of figures to the right-hand side of the decimal point as there are significant figures in the number.

For example, the number 0.003 21 has three significant figures (this becomes more obvious when it is written in scientific notation as 3.21×10^{-3}). Using a calculator to determine the logarithm gives −2.493 494 967 6. This should then be quoted as −2.493.

The reverse is also true; given a logarithm, you determine the correct number of significant figures in the antilogarithm from the number of figures to the right of the decimal point. Hence, using a calculator, we find that the antilogarithm of −4.18 is $6.606\,934\,48 \times 10^{-5}$. This should be quoted to two significant figures, namely 6.6×10^{-5}.

The above rules for logarithms apply to both common ($\log_{10}$) and natural (ln or $\log_e$) logarithms.

While significant figures and uncertainties are related, the rules for determining the uncertainty in the answer of a calculation are somewhat different from those for significant figures on the previous pages. They are as follows.

- When two or more numerical measurements are added or subtracted, the uncertainty in the final answer is the sum of the *absolute* uncertainties in the measurements.

  Before we see an example of this, we should recall that an absolute uncertainty is one that has the same units as the measurement. We will now consider adding two volumes, 21.05 mL and 18.55 mL. If the absolute uncertainty in each of the measurements is ±0.05 mL, then the uncertainty in the final answer is ±0.1 mL. We can explain this as follows. The range of possible values for the two measurements, as dictated by the uncertainty is 21.00 mL to 21.10 mL for the first, and 18.50 mL to 18.60 mL for the second. Thus when we add the two measurements, the smallest possible answer we could obtain is (21.00 mL + 18.50 mL) = 39.50 mL and the largest possible answer we could obtain is (21.10 mL + 18.60 mL) = 39.70 mL. Adding 21.05 mL and 18.55 mL gives 39.60 mL, and so we would quote the final answer as (39.6 ± 0.1) mL — in other words it could range from 39.5 mL to 39.7 mL, owing to the uncertainty in each reading.

  Notice that our final answer appears to have dropped a significant figure — it has 3 significant figures, where the initial measurements contained 4. This is a result of the magnitude of the uncertainty. As it is generally only the final figure in a measurement that has any uncertainty, we usually only quote uncertainties to a single figure. Therefore, the uncertainty we obtain from adding 0.05 and 0.05 is not 0.10, but 0.1. As this must refer to the final figure in the answer, we cannot quote the answer as (39.60 ± 0.1) mL — it must be (39.6 ± 0.1) mL. You may quite often find that the uncertainty dictates the precision of the answer.

- When two or more numerical measurements are multiplied or divided, the uncertainty in the final answer is the sum of the percentage uncertainties in the measurements.

As an example of this, consider calculating the density (ρ) of gold by dividing its mass (m) by its volume (V). Suppose that a (5.000 ± 0.005) g sample of gold occupies a volume of (0.2588 ± 0.0005) mL. We would calculate the density using the equation:

$$\rho = \frac{m}{V} = \frac{5.000 \text{ g}}{0.2588 \text{ mL}} = 19.32 \text{ g mL}^{-1}$$

To calculate the uncertainty in the answer, we must calculate the percentage uncertainties in the two measurements, and then add them. The reasoning for this is similar to that given above — the mass spans the possible values 4.995 g to 5.005 g, and the volume ranges from 0.2583 mL to 0.2593 mL. Therefore, the smallest and largest possible values of the density, as dictated by these uncertainties are:

$$\rho = \frac{m}{V} = \frac{4.995 \text{ g}}{0.2593 \text{ mL}} = 19.26 \text{ g mL}^{-1} \quad \text{and} \quad \rho = \frac{m}{V} = \frac{5.005 \text{ g}}{0.2583 \text{ mL}} = 19.38 \text{ g mL}^{-1}$$

However, treatment of the uncertainties here is not as simple as in the case of addition and subtraction. When measurements are added or subtracted, they must have the same units, and so the absolute uncertainties, which must also therefore have the same units, can be added. When measurements are multiplied or divided, then they do not necessarily need to have the same units, and we cannot then add the absolute uncertainties. We solve this problem by converting the absolute uncertainties to percentage uncertainties, which are dimensionless. These can then be added, and in the final step the total percentage uncertainty is converted back to an absolute uncertainty. Thus, in the above example, we convert the absolute uncertainties to percentage uncertainties as follows.

$$\% \text{ uncertainty in measurement} = \frac{\text{absolute uncertainty in measurement}}{\text{value of measurement}} \times 100$$

$$\text{Therefore, } \% \text{ uncertainty in mass} = \frac{0.005 \text{ g}}{5.000 \text{ g}} \times 100 = 0.1\%$$

$$\text{and } \% \text{ uncertainty in volume} = \frac{0.0005 \text{ mL}}{0.2588 \text{ mL}} = 0.2\%$$

Therefore, the total % uncertainty is 0.1% + 0.2% = 0.3%.

We convert the total % uncertainty to an absolute uncertainty as follows.

$$\text{absolute uncertainty in calculated value} = \frac{\text{total } \% \text{ uncertainty} \times \text{calculated value}}{100}$$

$$\text{Therefore, the absolute uncertainty in density} = \frac{0.3 \times 19.32 \text{ g mL}^{-1}}{100} = 0.06 \text{ g mL}^{-1}$$

You will note that ±0.06 g mL^{-1} represents the spread of values in the final answer that we calculated previously.

PRACTICE EXERCISE 2.8

Give the answers to the following calculations to the correct number of significant figures.

(a) $4.196 + 8.3492 + 14.73$
(b) 5.36×1.259
(c) $6.38 \times (2.514 + 5.4)$
(d) the total mass of two beakers of water, each of which weighs 37.50 g
(e) $\log(2.185 \times 5.48)$
(f) $\frac{5.39}{14.1} \times (268 + 64.1)$

PRACTICE EXERCISE 2.9

Perform the following calculations involving measurements. Give the answers to the correct number of significant figures, and use the appropriate units.

(a) 38.1931 g + 18.0 g
(b) 14.77 mL − 4.3 mL
(c) 77.371 g − 0.0446 g
(d) 13.4 g + 2.66 g + 0.926 g
(e) $\frac{10.4198 \text{ g}}{14.2 \text{ mL}}$
(f) 1.52 m × 1.029 m × 1.6 m

PRACTICE EXERCISE 2.10

A student wanted to prepare 100 mL of a 1.000 g L^{-1} solution of NaCl and devised the following method. The student weighed 0.100 g of solid NaCl on a balance and transferred this to a 100 mL volumetric flask. Sufficient water was then added, with stirring, to give a final volume of 100 mL according to the mark on the volumetric flask.

Given that the uncertainties in the balance and volumetric flask were ±0.001 g and ±1 mL, respectively, calculate the final uncertainty in the concentration of the NaCl solution. Did the student succeed in preparing a 1.000 g L^{-1} solution? If not, what was the concentration, including the uncertainty, of the solution?

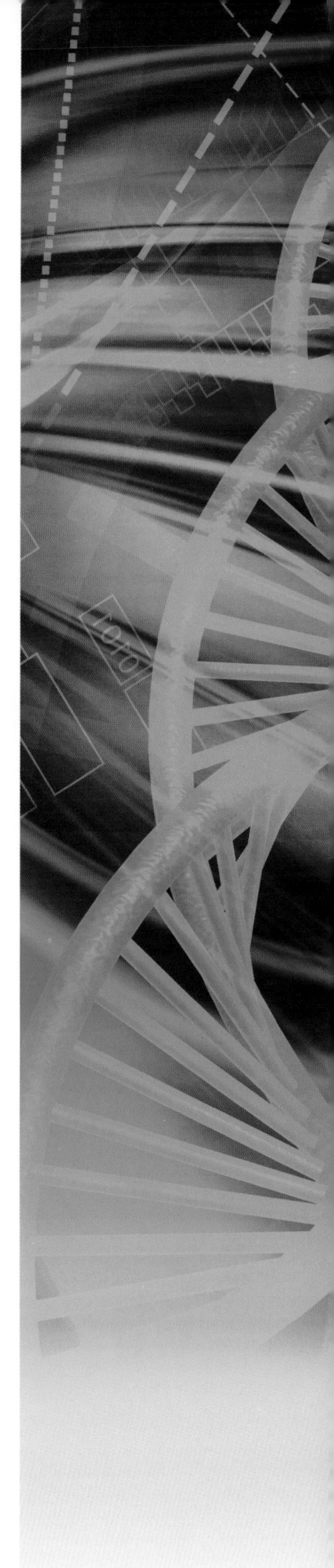

Chemical Connections

The National Measurement Institute

The National Measurement Institute (NMI) is the organisation responsible for providing unit and measurement standards and services in Australia. It has four main sections:

- legal metrology and business services
- physical metrology
- analytical services
- chemical and biological metrology.

We mentioned earlier in this chapter that the primary standard of mass is the international prototype of the kilogram, which is a platinum–iridium cylinder held at the International Bureau of Weights and Measures. The NMI holds 'Copy No. 44' (figure 2.6) of the international prototype of the kilogram. It serves as the Australian standard of mass and the NMI uses it to calibrate a set of 1 kg stainless steel standards. Those standards are then used to calibrate standards with masses from 0.5 mg to 20 kg, which are used for all calibration work on mass and the measurement of various other quantities.

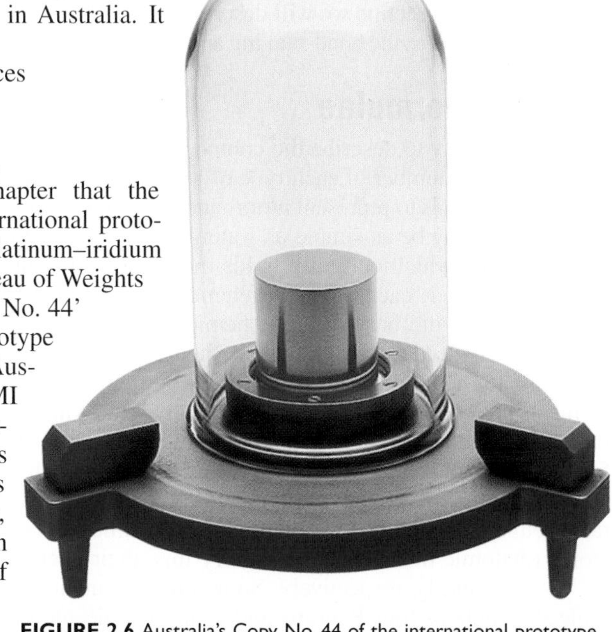

FIGURE 2.6 Australia's Copy No. 44 of the international prototype of the kilogram. This is made of a 90% platinum – 10% iridium alloy.

While measurement might seem a somewhat esoteric field, it is of great importance in many disciplines, including chemistry and business. For example, one of the key aims of the NMI is to promote uniform measurement policy and practice to help international trade. One area in which the NMI has been active is the wine industry.

Australia exports hundreds of millions of litres of wine each year. A wine connoisseur will tell you that taste, bouquet and colour are the qualities that characterise a wine. A metrologist (someone who specialises in measurement) might talk more about the levels of residues, contaminants and preservatives. In addition to health and quality concerns, many international markets have strict regulations on acceptable levels of ethanol, sulfites and heavy metals in wines. Exporters must comply with these regulations to ensure they can sell their wines into those markets.

NMI provides the wine industry with analytical chemistry services to determine pesticide, insecticide and herbicide residues, foreign matter, alcohol content and other important information. The techniques used have been validated against those used in other countries to ensure the measurements are accepted by export markets.

FIGURE 2.7 Chemical analysis can quantify what a wine expert detects using the senses.

2.2 Representations of molecules and reactions

(a) CH_4

(b)
```
    H
    |
H — C — H
    |
    H
```

(c) 3-D structural formula of CH_4

(d)

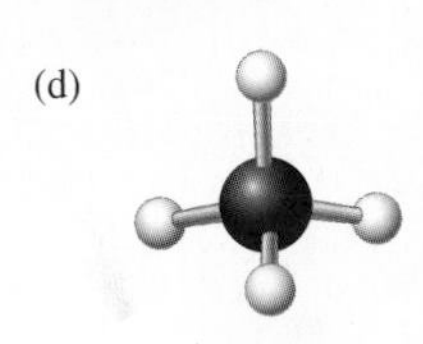

(e)

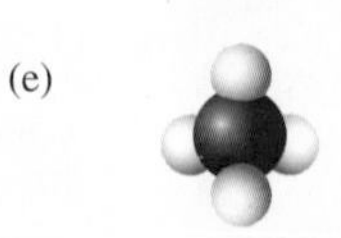

FIGURE 2.8 Different ways of representing the chemical species methane: **(a)** chemical formula, **(b)** structural formula, **(c)** 3-D structural formula, **(d)** ball-and-stick model and **(e)** space-filling model.

The language of chemistry is not only about the meaning of words, but also about how we can represent chemical species using particular combinations of letters, numbers, symbols and pictures (see figure 2.8). In this section we will describe some of the many different ways in which we can depict molecules, as well as the bond-making and bond-breaking processes that occur in chemical reactions.

Chemical formulae

The simplest way to describe the composition of any substance is to write its **chemical formula**. This shows the number of each type of atom present in a substance. A chemical formula contains elemental symbols to represent atoms and subscripted numbers to indicate the number of atoms of each type. It may be as simple as water, H_2O, or as complicated as $C_{30}H_{34}AuBClF_3N_6O_2P_2PtW$, the compound which currently holds the record for containing the most elements. As we have seen in chapter 1, each chemical element is described by a one-, two- or three-letter symbol, and these form the basis for all chemical formulae. The simplest chemical formulae describe pure elements and are usually just their elemental symbols as listed on the periodic table; for example, helium is He, silicon is Si and copper is Cu. Such formulae imply that the bulk elements are composed of either individual atoms which do not interact (such as He) or 'infinite' three-dimensional arrays of bonded atoms (such as Si and Cu). However, this is not the case for some elements and we must write more complex chemical formulae for these. For example, seven elements (hydrogen, nitrogen, oxygen, fluorine, chlorine, bromine and iodine) occur naturally neither as individual atoms nor as three-dimensional arrays but as molecules containing two atoms (diatomic molecules). To signify this, their chemical formulae are written as H_2, N_2, O_2, F_2, Cl_2, Br_2 and I_2, respectively. Some other elements occur as larger molecules; a phosphorus molecule contains four P atoms and a sulfur molecule contains eight S atoms so their chemical formulae are P_4 and S_8, respectively. A chemical formula that refers to a discrete molecule is often called a **molecular formula** as it describes the types and numbers of atoms present in the molecule. Note that a subscripted numeral refers either only to the atom immediately preceding it, or, if it follows a group of atoms enclosed in parentheses, to the entire group of enclosed atoms. For example, the chemical formula $B(OH)_3$ shows that the compound contains one boron atom, three oxygen atoms and three hydrogen atoms.

Chemical compounds contain more than one element. Therefore, there is potentially more than one way to write the formula of any compound. For example, hydrogen chloride is a diatomic molecule with one atom each of hydrogen and chlorine. Its chemical formula might therefore be written as HCl or ClH. To avoid possible confusion, chemists have standardised the writing of chemical formulae. For **binary compounds** (compounds containing only two elements) the following guidelines apply.

1. With the exception of hydrogen, the element further to the left in the periodic table appears first. Examples are KCl, PCl_3, Al_2S_3 and Fe_3O_4.
2. If hydrogen is present, it appears last except when the other element is from group 16 or 17 of the periodic table. For example, we write hydrogen last in LiH, NH_3, B_2H_6 and CH_4, but first in H_2O_2, H_2S, HCl and HI.
3. If both elements are from the same group of the periodic table, the lower one appears first, for example SiC and BrF_3.
4. If the compound is ionic, we write the cation (the positively charged ion) first, followed by the anion (the negatively charged ion). Examples are NaBr and $MgCl_2$.

Writing chemical formulae for compounds containing more than two elements requires some knowledge of the bonding within the compound. As we will discuss in greater detail in chapter 5, chemical compounds can be divided broadly into two classes — ionic compounds and covalent compounds — which are distinguished by the type of bonding. Ionic compounds are composed solely of ions, and the attractive forces between ions of opposite charge result in the formation of 'infinite' three-dimensional lattices. On the other hand, covalent compounds are characterised by bonding involving the sharing of electrons between adjacent atoms. To write the chemical formula of an ionic compound containing more than two elements, we again write the cation followed by the anion. For example, the formula of sodium nitrate, which contains positively charged sodium ions, Na^+, and negatively charged nitrate ions, NO_3^-, is $NaNO_3$. Note that the ordering of the elements in the NO_3^- ion conforms to the rules given above. Note also that we must always ensure that the total charge on the ionic compound is zero, so the number of Na^+ and NO_3^- ions must be the same, and we never include the individual ionic charges in the overall formula. Writing the chemical formula for calcium nitrate is a little more complicated. The formula of the calcium ion is Ca^{2+}, while that of the nitrate ion is NO_3^-. We can see that, to achieve an overall charge of zero, we will require one Ca^{2+} ion and two NO_3^- ions, and it is with the latter that a potential problem arises. We signify the presence of two NO_3^- ions, as we do for any ion, by using a subscripted '2' immediately following the ion. However, you can see that, if we do this, we will have two subscripted numbers immediately adjacent to each other, thereby leading to possible confusion. In cases such as this, we enclose the ion in parentheses, and place the subscripted '2' outside the parentheses. Putting the cation first, we therefore write the chemical formula as $Ca(NO_3)_2$.

$Ca(NO_3)_2$ ✓ $CaNO_{32}$ ✗

As stated on the previous page, it is understood that the subscripted '2' refers to the entire group of atoms within the parentheses; that is, the correct chemical formula $Ca(NO_3)_2$, implies one Ca^{2+} ion and two NO_3^- ions. The incorrect chemical formula, $CaNO_{32}$, implies one Ca, one N and 32 O atoms.

To illustrate the difference between the use of subscripted numbers and numbers of normal case, consider the following chemical reaction, which describes solid $Ca(NO_3)_2$ dissolving in water to form its constituent ions.

$$Ca(NO_3)_2(s) \rightarrow Ca^{2+}(aq) + 2NO_3^-(aq)$$

Note that subscripts are used only when we are referring to the number of constituent atoms or ions *within an element or compound*; in other words, they appear in chemical formulae only. Conversely, normal case numbers are used as coefficients to describe the number of atoms, ions or molecules present *in a particular chemical reaction*. This will be covered in much greater detail in chapter 3.

PRACTICE EXERCISE 2.11

Write the chemical formulae for the compounds resulting from all possible combinations of the following cations and anions (16 compounds in total).

Cations: NH_4^+, K^+, Ca^{2+}, Al^{3+} Anions: Br^-, ClO_4^-, CO_3^{2-}, PO_4^{3-}

If calcium nitrate, a white crystalline solid, is exposed to moist air, it absorbs water and forms a **hydrate**. This hydrate is also a white crystalline solid but it has the chemical formula $Ca(NO_3)_2{\cdot}4H_2O$. This process can be reversed by heating the hydrate under vacuum to remove the water molecules, which results in the formation of **anhydrous** $Ca(NO_3)_2$ (anhydrous means that the compound contains no water molecules). The formation of hydrates is relatively common among ionic compounds, and it is usual to write the water molecules in such compounds at the end of the chemical formula, with the number of water molecules indicated by a normal case number. This provides an exception to the rules given above.

The majority of covalent compounds are carbon-based organic compounds, the chemical formulae of which are often written with carbon first, followed by hydrogen and then the remaining elements in alphabetical order (e.g. C_2H_6O, C_4H_9BrO, CH_3Cl). The major shortcoming of a chemical formula is that it tells us little or nothing about the *structure* of the compound in question. To obtain this information, we require structural formulae.

Structural formulae

Chemists are interested in the way in which atoms are bonded together to form molecules. While molecular formulae are useful in telling us the chemical makeup of molecules, they only rarely give us any idea of which atom is bonded to which. For example, the molecular formula of water is always written as H_2O, but this tells us nothing about how the atoms are arranged in the molecule. We cannot tell from the molecular formula alone which of the three possible arrangements of atoms below is the correct one for H_2O.

```
                               O
                              / \
H—H—O        H—O—H           H—H
```

Structural formulae attempt to show the way in which the atoms in a molecule are bonded together, thereby giving us some structural information. The chemical symbols are still used for each element present, but now the constituent atoms are placed in the order in which they are bonded together and the bonds between neighbouring atoms are represented as lines. A single line represents a single bond, which, as we will see in chapter 5, consists of a pair of electrons. Consider for example the ammonia molecule, NH_3. We can show the way in which the four atoms are bonded together in this molecule by drawing the structural formula as follows.

```
H   H
 \ /
  N
  |
  H
```

From this depiction, it is obvious that the nitrogen atom is bonded to three H atoms by single bonds. It is also common to see ammonia depicted as:

```
H—N—H
  |
  H
```

which gives us the same structural information as the previous diagram (the N atom is bonded to three H atoms by single bonds), but shows the molecule apparently adopting a slightly different geometry. This illustrates an important point about structural formulae: *structural formulae do not necessarily show the correct geometry of a compound*, simply because it is difficult to accurately represent three-dimensional molecules in two dimensions. While both of the depictions on the previous page correctly show all of the bonds in the molecule, neither is strictly correct from a structural point of view as they do not show the actual three-dimensional arrangement of the atoms. Both of them also neglect the fact that there is a pair of electrons on the N atom which does not participate in bonding. This is termed a lone pair of electrons. While such electrons are often ignored in depictions of molecules (their presence is usually implied by the chemical formula), they can have significant structural consequences. Where necessary, they can be shown as follows.

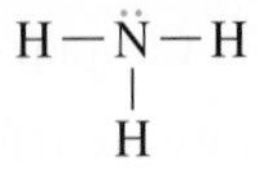

```
    ..
  H—N—H
    |
    H
```

In later chapters, you will find it can be important to show lone pairs on atoms, especially when you draw organic reaction mechanisms, which show the movement of electrons in bond-making and bond-breaking processes. Even with the inclusion of the lone pair, this is still not an accurate depiction of the ammonia molecule in a three-dimensional sense, but it shows the correct connections between the constituent atoms of the molecule. We will see later how to introduce three-dimensional elements into structural formulae.

It is in the world of organic chemistry that structural formulae are particularly useful. While many covalent inorganic compounds tend to be small molecules in which a single central atom is joined to 2, 3, 4, 5 or 6 surrounding atoms, organic molecules tend to exist as rings and chains, and have a huge variety of possible geometries. This is due to the unusual propensity of carbon atoms to bond to themselves, a property called **catenation**, which can lead to the formation of massive molecules. Each carbon atom in a molecule can form bonds to as many as four other carbon atoms, which means that even relatively small organic molecules have many different possible ways in which the constituent atoms can be bonded together. Structural formulae allow us to depict these possibilities.

Carbon is a tetravalent element, meaning that it prefers to form a total of four bonds within a molecule. These bonds may be single, double or triple. This means that there are four different ways that a carbon atom can form a total of four bonds: four single bonds, two double bonds, a double bond and two single bonds, or a triple bond and a single bond. Hydrogen, on the other hand, is monovalent, meaning that it usually forms only one single bond to another atom and, in an organic molecule, this is most often to a carbon atom. This can be illustrated by the propane molecule, which has the chemical formula C_3H_8. Given that each C atom must form four bonds and each hydrogen atom must form one, there is only one possible way of attaching all the elements; this is shown in a structural formula below.

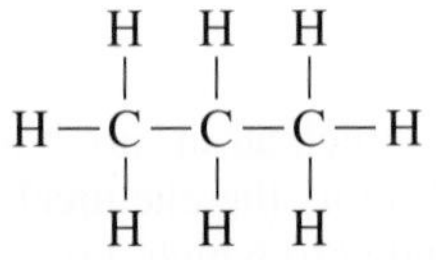

```
    H  H  H
    |  |  |
  H—C—C—C—H
    |  |  |
    H  H  H
```

propane, C_3H_8

We can see from this that the three carbon atoms of a propane molecule link to form a chain in which each carbon at the end of the chain (called a **terminal carbon**) is singly bonded to three hydrogen atoms, and the inner carbon is singly bonded to two hydrogen atoms. Notice that the structural formula of propane contains more information than the chemical formula. Both formulae identify the number of atoms (three C atoms and eight H atoms), but the structural formula also shows how the atoms are connected.

The propane molecule contains only C—C and C—H single bonds. It is also possible to have double bonds and triple bonds in organic molecules, which we designate by two and three lines respectively, as shown in figure 2.9.

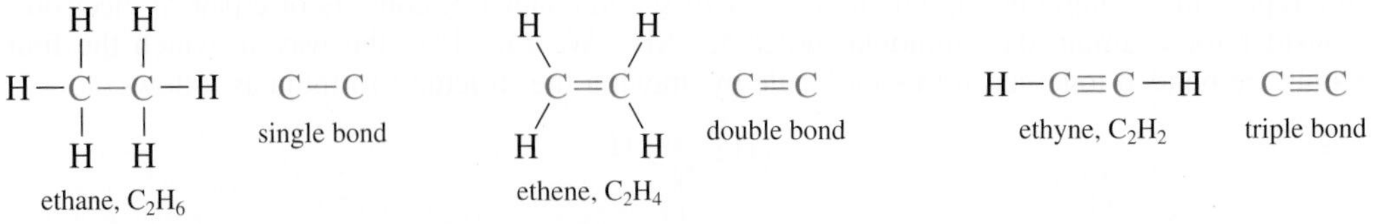

```
    H  H                    H     H
    |  |                     \   /
  H—C—C—H     C—C             C=C        C=C       H—C≡C—H      C≡C
    |  |                     /   \
    H  H    single bond     H     H    double bond  ethyne, C2H2  triple bond
 ethane, C2H6              ethene, C2H4
```

FIGURE 2.9 Depictions of single, double and triple bonds between carbon atoms.

Structural formulae can remove ambiguities inherent in chemical formulae. For example, the ethanol and dimethyl ether molecules have the same chemical formula, C_2H_6O. However, the structural formulae depicted in figure 2.10 clearly show the differences in the way the atoms are connected within the molecules.

```
    H       H              H  H
    |       |              |  |
H — C — O — C — H      H — C — C — O — H
    |       |              |  |
    H       H              H  H
  dimethyl ether            ethanol
```

FIGURE 2.10 Structural formulae of dimethyl ether and ethanol. Both molecules have the same chemical formula, C_2H_6O.

Molecules such as these that have the same chemical formula but different chemical structures are called **isomers**. Structural formulae are a convenient way to distinguish between isomers.

Two types of shorthand structural formulae are commonly used. The constituent atoms in **condensed structural formulae** are arranged in bonded groups, and the actual bonds are not drawn. For example, we would write the condensed structural formulae of dimethyl ether and ethanol as CH_3OCH_3 and CH_3CH_2OH respectively, which essentially gives us the same structural information as the structural formulae, but with a significant saving of space. Note that the condensed structural formulae attempt to show the order in which the atoms are bonded together, and differ from the chemical formula, which is C_2H_6O in both cases. Condensed structural formulae can also be drawn for more complex molecules such as 2-methylpropane, C_4H_{10}.

```
    H  Me  H
    |  |   |
H — C — C — C — H      Me = methyl, —CH3
    |  |   |
    H  H   H
```

The condensed structural formula of this molecule can be written as $CH_3CH(CH_3)CH_3$, $(CH_3)_2CHCH_3$ or $(CH_3)_3CH$, all of which are equivalent. In these cases, any $-CH_3$ group in parentheses is understood to be bonded to the middle carbon of the longest carbon chain.

We can also use **line structures**, in which C atoms are not drawn explicitly, to depict molecules. Line structures are constructed according to the following guidelines.

1. All bonds except C—H bonds are shown as lines.
2. C—H bonds and H atoms attached to carbon are not shown in the line structure.
3. Single bonds are shown as one line; double bonds are shown as two lines; triple bonds are shown as three lines.
4. Carbon atoms are not labelled. All other atoms are labelled with their elemental symbols.

Following these guidelines, we would then write the line structure of propane, C_3H_8, as:

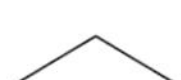

In this diagram a carbon atom is implied at each end of the chain and at the kink in the chain. The carbon atoms are singly bonded to each other and, as there are no other elemental symbols in the structure, it is assumed that each carbon atom bonds to the appropriate number of hydrogen atoms so that it forms a total of four bonds. This means the terminal carbon atoms will bond to three hydrogen atoms and the central carbon atom will bond to two hydrogen atoms. We illustrate how to draw line structures further in worked example 2.8.

WORKED EXAMPLE 2.8

Drawing line structures

Construct line structures for compounds with the following structural formulae.

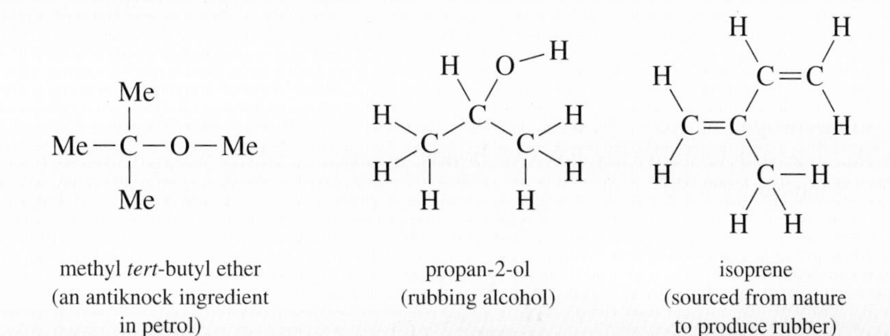

Analysis

We are asked to convert the structural formulae above to line structures. To do this, we simply apply the guidelines given previously.

Solution

Using guidelines 1 and 2, we remove all C—H bonds and H atoms attached to C atoms from the structure, thereby leaving the carbon-based framework of each molecule (remember that Me = —CH_3). Guideline 3 states that double bonds remain as two lines. This gives us the following.

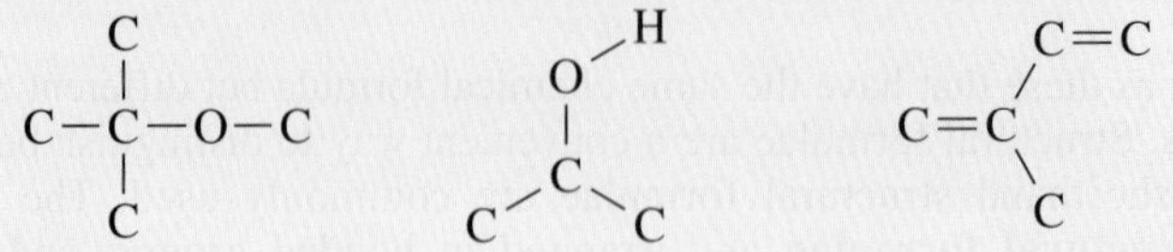

The final guideline tells us to remove the labels for all C atoms. This gives the answer:

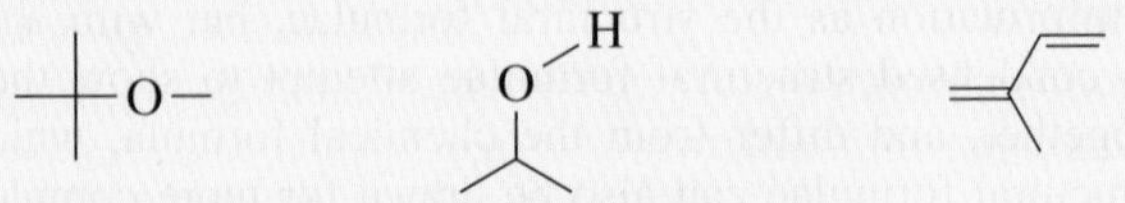

Is our answer reasonable?

To check that the line structures represent the correct substances, count the number of intersections and line ends, which should equal the number of C atoms in the compound. The first line structure has five, the second has three and the third has five, matching the chemical formulae. Therefore, our answers appear correct.

You should be familiar with line structures, and you should be able to convert from a structural formula to a line structure, and from a line structure to a structural formula. To do the latter, keep in mind that carbon atoms are not shown in a line structure, so the first step is to place a C at every line intersection or kink and at the end of every line. Then add singly bonded hydrogen atoms (—H) until every carbon atom has four bonds.

WORKED EXAMPLE 2.9

Converting line structures

Draw the structural formulae and determine the chemical formulae of the molecules in the following line structures.

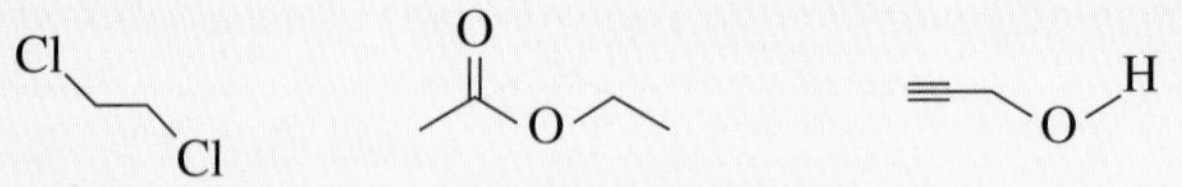

Analysis

We know that line structures show all structural features except carbon atoms and C—H bonds. We can therefore convert a line structure into a structural formula in two steps. First, place a C at any unlabelled line end and at each line intersection or kink. Second, add hydrogen atoms until each carbon atom has four bonds. The chemical formula is then obtained by counting the number of atoms of each element.

Solution

Begin by placing a C at each intersection, kink and line end.

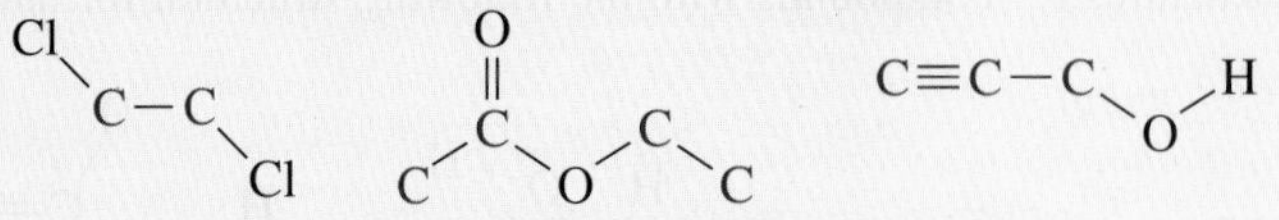

Now add hydrogen atoms until each carbon atom has a total of four bonds. Each carbon atom in the first structure already has two bonds, so each needs two hydrogen atoms. This gives the structural formula:

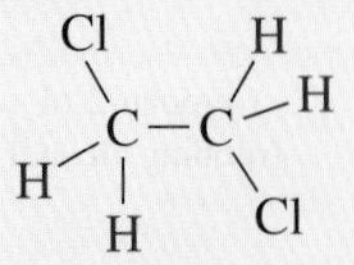

and the chemical formula $C_2H_4Cl_2$.

The second structure has four carbon atoms. The two terminal carbon atoms have just one bond, so each needs three hydrogen atoms. The carbon with the double bond to oxygen already has its complete set of four bonds. The other inner carbon atom has a bond to C and a bond to O, so it needs two hydrogen atoms. This gives the structural formula:

and the chemical formula $C_4H_8O_2$.

The third structure contains a triple bond. The terminal carbon atom needs one hydrogen atom, but the other carbon atom of the triple bond already has four bonds. The next carbon atom has two bonds, one to carbon and one to oxygen. Two hydrogen atoms are needed to give this atom four bonds, and this gives the structural formula:

$$H-C\equiv C-CH_2-O-H$$

and the chemical formula C_3H_4O.

Is our answer reasonable?

Check the consistencies of the structural formulae by counting the number of bonds associated with each carbon atom. If you have converted the line structure correctly, each carbon atom will have four bonds.

PRACTICE EXERCISE 2.12

Convert the following line structures into structural and chemical formulae.

(a) (b) (c)

While line structures are the basis for the depiction of most molecules in the chemical literature, it is common to see variations on the rules we have outlined previously. For example, we might see butane, C_4H_{10}, written as any of the following.

All of these are acceptable alternatives for the 'strictly correct' depiction shown below.

Notice also that in molecules with no double or triple bonds, the carbon chain can be drawn in any orientation. The following equivalent line structures can be drawn for pentane, C_5H_{12}:

but the first two of these would be most commonly seen.

The depiction of molecules as line structures can give rise to some very interesting shapes, and some of these have served as inspirations for the synthesis of particular molecules. Consider, for example, the molecules below, all of which have been synthesised.

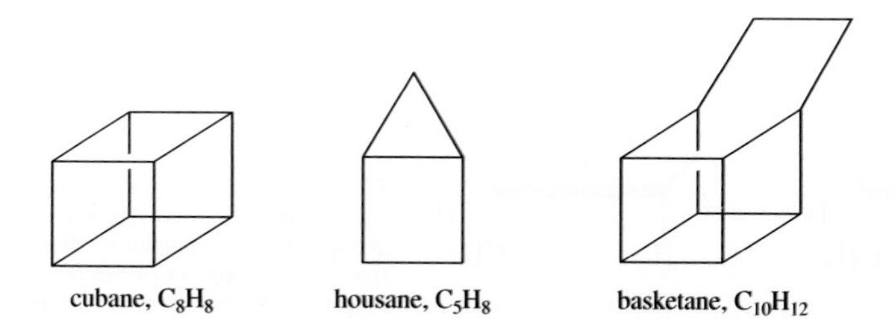
cubane, C_8H_8 housane, C_5H_8 basketane, $C_{10}H_{12}$

Probably the most remarkable example of molecular synthesis inspired by line structures is the series of molecules called Nanoputians, which were reported in 2003 by Dr Stephanie Chanteau and Professor James Tour from Rice University in the USA. These are molecules that, when drawn as line structures, resemble the human form. Two examples of Nanoputians (named after the Lilliputians in Jonathan Swift's *Gulliver's Travels*) are shown in figure 2.11. Had we chosen to describe these with their chemical formulae ($C_{39}H_{42}O_2$ and $C_{38}H_{44}O_2$ for NanoKid and NanoBalletDancer, respectively) we would have no idea of the delightful shapes that representations of these molecules can display.

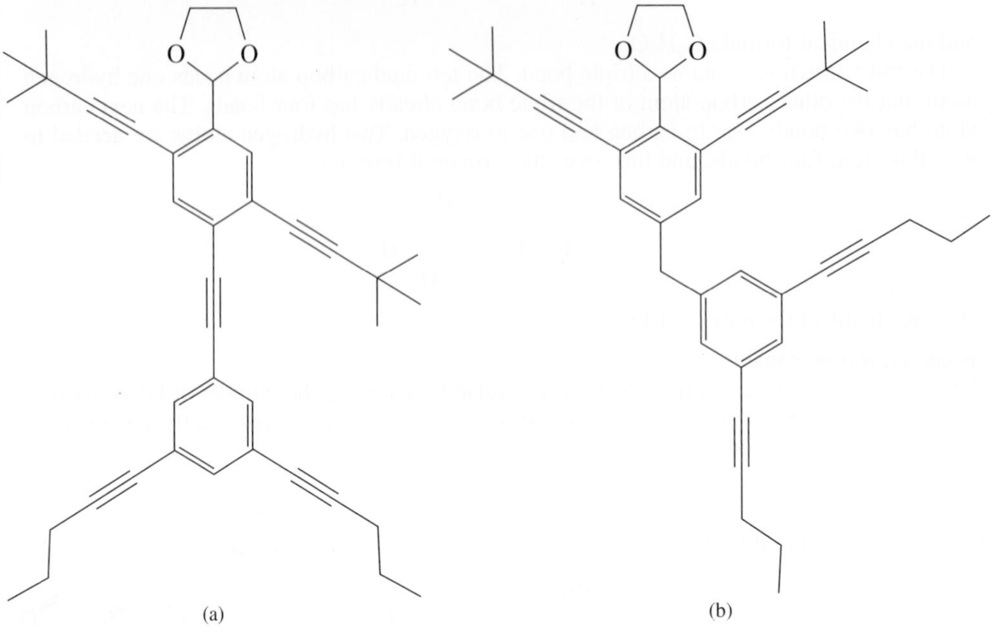

(a) (b)

FIGURE 2.11 Line structures of: **(a)** NanoKid **(b)** NanoBalletDancer.

Three-dimensional structures

While the structural formulae encountered so far give more structural information than chemical formulae, they do not necessarily give a complete representation of a molecule. For example, none of the molecules shown previously are flat, but you would not know this from looking at the structural formulae. We must therefore attempt to introduce some aspects of three-dimensionality in our representations. There are a number of ways this can be done.

Three-dimensional structural formulae

The simplest way of drawing a three-dimensional structure in two dimensions is to use a structural formula or line structure as a base, and then add some perspective. This is illustrated below using the molecule 1,2-dimethylcyclopentane. This is a cyclic molecule consisting of a pentagon of carbon atoms, with —CH_3 groups attached to two adjacent carbons. We can draw the line structure as follows.

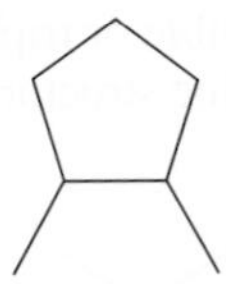

However, when this molecule is viewed in three dimensions from side-on, we can immediately see that the flat line structure is not an adequate representation; each carbon atom in the ring has two attached atoms, either two H atoms or a H atom and the C atom of a —CH_3 group, which point either above or below the plane of the ring. This leads to two possible relative arrangements of the —CH_3 groups within the molecule — either both on the same side of the ring, or on opposite sides — and hence two possible isomers named using the prefixes *cis* and *trans* (see figure 2.12).

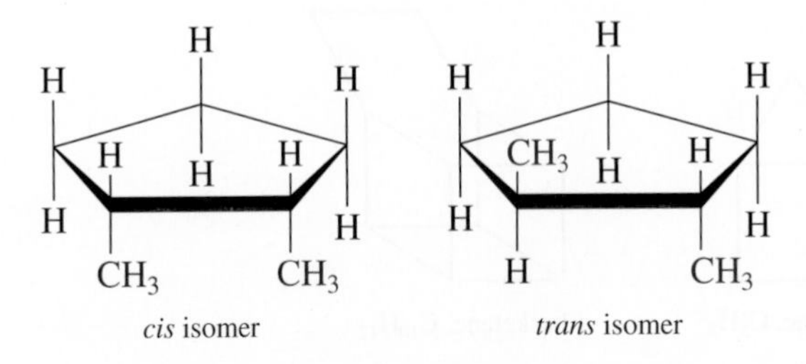

FIGURE 2.12 Representations of the two possible isomers of 1,2-dimethylcyclopentane. The two —CH_3 groups may be arranged so that they are on either the same or opposite sides of the ring. In reality, the cyclopentane rings are not flat but slightly puckered.

The *cis* isomer has both $-CH_3$ groups on the same side of the ring. (Note that it does not matter whether they are both on the top or bottom side of the ring — the same compound results.) The $-CH_3$ groups are on opposite sides of the ring in the *trans* isomer. We can show this in our line structure by drawing the bonds to these groups as wedges. A **solid wedge** (▬) represents a bond coming out of the page towards the observer, while a **hashed wedge** (⋯) represents a bond going back into the page. A normal line is used for bonds in the plane of the page. Therefore, the two isomers can be drawn as line structures as follows:

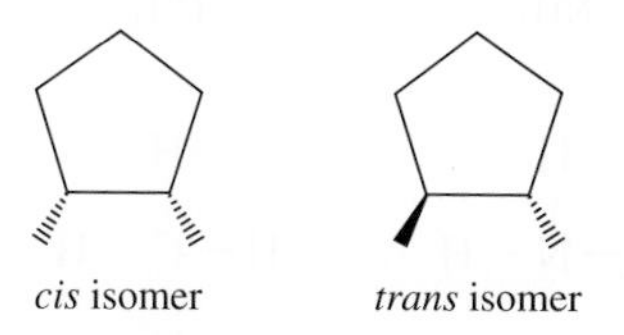

and it can be easily seen that they are different compounds.

When carbon is bonded to four atoms, those atoms generally adopt a tetrahedral arrangement around the carbon atom, as this arrangement places them as far as possible away from each other. It is conventional to draw the three-dimensional nature of the four bonds around a single carbon atom as follows:

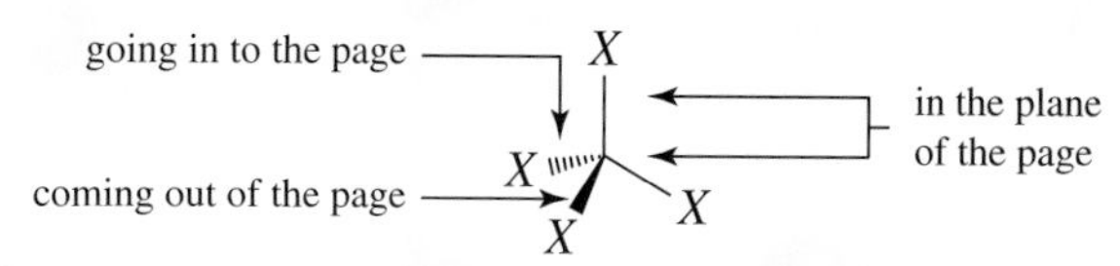

where X is any atom. In this representation, one bond goes into the plane of the page, one comes out of the plane of the page, and two lie in the plane of the page. We can show the three-dimensionality of molecules containing more than one carbon atom by joining these individual units together. Such three-dimensional depictions are not limited to organic compounds; we will see in chapter 13 that transition metal complexes can display a wide variety of geometries, and we can use analogous diagrams to portray these. For example, it is common to have six atoms bound to a transition metal ion to give a complex of general formula ML_6. Each atom sits at the corner of an imaginary octahedron in what is called an octahedral geometry. We would draw such an arrangement as follows.

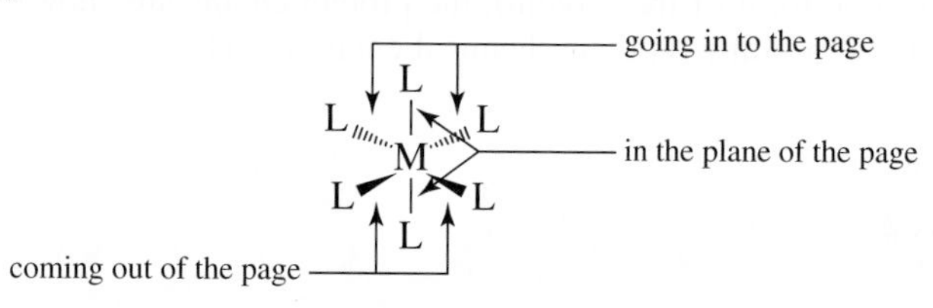

In this case we have the two 'vertical' bonds in the plane of the page, while two bonds go into the plane and two bonds come out of the plane.

While it is possible to make remarkably detailed three-dimensional drawings using only a pen and paper, it is easier to use computers to draw somewhat more realistic depictions of molecules.

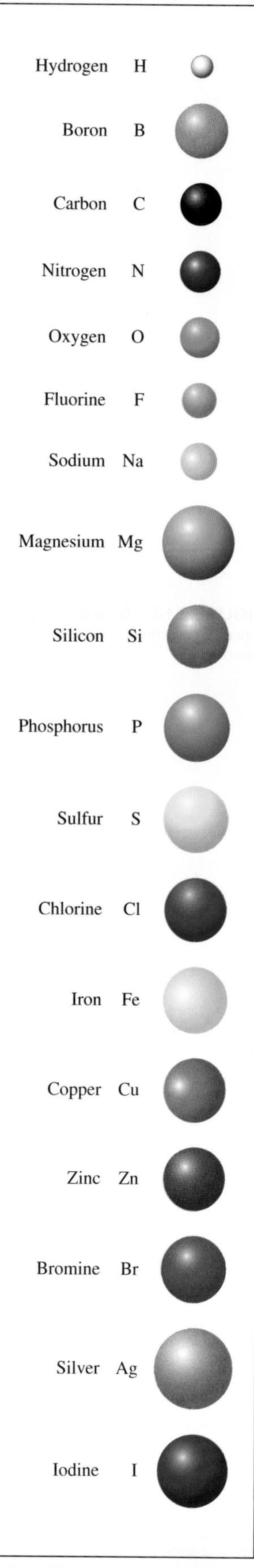

FIGURE 2.14 The colours and sizes of balls used to represent atoms in this book.

Ball-and-stick models

In a **ball-and-stick model**, balls of arbitrary size represent atoms and sticks represent chemical bonds. Figure 2.13 shows a ball-and-stick model of propane. The balls are usually drawn in different colours to distinguish between different elements present in the molecule (see figure 2.14). Such models generally enable the easy visualisation of three-dimensional features of the molecule.

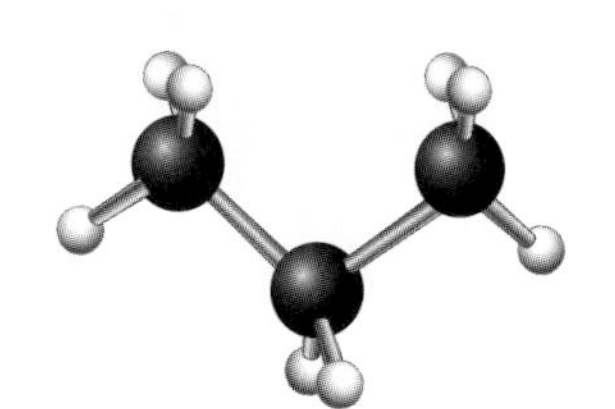

FIGURE 2.13 Ball-and-stick model of propane, C_3H_8.

Space-filling models

While ball-and-stick models are useful, they are far removed from reality — chemical bonds are not sticks, and the size of atoms is by no means arbitrary. A **space-filling model** recognises that a molecule is defined by the space occupied by its electrons, and attempts to represent the relative size of atoms within a molecule by showing the size of each atom's electron cloud. Recall from Rutherford's gold foil experiment (p. 10) that electron clouds make up nearly the entire volume of any atom. Each atom in a space-filling model is shown as a distorted sphere representing the volume occupied by its electrons. These spheres merge into one another to build up the entire molecule. A space-filling model can tell us at a glance whether the electron clouds of atoms in a molecule overlap, and therefore whether those atoms are bonded together or not. Figure 2.15 shows a space-filling model of propane; notice that

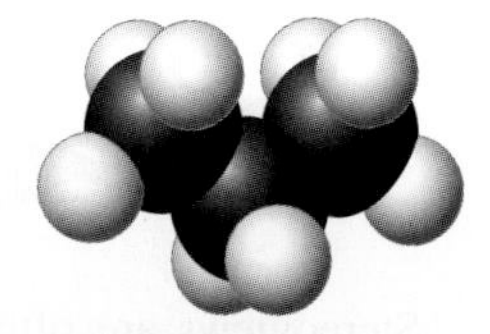

FIGURE 2.15 Space-filling model of propane, C_3H_8.

the spheres of the neighbouring C atoms overlap to a significant extent, implying these atoms are bonded together. The presence of C—H bonds is also obvious from the overlap of the spheres representing C and H. Figure 2.16 shows ball-and-stick and space-filling models of several chemical compounds common in everyday life.

<table>
<tr><th>Molecule</th><th>water</th><th>ammonia</th><th>methane</th><th>ethanol</th></tr>
<tr><td>Chemical formula</td><td>H_2O</td><td>NH_3</td><td>CH_4</td><td>C_2H_6O</td></tr>
<tr><td>Structural formula</td><td>H—O—H</td><td>H
|
H—N—H</td><td>H
|
H—C—H
|
H</td><td>H H
| |
H—C—C—O—H
| |
H H</td></tr>
<tr><td>Ball-and-stick model</td><td></td><td></td><td></td><td></td></tr>
<tr><td>Space-filling model</td><td></td><td></td><td></td><td></td></tr>
</table>

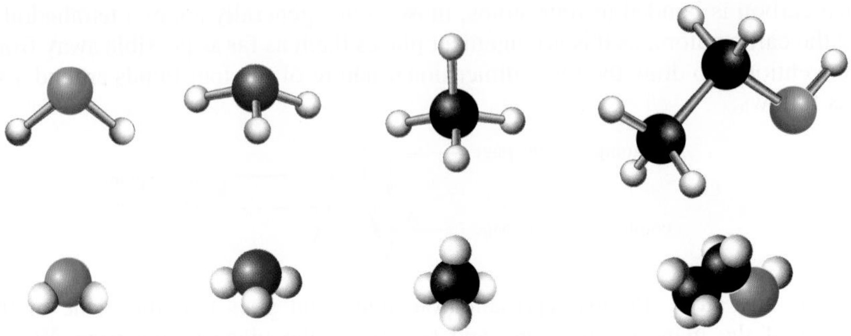

FIGURE 2.16 Different representations of some common compounds.

Other representations

It is often inconvenient, if not impossible, to show every atom in a large molecule such as a protein or nucleic acid, and it is usual to draw 'cartoon' type structures of such molecules. Figure 2.17 shows such a view of haemoglobin, the iron-containing protein that carries oxygen in human blood. Instead of showing each individual atom in the molecule ($C_{2952}H_{4664}N_{812}O_{832}S_8Fe_4$ is the molecular formula of one form of the protein), the protein chains are shown as 'ribbons' which fold in particular ways depending on their chemical environment.

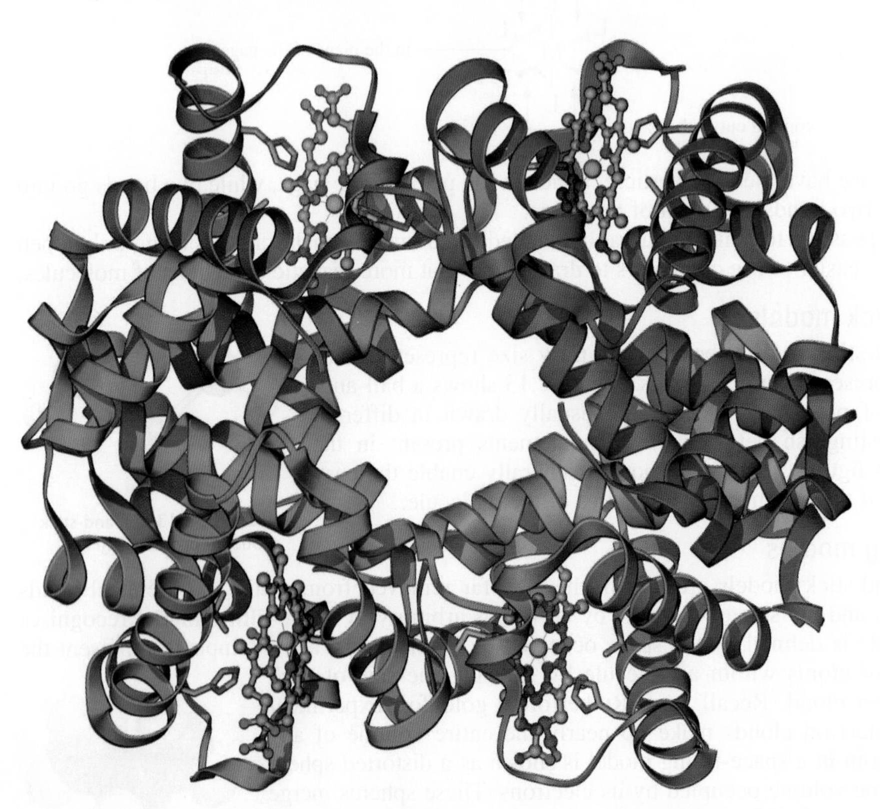

FIGURE 2.17 A representation of the structure of haemoglobin showing four haem groups (the ball-and-stick models). The different coloured ribbons denote different parts of the protein.

Stereoviews are often helpful in visualising large molecules, as they allow the image to be viewed in three dimensions, although viewing them properly requires some practice. An example of a stereoview of DNA is shown in figure 2.18.

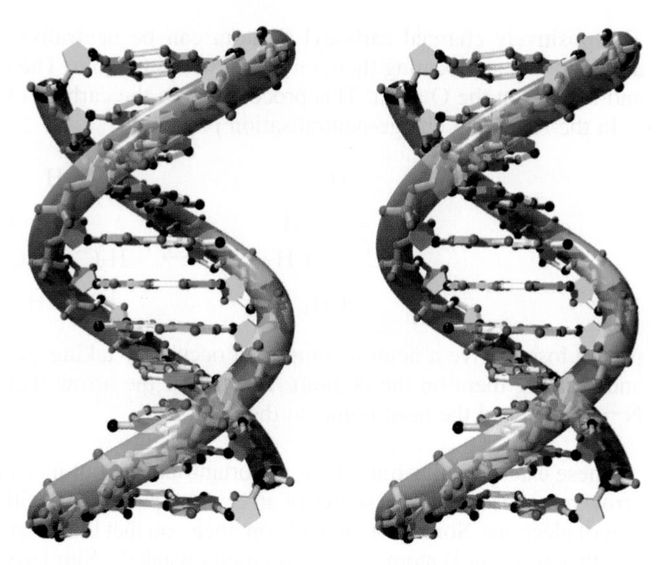

FIGURE 2.18 Stereoview of DNA.

To view stereoviews (which are three-dimensional images), you must be able to 'go cross-eyed'. You will eventually find that, with lots of practice, a three-dimensional image appears in the middle of the two structures in figure 2.18. If you do not get it the first time, keep trying!

Which of our various depictions is the 'correct' one? The answer is 'none of them'. Atoms are not finite coloured spheres and bonds are not sticks attached to atoms. The representations we use in chemistry are merely attempts to visualise things that, in all probability, are impossible to depict. However, this is not to say that our representations are useless. As we will see throughout this book, we can understand many important facets of molecular structure through studying these representations. But keep in mind at all times that the representations are models; nothing more and nothing less.

Mechanistic arrows in chemical reactions

In addition to representing the structures of molecules, we also want to be able to depict the way in which chemical bonds are broken and formed within and between chemical species as chemical reactions occur. We do this by means of **mechanistic arrows**, which show the movement of electrons in bond-breaking and bond-making processes.

A two-headed mechanistic arrow shows the movement of a pair of electrons. The electrons move from the tail to the head of the arrow, resulting in a number of possible chemical processes, some of which are detailed below. *Note*: In order to better understand the charges on the atoms in the following examples, you should read the section on formal charges in chapter 5.

Bond breaking

Consider the following bond-breaking process.

$$H{-}Cl \longrightarrow H^+ + Cl^-$$

In this case, the arrow starts at the middle of the bond. The bond breaks in a **heterolytic** fashion, with both electrons ending up on the Cl atom. This gives rise to the formation of an H^+ cation (proton) and a Cl^- anion.

Bond making

Consider the following bond-making process.

$$(H_3C)_2C{=}\ddot{O}: \;+\; H^+ \longrightarrow (H_3C)_2C{=}\overset{+}{\ddot{O}}{-}H$$

Here, a bond is formed between the carbonyl O atom and the H^+ ion, with the two electrons in the bond coming from one of the lone pairs of electrons on the carbonyl O atom. This leaves a formal positive charge on the O atom.

Charge neutralisation

Consider the following charge-neutralisation process.

$$(H_3C)_2C{=}\overset{+}{\ddot{O}}{-}H \longrightarrow (H_3C)_2\overset{+}{C}{-}\ddot{O}{-}H$$

The positively charged carbonyl O atom can be neutralised by taking two electrons from the double bond and locating them on the carbonyl O atom. The arrow starts halfway along the bond and finishes at the O atom. This process leaves the carbonyl C atom positively charged.

In the following charge-neutralisation process:

$$(H_3C)_3C{-}\overset{+}{N}H_3 \longrightarrow (H_3C)_3C{-}\ddot{N}H_2 + H^+$$

proton loss to give a neutral compound occurs by taking the two electrons from an N—H bond and locating them on the N atom. The tail of the arrow therefore is situated halfway along the N—H bond and the head points to the N atom.

These examples illustrate some important facts about mechanistic arrows. Firstly, the tail of the arrow must be situated at a source of an electron pair: that is, either halfway along a bond or on a lone pair of electrons. Some thought will convince you that this means a mechanistic arrow can never start at either H^+ or an H atom (a very common mistake!). Similarly, the head of the arrow must be situated in a region able to accept a pair of electrons. Thus it would be unlikely, for example, for the head of an arrow to be at an atom bearing a partial negative charge, or a negatively charged ion. It is important to remember that total charge must be conserved in any mechanistic process. Thus, in all the examples above, both sides of the equations have the same total charge. It should also be noted that we can use more than one arrow to represent multiple simultaneous bond-making and bond-breaking processes. This is illustrated below.

$$HO^- + (H_3C)_2CBr{-}CH_3 \longrightarrow (H_3C)_2C{=}CH_2 + H_2O + Br^-$$

Here, three processes occur essentially simultaneously.

1. OH^- removes a proton from a CH_3 group, resulting in the formation of H_2O.
2. The pair of electrons which were in the C—H bond move so as to give a C=C double bond.
3. The C—Br bond breaks heterolytically, with both electrons in the bond moving to the Br, to give a Br^- ion.

A single-headed arrow shows the movement of a single electron. Such arrows are usually found only in reactions involving radicals, chemical species which contain one or more unpaired electrons. An example is given below, in which the Br—Br single bond is broken in a **homolytic** fashion, such that each of the bonded atoms receives an electron, giving two bromine radicals.

$$Br{-}Br \longrightarrow Br^{\bullet} + Br^{\bullet}$$

You will make extensive use of mechanistic arrows in the later chapters of this book, especially in writing organic reaction mechanisms.

2.3 Nomenclature

Representing molecules as images allows us to impart a great deal of information concerning molecular structure. But molecules can also be named, and there are occasions when this is more convenient than drawing a picture. In the early days of chemistry, the list of known compounds was short, and chemists could memorise the names of all of them. New compounds were often named by their discoverer after their place of origin, physical appearance or properties. Today, more than 70 million compounds are known and millions more are synthesised each year. Chemists consequently need systematic procedures for naming chemical compounds. The International Union of Pure and Applied Chemistry (IUPAC) has established uniform guidelines for naming various types of chemical substances, and chemists increasingly use IUPAC-approved names rather than their common counterparts. In this section we will introduce the rules for **nomenclature**, the system for the naming of compounds. You will encounter many chemical names in subsequent chapters, and the basics that you learn in this section will aid your interpretation of these names. We will investigate the naming of transition metal complexes in chapter 13. You should be aware that there are some compounds that are better known by their common, unsystematic name, and not their systematic IUPAC name. The best example of this is water, the systematic name of which is oxidane. Unsystematic names for a small number of common compounds are accepted by IUPAC.

Chemical Connections

IUPAC

The International Union of Pure and Applied Chemistry (IUPAC) is the organisation responsible for helping chemists talk to each other about their science. It does this by developing and systematising the nomenclature, symbols and terminology that are required for chemistry. The names and formulae of chemical compounds together with the units and symbols used in chemistry are central to the language of chemists, and rules are required to tell people how to put them together and work out what they mean. The rules are regularly updated, both to make them simpler, and to provide ways of naming new kinds of molecules that were not envisaged when the rules were first written.

The different sets of rules for nomenclature, symbols and terminology appear in IUPAC's 'colour' books. These are:

- the Red Book (*Nomenclature of Inorganic Chemistry*) (figure 2.19)
- the Blue Book (*Nomenclature of Organic Chemistry*) (figure 2.19)
- the Green Book (*Quantities, Units and Symbols in Physical Chemistry*)
- the Purple Book (*Macromolecular Nomenclature*)
- the Gold Book (*Chemical Terminology*)
- the Orange Book (*Analytical Terminology*)
- the White Book (*Biochemical Nomenclature*).

These books are all available online, and links to them can be found at www.iupac.org/home/publications/e-resources/nomenclature-and-terminology.html.

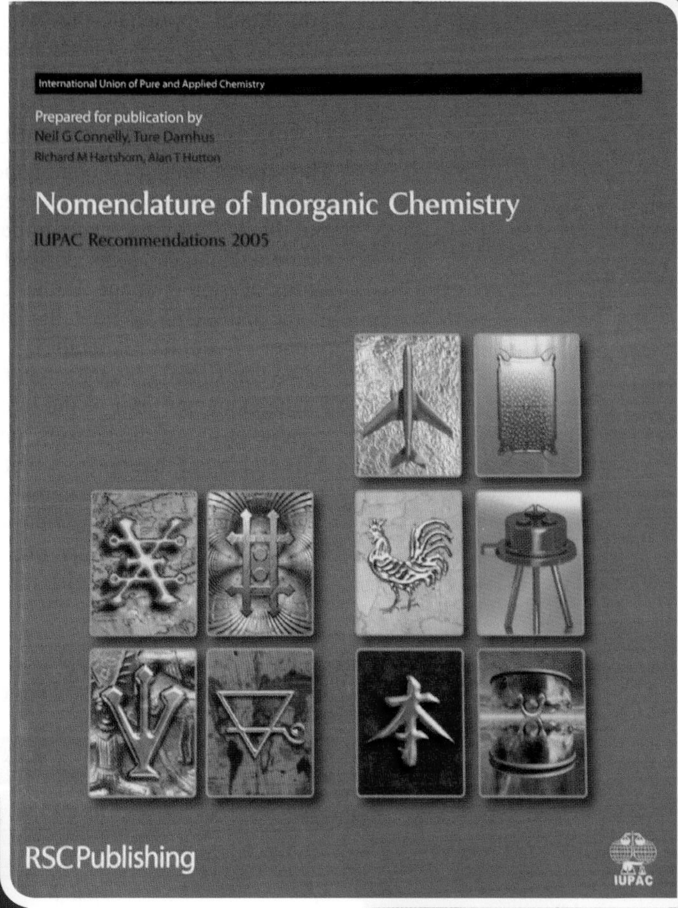

International Union of Pure and Applied Chemistry
Prepared for publication by
Neil G Connelly, Ture Damhus
Richard M Hartshorn, Alan T Hutton
Nomenclature of Inorganic Chemistry
IUPAC Recommendations 2005
RSCPublishing
IUPAC

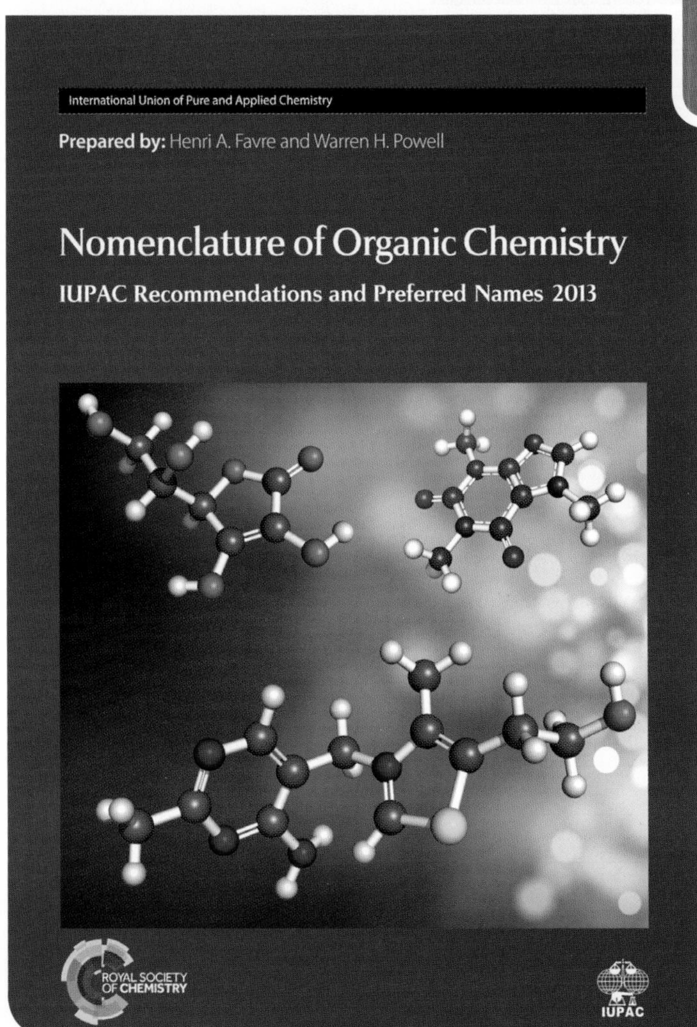

International Union of Pure and Applied Chemistry
Prepared by: Henri A. Favre and Warren H. Powell
Nomenclature of Organic Chemistry
IUPAC Recommendations and Preferred Names 2013
ROYAL SOCIETY OF CHEMISTRY
IUPAC

FIGURE 2.19 The Red Book and the Blue Book are IUPAC's 'rule books' for nomenclature, symbols and terminology.

Naming inorganic compounds

The way in which we name inorganic compounds (i.e. those not composed primarily of carbon and hydrogen) depends to some extent on the exact nature of the compound in question. We will examine the different possible types of inorganic compounds in the following pages. You should appreciate that what follows is, of necessity, an abbreviated version of the IUPAC rules for naming inorganic compounds. The full version may be found at the IUPAC website (www.iupac.org).

Nonmetallic binary compounds

The written name of a compound includes the names of the elements it contains and information about the number of atoms of each element present. The elements have to occur in some order, and this is set by the same guidelines as for the chemical formula (see p. 40). Names can contain elemental names, roots derived from elemental names and prefixes indicating the number of atoms of each element. Tables 2.4 and 2.5 list the more important roots and prefixes that appear in the names of nonmetallic binary compounds, compounds that contain only two elements, neither of which are metals.

We can summarise the rules for naming nonmetallic binary compounds in four guidelines.

1. The element closer to the left of the periodic table appears first. If both elements are from the same group of the periodic table, the lower one appears first.
2. The element that appears first retains its elemental name.
3. The second element begins with a root derived from its elemental name and ends with the suffix *-ide*. Some common roots are listed in table 2.4.
4. When there is more than one atom of a given element in the formula, the name of the element usually contains a prefix that specifies the number of atoms present. Common prefixes are given in table 2.5.

TABLE 2.4 Common roots for naming nonmetallic binary compounds.

Element	Full name	Root
As	arsenic	arsen-
Br	bromine	brom-
C	carbon	carb-
Cl	chlorine	chlor-
F	fluorine	fluor-
H	hydrogen	hydr-
I	iodine	iod-
N	nitrogen	nitr-
O	oxygen	ox-
P	phosphorus	phosph-
S	sulfur	sulf-

TABLE 2.5 Number prefixes for chemical names.

Number	Prefix	Example	Name
1	mono-	CO	carbon monoxide[(a)]
2	di-	SiO_2	silicon dioxide
3	tri-	NI_3	nitrogen triiodide
4	tetra-	$SnCl_4$	tin tetrachloride
5	penta-	PCl_5	phosphorus pentachloride
6	hexa-	SF_6	sulfur hexafluoride
7	hepta-	IF_7	iodine heptafluoride

(a) The final 'o' of the prefix is omitted in this case.

Numerical prefixes are essential in naming similar binary compounds. For example, nitrogen and oxygen form the six different compounds shown as space-filling models in figure 2.20: NO, nitrogen monoxide; NO_2, nitrogen dioxide; N_2O, dinitrogen oxide; N_2O_3, dinitrogen tri-oxide; N_2O_4, dinitrogen tetraoxide; and N_2O_5, dinitrogen pentaoxide.

FIGURE 2.20 Space-filling models of the six compounds formed between nitrogen and oxygen.

WORKED EXAMPLE 2.10

Naming binary compounds

Name the following binary compounds: SO_2, CS_2, BCl_3 and BrF_5.

Analysis

None of these compounds contains a metallic element, so we apply the guidelines for non-metallic binary compound nomenclature.

Solution

Name the first element, use a root plus *-ide* for the second element and indicate the number of atoms with prefixes. Therefore, we obtain:

SO_2: sulfur dioxide
CS_2: carbon disulfide
BCl_3: boron trichloride
BrF_5: bromine pentafluoride.

PRACTICE EXERCISE 2.13

Name the neutral compounds that contain:
(a) four atoms of bromine and one atom of silicon
(b) three atoms of oxygen and one atom of sulfur
(c) three atoms of fluorine and one atom of chlorine.

Binary compounds of hydrogen

Hydrogen requires special consideration because, as we saw earlier, it may appear first or second in the chemical formula of a compound and, as a result, it may appear first or second in the name. This is particularly evident in the diatomic molecules that hydrogen forms with elements from groups 1 and 17, which are named according to the guidelines on the previous page. For example, LiH is lithium hydride, and HF is hydrogen fluoride. With elements from groups 2 and 16, hydrogen forms compounds containing two atoms of hydrogen. Except for oxygen, there is only one commonly occurring binary compound for each of these elements, so the prefix *di-* is omitted. Examples are H_2S, hydrogen sulfide, and CaH_2, calcium hydride. Oxygen forms two binary compounds with hydrogen. These have unsystematic names: one is water, H_2O (systematic name oxidane), and the other is hydrogen peroxide, H_2O_2 (systematic name dioxidane). Binary compounds of hydrogen with elements from groups 13, 14 and 15 have unsystematic names in common use; B_2H_6 is diborane (systematic name diborane (6)), CH_4 is methane (systematic name carbane), NH_3 is ammonia (systematic name azane) and PH_3 is phosphine (systematic name phosphane). In fact, carbon, boron and silicon form many different binary compounds with hydrogen, and only the simplest of these are listed here.

Ionic compounds

Binary compounds which contain metal ions are often ionic, consisting of a cation and an anion (see chapter 1, p. 4). We name binary ionic compounds with the cation first and the anion, which takes the suffix *-ide*, last. For example, NaCl is named sodium chloride, while KI is named potassium iodide. Compounds such as CaF_2, in which the numbers of cations and anions are not the same, do not require the actual number of cations and anions to be specified. Thus, CaF_2 is named calcium fluoride, not calcium difluoride. The reason for this is that calcium is a group 2 element, and will therefore form only a 2+ cation. This will always combine with two 1− anions to form a neutral ionic compound, and therefore we need not specify *di*fluoride. Using the same reasoning, we can see that Na_2S and MgI_2 are named sodium sulfide and magnesium iodide, respectively. Ionic compounds containing polyatomic ions such as NH_4^+, NO_3^- and SO_4^{2-} are again named with the cation followed by the anion, but you will need to learn both the names and the charges of the common polyatomic ions to be able to name, and write chemical formulae for, compounds containing these ions. Table 2.6 (overleaf) lists the more common polyatomic ions.

TABLE 2.6 Names and chemical formulae of some common polyatomic ions.

Formula	Name	Formula	Name
Cations		**Oxoanions**	
NH_4^+	ammonium	SO_4^{2-}	sulfate
H_3O^+	hydronium (oxonium)	SO_3^{2-}	sulfite
Hg_2^{2+}	dimercury (2+)	NO_3^-	nitrate
Diatomic anions		NO_2^-	nitrite
OH^-	hydroxide	PO_4^{3-}	phosphate
CN^-	cyanide	MnO_4^-	permanganate
Anions containing carbon		CrO_4^{2-}	chromate
CO_3^{2-}	carbonate	$Cr_2O_7^{2-}$	dichromate
HCO_3^-	hydrogencarbonate (bicarbonate)	ClO_4^-	perchlorate
CH_3COO^-	acetate	ClO_3^-	chlorate
$C_2O_4^{2-}$	oxalate	ClO_2^-	chlorite
		ClO^-	hypochlorite

The **oxoanions** listed in table 2.6 consist of a central atom surrounded by oxygen atoms. Their names can be deduced using the following rules.

1. The name has a root taken from the name of the central atom (e.g. chlorate, ClO_3^-, and nitrite, NO_2^-).
2. When an element forms two different oxoanions, the one with fewer oxygen atoms ends in *-ite*, and the other ends in *-ate* (e.g. SO_3^{2-}, sulfite, and SO_4^{2-}, sulfate).
3. Chlorine, bromine and iodine each form four different oxoanions that are distinguished by prefixes and suffixes. The nomenclature of these ions is illustrated for bromine, but it applies to chlorine and iodine as well: BrO^-, hypobromite; BrO_2^-, bromite; BrO_3^-, bromate; and BrO_4^-, perbromate.
4. A polyatomic anion with a charge more negative than 1− may add H^+ to give another anion. These anions are named from the parent anion by adding the word hydrogen. For example, HCO_3^- is hydrogencarbonate, HPO_4^{2-} is hydrogenphosphate and $H_2PO_4^-$ is dihydrogenphosphate.

Naming organic compounds

Organic compounds are composed primarily of carbon and hydrogen atoms, and the naming system used is based on the number of carbon atoms in the particular molecule. Not surprisingly, there are an enormous number of rules for naming organic compounds and at this stage we will restrict ourselves to the basics. Indeed, computer programs are now able to generate names from nearly any structural diagram, providing a helpful supplement to knowledge of the rules of IUPAC nomenclature.

Functional groups

The concept of functional groups underpins all of organic chemistry and makes it the systematic discipline that it is. A **functional group** is simply a group of one or more atoms within a molecule, bonded together in a particular fashion, and is usually the point of reaction within a molecule. Organic molecules are named primarily according to the functional group or groups they contain. The power of the functional group concept is that molecules containing the same functional group tend to behave in chemically similar ways, and this allows us to predict their reactivity towards particular reagents with some confidence. For example, although the two molecules below look quite different, they both contain a carbonyl functional group, C═O, as part of an aldehyde group, —CHO.

We can therefore predict that both molecules will undergo a reaction called oxidation, in which the —CHO group is converted to a —COOH (carboxylic acid) functional group. Similarly, both compounds will react with species called reducing agents, which will convert the —CHO group to a —CH_2OH (primary alcohol) group. (We will learn a lot more about the specific reactions involved later in this textbook.)

It therefore makes sense to name organic compounds according to their functional groups, and we will learn how this is done on the following pages. But first, the common functional groups that you will encounter are outlined in table 2.7. Note that R is a commonly used symbol to denote either a hydrogen atom, or an alkyl group (table 2.9, p. 59). You should be able to identify the presence of each of these functional groups in any molecule.

TABLE 2.7 Common functional groups.

Functional group	Name of group	Found in	R =
$R{-}O{-}H$	hydroxyl	alcohols	C
$R{-}C({=}O){-}H$	carbonyl	aldehydes	C or H
$R{-}C({=}O){-}R$	carbonyl	ketones	C
$R{-}C({=}O){-}OH$	carboxyl	carboxylic acids	C or H

Alcohols

The —OH (hydroxyl) group is present in all alcohols. It is attached to a carbon atom, which itself may be attached to one, two or three other carbon atoms (the exception is methanol, CH_3OH, which contains only one carbon atom). Alcohols are classified as primary (1°), secondary (2°) or tertiary (3°) according to this number, as shown in figure 2.21.

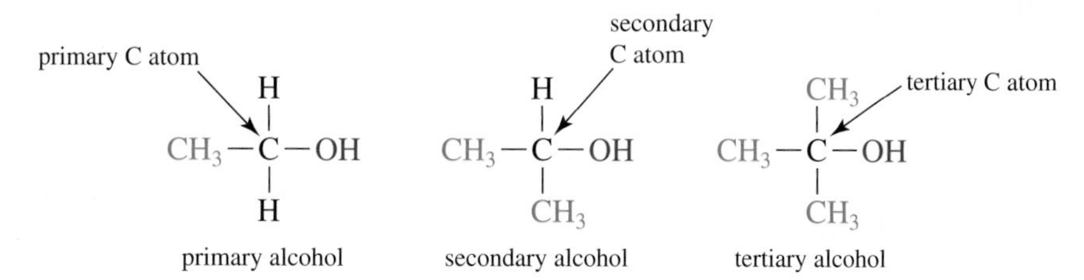

FIGURE 2.21 Structural formulae of primary, secondary and tertiary alcohols.

We can also classify carbon atoms within molecules according to this scheme. Thus a carbon atom attached to one carbon atom (as in a primary alcohol) is called a **primary carbon atom**, a carbon atom attached to two carbon atoms (as in a secondary alcohol) is called a **secondary carbon atom**, and a carbon atom attached to three carbon atoms (as in a tertiary alcohol) is called a **tertiary carbon atom** (figure 2.21).

WORKED EXAMPLE 2.11

Writing possible structures for alcohols

Write condensed structural formulae for the two alcohols with the molecular formula C_3H_8O. Classify each as primary, secondary or tertiary.

Analysis

We know that alcohols contain the —OH group. We must therefore draw the two possible structures of the formula C_3H_8O which contain an —OH group, remembering that each carbon atom in the molecule must have four bonds. In problems like this, it is generally easiest to draw the structural formula, and then convert it to the condensed form.

Solution

We have a three-carbon chain in the molecule, and so the —OH group can be attached only to either the terminal carbon (it does not matter which one; the same compound results regardless of which end we attach the functional group to) or the central carbon. Therefore, we begin by drawing the carbon chain, then attach the —OH group in the two possible positions, and fill in the hydrogen atoms. Doing this gives us the following two compounds.

$H{-}CH_2{-}CH_2{-}CH_2{-}O{-}H$ (each C bearing two H) or $CH_3CH_2CH_2OH$

1° alcohol

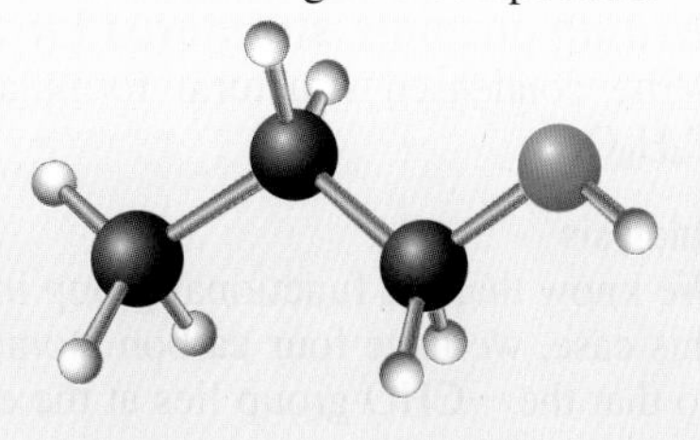

The compound above is a primary alcohol as the —OH group is attached to a carbon atom directly bonded to only one other carbon.

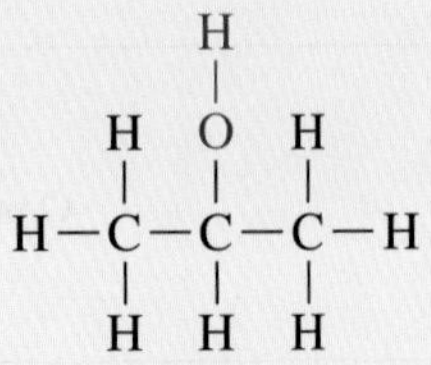

$$\begin{array}{ccccc} & & H & & \\ & & | & & \\ & H & O & H & \\ & | & | & | & \\ H- & C- & C- & C- & H \\ & | & | & | & \\ & H & H & H & \end{array} \quad \text{or} \quad CH_3CH(OH)CH_3$$

2° alcohol

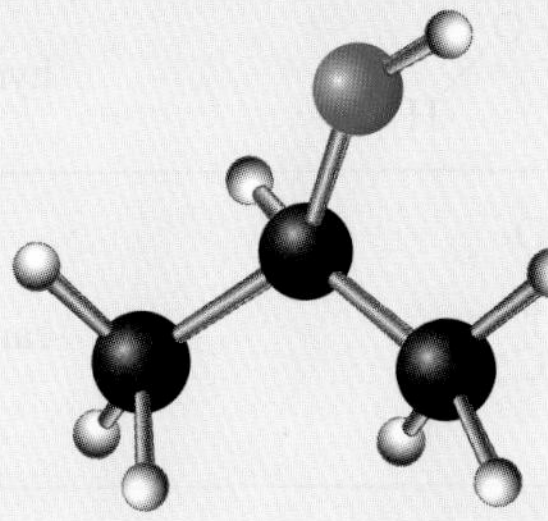

In contrast to the first structure, the above compound is a secondary alcohol, the central carbon in the chain being bonded to two other carbon atoms.

PRACTICE EXERCISE 2.14 Write condensed structural formulae for the eight alcohols that have the molecular formula $C_5H_{12}O$. Classify each as primary, secondary or tertiary.

Aldehydes and ketones

Aldehydes and ketones both contain the same functional group, the carbonyl group (C=O), but they differ in the way that the carbonyl group is bonded to the rest of the molecule. Aldehydes *always* have the carbon atom of the carbonyl group bonded to at least one hydrogen atom, and this means that the carbonyl group of an aldehyde must always be at the *end* of a carbon chain. Conversely, the carbon atom of the carbonyl group in ketones is *always* bonded to two other carbon atoms, and therefore can never be at the end of a carbon chain. We can write the aldehyde either explicitly as:

$$\overset{O}{\underset{RCH}{\|}}$$

or as RCHO in a condensed structural formula. The differences between aldehydes and ketones are shown in figure 2.22.

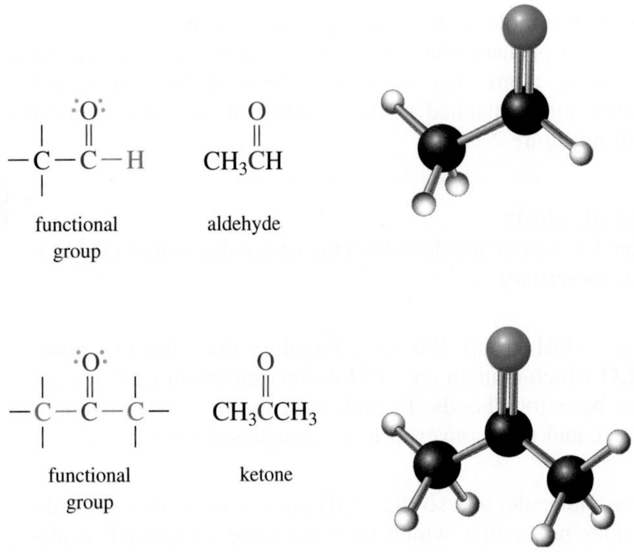

functional group: $-\overset{|}{\underset{|}{C}}-\overset{\ddot{O}}{\overset{\|}{C}}-H$ — aldehyde: $CH_3\overset{O}{\overset{\|}{C}}H$

functional group: $-\overset{|}{\underset{|}{C}}-\overset{\ddot{O}}{\overset{\|}{C}}-\overset{|}{\underset{|}{C}}-$ — ketone: $CH_3\overset{O}{\overset{\|}{C}}CH_3$

FIGURE 2.22 Structural formulae and ball-and-stick models of an aldehyde and a ketone.

WORKED EXAMPLE 2.12

Writing possible structures for aldehydes

Write condensed structural formulae for the two aldehydes with the chemical formula C_4H_8O.

Analysis

We know that the functional group in aldehydes can only be at the end of a carbon chain. In this case, we have four carbon atoms, and there are two possible ways of connecting these so that the —CHO group lies at the end of a chain.

Solution

Begin by writing the —CHO group and then attaching the remaining three carbon atoms. We can connect the four carbon atoms in a straight chain or a branched chain. The two possibilities are shown below.

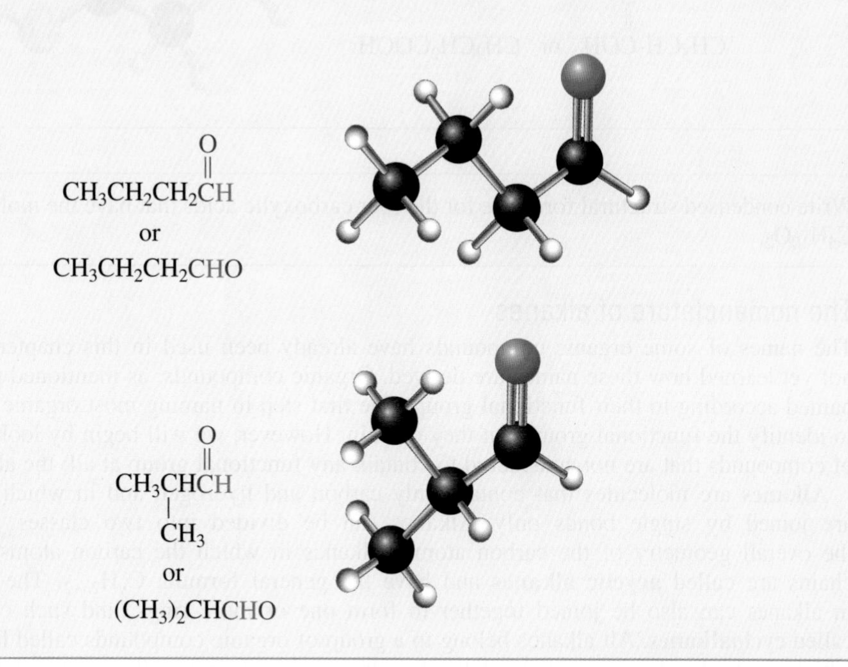

PRACTICE EXERCISE 2.15

Write condensed structural formulae for the three ketones with the molecular formula $C_5H_{10}O$.

Carboxylic acids

The functional group found in all carboxylic acids is the carboxyl group, —COOH. We can write this explicitly as:

$$\overset{\displaystyle O}{\overset{\|}{\text{RCOH}}}$$

or, more commonly, as RCOOH. As the carbon atom of a carboxyl group can bond to only one other atom, carboxyl groups are *always* found at the end of a carbon chain.

$$\text{R}-\overset{\displaystyle :\ddot{O}:}{\overset{\|}{C}}-\ddot{\underset{\cdot\cdot}{O}}-\text{H} \qquad \overset{\displaystyle O}{\overset{\|}{\text{CH}_3\text{COH}}}$$

functional group — acetic acid

WORKED EXAMPLE 2.13

Writing possible structures for carboxylic acids

Write a condensed structural formula for the carboxylic acid with the molecular formula $C_3H_6O_2$.

Analysis

There is only one possible compound we can draw here. It must contain a three-carbon chain with a —COOH group on the end.

Solution

Begin by drawing the —COOH group and then attach the remaining carbon atoms. Finish by adding hydrogen atoms so that all the carbon atoms have four bonds. This gives us the compound shown on the next page.

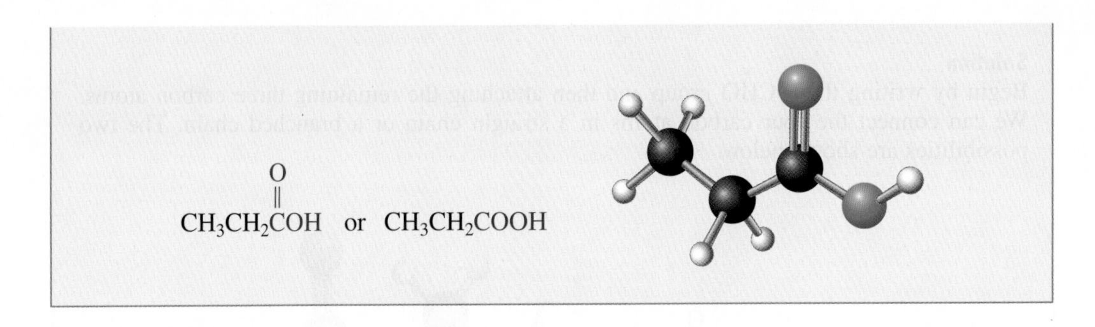

PRACTICE EXERCISE 2.16 Write condensed structural formulae for the four carboxylic acids that have the molecular formula $C_5H_{10}O_2$.

The nomenclature of alkanes

The names of some organic compounds have already been used in this chapter, but we have not yet learned how these names are derived. Organic compounds, as mentioned previously, are named according to their functional group. The first step in naming most organic compounds is to identify the functional group that they contain. However, we will begin by looking at a series of compounds that are not considered to contain any functional group at all: the alkanes.

Alkanes are molecules that contain only carbon and hydrogen and in which carbon atoms are joined by single bonds only. Alkanes can be divided into two classes, depending on the overall geometry of the carbon atoms. Alkanes in which the carbon atoms are joined in chains are called **acyclic alkanes** and have the general formula C_nH_{2n+2}. The carbon atoms in alkanes can also be joined together to form one or more rings, and such compounds are called **cycloalkanes**. All alkanes belong to a group of organic compounds called **hydrocarbons**, which are molecules composed solely of carbon and hydrogen. Alkanes are sometimes called **saturated hydrocarbons**, to denote the fact that all the carbon–carbon bonds are single. As there are no functional groups in alkanes, we name them according to the length of the longest carbon chain. There are two parts to the name of an alkane: a prefix, indicating the number of carbon atoms in the chain, and the ending *–ane*. The prefixes for carbon chain lengths up to 10 carbon atoms are shown in table 2.8.

TABLE 2.8 Prefixes used in the IUPAC system for chain lengths of 1 to 10 carbon atoms.

Prefix	Number of carbon atoms
meth-	1
eth-	2
prop-	3
but-	4
pent-	5
hex-	6
hept-	7
oct-	8
non-	9
dec-	10

The simplest alkane is methane, which contains one carbon atom and has the formula CH_4. Note that the suffix and prefix are not hyphenated, but are written together as one word. The alkane with six carbon atoms in a straight chain ($CH_3CH_2CH_2CH_2CH_2CH_3$) is likewise called hexane. Names for the straight-chain alkanes are straightforward, but things become more complex for alkanes in which the carbon chain is branched. In order to name such alkanes, identify the longest straight carbon chain (the **parent chain**), and then treat any groups which branch off this chain as **substituents**. The IUPAC name of an alkane with a branched chain consists of a root name that indicates the longest chain of carbon atoms in the compound, substituent names that indicate the groups bonded to the parent chain, and numbers that indicate the carbon atom to which the substituents are attached. This is illustrated opposite for 4-methyloctane, C_9H_{20}, in which the longest carbon chain contains eight carbon atoms (hence *octane*) and the $—CH_3$ (methyl) group bonded to C(4) is a substituent (hence *4-methyl*).

substituent

parent chain

$$\begin{array}{c} \quad\quad\quad CH_3 \\ \quad\quad\quad | \\ CH_3CH_2CH_2CHCH_2CH_2CH_2CH_3 \\ 1\ \ 2\ \ 3\ \ 4\ \ 5\ \ 6\ \ 7\ \ 8 \end{array}$$

4-methyloctane

The substituents in alkanes are named according to the alkane from which they are derived. For example, a $—CH_3$ substituent is derived from methane, CH_4, through the removal of one H atom. Similarly, a $—C_2H_5$ substituent is derived from ethane, C_2H_6, by removal of one H atom. A substituent derived from an alkane by the removal of one H atom is called an **alkyl group** and is commonly represented by the symbol R. Alkyl groups are named by dropping the *-ane* from the name of the parent alkane and adding the suffix *-yl*. Therefore, a $—CH_3$ substituent is called *methyl*, and a $—C_2H_5$ substituent *ethyl*. Table 2.9 gives names and structural formulae for eight of the most common alkyl groups. Note that, when alkyl substituents contain three or more carbon atoms, there is more than one possible arrangement of the carbon atoms in the substituent. We differentiate between these possibilities using the prefixes iso, *sec-* and *tert-*. Iso implies the presence of a $—CH(CH_3)_2$ group at the end of the substituent carbon chain (the only possibility for a three-atom chain), while the prefixes *sec-* and *tert-* imply that the substituent is bonded to the longest carbon chain via a secondary or tertiary carbon atom, respectively. When *sec-* and *tert-* are part of a name, they are always italicised and hyphenated. However, iso, as used in both isopropyl and isobutyl, is not treated as a prefix like *sec-* and *tert-*, and is neither italicised nor hyphenated.

TABLE 2.9 Names, formulae and abbreviations of the most common alkyl groups.

Name	Condensed structural formula	Abbreviation
methyl	$—CH_3$	Me
ethyl	$—CH_2CH_3$	Et
propyl	$—CH_2CH_2CH_3$	Pr
isopropyl	$—CHCH_3$ with CH_3 on C1	i-Pr
butyl	$—CH_2CH_2CH_2CH_3$	Bu
isobutyl	$—CH_2CHCH_3$ with CH_3 on C2	i-Bu
sec-butyl	$—CHCH_2CH_3$ with CH_3 on C1	*s*-Bu
tert-butyl	$—CCH_3$ with two CH_3 on C1	*t*-Bu

The complete IUPAC rules for naming alkanes are as follows.

1. The name for an alkane with an unbranched chain of carbon atoms consists of a prefix showing the number of carbon atoms in the chain (table 2.8) and the ending *–ane*.
2. For branched-chain alkanes, the longest chain of carbon atoms is the parent chain, and its name becomes the root name.
3. For an alkane with one substituent, number the parent chain so that the carbon atom bearing the substituent is given the lowest possible number. For example, we number the five-carbon chain in the following compound from the right-hand end.

$$\begin{array}{c} \quad\quad\quad\quad CH_3 \\ \quad\quad\quad\quad | \\ CH_3CH_2CH_2CHCH_3 \end{array}$$

4. Give the substituent on the parent chain a name and a number. The number shows the carbon atom of the parent chain to which the substituent is bonded. Use a hyphen to connect the number to the name. For example, the methyl group in 3-methylpentane is attached to C(3), as shown below.

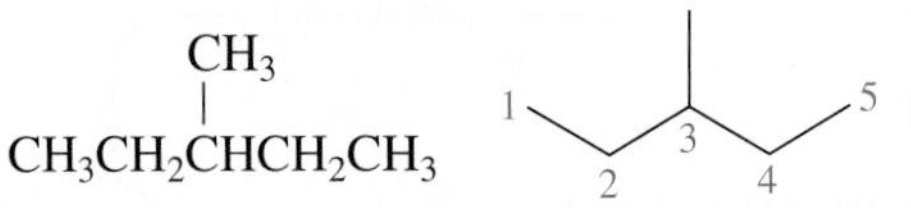

$$\begin{array}{c} CH_3 \\ | \\ CH_3CH_2CHCH_2CH_3 \end{array}$$

3-methylpentane

5. If there are two or more identical substituents, number the parent chain from the end that gives the lower number to the substituent closest to the end of the chain. If there are two or more identical substituents, the number of times they occur is indicated by the prefixes *di-*, *tri-*, *tetra-*, *penta-*, *hexa-* and so on. A comma is used to separate position numbers. Of the two possibilities below, we choose the one that numbers the carbon atoms to which the methyl groups are attached as 2 and 4, rather than 3 and 5.

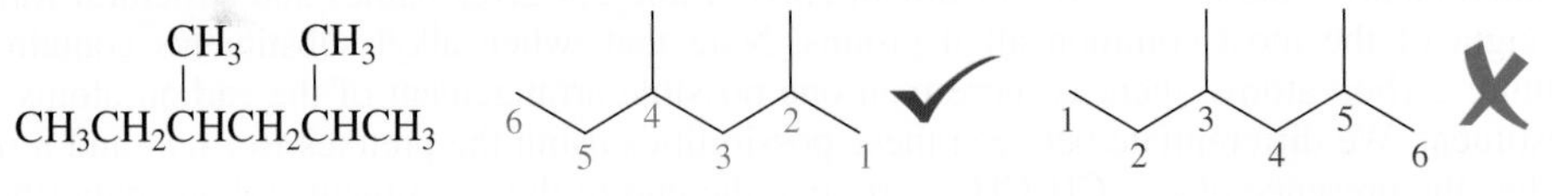

$$\begin{array}{c} CH_3 \quad CH_3 \\ | \qquad | \\ CH_3CH_2CHCH_2CHCH_3 \end{array}$$

2,4-dimethylhexane ✓ (not 3,5-dimethylhexane) ✗

6. If step 5 leads to more than one possibility, number the parent chain such that the first point of difference has the lowest possible number. Of the two possibilities below, we choose the one that numbers the methyl group shown in red 3 rather than 4.

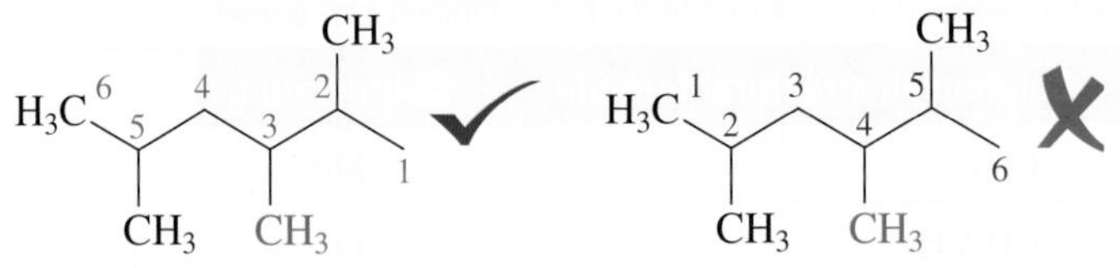

2,3,5-trimethylhexane ✓ (not 2,4,5-trimethylhexane) ✗

7. If there are two or more different substituents, list them in alphabetical order, and (as usual) number the chain from the end that gives the lower number to the substituent encountered first or that provides the first point of difference. If there are different substituents in equivalent positions on opposite ends of the parent chain, the substituent of lower alphabetical order is given the lower number. The importance of alphabetically ordering the substituents is shown in the following example.

$$\begin{array}{c} CH_3 \\ | \\ CH_3CH_2CHCH_2CHCH_2CH_3 \\ | \\ CH_2CH_3 \end{array}$$

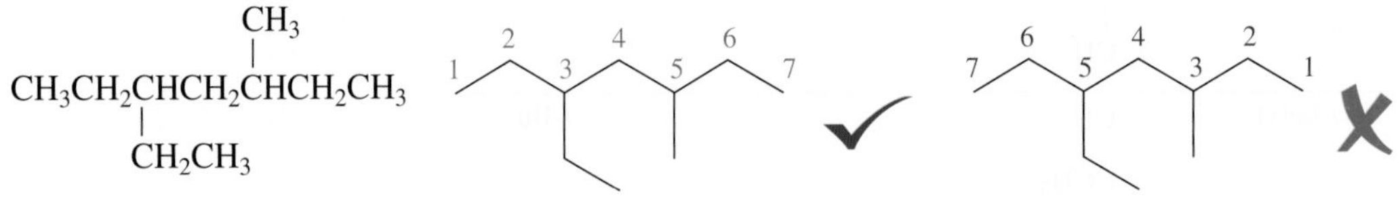

3-ethyl-5-methylheptane ✓ (not 3-methyl-5-ethylheptane) ✗

8. The prefixes *di-*, *tri-*, *tetra-* and so on, and the hyphenated prefixes *sec-* and *tert-* are disregarded for the purposes of placing the substituents in alphabetical order. Note that both isobutyl and isopropyl substituents are treated as beginning with the letter 'i'. Put the names of the substituents in alphabetical order first, and then insert the prefix. In the following example, the substituents are ordered as ethyl and methyl, not dimethyl and ethyl.

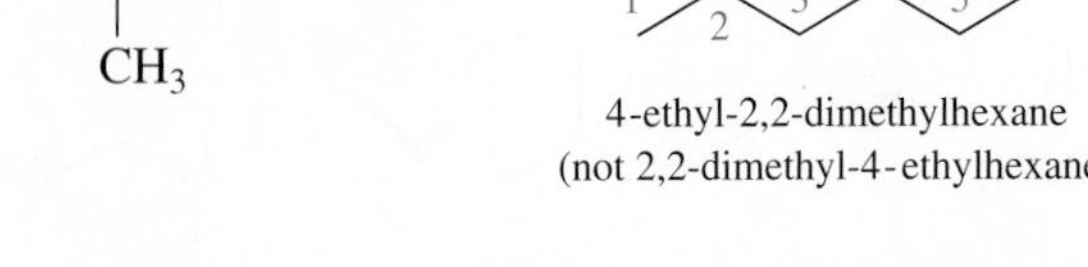

$$\begin{array}{c} CH_3 \quad CH_2CH_3 \\ | \qquad | \\ CH_3CCH_2CHCH_2CH_3 \\ | \\ CH_3 \end{array}$$

4-ethyl-2,2-dimethylhexane
(not 2,2-dimethyl-4-ethylhexane)

WORKED EXAMPLE 2.14

Writing IUPAC names for alkanes

Write IUPAC names for the following alkanes.

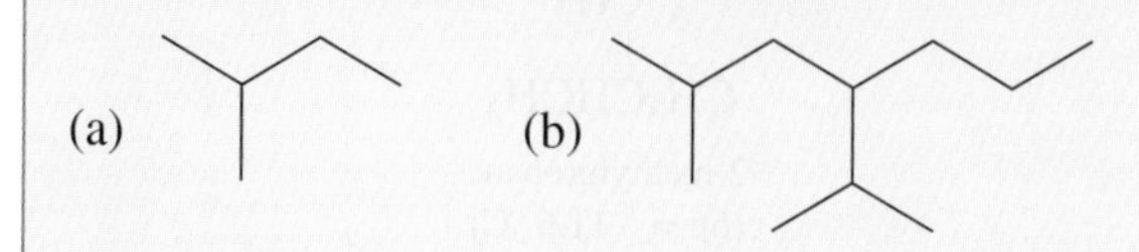

Analysis

We need to follow the rules for naming alkanes outlined previously. When naming alkanes, start by identifying the longest carbon chain, and then identify and number the substituents.

Solution

We will consider each alkane in turn. The longest carbon chain in (a) contains four carbon atoms and therefore will be named as a substituted butane. We then number the parent chain to give the methyl substituent the lowest possible number. This is done in either of the two equivalent ways shown below (note that the substituent is at C(2) in both cases).

(a)

We now name the substituent and specify its position on the parent chain to obtain the full name. This alkane is therefore named 2-methylbutane.

The longest carbon chain in (b) contains seven carbon atoms, and therefore it will be named as a substituted heptane. There are two substituents in this molecule: a methyl group and an isopropyl group, and therefore we must number the chain to give the substituent closest to the end of the chain the lowest number. This is shown in the following diagram, where we again have two equivalent possibilities.

(b)

To name the alkane, list the substituents in alphabetical order and specify their positions on the parent chain. As isopropyl comes before methyl, this alkane is named 4-isopropyl-2-methylheptane.

Is our answer reasonable?

The easiest way to check the answers to this type of question is to work backwards from the name we have deduced, and derive the chemical structure. If we do that in this case, we find that our answers are correct.

PRACTICE EXERCISE 2.17

Write the IUPAC names for the following alkanes.

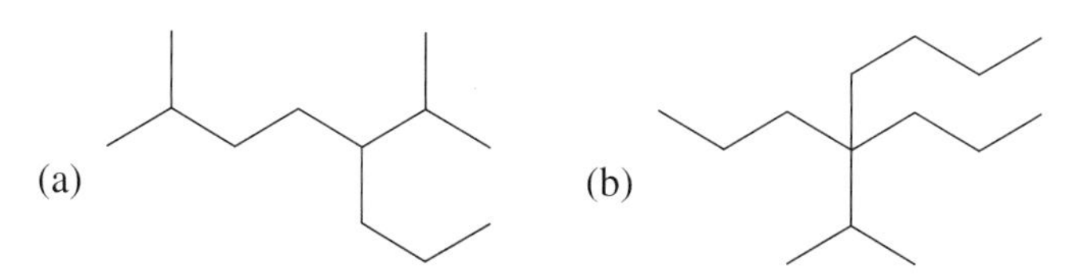

Constitutional isomerism in alkanes

The concept of isomerism was mentioned previously with respect to the relative positioning of the two methyl groups in 1,2-dimethylcyclopentane. The alkanes provide an excellent example of the different ways in which a particular number of carbon and hydrogen atoms may be joined. There is only one order of attachment, and hence one way of writing CH_4 (methane), C_2H_6 (ethane) and C_3H_8 (propane). However, when we come to C_4H_{10} two orders of attachment of atoms are possible.

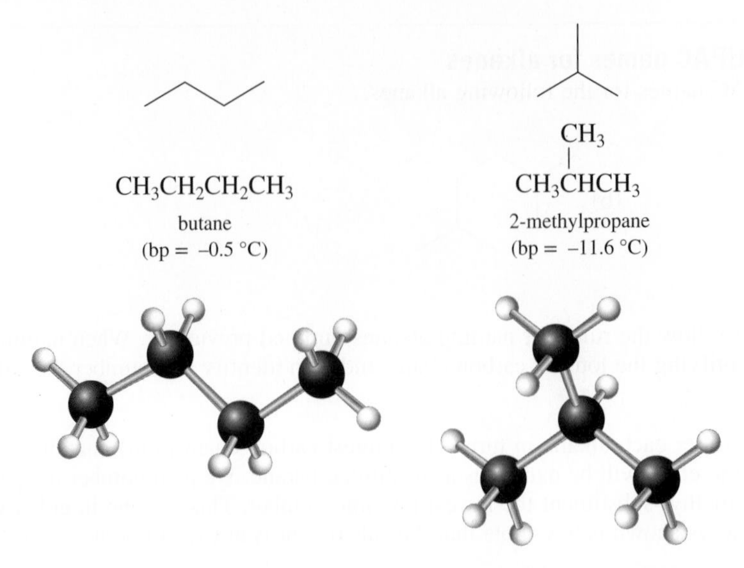

In one of these, butane, the four carbon atoms are bonded in a straight chain; in the other, 2-methylpropane, three carbon atoms are bonded in a straight chain, with the fourth carbon atom as a branch. Butane and 2-methylpropane are constitutional isomers. **Constitutional isomers** are compounds with the same chemical formula but a different order of attachment of the constituent atoms. Constitutional isomers can usually be distinguished by their differing physical properties; butane and 2-methylpropane provide a good example of this; butane boils at −0.5 °C, while 2-methylpropane boils at −11.6 °C. If we consider C_5H_{12}, we can obtain the three constitutional isomers shown below.

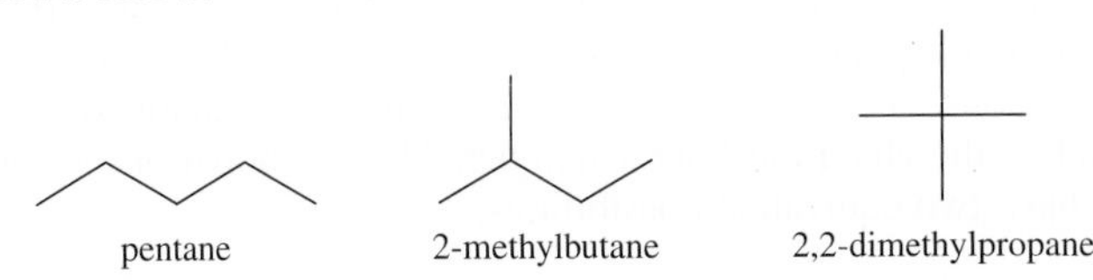

The number of possible constitutional isomers becomes very large very quickly. For example, the alkane with the chemical formula $C_{25}H_{52}$ has a staggering 36 797 588 possible constitutional isomers. Constitutional isomers are not unique to alkanes, and can be found throughout organic chemistry; indeed, worked examples 2.11 and 2.12 involved drawing constitutional isomers of alcohols and aldehydes, respectively.

WORKED EXAMPLE 2.15

Recognising constitutional isomers

Do the structural formulae in each of the following pairs of molecules represent the same compound or constitutional isomers?

(a) $CH_3CH_2CH_2CH_2CH_2CH_3$ and $CH_3CH_2CH_2$ (with $CH_2CH_2CH_3$ attached below the last CH_2) (each is C_6H_{14})

(b) CH_3CHCH_2CH (with CH_3 above the first CH, and CH_3 above and CH_3 below the last CH) and $CH_3CH_2CHCHCH_3$ (with CH_3 above the first CH and CH_3 below the second CH) (each is C_7H_{16})

Analysis

To determine whether these structural formulae represent the same compound or constitutional isomers, we must first find the longest chain of carbon atoms. Note that it makes no difference whether the chain is drawn straight or bent. We then number the longest chain from the end nearest the first branch and compare the lengths of each chain, and the sizes and locations of any branches. Structural formulae that have the same order of attachment of atoms represent the same compound; those which have different orders of attachment of atoms represent constitutional isomers. Note that similarities or differences between structural formulae are sometimes made clearer if they are drawn as line structures.

Solution

(a) Each structural formula has an unbranched chain of six carbon atoms. The two structures are identical and represent the same compound.

$$\overset{1}{CH_3}\overset{2}{CH_2}\overset{3}{CH_2}\overset{4}{CH_2}\overset{5}{CH_2}\overset{6}{CH_3} \text{ and } \overset{1}{CH_3}\overset{2}{CH_2}\overset{3}{CH_2}$$

$$\overset{4}{CH_2}\overset{5}{CH_2}\overset{6}{CH_3}$$

(b) Each structural formula has a chain of five carbon atoms with two branches. Although the branches are identical, they are at different locations on the chains. Therefore, these structural formulae represent constitutional isomers.

$CH_3CHCH_2CH(CH_3)_2$ — 2,4-dimethylpentane (bp = 80.6 °C) and $CH_3CH_2CH(CH_3)CHCH_3(CH_3)$ — 2,3-dimethylpentane (bp = 89.8 °C)

2,4-dimethylpentane
(bp = 80.6 °C)

2,3-dimethylpentane
(bp = 89.8 °C)

PRACTICE EXERCISE 2.18

Do the structural formulae in each of the following pairs of molecules represent the same compound or constitutional isomers?

(a) and (b) and

WORKED EXAMPLE 2.16

Drawing constitutional isomers

Draw structural formulae for the five constitutional isomers with the molecular formula C_6H_{14}.

Analysis

There are six carbon atoms in the molecule, so we can expect constitutional isomers with longest carbon chain lengths of six, five, four and possibly three. We must draw all possible isomers with these chain lengths.

Solution

The constitutional isomers of C_6H_{14} are as follows.

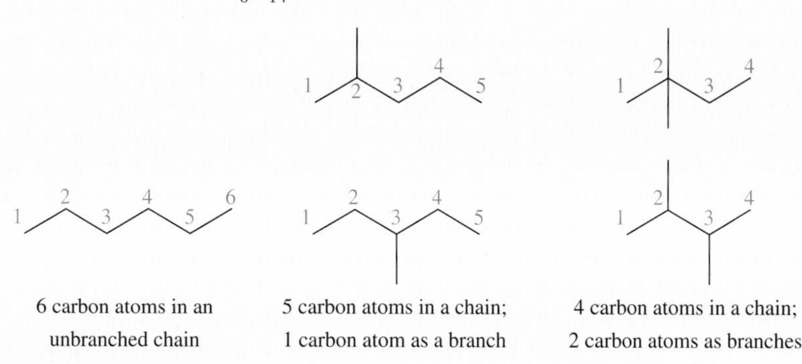

6 carbon atoms in an unbranched chain

5 carbon atoms in a chain; 1 carbon atom as a branch

4 carbon atoms in a chain; 2 carbon atoms as branches

Note that we find that there is no way of writing C_6H_{14} with only three carbon atoms in the longest chain.

Is our answer reasonable?

There are no other possible arrangements of six carbon atoms and fourteen hydrogen atoms, each singly bonded to each other. Therefore, we can be confident our answers are correct.

PRACTICE EXERCISE 2.19

Draw the nine constitutional isomers with the formula C_7H_{16}.

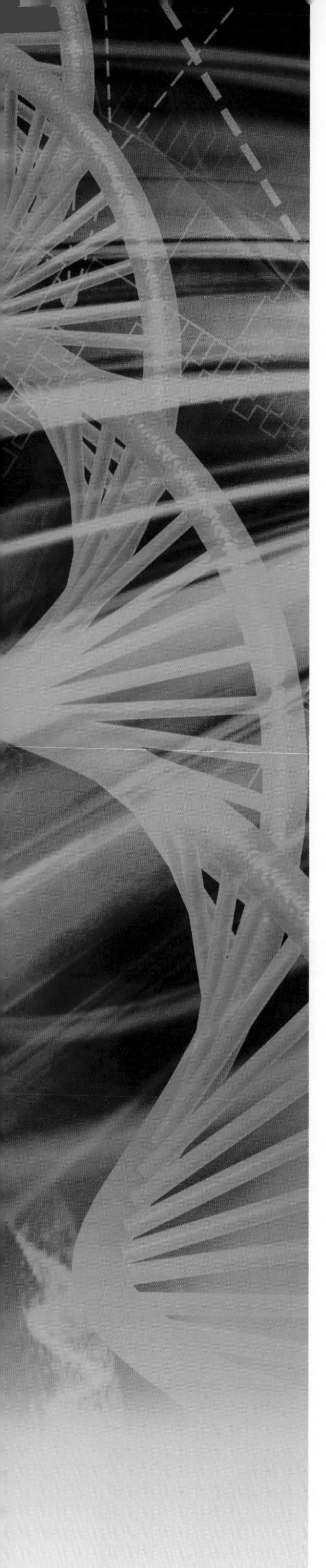

Chemical Connections

Bosutnib

In this chapter, we have looked at how to name some small organic molecules. In the case of large molecules, IUPAC names start becoming unwieldy, and we often give them trivial names instead. This is particularly true of drug molecules — for example, when we take a tablet containing *N*-(4-hydroxyphenyl)ethanamide to relieve pain, we refer to it as paracetamol. It is also generally much easier to draw the structure of a large molecule than to name it.

The molecule drawn below has the IUPAC name 4-[(2,4-dichloro-5-methoxyphenyl)amino]-6-methoxy-7-[3-(4-methylpiperazin-1-yl)propoxy]quinoline-3-carbonitrile, but is better known by its trivial name 'bosutnib'. In 2012 it was approved by the US FDA (the US Food and Drug Administration) for use as a drug against a particular type of leukaemia.

However, prior to this approval, scientists in England and the USA investigating the efficacy of this drug found some unusual results when they looked at its structure. They found that their 'bosutnib', which they had bought from a particular chemical supplier, was in fact not bosutnib at all, but a constitutional isomer of bosutnib which had the following structure.

You can see that the parts of the molecules highlighted in red are not the same. Sadly, this meant they had wasted a significant amount of time and money studying the wrong drug! This was presumed to have arisen from the use of the wrong molecule in the synthesis of bosutnib by the chemical supplier. In order to prepare authentic bosutnib, the molecule 2,4-dichloro-5-methoxyaniline is required, but the chemical supplier must have mistakenly used the constitutional isomer 3,5-dichloro-4-methoxyaniline.

2,4-dichloro-5-methoxyaniline

3,5-dichloro-4-methoxyaniline

How this mistake occurred is not known, but this story provides a very good chemical lesson that you should remember regardless of which field of science you are in — always check to make sure that the stuff in the bottle is in fact what it says on the label!

General organic nomenclature

The principles of IUPAC nomenclature as applied to alkanes can be extended to the naming of any organic compound. The name we give to any compound with a chain of carbon atoms consists of three parts: a prefix, an infix and a suffix. Each part provides specific information about the structural formula of the compound.

1. The prefix shows the number of carbon atoms in the parent chain. Prefixes that show the presence of 1 to 10 carbon atoms in a chain were given in table 2.8.
2. The infix shows the nature of the carbon–carbon bonds in the parent chain. The three possibilities are given in table 2.10.

TABLE 2.10 The three possible infixes and their meanings.

Infix	Nature of carbon–carbon bonds in the parent chain
-an-	all single bonds
-en-	one or more double bonds
-yn-	one or more triple bonds

3. The suffix shows the class of compound to which the substance belongs, and therefore the functional group(s) present in the compound. The suffixes for the classes of compound we have already met are as shown in table 2.11.

TABLE 2.11 Some suffixes and the class of compound that they designate.

Suffix	Class of compound
-e	hydrocarbon
-ol	alcohol
-al	aldehyde
-one	ketone
-oic acid	carboxylic acid

We can illustrate the use of prefixes, infixes and suffixes with respect to the following four organic compounds. We will divide each name into a prefix, an infix and a suffix, and specify the information about the structural formula that is contained in each part of the name.

(a)

$$CH_2{=}CHCH_3$$

propene

The prefix *prop-* means three carbon atoms in the parent chain. The infix *-en-* refers to the presence of one or more C=C double bonds in the molecule, while the suffix *-e* means that the compound is a hydrocarbon, composed of only carbon and hydrogen atoms.

(b)

$$CH_3CH_2OH$$

ethanol

The prefix *eth-* means two carbon atoms in the parent chain. The infix *-an-* means there are no carbon–carbon multiple bonds in the molecule, and the suffix *-ol* means that there is an alcohol functional group in the molecule.

(c)

$$CH_3CH_2CH_2CH_2\overset{\overset{\displaystyle O}{\|}}{C}OH$$

pentanoic acid

The prefix *pent-* means five carbon atoms in the parent chain. The infix *-an-* means there are no carbon–carbon multiple bonds in the molecule, and the suffix *-oic acid* refers to the presence of a carboxylic acid in the molecule.

(d)

$$HC{\equiv}CH$$

ethyne

The prefix *eth-* means two carbon atoms in the parent chain. The infix *-yn-* means that there are one or more C≡C triple bonds in the molecule. The suffix *-e* means that the compound is a hydrocarbon, composed of only carbon and hydrogen atoms.

We can summarise this information in the following diagram.

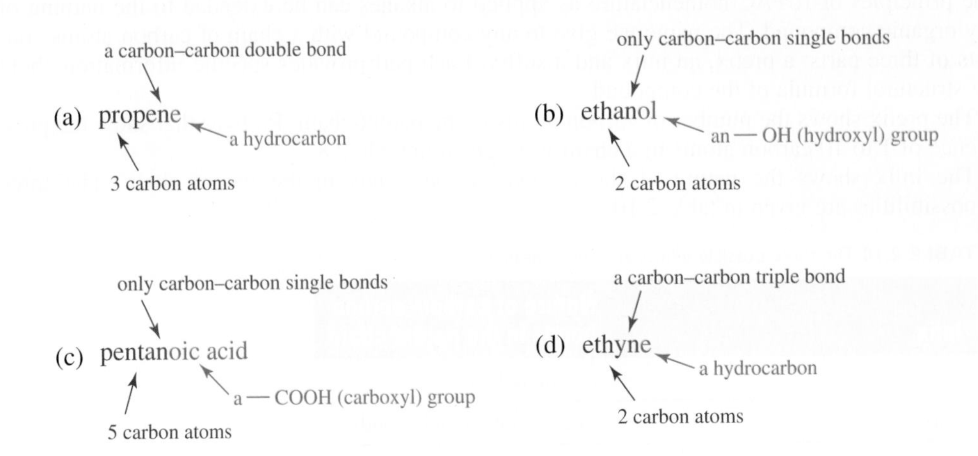

Notice that, in each of these examples, there is no ambiguity about the position of the functional group. We will deal later with examples where the position is not as clear.

PRACTICE EXERCISE 2.20 Combine the proper prefix, infix and suffix to write the IUPAC name for each of the following compounds.

(a) $CH_3CH_2CH_2C(=O)-OH$

(b) $HC\equiv C-CH_3$

The examples given in this section serve only as an introduction to organic nomenclature; you will find more detailed examples in the appropriate chapters later in the book.

We have seen in this chapter that such apparently disparate topics as units, nomenclature and representations of molecules are all part of the language of chemistry. There is still much of this language for you to learn, but this chapter has introduced many of the fundamentals that you will need in order to understand and appreciate the following chapters. Indeed, it is likely that you will often need to refer back to this chapter when you come across examples of measurement, representations of molecules or nomenclature later in this book.

SUMMARY

LO 1 Measurement

All measurements have an associated unit and uncertainty. The units used for measurements are based on the set of seven SI base units, which can be combined to give various derived units. The SI base units are the metre, kilogram, second, kelvin, mole, ampere and candela. Prefixes can be used to denote multiplication of these units by powers of 10. Dimensional analysis, which uses the fact that units undergo the same kinds of mathematical operations as the numbers to which they are attached, can be used as an aid to determining the correct form of an equation. Some non-SI units are in common usage but can be converted to SI units using a conversion factor.

Uncertainties arise in any measurement from the limitations of the apparatus used to make the measurement. Some idea of the uncertainty and therefore precision in a measurement can be obtained from the number of significant figures to which the measurement is quoted.

Expressing a number in scientific notation can make it easier to determine the number of significant figures. There will always be uncertainty in the final figure of any measurement. The accuracy of a measurement shows how close to the true value the measurement is, while the precision of a measurement represents how reproducible it is.

When measurements are manipulated in equations, the following rules apply.

- When measurements are multiplied or divided, the number of significant figures in the answer should not be greater than the number of significant figures in the least precise measurement.
- When measurements are added or subtracted, the answer should have the same number of decimal places as the quantity with the fewest number of decimal places.

LO 2 Representations of molecules and reactions

The chemical formula of a substance shows the relative number of each type of atom present, but gives no structural information. The rules for writing chemical formulae of binary compounds are as follows.

1. With the exception of hydrogen, the element further to the left in the periodic table appears first.
2. If hydrogen is present, it appears last except when the other element is from group 16 or 17 of the periodic table.

3. If both elements are from the same group of the periodic table, the lower one appears first.

The chemical formulae of ionic compounds are written with the cation followed by the anion, while organic compounds have C and H first, followed by the rest of the elements in alphabetical order.

Structural formulae attempt to show particular aspects of molecular structure in both two and three dimensions. Such depictions are especially important in organic chemistry where catenation leads to an enormous number of possible carbon-based molecules. Structural formulae are drawn using one line to represent single bonds, two lines for double bonds and three lines for triple bonds. Carbon atoms at the end of a carbon chain are called terminal carbon atoms. Structural formulae can help distinguish between isomers. Two types of shorthand structural formulae are widely used; condensed structural formulae do not include explicit depictions of bonds between atoms, while line structures remove atom labels for carbon atoms and all hydrogen atoms that are bonded to carbon atoms. Three-dimensional aspects can be introduced into structural formulae through the use of hashed and solid wedges, which show bonds going away from and coming towards the observer, respectively.

Ball-and-stick models and space-filling models are two other methods where three-dimensionality may be shown.

Mechanistic arrows can be used to show the movement of electrons in chemical processes. The tail of the arrow is situated at the electron source and the arrow head shows the destination.

LO 3 Nomenclature

Chemical species can be named according to sets of rules prescribed by IUPAC. The rules for naming nonmetallic binary compounds are as follows.

1. The element closer to the left of the periodic table appears first. If both elements are from the same group of the periodic table, the lower one appears first.
2. The element that appears first retains its elemental name.
3. The second element begins with a root derived from its elemental name and ends with the suffix *-ide*.
4. When there is more than one atom of a given element in the formula, the name of the element usually contains a prefix that specifies the number of atoms present.

Binary compounds containing hydrogen can be named with hydrogen either first or second, depending on the position of the other element in the periodic table, according to the above rules.

Ionic compounds are named with the cation first, followed by the anion, while oxoanions are named according to an extensive set of rules.

Organic compounds are named primarily according to their functional group; those of importance in this chapter are the —OH group found in alcohols, the —CHO group found in aldehydes, the C═O group found in ketones and the —COOH group found in carboxylic acids. Alkanes belong to the group of compounds called hydrocarbons and are sometimes called saturated hydrocarbons. Those consisting solely of chains of carbon atoms are called acyclic alkanes, while those containing one or more rings of carbon atoms are called cycloalkanes. Their nomenclature is based on the longest carbon chain within the molecule, with a prefix denoting the length of that chain. Names of organic compounds generally comprise a prefix, infix and suffix. The prefix shows the number of carbon atoms in the parent chain, the infix shows the nature of the carbon–carbon bonds in the parent chain, and the suffix shows the class of compound to which the substance belongs, and therefore the functional group(s) present in the compound.

Constitutional isomers of alkanes have the same chemical formulae but different structural formulae and hence different names. These isomers generally have different physical properties.

KEY CONCEPTS AND EQUATIONS

SI units
(section 2.1)

The base SI units: metre, kilogram, second, kelvin, mole, ampere and candela are used either directly or to derive units for any measurement in chemistry.

SI prefixes
(section 2.1)

SI prefixes are used to create larger and smaller units. The prefixes help us to convert between differently sized units.

Dimensional analysis
(section 2.1)

Dimensional analysis assists in determining the correct form of an equation through knowledge of the units of all components of the equation.

Rules for significant figures in calculations
(section 2.1)

The following rules are used to round answers to the correct number of significant figures.

- When measurements are multiplied or divided, the number of significant figures in the answer should not be greater than the number of significant figures in the *least* precise measurement.
- When measurements are added or subtracted, the answer should have the same number of decimal places as the quantity with the *fewest* number of decimal places.

Rules for writing chemical formulae of binary compounds
(section 2.2)

The following rules are used to write the correct chemical formulae for compounds containing only two elements.

1. With the exception of hydrogen, the element further to the left in the periodic table appears first.
2. If hydrogen is present, it appears last except when the other element is from group 16 or 17 of the periodic table.
3. If both elements are from the same group of the periodic table, the lower one appears first.
4. If the compound is ionic, we write the cation first, followed by the anion.

Rules for drawing line structures
(section 2.2)

These rules are used for converting structural formulae to line structures, and can also be used to carry out the reverse operation.

1. All bonds except C—H bonds are shown as lines.
2. C—H bonds and H atoms attached to carbon are not shown in the line structure.
3. Single bonds are shown as one line; double bonds are shown as two lines; triple bonds are shown as three lines.
4. Carbon atoms are not labelled. All other atoms are labelled with their elemental symbols.

Hashed and solid wedges
(section 2.2)

These are used to introduce three-dimensional aspects into structural formulae.

A solid wedge (▬) represents a bond coming out of the page, while a hashed wedge (⋯) represents a bond going back into the page. A normal line is used for bonds in the plane of the page.

Mechanistic arrows
(section 2.2)

These are used to show the movement of electrons in chemical processes. Single-headed arrows show the movement of a single electron, while two-headed arrows show the movement of electron pairs. The direction of electron movement is from the tail of the arrow to the head.

Naming nonmetallic binary compounds
(section 2.3)

These rules are used to name compounds containing only two nonmetallic elements.

1. The element closer to the left of the periodic table appears first. If both elements are from the same group of the periodic table, the lower one appears first.
2. The element that appears first retains its elemental name.
3. The second element begins with a root derived from its elemental name and ends with the suffix *-ide*.
4. When there is more than one atom of a given element in the formula, the name of the element usually contains a prefix that specifies the number of atoms present.

Naming oxoanions
(section 2.3)

These rules are used to name anions containing oxygen and at least one other element.

1. The name has a root taken from the name of the central atom.
2. When an element forms two different oxoanions, the one with fewer oxygen atoms ends in *-ite* and the other ends in *-ate*.
3. Chlorine, bromine and iodine each form four different oxoanions that are distinguished by prefixes and suffixes.
4. A polyatomic anion with a charge more negative than 1– may add H^+ to give another anion. These anions are named from the parent anion by adding the word *hydrogen*.

Functional groups
(section 2.3)

Functional groups are the basis for naming organic compounds.

- Alcohols contain a hydroxyl group.
- Aldehydes contain a terminal carbonyl group.
- Ketones contain a non-terminal carbonyl group.
- Carboxylic acids contain a carboxyl group.

Naming alkanes
(section 2.3)

There are two parts to the name of an alkane: a prefix, indicating the number of carbon atoms in the chain, and the suffix *-ane*.

Prefixes for carbon chain length
(section 2.3)

Knowledge of these prefixes is necessary for naming any organic compound: 1, meth-; 2, eth-; 3, prop-; 4, but-; 5, pent-; 6, hex-; 7, hept-; 8, oct-; 9, non-; 10, dec-.

General organic nomenclature
(section 2.3)

The prefix, infix and suffix of the name of an organic compound reflect the structure of that compound.

KEY TERMS

absolute uncertainty *p. 35*
accuracy *p. 32*
acyclic alkanes *p. 58*
alkane *p. 58*
alkyl group *p. 59*
ampere (A) *p. 27*
anhydrous *p. 41*
ball-and-stick model *p. 47*
binary compound *p. 40*
candela *p. 27*
catenation *p. 42*
chemical formula *p. 40*
condensed structural formula *p. 43*
constitutional isomers *p. 62*
cycloalkane *p. 58*
dimensional analysis *p. 31*
functional group *p. 54*
hashed wedge *p. 47*
heterolytic *p. 49*
homolytic *p. 50*
hydrate *p. 41*
hydrocarbon *p. 58*
isomers *p. 43*
kelvin *p. 27*
kilogram *p. 27*
line structure *p. 43*
mechanistic arrows *p. 49*
metre *p. 27*
mole *p. 27*
molecular formula *p. 40*
nomenclature *p. 50*
oxoanion *p. 54*
parent chain *p. 58*
percentage uncertainty *p. 35*
precision *p. 32*
primary carbon atom *p. 55*
saturated hydrocarbon *p. 58*
scientific notation *p. 34*
second *p. 27*
secondary carbon atom *p. 55*
SI (Système International) units *p. 26*
significant figures *p. 33*
solid wedge *p. 47*
space-filling model *p. 47*
structural formula *p. 41*
substituent *p. 58*
terminal carbon *p. 42*
tertiary carbon atom *p. 55*
unit *p. 26*

REVIEW QUESTIONS

LO 1 Measurement

2.1 Why must measurements always be written with a unit?

2.2 What does the abbreviation SI stand for?

2.3 Which SI base unit is defined in terms of a physical object?

2.4 What is the only SI base unit that includes a decimal prefix?

2.5 Einstein's famous equation states that $E = mc^2$, where E is energy, m is the mass of an object in kg, and c is the speed of light in m s^{-1}. What is the SI derived unit of energy?

2.6 Give the prefixes that correspond to the following powers of 10.

(a) 10^{-2} (b) 10^{-3} (c) 10^{3} (d) 10^{-6} (e) 10^{-9} (f) 10^{-12} (g) 10^{9} (h) 10^{-1} (i) 10^{2}

2.7 What abbreviation is used for each of the powers of 10 given in question 2.6?

2.8 Chemists often use the non-SI unit litre (L) as the unit of volume in the laboratory. What is the derived SI unit of volume that is equal in volume to 1 L?

2.9 Why does dimensional analysis allow you to determine the correct form of an equation from knowledge of the units of the measurements involved?

2.10 What is the difference between 'accuracy' and 'precision'?

2.11 Analysis of a blood sample showed a concentration of 3.0 μmol mL^{-1} of a particular toxin. How high might the concentration actually have been? A more precise technique gave the concentration of the toxin as 3.027 μmol mL^{-1} of blood. How high might the concentration have been in this case?

2.12 There are 3600 s in 1 hour. By what conversion factor would you multiply 250 s to convert it to hours? By what conversion factor would you multiply 3.84 h to convert it to seconds?

LO 2 Representations of molecules and reactions

2.13 Subscripts are often used in chemical formulae. How does a subscript that follows a group of atoms enclosed in parentheses differ from one that does not?

2.14 What must be true about two substances if they are to be called isomers of each other?

2.15 Write the chemical formula of each of the following substances.
(a) trinitrotoluene (TNT), which contains 5 hydrogen atoms, 7 carbon atoms, 6 oxygen atoms and 3 nitrogen atoms
(b) barium phosphate, which contains 3 Ba^{2+} ions and 2 PO_4^{3-} ions
(c) ibuprofen, which contains 13 carbon atoms, 2 oxygen atoms and 18 hydrogen atoms

2.16 The line structures that follow represent starting materials for making plastics. For each of them, draw the structural formula and give the chemical formula.

(a) isoprene (sourced from nature to produce rubber)

(b) styrene (used to make polystyrene foam)

(c) O methyl methacrylate (used to make perspex)

2.17 Write line structures for compounds (a) to (f).

(a) $CH_3CH(CH_3)CH(CH_3)CH_2CH(CH_3)CH_3$ — printed as:

```
      CH3
      |
CH3CHCHCH2CHCH3
        |      |
       CH3    CH3
```

(b)

```
                 CH3   CH2CH2CH3
                  |     |
CH3CH2CHCH2CHCHCHCH3
         |           |
        CH3        CH2CH3
```

(c)

```
CH3        CH3
|          |
CHCH2CH2CH
|          |
CH3        CH2CH3
```

(d) $CH_3CH(CH_3)CH_2C(CH_3)_2CH_3$

(e) $(CH_3)_3CC(CH_3)_3$

(f) $(CH_3CH_2)_2CHCH_2CH(CH_3)CH_3$

2.18 Write the chemical formula and the condensed structural formula for each of the following alkanes.

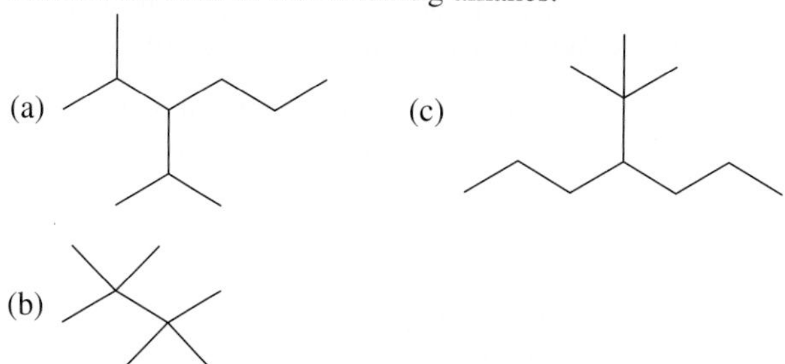

2.19 Provide an even more abbreviated formula for each of the following condensed structural formulae, using parentheses and subscripts.

(a)

```
   CH3      CH3
   |        |
CH3CHCH2CH2CCH3
            |
            CH3
```

(b)

```
   CH3            CH2CH2CH3
   |              |
CH3CHCHCH2CH2CHCHCH3
      |          |
      CH3        CH2CH3
```

(c)

```
   CH3       CH3
   |         |
CH3CCH2CH2CCH3
   |         |
   CH3       CH2CH3
```

2.20 Describe the process represented by a two-headed mechanistic arrow. How does this differ from that represented by a single-headed mechanistic arrow?

LO 3 Nomenclature

2.21 Write chemical formulae for the following compounds.
(a) chlorine monofluoride
(b) selenium trioxide
(c) hydrogen bromide
(d) silicon tetrachloride
(e) sulfur dioxide
(f) hydrogen peroxide

2.22 Name the following compounds.
(a) BrF_3 (b) H_2S (c) OF_2 (d) SbF_5 (e) PCl_3 (f) N_2O_3 (g) SO_3 (h) BBr_3

2.23 Draw the structural formulae of the following molecules.
(a) butane
(b) 3-methylpentane
(c) 2,3-dimethylhexane

2.24 Why is there no such thing as a primary ketone?

REVIEW PROBLEMS

2.25 What number should replace the question mark in each of the following? **LO1**

(a) 1 cm = ? m
(b) 1 km = ? m
(c) 1 m = ? pm
(d) 1 dm = ? m
(e) 1 g = ? kg
(f) 1 cg = ? g
(g) 1 nm = ? m
(h) 1 μg = ? g
(i) 1 kg = ? g
(j) 1 Mg = ? g
(k) 1 mg = ? g
(l) 1 dg = ? g

2.26 Convert each of the following measurements to the given units. **LO1**

(a) 100 °C to K (the boiling point of water)
(b) −78 °C to K (the temperature of a dry-ice/acetone bath)
(c) 273 K to °C (the freezing point of water)
(d) 298 K to °C (room temperature)
(e) 37 °C to K (body temperature)

2.27 Natural gas is mostly methane, a substance that boils at a temperature of 111 K. What is its boiling point in °C? **LO1**

2.28 Tungsten is often used as the filament in lightbulbs because of its melting point of 3410 °C, the highest of any pure metal. What is its melting point in K? **LO1**

2.29 How many significant figures do the following measured quantities have? **LO1**

(a) 37.53 cm
(b) 37.240 cm
(c) 202.0 g
(d) 0.000 24 kg
(e) 0.070 80 m
(f) 2400 mL

2.30 How many significant figures do the following measured quantities have? **LO1**

(a) 0.5070 m
(b) 10.024 g
(c) 0.002 mol
(d) 2.00 km
(e) 10 kg
(f) 3.8005 s

2.31 How many significant figures do the following measured quantities have? **LO1**

(a) 1.0230 kg
(b) 3.0200 m
(c) 0.0030 L
(d) 27.300 g
(e) 0.043 20 mm

2.32 How many significant figures do the following measured quantities have? **LO1**

(a) 2.777 μmol
(b) 0.040 00 m^3
(c) 21 740 m
(d) 8100 L
(e) 0.0206 V

2.33 Carry out the following calculations and round the answers to the correct number of significant figures. **LO1**

(a) 0.0023 m × 315 m
(b) 84.25 kg − 0.010 75 kg
(c) (94.45 g − 84.45 g)/(31.4 mL − 9.9 mL)
(d) (23.4 g + 102.4 g + 0.003 g)/(6.478 mL)
(e) (313.44 cm − 209.1 cm) × 8.2234 cm

2.34 Carry out the following calculations and round the answers to the correct number of significant figures. **LO1**

(a) 2.60 mol/4.100 L
(b) 3.88 kg + 0.001 kg
(c) (14.2 g − 6.9 g)/(1.501 L − 0.002 L)
(d) (2.918 kJ + 0.015 kJ)/(13.77 mol − 5.1084 mol)
(e) (10.625 m − 10.622 m)/(5.10 s × 1.001 s)

2.35 Express the following numbers in scientific notation to three significant figures. **LO1**

(a) 4340
(b) 32 000 000
(c) 0.003 287
(d) 42 000
(e) 0.000 008 00
(f) 324 300

2.36 Express the following numbers in scientific notation. Assume, in this problem, that only the non-zero digits are significant figures. **LO1**

(a) 866
(b) 0.003 79
(c) 42 300
(d) 0.055 38
(e) 7.91
(f) 16 240

2.37 Write the following numbers in standard (non-scientific) form. **LO1**

(a) 3.1×10^5
(b) 4.35×10^{-6}
(c) 3.9×10^3
(d) 4.4×10^{-12}
(e) 35.6×10^{-7}
(f) 8.8×10^8

2.38 Write the following numbers in standard (non-scientific) form. **LO1**

(a) 8.26×10^{-1}
(b) 9.79×10^4
(c) 23.8×10^{-5}
(d) 6.75×10^{-7}
(e) 2.0000×10^3
(f) 88.1×10^9

2.39 Carry out the following calculations and express the answers in scientific notation. **LO1**

(a) $(4.0 \times 10^7) - (2.1 \times 10^5)$
(b) $(3.0 \times 10^{-2}) + (3.21 \times 10^{-5})$
(c) $(2.1 \times 10^7) \times (2.1 \times 10^5)$
(d) $(3.0 \times 10^4) \times (8.2 \times 10^{-5})$
(e) $(9.10 \times 10^{12})/(2.0 \times 10^{-3})^2$

2.40 Carry out the following calculations and express the answers in scientific notation. **LO1**

(a) $(2.8 \times 10^4) \times (9.8 \times 10^2)$
(b) $(7.4 \times 10^{13})/(5.6 \times 10^{-1})^2$
(c) $(4.3 \times 10^{-3}) \times (8.8 \times 10^{-6})$
(d) $(3.7 \times 10^4) - (2.6 \times 10^3)$
(e) $(7.2 \times 10^{-4}) + (5.12 \times 10^{-2})$

2.41 Convert each of the following measurements to the given units. **LO1**

(a) 32.0 dm to km
(b) 8.2 mg to μg
(c) 75.3 mg to kg
(d) 137.5 mL to L
(e) 0.025 L to mL
(f) 342 pm to dm

2.42 Convert each of the following measurements to the given units. **LO1**

(a) 55 L to m^3
(b) 4.0 kg to mg
(c) 44 mmol to mol
(d) 43.8 m to km
(e) 2.31 mL to L
(f) 3.2 kJ to J

2.43 Convert each of the following measurements to the given units and express your answers in scientific notation. **LO1**

(a) 230 km to cm
(b) 423 kg to mg
(c) 423 kg to Mg
(d) 430 μL to mL
(e) 27 ng to kg
(f) 730 nL to kL

2.44 Convert each of the following measurements to the given units and express your answers in scientific notation. **LO1**

(a) 326 m to dm
(b) 5.21 kg to g
(c) 5.39 km to μm
(d) 51 nm to pm
(e) 0.55 m^3 to mm^3
(f) 47.1 g to mg

2.45 Which functional group does each of the following compounds contain? **LO3**

(a) $CH_3CH{=}CH_2$
(b) CH_3CH_2OH
(c) $CH_3CH_2CH_2COOH$
(d) $HOCH_2CH_2CH_3$

2.46 Which functional group does each of the following compounds contain? **LO3**

(a) $CH_2{=}CHCH_2CH_3$

(b) CH_3CH_2COOH

(c) $CH_3CH_2CH(OH)CH_2CH_3$ (OH attached to the third carbon)

2.47 Decide whether the members of each pair are identical, isomers or unrelated. **LO 2,3**

(a) CH_3-CH_3 and CH_3 bonded vertically to CH_3

(b) $H_3C-CH_2-CH_3$ (drawn as a zigzag) and $H_3C-CH_2-CH_3$ (drawn with CH_2 at the top)

(c) CH_3CH_2OH and $CH_3CH_2CH_2OH$

(d) $CH_3CH=CH_2$ and a three-membered ring of H_2C, CH_2 and CH_2

(e) $H-C(=O)-CH_3$ and $CH_3-C(=O)-H$

(f) $CH_3CH(CH_3)CH_3$ (the CH_3 branch drawn above) and $CH_3CH(CH_3)$ with a second CH_3 drawn below the CH

(g) $CH_3CH_2CH_2NH_2$ and $CH_3CH_2-NH-CH_3$

2.48 Examine each pair of molecules and decide whether they are identical, isomers or unrelated. **LO 2,3**

(a) $CH_3CH_2C(=O)CH_3$ and $CH_3C(=O)CH_2CH_3$

(b) $HC(=O)CH_2CH_2OH$ and $HOCH_2CH_2CH_2C(=O)H$

(c) $HC(=O)CH_2CH_2C(=O)OH$ and $HOC(=O)CH_2CH_2C(=O)H$

(d) $CH_3CH_2C(=O)CH_2CH_3$ and $CH_3C(=O)CH_2CH_2CH_3$

2.49 Write the IUPAC names of the following hydrocarbons. **LO 2,3**

(a) $CH_3CH_2CH_2CH_2CH_3$

(b) $CH_3CH_2CH_2CH(CH_3)CH_3$

(c) $CH_3CH(CH_3)CH_2CH(CH_3)CH_2CH_3$

(d) $CH_3CH_2CH(CH_3)CH_2CH(CH_3)CH_3$

2.50 There are only three saturated alcohols that contain three carbon atoms. Draw the structural formulae of these molecules. **LO 2**

2.51 Write structural formulae for all of the possible alcohols with the general formula $C_4H_{10}O$. **LO 2**

2.52 Write chemical formulae for the molecules whose ball-and-stick models follow. **LO 2**

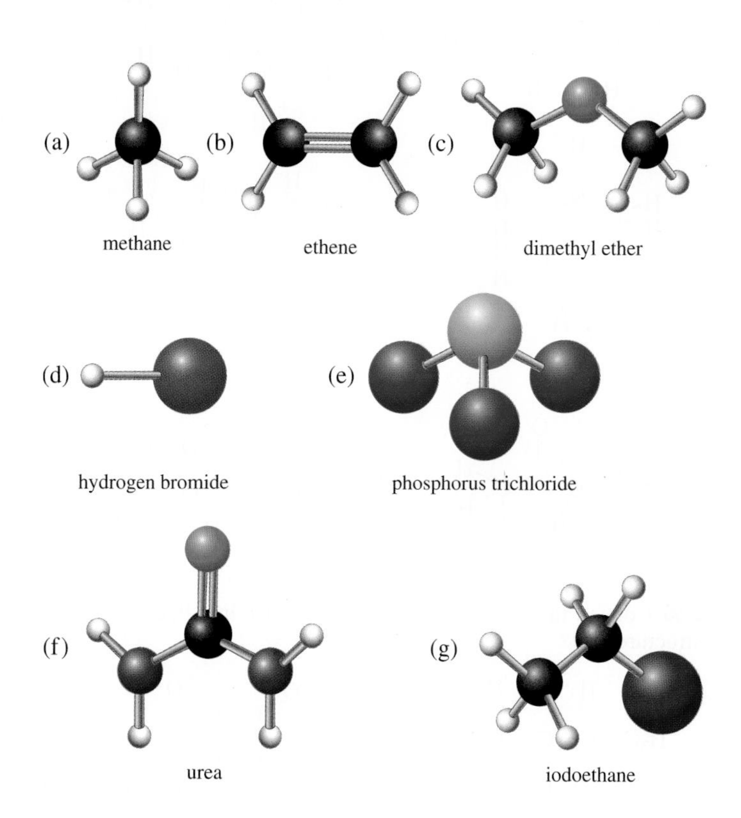

2.53 Write chemical formulae for the molecules whose ball-and-stick models follow. **LO 2**

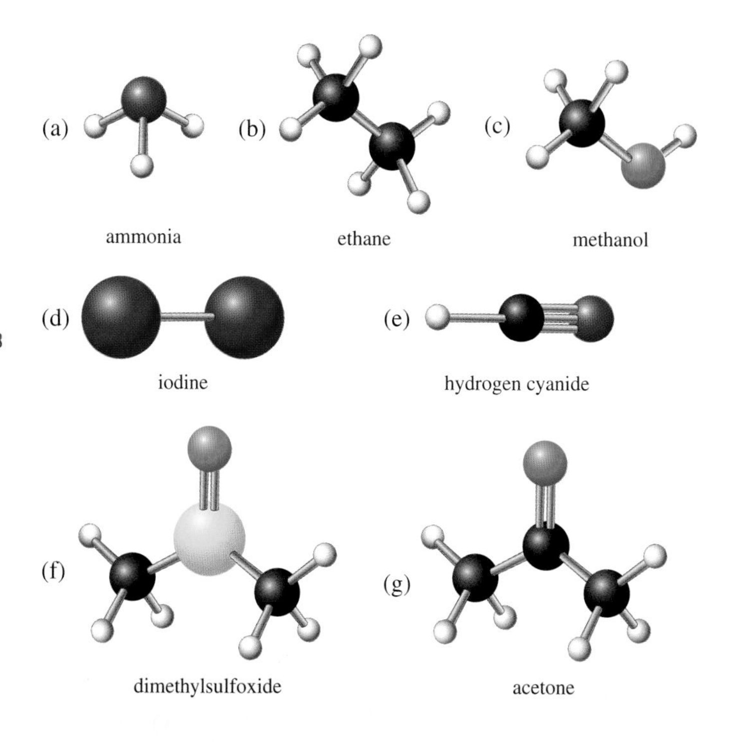

2.54 Write structural formulae for the molecules in questions 2.52 and 2.53. **LO 2**

2.55 Convert the following structural formulae into line structures. **LO2**

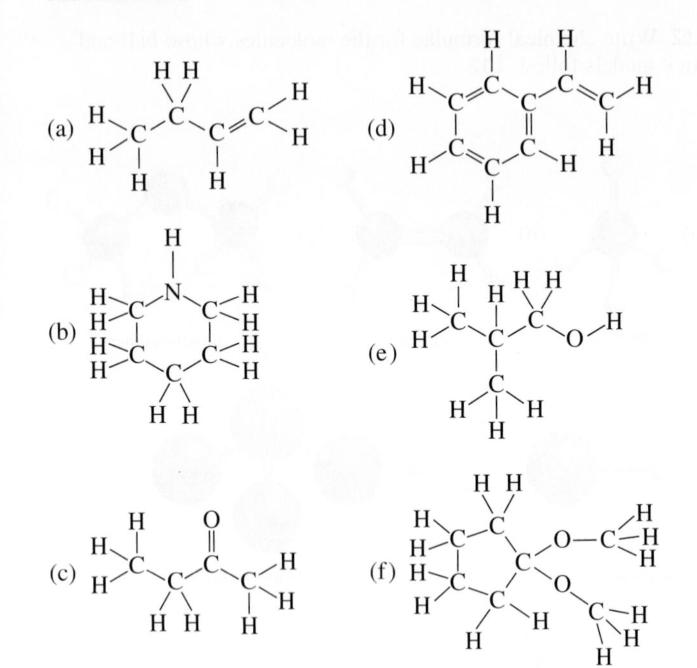

2.56 Convert the following structural formulae into line structures. **LO2**

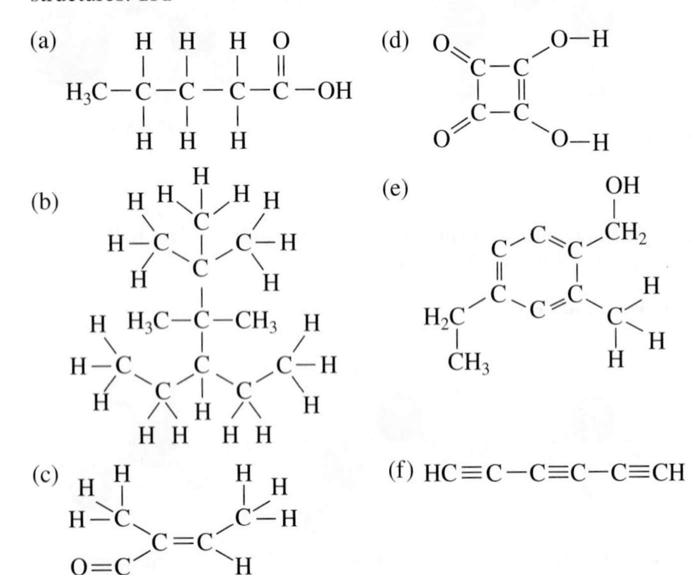

2.57 Convert the following line structures into structural formulae. **LO2**

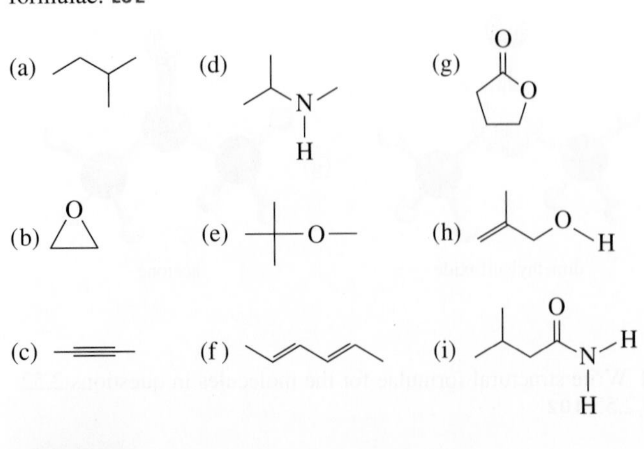

2.58 Convert the following line structures into structural formulae. **LO2**

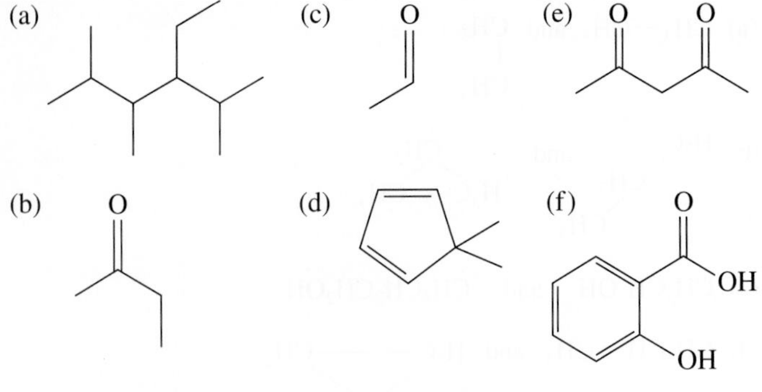

2.59 Which of the following statements are true about constitutional isomers? **LO3**

(a) They have the same chemical formula.
(b) They have the same atomic mass.
(c) They have the same order of attachment of atoms.
(d) They have the same physical properties.

2.60 Each member of the following set of compounds is an alcohol. **LO2,3**

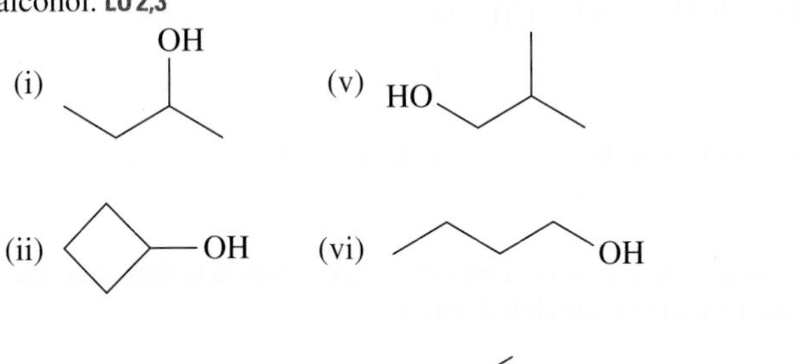

(iii) (vii) (iv) (viii)

Which structural formulae represent:

(a) the same compound?
(b) compounds that are constitutional isomers?
(c) compounds that are not constitutional isomers?

2.61 Each member of the following set of compounds is either an aldehyde or a ketone. **LO2,3**

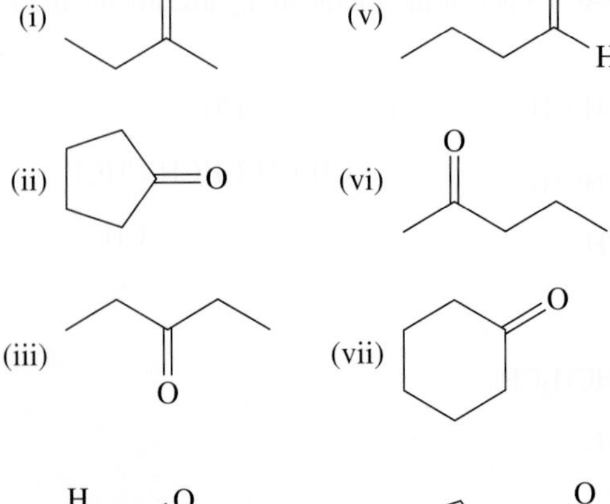

(iv) (viii)

Which structural formulae represent:
(a) the same compound?
(b) compounds that are constitutional isomers?
(c) compounds that are not constitutional isomers?

2.62 For each of the following pairs of compounds, determine whether the structural formulae shown represent: **LO2,3**
(i) the same compound
(ii) different compounds that are constitutional isomers
(iii) different compounds that are not constitutional isomers.

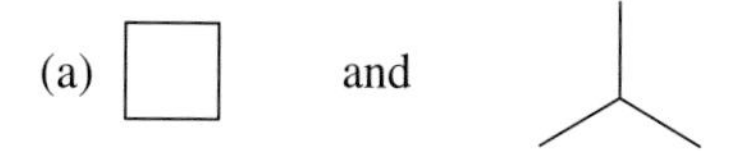

(b) and

(c) and

(d) and

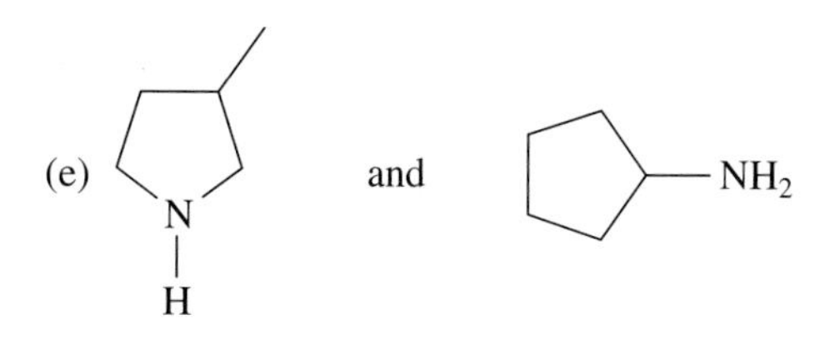

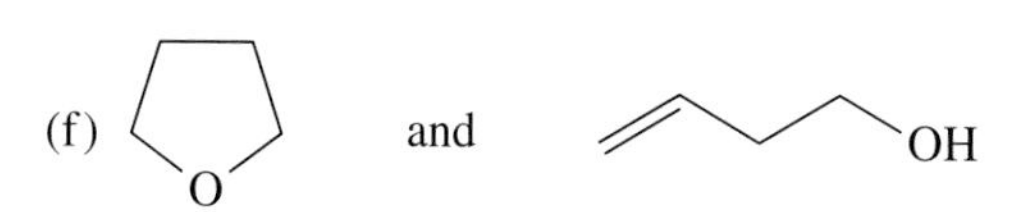

2.63 Which sets of compounds are constitutional isomers? **LO2,3**

(a) CH_3CH_2OH and CH_3OCH_3

(b) $CH_3\overset{O}{\overset{\|}{C}}CH_3$ and $CH_3CH_2\overset{O}{\overset{\|}{C}}H$

(c) $CH_3\overset{O}{\overset{\|}{C}}OCH_3$ and $CH_3CH_2\overset{O}{\overset{\|}{C}}OH$

(d) $CH_3\overset{OH}{\overset{|}{C}}HCH_2CH_3$ and $CH_3\overset{O}{\overset{\|}{C}}CH_2CH_3$

(e) [cyclopentane] and $CH_3CH_2CH_2CH_2CH_3$

(f) [cyclopentane] and $CH_2{=}CHCH_2CH_2CH_3$

2.64 Draw the product(s) that result(s) from the following mechanistic processes. **LO2**

(a) H–C(H)(H)–C(=O)–O–H + ⁻O–H

(b) I⁻ + H–C(CH₃)(H)–Br

(c) H_3C–C(=O)–CH_3 + ⁻C≡N

2.65 Draw line structures for: **LO2**
(a) four aldehydes with chemical formula $C_5H_{10}O$
(b) one ketone with chemical formula C_4H_6O
(c) three ketones with chemical formula $C_5H_{10}O$
(d) four carboxylic acids with chemical formula $C_5H_{10}O_2$.

2.66 Write chemical formulae for these compounds. **LO3**
(a) methane
(b) hydrogen iodide
(c) calcium hydride
(d) phosphorus trichloride
(e) dinitrogen pentaoxide
(f) sulfur hexafluoride
(g) boron trifluoride

2.67 Write chemical formulae for these compounds. **LO3**
(a) aluminium hydride
(b) hydrogen selenide
(c) butanone
(d) beryllium oxide
(e) molecular oxygen
(f) oxygen difluoride
(g) nitrogen triiodide

2.68 Name the following compounds. **LO3**
(a) S_2Cl_2
(b) IF_7
(c) HBr
(d) N_2O_3
(e) SiC
(f) CH_3OH

2.69 Name the following compounds. **LO3**
(a) $SiCl_4$
(b) N_2O_5
(c) CsH
(d) PbO_2
(e) CH_3CH_2COOH

2.70 Write the chemical formula of each compound. **LO3**
(a) hydrogen fluoride
(b) calcium fluoride
(c) aluminium sulfate
(d) ammonium sulfide

2.71 Write the chemical formula of each compound. **LO3**
(a) sodium chlorite
(b) potassium permanganate
(c) sodium hydrogencarbonate

2.72 Name each of the following compounds. **LO3**
(a) CaO
(b) K_2CO_3
(c) HI
(d) Na_2HPO_4

2.73 Name each of the following compounds. **LO3**
(a) Na_2SO_3
(b) K_3PO_4
(c) $LiNO_2$
(d) $K_2Cr_2O_7$

2.74 Write chemical formulae for these compounds. **LO3**
(a) sodium sulfate
(b) potassium sulfide
(c) potassium dihydrogenphosphate
(d) sodium carbonate
(e) lithium perbromate

2.75 Write chemical formulae for these compounds. **LO3**
(a) potassium perchlorate
(b) sodium hydrogensulfate
(c) aluminium oxide
(d) potassium oxalate

2.76 Name each of the following compounds. **LO 3**
(a) K_2SO_4
(b) $NaClO$
(c) $BeCl_2$
(d) SrO
(e) $NaNO_3$
(f) $CaSO_3$
(g) K_2CrO_4

2.77 Name each of the following compounds. **LO 3**
(a) $NaHSO_3$
(b) $Mg(OOCCH_3)_2$
(c) Rb_2CO_3
(d) CsI
(e) $AlPO_4$

2.78 Write IUPAC names for these alkanes. **LO 2,3**

(a) $CH_3CHCH_2CH_2CH_3$ with CH_3 on C-2

(b) 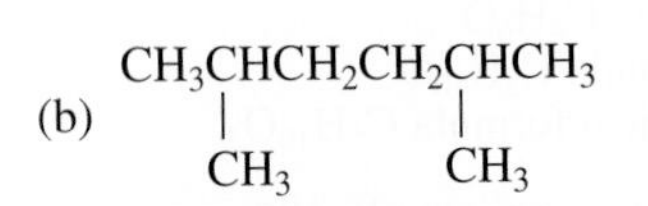

(c) $CH_3(CH_2)_4CHCH_2CH_3$ with CH_2CH_3 on the CH

(d)

2.79 Draw line structures for these alkanes. **LO 2,3**
(a) 2,2-dimethylpropane
(b) 2,3-dimethylbutane
(c) 2,3,4-trimethylnonane
(d) 4-isopropylheptane
(e) 3-ethyl-2,3,5-trimethylnonane
(f) 2,3,4,5-tetramethylhexane

2.80 Explain why each of the following names is an incorrect IUPAC name, and write the correct IUPAC name for the intended compound. **LO 2,3**
(a) 1,3-dimethylbutane
(b) 4-methylpentane
(c) 2,2-diethylbutane
(d) 2-ethyl-3-methylpentane
(e) 2-propylpentane
(f) 2,2-diethylheptane

2.81 Draw a structural formula for each of the following compounds. **LO 2,3**
(a) methanol
(b) methanal
(c) methanoic acid
(d) propanone
(e) propanal
(f) propanoic acid
(g) butanal

2.82 Write the IUPAC name for each compound. **LO 2,3**

(a) 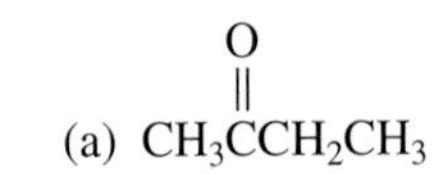

(b) 

(c) 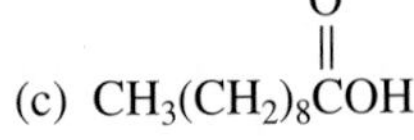

ADDITIONAL EXERCISES

2.83 As stated earlier in this chapter, chemists tend to use the litre (L), a non-SI unit, and its derivative, the millilitre (mL) as units of volume. A disadvantage of this is that there are no obvious conversion factors between these units and the derived SI unit of volume, the cubic metre (m^3). However, such conversion factors may be worked out using the following method. Consider a cube having edge lengths of 1 m. This has a volume of 1 m × 1 m × 1 m = 1 m^3.

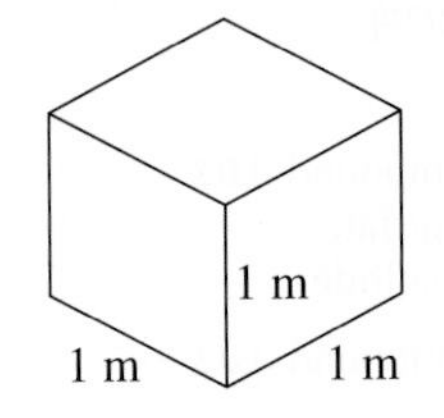

volume = 1 m × 1 m × 1 m
= 1 m^3

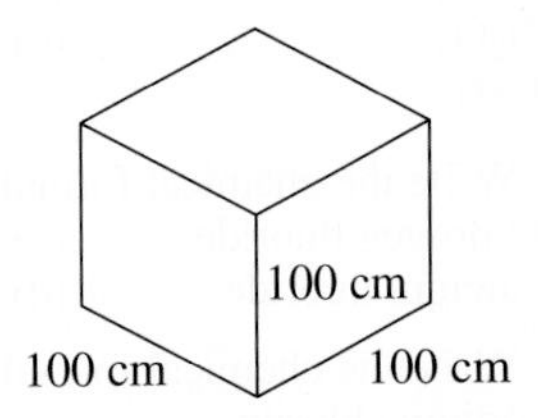

volume = 100 cm × 100 cm × 100 cm
= 1 000 000 cm^3

We now take the same cube and, knowing there are 100 cm in 1 m, we express its edge lengths as 100 cm. Therefore, the volume of this same cube is now 100 cm × 100 cm × 100 cm = 1 000 000 cm^3. Thus, there are 1 000 000 cm^3 in 1 m^3. Use this method to calculate the following conversion factors. **LO 1**
(a) How many dm^3 are there in 1 m^3?
(b) How many cm^3 are there in 1 dm^3?
(c) How many mm^3 are there in 1 m^3?
(d) How many mm^3 are there in 1 dm^3?
Note that 1 dm^3 = 1 L.

2.84 The IUPAC system divides the name of an organic compound into a prefix (showing the number of carbon atoms), an infix (showing the presence of carbon–carbon single, double or triple bonds), and a suffix (showing the presence of an alcohol, an aldehyde, a ketone or a carboxylic acid). Assume for the purposes of this problem that, to be an alcohol (-ol), the hydroxyl group must be bonded to a carbon atom which has only single bonds to other attached atoms.

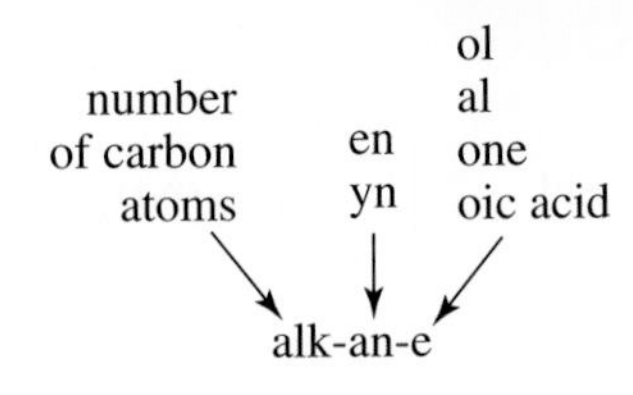

Given this information, write the structural formula of a compound with an unbranched chain of four carbon atoms that is an:
(a) alkane
(b) alkene
(c) alkyne
(d) alkanol
(e) alkenol
(f) alkynol
(g) alkanal
(h) alkenal
(i) alkynal
(j) alkanone
(k) alkenone
(l) alkynone
(m) alkanoic acid
(n) alkenoic acid
(o) alkynoic acid.

(*Note:* There is only one structural formula possible for some parts of this problem. For other parts, two or more structural formulae are possible. Where two or more are possible, we will deal with how the IUPAC system distinguishes between them when we come to the chapters on those particular functional groups.) **LO 2,3**

2.85 In everyday life, we encounter chemicals with unsystematic names. What is the name a chemist would use for each of the following substances? **LO 3**
(a) limestone, $CaCO_3$
(b) laughing gas, N_2O
(c) potash, K_2CO_3
(d) silica, SiO_2
(e) caustic soda, $NaOH$
(f) bleach, $NaClO$
(g) titanium white, TiO_2
(h) cadmium yellow, CdS

2.86 A mineral is a chemical compound found in the Earth's crust. What are the chemical names of the following minerals? **LO 3**
(a) PbS (galena)
(b) Al_2O_3 (bauxite)
(c) $BaSO_4$ (barite)
(d) HgS (cinnabar)

2.87 The line structures of three different vitamins are given below. Write the chemical formula for each compound. **LO2**

(a) 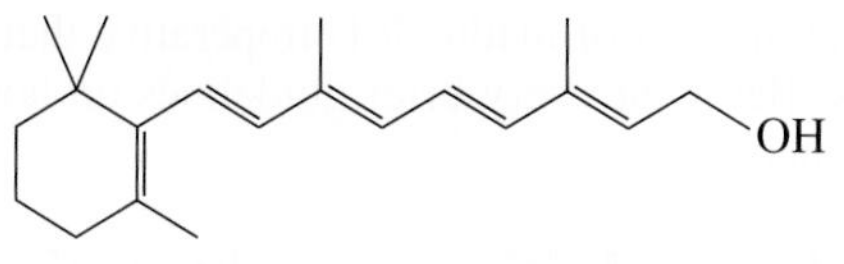 vitamin A

(b) 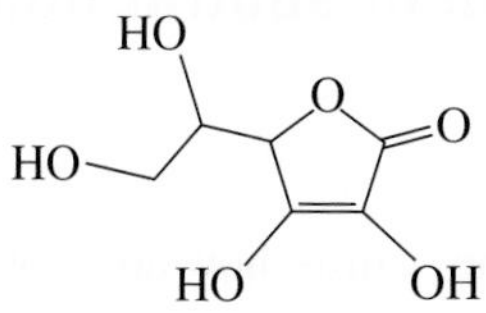 vitamin C

(c) 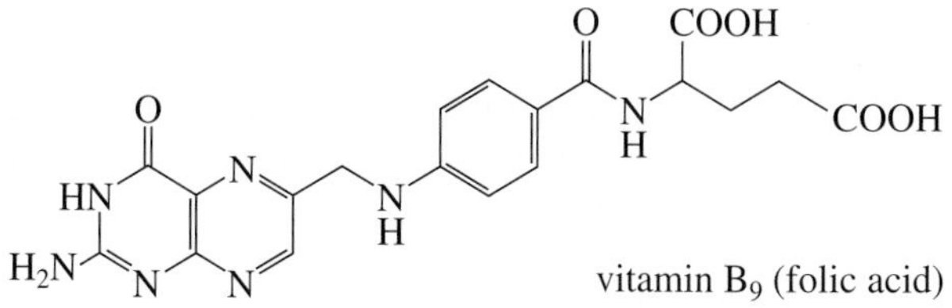 vitamin B_9 (folic acid)

2.88 The following molecules are known for their characteristic fragrances. Convert each line structure into a complete structural formula. **LO2**

(a) 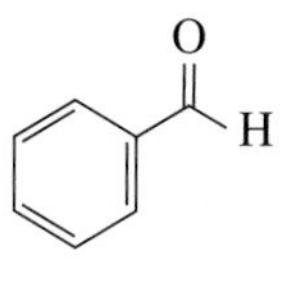 benzaldehyde (almond)

(b) 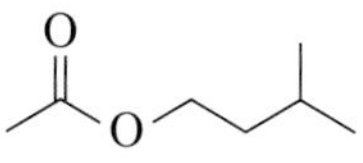 methylbutyl acetate (banana)

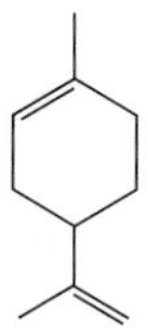

(c) 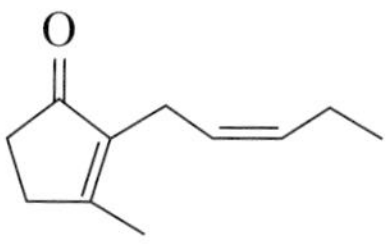 jasmone (jasmine)

(d) limonene (lemon)

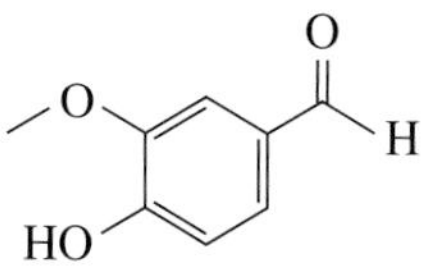

(e) vanillin (vanilla)

2.89 The compounds in the following pairs of substances are quite different from each other despite having similar names. Write correct formulae for each pair. **LO3**

(a) sodium nitrite and sodium nitrate
(b) potassium carbonate and potassium hydrogencarbonate
(c) iodine and iodide ion
(d) sodium chloride and sodium hypochlorite
(e) nitrogen oxide and nitrogen dioxide
(f) potassium chlorate and potassium perchlorate
(g) ammonia and ammonium ion

2.90 Provide IUPAC-approved names for each of the following compounds. **LO3**

(a) NH_4Cl (b) BrF_5 (c) SO_2

2.91 Provide the IUPAC names of each of the following compounds. **LO3**

(a) Li_3PO_4 (b) Li_2HPO_4 (c) LiH_2PO_4 (d) $NaIO_3$ (e) $NaIO_4$

2.92 Express the results of the following logarithmic calculations to the correct number of significant figures. **LO1**

(a) $\log(2.75 \times 10^{-5})$
(b) antilog(−2.53)
(c) ln(2.0)
(d) log(1.56 + 0.254)
(e) $e^{7.21}$
(f) $\ln\left(\frac{6.91\times10^{-2}\times4.667\times10^{-4}}{2.0\times10^{-3}}\right)$

2.93 As we will see in chapter 6, the root mean square velocity $\bar{\mu}_{rms}$, of a gas molecule is given by the equation:

$$\bar{\mu}_{rms} = \left(\frac{3RT}{M}\right)^{\frac{1}{2}}$$

where the units of R are J mol^{-1} K^{-1}, T are K, and M are kg mol^{-1}. Given that the units of $\bar{\mu}_{rms}$ are m s^{-1}, show that the units on both sides of the equation are the same. **LO1**

2.94 In chapter 15, we will learn about rates of chemical reactions and how they can be described by rate laws. The rate law for the gas-phase reaction:

$$2NO + O_2 \rightarrow 2NO_2$$

is:

$$\text{rate of reaction} = k[NO]^2[O_2]$$

where the square brackets denote the concentrations of NO and O_2 in mol L^{-1}. Determine the units of the rate constant k, given that the rate of reaction has units of mol L^{-1} s^{-1}. **LO1**

2.95 From knowledge of their given units, determine an equation relating each of the following groups of variables and constants. **LO1**

(a) volume (V, L), concentration (c, mol L^{-1}) and amount (n, mol). Make amount the subject.
(b) speed of light (c, m s^{-1}), energy (E, J), Planck's constant (h, J s) and wavelength (λ, m). Make energy the subject.
(c) specific heat (c, J g^{-1} K^{-1}), temperature change (ΔT, K), mass (m, g) and heat (q, J). Make mass the subject.
(d) gas density (ρ, kg m^{-3}), pressure (p, Pa), molar mass (M, kg mol^{-1}), the gas constant (R, J mol^{-1} K^{-1}) and temperature (T, K). Make pressure the subject. (*Hint:* Recall that Pa is a derived unit.)

2.96 A student wished to use mechanistic arrows to depict the electron movements in the following process.

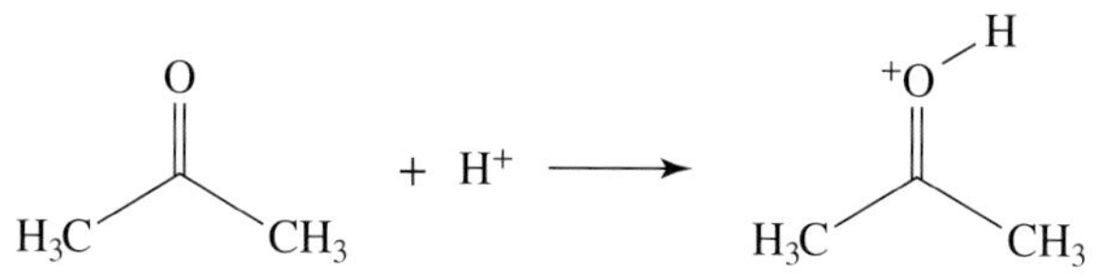

$$CH_3COCH_3 + H^+ \longrightarrow CH_3C(=O^+H)CH_3$$

Explain why the student's use of arrows in each answer below is incorrect. **LO2**

(a)

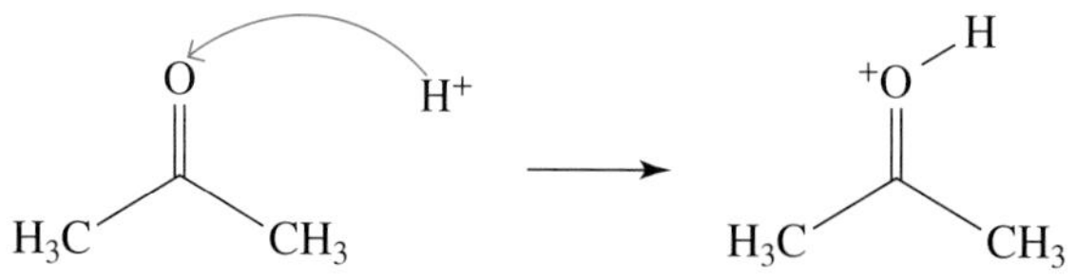

(b)

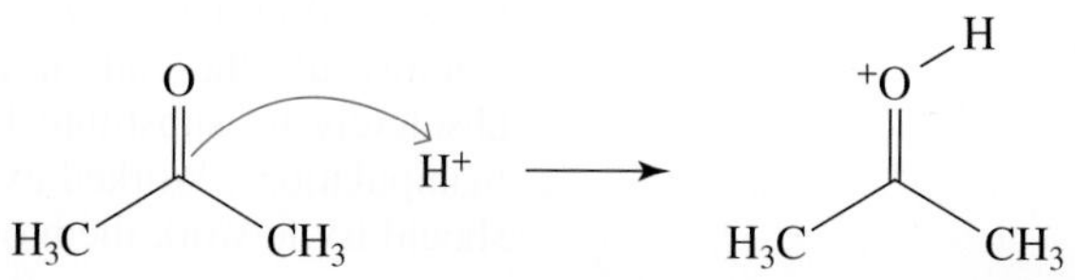

In chemistry, as in any science involving a significant amount of maths, it is imperative that you are confident in rearranging mathematical equations. Below are very brief guidelines on how to do this for two major classes of equations you will encounter.

The equation to be rearranged involves addition or subtraction

Consider the equation:

$$a + b = c$$

How do we rearrange this to make b the subject of the equation? In other words, how do we write the equation in the form:

$$b = ?$$

We start by knowing that b is going to be the only term on the left-hand side of the equation, so both a and c must appear on the right-hand side of the equation. As c is already on the right-hand side of the equation, we only have to move a. But how do we do this?

We make use of the fact that *we can make the same change to both sides of any equation without altering the overall equation*. For example, consider the equation:

$$2 + 4 = 6$$

We can make any change to the equation that we want to, as long as we make the exact same change on both sides of the equals sign. For example, if we added 5 to both sides of the equation, we would have:

$$2 + 4 + 5 = 6 + 5$$

and both sides of the equation are still equal.

We can now use this fact to rearrange the above equation $a + b = c$ to make b the subject. We want to remove a from the left-hand side of the equation, so we need to subtract a from both sides of the equation. This will then give:

$$a + b - a = c - a$$

On the left-hand side, we know that $a + b - a = b$, and therefore we rewrite the above equation as:

$$b = c - a$$

and there we have our rearranged equation.

The equation to be rearranged involves multiplication or division

Consider the equation:

$$a \times b = c$$

How do we rearrange this to make b the subject of the equation? In other words, how do we write the equation in the form:

$$b = ?$$

In this case, we again want to move a to the right-hand side of the equation. We now do this by dividing both sides of the equation by a. This will then give:

$$\frac{a \times b}{a} = \frac{c}{a}$$

The a terms on the left-hand side of the equation can be cancelled and this then gives:

$$b = \frac{c}{a}$$

The examples above are the simplest possible, but the principles involved are identical to those used in rearranging much more complicated equations. As we outlined in the chapter, you can also use the units in an equation as a check that you have rearranged it correctly. There is absolutely no substitute for practice when it comes to becoming expert in carrying out these manipulations. Worked example 2.17 presents some rearrangements to assist you with this. You should try to work them out yourself before looking at the solutions.

WORKED EXAMPLE 2.17

Rearranging equations

(a) Rearrange the First Law of Thermodynamics, $\Delta U = q + w$, to make q the subject.

(b) Rearrange the ideal gas equation $pV = nRT$ to make T the subject.

(c) Rearrange Faraday's Law, amount of product $= \dfrac{It}{zF}$ to make F the subject.

(d) Rearrange the expression for enthalpy, $H = U + pV$, to make p the subject.

(e) Rearrange the expression for the energy of a photon, $E = \dfrac{hc}{\lambda}$, to make c the subject.

(f) Rearrange the expression for cell potential, $E^{o} = \dfrac{RT}{zF}\ln K_c$, to make R the subject.

(g) Rearrange the van der Waals equation $\left(p + \dfrac{n^2 a}{V^2}\right)(V - nb) = nRT$ to make b the subject.

Analysis

These examples are progressively more complicated, but they all make use of the same two principles we described above. We can make any change we want to any equation as long as we make the same change on both sides of the equals sign.

Solution

(a) $\Delta U = q + w$

We want q as the subject, so subtract w from both sides of the equation. This then gives:

$$\Delta U - w = q + w - w$$

which can then be rewritten as:

$$\Delta U - w = q$$

or:

$$q = \Delta U - w$$

(b) $pV = nRT$

We want T as the subject, so divide both sides of the equation by nR. This then gives:

$$\frac{pV}{nR} = \frac{nRT}{nR}$$

The nR terms on the right-hand side of the equation will cancel out, to give:

$$\frac{pV}{nR} = T$$

or:

$$T = \frac{pV}{nR}$$

(c) amount of product $= \dfrac{It}{zF}$

We want F as the subject. We do this in two steps. Firstly, we multiply both sides of the equation by F, to get F on the left-hand side of the equation. This will give:

$$(\text{amount of product}) \times F = \frac{ItF}{zF}$$

The F terms on the right hand side of the equation will cancel out, to give:

$$(\text{amount of product}) \times F = \frac{It}{z}$$

Now we move (amount of product) to the right-hand side of the equation by dividing both sides of the equation by (amount of product). This then gives:

$$\frac{(\text{amount of product}) \times F}{(\text{amount of product})} = \frac{\left(\dfrac{It}{z}\right)}{(\text{amount of product})}$$

The (amount of product) terms on the left-hand side will cancel out, to give:

$$F = \frac{\left(\frac{It}{z}\right)}{\text{(amount of product)}}$$

You can leave the equation in this form, or carry out one last simplification, which gives:

$$F = \left(\frac{It}{z \times \text{(amount of product)}}\right)$$

(d) $H = U + pV$

We want p to be the subject. We do this in two steps. Firstly, subtract U from both sides of the equation, to give:

$$H - U = U + pV - U$$

which can then be rewritten as:

$$H - U = pV$$

We now divide both sides of the equation by V, which gives:

$$\frac{H - U}{V} = \frac{pV}{V}$$

The V terms on the right-hand side cancel out, to give:

$$\frac{H - U}{V} = p$$

or:

$$p = \frac{H - U}{V}$$

(e) $E = \frac{hc}{\lambda}$

We want c to be the subject. We could do this the way that we have learned to above, in two steps, by firstly multiplying both sides of the equation by λ, then dividing both sides of the equation by h. However, there are two equivalent ways in which we can do this in a single step. We could divide both sides of the equation by $\frac{h}{\lambda}$, or multiply both sides of the equation by $\frac{\lambda}{h}$. These both lead to the same answer, namely:

$$\frac{E\lambda}{h} = c$$

or:

$$c = \frac{E\lambda}{h}$$

(f) $E^{\text{o}} = \frac{RT}{zF} \ln K_{\text{c}}$

We want R to be the subject. There are a number of equivalent ways we could do this. We will begin by dividing both sides of the equation by $\ln K_{\text{c}}$, which will give:

$$\frac{E^{\text{o}}}{\ln K_{\text{c}}} = \frac{RT}{zF} \frac{\ln K_{\text{c}}}{\ln K_{\text{c}}}$$

The $\ln K_{\text{c}}$ terms on the right hand side of the equation will cancel out to give:

$$\frac{E^{\text{o}}}{\ln K_{\text{c}}} = \frac{RT}{zF}$$

We can then either divide both sides of the equation by $\frac{T}{zF}$ or multiply both sides of the equation by $\frac{zF}{T}$. Both approaches give the same answer, namely:

$$\frac{E^{\text{o}}}{\ln K_{\text{c}}} \frac{zF}{T} = \frac{RT}{zF} \frac{zF}{T}$$

On the right hand side of the equation, $\frac{T}{zF} \times \frac{zF}{T} = 1$ and therefore, the above equation becomes:

$$\frac{E^{o}}{\ln K_c}\frac{zF}{T} = R$$

or:

$$R = \frac{E^{o}}{\ln K_c}\frac{zF}{T}$$

(g) $\left(p + \frac{n^2 a}{V^2}\right)(V - nb) = nRT$

We want b to be the subject. Again, there are a number of equivalent ways of doing this. We will start by trying to isolate b on the left-hand side of the equation. Therefore, dividing both sides of the equation by $\left(p + \frac{n^2 a}{V^2}\right)$ will give:

$$\frac{\left(p + \frac{n^2 a}{V^2}\right)(V - nb)}{\left(p + \frac{n^2 a}{V^2}\right)} = \frac{nRT}{\left(p + \frac{n^2 a}{V^2}\right)}$$

The $\left(p + \frac{n^2 a}{V^2}\right)$ terms on the left-hand side of the equation will cancel out, leaving:

$$(V - nb) = \frac{nRT}{\left(p + \frac{n^2 a}{V^2}\right)}$$

Now we will subtract V from both sides of the equation:

$$(V - nb) - V = \left(\frac{nRT}{\left(p + \frac{n^2 a}{V^2}\right)}\right) - V$$

and simplifying this gives:

$$-nb = \left(\frac{nRT}{\left(p + \frac{n^2 a}{V^2}\right)}\right) - V$$

Finally, if we divide both sides of the equation by $-n$, we obtain:

$$\frac{-nb}{-n} = \frac{\left(\frac{nRT}{\left(p + \frac{n^2 a}{V^2}\right)}\right) - V}{-n}$$

which simplifies to:

$$b = \frac{\left(\frac{nRT}{\left(p + \frac{n^2 a}{V^2}\right)}\right) - V}{-n}$$

This is probably sufficient, but can be simplified further if you're feeling brave.

3 Chemical reactions and stoichiometry

Chemical plants produce large quantities of basic chemicals used in everyday consumer goods, industrial processes and research laboratories. However, if you manage a pharmaceutical manufacturing plant like the one pictured and want to run it efficiently, you must know how much starting material is needed to synthesise a product with the best possible yield.

Some of the information comes from chemical equations. Chemical equations conveniently and succinctly summarise information about chemical reactions. They show the substances you start with and the substances you end up with. They also show the relative amounts of each substance involved.

In this chapter, you will learn how to use balanced chemical equations and the rules of stoichiometry to work out the mass of the substances required to form a desired mass of a product. A good understanding of stoichiometry is essential for all areas of chemistry, including industrial-scale production, pharmaceutical manufacture and controlling reactions in research laboratories. It is also essential for students wanting to understand basic chemistry.

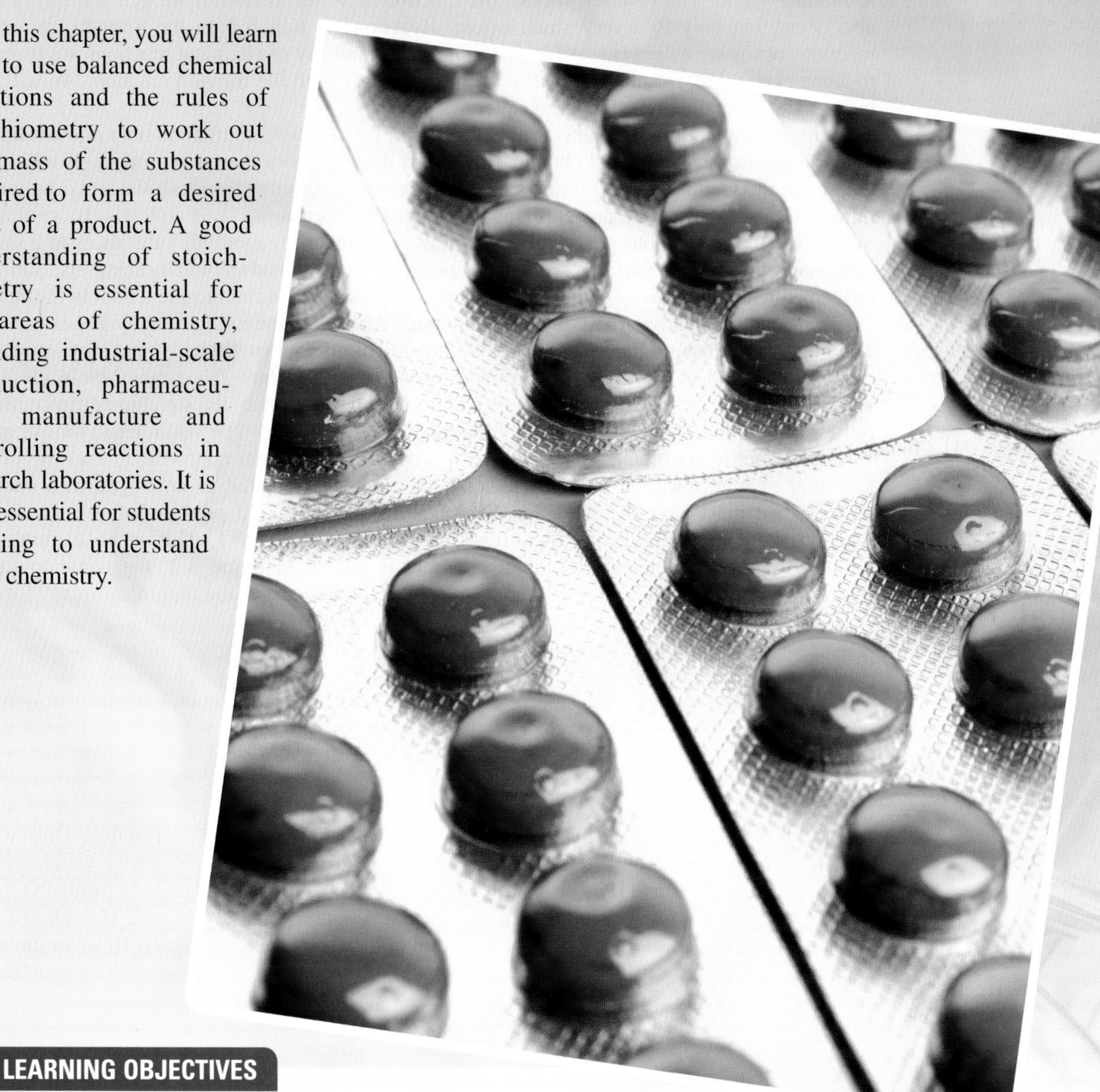

LEARNING OBJECTIVES

After studying this chapter, you should be able to:

3.1 explain how a chemical equation represents a chemical reaction

3.2 balance chemical equations

3.3 use the mole concept in stoichiometric calculations

3.4 determine empirical formulae from quantitative analysis data

3.5 calculate theoretical and actual yields based on balanced equations

3.6 apply stoichiometric concepts to reactions in solutions.

3.1 Chemical equations

The production of new substances from one or more chemical species is called a **chemical reaction**. A **chemical equation**, a concept we briefly introduced in chapter 1, may be used to describe what happens when a chemical reaction occurs. It uses chemical formulae to provide a before-and-after picture of the chemical substances involved. Consider, for example, the reaction between hydrogen and oxygen to give water. The chemical equation that describes this reaction is:

$$2H_2 + O_2 \rightarrow 2H_2O$$

The two formulae that appear to the left of the arrow are those of the **reactants**, hydrogen and oxygen, that is, the substances that react to form the product. The formula of the product, water, is written to the right of the arrow. **Products** are the substances that are formed in the reaction. The arrow means 'react to produce'. Thus, this equation tells us that *hydrogen and oxygen react to produce water.* The reverse process, the reaction of water to form hydrogen and oxygen, can also occur, but only to a very small extent. Reactions in which both the forward and reverse reactions occur are called **reversible reactions**. The majority of chemical reactions are reversible to some extent, but, for now, we will be concerned only with the forward reaction. We will learn more about reversible reactions in chapter 9.

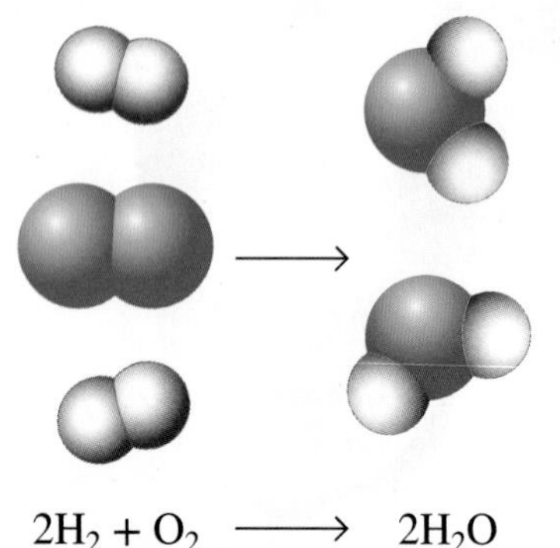

FIGURE 3.1 The reaction between 2 molecules of hydrogen and 1 molecule of oxygen gives 2 molecules of water, as depicted by the space-filling models and chemical equation shown here.

In this example, only one substance is formed in the reaction, so there is only one product. As we will see, however, in most chemical reactions there is more than one product. If we want to determine the identity of the products, we are engaged in **qualitative analysis**; we simply determine which substances are present in a sample without measuring their amounts. By contrast, in **quantitative analysis**, our goal is to measure the amounts of the various substances in a sample.

Stoichiometry (from the Greek *stoicheion* meaning 'element' and *metrein* meaning 'measuring') is concerned with the relative amounts of reactants and products in a chemical reaction. In the equation for the reaction of hydrogen and oxygen, the number 2 precedes the formulae of hydrogen and water. The numbers in front of the formulae are called **stoichiometric coefficients**, and they indicate the ratio of species of each kind among the reactants and products. Thus, $2H_2$ means 2 molecules of H_2, and $2H_2O$ means 2 molecules of H_2O. When no number is written, the stoichiometric coefficient is 1 (so the coefficient of O_2 equals 1). Stoichiometric coefficients can also refer to ions or atoms.

Stoichiometric coefficients are required to ensure the equation conforms to the law of conservation of mass (section 1.2). Because atoms cannot be created or destroyed in a chemical reaction, we must have the same number of atoms of each kind present before and after the reaction (i.e. on both sides of the arrow), as shown in figure 3.1. This is an example of a **balanced chemical equation**. Another example is shown in figure 3.2: the equation for the reaction of butane, C_4H_{10}, with oxygen, O_2. Butane can be used as the fuel in camping stoves.

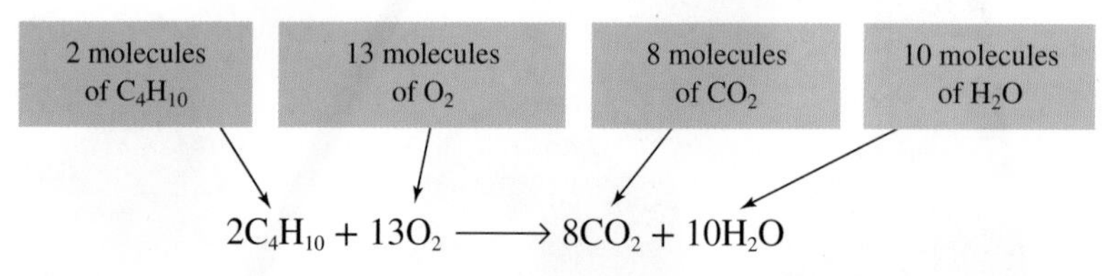

FIGURE 3.2 The chemical equation for the combustion of butane, C_4H_{10}. The products are carbon dioxide and water.

The 2 before C_4H_{10} tells us that 2 molecules of butane are involved in the reaction. This involves a total of 8 carbon atoms and 20 hydrogen atoms (see figure 3.3). Notice that we have multiplied the numbers of atoms of C and H in 1 molecule of C_4H_{10} by the coefficient 2. The 13 in front of O_2 means that 13 molecules of O_2 are required for complete reaction of 2 molecules of C_4H_{10}. On the right, we find 8 molecules of CO_2, which contain a total of 8 carbon atoms. Similarly, 10 water molecules contain 20 hydrogen atoms. Finally, we can count 26 oxygen atoms on both sides of the equation.

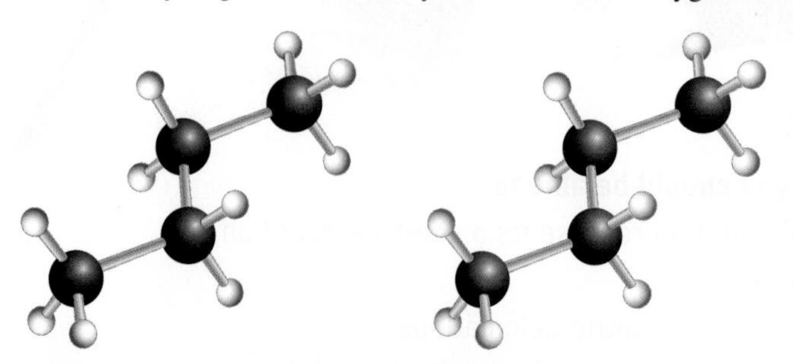

FIGURE 3.3 Understanding coefficients in an equation. The expression $2C_4H_{10}$ describes 2 molecules of butane (depicted here as ball-and-stick models), each of which contains 4 carbon and 10 hydrogen atoms. This gives a total of 8 carbon and 20 hydrogen atoms.

Specifying states of matter

In a chemical equation, it is useful to specify the physical states of the reactants and products, that is, whether they are solids, liquids or gases. This is done by writing (s) for solid, (l) for liquid or (g) for gas after the chemical formula. For example, the equation for the combustion of carbon in a charcoal briquette can be written as:

$$C(s) + O_2(g) \rightarrow CO_2(g)$$

At times, we will also find it useful to indicate that a particular substance is dissolved in water. We do this by writing (aq), meaning 'aqueous solution', after the formula. For instance, the reaction between stomach acid (an aqueous solution of HCl) and $CaCO_3$, the active ingredient in some antacids, is:

$$2HCl(aq) + CaCO_3(s) \rightarrow CaCl_2(aq) + H_2O(l) + CO_2(g)$$

This type of chemical equation is called a molecular equation. It shows all reactants and products, with molecular substances written as discrete molecules (i.e. HCl, H_2O and CO_2) and ionic compounds written in terms of their empirical formulae (i.e. $CaCO_3$ and $CaCl_2$; see section 3.4). If ionic compounds are involved we can also write net ionic equations that only show the ions involved in the actual reaction (see p. 107).

PRACTICE EXERCISE 3.1

How many atoms of each element appear on each side of the arrow in the following equation?

$$Mg(OH)_2 + 2HCl \rightarrow MgCl_2 + 2H_2O$$

PRACTICE EXERCISE 3.2

Rewrite the equation in practice exercise 3.1 to show that $Mg(OH)_2$ is a solid, HCl and $MgCl_2$ are dissolved in water, and H_2O is a liquid.

3.2 Balancing chemical equations

We have learned that a chemical equation is a shorthand, quantitative description of a chemical reaction. An equation is *balanced* when the atoms present among the reactants are also present in the same number among the products. As we learned, stoichiometric coefficients (the numbers in front of formulae) are multiplier numbers for their respective formulae, and the values of the coefficients determine whether an equation is balanced.

Always approach the balancing of an equation as a two-step process.

Step 1: Write the unbalanced 'equation'. Organise the formulae in the pattern of an equation with plus signs and an arrow.

Step 2: Adjust the coefficients so that there are equal numbers of each kind of atom on each side of the arrow. Apply the following guidelines in sequence.

(a) Balance elements other than H and O.

(b) Balance as a group those polyatomic ions (ions consisting of two or more covalently bonded atoms; see chapter 2) that appear unchanged on both sides of the arrow.

(c) Balance ions so that the overall charge is the same on both sides of the arrow.

(d) Balance those species that appear on their own (as elements or ions).

(e) Balance the number of H and O atoms on both sides (e.g. by using H_2O or OH^-).

The guidelines are derived from experience, but are not hard and fast rules. The sequence of rules is chosen so that the species that are most difficult to balance are handled first and the easiest ones last (H and O are easily balanced in aqueous solutions).

When you carry out step 2, make no changes in the formulae, either in the atomic symbols or their subscripts. If you do, the equation will involve different substances from those intended. It may still balance, but the equation will not be for the reaction you want.

We will begin with simple equations that can be balanced by inspection. An example is the reaction of aluminium metal with hydrochloric acid (see figure 3.4). First, we need the correct formulae and the correct physical states. In this example, we use the molecular equation (see above) even though both HCl and $AlCl_3$ exist as ions in aqueous solution. The reactants are aluminium, Al(s), and hydrochloric acid, HCl(aq). We also need formulae for the products. The reaction results in the formation of aqueous aluminium chloride, which we write as $AlCl_3(aq)$, and hydrogen gas, $H_2(g)$. Recall that hydrogen normally exists as diatomic molecules and not as individual atoms.

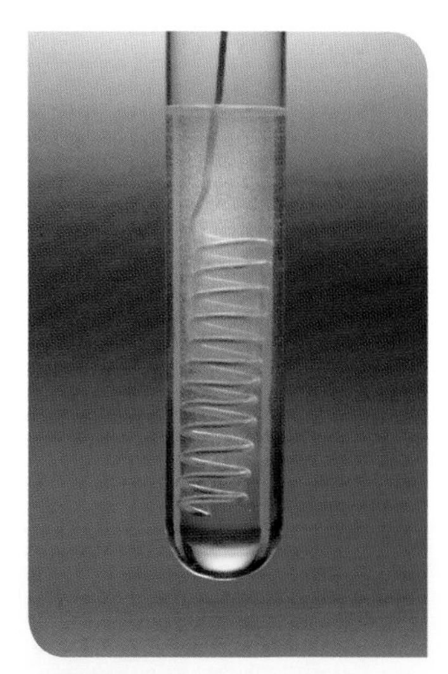

FIGURE 3.4 Aluminium and hydrochloric acid react to form aluminium chloride and hydrogen gas.

Step 1: Write an unbalanced equation.

$$Al(s) + HCl(aq) \rightarrow AlCl_3(aq) + H_2(g) \quad \text{(unbalanced)}$$

Step 2: Adjust the coefficients to get equal numbers of each kind of atom on both sides of the arrow by applying the guidelines in the order given above. Using the guidelines, we will look at Cl first in our example. Because there are 3 Cl atoms to the right of the arrow but only 1 to the left, we put a 3 in front of the HCl on the left side. The result is:

$$Al(s) + 3HCl(aq) \rightarrow AlCl_3(aq) + H_2(g)$$

Since there are no polyatomic ions in this example, we look next at species that appear on their own (Al and H_2, in this example). You can see that Al looks balanced, but there are 3 H atoms on the left and only 2 on the right. We can balance these by putting a coefficient of 3 in front of H_2 and then doubling the coefficient for HCl.

$$Al(s) + 6HCl(aq) \rightarrow AlCl_3(aq) + 3H_2(g)$$

However, Cl is now not balanced; there are 6 Cl atoms on the left and only 3 on the right. Therefore, we need a coefficient of 2 in front of $AlCl_3$.

$$Al(s) + 6HCl(aq) \rightarrow 2AlCl_3(aq) + 3H_2(g)$$

Everything is now balanced except for Al; there is 1 Al atom on the left but 2 on the right. We put a coefficient of 2 in front of Al to give:

$$2Al(s) + 6HCl(aq) \rightarrow 2AlCl_3(aq) + 3H_2(g)$$

Now there are 2 Al, 6 H and 6 Cl atoms on each side of the equation and everything is balanced.

Note that the equation $4Al(s) + 12HCl(aq) \rightarrow 4AlCl_3(aq) + 6H_2(g)$ is also balanced, but it is usual to write balanced equations using the *smallest* whole-number coefficients.

WORKED EXAMPLE 3.1

Writing a balanced equation

Aqueous solutions of calcium hydroxide, $Ca(OH)_2$, and phosphoric acid, H_3PO_4, react to give calcium phosphate, $Ca_3(PO_4)_2$, and water. The calcium phosphate precipitates from solution. Write the balanced equation for this reaction.

Analysis

First, write an unbalanced equation that includes the reactant formulae on the left-hand side and the product formulae on the right. Include the designation (aq) for all substances dissolved in water and (l) for H_2O itself when it is in its liquid state. Then adjust stoichiometric coefficients (*never subscripts!*) until there is the same number of each type of atom on the left and right sides of the equation.

Solution

$$Ca(OH)_2(aq) + H_3PO_4(aq) \rightarrow Ca_3(PO_4)_2(s) + H_2O(l) \qquad \text{(unbalanced)}$$

There are several things not in balance, but our guidelines suggest that we work with Ca first, rather than with O and H atoms, or the PO_4^{3-} polyatomic ions, which remain unchanged on both sides of the equation. There are 3 Ca atoms on the right side, so we put a 3 in front of $Ca(OH)_2$ on the left, as a trial.

$$3Ca(OH)_2(aq) + H_3PO_4(aq) \rightarrow Ca_3(PO_4)_2(s) + H_2O(l) \qquad \text{(unbalanced)}$$

Now the Ca atoms are in balance. The PO_4^{3-} anions are not balanced; there are 2 on the right side and only 1 on the left side. We put a 2 in front of H_3PO_4 as the next step.

$$3Ca(OH)_2(aq) + 2H_3PO_4(aq) \rightarrow Ca_3(PO_4)_2(s) + H_2O(l) \qquad \text{(unbalanced)}$$

Not counting the PO_4^{3-} anions, we have on the left 6 O and 6 H atoms in $3Ca(OH)_2$ plus 6 H atoms in H_3PO_4, for a net total of 6 O and 12 H atoms on the left. On the right, in H_2O, we have 1 O and 2 H atoms. The ratio of 6 O to 12 H atoms on the left is equivalent to the ratio of 1 O to 2 H atoms on the right, so we write the multiplier (coefficient) 6 in front of H_2O.

$$3Ca(OH)_2(aq) + 2H_3PO_4(aq) \rightarrow Ca_3(PO_4)_2(s) + 6H_2O(l) \qquad \text{(balanced)}$$

We now have a balanced equation.

Is our answer reasonable?

A count of the total number of atoms on each side of the equation shows that they are the same. On each side we have 3 Ca, 12 H, 2 P and 14 O atoms. The coefficients used cannot be reduced to smaller whole numbers and so we have an equation balanced with the smallest whole-number coefficients.

PRACTICE EXERCISE 3.3

When aqueous solutions of barium chloride, $BaCl_2$, and aluminium sulfate, $Al_2(SO_4)_3$, are mixed, a reaction occurs in which solid barium sulfate, $BaSO_4$, precipitates from the solution. The other product is $AlCl_3(aq)$. Write the balanced equation for this reaction.

3.3 The mole

In the preceding sections, you have seen that chemical formulae tell us about the relative numerical proportions of atoms in molecules or ionic compounds, and that chemical equations are a convenient way to summarise what happens in chemical reactions. In this and the following sections, you will investigate the quantitative relationships further and find that these are governed by the rules of stoichiometry.

In a chemical reaction, the relative ratios of reactants resulting in a given ratio of products are given by the chemical equations that we looked at in earlier sections of this chapter. That, however, is on an atomic or molecular level. For example:

$$A + 2B \rightarrow AB_2$$

tells you that 1 molecule of A reacts with 2 molecules of B to form 1 molecule of AB_2. Suppose that we wanted to carry out this reaction in the laboratory. Given that we cannot see or count individual atoms, ions or molecules, we need to weigh out and work with larger quantities, but how do we know what masses of A and B to use? It turns out that some of the information we need to answer this question is given in the balanced chemical equation.

Think about it this way. If you need 50 000 sheets of white photocopy paper and you order them from a paper mill, the mill will weigh them out rather than counting the individual sheets. This procedure is much simpler and quicker and relies on the fact that we know the mass of one sheet of paper, and that every sheet of paper weighs the same (at least to a very good approximation). It is, therefore, easy to calculate the weight that corresponds to 50 000 sheets of paper. The same principle applies to counting large numbers of coins if you want to deposit them at a bank.

We can take a similar approach to 'counting' ions, atoms and molecules. Every atom has a certain mass. As we saw in chapter 1, the masses of individual atoms can be measured in atomic mass units (u), where $1\ \text{u} = 1.660\,54 \times 10^{-24}$ g, exactly $\frac{1}{12}$ of the mass of a ^{12}C atom. If we reverse this definition, we see that $6.022\,14 \times 10^{23}\ \text{u} = 1$ g, so 12 g of ^{12}C contains exactly $6.022\,14 \times 10^{23}$ atoms. It is clear that even a small mass of a substance contains a very large number of atoms. Therefore, we use a quantity called the **mole**. The mole (unit symbol: mol) is the SI unit of **amount of substance**. One mole is the amount of substance that contains the same number of specified entities as there are atoms in exactly 12 g of ^{12}C. This number is called the **Avogadro constant** (N_A), and it underpins all of chemistry.

$$N_A = 6.022 \times 10^{23}\ \text{mol}^{-1}$$

FIGURE 3.5 One mole of different elements (clockwise from top left: sulfur, aluminium, mercury and copper). Each of these element samples contains 6.022×10^{23} atoms.

We will generally quote the Avogadro constant to four significant figures. When we talk about an amount, we mean an amount of substance in moles. The numerical value of the Avogadro constant is almost beyond comprehension. For example, 6.022×10^{23} grains of sand would cover Australia and New Zealand to a depth of about 1.3 metres. However, because atoms are so incredibly small, a mole of atoms, molecules or ions takes up comparatively little space (see figures 3.5 and 3.6). One mole of anything contains 6.022×10^{23} entities. In chemistry terms, these entities are usually atoms, ions or molecules, but the definition of a mole does not restrict the nature of these entities. For example, one mole of eggs would contain 6.022×10^{23} eggs. We can in fact think of the mole as being the 'chemist's dozen' — just as 12 eggs constitute a dozen, 6.022×10^{23} eggs constitute a mole. Most importantly, the mole conveniently links atomic mass and macroscopic amounts; whatever anything weighs in atomic mass units, a mole of it will weigh in grams.

It is important that we specify exactly which particular species we are discussing when we use the mole. For example, the phrase '1 mole of oxygen' is ambiguous, because we have not specified whether we are talking about 1 mole of oxygen atoms, O, or oxygen molecules, O_2, so it is usual to include a chemical formula whenever the mole is used.

Of particular interest to us in chemistry is the mass of 1 mole. We saw in the definition of the mole above that 1 mol of ^{12}C weighs 12 g, but it stands to reason that 1 mol of eggs is going to weigh rather more than this. Therefore, while the number of specified entities in a mole is constant, the mass of 1 mole of those entities depends on the mass of the individual entities. We call the mass of 1 mol of any specified entity the **molar mass (M)**. This serves as the link between mass (m) and amount of substance (n) through the equation:

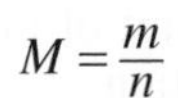

$$M = \frac{m}{n}$$

FIGURE 3.6 One mole of different compounds (clockwise from top left: copper sulfate pentahydrate, water, sodium chromate and sodium chloride).

The SI unit of molar mass is actually kg mol^{-1}, but this is rarely used in practice. Molar mass values are almost always quoted in units of g mol^{-1}; that is, the number of grams per mole of substance.

In chapter 1, we introduced the concept of atomic mass, the mass of a single atom of an element, which we measured in atomic mass units. The molar mass, which as the mass of 1 mole of atoms of an element, is numerically equal to the atomic mass, but the units differ. Values of molar mass for the elements are tabulated inside the front cover of this book, and it is informative to look at some of these. Note that these values refer to a mole of atoms of the elements, regardless of whether the elements themselves exist in atomic form. For example, the tabulated molar mass of H is 1.007 94 g mol^{-1} even though elemental hydrogen exists as H_2 molecules, not H atoms. The molar mass of C is 12.0107 g mol^{-1}, a value that might surprise you, given that we stated previously that 1 mole of ^{12}C weighs exactly 12 g. We need to appreciate that the molar mass of an element is determined by the isotopic make-up of that element. Carbon, for example, has two major naturally occurring isotopes, namely ^{12}C and ^{13}C, and any sample of carbon will contain these isotopes in 98.89% and 1.11% abundance, respectively. This means that the molar mass of *naturally occurring* carbon is not exactly 12 g mol^{-1}, but just a little greater, reflecting the small amount of the heavier ^{13}C isotope present. The molar masses of all elements with more than one naturally occurring isotope reflect the relative mass and abundance of each isotope as discussed in chapter 1.

Because no element has a molar mass exactly equal to an integer (even those with a single isotope, as we saw on p. 12), for simplicity of calculations in this book we will usually use molar masses to four significant figures. There is no need to learn the values of molar mass (they will always be available), but it is important to learn how to use them. The equation:

$$M = \frac{m}{n}$$

which defines molar mass, is one of the most important in all of chemistry, and we will use it often. With it, and just one other equation (which we will encounter later in this chapter), we have everything we need to solve *any* stoichiometry problem. The approach to doing stoichiometry problems is deciding how to apply these two equations.

WORKED EXAMPLE 3.2

Calculating an amount from a mass

The *Golden Jubilee* diamond is the largest faceted diamond in the world. It has a mass of 109.13 g. If the stone consists of pure carbon, what amount of C does the stone contain, given that the molar mass of C is 12.01 g mol^{-1}?

Analysis

A good method to follow when beginning any stoichiometry problem is to ask two questions: 'What value does the question want me to determine?', and 'What data have I been given?'. Write down the answers to these questions and then look for an equation that contains as many as possible of these. In this case, we are asked to determine an amount (n), and we have been given a molar mass (M) and a mass (m). The equation that contains all of these is $M = \frac{m}{n}$.

Solution

We rearrange $M = \frac{m}{n}$ to solve for n: $n = \frac{m}{M}$

Hence:
$$n = \frac{109.13\ \text{g}}{12.01\ \text{g mol}^{-1}} = 9.087\ \text{mol}$$

Therefore, this diamond contains 9.087 moles of carbon.

Is our answer reasonable?

We have 109.13 g of carbon in the diamond. Given that 1 mole of C has a mass of 12.01 g, 10 moles would have a mass of 120.1 g. We should therefore expect an answer of slightly fewer than 10 moles, which we have. Our answer therefore seems reasonable.

PRACTICE EXERCISE 3.4 What amount of sulfur molecules, S_8, is present in 35.6 g of sulfur? The molar mass of atomic sulfur, S, is 32.06 g mol^{-1}.

When solving this type of problem, always check the units of the answer. This is especially important if it is necessary to rearrange the equation to solve the problem. The units will *always* be correct if you have rearranged the equation correctly. In worked example 3.2, we are dividing mass by molar mass. Grams in the numerator (the top line) will cancel grams in the denominator (the bottom line), leaving $\frac{1}{\text{mol}^{-1}}$ as our final unit. This is the same as $\frac{1}{\frac{1}{\text{mol}}}$, which is mol, meaning that our units are correct. As we saw on p. 30–2, units are also very powerful in helping you to remember equations. The equation $M = \frac{m}{n}$ is easy to remember if you know that the units of molar mass are g mol^{-1}: in other words, grams (the unit of mass) divided by moles (the unit of amount). This then gives the correct form of the equation.

WORKED EXAMPLE 3.3

Calculating a mass from an amount

Calcium phosphate, $Ca_3(PO_4)_2$, is often used to coat some of the surfaces of bone or dental implants to permit bone to bond with the implant surface. A schematic is shown in figure 3.7.

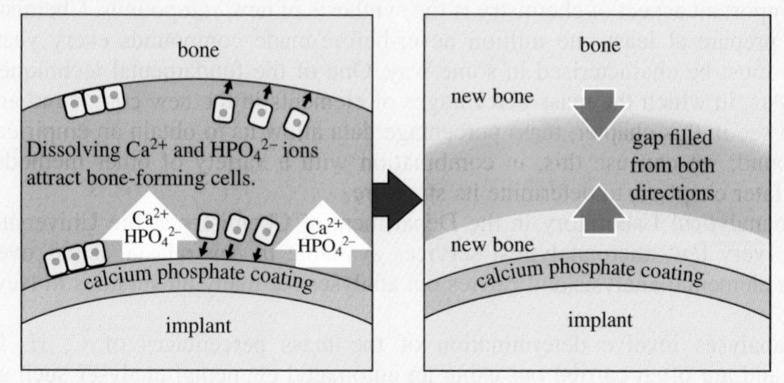

FIGURE 3.7 Some surfaces on bone implants are coated with calcium phosphate to permit bone to bond to the surface.

If a coating procedure can deposit 0.115 mol of pure $Ca_3(PO_4)_2$ on an implant, what is the mass of the coating? The molar mass of Ca is 40.08 g mol^{-1}, of P is 30.97 g mol^{-1} and of O is 16.00 g mol^{-1}.

Analysis

We are asked to calculate a mass (m) and we are given an amount (n) and the molar masses (M) of Ca, P and O. We will therefore use $M = \frac{m}{n}$ again, but we will have to calculate the molar mass of $Ca_3(PO_4)_2$ from the individual molar masses of the elements before we can substitute the data into the equation.

Solution

$$M_{Ca_3(PO_4)_2} = (3 \times 40.08 \text{ g mol}^{-1}) + (2 \times 30.97 \text{ g mol}^{-1}) + (8 \times 16.00 \text{ g mol}^{-1})$$
$$= 310.2 \text{ g mol}^{-1}$$

We now rearrange $M = \frac{m}{n}$ to make m the subject:

$$m = Mn$$

Substitution of the data gives:

$$m = 310.2 \text{ g mol}^{-1} \times 0.115 \text{ mol}$$
$$= 35.7 \text{ g}$$

Therefore, the coating of $Ca_3(PO_4)_2$ weighs 35.7 g.

Is our answer reasonable?

The coating contains a little over 0.1 of a mole of $Ca_3(PO_4)_2$. Since 0.1 of 310 g is 31 g, and our answer of 35.7 g is slightly more than 31 g, it makes sense.

What mass of sodium carbonate, Na_2CO_3, corresponds to 0.125 mol of Na_2CO_3? **PRACTICE EXERCISE 3.5**

Determine the amount of H_2SO_4 in a 45.8 g sample of H_2SO_4. **PRACTICE EXERCISE 3.6**

3.4 Empirical formulae

Whenever a new compound is prepared, it is usually analysed to determine the mass percentages of the elements it contains. From these mass percentages, it is possible to obtain an empirical formula of the compound. The **empirical formula** is the simplest whole-number ratio of atoms within that compound. For example, the empirical formula of butane (chemical formula C_4H_{10}) is C_2H_5. If we know the ratio by mass of each element within a compound, we can determine its empirical formula using the relationship $M = \frac{m}{n}$.

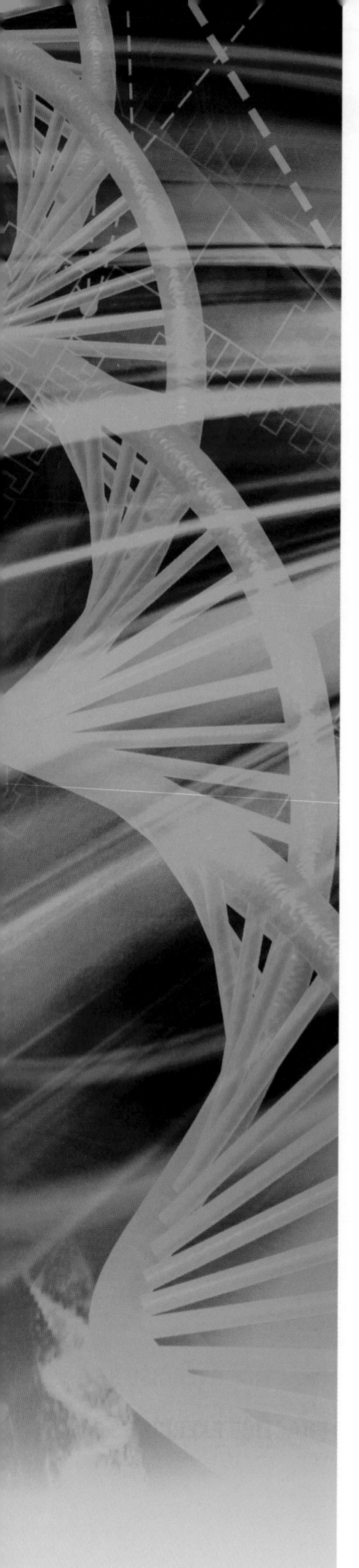

Chemical Connections

Elemental analysis at the Campbell Microanalytical Laboratory

Arguably the most important aspect of chemistry is the synthesis of new compounds. Chemists around the world prepare at least one million never-before-made compounds every year, and every one of these must be characterised in some way. One of the fundamental techniques used is elemental analysis, in which the mass percentages of elements in the new compound are determined. As we will see in this chapter, mass percentage data allow us to obtain an empirical formula for the compound; we can use this, in combination with a variety of other methods, which we will meet in later chapters, to determine its structure.

The Campbell Microanalytical Laboratory in the Department of Chemistry at the University of Otago is one of the very few microanalytical services available in Australasia. It has over 50 years of expertise in elemental analysis and carries out analyses for many universities in New Zealand and Australia.

Routine elemental analyses involve determination of the mass percentages of C, H, N and S in a compound, and are often carried out using an automated elemental analyser such as the one shown in the schematic diagram in figure 3.8. The samples are combusted with oxygen at high temperature (oxidation) in the presence of a catalyst. The resulting gases are then passed over a catalyst layer and through copper to remove excess O_2 and to reduce nitrogen oxides to nitrogen (reduction). The gases, CO_2, H_2O, SO_2 and N_2, are then separated by injection onto a chromatography column, and the amount of each is measured using a very sensitive thermal conductivity detector. These data are then converted to mass percentages of C, H, N and S, respectively.

Elements other than C, H, N or S in a sample must be analysed manually, and standard methods are available for these. Note that O is not usually analysed directly, and its mass percentage in an organic compound is generally obtained by subtracting the total mass percentages of all the other elements from 100.

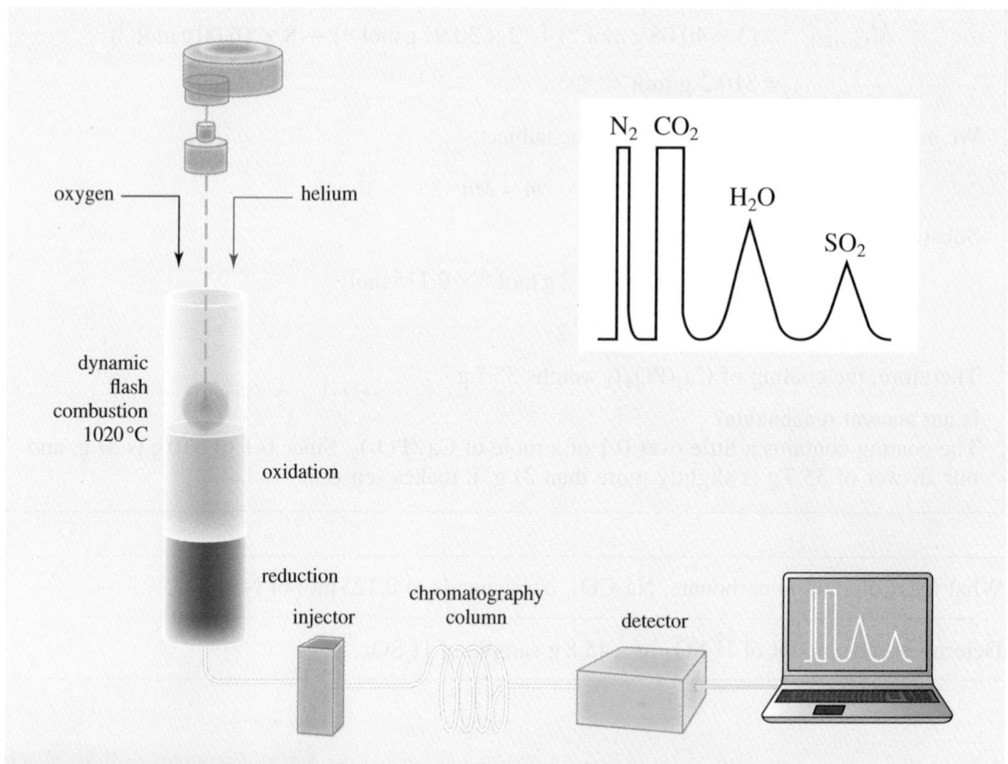

FIGURE 3.8 Diagram of an automated elemental analysis system for carbon, hydrogen, nitrogen and sulfur. The diagram illustrates the combustion chamber, the chromatography column, a thermal conductivity detector, and a sample of a chromatogram with peaks for N_2, CO_2, H_2O and SO_2.

Mole ratios from chemical formulae

Consider the chemical formula for water, H_2O.

- 1 molecule of water contains 2 H atoms and 1 O atom.
- 2 molecules of water contain 4 H atoms and 2 O atoms.
- 1 dozen molecules of water contain 2 dozen H atoms and 1 dozen O atoms.
- 1 mole of molecules of water contains 2 moles of H atoms and 1 mole of O atoms.

Whether we are dealing with atoms, dozens of atoms or moles of atoms, the chemical formula tells us that the ratio of H atoms to O atoms in H_2O is always 2 to 1. *For chemical compounds, mole ratios are the same as the ratios of the individual atoms.*

This fact lets us prepare mole-to-mole conversion factors involving elements in compounds as we need them. For example, in the formula P_4O_{10}, the subscripts mean that there are 4 moles of P for every 10 moles of O in this compound. We also know that 1 mole of P_4O_{10} contains 4 moles of P and 10 moles of O.

In the laboratory, stoichiometry is commonly used to relate the masses of two starting materials that are needed to synthesise a compound and to determine reaction yields, as we will see later in this chapter.

WORKED EXAMPLE 3.4

Calculating amounts of elements within a compound

Leaves have a characteristic green colour (figure 3.9) due to the presence of the green pigment chlorophyll, which has the formula $C_{55}H_{72}MgN_4O_5$. A particular sample of chlorophyll was found to contain 0.0011 g of Mg. What mass of carbon was present in the sample? ($M_C = 12.01\ \text{g mol}^{-1}$; $M_{Mg} = 24.31\ \text{g mol}^{-1}$)

Analysis

We are asked to calculate a mass of carbon (m_C) and are given a mass of magnesium (m_{Mg}) and molar masses of both carbon and magnesium. The data given allow us to calculate the amount of magnesium (n) in the sample using $M = \frac{m}{n}$; we then use the mole ratio between Mg and C, which is implicit in the formula of chlorophyll, to determine the amount of C present. Once we have this amount, we then use $M = \frac{m}{n}$ again to convert it to mass.

FIGURE 3.9 The characteristic green colour of leaves is imparted by the presence of the pigment chlorophyll.

Solution

First, we calculate the amount of Mg corresponding to 0.0011 g of Mg.

$$M = \frac{m}{n}$$

$$n = \frac{m}{M} = \frac{0.0011\ \text{g}}{24.31\ \text{g mol}^{-1}} = 4.5 \times 10^{-5}\ \text{mol}$$

We now look at the formula of chlorophyll, $C_{55}H_{72}MgN_4O_5$, to obtain the mole ratio between Mg and C. In 1 molecule of chlorophyll, we have 1 atom of Mg and 55 atoms of C. Therefore, in 1 mole of chlorophyll, we have 1 mole of Mg and 55 moles of C. The mole ratio between Mg and C is therefore 1 : 55, meaning that there is 55 times the amount of C as there is Mg in chlorophyll. In this problem, we have 4.5×10^{-5} mol Mg, so we obtain the amount of C from:

$$n_C = n_{Mg} \times 55 = 4.5 \times 10^{-5}\ \text{mol} \times 55 = 2.5 \times 10^{-3}\ \text{mol}$$

Now we know the amount of carbon, we can convert this to mass.

$$m = Mn = 12.01\ \text{g mol}^{-1} \times (2.5 \times 10^{-3}\ \text{mol}) = 3.0 \times 10^{-2}\ \text{g}$$

Therefore, 0.030 g C was present in the sample.

Is our answer reasonable?

The molar mass of magnesium is about twice that of C. Since we need 55 mol C for every 1 mol Mg, the mass of C we require should be about 30 times that of Mg. Our answer was 0.030 g C for 0.0011 g Mg, so the mass of C we got is about 30 times the mass of the magnesium we started with. This seems reasonable.

PRACTICE EXERCISE 3.7 What mass of iron is in a 15.0 g sample of Fe_2O_3?

PRACTICE EXERCISE 3.8 What mass of iron is required to react with 25.6 g of O to form Fe_2O_3? Assume that the reaction proceeds to completion.

Determination of chemical formulae

In research laboratories, chemists often synthesise entirely new compounds or isolate previously unknown compounds from plant and animal tissues. They must then determine the formula and structure of the compound. This usually involves mass spectrometry (see chapter 20), which gives an experimental value for the molar mass. The compound can also be decomposed chemically to find the masses of elements within a given amount of compound, a process called **elemental analysis** (see Chemical Connections on p. 88). Let us see how experimental mass measurements can be used to determine the compound's formula.

The usual form for describing the relative masses of the elements in a compound is a list of percentages by mass called the compound's **mass percentage composition**. In general, a **percentage by mass** is found by using the following equation.

$$\% \text{ by mass of element} = \frac{\text{mass of element present in the sample}}{\text{mass of whole sample}} \times 100\%$$

WORKED EXAMPLE 3.5

Calculating a percentage composition from chemical analysis

A sample of a liquid with a mass of 8.657 g was found to contain 5.217 g of carbon, 0.9620 g of hydrogen and 2.478 g of oxygen. What is the percentage composition of this compound?

Analysis

We must apply the percentage composition equation for each element. The 'mass of whole sample' here is 8.657 g, so we take each element in turn and do the calculations.

Solution

$$\text{mass \% (C)} = \frac{5.217\text{ g}}{8.657\text{ g}} \times 100\% = 60.26\%$$

$$\text{mass \% (H)} = \frac{0.9620\text{ g}}{8.657\text{ g}} \times 100\% = 11.11\%$$

$$\text{mass \% (O)} = \frac{2.478\text{ g}}{8.657\text{ g}} \times 100\% = 28.62\%$$

$$\text{Sum of percentages} = 99.99\%$$

One of the useful things about a percentage composition is that it allows us to work out the mass of each of the elements in 100 g of the substance without further calculation. For example, the results in this problem tell us that in 100.00 g of the liquid there is 60.26 g of carbon, 11.11 g of hydrogen and 28.62 g of oxygen.

Is our answer reasonable?

The 'check' is that the percentages must add up to 100% if the individual masses add up to the total mass, allowing for small differences caused by rounding and experimental error.

PRACTICE EXERCISE 3.9 From 0.5462 g of a compound, 0.1417 g of nitrogen and 0.4045 g of oxygen was isolated. What is the percentage composition of this compound? Are any other elements present?

Elements can combine in many different ways. Nitrogen and oxygen, for example, form all of the following compounds: N_2O, NO, NO_2, N_2O_3, N_2O_4 and N_2O_5. To identify an unknown sample of a compound of nitrogen and oxygen, we compare the percentage composition found by experiment with the calculated percentages for each possible formula.

WORKED EXAMPLE 3.6

Calculating a theoretical percentage composition from a chemical formula

Are the mass percentages of 25.92% N and 74.09% O consistent with the formula N_2O_5? ($M_N = 14.01\ g\ mol^{-1}$; $M_O = 16.00\ g\ mol^{-1}$)

Analysis

Essentially, we are asked to calculate the mass percentages of N and O in the compound N_2O_5. The data we are given are the molar masses (M) of N and O. The only equation we know that contains molar mass is $M = \frac{m}{n}$ so we will have to use this at some stage. To calculate the theoretical percentages by mass of N and O in N_2O_5, we need the masses of N and O in a specific sample of N_2O_5. If we choose 1 mole of the given compound as the sample, calculating the rest of the data will be easier.

Solution

We know that 1 mole of N_2O_5 contains 2 moles of N and 5 moles of O. The corresponding masses of N and O are found using $M = \frac{m}{n}$ as follows.

$$m_N = Mn = 14.01\ g\ mol^{-1} \times 2\ mol = 28.02\ g$$

$$m_O = Mn = 16.00\ g\ mol^{-1} \times 5\ mol = 80.00\ g$$

The mass of 1 mole of N_2O_5 will, therefore, be the sum of the mass of N and the mass of O in 1 mole of the molecule; that is, 28.02 g + 80.00 g = 108.02 g.

We can now calculate the percentages by mass as we did in worked example 3.5.

$$\text{mass \% (N)} = \frac{28.02\ g}{108.02\ g} \times 100 = 25.94\%$$

$$\text{mass \% (O)} = \frac{80.00\ g}{108.02\ g} \times 100 = 74.06\%$$

Therefore, the experimental data are consistent with the formula N_2O_5.

Is our answer reasonable?

We get a good match between the calculated and experimental values, which is always encouraging! In problems like this, you should double-check your calculations and make sure that the percentages of all the elements present add up to 100% if the individual masses add up to the total mass, within rounding and experimental error.

PRACTICE EXERCISE 3.10

Calculate the percentage composition of N_2O_4.

Determination of empirical formulae

The compound that forms when phosphorus burns in oxygen consists of molecules with the formula P_4O_{10}. When a formula gives the composition of one *molecule*, it is called the **molecular formula**. Notice, however, that both the subscripts 4 and 10 are divisible by 2, so the *smallest* numbers that tell us the *ratio* of P to O is 2 to 5. A simpler (but less informative) formula that expresses this ratio is P_2O_5. This is sometimes called the *simplest formula* for the compound. It is also called the empirical formula because it can be obtained from an experimental analysis of the compound.

To obtain an empirical formula experimentally, we need to determine the mass of each element in a sample of the compound. From mass, we then calculate amount, from which we obtain the mole ratios of the elements. Because the ratio by moles is the same as the ratio by atoms, we can construct the empirical formula.

Only rarely is it possible to obtain the masses of every element in a compound by the use of just one weighed sample; two or more analyses on different samples may be required. For example, suppose an analyst is given a compound known to consist exclusively of calcium, chlorine and oxygen. The mass of calcium in one weighed sample and the mass of chlorine in another sample would be determined in separate experiments. Then the mass data for calcium and chlorine would be converted to percentages by mass (equivalent to the mass of each of these elements in a sample of 100 g). This allows the data from different samples to be compared directly. The mass percentage of oxygen would then be calculated by difference because %Ca + %Cl + %O = 100%. Each mass percentage can be related to the corresponding amount of the element. The mole proportions are converted to whole numbers as we have just studied, giving us the subscripts for the empirical formula.

Therefore, the three steps necessary to determine an empirical formula are as follows.

1. Assume we are studying 100 g of the compound. Therefore, the individual mass percentages correspond to the actual masses of the elements in 100 g of the compound.
2. Convert the ratio of elements by mass to a ratio by amount, by dividing the mass of each element by its molar mass.
3. Divide the resulting numbers by the smallest one, which will give the smallest ratio of the elements in the compound. If this gives any numbers that are not integers, multiply all numbers in the ratio by the smallest factor that will result in all numbers being integers.

WORKED EXAMPLE 3.7

Calculating an empirical formula from percentage composition

A white powder used in paints, enamels and ceramics has the following mass percentage composition: Ba, 69.6%; C, 6.09%; and O, 24.3%. What is its empirical formula? (M_{Ba} = 137.3 g mol^{-1}; M_C = 12.01 g mol^{-1}; M_O = 16.00 g mol^{-1})

Analysis

We are asked to calculate an empirical formula, and we are given mass percentage compositions and molar mass (*M*) values. We can convert mass percentage to mass (*m*) simply by choosing 100 g of compound to study. We can then calculate amounts of Ba, C and O by using $M = \frac{m}{n}$, and use these to obtain the necessary mole ratios for the empirical formula.

Solution

A 100 g sample of the compound will contain 69.6 g Ba, 6.09 g C and 24.3 g O. We first relate these masses to amounts.

$$n_{Ba} = \frac{m}{M} = \frac{69.6\ \text{g}}{137.3\ \text{g mol}^{-1}} = 0.507\ \text{mol}$$

$$n_{C} = \frac{m}{M} = \frac{6.09\ \text{g}}{12.01\ \text{g mol}^{-1}} = 0.507\ \text{mol}$$

$$n_{O} = \frac{m}{M} = \frac{24.3\ \text{g}}{16.00\ \text{g mol}^{-1}} = 1.52\ \text{mol}$$

Our ratio of Ba : C : O is therefore 0.507 : 0.507 : 1.52. However, chemical formulae contain whole numbers, so we divide each amount by the smallest amount. The ratio then becomes:

$$\frac{0.507\ \text{mol}}{0.507\ \text{mol}} : \frac{0.507\ \text{mol}}{0.507\ \text{mol}} : \frac{1.52\ \text{mol}}{0.507\ \text{mol}} = 1:1:3$$

Therefore, our mole ratio is 1 Ba : 1 C : 3 O, and the empirical formula is $BaCO_3$.

Is our answer reasonable?

If the formula is correct, there are 3 moles of oxygen for every 1 mole of carbon in the compound. So, for every 1 mole (12 g) of carbon in the compound, there should be 3 moles (48 g) of oxygen. The carbon-to-oxygen mass ratio is 12 : 48, or 1 : 4. This is consistent with the ratio of carbon and oxygen percentages (6.09 : 24.3 ≈ 1 : 4).

PRACTICE EXERCISE 3.11

A white solid used to whiten paper has the following mass percentage composition: Na, 32.4%; S, 22.6%. The rest of the mass is oxygen. What is the empirical formula of the compound?

In practice, a compound is seldom broken down completely to its elements in a quantitative analysis. Instead, the compound is changed into other compounds. The reactions separate the elements by capturing each one entirely (quantitatively) in a separate compound with a known formula.

In the following example, we illustrate an indirect analysis of a compound made entirely of carbon, hydrogen and oxygen. Such compounds burn completely in pure oxygen — the reaction is called **combustion** — and the sole products are carbon dioxide and water vapour. The complete combustion of ethanol, CH_3CH_2OH, for example, occurs according to the following equation.

$$CH_3CH_2OH(l) + 3O_2(g) \rightarrow 2CO_2(g) + 3H_2O(g)$$

The carbon dioxide and water can be separated and then their masses determined. Notice that all of the carbon atoms in the original compound end up among the CO_2 molecules and all of the hydrogen atoms are in H_2O molecules. In this way at least two of the original elements, C and H, are entirely separated.

Modern microanalytical techniques allow the simultaneous determination of C, H, N and S, as described in the Chemical Connections feature on page 88. Initial combustion of the sample and subsequent passage of the resulting gas mixture over a catalyst results in the conversion of all C in the sample to CO_2, all the H to H_2O, all the N to N_2 and all the S to S_2. The mass of each gas, and hence the mass of each element (C, H, N and S), is then determined. We convert the masses of the elements into percentages of the element by mass in the compound. We then add up the percentages and subtract from 100 to get the percentage of O. We then calculate the empirical formula from the percentage composition as described earlier.

The empirical formula is all that we need for ionic compounds. For molecular compounds, however, chemists prefer *molecular* formulae because they give the number of atoms of each type in a molecule, rather than just the ratio of the elements in a compound, as the empirical formula does.

Sometimes an empirical formula and a molecular formula are the same. Two examples are H_2O and NH_3. More often, however, the subscripts of a molecular formula are whole-number multiples of those in the empirical formula. The subscripts of the molecular formula P_4O_{10}, for example, are each two times those in the empirical formula, P_2O_5, as we saw earlier. The molar mass of P_4O_{10} is likewise two times the molar mass of P_2O_5. This observation helps us find the molecular formula for a compound, provided we have a way of experimentally determining the molar mass of the compound. If the experimental molar mass *equals* the calculated mass of an empirical formula, the empirical formula itself is also a molecular formula. Otherwise, the experimental molar mass will be some whole-number multiple of the value calculated from the empirical formula. Whatever the whole number is, it is a common multiplier for the subscripts of the empirical formula.

3.5 Stoichiometry, limiting reagents and percentage yield

So far we have focused on relationships between elements within a single compound. We have seen that the critical link between amounts of elements within a compound is the mole-to-mole ratio obtained from the formula of the compound. In this section, we will see that the same techniques can be used to relate amounts of substances involved in a chemical reaction. The critical link between amounts of substances involved in a reaction is a mole-to-mole ratio obtained from the chemical equation that describes the reaction.

To see how a chemical equation can be used to obtain mole-to-mole relationships, consider the equation that describes the burning of octane, C_8H_{18}, in oxygen, O_2, to give carbon dioxide and water vapour.

$$2C_8H_{18}(l) + 25O_2(g) \rightarrow 16CO_2(g) + 18H_2O(g)$$

This equation can be interpreted on a molecular scale as follows: for every 2 molecules of liquid octane that react with 25 molecules of oxygen gas, 16 molecules of carbon dioxide gas and 18 molecules of water vapour are produced.

We can also interpret the equation on a molar scale as follows; 2 moles of liquid octane react with 25 moles of oxygen gas to produce 16 moles of carbon dioxide gas and 18 moles of water vapour.

To use these relationships in a stoichiometry problem, the equation must be balanced. We must always check to see whether this is so for a given equation before we can use the coefficients in the equation to calculate stoichiometric ratios.

First, let us see how mole-to-mole relationships obtained from a chemical equation can be used to relate the amount of one substance to the amount of another when both substances are involved in a chemical reaction.

WORKED EXAMPLE 3.8

Stoichiometry of chemical reactions

What amount of sodium phosphate, Na_3PO_4, can be made from 0.240 mol of NaOH by the following reaction?

$$3NaOH(aq) + H_3PO_4(aq) \rightarrow Na_3PO_4(aq) + 3H_2O(l)$$

Analysis

We use a similar approach to that used in worked example 3.4 but, in this case, we require the mole ratio between different compounds. We obtain this mole ratio directly from the stoichiometry of the balanced chemical equation.

Solution

We can see from the stoichiometric coefficients that 3 moles of NaOH will react to give 1 mole of Na_3PO_4. Therefore, if we start with 0.240 mol NaOH, we will obtain:

$$\frac{0.240\text{ mol}}{3} = 0.0800\text{ mol }Na_3PO_4$$

Is our answer reasonable?

There are really only two possible answers we could get to this problem — the correct one (0.0800 mol) and the incorrect one (0.720 mol), which would be obtained by multiplying rather than dividing by 3. Looking at the balanced chemical equation, we can see that the amount of Na_3PO_4 produced is less than the original amount of NaOH. In this case, our answer should be less than 0.240 mol, and it is.

PRACTICE EXERCISE 3.12 What amount of sulfuric acid, H_2SO_4, is needed to react completely with 0.366 mol of NaOH according to the following reaction?

$$2NaOH(aq) + H_2SO_4(aq) \rightarrow Na_2SO_4(aq) + 2H_2O(l)$$

PRACTICE EXERCISE 3.13 If 0.575 mol of CO_2 is produced by the combustion of propane, C_3H_8, what amount of oxygen is consumed? The balanced chemical equation is as follows.

$$C_3H_8(g) + 5O_2(g) \rightarrow 3CO_2(g) + 4H_2O(g)$$

Mole ratios in chemical reactions

We often need to relate mass of one substance to mass of another in a chemical reaction. Consider, for example, the reaction of glucose, $C_6H_{12}O_6$, one of the body's primary energy sources, with oxygen. The body combines glucose and oxygen to give carbon dioxide and water. The balanced equation for the overall reaction is:

$$C_6H_{12}O_6(aq) + 6O_2(aq) \rightarrow 6CO_2(aq) + 6H_2O(l)$$

What mass of oxygen must the body take in to completely process 1.00 g of glucose?

The first thing we should notice about this problem is that we are relating amounts of *two different substances* in a reaction. The link between the amounts of substances is the mole-to-mole relationship between glucose and O_2 given by the chemical equation. In this case, the equation tells us that 1 mole of $C_6H_{12}O_6$ will react completely with 6 moles of O_2.

If we insert the mole-to-mole conversion between our starting point (1.00 g $C_6H_{12}O_6$) and the desired quantity (g O_2), we cut the problem into three simple steps.

$$1.00\text{ g }C_6H_{12}O_6 \rightarrow \text{amount }C_6H_{12}O_6 \rightarrow \text{amount }O_2 \rightarrow \text{mass }O_2$$

We obtain the amount of $C_6H_{12}O_6$ from the mass of $C_6H_{12}O_6$ using the molar mass of $C_6H_{12}O_6$. We then obtain the amount of O_2 from the amount of $C_6H_{12}O_6$, using the stoichiometry of the balanced equation. Finally, we obtain the mass of O_2 from the amount of O_2 using the molar mass of O_2. If we have carried out these manipulations correctly, the answer should be 1.07 g O_2.

One of the most common mistakes made in stoichiometry problems like this is the incorrect determination and use of the mole ratios for substances in the reaction.

Consider the general chemical reaction:

$$aA + bB \rightarrow cC + dD$$

where *A*, *B*, *C* and *D* are reactants and products in the reaction, and *a*, *b*, *c* and *d* are the stoichiometric coefficients of each. For this reaction, the following *always* holds.

$$\frac{n_A}{a} = \frac{n_B}{b} = \frac{n_C}{c} = \frac{n_D}{d}$$

Applying this to the example above, we find that:

$$\frac{n_{C_6H_{12}O_6}}{1} = \frac{n_{O_2}}{6} = \frac{n_{CO_2}}{6} = \frac{n_{H_2O}}{6}$$

and, if we limit ourselves to the mole ratio between glucose and oxygen, we find:

$$\frac{n_{C_6H_{12}O_6}}{1} = \frac{n_{O_2}}{6}$$

Having calculated the amount of glucose, we know from the previous equation that this is equal to $n_{O_2}/6$. We then rearrange this equation to make n_{O_2} the subject:

$$n_{O_2} = 6n_{C_6H_{12}O_6}$$

and we find that we must *multiply* $n_{C_6H_{12}O_6}$ by 6 to obtain our answer.

WORKED EXAMPLE 3.9

Stoichiometric mass calculations

Metallic iron can be made by the thermite reaction, the spectacular reaction of aluminium with iron oxide, Fe_2O_3. So much heat is generated that the iron forms in the liquid state. The equation is:

$$2Al(s) + Fe_2O_3(s) \rightarrow Al_2O_3(s) + 2Fe(l)$$

Assume that you need to produce 86.0 g of Fe in a welding operation.

What mass of both Fe_2O_3 and aluminium must be used for this operation, assuming all the Fe_2O_3 is converted to Fe? ($M_{Fe} = 55.85$ g mol^{-1}; $M_O = 16.00$ g mol^{-1}; $M_{Al} = 26.98$ g mol^{-1})

Analysis

We are asked to find the masses of Al and Fe_2O_3 required to prepare 86.0 g of Fe. We are given a mass of Fe (m) and the molar mass of Fe (M) so we can calculate the amount of Fe (n) using $M = \frac{m}{n}$. We then use the mole ratios in the balanced chemical equation to calculate the amounts of Al and Fe_2O_3. The final step is to use $M = \frac{m}{n}$ to obtain the masses from these amounts.

In summary:

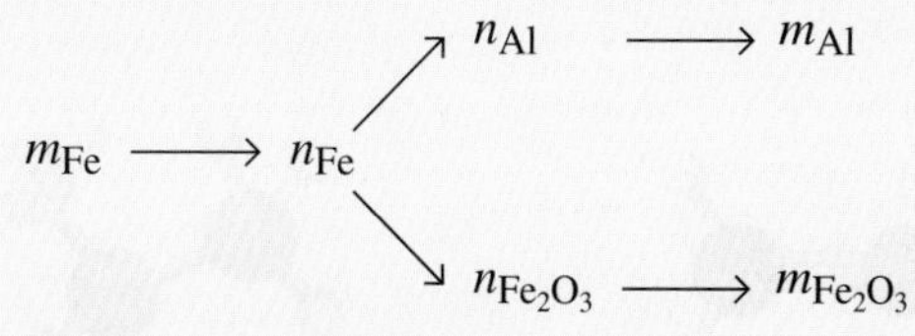

Solution

Firstly, we calculate the amount of Fe in 86.0 g:

$$n = \frac{m}{M} = \frac{86.0\text{ g}}{55.85\text{ g mol}^{-1}} = 1.54\text{ mol}$$

Using the stoichiometric coefficients from the balanced chemical equation, we can see:

$$\frac{n_{Al}}{2} = \frac{n_{Fe_2O_3}}{1} = \frac{n_{Al_2O_3}}{1} = \frac{n_{Fe}}{2}$$

$$n_{Fe_2O_3} = \frac{n_{Fe}}{2} = \frac{n_{Al}}{2}$$

Therefore, the formation of 2 moles of Fe requires 2 moles of Al and 1 mole of Fe_2O_3. In other words, we will require the same amount of Al, but half the amount of Fe_2O_3.

Therefore:

$$n_{Al} = n_{Fe} = 1.54\text{ mol}$$

$$n_{Fe_2O_3} = \frac{n_{Fe}}{2} = 0.770\text{ mol}$$

Thus, $n_{Al} = n_{Fe}$. We then use $M = \frac{m}{n}$ to obtain the masses from these amounts.

$$m_{Al} = Mn = 26.98\text{ g mol}^{-1} \times 1.54\text{ mol} = 41.5\text{ g}$$

$$m_{Fe_2O_3} = Mn = 159.70\text{ g mol}^{-1} \times 0.770\text{ mol} = 123\text{ g}$$

Is our answer reasonable?

As the formation of Fe from Fe_2O_3 involves loss of oxygen from Fe_2O_3, we would expect the mass of Fe_2O_3 required to be greater than the mass of Fe produced. Given that 2 moles of Al give 2 moles of Fe, we would expect the mass of Al required to be less than the mass of Fe produced (86.0 g), as the molar mass of Al is less than that of Fe. Hence, our answers appear reasonable.

PRACTICE EXERCISE 3.14

What mass of aluminium oxide is also produced when 86.0 g of Fe is formed by the reaction described in worked example 3.9?

Limiting reagents

We have seen that balanced chemical equations can tell us how to mix reactants together in just the right proportions to get a certain amount of product. For example, ethanol, CH_3CH_2OH, is prepared industrially as follows.

$$\underset{\text{ethene}}{C_2H_4} + H_2O \rightarrow \underset{\text{ethanol}}{CH_3CH_2OH}$$

The equation tells us that 1 mole of ethene will react with 1 mole of water to give 1 mole of ethanol. We can also interpret the equation on a molecular level; every molecule of ethene that reacts requires 1 molecule of water to produce 1 molecule of ethanol.

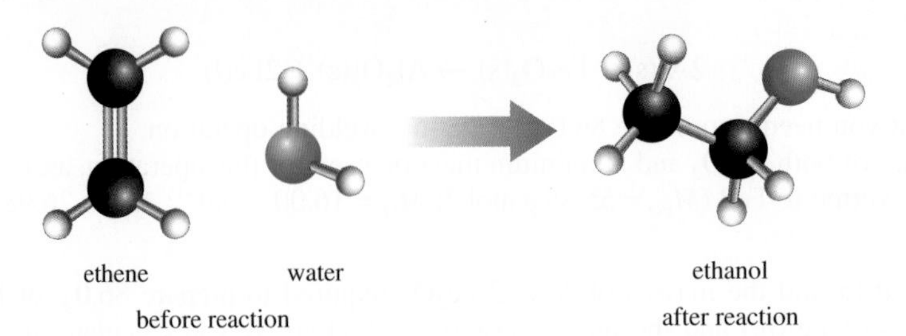

If we have 3 molecules of ethene reacting with 3 molecules of water, 3 ethanol molecules are produced.

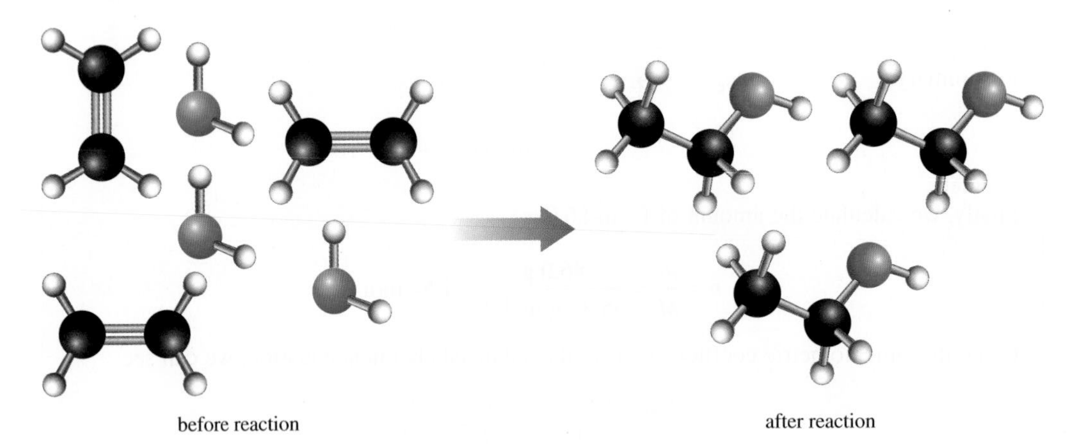

What happens if we mix 3 molecules of ethene with 5 molecules of water? The ethene will be completely used up before all the water, and the product mixture will contain 2 unreacted water molecules, as shown below.

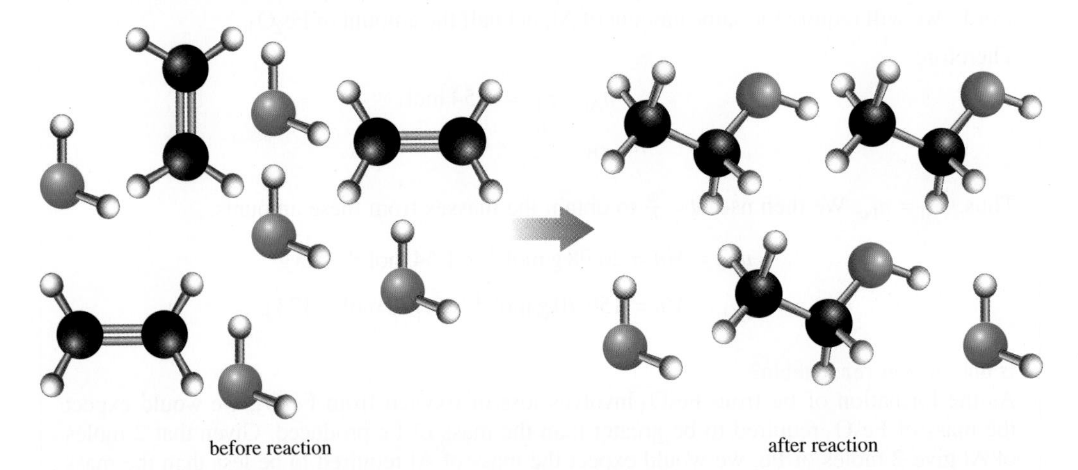

This situation can be a problem in the manufacture of chemicals because we waste one of our reactants (water, in this case) and also obtain a product that is contaminated with unused reactant.

In the reaction mixture above, ethene is called the **limiting reagent** because it limits the amount of product (ethanol) that can form. Water is called an **excess reagent** because we have more of it than is needed to completely consume all the ethene. It is important to realise that in any reaction one or more reactants will be limiting. If all reactants are present in the exact amounts required by the stoichiometry of the reaction, then in essence *all* are limiting the amount of product that can be formed. They are not usually referred to as limiting reagents in this case.

To predict the amount of product we may theoretically obtain in a reaction, we need to identify the limiting reagent(s). In the previous example, we saw that we needed only three H_2O molecules to react with three C_2H_4 molecules; we had five H_2O molecules, so H_2O was present in excess and C_2H_4 was the limiting reagent.

Once we have identified the limiting reagent(s), it is possible to calculate the amount of product that might form and the amount of excess reagent(s), if any, that would be left over.

WORKED EXAMPLE 3.10

Limiting reagent

Gold(III) hydroxide, $Au(OH)_3$, is used for electroplating gold onto other metals. It can be made by the following reaction.

$$2KAuCl_4(aq) + 3Na_2CO_3(aq) + 3H_2O(l) \rightarrow 2Au(OH)_3(s) + 6NaCl(aq) + 2KCl(aq) + 3CO_2(g)$$

To prepare a fresh supply of $Au(OH)_3$, a chemist at an electroplating plant mixed 20.00 g of $KAuCl_4$ with 25.00 g of Na_2CO_3 (both dissolved in a large excess of water).

What is the maximum mass of $Au(OH)_3$ that can form? ($M_{Na_2CO_3} = 105.99\ g\ mol^{-1}$; $M_{KAuCl_4} = 377.88\ g\ mol^{-1}$; $M_{Au(OH)_3} = 247.99\ g\ mol^{-1}$)

Analysis

We are asked to determine the maximum mass of $Au(OH)_3$ that can be formed. We are given masses (m) of both $KAuCl_4$ and Na_2CO_3 and their molar masses (M). This allows us to calculate the amounts of both $KAuCl_4$ and Na_2CO_3 using $M = \frac{m}{n}$. We then have to use the ratios of amounts to stoichiometric coefficients in the balanced chemical equation to determine if either $KAuCl_4$ or Na_2CO_3 is present in excess. The other substance would then be the limiting reagent.

Solution

Begin by calculating the amounts of $KAuCl_4$ and Na_2CO_3 using $M = \frac{m}{n}$:

$$M = \frac{m}{n}$$

$$n = \frac{m}{M}$$

$$n_{Na_2CO_3} = \frac{m}{M} = \frac{25.00\ g}{105.99\ g\ mol^{-1}} = 2.359 \times 10^{-1}\ mol$$

$$n_{KAuCl_4} = \frac{m}{M} = \frac{20.00\ g}{377.88\ g\ mol^{-1}} = 5.293 \times 10^{-2}\ mol$$

We know, from the stoichiometry of the reaction, that 2 moles of $KAuCl_4$ require 3 moles of Na_2CO_3 for complete reaction. In other words, the amount of $KAuCl_4$ divided by 2 should equal the amount of Na_2CO_3 divided by 3 for the reactants to be present in the quantities required by the stoichiometry of the reaction. This can be represented as:

$$\frac{n_{Na_2CO_3}}{3} = \frac{n_{KAuCl_4}}{2}$$

If the ratios are unequal, the *smaller* number indicates the limiting reagent. In the experiment, we use:

$$\frac{n_{Na_2CO_3}}{3} = \frac{2.359 \times 10^{-1}\ mol}{3} = 0.078\,63\ mol$$

$$\frac{n_{KAuCl_4}}{2} = \frac{5.293 \times 10^{-2}\ mol}{2} = 0.026\,47\ mol$$

This means that we have too much Na_2CO_3, so $KAuCl_4$ is the limiting reagent. The amount of $KAuCl_4$ now determines the amount of $Au(OH)_3$ that can be formed. To calculate how much $Au(OH)_3$ will be formed, we again use the stoichiometric coefficients from the balanced chemical equation. We see that 2 moles of $KAuCl_4$ will form 2 moles of $Au(OH)_3$ so:

$$n_{Au(OH)_3} = n_{KAuCl_4} = 5.293 \times 10^{-2}\ mol$$

Converting this amount of $Au(OH)_3$ to mass gives:

$$m = Mn = 247.99\ g\ mol^{-1} \times (5.293 \times 10^{-2}\ mol) = 13.13\ g$$

Therefore, using the reaction conditions given, we obtain a maximum of 13.13 g of $Au(OH)_3$.

Is our answer reasonable?

The main thing in calculations of this type is to get the mole ratios correct and, indeed, to interpret them correctly. We can see that the molar mass of $KAuCl_4$ is more than three times that of Na_2CO_3. Since we are using a larger mass of Na_2CO_3 in the reaction, we will have a significantly larger amount of Na_2CO_3 than $KAuCl_4$, and our calculations bear this out. The molar mass of $Au(OH)_3$ is about two-thirds that of $KAuCl_4$ and, given that their mole ratio in the balanced chemical equation is 1 : 1, we would expect the mass of $Au(OH)_3$ to be about two-thirds the initial mass of $KAuCl_4$, which it is. The answer is reasonable.

PRACTICE EXERCISE 3.15

In an industrial process for producing nitric acid, the first step is the reaction of ammonia with oxygen at high temperature in the presence of a platinum gauze to form nitrogen monoxide as follows.

$$4NH_3(g) + 5O_2(g) \rightarrow 4NO(g) + 6H_2O(g)$$

What is the maximum mass of nitrogen monoxide that can form from an initial mixture of 30.00 g of NH_3 and 40.00 g of O_2?

Percentage yield

The balanced chemical equation for a reaction allows you to calculate the theoretical maximum amount of product that can be synthesised from a given amount of starting material. In most chemical syntheses, the amount of a product actually isolated falls short of the calculated maximum amount. Losses occur for several reasons. Some are mechanical, such as materials sticking to glassware. In some reactions, losses occur by the evaporation of a volatile product. In others, a solid product separates from the solution as it forms because it is largely insoluble. The solid is removed by filtration. What stays in solution, although relatively small, contributes to some loss of product.

One of the common causes of obtaining less than the stoichiometric amount of a product is the occurrence of a reaction or reactions that compete with the desired reaction. A competing reaction produces a **by-product**, that is, a substance other than the product that we want to obtain. A second product of the desired reaction is also called a by-product; however, it does not reduce the yield. The synthesis of phosphorus trichloride from P and Cl_2, for example, gives some phosphorus pentachloride as well, because PCl_3 can react further with Cl_2.

$$2P(s) + 3Cl_2(g) \rightarrow 2PCl_3(l) \qquad \text{(desired reaction)}$$

$$PCl_3(l) + Cl_2(g) \rightarrow PCl_5(s) \qquad \text{(competing reaction)}$$

The competition is between newly formed PCl_3 and unreacted phosphorus for chlorine. The production of TiO_2 is an example of the formation of a by-product that does not reduce the yield.

$$TiCl_4(l) + 2H_2O(l) \rightarrow TiO_2(s) + 4HCl(g)$$

Here HCl is the by-product, but it does not reduce the yield of TiO_2.

Another cause of lower than theoretical yields is a reversible reaction, where the forward and backward reactions occur concurrently.

The **theoretical yield** of a product is what would be obtained if the maximum amount of product was formed and no losses occurred. It is calculated based on the balanced chemical equation and the amounts of the reactants available. The **actual yield** of the product is simply how much has been isolated in practice. You can calculate the **percentage yield** of product to describe how successful the preparation was. The percentage yield is the actual yield calculated as a percentage of the theoretical yield.

$$\text{percentage yield} = \frac{\text{actual yield}}{\text{theoretical yield}} \times 100\%$$

Both the actual and theoretical yields must, of course, be in the same units.

It is important to realise that the actual yield is an experimentally determined quantity. The theoretical yield is always a calculated quantity based on a chemical equation and the amounts of the reactants available.

WORKED EXAMPLE 3.11

Calculating a percentage yield

A chemist set up a synthesis of phosphorus trichloride, PCl_3, by mixing 12.0 g of P with 35.0 g of Cl_2 and obtained 42.4 g of PCl_3. The equation for the reaction is:

$$2P(s) + 3Cl_2(g) \rightarrow 2PCl_3(l)$$

Calculate the percentage yield of this compound. ($M_P = 30.97\ \text{g mol}^{-1}$; $M_{Cl_2} = 70.90\ \text{g mol}^{-1}$; $M_{PCl_3} = 137.32\ \text{g mol}^{-1}$)

Analysis

We are asked to calculate a percentage yield. We are given masses of all three components in the reaction and their molar masses (M). This means we can calculate the amounts of all three using $M = \frac{m}{n}$. To calculate the percentage yield, we must first decide on which reactant to base our calculations. We must calculate the amounts of each before we can make that decision. If we find that a reactant is a limiting reagent, the percentage yield is based on that reactant. To calculate the percentage yield, we must first calculate the theoretical yield: in other words, the maximum amount of product that can theoretically be formed according to the balanced chemical equation. In addition to the chemical equation, we only need $M = \frac{m}{n}$.

Solution

We begin by calculating the amounts of both P and Cl_2 used.

$$n_P = \frac{m}{M} = \frac{12.0\ \text{g}}{30.97\ \text{g mol}^{-1}} = 3.87 \times 10^{-1}\ \text{mol}$$

$$n_{Cl_2} = \frac{m}{M} = \frac{35.0\ \text{g}}{70.90\ \text{g mol}^{-1}} = 4.94 \times 10^{-1}\ \text{mol}$$

From the balanced chemical equation, we know that P reacts with Cl_2 in the mole ratio 2:3. Therefore, for complete reaction, theoretically:

$$\frac{n_{Cl_2}}{3} = \frac{n_P}{2}$$

The experimental ratios are 0.494 mol/3 = 0.165 for Cl_2 and 0.387 mol/2 = 0.194 for P. This means that we do not have sufficient Cl_2 to react with all the P present, so Cl_2 is the limiting reagent. We now base our subsequent calculations on this. From the balanced chemical equation, we know that:

$$\frac{n_{Cl_2}}{3} = \frac{n_{PCl_3}}{2}$$

Therefore, 0.165 mol $Cl_2 = n_{PCl_3}/2$, so $n_{PCl_3} = 2(0.165\ \text{mol}) = 0.330\ \text{mol}$.
Converting this to mass, we obtain:

$$m = Mn = 137.32\ \text{g mol}^{-1} \times 0.330\ \text{mol} = 45.3\ \text{g}$$

The actual mass obtained was 42.4 g, so the percentage yield is given by:

$$\text{percentage yield} = \frac{\text{actual yield}}{\text{theoretical yield}} \times 100\% = \frac{42.4\ \text{g}}{45.3\ \text{g}} \times 100\% = 93.6\%$$

Is our answer reasonable?

The obvious check is that the actual yield can never be more than the theoretical yield. So, on this score, our answer is reasonable. Again, the most common place where errors are made is in the use of the mole ratios. In this example, we must end up with *fewer* moles of PCl_3 than Cl_2, so we must multiply n_{Cl_2} by $\frac{2}{3}$, not $\frac{3}{2}$.

PRACTICE EXERCISE 3.16

Ethanol, CH_3CH_2OH, can be converted to acetic acid (the acid in vinegar), CH_3COOH, by the action of sodium dichromate, $Na_2Cr_2O_7$, in aqueous sulfuric acid according to the following equation.

$$3CH_3CH_2OH(aq) + 2Na_2Cr_2O_7(aq) + 8H_2SO_4(aq) \rightarrow 3CH_3COOH(aq) + 2Cr_2(SO_4)_3(aq) + 2Na_2SO_4(aq) + 11H_2O(l)$$

In one experiment, 24.0 g of CH_3CH_2OH, 90.0 g of $Na_2Cr_2O_7$ and a known excess of sulfuric acid was mixed, and 26.6 g of acetic acid was isolated. Calculate the percentage yield of CH_3COOH.

Another way to look at the yield of a reaction is to take into account the so-called **atom efficiency**. Atom efficiency takes into account all atoms in the starting materials and compares them to the number of atoms in the desired product. There is a difference between a reaction where most of the atoms of the starting material(s) end up forming the desired product and a reaction where a large fraction of the atoms in the starting material(s) end up as unwanted by-products. There may be a trade-off between the percentage yield of a reaction and the atom efficiency on occasion. Concerns about pollution and cost certainly favour reactions (at least on the industrial scale) that produce the least amount of waste material.

3.6 Solution stoichiometry

Chemical reactions are most often carried out in solution because this allows intimate mixing of the reactants, which leads to more rapid reaction. A **solution** is a homogeneous mixture in which the atoms, molecules or ions of the components freely intermingle (see figure 3.10). When a solution forms, at least two substances are involved. One is called the solvent and all of the others are called solutes. The **solvent** is usually taken to be the component present in largest amount and is the medium into which the solutes are mixed or dissolved. Liquid water is a typical and very common solvent, but the solvent can actually be in any physical state — solid, liquid, or gas. Unless stated otherwise, we will assume that any solutions we mention are aqueous solutions, so that liquid water is understood to be the solvent.

A **solute** is any substance dissolved in the solvent. It might be a gas, like the carbon dioxide dissolved in carbonated drinks. Some solutes are liquids, like the ethylene glycol dissolved in water to stop it from freezing in a car's radiator. Solids, of course, can be solutes, like the sugar dissolved in lemonade or the salt dissolved in sea water.

solvent molecule (ethanol)
solute molecule (iodine)
A crystal of solute is placed in the solvent.
A solution. Solute molecules are dispersed throughout the solvent.

FIGURE 3.10 In the photo on the left, a crystal of iodine, I_2, that was added to a beaker containing ethanol started dissolving on its way to the bottom of the beaker, the purplish iodine crystal forming a reddish-brown solution. The photo on the right shows how after stirring the mixture with a spatula the solution takes on a homogeneous reddish-brown colour, due to the iodine molecules dissolved in the solvent as indicated in the schematic below the picture. Called 'tincture of iodine', such solutions were once widely used as antiseptics.

The concentration of solutions

We use the concentration of a solution to describe its composition. The **concentration** is defined as the amount of solute dissolved in a particular volume of solution. The concentration of a substance X is often represented as $[X]$. When the amount is given in moles and the volume in litres, it is called the **molarity** or molar concentration and has the units mol L^{-1} (often abbreviated M, not to be confused with the abbreviation for molar mass, M). The equation that defines concentration (c) is:

$$c = \frac{n}{V}$$

We stated earlier that there were only two equations needed to solve any stoichiometric problem. $M = \frac{m}{n}$ was the first, and this is the second. Again, we must commit this to memory, but it is easy to remember when we consider the units of concentration; mol L^{-1} implies amount divided by volume (in L).

Thus, a 1.00 L solution that contains 0.100 mol of NaCl has a molarity of 0.100 M, and we would refer to this solution as 0.100 *molar* NaCl or as 0.100 M NaCl. A 0.100 L solution containing 0.0100 mol of NaCl would have the same concentration because the *ratio* of amount of solute to volume of solution is the same.

$$\frac{0.100\ \text{mol NaCl}}{1.00\ \text{L NaCl}} = \frac{0.0100\ \text{mol NaCl}}{0.100\ \text{L NaCl}} = 0.100\ \text{M NaCl}$$

Concentration is particularly useful because it lets us obtain a given amount of a substance simply by measuring a volume of a previously prepared solution, and this measurement is quick and easy to do in the laboratory. If we had a stock supply of 0.100 M NaCl solution, for example, and we needed 0.100 mol of NaCl for a reaction, we would simply measure out 1.00 L of the solution because there is 0.100 mol of NaCl in this volume.

We do not have to work with whole litres of solutions. It is the *ratio* of the amount of solute to the volume of solution, not the total volume, that matters. Suppose, for example, that we dissolved 0.0200 mol of sodium chromate, Na_2CrO_4, in sufficient solvent to give a final volume of 0.250 L. The concentration of the solution would be found by:

$$c = \frac{n}{V} = \frac{0.0200\ \text{mol}}{0.250\ \text{L}} = 0.0800\ \text{M}$$

Such a solution can be accurately prepared in a *volumetric flask*, which is a narrow-necked container with an etched mark high on its neck. When filled 'up to the mark', the flask contains the volume given by its label. Figure 3.11 shows how to use a 250 mL volumetric flask to prepare a solution of known concentration.

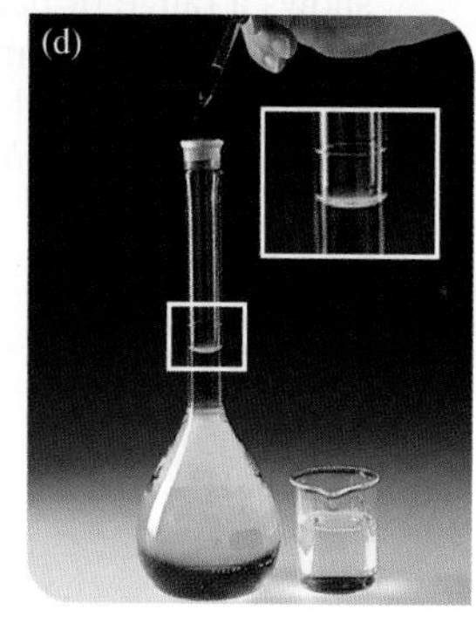

FIGURE 3.11 The preparation of a solution of known molarity: **(a)** A 250 mL volumetric flask, one of a number of sizes available for preparing solutions. When filled to the line etched around its neck, this flask contains exactly 250 mL of solution. The flask here already contains a weighed amount of solute. **(b)** Water is being added. **(c)** The solute is brought completely into solution before the level is brought up to the narrow neck of the flask. **(d)** More water is added to bring the level of the solution to the etched line or 'up to the mark'. **(e)** The flask is stoppered and then inverted many times to mix its contents thoroughly.

WORKED EXAMPLE 3.12

Calculating the concentration of a solution

A student prepared a solution of NaCl by dissolving 1.461 g of NaCl in water and making up to the final volume of 250.0 mL in a volumetric flask to study the effect of dissolved salt on the rusting of an iron sample. What is the concentration of this solution in mol L^{-1}?

Solution

Firstly, we convert the mass of NaCl to amount using:

$$M = \frac{m}{n}$$

$$n = \frac{m}{M} = \frac{1.461\ \text{g}}{58.44\ \text{g mol}^{-1}}$$
$$= 2.500 \times 10^{-2}\ \text{mol}$$

We then substitute the data into $c = \frac{n}{V}$ and solve for c.

$$c = \frac{n}{V} = \frac{2.500 \times 10^{-2} \text{ mol}}{0.2500 \text{ L}}$$
$$= 1.000 \times 10^{-1} \text{ mol L}^{-1}$$

Therefore, the concentration of the NaCl solution is 1.000×10^{-1} mol L^{-1}.

Is our answer reasonable?

A 1 mol L^{-1} solution of NaCl would contain 58.44 g of NaCl in 1 litre, so a 0.1 mol L^{-1} solution would contain about 6 g in 1 L. We have 0.25 L, so we would expect about $(0.25 \times 6 \text{ g}) = 1.5$ g in 0.25 L, which is roughly what we have. The answer is certainly reasonable. An easy mistake to make in problems of this type is to forget to convert mL to L. As the concentration was required in mol L^{-1} and the solution volume is given in mL, we must *divide* by 1000 to convert to L.

PRACTICE EXERCISE 3.17

As part of a study to determine if just the chloride ion in NaCl accelerates the rusting of iron, a student decided to see if iron rusted as rapidly when exposed to a solution of Na_2SO_4. This solution was made by dissolving 3.550 g of Na_2SO_4 in water and making up the final volume to 100.0 mL in a volumetric flask. What is the concentration of this solution?

Notice that worked example 3.12 uses both $M = \frac{m}{n}$ and $c = \frac{n}{V}$. As we have stressed, with these two equations, there is no stoichiometric problem we cannot solve!

In the laboratory, we often must prepare a solution with a specific concentration. Usually, we select the volume of solution that we need and then calculate how much solute must be used. Depending on the accuracy that you require for the concentration, you will need to use a balance that allows you to weigh your sample to the required number of significant figures. Figure 3.12 shows a range of balances and glassware used in the preparation of solutions of known concentration. Worked example 3.13 illustrates the kind of calculation required and also demonstrates the following useful relationship: *whenever we know both the volume and the concentration of a solution, we can calculate the amount of solute from* $c = \frac{n}{V}$.

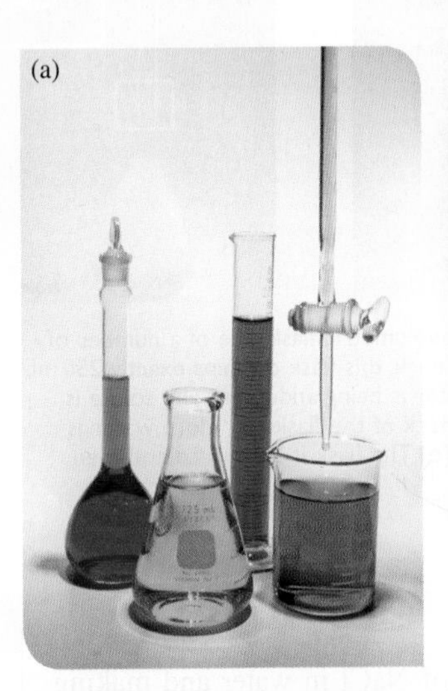

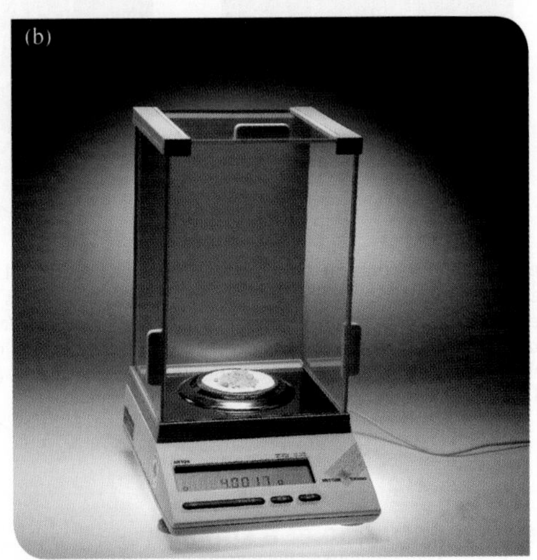

FIGURE 3.12 (a) Glassware that can be used for measuring volumes to different accuracies and **(b)** an analytical balance that can be read to 1×10^{-4} g.

WORKED EXAMPLE 3.13

Preparing a solution of known concentration

Strontium nitrate, $Sr(NO_3)_2$, is used in fireworks to produce brilliant red colours like those shown in figure 3.13. What mass of strontium nitrate would a chemist need to prepare 250.0 mL of a 0.100 M $Sr(NO_3)_2$ solution? ($M_{Sr(NO_3)_2} = 211.64$ g mol^{-1})

Analysis

We are asked to calculate a mass (m) of strontium nitrate and we are given the volume (V) and concentration (c) of a strontium nitrate solution, as well as the molar mass (M) of strontium nitrate. The concentration and volume will allow us to calculate the amount (n)

of strontium nitrate using $c = \frac{n}{V}$, and we can then use this amount in the equation $M = \frac{m}{n}$ to calculate the mass of strontium nitrate.

FIGURE 3.13 The red colour in some fireworks is a result of the presence of strontium nitrate, $Sr(NO_3)_2$.

Solution

We rearrange $c = \frac{n}{V}$ to solve for n.

$$\begin{aligned} n &= cV \\ &= 0.100\ \text{M} \times 0.2500\ \text{L} \\ &= 0.0250\ \text{mol} \\ &= 2.50 \times 10^{-2}\ \text{mol} \end{aligned}$$

We can now use $M = \frac{m}{n}$ to obtain the mass.

$$\begin{aligned} m = Mn &= 211.64\ \text{g mol}^{-1} \times (2.50 \times 10^{-2}\ \text{mol}) \\ &= 5.29\ \text{g} \end{aligned}$$

Therefore, the chemist would require 5.29 g of $Sr(NO_3)_2$ to prepare the solution.

Is our answer reasonable?

In 1 litre of 0.100 M $Sr(NO_3)_2$, there would be 0.1 mol of $Sr(NO_3)_2$, about 20 g. Therefore, in 0.25 L of this solution, we would have roughly (0.25 × 20 g) = 5 g. The answer, 5.29 g, is close to this, so it makes sense.

PRACTICE EXERCISE 3.18

What mass of $AgNO_3$ is needed to prepare 250 mL of 0.0125 M $AgNO_3$ solution?

Diluting a solution

It is not always necessary to begin with a solute in pure form to prepare a solution with a known concentration. The solute can already be dissolved in a solution of relatively high concentration, often called the stock solution, and this can be *diluted* with the same pure solvent to make a solution of the desired lower concentration. Dilution is accomplished by adding more solvent to the solution, which causes the concentration (the amount per unit volume) to decrease (figure 3.14). The choice of apparatus used depends on the precision required. If high precision is necessary, pipettes and volumetric flasks are used (see figure 3.12a). If we can do with less precision, we might use graduated cylinders instead. Worked example 3.14 shows how we can use $c = \frac{n}{V}$ to calculate the final concentration of a diluted solution.

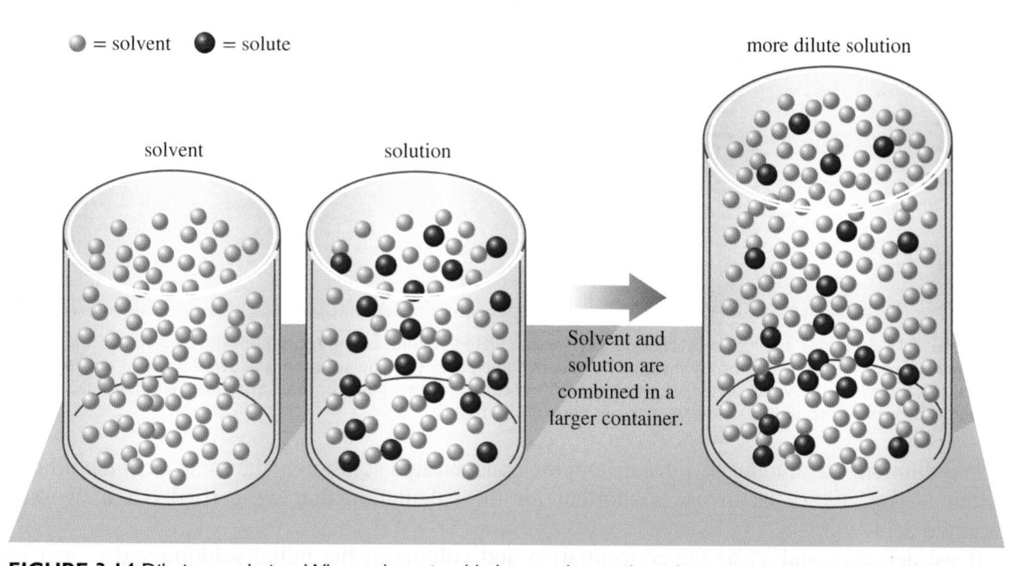

FIGURE 3.14 Diluting a solution. When solvent is added to a solution, the solute particles become more spread out and the solution becomes more dilute. The concentration of the solute in the solution becomes lower, while the total amount of solute remains constant.

WORKED EXAMPLE 3.14

Preparing a solution of known concentration by dilution

How could we prepare 100.0 mL of 0.0400 M $K_2Cr_2O_7$ from a solution of 0.200 M $K_2Cr_2O_7$?

Analysis

We are asked how to dilute a concentrated solution to give a solution of a particular concentration. We know both the volume (V) and the concentration (c) of the final solution, so use of $c = \frac{n}{V}$ will allow us to calculate n from these data. As we know the concentration of the initial solution, and we have calculated the required n, $c = \frac{n}{V}$ will allow us to calculate V, the volume of the initial solution required.

Solution

We are going to use $c = \frac{n}{V}$ twice, to calculate two different things. Firstly, we calculate $n_{K_2Cr_2O_7}$ in our final solution.

$$n = cV = 0.0400 \text{ mol L}^{-1} \times 0.1000 \text{ L}$$
$$= 4.00 \times 10^{-3} \text{ mol}$$

This is the amount of $K_2Cr_2O_7$ that, when dissolved in 100.0 mL, will give a concentration of 0.0400 M. We now turn our attention to the initial solution and use this calculated amount, in conjunction with the concentration, to calculate V, the volume of the initial solution required.

$$V = \frac{n}{c} = \frac{4.00 \times 10^{-3} \text{ mol}}{0.200 \text{ mol L}^{-1}}$$
$$= 0.0200 \text{ L}$$

Therefore, we take 20.0 mL of the original solution, pour it into a 100.0 mL volumetric flask and make it up to the mark with water to get a 0.0400 M solution of $K_2Cr_2O_7$ (figure 3.15).

Is our answer reasonable?

Think about the magnitude of the dilution, from 0.2 M to 0.04 M. Notice that the concentrated solution is five times as concentrated as the dilute solution (5 × 0.04 = 0.2). To reduce the concentration by a factor of five requires that we increase the volume by a factor of five, and we see that 100 mL is 5 × 20 mL. The answer appears to be correct.

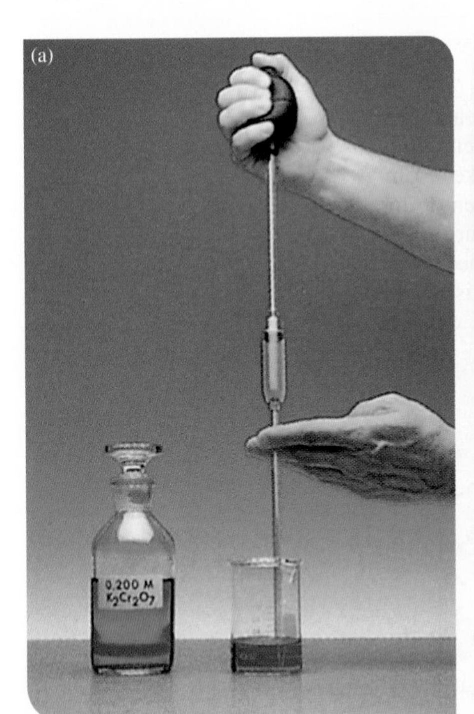

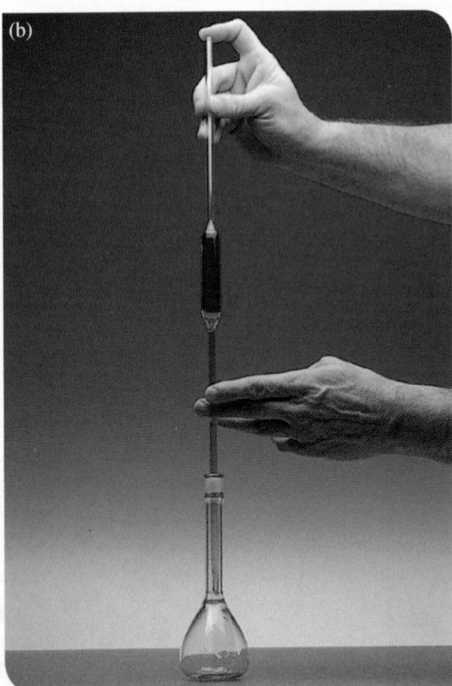

FIGURE 3.15 Preparing a solution by dilution: **(a)** The calculated volume of the more concentrated solution is withdrawn from the stock solution by means of a volumetric pipette. **(b)** The solution is allowed to drain entirely from the pipette into the volumetric flask. **(c)** Water is added to the flask, the contents are mixed and the final volume is brought up to the mark on the narrow neck of the flask.

In a dilution, the amount of substance in the concentrated solution is the same as in the diluted solution. This allows us to use a shortcut for the calculation that we carried out in worked example 3.14.

If we define c_1 and V_1 as the concentration and volume of the initial solution and c_2 and V_2 as those of the final solution, because n is the same in both solutions we can write $n = c_1V_1$ and $n = c_2V_2$. As the amount, n, is the same for both solutions, we can then cancel out n, to give the equation $c_1V_1 = c_2V_2$. If we know the initial concentration and the final volume and concentration we can solve for the initial volume.

$$V_1 = \frac{c_2}{c_1} \times V_2$$

That is, if we multiply the final volume by the ratio of final to initial concentration, we obtain the required volume of the initial solution. For worked example 3.14, we could write:

$$\begin{aligned} V_1 &= \frac{c_2}{c_1} \times V_2 \\ &= \frac{0.0400\ \text{M}}{0.200\ \text{M}} \times 0.1000\ \text{L} \\ &= 0.0200\ \text{L} \end{aligned}$$

PRACTICE EXERCISE 3.19

How could 0.1 L of 0.125 M H_2SO_4 solution be made from a 0.500 M H_2SO_4 solution?

Applications of solution stoichiometry

When we deal quantitatively with reactions in solution, we often work with volumes of solutions and concentrations.

WORKED EXAMPLE 3.15

Stoichiometry involving reactions in solution

Before the advent of digital cameras, silver bromide, AgBr, was used extensively in photographic film. This compound is essentially insoluble in water, and one way to prepare it is to mix solutions of the water-soluble compounds, silver nitrate and calcium bromide (see figure 3.16).

Suppose we wished to prepare AgBr by the following precipitation reaction.

$$2AgNO_3(aq) + CaBr_2(aq) \rightarrow 2AgBr(s) + Ca(NO_3)_2(aq)$$

What volume of 0.125 M $CaBr_2$ solution would be required to react completely with 50.0 mL of 0.115 M $AgNO_3$?

FIGURE 3.16 Addition of $CaBr_2$(aq) to $AgNO_3$(aq) results in precipitation of creamy white AgBr(s).

Analysis

We are asked to calculate a volume of $CaBr_2$ solution, and we are given the volume (V) and concentration (c) of a $AgNO_3$ solution. We can calculate the amount (n) of $AgNO_3$ using $c = \frac{n}{V}$. We are also given a balanced chemical equation, and we will need to use this to obtain the mole ratio between $AgNO_3$ and $CaBr_2$. We then use the stoichiometric coefficients from the balanced chemical equation to calculate n_{CaBr_2}. Finally, we will use $c = \frac{n}{V}$ along with the concentration of the $CaBr_2$ solution to calculate the volume of the initial solution required.

Solution

We start by calculating n_{AgNO_3}.

$$\begin{aligned} c &= \frac{n}{V} \\ n &= cV = 0.115\ \text{mol L}^{-1} \times (50.0 \times 10^{-3}\ \text{L}) \\ &= 5.75 \times 10^{-3}\ \text{mol} \end{aligned}$$

From the balanced chemical equation, we know that 2 moles of $AgNO_3$ will react with 1 mole of $CaBr_2$. Therefore:

$$n_{CaBr_2} = \frac{n_{AgNO_3}}{2} = \frac{5.75 \times 10^{-3}\ \text{mol}}{2} = 2.88 \times 10^{-3}\ \text{mol}$$

We now know n_{CaBr_2} and the concentration of the $CaBr_2$ solution, so we can substitute these values into $c = \frac{n}{V}$ to calculate V, the volume of the $CaBr_2$ solution required.

$$V = \frac{n}{c} = \frac{2.88 \times 10^{-3}\ \text{mol}}{0.125\ \text{mol L}^{-1}} = 0.0230\ \text{L} = 2.30 \times 10^{-2}\ \text{L}$$

Therefore, 23.0 mL of the $CaBr_2$ solution will react completely with 50.0 mL of the 0.115 M $AgNO_3$ solution.

Is our answer reasonable?

The molarities of the two solutions are about the same, but only 1 mole of $CaBr_2$ is needed for every 2 moles of $AgNO_3$. Therefore, the volume of $CaBr_2$ solution needed (23.0 mL) should be about half the volume of $AgNO_3$ solution taken (50.0 mL), which it is.

PRACTICE EXERCISE 3.20 What volume of 0.124 M NaOH is required to react completely with 15.4 mL of 0.108 M H_2SO_4 according to the following equation?

$$2NaOH(aq) + H_2SO_4(aq) \rightarrow Na_2SO_4(aq) + 2H_2O(l)$$

Stoichiometry of solutions containing ions

In worked example 3.15, we used an equation involving ionic compounds to solve a stoichiometry problem. Ionic compounds dissociate into their constituent ions if they dissolve in water and this can have important consequences in stoichiometric calculations involving such solutions. For example, in worked example 3.15, we used calcium bromide, $CaBr_2$, as a reactant. Suppose we are working with an aqueous solution labelled '0.10 M $CaBr_2$'. This means there is 0.10 mole of $CaBr_2$ dissolved in each litre of this solution **dissociated** into Ca^{2+} and Br^- ions.

$$CaBr_2(s) \rightarrow Ca^{2+}(aq) + 2Br^-(aq)$$

From the stoichiometry of the dissociation, we see that 1 mole of $Ca^{2+}(aq)$ and 2 moles of $Br^-(aq)$ are formed from each mole of $CaBr_2(s)$. Therefore, 0.10 mole of $CaBr_2(s)$ will yield 0.10 mole of $Ca^{2+}(aq)$ and 0.20 moles of $Br^-(aq)$. In 0.10 M $CaBr_2$, then, the concentration of $Ca^{2+}(aq)$ is 0.10 M and the concentration of $Br^-(aq)$ is 0.20 M. Notice that the concentration of a particular ion equals the concentration of the salt multiplied by the number of ions of that kind in the formula of the salt.

We use the term 'dissolution' to describe the process of a solute dissolving in a solvent, and it is important that you understand the difference between the terms 'dissolution' and 'dissociation'. When an ionic compound undergoes dissolution (dissolves) in a solvent, it undergoes dissociation into its constituent ions. We showed the equation for this process with reference to $CaBr_2(s)$ above, and assumed that the salt dissociates fully. However, dissociation reactions do not necessarily do that, and the extent of such reactions depends on the nature of the ionic compound. We will assume 100% dissociation of all ionic compounds in the remaining examples in this chapter. We will investigate partial dissociation in chapter 10.

Initially you may find it difficult to know exactly which ionic species dissolve in water and which ones do not, but with experience you will be able to make reasonable assessments. In the first instance a table of solubility rules can be a great help. Table 3.1 summarises the rules for the solubilities of many common ionic compounds and you may find it helpful to be familiar with these.

TABLE 3.1 Solubility of common binary ionic compounds in water.

If the anion is	the compound is usually	except for
F^-	soluble	Mg^{2+}, Ca^{2+}, Sr^{2+}, Ba^{2+} and Pb^{2+} (Al^{3+})
Cl^-	soluble	Ag^+ and Hg_2^{2+} (Pb^{2+})
Br^-	soluble	Ag^+ and Hg_2^{2+} (Hg^{2+} and Pb^{2+})
I^-	soluble	Ag^+, Hg^{2+}, Hg_2^{2+} and Pb^{2+}
NO_3^-	soluble	none
SO_4^{2-}	soluble	Sr^{2+}, Ba^{2+}, Hg_2^{2+} and Pb^{2+} (Ag^+ and Ca^{2+})
CH_3COO^-	soluble	none (Ag^+ and Hg_2^{2+})
OH^-	insoluble	Li^+, Na^+, K^+ and Ba^{2+} (Ca^{2+} and Sr^{2+})
SO_3^{2-}	insoluble	Li^+, Na^+, K^+, NH_4^+ and Mg^{2+}
PO_4^{3-}	insoluble	Na^+, K^+ and NH_4^+
CO_3^{2-}	insoluble	Li^+, Na^+, K^+ and NH_4^+
$C_2O_4^{2-}$	insoluble	Li^+, Na^+, K^+ and NH_4^+

Note: For the purposes of this table, insoluble compounds are deemed to have molar solubilities of less than 1×10^{-2} M in water, slightly soluble compounds have molar solubilities in water between 1×10^{-1} M and 1×10^{-2} M, while soluble compounds have molar solubilities greater than 1×10^{-1} M in water. Cations in parentheses form slightly soluble compounds with the specified anion. Most oxides and sulfides are insoluble in water; apparent exceptions to this rule (e.g. Li_2O and MgS) are generally complicated by hydrolysis reactions and are therefore not listed.

Note that, in worked example 3.15, we wrote the balanced chemical equation as the molecular equation.

$$2AgNO_3(aq) + CaBr_2(aq) \rightarrow 2AgBr(s) + Ca(NO_3)_2(aq)$$

We can also write this in terms of the ions present as the ionic equation.

$$2Ag^+(aq) + 2NO_3^-(aq) + Ca^{2+}(aq) + 2Br^-(aq) \rightarrow 2AgBr(s) + Ca^{2+}(aq) + 2NO_3^-(aq)$$

The net reaction that occurred was:

$$2Ag^+(aq) + 2Br^-(aq) \rightarrow 2AgBr(s)$$

and the NO_3^- and Ca^{2+} took no part in the reaction, remaining simply as aqueous ions throughout. We call such ions **spectator ions**. Balanced ionic equations written without the inclusion of spectator ions are called **net ionic equations**. That is, net ionic equations show only those ions that participate in the reaction.

WORKED EXAMPLE 3.16

Calculating the concentrations of ions in a solution

What are the concentrations of the ions in a 0.20 M solution of $Al_2(SO_4)_3(aq)$?

Analysis

The concentrations of the ions are determined by the concentration of the salt and its stoichiometry. We need to write a balanced chemical equation for the dissolution of the salt in water; from this, we can determine the concentration of each ion in the solution using the given concentration of the salt.

Solution

When $Al_2(SO_4)_3$ dissolves, it dissociates as follows:

$$Al_2(SO_4)_3(s) \rightarrow 2Al^{3+}(aq) + 3SO_4^{2-}(aq)$$

Each mole of $Al_2(SO_4)_3$ yields 2 moles of Al^{3+} ions and 3 moles of SO_4^{2-} ions on dissolution, assuming complete dissociation. Therefore, if we consider 1 litre of solution, the 0.20 mol of $Al_2(SO_4)_3$ will yield 0.40 mol Al^{3+} and 0.60 mol SO_4^{2-}. Hence, the concentrations of the ions in the solution will be 0.40 M Al^{3+} and 0.60 M SO_4^{2-}.

PRACTICE EXERCISE 3.21

What are the molar concentrations of the ions in 0.40 M $FeCl_3(aq)$?

PRACTICE EXERCISE 3.22

In an aqueous solution of Na_3PO_4, the PO_4^{3-} concentration was determined to be 0.250 M. What was the sodium ion concentration in the solution?

We have seen that a net ionic equation is convenient for focusing on the net chemical change in reactions involving ions. Let us study some examples that illustrate how such equations can be used in stoichiometric calculations.

WORKED EXAMPLE 3.17

Stoichiometry calculations using net ionic equations

When aqueous solutions of $AgNO_3$ and $CaCl_2$ are mixed, a white precipitate of AgCl(s) is formed (see figure 3.17). The net ionic equation for the formation of this precipitate is:

$$Ag^+(aq) + Cl^-(aq) \rightarrow AgCl(s)$$

What volume of 0.100 M $AgNO_3$ solution is needed to react completely with 25.0 mL of 0.400 M $CaCl_2$ solution to form AgCl(s)?

Analysis

We are asked to calculate a volume of $AgNO_3$ solution. The volume (V) and concentration (c) of the $CaCl_2$ solution given in the question allow us to calculate n_{CaCl_2} and hence n_{Cl^-} using $c = \frac{n}{V}$. From the stoichiometry of the reaction, we can then determine the required n_{Ag^+}. We can then calculate the required volume using the given concentration of the $AgNO_3$ solution.

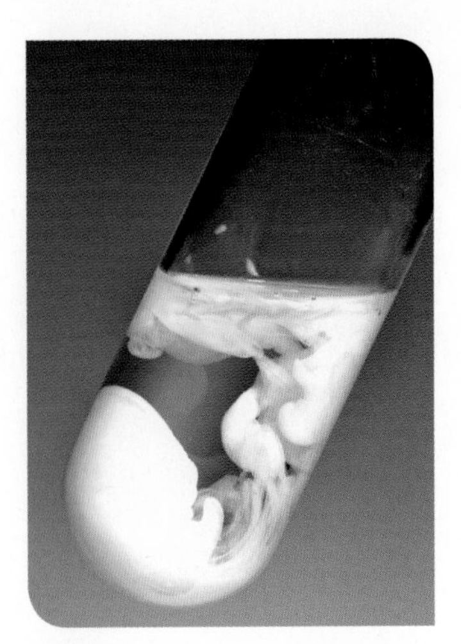

FIGURE 3.17 A solution of $AgNO_3(aq)$ is added to a solution of $CaCl_2(aq)$, producing a precipitate of $AgCl(s)$.

Solution

We are given a volume and a concentration of the $CaCl_2$ solution, so we substitute these values into $c = \frac{n}{V}$.

$$n_{CaCl_2} = cV = 0.400\text{ mol L}^{-1} \times (25.0 \times 10^{-3}\text{ L}) = 1.00 \times 10^{-2}\text{ mol}$$

We know that 1 mole of $CaCl_2$ will give 2 moles of Cl^- on dissolution in water (assuming complete dissociation), so:

$$n_{Cl^-} = 2 \times (1.00 \times 10^{-2}\text{ mol}) = 2.00 \times 10^{-2}\text{ mol}$$

The stoichiometry of the balanced chemical equation for the reaction of Ag^+ and Cl^- shows that 1 mole of Cl^- will react with 1 mole of Ag^+. Therefore, we can say that:

$$n_{Ag^+} = n_{Cl^-} = 2.00 \times 10^{-2}\text{ mol}$$

Before we calculate the required volume, we need to look at the stoichiometry of $AgNO_3$. Because 1 mole of $AgNO_3$ will give us 1 mole of Ag^+ when dissolved in solution:

$$n_{Ag^+} = n_{AgNO_3}$$

Now we obtain:

$$V = \frac{n}{c} = \frac{2.00 \times 10^{-2}\text{ mol}}{0.100\text{ mol L}^{-1}} = 0.200\text{ L}$$

Thus, we require 200 mL of 0.100 M $AgNO_3$ solution.

Is our answer reasonable?

Silver ions react with chloride ions in a 1 : 1 ratio. $CaCl_2$ solution is four times the concentration of the $AgNO_3$ solution and, as 1 mole of $CaCl_2$ dissolves to give 2 moles of Cl^-, the Cl^- concentration in this solution is eight times the concentration of the $AgNO_3$ solution. Therefore, we would expect a volume of $AgNO_3$ solution eight times the volume of the $CaCl_2$ solution. As $8 \times 25 = 200$, our answer appears correct.

PRACTICE EXERCISE 3.23

What volume of 0.500 M KOH is needed to react completely with 60.0 mL of a 0.250 M $FeCl_2$ solution to precipitate $Fe(OH)_2$?

Earlier in this chapter (pp. 96–8), we learned how to recognise and solve limiting reagent problems. Similar problems can be encountered when working with solutions.

WORKED EXAMPLE 3.18

Calculation involving the stoichiometry of an ionic reaction

A suspension of $Mg(OH)_2$ in water is sometimes used as an antacid. It can be made by adding NaOH to a solution containing Mg^{2+}. Suppose that 40.0 mL of a 0.200 M NaOH solution is added to 25.0 mL of a 0.300 M $MgCl_2$ solution. The net ionic equation for the reaction is:

$$Mg^{2+}(aq) + 2OH^-(aq) \rightarrow Mg(OH)_2(s)$$

What mass of $Mg(OH)_2$ will be formed, and what will the concentrations of the ions in the solution be after the reaction is complete? ($M_{Mg(OH)_2} = 58.33\text{ g mol}^{-1}$)

Analysis

We are asked to calculate the mass of $Mg(OH)_2$ and the concentrations of ions in solution at the end of the reaction. We are given concentrations (c) and volumes (V) of both the NaOH and $MgCl_2$ solutions, which allow us to calculate amounts (n) of these using $c = \frac{n}{V}$. We are also given the molar mass (M) of $Mg(OH)_2$, which allows us to use $M = \frac{m}{n}$ to calculate the mass (m) of $Mg(OH)_2$. This problem is similar to worked example 3.17, but slightly more complex. We must calculate the amounts of Mg^{2+} and OH^- present initially, and work out which of these will be limiting. We then proceed in the same way as we have done in other limiting reagent problems, using $M = \frac{m}{n}$ to calculate the mass of $Mg(OH)_2$ formed. We obtain the concentrations of the ions left in solution by difference.

Solution

We begin by calculating the amounts of both NaOH(aq) and $MgCl_2(aq)$, and then calculating $n_{Mg^{2+}}$ and n_{OH^-} from these.

$$n_{NaOH} = cV = 0.200\text{ mol L}^{-1} \times (40.0 \times 10^{-3}\text{ L}) = 8.00 \times 10^{-3}\text{ mol}$$

As 1 mole of NaOH gives 1 mole of OH^- on dissolution in water:

$$n_{OH^-} = n_{NaOH} = 8.00 \times 10^{-3}\text{ mol}$$

Similarly:

$$n_{MgCl_2} = cV = 0.300 \text{ mol L}^{-1} \times (25.0 \times 10^{-3} \text{ L}) = 7.50 \times 10^{-3} \text{ mol}$$

Since 1 mole of $MgCl_2$ gives 1 mole of Mg^{2+} on dissolution in water:

$$n_{Mg^{2+}} = n_{MgCl_2} = 7.50 \times 10^{-3} \text{ mol}$$

Looking at the balanced chemical equation, we see that 1 mole of Mg^{2+} will react with 2 moles of OH^-; so $n_{Mg^{2+}} = n_{OH^-}/2$ for complete reaction. Therefore, we would require 1.50×10^{-2} mol of OH^-, but we have only 8.00×10^{-3} mol, meaning that OH^- is the limiting reagent. Therefore, from the balanced chemical equation, 8.00×10^{-3} mol OH^- will react with 4.00×10^{-3} mol Mg^{2+} to give 4.00×10^{-3} mol $Mg(OH)_2$. We can obtain the mass of this amount using $M = \frac{m}{n}$ and the given molar mass of $Mg(OH)_2$.

$$m = Mn = 58.33 \text{ g mol}^{-1} \times (4.00 \times 10^{-3} \text{ mol}) = 0.233 \text{ g}$$

Therefore, the reaction forms 0.233 g $Mg(OH)_2(s)$.

We still have to calculate the concentrations of the ions in solution at the end of the reaction. The ions present will be Mg^{2+}, Cl^- and Na^+. We have already calculated the amount of NaOH as 8.00×10^{-3} mol, so $n_{Na^+} = 8.00 \times 10^{-3}$ mol. The final volume of the solution is 65.0 mL (40.0 mL of NaOH solution plus 25.0 mL of $MgCl_2$ solution), so:

$$c_{Na^+} = \frac{n}{V} = \frac{8.00 \times 10^{-3} \text{ mol}}{65.0 \times 10^{-3} \text{ L}} = 0.123 \text{ mol L}^{-1}$$

We have also calculated the amount of $MgCl_2$ as 7.50×10^{-3} mol, and, as 1 mole of $MgCl_2$ gives 2 moles of Cl^- on dissolution in water, $n_{Cl^-} = 2n_{MgCl_2} = 1.50 \times 10^{-2}$ mol. We can then obtain the concentration from:

$$c_{Cl^-} = \frac{n}{V} = \frac{1.50 \times 10^{-2} \text{ mol}}{65.0 \times 10^{-3} \text{ L}} = 0.231 \text{ mol L}^{-1}$$

We will assume that all of the OH^- in solution has precipitated as $Mg(OH)_2$ so the final concentration of OH^- will be 0.00 mol L^{-1} (in fact, in chapter 10, we will see that a very small amount of OH^- will remain in solution, but not enough to make a significant difference to our calculation here). The final concentration we need to calculate is that of Mg^{2+}. This is slightly more involved than the others, as we need to remember that some of the Mg^{2+} has been precipitated as $Mg(OH)_2$. Earlier, we calculated the initial amount of Mg^{2+} as 7.50×10^{-3} mol. We also calculated that 4.00×10^{-3} mol of Mg^{2+} has precipitated as $Mg(OH)_2$, so we can obtain the amount of Mg^{2+} in solution by difference.

$$n_{Mg^{2+} \text{ in solution}} = (7.50 \times 10^{-3} \text{ mol}) - (4.00 \times 10^{-3} \text{ mol}) = 3.50 \times 10^{-3} \text{ mol}$$

Therefore, the concentration of Mg^{2+} in solution is:

$$c_{Mg^{2+}} = \frac{n}{V} = \frac{3.50 \times 10^{-3} \text{ mol}}{65.0 \times 10^{-3} \text{ L}} = 5.38 \times 10^{-2} \text{ mol L}^{-1}$$

Is our answer reasonable?

This was a fairly complex problem and there are no simple calculations we can make to verify our answers. However, there are a couple of key steps we might double-check. First, we can check that we have identified the limiting reagent correctly. In the previous calculation, we selected Mg^{2+} and determined if there was enough OH^- to react with it. As a check, let us look at OH^- to see if there is enough Mg^{2+}. We began with 8×10^{-3} mol of OH^-, which requires half that amount of Mg^{2+} (4×10^{-3} mol Mg^{2+}) to react completely. The amount of Mg^{2+} available (7.5×10^{-3} mol) is more than enough, so Mg^{2+} will be left over and OH^- is limiting. Our analysis confirms that we have selected the correct limiting reagent.

Another place to be careful is in the calculation of the final concentrations of the ions in the reaction mixture. Whenever we add one aqueous solution to another, the mixture will have a final volume that is the sum of the volumes of the two solutions combined. When calculating the concentrations of anything in the final mixture, we have to use the final combined volume of the mixture. We see that we have taken this into account, so we can feel confident we've done this part of the calculation correctly.

PRACTICE EXERCISE 3.24

What amount of $BaSO_4$ will form if 20.0 mL of 0.600 M $BaCl_2$ is mixed with 30.0 mL of 0.500 M $MgSO_4$? Determine the concentration of each ion in the final reaction mixture.

The worked examples in this chapter so far have specified the reactants and products involved. There will be occasions when we do not know the composition of one of these, but stoichiometry can still help to determine the amount of a particular element in an unknown compound. The calculations required for these kinds of problems are not new; they are simply applications of the stoichiometric calculations we have already learned.

WORKED EXAMPLE 3.19

Calculation involving an unknown compound

A compound used as an insecticide contains carbon, hydrogen and chlorine. Reactions were carried out on a 1.340 g sample of the compound that converted all of its chlorine to chloride ions dissolved in water. This aqueous solution was treated with an excess of $AgNO_3$ solution, and the AgCl precipitate was collected and then weighed. Its mass was 2.709 g. What was the percentage by mass of Cl in the original insecticide sample? ($M_{Cl} = 35.45\ g\ mol^{-1}$; $M_{AgCl} = 143.32\ g\ mol^{-1}$)

Analysis

We are asked to calculate a percentage by mass. We are given the mass (m) and molar mass (M) of AgCl so we can calculate the amount of AgCl using $M = \frac{m}{n}$. We can obtain the rest of the necessary data using mole ratios. The important point to understand in this problem is that all the chlorine in the AgCl that was collected originated from the insecticide sample. The strategy, therefore, is to determine the mass of Cl in 2.709 g of AgCl. We then assume that this is the mass of chlorine in the original 1.340 g sample of insecticide and calculate the percentage Cl as follows.

$$\%\text{Cl by mass} = \frac{\text{mass of Cl in sample}}{\text{mass of sample}} \times 100\%$$

Solution

We begin by calculating the amount of AgCl using $M = \frac{m}{n}$.

$$n_{AgCl} = \frac{m}{M} = \frac{2.709\ \text{g}}{143.32\ \text{g mol}^{-1}} = 1.890 \times 10^{-2}\ \text{mol}$$

As 1 mole of AgCl contains 1 mole of Cl, the amount of Cl in the original insecticide sample must be 1.890×10^{-2} mol. We calculate the mass of this amount using $M = \frac{m}{n}$.

$$m = Mn = 35.45\ \text{g mol}^{-1} \times 1.890 \times 10^{-2}\ \text{mol} = 0.6700\ \text{g}$$

We can now calculate the %Cl by mass.

$$\%\text{Cl by mass} = \frac{\text{mass of Cl in sample}}{\text{mass of sample}} = \frac{0.6700\ \text{g}}{1.340\ \text{g}} \times 100\% = 50.00\%$$

Is our answer reasonable?

Let us consider the molar masses of AgCl and Cl (143.32 g mol^{-1} and 35.45 g mol^{-1}, respectively). If we round these to 140 and 35, we can say that AgCl is approximately 25% of Cl by mass (35/140 = 0.25). Approximately 3 g of AgCl was obtained; 25% of 3 g equals 0.75 g of Cl, which is close to the value we obtained. The original sample weighed 1.34 g, which is nearly twice 0.75 g (the mass of Cl), so an answer of 50% Cl by mass seems reasonable.

PRACTICE EXERCISE 3.25

A sample of a mixture containing $CaCl_2$ and $MgCl_2$ weighed 2.000 g. The sample was dissolved in water, and H_2SO_4 was added until the precipitation of $CaSO_4$ was complete. The $CaSO_4$ was filtered, dried completely and weighed. A total of 0.736 g of $CaSO_4$ was obtained. What amount of Ca^{2+} was in the original 2.000 g sample?

SUMMARY

LO 1 Chemical equations

A chemical reaction is the formation of new substances (products) from one or more starting materials (reactants). A chemical equation is a before-and-after description of a chemical reaction. A qualitative analysis is concerned only with which substances are present. A quantitative analysis is concerned with the amounts of all the various substances. Stoichiometry is concerned with the relative amounts of products and reactants in a reaction.

Stoichiometric coefficients are written in front of formulae in chemical equations to indicate the number of specified entities of each kind among the reactants and products. A balanced chemical equation has the same number of atoms of each kind in the products and reactants, thus conforming to the law of conservation of mass. The physical states of reactants and products can be shown in chemical equations using (s) for solid, (l) for liquid and (g) for gas. We use (aq) to indicate an aqueous solution (i.e. when a substance is dissolved in water).

LO 2 Balancing chemical equations

A chemical equation is balanced when all atoms present among the reactants are also somewhere among the products. To balance an equation, we first write the unbalanced equation and then adjust the stoichiometric coefficients to get equal numbers of each kind of atom on each side of the arrow.

LO 3 The mole

The mole (mol) is the base SI unit for amount. It is the amount of any substance with the same number of atoms or molecules as there are atoms in exactly 12 g of ^{12}C. The Avogadro constant, N_A, is 6.022×10^{23} mol^{-1} and gives the number of specified entities in 1 mol of a substance. The molar mass, M, is the mass of 1 mol of a substance.

LO 4 Empirical formulae

The actual composition of a molecule is given by its molecular formula. An empirical formula gives the ratio of atoms, but in the smallest whole numbers, and it is generally the only formula we write for ionic compounds. The molar mass of a molecular compound is equal to that calculated from the compound's empirical formula or to a simple multiple of it. When the two calculated masses are the same, the molecular formula is the same as the empirical formula.

An empirical formula can be found from the masses of the elements obtained by the quantitative analysis of a known sample of the compound, or it can be calculated from the percentage composition (the percentage of an element in a compound is the same as the number of grams of the element in 100 g of the compound). If there is J% of X in $X_x Z_z$, then 100 g of $X_x Z_z$ contains J g of X. The amounts of X and Z can be calculated from the masses of X and Z. The subscripts in the empirical formula are obtained by adjusting these amounts to their corresponding whole-number ratios.

LO 5 Stoichiometry, limiting reagents and percentage yield

A chemical formula is a tool for stoichiometric calculations, because its subscripts tell us the mole ratios in which the various elements are combined. In 1 mole of $X_x Z_z$, x moles of X are combined with z moles of Z.

A balanced equation is a tool for reaction stoichiometry because its coefficients disclose the stoichiometric relationships. All problems of reaction stoichiometry must be solved at the mole level of the substances involved. If masses are given, they must first be changed to amounts.

From the generic balanced equation $aA + bB \rightarrow cC + dD$, it follows that the mole ratios for exact stoichiometric reaction are:

$$\frac{n_A}{a} = \frac{n_B}{b} = \frac{n_C}{c} = \frac{n_D}{d}$$

In cases where this equation does not hold, the reactant present in the smallest mole ratio is called the limiting reagent. The theoretical yield of a product can be no more than that permitted by the limiting reagent. Sometimes, competing reactions (side reactions) that produce by-products reduce the actual yield. The ratio of the actual yield to the theoretical yield, expressed as a percentage, is the percentage yield.

LO 6 Solution stoichiometry

A solution is a homogeneous mixture in which one or more solutes are dissolved in a solvent. Concentration is the ratio of the amount of solute to the volume of solution. Solutions of known concentration can be prepared by weighing accurate masses of compounds using an analytical balance and then making them up to a particular volume with solvent in a volumetric flask. Concentrated solutions of known concentration can be diluted quantitatively using volumetric glassware such as pipettes and volumetric flasks. Molarity is a useful concentration unit for any calculation involving the stoichiometry of reactions in solution. In ionic reactions, the concentrations of the ions in a solution of a salt can be derived from the molar concentration of the salt, taking into account the number of ions formed on dissolution of the salt. For reactions involving ions there are three possible ways to write reaction equations. A molecular equation uses the empirical formula for each compound to aid stoichiometric calculations. The ionic equation shows all the ions that are formed in solution, while the net ionic equation shows only those ions that are actually involved in the net reaction (i.e. excluding spectator ions).

KEY CONCEPTS AND EQUATIONS

Chemical formula *(section 3.1)*	We use subscripts in a formula to establish atom ratios and mole ratios between the elements in the substance.
Balanced chemical equation *(section 3.2)*	We use coefficients to establish stoichiometric equivalencies that relate the amount of one substance to the amount of another in a chemical reaction. The coefficients also establish the ratios of reactants and products involved in the reaction.
The mole *(section 3.3)*	The mole provides the basis for determination of the amount of substance.
Avogadro constant (N_A) *(section 3.3)*	The Avogadro constant relates macroscopic quantities used in a laboratory to numbers of individual atomic-sized particles, such as atoms, molecules and ions.
Molar mass (M) *(section 3.3)*	Molar mass can be used to calculate the mass of 1 mole of any substance or to convert between amount and mass using $M = \frac{m}{n}$.
$M = \frac{m}{n}$ *(section 3.3)*	This equation represents the relationship between mass (m), amount (n) and molar mass (M).

Percentage composition *(section 3.4)*

We use percentage composition to represent the composition of a compound and as the basis for calculating the empirical formula. Comparing experimental and theoretical percentage compositions can help to establish the identity of a compound.

Theoretical, actual and percentage yields *(section 3.5)*

We use these values to estimate the efficiency of a reaction. Remember that theoretical yield is calculated from the limiting reagent using the balanced chemical equation. The percentage yield is given by the equation: percentage yield = (actual yield/theoretical yield) × 100%.

Balanced equations and mole ratios *(section 3.5)*

For the generic balanced equation:

$$aA + bB \rightarrow cC + dD$$

the mole ratios for exact stoichiometric reaction are:

$$\frac{n_A}{a} = \frac{n_B}{b} = \frac{n_C}{c} = \frac{n_D}{d}$$

$c = \frac{n}{V}$ *(section 3.6)*

This equation represents the relationship between concentration (c), amount (n) and volume (V).

KEY TERMS

actual yield *p. 98*
amount of substance *p. 85*
atom efficiency *p. 99*
Avogadro constant (N_A) *p. 85*
balanced chemical equation *p. 82*
by-product *p. 98*
chemical equation *p. 82*
chemical reaction *p. 82*
combustion *p. 92*
concentration *p. 100*
dissociate *p. 106*
elemental analysis *p. 90*
empirical formula *p. 87*
excess reagent *p. 96*
limiting reagent *p. 96*
mass percentage composition *p. 90*
molarity *p. 100*
molar mass (M) *p. 85*
mole *p. 85*
molecular formula *p. 91*
net ionic equation *p. 107*
percentage by mass *p. 90*
percentage yield *p. 98*
product *p. 82*
qualitative analysis *p. 82*
quantitative analysis *p. 82*
reactant *p. 82*
reversible reaction *p. 82*
solute *p. 100*
solution *p. 100*
solvent *p. 100*
spectator ion *p. 107*
stoichiometric coefficients *p. 82*
stoichiometry *p. 82*
theoretical yield *p. 98*

REVIEW QUESTIONS

LO 1 Chemical equations

3.1 What do we mean when we say a chemical equation is 'balanced'? Why do we balance chemical equations?

3.2 In a chemical reaction, what do we mean by the term 'reactants'? What do we mean by the term 'products'?

3.3 What do we call the numbers that are written in front of the formulae in a balanced chemical equation?

3.4 The combustion of a thin wire of magnesium metal, Mg, in an atmosphere of pure oxygen produces the brilliant light of a flashbulb, once commonly used in photography. After the reaction, a thin film of magnesium oxide is seen on the inside of the bulb. The chemical reaction that occurs can be represented by the equation:

$$2Mg + O_2 \rightarrow 2MgO$$

(a) State in words how this equation is read.
(b) Give the formulae of the reactant(s).
(c) Give the formula of the product(s).
(d) Rewrite the equation to show that Mg and MgO are solids and O_2 is a gas.

LO 2 Balancing chemical equations

3.5 When given the *unbalanced* equation:

$$Na(s) + Cl_2(g) \rightarrow NaCl(s)$$

and asked to balance it, student A wrote:

$$Na(s) + Cl_2(g) \rightarrow NaCl_2(s)$$

and student B wrote:

$$2Na(s) + Cl_2(g) \rightarrow 2NaCl(s)$$

(a) Both equations are balanced, but which student is correct?
(b) Explain why the other student's answer is incorrect.

LO 3 The mole

3.6 Define the term 'mole'.

3.7 Why are stoichiometry problems solved using amounts (in mol) rather than masses of reagents?

3.8 What information is required to calculate the amount of a substance from a given mass of that same substance?

3.9 Give the equation that relates molar mass, amount and mass.

3.10 How would you estimate the number of atoms in a gram of aluminium, using atomic mass units?

3.11 Which contains more molecules: 2.5 moles of H_2O or 2.5 moles of H_2?

3.12 What amount of iron atoms is in 1 mol of Fe_2O_3? How many iron atoms are in 1 mol of Fe_2O_3?

3.13 Why is the expression '1.0 mol of nitrogen' ambiguous?

LO 4 Empirical formulae

3.14 In general, what fundamental information, obtained from experimental measurements, is required to calculate the empirical formula of a compound?

3.15 Why is the changing of subscripts not allowed when balancing a chemical equation?

3.16 Give a step-by-step procedure for estimating the mass of *A* required to completely react with 10 moles of *B*, given the following information.
(i) *A* and *B* react to form A_5B_2.
(ii) *A* has a molar mass of 100.0 g mol^{-1}.
(iii) *B* has a molar mass of 200.0 g mol^{-1}.
(iv) There are 6.022×10^{23} molecules of *A* in 1 mole of *A*.
Which of these pieces of information was/were not needed?

LO 5 Stoichiometry, limiting reagents and percentage yield

3.17 What information is required to determine the mass of oxygen that would react with 1 gram of arsenic?

3.18 A mixture of 0.020 mol of Mg and 0.020 mol of Cl_2 reacted completely to form $MgCl_2$ according to the equation:

$$Mg(s) + Cl_2(g) \rightarrow MgCl_2(s)$$

(a) What information describes the *stoichiometry* of this reaction?
(b) What information gives the *scale* of the reaction?

3.19 In a report to a supervisor, a chemist described an experiment in the following way: '0.0800 mol of H_2O_2 decomposed into 0.0800 mol of H_2O and 0.0400 mol of O_2'. Express the chemistry and stoichiometry of this reaction by a conventional chemical equation.

LO 6 Solution stoichiometry

3.20 Give the equation that relates concentration, volume and amount.

3.21 What is the definition of 'molarity'? Show that the ratio of millimoles (mmol) to millilitres (mL) is equivalent to the ratio of moles to litres.

3.22 What are spectator ions?

3.23 When a solution labelled 0.75 M HNO_3 is diluted with water to give 0.15 M HNO_3, what happens to the amount of HNO_3 in the solution?

3.24 Solutions A and B are labelled '0.15 M $CaCl_2$' and '0.25 M $CaCl_2$', respectively. Both solutions contain the same amount of $CaCl_2$. If solution A has a volume of 50 mL, what is the volume of solution B?

3.25 What is the difference between a qualitative analysis and a quantitative analysis?

3.26 The following equation represents the formation of cobalt(II) hydroxide.

$$Co^{2+}(aq) + 2Cl^-(aq) + 2K^+(aq) + 2OH^-(aq) \rightarrow Co(OH)_2(s) + 2K^+(aq) + 2Cl^-(aq)$$

Identify the spectator ions and write the net ionic equation.

3.27 Why is the following equation not balanced? Find the errors and fix them.

$$3Co^{3+} + 2HPO_4^{2-} \rightarrow Co_3(PO_4)_2 + 2H^+$$

REVIEW PROBLEMS

3.28 Consider the balanced equation: **LO1**

$$2Fe(NO_3)_3 + 3Na_2CO_3 \rightarrow Fe_2(CO_3)_3 + 6NaNO_3$$

(a) How many atoms of Na are on each side of the equation?
(b) How many atoms of C are on each side of the equation?
(c) How many atoms of O are on each side of the equation?

3.29 When sulfur impurities in fuels burn, they produce pollutants such as sulfur dioxide, a major contributor to acid rain. The following is a ball-and-stick model of a typical reaction. **LO1**

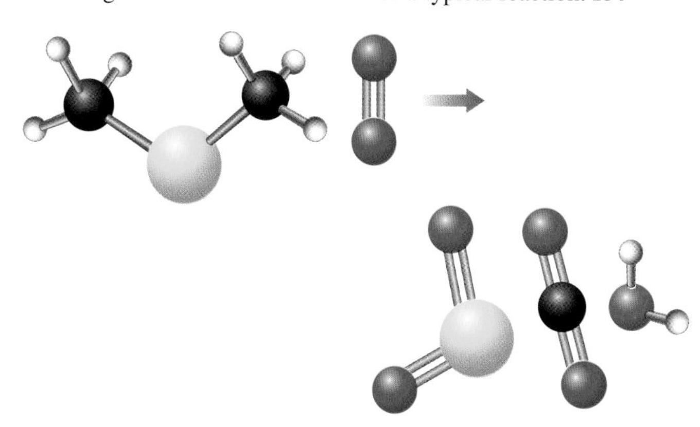

On the left are reactant molecules and on the right are product molecules in a chemical reaction. Write the balanced chemical equation for this reaction, using the stoichiometric coefficients that give the smallest whole number of product molecules

3.30 Write the equation that expresses in acceptable chemical shorthand the following statement: 'Iron can be made to react with molecular oxygen, O_2, to give iron oxide with the formula Fe_2O_3.' **LO1**

3.31 Racing car drivers can get extra power by burning methyl alcohol with nitrous oxide. Below on the left are the reactant molecules and on the right are product molecules of the chemical reaction. Write the balanced chemical equation for this reaction, using the stoichiometric coefficients that give the smallest whole number of product molecules. **LO1,2**

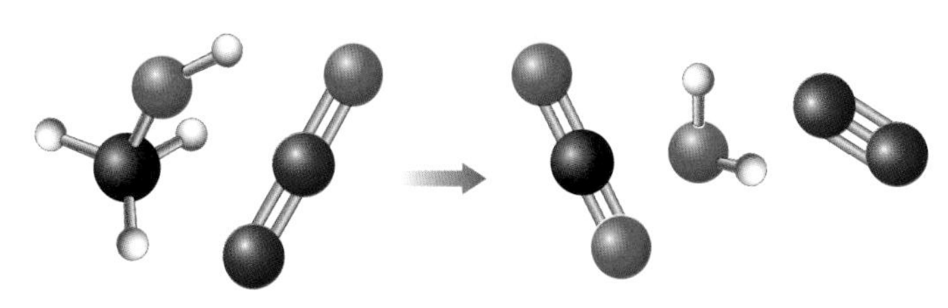

3.32 Is the following chemical equation balanced? This reaction is used for the production of nitric acid, HNO_3, and is one of the reactions responsible for acid rain. **LO1,2**

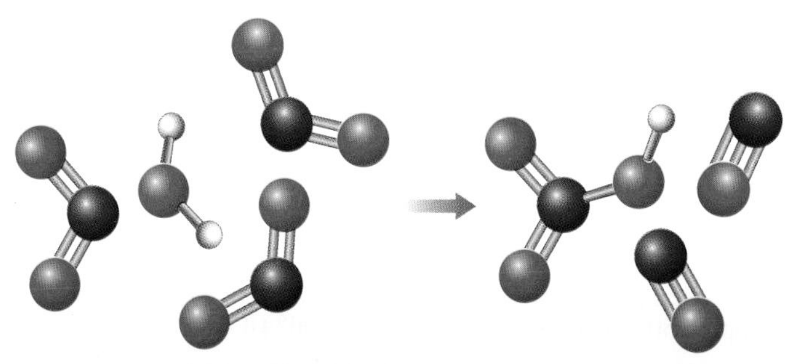

$$3NO_2 + H_2O \rightarrow HNO_3 + 2NO$$

If the equation is not balanced, find coefficients that would make it balanced.

3.33 Write the balanced chemical equation for the chemical reaction depicted by the space-filling models below. **LO1,2**

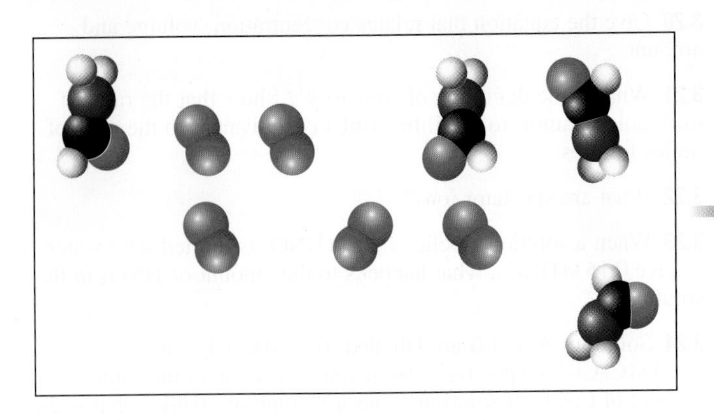

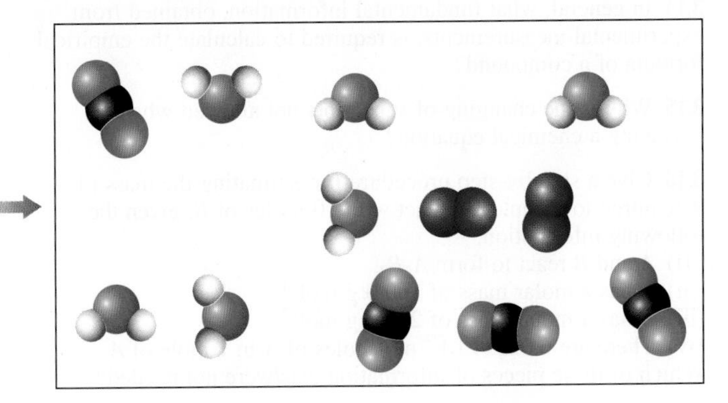

3.34 Balance the following equations. **LO2**
(a) $Ca(OH)_2 + HCl \rightarrow CaCl_2 + H_2O$
(b) $NaHCO_3 + H_2SO_4 \rightarrow Na_2SO_4 + H_2O + CO_2$
(c) $AgNO_3 + CaCl_2 \rightarrow Ca(NO_3)_2 + AgCl$
(d) $C_4H_{10} + O_2 \rightarrow CO_2 + H_2O$
(e) $Fe_2O_3 + C \rightarrow Fe + CO_2$

3.35 In what smallest whole-number ratio must N and O atoms combine to make dinitrogen tetroxide, N_2O_4? What is the mole ratio of the elements in this compound? **LO3,4**

3.36 What amount of sodium atoms corresponds to 1.56×10^{21} atoms of sodium? **LO3**

3.37 What amount of Al atoms are needed to combine with 2.00 mol of O atoms to produce the maximum amount of aluminium oxide, Al_2O_3? **LO4,5**

3.38 What amount of H_2 and N_2 can be formed by the decomposition of 0.145 mol of ammonia, NH_3? **LO2,4**

3.39 What amount of UF_6 would have to be decomposed to provide enough fluorine to prepare 1.25 mol of CF_4? **LO2,4**

3.40 Determine the mass in grams of each of the following. **LO3**
(a) 1.85 mol Fe (b) 1.85 mol O (c) 1.85 mol Ca

3.41 What is the mass, in grams, of 2.00×10^{12} atoms of magnesium? **LO3**

3.42 What amount of nickel is in 17.7 g of Ni? **LO3**

3.43 Calculate the molar mass of each of the following. **LO3**
(a) $NaHCO_3$
(b) $K_2Cr_2O_7$
(c) $(NH_4)_2CO_3$
(d) $Al_2(SO_4)_3$
(e) $CuSO_4{\cdot}5H_2O$

3.44 Calculate the mass in grams of each of the following. **LO3**
(a) 1.25 mol $Ca_3(PO_4)_2$
(b) 0.625 mol $Fe(NO_3)_3$
(c) 0.600 mol C_4H_{10}
(d) 1.45 mol $(NH_4)_2CO_3$

3.45 Calculate the amount of the compound in each of the following samples. **LO3**
(a) 21.5 g $CaCO_3$
(b) 1.56 g NH_3
(c) 16.8 g $Sr(NO_3)_2$
(d) 6.98 μg Na_2CrO_4

3.46 What mass of $(NH_4)_2CO_3$ fertiliser would be required to supply 250 g of atomic nitrogen to the soil? **LO3**

3.47 What mass of O is combined with 6.35×10^{21} atoms of N in the compound N_2O_5? **LO3**

3.48 The incandescent white of a fireworks display is caused by the reaction of phosphorus with O_2 to give P_4O_{10}. **LO2,3**
(a) Write the balanced chemical equation for the reaction.
(b) What mass of O_2 is needed to combine with 6.85 g of P?
(c) What mass of P_4O_{10} can be made from 8.00 g of O_2?
(d) What mass of P is needed to make 7.46 g of P_4O_{10}?

3.49 Radioactive sodium pertechnetate is used as a brain-scanning agent in medicine. Quantitative analysis of a 0.896 g sample of sodium pertechnetate found 0.111 g of sodium and 0.477 g of technetium. The remainder was oxygen. Calculate the empirical formula of sodium pertechnetate. **LO4**

3.50 The composition of a dry-cleaning fluid, composed of only carbon and chlorine, was found to be 14.5% C and 85.5% Cl (by mass). What is the empirical formula of this compound? **LO4**

3.51 When 0.684 g of an organic compound containing only carbon, hydrogen and oxygen was burned in oxygen, 1.312 g CO_2 and 0.805 g H_2O was obtained. What is the empirical formula of the compound? **LO4**

3.52 A sample of a compound of mercury and bromine with a mass of 0.389 g was found to contain 0.111 g bromine. Its molar mass was found to be 561 g mol^{-1}. What are its empirical and molecular formulae? **LO4**

3.53 Calculate the percentage composition by mass of all elements in each of the following. **LO4**
(a) NaH_2PO_4
(b) $NH_4H_2PO_4$
(c) $(CH_3)_2CO$
(d) $CaSO_4$
(e) $CaSO_4{\cdot}2H_2O$

3.54 Chlorine is used by textile manufacturers to bleach cloth. Excess chlorine is destroyed by its reaction with sodium thiosulfate, $Na_2S_2O_3$, as follows. **LO5**

$$Na_2S_2O_3(aq) + 4Cl_2(g) + 5H_2O(l) \rightarrow 2NaHSO_4(aq) + 8HCl(aq)$$

(a) What amount of $Na_2S_2O_3$ is needed to react with 0.12 mol of Cl_2?
(b) What amount of HCl can form from 0.12 mol of Cl_2?
(c) What amount of H_2O is required for the reaction of 0.12 mol of Cl_2?
(d) What amount of H_2O reacts if 0.24 mol HCl is formed?

3.55 The following reaction is used to extract gold from pretreated gold ore. **LO5**

$$2Au(CN)_2^-(aq) + Zn(s) \rightarrow 2Au(s) + Zn(CN)_4^{2-}(aq)$$

(a) What mass of Zn is needed to react with 0.11 mol of $Au(CN)_2^-$?
(b) What mass of Au can form from 0.11 mol of $Au(CN)_2^-$?
(c) What mass of $Au(CN)_2^-$ is required for the reaction of 0.11 mol of Zn?

3.56 Oxygen gas can be produced in the laboratory by decomposition of hydrogen peroxide, H_2O_2. **LO5**

$$2H_2O_2(l) \rightarrow 2H_2O(l) + O_2(g)$$

What mass of O_2 can be produced from 255 g of H_2O_2?

3.57 Given the balanced chemical equation:

$$2C_2H_5SH(g) + 9O_2(g) \rightarrow 4CO_2(g) + 2SO_2(g) + 6H_2O(g)$$

determine the number of molecules of SO_2 that would be formed starting from the actual number of molecules of C_2H_5SH and O_2 shown below. **LO5**

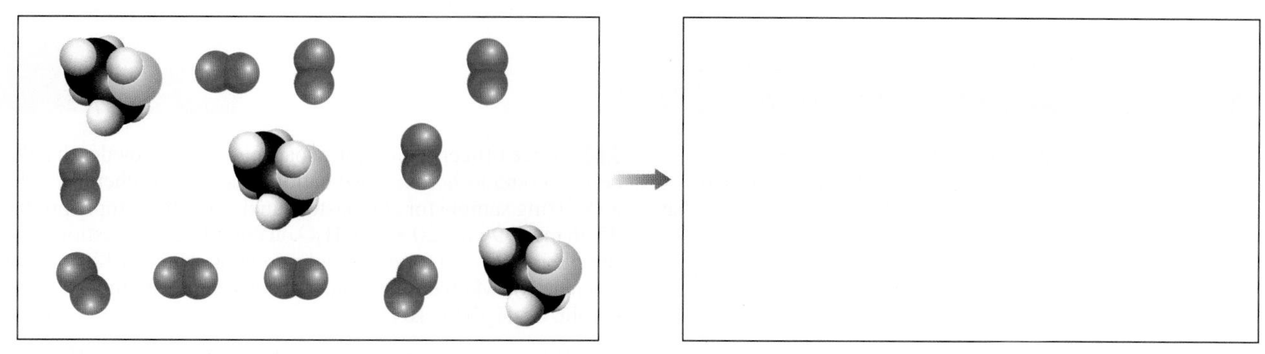

3.58 Using the balanced chemical equation from question 3.57, which of C_2H_5SH or O_2 was the limiting reagent if the product distribution for the reaction was that shown below? **LO5**

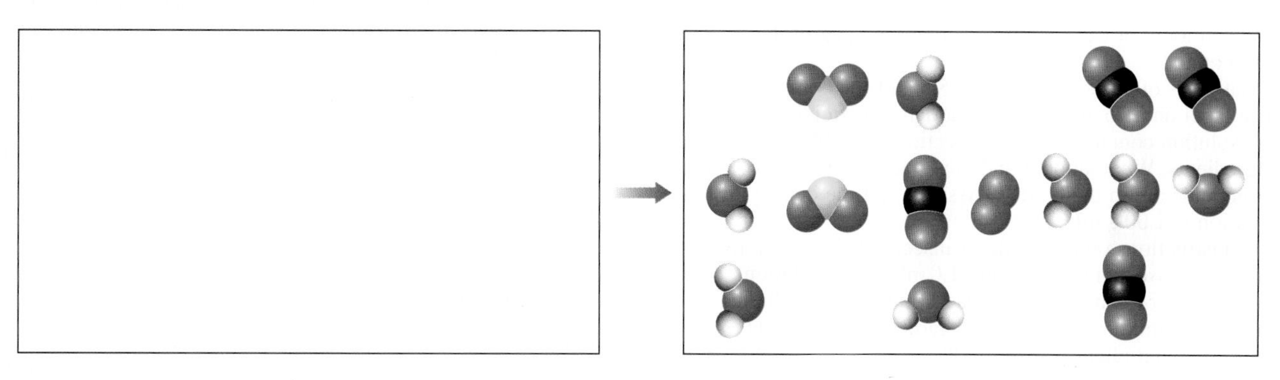

3.59 The reaction of powdered aluminium and iron(III) oxide: **LO5**

$$2Al(s) + Fe_2O_3(s) \rightarrow Al_2O_3(s) + 2Fe(l)$$

produces so much heat that the iron that forms is molten. Because of this, the reaction is used when laying railway tracks to provide molten steel to weld steel rails together. Suppose that, in one batch of reactants, 4.00 mol of Al was mixed with 1.50 mol of Fe_2O_3.
(a) Which reactant, if either, was the limiting reagent?
(b) Calculate the mass of iron that can be formed from this mixture of reactants.

3.60 Silver nitrate, $AgNO_3$, reacts with iron(III) chloride, $FeCl_3$, to give silver chloride, AgCl, and iron(III) nitrate, $Fe(NO_3)_3$. A solution containing 18.0 g of $AgNO_3$ was mixed with a solution containing 32.4 g of $FeCl_3$. What mass of which reactant remains after the reaction is complete? **LO5**

3.61 Barium sulfate, $BaSO_4$, is made by the following reaction. **LO5**

$$Ba(NO_3)_2(aq) + Na_2SO_4(aq) \rightarrow BaSO_4(s) + 2NaNO_3(aq)$$

An experiment was begun with 75.00 g of $Ba(NO_3)_2$ and an excess of Na_2SO_4. After collecting and drying the product, 64.45 g of $BaSO_4$ was obtained. Calculate the theoretical yield and percentage yield of $BaSO_4$.

3.62 The potassium salt of benzoic acid (potassium benzoate, C_6H_5COOK), can be made by the action of potassium permanganate, $KMnO_4$, on toluene, $C_6H_5CH_3$, as follows. **LO5**

$$C_6H_5CH_3 + 2KMnO_4 \rightarrow KC_6H_5COO + 2MnO_2 + KOH + H_2O$$

The maximum practical yield for this reaction is 71%. Given that, what is the minimum mass of toluene required to produce 14.5 g of potassium benzoate?

3.63 Calculate the concentration of a solution prepared by dissolving: **LO6**
(a) 4.00 g of NaOH in 100.0 mL of solution
(b) 16.0 g of $CaCl_2$ in 250.0 mL of solution
(c) 14.0 g of KOH in 75.0 mL of solution
(d) 6.75 g of HO_2CCO_2H in 500 mL of solution.

3.64 Calculate the mass of each solute needed to make each of the following solutions. **LO6**
(a) 125 mL of 0.200 M NaCl(aq)
(b) 250 mL of 0.360 M $C_6H_{12}O_6$(aq) (glucose)
(c) 250 mL of 0.250 M H_2SO_4(aq)

3.65 If 25.0 mL of 0.56 M H_2SO_4 is diluted to a volume of 125 mL, what is the concentration of the resulting solution? **LO6**

3.66 To what volume must 25.0 mL of 18.0 M H_2SO_4 be diluted to produce 1.50 M H_2SO_4(aq)? **LO6**

3.67 Calculate the amounts of each of the ions in the following solutions. **LO6**
(a) 35.0 mL of 1.25 M KOH(aq)
(b) 32.3 mL of 0.45 M $CaCl_2$(aq)
(c) 50.0 mL of 0.40 M $AlCl_3$(aq)

3.68 Calculate the concentrations of each of the ions in the following solution? **LO6**
(a) 0.25 M $Cr(NO_3)_2$(aq)
(b) 0.10 M $CuSO_4$(aq)
(c) 0.16 M Na_3PO_4(aq)
(d) 0.075 M $Al_2(SO_4)_3$(aq)

3.69 In a solution of $Al_2(SO_4)_3$(aq), the Al^{3+} concentration is 0.12 M. What mass of $Al_2(SO_4)_3$ is in 50.0 mL of this solution? **LO6**

3.70 What volume of 0.25 M $NiCl_2$(aq) solution is needed to react completely with 20.0 mL of 0.15 M Na_2CO_3(aq) solution? What mass of $NiCO_3$ will be formed? The reaction can be represented by the following equation. **LO6**

$$Na_2CO_3(aq) + NiCl_2(aq) \rightarrow NiCO_3(s) + 2NaCl(aq)$$

3.71 What volume of 0.150 M $FeCl_3$(aq) solution is needed to react completely with 20.0 mL of 0.0450 M $AgNO_3$(aq) solution? What mass of AgCl will be formed? The net ionic equation for the reaction is: **LO6**

$$Ag^+(aq) + Cl^-(aq) \rightarrow AgCl(s)$$

3.72 Consider the reaction of aluminium chloride with silver acetate. What volume of 0.200 M $AlCl_3$(aq) would be needed to react completely with 40.0 mL of 0.500 M $AgOOCCH_3$(aq) solution? The net ionic equation for the reaction is: **LO6**

$$Ag^+(aq) + Cl^-(aq) \rightarrow AgCl(s)$$

3.73 Suppose that 25.0 mL of 0.440 M NaCl(aq) is added to 25.0 mL of 0.320 M $AgNO_3$(aq). **LO 6**
(a) What amount of AgCl would precipitate?
(b) Determine the concentration of each of the ions in the reaction mixture after the reaction.

3.74 A mixture is prepared by adding 25.0 mL of 0.185 M Na_3PO_4(aq) to 34.0 mL of 0.140 M $Ca(NO_3)_2$(aq). **LO 6**
(a) What mass of $Ca_3(PO_4)_2$ will be formed?
(b) Determine the concentration of each of the ions in the mixture after reaction.

ADDITIONAL EXERCISES

3.75 Write the net ionic equations for the following reactions. Include the designations for the states — solid, liquid, gas and aqueous solution. (For any substance soluble in water, assume that it is used as an aqueous solution.) **LO 6**
(a) $CaCO_3 + 2HNO_3 \rightarrow Ca(NO_3)_2 + CO_2 + H_2O$
(b) $CaCO_3 + H_2SO_4 \rightarrow CaSO_4 + CO_2 + H_2O$
(c) $FeS + 2HBr \rightarrow H_2S + FeBr_2$
(d) $2KOH + SnCl_2 \rightarrow 2KCl + Sn(OH)_2$

3.76 Answer the following questions about an insecticide with the chemical formula $C_{12}H_{11}NO_2$. **LO 3,6**
(a) What amount of carbon atoms is there in 8.3 g of the compound?
(b) What mass of oxygen is there in 8.3 g of the compound?
(c) The label on a 75 mL bottle of garden insecticide states that the solution contains 0.010% insecticide and 99.99% inert ingredients. What amount and how many molecules of the insecticide are in the bottle? (Assume that the density of the solution is 1.00 g mL^{-1}.)
(d) The instructions on the bottle of insecticide in (c) are to dilute the insecticide by adding 1.0 mL of the solution to 4.0 litres of water. If you spray 60 litres of the diluted insecticide mixture on your rose garden, what amount of the active ingredient is dispersed?

3.77 The mineral turquoise has the following formula. **LO 3**

$$CuAl_6(PO_4)_4(OH)_8{\cdot}4H_2O$$

(a) What mass of aluminium is in a 6.75 g sample of turquoise?
(b) How many phosphate ions are in a sample of turquoise that contains 5.00×10^{-3} g of oxygen?

3.78 Adenosine triphosphate (ATP) is used to generate chemical energy in plant and animal cells. The molecular formula of ATP is $C_{10}H_{16}N_5O_{13}P_3$. Answer the following questions about ATP. **LO 3,5**
(a) What is the percentage composition by mass of each element in ATP?
(b) How many P atoms are in 1.75 μg of ATP?
(c) If a cell consumes 3.0 pmol of ATP, what mass has it consumed?
(d) What mass of ATP contains the same number of atoms of H as the number of N atoms in 37.5 mg of ATP?

3.79 The thyroid gland produces hormones that help regulate body temperature, metabolic rate, reproduction, synthesis of red blood cells and more. Iodine must be present in the diet to produce these thyroid hormones. Iodine deficiency leads to sluggishness and weight gain, and can cause severe problems in the development of a foetus. One thyroid hormone is thyroxine, with the chemical formula $C_{15}H_{11}I_4NO_4$. What mass of thyroxine can be produced from 210 mg of iodine atoms, the amount a typical adult consumes per day? How many molecules is this? **LO 3,5**

thyroxine
$C_{15}H_{11}I_4NO_4$

3.80 Police officers confiscate a packet of white powder that they believe contains heroin. Purification by a forensic chemist yields a 38.70 mg sample for combustion analysis. This sample produces 97.46 mg CO_2 and 20.81 mg H_2O on complete combustion. A second sample is analysed for its nitrogen content, which is 3.8%. Show by calculations whether these data are consistent with the formula for heroin, $C_{21}H_{22}NO_5$. **LO 4,5**

3.81 When an unknown compound containing only carbon and hydrogen is burned completely in O_2, 1.45 g of CO_2 and 0.47 g of H_2O is recovered. What additional information is needed before the molecular formula of the unknown compound can be determined? **LO 4,5**

3.82 A sulfur-containing ore of copper releases sulfur dioxide when heated in air. A 5.26 g sample of the ore releases 2.12 g of SO_2 on heating. Assuming that the ore contains only copper and sulfur, what is the empirical formula? **LO 4**

3.83 Vitamin B_{12} is a large molecule called cobalamin. There is 1 atom of cobalt in each molecule of vitamin B_{12}, and the mass percentage of cobalt is 4.34%. Calculate the molar mass of cobalamin. **LO 4**

3.84 Element *E* forms a compound with the formula ECl_5. Elemental analysis shows that the compound is 58.28% chlorine by mass. Identify element *E*. **LO 4**

3.85 A sample of a component of petroleum was subjected to combustion analysis. An empty vial of mass 2.7534 g was filled with the sample, after which the vial plus sample had a mass of 2.8954 g. The sample was burned and the resulting CO_2 and H_2O were collected in separate traps. Before combustion, the CO_2 trap had a mass of 54.4375 g and the H_2O trap had a mass of 47.8845 g. At the end of the analysis, the CO_2 trap had a new mass of 54.9140 g and the H_2O trap had a new mass of 47.9961 g. Determine the empirical formula of this component of petroleum. **LO 4**

3.86 In the Haber synthesis of ammonia, N_2 and H_2 react at high temperature, but they never react completely. In a typical reaction, 24.0 kg of H_2 and 84.0 kg of N_2 react to produce 68 kg of NH_3. Find the theoretical yield, the percentage yield and the masses of H_2 and N_2 that remain unreacted, assuming that no other products form. **LO 5**

3.87 Silicon tetrachloride is used in the electronics industry to make elemental silicon for computer chips. Silicon tetrachloride is prepared from silicon dioxide, graphite and chlorine gas according to the equation: **LO 5**

$$SiO_2(s) + 2C(s) + 2Cl_2(g) \rightarrow SiCl_4(l) + 2CO(g)$$

If the reaction achieves 95.7% yield, how much silicon tetrachloride can be prepared from 75.0 g of each starting material, and how much of each reactant remains unreacted?

3.88 A 0.113 g sample of a mixture of sodium chloride and sodium nitrate was dissolved in water, and enough silver nitrate solution was added to precipitate all of the chloride ions. The mass of silver chloride obtained was 0.250 g. What mass percentage of the sample was sodium chloride? **LO 6**

3.89 In an experiment, 40.0 mL of 0.27 0 M $Ba(OH)_2$(aq) was mixed with 25.0 mL of 0.330 M $Al_2(SO_4)_3$(aq). **LO 6**
(a) Write the net ionic equation for the reaction that takes place.
(b) What is the total mass of precipitate that forms?
(c) What are the molar concentrations of the ions that remain in the solution after the reaction is complete?

3.90 What volume of 0.10 M HCl(aq) must be added to 40.0 mL of 0.40 M HCl(aq) to give a final solution that has a molarity of 0.25 M? **LO 6**

3.91 Solution A is prepared by dissolving 90.0 g of Na_3PO_4 in enough water to make 1.5 L of solution. Solution B is 2.5 L of 0.705 M Na_2SO_4(aq). **LO 6**
(a) What is the molar concentration of Na_3PO_4 in solution A?
(b) What volume of solution A contains 2.50 g of Na_3PO_4?
(c) A 50.0 mL sample of solution B is mixed with a 75.00 mL sample of solution A. Calculate the concentration of Na^+ ions in the final solution.

3.92 A chemist needs a solution that contains aluminium ions, sodium ions and sulfate ions. In the laboratory she finds a large volume of 0.355 M sodium sulfate solution and a bottle of solid $Al_2(SO_4)_3 \cdot 18H_2O$. The chemist puts 250 mL of the sodium sulfate solution and 5.13 g of aluminium sulfate hydrate in a 500 mL volumetric flask. The flask is made up to the mark with water. Determine the molarity of aluminium ions, sodium ions and sulfate ions in the solution. **LO 6**

3.93 The waters of the oceans contain many elements in trace amounts. Rubidium, for example, is present at the level of 2.2 nM (nanomolar). How many ions of rubidium are present in 1.00 L of sea water? How many litres would have to be processed to recover 1.00 kg of rubidium, assuming the recovery process was 100% efficient? **LO 6**

3.94 Pure acetic acid, CH_3COOH, has a concentration of 17.4 M. A laboratory worker measured out 100.0 mL of pure acetic acid and added enough water to make 500.0 mL of solution. A 75.0 mL portion of the acetic acid solution was then mixed with enough water to make 1.50 L of dilute solution. What was the final molarity of acetic acid in the dilute solution? **LO 6**

3.95 A worker in a biological laboratory needed a solution that was 0.30 M in sodium acetate, $NaOOCCH_3$, and 0.15 M in acetic acid, CH_3COOH. On hand were stock solutions of 2.5 M sodium acetate and 2.5 M acetic acid. Describe how the worker prepared 1.5 L of the desired solution. **LO 6**

3.96 Silver jewellery is usually made from silver and copper alloys. The amount of copper in an alloy can vary considerably. The finest quality alloy is sterling silver, which is 92.5% silver by mass. To determine the composition of a silver–copper alloy, a jeweller dissolved 0.135 g of metal shavings in 50 mL of concentrated nitric acid and then added 1.00 M KCl solution until no more precipitate formed. Filtration and drying yielded 0.156 g of AgCl precipitate. What was the mass composition of the silver alloy? **LO 6**

3.97 Aluminium sulfate is used in the manufacture of paper and in the water purification industry. In the solid state, aluminium sulfate is a hydrate with the formula $Al_2(SO_4)_3 \cdot 18H_2O$. **LO 3,6**
(a) What mass of sulfur is there in 0.500 moles of solid aluminium sulfate?
(b) How many water molecules are there in a 4.81 g sample of solid aluminium sulfate?
(c) What amount of sulfate ions is there in a sample of solid aluminium sulfate that contains 12.5 moles of oxygen atoms?
(d) An aqueous solution of aluminium sulfate has a density of 1.05 g mL^{-1}. What is the molarity of aluminium ions in the solution?

4 Atomic energy levels

In the right combination, atoms and light combine to create one of the most remarkable tools of modern technology, the laser. Laser light is more highly organised than normal light. Laser light is monochromatic (has a single wavelength) and is highly directional, whereas conventional light sources typically produce light of many wavelengths. The pictures on these pages demonstrate these differences. In the picture opposite, we see red laser light with wavelength of 650 nm pass through a prism. While the light is diffracted, there is no splitting into different colours, as we see for white light in the bottom picture. This confirms that laser light consists of only one wavelength while normal white light comprises many wavelengths. Many of the applications of lasers take advantage of the high degree of organisation.

Lasers are used in, for example, DVD scanners, eye surgery and fibre optics communications. Lasers have also become versatile tools for scientific research. For example, finely tuned lasers have been used to deposit vapour-phase atoms on solid surfaces in regular patterns. The ability to manipulate individual atoms is likely to have important applications in nanotechnology.

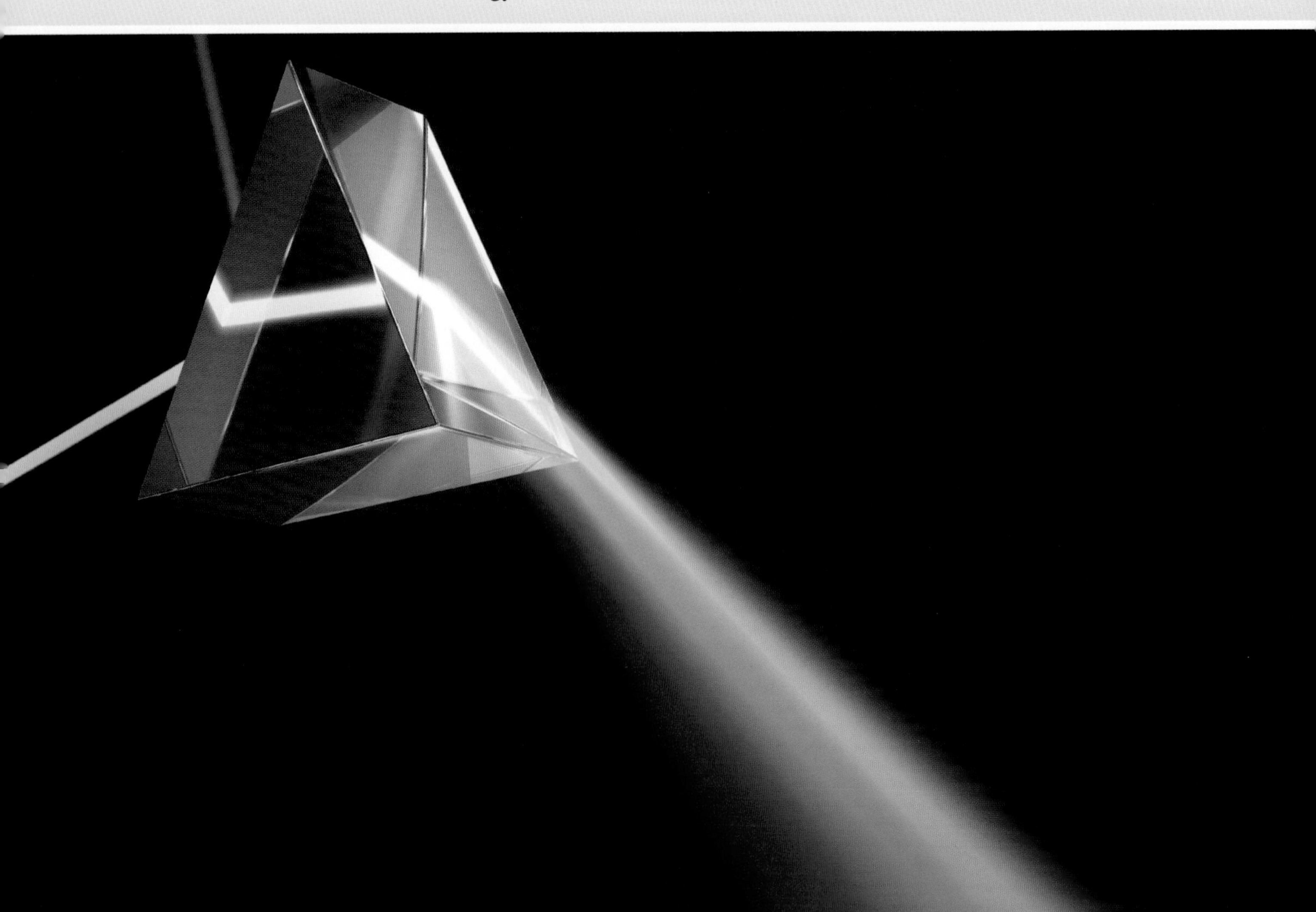

Every laser is based on the interactions between light and atoms or molecules. In this chapter, we describe the properties of atoms and light and the energy changes that accompany the interactions between them. Then we describe the properties of electrons bound to atoms and how these contribute to atomic structure. We explore the details of orbital energies and relate them to the ordering of atoms that leads to the familiar form and structure of the periodic table. Orbital energy levels have consequences that are far reaching, for example, they determine the stability and reactivity of atoms. The periodic table is based on orbital energy levels and provides the foundation for interpreting patterns of chemical behaviour related to an element's position in the table.

LEARNING OBJECTIVES

After studying this chapter, you should be able to:

4.1 list some basic properties of atoms

4.2 explain the wave-like and particle-like properties of light, and how light interacts with atoms

4.3 describe the properties of electrons

4.4 explain the properties of bound electrons

4.5 describe orbitals in terms of electron distributions and energies

4.6 explain the structure of the periodic table in terms of electrons

4.7 write electron configurations

4.8 describe periodic trends in chemical and physical properties

4.9 describe periodic trends in relation to ions.

4.1 Characteristics of atoms

We introduced the concept of the atom in chapter 1, and stated that it was the fundamental building block of all matter. In this chapter, we will investigate the properties of atoms in more detail. We will start by outlining the basic characteristics of atoms; we read about many of these characteristics in chapter 1.

Atoms possess mass, most of which, as Rutherford showed, is concentrated in the nucleus. The nucleus of an atom is small and positively charged, and all nuclei, except that of the hydrogen atom ^{1}H, consist of both positively charged protons and neutral neutrons. For neutral atoms, the positive charge of the nucleus is exactly balanced by negatively charged electrons, which occupy the region of space around the nucleus, that is, a neutral atom contains equal numbers of protons and electrons. Atoms occupy volume, the majority of which is taken up by the electron cloud. The chemical properties of elements are determined to a large extent by both their atomic size and their number of accessible (valence) electrons. Finally, atoms can attract one another, and, as a result, can combine to form molecules.

This chapter examines in detail how both the number of electrons and their specific arrangement influence chemical properties. Since there are similarities between the properties of light and electrons, and light is an essential tool for probing the properties of electrons, we will begin our discussion by describing light and its interaction with atoms.

4.2 Characteristics of light

The most useful tool for studying the structure of atoms is **electromagnetic radiation**. What we call **light** is one form of this radiation, with other forms including radiowaves, microwaves and X-rays. We need to know about the fundamental properties of light to understand what electromagnetic radiation reveals about atomic structure.

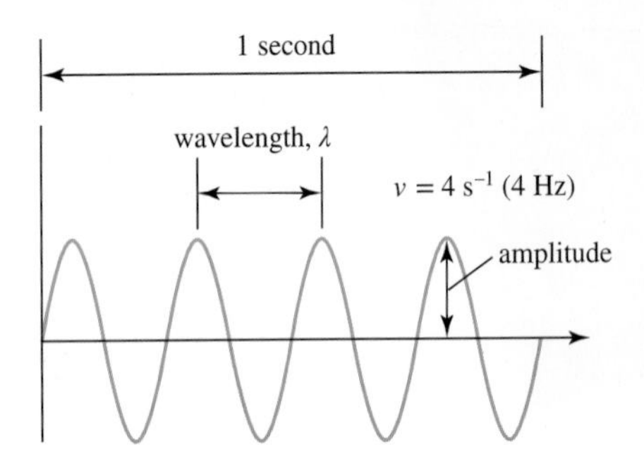

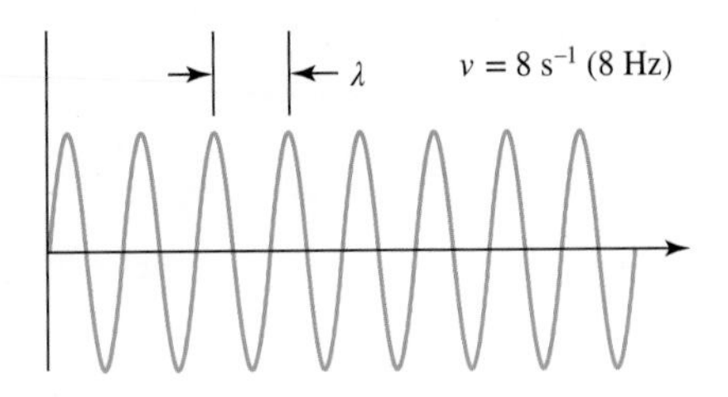

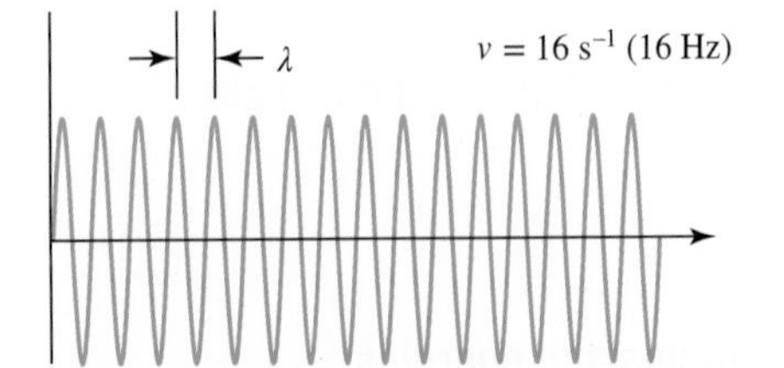

FIGURE 4.1 A light wave can be described by its wavelength or frequency. As wavelength increases, frequency decreases, and vice versa.

Wave-like properties of light

Light has wave-like properties. A wave is a regular oscillation in some particular property, such as the up-and-down variation in position of water waves. Water waves vary with time. A surfer waiting for a 'big one' bobs up and down as 'small ones' pass by. Light waves vary with time too, in a more regular manner than water waves. This variation is characterised by the wave's **frequency** (ν), which is the number of wave crests passing a point in space in 1 second (so the unit is s^{-1}, also designated hertz or Hz). Water waves also vary in height as you move away from the beach; that is, the height of the wave varies with position. Light waves vary in space in a manner illustrated in figure 4.1. This variation in space is characterised by the **wavelength** (λ), which is the distance between successive wave crests. Wavelengths are measured in units of length, such as metres or nanometres. As we will see below, frequency and wavelength are not independent variables, but are inversely proportional to each other.

Amplitude is the maximum displacement of a wave from its centre. The amplitude of a light wave determines the **intensity** of the light. As figure 4.2 shows, a bright light is more intense than a dim one as its waves have higher amplitudes. It is important to note that the intensity of light is proportional to the square of its amplitude. The amplitude itself has no physical meaning because at any moment in time the amplitude of the wave can be negative or positive, while the square is always positive and equivalent to photon density.

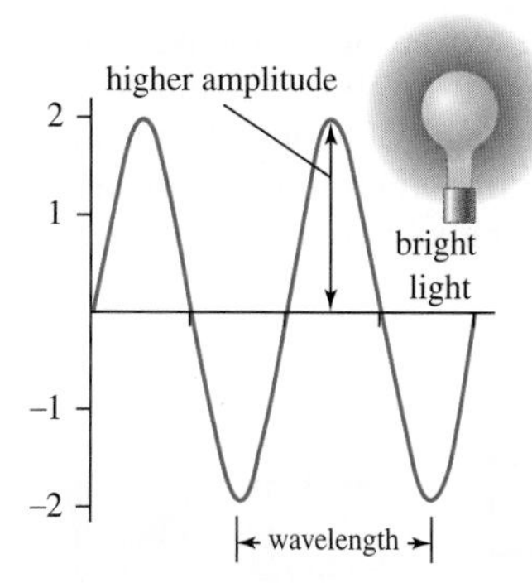

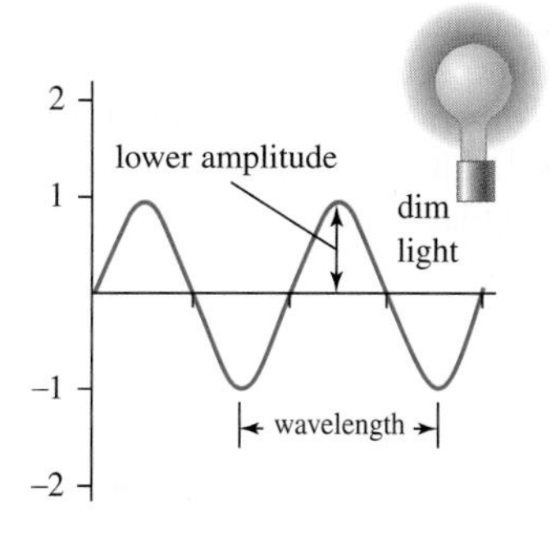

FIGURE 4.2 The amplitude of a light wave determines the intensity of the light. A bright light is more intense, that is, has a higher amplitude than a dim light.

Waves can also be described in terms of their phase. **Phase** refers to the starting position of a wave with respect to one wavelength. We can see that waves in figures 4.3a and 4.3b have the same amplitude and wavelengths, but different starting points. We say that the two waves have different phases. We can see this more clearly in figure 4.3c which shows the two waves superimposed. If they had the same phase (and the same amplitude and wavelength), they would line up perfectly. As we will see on p. 196, when waves interact, the outcome is influenced, among other things, by their relative phases.

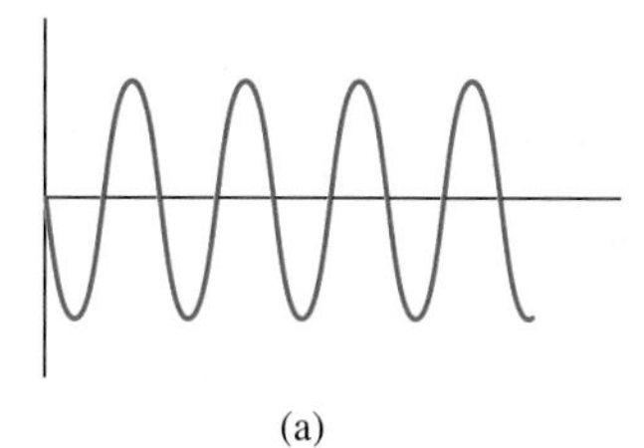
(a)

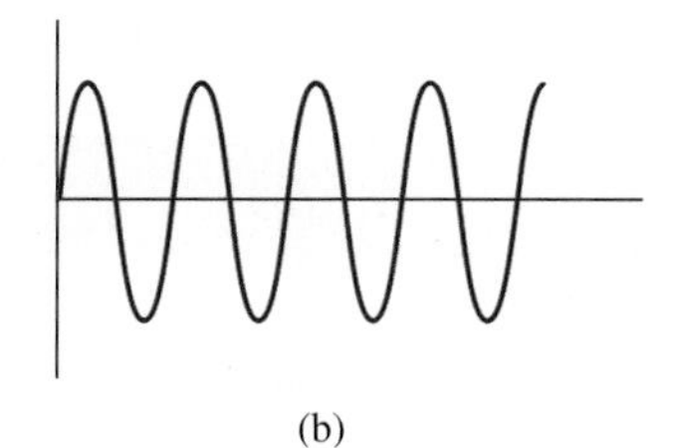
(b)

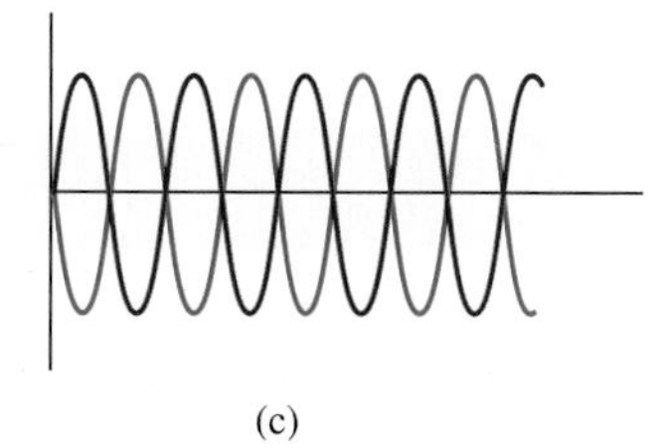
(c)

FIGURE 4.3 Phase refers to the starting position of a wave with respect to one wavelength. The waves in **(a)** and **(b)** have different phases; **(c)** shows an overlap of (a) and (b).

Light waves, and all other types of electromagnetic radiation, always move through a vacuum at the same speed. The speed of light is a fundamental constant, denoted by the symbol c: $c = 2.997\,924\,58 \times 10^8\,\text{m s}^{-1}$ (we will round it to $2.998 \times 10^8\,\text{m s}^{-1}$). For any wave, its frequency (ν) (in units of s^{-1}) multiplied by its wavelength (λ) (in units of m) equals its speed c (m s^{-1}), that is:

$$c = \nu\lambda$$

WORKED EXAMPLE 4.1

Wavelength–frequency conversion

An FM radio station transmits its signal at 88.1 MHz. What is the wavelength of the signal?

Analysis

The link between wavelength (λ) and frequency (ν) is given by $c = \nu\lambda$.

Solution

First, we summarise the data.

$$c = 2.998 \times 10^8\,\text{m s}^{-1} \qquad \nu = 88.1\,\text{MHz}$$

Next, we rearrange the equation to solve for wavelength.

$$c = \nu\lambda \quad \text{so} \quad \lambda = \frac{c}{\nu}$$

To obtain equivalent units, convert the frequency units from MHz to Hz. The prefix 'M' stands for 'mega', which is a factor of 10^6. Remember that Hz is equivalent to s^{-1}.

$$\lambda = \frac{2.998 \times 10^8\,\text{m s}^{-1}}{88.1 \times 10^6\,\text{s}^{-1}} = 3.40\,\text{m}$$

Is our answer reasonable?

The wavelength of 3.40 m may seem rather long, but radio waves are known as long wavelength radiation. See figure 4.4 for a sense of the wavelengths of electromagnetic radiation.

PRACTICE EXERCISE 4.1

What is the frequency of electromagnetic radiation that has a wavelength of 1.40 cm?

The wavelengths or frequencies of electromagnetic radiation cover an immense range. Figure 4.4 shows that the visible spectrum covers the wavelength range from about 380 nm (violet) to 780 nm (red). The centre of this range is yellow light, with a wavelength of about 580 nm and a frequency around $5.2 \times 10^{14}\,s^{-1}$. Although visible light is extremely important to living creatures for vision, the whole, electromagnetic spectrum from gamma rays to radio waves also has diverse effects on, and uses in, our lives.

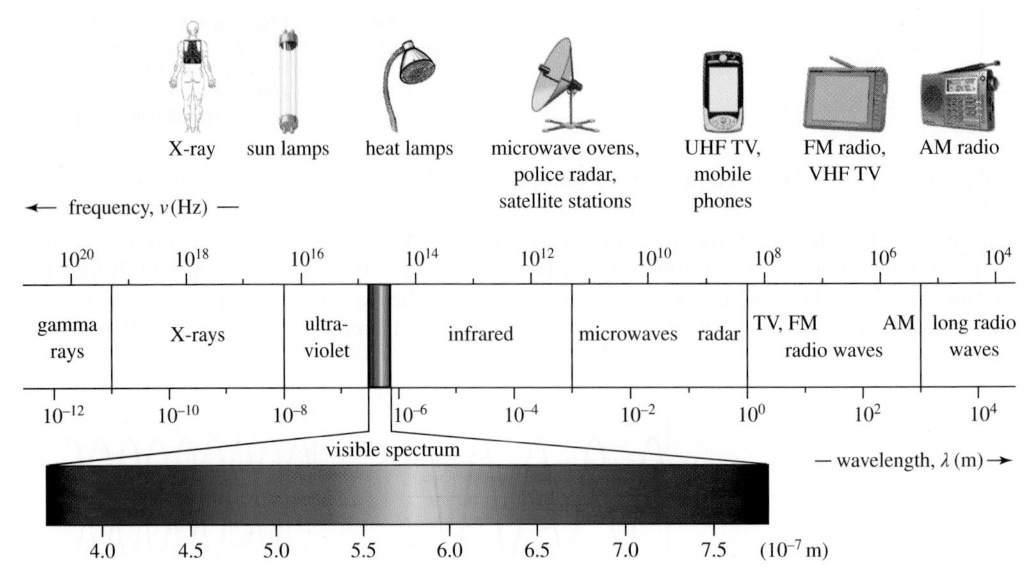

FIGURE 4.4 The electromagnetic spectrum, showing its various regions and the associated wavelengths and frequencies.

Radiation with short wavelengths, in the X-ray and gamma ray regions, can generate ions by removing electrons from atoms and molecules. These ions are highly reactive and can cause serious damage to the material that absorbs the radiation. However, under closely controlled conditions, X-rays are used in medical imaging, and gamma rays are used to treat cancer. The wavelength of ultraviolet (UV) radiation lies between that of X-rays and visible light. Ultraviolet radiation can also damage materials, especially in high doses. High rates of skin cancer in Australia and New Zealand are a direct result of exposure to damaging UV radiation in sunlight.

Radiation with long wavelengths falls in the infrared, microwave or radio frequency regions. Heat lamps make use of infrared radiation, microwave ovens cook with microwave radiation, and radio and television signals are transmitted by radio waves.

What we perceive as white light actually contains a range of wavelengths. This becomes apparent when white light passes through a prism (see p. 118) or through raindrops (which creates a rainbow as shown in figure 4.5). These objects diffract different wavelengths of light through different angles, so the light that passes through spreads out in space, with each wavelength appearing at its own characteristic angle.

FIGURE 4.5 Light rays bend as they pass through raindrops, causing white light to separate into its rainbow of colours.

Particle properties of light

Light carries energy. When our bodies absorb sunlight, for example, we feel warm because energy from the sunlight has been transferred to our skin. A phenomenon known as the **photoelectric effect** shows how the energy of light depends on its frequency and intensity. The photoelectric effect is the basis for many light-sensing devices, such as automatic door openers and camera exposure meters. Figure 4.6 illustrates a photoelectric experiment in which a beam of light strikes the surface of a metal. Under the right conditions, the light causes electrons to be ejected from the metal surface. A detailed study of the photoelectric effect reveals how the behaviour of these electrons is related to the characteristics of the light.

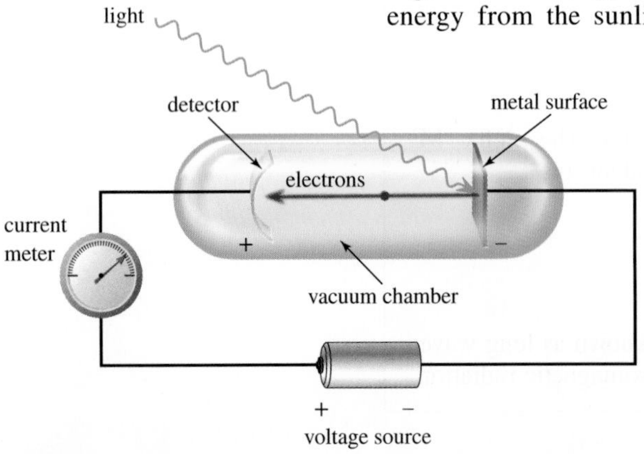

1. Below a characteristic threshold frequency, ν_0, no electrons are observed, regardless of the light's intensity.
2. Above the threshold frequency, the maximum kinetic energy of ejected electrons increases linearly with the frequency of the light, as shown in figure 4.7. (Kinetic energy is a function of the electron's speed. We will learn more about kinetic energy in chapter 6.)

FIGURE 4.6 A diagrammatic view of the photoelectric effect. When light of a high enough frequency strikes a metal, electrons are ejected from the surface.

3. Above the threshold frequency, the *number* of emitted electrons increases with the light's intensity, but the *kinetic energy* per electron does not depend on the light's intensity.
4. All metals exhibit the same behaviour, but, as figure 4.7 indicates, each metal has a different threshold frequency.

In 1905, Albert Einstein postulated that light comes in 'packets' or 'bundles', called **photons**. Each photon has an energy that is directly proportional to its frequency.

$$E_{\text{photon}} = h\nu_{\text{photon}}$$

In this equation, E is the energy of a photon of light and ν is its frequency. The proportionality constant between energy and frequency is known as **Planck's constant (h)** and has a value of $6.626\,069\,57(29) \times 10^{-34}$ J s (we will use 6.626×10^{-34} J s). The 29 in parentheses refers to the uncertainty in the final digits; therefore, $h = (6.626\,069\,57 \pm 0.000\,000\,29) \times 10^{-34}$ J s.

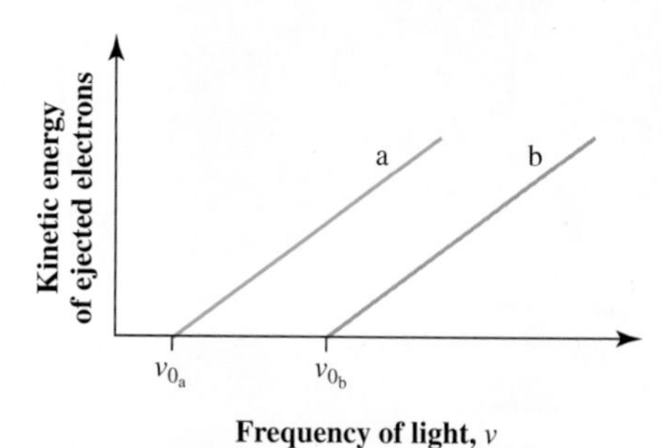

FIGURE 4.7 Variation in the maximum kinetic energy of electrons ejected from two different metal surfaces (a and b) as a function of the frequency of incoming light.

WORKED EXAMPLE 4.2

The energy of light

What is the energy of a photon of red light with a wavelength of 655 nm?

Analysis

This conversion problem requires two steps. We do not have an equation that will calculate E based on the data we have, but, by combining the two equations we do know, we can relate the energy of a photon to its frequency and wavelength.

Solution

We summarise the data.

$$h = 6.626 \times 10^{-34}\ \text{J s} \qquad \lambda = 655\ \text{nm} \qquad c = 2.998 \times 10^{8}\ \text{m s}^{-1}$$

We combine the equations into an equation that relates energy to wavelength.

$$E = h\nu \quad \text{and} \quad \nu = \frac{c}{\lambda} \quad \text{so} \quad E = \frac{hc}{\lambda}$$

Substituting our data, we find:

$$E_{\text{photon}} = \frac{(6.626\times10^{-34}\ \text{J s})\,(2.998\times10^{8}\ \text{m s}^{-1})}{655\times10^{-9}\ \text{m}}$$

$$= \frac{(6.626\times10^{-34}\ \text{J}\cancel{\text{s}})\,(2.998\times10^{8}\ \cancel{\text{m}}\ \cancel{\text{s}^{-1}})}{655\times10^{-9}\ \cancel{\text{m}}} = 3.03\times10^{-19}\ \text{J}$$

Is our answer reasonable?

An energy of 10^{-19} J seems very small, but we must remember that this is the energy of a single photon. If we want to calculate the energy of a mole of these photons, we need to multiply our result by the Avogadro constant, that is, $E_{\text{one mole of photons}} = E_{\text{photon}} \times N_A = (3.03 \times 10^{-19}\ \text{J}) \times (6.022 \times 10^{23}\ \text{mol}^{-1}) = 182 \times 10^{3}\ \text{J mol}^{-1} = 182\ \text{kJ mol}^{-1}$. This energy is comparable to the energies of chemical bonds.

PRACTICE EXERCISE 4.2

What is the energy of a photon with a wavelength of 254 nm (UV radiation)?

Einstein concluded that the energy of a photon that has the threshold frequency (ν_0) corresponds to the binding energy of the electron. Energy beyond the threshold frequency increases the electron's kinetic energy as shown in figure 4.8. This can be described by:

$$\text{electron kinetic energy} = \text{photon energy} - \text{binding energy}$$

$$E_{\text{kinetic(electron)}} = h\nu - h\nu_0$$

Einstein's explanation accounts for the observed properties of the photoelectric effect. When the energy of the photon is less than $h\nu_0$ (low-frequency light), no electrons can escape from the metal surface, no matter how intense the light. When the energy of the photon is greater than $h\nu_0$, an electron is ejected and any excess energy is transferred to that electron as kinetic energy. The intensity of a light beam is a measure of the number of photons per unit time; light of a higher amplitude carries more photons than light of lower amplitude. The intensity of the light *does not* determine the amount of energy per photon. More photons striking the metal result in more electrons being ejected, but the energy of each photon and each electron is unchanged. Finally, the fact that each metal has its own characteristic threshold frequency suggests that electrons are bound more tightly to some metals than to others.

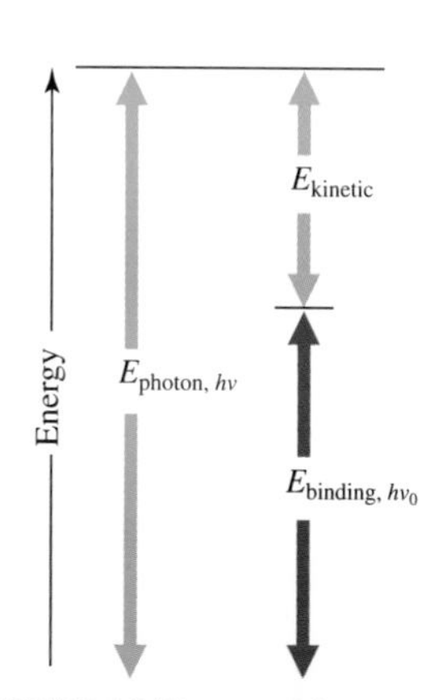

FIGURE 4.8 Diagram of the energy balance for the photoelectric effect.

Chemical Connections

The Australian Synchrotron

Electromagnetic radiation in its many forms is frequently used to probe the nature of chemical compounds, whether they are employed as materials for electronic devices, newly developed drugs or anything in between. The intensity, coherence and wavelength of the available radiation strongly influence the detailed information that can be extracted and which is needed to advance chemistry in all its forms. Important properties of electromagnetic radiation can be maximised using a synchrotron source. The Australian Synchrotron in Melbourne is one of about 50 of these facilities in the world.

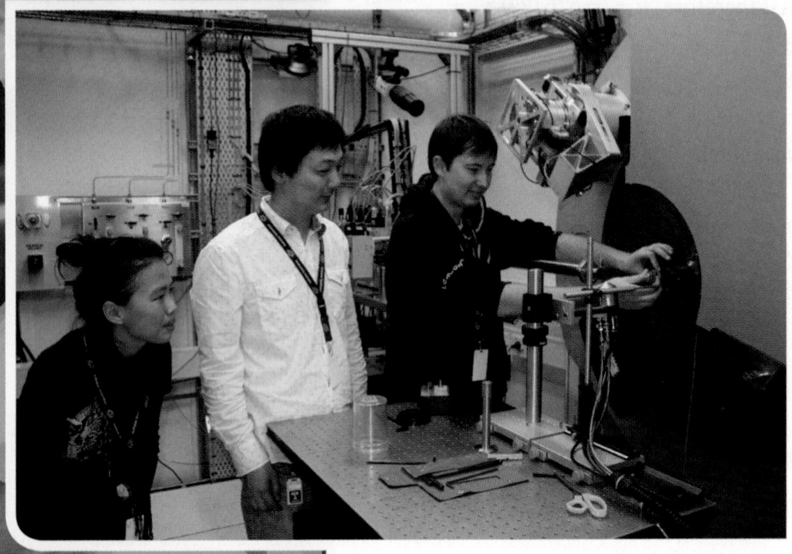

FIGURE 4.9 Beamline scientist (middle) helping postgraduate students using the Australian Synchrotron facilities to assist their studies.

Figure 4.9 shows a beamline scientist and two students from the University of Sydney aligning a sample on the powder diffraction beamline at the Australian Synchrotron. So far about 4200 scientists, mainly from Australia and New Zealand, have used beamlines at the Australian Synchrotron for their research. Many of these users are students working on their Honours or PhD theses, and one day you might use the facility yourself.

Australian scientists are able to use synchrotron light radiation — x-rays and infrared — to understand in precise detail, the structure of materials or the workings of whole systems, and then apply that valuable information to their work. For example, researchers have: 'seen' the structure of milk fat when it's digested and how it might be used for drug delivery; discovered the trigger for coeliac disease to help develop a vaccine; observed how some native plants immobilise heavy metals and how they might remediate contaminated soils; better understood complex viruses and disease to develop more effective and affordable drugs; and determined under which conditions fat is absorbed into potatoes, to create a low-fat chip.

Synchrotron light is also used in archeology and cultural studies in: determining the source of ochre in rock painting in the Northern Territory; digitally reproducing hidden paintings by famous artists such as Australian Arthur Streeton, Picasso and Degas; or even discovering what a Victorian dinosaur had for its last meal without breaking bedrock.

In order to understand synchrotron science we need to discuss the origin of synchrotron radiation. When electrons are forced to travel in a circular orbit at speeds approaching the speed of light, they emit radiation in the form of photons of various wavelengths, ranging from infrared to hard X-rays (wavelengths of 10^{-6} to 10^{-11} m, respectively, see figure 4.4). This radiation is called synchrotron radiation and is characterised by its high intensity (figure 4.10), making it very useful for studying the absorption, reflection and diffraction properties of matter.

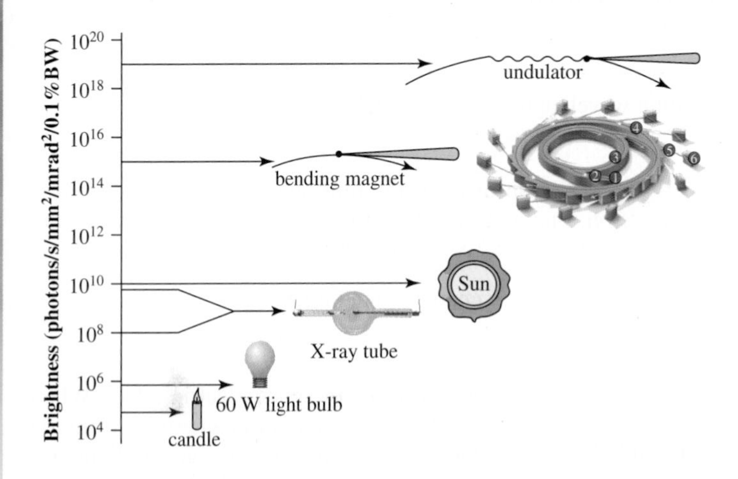

FIGURE 4.10 Comparison of brightness (intensity) of radiation from a range of sources.

Figure 4.11 shows a schematic of the Australian Synchrotron. Electrons are generated by a central electron gun and are accelerated to almost the speed of light by the linear accelerator (linac) and the booster ring. They are then transferred to the outer storage ring. The electrons are confined to the circular orbit by a series of powerful bending magnets. As the electrons are deflected through the magnetic field created by the bending magnets, they give off electromagnetic radiation, so that a beam of synchrotron radiation is produced at each bending magnet.

These intense beams of white radiation (i.e. with a spectrum of wavelengths from infrared to hard X-rays) can be captured and directed to a beamline end station where a specific wavelength appropriate for a particular technique or specific experiment can be selected.

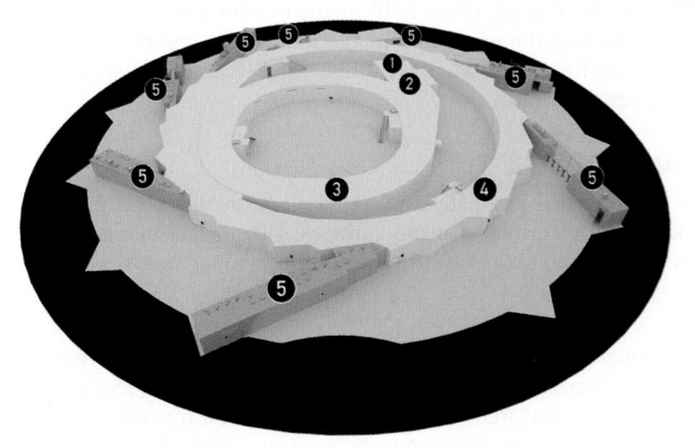

1. electron gun
2. linear accelerator (Linac)
3. booster ring
4. storage ring
5. beamline

Source: Illustrator Michael Payne. Image supplied courtesy of Australian Synchrotron.

FIGURE 4.11 Schematic of the Australian Synchrotron.

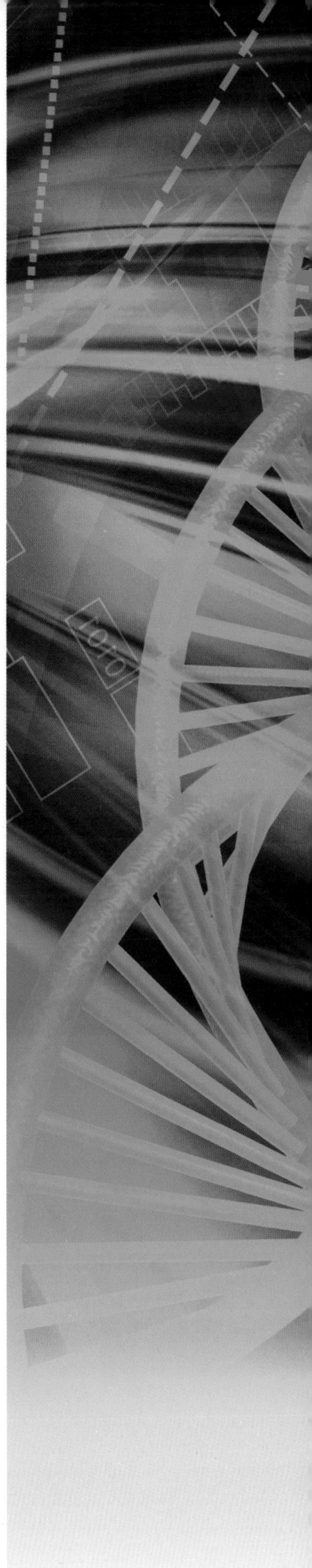

At this stage the Australian Synchrotron has nine beamlines but could operate more than 30. Each of these beamlines is dedicated to the specific purposes outlined in the table below.

Beamline	Capabilities
High-throughput protein crystallography	Dedicated facility for crystallography of large protein crystals, set up with robotic loading and centring, and for remote operation
Protein microcrystal and small molecule diffraction	Facility with finely focused X-ray beam for determining the crystal structure and electron density maps of small, hard-to-crystallise proteins, nucleic acids, and for small molecules
Powder diffraction	Versatile, high-resolution powder diffraction facility equipped with sample chambers for a wide range of in-situ experiments
Small and wide angle X-ray scattering	Measurement of long-range order in complex molecules and materials
X-ray absorption spectroscopy	Measurement of short and medium range order, bond lengths, coordination numbers and local coordination geometry, and the oxidation state of atoms from atomic number $Z = 20$ upwards
Soft X-ray spectroscopy	As for beamline 5, for the light elements, $Z < 20$; also for the analysis of surfaces and thin films
Infrared spectroscopy	Analysis of bonds in complex molecules, biological materials, minerals and band structures in certain semiconductors
Microspectroscopy	For production of high-resolution maps of elemental distribution in a sample; also for determination of the oxidation state and coordination geometry of atoms in particles down to submicron size
Imaging and medical therapy	Very flexible beamline for research into high-contrast imaging of objects from small animals through to engineering components; also for research into the physics and biophysics of cancer therapy techniques

The intensity of the radiation can be increased by the use of 'insertion devices' in the straight sections of the ring. There are two classes of insertion devices, multipole wigglers and undulators. Current 'third-generation' designs of synchrotrons aim to optimise the intensity that can be obtained from insertion devices. In particular, attention is given to the size and positioning of the straight sections that accommodate the insertion devices. The Australian Synchrotron is an advanced third-generation design. It accommodates the three different types of light sources (bending magnets, multipole wigglers and undulators) to enable a wide range of advanced experiments and measurements to be carried out.

Einstein's explanation of the photoelectric effect showed that light has some properties of particles. A complete description of light includes both wave-like and particle-like properties. When light interacts with a relatively large body such as a raindrop or a prism, its wave properties dominate the interaction. On the other hand, when light interacts with a small body such as an atom or an electron, particle properties dominate the interaction. Each view provides different information about the properties of light, and, when we think about light, we must think of *particle–waves* that combine both types of features.

Absorption and emission spectra

In the photoelectric effect, when photons strike a metal surface, the energy absorbed provides information about the binding energies of electrons at metal surfaces. When light interacts with free atoms, the interaction reveals information about electrons within individual atoms.

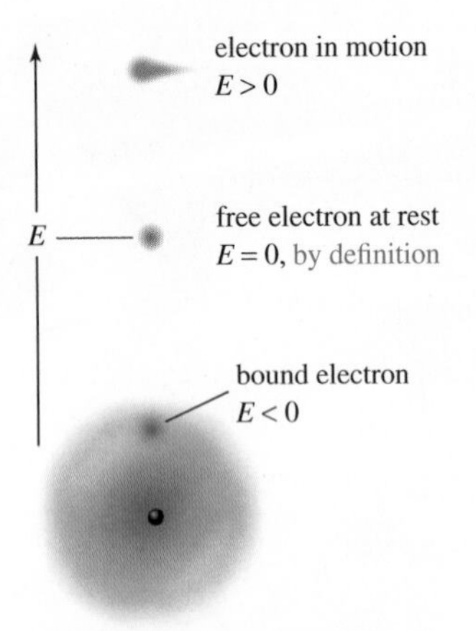

FIGURE 4.12 By convention, a free stationary electron has zero energy, and bound electrons have negative energies.

Light and atoms

Attractive electrostatic forces hold an electron within an atom. Energy must be supplied to remove the electron from the atom. The lower the atom's energy state, the more energy is required to remove its electron. The energy required is measured relative to the energy of a free electron. We define the energy of a hypothetical free stationary electron to be zero. As figure 4.12 indicates, the kinetic energy of a freely moving electron is positive relative to this conventional zero point. In contrast, bound electrons are lower in energy than the hypothetical free stationary electron, so we assign them negative energy values.

The absorption of photons by free atoms has two possible results, depending on the energy of the photon. When an atom absorbs a photon of sufficiently high energy, an electron is ejected (i.e. the atom ionises). This process is described later in this chapter. Here, we focus on the second type of result, in which the atom gains energy, but does not ionise. Instead, the atom moves from the lowest energy or **ground state** to a higher energy state called an **excited state**. As we saw in chapter 1, this process is called an electronic transition. Excited states are not stable and atoms in excited states subsequently give up their excess energy to return to lower energy states, and eventually the ground state. Atoms lose their excess energy in collisions with other atoms or by emitting photons.

The key feature in the exchange of energy between atoms and light (photons) is that energy is conserved. This requires that the change in energy of the atom exactly equals the energy of the photon.

$$\Delta E_{\text{atom}} = \pm h\nu_{\text{photon}}$$

When an atom *absorbs* a photon, the atom gains the photon's energy, so ΔE_{atom} is positive. When an atom *emits* a photon, the atom loses the photon's energy, so ΔE_{atom} is negative. As an atom returns from an excited state to the ground state, it must lose exactly the amount of energy that it originally gained. However, excited atoms usually lose excess energy in several steps involving small energy changes, so the frequencies of emitted photons are often lower than those of absorbed photons.

WORKED EXAMPLE 4.3

Emission energies

A sodium-vapour street lamp emits yellow light (figure 4.13) at a wavelength of 589 nm. What is the energy change for a sodium atom involved in this emission? How much energy is released per mole of sodium atoms?

Analysis

This problem relates energies of photons to energy changes of atoms. The solution requires a conversion involving wavelength and energy.

Solution

Sodium atoms lose energy by emitting photons of light ($\lambda = 589$ nm). We can relate the wavelength of one of these photons to its energy.

$$E_{\text{photon}} = h\nu = \frac{hc}{\lambda}$$

Energy is conserved, so the energy of the emitted photon must exactly equal the energy lost by the atom.

$$\Delta E_{\text{atom}} = -E_{\text{photon}} = -\frac{hc}{\lambda}$$

The equation includes a negative sign because the atom *loses* energy.

FIGURE 4.13 Sodium-vapour lamps emit yellow light.

To calculate the energy of a photon, we need the following data.

$h = 6.626 \times 10^{-34}\,\text{J s} \qquad c = 2.998 \times 10^{8}\,\text{m s}^{-1} \qquad \lambda = 589\,\text{nm} = 589 \times 10^{-9}\,\text{m}$

$$\Delta E_{\text{atom}} = -E_{\text{photon}} = -\frac{hc}{\lambda} = -\frac{(6.626\times10^{-34}\,\text{J s})(2.998\times10^{8}\,\text{m s}^{-1})}{589\times10^{-9}\,\text{m}}$$

$$= -3.37 \times 10^{-19}\,\text{J}$$

This calculation gives the energy change for one sodium atom emitting one photon. We multiply this value by the Avogadro constant to convert from energy per atom to energy per mole.

$$\Delta E_{\text{mol}} = (\Delta E_{\text{atom}})(N_A) = (-3.37 \times 10^{-19}\,\text{J})(6.022 \times 10^{23}\,\text{mol}^{-1})$$

$$= -2.03 \times 10^{5}\,\text{J mol}^{-1} = -203\,\text{kJ mol}^{-1}$$

Remembering that the negative sign simply means that the atom is losing energy, the actual energy released is the absolute value of this, that is, 203 kJ mol^{-1}.

PRACTICE EXERCISE 4.3

Mercury lamps emit photons with a wavelength of 436 nm. Calculate the energy change of the mercury atoms in J atom^{-1} and kJ mol^{-1}.

Atomic spectra

When a light beam passes through a tube containing a gas consisting of atoms of a single element, the atoms absorb specific frequencies of light characteristic of that element. As a result, the beam emerging from the sample tube has fewer photons at these specific frequencies. The frequencies with diminished intensity in the visible portion of the spectrum can be detected by passing the emerging light through a prism. The prism diffracts the light, with different frequencies diffracting through different angles. After leaving the prism, the beam strikes a screen, where the frequencies with diminished intensity appear as gaps or dark bands called 'lines', which are images of the slit. These are the frequencies of light absorbed by the atoms in the sample tube. The resulting pattern, shown schematically in figure 4.14, is called an **absorption spectrum**.

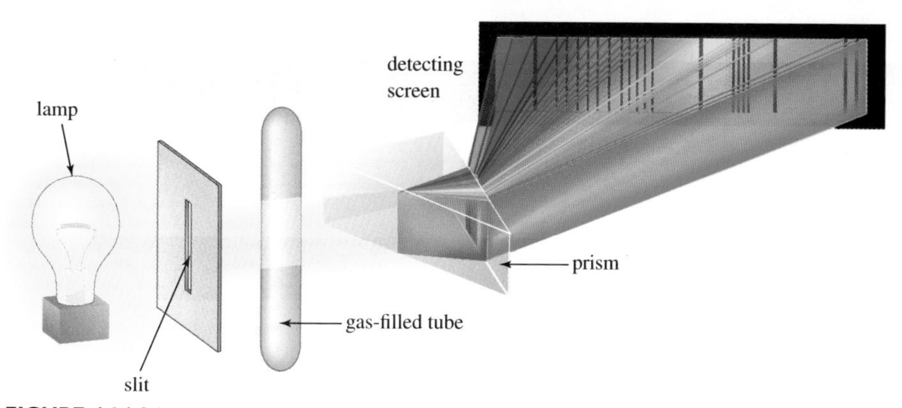

FIGURE 4.14 Schematic representation of an apparatus that measures the absorption spectrum of a monatomic gas. The gas in the tube absorbs light at specific wavelengths, called lines, so the intensity of transmitted light is low and, therefore, dark lines appear at these particular wavelengths.

An absorption spectrum is used to determine the frequencies of the photons that an atom *absorbs*. A similar experiment can measure the energies of the photons *emitted* by atoms in excited states. Figure 4.15 (overleaf) outlines the features of an apparatus that measures these emitted photons. An electrical discharge excites a collection of atoms from their ground state into higher energy states. These excited atoms then lose all, or part of, their excess energy by emitting photons. This emitted light can be analysed by passing it through a prism to give an **emission spectrum**. This is a plot of the intensity of light emitted as a function of frequency. The emission spectrum for atomic hydrogen, shown in figure 4.18, shows several sharp emission lines of high intensity. The frequencies of these lines correspond to photons emitted by the hydrogen atoms as they return to their ground state.

Each element has unique absorption and emission spectra. That is, each element has a set of characteristic frequencies of light that it can absorb or emit. It is important to note that figures 4.14, 4.15, 4.16 and 4.18 show only the visible portions of the absorption and emission spectra. Electronic transitions also take place in regions of the electromagnetic spectrum that the human eye cannot detect. Instruments allow scientists to observe electromagnetic radiation in these regions.

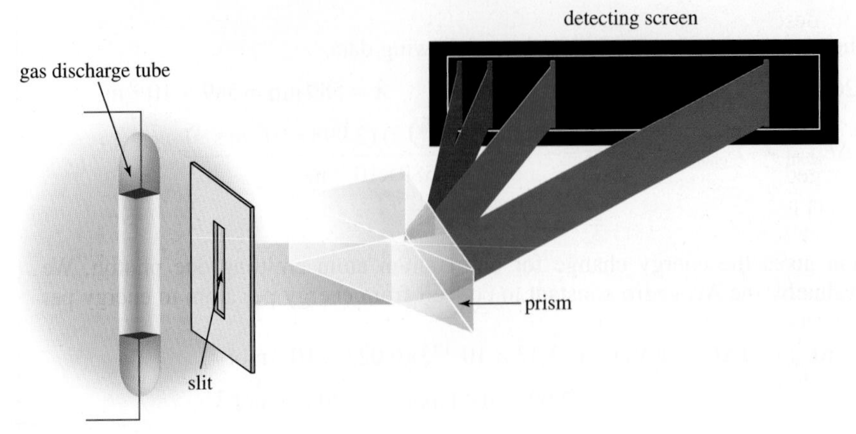

FIGURE 4.15 Schematic representation of an apparatus that measures the emission spectrum of a gaseous element. Emission lines appear bright against a dark background. The spectrum shown is the emission spectrum for hydrogen atoms.

Each frequency absorbed or emitted by an atom corresponds to a particular energy change for the atom. These characteristic patterns of energy gains and losses provide information about atomic structure. Figure 4.16 shows the emission spectra for several selected atoms.

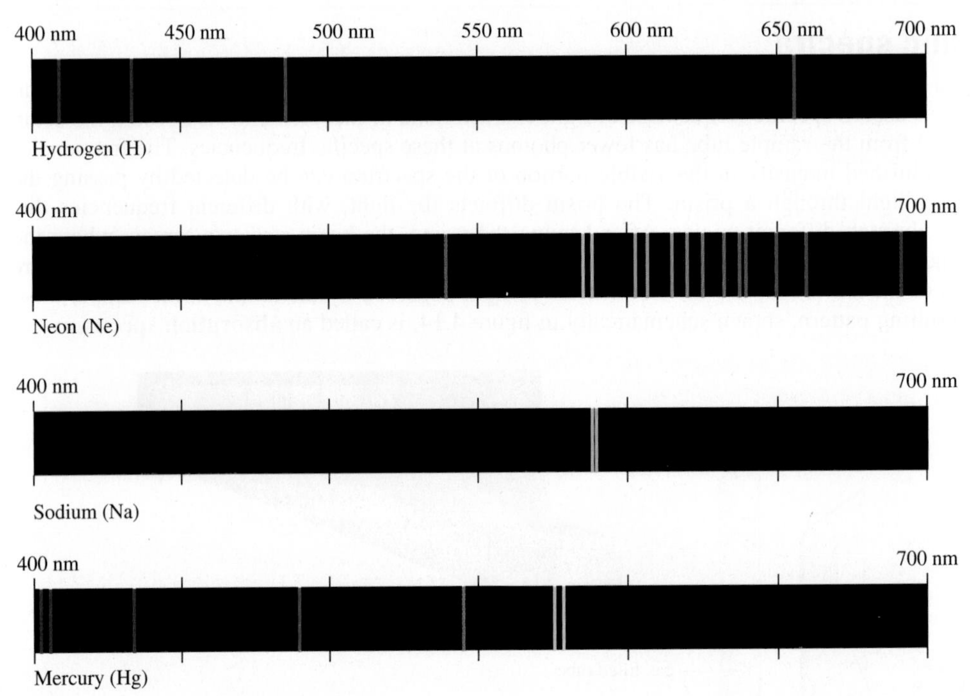

FIGURE 4.16 Emission patterns for hydrogen, neon, sodium and mercury. Each element has a unique emission pattern that provides valuable clues about its atomic structure.

Quantisation of energy

When an atom absorbs a photon of light of frequency ν, the light beam loses energy equal to $h\nu$, and the atom gains that amount of energy. What happens to the energy that the atom gains? A clue is that, when the frequency of the incoming (or absorbed) light is high enough, it produces cations and free electrons. In other words, a photon with high enough energy can cause an atom to lose one of its electrons, that is, ionise. This implies that absorption of a photon results in an energy gain for an electron in the atom. Stated as an equation:

$$\Delta E_{\text{atom}} = \Delta E_{\text{electron}} = h\nu$$

The atomic spectra of most elements are complex and show little apparent regularity. However, in 1885, the Swiss mathematician and physicist Johann Balmer was able to propose a single equation that could describe the emission spectrum of the hydrogen atom. The equation was:

$$\nu_{\text{emission}} = (3.29 \times 10^{15}\ \text{s}^{-1})\left(\frac{1}{n_1^2} - \frac{1}{n_2^2}\right)$$

in which n_1 and n_2 are integers (1, 2, 3 etc.). In 1913, the Danish physicist Niels Bohr used Einstein's postulate ($E = h\nu$) to interpret Balmer's observations. Combining the two equations,

Bohr was able to describe the energy levels of the electron in the hydrogen atom with the following equation, where n is a positive integer.

$$E_n = -\frac{2.18 \times 10^{-18}\ \text{J}}{n^2}$$

As previously stated, the negative sign in the equation reflects the fact that bound electrons are lower in energy than a stationary free electron (with its energy defined as zero).

Bohr realised that the emission frequencies have specific values because the electron in a hydrogen atom is restricted to specific energies. Bohr's idea of restricted energy levels was revolutionary because scientists at that time thought that the electron in a hydrogen atom could have any energy. In contrast, Bohr interpreted the hydrogen emission spectrum to mean that electrons bound to atoms can have only certain specific energy values. He was awarded the Nobel Prize in physics in 1922 for this interpretation. A property that cannot change continuously, and is restricted to specific values, is said to be **quantised**. The atomic energy levels of hydrogen (and other elements) are quantised. Each integer value of n describes one of the allowed energy levels of the hydrogen atom. For example, the energy of an electron in hydrogen's fourth level is:

$$E_4 = -\frac{2.18 \times 10^{-18}\ \text{J}}{4^2} = -1.36 \times 10^{-19}\ \text{J}$$

When an electron changes energy levels, the change is an electronic transition between quantised levels. When a hydrogen atom absorbs or emits a photon, its electron moves from one energy level to another. Thus, the change in energy (ΔE) of the atom for this process is the energy difference between the two levels.

$$\Delta E_{\text{atom}} = E_{\text{final}} - E_{\text{initial}}$$

Photons always have positive energies, but energy changes (ΔE) can be positive or negative. When a photon is absorbed, an electron is promoted to a higher energy level, the atom gains energy and ΔE for the atom is positive.

$$E_{\text{absorbed photon}} = \Delta E_{\text{atom}}$$

When a photon is emitted, an electron falls from a higher energy level to a lower one, the atom loses energy and ΔE for the atom is negative.

$$E_{\text{emitted photon}} = -\Delta E_{\text{atom}}$$

We can combine these two equations by using absolute values.

$$E_{\text{photon}} = \left|\Delta E_{\text{atom}}\right|$$

WORKED EXAMPLE 4.4

Energy levels of the hydrogen atom

What is the energy change when the electron in a hydrogen atom undergoes a transition from the fourth energy level to the second energy level? What is the wavelength of the photon emitted?

Analysis

The question asks about energy and the wavelength of a photon emitted by a hydrogen atom. Wavelength is related to energy, and photon energy is determined by the difference in energy between the two levels involved in the transition. In this case, the electron moves from the fourth to the second energy level. The energy difference between these two states is given by:

$$\Delta E_{\text{atom}} = E_{\text{final}} - E_{\text{initial}} = E_2 - E_4$$

Solution

As shown above, $E_4 = -1.36 \times 10^{-19}$ J. A similar calculation gives the energy of the second level.

$$E_2 = -\frac{2.18 \times 10^{-18}\ \text{J}}{2^2} = -5.45 \times 10^{-19}\ \text{J}$$

This energy change is negative because the atom loses energy. The energy difference between the two levels is:

$$\begin{aligned}\Delta E_{\text{atom}} = E_{\text{final}} - E_{\text{initial}} &= (-5.45 \times 10^{-19}\ \text{J}) - (-1.36 \times 10^{-19}\ \text{J}) \\ &= -4.09 \times 10^{-19}\ \text{J}\end{aligned}$$

The negative sign for the energy change is consistent with the atom losing energy. This lost energy appears as a photon with energy of:

$$E_{\text{photon}} = \left|\Delta E_{\text{atom}}\right| = \left|-4.09 \times 10^{-19}\ \text{J}\right| = 4.09 \times 10^{-19}\ \text{J}$$

To determine the wavelength of the photon, we use:

$$E_{\text{photon}} = h\nu = \frac{hc}{\lambda}$$

Solving for λ gives:

$$\lambda_{\text{photon}} = \frac{hc}{E_{\text{photon}}}$$

$$= \frac{(6.626 \times 10^{-34}\ \text{J s})\ (2.998 \times 10^{8}\ \text{m s}^{-1})}{4.09 \times 10^{-19}\ \text{J}} = 486 \times 10^{-9}\ \text{m} = 486\ \text{nm}$$

Is our answer reasonable?

The signs of the energies are consistent — the atom loses energy and the photon emitted has positive energy. From figure 4.4 (p. 122) we can see that light with a wavelength of 486 nm has a blue-green colour. Cross-checking against the emission spectrum of hydrogen (figure 4.16), we can see that it has such a blue-green line. Therefore, we can be confident that our answer is correct.

PRACTICE EXERCISE 4.4 What minimum energy must a photon have to excite a hydrogen atom from its ground state, that is, $n = 1$, to the $n = 4$ level? What is the corresponding wavelength?

Energy level diagrams

Electronic transitions in atoms can be represented using an **energy level diagram** such as the one shown in figure 4.17b. Energy level diagrams are a concise way to summarise information about atomic energies. They show energy along the vertical axis and each energy state of the atom is represented by a horizontal line.

We can begin to understand the quantum levels of an electron bound to an atom by picturing a ball on a staircase. This analogy is illustrated in figure 4.17. A ball may sit on any of the steps. To move a ball from the bottom of the staircase to step 5 requires the addition of a specific amount of energy, $\Delta E = E_5 - E_1$. If too little energy is supplied, the ball cannot reach this step. Conversely, if a ball moves down the staircase, it releases specific amounts of energy. If a ball moves from step 5 to step 3, it loses energy, $\Delta E = E_3 - E_5$. Although a ball may rest squarely on any step, it cannot be suspended between the steps. Electrons in atoms, like balls on steps, cannot exist 'between steps', but must occupy one of the specific, quantised energy levels. *Remember that this is an analogy; atomic energy levels are not at all similar to staircases except in being quantised.*

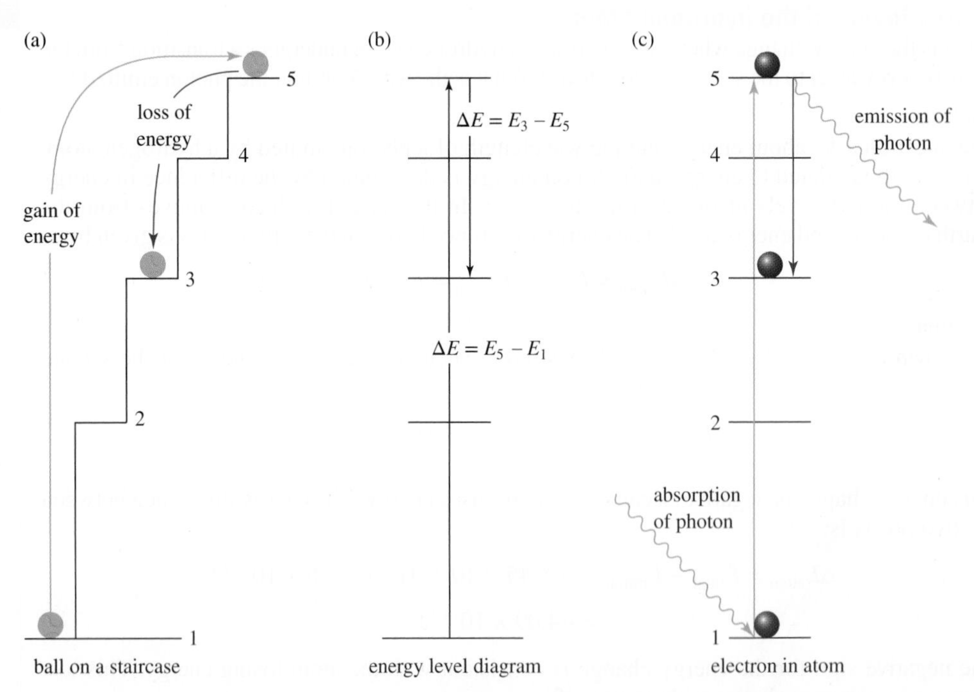

FIGURE 4.17 (a) A ball on a staircase shows some of the properties of quantised energy levels. **(b)** Quantised energy levels can be depicted using an energy level diagram. **(c)** Electronic transitions occur between quantised energy levels through either absorption or emission of photons.

A hydrogen atom has a regular progression of quantised energy levels. Figure 4.18 shows the energy level diagram for hydrogen atoms; arrows represent some of the possible absorption and emission transitions. Notice that the energies of absorption from the lowest energy level are identical to the energies of emission to the lowest energy level. This means that the wavelengths of light absorbed in these upward transitions are identical to the wavelengths of light emitted in the corresponding downward transitions.

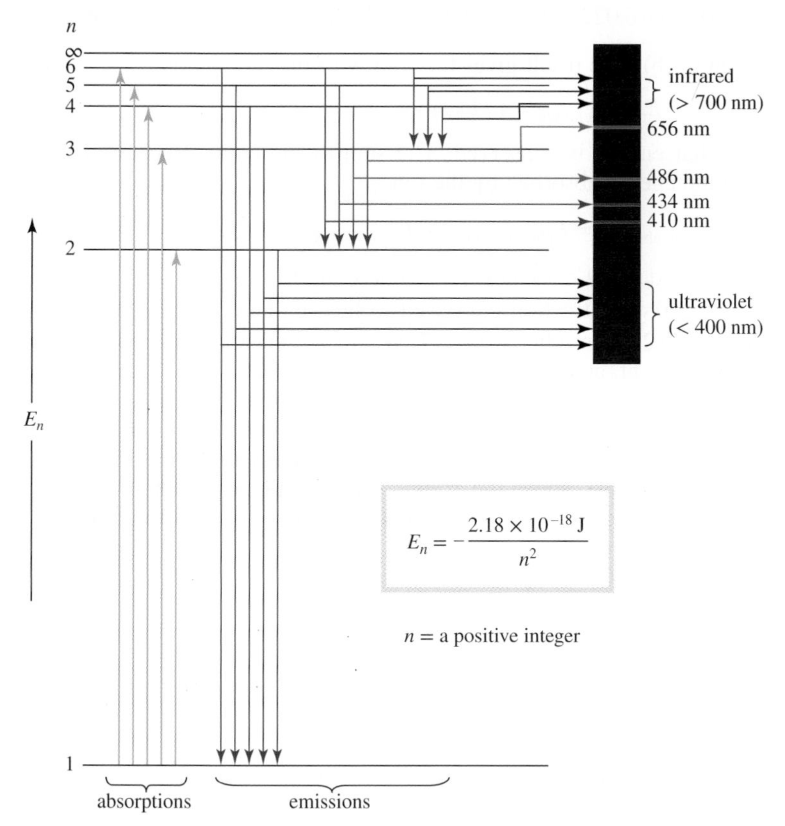

FIGURE 4.18 Energy levels for the hydrogen atom and some of the transitions that occur between levels, as well as the resulting emission spectrum. Upward arrows represent absorption transitions, and downward arrows represent emission transitions.

Elements other than hydrogen also have quantised energy levels, but as we will see later in this chapter they cannot be so simply described because they have more than one electron. In these cases, scientists use experimental values for observed absorption and emission lines to calculate the allowed energy levels for each different element.

WORKED EXAMPLE 4.5

Energy level diagrams

Ruby lasers use crystals of Al_2O_3 that contain small amounts of Cr^{3+} ions, which absorb light between 400 and 560 nm. The excited-state ions lose some energy as heat. After losing heat, the Cr^{3+} ions return to the ground state by emitting red light with a wavelength of 694 nm.

(a) Calculate the energy in kJ mol^{-1} of the 500 nm radiation used to excite the Cr^{3+} ions.
(b) Calculate the energy in kJ mol^{-1} of the emitted light.
(c) Calculate the fraction of the excitation energy emitted as red photons and the fraction lost as heat.
(d) Draw an energy level diagram that summarises these processes.

Analysis

This problem asks about energies and light. With multipart problems, the best strategy is to work through the parts one at a time. Parts (a) and (b) concern the link between light and energy as discussed earlier in this section. Once the transition energies have been determined, we can calculate the fraction of the excited-state energy lost as heat for part (c), and we can also draw an energy level diagram that shows how the levels are related for part (d).

Solution

We solve (a) using the following equations and the Avogadro constant.

$$E_{\text{photon}} = h\nu \quad \text{and} \quad c = \nu\lambda \quad \text{so} \quad E_{\text{photon}} = \frac{hc}{\lambda}$$

$$E_{\text{photon absorbed}} = \frac{(6.626\times10^{-34}\ \text{J s})\ (2.998\times10^{8}\ \text{m s}^{-1})}{500\times10^{-9}\ \text{m}} = 3.97\times10^{-19}\ \text{J}$$

This is the energy change per ion. To calculate the change per mole, we multiply by the Avogadro constant. We also need to convert from J to kJ.

$$E_{\text{photon absorbed}} = (3.97 \times 10^{-19}\,\text{J})(6.022 \times 10^{23}\,\text{mol}^{-1}) = 239 \times 10^{3}\,\text{J mol}^{-1} = 239\,\text{kJ mol}^{-1}$$

Apply the same approach to (b). The result gives the energy of the red photon.

$$E_{\text{photon emitted}} = 172\,\text{kJ mol}^{-1}$$

To solve (c), remember that energy is conserved. The sum of the emitted heat and the emitted photon must equal the energy absorbed by the ion.

$$E_{\text{photon absorbed}} = E_{\text{photon emitted}} + E_{\text{heat emitted}}$$

Because 239 kJ mol^{-1} is absorbed and 172 kJ mol^{-1} is emitted, the fraction of the excitation energy re-emitted is $\frac{172}{239} = 0.720$. The fraction converted to heat is the difference between this value and 1.000, which is 0.280. In other words, 72.0% of the energy absorbed by the chromium ion is emitted as red light, and the other 28.0% is lost as heat.

Part (d) asks for an energy level diagram for this process. The electron starts in the ground state. On absorption of a photon, the electron moves to an energy level that is higher by 239 kJ mol^{-1}. The chromium ion loses 28.0% of its excited-state energy as heat as the electron moves to a level that is 172 kJ mol^{-1} above the ground state. Finally, emission of the red photon returns the Cr^{3+} ion to the ground state. The numerical values allow us to construct an accurate diagram.

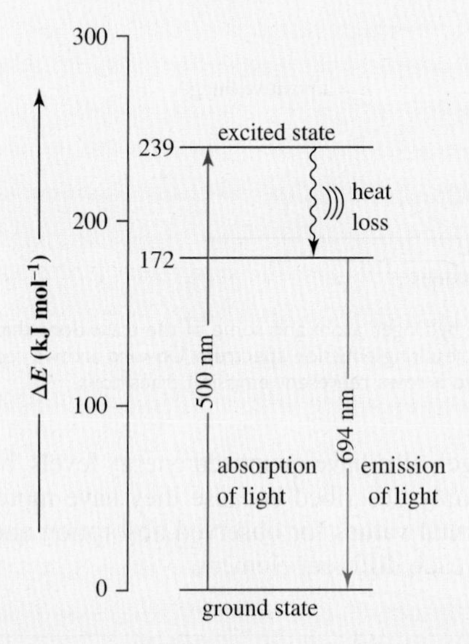

Is our answer reasonable?

When Cr^{3+} ions absorb light, they are promoted to an energy level higher than the energy that they later emit. If calculations had shown that the emitted light had a larger energy than the absorbed light, the result would have been unreasonable because the system would violate the law of conservation of energy.

PRACTICE EXERCISE 4.5 After excitation in an electric discharge, an atom of Hg returns to the ground state by emitting two photons with wavelengths of 436 nm and 254 nm. Calculate the excitation energy of the excited state in kJ mol^{-1}.

4.3 Properties of electrons

The energies of electrons in atoms play a central role in determining chemical behaviour. Several other properties of electrons also influence the physical and chemical characteristics of atoms and molecules. Some properties are characteristic of all electrons, but others arise only when electrons are bound to atoms or molecules. In this section, we describe the properties characteristic of all electrons.

Every electron has a mass of 9.109×10^{-31} kg and a charge of 1.602×10^{-19} C. Electrons have magnetic properties that arise from a property called spin, which we met briefly in chapter 1 and will describe in more detail in section 4.4.

The French physicist Louis de Broglie (1892–1987, Nobel Prize in physics, 1929) was the first to suggest that, like photons, electrons also display wave–particle duality, which means that they have both particle and wave properties.

Experiments had shown that a beam of light shining on an object exerts a pressure, which implies that a photon has momentum. Quantitative measurements of the pressure exerted by light showed that a simple equation relates the momentum of light (p) to its energy.

$$E = pc$$

As we have already described, light energy is also related to its wavelength.

$$E = h\nu = \frac{hc}{\lambda}$$

Combining these two energy expressions gives an expression relating p to λ.

$$pc = \frac{hc}{\lambda}$$

The speed of light cancels, leaving an equation that de Broglie suggested should apply to electrons and other particles as well as to photons.

$$p = \frac{h}{\lambda}$$

The momentum of a particle is the product of its mass and velocity (u), $p = mu$. Substituting this and solving for λ gives a form of the de Broglie equation that links the wavelength of a particle with its mass and velocity.

$$\lambda_{\text{particle}} = \frac{h}{mu}$$

De Broglie's theory predicted that electrons are wave-like. How might this be confirmed? Figure 4.19 shows examples of the characteristic intensity patterns displayed by waves. In figure 4.19a, water waves radiate away from two bobbing floats and form a standing pattern. In figure 4.19b, diffracted X-rays form a similar wave pattern. High-energy photons have passed through the regular array of atoms in a crystal, whose electron clouds scatter the photon waves. If electrons have wave properties, they should display regular wave patterns like these.

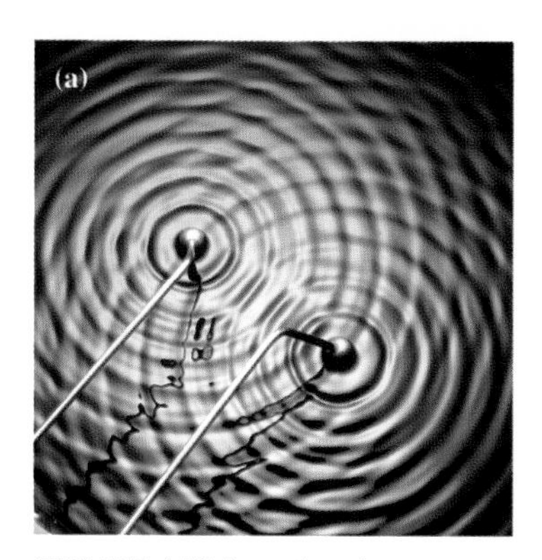

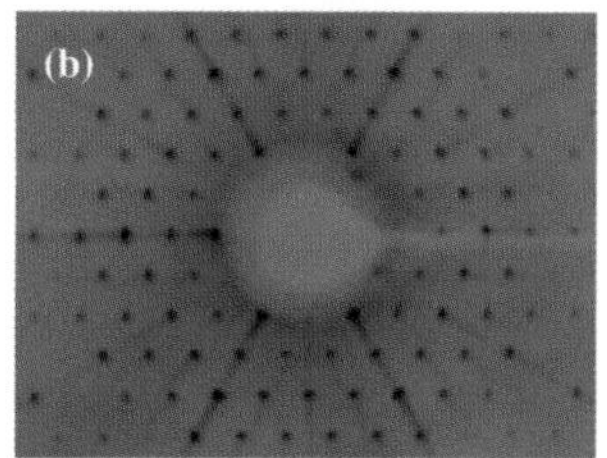

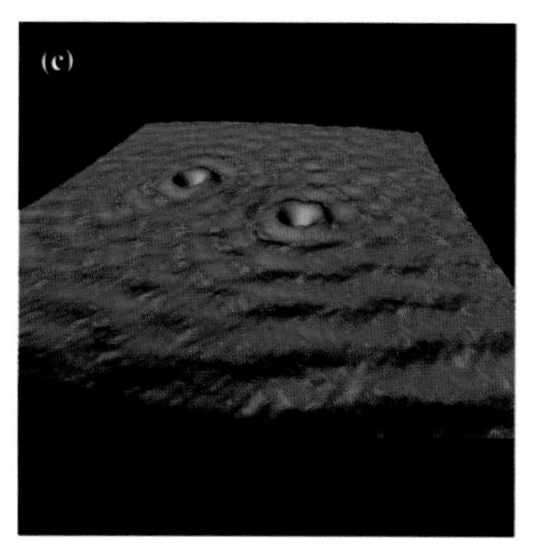

FIGURE 4.19 Examples of wave patterns: **(a)** Floats produce standing water waves. **(b)** X-rays generate wave interference patterns. **(c)** Protruding atoms on a metal surface generate standing electron waves.

In 1927, American physicists Clinton Davisson and Lester Germer, and British physicist George Thomson, separately carried out experiments in which they exposed metal films to electron beams with well-defined kinetic energies. Both experiments generated patterns like those shown in figure 4.19b, confirming the validity of the de Broglie equation for electron wavelengths. This established the wave nature of electrons. Davisson and Thomson were awarded the Nobel Prize in physics in 1937 for this discovery.

In recent years, scanning tunnelling electron microscopes have produced images of electron waves, an example of which appears in figure 4.19c. Here, two atoms on an otherwise smooth metal surface act like the floats in figure 4.19a, and cause the electrons in the metal to set up a standing wave pattern.

Both photons and electrons have wave and particle properties, but different equations describe their properties. Table 4.1 (overleaf) summarises the properties of photons and free electrons.

TABLE 4.1 Equations for photons and free electrons.

Property	Photon equation	Electron equation
energy	$E = h\nu$	$E_{\text{kinetic}} = \frac{1}{2}mu^2$
wavelength	$\lambda = \frac{hc}{E}$	$\lambda = \frac{h}{mu}$
speed	$c = 2.998 \times 10^8 \text{ m s}^{-1}$	$u = \sqrt{\frac{2E_{\text{kinetic}}}{m}}$

h: Planck's constant; ν: frequency; m: mass; u: velocity

WORKED EXAMPLE 4.6

Wavelengths

The structure of a crystal can be studied by observing the wave interference patterns that result from passing particle–waves through the crystal. To generate well-defined patterns, the wavelength of the particle–wave must be similar to the distance between atoms in the crystal. Determine the energy of a photon particle–wave beam with a wavelength of 0.25 nm and the energy of an electron particle–wave beam with this wavelength.

Analysis

This problem has two parts, one dealing with photons and the other with electrons. We are asked to relate the wavelengths of the particle–waves to their corresponding energies. Table 4.1 emphasises that photons and electrons have different relationships between energy and wavelength. Thus, we use different equations for the two calculations.

Solution

Photon energy is given by $E = h\nu = hc/\lambda$. We substitute and evaluate, being careful with units.

$$E_{\text{photon}} = \frac{(6.626\times10^{-34}\text{ J s})\,(2.998\times10^{8}\text{ m s}^{-1})}{0.25\times10^{-9}\text{ m}} = 7.9\times10^{-16}\text{ J}$$

For an electron, we need to work with two equations. The de Broglie equation links the velocity of an electron with its wavelength.

$$\lambda_{\text{particle}} = \frac{h}{mu}$$

The kinetic energy equation links the velocity of an electron with its kinetic energy.

$$E_{\text{kinetic}} = \tfrac{1}{2}mu^2$$

Begin by determining the velocity of the electron, recalling from chapter 2 that $1 \text{ J} = 1 \text{ kg m}^2 \text{ s}^{-2}$.

$$u_{\text{electron}} = \frac{h}{m\lambda} = \frac{6.626\times10^{-34}\text{ kg m}^2\text{ s}^{-1}}{(9.109\times10^{-31}\text{ kg})\,(0.25\times10^{-9}\text{ m})} = 2.91\times10^{6}\text{ m s}^{-1}$$

Next, use the velocity to find the kinetic energy of the electron.

$$E_{\text{kinetic}} = \tfrac{1}{2}mu^2 = \frac{(9.109\times10^{-31}\text{ kg})\,(2.91\times10^{6}\text{ m s}^{-1})^2}{2}$$

$$= 3.9 \times 10^{-18}\text{ kg m}^2\text{ s}^{-2} = 3.9 \times 10^{-18}\text{ J}$$

Is our answer reasonable?

The energy of the photon is higher than the kinetic energy of the electron, which is reasonable since the electron has a mass.

PRACTICE EXERCISE 4.6 In a photoelectric-effect experiment, a photon with an energy of 1.25×10^{-18} J is absorbed, causing the ejection of an electron with a kinetic energy of 2.5×10^{-19} J. Calculate the wavelength (in nm) associated with each.

The de Broglie equation predicts that every particle has wave characteristics. The wave properties of subatomic particles such as electrons and neutrons play important roles in their behaviour, but for larger objects, such as table tennis balls or cars, they do not. The reason is the scale of the waves. For all except subatomic particles, the wavelengths involved are so short that we are unable to detect the wave properties.

WORKED EXAMPLE 4.7

Matter waves

Calculate the wavelengths of an electron travelling at $1.00 \times 10^5\ \mathrm{m\,s^{-1}}$ and a table tennis ball with a mass of 11 g travelling at $2.5\ \mathrm{m\,s^{-1}}$.

Analysis

This problem deals with particle–waves that have mass. The de Broglie equation relates the mass and speed of an object to its wavelength.

Solution

For the electron: $m_{\text{electron}} = 9.109 \times 10^{-31}$ kg, $u = 1.00 \times 10^5\ \mathrm{m\,s^{-1}}$

$$\lambda_{\text{electron}} = \frac{h}{mu}$$

$$= \frac{6.626 \times 10^{-34}\ \mathrm{kg\,m^2\,s^{-1}}}{(9.109 \times 10^{-31}\ \mathrm{kg})\,(1.00 \times 10^5\ \mathrm{m\,s^{-1}})}$$

$$= 7.27 \times 10^{-9}\ \mathrm{m}$$

For the table tennis ball: $m_{\text{ball}} = 11$ g, $u = 2.5\ \mathrm{m\,s^{-1}}$

$$\lambda_{\text{ball}} = \frac{h}{mu} = \frac{6.626 \times 10^{-34}\ \mathrm{kg\,m^2\,s^{-1}}}{(11 \times 10^{-3}\ \mathrm{kg})\,(2.5\ \mathrm{m\,s^{-1}})} = 2.4 \times 10^{-32}\ \mathrm{m}$$

Is our answer reasonable?

The wavelength of the electron is in the order of the atomic size, which is consistent with the fact that electrons have particle–wave duality. The wavelength of the table tennis ball is extremely small, which is expected as we do not observe wave characteristics for the table tennis ball.

PRACTICE EXERCISE 4.7

Calculate the wavelength associated with a proton that is moving at a speed of $2.85 \times 10^5\ \mathrm{m\,s^{-1}}$.

The Heisenberg uncertainty principle

A particle occupies a particular location, but a wave has no exact position. A wave extends over some region of space. Because of their wave properties, electrons are always spread out rather than located in one particular place. As a result, *the position of a moving electron cannot be precisely defined.* We describe electrons as delocalised because their waves are spread out rather than pinpointed.

Mathematically, the position and momentum of a particle–wave are linked. Werner Heisenberg (1901–1976, Nobel Prize in physics, 1932), a German physicist, showed in the 1920s that the momentum and position of a particle–wave cannot be simultaneously determined. If a particular particle–wave can be pinpointed in a specific location, its momentum cannot be known accurately. Conversely, if the momentum of a particle–wave is known precisely, its location cannot be identified. Heisenberg summarised this uncertainty in what has become known as the **uncertainty principle**: the more accurately we know position, the more uncertain we are about momentum, and vice versa. Uncertainty is a feature of all objects, but it becomes important only for very tiny objects like electrons.

4.4 Quantisation and quantum numbers

The properties of electrons mentioned so far (mass, charge, spin and wave nature) apply to all electrons. Electrons travelling freely in space, electrons moving in a copper wire and electrons bound to atoms all have these characteristics. Bound electrons, those held in a specific region of an atom by electrostatic forces, have additional important properties relating to their energies and the shapes of their waves. These additional properties can have only certain specific values, that is, they are quantised.

As described in section 4.2, atoms of each element have unique, quantised electronic energy levels (see figures 4.17 and 4.18). This quantisation of energy is a property of bound electrons. The absorption and emission spectra of atoms consist of specific discrete energies because electrons undergo transitions from one bound state to another. In contrast, if an atom absorbs enough energy to remove an electron completely, the electron is no longer bound and can take on any amount of kinetic energy. *Bound* electrons have quantised energies; *free* electrons can have any amount of energy.

Experimental values for the quantised energies of atomic electrons can be calculated from absorption and emission spectra. The theory of quantum mechanics provides a mathematical explanation that links quantised energies to the wave characteristics of electrons. These wave properties of atomic electrons are described by the Schrödinger equation:

$$\hat{H}\psi = E\psi$$

where $\hat{H}$ is the Hamiltonian operator for the system (containing terms for the kinetic and potential energy), E is the energy of the system and ψ is the wavefunction for the system. In particular, ψ is the amplitude of the electron's matter wave. As with light, the amplitude itself does not have any physical meaning, but its square represents the electron density distribution or probability of finding an electron. A wavefunction is a mathematical function that gives us information about an electron's position in an atom. The solution of the Schrödinger equation gives us both the energies and the associated wavefunctions of a chemical system. Despite its concise form, the Schrödinger equation can be solved *exactly* only for systems containing one electron.

The Schrödinger equation has solutions only for specific, quantised energy values. For each quantised energy value, the Schrödinger equation generates a wavefunction that describes how the electrons are distributed in space. A one-electron wavefunction is called an **orbital**. We will describe the properties of orbitals in section 4.5.

Each described property can be identified, or indexed, using a **quantum number**. These are numbers that specify the values of the electron's quantised properties. Each electron in an atom has three quantum numbers that specify its three variable properties: energy (or orbital size), angular momentum (or orbital shape) and orbital orientation. A fourth quantum number describes the spin of an electron. To describe an atomic electron completely, chemists specify a value for each of its four quantum numbers: the principal quantum number, azimuthal quantum number, magnetic quantum number and spin quantum number.

Principal quantum number (n)

The most important quantised property of an atomic electron is its energy. The quantum number that indexes energy for an atom or ion containing only a single electron is the **principal quantum number** (n). For the simplest atom, hydrogen, we can use $E_n = \frac{-2.18 \times 10^{-18}\text{ J}}{n^2}$ to calculate the energy of the electron if we know n (as seen in section 4.2). However, that equation applies only to the hydrogen atom and, in slightly modified form, to other one-electron such as He^+ and Li^{2+} systems.

No known equation provides the exact energies of an atom that has more than one electron. Nevertheless, each electron in a multi-electron atom can be assigned a value of n that is a positive integer and which broadly correlates with the energy of the electron. The lowest energy for an atomic electron corresponds to $n = 1$, and each successively higher value of n describes a higher energy state. *The principal quantum number must be a positive integer:* $n = 1, 2, 3$ etc.

The principal quantum number also tells us something about the size of an atomic orbital, because the energy of an electron is correlated with its distribution in space. The higher the principal quantum number, the more energy the electron has and the greater is its average distance from the nucleus.

In summary, the principal quantum number (n) can have any positive integer value. It indexes the energy of the electron and is correlated with orbital size. As n increases, the energy of the electron increases, its orbital gets bigger and the electron is less tightly bound to the atom.

Azimuthal quantum number (l)

A second quantum number indexes the angular momentum of an atomic orbital. This quantum number is the **azimuthal quantum number** (l). The solutions for the Schrödinger equation and experimental evidence show that the electron distribution associated with orbitals can be described by a variety of shapes. Note that it is technically incorrect to talk about the 'shape' of an orbital itself. An orbital is a mathematical function that describes the amplitude of an electron's three-dimensional matter wave. Because the amplitude can fluctuate between positive and negative values (just like any other wave), it has no physical meaning by itself. We can, however, talk about the shape of the electron distribution associated with a particular orbital, and this tells us where we are most likely to find an electron within this orbital.

We can categorise the shapes of objects, such as the soccer ball, rugby ball and four-leaf clover shown in figure 4.20, according to their preferred axes. A preferred axis is a line through the centre of mass of an object or shape, about which the shape can be aligned or distributed. A soccer ball has no preferred axis because its mass is distributed equally in all directions about its centre. A rugby ball has one preferred axis, with more mass along this axis than in any other direction. A four-leaf clover has two preferred axes, at right angles to each other. In an analogous fashion, electron density in an orbital can be concentrated along preferred axes.

FIGURE 4.20 Some everyday objects have their masses concentrated along preferred axes. A soccer ball has no such axis, but a rugby ball has one and a four-leaf clover has two.

The value of l correlates with the number of preferred axes of a particular orbital and thereby identifies the shape of the electron distribution for that orbital. According to quantum theory, these shapes are highly restricted. These restrictions are linked to energy, so the value of the principal quantum number (n) limits the possible values of l. The smaller the value of n is, the more compact the orbital and the more restricted its possible electron distributions. *The azimuthal quantum number (l) can be zero or any positive integer smaller than n:* $l = 0, 1, 2, \ldots, (n - 1)$; that is, there are n values of l.

Historically, orbitals have been identified with letters rather than numbers. These letter designations correspond to the values of l as shown below.

Value of l	0	1	2	3	4
Orbital designation	*s*	*p*	*d*	*f*	*g*

An orbital is named by listing the numerical value for n, followed by the letter that corresponds to the numerical value for l. Thus, a 3*s* orbital has quantum numbers $n = 3$, $l = 0$. A 5*f* orbital has $n = 5$, $l = 3$. Notice that the restrictions on l mean that, when $n = 1$, l can only be zero. In other words, 1*s* orbitals exist, but there are no 1*p*, 1*d*, 1*f* or 1*g* orbitals. Similarly, there are 2*s* and 2*p* orbitals but no 2*d*, 2*f* or 2*g* orbitals.

Magnetic quantum number (m_l)

A sphere has no preferred axis, so it has no directionality in space. When there is a preferred axis, as for a rugby ball, figure 4.21 shows that the axis can point in many different directions relative to an *xyz* coordinate system. Thus, objects with preferred axes have directionality as well as shape.

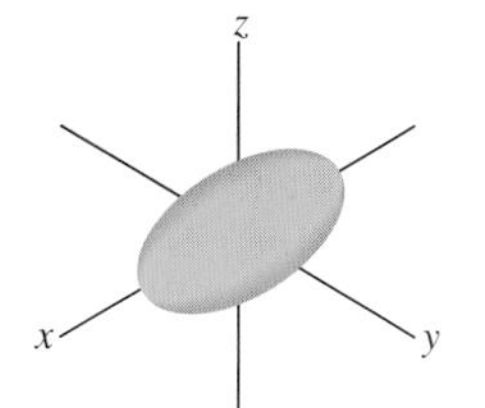

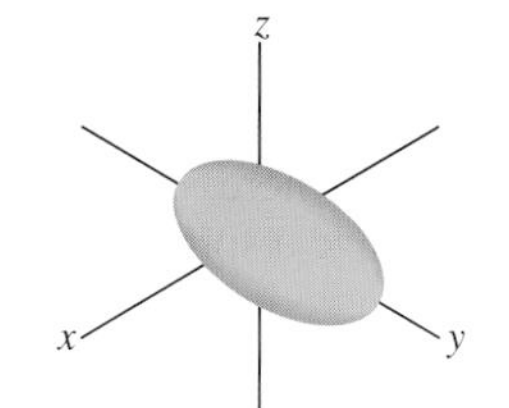

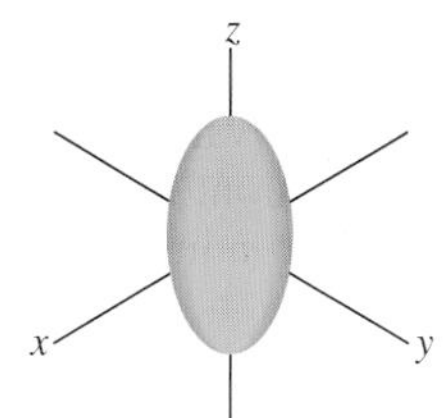

FIGURE 4.21 A rugby ball has directionality and shape. The figure shows three of the many ways in which a rugby ball can be oriented relative to a set of *x*, *y* and *z* axes.

The electron distribution within an *s* orbital is spherical and has no directionality. Electron distributions in other orbitals are non-spherical and therefore have a directional dependence. Like energy and orbital electron distribution, this directional dependence is quantised. Unlike rugby balls, the electron distributions within *p*, *d* and *f* orbitals have restricted numbers of possible orientations. The **magnetic quantum number (m_l)** indexes these restrictions.

The number of preferred axes (l) limits the possible orientations of the preferred axes (m_l). When $l = 0$, there is no preferred axis and therefore no preferred orientation, so $m_l = 0$ for *s* orbitals. One preferred axis ($l = 1$) can orient in any of three directions, giving three possible values for m_l: +1, 0 and −1. Two preferred axes ($l = 2$) can orient in any of five ways, giving five possible values for m_l: +2, +1, 0, −1 and −2. Each time l increases in value by one unit, two additional values of m_l become possible, and the number of possible orientations increases by two. *The magnetic quantum number (m_l) can be any positive or negative integer between 0 and l:* $m_l = 0, \pm1, \pm2, \ldots, \pm l$; that is, there are $(2l + 1)$ values of m_l.

Spin quantum number (m_s)

As we saw in section 1.5, all electrons have a property called **spin**. When a beam of silver atoms is passed through a magnetic field (similar to the experiment carried out by Otto Stern and Walter Gerlach in 1921, see figure 4.22), the atom beam is split, some atoms are deflected in one direction and the rest are deflected in the opposite direction. Since classical physics associates a spinning electric charge with a magnetic field, the experimental observation was explained by the Dutch physicists George Uhlenbeck and Samuel Gouldsmit with a property they called electron spin. (*Note:* There is no physical evidence that an electron actually spins.) The fact that only two responses are observed demonstrates that spin is quantised. The **spin quantum number** (m_s) indexes this behaviour. The two possible values of m_s are $+\frac{1}{2}$ and $-\frac{1}{2}$.

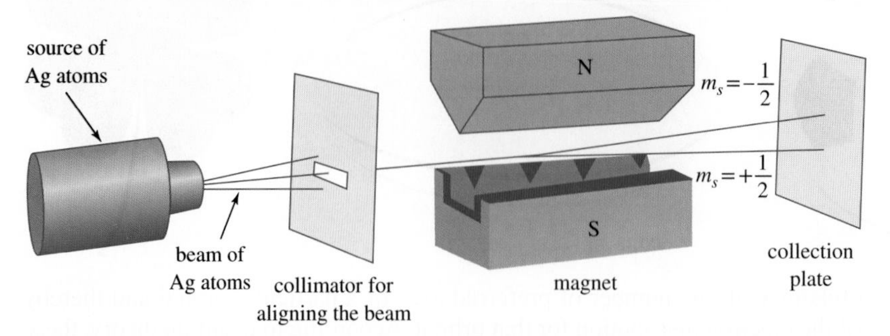

FIGURE 4.22 Schematic diagram of the set-up for the Stern–Gerlach magnetic field experiment.

The Pauli exclusion principle

A complete description of an atomic electron requires a set of four quantum numbers: n, l, m_l and m_s, which must meet all the restrictions mentioned earlier and summarised in table 4.2. Each electron in an atom has a unique set of quantum numbers; no two electrons in an atom have exactly the same set of quantum numbers. This was first postulated by the Austrian physicist Wolfgang Pauli (1900–1958, Nobel Prize in physics, 1945) and is known as the **Pauli exclusion principle**. The principle is derived from quantum mechanics and supported by all experimental evidence.

TABLE 4.2 Restrictions on quantum numbers for electrons in atoms.

Quantum number	Restrictions	Range
n	positive integers	$1, 2, \ldots, \infty$
l	positive integers less than n	$0, 1, \ldots, (n-1)$
m_l	integers between $-l$ and l	$-l, \ldots, -1, 0, +1, \ldots, +l$
m_s	$-\frac{1}{2}$ or $+\frac{1}{2}$	$-\frac{1}{2}, +\frac{1}{2}$

The number of possible sets of quantum numbers increases rapidly as n increases. An atomic orbital is designated by its n and l values, such as 1s, 3d, 4p and so on. When $l > 0$, there is more than one orbital of each designation; that is, when $l = 1$, there are three p orbitals and for $l = 2$ there are five d orbitals. An electron in any orbital can have a spin quantum number (m_s) of either $+\frac{1}{2}$ or $-\frac{1}{2}$. Thus, there are many sets of valid quantum numbers. An electron in a 3p orbital, for example, can have any one of six valid sets of quantum numbers.

$$n = 3, l = 1, m_l = +1, m_s = +\tfrac{1}{2} \qquad n = 3, l = 1, m_l = +1, m_s = -\tfrac{1}{2}$$

$$n = 3, l = 1, m_l = 0, \ m_s = +\tfrac{1}{2} \qquad n = 3, l = 1, m_l = 0, \ m_s = -\tfrac{1}{2}$$

$$n = 3, l = 1, m_l = -1, m_s = +\tfrac{1}{2} \qquad n = 3, l = 1, m_l = -1, m_s = -\tfrac{1}{2}$$

A direct consequence of the Pauli exclusion principle is that any orbital can contain a maximum of two electrons (we say that an orbital containing two electrons is full) and that the two electrons in a full orbital must be of opposite spin.

WORKED EXAMPLE 4.8

Valid quantum numbers

How many valid sets of quantum numbers exist for 4d orbitals? Give two examples.

Analysis

The question asks for the sets of quantum numbers that have $n = 4$ and $l = 2$. Each set must meet all the restrictions listed in table 4.2. The easiest way to see how many valid sets there are is to list all the valid quantum numbers.

Solution

Because this is a 4d orbital, n and l are specified and cannot vary. The other two quantum numbers, however, have several acceptable values. For each value of m_l, either value of m_s is acceptable, so the total number of possibilities is the product of the number of possible values for each quantum number. This is shown on the next page.

Quantum number	n	l	m_l	m_s
Possible values	4	2	2, 1, 0, −1, −2	$+\frac{1}{2}, -\frac{1}{2}$
Number of possible values	1	1	5	2

Possible sets of values for a $4d$ electron: (1)(1)(5)(2) = 10

There are 10 sets of quantum numbers that can describe a $4d$ orbital. Two of them are shown below.

$$n = 4, \quad l = 2, \quad m_l = 1, \quad m_s = +\tfrac{1}{2}$$

$$n = 4, \quad l = 2, \quad m_l = -2, \quad m_s = -\tfrac{1}{2}$$

You should be able to list the other eight sets.

Is our answer reasonable?

If you write down all the possible sets, you will find that there are 10. Notice that there are always $2l + 1$ possible values of m_l and 2 possible values of m_s, so every nd set of orbitals has 10 possible sets of quantum numbers.

PRACTICE EXERCISE 4.8

Write all the valid sets of quantum numbers of the $5p$ orbitals.

4.5 Atomic orbital electron distributions and energies

The chemical properties of atoms are determined by the behaviour of their electrons. Because atomic electrons are described by orbitals, the interactions of electrons can be described in terms of orbital interactions. The two characteristics of orbitals that determine how they interact are their electron distributions in three-dimensional space and their energies.

Orbital electron distributions

Wave-like properties cause electrons to be smeared out rather than localised at an exact position. This smeared-out distribution can be described using electron density. Where electrons are most likely to be found, there is high electron density. Low electron density occurs in regions where electrons are less likely to be found. Each electron, rather than being a point charge, is a three-dimensional particle–wave that is distributed over space as an orbital. Orbitals describe the delocalisation of electrons. Moreover, when the energy of an electron changes, the size and shape of its distribution in space change as well. Each atomic energy level is associated with a specific three-dimensional atomic orbital.

An atom that contains many electrons can be described by superimposing (adding together) the orbitals for all of its electrons to obtain the overall size and 'shape' of the atom.

The quantum numbers n and l determine the size and electron distribution of an orbital. As n increases, the size of the orbital increases, and, as l increases, the electron distribution of the orbital becomes more elaborate. Orbitals can have nodal planes or radial nodes, and the number of these depends on n and l; that is, for every orbital, there are $n - 1$ nodes, of which l are nodal planes and the remainder are spherical (radial) nodes.

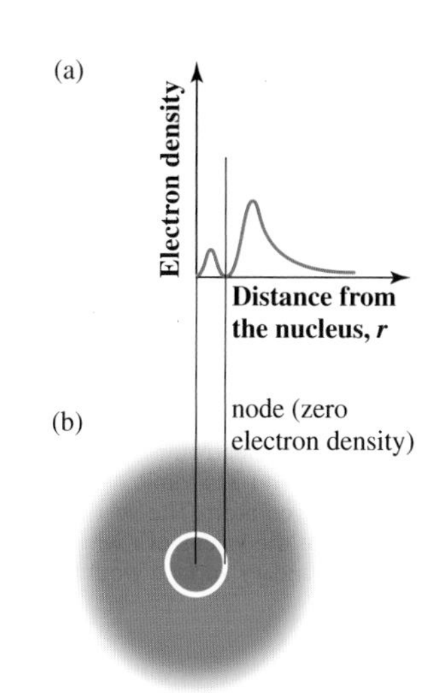

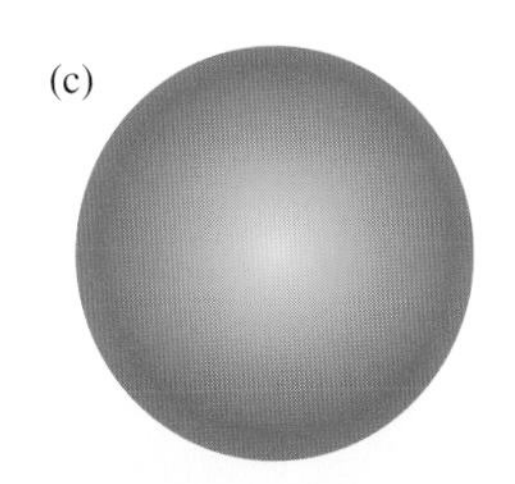

FIGURE 4.23 Different depictions of the $2s$ orbital: **(a)** a plot of electron density versus distance from the nucleus, **(b)** an electron density picture and **(c)** a boundary surface diagram.

Orbital depictions

We need ways to visualise electrons as particle–waves delocalised in three-dimensional space. Orbital pictures provide maps of how an electron wave is distributed in space. There are several ways to represent these three-dimensional maps. Each one shows some important orbital features, but none shows all of them. We use three different representations: electron density plots, electron density pictures and boundary surface diagrams.

An **electron density plot** represents the electron distribution in an orbital as a two-dimensional graph. These graphs show electron density along the y-axis and distance from the nucleus, r, along the x-axis. Figure 4.23a shows an electron density plot for the $2s$ orbital.

Electron density plots are useful because plots for several orbitals can be superimposed to indicate the relative sizes of various orbitals. Electron density plots do not, however, show the three-dimensionality of an orbital.

Electron density pictures can indicate the three-dimensional nature of orbitals. One type of orbital picture is a two-dimensional colour pattern in which the density of colour represents electron density. Figure 4.23b shows such an electron density picture of the 2*s* orbital. This two-dimensional pattern of colour density shows a cross-sectional slice through the middle of the orbital. Figures 4.23a and b show that the 2*s* orbital has one spherical node, as we discussed on the previous page.

A **boundary surface diagram** provides a simplified orbital picture. In this representation, we draw a solid surface that encloses most (usually 90%) of the electron density. Thus, the electron density is high inside the boundary surface but very low outside. Figure 4.23c shows a boundary surface diagram of the 2*s* orbital. One drawback of boundary surface diagrams is that they do not show details of electron density below the surface, including spherical nodes. They do, however, show all planar nodes, and from those it is possible to deduce the number of hidden spherical nodes.

A useful analogy for understanding the value of boundary surfaces is a swarm of bees around a hive. At any one time, some bees will be off foraging for nectar, so a boundary surface drawn around all the bees might cover several hectares. This would not be a very useful map of bee density. A boundary surface containing 90% of the bees, on the other hand, would be only slightly bigger than the hive itself. This would be a very useful map of bee density, because anyone inside that boundary surface would surely interact with bees.

Another drawback of boundary surface diagrams is that all details of electron density inside the surface are lost. Thus, if we want to convey the maximum information about orbitals, we must use combinations of the various depictions.

The advantages and disadvantages of the three types of plots can be seen by how they show one characteristic feature of orbitals. Figure 4.23a shows clearly that there is a value for r where the electron density falls to zero. A place where electron density is zero is called a **node**. Figure 4.23b shows the node for the 2*s* orbital as a white ring. In three dimensions, this node is a spherical surface. Figure 4.23c does not show the node, because the spherical nodal surface is hidden inside the 90% boundary surface. The two-dimensional graph shows the location of the node most clearly, the electron density picture gives the best sense of the shape of the node, while the boundary surface diagram provides the best three-dimensional view of an orbital.

Orbital size

Experiments that determine atomic radii provide information about the sizes of orbitals. In addition, theoretical models of the atom can predict how the electron density of a particular orbital changes with distance from the nucleus.

In any particular atom, orbitals get larger as the value of n increases. The $n = 2$ orbitals are larger than the $n = 1$ orbital, the $n = 3$ orbitals are larger than the $n = 2$ orbitals, and so on. The electron density plots in figure 4.24 show this trend for the first three *s* orbitals of the hydrogen atom. This plot also shows that the number of nodes increases as n increases.

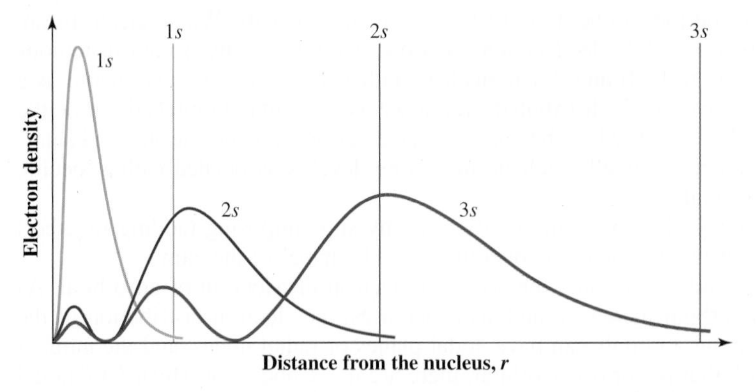

FIGURE 4.24 Electron density plots for the 1*s*, 2*s* and 3*s* atomic orbitals of the hydrogen atom. The vertical lines indicate the values of r where the 90% boundary surface would be located.

In any particular atom, all orbitals with the same principal quantum number are similar in size. As an example, figure 4.25 shows that the $n = 3$ orbitals of the copper atom have their maximum electron densities at similar distances from the nucleus. The same regularity holds for all other atoms. The quantum numbers other than n affect orbital size only slightly. We describe these small effects in the context of orbital energies later in this section.

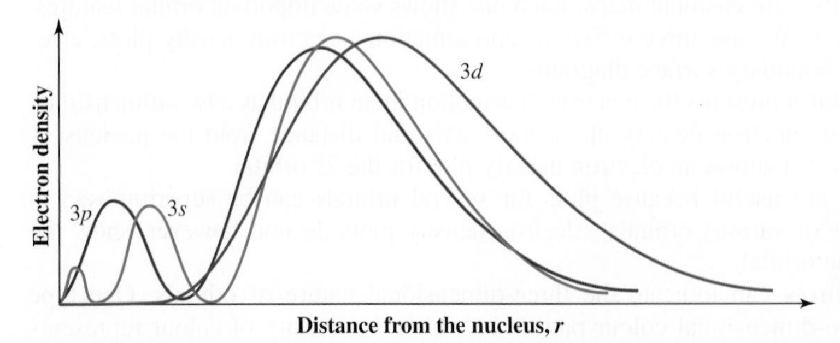

FIGURE 4.25 Electron density plots for the 3*s* (red line), 3*p* (blue line), and 3*d* (green line) orbitals for the copper atom. All three orbitals are nearly the same size.

A specific orbital becomes smaller as the atomic nuclear charge increases. As the positive charge of the nucleus increases, the electrostatic force exerted by the nucleus on the negatively charged electrons increases and electrons become more tightly bound. This reduces the radius of the orbital. As a result, each orbital shrinks in size as atomic number increases. For example, figure 4.26 shows that the 2*s* orbital steadily decreases in size across the second row of the periodic table from Li ($Z = 3$) to Ne ($Z = 10$). The atomic number, Z, is equal to the number of protons in the nucleus, so increasing Z means increasing nuclear charge.

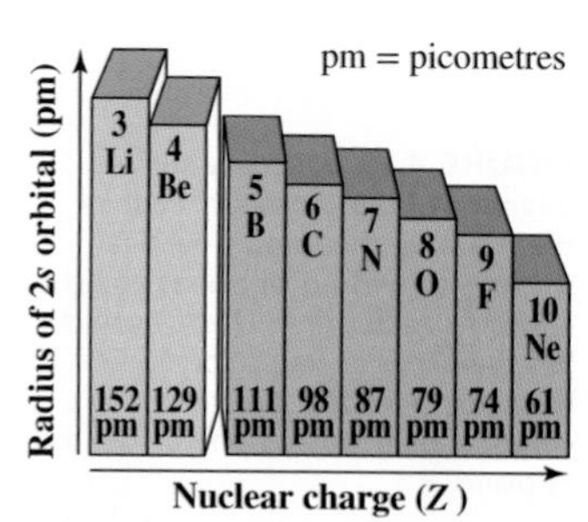

FIGURE 4.26 The radius of the 2*s* orbital decreases as Z increases. The number at the bottom of each column is the radius of the 2*s* orbital in pm (picometres).

Details of orbital electron distributions

The electron distributions of orbitals strongly influence chemical interactions. Hence, we need to have detailed pictures of these electron distributions to understand the chemistry of the elements.

The quantum number $l = 0$ corresponds to an *s* orbital. According to the restrictions on quantum numbers, there is only one *s* orbital for each value of the principal quantum number. The electron distribution of *s* orbitals is spherical, with radii and number of nodes that increase as n increases. Figure 4.27 shows a boundary surface diagram of the 1*s* orbital.

We introduced the concept of the phase of a wave in section 4.2, and this concept also applies to orbitals. An *s* orbital has a single phase, which we refer to as either positive (+) or negative (–). We generally depict different phases using + and – signs superimposed on the orbital electron distribution, or by using different colours corresponding to + and – phases.

FIGURE 4.27 Boundary surface diagram of the 1*s* orbital.

The quantum number $l = 1$ corresponds to a *p* orbital. An electron in a *p* orbital can have any of three values for m_l, so for each value of n there are three different *p* orbitals. The nonspherical electron distribution of *p* orbitals can be shown in various ways. The most convenient representation shows the three orbitals with identical electron distributions but pointing in three different directions. Figure 4.28 shows boundary surface diagrams of the 2*p* orbitals. Each *p* orbital has high electron density in one particular direction, perpendicular to the other two orbitals, with the nucleus at the centre of the system. The three different orbitals can be represented so that each has its electron density concentrated on both sides of the nucleus along a preferred axis. We can write subscripts on the orbitals to distinguish the three distinct orientations: p_x, p_y and p_z. Each *p* orbital also has a nodal plane that passes through the nucleus. The nodal plane for the p_x orbital is the *yz* plane, for the p_y orbital the nodal plane is the *xz* plane and for the p_z orbital it is the *xy* plane. The two lobes of a single *p* orbital have opposite phases; this is shown in figure 4.28 by using a different colour for each lobe.

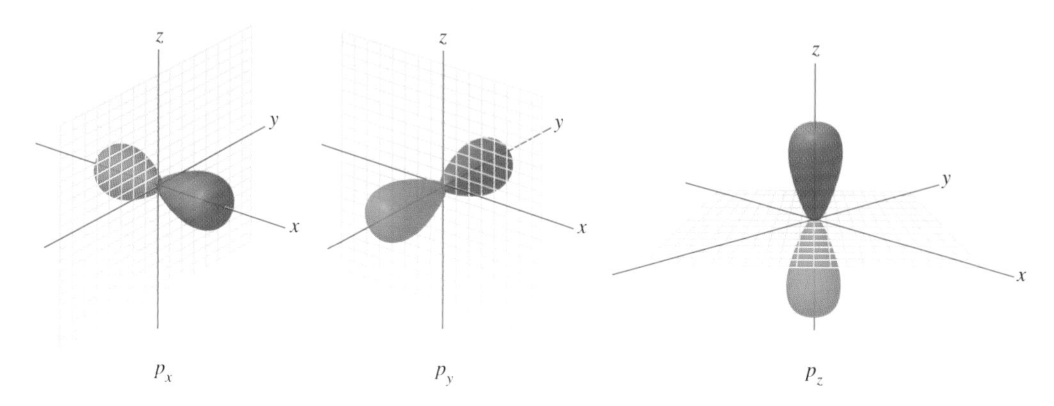

FIGURE 4.28 Boundary surface diagrams of the three 2*p* orbitals. The three orbitals have the same overall electron distribution, but each is oriented perpendicular to the other two. The nodal plane in each case is illustrated by the grey hatched surface.

As n increases, the number of nodes within each *p* orbital increases, just as for *s* orbitals. Nevertheless, the directionality of the electron distribution does not change. Each *p* orbital is perpendicular to the other two in its set and has its lobes along its preferred axis, where electron density is high. To an approaching atom, therefore, an electron in a 3*p* orbital presents the same characteristics as one in a 2*p* orbital, except that the 3*p* orbital is bigger. Consequently, the electron distributions and relative orientations of the 2*p* orbitals in figure 4.28 represent the prominent spatial features of all *p* orbitals.

The quantum number $l = 2$ corresponds to a *d* orbital. An electron in a *d* orbital can have any of five values for m_l (–2, –1, 0, +1 and +2), so there are five different orbitals in each set. Each *d* orbital has two nodal planes. Consequently, the electron distributions of the *d* orbitals are more complicated than their *s* and *p* counterparts. The boundary surface diagrams in figure 4.29 show these orbitals in the most convenient way. In these drawings, three orbitals have electron distributions that look like three-dimensional 'clover leaves' lying in a plane with the lobes pointed between the axes. A subscript identifies the plane in which each lies: d_{xy}, d_{xz} and d_{yz}. A fourth orbital has an electron distribution that is a clover leaf in the *xy* plane, but its lobes point along the *x* and *y* axes. This orbital is designated $d_{x^2-y^2}$. In each of these 'cloverleaf' orbitals, the lobes situated opposite each other have the same phases, as shown in figure 4.29 (overleaf). The electron distribution in the fifth orbital is quite different. Its major lobes point along the *z*-axis, but there is also a 'doughnut' of electron density in the *xy* plane. This orbital is designated d_{z^2}. The two lobes have the same phase, while the 'doughnut' is of opposite phase.

FIGURE 4.29 Boundary surface diagrams of the *d* orbitals. Four of the five have the same cloverleaf shape with two nodal planes at right angles to each other. The d_{z^2} orbital has the z-axis as its preferred axis. It has two nodes that are shaped like cones, one above and one below the *xy* plane.

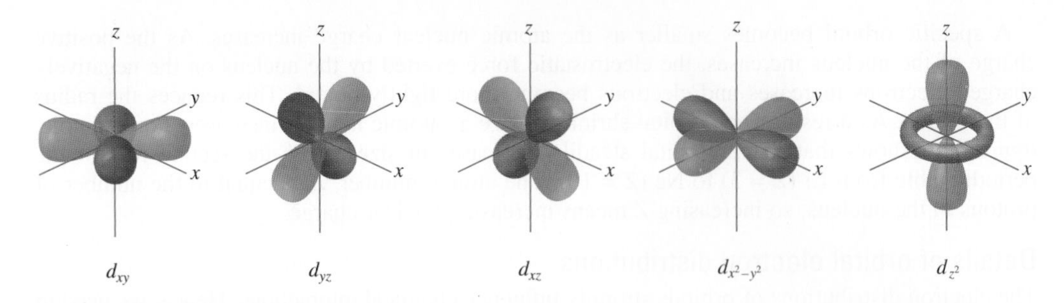

The quantum number $l = 3$ corresponds to an *f* orbital. The possible values of m_l (−3, −2, −1, 0, +1, +2 and +3) mean that there are seven *f* orbitals, but as they become important only for the lanthanoid elements and beyond, we will not detail their properties here.

Orbital energies

A hydrogen atom can absorb a photon and change from its most stable, lowest energy state (ground state) to a less stable, higher energy state (excited state), as described in section 4.2. We can explain this process using atomic orbitals. When a hydrogen atom absorbs a photon, its electron can undergo a transition to an orbital that has a larger principal quantum number, i.e. a higher energy. Figure 4.30 illustrates this process for the $1s \rightarrow 2p$ transition in the hydrogen atom.

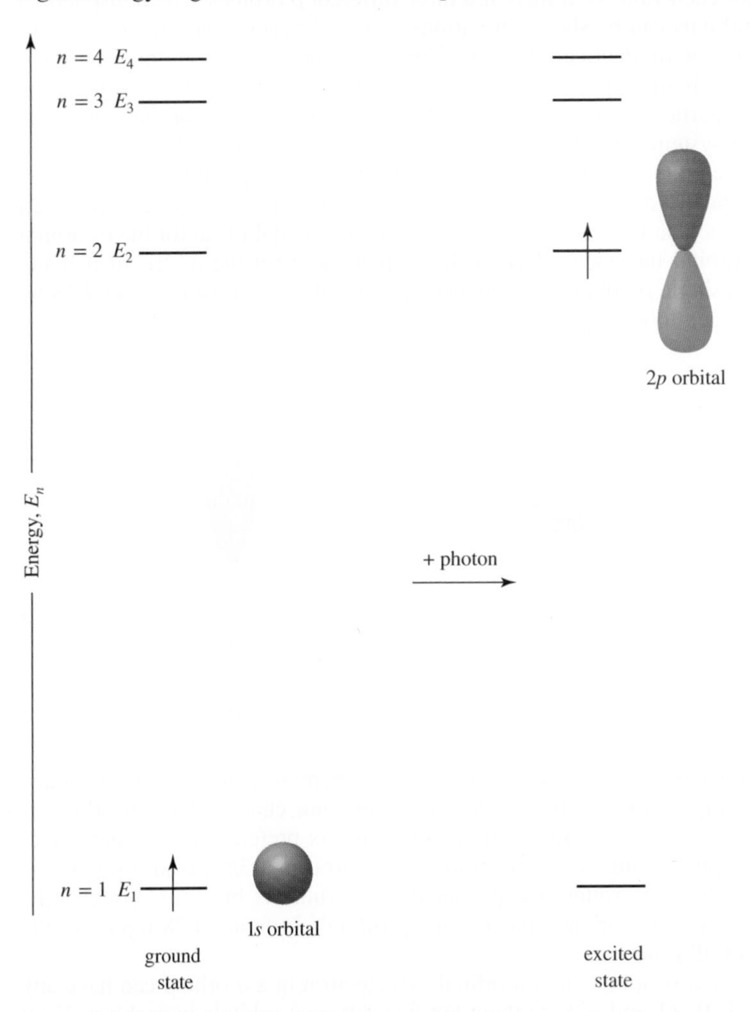

FIGURE 4.30 When a hydrogen atom in its ground state absorbs light of wavelength 121.6 nm, it is converted to an excited state, in which the electron occupies a 2*p* orbital.

Electronic transitions cannot necessarily occur between any set of orbitals; in the hydrogen atom, and other one-electron systems, they are governed by the following selection rules.

1. Δn = anything. There is no restriction on the values of the initial and final principal quantum numbers.
2. $\Delta l = \pm 1$. The azimuthal quantum number must increase or decrease by 1. This means that we cannot have a transition from an *s* orbital to another *s* orbital, or from a *p* orbital to another *p* orbital.
3. $\Delta m_l = 0, \pm 1$. The magnetic quantum number may stay the same, or it can increase or decrease by 1.

The atomic orbital model explains the spectra and energy levels of the hydrogen atom perfectly. For a one-electron system like the hydrogen atom, the energy levels for all orbitals depend only on the principal quantum number, *n*. All orbitals having the same value of *n* have the same energy and are said to be **degenerate**. For multi-electron systems, experiments show that the underlying principles are the same, although the details for each kind of atom are different. Variations in nuclear charge and the number of electrons change the magnitudes of the electrical forces that hold electrons in their orbitals. Differences in forces cause changes in orbital energies that can be understood qualitatively using forces of electrical attraction and repulsion, as we describe later in this chapter. As a consequence, the degeneracy of orbitals for each level of n is partially removed.

The effect of nuclear charge

A helium +1 cation, like a hydrogen atom, has one electron. Absorption and emission spectra show that He^+ has energy levels that depend on n only, just like those of the hydrogen atom. Nevertheless, figure 4.31 shows that the emission spectra of He^+ and H differ, which means that these two species must have different energy levels. We conclude that something besides n influences orbital energy.

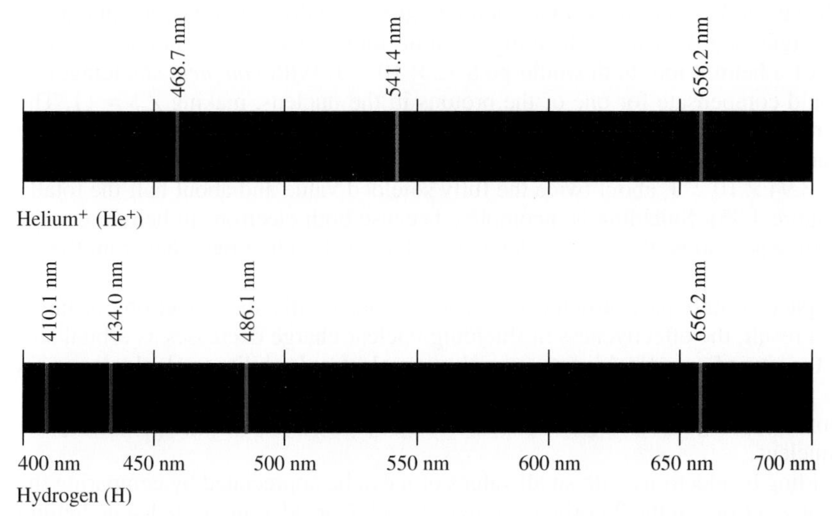

FIGURE 4.31 The emission spectra of He^+ and H reveal transitions at characteristic energies. Most of the emitted photons have different energies because He^+ has quantised energy levels that are different from those of H.

The difference between He^+ and H is in their nuclei. A hydrogen nucleus is a single proton with a +1 charge, whereas a helium nucleus contains two protons (and two neutrons) and has a charge of +2. The larger nuclear charge of He^+ attracts its single electron more strongly than the smaller charge of H. As a result, He^+ binds the electron with a stronger force. Thus, any given energy level in the helium ion is lower in energy than the corresponding level in the hydrogen atom.

The energy of an orbital can be determined by measuring the amount of energy required to remove an electron completely from that orbital. This is the **ionisation energy (E_i)**.

$$H \longrightarrow H^+ + e^- \qquad E_{i_H} = 2.18 \times 10^{-18}\ J$$

$$He^+ \longrightarrow He^{2+} + e^- \qquad E_{i_{He^+}} = 8.72 \times 10^{-18}\ J$$

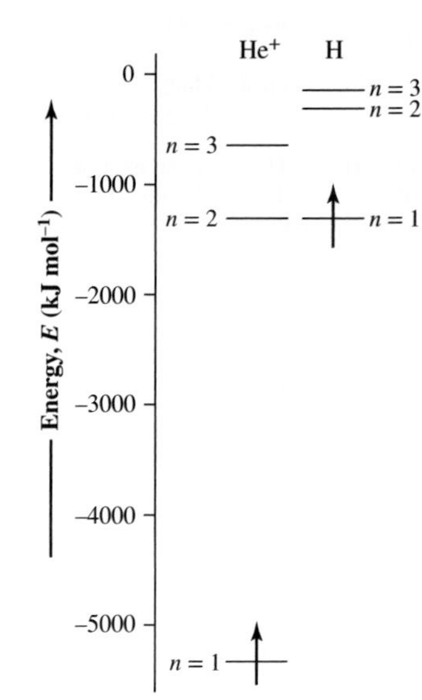

FIGURE 4.32 An energy level diagram for He^+ and H. Each species has one electron, but the different nuclear charges make each He^+ orbital four times lower in energy than the corresponding H orbital.

The ionisation energy of He^+ is four times that of H. Thus, the ground state orbital for He^+ must be four times lower in energy than the ground state orbital for H. Spectral analysis shows that each orbital of a helium cation is four times lower in energy than its counterpart orbital in a hydrogen atom, showing that orbital energy depends on Z^2 in a one-electron system. While the energy levels in H and He^+ are different for corresponding orbitals due to the Z^2 relationship, there are some orbital energies that coincide and hence we can get emission energies that are identical for both species (see figure 4.31). Figure 4.32 shows the relationship between the energy levels of He^+ and H. The diagram is in exact agreement with calculations based on the Schrödinger equation.

The effect of other electrons

A hydrogen atom or a helium cation are rare examples of one-electron systems; most other atoms and ions contain collections of electrons. In a multi-electron atom, electrons affect each other's properties. These electron–electron interactions make the orbital energies of every element unique.

A given orbital is of higher energy in a multi-electron atom than it is in the single-electron ion with the same nuclear charge. For instance, figure 4.33 shows that it takes more than twice as much energy to remove the electron from He^+ (one electron) as it does to remove one of the electrons from a neutral He atom (two electrons). This demonstrates that the 1s orbital in He^+ is more than two times lower in energy than the 1s orbital in neutral He. The nuclear charge of both species is +2, so the smaller ionisation energy for He must result from the presence of the second electron. A negatively charged electron in a multi-electron atom is attracted to the positively charged nucleus, but it is repelled by the other negatively charged electrons. This electron–electron repulsion contributes to the lower ionisation energy of the helium atom.

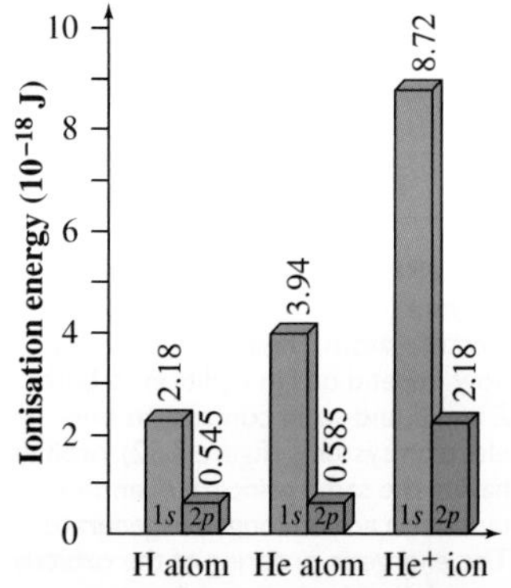

FIGURE 4.33 Ionisation energies.

Shielding

Figure 4.34 shows a free electron approaching a He^+ cation. The +2 charge of the nucleus attracts the incoming electron, but the negative charge of the He^+ 1*s* electron repels it. As electrons have a like charge, electron–electron repulsion cancels a portion of the attraction between the nucleus and the incoming electron. Chemists call this partial cancellation **shielding**.

With its –1 charge, a bound electron could reduce the total charge by a maximum of 1 charge unit. Indeed, when an approaching electron is far enough away from a He^+ ion, it feels an attraction due to the net charge on the ion, +1. However, the 1*s* orbital is spread out all around the nucleus. This means that, as an approaching electron gets close enough, the 1*s* electron shields only part of the total nuclear charge. Consequently, an approaching electron is subject to a net attraction resulting from some **effective nuclear charge** (Z_{eff}) less than +2 but greater than +1.

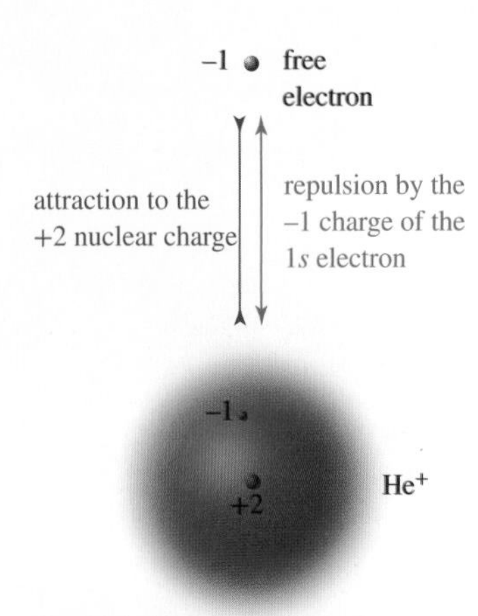

FIGURE 4.34 As a free electron approaches a He^+ cation, it is attracted to the +2 charge on the nucleus but repelled by the –1 charge on the 1*s* electron. When it is far from the cation, the free electron experiences a net charge of +1.

Incomplete shielding can be seen in the ionisation energies of hydrogen atoms, helium atoms and helium ions (see figure 4.33). Without shielding, the ionisation energy of a helium atom would be the same as that of a helium ion; both would be 8.72×10^{-18} J. With *complete* shielding, one helium electron would compensate for one of the protons in the nucleus, making $Z_{eff} = +1$. The energy required to remove an electron from a helium atom would then be the same as the energy required to remove an electron from a hydrogen atom, 2.18×10^{-18} J. The actual ionisation energy of a helium atom is 3.94×10^{-18} J, about twice the fully shielded value and about half the totally unshielded value (figure 4.35). Shielding is incomplete because both electrons in helium occupy an extended region of space, so neither is completely effective at shielding the other from the +2 charge of the nucleus.

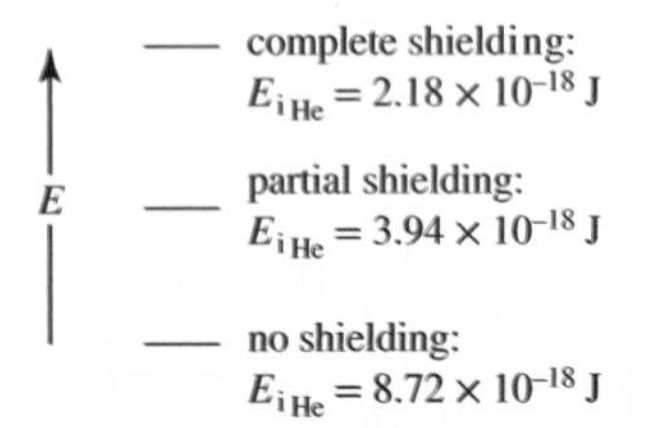

FIGURE 4.35 Ionisation energies of a helium atom with different amounts of shielding.

Electrons in compact orbitals pack around the nucleus more tightly than electrons in large, diffuse orbitals. As a result, the effectiveness in shielding nuclear charge decreases as orbital size increases. Because the size of an orbital increases with *n*, an electron's ability to shield decreases as *n* increases. In a multi-electron atom, lower *n* electrons are concentrated between the nucleus and higher *n* electrons. The negative charges of these inner electrons counteract most of the positive charge of the nucleus.

The efficient shielding by electrons with small values of *n* can be appreciated by comparing the ionisation energies of electrons in the 2*p* orbitals in figure 4.33. Consider an excited-state helium atom that has one of its electrons in the 1*s* orbital and its other electron in a 2*p* orbital. It takes 0.585×10^{-18} J to remove the 2*p* electron from this excited-state helium atom. This value is close to that of an excited hydrogen atom with its lone electron in a 2*p* orbital, 0.545×10^{-18} J. It is much less than the 2*p* orbital ionisation energy of an excited He^+ ion, 2.18×10^{-18} J. These data show that Z_{eff} is quite close to +1 for the 2*p* orbital of an excited atom. In the excited He atom, the electron in the 1*s* orbital is very effective at shielding the electron in the 2*p* orbital from the full +2 charge of the nucleus.

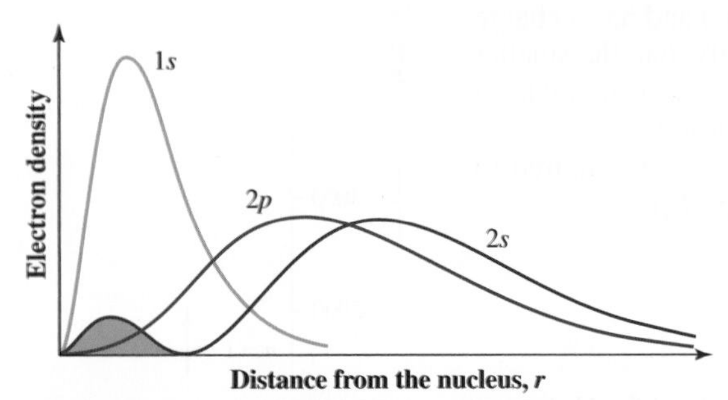

FIGURE 4.36 Electron density plots for the 1*s*, 2*s* and 2*p* orbitals. Unlike the 2*p* orbital, the 2*s* orbital has significant electron density very near the nucleus (shaded region).

In multi-electron atoms, electrons with a given value of *n* generally provide effective shielding for orbitals with a larger value of *n*. That is, $n = 1$ electrons shield the $n = 2$, $n = 3$ and larger orbitals, whereas $n = 2$ electrons provide effective shielding for the $n = 3$, $n = 4$ and larger orbitals but provide little shielding for the $n = 1$ orbital. The amount of shielding also depends on the electron distribution of the shielded orbital and decreases with increasing azimuthal quantum number, *l*. The shaded area of the electron density plot in figure 4.36 emphasises that the 2*s* orbital has a region of significant electron density near the nucleus. A 2*p* orbital lacks this inner layer, so virtually all of its electron density lies outside the region occupied by the 1*s* orbital. Consequently, a 1*s* orbital shields the 2*p* orbital more effectively than the 2*s* orbital, even though both $n = 2$ orbitals are about the same size. Thus, a 2*s* electron is subject to a larger effective nuclear charge than a 2*p* electron. This results in stronger electrostatic attraction to the nucleus, which makes the 2*s* orbital lower in energy than the 2*p* orbitals. The 2*p* orbitals of any multi-electron atom are always slightly higher in energy than the 2*s* orbital.

The shielding differences experienced by the 2*s* and 2*p* orbitals also extend to larger values of *n*. The 3*s* orbital is of lower energy than the 3*p*, the 4*s* orbital is of lower energy than the 4*p*, and so on. Orbitals with higher *l* values show similar effects. The 3*d* orbitals are always higher in energy than the 3*p* orbitals, and the 4*d* orbitals are higher in energy than the 4*p* orbitals. These effects can be summarised in a single general statement: *The higher the value of the l quantum number, the more that orbital is shielded by electrons in smaller, lower energy orbitals.*

FIGURE 4.37 An energy level diagram (not to scale) for a multi-electron atom. The orbital energies now depend on *l* in addition to both *Z* and *n*, and so, in contrast to one-electron systems (figure 4.32), orbitals having the same principal quantum number *n* are no longer degenerate. The energetic ordering of the orbitals is explained in section 4.6.

In a one-electron system (H, He^+, Li^{2+} and so on) the energy of the orbitals depends only on *Z* and *n*. In multi-electron systems, orbital energy depends primarily on *Z* and *n*, but it also depends significantly on *l*. In a sense, *l* finetunes orbital energies and as a consequence there are fewer degenerate orbitals.

The relative orbital energies for a multi-electron atom are shown in figure 4.37. It can be seen that, whereas in a one-electron system the *ns*, *np* and *nd* orbitals are degenerate, this is not the case in a multi-electron atom.

Electrons with the same l value but different values of m_l do not shield one another effectively. For example, when electrons occupy different p orbitals, the amount of mutual shielding is slight. This is because shielding is effective only when much of the electron density of one orbital lies between the nucleus and the electron density of another. Recall from figure 4.28 (p. 141) that the p orbitals are perpendicular to one another, with high electron densities in different regions of space. The electron density of the $2p_x$ orbital does not lie between the $2p_y$ orbital and the nucleus, so there is little shielding. The d orbitals also occupy different regions of space from one another, so mutual shielding among electrons in these orbitals is small as well.

WORKED EXAMPLE 4.9

Shielding

Draw a qualitative electron density plot showing the 1s, 2p and 3d orbitals to scale. Label the plot in a way that summarises the shielding properties of these orbitals.

Analysis

This is a qualitative problem that asks us to combine information about three different orbitals on a single plot. We need to find electron density information and draw a single graph to scale.

Solution

Figures 4.24, 4.25 and 4.36 show electron density plots of the n = 1, n = 2 and n = 3 orbitals. We extract the electron distributions of the 1s, 2p and 3d orbitals from these graphs. Then we add labels that summarise the shielding properties of these orbitals. Shielding is provided by small orbitals whose electron density is concentrated closer to the nucleus than that of larger orbitals. In this case, 1s shields both 2p and 3d; 2p shields 3d, but not 1s; and 3d shields neither 1s nor 2p. The shielding patterns can be labelled as shown.

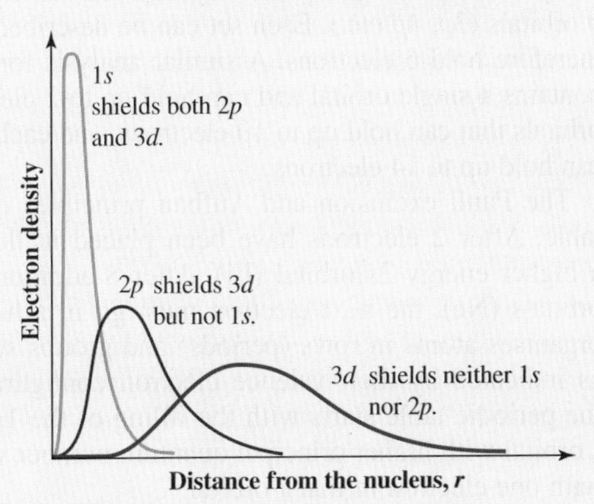

Is our answer reasonable?

We know that the most important factor for orbital size is the value of n and that small orbitals shield better than large ones, so the shielding sequence makes sense.

PRACTICE EXERCISE 4.9

Construct and label a graph illustrating that the 2s orbital shields the 3s orbital more effectively than the 3s shields the 2s.

4.6 Structure of the periodic table

As we have seen in chapter 1, the periodic table lists the elements in order of increasing atomic number. Because the number of electrons in a neutral atom is the same as its atomic number, this list is also in order of increasing number of atomic electrons. The periodic table organises atoms in rows (periods) and groups such that atoms with the same valence electron configuration, leading to similar chemical properties, are located in the same group. We have since learned that the electrons are mainly responsible for an atom's chemical properties, and equipped with the knowledge of orbital electron distributions and energies we can now start to understand the reasons for the placement of atoms in the periodic table.

The Aufbau principle and order of orbital filling

The most stable arrangement of an atom's electrons is called its ground state. *Most stable* means that the electrons occupy the lowest energy orbitals available. We construct the ground-state configuration of an atom by placing electrons in the orbitals starting with the lowest in energy and moving progressively upward. In accordance with the Pauli exclusion principle, each successive electron is placed in the lowest energy orbital whose quantum numbers are not already assigned to another electron. This is called the **Aufbau principle** (Aufbau is a German word literally meaning 'build-up'). We have seen earlier that for multi-electron atoms shielding causes the orbitals with the same principal quantum number to increase in energy as l increases. Consequently, the 2s orbital, being of lower energy than the 2p orbital, fills first. Similarly, 3s fills before 3p. The energy differences between orbitals become less pronounced the further away the orbitals are from the nucleus. The consequence of this is that, for orbitals with principal quantum numbers larger than $n = 2$, the order of orbital energies is not determined by n alone. For example, the 4s orbital fills before the 3d orbitals. Therefore, the order of orbital filling is as given in figure 4.38a. The sequence of orbital energies is determined by a combination of $n + l$, and figure 4.38b shows a way to remember that sequence. The orbitals are listed in increasing order of n (bottom to top) and l (right to left) with the red arrows indicating the directions where $n + l$ is constant. For equal values of $n + l$, lower values of n are filled first. When orbitals are close in energy, anomalies may occur in the order of filling suggested by figure 4.38. While the 4*s* orbital is lower in energy than the empty 3*d* orbital, once 3*d* states are populated the energy order switches and now the 4*s* electrons are higher in energy than the 3*d* electrons. We shall look at this in more detail in chapter 13.

In applying the Aufbau principle, remember that a full description of an electron requires four quantum numbers: n, l, m_l and m_s. Each combination of n and l describes one quantised energy level. Moreover, each level with $l > 1$ includes multiple orbitals, each with a different value of m_l. For any combination of n and l, all orbitals with the same value of l are degenerate, i.e. in an atom they have the same energy. For example, the 2*p* energy level ($n = 2$, $l = 1$) consists of three distinct *p* orbitals ($m_l = -1$, 0 and +1), all with the same energy. In addition, the spin quantum number, m_s, describes the two possible different spin orientations of an electron.

In other words, the 2*p* energy level consists of 3 orbitals with different m_l values that can hold as many as 6 electrons without violating the Pauli exclusion principle. The same is true of every set of *p* orbitals (3*p*, 4*p* etc.). Each set can be described by 6 different sets of quantum numbers and can therefore hold 6 electrons. A similar analysis for other values of l shows that each *s* energy level contains a single orbital and can hold up to 2 electrons, each *d* energy level consists of 5 different orbitals that can hold up to 10 electrons, and each *f* energy level consists of 7 different orbitals that can hold up to 14 electrons.

The Pauli exclusion and Aufbau principles dictate the length of the periods in the periodic table. After 2 electrons have been placed in the 1*s* orbital (He), the next electron must go in a higher energy 2*s* orbital (Li). After 8 additional electrons have been placed in the 2*s* and 2*p* orbitals (Ne), the next electron must go in a higher energy 3*s* orbital (Na). The periodic table organises atoms in rows (periods) and groups such that atoms with similar chemical properties, as indicated by their valence electron configurations, are located in the same group. Because the periodic table starts with the filling of the 1*s* orbital, every row has to finish before the next *s* orbital with higher principal quantum number will be filled. The start of the next row is an atom with one electron in that *s* orbital.

Here is a summary of the conditions for atomic ground states.

1. The electrons in the atom occupy the lowest energy orbitals (see figure 4.38a).
2. No two electrons can have identical sets of quantum numbers.
3. Orbital capacities are as follows: *s* has 2 electrons; *p* set has 6 electrons; *d* set has 10 electrons; *f* set has 14 electrons.

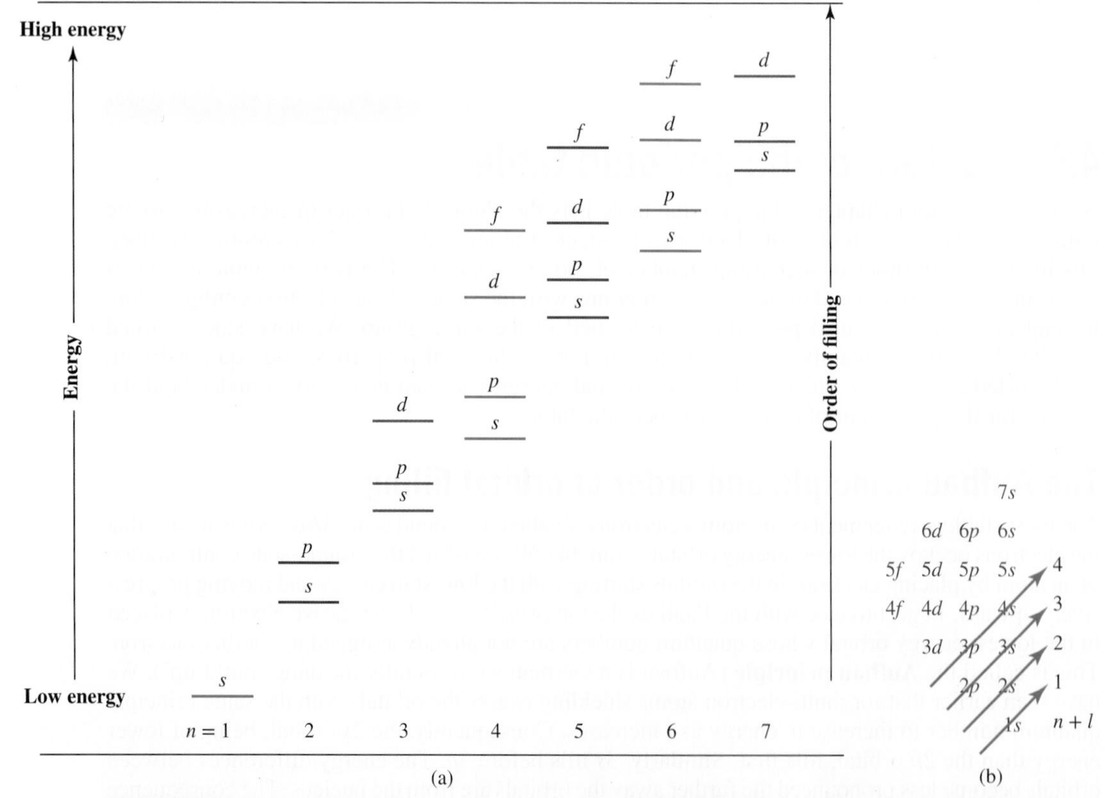

FIGURE 4.38 (a) The order of filling orbitals with electrons. **(b)** An easy way to remember the general order of electron filling. Orbitals fill in order of increasing ($n + l$). Where orbitals have the same value of ($n + l$), the one with the lowest n will fill first, followed by the others in increasing order of n. Note, however, for higher values of n, there are a number of exceptions to this general order (see the electron configurations given in figure 4.40).

Armed with these conditions, we can correlate the rows and columns of the periodic table with values of the quantum numbers n and l. This is shown in the periodic table in figure 4.39. Remember that the elements are arranged so that Z increases one unit at a time from left to right across a row. At the end of each row, we move to the next higher value of Z one row down on the left side of the periodic table.

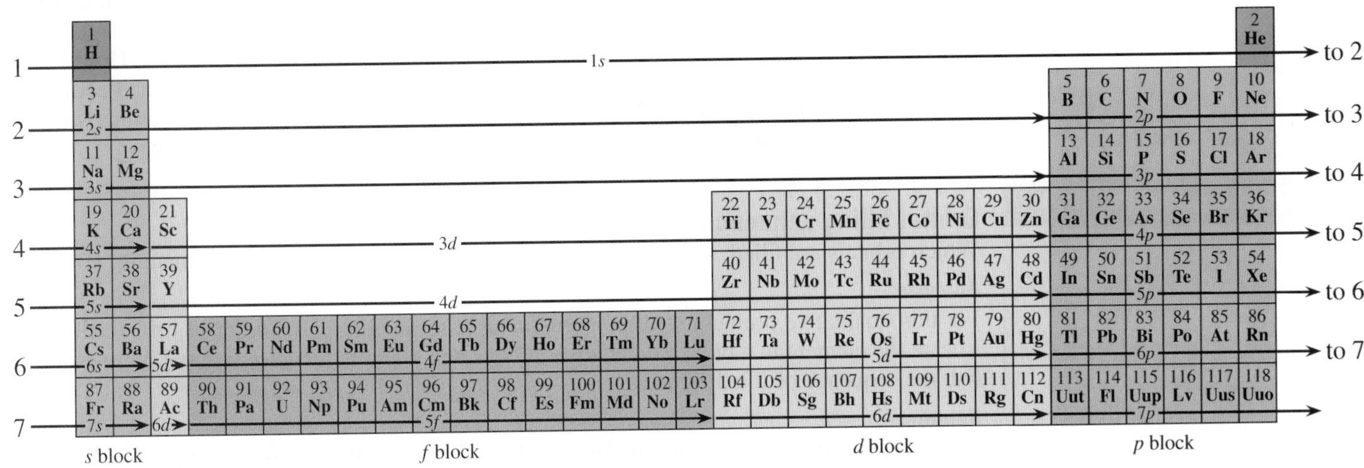

FIGURE 4.39 The periodic table in block form, showing the filling sequence of the atomic orbitals. Filling proceeds from left to right across each row and from the right end of each row to the left end of the succeeding row.

The number of elements per row increases as we increase n. The first period contains only hydrogen and helium. Then there are two 8-element periods, followed by two 18-element periods, and finally two 32-element periods.

In figure 4.40, each row is labelled with the highest principal quantum number of any of its occupied orbitals. For example, elements of the third row (Na to Ar) have electrons in orbitals with $n = 3$ (in addition to electrons with $n = 1$ and 2). Each column is labelled with its group number, starting with group 1 on the left and proceeding to group 18 on the right (the f block does not have group numbers). In general, elements in the same group of the periodic table have the same arrangement of electrons in their highest energy occupied orbitals, giving rise to their similar chemical properties.

Group number

Row number	1	2	3	4	5	6	7	8	9	10	11	12	13	14	15	16	17	18
1	1 H $1s^1$																	2 He $1s^2$
	main group (s block)												main group (p block)					
2	3 Li $2s^1$	4 Be $2s^2$											5 B $2s^22p^1$	6 C $2s^22p^2$	7 N $2s^22p^3$	8 O $2s^22p^4$	9 F $2s^22p^5$	10 Ne $2s^22p^6$
3	11 Na $3s^1$	12 Mg $3s^2$	transition metals (d block)										13 Al $3s^23p^1$	14 Si $3s^23p^2$	15 P $3s^23p^3$	16 S $3s^23p^4$	17 Cl $3s^23p^5$	18 Ar $3s^23p^6$
4	19 K $4s^1$	20 Ca $4s^2$	21 Sc $4s^23d^1$	22 Ti $4s^23d^2$	23 V $4s^23d^3$	24 Cr $4s^13d^5$	25 Mn $4s^23d^5$	26 Fe $4s^23d^6$	27 Co $4s^23d^7$	28 Ni $4s^23d^8$	29 Cu $4s^13d^{10}$	30 Zn $4s^23d^{10}$	31 Ga $4s^24p^1$	32 Ge $4s^24p^2$	33 As $4s^24p^3$	34 Se $4s^24p^4$	35 Br $4s^24p^5$	36 Kr $4s^24p^6$
5	37 Rb $5s^1$	38 Sr $5s^2$	39 Y $5s^24d^1$	40 Zr $5s^24d^2$	41 Nb $5s^14d^4$	42 Mo $5s^14d^5$	43 Tc $5s^24d^5$	44 Ru $5s^14d^7$	45 Rh $5s^14d^8$	46 Pd $4d^{10}$	47 Ag $5s^14d^{10}$	48 Cd $5s^24d^{10}$	49 In $5s^25p^1$	50 Sn $5s^25p^2$	51 Sb $5s^25p^3$	52 Te $5s^25p^4$	53 I $5s^25p^5$	54 Xe $5s^25p^6$
6	55 Cs $6s^1$	56 Ba $6s^2$	57–71 *	72 Hf $6s^25d^2$	73 Ta $6s^25d^3$	74 W $6s^25d^4$	75 Re $6s^25d^5$	76 Os $6s^25d^6$	77 Ir $6s^25d^7$	78 Pt $6s^15d^9$	79 Au $6s^15d^{10}$	80 Hg $6s^25d^{10}$	81 Tl $6s^26p^1$	82 Pb $6s^26p^2$	83 Bi $6s^26p^3$	84 Po $6s^26p^4$	85 At $6s^26p^5$	86 Rn $6s^26p^6$
7	87 Fr $7s^1$	88 Ra $7s^2$	89–103 **	104 Rf $7s^26d^2$	105 Db $7s^26d^3$	106 Sg $7s^26d^4$	107 Bh $7s^26d^5$	108 Hs $7s^26d^6$	109 Mt $7s^26d^7$	110 Ds $7s^26d^8$	111 Rg $7s^16d^{10}$	112 Cn $7s^26d^{10}$	113 Uut $7s^27p^1$	114 Fl $7s^27p^2$	115 Uup $7s^27p^3$	116 Lv $7s^27p^4$	117 Uus $7s^27p^5$	118 Uuo $7s^27p^5$

lanthoids and actinoids (f block)

6	lanthanoid series *	57 La $6s^25d^1$	58 Ce $6s^24f^15d^1$	59 Pr $6s^24f^3$	60 Nd $6s^24f^4$	61 Pm $6s^24f^5$	62 Sm $6s^24f^6$	63 Eu $6s^24f^7$	64 Gd $6s^24f^75d^1$	65 Tb $6s^24f^9$	66 Dy $6s^24f^{10}$	67 Ho $6s^24f^{11}$	68 Er $6s^24f^{12}$	69 Tm $6s^24f^{13}$	70 Yb $6s^24f^{14}$	71 Lu $6s^24f^{14}5d^1$
7	actinoid series **	89 Ac $7s^26d^1$	90 Th $7s^26d^2$	91 Pa $7s^25f^26d^1$	92 U $7s^25f^36d^1$	93 Np $7s^25f^46d^1$	94 Pu $7s^25f^6$	95 Am $7s^25f^7$	96 Cm $7s^25f^76d^1$	97 Bk $7s^25f^9$	98 Cf $7s^25f^{10}$	99 Es $7s^25f^{11}$	100 Fm $7s^25f^{12}$	101 Md $7s^25f^{13}$	102 No $7s^25f^{14}$	103 Lr $7s^25f^{14}6d^1$

FIGURE 4.40 The periodic table with its rows and blocks labelled to show the relationship between sectors of the table and ground-state configurations. Rows are labelled with the highest principal quantum number of the occupied orbitals, and each block is labelled with the letter (s, p, d, f) indicating the orbital set that is filling. Electron configurations for elements 103–118 are tentative, and are based on the electron configurations of other elements in the same group.

Every period ends at the end of the p block. This indicates that, when the np orbitals are full, the next orbital to accept electrons is the $(n + 1)s$ orbital. For example, after filling the $3p$ orbitals from Al ($Z = 13$) to Ar ($Z = 18$), the next element, potassium, has its final electron in the $4s$ orbital rather than in one of the $3d$ orbitals. This confirms (see figure 4.38a) that the $4s$ orbital of the potassium atom is of lower energy than the $3d$ orbital. According to the Aufbau principle, the $3d$ orbitals fill after the $4s$ orbital is full, starting with scandium ($Z = 21$).

A similar situation exists at the end of the next row. When the $4p$ orbital is full (Kr, $Z = 36$), the next element (Rb, $Z = 37$) has an electron in the $5s$ orbital rather than either a $4d$ or $4f$ orbital. In fact, electrons are not added to the $4f$ orbitals until element 58, after the $5s$, $5p$ and $6s$ orbitals have filled.

The seven f orbitals can hold 14 electrons, and therefore there should be 14 f-block elements. However, the periodic table shown in figure 4.40 shows 15 elements in both the lanthanoid and actinoid series. This reflects the current debate over whether it is the first element or the last in each row (La/Ac and Lu/Lr, respectively) that is part of the d block with the other then being part of the f block. Older versions of the periodic table place La and Ac directly beneath Sc and Y, followed by 14 lanthanoid and actinoid elements. However, it is argued that the chemistry of both La and Ac resembles that of the lanthanoids and actinoids, respectively, more than that of transition elements, despite their f^0 electron configurations (indeed, the terms lanthanoids and actinoids derive from lanthanum and actinium, respectively).

WORKED EXAMPLE 4.10

Orbital filling sequence

Which orbitals are filled, and which set of orbitals is partially filled, in a germanium atom?

Analysis

For this qualitative problem, use the periodic table to determine the order of orbital filling. Locate the element in a block and identify its row and column, then use figure 4.39 to establish the sequence of filled orbitals.

Solution

Germanium is element 32. Consult figure 4.39 to determine that Ge is in group 14, row 4 of the p block.

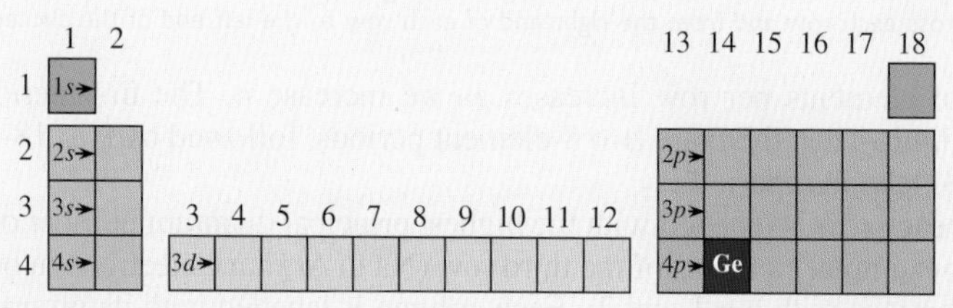

Starting from the top left of the periodic table and working left to right across the rows until we reach Ge, we identify the filled orbitals: $1s$, $2s$, $2p$, $3s$, $3p$, $4s$ and $3d$. Germanium is in row 4 of the p block, so the $4p$ set of orbitals is partially filled.

Is our answer reasonable?

Ge has $Z = 32$, meaning its neutral atoms contain 32 electrons. We can count how many electrons each orbital can hold: 2 for each s orbital, 6 for each p orbital set and 10 for each d orbital set. $2(1s) + 2(2s) + 6(2p) + 2(3s) + 6(3p) + 2(4s) + 10(3d) = 30$ electrons, leaving 2 in the partially filled $4p$ orbital set.

PRACTICE EXERCISE 4.10

Determine which orbitals are filled and which set is partially filled for the element Zr.

Valence electrons

The chemical behaviour of an atom is determined by its size and the electrons that are accessible to an approaching chemical reagent. Accessibility can be considered from a spatial or an energetic point of view. An electron is accessible *spatially* when it occupies one of the largest orbitals of the atom. Electrons on the perimeter of the atom, farthest from the nucleus, are the first ones encountered by an incoming chemical reagent. An electron is accessible *energetically* when it occupies one of the highest energy occupied orbitals of the atom. Electrons in high-energy orbitals are more chemically active than electrons in lower energy orbitals.

Similar electron accessibility generates similar chemical behaviour. For example, iodine has many more electrons than chlorine, but these two elements display similar chemical behaviour, as reflected by their placement in the same group of the periodic table. This is because the chemistry of chlorine and iodine is determined by the number of electrons in their largest and highest energy occupied orbitals: $3s$ and $3p$ for chlorine, and $5s$ and $5p$ for iodine. Each of these elements has seven accessible electrons, and this accounts for their chemical similarities.

Accessible electrons are called **valence electrons**, and inaccessible electrons are called **core electrons**. Valence electrons may participate in chemical reactions, while core electrons do not. In general, valence electrons are those that have been added in the period of the periodic table (see figure 4.40) that the atom is located in. For example, carbon with $Z = 6$ is located in period 2, so its valence electrons are those in the $2s$ and $2p$ orbitals (totalling 4). The situation is similar for phosphorus in period 3, which has $3s$ and $3p$ valence electrons (totalling 5). Manganese in period 4 has valence electrons in the $4s$ and $3d$ orbitals (totalling 7).

The nearly equal energies of *ns* and $(n - 1)d$ orbitals create some ambiguity about valence and core electrons for elements in the *d* and *f* blocks. For example, titanium forms a chloride and an oxide with chemical formulae consistent with four valence electrons: $TiCl_4$ and TiO_2. This shows that the two 3*d* electrons of titanium participate in chemical reactions. On the other hand, zinc forms compounds, such as $ZnCl_2$ and ZnO, with only two electrons involved in its chemistry. This chemical behaviour indicates that zinc's ten 3*d* electrons, which completely fill the 3*d* orbitals, are inaccessible for chemical reactions. When the *d* orbitals are partially filled, the *d* electrons participate in chemical reactions, but, when the *d* orbitals are completely filled, the *d* electrons do not generally participate in reactions (see below for a few exceptions). The predominant oxidation state of 3+ throughout the *f*-block elements also suggests that generally only one *d* or *f* electron (see ground state configurations in figure 4.40) is accessible for chemical reactions. The definitions of core and valence electrons do not depend on whether the electrons are used in chemical reactions, but rather whether they are in an accessible energy level or not.

We find that the number of valence electrons can be determined easily from group numbers in the periodic table. For groups 1–10, the number of valence electrons equals the group number. As examples, potassium and rubidium, members of group 1, have just one valence electron each. Tungsten, in group 6, has six valence electrons: two 6*s* electrons and four 5*d* electrons. For groups 12–18, the number of valence electrons equals the group number minus 10 (the number of electrons it takes to fill the *d* orbitals). Thus, antimony and nitrogen, in group 15, have $15 - 10 = 5$ valence electrons each (two *s* and three *p*). For group 11, the number of valence electrons is expected to be 1 (the one *s* electron). However, often compounds of these elements involve their *d* electrons.

4.7 Electron configurations

A complete specification of which energy levels an atom's electrons populate is called an **electron configuration**. There are three common ways to represent electron configurations. One is a complete specification of quantum numbers. The second is a shorthand notation from which the quantum numbers can be inferred. The third is a diagrammatic representation of orbital energy levels and their occupancies.

A list of the values of all quantum numbers is easy for the single electron in a hydrogen atom.

- $n = 1, l = 0, m_l = 0, m_s = +\frac{1}{2}$
- $n = 1, l = 0, m_l = 0, m_s = -\frac{1}{2}$

Either designation is equally valid, because under normal conditions these two states are equal in energy. In a large collection of hydrogen atoms, half the atoms have one designation and the other half have the other designation.

As the number of electrons in an atom increases, a listing of all quantum numbers quickly becomes tedious. For example, iron, with 26 electrons, would require the specification of 26 sets of four quantum numbers. To save time and space, chemists have devised a shorthand notation to write electron configurations. The orbital symbols (1*s*, 2*p*, 4*d* etc.) are followed by superscripts designating how many electrons are in each set of orbitals. The compact configuration for a hydrogen atom is $1s^1$, indicating 1 electron in the 1*s* orbital. For an iron atom the compact configuration is $1s^22s^22p^63s^23p^64s^23d^6$.

The third way to represent an electron configuration uses an energy level diagram to designate orbitals. Orbital energy levels are indicated by a horizontal line, and these are arranged vertically in order of increasing energy. Each electron is represented by an arrow and is placed on the appropriate horizontal line. The direction of the arrow indicates the value of m_s. By convention, we fill orbitals starting from the left-hand side, and the arrow points upward for $m_s = +\frac{1}{2}$ and downward for $m_s = -\frac{1}{2}$. The configuration of hydrogen can be represented by a single arrow in a 1*s* orbital.

1*s* ↑ or 1*s* ↓

A neutral helium atom has two electrons. To write the ground-state electron configuration of He, we apply the Aufbau principle. One unique set of quantum numbers is assigned to each electron, moving from the lowest energy orbital upward until all electrons have been assigned. The lowest energy orbital is always 1*s* ($n = 1, l = 0, m_l = 0$). Both helium electrons can occupy the 1*s* orbital, provided one of them has $m_s = +\frac{1}{2}$ and the other has $m_s = -\frac{1}{2}$. Below are the three representations of helium's ground-state electron configuration.

$n = 1, l = 0, m_l = 0, m_s = +\frac{1}{2}$
$n = 1, l = 0, m_l = 0, m_s = -\frac{1}{2}$ $1s^2$ 1*s* ↑↓

The two electrons in this configuration are said to be paired electrons, meaning that they are in the same orbital, with opposing spins. Opposing spins cancel, so paired electrons have zero net spin.

A lithium atom has three electrons. The first two electrons fill lithium's lowest possible energy level, the 1*s* orbital, and the third electron occupies the 2*s* orbital. The three representations for the ground-state electron configuration of a lithium atom are shown below.

$$n = 2, l = 0, m_l = 0, m_s = +\tfrac{1}{2}$$
$$n = 1, l = 0, m_l = 0, m_s = +\tfrac{1}{2} \qquad 1s^22s^1$$
$$n = 1, l = 0, m_l = 0, m_s = -\tfrac{1}{2}$$

E: 2s ↑; 1s ↑↓

The set $n = 2$, $l = 0$, $m_l = 0$, $m_s = -\frac{1}{2}$ is equally valid for the third electron.

The next atoms of the periodic table are beryllium and boron. You should be able to write the three different representations (the set of quantum numbers, the shorthand notation and the energy level diagram) for the ground-state configurations of these elements. The filling principles are the same as we move to higher atomic numbers.

WORKED EXAMPLE 4.11

An electron configuration

Construct an energy level diagram and the compact notation of the ground-state configuration of aluminium. Provide one set of valid quantum numbers for the highest-energy electron.

Analysis

First consult the periodic table to locate aluminium and determine how many electrons are present in a neutral atom. Then construct the electron configuration using the patterns of the periodic table.

Solution

Aluminium has $Z = 13$, so a neutral atom of Al has 13 electrons. Place the 13 electrons sequentially, using arrows, into the lowest energy orbitals available. The $n = 1$ orbital is filled by two electrons, eight electrons fill the $n = 2$ orbitals, two electrons fill the 3*s* orbital, and one electron goes in a 3*p* orbital.

The last electron could be placed in any of the 3*p* orbitals, because these three orbitals are equal in energy. The final electron could also be given either spin orientation. By convention, we place electrons in unfilled orbitals starting with the left-hand side, with spins pointed up.

The compact notation is $1s^22s^22p^63s^23p^1$.

E: 3p ↑ — —; 3s ↑↓; 2p ↑↓ ↑↓ ↑↓; 2s ↑↓; 1s ↑↓

The highest energy electron is in a 3*p* orbital, meaning $n = 3$ and $l = 1$. The value of m_l can be any of three values: +1, −1 or 0. The spin quantum number, m_s, can be $+\frac{1}{2}$ or $-\frac{1}{2}$. One valid set of quantum numbers is:

$$n = 3, l = 1, m_l = 1, m_s = +\tfrac{1}{2}$$

You should be able to write the other five possible sets.

Is our answer reasonable?

Aluminium, in group 13, has three valence electrons. The configurations show three electrons with $n = 3$, so the configuration is consistent with the valence electron count.

PRACTICE EXERCISE 4.11 Determine the energy level diagram and compact notation for the electron configuration of the fluorine atom.

Electron configurations become longer as the number of electrons increases. To avoid the need to write long configurations, chemists make use of the regular pattern for the electrons with lower principal quantum numbers. Compare the compact electron configurations of neon and aluminium shown below.

Ne (10 electrons) $1s^22s^22p^6$

Al (13 electrons) $1s^22s^22p^63s^23p^1$

The description of the first 10 electrons in the configuration of aluminium is identical to that of neon. We can take advantage of this pattern and represent that portion as [Ne]. With this notation, we can write the abbreviated compact electron configuration of Al as [Ne]$3s^23p^1$. The element at the end of each row of the periodic table has a **noble gas configuration**. These configurations can be written in the following notation.

Notation	Configuration	Element
[He]	$1s^2$	He (2 electrons)
[Ne]	[He]$2s^22p^6$	Ne (10 electrons)
[Ar]	[Ne]$3s^23p^6$	Ar (18 electrons)
[Kr]	[Ar]$4s^23d^{10}4p^6$	Kr (36 electrons)
[Xe]	[Kr]$5s^24d^{10}5p^6$	Xe (54 electrons)
[Rn]	[Xe]$6s^25d^{10}4f^{14}6p^6$	Rn (86 electrons)

To write the abbreviated compact electron configuration of any other element, we first consult the periodic table to find its location relative to the noble gases. Then we specify the configuration of the preceding noble gas and build the remaining portion of the configuration according to the Aufbau principle.

WORKED EXAMPLE 4.12

A shorthand electron configuration

Determine the electron configuration of indium, first in abbreviated compact form and then in compact form, and draw the energy level diagram.

Analysis

Locate the element in the periodic table, and find the nearest noble gas with a smaller atomic number. Start with the electron configuration of that noble gas, and add enough additional electrons to the next-filling orbitals to give the neutral atom.

Solution

Indium, In, ($Z = 49$) is in row 5, group 13. The nearest noble gas of smaller Z is Kr ($Z = 36$). Thus, the configuration of In has 36 electrons in the Kr configuration and 13 additional electrons. The last orbital to fill in Kr is $4p$, and the periodic table shows that the next orbitals to fill are the $5s$, $4d$ and $5p$ orbitals.

Abbreviated compact electron configuration of indium: [Kr]$5s^24d^{10}5p^1$

To write the compact configuration, expand the krypton configuration.

[Kr] = [Ar]$4s^23d^{10}4p^6$ = $1s^22s^22p^63s^23p^64s^23d^{10}4p^6$

Compact configuration of In (49 electrons): $1s^22s^22p^63s^23p^64s^23d^{10}4p^65s^24d^{10}5p^1$

Energy level diagram:

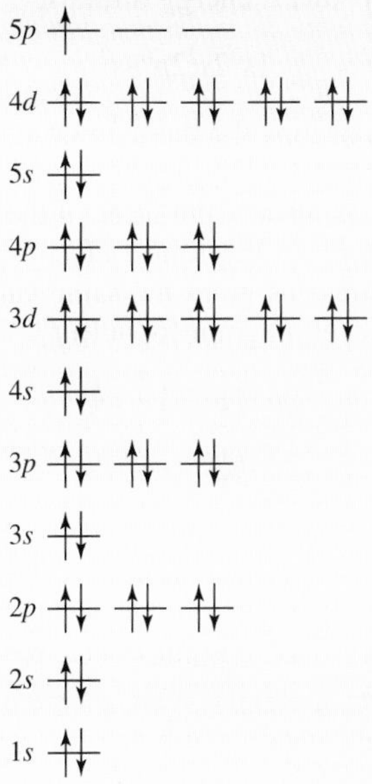

5p ↑ __ __

4d ↑↓ ↑↓ ↑↓ ↑↓ ↑↓

5s ↑↓

4p ↑↓ ↑↓ ↑↓

3d ↑↓ ↑↓ ↑↓ ↑↓ ↑↓

4s ↑↓

3p ↑↓ ↑↓ ↑↓

3s ↑↓

2p ↑↓ ↑↓ ↑↓

2s ↑↓

1s ↑↓

Note that it is usual to indicate empty orbitals in an energy level diagram when they are part of a set of degenerate orbitals. For example, in the energy level diagram above, we show the two empty p orbitals of the $5p$ set because one of the set contains a single electron.

Is our answer reasonable?

Indium, in group 13, has three valence electrons. The configurations show three electrons with $n = 5$, so the configuration is consistent with the valence electron count.

PRACTICE EXERCISE 4.12

Determine the abbreviated compact notation for the electron configuration of the cadmium atom.

Electron–electron repulsion

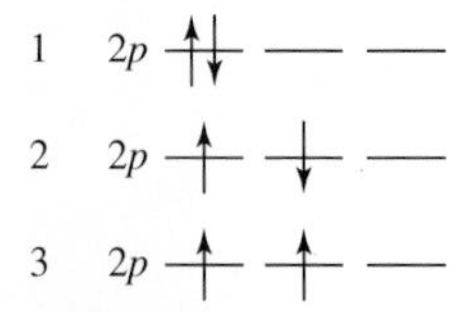

FIGURE 4.41 Three different arrangements of two 2*p* electrons that obey the Pauli exclusion principle.

The Aufbau principle allows us to assign quantum numbers to aluminium's 13 electrons without ambiguity. The first 12 electrons fill the 1*s*, 2*s*, 2*p* and 3*s* energy levels, and the last electron can occupy any 3*p* orbital with either spin orientation. But what happens when more than one electron must be placed in a *p* energy level? Carbon atoms, for example, have six electrons, two of which occupy 2*p* orbitals. How should these two electrons be arranged in the 2*p* orbitals? Figure 4.41 shows three arrangements of the two 2*p* elements that obey the Pauli exclusion principle.

1. The electrons could be paired in the same 2*p* orbital (same m_l value but different m_s values).
2. The electrons could occupy different 2*p* orbitals with opposite spin orientations (different m_l values and different m_s values).
3. The electrons could occupy different 2*p* orbitals with the same spin orientation (different m_l values but the same m_s value).

Note that there are a total of 15 possible arrangements. These three arrangements have different energies because electrons that are close together repel each other more than electrons that are far apart. As a result, for two or more degenerate orbitals, the lowest energy situation results when electrons occupy the orbitals that keep them furthest apart. Placing two electrons in different *p* orbitals keeps them relatively far apart, so an atom is of lower energy with the two electrons in different *p* orbitals. Thus, arrangements 2 and 3 are of lower energy than arrangement 1.

Arrangements 2 and 3 look spatially equivalent, but experiments show that a configuration that gives unpaired electrons the same spin orientation is always of lower energy than one that gives them opposite orientations. **Hund's rule** summarises the way in which electrons occupy orbitals of equal energies. *The lowest energy configuration involving orbitals of equal energies is the one with the maximum number of unpaired electrons with the same spin orientation.*

According to Hund's rule, arrangement 3 of figure 4.41 is consistent with the ground-state configuration for carbon atoms.

WORKED EXAMPLE 4.13

Applying Hund's rule

Write the compact electron configuration and draw the ground-state orbital energy level diagram for the valence electrons in a sulfur atom.

Analysis

From the periodic table, we see that sulfur has 16 electrons and is in the *p* block, group 16. To build the ground-state configuration, apply the Aufbau and the Pauli exclusion principles and then apply Hund's rule if needed.

Solution

The first 12 electrons fill the four lowest energy orbitals.

$$1s^22s^22p^63s^2$$

Sulfur's remaining four electrons occupy the three 3*p* orbitals. The complete configuration is $1s^22s^22p^63s^23p^4$ or, in abbreviated form, $[Ne]3s^23p^4$.

To minimise electron–electron repulsion, put three of the 3*p* electrons in different orbitals, all with the same spin, and then place the fourth electron, with opposite spin, in the first orbital. In accord with Hund's rule, this gives the same value of m_s to all electrons that are not paired. Below is the energy level diagram for sulfur's valence electrons.

3p ↑↓ ↑ ↑

3s ↑↓

Is our answer reasonable?

Sulfur, in row 3 and group 16, has six valence electrons. The configurations show six electrons with $n = 3$, so the configuration is consistent with the valence electron count. The electrons are spread among the three 3*p* orbitals, which minimises electron–electron repulsion.

PRACTICE EXERCISE 4.13

Determine the abbreviated compact electron configuration for the nitrogen atom and write one valid set of quantum numbers for its valence electrons.

Orbitals with nearly equal energies

The filling order embodied in the periodic table predicts a regular progression of ground-state configurations. Experiments show, however, that some elements have ground-state configurations different from the predictions of the regular progression. Among the first 40 elements, there are only two exceptions: copper and chromium. Chromium ($Z = 24$) is in group 6, four elements into the *d* block. We would predict that the ground-state valence configuration of chromium should be $4s^23d^4$.

Instead, experiments show that the ground-state configuration of this element is $4s^13d^5$. Likewise, the configuration of copper ($Z = 29$) is $4s^13d^{10}$ rather than the predicted $4s^23d^9$.

Look again at figure 4.38a, which shows that these two sets of orbitals are nearly the same in energy. Each $(n - 1)d$ orbital has nearly the same energy as its ns counterpart. In addition, each $(n - 2)f$ orbital has nearly the same energy as its $(n - 1)d$ counterpart. Table 4.3 lists these orbitals and the atomic numbers for which the filling sequence differs from the expected pattern. These configurations are also indicated in figure 4.40 on page 147.

TABLE 4.3 Atomic orbitals with nearly equal energies.

Orbitals	Atomic numbers affected	Example
$4s$, $3d$	24, 29	Cr: [Ar]$4s^13d^5$
$5s$, $4d$	41, 42, 44, 45, 46, 47	Ru: [Kr]$5s^14d^7$
$6s$, $5d$, $4f$	57, 58, 64, 78, 79	Au: [Xe]$6s^14f^{14}5d^{10}$
$6d$, $5f$	89, 91–93, 96	U: [Rn]$7s^25f^36d^1$

Often, an s orbital contains only one electron rather than two. Five of the exceptional ground-state configurations have a common pattern and are easy to remember: Cr and Mo are s^1d^5, and Cu, Ag and Au are s^1d^{10}. The other exceptional cases follow no recognisable patterns, because they are generated by subtle interactions between all the electrons. Among elements with valence electrons filling orbitals with nearly equal energies, several factors help to determine the ground-state configuration. The details are beyond the scope of this book, except that you should recognise that even a subtle change can cause variations in the filling pattern predicted by the periodic table.

Configurations of ions

The electron configurations of atomic ions are written using the same procedure as for neutral atoms, taking into account the proper number of electrons. An anion has one additional electron for each unit of negative charge. A cation has one fewer electron for each unit of positive charge.

For most atomic ions, the electron configuration can be deduced easily from that of the corresponding neutral atom. For example, Na^+, Ne and F^- all contain 10 electrons, and each has the configuration $1s^22s^22p^6$. Atoms and ions that have the same number of electrons are said to be **isoelectronic**.

In general, you would expect that the last electron to be added would be the first electron to be lost on ionisation. The nearly equal energies of ns and $(n - 1)d$ orbitals causes the configurations of transition metal cations to differ from the configurations predicted by the filling pattern of the periodic table. Remember that the energy ranking of orbitals such as $4s$ and $3d$ depends on a balance of several factors, and that even a small variation in that balance can change the order of the orbitals. This feature is particularly important for the transition metals, for which the d orbitals are occupied. Experiments show that in transition metal cations the $(n - 1)d$ orbitals are *always* of lower energy than the ns orbitals. For example, an Fe^{3+} cation contains 23 electrons. The first 18 electrons fill the $1s$, $2s$, $2p$, $3s$ and $3p$ orbitals, as predicted by the periodic table. However, the five remaining electrons populate the $3d$ set, leaving the $4s$ orbital empty. Thus the configuration of the Fe^{3+} cation is [Ar]$3d^5$.

Vanadium, V, atoms ([Ar]$4s^23d^3$) and Fe^{3+} cations ([Ar]$3d^5$) have different configurations, even though each has 23 electrons.

WORKED EXAMPLE 4.14

Configuration of a cation

What is the ground-state electron configuration of a Cr^{3+} cation?

Analysis

Use the Aufbau principle, remembering that for transition metals, the $4s$ orbital and $3d$ orbitals swap positions once the levels are populated. Thus $4s$ electrons will be lost before $3d$ electrons.

Solution

A neutral chromium atom has 24 electrons, so the corresponding Cr^{3+} cation has 21 electrons. The first 18 electrons follow the usual filling order to give the argon core configuration: $1s^22s^22p^63s^23p^6$, or [Ar]. Place the remaining three electrons in the $3d$ set of orbitals, following Hund's rule: [Ar]$3d^3$.

[Ar] $4s$ —
$3d$ ↑ ↑ ↑ — —

PRACTICE EXERCISE 4.14 Determine the ground-state electron configuration (abbreviated compact form) of a Ru^{3+} cation.

We can summarise the discussion in this section into guidelines for determining electron configurations for atoms and ions.

1. Count the total number of electrons for a neutral atom.
   (a) Add electrons for anions.
   (b) Subtract electrons for cations.
2. Fill orbitals to match the nearest noble gas of smaller atomic number.
3. Add remaining electrons to the next orbitals to be filled according to the Aufbau and Pauli exclusion principles, as well as Hund's rule.
   (a) For neutral atoms and anions, place electrons in *ns* before $(n - 1)d$ and $(n - 2)f$, keeping in mind that the order changes once the *d* and *f* levels are at least partially occupied.
   (b) For cations, place electrons in the $(n - 1)d$ before the *ns* orbitals.
4. Look for exceptions and correct the configuration if necessary.

Magnetic properties of atoms

How do we know that an Fe^{3+} ion in its ground state has the configuration $[Ar]3d^5$ rather than the $[Ar]4s^23d^3$ configuration predicted by the orbital filling sequence? Remember that electron spin gives rise to magnetic properties. Consequently, any atom or ion with unpaired electrons has non-zero net spin and is attracted by a strong magnet. We can divide the electrons of an atom or ion into two categories with different spin characteristics. In filled orbitals, all the electrons are paired. Each electron with spin orientation $+\frac{1}{2}$ has a partner with spin orientation $-\frac{1}{2}$. The spins of these electrons cancel each other, giving a net spin of zero. An atom or ion with all electrons paired is not attracted by strong magnets and is termed **diamagnetic**. In contrast, spins do not cancel when unpaired electrons are present with spins aligned in the same direction. An atom or ion with unpaired electrons is attracted to strong magnets and is called **paramagnetic**. Moreover, the spins of all the unpaired electrons are additive, so the amount of paramagnetism shown by an atom or ion is proportional to the number of unpaired spins.

In Fe^{3+}, Hund's rule dictates that the five *d* electrons all have the same spin orientation. For these five electrons, the spins all act together, giving a net spin of $5 \times \frac{1}{2} = \frac{5}{2}$. The alternative configuration for Fe^{3+}, $[Ar]4s^23d^3$, is paramagnetic, too, but its net spin is $3 \times \frac{1}{2} = \frac{3}{2}$ as it has only three unpaired electrons. Experiments show that Fe^{3+} has a net spin of $\frac{5}{2}$. In fact, magnetic measurements on a wide range of transition metal cations are all consistent with the $(n - 1)d$ orbitals being occupied rather than the *ns* orbital.

WORKED EXAMPLE 4.15

Unpaired electrons

Which of these species is paramagnetic: F^-, Zn^{2+} or Ti?

Analysis

Paramagnetism results from unpaired spins, which exist only in partially filled sets of orbitals. We need to build the configurations and then look for any orbitals that are partially filled.

Solution

F^-: A fluorine atom has 9 electrons, so F^- has 10 electrons. The configuration is $1s^22s^22p^6$. There are no partially filled orbitals, so the fluoride ion is diamagnetic.

Zn^{2+}: The parent zinc atom has 30 electrons, and the cation has 28, so the configuration for Zn^{2+} is $[Ar]3d^{10}$. Again, there are no partially filled orbitals, so this ion is also diamagnetic.

Ti: A neutral titanium atom has 22 electrons. The ground-state configuration is $[Ar]4s^23d^2$. The spins of the 4*s* electrons cancel. The two electrons in 3*d* orbitals have the same spin orientation, so their effect is additive.

4s ↑↓

3d ↑ ↑ — — —

This atom is paramagnetic, with a net spin of $\frac{1}{2} + \frac{1}{2} = 1$.

Is our answer reasonable?

Filled orbitals always have all electrons paired, and two of these three species have no partially filled orbitals. Only Ti has a partially filled orbital set.

PRACTICE EXERCISE 4.15 Most transition metal cations are paramagnetic. Which cations in the 3*d* transition metal series have net charges less than +4 and are exceptions to this generalisation?

The ground-state configurations of most neutral atoms and many ions contain unpaired electrons, so we might expect most materials to be paramagnetic. On the contrary, most substances are diamagnetic. This is because stable substances seldom contain free atoms. Instead, in molecular substances, atoms are bonded together through pairing of electrons to give molecules (see chapter 5), and such bonding results in the cancellation of overall spin. This is different for compounds of transition metals and lanthanoids and actinoids, since they often have unpaired electrons left over (in partially filled *d* and *f* orbitals). We will examine the magnetic properties of transition metal complexes in detail in chapter 13.

Excited states

The ground-state configuration is the lowest energy arrangement of electrons, so an atom or ion will usually have this configuration. When an atom absorbs energy, however, it can reach an excited state with a new electron configuration. For example, sodium atoms normally have the ground-state configuration [Ne]$3s^1$, but when sodium atoms are in the gas phase, an electrical discharge can induce transfer of the $3s$ electron to a higher energy orbital, such as $4p$. Excited atoms are unstable and spontaneously return to the ground-state configuration, giving up their excess energy in the process.

Excited-state configurations are perfectly valid as long as they meet the restrictions given in table 4.2. In the electrical discharge of a sodium-vapour lamp, for instance, we find some sodium atoms in excited states with configurations such as $1s^22s^22p^63p^1$ or $1s^22s^22p^53s^2$. These configurations use valid orbitals and are in accord with the Pauli exclusion principle, but they describe atoms that are of higher energy than those in the ground state. It should be noted that both the Aufbau principle and Hund's rule can be violated when writing excited-state electronic configurations, but the Pauli exclusion principle must *always* be obeyed.

Excited states play important roles in chemistry. Properties of atoms can be studied by observing excited states. In fact, chemists and physicists use the characteristics of excited states extensively to probe the structure and reactivity of atoms, ions and molecules. Excited states also have practical applications. For example, sodium-vapour lamps, which are used for street lighting, use the emissions from excited sodium atoms returning to their ground states, and the dazzling colours of a fireworks display come from photons emitted by various metal ions in excited states.

4.8 Periodicity of atomic properties

The periodicity of physical and chemical properties, which is summarised in the periodic table of the elements, is one of the most useful organising principles in chemistry. The regular periodic trends can be explained using electron configurations and effective nuclear charges. The most important of those trends is the atomic radius, which we will discuss first, since it influences all other periodic trends.

Atomic radii

The size of an atom is determined by its electron cloud and therefore by the sizes of its orbitals. As we have seen earlier, these sizes are determined by a number of factors, including effective nuclear charge (Z_{eff}), orbital energy and electron distribution.

Assume you are moving from left to right across period 2 of the periodic table ($n = 2$). While Z increases, the electrons that are added do not shield the increasing nuclear charge effectively, so Z_{eff} increases. A larger Z_{eff} exerts a stronger electrostatic attraction on the electron cloud, and this results in smaller orbitals. Moving from *left to right* across that period, orbitals become *smaller*, that is, *have lower energy*.

Now assume you proceed down group 1 of the periodic table. As n increases, we would expect orbitals to increase in energy and become larger. However, as Z increases, we would expect orbitals to become smaller and decrease in energy. Which trend dominates here? Recall that the number of core electrons increases as we move down any group. For example, sodium ($Z = 11$) has 10 core electrons and 1 valence electron. In the next period, potassium ($Z = 19$) has 18 core electrons and 1 valence electron. The shielding provided by potassium's additional 8 core electrons largely cancels the effect of the additional 8 protons in its nucleus. Consequently, increased shielding largely offsets the increase in Z value from one period to the next and, therefore, valence orbitals become larger and increase in energy from top to bottom of a group.

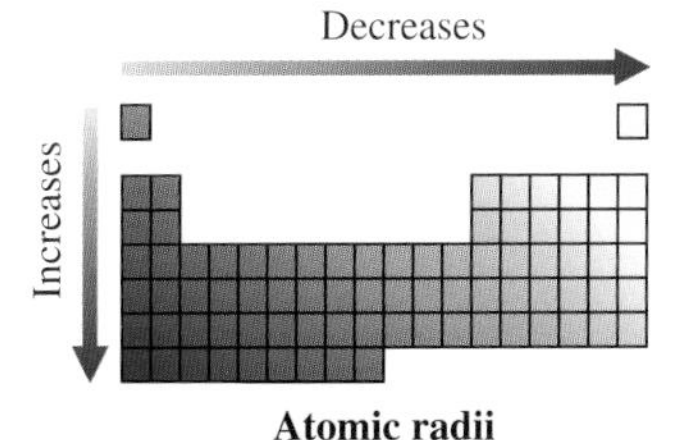

FIGURE 4.42 Representation of the periodic table showing that atomic radius decreases from left to right within a row and increases from top to bottom within a group.

In summary we can say that *atomic size decreases from left to right and increases from top to bottom of the periodic table* (figure 4.42).

A convenient measure of atomic size is the radius of the atom. Figure 4.43 (overleaf) shows the trends in **atomic radii**. For example, the atomic radius decreases smoothly across the third

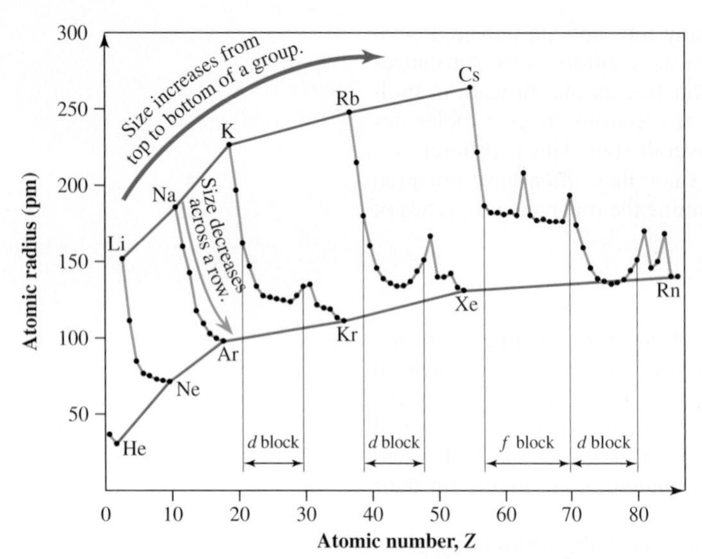

FIGURE 4.43 The radii of atoms vary in periodic fashion (data are for atoms in the gas phase). Atomic radius decreases from left to right within any row (blue lines) and increases from top to bottom within any group (red lines).

period, from 186 pm for sodium to 97 pm for argon (blue arrow). The atomic radius increases smoothly down group 1, from 152 pm for lithium to 265 pm for caesium (red arrow). Notice, however, that the atomic radius changes very little across the *d* and *f* blocks. This is due to shielding. For these elements, the largest orbital is the filled *ns* orbital. Moving from left to right across a row, *Z* increases, but electrons add to the smaller $(n-1)d$ or $(n-2)f$ orbitals. An increase in *Z* by 1 unit is matched by the addition of 1 shielding electron. From the perspective of the outlying *s* orbital, the increases in *Z* are balanced by increased shielding from the added *d* or *f* electrons. Thus, the electron in the outermost occupied orbital, *ns*, is subject to an effective nuclear charge that changes very little across these blocks. As a consequence, the relative changes in atomic size across the *d* and *f* blocks are much smaller than those found in the *s* and *p* blocks.

A knowledge of periodic trends in physical and chemical properties and an understanding of the principles that give rise to these trends are important since they enable you, for example, to predict certain chemical behaviours, even if you are not familiar with a particular element.

WORKED EXAMPLE 4.16

Trends in atomic radii

For each of the following pairs, predict which atom is larger and explain why. (a) Si or Cl (b) S or Se (c) Mo or Ag

Analysis

Qualitative predictions about atomic size can be made on the basis of an atom's position in the periodic table.

Solution

(a) Silicon and chlorine are in the third row of the periodic table.

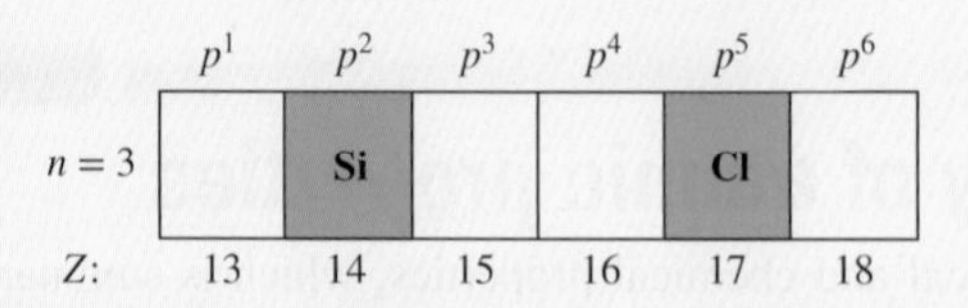

Chlorine's nuclear charge (+17) is larger than silicon's (+14), so chlorine's nucleus exerts a stronger pull on its electron cloud. Chlorine also has three more electrons than silicon, which increases the possibility that shielding effects could counter the extra nuclear charge. Remember, however, that electrons in the same type of orbital do a poor job of shielding one another from the nuclear charge. For Si and Cl, shielding comes mainly from the core electrons, not from the electrons in the 3*p* orbitals. Because shielding effects are similar for these elements, the effective nuclear charge increases from Si to Cl. Therefore, we conclude that chlorine, with its greater nuclear attraction for the electron cloud, is the smaller atom.

(b) Sulfur and selenium are in group 16 of the periodic table. Although they both have the s^2p^2 valence configurations, selenium's least stable electrons are in orbitals with a larger *n* value. Orbital size increases with *n*. Selenium also has a greater nuclear charge than sulfur, which raises the possibility that nuclear attraction could offset increased *n*.

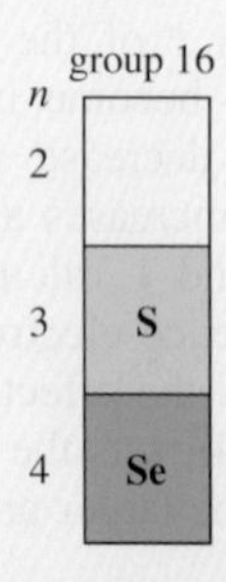

Remember, however, that much of this extra nuclear charge is offset by the shielding influence of the core electrons. Selenium has 18 core electrons and sulfur has 10. Thus, we conclude that selenium, with its larger *n* value, is larger than sulfur.

(c) Molybdenum and silver are in the same row of the d block.

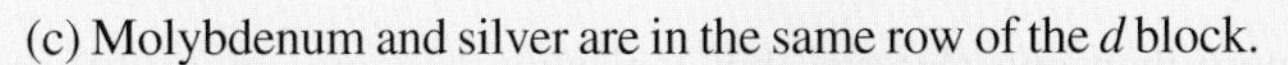

They have the following configurations.

$$\text{Mo} = [\text{Kr}]5s^1 4d^5 \quad \text{Ag} = [\text{Kr}]5s^1 4d^{10}$$

In each case, $5s$ is the largest occupied orbital. The $4d$ orbitals are smaller, with their electron density located mostly inside the $5s$ orbital. Consequently, $4d$ is effective at shielding $5s$. The nuclear charge of silver is five units larger than that of molybdenum, but silver also has five extra shielding electrons in d orbitals. These offset the extra nuclear charge, making Mo and Ag nearly the same size.

Is our answer reasonable?

The trends in figure 4.43 confirm the results. Chlorine lies to the right of silicon in the same row of the periodic table. Size decreases from left to right in any row; thus, chlorine is smaller than silicon. Selenium is immediately below sulfur in the same column of the periodic table. Size increases down a column; thus, selenium is larger than sulfur. Molybdenum and silver occupy the same row of the d block of the periodic table, across which size changes relatively little; thus, molybdenum and silver are nearly the same size.

PRACTICE EXERCISE 4.16

Use periodic trends to determine which of the following are smaller than As and which are larger: P, Ge, Se and Sb.

Ionisation energy

When an atom absorbs a photon, the gain in energy promotes an electron to a higher energy orbital in which the electron is, on average, further from the nucleus and therefore experiences less electrical attraction to the nucleus. If the absorbed photon has enough energy, an electron can be ejected from the atom.

The minimum amount of energy needed to remove an electron from a neutral atom is the first ionisation energy (E_{i1}). Ionisation energies are measured for gaseous elements to ensure that the atoms are isolated from one another.

Variations in ionisation energy mirror variations in orbital energy, because an electron in a higher energy orbital is easier to remove than one in a lower energy orbital.

Figures 4.44 and 4.45 show how the first ionisation energies of gaseous atoms vary with atomic number. Notice the general trends in ionisation energy. Ionisation energy increases from left to right across each period (third period: 496 kJ mol^{-1} for Na to 1520 kJ mol^{-1} for Ar) and decreases from top to bottom of each group (group 18: 2372 kJ mol^{-1} for He to 1037 kJ mol^{-1} for Rn). As with atomic radius, ionisation energy changes less for elements in the d and f blocks, because increased shielding from the d and f orbitals offsets increases in Z. As a rule of thumb, the trend in the ionisation energies is inverse to that of the atomic radii, that is, smaller atoms have higher ionisation energies. Therefore, the trend in the ionisation energies can be rationalised in similar fashion.

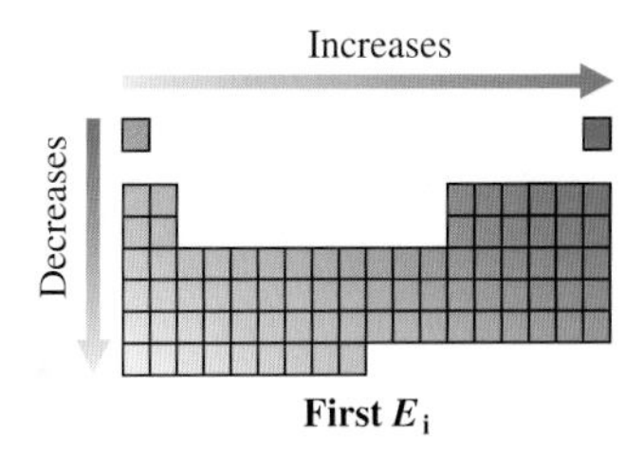

FIGURE 4.44 Representation of the periodic table showing that first ionisation energy increases from left to right within a row and decreases from top to bottom within a group.

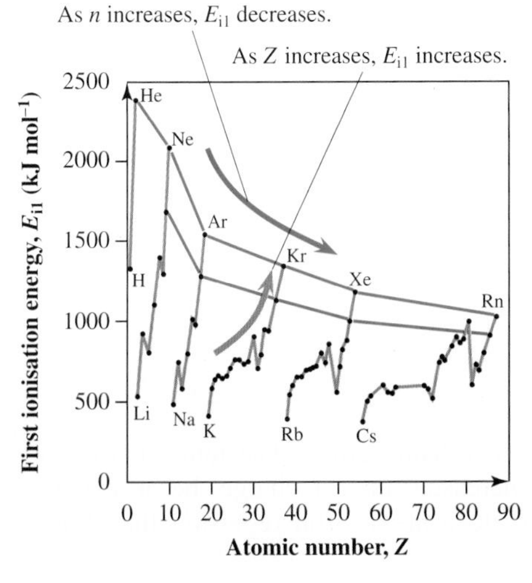

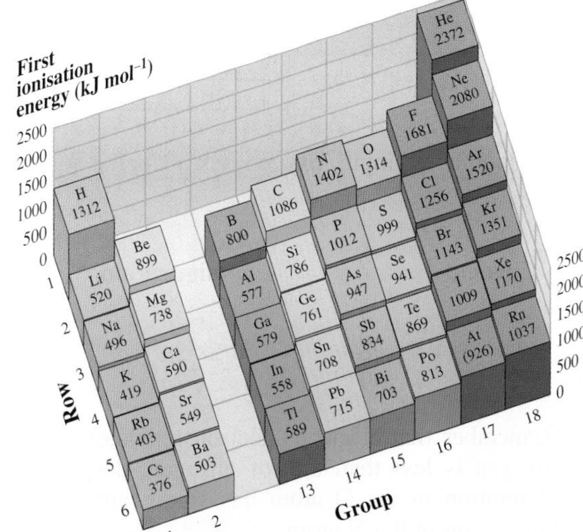

FIGURE 4.45 The first ionisation energy generally increases from left to right across a period (blue arrow) and decreases from top to bottom down a group (red arrow) of the periodic table.

Higher ionisations

A multi-electron atom can lose more than one electron, but ionisation becomes more difficult as cationic charge increases. The first three ionisation energies for a magnesium atom in the gas phase provide an illustration.

Process	Configurations	E_i
$Mg(g) \longrightarrow Mg^+(g) + e^-$	$[Ne]3s^2 \longrightarrow [Ne]3s^1$	738 kJ mol^{-1}
$Mg^+(g) \longrightarrow Mg^{2+}(g) + e^-$	$[Ne]3s^1 \longrightarrow [Ne]$	1450 kJ mol^{-1}
$Mg^{2+}(g) \longrightarrow Mg^{3+}(g) + e^-$	$[Ne] \longrightarrow [He]2s^22p^5$	7730 kJ mol^{-1}

Notice that the second ionisation energy of magnesium is almost twice as large as the first, even though each electron is removed from a 3s orbital. This is because Z_{eff} increases as the number of electrons decreases. That is, the positive charge on the magnesium nucleus remains the same throughout the ionisation process, but the net charge of the electron cloud decreases with each successive ionisation. As the number of electrons decreases, each electron feels a greater electrostatic attraction to the nucleus (due to minimised electron–electron repulsion), resulting in a larger ionisation energy.

The third ionisation energy of magnesium is more than 10 times the first ionisation energy. This large increase occurs because the third ionisation removes a core electron (2p) rather than a valence electron (3s). Removing core electrons from any atom requires much more energy than removing valence electrons. The second ionisation energy of any group 1 metal is substantially larger than the first ionisation energy; the third ionisation energy of any group 2 metal is substantially larger than the first or second ionisation energy, and so on. Appendix G gives the first three ionisation energies for the first 36 elements.

Irregularities in ionisation energies

Ionisation energies deviate somewhat from smooth periodic behaviour. These deviations can be attributed to shielding effects and electron–electron repulsion. Aluminium, for example, has a smaller first ionisation energy than either of its neighbours in row 3 (see figure 4.39).

Element	Z	Atom configuration	E_{i1}	Cation configuration
Mg	12	$[Ne]3s^2$	738 kJ mol^{-1}	$[Ne]3s^1$
Al	13	$[Ne]3s^23p^1$	577 kJ mol^{-1}	$[Ne]3s^2$
Si	14	$[Ne]3s^23p^2$	786 kJ mol^{-1}	$[Ne]3s^23p^1$

The configurations of these elements show that a 3s electron must be removed to ionise magnesium, whereas a 3p electron must be removed to ionise aluminium or silicon. Shielding makes the 3p orbitals of significantly higher energy than a 3s orbital, and this difference in energy more than offsets the increase in nuclear charge in going from magnesium to aluminium. A second electron in a different 3p orbital does not contribute to the shielding of the other 3p electron and thus the increased Z_{eff} means that it is harder to remove an electron from Si than from Al.

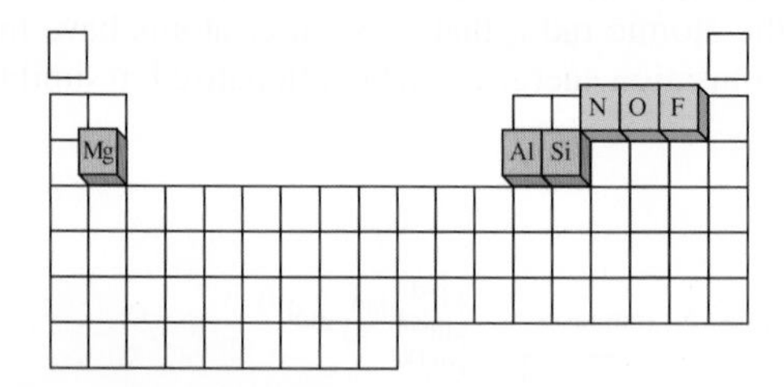

As another example, oxygen has a smaller ionisation energy than either of its neighbours in row 2.

Element	Z	Atom configuration	E_{i1}	Cation configuration
N	7	$1s^22s^22p^3$	1402 kJ mol^{-1}	$1s^22s^22p^2$
O	8	$1s^22s^22p^4$	1314 kJ mol^{-1}	$1s^22s^22p^3$
F	9	$1s^22s^22p^5$	1681 kJ mol^{-1}	$1s^22s^22p^4$

Remember that electron–electron repulsion has a destabilising effect. The ionisation energy of oxygen is less than that of nitrogen, despite the increased nuclear charge, because the p^4 configuration in the O atom has significantly greater electron–electron repulsion than the p^3 configuration in the N atom.

N: 2p ↑ ↑ ↑; 2s ↑↓

O: 2p ↑↓ ↑ ↑; 2s ↑↓

Electron affinity

A neutral atom can add an electron to form an anion. The energy change when an electron is added to an atom in the gas phase is called the **electron affinity (E_{ea})**. Both ionisation energy (E_i) and electron affinity measure the stability of a bound electron, but for different species. Below, for example, are the values for fluorine.

$$F(g) \longrightarrow F^+(g) + e^- \quad E_{i1} = 1681\ \text{kJ mol}^{-1}$$

$$F(g) + e^- \longrightarrow F^-(g) \quad E_{ea} = -322\ \text{kJ mol}^{-1}$$

Energy is released when an electron is added to a fluorine atom to form a fluoride anion. This means that a fluoride anion is of lower energy than a fluorine atom plus a free electron. Another way of saying this is that fluorine atoms have an affinity for electrons.

The energy associated with removing an electron to convert an anion to a neutral atom (that is, the reverse of electron attachment) has the same magnitude as the electron affinity, but the opposite sign. Removing an electron from F^-, for example, requires energy, giving a positive energy change.

$$F^-(g) \longrightarrow F(g) + e^- \quad \Delta E = 322\ \text{kJ mol}^{-1}$$

The Aufbau principle must be obeyed when an electron is added to a neutral atom, so the electron goes into the lowest energy orbital available. Hence, we expect trends in electron affinity to parallel trends in orbital energy. However, electron–electron repulsion and shielding are more important for negative ions than for neutral atoms, so there is no clear trend in electron affinities as n increases. Thus, there is only one general pattern: *Electron affinity tends to become more negative from left to right across a period of the periodic table.*

The plot in figure 4.46 shows how electron affinity changes with atomic number. The blue lines reveal the trend: Electron affinity increases in magnitude across each period of the periodic table. This trend is due to increasing effective nuclear charge, which binds the added electron more tightly to the nucleus. Notice that, in contrast to the pattern for ionisation energies (see figure 4.45), values for electron affinities remain nearly constant among elements occupying the same group of the periodic table.

The electron affinity values for many of the elements shown in figure 4.46 appear to lie on the x-axis. Actually, these elements have positive electron affinities, meaning the resulting anion is of higher energy than the neutral atom. Moreover, the second electron affinity of every element is large and positive. Positive electron affinities cannot be measured directly. Instead, these values are estimated using other methods.

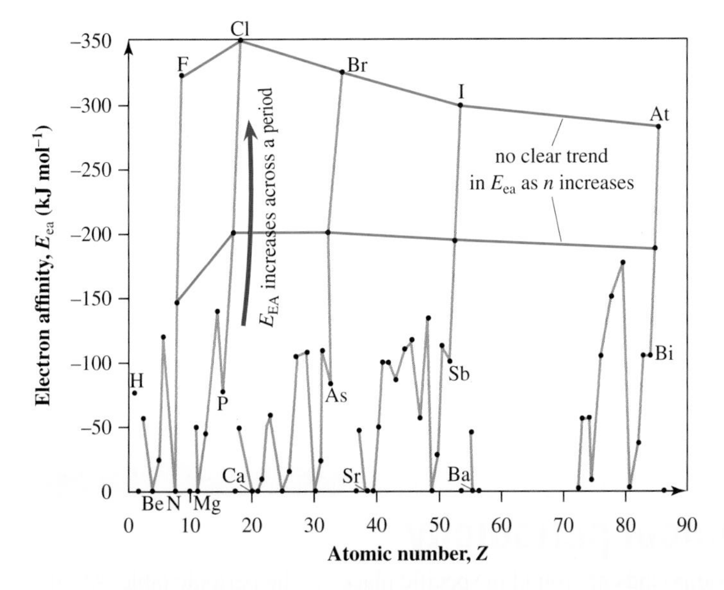

FIGURE 4.46 The electron affinity of atoms varies with atomic number. In moving across any period (blue lines), the electron affinity becomes more negative, but this is the only clear trend.

Although electron affinity values show only one clear trend, there is another recognisable pattern in the values that are positive. When the electron that is added must occupy a new orbital, the resulting anion is unstable. Thus, all the elements of group 2 have positive electron affinities because their valence *ns* orbitals are filled. Similarly, all the noble gases have positive electron affinities because their valence *np* orbitals are filled. Elements with half-filled orbitals also have less negative electron affinities than their neighbours. As examples, N (half-filled 2*p* orbital set) has a positive electron affinity, and so does Mn (half-filled 3*d* orbital set).

Sizes of ions

An atomic cation is always smaller than the corresponding neutral atom. Conversely, an atomic anion is always larger than the neutral atom. Figure 4.47 illustrates these trends, and electron–electron repulsion explains them. A cation has fewer electrons than its parent neutral atom. This reduction in the number of electrons means that the cation's remaining electrons experience less electron–electron repulsion. An anion has more electrons than its parent neutral atom. This increase means that there is more electron–electron repulsion in the anion than in the parent neutral atom.

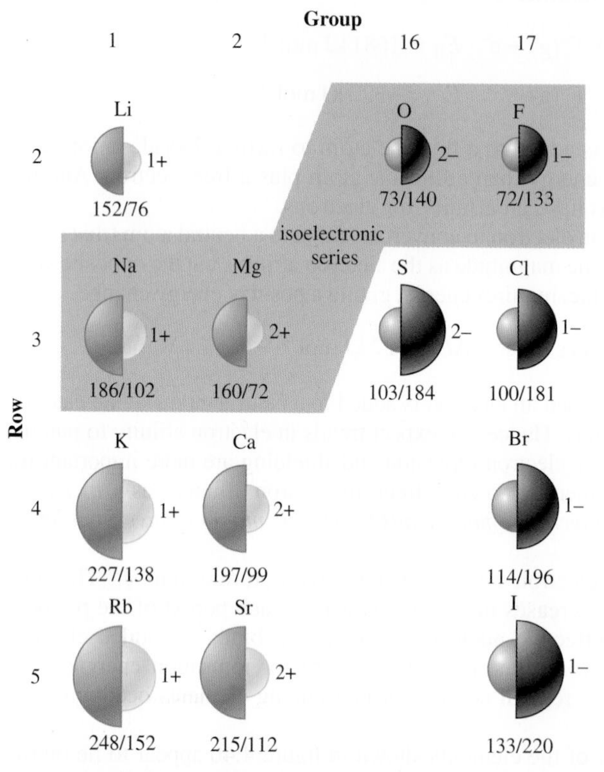

FIGURE 4.47 Comparison of the sizes (radius, *r*, in pm) of neutral atoms (brown) and their ions (cations yellow; anions black) for some elements. The blue highlight indicates isoelectronic ions.

Figure 4.47 also highlights the relationships between isoelectronic species, i.e. those that possess equal numbers of electrons. As noted earlier, the F^- anion and the Na^+ cation are isoelectronic, each having 10 electrons and the configuration [He]$2s^22p^6$. For isoelectronic species, properties change regularly with *Z*. For example, table 4.4 shows two properties of the 10-electron isoelectronic sequence. A progressive increase in nuclear charge results in a corresponding decrease in ionic radius, a result of stronger electrical force between the nucleus and the electron cloud. For the same reason, as *Z* increases, it becomes progressively more difficult to remove an electron.

TABLE 4.4 Trends in an isoelectronic sequence.

	Species			
Property	O^{2-}	F^-	Na^+	Mg^{2+}
Z	8	9	11	12
radius (pm)	140	133	102	72
ionisation energy (kJ mol^{-1})	< 0 ($-E_{ea_2}$)	322 ($-E_{ea}$)	4560 (E_{i2})	7730 (E_{i3})

4.9 Ions and chemical periodicity

The elements that form ionic compounds are found in specific places in the periodic table. Atomic anions are restricted to elements on the right side of the table: the halogens, oxygen and sulfur. All the elements in the *s*, *d* and *f* blocks form compounds containing atomic cations.

Cation stability

Knowing that the energy cost of removing core electrons is always very large, we can predict that the ionisation process will stop when all valence electrons have been removed. Thus, a knowledge of ground-state configurations is all that we need to make qualitative predictions about cation stability.

Each element in group 1 of the periodic table has one valence electron. These elements form ionic compounds containing M^+ cations. Examples are KCl and Na_2CO_3. Each element in group 2 of the periodic table has two valence electrons and forms ionic compounds containing M^{2+} cations. Examples are $CaCO_3$ and $MgCl_2$.

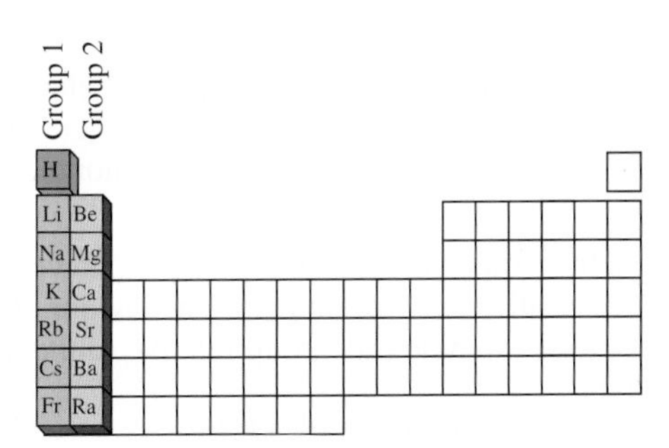

Beyond these two groups, the removal of all valence electrons is usually not energetically possible. For example, iron has eight valence electrons but forms only two stable cations, Fe^{2+} and Fe^{3+}. Compounds of iron containing these ions are abundant in the Earth's crust.

Other metallic elements form ionic compounds with cation charges ranging from +1 to +3. Aluminium nitrate nonahydrate, $Al(NO_3)_3{\cdot}9H_2O$, (the word 'nonahydrate' means nine water molecules are associated with the compound) is composed of Al^{3+} cations, NO_3^- anions, and water molecules. Silver nitrate, $AgNO_3$, which contains Ag^+ cations, is a water-soluble silver salt that is used in silver plating.

Anion stability

Halogens, the elements in group 17 of the periodic table, have the largest electron affinities of all the elements. So halogen atoms (ns^2np^5) readily accept electrons to produce halide anions (ns^2np^6). This allows halogens to react with many metals to form binary compounds, called halides, which contain metal cations and halide anions. Examples include NaCl (chloride anion), CaF_2 (fluoride anion), AgBr (bromide anion) and KI (iodide anion).

Isolated atomic anions with charges more negative than −1 are always unstable, but oxide (O^{2-}, $1s^22s^22p^6$) and sulfide (S^{2-}, $[Ne]3s^23p^6$) are found in many ionic solids, such as CaO and Na_2S. The lattice energies (see pp. 176–7) of these solids are large enough to make the overall reaction energy releasing despite the large positive second electron affinity of the anions. In addition, three-dimensional arrays of surrounding cations stabilise the −2 anions in these solids.

Metals, nonmetals and metalloids

Ion formation is only one pattern of chemical behaviour. Many other chemical trends can be traced to valence electron configurations, but we need the description of chemical bonding from chapter 5 to explain such periodic properties. Nevertheless, we can relate important patterns in chemical behaviour to the ability of some elements to form ions. One example is the subdivision of the periodic table into metals, nonmetals and metalloids, first introduced in chapter 1.

The elements that can form cations relatively easily are metals. All metals have similar properties, in part because their outermost *s* electrons are relatively easy to remove. All elements in the *s* block have ns^1 or ns^2 valence configurations. The *d*-block elements have one or two *ns* electrons and various numbers of $(n-1)d$ electrons. Examples are titanium ($4s^23d^2$) and silver ($5s^14d^{10}$). Elements in the *f* block have two *ns* electrons and a number of $(n-2)f$ electrons. Samarium, for example, has the valence configuration $6s^24f^6$. The metallic behaviour of these elements occurs partly because the *s* electrons are shared readily among all atoms. Metals form ionic salts because *s* electrons (and some *p*, *d* and *f* electrons) can be readily removed from the metal atoms to form cations.

Whereas the other blocks contain only metals, elemental properties vary widely within the *p* block. We have already noted that aluminium ($3s^23p^1$) can lose its three valence electrons to form Al^{3+} cations. The six elements within the triangle in figure 4.48 also lose *p* electrons easily and therefore have metallic properties. Examples are tin ($5s^25p^2$) and bismuth ($6s^26p^3$).

In contrast, the halogens and noble gases on the right of the *p* block are distinctly nonmetallic. The noble gases, group 18 of the periodic table, are monatomic gases that resist chemical attack because their electron configurations involve completely filled *s* and *p* orbitals.

Elements in any intermediate column of the *p* block display a range of chemical properties even though they have the same valence configurations. Carbon, silicon, germanium and tin all

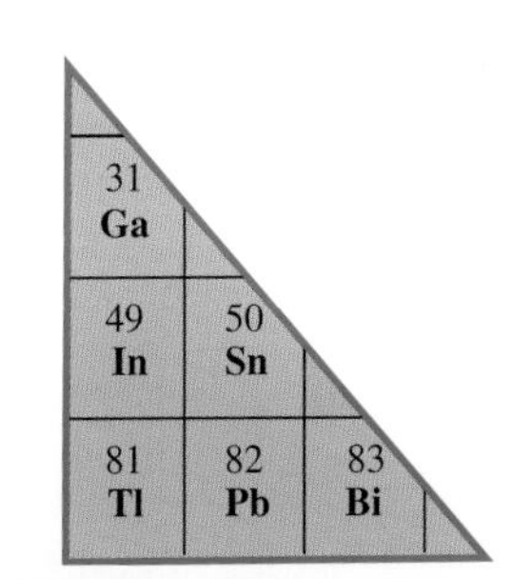

FIGURE 4.48 These six elements lose *p* electrons easily and so have metallic properties.

have ns^2np^2 valence configurations; yet carbon is a nonmetal, silicon and germanium are metalloids, and tin is a metal.

Qualitatively, we can understand this variation by recalling that, as the principal quantum number increases, the valence orbitals increase in energy. In tin, the four $n = 5$ valence electrons are bound relatively loosely to the atom, resulting in the metallic properties associated with electrons that are easily removed. In carbon, the four $n = 2$ valence electrons are bound relatively tightly to the atom, resulting in nonmetallic behaviour. Silicon ($n = 3$) and germanium ($n = 4$) fall in between these two extremes.

WORKED EXAMPLE 4.17

Classifying elements

Nitrogen is a colourless diatomic gas. Phosphorus has several elemental forms, one being a red solid that is used in match tips. Arsenic and antimony are grey solids, and bismuth is a silvery solid. Classify these elements of group 15 as metals, nonmetals or metalloids.

Analysis

All elements except those in the *p* block are metals. Group 15, however, is part of the *p* block, within which elements display all forms of elemental behaviour. To decide the classifications of these elements, we must examine this group relative to the diagonal arrangement of the metalloids (see chapter 1).

Solution

We see that group 15 passes through all three classes of elements. The elements with the lowest *Z* values, nitrogen and phosphorus, are nonmetals. The element with highest *Z* value, bismuth, is a metal, and the two elements with intermediate *Z* values, arsenic and antimony, are metalloids.

Is our answer reasonable?

As the principal quantum number increases, valence electrons become progressively easier to remove, and metals are those elements with valence electrons that are easily removed.

PRACTICE EXERCISE 4.17 Classify the 4*p* set of elements, from Ga to Kr as metals, nonmetals or metalloids.

s-block elements

The electron configuration of any element in groups 1 and 2 of the periodic table contains a core of tightly bound electrons and one or two *s* electrons that are loosely bound. The group 1 metals (ns^1 configuration) and the group 2 metals (ns^2 configuration) form stable ionic salts because their valence electrons are easily removed. Nearly all salts of group 1 metals and many salts of group 2 metals dissolve readily in water, so naturally occurring sources of water frequently contain these ions.

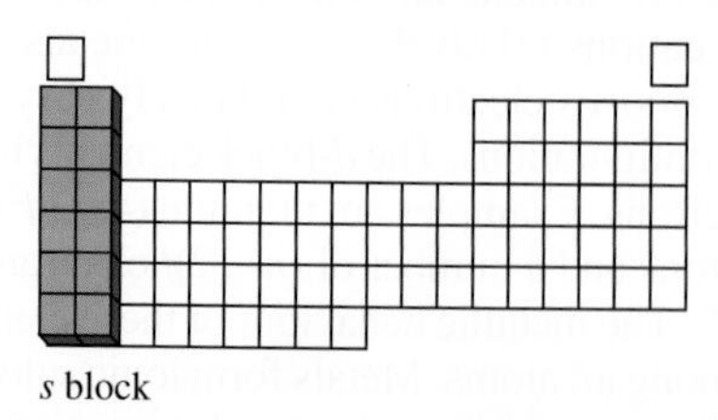

The four most abundant *s*-block elements in the Earth's crust are sodium, potassium, magnesium and calcium (table 4.5). These elements are found in nature in salts such as NaCl, KNO_3, $MgCl_2$, $MgCO_3$ and $CaCO_3$. Portions of these solid salts dissolve in rainwater as it percolates through the Earth's crust. The resulting solution of anions and cations eventually finds its way to the oceans. When water evaporates from the oceans, the ions are left behind. Over many aeons the continual influx of river water containing these ions has built up the substantial salt concentrations found in the oceans.

Table 4.5 shows that each of the four common *s*-block ions is abundant not only in sea water but also in body fluids, where they play essential biochemical roles. Sodium is the most abundant cation in fluids that are outside of cells, and proper functioning of body cells requires that sodium concentrations be maintained within a narrow range. One of the main functions of the kidneys is to control the excretion of sodium. Whereas sodium cations are abundant in the fluids outside cells, potassium cations are the most abundant ions in the fluids inside cells. The difference in ion concentration across cell membranes is responsible for the generation of nerve impulses that drive muscle contraction. If the difference in potassium ion concentration across cell membranes deteriorates, muscular activity, including the regular muscle contractions of the heart, can be seriously disrupted.

TABLE 4.5 Abundance of s-block elements.

Element	Abundance in Earth's crust (% by mass)	Abundance in sea water (mol L^{-1})	Abundance in human blood plasma (mol L^{-1})
Na	2.27	0.462	0.142
K	1.84	0.097	0.005
Mg	2.76	0.053	0.003
Ca	4.66	0.100	0.005

The cations Mg^{2+} and Ca^{2+} are major components of bones. Calcium occurs in hydroxyapatite, $Ca_5(PO_4)_3(OH)$. The structural function of magnesium in bones is not fully understood. In addition to being essential ingredients of bone, these two cations play key roles in various biochemical reactions, including photosynthesis, transmission of nerve impulses, muscle contraction and the formation of blood clots.

Beryllium behaves differently from the other *s*-block elements because its $n = 2$ orbitals are more compact than orbitals with higher principal quantum numbers. The first ionisation energy of beryllium, 899 kJ mol^{-1}, is comparable with those of nonmetals, so beryllium does not form compounds that are clearly ionic.

Some compounds of the *s*-block elements are important industrial and agricultural chemicals. For example, K_2CO_3 (potassium carbonate or potash) is obtained from mineral deposits and is the most common source of potassium for fertilisers. Potassium is essential for healthy plant growth. However, potassium salts are highly soluble in water, so potassium quickly becomes depleted from the soil. Consequently, agricultural land requires frequent addition of potassium fertilisers.

Three other compounds of *s*-block elements — calcium oxide, CaO or 'lime', sodium hydroxide, NaOH, and sodium carbonate, Na_2CO_3 — are major industrial chemicals. For example, lime is the key ingredient in construction materials such as concrete, cement, mortar and plaster. Two other compounds, calcium chloride, $CaCl_2$, and sodium sulfate, Na_2SO_4, are also of industrial importance.

Many industrial processes make use of anions such as hydroxide, OH^-, carbonate, CO_3^{2-}, and chlorate, ClO_3^-. These anions must be supplied as chemical compounds that include cations. Sodium is most frequently used as this spectator cation because it is abundant, inexpensive and relatively non-toxic. The hydroxide ion is industrially important because it is a strong base (see chapter 11). Sodium hydroxide is used to manufacture other chemicals, textiles, paper, soaps and detergents. Sodium carbonate and sand are the major starting materials in the manufacture of glass. Glass contains sodium and other cations embedded in a matrix of silicate, SiO_3^{2-}, anions. About half the sodium carbonate produced in the world is used in glass making.

p-block elements

The properties of elements in the *p* block vary across the entire spectrum of chemical possibilities. The elements in group 13, with a single electron in a *p* orbital as well as two valence *s* electrons, display chemical reactivity characteristic of three valence electrons. Except for boron, the elements of this group are metals that form stable +3 cations. Examples are $Al(OH)_3$ and GaF_3. Metallic character diminishes rapidly as additional *p* electrons are added. This change culminates in the elements in group 18. With filled *p* orbitals, these elements are so unreactive that for many years they were thought to be completely inert. Xenon is now known to form compounds with the most reactive nonmetals: oxygen, fluorine and chlorine; krypton also forms a few highly unstable compounds with these elements.

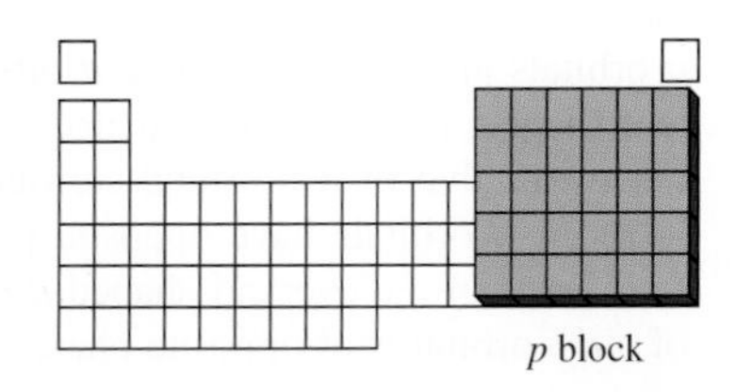

Although the nonmetals do not readily form cations, many of them combine with oxygen to form polyatomic oxoanions. These anions have various stoichiometries, but there are some common patterns. Two second-row elements form oxoanions with three oxygen atoms: carbon (four valence electrons) forms carbonate, CO_3^{2-}, and nitrogen (five valence electrons) forms nitrate, NO_3^-. In the third row, the most stable oxoanions contain four oxygen atoms: SiO_4^{4-}, PO_4^{3-}, SO_4^{2-} and ClO_4^-.

Many of the minerals that form the Earth's crust contain oxoanions. Examples of carbonates are $CaCO_3$ (limestone) and $MgCa(CO_3)_2$ (dolomite). Barite, $BaSO_4$, is a sulfate mineral. An important phosphate is $Ca_5(PO_4)_3F$ (apatite). Two silicates are zircon, $ZrSiO_4$, and olivine, a mixture of $MgSiO_4$ and $FeSiO_4$.

Several leading industrial chemicals contain oxoanions or are acids resulting from addition of H^+ to the anions. Sulfuric acid, H_2SO_4, nitric acid, HNO_3, and phosphoric acid, H_3PO_4, are all industrially important. Heating $CaCO_3$ drives off CO_2 to form CaO, the essential ingredient of building materials mentioned earlier. Other industrially important salts include ammonium sulfate, $(NH_4)_2SO_4$, aluminium sulfate, $Al_2(SO_4)_3$, sodium carbonate, Na_2CO_3, and ammonium nitrate, NH_4NO_3. We will learn more about the *p*-block elements in chapter 14.

SUMMARY

LO 1 Characteristics of atoms

All atoms display certain basic characteristics. They possess mass, occupy volume, attract one another and can combine with one another. Atoms contain positively charged nuclei and negatively charged electrons, and the properties of elements are determined by variations in nuclear charges and numbers of electrons.

LO 2 Characteristics of light

Electromagnetic radiation in the form of light may be used to probe the structure of atoms. Light has both wave- and particle-like properties.

The wave properties are characterised by:

- frequency (ν): the number of wave crests passing a given point in 1 second
- wavelength (λ): the distance between successive wave crests
- amplitude: the height of the wave
- phase: the starting position of the wave with respect to one wavelength.

The intensity (or brightness) of light is proportional to the square of the amplitude.

Light waves move at the speed of light, *c* (approximately 2.998×10^8 m s^{-1}) in a vacuum. The relationship between speed, frequency and wavelength of light is $c = \nu\lambda$. Visible light has wavelengths between 380 and 780 nm. The energy of light is inversely related to its wavelength — the shorter the wavelength, the higher the energy.

The particle properties of light are demonstrated by the photoelectric effect, where light striking the surface of a metal causes electrons to be ejected from the metal. Einstein postulated that this behaviour could be explained by 'packets' or 'bundles' of light called photons. The energy of a photon is given by the equation $E_{photon} = h\nu_{photon}$, where *h* is Planck's constant (approximately 6.626×10^{-34} J s). An atom can absorb a photon of a particular energy and form a higher energy excited state. The excited-state atom can then release the excess energy by emitting a photon and returning to its ground state. These photon absorption and emission processes can be observed in atomic spectra. An absorption spectrum gives dark lines in the visible spectrum where light of a particular wavelength has been absorbed, while an emission spectrum gives coloured lines on a dark background which correspond to emission of photons. Atoms can absorb or emit photons of only certain definite energies, meaning that the electrons in atoms have only certain specific energy values — that is, their energies are quantised. We show the quantised energy levels within an atom by means of an energy level diagram.

LO 3 Properties of electrons

Like atoms, all electrons display certain basic characteristics. Every electron has a mass of 9.109×10^{-31} kg and a charge of 1.602×10^{-19} C. Electrons have magnetic properties and, like light, electrons also display particle–wave duality. The wavelength associated with an electron is given by the equation $\lambda = \frac{h}{mu}$. The Heisenberg uncertainty principle states that we cannot simultaneously know both the position and the momentum of any particle, and this is particularly important for electrons.

LO 4 Quantisation and quantum numbers

Electrons bound to atoms have quantised energies, while free electrons can have any energy. The quantised energies of bound electrons are linked to their wave properties by the Schrödinger equation $\hat{H}\psi = E\psi$, which gives both the energies and wavefunctions of a chemical system. This equation can be solved exactly only for systems containing one electron, and a one-electron wavefunction is called an orbital.

Quantised properties of an atom can be indexed using quantum numbers. The principal quantum number (*n*) can have only positive integer values, and indexes both the energy of the electron and orbital size. The size of an orbital increases as *n* increases. The azimuthal quantum number (*l*) indexes the angular momentum of an atomic orbital and identifies the shape of the electron distribution of an orbital; it can be zero or any positive integer value smaller than *n*. Atomic orbitals are labelled according to their *l* value; *s* orbitals have $l = 0$, *p* orbitals have $l = 1$, *d* orbitals have $l = 2$ and *f* orbitals have $l = 3$. The magnetic quantum number (m_l) indexes the different possible orientations of the electron distribution of an orbital; it can have any positive or negative integer value from 0 to *l*. The spin quantum number (m_s) has possible values of $+\frac{1}{2}$ and $-\frac{1}{2}$, corresponding to the two possible spin states of an electron. All electrons within an atom can be described using sets of these four quantum numbers, and the Pauli exclusion principle states that each electron has a unique set.

LO 5 Atomic orbital electron distributions and energies

The electron distribution of an orbital can be depicted in a variety of ways. An electron density plot shows the distribution graphically in two dimensions, an electron density picture shows a cross-section of the distribution, while a boundary surface diagram shows the surface which encompasses (usually) 90% of the electron density. Areas of zero electron density within an orbital are called nodes, and the number of these increase as *n* increases. The electron distribution of an *s* orbital is spherical, while those of *p* orbitals and four of the five *d* orbitals are dumbbell and cloverleaf shaped, respectively. The electron distribution of the d_{z^2} orbital resembles that of a *p* orbital surrounded by a doughnut. The two lobes of *p* orbitals have opposite phases, while the phases of the four lobes of a cloverleaf-shaped *d* orbital alternate. The doughnut of a d_{z^2} orbital is of opposite phase to the rest of the orbital.

The energy of an orbital in a one-electron system is determined by both the atomic number Z and the principal quantum number n. All orbitals in a one-electron system with the same value of n have the same energy — they are degenerate. This degeneracy is destroyed in multi-electron atoms. The energy of an orbital can be determined by measuring the ionisation energy, which is the energy required to remove an electron completely from that (usually valence) orbital. Ionisation energies are dependent on many factors, one of which is the shielding of the valence electrons from the nuclear charge by core electrons. Consequently, a valence electron in a multi-electron atom experiences a charge less than the full nuclear charge; this is called the effective nuclear charge (Z_{eff}). Electrons with a lower n shield electrons of higher n very efficiently, while, within groups of electrons of the same n, those having the lowest value of l shield those of higher l.

LO 6 Structure of the periodic table

The ground-state configuration of an atom is the lowest possible energy arrangement of the electrons. The order of filling atomic orbitals with electrons is dictated by: the Aufbau principle, which states that orbitals are filled in order of increasing energy; the Pauli exclusion principle, which states that no two electrons in an atom can have identical sets of the four quantum numbers; and Hund's rule, which states that electrons occupy sets of degenerate orbitals so as to give the maximum number of unpaired parallel spins. An s orbital can accommodate 2 electrons, a set of p orbitals 6 and a set of d orbitals 10.

The shape of the periodic table of the elements derives from these orbital occupancies and the order in which the orbitals are filled. Elements with the same valence-electron configurations lie in the same group of the periodic table and have similar chemical properties. Valence electrons largely determine the chemistry of an element, while core electrons do not participate in chemical reactions.

LO 7 Electron configurations

The electron configuration of an atom can be written by recording quantum numbers for all electrons in the atom, using a compact notation which shows the electron occupancy of each orbital type, or through an energy level diagram which shows the energy and occupancy of each orbital. In the last, electrons are depicted as arrows pointing either up ($m_s = +\frac{1}{2}$) or down ($m_s = -\frac{1}{2}$). We can also use an abbreviated compact notation that does not explicitly include the core electrons, but abbreviates them using the noble gas configuration of the nearest noble gas with lower Z. Some exceptions to the order of orbital filling occur, most notably for Cr and Cu. For anions electrons are added to the electron configuration of the neutral atom, while for cations electrons are removed from that configuration. Atoms or ions having unpaired electrons are paramagnetic and will be attracted to a magnetic field, while species having all electrons paired are diamagnetic and are not attracted to a magnetic field. The electron configurations of excited states can violate Hund's rule and the Aufbau principle but not the Pauli exclusion principle.

LO 8 Periodicity of atomic properties

The size of an atom is determined by factors such as nuclear charge, orbital energy and electron distribution, and shielding. Atomic radii give us a measure of the sizes of atoms. Atomic radii become smaller across a period and larger down a group of the periodic table. The first ionisation energies of the elements generally increase across a period and decrease down a group of the periodic table. Exceptions to these general trends occur when p orbitals are occupied by 1 or 4 electrons. Electron affinities become more negative across a period, but there is no clear trend down a group of the periodic table. Atomic cations are always smaller than the neutral atom and atomic anions are always larger than the neutral atom.

LO 9 Ions and chemical periodicity

The nature of cations and anions is determined by the electron configuration of the neutral atom. Group 1 and group 2 elements form 1+ and 2+ cations respectively, while group 17 elements form 1− anions. Isolated atomic anions with charges more negative than 1− are always unstable, and O^{2-} and S^{2-} are found only in combination with cations in ionic solids. Metals generally form cations readily through loss of s or d electrons. The p-block elements may be metals, metalloids or nonmetals. The s-block elements are found in the Earth's crust and also in living systems, and p-block elements display a wide variety of chemical properties.

KEY CONCEPTS AND EQUATIONS

Wavelength–frequency relationship
(section 4.2)

The product of wavelength (λ) and frequency (ν) is the speed of light (c).

$$c = \nu\lambda$$

Energy of a photon
(section 4.2)

This equation is used to calculate the energy carried by a photon of frequency ν. Also, ν can be calculated if E is known.

$$E = h\nu$$

de Broglie equation
(section 4.3)

This links the wavelength of a particle with its speed and mass.

$$p = \frac{h}{\lambda}$$

$$\lambda_{\text{particle}} = \frac{h}{mu}$$

Quantum numbers for atoms
(section 4.4)

Each electron in an atom is described by a unique set of four quantum numbers. The allowed values of these quantum numbers are as follows.

Quantum number	Range
principal quantum number (n)	$1, 2, \ldots, \infty$
azimuthal quantum number (l)	$0, 1, \ldots, (n-1)$
magnetic quantum number (m_l)	$-l, \ldots, -1, 0, +1, \ldots, +l$
spin quantum number (m_s)	$-\frac{1}{2}, +\frac{1}{2}$

Periodic table *(section 4.6)*

The periodic table is an aid in writing electron configurations of the elements. An element's location in the table is related to properties such as atomic radius, ionisation energy and electron affinity.

Electron configurations *(section 4.7)*

These guidelines help us to write electron configurations for atoms or ions.

1. Count the total number of electrons for a neutral atom.
   (a) Add electrons for anions.
   (b) Subtract electrons for cations.
2. Fill orbitals to match the nearest noble gas of smaller atomic number.
3. Add remaining electrons to the next orbitals to be filled according to the Aufbau and Pauli exclusion principles, as well as Hund's rule.
   (a) For neutral atoms and anions, place electrons in ns before $(n-1)d$ and $(n-2)f$, keeping in mind that the order changes once the d and f levels are at least partially occupied.
   (b) For cations, place electrons in the $(n-1)d$ before the ns orbitals.
4. Look for exceptions and correct the configuration if necessary.

Periodic trends in atomic and ionic size *(section 4.8)*

Periodic trends are evident in, for example:

- atomic and ionic size
- ionisation energy
- electron affinity.

KEY TERMS

absorption spectrum *p. 127*
amplitude *p. 120*
atomic radius *p. 155*
Aufbau principle *p. 145*
azimuthal quantum number (l) *p. 136*
boundary surface diagram *p. 140*
core electrons *p. 148*
degenerate *p. 143*
diamagnetic *p. 154*
effective nuclear charge (Z_{eff}) *p. 144*
electromagnetic radiation *p. 120*
electron affinity (E_{ea}) *p. 159*
electron configuration *p. 149*
electron density picture *p. 139*
electron density plot *p. 139*
emission spectrum *p. 127*
energy level diagram *p. 130*
excited state *p. 126*
frequency (ν) *p. 120*
ground state *p. 126*
Hund's rule *p. 152*
intensity *p. 120*
ionisation energy (E_i) *p. 143*
isoelectronic *p. 153*
light *p. 120*
magnetic quantum number (m_l) *p. 137*
noble gas configuration *p. 151*
node *p. 140*
orbital *p. 136*
paramagnetic *p. 154*
Pauli exclusion principle *p. 138*
phase *p. 121*
photoelectric effect *p. 122*
photons *p. 123*
Planck's constant (h) *p. 123*
principal quantum number (n) *p. 136*
quantised *p. 129*
quantum number *p. 136*
shielding *p. 144*
spin *p. 137*
spin quantum number (m_s) *p. 138*
uncertainty principle *p. 135*
valence electrons *p. 148*
wavelength (λ) *p. 120*

REVIEW QUESTIONS

LO 1 Characteristics of atoms

4.1 The density of silver is 1.050×10^4 kg m^{-3}, and the density of lead is 1.134×10^4 kg m^{-3}. For each metal: (a) calculate the volume occupied per atom, (b) estimate the atomic diameter and (c) using this estimate, calculate the thickness of a metal foil containing 9.3×10^6 atomic layers of the metal.

4.2 Describe evidence that indicates that atoms have mass.

4.3 Describe evidence that indicates that atoms have volume.

LO 2 Characteristics of light

4.4 Calculate the frequencies (Hz) corresponding to the following wavelengths, using power-of-ten notation. (a) 5.00 nm (b) 2.50×10^{-10} m (c) 801 mm (d) 3.57 μm

4.5 Calculate the wavelengths corresponding to the following frequencies, expressing the result in the indicated units.
(a) 3.77 GHz (m) (b) 25.0 kHz (cm) (c) 84 Hz (mm)
(d) 2.22 MHz (μm)

4.6 Calculate the energy in joules per photon and in kilojoules per mole of the following. (a) blue-green light with a wavelength of 490.6 nm (b) X-rays with a wavelength of 25.5 nm
(c) microwaves with a frequency of 2.5437×10^{10} Hz

4.7 What are the wavelength and frequency of photons with the following energies? (a) 828 kJ mol^{-1} (b) 4.35×10^{-19} J photon^{-1}

4.8 When light of frequency 1.30×10^{15} s^{-1} shines on the surface of caesium metal, electrons are ejected with a maximum kinetic energy of 5.2×10^{-19} J. Calculate: (a) the wavelength of this light, (b) the binding energy of electrons to caesium metal and (c) the longest wavelength of light that will eject electrons.

4.9 A phototube delivers an electrical current when a beam of light strikes a metal surface inside the tube. Phototubes do not respond to infrared photons. Draw an energy level diagram for electrons in the metal of a phototube and use it to explain why phototubes do not respond to infrared light.

4.10 Refer to figure 4.4 to answer the following questions.
(a) What is the wavelength range for X-rays?
(b) What colour is light with a wavelength of 6.8×10^{-7} m?
(c) In what region does radiation with frequency of 1.0×10^{12} Hz lie?

4.11 Determine the wavelengths of radiation that hydrogen atoms absorb to reach the $n = 8$ and $n = 9$ states from the ground state. In what region of the electromagnetic spectrum do these photons lie?

LO 3 Properties of electrons

4.12 What is the charge in coulombs of one mole of electrons?

4.13 Determine the wavelengths of electrons with the following kinetic energies. (a) 1.15×10^{-19} J (b) 3.55 kJ mol^{-1}
(c) 7.45×10^{-3} J mol^{-1}

4.14 Determine the kinetic energies in joules of electrons with the following wavelengths. (a) 4.88 m (b) 7.93 mm (c) 3.25 nm

LO 4 Quantisation and quantum numbers

4.15 List all the valid sets of quantum numbers for a $5p$ electron.

4.16 If you know that an electron has $n = 4$, what are the possible values for its other quantum numbers?

4.17 For the following sets of quantum numbers, determine which describe actual orbitals and which are nonexistent. For each one that is nonexistent, list the restriction that forbids it.

	n	l	m_l	m_s
(a)	4	1	2	$-\frac{1}{2}$
(b)	3	1	0	$+\frac{1}{2}$
(c)	4	2	1	$-\frac{1}{2}$
(d)	2	1	−1	$+\frac{1}{2}$

LO 5 Atomic orbital electron distributions and energies

4.18 Refer to figure 4.28. Draw the analogous set of three depictions for an orbital that has $n = 2$, $l = 1$.

4.19 Shown below are electron density pictures and electron density plots for the 1*s*, 2*s*, 2*p* and 3*p* orbitals. Assign the various depictions to their respective orbitals.

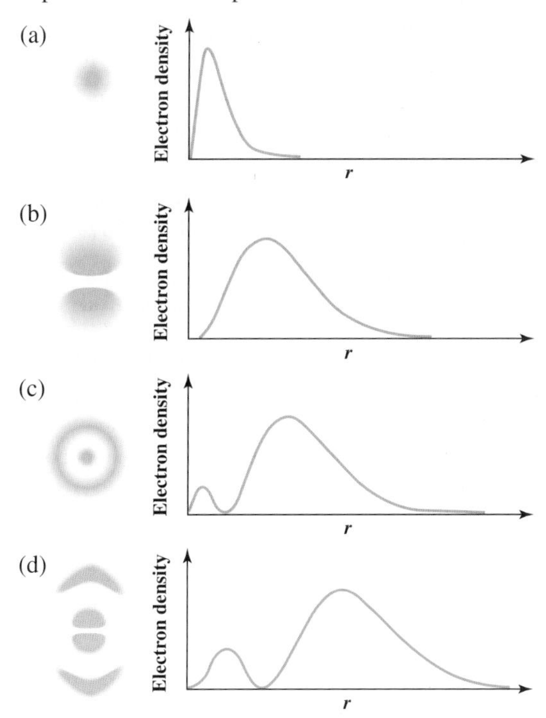

4.20 What are the limitations of plots of electron density versus *r*?

4.21 The conventional method of showing the three-dimensional electron distribution of an orbital is a boundary surface diagram. What are the limitations of this representation?

4.22 For each of the following pairs of orbitals, determine which is more stable and explain why. (a) He 1*s* and He 2*s* (b) Kr 5*p* and Kr 5*s* (c) He 2*s* and He^+ 2*s*

4.23 In a hydrogen atom the 3*s*, 3*p* and 3*d* orbitals all have the same energy. In a helium atom, however, the 3*s* orbital is lower in energy than the 3*p* orbital, which in turn is lower in energy than the 3*d* orbital. Explain why the energy rankings of hydrogen and helium are different.

4.24 The energy of the $n = 2$ orbital of the He^+ ion is the same as the energy of the $n = 1$ orbital of the H atom. Explain this fact.

4.25 Refer to figure 4.33 to answer the following questions. In each case, provide a brief explanation of your choice. (a) Which ionisation energies show that an electron in a 1*s* orbital provides nearly complete shielding of an electron in a 2*p* orbital? (b) Which ionisation energies show that the stability of $n = 2$ orbitals increases with Z^2?

LO 6 Structure of the periodic table

4.26 Draw the periodic table in block form, and outline and label each of the following sets. (a) elements that have half-filled *p* orbitals (b) elements for which $n = 4$ orbitals are filling (c) elements that are two electrons short of filled *d* orbitals (d) the first element that contains a 6*s* electron

4.27 How many valence electrons does each of the following atoms have? (a) N (b) Ti (c) Cs (d) Pb (e) Hg

LO 7 Electron configurations

4.28 Write the compact electron configuration, and list a correct set of quantum numbers for each of the valence electrons in the ground-state configurations of the following. (a) B (b) S (c) Ne (d) P

4.29 Which of the atoms from question 4.28 are paramagnetic? Draw orbital energy level diagrams to support your answer.

4.30 The following are hypothetical configurations for a beryllium atom. Which use nonexistent orbitals, which are forbidden by the Pauli principle, which are excited states, and which one is the ground-state configuration? (a) $1s^32s^1$ (b) $1s^12s^3$ (c) $1s^12p^3$ (d) $1s^22s^12p^1$ (e) $1s^22s^2$ (f) $1s^21p^2$ (g) $1s^22s^12d^1$

4.31 Write the correct ground-state electron configurations, in abbreviated compact form, for F, Cr, As and I.

LO 8 Periodicity of atomic properties

4.32 Arrange the following atoms in order of decreasing first ionisation energy (smallest last): Ar, Cl, Cs and K.

4.33 One element has these ionisation energies and electron affinity (all in kJ mol^{-1}): $E_{i1} = 376$, $E_{i2} = 2420$, $E_{i3} = 3400$, $E_{ea} = -45.5$. In which column of the periodic table is this element found? Give your reasoning.

4.34 According to appendix G, each of the following elements has a positive electron affinity. For each one, construct its valence orbital energy level diagram and use it to explain why the anion is unstable. (a) N (b) Mg (c) Zn

4.35 From the following list, select the elements that form ionic compounds: Ca, C, Cu, Cs, Cl and Cr. Indicate whether each forms a stable cation or a stable anion.

LO 9 Ions and chemical periodicity

4.36 Classify each of the elements from group 15 of the periodic table as a metal, a nonmetal or a metalloid.

4.37 Classify each of the elements listed in question 4.35 as a metal, a nonmetal or a metalloid.

REVIEW PROBLEMS

4.38 Redraw the first light wave in figure 4.3 to show a wave whose amplitude is twice as large as that shown in the figure, but whose frequency is the same. **LO 2**

4.39 A minimum of 216.4 kJ mol^{-1} is required to remove an electron from a potassium metal surface. What is the longest wavelength of light that can do this? **LO 2**

4.40 In a photoelectric effect experiment, the minimum frequency needed to eject electrons from a metal is 7.5×10^{14} s^{-1}. Suppose that a 366 nm photon from a mercury discharge lamp strikes the metal. Calculate: (a) the binding energy of the electrons in the metal, (b) the maximum kinetic energy of the ejected electrons and (c) the wavelength associated with those electrons. **LO 2**

4.41 Microwave ovens use radiation with a wavelength of 20.0 cm. What is the frequency, and energy in kJ mol^{-1}, of this radiation? **LO 2**

4.42 Barium salts in fireworks generate a yellow-green colour. Ba^{2+} ions emit light with $\lambda = 487$, 514, 543, 553 and 578 nm. Convert these wavelengths into frequencies and into energies in kJ mol^{-1}. **LO 2**

4.43 A hydrogen atom emits a photon as its electron changes from $n = 5$ to $n = 1$. What is the wavelength of the photon? In what region of the electromagnetic spectrum is this photon found? **LO 2**

4.44 List the properties of electrons and of photons, including the equations used to describe each. **LO 2,3**

4.45 Write a short description of: (a) the photoelectric effect, (b) particle–wave duality, (c) electron spin and (d) the uncertainty principle. **LO 2,3**

4.46 Write brief explanations of: (a) shielding, (b) the Pauli exclusion principle, (c) the Aufbau principle, (d) Hund's rule and (e) valence electrons. **LO 4,5,6**

4.47 Describe an atomic energy level diagram and the information it incorporates. **LO 5**

4.48 Construct an orbital energy level diagram for all orbitals with $n < 6$ and $l < 4$. Use the periodic table to help determine the correct order of the energy levels. **LO 5**

4.49 Describe periodic variations in electron configurations; explain how they affect ionisation energy and electron affinity. **LO 7,8**

4.50 Write the abbreviated compact electron configuration for the Mn^{3+} ground state, and give a set of quantum numbers for all electrons in the least stable occupied orbital. **LO 7**

4.51 Which has the most unpaired electrons, S^+, S or S^{2-}? Use electron configurations to support your answer. **LO 7**

4.52 None of the following electron energy diagrams describes the ground state of a sulfur atom. For each, state the reason why it is not correct. **LO 7**

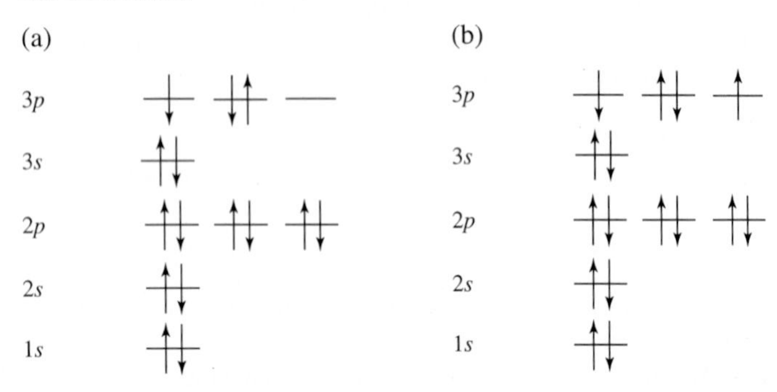

(c) (d)

3p
3s
2p
2s
1s

4.53 Identify the ionic compounds that best fit the following descriptions. **LO 8,9**

(a) the group 2 cation with the second smallest radius combined with a group 16 anion that is isoelectronic with the noble gas from row 3
(b) a +1 ion that is isoelectronic with the noble gas from row 3, combined with the anion formed from the row 2 element with the highest electron affinity
(c) the group 2 metal with the highest second ionisation energy that combines in a 1 : 2 ratio with an element from row 3.

4.54 Arrange the following in order of decreasing size (radius): Cl^-, K^+, Cl and Br^-. Explain your rankings in terms of quantum numbers and electrical interactions. **LO 8**

4.55 Draw energy level diagrams that show the ground-state valence electron configurations for Cu^{2+}, Mn^{3+} and Ag^+. **LO 7**

4.56 Write a brief explanation for each of the following. **LO 8**

(a) In a hydrogen atom, the 2*s* and 2*p* orbitals have identical energy.
(b) In a helium atom, the 2*s* and 2*p* orbitals have different energies.

4.57 Refer to figure 4.45 to answer the following questions about first ionisation energies. **LO 8**

(a) Which element shows the greatest decrease from its neighbour of next lower *Z*?
(b) What is the atomic number of the element with the lowest value?
(c) Identify three ranges of *Z* across which the value changes the least.
(d) List the atomic numbers of all elements whose values are between 925 and 1050 kJ mol^{-1}.

ADDITIONAL EXERCISES

4.58 The graph below shows the results of photoelectron experiments on two metals, using light of the same energy. **LO 2**

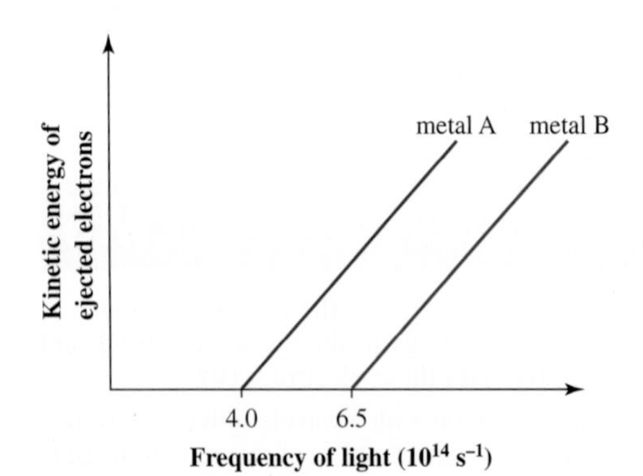

(a) Calculate the binding energy of each metal. Which has the higher binding energy? Explain.
(b) Calculate the kinetic energies of electrons ejected from each metal by photons with wavelength of 125 nm.
(c) Calculate the wavelength range over which photons can eject electrons from one metal but not from the other.

4.59 Energetic free electrons can transfer their energy to bound electrons in atoms. In 1913, James Franck and Gustav Hertz passed electrons through mercury vapour at low pressure to determine the minimum kinetic energy required to produce the excited state that emits ultraviolet light at 253.7 nm. What is that minimum kinetic energy? What wavelength is associated with electrons of this energy? **LO 2**

4.60 The diameter of a typical atom is 10^{-10} m, and the diameter of a typical nucleus is 10^{-15} m. Calculate typical atomic and nuclear volumes and determine what fraction of the volume of a typical atom is occupied by its nucleus. **LO 5**

4.61 Neutrons, like electrons and photons, are particle–waves with diffraction patterns that can be used to determine the structures of molecules. Calculate the kinetic energy of a neutron with a wavelength of 68 pm. **LO 3**

4.62 The human eye can detect as little as 2.35×10^{-18} J of green light with a wavelength of 510 nm. Calculate the minimum number of photons of green light that can be detected by the human eye. **LO2**

4.63 Gaseous lithium atoms absorb light with a wavelength of 323 nm. The resulting excited lithium atoms lose some energy through collisions with other atoms. The atoms then return to their ground state by emitting two photons with $\lambda = 812.7$ and 670.8 nm. Draw an energy level diagram that shows this process. What fraction of the energy of the absorbed photon is lost in collisions? **LO2**

4.64 Calculate the wavelengths associated with an electron and a proton, each travelling at 6.500% of the speed of light. **LO2**

4.65 One hydrogen emission line has a wavelength of 486 nm. Identify the values for n_{final} and $n_{initial}$ for the transition giving rise to this line. **LO2**

4.66 An atomic energy level diagram, shown to scale, follows. **LO2**

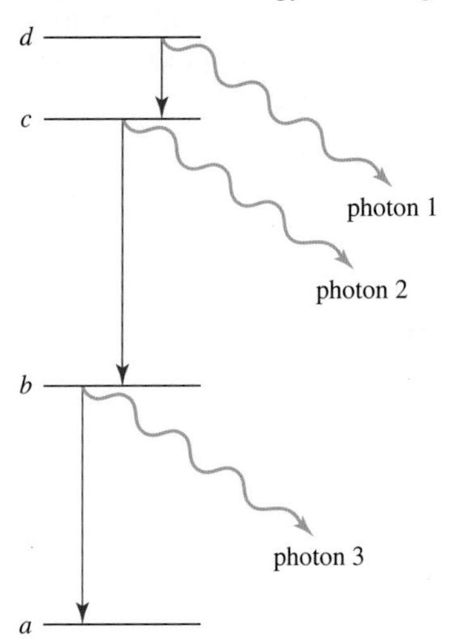

An excited-state atom emits photons when the electron moves in succession from level d to level c, from level c to level b, and from level b to the ground state (level a). The wavelengths of the emitted photons are 565 nm, 152 nm and 121 nm (not necessarily in the proper sequence). Match each emission with the appropriate wavelength and calculate the energies of levels b, c and d relative to level a.

4.67 Small helium–neon lasers emit 1.0 mJ s^{-1} of light at 634 nm. How many photons does such a laser emit in one minute? **LO2**

4.68 The argon-ion laser has major emission lines at 488 and 514 nm. Each of these emissions leaves the Ar^+ ion in an energy level that is 2.76×10^{-18} J above the ground state. **LO2**

(a) Calculate the energies of the two emission wavelengths in joules.
(b) Draw an energy level diagram (in joules per atom) that illustrates these facts.
(c) What frequency and wavelength radiation is emitted when the Ar^+ ion returns to its lowest energy level?

4.69 The series of emission lines that results from excited hydrogen atoms undergoing transitions to the $n = 3$ level is called the Paschen series. Calculate the energies of the first five lines in this series of transitions, and draw an energy level diagram that shows them to scale. **LO2**

4.70 It takes 486 kJ mol^{-1} to remove electrons completely from sodium atoms. Sodium atoms absorb and emit light of wavelengths 589.6 and 590.0 nm. **LO2**

(a) Calculate the energies of these two wavelengths in kJ mol^{-1}.
(b) Draw an energy level diagram for sodium atoms that shows the levels involved in these transitions and the ionisation energy.
(c) If a sodium atom has already absorbed a 590.0 nm photon, what is the wavelength of the second photon a sodium atom must absorb in order to remove an electron completely?

4.71 The photoelectric effect for magnesium metal has a threshold frequency of 8.95×10^{14} s^{-1}. Can Mg be used in photoelectric devices that sense visible light? Do a calculation to support your answer. **LO2**

4.72 Make an electron density plot that shows how the $3s$ and $3p$ orbitals are shielded effectively by the $2p$ orbitals. Provide a brief explanation of your plot. **LO5**

4.73 The idea of an atomic radius is inherently ambiguous. Explain why this is so. **LO5**

4.74 Write the electron configuration for the lowest energy excited state of each of the following: Be, O^{2-}, Br^-, Ca^{2+} and Sb^{3+}. **LO7**

4.75 Write the ground-state configurations for the isoelectronic species Ce^{2+}, La^+ and Ba. Are they the same? What features of orbital energies account for this? **LO7**

4.76 The ionisation energy of lithium atoms in the gas phase is about half as large as the ionisation energy of beryllium atoms in the gas phase. In contrast, the ionisation energy of Li^+ is about four times larger than the ionisation energy of Be^+. Explain the difference between the atoms and the ions. **LO8**

4.77 The figures below show Br^-, Kr, and Rb^+ drawn to scale. Decide which figure corresponds to which species and explain your reasoning. **LO8**

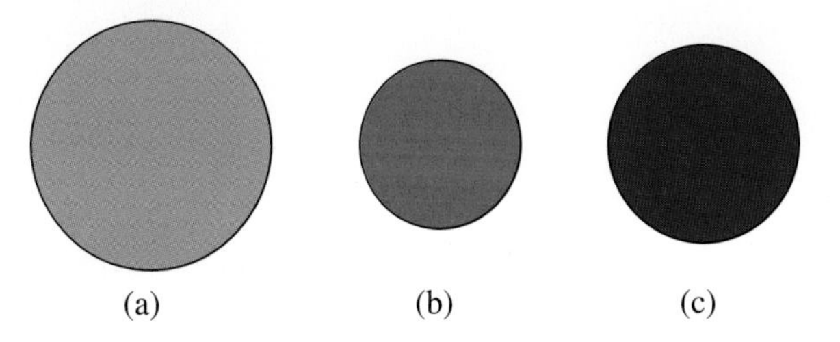

4.78 Use the data in appendix G to explain why the noble gases seldom take part in chemical reactions. **LO8,9**

4.79 From its location in the periodic table, predict some of the physical and chemical properties of francium. What element does it most closely resemble? **LO8**

4.80 What would be the next two orbitals to fill after the $5p$ orbital? **LO6**

4.81 Which has the more stable $2s$ orbital, a lithium atom or a Li^+ cation? Explain your reasoning. **LO8**

4.82 The first four ionisation energies of aluminium are as follows. $E_{i1} = 577$ kJ mol^{-1}, $E_{i2} = 1817$ kJ mol^{-1}, $E_{i3} = 2745$ kJ mol^{-1} and $E_{i4} = 11\,578$ kJ mol^{-1}. **LO8**

(a) Explain the trend in ionisation energies.
(b) Rank ions of aluminium in order of ionic radius, from largest to smallest.
(c) Which ion of aluminium has the largest electron affinity?

4.83 Use electron–electron repulsion and orbital energies to explain the following irregularities in first ionisation energies. **LO5,8**

(a) Boron has a lower ionisation energy than beryllium.
(b) Sulfur has a lower ionisation energy than phosphorus.

5 Chemical bonding and molecular structure

Australia's Samantha Stosur serving at Wimbledon — the high-tech fibres in her tennis racquet that are responsible for some of the power are held together by chemical bonds. Producing such high-tech fibres requires a thorough understanding of chemical bonding, because the fibres are constructed of polymers, large molecules that are held together by covalent bonds in which electrons are shared between atoms. The ways in which chemical bonds are formed depend on the nature of the elements involved, and chemical bonds show a variety of forms and strengths.

This chapter describes chemical bonding and its relationship to molecular structure. We discuss the two extremes of chemical bonding: ionic, in which electrostatic forces hold ionic arrangements together, and covalent, where shared electron pairs dictate the bonding and geometry of individual molecules. We show how simple Lewis theory gives a good approximation of the distribution of electrons within molecules, and how we can use this to predict molecular geometry using valence-shell-electron-pair repulsion theory. We then consider in detail the two predominant theories used to describe covalent bonding: valence bond theory and molecular orbital theory.

LEARNING OBJECTIVES

After studying this chapter, you should be able to:

5.1 describe qualitatively the bonding in molecules

5.2 describe qualitatively the bonding in ionic compounds

5.3 determine Lewis structures for molecular compounds

5.4 apply the VESPR concept to determine the geometry of molecular compounds

5.5 identify and explain properties of covalent bonds

5.6 analyse bonding in molecules using valence bond theory

5.7 analyse bonding in molecules using molecular orbital theory.

5.1 Fundamentals of bonding

We learned in chapter 4 that the size, energy levels and electron configuration of an atom determine its chemical properties. In molecular compounds, electrons on atoms interact and are shared between atoms. In ionic compounds, electrons are transferred completely between atoms to form positively and negatively charged ions. Since electrons have both particle- and wave-like properties, bonding interactions can be described from either viewpoint. Keep in mind, though, that these models do not necessarily reflect reality exactly.

The electrostatic energy between two charged species is proportional to the magnitudes of the charges and inversely proportional to the distance between them. Charges of opposite sign attract one another, but like charges repel. The relationship is given by Coulomb's law as:

$$E_{\text{electrostatic}} = k\frac{q_1 q_2}{r}$$

where:

$E_{\text{electrostatic}}$ = electrostatic potential energy (J)

q_1 = charge on species 1 (C)

q_2 = charge on species 2 (C)

r = separation distance between the pair of charges (m)

$k = 9.00 \times 10^9\ \text{N m}^2\ \text{C}^{-2}$ (Coulomb's constant in vacuum)

The equation describes the potential energy of one pair of charges. Molecules, however, contain two or more nuclei and two or more electrons. To obtain the total potential energy of a molecule, the equation must be applied to every possible pair of charged species. These pair-wise interactions are of three types (see figure 5.1).

1. Electrons and nuclei attract one another. Attractive interactions are energetically favourable, so an electron attracted to a nucleus is at a lower energy than a free electron.
2. Electrons repel each other, raising the energy and reducing the stability of a molecule.
3. Nuclei repel each other, so these interactions also reduce the stability of a molecule.

In any molecule, these three interactions are balanced to give the molecule its greatest possible stability. This balance is achieved when the electron density is situated between the nuclei of bonded atoms. We view the electrons as being shared between the nuclei and call this shared electron density a **covalent bond**. In any covalent bond, the attractive energy between nuclei and electrons exceeds the repulsive energy arising from nuclear–nuclear and electron–electron interactions.

The hydrogen molecule

These concepts are demonstrated below for the simplest stable neutral molecule, molecular hydrogen. A hydrogen molecule, H_2, contains just two nuclei and two electrons.

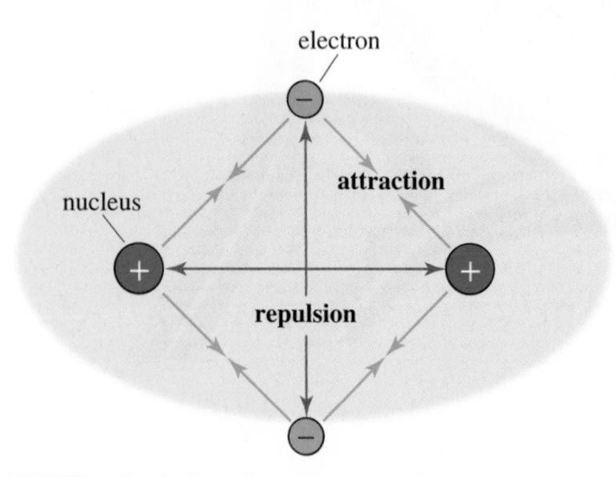

FIGURE 5.1 When electrons are in the region between two hydrogen nuclei, attractive electrostatic forces exceed repulsive electrostatic forces, leading to the stable arrangement of a chemical bond.

Consider what happens when two hydrogen atoms come together and form a covalent bond. As the atoms approach, each nucleus attracts the opposite electron, pulling the two atoms closer together. At the same time, the two nuclei repel each other, as do the two electrons. These repulsive interactions tend to drive the atoms apart.

For H_2 to be a stable molecule, the sum of the attractive energies must exceed the sum of the repulsive energies. Figure 5.1 shows a static arrangement of electrons and nuclei in which the electron–nucleus distances are shorter than the electron–electron and nucleus–nucleus distances. As the charges on the hydrogen nuclei and electrons are 1+ and 1− respectively, the distances between them become crucial in determining the energy of this arrangement. In this case, attractive interactions exceed repulsive interactions, leading to a stable molecule. Notice that the two electrons occupy the region between the two nuclei where they can interact with both nuclei at once; that is, the atoms share the electrons in a covalent bond.

An actual molecule is dynamic, not static, and both the electrons and nuclei move continuously. In a covalent bond, the most probable electron locations are between the nuclei, where, in this model, they can be viewed as being shared between the bonded atoms.

Bond length and bond energy

As two hydrogen atoms come together to form a molecule, attractive forces between nuclei and electrons make the hydrogen molecule more stable than the individual hydrogen atoms. The amount of increased stability depends on the distance between the nuclei, as shown in figure 5.2. At distances greater than 300 pm (300×10^{-12} m), there is almost no interaction between the atoms, so the total energy is the sum of the energies of the two individual atoms. At closer distances, the attraction between the electron of one atom and the nucleus of the other atom increases, and the combined atoms become more stable. Moving the atoms closer together generates greater stability until the nuclei are 74 pm apart. At this distance, we obtain the best possible balance between the attractive and repulsive forces in the molecule. At distances closer than 74 pm, the nucleus–nucleus repulsion begins to dominate and the energy of the system increases rapidly. Thus, at a separation distance of 74 pm, the hydrogen molecule has the maximum energetic advantage (7.22×10^{-19} J) over two separated hydrogen atoms. The atoms in the molecule are not static; instead they move continuously, oscillating about their lowest energy separation distance like two spheres attached to opposite ends of a spring.

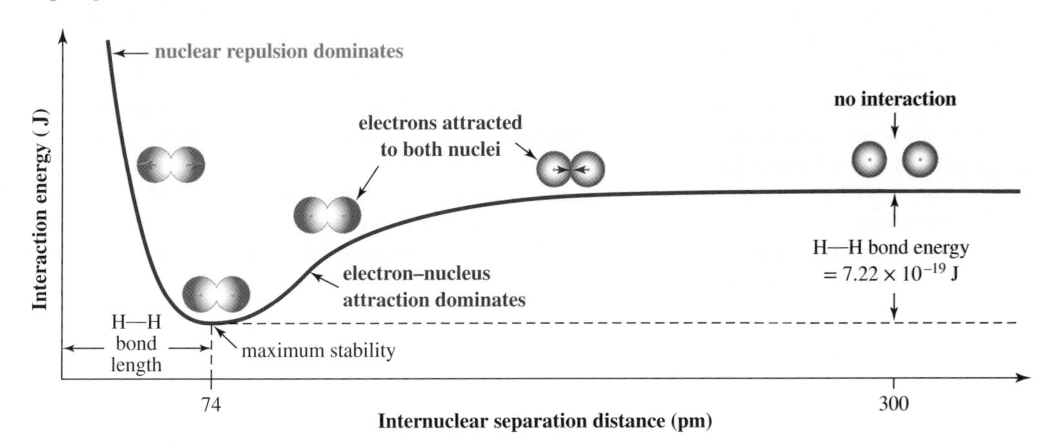

FIGURE 5.2 The interaction energy of two hydrogen atoms depends on the distance between the nuclei.

Figure 5.2 shows two characteristic features of covalent bonds. The separation distance at which the molecule has the maximum energetic advantage over the separated atoms (74 pm for H_2) is known as the **bond length**. The energy difference between the molecule and the separated atoms at this separation distance is called the bond energy. **Bond energy** is defined as the energy required to break the bond and therefore is always positive. Bond lengths and energies are important properties of chemical bonds and, as we will see, they are related. It is usual to quote bond energies in units of kJ mol^{-1}. We can obtain this value for the hydrogen molecule by multiplying the energy of a single H—H bond (7.22×10^{-19} J) by N_A (6.022×10^{23} mol^{-1}). This gives a H—H bond energy of 435 kJ mol^{-1}. Each different chemical bond has a characteristic bond length and bond energy.

The most probable position of the electrons in the hydrogen molecule is between the nuclei. If we rotate the molecule about its internuclear axis (an imaginary line joining the two nuclei) we find that the electron distribution between the nuclei looks exactly the same, regardless of the angle through which it was rotated. In other words, the bond is symmetric with respect to rotation about the internuclear axis. Bonds for which this is true are called **sigma (σ) bonds**.

Other diatomic molecules: F_2

Bond formation in H_2 is relatively easy to describe because we need to consider just two electrons. However, even when the atoms within a molecule contain many electrons, we can still consider bond formation as the sharing of only two electrons. We will illustrate this with the fluorine molecule, F_2.

As two fluorine atoms approach each other, the electrons of each atom feel the attraction of the nucleus of the other atom, just as in H_2. In this case, the valence $2p$ electrons are closest to the neighbouring nucleus and experience the strongest attraction. The single electron in one of the $2p$ orbitals approaches the opposite nucleus. Consequently, we can describe the F—F bond as the sharing of two electrons from the fluorine $2p$ orbitals that point directly towards each other (see figure 5.3). Note that the resulting bond is a σ bond.

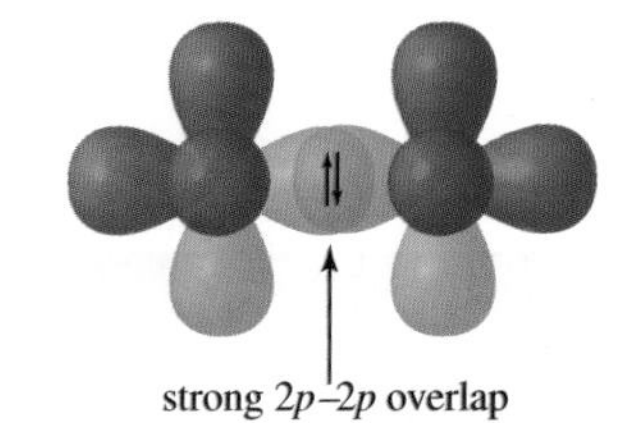

FIGURE 5.3 The chemical bond in F_2 forms from strong electrostatic attraction of the electron in the fluorine $2p$ orbital that points directly at the neighbouring nucleus.

Unequal electron sharing

Each nucleus within a hydrogen molecule and each within a fluorine molecule has the same charge. Consequently, both nuclei attract electrons equally. The result is a symmetrical distribution of the electron density around the atoms. In a chemical bond between any two identical atoms, the nuclei share the bonding electrons equally.

In contrast to the symmetrical forces in H_2 and F_2, the bonding electrons in HF experience *unsymmetrical* attractive forces. The effective nuclear charge of a fluorine atom is significantly greater than the effective charge of a hydrogen atom due to the greater number of protons, so the electrons shared between H and F feel a stronger attraction to the F atom. Unequal attractive forces lead to an unsymmetrical distribution of the bonding electrons. The HF molecule is most stable when its bonding electrons are concentrated closer to the fluorine atom and away from the hydrogen atom. This unequal distribution of electron density gives the fluorine end of the molecule a small negative charge and the hydrogen end a small positive charge; however, the molecule as a whole remains electrically neutral. These partial charges are less than one charge unit and are equal in magnitude. To indicate such partial charges, we use the symbols $\delta+$ and $\delta-$ (δ is the lower-case Greek letter delta) or an arrow pointing from the negative end towards the positive end of the molecule, or an electron density surface, which shows the spread of charge using colours, with blue being positive and red being negative as shown for HF in figure 5.4. This unequal sharing of electrons results in a **polar covalent bond**. We describe bond polarity and its consequences in section 5.5.

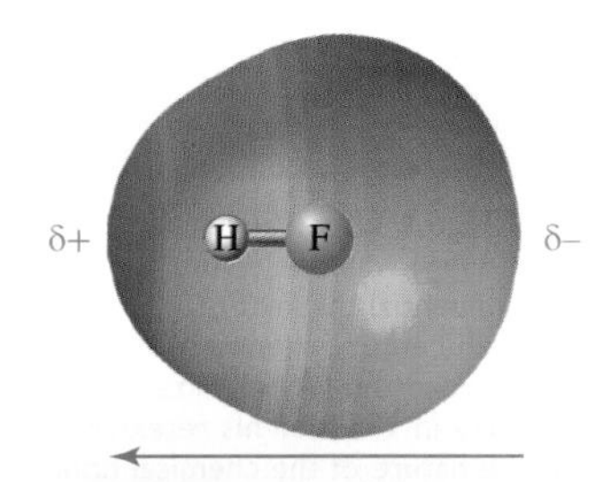

FIGURE 5.4 Unequal sharing of electron density in HF results in a polar covalent bond. The colour gradient represents the variation in electron density shared between the atoms.

Note that, despite being polar, the H—F bond is still a σ bond, because it remains unchanged as we rotate the molecule about the H—F bond axis.

As elements differ in their effective nuclear charges, atoms of each element have a characteristic ability to attract electrons within a covalent bond. This ability is called **electronegativity** and is symbolised by χ (the lower-case Greek letter chi). A bond between two elements of different electronegativities will be polar, and the greater the difference ($\Delta\chi$), the more polar the bond.

Electronegativity gives a numerical value to how strongly an atom attracts the electrons in a chemical bond. This property of an atom involved in a bond is related to, but distinct from, ionisation energy and electron affinity. (As described in chapter 4, ionisation energy measures how strongly an atom attracts one of its own electrons. Electron affinity specifies how strongly an atom attracts a free electron.)

Electronegativities, which have no units, are estimated by using combinations of atomic and molecular properties. The periodic table shown in figure 5.5 presents commonly used values, developed by Linus Pauling (figure 5.6), based on bond energies. These values are often called the Pauling scale of electronegativity.

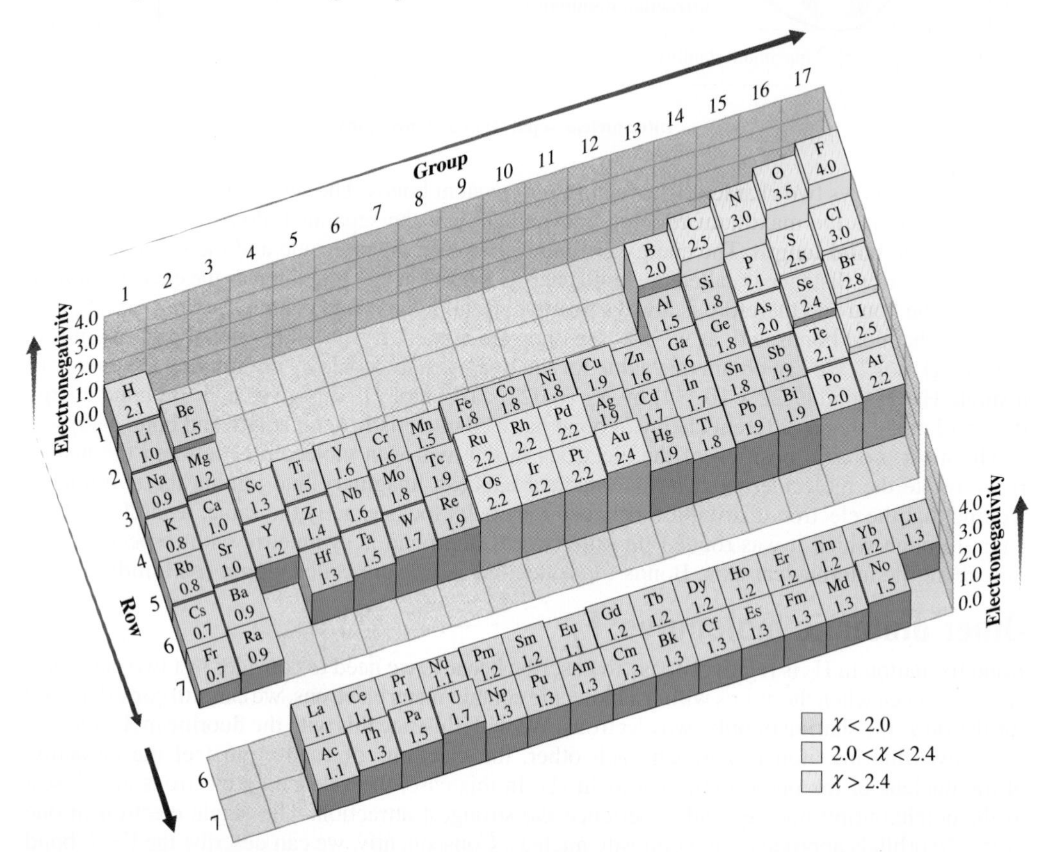

FIGURE 5.5 Electronegativities of the elements vary periodically. While there are some exceptions, the general trend is that electronegativity increases from left to right and decreases from top to bottom of the periodic table. The Pauling scale omits group 18 elements (the noble gases) because they are very unreactive and, with very few exceptions, do not form covalent bonds. Some other electronegativity scales do assign values to these elements.

FIGURE 5.6 Linus Pauling (1901–1994) was one of the most influential thinkers of twentieth-century chemistry and was awarded the Nobel Prize in chemistry in 1954 for his research into the nature of the chemical bond. He was also one of the few people to be awarded two Nobel Prizes, receiving the peace prize in 1962 for his stand against nuclear weapons, which led the US government to revoke his passport in 1952.

Figure 5.5 shows that electronegativity is a periodic property. Electronegativities generally increase from the lower left to the upper right of the periodic table. The elements francium and caesium ($\chi = 0.7$) have the lowest value, and fluorine ($\chi = 4.0$) has the highest value. Electronegativities also generally decrease down most groups and increase from left to right across the *s* and *p* blocks. As with ionisation energies and electron affinities, variations in effective nuclear charge and principal quantum number explain electronegativity trends. Metals generally have low electronegativities ($\chi = 0.7$ to 2.4) and nonmetals have high electronegativities ($\chi = 2.1$ to 4.0). As expected from their intermediate character between metals and nonmetals, metalloids have electronegativities that are larger than the values for most metals and smaller than those for most nonmetals.

Electronegativity differences ($\Delta\chi$) between bonded atoms indicate where a particular bond lies on the continuum of bond polarities. Three fluorine-containing substances, F_2, HF and CsF, represent the range of variation. (Note that, as no weighable quantity of francium has ever been isolated, the compound FrF is unknown.) At one end of the continuum, the bonding electrons in F_2 are shared equally between the two fluorine atoms ($\Delta\chi = 4.0 - 4.0 = 0$). At the other limit, CsF ($\Delta\chi = 4.0 - 0.7 = 3.3$) is a compound in which electrons have been fully transferred to give Cs^+ cations and F^- anions. We call such compounds **ionic compounds**, and they will be discussed in more detail in the next section. Most bonds, including the bond in HF ($\Delta\chi = 4.0 - 2.1 = 1.9$), fall between these extremes. These are polar covalent bonds, in which two atoms share electrons unequally.

Chemical Connections

Electromagnetic radiation and cancer

The ability of light to provide the energy for chemical reactions enables life to exist on our planet. Green plants absorb sunlight and, with the help of chlorophyll, convert carbon dioxide and water into carbohydrates (e.g. sugars and cellulose), which are essential constituents of the food chain. However, not all effects of sunlight are so beneficial.

As we saw in chapter 4, light has energy that is proportional to its frequency, and, if the photons that are absorbed by a substance have enough energy, they can rupture chemical bonds and initiate chemical reactions. The carbon–carbon and carbon–hydrogen covalent bonds in organic molecules typically have energies of at least 350 kJ mol^{-1}. Visible light, which ranges from 780 nm to 380 nm in wavelength, has a corresponding energy range of approximately 153 to 315 kJ mol^{-1}, and therefore photons of visible light are not sufficiently energetic to break such bonds. However, ultraviolet radiation, which has shorter wavelengths and hence higher energies, can be dangerous in this respect, particularly light of wavelength less than 340 nm, which has energy greater than 350 kJ mol^{-1}. The sunlight illuminating the Earth contains substantial amounts of UV radiation. Fortunately, a layer of ozone (O_3) in the stratosphere, a region of the atmosphere extending from about 45 to 55 km altitude, absorbs most of the incoming UV, protecting life on the surface. However, some UV radiation does get through, and the part of the spectrum of most concern is called UV-B with wavelengths between 280 and 320 nm.

What makes UV-B so dangerous is its ability to affect the DNA in our cells (the structure of DNA and its replication are discussed in chapter 25). Absorption of UV radiation causes constituents of the DNA, called *pyrimidine bases*, to undergo reactions that form bonds between them. This causes transcription errors when the DNA replicates during cell division, giving rise to genetic mutations that can lead to skin cancers. These skin cancers fall into three classes: basal cell carcinomas, squamous cell carcinomas and melanomas, the last being the most dangerous. Each year more than 1 million cases of skin cancer are diagnosed worldwide. It is estimated that more than 90% of skin cancers are due to absorption of UV-B radiation.

In Australia, public health campaigns encourage people to avoid exposure to the sun when the UV index is high. They are further encouraged to protect themselves with clothing, hats, sunglasses and sunblock creams (figure 5.7). Sunblock creams work by either reflecting or absorbing UV radiation before it reaches the skin.

FIGURE 5.7 Sunlight carries the risk of skin cancer to those particularly susceptible. Understanding the risk allows us to protect ourselves with clothing and sunblock creams.

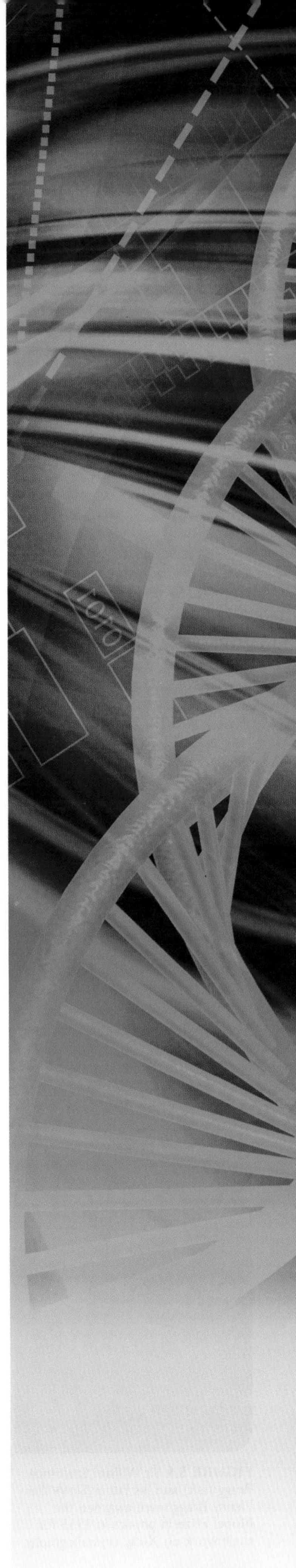

5.2 Ionic bonding

Compounds formed between elements of very different electronegativities are often predominantly ionic in character. For example, ionic compounds tend to form between cations from groups 1 or 2 and anions from groups 16 or 17. Most ionic compounds are solids with relatively high melting points, particularly if the ionic solids are composed of atomic or small molecular ions. Apart from these high-melting-point ionic solids, there is also a class of compounds called ionic liquids, compounds of large molecular ions. These are ionic compounds that have melting points below 100 °C. There are some ionic liquids that are liquid even at room temperature, and those have important applications. Ionic liquids are not further discussed in this section.

The bonding in ionic compounds differs fundamentally from that in covalent compounds; it does not involve the sharing of pairs of electrons. Instead, ionic compounds are held together by the non-directional attractive forces between oppositely charged ions, leading to three-dimensional arrangements rather than distinct molecules.

Consider, sodium chloride, NaCl. The structure of this compound has each Na^+ cation surrounded by six Cl^- anions and each Cl^- anion surrounded by six Na^+ cations, as shown in figure 5.8.

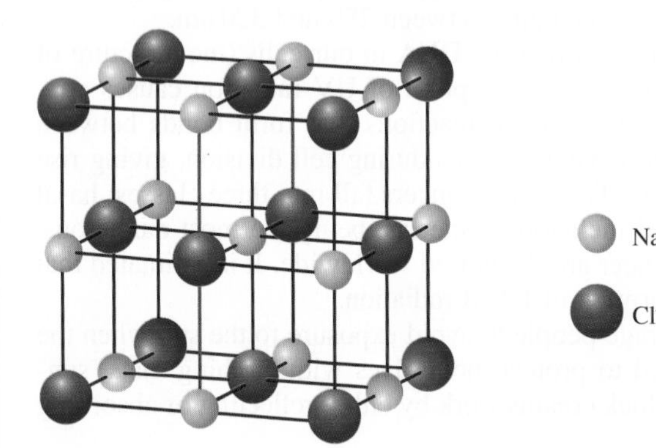

FIGURE 5.8 Representation of part of the structure of sodium chloride, NaCl. Note that each Na^+ cation is surrounded by six Cl^- anions, and each Cl^- anion is surrounded by six Na^+ cations.

This arrangement of Na^+ cations and Cl^- anions was first determined in 1913 by Australian-born Sir William Lawrence Bragg, using X-ray equipment designed by his father, Sir William Henry Bragg (figure 5.9). NaCl was the first compound whose structure was determined by X-ray crystallography (see chapter 7), a field in which the Braggs were pioneers. They were jointly awarded the Nobel Prize in physics in 1915, and William Lawrence Bragg became the youngest ever awardee at the age of 25. Crystal structures determined through analysis of X-ray diffraction data show the precise locations of atoms within a chemical species and give atomic separations accurate to about 1×10^{-12} m. X-ray crystallography is a vital part of modern chemistry and is the source of much of the structural information that you will encounter in later chapters.

The distance between the ions in NaCl is determined by a balance of the Coulombic (electrostatic) attractions and repulsions and the short-range repulsive interactions between both the electron densities and nuclei of neighbouring ions. At room temperature, the distance between the centres of neighbouring ions is 2.82×10^{-10} m. Using this number, we can calculate the magnitude of the attractive forces between two neighbouring Na^+ and Cl^- ions using the equation introduced at the start of this chapter.

FIGURE 5.9 Sir William Lawrence Bragg (left) and his father Sir William Henry Bragg were awarded the Nobel Prize in physics in 1915 for their work on X-ray crystallography.

$$E_{\text{electrostatic}} = k\frac{q_1 q_2}{r}$$

Given that q_1 is $+1.60 \times 10^{-19}$ C, q_2 is -1.60×10^{-19} C (-1.60×10^{-19} C is the fundamental charge of a single electron) and $k = 9.00 \times 10^9$ N m^2 C^{-2}, we obtain a value of -8.17×10^{-19} J, which equates to -492 kJ mol^{-1} when multiplied by the Avogadro constant. This corresponds to the energy released (hence the negative sign) when 1 mole of gaseous Na^+ and gaseous Cl^- ions are brought from an infinite distance to a separation of 2.82×10^{-10} m. However, this value takes into account only the attractive forces to one nearest neighbour, whereas each ion in the crystal has six. If we take all six interactions into account, as well as repulsive interactions between ions of the same charge in the structure, we obtain a value of -769 kJ mol^{-1}. This is a substantial amount of energy, and means that, in order to break 1 mole of the compound into its constituent gaseous ions, we would have to supply 769 kJ of energy. This is called the **lattice energy** of the compound. Among other factors, the lattice energy of a compound is dependent on the charges of the ions; as the charges increase, the lattice energy becomes more positive and the ions become harder to separate from one another. We also expect lattice energy values to decrease as the ions get larger, because *r*, the distance between them, will increase. These trends are illustrated in table 5.1, which gives some lattice energy values for ionic species.

TABLE 5.1 Lattice energies for a number of ionic compounds. All values are in kJ mol^{-1}.

	Anion				
Cation	F^-	Cl^-	Br^-	I^-	O^{2-}
Li^+	1030	834	788	730	2799
Na^+	910	769	732	682	2481
K^+	808	701	671	632	2238
Rb^+	774	680	651	617	2163
Mg^{2+}	2913	2326	2097	1944	3795
Ca^{2+}	2609	2223	2132	1905	3414
Sr^{2+}	2476	2127	2008	1937	3217
Ba^{2+}	2341	2033	1950	1831	3029

The effect of ionic size on lattice energy is shown by the fact that the lattice energies of the group 1 chlorides decrease as the size of the cation increases from Li^+ to Rb^+, and the lattice energies of the sodium halides decrease as the size of the anion increases from F^- to I^-. The effect of ionic charge is seen by comparing the lattice energies of the compounds formed from the 2+ ions with halide anions. In all cases, these lattice energies are much larger than those for the 1+/1− combinations. The values for compounds containing 2+ and 2− ions are even larger, with the lattice energy of MgO being almost four times the lattice energy of LiF. However, it is important to note that meaningful comparisons of lattice energy values can be made only between compounds with the same structure type.

While the ionic model describes a number of metal halides, oxides and sulfides well, it does not describe most other chemical substances adequately. Compounds such as CaO, NaCl and MgF_2 behave like simple cations and anions held together by electrostatic attraction, but compounds such as CO, Cl_2 and HF do not. We must describe the bonding in these latter compounds in terms of the sharing of electrons between atoms. There are two predominant theories of bonding which do this. Before we study these, we will look at a simple way of determining the electron distribution in molecules and how we can use this to predict the shape of a particular molecule.

5.3 Lewis structures

In this section, we develop a process for making schematic drawings that show how the atoms in a molecule are bonded. In addition, these drawings, known as **Lewis structures**, differentiate between bonding and nonbonding valence electrons in a molecule. From this perspective, writing a Lewis structure is the first step in developing a bonding description of a molecule. Lewis structures are named after their originator, Gilbert Newton Lewis (1875–1946), former Professor of Chemistry at the University of California, Berkeley, USA.

The conventions

Lewis structures are drawn according to the following conventions.

1. *Each atom is represented by its elemental symbol.* In this respect, a Lewis structure is an extension of the chemical formula.
2. *Only the valence electrons appear in a Lewis structure.* Recall from chapter 4 that only valence electrons are accessible for bonding.
3. *A line joining two elemental symbols represents one pair of electrons that is shared between two atoms.* Two atoms may share electrons in a **single bond** (2 shared electrons, 1 line), a **double bond** (4 shared electrons, 2 lines) or a **triple bond** (6 shared electrons, 3 lines).
4. *Dots placed next to an elemental symbol represent nonbonding electrons on that atom.* Nonbonding electrons usually occur in pairs.

The use of these conventions is shown, using the HF molecule as an example, in figure 5.10.

1. Elemental symbols represent atoms.	H	F
2. Only valence electrons appear.	1s ↑	2s ↑↓ ↑↓ ↑↓ ↑ 2p
3. Lines represent bonds.	H—F	
4. Dots represent nonbonding electrons.	H—F: (with lone pairs)	

FIGURE 5.10 The Lewis structure conventions for hydrogen fluoride.

These conventions divide electrons in molecules into three groups. Core electrons do not appear in Lewis structures. Bonding valence electrons are shared between atoms and are represented by lines. Nonbonding valence electrons are localised on atoms and are represented by dots.

We use a five-step procedure to draw Lewis structures.

Step 1: Count the valence electrons.
Step 2: Assemble the bonding framework using single bonds.
Step 3: Place three nonbonding pairs of electrons on each outer atom except H. (*Note:* Outer atoms are those that are bonded to only one other atom, while inner atoms are those bonded to more than one other atom.)
Step 4: Assign the remaining valence electrons to inner atoms.
Step 5: Minimise formal charges on all atoms.

Building Lewis structures

Learning how to draw Lewis structures is best done through examples. We will work through the five steps using the sulfur dioxide molecule, SO_2.

Step 1: Count the valence electrons. If the species is an ion, add or subtract one electron for each negative or positive charge, respectively. Sulfur (group 16) has the valence electron configuration $3s^23p^4$, while the configuration of oxygen (also group 16) is $2s^22p^4$. Hence, there is a total of $(6 + [2 \times 6]) = 18$ valence electrons from the atoms themselves. The molecule is uncharged, so the total number of valence electrons is 18.

Step 2: Assemble the bonding framework using single bonds. This step is straightforward for organic molecules, which consist predominantly of C and H atoms, as there is no ambiguity as to the possible placement of atoms. However, care must be taken when drawing structures of inorganic molecules. Structures will generally consist of an inner (central) atom connected to two or more other (outer) atoms. The outer atoms will generally be either hydrogen or the most electronegative of the atoms present, and will be bonded only to the inner atom. In the case of sulfur dioxide, oxygen is the most electronegative element present, so sulfur will be the inner atom. The bonding framework will therefore be as follows.

O—S—O

As each single bond in the framework consists of two electrons, we have used four of the original 18 valence electrons so far.

Step 3: Place three nonbonding pairs of electrons on each outer atom except H. With the exception of hydrogen, all outer atoms are associated with eight electrons (four pairs in the valence *s* and *p* orbitals), which may be bonding or nonbonding. Such a set of four pairs of electrons associated with an atom is often called an octet.

:Ö—S—Ö: (each O with three lone pairs)

The nonbonding pairs are called **lone pairs**. Each O atom is associated with three lone pairs and one bonding pair and therefore has a complete octet. We used $(2 \times 6) = 12$ electrons in this step. Combined with the four used in the previous step, this means 16 of the original 18 electrons have been used.

Step 4: Assign the remaining valence electrons to inner atoms. There are two remaining valence electrons. These are placed on the inner S atom as a lone pair as follows.

:Ö—S̈—Ö: (each O with three lone pairs, S with one lone pair)

We have now accounted for all the valence electrons of SO_2.

Step 5: Minimise formal charges on all atoms. At this point, we need to analyse the structure obtained and calculate the **formal charge** on each atom. We then try to minimise these by adjusting the electron distribution, since unnecessarily large formal charges do not make chemical sense. Formal charge is defined as the difference between the number of valence electrons in the free atom and the number of electrons assigned to that atom in the Lewis structure.

$$\text{formal charge} = (\text{valence electrons of free atom}) - (\text{lone pair electrons}) - \tfrac{1}{2}(\text{shared electrons})$$

Assume that lone pair electrons 'belong' to the atom on which they are located, while electrons in bonds are shared equally between the two atoms. A useful feature of formal charges is that the sum of the formal charges on all atoms equals the charge of the species. For a neutral molecule, the sum of the formal charges must be zero. For a cation or anion, the sum of the formal charges equals the charge on the ion.

The formal charges of the atoms in the structure obtained in step 4 are calculated as follows.

S	Valence electrons of free atom	6
	Lone pair electrons	2
	Shared electrons	4
	Formal charge	$6 - 2 - \frac{1}{2}(4) = +2$
O	Valence electrons of free atom	6
	Lone pair electrons	6
	Shared electrons	2
	Formal charge	$6 - 6 - \frac{1}{2}(2) = -1$

We can minimise the formal charges by converting a lone pair of electrons on each O atom to a bonding pair. We can represent this process diagrammatically as follows.

$$:\ddot{\underset{..}{O}}-\ddot{S}-\ddot{\underset{..}{O}}: \longrightarrow \ddot{\underset{..}{O}}=\ddot{S}=\ddot{\underset{..}{O}}$$

If we analyse the formal charges on the right-hand structure, we find that they are all zero. Therefore, the right-hand structure is the best Lewis structural representation of the SO_2 molecule.

In this structure each oxygen atom is surrounded by eight electrons, and therefore has an octet, while the inner sulfur atom is surrounded by 10 electrons. We generally find octets surrounding atoms from the second period of the periodic table, especially C, N, O and F, as such atoms can accommodate a maximum of eight electrons in the available orbitals. However, the larger atoms from the third and subsequent periods often accommodate more than eight electrons, as observed for the sulfur atom. This is due to the fact that these atoms have more than four orbitals available for valence electrons.

WORKED EXAMPLE 5.1

Lewis structure of ClF_3

Chlorine trifluoride, ClF_3, is used to recover uranium from nuclear fuel rods in a high-temperature reaction that produces gaseous uranium hexafluoride.

$$2ClF_3(g) + U(s) \longrightarrow UF_6(g) + Cl_2(g)$$

Determine the Lewis structure of ClF_3.

Analysis

We follow the five steps outlined previously for drawing a Lewis structure.

Solution

1. All four atoms are halogens (group 17, ns^2np^5), so ClF_3 has 28 valence electrons.
2. Chlorine, with lower electronegativity than fluorine, is the inner atom. Make a single bond to each of the three fluorine atoms.

```
      F
      |
  F — Cl — F
```

3. Add three nonbonding pairs to each fluorine atom.

```
      :F̤̈:
       |
  :F̤̈ — Cl — F̤̈:
```

4. Four electrons remain unassigned. Place two electron pairs on the chlorine atom.

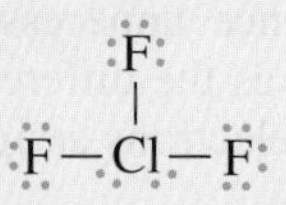

5. Determine the formal charges on all the atoms and adjust the electron distribution if necessary. In this case, each atom has a formal charge of zero, and so no redistribution of electrons is required.

Is our answer reasonable?

Step 4 results in 10 electrons around the chlorine atom. However, chlorine is a third-row element and is therefore able to accommodate more than eight electrons. We have followed the rules and obtained the structure with formal charges of zero on every atom, so our answer is reasonable.

PRACTICE EXERCISE 5.1 Determine the Lewis structure of sulfur tetrafluoride, SF_4.

Note that some molecules and all ions will have preferred Lewis structures in which the formal charges are not all zero. In these cases, ensure that negative formal charges are situated on the most electronegative atoms and positive formal charges on the least electronegative atoms, if possible (for example, the nitrate ion in figure 5.11, where the more electronegative O atoms, rather than the N atom, have a negative formal charge). It is also important to realise that formal charges are not the same as the partial charges induced as a result of bond polarity. For example, in ClF_3, the chlorine atom would have a $\delta+$ charge, despite the fact that its formal charge is zero.

Resonance structures

In completing step 5 of the Lewis structure procedure, there might be more than one way to minimise the formal charges on the atoms; that is, more than one Lewis structure is possible for a particular molecule or ion. This is shown below with the nitrate anion, NO_3^-.

The anion has 24 valence electrons, and nitrogen is the inner atom. Three N—O bonds use six electrons, and assigning six electrons to each of the three outer oxygen atoms uses the remaining 18 electrons. This gives a Lewis structure in which the O atoms have formal charges of −1 and the N atom has a formal charge of +2. We can minimise the formal charges by converting the lone pair on one oxygen atom to a bonding pair. However, as figure 5.11 shows, a lone pair from any of the three oxygen atoms can be moved to form this double bond.

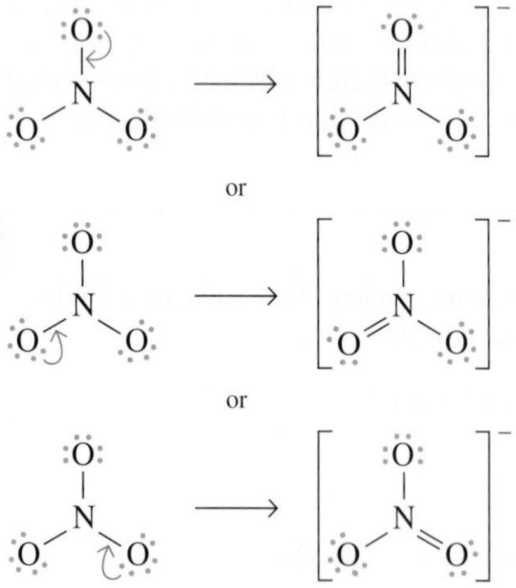

or

or

FIGURE 5.11 There are three possible ways to minimise the formal charges in the nitrate ion; any of the three oxygen atoms can supply a pair of electrons.

Which of these options is the correct Lewis structure? Actually, no single Lewis structure is an accurate representation of NO_3^-. Any single structure of the anion shows nitrate with one N═O double bond and two N—O single bonds. Experiments, however, show that all three N—O bonds are identical (we cannot form two double bonds since nitrogen can accommodate only eight electrons in its orbitals). To show that the nitrate N—O bonds are all alike, we use a composite of the three equivalent Lewis structures. These are then called **resonance structures**. Resonance structures are connected by double-headed arrows to emphasise that a complete depiction requires all of them.

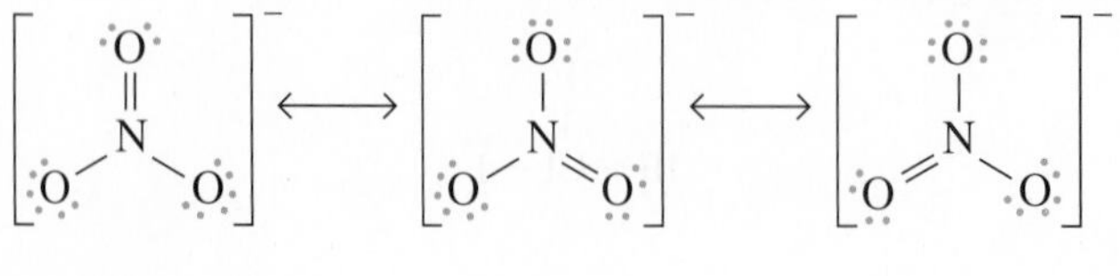

It is essential to realise that electrons in the nitrate anion *do not* 'flip' back and forth among the three bonds, as implied by separate structures. None of these structures actually exist. The true character of the anion is a blend of the three, in which all three nitrogen–oxygen bonds are equivalent. Note also that resonance structures differ *only* in the position of the electrons; you cannot move atoms when drawing resonance structures. The need to show several equivalent structures for species such as the nitrate ion reflects the fact that Lewis structures are approximate representations of structures with delocalised electrons. While they reveal much about how electrons are distributed in a molecule or ion, they cannot describe the entire story of chemical bonding.

WORKED EXAMPLE 5.2

Resonance structures of an oxoanion

Determine the possible resonance structures of the phosphate anion, PO_4^{3-}.

Analysis

We must first determine the Lewis structure by following the five-step procedure outlined previously. In step 5, we determine the different ways we can minimise the formal charges within the ion, thereby obtaining the resonance structures.

Solution

We have 32 electrons to be used (remember that the ion has a 3– charge). Following the rules outlined previously, we arrive at the following structure, in which all the electrons are used after step 4.

```
      :Ö:
       |
  :Ö — P — Ö:
       |
      :Ö:
```

In this structure, the O atoms all have formal charges of –1, while the P atom has a formal charge of +1. Turning one lone pair from an oxygen atom into a bonding pair makes the formal charge on the P atom zero and also makes the formal charge on the O atom involved equal to zero, thereby minimising the formal charges in the ion. We can shift electrons from any of the outer O atoms, so there are four possible resonance structures as shown below. (Note that we have omitted the 3– charge on each structure for clarity.)

```
     :Ö:              :Ö:              :Ö:              :Ö:
      |                |                |                |
 :Ö — P — Ö:      :Ö — P — Ö:      :Ö — P — Ö:      :Ö — P — Ö:
      |                |                |                |
     :Ö:              :Ö:              :Ö:              :Ö:
      ↓                ↓                ↓                ↓
     :Ö:              :Ö:              Ö                :Ö:
      |                |               ||                |
 :Ö — P — Ö:  ⟷  :Ö — P = Ö   ⟷  :Ö — P — Ö:  ⟷   Ö = P — Ö:
      ||               |                |                |
      Ö               :Ö:              :Ö:              :Ö:
```

Is our answer reasonable?

There are four possible ways to minimise the formal charges, so we must have four resonance structures. The four structures differ only in the position of a double bond, and therefore we can be confident that our answer is correct.

PRACTICE EXERCISE 5.2

Determine the Lewis structure of the acetate anion, CH_3COO^-.

In the examples presented so far, all the resonance structures are equivalent in energy, but resonance structures are not always equivalent. Resonance structures that are not equivalent occur when step 5 requires shifting electrons from atoms of different elements. In such cases, different possible structures may have different formal charge distributions, and the optimal set of resonance structures includes those forms with the lowest possible formal charge on each atom.

WORKED EXAMPLE 5.3

Resonance structures of N_2O

Determine the possible resonance structures of dinitrogen oxide, N_2O, a gas used as an anaesthetic, a foaming agent and a propellant for whipped cream.

Analysis

Once again, we first determine the Lewis structure, and then inspect any possible resonance structures.

Solution

We must first choose whether N or O will be the inner atom. As O is more electronegative than N, the structure will have an inner N atom bonded to an O atom and another N atom. The structure we obtain following the first four steps is:

$$:\ddot{N}-N-\ddot{O}:$$

This structure has formal charges of −2 on the outer N atom, +3 on the inner N atom and −1 on the O atom. We can convert lone pairs on the end N atom or the O atom into bonding pairs to reduce the formal charges, but we must ensure that the inner N atom has no more than four electron pairs. This leaves us with the following possibilities, in which the numbers denote the formal charges.

$$\underset{-1}{N}=\overset{+1}{N}=\underset{0}{O} \quad \underset{0}{:N}\equiv\overset{+1}{N}-\underset{-1}{O:} \quad \underset{-2}{:N}-\overset{+1}{N}\equiv\underset{+1}{O:}$$

The formal charges of the first two structures are identical, and less than those on the third structure. Thus, the optimal Lewis structure of the N_2O molecule is a composite of the first two structures, which are formally inequivalent, but not the third.

$$N=N=O \longleftrightarrow :N\equiv N-O:$$

Is our answer reasonable?

We have chosen the two structures in which the formal charges are minimised, and so our answer should be correct. In fact, experimental studies show that the nitrogen–oxygen bond in N_2O is somewhere between an N—O single and an N=O double bond.

PRACTICE EXERCISE 5.3 Determine the Lewis structure of ozone, O_3, in which the three oxygen atoms are linked in a row.

5.4 Valence-shell-electron-pair repulsion (VSEPR) theory

The Lewis structure of a molecule shows how the valence electrons are distributed among the atoms. It does not, however, show the three-dimensional structure, which plays an essential role in determining chemical reactivity. However, we can use the Lewis structure as a starting point to determine the shape of a molecule or ion. The **valence-shell-electron-pair repulsion (VSEPR)** theory considers that molecular shape is determined primarily by the repulsions between pairs of electrons in the molecule or molecular ion, be they bonding pairs or lone pairs. Therefore, to minimise these repulsions, electron pairs around an inner atom within a molecule will be situated as far apart as possible in the preferred three-dimensional structure.

To determine the shape of a molecule using VSEPR theory, we use the following procedure.

1. Determine and draw the Lewis structure of the molecule.
2. Count the number of sets of bonding pairs and lone pairs of electrons around any inner atom, and use table 5.2 to determine the optimum geometry of these sets.
   Note that VSEPR theory makes no distinction between electron pairs in single, double and triple bonds. Each is treated as being one set.

TABLE 5.2 Optimum geometry of sets of electron pairs.

Number of sets of electron pairs	Geometry of sets of electron pairs
2	linear
3	trigonal planar
4	tetrahedral
5	trigonal bipyramidal
6	octahedral

3. Modify the geometry, if necessary, to take account of the fact that the magnitudes of repulsions between sets of electron pairs depend on whether the electron pairs involved are bonding pairs (BP) or lone pairs (LP). The repulsions are in the order:

$$\text{LP–LP} > \text{BP–LP} > \text{BP–BP}$$

Therefore, two neighbouring lone pairs on an inner atom will repel each other to a greater extent than will two neighbouring bonding pairs. This is because lone pairs occupy a larger volume than bonding pairs. Note that double bonds occupy a larger volume than single bonds, and some structures may have to be modified for this.

It is important to make the distinction between the geometry of the sets of electron pairs, and the shape of the molecule. The former refers to the way in which sets of electron pairs are arranged around an inner atom of the molecule, while the latter refers to the arrangement of the atoms bonded to that inner atom. As we will see, when lone pairs are present, the two are not the same.

We will now illustrate VSEPR using examples of each different geometry of sets of electron pairs.

Two sets of electron pairs: linear geometry

Beryllium hydride, BeH_2

The Lewis structure of BeH_2 is:

H—Be—H

In this case, we have two sets of electron pairs around the inner atom that we need to situate as far away from each other as possible. This is satisfied by placing the two sets in a **linear** arrangement (see table 5.2) and results in a linear shape for this molecule, with a H—Be—H bond angle of 180°. Note that we do not adjust the optimum geometry of the electron pairs, since there are only two single bonding pairs and no lone pairs.

Carbon dioxide, CO_2

The Lewis structure of CO_2 shows an inner C atom bonded to two outer O atoms.

O=C=O

two sets of electron pairs around the C atom

linear shape
bond angle = 180°

This is similar to BeH_2 because we have two sets of electron pairs again. Even though each set comprises the two electron pairs of a double bond, the principle is still the same. We place the two sets as far apart as possible, that is, in a linear arrangement, and this leads to a linear shape. Notice that the lone pairs on the O atoms do not affect the overall shape of the molecule; only sets of electrons that are either bonded to, or lie completely on, an inner atom determine molecular shape.

Three sets of electron pairs: trigonal planar geometry

Boron trifluoride, BF_3

The Lewis structure of BF_3 shows an inner B atom singly bonded to three F atoms. We have three sets of electron pairs about the inner atom.

:F:
|
B
:F: :F:

The optimal geometry for three sets of electron pairs is **trigonal planar** (see table 5.2). In this arrangement, the sets are oriented at 120° to each other and are coplanar. This leads to a trigonal planar shape for the BF_3 molecule, with all atoms coplanar and all F—B—F bond angles of 120°. Note that, again, as all the sets of electron pairs involved are equivalent, we do not have to adjust their optimal geometry.

The nitrite ion, NO_2^-

Two possible resonance structures can be drawn for the nitrite ion, which are as follows.

O=N—O⁻ ⟷ ⁻O—N=O

This is the first example we have encountered in which the inner atom bears a lone pair, and this has important implications when applying VSEPR theory. Note that we can choose either resonance structure and end up with the same result. There are three sets of electron pairs, two sets of bonding pairs and a lone pair. The idealised geometry is trigonal planar. However, in this case, not all the repulsions between the sets of electron pairs are the same; there are BP–BP repulsions and BP–LP repulsions. The BP–LP repulsions are stronger, which means that the two bonding pairs will be pushed slightly closer together than an exact trigonal planar arrangement. The result is that the O—N—O bond angle is about 115°, rather than 120°. This example also illustrates the difference between geometry and shape. The sets of electron pairs have an approximately trigonal planar geometry, but we would describe the shape of the NO_2^- ion as bent, because shape refers only to the positions of the atoms and not the electrons.

Four sets of electron pairs: tetrahedral geometry

Methane, CH_4

The Lewis structure of methane, CH_4, shows that the molecule contains four C—H single covalent bonds, and therefore four sets of equivalent electron pairs.

```
     H
     |
 H — C — H
     |
     H
```

The optimal geometry for four sets of electron pairs is **tetrahedral** (see table 5.2). This means that the methane molecule is tetrahedral in shape, with H—C—H bond angles of 109.5°, and the H atoms situated at the four vertices of a tetrahedron.

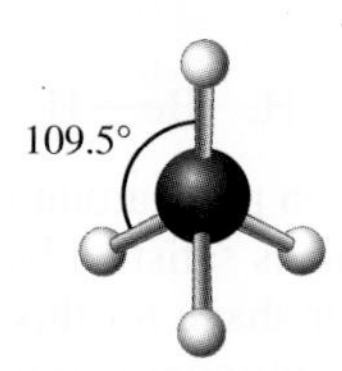

Ammonia, NH_3

The Lewis structure of ammonia, NH_3, shows a total of four sets of electron pairs, three bonding pairs and a lone pair.

```
     ..
 H — N — H
     |
     H
```

The idealised arrangement of four sets of electron pairs is tetrahedral. But we now have BP–BP and BP–LP repulsions to consider. The single lone pair on the N atom takes up more space than each of the three bonding pairs, and this distorts the idealised arrangement such that the H—N—H angles in ammonia are about 107°. Ammonia adopts a trigonal pyramidal shape, with the N atom lying out of the plane of the three H atoms.

Water, H_2O

Like methane and ammonia, water has four sets of electron pairs, as shown in its Lewis structure, and the idealised arrangement is again tetrahedral.

```
     ..
 H — O — H
     ..
```

However, in this case we have two sets of bonding pairs and, significantly, two lone pairs. The LP–LP repulsion dominates, and there are also BP–LP repulsions to consider. The result is that the idealised tetrahedral geometry of the four sets of electron pairs is significantly distorted, such that the H—O—H bond angle in water is about 104.5°. Water adopts a bent shape.

Figure 5.12 shows the different shapes of methane, ammonia and water. The important effect of lone pairs on the shapes of the molecules is evident in these three diagrams.

tetrahedral

109.5° 109.5° 109.5° 109.5°

CH_4
lone pairs = 0

trigonal pyramidal

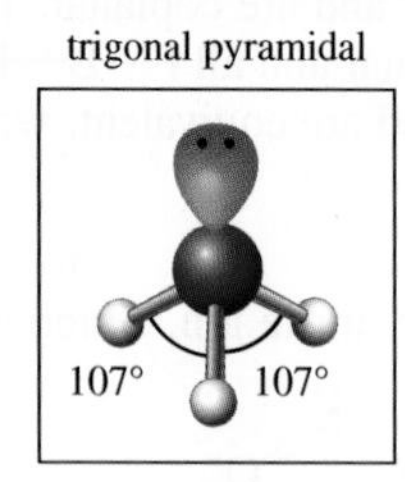

NH_3
lone pairs = 1

bent

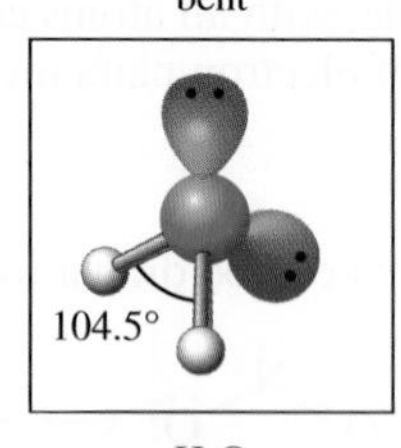

H_2O
lone pairs = 2

FIGURE 5.12 Representations of the different shapes of methane, ammonia and water molecules, each with four sets of electron pairs.

WORKED EXAMPLE 5.4

Shape of the hydronium ion

Determine the shape of the hydronium ion, H_3O^+, using VSEPR theory. Make a sketch of the ion that shows the three-dimensional shape, including any lone pairs that may be present.

Analysis

Follow the three-step process described previously. Begin with the Lewis structure, and then use this to determine the number of sets of electron pairs and therefore the geometry of these. Finally, take into account any lone pairs to deduce the molecular shape.

Solution

1. Determine the Lewis structure. A hydronium ion has eight valence electrons. Six are used to make three O—H single bonds, and two are placed as a lone pair on the oxygen atom.

$$\left[\begin{array}{c} \mathrm{H} \\ | \\ \mathrm{H{-}\ddot{O}{-}H} \end{array}\right]^{+}$$

2. There are four sets of electron pairs, three bonding pairs and one lone pair. The idealised geometry is therefore tetrahedral.
3. The presence of a lone pair means that the idealised tetrahedral geometry of the sets of electron pairs will be distorted, owing to LP–BP repulsions. We would therefore expect the H—O—H bond angles to be somewhat less than the 109.5°. The shape of the hydronium ion is trigonal pyramidal, as outlined in the following diagram.

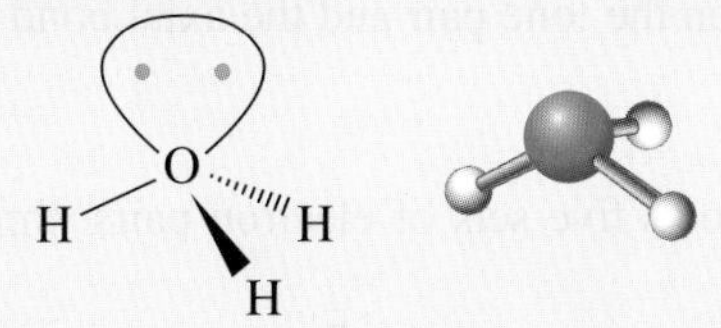

Is our answer reasonable?

The hydronium ion has the same number of electrons and atoms as the ammonia molecule, so it is reasonable for these species to have a similar shape.

PRACTICE EXERCISE 5.4

Determine the shape of chloromethane, CH_3Cl. Make a sketch of the molecule that shows its three-dimensional shape.

Five sets of electron pairs: trigonal bipyramidal geometry

Phosphorus pentachloride, PCl_5

The Lewis structure of PCl_5 shows five sets of equivalent electron pairs around a phosphorus atom.

```
         ..
        :Cl:
   ..    |    ..
  :Cl—   P   —Cl:
   ..   / \   ..
      :Cl: :Cl:
       ..   ..
```

The geometry which places these as far apart as possible is **trigonal bipyramidal** (see table 5.2). A trigonal bipyramid is so called because it can be viewed as two pyramids that share a triangular base, as shown in figure 5.13a. In contrast to the geometries encountered so far, the five positions in a trigonal bipyramid are not all equivalent. Three positions lie at the corners of an equilateral triangle around the inner atom, separated by 120° bond angles. Atoms in the trigonal plane are in *equatorial positions*. The other two positions lie along an axis above and below the trigonal plane, separated from equatorial positions by 90° bond angles. Atoms in these sites are in *axial positions*. These differences are outlined in figure 5.13b.

As PCl_5 contains only bonding pairs of electrons, we would expect it to adopt a regular trigonal bipyramidal shape, as shown in figure 5.13.

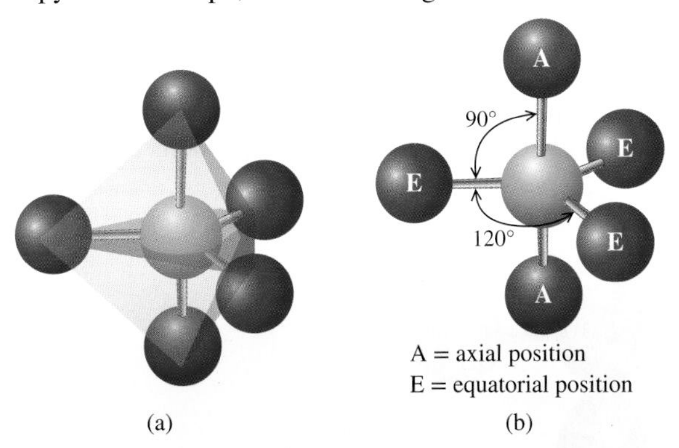

FIGURE 5.13 (a) Model representing the equatorial plane and opposing pyramids that make up a trigonal bipyramid. PCl_5 adopts a regular trigonal bipyramidal shape. **(b)** Ball-and-stick model representing a trigonal bipyramid with axial and equatorial sites labelled.

When lone pairs are present in molecules with five sets of electron pairs, the consequences of the inequivalence of the axial and equatorial sites become more apparent. We will illustrate this with SF_4.

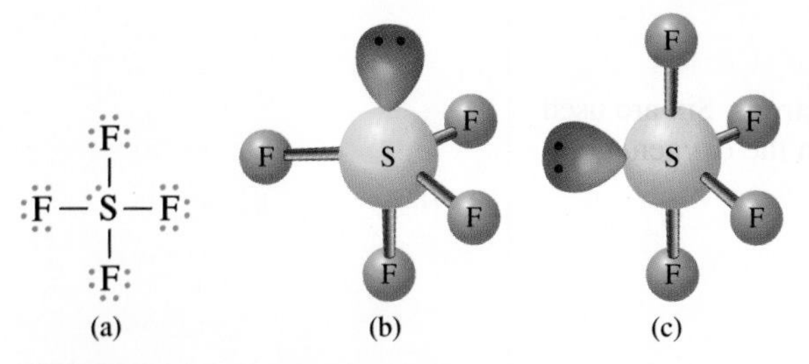

FIGURE 5.14 Views of sulfur tetrafluoride: **(a)** Lewis structure, **(b)** ball-and-stick model of the trigonal pyramid and **(c)** ball-and-stick model showing the seesaw form.

Sulfur tetrafluoride, SF_4

The Lewis structure of SF_4 (figure 5.14) shows four S—F bonds and one lone pair of electrons on the sulfur atom. We would expect a trigonal bipyramidal geometry for these five sets of electron pairs, but the presence of the lone pair means there will be some slight distortion.

In addition, because the equatorial and axial positions of a trigonal bipyramid differ, we can draw two possible structures for SF_4 with the lone pair in either axial or equatorial position. As figure 5.14 shows, placing the lone pair in an axial position gives a trigonal pyramidal shape, whereas placing the lone pair in an equatorial position gives a **seesaw shape**.

Experiments show that SF_4 has the seesaw shape, which means that this is more stable than the trigonal pyramid. This can be explained by noting that the trigonal pyramid structure has three LP–BP repulsions at 90°, while the seesaw has two at 90° and two at 120°. Minimising the number of lowest angle repulsive interactions involving lone pairs gives the lowest energy structure and consequently the seesaw shape is preferred. Note that, because of the LP–BP repulsions between the lone pair and the axial bond pairs, the axial F—S—F bond angle is slightly less than 180°.

Chlorine trifluoride, ClF_3

The Lewis structure of ClF_3 shows five sets of electron pairs: three bonding pairs and two lone pairs.

:F:
|
:F—Cl:
|
:F:

We again expect a distorted trigonal bipyramidal geometry of the five sets of electron pairs, but now we must determine whether the two lone pairs occupy axial or equatorial positions.

We can immediately discount placing one lone pair axial and the other equatorial, as this would lead to a LP–LP repulsion, the largest possible, at 90°. You might imagine that placing the lone pairs in the axial positions would be preferred, as it places them furthest away, but it also leads to a total of six LP–BP repulsions at 90°. In fact, experiments show that the two lone pairs are equatorial; this leads to four LP–BP repulsions at 90° and is energetically preferred, despite the fact that there is a LP–LP repulsion at 120°. Thus, ClF_3 is T-shaped with two equatorial lone pairs.

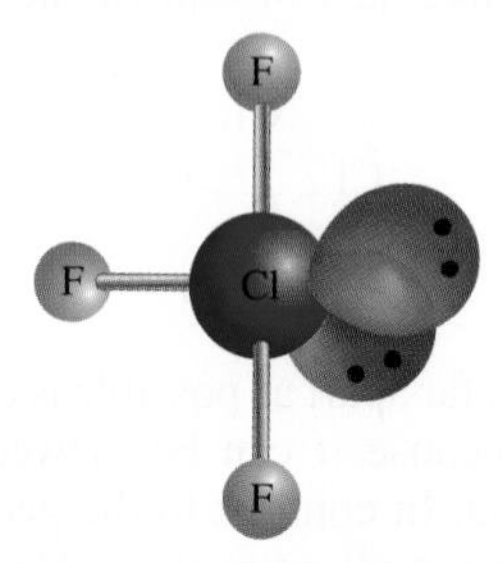

It is sometimes difficult to determine which of the possible structures is correct using simple VSEPR theory, and we must then resort to experimental results for definitive answers.

The triiodide ion, I_3^-

The Lewis structure shows five sets of electron pairs around the inner I atom: two bonding and three lone pairs.

$$\left[:\ddot{I}-\ddot{I}-\ddot{I}:\right]^-$$

We have seen in the previous examples that lone pairs prefer to be in equatorial positions in trigonal bipyramidal geometries, and the same principle applies here. All three lone pairs are equatorial, as this is the only way to avoid any highly destabilising LP–LP repulsions at 90°.

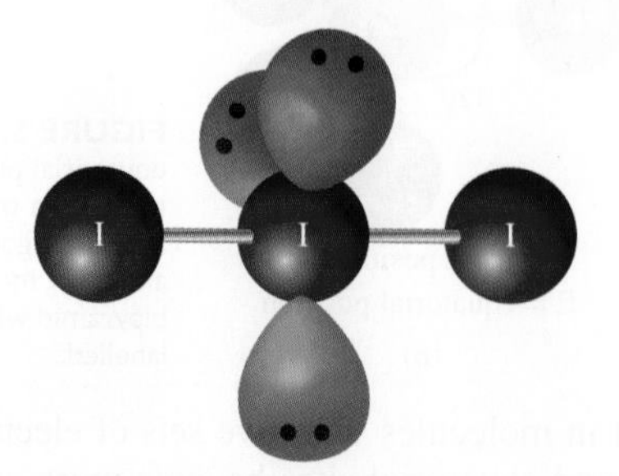

Figure 5.15 summarises the different possible shapes of species with five sets of electron pairs.

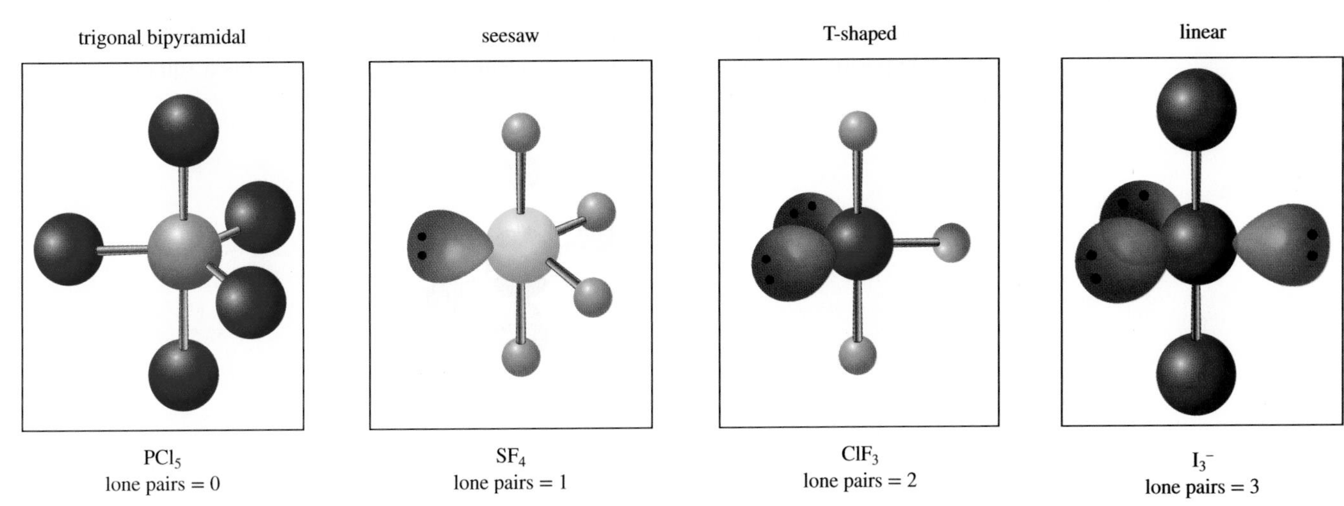

FIGURE 5.15 The different possible shapes of species with five sets of electron pairs.

Six sets of electron pairs: octahedral geometry

Sulfur hexafluoride, SF_6

The Lewis structure of SF_6, shown in figure 5.16a, indicates that sulfur has six S—F bonds and no lone pairs, and therefore has six sets of electron pairs around the inner S atom. An **octahedral** geometry of the six sets (see table 5.2) places them as far apart as possible, as shown in figure 5.16b. This gives an octahedral shape; each F atom lies at one of the six vertices of an octahedron, with possible F—S—F bond angles of 90° and 180°. An octahedron is so called because it has eight triangular faces, as shown in figure 5.16c.

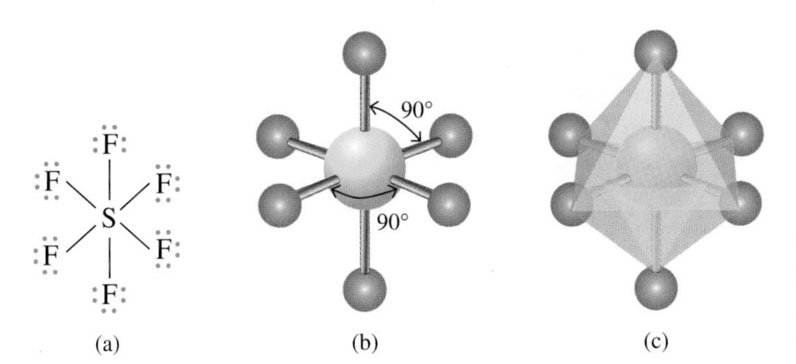

FIGURE 5.16 Views of sulfur hexafluoride: **(a)** Lewis structure, **(b)** ball-and-stick model and **(c)** ball-and-stick model showing the triangular faces of the octahedron.

All six positions around an octahedron are equivalent, as figure 5.17 demonstrates. Replacing one fluorine atom in SF_6 with a chlorine atom gives SF_5Cl. No matter which fluorine is replaced, the SF_5Cl molecule has four fluorine atoms in a square, with the fifth fluorine and the chlorine atom on opposite sides, at right angles to the plane of the square.

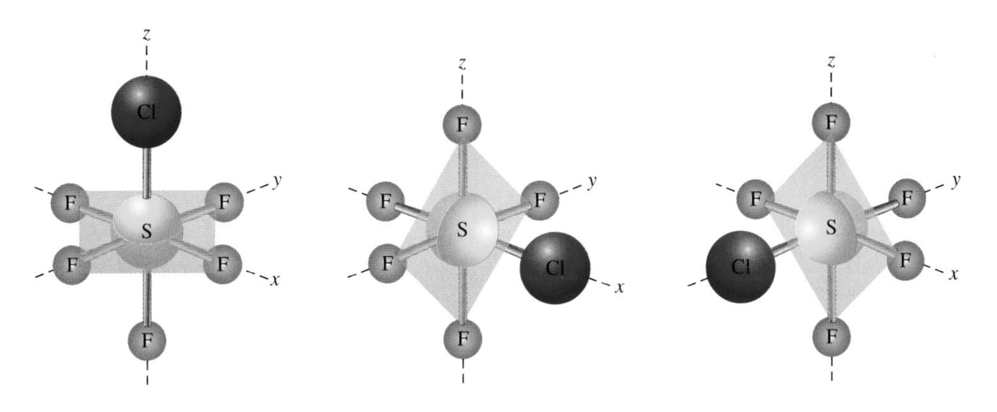

FIGURE 5.17 Replacing any of the fluorine atoms of SF_6 with a chlorine atom gives a molecule with a square of fluorine atoms capped by one fluorine and one chlorine. This shows that all six positions of the octahedron are equivalent.

Chlorine pentafluoride, ClF_5

The Lewis structure of ClF_5 shows a total of six sets of electron pairs: five bonding pairs and one lone pair.

We again expect an octahedral geometry of the six sets of electron pairs, and, in this case, all six possible positions of the lone pair are equivalent. Therefore, ClF_5 adopts a square pyramidal shape.

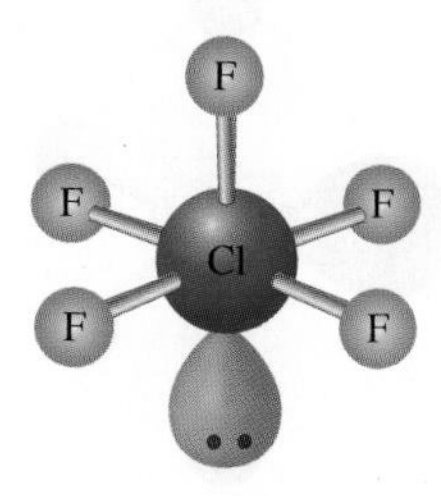

Xenon tetrafluoride, XeF_4

The Lewis structure of this molecule shows six sets of electron pairs: four bonding pairs and two lone pairs.

```
      :F:
       |
  :F — Xe — F:
       |
      :F:
```

We therefore expect an octahedral geometry of the sets of electron pairs, but, as always, we must determine how to arrange the lone pairs. In XeF_4, both lone pairs *must* be opposite each other because any other arrangement would lead to highly destabilising LP–LP repulsions at 90°. This leaves the four fluorine atoms in a square plane around xenon.

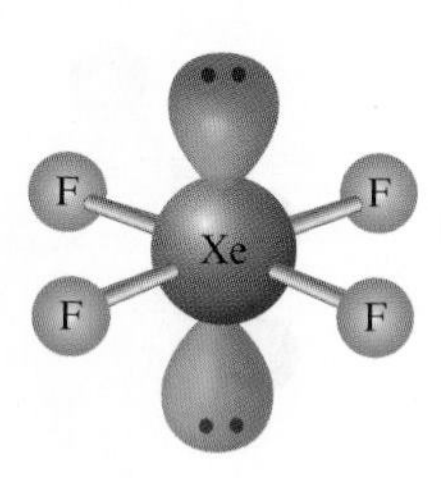

Figure 5.18 summarises the shapes of molecules with six sets of electron pairs. This completes our inventory of molecular shapes.

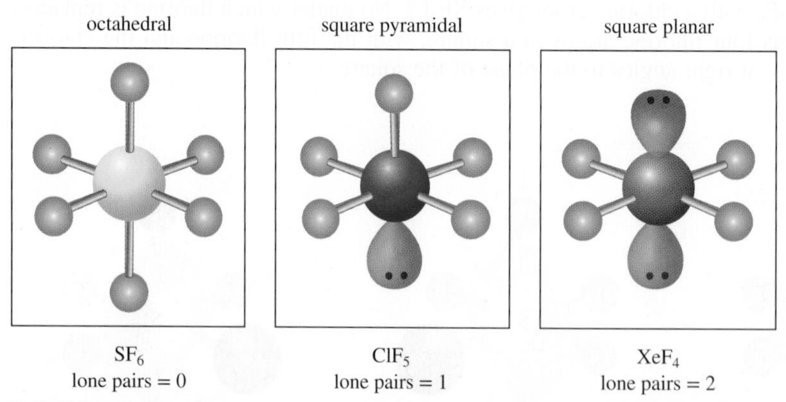

FIGURE 5.18 The different possible shapes of some molecules with six sets of electron pairs.

The most common mistake made when determining a molecular shape using VSEPR theory is to count the number of *atoms* surrounding an inner atom, rather than the number of *sets of electron pairs*. For example, a molecule such as SF_4 may be easily mistaken as tetrahedral, by analogy with tetrahedral CH_4, as both molecules have four outer atoms surrounding an inner atom. However, we have shown previously that SF_4 has five sets of electron pairs and its shape is based on a trigonal bipyramidal arrangement of these. While VSEPR theory often gives remarkably accurate results, it does not necessarily work for all molecules. In cases where repulsion between the electron pairs is not the dominant factor in determining molecular geometry, VSEPR theory fails. You will encounter some examples of this in chapter 13.

5.5 Properties of covalent bonds

Having developed ideas about Lewis structures and shapes of molecules, we can now explore some of the important features of covalent bonds.

Dipole moments

As described in section 5.1, most chemical bonds are polar, meaning that one end of the bond is slightly negative, and the other is slightly positive. Bond polarities can lead to molecules with negative ends and positive ends. A molecule with this type of electron density distribution is said to have a **dipole moment**, μ (the lowercase Greek letter mu).

The dipole moment of a polar molecule can be measured by placing a sample in an electric field. For example, figure 5.19 shows a schematic diagram of hydrogen fluoride, HF, molecules between a pair of metal plates. In the absence of an applied field, the molecules are oriented randomly throughout the volume of the device. When an electric potential is applied across the plates, the HF molecules align spontaneously according to the principles of electrostatic force. The positive ends of the molecules (the hydrogen atoms) point towards the negative plate, and the negative ends (the fluorine atoms) point towards the positive plate. The extent of alignment depends on the magnitude of the dipole moment. The SI unit for the dipole moment is the coulomb metre (C m). Bond dipole moments span the range from 0 to about 7×10^{-30} C m. Hydrogen fluoride has $\mu = 5.95 \times 10^{-30}$ C m, while a non-polar molecule like F_2 has no dipole moment, $\mu = 0$ C m. Experimental values are also often reported in units of debyes, D (1 D $= 3.34 \times 10^{-30}$ C m).

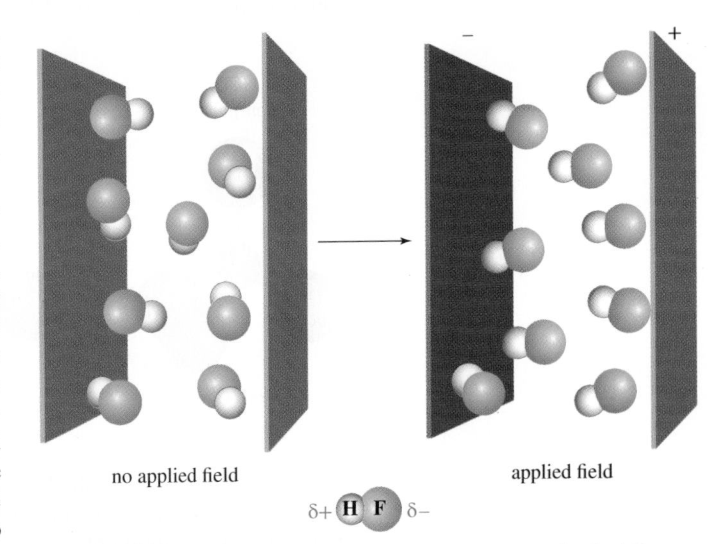

FIGURE 5.19 An applied electric field causes an alignment of polar HF molecules. The extent of alignment depends on the magnitude of the dipole moment.

Dipole moments depend on bond polarities. For example, the trend in dipole moments for the hydrogen halides follows the trend in electronegativity differences; the more polar the bond (indicated by the difference in electronegativity, $\Delta\chi$), the larger the dipole moment, μ.

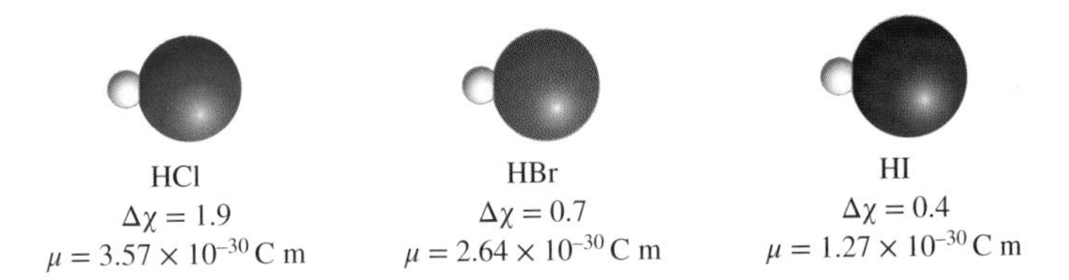

Dipole moments also depend on molecular shape. Any diatomic molecule composed of atoms with different electronegativities has a dipole moment. For more complex molecules, we must evaluate dipole moments using both bond polarity and molecular shape. A molecule with polar bonds will have no dipole moment if its shape causes the dipole moments of individual bonds to cancel one another, which is more likely for small symmetrical molecules.

Figure 5.20 illustrates the dramatic effect of shape on the dipole moments of triatomic molecules. Recall from section 5.1 that an arrow can be used to represent a dipole moment. The head of the arrow points to the partial positive end (δ+) of the polar bond. Figure 5.20a shows that, although both bonds in linear CO_2 are polar ($\Delta\chi = 1.0$), the two arrows indicating individual bond polarities point in exactly opposite directions. Thus, the effect of one polar bond exactly cancels the effect of the other. For bent H_2O, in contrast, figure 5.20b shows that the effects of the two polar bonds do not cancel. Water has a partial negative charge on its oxygen atom and partial positive charges on its hydrogen atoms. The resulting dipole moment of the molecule is $\mu = 6.18 \times 10^{-30}$ C m, with the direction of the dipole moment shown by the coloured arrow in figure 5.20b.

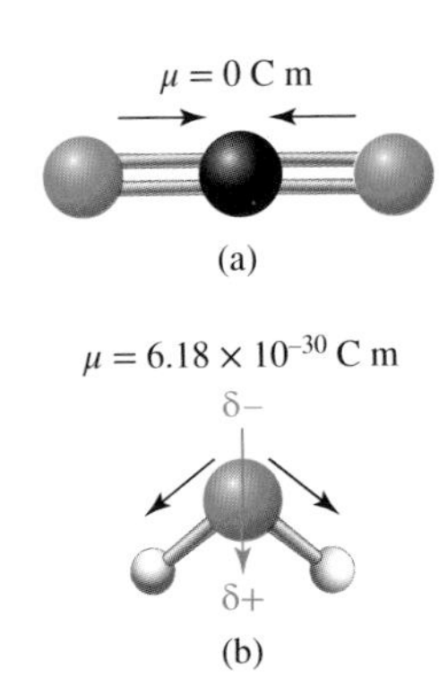

FIGURE 5.20 (a) When identical polar bonds point in opposite directions, as in CO_2, their polarity effects cancel, giving zero net dipole moment. **(b)** When two identical polar bonds do not point in exactly opposite directions, as in H_2O, there is a net dipole moment.

Molecules containing polar bonds whose dipole moments cancel each other do not have overall dipole moments. Such molecules tend to have highly symmetrical geometries. Phosphorus pentachloride, for example, has $\mu = 0$ C m. The two axial P—Cl bonds point in opposite directions, and, although three P—Cl bonds arranged in a trigonal plane have no counterparts pointing in opposite directions, trigonometric analysis shows that the polar effects of three identical bonds in a trigonal plane cancel exactly. Likewise, the bonds in a tetrahedron are arranged so that their polarities cancel exactly; i.e. the tetrahedral molecule CCl_4 has no dipole moment.

The perfect symmetry of these geometric forms is disrupted when a lone pair replaces a bond, giving a molecule with a dipole moment. Examples include SF_4 (seesaw), ClF_3 (T-shape), NH_3 (trigonal pyramid) and H_2O (bent), all of which have dipole moments. Replacing one or more bonds with a bond to a different kind of atom also introduces a dipole moment because the symmetry is lowered. Thus, chloroform, $CHCl_3$, has a dipole moment but CCl_4 does not. The carbon atom of chloroform has four bonds in a near-regular tetrahedron, but the four bonds are not identical. The C—Cl bonds are more polar than the C—H bond, so the polarities of the four bonds do not cancel. Figure 5.21 shows the magnitude and direction of the dipole moment in chloroform ($\mu = 3.47 \times 10^{-30}$ C m).

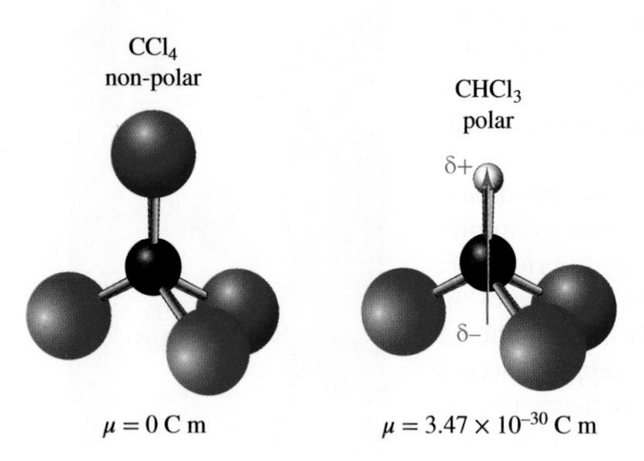

FIGURE 5.21 Carbon tetrachloride, CCl_4, is a symmetrical tetrahedral molecule, so the individual bond polarities cancel. Chloroform, $CHCl_3$, is also a tetrahedral molecule, but the four bonds are not identical, so the bond polarities do not cancel.

In a symmetrical octahedral system such as SF_6, each polar S—F bond has a counterpart pointing in the opposite direction. The bond polarities cancel in pairs, leaving this molecule without a dipole moment.

WORKED EXAMPLE 5.5

Predicting dipole moments

Does either ClF_5 or XeF_4 have a dipole moment?

Analysis

Molecules composed of atoms with different electronegativities have dipole moments unless their symmetries are sufficient to cancel their bond polarities. Therefore, we must examine each molecular structure and the orientation of the bonds in each molecule.

Solution

The molecular structures of chlorine pentafluoride and xenon tetrafluoride are shown in figure 5.18. Each has six sets of electron pairs around an inner atom, but these include lone pairs. As a result, ClF_5 has a square pyramidal shape, whereas XeF_4 has a square planar shape. Pictures can help us determine whether or not the bond polarities cancel.

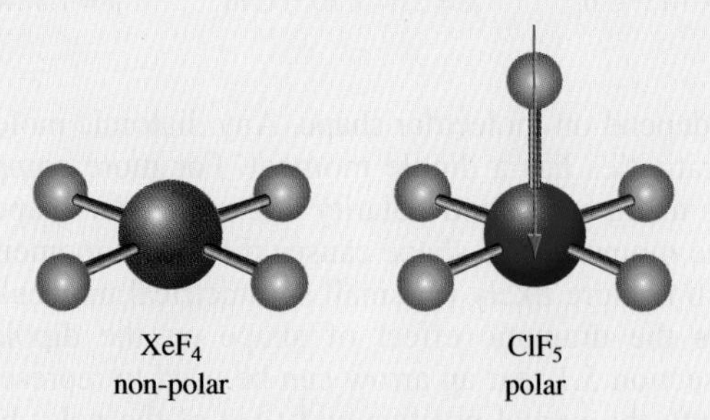

Each molecule has four fluorine atoms at the corners of a square. The Xe—F bond polarities cancel in pairs, leaving XeF_4 without a dipole moment. Four bond polarities also cancel in ClF_5, but the fifth Cl—F bond has no counterpart in the opposing direction, so ClF_5 has a dipole moment that points along the axis containing the lone pair and the fifth Cl—F bond.

Is our answer reasonable?

Any molecule in which individual dipole moments do not cancel will have an overall dipole moment. It is therefore reasonable that XeF_4 will have no dipole moment while the less symmetric ClF_5 molecule will.

PRACTICE EXERCISE 5.5 Determine whether ethane, C_2H_6, and ethanol, C_2H_5OH, have dipole moments.

Bond length

As we described in section 5.1, the bond length of a covalent bond is the nuclear separation distance at which the molecule is most stable. The H—H bond length in molecular hydrogen is 74 pm. At this distance, attractive interactions are maximised relative to repulsive interactions (see figure 5.2). Having developed ideas about Lewis structures and molecular shapes, we can now examine bond lengths in more detail.

Table 5.3 lists average bond lengths for the most common chemical bonds. The table displays several trends. One trend is that bonds become longer as atom size increases, as shown by bond lengths of the diatomic halogens.

	F_2	Cl_2	Br_2	I_2
Bond length (pm)	142	199	228	267
Atomic radius (pm)	72	100	114	133

The trend is consistent with our introduction to orbital overlap. Recall that bonding involves valence orbitals, and it is the occupied valence orbitals that determine the size of an atom.

TABLE 5.3 Average bond lengths (all values in picometres, 1 pm = 10^{-12} m).

n_a[a]	n_b[a]	Bond lengths							
H—X bonds									
1	1	H—H	74						
1	2	H—C	109	H—N	101	H—O	96	H—F	92
1	3	H—Si	148	H—P	144	H—S	134	H—Cl	127
1	4							H—Br	141
Second-row elements									
2	2	C—C	154	C—N	147	C—O	143	C—F	135
2	2			N—N	145	O—O	148	F—F	142
2	3	C—Si	185	C—P	184	C—S	182	C—Cl	177
2	3	O—Si	166	O—P	163	O—S	158	N—Cl	175
2	3	F—Si	157	F—P	157	F—S	156		
2	4, 5			F—Xe	190	C—Br	194	C—I	214
Larger elements									
3	3	Si—Si	235	P—P	221	S—S	205	Cl—Cl	199
3	3	Si—Cl	202	P—Cl	203	S—Cl	207		
4	4							Br—Br	228
5	5							I—I	267
Multiple bonds									
		C=C	133	C=N	138	C=O	120	O=O	121
		P=O	150	S=O	143				
		C≡C	120	C≡N	116	C≡O	113	N≡N	110

(a) n_a is the principal quantum number for the first named atom and n_b is for the second atom.

Bond polarity also contributes to bond length because partial charges generate electrostatic attraction that pulls the atoms closer together. For example, notice in table 5.3 that C—N bonds are slightly shorter than the average of C—C and N—N bonds. This is a result of the polarity of the C—N bond.

The bond lengths in table 5.3 show one final feature: a multiple bond is shorter than the single bond between the same two atoms. This is because placing additional electrons between atoms decreases internuclear repulsion, leading to an increase in net attraction, thereby allowing the atoms to come closer together. Thus, triple bonds are the shortest of all bonds among second-row elements.

Because bond lengths depend on the number of electrons involved in the bond, we can use bond lengths to decide which Lewis structure best represents the actual electron distribution in a molecule. Sulfur–oxygen bonds provide a good example. Figure 5.22 shows two possible sets of Lewis structures for species that contain sulfur–oxygen bonds. One set, the 'unoptimised structures', places octets of electrons on the inner sulfur atom, while the other set, the optimised structures, reduces the formal charge on sulfur to zero at the expense of exceeding the octet on the S atom. As the figure shows, the two sets of structures predict different bond types.

	Sulfuric acid	Sulfate anion	Sulfur dioxide	Sulfur trioxide
Octet structures		$2-$	plus other resonance structures	plus other resonance structures
Bond types	4 single bonds	4 single bonds	2 intermediate bonds	3 intermediate bonds
Optimised structures		$2-$ plus other resonance structures		
Bond types	2 single bonds, 2 double bonds	4 intermediate bonds	2 double bonds	3 double bonds

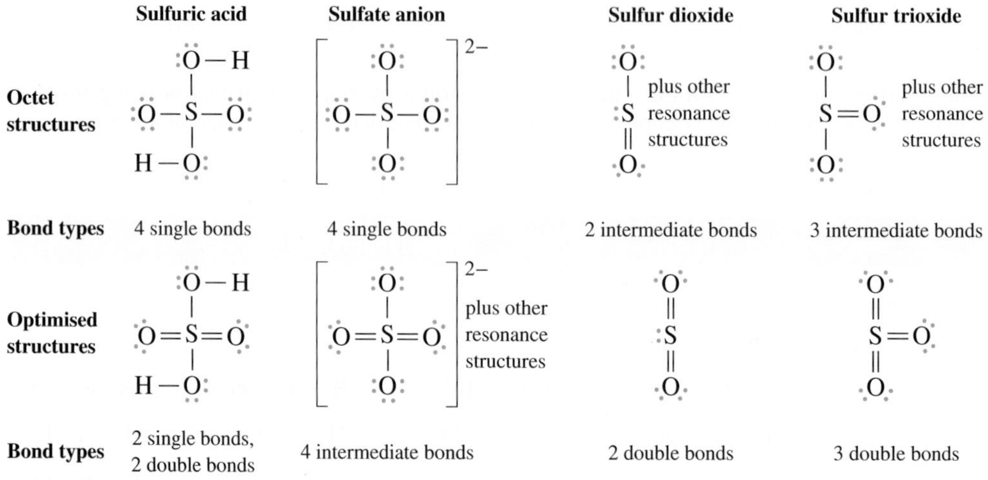

FIGURE 5.22 'Unoptimised structures' and optimised structures predict different bond types for sulfur–oxygen bonds.

We can use experimental bond length values to support the optimised Lewis structures. In sulfuric acid, there are two distinctly different bond types with bond lengths of 157 pm and 142 pm. In the sulfate anion, all the bond lengths are 147 pm, while the bond lengths in SO_2 and SO_3 are 143 pm and 142 pm, respectively. These values indicate that a bond length of 157 pm is consistent with an S—O single bond, and 142 pm with an S═O double bond. A bond length of 147 pm indicates an intermediate bond character (see table 5.3). These values support the optimised Lewis structures, as shown in figure 5.22, which predict double bonds only for SO_2 and SO_3, double and single bonds for H_2SO_4 and intermediate bonds for SO_4^{2-}.

To summarise, generally the following factors influence bond lengths.

1. For a given pair of atoms, the more electrons in the bond between them, the shorter the bond.
2. For two atoms joined by the same type of bond (single, double or triple), the larger the electronegativity difference between the bonded atoms, the shorter the bond.

WORKED EXAMPLE 5.6

Bond lengths

What factor accounts for each of the following differences in bond length?

(a) I_2 has a longer bond than Br_2.

(b) C—N bonds are shorter than C—C bonds.

(c) H—C bonds are shorter than the C≡O bond.

(d) The carbon–oxygen bond in formaldehyde, $H_2C{=}O$, is longer than the bond in carbon monoxide, C≡O.

Analysis

Bond lengths are controlled by three factors, some of which are more influential than others. To explain a difference in bond length, we need to determine the way that the factors are balanced.

Solution

(a) I—I > Br—Br. Iodine is just below bromine in the periodic table, so the valence orbitals of iodine are larger than the valence orbitals of bromine. Thus, the I_2 bond is longer than the Br_2 bond because iodine has a larger atomic radius.

(b) C—N < C—C. Carbon and nitrogen are both second-row elements. However, nitrogen has a higher effective nuclear charge than carbon, so nitrogen has the smaller radius. This makes C—N bonds shorter than C—C bonds. In addition, a C—N bond is polar, which contributes to the shortening of the C—N bond.

(c) H—C < C≡O. Here, we are comparing bonds in which the atomic radii and the amount of multiple bonding influence bond length. The experimental fact that H—C bonds are shorter than the triple C≡O bond indicates that the size of the hydrogen orbital is a more important factor than the presence of multiple bonding in this case.

(d) C=O > C≡O. Both bonds are between carbon and oxygen, so n (the principal quantum number, which determines size), atomic number (Z) and electronegativity difference ($\Delta\chi$) are the same. However, carbon monoxide contains a triple bond, whereas formaldehyde has a double bond. The triple bond in CO is shorter than the double bond in H_2CO because more shared electrons means a shorter bond.

Is our answer reasonable?

When factors work in the same direction (effective nuclear charge and bond polarity in the case of C—N < C—C), we can make confident predictions. When factors work in the opposite direction (n value and bond multiplicity in the case of H—C < C≡O), we can explain the observed values but cannot confidently predict them.

PRACTICE EXERCISE 5.6

What factor accounts for each of the following differences in bond length? (a) The C=C bond is longer than the C≡C bond. (b) The C—Cl bond is shorter than the Si—Cl bond. (c) The C—C bond is longer than the O—O bond.

Bond energy

Bond energies are defined as the amount of energy that must be supplied to break a particular chemical bond. Bond energies, like bond lengths, vary in ways that can be traced to atomic properties, and there are three consistent trends in bond energies.

1. *Bond energies increase as more electrons are shared between the atoms.* Shared electrons are the 'glue' of chemical bonding, so sharing more electrons strengthens the bond.
2. *Bond energies increase as the electronegativity difference ($\Delta\chi$) between bonded atoms increases.* Polar bonds gain stability from the electrostatic attraction between the negative and positive partial charges around the bonded atoms. Bonds between oxygen and other second-period elements exemplify this trend (note that the bond lengths also change slightly).

Bond	Difference in electronegativity ($\Delta\chi$)	Bond energy (kJ mol^{-1})
O—O	0.0	145
O—N	0.5	200
O—C	1.0	360

3. *Bond energies decrease as bonds become longer.* As atoms become larger, the electron density of a bond is spread over a wider region. This decreases the net attraction between the electrons and the nuclei. The following table of bond energies illustrates this effect (note that the electronegativity difference also changes, strengthening the trend).

Bond	Bond length (pm)	Bond energy (kJ mol^{-1})
H—F	92	565
H—Cl	127	430
H—Br	141	360
H—I	161	295

Like bond lengths, bond energies result from the interplay of several factors, including effective nuclear charge, principal quantum number, electrostatic forces and electronegativity. Thus, it is not surprising that there are numerous exceptions to these three trends in bond energies as, in most cases, more than one of the determining factors changes at the same time (see tables above) and often not with the same influence on bond energies. Although it is possible to *explain* many differences in bond energy, it frequently is not possible to *predict* differences with confidence.

Chemical Connections

Making use of the energy stored in covalent bonds

Australian industries use large amounts of explosives, estimated at about 2.5 million tonnes each year. Most explosives are used in the mining and quarrying industries (figure 5.23), but they are also used in the manufacturing and construction industries, the demolition industry and, to a smaller extent, in the entertainment industry (for fireworks and special effects).

FIGURE 5.23 Australian industry uses about 2.5 million tonnes of explosives a year. This blast is at the Argyle Diamond Mine in Western Australia (see p. 696).

The difference in the strengths of chemical bonds between the starting material and the products is responsible for the characteristics of an explosive. Ideally the explosive is a compound that has weak chemical bonds with the reaction producing molecules with strong chemical bonds, thereby ensuring that the reaction releases a lot of energy. However, of course the explosive must be stable enough so that it can be handled safely. The products of the explosive reaction should be gaseous, so that a large pressure is created accompanying the explosion, and the reaction must also be very fast.

One of the best-known explosives is liquid nitroglycerin (figure 5.24), also known as trinitroglycerine (with the systematic IUPAC name 1,2,3-trinitroxypropane). Nitroglycerine is very unsafe to handle — merely shaking it can lead to an explosion. Alfred Nobel (1833–1896) discovered that nitroglycerine can be stabilised by cellulose, and it was this discovery that has allowed the compound to play an important role in the explosives industry. Nobel's accidental discovery has had a major impact on society and — along with the many other accidental scientific discoveries — strongly supports the need for fundamental scientific research. The fortune Nobel made from his discovery eventually found its way into the funding of the Nobel Prize in chemistry, amongst others, thereby creating an ongoing connection between Alfred Nobel and scientific discovery.

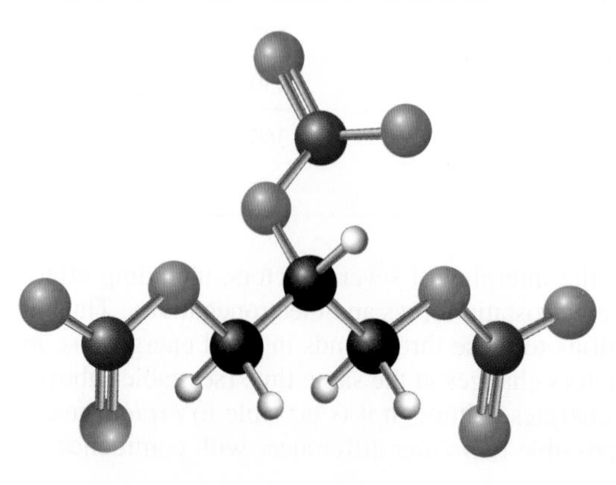

FIGURE 5.24 A ball-and-stick model of a molecule of nitroglycerin (o 1,2,3-trinitroxypropane).

Summary of molecular shapes

Table 5.4 summarises the relationships between number of sets of electron pairs, the geometry of the sets of electron pairs and molecular shape. If you remember the geometry associated with each number of sets of electron pairs, you can deduce molecular shapes, bond angles and the existence of dipole moments.

TABLE 5.4 Features of molecular geometries.

Number of sets of electron pairs	Number of outer atoms	Lone pairs	Geometry of sets of electron pairs	Molecular shape	Bond angles	Dipole moment[(a)]	Example
2	2	0	linear	linear	180°	no	CO_2
3	3	0	trigonal planar	trigonal planar	120°	no	BF_3
	2	1	trigonal planar	bent	<120°	yes	NO_2^- (plus other resonance structures)
4	4	0	tetrahedral	tetrahedral	109.5°	no	CH_4
	3	1	tetrahedral	trigonal pyramidal	<109.5°	yes	NH_3
	2	2	tetrahedral	bent	<109.5°	yes	H_2O
5	5	0	trigonal bipyramidal	trigonal bipyramidal	90°, 120°	no	PCl_5
	4	1	trigonal bipyramidal	seesaw	<90°, <120°	yes	SF_4
	3	2	trigonal bipyramidal	T-shaped	<90°, <120°	yes	ClF_3
	2	3	trigonal bipyramidal	linear	180°	no	I_3^-
6	6	0	octahedral	octahedral	90°	no	SF_6
	5	1	octahedral	square pyramidal	<90°	yes	ClF_5
	4	2	octahedral	square planar	90°	no	XeF_4

(a) Applies only to molecules with identical outer atoms.

This relatively small catalogue of molecular shapes accounts for a remarkable number of molecules. Even complicated molecules such as proteins and other polymers have shapes that can be traced back to these relatively simple templates. The overall shape of a large molecule is a composite of the shapes associated with its inner atoms, with the shape around each inner atom determined by the number of sets of surrounding bonding electron pairs and the number of lone pairs.

5.6 Valence bond theory

Lewis structures are blueprints that show the distribution of valence electrons in molecules. However, the dots and lines of a Lewis structure do not show how bonds form, how molecules react or the shape of a molecule. In this respect, a Lewis structure is like the electron configuration of an atom; both tell us about electron distributions, but neither provides detailed descriptions. Just as we need atomic orbitals to understand how electrons are distributed in an atom, we need an orbital view to understand how electrons are distributed in a molecule.

In the remainder of this chapter we describe two ways to think about covalent bonding: valence bond theory, which uses **localised bonds**, and molecular orbital theory, which uses **delocalised bonds**. The valence bond approach to molecules assumes that electrons are either localised in bonds between two atoms or localised on a single atom, usually in pairs. Accordingly, the localised bonding model develops orbitals of two types. One type is a bonding orbital that has high electron density between two atoms. The other type is an orbital located on a single atom. In other words, any electron is restricted to the region around a single atom, either in a bond to another atom or in a nonbonding orbital.

Localised bonds are easy to apply, even to very complex molecules, and they do an excellent job of explaining much chemical behaviour. In many instances, however, localised bonds are insufficient to explain molecular properties and chemical reactivity. We therefore also show how to construct delocalised bonds that spread over several atoms using molecular orbital theory. Delocalisation requires a more complicated analysis, but it explains chemical properties that localised bonds cannot.

Orbital overlap

As described earlier in this chapter, the electrons in a hydrogen molecule are located between the two nuclei in a way that maximises electron–nucleus attraction. To understand chemical bonding, we must develop a new orbital model that accounts for shared electrons. In other words, we need to develop a set of **bonding orbitals**.

The model of bonding used by most chemists is an extension of the atomic theory that incorporates a number of important ideas from chapters 1 and 4. Bonding orbitals are created by combining atomic orbitals. Remember that orbitals have wave-like properties. Waves interact, resulting in addition of their amplitudes. Two waves that occupy the same region of space will be superimposed, generating a new wave that is a composite of the original waves. In regions where the amplitudes of the superimposed waves have the same phase, addition of the amplitudes of the waves leads to constructive overlap. This gives a new wave amplitude that is larger than either original wave in that region, as figure 5.25a shows. In regions where the amplitudes have opposite phases, addition of the amplitudes of the waves leads to destructive overlap, giving a new wave amplitude that is smaller than either of the original waves, as illustrated in figure 5.25b.

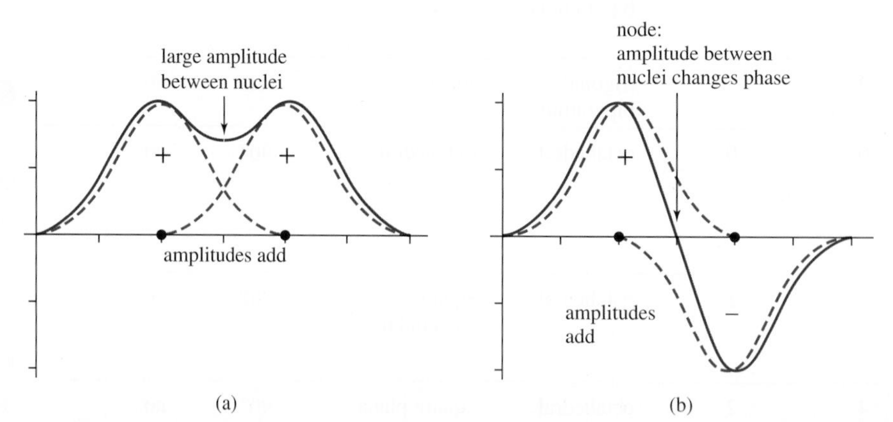

FIGURE 5.25 (a) When two waves (dashed lines) have amplitudes of the same phase, addition gives a new wave (solid line) with a large amplitude in the overlap region. **(b)** When two waves (dashed lines) have amplitudes of opposite phases, addition gives a new wave (solid line) with a small amplitude in the overlap region.

Because electrons have wave-like properties, orbital interactions involve similar addition or subtraction of wavefunctions. When two orbitals of the same phase are superimposed, the result is a new orbital that is a composite of the originals, as shown for molecular hydrogen in figure 5.26. This interaction is called **orbital overlap**, and is the foundation of the bonding models we will describe.

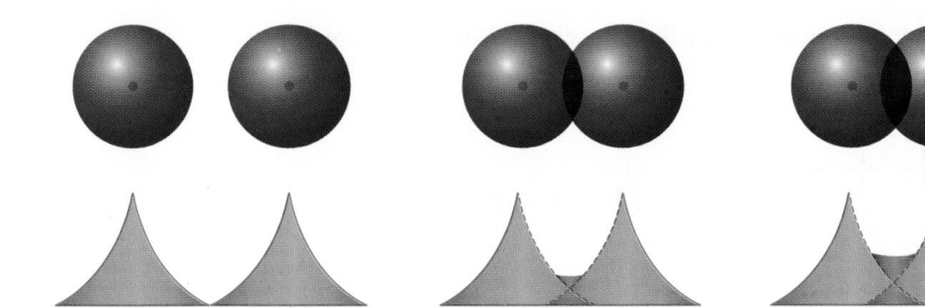

FIGURE 5.26 As two hydrogen atoms approach each other, the overlap of their 1s atomic orbitals increases. The wave amplitudes add to generate a new orbital with high electron density between the nuclei.

Conventions of the orbital overlap model

Orbital overlap models proceed from the following assumptions.

1. Each electron in a molecule is assigned to a specific orbital.
2. No two electrons in a molecule have identical descriptions because the Pauli exclusion principle (chapter 4) applies to electrons in molecules as well as in atoms.
3. The electrons in molecules obey the Aufbau principle (chapter 4), meaning that, in the ground state, they occupy the lowest energy orbitals available to them.
4. Only the valence orbitals are needed to describe bonding.

Bonding involves the valence orbitals exclusively in this model, because these orbitals have the appropriate sizes and energies to interact strongly. We have already seen examples of the overlap of valence atomic orbitals in the H_2 and F_2 molecules earlier in this chapter, and this simple approach is perfectly adequate for these small diatomic molecules. There are, however, instances where this simple model does not reproduce the experimentally determined structures of certain molecules. For example, if we consider the bonding in a water molecule to arise from overlap of the valence atomic orbitals on O and H, we find that we cannot reproduce the 104.5° H—O—H bond angle. The two singly occupied *p* orbitals on oxygen lie at 90° to each other, and overlap of these with the H 1*s* orbitals should give a bond angle of 90°.

Valence bond theory recognises this problem, and attempts to solve it by constructing hybrid orbitals from valence atomic orbitals. This approach gives much better agreement with experiments, and is outlined in detail on the following pages, with particular emphasis on the tetrahedral molecule methane.

Hybridisation of atomic orbitals

The experimentally determined structure of methane is shown in figure 5.27. Methane adopts a tetrahedral molecular geometry with H—C—H bond angles of 109.5° and the H atoms situated at the vertices of a tetrahedron. If we consider the bonding in methane solely in terms of the overlap of atomic valence orbitals, we quickly come to the conclusion that such a model could not reproduce these bond angles; the three 2*p* orbitals on C are at 90° to each other, while the 2*s* orbital has no directional preference, and we would therefore expect a structure with three of the four C—H bonds at 90° to each other. Obviously this simple model is inadequate to describe the bonding in methane. We can, however, use the individual valence atomic orbitals as a starting point to construct **hybrid orbitals**. These are combinations of atomic orbitals and the process by which we combine them is called hydridisation. We illustrate this with methane.

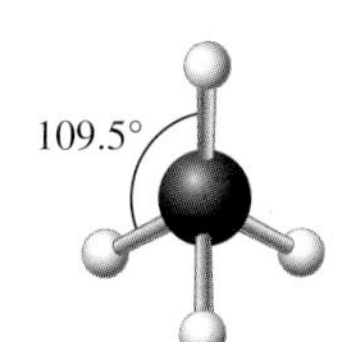

FIGURE 5.27 The experimentally determined structure of methane. The molecule has a tetrahedral geometry with 109.5° bond angles.

Methane: sp^3 hybrid orbitals

To construct hybrid orbitals, we first need to identify the valence atomic orbitals that we will use. Carbon has the electron configuration $1s^2 2s^2 2p^2$, which means that the valence orbitals of interest are the one 2*s* and three 2*p* orbitals. We then 'mix' these four atomic orbitals to form four new hybrid orbitals that are called **sp^3 hybrid orbitals**. Recall from chapter 4 that a *p* orbital has two lobes of opposite phase (indicated in figure 5.28a as pink and blue lobes). If we imagine placing an *s* and a *p* orbital such that their centres coincide, one of these lobes will interact constructively with the *s* orbital and will therefore be enhanced, while the other will interact destructively and be diminished. The name sp^3 refers to the fact that the four hybrid orbitals are constructed from one *s* and three *p* atomic orbitals. Equally, and possibly more instructively, it can also be interpreted as each individual sp^3 hybrid orbital having $\frac{1}{4}s$ character and $\frac{3}{4}p$ character. Figure 5.28a shows the shape of sp^3 hybrid orbitals, and also shows diagrammatically the process by which they are formed.

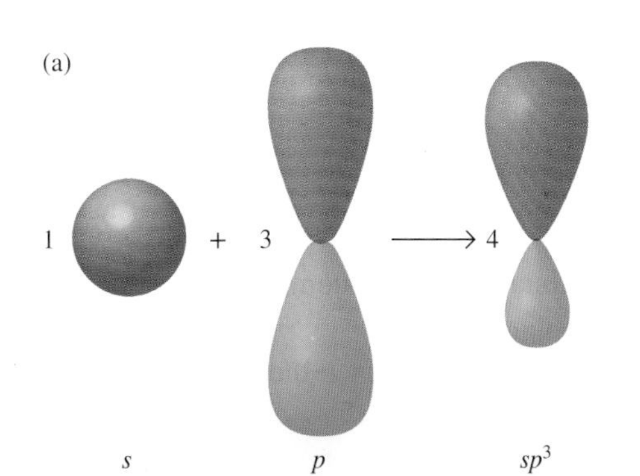

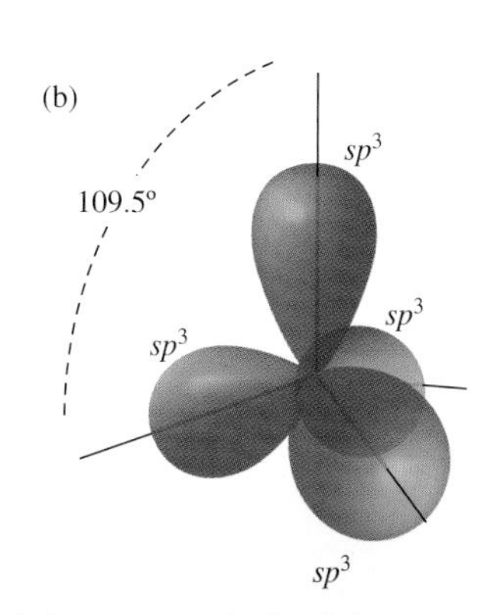

FIGURE 5.28 (a) The formation of four sp^3 hybrid orbitals from one *s* orbital and three *p* orbitals. **(b)** The tetrahedral arrangement of the four sp^3 hybrid orbitals. The smaller lobes of opposite phase have been omitted for clarity as they are not significantly involved in any major bonding interactions.

The energy level diagram for this process is shown in figure 5.29.

E

$2p$ ↑ ↑ — hybridise → ↑ ↑ ↑ ↑ sp^3

$2s$ ↑↓

atomic orbitals — sp^3 hybrid orbitals

FIGURE 5.29 Energy level diagram for the formation of four sp^3 hybrid orbitals from one 2*s* and three 2*p* atomic valence orbitals.

Because we can neither create nor destroy energy, the total energy of the four atomic orbitals must equal the total energy of the four hybrid orbitals. This means that the energy of the sp^3 hybrid orbitals must lie between the energies of the 2*s* and 2*p* atomic orbitals. Specifically, the energy of each sp^3 hybrid orbital must be $\frac{1}{4}E_{2s} + \frac{3}{4}E_{2p}$. As the four hybrid orbitals are degenerate, each of them accommodates one valence electron.

The four sp^3 hybrid orbitals are arranged such that their electrons undergo the minimum repulsion. This results in a tetrahedral arrangement in which each hybrid orbital points towards a different corner of a tetrahedron. Every hybrid orbital is directional, with a lobe of high electron density pointing in one specific direction, as shown in figure 5.28b.

We can now complete our model of methane. We overlap each singly occupied sp^3 hybrid orbital with one singly occupied 1*s* orbital of a H atom. This leads to the formation of four identical σ bonds and a regular tetrahedral shape for the molecule, as shown in figure 5.30.

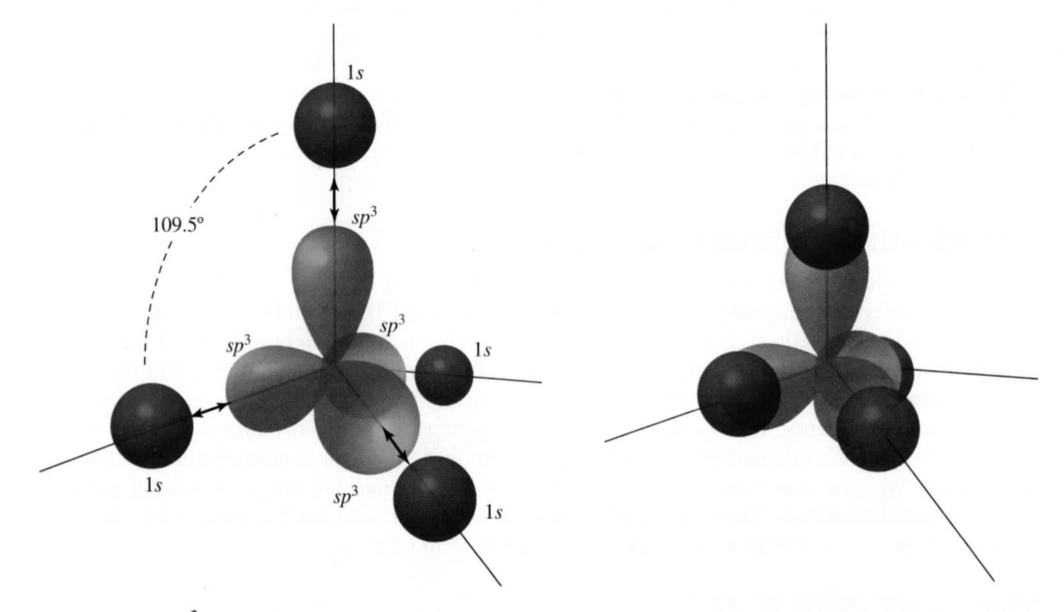

FIGURE 5.30 Methane forms from orbital overlap between the hydrogen 1*s* orbitals and the sp^3 hybrid orbitals of the carbon atom. *Note:* The large lobes of the sp^3 orbitals and the hydrogen *s* orbitals shown are of the same phase. The different colours are used for clarity only. Diminished lobes of opposite phase for the sp^3 hybrid orbitals have been omitted.

Note that sp^3 hybridisation is not limited to C atoms. In fact, any atom which exhibits a tetrahedral (or nearly so) arrangement of electron pairs can be considered to be sp^3 hybridised. This is illustrated in worked example 5.7 for the hydronium ion in which the O atom is sp^3 hybridised.

WORKED EXAMPLE 5.7

Bonding in the hydronium ion

Describe the bonding of the hydronium ion, H_3O^+, in terms of hybrid orbitals.

Analysis

Determine the Lewis structure to confirm that the inner O atom is surrounded by four sets of electron pairs, which will adopt a nearly tetrahedral geometry. This suggests sp^3 hybridisation of the O atom. Then construct the four sp^3 hybrid orbitals from the appropriate atomic orbitals on the O atom.

Solution

We have determined the Lewis structure of the hydronium ion before (worked example 5.4).

```
[    H    ]+
     |
[ H—O—H  ]
     ..
```

As the inner atom is O, we use the $2s$ and $2p$ atomic orbitals to construct the sp^3 hybrid orbitals. These will contain a total of six electrons.

$$\begin{array}{llll} & 2p\ \uparrow\downarrow\ \uparrow\ \uparrow & \xrightarrow{\text{hybridise}} & \uparrow\downarrow\ \uparrow\downarrow\ \uparrow\ \uparrow\ sp^3 \\ E & & & \\ & 2s\ \uparrow\downarrow & & \\ & \text{atomic orbitals} & & sp^3 \text{ hybrid orbitals} \end{array}$$

We have two half-occupied sp^3 hybrid orbitals that can overlap with two half-occupied $1s$ orbitals on H to form two O—H σ bonds. We form the third bond by overlapping one of the *full* sp^3 orbitals with an empty $1s$ orbital on an H^+ ion. Although both electrons in this bond derive formally from the O atom, the bond is identical to the other two O—H bonds. This leaves an sp^3 hybrid orbital not involved in bonding, which contains two electrons. These electrons constitute a lone pair on the O atom. We can visualise the bonding situation as:

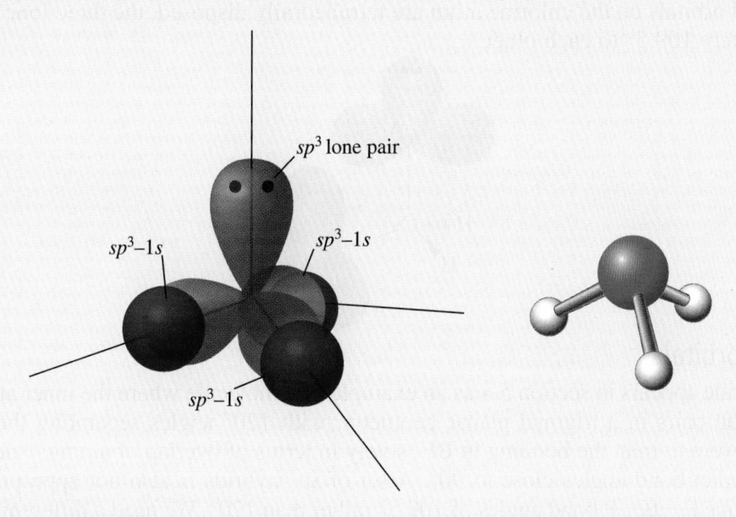

Note: The large lobes of the sp^3 orbitals and the hydrogen *s* orbitals shown are of the same phase. The different colours are used for clarity only. Diminished lobes of opposite phase for the sp^3 hybrid orbitals have been omitted.

Is our answer reasonable?

Our bonding description accounts for the trigonal pyramidal shape of the hydronium ion. This example provides an excellent illustration of how valence bond theory can account for the presence of lone pairs in a molecule. The actual H—O—H bond angles in the hydronium ion will be slightly less than the idealised 109.5° due to the presence of the lone pair.

PRACTICE EXERCISE 5.7

Describe the bonding in a water molecule in terms of hybrid orbitals.

Note that, in methane, the number of hybridised valence atomic orbitals equals the number of valence atomic orbitals participating in hybridisation. This is always the case, regardless of the nature of the hybridisation. For example, as we will see later, hybridisation of three atomic orbitals gives three sp^2 hybrid orbitals, while hybridisation of two atomic orbitals gives two *sp* hybrid orbitals.

To this point, we have considered only molecules in which hybrid orbitals on the inner atom overlap with H $1s$ atomic orbitals. In cases where the outer atoms are not hydrogen, we also hybridise these atoms to obtain a better description of the geometry of any lone pairs present. We can illustrate this with reference to the dichloromethane molecule, CH_2Cl_2, the Lewis structure of which is:

```
        H
        |
  :Cl — C — H
        |
       :Cl:
```

We hybridise the inner C atom to generate four singly occupied sp^3 hybrid orbitals, and we can generate two C—H σ bonds by overlapping these with two singly occupied H 1*s* orbitals. We could also form the C—Cl σ bonds by overlapping the remaining two hybrid orbitals with the singly occupied *p* orbital on each of two Cl atoms. However, the problem with this approach is the same as that which led us to hybridise atomic orbitals in the first place — it leads to an incorrect geometry of electron pairs (in this case, the three lone pairs on each Cl atom). We can hybridise the valence orbitals on Cl in the same way as we have done previously for C, as shown in figure 5.31.

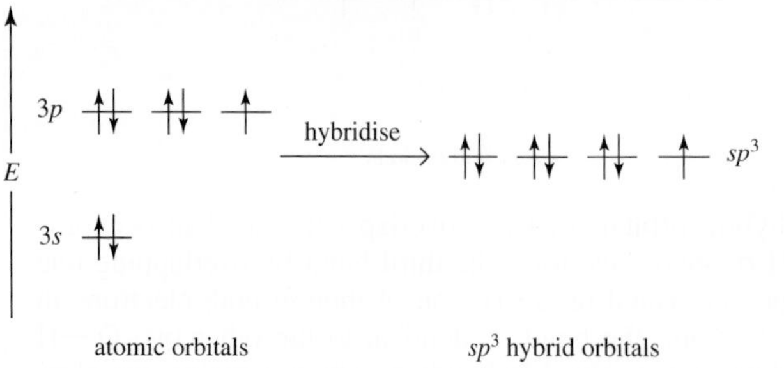

E

3*p* ↑↓ ↑↓ ↑ —hybridise→ ↑↓ ↑↓ ↑↓ ↑ sp^3

3*s* ↑↓

atomic orbitals sp^3 hybrid orbitals

FIGURE 5.31 Energy level diagram for the formation of sp^3 hybrid orbitals on Cl from the 3*s* and 3*p* valence atomic orbitals.

This hybridisation gives one singly occupied sp^3 hybrid orbital on each Cl atom, which can overlap with the sp^3 hybrid orbitals on the carbon atom to form two C—Cl σ bonds. Given that the sp^3 hybrid orbitals on the chlorine atom are tetrahedrally disposed, the three lone pairs will be at approximately 109.5° to each other.

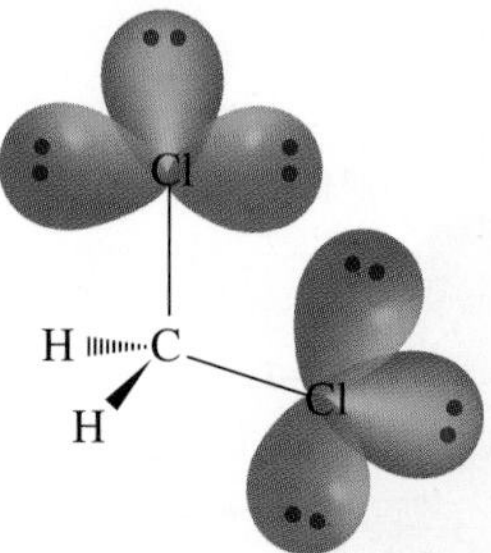

sp^2 hybrid orbitals

Boron trifluoride appears in section 5.4 as an example of a molecule where the inner atom has three sets of electron pairs in a trigonal planar geometry, with 120° angles separating the three B—F bonds. If we were to treat the bonding in BF_3 solely in terms of overlap of atomic valence orbitals, we would predict bond angles close to 90°. A set of sp^3 hybrids is also not appropriate, because these would give F—B—F bond angles of 109.5° rather than 120°. We need a different set of hybrid orbitals to represent an atom with trigonal planar geometry. In this case we can generate a set of three **sp^2 hybrid orbitals** by mixing the 2*s* orbital with two 2*p* orbitals on the boron atom. We represent this process in figure 5.32, in which each resulting sp^2 hybrid orbital has $\frac{1}{3}s$ and $\frac{2}{3}p$ character.

(a) (b)

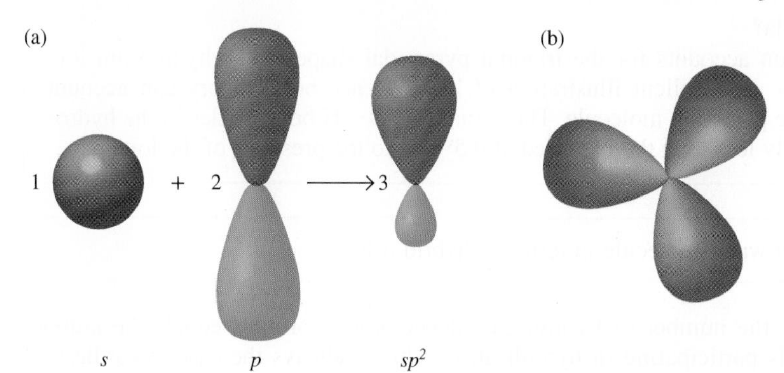

FIGURE 5.32 (a) The formation of three sp^2 hybrid orbitals from one *s* orbital and two *p* orbitals. **(b)** The orientation of the three sp^2 hybrid orbitals. They lie in the same plane at an angle of 120° to each other. The smaller lobes of opposite phase have been omitted for clarity as they are not involved in the major bonding interaction.

Although an individual sp^2 hybrid orbital looks very much like its sp^3 counterpart, the energetics of the hybridisation process are quite different, as shown in figure 5.33.

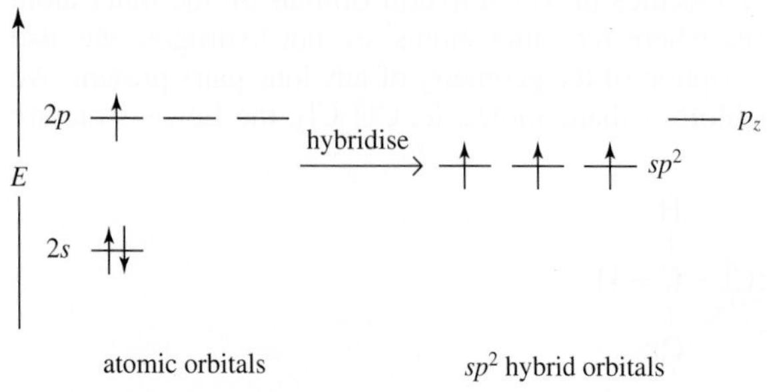

E

2*p* ↑ — — —hybridise→ ↑ ↑ ↑ sp^2 — p_z

2*s* ↑↓

atomic orbitals sp^2 hybrid orbitals

FIGURE 5.33 Energy level diagram for the formation of sp^2 hybrid orbitals from the 2*s* and 2*p* valence atomic orbitals on a B atom.

The three sp^2 hybrid orbitals are coplanar, and are positioned at angles of 120° to each other around the B atom. As we use only two of the three $2p$ orbitals in constructing the sp^2 hybrid orbitals, this leaves one orbital (by convention, the p_z) unhybridised. Figure 5.34 shows the arrangement of all the orbitals.

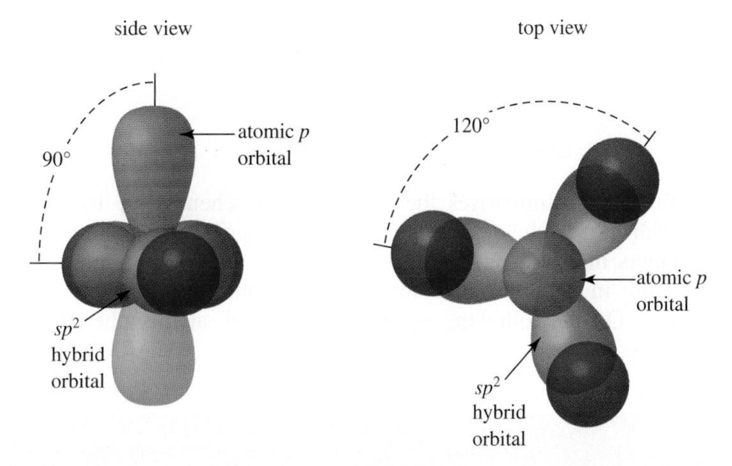

FIGURE 5.34 An sp^2 hybridised atom has three coplanar hybrid orbitals separated by 120° angles. One unchanged p orbital is perpendicular to the plane of the hybrid orbitals. *Note:* Apart from one half (pink or blue) of the unhybridised p_z atomic orbital, all orbitals are of the same phase. The different colours are used for clarity only. Diminished lobes of opposite phase have been omitted.

We also hybridise the F atoms. Each F atom is surrounded by four electron pairs, which means the atoms will be sp^3 hybridised, and the process is entirely analogous to that used previously for the Cl atoms in CH_2Cl_2. We now overlap each of the three singly occupied sp^2 hybrid orbitals with a singly occupied sp^3 hybrid orbital on a fluorine atom to form the three identical σ B—F bonds.

The presence of the unoccupied p_z orbital on the B atom has significant consequences for the reactivity of BF_3. BF_3 often reacts very easily with molecules that have a lone pair of electrons, such as NH_3. The p_z orbital serves to accept the pair of electrons from such molecules, thereby giving the B atom an octet and converting it to sp^3 hybridisation in the process.

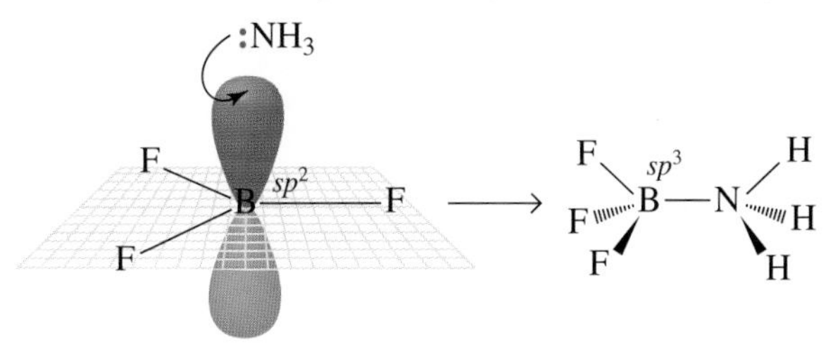

sp hybrid orbitals

Beryllium hydride appears in section 5.4 as an example of a molecule with linear geometry. There are two bonds to the beryllium atom and a H—Be—H bond angle of 180°. To describe linear orbital geometry, we need a hybridisation scheme that generates two orbitals pointing in opposite directions. We do this by hybridising the $2s$ orbital of Be with one of the $2p$ orbitals to generate a pair of ***sp* hybrid orbitals**. We represent this process in figure 5.35, in which each sp hybrid orbital has $\frac{1}{2}s$ character and $\frac{1}{2}p$ character.

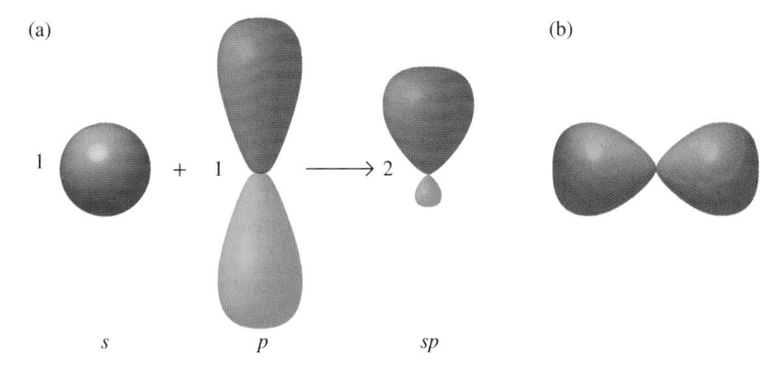

FIGURE 5.35 (a) The formation of two sp hybrid orbitals from one s orbital and one p orbital. **(b)** The orientation of the two sp hybrid orbitals. They lie at an angle of 180° to each other. The smaller lobes have been omitted for clarity as they are not involved in the major bonding interaction.

Figure 5.36 (overleaf) shows the energetics of the hybridisation process. In this case, we use only one of the p orbitals, which leaves two of them unhybridised.

The two sp hybrid orbitals are oriented at 180° from each other, which leads to a linear shape for the molecule. We form two identical Be—H σ bonds by overlapping each half-filled hybrid orbital with one half-filled $1s$ orbital on a H atom. Note that we cannot hybridise the outer H atoms, as they have only a $1s$ orbital each. Figure 5.37 (overleaf) shows a diagram of the bonding for BeH_2. Note that the two unoccupied and unhybridised p orbitals lie perpendicular to each other. We will see how such orbitals can participate in bonding when we consider triply bonded molecules later in this chapter.

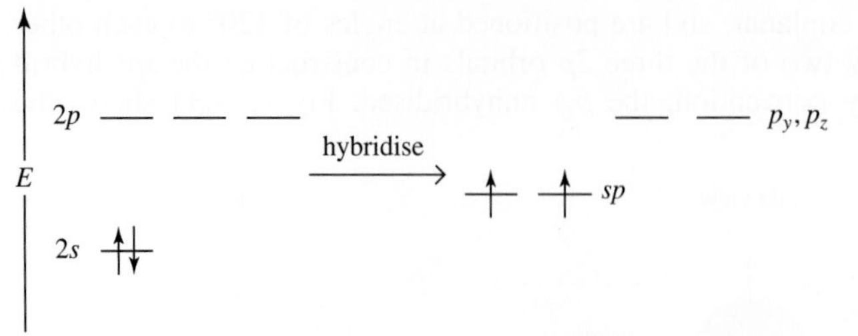

FIGURE 5.36 Energy level diagram for the formation of *sp* hybrid orbitals from the 2*s* and 2*p* valence atomic orbitals on Be.

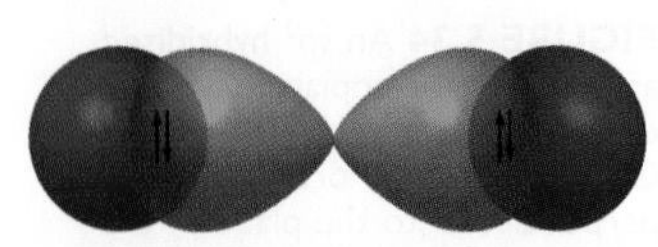

FIGURE 5.37 The bonding in BeH_2. *Note:* The large lobes of the sp^3 orbitals and the hydrogen *s* orbitals shown are of the same phase. The different colours are used for clarity only. Diminished lobes of opposite phase in the *sp* hybrid orbitals have been omitted.

Table 5.5 summarises the hybridisation schemes we have encountered so far. While other hybridisation schemes are required to describe the bonding in molecules with coordination numbers higher than 4, *sp*, sp^2 and sp^3 are by far the most useful, particularly to describe the bonding in organic molecules. Multiple bonding is one of the features of many organic molecules. On the following pages, we will look at this phenomenon from a valence bond point of view.

TABLE 5.5 A summary of valence orbital hybridisation.

Number of sets of electron pairs	Electron group geometry	Hybridisation	Number of hybrid orbitals	Number of unused *p* orbitals	Diagram[(a)]
2	linear	*sp*	2	2	
3	trigonal planar	sp^2	3	1	
4	tetrahedral	sp^3	4	0	

(a) Apart from one half (pink or blue) of the unhybridised p_z atomic orbital, all orbitals are of the same phase. The different colours are used for clarity only. Diminished lobes of opposite phase have also been omitted.

Multiple bonds

Many of the Lewis structures in this book represent molecules that contain double and triple bonds. From simple molecules such as ethene and ethyne, to complex compounds such as chlorophyll and vitamin B_{12}, multiple bonds are abundant in chemistry. Double and triple bonds can be described by extending the valence bond model. We begin with ethene, a simple hydrocarbon with the formula C_2H_4.

Bonding in ethene

Ethene (ethylene) is a colourless, flammable gas with a boiling point of −104 °C that is predominantly used in the manufacture of plastics such as polyethylene. Because ethene stimulates the breakdown of cell walls, it is used commercially to speed up the ripening of fruit, particularly bananas.

Every description of bonding starts with a Lewis structure. Ethene has 12 valence electrons. The bond framework of the molecule has one C—C bond and four C—H bonds, requiring 10 of these electrons. If we place the final two electrons as a lone pair on one of the carbon atoms, this gives a formal charge of −1 on one carbon atom and +1 on the other. These charges can be minimised by forming a double bond between the carbon atoms, leaving them both with an octet of electrons.

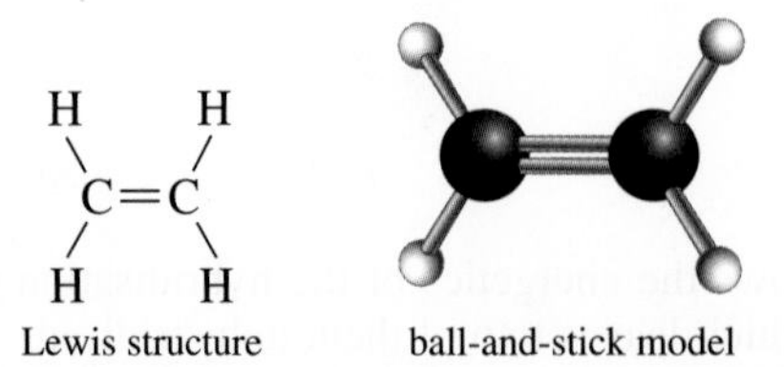

Each carbon atom is thus surrounded by three sets of electron pairs, which implies that the carbon atoms are sp^2 hybridised, with a trigonal planar geometry. (Recall from page 182 that double bonds and triple bonds are treated as being a single set of electron pairs.)

To develop a bonding picture for molecules with multiple bonds, we start by considering *only* the singly bonded σ framework of the molecule, which we construct using hybrid orbitals. The orbital diagram for an sp^2 hybridised C atom is shown in figure 5.38 and the σ framework is shown in figure 5.39. Note that the electron configuration on the right-hand side of figure 5.38 violates the Aufbau principle. This is justified as the energy difference between the sp^2 orbitals and the p_z orbital is small, and therefore the energy required to populate the p_z orbital is smaller than the spin-pairing energy that would otherwise be required to occupy one of the hybrid orbitals with two electrons. Such an electron configuration provides us with results that agree with experiments.

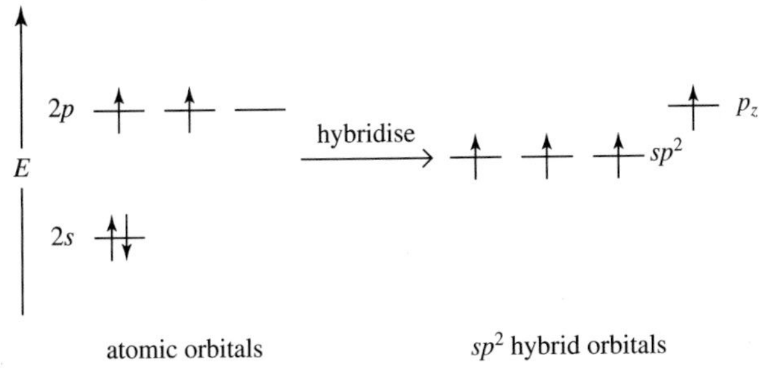

E

2p

2s

hybridise

p_z

sp^2

atomic orbitals

sp^2 hybrid orbitals

FIGURE 5.38 Energy level diagram for the formation of sp^2 hybrid orbitals from the 2*s* and 2*p* valence atomic orbitals on C.

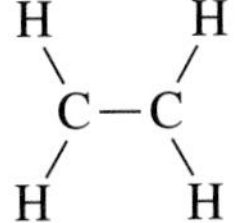

FIGURE 5.39 The σ framework of the ethene molecule.

For the σ framework, we must form four C—H bonds and one C—C bond. We can do this, as shown in figure 5.40, by using two sp^2 hybrid orbitals on each carbon to form two C—H σ bonds, with the C—C bond formed by overlap of the remaining sp^2 hybrid orbital on each C atom.

Having constructed the σ framework, we are left with one half-occupied p_z orbital on each C atom, which is used to construct the double bond. As we have seen earlier, a σ bond has its maximum electron density along the internuclear axis between the two nuclei. To construct the double bond, we overlap the two p_z orbitals in a side-by-side fashion to form a **pi (π) bond**, as shown in figure 5.41. A π bond has its maximum electron density above and below the plane of the nuclei, and has zero electron density along the internuclear axis between the nuclei. A double bond always consists of one σ bond and one π bond. The σ bond forms from the end-on overlap of two hybrid orbitals, and the π bond forms from the side-by-side overlap of two atomic *p* orbitals. As the side-by-side overlap is not as efficient as the end-on overlap, double bonds are not twice as strong as single bonds between the same atoms. Figure 5.41 shows the complete orbital picture of the bonding in ethene. Ethene is the simplest of a class of molecules, the alkenes, which contain C=C double bonds.

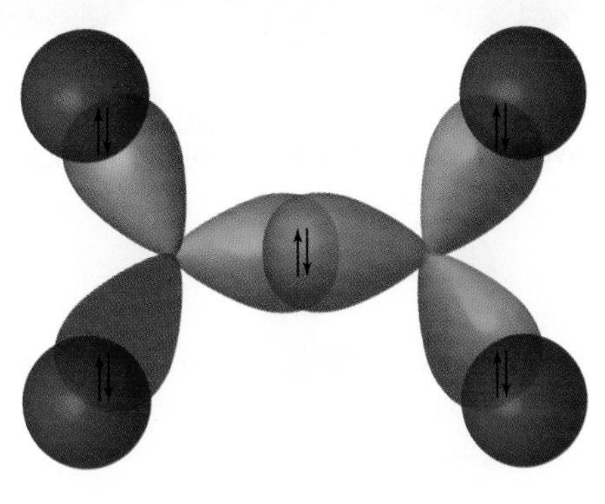

FIGURE 5.40 Orbital overlaps used to form the σ framework of ethene. Note that all orbitals shown are of the same phase and colours are used for clarity only.

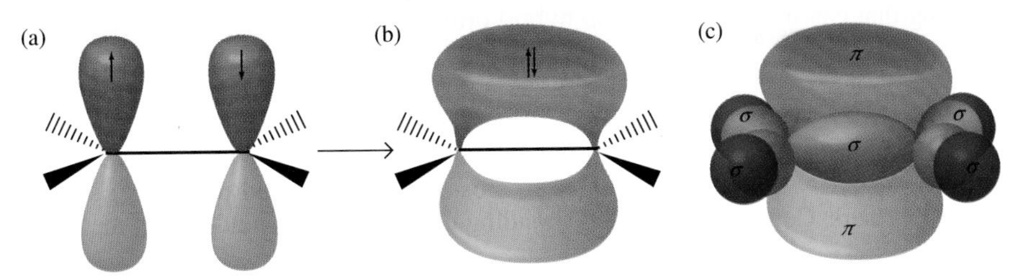

FIGURE 5.41 Orbital pictures of the bonding in ethene from three perspectives: **(a)** view of the p_z atomic orbitals which overlap to form the π bond, **(b)** view of the π bond formed from the overlap of the p_z atomic orbitals and **(c)** the π bond superimposed on the σ framework. Note that the pink and blue lobes together represent a single π bond, which is occupied by two electrons.

The availability of an unhybridised valence *p* orbital to form π bonds is characteristic of sp^2 hybridisation and is not restricted to carbon atoms. Many organic molecules contain C=N and C=O double bonds and we can describe the bonding in such compounds in a similar fashion to the above description of a C=C double bond.

To summarise, we can construct the valence bond description of any double bond using the following four-step procedure.

1. Determine the Lewis structure.
2. Use the Lewis structure to determine the type of hybridisation.
3. Construct the σ bond framework.
4. Add the π bonds.

Bonding in ethyne

The ethyne molecule may be familiar to you by its more common name acetylene, the gas used in oxyacetylene welding torches. Ethyne has 10 valence electrons, and we use six of these to form one C—C bond and two C—H bonds. This leaves each C atom in the molecule with a share of only four electrons. We can complete the octet of one of the C atoms by allocating the remaining four electrons to it in the form of two lone pairs. However, this leads to a –2 charge on this and a formal charge of +2 on the other C atom. Converting the two lone pairs on the C atom to two bonding pairs situated between the two C atoms, we obtain formal charges of zero for all atoms in the molecule. This produces the following Lewis structure in which the carbon atoms are joined by a triple bond.

H—C—C—H

Each carbon atom is surrounded by two sets of electron pairs, and we would predict a linear geometry for this molecule. We can therefore use *sp* hybrid orbitals to describe the bonding in ethyne.

As we did for ethene, we first construct the singly bonded σ framework using the hybrid orbitals. The orbital diagram for an *sp* hybridised C atom is shown in figure 5.42, and the σ framework of ethyne is shown in figure 5.43.

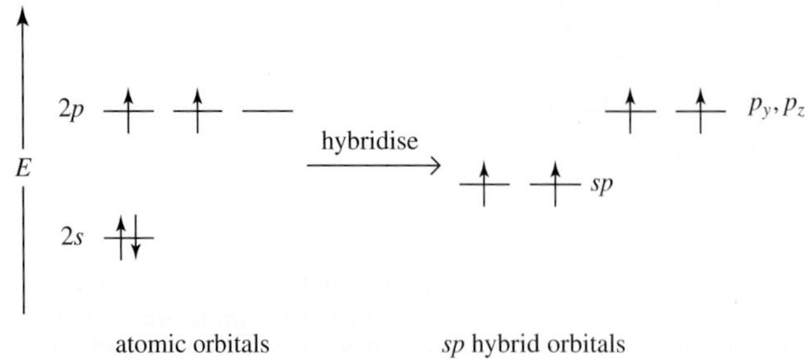

E

2*p* ↑ ↑ —

2*s* ↑↓

hybridise →

↑ ↑ p_y, p_z

↑ ↑ *sp*

atomic orbitals | *sp* hybrid orbitals

FIGURE 5.42 Energy level diagram for the formation of *sp* hybrid orbitals from the 2*s* and 2*p* valence atomic orbitals on C.

H—C—C—H

FIGURE 5.43 The σ framework of the ethyne molecule.

We form the σ framework by forming two C—H bonds and one C—C bond using the *sp* hybrid orbitals, as shown in figure 5.44.

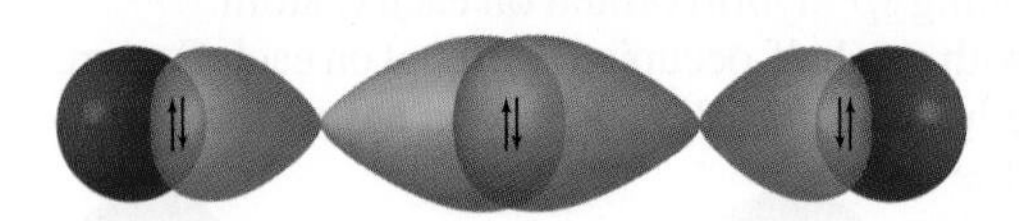

FIGURE 5.44 Orbital overlaps used to form the σ framework of ethyne. All orbitals are in the same phase; different colours are used for clarity only.

This then leaves two singly occupied *p* orbitals on each C atom. We can form two π bonds by overlapping the p_y and p_z orbitals on each C atom in a side-by-side fashion, as shown in figure 5.45.

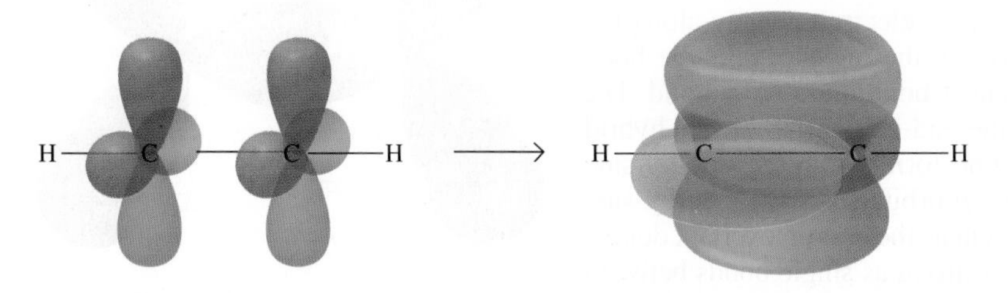

FIGURE 5.45 Overlapping non-hybridised *p* orbitals form the π framework of ethyne.

Thus, each triple bond has one C—C σ bond and two C—C π bonds perpendicular to each other, which is always the case in any alkyne. Worked example 5.8 treats another molecule with bonding that can be described by *sp* hybrid orbitals.

WORKED EXAMPLE 5.8

Orbital overlap in triple bonds

Hydrogen cyanide, HCN, is an extremely poisonous gas. Approximately half a million tonnes of HCN is produced each year, most of which is used to prepare starting materials for polymers. Construct a complete bonding picture for HCN and sketch the various orbitals.

Analysis

Use the four-step procedure described previously. Begin with the Lewis structure for the molecule, and then identify the appropriate hybrid orbitals. Construct a σ bond framework, and complete the bonding picture by assembling the π bonds from the unhybridised *p* orbitals.

Solution

The five steps of our procedure for writing Lewis structures lead to a triple bond between carbon and nitrogen.

$$H-C\equiv N:$$

The inner C atom is surrounded by two sets of electron pairs, as is the N atom. These atoms can therefore be considered to be *sp* hybridised. We have derived the orbital diagram for an *sp* hybridised C atom previously, while that for an *sp* N atom is shown below.

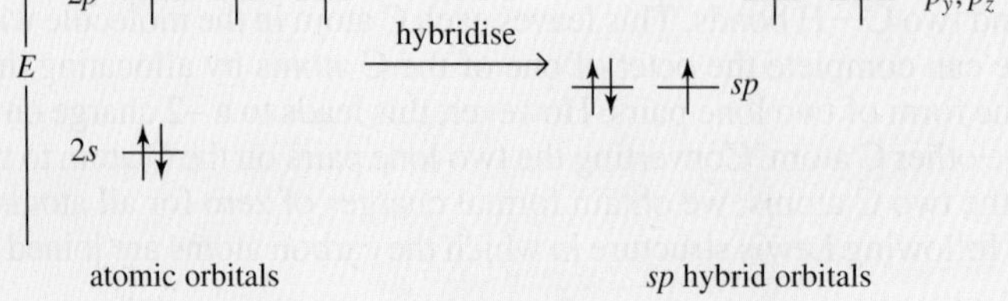

E

2*p* ↑ ↑ ↑

2*s* ↑↓

hybridise →

↑ ↑ p_y, p_z

↑↓ ↑ *sp*

atomic orbitals | *sp* hybrid orbitals

We form the linear singly bonded σ framework shown below by overlapping one of the *sp* hybrid orbitals on C with the 1*s* orbital on the H atom to give a C—H σ bond. Similarly, we form a C—N σ bond by overlapping the other *sp* hybrid orbital on the C atom with the singly occupied *sp* hybrid orbital on the N atom.

$$\mathrm{H{-}C{-}N}$$

This leaves two singly occupied *p* orbitals on C and two singly occupied *p* orbitals on the N atom with which we can construct the C≡N triple bond, in the same way as described previously for ethyne. We are then left with an *sp* hybrid orbital containing two electrons situated on the N atom. This is a non-bonding lone pair and is shown in the diagram below in gold, for clarity.

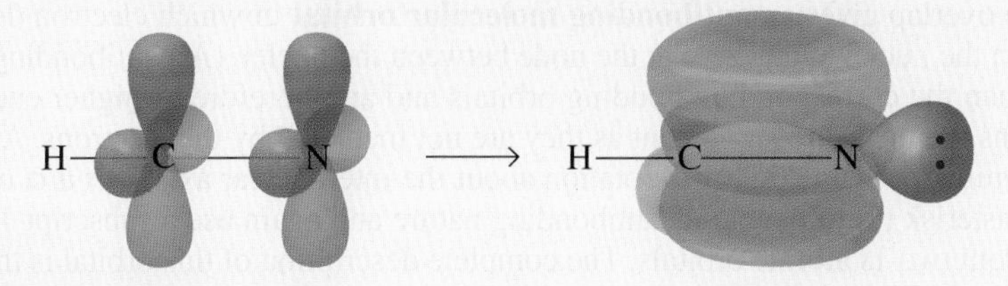

Is our answer reasonable?

We have already shown that a triple bond consists of one σ and two π bonds, which is what we have derived. Our bonding scheme also accounts for the presence of a lone pair on the N atom, so it is consistent with the experimentally determined structure.

PRACTICE EXERCISE 5.8

Describe the bonding in the but-2-yne molecule, $CH_3C{\equiv}CCH_3$, using valence bond theory, and sketch the orbitals involved.

5.7 Molecular orbital theory: diatomic molecules

Hybrid orbitals and localised bonds provide a model of bonding that can be easily applied to a wide range of molecules. Using this model we do an excellent job of rationalising and predicting chemical structures, but localised bonds cannot predict or interpret all aspects of bonding and reactivity. For example, it fails to explain why there are unpaired electrons in a molecule of O_2. In this section we introduce **molecular orbital theory**, a powerful theory of bonding that is more successful than valence bond theory in predicting and explaining molecular properties. Molecular orbital theory differs fundamentally from valence bond theory in that the electrons within a molecule are not localised either on one atom or between two atoms; instead they occupy **molecular orbitals (MOs)** that can cover the entire molecule.

Although molecular orbital theory is more complex than the hybrid orbital approach, we can illustrate the fundamentals of the theory with reference to simple diatomic molecules.

Molecular orbitals of H_2 and He_2

Molecular orbital theory considers the formation of molecular orbitals from the overlap of atomic orbitals on all of the individual atoms in the molecule. The overlap of *N* atomic orbitals will always lead to the formation of *N* molecular orbitals. Because of the large number of possible overlaps in even quite small molecules, we will restrict our discussion to diatomic molecules, and will first consider the molecular orbital treatment of the H_2 molecule.

We can consider the MOs of H_2 to be formed from the overlap of the 1*s* orbitals on the H atoms. As each atomic orbital can have either a positive or negative phase, there are two possible ways in which these atomic orbitals can overlap — both orbitals can have the same phase (in-phase overlap) or they can have opposite phases (out-of-phase overlap). In-phase overlap is constructive and results in the formation of a molecular orbital with a large amplitude, and consequently high electron density, between the nuclei. Out-of-phase overlap is destructive and gives a molecular orbital with zero amplitude, and therefore zero electron density, exactly halfway between the nuclei. As for atomic orbitals, this position where the amplitude changes phase is called a **node**. These overlaps and the resulting molecular orbitals are shown in figures 5.25 and 5.46 (overleaf).

Constructive overlap gives a **bonding molecular orbital**, in which electron density is maximised between the two nuclei. Such an arrangement of electrons minimises internuclear repulsions. The orbital is completely symmetric with respect to rotation around the internuclear axis and is therefore a σ orbital. We use a 1*s* subscript to show that it derives from the overlap of two 1*s* atomic orbitals, and the complete description of this orbital is therefore σ_{1s}.

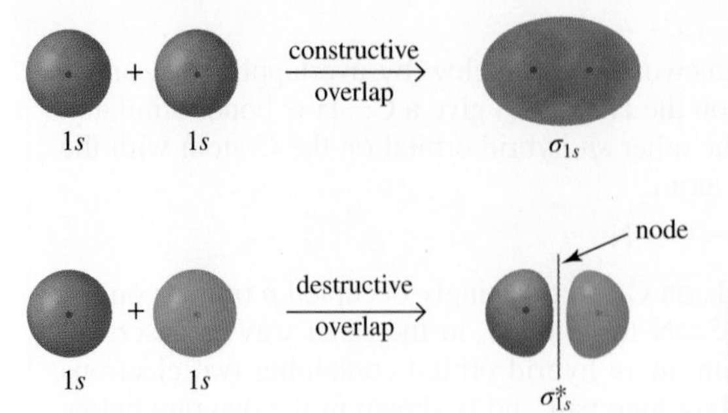

FIGURE 5.46 When two hydrogen 1s atomic orbitals interact, they generate two molecular orbitals, one bonding (σ_{1s}) and one antibonding (σ^*_{1s}).

Destructive overlap gives an **antibonding molecular orbital** in which electron density is minimised between the nuclei and is zero at the node between the nuclei. (All antibonding orbitals have more nodes than the corresponding bonding orbitals and are therefore of higher energy.) Internuclear repulsions are therefore significant as they are not mediated by the electrons. As the orbital is completely symmetric with respect to rotation about the internuclear axis, it is a σ orbital. We use a superscript asterisk (*) to denote its antibonding nature and again use a subscript 1*s* to show that it is derived from two 1*s* atomic orbitals. The complete description of this orbital is therefore σ^*_{1s}.

The relative energies of atomic orbitals and the molecular orbitals they form are shown in a **molecular orbital diagram**. Figure 5.47 shows the molecular orbital diagram for H_2. It shows that the bonding orbital is of lower energy than the orbitals from which it is formed, whereas the antibonding orbital is of higher energy. When a hydrogen molecule forms, its two electrons obey the same principles for distributing electrons as described in chapter 4. The lowest energy arrangement of electrons requires adherence to the Aufbau principle, the Pauli exclusion principle and Hund's rule, no matter what types of orbitals they occupy. The two valence electrons of molecular hydrogen therefore fill the lowest energy orbital, the σ_{1s} orbital, with their spins paired, leaving the σ^*_{1s} orbital empty.

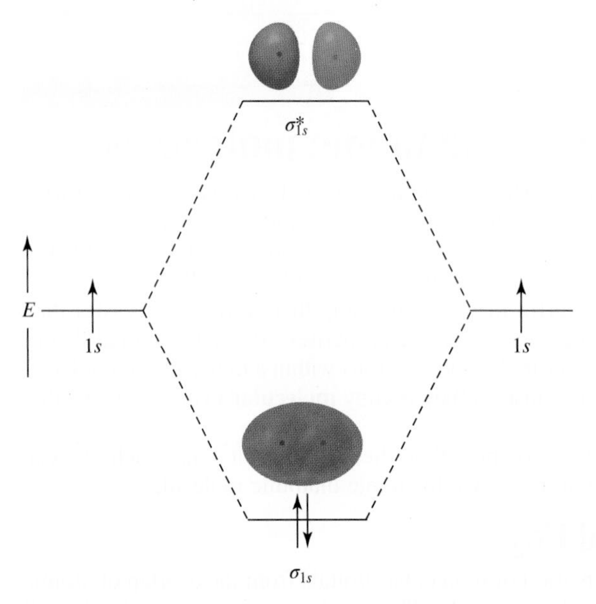

FIGURE 5.47 Molecular orbital diagram for H_2.

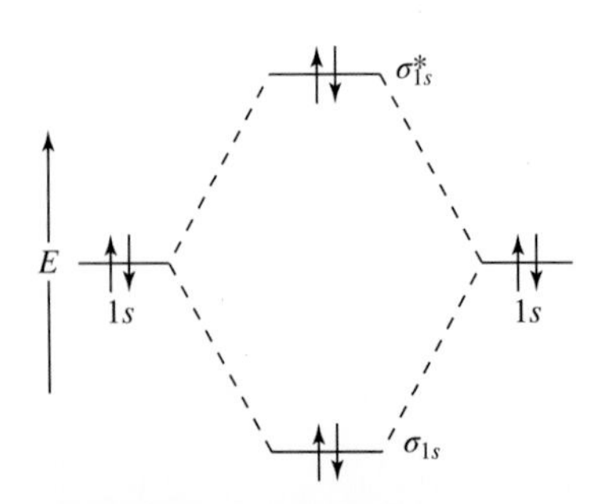

FIGURE 5.48 Diatomic He_2 does not exist because the stability imparted by the electrons in bonding orbitals is offset by the destabilisation due to the electrons in antibonding orbitals.

The molecular orbital diagram shown in figure 5.47 applies to all diatomic species from the first row of the periodic table, except that the electron occupancy and the absolute energies involved are different. We can, for instance, use molecular orbital theory to rationalise why two helium atoms do not combine to form a molecule of He_2. The electron configuration of He is $1s^2$ and therefore the hypothetical He_2 molecule would have four electrons. Two of these would populate the bonding σ_{1s} orbital, while the remaining two electrons would have to occupy the antibonding σ^*_{1s} orbital, as shown in figure 5.48. The total electronic energy of the He_2 molecule would therefore be exactly the same as that of two isolated He atoms, as the destabilising effect of the electrons in the antibonding orbital cancels out any stabilisation due to the electrons in the bonding orbital. As there is no energetic advantage, a molecule is not formed (see the discussion of entropy in chapter 8 for an explanation). A convenient way to summarise this argument is by calculating the **bond order**, which represents the net amount of bonding between two atoms.

$$\text{Bond order} = \tfrac{1}{2}(\text{number of electrons in bonding molecular orbitals} - \text{number of electrons in antibonding molecular orbitals})$$

Bond orders usually have integer or half-integer values: the larger the bond order, the stronger the bond (for the same type of atom).

For molecular hydrogen, with two bonding electrons and no electrons in the antibonding orbitals, the bond order is $\frac{1}{2}(2 - 0) = 1$. A bond order of 1 corresponds to a single bond, consistent with the Lewis structure of H_2. For the hypothetical He_2 molecule, two bonding and two antibonding electrons give a bond order of $\frac{1}{2}(2 - 2) = 0$, so no bond forms and helium does not form stable diatomic molecules.

WORKED EXAMPLE 5.9

Does He_2^+ exist?

Use a molecular orbital diagram to predict if it is possible to form the He_2^+ cation.

Analysis

The schematic MO diagram shown in figure 5.47 can be applied to any of the possible diatomic molecules or ions formed from the first-row elements, hydrogen and helium. Count the electrons of He_2^+, place the electrons in the molecular orbital diagram and calculate the bond order. If the bond order is greater than zero, the species can potentially form under the right conditions.

Solution

One He atom has two electrons, so a He_2^+ cation has three electrons. Following the Aufbau principle, two electrons fill the lower-energy σ_{1s} orbital, so the third must be placed in the antibonding σ_{1s}^* orbital in either spin orientation.

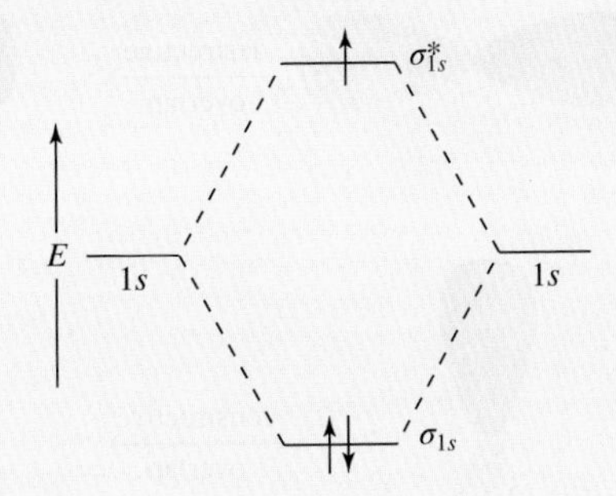

The bond order of the cation is therefore $\frac{1}{2}(2 - 1) = \frac{1}{2}$. Although a bond order of $\frac{1}{2}$ might sound unusual, it is greater than zero and therefore we predict that He_2^+, with a net bond order of $\frac{1}{2}$, can be prepared in the laboratory under appropriate conditions.

Is our answer reasonable?

Up until this point, we have discussed only single, double and triple bonds, and have not encountered bonds with non-integer orders. In fact, such bonds are quite common in chemistry. He_2^+ can be prepared by passing an electrical discharge through a sample of helium gas. The He_2^+ ion is unstable but survives long enough to allow its study. The bond dissociation energy of He_2^+ is 250 kJ mol^{-1}, approximately 60% as strong as the bond in the H_2 molecule, which has a bond order of 1.

PRACTICE EXERCISE 5.9

Which species has the stronger bond, H_2 or H_2^-? Use a molecular orbital diagram to support your answer.

Molecular orbitals of O_2

The next step in the development of molecular orbital theory is to consider the molecular orbitals formed by constructive and destructive overlap of atomic *p* orbitals. We use molecular oxygen, O_2, as a case study.

The Lewis structure of O_2 shows the two atoms connected by a double bond, with two nonbonding electron pairs on each atom. Note that there are no unpaired electrons in the Lewis structure.

$$\ddot{\text{O}}=\ddot{\text{O}}$$

We derive the molecular orbital diagram for the O_2 molecule by considering overlap of only the valence atomic orbitals on the two oxygen atoms; an exact treatment would also include the core 1*s* orbitals, but we will ignore these in the interests of clarity. Each oxygen atom has four valence orbitals, so the molecular orbital diagram for O_2 will contain eight molecular orbitals. The first interaction involves the 2*s* atomic orbitals. This pair interacts in exactly the same way as the 1*s* orbitals described previously for H_2. The constructive overlap generates a bonding σ_s orbital, and the destructive overlap forms an antibonding σ_s^* orbital. Because of the larger size of the 2*s* atomic orbitals, these two molecular orbitals are larger than their hydrogen σ_{1s} and σ_{1s}^* counterparts, but their overall appearances are similar.

Figure 5.49 shows the construction of the 2p-based molecular orbitals. One pair of molecular orbitals forms from end-on constructive and destructive overlap of the p_z orbitals that point towards each other along the bond axis. (By convention, the bond axis is chosen to be the z-axis.) These end-on overlaps give a pair of σ orbitals, the bonding σ_p and the antibonding σ_p^*, as shown in figure 5.49a. (Note that here we have simply labelled the molecular orbitals as σ_p and σ_p^*, rather than σ_{2p} and σ_{2p}^* for reasons we will explain on p. 210.) The remaining sets of p orbitals undergo constructive and destructive overlap in a side-by-side fashion to form pairs of molecular orbitals with electron density located primarily above and below the internuclear axis. Such molecular orbitals are called orbitals. Constructive overlap gives a bonding π orbital, while destructive overlap gives an antibonding π^* orbital. One of these pairs comes from overlap of the p_y orbitals, while the other comes from overlap of the p_x orbitals. Figure 5.49b shows only the p_y pair of π orbitals, which we label π_y — the π_x pair of orbitals derived from the p_x atomic orbitals has the same appearance, but is perpendicular to the π_y orbitals. Note that the antibonding π^* orbital contains a node between the two O atoms.

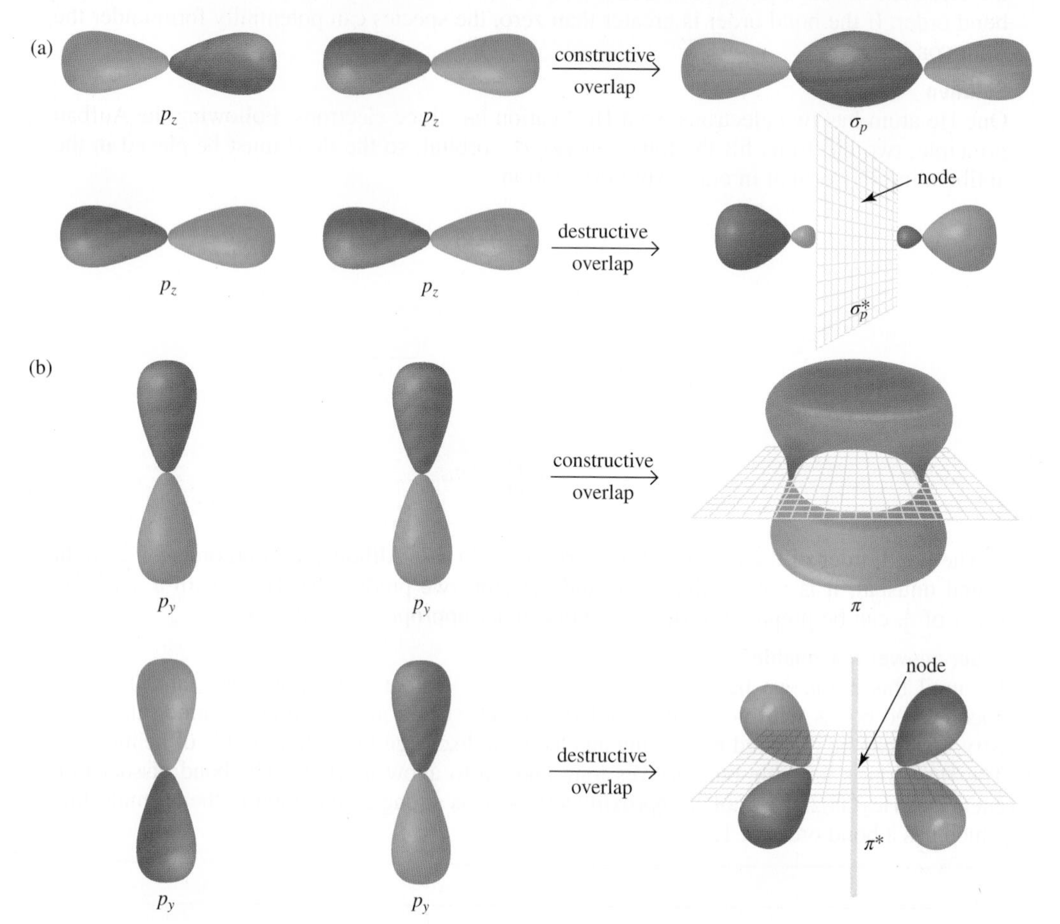

FIGURE 5.49 Constructive and destructive overlap of p orbitals lead to bonding and antibonding orbitals: **(a)** End-on overlap of the p_z orbitals gives σ orbitals. **(b)** Side-by-side overlap of the p_y atomic orbitals gives π_y and π_y^* molecular orbitals. Similar combinations of p_x orbitals would give π_x and π_x^* MOs.

To draw the molecular orbital diagram for O_2, we must first determine the relative energies of the molecular orbitals. We can do this with the following rules.

1. The core σ_s and σ_s^* orbitals derived from the overlap of the 1s atomic orbitals are of lowest energy and contribute little to bonding. For this reason, they are not shown explicitly.
2. The bonding σ_s and antibonding σ_s^* orbitals derived from the overlap of the 2s orbitals are of lower energy than any of the six molecular orbitals derived from the 2p orbitals. This is because the 2s orbitals that generate the σ_s and σ_s^* orbitals are of lower energy than the 2p atomic orbitals.
3. The π_x and π_y bonding orbitals are degenerate, because the corresponding atomic p orbitals from which they are derived are degenerate. The π_x^* and π_y^* orbitals are degenerate for the same reason.
4. The antibonding orbitals formed from the atomic 2p orbitals are highest in energy, with the σ_p^* orbital higher than the degenerate π_x^* and π_y^* pair of orbitals.

After applying these features, we are left with the bonding σ_p and π molecular orbitals. These must be placed between σ_s^* and π^*, but which of them is lower in energy? It can be shown mathematically that end-on overlap is more efficient than side-by-side overlap, which suggests that the σ_p orbital should be of lower energy than the π orbitals. This is the case for the O_2 molecule and figure 5.50 shows the complete diagram. To obtain the ground state electron configuration, we place the 12 valence electrons in the molecular orbitals in accordance with the Pauli and Aufbau principles and Hund's rule. We can describe the resulting electron configuration by naming the orbitals (σ_s, π_x etc.), and using superscripts to show how many electrons each orbital contains. Therefore, the configuration for the ground state of O_2 is:

$$(\sigma_s)^2\,(\sigma_s^*)^2\,(\sigma_p)^2\,(\pi_x)^2\,(\pi_y)^2\,(\pi_x^*)^1\,(\pi_y^*)^1$$

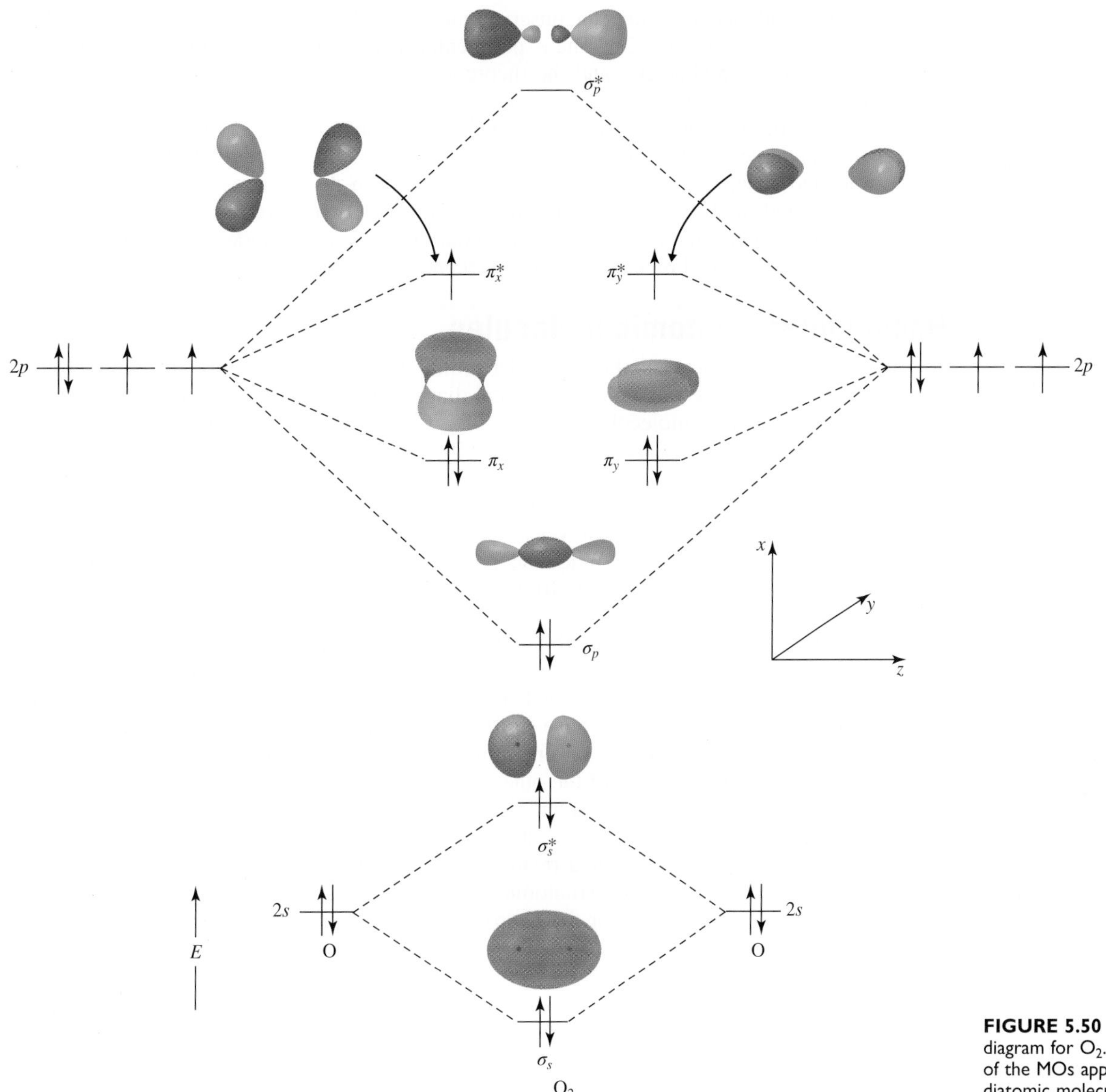

FIGURE 5.50 The molecular orbital diagram for O_2. The energy ordering of the MOs applies to second-row diatomic molecules with $Z > 7$.

Note that, to satisfy Hund's rule, we must place one electron in each of the two π^* orbitals with parallel spins. Therefore, this model of bonding predicts that O_2 should contain two unpaired electrons, in contrast to both the Lewis structure and the valence bond theory. Which model agrees with experiment?

Figure 5.51 shows that liquid oxygen adheres to the poles of a magnet. Attraction to a magnetic field shows that molecular oxygen is paramagnetic. Recall from chapter 4 that paramagnetism arises when the electron configuration includes unpaired electrons. Neither the Lewis structure of O_2 nor the valence bond model predicts the presence of unpaired electrons. However, the molecular orbital description shows that the two highest energy electrons of the oxygen molecule occupy the two degenerate π_x^* and π_y^* antibonding orbitals (figure 5.51). So, molecular orbital theory gives us a more appropriate model of the bonding in O_2 than either the Lewis or the valence bond theories. This does not necessarily mean that the molecular orbital theory is 'correct', just that it gives better agreement with experiment than the other models.

FIGURE 5.51 Molecular oxygen is paramagnetic, so it clings to the poles of a magnet.

Bond length and bond energy measurements on oxygen and its cation provide evidence that the π^* orbitals are indeed antibonding. The bond order of the O_2 molecule is $\frac{1}{2}(8 - 4) = 2$. If we remove an electron from O_2, we form the O_2^+ cation. Experimental measurements show that this species has a larger bond energy and shorter bond length than O_2, both observations consistent with an increased bond order. This indicates that the electron removed must be antibonding in character, as its removal has reduced the amount of antibonding and strengthened the bond.

Species	Bond length	Bond energy	Configuration	Bond order
O_2	121 pm	496 kJ mol^{-1}	$\ldots (\sigma_p)^2 (\pi_x)^2 (\pi^*_x)^1 (\pi^*_y)^1$	2
O_2^+	112 pm	643 kJ mol^{-1}	$\ldots (\sigma_p)^2 (\pi_x)^2 (\pi_y)^2 (\pi^*_{x,y})^1$	2.5

Molecular orbital theory predicts the observed increase in bond order, with the calculated bond order of O_2^+ being $\frac{1}{2}(8 - 3) = 2.5$. The experimental data show that the highest energy orbital of O_2 is antibonding in character and the theoretical molecular orbital predictions are in good agreement.

Our treatment of O_2 shows that the extra complexity of the molecular orbital approach explains features that a simpler description of bonding cannot. The Lewis structure of O_2 does not reveal its two unpaired electrons, but a molecular orbital approach does. The simple σ–π description of the double bond in O_2 does not predict that the bond in O_2^+ is stronger than that in O_2, but a molecular orbital approach does. We now extend our discussion to include all of the homonuclear diatomic molecules formed by elements in the second row of the periodic table.

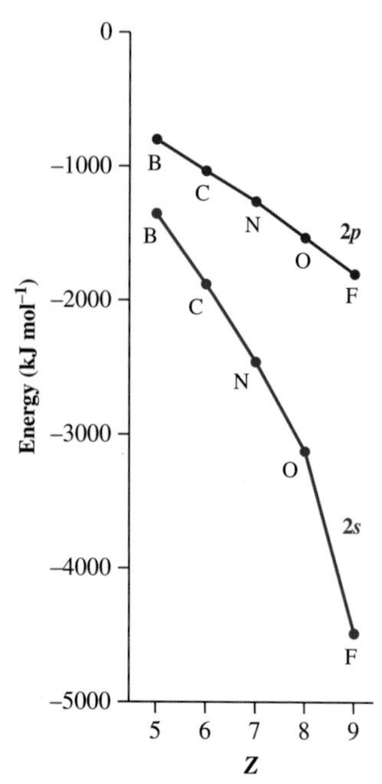

FIGURE 5.52 Energies of the $n = 2$ valence orbitals as a function of Z.

Homonuclear diatomic molecules

We saw previously that very similar molecular orbital diagrams can describe the bonding in H_2, He_2^+ and He_2. Can we use the molecular orbital diagram for O_2 to explain the bonding in all of the second-row diatomic molecules (Li_2, Be_2, B_2, C_2, N_2, O_2, F_2 and Ne_2)? Consider the B_2 molecule, which can be formed in the gas phase by strong heating of solid boron. If we use the molecular orbital diagram for O_2 and insert the six valence electrons of B_2, we would predict the electron configuration $(\sigma_s)^2(\sigma_s^*)^2(\sigma_p)^2$ for this molecule. With all its electrons paired, B_2 should be diamagnetic. However, experiments show that B_2 molecules are paramagnetic, with two unpaired electrons. When theory and experiment conflict, theory must be revised. The molecular orbital diagram we used for O_2 must be revised to account for the paramagnetism of B_2.

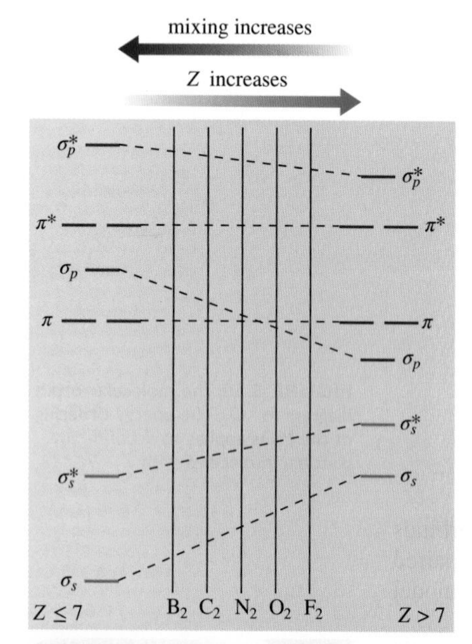

FIGURE 5.53 Mixing of the 2s and $2p_z$ orbitals causes the σ_s MOs to become lower in energy and the σ_p MOs to increase in energy. The amount of mixing decreases across the second row of the periodic table. For B_2, C_2 and N_2, the π MOs are of lower energy than the σ_p orbitals. For O_2 and F_2, the opposite is true.

The view of molecular orbital theory that leads to the molecular orbital diagram in figure 5.50 assumes that the 2*s* and 2*p* orbitals act independently. A more refined treatment considers interactions between the 2*s* and 2*p* orbitals. Recall from chapter 4 that the 2*s* and 2*p* orbitals have similar radii. Consequently, when two atoms approach each other, the 2*s* orbital on one atom will overlap not only with the 2*s* orbital, but also with the $2p_z$ orbital of the other atom. This mixed interaction of 2*s* and $2p_z$ stabilises the σ_s molecular orbital and destabilises the σ_p molecular orbital. In other words, **orbital mixing** causes the σ_s and σ_p molecular orbitals to move further apart in energy. In this more refined treatment, the resulting molecular orbitals are therefore not derived solely from two atomic orbitals, and hence we do not use specific labels, such as σ_{1s}, as we did for the hydrogen and helium cases. Labels of the type σ_s and π_y show the *predominant* type of overlap that contributes to the molecular orbital. The amount of mixing depends on the difference in energy between the 2*s* and 2*p* atomic orbitals. Mixing is largest when the energies of the orbitals are nearly the same. As figure 5.52 shows, the energies of the 2*s* and 2*p* atomic orbitals diverge as *Z* increases across the second row, so mixing is large for B_2 but small for F_2. Figure 5.53 shows how mixing affects the energies of the σ_s and σ_p molecular orbitals.

The main effect of mixing is to make the σ_p molecular orbital of higher energy than the degenerate π orbitals for the B_2, C_2 and N_2 molecules. Therefore, according to the Aufbau principle, we first fill the π orbitals and then the σ_p orbital for these molecules. Following Hund's rule, this means that B_2 is paramagnetic, as it has the electron configuration $(\sigma_s)^2(\sigma_s^*)^2(\pi_x)^1(\pi_y)^1$, in agreement with experiment. Notice that crossover of the σ_p and π energy levels takes place between molecular nitrogen and molecular oxygen. Because of orbital mixing, we therefore require two generalised molecular orbital diagrams for diatomic molecules, one for B_2, C_2 and N_2, and the other for O_2 and F_2. Figure 5.53 shows both diagrams.

These molecular orbital diagrams help to rationalise experimental observations about molecules.

WORKED EXAMPLE 5.10

Trends in bond energy

Use molecular orbital diagrams to explain the trend in the following bond energies. $B_2 = 290\ kJ\ mol^{-1}$, $C_2 = 600\ kJ\ mol^{-1}$ and $N_2 = 942\ kJ\ mol^{-1}$.

Analysis

The data show that bond energies for these three diatomic molecules increase as we move across the second row of the periodic table. We must construct molecular orbital diagrams for the three molecules and use the results to interpret the trend.

Solution

The crossover point for the σ_p/π energy levels takes place between N_2 and O_2, so the general diagram for $Z \leq 7$ in figure 5.53 applies to all three molecules. The valence electron counts are $B_2 = 6e^-$, $C_2 = 8e^-$ and $N_2 = 10e^-$. Place these electrons in the MO diagram following the Pauli exclusion and Aufbau principles and Hund's rule. Here are the results.

B_2: σ_p^* (empty); π_x^*, π_y^* (empty); σ_p (empty); π_x, π_y (↑ ↑); σ_s^* (↑↓); σ_s (↑↓)

C_2: σ_p^* (empty); π_x^*, π_y^* (empty); σ_p (empty); π_x, π_y (↑↓ ↑↓); σ_s^* (↑↓); σ_s (↑↓)

N_2: σ_p^* (empty); π_x^*, π_y^* (empty); σ_p (↑↓); π_x, π_y (↑↓ ↑↓); σ_s^* (↑↓); σ_s (↑↓)

The diagrams show that the increasing electron counts as we progress from boron to carbon to nitrogen result in the filling of bonding molecular orbitals. This is revealed most clearly by calculating the bond orders for the three molecules: $B_2 = \frac{1}{2}(4 - 2) = 1$, $C_2 = \frac{1}{2}(6 - 2) = 2$ and $N_2 = \frac{1}{2}(8 - 2) = 3$. Increasing bond order corresponds to stronger bonding between the atoms and therefore greater bond energies.

Is our answer reasonable?

Our predicted bond orders correlate perfectly with the experimental bond energies and therefore our answer is reasonable. Of these three diatomic molecules, only N_2 exists under normal conditions. Boron and carbon form nonmolecular solids rather than isolated diatomic molecules. However, molecular orbital theory predicts that B_2 and C_2 may exist under the right conditions, and in fact both molecules can be generated in the gas phase by vaporising solid boron or solid carbon.

PRACTICE EXERCISE 5.10

Use molecular orbital diagrams to explain the trend in bond energies for the following diatomic molecules: $N_2 = 942\ \text{kJ mol}^{-1}$, $O_2 = 495\ \text{kJ mol}^{-1}$ and $F_2 = 155\ \text{kJ mol}^{-1}$.

Heteronuclear diatomic molecules

Nitrogen oxide (NO) is an example of a heteronuclear diatomic molecule. This interesting molecule has been in the news in recent years because of important discoveries about the role of NO as a biological messenger. A molecular orbital diagram of NO is shown in figure 5.54.

Because the qualitative features of orbital overlap do not depend on the identity of the atoms, the bonding in NO can be described by the same sets of orbitals that describe homonuclear diatomic molecules. Which of the two general MO diagrams for diatomic molecules applies to nitrogen oxide? The crossover point for the energy rankings of the σ_p and π molecular orbitals falls between N and O, so we expect the orbital energies to be nearly equal. Experiments have shown that the σ_p MO is slightly more stable than the π MO. Consequently, the MO configuration for the 11 valence electrons of NO mirrors that for the 12 valence electrons of O_2, except that there is a single electron in the π^* orbital.

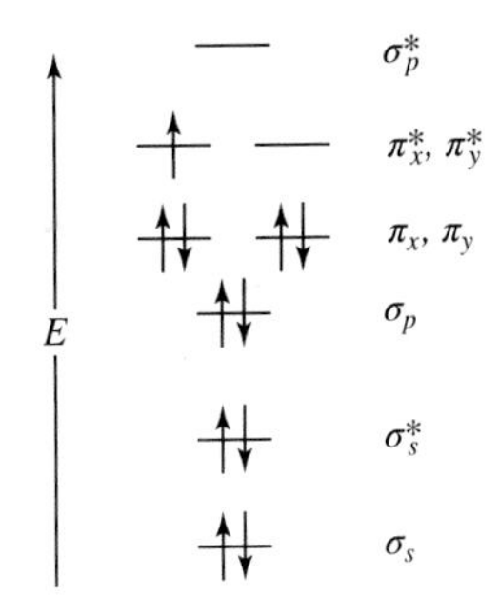

FIGURE 5.54 Molecular orbital diagram of NO.

$$(\sigma_s)^2 (\sigma_s^*)^2 (\sigma_p)^2 (\pi_x)^2 (\pi_y)^2 (\pi_x^*)^1$$

There are eight bonding electrons and three antibonding electrons, so the bond order of NO is 2.5. This bond is stronger and shorter than the double bond of O_2 but weaker and longer than the triple bond of N_2.

Species	Bond length	Bond energy	Configuration	Bond order
O_2	121 pm	495 kJ mol^{-1}	$\ldots (\sigma_p)^2 (\pi_x)^2 (\pi_y)^2 (\pi_x^*)^1 (\pi_y^*)^1$	2
NO	115 pm	605 kJ mol^{-1}	$\ldots (\sigma_p)^2 (\pi_x)^2 (\pi_y)^2 (\pi_{x,y}^*)^1$	2.5
N_2	110 pm	945 kJ mol^{-1}	$\ldots (\pi_x)^2 (\pi_y)^2 (\sigma_p)^2$	3

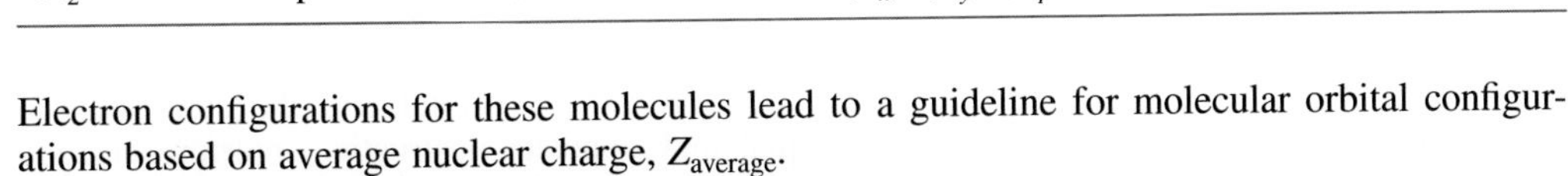

Electron configurations for these molecules lead to a guideline for molecular orbital configurations based on average nuclear charge, $Z_{average}$.

For second-row diatomic molecules and ions:

- σ_p *is lower in energy than* π_x, π_y *when* $Z_{average} > 7$
- π_x, π_y *is lower in energy than* σ_p *when* $Z_{average} \le 7$.

Worked example 5.11 compares the ionisation energies of three diatomic molecules and uses MO theory to explain the differences.

WORKED EXAMPLE 5.11

Comparing ionisation energies

The first ionisation energy of NO is 891 kJ mol^{-1}, that of N_2 is 1500 kJ mol^{-1}, and that of CO is 1350 kJ mol^{-1}. Use electron configurations to explain why NO ionises more easily than either N_2 or CO.

Analysis

Recall from chapter 4 that ionisation energy refers to the removal of an electron from an atom, or, in this case, from a molecule. We must count the valence electrons, choose the correct MO diagram, follow the Aufbau principle in placing the electrons, and then use the configurations to explain the ionisation energy data.

Solution

Begin with a summary of the important information for each molecule: NO has 11 valence electrons and $Z_{average} = 7.5$; N_2 has 10 valence electrons and $Z_{average} = 7$; CO has 10 valence electrons and $Z_{average} = 7$. Applying the general diatomic MO diagrams of figure 5.53 gives the following configurations:

$$NO = (\sigma_s)^2 (\sigma_s^*)^2 (\pi_x)^2 (\pi_y)^2 (\sigma_p)^2 (\pi_{x,y}^*)^1$$

$$N_2 = (\sigma_s)^2 (\sigma_s^*)^2 (\sigma_p)^2 (\pi_x)^2 (\pi_y)^2 (\sigma_p)^2$$

$$CO = (\sigma_s)^2 (\sigma_s^*)^2 (\pi_x)^2 (\pi_y)^2 (\sigma_p)^2$$

CO and N_2 are isoelectronic species, meaning they have the same number of electrons (see p. 160), and each has a bond order of 3. It makes sense that these two molecules have comparable ionisation energies. NO has one electron more than N_2 and CO, and this electron occupies one of the antibonding π^* orbitals. It is much easier to remove this antibonding electron from NO than it is to remove a bonding electron from either of the other two species.

PRACTICE EXERCISE 5.11 Use electron configurations to predict which of the following is the most stable diatomic combination of carbon and nitrogen: CN^-, CN or CN^+.

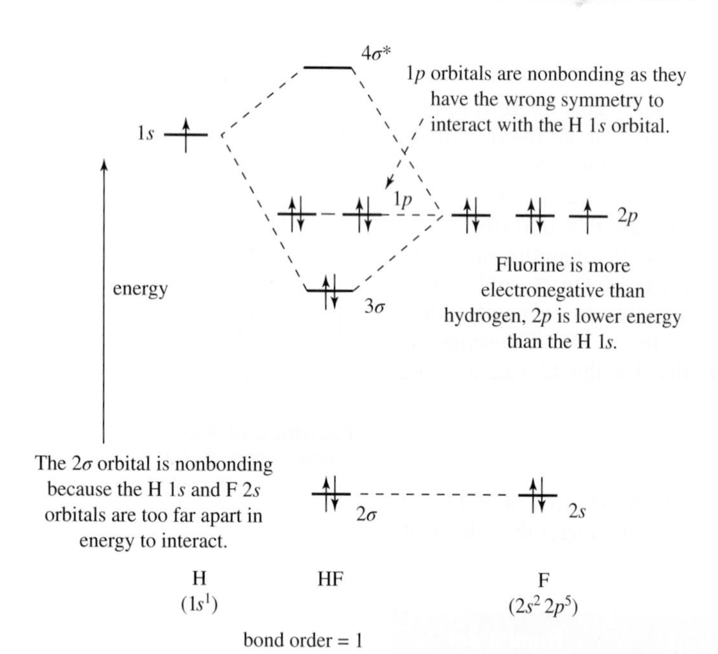

4σ*
1p orbitals are nonbonding as they have the wrong symmetry to interact with the H 1s orbital.
1s
1p
2p
energy
3σ
Fluorine is more electronegative than hydrogen, 2p is lower energy than the H 1s.
The 2σ orbital is nonbonding because the H 1s and F 2s orbitals are too far apart in energy to interact.
2σ
2s
H
(1s¹)
HF
F
(2s² 2p⁵)
bond order = 1

FIGURE 5.55 The molecular orbital diagram of HF.

For diatomic molecules composed of atoms that have very different energies of their atomic orbitals, the MO diagram becomes slightly more complicated. We will use HF as an example and show the energy levels for both atomic orbitals and molecular orbitals in figure 5.55.

Generally, orbitals are of lower energy in a more electronegative atom and orbitals with the same designation are lower in energy if the nuclear charge is larger. Armed with these rules we can understand that the 1*s* orbital in the F atom is much lower in energy than the 1*s* orbital of H.

Molecular orbitals can form only when there is net overlap between the constituent atomic orbitals and when the atomic orbitals are relatively close in energy. We can illustrate this by considering the approach of the hydrogen 1*s* orbital with either the $2p_x$ or $2p_y$ orbital of the fluorine atom.

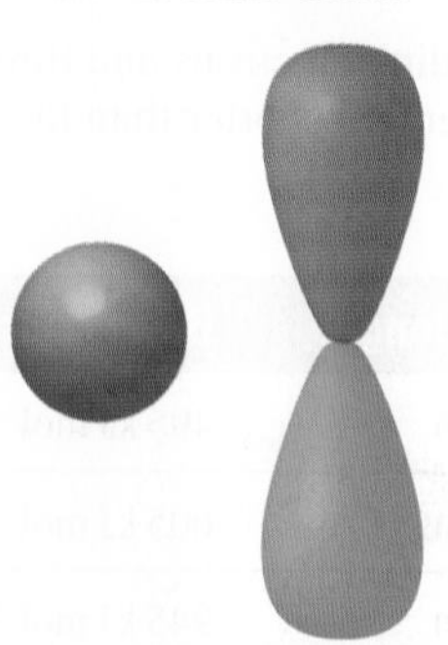

As these orbitals approach each other in a 'side-on' manner, there will be constructive interference, and hence a bonding interaction, between the *s* orbital and the same phased lobe of the *p* orbital, but there will be an equal and opposite destructive interference, and hence an antibonding interaction, between the *s* orbital and the opposite phased lobe of the *p* orbital (figure 5.55). Therefore, there is no net overlap between an *s* orbital and a p_x or p_y orbital when they approach in this manner, and no molecular orbitals are formed. Instead, the $2p_x$ and $2p_y$ orbitals are located primarily on the fluorine atom as essentially nonbonding molecular orbitals. While overlap of the hydrogen 1*s* and fluorine 2*s* orbitals can, in theory, occur, the energies of the orbitals are too different to permit effective overlap, and for this reason, the fluorine $2s$ orbital remains essentially nonbonding. The $1s$ orbital of hydrogen and the $2p_z$ orbital of fluorine overlap effectively and form two MOs. The 3σ orbital is much closer in energy to the $2p_y$ orbital of fluorine and therefore has more $2p_y$ character and is closer to the fluorine nucleus, while the antibonding $4\sigma^*$ orbital has more hydrogen $1s$ character. The MO diagram in figure 5.55 shows the sequence of the valence molecular orbitals.

The CO molecule is another heteronuclear diatomic molecule that can serve to illustrate features of MO theory. Both C and O have $2s$ and $2p$ valence orbitals and therefore can form π bonds. Given that O is more electronegative than C, the valence orbitals on O will be of lower energy than the corresponding orbitals on C.

Figure 5.56 shows the energy level diagram for CO. The $2s$ and the $2p_z$ orbitals on both C and O overlap to form four σ molecular orbitals, while the other two p orbitals on C and O overlap to form two bonding and two antibonding π orbitals. The bonding π orbitals are lower in energy than the σ bonding orbital with mainly p character due to s–p mixing, which is generally large in heteronuclear diatomic molecules.

We have barely scratched the surface of molecular orbital theory. More complicated systems (three or more atoms) require the use of powerful computers to adequately treat the bonding. Indeed, even the simple diatomic systems we have discussed have involved huge simplifications; it is, in fact, currently impossible to exactly describe a molecule containing more than one electron, as the mathematics of such a system have so far defied solution. We have discussed a number of ways to describe the bonding in molecules; although each of them has particular shortcomings, if applied appropriately, we can use them to satisfactorily explain the bonding in virtually all chemical systems.

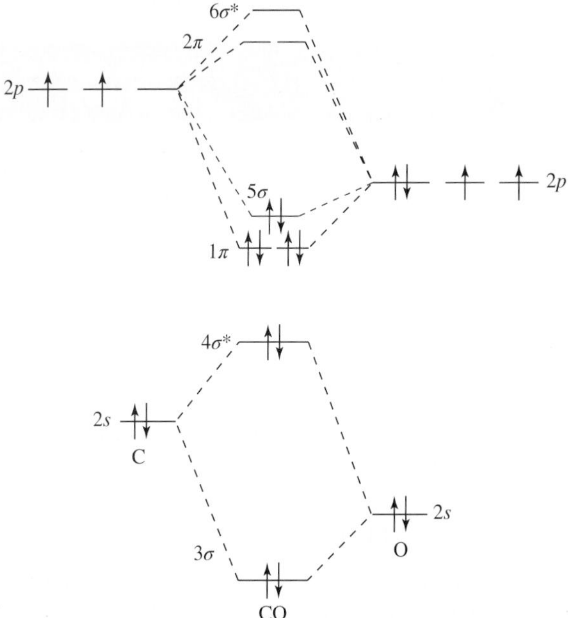

FIGURE 5.56 The molecular orbital diagram of CO.

SUMMARY

LO 1 Fundamentals of bonding

Covalent bonds are formed as a result of the sharing of electrons between nuclei in a way that balances attractive and repulsive forces. In the hydrogen molecule, H_2, this leads to a lowest energy internuclear separation of 74 pm. This distance is called the bond length, and the energy difference between the molecule at this distance and the separated atoms is called the bond energy. The single bond in H_2 is an example of a σ bond, a bond that is totally symmetric with respect to rotation about the internuclear axis. In simple diatomic molecules such as H_2 and F_2 we can describe bonding in terms of overlap of atomic orbitals containing single electrons. When the two atoms in a diatomic molecule are different, the electron pair will tend to be attracted more to one atom, resulting in unequal sharing of the electrons and a polar covalent bond. The unequal sharing is a consequence of the different electronegativity values of the atoms in the molecule; the greater the electronegativity difference between the atoms, the more polar the bond. Electronegativities generally increase across a period and decrease down a group in the periodic table.

LO 2 Ionic bonding

Compounds formed between elements with very different electronegativities are often ionic in character. Ionic compounds generally consist of alternating positively and negatively charged ions that are held together by attractive electrostatic forces between the oppositely charged ions. The magnitude of the attractive forces is dependent on the charge of the ions and the separation between them, as well as a number of other factors. The total energy that must be supplied to break the crystal lattice into its gaseous ions is called the lattice energy. Trends in lattice energy values for ionic compounds can be explained in terms of ionic charge and separation.

LO 3 Lewis structures

Lewis structures show the distribution of valence electrons within a molecule. Bonding electrons are found in single bonds (1 electron pair), double bonds (2 electron pairs) or triple bonds (3 electron pairs) and these are represented as 1, 2 and 3 lines respectively between the bonded atoms.

Nonbonding electrons, called lone pairs, are indicated as dots next to the elemental symbol for the atom on which they are located. Lewis structures may be drawn using the following five-step procedure.

Step 1: Count the valence electrons.
Step 2: Assemble the bonding framework using single bonds.
Step 3: Place three nonbonding pairs of electrons on each outer atom, except H.
Step 4: Assign the remaining valence electrons to inner atoms.
Step 5: Minimise formal charges on all atoms.

An atom has an octet if it is surrounded by four pairs of electrons. This arrangement is often found for elements in the second period of the periodic table. When there is a choice of electron arrangement, the one which minimises the formal charge on all atoms will be preferred.

In some cases, two or more energetically equivalent Lewis structures for a molecule can be drawn which differ only in the position of one electron pair. Such structures are called resonance structures. We can also have energetically inequivalent resonance structures; in these cases, the preferred structure will again be that with the lowest possible formal charge on each atom.

LO 4 Valence-shell-electron-pair repulsion (VSEPR) theory

VSEPR theory states that a molecule will adopt a shape in which electron-pair repulsions are minimised. This is achieved by placing electron pairs as far apart as possible.

We can predict the structure of a molecule by using the following three-step procedure.

1. Draw the Lewis structure of the molecule.
2. Count the number of sets of bonding pairs and lone pairs of electrons around any inner atom, and use the following table to determine the optimum geometry of these sets.

Number of sets of electron pairs	Geometry of sets of electron pairs
2	linear
3	trigonal planar
4	tetrahedral
5	trigonal bipyramidal
6	octahedral

3. Modify the geometry, if necessary, to take account of the fact that the magnitudes of repulsions between sets of electron pairs depend on whether the electron pairs involved are bonding pairs (BP) or lone pairs (LP). The repulsions are in the order:

$$\text{LP–LP} > \text{BP–LP} > \text{BP–BP}$$

The idealised electron pair geometries will be slightly distorted whenever lone pairs are present, because lone pairs occupy more space than bonding pairs.

LO 5 Properties of covalent bonds

Molecules where the centres of partial negative and partial positive changes do not coincide have a dipole moment. The magnitudes of dipole moments are measured in C m (coulomb metres). Molecules containing polar bonds do not necessarily have a dipole moment, as the polarities of such bonds can cancel each other. Carbon dioxide provides such an example.

Bond lengths in molecules are dependent on the atomic radii of the bonded atoms, which themselves are dependent on effective nuclear charge. A multiple bond between two atoms is shorter than a single bond between the same atoms. Bond energies increase as the number of electrons in the bond increases, and also as the electronegativity difference between the bonded atoms increases. The length of a bond is inversely proportional to its bond energy — as a bond lengthens, its energy becomes less.

LO 6 Valence bond theory

Valence bond theory describes bonding in molecules using localised bonds formed from orbital overlap of hybrid orbitals. As the name suggests, valence bond theory considers only valence orbitals in the formation of bonds. Valence atomic orbitals undergo hybridisation to form hybrid orbitals, with the orbitals involved depending on the type of hybrid orbital required. Hybridisation of one s and three p orbitals gives four sp^3 hybrid orbitals, hybridising one s and two p orbitals gives three sp^2 hybrid orbitals, and hybridising one s and one p orbital gives two sp hybrid orbitals. These processes leave zero, one and two p orbitals, respectively, unhybridised. The sp^3 hybrid orbitals are positioned tetrahedrally, sp^2 hybrid orbitals are arranged in a trigonal plane, and sp hybrid orbitals adopt a linear arrangement. The hybrid orbitals form σ bonds by overlapping with either atomic orbitals or hybrid orbitals on adjacent atoms. Multiple bonds can be described in terms of one σ bond plus one π bond (double bond) or one σ bond plus two π bonds (triple bond); π bonds are formed by overlap of the unhybridised p orbitals on adjacent sp^2 or sp hybridised atoms.

LO 7 Molecular orbital theory: diatomic molecules

Molecular orbital theory considers all possible overlaps between atomic orbitals in a molecule and describes bonding in terms of delocalised bonds. Overlap of two atomic orbitals gives the formation of molecular orbitals that can cover the entire molecule. Atomic orbitals can overlap in phase (constructive overlap) to give a low energy bonding molecular orbital, or out of phase (destructive overlap) to give a high energy antibonding molecular orbital. Electron density is maximised between the nuclei in bonding MOs, while antibonding MOs contain a node, a position where the amplitude changes sign, between the nuclei. The relative energies of the resulting orbitals are shown on a molecular orbital diagram. Electrons are placed into a molecular orbital diagram following the same rules as used in an atomic energy level diagram. The bond order of a simple diatomic molecule can be calculated from the expression:

$$\text{bond order} = \tfrac{1}{2}(\text{number of electrons in bonding molecular orbitals} - \text{number of electrons in antibonding molecular orbitals})$$

While bond orders are generally 1, 2 or 3, it is possible to have non-integer bond orders. Molecular orbital theory can account for the observed paramagnetism of O_2 while valence bond theory cannot. However, to account for the observed properties of other diatomic molecules, orbital mixing, the overlap of s atomic orbitals with p_z atomic orbitals, must be considered.

KEY CONCEPTS AND EQUATIONS

Periodic trends in electronegativity *(section 5.1)*

These trends allow us to use the positions of elements in the periodic table to estimate the degree of bond polarity and to estimate which atom in a bond is the most electronegative.

Method for drawing Lewis structures *(section 5.3)*

Lewis structures show the distribution of valence electrons in a molecule or ion and are drawn according to the following conventions.

1. Each atom is represented by its elemental symbol.
2. Only the valence electrons appear in a Lewis structure.
3. A line joining two elemental symbols represents one pair of electrons shared between two atoms.
4. Dots placed next to an elemental symbol represent nonbonding electrons on that atom.

Formal charges *(section 5.3)*

By assigning formal charges, we can select the best Lewis structure for a molecule or polyatomic ion.

Method for determining resonance structures *(section 5.3)*

By distributing multiple bonds over atoms in a molecule, we obtain a better description of the bonding.

VSEPR theory *(section 5.4)*

This theory enables us to predict the shape of a molecule or polyatomic ion when its Lewis structure is known.

1. Draw the Lewis structure of the molecule.
2. Count the number of sets of bonding pairs and lone pairs of electrons around any inner atom, and use the following table to determine the optimum geometry of these sets.

Number of sets of electron pairs	Geometry of sets of electron pairs
2	linear
3	trigonal planar
4	tetrahedral
5	trigonal bipyramidal
6	octahedral

3. Modify the geometry, if necessary, to take account of the fact that the magnitudes of repulsions between sets of electron pairs depend on whether the electron pairs involved are bonding pairs (BP) or lone pairs (LP). The repulsions are in the order:

$$LP–LP > BP–LP > BP–BP$$

Hybrid orbitals *(section 5.6)*

Hybrid orbitals help to explain the bonding in molecules, and in particular can be used to rationalise the shapes of molecules.

Molecular orbital diagrams *(section 5.7)*

By placing electrons in molecular orbital diagrams we are able to obtain bond orders that agree with experiments for simple diatomic molecules.

KEY TERMS

antibonding molecular orbital *p. 206*
bond energy *p. 173*
bond length *p. 173*
bond order *p. 206*
bonding molecular orbital *p. 205*
bonding orbital *p. 196*
covalent bond *p. 172*
delocalised bond *p. 196*
dipole moment *p. 189*
double bond *p. 177*
electronegativity *p. 174*
formal charge *p. 179*
hybrid orbital *p. 197*
ionic compound *p. 174*
lattice energy *p. 176*
Lewis structure *p. 177*
linear *p. 183*
localised bond *p. 196*
lone pair *p. 178*
molecular orbital (MO) *p. 205*
molecular orbital diagram *p. 206*
molecular orbital theory *p. 205*
node *p. 205*
octahedral *p. 187*
orbital mixing *p. 210*
orbital overlap *p. 196*
pi (π) bond *p. 203*
polar covalent bond *p. 173*
resonance structures *p. 180*
seesaw shape *p. 186*
sigma (σ) bond *p. 173*
single bond *p. 177*
sp hybrid orbitals *p. 201*
sp^2 hybrid orbitals *p. 200*
sp^3 hybrid orbitals *p. 197*
tetrahedral *p. 184*
trigonal bipyramidal *p. 185*
trigonal planar *p. 183*
triple bond *p. 177*
valence-shell-electron-pair repulsion (VSEPR) *p. 182*

REVIEW QUESTIONS

LO 1 Fundamentals of bonding

5.1 For the following atoms, write the complete electron configuration and identify which of the electrons will be involved in bond formation: (a) N (b) S (c) Be (d) Br.

5.2 Describe bond formation between a hydrogen atom and a bromine atom to form a molecule of HBr, and include a diagram of the overlapping orbitals.

5.3 Give the group number and the number of valence electrons for the following elements: (a) boron (b) phosphorus (c) iodine (d) germanium.

5.4 For each of the following pairs, identify which element tends to attract electron density from the other in a covalent bond: (a) C and N (b) O and H (c) Zn and I (d) P and Ga.

5.5 Show the direction of bond polarity for the following bonds using δ+/δ– notation: (a) Si—N (b) O—C (c) Br—F (d) C—P.

5.6 Arrange the following molecules in order of increasing bond polarity: H_2Se, AsH_3, PH_3 and H_2S.

LO 2 Ionic bonding

5.7 From the following list, select the elements that form ionic compounds: Ba, B, Cu, Rb, Cl and V. Indicate whether each forms a stable cation or a stable anion.

5.8 From the following list, select the elements that form ionic compounds: C, Mg, K, Bi and Br. Indicate whether each forms a stable cation or a stable anion.

5.9 Consider three possible ionic compounds formed by barium and oxygen: Ba^+O^-, $Ba^{2+}O^{2-}$ and $Ba^{3+}O^{3-}$. (a) Which would have the greatest lattice energy? (b) Which would require the least energy to form the ions? (c) Which compound actually exists, and why?

LO 3 Lewis structures

5.10 Count the total number of valence electrons in the following species: (a) $H_2PO_4^-$ (b) $(C_6H_5)_3CH$ (c) $(NH_2)_2CO$ (d) HSO_4^-.

5.11 Convert the following formulae into molecular frameworks. For each molecule, calculate how many valence electrons are required to construct the framework: (a) $(CH_3CH_2)_3CBr$ (b) $(CH_3CH_2)_2NH$ (c) $HClO_4$ (d) $OP(OCH_3)_3$.

5.12 Determine the Lewis structures of: (a) NH_3 (b) NH_4^+ (c) H_2N^-.

5.13 Determine the Lewis structures of: (a) PBr_3 (b) SiF_4 (c) BF_4^-.

5.14 Use the standard procedures to determine the Lewis structures of: (a) H_3CNH_2 (b) CF_2Cl_2 (c) OF_2.

5.15 Determine the Lewis structures of: (a) IF_5 (b) SO_3 (c) $OPCl_3$ (d) XeF_2.

5.16 Determine the Lewis structure of each of the following polyatomic ions. Include all resonance structures and formal charges where appropriate. (a) NO_3^- (b) HSO_4^- (c) CO_3^{2-} (d) ClO_2^-

LO 4 Valence-shell-electron-pair repulsion (VSEPR) theory

5.17 Sketch the following molecular shapes and give the various bond angles in the structures: (a) trigonal planar (b) tetrahedral (c) octahedral.

5.18 Sketch and name the shapes of the following molecules: (a) CH_2Cl_2 (b) BF_4 (c) PBr_3.

5.19 Draw a ball-and-stick model that shows the geometry of 1,2-dichloroethane, ClH_2CCH_2Cl.

5.20 Write the Lewis structure of dimethylamine, $(CH_3)_2NH$. Determine its geometry.

5.21 Iodine forms three compounds with chlorine: (a) IF (b) IF_3 (c) IF_5. Determine the Lewis structure, describe the shape and draw a ball-and-stick model of each compound.

5.22 Determine the molecular shape and the ideal bond angles for each of the following: (a) SO_2 (b) SbF_5 (c) ClF_4^+ (d) ICl_4^-.

LO 5 Properties of covalent bonds

5.23 Determine the Lewis structures of the following compounds, and determine which have dipole moments. For each molecule that has a dipole moment, draw a ball-and-stick model and include an arrow to indicate the direction of the dipole moment. (a) SiF_4 (b) H_2S (c) XeF_2 (d) $GaCl_3$ (e) NF_3

5.24 Carbon dioxide has no dipole moment, but sulfur dioxide has $\mu = 5.44 \times 10^{-30}$ C m. Use Lewis structures to account for this difference in dipole moments.

5.25 Which of the following molecules would you expect to have bond angles that deviate from the ideal VSEPR values? For the molecules that do, make sketches that illustrate the deviations. (a) PCl_5 (b) CH_3I (c) ICl_5

5.26 Use table 5.3 to arrange the following bonds in order of increasing bond strength (weakest first). List the single most important factor for each successive increase in strength. C═C, H—N, C═O, N≡N and C—C

LO 6 Valence bond theory

5.27 Describe the bonding between fluorine and chlorine atoms in the mixed halogen ICl.

5.28 Describe the bonding between hydrogen and fluorine atoms in HF and include a picture of the overlapping orbitals.

5.29 The bond angles in antimony trifluoride are 87°. Describe the bonding in SbF_3, including a picture of the orbital overlap interaction that creates the Sb—F bonds.

5.30 Determine the hybridisation of an inner atom in a molecule that has each of the following characteristics: (a) 2 lone pairs and 2 ligands (b) 3 ligands and 1 lone pair (c) 3 ligands and no lone pairs.

5.31 Name the hybrid orbitals formed by combining each of the following sets of atomic orbitals: (a) 2*s* and three 2*p* orbitals (b) 3*s* and two 3*p* orbitals.

5.32 Identify the hybridisation of the bolded atom in each of the following species: (a) $(CH_3)_2NH$ (b) SO_2 (c) CO_2.

5.33 Describe the bonding in the chloroform molecule, $CHCl_3$. Sketch an orbital overlap diagram of the molecule.

5.34 Describe the bonding in the hydrazine molecule, H_2NNH_2. Sketch an orbital overlap diagram of the molecule.

5.35 Describe the bonding in the common solvent acetone, $(CH_3)_2CO$, and include sketches of all the bonding orbitals.

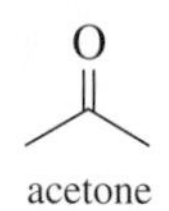

acetone

5.36 The carbon compounds penta-1,4-diene, pent-1-yne and cyclopentene all have the molecular formula C_5H_8. Use the number of sets of electron pairs and hybridisation to develop bonding pictures of these three molecules.

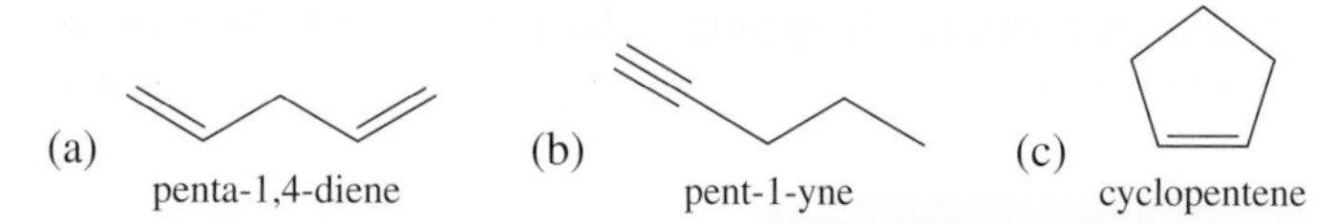

5.37 Decide if the following pairs of orbitals overlap to form a σ bond, π bond, or no bond at all. Explain your reasoning in each case, including a sketch of the orbitals. Assume the bond lies along the *z*-axis. (a) $2p_z$ and $2p_z$ (b) $2p_y$ and $2p_x$ (c) sp^3 and $2p_z$ (d) $2p_y$ and $2p_y$

LO 7 Molecular orbital theory: diatomic molecules

5.38 Use molecular orbital diagrams to rank the bond energies of the following diatomic species from weakest to strongest: H_2, H_2^- and H_2^+.

5.39 Which of N_2 or N_2^+ has the stronger bond? Use orbital configurations to justify your selection.

5.40 For each of the following interactions between orbitals of two different atoms, sketch the resulting molecular orbitals. Assume that the nuclei lie along the *z*-axis, and include at least two coordinate axes in your drawing. Label each MO as bonding or antibonding and σ or π. (a) 2*s* and $2p_z$ (b) $2p_x$ and $2p_x$

5.41 Below is an illustration showing two 3*d* orbitals about to overlap. The drawings also show the algebraic signs of the wavefunctions for both orbitals in this combination. Will this combination of orbitals produce a bonding or an antibonding MO?

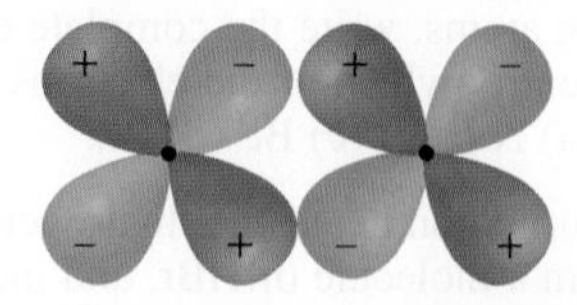

REVIEW PROBLEMS

5.42 List the following *X*—H bonds from smallest bond polarity to largest bond polarity: C—H, F—H, N—H, O—H and Si—H. **LO1**

5.43 Using information in table 5.1, determine the average difference in lattice energy between group 1 fluorides and group 1 chlorides, and the average difference in lattice energy between group 1 bromides and group 1 iodides. Is there a trend in the individual values? If so, describe and explain it. **LO2**

5.44 Determine the molecular geometries of the following molecules. (a) $SiCl_4$ (b) SeF_4 (c) CI_4 **LO4**

5.45 Determine the molecular geometries of the following ions. (a) ClF_2^- (b) BF_4^- (c) PF_4^+ **LO4**

5.46 How many different structural isomers are there for octahedral molecules with the general formula AX_3Y_3? Draw three-dimensional structures of each. **LO4**

5.47 Carbon, nitrogen and oxygen form two different polyatomic ions: the cyanate ion, NCO^-, and the isocyanate ion, CNO^-. Write Lewis structures for each anion; include near-equivalent resonance structures and indicate formal charges. **LO4**

5.48 Species with the general chemical formula XY_4 can have the shapes shown in the following diagrams. For each, name the molecular geometry, identify the ideal VSEPR bond angles, give the number of lone pairs present in the structure and provide a specific example. **LO4**

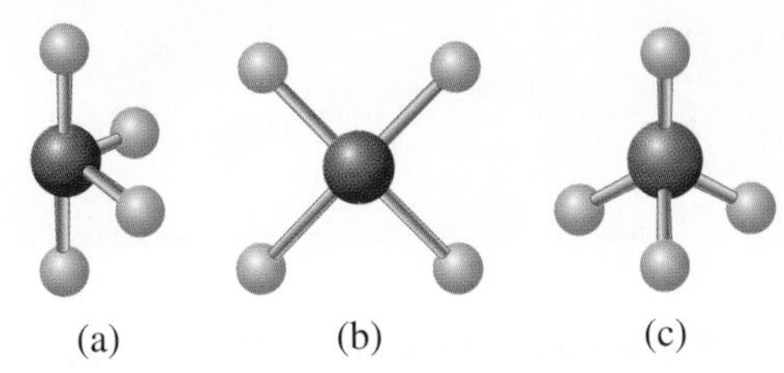

(a) (b) (c)

5.49 Species with the general chemical formula XY_3 can have the shapes shown in the following diagrams. For each, name the molecular geometry, identify the ideal VSEPR bond angles, give the number of lone pairs present in the structure and provide a specific example. **LO4**

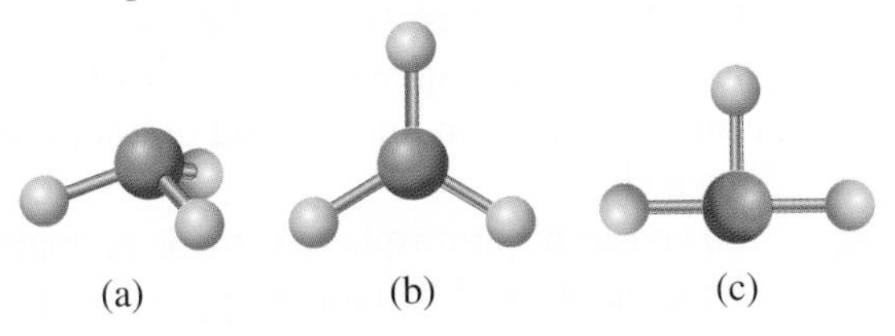

(a) (b) (c)

5.50 In the lower atmosphere, NO_2 participates in a series of reactions in air that is also contaminated with unburned hydrocarbons. One product of these reactions is peroxyacetyl nitrate (PAN). The skeletal arrangement of the atoms in PAN appears below. **LO4**

```
     H   O            O
     |   |↙          /
H —*C —*C —*O — O —*N←
   ↗ |                \
     H                 O
```

(a) Complete the Lewis structure of this compound.
(b) Determine the geometry around each atom marked with an asterisk.
(c) Give the approximate values of the bond angles indicated with arrows.

5.51 The H—O—H bond angle in a water molecule is 104.5°. The H—S—H bond angle in hydrogen sulfide is only 92.2°. Explain these variations in bond angles using orbital sizes and electron–electron repulsion arguments. Draw space-filling models to illustrate your explanation. **LO4**

5.52 Sulfur forms two stable oxides, SO_2 and SO_3. Describe the bonding and geometry of these compounds. **LO4**

5.53 Capsaicin is the molecule responsible for the hot spiciness of chillies: **LO4**

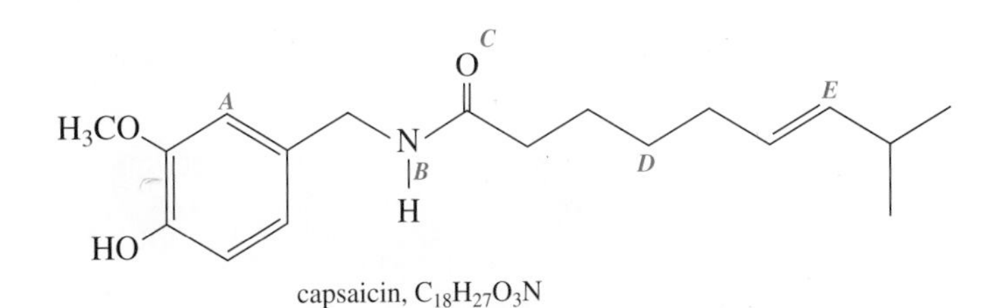

capsaicin, $C_{18}H_{27}O_3N$

(a) How many π bonds does capsaicin have?
(b) Which orbitals are used for bonding by each of the labelled atoms?
(c) What are the bond angles around each of the labelled atoms?
(d) Redraw the structure of capsaicin adding the lone pairs.

5.54 Determine the Lewis structures for the two possible arrangements of the N_2O molecule: N—N—O and N—O—N. Experiments show that the molecule is linear and has a dipole moment. What is the arrangement of atoms? Justify your choice. **LO4**

5.55 The inner atom of a triatomic molecule can have any of four different geometries for the surrounding sets of electron pairs. Identify the four, describe the shape associated with each and give a specific example of each. **LO4**

ADDITIONAL EXERCISES

5.56 Phosphonic acid, H_3PO_3, is an exception to the rule that hydrogen always bonds to oxygen in oxoacids. In this compound, one of the hydrogen atoms bonds to phosphorus. Determine the Lewis structure of phosphonic acid and determine the geometry around the phosphorus atom. Draw a ball-and-stick model of the molecule. **LO4**

5.57 When an oxalate anion, $C_2O_4^{2-}$, adds two protons to form oxalic acid, two C—O bonds become longer and two become shorter than the bonds in oxalate anions. Which bonds get longer and which shorter? Use bonding principles to explain the changes. **LO5**

5.58 Determine the type of orbitals (atomic, sp^3 or sp^2) used by each atom in the molecules shown below. **LO6**

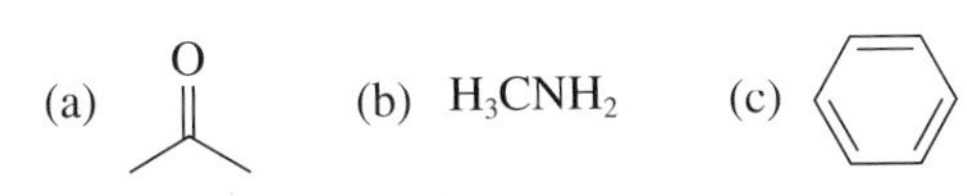

5.59 Both PF_3 and PF_5 are known compounds. NF_3 also exists, but NF_5 does not. Why is there no molecule with the formula NF_5? **LO6**

5.60 Chlorine forms the neutral oxide, ClO_2. Describe the bonding in this unusual compound. Explain why it is considered unusual. **LO6**

5.61 Use electron configurations to decide if the following species are paramagnetic or diamagnetic: (a) N_2^+ (b) O_2^+. **LO7**

5.62 Nitrogen molecules can absorb photons to generate excited-state molecules. Construct an energy level diagram and place the valence electrons so that it describes the most stable excited state of an N_2 molecule. Is the N—N bond in this excited-state N_2 molecule stronger or weaker than the N—N bond in ground-state nitrogen? Explain your answer. **LO7**

5.63 Dilithium molecules can be generated by vaporising lithium metal at very low pressure. Do you think it is possible to prepare diberyllium? Explain your reasoning using MO diagrams for Li_2 and Be_2. **LO7**

5.64 Consider the bond lengths of the following diatomic molecules: N_2, 110 pm; O_2, 121 pm; F_2, 143 pm. Explain the variation in length in terms of the molecular orbital descriptions of these molecules. **LO7**

6 Gases

If you take a look over all the elements on the periodic table you will find that only a few occur naturally as gases. This is mirrored in the fact that overall not many compounds occur as gases under ambient conditions. However, we should not underestimate the importance of gases, since we spend almost all of our time immersed in them and depend on them for our survival.

The Earth's atmosphere is made up mainly of nitrogen, N_2, (≈78%) and oxygen, O_2, (≈21%) with some other species in trace amounts. The exact composition and temperature of this gas mixture is responsible for our weather. The photo on this page shows hot-air balloons during the annual balloon festival over Canberra's Lake Burley Griffin and reminds us that hot-air balloonists exploit the fact that hot air rises.

We begin this chapter with a discussion of the variables that characterise gases. Then we develop a molecular description that explains gas behaviour. Next, we explore additional gas properties and show how to carry out stoichiometric calculations for reactions involving gas-phase species. We then explore the kinds of attractions that exist between molecules and how these forces affect the physical properties of gases.

LEARNING OBJECTIVES

After studying this chapter, you should be able to:

6.1 list the three common states of matter

6.2 describe qualitatively the macroscopic characteristics of gases

6.3 outline and apply the molecular description of gases

6.4 describe the behaviour of gas mixtures

6.5 apply the ideal gas equation to pure gases and mixtures

6.6 complete stoichiometric calculations involving gases

6.7 extend the description of gases to those that deviate from ideal behaviour

6.8 identify and explain intermolecular forces.

6.1 The states of matter

It is remarkable that essentially all of the matter in the universe exists in only three states: solid, liquid or gas. A fourth state of matter, plasma, exists only under extreme conditions such as those found in the interiors of stars, while the Bose–Einstein condensate, a recently discovered state of matter, is found only at temperatures approaching absolute zero. We will also encounter supercritical fluids, which have properties of both liquids and gases, in chapter 7.

It is also interesting that most substances can exist in any of these three states, and that we can change the state that a particular substance is in by varying the temperature and/or pressure; for example, cooling liquid water to below 0 °C at ambient pressure will produce solid water (ice).

Why, at 25 °C and atmospheric pressure, does water exist as a liquid, while oxygen is a gas and gold is a solid? The answer lies in the forces that occur between individual atoms, molecules or ions in a substance. Solids are held together by relatively strong forces between their components, while the corresponding forces in gases are relatively weak. This chapter and chapter 7 focus on the different possible states of matter. In this chapter, we concentrate on the gaseous state of matter, and investigate the different types of forces that can occur between atoms, ions and molecules. Chapter 7 is primarily concerned with liquids and solids (the so-called condensed phases), and also looks at the processes involved when substances change from one state to another.

6.2 Describing gases

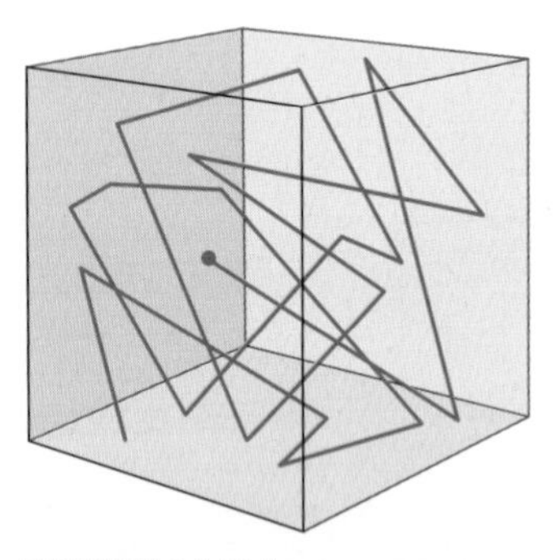

FIGURE 6.1 Molecules or atoms of a gas move freely throughout the entire volume of a container, changing direction whenever they collide with other molecules or atoms or with a wall. The line traces a possible path of a single molecule or atom.

Gases are characterised by the fact that, in contrast to liquids and solids, they expand to occupy all the space of their container. For example, gaseous water (water vapour) will be found in all parts of any flask it is in, but, if the temperature is lowered, the resulting liquid water occupies only the bottom of the flask. Lowering the temperature further to below the freezing point produces ice, which also occupies only the bottom of the flask. The fact that gases fill all available space implies that the individual gas atoms or molecules are free to move anywhere within their container (figure 6.1), and therefore the forces between them must be very weak. One of the defining characteristics of gases is the pressure they exert. This results from the rapid motion of individual gas atoms or molecules and their collisions with the walls of the container. The pressure (p) exerted by a gas is dependent on the amount of gas present (n), the volume in which it is contained (V) and its temperature (T). On the following pages, we derive a relationship relating all of these variables.

Pressure (p)

Any object that strikes a surface exerts a force against that surface. In the same way that a ball exerts a force against the ground when it strikes it, the gas molecules inside the ball also exert pressure through collisions with the walls of the ball. At any temperature above absolute zero, atoms and molecules are always in motion. At the molecular level, atoms and molecules exert forces through never-ending collisions with each other and with the walls of their container (e.g. the inside surface of the ball). The collective result of these collisions is what we call **pressure**. We experience pressure as a macroscopic property.

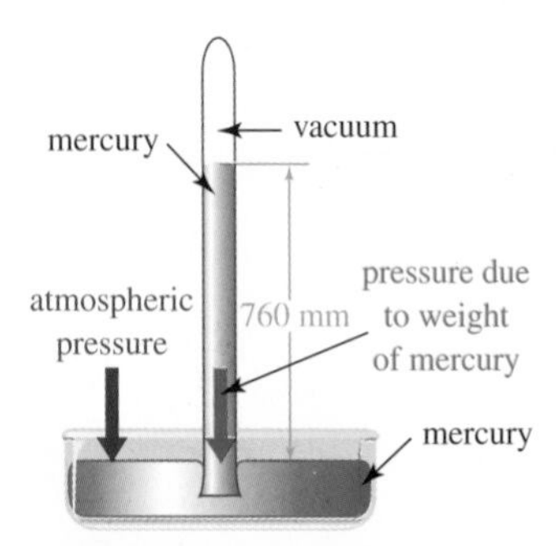

FIGURE 6.2 A mercury barometer. The pressure of the atmosphere on the surface of the mercury balances the pressure of the column of mercury.

We can get a feel for the macroscopic characteristics of gas pressure by examining the Earth's atmosphere. The atmosphere is a huge reservoir of gas that, due to Earth's gravity, exerts pressure on the Earth's surface. The pressure of the atmosphere can be measured with an instrument called a **barometer**. Figure 6.2 shows a schematic view of a simple mercury barometer. A long glass tube, closed at one end, is filled with liquid mercury. The filled tube is inverted carefully into a dish that is partially filled with more mercury. The force of gravity pulls downwards on the mercury in the tube. With no opposing force, all the mercury would run out of the tube and mix with the mercury in the dish. The mercury does fall, but the flow stops at a certain height. The column of mercury stops falling because the atmosphere exerts pressure on the mercury in the dish, pushing the column up the tube. When the height of the mercury column generates a downward force on the inside of the tube that exactly balances the force exerted by the atmosphere on the outside of the tube, the column of mercury will remain steady.

At sea level, atmospheric pressure supports a mercury column approximately 760 mm high. Changes in altitude and weather cause fluctuations in atmospheric pressure. At sea level the height of the mercury column seldom varies by more than 10 mm, except under extreme conditions, such as in the eye of a cyclone, when it may fall below 740 mm.

A **manometer** is similar to a barometer, but in a manometer gases exert pressure on both liquid surfaces. Consequently, a manometer measures the difference in pressures exerted by two gases. A simple manometer, shown in figure 6.3, is a U-shaped glass tube containing mercury. One side of the tube is exposed to the atmosphere and the other to a gas whose pressure we want to measure. In figure 6.3, the pressure exerted by the atmosphere (p_{atm}) is less than the pressure exerted by the gas in the bulb (p_{gas}). The difference in heights of mercury between the two sides of the manometer depends on the difference in the pressures exerted by the gas and the atmosphere (equal to p_{Hg}).

As we learned in chapter 2, the SI unit for pressure is the **pascal (Pa)**. Pressure is defined as force per unit area, so the pascal is expressed by combining the SI units for these two variables. The SI unit of force is the newton (N), and the unit of area is square metres (m^2). Thus, $1\ Pa = 1\ N\ m^{-2} = 1\ kg\ m^{-1}\ s^{-2} = 1\ J\ m^{-3}$ (with $1\ N = 1\ kg\ m\ s^{-2}$ and $1\ J = 1\ kg\ m^2\ s^{-2}$).

A number of non-SI units for pressure are also commonly used.

- One standard atmosphere (1 atm) is the pressure that will support a column of mercury 760 mm in height (i.e. average atmospheric pressure at sea level); $1\ atm = 1.01325 \times 10^5\ Pa$.
- One torr is the pressure exerted by a column of mercury 1 mm in height; 1 atm = 760 torr.
- $1\ bar = 1 \times 10^5\ Pa$.

In this textbook we will always express pressure in the SI unit: the pascal (Pa). Standard pressure, $p^{\ominus}$ is defined as is 1×10^5 Pa, and we will encounter that in chapter 8.

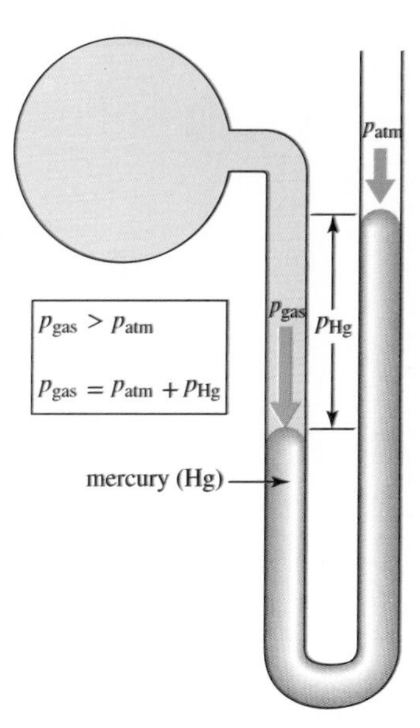

FIGURE 6.3 The difference in heights of liquid on the two sides of a manometer is a measure of the difference in gas pressures applied to the two sides.

The gas laws

The volume occupied by a gas changes in response to changes in pressure, temperature and amount of gas. The relationships among these variables are known as the gas laws.

Boyle's Law: the relationship between volume and pressure

Figure 6.4a illustrates experiments conducted at constant temperature by the English scientist Robert Boyle in about 1660. Boyle added liquid mercury to the open end of a J-shaped glass tube, trapping a fixed amount of air on the closed side. When Boyle added additional mercury to the open end of the tube, the pressure exerted by the weight of the additional mercury compressed the trapped gas into a smaller volume.

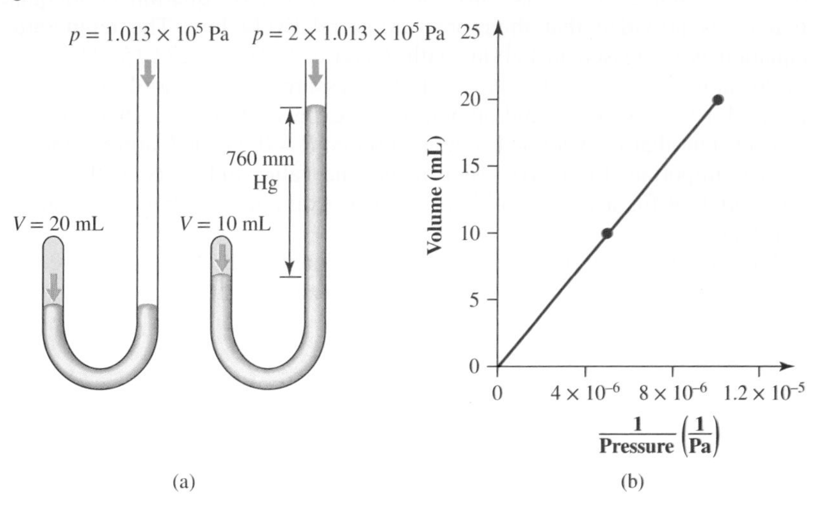

FIGURE 6.4 (a) Schematic illustration of Boyle's experiments on air trapped in J-shaped tubes. The pressure on the enclosed gas in the right-hand tube is twice that in the left-hand tube, and the volume of the enclosed gas is halved. **(b)** Graph showing the linear variation of V versus $\frac{1}{p}$.

For example, as shown in figure 6.4a, Boyle found that doubling the total pressure on the trapped air by the addition of mercury halved its volume. In more general terms, Boyle's experiments showed that the volume of the trapped air was inversely proportional to the total pressure applied by the mercury plus the atmosphere.

$$V_{gas} \propto \frac{1}{p_{gas}} \quad \text{(fixed temperature and amount of gas)}$$

or, alternatively:

$$p_{gas}V_{gas} = k \quad \text{(fixed temperature and amount of gas)}$$

where k is a constant. This is shown graphically in figure 6.4b.

Charles' Law: the relationship between volume and temperature

Boyle also observed that heating a gas causes it to expand in volume, but more than a century passed before Jacques-Alexandre-César Charles reported the first quantitative studies of gas volume as a function of temperature. Charles found that, for a fixed amount of gas, a graph of gas volume versus temperature gives a straight line, as shown in figure 6.5. In other words, the volume of a gas is directly proportional to its temperature.

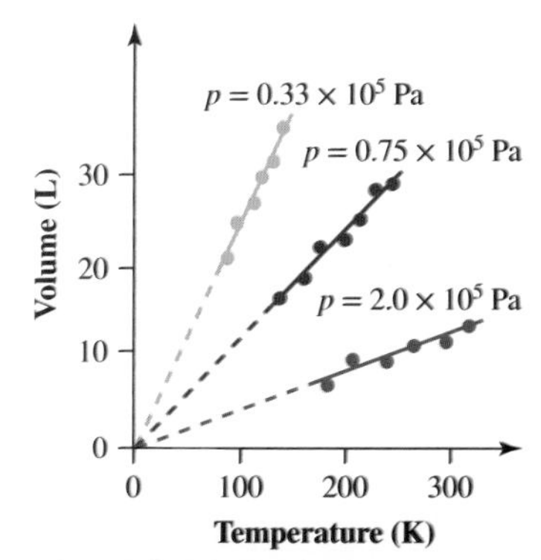

FIGURE 6.5 Plots of volume versus temperature for 1 mol of air at three constant pressures. The dots represent experimental data, with dashed lines of best fit extrapolated to 0 K.

$$V_{gas} \propto T_{gas} \quad \text{(fixed pressure and amount of gas)}$$

or, alternatively:

$$V_{gas} = k' \, T_{gas} \quad \text{(fixed pressure and amount of gas)}$$

where k′ is a constant.

Avogadro's Law: the relationship between volume and amount

The volume of a gas also changes when the amount of the gas changes. If we double the amount of a gas while keeping the temperature and pressure fixed, the gas volume doubles. In other words, gas volume is proportional to the amount of gas.

$$V_{gas} \propto n_{gas} \quad \text{(fixed pressure and temperature of gas)}$$

or, alternatively:

$$V_{gas} = k''n_{gas} \quad \text{(fixed pressure and temperature of gas)}$$

where k″ is a constant.

The ideal gas equation

The work of Boyle, Charles, Avogadro and other early scientists showed that the volume of any gas is proportional to the temperature and amount of the gas, and inversely proportional to the pressure of the gas. Combining these observations, we find that:

$$V \propto \frac{nT}{p}$$

This proportionality is converted into an equality by introducing the **gas constant**, represented by the symbol R. Introducing R and multiplying both sides of the equation by p gives:

$$pV = nRT$$

This is known as the **ideal gas equation**, because it describes the behaviour of an **ideal gas**. While no gas is truly ideal, the ideal gas equation is adequate to describe most real gases provided that their pressure is relatively low. The temperature in the ideal gas equation is expressed in kelvin, with T (K) = T (°C) + 273.15. The gas constant is defined in SI units as R = 8.314 J mol^{-1} K^{-1}. Pressure is expressed in Pa, volume is expressed in m^3 (1 m^3 = 1 × 10^3 L) and amount (n) is expressed in mol. There is a second version of the gas constant that is expressed in non-SI units: R = 0.082 06 L atm K^{-1} mol^{-1}. This is still in use, so it is important to always confirm that the value and units of the gas constant are chosen appropriately for a given problem. Worked example 6.1 shows the calculations for both gas constants.

It follows naturally from the ideal gas equation that all ideal gases have the same molar volume under a given set of standard conditions. For example, at 0 °C and 1 × 10^5 Pa, 1 mol of an ideal gas occupies 22.71 L.

WORKED EXAMPLE 6.1

Calculation of gas pressure

A 1.000 × 10^3 L steel storage tank contains 88.5 kg of methane, CH_4. If the temperature is 25 °C, what is the pressure inside the tank?

Analysis

The problem asks for the pressure exerted by a gas. The gas is stored in a steel tank, so the volume of the gas cannot change. No chemical changes are described in the problem. We can therefore solve this problem using the ideal gas equation.

Solution

We are told that:

$$V = 1.000 \times 10^3\,\text{L} = 1.000\,\text{m}^3 \qquad T = 25\,°\text{C} \qquad m = 88.5\,\text{kg of methane}$$

We can calculate the pressure of the gas using the ideal gas equation, but we need to make sure all the variables are expressed in appropriate units. Temperature must be in kelvin, amount of methane in mol and volume in m^3.

We rearrange the ideal gas equation by dividing both sides by V, thus making pressure the subject of the equation.

$$pV = nRT$$

$$p = \frac{nRT}{V}$$

Add 273.15 to the temperature in °C to convert to kelvin.

$$T = 25 + 273.15 = 298\,\text{K}$$

Note that, when we add numbers, the result has the same number of decimal places as the number with the fewest decimal places (zero in this example).

Use the molar mass of methane to calculate the amount of gas.

$$n_{CH_4} = \frac{88.5 \times 10^3\,g}{16.04\,g\,mol^{-1}} = 5.52 \times 10^3\,mol$$

Now substitute this into the rearranged ideal gas equation and calculate the pressure.

$$p = \frac{(5.52 \times 10^3\,mol)(8.314\,J\,mol^{-1}\,K^{-1})(298\,K)}{1.000\,m^3} = 1.37 \times 10^7\,Pa$$

If we use the non-SI version of the gas constant, 0.082 06 L atm K^{-1} mol^{-1}, we need to express the volume in litres (L).

$$p = \frac{(5.52 \times 10^3\,mol)(0.082\,06\,L\,atm\,K^{-1}\,mol^{-1})(298\,K)}{1000\,L} = 135\,atm$$

Is our answer reasonable?

It is important to check that the units cancel properly, giving a result in pressure units (recall from p. 221 that 1 Pa = 1 J m^{-3}). The problem describes a large amount of methane in a relatively small volume, so a high pressure (over 100 times atmospheric pressure) is reasonable. This high value indicates why gases must be stored in tanks made of materials such as steel that can withstand high pressures.

PRACTICE EXERCISE 6.1

Determine the new pressure if the tank in worked example 6.1 is stored in a hot shed where the temperature reaches 42 °C.

During chemical and physical transformations, any of the four variables in the ideal gas equation (p, V, n, T) may change, and any of them may remain constant. The ideal gas equation can be rearranged as:

$$R = \frac{pV}{nT}$$

Then, for different initial and final conditions, we have:

$$\frac{p_i V_i}{n_i T_i} = \frac{p_f V_f}{n_f T_f}$$

WORKED EXAMPLE 6.2

Pressure–volume variations

A sample of helium gas is held at constant temperature inside a cylinder with a volume of 0.80 L when a piston exerts a pressure of 1.5 × 10^5 Pa. If the external pressure on the piston is increased to 2.1 × 10^5 Pa, what will the new volume be?

Analysis

We can visualise the conditions by drawing a schematic diagram of the initial and final conditions (p_i = initial pressure, p_f = final pressure, V_i = initial volume, V_f = final volume).

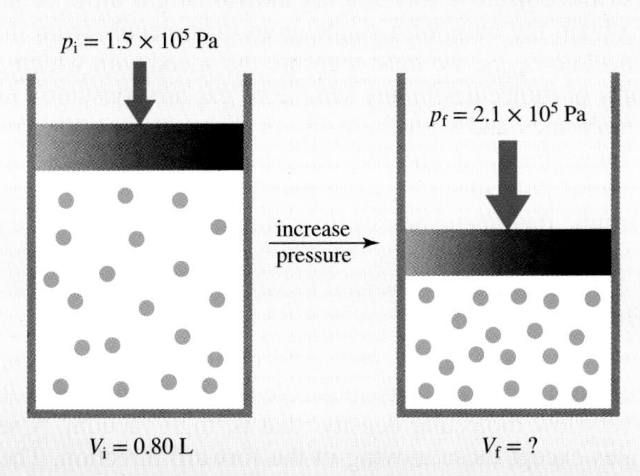

Solution

Gas behaviour is involved, so the equation that applies is the ideal gas equation, $pV = nRT$. Rearranging the gas equation to solve for V will not help because we do not know n, the amount of He present, or T, the temperature of the gas. We do know, however, that n and T remain unchanged as the pressure increases.

To determine the final volume of the helium gas, apply the ideal gas equation to the initial and final conditions (shown with subscripts i and f, respectively).

$$p_iV_i = n_iRT_i \qquad \text{and} \qquad p_fV_f = n_fRT_f$$

In this problem, the quantity of He inside the cylinder and the temperature of the gas are constant.

$$n_i = n_f \qquad \text{and} \qquad T_i = T_f \qquad \text{so} \qquad n_iRT_i = n_fRT_f$$

Therefore:

$$p_iV_i = p_fV_f \qquad \text{(constant } n \text{ and } T\text{)}$$

Notice that we can solve this for V_f without knowing the values of n and T.

Now we rearrange the above equation, substitute the appropriate values and calculate the final volume.

$$V_f = \frac{p_iV_i}{p_f} = \frac{(1.5\times10^5\ \text{Pa})(0.80\times10^{-3}\ \text{m}^3)}{2.1\times10^5\ \text{Pa}} = 0.57\times10^{-3}\ \text{m}^3$$
$$= 0.57\ \text{L}$$

Is our answer reasonable?

This answer is reasonable because an external pressure increase has caused a volume decrease.

PRACTICE EXERCISE 6.2 The piston in worked example 6.2 is withdrawn until the gas volume is 2.55 L. Calculate the final pressure of the gas.

In worked example 6.2, the quantities on the right-hand side of the ideal gas equation ($pV = nRT$) are fixed, whereas the quantities on the left are changing. A useful strategy for organising gas calculations is to rearrange the gas equation to place all variables that do not change on one side.

6.3 Molecular view of gases

To understand why all gases can be described by a single equation, we need to explore how gases behave at the molecular level. As gases are characterised by rapid motion of their constituent atoms or molecules, the most important energy component to consider is their kinetic energy, E_{kinetic}. The kinetic energy of an object is given by the equation:

$$E_{\text{kinetic}} = \tfrac{1}{2}mu^2$$

where m is the mass of the object (in this case an individual gas atom or molecule) and u is its speed. We can easily obtain the mass of a single atom or molecule from the molar mass of the gas, but in order to calculate E_{kinetic} we must measure the speed with which an atom or molecule is moving. As the atoms or molecules in any sample of gas are constantly undergoing collisions with other atoms or molecules and with the walls of the container, they might not necessarily all be moving at the same speed. On the following pages, we describe a method of measuring the speeds of gas atoms or molecules to give an overall distribution of speeds for a gas sample. This allows us to determine the kinetic energy distribution of the atoms or molecules in a sample of gas.

Molecular speeds

The speeds of molecules in a gas can be measured using a molecular beam apparatus, which is shown schematically in figure 6.6a. Gas molecules in an oven escape through a small hole into a chamber with very low molecular density: that is, high vacuum. A set of slits blocks the passage of all molecules except those moving in the forward direction. The result is a beam of molecules, all moving in the same direction. A rotating disc blocks the beam path except for a small slit that allows a packet of molecules to pass through. Each molecule moves down the beam axis at its own speed undergoing minimal collisions with other molecules in the packet. The faster a molecule moves, the less time it takes to travel the length of the chamber. A detector at the end of the chamber measures the number of molecules arriving as a function of time, giving a profile of speeds.

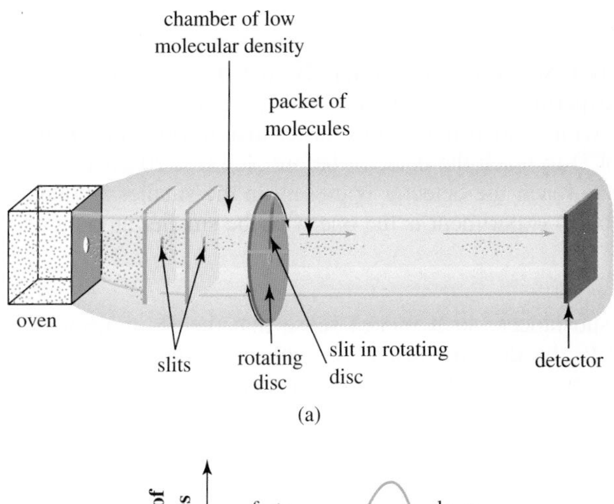

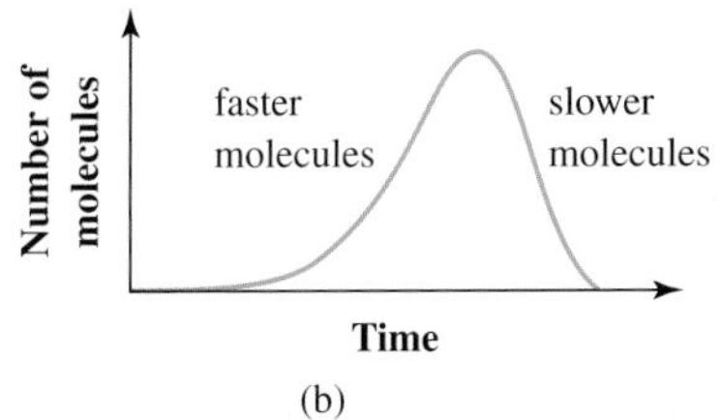

FIGURE 6.6 (a) Schematic diagram of a molecular beam apparatus designed to measure the speeds of gas molecules. **(b)** Distribution of molecules observed by the detector as a function of the time after a packet of molecules reaches the slit in the disc.

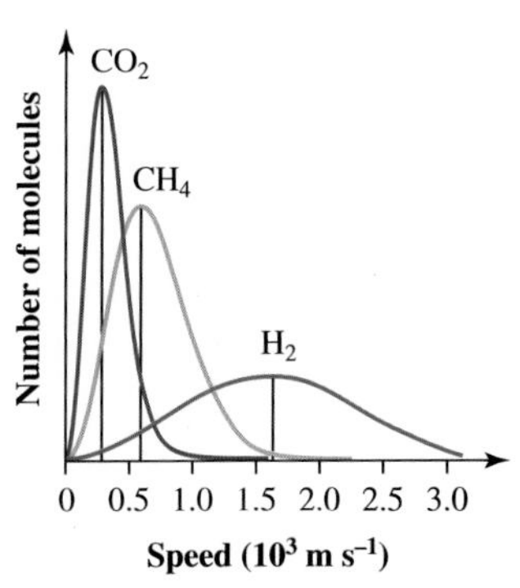

FIGURE 6.7 Graph showing molecular speed distributions for CO_2, CH_4 and H_2 at a temperature of 300 K.

When the speed profile of a gas is measured in this way, the results give a distribution like the one shown in figure 6.6b. If all the molecules travelled at the same speed, they would all reach the detector at the same time, in a single clump. Instead, faster molecules move ahead of the main packet, and slower molecules fall behind. This experiment shows that molecules in a gas have a distribution of speeds.

A pattern emerges when this molecular beam experiment is repeated for various gases at a common temperature; molecules with small masses move, on average, faster than those with large masses. Figure 6.7 shows this for H_2, CH_4 and CO_2. Of these molecules, H_2 has the smallest mass and CO_2 the largest. The vertical line drawn for each gas shows the speed at which there is the largest number of molecules. More molecules have this speed than any other, so this is the most probable speed for molecules of that gas. The most probable speed for a molecule of hydrogen at 300 K is 1.57×10^3 m s^{-1} or 5.65×10^3 km h^{-1}.

WORKED EXAMPLE 6.3

A molecular beam experiment

The figures below represent mixtures of neon atoms and hydrogen molecules.

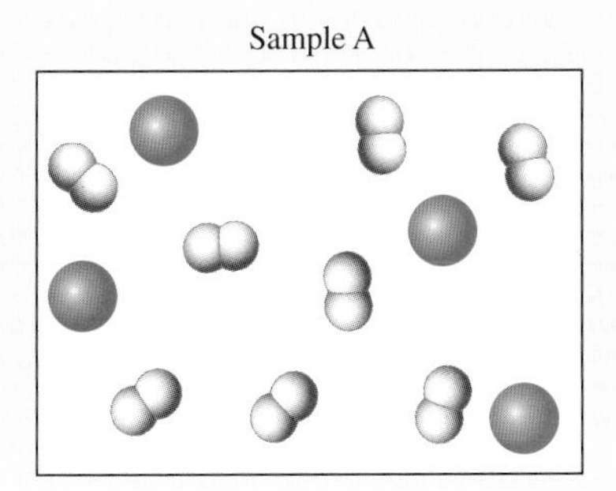

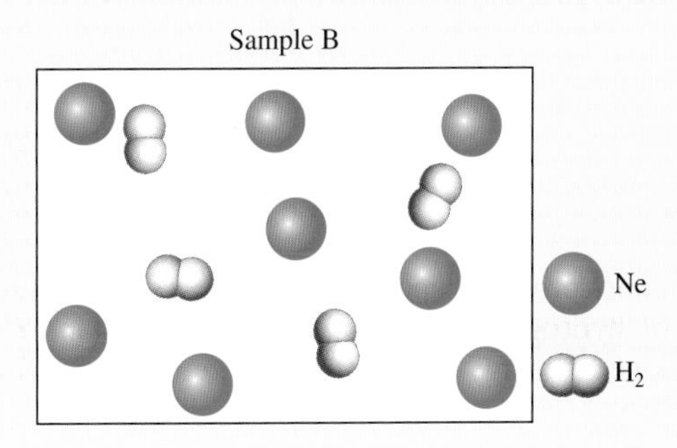

One of the gas mixtures was used in a molecular beam experiment. The result of the experiment is shown below. Which of the two gas samples, A or B, was used for this experiment?

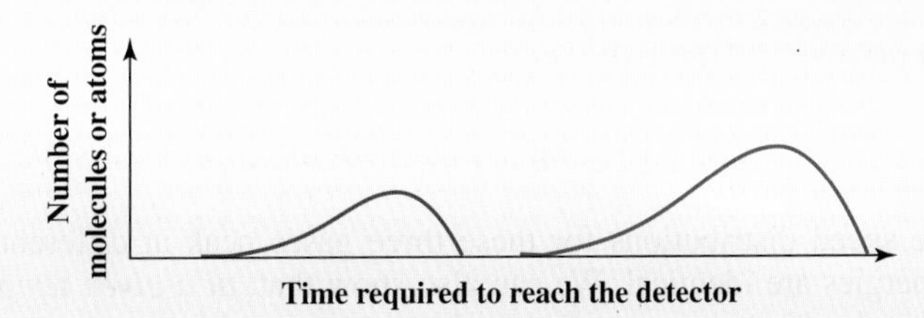

Analysis

Examine the two samples, and then decide how each of them would behave in a beam experiment.

Solution

Sample A contains 8 H_2 molecules and 4 Ne atoms, and sample B contains 4 H_2 molecules and 8 Ne atoms. In a molecular beam experiment, both samples would give two peaks in relative areas of 2 : 1. Atoms or molecules with a small mass move faster than those with a large mass, so we expect H_2 ($M = 2.02\,g\,mol^{-1}$) to reach the detector before Ne ($M = 20.2\,g\,mol^{-1}$). The data show that the first substance to reach the detector is present in the smaller amount. Consequently, the sample used in the beam experiment is the one with the smaller amount of hydrogen molecules: sample B.

Is our answer reasonable?

H_2 moves faster than other, heavier substances, so it makes sense that the H_2 molecules, which are lighter than Ne atoms, get to the detector faster. H_2 molecules are present in a smaller amount than Ne atoms, and the experimental data are consistent with that.

PRACTICE EXERCISE 6.3

Suppose that, in worked example 6.3, sample A is used and that the second gas is helium instead of neon. Sketch a graph, similar to the graph shown in worked example 6.3, showing the number of molecules/atoms versus time.

Speed and energy

The energy of a molecule is related to its speed. As we learned on page 224, any moving object has kinetic energy ($E_{kinetic}$) with a magnitude given by:

$$E_{kinetic} = \frac{1}{2}mu^2$$

where m is the object's mass and u is its speed. The most probable speed of hydrogen molecules at 300 K is $1.57 \times 10^3\,m\,s^{-1}$ (see p. 225). We need the mass of one H_2 molecule to be able to calculate its most probable kinetic energy. The SI unit of energy is the joule (J), which equals $1\,kg\,m^2\,s^{-2}$. Thus, we need the mass (m) in kilograms. The molar mass gives the mass of 1 mole of molecules, so dividing molar mass by the Avogadro constant (N_A) gives the mass of 1 molecule.

$$m = \frac{M}{N_A} = \frac{2.016\,g\,mol^{-1}}{6.022 \times 10^{23}\,mol^{-1}}$$

$$m = 3.348 \times 10^{-24}\,g$$

$$= 3.348 \times 10^{-27}\,kg$$

Now apply:

$$E_{kinetic} = \frac{1}{2}mu^2$$

$$E_{kinetic}\,(\text{most probable}) = \frac{1}{2}(3.348 \times 10^{-27}\,kg)(1.57 \times 10^3\,m\,s^{-1})^2$$

$$= 4.13 \times 10^{-21}\,kg\,m^2\,s^{-2}$$

$$E_{kinetic}\,(\text{most probable}) = 4.13 \times 10^{-21}\,J$$

The most probable speeds of methane and carbon dioxide are slower than the most probable speed of hydrogen (see figure 6.7), but CH_4 and CO_2 molecules have larger masses than H_2. When kinetic energy calculations are repeated for these gases, they show that the most probable kinetic energy is the same for all three gases.

For a CH_4 molecule ($m = 2.664 \times 10^{-26}\,kg$):

$$E_{kinetic} = \frac{1}{2} \times 2.664 \times 10^{-26}\,kg \times (5.57 \times 10^2\,m\,s^{-1})^2 = 4.13 \times 10^{-21}\,J$$

For a CO_2 molecule ($m = 7.308 \times 10^{-26}\,kg$):

$$E_{kinetic} = \frac{1}{2} \times 7.308 \times 10^{-26}\,kg \times (3.36 \times 10^2\,m\,s^{-1})^2 = 4.13 \times 10^{-21}\,J$$

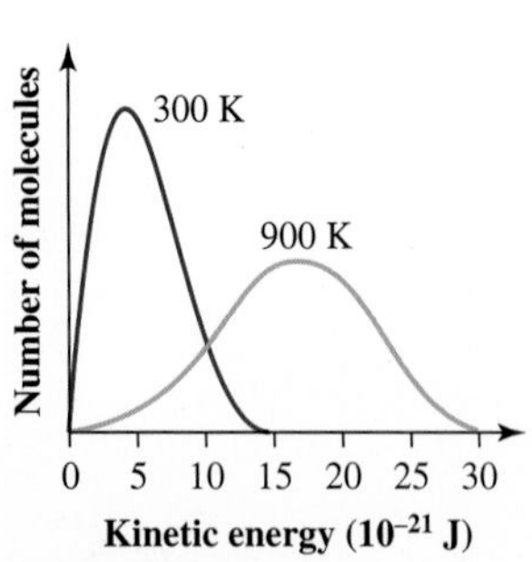

FIGURE 6.8 Graph of distribution of molecular energies for any gas at 300 K and 900 K.

Even though the speed distributions for these three gases peak at different values, the most probable kinetic energies are identical. We can also show that, *at a given temperature, all gases have the same molecular kinetic energy distribution.*

Molecular beam experiments show that molecules move faster as the temperature increases. Molecules escaping from the oven at 900 K take less time to reach the detector than molecules escaping at 300 K. As the molecular speed increases, so does the kinetic energy. Figure 6.8 shows the molecular kinetic energy distributions at 300 K and 900 K. By comparing this figure with figure 6.7, we see that the molecular energies show a wide distribution that is similar to the distribution in molecular speeds. Unlike speeds, however, the molecular energy distributions are the same for all gases at any particular temperature. The distributions shown in figure 6.8 apply to any gas.

Average kinetic energy and temperature

The most probable kinetic energy is not the same as the average kinetic energy. To find the average kinetic energy per molecule ($\bar{E}_{kinetic}$; the bar over a variable indicates that it is an average value), we must add all the individual molecular energies and divide by the total number of molecules. The result describes how the average kinetic energy of gas molecules is related to the temperature of the gas.

$$\bar{E}_{kinetic} = \frac{3RT}{2N_A}$$

In this equation, T is the temperature in kelvin, N_A is the Avogadro constant and R is the gas constant (8.314 J mol^{-1} K^{-1}).

The average kinetic energy expressed by the equation is kinetic energy per molecule. We find the total kinetic energy of 1 mol ($E_{kinetic,\ molar}$) of gas molecules by multiplying by the Avogadro constant.

$$E_{kinetic,\ molar} = N_A \frac{3RT}{2N_A} = \frac{3}{2}RT$$

Thus, 1 mol of a gas has a total molecular kinetic energy of $\frac{3}{2}RT$ independent of the nature of that gas. While the equation is usually stated in this way, it is worthwhile to note that it is the temperature that is a measure of average kinetic energy.

WORKED EXAMPLE 6.4

Molecular kinetic energies

Determine the average molecular kinetic energy and molar kinetic energy of gaseous sulfur hexafluoride, SF_6, at 150 °C.

Analysis

The problem asks for a calculation of kinetic energies and provides a temperature. We have two equations for the kinetic energy of a gas.

$$\bar{E}_{kinetic} = \frac{3RT}{2N_A} \quad \text{and} \quad E_{kinetic,\ molar} = \frac{3}{2}RT$$

To carry out the calculations, we need to know the constants R (8.314 J mol^{-1} K^{-1}) and N_A (6.022×10^{23} mol^{-1}) as well as T in kelvin.

Solution

Convert the temperature from °C to K by adding 273.15.

$$T = 150\ °C + 273.15 = 423\ K$$

Now substitute this into the kinetic energy equation and calculate.

For a single molecule:

$$\bar{E}_{kinetic} = \frac{3(8.314\ J\ mol^{-1}\ K^{-1})(423\ K)}{2(6.022 \times 10^{23}\ mol^{-1})} = 8.76 \times 10^{-21}\ J$$

For 1 mol of molecules:

$$E_{kinetic,\ molar} = \frac{3}{2}(8.314\ J\ mol^{-1}\ K^{-1})(423\ K) = 5.28 \times 10^{3}\ J\ mol^{-1}$$

Is our answer reasonable?

We have shown that the average kinetic energy is independent of the nature of the gas. Figure 6.8 shows the molecular energy distributions for a temperature of 300 K and 900 K. The temperature of 423 K is in between these and the molecular energy that we calculated is also between the values at 300 K and 900 K. This answer, therefore, seems reasonable.

PRACTICE EXERCISE 6.4

The atmospheric temperature at the altitudes where commercial jets fly is around −35 °C. Calculate the average molecular and molar kinetic energies of N_2 at this temperature.

Rates of gas movement

The average speed of the molecules in a gas determines its temperature. To state this dependence quantitatively, we start with:

$$\overline{E}_{\text{kinetic}} = \tfrac{1}{2}m\overline{u}^2$$

and:

$$\overline{E}_{\text{kinetic}} = \frac{3RT}{2N_A}$$

We set the two expressions for kinetic energy equal to each other.

$$\tfrac{1}{2}m\overline{u}^2 = \frac{3RT}{2N_A}$$

Now solve this equality for u, noting that the product mN_A is equal to the molar mass of the gas in kg mol^{-1}.

$$\overline{u}^2 = \frac{3RT}{mN_A} = \frac{3RT}{M} \quad \text{so} \quad \overline{u}_{\text{rms}} = \left(\frac{3RT}{M}\right)^{\frac{1}{2}}$$

This speed is called the **root-mean-square speed** because it is found by taking the square root of $\overline{u}^2$. The average speed of gas molecules is directly proportional to the square root of the temperature and is inversely proportional to the square root of the molar mass of the gas in kg mol^{-1}. This equation can be applied directly to the movement of molecules escaping from a container into a vacuum. This process is one example of **effusion** (more generally, effusion is the flow of molecules through an opening without collisions between molecules). Effusion is exemplified by the escape of molecules from the oven shown in figure 6.6a.

A second type of gas movement is **diffusion**, the movement of one type of molecule through molecules of the same or another type. Diffusion is exemplified by air escaping from a punctured tyre. The escaping molecules must diffuse among the molecules already present in the atmosphere. Diffusing molecules undergo frequent collisions, so their paths are similar to that shown in figure 6.1. Nevertheless, their average rate of movement depends on temperature and molar mass. Figure 6.9 shows a molecular-level comparison of effusion and diffusion.

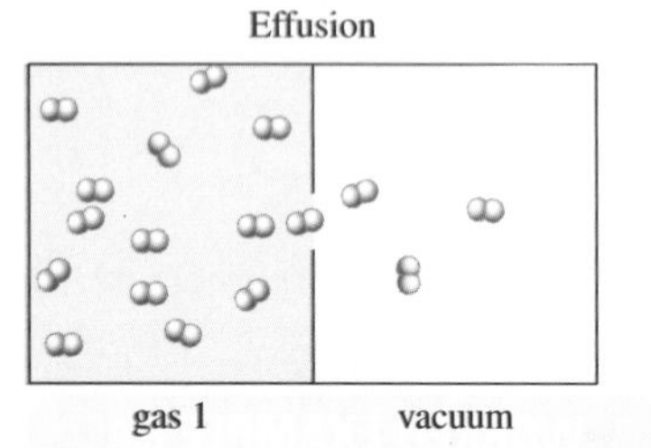

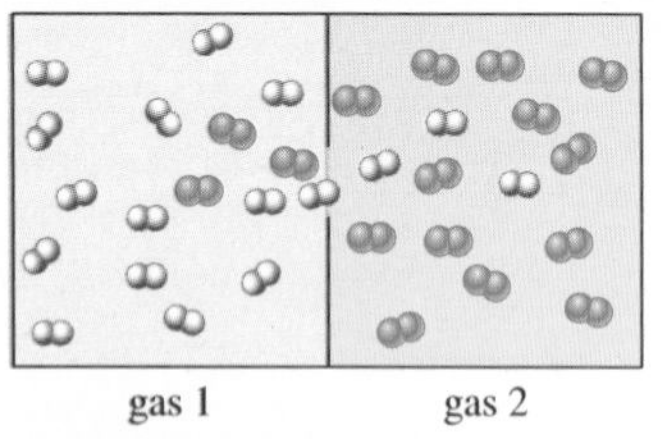

FIGURE 6.9 Effusion is the movement of a gas into a vacuum. Diffusion is the mixing of two or more gases.

An example of diffusion is shown in figure 6.10, in which concentrated aqueous solutions of hydrochloric acid, HCl, and ammonia, NH_3, are placed at opposite ends of a glass tube. Molecules of HCl gas and NH_3 gas escape from the solutions and diffuse through the air in the tube. When the two gases meet, they undergo a reaction that produces ammonium chloride, a white solid salt.

$$HCl(g) + NH_3(g) \longrightarrow NH_4Cl(s)$$

The lighter NH_3 molecules diffuse more rapidly than the heavier HCl molecules, so the white band of salt forms closer to the HCl end of the tube, as can be seen in figure 6.10.

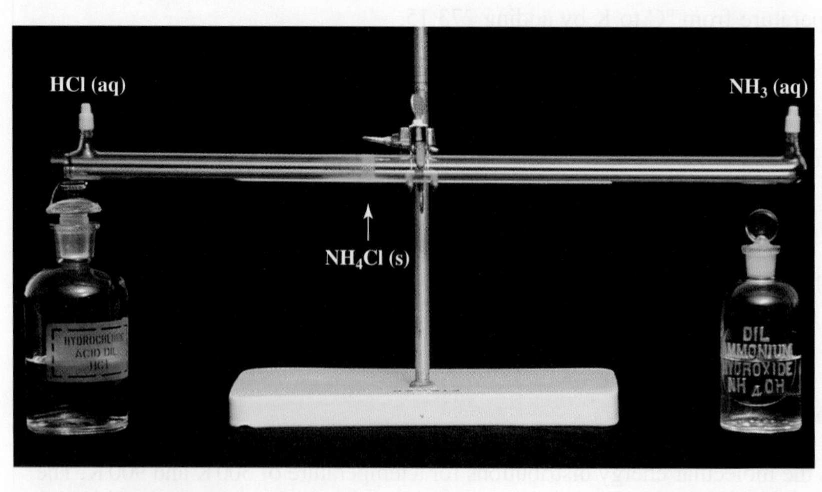

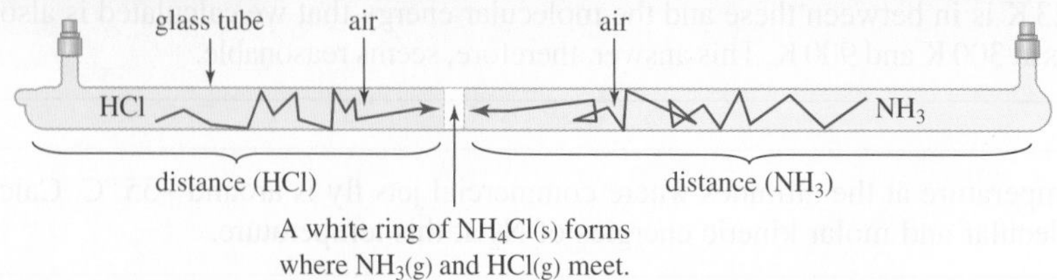

FIGURE 6.10 NH_3 diffuses through a glass tube faster than HCl. When the two gases meet, they form solid ammonium chloride, $NH_4Cl(s)$, which appears as a white band closer to the end of the tube that contains HCl.

From the discussion of root-mean-square speed on the previous page, it follows that the ratio of distances that HCl and NH_3 travel, respectively, is inversely related to the square root of the molar masses:

$$\frac{d_{NH_3}}{d_{HCl}} = \sqrt{\frac{M_{HCl}}{M_{NH_3}}}$$

Rates of molecular motion are directly proportional to molecular speeds, so, for any gas, rates of effusion and diffusion increase with the square root of the temperature in kelvin. Also, at any particular temperature, ratios of effusion and diffusion are faster for molecules with small molar masses.

Ideal gases

As we have seen in this chapter, gases can consist of either atoms or molecules. In what follows, we will use the term 'molecules' exclusively for brevity's sake, but you must remember that everything said about molecules in this section applies equally to atoms.

The behaviour of gases suggests that gas molecules have little effect on one another. Gases are easy to compress, showing that there is lots of empty space between molecules. Gases also escape easily through any opening, indicating that their molecules are not strongly attracted to one another. An ideal gas is defined as a gas for which the ideal gas equation holds. This requires that both the volume of molecules and the forces between the molecules are so small that they have no effect on the behaviour of the gas. We can assume that this is the case when molecular sizes are negligible compared with the volume of the container, and the energies generated by forces between molecules are negligible compared with molecular kinetic energies.

As we describe later in this chapter, small molecules usually have small attractive forces with respect to their kinetic energies, or with respect to RT, often called the *thermal energy*. Table 6.1 lists substances that exist as gases under ambient conditions (note that a binary gas is one that contains molecules formed from two elements). These gases are mostly small molecules with molar masses less than 50 g mol^{-1}. There are some notable exceptions to this, however. Tungsten hexafluoride, WF_6 (M = 297.8 g mol^{-1}), and the element radon, Rn (M = 222 g mol^{-1}), are examples of gases having molar masses greater than 200 g mol^{-1}.

TABLE 6.1 Molar masses of some gaseous substances (at 298 K and 1.013 x 10^5 Pa).

Elemental gases			Binary gases		
Substance	**Formula**	**M (g mol^{-1})**	**Substance**	**Formula**	**M (g mol^{-1})**
hydrogen	H_2	2.016	methane	CH_4	16.04
helium	He	4.003	ammonia	NH_3	17.03
neon	Ne	20.18	carbon monoxide	CO	28.01
nitrogen	N_2	28.02	nitrogen oxide	NO	30.01
oxygen	O_2	32.00	ethane	C_2H_6	30.07
fluorine	F_2	38.00	hydrogen sulfide	H_2S	34.09
argon	Ar	39.95	hydrogen chloride	HCl	36.46
ozone	O_3	48.00	carbon dioxide	CO_2	44.01
chlorine	Cl_2	70.90	nitrogen dioxide	NO_2	46.01

How does an ideal gas behave? We can answer this question by considering, for example, how changes in V, T or n affect the pressure, p. Each time a molecule strikes a wall, it exerts a force on the wall. During each second, many collisions exert many such forces. Pressure is the sum of all these forces per unit area and unit time. In an ideal gas, each molecule is independent of all others. This independence means that the total pressure is the sum of the pressure created by each individual molecule.

To see how pressure depends on V, T or n, we consider the effect of changing one property while holding the other properties constant. We analyse what happens to the molecular collisions as each property changes.

First, consider increasing the amount of gas while keeping the temperature and volume fixed. Doubling the amount of gas in a fixed volume (figure 6.11, overleaf) doubles the number of collisions with the walls. Thus, as pressure is essentially a measure of the number of collisions of the molecules with the walls of the container, pressure is directly proportional to the amount of gas. This agrees with the ideal gas equation.

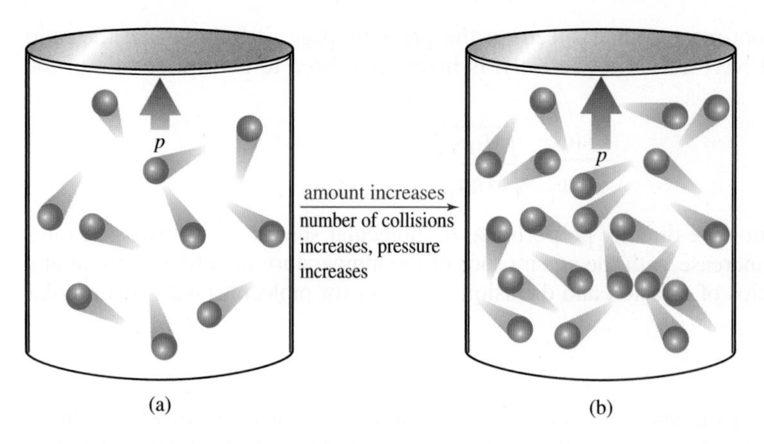

FIGURE 6.11 Schematic view of the effect of doubling the number of gas molecules in a fixed volume. The container in (b) has twice as many molecules as it does in (a). Consequently, the molecular density is twice as large in (b), with twice as many collisions per second with the walls.

Next, consider keeping the temperature and amount fixed and changing the volume of the gas. Figure 6.12 shows that compressing a gas into a smaller volume has the same effect as adding more molecules. The result is more collisions with the walls. If the molecules act independently, cutting the volume in half will double the pressure. In other words, pressure is inversely proportional to volume, again in agreement with the ideal gas equation.

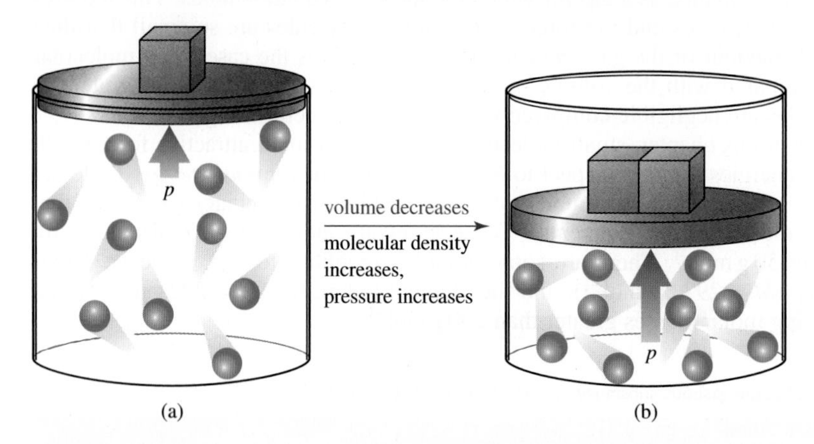

FIGURE 6.12 Schematic view of the effect of compressing a fixed quantity of gas into a smaller volume at constant *T*. Decreasing the volume increases the molecular density, which increases the number of collisions per second with the walls.

To complete our analysis, we must determine the effect of a change in temperature, keeping volume and amount fixed. Recall that kinetic energy is proportional to the square of the molecular speed (a microscopic variable), and temperature (a macroscopic variable) is a measure of average molecular kinetic energy. Thus, the square of the molecular speed is proportional to temperature. We can summarise these relationships by:

$$T \propto E_{\text{kinetic}} \quad \text{and} \quad E_{\text{kinetic}} = \tfrac{1}{2}mu^2 \quad \text{so} \quad T \propto u^2$$

Molecular speed affects pressure in two ways, which are illustrated in figure 6.13. First, faster moving molecules hit the walls more often than slower moving molecules. The number of collisions each molecule makes with the wall is proportional to the molecule's speed. Second, the force exerted when a molecule strikes the wall also depends on the molecule's speed. A fast-moving molecule exerts a larger force than the same molecule moving more slowly. Force per collision and number of collisions increase with speed, so the total effect of a single molecule on the pressure of a gas is proportional to the square of its speed.

Pressure is proportional to molecular speed squared, which in turn is proportional to temperature. For an ideal gas, then, the pressure is directly proportional to temperature, and a plot of p versus T yields a straight line. Again, this agrees with the ideal gas equation.

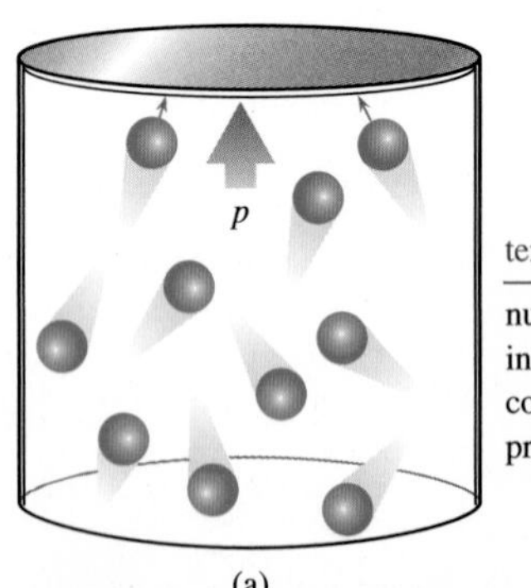

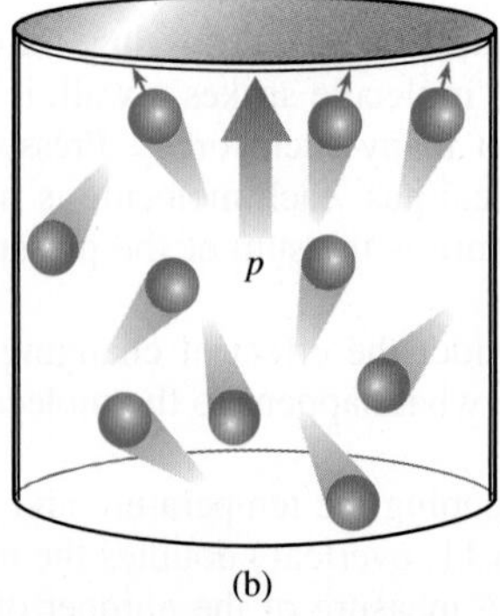

FIGURE 6.13 Schematic view of the effect of increasing molecular speeds (as measured by an increase in temperature), resulting in more wall collisions and more force per collision, and therefore an increase in pressure.

A gas will obey the ideal gas equation if its molecular sizes are negligible compared with the volume of the container, and the energies generated by forces between molecules are negligible compared with molecular kinetic energies. The behaviour of any real gas departs somewhat from ideality because real molecules occupy volume and exert forces on one another. Nevertheless, departures from ideality are small enough to neglect under many circumstances, and real gases approximate ideal behaviour under conditions of low pressure and high temperature. We consider departures from ideal gas behaviour later in this chapter.

6.4 Gas mixtures

The atmosphere, which contains nitrogen, oxygen and various trace gases, is one obvious example of a gas mixture. Another example is the gas used by deep-sea divers, which contains a mixture of helium and oxygen (figure 6.14). The ideal gas model provides guidance as to how to describe mixtures of gases.

All constituents of an ideal gas, whether atoms or molecules, act independently. This notion applies to gas mixtures as well as to pure gases. Gas behaviour depends on the *number* of gas atoms or molecules but not on their *identity*. The ideal gas equation applies to each gas in the mixture, as well as to the entire collection of atoms or molecules.

Suppose we introduce 0.1 mol of helium into an evacuated 20 L flask. If we add another 0.1 mol of He, the container now contains 0.2 mol of gas. The pressure can be calculated using the ideal gas equation, with n = 0.1 mol + 0.1 mol = 0.2 mol. Suppose that we add 0.1 mol of molecular oxygen. Now the container holds a total of 0.3 mol of gas. According to the ideal gas model, it does not matter whether we add the same gas or a different gas because all atoms or molecules in a sample of an ideal gas behave independently. The pressure increases in proportion to the increase in the total amount of gas. Thus, we can calculate the total pressure from the ideal gas equation, using n = 0.2 mol + 0.1 mol = 0.3 mol.

How does a 1 : 2 mixture of O_2 and He appear on the molecular level? As O_2 is added to the container, its molecules move throughout the volume and become distributed uniformly. Diffusion causes gas mixtures to become homogeneous. Figure 6.15 illustrates this.

FIGURE 6.14 Deep-sea divers use mixtures of helium and molecular oxygen in their breathing tanks.

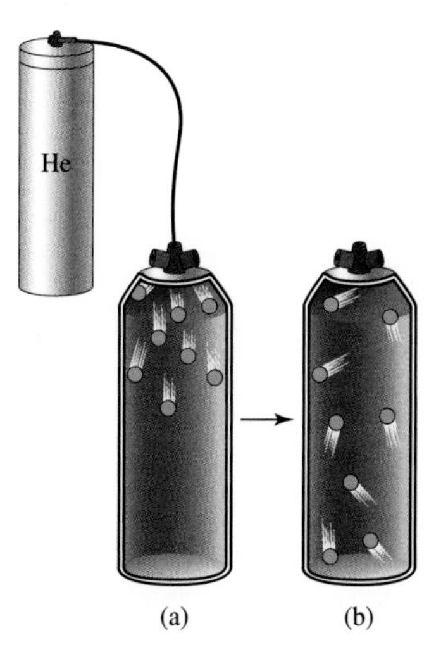

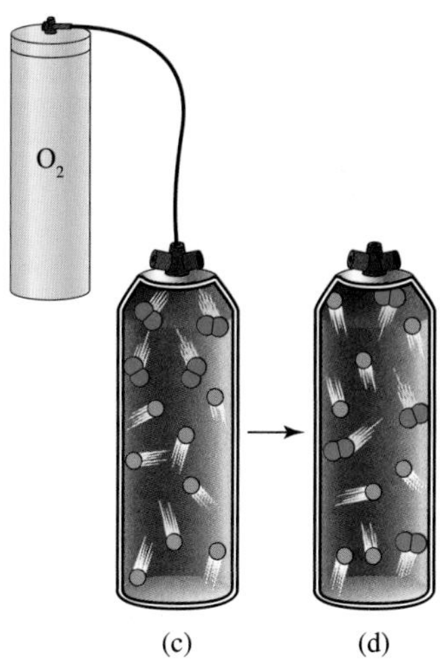

FIGURE 6.15 (a) 0.2 mol of He are added to a container. **(b)** The atoms quickly become distributed uniformly throughout the container. **(c)** 0.1 mol of O_2 are added to this container. **(d)** The molecules move about independently of the He atoms, causing the gases to mix uniformly.

Dalton's law of partial pressures

The pressure exerted by an ideal gas mixture is determined by the total amount (n_{total}) of gas.

$$p = \frac{n_{total}RT}{V}$$

We can express the total amount of gas as the sum of the amounts of the individual gases. For the He and O_2 mixture:

$$n_{total} = n_{He} + n_{O_2}$$

Substitution gives a two-term equation for the total pressure.

$$p = \frac{(n_{He} + n_{O_2})RT}{V}$$

$$= \frac{n_{He}RT}{V} + \frac{n_{O_2}RT}{V}$$

Notice that each term on the right resembles the ideal gas equation rearranged to express pressure. Each term therefore represents the **partial pressure** of one of the gases. As figure 6.16 illustrates, partial pressure is the pressure that would be present in a gas container if one gas were present by itself.

$$p_{He} = \frac{n_{He}RT}{V} \quad \text{and} \quad p_{O_2} = \frac{n_{O_2}RT}{V}$$

The total pressure in the container is the sum of the partial pressures.

$$p_{total} = p_{He} + p_{O_2}$$

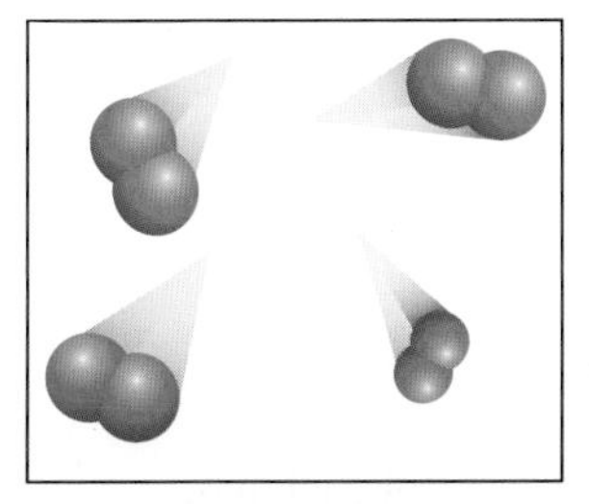

20.0 L at 273 K
0.100 mol O_2
$p_{O_2} = 1.13 \times 10^4$ Pa

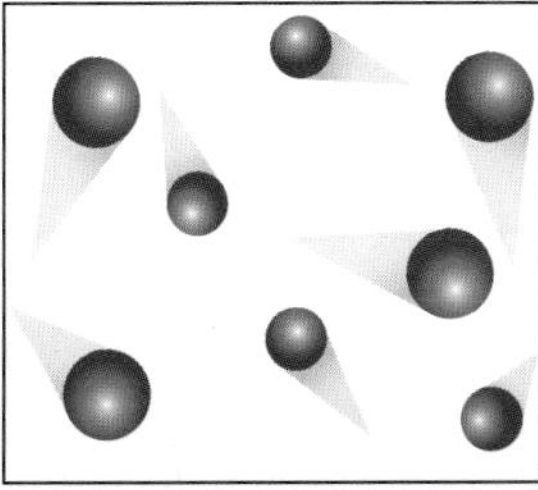

20.0 L at 273 K
0.200 mol He
$p_{He} = 2.27 \times 10^4$ Pa

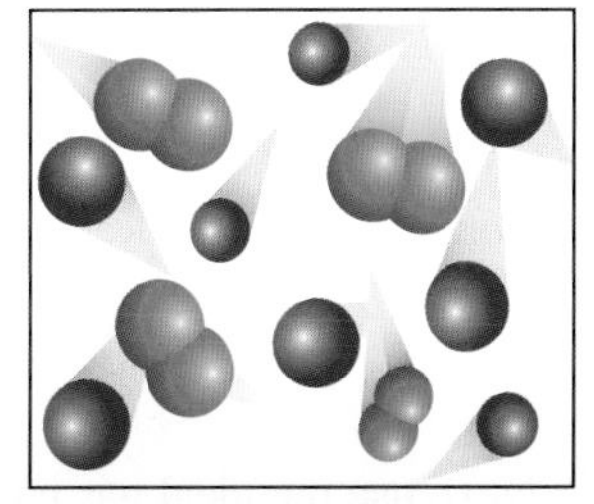

20.0 L at 273 K
0.300 mol gas mixture
$p_{O_2} = 1.13 \times 10^4$ Pa
$p_{He} = 2.27 \times 10^4$ Pa
$p_{total} = 3.40 \times 10^4$ Pa

FIGURE 6.16 Schematic representation of a sample of O_2, a sample of He and a mixture of the two gases. Both components are distributed uniformly throughout the gas volume. Each gas exerts the same pressure, whether it is pure or part of a mixture.

We have used He and O_2 to illustrate the behaviour of a mixture of ideal gases, but this relationship can be extended from a mixture of two ideal gases to mixtures with three or more substances behaving as ideal gases. *In a mixture of gases in which no chemical reaction occurs, each gas contributes to the total pressure the amount that it would exert if the gas were present in the container by itself.* This is **Dalton's law of partial pressures**. To obtain a total pressure, simply add the contributions from all gases present.

$$p_{total} = p_1 + p_2 + p_3 + \cdots + p_i$$

When doing calculations on a mixture of gases, we can apply the ideal gas equation to each component to find its partial pressure (p_i). Alternatively, we can treat the entire gas as a unit, using the total amount of gases present to determine the total pressure of the mixture (p).

Describing gas mixtures

There are several ways to describe the chemical composition of a mixture of gases. The simplest method is to list each component with its partial pressure or amount. Two other descriptions, mole fractions and parts per million, are also used frequently.

Chemists often express chemical composition in fractional terms, stating the amount of a substance as a fraction of the amount of all substances in the mixture. This way of stating composition is the **mole fraction (x)**.

$$\text{mole fraction of A} = x_A = \frac{n_A}{n_{total}}$$

Mole fractions provide a simple way to relate the partial pressure of one component to the total pressure of the gas mixture.

$$p_A = \frac{n_A RT}{V} \quad \text{and} \quad p_{total} = \frac{n_{total} RT}{V}$$

Dividing p_A by p_{total} gives:

$$\frac{p_A}{p_{total}} = \frac{\left(\frac{n_A RT}{V}\right)}{\left(\frac{n_{total} RT}{V}\right)} = \frac{n_A}{n_{total}} = x_A$$

or:

$$p_A = x_A p_{total}$$

The partial pressure of a component in a gas mixture is its mole fraction multiplied by the total pressure.

WORKED EXAMPLE 6.5

Gas mixtures

Exactly 8.00 g of O_2 and 2.00 g of He was placed in a 5.00 L tank at 298 K. Determine the total pressure of the mixture, and find the partial pressures and mole fractions of the two gases.

Analysis

We have a mixture of two gases in a container with known volume and temperature. The problem asks for pressures and mole fractions. Assuming that molecular interactions are negligible, each gas can be described independently by the ideal gas equation. As usual, we need molar amounts for the calculations.

Solution

Begin with the data provided.

$$V = 5.00\ \text{L} = 5.00 \times 10^{-3}\ \text{m}^3 \qquad T = 298\ \text{K} \qquad m_{He} = 2.00\ \text{g} \qquad m_{O_2} = 8.00\ \text{g}$$

Convert the mass of each gas into an amount in moles.

$$n = \frac{m}{M} \qquad n_{He} = 0.500\ \text{mol} \qquad n_{O_2} = 0.250\ \text{mol}$$

Next, use the ideal gas equation to calculate the partial pressure of each gas.

$$p_{He} = \frac{(0.500\ \text{mol})(8.314\ \text{J mol}^{-1}\ \text{K}^{-1})(298\ \text{K})}{5.00 \times 10^{-3}\ \text{m}^3} = 2.48 \times 10^5\ \text{Pa}$$

The same calculation for O_2 gives $p_{O_2} = 1.24 \times 10^5$ Pa.
The total pressure is the sum of the partial pressures.

$$p_{\text{total}} = p_{\text{He}} + p_{O_2} = (2.48 \times 10^5 \text{ Pa}) + (1.24 \times 10^5 \text{ Pa}) = 3.72 \times 10^5 \text{ Pa}$$

Mole fractions can be calculated from amounts or partial pressures.

$$x_{\text{He}} = \frac{n_{\text{He}}}{n_{\text{total}}} = \frac{0.500 \text{ mol}}{0.750 \text{ mol}} = 0.667 \quad \text{and} \quad x_{O_2} = \frac{p_{O_2}}{p_{\text{total}}} = \frac{1.24 \times 10^5 \text{ Pa}}{3.72 \times 10^5 \text{ Pa}} = 0.333$$

Is our answer reasonable?

The mole fractions calculated from the partial pressures must sum to 1.00. They do in this case, suggesting that our answer is reasonable.

PRACTICE EXERCISE 6.5

A 5.00 L tank is charged with 7.50 g of O_2 and 2.50 g of He. Calculate the mole fractions and the partial pressures of the gases at 25 °C.

When referring to lower concentrations in a gas mixture, scientists typically use **parts per million (ppm)** or **parts per billion (ppb)**. Mole fractions, parts per million and parts per billion can all describe ratios of amount of a particular substance to total amount of a sample. Mole fraction is moles per amounts, ppm is moles per million moles and ppb is moles per billion moles. A concentration of 1 ppm is a mole fraction of 1×10^{-6}, and 1 ppb is a mole fraction of 1×10^{-9}.

WORKED EXAMPLE 6.6

Working with parts per million (ppm)

The exhaust gas from an average car contains 206 ppm of the pollutant nitrogen oxide, NO. If a car emits 0.125 m^3 of exhaust gas at 1.00×10^5 Pa and exactly 350 K, what mass of NO has been added to the atmosphere?

Analysis

The question asks for the mass of NO. Information about concentration in ppm tells us the amount of NO present in 1 mole of exhaust gas. We can use the ideal gas equation to determine the total amount of gas emitted, then use the concentration information to find the amount of NO, and use the molar mass of NO to obtain the required mass.

Solution

$$[\text{NO}] = 206 \text{ ppm} \qquad V = 0.125 \text{ m}^3 \qquad p = 1.00 \times 10^5 \text{ Pa} \qquad T = 350 \text{ K}$$

$$n_{\text{gas}} = \frac{pV}{RT} = \frac{(1.00 \times 10^5 \text{ Pa})(0.125 \text{ m}^3)}{(8.314 \text{ J mol}^{-1} \text{ K}^{-1})(350 \text{ K})} = 4.30 \text{ mol}$$

$$x_{\text{NO}} = 206 \text{ ppm} = 206 \times 10^{-6} = 2.06 \times 10^{-4}$$

$$n_{\text{NO}} = (x_{\text{NO}})(n_{\text{gas}}) = (2.06 \times 10^{-4})(4.30 \text{ mol}) = 8.86 \times 10^{-4} \text{ mol}$$

$$m = nM = (8.86 \times 10^{-4} \text{ mol})(30.0 \text{ g mol}^{-1}) = 2.66 \times 10^{-2} \text{ g}$$

Is our answer reasonable?

The mass of NO is rather small, only 27 mg, but the mole fraction is also small, so this is a reasonable result.

PRACTICE EXERCISE 6.6

If the maximum NO emission allowed in worked example 6.6 is 762 ppm, what mass of NO is allowed per litre of exhaust emitted at a temperature of exactly 50 °C?

The description presented in this section applies to a gas mixture that is not undergoing chemical reactions. As long as reactions do not occur, the amount of each gas is determined by the amount of that substance initially present. When reactions occur, the amounts of reactants and products change as predicted by the principles of stoichiometry. Changes in composition must be taken into account before the properties of the gas mixture can be calculated. Gas stoichiometry is described in section 6.6.

6.5 Applications of the ideal gas equation

The ideal gas equation and the molecular view of gases lead to several useful applications. We have already described how to carry out calculations involving p–V–n–T relationships. In this section, we examine the use of the ideal gas equation to determine molar masses and gas density.

Determination of molar mass

The ideal gas equation can be combined with the equation $n = \frac{m}{M}$ (from chapter 3) to find the molar mass of an unknown gas.

$$pV = nRT \quad \text{and} \quad n = \frac{m}{M}$$

If we know the pressure, volume and temperature of a gas sample, we can use this information to calculate the amount in moles of the ideal unknown gas in the sample.

$$n = \frac{pV}{RT}$$

If we also know the mass of the gas sample, we can use that information to determine the molar mass of the gas.

$$M = \frac{m}{n}$$

WORKED EXAMPLE 6.7

Molar mass determination

Pure calcium carbide, CaC_2, is a hard, colourless solid that was used in caving lamps. It has a melting point of 2160 °C, and it reacts vigorously with water to produce a gas and a solution containing OH^- ions.

A 12.8 g sample of CaC_2 was treated with excess water and the resulting gas was collected in an evacuated 5.00 L glass bulb with a mass of 1054.49 g. The filled bulb had a mass of 1059.70 g and the pressure of the gas inside was 1.00×10^5 Pa when the temperature was 26.8 °C. Calculate the molar mass of the gas and determine its formula. (Assume that the product gas is insoluble in water and that the vapour pressure of water is negligible in relation to the pressure of the product gas.)

Analysis

We can use the ideal gas equation to calculate the molar mass. Then we can use the molar mass to identify the correct molecular formula from a group of possible candidates, as the products may contain only the same elements as the reactants. The problem involves a chemical reaction, so we must make a connection between the gas measurements and the chemistry that takes place. Because the reactants and one product are known, we can write a partial equation that describes the chemical reaction.

$$CaC_2(s) + H_2O(l) \longrightarrow OH^-(aq) + ?(?)$$

In any chemical reaction, atoms must be conserved, so the gas molecules can contain only H, O, C and Ca atoms. To determine the chemical formula of the gas, we must find the combination of these elements that gives the observed molar mass.

Solution

Use the data given to determine the molar mass of the unknown gas.

$$V_{bulb} = V_{gas} = 5.00\,L = 5.00 \times 10^{-3}\,m^3 \qquad T = 26.8\,°C = 300.0\,K \qquad p = 1.00 \times 10^5\,Pa$$

$$m_{bulb+gas} = 1059.70\,g \qquad m_{bulb} = 1054.49\,g \qquad m_{gas} = m_{bulb+gas} - m_{bulb} = 5.21\,g$$

We use V, T, p and the ideal gas equation to find the amount of gas in moles. Then, with the mass of the gas sample, we can determine the molar mass.

$$n_{gas} = \frac{pV}{RT} = \frac{m_{gas}}{M}$$

We could rearrange this expression and solve the equality for M directly, but to reduce the likelihood of introducing errors, it is often better to carry out each step explicitly. Therefore, we will first determine n and then solve for M.

$$n_{gas} = \frac{pV}{RT} = \frac{(1.00 \times 10^5\,\text{Pa})(5.00 \times 10^{-3}\,\text{m}^3)}{(8.314\,\text{J mol}^{-1}\,\text{K}^{-1})(300.0\,\text{K})} = 2.00 \times 10^{-1}\,\text{mol}$$

$$M = \frac{m_{gas}}{n_{gas}} = \frac{5.21\,\text{g}}{2.00 \times 10^{-1}\,\text{mol}} = 26.0\,\text{g mol}^{-1}$$

To identify the gas, we examine the formulae and molar masses of known compounds that contain only H, O, C or Ca.

Formula	*M*	Comment
Ca	40	A gas with $M = 26.0\,\text{g mol}^{-1}$ cannot contain Ca.
CO	28	This is too high.
O_2	32	This is also too high.
H_2O	18	This is too low, and H_2O is a reactant not a product.
CH_4	16	This is too low.
C_2H_2	26	This substance has the observed molar mass.

Consideration of these possibilities leads to the molecular formula C_2H_2, ethyne (commonly known as acetylene).

Is our answer reasonable?

Knowing the formula of the gaseous product and that OH^- is another product, we can write a balanced equation for the generation of ethyne.

$$CaC_2(s) + 2H_2O(l) \longrightarrow Ca^{2+}(aq) + 2OH^-(aq) + C_2H_2(g)$$

The balanced equation suggests that the result is reasonable. A systematic evaluation of all possible gases from the reagents that involve only one or two atoms other than hydrogen shows that there are no other possibilities for molar mass $26\,\text{g mol}^{-1}$.

PRACTICE EXERCISE 6.7

The glass bulb in worked example 6.7 is filled with an unknown gas until the pressure is 1.03×10^5 Pa at a temperature of 24.5 °C. The mass of the bulb plus contents is 1260.33 g. Determine the molar mass of the unknown gas.

Determination of gas density

One characteristic of gases is that the density of a gas varies significantly with the conditions. To see this, we combine the ideal gas equation and $n = \frac{m}{M}$ and rearrange to obtain an equation for density ($\rho = \frac{m}{V}$).

$$n = \frac{pV}{RT} \quad \text{and} \quad n = \frac{m}{M}$$

Set the two expressions for n equal to each other.

$$\frac{m}{M} = \frac{pV}{RT}$$

Now, multiply both sides of the equality by M and divide both sides by V.

$$\rho_{gas} = \frac{m}{V} = \frac{pM}{RT}$$

This equation reveals three features of gas density.

1. The density of an ideal gas increases linearly with increasing pressure at fixed temperature. The reason is that increasing the pressure compresses the gas into a smaller volume without changing its mass.

$$\rho_{gas} \propto p$$

2. The density of an ideal gas decreases linearly with increasing temperature at fixed pressure. The reason is that increasing the temperature causes the gas to expand without changing its mass.

$$\rho_{gas} \propto \frac{1}{T}$$

FIGURE 6.17 A helium-filled balloon floats because helium is less dense than air.

3. The density of ideal gases increases linearly with increasing molar mass at a given temperature and pressure. The reason is that equal amounts of different gases occupy equal volumes at a given temperature and pressure.

$$\rho_{gas} \propto M$$

There are practical applications of all these features.

- SCUBA divers use high-pressure cylinders, where the high density of the gas gives them more dive time per volume.
- Balloons inflated with helium rise in the atmosphere because the molar mass of helium is substantially lower than the average molar mass of air (figure 6.17). Consequently, the density of a helium-filled balloon is less than the density of air, and the balloon rises, just as a cork released under water rises to the surface.
- When the air within a hot-air balloon is heated, its density decreases below that of the outside air. With sufficient heating, the balloon rises and floats. In contrast, cold air is denser than warm air, so cold air sinks. For this reason, valleys often are colder than the surrounding hillsides.

When a gas is released into the atmosphere, whether it rises or sinks depends on its molar mass. If the molar mass of the gas is greater than the average molar mass of air, the gas remains near the ground. Carbon dioxide fire extinguishers are effective for certain fires because of this feature. The molar mass of CO_2 (44.0 g mol^{-1}) is greater than that of N_2 (28.0 g mol^{-1}) or O_2 (32.0 g mol^{-1}), and so a CO_2 fire extinguisher lays down a blanket of this gas that excludes oxygen from the fire.

WORKED EXAMPLE 6.8

Gas density

A certain hot-air balloon will rise when the density of its air is 15.0% lower than that of the surrounding atmospheric air. Calculate the density of air at 295 K and 1.00×10^5 Pa (assume that dry air is 78.0% N_2 and 22.0% O_2), and determine the minimum temperature of air inside the balloon that will cause the balloon to rise.

Analysis

The problem has two parts. First, we must calculate the density of atmospheric air. To do this, we need to determine the molar mass of dry air, which is the weighted average of the molar masses of its components. Then we must calculate the temperature needed to reduce the density by 15.0%. For both calculations, we use the ideal gas equation as rearranged to give gas density.

$$\rho_{gas} = \frac{pM}{RT}$$

Solution

Begin by calculating the molar mass of dry air. Multiply the fraction of each component by its molar mass.

$$M_{air} = \left(\frac{0.78}{1.00}\right)(28.02\,\text{g mol}^{-1}) + \left(\frac{0.22}{1.00}\right)(32.00\,\text{g mol}^{-1}) = 28.9\,\text{g mol}^{-1}$$

The other conditions are stated in the problem.

$$T = 295\,\text{K} \qquad p = 1.00 \times 10^5\,\text{Pa}$$

Recall that, because the unit Pa is defined in terms of N m^{-2}, and that 1 N = 1 kg m s^{-2}, we must use a molar mass in kg mol^{-1} in this equation. Substituting these into the equation for gas density gives:

$$\rho_{gas} = \frac{pM}{RT} = \frac{(1.00 \times 10^5\,\text{Pa})(28.9 \times 10^{-3}\,\text{kg mol}^{-1})}{(8.314\,\text{J mol}^{-1}\,\text{K}^{-1})(295\,\text{K})} = 1.18\,\text{kg m}^{-3}$$

The balloon will rise when the density of its contents is 15% less than the density of the surrounding air.

$$\rho_{gas} = 1.18\,\text{kg m}^{-3} - \left(\frac{0.15}{1.00}\right)(1.18\,\text{kg m}^{-3}) = 1.00\,\text{kg m}^{-3}$$

Rearrange the density equation to solve for temperature, then substitute and calculate.

$$\rho_{gas} = \frac{pM}{RT} \quad \text{so} \quad T = \frac{pM}{R\rho}$$

$$T = \frac{(1.00 \times 10^5\,\text{Pa})(28.9 \times 10^{-3}\,\text{kg mol}^{-1})}{(8.314\,\text{J mol}^{-1}\,\text{K}^{-1})(1.00\,\text{kg m}^{-3})} = 348\,\text{K}$$

Is our answer reasonable?

A temperature of 348 K or 75 °C is a reasonable result as it is higher than the temperature of the surrounding air. Notice that the molar mass and gas constant appear in both calculations, so we can also find the temperature requirement by simple proportions (comparing the air inside and outside the balloon).

$$T_{\text{inside}} = T_{\text{outside}}\left(\frac{\rho_{\text{outside}}}{\rho_{\text{inside}}}\right) = \frac{(295\,\text{K})(1.18\,\text{kg m}^{-3})}{1.00\,\text{kg m}^{-3}} = 348\,\text{K, which rounds to } 350\,\text{K.}$$

PRACTICE EXERCISE 6.8

Helium is used for lighter-than-air blimps, whereas argon is used to exclude air from flasks in which air-sensitive syntheses are performed. Calculate the densities of these two group 18 gases at 295 K and 1.00×10^5 Pa, and explain why the two gases have different uses.

Gas density has a significant effect on the interactions between molecules of a gas. As molecules move about, they collide regularly with one another and with the walls of their container. Figure 6.18 shows that the frequency of collisions depends on the density of the gas. At low density, a molecule may move a significant distance before it encounters another molecule. At high density, a molecule travels only a short distance before it collides with another molecule.

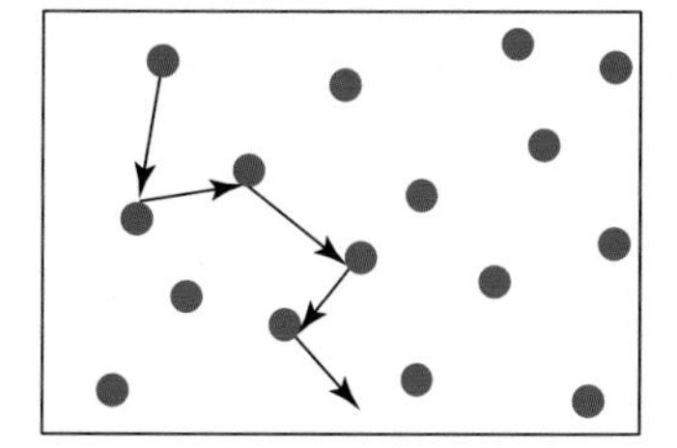

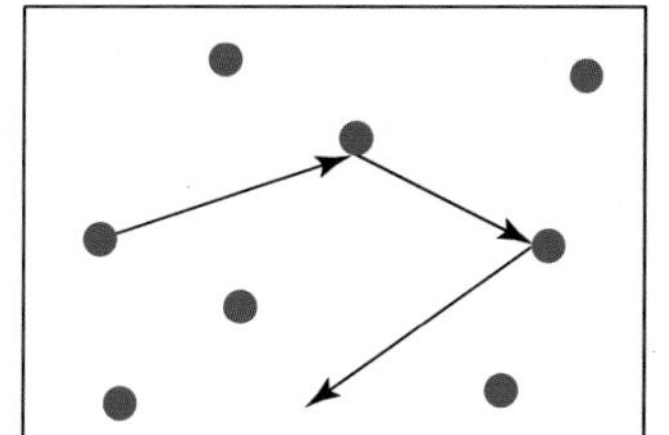

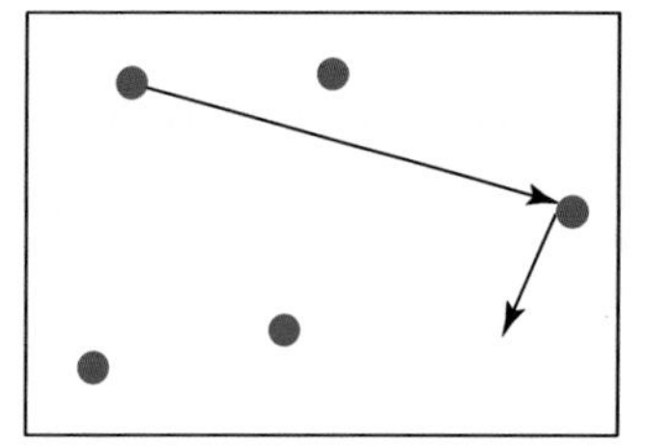

FIGURE 6.18 As the molecular density decreases, the average distance travelled between molecular collisions increases.

6.6 Gas stoichiometry

The principles of stoichiometry apply equally to solids, liquids and gases. The ideal gas equation relates the amount of gas to the physical properties of that gas. When a chemical reaction involves a gas, the ideal gas equation provides the link between p–V–T data and molar amounts.

$$n_i = \frac{p_i V}{RT}$$

Stoichiometric calculations always require amounts and, for gases, these are usually calculated from the ideal gas equation.

WORKED EXAMPLE 6.9

Gas stoichiometry

Worked example 6.7 described the formation of ethyne, C_2H_2, from calcium carbide, CaC_2. Modern industrial production of ethyne is based on a reaction of methane, CH_4, under carefully controlled conditions. At temperatures greater than 1600 K, two methane molecules rearrange to give three molecules of hydrogen and one molecule of ethyne.

$$2CH_4(g) \xrightarrow{1600\text{ K}} C_2H_2(g) + 3H_2(g)$$

A 50.0 L steel vessel, filled with CH_4 to a pressure of 10.0×10^5 Pa at 298 K, is heated to exactly 1600 K to convert CH_4 into C_2H_2. What is the maximum possible mass of C_2H_2 that can be produced? What pressure does the reactor reach at 1600 K? Assume that both CH_4 and C_2H_2 behave as ideal gases under the conditions of the reaction.

Analysis

This is a stoichiometry problem involving gases. The mass of a product and the final pressure must be calculated. We can draw a diagram showing the process and listing the data.

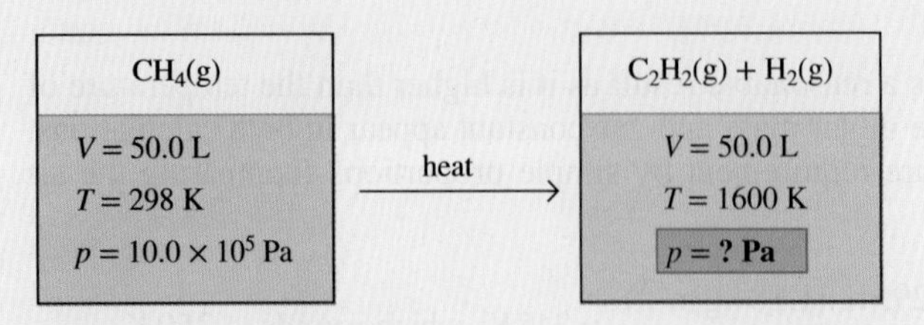

Solution

In any stoichiometry problem, we work with amounts. This problem involves gases, so we use the ideal gas equation to obtain amounts in moles from the p–V–T information. Use the ideal gas equation to calculate the amount of CH_4 present initially.

$$n = \frac{p_i V}{RT_i} = \frac{(10.0 \times 10^5\,\text{Pa})(0.0500\,\text{m}^3)}{(8.314\,\text{J mol}^{-1}\,\text{K}^{-1})(298\,\text{K})}$$

$$= 20.18\,\text{mol CH}_4$$

From the stoichiometry of the balanced chemical equation, we can determine that complete conversion of 20.18 mol of CH_4 into C_2H_2 and H_2 will give ($\frac{1}{2} \times 20.18$) = 10.09 mol C_2H_2 and ($\frac{3}{2} \times 20.18$) = 30.27 mol H_2.

Now we determine the mass of ethyne formed.

$$m = nM = (10.09\,\text{mol})(26.04\,\text{g mol}^{-1})$$
$$= 263\,\text{g C}_2\text{H}_2$$

Using the ideal gas equation and the total amount present at the end of the reaction, and assuming all the CH_4 has reacted, we can calculate the final pressure.

$$n_{\text{total}} = 10.09 + 30.27 = 40.36\,\text{mol}$$

$$pV = nRT, \text{ and so } p = \frac{nRT}{V}$$

$$p = \frac{(40.36\,\text{mol})(8.314\,\text{J mol}^{-1}\,\text{K}^{-1})(1600\,\text{K})}{0.0500\,\text{m}^3} = 1.07 \times 10^7\,\text{Pa}$$

Is our answer reasonable?

From the initial amount of methane, we can calculate the initial mass of methane, which is 323 g. The mass of ethyne, 263 g, is less than this, a reasonable result given that H_2 is also produced. The final pressure, 1.07×10^7 Pa, seems high, but the temperature of the reactor has increased substantially and the amount of products is twice the amount of reactants, so a large pressure increase is to be expected.

PRACTICE EXERCISE 6.9

Calculate the volume of hydrogen gas generated when 1.52 g of Mg reacts with an excess of aqueous HCl, if the gas is collected at 1.00×10^5 Pa and 22.5 °C. The balanced chemical equation for the reaction is:

$$\text{Mg(s)} + 2\text{HCl(aq)} \rightarrow \text{MgCl}_2\text{(aq)} + \text{H}_2\text{(g)}$$

Any of the types of problems discussed in chapter 3 can involve gases. The strategy for doing stoichiometric calculations is the same whether the species involved are solids, liquids or gases. In this chapter, we add the ideal gas equation to our equations for relating measured quantities to amounts.

WORKED EXAMPLE 6.10

Limiting reagents in a gas mixture

Margarine can be made from natural oils such as coconut oil by hydrogenation.

$$\underset{\text{oil}}{\text{C}_{57}\text{H}_{104}\text{O}_6\text{(l)}} + 3\text{H}_2\text{(g)} \xrightarrow{200\,°\text{C},\,7\times10^5\,\text{Pa, Ni catalyst}} \underset{\text{margarine}}{\text{C}_{57}\text{H}_{110}\text{O}_6\text{(s)}}$$

A hydrogenator (figure 6.19) with a volume of 2.50×10^2 L is charged with 12.0 kg of oil. The remaining volume is then filled with 7.00×10^5 Pa of hydrogen, H_2, at 473 K. The reaction produces the maximum amount of margarine. What is the final pressure of H_2, and what mass of margarine will be produced? Assume that H_2 does not dissolve in coconut oil or margarine, that it behaves as an ideal gas under the reaction conditions and that the coconut oil and margarine occupy a negligible volume in the hydrogenator.

Analysis

We have data from which we can calculate the amounts of both starting materials. Given the chemical equation, the first step in a stoichiometry problem is to determine the initial amount of each starting material. Next, calculate ratios of amounts to identify the limiting reagent. After that, use the ideal gas equation and the stoichiometric principles we learned in chapter 3 to obtain the final answers.

Solution

The ideal gas equation is used for the gaseous starting material.

$$n_{H_2} = \frac{pV}{RT} = \frac{(7.00 \times 10^5\,\text{Pa})(2.50 \times 10^{-1}\,\text{m}^3)}{(8.314\,\text{J mol}^{-1}\,\text{K}^{-1})(473\,\text{K})} = 44.5\,\text{mol}$$

We obtain the amount of oil from the mass and the molar mass.

$$n_{oil} = \frac{m}{M} = \frac{1.20 \times 10^4\,\text{g}}{885.4\,\text{g mol}^{-1}} = 13.6\,\text{mol}$$

Now we divide each amount by the stoichiometric coefficient to identify the limiting reagent.

$$\text{For } H_2, \frac{44.50\,\text{mol}}{3\,\text{mol}} = 14.83$$

$$\text{For oil, } \frac{13.55\,\text{mol}}{1\,\text{mol}} = 13.55$$

The reactant with the smaller ratio, oil, is limiting. (Excess hydrogen is easily recovered from a gas-phase reactor, so margarine manufacturers make the oil the limiting reagent to ensure complete conversion of oil into margarine. The excess H_2 gas is recovered and reused.)

The balanced chemical equation shows that 13.6 mol of oil will react with 40.8 mol (3×13.6) of H_2 to give 13.6 mol of margarine. Therefore, the amount of unreacted H_2 will be $(44.5 - 40.8) = 3.7$ mol.

Using the ideal gas equation, we calculate the pressure of hydrogen at the end of the reaction.

$$p = \frac{nRT}{V} = \frac{(3.7\,\text{mol})(8.314\,\text{J mol}^{-1}\,\text{K}^{-1})(473\,\text{K})}{2.50 \times 10^{-1}\,\text{m}^3}$$

$$= 0.61 \times 10^5\,\text{Pa}$$

Now we calculate the mass of margarine.

$$m = nM = (13.6\,\text{mol})(891.5\,\text{g mol}^{-1})$$

$$= 1.21 \times 10^4\,\text{g} = 12.1\,\text{kg}$$

Is our answer reasonable?

In this example, oil is the limiting reagent and not all the H_2 will react. Therefore, some H_2 will be left, but we expect the pressure of H_2 at the end of the reaction to be lower than at the start, which it is. Notice that the mass of margarine produced (12.1 kg) is only slightly greater than the mass of oil we started with (12.0 kg). This is to be expected, as the molar masses of coconut oil and margarine are very similar.

FIGURE 6.19 A hydrogenator available commercially for small-scale use, e.g. in a research laboratory.
Image courtesy Armfield Ltd. FT67 Hydrogenation Unit.

PRACTICE EXERCISE 6.10

In the industrial production of nitric acid, one step is the oxidation of nitrogen oxide: $2NO(g) + O_2(g) \rightarrow 2NO_2(g)$. A reaction chamber is charged with 5.00×10^5 Pa each of NO and O_2. If the reaction gives the maximum possible amount of $NO_2(g)$ at constant temperature and volume, calculate the final pressures of each reagent.

Summary of mole conversions

On the microscopic level, moles are the currency of chemistry; therefore, all stoichiometric calculations require amounts. On the macroscopic level, we measure mass, volume, temperature and pressure. Table 6.2 (overleaf) lists three equations that can be used to convert between the microscopic and macroscopic levels. Each of these equations applies to a particular category of chemical substances. You should have a working knowledge of these three equations, along with the category of substances to which they apply.

TABLE 6.2 Summary of stoichiometric relationships.

Substance	Relationship	Equation
pure solid, liquid or gas	amount (mol) $= \dfrac{\text{mass (g)}}{\text{molar mass (g mol}^{-1})}$	$n = \dfrac{m}{M}$
solution	amount (mol) = concentration (mol L^{-1}) × volume (L)	$n = cV$
gas	amount (mol) $= \dfrac{\text{pressure (Pa)} \times \text{volume (m}^3)}{\text{constant (J mol}^{-1}\text{ K}^{-1}) \times \text{temperature (K)}}$	$n = \dfrac{pV}{RT}$

Worked example 6.11 uses all three relationships. Viewed as a whole, the example may seem complicated. However, as the solution illustrates, breaking the problem into separate parts allows each part to be solved using simple chemical and stoichiometric principles. Complicated problems are often simplified considerably by looking at them one piece at a time.

WORKED EXAMPLE 6.11

FIGURE 6.20 Magnesium reacting with aqueous HCl.

General stoichiometry

Reaction of magnesium metal with acid (figure 6.20) generates hydrogen gas and an aqueous solution of ions. Suppose that 3.50 g of magnesium metal is dropped into 0.150 L of 6.00 M HCl in a 5.00 L cylinder at 25.0 °C. The initial gas pressure of the cylinder is 1.00×10^5 Pa, and it is immediately sealed. Find the final partial pressure of hydrogen, the total pressure in the container and the concentration of Mg^{2+} in solution.

Analysis

Data are given for all reactants. We must first write a balanced chemical equation and then use the equations in table 6.2, along with the stoichiometry of the reaction, to calculate the required pressures and concentrations.

Solution

We begin by writing the balanced chemical equation, which we have seen previously in practice exercise 6.9.

$$Mg(s) + 2HCl(aq) \rightarrow MgCl_2(aq) + H_2(g)$$

The problem asks for pressures and ion concentrations. The final pressure can be determined from *p–V–T* data and n_{H_2}. The amount of hydrogen can be found from the mass of magnesium and the stoichiometric ratio. Here is a summary of the data.

$$V_{container} = 5.00\,L = 5.00 \times 10^{-3}\,m^3 \qquad T = 298\,K \qquad m_{Mg} = 3.50\,g$$

$$V_{solution} = 0.150 \times 10^{-3}\,m^3 \qquad p_{air} = 1.00 \times 10^5\,Pa \qquad [HCl] = 6.00\,M$$

Now we analyse the stoichiometry of the reaction. The starting mass of Mg and the volume and concentration of HCl are given, and we can use these data to find the limiting reactant.

$$n_{Mg} = \frac{m}{M} = \frac{3.50\,g}{24.31\,g\,mol^{-1}} = 0.144\,mol$$

$$n_{HCl} = cV = (6.00\,mol\,L^{-1})(0.150\,L) = 0.900\,mol$$

We divide by the stoichiometric coefficients to show that magnesium is the limiting reagent.

$$\text{For Mg,}\ \frac{0.144\,mol}{1\,mol} = 0.144 \qquad \text{For HCl,}\ \frac{0.900\,mol}{2\,mol} = 0.450$$

From the stoichiometry of the balanced chemical equation, we can show that 0.144 mol of Mg will form 0.144 mol H_2. We therefore use this amount to calculate the pressure of H_2. However, before doing this, we must visualise the reaction vessel. The container's total volume is $5.00 \times 10^{-3}\,m^3$, but $0.150 \times 10^{-3}\,m^3$ is occupied by the aqueous solution. This leaves $4.85 \times 10^{-3}\,m^3$ for the gas mixture. The partial pressure of hydrogen is calculated using the ideal gas equation and assuming that no H_2 remains in solution; this is a good approximation because hydrogen gas is not very soluble in water.

$$p_{H_2} = \frac{nRT}{V} = \frac{(0.144\,mol)(8.314\,J\,mol^{-1}K^{-1})(298\,K)}{4.85 \times 10^{-3}\,m^3} = 0.736 \times 10^5\,Pa$$

Assuming that the amount of air originally present does not change in the reaction, the pressure exerted by the air remains constant at 1.00×10^5 Pa. The final total pressure is the sum of the partial pressures.

$$p_{\text{total}} = p_{H_2} + p_{\text{initial}} = 0.736 \times 10^5\ \text{Pa} + 1.00 \times 10^5\ \text{Pa}$$
$$= 1.74 \times 10^5\ \text{Pa}$$

We can calculate the final concentration of Mg^{2+} using the initial amount of Mg and the volume of the solution.

$$[Mg^{2+}] = \frac{0.144\ \text{mol}}{0.150\ \text{L}} = 0.960\ \text{M}$$

PRACTICE EXERCISE 6.11

Repeat the calculations from worked example 6.11 with all conditions the same except that 14.0 g of Mg is added to the HCl solution.

6.7 Real gases

In chapter 5 we looked at what holds atoms and ions together, and we listed three types of bonding forces: ionic, covalent and metallic. In addition to these relatively strong bonding forces, there are much weaker forces that exist between atoms, ions and molecules. In the previous sections of this chapter we looked at ideal gases, assuming that there are no attractive forces between either atoms or molecules. However, weak attractive forces do occur in real gases, and much stronger forces are present in liquids and solids. In the absence of these **intermolecular forces**, all molecules would move independently, and all molecular substances would be gases. These intermolecular forces are essentially due to attractions between polarised molecules, but they also include polar interactions between ions and molecules as well as atoms. The name, intermolecular forces, is therefore somewhat inaccurate, but it can be justified by the fact that an overwhelming number of all chemical compounds are molecules.

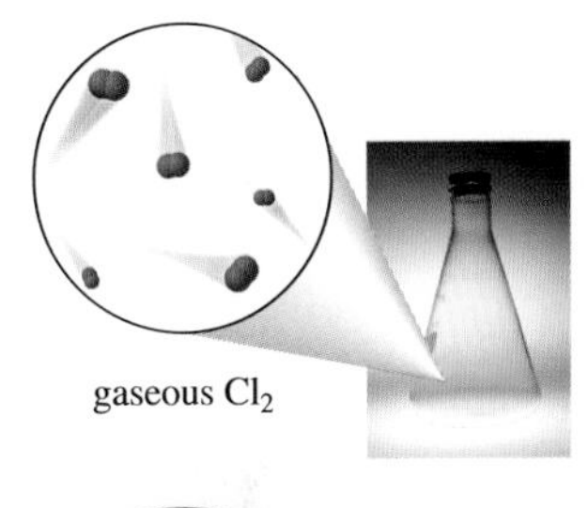

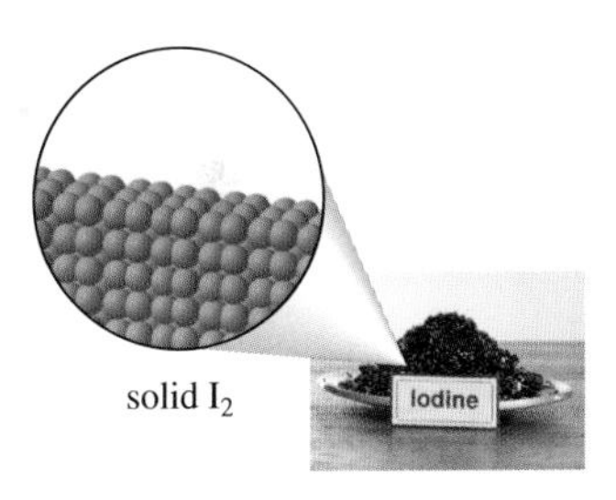

FIGURE 6.21 Under normal conditions, chlorine is a pale yellow-green gas, bromine is a dark red liquid, and iodine is a dark crystalline solid.

The halogens

The halogens, the elements from group 17 of the periodic table, provide an introduction to intermolecular forces. These elements exist as diatomic molecules: F_2, Cl_2, Br_2 and I_2. Each molecule contains two atoms held together by a single covalent bond that can be described by end-on overlap of valence *p* orbitals.

Although they have similar covalent bonding, bromine and iodine differ from chlorine and fluorine in their macroscopic physical appearance and in their molecular behaviour. As figure 6.21 shows, at room temperature and pressure, chlorine (like fluorine) is a gas, bromine is a liquid and iodine is a solid.

Gases and condensed phases (a term used to collectively describe liquids and solids) look very different at the molecular level. Molecules of F_2 or Cl_2 move freely throughout their gaseous volume, travelling many molecular diameters before colliding with one another or with the walls of their container. Because much of the volume of a gas is empty space, samples of gaseous F_2 and Cl_2 readily expand or contract in response to changes in pressure. This freedom of motion exists because the intermolecular forces between these molecules are small with respect to their kinetic energy.

Molecules of liquid bromine also move about relatively freely, but there is not much empty space between molecules. A liquid cannot be compressed significantly by increasing the pressure, because its molecules are already in close contact with one another. Neither does a liquid expand significantly if the pressure is reduced, because the intermolecular forces in a liquid are strong enough to prevent the molecules from breaking away from one another. This is no longer the case, however, if the pressure is reduced enough to reach the boiling point of the liquid.

Solid iodine, like liquid bromine, has little empty space between molecules. Like liquids, solids have sufficiently strong intermolecular forces that they neither expand nor contract significantly when the pressure changes. In solids, the intermolecular forces are strong enough to prevent molecules from moving freely past one another. Instead, the I_2 molecules in the solid phase are arranged in ordered arrays. Each molecule vibrates back and forth around a single most stable position, but it cannot slide easily past its neighbours.

The balance between molecular kinetic energies and intermolecular attractive forces accounts for these striking differences (figure 6.22). Recall from section 6.2 that molecules

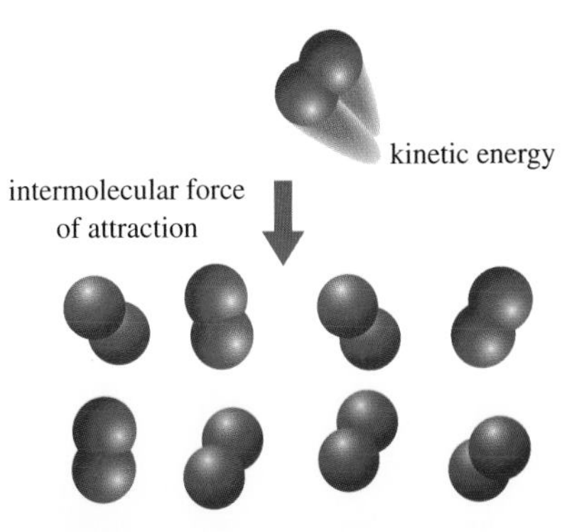

FIGURE 6.22 A substance exists in a condensed phase when its molecules have too little average kinetic energy to overcome intermolecular forces of attraction.

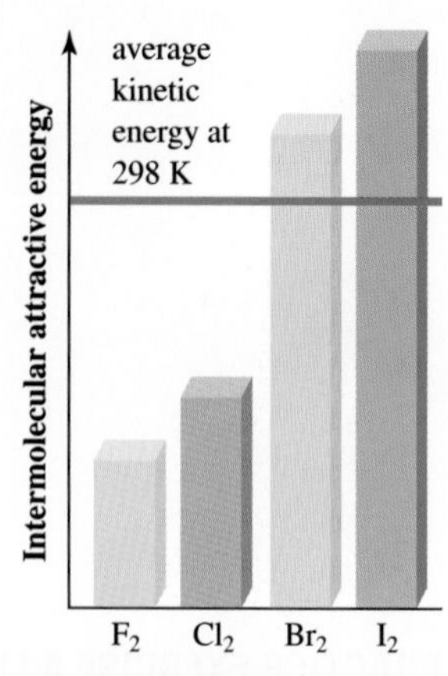

FIGURE 6.23 The balance of average kinetic energy (red line) and intermolecular energies of attraction favours the gas phase for F_2 and Cl_2, and the condensed phases for Br_2 and I_2.

are always moving. Intermolecular attractive forces tend to hold molecules together in a condensed phase, but molecules that are moving fast enough can overcome these forces and move freely in the gas phase. When the average kinetic energy is large enough, molecules remain separated from one another, and the substance is a gas. Conversely, when intermolecular attractive forces are large enough, molecules remain close to one another, and the substance is a liquid or solid.

The bars in figure 6.23 compare the attractive energies arising from intermolecular forces acting on the halogens. The figure also shows the average molecular kinetic energy at room temperature. For fluorine and chlorine, the attractive energy generated by intermolecular forces is smaller than the average molecular kinetic energy at room temperature. Thus, these elements are gases under these conditions. In contrast, bromine molecules have enough kinetic energy to move freely about, but their energy is insufficient to overcome the intermolecular forces of the liquid phase. Finally, attractive forces in iodine are strong enough to lock the I_2 molecules in position in the solid state.

Properties of real gases

As we described in section 6.3, the ideal gas model makes two assumptions: a gas has negligible forces between its constituent atoms or molecules, and gas atoms or molecules have negligible volumes. Neither of these assumptions is true for a real gas. At sufficiently high pressures and sufficiently low temperatures, all gases can be liquefied.

How close do real gases come to ideal behaviour? To answer this question, we rearrange the ideal gas equation to examine the ratio $\frac{pV}{nRT}$. Figure 6.24 shows how the $\frac{pV}{nRT}$ ratio varies with pressure for chlorine gas at room temperature. If chlorine were an ideal gas, the $\frac{pV}{nRT}$ ratio would always be 1, as shown by the red line on the graph. Instead, chlorine shows deviations from 1 as the pressure increases.

Notice in the inset of figure 6.24 that chlorine is nearly ideal at pressures around 1×10^5 Pa. In fact, $\frac{pV}{nRT}$ deviates from 1.0 by less than 4% at pressures below 4×10^5 Pa. As the pressure increases, however, the deviations become increasingly significant. With increasing pressure, the $\frac{pV}{nRT}$ ratio for chlorine drops below 1. This is because the chlorine molecules are close enough together for attractive forces to play a significant role. These intermolecular attractions hold molecules together and reduce the forces exerted when the molecules strike the walls of the container. Intermolecular attractions tend to make the pressure of a real gas lower than the ideal value.

Figure 6.24 also shows that at pressures greater than 375×10^5 Pa, $\frac{pV}{nRT}$ becomes *larger* than 1. This is due to the effect of molecular size. At high enough pressure, the molecules are packed so close together that the total volume of the molecules is no longer negligible compared with the overall volume of the container. The result is that the container volume is effectively decreased by the molecules' volume and the pressure of the real gas tends to be greater than that of the ideal gas.

Every gas shows deviations from ideal behaviour at high pressure. Figure 6.25 shows $\frac{pV}{nRT}$ for He, F_2, CH_4 and N_2, all of which are gases at room temperature. Notice that $\frac{pV}{nRT}$ for helium increases steadily as pressure increases. Interatomic forces for helium are too small to reduce the ratio below 1, but the finite size of the helium atom generates deviations from ideal behaviour that become significant at pressures above 100×10^5 Pa.

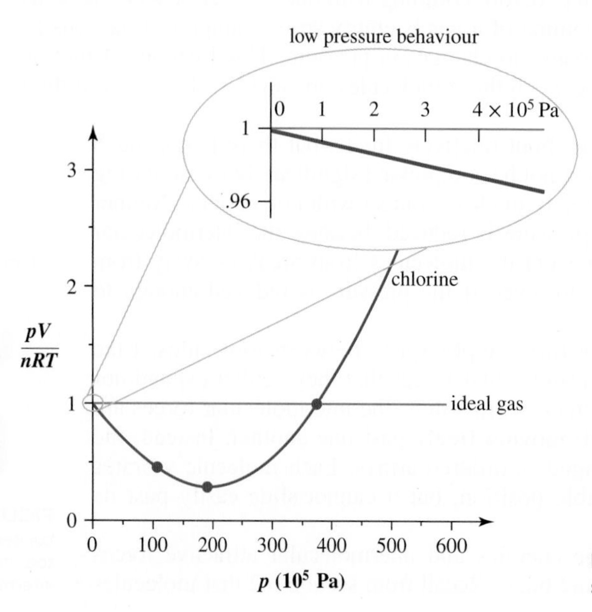

FIGURE 6.24 The variation in $\frac{pV}{nRT}$ with pressure shows that chlorine is not an ideal gas.

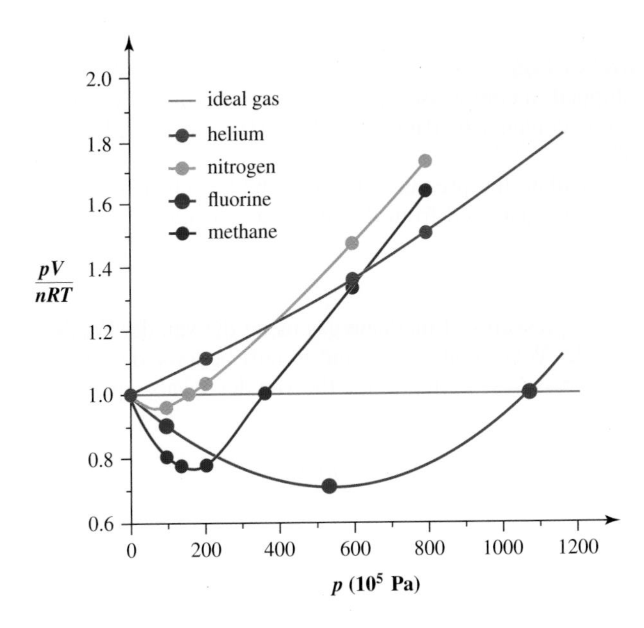

FIGURE 6.25 Variations in $\frac{pV}{nRT}$ for He, N_2, F_2 and CH_4 at 300 K.

Given that every gas deviates from ideal behaviour to some degree, can we use the ideal gas model to discuss the properties of real gases? The answer is yes, as long as conditions do not become too extreme. The gases with which chemists often work, such as chlorine, helium and nitrogen, are nearly ideal at room temperature at pressures below about 10×10^5 Pa, but, at very high pressures or near the condensation points of substances, deviations from the ideal gas equation become large.

The van der Waals equation

It would be useful to have an equation that describes the relationship between pressure and volume for a real gas, just as $\frac{pV}{nRT}$ describes an ideal gas. One way to approach real gas behaviour is to modify the ideal gas equation to account for attractive forces and molecular volumes. The result is the **van der Waals equation**, named after the scientist who first proposed it in 1873, Johannes van der Waals (1837–1923, Nobel Prize in physics, 1910).

$$\left(p + \frac{n^2 a}{V^2}\right)(V - nb) = nRT$$

The van der Waals equation adds two correction terms to the ideal gas equation. Each correction term includes a constant that has a specific value for every gas. The first correction term, $\frac{n^2 a}{V^2}$, adjusts for attractive intermolecular forces. The van der Waals constant a reflects the strength of intermolecular forces for the gas; the stronger the forces, the larger the value of a. The second correction term, nb, adjusts for molecular sizes. The van der Waals constant b reflects the size of gas atoms or molecules; the larger they are, the larger the value of b.

The van der Waals constants for a number of gases are given in table 6.3, and the magnitude of van der Waals corrections is explored in worked example 6.12.

TABLE 6.3 Values of van der Waals constants for a variety of gases.

Substance	a (m^6 Pa mol^{-2})	b (m^3 mol^{-1})	Substance	a (m^6 Pa mol^{-2})	b (m^3 mol^{-1})
He	3.457×10^{-3}	2.37×10^{-5}	Cl_2	6.579×10^{-1}	5.622×10^{-5}
Ne	2.135×10^{-2}	1.709×10^{-5}	CO_2	3.640×10^{-1}	4.267×10^{-5}
Ar	1.363×10^{-1}	3.219×10^{-5}	H_2O	5.536×10^{-1}	3.049×10^{-5}
H_2	2.476×10^{-2}	2.661×10^{-5}	NH_3	4.225×10^{-1}	3.707×10^{-5}
N_2	1.408×10^{-1}	3.913×10^{-5}	CH_4	2.283×10^{-1}	4.278×10^{-5}
O_2	1.378×10^{-1}	3.183×10^{-5}	C_2H_6	5.562×10^{-1}	6.38×10^{-5}
CO	1.505×10^{-1}	3.985×10^{-5}	C_6H_6	1.824×10^{-1}	1.154×10^{-5}
F_2	1.156×10^{-1}	2.90×10^{-5}			

WORKED EXAMPLE 6.12

Magnitudes of van der Waals corrections

Gases such as methane are sold and shipped in compressed gas cylinders. A typical cylinder has a volume of 15.0 L and, when full, contains 62.0 mol of CH_4. After prolonged use, 0.620 mol of CH_4 remains in the cylinder.

Use the van der Waals equation to calculate the pressures in the cylinder when full and after use, and compare the values with those obtained from the ideal gas equation. Assume a constant temperature of 25 °C.

Analysis

We are asked to calculate and compare the pressures of methane gas using the van der Waals equation and the ideal gas equation. Van der Waals constants a and b must be looked up in a data table such as table 6.3. To calculate pressures, we rearrange the van der Waals equation and the ideal gas equation.

$$p_{\text{real}} = \frac{nRT}{V - nb} - \frac{n^2 a}{V^2} \quad \text{and} \quad p_{\text{ideal}} = \frac{nRT}{V}$$

Solution

The van der Waals constants for CH_4 are:

$$a = 2.283 \times 10^{-1}\,\text{m}^6\,\text{Pa mol}^{-2} \qquad b = 4.278 \times 10^{-5}\,\text{m}^3\,\text{mol}^{-1}$$

$$V = 15.0 \times 10^{-3}\,\text{m}^3 \qquad T = (25 + 273.15)\,\text{K} = 298\,\text{K}$$

When the tank is full, $n = 62.0$ mol.

$$p_{\text{real}} = \frac{(62.0\,\text{mol})(8.314\,\text{J mol}^{-1}\,\text{K}^{-1})(298\,\text{K})}{15.0 \times 10^{-3}\,\text{m}^3 - (62.0\,\text{mol})(4.278 \times 10^{-5}\,\text{m}^3\,\text{mol}^{-1})} - \frac{(62.0\,\text{mol})^2 (2.283 \times 10^{-1}\,\text{m}^6\,\text{Pa mol}^{-2})}{(15.0 \times 10^{-3}\,\text{m}^3)^2}$$

$$p_{\text{real}} = \frac{1.537 \times 10^5\,\text{J}}{1.24 \times 10^{-2}\,\text{m}^3} - 3.900 \times 10^6\,\text{Pa} = 85.5 \times 10^5\,\text{Pa} \quad \text{(recall that } 1\,\text{Pa} = 1\,\text{J m}^{-3}\text{)}$$

$$p_{\text{ideal}} = \frac{nRT}{V} = \frac{(62.0\,\text{mol})(8.314\,\text{J mol}^{-1}\,\text{K}^{-1})(298\,\text{K})}{15.0 \times 10^{-3}\,\text{m}^3} = 102 \times 10^5\,\text{Pa}$$

After use, $n = 0.620$ mol.

$$p_{\text{real}} = \frac{(0.620\,\text{mol})(8.314\,\text{J mol}^{-1}\,\text{K}^{-1})(298\,\text{K})}{15.0 \times 10^{-3}\,\text{m}^3 - (0.620\,\text{mol})(4.278 \times 10^{-5}\,\text{m}^3\,\text{mol}^{-1})} - \frac{(0.620\,\text{mol})^2 (2.283 \times 10^{-1}\,\text{m}^6\,\text{Pa mol}^{-2})}{(15.0 \times 10^{-3}\,\text{m}^3)^2}$$

$$p_{\text{real}} = \frac{1.537 \times 10^3\,\text{J}}{1.50 \times 10^{-2}\,\text{m}^3} - 3.90 \times 10^2\,\text{Pa} = 1.02 \times 10^5\,\text{Pa}$$

$$p_{\text{ideal}} = \frac{nRT}{V} = \frac{(0.620\,\text{mol})(8.314\,\text{J mol}^{-1}\,\text{K}^{-1})(298\,\text{K})}{15.0 \times 10^{-3}\,\text{m}^3} = 1.02 \times 10^5\,\text{Pa}$$

Organise the results to make a comparison.

$$p_{\text{full, real}} = 85.5 \times 10^5\,\text{Pa} \quad \text{and} \quad p_{\text{full, ideal}} = 102 \times 10^5\,\text{Pa}$$

$$p_{\text{used, real}} = 1.02 \times 10^5\,\text{Pa} \quad \text{and} \quad p_{\text{used, ideal}} = 1.02 \times 10^5\,\text{Pa}$$

Notice that the van der Waals correction is appreciable (15.4%) at high pressure, but is negligible at atmospheric pressure (1.013×10^5 Pa).

Is our answer reasonable?

Methane is a low molar mass gas that is non-polar and has only dispersion forces acting between the molecules. It is therefore reasonable to expect that it behaves like an ideal gas under ambient conditions.

Methane boils at −164 °C. Use the ideal gas equation to calculate the molar volume of methane gas at this temperature at 1.013×10^5 Pa pressure. Then use the van der Waals equation to calculate the pressure exerted by 1.00 mole of the gas if it occupies this volume at −164 °C.

PRACTICE EXERCISE 6.12

Melting and boiling points

Melting points and boiling points can be used as indicators of the strengths of intermolecular forces. Remember that the temperature increases with the average kinetic energy of molecular motion. The boiling point of a substance is the temperature at which the average kinetic energy of molecular motion balances the attractive energy of intermolecular attractions. When the pressure is 1.013×10^5 Pa, that temperature is the **normal boiling point**. For example, the normal boiling point of bromine is 332 K (59 °C). Above this temperature, the average kinetic energy exceeds the attractive energies created by intermolecular forces of attraction, and bromine exists as vapour. Under atmospheric pressure, bromine is a liquid at temperatures below 332 K. Note that, because boiling points are determined primarily by intermolecular attractive forces, factors other than the molar mass of the molecule may be important. For example, carbon tetrachloride, CCl_4, (M = 153.81 g mol^{-1}) has a lower boiling point (77 °C) than the lighter molecule phenol, C_6H_5OH, (M = 94.12 g mol^{-1}, bp = 182 °C) because of the much stronger intermolecular forces in phenol.

The conversion of a liquid into a gas is called **vaporisation**. A liquid vaporises when molecules leave the liquid phase faster than they are captured from the gas. **Condensation** is the reverse process. A gas condenses when molecules leave the gas phase more rapidly than they escape from the liquid. We explore these processes in more detail in chapter 7.

The molecules in a liquid are able to move about freely. When a liquid is cooled, however, the kinetic energies decrease. At temperatures below the freezing point, the molecules become locked in place and the liquid solidifies. When the pressure is 1.013×10^5 Pa, that temperature is the **normal freezing point**. A liquid freezes when liquid molecules have too little kinetic energy to slide past one another. Conversely, a solid melts when its molecules have enough kinetic energy to move freely past one another. Just as intermolecular forces determine the normal boiling point, these forces also determine the freezing point. The stronger the intermolecular forces, the higher the freezing point; fluorine freezes at 53.5 K, chlorine at 172 K and bromine at 266 K.

Boiling points and melting points depend on the strengths of intermolecular forces. This is because the rates of escape and capture from a phase depend on the balance between molecular kinetic energies and intermolecular forces of attraction. A substance with large intermolecular forces must be raised to a high temperature before its molecules have sufficient kinetic energies to overcome those forces. A substance with small intermolecular forces must be cooled to a low temperature before its molecules have small enough kinetic energies to coalesce into a condensed phase. Table 6.4 lists the boiling and melting points of some elemental substances.

TABLE 6.4 Normal melting and boiling points of selected elements.

Substance	mp (K)	bp (K)	Substance	mp (K)	bp (K)
He	0.95[(a)]	4.2	Br_2	266	332
H_2	14.0	20.3	I_2	387	458
N_2	63.3	77.4	P_4	317	553
F_2	53.5	85.0	Na	371	1156
Ar	83.8	87.3	Mg	922	1363
O_2	54.8	90.2	Si	1683	2628
Cl_2	172	239	Fe	1808	3023

(a) Under high pressure, as helium cannot be solidified at atmospheric pressure.

6.8 Intermolecular forces

In chapter 5 we encountered ionic and covalent bonding that hold atoms and ions together. In addition, so-called intermolecular forces exist between all molecules (hence the name), but they also include attractive forces between atoms, ions and molecules other than bonding forces. Intermolecular forces are typically much weaker than bonding forces and generally influence physical properties of compounds (e.g. whether they are gaseous, liquid or solid at ambient conditions) rather than their chemical properties. As stated earlier in this chapter, intermolecular forces are caused by differences in polarity (permanent or temporary) between entities (see below). In this section, we will talk only about molecules, but the discussion also encompasses ions and atoms.

Different intermolecular forces can be thought of as points along a bonding continuum; all of them rely on the attraction between positively and negatively charged parts of neighbouring molecules (i.e. electron clouds and nuclei), and the distinction between them is in some cases rather fluid. Forces in order of increasing strength are:

- instantaneous dipole-induced dipole forces (also called dispersion forces)
- dipole-induced dipole forces
- dipole–dipole forces (including hydrogen bonds, which are typically relatively strong dipole–dipole interactions). These are also called dipolar forces.

Analogous forces also exist between ions and both dipoles and induced dipoles. Ion–dipole forces and ion-induced dipole forces are stronger than the intermolecular forces above.

Dispersion forces are the attractions between the negatively charged electron cloud of one molecule and the positively charged nuclei of neighbouring molecules. All substances display dispersion forces. They are caused by the usually symmetrical electron distribution around a molecule not being symmetrical at every instant of time. This creates an instantaneous dipole, which in turn can induce a dipole in a neighbouring molecule.

Dipole-induced dipole forces are somewhat stronger than dispersion forces; the difference is that, in this case, a molecule with a permanent dipole induces a dipole in a neighbouring molecule.

Dipole–dipole forces are the attractions between the negatively charged end of a polar molecule and the positively charged ends of neighbouring polar molecules. Dipole–dipole forces exist only for compounds that possess permanent dipole moments (see section 5.5).

These dipole–dipole forces are called hydrogen bonds if the positively charged end is formed by a hydrogen atom. They occur mainly between a lone pair of electrons on a small, highly electronegative atom (usually N, O or F) and a hydrogen atom bonded to a highly electronegative atom. Strong hydrogen bonds occur primarily with molecules that contain O—H, N—H and F—H covalent bonds.

Dispersion forces, dipole–dipole forces and hydrogen bonds are all much weaker than intramolecular covalent bonds. For example, the average C—C bond energy is 345 kJ mol^{-1}, whereas dispersion forces are just 0.1 to 5 kJ mol^{-1} for small alkanes such as propane. Dipole–dipole forces between polar molecules such as acetone range between 5 and 20 kJ mol^{-1}, and hydrogen bonds range between 5 and 50 kJ mol^{-1}. This generalisation, however, needs to be treated with care. In a large molecule, the sum of all the weak dispersion forces may be much stronger than a dipole–dipole interaction in a different molecule, leading to a higher melting or boiling point.

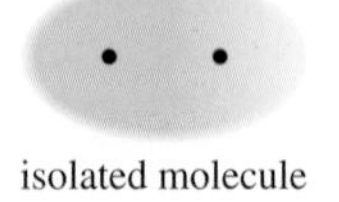

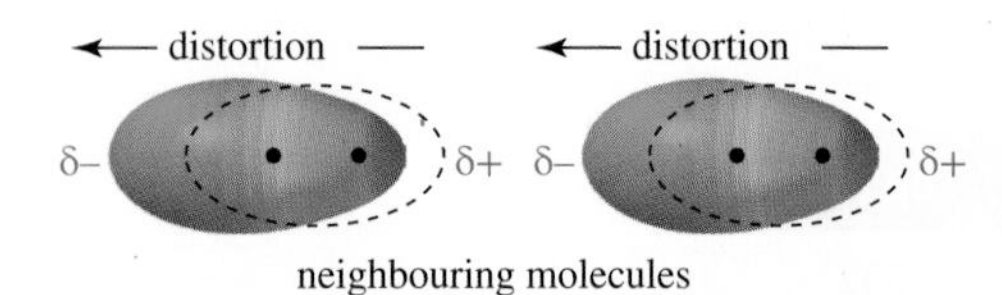

FIGURE 6.26 Exaggerated view of how dispersion forces arise.

Dispersion forces

Dispersion forces exist because the electron clouds of molecules can be distorted. For example, consider what happens when two halogen molecules approach each other. Each molecule contains positive nuclei surrounded by a cloud of negative electrons. As two molecules approach, the nucleus of one molecule attracts the electron cloud of the other. Electrons are highly mobile, so the electron clouds change shape in response to this attraction. At the same time, the two electron clouds repel each other, which leads to further distortion to minimise electron–electron repulsion. As figure 6.26 indicates, this distortion of the electron clouds creates a charge imbalance, giving the molecule a slight positive charge at one end and a slight negative charge at the other. Dispersion forces are the net attractive forces between molecules generated by all these induced charge imbalances.

The magnitude of dispersion forces depends on how easy it is to distort the electron cloud of a molecule. This ease of distortion is called the **polarisability** because distortion of an electron cloud generates a temporary polarity, called an instantaneous dipole, within the molecule. We can explore how polarisability varies by examining boiling points of the halogens, as shown in figure 6.27. The data reveal that boiling points increase with the total number of electrons. Fluorine, with 18 total electrons, has the lowest boiling point (85 K). Iodine, with 106 electrons, has the highest boiling point (458 K). The large electron cloud of I_2 distorts more readily than the small electron cloud of F_2. Figure 6.28 shows schematically how a large electron cloud distorts more than a small one, generating larger dispersion forces and leading to a higher boiling point.

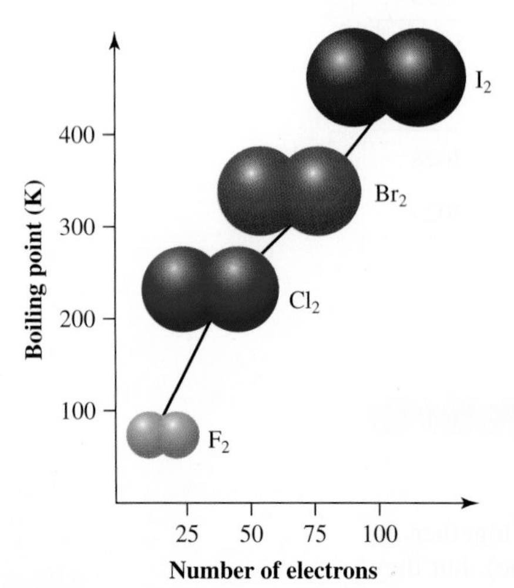

FIGURE 6.27 Boiling points of the halogens increase with the number of electrons.

Molecular size increases with the number as well as with the size of individual atoms. Figure 6.29 shows how the boiling points of alkanes increase as the carbon chain gets longer. As alkanes get longer, their electron clouds become larger and more polarisable, making dispersion forces larger and raising the boiling point. As examples, methane (CH_4, 10 electrons) is a gas at 298 K, pentane (C_5H_{12}, 42 electrons) is a low-boiling-point liquid, decane ($C_{10}H_{22}$, 82 electrons) is a high-boiling-point liquid and icosane ($C_{20}H_{42}$, 162 electrons) is a waxy solid. The boiling point increases progressively as molecules become larger and more polarisable.

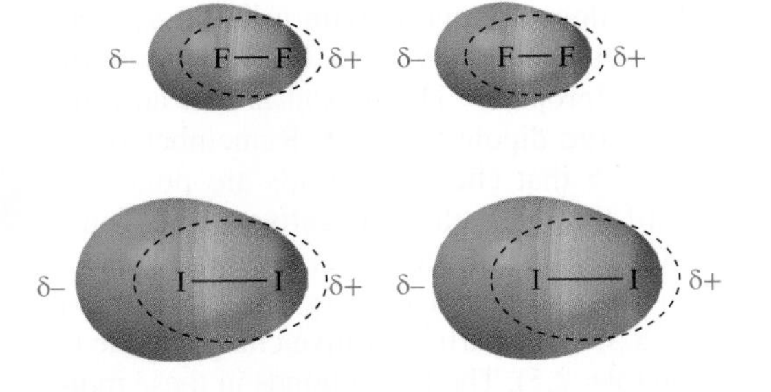

FIGURE 6.28 With many more electrons, the electron cloud of iodine is much larger and more polarisable than that of fluorine.

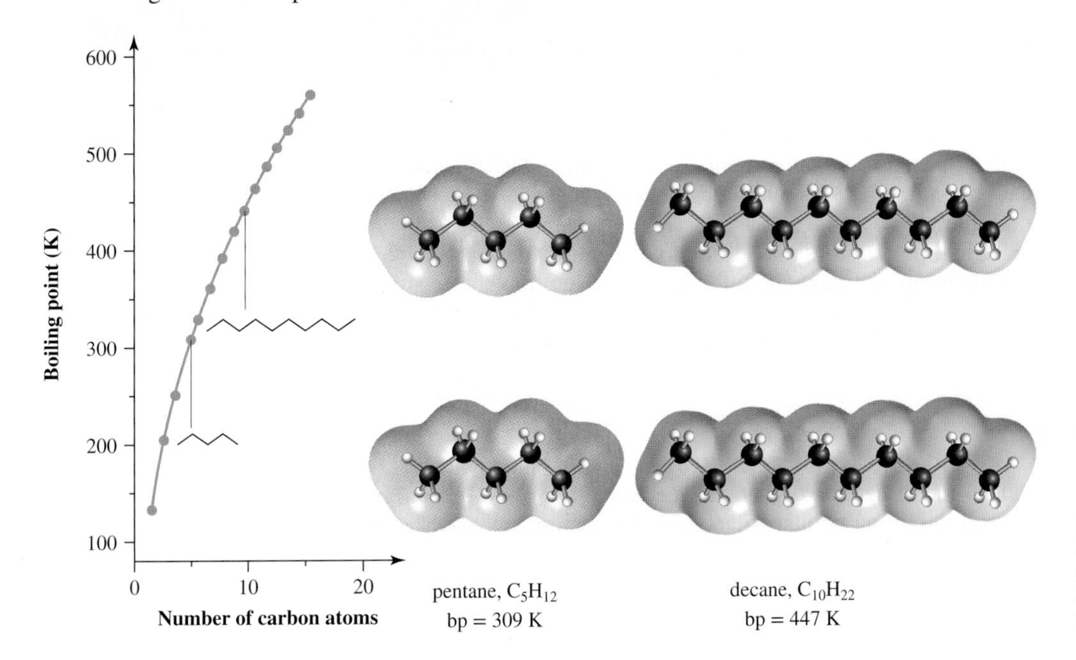

FIGURE 6.29 The boiling points of alkanes increase with the length of the carbon chain, because a large electron cloud is more polarisable than a small one.

Dispersion forces increase in strength with the number of electrons because larger electron clouds are more polarisable than smaller electron clouds. For molecules with comparable numbers of electrons, the shape of the molecule makes an important secondary contribution to the magnitude of dispersion forces. For example, figure 6.30 shows the shapes of pentane and 2,2-dimethylpropane. Both of these molecules have the formula C_5H_{12}, with 72 total electrons. Notice that 2,2-dimethylpropane has a more compact structure than pentane. This compactness results in a less polarisable electron cloud and smaller dispersion forces. Accordingly, pentane has a boiling point of 309 K, while 2,2-dimethylpropane boils at 283 K.

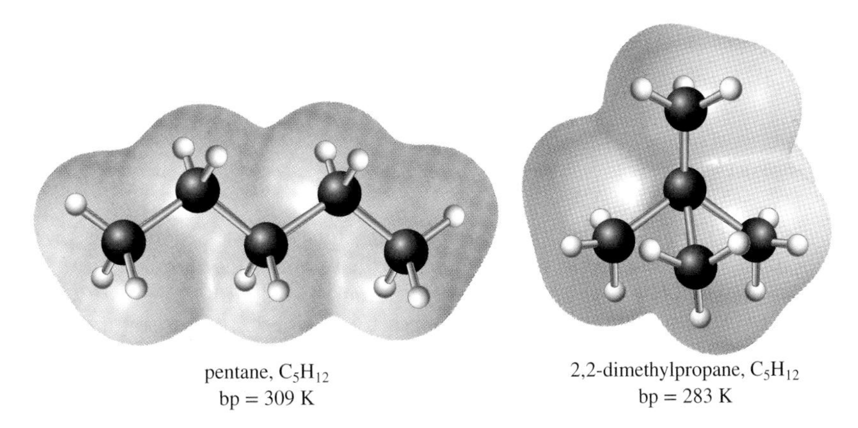

FIGURE 6.30 The boiling point of pentane is higher than the boiling point of 2,2-dimethylpropane because an extended electron cloud is more polarisable than a compact one.

Dipolar forces

Dispersion forces exist between all molecules, but some substances remain liquid at much higher temperatures than can be accounted for by dispersion forces alone. Consider 2-methylpropane and propanone (acetone), whose structures are shown in figure 6.31 (overleaf). These molecules have similar shapes and nearly the same number of electrons (34 versus 32). They are so similar that we might expect the two to have nearly equal boiling points, but acetone is a liquid at room temperature, whereas 2-methylpropane is a gas.

Why does acetone remain a liquid at temperatures well above the boiling point of 2-methylpropane? The reason is that acetone has a large dipole moment. Remember from chapter 5 that chemical bonds are polarised toward the more electronegative atom. Thus, the C=O bond of acetone is highly polarised, with a partial negative charge on the O atom ($\chi = 3.5$) and a partial positive charge on the C atom ($\chi = 2.5$). The C—H bonds in these molecules, in contrast, are only very slightly polar, because the electronegativity of hydrogen ($\chi = 2.1$) is only slightly smaller than that of carbon.

When two polar acetone molecules approach each other, they align so that the δ+ end of one molecule is close to the δ− end of the other (see figure 6.31). In a liquid array, this repeating pattern of head-to-tail alignment creates significant net attractive dipolar forces between the molecules.

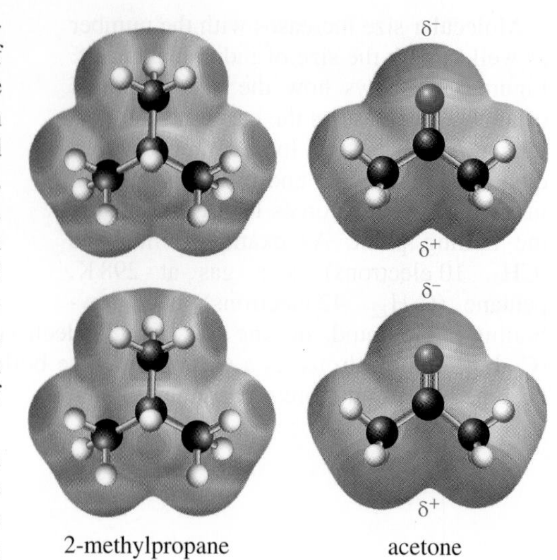

FIGURE 6.31 Acetone and 2-methylpropane have similar molecular shapes, but acetone has a large dipole moment resulting from its polar C=O bond.

These dipolar forces in addition to dispersion forces, which are nearly the same in acetone and 2-methylpropane, make the total amount of intermolecular attraction between acetone molecules substantially greater than between molecules of 2-methylpropane. Consequently, acetone boils at a considerably higher temperature than 2-methylpropane.

WORKED EXAMPLE 6.13

Boiling points and structure

The line structures of butane, methoxyethane and acetone are shown below. Explain the trend in boiling points: butane (273 K), methoxyethane (281 K) and acetone (329 K).

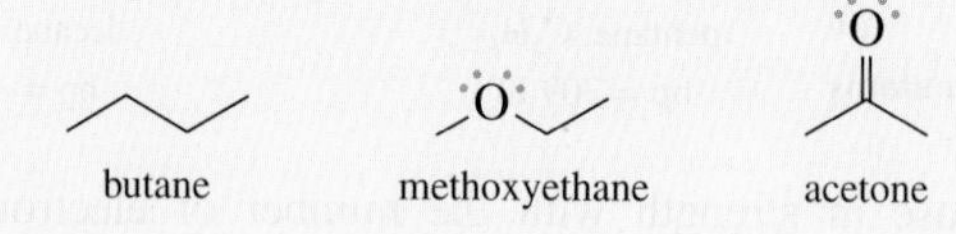

Analysis

These boiling points can be explained in terms of dispersion and dipolar forces. First, assess the magnitudes of dispersion forces, which are present in all substances, and then look for polarity within the molecules.

Solution

A table helps to organise the available information.

Substance	Boiling point	Total electrons
butane	273 K	34
methoxyethane	281 K	34
acetone	329 K	32

The table shows that dispersion forces alone cannot account for the range in boiling temperatures. Methoxyethane and butane have the same number of electrons and similar shapes; yet their boiling points are different. In addition, acetone, which has two fewer electrons than the other compounds, boils at the highest temperature. The order of boiling points indicates that acetone is a more polar molecule than methoxyethane, which in turn is more polar than butane.

We expect butane to be non-polar because of the small electronegativity difference between carbon and hydrogen. Acetone and methoxyethane, on the other hand, both contain polar carbon–oxygen bonds. Molecular geometry reveals why acetone is more polar than methoxyethane. The full structures of these molecules show that the oxygen atom in the ether has four sets of electron pairs and a bent geometry. The two C—O bond dipoles in methoxyethane partially cancel each other, leaving a relatively small molecular dipole moment. On the other hand, the polar C=O bond in acetone is unopposed, so acetone has a larger dipole moment and is more polar than methoxyethane.

Is our answer reasonable?

For the molecules in this example, we would expect those containing electronegative oxygen atoms to have higher boiling points than butane because of the presence of significant permanent dipoles. An O atom bonded to only one atom results in a large dipole moment for the molecule as explained previously, and therefore we expect acetone to have a higher boiling point than methoxyethane. Experimental values for the dipole moments are: butane, 0 C m; methoxyethane, 3.74×10^{-30} C m; and acetone, 9.62×10^{-30} C m.

PRACTICE EXERCISE 6.13

Acetaldehyde, CH_3CHO, has a structure similar to acetone but with a CH_3 group replaced by H. This substance boils at 294 K. Explain its boiling point relative to the three compounds described in worked example 6.13.

Although not strictly classed as an intermolecular force, the interaction between an ion and the permanent dipole of a molecule can be significant. Such interactions are called ion–dipole interactions, and these can be viewed as lying between dipole–dipole interactions and pure ionic interactions. In addition, ions polarise neighbouring molecules, leading to induced dipole interactions, which are much weaker than ion–dipole interactions.

Hydrogen bonds

Methoxyethane is a gas at room temperature (boiling point = 281 K), but propan-1-ol (see figure 6.32) is a liquid (boiling point = 370 K). The compounds have the same molecular formula, C_3H_8O, and each has a chain of four inner atoms, C—O—C—C and O—C—C—C. Consequently, the electron clouds of these two molecules are about the same size, and their dispersion forces are comparable. Each molecule has an sp^3-hybridised oxygen atom with two polar single bonds, so their dipolar forces should be similar. The very different boiling points of propan-1-ol and methoxyethane make it clear that dispersion and dipolar forces do not reveal the entire story of intermolecular attractions.

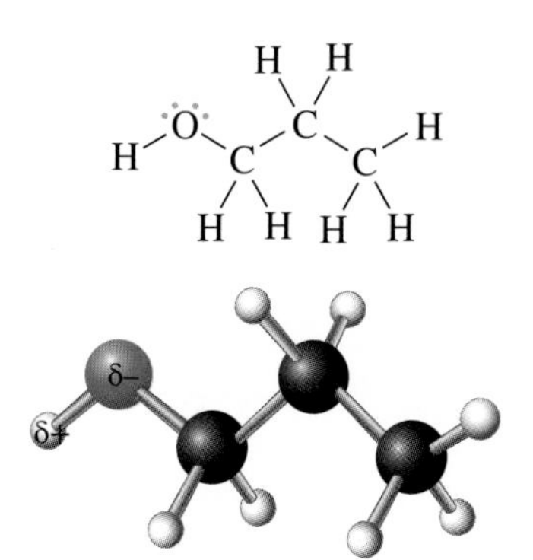

FIGURE 6.32 The structural formula and a ball-and-stick model of propan-1-ol.

The forces of attraction between propan-1-ol molecules are stronger than those between methoxyethane molecules because of an intermolecular interaction called a hydrogen bond. A hydrogen bond occurs between a small, highly electronegative atom with a lone pair of electrons and a positively polarised hydrogen atom. These interactions can be considered a special example of dipole–dipole interactions, and their strengths generally lie between those of dispersion forces and covalent bonds.

There are two requirements for hydrogen bond formation. First, there must be an electron-deficient hydrogen atom that can be attracted to an electron pair. Hydrogen atoms in O—H, F—H and N—H bonds meet this requirement. Second, there must be a small, highly electronegative atom with an electron pair that can interact with the electron-deficient hydrogen atom. Three second-row elements, O, N and F, meet this requirement, and there is evidence that S and Cl atoms can also form hydrogen bonds, but these are much weaker than those involving F, O and N atoms. Figure 6.33 shows representative examples of hydrogen bonding. The hydrogen bonds are shown as dotted lines to indicate the weak nature of these dipolar interactions.

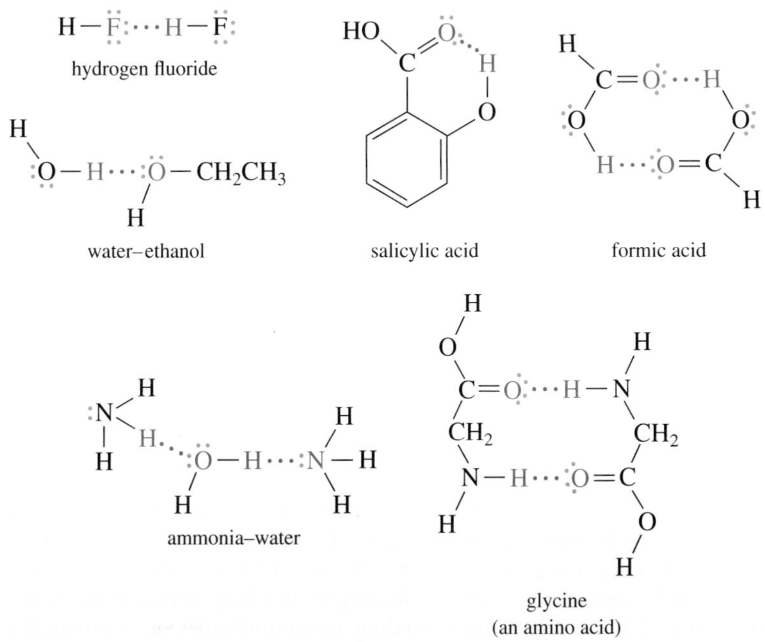

H—F:···H—F:

hydrogen fluoride

water–ethanol

salicylic acid

formic acid

ammonia–water

glycine
(an amino acid)

FIGURE 6.33 Examples of hydrogen bonding.

Notice from the examples shown in figure 6.34 that hydrogen bonds can form between different molecules (e.g. $H_3N\cdots H_2O$) or between identical molecules (e.g. $HF\cdots HF$). Also notice that molecules can form more than one hydrogen bond (e.g. glycine) and that hydrogen bonds can form between molecules as well as within a molecule (e.g. salicylic acid).

WORKED EXAMPLE 6.14

Formation of hydrogen bonds

In which of the following systems will hydrogen bonding play an important role: CH_3F, $(CH_3)_2CO$ (acetone), CH_3OH, and NH_3 dissolved in $(CH_3)_2CO$?

Analysis

Hydrogen bonds require electron-deficient hydrogen atoms in polar H—X bonds and small, highly electronegative atoms with nonbonding pairs of electrons. Use structural formulae to determine whether these requirements are met.

Solution

Here are the structural formulae of the four molecules with nonbonding electron pairs indicated.

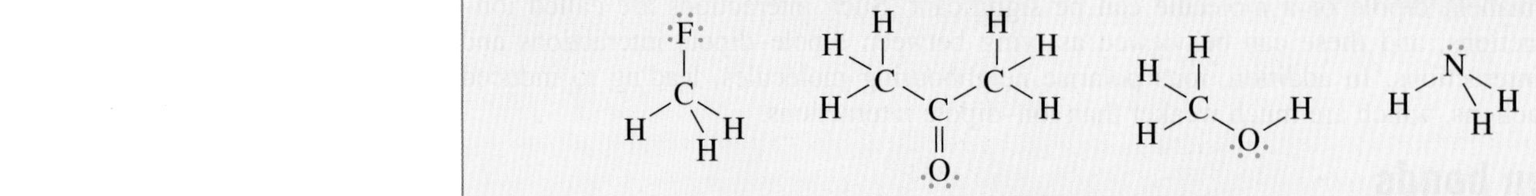

Acetone and CH_3F contain electronegative atoms with nonbonding pairs, but neither has any highly polar H—X bonds. Thus, the magnitude of any hydrogen bonding between molecules of these substances will be small.

The O—H bond in CH_3OH meets both of the requirements for hydrogen bonding. The O—H hydrogen atom on one molecule interacts with the oxygen atom of a neighbouring molecule.

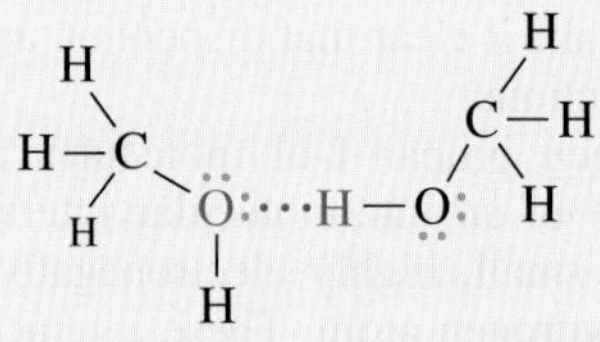

For a solution of ammonia in acetone, we must examine both components. Acetone has an electronegative oxygen atom with nonbonding pairs, whereas NH_3 has a polar N—H bond. Consequently, a mixture of these two compounds displays hydrogen bonding between the hydrogen atoms of ammonia and oxygen atoms of acetone, and, depending on the concentration, there will also be hydrogen bonds between ammonia molecules.

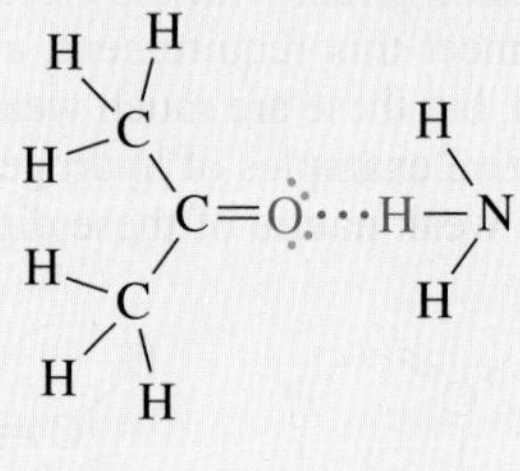

Is our answer reasonable?

Methanol has a considerably higher boiling point than the alkanes, methane and ethane, consistent with significant intermolecular forces. Ammonia dissolves readily in acetone, also consistent with significant intermolecular forces.

PRACTICE EXERCISE 6.14

Draw a picture that shows the hydrogen-bonding interactions of an acetone molecule dissolved in water.

Hydrogen bonding is particularly important in biochemical systems, because biomolecules contain many oxygen and nitrogen atoms that participate in hydrogen bonding. The most important example of biological hydrogen bonding is seen in DNA, where hydrogen bonds hold the two strands of DNA together in a double helix geometry. Likewise, the amino acids from which proteins are made contain $—NH_2$ (amino) and $—CO_2H$ (carboxylic acid) groups, and four different types of hydrogen bonds exist in these systems: O···H—N, N···H—O, O···H—O and N···H—N. When biological molecules contain S atoms, S···H—O and S···H—N hydrogen bonds can also form. Figure 6.33 includes a view of hydrogen bonding between molecules of the amino acid glycine. More details of hydrogen bonding in biomolecules are examined in chapters 24 and 25.

Binary hydrogen compounds

Boiling points of binary hydrogen compounds illustrate the interplay between different types of intermolecular forces. Figure 6.34 shows that there are periodic trends in the boiling points of these compounds. In general, the boiling points of the binary hydrogen compounds increase from top to bottom of each column of the periodic table. This trend is due to increasing dispersion forces; the more electrons the molecule has, the stronger are the dispersion forces and the higher is the boiling point. In group 16, for example, H_2S (18 electrons) boils at 213 K, H_2Se (36 electrons) at 232 K and H_2Te (54 electrons) at 269 K.

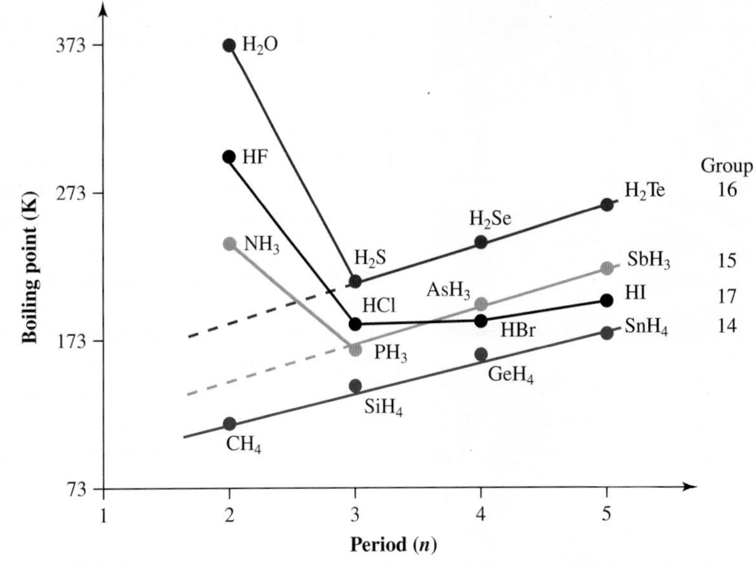

FIGURE 6.34 Periodic trends in the boiling points of binary hydrogen compounds. The expectation that the boiling point for substances decreases with molar mass is not followed by all of the binary hydrogen compounds. NH_3, H_2O and HF are clear exceptions to that trend.

As figure 6.34 shows, ammonia, water and hydrogen fluoride depart dramatically from the periodic behaviour. This is because their molecules experience particularly large intermolecular forces resulting from hydrogen bonding. In hydrogen fluoride, for instance, hydrogen bonds form between the highly electronegative fluorine atom of one HF molecule and the electron-deficient hydrogen atom of another HF molecule. Similar interactions among many HF molecules result in chains of hydrogen-bonded HF molecules leading to a boiling point much higher than those of HCl, HBr and HI. Note, however, that even the boiling point of HCl is higher than if it were on a trendline parallel to those for groups 15 and 16, indicating that weak hydrogen bonds may be formed.

Fluorine has the highest electronegativity of any element, so the strongest *individual* hydrogen bonds are those in HF. Every hydrogen atom in liquid HF is involved in a hydrogen bond, but there is only one polar hydrogen atom per molecule. Thus, each HF molecule may participate in two hydrogen bonds with two other HF partners. There is one hydrogen bond involving the partially positive hydrogen atom and a second involving the partially negative fluorine atom, leading to the formation of HF chains.

Water has a substantially higher boiling point than hydrogen fluoride, which indicates that the *overall* hydrogen bonding in H_2O is stronger than that in HF, even though the individual hydrogen bonds in HF are stronger. The higher boiling point of water reflects the fact that it forms more hydrogen bonds of significant strength per molecule than hydrogen fluoride. A water molecule has two hydrogen atoms that can form hydrogen bonds and two nonbonding electron pairs on each oxygen atom. This permits every water molecule to be involved in four hydrogen bonds to four other H_2O partners, as shown in figure 6.35a.

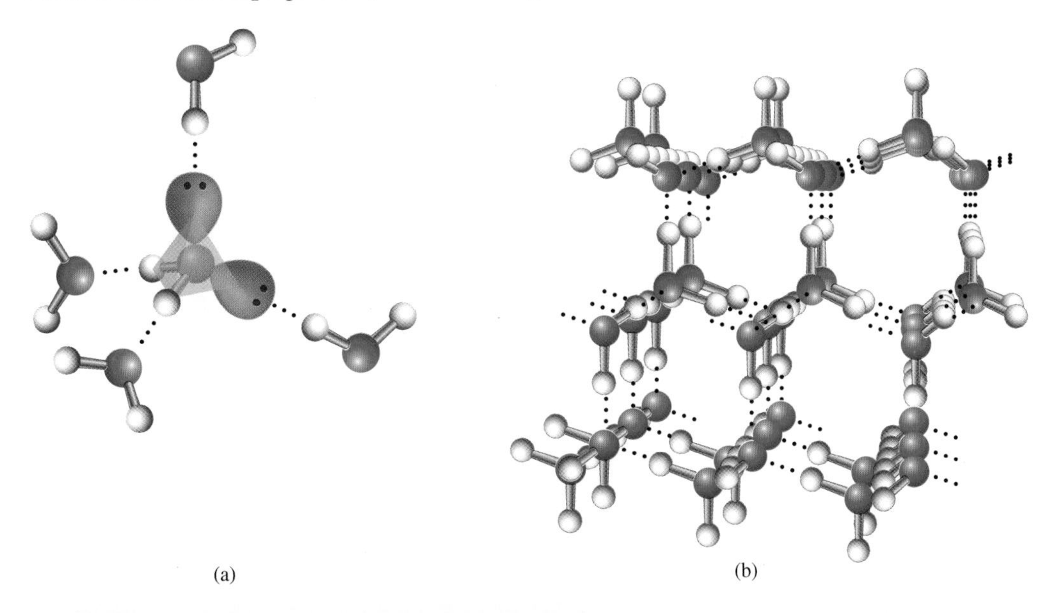

FIGURE 6.35 Two representations of the structure of ice: **(a)** Each oxygen atom is at the centre of a distorted tetrahedron of hydrogen atoms. The tetrahedron is composed of two short covalent O—H bonds and two long H···O hydrogen bonds. **(b)** Water molecules in ice are held in a network of these tetrahedra.

Hydrogen bonding in ice creates a three-dimensional network (figure 6.35b) that puts each oxygen atom at the centre of a distorted tetrahedron. Two arms of the tetrahedron are regular covalent O—H bonds, whereas the other two arms of the tetrahedron are hydrogen bonds to two different water molecules.

Therefore, both the strength *and* number of hydrogen bonds that a binary hydrogen compound can form determine its boiling point. We will discuss the role of hydrogen bonding in solids in the next chapter.

Chemical Connections

Atmospheric carbon dioxide

In the periodic table, elemental gases are relatively rare — and that translates to the fact that not many compounds overall are gases. This does, however, not indicate that they are not important for our everyday lives, and indeed it might be argued they are what we depend on most.

We are immersed in a gaseous atmosphere that has an approximate composition of 78% N_2 and 21% O_2, along with a number of other gaseous substances in very small concentrations. The concentration of one of the minor species, CO_2, is creating much controversy in the debate on climate change.

Figure 6.36 shows the increase in the concentration of CO_2 in the atmosphere over the past 250 years. There is clearly a very strong increase since the start of the industrial revolution in the mid 1800s. While the overall levels of CO_2 might be small, it is the rapid relative change that threatens to bring Earth to the brink of disaster. The ability for CO_2 to act as a greenhouse gas has attracted most attention and controversy among politicians. What has been in the media much less, is the negative effects of an increasing CO_2 concentration on the acidity of our oceans and consequently on the health of World Heritage listed natural features such as the Great Barrier Reef (figure 6.37). As you will see in later chapters, there is a complex interplay between the increasing concentration of CO_2 in the atmosphere, the solubility of CO_2 in sea water and the resulting pH — and its effect on the oceans' health.

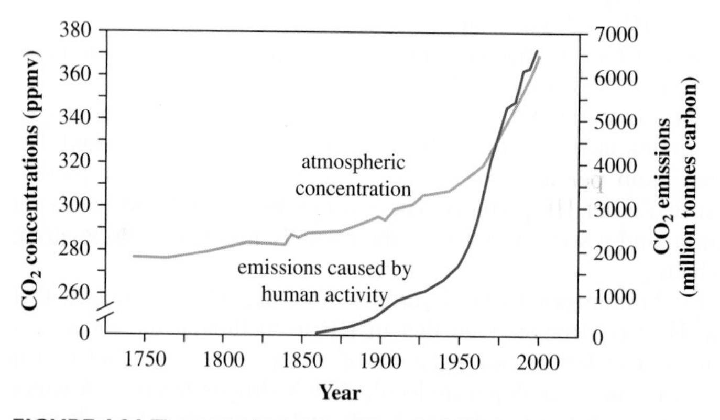

FIGURE 6.36 The atmospheric concentration of CO_2 has increased over the past 250 years, most notably as CO_2 emissions associated with human activity have increased.

FIGURE 6.37 There is a complex relationship between the concentration of CO_2 in the atmosphere and the solubility of CO_2 in the ocean. Changing amounts of CO_2 in the ocean affects the water's pH, with consequences for the health of marine life. Corals like those of the Great Barrier Reef are particularly sensitive to such changes.

LO 1 The states of matter

The three most important states of matter are solid, liquid and gas. Most substances can exist in any of these states, and the state adopted by a particular substance under defined conditions is determined by the magnitude of the forces between individual atoms, molecules or ions in the substance.

LO 2 Describing gases

Gases occupy all of the space in which they are contained. The pressure exerted by a gas is due to the collisions of rapidly moving gas atoms or molecules with the walls of the container. The SI unit of pressure is the pascal (Pa). Atmospheric pressure is measured with a barometer in which a pressure of 1 standard atmosphere (1 atm) will support a column of mercury 760 mm high. This is a pressure of 760 torr. By definition, 1 atm = $1.013\,25 \times 10^5$ pascals (Pa), and standard pressure ($p^{\ominus}$) = 1×10^5 Pa. Manometers are used to measure the pressure of trapped gases. Boyle's Law, Charles' Law and Avogadro's Law describe the relationships between volume and pressure, volume and temperature, and volume and amount of a gas, respectively. The combination of these three laws gives the ideal gas equation:

$$pV = nRT$$

where p = pressure (Pa), V = volume (m^3), n = amount (mol), R = the gas constant (8.314 J mol^{-1} K^{-1}) and T is temperature (K).

LO 3 Molecular view of gases

In order to determine the kinetic energy of a gas molecule, it is necessary to measure the speed with which it is moving. This can be done using a molecular beam apparatus. This shows that all molecules within a particular gas sample do not move with the same speed and that small molecules move, on average, faster than large molecules. However, the most probable kinetic energy of a gas molecule is the same, regardless of its mass, and all gases have an identical molecular kinetic energy distribution. The average kinetic energy of a mole of any gas is related to the temperature in kelvin through the equation:

$$E_{\text{kinetic, molar}} = \tfrac{3}{2}RT$$

An ideal gas is one for which the volume of the molecules and the forces between the molecules are so small as to be insignificant. An ideal gas obeys the ideal gas equation exactly. Real gases approximate ideal behaviour under conditions of low pressure and high temperature.

The root-mean-square speed of a gas ($\overline{u}_{\text{RMS}}$) can be found from the equation:

$$\overline{u}_{\text{RMS}} = \left(\frac{3RT}{M}\right)^{\frac{1}{2}}$$

The movement of gas molecules can be described as either effusion (the movement of molecules into a vacuum) or diffusion (the movement of one type of molecule through molecules of the same or another type).

LO 4 Gas mixtures

Each gaseous component in a mixture of ideal gases exerts a partial pressure. The sum of partial pressures of all gases in a mixture equals the total pressure. This is a statement of Dalton's law of partial pressures, which can also be expressed as:

$$p_{\text{total}} = p_1 + p_2 + p_3 + \cdots + p_i$$

The mole fraction (x_A) of a substance A equals the ratio of the amount of A (n_A), to the total amount (n_{total}) of all the components of a mixture, that is, $x_A = \frac{n_A}{n_{\text{total}}}$. In terms of mole fractions, $p_A = x_A p_{\text{total}}$. The composition of a gas mixture can also be expressed in terms of parts per million or parts per billion.

LO 5 Applications of the ideal gas equation

Combination of the ideal gas equation with the stoichiometric equation, $n = \frac{m}{M}$, allows determination of the molar mass and density of an unknown gas.

LO 6 Gas stoichiometry

The ideal gas equation can be used to determine amounts, through rearrangement into the form $n_i = \frac{p_i V}{RT}$, and can therefore be used in stoichiometric calculations involving pure gases and gas mixtures.

LO 7 Real gases

Intermolecular forces are partially responsible for the fact that real gases do not exactly obey the ideal gas laws. The variation from ideal behaviour can be shown by plotting $\frac{pV}{nRT}$ versus pressure; for an ideal gas, $\frac{pV}{nRT}$ should always equal 1. Values less than 1 are a result of significant intermolecular interactions between individual gas molecules, while values greater than 1 arise from the nonzero volume of the gas molecules. Every gas shows deviation from ideal behaviour at high pressures and at temperatures near the condensation point. The van der Waals equation for a real gas:

$$\left(p + \frac{n^2a}{V^2}\right)(V - nb) = nRT$$

makes corrections for the volume of the gas molecules and for the attractive force between them. The van der Waals constant a reflects the attractive forces between molecules, whereas the constant b reflects the relative size of the gas molecules. Melting points and, especially, boiling points give good indications of the strengths of intermolecular forces. The normal boiling point is the temperature at which the average kinetic energy of molecular motion balances the attractive energy of intermolecular attractions at a pressure of 1.013×10^5 Pa. Under these conditions, the substance is converted from a liquid to a gas in a process called vaporisation. The reverse process is called condensation. The normal freezing point is the temperature at which a liquid solidifies at a pressure of 1.013×10^5 Pa.

LO 8 Intermolecular forces

There are three general types of intermolecular forces: dispersion forces, dipole-induced dipole forces and hydrogen bonds. These range in strength from 0.1 to 5 kJ mol^{-1}, 5 to 20 kJ mol^{-1} and 5 to 50 kJ mol^{-1}, respectively.

Dispersion forces result from the asymmetrical distribution of a molecule's electrons at any instant of time, leading to the formation of instantaneous dipoles within the molecule. The ease of distortion of an electron cloud is called the polarisability of the molecule; molecules with large electron clouds have high polarisabilities and consequently display large dispersion forces. Dispersion forces occur in all molecules.

Polar molecules attract each other by dipolar forces, which arise because the positive end of a permanent dipole attracts the negative end of a dipole in a neighbouring molecule.

Hydrogen bonding, a special case of dipole–dipole attractions, occurs between molecules in which hydrogen is covalently bonded to a small, very electronegative atom — principally, nitrogen, oxygen or fluorine — and molecules containing atoms that have lone pairs of electrons (again mainly nitrogen, oxygen or fluorine). Hydrogen bonding is much stronger than the other types of intermolecular attractions and is the reason for the anomalous behaviour of water, ammonia and hydrogen fluoride when compared with other binary hydrogen compounds.

KEY CONCEPTS AND EQUATIONS

Ideal gas equation
(section 6.2)

This can be used when any three of the four variables p, V, T or n, are known and we wish to calculate the value of the fourth.

$$pV = nRT$$

Dalton's law of partial pressures
(section 6.4)

We use this law to calculate the partial pressure of one gas in a mixture of gases. This requires the total pressure and either the partial pressures of the other gases or their mole fractions. If the partial pressures are known, their sum is the total pressure.

$$p_{\text{total}} = p_1 + p_2 + p_3 + \cdots + p_{\text{i}}$$

Mole fractions
(section 6.4)

Given the composition of a gas mixture, we can calculate the mole fraction of a component. The mole fraction can then be used to find the partial pressure of the component given the total pressure. If the total pressure and partial pressure of a component are known, we can calculate the mole fraction of the component.

$$p_A = x_A p_{\text{total}}$$

Relationship between intermolecular forces and molecular structure
(section 6.8)

From the molecular structure, we can determine whether a molecule is polar and whether it has N—H or O—H bonds. This lets us predict and compare the strengths of intermolecular attractions. You should be able to identify when dispersion forces, dipolar forces or hydrogen bonding occur.

van der Waals equation
(section 6.8)

$$\left(p + \frac{n^2 a}{V^2}\right)(V - nb) = nRT$$

Boiling points of substances
(section 6.8)

These allow us to compare the strengths of intermolecular forces in substances based on their boiling points.

KEY TERMS

REVIEW QUESTIONS

LO 2 Describing gases

6.1 Describe what would happen to the barometer in figure 6.2 if the tube holding the mercury had a pinhole at its top.

6.2 Explain how the difference between an empty and an inflated and balloon shows that gas molecules exert pressure.

6.3 Express the following in units of pascal: (a) 366 torr (b) 3.01 atm (c) 0.532 torr (d) 1.41×10^{-2} atm.

6.4 A sample of air was compressed to a volume of 15×10^{-3} m^3. The temperature was 298 K and the pressure was 6.67×10^5 Pa. What amount of gas was in the sample? If the sample was collected from air at a pressure of 1.00×10^5 Pa with a temperature of 298 K, what was the original volume of the gas?

6.5 Rearrange the ideal gas equation to give the following expressions.
(a) an equation that relates p_{i}, T_{i}, p_{f} and T_{f} when n and V are constant
(b) V = ?
(c) an equation that relates p_{i}, V_{i}, p_{f} and V_{f} when n and T are constant

6.6 It requires 0.355 L of air to fill a metal foil balloon to 1.000×10^5 Pa pressure at 25 °C. The balloon is tied off and placed in a freezer at −15 °C. What is the new volume of air in the balloon?

6.7 Under which of the following conditions could you use the equation $p_{\text{f}}V_{\text{f}} = p_{\text{i}}V_{\text{i}}$?

(a) A gas is compressed at constant *T*.
(b) A gas-phase chemical reaction occurs.
(c) A container of gas is heated.
(d) A container of liquid is compressed at constant *T*.

LO 3 Molecular view of gases

6.8 Redraw figure 6.6b to show the distribution of molecules if the temperature of the oven is doubled.

6.9 Draw a single graph that shows the speed distributions of N_2 at 200 K, N_2 at 300 K and He at 300 K.

6.10 Calculate the root-mean-square speed and average kinetic energy per mole for each of the following.
(a) He at 527 °C
(b) O_2 at 127 °C
(c) SF_6 at 27 °C

6.11 Explain in molecular terms why each of the following statements is true.
(a) At very high pressure, no gas behaves ideally.
(b) At very low temperature, no gas behaves ideally.

6.12 The figure below represents an ideal gas in a container with a movable friction-free piston.

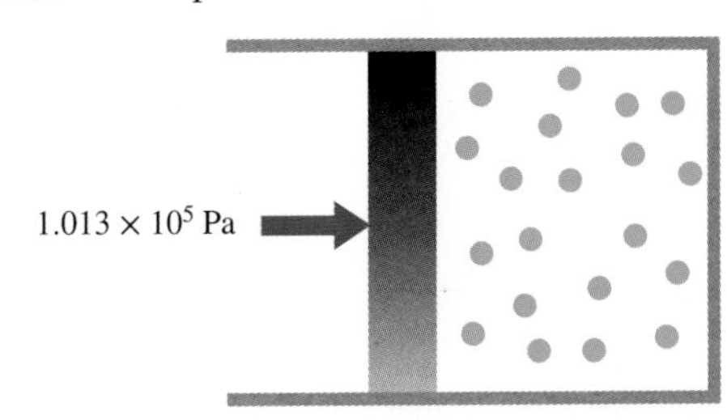

(a) The external pressure on the piston exerted by the atmosphere is 1.013×10^5 Pa. If the piston is not moving, what is the pressure inside the container? Explain in terms of molecular collisions.
(b) Redraw the sketch to show what would happen if the temperature of the gas in the container is doubled. Explain in terms of molecular collisions.

6.13 Describe the molecular changes that account for the result in question 6.6.

6.14 Determine the root-mean-square speed of SF_6 molecules at 1.01×10^5 Pa and 25 °C.

6.15 A student proposes to separate CO from N_2 by an effusion process. Is this likely to work? Why or why not?

LO 4 Gas mixtures

6.16 The concentration of NO_2 in a smoggy atmosphere was measured as 0.87 ppm. The barometric pressure was 1.011×10^5 Pa. Calculate the partial pressure of NO_2 in Pa.

6.17 In dry atmospheric air, the four most abundant components are N_2, $x = 0.7808$; O_2, $x = 0.2095$; Ar, $x = 9.34 \times 10^{-3}$; and CO_2, $x = 3.25 \times 10^{-4}$. Calculate the partial pressures of these four gases, in Pa, under standard atmospheric conditions.

6.18 The figures shown below represent very small portions of three gas mixtures, all at the same volume and temperature.

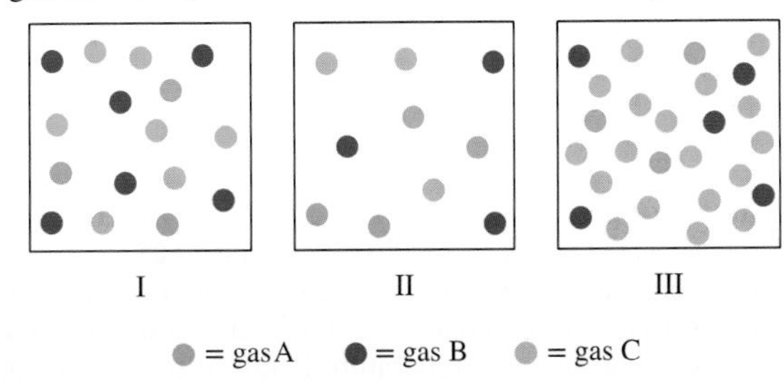

(a) Which sample has the highest partial pressure of gas A?
(b) Which sample has the highest mole fraction of gas B?
(c) In sample III, what is the concentration of gas A in ppm?

6.19 A sample of car exhaust is analysed. The gas contains 487.4 ppm CO_2, 10.3 ppb NO and 4.2 ppb CO. Calculate the partial pressures of these gases if the exhaust is emitted at 111.1 kPa pressure.

LO 5 Applications of the ideal gas equation

6.20 What is the density (g L^{-1}) of SF_6 gas at 1.013×10^5 Pa and 25 °C?

LO 6 Gas stoichiometry

6.21 Humans consume glucose to produce energy. The products of glucose consumption are CO_2 and H_2O.

$$C_6H_{12}O_6(s) + O_2(g) \rightarrow CO_2(g) + H_2O(l) \qquad \text{(unbalanced)}$$

What volume of CO_2 is produced under body conditions (37 °C, 1.013×10^5 Pa) during the consumption of 3.33 g of glucose?

6.22 Sodium metal reacts with molecular chlorine gas to form sodium chloride. A closed container of volume 3.00×10^3 mL contains chlorine gas at 27 °C and 1.67×10^5 Pa. Then 6.90 g of solid sodium is introduced, and the reaction goes to completion. What is the final pressure if the temperature rises to 47 °C?

6.23 Ammonia is produced industrially by reacting N_2 with H_2 at elevated pressure and temperature in the presence of a catalyst.

$$N_2(g) + H_2(g) \rightarrow NH_3(g) \qquad \text{(unbalanced)}$$

Assuming 100% yield, what mass of ammonia would be produced from a 1 : 1 mole ratio mixture in a reactor that has a volume of 8.75×10^3 L, under a total pressure of 275×10^5 Pa at 455 °C?

6.24 In practice, the reaction of question 6.23 gives a yield of only 17%. Repeat the calculation taking this into account.

LO 7 Real gases

6.25 Predict whether the effects of molecular volume and intermolecular attractions become more or less significant when the following changes are imposed.
(a) A gas expands into a larger volume at constant temperature.
(b) More gas is introduced into a container of constant volume at constant temperature.
(c) The temperature of a gas increases at constant pressure.

6.26 From the following experimental data, calculate the percentage deviation from ideal behaviour: 1.00 mol CO_2 in a 1.20 L container at 25 °C exerts 19.7×10^5 Pa pressure.

6.27 Chlorine gas is commercially produced by electrolysis of sea water and is stored under pressure in metal tanks. A typical tank has a volume of 15.0 L and contains 1.25 kg of Cl_2. Use the van der Waals equation to calculate the pressure in this tank if the temperature is 295 K, and compare the result with the ideal gas value.

LO 8 Intermolecular forces

6.28 Arrange the following in order of ease of liquefaction: CCl_4, CH_4 and CF_4. Explain your ranking.

6.29 Arrange the following in order of increasing boiling point: Cl_2, I_2, Br_2 and F_2. Explain your ranking.

6.30 List ethanol, CH_3CH_2OH, propane, $CH_3CH_2CH_3$, and pentane, $CH_3CH_2CH_2CH_2CH_3$, in order of increasing boiling point, and explain what features determine this order.

6.31 Which of the following will form hydrogen bonds with another molecule of the same substance? Draw molecular pictures illustrating these hydrogen bonds.
(a) CH_2Cl_2 (b) H_2SO_4 (c) H_3COCH_3 (d) H_2NCH_2COOH

6.32 Which of the following molecules form hydrogen bonds with water? Draw molecular pictures illustrating these hydrogen bonds.
(a) CH_4
(b) I_2
(c) HF
(d) H_3COCH_3
(e) $(CH_3)_3COH$

REVIEW PROBLEMS

6.33 The figure below shows three chambers with equal volumes, all at the same temperature, connected by closed valves. Each chamber contains helium gas, with amounts proportional to the number of dots representing atoms. **LO2**

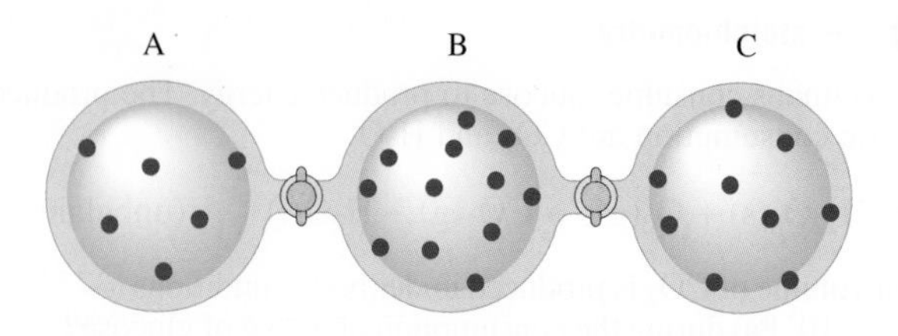

Answer each of the following questions, briefly stating your reasoning.

(a) Which of the three has the highest pressure?
(b) If the pressure in B is 1.0×10^5 Pa, what is the pressure in A?
(c) If the pressure in A starts at 1.0×10^5 Pa, and then all of the atoms from B and C are transferred to A, what will be the new pressure?
(d) If the pressure in B is 0.50×10^5 Pa, what will the pressure be after the valves are opened?

6.34 In the two figures shown below, the green molecule is about to strike the wall of its container. Assuming all other conditions are identical, which collision will exert greater pressure on the wall? Explain in terms of intermolecular interactions. **LO2**

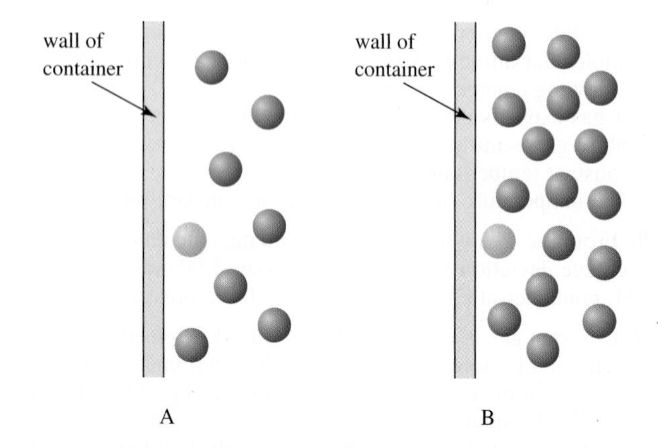

6.35 People often remark that 'the air is thin' at higher elevation. Explain this comment in molecular terms using the fact that the atmospheric pressure at the top of Mount Everest is about 3.33×10^4 Pa. **LO2**

6.36 At an altitude of 40 km above the Earth's surface, the temperature is about −32 °C, and the pressure is about 4.0×10^2 Pa. Calculate the average molecular speed of ozone, O_3, at this altitude. **LO3**

6.37 Consider two gas bulbs of equal volume, one filled with H_2 gas at 0 °C and 2×10^5 Pa, the other containing O_2 gas at 25 °C and 1×10^5 Pa. Which bulb has (a) more molecules (b) a greater mass (c) a higher average kinetic energy of molecules and (d) a higher average molecular speed? **LO3**

6.38 Molecular beam experiments on ammonia at 425 K give the speed distribution shown in the figure. **LO3**

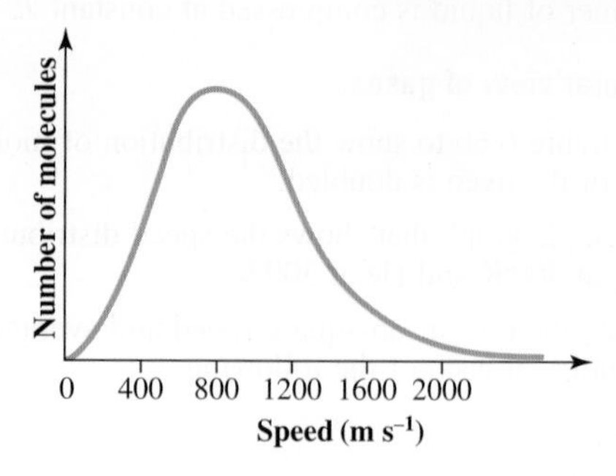

(a) What is the most probable speed?
(b) What is the most probable kinetic energy?

6.39 The figures shown below represent mixtures of argon atoms and hydrogen molecules. The volume of container B is twice the volume of container A. **LO2,3**

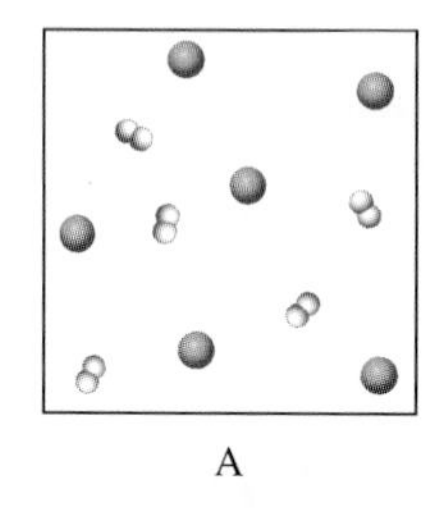

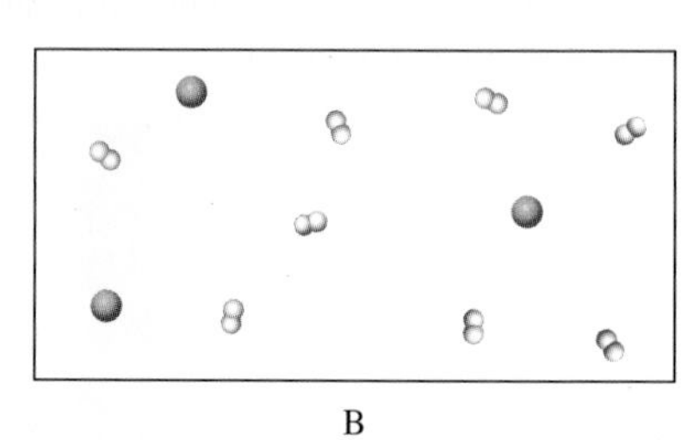

(a) Which container has a higher total gas pressure? Explain.
(b) Which container has a higher partial pressure of molecular hydrogen? Explain.
(c) One of the two gas mixtures (A or B) was used in a pulsed molecular beam experiment. The result of the experiment is shown below. Which of the two gas samples, A or B, was used for this experiment? Explain.

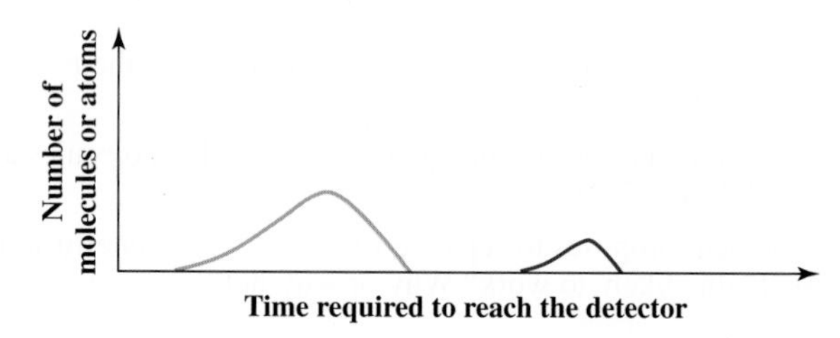

6.40 Nitrogen gas is available commercially in pressurised 9.50 L steel cylinders. If a tank has a pressure of 115×10^5 Pa at 298 K, what amount of N_2 is in the tank? What is the mass of N_2 in the tank? **LO5**

6.41 Molecular clouds composed mostly of hydrogen molecules have been detected in interstellar space. The molecular density in these clouds is about 10^{10} molecules per m^3, and their temperature is around 25 K. What is the pressure in such a cloud? **LO5**

6.42 Describe a gas experiment that would show that the element oxygen exists naturally as diatomic molecules. **LO5**

6.43 A mixture of cyclopropane gas, C_3H_6, and oxygen, O_2, in a 1.00 : 4.00 mole ratio is used as an anaesthetic. What mass of each of these gases is present in a 2.00 L bulb at 23.5 °C if the total pressure is 1.00×10^5 Pa? **LO5**

6.44 A sample of gas is found to exert a pressure of 7.00×10^4 Pa when it is in a 3.00 L flask at 0.00 °C. Calculate: **LO5**

(a) the new volume if p becomes 1.53×10^5 Pa and T is unchanged
(b) the new pressure if V becomes 2.50 L and T is unchanged
(c) the new pressure if the temperature is raised to 88.3 °C and V is unchanged.

6.45 Determine whether each of the following statements is true or false. If false, rewrite the statement so that it is true. **LO5**

(a) At constant T and V, p is inversely proportional to the amount of gas.
(b) At constant V, the pressure of a fixed amount of gas is directly proportional to T.
(c) At fixed n and V, the product of p and T is constant.

6.46 A 3.00 g sample of an ideal gas at 22 °C and 0.969×10^5 Pa occupies 0.963 L. What is its volume at 15 °C and 1.00×10^5 Pa? **LO5**

6.47 Two chambers are connected by a valve. One chamber has a volume of 15 L and contains N_2 gas at a pressure of 2.0×10^5 Pa. The other has a volume of 1.5 L and contains O_2 gas at 3.0×10^5 Pa. The valve is opened, and the two gases are allowed to mix thoroughly. The temperature is constant at 300 K throughout this process. **LO5**

(a) What amount of each of N_2 and O_2 is present?
(b) What are the final pressures of N_2 and O_2, and what is the total pressure?
(c) What fraction of the O_2 is in the smaller chamber after mixing?

ADDITIONAL EXERCISES

6.48 Recently, carbon dioxide levels at the South Pole reached 374.6 parts per million by volume. (The 1958 reading was 314.6 ppm by volume.) Convert this reading to a partial pressure in Pa. At this level, how many CO_2 molecules are there in 1.0 L of dry air at −45 °C? **LO6**

6.49 Liquid oxygen, used in some large rockets, is produced by cooling dry air to −183 °C. How many litres of dry air at 25 °C and 1.00×10^5 Pa have to be processed to produce 175 L of liquid oxygen (density = 1.14 g mL^{-1})? Assume that air is 21% O_2 by volume. **LO6**

6.50 In an explosion, a compound that is a solid or a liquid decomposes very rapidly, producing large volumes of gas. The force of the explosion results from the rapid expansion of the hot gases. For example, TNT (trinitrotoluene) explodes according to the following balanced equation. **LO6**

$$2C_7H_5(NO_2)_3(s) \rightarrow 12CO(g) + 2C(s) + 5H_2(g) + 3N_2(g)$$

TNT
$C_7H_5(NO_2)_3$

(a) What amount of gas is produced in the explosion of 1.0 kg of TNT?
(b) What volume will these gases occupy if they expand to a total pressure of 1.0×10^5 Pa at 25 °C?
(c) Calculate the partial pressure of each gas at 1.0×10^5 Pa total pressure.

6.51 The Haber synthesis of ammonia occurs in the gas phase at high temperature (400 to 500 °C) and pressure (100 to 300 × 10^5 Pa). The starting materials for the Haber synthesis are placed inside a container, in proportions shown in the figure. **LO6**

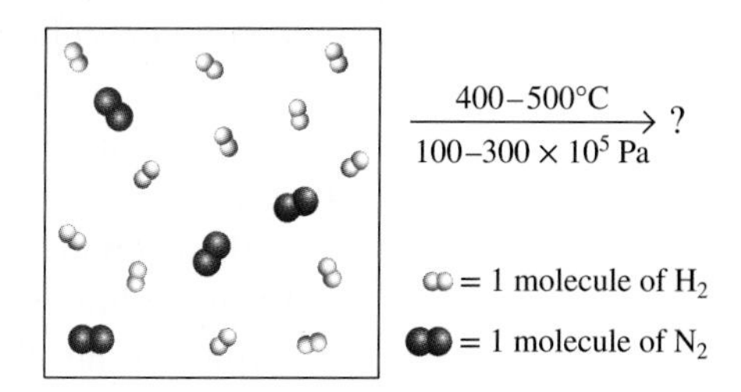

Assuming 100% yield, sketch the system at the end of the reaction.

6.52 A 0.1054 g mixture of $KClO_3$ and a catalyst was placed in a quartz tube and heated vigorously to drive off all the oxygen as O_2. The O_2 was collected at 25.17 °C and a pressure of 1.012×10^5 Pa. The volume of gas collected was 22.96 mL. **LO6**

(a) What amount of O_2 was produced?
(b) What amount of $KClO_3$ was in the original mixture?
(c) What was the mass percent of $KClO_3$ in the original mixture?

6.53 Which gas deviates more from ideal $\frac{pV}{nRT}$ behaviour, F_2 or Cl_2? Explain your choice. **LO7**

6.54 List the different kinds of forces that must be overcome to convert each of the following from a liquid to a gas. **LO8**

(a) NH_3 (b) $CHCl_3$ (c) CCl_4 (d) CO_2

6.55 For each of the following pairs, identify which has the higher boiling point, and identify the type of force that is responsible. **LO8**

(a) H_3COCH_3 and CH_3OH
(b) SO_2 and SiO_2
(c) HF and HCl
(d) Br_2 and I_2

6.56 Does the boiling point of HCl (see figure 6.34) suggest that it may form hydrogen bonds? Explain your answer and draw a molecular picture that shows the possible hydrogen bonds between HCl molecules. **LO8**

6.57 Molecular hydrogen and atomic helium both have two electrons, but He boils at 4.2 K, whereas H_2 boils at 20 K. Neon boils at 27.1 K, whereas methane, which has the same number of electrons, boils at 114 K. Explain why molecular substances boil at a higher temperature than atomic substances with the same number of electrons. **LO8**

7 Condensed phases: liquids and solids

In the previous chapter, we learned about gases, one of the three common states of matter. Icebergs floating in the ocean exemplify two common states of matter: liquid and solid. Ice is quite unusual because it is *less* dense than its liquid phase and, therefore, floats on the surface of water. This has wide-ranging consequences for life on Earth; it means that ice will accumulate at the surface of a body of water, rather than at the bottom, ensuring that aquatic life can survive even in cold winters. It also means that a full bottle of drink that you put into the freezer might crack because water expands when it changes phase from liquid to solid.

In this chapter we will investigate liquids and solids, the two condensed phases of matter. We will also look at what happens when substances change phase between solid, liquid and gas, and how these phase changes depend on both temperature and pressure. Finally we will examine order in solids and some applications of modern materials.

LEARNING OBJECTIVES

After studying this chapter, you should be able to:

7.1 describe qualitatively the macroscopic characteristics of liquids
7.2 describe qualitatively the macroscopic characteristics of solids
7.3 explain the phase changes between the common states of matter
7.4 describe order in solid structures
7.5 explain how X-ray diffraction helps to analyse order in solid structures
7.6 describe the characteristics of amorphous solids
7.7 describe the consequences of imperfections in crystal structures
7.8 outline some modern ceramics.

7.1 Liquids

In the previous chapter, we saw that, although many gases behave like an ideal gas under ambient conditions, all of them have intermolecular forces which may cause deviations from ideal behaviour. One consequence of these intermolecular forces is that a gas condenses to a liquid if it is cooled sufficiently. Condensation occurs when the average kinetic energy of the molecules falls below the value needed for them to move about independently. In a liquid, intermolecular forces are strong enough to confine the molecules to a specific volume, but not strong enough to keep molecules from moving freely within the liquid. As a consequence, like gases, liquids are fluid, and most flow easily from place to place. Unlike gases, however, liquids are compact, so they cannot expand or contract much.

Properties of liquids

Intermolecular forces in liquids lead to three important properties: surface tension, capillary action and viscosity.

Surface tension is a measure of the resistance of a liquid to an increase in its surface area. This property is caused by attractive intermolecular forces between the molecules in a liquid. These attractive forces are known as cohesive forces.

Figure 7.1 illustrates at the molecular level why liquids exhibit surface tension. A molecule in the interior of a liquid is completely surrounded by other molecules. A molecule at a liquid surface, on the other hand, has other molecules beside it and beneath it, but very few above it in the gas phase. This difference means that there is a net attractive force on molecules at the surface that pulls them towards the interior of the liquid, resulting in as few particles as possible at the surface. A liquid will therefore adopt a shape with the minimum possible surface area. Small amounts of a liquid will adopt a spherical shape because spheres have less surface area per unit volume than any other shape. For example, water drips from a tap in nearly spherical liquid droplets. Large drops are distorted from ideal spheres by the force of gravity.

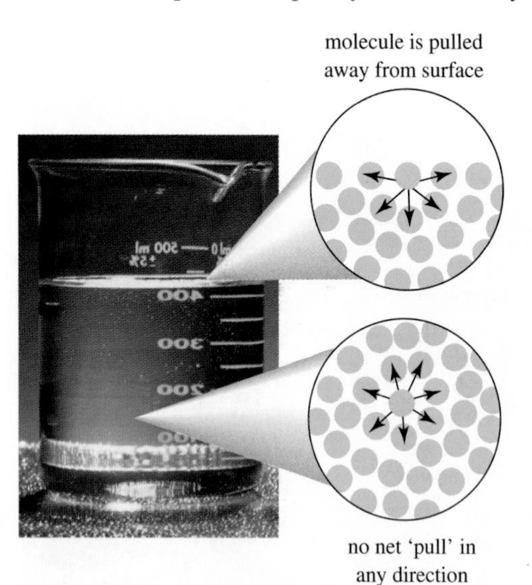

FIGURE 7.1 In the interior of a liquid (*bottom*), each molecule experiences equal forces in all directions (represented by the arrows). A molecule at the surface of a liquid (*top*) is pulled back into the liquid by intermolecular forces.

Molecules in contact with the surface of their container experience two sets of intermolecular forces. *Cohesive forces* attract molecules in the liquid to one another. *Adhesive forces* attract molecules in the liquid to the walls of the container.

One result of adhesive forces is the curved surface of a liquid, called a **meniscus**. Water in a glass tube forms a concave meniscus that increases the number of water molecules in contact with the walls of the tube. This is because the adhesive forces between water and glass are stronger than the cohesive forces between water molecules.

Figure 7.2 shows another result of adhesive forces. In a tube of small enough diameter, water actually climbs the walls because the adhesive forces between water and glass are stronger than the force of gravity. This upward movement of water against the downward force of gravity is called **capillary action** and is due to attractions between polar water molecules and oxygen atoms in glass (mainly SiO_2). Similarly, capillary action involving sap and the cellulose walls of wood fibre plays a role in how trees transport sap from their roots to their highest branches.

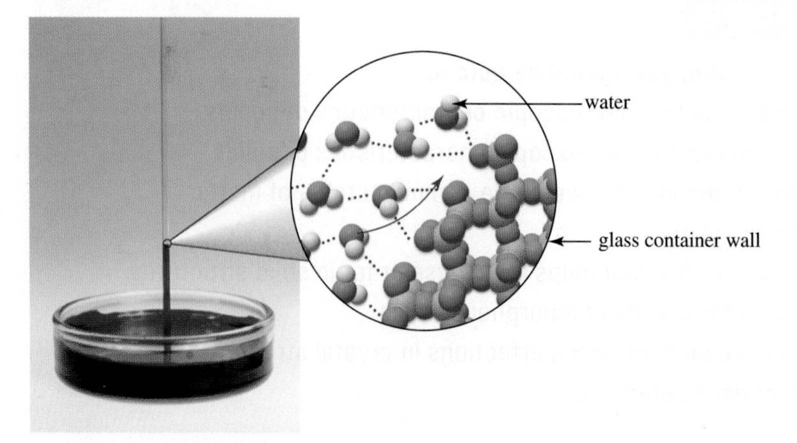

FIGURE 7.2 Water rises inside a small-diameter tube because of capillary action. The water in this photo has been coloured to show the effect of capillary action more clearly.

Water can be poured very quickly from one container to another, while oil pours more slowly, and honey sometimes seems to take forever. A liquid's resistance to flow is called its **viscosity**; the greater the viscosity, the more difficult the liquid is to pour. Viscosity is a measure of how easily molecules slide past one another. Viscosity is, therefore, affected by a combination of molecular shape and the strength of the intermolecular forces. The greater the contact area of each molecule in a liquid, the higher the viscosity of the liquid. The molecules in liquids such as water, acetone and benzene are small and compact; they have small contact areas and thus low viscosity. In contrast, molecules with a large surface area, such as the sugars in honey and the hydrocarbons found in oils, have large areas of contact and thus high viscosity.

Viscosity is affected by temperature. This dependence is quite noticeable for highly viscous substances such as honey and syrup, which are much easier to pour when hot than when cold. At higher temperature, the average kinetic energy of the molecules in the liquid is higher, which allows them to overcome the intermolecular forces more easily. Thus, viscosity decreases as temperature increases.

Vapour pressure

Our senses tell us that molecules escape from a liquid. For example, both the smell of petrol around an open tank and the evaporation of a rain puddle in the sunshine suggest that molecules escape from the liquid into the vapour phase. The red vapour phase that can be observed above liquid bromine suggests that bromine molecules are present in both phases.

Recall from section 6.3 that any collection of molecules has a distribution of kinetic energies. In a liquid, the distribution of kinetic energies is such that molecules can move about within the liquid, but on average they do not have sufficient kinetic energy to escape into the gas phase. Nevertheless, the distribution of molecular energies guarantees that some of the molecules in any liquid have enough kinetic energy to overcome the intermolecular forces that confine the liquid. These molecules escape into the vapour phase whenever they are at the surface of the liquid. If the surface of a liquid is in contact with a gas phase, some of its molecules will escape into the gas phase.

The number of molecules of a liquid that have enough energy to escape into the vapour phase depends both on the strength of intermolecular forces within the liquid and the temperature. As figure 7.3 shows, at a particular temperature more molecules can escape from liquid bromine than from liquid water. This is because the intermolecular forces between water molecules are stronger than those between bromine molecules. A water molecule needs more kinetic energy to escape the liquid phase. Figure 7.3b illustrates that, when the temperature rises, the fraction of molecules having enough energy to escape increases.

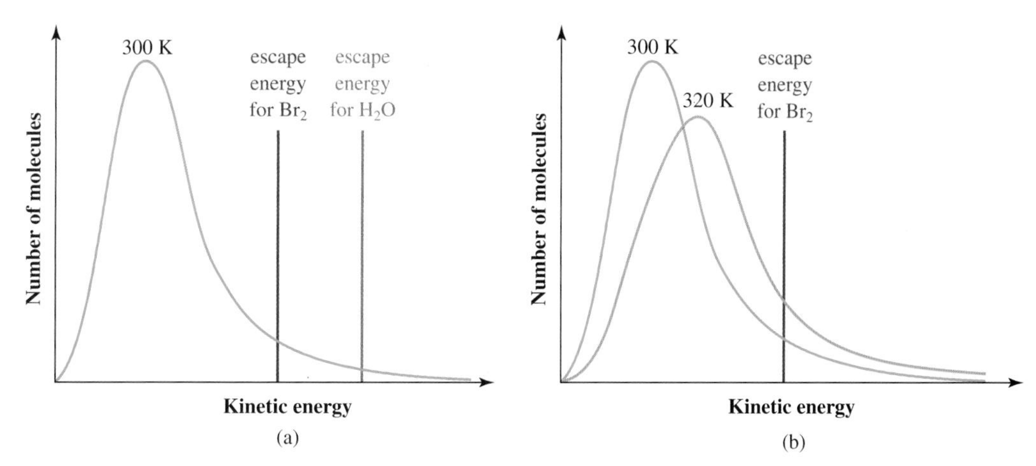

FIGURE 7.3 The fraction of molecules with enough kinetic energy to escape a liquid depends on the strength of intermolecular forces and temperature. **(a)** At 300 K, more bromine molecules can escape than water molecules because bromine has weaker intermolecular forces than water. **(b)** More bromine molecules can escape at 320 K (area under orange curve to the right of escape energy) than at 300 K (area under the blue curve to the right of escape energy) because there are more molecules with at least the escape energy.

A liquid in an open container continually loses molecules until eventually it has evaporated completely. However, in a closed container such as the one shown in figure 7.4 (overleaf), the partial pressure of the vapour increases as more and more molecules enter the gas phase. As this partial pressure builds, increasing numbers of molecules from the vapour strike the liquid surface and return to the liquid. Eventually, as figure 7.4 illustrates, the number of molecules escaping from the liquid exactly matches the number of molecules being captured by the liquid. There is no net change in the total number of molecules in the gas and the liquid, and we call this a dynamic equilibrium. The pressure at which this equilibrium exists is the **vapour pressure** of the liquid. The vapour pressure of any liquid rises with increasing temperature because more molecules have sufficient kinetic energy to escape the liquid phase into the gas phase.

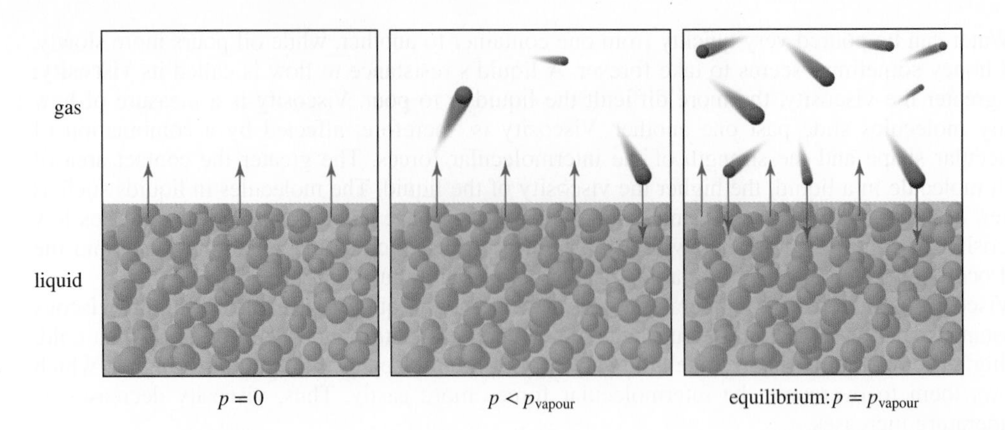

FIGURE 7.4 As the partial pressure of a substance in the gas phase above a liquid in a closed container increases, the difference between the number of molecules escaping into the gas phase (represented by the upward red arrows) and the number re-entering the liquid phase (represented by the downward red arrows) decreases until, at the vapour pressure of the substance, equilibrium is established.

Once the vapour pressure of the liquid in an open container reaches the external pressure, the liquid starts to boil. Water, for instance, boils at 100 °C at average atmospheric pressure at sea level (1.013×10^5 Pa). As we learned in chapter 6, the temperature at which a liquid boils at this pressure is known as the normal boiling point. At lower pressure, such as we would find on top of a high mountain, the boiling point is lower because the required vapour pressure is reached at a lower temperature. For example, on top of New Zealand's Aoraki/Mount Cook, which is 3754 m high, average atmospheric pressure is 0.637×10^5 Pa and water boils at about 85 °C. Conversely, a pressure cooker works on the principle that if you increase the external pressure the boiling point increases. For instance, at normal atmospheric pressure, you cannot heat liquid water to above 100 °C, whereas in a household pressure cooker temperatures of up to 120 °C can be reached. Figure 7.5 shows the dependence of vapour pressure on temperature for a range of liquids. The variation between vapour pressures of different substances is due to the different strengths of intermolecular forces.

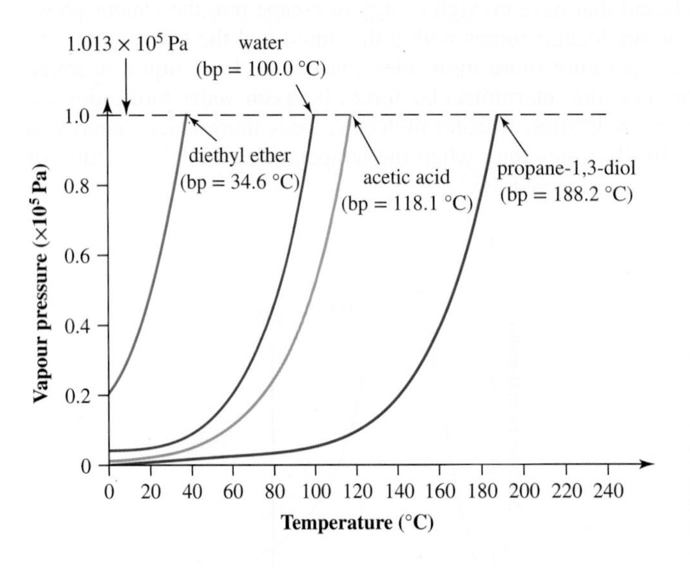

FIGURE 7.5 Vapour pressure as a function of temperature for diethyl ether, water, acetic acid and propane-1,3-diol.

7.2 Solids

The vast majority of compounds are solids under ambient conditions. A substance is a solid when its ions, atoms or molecules are held together so strongly that they cannot easily move past each other. Unlike gases and liquids, solids are characterised by rigidity, which gives them the stable shapes that we see in structures ranging from bones to aeroplane wings. One of the most active areas of research in chemistry, physics and engineering is the development of new or improved solid materials. Solids continue to play a large role in society, from capacitors and solid-state battery components in mobile phones and laptops to new tissue-compatible solids for surgical implants. In this section, we describe the various types of solids.

Magnitudes of forces

In chapter 6, we saw that the melting points of solids span an immense range, for example, from 0.95 K (He) to 1808 K (Fe) for those presented in table 6.4 (p. 245). These values indicate that the forces in solids range from very small to very large. This is because the ions, atoms or molecules in solids can be bound together by various attractive forces: intermolecular forces, metallic bonding, covalent bonds and ionic interactions.

The molecules of a **molecular solid** are held in place by intermolecular forces: dispersion forces and dipolar interactions, including hydrogen bonds. The atoms of a **metallic solid** are held in place by delocalised bonding involving mobile electrons (pp. 265–6). A **network solid** contains an array of covalent bonds linking every atom to its neighbours. An **ionic solid** contains cations and anions, attracted to one another by electrostatic forces as described in section 5.2. Table 7.1 contrasts the forces and energies associated with these four types of solids.

TABLE 7.1 Characteristics of different types of solids.

Solid type	atomic/molecular	molecular	molecular	metallic	network	ionic
Attractive forces	dispersion	dispersion + dipolar	dispersion + dipolar + hydrogen bonding	delocalised bonding	covalent	electrostatic
Energy ($kJ\,mol^{-1}$)	0.05–40	5–25	5–50	75–1000	150–500	400–4000
Example	Ar	HCl	H_2O	Cu	SiO_2	NaCl
Melting point (°C)	–189	–114	0	1088	1713	801
Schematic diagram	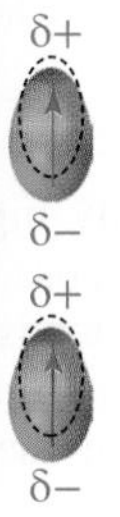 δ+ δ– δ+ δ–	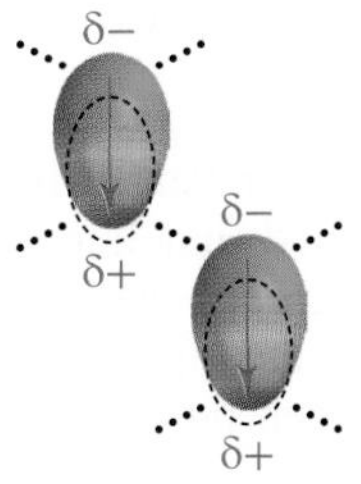 δ– δ+ δ– δ+	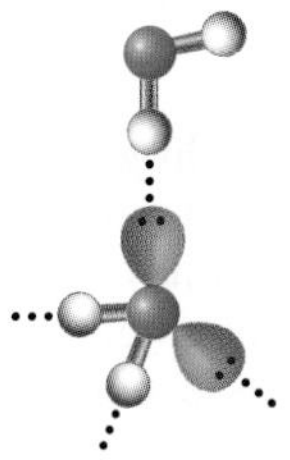	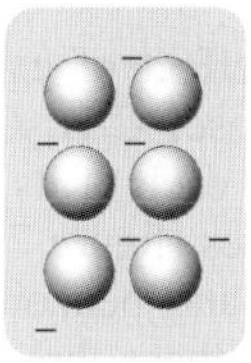	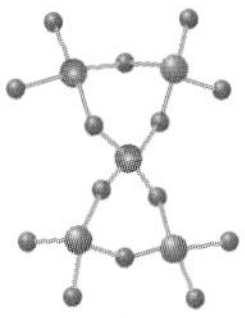	

Molecular solids

Molecular solids are aggregates of molecules bound together by intermolecular forces. The forces can be dispersion forces, dipolar forces, hydrogen bonding or a combination of these.

Many larger molecules have sufficient dispersion forces to exist as solids at room temperature. One example is naphthalene, $C_{10}H_8$, the active substance in mothballs. Naphthalene is a white solid that melts at 80 °C. Naphthalene has a planar structure with a cloud of 10 π electrons delocalised above and below the molecular plane (figure 7.6). Naphthalene molecules are held in the solid state by strong dispersion forces involving these highly polarisable π electrons. The molecules in crystalline naphthalene are arranged to maximise these dispersion forces, thus leading to the formation of plate-like macroscopic crystals.

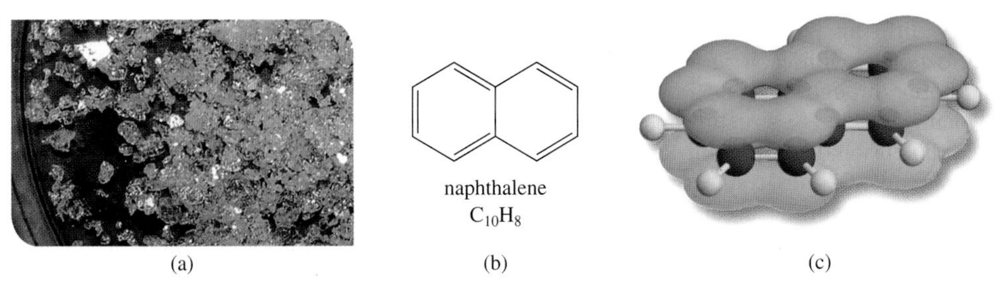

FIGURE 7.6 (a) Naphthalene, **(b)** the line structure of naphthalene and **(c)** a ball-and-stick model, overlaid with the lowest energy π bonding molecular orbital.

Dimethyl oxalate, $CH_3OC(O)C(O)OCH_3$, (figure 7.7) is an example of a molecular solid (mp = 52 °C) in which neighbouring molecules are held together predominantly by dipolar interactions. The two carbonyl C atoms have significant $\delta+$ character, due to the fact that each is bonded to two electronegative O atoms. The $\delta+$ C atoms are involved in attractive dipolar interactions with the $\delta-$ O atoms of neighbouring molecules.

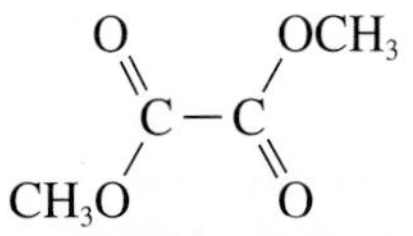

FIGURE 7.7 Structural formula of dimethyl oxalate.

In addition to dispersion and dipolar forces, molecular solids often involve hydrogen bonding. Benzoic acid, C_6H_5COOH, whose sodium salt is a common food preservative, provides a good example, as illustrated in figure 7.8. Molecules of benzoic acid are held in place by a combination of dispersion forces between the π electrons and hydrogen bonding between the —COOH groups. With fewer π electrons, benzoic acid has weaker dispersion forces than naphthalene, but its hydrogen bonding gives benzoic acid a higher melting point of 122 °C.

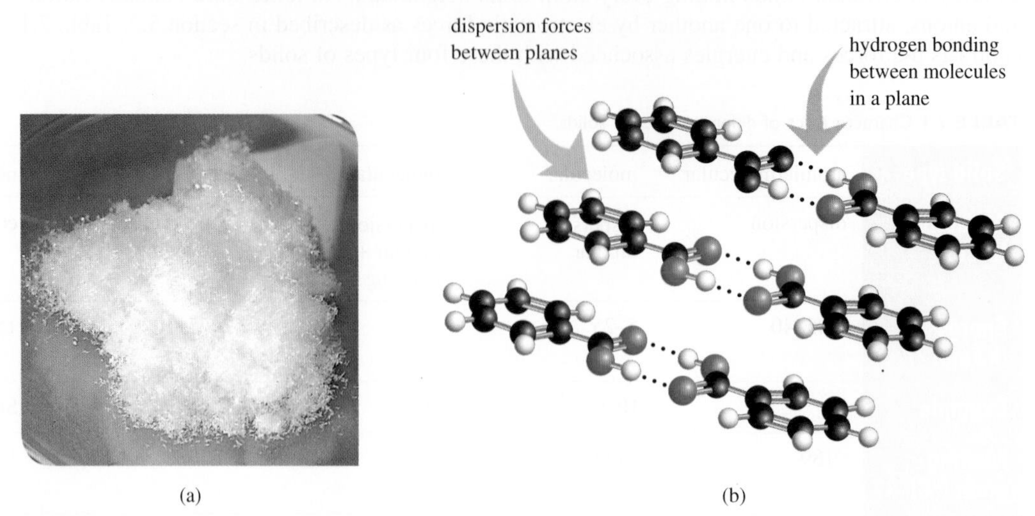

FIGURE 7.8 (a) Crystals of benzoic acid contain **(b)** pairs of molecules held together head-to-head by hydrogen bonds. These pairs then stack in planes that are held together by dispersion forces.

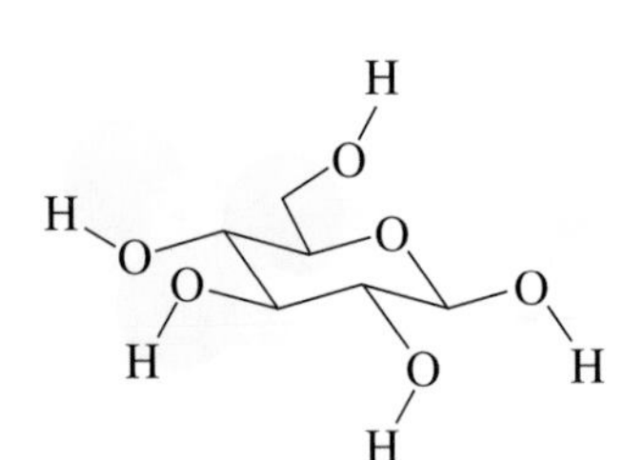

FIGURE 7.9 The structural formula of glucose, $C_6H_{12}O_6$.

The effect of extensive hydrogen bonding is revealed by the relatively high melting point of glucose, $C_6H_{12}O_6$, the sugar found in blood and human tissue. Glucose melts at 155 °C because each of its molecules has five —OH groups that form hydrogen bonds to neighbouring molecules (figure 7.9). Although glucose lacks the highly polarisable π electrons found in naphthalene and benzoic acid, its extensive hydrogen bonding gives this sugar the highest melting point of these three compounds.

Network solids

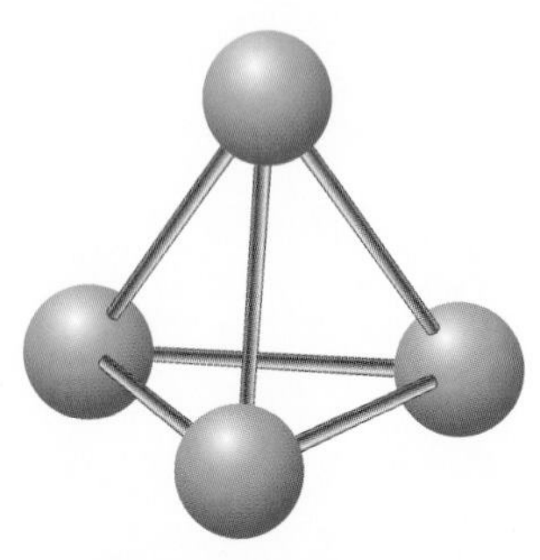

FIGURE 7.10 Representation of a P_4 molecule in white phosphorus.

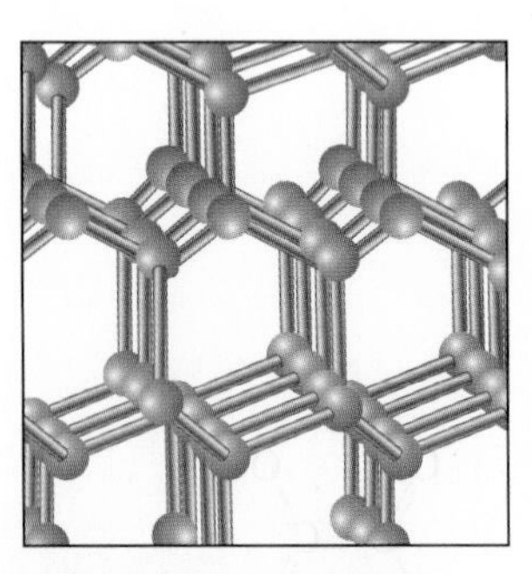

FIGURE 7.11 Part of the structural model of silicon, showing silicon atoms connected to each other to form a three-dimensional network.

In sharp contrast to molecular solids, network solids have very high melting points. Compare the behaviour of phosphorus and silicon, third-row neighbours in the periodic table. White phosphorus melts at 44 °C, but silicon melts at 1410 °C. White phosphorus is a molecular solid that contains individual P_4 molecules (see figure 7.10), but silicon is a network solid in which covalent bonds connect every Si atom to each of its four neighbours (see figure 7.11).

The cause of the great difference in the melting points of these two elements is evident from table 7.1. Covalent bonds are much stronger than intermolecular forces. For solid silicon to melt, a significant fraction of its Si—Si covalent bonds must break. The average Si—Si bond energy is 225 kJ mol^{-1}, whereas attractive energies due to intermolecular forces in P_4 are much lower (as shown in table 7.1, intermolecular forces are generally less than 50 kJ mol^{-1}) so it takes a much higher temperature to melt silicon than white phosphorus.

Diamond and graphite, the two forms of elemental carbon that occur naturally on Earth, have very different physical and chemical properties. Diamond contains a three-dimensional array of σ bonds, with each sp^3-hybridised carbon atom linked to each of its four neighbouring carbon atoms through covalent bonds resulting in a tetrahedral geometry around each carbon atom (identical to the Si structure mentioned above, see figure 7.12a). It is thus a network solid. Its three-dimensional network of strong covalent bonds makes diamond extremely strong and abrasive. Covalent bonds connecting atoms in three dimensions make network solids extremely durable. The sp^2 hybridised carbon atoms in graphite, in contrast, are connected to only three neighbouring carbon atoms via covalent bonds in a planar arrangement. The bonding is supplemented by delocalised π bonding above and below the plane of the σ bonds (see figure 7.12b). Each two-dimensional layer is attracted to its neighbouring layers only by dispersion forces between the π electrons. As a result, planes of carbon atoms easily slide past one another, making graphite a brittle lubricant.

Compounds such as silica (silicon dioxide, SiO_2) may also exist as network solids. Opals are composed of nanosized silica spheroids. Their prized colour is due to the way they diffract light, rather than the impurities that give most other gemstones their colour. Another example is silicon carbide, SiC, which has a similar structure to diamond and is used as an abrasive in sandpaper and as an edge on cutting tools as it is less expensive to produce. These substances have very high melting points because their atoms are held together by networks of strong σ covalent bonds.

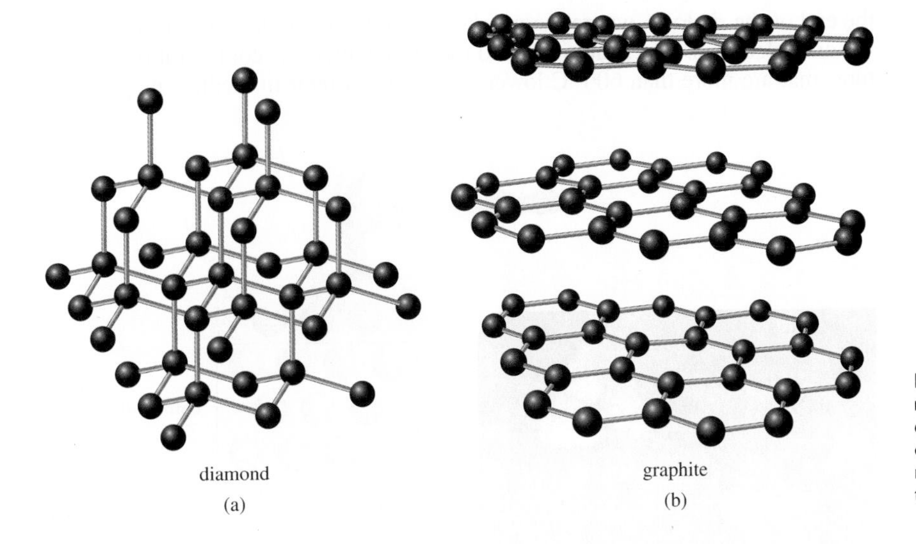

FIGURE 7.12 Part of the structural models of: **(a)** diamond, showing carbon atoms connected to each other to form a three-dimensional network and **(b)** graphite, showing the two-dimensional layers.

Metallic solids

The bonding in solid metals differs from that in other types of solids because it derives primarily from electrons in highly delocalised valence orbitals. Recall from the molecular orbital theory discussed in chapter 5 that overlap of n valence atomic orbitals gives rise to n molecular orbitals. When n is very large, as is the case in 1 mole of a solid metal, then a large number of molecular orbitals are formed. The energies of these molecular orbitals are so close together that they essentially form a continuum of energies, called a band, which extends over all of the atoms in the metal. Therefore, an electron in a partially filled orbital is delocalised over all the atoms, and is consequently able to move throughout the entire metal. Hence, we can view a metal as consisting of a regular array of metal atom cores embedded in a 'sea' of mobile valence electrons (see figure 7.13). The properties of metals, such as electrical and thermal conductivity, can be explained on the basis of this model.

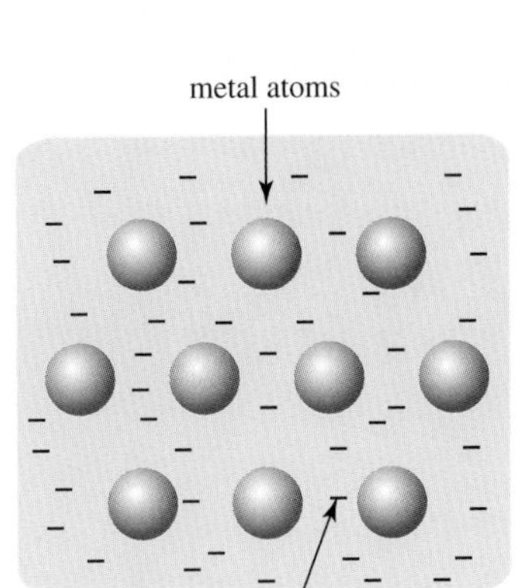

FIGURE 7.13 A diagrammatic representation of metallic bonding. The metal atoms are embedded in a 'sea' of mobile valence electrons.

Metals display a wide range of melting points, indicating that the strength of metallic bonding is variable. The group 1 metals are quite soft and melt at relatively low temperatures (see figure 7.14a and c). Sodium, for example, melts at 98 °C and caesium melts at 28.5 °C. Bonding is weak in these metals because each atom of a group 1 metal contributes only one valence electron to the bond-forming energy band. Metals near the middle of the *d* block, on the other hand, are very hard and have some of the highest known melting points: tungsten melts at 3407 °C (figure 7.14b), rhenium at 3180 °C and chromium at 1857 °C. Atoms of these metals contribute several *d* electrons to bond formation, leading to extremely strong metallic bonding.

Metals are ductile, meaning they can be drawn into wires, and malleable, meaning they can be hammered into thin sheets. When a piece of metal forms a new shape, its atoms change position. However, because the bonding electrons are fully delocalised, changing the positions of the atoms does not cause corresponding changes in the energy levels of the electrons. The 'sea' of electrons is largely unaffected by the pattern of metal atoms, as figure 7.15 (overleaf) illustrates. Thus, metals can be forced into many shapes, including sheets and wires, without destroying their bonding nature.

The *d*-block metals display a range of properties. Copper and silver are much better electrical conductors than chromium. Tungsten has very low ductility. Mercury is a liquid at room temperature. These differences arise in part because of variations in the number of valence electrons. The *d*-block metals vanadium and chromium have five or six valence electrons per atom, respectively, all of which occupy bonding orbitals in the metal atoms. As a result, there are strong attractive forces between the metal atoms, and vanadium and chromium are strong and hard. Beyond the middle of the *d*-block series, the additional valence electrons occupy antibonding orbitals, which reduces the net bonding. This effect is most pronounced at

(a)

(b)

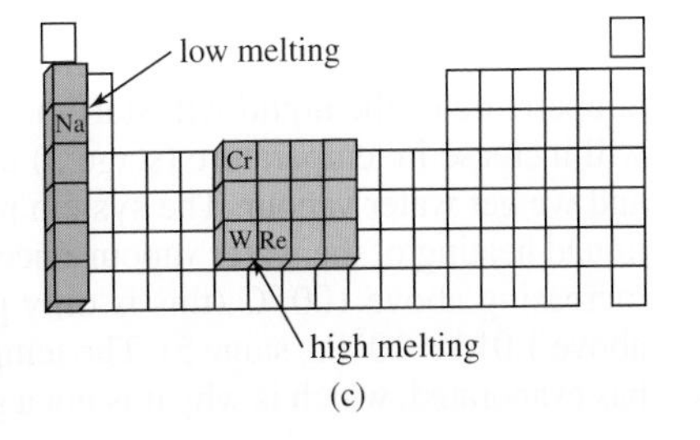

(c)

FIGURE 7.14 **(a)** Sodium, Na, can be cut with a knife. **(b)** Tungsten, W, can be heated to incandescence without melting, which is why it is used as the filament in light bulbs. **(c)** Group 1 metals melt at relatively low temperatures whereas metals near the middle of the *d* block have some of the highest known melting points.

the end of the *d* block, where the number of antibonding electrons nearly matches the number of bonding electrons. Zinc, cadmium and mercury, with $d^{10}s^2$ configurations, have melting temperatures that are more than 600 °C lower than those of their immediate neighbours.

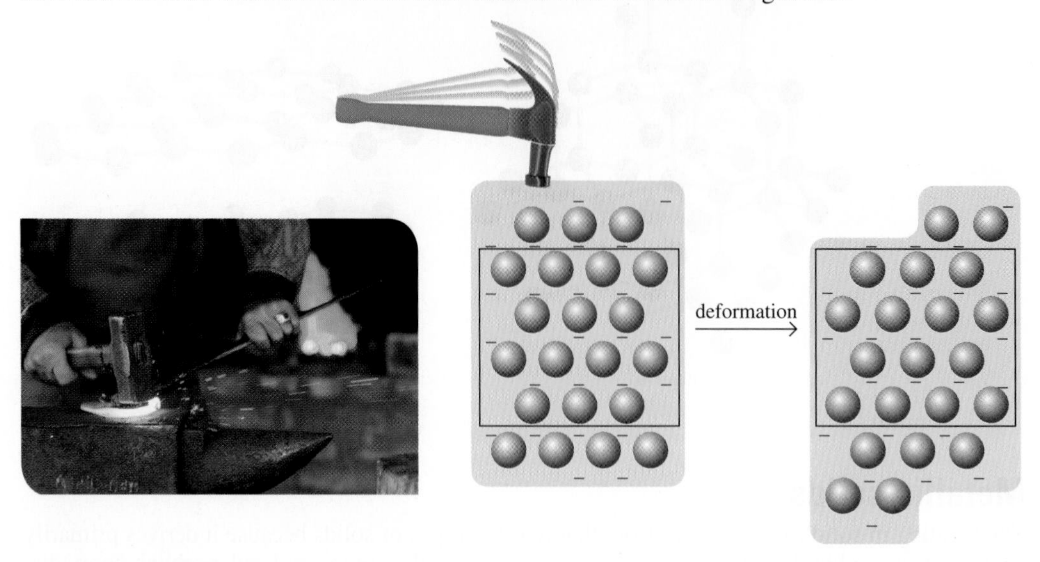

FIGURE 7.15 When a metal changes shape, its atoms shift position. However, because the valence electrons are fully delocalised, the energy of these electrons is unaffected.

Ionic solids

As described in chapter 5, ionic solids contain cations and anions strongly attracted to each other by electrostatic forces. Ionic solids must be electrically neutral, so their stoichiometries are determined by the charges carried by the positive and negative ions.

Many ionic solids contain metal cations and polyatomic anions. Here again, the stoichiometry of the solid is dictated by the charges on the ions and the need for the solid to maintain electrical neutrality. Some examples of 1 : 1 ionic solids containing polyatomic anions are NaOH, KNO_3, $CuSO_4$, $BaCO_3$ and $NaClO_3$. Some metallic ores have 1 : 1 stoichiometry, such as scheelite, $CaWO_4$ (contains WO_4^{2-}), zircon, $ZrSiO_4$ (contains SiO_4^{4-}) and ilmenite, $FeTiO_3$ (contains TiO_3^{2-}). Ilmenite is the source of TiO_2, which is used as a white pigment in paper and paints.

Minerals often contain more than one cation or anion. For example, apatite, $Ca_5(PO_4)_3OH$, the main substance in tooth enamel, contains both phosphate and hydroxide anions. Beryl, $Be_3Al_2Si_6O_{18}$, contains beryllium and aluminium cations as well as the $Si_6O_{18}^{12-}$ polyatomic anion. Beryl is the substance that makes up emeralds. An even more complicated mineral is garnierite, $(Ni, Mg)_6Si_4O_{10}(OH)_2$, which has a variable composition of Ni^{2+} and Mg^{2+} cations. Although the relative proportions of Mg^{2+} and Ni^{2+} vary, garnierite always has six cations and two anions of OH^- for every anion of $Si_4O_{10}^{10-}$.

Some mixed oxides that contain rare earth metals are *superconductors*. Below a certain temperature, a superconductor can carry an immense electrical current without incurring losses from resistance. Optimal superconducting behaviour often requires a slight deviation from stoichiometric composition, which we will learn more about in section 7.8.

7.3 Phase changes

In chapter 6 and the first two sections of this chapter, we have examined the three most common phases of matter (gas, liquid and solid) in isolation. We will now investigate what governs a **phase change** — the transition of a substance from one phase to another. Substances can undergo phase changes given the appropriate conditions. Phase changes depend on temperature, pressure, and the magnitudes of bonding and intermolecular forces.

We are all familiar with the phase changes of water at ambient pressure, which are illustrated in figure 7.16. Consider taking ice cubes from the freezer at −18 °C and placing them in a container at room temperature. At first the water is in the solid phase (stage 1). Outside the freezer, the temperature of the ice cubes begins to increase. When the temperature of the ice cubes reaches 0 °C, they begin to melt and we get a mixture of solid and liquid water (stage 2). The temperature of this mixture does not increase until all the ice has melted. Once the ice has completely melted, the temperature of the liquid will start increasing (stage 3). If we apply heat to the container the liquid will increase in temperature (stage 3) until it reaches 100 °C. The water then starts to boil (stage 4) and we get water vapour. The system remains at 100 °C until all of the water has evaporated. Continued heating of the water vapour once all of the water has evaporated will result in its temperature increasing above 100 °C (this is only possible in a confined space and will increase the pressure above 1.013×10^5 Pa, stage 5). The temperature of the container will also increase once all the water has evaporated, which is why it is not a good idea to allow a saucepan to boil dry on a hotplate.

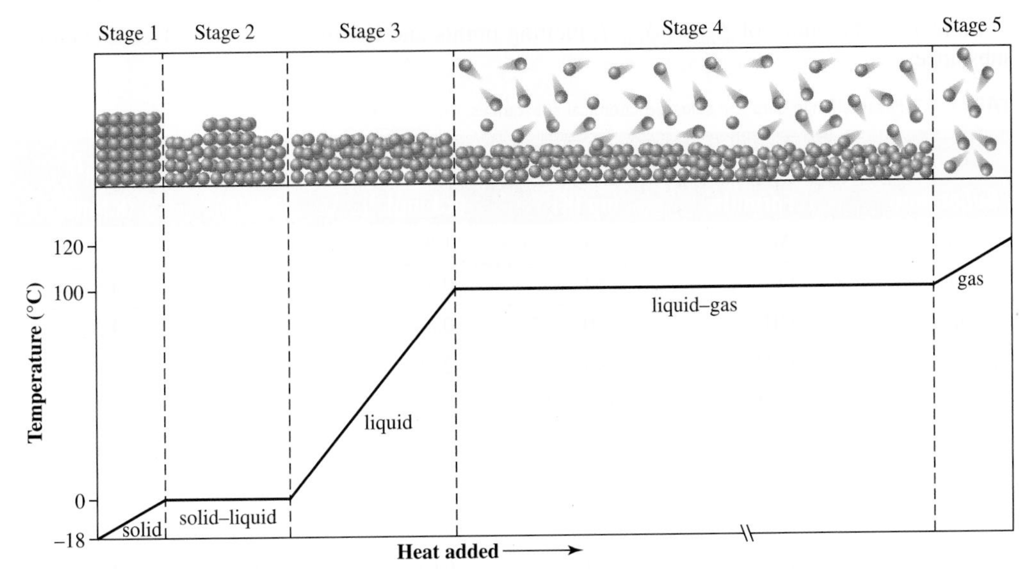

Adapted from: *Chemistry: The molecular nature of matter and change*, 3rd edition, Martin S Silberberg, p. 424, © 2003 The McGraw-Hill Companies, Inc.

FIGURE 7.16 When heat is supplied to a sample of H_2O in a sealed container at a constant pressure of 1.013×10^5 Pa, phase changes occur at 0 °C and 100 °C. The *x*-axis represents the heat required for each individual process. It should be noted that there are molecules in the vapour phase above both solid and liquid water, but these have been omitted in stages 1, 2 and 3 for clarity.

Phase changes require that energy (usually in the form of heat) be either supplied to or removed from the substance undergoing the phase change. Restricting ourselves to molecular substances, but recognising that similar arguments hold for atomic and ionic substances, a molecular perspective reveals why this is so. Any phase change that results in increased molecular mobility requires that intermolecular forces be overcome. For example, hydrogen bonds in ice must be broken to change the ice to liquid water, and hydrogen bonds in liquid water must be broken to convert it to water vapour. About 41 kJ of heat must be supplied to transfer 1 mole of water molecules from the liquid phase into the vapour phase. Similarly, 41 kJ of heat must be removed to liquefy 1 mole of water vapour. As figure 7.16 indicates, the heat required to change between the liquid and solid phases is considerably less.

As you will see in more detail in chapter 8, the amount of heat transferred at constant pressure is equal to a quantity called the enthalpy change (ΔH). The magnitude of the change depends on the strength of intermolecular forces in the substance undergoing the phase change.

The amount of heat required to vaporise a substance also depends on the amount in the sample. The energy required to vaporise 2 moles of water is twice that needed for 1 mole. The heat needed to vaporise 1 mole of a substance at its normal boiling point is called the **molar enthalpy of vaporisation ($\Delta_{vap}H$)**. For example, $\Delta_{vap}H$ of water is about 41 kJ mol^{-1}.

Energy must also be provided to melt a solid substance. This energy is used to overcome the intermolecular forces that hold molecules in fixed positions in the solid phase. The heat needed to melt 1 mole of a substance at its normal melting point is called the **molar enthalpy of fusion ($\Delta_{fus}H$)**. For example, $\Delta_{fus}H$ of water is about 6 kJ mol^{-1}.

Phase changes between solid and liquid, and between liquid and gas, are the most common, but a phase change in which a solid converts directly to a gas without passing through the liquid phase is also possible. This transition is known as **sublimation** (**deposition** is the opposite transition). Dry ice (solid CO_2) sublimes at 195 K with the **molar enthalpy of sublimation ($\Delta_{sub}H$)** = 25.2 kJ mol^{-1}. Mothballs contain naphthalene, $C_{10}H_8$ ($\Delta_{sub}H = 73$ kJ mol^{-1}), a crystalline white solid that sublimes to produce a vapour that repels moths. The purple colour of the gas above iodine crystals in a closed container (figure 7.17) provides visible evidence that this solid also sublimes at room temperature ($\Delta_{sub}H = 62.4$ kJ mol^{-1}). Both naphthalene and iodine melt at ambient pressure (at 80 °C and 114 °C, respectively), indicating that, while a small amount can sublime at ambient pressure, it is not a thermodynamic equilibrium.

FIGURE 7.17 The purple colour of the vapour above solid iodine is due to I_2 molecules in the gas phase.

Phase changes can go in either direction; ice melts upon adding heat, and liquid water freezes on removing heat from the system. Heat is absorbed as a solid melts to a liquid and is released as a liquid freezes to a solid. To make ice cubes, for instance, water is placed in a freezer that absorbs the heat released during the formation of ice. The heat released when liquid water changes to ice is equal in magnitude to the heat required to change ice to liquid water.

By convention, tabulated values of enthalpies of phase changes are always specified in terms of the phase change that requires the addition of heat. The reverse processes have the same magnitude but the opposite sign.

$$\text{solid} \longrightarrow \text{liquid} \quad \Delta_{fus}H \qquad \text{liquid} \longrightarrow \text{solid} \quad \Delta H_{solidification} = -\Delta_{fus}H$$

$$\text{liquid} \longrightarrow \text{gas} \quad \Delta_{vap}H \qquad \text{gas} \longrightarrow \text{liquid} \quad \Delta H_{condensation} = -\Delta_{vap}H$$

$$\text{solid} \longrightarrow \text{gas} \quad \Delta_{sub}H \qquad \text{gas} \longrightarrow \text{solid} \quad \Delta H_{deposition} = -\Delta_{sub}H$$

Table 7.2 lists values of $\Delta_{fus}H$, $\Delta_{vap}H$, melting points and boiling points for selected chemical substances.

TABLE 7.2 Phase change data for selected chemical substances.

Substance	Formula	mp (K)	$\Delta_{fus}H$ (kJ mol^{-1})	bp (K)	$\Delta_{vap}H$ (kJ mol^{-1})
argon	Ar	83	1.3	87	6.3
oxygen	O_2	54	0.45	90	9.8
methane	CH_4	90	0.84	112	9.2
ethane	C_2H_6	90	2.85	184	15.5
diethyl ether	$(C_2H_5)_2O$	157	6.90	308	26.0
bromine	Br_2	266	10.8	332	30.5
ethanol	C_2H_5OH	156	7.61	351	39.3
benzene	C_6H_6	278	10.9	353	31.0
water	H_2O	273	6.01	373	40.79
mercury	Hg	234	23.4	630	59.0

WORKED EXAMPLE 7.1

Enthalpy of phase change

A swimmer emerging from a pool is covered with a film containing about 75 g of water. How much heat must be supplied to evaporate this water?

Analysis

Energy in the form of heat is required to evaporate the water from the swimmer's skin. The energy needed to vaporise the water can be found using the molar enthalpy of vaporisation and the amount of water.

Solution

$\Delta_{vap}H$ of water is 40.79 kJ mol^{-1} (see table 7.2). The molar mass of water is 18.02 g mol^{-1}, so 75 g of water is 4.16 mol. Therefore, the heat that must be supplied is:

$$n\Delta_{vap}H = (4.16\ \text{mol})(40.79\ \text{kJ mol}^{-1}) = 1.7 \times 10^2\ \text{kJ}$$

Is our answer reasonable?

If the swimmer's body must supply all this heat, a substantial chilling effect occurs. Thus, swimmers usually towel off (to reduce the amount of water that must be evaporated) or lie in the sun (to let the sun provide most of the heat required).

PRACTICE EXERCISE 7.1

Determine how much heat is involved in freezing 125 g of water in an ice-cube tray. What is the direction of heat flow in this process?

Recall from our discussion of figure 7.16 that adding heat to boiling water does not cause the temperature of the water to increase. Instead, the added energy is used to overcome intermolecular attractions as more molecules leave the liquid phase and enter the gas phase. Other two-phase systems show similar behaviour. This can be used to hold a chemical system at a fixed temperature. A temperature of 100 °C can be maintained by a boiling water bath, and an ice bath holds a system at 0 °C. Lower temperatures can be achieved with other substances. Dry ice (solid CO_2) suspended in liquid acetone maintains a temperature of −78 °C (195 K); a bath of liquid nitrogen has a constant temperature of −196 °C (77 K); and liquid helium, which boils at 4.2 K, is used for research requiring ultracold temperatures.

At the start of this section we mentioned that pressure is also involved in phase changes. The effect of pressure is mainly seen for phase transitions involving gases. Recall from chapter 6 that gas density increases with pressure. A gas at constant temperature can be liquefied by increasing the pressure. This is important in the storage of liquefied petroleum gas (LPG), a mixture of propane and butane that is used as a fuel for cars, barbecues and heaters. Propane and butane are gases at room temperature and normal atmospheric pressure. To make the storage and transport of the fuel practicable, it is liquefied by placing it under pressure. The pressure at which a gas liquefies at a specific temperature is known as the condensation point.

Supercritical fluids

We have learned about the three phases we are familiar with from everyday life: solids, liquids and gases. We have seen that we can liquefy gases by cooling or increasing the pressure. As the pressure on a gas increases, the gas is compressed into an ever-smaller volume. If the temperature is low enough, this compression eventually results in liquefaction. At high enough temperature, however, no amount of compression can cause liquefaction; that is, the liquid–gas transition is no longer possible. Instead, the substance becomes a **supercritical fluid**. A supercritical fluid is a fluid that has certain characteristics of both liquids and gases but is neither.

Imagine that you have a liquid in a sealed container. On heating, the vapour pressure increases as more molecules from the liquid escape into the gas phase, thus increasing the density in the gas phase while also decreasing the density in the liquid phase. As we continue heating the container, the densities in the gas phase and the liquid phase approach the same value. At the point where they become equal we can no longer distinguish between the liquid and gas phases, i.e. we cannot observe a phase boundary. The temperature at which this occurs is called the **critical temperature** (T_c); the associated pressure is called the **critical pressure**. The combination of the critical temperature and the critical pressure is called the **critical point**. The substance is now in its supercritical state (figure 7.18). It has the viscosity typical of a liquid, but it is able to expand or contract like a gas.

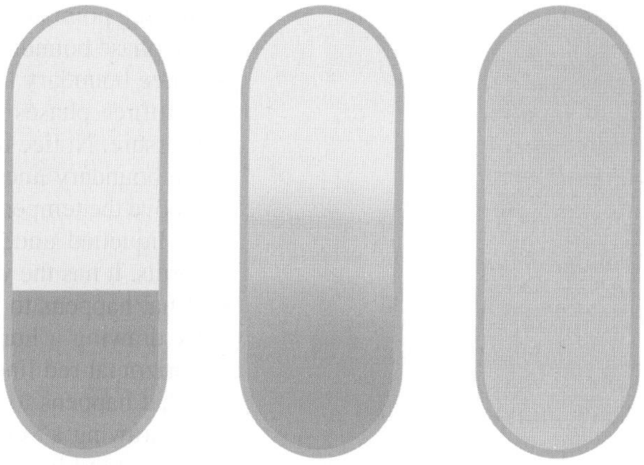

$T < T_c$ (a) $T \sim T_c$ (b) $T > T_c$ (c)

FIGURE 7.18(a) Below the critical temperature for CO_2 you can clearly see the phase boundary between the liquid and the gas phase. **(b)** As the temperature rises the liquid becomes less dense, the vapour phase becomes more dense, and the phase boundary blurs. **(c)** Above the critical temperature the phase boundary has disappeared and only a supercritical fluid exists.

The critical point of water is $T = 647$ K, $p = 221 \times 10^5$ Pa, that of CO_2 is $T = 304$ K, $p = 73.9 \times 10^5$ Pa, and that of N_2 is $T = 126$ K, $p = 33.9 \times 10^5$ Pa. Although critical pressures are many times greater than atmospheric pressure, supercritical fluids have important commercial applications. The most important of these is the use of supercritical carbon dioxide as a solvent. Supercritical CO_2 diffuses through a solid matrix rapidly, and it transports materials well because it has a lower viscosity than liquids. Supercritical CO_2 is currently used as a solvent for dry cleaning, for petroleum extraction, for decaffeination and for polymer synthesis. Compared with solvents used previously for some of these processes, such as dichloromethane, CH_2Cl_2, and tetrachloromethane, CCl_4, supercritical CO_2 is a much better choice.

Phase diagrams

The phase behaviour of a substance as a functional temperature and pressure can be conveniently summarised in a **phase diagram**. An example is provided in figure 7.19. Pressure is plotted along the *y*-axis, and temperature is plotted along the *x*-axis. In the region of low *T* and high *p*, the substance exists as a solid. In the region of high *T* and low *p*, the substance exists as a gas. In some intermediate range of *T* and *p*, the substance exists as a liquid.

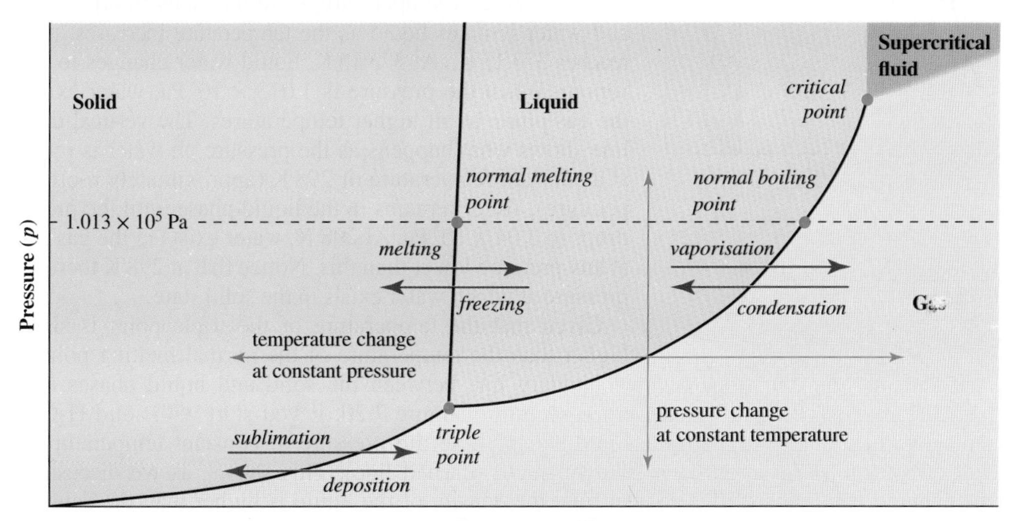

FIGURE 7.19 The general form of a phase diagram. Any point on the diagram corresponds to a specific temperature and pressure. Boundary lines trace conditions where neighbouring phases exist in equilibrium, and the blue arrows show six types of phase transitions. The red arrows highlight a sequence of phase changes on changing temperature at constant pressure, or on changing pressure at constant temperature.

Figure 7.19 illustrates many of the characteristic features of phase diagrams.

1. Boundary lines between phases separate the regions where each phase is thermodynamically stable. Substances can exist as a metastable phase outside their thermodynamically determined boundaries, e.g. carbon can exist as diamond, rather than graphite, under ambient conditions.
2. Movement across a boundary line corresponds to a phase change. The blue arrows on the figure show six different phase changes: sublimation and its reverse, deposition; melting and its reverse, freezing; and vaporisation and its reverse, condensation.
3. On a boundary line, the two neighbouring phases coexist in a dynamic equilibrium. In addition, at a given pressure, for example, when heat is added or removed, the temperature of this two-phase system does not change until all of one phase has converted to the other. The normal

melting point and normal boiling point of a substance (shown by red dots) are the points where the phase boundary lines intersect the horizontal line that represents $p = 1.013 \times 10^5$ Pa.

4. Three boundary lines meet at a single point (shown by another red dot), called a **triple point**. All three phases are present simultaneously at this unique combination of temperature and pressure. Notice that, although two phases are stable under any of the conditions specified by the boundary lines, three phases can be simultaneously stable only at a triple point.
5. Above the temperature specified by the critical point (again shown by a red dot), the gas cannot be liquefied under any pressure. Instead if the pressure is high enough a supercritical fluid forms. It has the viscosity typical of a liquid, but it is able to expand or contract like a gas.
6. What happens to a substance as temperature changes at constant pressure can be determined by drawing a horizontal line at the appropriate pressure on the phase diagram (shown as a horizontal red line).
7. What happens to a substance as pressure changes at constant temperature can be determined by drawing a vertical line at the appropriate temperature on the phase diagram (shown as a vertical red line).
8. The temperature for conversion between the gas phase and a condensed phase depends strongly on pressure. Qualitatively, this is because compressing a gas increases the collision rate and makes condensation more favourable.
9. The melting temperature is almost independent of pressure, making the boundary line between solid and liquid nearly vertical. Qualitatively, this is because moderate pressure has hardly any effect on the condensed liquid and solid phases.
10. The solid–gas boundary line extrapolates to $p = 0$ Pa and $T = 0$ K. This is a consequence of the direct link between temperature and energy. At 0 K, atoms, ions and molecules have minimum energy, so they cannot escape from the solid lattice. At 0 K, the vapour pressure of every solid substance would be 0 Pa.

The phase diagram for water, shown in figure 7.20, illustrates these features for a familiar substance. The figure shows that liquid water and solid ice coexist at the normal melting point (mp), $T = 273.15$ K and $p = 1.013 \times 10^5$ Pa. Liquid water and water vapour coexist at the normal boiling point (bp), $T = 373.15$ K and $p = 1.013 \times 10^5$ Pa. The triple point (tp) of water occurs at $T = 273.16$ K and $p = 0.0061 \times 10^5$ Pa. The figure shows that when p is lower than 0.0061×10^5 Pa, there is no temperature at which water is stable as a liquid. At those low pressures, ice sublimes but does not melt.

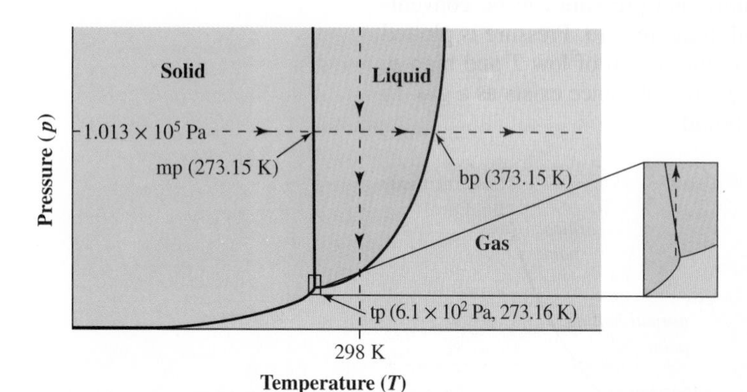

FIGURE 7.20 The phase diagram for water. The critical point is at $T = 647$ K, $p = 221 \times 10^5$ Pa, well outside this diagram.

The dashed lines on figure 7.20 show two paths that involve phase changes for water. The horizontal dashed line shows what happens as the temperature increases at a constant pressure of 1.013×10^5 Pa. As ice warms from a low temperature, it remains in the solid phase until the temperature reaches 273.15 K. At that temperature, solid ice melts to liquid water, and water remains liquid as the temperature increases, until it reaches 373.15 K. At 373.15 K, liquid water changes to water vapour. When the pressure is 1.013×10^5 Pa, water exists in the gas phase at all higher temperatures. The vertical dashed line shows what happens as the pressure on water is reduced at a constant temperature of 298 K (approximately room temperature). Water remains in the liquid phase until the pressure drops to 3.04×10^3 Pa. At 298 K, water exists in the gas phase at any pressure lower than this. Notice that at 298 K there is no pressure at which water exists in the solid state.

Given that the temperature of the triple point is slightly higher than the temperature of the normal melting point, we can conclude that the slope of the boundary line between the solid and liquid phases has to be negative. As shown in the magnified section of figure 7.20, if you start with solid H_2O at a temperature just below the triple point and increase the pressure at constant temperature you will effect a phase change from solid to liquid (dashed line). This shows, as we discussed in the opening paragraph of this chapter, that the density of the liquid is higher than the density of the solid, which is why ice floats on water. Ice-skaters make use of this pressure-induced phase transition. The thin blades of their ice skates put pressure on the ice, changing it to liquid water which refreezes once the pressure is removed. This property of water is relatively uncommon among substances. Most solids have higher density than their corresponding liquids. For water, the lower density is caused by the larger number of hydrogen bonds per water molecule in ice compared with liquid water.

All phase diagrams of pure substances share the 10 common features listed previously. However, the detailed appearance of a phase diagram is different for each substance, as determined by the strength of the interactions between its constituents. Figure 7.21 shows two examples, the phase diagrams for nitrogen and carbon dioxide. Both of these substances are gases under normal conditions. In contrast to H_2O, which has a triple point close to 298 K, N_2 and CO_2 have triple points that are well below room temperature. Although both are gases at ambient temperature and pressure, they behave differently when cooled at atmospheric pressure. Molecular nitrogen liquefies at 77.4 K and then solidifies at 63.29 K, whereas carbon dioxide condenses directly to the solid phase at 195 K. This difference in behaviour arises because the triple point of CO_2, in contrast to the triple points of H_2O and N_2, occurs at a pressure greater than 1.013×10^5 Pa. The phase diagram of CO_2 shows that at a pressure of 1.013×10^5 Pa, there is no temperature at which CO_2 exists in the liquid phase.

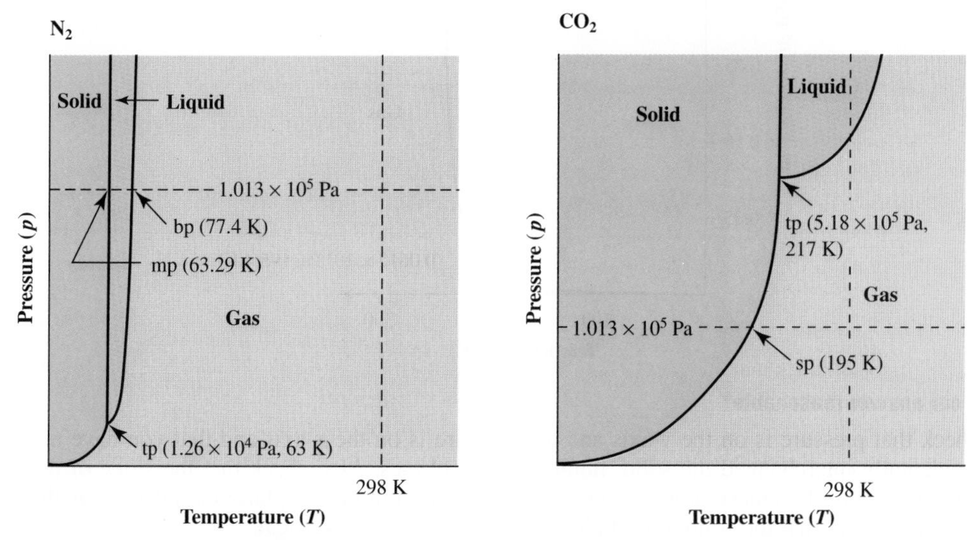

FIGURE 7.21 Phase diagrams for nitrogen and carbon dioxide, two substances that are gases at room temperature and pressure (sp = normal sublimation point).

Phase diagrams are constructed by measuring the temperatures and pressures at which phase changes occur. Approximate phase diagrams such as those shown in figures 7.20 and 7.21 can be constructed from the triple point, normal melting point and normal boiling point of a substance.

WORKED EXAMPLE 7.2

Constructing a phase diagram

Ammonia is a gas under ambient conditions. Its normal boiling point is 239.8 K, and its normal melting point is 195.5 K. The triple point for NH_3 is $p = 0.0612 \times 10^5$ Pa and $T = 195.4$ K. Use this information to construct an approximate phase diagram for NH_3.

Analysis

The normal melting and boiling points and the triple point give three points on the phase boundary curves. To construct the curves from knowledge of these three points, use the common features of phase diagrams: the gas–liquid and gas–solid boundaries of phase diagrams slope upwards, the liquid–solid line is nearly vertical, and the gas–solid line begins at $T = 0$ K and $p = 0$ Pa.

Solution

Begin by choosing appropriate scales, drawing the $p = 1.013 \times 10^5$ Pa line and locating the given data points. An upper temperature limit of 300 K encompasses all the data.

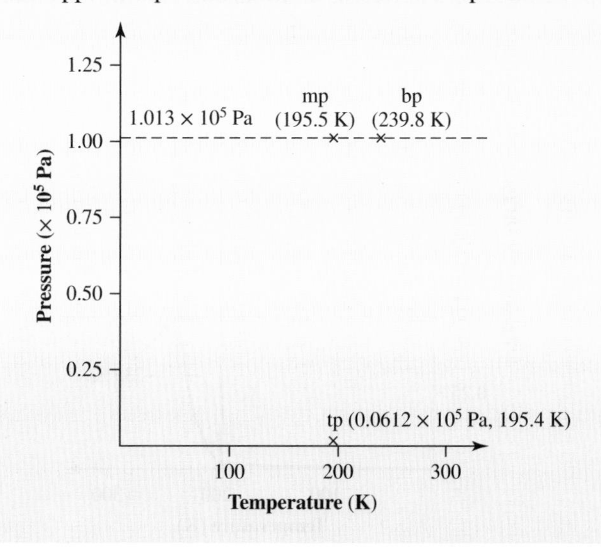

Next, connect the points and label the domains.

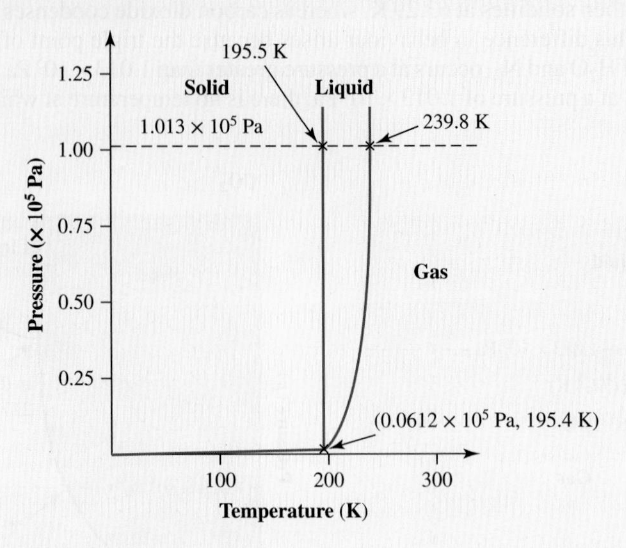

Is our answer reasonable?

Check that pressure is on the y-axis and temperature is on the x-axis and that you have not accidentally interchanged the solid, liquid and gas phases. Also check that the slope of the solid–gas phase boundary is less steep than that of the liquid–gas phase boundary near the triple point and that the solid–liquid phase boundary is nearly vertical.

PRACTICE EXERCISE 7.2

Molecular chlorine melts at 172 K, boils at 239 K and has its triple point at 0.014×10^5 Pa and 172 K. Sketch the phase diagram for this compound.

Phase diagrams can be used to determine which phase of a substance is stable at any particular pressure and temperature. They also summarise what phase changes occur as either condition is varied.

WORKED EXAMPLE 7.3

Interpreting a phase diagram

A chemist wants to perform a synthesis in a vessel at $p = 0.50 \times 10^5$ Pa using liquid NH_3 as the solvent. What temperature range would be suitable? When the synthesis is complete, the chemist wants to boil off the solvent without raising T above 220 K. Is this possible?

Analysis

The phase diagram for NH_3 (refer to worked example 7.2) shows the boundary lines for the liquid domain. These boundary lines can be used to determine the conditions under which phase changes occur.

Solution

Because the chemist wants to work at $p = 0.50 \times 10^5$ Pa, draw a horizontal line across the phase diagram at $p = 0.50 \times 10^5$ Pa. Here is an expanded view of the phase diagram between 150 and 300 K.

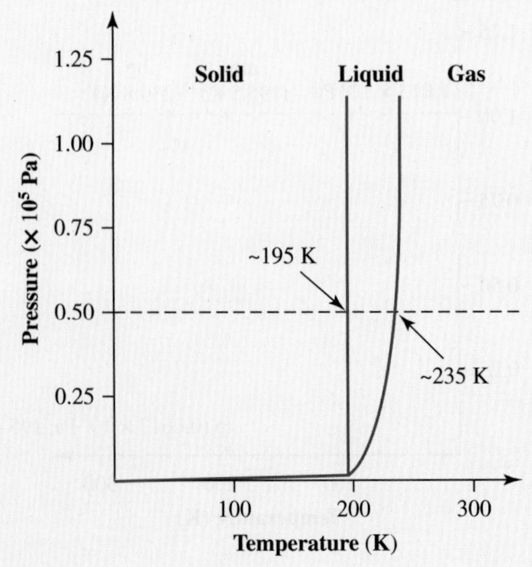

The horizontal line intersects the boundary lines at about 235 K and 195 K. Liquid NH_3 is stable between these temperatures at this pressure.

At the completion of the synthesis, the chemist wants to remove the solvent without raising T above 220 K. A vertical line at 220 K on the phase diagram represents this condition.

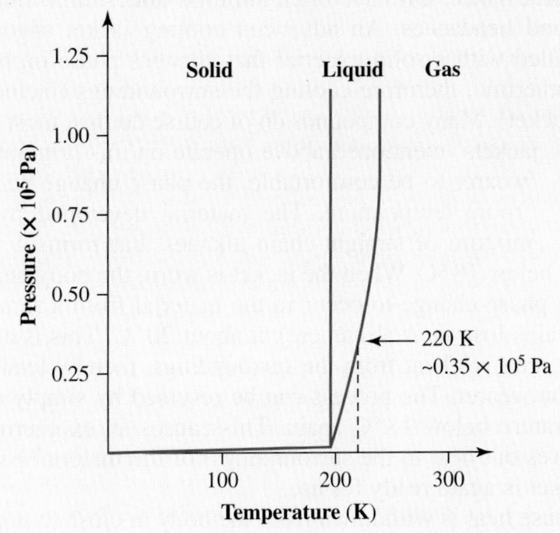

The line intersects the liquid–gas boundary at about 0.35×10^5 Pa. A vacuum pump capable of reducing p below 0.35×10^5 Pa can be used to vaporise and remove the NH_3 while keeping the temperature below 220 K.

Is our answer reasonable?

Check that the lines for pressure and temperature are entered correctly. Since NH_3(l) boils at 235 K and 0.5×10^5 Pa, for NH_3(l) to boil at the lower temperature of 220 K we need a reduced pressure, as we have determined.

PRACTICE EXERCISE 7.3

A sample of N_2 gas is at $p = 0.1 \times 10^5$ Pa and $T = 63.1$ K. Use figure 7.21 to determine what will happen to this sample if the temperature remains fixed while the pressure slowly increases to 1×10^5 Pa.

The phase diagrams we have studied are the simplest ones that occur for pure substances. However, many substances have more than one solid phase. (There are 14 different known forms of ice!) Phase diagrams of solids are particularly useful in geology, because many minerals undergo solid–solid phase transitions at the very high temperatures and pressures found deep in the Earth's crust. Figure 7.22 is the phase diagram for silica (silicon dioxide, SiO_2), an important geological substance. Notice that the scale reflects geological pressures and temperatures. Notice also that there are six different forms of crystalline silica, each stable in a different temperature–pressure region. A geologist who encounters a sample of stishovite can be confident that it solidified under extremely high pressure conditions.

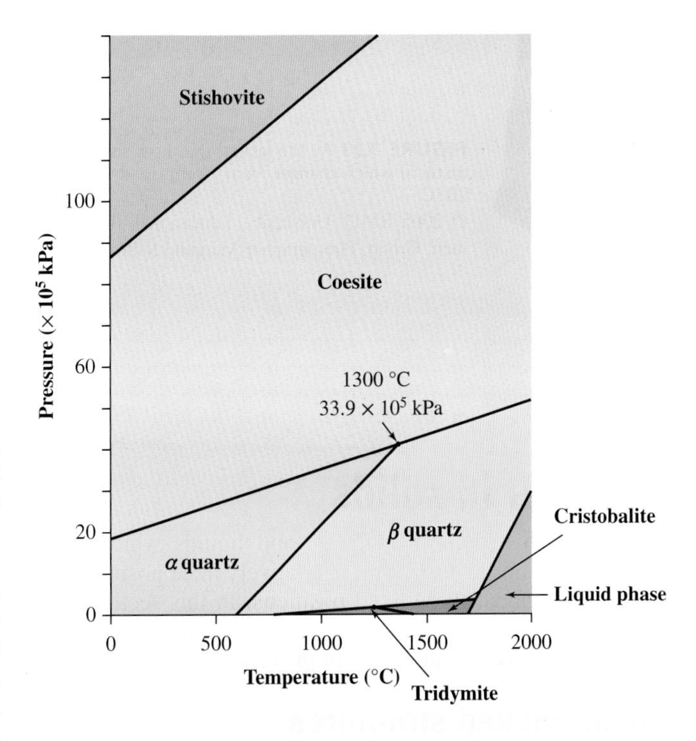

FIGURE 7.22 The phase diagram for silica shows six different forms of the solid, each stable under different temperature and pressure conditions.

If we have more than one component in a mixture, phase diagrams become even more complicated, and treatment of these is beyond the scope of this book.

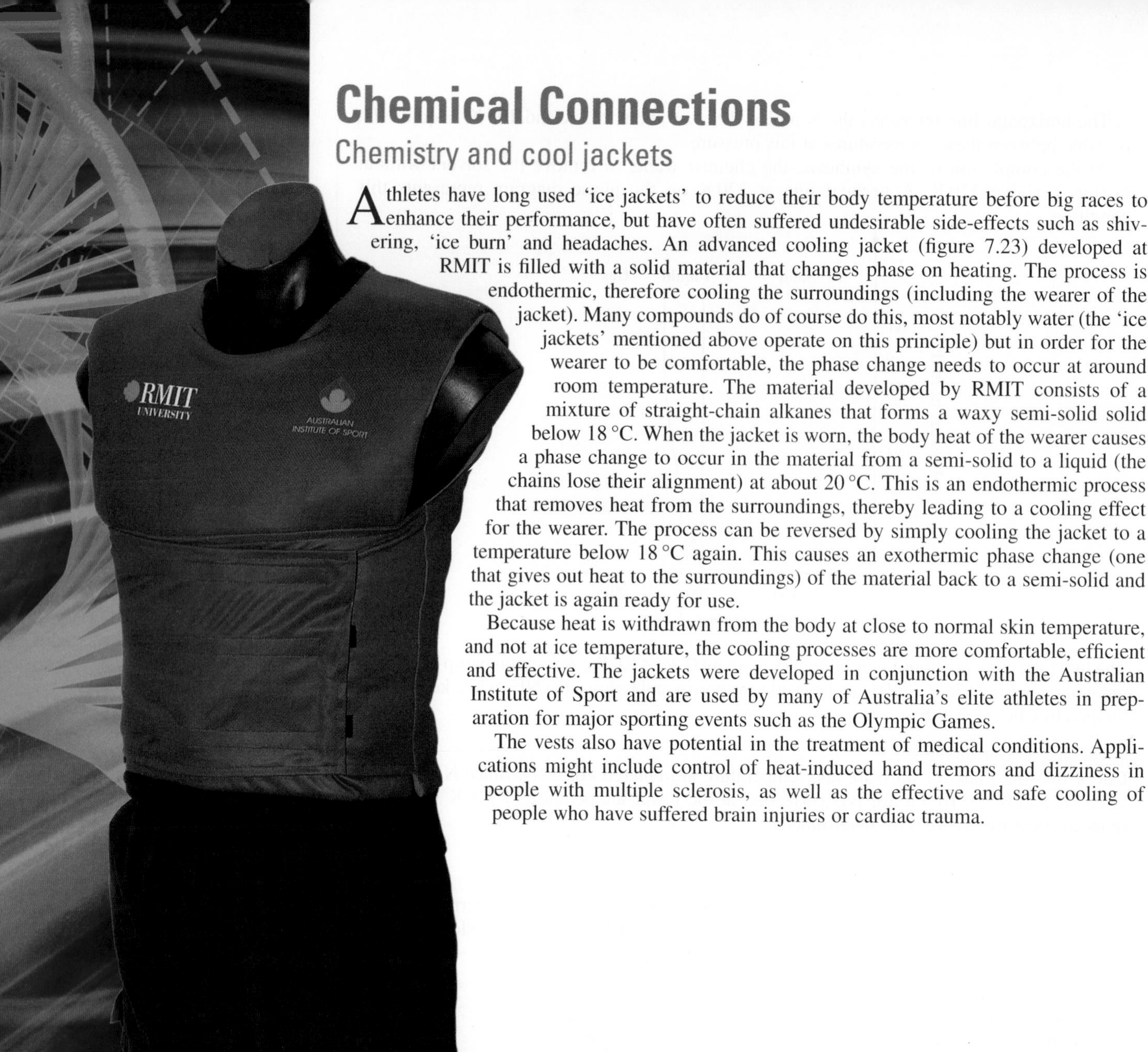

Chemical Connections

Chemistry and cool jackets

Athletes have long used 'ice jackets' to reduce their body temperature before big races to enhance their performance, but have often suffered undesirable side-effects such as shivering, 'ice burn' and headaches. An advanced cooling jacket (figure 7.23) developed at RMIT is filled with a solid material that changes phase on heating. The process is endothermic, therefore cooling the surroundings (including the wearer of the jacket). Many compounds do of course do this, most notably water (the 'ice jackets' mentioned above operate on this principle) but in order for the wearer to be comfortable, the phase change needs to occur at around room temperature. The material developed by RMIT consists of a mixture of straight-chain alkanes that forms a waxy semi-solid solid below 18 °C. When the jacket is worn, the body heat of the wearer causes a phase change to occur in the material from a semi-solid to a liquid (the chains lose their alignment) at about 20 °C. This is an endothermic process that removes heat from the surroundings, thereby leading to a cooling effect for the wearer. The process can be reversed by simply cooling the jacket to a temperature below 18 °C again. This causes an exothermic phase change (one that gives out heat to the surroundings) of the material back to a semi-solid and the jacket is again ready for use.

Because heat is withdrawn from the body at close to normal skin temperature, and not at ice temperature, the cooling processes are more comfortable, efficient and effective. The jackets were developed in conjunction with the Australian Institute of Sport and are used by many of Australia's elite athletes in preparation for major sporting events such as the Olympic Games.

The vests also have potential in the treatment of medical conditions. Applications might include control of heat-induced hand tremors and dizziness in people with multiple sclerosis, as well as the effective and safe cooling of people who have suffered brain injuries or cardiac trauma.

FIGURE 7.23 An 'ice jacket' that uses solid phase change material which changes from a semi-solid to a fluid at about 20 °C.

7.4 Order in solids

In contrast to molecules in gases and liquids which can move relatively freely, the atoms, molecules or ions in a solid are in relatively fixed positions. Generally, their motions are restricted to vibrations about these fixed positions. In this section, we examine the principles that govern the relative arrangements of atoms, molecules and ions in solids. To start our exploration of order in solids we will look at some simple cases.

Close-packed structures

Most elements in the periodic table are metals and, apart from mercury, all metals are solids at standard conditions. In order to develop a picture of metal structures, we will assume that we can approximate a metal atom with a sphere. Figure 7.24 shows both square and hexagonal arrangements of identical spheres. We can see that the hexagonal arrangement is able to fit more spheres into the same area; every sphere in the hexagonal arrangement has six direct neighbours, but only four in the square arrangement. This hexagonal arrangement of identical spheres is the most dense packing that you can achieve in one layer of spheres. You can see this when you place pool balls in the triangle at the start of the game (figure 7.25). Only if you arrange them this way can all of the balls fit into the triangle.

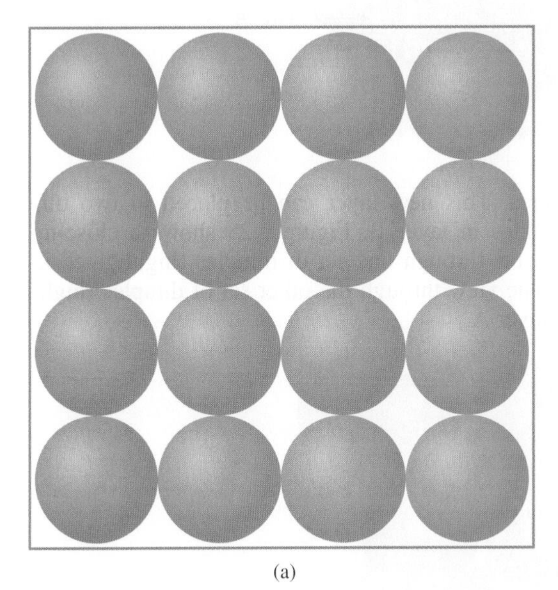

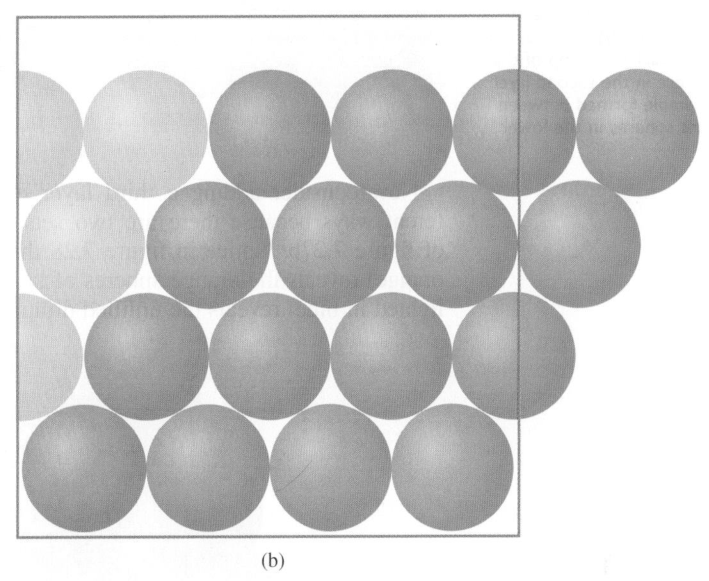

FIGURE 7.24 Comparison of **(a)** square and **(b)** hexagonal arrangement of identical spheres. You can see that both squares fit 16 spheres (moving the spheres outside the square in (b) into the empty space at the other side of the square (lighter shaded spheres). In (b), there is space left over at the top of the square, while in (a) there is no space left over, indicating that the spheres in the hexagonal arrangement are closer together.

Now add a second layer of spheres to the hexagonal arrangement. Let's call the first layer 'A' and the second layer 'B'. To achieve the most compact arrangement, each sphere in layer B sits in one of the 'dimples' between a triangle of spheres in layer A. There are six dimples around any sphere in layer A. You can see that the centres of two neighbouring dimples are closer together than the diameter of the spheres. Hence, it is impossible to place a sphere over every dimple. For example, if we place spheres as indicated by the red rings in figure 7.26, we can see that it is impossible to place spheres in the positions indicated by the blue rings at the same time.

FIGURE 7.25 Pool balls fit into the triangle in a hexagonal arrangement.

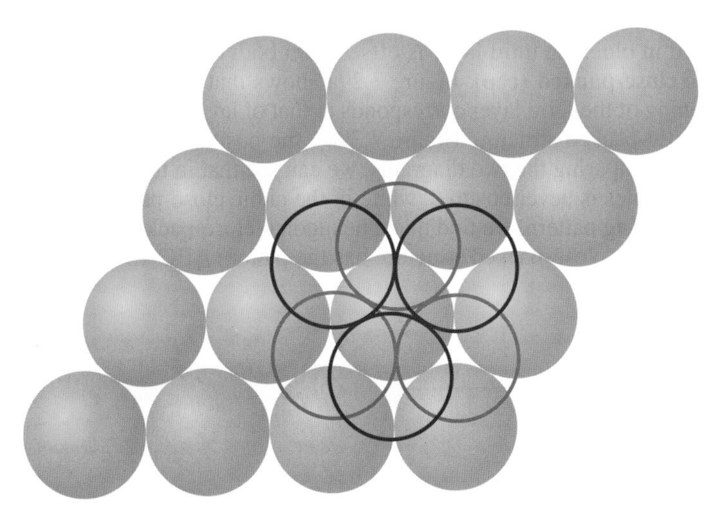

FIGURE 7.26 It is impossible to place a sphere over every dimple. We can place spheres only in the red positions or blue positions, not both at the same time.

If we fill all the dimples corresponding to the red rings in figure 7.27a (overleaf), we create a second layer as shown in figure 7.27b (overleaf). This adds three more adjoining spheres for each sphere in layer A. As additional spheres are added, layer B eventually looks identical to layer A, except that it is offset slightly to allow the spheres to nestle in the dimples formed by the layer below. There is no difference between filling all dimples corresponding to the red rings or the blue rings; the resulting arrangements of spheres are indistinguishable.

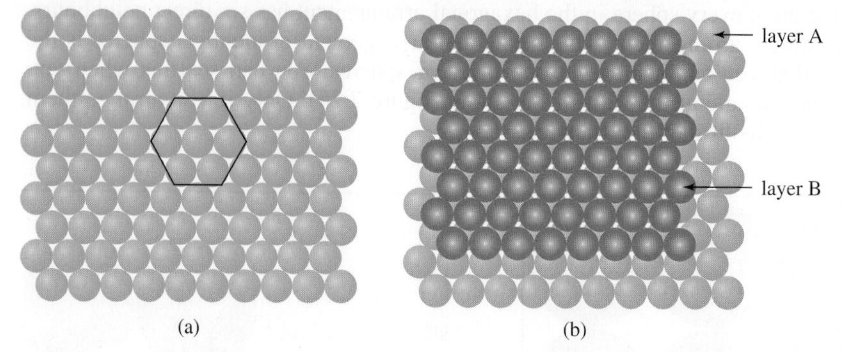

FIGURE 7.27 (a) Spheres close-packed in a layer generate a hexagonal pattern. **(b)** When a second layer is packed on top of the first, each sphere in the second layer nestles in a dimple formed between three adjacent spheres in the lower layer.

Now consider adding a third layer of spheres. This new layer can be placed in two different ways because there are two sets of dimples in layer B. Figure 7.28 shows a close-up of figure 7.27b. Notice in figure 7.28 that the view through one set of dimples (highlighted in orange) reveals the orange spheres of layer A. The view through the other set of dimples (highlighted in blue) reveals the unfilled dimples of layer A.

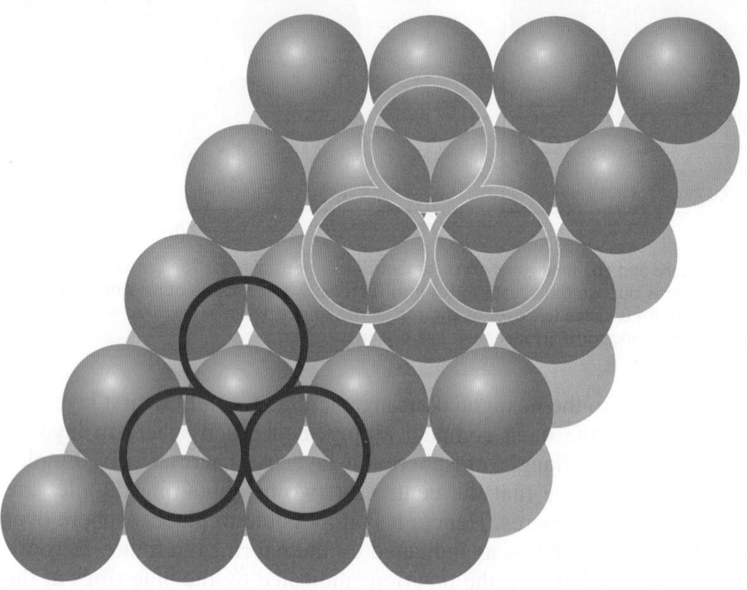

FIGURE 7.28 A third layer of spheres can be placed in two different ways, as indicated by the orange and blue rings.

If spheres in the third layer lie in the dimples highlighted in orange, the third layer is directly above the first, and the structure is called **hexagonal close-packed (hcp)**. If the spheres of the third layer lie in the dimples highlighted in blue, the third layer is offset from both of the lower layers. This arrangement is called **cubic close-packed (ccp)**. Figure 7.29a shows that in the hexagonal close-packed structure the positioning of the third layer corresponds to that of layer A and therefore we also label it A. In the cubic close-packed structure (figure 7.29b), the positioning is different and so we label the third layer 'C'. Comparing the two diagrams, we can see that in the hexagonal close-packed structure one set of dimples is never covered, while in the cubic close-packed structure, they are. If we continue this layering pattern, we find that a hexagonal close-packed structure has the pattern ABAB etc., while a cubic close-packed structure has the pattern ABCABC etc.

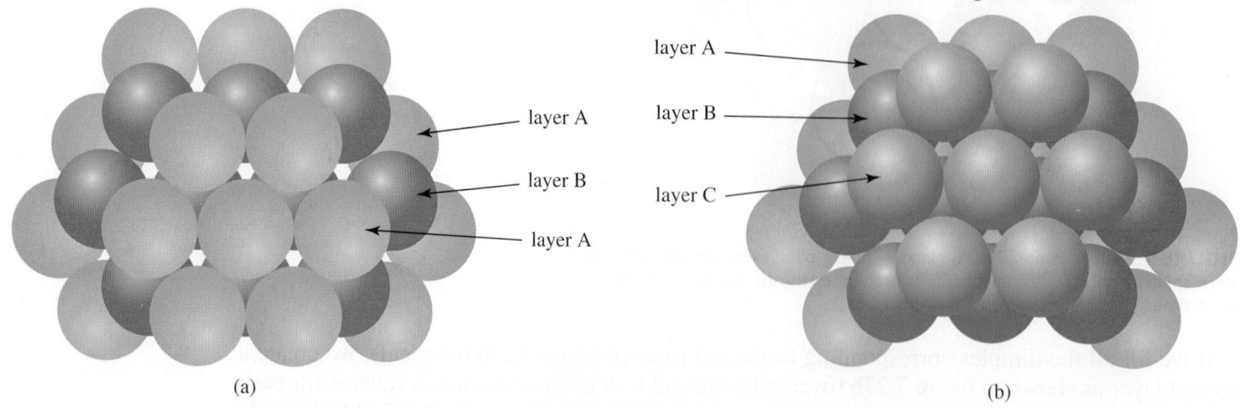

FIGURE 7.29 Top-down views of hexagonal and cubic close-packed arrangements: **(a)** The hexagonal close-packed structure has a layer repeating pattern ABAB etc. **(b)** The cubic close-packed structure has a layer repeating pattern ABCABC etc.

The two structure types described are called **close-packed structures** because they achieve maximum space filling for identical spheres. In either of these arrangements, 74% of the total space is filled, and each sphere has 12 nearest neighbours (i.e. neighbours with the same shortest distance to that sphere): six in the same plane, three above and three below. This is shown for the hexagonal and cubic close-packed structures in figure 7.30. The number of nearest neighbours is also called the coordination number, and therefore spheres in close-packed structures have a coordination number of 12.

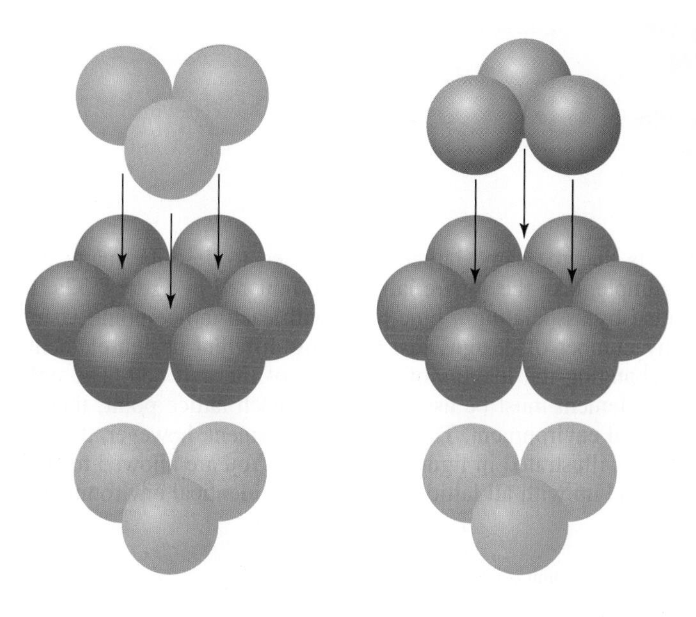

FIGURE 7.30 In close-packed structures, each sphere has 12 nearest neighbours: six in the same plane, three above and three below.

Many metals form crystals with hexagonal or cubic close-packed geometries, since metal bonding that involves sharing valence electrons benefits from close contact between participating metal atoms. For instance, magnesium and zinc crystallise with their atoms in a hexagonal close-packed array, while silver, aluminium and gold crystallise in the cubic close-packed arrangement. Other spherical entities, such as argon atoms, and methane and C_{60} molecules, also form close-packed arrangements. Argon solidifies at low temperature as a cubic close-packed crystal, and neon can solidify in either the hcp or ccp form.

The crystal lattice and the unit cell

In a solid, the number of atoms, ions or molecules is enormous. If you could imagine being at the centre of even a small sample of a solid, you would find that the atoms, ions or molecules go on as far as you can see in every direction. Describing all the individual positions is impossible and, fortunately, unnecessary if we are able to identify a repeating pattern within the solid.

There are a number of rules to consider when defining such a repeating pattern, and to start off we will look at examples in two dimensions. We can recognise a repeating motif in the wallpaper design of figure 7.31a (overleaf). How can we describe this pattern? In order to concentrate on the repeating features of the pattern, we start by randomly selecting a point and then mark every other point on the wallpaper that has an identical environment (to make that easier to see we choose the eye of the rearmost of the three fish).

Figure 7.31b (overleaf) shows this set of points, which have been connected by lines. In figure 7.31c (overleaf) we take away the wallpaper, leaving just the connected points. We call this pattern of points a **lattice**, and every individual point is a **lattice point**. Since we selected a set of points by requiring each one to have an identical environment, this set of points represents the repeating feature of our wallpaper design, and every lattice point represents one repeating motif.

In the next step, we must find the smallest repeating unit within our lattice, which in this case is one of the rectangles in figure 7.31c. We call the smallest repeating unit of a lattice the **unit cell** (there are a number of rules for the correct choice of unit cell, but they are beyond the scope of this book). Repeating the unit cell in two dimensions will create the lattice. Therefore, all we need to describe any pattern is the lattice and the content of the unit cell.

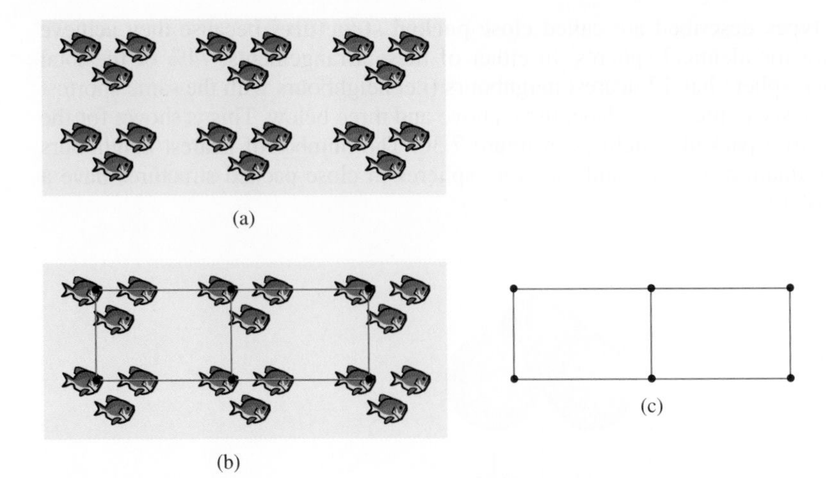

FIGURE 7.31 Defining a repeating pattern: **(a)** There is a repeating pattern in the wallpaper design. **(b)** The pattern can be described by a set of points with identical environments. **(c)** The resulting pattern is a lattice and each black dot is a lattice point.

An important fact about lattices is that the same lattice can be used to describe many different designs or structures. For example, take the square lattice and unit cell in figure 7.32a. In figure 7.32b, we see a design formed by associating a blue heart with each lattice point. Using a square lattice, we could form any number of designs just by using different design elements (e.g. a rose or a diamond), or by changing the lengths of the edges of the unit cell. The only requirement is that the same design element must be associated with each lattice point; that is, each lattice point must have an identical environment. There is no requirement, however, that the lattice point is centred on the object, as illustrated in figure 7.32c and d. You are allowed to shift the entire lattice as long as you still end up with all lattice points having identical environments.

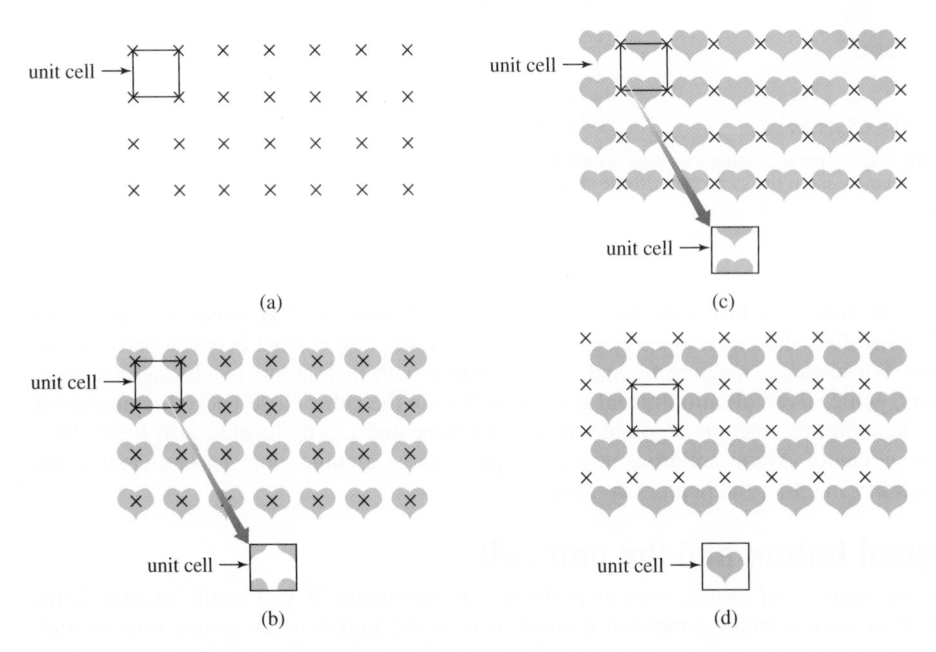

FIGURE 7.32 A two-dimensional lattice: **(a)** a simple square lattice, for which the unit cell is a square with lattice points at the corners, **(b)** a wallpaper pattern formed by associating a design element (blue heart) with each lattice point. The x centred on each heart corresponds to a lattice point. Notice that the unit cell contains portions of a heart at each corner. (The rest of each heart lies in adjacent unit cells.) **(c)** and **(d)** There is no requirement that the lattice point is centred on any object. In (a), (b), (c) and (d), all the information needed to generate the entire array is present in the unit cell, which is repeated over and over.

Another example is shown in the picture by the Dutch artist MC Escher (1898–1972) reproduced in figure 7.33. Escher often used symmetrical patterns aligned together to create an overall design. The repeating units can be visualised as tiles placed edge to edge. The faces of unit cells can be squares, rectangles or parallelograms. In figure 7.33a each tile is a parallelogram. Once you have chosen the size and shape of the unit cell, you can move its origin and still create the same overall pattern (figure 7.33b).

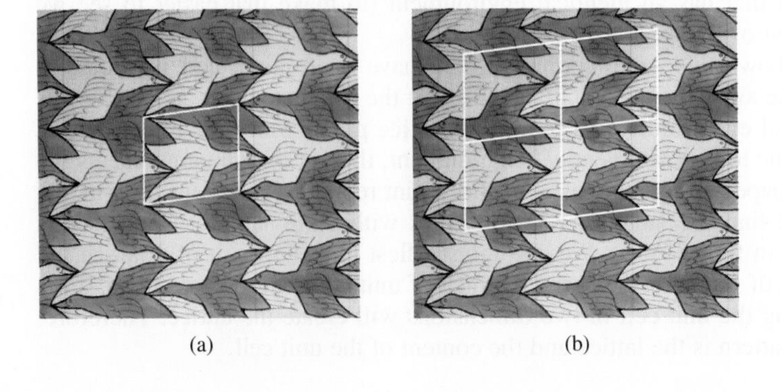

FIGURE 7.33 A drawing by MC Escher that contains a repeating pattern. A two-dimensional unit cell is highlighted in **(a)**. In **(b)**, the unit cell has been shifted and it is shown how repeating it in two dimensions will produce the entire pattern. (*Source:* M.C. Escher's 'Symmetry Drawing E18', © 2015 The M.C. Escher Company–The Netherlands. All rights reserved. www.mcescher.com)

For the sake of simplicity we have looked at lattices and unit cells in two dimensions. We can also apply these concepts to three-dimensional patterns. The ions, atoms or molecules in many solids are arranged in regular three-dimensional patterns. Such solids are classified as **crystalline solids**. Diamonds are an example. We can describe crystalline solids using the concepts of a **crystal lattice**, that is, a three-dimensional lattice and a unit cell. It is possible to describe limitless numbers of different compounds with just a limited set of three-dimensional lattices. In fact, there are only 14 different types of three-dimensional lattices possible, which means that essentially all crystalline chemical substances must form crystals with one of these 14 lattice types, albeit with different unit cell dimensions. On the following pages, we describe some metal structures with cubic unit cells. These crystal structures can be described such that all the atoms in a unit cell are associated with the lattice points, that is, the number of atoms in the unit cell and the number of lattice points will be identical. This is because all metal atoms in each of these very simple structures have identical surroundings and therefore all can be associated with lattice points. This is not so for most other crystalline solid structures.

Cubic structures

The easiest three-dimensional structure to visualise has a simple cube as the unit cell. In a simple cubic crystal, layers of atoms stack one directly above another, so that all atoms lie along straight lines at right angles, as figure 7.34 shows. Each atom in this structure has a coordination number of 6; there are four nearest neighbours within the same plane, one above the plane and one below the plane. Taking four atoms that form a square, adding an atom on top of (or below) each forms a cube. This structure is called the **primitive cubic structure**, because it has a primitive cubic lattice. (In this context, primitive means that there is only one lattice point per unit cell. A method for determining the number of lattice points is described below.) The unit cell of the primitive cubic structure is shown in the cutaway portion of figure 7.34. To calculate how many atoms are in the unit cell, we count only that proportion of the atom that is within the unit cell boundaries. An atom that is located on the corner of a cube is shared by eight cubes in total. Therefore, only $\frac{1}{8}$ of that atom counts for any one of those cubes. The unit cell contains $\frac{1}{8}$ of an atom at each of the eight corners of a cube. The total number of atoms in the unit cell is thus $8(\frac{1}{8}) = 1$. The same applies to counting lattice points; a unit cell with lattice points only on the corners ends up with one lattice point, that is, a primitive lattice. The fraction of space that is filled by the atoms in this primitive cubic unit cell is only 52%. This is much less than the most efficient packing, which achieves 74% space filling (see p. 277). Consequently, this structure is adopted by only one metal: polonium.

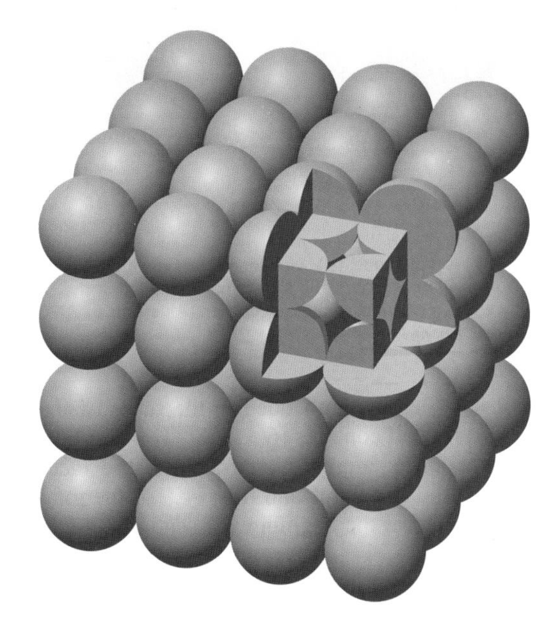

FIGURE 7.34 The primitive cubic structure is built from square layers of spheres stacked one directly above another. The cutaway view shows one unit cell.

When eight atoms form a cube, there is a cavity at the centre of the cube. The cavity is not large enough to hold an additional atom, but moving the eight corner atoms slightly away from each other does allow the structure to accommodate another atom of the same type in the centre. The result is the body-centred cubic structure, described by a **body-centred cubic lattice (bcc)**, which now has two lattice points: one on a corner and one in the centre of the cube. Each atom in a body-centred cubic lattice is at the centre of a cube and contacts eight neighbouring corner atoms (see figure 7.35, overleaf). In other words, every atom in the structure can be viewed as the centre atom of one body-centred cube or as a corner atom in another cube. The unit cell of a body-centred cube contains a total of $[1 + 8(\frac{1}{8})] = 2$ atoms. The views of the iron crystal structure in figure 7.35 on the next page illustrate this arrangement.

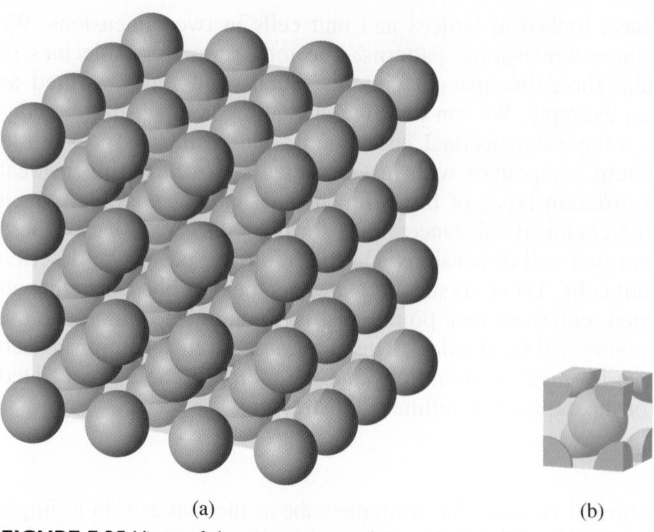

(a) (b)

FIGURE 7.35 Views of the arrangement of atoms in iron: **(a)** a view of the body-centred crystal lattice and **(b)** the unit cell.

Although the corner atoms are further apart in a body-centred cube than in a primitive cube, the extra atom in the centre of the structure makes the body-centred cubic structure more compact than the primitive cubic structure. It fills 68% of the total space, which is close to the maximum possible. All of the group 1 metals, as well as iron and the transition metals from groups 5 and 6, form crystals with body-centred cubic structures.

Another common cubic crystal structure is the **face-centred cubic structure (fcc)**, which is the same as the cubic close-packed structure that we encountered on page 276. When four atoms form a square, there is open space at the centre of the square. A fifth atom can fit into this space by moving the other four atoms away from one another. Additional atoms can be placed in the centres of all six faces of the simple cube, as figure 7.36 shows.

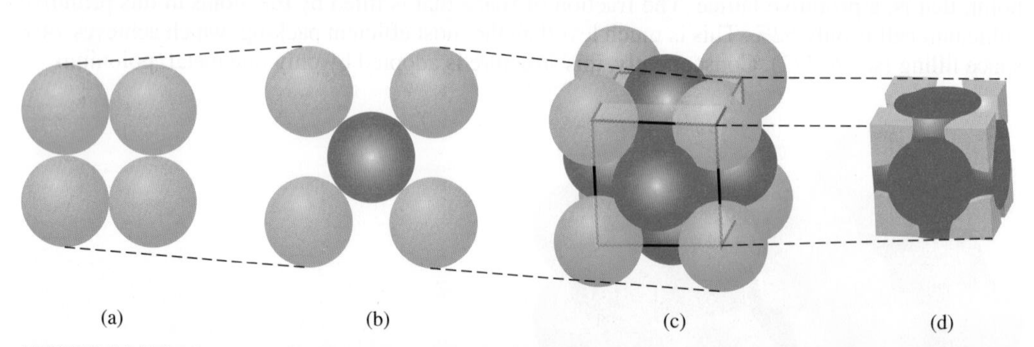

(a) (b) (c) (d)

FIGURE 7.36 The face-centred cubic lattice can be viewed as: **(a)** a simple cube, **(b)** with each face expanded just enough to fit an additional atom in the centre, **(c)** giving eight atoms at the corners and six atoms embedded in the faces. **(d)** The unit cell has $\frac{1}{2}$ of an atom at the centre of each face and $\frac{1}{8}$ of an atom at each corner. (All atoms in this arrangement are the same. Different colours are used only for clarity.)

The unit cell of the face-centred cube consists of six $\frac{1}{2}$-atoms embedded in the faces of the cube and eight $\frac{1}{8}$-atoms at the corners of the cube. This unit cell contains $[6(\frac{1}{2}) + 8(\frac{1}{8})] = 4$ atoms. Each atom in a face-centred cube contacts 12 neighbouring atoms. Consequently, the face-centred cube is more compact than either of the other cubic structures and fills 74% of the total space. The cubic close-packed structure that we encountered at the beginning of section 7.4 can be described by this face-centred cubic unit cell.

Figure 7.37 shows the relationship between a set of hexagonal layers and a face-centred cubic unit cell. Figure 7.37a shows a top-down view of the hexagonal layers. In figure 7.37b we have removed all but one of the atoms from the top and bottom layers, and have removed all but six atoms in each of the two middle layers. We can see from figure 7.37c that a line drawn between the orange atoms (from the top and bottom layers) represents the body diagonal of a cube. Each face of the cube has an atom at each corner and one atom in the centre, as shown by figure 7.37c. Since all the atoms are identical, this is a face-centred cubic arrangement as revealed in the perspectives of figures 7.37c and 7.37d.

We can now see that the arrangement of hexagonal layers in an ABCABC fashion is called a cubic close-packed structure, because the structure can be described by a face-centred cubic unit cell (the terms 'cubic close-packed structure' and 'face-centred cubic structure' are often used synonmously). Equally, the hexagonal close-packed structure gets its name because it can be described by a hexagonal unit cell as shown in figure 7.38.

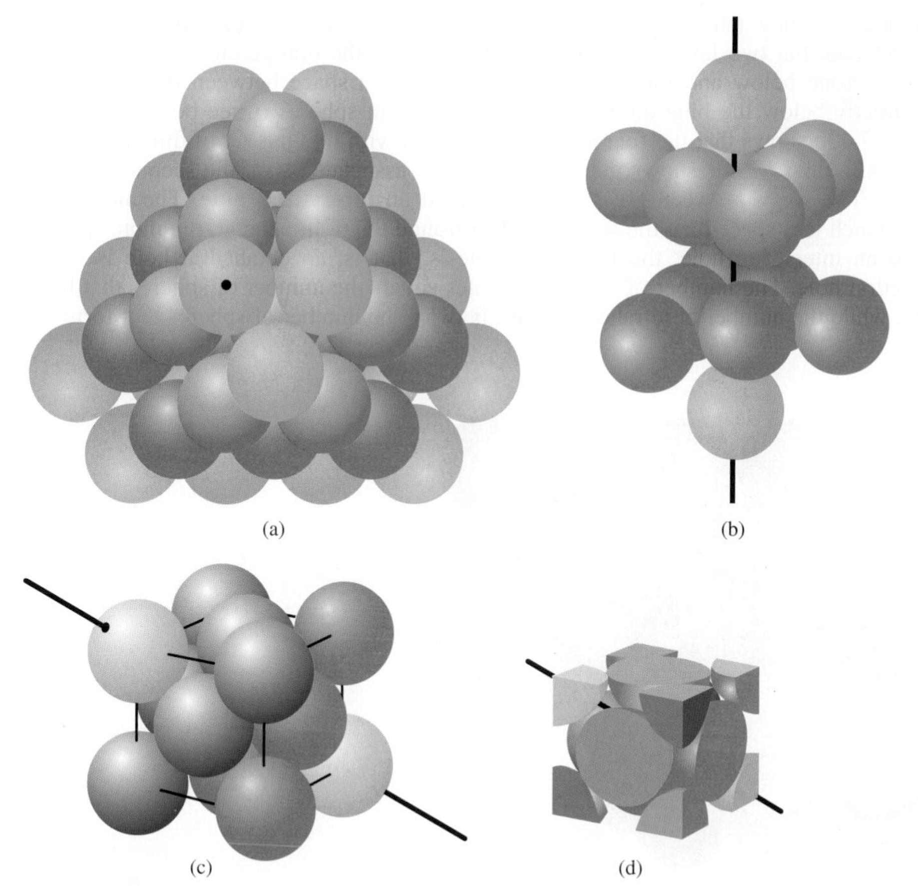

FIGURE 7.37 Four views of the hexagonal layers contained within a face-centred cubic arrangement: **(a)** a top-down view of the hexagonal layers, with an axis shown as a black dot, **(b)** a side view showing four layers along the axis, **(c)** tilting (b) leads to the arrangement shown here, with the vertical black line in (b) now the body diagonal of this cube, and **(d)** the unit cell. (All atoms in this arrangement are the same. Different colours are used for clarity only.)

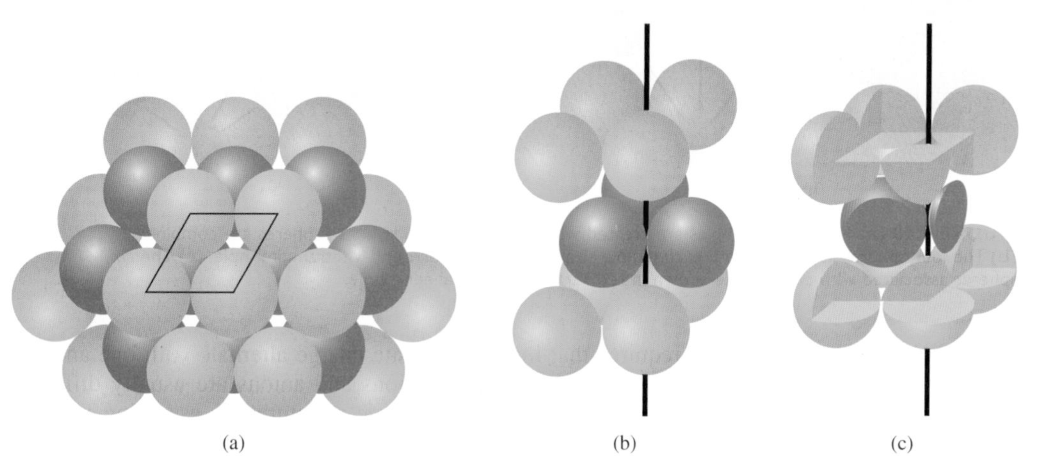

FIGURE 7.38 Three views of the hexagonal layers contained within a hexagonal close-packed structure: **(a)** a top-down view of the hexagonal layers with the unit cell shown, **(b)** a side view showing three layers and **(c)** the unit cell. This unit cell contains two atoms and fills 74% of the space. (All atoms in this arrangement are of the same type. Different colours are used for clarity only.)

Up to now, we have described the crystalline arrangements favoured by spherical objects such as atoms, but most molecules are far from spherical. Nonspherical objects require different arrays to achieve maximal stability. To illustrate, compare a stack of bananas with a stack of oranges (figure 7.39). Just as the stacking pattern for bananas is less symmetrical than that for oranges, the structural patterns for most molecular crystals are less symmetrical than those for crystals of spherical atoms, reflecting the lower symmetry of the molecules.

FIGURE 7.39 Nonspherical bananas require different packing schemes from spherical oranges.

Ionic solids

We will now show how the principles discussed previously relate to the actual structures of ionic solids. Recall from figure 7.28 that two sets of dimples are created when we add two hexagonal layers of spheres. This is shown again in figure 7.40a (overleaf). One set of dimples has an atom directly below (shown in orange), while the other set does not (shown in blue). The open space between the two layers of spheres directly below the orange circle is surrounded by four spheres (one below and three above), while the open space between the two layers of spheres directly below the blue circle is surrounded by six spheres (three below and three above). Figure 7.40b shows that the four spheres associated with the first set of dimples form a tetrahedron, while figure 7.40c shows that the six spheres associated with the second set form an octahedron. The space between the spheres is called an **interstitial hole**. The interstitial holes take up much less space than the spheres that surround them. In general, the more spheres that surround an interstitial hole, the larger the hole is, that is, an octahedral hole is larger than a tetrahedral hole. The number of octahedral holes equals the number of spheres that form the structure, while the number of tetrahedral holes is twice the number of spheres that form the structure.

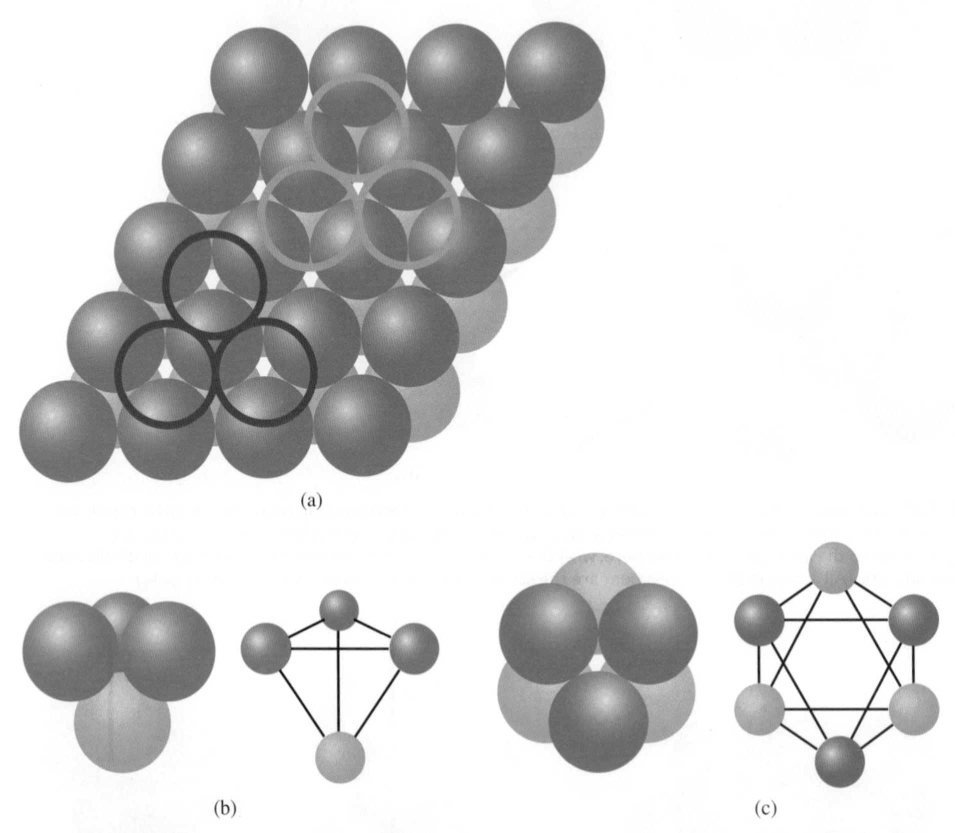

FIGURE 7.40 (a) Two sets of dimples are created when two hexagonal layers of spheres are combined. **(b)** The four spheres associated with each dimple shown in orange surround a tetrahedral hole. **(c)** The six spheres associated with each dimple shown in red surround an octahedral hole.

The packing in ionic crystals requires that ions of opposite charge alternate with one another to maximise attractions and minimise repulsions. The cations and anions are usually different sizes (often the cations are smaller). We can understand ionic structures by assuming that the close-packed structure is formed by the larger ions, while the smaller ions fill the interstitial sites. We can identify a crystal lattice for the larger ions, and then describe how the smaller ions pack within that lattice. Because the crystal is made up of a huge number of identical unit cells, the stoichiometry within the unit cell must match the overall stoichiometry of the compound.

Ionic compounds adopt a variety of structures that depend on the stoichiometry and relative sizes of the ions. For many 1 : 1 ionic crystals such as NaCl, the most stable arrangement is a face-centred cubic array of anions (Cl^-) with the cations (Na^+) packed into the octahedral holes between the anions. This structure appears in figure 7.41a. Notice that every cation is surrounded by six anions and every anion is surrounded by six cations. The unit cell based on close packing of Cl^- is shown. In addition to the face-centred cubic arrangement of chloride ions, the unit cell in figure 7.41b shows $\frac{1}{4}$ of a sodium cation along each edge of the cell, and there is an additional sodium cation in the centre of the unit cell. Summing the components reveals that each unit cell contains four complete NaCl units, thus giving the correct 1 : 1 stoichiometry.

$$\text{chloride anions: 6 faces } (\tfrac{1}{2}\text{ anion/face}) + 8 \text{ corners } (\tfrac{1}{8}\text{ anion/corner}) = 4\ Cl^- \text{ anions}$$
$$\text{sodium cations: 12 edges } (\tfrac{1}{4}\text{ cation/edge}) + 1 \text{ in centre} = 4\ Na^+ \text{ cations}$$

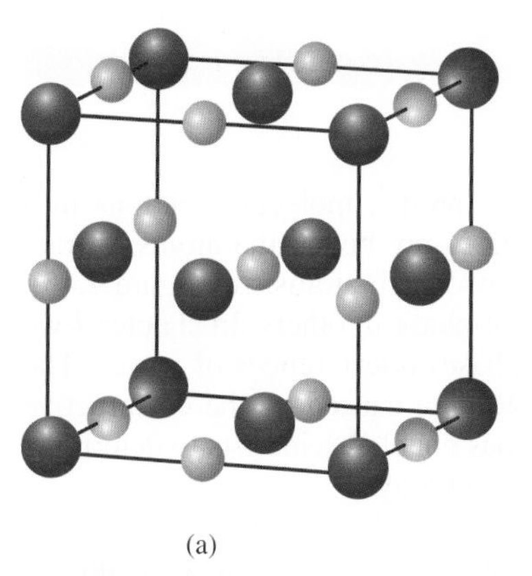

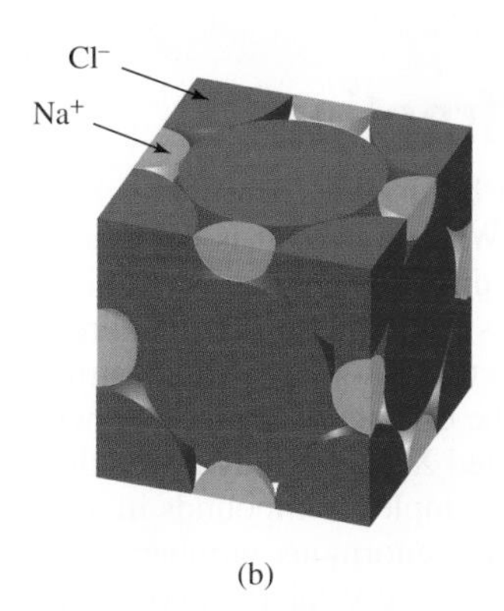

(a) (b)

FIGURE 7.41 (a) The structure of sodium chloride can be described as a face-centred cubic arrangement of Cl^- anions (green), with the Na^+ cations (orange) occupying the interstitial octahedral sites. **(b)** Because Cl^- anions are considerably larger than Na^+ cations, the commonly used unit cell places anions at the corners. There is a cation at the centre of this unit cell.

Many ionic solids have the same structure type as sodium chloride. Examples are all the halides of Li^+, Na^+, K^+ and Rb^+, as well as the oxides of Mg^{2+}, Ca^{2+}, Sr^{2+} and Ba^{2+}.

Caesium chloride, in which the relative size of the cation is much larger than in sodium chloride, forms a different type of structure. Because its cation and anion are close to the same size, caesium chloride is most stable in a primitive cubic lattice. There are Cl^- anions at the corners of the cube, with a Cs^+ in the centre, as figure 7.42 shows. The unit cell contains $8(\frac{1}{8}) = 1$ chloride ion and 1 caesium ion. The caesium chloride structure type is also found for CsBr, CsI and several other 1 : 1 ionic crystals in which the cations and anions have similar sizes.

In zinc sulfide, ZnS, the relative size ratios are such that the zinc cation is too small for octahedral sites and instead occupies tetrahedral holes in the structure. Since there are twice as many tetrahedral sites as spheres that form the structure, only half will be occupied by Zn^{2+} cations in a regular fashion. We have seen previously that a face-centred arrangement of anions leads to four anions in the unit cell. We can see in figure 7.43 that there are also four cations within the unit cell, resulting in the correct 1 : 1 stoichiometry. In this structure, as seen in figure 7.43b, every cation is surrounded by four anions, and every anion is surrounded by four cations in a tetrahedral arrangement.

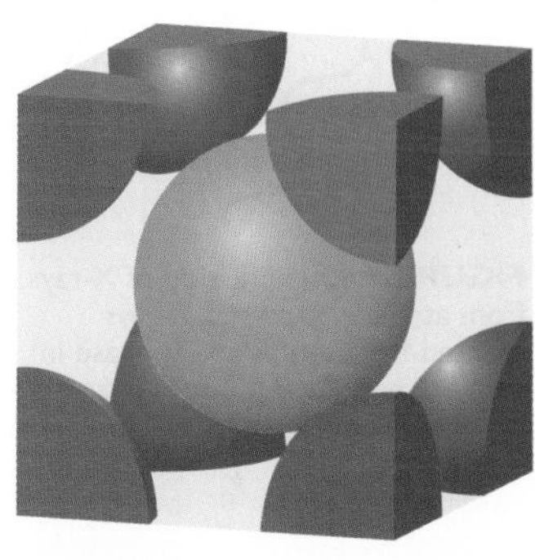

FIGURE 7.42 The caesium chloride structure consists of a primitive cubic array of chloride anions with a caesium cation at the centre of each unit cell.

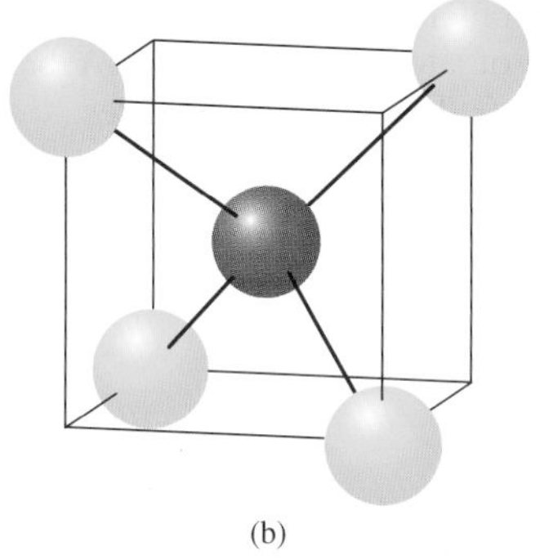

(a) (b)

FIGURE 7.43 Zinc sulfide. **(a)** The unit cell of cubic zinc sulfide has anions in a face-centred cubic arrangement. **(b)** Each cation is surrounded by four anions in the form of a tetrahedron.

Many ionic compounds have stoichiometries that differ from 1 : 1. To give an example, the fluorite structure is named for the naturally occurring form of CaF_2 and is common for salts of the general formula MX_2. It is easiest to describe this structure by thinking of the Ca^{2+} cations arranged in a face-centred cubic structure. In this arrangement, the F^- anions fill all the tetrahedral holes in the lattice, forming a simple cube of eight anions in the interior of the face-centred cube. This is in contrast to zinc sulfide, where only half of the tetrahedral holes were filled. Figure 7.44a shows the unit cell of the fluorite structure, which contains a total of four Ca^{2+} cations and eight F^- anions (consistent with the number of tetrahedral holes being twice the number of spheres that form the structure).

$$\text{calcium anions: 6 faces } (\tfrac{1}{2}\text{ cation/face}) + 8\text{ corners } (\tfrac{1}{8}\text{ cation/corner}) = 4\ Ca^{2+}\text{ cations}$$

$$\text{fluoride anions: 8 complete anions within the face-centred cube} = 8\ F^-\text{ anions}$$

This gives the correct 1 : 2 stoichiometry, resulting in electrical neutrality.

The calcium ions are arranged in a tetrahedron about the fluoride ions as shown in figure 7.44b. We have briefly mentioned four simple ionic lattices, but there are many others.

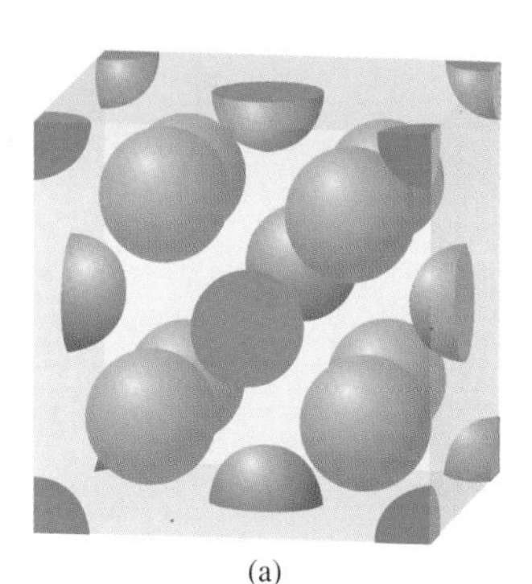

(a)

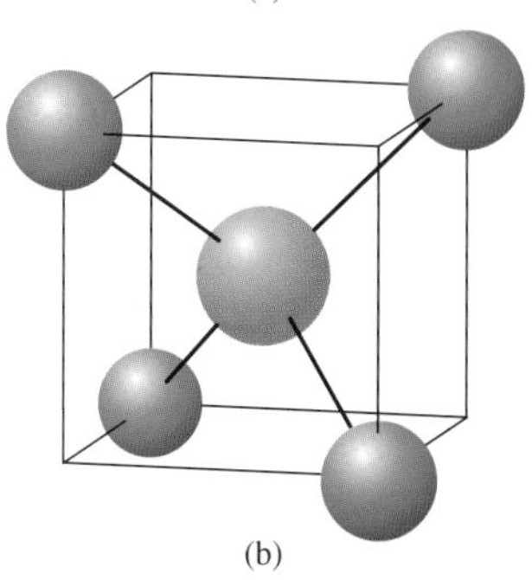

(b)

FIGURE 7.44 (a) The unit cell of the fluorite structure has cations in a face-centred cubic arrangement, with a simple cube of anions within the unit cell. **(b)** All fluoride ions are surrounded by four calcium ions in a tetrahedral arrangement.

7.5 X-ray diffraction

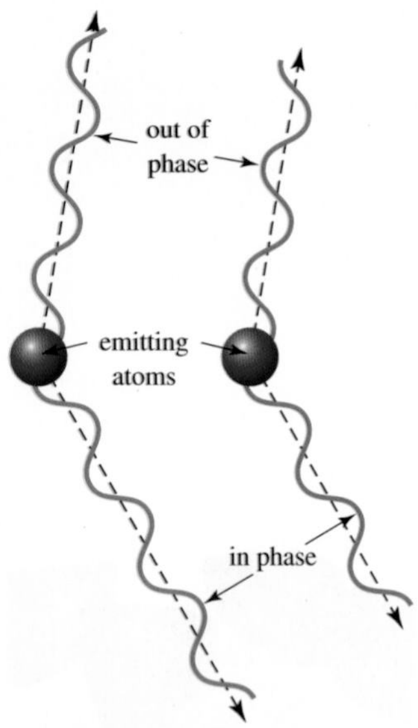

X-ray diffraction can be used to determine the arrangement of atoms, molecules or ions in a crystalline structure. When a crystal is exposed to an incoming X-ray beam, the atoms diffract the X-rays into many directions. If we look at diffraction from two such atoms (see figure 7.45), we find that the X-rays are in phase in some directions but out of phase in others. In chapter 4 we learned about constructive (in-phase) and destructive (out-of-phase) interferences of waves. The dots (reflections) in the diffraction pattern shown in figure 7.46b are caused by the constructive interference of diffracted X-rays. X-ray diffraction by crystals has enabled scientists to determine the structures of very complex compounds in a particularly elegant way.

In a crystal, there are enormous numbers of atoms, and these can be described by a lattice and a unit cell. When the crystal is exposed to X-rays, diffracted beams due to constructive interference appear only in specific directions. In other directions, no X-rays appear because of destructive interference. The X-rays coming from the crystal are recorded and form a **diffraction pattern** (see figure 7.46).

FIGURE 7.45 Diffraction of X-rays from atoms in a crystal. X-rays emitted from atoms are in phase in some directions and out of phase in other directions.

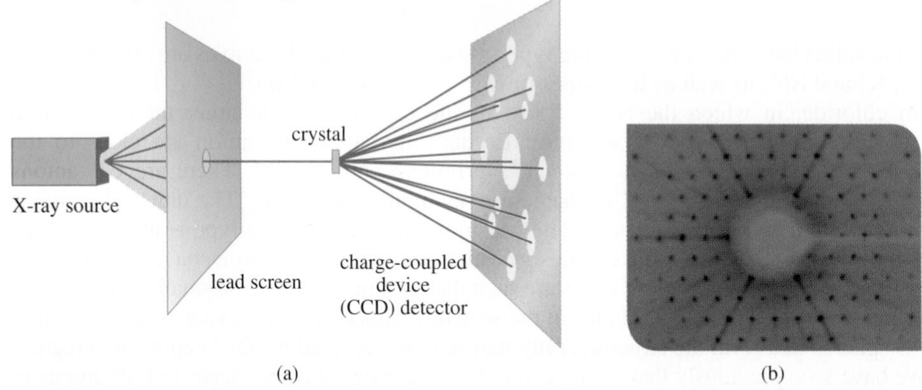

FIGURE 7.46 X-ray diffraction: **(a)** schematic set-up for recording an X-ray diffraction pattern and **(b)** an X-ray diffraction pattern.

In 1913, William Henry Bragg and his Australian-born son William Lawrence Bragg discovered that just a few variables control the appearance of an X-ray diffraction pattern. Figure 7.47 illustrates the conditions necessary to obtain constructive interference of the X-rays from successive layers (planes) of atoms in a crystal. A beam of X-rays with a wavelength λ strikes the layers at an angle θ. For certain distances, d, between the layers, constructive interference causes an intense diffracted beam to emerge at the same angle, θ.

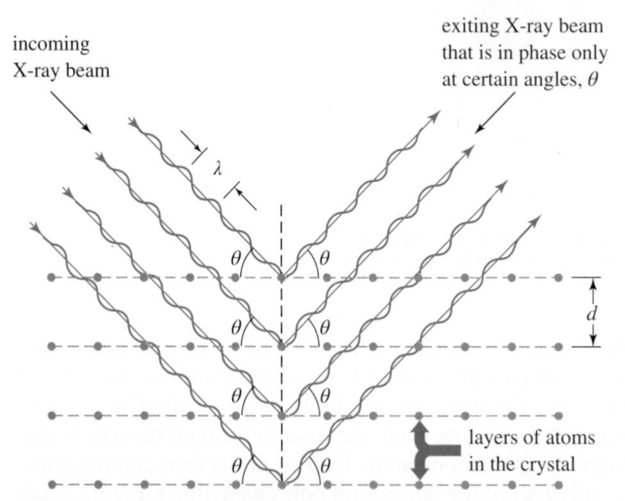

FIGURE 7.47 Diffraction of X-rays from successive layers of atoms in a crystal. The layers of atoms are separated by a distance d. The X-rays of wavelength λ enter and emerge at an angle θ relative to the layers of the atoms. For the emerging beam of X-rays to have any intensity, the condition $n\lambda = 2d \sin \theta$ must be fulfilled, where n is an integer.

The **Bragg equation** relates λ, θ and the distance between the planes of atoms, d:

$$n\lambda = 2d \sin \theta$$

where n is any integer. The Bragg equation is the basic tool used by scientists in the study of solid structures. Let us briefly see how it is used.

In any crystal, many different sets of imaginary planes can be passed through the structure. Figure 7.48 illustrates this idea in two dimensions for a simple pattern of points. When a crystal is exposed to X-rays, many diffracted beams are produced because of the diffraction from the many possible sets of planes. An apparatus called a diffractometer (figure 7.49) is used to record the diffraction pattern and allows the determination of the angles at which the diffracted beams emerge from each distinct set of planes. Using the Bragg equation, we can calculate the distances between planes of atoms from the measured angles of diffraction, the wavelength of the X-rays (source dependent) and the value of n (usually 1). The next step is to use the calculated interplanar distances to deduce the unit cell of the lattice. The locations of the atoms within the unit cell are then determined from the relative intensities of the diffracted beams. If this sounds like a difficult task, it is! For all but the simplest crystalline structures, sophisticated mathematics and computers are needed to accomplish it. The efforts, however, are well rewarded because the calculations give very accurate locations for the atoms within the unit cell. This enables us to understand the bonding, intermolecular interactions and properties of crystalline materials.

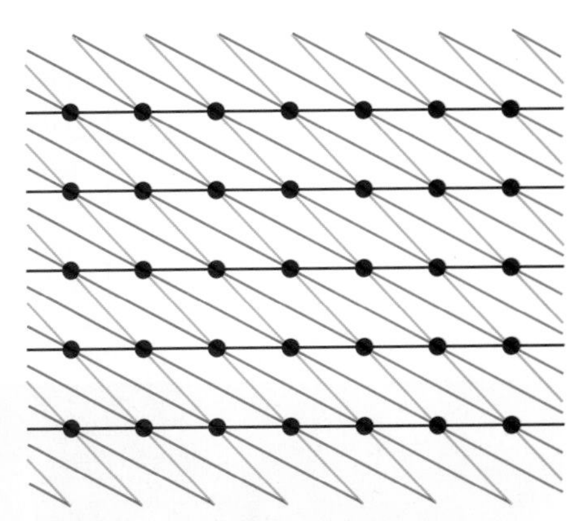

FIGURE 7.48 A two-dimensional pattern of points with many possible sets of parallel lines (three are shown here). In a three-dimensional crystal lattice there are many sets of parallel planes.

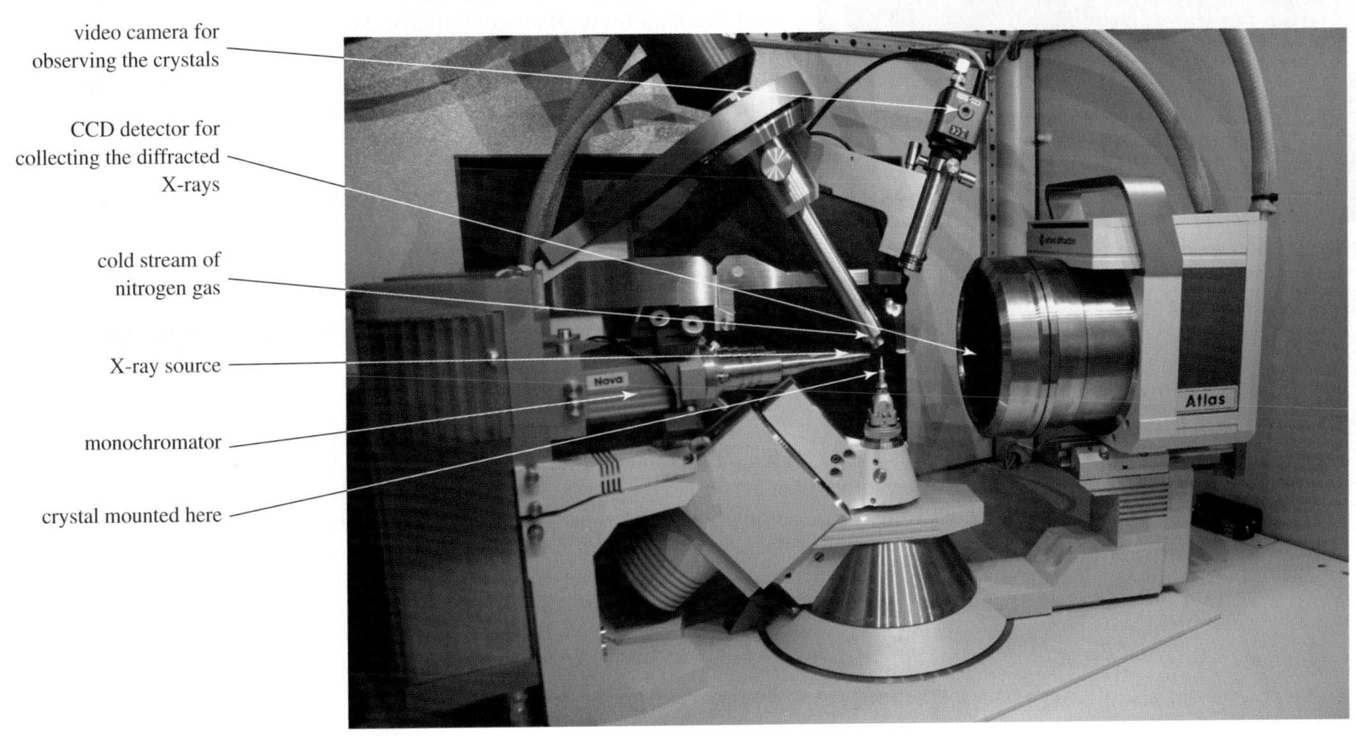

FIGURE 7.49 A modern X-ray diffractometer.

WORKED EXAMPLE 7.4

Using crystal structure data to calculate atomic sizes

X-ray diffraction measurements reveal that copper crystallises with a face-centred cubic lattice in which the unit cell length is 3.62×10^{-10} m. What is the radius of a copper atom? What is the copper–copper bond length?

Analysis

Copper atoms are in contact along a face diagonal (the dashed line at right) that runs from one corner of a face to another corner.

Using geometry, we can calculate the length of this diagonal, which equals four times the radius of a copper atom.

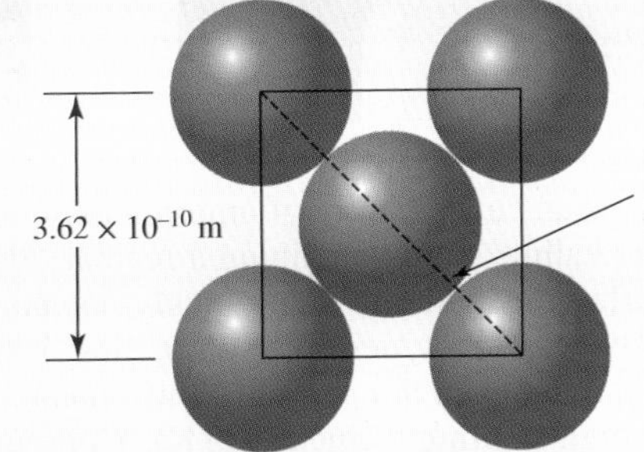

Solution

From geometry, the length of the diagonal is $\sqrt{2}$ times the length of the edge of the unit cell.

$$\text{diagonal} = \sqrt{2} \times (3.62 \times 10^{-10}) = 5.12 \times 10^{-10}\,\text{m}$$

If we call the radius of the copper atom r_{Cu}, then the diagonal equals $4 \times r_{Cu}$. Therefore:

$$4 \times r_{Cu} = 5.12 \times 10^{-10}\,\text{m}$$

$$r_{Cu} = 1.28 \times 10^{-10}\,\text{m}$$

The calculated radius of the copper atom is 1.28×10^{-10} m.
The copper–copper bond length is twice this number, that is, 2.56×10^{-10} m.

Is our answer reasonable?

You will know from chapter 5 that bond distances between atoms are about $1\text{–}3 \times 10^{-10}$ m and atomic radii are about half of this. The calculated Cu radius and Cu—Cu bond length fall within this range, so our answer is reasonable.

FIGURE 7.50 A computer-generated model of the DNA double helix.

X-ray diffraction has had a profound impact on the study of biochemistry. Biochemists often use X-ray diffraction to determine the structures of enzymes and other large molecules produced by living systems. The most famous example is the experimental determination of the molecular structure of DNA. DNA is found in the nuclei of cells, and serves to carry an organism's genetic information. In 1953, using X-ray diffraction photographs of DNA fibres obtained by Rosalind Franklin and New Zealand-born Maurice Wilkins, James Watson and Francis Crick came to the conclusion that the DNA structure consists of the now-famous double helix (see figure 7.50). Watson, Crick and Wilkins shared the 1962 Nobel Prize in physiology or medicine for their discovery (Franklin had died of cancer in 1958, and it was not possible to nominate her for the Nobel Prize posthumously).

X-ray diffraction continues to be one of the main tools used to determine the structures of complex proteins and enzymes.

7.6 Amorphous solids

When a pure liquid or a melt is cooled slowly, it often solidifies as a well-ordered crystalline solid. When solids form rapidly, on the other hand, their atoms, molecules or ions may become locked into positions other than those of a regular crystal, giving materials that are **amorphous**, meaning 'without form'. For example, ordinary cane sugar is crystalline, but rapidly cooling melted sugar gives fairy (or candy) floss, which contains long threads of amorphous sugar (figure 7.51). While amorphous solids can be formed in a variety of ways and by different mechanisms, one feature that clearly distinguishes them from crystalline solids is the fact that they do not diffract X-rays (see section 7.5).

(a)

(b)

FIGURE 7.51 (a) Fairy floss is amorphous, while **(b)** ordinary cane sugar is crystalline.

What we call **glass** is an entire family of amorphous solids based on silica, SiO_2. Pure SiO_2 is usually found as a crystalline solid containing a regular array of Si and O atoms. This is known as quartz. When quartz is melted and then quickly cooled, it forms fused silica, an amorphous solid glass. Silica glass has many desirable properties. It resists corrosion, transmits light well and withstands wide variations in temperature. Unfortunately, pure silica is very difficult to work with because of its high melting point (1983 K). Consequently, silica glass is used only for special applications.

Mixing sodium oxide, Na_2O, with silica results in a glass that can be shaped at a lower temperature. Sodium—oxygen bonds are ionic, and incorporating sodium ions into the mixture breaks the Si—O—Si chain of covalent bonds. This weakens the lattice strength of the glass, lowers its melting point and reduces the viscosity of the resulting liquid. However, the weakened lattice also means that glass made from mixed sodium and silicon oxides is vulnerable to chemical attack.

A desirable glass melts at a reasonable temperature, is easy to work with and yet is chemically inert. Such a glass can be prepared by adding a third component that has bonding characteristics between those of purely ionic sodium oxide and purely covalent silicon dioxide. Several different components are used, depending on the properties required for the glass.

The glass used for windowpanes and bottles is soda–lime–silica glass, a mixture of sodium oxide, calcium oxide and silicon dioxide. The addition of CaO strengthens the lattice enough to make the glass chemically inert to most common substances (strong bases and HF, however, attack this glass). Pyrex®, the glass used in coffeepots and laboratory glassware, is a mixture of B_2O_3, CaO and SiO_2. This glass can withstand rapid temperature changes that would crack soda–lime–silica glass. Lenses and other optical components are made from glass that contains PbO. Light rays are strongly bent as they pass through lenses made of this glass. Coloured glasses contain small amounts of coloured metal oxides such as Cr_2O_3 (amber), NiO (green) or CoO (brown).

7.7 Crystal imperfections

The two extremes of ordering in solids are perfect crystals with complete regularity and amorphous solids that have little order. Many solid materials are crystalline but contain defects. **Crystalline defects** can profoundly alter the properties of a solid material, often in ways that have useful applications. Doped semiconductors are solids into which impurity 'defects' are introduced deliberately in order to modify electrical conductivity. Some gemstones are crystals containing impurities that give them their colour. Sapphires and rubies are imperfect crystals of colourless Al_2O_3; Ti^{3+} or Fe^{3+} makes sapphires blue, and Cr^{3+} makes rubies red (figure 7.52).

FIGURE 7.52 The vivid colours of sapphires and rubies arise due to imperfections in their crystal structures.

Yttrium barium copper oxide superconductors have optimal properties when they have a slight deficiency of oxygen, as indicated by the δ sign in the formula $YBa_2Cu_3O_{7-\delta}$. This departure from stoichiometric composition is accommodated by defects in the crystal structure. The superconductor crystal has oxygen anions missing from some positions in the crystal lattice, and the number of missing anions can vary sufficiently to give the material a variable composition. The solid remains electrically neutral when its anion content is varied because some of the cations take on different charges. The copper cations in the superconductor can have a +2 or +3 charge. The relative number of Cu^{3+} ions decreases as oxygen anions are removed from the structure.

Substitutional impurities replace one atom with another, while interstitial impurities occupy the spaces between regular atoms. Interstitial impurities create imperfections that play important roles in the properties of metals. For example, small amounts of impurities are deliberately added to iron to improve its mechanical properties. Pure iron is relatively soft, easily deformed and corrodes readily, but the addition of a small amount of carbon creates steel, a much harder material. Carbon atoms fill some of the open spaces between iron atoms in the crystalline structure of steel. Although carbon atoms fit easily into these spaces, their presence reduces the ability of adjacent layers of iron atoms to slide past each other, as figure 7.53 illustrates.

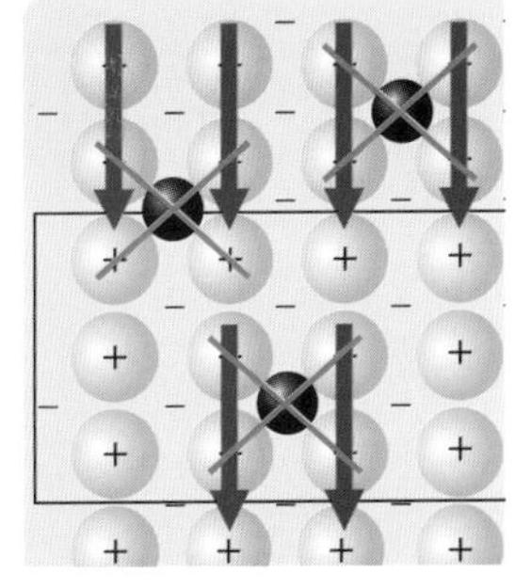

FIGURE 7.53 The presence of even a few carbon atoms (black spheres) in the interstitial holes in an iron lattice prevents adjacent layers of iron atoms from sliding past one another and hardens the iron into steel.

7.8 Modern ceramics

Ceramics are materials composed of inorganic components that have been heat treated. They have a long history, dating to prehistoric times. Examples of pottery about 13 000 years old have been found in several parts of the world. Today, ceramics include common inorganic building materials such as brick, cement and glass. We find ceramics around the home as porcelain dinnerware, tiles, sinks, toilets, and artistic pottery and figurines. We also find ceramics in places we might not expect, such as mobile phones, diesel engines and bullet-proof vests.

Many manufactured ceramics are made from inorganic minerals such as clay, silica (sand) and other silicates (compounds containing anions composed of silicon and oxygen), which are taken from the Earth's crust. In recent times, an entirely new set of materials with high-tech applications, generally referred to as advanced ceramics, has been prepared by chemists. We will focus our discussions mainly on these.

Properties of ceramics

There are some properties that ceramics have in common, and there are others that can be tailored by controlling the ceramic's composition and method of preparation. For example, many ceramic materials are very hard and have very high melting points. Table 7.3 contains a list of some advanced ceramic materials and their properties, along with the properties of some metals in common use. The hardest substance known is diamond, and some ceramics have hardnesses that approach diamond. Silicon carbide, which is relatively inexpensive to make in bulk, has long been used as an abrasive in sandpaper and grinding wheels.

TABLE 7.3 Some properties of ceramic materials, diamond, steel and aluminium.

Ceramic material	Hardness[(a)] (GPa)	Melting point (°C)	Elastic modulus[(b)] (GPa)	Density ($g\ mL^{-1}$)
diamond, C	80	3800	910	3.52
boron nitride, BN	50	2730	660	3.48
titanium carbide, TiC	28	3067	470	4.93
silicon carbide, SiC	26	2760	480	3.22
zirconium carbide, ZrC	25	3445	400	6.63
tungsten carbide, WC	23	2776	720	15.72
titanium nitride, TiN	21	2950	590	5.40
aluminium oxide, Al_2O_3 (alumina)	21	2047	400	3.98
beryllium oxide, BeO (beryllia)	15	2550	390	3.03
zirconium oxide, ZrO_2 (zirconia)	12	2677	190	5.76
aluminium nitride, AlN	12	2250	350	3.36
titanium dioxide, TiO_2 (titania)	11	1867	205	4.25
stainless steel	2.35	1420	200	7.89
aluminium	1.51	658	69	2.70

(a) Pressure required to dent the material: the larger the value, the harder the material.
(b) Force per unit cross-sectional area required to elongate (stretch) the solid: the larger the value, the stronger the material and the more it resists stretching.

Notice that all of the listed ceramics are harder than steel and also have lower densities and very high melting points. Those with high strength and relatively low density are useful, for example, for applications in space.

Because of their high melting points, we often find ceramics used as **refractories** (heat-resistant materials) for lining furnaces and rocket engine exhausts. When given a glassy porcelain surface, ceramics are impervious to water, but under the right conditions they can be made porous and used as filters in applications where high temperatures would destroy other materials.

Most ceramics do not conduct electricity, so they are used as insulators for high-voltage power lines, car spark plugs and in TV sets. However, some ceramic materials become excellent conductors of electricity when cooled to very low temperatures. There is a whole class of these materials that are superconductors with many potential applications (see pp. 289–90).

Table 7.3 shows that ceramics generally contain metals in relatively high positive oxidation states, combined with small nonmetals (e.g. O, N and C) with high negative oxidation states. As we discussed earlier, the assignment of a formal oxidation state does not mean that the electrons have completely transferred from one bonding partner to another; that is, it allows for all forms of chemical bonding.

Actually, the ceramic compounds listed in table 7.3 possess substantial covalent bonding between the atoms. Imagining that the compounds were purely ionic, small cations with high charges would be able to pull electron density from highly charged anions into the region between the ions, causing the bonds to become substantially covalent. Furthermore, elements such as oxygen, nitrogen and carbon are able to form covalent bonds between two or more atoms. Therefore, a mixture of ionic and covalent bonding between the atoms, that is, very polar covalent bonds, causes both the strength and high melting points of ceramic materials.

Applications of advanced ceramics

The field of high-tech ceramics is very competitive new materials with new applications are continually being developed. Their uses are extremely varied and a few examples are given on the following pages.

Thin ceramic films are used as antireflective coatings on optical surfaces and as filters for lighting. Tools such as drill bits are given thin coatings of titanium nitride, TiN, to make them more wear resistant (figure 7.54).

FIGURE 7.54 Drill bits with a thin golden coating of titanium nitride, TiN, will retain their sharpness longer than steel drill bits.

Partially stabilised zirconia (ZrO_2 with small amounts of other metal oxides) is used to make portions of hip-joint replacements. This application is possible because of the toughness of the ceramic.

One form of boron nitride powder, BN, is composed of flat, platelike crystals that can easily slide over one another. It is used in cosmetics, where it gives a silky texture. Another boron-containing ceramic, boron carbide, is used along with Kevlar™ polymer to make bullet-proof vests. When a bullet strikes the vest, it is either shattered by the ceramic, or it cracks the ceramic, which then absorbs most of the kinetic energy. Residual energy is absorbed by the Kevlar backing.

Silicon nitride, Si_3N_4, is used to make engine components for diesel engines because it is wear resistant and extremely hard. It has a high stiffness with low density, and can withstand extremely high temperatures and harsh chemical environments.

Piezoelectric ceramics produce an electric potential when their shape is deformed. They also deform when an electric potential is applied to them. A company makes 'smart skis' that incorporate piezoelectric devices that use both properties. Vibrations in the skis are detected by the potential developed when they deform. A potential is then applied to cancel the vibration. Smart materials possess both sensing and action capability. They adaptively respond to changing stimuli.

High-temperature superconductors

A **superconductor** is a material which offers no resistance to the flow of electricity. Many compounds are superconductors at very low temperatures, which can be reached only by using liquid helium for cooling. Because liquid helium is very expensive, materials have been sought with critical temperatures, T_c (the temperature at which they become superconductors), above 77.4 K, the boiling point of liquid nitrogen. The refrigeration technology for producing and handling liquid nitrogen is well known and relatively simple, and liquid nitrogen costs about as much as milk or orange juice. In 1987, a ceramic mixed metal oxide, $YBa_2Cu_3O_7$, was discovered with a T_c of 93 K. Materials that have T_c values above about 30 K are called *high-temperature superconductors*. Materials with T_c values above 130 K (at atmospheric pressure) have been found.

The chief problem with ceramic superconductors is that they are brittle, but during the 1990s scientists and engineers successfully made wire-like superconducting materials that could be used to build electrical cables. The cables can carry up to three times more electricity than standard electrical cables. New-generation magnets made with high-temperature superconducting wires are now commercially available.

One of the most fascinating properties of superconducting materials is that they can be levitated by a magnetic field. Besides zero electrical resistance, a substance in its superconducting state permits no magnetic field within itself. A weak magnetic field is actually *repelled* by a superconductor (the Meisner effect), and this is what makes the levitation of superconductors possible (figure 7.55).

FIGURE 7.55 Levitation of a magnet above a superconductor. When a ceramic is cooled below T_c, it becomes superconducting and repels a magnetic field. That repulsion is sufficient to hold a magnet suspended above it.

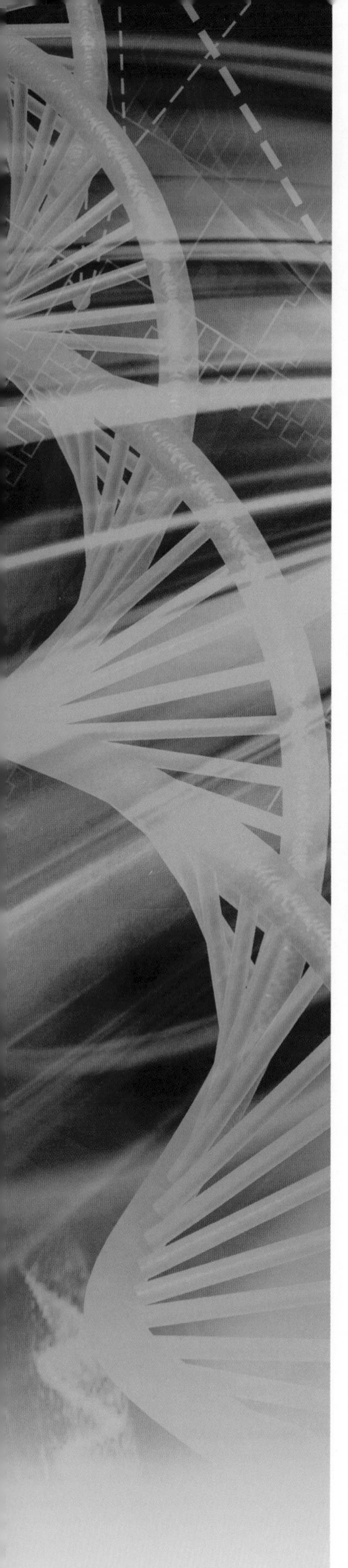

Chemical Connections

Fast trains on superconductors

The Dutch scientist Heike Kamerlingh-Onnes, was awarded the Nobel Prize in physics for his 1911 discovery of superconductors. These become perfect electrical conductors when cooled below a critical temperature, T_c, where they show no resistance to the flow of electrical current. Therefore, superconducting wires can potentially be used to carry electricity with no loss. Until 1986, all known superconductors had to be cooled using liquid helium. The record T_c at the time was 23 K. That year, physicists Georg Bednorz and Alex Müller discovered the first of the so-called high-temperature superconductors (HTS) with $T_c \approx 30$ K. This set in motion an international race to discover HTS with higher T_c values. Over the next few years, T_c increased an astounding sevenfold. Today there are more than 70 known HTS.

Most HTS are brittle ceramics that cannot be easily manufactured into wires. The most useful is $Bi_2Sr_2Ca_2Cu_3O_{10}$, or BSCCO (figure 7.56). It was first identified by a research group working in New Zealand and its T_c of 108 K means that it can be cooled using liquid nitrogen, which is much cheaper than liquid helium. Products made using this material include magnets for ion implantation, research, testing of computer hard drives, synchrotron dipole magnets, coils for motors and generators, and nuclear magnetic resonance instruments.

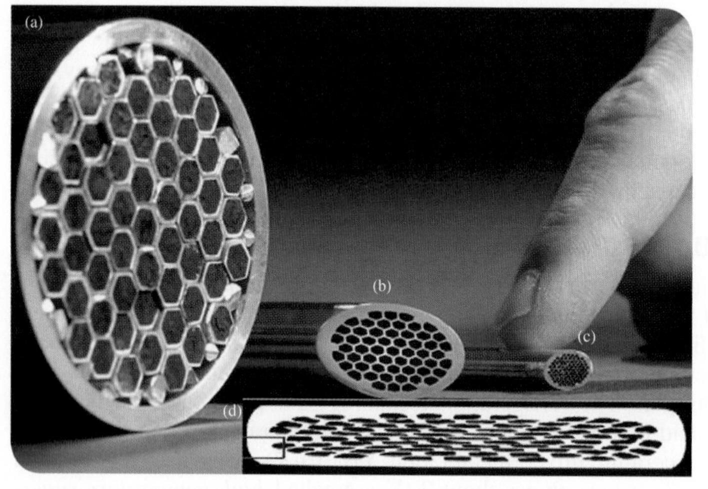

FIGURE 7.56 BSCCO wires: **(a)** precursor materials are loaded into a silver billet, extruded, **(b)** rebundled into 55 hexagonal tubes, re-extruded, **(c)** drawn down to about 1 mm diameter, and then **(d)** rolled flat to form a multifilamentary tape. This is then reacted at high temperature to complete the formation of the final BSCCO product.

Another aspect of superconductors is that they expel magnetic fields perfectly, which can be used to float a magnet over a superconductor, or vice versa (see figure 7.57). One large-scale application of this phenomenon is a magnetic levitation train. It was announced in 2014 that Japan is going to build a train line based on this principle that will connect Tokyo with Nagoya and allow trains to travel at speeds of 500 km h^{-1} in standard operation. The 285 km track is estimated to cost A$100 billion and will be finished in 2027.

There are many aspects of HTS science yet to be understood and this is an ongoing research focus across the globe. The ultimate goal is the discovery of room temperature superconductors, which would revolutionise the production and use of electricity on our planet.

FIGURE 7.57 Magnetic levitation (maglev) trains use superconducting magnets to create a magnetic field that enables the train to levitate over its track.

LO 1 Liquids

Liquids, like gases, are fluid, but cannot expand or contract much. Three properties of liquids, viscosity, surface tension and capillary action, depend mostly on the strengths of intermolecular attractions within the liquid. Cohesive forces occur between molecules in a liquid and are responsible for surface tension. Adhesive forces occur between the liquid molecules and the walls of the container and give rise to a curved meniscus at the surface of a liquid. Adhesive forces are also responsible for capillary action. Viscosity is related to both the shape of the molecules and the magnitude of the intermolecular forces in a liquid.

Liquids evaporate when their molecules have sufficient energy to overcome attractive intermolecular forces. The equilibrium vapour pressure of a liquid is the pressure at which the number of molecules escaping from the liquid exactly matches the number being captured by the liquid. Vapour pressure increases with increasing temperature until the normal boiling point, the point where the vapour pressure equals the external atmospheric pressure, is reached.

LO 2 Solids

Solids are rigid, meaning that their constituent atoms, molecules or ions occupy well-defined positions relative to one another. Solids may be classified as molecular, metallic, network or ionic solids according to the nature of their constituents. Each category has characteristic interactions between its constituents. Molecular solids are held together by dispersion forces. In some solids and liquids, dipolar and hydrogen bonding forces may also occur depending on the nature of the individual molecules. In network solids, the constituents are covalently bonded together, while ionic solids are held together by electrostatic attractions between oppositely charged ions. The bonding in metallic solids is due to a delocalised 'sea' of electrons.

LO 3 Phase changes

A phase change occurs when a substance undergoes a transition from one phase to another. Energy must be either supplied to, or removed from, a substance undergoing a phase change. This can be quantified by considering the values of the molar enthalpy of vaporisation $\Delta_{vap}H$ or the molar enthalpy of fusion $\Delta_{fus}H$, which are the enthalpy changes that occur on vaporising or melting 1 mole of substance, respectively. In addition to these phase changes, sublimation, the direct conversion of a solid to a gas, can also occur, and it has an associated molar enthalpy of sublimation, $\Delta_{sub}H$. The reverse of this process is called deposition.

Heating a liquid in a closed container results in gradual evaporation and a corresponding increase in pressure. Eventually, a point may be reached at which the liquid and gas phases cannot be distinguished and the substance exists as a supercritical fluid. The temperature and pressure at which this occurs are called the critical temperature and critical pressure.

Temperatures and pressures at which equilibria can exist between phases are shown graphically in a phase diagram. Three equilibrium lines intersect at the triple point. The liquid–gas line terminates at the critical point. Above the critical point, a liquid phase cannot be formed; the single phase that exists is a supercritical fluid. The equilibrium lines also divide a phase diagram into temperature–pressure regions in which a substance exists in just a single phase. Water is different from most substances in that its melting point decreases with increasing pressure, albeit very slightly.

LO 4 Order in solids

Approximating atoms as spheres allows us to determine their possible close-packed arrangements. A hexagonal arrangement is more efficient than a square arrangement for a single layer of spheres. Additional layers may be added to give either an ABABAB or ABCABC arrangement, depending on how the additional atoms are arranged relative to the original layer. The former is called a hexagonal close-packed (hcp) arrangement and the latter is a cubic close-packed (ccp) arrangement. Both are called close-packed structures as they fill 74% of the space, which is the maximum possible for identical spheres.

The overall structure of any crystalline solid can be described in terms of a repeating three-dimensional array of lattice points, which is called a lattice. The simplest portion of a lattice is its unit cell. Many structures can be described by the same lattice by changing the contents and dimensions of the unit cell. Three possible cubic unit cells are: primitive cubic, face-centred cubic and body-centred cubic. In a lattice of spheres, the spaces between the spheres are called interstitial holes.

Sodium chloride and many other alkali metal halides crystallise in a face-centred cubic arrangement, which contains four NaCl units per unit cell. The relative sizes of the cations and anions in ionic crystals determine the structure adopted; hence CsCl, in which the cations and anions are nearly the same size, crystallises in a primitive cubic structure. The ratio of cations to anions is also important; for example, CaF_2 adopts a structure in which the Ca^{2+} ions are arranged in a face-centred cubic array, with the F^- ions filling all the tetrahedral holes.

LO 5 X-ray diffraction

X-ray diffraction can be used to determine the structure of crystalline solids. The diffraction of X-rays from a crystalline solid results in constructive interference of the X-rays in particular directions as determined by the Bragg equation. This gives a diffraction pattern, which can be analysed to allow the calculation of the structure of the solid.

LO 6 Amorphous solids

Some solids, when formed by rapid cooling, do not crystallise with their constituent atoms, molecules or ions in a regular arrangement. These are known as amorphous solids. The most important example is glass, a family of amorphous solids based on silica. Rapid cooling of molten SiO_2 gives an amorphous solid glass, while the addition of compounds such as Na_2O, B_2O_3 and CaO gives different glasses with particular physical and chemical properties.

LO 7 Crystal imperfections

Crystalline defects can significantly alter the properties of a solid. The introduction of small impurity 'defects' can change the electrical conductivity of a solid, and even its colour. Substitutional impurities arise from the replacement of one atom with another of a different type, while interstitial impurities result from the placement of atoms in interstitial holes.

LO 8 Modern ceramics

Ceramics are composed of inorganic components such as metal oxides or nitrides that have been heat-treated. Most ceramics have high melting points and are very hard. This is a result of significant covalent bonding, which gives rise to network-type solids. Modern ceramics are used in such diverse areas as coatings, filters, cosmetics and bullet-proof vests. Piezoelectric ceramics can deform in response to an electric potential and can be used in smart materials. High-temperature superconductors are ceramics and are used in rapid maglev trains.

KEY CONCEPTS AND EQUATIONS

Boiling points of substances *(section 7.1)*
The strengths of intermolecular forces in substances can be compared based on their boiling points.

Phase diagram *(section 7.3)*
We use a phase diagram to identify temperatures and pressures at which equilibrium can exist between phases of a substance and to identify conditions under which only a single phase can exist.

Unit cell *(section 7.4)*
A unit cell shows how the constituent atoms, molecules or ions are arranged in three dimensions in a crystalline solid. It is the smallest repeating unit of a crystalline structure.

The Bragg equation *(section 7.5)*
The Bragg equation relates λ, θ and the distance between the planes of atoms, d, and allows the determination of the structures of solids from X-ray diffraction patterns.

$$n\lambda = 2d \sin \theta$$

KEY TERMS

REVIEW QUESTIONS

LO 1 Liquids

7.1 Pentane is a C_5 hydrocarbon; gasoline contains mostly C_8 hydrocarbons; and fuel oil contains hydrocarbons in the C_{12} range. List these three hydrocarbons in order of increasing viscosity, and explain what molecular feature accounts for the variation.

7.2 Aluminium tubing has a thin surface layer of aluminium oxide. What shape meniscus would you expect to find for water and for mercury inside aluminium tubing? Explain your answers in terms of intermolecular forces.

7.3 Water forms 'beads' on the surface of a freshly waxed car, but forms a film on the clean windscreen. Why is this?

7.4 For each of the following pairs of liquids, choose which has the lower vapour pressure at room temperature and explain your reasoning.
(a) water, H_2O, or ethanol, CH_3CH_2OH
(b) pentan-1-ol, $C_5H_{12}OH$, or hexan-1-ol, $C_6H_{13}OH$
(c) chloromethane, CH_3Cl, or chloroform, $CHCl_3$

LO 2 Solids

7.5 Classify each of the following as an ionic, network, molecular or metallic solid: Bi, S_8, Se, SiO_2 and $Ba(NO_3)_2$.

7.6 Describe the differences between molecular and ionic solids in relation to: (a) interparticle forces and (b) macroscopic properties.

7.7 Indicate which type of solid (ionic, network, metallic or molecular/atomic) each of the following forms on solidification.
(a) Cl_2
(b) NaI
(c) Cs
(d) SiO_2
(e) Xe

LO 3 Phase changes

7.8 From table 7.2, determine which of each of the following pairs has the higher value. Use intermolecular forces to explain the data.
(a) enthalpies of vaporisation of ethane and propane
(b) enthalpies of vaporisation of ethanol and diethyl ether
(c) enthalpies of fusion of argon and methane

7.9 Sketch the approximate phase diagram for Br_2 from the following information: normal melting point is 265.9 K, normal boiling point is 331.9 K, triple point is at $p = 5.87 \times 10^3$ Pa and $T = 265.7$ K. Label the axes and the area where each phase is stable.

7.10 Using the phase diagram from question 7.9, describe what happens to a sample of Br_2 as the following processes take place. Draw lines on the phase diagram to represent each process.
(a) A sample at $T = 400$ K is cooled to 250 K at constant $p = 1.013 \times 10^5$ Pa.
(b) A sample is compressed from $p = 1.013 \times 10^2$ Pa to $p = 1.013 \times 10^8$ Pa at constant $T = 265.8$ K.
(c) A sample is cooled from 350 K to 250 K at constant $p = 2.026 \times 10^3$ Pa.

LO 4 Order in solids

7.11 The unit cell of a perovskite-type compound is shown below. What is the empirical formula of this perovskite?

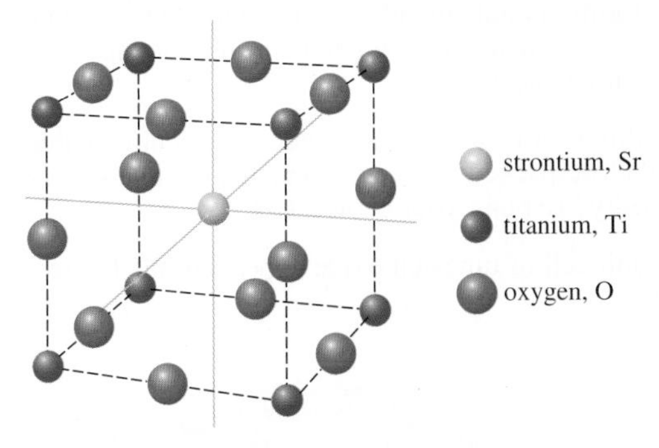

7.12 For the pattern shown below, draw a tile that represents a unit cell for the pattern and contains only complete fish. Draw a second tile that is a unit cell but contains no complete fish.

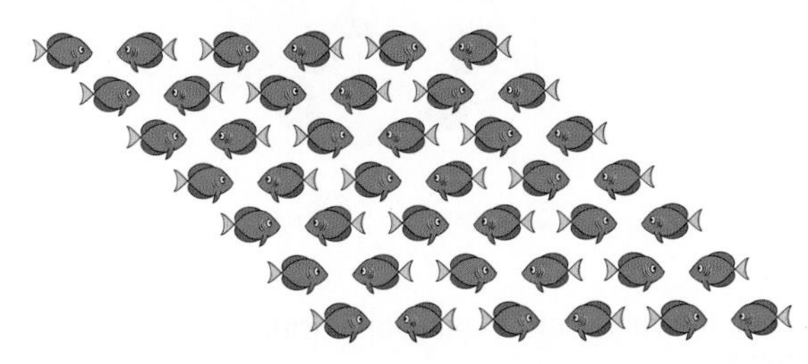

7.13 In the unit cell of lithium oxide, the oxide ions adopt a face-centred cubic arrangement. The lithium ions are in the holes within the face-centred cubic unit cell. How many complete lithium ions must be packed within each face-centred cube for the unit cell to be charge-neutral?

LO 5 X-ray diffraction

7.14 Write the Bragg equation and define the symbols used.

7.15 Explain in general terms how an X-ray diffraction pattern of a crystal and the Bragg equation provide information that allows chemists to determine the structures of molecules.

LO 6 Amorphous solids

7.16 What does the word 'amorphous' mean?

7.17 What is an amorphous solid? Compare what happens when crystalline and amorphous solids are broken into pieces.

7.18 Amorphous silica has a density of around $2.3\ g\ mL^{-1}$ and crystalline quartz has a density of $2.65\ g\ mL^{-1}$. Describe the bonding features that cause these two forms of the same substance to have different densities.

LO 7 Crystal imperfections

7.19 What are 'substitutional impurities'?

7.20 What are 'interstitial impurities'?

LO 8 Modern ceramics

7.21 What are 'ceramics'?

7.22 What are two physical properties that make ceramics useful materials?

7.23 What are 'refractories'? Give two uses for refractories.

7.24 What are 'superconductors'? Why is it very expensive to use typical superconductors in practical applications?

REVIEW PROBLEMS

7.25 List the different kinds of forces that must be overcome to convert each of the following compounds from a liquid to a gas. **LO 1**
(a) H_2O (c) CF_4
(b) CH_3Cl (d) CO_2

7.26 List all the intermolecular forces that stabilise the liquid phase of each of the following. **LO 1**
(a) Xe (c) CCl_4
(b) SF_4 (d) CH_3CH_2OH

7.27 For each of the following pairs, identify which has the higher boiling point, and identify the type of force that is responsible. **LO 1**
(a) H_3COCH_3 and CH_3OH (c) HF and HCl
(b) SO_2 and SiO_2 (d) Br_2 and I_2

7.28 Why do the two isomers represented below have very different melting points? **LO 2**

CH_3O, O, OH
$C_8H_8O_3$
methyl 2-hydroxybenzoate
(oil of wintergreen)
mp = −8 °C

CH_3O, O, OH
$C_8H_8O_3$
methyl 4-hydroxybenzoate
mp = 127 °C

7.29 Solid sodium metal can exist in two different crystalline forms, body-centred cubic and hexagonal. One is stable only below 5 K at relatively low pressure; the other is stable under all other conditions. Draw atomic pictures of both phases, identify the phase that is stable only at low temperature and pressure, and explain your choice. **LO 2**

7.30 Tungsten does not melt even when heated to incandescence. How do interatomic forces in tungsten compare with those in more typical metals such as gold or nickel? **LO 2**

7.31 Tin(IV) chloride, $SnCl_4$, has soft crystals with a melting point of −30.2 °C. The liquid is nonconducting. What type of solid (ionic, molecular, covalent or metallic) is formed by $SnCl_4$? **LO 2**

7.32 Elemental boron is a semiconductor, is very hard and has a melting point of about 2250 °C. What type of solid is formed by boron? **LO 2**

7.33 Gallium crystals are shiny and conduct electricity. Gallium melts at 29.8 °C. What type of solid does gallium form? **LO 2**

7.34 Titanium(IV) bromide forms soft orange-yellow crystals that melt at 39 °C to give a liquid that does not conduct electricity. The liquid boils at 230 °C. What type of solid does $TiBr_4$ form? **LO 2**

7.35 The element niobium is shiny, soft and ductile. It melts at 2468 °C and the solid conducts electricity. What type of solid does it form? **LO 2**

7.36 Elemental phosphorus consists of soft, white, 'waxy' crystals that are easily crushed and melt at 44 °C. The solid does not conduct electricity. What type of solid does phosphorus form? **LO 2**

7.37 Indicate which type of solid each of the following would form when it solidifies. **LO 2**
(a) Cl_2 (c) Li_2O (e) PH_3 (g) Si
(b) CsF (d) W (f) NaOH

7.38 Indicate which type of solid each of the following would form when it solidifies. **LO2**
(a) Na_2SO_4 (b) Pd (c) Ar (d) H_2Se (e) Ge (f) CsCl (g) N_2

7.39 Molecular hydrogen and atomic helium both have two electrons, but He boils at 4.2 K, whereas H_2 boils at 20 K. Neon boils at 27.1 K, whereas methane, which has the same number of electrons, boils at 114 K. Explain why molecular substances boil at a higher temperature than atomic substances with the same number of electrons. **LO2**

7.40 Refer to figure 7.21. Describe in detail what occurs when each of the following processes is carried out. **LO3**
(a) A sample of CO_2 gas is compressed from 1.013×10^5 Pa to 5.065×10^6 Pa at $T = 298$ K.
(b) Dry ice at 195 K is heated to 350 K at $p = 6.078 \times 10^5$ Pa.
(c) A sample of CO_2 gas at $p = 1.013 \times 10^5$ Pa is cooled from 298 K to 50 K.

7.41 With reference to figure 7.22, describe what happens to silica as it is slowly heated from room temperature to 2000 °C at atmospheric pressure (1.013×10^5 Pa). **LO3**

7.42 How many triple points appear on the phase diagram in figure 7.22? For each one, describe the conditions and name the three phases that coexist under these conditions. **LO3**

7.43 The figure below shows the phase diagram for elemental sulfur (not to scale). Use it to answer the following questions. **LO3**

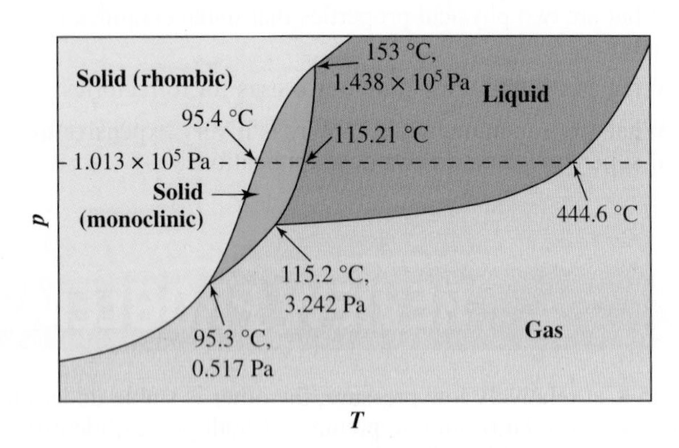

(a) Under what conditions does rhombic sulfur melt?
(b) Under what conditions does rhombic sulfur convert to monoclinic sulfur?
(c) Under what conditions does rhombic sulfur sublime?

7.44 Use the figure in question 7.43 to answer the following. **LO3**
(a) How many triple points are there in the diagram?
(b) Which phases coexist at each triple point?
(c) If sulfur is at the triple point at 153 °C, what happens if the pressure is reduced?

7.45 Sketch the phase diagram for a substance that has a triple point at −15.0 °C and 0.30×10^5 Pa, melts at −10.0 °C at 1×10^5 Pa, and has a normal boiling point of 90 °C. **LO3**

7.46 Based on the phase diagram in question 7.43, below what pressure will the substance undergo sublimation? How does the density of the liquid compare with the density of the solid? **LO3**

7.47 Construct a qualitative graph similar to the one in figure 7.16 that summarises the energy changes that accompany the following process: A sample of water at 40 °C is placed in a freezer until the temperature reaches −30 °C. Plot temperature along the y-axis and kilojoules of heat removed along the x-axis. **LO3**

7.48 Iron forms crystals in either body-centred cubic or face-centred cubic geometry. Which geometry has the higher density? Explain your choice. **LO4**

7.49 Beryllium forms crystals in either body-centred cubic or hexagonal close-packed geometry. Which solid phase has the higher density? Explain your choice. **LO4**

7.50 The unit cell of tungsten oxide appears in the figure below. **LO4**

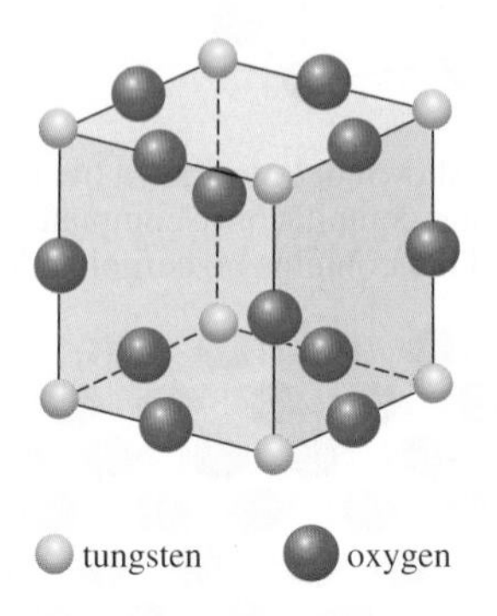

(a) Name the cubic lattice compatible with this arrangement of the atoms.
(b) Calculate the number of rhenium and oxygen atoms in the unit cell and identify the formula of rhenium oxide (the chemical symbol for rhenium is Re).

7.51 One form of niobium oxide crystallises in the unit cell shown below. Calculate the number of niobium and oxygen atoms in the unit cell, and identify the formula of niobium oxide (the chemical symbol for niobium is Nb). **LO4**

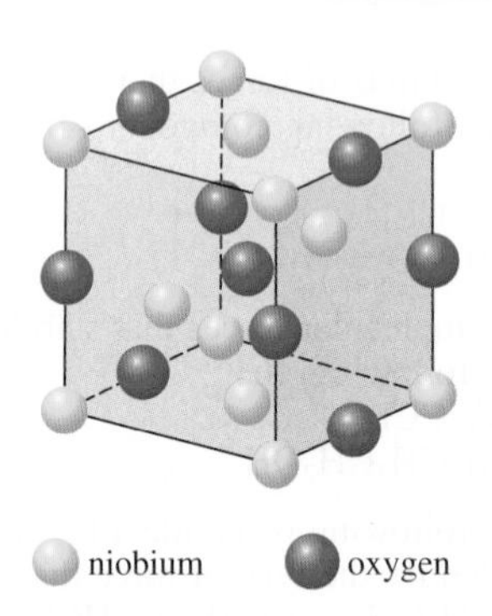

7.52 Silver forms face-centred cubic crystals. The atomic radius of a silver atom is 144 pm. Draw the face of a unit cell with the nuclei of the silver atoms at the lattice points. The atoms are in contact along the diagonal. Calculate the length of an edge of this unit cell. **LO4**

7.53 The atomic radius of nickel is 124 pm. Nickel crystallises in a face-centred cubic lattice. What is the length of the edge of the unit cell expressed in picometres? **LO4**

ADDITIONAL EXERCISES

7.54 Potassium ions have a radius of 133 pm, and bromide ions have a radius of 195 pm. The crystal structure of potassium bromide is the same as sodium chloride. Estimate the length of the edge of the unit cell in potassium bromide. **LO4**

7.55 The unit cell in sodium chloride has an edge length of 564.0 pm. The sodium ion has a radius of 95 pm. What is the diameter of a chloride ion? **LO4**

7.56 Caesium chloride, CsCl, crystallises with a cubic unit cell of edge length 412.3 pm. The density of CsCl is 3.99 g mL^{-1}. Show that the unit cell cannot be body-centred cubic or face-centred cubic. **LO 4**

7.57 Draw the unit cell of the NaCl crystal and determine the number of nearest neighbours of opposite charge for each ion in this unit cell. **LO 4**

7.58 Below is an illustration of a 'herringbone' pattern for patio pavers. Identify the unit cell for this pattern. **LO 4**

7.59 Gold crystallises in a face-centred cubic lattice. The edge of the unit cell has a length of 407.86 pm. The density of gold is 19.31 g mL^{-1}. **LO 4**

(a) Use these data and the atomic mass of gold to calculate the value of the Avogadro constant.
(b) Calculate the atomic radius of gold in units of picometres.

7.60 Silver has an atomic radius of 144 pm. What would the density of silver be in g mL^{-1} if it were to crystallise in: **LO 4**

(a) a simple cubic lattice
(b) a body-centred cubic lattice
(c) a face-centred cubic lattice?

The actual density of silver is 10.6 g mL^{-1}. Which cubic lattice does silver have?

7.61 Which of the following molecules would you expect to wet a greasy glass surface more effectively? **LO 1,2**

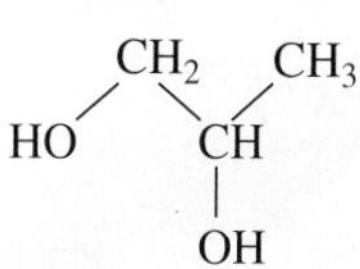

propylene glycol (propane-1,2-diol)

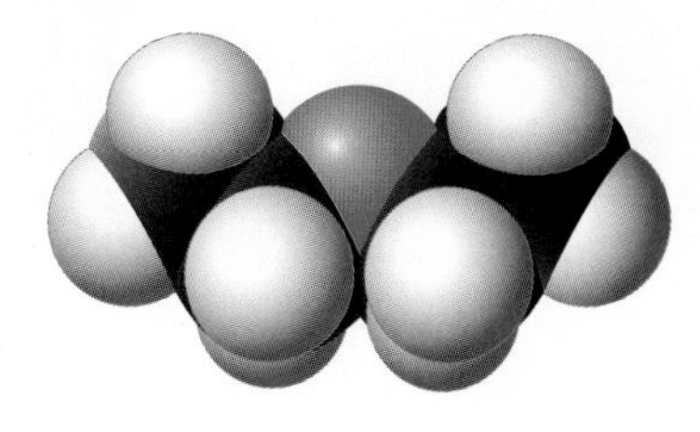

CH_2 CH_2
H_3C O CH_3

diethyl ether

8 Chemical thermodynamics

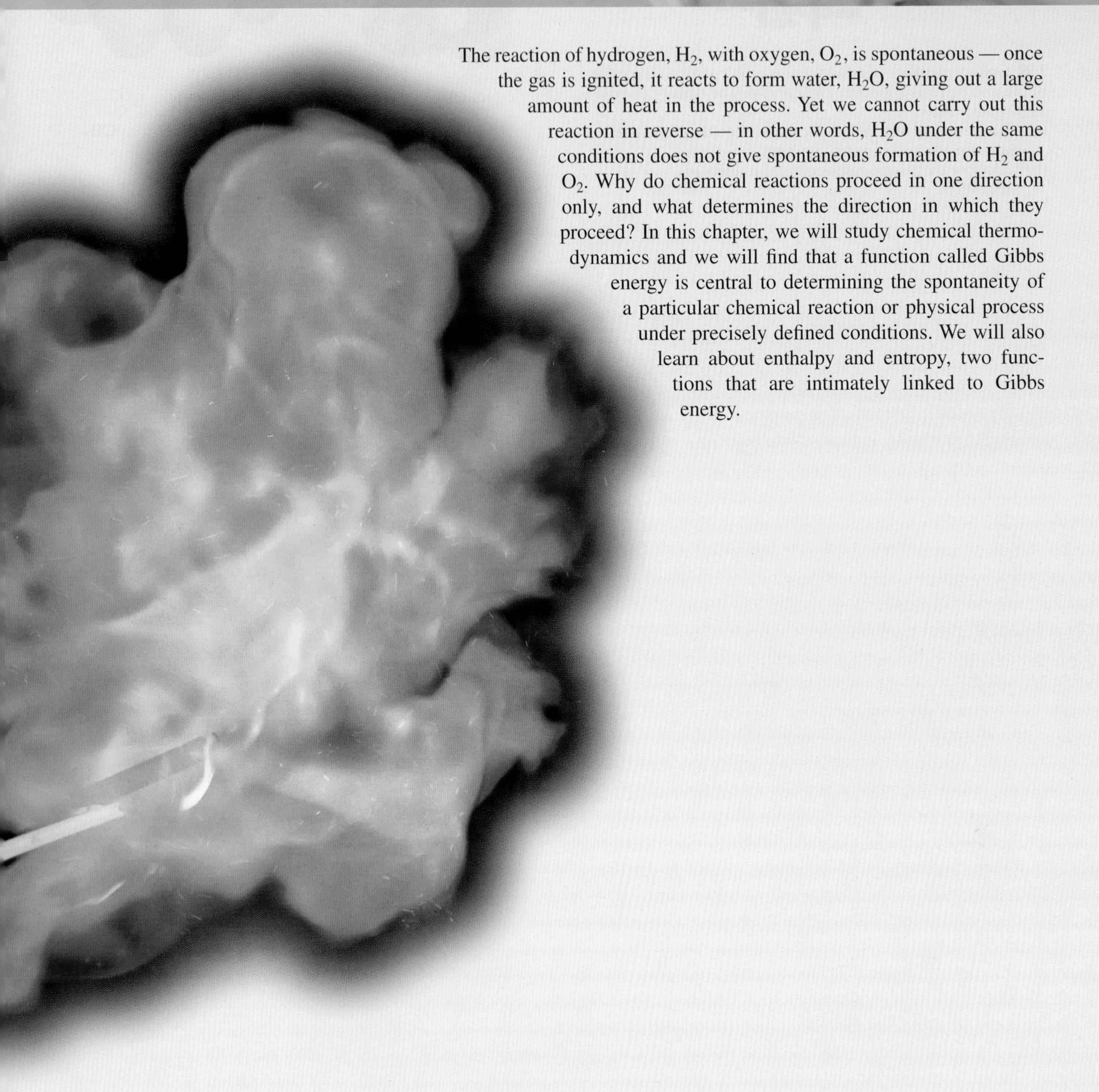

The reaction of hydrogen, H_2, with oxygen, O_2, is spontaneous — once the gas is ignited, it reacts to form water, H_2O, giving out a large amount of heat in the process. Yet we cannot carry out this reaction in reverse — in other words, H_2O under the same conditions does not give spontaneous formation of H_2 and O_2. Why do chemical reactions proceed in one direction only, and what determines the direction in which they proceed? In this chapter, we will study chemical thermodynamics and we will find that a function called Gibbs energy is central to determining the spontaneity of a particular chemical reaction or physical process under precisely defined conditions. We will also learn about enthalpy and entropy, two functions that are intimately linked to Gibbs energy.

Chemical thermodynamics is crucial to understanding chemical equilibria and forms the basis of this chapter and chapters 9, 10, 11 and 12. It is an essential part not only of chemistry, but also of life itself.

LEARNING OBJECTIVES

After studying this chapter, you should be able to:

8.1 explain the difference between spontaneous and non-spontaneous chemical processes

8.2 use the terminology and units of chemical thermodynamics

8.3 carry out calculations related to the first law of thermodynamics

8.4 explain the concept of enthalpy and use Hess's Law

8.5 explain the concept of entropy

8.6 state the second law of thermodynamics, and appreciate that any spontaneous process involves an increase in the entropy of the universe

8.7 calculate entropies of reaction

8.8 determine the spontaneity of a process using the change in Gibbs energy.

8.1 Introduction to chemical thermodynamics

FIGURE 8.1 Metallic sodium reacts violently with water. In the reaction, sodium is converted to Na^+, and water gives hydrogen gas and hydroxide ions. When the reaction is over, the solution contains sodium hydroxide.

When a lump of solid sodium metal is placed in water, a vigorous chemical reaction occurs that forms hydrogen gas and aqueous sodium hydroxide (figure 8.1).

$$\text{Na(s)} + \text{H}_2\text{O(l)} \rightarrow \text{NaOH(aq)} + \tfrac{1}{2}\text{H}_2\text{(g)}$$

However, if we try to carry out the reverse reaction by bubbling hydrogen gas through an aqueous solution of sodium hydroxide, no chemical reaction is observed.

$$\text{NaOH(aq)} + \tfrac{1}{2}\text{H}_2\text{(g)} \rightarrow \text{no reaction}$$

Similarly, if a block of ice is left at room temperature and atmospheric pressure, it will eventually melt to give liquid water:

$$\text{H}_2\text{O(s, 25 °C, } 1.013\times10^5\text{ Pa)} \rightarrow \text{H}_2\text{O(l, 25 °C, } 1.013\times10^5\text{ Pa)}$$

but liquid water under the same conditions will never solidify into ice:

$$\text{H}_2\text{O(l, 25 °C, } 1.013\times10^5\text{ Pa)} \rightarrow \text{no change}$$

These examples show that both chemical reactions and physical changes occur in one particular direction only under particular conditions of temperature and pressure, and we say that such changes are **spontaneous**; in other words, once started they proceed without the involvement of any outside factors. The reaction of sodium with water and the melting of ice described above are examples of spontaneous processes that proceed to completion. However, many spontaneous chemical reactions do not give complete conversion of reactants to products. An example of a spontaneous reaction that does not go to completion is the reaction of N_2O (laughing gas) with oxygen at room temperature and atmospheric pressure to give $NO_2(g)$. Although we can write the balanced chemical equation for this reaction as:

$$2\text{N}_2\text{O(g)} + 3\text{O}_2\text{(g)} \rightarrow 4\text{NO}_2\text{(g)}$$

this does not mean that, if we mixed 2 moles of $N_2O(g)$ with 3 moles of $O_2(g)$ in a container at room temperature and pressure, we would obtain 4 moles of $NO_2(g)$. In fact, we would find that, no matter how long we waited, no more than about 2 moles of $NO_2(g)$ would form and there would always be significant amounts of the starting materials present in the reaction mixture. It is important for you to appreciate that *a balanced chemical equation tells us nothing about whether or not a particular reaction proceeds to completion; it simply tells us the mole ratios in which reactants react to give products.*

We can force nonspontaneous reactions to proceed provided that we couple them to a spontaneous process. For example, water does not spontaneously decompose into hydrogen gas and oxygen gas, but, by passing an electric current (a spontaneous flow of electrons) through water, we can force the reaction:

$$2\text{H}_2\text{O(l)} \rightarrow 2\text{H}_2\text{(g)} + \text{O}_2\text{(g)}$$

to proceed (figure 8.2). This process is called electrolysis. Production of hydrogen and oxygen will continue, however, only as long as the electric current is maintained. As soon as the supply of electricity is cut off, the decomposition ceases. This example demonstrates the difference between spontaneous and nonspontaneous changes. Once a spontaneous event begins, it tends to continue until it stops of its own accord (figure 8.3). A nonspontaneous event, on the other hand, can continue only as long as it receives some sort of outside assistance. You should also note that the electrolysis of water requires some sort of spontaneous mechanical or chemical change to generate the needed electricity. In short, *all nonspontaneous events occur at the expense of spontaneous ones.* Everything that happens can be traced, either directly or indirectly, to spontaneous changes.

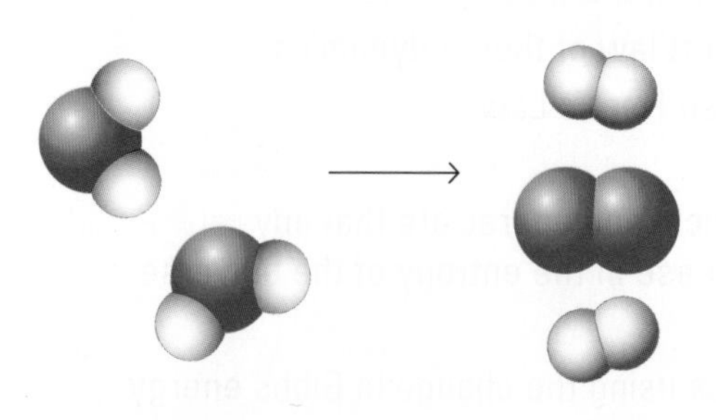
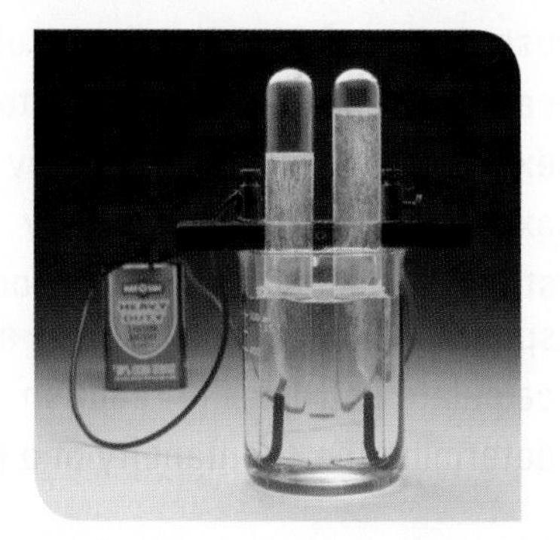
FIGURE 8.2 The electrolysis of water produces H_2 and O_2 gases. It is a nonspontaneous change that continues only as long as electricity is supplied.

FIGURE 8.3 Three common spontaneous events — iron rusts, fuel burns and an ice cube melts at room temperature.

As chemists, we are interested in predicting whether a reaction is spontaneous under our conditions of interest, and, if it is, how far it will proceed towards completion under those conditions. The science of **chemical thermodynamics** allows us to predict both the direction and extent of spontaneous chemical and physical change under particular conditions. In order to do this, we need to study four important thermodynamic functions: internal energy (U, section 8.3), enthalpy (H, section 8.4), entropy (S, section 8.5) and Gibbs energy (G, section 8.8). Very brief definitions of these are as follows.

- **Internal energy** (U) is the sum of all the energies of all of the individual particles in a sample of matter.
- **Enthalpy** (H) is a function related to the heat absorbed or evolved by a chemical system during a chemical reaction or physical change under conditions of constant pressure.
- **Entropy** (S) is a measure of the number of ways energy is distributed throughout a chemical system.
- **Gibbs energy** (G), named after the American mathematical physicist Josiah Willard Gibbs (1839–1903), is defined as:

$$G = H - TS$$

where T is the thermodynamic temperature, measured in kelvin.

We will look at these thermodynamic functions in greater detail, and provide exact definitions, in the appropriate sections of this chapter. For now, it is sufficient to state that Gibbs energy, G, allows us to predict spontaneity. We can liken a spontaneous change to a ball at the top of a hill — once pushed, it will roll spontaneously down the hill until it reaches the bottom, the point of minimum energy. In the same way, we will see that chemical reactions proceed in the direction that leads to a decrease in the Gibbs energy of the system. We will also see that overall chemical and physical change will cease once the Gibbs energy of the system is minimised, and that, under these conditions, the system is at equilibrium.

8.2 Thermodynamic concepts

Before we begin our study of chemical thermodynamics, we need to discuss some important thermodynamic concepts.

Heat and temperature

Arguably the most important and indeed most familiar thermodynamic term, **heat**, is perhaps the most conceptually difficult. Heat is a transfer of energy due to a temperature difference. If two bodies having different temperatures are brought into direct contact, there will be heat flow from the hotter to the colder body, until both are at the same temperature. As we will see, we cannot actually measure heat directly. The definition of temperature also follows from the above; two bodies have the same temperature if they are in thermal equilibrium; i.e. there is no heat flow between them when they are in direct contact. It is important to note that the **thermodynamic temperature** (sometimes called the absolute temperature) scale is used in nearly all thermodynamic calculations.

The thermodynamic temperature is measured in kelvin (K), and a thermodynamic temperature can be converted to a temperature in degrees Celsius by subtracting 273.15. You should note that a temperature *difference* of 1 K is numerically equal to a temperature difference of 1 °C, so, for calculations involving a temperature *change*, the numerical value of this change will be the same, regardless of whether the individual temperatures are expressed in K or °C.

System, surroundings and universe

We will use the word **system** to refer to the particular part of the universe (generally one or more chemical species) that we are studying, while everything else is the **surroundings**. Together, the system and surroundings constitute the **universe** (figure 8.4). As we are usually interested in heat flow between the system and surroundings, it is very important to specify the **boundary** across which heat flows. The boundary might be visible (such as the walls of a beaker) or invisible (such as the boundary that separates warm air from cold air along a weather front). Three types of system are possible, depending on whether matter or energy can cross the boundary.

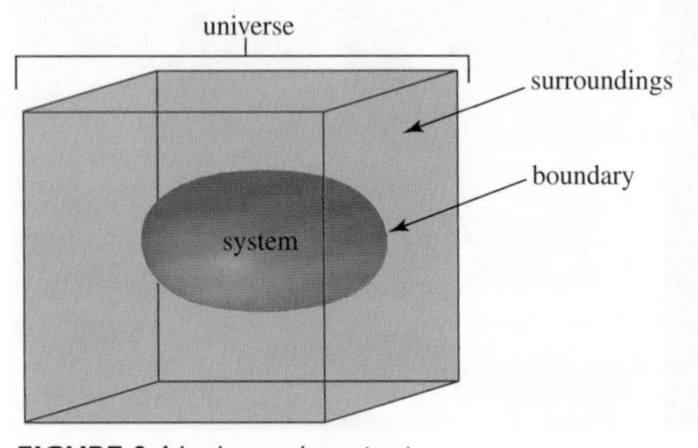

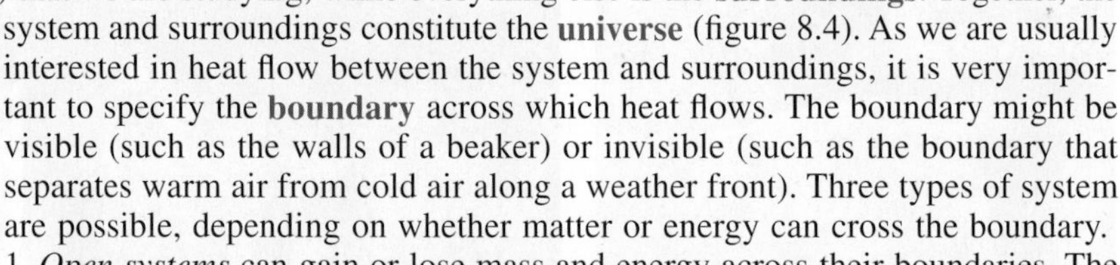

FIGURE 8.4 In thermodynamics, it is important to define the system, surroundings, universe and the boundary.

1. *Open systems* can gain or lose mass and energy across their boundaries. The human body is an example of an open system.
2. *Closed systems* can absorb or release energy, but not mass, across the boundary. The mass of a closed system is constant, no matter what happens inside. A light bulb is an example of a closed system as it can release energy as both heat and light, but its mass is constant.
3. *Isolated systems* cannot exchange matter or energy with their surroundings. Because energy cannot be created or destroyed, the energy of an isolated system is constant, no matter what happens inside. A stoppered vacuum flask is a good approximation of an isolated system. Processes that occur within an isolated system with no heat transfer to the surroundings are called *adiabatic*, from the Greek words *a* and *diabatos* meaning 'not passable'.

Units

We must be aware of the units used in chemical thermodynamics. The SI unit of energy, work (motion against an opposing force) and heat is the **joule (J)**. Recall that the joule is a derived unit (table 2.2, p. 27) and can be expressed in terms of SI base units as:

$$1\,J = 1\,kg\,m^2\,s^{-2}$$

A joule is the amount of kinetic energy possessed by a 2-kilogram object moving at a speed of 1 metre per second, or, to use another example, 1 joule is approximately the energy of a single human heartbeat. The joule is actually a rather small amount of energy and in most cases we will use the larger unit, the **kilojoule (kJ)**.

$$1\,kJ = 10^3\,J$$

The use of both J and kJ in thermodynamics means that you must be very careful when carrying out calculations. For example, as we shall see, values of ΔH are commonly reported in $kJ\,mol^{-1}$ while ΔS values are usually given in $J\,mol^{-1}\,K^{-1}$. Therefore, when using these data in the important equation:

$$\Delta G = \Delta H - T\Delta S$$

it is necessary to first ensure the values have consistent units. This will involve either converting ΔH to $J\,mol^{-1}$, by multiplying by 1000, or converting ΔS to $kJ\,mol^{-1}\,K^{-1}$, by dividing by 1000. This will then ensure that the term $\Delta H - T\Delta S$, and hence ΔG, will have consistent units of either:

$$J\,mol^{-1} - (K)(J\,mol^{-1}\,K^{-1}) = J\,mol^{-1}$$

or:

$$kJ\,mol^{-1} - (K)(kJ\,mol^{-1}\,K^{-1}) = kJ\,mol^{-1}$$

Δ*X*: the change in *X*

We are generally interested in the *change* in the values of *U*, *H*, *S* and *G* as the result of either a chemical reaction or a change in phase, rather than their absolute values. We denote such a change using Δ (the uppercase Greek letter Δ delta). Therefore, for any thermodynamic quantity *X*:

$$\Delta X = \text{final value of } X - \text{initial value of } X$$

We often use the Δ terminology when discussing changes in temperature. Thus, any change in temperature, ΔT, is defined as:

$$\Delta T = T_{final} - T_{initial}$$

Note that, if T_{final} is less than $T_{initial}$, ΔT is then, by definition, negative.

The thermodynamic functions ΔU, ΔH, ΔS and ΔG, can refer to both chemical and physical changes. Chemical changes generally occur as a result of chemical reactions; in these, the reactants and products are chemically different, as the result of the making and/or breaking of chemical bonds, or the transfer of electrons. Physical changes refer primarily to changes of phase. Here, the chemical identities of the reactants and products are the same, but their physical states are different; for example, liquid water freezing to give ice, and solid iodine subliming to give gaseous iodine. On the occasions when we are referring to the values of these thermodynamic functions *specifically* for chemical reactions, we denote this using the symbols $\Delta_r U$, $\Delta_r H$, $\Delta_r S$ and $\Delta_r G$.

Whenever we make any measurement, we do so relative to some reference. This reference can be either arbitrary or absolute. For example, the zero point on the °C temperature scale has been arbitrarily set to the point at which water freezes or ice melts, and all temperatures in °C are measured relative to this. Conversely, any temperature in K is measured relative to absolute zero (0 K), the point at which all atomic and molecular motion ceases, and which is therefore the coldest possible temperature. For this reason, the temperature in K is sometimes called the absolute temperature.

The necessity for a reference is also true of thermodynamic measurements. In this chapter, we will encounter thermodynamic functions such as H (enthalpy) and G (Gibbs energy) for which absolute values cannot be measured, and for which we must therefore assign an arbitrary zero, and S (entropy), for which absolute values can be measured relative to a true zero.

As we shall see, our arbitrary zero for thermodynamic functions such as enthalpy and Gibbs energy involves standard states, the definitions of which depend on the phase of the substance in question. These standard states are given in table 8.1.

TABLE 8.1 Standard states

Phase of substance	Standard state
pure solid	the pure solid at a pressure of $p^{\ominus}$ (1×10^5 Pa)
pure liquid	the pure liquid at a pressure of $p^{\ominus}$
pure gas	the pure ideal gas at a pressure of $p^{\ominus}$
species in solution	the ideal solution at a pressure of $p^{\ominus}$, having either a molality of $m^{\ominus}$ (1 mol kg^{-1}) or a concentration of $c^{\ominus}$ (1 mol L^{-1})

It should be noted that temperature is not part of the definition of any standard state and it must therefore always be specified. However, it is very common to quote thermodynamic measurements at 25 °C. It should also be noted that the standard states for pure gases and species in solution are unattainable and therefore hypothetical; as we have seen in chapter 6 and will see in chapter 10, no gases or solutions exhibit ideal behaviour.

Standard thermodynamic functions for physical processes (i.e. phase changes) or chemical reactions refer to processes or reactions in which the unmixed reactants in their standard states are converted to the unmixed products in their standard states. Such functions are denoted by a superscript plimsoll ($^{\ominus}$), and such processes or reactions are said to be carried out under standard conditions.

State functions

As we will see, ΔU, ΔH, ΔS and ΔG are all examples of a **state function**. This means that the value of each depends only on the current state of the system and it does not matter how the system came to be in its current state or how the system will behave in the future. We can usually define the state of a chemical system by specifying the amount and type of substances present in the system, and the pressure, temperature and volume of the system. When we then change the state of the system by changing one or more of the above variables, the change in the value of any state function X of the system is given by:

$$\Delta X = X_{\text{final}} - X_{\text{initial}}$$

and *depends only on the initial and final states of the system* and not on the path taken during the change in the state of the system. This concept is illustrated in figure 8.5.

As we will see later, the advantage of recognising that a particular property is a state function is that many calculations become much easier.

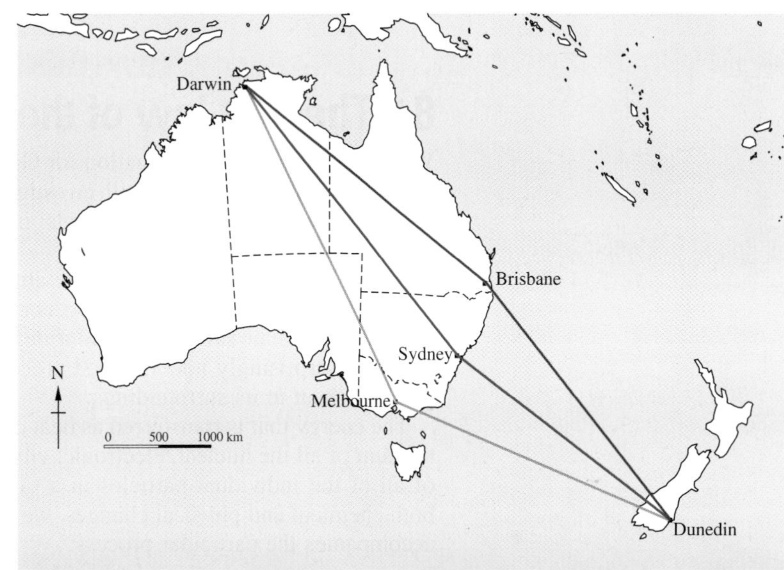

FIGURE 8.5 The latitude of Darwin airport is 12°24'S, while that of Dunedin airport is 45°55'S. As there are no direct flights between Darwin and Dunedin, if we wish to travel between them, we must fly via Brisbane (blue line, a total of 5401 km), Sydney (green line, 5244 km) or Melbourne (orange line, 5410 km). The actual distance travelled is different in each case, and is therefore not a state function as it depends on the path taken. However, the change in latitude that occurs, 33°31', is the same, regardless of which route we take, and is therefore a state function.

ΔG and spontaneity

The primary aim of this chapter is to show that the spontaneity of a chemical or physical process at constant temperature and pressure can be predicted from the sign of the Gibbs energy change for the process (ΔG), which itself depends on the values of ΔH and ΔS via the equation:

$$\Delta G = \Delta H - T\Delta S$$

This is one of the most important equations in all of chemistry. A negative value of ΔG means that the products are of lower G than the reactants; the process must proceed in the direction of decreasing G so it is spontaneous. Conversely, if ΔG is positive, the products are of higher G than the reactants; the process will not be spontaneous as it has to proceed in the direction of increasing G (figure 8.6).

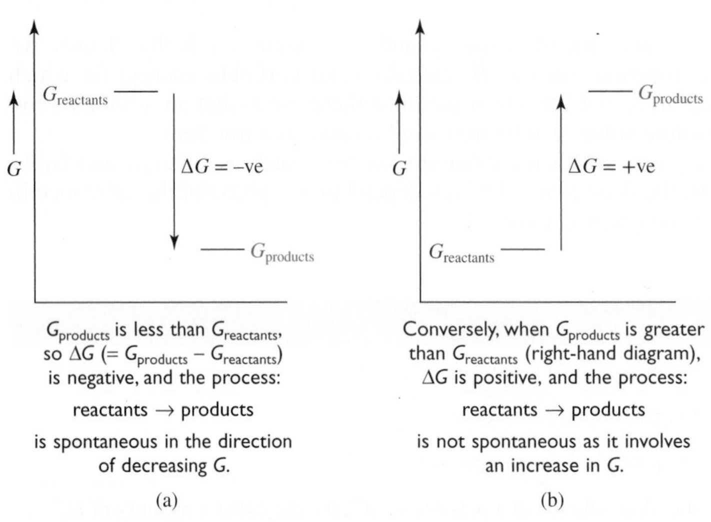

$G_{reactants}$
G
$\Delta G = -ve$
$G_{products}$
$G_{products}$ is less than $G_{reactants}$, so ΔG (= $G_{products} - G_{reactants}$) is negative, and the process:
reactants → products
is spontaneous in the direction of decreasing G.
(a)
$G_{products}$
G
$\Delta G = +ve$
$G_{reactants}$
Conversely, when $G_{products}$ is greater than $G_{reactants}$ (right-hand diagram), ΔG is positive, and the process:
reactants → products
is not spontaneous as it involves an increase in G.
(b)

FIGURE 8.6 Diagrams showing the origin of the sign of ΔG.

If $\Delta G = 0$, the Gibbs energy of both the reactants and products is the same, and the system is at equilibrium, a state in which no overall chemical or physical change is observed. In short, we can say that, for a process occurring at constant temperature and pressure:

- if $\Delta G < 0$, the process is spontaneous
- if $\Delta G > 0$, the process is nonspontaneous
- if $\Delta G = 0$, the system is at equilibrium.

The magnitude of ΔG under standard conditions is also important in telling us how far a reaction goes towards completion, as we will see in chapter 9. Obviously, to determine the value of ΔG, we need to know the values of both ΔH and ΔS. We will define the former by way of the first law of thermodynamics. In so doing, we will introduce the concepts of heat (q), work (w) and internal energy (u).

8.3 The first law of thermodynamics

We stated earlier that the equation for Gibbs energy contained terms involving both enthalpy and entropy. In this section, we will investigate what is meant by the term 'enthalpy'. In doing this, we will introduce some new concepts: internal energy (U), work (w) and heat (q), as well as the first law of thermodynamics.

If you have ever dissolved a large amount of sodium chloride, NaCl, in water, you will have noticed that the resulting solution gets colder as the salt dissolves. However, if you dissolve the chemically similar salt lithium chloride, LiCl, in water, you will find that the resulting solution becomes surprisingly hot. The first process absorbs heat from its surroundings while the second gives out heat to its surroundings.

The energy that is transferred as heat comes from an object's internal energy. Internal energy is the sum of all the nuclear, electronic, vibrational, rotational, translational and interaction energies of all of the individual particles in a sample of matter and is given the symbol U. In studying both chemical and physical changes, we are interested in the *change* in internal energy (ΔU) that accompanies the particular process:

$$\Delta U = U_{final} - U_{initial}$$

For a chemical reaction, U_{final} corresponds to the internal energy of the products, so we will write it as $U_{products}$. Similarly, we will use the symbol $U_{reactants}$ for $U_{initial}$. So for a chemical reaction the change in internal energy is given by:

$$\Delta_r U = U_{products} - U_{reactants}$$

Thus, when we dissolve sodium chloride in water, the system *absorbs* energy from its surroundings; its final energy is therefore greater than its initial energy, so $\Delta_r U$ is positive. Conversely, when we dissolve lithium chloride in water, the system gives out energy to its surroundings; its final energy is therefore less than its initial energy, so $\Delta_r U$ is negative.

A chemical system can exchange energy with its surroundings in two ways. The first, as we have seen above, is by either absorbing heat from or emitting heat to the surroundings. The second is by doing work on the surroundings, or having the surroundings do work on it. **Work** may be defined simply as motion against an opposing force. For example, lifting a weight against the force of gravity is work, as is stretching or compressing a spring. While there are a number of different types of work possible in chemical systems (such as electrical work and osmotic work), we will consider only the work associated with the compression or expansion of a gas in a closed chemical system to illustrate a number of important thermodynamic concepts. This type of work is often called pressure–volume or pV work, as it involves a change in pressure and volume of a gas sample. When a gas is compressed at constant temperature, the surroundings are doing work on the system, while the expansion of a gas at constant temperature is an example of a system doing work on the surroundings (figure 8.7).

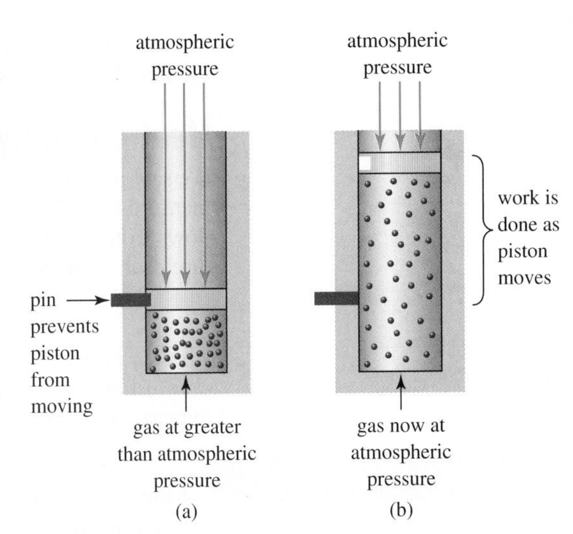

FIGURE 8.7 Pressure–volume work: **(a)** A gas is confined under pressure in a cylinder fitted with a piston that is held in place by a sliding pin. **(b)** When the piston is released, the gas inside the cylinder expands and pushes the piston upwards against the opposing pressure of the atmosphere. As it does so, the gas does some pressure–volume work on the surroundings.

The magnitude of the work done during a volume change, ΔV, against an opposing pressure, p, is given by the equation:

$$w = -p\Delta V$$

As ΔV ($V_{final} - V_{initial}$) is positive for the expansion of a gas, work is negative for this process at constant temperature where the system does work on the surroundings, while similar reasoning shows that work is positive for the compression of a gas where the surroundings do work on the system.

As heat and work are the only ways by which a closed chemical system can exchange energy with its surroundings, it follows that the change in internal energy of a closed chemical system during a chemical or physical change must be equal to the sum of the heat absorbed or emitted by the system and the work done on or by the system. In mathematical terms, this becomes:

$$\Delta U = q + w$$

where q = heat and w = work. This equation is known as the **first law of thermodynamics**. In simple terms, it means that energy can be transferred between a system and its surroundings as either heat or work, but it can never be created or destroyed. An alternative statement of the first law of thermodynamics is that the energy of an isolated system is constant. Both formulations are essentially alternative versions of the law of conservation of energy, which says that energy cannot be created or destroyed. As with H, S and G, we are not interested in the absolute value of U, but rather the change in U (ΔU) that occurs as the result of a chemical or physical change.

While ΔU is a state function, the values of q and w depend on what happens between the initial and final states of any change; in other words, q and w are not state functions. For example, consider the discharge of a car battery by two different paths (figure 8.8). Both paths take us between the same two states, one being the fully charged state and the other the fully discharged state. Because ΔU is a state function and because both paths have the same initial and final states, ΔU must be the same for both paths. But what about q and w?

In path 1 of figure 8.8, we simply short-circuit the battery by placing a heavy spanner across the terminals. If you have ever done this, even by accident, you know how violent the result can be. Sparks fly and the spanner becomes very hot as the battery quickly discharges. Lots of heat is given off, but *the system does no work* ($w = 0$). The heat loss to the surroundings is ΔU, the difference in U between the fully discharged and fully charged battery.

In path 2, we discharge the battery more slowly by using it to operate a motor. Along this path, most of the energy represented by ΔU appears as work (running the motor) and only a relatively small amount appears as heat (from the friction within the motor and the electrical resistance of the wires).

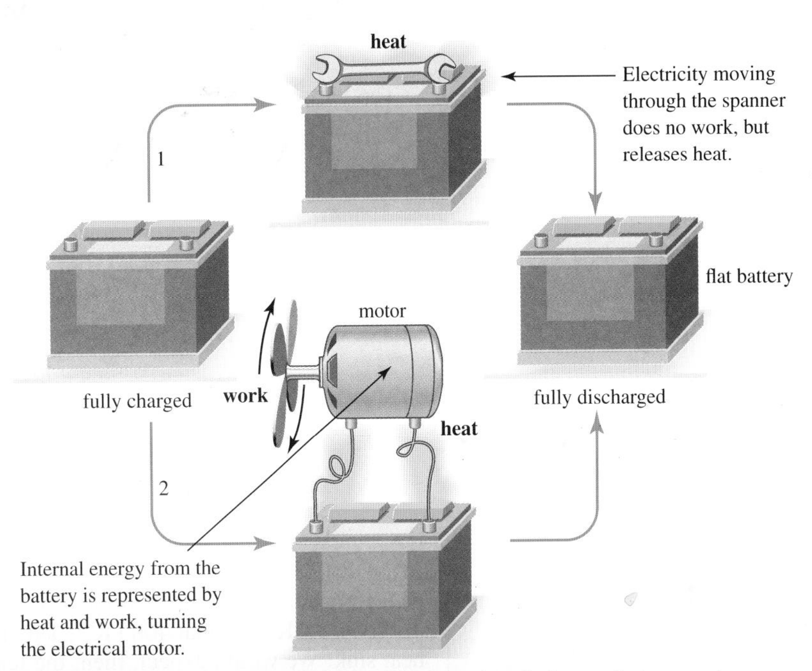

FIGURE 8.8 Internal energy, heat and work. The complete discharge of a battery along two different paths involves the same total amount of internal energy, ΔU. However, if the battery is simply short-circuited with a heavy spanner, as shown in path 1, this energy appears entirely as heat. Path 2 gives part of the total internal energy as heat, but most of the internal energy appears as work done by the motor.

Therefore, neither q nor w is a state function. Their values depend *entirely* on the path between the initial and final states.

Heat capacity and specific heat

We saw in section 8.2 that heat is a transfer of energy due to a temperature difference. We cannot measure heat directly, but we can calculate it using the temperature change that occurs when heat flows from one body to another. There is, in fact, a linear relationship between heat and temperature change given by the equation:

$$q = C\Delta T$$

where q is heat and C is the **heat capacity** of the object in question. The units of heat capacity ($J\,K^{-1}$) help us to understand what it is; it can be thought of as the amount of heat required to increase the temperature of a substance by 1 K.

WORKED EXAMPLE 8.1

Experimental determination of heat capacity

Central processing chips in computers generate a tremendous amount of heat — enough to damage themselves permanently if the chip is not cooled somehow. Aluminium 'heat sinks' are often attached to the chips to carry away excess heat. We can measure the heat capacity of such heat sinks by placing them (hot) into a known volume of water and measuring the temperature rise of the water. Suppose that a heat sink at 71.3 °C is dropped into a polystyrene cup containing 100.0 g of water at 25.0 °C. The temperature of the water rises to 27.4 °C. Given that it takes 4.18 J to raise the temperature of 1 g of water by 1 K, what is the heat capacity of the heat sink?

Analysis

We want to find the heat capacity, C. We can solve $q = C\Delta T$ for C by dividing both sides by the temperature change, ΔT.

$$C = \frac{q}{\Delta T}$$

We must find the temperature change of the heat sink and the amount of heat, q, that it exchanges with the water in the cup.

Solution

To find q, we can assume that the heat lost by the heat sink will be gained by the much cooler water, so we start by calculating the heat gained by the water.

$$\text{heat lost by heat sink} = -\text{heat gained by the water}$$

The temperature of the water rises from 25.0 °C to 27.4 °C, a rise of 2.4 K (recall that a temperature *difference* has the same numerical value in both K and °C). It would take $2.4\,K \times 4.18\,J\,K^{-1} = 1.0 \times 10^{1}\,J$ of heat to raise the temperature of each gram of water by 2.4 K, so it would take $100.0 \times 1.0 \times 10^{1}\,J = 1.0 \times 10^{3}\,J$ to raise the temperature of 100.0 g of water by 2.4 K.

The temperature of the heat sink drops from 71.3 °C to 27.4 °C, so:

$$\begin{aligned}\Delta T &= T_{\text{final}} - T_{\text{initial}} \\ &= 27.4\,°C - 71.3\,°C \\ &= -43.9\,°C \\ &= -43.9\,K\end{aligned}$$

The heat gained by the water is $1.0 \times 10^{3}\,J$. The heat lost by the heat sink will be $-1.0 \times 10^{3}\,J$ (notice that the negative sign means heat was lost by the object). The heat capacity is:

$$\begin{aligned}C &= \frac{q}{\Delta T} \\ &= \frac{-1.0 \times 10^{3}\,J}{-43.9\,K} \\ &= 23\,J\,K^{-1}\end{aligned}$$

Is our answer reasonable?

In any calculation involving energy transfer, we first check to see that all quantities have the correct signs. The water gained about 1000 J and warmed up by 2.4 K, so the water's heat capacity is 1000 J/2.4 K, or about 400 $J\,K^{-1}$. This is about 20 times the calculated heat capacity of the heat sink. We would expect, then, the temperature rise of the water to be about 1/20 of the temperature drop of the heat sink, since it absorbs 20 times as much heat as the heat sink before its temperature will rise 1 degree. The temperature of the heat sink dropped by 43.9 K, and the temperature of the water rose by 2.4 K, which agree with this prediction.

PRACTICE EXERCISE 8.1

A ball bearing having a temperature of 220 °C is dropped into a polystyrene cup containing 100 g of water at a temperature of 20 °C. The final temperature of both the ball bearing and the water is 32.5 °C. Assuming no heat is lost to the surroundings, calculate the heat capacity of the ball bearing, in $J\ K^{-1}$.

Heat capacity depends on the size of the sample. If it takes 4.18 J to raise the temperature of 1 g of water by 1 K, it will take twice that amount of energy (8.36 J) to obtain the same temperature rise in 2 g of water. Any property with a value that depends on the size of the sample is called an **extensive property**, while one with a value that is the same regardless of the size of the sample is called an **intensive property**. For example, the volume of a system is an extensive property, whereas its temperature is an intensive property. We can turn heat capacity, an extensive property, into a new intensive property called **specific heat capacity** by dividing by the mass of the sample. Specific heat capacity (often called **specific heat**) is simply the heat capacity per gram of substance. It has the symbol c and is defined as:

$$c = \frac{C}{m}$$

giving c the units of $J\ g^{-1}\ K^{-1}$. The advantage of using the intensive property specific heat is that we can easily compare values for the same mass of different substances by inspection. As we can see in table 8.2, the specific heat of water is approximately nine times that of iron, meaning that it takes nine times as much heat to raise the temperature of water by a specified amount as it does for the same mass of iron. The heat required to raise the temperature of 1 mole of a substance by 1 K is called the **molar heat capacity** and has units of $J\ mol^{-1}\ K^{-1}$.

TABLE 8.2 Specific heats of selected substances at 25 °C.

Substance	Specific heat ($J\ g^{-1}\ K^{-1}$)	Molar heat capacity ($J\ mol^{-1}\ K^{-1}$)
lead	0.128	26.5
gold	0.129	25.4
silver	0.235	25.4
copper	0.387	24.6
iron	0.4498	25.1
carbon (graphite)	0.711	8.5
oxygen (gas)	0.920	29.4
neon (gas)	1.03	20.8
nitrogen (gas)	1.04	29.1
ethanol	2.45	113
water (liquid)	4.18	75.3

We can calculate the heat absorbed or emitted by an object given its mass (m), temperature change (ΔT) and specific heat (c) by:

$$q = cm\Delta T$$

WORKED EXAMPLE 8.2

Calculating heat from a temperature change, mass and specific heat

If a gold ring with a mass of 5.50 g changes in temperature from 25.0 to 28.0 °C, how much heat has it absorbed?

Analysis

The question asks us to connect the heat absorbed by the ring with its temperature change, ΔT. We don't know the heat capacity of the ring. We do know the mass of the ring and that it is made of gold, so we can look up the specific heat. We will use the value of the specific heat of gold given for 25 °C in table 8.2, that is, 0.129 $J\ g^{-1}\ K^{-1}$.

Solution

The mass, m, of the ring is 5.50 g, the specific heat, c, is 0.129 $J\ g^{-1}\ K^{-1}$, and the temperature increases from 25.0 to 28.0 °C, so $\Delta T = 3.0\ °C = 3.0\ K$. Using these values gives:

$$\begin{aligned} q &= cm\Delta T \\ &= (0.129\ J\ g^{-1}\ K^{-1}) \times (5.50\ g) \times (3.0\ K) \\ &= 2.1\ J \end{aligned}$$

Thus, just 2.1 J raises the temperature of 5.50 g of gold by 3.0 °C. Because ΔT is positive, so is the heat, 2.1 J. Thus, the *sign* of q signifies that the ring *absorbs* heat.

Is our answer reasonable?

If the ring had a mass of only 1 g and its temperature increased by 1 °C, we'd know from the specific heat of gold (let's round it to 0.13 J g^{-1} K^{-1}) that the ring would absorb 0.13 J. For a 3-degree temperature increase, the answer would be three times as much, or 0.39 J, nearly 0.4 J. For a ring a little heavier than 5 g, the heat absorbed would be five times as much, or about 2.0 J. So our answer (2.1 J) is clearly reasonable. Notice also that analysis of units can be very helpful in problems like these. We want our final answer in joules, and there is only one possible way of combining specific heat (J g^{-1} K^{-1}), mass (g) and ΔT (K) to give this unit — namely multiplying them together. If you cannot remember the form of the equation, you can always figure it out using units (see section 2.1).

PRACTICE EXERCISE 8.2 The temperature of 250 g of water was changed from 25.0 to 30.0 °C. How much heat was transferred into the water?

Determination of heat

Now that we know the relationship between temperature change and heat, we can use this to determine the heat lost or gained in a chemical reaction or physical process. We carry out the reaction or process in a **calorimeter**, which is a piece of apparatus especially designed to minimise heat loss between the system and surroundings. There are two main types of calorimeter — those that operate under conditions such that the system remains at constant volume and those that contain the system at constant pressure. While the distinction may appear small, we will see that there are important implications. Let us first consider **combustion reactions** where a substance undergoes reaction with oxygen in a constant volume **bomb calorimeter**, an example of which is shown in figure 8.9. Because the volume of the system cannot change during the reaction, ΔV for the system will be 0. This means that $p\Delta V$ must also be 0. Remember that $w = -p\Delta V$; therefore, no work can be done by or on the system. Recall that:

$$\Delta U = q + w$$

so, for a reaction carried out under constant volume conditions in a bomb calorimeter, it follows that:

$$\Delta_r U = q_V$$

where q_V is the **heat of reaction at constant volume**.

Food scientists determine the internal energy of foods and food ingredients by burning them in a bomb calorimeter. The reactions that break down foods in the body are complex, but they produce the same products as the combustion reaction for the food.

It is important that you use the correct sign of the heat change in any calculations based on calorimetric measurements. Consider, for example, a combustion reaction carried out in a bomb calorimeter of the type shown in figure 8.9. The heat given out by the chemical reaction

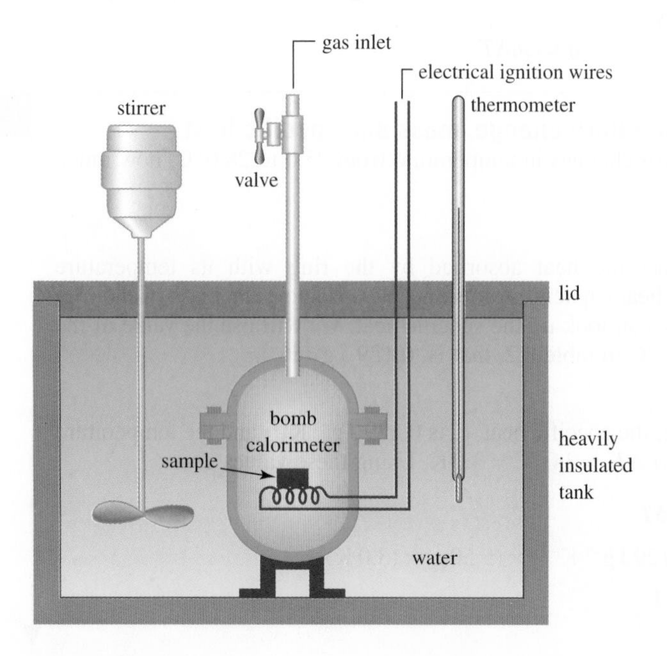

FIGURE 8.9 A bomb calorimeter. The sample is placed in the steel bomb calorimeter which is then filled with $O_2(g)$. The initial temperature of the stirred water is measured, and the combustion reaction is then initiated electrically. The reaction gives out heat, which is absorbed by the bomb calorimeter and the surrounding water. Measurement of the temperature rise of the water, together with knowledge of the heat capacities of the bomb calorimeter and the water, allow the heat of reaction at constant volume, q_V, and therefore the internal energy of the reaction, $\Delta_r U$, to be calculated.

is absorbed by the calorimeter and the surrounding water, leading to a measured temperature rise of the water. In simple terms, the chemical reaction gives out an amount of heat ($q_{reaction}$), which is gained by the calorimeter and its surroundings ($q_{calorimeter}$). These two amounts of heat are numerically equal, but opposite in sign. It is always true that:

$$q_{calorimeter} = -q_{reaction}$$

regardless of whether the reaction gives out, or takes in, heat.

WORKED EXAMPLE 8.3

Bomb calorimetry

(a) When 1.000 g of olive oil was completely burned in pure oxygen in a bomb calorimeter like the one shown in figure 8.9, the temperature of the water bath, which contained 0.750 kg of water, increased from 22.000 °C to 22.241 °C. The heat capacity of the bomb calorimeter was 5.896 kJ K^{-1}. How much heat is produced on burning 1.000 g of olive oil?

(b) Olive oil is almost pure glyceryl trioleate, $C_{57}H_{104}O_6$. The equation for the combustion of glyceryl trioleate is:

$$C_{57}H_{104}O_6\ (l) + 80O_2\ (g) \rightarrow 57CO_2\ (g) + 52H_2O(l)$$

What is the change in internal energy, $\Delta_r U$, in kJ for the combustion of 1 mole of glyceryl trioleate? Assume the olive oil burned in (a) was pure glyceryl trioleate.

Analysis

In (a), we are given the heat capacity of the bomb calorimeter, the specific heat and mass of water, and the temperature change observed produced on burning 1.000 g of olive oil, and we are asked to calculate the heat evolved. Both the bomb calorimeter and the water bath absorb the heat released by the burning olive oil. We can calculate the heat absorbed by both the bomb calorimeter ($q_{calorimeter}$) and the water bath (q_{water}) from the temperature change, heat capacity of the bomb calorimeter and specific heat and mass of water. The heat released by the oil ($q_{reaction}$) will be the negative of the sum of these. We will have to take care with units, as the heat capacity of the bomb calorimeter is given in kJ K^{-1}, and the specific heat of water is given in J g^{-1} K^{-1}.

In (b), we are asked for the change in internal energy, $\Delta_r U$. Bomb calorimetry measures heat at constant volume, q_v, which is equal to the internal energy change for the reaction. Thus, the heat calculated in (a) is equal to $\Delta_r U$ for 1 g of glyceryl trioleate. We can use the molar mass of glyceryl trioleate to convert $\Delta_r U$ per gram to $\Delta_r U$ per mole.

Solution

(a) We firstly calculate the heat absorbed by the water bath when 1.000 g of olive oil is burned using:

$$\begin{aligned} q_{water} &= cm\Delta T \\ &= 4.18\ \text{J g}^{-1}\ \text{K}^{-1} \times 750\ \text{g} \times (22.241 - 22.000)\ \text{K} \\ &= 756\ \text{J} \end{aligned}$$

The heat absorbed by the bomb calorimeter can then be calculated using:

$$\begin{aligned} q_{calorimeter} &= C\Delta T \\ &= 5.896\ \text{kJ K}^{-1} \times (22.241 - 22.000)\ \text{K} \\ &= 1.42\ \text{kJ} \end{aligned}$$

The total heat given out in the combustion reaction, $q_{reaction}$, will be the negative of the sum of these, that is:

$$q_{calorimeter} + q_{water} = -q_{reaction}$$

Therefore, $q_{reaction}$ for burning 1.000 g of olive oil is –(1.42 + 0.756) kJ = –2.17 kJ, and the energy content of the olive oil on a mass basis is thus –2.17 kJ g^{-1}.

(b) We convert the heat produced per gram to the heat produced per mole, using the molar mass of $C_{57}H_{104}O_6$, 885.4 g mol^{-1}.

$$-2.17\ \text{kJ g}^{-1} \times 885.4\ \text{g mol}^{-1} = -1.92 \times 10^3\ \text{kJ mol}^{-1}$$

Since this is heat at constant volume, q_v, we have $\Delta_r U = q_v = -1.92 \times 10^3$ kJ for combustion of 1 mole of $C_{57}H_{104}O_6$.

Is our answer reasonable?

Always check the signs of calculated heats in calorimetry. The combustion reaction releases heat, so the sign must be negative. We appear to have done the numerical calculations correctly, so we can assume our answer is correct.

PRACTICE EXERCISE 8.3 A sample of carbon weighing 1.07 g is completely burned in an atmosphere of excess oxygen in a bomb calorimeter having a heat capacity of $4.504\,\text{kJ}\,\text{K}^{-1}$. The temperature of the 0.655 kg of water surrounding the calorimeter rises from 20.00 °C to 24.84 °C. Use these data to calculate $\Delta_r U$ for the combustion of 1 mole of carbon.

8.4 Enthalpy

In chemistry, we are most often concerned with reactions that occur in solution, and it is more convenient to carry out such reactions in a beaker, flask or test tube open to the atmosphere than inside a metal 'bomb'. In these cases, we are working under conditions of constant atmospheric pressure. Hence, the system is free to expand or contract and this means that it can potentially do pV work. Thus we can see that we will not be able to make the same simplifications as we did under constant volume conditions. As we have already seen, when a system expands against a constant pressure, p, the pV work done is given by:

$$w = -p\Delta V$$

where:

$$\Delta V = V_{\text{final}} - V_{\text{initial}}$$

Under conditions where the system expands, ΔV is positive and so w is negative, meaning that the system has done work on the surroundings and consequently lowered its internal energy. The opposite is true if the system is compressed. We now rewrite:

$$\Delta U = q + w$$

under conditions of constant pressure as:

$$\Delta U = q_p - p\Delta V$$

where q_p is the **heat of reaction at constant pressure**. However, this equation is inconvenient as we need to know the value of ΔV if we wish to calculate ΔU. At this point, we define a new thermodynamic function called enthalpy, which has the symbol H, as:

$$H = U + pV$$

Thus, under conditions of constant pressure:

$$\Delta H = \Delta U + p\Delta V$$

Substituting $\Delta U = q_p - p\Delta V$ into this gives:

$$\Delta H = q_p - p\Delta V + p\Delta V$$

so:

$$\Delta H = q_p$$

Thus, the heat of reaction at constant pressure is equal to $\Delta_r H$, in the same way that the heat of reaction at constant volume is equal to $\Delta_r U$. Like $\Delta_r U$, $\Delta_r H$ is a state function, meaning that the enthalpy change for a reaction depends only on the initial and final states of the system. If the final enthalpy of the system is greater than the initial, the system has absorbed heat from its surroundings, so $\Delta_r H$ will be positive and the reaction is said to be **endothermic**. Conversely, if the system has lost heat to the surroundings, the enthalpy of the system has decreased, so $\Delta_r H$ will be negative and the reaction is said to be **exothermic**.

The difference between the enthalpy change and the internal energy change for a reaction is $p\Delta V$. The difference between $\Delta_r U$ and $\Delta_r H$ can be very large for reactions that produce or consume gases, because these reactions can have very large volume changes. If a reaction involves only solids and liquids, though, the values of ΔV are tiny, so $\Delta_r U$ and $\Delta_r H$ for these reactions are nearly identical.

A very simple constant pressure calorimeter, dubbed the coffee cup calorimeter, comprises two nested and capped cups made of polystyrene, a very good insulator (figure 8.10). A reaction occurring in such a calorimeter exchanges very little heat with the surroundings, particularly if the reaction is fast, under which conditions the temperature change is rapid and easily measured. We can find the heat of reaction if we have determined the heat capacity of the calorimeter and its contents before the reaction. As the styrofoam cup and the thermometer absorb only a tiny amount of heat, we can usually ignore them in our calculations. Therefore, in our calculations of $q_{\text{calorimeter}}$, we need only use the specific heat of the solution.

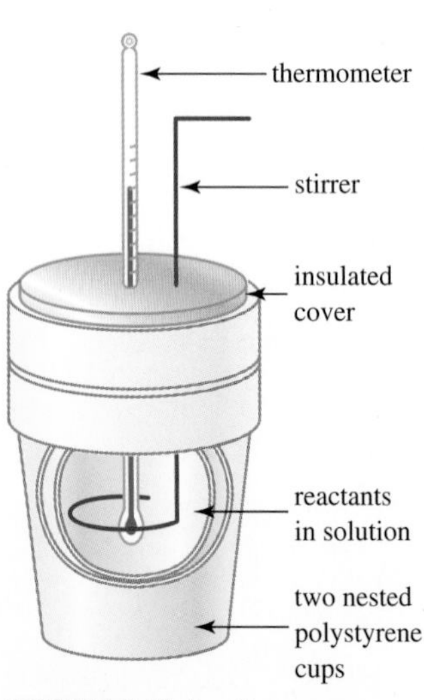

FIGURE 8.10 A coffee cup calorimeter used to measure heats of reaction at constant pressure.

WORKED EXAMPLE 8.4

Constant pressure calorimetry

The reaction of hydrochloric acid and sodium hydroxide is very rapid and exothermic. The balanced chemical equation is:

$$\mathrm{HCl(aq) + NaOH(aq) \longrightarrow NaCl(aq) + H_2O(l)}$$

In one experiment, 50.0 mL of 1.00 M HCl at 25.5 °C was placed in a coffee cup calorimeter. To this was added 50.0 mL of 1.00 M NaOH solution, also at 25.5 °C. The mixture was stirred, and the temperature quickly increased to a maximum of 32.2 °C. The density of 1.00 M HCl is 1.02 g $\mathrm{mL^{-1}}$ and that of 1.00 M NaOH is 1.04 g $\mathrm{mL^{-1}}$. What is the heat evolved in joules per mole of HCl? Assume that the specific heats of the solutions are the same as that of water, 4.18 J $\mathrm{g^{-1}\,K^{-1}}$. (We will neglect the heat lost to the polystyrene, the thermometer and the surrounding air.)

Analysis

Before we can use $q = cm\Delta T$ to find the heat evolved, we have to calculate the system's total mass and the temperature change. The mass here refers to the *total* mass of the combined solutions, but we were given volumes. So we have to use their densities to calculate their masses, using the equation:

$$\text{mass} = \text{density} \times \text{volume (table 2.2, p. 27)}$$

For the HCl solution, the density is 1.02 g $\mathrm{mL^{-1}}$ so:

$$\text{mass}_{\mathrm{HCl}} = 1.02\ \mathrm{g\ mL^{-1}} \times 50.0\ \mathrm{mL} = 51.0\ \mathrm{g}$$

For the NaOH solution, the density is 1.04 g $\mathrm{mL^{-1}}$ so:

$$\text{mass}_{\mathrm{NaOH}} = 1.04\ \mathrm{g\ mL^{-1}} \times 50.0\ \mathrm{mL} = 52.0\ \mathrm{g}$$

Thus, the total mass = 51.0 g + 52.0 g = 103.0 g.

Solution

The reaction changes the temperature of the calorimeter by $T_{\text{final}} - T_{\text{initial}}$, so:

$$\Delta T = (32.2 - 25.5)\ \mathrm{K} = 6.7\ \mathrm{K}$$

Now we can calculate the heat absorbed by the calorimeter, $q_{\text{calorimeter}}$.

$$\begin{aligned} q_{\text{calorimeter}} &= cm\Delta T \\ &= 4.18\ \mathrm{J\ g^{-1}\ K^{-1}} \times 103.0\ \mathrm{g} \times 6.7\ \mathrm{K} \\ &= 2.9 \times 10^3\ \mathrm{J} \end{aligned}$$

Knowing that $q_{\text{reaction}} = -q_{\text{calorimeter}}$, we obtain $q_{\text{reaction}} = -2.9 \times 10^3$ J.

This is the heat evolved for the specific mixture prepared. However, the problem calls for joules *per mole of HCl*. We calculate the amount of HCl using the equation:

$$\begin{aligned} n_{\mathrm{HCl}} &= cV \\ &= 1.00\ \mathrm{mol\ L^{-1}} \times 0.0500\ \mathrm{L} \\ &= 0.0500\ \mathrm{mol} \end{aligned}$$

Neutralising 0.0500 mol of acid gave $q_{\text{reaction}} = -2.9 \times 10^3$ J. So, on a per mole basis:

$$\begin{aligned} \text{heat evolved} &= \frac{-2.9 \times 10^3\ \mathrm{J}}{0.0500\ \mathrm{mol}} \\ &= -58 \times 10^3\ \mathrm{J\ mol^{-1}} = -58\ \mathrm{kJ\ mol^{-1}} \end{aligned}$$

Thus, the enthalpy change for this reaction is −58 kJ $\mathrm{mol^{-1}}$. (The negative sign shows that the reaction is exothermic.)

Is our answer reasonable?

Let's first review the logic of the steps we used. *Notice how the logic is driven by definitions, which carry specific units.* Working backwards, knowing that we want units of joules per mole in the answer, we must calculate separately the amount of acid neutralised and the amount of heat evolved. The latter will emerge when we multiply the solution's mass (g) and specific heat (J $\mathrm{g^{-1}\,K^{-1}}$) by the temperature increase (K).

The step in the calculation that involves the mass of the final sample is straightforward. Because the densities are close to 1 g mL^{-1}, the combined masses are about 50 g per solution or a total of 100 g.

The specific heat is around 4.2 J g^{-1} K^{-1}, which means that, if the final solution weighed 1 g, its heat capacity would be 4.2 J K^{-1}. If the mass is 100 g, the heat capacity is 100 times as much, or 420 J K^{-1}.

The temperature increased by about 7 K so, when we multiply 420 J K^{-1} by 7, we get the heat absorbed by the calorimeter, about 2940 J or 2.9 × 10^3 J, which is the same as 2.9 kJ; $q_{reaction}$ is the negative of this.

This much heat is associated with neutralising 0.0500 mol HCl, so the heat per mole is −2.9 kJ divided by 0.0500 mol, or −58 kJ mol^{-1}.

PRACTICE EXERCISE 8.4 When pure sulfuric acid dissolves in water, a large amount of heat is given off. To measure it, 175 g of water was placed in a coffee cup calorimeter and chilled to 10.0 °C. Then 4.90 g of sulfuric acid, H_2SO_4, also at 10.0 °C, was added, and the mixture was quickly stirred with a thermometer. The temperature rose rapidly to 14.9 °C. Assume that the value of the specific heat of the solution is 4.18 J g^{-1} K^{-1} and that the solution absorbs all the heat evolved. Calculate the heat evolved in kilojoules on formation of this solution. (Remember to use the *total* mass of the solution: the water plus the acid.) Calculate also the heat evolved *per mole* of sulfuric acid.

Standard enthalpy of reaction

The amount of heat a reaction produces or absorbs depends on the amounts of reactants we combine. It makes sense that, if we burn 2 moles of carbon, we will get twice as much heat as if we had burned 1 mole. For heats of reaction to have meaning, we must describe the system completely. Our description must include amounts and concentrations of reactants, amounts and concentrations of products, temperature and pressure, because all of these things can influence heats of reaction.

The **standard enthalpy of reaction** ($\Delta_r H^{\ominus}$) at temperature T is the value of ΔH for a reaction in which the pure, separated reactants in their standard states are converted to the pure separated products in their standard states, all at temperature T, and in which the stoichiometric coefficients in the balanced chemical equation refer to actual numbers of moles. While the SI unit of $\Delta_r H^{\ominus}$ is J mol^{-1}, values are usually quoted in kJ mol^{-1}.

To illustrate clearly what we mean by $\Delta_r H^{\ominus}$, let us use the reaction between gaseous nitrogen and hydrogen that produces gaseous ammonia.

$$N_2(g) + 3H_2(g) \longrightarrow 2NH_3(g)$$

When 1.000 mol of $N_2(g)$ at 25 °C and 10^5 Pa and 3.000 mol of $H_2(g)$ at 25 °C and 10^5 Pa react to form 2.000 mol of $NH_3(g)$ at 25 °C and 10^5 Pa, the reaction releases 92.38 kJ. Hence, for the reaction as given by the preceding equation, $\Delta_r H^{\ominus} = -92.38$ kJ mol^{-1}. Often, the enthalpy change is given immediately after the equation, for example:

$$N_2(g) + 3H_2(g) \longrightarrow 2NH_3(g) \qquad \Delta_r H^{\ominus} = -92.38 \text{ kJ mol}^{-1}$$

A chemical equation that also shows the value of $\Delta_r H^{\ominus}$ is called a **thermochemical equation**. Such equations always specify the physical states of the reactants and products, and their associated $\Delta_r H^{\ominus}$ value applies to the situation where the stoichiometric coefficients refer to actual numbers of moles. The equation above, for example, shows a release of 92.38 kJ if 2 moles of NH_3 form. If we were to make twice as much, or 4.000 mol, of NH_3 (from 2.000 mol of N_2 and 6.000 mol of H_2), then twice as much heat (184.8 kJ) would be released. On the other hand, if only 0.5000 mol of N_2 and 1.500 mol of H_2 were to react to form just 1.000 mol of NH_3, then only half as much heat (46.19 kJ) would be released. For the various reactions just described, for example, we would have the following thermochemical equations.

$$N_2(g) + 3H_2(g) \longrightarrow 2NH_3(g) \quad \Delta_r H^{\ominus} = -92.38 \text{ kJ mol}^{-1}$$
$$2N_2(g) + 6H_2(g) \longrightarrow 4NH_3(g) \quad \Delta_r H^{\ominus} = -184.8 \text{ kJ mol}^{-1}$$
$$\tfrac{1}{2}N_2(g) + \tfrac{3}{2}H_2(g) \longrightarrow NH_3(g) \quad \Delta_r H^{\ominus} = -46.19 \text{ kJ mol}^{-1}$$

Because the coefficients of a thermochemical equation always mean *moles*, not molecules, we may use fractional coefficients.

Chemical Connections

Cold fusion — a nuclear reaction at room temperature

On 23 March 1989, an extraordinary press conference was given by two chemists, Professors Martin Fleischmann and Stanley Pons, at the University of Utah, in the US, in which they announced their discovery of 'cold fusion'. This process, if real, would have solved the energy problems of the world. Fleischmann and Pons (figure 8.11) claimed that they had developed a method by which nuclear fusion, the process that powers our sun, could be carried out in a controlled fashion in a chemistry laboratory, using nothing more than D_2O (water in which the protons, ^1_1H, have been replaced by deuterons, ^2_1H), the element palladium and an electrolysis setup very similar to that shown in figure 8.2 (p. 298).

The reaction to this news was huge. Laboratories around the world set about replicating Pons and Fleischmann's results, and it did not take long for serious doubts about their work to be raised. On the basis of calorimetry measurements, Pons and Fleischmann had claimed that their apparatus produced more energy than it used up. However, two major flaws in their experimental setup were apparent once they published their work. Firstly, they were working with an open system (see p. 300) in which matter, in the form of D_2 and O_2 gas, and therefore heat, could escape. Secondly, they did not have a stirrer in their calorimeter, relying instead on the bubbles of D_2 and O_2 gas to stir the solution. This meant that the temperature throughout their solution was not uniform, and the excess heat they apparently measured could simply have come from local hotspots in the solution.

A number of other problems with their results meant that cold fusion was eventually shown to be false, although a few diehard believers can be found on the internet. The driving force for the announcement of cold fusion was not science but money — the potential windfall that would have come the way of Pons, Fleischmann and the University of Utah (not to mention the certain Nobel Prizes) would have been truly mind-boggling, and these riches would only go to the first people to proclaim the discovery — there are no prizes in science for coming second. Sadly, their names will now be forever associated with a scientific fiasco. Maybe their embarrassment would have been spared had they paid more attention to their first-year chemistry.

FIGURE 8.11 Fleischmann and Pons claimed their cold fusion apparatus produced more energy than it used. Unfortunately, they were wrong.

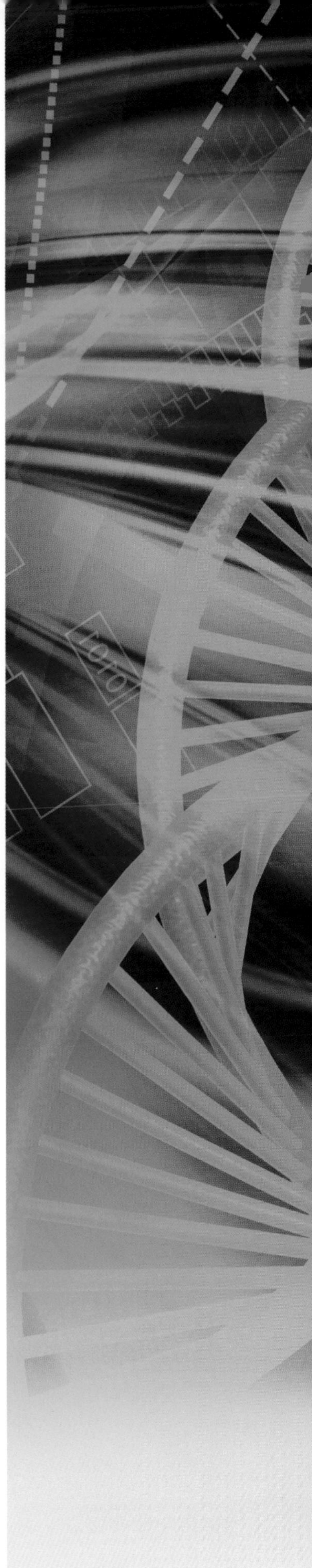

You must write down physical states for all reactants and products in thermochemical equations. The combustion of 1 mole of methane, for example, has different values of $\Delta_r H^\ominus$ if the water produced is in its liquid or its gaseous state.

$$CH_4(g) + 2O_2(g) \longrightarrow CO_2(g) + 2H_2O(l) \quad \Delta_r H^\ominus = -890.5\ \text{kJ mol}^{-1}$$

$$CH_4(g) + 2O_2(g) \longrightarrow CO_2(g) + 2H_2O(g) \quad \Delta_r H^\ominus = -802.3\ \text{kJ mol}^{-1}$$

The difference in $\Delta_r H^\ominus$ values for these two reactions is the quantity of energy that would be released by the physical change of 2 moles of water vapour at 25 °C to 2 moles of liquid water at 25 °C.

WORKED EXAMPLE 8.5

Writing a thermochemical equation

The following thermochemical equation is that for the exothermic reaction of hydrogen and oxygen that produces water (see figure 8.12).

$$2H_2(g) + O_2(g) \rightarrow 2H_2O(g) \quad \Delta_r H^\ominus = -483.6\ \text{kJ mol}^{-1}$$

What is the thermochemical equation for this reaction when it is carried out to produce 1.000 mol H_2O?

Analysis

The given equation is for the production of 2.000 mol of H_2O, and any changes in the coefficient for water must be made identically to all other coefficients, as well as to the value of $\Delta_r H^\ominus$.

Solution

We divide everything by 2 to obtain:

$$H_2(g) + \tfrac{1}{2}O_2(g) \rightarrow H_2O(g) \quad \Delta_r H^\ominus = -241.8\ \text{kJ mol}^{-1}$$

Is our answer reasonable?

Compare the equation just found with the initial equation to see that the coefficients and the value of $\Delta_r H^\ominus$ are all divided by 2.

FIGURE 8.12 The airship *Hindenburg* used hydrogen, a gas that is lighter than air but also flammable, to provide lift. On 6 May 1937, at Lakehurst, New Jersey, USA, it caught fire while mooring, probably as a result of a build-up of static electricity, and the explosive reaction of hydrogen with oxygen destroyed the airship in minutes.

PRACTICE EXERCISE 8.5

Use the data given in worked example 8.5 to write the thermochemical equation for the formation of 3.200 mol of H_2O.

Once we have the thermochemical equation for a given reaction, we can write the equation for the reverse reaction, regardless of how hard it might actually be to make it happen. For example, the thermochemical equation for the combustion of carbon in oxygen to give carbon dioxide is:

$$C(s) + O_2(g) \longrightarrow CO_2(g) \quad \Delta_r H^\ominus = -393.5\ \text{kJ mol}^{-1}$$

The reverse reaction, which is extremely difficult to carry out, would be the decomposition of carbon dioxide to carbon and oxygen.

$$CO_2(g) \longrightarrow C(s) + O_2(g) \quad \Delta_r H^\ominus = ?$$

Despite the difficulty of carrying out this reaction, we can still determine its value of $\Delta_r H^\ominus$. This is because ΔH is a state function, and therefore the absolute difference between the enthalpies of the reactants and products does not depend on the direction of the reaction. Therefore, $\Delta_r H^\ominus$ for the decomposition reaction must be +393.5 kJ mol^{-1}, that is, equal in size but opposite in sign to $\Delta_r H^\ominus$ for the reaction written in the opposite direction. The law of conservation of energy requires this remarkable result. If the values of $\Delta_r H^\ominus$ for the forward and the reverse reactions were not equal but opposite in sign, then perpetual-motion machines would be possible. (Despite numerous claims, such machines will always be an unattainable dream.) Thus, regardless of the difficulty of directly decomposing CO_2 into its elements, we can still write a thermochemical equation for it.

$$CO_2(g) \longrightarrow C(s) + O_2(g) \quad \Delta_r H^\ominus = +393.5\ \text{kJ mol}^{-1}$$

To repeat, if we know $\Delta_r H^\ominus$ for a given reaction, then we also know $\Delta_r H^\ominus$ for the reverse reaction; it has the same numerical value, but the opposite algebraic sign. This extremely useful fact makes thermochemical data available that would otherwise be impossible to measure.

Hess's law

Hess's law is a method for combining known thermochemical equations in a way that allows us to calculate $\Delta_r H^\ominus$ for another reaction. This requires experience in other kinds of manipulations of equations. We will illustrate this using the combustion of carbon.

We can imagine two paths leading from 1 mole each of carbon and oxygen to 1 mole of carbon dioxide.

One-step path

Let C and O_2 react to give CO_2 directly.

$$C(s) + O_2(g) \longrightarrow CO_2(g) \qquad \Delta_r H^\ominus = -393.5 \text{ kJ mol}^{-1}$$

Two-step path

Let C and O_2 react to give CO, and then let CO react with O_2 to give CO_2.

Step 1: $$C(s) + \tfrac{1}{2}O_2(g) \longrightarrow CO(g) \qquad \Delta_r H^\ominus = -110.5 \text{ kJ mol}^{-1}$$

Step 2: $$CO(g) + \tfrac{1}{2}O_2(g) \longrightarrow CO_2(g) \qquad \Delta_r H^\ominus = -283.0 \text{ kJ mol}^{-1}$$

Overall, the two-step path consumes 1 mole each of C and O_2 to make 1 mole of CO_2, just like the one-step path. The initial and final states for the two routes to CO_2 are identical.

If $\Delta_r H^\ominus$ is a state function dependent only on the initial and final states and independent of path, the values of $\Delta_r H^\ominus$ for both routes should be identical. We can see that this is true simply by adding the equations for the two-step path and comparing the result with the equation for the one-step path.

$$\begin{array}{llll}
\textit{Step 1:} & C(s) + \tfrac{1}{2}O_2(g) \longrightarrow \cancel{CO(g)} & \Delta_r H^\ominus = -110.5 \text{ kJ mol}^{-1} \\
\textit{Step 2:} & \cancel{CO(g)} + \tfrac{1}{2}O_2(g) \longrightarrow CO_2(g) & \Delta_r H^\ominus = -283.0 \text{ kJ mol}^{-1} \\
\hline
 & C(s) + O_2(g) \longrightarrow CO_2(g) & \Delta_r H^\ominus = -110.5 \text{ kJ mol}^{-1} + (-283.0 \text{ kJ mol}^{-1}) \\
 & & \quad = -393.5 \text{ kJ mol}^{-1}
\end{array}$$

Note that we can cancel CO(g) as it appears *identically* on opposite sides of the arrow in steps 1 and 2. Such a cancellation is permitted only when both the formula and the physical state of a species are identical on opposite sides of the arrow. The net thermochemical equation for the two-step process, therefore, is:

$$C(s) + O_2(g) \longrightarrow CO_2(g) \qquad \Delta_r H^\ominus = -393.5 \text{ kJ mol}^{-1}$$

The results, chemically and thermochemically, are identical for both routes to CO_2, demonstrating that $\Delta_r H^\ominus$ is a state function.

The above example is a manifestation of **Hess's law**, which states that *the overall enthalpy change for any chemical reaction is constant, regardless of how the reaction is carried out.*

The chief use of Hess's law is to calculate the enthalpy change for a reaction for which such data cannot be determined experimentally or are otherwise unavailable. Because this requires that we manipulate equations, let's restate the few rules that govern these operations.

Rules for manipulating thermochemical equations

1. When an equation is reversed — written in the opposite direction — the sign of $\Delta_r H^\ominus$ must also be reversed. To illustrate, the reverse of the equation:

$$C(s) + O_2(g) \longrightarrow CO_2(g) \qquad \Delta_r H^\ominus = -393.5 \text{ kJ mol}^{-1}$$

is the following equation:

$$CO_2(g) \longrightarrow C(s) + O_2(g) \qquad \Delta_r H^\ominus = +393.5 \text{ kJ mol}^{-1}$$

2. Substances can be cancelled from both sides of an equation only if the substance is in an identical physical state.
3. If all the coefficients of an equation are multiplied or divided by the same factor, the value of $\Delta_r H^\ominus$ must likewise be multiplied or divided by that factor.

WORKED EXAMPLE 8.6

Using Hess's law

Carbon monoxide is often used in metallurgy to remove oxygen from metal oxides to give the free metal. The thermochemical equation for the reaction of CO with iron(III) oxide, Fe_2O_3, is:

$$Fe_2O_3(s) + 3CO(g) \longrightarrow 2Fe(s) + 3CO_2(g) \qquad \Delta_r H^\ominus = -26.7 \text{ kJ mol}^{-1}$$

Use this equation and the equation for the combustion of CO:

$$CO(g) + \frac{1}{2}O_2(g) \longrightarrow CO_2(g) \quad \Delta_r H^{\ominus} = -283.0\,\text{kJ mol}^{-1}$$

to calculate the value of $\Delta_r H^{\ominus}$ for the following reaction:

$$2Fe(s) + \frac{3}{2}O_2(g) \longrightarrow Fe_2O_3(s)$$

Analysis

We cannot simply add the two given equations, because this will not produce the equation we want. We first have to manipulate these equations so that when we add them we will obtain the target equation.

Solution

We can manipulate the two given equations as follows:

Step 1: We begin with the iron atoms. The target equation must have 2Fe on the left, but the first equation in this worked example has 2Fe to the right of the arrow. To move it to the left, we must reverse the entire equation, remembering also to reverse the sign of $\Delta_r H^{\ominus}$. This puts Fe_2O_3 to the right of the arrow, which is where it has to be after we add our adjusted equations. After these manipulations, and reversing the sign of $\Delta_r H^{\ominus}$, we have:

$$2Fe(s) + 3CO_2(g) \longrightarrow Fe_2O_3(s) + 3CO(g) \quad \Delta_r H^{\ominus} = +26.7\,\text{kJ mol}^{-1}$$

Step 2: There must be $\frac{3}{2}O_2$ on the left, and we must be able to cancel 3CO and $3CO_2$ when the equations are added. If we multiply the second equation in this worked example by 3, we will obtain the necessary coefficients. We must also multiply the value of $\Delta_r H^{\ominus}$ for this equation by 3, because three times the amount of substances are now involved in the reaction. When we have done this, we have:

$$3CO(g) + \frac{3}{2}O_2(g) \longrightarrow 3CO_2(g) \quad \Delta_r H^{\ominus} = 3 \times (-283.0\,\text{kJ}) = -849.0\,\text{kJ mol}^{-1}$$

We now put our two equations together and find the answer:

$$2Fe(s) + \cancel{3CO_2(g)} \longrightarrow Fe_2O_3(s) + \cancel{3CO(g)} \quad \Delta_r H^{\ominus} = +26.7\,\text{kJ mol}^{-1}$$

$$\cancel{3CO(g)} + \frac{3}{2}O_2(g) \longrightarrow \cancel{3CO_2(g)} \quad \Delta_r H^{\ominus} = -849.0\,\text{kJ mol}^{-1}$$

$$\text{Sum:} \quad 2Fe(s) + \frac{3}{2}O_2(g) \longrightarrow Fe_2O_3(s) \quad \Delta_r H^{\ominus} = -822.3\,\text{kJ mol}^{-1}$$

Thus, the value of $\Delta_r H^{\ominus}$ for the oxidation of 2 mol Fe(s) to 1 mol $Fe_2O_3(s)$ is $-822.3\,\text{kJ mol}^{-1}$. (The reaction is *very* exothermic.)

Is our answer reasonable?

There is no shortcut to check that we are right. But, for each step, double-check that you have followed the rules for manipulating thermochemical equations.

PRACTICE EXERCISE 8.6 Ethanol, C_2H_5OH, is made industrially by the reaction of water with ethene, C_2H_4. Calculate the value of $\Delta_r H^{\ominus}$ for the reaction:

$$C_2H_4(g) + H_2O(l) \longrightarrow C_2H_5OH(l)$$

given the following thermochemical equations:

$$C_2H_4(g) + 3O_2(g) \longrightarrow 2CO_2(g) + 2H_2O(l) \quad \Delta_r H^{\ominus} = -1411.1\,\text{kJ mol}^{-1}$$

$$C_2H_5OH(l) + 3O_2(g) \longrightarrow 2CO_2(g) + 3H_2O(l) \quad \Delta_r H^{\ominus} = -1367.1\,\text{kJ mol}^{-1}$$

Standard enthalpy of formation

Enormous databases of thermochemical equations have been compiled to allow the calculation of any enthalpy of reaction using Hess's law. The most frequently tabulated data are those for formation reactions.

The **standard enthalpy of formation** ($\Delta_f H^{\ominus}$) of a substance is the enthalpy change when 1 mole of the substance is formed at 10^5 Pa and the specified temperature from its elements in their standard states. An element is in its standard state when it is in its most stable form and physical state (solid, liquid or gas) at 10^5 Pa and the specified temperature. Oxygen, for example, is in its standard state at a temperature of 25 °C only as a gas at 10^5 Pa and only as O_2 molecules, not as O atoms or O_3 (ozone) molecules. Carbon must be in the form of graphite, not diamond, to be in its standard state at a temperature of 25 °C because the graphite form of carbon is the most stable form at this temperature under standard conditions (10^5 Pa).

Standard enthalpies of formation for a variety of substances are given in table 8.3, and a more extensive table can be found in appendix A. Notice in particular that all values of $\Delta_f H^\ominus$ for the elements in their standard states are 0; forming an element from itself, of course, would yield no change in enthalpy. For this reason, values of $\Delta_f H^\ominus$ for the elements are generally not included in tables.

TABLE 8.3 Standard enthalpies of formation of selected substances at 25 °C.

Substance	$\Delta_f H^\ominus$ (kJ mol^{-1})	Substance	$\Delta_f H^\ominus$ (kJ mol^{-1})	Substance	$\Delta_f H^\ominus$ (kJ mol^{-1})
Ag(s)	0	$CaCl_2$(s)	–795.0	KCl(s)	–435.89
AgBr(s)	–100.4	CaO(s)	–635.5	K_2SO_4(s)	–1433.7
AgCl(s)	–127.0	$Ca(OH)_2$(s)	–986.59	N_2(g)	0
Al(s)	0	$CaSO_4$(s)	–1432.7	NH_3(g)	–46.19
Al_2O_3(s)	–1669.8	$CaSO_4 \cdot \frac{1}{2}H_2O$(s)	–1575.2	NH_4Cl(s)	–315.4
C(s, C_{60})	2320	$CaSO_4 \cdot 2H_2O$(s)	–2021.1	NO(g)	90.37
C(s, diamond)	1.9	Cl_2(g)	0	NO_2(g)	33.8
C(s, graphite)	0	Fe(s)	0	N_2O(g)	81.57
CH_3Cl(g)	–82.0	Fe_2O_3(s)	–822.3	N_2O_4(g)	9.67
CH_3I(g)	14.2	H_2O(g)	–241.8	N_2O_5(g)	11
CH_3OH(l)	–238.6	H_2O(l)	–285.9	Na(s)	0
CH_3COOH(l)	–487.0	H_2(g)	0	$NaHCO_3$(s)	–947.7
CH_4(g)	–74.848	H_2O_2(l)	–187.6	Na_2CO_3(s)	–1131
C_2H_2(g)	226.75	HBr(g)	–36	NaCl(s)	–411.0
C_2H_4(g)	52.284	HCl(g)	–92.30	NaOH(s)	–426.8
C_2H_6(g)	–84.667	HI(g)	26.6	Na_2SO_4(s)	–1384.5
C_2H_5OH(l)	–277.63	HNO_3(l)	–173.2	O_2(g)	0
CO(g)	–110.5	H_2SO_4(l)	–811.32	Pb(s)	0
CO_2(g)	–393.5	Hg(l)	0	PbO(s)	–219.2
$CO(NH_2)_2$(s)	–333.19	Hg(g)	60.84	S(s, rhombic)	0
Ca(s)	0	I_2(s)	0	SO_2(g)	–296.9
$CaBr_2$(s)	–682.8	K(s)	0	SO_3(g)	–395.2
$CaCO_3$(s)	–1207				

It is important to remember the meaning of the subscript f in the symbol $\Delta_f H^\ominus$. It is applied only when *1 mole* of the substance is formed *from its elements in their standard states.* Consider, for example, the following four thermochemical equations and their corresponding values of $\Delta H^\ominus$.

$$H_2(g) + \tfrac{1}{2}O_2(g) \longrightarrow H_2O(l) \quad \Delta_f H^\ominus = -285.9\ \text{kJ mol}^{-1}$$

$$2H_2(g) + O_2(g) \longrightarrow 2H_2O(l) \quad \Delta_r H^\ominus = -571.8\ \text{kJ mol}^{-1}$$

$$H_2O(g) \longrightarrow H_2O(l) \quad \Delta_r H^\ominus = -44.1\ \text{kJ mol}^{-1}$$

$$2H(g) + O(g) \longrightarrow H_2O(l) \quad \Delta_r H^\ominus = -971.1\ \text{kJ mol}^{-1}$$

Only in the first equation is ΔH° given the subscript f. It is the only reaction that satisfies both of the conditions specified for standard enthalpies of formation. The second equation shows the formation of 2 moles of water, rather than 1 mole. The third involves a compound as the reactant. The fourth involves the elements as atoms, which are not standard states for these elements. We can obtain $\Delta_r H^\circ$ for the second equation simply by multiplying the $\Delta_f H^\circ$ value of the first equation by 2.

WORKED EXAMPLE 8.7

Writing an equation for a standard enthalpy of formation

Write the equation to which $\Delta_f H^\circ_{HNO_3(l)}$ refers.

Analysis

The equation must show only 1 mole of the product. We begin with its formula and take whatever fractions of moles of the elements are needed to make it. We also remember to include the physical states. Table 8.3 gives the value of $\Delta_f H^\circ$ for $HNO_3(l)$ as $-173.2\ kJ\ mol^{-1}$ at 25 °C.

Solution

The three elements, H, N and O, all occur as diatomic molecules in the gaseous state, so the following fractions of moles supply exactly enough to make 1 mole of HNO_3.

$$\tfrac{1}{2}H_2(g) + \tfrac{1}{2}N_2(g) + \tfrac{3}{2}O_2(g) \longrightarrow HNO_3(l) \quad \Delta_f H^\circ = -173.2\ kJ\ mol^{-1}$$

Is our answer reasonable?

The answer correctly shows only 1 mole of HNO_3, and this governs the coefficients for the reactants. So simply check that the equation is balanced.

PRACTICE EXERCISE 8.7

Write the thermochemical equations corresponding to the standard enthalpies of formation of:

(a) solid sodium hydroxide, $NaOH(s)$
(b) liquid ethanol, $C_2H_5OH(l)$
(c) gaseous dinitrogen pentaoxide, $N_2O_5(g)$.

Standard enthalpies of formation are useful because they provide a convenient method for applying Hess's law without having to manipulate thermochemical equations. This is possible because, as we will demonstrate, $\Delta_r H^\circ$ for a reaction equals the sum of the standard enthalpies of formation of the products minus the sum of the standard enthalpies of formation of the reactants, with each $\Delta_f H^\circ$ value multiplied by the appropriate stoichiometric coefficient given by the thermochemical equation. In other words, for the reaction:

$$aA + bB \longrightarrow cC + dD$$

$$\Delta_r H^\circ = c\Delta_f H^\circ_C + d\Delta_f H^\circ_D - (a\Delta_f H^\circ_A + b\Delta_f H^\circ_B)$$

or, in more general terms:

$$\Delta_r H^\circ = \sum_i^{\text{products}} v_i \Delta_f H^\circ_i - \sum_j^{\text{reactants}} v_j \Delta_f H^\circ_j$$

where Σ means 'the sum over' and v_i and v_j are the stoichiometric coefficients for the products and reactants, respectively, in the balanced chemical equation for the reaction.

Although the two expressions above for $\Delta_r H^\circ$ look very different, they are merely alternative formulations of the Hess's law equation. While both are given in terms of standard enthalpies of formation, $\Delta_f H^\circ$, they also apply to standard enthalpies of combustion, $\Delta_c H^\circ$ (see p. 318). However, you need to ensure you use only either standard enthalpies of formation or standard enthalpies of combustion when calculating $\Delta_r H^\circ$ using the above equations — you must never mix them.

We will now demonstrate that the Hess's law equation works. Consider the reaction given by the following equation.

$$SO_3(g) \longrightarrow SO_2(g) + \tfrac{1}{2}O_2(g) \qquad \Delta_r H^\circ = ?$$

We wish to calculate the enthalpy of reaction using standard enthalpies of formation.

If we use the first method we learned, namely, the manipulation of thermochemical equations, we would need to imagine a path from the reactant to the products that involves first decomposing $SO_3(g)$ into its elements in their standard states and then recombining the elements to form the products. This path is shown in figure 8.13. The first step with the enthalpy change indicated as $\Delta_r H^\circ_1$, corresponds to the decomposition of $SO_3(g)$ into sulfur and oxygen. This is just the reverse of the equation for the formation of $SO_3(g)$, so we can write:

$$\Delta_r H_1^\ominus = -\Delta_f H^\ominus$$

We use a negative sign because, when we reverse a process, we change the sign of its $\Delta_r H$.

The second step in figure 8.13, with the enthalpy change indicated as $\Delta_r H_2^\ominus$, is the formation of $SO_2(g)$ plus $\frac{1}{2}$ mole of $O_2(g)$ from sulfur and oxygen. Therefore, we can write:

$$\Delta_r H_2^\ominus = \Delta_f H^\ominus_{SO_2(g)} + \tfrac{1}{2}\Delta_f H^\ominus_{O_2(g)}$$

The sum of these two steps gives the net change we want, so the sum of $\Delta_r H_1^\ominus$ and $\Delta_r H_2^\ominus$ must equal the desired $\Delta_r H^\ominus$.

$$\Delta_r H^\ominus = \Delta_r H_1^\ominus + \Delta_r H_2^\ominus$$

By substitution:

$$\Delta_r H^\ominus = (-\Delta_f H^\ominus_{SO_3(g)}) + (\Delta_f H^\ominus_{SO_2(g)} + \tfrac{1}{2}\Delta_f H^\ominus_{O_2(g)})$$

This can be rewritten as:

$$\Delta_r H^\ominus = (\Delta_f H^\ominus_{SO_2(g)} + \tfrac{1}{2}\Delta_f H^\ominus_{O_2(g)}) + (-\Delta_f H^\ominus_{SO_3(g)})$$

The change of sign gives:

$$\Delta_r H^\ominus = (\Delta_f H^\ominus_{SO_2(g)} + \tfrac{1}{2}\Delta_f H^\ominus_{O_2(g)}) - (\Delta_f H^\ominus_{SO_3(g)})$$

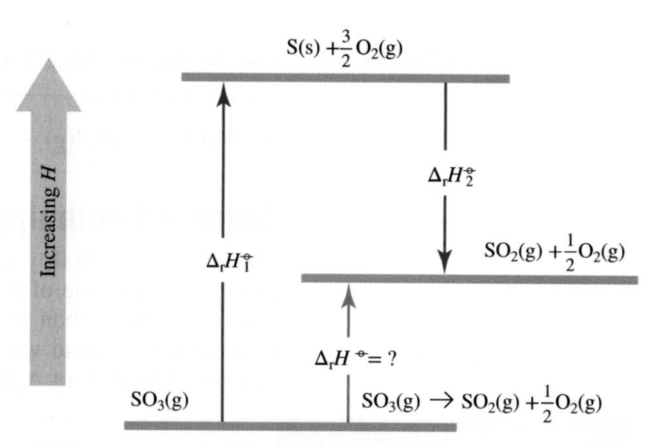

FIGURE 8.13 Enthalpy diagram for the reaction:

$$SO_3(g) \longrightarrow SO_2(g) + \tfrac{1}{2}O_2(g)$$

The path of the reaction in this diagram involves the reactant being decomposed into its elements in their standard states (upward blue arrow, $\Delta_r H_1^\ominus$), and then the elements being recombined to form the products (downward blue arrow, $\Delta_r H_2^\ominus$). The difference in the lengths of these two arrows is proportional to the net enthalpy change (upward red arrow $\Delta_r H^\ominus$).

Note the net result. $\Delta_r H^\ominus$ for the reaction equals the sum of the enthalpies of formation of the products minus the enthalpy of formation of the reactant, each multiplied by the appropriate coefficient. So, we could have obtained the identical result by using the Hess's law equation directly, instead of by the more laborious method of manipulating thermochemical equations.

WORKED EXAMPLE 8.8

Using Hess's law and standard enthalpies of formation

Some chefs keep baking soda, $NaHCO_3$, handy to put out fat and oil fires. When thrown on the fire, baking soda partly smothers the fire and the heat decomposes it to give CO_2, which further smothers the flame. The equation for the decomposition of $NaHCO_3$ is:

$$2NaHCO_3(s) \longrightarrow Na_2CO_3(s) + H_2O(g) + CO_2(g)$$

Use the data in table 8.3 to calculate $\Delta_r H^\ominus$ for this reaction.

Analysis

The Hess's law equation is now our basic tool for calculating values of $\Delta_r H^\ominus$. So we calculate the sum of the $\Delta_f H^\ominus$ values for the products (table 8.3), each multiplied by the appropriate stoichiometric coefficient, and do the same for the reactants. Then we subtract the latter from the former to calculate $\Delta_r H^\ominus$.

Solution

The Hess's law equation gives us:

$$\Delta_r H^\ominus = (1 \times \Delta_f H^\ominus_{Na_2CO_3(s)}) + (1 \times \Delta_f H^\ominus_{H_2O(g)}) + (1 \times \Delta_f H^\ominus_{CO_2(g)}) - (2 \times \Delta_f H^\ominus_{NaHCO_3(s)})$$

We now use table 8.3 to find the values of $\Delta_f H^\ominus$ for each substance in its given physical state.

$$\begin{aligned}\Delta_r H^\ominus &= 1 \times (-1131\,\text{kJ mol}^{-1}) + 1 \times (-241.8\,\text{kJ mol}^{-1}) + \\ &\qquad 1 \times (-393.5\,\text{kJ mol}^{-1}) - 2 \times (-947.7\,\text{kJ mol}^{-1}) \\ &= (-1766\,\text{kJ mol}^{-1}) - (-1895\,\text{kJ mol}^{-1}) \\ &= +129\,\text{kJ mol}^{-1}\end{aligned}$$

Thus, under standard conditions, the reaction is endothermic by $129\,\text{kJ mol}^{-1}$. (Notice that we did not have to manipulate any equations.)

Is our answer reasonable?

It is difficult to know if the answer is reasonable without checking your calculations. Make sure you have multiplied values by the appropriate stoichiometric coefficients, subtracted values for reactants from those for products, and kept track of all signs.

PRACTICE EXERCISE 8.8 Use the data in table 8.3 to calculate $\Delta_r H^\ominus$ for the following reactions.

(a) $2NO(g) + O_2(g) \longrightarrow 2NO_2(g)$

(b) $NaOH(s) + HCl(g) \longrightarrow NaCl(s) + H_2O(l)$

Standard enthalpy of combustion

The **standard enthalpy of combustion** ($\Delta_c H^\ominus$) of a substance at temperature T is the enthalpy change when 1 mole of the substance is completely burned in pure oxygen gas, under standard conditions. All carbon in the substance is converted to carbon dioxide gas, and all hydrogen is converted to liquid water. *Combustion reactions are always exothermic,* so $\Delta_c H^\ominus$ is always negative. The units of $\Delta_c H^\ominus$ are usually kJ mol^{-1}.

WORKED EXAMPLE 8.9

Writing an equation for a standard enthalpy of combustion

What amount of carbon dioxide gas is produced by a gas-fired power plant for every 1.0 MJ (megajoule) of energy it produces? The plant burns methane, $CH_4(g)$, for which $\Delta_c H^\ominus$ is –890 kJ mol^{-1}.

Analysis

We need to link the amount of carbon dioxide with the amount of heat produced. To do this, we require a balanced thermochemical equation. As this is a combustion reaction, the reactants will be the fuel, $CH_4(g)$, and oxygen, $O_2(g)$, and the products will be carbon dioxide gas (because the fuel contains carbon) and liquid water (because the fuel contains hydrogen). We must therefore balance the following equation for combustion of 1 mole of $CH_4(g)$.

$$CH_4(g) + ?O_2(g) \longrightarrow ?CO_2(g) + ?H_2O(l) \quad \Delta_c H^\ominus = -890 \text{ kJ mol}^{-1}$$

The coefficient in front of $CO_2(g)$ will be the amount of CO_2 produced when 890 kJ of heat is released. Using this, we can then obtain a conversion factor to relate megajoules of heat to the amount of $CO_2(g)$. To link megajoules with kilojoules, we'll need to remember the meanings of the SI prefixes *kilo* and *mega*.

$$1 \text{ kJ} = 10^3 \text{ J}$$

$$1 \text{ MJ} = 10^6 \text{ J}$$

These three relations are all we need to link megajoules of heat with the amount of $CO_2(g)$.

Solution

We balance the equation for 1 mole of CH_4 as follows:

$$CH_4(g) + 2O_2(g) \longrightarrow CO_2(g) + 2H_2O(l) \quad \Delta_c H^\ominus = -890 \text{ kJ mol}^{-1}$$

so 1 mol CO_2 is released for every 890 kJ of heat released. Knowing that 1 MJ = 1000 kJ, the amount of CO_2 produced when 1 MJ of heat is released is:

$$\frac{1.0 \times 10^3 \text{ kJ}}{890 \text{ kJ mol}^{-1}} = 1.1 \text{ mol } CO_2$$

Is our answer reasonable?

We can see that the equation has been correctly balanced for 1 mol CH_4; each side has one C atom, four H atoms and four O atoms. We can also see that it makes sense that only slightly more than 1 mole of CO_2 is produced with 1 MJ (1000 kJ) of heat, because 1000 kJ is only slightly more than 890 kJ, the amount of heat produced on formation of 1 mole of CO_2.

PRACTICE EXERCISE 8.9 A 9 kg cylinder of LPG (liquefied petroleum gas) can hold about 204 moles of propane, C_3H_8, which has a standard enthalpy of combustion of –2220 kJ mol^{-1}. How much heat could be produced by burning the propane gas contained in a full cylinder of LPG at atmospheric pressure?

Bond enthalpies

Chemical reactions generally involve the breaking of chemical bonds in the reactants and the making of chemical bonds in the products. The observed enthalpy change for any chemical reaction is predominantly a result of these bond-breaking and bond-making processes; energy is required to break a chemical bond and energy is liberated on formation of a chemical bond. Therefore, a knowledge of individual bond energies can provide us with a means to estimate the overall enthalpy change in a chemical reaction.

A **bond enthalpy** is the enthalpy change on breaking 1 mole of a particular chemical bond to give electrically neutral fragments in the gas phase at temperature T. You should appreciate that the bond energies we have encountered in previous chapters are associated with ΔU and refer to conditions of constant volume. Bond enthalpies are useful in rationalising chemical reactivity, as they provide a measure of the ease with which a particular chemical bond can be broken. Elemental nitrogen, for example, has a very low degree of reactivity, which is generally attributed to the very strong triple bond in N_2. Reactions that involve breaking this bond in a single step simply do not occur. When N_2 does react, it is by a stepwise breaking of its three bonds, one at a time.

Bond enthalpies and Hess's law

The bond enthalpies of simple diatomic molecules such as H_2, O_2 and Cl_2 are usually measured spectroscopically, using techniques similar to those we discussed in chapter 4 (p. 131). An electric discharge is used to excite the molecules, causing them to emit light. An analysis of the spectrum of the emitted light allows calculation of the amount of energy needed to break the bond.

For more complex molecules, thermochemical data can be used to calculate bond enthalpies using Hess's law. We will use the standard enthalpy of formation of methane to illustrate how this is accomplished. However, before we can attempt such a calculation, we must first define a thermochemical quantity called the **atomisation enthalpy ($\Delta_{at}H$)**. This is the enthalpy change that occurs on breaking all the chemical bonds in 1 mole of gaseous molecules to give gaseous atoms as products. For example, the equation for the atomisation of methane is:

$$CH_4(g) \longrightarrow C(g) + 4H(g)$$

and the enthalpy change for the process is $\Delta_{at}H^{\ominus}$ under standard conditions. For this particular molecule, $\Delta_{at}H^{\ominus}$ corresponds to the total enthalpy change on breaking all the C—H bonds in 1 mole of CH_4 under standard conditions; therefore, division of $\Delta_{at}H^{\ominus}$ by 4 would give the average C—H bond enthalpy in methane.

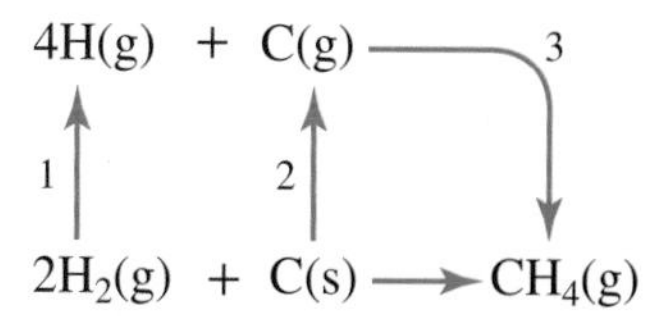

FIGURE 8.14 Two paths (equations) for the formation of methane from its elements in their standard states. As described in the text, steps 1, 2 and 3 of the upper path are the formation of gaseous atoms of the elements and then the formation of the bonds in CH_4. The lower path is the direct combination of the elements in their standard states to give CH_4. Because ΔH is a state function, the sum of the enthalpy changes along the upper path must equal the enthalpy change for the lower path, $\Delta_f H^{\ominus}$.

Figure 8.14 shows how we can use the standard enthalpy of formation, $\Delta_f H^{\ominus}$, to calculate the atomisation enthalpy. Across the bottom we have the chemical equation for the formation of 1 mole of CH_4 from its elements in their standard states. The enthalpy change for this reaction is $\Delta_f H^{\ominus}_{CH_4(g)}$. In this figure, we also can see an alternative three-step path that leads to $CH_4(g)$. One step is the breaking of H—H bonds in the H_2 molecules to give 4 moles of gaseous hydrogen atoms; another is the vaporisation of carbon to give 1 mole of gaseous carbon atoms, and the third is the combination of the gaseous atoms to form 1 mole of gaseous CH_4 molecules. These changes are labelled 1, 2 and 3 in the figure.

Since enthalpy is a state function, the net enthalpy change from one state to another is the same regardless of the path that we follow. This means that the sum of the enthalpy changes along the upper path must be the same as the enthalpy change along the lower path, $\Delta_f H^{\ominus}$. Perhaps this can be seen more easily in Hess's law terms if we write the changes along the upper path in the form of thermochemical equations.

Steps 1 and 2 have enthalpy changes that are standard enthalpies of formation of gaseous atoms. Values for these quantities have been measured for many of the elements, and some are given in table 8.4. Step 3 is the opposite of atomisation, and its enthalpy change will, therefore, be the negative of $\Delta_{at}H^{\ominus}$. (Recall that if we reverse a reaction, we change the sign of its ΔH.)

Step 1: $$2H_2(g) \longrightarrow 4H(g) \quad \Delta_r H^{\ominus}_1 = 4\Delta_f H^{\ominus}_{H(g)}$$

Step 2: $$C(s) \longrightarrow C(g) \quad \Delta_r H^{\ominus}_2 = \Delta_f H^{\ominus}_{C(g)}$$

Step 3: $$4H(g) + C(g) \longrightarrow CH_4(g) \quad \Delta_r H^{\ominus}_3 = -\Delta_{at}H^{\ominus}$$

$$2H_2(g) + C(s) \longrightarrow CH_4(g) \quad \Delta_r H^{\ominus} = \Delta_f H^{\ominus}_{CH_4(g)}$$

Notice that, by adding the first three equations, we get the equation for the formation of CH_4 from its elements in their standard states. This means that adding the $\Delta_r H^{\ominus}$ values of the first three equations should give $\Delta_f H^{\ominus}$ for CH_4:

$$\Delta_r H^{\ominus}_1 + \Delta_r H^{\ominus}_2 + \Delta_r H^{\ominus}_3 = \Delta_f H^{\ominus}_{CH_4(g)}$$

TABLE 8.4 Standard enthalpies of formation of selected gaseous atoms from the elements in their standard states at 25 °C.

Atom	$\Delta_f H^{\ominus}$ (kJ mol^{-1})[(a)]	Atom	$\Delta_f H^{\ominus}$ (kJ mol^{-1})[(a)]
B	560	I	107.48
Be	324.3	Li	161.5
Br	112.38	N	472.68
C	716.67	O	249.17
Cl	121.47	P	332.2
F	79.14	S	276.98
H	217.89	Si	450

(a) All values are positive because formation of the gaseous atoms from the elements involves bond breaking, which requires an input of energy.

We will substitute for $\Delta_r H_1^\ominus$, $\Delta_r H_2^\ominus$ and $\Delta_r H_3^\ominus$, and then solve for $\Delta_{at}H^\ominus$. First, we substitute for the $\Delta_r H^\ominus$ quantities.

$$4\Delta_f H^\ominus_{H(g)} + \Delta_f H^\ominus_{C(g)} + (-\Delta_{at}H^\ominus) = \Delta_f H^\ominus_{CH_4(g)}$$

Next, we solve for $(-\Delta_{at}H^\ominus)$.

$$-\Delta_{at}H^\ominus = \Delta_f H^\ominus_{CH_4(g)} - 4\Delta_f H^\ominus_{H(g)} - \Delta_f H^\ominus_{C(g)}$$

Changing signs and rearranging the right-hand side of the equation gives:

$$\Delta_{at}H^\ominus = 4\Delta_f H^\ominus_{H(g)} + \Delta_f H^\ominus_{C(g)} - \Delta_f H^\ominus_{CH_4(g)}$$

Now all we need are the $\Delta_f H^\ominus$ values on the right-hand side. We obtain $\Delta_f H^\ominus_{H(g)}$ and $\Delta_f H^\ominus_{C(g)}$ from table 8.4 and $\Delta_f H^\ominus_{CH_4(g)}$ from table 8.3 and round these to the nearest 0.1 kJ mol^{-1}.

$$\Delta_f H^\ominus_{H(g)} = +217.9 \text{ kJ mol}^{-1}$$

$$\Delta_f H^\ominus_{C(g)} = +716.7 \text{ kJ mol}^{-1}$$

$$\Delta_f H^\ominus_{CH_4(g)} = -74.8 \text{ kJ mol}^{-1}$$

Substituting these values gives:

$$\Delta_{at}H^\ominus = 1663.1 \text{ kJ mol}^{-1}$$

Given that each CH_4 molecule contains four C—H bonds, division by 4 gives an estimate of the average C—H bond enthalpy in this molecule.

$$\text{bond enthalpy} = \frac{1663.1 \text{ kJ mol}^{-1}}{4}$$

$$= 415.8 \text{ kilojoules per mole of C—H bonds}$$

This value is quite close to the one in table 8.5, which is an average of the C—H bond enthalpies in many different compounds. The other bond enthalpies in table 8.5 are also based on thermochemical data and were obtained by similar calculations.

TABLE 8.5 Some average bond enthalpies at 25 °C.

Bond	Bond enthalpy (kJ mol^{-1})	Bond	Bond enthalpy (kJ mol^{-1})	Bond	Bond enthalpy (kJ mol^{-1})
C—C	348	C—F	484	H—H	436
C═C	612	C—Cl	338	H—F	565
C≡C	960	C—Br	276	H—Cl	431
C—H	412	C—I	238	H—Br	366
C—N	305			H—I	299
C═N	613			H—N	388
C≡N	890			H—O	463
C—O	360			H—S	338
C═O	743			H—Si	376

A remarkable thing about many covalent bond enthalpies is that they are very nearly the same in many different compounds. This suggests, for example, that a C—H bond is very nearly the same in CH_4 as it is in a large number of other compounds that contain this kind of bond. Also note that, in general, triple bond enthalpies (e.g. C≡C) are greater than double bond enthalpies (C═C), which in turn are greater than single bond enthalpies (C—C).

Because the bond enthalpy does not vary much from compound to compound, we can use tabulated bond enthalpies to estimate the enthalpies of formation of substances. We will illustrate this by calculating the standard enthalpy of formation of gaseous methanol, $CH_3OH(g)$. The structural formula of methanol is:

```
    H
    |
H—C—O—H
    |
    H
```

To carry out this calculation, we set up two paths from the elements to the compound, as shown in figure 8.15. The lower path has an enthalpy change corresponding to $\Delta_f H^{\ominus}_{CH_3OH(g)}$ while the upper path takes us to the gaseous elements and then through the enthalpy change when the bonds in the molecule are formed. This latter enthalpy can be calculated from the bond enthalpies in table 8.5. As before, the sum of the enthalpy changes along the upper path must be the same as the enthalpy change along the lower path, and this allows us to determine $\Delta_f H^{\ominus}_{CH_3OH(g)}$.

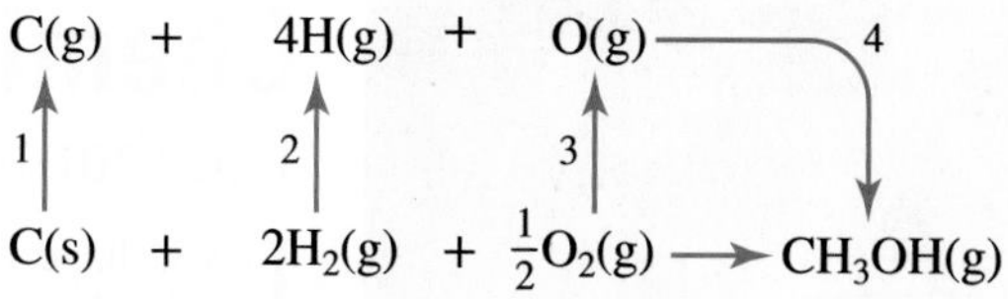

Steps 1, 2 and 3 in figure 8.15 involve the formation of the gaseous atoms from the elements, and their enthalpy changes are obtained from table 8.4.

$$C(g) + 4H(g) + O(g) \xrightarrow{4} CH_3OH(g)$$
$$C(s) + 2H_2(g) + \tfrac{1}{2}O_2(g) \longrightarrow CH_3OH(g)$$

FIGURE 8.15 Two paths for the formation of methanol vapour from its elements in their standard states. The numbered steps are discussed in the text.

$$\Delta_r H^{\ominus}_1 = \Delta_f H^{\ominus}_{C(g)} = 1 \times 716.7\ \text{kJ mol}^{-1} = 716.7\ \text{kJ mol}^{-1}$$
$$\Delta_r H^{\ominus}_2 = 4\Delta_f H^{\ominus}_{H(g)} = 4 \times 217.9\ \text{kJ mol}^{-1} = 871.6\ \text{kJ mol}^{-1}$$
$$\Delta_r H^{\ominus}_3 = \Delta_f H^{\ominus}_{O(g)} = 1 \times 249.2\ \text{kJ mol}^{-1} = 249.2\ \text{kJ mol}^{-1}$$

The sum of these values, +1837.5 kJ mol^{-1}, is the net $\Delta_r H^{\ominus}$ for the first three steps.

The formation of the CH_3OH molecule from the gaseous atoms is exothermic because energy is always released when atoms become joined by covalent bonds. In this molecule, we can count three C—H bonds, one C—O bond and one O—H bond. Their formation releases energy equal to the sum of their bond enthalpies, which we obtain from table 8.5.

Bond	Bond enthalpy (kJ mol^{-1})
3(C—H)	3 × 412 = 1236
C—O	360
O—H	463

Adding these gives a total of 2059 kJ mol^{-1}. Therefore, $\Delta_r H^{\ominus}_4$ is −2059 kJ mol^{-1} because it involves bond formation, and so it is exothermic. Now we can calculate the total enthalpy change for the upper path.

$$\Delta_r H^{\ominus} = (+1837.5\ \text{kJ mol}^{-1}) + (-2059\ \text{kJ mol}^{-1}) = -222\ \text{kJ mol}^{-1}$$

The value just calculated should be equal to $\Delta_f H^{\ominus}$ for $CH_3OH(g)$. For comparison, the experimental value of $\Delta_f H^{\ominus}$ for $CH_3OH(g)$ is −201 kJ mol^{-1}. At first glance, the agreement does not seem very good, but on a relative basis the calculated value (−222 kJ) differs from the experimental value by only about 10%.

Lattice enthalpies and Hess's law

Hess's law can be used in combination with particular thermodynamic data to calculate the values of lattice enthalpies for ionic solids. We will illustrate this using NaCl(s). Recall from chapter 5 (p. 176) that the lattice enthalpy (or, as we called it then, lattice energy) of an ionic solid is the enthalpy change for the conversion of 1 mole of the ionic solid into its constituent gaseous ions. For NaCl(s) we would represent this process by the equation:

$$NaCl(s) \rightarrow Na^+(g) + Cl^-(g) \quad \Delta_r H^{\ominus} = \text{lattice enthalpy (under standard conditions)}$$

We can break this process into smaller steps, each with an associated enthalpy change under standard conditions, as shown in figure 8.16.

$$NaCl(s) \xrightarrow{\text{−(enthalpy of formation of NaCl(s))}} Na(s) + \tfrac{1}{2}Cl_2(g) \quad \Delta H_1 = -\Delta_f H^{\ominus}_{NaCl(s)}$$
$$Na(s) \xrightarrow{\text{enthalpy of formation of Na(g)}} Na(g) \quad \Delta H_2 = \Delta_f H^{\ominus}_{Na(g)}$$
$$Na(g) \xrightarrow{\text{ionisation energy of Na(g)}} Na^+(g) + e^- \quad \Delta H_3 = E_{i1\ Na(g)}$$
$$\tfrac{1}{2}Cl_2(g) \xrightarrow{\text{enthalpy of formation of Cl(g)}} Cl(g) \quad \Delta H_4 = \Delta_f H^{\ominus}_{Cl(g)}$$
$$Cl(g)^{+e^-} \xrightarrow{\text{electron affinity of Cl(g)}} Cl^-(g) \quad \Delta H_5 = E_{EA\ Cl(g)}$$
$$NaCl(s) \xrightarrow{\text{lattice enthalpy of NaCl(s)}} Na^+(g) + Cl^-(g) \quad \Delta H_6$$

FIGURE 8.16 An enthalpy diagram that can be used to calculate the lattice enthalpy of NaCl(s).

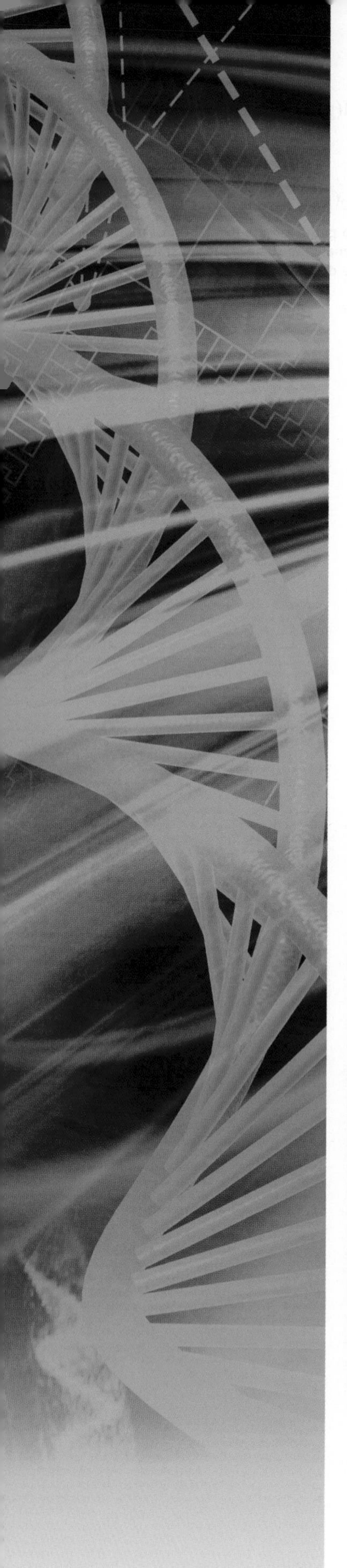

Chemical Connections

The thermodynamics of steam burns

If you have ever been unfortunate enough to accidentally place any of your bare skin over a boiling jug of water, you will know the enormous pain that ensues. In fact, a steam burn causes more damage to the skin than does a burn caused by boiling water. And the reason why this is so can be found in thermodynamics.

Let's consider what happens when we heat up a sample of water at constant pressure. You know from what you have learned in this chapter that it takes 4.18 J to increase the temperature of 1 g of water by 1 °C. So if we heat 1 kg of water, initially at 25 °C, it will absorb 4.18 kJ for every degree Celsius by which its temperature increases. By the time the water reaches 100 °C and starts boiling, it will have absorbed (75 × 4.18 kJ) = 313.5 kJ. Water at this temperature contains a large amount of energy, and will, as you know, cause severe burns.

We could take the boiling water at 100 °C and continue heating it, but the temperature of the water would not increase. The added energy simply vaporises the water and turns it into water vapour — or, as we better know it, steam. This process will continue until the liquid water at 100 °C has been completely converted to steam at 100 °C. In thermodynamic terms, the amount of energy (or, more correctly, enthalpy) required to carry out this phase change is the enthalpy of vaporisation, $\Delta_{vap}H$. For water at 100 °C, $\Delta_{vap}H = 40.66$ kJ mol^{-1}. On a mass basis this means that we would require 2260 kJ to convert our 1 kg of water completely to steam.

Let's now compare the numbers. It takes 313.5 kJ to heat 1 kg of water from 25 °C to 100 °C, and 2260 kJ, or around 7 times as much energy, to convert the boiling water to steam at 100 °C. You can see that steam therefore contains significantly more energy than does the same amount of liquid water at the same temperature. This energy will be rapidly released on contact with skin, and this will therefore lead to substantially more tissue damage than a burn caused by boiling water.

So be careful when making those cappuccinos!

FIGURE 8.17 Steam contains significantly more energy than water at the same temperature and can cause very severe burns.

The unknown lattice enthalpy, ΔH_6, is shown at the bottom of figure 8.15. This is an endothermic process that requires energy to turn NaCl(s) into its gaseous ions $Na^+(g)$ and $Cl^-(g)$. From Hess's law, we know that $\Delta H_6 = \Delta H_1 + \Delta H_2 + \Delta H_3 + \Delta H_4 + \Delta H_5$. As each of ΔH_1 to ΔH_5 is a tabulated thermodynamic quantity, we can calculate ΔH_6, the lattice enthalpy of NaCl(s). Notice that we have defined ΔH_1 as the enthalpy change for the process:

$$\mathrm{NaCl(s)} \rightarrow \mathrm{Na(s)} + \tfrac{1}{2}\mathrm{Cl_2(g)}$$

This is the reverse of the equation defining the enthalpy of formation of NaCl(s):

$$\mathrm{Na(s)} + \tfrac{1}{2}\mathrm{Cl_2(g)} \rightarrow \mathrm{NaCl(s)}$$

so we must use the *negative* of the enthalpy of formation of NaCl(s) in our calculations.

The calculation is then:

$$\begin{aligned}\Delta H_6 &= -\Delta_f H^{\ominus}_{\mathrm{NaCl(s)}} + \Delta_f H^{\ominus}_{\mathrm{Na(g)}} + E_{\mathrm{i1\ Na(g)}} + \Delta_f H^{\ominus}_{\mathrm{Cl(g)}} + E_{\mathrm{EA\ Cl(g)}} \\ &= 411\,\mathrm{kJ\,mol^{-1}} + 107\,\mathrm{kJ\,mol^{-1}} + 496\,\mathrm{kJ\,mol^{-1}} + 121\,\mathrm{kJ\,mol^{-1}} + (-349\,\mathrm{kJ\,mol^{-1}}) \\ &= 786\,\mathrm{kJ\,mol^{-1}}\end{aligned}$$

The lattice enthalpy of NaCl(s) is thus 786 kJ mol^{-1}.

This approach gives values close to the experimentally measured lattice enthalpies for ionic solids containing relatively small ions. As the sizes of the ions increase, the values become less accurate because the bonding in such solids is no longer purely ionic.

8.5 Entropy

In the previous section, we defined enthalpy, showing how it is related to heat. We will now discuss the thermodynamic function entropy. While there are many different textbook definitions of entropy, here we will concentrate on its statistical basis.

Entropy and probability

Earlier in this chapter, you learned that, when a hot object is placed in contact with a colder one, heat will flow spontaneously from the hot object to the colder one. But why? Energy will be conserved no matter what the direction of the heat flow is. And if we believe that the lowering of energy is the driving force behind spontaneous change, we might wonder why the lowering of the energy of the hot object is favoured over the gain in energy that results for the colder one.

We will build a simple model to explain the direction of heat flow. Imagine that we have two objects made of molecules that have a low-energy ground state and a high-energy excited state. They will actually have many different excited states, but we will keep things simple and assume that the molecules can be either 'low energy' in the ground state or 'high energy' in just one excited state. We will use a blue dot to represent a low-energy cold molecule, and a red dot to represent a high-energy hot molecule. If we place three high-energy molecules in contact with three low-energy molecules, we initially have a configuration like the one shown on the right, where the three molecules on the top represent an object that is 'hot' and the three on the bottom represent a 'cold' object. After the objects are placed in contact, energy can be transferred between molecules in the two objects. The sum of the energies of the two objects must be the same before and after contact. Because our molecules can have only high-energy (red) or low-energy (blue) states, the total number of red molecules must be the same before and after contact. Here are all possible distributions of energy among the six molecules after the objects have been placed into contact:

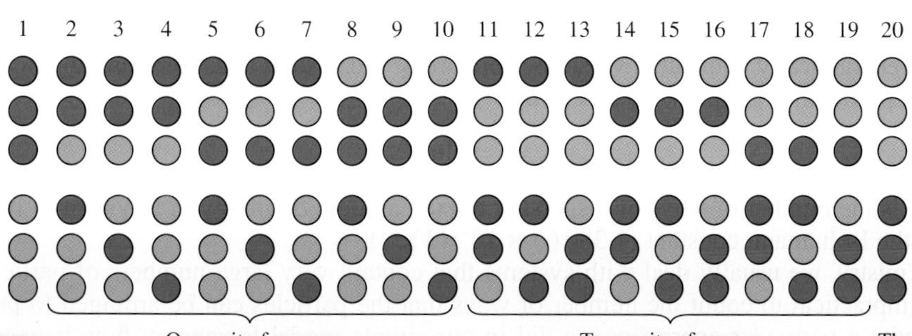
1 2 3 4 5 6 7 8 9 10 11 12 13 14 15 16 17 18 19 20
No energy is transferred.
One unit of energy is transferred.
Two units of energy are transferred.
Three units of energy are transferred.

We can see four possible outcomes; 0, 1, 2 or 3 units of energy can be transferred from the hot body to the colder one. Which is most likely to occur?

Notice that some of the outcomes can be produced in a number of different ways. For example, arrangements 2 to 10 represent ways energy can be distributed among the molecules after 1 unit of energy is transferred. The more ways that a state can be produced, the more likely it is to occur. We can use this fact to estimate the probability of the state occurring. There are 20 possible arrangements of energy among the particles after the hot and cold objects are brought into contact. Let's assume that all of these arrangements are equally probable. The probability of a particular outcome can be calculated as:

$$\text{probability of an outcome} = \frac{\text{number of ways the outcome can be produced}}{\text{total number of ways all outcomes can be produced}}$$

For example, there are nine arrangements that correspond to a transfer of 1 energy unit and 20 arrangements in all, so the probability that 1 energy unit will be transferred is 9/20 = 0.45 or 45%.

Units of energy transferred	Number of equivalent ways to produce the energy transfer	Probability of energy transfer
0	1	1/20 = 5%
1	9	9/20 = 45%
2	9	9/20 = 45%
3	1	1/20 = 5%

In this model system, there is a 19/20 or 95% chance that some amount of energy transfer will occur. This agrees with our expectation that heat should flow spontaneously from the hot object to the colder one.

To show the importance of probability in determining the direction of heat flow, consider the possible outcomes if the initial state of the system was given by arrangement 2 on the previous page, as shown on the right. Here, the hot object contains two high-energy molecules and one low-energy molecule, while the cold object contains two low-energy molecules and one high-energy molecule. If we bring the objects into contact, the final possible states are still given by arrangements 1 to 20 on the previous page. Notice though that, if the final state of the system is arrangement 1, as shown below right, then this formally involves the 'hot' object getting hotter and the 'cold' object getting colder, something alien to our everyday experience. However, as this is only one of 20 possible final arrangements, the probability of this occurring is only 5%. If we repeat this analysis with objects that contain more and more particles, the probability of energy transfer becomes so high that we can be quite certain that heat will flow from a hot object to a colder one. If we are dealing with hot and cold objects that contain moles of particles, the chance that no energy transfer would occur once they were placed into contact is negligible.

Although our model for heat transfer is very simple, it demonstrates the role of probability in determining the direction of a spontaneous process. *Spontaneous processes tend to proceed from states of low probability to states of higher probability*. The higher probability states are those that allow more options for distributing energy among the molecules, so we can also say that *spontaneous processes tend to disperse energy.*

Entropy and entropy change

Because statistical probability is so important in determining the outcome of chemical and physical events, thermodynamics defines a quantity, called entropy (S), that describes the number of equivalent ways that energy can be distributed in the system. The larger the value of the entropy, the more energetically equivalent versions there are of the system and, therefore, the higher is its statistical probability. In mathematical terms:

$$S = k \ln W$$

where W is the number of ways of dispersing a fixed amount of energy in a system of fixed size, and k is the Boltzmann constant ($1.380\,65 \times 10^{-23}\,\text{J K}^{-1}$).

In chemistry, we usually deal with systems that contain very large numbers of particles. It is usually impractical to count the number of ways that the particles can be arranged to produce a system with a particular energy, as we did in our simple model for energy flow between a hot and cold object. Fortunately, we do not need to do so. The entropy of a system can be related to experimental heat and temperature measurements. We can also recognise increases and decreases in entropy without explicitly counting the number of ways the system can be realised.

Like enthalpy, entropy is a state function. It depends only on the state of the system, so a change in entropy, ΔS, is independent of the path from start to finish. As with other thermodynamic quantities, ΔS is defined as 'final minus initial' or 'products minus reactants'. Thus:

$$\Delta S = S_{final} - S_{initial}$$

or, for a chemical reaction:

$$\Delta_r S = S_{products} - S_{reactants}$$

The units of $\Delta_r S$ are J mol^{-1} K^{-1}.

As you can see, when S_{final} is larger than $S_{initial}$ (or when $S_{products}$ is larger than $S_{reactants}$), the value of ΔS is positive. A positive value for ΔS means an increase in the number of energy-equivalent ways the system can be produced; we have seen that this kind of change tends to be spontaneous (see figure 8.18). This leads to a general statement about entropy: *Any event that is accompanied by an increase in the entropy of an isolated system has a tendency to occur spontaneously.*

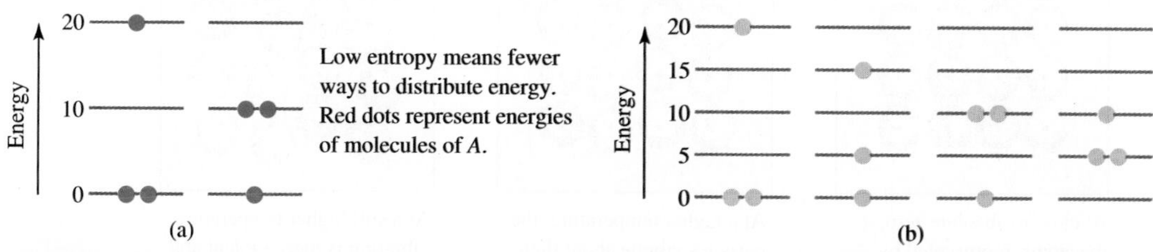

FIGURE 8.18 A positive value for $\Delta_r S$ means an increase in the number of ways energy can be distributed among a system's molecules. Consider the reaction $3A \rightarrow 3B$, where A molecules can take on energies that are multiples of 10 energy units, and B molecules can take on energies that are multiples of 5 units. Suppose that the total energy of the reacting mixture is 20 units. **(a)** There are two ways to distribute 20 units of energy among three molecules of A. **(b)** There are four ways to distribute 20 units of energy among three molecules of B. The entropy of B is higher than the entropy of A because there are more ways to distribute the same amount of energy in B than in A. The reaction $3A \rightarrow 3B$ is therefore spontaneous.

Factors that affect entropy

It is often possible to predict whether ΔS is positive or negative for a particular change. This is because several factors influence the magnitude of the entropy in predictable ways.

Volume

For gases, the entropy increases with increasing volume, as illustrated in figure 8.19. At the left we see a gas confined to one side of a container, separated from a vacuum by a removable partition. Let's suppose the partition could be pulled away in an instant, as shown in figure 8.19b. Now we find a situation in which all the molecules of the gas are at one end of a larger container. There are many more possible ways that the total kinetic energy can be distributed if we give the molecules greater freedom of movement and spread them throughout the larger volume. That makes the configuration in figure 8.19b extremely unlikely. Therefore, the gas expands spontaneously to achieve a more probable (higher entropy) particle distribution (figure 8.19c).

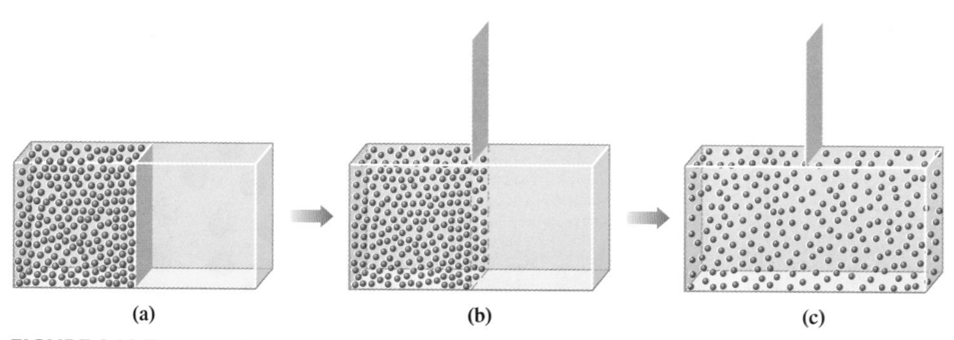

FIGURE 8.19 The expansion of a gas into a vacuum: **(a)** a gas in a container separated from a vacuum by a partition and **(b)** the gas at the moment the partition is removed. **(c)** The gas expands to achieve a more probable (higher entropy) particle distribution.

Temperature

Entropy is affected by temperature: the higher the temperature, the higher the entropy. For example, when a substance is a solid at temperatures close to absolute zero, its particles are essentially motionless. There is relatively little kinetic energy; with few ways to distribute kinetic energy among the particles, the entropy of the solid is relatively low (figure 8.20a). If some heat is added to the solid, the kinetic energy of the particles increases along with the temperature. This causes the particles to vibrate within the crystal, so at a particular moment (figure 8.20b) the particles are not found exactly at their equilibrium lattice sites. There is more kinetic energy and more freedom of movement, so there are more ways to distribute the energy among the molecules. At a still higher temperature, there is more kinetic energy and more possible ways to distribute it, so the solid has a still higher entropy (figure 8.20c).

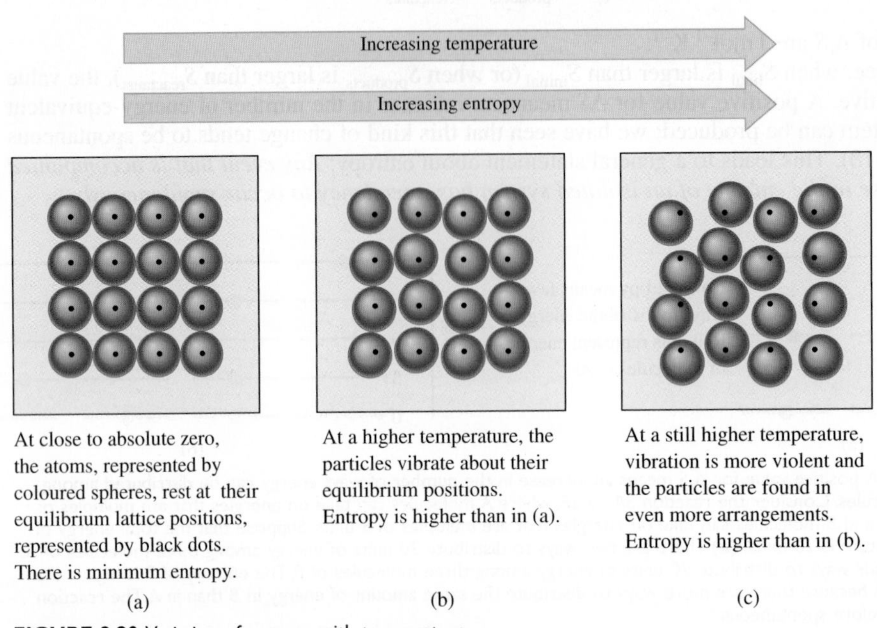

Increasing temperature
Increasing entropy
At close to absolute zero, the atoms, represented by coloured spheres, rest at their equilibrium lattice positions, represented by black dots. There is minimum entropy.
(a)
At a higher temperature, the particles vibrate about their equilibrium positions. Entropy is higher than in (a).
(b)
At a still higher temperature, vibration is more violent and the particles are found in even more arrangements. Entropy is higher than in (b).
(c)

FIGURE 8.20 Variation of entropy with temperature.

Physical state

One of the major factors that affects the entropy of a system is its physical state, which is demonstrated in figure 8.21. Suppose that the contents of the three containers are ice, liquid water and water vapour at the same temperature. There is greater freedom of molecular movement in liquid water than in ice at the same temperature, and so there are more ways to distribute kinetic energy among the molecules of liquid water than there are in ice. Water vapour molecules are free to move through the entire container, so there are many more possible ways to distribute the kinetic energy among the gas molecules than there are in liquid or solid water. In fact, any gas has such a large entropy compared with a liquid or solid that changes which produce gases from liquids or solids are almost always accompanied by increases in entropy.

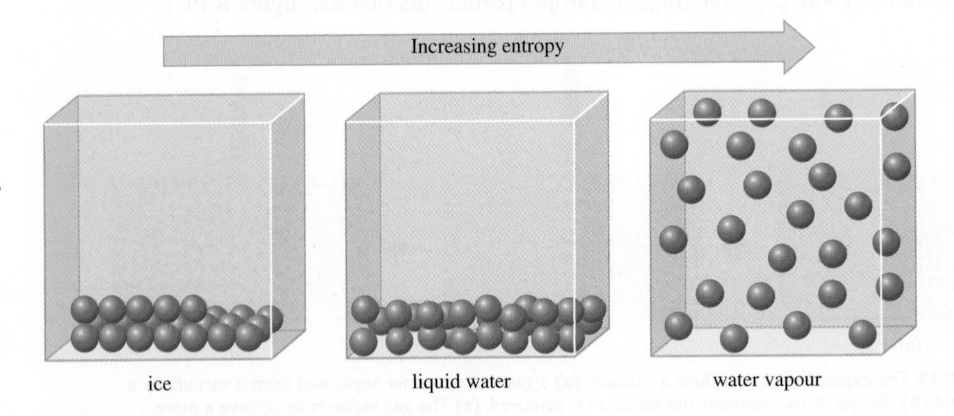

FIGURE 8.21 Comparison of the entropies of the solid, liquid, and gaseous states of water. The crystalline solid has a very low entropy. The liquid has a higher entropy because its molecules can move more freely and, therefore, there are more ways to distribute kinetic energy among them. All the molecules are still found at the bottom of the container. The gas has the highest entropy because the molecules are randomly distributed throughout the entire container, so there are many ways to distribute the kinetic energy.

For reactions that involve gases, we can simply calculate the change in the amount of gas, Δn_{gas}, on going from reactants to products.

$$\Delta n_{gas} = n_{gaseous\ products} - n_{gaseous\ reactants}$$

When Δn_{gas} is positive, so is the entropy change.

When a chemical reaction produces or consumes gases, the sign of its entropy change is usually easy to predict. This is because the entropy of a gas is so much larger than that of either a liquid or solid. For example, the thermal decomposition of sodium bicarbonate produces two gases, CO_2 and H_2O.

$$2NaHCO_3(s) \xrightarrow{\text{heat}} Na_2CO_3(s) + CO_2(g) + H_2O(g)$$

Because the amount of gaseous products is larger than the amount of gaseous reactants, we can predict that $\Delta_r S$ for the reaction is positive. On the other hand, the reaction:

$$CaO(s) + SO_2(g) \longrightarrow CaSO_3(s)$$

which can be used to remove sulfur dioxide, an atmospheric pollutant, from a gas mixture, has a negative entropy change as $\Delta n_{gas} < 0$.

Number of particles

For chemical reactions, another major factor that affects the sign of $\Delta_r S$ is an increase in the total number of molecules as the reaction proceeds. When more molecules are produced during a reaction, more ways of distributing the energy among the molecules are possible. *When all other things are equal, reactions that increase the number of particles in the system tend to have a positive entropy change*, as shown in figure 8.22. This is particularly true when the products are gases, owing to their inherently large, positive entropies.

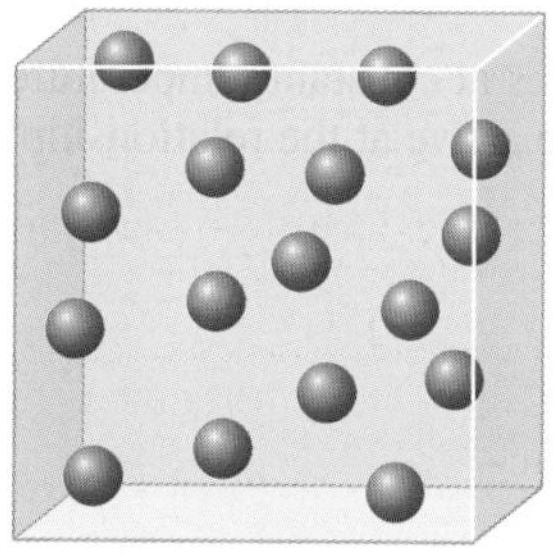
lower entropy

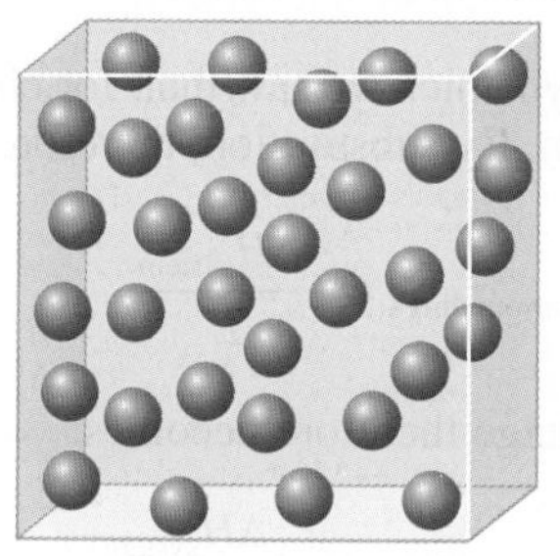
higher entropy

FIGURE 8.22 Entropy is affected by the number of particles. Adding particles to a system increases the number of ways that energy can be distributed in the system so, with all other things being equal, a reaction that produces more particles has a positive value of $\Delta_r S$.

WORKED EXAMPLE 8.10

Predicting the sign of $\Delta_r S$

Predict the algebraic sign of $\Delta_r S$ for the following reactions.

(a) $2NO_2(g) \rightarrow N_2O_4(g)$

(b) $C_3H_8(g) + 5O_2(g) \rightarrow 3CO_2(g) + 4H_2O(g)$

(c) $CO(g) + H_2O(g) \rightarrow H_2(g) + CO_2(g)$

Analysis

As we examine the equations, we look for changes in the number of molecules of gas and changes in the number of particles on going from reactants to products.

Solution

In reaction (a), we are forming one N_2O_4 molecule from the two NO_2 molecules. Since we are forming fewer molecules, there are fewer ways to distribute energy among them, which means that the entropy must be decreasing. Therefore, $\Delta_r S$ must be negative.

For reaction (b), we can count the number of molecules on both sides. On the left of the equation we have six molecules; on the right there are seven. There are more ways to distribute kinetic energy among seven molecules than among six, so for reaction (b) we expect $\Delta_r S$ to be positive.

For reaction (c), there is the same number of molecules on both sides. We cannot therefore predict the sign of $\Delta_r S$ with confidence, but we expect its magnitude to be small.

PRACTICE EXERCISE 8.10

Predict the sign of the entropy change for the following phase changes: (a) the condensation of steam to liquid water and (b) the sublimation of a solid.

8.6 The second law of thermodynamics

The **second law of thermodynamics** can help us appreciate the importance of entropy. This law states that, *whenever a spontaneous event takes place in our universe, the total entropy of the universe increases* ($\Delta S_{total} > 0$). Notice that the increase in entropy referred to here is for the total entropy of the universe (system plus surroundings), not just the system alone. This means that the entropy of a system can decrease, as long as there is a larger increase in the entropy of the surroundings so that the overall entropy change is positive. The consequence of this is that we cannot use the entropy change of the system alone as a criterion for spontaneity of a particular chemical reaction. Because everything that happens relies on spontaneous changes of some sort, the entropy of the universe is constantly increasing.

Now let's examine the entropy change for the universe more closely. As we have suggested, this quantity equals the sum of the entropy change for the system plus the entropy change for the surroundings.

$$\Delta S_{total} = \Delta S_{system} + \Delta S_{surroundings}$$

It can be shown that the entropy change for the surroundings is equal to the heat transferred *to* the surroundings *from* the system, $q_{surroundings}$, divided by the thermodynamic temperature, T, at which it is transferred.

$$\Delta S_{surroundings} = \frac{q_{surroundings}}{T}$$

The law of conservation of energy requires that the heat gained by the surroundings equals the negative of the heat lost by the system, so we can write:

$$q_{surroundings} = -q_{system}$$

In our study of the first law of thermodynamics we saw that, for changes at constant temperature and pressure, $q_{system} = \Delta H$ for the system. By substitution, therefore, we arrive at the relationship:

$$\Delta S_{surroundings} = \frac{-\Delta H_{system}}{T}$$

and our expression for total entropy change therefore becomes:

$$\Delta S_{total} = \Delta S_{system} - \frac{\Delta H_{system}}{T}$$

Multiplying through by T gives:

$$T\Delta S_{total} = T\Delta S_{system} - \Delta H_{system}$$

or:

$$T\Delta S_{total} = -(\Delta H_{system} - T\Delta S_{system})$$

We stated above that ΔS_{total} is > 0 for a spontaneous process, and, because T is always positive, the left-hand side of this equation will be positive for a spontaneous process. To make the right-hand side of this equation positive requires that the quantity in parentheses is negative. In other words, for a spontaneous process:

$$\Delta H_{system} - T\Delta S_{system} < 0$$

We now formally define the thermodynamic function Gibbs energy (G) as:

$$G = H - TS$$

For changes at constant T and p, we can therefore write:

$$\Delta G = \Delta H - T\Delta S$$

where $\Delta G = G_{final} - G_{initial}$

We showed above that $\Delta H_{system} - T\Delta S_{system} < 0$ for a spontaneous process. As $\Delta G = \Delta H - T\Delta S$, it follows from this that, if $\Delta G < 0$ for a chemical process in a system at constant T and p, that process will be spontaneous. We will investigate this concept further in section 8.8.

8.7 The third law of thermodynamics

Earlier we described how the entropy of a substance depends on temperature, and we noted that at temperatures approaching absolute zero the order within a crystal increases and the entropy decreases. The **third law of thermodynamics** goes one step further by stating that, *at absolute zero, the entropy of a perfectly ordered pure crystalline substance is zero.*

$$S = 0 \quad \text{at} \quad T = 0\,\text{K}$$

Because we know the point at which entropy has a value of 0, it is possible by *experimental measurement* and calculation to determine the entropy of a substance at temperatures above 0 K. If the entropy of 1 mole of a substance is determined under standard conditions, we call it the **standard entropy** ($S^\ominus$). Table 8.6 lists the standard entropies of a number of substances at 25 °C.

TABLE 8.6 Standard entropies of selected substances at 25 °C.

Substance	$S^\ominus$(J mol^{-1} K^{-1})	Substance	$S^\ominus$(J mol^{-1} K^{-1})	Substance	$S^\ominus$(J mol^{-1} K^{-1})
Ag(s)	42.55	$CaCl_2$(s)	114	K_2SO_4(s)	176
AgCl(s)	96.2	CaO(s)	40	N_2(g)	191.5
Al(s)	28.3	$Ca(OH)_2$(s)	76.1	NH_3(g)	192.5
Al_2O_3(s)	51.00	$CaSO_4$(s)	107	NH_4Cl(s)	94.6
C(s, diamond)	2.4	$CaSO_4 \cdot \frac{1}{2}H_2O$(s)	131	NO(g)	210.6
C(s, graphite)	5.69	$CaSO_4 \cdot 2H_2O$(s)	194.0	NO_2(g)	240.5
CH_3Cl(g)	234.2	Cl_2(g)	223.0	N_2O(g)	220.0
CH_3OH(l)	126.8	Fe(s)	27	N_2O_4(g)	304
CH_3COOH(l)	160	Fe_2O_3(s)	90.0	Na(s)	51.0
CH_4(g)	186.2	H_2(g)	130.6	Na_2CO_3(s)	136
C_2H_2(g)	200.8	H_2O(g)	188.7	$NaHCO_3$(s)	102
C_2H_4(g)	219.8	H_2O(l)	69.96	NaCl(s)	72.38
C_2H_6(g)	229.5	HCl(g)	186.7	NaOH(s)	64.18
C_2H_5OH(l)	161	HNO_3(l)	155.6	Na_2SO_4(s)	149.4
C_8H_{18}(l)	466.9	H_2SO_4(l)	157	O_2(g)	205.0
CO(g)	197.9	Hg(l)	76.1	PbO(s)	67.8
CO_2(g)	213.6	Hg(g)	175	S(s, rhombic)	31.9
$CO(NH_2)_2$(s)	104.6	K(s)	64.18	SO_2(g)	248.5
Ca(s)	41.4	KCl(s)	82.59	SO_3(g)	256.2
$CaCO_3$(s)	92.9				

Once we have the entropies of a variety of substances, we can calculate the **standard entropy of reaction** ($\Delta_r S^\ominus$) for chemical reactions in much the same way as we calculated $\Delta_r H^\ominus$ earlier. For the reaction:

$$aA + bB \rightarrow cC + dD$$

$$\Delta_r S^\ominus = cS_C^\ominus + dS_D^\ominus - (aS_A^\ominus + bS_B^\ominus)$$

or:

$$\Delta_r S^\ominus = \sum_i^{\text{products}} \nu_i S_i^\ominus - \sum_j^{\text{reactants}} \nu_j S_j^\ominus$$

If the reaction we are working with happens to correspond to the formation of 1 mole of a compound from its elements, the $\Delta_r S^\ominus$ that we calculate can be referred to as the **standard entropy of formation** ($\Delta_f S^\ominus$). Values of $\Delta_f S^\ominus$ are not tabulated, however; if we need them for some purpose, we can calculate them from tabulated values of $S^\ominus$.

WORKED EXAMPLE 8.11

Calculating $\Delta_r S^\ominus$ from standard entropies

Urea (a compound found in urine) is manufactured commercially from CO_2 and NH_3. One of its uses is as a fertiliser, where it reacts slowly with water in the soil to produce ammonia and carbon dioxide:

$$\underset{\text{urea}}{CO(NH_2)_2(s)} + H_2O(l) \longrightarrow CO_2(g) + 2NH_3(g)$$

The ammonia provides nitrogen for growing plants. What is the standard entropy of reaction when 1 mole of urea reacts with water at 298 K?

Analysis

We can calculate the standard entropy of reaction for the reaction using the standard entropies, $S^\ominus$, of each reactant and product. The data we need can be found in table 8.6 and are listed below.

Substance	$S^\ominus$ (J mol^{-1} K^{-1})
$CO(NH_2)_2(s)$	104.6
$H_2O(l)$	69.96
$CO_2(g)$	213.6
$NH_3(g)$	192.5

Solution

The standard entropy of reaction can be calculated from the equation:

$$\begin{aligned}\Delta S^\ominus &= (S^\ominus_{CO_2(g)} + 2S^\ominus_{NH_3(g)}) - (S^\ominus_{CO(NH_2)_2(s)} + S^\ominus_{H_2O(l)}) \\ &= (1 \times 213.6\,\text{J mol}^{-1}\,\text{K}^{-1}) + (2 \times 192.5\,\text{J mol}^{-1}\,\text{K}^{-1}) \\ &\quad - [(1 \times 104.6\,\text{J mol}^{-1}\,\text{K}^{-1}) + (1 \times 69.96\,\text{J mol}^{-1}\,\text{K}^{-1})] \\ &= (598.6\,\text{J mol}^{-1}\,\text{K}^{-1}) - (174.6\,\text{J mol}^{-1}\,\text{K}^{-1}) \\ &= 424.0\,\text{J mol}^{-1}\,\text{K}^{-1}\end{aligned}$$

Thus, the standard entropy of reaction for this reaction is +424.0 J mol^{-1} K^{-1}.

Is our answer reasonable?

In the reaction, gases are formed from liquid reactants. Since gases have much larger entropies than liquids, we expect $\Delta_r S^\ominus$ to be positive, which agrees with our answer.

PRACTICE EXERCISE 8.11

Calculate the standard entropy of reaction, $\Delta_r S^\ominus$, for each of the following reactions.

(a) $CaCO_3(s) \rightarrow CaO(s) + CO_2(g)$

(b) $C_2H_2(g) + 2H_2(g) \rightarrow C_2H_6(g)$

As there is no standard temperature, the temperature is often specified as a subscript on the symbol for standard thermodynamic quantities: for example, $\Delta S^\ominus_{298\,K}$. As you will see later, there are times when it is desirable to indicate the temperature explicitly.

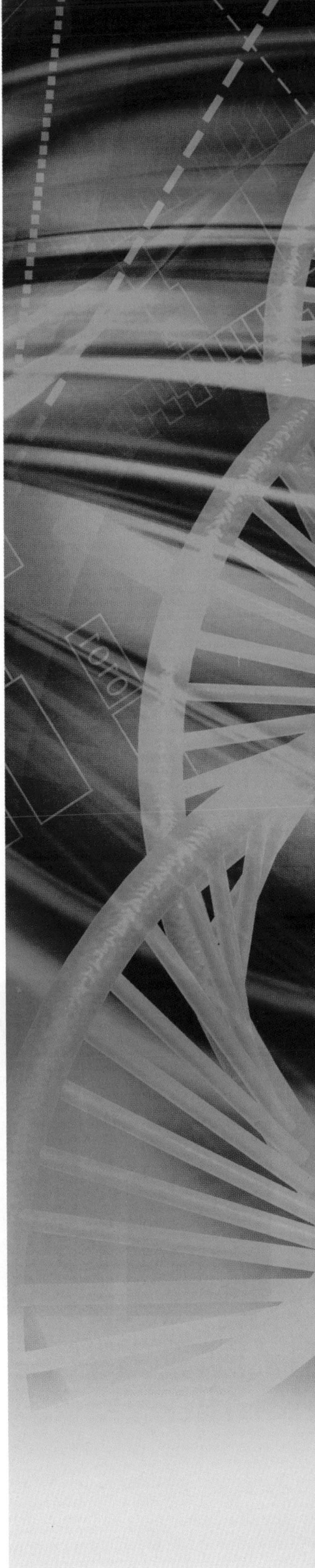

Chemical Connections

Airconditioning and the second law of thermodynamics

Regardless of where you live, Queensland, Queenstown or beyond, chances are it will be either too hot or too cold inside your house at some stage of the day or night. While it is easy to heat the interior of a house using central heating or a radiant heater, cooling it presents us with a more difficult problem — somehow we need to get rid of unwanted heat, and this is difficult when it's hotter outside than inside. Of course, we all know that the interior of a house can be cooled by using an airconditioner (figure 8.23), but let's stop and think about what actually happens when we do this.

Let's assume that it's a scorching 45 °C outside while the inside temperature of our house is a tropical 35 °C. The normal direction of heat flow will be from the outside to the inside; if we were to open all the windows, the inside temperature would increase and eventually become the same as that outside, a spontaneous process precisely in accord with what we have learned in this chapter. The reverse process, heat flowing from the 35 °C interior of the house (the system) to the 45 °C exterior (the surroundings), is nonspontaneous, and we must therefore use an airconditioner to drive this. In terms of entropy, we can show, from the equation $\Delta S = \frac{\Delta H}{T}$, that adding heat to, or removing heat from, a cool system gives a greater entropy change than the analogous process with a hotter system. This means that cooling the house will lower the entropy of the inside by more than it will increase the entropy of the outside; it may therefore appear that the overall process leads to a *decrease* in entropy of the universe, violating the second law of thermodynamics. Of course, this is not the case, as we have neglected to include the airconditioner itself in this discussion. It uses electrical energy to do work by compressing and expanding a fluid that carries the heat from the inside to the outside. In so doing, the airconditioner itself provides an entropy contribution, and the sum of this and the entropy increase of the surroundings is greater than the entropy decrease of the system, thereby ensuring that the overall entropy of the universe increases and that the second law of thermodynamics remains intact.

Heat pumps operate on the same principle but in the opposite direction, taking heat from the chilly outside and delivering it to the warmer inside, and the same arguments concerning the total entropy change apply.

FIGURE 8.23 Airconditioners use electrical energy to compress and expand a fluid that carries heat from the inside of a building to the outside. In doing so, the airconditioner provides an entropy contribution to the system, and the overall entropy of the universe increases.

8.8 Gibbs energy and reaction spontaneity

Now that we have defined enthalpy and entropy, we are ready to look in detail at Gibbs energy. In section 8.2, we stated that a process is spontaneous if ΔG for the process is negative at constant temperature and pressure. We will begin by looking at the factors that determine the sign of ΔG.

The sign of ΔG

We might naively expect, as did a number of nineteenth-century scientists, that any chemical process that evolves heat will be spontaneous, as it will lead to a lowering in energy. However, we can show that this is not the case by considering the melting of ice under standard conditions at 25.0 °C, a process that is spontaneous. $\Delta H^\ominus$ for this process is approximately 6 kJ mol^{-1}, meaning it is endothermic. Given the expression for $\Delta G^\ominus$:

$$\Delta G^\ominus = \Delta H^\ominus - T\Delta S^\ominus$$

and remembering that $\Delta G^\ominus$ must be negative for the process to be spontaneous under standard conditions, we can see that, in this case, if $\Delta H^\ominus$ is positive, the product of T and $\Delta S^\ominus$ must be *more* positive to ensure that $\Delta G^\ominus$ is negative. As T is always positive, $\Delta S^\ominus$ *must* be positive for the melting of ice to be spontaneous. (We have seen in section 8.5 that liquids generally have larger entropies than solids, so we would expect $\Delta S^\ominus$ for the melting of ice to be positive.) You can see from this example that the signs of $\Delta H^\ominus$ and $\Delta S^\ominus$ are crucial in determining the spontaneity of physical and chemical processes under standard conditions.

The combustion of octane, like all combustion reactions (p. 318), is exothermic. There is also a large increase in entropy since the number of particles in the system increases. For this change, ΔH is negative and ΔS is positive, both of which favour spontaneity.

$$\begin{aligned}&\Delta H \text{ is negative } (-\text{ve})\\ &\Delta S \text{ is positive } (+\text{ve})\\ &\Delta G = \Delta H - T\Delta S\\ &\quad\;\; = (-\text{ve}) - (+\text{ve})\\ &\quad\;\; = (-\text{ve})\end{aligned}$$

Notice that, regardless of the thermodynamic temperature, which must have a positive value, ΔG will be negative. This means that such a change must be spontaneous at any temperature. In fact, once started, fires continue to consume all available fuel or oxygen because combustion reactions are always spontaneous (see figure 8.24a).

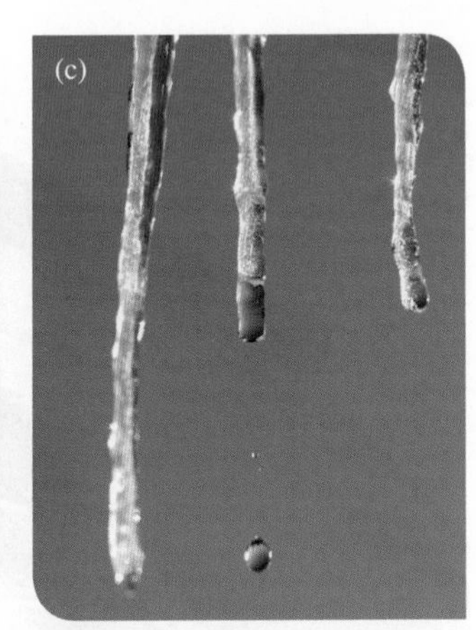

FIGURE 8.24 Process spontaneity can be predicted if ΔH, ΔS and T are known. **(a)** When ΔH is negative and ΔS is positive, as in any combustion reaction, the reaction is spontaneous at any temperature. **(b)** When ΔH is positive and ΔS is negative, the process is not spontaneous. Wood does not spontaneously re-form from ash, carbon dioxide and water. **(c)** When ΔH and ΔS have the same sign, temperature determines whether the process is spontaneous. Water spontaneously becomes ice below 0 °C, and ice spontaneously melts into liquid water above 0 °C.

When a change is endothermic and is accompanied by a lowering of the entropy, both factors work against spontaneity.

$$\begin{aligned}&\Delta H \text{ is positive } (+\text{ve})\\ &\Delta S \text{ is negative } (-\text{ve})\\ &\Delta G = \Delta H - T\Delta S\\ &\quad\;\; = (+\text{ve}) - (-\text{ve})\\ &\quad\;\; = (+\text{ve})\end{aligned}$$

Now, no matter what the temperature is, ΔG will be positive and the change must be nonspontaneous. An example would be carbon dioxide and water recombining to form wood and oxygen after a fire. If you saw such a thing happen on a film, experience would tell you that the film was being played backwards.

When ΔH and ΔS have the same algebraic sign, the temperature becomes the determining factor in controlling spontaneity. If ΔH and ΔS are both positive:

$$\begin{aligned}\Delta G &= (+\text{ve}) - (+\text{ve}) \\ &= (+\text{ve}) \text{ or } (-\text{ve})\end{aligned}$$

ΔG is the difference between two positive quantities, ΔH and $T\Delta S$. This difference will be negative only if the term $T\Delta S$ is larger than ΔH, and this will be true only when the temperature is high. (*Note:* We assume that the values of ΔH and ΔS change negligibly with change in temperature.) In other words, when ΔH and ΔS are both positive, the change will be spontaneous at high temperature but not at low temperature. An example we have already seen is the melting of ice.

$$H_2O(s) \longrightarrow H_2O(l)$$

This is an endothermic change that is accompanied by an increase in entropy. We know that at high temperature (above 0 °C) melting is spontaneous, but at low temperature (below 0 °C) it is not.

For similar reasons, when ΔH and ΔS are both negative, ΔG will be negative (and the change spontaneous) only at low temperature.

$$\begin{aligned}\Delta G &= (-\text{ve}) - (-\text{ve}) \\ &= (-\text{ve}) \text{ or } (+\text{ve})\end{aligned}$$

Only when the negative value of ΔH is larger than the negative value of $T\Delta S$ will ΔG be negative. Such a change is spontaneous only at low temperature. An example is the freezing of water.

$$H_2O(l) \longrightarrow H_2O(s)$$

This is an exothermic change that is accompanied by a decrease in entropy; it is spontaneous only at low temperatures (below 0 °C).

Figure 8.25 summarises the effects of the signs of ΔH and ΔS on that of ΔG, and hence on the spontaneity of physical and chemical processes. It should be noted at this point that processes having positive values of ΔG are sometimes termed **endergonic**, while those having negative values are called **exergonic**.

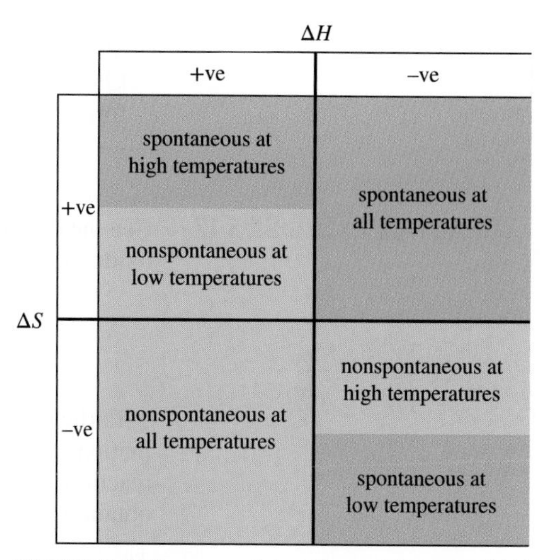

FIGURE 8.25 Summary of the effects of the signs of ΔH and ΔS on spontaneity.

Standard Gibbs energy change

When ΔG is determined at 10^5 Pa, we call it the **standard Gibbs energy change** ($\Delta G^\ominus$). There are several ways of obtaining $\Delta_r G^\ominus$ for a reaction. One is to calculate $\Delta_r G^\ominus$ from $\Delta_r H^\ominus$ and $\Delta_r S^\ominus$.

$$\Delta_r G^\ominus = \Delta_r H^\ominus - T\Delta_r S^\ominus$$

WORKED EXAMPLE 8.12

Calculating $\Delta_r G^\ominus$ from $\Delta_r H^\ominus$ and $\Delta_r S^\ominus$

Calculate $\Delta_r G^\ominus$ for the reaction of urea with water at 25.0 °C from values of $\Delta_r H^\ominus$ and $\Delta_r S^\ominus$ at this temperature.

$$CO(NH_2)_2(s) + H_2O(l) \longrightarrow CO_2(g) + 2NH_3(g)$$

Analysis

We can calculate $\Delta_r G^\ominus$ using the equation.

$$\Delta_r G^\ominus = \Delta_r H^\ominus - T\Delta_r S^\ominus$$

The data needed to calculate $\Delta_r H^\ominus$ come from table 8.3 and require a Hess's law calculation. To obtain $\Delta_r S^\ominus$, we would normally need to do a similar calculation with data from table 8.6. However, we already calculated this in worked example 8.11.

Solution

First we calculate $\Delta_r H^\ominus$ from data in table 8.3.

$$\begin{aligned}\Delta H^\ominus &= (\Delta_f H^\ominus_{CO_2(g)} + 2\Delta_f H^\ominus_{NH_3(g)}) - (\Delta_f H^\ominus_{CO(NH_2)_2(s)} + \Delta_f H^\ominus_{H_2O(l)}) \\ &= [1 \times (-393.5 \text{ kJ mol}^{-1}) + 2 \times (-46.19 \text{ kJ mol}^{-1})] \\ &\quad - [1 \times (-333.19 \text{ kJ mol}^{-1}) + 1 \times (-285.9 \text{ kJ mol}^{-1})] \\ &= (-485.9 \text{ kJ mol}^{-1}) - (-619.1 \text{ kJ mol}^{-1}) \\ &= +133.2 \text{ kJ mol}^{-1}\end{aligned}$$

In worked example 8.11, we found $\Delta_r S^\ominus$ to be +424.0 J mol^{-1} K^{-1}. To calculate $\Delta_r G^\ominus$, we also need to express 25.0 °C as a thermodynamic temperature by adding 273 K. Thus, 25.0 °C = 298 K. Also, we must be careful to express $\Delta_r H^\ominus$ and $T\Delta_r S^\ominus$ in the same energy units, so we change the units of the enthalpy change to J mol^{-1} to give $\Delta_r H^\ominus = +133.2 \times 10^3$ J mol^{-1}. Substituting into the equation for $\Delta_r G^\ominus$:

$$\begin{aligned}\Delta G^\ominus &= +133.2\times10^3\ \text{J mol}^{-1} - [298\ \text{K}\times(+424.0\ \text{J mol}^{-1}\ \text{K}^{-1})]\\ &= +133.2\times10^3\ \text{J mol}^{-1} - 126\times10^3\ \text{J mol}^{-1}\\ &= +7.2\times10^3\ \text{J mol}^{-1} = 7.2\ \text{kJ mol}^{-1}\end{aligned}$$

Therefore, for this reaction, $\Delta_r G^\ominus = +7.2$ kJ mol^{-1}.

Is our answer reasonable?

There's not much we can do to check the reasonableness of the answer. We just need to be sure we've done the calculations correctly and that the units are consistent. Note that, instead of changing the units of $\Delta_r H^\ominus$ to J mol^{-1}, we could also have changed the units of $\Delta S^\ominus$ to kJ mol^{-1} K^{-1} by dividing by 10^3. The most common mistake made in calculations like this is forgetting to change the units of either $\Delta_r H^\ominus$ or $\Delta_r S^\ominus$ to ensure the units are consistent.

PRACTICE EXERCISE 8.12

Use the data in tables 8.3 and 8.6 to calculate $\Delta_r G^\ominus$ for the formation of iron(III) oxide (the iron oxide in rust). The equation for the reaction is:

$$4Fe(s) + 3O_2(g) \longrightarrow 2Fe_2O_3(s)$$

Earlier in this chapter you learned that it is useful to have tabulated standard enthalpies of formation, $\Delta_f H^\ominus$, because they can be used with Hess's law to calculate $\Delta_r H^\ominus$ for many different reactions. Standard Gibbs energies of formation, $\Delta_f G^\ominus$, can be used in similar calculations to obtain $\Delta_r G^\ominus$. For the reaction:

$$aA + bB \rightarrow cC + dD$$

$$\Delta_r G^\ominus = c\Delta_f G^\ominus_C + d\Delta_f G^\ominus_D - (a\Delta_f G^\ominus_A + b\Delta_f G^\ominus_B)$$

or:

$$\Delta_r G^\ominus = \sum_i^{\text{products}} v_i \Delta_f G^\ominus_i - \sum_j^{\text{reactants}} v_j \Delta_f G^\ominus_j$$

The $\Delta_f G^\ominus$ values for some typical substances are found in table 8.7. Worked example 8.13 shows how we can use them to calculate $\Delta_r G^\ominus$ for a reaction.

WORKED EXAMPLE 8.13

Calculating $\Delta_r G^\ominus$ from $\Delta_f G^\ominus$

Ethanol, C_2H_5OH, is made from grain by fermentation and can be added to petrol to produce a product called gasohol or E10. What is $\Delta_r G^\ominus$ of the combustion of liquid ethanol to give $CO_2(g)$ and $H_2O(g)$?

Analysis

We can use the equation:

$$\Delta_r G^\ominus = c\Delta_f G^\ominus_C + d\Delta_f G^\ominus_D - (a\Delta_f G^\ominus_A + b\Delta_f G^\ominus_B)$$

to calculate the standard Gibbs energy change for the reaction. We will need the standard Gibbs energy changes of formation for each reactant and product. These data can be found in table 8.7 on the next page.

Solution

First, we need the balanced equation for the reaction:

$$C_2H_5OH(l) + 3O_2(g) \longrightarrow 2CO_2(g) + 3H_2O(g)$$

$$\Delta_r G^\ominus = (2\Delta_f G^\ominus_{CO_2(g)} + 3\Delta_f G^\ominus_{H_2O(g)}) - (\Delta_f G^\ominus_{C_2H_5OH(l)} + 3\Delta_f G^\ominus_{O_2(g)})$$

As with $\Delta_f H^{\ominus}$, $\Delta_f G^{\ominus}$ for any element in its standard state is 0. Therefore, using the data from table 8.6:

$$\Delta_r G^{\ominus} = [2 \times -(394.4)\ \text{kJ mol}^{-1} + 3 \times (-228.6)\ \text{kJ mol}^{-1}]$$
$$- [1 \times (-174.8)\ \text{kJ mol}^{-1} + 3 \times 0\ \text{kJ mol}^{-1}]$$
$$= (-1474.6\ \text{kJ mol}^{-1}) - (-174.8\ \text{kJ mol}^{-1})$$
$$= -1299.8\ \text{kJ mol}^{-1}$$

The standard Gibbs energy change for the reaction is $-1299.8\ \text{kJ mol}^{-1}$.

Is our answer reasonable?

There's no quick way to estimate the answer, although we could calculate $\Delta_r G^{\ominus}$ from $\Delta_r H^{\ominus}$ and $\Delta_r S^{\ominus}$ following the method used in worked example 8.12 for a thorough check. We know that ethanol is quite flammable under standard conditions, so we expect this combustion reaction to proceed spontaneously. The large negative standard Gibbs energy change makes sense.

PRACTICE EXERCISE 8.13

Calculate $\Delta_r G^{\circ}$ for the following reactions using the data in table 8.7.

(a) $CH_4(g) + 2O_2(g) \rightarrow CO_2(g) + {}_2H_2O(l)$

(b) $NH_3(g) + HCl(g) \rightarrow NH_4Cl(s)$

TABLE 8.7 Standard Gibbs energies of formation of selected substances at 25 °C.

Substance	$\Delta_f G^{\ominus}$ (kJ mol^{-1})	Substance	$\Delta_f G^{\ominus}$ (kJ mol^{-1})	Substance	$\Delta_f G^{\ominus}$ (kJ mol^{-1})
Ag(s)	0	$CaCl_2$(s)	−750.2	K_2SO_4(s)	−1316.4
AgCl(s)	−109.7	CaO(s)	−604.2	N_2(g)	0
Al(s)	0	$Ca(OH)_2$(s)	−896.76	NH_3(g)	−16.7
Al_2O_3(s)	−1576.4	$CaSO_4$(s)	−1320.3	NH_4Cl(s)	−203.9
C(s, diamond)	2.9	$CaSO_4\cdot\frac{1}{2}H_2O$(s)	−1435.2	NO(g)	+86.69
C(s, graphite)	0	$CaSO_4\cdot 2H_2O$(s)	−1795.7	NO_2(g)	+51.84
CH_3Cl(g)	−58.6	Cl_2(g)	0	N_2O(g)	+103.6
CH_3OH(l)	−166.2	Fe(s)	0	N_2O_4(g)	+98.28
CH_3COOH(l)	−392.5	Fe_2O_3(s)	−741.0	Na(s)	0
CH_4(g)	−50.79	H_2(g)	0	Na_2CO_3(s)	−1048
C_2H_2(g)	+209	H_2O(g)	−228.6	$NaHCO_3$(s)	−851.9
C_2H_4(g)	+68.12	H_2O(l)	−237.2	NaCl(s)	−384.0
C_2H_6(g)	−32.9	HCl(g)	−95.27	NaOH(s)	−382
C_2H_5OH(l)	−174.8	HNO_3(l)	−79.91	Na_2SO_4(s)	−1266.8
C_8H_{18}(l)	+17.3	H_2SO_4(l)	−689.9	O_2(g)	0
CO(g)	−137.3	Hg(l)	0	PbO(s)	−189.3
CO_2(g)	−394.4	Hg(g)	+31.8	S(s, rhombic)	0
$CO(NH_2)_2$(s)	−197.2	K(s)	0	SO_2(g)	−300.4
Ca(s)	0	KCl(s)	−408.3	SO_3(g)	−370.4
$CaCO_3$(s)	−1128.8				

Gibbs energy and work

One of the chief uses of spontaneous chemical reactions is the production of useful work. For example, fuels are burned in petrol and diesel engines to power cars and heavy machinery, and chemical reactions in batteries power everything from mobile phones to laptop computers.

When chemical reactions occur, however, their energy is not always harnessed to do work. For instance, if petrol is burned in an open dish, the energy evolved is lost entirely as heat and no useful work is accomplished. Engineers, therefore, seek ways to capture as much energy as possible in the form of work. One of their primary goals is to maximise the efficiency with which chemical energy is converted to work and to minimise the amount of energy transferred unproductively to the surroundings as heat.

The maximum conversion of chemical energy to work occurs if a reaction is carried out under conditions that are said to be thermodynamically reversible. A process is defined as thermodynamically reversible if its driving force is opposed by another force that is just slightly weaker, so that the slightest increase in the opposing force will cause the direction of the change to be reversed. An example of a nearly reversible process is illustrated in figure 8.26, where we have a compressed gas in a cylinder pushing against a piston that is held in place by liquid water above it. If a water molecule evaporates, the external pressure drops slightly and the gas can expand just a bit. Gradually, as one water molecule after another evaporates, the gas inside the cylinder slowly expands. At any time, however, the process can be reversed by the condensation of a water molecule.

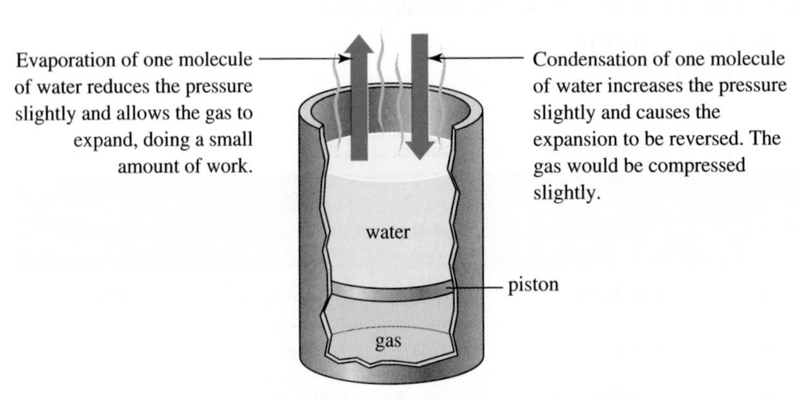

Evaporation of one molecule of water reduces the pressure slightly and allows the gas to expand, doing a small amount of work.
Condensation of one molecule of water increases the pressure slightly and causes the expansion to be reversed. The gas would be compressed slightly.
water
piston
gas

FIGURE 8.26 A reversible expansion of a gas. As water molecules evaporate one at a time, the external pressure gradually decreases and the gas slowly expands. The process would be reversed if a molecule of water were to condense into the liquid. The ability of the expansion to be reversed by the slightest increase in the opposing pressure is what makes this a reversible process.

As we will see in chapter 9, thermodynamic reversibility implies that the system is very close to equilibrium (a state in which there is no driving force for chemical or physical change) at every stage of a process. While we sometimes say that a reaction such as dissociation of a weak acid in water (chapter 11) is 'reversible', simply because it runs in both the forward and reverse directions, we cannot say the reaction is *thermodynamically reversible* unless the concentrations are only infinitesimally different from their equilibrium values as the reaction occurs.

Although we could obtain the maximum work by carrying out a change reversibly, a thermodynamically reversible process requires so many steps that it proceeds at an extremely slow speed. If the work cannot be done at a reasonable rate, it is of little value to us. Our goal, then, is to approach thermodynamic reversibility for maximum efficiency, but to carry out the change at a pace that will deliver work at acceptable rates. Note that, irrespective of the rate that we carry out the process shown in figure 8.26, we will always lose some energy as heat, due to friction. In fact, it is impossible to extract all the energy from any chemical or physical process as work, and therefore there is no such thing as a truly thermodynamically reversible process.

The relationship of useful work to thermodynamic reversibility was illustrated earlier (section 8.3, p. 303) in our discussion of the discharge of a car battery. Recall that, when the battery is short-circuited with a heavy spanner, no work is done and all the energy appears as heat. In this case, there is nothing opposing the discharge, and it occurs in the most thermodynamically irreversible manner possible. However, when the current is passed through a small electric motor, the motor itself offers resistance to the passage of the electricity and the discharge takes place slowly. In this instance, the discharge occurs in a more nearly thermodynamically reversible manner because of the opposition provided by the motor, and a relatively large amount of the available energy appears in the form of work accomplished by the motor.

The preceding discussion leads naturally to the question, 'Is there a limit to the amount of the available energy in a reaction that can be harnessed as useful (i.e. non-pV) work?' The answer to this question is yes. The limit is the Gibbs energy for the reaction.

The maximum amount of energy produced by a reaction that can be *theoretically* harnessed as non-pV work is equal to $\Delta_r G$. This is the energy that need not be lost to the surroundings as heat and is, therefore, available to be used for work. Thus, by determining the value of $\Delta_r G$, we can find out whether a given reaction will be an effective source of useful energy. Also, by comparing the actual amount of non-pV work derived from a given system with the $\Delta_r G$ values for the reactions involved, we can measure the efficiency of the system.

WORKED EXAMPLE 8.14

Calculating maximum non-pV work

Calculate the maximum non-pV work available from the oxidation of 1 mole of octane, $C_8H_{18}(l)$, by oxygen to give $CO_2(g)$ and $H_2O(l)$ at 25 °C and 10^5 Pa.

Analysis

The maximum non-pV work is equal to $\Delta_r G$ for the reaction. Standard thermodynamic conditions are specified, so we need to calculate $\Delta_r G^{\ominus}$.

Solution

First we need a balanced equation for the reaction. For the combustion of 1 mole of C_8H_{18} we have:

$$C_8H_{18}(l) + 12\tfrac{1}{2}O_2(g) \longrightarrow 8CO_2(g) + 9H_2O(l)$$

$$\Delta_r G^{\ominus} = (8\Delta_f G^{\ominus}_{CO_2(g)} + 9\Delta_f G^{\ominus}_{H_2O(l)}) - (\Delta_f G^{\ominus}_{C_8H_{18}(l)} + 12\tfrac{1}{2}\Delta_f G^{\ominus}_{O_2(g)})$$

Referring to table 8.7:

$$\begin{aligned}\Delta_r G^{\ominus} &= 8 \times (-394.4\ \text{kJ mol}^{-1}) + 9 \times (-237.2\ \text{kJ mol}^{-1}) \\ &\quad - [1 \times (+17.3\ \text{kJ mol}^{-1}) + 12\tfrac{1}{2} \times 0\ \text{kJ mol}^{-1}] \\ &= (-5290\ \text{kJ mol}^{-1}) - (+17.3\ \text{kJ mol}^{-1}) \\ &= -5307\ \text{kJ mol}^{-1}\end{aligned}$$

Thus, at 25 °C and 10^5 Pa, we can expect no more than 5307 kJ of non-pV work from the oxidation of 1 mole of C_8H_{18}.

Is our answer reasonable?

Be sure to check the algebraic signs of each of the terms in the calculation. This is a combustion reaction, so we would expect it to be spontaneous and hence have a negative $\Delta_r G$ value, which it does.

PRACTICE EXERCISE 8.14

Calculate the maximum non-pV work that could be obtained at 25 °C and 10^5 Pa from the oxidation of 1.00 mole of aluminium by $O_2(g)$ to give $Al_2O_3(s)$. (Aluminium was one of the components of the booster rocket that was used to launch the world's first successful scramjet flight in Australia.)

Gibbs energy and equilibrium

We have seen that, when the value of ΔG for a given change is negative, the change occurs spontaneously. We have also seen that a change is nonspontaneous when ΔG is positive. However, when ΔG is neither positive nor negative, the change is neither spontaneous nor nonspontaneous — the system is in a state of equilibrium. This occurs when ΔG is equal to 0. When a system is in a state of equilibrium in which two opposing physical processes or chemical reactions are occurring at the same rate (we call this a dynamic equilibrium) at constant temperature and pressure.

$$G_{\text{products}} = G_{\text{reactants}} \text{ and } \Delta G = 0$$

Let's again consider the freezing of water at $p = 1.013 \times 10^5$ Pa.

$$H_2O(l) \rightleftharpoons H_2O(s)$$

Below 0 °C, ΔG for this change is negative and freezing is spontaneous, that is, we would write $H_2O(l) \rightarrow H_2O(s)$. On the other hand, above 0 °C we find that ΔG is positive and freezing is nonspontaneous, that is, we would write $H_2O(s) \rightarrow H_2O(l)$. When the temperature is exactly 0 °C, $\Delta G = 0$ and an ice–water mixture exists in a condition of equilibrium. As long as heat is not added or removed from the system, neither overall freezing nor melting is spontaneous and the ice and liquid water can exist together indefinitely. In this state of dynamic equilibrium, the rate at which liquid water freezes to give ice is exactly equal to the rate at which ice melts to give liquid water.

We have identified ΔG as a quantity that specifies the amount of non-pV work available from a system. Since $\Delta G = 0$ at equilibrium, the amount of non-pV work available is 0 also. Therefore, when a system is at equilibrium, no non-pV work can be extracted from it. As an example, we will consider again the common lead storage battery that we use to start a car.

When a battery is fully charged, there are virtually no products of the discharge reaction present. The chemical reactants, however, are present in large amounts. Therefore, the total Gibbs energy of the reactants far exceeds the total Gibbs energy of products and, since:

$$\Delta_r G = G_{\text{products}} - G_{\text{reactants}}$$

the $\Delta_r G$ of the system has a large negative value. This means that a lot of energy is available to do non-pV work. As the battery discharges, the reactants are converted to products and G_{products} gets larger while $G_{\text{reactants}}$ gets smaller, so $\Delta_r G$ becomes less negative, and less energy is available to do non-pV work. Finally, the battery reaches equilibrium. The total free energies of the reactants and the products have become equal, so:

$$G_{\text{products}} - G_{\text{reactants}} = 0 \text{ and } \Delta_r G = 0$$

No further non-pV work can be extracted and we say the battery is flat.

When we have equilibrium in any system, we know that $\Delta G = 0$. For a phase change such as $H_2O(l) \rightarrow H_2O(s)$, equilibrium can be established only at one particular temperature at atmospheric pressure. In this instance, that temperature is 0 °C. Above 0 °C, only liquid water can exist, and below 0 °C all the liquid will freeze to give ice. This yields an interesting relationship between ΔH and ΔS that applies not only to phase changes but also to chemical reactions. Since at equilibrium, $\Delta G = 0$:

$$\Delta G = 0 = \Delta H - T\Delta S$$

Therefore:

$$\Delta H = T\Delta S$$

and:

$$\Delta S = \frac{\Delta H}{T}$$

Thus, if we know ΔH for a chemical reaction or phase change and the temperature at which equilibrium is established, we can calculate ΔS. We can also rearrange the above equation to give:

$$T = \frac{\Delta H}{\Delta S}$$

Thus, if we know ΔH and ΔS, we can calculate the temperature at which equilibrium will be established.

WORKED EXAMPLE 8.15

Calculating the equilibrium temperature for a phase change

For the phase change $Br_2(l) \rightarrow Br_2(g)$, $\Delta H^\circ = +30.9\,\text{kJ mol}^{-1}$ and $\Delta S^\circ = 93.2\,\text{J mol}^{-1}\,\text{K}^{-1}$ at 25 °C. Assuming that ΔH and ΔS are temperature independent, calculate the approximate temperature at which $Br_2(l)$ will be in equilibrium with $Br_2(g)$ at 10^5 Pa (i.e. the normal boiling point of liquid Br_2).

Analysis

The temperature at which equilibrium exists is given by:

$$T = \frac{\Delta H}{\Delta S}$$

We are working under standard conditions (10^5 Pa), so we can use ΔH° and ΔS° in this equation. That is:

$$T = \frac{\Delta H^\circ}{\Delta S^\circ}$$

Solution

Substituting the data given in the problem:

$$T = \frac{30.9 \times 10^3\,\text{J mol}^{-1}}{93.2\,\text{J mol}^{-1}\,\text{K}^{-1}}$$

$$= 332\,\text{K}$$

$$= (332 - 273)\,°\text{C} = 59\,°\text{C}$$

Notice that we must express ΔH° in $J\,mol^{-1}$, not $kJ\,mol^{-1}$, so the units cancel correctly. It is also interesting that the boiling point we calculate is quite close to the measured normal boiling point of 58.8 °C.

Is our answer reasonable?

ΔH° equals 30 900 and ΔS° equals approximately 100, which means the temperature should be about 310 K. Our value, 332 K, is not far from that, so the answer is reasonable.

PRACTICE EXERCISE 8.15

The standard enthalpy of vaporisation of silicon tetrachloride ($SiCl_4$) is $28.7\,kJ\,mol^{-1}$. For $SiCl_4(l)$, $S^\circ_{298\,K} = 239.3\,J\,mol^{-1}\,K^{-1}$ and for $SiCl_4(g)$, $S^\circ_{298\,K} = 331.4\,J\,mol^{-1}\,K^{-1}$. Estimate the normal boiling point of liquid $SiCl_4$.

We can also use the equation $T = \frac{\Delta H}{\Delta S}$ to determine the temperature at which a nonspontaneous reaction first becomes spontaneous. To illustrate this, consider the decomposition of calcium carbonate under standard conditions at 25 °C, which occurs according to the equation:

$$CaCO_3(s) \rightarrow CaO(s) + CO_2(g)$$

Using the $\Delta_f H^\circ$ and S° data in tables 8.3 and 8.6, or the $\Delta_f G^\circ$ data in table 8.7, we can calculate that $\Delta G^\circ = 130\,kJ\,mol^{-1}$ for this reaction at 25 °C. Therefore, the reaction is nonspontaneous at this temperature under standard conditions. However, we showed on page 332 that the spontaneity of reactions with positive ΔH and ΔS values depends on the temperature, and that such reactions are generally spontaneous at high temperature. We can substitute the values of $\Delta_r H^\circ = +178\,kJ\,mol^{-1}$ and $\Delta S^\circ = 160.7\,J\,mol^{-1}\,K^{-1}$ for this reaction into the equation:

$$T = \frac{\Delta_r H^\circ}{\Delta_r S^\circ} = \frac{178 \times 10^3\,J\,mol^{-1}}{160.7\,J\,mol^{-1}\,K^{-1}} = 1.11 \times 10^3\,K$$

This shows that the system will be in equilibrium at 1110 K under standard conditions (assuming that $\Delta_r H^\circ$ and $\Delta_r S^\circ$ are independent of temperature) and that the forward reaction will be spontaneous at any temperature greater than this. We summarise this in figure 8.27.

In this chapter, we have shown that ΔG can be used as the criterion for spontaneity of a chemical reaction or phase change. Furthermore, we have shown that the sign of ΔG depends on the signs of both ΔH and ΔS and that, when $\Delta G = 0$, there is no driving force for chemical or physical change and the system is at equilibrium. We will investigate chemical equilibrium in more detail in the next chapter, and we will see how the value of $\Delta_r G$ measured under standard conditions, $\Delta_r G^\circ$, is related to the extent of reaction.

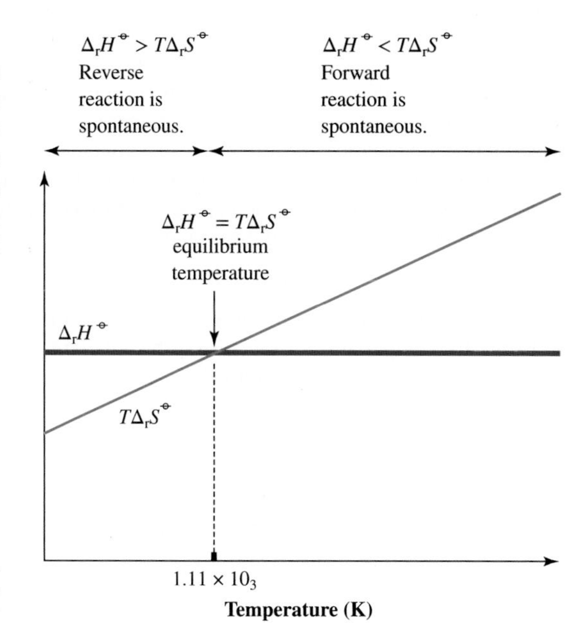

FIGURE 8.27 Diagram showing the relative values of $\Delta_r H^\circ$ and $T\Delta_r S^\circ$ for the decomposition of calcium carbonate. At low temperatures, $\Delta_r H^\circ$ (blue line) is greater than $T\Delta_r S^\circ$ (red line). This means that $\Delta_r G^\circ$ will be positive under these conditions and the reaction will be nonspontaneous. As the temperature increases, the value of $T\Delta_r S^\circ$ increases and eventually, at $T = 1.11 \times 10^3\,K$, $\Delta_r H^\circ = T\Delta_r S^\circ$ and the system is at equilibrium. At any temperature greater than $1.11 \times 10^3\,K$, $\Delta_r H^\circ$ is less than $T\Delta_r S^\circ$. Under these conditions, $\Delta_r G^\circ$ will be negative and the decomposition reaction will be spontaneous.

SUMMARY

LO 1 Introduction to chemical thermodynamics

A spontaneous change occurs without outside assistance, whereas a nonspontaneous change requires continuous help and can occur only if it is accompanied by, and linked to, some other spontaneous event.

The science of chemical thermodynamics allows us to predict both the direction and extent of spontaneous chemical and physical change under particular conditions.

LO 2 Thermodynamic concepts

We call the object we are interested in the 'system', which is enclosed by a 'boundary'. Everything else in the 'universe' is the system's 'surroundings'.

Chemical thermodynamics uses the SI unit of energy, work and heat, the joule (J) or the kilojoule (kJ). Many thermodynamic equations require the temperature to be expressed in kelvin. Standard states for chemical substances depend on the phase of the substance, all standard states involve $p^{\ominus}$ (1×10^5 pa). Standard thermodynamic functions refer to conversion of unmixed reactants in their standard states to unmixed products in their standard states. Such functions are denoted by a superscript plimsoll ($^{\ominus}$) and are said to be determined under standard conditions. Note there is no standard temperature.

Gibbs energy, G, is defined by the equation $G = H - TS$, where H is the enthalpy of the system, S is the entropy of the system and T is the thermodynamic temperature. We use ΔG to symbolise the change in Gibbs energy that occurs when a chemical system changes from one state to another due to a reaction or physical change at constant temperature and pressure. The sign of ΔG allows us to determine whether a particular chemical reaction or physical change is spontaneous.

LO 3 The first law of thermodynamics

The internal energy U, is the sum of all the nuclear, electronic, vibrational, rotational, translational and interaction energies of all of the individual particles in a sample of matter. It is an extensive property, and the change in internal energy, $\Delta_r U$, for a chemical reaction is given by the equation:

$$\Delta_r U = U_{\text{products}} - U_{\text{reactants}}$$

The value of the state function, internal energy, depends only on the current condition or state of the system. Temperature is an intensive property that determines the direction of spontaneous heat flow. Heat, q, is a transfer of energy from an object at a high temperature to one at a lower temperature and can be determined by measuring temperature changes, ΔT, using the equation:

$$q = C\Delta T$$

where C is the heat capacity of the system: that is, the heat needed to change the temperature of the system by 1 kelvin. The specific heat (c) of a pure substance is defined as:

$$c = \frac{C}{m}$$

Heat capacity is specific heat multiplied by mass. Heat capacity calculated on a per mole basis is called molar heat capacity. When the mass and specific heat of an object are known, heat can be calculated using the equation:

$$q = cm\Delta T$$

Heats of reaction are measured at constant volume or at constant pressure. When a gas evolves and pushes against the atmosphere or causes a piston to move in a cylinder, pressure–volume work is being done. When the volume change, ΔV, occurs at constant opposing pressure, p, the associated pressure–volume work is given by $w = -p\Delta V$. The energy expended in doing this pressure–volume work causes heats of reaction measured at constant volume, q_V, to differ numerically from heats measured at constant pressure, q_p.

The first law of thermodynamics states that:

$$\Delta U = q + w$$

The algebraic signs for q and w are negative when the system gives heat to or does work on the surroundings. The signs are positive when the system absorbs heat or receives work energy done to it. When a closed system has rigid and immovable walls, no pressure–volume work can be done to or by the system so $\Delta U = q_V$. Values of q_V are determined using a bomb calorimeter.

LO 4 Enthalpy

When a system is under constant pressure (e.g. open to the atmosphere), pressure–volume work is possible and a new thermodynamic property called enthalpy, H, can be defined.

$$H = U + pV$$

Like U, absolute values of H cannot be easily measured or calculated, but the difference in enthalpy between reactants and products can be determined. The heat of reaction is now called the enthalpy change, ΔH. Under constant pressure:

$$\Delta H = \Delta U + p\Delta V = q_p$$

While ΔH is a state function, its value differs from that of ΔU by the work involved in interacting with the surroundings when the change occurs at constant pressure. In general, the difference between ΔU and ΔH is quite small.

Exothermic reactions have negative values of $\Delta_r H$; endothermic changes have positive values. A balanced chemical equation that includes both the enthalpy change and the physical states of the substances is called a thermochemical equation. These can be added, reversed (reversing also the sign of the enthalpy change) or multiplied by a constant multiplier (doing the same to the enthalpy change). If substances are cancelled or added, they must be in identical physical states.

Thermochemical equations can be added because enthalpy is a state function. Values of $\Delta_r H^{\ominus}$ can be determined by the manipulation of any combination of thermochemical equations that add up to the final net equation.

An enthalpy change measured under standard conditions at the specified temperature T is called the standard enthalpy of reaction, $\Delta_r H^{\ominus}$. The value of $\Delta_r H^{\ominus}$ is a function of the amounts of substances involved in the reaction. If the amount of reactants is doubled, the standard enthalpy of the reaction doubles.

The enthalpy change for the formation of 1 mole of a substance under standard conditions from its elements in their standard states is called the standard enthalpy of formation, $\Delta_f H^{\ominus}$, and it is generally given in units of kilojoules per mole (kJ mol^{-1}). Hess's law is possible because enthalpy is a state function. The value of $\Delta_r H^{\ominus}$ for a reaction can be calculated as the sum of the $\Delta_f H^{\ominus}$ values of the products minus the sum of the $\Delta_f H^{\ominus}$ values of the reactants, each multiplied by the corresponding stoichiometric coefficient in the equation.

The enthalpy change for the complete combustion of 1 mole of a pure substance under standard conditions is called the standard enthalpy of combustion, $\Delta_c H^{\ominus}$.

The bond enthalpy is the enthalpy change on breaking 1 mole of a particular chemical bond to give electrically neutral fragments. The sum of all the bond enthalpies in a molecule is the atomisation enthalpy, $\Delta_{at} H$, and, on a mole basis, it corresponds to the energy needed to break 1 mole of molecules into individual atoms.

LO 5 Entropy

Spontaneity is associated with statistical probability; spontaneous processes tend to proceed from lower probability to higher probability states. The thermodynamic quantity associated with the probability of a state is entropy, S. Entropy is a measure of the number of energetically equivalent ways a state can be realised.

LO 6 The second law of thermodynamics

The second law of thermodynamics states that the entropy of the universe increases whenever a spontaneous change occurs.

In general, gases have much higher entropies than liquids, which have somewhat higher entropies than solids. Entropy increases with volume for a gas and with the temperature. During a chemical reaction, the entropy tends to increase if the number of molecules increases on going from reactants to products.

LO 7 The third law of thermodynamics

The third law of thermodynamics states that the entropy of a pure crystalline substance is equal to 0 at absolute zero (0 K). Because we know where the zero point is on the entropy scale, it is possible to measure absolute entropies. Standard entropies, $S^\ominus$, are calculated at 10^5 Pa (table 8.6) and can be used to calculate $\Delta S^\ominus$ for chemical reactions.

LO 8 Gibbs energy and reaction spontaneity

Gibbs energy is defined as $G = H - TS$. A chemical or physical change at constant temperature and pressure is spontaneous only if the Gibbs energy of the system decreases (ΔG is negative). When ΔH and ΔS have the same algebraic sign, the temperature becomes the critical factor in determining spontaneity.

When ΔG is measured under standard conditions, it is the standard Gibbs energy change, $\Delta G^\ominus$. As with enthalpy changes, the standard Gibbs energies of formation, $\Delta_f G^\ominus$, (table 8.7) can be used to obtain $\Delta_r G^\ominus$ for chemical reactions by a Hess's law type of calculation.

For any system, the value of ΔG is equal to the maximum amount of energy that can be obtained in the form of useful non-pV work. This maximum work can be obtained only if the change takes place reversibly. All real changes are irreversible and we always obtain less work than is theoretically available; the rest is lost as heat.

When a system reaches equilibrium, $\Delta G = 0$ and no useful work can be obtained from it. At any particular pressure, an equilibrium between two phases of a substance (e.g. liquid–solid or solid–vapour) can occur at only one temperature. The entropy change can be calculated as $\Delta S = \frac{\Delta H}{T}$. The temperature at which the equilibrium occurs can be calculated from $T = \frac{\Delta H}{\Delta S}$.

KEY CONCEPTS AND EQUATIONS

Heat capacity and specific heat *(section 8.3)*

The equation:

$$q = C\Delta T$$

(q = heat, C = heat capacity, ΔT = temperature change) can be used to calculate the heat of a reaction. Specific heat (c) is defined as:

$$c = C/m$$

Thermochemical equations *(section 8.4)*

We can use thermochemical equations for one set of reactions to write a thermochemical equation for some net reaction.

Hess's law *(section 8.4)*

Hess's law states that the overall enthalpy change for a reaction is the same regardless of how it is carried out. This enables us to use standard enthalpies of formation to calculate the enthalpy of a reaction.

$\Delta_f H^\ominus$ and $S^\ominus$ *(sections 8.4, 8.7)*

Standard enthalpies of formation and standard entropies can be used to calculate, respectively, $\Delta_r H^\ominus$ and $\Delta_r S^\ominus$ for a reaction.

ΔG as a predictor of spontaneity *(section 8.8)*

If $\Delta_r G$ for a reaction is negative, the reaction is spontaneous under the defined conditions.

Calculating $\Delta G^\ominus$ from $\Delta H^\ominus$ and $\Delta S^\ominus$ *(section 8.8)*

$\Delta G^\ominus$ values can be calculated from values of $\Delta H^\ominus$, $\Delta S^\ominus$ and T through the equation:

$$\Delta G^\ominus = \Delta H^\ominus - T\Delta S^\ominus$$

Using $\Delta_f G^\ominus$ values *(section 8.8)*

Standard Gibbs energy of formation values can be used to calculate $\Delta_r G^\ominus$ for a reaction.

KEY TERMS

atomisation enthalpy ($\Delta_{at}H$) *p. 319*
bomb calorimeter *p. 306*
bond enthalpy *p. 318*
boundary *p. 300*
calorimeter *p. 306*
chemical thermodynamics *p. 299*
combustion reaction *p. 306*
endergonic *p. 333*
endothermic *p. 308*
enthalpy (H) *p. 299*
entropy (S) *p. 299*
exergonic *p. 333*
exothermic *p. 308*
extensive property *p. 305*
first law of thermodynamics *p. 303*
Gibbs energy (G) *p. 299*
heat (q) *p. 299*
heat capacity (C) *p. 304*
heat of reaction at constant pressure (q_p) *p. 308*
heat of reaction at constant volume (q_v) *p. 306*
Hess's law *p. 313*
intensive property *p. 305*
internal energy (U) *p. 299*
joule (J) *p. 300*
kilojoule (kJ) *p. 300*
molar heat capacity *p. 305*
second law of thermodynamics *p. 327*
specific heat (c) *p. 305*
specific heat capacity (c) *p. 305*
spontaneous *p. 298*
standard enthalpy of combustion ($\Delta_c H^\ominus$) *p. 318*
standard enthalpy of formation ($\Delta_f H^\ominus$) *p. 314*
standard enthalpy of reaction ($\Delta_r H^\ominus$) *p. 310*
standard entropy ($S^\ominus$) *p. 329*
standard entropy of formation ($\Delta_f S^\ominus$) *p. 329*
standard entropy of reaction ($\Delta_r S^\ominus$) *p. 329*
standard Gibbs energy change ($\Delta G^\ominus$) *p. 333*
state function *p. 301*
surroundings *p. 300*
system *p. 300*
thermochemical equation *p. 310*
thermodynamic temperature *p. 299*
third law of thermodynamics *p. 328*
universe *p. 300*
work (w) *p. 303*

REVIEW QUESTIONS

LO 1 Introduction to chemical thermodynamics

8.1 What is a spontaneous change? How can nonspontaneous processes be made to occur?

LO 2 Thermodynamic concepts

8.2 Define the terms *system, surroundings* and *universe,* as they are used in thermodynamics.

8.3 What is a state function? Give two examples.

LO 3 The first law of thermodynamics

8.4 What kinds of energies contribute to the internal energy of a system?

8.5 Write the equation for the first law of thermodynamics. To what type of systems does this equation apply? What are the only two ways in which energy can be transferred between these systems and their surroundings?

8.6 Define heat capacity in words.

8.7 If the same amount of heat is applied to 1 kg of water and 1 kg of iron, which are initially at the same temperature, for the same amount of time, will they still have the same temperature after the heat has been applied? If not, which will have the higher temperature? Explain your answer in terms of specific heat.

8.8 How do heat capacity and specific heat differ?

8.9 If the specific heat values in table 8.2 were in units of $kJ\,kg^{-1}\,K^{-1}$, would the values be numerically different? Explain.

8.10 Is the density of a pure chemical substance an extensive or intensive property? Explain your answer.

8.11 Why can no pV work be done in a bomb calorimeter?

LO 4 Enthalpy

8.12 Give the equation that defines ΔH in terms of internal energy, pressure and volume.

8.13 Why does the calorimeter in figure 8.10 consist of polystyrene cups rather than, for instance, a glass beaker?

8.14 What is the sign of ΔH for an endothermic change?

8.15 In numerical terms, what is the major difference between $\Delta_r H$ and $\Delta_r H^{\ominus}$ for a particular chemical reaction at temperature T?

8.16 What distinguishes a thermochemical equation from an ordinary chemical equation?

8.17 What fundamental property of $\Delta_r H$ makes Hess's law possible?

8.18 If $\Delta_r H^{\ominus}$ for a particular chemical reaction is x kJ mol^{-1}, what will be the value of $\Delta_r H^{\ominus}$ if the balanced chemical equation is reversed and the stoichiometric coefficients are multiplied by 3?

8.19 What conditions must be met by a thermochemical equation so that its standard enthalpy change can be given the symbol $\Delta_f H^{\ominus}$?

8.20 How can Hess's law be expressed in terms of standard enthalpies of formation?

8.21 Write Hess's law in terms of standard enthalpies of combustion.

8.22 Define atomisation enthalpy, $\Delta_{at}H$.

8.23 Why are lattice enthalpies always positive?

LO 5 Entropy

8.24 Give a definition of entropy in terms of the distribution of energy in a system.

8.25 Will the entropy change for each of the following be positive or negative?
(a) Moisture condenses on the outside of a cold glass.
(b) Raindrops form in a cloud.
(c) Air is pumped into a tyre.
(d) Frost forms on the windscreen of your car.
(e) Sugar dissolves in coffee.

8.26 Why should an increase in the temperature of a system lead to an increase in entropy of the system?

8.27 Consider two systems, one of which comprises 1 mol of ice, and the other, 1 mol of liquid water, at exactly 0 °C. Would you expect the entropies of the two systems to be the same? Explain your answer.

LO 6 The second law of thermodynamics

8.28 State the second law of thermodynamics.

8.29 How can a chemical or physical process lead to a negative entropy change for the system under study and still be spontaneous?

LO 7 The third law of thermodynamics

8.30 State the third law of thermodynamics.

8.31 Would you expect the entropy of an alloy (a solution of two metals) to be 0 at 0 K? Explain your answer.

8.32 Does $S = 0$ at 0 K for glass? Explain.

LO 8 Gibbs energy and reaction spontaneity

8.33 What is the equation expressing the change in the Gibbs energy for a reaction occurring at constant temperature and pressure?

8.34 For each of the following conditions, state in terms of the algebraic signs of ΔH and ΔS the circumstances under which a change will be spontaneous.
(a) at all temperatures
(b) at low temperatures but not at high temperatures
(c) at high temperatures but not at low temperatures

8.35 Under what circumstances will a change be nonspontaneous regardless of the temperature?

8.36 Define the terms *exergonic* and *endergonic*.

8.37 How is Gibbs energy related to useful work?

8.38 What is a thermodynamically reversible process? Why are such processes non-existent?

8.39 What is a dynamic equilibrium?

8.40 What is the difference in Gibbs energy between reactants and products when a chemical process is at equilibrium?

REVIEW PROBLEMS

Note: Unless otherwise stated, a temperature of 25 °C is assumed in all problems requiring calculations. Thermodynamic data required for calculations can be found in the appropriate tables in this chapter or in the appendices.

8.41 The value of ΔU for a certain process is −1455 J. During the process, the system absorbed 812 J of heat. Did the system do work, or was work done on the system? How much work was involved? **LO 3**

8.42 How much heat must be removed from 130 g of water to lower its temperature from 22.9 °C to 16.1 °C? **LO3**

8.43 How much heat is needed to increase the temperature of 1.00 kg of water from 25.0 °C to 100.0 °C? This amount of heat would be comparable to making four cups of coffee. **LO3**

8.44 If 500 J of heat is added to 100.0 g of copper, initially at 22.0 °C, what is its final temperature? **LO3**

8.45 A sample of copper was heated to 120 °C and then plunged into an insulated vessel containing 200 g of water at 25.00 °C. The final temperature of the mixture was 26.50 °C.
(a) How much heat was absorbed by the water?
(b) How much heat was lost by the copper sample?
(c) What was the mass of the copper sample? **LO3**

8.46 Use the value of the specific heat of silver in table 8.2 to calculate its molar heat capacity. **LO3**

8.47 Nitric acid, HNO_3, reacts with potassium hydroxide, KOH, according to the equation:

$$HNO_3(aq) + KOH(aq) \rightarrow KNO_3(aq) + H_2O(l)$$

A student poured 72.0 mL of 1.0 M HNO_3 into a coffee cup calorimeter, noted that the temperature was 25.0 °C and added 72.0 mL of 1.0 M KOH, also at 25.0 °C. The mixture was stirred quickly with a thermometer, and its temperature rose to 31.3 °C. Assuming that the specific heats of all solutions are 4.18 J g^{-1} K^{-1} and that all densities are 1.00 g mL^{-1}, calculate both the enthalpy of reaction, and the enthalpy of reaction per mole of acid. **LO3,4**

8.48 A 0.100 mol sample of propane, a gas used in portable gas cookers, was placed in a bomb calorimeter where it was burned in the presence of excess oxygen. The reaction was:

$$C_3H_8(g) + 5O_2(g) \rightarrow 3CO_2(g) + 4H_2O(l)$$

The water bath surrounding the bomb calorimeter contained 1.000 kg of water at an initial temperature of 25.000 °C. The heat capacity of the bomb calorimeter was 93.1 kJ K^{-1}. The reaction raised the temperature of the water bath to 27.282 °C. The specific heat of water is 4.18 J g^{-1} K^{-1}. **LO 3,4**
(a) How much heat was liberated in this reaction?
(b) What is the enthalpy of reaction of propane with oxygen?

8.49 Toluene, C_7H_8, is used in the manufacture of explosives such as TNT (trinitrotoluene). A 1.500 g sample of liquid toluene was placed in a bomb calorimeter along with excess oxygen. When the combustion of the toluene was initiated, the temperature of the 2.500 kg of water in the water bath surrounding the bomb calorimeter rose from 25.000 °C to 26.413 °C. The products of the combustion were $CO_2(g)$ and $H_2O(l)$, and the heat capacity of the calorimeter was 34.61 kJ K^{-1}. The reaction was: **LO3,4**

$$C_7H_8(l) + 9O_2(g) \rightarrow 7CO_2(g) + 4H_2O(l)$$

(a) How much heat was liberated by the reaction?
(b) How much heat would be liberated under similar conditions if 1.00 mol of toluene was burned?

8.50 One thermochemical equation for the reaction of carbon monoxide with oxygen is: **LO4**

$$3CO(g) + \tfrac{3}{2}O_2(g) \rightarrow 3CO_2(g) \qquad \Delta_r H^\ominus = -849\ \text{kJ mol}^{-1}$$

(a) Write the thermochemical equation for the reaction using 2 mol of CO.
(b) What is $\Delta_r H(^\ominus)$ for the formation of 1 mol of CO_2 from this reaction?

8.51 Ammonia reacts with oxygen as follows: **LO4**

$$4NH_3(g) + 7O_2(g) \rightarrow 4NO_2(g) + 6H_2O(g) \qquad \Delta_r H^\ominus = -1132\ \text{kJ mol}^{-1}$$

(a) Calculate the enthalpy change for the combustion of 1 mol of NH_3.
(b) Write the thermochemical equation for the reaction in which 1 mol of H_2O is formed.

8.52 Aluminium and iron(III) oxide, Fe_2O_3, react to form aluminium oxide, Al_2O_3, and iron. For each mole of aluminium used, 426.9 kJ of energy is released under standard conditions. Write the thermochemical equation that shows the consumption of 4 mol of Al. (All of the substances are solids.) **LO4**

8.53 Magnesium burns in air to produce a bright light and is often used in fireworks displays. The thermochemical equations for this process is: **LO4**

$$2Mg(s) + O_2(g) \longrightarrow 2MgO(s) \qquad \Delta_r H^\ominus = -1203\ \text{kJ mol}^{-1}$$

What is the enthalpy change for the combustion of 4.27 g of magnesium under standard conditions?

8.54 Show how the thermochemical equations: **LO4**

$$N_2O_4(g) \longrightarrow 2NO_2(g) \qquad \Delta_r H^\ominus = +57.93\ \text{kJ mol}^{-1}$$

$$2NO(g) + O_2(g) \longrightarrow 2NO_2(g) \qquad \Delta_r H^\ominus = -113.14\ \text{kJ mol}^{-1}$$

can be manipulated to give $\Delta_r H^\ominus$ for the following reaction.

$$2NO(g) + O_2(g) \longrightarrow N_2O_4(g)$$

8.55 Hydrogen chloride can be generated by heating a mixture of sulfuric acid and potassium chloride according to the equation:

$$2KCl(s) + H_2SO_4(l) \longrightarrow 2HCl(g) + K_2SO_4(s)$$

Calculate $\Delta_r H^\ominus$ for this reaction from the following thermochemical equations. **LO4**

$$HCl(g) + KOH(s) \longrightarrow KCl(s) + H_2O(l) \qquad \Delta_r H^\ominus = -203.6\ \text{kJ mol}^{-1}$$

$$H_2SO_4(l) + 2KOH(s) \longrightarrow K_2SO_4(s) + 2H_2O(l) \qquad \Delta_r H^\ominus = -342.4\ \text{kJ mol}^{-1}$$

8.56 Calculate $\Delta_r H^\ominus$ for the following reaction. **LO4**

$$HCl(g) + NaNO_2(s) \longrightarrow HNO_2(l) + NaCl(s)$$

Using the following thermochemical equations.

$$2NaCl(s) + H_2O(l) \longrightarrow 2HCl(g) + Na_2O(s) \qquad \Delta_r H^\ominus = +507.31\ \text{kJ mol}^{-1}$$

$$NO(g) + NO_2(g) + Na_2O(s) \longrightarrow 2NaNO_2(s) \qquad \Delta_r H^\ominus = -427.14\ \text{kJ mol}^{-1}$$

$$NO(g) + NO_2(g) \longrightarrow N_2O(g) + O_2(g) \qquad \Delta_r H^\ominus = -42.68\ \text{kJ mol}^{-1}$$

$$2HNO_2(l) \longrightarrow N_2O(g) + O_2(g) + H_2O(l) \qquad \Delta_r H^\ominus = +34.35\ \text{kJ mol}^{-1}$$

8.57 Barium oxide reacts with sulfuric acid according to the equation: **LO4**

$$BaO(s) + H_2SO_4(l) \longrightarrow BaSO_4(s) + H_2O(l)$$

Calculate $\Delta_r H^\ominus$ for this reaction. The following thermochemical equations can be used.

$$SO_3(g) + H_2O(l) \longrightarrow H_2SO_4(l) \qquad \Delta_r H^\ominus = -78.2\ \text{kJ mol}^{-1}$$

$$BaO(s) + SO_3(g) \longrightarrow BaSO_4(s) \qquad \Delta_r H^\ominus = -213\ \text{kJ mol}^{-1}$$

8.58 Use the following thermochemical equations to calculate the standard enthalpy of formation of CuO(s). **LO 4**

$$2Cu(s) + S(s) \longrightarrow Cu_2S(s) \qquad \Delta_r H^\ominus = -79.5\,kJ\,mol^{-1}$$

$$S(s) + O_2(g) \longrightarrow SO_2(g) \qquad \Delta_r H^\ominus = -297\,kJ\,mol^{-1}$$

$$Cu_2S(s) + 2O_2(g) \longrightarrow 2CuO(s) + SO_2(g) \qquad \Delta_r H^\ominus = -527.5\,kJ\,mol^{-1}$$

8.59 Use the following thermochemical equations to calculate the standard enthalpy of formation of $Mg(NO_3)_2(s)$. **LO 4**

$$8Mg(s) + Mg(NO_3)_2(s) \longrightarrow Mg_3N_2(s) + 6MgO(s) \qquad \Delta_r H^\ominus = -3884\,kJ\,mol^{-1}$$

$$Mg_3N_2(s) \longrightarrow 3Mg(s) + N_2(g) \qquad \Delta_r H^\ominus = +463\,kJ\,mol^{-1}$$

$$2MgO(s) \longrightarrow 2Mg(s) + O_2(g) \qquad \Delta_r H^\ominus = +1203\,kJ\,mol^{-1}$$

8.60 Which of the following equations has a value of $\Delta_r H^\ominus$ that would properly be labelled as $\Delta_f H^\ominus$? **LO 4**

(a) $CaCO_3(s) \rightarrow CaO(s) + CO_2(g)$
(b) $Ca(s) + \frac{1}{2}O_2(g) \rightarrow CaO(s)$
(c) $2Cu(s) + O_2(g) \rightarrow 2CuO(s)$
(d) $2Fe(s) + O_2(g) \rightarrow 2FeO(s)$
(e) $SO_2(g) + \frac{1}{2}O_2(g) \rightarrow SO_3(g)$
(f) $N_2(g) + \frac{5}{2}O_2(g) \rightarrow N_2O_5(g)$

8.61 Write the thermochemical equations for the standard enthalpy of formation of the following compounds at 25 °C.

(a) C(s, diamond)
(b) $CH_3Cl(g)$
(c) $HNO_3(l)$
(d) $CaSO_4{\cdot}2H_2O(s)$
(e) $Na_2CO_3(s)$
(f) Hg(g)

8.62 Using the data in table 8.3, calculate $\Delta_r H^\ominus$ for each of the following reactions. **LO 4**

(a) $2H_2O_2(l) \rightarrow 2H_2O(l) + O_2(g)$
(b) $HCl(g) + NaOH(s) \rightarrow NaCl(s) + H_2O(l)$
(c) $CH_4(g) + Cl_2(g) \rightarrow CH_3Cl(g) + HCl(g)$
(d) $2NH_3(g) + CO_2(g) \rightarrow CO(NH_2)_2(s) + H_2O(l)$

8.63 The standard enthalpy of combustion, $\Delta_c H^\ominus$, of glucose ($C_6H_{12}O_6$) is -2.80×10^3 kJ mol^{-1}. The sole products of the combustion of glucose are $CO_2(g)$ and $H_2O(l)$. Write the thermochemical equation for the combustion of 1 mol of glucose and calculate $\Delta_f H^\ominus$ for this compound. **LO 4**

8.64 The thermochemical equation for the combustion of ethyne (acetylene) gas, $C_2H_2(g)$, is: **LO 4**

$$2C_2H_2(g) + 5O_2(g) \longrightarrow 4CO_2(g) + 2H_2O(l) \qquad \Delta_r H^\ominus = -2599.3\,kJ\,mol^{-1}$$

Using the data in table 8.3, determine the value of $\Delta_f H^\ominus$ for ethyne (acetylene) gas.

8.65 Use the data in tables 8.3 and 8.4 to calculate the approximate atomisation enthalpy of CO_2. Use your answer to obtain an approximate value for a C═O bond enthalpy. **LO 4**

8.66 The standard enthalpy of formation of ethanol vapour, $C_2H_5OH(g)$, is -235.3 kJ mol^{-1}. Use the data in table 8.4 and the average bond enthalpies for C—C, C—H and O—H bonds to estimate the C—O bond enthalpy in this molecule. The structure of the molecule is: **LO 4**

```
    H   H
    |   |  ..
H — C — C — O — H
    |   |  ..
    H   H
```

8.67 Gaseous hydrogen sulfide, $H_2S(g)$, has $\Delta_f H^\ominus = -20.15$ kJ mol^{-1}. Use the data in table 8.4 to calculate the average S—H bond enthalpy in this molecule. **LO 4**

8.68 For $SF_6(g)$, $\Delta_f H^\ominus = -1209$ kJ mol^{-1}. Use the data in table 8.4 to calculate the average S—F bond enthalpy in this molecule. **LO 4**

8.69 Use the results of question 8.68 and the data in table 8.3 to calculate the standard enthalpy of formation of $SF_4(g)$. The measured value of $\Delta_f H^\ominus$ for $SF_4(g)$ is -718.4 kJ mol^{-1}. What is the percentage difference between your calculated value of $\Delta_f H^\ominus$ and the experimentally determined value? **LO 4**

8.70 Use the data in tables 8.4 and 8.5 to estimate the standard enthalpy of formation of $C_2H_4(g)$. **LO 4**

8.71 Calculate the approximate enthalpy of formation of CCl_4 vapour at 25 °C and 10^5 Pa. **LO 4**

8.72 Which gaseous compound should have the more exothermic enthalpy of formation, HBr or HCl? **LO 4**

8.73 Use the data from table 8.3 to calculate $\Delta_r H^\ominus$ for each of the following reactions. **LO 4**

(a) $CaO(s) + CO_2(g) \rightarrow CaCO_3(s)$
(b) $C_2H_2(g) + 2H_2(g) \rightarrow C_2H_6(g)$
(c) $3CaO(s) + 2Fe(s) \rightarrow 3Ca(s) + Fe_2O_3(s)$
(d) $Ca(OH)_2(s) \rightarrow CaO(s) + H_2O(l)$
(e) $2NaCl(s) + H_2SO_4(l) \rightarrow Na_2SO_4(s) + 2HCl(g)$
(f) $2C_2H_2(g) + 5O_2(g) \rightarrow 4CO_2(g) + 2H_2O(g)$
(g) $C_2H_2(g) + 5N_2O(g) \rightarrow 2CO_2(g) + H_2O(g) + 5N_2(g)$
(h) $Fe_2O_3(s) + 2Al(s) \rightarrow Al_2O_3(s) + 2Fe(s)$
(i) $NH_4Cl(s) \rightarrow NH_3(g) + HCl(g)$
(j) $Ag(s) + KCl(s) \rightarrow AgCl(s) + K(s)$

8.74 Predict the algebraic sign of the entropy change for each of the following reactions. **LO 5**

(a) $PCl_3(g) + Cl_2(g) \rightarrow PCl_5(g)$
(b) $SO_2(g) + CaO(s) \rightarrow CaSO_3(s)$
(c) $CO_2(g) + H_2O(l) \rightarrow H_2CO_3(aq)$
(d) $Ni(s) + 2HCl(aq) \rightarrow H_2(g) + NiCl_2(aq)$
(e) $I_2(s) \rightarrow I_2(g)$
(f) $Br_2(g) + 3Cl_2(g) \rightarrow 2BrCl_3(g)$
(g) $NH_3(g) + HCl(g) \rightarrow NH_4Cl(s)$
(h) $CaO(s) + H_2O(l) \rightarrow Ca(OH)_2(s)$

8.75 Calculate $\Delta_r S^\ominus$ for each of the following reactions from the data in table 8.6. **LO 7**

(a) $N_2(g) + 3H_2(g) \rightarrow 2NH_3(g)$
(b) $CO(g) + 2H_2(g) \rightarrow CH_3OH(l)$
(c) $2C_2H_6(g) + 7O_2(g) \rightarrow 4CO_2(g) + 6H_2O(g)$
(d) $Ca(OH)_2(s) + H_2SO_4(l) \rightarrow CaSO_4(s) + 2H_2O(l)$
(e) $S(s) + 2N_2O(g) \rightarrow SO_2(g) + 2N_2(g)$
(f) $Ag(s) + \frac{1}{2}Cl_2(g) \rightarrow AgCl(s)$
(g) $H_2(g) + \frac{1}{2}O_2(g) \rightarrow H_2O(g)$
(h) $H_2(g) + \frac{1}{2}O_2(g) \rightarrow H_2O(l)$
(i) $CaCO_3(s) + H_2SO_4(l) \rightarrow CaSO_4(s) + H_2O(g) + CO_2(g)$
(j) $NH_3(g) + HCl(g) \rightarrow NH_4Cl(s)$

8.76 Calculate $\Delta_f S^\ominus$ for each of the following compounds. **LO 7**

(a) $K_2SO_4(s)$
(b) $C_8H_{18}(l)$
(c) KCl(s)
(d) $CH_3OH(l)$
(e) $Ca(OH)_2(s)$
(f) $HNO_3(l)$

8.77 Good wine will turn to vinegar if it is left exposed to air because the alcohol is oxidised to acetic acid. The equation for the reaction is: **LO 7**

$$C_2H_5OH(l) + O_2(g) \longrightarrow CH_3COOH(l) + H_2O(l)$$

Calculate $\Delta_r S^{\ominus}$ for this reaction.

8.78 Calculate $\Delta_r G^{\ominus}$ for each of the following reactions, using the data in table 8.7. **LO 8**

(a) $SO_3(g) + H_2O(l) \rightarrow H_2SO_4(l)$
(b) $2NH_4Cl(s) + CaO(s) \rightarrow CaCl_2(s) + H_2O(l) + 2NH_3(g)$
(c) $CaSO_4(s) + 2HCl(g) \rightarrow CaCl_2(s) + H_2SO_4(l)$
(d) $C_2H_4(g) + H_2O(g) \rightarrow C_2H_5OH(l)$
(e) $Ca(s) + 2H_2SO_4(l) \rightarrow CaSO_4(s) + SO_2(g) + 2H_2O(l)$
(f) $2HCl(g) + CaO(s) \rightarrow CaCl_2(s) + H_2O(g)$
(g) $H_2SO_4(l) + 2NaCl(s) \rightarrow 2HCl(g) + Na_2SO_4(s)$
(h) $3NO_2(g) + H_2O(l) \rightarrow 2HNO_3(l) + NO(g)$
(i) $2AgCl(s) + Ca(s) \rightarrow CaCl_2(s) + 2Ag(s)$
(j) $NH_3(g) + HCl(g) \rightarrow NH_4Cl(s)$

8.79 Given the following information: **LO 8**

$$4NO(g) \longrightarrow 2N_2O(g) + O_2(g) \qquad \Delta_r G^{\ominus} = -139.56\ \text{kJ mol}^{-1}$$

$$2NO(g) + O_2(g) \longrightarrow 2NO_2(g) \qquad \Delta_r G^{\ominus} = -69.70\ \text{kJ mol}^{-1}$$

calculate $\Delta_r G^{\ominus}$ for the reaction:

$$2N_2O(g) + 3O_2(g) \longrightarrow 4NO_2(g)$$

8.80 Given these reactions and their $\Delta_r G^{\ominus}$ values: **LO 8**

$$COCl_2(g) + 4NH_3(g) \longrightarrow CO(NH_2)_2(s) + 2NH_4Cl(s) \qquad \Delta_r G^{\ominus} = -332.0\ \text{kJ mol}^{-1}$$

$$COCl_2(g) + H_2O(l) \longrightarrow CO_2(g) + 2HCl(g) \qquad \Delta_r G^{\ominus} = -141.8\ \text{kJ mol}^{-1}$$

$$NH_3(g) + HCl(g) \longrightarrow NH_4Cl(s) \qquad \Delta_r G^{\ominus} = -91.96\ \text{kJ mol}^{-1}$$

calculate the value of $\Delta_r G^{\ominus}$ for the reaction:

$$CO(NH_2)_2(s) + H_2O(l) \longrightarrow CO_2(g) + 2NH_3(g)$$

8.81 What is the maximum amount of useful work that could theoretically be obtained at 25 °C and 10^5 Pa from the combustion of 62.0 g of ethane, $C_2H_6(g)$, to give $CO_2(g)$ and $H_2O(g)$? **LO 8**

8.82 Chloroform, $CHCl_3$, was formerly used as an anaesthetic and is now believed to be a carcinogen (cancer-causing agent). It has an enthalpy of vaporisation ($\Delta_{vap}H$) of 31.4 kJ mol^{-1}. The change, $CHCl_3(l) \rightarrow CHCl_3(g)$, has $\Delta S^{\ominus} = 94.2$ J mol^{-1} K^{-1}. At what temperature will $CHCl_3$ boil (i.e. at what temperature will liquid and vapour be in equilibrium at 10^5 Pa)? **LO 8**

8.83 Isooctane, an important constituent of petrol, has a boiling point of 99.3 °C and an enthalpy of vaporisation of 37.7 kJ mol^{-1}. What is ΔS for the vaporisation of 1 mol of isooctane? **LO 8,7**

8.84 Acetone, $(CH_3)_2CO$, has a boiling point of 56.2 °C. The change $(CH_3)_2CO(l) \rightarrow (CH_3)_2CO(g)$ has $\Delta H^{\ominus} = 31.9$ kJ mol^{-1}. What is $\Delta S^{\ominus}$ for this change? **LO 8,7**

8.85 Which of the following equations (which are not necessarily balanced) represent a reaction that would be expected to be spontaneous at 25 °C and 10^5 Pa? **LO 8**

(a) $PbO(s) + NH_3(g) \rightarrow Pb(s) + N_2(g) + H_2O(g)$
(b) $NaOH(s) + HCl(g) \rightarrow NaCl(s) + H_2O(l)$
(c) $Al_2O_3(s) + Fe(s) \rightarrow Fe_2O_3(s) + Al(s)$
(d) $2CH_4(g) \rightarrow C_2H_6(g) + H_2(g)$

8.86 How much work is accomplished by the following chemical reaction if it occurs inside a bomb calorimeter? **LO 3**

$$2C_6H_{14}(l) + 19O_2(g) \longrightarrow 12CO_2(g) + 14H_2O(l)$$

What is the maximum amount of work that could be done if the reaction is carried out at 10^5 Pa, with all reactants and products at 25 °C? (For $C_6H_{14}(l)$, $\Delta_f H^{\ominus} = -199$ kJ mol^{-1} and $\Delta S^{\ominus} = 295$ J mol^{-1} k^{-1}.)

ADDITIONAL EXERCISES

8.87 A cylinder fitted with a piston contains 5.00 L of a gas at a pressure of 4×10^5 Pa. The entire apparatus is maintained at a constant temperature of 25 °C. The piston is released and the gas expands until the pressure inside the cylinder equals the atmospheric pressure outside, which is 10^5 Pa. Assume ideal gas behaviour and calculate the amount of work done by the gas as it expands at constant temperature. **LO 3**

8.88 The experiment described in question 8.87 is repeated, but this time a weight, which exerts a pressure of 2×10^5 Pa, is placed on the piston. When the gas expands, its pressure drops to this 2×10^5 Pa pressure. Then the weight is removed and the gas is allowed to expand again to a final pressure of 10^5 Pa. Throughout both expansions, the temperature of the apparatus was held at a constant 25 °C. Calculate the amount of work done by the gas in each step. How does the combined total amount of work in this two-step expansion compare with the amount of work done by the gas in the one-step expansion described in question 8.87? **LO 3**

8.89 A body of water with a mass of 750 g changed in temperature from 25.50 °C to 19.50 °C. **LO 3**

(a) What would have to be done to cause such a change?
(b) How much energy is involved in this change?

8.90 1.000 kg of gold at a temperature of 120.00 °C was plunged into an insulated vat of water. The mass of the water was 2.000 kg, and its initial temperature was 25.00 °C. What was the temperature of the resulting system when it stabilised? **LO 3**

8.91 A dilute solution of hydrochloric acid with a mass of 610.29 g and containing 0.331 83 mol of HCl was exactly neutralised in a calorimeter by the sodium hydroxide in 615.31 g of a comparably dilute solution. The temperature increased from 16.784 °C to 20.610 °C. The specific heat of the HCl solution was 4.031 J g^{-1} K^{-1} and that of the NaOH solution was 4.046 J g^{-1} K^{-1}. The heat capacity of the calorimeter was 77.99 J K^{-1}. Use this information to calculate the enthalpy change for the following reaction.

$$HCl(aq) + NaOH(aq) \longrightarrow NaCl(aq) + H_2O(l)$$

What is the enthalpy of neutralisation per mole of HCl? Assume that the original solutions made independent contributions to the total heat capacity of the system following their mixing.

8.92 In the recovery of iron from iron ore, the reduction of the ore is actually accomplished by reactions involving carbon monoxide. Use the following thermochemical equations: **LO4**

$$Fe_2O_3(s) + 3CO(g) \longrightarrow 2Fe(s) + 3CO_2(g) \quad \Delta_rH^\ominus = -28\ \text{kJ mol}^{-1}$$

$$3Fe_2O_3(s) + CO(g) \longrightarrow 2Fe_3O_4(s) + CO_2(g) \quad \Delta_rH^\ominus = -59\ \text{kJ mol}^{-1}$$

$$Fe_3O_4(s) + CO(g) \longrightarrow 3FeO(s) + CO_2(g) \quad \Delta_rH^\ominus = +38\ \text{kJ mol}^{-1}$$

to calculate $\Delta_rH^\ominus$ for the reaction:

$$FeO(s) + CO(g) \longrightarrow Fe(s) + CO_2(g)$$

8.93 Use the results of question 8.92 and the data in table 8.3 to calculate the value of $\Delta_fH^\ominus$ for FeO. **LO4**

8.94 Phosphorus burns in air to give tetraphosphorus decaoxide.

$$4P(s) + 5O_2(g) \longrightarrow P_4O_{10}(s) \quad \Delta_rH^\ominus = -3062\ \text{kJ mol}^{-1}$$

The product combines with water to give phosphoric acid, H_3PO_4.

$$P_4O_{10}(s) + 6H_2O(l) \longrightarrow 4H_3PO_4(l) \quad \Delta_rH^\ominus = -257.2\ \text{kJ mol}^{-1}$$

Using these equations (and any others in the chapter, as needed), write the thermochemical equation for the formation of 1 mol of $H_3PO_4(l)$ from the elements, and calculate $\Delta_fH^\ominus$. **LO4**

8.95 The amino acid alanine, $H_2NCH(CH_3)COOH$, is one of the compounds used by the body to make proteins. The equation for its combustion is: **LO4**

$$2H_2NCH(CH_3)COOH(s) + 7.5O_2(g) \rightarrow 6CO_2(g) + 7H_2O(l) + N_2(g)$$

For each mole of alanine that burns, 1620 kJ of heat is liberated. Use this information plus values of $\Delta_fH^\ominus$ for the products of combustion to calculate $\Delta_fH^\ominus$ for alanine.

8.96 The value of $\Delta_fH^\ominus$ for HBr(g) was first evaluated using the following standard enthalpy values obtained experimentally.

$$Cl_2(g) + 2KBr(aq) \longrightarrow Br_2(aq) + 2KCl(aq) \quad \Delta_rH^\ominus = -96.2\ \text{kJ mol}^{-1}$$

$$H_2(g) + Cl_2(g) \longrightarrow 2HCl(g) \quad \Delta_rH^\ominus = -184\ \text{kJ mol}^{-1}$$

$$HCl(aq) + KOH(aq) \longrightarrow KCl(aq) + H_2O(l) \quad \Delta_rH^\ominus = -57.3\ \text{kJ mol}^{-1}$$

$$HBr(aq) + KOH(aq) \longrightarrow KBr(aq) + H_2O(l) \quad \Delta_rH^\ominus = -57.3\ \text{kJ mol}^{-1}$$

$$HCl(g) \longrightarrow HCl(aq) \quad \Delta_rH^\ominus = -77.0\ \text{kJ mol}^{-1}$$

$$Br_2(l) \longrightarrow Br_2(aq) \quad \Delta_rH^\ominus = -4.2\ \text{kJ mol}^{-1}$$

$$HBr(g) \longrightarrow HBr(aq) \quad \Delta_rH^\ominus = -79.9\ \text{kJ mol}^{-1}$$

Use this information to calculate the value of $\Delta_fH^\ominus$ for HBr(g). **LO4**

8.97 The reaction for the metabolism of sucrose, $C_{12}H_{22}O_{11}$, in the body is the same as for its combustion in oxygen to yield $CO_2(g)$ and $H_2O(l)$. The standard enthalpy of formation of sucrose is −2230 kJ mol^{-1}. Use data in table 8.3 to calculate the amount of energy released by metabolising 50.0 g of sucrose. **LO4**

8.98 Consider the following thermochemical equations. **LO4**

$$CH_3OH(l) + O_2(g) \longrightarrow HCOOH(l) + H_2O(l) \quad \Delta_rH^\ominus = -411\ \text{kJ mol}^{-1} \quad (1)$$

$$CO(g) + 2H_2(g) \longrightarrow CH_3OH(l) \quad \Delta_rH^\ominus = -128\ \text{kJ mol}^{-1} \quad (2)$$

$$HCOOH(l) \longrightarrow CO(g) + H_2O(l) \quad \Delta_rH^\ominus = -33\ \text{kJ mol}^{-1} \quad (3)$$

Suppose equation 1 is reversed and divided by 2, equations 2 and 3 are multiplied by $\frac{1}{2}$, and then the three adjusted equations are added. What is the net reaction, and what is the value of $\Delta_rH^\ominus$ for the net reaction?

8.99 The reaction $Cl_2(g) + Br_2(g) \rightarrow 2BrCl(g)$ has a very small value for $\Delta_rS^\ominus$ (+11.6 J K^{-1}). Why? **LO5**

8.100 The enthalpy of combustion, $\Delta_cH^\ominus$, of oxalic acid, $H_2C_2O_4(s)$, is −246.05 kJ mol^{-1}. Consider the following data (all at 25 °C). **LO4,7,8**

Substance	$\Delta_fH^\ominus$ (kJ mol^{-1})	$S^\ominus$ (J mol^{-1} K^{-1})
C(s, graphite)	0	5.69
$CO_2(g)$	−393.5	213.6
$H_2(g)$	0	130.6
$H_2O(l)$	−285.9	69.96
$O_2(g)$	0	205.0
$H_2C_2O_4(s)$	?	120.1

(a) Write the balanced thermochemical equation for the combustion of 1 mol of oxalic acid.
(b) Write the balanced thermochemical equation for the formation of 1 mol of oxalic acid from its elements in their standard states at 25 °C.
(c) Use the information in the table and your equations in (a) and (b) to calculate $\Delta_fH^\ominus$ for oxalic acid.
(d) Calculate $\Delta_fS^\ominus$ for oxalic acid and $\Delta_rS^\ominus$ for the combustion of 1 mol of oxalic acid.
(e) Calculate $\Delta_fG^\ominus$ for oxalic acid and $\Delta_rG^\ominus$ for the combustion of 1 mol of oxalic acid at 25 °C.

8.101 Many biochemical reactions have positive values for Δ_rG under biological conditions and so should not be expected to be spontaneous. They occur, however, because they are chemically coupled with other reactions that have negative values of Δ_rG. An example is the set of reactions that begins the sequence of reactions involved in the metabolism of glucose, a sugar. Given these reactions and their corresponding Δ_rG values under biological conditions: **LO8**

$$\text{glucose} + \text{phosphate} \longrightarrow \text{glucose-6-phosphate} + H_2O \quad \Delta_rG = +13.13\ \text{kJ mol}^{-1}$$

$$\text{ATP} + H_2O \longrightarrow \text{ADP} + \text{phosphate} \quad \Delta_rG = -32.22\ \text{kJ mol}^{-1}$$

calculate Δ_rG for the coupled reaction:

$$\text{glucose} + \text{ATP} \longrightarrow \text{glucose-6-phosphate} + \text{ADP}$$

8.102 Cars, trucks and other machines that use petrol or diesel engines for power have cooling systems. In terms of thermodynamics, what makes these cooling systems necessary? **LO8**

8.103 Ethanol, C_2H_5OH, has been suggested as an alternative to petrol as a fuel. In worked example 8.13, we calculated $\Delta_r G^\ominus$ for combustion of 1 mol of C_2H_5OH; in worked example 8.14, we calculated $\Delta_r G^\ominus$ for combustion of 1 mol of octane. We will assume that petrol has the same properties as octane (one of its constituents). The density of C_2H_5OH is 0.7893 g mL^{-1}; the density of octane, C_8H_{18}, is 0.7025 g mL^{-1}. Calculate the maximum work that could be obtained by burning 1 L each of C_2H_5OH and C_8H_{18}. On a volume basis, which is a better fuel? Explain your answer. **LO 8**

8.104 Use the data in table 8.4 to calculate the bond enthalpy in the nitrogen molecule and in the oxygen molecule. **LO 4**

8.105 The enthalpy of vaporisation of carbon tetrachloride, CCl_4, is 29.9 kJ mol^{-1}. Using this information and data in tables 8.4 and 8.5, estimate the standard enthalpy of formation of liquid CCl_4. **LO 4**

8.106 Ammonium nitrate, NH_4NO_3, is a white, crystalline substance manufactured on an enormous scale, for use as a fertiliser, from the reaction of anhydrous ammonia with concentrated nitric acid. However, ammonium nitrate must be handled with care as it is potentially explosive. **LO 3,4**

(a) The standard enthalpy of formation ($\Delta_f H^\ominus$) of $NH_4NO_3(s)$ is −365.56 kJ mol^{-1}. Write the balanced chemical equation to which this value refers. Specify the standard state for each species in this reaction.
(b) Write the balanced chemical equation for the formation of $NH_4NO_3(s)$ from $NH_3(g)$ and $HNO_3(l)$. Calculate $\Delta_r H^\ominus$ for this reaction.
(c) When gently heated, $NH_4NO_3(s)$ can decompose according to the following equation.

$$NH_4NO_3(s) \rightarrow N_2O(g) + 2H_2O(g)$$

Calculate $\Delta_r H^\ominus$ for this reaction.
(d) When vigorously heated, NH_4NO_3 can decompose explosively according to the following equation.

$$NH_4NO_3(s) \rightarrow N_2(g) + 2H_2O(g) + \tfrac{1}{2}O_2(g)$$

Calculate the enthalpy change if 5.000 kg of NH_4NO_3 decomposes completely according to this equation under standard conditions.
(e) If the energy released from the reaction in (d) was used to heat 50.0 kg of water, initially at 25.0 °C, what would be the final temperature of the water?

9 Chemical equilibrium

The Earth's oceans are extremely complex chemical systems, comprising vast quantities of dissolved inorganic salts, dissolved organic matter and dissolved gases. Of the gases, carbon dioxide is very important because it provides a source of carbonate ions for the growth of coral, the exoskeleton of which is composed of calcium carbonate. The rate at which coral reefs grow, therefore, depends critically on the amount of dissolved carbon dioxide in the ocean, and any process which alters this amount will affect coral reefs.

We know that carbon dioxide levels in the Earth's atmosphere are currently increasing and the pH of the Earth's oceans is decreasing. It is also becoming increasingly apparent that the Earth's average temperature is rising. All of these factors can influence the chemical equilibrium between solid and dissolved calcium carbonate in the oceans, thereby potentially putting coral reefs, such as the Great Barrier Reef, at risk.

This chapter will introduce the concept of chemical equilibrium, the factors that can affect the behaviour of chemical systems at equilibrium and how a system in which equilibrium has been perturbed must respond to re-establish equilibrium. An understanding of chemical equilibrium is necessary in all forms of chemistry, and we will study important applications of chemical equilibrium in detail in chapters 10, 11 and 12.

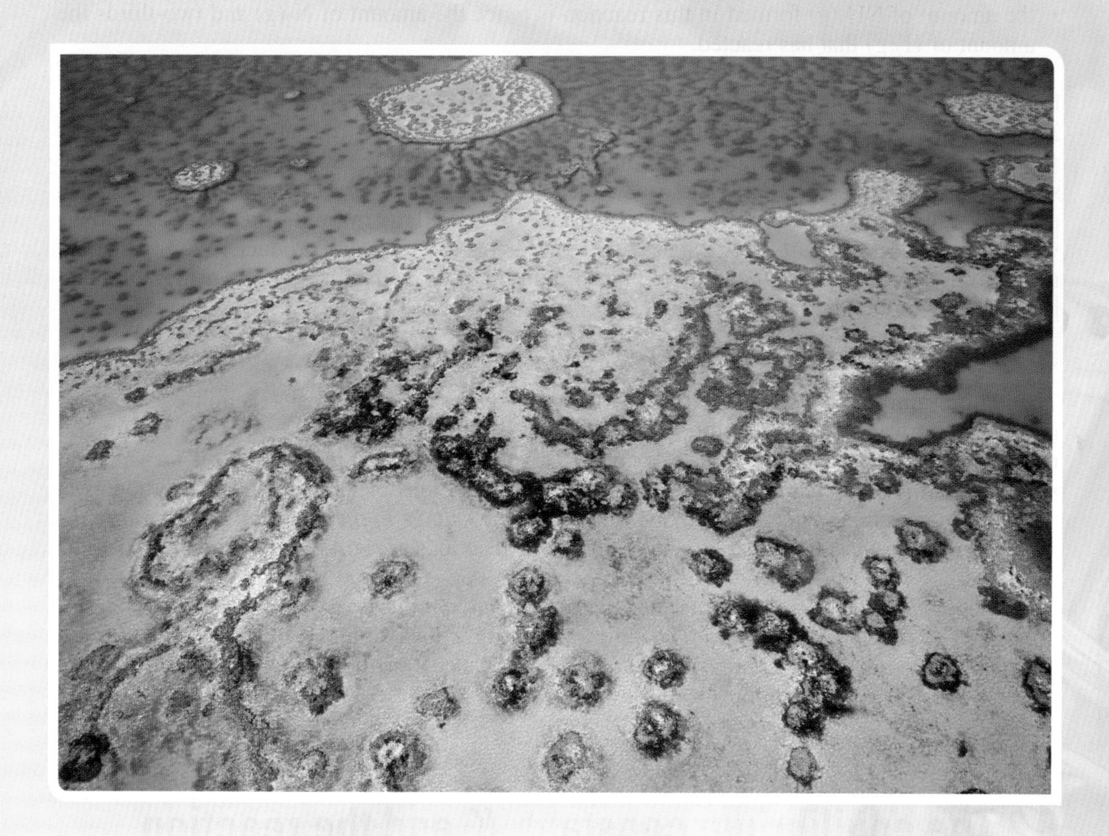

LEARNING OBJECTIVES

After studying this chapter, you should be able to:

9.1 define chemical equilibrium

9.2 use the equilibrium constant, K, and the reaction quotient, Q

9.3 show how equilibrium and Gibbs energy are related

9.4 predict how an equilibrium will respond to change

9.5 carry out calculations involving equilibria.

9.1 Chemical equilibrium

In chapter 8, we learned how to use the value of ΔG to predict whether a particular chemical reaction or physical change would be spontaneous under the specified conditions. We also briefly introduced the concept of equilibrium and showed that a system is at equilibrium when $\Delta G = 0$; under these conditions, no non-pV work can be extracted from the system, and there is no driving force for chemical or physical change. In this chapter, we will develop the idea of equilibrium further; most importantly, we will show that equilibria in chemical reactions can be quantified using the equilibrium constant, K, and that this is related to the thermodynamic functions we encountered in chapter 8.

It is important to realise that a negative value of ΔG for a particular chemical reaction under specific conditions tells us only that spontaneous chemical change will occur in the forward direction — it does *not* necessarily mean that *complete* conversion of reactants to products will occur. In fact, many chemical reactions proceed only part way to completion and give a mixture of reactants and products. We will illustrate this using the Haber–Bosch process, the catalysed formation of gaseous ammonia, NH_3, from the reaction of nitrogen gas, N_2, and hydrogen gas, H_2.

$$N_2(g) + 3H_2(g) \rightarrow 2NH_3(g)$$

The stoichiometric coefficients tell us that:

- N_2(g) always reacts with H_2(g) in a 1 : 3 mole ratio
- the amount of NH_3(g) formed in this reaction is twice the amount of N_2(g) and two-thirds the amount of H_2(g) that has reacted.

We cannot, however, conclude from the balanced chemical equation that mixing 1 mole of N_2(g) with 3 moles of H_2(g) will produce 2 moles of NH_3(g). If we mixed 1 mole of N_2 with 3 moles of H_2 at elevated temperature and pressure over a catalyst (the uncatalysed reaction is very slow) and monitored the composition of the reaction mixture over time, we would indeed see the concentration of NH_3(g) increase and the concentrations of N_2(g) and H_2(g) decrease; eventually, however, the concentrations of N_2(g), H_2(g) and NH_3(g) would stop changing, and we would end up not with pure NH_3(g) but with a mixture of reactants and products. We say that the reaction mixture has come to **chemical equilibrium** and we distinguish this by using a double-headed arrow $\rightleftharpoons$ (called an equilibrium arrow).

$$N_2(g) + 3H_2(g) \rightleftharpoons 2NH_3(g)$$

The equilibrium arrow highlights a very important point — a chemical reaction can proceed in both the forward and reverse directions at the same time. At equilibrium, N_2 and H_2 are still reacting to give NH_3, and NH_3 is also reacting to give N_2 and H_2, but *the rates of the forward and reverse reactions are equal*. This means that the concentrations of the reactants and products in the reaction mixture remain constant and there is *no net change in the overall composition of the reaction mixture*. Since both the forward and reverse reactions are still occurring, we say that the system is in a state of **dynamic equilibrium**.

In a chemical equilibrium, the terms 'reactants' and 'products' do not have the usual significance because the reaction is proceeding in both directions simultaneously. Instead, we use 'reactants' and 'products' simply to identify the substances on the left- and right-hand sides, respectively, of the equation for the equilibrium.

FIGURE 9.1 Equilibrium mixtures of N_2O_4(g) and NO_2(g) at different temperatures. In ice (right-hand side), colourless N_2O_4 predominates and the mixture is pale. At high temperature (left-hand side), dark brown NO_2 predominates. At room temperature (middle), there is a mixture of the gases, with N_2O_4 in the larger amount.

9.2 The equilibrium constant, *K*, and the reaction quotient, *Q*

A classic experiment that demonstrates many important concepts of equilibrium is the decomposition of N_2O_4(g) (dinitrogen tetroxide) to give NO_2(g) (nitrogen dioxide) (figure 9.1).

$$N_2O_4(g) \rightleftharpoons 2NO_2(g)$$

This reaction is easily monitored by observing the colour of the reaction mixture. Pure N_2O_4(g) is colourless, while pure NO_2(g) is brown; as N_2O_4 decomposes, the mixture becomes more and more brown. Figure 9.2 shows how the concentrations of the two gases change over time as pure N_2O_4(g) is converted to an equilibrium mixture of N_2O_4(g) and NO_2(g).

Suppose we set up the two experiments shown in figure 9.3 at constant temperature. In the first 1-litre flask, we put 0.0350 mol N_2O_4. Since no NO_2 is present, some N_2O_4 must decompose for the mixture to reach equilibrium, so the reaction will proceed in the forward direction (i.e. from left to right). When equilibrium is reached, we find the concentration of N_2O_4 has dropped to 0.0292 mol L^{-1} and the concentration of NO_2 has become 0.0116 mol L^{-1}.

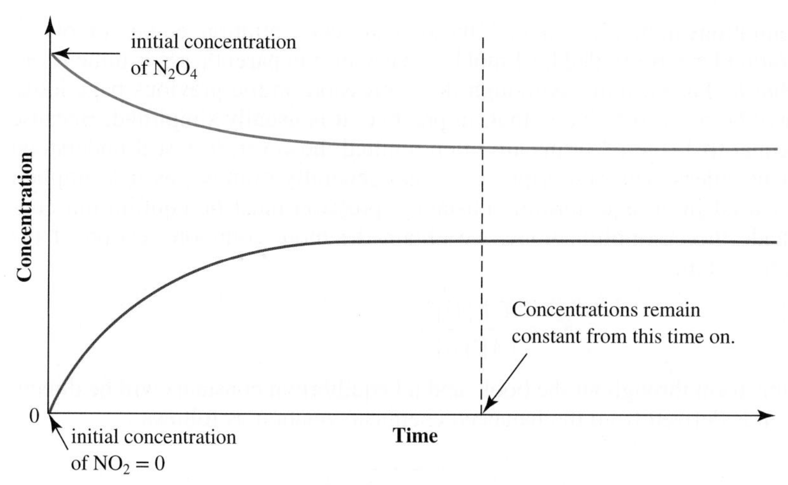

FIGURE 9.2 The approach to equilibrium. In the decomposition of $N_2O_4(g)$ into $NO_2(g)$, $N_2O_4(g) \rightleftharpoons 2NO_2(g)$, the concentrations of N_2O_4 and NO_2 change relatively quickly at first. As time passes, the concentrations change more and more slowly. When equilibrium is reached, the concentrations of N_2O_4 and NO_2 no longer change with time; they remain constant. Note that the concentration change of NO_2 over the course of the reaction is twice the concentration change of N_2O_4.

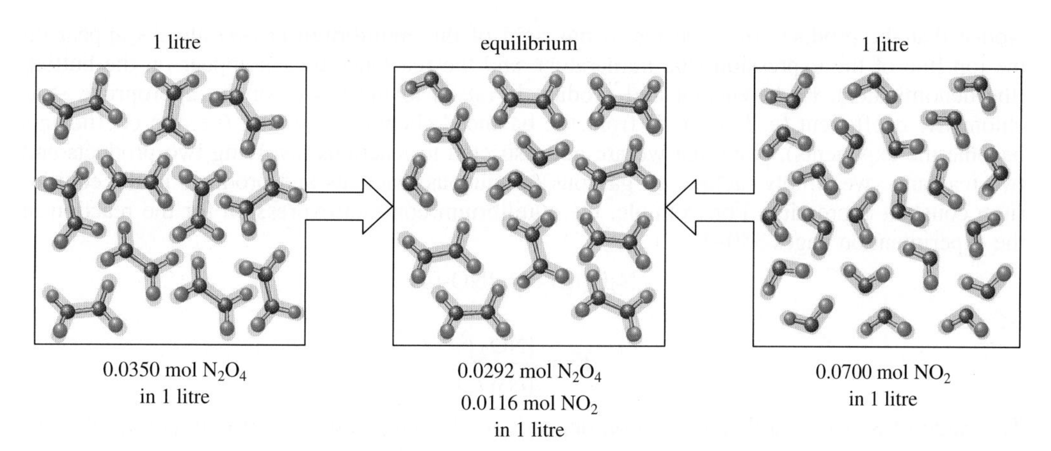

FIGURE 9.3 Reaction reversibility for the equilibrium:

$$N_2O_4(g) \rightleftharpoons 2NO_2(g)$$

The same equilibrium composition is reached starting from either pure $N_2O_4(g)$ or $NO_2(g)$, provided the *overall* system composition is the same. Because pure NO_2 is brown and pure N_2O_4 is colourless, the amber colour of the equilibrium mixture indicates that both species are present at equilibrium.

In the second 1-litre flask we put 0.0700 mol of NO_2 (*precisely* the amount of NO_2 that would form if 0.0350 mol of N_2O_4 — the amount placed in the first flask — decomposed completely). In this second flask, there is no N_2O_4 present initially, so NO_2 molecules must combine, following the reverse reaction (right to left) shown above, to give N_2O_4. When we measure the concentrations at equilibrium in the second flask, we find, once again, 0.0292 mol L^{-1} of N_2O_4 and 0.0116 mol L^{-1} of NO_2. (In practice, these experiments are difficult, if not impossible, to carry out, because $N_2O_4(g)$ and $NO_2(g)$ are always in equilibrium, so they cannot be obtained absolutely pure.)

We see here that the same equilibrium composition is reached, whether we begin with pure NO_2 or pure N_2O_4, as long as the *total* amount of N and O to be divided between these two substances is the same. Similar observations apply to other chemical systems as well, which leads to the following generalisation: for a given *overall* system composition at constant temperature, we always reach the same equilibrium concentrations regardless of the direction from which equilibrium is approched.

For any reaction involving only gases or species in solution (generally aqueous) in which a moles of substance A react with b moles of substance B to give c moles of substance C and d moles of substance D:

$$aA + bB \rightleftharpoons cC + dD$$

the following holds when equilibrium is established:

$$K_c = \frac{\left(\frac{[C]_e}{c^{\ominus}}\right)^c \left(\frac{[D]_e}{c^{\ominus}}\right)^d}{\left(\frac{[A]_e}{c^{\ominus}}\right)^a \left(\frac{[B]_e}{c^{\ominus}}\right)^b}$$

K_c is called the **equilibrium constant** (or, occasionally, the concentration-based equilibrium constant) and the above expression is called the **equilibrium constant expression**. The square brackets refer to concentrations in mol L^{-1}, and $c^{\ominus}$, the standard concentration, equals 1 mol L^{-1}. Because each concentration term is divided by 1 mol L^{-1}, each term in parentheses is dimensionless, and this means that K_c has no units. Although the expression on the previous page looks rather daunting, you will be pleased to know that, in practice, it is usually simplified. Because they are numerically equal to 1, the $c^{\ominus}$ terms are often omitted; however, it is still understood that each term is dimensionless. The subscript 'e' is also generally omitted, as it is implicit that the concentrations used in an equilibrium constant expression must be equilibrium concentrations. Having made these simplifications, we obtain the more common version of the equilibrium constant expression:

$$K_c = \frac{[C]^c[D]^d}{[A]^a[B]^b}$$

Note that we will use this form throughout the book, and all equilibrium constants will be dimensionless. This expression is derived from the balanced chemical equation as follows.

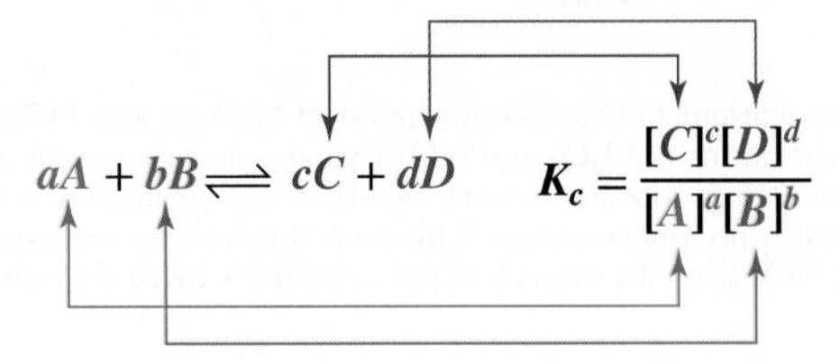

Notice that the products (the species to the right of the equilibrium arrow) always appear on the top line of the expression (the numerator), and the reactants always appear on the bottom (the denominator). Each reactant and product is raised to the power of the appropriate stoichiometric coefficient (a, b, c or d) from the balanced chemical equation (i.e. the coefficients become the exponents). Note that we are not restricted to reactions involving two products and two reactants; we simply include all gaseous or aqueous reactants and products in the equilibrium constant expression. For example, the equilibrium constant expression for the reaction in the experiment on pages 350–1:

$$N_2O_4(g) \rightleftharpoons 2NO_2(g)$$

is:

$$K_c = \frac{[NO_2]^2}{[N_2O_4]}$$

The value of K_c for a particular reaction depends on the temperature, so it is important that the temperature is always specified when K_c is reported. Tabulated values of K_c generally refer to 25.0 °C and concentrations in mol L^{-1}.

Table 9.1 gives the results of four further experiments using differing initial concentrations of $N_2O_4(g)$ and $NO_2(g)$. The final column shows that the value of K_c is constant (within experimental error). We can, therefore, say that the equilibrium constant for this equilibrium:

$$N_2O_4(g) \rightleftharpoons 2NO_2(g)$$

is:

$$K_c = \frac{[NO_2]^2}{[N_2O_4]} = 4.61 \times 10^{-3} \text{ at } 25.0\,°C$$

TABLE 9.1 Equilibrium mixture compositions for various initial concentrations of $NO_2(g)$ and $N_2O_4(g)$.

Initial concentration		Equilibrium concentration		
$[N_2O_4]$ (mol L^{-1})	$[NO_2]$ (mol L^{-1})	$[N_2O_4]$ (mol L^{-1})	$[NO_2]$ (mol L^{-1})	$K_c = \frac{[NO_2]^2}{[N_2O_4]}$
0.0450	0	0.0384	0.0133	$\frac{(0.0133)^2}{0.0384} = 4.61 \times 10^{-3}$
0.0150	0	0.0114	0.007 24	$\frac{(0.007\,24)^2}{0.0114} = 4.60 \times 10^{-3}$
0	0.0600	0.0248	0.0107	$\frac{(0.0107)^2}{0.0247} = 4.62 \times 10^{-3}$
0	0.0500	0.0202	0.009 64	$\frac{(0.009\,64)^2}{0.0202} = 4.60 \times 10^{-3}$

The values of K_c are equal within experimental error.

Similarly, if we were to study the reaction:

$$H_2(g) + I_2(g) \rightleftharpoons 2HI(g)$$

at different initial concentrations of $H_2(g)$, $I_2(g)$ and HI(g) at 400 °C, we would find that, when the system came to equilibrium:

$$K_c = \frac{[HI]^2}{[H_2][I_2]} = 49.5 \text{ at } 400\,°C$$

We can apply the equilibrium constant expression to any system at equilibrium. We can also write an expression for systems that are not at equilibrium. We call this the **reaction quotient (Q)**. For:

$$aA + bB \rightleftharpoons cC + dD$$

we can write the **reaction quotient expression**:

$$Q_c = \frac{[C]^c[D]^d}{[A]^a[B]^b}$$

(A 'quotient' is the number obtained from a division.) The expression takes a similar form to that for K_c, and we have made the same simplifications for Q_c as for K_c. Q_c and K_c differ in that:

- K_c involves equilibrium concentrations and so refers to a system *at equilibrium*
- Q_c refers to systems that are not necessarily at equilibrium.

K_c can have only one positive value at a particular temperature, while Q_c can have any positive value.

It is important to realise that *all chemical systems will eventually come to equilibrium.* When equilibrium is established, $Q_c = K_c$. In systems where $Q_c \neq K_c$, a comparison of their values can tell us how a system must change to establish equilibrium, as shown in figure 9.4.

$$aA + bB \rightleftharpoons cC + dD$$

$$Q_c = \frac{[C]^c[D]^d}{[A]^a[B]^b}$$

$Q_c < K_c$ equilibrium shifts →	$Q_c = K_c$ equilibrium	$Q_c > K_c$ equilibrium shifts ←
• If $Q_c < K_c$, we must make the value of Q_c larger so that $Q_c = K_c$. • We make the value of Q_c larger by either increasing the value of the top line or decreasing the value of the bottom line of the expression for Q_c. • The value of the top line is increased if more products are formed. • The value of the bottom line is decreased if more reactants are used up. • Therefore, the reaction mixture composition will move towards products.		• If $Q_c > K_c$, we must make the value of Q_c smaller so that $Q_c = K_c$. • We make the value of Q_c smaller by either decreasing the value of the top line or increasing the value of the bottom line of the expression for Q_c. • The value of the top line is decreased if more products are used up. • The value of the bottom line is increased if more reactants are formed. • Therefore, the reaction mixture composition will move towards reactants.

FIGURE 9.4 How a system must change to establish equilibrium.

If $Q_c > K_c$, the system reacts to use up products and generate more reactants, thereby decreasing Q_c to the point where it equals K_c; that is, the system is at equilibrium. If $Q_c < K_c$, the opposite occurs; the system reacts to use up reactants and form more products, thereby increasing Q_c to the point where it equals K_c, again meaning the system is at equilibrium (see section 9.4).

The subscript 'c' in both Q_c and K_c refers to the fact that we are expressing the composition of the reaction mixtures in terms of concentration. For equilibria involving gases, we can also express the composition of the equilibrium mixture in terms of partial pressure (p. 231), rather than concentration. When we do this, the equilibrium constant expression takes the form:

$$K_p = \frac{\left(\frac{p_C}{p^\ominus}\right)^c \left(\frac{p_D}{p^\ominus}\right)^d}{\left(\frac{p_A}{p^\ominus}\right)^a \left(\frac{p_B}{p^\ominus}\right)^b}$$

where p_A, p_B, p_C and p_D are the equilibrium partial pressures of A, B, C and D, respectively, and $p^\ominus$ is the standard pressure.

Because $p^{\ominus} = 1 \times 10^5$ Pa, we cannot make the simplification that we made earlier for K_c as, in this case, the $p^{\ominus}$ terms affect the actual value of K_p. Therefore, expressions for K_p must be written in full, including the $p^{\ominus}$ terms when pressures are measured in Pa. Like K_c, K_p has no units.

WORKED EXAMPLE 9.1

Writing expressions for K_c

Interest is growing in the potential use of hydrogen as an alternative fuel to replace fossil fuels. Ironically, much of the hydrogen produced in the world is derived from methane, CH_4, in natural gas using the forward reaction of the equilibrium:

$$CH_4(g) + H_2O(g) \rightleftharpoons CO(g) + 3H_2(g)$$

Write the expression for K_c for this equilibrium.

Analysis

To write the equilibrium constant expression, we place the concentrations of the products in the numerator and the concentrations of the reactants in the denominator, each raised to the value of the appropriate stoichiometric coefficient. As stated earlier, it is usual to ignore the $c^{\ominus}$ terms as these do not affect the value of K_c.

$$\mathbf{CH_4 + H_2O \rightleftharpoons CO + 3H_2 \qquad K_c = \frac{[CO][H_2]^3}{[CH_4][H_2O]}}$$

Solution

The equilibrium constant expression is:

$$K_c = \frac{[CO][H_2]^3}{[CH_4][H_2O]}$$

Is our answer reasonable?

Check to see that products are on the top line and that reactants are on the bottom. Also check the superscripts and be sure that they are the same as those in the balanced chemical equation. Notice that we omit the exponent when it is equal to 1.

PRACTICE EXERCISE 9.1

Write the equilibrium constant expression K_c for each of the following equilibria.

(a) $2NO(g) \rightleftharpoons N_2(g) + O_2(g)$

(b) $SO_3(g) + NO(g) \rightleftharpoons NO_2(g) + SO_2(g)$

WORKED EXAMPLE 9.2

Writing expressions for K_p

Much of the world's supply of methanol, CH_3OH, is produced by the reaction of CO(g) with $H_2(g)$.

$$CO(g) + 2H_2(g) \rightleftharpoons CH_3OH(g)$$

Write the expression for K_p for this equilibrium.

Analysis

For K_p, we use partial pressures in the equilibrium constant expression. We put the equilibrium partial pressures of the products in the numerator and the equilibrium partial pressures of the reactants in the denominator. The coefficients in the equation become exponents of the pressures. Because we are working in SI units, in which $p^{\ominus} = 1 \times 10^5$ Pa, we must explicitly divide each term by $p^{\ominus}$ to calculate K_p.

Solution

The expression for K_p for this reaction is:

$$K_p = \frac{\left(\frac{p_{CH_3OH}}{p^{\ominus}}\right)}{\left(\frac{p_{CO}}{p^{\ominus}}\right)\left(\frac{p_{H_2}}{p^{\ominus}}\right)^2}$$

Is our answer reasonable?
Check to make sure you have written the expression in terms of products over reactants, and not the other way around. Check each superscript against the coefficients in the balanced chemical equation.

PRACTICE EXERCISE 9.2

Using partial pressures, write the equilibrium constant expression for the equilibrium.

$$H_2(g) + I_2(g) \rightleftharpoons 2HI(g)$$

The values of K_p and K_c are related through the equation

$$K_p = K_c\left(\frac{1000\,RT}{p^{\circ}}\right)^{\Delta n_{gas}}$$

where Δn_{gas} is the change in the *number of moles of gas* in going from the reactants to the products:

$$\Delta n_{gas} = (\text{number of moles of } \textit{gaseous} \text{ products}) - (\text{number of moles of } \textit{gaseous} \text{ reactants})$$

For example, the reaction:

$$N_2(g) + 3H_2(g) \rightleftharpoons 2NH_3(g)$$

has $\Delta n_{gas} = 2 - 4 = -2$.

The exact thermodynamic treatment of equilibrium uses **activities**, rather than concentrations or pressures, in the equilibrium expression. Activities are dimensionless quantities that take account of the fact that neither gases in gas mixtures nor species in solution behave ideally; atoms, molecules and ions in the gas or solution phase tend to interact with each other, with the magnitude of the interaction increasing as the concentration increases. Activities can, therefore, be thought of as 'effective' concentrations and, as they are dimensionless, this leads to equilibrium constants having no units. Providing that we work at low gas pressures or with dilute solutions, activities approximate concentrations to an acceptable degree. Equilibrium constants obtained using activities, rather than pressures or concentrations, are often called **thermodynamic equilibrium constants**.

Manipulating equilibrium constant expressions

Sometimes it is useful to be able to combine chemical equilibria to obtain the equation for some other reaction of interest. In doing this, we perform various operations such as reversing an equation, multiplying the coefficients by some factor and adding the equations to give the desired equation. In our discussion of thermochemistry, you learned how such manipulations affect ΔH, ΔS and ΔG values. Some different rules apply to equilibrium constant expressions.

Changing the direction of an equilibrium

When the direction of an equation is reversed, the new equilibrium constant is the reciprocal of the original. As an example, when we reverse the equilibrium:

$$PCl_3(g) + Cl_2(g) \rightleftharpoons PCl_5(g) \qquad K_c = \frac{[PCl_5]}{[PCl_3][Cl_2]}$$

we obtain:

$$PCl_5(g) \rightleftharpoons PCl_3(g) + Cl_2(g) \qquad K_c' = \frac{[PCl_3][Cl_2]}{[PCl_5]}$$

The equilibrium constant expression for the second reaction is the reciprocal of that for the first, so $K_c' = \frac{1}{K_c}$.

Multiplying the coefficients by a factor

When the coefficients in an equation are multiplied by a factor, the equilibrium constant is raised to a power equal to that factor. For example, if we multiply the coefficients of the equation:

$$PCl_3(g) + Cl_2(g) \rightleftharpoons PCl_5(g) \qquad K_c = \frac{[PCl_5]}{[PCl_3][Cl_2]}$$

by 2, this gives:

$$2PCl_3(g) + 2Cl_2(g) \rightleftharpoons 2PCl_5(g) \qquad K_c'' = \frac{[PCl_5]^2}{[PCl_3]^2[Cl_2]^2}$$

Comparing equilibrium constant expressions, we see that $K_c'' = K_c^2$.

Adding chemical equilibria

When chemical equilibria are added, their equilibrium constants are multiplied. For example, consider the following.

$$2N_2(g) + O_2(g) \rightleftharpoons 2N_2O(g) \quad K_{c1} = \frac{[N_2O]^2}{[N_2]^2[O_2]}$$

$$2N_2O(g) + 3O_2(g) \rightleftharpoons 4NO_2(g) \quad K_{c2} = \frac{[NO_2]^4}{[N_2O]^2[O_2]^3}$$

$$2N_2(g) + 4O_2(g) \rightleftharpoons 4NO_2(g) \quad K_{c3} = \frac{[NO_2]^4}{[N_2]^2[O_2]^4}$$

We have numbered the equilibrium constants just to distinguish between them. If we multiply the equilibrium constant expression for K_{c1} by that for K_{c2}, we obtain the equilibrium constant expression for K_{c3}.

$$\frac{[N_2O]^2}{[N_2]^2[O_2]} \times \frac{[NO_2]^4}{[N_2O]^2[O_2]^3} = \frac{[NO_2]^4}{[N_2]^2[O_2]^4}$$

Therefore, $K_{c3} = K_{c1} \times K_{c2}$.

PRACTICE EXERCISE 9.3

Under particular conditions of temperature, $K_c = 8.7 \times 10^4$ for the equilibrium:

$$HCOOH(g) \rightleftharpoons CO(g) + H_2O(g)$$

What is the value of K_c for the equilibrium:

$$CO(g) + H_2O(g) \rightleftharpoons HCOOH(g)$$

under these conditions?

PRACTICE EXERCISE 9.4

At 25 °C, the equilibrium constants for the following equilibria are as shown.

$$2CO(g) + O_2(g) \rightleftharpoons 2CO_2(g) \quad K_c = 3.3 \times 10^{91}$$

$$2H_2(g) + O_2(g) \rightleftharpoons 2H_2O(g) \quad K_c = 9.1 \times 10^{80}$$

Use these data to calculate K_c for the reaction:

$$H_2O(g) + CO(g) \rightleftharpoons CO_2(g) + H_2(g)$$

The magnitude of the equilibrium constant

Because the concentrations of the products are always on the top line of the equilibrium constant expression K_c, the value of the equilibrium constant gives us a measure of how far the reaction has proceeded towards completion when equilibrium is reached. For example, the equilibrium:

$$2H_2(g) + O_2(g) \rightleftharpoons 2H_2O(g)$$

has $K_c = 9.1 \times 10^{80}$ at 25 °C. This means that when there is an equilibrium between these gases:

$$K_c = \frac{[H_2O]^2}{[H_2]^2[O_2]} = 9.1 \times 10^{80} = \frac{9.1 \times 10^{80}}{1}$$

By writing K_c as a fraction, $\frac{9.1 \times 10^{80}}{1}$, we see that the top line (numerator) of the equilibrium constant expression is enormous compared with the bottom line (denominator), which means that, at equilibrium, the concentration of H_2O has to be enormous compared with the concentrations of H_2 and O_2. At equilibrium, therefore, most of the hydrogen and oxygen atoms in the system are found in H_2O and very few are present in H_2 and O_2. The enormous value of K_c tells us that the reaction between H_2 and O_2 goes essentially to completion and that the products are strongly favoured at equilibrium.

The reaction between N_2 and O_2 to give NO:

$$N_2(g) + O_2(g) \rightleftharpoons 2NO(g)$$

has a very small equilibrium constant: $K_c = 4.8 \times 10^{-31}$ at 25 °C. The equilibrium constant expression for this reaction is:

$$K_c = \frac{[NO]^2}{[N_2][O_2]} = 4.8 \times 10^{-31}$$

Since $10^{-31} = \frac{1}{10^{31}}$, we can write this as:

$$K_c = \frac{[NO]^2}{[N_2][O_2]} = 4.8 \times \frac{1}{10^{31}} = \frac{4.8}{10^{31}}$$

Here the bottom line is huge compared with the top line, so the concentrations of N_2 and O_2 must be very much larger than the concentration of NO at equilibrium. This means that, in a mixture of N_2 and O_2 at this temperature, the amount of NO that is formed is negligible. The reaction hardly proceeds at all towards completion before equilibrium is reached, so the reactants are favoured at equilibrium.

The relationship between the equilibrium constant and the position of equilibrium is summarised in figure 9.5.

reactant   product

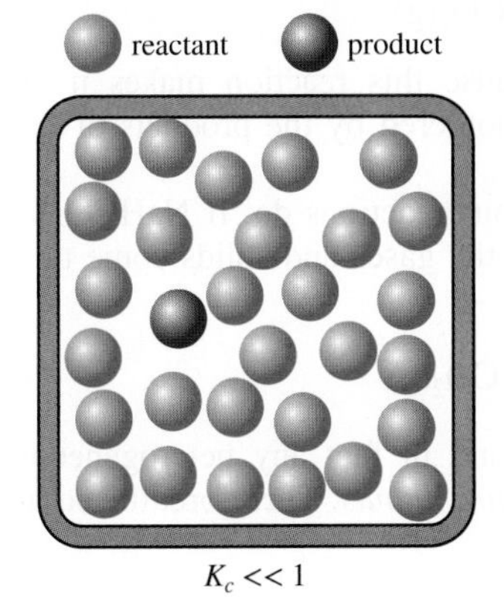

$K_c << 1$

Extremely small amounts of products are formed. The position of equilibrium lies strongly towards the reactants.

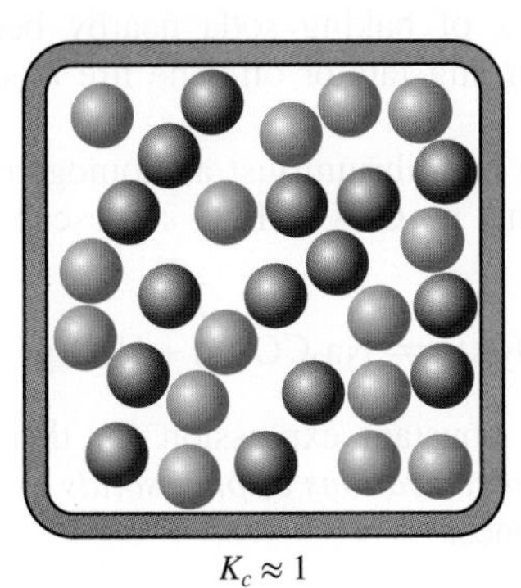

$K_c \approx 1$

The concentrations of reactants and products are similar at equilibrium. The position of equilibrium lies approximately midway between reactants and products.

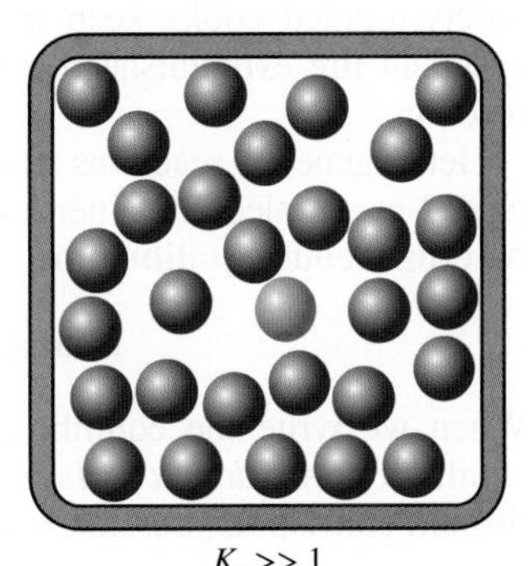

$K_c >> 1$

The reaction proceeds to give essentially complete formation of products. The position of equilibrium lies strongly towards the products.

FIGURE 9.5 The magnitude of K_c and the position of equilibrium.

Consider the reaction:

$$\text{reactant} \rightleftharpoons \text{product}$$

When K_c is very large ($K_c >> 1$), there is a large amount of product and very little reactant in the reaction mixture at equilibrium, so we say *the position of equilibrium lies to the right*. When $K_c \approx 1$, similar amounts of reactant and product are present at equilibrium. When $K_c << 1$, the reaction mixture contains a large amount of reactant at equilibrium and very little product, so we say *the position of equilibrium lies to the left*.

One of the ways that we can use equilibrium constants is to compare the extents to which two or more reactions proceed to completion. Take care in making such comparisons, however, because, unless the K values are greatly different, the comparison is valid *only if both reactions have the same number of reactant and product molecules appearing in their balanced chemical equations*. Note also that the magnitude of the equilibrium constant tells us nothing about how rapidly a system reaches equilibrium. For example, we saw on the previous page that K_c for the formation of water from H_2 and O_2 is enormous, yet a mixture of H_2 and O_2 is stable almost indefinitely at room temperature. However, once initiated by a spark, the reaction proceeds rapidly to completion with explosive force.

PRACTICE EXERCISE 9.5

Values of the equilibrium constant for the reaction:

$$2NO(g) + Cl_2(g) \rightleftharpoons 2NOCl(g)$$

at different temperatures are as follows.

T/K	K_c
300	2.06×10^8
400	1.34×10^5
500	1.73×10^3

If we begin with a closed flask containing NO(g), Cl_2(g) and NOCl(g) at equilibrium at room temperature and heat it, at which temperature, 300 K, 400 K or 500 K, will the greatest amount of NOCl be present in the equilibrium mixture?

Equilibrium constant expressions for heterogeneous systems

In a **homogeneous reaction** — or a **homogeneous equilibrium** — all of the reactants and products are in the same phase. Equilibria among gases are homogeneous because all gases mix freely with each other, so a single phase exists. There are also many equilibria in which reactants and products are dissolved in the same liquid phase.

When more than one phase exists in a reaction mixture, we call it a **heterogeneous reaction**. A common example is the combustion of wood, in which a solid fuel reacts with gaseous oxygen. Another is the thermal decomposition of sodium bicarbonate (baking soda), which occurs when the compound is sprinkled on a fire.

$$2NaHCO_3(s) \rightarrow Na_2CO_3(s) + H_2O(g) + CO_2(g)$$

Safety-minded cooks keep a box of baking soda nearby because this reaction makes it an excellent fire extinguisher for burning fats or oil. The fire is smothered by the products of the reaction.

Heterogeneous reactions reach equilibrium just as homogeneous reactions do. If $NaHCO_3$ is placed in a sealed container so that no CO_2 or H_2O can escape, the gases and solids come to a **heterogeneous equilibrium**.

$$2NaHCO_3(s) \rightleftharpoons Na_2CO_3(s) + H_2O(g) + CO_2(g)$$

When we write the equilibrium constant expression for this, and in fact any heterogeneous equilibrium, *we do not include concentrations of pure solids or pure liquids*. Therefore, the equilibrium constant expression becomes:

$$K_c = [H_2O(g)][CO_2(g)]$$

If we consider the reverse reaction:

$$Na_2CO_3(s) + H_2O(g) + CO_2(g) \rightleftharpoons 2NaHCO_3(s)$$

following the rules we learned earlier about changing the direction of an equilibrium, we write the equilibrium constant expression as:

$$K_c = \frac{1}{[H_2O(g)][CO_2(g)]}$$

The reason we do not include pure solids and pure liquids is that the concentration of a pure liquid or solid is unchangeable at a given temperature. *For any pure liquid or solid at constant temperature, the ratio of amount of substance to volume of substance is constant.* For example, at 25.0 °C and 1.013×10^5 Pa, 1 mole of $NaHCO_3$ will occupy a volume of 38.9 mL and 2 moles of $NaHCO_3$ will occupy twice this volume, 77.8 mL, but the *ratio* of the amount to volume (i.e. the molar concentration) remains the same (figure 9.6).

For $NaHCO_3$, the concentration of the solid at 25.0 °C is:

$$\frac{1\,\text{mol}}{0.0389\,\text{L}} = \frac{2\,\text{mol}}{0.0778\,\text{L}}$$
$$= 25.7\,\text{mol L}^{-1}$$

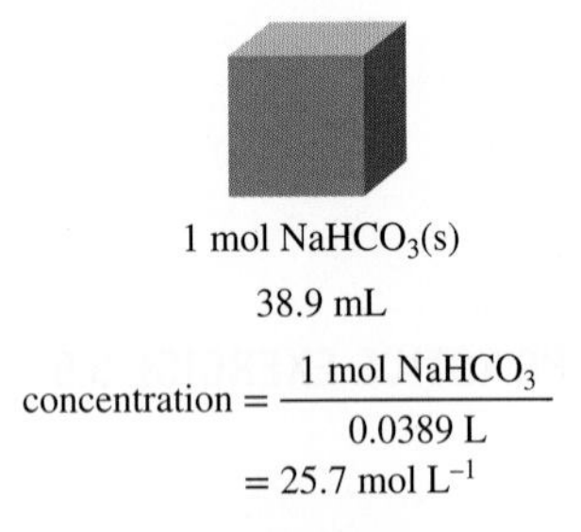

1 mol $NaHCO_3(s)$

38.9 mL

$$\text{concentration} = \frac{1\text{ mol NaHCO}_3}{0.0389\text{ L}} = 25.7\text{ mol L}^{-1}$$

This is the concentration of $NaHCO_3$ in the solid, regardless of the size of the solid sample. In other words, the concentration of $NaHCO_3(s)$ is constant at constant temperature, provided there is some of it present in the reaction mixture. Even though neither $NaHCO_3(s)$ nor $Na_2CO_3(s)$ appears in the equilibrium constant expression for the reaction:

$$2NaHCO_3(s) \rightleftharpoons Na_2CO_3(s) + H_2O(g) + CO_2(g)$$

it is important to realise that some $NaHCO_3(s)$ and $Na_2CO_3(s)$ *must* be present in the reaction mixture in order for equilibrium to be established. In any equilibrium, the presence of *all* reactants and products in the balanced chemical equation is necessary at all times after the reaction is initiated in order for the system to reach equilibrium.

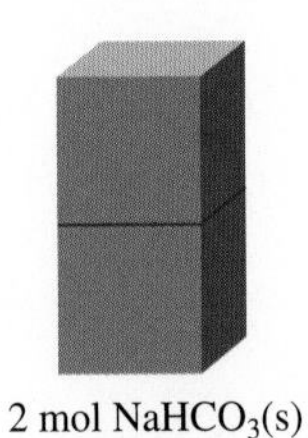

2 mol $NaHCO_3(s)$

77.8 mL

$$\text{concentration} = \frac{2\text{ mol NaHCO}_3}{0.0778\text{ L}} = 25.7\text{ mol L}^{-1}$$

FIGURE 9.6 The concentration of a substance in the solid state is constant. Doubling the amount also doubles the volume, but the ratio of amount to volume remains the same.

WORKED EXAMPLE 9.3

Writing the equilibrium constant expression for a heterogeneous reaction

The air pollutant sulfur dioxide can be removed from a gas mixture by passing it over calcium oxide. The equation for this reaction is:

$$CaO(s) + SO_2(g) \rightleftharpoons CaSO_3(s)$$

Write the equilibrium constant expression for K_c for this reaction.

Analysis

The concentrations of the two solids, CaO and $CaSO_3$, are incorporated into the equilibrium constant, K_c, for the reaction. The only concentration that should appear in the equilibrium constant expression is that of $SO_2(g)$.

Solution

The equilibrium constant expression is simply:

$$K_c = \frac{1}{[SO_2]}$$

Is our answer reasonable?

The only gaseous component in the reaction mixture is a reactant, and must therefore appear on the bottom line of the expression for K_c, which it does.

PRACTICE EXERCISE 9.6

Write the equilibrium constant expression K_c for each of the following heterogeneous equilibria.

(a) $2Hg(l) + Cl_2(g) \rightleftharpoons Hg_2Cl_2(s)$

(b) $NH_3(g) + HCl(g) \rightleftharpoons NH_4Cl(s)$

(c) $Na(s) + H_2O(l) \rightleftharpoons Na^+(aq) + OH^-(aq) + H_2(g)$

(d) $Ag_2CrO_4(s) \rightleftharpoons 2Ag^+(aq) + CrO_4^{2-}(aq)$

(e) $CaCO_3(s) + H_2O(l) + CO_2(aq) \rightleftharpoons Ca^{2+}(aq) + 2HCO_3^-(aq)$

9.3 Equilibrium and Gibbs energy

In chapter 8 we introduced the concept of Gibbs energy. In this section, we will detail the relationship between Gibbs energy and chemical equilibrium.

Gibbs energy diagrams

We have seen in chapter 8 that the sign of ΔG tells us the direction of spontaneous change in a chemical or physical system, and in this chapter we have seen that the equilibrium constant, K, gives us a measure of how far a spontaneous process will proceed towards completion. What then is the relationship between Gibbs energy and K? To answer this question, we will first consider phase changes and then look at chemical reactions.

Phase changes

In a phase change such as $H_2O(l) \rightarrow H_2O(s)$, equilibrium can exist for a given pressure only at one particular temperature; for water at a pressure of 1.013×10^5 Pa (atmospheric pressure), this temperature is 0 °C, the normal freezing point. At other temperatures, the phase change proceeds to completion in one direction or the other. One way to gain a better understanding of this is by studying **Gibbs energy diagrams**, which depict how the Gibbs energy changes as we proceed from the 'reactants' to the 'products'.

Let us consider the phase change:

$$H_2O(l) \rightarrow H_2O(s)$$

at three different temperatures: below 0 °C, at 0 °C and above 0 °C. We know, from experience that below 0 °C this process will occur spontaneously, while above 0 °C this process will not occur — in fact, the reverse process will be spontaneous. At 0 °C, we will obtain an equilibrium mixture of ice and water. Why is this? Inspection of the three Gibbs energy diagrams in figure 9.7 (overleaf), which correspond to these three situations, provides an answer. Each diagram shows how the Gibbs energy, G, of the system varies with the composition of the system, from pure liquid water on the left of each diagram to pure solid water (ice) on the right. In each case, ΔG for the phase change $H_2O(l) \rightarrow H_2O(s)$ is equal to $G(H_2O, s) - G(H_2O, l)$.

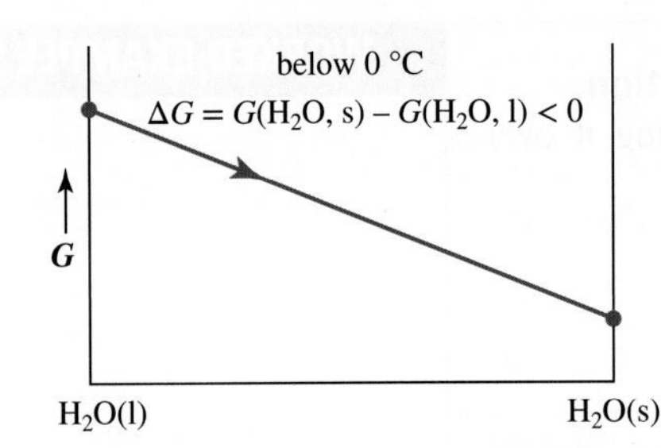

Gibbs energy decreases in the direction $H_2O(l) \longrightarrow H_2O(s)$ so the mixture freezes spontaneously to give complete formation of $H_2O(s)$.

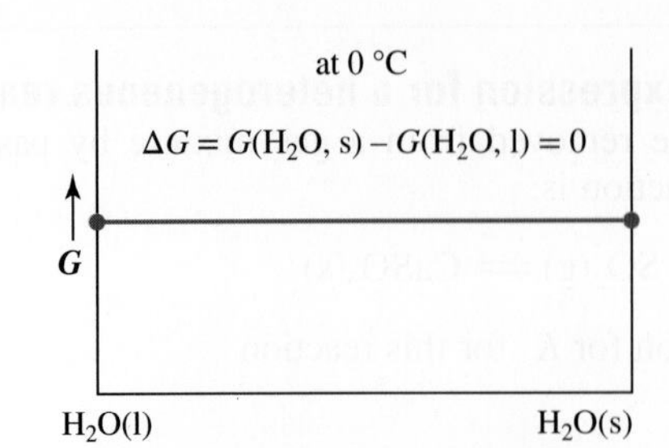

Ice–liquid equilibrium can exist for any ratio of solid to liquid. The Gibbs energy is constant across the entire range of compositions of the system.

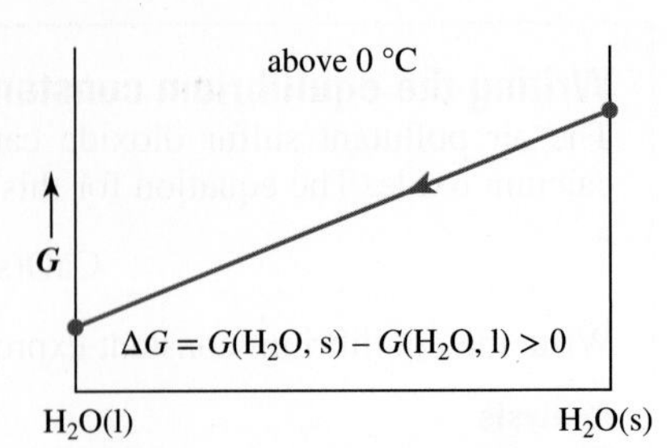

Gibbs energy decreases in the direction $H_2O(s) \longrightarrow H_2O(1)$ so the mixture melts spontaneously to give complete formation of $H_2O(1)$.

FIGURE 9.7 Gibbs energy diagrams (not to scale) for the phase change between liquid H_2O and solid H_2O at 1.013×10^5 Pa. The horizontal axis represents the amount of $H_2O(l)$ and $H_2O(s)$ present. At the left of each diagram the system consists entirely of $H_2O(l)$, while at the right of each diagram only $H_2O(s)$ is present. The green line shows the direction of decreasing G, and hence spontaneous change, in each case.

Below 0 °C, we see that the Gibbs energy of the liquid is higher than that of the solid. You have learned that a spontaneous change occurs if the Gibbs energy can decrease. Therefore, the first diagram tells us that the liquid phase, or any mixture of liquid or solid, below 0 °C freezes until only the solid is present. This is because the Gibbs energy decreases continuously until all the liquid has frozen. At this point, G is minimised. In other words, ΔG for the reaction $H_2O(l) \rightarrow H_2O(s)$ is negative at this temperature.

Above 0 °C, we have the opposite situation. The Gibbs energy of the solid is greater than that of the liquid. It therefore decreases in the direction of $H_2O(s) \rightarrow H_2O(l)$, and it continues to drop until all the solid has melted. This means that ice, or any mixture of ice and liquid water, above 0 °C continues to melt until only the liquid is present. ΔG for the melting of ice at this temperature is therefore negative.

Above or below 0 °C, the system is unable to establish an equilibrium mixture of liquid and solid. Melting or freezing occurs spontaneously until only one phase is present. However, at 0 °C, the Gibbs energies of both $H_2O(l)$ and $H_2O(s)$ are the same. Therefore, there is no change in Gibbs energy if either melting or freezing occurs, so there is no driving force for either change. Therefore, as long as a system of ice and liquid water is insulated from warmer or colder surroundings, any particular mixture of the two phases is stable and a state of equilibrium exists. Remember that this is a dynamic equilibrium; the ice melts and the water freezes, but both processes occur at the same rate, so no overall change is observed.

Chemical reactions

The Gibbs energy changes that occur in most chemical reactions are more complex than those in phase changes. As an example, let's study a reaction you've seen before — the decomposition of N_2O_4 into NO_2.

$$N_2O_4(g) \rightleftharpoons 2NO_2(g)$$

In our previous discussion of chemical equilibrium, we noted that equilibrium in this system can be approached from *either* direction, with the same equilibrium concentrations being achieved provided we begin with the same overall system composition.

Figure 9.8 shows the Gibbs energy diagram for the reaction. Notice that, in going from reactant to product, the Gibbs energy has a minimum, in contrast to the diagrams in figure 9.7; it drops below that of both pure N_2O_4 and pure NO_2.

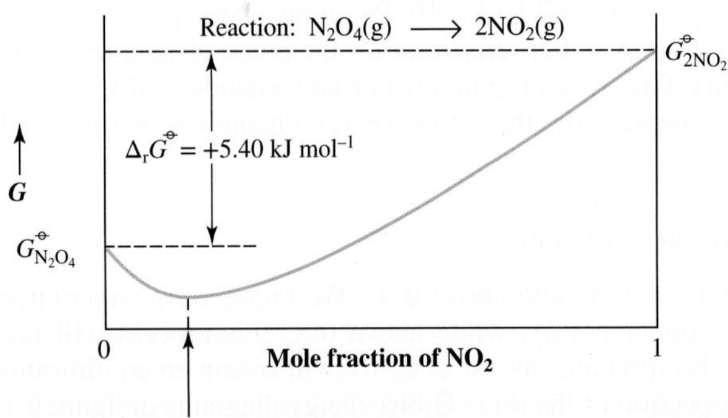

FIGURE 9.8 Gibbs energy diagram for the decomposition of $N_2O_4(g)$. The minimum on the curve indicates the composition of the reaction mixture at equilibrium. Because $\Delta_r G^{\ominus}$ is positive, the position of equilibrium lies close to the reactants. Not much product forms by the time the system reaches equilibrium.

Why is this? Let's suppose we start with pure $N_2O_4(g)$ and allow the reaction to come to equilibrium. As the reaction proceeds, $NO_2(g)$ molecules start to form, and they mix spontaneously with the remaining $N_2O_4(g)$ molecules so that the reaction mixture is homogeneous at all times. This spontaneous mixing has an associated Gibbs energy change, $\Delta_{mix}G$, that is *negative* at all possible reaction mixture compositions — this simply means that, no matter what the proportions of NO_2 and N_2O_4 are in the reaction mixture, the two gases *always* mix spontaneously as this leads to an increase in entropy of the system relative to the unmixed gases. Hence, the total Gibbs energy change for this gaseous system becomes the Gibbs energy change for the chemical reaction *plus* the Gibbs energy change of mixing. Surprisingly, it is the contribution from $\Delta_{mix}G$ that leads to the minimum in the Gibbs energy diagram (figure 9.9).

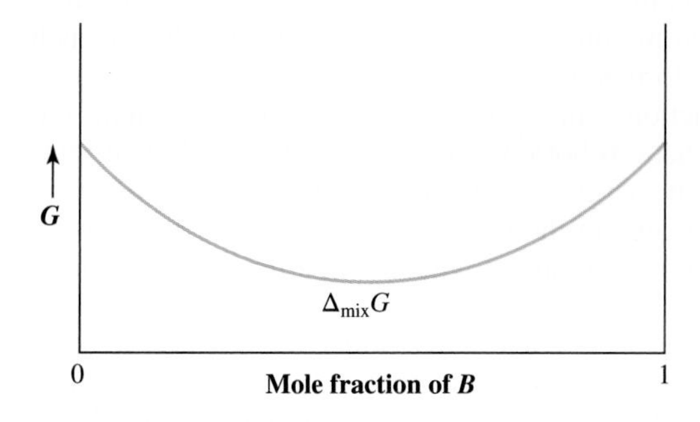

FIGURE 9.9 Gibbs energy diagram for mixing *A*(g) and *B*(g) in the hypothetical reaction *A*(g) → *B*(g). The minimum corresponds to an equimolar mixture of *A*(g) and *B*(g).

If the $NO_2(g)$ molecules formed in the reaction did not mix with the $N_2O_4(g)$ molecules, we would find that the Gibbs energy diagram would simply be a straight line joining $G^{\ominus}_{N_2O_4(g)}$ with $G^{\ominus}_{NO_2(g)}$, similar to that shown on the right of figure 9.7. ($H_2O(l)$ and $H_2O(s)$ are different phases and hence cannot mix.) However, when we add the contribution from $\Delta_{mix}G$, we always find that a minimum exists in the Gibbs energy curve, with its position depending primarily on the sign and magnitude of $\Delta_rG^{\ominus}$, the standard Gibbs energy change for the reaction.

We said in the previous chapter that $\Delta G = 0$ at equilibrium. Let's now look at this statement with reference to figure 9.10.

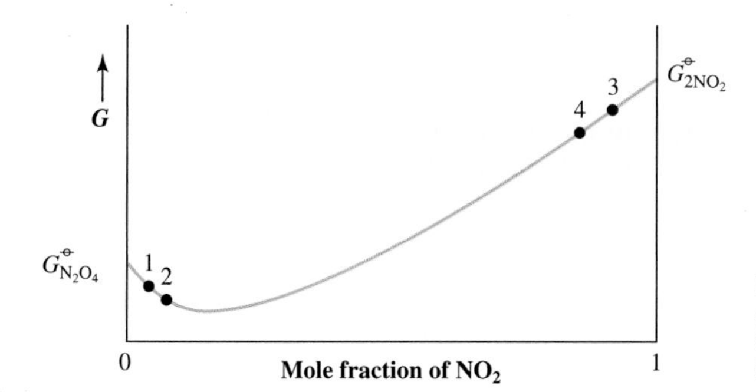

FIGURE 9.10 Gibbs energy diagram for the decomposition of $N_2O_4(g)$. ΔG for any change in composition of the reaction mixture is given by $G_{final} - G_{initial}$.

Again, let us imagine a reaction mixture having the composition indicated by point 1 on the curve and see what happens when the system changes to the composition indicated by point 2. Δ_rG for this change is $G_2 - G_1$ and is negative — thus, we expect the system at point 1 to undergo a spontaneous change to use up $N_2O_4(g)$ and give more $NO_2(g)$. The reverse change, where the system at point 2 changes to give the system at point 1 does not occur as Δ_rG for this change, $G_1 - G_2$, is positive. Similarly, we can see that the system at point 3 will change spontaneously to point 4 as Δ_rG for this change, $G_4 - G_3$, is negative — in other words, $NO_2(g)$ is consumed and $N_2O_4(g)$ is produced. As we approach the minimum in the curve from either direction, Δ_rG for any small change in composition of the reaction mixture starts becoming less negative, until, when we reach the minimum, Δ_rG for any infinitesimally small change in composition of the reaction mixture becomes 0, so the system is at equilibrium. Obviously, when the system has a composition corresponding to the minimum in the Gibbs energy diagram, Δ_rG for *any* change in composition is positive, so no change in composition occurs.

It is very important that you appreciate the difference between Δ_rG and $\Delta_rG^{\ominus}$. You will notice from figure 9.8 that $\Delta_rG^{\ominus}$ for the reaction $N_2O_4(g) \rightarrow 2NO_2(g)$ is positive so you might ask why the reaction proceeds in the forward direction at all. To answer this question, we need to remember that $\Delta_rG^{\ominus}$ is the Gibbs energy change for the conversion of reactants *in their standard states* to products *in their standard states* — in this case, the standard state is the pure gas, each at $p^{\ominus}$. A sample of $N_2O_4(g)$ at $p^{\ominus}$ reacts spontaneously to give some $NO_2(g)$; however, the pressure of $NO_2(g)$ at equilibrium is much less than $p^{\ominus}$ so $N_2O_4(g)$ predominates in the equilibrium mixture. A more familiar example is the vaporisation of water, for which $\Delta G^{\ominus}$ is +8.6 kJ mol^{-1} at 25.0 °C. You know from experience that water evaporates spontaneously at 25.0 °C; in this case, the positive value of $\Delta G^{\ominus}$ simply means that the partial pressure of water never reaches $p^{\ominus}$ under these conditions when the system is at equilibrium. Similarly, a negative value of $\Delta G^{\ominus}$ simply means that the products are present in a large amount at equilibrium, but there are also some reactants in the equilibrium mixture. Hence, we can see that the sign of $\Delta_r G^\ominus$ tells us about *the composition of the reaction mixture at equilibrium*; $\Delta_r G^\ominus$ should not be used as a criterion of spontaneity *except in the very rare case when the reaction is carried out under standard conditions*. It is the *sign* of $\Delta_r G$ that tells us whether a particular change in composition of the reaction mixture is spontaneous.

You might also ask why both the forward and reverse reactions occur spontaneously at equilibrium when the fact that $\Delta_r G = 0$ might suggest that no spontaneous reaction should occur. This apparent contradiction is possibly more a matter of semantics than anything else. Although it is customary to talk of the value of $\Delta_r G$ for a particular reaction, it is more correct, and indeed probably more instructive, to talk of $\Delta_r G$ for a particular *change in the composition of the reaction mixture*. When $\Delta_r G$ is negative, it means that the reaction mixture undergoes a spontaneous change such that reactants are used up and more products are formed in order to minimise the Gibbs energy of the system. Likewise, the term 'spontaneous reaction' can be confusing — when any system is at dynamic equilibrium, the forward and reverse reactions both occur spontaneously, with their rates being governed by both their respective activation energies and the temperature, as we will see in chapter 15. However, there is no spontaneous change in the *composition of the reaction mixture* at equilibrium, and we are referring to this when we talk about whether a particular 'reaction' is spontaneous.

The extent to which a reaction proceeds is very sensitive to the magnitude of $\Delta_r G^\ominus$. If the $\Delta_r G^\ominus$ value for a reaction is reasonably large — about 20 kJ mol^{-1} or more — and positive, almost no reaction will occur at room temperature. On the other hand, the reaction will eventually go almost to completion under the same conditions if $\Delta_r G^\ominus$ is both large and negative. From a practical standpoint, then, *the magnitude and sign of $\Delta_r G^\ominus$ indicate whether spontaneous reaction will occur.* However, even a large negative value of $\Delta_r G^\ominus$ is no guarantee that the predicted spontaneous reaction will actually be observed. Consider, for example the equilibrium:

$$C(s, graphite) + O_2(g) \rightleftharpoons CO_2(g)$$

which has $\Delta_r G^\ominus = -394$ kJ mol^{-1} at 25 °C. The fact that the lead in pencils (which is graphite) can exist in air, which is 21% $O_2(g)$, suggests that another factor is important in determining whether or not a spontaneous reaction will be observed. That factor is kinetics, which we will learn more about in chapter 15.

WORKED EXAMPLE 9.4

Using $\Delta_r G^\ominus$ as a predictor of the outcome of a reaction

Would we expect to be able to observe formation of products in the following reaction at 25 °C?

$$NH_4Cl(s) \rightarrow NH_3(g) + HCl(g)$$

Analysis

We need to determine the magnitude and sign of $\Delta_r G^\ominus$ for the reaction. If $\Delta_r G^\ominus$ is reasonably large and positive, the reaction won't be observed. If it is reasonably large and negative, we can expect to see the reaction go nearly to completion.

Solution

First, let's calculate $\Delta_r G^\ominus$ for the reaction using the data in table 8.7 on page 335. The procedure is the same as that discussed on pages 333–4.

$$\begin{aligned}\Delta_r G^\ominus &= (\Delta_f G^\ominus_{NH_3(g)} + \Delta_f G^\ominus_{HCl(g)}) - (\Delta_f G^\ominus_{NH_4Cl(s)}) \\ &= [1\times(-16.7\,\text{kJ mol}^{-1}) + (1\times(-95.27\,\text{kJ mol}^{-1})] - [1\times(-203.9\,\text{kJ mol}^{-1})] \\ &= +91.9\,\text{kJ mol}^{-1}\end{aligned}$$

Because $\Delta_r G^\ominus$ is large and positive, we expect to observe the spontaneous formation of only extremely small amounts of products at this temperature.

PRACTICE EXERCISE 9.7

Use the data in table 8.7 to show that the reaction:

$$4NH_3(g) + 7O_2(g) \rightarrow 4NO_2(g) + 6H_2O(g)$$

should produce substantial amounts of $NO_2(g)$ at 25 °C. Given that $NO_2(g)$ is red-brown in colour, why do you think that bottles of aqueous ammonia ($NH_3(g)$ dissolved in water) do not show a red-brown colour above the liquid?

The relationship between $\Delta_r G^\ominus$ and K

In the preceding discussion, we learned in a qualitative way that the position of equilibrium in a reaction is determined by the sign and magnitude of $\Delta_r G^\ominus$. We also learned that the direction in which a reaction proceeds depends on where the system composition stands relative to the minimum on the Gibbs energy curve. Thus, the reaction proceeds spontaneously in the forward direction only if it lowers the Gibbs energy (i.e. if $\Delta_r G$ for the change in question is negative).

Quantitatively, the relationship between $\Delta_r G$ and $\Delta_r G^\ominus$ is expressed by the following equation, which is derived from the ideal gas equation (chapter 6, p. 222):

$$\Delta_r G = \Delta_r G^\ominus + RT \ln Q$$

where R is the gas constant (8.314 J mol^{-1} K^{-1}), T is the temperature in kelvin and $\ln Q$ is the natural logarithm of the reaction quotient. For gaseous reactions, Q_p is calculated using partial pressures expressed in Pa, while, for reactions in solution, Q_c is calculated from molar concentrations. This equation allows us to predict the direction of the spontaneous change in a reaction mixture if we know $\Delta_r G^\ominus$ and the composition of the mixture, as illustrated in worked example 9.5.

WORKED EXAMPLE 9.5

Determining the direction of a spontaneous reaction

The reaction $2NO_2(g) \rightleftharpoons N_2O_4(g)$ has $\Delta_r G^\ominus = -5.40$ kJ mol^{-1} at 298 K. At the instant of mixing samples of the two gases, the partial pressure of NO_2 is 0.25×10^5 Pa and the partial pressure of N_2O_4 is 0.60×10^5 Pa. How must the composition of the reaction mixture change to reach equilibrium?

Analysis

Since we know that reactions proceed spontaneously *towards equilibrium,* we are really being asked to determine whether the reaction proceeds spontaneously in the forward or reverse direction. We can calculate $\Delta_r G$ for the forward reaction. If $\Delta_r G$ is negative, the forward reaction is spontaneous. However, if the calculated $\Delta_r G$ is positive, the reverse reaction is spontaneous.

Solution

First, we need the correct form for the reaction quotient expression. Expressed in terms of partial pressures in Pa, this is:

$$Q_p = \frac{\dfrac{p_{N_2O_4}}{p^\ominus}}{\left(\dfrac{p_{NO_2}}{p^\ominus}\right)^2}$$

Therefore, the equation we use is:

$$\Delta_r G = \Delta_r G^\ominus + RT \ln\left(\frac{\dfrac{p_{N_2O_4}}{p^\ominus}}{\left(\dfrac{p_{NO_2}}{p^\ominus}\right)^2}\right)$$

Now we assemble the data:

$\Delta_r G^\ominus = -5.40$ kJ mol^{-1} $= -5.40 \times 10^3$ J mol^{-1} $\quad R = 8.314$ J mol^{-1} K^{-1}

$T = 298$ K $\quad p_{N_2O_4} = 0.60 \times 10^5$ Pa $\quad p_{NO_2} = 0.25 \times 10^5$ Pa $\quad p^\ominus = 1 \times 10^5$ Pa

Notice that we must change the units of $\Delta_r G^\ominus$ to J mol^{-1} so they are compatible with those of R. Substituting quantities gives:

$$\Delta_r G = -5.40 \times 10^3\ \text{J mol}^{-1} + (8.314\ \text{J mol}^{-1}\ \text{K}^{-1})(298\ \text{K}) \ln\left(\frac{\dfrac{0.60\times10^5}{1\times10^5}}{\left(\dfrac{0.25\times10^5}{1\times10^5}\right)^2}\right)$$

$$= -5.40 \times 10^3\ \text{J mol}^{-1} + (8.314\ \text{J mol}^{-1}\ \text{K}^{-1})(298\ \text{K})(2.26)$$

$$= -5.40 \times 10^3\ \text{J mol}^{-1} + 5.60 \times 10^3\ \text{J mol}^{-1}$$

$$= +2.0 \times 10^2\ \text{J mol}^{-1}$$

Since $\Delta_r G$ is positive, the forward reaction is nonspontaneous so the reverse reaction must occur for the system to reach equilibrium. The composition of the reaction mixture changes to increase the amount of NO_2 and decrease the amount of N_2O_4.

Is our answer reasonable?

There is no simple check. However, we can check that the energy units in both terms on the right are the same. Notice that here we have changed the units for $\Delta_r G^\ominus$ to J mol^{-1}, to match those of R. Also notice that the temperature is expressed in kelvin, to match the temperature units in R.

PRACTICE EXERCISE 9.8 $\Delta_r G^{\ominus}_{298\,K} = 15\,kJ\,mol^{-1}$ for the forward reaction in the equilibrium:

$$2HI \rightleftharpoons H_2 + I_2$$

The partial pressures of HI, H_2 and I_2 in a mixture of the gases were 1.00×10^4 Pa, 6.50×10^4 Pa and 4.25×10^4 Pa, respectively. In which direction would you expect spontaneous change to occur in this system?

At this point, we have two seemingly separate criteria for determining whether or not a system is at equilibrium. We know that, when $\Delta_r G = 0$ for an infinitesimal change in composition of a reaction mixture, or when $Q = K$ for a particular set of reaction conditions, the system is at equilibrium. We will now substitute these conditions into the equation:

$$\Delta_r G = \Delta_r G^{\ominus} + RT\ln Q$$

If $\Delta_r G = 0$, $Q = K$, and thus:

$$0 = \Delta_r G^{\ominus} + RT\ln K$$

and therefore:

$$\Delta_r G^{\ominus} = -RT\ln K$$

This is one of the most important equations in thermodynamics, if not all of chemistry. It allows us to use tabulated values of $\Delta_r G^{\ominus}$ to calculate the equilibrium constant for a particular reaction, and, conversely, we can use measured values of the equilibrium constant to determine $\Delta_r G^{\ominus}$ for a reaction. *K in the reaction as written corresponds to K_p for reactions involving gases and K_c for reactions involving species in solution.*

From this equation you can see that, if $\Delta_r G^{\ominus}$ for a reaction is large and negative, ln$K$ is large and positive, and hence K is large — the equilibrium position lies towards products. Similarly, if $\Delta_r G^{\ominus}$ is large and positive, ln$K$ is negative, and hence K is small — the equilibrium position lies towards reactants.

WORKED EXAMPLE 9.6

Equilibrium constants

The brownish haze associated with air pollution is caused by nitrogen dioxide, NO_2, a red-brown gas (figure 9.11). Nitric oxide, NO, is formed in car engines and some of it escapes into the air, where it is oxidised to NO_2 by oxygen.

$$2NO(g) + O_2(g) \rightleftharpoons 2NO_2(g)$$

The value of K_p for this reaction is 1.7×10^{12} at 25 °C. What is $\Delta_r G^{\ominus}$ for the forward reaction at this temperature?

FIGURE 9.11 The atmospheric pollutant nitrogen dioxide, NO_2, produces a characteristic brown haze.

Analysis

We can calculate $\Delta_r G^{\ominus}$ from the given value of K_p using the following relationship.

$$\Delta_r G^{\ominus} = -RT\ln K_p$$

We'll need the following data.

$$R = 8.314\,J\,mol^{-1}\,K^{-1} \qquad T = 298\,K \qquad K_p = 1.7 \times 10^{12}$$

Solution

Substituting these values into the equation, we have:

$$\begin{aligned}\Delta_r G^{\ominus} &= -(8.314\,J\,mol^{-1}\,K^{-1} \times 298\,K)\ln(1.7 \times 10^{12})\\ &= -(8.314\,J\,mol^{-1}\,K^{-1} \times 298\,K)(28.16)\\ &= -7.0 \times 10^4\,J\,mol^{-1}\end{aligned}$$

Expressed in kilojoules per mole, $\Delta_r G^{\ominus} = -7.0 \times 10^1\,kJ\,mol^{-1}$.

Is our answer reasonable?

The large value of K_p tells us that the position of equilibrium lies far to the right, which means $\Delta_r G^{\ominus}$ must be large and negative. Therefore, the answer, $-70\,kJ\,mol^{-1}$, seems reasonable.

PRACTICE EXERCISE 9.9

The equilibrium $N_2(g) + 3H_2(g) \rightleftharpoons 2NH_3(g)$ has $K_p = 6.9 \times 10^5$ at 25.0 °C. Calculate $\Delta_r G^\ominus$ for the forward reaction.

WORKED EXAMPLE 9.7

Equilibrium constants

Sulfur dioxide, which is sometimes present in polluted air, reacts with oxygen when it passes over the catalyst in a car's catalytic converter, producing the acidic oxide SO_3.

$$2SO_2(g) + O_2(g) \rightleftharpoons 2SO_3(g)$$

$\Delta_r G^\ominus = -1.40 \times 10^2\,\text{kJ mol}^{-1}$ for the forward reaction at 25 °C. What is the value of K_p at this temperature?

Analysis

As in worked example 9.6, we use the equation:

$$\Delta_r G^\ominus = -RT\ln K_p$$

Our data are:

$R = 8.314\,\text{J mol}^{-1}\,\text{K}^{-1}$ $\quad T = 298\,\text{K}$ $\quad \Delta_r G^\ominus = -1.40 \times 10^2\,\text{kJ mol}^{-1} = -1.40 \times 10^5\,\text{J mol}^{-1}$

To calculate K_p, it is easiest to rearrange the equation and solve for $\ln K_p$.

$$\ln K_p = \frac{-\Delta_r G^\ominus}{RT}$$

Solution

Substituting values gives:

$$\ln K_p = \frac{-(-1.40 \times 10^5\,\text{J mol}^{-1})}{(8.314\,\text{J mol}^{-1}\,\text{K}^{-1})(298\,\text{K})}$$

$$= +56.5$$

To calculate K_p, we take the antilogarithm:

$$K_p = e^{56.5}$$

$$= 3 \times 10^{24}$$

Notice that we have expressed the answer to only one significant figure. This is because, as we saw in chapter 2, when taking a logarithm, the number of digits written after the decimal place equals the number of significant figures in the number. Conversely, the number of significant figures in the antilogarithm equals the number of digits after the decimal in the logarithm.

Is our answer reasonable?

The value of $\Delta_r G^\ominus$ is large and negative, so the position of equilibrium should favour the products. The large value of K_p is therefore reasonable.

PRACTICE EXERCISE 9.10

The forward reaction in the equilibrium:

$$CO(g) + \frac{1}{2}O_2(g) \rightleftharpoons CO_2(g)$$

has $\Delta_r G^\ominus_{298\,\text{K}} = -257.1\,\text{kJ mol}^{-1}$. What is the value of K_p at this temperature?

We can use thermodynamic data collected at 25 °C to calculate good approximations of the values of equilibrium constants at temperatures other than 25 °C. This is because the values of $\Delta_r H^\ominus$ and $\Delta_r S^\ominus$ do not change much with temperature, so we can use these to calculate $\Delta_r G^\ominus$ and, hence, K at the new temperature.

WORKED EXAMPLE 9.8

Calculating *K* at temperatures other than 25 °C

Nitrous oxide, N_2O, is an anaesthetic known as laughing gas because it sometimes relieves patients of their inhibitions. The decomposition of nitrous oxide has $K_p = 1.8 \times 10^{36}$ at 25 °C. The equation is:

$$2N_2O(g) \rightleftharpoons 2N_2(g) + O_2(g)$$

For this reaction, $\Delta_r H^\circ = -163\,\text{kJ mol}^{-1}$ and $\Delta_r S^\circ = +148\,\text{J mol}^{-1}\,\text{K}^{-1}$. What is the approximate value of K_p for this reaction at 40 °C?

Analysis

We need a value of $\Delta_r G^\circ$ at 40 °C (313 K), which we can represent as $\Delta_r G^\circ_{313\,\text{K}}$. We can estimate $\Delta_r G^\circ_{313\,\text{K}}$ using the values of $\Delta_r H^\circ$ and $\Delta_r S^\circ$ measured at 25 °C and the equation $\Delta_r G^\circ = \Delta_r H^\circ - T\Delta_r S^\circ$.

Hence:

$$\Delta_r G^\circ_{313\,\text{K}} \approx \Delta_r H^\circ_{298\,\text{K}} - (313\,\text{K})\Delta_r S^\circ_{298\,\text{K}}$$

Solution

Substituting the values of $\Delta_r H^\circ$ and $\Delta_r S^\circ$ provided in the question gives:

$$\Delta_r G^\circ_{313\,\text{K}} \approx -1.63 \times 10^5\,\text{J mol}^{-1} - (313\,\text{K})(+148\,\text{J mol}^{-1}\,\text{K}^{-1})$$

Notice that we have converted kJ mol^{-1} to J mol^{-1}. Performing the arithmetic gives:

$$\Delta_r G^\circ_{313\,\text{K}} \approx -2.09 \times 10^5\,\text{J mol}^{-1}$$

The next step is to use this value of $\Delta_r G^\circ_{313\,\text{K}}$ to calculate K_p. First, let's solve for $\ln K_p$.

$$\ln K_p = \frac{-\Delta_r G^\circ_{313\,\text{K}}}{RT}$$

Substituting with $R = 8.314\,\text{J mol}^{-1}\,\text{K}^{-1}$ and $T = 313\,\text{K}$ gives:

$$\begin{aligned}\ln K_p &= \frac{2.09 \times 10^5\,\text{J mol}^{-1}}{(8.314\,\text{J mol}^{-1}\,\text{K}^{-1})(313\,\text{K})} \\ &= +80.3\end{aligned}$$

Taking the antilogarithm:

$$\begin{aligned}K_p &= e^{80.3} \\ &= 7 \times 10^{34}\end{aligned}$$

Notice that, at this higher temperature, $N_2O(g)$ is actually slightly more stable with respect to dissociation into its elements than at 25 °C, as reflected in the slightly smaller value for the equilibrium constant for its decomposition.

Is our answer reasonable?

As we will learn in the next section, the equilibrium constant for an exothermic reaction decreases with increasing temperature. Therefore, the answer is reasonable.

PRACTICE EXERCISE 9.11

The forward reaction in the equilibrium $N_2(g) + 3H_2(g) \rightleftharpoons 2NH_3(g)$ has a standard enthalpy of reaction of $-92.4\,\text{kJ mol}^{-1}$ and a standard entropy of reaction of $-198.3\,\text{J mol}^{-1}\,\text{K}^{-1}$ at 25 °C. Estimate the value of K_p for this reaction at 50 °C.

9.4 How systems at equilibrium respond to change

We have seen that the composition of a reaction mixture changes in the direction of decreasing Gibbs energy until it reaches equilibrium. What we have not yet discussed is how a system at equilibrium behaves when a change is made that perturbs the system. Such a change might be a change in temperature, volume or pressure of the system, addition or removal of chemical species to or from the system, or addition of a catalyst. Such knowledge is important as it may allow us to optimise reaction conditions to obtain the maximum amount of a desired product.

Le Châtelier's principle

One of the first to study in detail the effect of change on a system at equilibrium was the French chemist Henri Le Châtelier (1850–1936) who proposed what came to be known as **Le Châtelier's principle**. This states that, *if an outside influence upsets an equilibrium, the system undergoes a change in a direction that counteracts the disturbing influence and, if possible, returns the* *system to equilibrium*. This is often interpreted as meaning that, if a chemical species is added to one side of an equilibrium, the equilibrium position will shift to use up the added species. Le Châtelier's principle is useful as a predictive tool in many circumstances. Consider, for example, the equilibrium:

$$2NO_2(g) \rightleftharpoons N_2O_4(g)$$

for which the forward reaction is exothermic under standard conditions. According to Le Châtelier's principle, addition of $NO_2(g)$ to the reaction mixture at equilibrium will result in the production of more $N_2O_4(g)$, as the equilibrium position will move to the right to use up the added $NO_2(g)$. Using similar reasoning, we can see that addition of $N_2O_4(g)$ to the reaction mixture at equilibrium will result in the production of more $NO_2(g)$. If we take the above system at equilibrium and increase the volume of the container in which the reaction is occurring, this will have the effect of decreasing the pressure in the container. To counteract this disturbing influence, we would predict, from Le Châtelier's principle, that the equilibrium position will move to the left to produce more $NO_2(g)$, as this will minimise the pressure decrease by maximising the amount of gas in the container. Finally, as the forward reaction is exothermic, it gives out heat, which we can therefore consider as being a 'product' of the reaction; in other words, we can write the equation:

$$2NO_2(g) \rightleftharpoons N_2O_4(g) + \text{heat}$$

If we now take the system at equilibrium and increase its temperature, the equilibrium position will shift to the left to use up the added heat, resulting in the production of more $NO_2(g)$.

However, predictions made on the basis of Le Châtelier's principle can be incorrect if it is not applied correctly; in fact, adding a reactant or product to a system at equilibrium does not necessarily result in a shift of the equilibrium position. For example, addition of AgCl(s) to a saturated aqueous solution of AgCl in equilibrium with excess AgCl(s) does not result in an increase in $[Ag^+(aq)]$ and $[Cl^-(aq)]$ in the solution (we will see the reason why below). Most importantly, Le Châtelier's principle does not explain *why* the equilibrium position does or does not change on perturbation of the equilibrium. Therefore, it is better to consider the perturbation of systems at equilibrium by comparison of the values of the equilibrium constant, K, and the reaction quotient, Q, for a particular chemical process. We start with a system at equilibrium for which $Q = K$. We then perturb the system and consider how this alters the value of Q and how the system must alter to make Q again equal to K. We will consider a number of different scenarios using this method.

Adding or removing a product or reactant

The addition or removal of a product or reactant instantaneously alters the concentration of that species in the reaction mixture, provided that the reactant or product in question is not a pure solid or liquid. When this happens, the value of Q changes so that $Q \neq K$ and the system is no longer at equilibrium. We will illustrate this using an equilibrium involving coordination complexes (chapter 13).

$$\underset{\text{blue}}{[Cu(OH_2)_4]^{2+}(aq)} + 4Cl^-(aq) \rightleftharpoons \underset{\text{yellow}}{[CuCl_4]^{2-}(aq)} + 4H_2O(l)$$

$[Cu(OH_2)_4]^{2+}$ is blue and $[CuCl_4]^{2-}$ is yellow; mixtures of the two have an intermediate colour and, therefore, appear blue-green, as illustrated in figure 9.12.

What happens when we increase the chloride ion concentration by adding a small volume of a concentrated aqueous solution of HCl to the solution at equilibrium? To answer this question, we need to write the expression for Q, the reaction quotient. Recalling from the section on heterogeneous systems (p. 358) that H_2O does not appear in the expression for Q because it is a pure liquid, we can write:

$$Q = \frac{[CuCl_4{}^{2-}]}{[Cu(OH_2)_4{}^{2+}][Cl^-]^4}$$

Only reactants and products that appear in the expression for Q can influence the position of the equilibrium. We can see that, immediately after adding the HCl solution, the $[Cl^-]$ in the solution increases. This has the effect of *decreasing* the value of Q, because $[Cl^-]$ appears on the bottom line of the reaction quotient expression. Thus, we now have the situation where the solution is no longer at equilibrium and $Q < K$. Obviously, to restore equilibrium, we must alter the concentrations of the reactants and products to increase Q so that it again equals K. (Remember that K is a constant at constant temperature, so we must alter the value of Q.) With reference to the expression for Q, there are two ways that we can increase its value; we can make the top line larger,

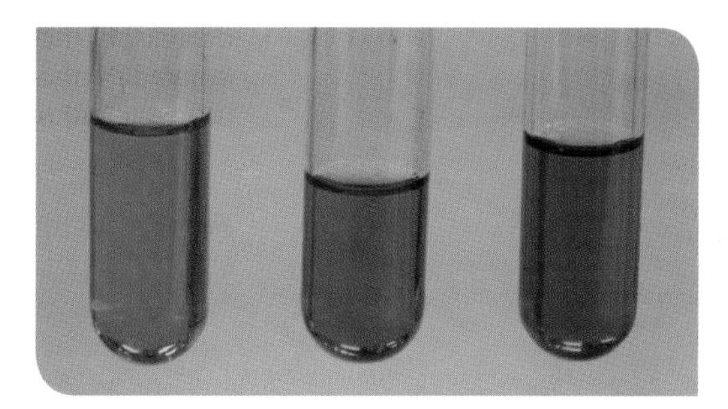

FIGURE 9.12 The effect of concentration changes on the position of equilibrium. The solution in the centre contains a mixture of blue $[Cu(OH_2)_4]^{2+}$ and yellow $[CuCl_4]^{2-}$, so it has a blue-green colour. The tube on the right contains some of the same mixture after the addition of concentrated HCl. It has a more pronounced green colour because the equilibrium is shifted towards $[CuCl_4]^{2-}$. At the left is some of the original mixture after removal of some Cl^- as insoluble AgCl and subsequent filtering. It is blue because the equilibrium has shifted towards $[Cu(OH_2)_4]^{2+}$.

by increasing the amount and, therefore, concentration of $[CuCl_4]^{2-}$, or we can make the bottom line smaller, by decreasing the amounts and, therefore, concentrations of $[Cu(OH_2)_4]^{2+}$ and Cl^-. These, in fact, equate to the same thing — a shift of the equilibrium position towards *products*. The pronounced green colour of the solution at the right of figure 9.12 shows that this is in fact what happens; there is more yellow $[CuCl_4]^{2-}$ and less blue $[Cu(OH_2)_4]^{2+}$ present in the solution after the system re-establishes equilibrium following addition of concentrated HCl(aq).

We can use the same reasoning to predict the effect of removing a reactant from the system. If we take the system at equilibrium and add a small volume of a concentrated solution of silver perchlorate, $AgClO_4$, the Ag^+ ions react with free Cl^- ions to form insoluble AgCl(s), thus effectively removing Cl^- ions from the mixture. Again, looking at our expression for Q, we can see that lowering $[Cl^-]$ makes the bottom line of the reaction quotient expression smaller, which leads to an instantaneous *increase* in the value of Q. The system is no longer at equilibrium, and Q is greater than K. The concentrations of reactants and products must, therefore, change to *decrease* the value of Q so that it again is equal to K. This may be achieved by either decreasing the value of the top line of the reaction quotient expression (lowering the amount and, therefore, concentration of $[CuCl_4]^{2-}$) or increasing the value of the bottom line (increasing the amounts and, therefore, concentrations of $[Cu(OH_2)_4]^{2+}$ and Cl^-). Again, figure 9.12 shows that the colour of the solution becomes more blue following removal of some Cl^- as insoluble AgCl (and subsequent filtering), consistent with the production of more $[Cu(OH_2)_4]^{2+}$.

This treatment is purely qualitative but allows us to predict the manner in which an equilibrium shifts *without having to know the actual values of K and Q*. See figure 9.4 for a summary of this treatment.

An important example already alluded to (p. 367) is that of a sparingly soluble ionic salt in equilibrium with its constituent aqueous ions. Consider, for instance, a saturated aqueous solution of silver chloride, AgCl, in the presence of solid AgCl. The relevant equilibrium is:

$$AgCl(s) \rightleftharpoons Ag^+(aq) + Cl^-(aq)$$

and we write the reaction quotient expression as:

$$Q = [Ag^+][Cl^-]$$

remembering that pure solids and pure liquids do not appear in the expressions for either K or Q. What would be the effect of adding more solid AgCl to the equilibrium mixture? AgCl(s) does not appear in the expression for Q so addition of AgCl(s) has *no effect* on the position of equilibrium. Le Châtelier's principle can lead to incorrect predictions in problems of this type (in this case, it would predict an increase in $[Ag^+]$ and $[Cl^-]$), and it is preferable to use the comparison of Q and K to determine the effect of adding or removing reactants or products.

We will see more of these types of problems in chapter 10.

Changing the pressure in gaseous reactions

There are two ways of changing the total pressure in a gaseous reaction mixture at equilibrium:

- changing the volume of the system
- adding an inert gas.

We will look at these in turn.

Changing the volume of the system

Changing the volume of a gas mixture at equilibrium changes both the concentrations and partial pressures of the reactant and product gases, and this can alter the value of Q. Increasing the volume will decrease the partial pressures of all gases and therefore decrease the total pressure, whereas decreasing the volume will have the opposite effect. We will consider the effect of increasing the volume of the reaction vessel on the equilibrium position for the formation of ammonia from nitrogen and hydrogen.

$$N_2(g) + 3H_2(g) \rightleftharpoons 2NH_3(g)$$

For this equilibrium, we would normally write Q_p as:

$$Q_p = \frac{\left(\frac{p_{NH_3}}{p^\ominus}\right)^2}{\left(\frac{p_{N_2}}{p^\ominus}\right)\left(\frac{p_{H_2}}{p^\ominus}\right)^3}$$

However, in this case, it is more instructive to use Q_c to determine the effect of changing the volume of the reaction mixture. Hence, we can write:

$$Q_c = \frac{[NH_3]^2}{[N_2][H_2]^3}$$

Remembering that $c = \frac{n}{V}$, we can rewrite this as:

$$Q_c = \frac{\frac{(n_{NH_3})^2}{V^2}}{\frac{n_{N_2}}{V} \times \frac{(n_{H_2})^3}{V^3}}$$

As $\frac{n}{V}$ is the same as nV^{-1}, we can simplify this expression as follows:

$$Q_c = \frac{\frac{(n_{NH_3})^2}{V^2}}{\frac{n_{N_2}}{V} \times \frac{(n_{H_2})^3}{V^3}}$$

$$= \frac{(n_{NH_3})^2 V^{-2}}{(n_{N_2})V^{-1} \times (n_{H_2})^3 V^{-3}}$$

$$= \frac{(n_{NH_3})^2}{(n_{N_2})(n_{H_2})^3} \times \frac{V^{-2}}{V^{-4}} = \frac{(n_{NH_3})^2}{n_{N_2}(n_{H_2})^3} \times V^2$$

While this may look somewhat daunting, the important point is that we have shown that the value of *Q_c is proportional to V^2* — in other words, when we increase V, we increase the value of Q. Using the same reasoning as on page 353, we can say that increasing the volume of the reaction vessel, V, will increase the value of Q, making $Q > K$. The system will then change in a way that makes the value of Q smaller: that is, making the value of the top line of the reaction quotient expression smaller by reducing n_{NH_3} or making the value of the bottom line larger by increasing n_{N_2} and n_{H_2}. This means that, when the volume of the reaction mixture is increased, the reaction mixture composition shifts towards reactants, resulting in the production of more N_2 and H_2. Increasing the volume of the reaction vessel in gas phase reactions which have a larger number of reactant molecules than product molecules will always result in a shift of the reaction mixture composition towards reactants.

If we consider the equilibrium:

$$H_2(g) + I_2(g) \rightleftharpoons 2HI(g)$$

a similar treatment to that above shows that:

$$Q_c = \frac{\frac{(n_{HI})^2}{V^2}}{\frac{n_{H_2}}{V} \times \frac{n_{I_2}}{V}}$$

$$= \frac{(n_{HI})^2 V^{-2}}{n_{H_2} V^{-1} \times n_{I_2} V^{-1}}$$

$$= \frac{(n_{HI})^2}{n_{H_2} n_{I_2}} \times \frac{V^{-2}}{V^{-2}} = \frac{(n_{HI})^2}{n_{H_2} n_{I_2}}$$

In this case, the value of *Q_c does not depend on V*. Thus, increasing or decreasing the volume of this reaction mixture has *no effect* on the position of the equilibrium. This is always the case when there is the same number of moles of gas on each side of the equation in a gas phase reaction.

The final example uses the familiar equilibrium:

$$N_2O_4(g) \rightleftharpoons 2NO_2(g)$$

To look at the effect of increasing the volume of the system at equilibrium, we use the same treatment as above to give:

$$Q_c = \frac{\frac{(n_{NO_2})^2}{V^2}}{\frac{n_{N_2O_4}}{V}}$$

$$= \frac{(n_{NO_2})^2 V^{-2}}{n_{N_2O_4} V^{-1}}$$

$$= \frac{(n_{NO_2})^2}{n_{N_2O_4}} \times \frac{V^{-2}}{V^{-1}} = \frac{(n_{NO_2})^2}{n_{N_2O_4}} \times \frac{1}{V}$$

and we can see that the value of *Q_c is proportional to the inverse of V* — in other words, as the volume increases, the value of Q_c decreases so $Q < K$. To re-establish equilibrium, the system will respond to increase the value of Q: that is, making the value of the top line of the reaction quotient expression larger by increasing n_{NO_2} or making the value of the bottom line smaller by

FIGURE 9.13 Jacobus Henricus van't Hoff (1852–1911), a Dutch chemist, was the first winner of the Nobel Prize in chemistry in 1901 for his significant contributions to chemical thermodynamics. However, possibly his most remarkable achievement was in organic chemistry, with his realisation that carbon atoms bearing four attached atoms adopt a tetrahedral, rather than a square planar, arrangement. What made this so remarkable was that he published this work at the age of 22 — not much older than most of you reading this.

decreasing $n_{N_2O_4}$. Both scenarios correspond to an equilibrium shift towards products following an increase in the volume of the reaction mixture. Increasing the volume of the reaction vessel in gas phase reactions which have a larger number of product molecules than reactant molecules will always result in a shift of the equilibrium towards products.

In summary, we have shown that Q_c is proportional to $V^{-\Delta n_{gas}}$, where Δn_{gas} is defined the same way as on page 355, namely as the number of moles of gaseous products minus the number of moles of gaseous reactants. Knowing this proportionality allows us to determine the effect of changing the volume of the reaction vessel on any system at equilibrium.

Adding an inert gas at constant volume

Adding an inert gas (which, by definition, is unreactive) to a gaseous reaction mixture at equilibrium increases the total pressure of the system, but *does not alter the position of equilibrium*. To prove this, let's consider adding the inert gas helium to the N_2O_4/NO_2 equilibrium mixture. As helium does not react with either the reactant or product, our expressions for both Q_c and Q_p for this equilibrium remain unchanged in the presence of helium. Hence, as there is no term for [He] or p_{He} in the expressions for Q_c or Q_p, respectively, addition of helium *cannot have any effect on the position of equilibrium* as it cannot alter the value of Q. This is true in all cases where the total pressure of the reaction mixture is altered by addition of an inert gas.

Changing the temperature of a reaction mixture

The value of the equilibrium constant, K, for any reaction can be changed only by altering the temperature of the reaction mixture. We can predict the effect of changing the temperature of a reaction mixture at equilibrium from the sign of $\Delta_r H^\ominus$ for the forward reaction. The **van't Hoff equation** (figures 9.13 and 9.14) states:

$$\frac{d\ln K}{dT} = \frac{\Delta_r H^\ominus}{RT^2}$$

and this means that the slope of the plot of $\ln K$ versus T has the same sign as that of $\Delta_r H^\ominus$.

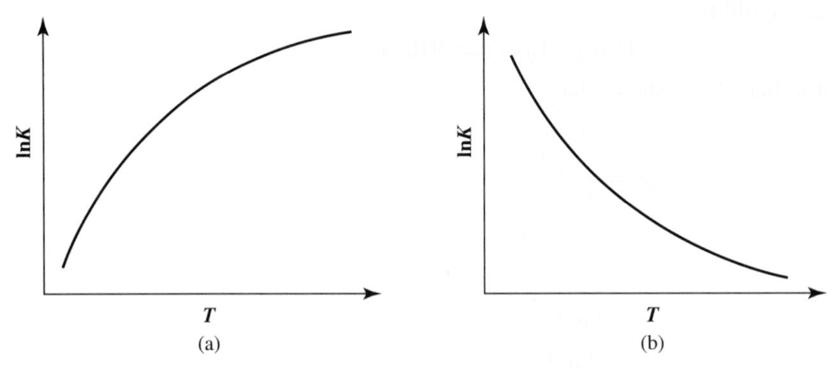

FIGURE 9.14 Plots of $\ln K$ versus T for situations where $\Delta_r H^\ominus$ is: **(a)** positive and **(b)** negative.

For example, the reaction of nitrogen and hydrogen to form ammonia is exothermic:

$$N_2(g) + 3H_2(g) \rightleftharpoons 2NH_3(g) \qquad \Delta_r H^\ominus = -92.38\ \text{kJ mol}^{-1}$$

so the graph of $\ln K$ versus T should have a negative slope. This means that, as we increase the temperature, $\ln K$, and hence K, should *decrease*. Thus, the production of ammonia becomes less favourable as we increase the temperature of the reaction mixture. We can make the following generalisations about the effect of temperature changes on the position of equilibrium.

- For an *endothermic* reaction ($\Delta_r H$ is positive), increasing the temperature *increases* the equilibrium constant, so products become more favoured (figure 9.14a).
- For an *exothermic* reaction ($\Delta_r H$ is negative), increasing the temperature *decreases* the equilibrium constant, so reactants become more favoured (figure 9.14b).

Figure 9.15 demonstrates the temperature dependence of the equilibrium involving complexes of Cu^{2+} discussed previously (p. 367), for which the forward reaction is endothermic. Cooling the reaction mixture *decreases* the equilibrium constant, giving a shift towards the reactants; heating the mixture *increases* the equilibrium constant, so more product is formed.

This illustrates the qualitative use of the van't Hoff equation. We can also use it quantitatively; if we know the value of an equilibrium constant, K, at a particular temperature, T_1, we can use the following form of the van't Hoff equation to determine the value of the equilibrium constant at another temperature, T_2.

$$\ln K_{T_2} - \ln K_{T_1} = \frac{-\Delta_r H^\ominus}{R}\left(\frac{1}{T_2} - \frac{1}{T_1}\right)$$

This assumes that $\Delta_r H^\ominus$ is constant over the temperature range T_1 to T_2.

FIGURE 9.15 The effect of temperature on the equilibrium:

$$[Cu(OH_2)_4]^{2+} + 4Cl^- \rightleftharpoons [CuCl_4]^{2-} + 4H_2O$$

The tube in the centre shows an equilibrium mixture of the two complexes. When the solution is cooled in ice (left), the equilibrium shifts towards the blue $[Cu(OH_2)_4]^{2+}$. When heated in boiling water (right), the equilibrium shifts towards the yellow $[CuCl_4]^{2-}$. This behaviour indicates that the reaction is endothermic in the forward direction.

Addition of a catalyst

Catalysts are substances that affect the rates of chemical reactions without actually being used up. While addition of a catalyst helps bring a system to chemical equilibrium more rapidly, the position of equilibrium is not affected. This is because catalysts are not included in the overall balanced chemical equation for the reaction, and so do not appear in the expression for the reaction quotient, Q.

WORKED EXAMPLE 9.9

Predicting equilibrium shifts

The reaction $N_2O_4(g) \rightleftharpoons 2NO_2(g)$ is endothermic, with $\Delta_r H^\ominus = +56.9\,kJ\,mol^{-1}$. If we start with an equilibrium mixture of $N_2O_4(g)$ and $NO_2(g)$, how is the amount of NO_2 in the mixture affected by: (a) adding N_2O_4, (b) lowering the pressure by increasing the volume of the container, (c) raising the temperature and (d) adding a catalyst to the system? Which of these changes alters the value of K_c?

Analysis

We are taking an equilibrium mixture of $N_2O_4(g)$ and $NO_2(g)$ and making various changes to the system. We must determine, via comparison of the values of Q and K, whether these changes alter the equilibrium position and, therefore, how this affects the amount of $NO_2(g)$.

Solution

To answer these questions, first write the expression for Q and then use it to determine the effect of the change made.

$$Q_c = \frac{[NO_2]^2}{[N_2O_4]}$$

(a) Adding $N_2O_4(g)$ instantaneously decreases the value of Q_c so $Q_c < K_c$. To make $Q_c = K_c$, the equilibrium shifts to the right to increase the value of Q_c, resulting in the production of more $NO_2(g)$. The amount of $NO_2(g)$ *increases*.

(b) We have seen on pages 369–70 that increasing the volume of the container reduces the value of Q_c because the value of Q_c is inversely proportional to V. Hence, to make $Q_c = K_c$, the equilibrium shifts to the right to increase the value of Q_c, resulting in the production of more $NO_2(g)$. The amount of $NO_2(g)$ *increases*.

(c) The forward reaction is endothermic, meaning that $\Delta_r H^\ominus$ is positive; by applying the van't Hoff equation, a plot of $\ln K_c$ versus T has a positive slope. Thus $\ln K_c$, and therefore K_c, increases with increasing temperature, so $Q_c < K_c$. To make $Q_c = K_c$, the value of Q_c must increase, resulting in the production of more $NO_2(g)$ at equilibrium. The amount of $NO_2(g)$ *increases*.

(d) As stated above, a catalyst can have no effect on the position of equilibrium. The amount of $NO_2(g)$ remains unchanged.

Finally, the only change that alters the value of K_c is a change in temperature.

Is our answer reasonable?

Compare each of our answers with examples given in this section to confirm that they are logically consistent.

PRACTICE EXERCISE 9.12

Consider the equilibrium:

$$S(s) + O_2(g) \rightleftharpoons SO_2(g)$$

For the forward reaction, $\Delta_r H^\ominus = -297\,kJ\,mol^{-1}$. How is the amount of SO_2 at equilibrium affected by: (a) adding $O_2(g)$, (b) adding $S(s)$, (c) raising the temperature, and (d) decreasing the volume of the container? How (if at all) does each of these changes affect K_p for the reaction?

Chemical Connections

Air pollution and equilibrium

Cars are one of the most serious sources of air pollution (figure 9.16). Every moment a car engine runs, potentially harmful compounds such as carbon monoxide, CO, carbon dioxide, CO_2, and a number of nitrogen oxides leave its engine as exhaust gases.

When air is drawn into a car's engine, both N_2 and O_2 are present. During combustion of the petrol or diesel, oxygen reacts with the hydrocarbons in the fuel to produce CO_2 through complete combustion, CO through incomplete combustion, and H_2O. However, N_2 and O_2 can also react to form nitric oxide, NO, at these elevated temperatures, according to the equation:

$$N_2(g) + O_2(g) \rightleftharpoons 2NO(g)$$

Although nitric oxide is an important biological molecule (it is involved in signalling in the body), it also reacts rapidly with oxygen in the atmosphere to form the toxic brown gas nitrogen dioxide, NO_2. At room temperature, K_c for the formation of NO from N_2 and O_2 is 4.8×10^{-31}. Its small value tells us that the equilibrium concentration of NO should be very small. Therefore, we don't find N_2 reacting with O_2 under ambient conditions.

Figure 9.16 Air pollution from car exhaust emissions can pose a health hazard. This can be particularly severe in large cities where heavy traffic and atmospheric conditions can combine to produce dangerous levels of smog.

The reaction of N_2 and O_2 to form NO is endothermic. Our discussion on the previous page tells us that at high temperatures, such as those found in the cylinders of a petrol or diesel engine during combustion, this equilibrium should be shifted to the right, and some NO does indeed form at high temperatures. Unfortunately, when the exhaust gases leave the engine, they cool so rapidly that the reaction rate becomes too slow for the NO to decompose back to N_2 and O_2. As a result, some NO is present among the exhaust gases. Once in the atmosphere, this NO reacts rapidly with oxygen to give NO_2, which is responsible for the brownish haze often associated with severe air pollution.

Various methods have been devised to minimise the amount of nitrogen oxides that enter the atmosphere. For instance, almost all modern cars are equipped with catalytic converters. These devices catalyse the decomposition of nitrogen oxides to N_2 and O_2. Catalytic converters are expensive, however, as the catalysts used in these are based on precious metals such as platinum, palladium and rhodium. Other methods of accomplishing the same goals are being studied.

One way to reduce the amount of NO_2 pollution of the atmosphere is to reduce the amount of NO that is formed in car engines. Since the extent to which the reaction proceeds towards the formation of NO increases as the temperature increases, the amount of NO formed can be reduced simply by running the combustion reaction at a lower temperature. This can be accomplished by lowering the compression ratio of the engine; this is the ratio between the volume of the cylinder when the piston is at the bottom of its stroke and its volume after the piston has compressed the air–fuel mixture. At high compression ratios, the air–fuel mixture is heated to a high temperature before it is ignited. After combustion, the gases are very hot, which favours the production of NO. Lowering the compression ratio lowers the maximum combustion temperature, which decreases the tendency for NO to be formed. Unfortunately, lowering the compression ratio also lowers the efficiency of the engine, which reduces fuel economy.

Another method for controlling NO emissions that has been trialled is to mix water with the air–fuel mixture. Some of the heat from the combustion is absorbed by the water vapour, so the mixture of exhaust gases does not get as hot as it would otherwise. At these lower temperatures, the concentration of NO in the exhaust is greatly reduced.

9.5 Equilibrium calculations

You have seen that the magnitude of an equilibrium constant gives us some feel for the composition of a reaction mixture at equilibrium. Sometimes, however, it is necessary to have more than merely a qualitative knowledge of equilibrium concentrations. This requires that we be able to use the equilibrium constant expression for quantitative calculations.

Equilibrium calculations for gaseous reactions can be performed using either K_p or K_c, but for reactions in solution we must use K_c. Whether we deal with concentrations or partial pressures, however, the same basic principles apply.

Overall, we can divide equilibrium calculations into two main categories:

1. calculating equilibrium constants from known equilibrium concentrations or partial pressures
2. calculating one or more equilibrium concentrations or partial pressures using a known value of K_c or K_p, given initial concentrations of reactants and products.

Calculating K_c from equilibrium concentrations

One way to determine the value of K_c is to carry out the reaction, measure the concentrations of reactants and products after equilibrium has been reached, and then substitute these equilibrium values into the equilibrium constant expression to calculate K_c. As an example, let's look again at the decomposition of N_2O_4.

$$N_2O_4(g) \rightleftharpoons 2NO_2(g)$$

In section 9.2, we saw that, if 0.0350 mol of N_2O_4 is placed into a 1 L flask at 25 °C, the concentrations of N_2O_4 and NO_2 at equilibrium are:

$$[N_2O_4] = 0.0292 \text{ mol L}^{-1}$$
$$[NO_2] = 0.0116 \text{ mol L}^{-1}$$

To calculate K_c for this reaction, we substitute the equilibrium concentrations into the equilibrium constant expression.

$$K_c = \frac{[NO_2]^2}{[N_2O_4]}$$

$$= \frac{(0.0116)^2}{0.0292}$$

Therefore:

$$K_c = 4.61 \times 10^{-3}$$

Although calculating an equilibrium constant in this way is usually straightforward, sometimes we have to do some stoichiometric manipulations, as shown in worked example 9.10. This example introduces the use of a **concentration table**, an extremely useful tool for solving equilibrium problems.

WORKED EXAMPLE 9.10

Calculating K_c from equilibrium concentrations

At a certain temperature, a mixture of H_2 and I_2 was prepared by placing 0.200 mol of H_2 and 0.200 mol of I_2 into a 2.00 L flask. After a period of time, the equilibrium:

$$H_2(g) + I_2(g) \rightleftharpoons 2HI(g)$$

was established. The purple colour of the I_2 vapour was used to monitor the reaction, and it was determined that, at equilibrium, the I_2 concentration had dropped to 0.020 mol L^{-1}. What is the value of K_c for this reaction at this temperature?

Analysis

The first step in any equilibrium problem is to write the balanced chemical equation and the equilibrium constant expression. The equation is already given, and its corresponding equilibrium constant expression is:

$$K_c = \frac{[HI]^2}{[H_2][I_2]}$$

To calculate the value of K_c, we must substitute the *equilibrium concentrations* of H_2, I_2 and HI into the equilibrium constant expression. But what are they? We have been given only one directly, the value of $[I_2]$. To obtain the others, we have to do some stoichiometric calculations.

The system starts with a set of initial concentrations, and a chemical change occurs to reach equilibrium. There is no HI present initially, so the reaction above must proceed to the right, because there has to be some HI present for the system to be at equilibrium. This change increases the HI concentration and decreases the concentrations of H_2 and I_2. If we can determine what the *changes* are, we can calculate the equilibrium concentrations. In fact, the key to solving almost all equilibrium problems is determining how the concentrations change as the system comes to equilibrium.

To help us in our analysis, we will construct a concentration table beneath the chemical equation. We will do this for most of the equilibrium problems we encounter from now on, so let's examine the general form of the table.

In the first row, we write the initial *molar concentrations* of the reactants and products. (Strictly speaking, we should use amounts, rather than concentrations, in these calculations. However, as long as the system remains at constant volume, the calculations will be the same, regardless of whether we use amounts or concentrations.) In the second row, we enter the changes in concentration, using a positive sign if the concentration is increasing and a negative sign if it is decreasing. Finally, adding the changes to the initial concentrations gives the equilibrium concentrations:

$$\begin{pmatrix}\text{equilibrium}\\\text{concentration}\end{pmatrix} = \begin{pmatrix}\text{initial}\\\text{concentration}\end{pmatrix} + \begin{pmatrix}\text{change in}\\\text{concentration}\end{pmatrix}$$

To set up the concentration table, it is essential that you realise that the changes in concentration must be in the same ratio as the coefficients of the balanced equation. This is because the changes are caused by the chemical reaction proceeding in one direction or the other. In this problem, the coefficients of H_2 and I_2 are the same, so, as the reaction proceeds to the right, the concentrations of H_2 and I_2 must decrease by the *same* amount. Similarly, because the coefficient of HI is twice that of H_2 or I_2, the concentration of HI must increase by *twice* as much. Now, let's look at the finished concentration table, how the individual entries are obtained and the solution to the problem.

Solution

To construct the concentration table, we begin by entering the data given in the statement of the problem. These are shown in the row labelled 'Initial concentration'. The other values are then derived as described below. Note that, in all concentration tables, we will highlight the known concentration data.

	$H_2(g)$	+	$I_2(g)$	$\rightleftharpoons$	$2HI(g)$
Initial concentration (mol L^{-1})	0.100		0.100		0
Change in concentration (mol L^{-1})	−0.080		−0.080		+2 × (0.080)
Equilibrium concentration (mol L^{-1})	0.020		0.020		0.160

Initial concentration: Notice that we have calculated the initial concentrations of both H_2 and I_2, that is, 0.200 mol/2.00 L = 0.100 M. Because no HI was placed into the reaction mixture, its initial concentration has been set to exactly 0 at the instant of mixing H_2 and I_2.

Change in concentration: We have been given both the initial and equilibrium concentrations of I_2, so, by difference, we can calculate the change for I_2 (−0.080 M). The other changes in concentration are then calculated from the mole ratios specified in the chemical equation. As noted above, because H_2 and I_2 have the same coefficients (i.e. 1), their changes in concentration are equal. The coefficient of HI is 2, so its change in concentration must be twice that of I_2. Reactant concentrations decrease, so the changes in concentrations of H_2 and I_2 are negative, while product concentrations increase, so the changes in HI concentration is positive.

Equilibrium concentration: For H_2 and HI, we just add the changes in concentration to the initial concentration value.

Now we can substitute the equilibrium concentrations into the equilibrium constant expression and calculate K_c.

$$K_c = \frac{[HI]^2}{[H_2][I_2]}$$

$$= \frac{(0.160)^2}{(0.020)(0.020)}$$

$$= 64$$

Is our answer reasonable?

First, carefully examine the equilibrium constant expression. As always, products must be placed in the numerator and reactants in the denominator. Be sure that each exponent corresponds to the correct coefficient in the balanced chemical equation for the reaction.

Now check the concentration table. Notice that the changes in concentration of both reactants have the same sign, and this sign is the opposite of that of the product. This relationship between the signs of the changes in concentration is always true and can serve as a useful check when you construct a concentration table.

Finally, check that each of the equilibrium concentrations is properly placed in the equilibrium constant expression. The equilibrium constant we obtained is greater than 1. That implies that a significant amount of product should be present at equilibrium, and examination of the concentration table shows that this is so.

The concentration table — a summary

Let us review some key points that apply not only to worked example 9.10 but also to others that deal with equilibrium calculations.

1. The only values that we can substitute into the equilibrium constant expression are equilibrium concentrations — the values that appear in the last row of the table.
2. When we enter initial concentrations into the table, they should be in units of moles per litre (mol L^{-1}). The initial concentrations are those present in the reaction mixture when it is prepared; we imagine that no reaction occurs until everything is mixed.
3. The changes in concentrations *always* occur in the same ratio as the coefficients in the balanced equation. For example, if we are dealing with the equilibrium:

$$3H_2(g) + N_2(g) \rightleftharpoons 2NH_3(g)$$

and we find that the $N_2(g)$ concentration decreases by 0.10 M during the approach to equilibrium, the entries in the 'Change in concentration' row would be as follows.

	$3H_2(g)$	+	$N_2(g)$	$\rightleftharpoons$	$2NH_3(g)$
Change in concentration (mol L^{-1})	$-3 \times (0.10)$ $= -0.30$		$-1 \times (0.10)$ $= -0.10$		$+2 \times (0.10)$ $= +0.20$

4. In constructing the 'Change in concentration' row, be sure that the reactant concentrations all change in the same direction and that the product concentrations all change in the opposite direction. If the concentrations of the reactants decrease, all the entries for the reactants in the 'Change in concentration' row should have a negative sign and all the entries for the products should be positive.

Keep these points in mind as we construct the concentration tables for other equilibrium problems.

PRACTICE EXERCISE 9.13

The water-gas shift reaction:

$$CO(g) + H_2O(g) \rightleftharpoons CO_2(g) + H_2(g)$$

is used in industry to prepare hydrogen. At equilibrium, the following concentrations were found for this reaction at 500 °C: [CO] = 0.180 M, $[H_2O]$ = 0.0411 M, $[CO_2]$ = 0.150 M and $[H_2]$ = 0.200 M. What is the value of K_c for this reaction?

PRACTICE EXERCISE 9.14

$NO_2(g)$ can decompose to give NO(g) and $O_2(g)$ according to the equilibrium:

$$2NO_2(g) \rightleftharpoons O_2(g) + 2NO(g)$$

When a mixture of the three gases was introduced into an empty vessel and the system was allowed to come to equilibrium, it was found that the NO concentration had decreased by 0.000 50 mol L^{-1}. How had the concentrations of O_2 and NO_2 changed?

PRACTICE EXERCISE 9.15

A student placed 0.200 mol of $PCl_3(g)$ and 0.100 mol of $Cl_2(g)$ into a 1.00 L container at 250 °C. After the reaction:

$$PCl_3(g) + Cl_2(g) \rightleftharpoons PCl_5(g)$$

came to equilibrium, it was found that the container contained 0.120 mol of PCl_3.

(a) What were the initial concentrations of the reactants and product?
(b) By how much had the concentrations of the reactants and product changed when the reaction reached equilibrium?
(c) What were the equilibrium concentrations of the reactants and product?
(d) What is the value of K_c for this reaction at this temperature?

Calculating equilibrium concentrations from initial concentrations

A more complex type of calculation involves the use of initial concentrations and K_c to calculate equilibrium concentrations. Although some of these problems can be rather complicated, we can learn the general principles involved by working on simple calculations. Even these, however, require a little applied algebra. This is where the concentration table can be very helpful.

WORKED EXAMPLE 9.11

Using K_c to calculate equilibrium concentrations — 1

The water-gas shift reaction:

$$CO(g) + H_2O(g) \rightleftharpoons CO_2(g) + H_2(g)$$

has $K_c = 4.06$ at 500 °C. If 0.100 mol of CO and 0.100 mol of $H_2O(g)$ are placed in a 1.00 L reaction vessel at this temperature, what are the concentrations of the reactants and products when the system reaches equilibrium?

Analysis

The key to solving this kind of problem is recognising that, at equilibrium:

$$K_c = \frac{[CO_2][H_2]}{[CO][H_2O]} = 4.06$$

We must find values for the concentrations that satisfy this condition.

Because we don't know what these concentrations are, we represent them algebraically in the concentration table as unknowns.

Initial concentration: The initial concentrations of CO and H_2O are both 0.100 mol L^{-1} = 0.100 M. Since no CO_2 or H_2 is initially placed into the reaction vessel, their initial concentrations are both exactly 0 at the instant of mixing CO and H_2O.

Change in concentration: Some CO_2 and H_2 must form for the reaction to reach equilibrium. This also means that some CO and H_2O must react. But how much? If we knew the answer, we could calculate the equilibrium concentrations. Therefore, the changes in concentration are our unknown quantities.

Let us allow x to represent the amount of CO per litre that reacts. Notice that we could just as easily have chosen x to be the amount of H_2O per litre that reacts or the amount of CO_2 or H_2 per litre that forms. There is nothing special about having chosen CO to define x. The change in the concentration of CO is then $-x$. (It is negative because the approach of the system to equilibrium decreases the CO concentration.) Because CO and H_2O react in a 1 : 1 mole ratio, the change in the H_2O concentration is also $-x$. Since 1 mole each of CO_2 and H_2 is formed from 1 mole of CO, the CO_2 and H_2O concentrations both increase by x (their changes are $+x$).

Equilibrium concentration: We obtain the equilibrium concentrations as:

$$\begin{pmatrix}\text{equilibrium}\\ \text{concentration}\end{pmatrix} = \begin{pmatrix}\text{initial}\\ \text{concentration}\end{pmatrix} + \begin{pmatrix}\text{change in}\\ \text{concentration}\end{pmatrix}$$

Solution

Here is the completed concentration table.

	CO(g)	+	$H_2O(g)$	$\rightleftharpoons$	$CO_2(g)$	+	$H_2(g)$
Initial concentration (mol L^{-1})	0.100		0.100		0		0
Change in concentration (mol L^{-1})	$-x$		$-x$		$+x$		$+x$
Equilibrium concentration (mol L^{-1})	$0.100 - x$		$0.100 - x$		x		x

Note that the last line in the table tells us that the equilibrium CO and H_2O concentrations are equal to the amount per litre that was present initially minus the amount per litre that has reacted. The equilibrium concentrations of CO_2 and H_2 equal the amount per litre of each that forms, since no CO_2 or H_2 is present initially.

Next, we substitute the quantities from the 'Equilibrium concentration' row into the equilibrium constant expression and solve for x.

$$K_c = \frac{[CO_2][H_2]}{[CO][H_2O]} = 4.06$$

$$= \frac{(x)(x)}{(0.100 - x)(0.100 - x)} = 4.06$$

which we can write as:

$$\frac{x^2}{(0.100 - x)^2} = 4.06$$

In this problem, we can solve the equation for x most easily by taking the square root of both sides.

$$\frac{x}{(0.100 - x)} = \sqrt{4.06} = 2.01$$

Rearranging gives:

$$x = 2.01(0.100 - x)$$
$$= 0.201 - 2.01x$$

Collecting terms in x gives:

$$x + 2.01x = 0.201$$
$$3.01x = 0.201$$
$$x = 0.0668$$

Now that we know the value of x, we can calculate the equilibrium concentrations from the last row of the table.

$$[CO] = 0.100 - x = 0.100 - 0.0668 = 0.033\ \text{M}$$
$$[H_2O] = 0.100 - x = 0.100 - 0.0668 = 0.033\ \text{M}$$
$$[CO_2] = x = 0.0668\ \text{M}$$
$$[H_2] = x = 0.0668\ \text{M}$$

Is our answer reasonable?

First, we should check that all concentrations are positive numbers. They are. One way to check the answers is to substitute the equilibrium concentrations we've found into the equilibrium constant expression and evaluate the reaction quotient. If our answers are correct, Q_c should equal K_c. Let us do this.

$$Q_c = \frac{[CO_2][H_2]}{[CO][H_2O]}$$

$$= \frac{(0.0668)^2}{(0.033)^2} = 4.1$$

Rounding K_c to two significant figures gives 4.1, so the calculated concentrations are correct.

WORKED EXAMPLE 9.12

Using K_c to calculate equilibrium concentrations — 2

In worked example 9.11, it was stated that the reaction:

$$CO(g) + H_2O(g) \rightleftharpoons CO_2(g) + H_2(g)$$

has $K_c = 4.06$ at 500 °C. Suppose 0.0600 mol of CO and 0.0600 mol of H_2O are mixed with 0.100 mol of CO_2 and 0.100 mol of H_2 in a 1.00 L reaction vessel. Determine the concentrations of all the substances when the mixture reaches equilibrium at this temperature.

Analysis

We will proceed in much the same way as in worked example 9.11. However, this time, determining the algebraic signs of x will not be quite so simple because none of the initial concentrations is 0. The best way to determine the algebraic signs is to use the initial concentrations to calculate the initial reaction quotient. Then we can compare the values of Q_c and K_c, and then by reasoning we will determine which way the reaction must proceed to make the value of Q_c equal that of K_c.

Solution

The equilibrium constant expression for the reaction is:

$$K_c = \frac{[CO_2][H_2]}{[CO][H_2O]}$$

Let's use the initial concentrations, shown in the concentration table below, to determine the value of the reaction quotient immediately after mixing.

$$Q_c = \frac{(0.100)(0.100)}{(0.0600)(0.0600)} = 2.78 < K_c$$

As indicated, the value of Q_c is less than that of K_c, so the system is not at equilibrium. To reach equilibrium, the value of Q_c must become larger, which requires an increase in the concentrations of CO_2 and H_2 and a corresponding decrease in the concentrations of CO and H_2O. This means that, for CO_2 and H_2, the change must be positive, and, for CO and H_2O, the change must be negative.

Here is the completed concentration table.

	CO(g)	+	H_2O(g)	⇌	CO_2(g)	+	H_2(g)
Initial concentration (mol L^{-1})	0.0600		0.0600		0.100		0.100
Change in concentration (mol L^{-1})	$-x$		$-x$		$+x$		$+x$
Equilibrium concentration (mol L^{-1})	$0.0600 - x$		$0.0600 - x$		$0.100 + x$		$0.100 + x$

Substituting equilibrium quantities into the equilibrium constant expression gives us:

$$K_c = \frac{[CO_2][H_2]}{[CO][H_2O]} = \frac{(0.100 + x)^2}{(0.0600 - x)^2} = 4.06$$

Taking the square root of both sides yields:

$$\frac{0.100 + x}{0.0600 - x} = 2.01$$

To solve for x, we first multiply each side by $(0.0600 - x)$ to obtain:

$$0.100 + x = 2.01(0.0600 - x) = 0.121 - 2.01x$$

Collecting terms in x to one side and the constants to the other gives:

$$x + 2.01x = 0.121 - 0.100$$
$$3.01x = 0.021$$
$$x = 0.0070$$

Now we can calculate the equilibrium concentrations:

$$[CO] = [H_2O] = (0.0600 - x) = 0.0600 - 0.0070 = 0.0530\,\text{M}$$
$$[CO_2] = [H_2] = (0.100 + x) = 0.100 + 0.0070 = 0.107\,\text{M}$$

Is our answer reasonable?

As a check, we can evaluate the reaction quotient using the calculated equilibrium concentrations.

$$Q = \frac{(0.107)^2}{(0.0530)^2} = 4.08$$

This is acceptably close to the value of K_c. (It is not *exactly* equal to K_c because of the rounding of answers during the calculations.)

In worked examples 9.11 and 9.12, we were able to simplify the solution by taking the square root of both sides of the algebraic equation obtained by substituting equilibrium concentrations into the equilibrium constant expression. Such simplifications are not always possible, however, as illustrated in worked example 9.13.

WORKED EXAMPLE 9.13

Using K_c to calculate equilibrium concentrations — 3

At a certain temperature, $K_c = 4.50$ for the reaction:

$$N_2O_4(g) \rightleftharpoons 2NO_2(g)$$

If 0.300 mol of N_2O_4 is placed into a 2.00 L container at this temperature, calculate the equilibrium concentrations of both gases.

Analysis

As in worked example 9.12, we first write the equilibrium constant expression:

$$K_c = \frac{[NO_2]^2}{[N_2O_4]} = 4.50$$

We will need to find algebraic expressions for the equilibrium concentrations and substitute them into the equilibrium constant expression. To obtain these, we set up the concentration table for the reaction.

Initial concentration: The initial concentration of N_2O_4 is 0.300 mol/2.00 L = 0.150 M. Since no NO_2 was placed in the reaction vessel, its initial concentration is exactly 0.

Change in concentration: There is initially no NO_2 in the reaction mixture, so we know its concentration must increase. This means the N_2O_4 concentration must decrease as some of the NO_2 is formed. Let's allow x to represent the amount of N_2O_4 per litre that reacts, so the change in the N_2O_4 concentration is $-x$. Because of the stoichiometry of the reaction, the NO_2 concentration must increase by $2x$, so its change in concentration is $2x$.

Equilibrium concentration: As before, we add the change to the initial concentration in each column to obtain expressions for the equilibrium concentrations.

Solution

Here is the concentration table.

	$N_2O_4(g)$	$\rightleftharpoons$	$2NO_2(g)$
Initial concentration (mol L^{-1})	0.150		0
Change in concentration (mol L^{-1})	$-x$		$+2x$
Equilibrium concentration (mol L^{-1})	$0.150 - x$		$2x$

Now we substitute the equilibrium quantities into the equilibrium constant expression:

$$K_c = \frac{[NO_2]^2}{[N_2O_4]} = 4.50$$

$$= \frac{(2x)^2}{(0.150 - x)} = 4.50$$

$$= \frac{4x^2}{(0.150 - x)} = 4.50$$

This time the left side of the equation is not a perfect square, so we cannot just take the square root of both sides as in worked examples 9.11 and 9.12. However, because the equation involves terms in x^2 and x and a constant, we can use the quadratic formula. For a quadratic equation of the form $ax^2 + bx + c = 0$:

$$x = \frac{-b \pm \sqrt{b^2 - 4ac}}{2a}$$

Expanding the above equation, we get:

$$4x^2 = 4.50\,(0.150 - x)$$
$$= 0.675 - 4.50x$$

Arranging terms in the standard order gives:

$$4x^2 + 4.50x - 0.675 = 0$$

Therefore, the quantities we will substitute into the quadratic formula are: $a = 4$, $b = 4.50$ and $c = -0.675$. Making these substitutions gives:

$$x = \frac{-4.50 \pm \sqrt{(4.50)^2 - 4(4)(-0.675)}}{2(4)}$$

$$= \frac{-4.50 \pm \sqrt{31.05}}{8}$$

$$= \frac{-4.50 \pm 5.57}{8}$$

Because of the ± term, there are two values of x that satisfy the equation: $x = 0.134$ and $x = -1.26$. However, only the first value, $x = 0.134$, makes any sense chemically (see below). Using this value, the equilibrium concentrations are:

$$[N_2O_4] = 0.150 - 0.134 = 0.016\,\text{M}$$
$$[NO_2] = 2(0.134) = 0.268\,\text{M}$$

Notice that, if we had used the negative root, −1.26, the equilibrium concentration of NO_2 would be negative. Negative concentrations are impossible, so $x = -1.26$ is not acceptable *for chemical reasons*. In general, whenever you use the quadratic equation in a chemical calculation, one root will be satisfactory and the other will lead to an answer that does not make sense chemically.

Is our answer reasonable?

Once again, we can evaluate the reaction quotient using the calculated equilibrium values. When we do this, we obtain $Q = 4.49$, which is acceptably close to the given value of K_c.

PRACTICE EXERCISE 9.16

0.0500 mol of $H_2(g)$ and 0.0500 mol of $I_2(g)$ were placed into a 0.500 L vessel and the following equilibrium was established.

$$H_2(g) + I_2(g) \rightleftharpoons 2HI(g)$$

If $K_c = 49.5$ for this equilibrium at the temperature of the experiment, determine the equilibrium concentrations of $H_2(g)$, $I_2(g)$ and $HI(g)$.

Equilibrium problems can be much more complex than worked example 9.13. However, when K is either very large or very small, simplifying assumptions can be made that allow an approximate solution to be obtained. This is illustrated in worked example 9.14.

WORKED EXAMPLE 9.14

Simplifying approximations in equilibrium calculations

The thermal decomposition of water to its elements occurs only to a very small extent, even at high temperatures; at 1000 °C, $K_c = 7.3 \times 10^{-18}$ for the equilibrium:

$$2H_2O(g) \rightleftharpoons 2H_2(g) + O_2(g)$$

If water vapour at an initial concentration of 0.100 M is allowed to react at 1000 °C, what will the H_2 concentration be when the reaction reaches equilibrium?

Analysis

We begin, as always, by writing the equilibrium constant expression, K_c:

$$K_c = \frac{[H_2]^2[O_2]}{[H_2O]^2} = 7.3 \times 10^{-18}$$

and we then set up the concentration table and proceed as we have done before. However, we will find that we can make simplifying assumptions in this problem because of the extremely small value of the equilibrium constant.

Solution

Here is the concentration table.

	$2H_2O(g)$	$\rightleftharpoons$	$2H_2(g)$	+	$O_2(g)$
Initial concentration (mol L^{-1})	0.100		0		0
Change in concentration (mol L^{-1})	$-2x$		$+2x$		$+x$
Equilibrium concentration (mol L^{-1})	$0.100 - 2x$		$2x$		x

We now substitute the equilibrium concentrations into the equilibrium constant expression.

$$K_c = \frac{[H_2]^2[O_2]}{[H_2O]^2} = \frac{(2x)^2(x)}{(0.100 - 2x)^2} = \frac{4x^3}{(0.100 - 2x)^2} = 7.3 \times 10^{-18}$$

This is a cubic equation and the mathematics required to solve such equations exactly are well beyond the scope of this book. However, in this case, the magnitude of K_c allows us to make simplifications to this equation that allow its straightforward solution. Because K_c is so incredibly small, we know that, at equilibrium, not much H_2O will have reacted. This means that, to a very good approximation, the concentration of H_2O at equilibrium equals the initial concentration of H_2O. In other words, the term x in the concentration table is tiny, so $0.100 - 2x \approx 0.100$. Making this assumption simplifies the cubic expression as follows.

$$K_c = \frac{4x^3}{(0.100 - 2x)^2} \approx \frac{4x^3}{(0.100)^2} = 7.3 \times 10^{-18}$$

Hence:

$$4x^3 = (0.100)^2 \times (7.3 \times 10^{-18})$$
$$= 7.3 \times 10^{-20}$$
$$x^3 = 1.8 \times 10^{-20}$$
$$x = \sqrt[3]{1.8 \times 10^{-20}}$$
$$= 2.6 \times 10^{-7}$$

We can now obtain the H_2 concentration, which is equal to $2x$ (from the concentration table).

$$[H_2] = 2 \times (2.6 \times 10^{-7})$$
$$= 5.2 \times 10^{-7}\ \text{M}$$

Is our answer reasonable?

We must first show that the assumption we made was justified. We do this by showing that $(0.100 - 2x)$ is indeed essentially equal to 0.100.

$$0.100 - 2x = 0.100 - (5.2 \times 10^{-7}) = 0.099\,999\,48$$

When we round 0.099 999 48 to three significant figures, it is 0.100. This shows that our assumption is justified.

We can now show that our answer is reasonable. The tiny value of K_c leads us to expect that the equilibrium concentration of hydrogen is very small, which it is. We can also substitute our equilibrium values into the expression for K_c (remember that $[O_2] = x$) and confirm that this gives a value acceptably close to 7.3×10^{-18}. It should also be noted that the exact solution of the original cubic equation using a calculator gives $x = 2.6 \times 10^{-7}$, again showing that our assumption leads to a reasonable answer.

PRACTICE EXERCISE 9.17

In air at 25 °C and 1.013×10^5 Pa, the N_2 concentration is 0.033 M and the O_2 concentration is 0.008 10 M. The reaction:

$$N_2(g) + O_2(g) \rightleftharpoons 2NO(g)$$

has $K_c = 4.8 \times 10^{-31}$ at 25 °C. Taking the N_2 and O_2 concentrations given above as initial values, calculate the equilibrium NO concentration that should exist in our atmosphere from this reaction at 25 °C.

The simplifying assumption made in worked example 9.14 is valid because a very small number was subtracted from a much larger one. We could also have neglected x (or $2x$) if it were a very small number that was being added to a much larger one. Remember that you can neglect x only if it is added or subtracted; you can never neglect x if it occurs as a multiplying or dividing factor.

As an arbitrary rule of thumb, you can expect these simplifying assumptions to be valid if the concentration from which x is subtracted, or to which x is added, is at least 400 times greater than K. For instance, in worked example 9.14, $2x$ was subtracted from 0.100. Since 0.100 is much larger than $400 \times (7.3 \times 10^{-18})$, we expect the assumption $0.100 - 2x \approx 0.100$ to be valid. However, even though the simplifying assumption is expected to be valid, always check to see if it really is after finishing the calculation. If the assumption proves invalid, then some other way to solve the algebra must be found.

The routine use of scientific calculators has rendered the above approximations essentially redundant, as exact solutions of the most complicated equations can be obtained at the touch of a button. However, this is not to say that the principles underlying the assumptions are unimportant. You should appreciate that the concentrations of some equilibrium components are going to be negligibly small when the equilibrium constant is either very large or very small.

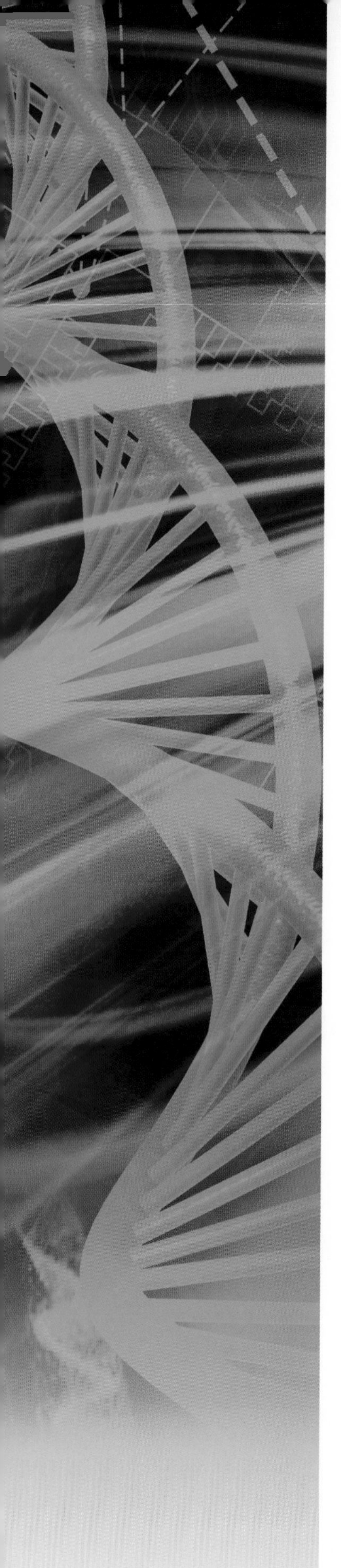

Chemistry Connections

Hyperventilation

Have you ever found yourself getting so excited or scared that you reach the point where you're breathing extremely rapidly? If so, you may have experienced *hyperventilation*, the symptoms of which are dizziness, palpitations and sweating, amongst other things. When you breathe out, you expel significant amounts of $CO_2(g)$ and when you breathe very fast, you breathe out more $CO_2(g)$ than normal. In the body, CO_2 dissolves in blood (an aqueous solution) to form 'carbonic acid', $H_2CO_3(aq)$, and any alteration of CO_2 levels affects the position of the equilibrium:

$$H_2CO_3(aq) + H_2O(l) \rightleftharpoons HCO_3^-\,(aq) + H_3O^+\,(aq)$$

which can then affect the pH of your blood by affecting $[H_3O^+]$. We can use the principles of equilibrium we have learned in this chapter to explain why breathing into a paper bag can be used to stop hyperventilation (figure 9.17). Consider the above equilibrium for which the reaction quotient expression is:

$$Q_c = \frac{[H_3O^+][HCO_3^-]}{[H_2CO_3]}$$

If we remove $H_2CO_3(aq)$ by breathing out lots of $CO_2(g)$, then the value of Q_c will increase, and the equilibrium will therefore move to the left to decrease the value of Q_c. This then alters the pH by decreasing $[H_3O^+]$ and it is this that causes the symptoms of hyperventilation. If you breathe into a paper bag, you are essentially breathing in the $CO_2(g)$ that you have just breathed out. This means that the amount of $CO_2(g)$, and hence $[H_2CO_3]$, will increase, counteracting the above process and resulting in the position of equilibrium moving back towards the right.

FIGURE 9.17 Breathing into a paper bag results in breathing in CO_2, re-establishing the equilibrium in the blood that was disrupted by hyperventilating.

SUMMARY

LO 1 Chemical equilibrium

Chemical equilibrium occurs when the chemical composition of a system does not change with time. When the forward and reverse reactions in a chemical system occur at equal rates, a dynamic equilibrium exists and the concentrations of the reactants and products remain constant. For a given overall chemical composition, the amounts of reactants and products that are present at equilibrium are the same regardless of whether the equilibrium is approached from the direction of pure 'reactants', pure 'products' or any mixture of them.

LO 2 The equilibrium constant, *K*, and the reaction quotient, *Q*

The equilibrium constant expression is a fraction. The concentrations of the products at equilibrium, each divided by $c^{\ominus}$ and raised to a power equal to its stoichiometric coefficient in the balanced chemical equation, are multiplied together in the top line (numerator). The bottom line (denominator) is constructed in the same way from the concentrations of the reactants at equilibrium, each divided by $c^{\ominus}$ and raised to a power equal to its stoichiometric coefficient. The numerical value of the equilibrium constant expression is the equilibrium constant, K_c. If partial pressures of gases, divided by $p^{\ominus}$ and raised to their stoichiometric coefficients, are used in the equilibrium constant expression, K_p is obtained. The thermodynamic equilibrium constant is obtained when activities, rather than pressures or concentrations, are used in the equilibrium constant expression. The equilibrium constant is constant at constant temperature. The magnitude of the equilibrium constant reflects the extent to which the forward reaction has proceeded to completion when equilibrium is reached.

The reaction quotient, Q, has the same form as K, but the concentrations or pressures are nonequilibrium values. Pure solids and pure liquids do not appear in the expressions for K and Q.

When an equation is multiplied by a factor n to obtain a new equation, we raise its K to the power of n to obtain K for the new equation. When two equations are added, we multiply their values of K to obtain the new K. When an equation is reversed, we take the reciprocal of its K to obtain the new K.

The values of K_p and K_c are equal only if the same number of moles of gas are represented on both sides of the chemical equation. When the numbers of moles of gas are different, K_p is related to K_c by the equation:

$$K_p = K_c\left(\frac{1000RT}{p^{\ominus}}\right)^{\Delta n_{\text{gas}}}$$

where $R = 8.314\ \text{J mol}^{-1}\ \text{K}^{-1}$, T = absolute temperature and Δn_{gas} is the difference between the number of moles of gaseous products and the number of moles of gaseous reactants in the balanced equation.

An equilibrium where all substances are in the same phase is called a homogeneous equilibrium, while one with substances in more than one phase is a heterogeneous equilibrium. The equilibrium constant expression for a heterogeneous equilibrium omits concentration terms for pure liquids and pure solids.

LO 3 Equilibrium and Gibbs energy

Gibbs energy diagrams for chemical reactions show the variation in Gibbs energy of a system as a reaction mixture proceeds from pure reactants to pure products. These diagrams show a minimum, due to the Gibbs energy of mixing of reactants and products, and this minimum corresponds to the equilibrium composition of the reaction mixture. The position of the minimum with respect to the proportion of products to reactants is related to the sign and magnitude of $\Delta_r G^{\ominus}$ for the forward reaction: if $\Delta_r G^{\ominus}$ is large and negative, the forward reaction proceeds essentially to completion and the equilibrium mixture consists predominantly of products; if $\Delta_r G^{\ominus}$ is large and positive, the forward reaction proceeds only to a very small extent and negligible amounts of products are present at equilibrium. $\Delta_r G^{\ominus}$ is related to the equilibrium constant for a chemical reaction by the equation:

$$\Delta_r G^{\ominus} = -RT\ln K$$

The spontaneity of any change in composition of a reaction mixture is given by the sign of $\Delta_r G$ for that change. $\Delta_r G$ is related to $\Delta_r G^{\ominus}$ by the equation:

$$\Delta_r G = \Delta_r G^{\ominus} + RT\ln Q$$

LO 4 How systems at equilibrium respond to change

The response of a system at equilibrium to a chemical or physical change can be predicted by determining the effect of such a change on the value of Q, the reaction quotient. Any change that affects the value of Q instantaneously removes the system from equilibrium because $Q \neq K$; the way in which the value of Q must change (increase or decrease) to again make $Q = K$ determines the direction of spontaneous change in the reaction mixture. Temperature alone can change the value of K for a chemical equilibrium, and the direction in which spontaneous change must occur to re-establish equilibrium in these cases may be determined by the sign of $\Delta_r H^{\ominus}$ for the forward reaction, via the van't Hoff equation.

$$\frac{d\ln K}{dT} = \frac{\Delta_r H^{\ominus}}{RT^2}$$

Pure solids, pure liquids, inert gases and catalysts do not appear in the expression for Q so addition of these can have no effect on the position of equilibrium.

LO 5 Equilibrium calculations

Given a set of equilibrium concentration or pressure conditions, we can calculate the value of the equilibrium constant for a reaction, and we can also use the given value of an equilibrium constant to calculate equilibrium concentrations or pressures. We can use a concentration table to assist in such calculations.

KEY CONCEPTS AND EQUATIONS

Equilibrium constant expression *(section 9.2)*

This expression defines the relationship between the equilibrium concentrations or partial pressures of the reactants and products in a chemical reaction. For the equilibrium:

$$aA + bB \rightleftharpoons cC + dD$$

$$K_c = \frac{[C]^c[D]^d}{[A]^a[B]^b} \quad \text{and} \quad K_p = \frac{\left(\frac{p_C}{p^{\ominus}}\right)^c\left(\frac{p_D}{p^{\ominus}}\right)^d}{\left(\frac{p_A}{p^{\ominus}}\right)^a\left(\frac{p_B}{p^{\ominus}}\right)^b}$$

Reaction quotient expression
(section 9.2)

This expression has the same form as the equilibrium constant expression, and is used to characterise systems that are not at equilibrium. The concentrations or partial pressures in the expression can have any value. For the equilibrium

$$aA + bB \rightleftharpoons cC + dD$$

$$Q_c = \frac{[C]^c[D]^d}{[A]^a[B]^b} \quad \text{and} \quad Q_p = \frac{\left(\frac{p_C}{p^\ominus}\right)^c \left(\frac{p_D}{p^\ominus}\right)^d}{\left(\frac{p_A}{p^\ominus}\right)^a \left(\frac{p_B}{p^\ominus}\right)^b}$$

Comparison of the values of Q and K
(section 9.2)

At equilibrium, $Q = K$. If the system is disturbed such that $Q > K$, the reaction must shift towards reactants to restore equilibrium; if $Q < K$, the reaction must shift towards products to restore equilibrium.

$K_p = K_c\left(\frac{1000RT}{p^\ominus}\right)^{\Delta n_{gas}}$
(section 9.2)

This equation allows interconversion of K_p and K_c values; Δn_{gas} is the change in the number of moles of gas, which is calculated using the coefficients of gaseous products and reactants.

Manipulation of equilibrium equations
(section 9.2)

1. When the direction of an equation is reversed, the new equilibrium constant is the reciprocal of the original.
2. When the coefficients are multiplied by a factor, the equilibrium constant is raised to a power equal to that factor.
3. When chemical equilibria are added, their equilibrium constants are multiplied.

Magnitude of the equilibrium constant
(section 9.2)

1. When K_c is very large, the equilibrium position lies towards the products.
2. When $K_c \approx 1$, the concentrations of reactants and products are similar at equilibrium.
3. When K_c is very small, the equilibrium position lies towards the reactants.

Value of $\Delta_r G^\ominus$ for a reaction
(section 9.3)

The magnitude and sign of $\Delta_r G^\ominus$ indicate qualitatively whether or not a significant amount of products will form. If $\Delta_r G^\ominus$ is reasonably large and negative, the reaction will proceed largely towards products. If $\Delta_r G^\ominus$ is reasonably large and positive, the equilibrium mixture will contain mainly reactants.

$\Delta_r G = \Delta_r G^\ominus + RT\ln Q$
(section 9.3)

This equation relates the reaction quotient, Q, to $\Delta_r G$, which enables us to determine whether a reaction is at equilibrium and, if not, the direction of spontaneous change required to reach equilibrium.

$\Delta_r G^\ominus = -RT\ln K$
(section 9.3)

This equation relates $\Delta_r G^\ominus$ to the equilibrium constant, K.

Concentration table
(section 9.5)

This may be used in problems involving equilibria to determine the concentrations of reactants and products at equilibrium.

KEY TERMS

REVIEW QUESTIONS

LO1 Chemical equilibrium

9.1 What information do the stoichiometric coefficients in a balanced chemical equation provide in terms of how far the reaction proceeds towards completion?

9.2 What is meant when we say that chemical reactions are reversible?

9.3 What is a dynamic equilibrium?

LO 2 The equilibrium constant, *K*, and reaction quotient, *Q*

9.4 Although the forms of the equilibrium constant and the reaction quotient appear similar, they differ in one crucial aspect. What is the difference between the two?

9.5 Under what conditions are the values of Q and K the same?

9.6 State, in words, the method for writing the equilibrium constant expression K_c for any equilibrium.

9.7 Give the equation which relates K_p to K_c and define all the terms used.

9.8 In the equation referred to in question 9.7, why is the factor of 1000 required? (*Hint:* Consider the concentration units that result from use of the ideal gas equation.)

9.9 At 225 °C, $K_p = 6.3 \times 10^{-3}$ for the equilibrium:

$$CO(g) + 2H_2(g) \rightleftharpoons CH_3OH(g)$$

Would we expect this reaction to go nearly to completion?

9.10 Arrange the following equilibria in order of their increasing tendency to proceed towards completion.

(a) $SO_2(g) + NO_2(g) \rightleftharpoons NO(g) + SO_3(g)$ $K_c = 8.50 \times 10^2$
(b) $N_2O(g) + NO_2(g) \rightleftharpoons 3NO(g)$ $K_c = 1.4 \times 10^{-13}$
(c) $HCOOH(g) \rightleftharpoons CO(g) + H_2O(g)$ $K_c = 4.3 \times 10^5$

9.11 What is the difference between a heterogeneous equilibrium and a homogeneous equilibrium?

9.12 Under what conditions can the concentrations of pure liquids and solids change at constant pressure?

LO 3 Equilibrium and Gibbs energy

9.13 Sketch a graph to show how Gibbs energy changes during a phase change such as the melting of a solid.

9.14 Sketch the shape of the Gibbs energy curve for a chemical reaction that has a negative $\Delta_r G^\ominus$. Indicate the composition of the reaction mixture when it is at equilibrium.

9.15 Many reactions that have large, negative values of $\Delta_r G^\ominus$ are not actually observed to occur at 25 °C and 10^5 Pa. Why?

9.16 Suppose a reaction has a positive $\Delta_r S^\ominus$. Will more or less product be present at equilibrium as the temperature is raised? Assume that $\Delta_r H^\ominus$ and $\Delta_r S^\ominus$ are independent of temperature.

9.17 Write the equation that relates the Gibbs energy change to the value of the reaction quotient for a reaction.

9.18 The equation that relates the equilibrium constant to the standard Gibbs energy change for a reaction is $\Delta_r G^\ominus = -RT\ln K$. Which equilibrium constant, K_p or K_c, must we use in this equation for a gas phase equilibrium?

9.19 What is a 'thermodynamic equilibrium constant'?

9.20 What is the value of K for a reaction for which $\Delta_r G^\ominus = 0$?

LO 4 How systems at equilibrium respond to change

9.21 How will the equilibrium:

$$CH_4(g) + 2H_2S(g) \rightleftharpoons CS_2(g) + 4H_2(g)$$

be affected by the following?

(a) addition of $CH_4(g)$
(b) addition of $H_2(g)$
(c) removal of $CS_2(g)$
(d) decreasing the volume of the container
(e) increasing the temperature (the forward reaction is endothermic)

9.22 The forward reaction in the equilibrium:

$$2NO(g) + O_2(g) \rightleftharpoons N_2O_4(g)$$

has $\Delta_r H^\ominus = -171.07\text{ kJ mol}^{-1}$. How will the amount of $N_2O_4(g)$ present in an equilibrium mixture of these gases be affected by the following?

(a) adding $NO(g)$
(b) removing $O_2(g)$
(c) increasing the volume of the reaction vessel
(d) adding a catalyst
(e) decreasing the temperature

9.23 The forward reaction in the equilibrium:

$$N_2O(g) + NO_2(g) \rightleftharpoons 3NO(g)$$

has $\Delta_r H^\ominus = 155.7\text{ kJ mol}^{-1}$. In which direction will the position of equilibrium be shifted by the following changes?

(a) addition of $N_2O(g)$
(b) removal of $NO_2(g)$
(c) addition of $NO(g)$
(d) increasing the temperature of the reaction mixture
(e) addition of helium gas to the reaction mixture while keeping the volume of the reaction vessel constant
(f) decreasing the volume of the container

9.24 Consider the equilibrium:

$$2NO(g) + Cl_2(g) \rightleftharpoons 2NOCl(g)$$

This has $\Delta_r H^\ominus = -77.07\text{ kJ mol}^{-1}$ for the forward reaction. How will the amount of NO in an equilibrium mixture of these gases be affected by the following?

(a) addition of $Cl_2(g)$
(b) removal of $NOCl(g)$
(c) decreasing the temperature
(d) increasing the volume of the reaction vessel

9.25 Under what circumstances will changing the volume of the reaction vessel not affect the position of equilibrium of a gas phase equilibrium?

LO 5 Equilibrium calculations

9.26 In order to study the equilibrium:

$$NO(g) + NO_2(g) + H_2O(g) \rightleftharpoons 2HNO_2(g)$$

a mixture of $NO(g)$, $NO_2(g)$ and $H_2O(g)$ was prepared in a 10.0 L glass bulb at 20 °C with the following initial concentrations: $[NO] = [NO_2] = 2.59 \times 10^{-3}$ M, $[H_2O] = 9.44 \times 10^{-4}$ M and $[HNO_2] = 0.0$ M. When equilibrium was reached, $[HNO_2]$ was 4.0×10^{-4} M. Use these data to prepare a concentration table for this system.

9.27 When carrying out equilibrium calculations, it is sometimes possible to make simplifying assumptions. Under what conditions can such assumptions be made?

9.28 Cl_2 reacts with Br_2 in CCl_4 solution to form BrCl according to the equation:

$$Br_2 + Cl_2 \rightleftharpoons 2BrCl \qquad K_c = 2$$

If the initial concentrations were $[Br_2] = 0.6$ M, $[Cl_2] = 0.4$ M and $[BrCl] = 0.0$ M, which of the following concentration versus time graphs could represent this reaction? Explain why you rejected each of the other four graphs.

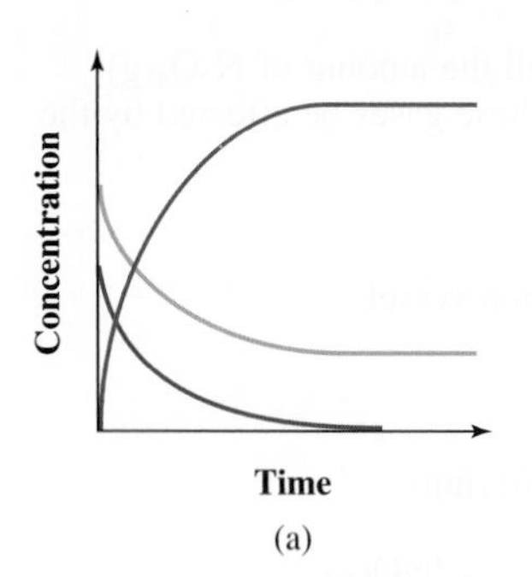

(a)

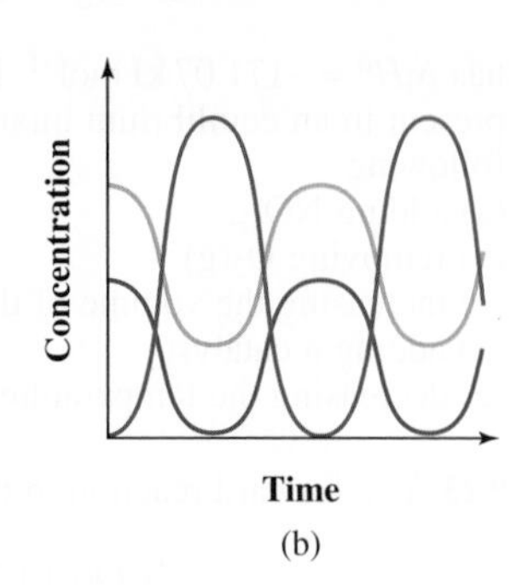

(b)

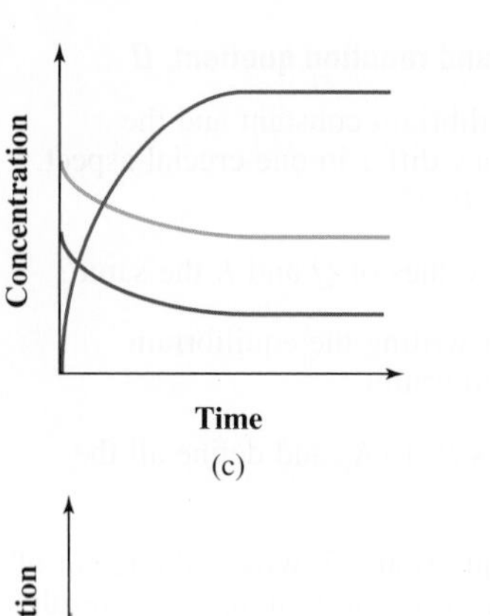

(c)

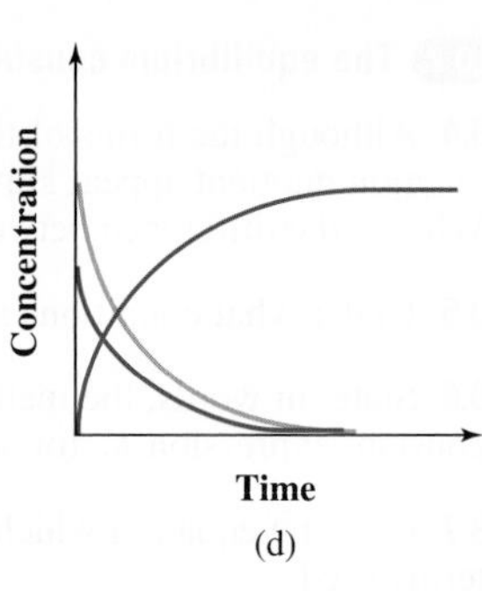

(d)

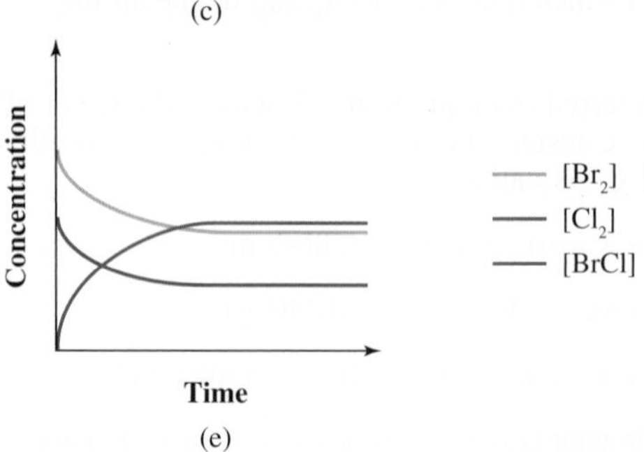

(e)

REVIEW PROBLEMS

9.29 Write the equilibrium constant expression for each of the following equilibria in terms of concentrations. **LO2**

(a) $2PCl_3(g) + O_2(g) \rightleftharpoons 2POCl_3(g)$
(b) $2SO_3(g) \rightleftharpoons 2SO_2(g) + O_2(g)$
(c) $N_2H_4(g) + 2O_2(g) \rightleftharpoons 2NO(g) + 2H_2O(g)$
(d) $N_2H_4(g) + 6H_2O_2(g) \rightleftharpoons 2NO_2(g) + 8H_2O(g)$
(e) $SOCl_2(g) + H_2O(g) \rightleftharpoons SO_2(g) + 2HCl(g)$
(f) $3Cl_2(g) + NH_3(g) \rightleftharpoons NCl_3(g) + 3HCl(g)$
(g) $PCl_3(g) + PBr_3(g) \rightleftharpoons PCl_2Br(g) + PClBr_2(g)$
(h) $NO(g) + NO_2(g) + H_2O(g) \rightleftharpoons 2HNO_2(g)$
(i) $H_2O(g) + Cl_2O(g) \rightleftharpoons 2HOCl(g)$
(j) $Br_2(g) + 5F_2(g) \rightleftharpoons 2BrF_5(g)$

9.30 Reverse the equilibrium equations in question 9.29 and write the K_p expressions for these. **LO2**

9.31 Write the equilibrium constant expression for each of the following equilibria in aqueous solution. **LO2**

(a) $Ag^+(aq) + 2NH_3(aq) \rightleftharpoons [Ag(NH_3)_2]^+(aq)$
(b) $Cd^{2+}(aq) + 4SCN^-(aq) \rightleftharpoons [Cd(SCN)_4]^{2-}(aq)$
(c) $HClO(aq) + H_2O(l) \rightleftharpoons H_3O^+(aq) + ClO^-(aq)$
(d) $CO_3^{2-}(aq) + HSO_4^-(aq) \rightleftharpoons HCO_3^-(aq) + SO_4^{2-}(aq)$

9.32 At 27 °C, $K_p = 1.5 \times 10^{18}$ for the following equilibrium. **LO2**

$$3NO(g) \rightleftharpoons N_2O(g) + NO_2(g)$$

What is the value of K_p for the equilibrium:

$$N_2O(g) + NO_2(g) \rightleftharpoons 3NO(g)$$

at the same temperature?

9.33 Use the following equilibria: **LO2**

$$2CH_4(g) \rightleftharpoons C_2H_6(g) + H_2(g) \qquad K_c = 9.5 \times 10^{-13}$$

$$CH_4(g) + H_2O(g) \rightleftharpoons CH_3OH(g) + H_2(g) \quad K_c = 2.8 \times 10^{-21}$$

to calculate K_c for the equilibrium:

$$2CH_3OH(g) + H_2(g) \rightleftharpoons C_2H_6(g) + 2H_2O(g)$$

9.34 Write the equilibrium constant expression for each of the following equilibria in terms of concentrations. **LO2**

(a) $H_2(g) + Cl_2(g) \rightleftharpoons 2HCl(g)$
(b) $\frac{1}{2}H_2(g) + \frac{1}{2}Cl_2(g) \rightleftharpoons HCl(g)$

How does the value of K_c for equilibrium (a) compare with that of K_c for equilibrium (b)?

9.35 Write the equilibrium constant expression for the equilibrium: **LO2**

$$2HCl(g) \rightleftharpoons H_2(g) + Cl_2(g)$$

How does the value of K_c for this equilibrium compare with that of K_c for equilibrium (a) in question 9.34?

9.36 A sample of $H_2(g)$ in a 740 mL container exhibited a pressure of 5.37×10^4 Pa at a temperature of 37 °C. Calculate the concentration of $H_2(g)$ in the container in units of mol L^{-1} assuming it behaves as an ideal gas. **LO2**

9.37 The concentration of water vapour is 0.0200 M in a certain container at 145 °C. What is the partial pressure of H_2O in the container? **LO2**

9.38 For which of the following equilibria does the value of K_p equal that of K_c? **LO2**

(a) $H_2O(g) + Cl_2O(g) \rightleftharpoons 2HOCl$
(b) $N_2(g) + 3H_2(g) \rightleftharpoons 2NH_3(g)$
(c) $CaO(s) + SO_2(g) \rightleftharpoons CaSO_3(s)$
(d) $H_2(g) + I_2(g) \rightleftharpoons 2HI(g)$
(e) $CO(g) + H_2O(g) \rightleftharpoons CO_2(g) + H_2(g)$

9.39 The equilibrium $CO(g) + 2H_2(g) \rightleftharpoons CH_3OH(g)$ has $K_p = 6.3 \times 10^{-3}$ at 225 °C. What is the value of K_c at this temperature? **LO2**

9.40 The equilibrium:

$$3NO(g) \rightleftharpoons N_2O(g) + NO_2(g)$$

has $K_c = 2.38 \times 10^3$ at 500 °C. What is the value of K_p at this temperature? **LO2**

9.41 One possible way of removing NO from the exhaust of a petrol engine is to cause it to react with CO in the presence of a suitable catalyst.

$$2NO(g) + 2CO(g) \rightleftharpoons N_2(g) + 2CO_2(g)$$

At 300 °C, this equilibrium has $K_c = 2.2 \times 10^{59}$. What is K_p at 300 °C? **LO2**

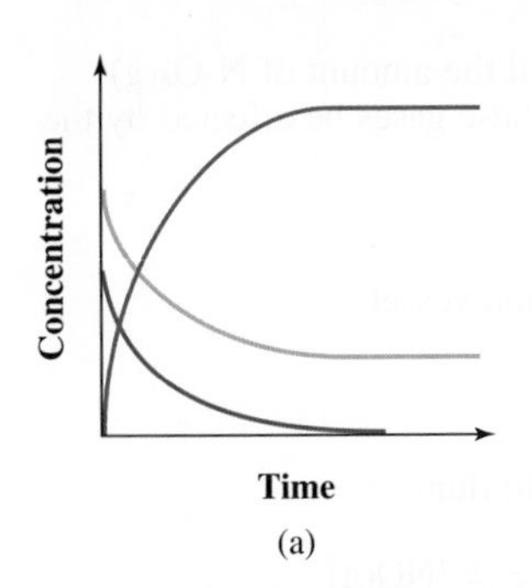

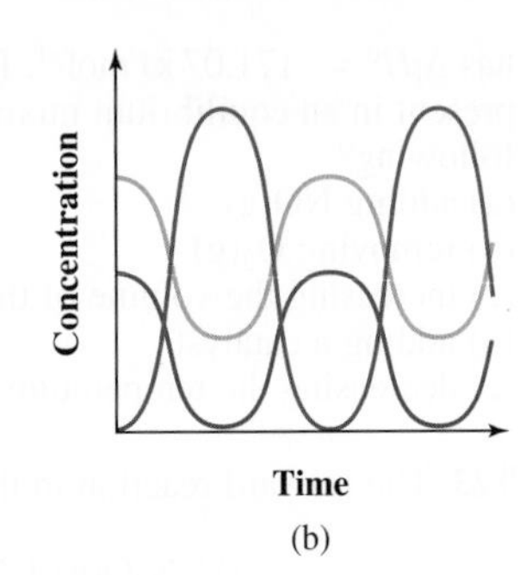

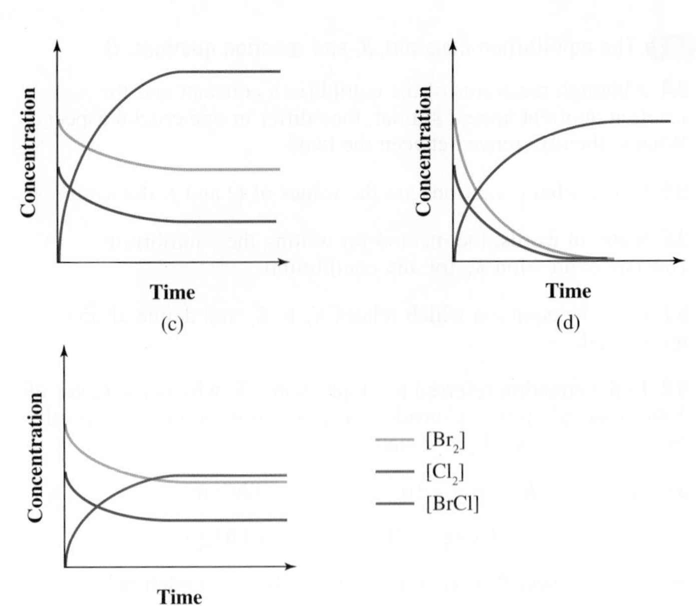

9.42 The equilibrium:

$$CO(g) + H_2O(g) \rightleftharpoons HCOOH(g)$$

has $K_p = 6.25 \times 10^{-7}$ at 400 °C. What is the value of K_c at this temperature? **LO2**

9.43 The equilibrium:

$$CO(g) + Cl_2(g) \rightleftharpoons COCl_2(g)$$

has $K_p = 2.17 \times 10^{1}$ at 395 °C. What is the value of K_c at this temperature? **LO2**

9.44 Calculate the concentration of water in: (a) 18.0 mL of H_2O, (b) 100.0 mL of H_2O and (c) 1.00 L of H_2O. Assume that the density of water is 1.00 g mL^{-1}. **LO2**

9.45 The density of solid silver bromide is 6.47 g cm^{-3}. What is the concentration of AgBr in a 15.0 cm^3 sample of pure AgBr? What is the concentration of AgBr in a 40.0 g sample of pure AgBr? **LO2**

9.46 Write the equilibrium constant expression, K_c, for each of the following heterogeneous equilibria. **LO2**

(a) $2C(s) + O_2(g) \rightleftharpoons 2CO(g)$

(b) $2NaHSO_3(s) \rightleftharpoons Na_2SO_3(s) + H_2O(g) + SO_2(g)$

(c) $2C(s) + 2H_2O(g) \rightleftharpoons CH_4(g) + CO_2(g)$

(d) $CaCO_3(s) + 2HF(g) \rightleftharpoons CaF_2(s) + H_2O(g) + CO_2(g)$

(e) $CuSO_4{\cdot}5H_2O(s) \rightleftharpoons CuSO_4(s) + 5H_2O(g)$

9.47 The heterogeneous equilibrium:

$$2HI(g) + Cl_2(g) \rightleftharpoons 2HCl(g) + I_2(s)$$

has $K_c = 6.25 \times 10^{33}$ at 25 °C. Suppose 0.200 mol each of HCl(g) and I_2(s) are placed in a 1.00 L reaction vessel. Calculate the equilibrium concentrations of HI(g) and Cl_2(g) in the reaction vessel. **LO5**

9.48 At 25 °C, $K_c = 360$ for the equilibrium:

$$AgCl(s) + Br^-(aq) \rightleftharpoons AgBr(s) + Cl^-(aq)$$

If solid AgCl is added to a solution containing 0.1000 M Br^-, calculate the equilibrium concentrations of Br^- and Cl^-. **LO2**

9.49 Use the $\Delta_f G^\circ$ data in appendix A to calculate the value of K_p for the following equilibria at 25 °C. **LO3**

(a) $2NO(g) + 2H_2O(g) \rightleftharpoons N_2H_4(g) + 2O_2(g)$

(b) $2POCl_3(g) \rightleftharpoons 2PCl_3(g) + O_2(g)$

9.50 The forward reaction in the equilibrium:

$$NO_2(g) + NO(g) \rightleftharpoons N_2O(g) + O_2(g)$$

has $\Delta_r G^\circ_{1273\ K} = -9.67$ kJ mol^{-1}. A 1.00 L reaction vessel at 1273 K contains 0.0200 mol NO_2, 0.040 mol NO, 0.015 mol N_2O and 0.0350 mol O_2. Is the reaction mixture at equilibrium? If not, in which direction will spontaneous change occur for the system to reach equilibrium? **LO3**

9.51 The forward reaction in the equilibrium:

$$HCOOH(g) \rightleftharpoons CO(g) + H_2O(g)$$

has $\Delta_r G^\circ_{673\ K} = -79.8$ kJ mol^{-1}. Is a gas mixture containing 0.047 mol HCOOH, 0.061 mol CO and 0.013 mol H_2O in a 1.75 L reaction vessel at 400 °C at equilibrium? If not, in which direction will spontaneous change occur for the system to reach equilibrium? **LO3**

9.52 The forward reaction in the equilibrium:

$$C(s) + 2H_2(g) \rightleftharpoons CH_4(g)$$

for which $\Delta_r G^\circ = -50.79$ kJ mol^{-1}; can be used to convert coal to methane, the major component of natural gas. What is the value of K_p for this reaction at 25 °C? Does this value of K_p suggest that it is worth studying this reaction as a means of methane production? **LO3**

9.53 One of the important reactions in living cells from which the organism draws energy is the reaction of adenosine triphosphate (ATP) with water to give adenosine diphosphate (ADP) and a free phosphate ion. **LO3**

$$ATP(aq) + H_2O(l) \rightleftharpoons ADP(aq) + PO_4^{3-}(aq)$$

The value of $\Delta_r G^\circ$ for the forward reaction at 37 °C (normal human body temperature) is −33 kJ mol^{-1}. Write the equilibrium constant expression and calculate the value of the equilibrium constant for the equilibrium at this temperature.

9.54 What is the value of the reaction quotient, Q, at the very beginning of a chemical reaction before any products have been formed? What happens to the value of Q as the reaction proceeds to give products? **LO3**

9.55 Methanol, a potential replacement for petrol as a fuel in cars, can be made from H_2 and CO by the forward reaction in the equilibrium:

$$CO(g) + 2H_2(g) \rightleftharpoons CH_3OH(g)$$

At 500 K, this equilibrium has $K_p = 6.25 \times 10^{-3}$. Calculate $\Delta_r G^\circ_{500\ K}$ for the forward reaction in this equilibrium. **LO3**

9.56 Refer to the data in table 8.7 on p. 335 to determine $\Delta_r G^\circ$ for the forward reaction in the equilibrium:

$$N_2(g) + 2CO_2(g) \rightleftharpoons 2NO(g) + 2CO(g)$$

What is the value of K_p for this equilibrium at 25 °C? **LO3**

9.57 Use the data in chapters to calculate $\Delta_r G^\circ_{773\ K}$ for the forward reaction in the equilibrium:

$$2NO(g) + 2CO(g) \rightleftharpoons N_2(g) + 2CO_2(g)$$

What is the value of K_p for the equilibrium at this temperature? **LO3**

9.58 The equilibrium:

$$PCl_3(g) + Cl_2(g) \rightleftharpoons PCl_5(g)$$

has $K_c = 0.18$ at a particular temperature. Analysis of a reaction vessel containing these three gases at this temperature showed them to be present in the following concentrations: $[PCl_3] = 0.0275$ M, $[Cl_2] = 0.0317$ M, $[PCl_5] = 0.00168$ M. **LO3**

(a) Is the system at equilibrium?

(b) If not, in which direction will a spontaneous change occur to restore equilibrium?

9.59 At 460 °C, the equilibrium:

$$SO_2(g) + NO_2(g) \rightleftharpoons NO(g) + SO_3(g)$$

has $K_c = 85.0$. A reaction flask at 460 °C contains these gases at the following concentrations: $[SO_2] = 0.002\,50$ M, $[NO_2] = 0.003\,50$ M, $[NO] = 0.0250$ M, $[SO_3] = 0.0400$ M. **LO5**

(a) Is the system at equilibrium?

(b) If not, in which direction will a spontaneous change occur to restore equilibrium?

9.60 At a certain temperature, the equilibrium:

$$CO(g) + 2H_2(g) \rightleftharpoons CH_3OH(g)$$

has $K_c = 0.521$. If a reaction mixture at equilibrium contains 0.247 M CO and 0.194 M H_2, what is the concentration of CH_3OH in the mixture? **LO5**

9.61 $K_c = 64$ for the equilibrium:

$$N_2(g) + 3H_2(g) \rightleftharpoons 2NH_3(g)$$

at a certain temperature. Suppose it was found that an equilibrium mixture of these gases contained 0.360 M NH_3 and 0.0192 M N_2. What was the equilibrium concentration of H_2 in the mixture? **LO5**

9.62 At 773 °C, a mixture of CO(g), H_2(g) and CH_3OH(g) was allowed to come to equilibrium. At equilibrium, the concentrations of the three gases were: [CO] = 0.105 M, $[H_2] = 0.250$ M, $[CH_3OH] = 0.002\,61$ M. Calculate K_c for the equilibrium: **LO5**

$$CH_3OH(g) \rightleftharpoons CO(g) + 2H_2(g)$$

9.63 Ethene, C_2H_4, and water react under appropriate conditions to give ethanol. **LO5**

An equilibrium mixture of these gases at a certain temperature had the following concentrations: $[C_2H_4] = 0.0148$ M, $[H_2O] = 0.0336$ M, $[C_2H_5OH] = 0.180$ M. What is the value of K_c for the equilibrium?

$$C_2H_4(g) + H_2O(g) \rightleftharpoons C_2H_5OH(g)$$

9.64 A mixture of $H_2(g)$, $Br_2(g)$ and HBr(g) will eventually come to equilibrium. At a high temperature, 2.00 mol of HBr was placed in a 4.00 L container. At equilibrium, the concentration of Br_2 was 0.0955 M. What is K_c for the equilibrium:

$$H_2(g) + Br_2(g) \rightleftharpoons 2HBr(g)$$

at this temperature? **LO5**

9.65 A 0.050 mol sample of formaldehyde vapour, HCHO, was placed in a heated 500 mL vessel and some of it decomposed to set up the equilibrium:

$$HCHO(g) \rightleftharpoons H_2(g) + CO(g)$$

At equilibrium, the HCHO(g) concentration was 0.066 mol L^{-1}. Calculate the value of K_c for this equilibrium. **LO5**

9.66 A reaction vessel at high temperature contained the following gases at the given initial concentrations: $[N_2O] = 0.184$ M, $[O_2] = 0.377$ M, $[NO_2] = 0.0560$ M, $[NO] = 0.294$ M. As NO_2 is the only coloured gas in the mixture, its concentration was monitored by following the intensity of the colour. At equilibrium, the NO_2 concentration was 0.118 M. What is the value of K_c for the equilibrium:

$$N_2O(g) + O_2(g) \rightleftharpoons NO_2(g) + NO(g)$$

at this temperature? **LO5**

9.67 At 25 °C, 0.0560 mol O_2 and 0.020 mol N_2O were placed in a 1.00 L container where the following equilibrium was then established. **LO5**

$$2N_2O(g) + 3O_2(g) \rightleftharpoons 4NO_2(g)$$

At equilibrium, the NO_2 concentration was 0.020 M. What is the value of K_c for this equilibrium?

9.68 At 25 °C, $K_c = 0.145$ for the following equilibrium in the solvent CCl_4. **LO5**

$$2BrCl \rightleftharpoons Br_2 + Cl_2$$

If the initial concentration of BrCl in the solution is 0.029 M, calculate the equilibrium concentrations of Br_2 and Cl_2.

9.69 The equilibrium constant, K_c, for the equilibrium:

$$SO_3(g) + NO(g) \rightleftharpoons NO_2(g) + SO_2(g)$$

was found to be 0.500 at a certain temperature. If 0.240 mol of SO_3 and 0.240 mol of NO are placed in a 2.00 L container and allowed to react, calculate the equilibrium concentration of each gas. **LO5**

9.70 For the equilibrium in question 9.69, a reaction mixture is prepared in which 0.120 mol NO_2 and 0.120 mol of SO_2 are placed in a 1.00 L vessel. What are the concentrations of all four gases at equilibrium? How do these equilibrium values compare with those calculated in question 9.69? Account for your observation. **LO5**

9.71 At a certain temperature, the equilibrium:

$$CO(g) + H_2O(g) \rightleftharpoons CO_2(g) + H_2(g)$$

has $K_c = 0.400$. If 1.50 mol of each gas was placed in a 100 L vessel and the mixture allowed to come to equilibrium, what will be the equilibrium concentration of each gas? **LO5**

9.72 The equilibrium $2HCl(g) \rightleftharpoons H_2(g) + Cl_2(g)$ has $K_c = 3.2 \times 10^{-34}$ at 25 °C. If a reaction vessel initially contains 0.0500 mol L^{-1} of HCl, what are the concentrations of H_2 and Cl_2 at equilibrium? **LO5**

9.73 At 200 °C, $K_c = 1.4 \times 10^{-10}$ for the equilibrium:

$$N_2O(g) + NO_2(g) \rightleftharpoons 3NO(g)$$

If 0.225 mol of N_2O and 0.450 mol of NO_2 are placed in a 4.00 L container, what is the NO concentration at equilibrium? **LO5**

9.74 At 2000 °C, the decomposition of CO_2:

$$2CO_2(g) \rightleftharpoons 2CO(g) + O_2(g)$$

has $K_c = 6.4 \times 10^{-7}$. If a 1.00 L container holding 1.0×10^{-2} mol of CO_2 is heated to 2000 °C, what is the concentration of CO at equilibrium? **LO5**

9.75 At 500 °C, the decomposition of $H_2O(g)$ into $H_2(g)$ and $O_2(g)$ proceeds according to the equilibrium:

$$2H_2O(g) \rightleftharpoons 2H_2(g) + O_2(g)$$

with $K_c = 6.0 \times 10^{-28}$. What amounts of $H_2(g)$ and $O_2(g)$ are present at equilibrium in a 2.00 L reaction vessel at this temperature if the container originally held 0.00721 mol H_2O? **LO5**

9.76 At a certain temperature, $K_c = 0.18$ for the equilibrium:

$$PCl_3(g) + Cl_2(g) \rightleftharpoons PCl_5(g)$$

If 0.026 mol of PCl_5 is placed in a 2.00 L vessel at this temperature, what is the concentration of PCl_3 at equilibrium? **LO5**

9.77 At 460 °C, the equilibrium:

$$SO_2(g) + NO_2(g) \rightleftharpoons NO(g) + SO_3(g)$$

has $K_c = 85.0$. Suppose 0.120 mol of SO_2, 0.0328 mol of NO_2, 0.0913 mol of NO and 0.159 mol of SO_3 are placed in a 10.0 L container at this temperature. What are the concentrations of all the gases when the system reaches equilibrium? **LO5**

9.78 At a certain temperature, $K_c = 0.500$ for the equilibrium:

$$SO_3(g) + NO(g) \rightleftharpoons NO_2(g) + SO_2(g)$$

If 0.100 mol SO_3 and 0.200 mol NO are placed in a 2.00 L container and allowed to come to equilibrium, what are the NO_2 and SO_2 concentrations? **LO5**

9.79 At 25 °C, $K_c = 6.90$ for the following equilibrium in the solvent CCl_4. **LO5**

$$Br_2 + Cl_2 \rightleftharpoons 2BrCl$$

A solution was prepared with the following initial concentrations: $[BrCl] = 0.0425$ M, $[Br_2] = 0.0305$ M, $[Cl_2] = 0.0243$ M. What are the equilibrium concentrations of all three compounds?

9.80 At a certain temperature, $K_c = 4.3 \times 10^5$ for the equilibrium:

$$HCOOH(g) \rightleftharpoons CO(g) + H_2O(g)$$

If 0.200 mol of HCOOH is placed in a 1.00 L vessel, what are the concentrations of CO and H_2O when the system reaches equilibrium? **LO5**

9.81 The equilibrium

$$H_2(g) + Br_2(g) \rightleftharpoons 2HBr(g)$$

has $K_c = 2.0 \times 10^9$ at 25 °C. If 0.112 mol of H_2 and 0.224 mol of Br_2 are placed in a 10.0 L container, what are the equilibrium concentrations of all components at 25 °C? **LO5**

ADDITIONAL EXERCISES

9.82 Show, by writing expressions for K_p and K_c, and making appropriate substitutions, that for the equilibrium: **LO2**

$$2NO_2(g) \rightleftharpoons N_2O_4(g)$$

$$K_p = K_c \left(\frac{1000\,RT}{p^{\ominus}} \right)^{-1}$$

9.83 If $K = 1$ for a particular equilibrium, does this necessarily mean that the concentrations of reactants and products are equal at equilibrium? Outline your reasoning. **LO2**

9.84 Consider the two equilibria:

$$X(g) \rightleftharpoons Y(g) \qquad \text{(equilibrium 1)}$$

and

$$2X(g) \rightleftharpoons Y(g) \qquad \text{(equilibrium 2)}$$

both of which have $K_c = 2$. Show that, at equilibrium, the ratio of products to reactants, [Y]/[X], in the second equilibrium increases as [Y] increases, while the same ratio remains constant in the first equilibrium. At what [Y] is the ratio [Y]/[X] = 1 for the second equilibrium? Discuss how this example is important with respect to determining the position of equilibrium in equilibria having similar values of K. **LO2**

9.85 With reference to figure 9.10 (p. 361), would you expect the change in composition 4 → 1 to occur spontaneously? Outline your reasoning. **LO3**

9.86 At a certain temperature, $K_c = 0.914$ for the equilibrium:

$$NO_2(g) + NO(g) \rightleftharpoons N_2O(g) + O_2(g)$$

A mixture was prepared containing 0.200 mol NO_2, 0.300 mol NO, 0.150 mol N_2O and 0.250 mol O_2 in a 4.00 L container. What are the equilibrium concentrations of each gas? **LO5**

9.87 At 27 °C, $K_p = 1.5 \times 10^{18}$ for the equilibrium: **LO5**

$$3NO(g) \rightleftharpoons N_2O(g) + NO_2(g)$$

If 0.030 mol of NO is placed in a 1.00 L vessel and equilibrium is established, what are the equilibrium concentrations of NO, N_2O and NO_2?

9.88 At a certain temperature, $K_c = 0.914$ for the equilibrium: **LO5**

$$NO_2(g) + NO(g) \rightleftharpoons N_2O(g) + O_2(g)$$

A mixture was prepared containing 0.200 mol of each gas in a 5.00 L container. Calculate the equilibrium concentration of each gas. How will the concentrations change if 0.050 mol of NO_2 is added to the equilibrium mixture?

9.89 A sample of liquid water was injected into an evacuated reaction vessel at 25.0 °C and the system was allowed to come to equilibrium at this temperature. If $\Delta_r G^\ominus = 8.60$ kJ mol^{-1} for the forward process of the equilibrium:

$$H_2O(l) \rightleftharpoons H_2O(g)$$

what is the pressure of $H_2O(g)$ in the reaction vessel at equilibrium at 25.0 °C? **LO3**

9.90 It is important that you obtain an appreciation for the values of K that correspond to particular values of $\Delta G^\ominus$, and vice versa. With this in mind, complete the following table for $T = 25.0$ °C. **LO3**

$\Delta G^\ominus$/kJ mol^{-1}	K	K	$\Delta G^\ominus$/kJ mol^{-1}
−100.00		1.00×10^{-50}	
−50.00		1.00×10^{-20}	
−20.00		1.00×10^{-10}	
−10.00		1.00×10^{-5}	
−5.00		1.00×10^{-2}	
0.00		1.00×10^{0}	
5.00		1.00×10^{1}	
10.00		1.00×10^{2}	
20.00		1.00×10^{3}	
50.00		1.00×10^{4}	
100.00		1.00×10^{5}	
		1.00×10^{10}	
		1.00×10^{20}	
		1.00×10^{50}	

MATHS FOR CHEMISTRY

The natural logarithm: ln or log$_e$

You are probably familiar with logarithms. You know, for example that $10^2 = 100$ and that therefore, the logarithm of 100 is 2. Likewise, 1 000 000 = 10^6, and so the logarithm of 1 000 000 is 6. However, it is important to realise that there is more than one type of logarithm. The logarithms discussed in the above sentences are logarithms to the base 10 — in other words, these are the powers to which 10 must be raised in order to obtain the desired number. We sometimes signify these types of logarithms by the symbol '$\log_{10}$', although we usually just use the symbol 'log'. In chemistry we use these logarithms extensively in problems involving pH, where pH = $-\log[H_3O^+]$ (chapter 11).

Another type of logarithm which is commonly used is the natural logarithm (ln), sometimes called the logarithm to the base e ($\log_e$). In this case, instead of using 10 as the base as we did above, the number e is used. e is an irrational number, whose first few digits are 2.718 281 828 4. Just as any number can be expressed in the form 10^y, where y is the logarithm ($\log_{10}$) of the number, so any number can be expressed in the form e^x, where x is the natural logarithm of the number. For example, $100 = e^{4.605\,170}$, and therefore the natural logarithm of 100 is 4.605 170. Natural logarithms are important in calculus, for the reason that $\int \frac{1}{x} = \ln x + C$. In chemistry, they are used most notably in the equation $\Delta G^\ominus = -RT\ln K$ and in the Nernst equation (see chapter 12).

It is vital that you recognise the difference between the two types of logarithm discussed above and use the appropriate logarithm in your calculations. Your calculator will have separate buttons for both types of logarithms — don't get them mixed up!

10 Solutions and solubility

The discovery of X-rays in 1895 enabled doctors to take images of the inside of the living human body for the first time, and this technique is still used over one hundred years later. It relies on the absorption of X-rays by constituents of the body. While objects such as bone give clear images, soft tissue is more difficult to see. Therefore, to create an X-ray image of areas such as the gastrointestinal tract, it is necessary to use a contrast agent that absorbs X-rays well and which can be delivered specifically to the required part of the body. Barium sulfate, $BaSO_4$, is one such contrast agent. This is ingested by the patient in the form of a barium meal, thereby coating the gastrointestinal tract, and an X-ray is taken shortly thereafter. This photo is an example of an X-ray of the gastrointestinal tract using $BaSO_4$ as a contrast agent.

It may surprise you to learn that the Ba^{2+} ion is highly toxic, and that swallowing a barium salt would generally end in death. However, barium sulfate is only slightly soluble in aqueous solution, meaning that the concentration of Ba^{2+} never approaches toxic levels.

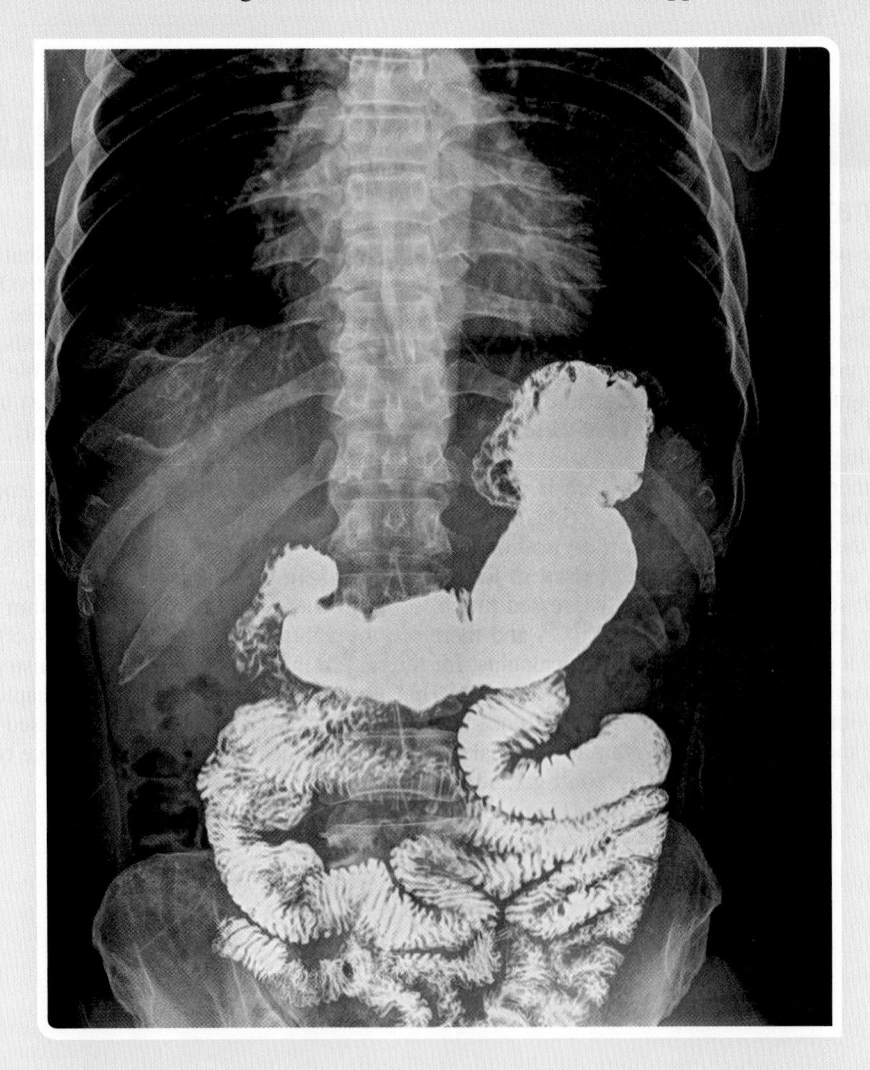

In this chapter, we will examine various aspects of solutions and solubility. In addition to outlining the different possible types of solutions, we will investigate the energetics behind the formation of solutions, quantify the solubility of slightly soluble salts using the solubility product, K_{sp}, and show that physical properties such as the freezing and boiling points of solutions depend on the number of solute particles in the solution.

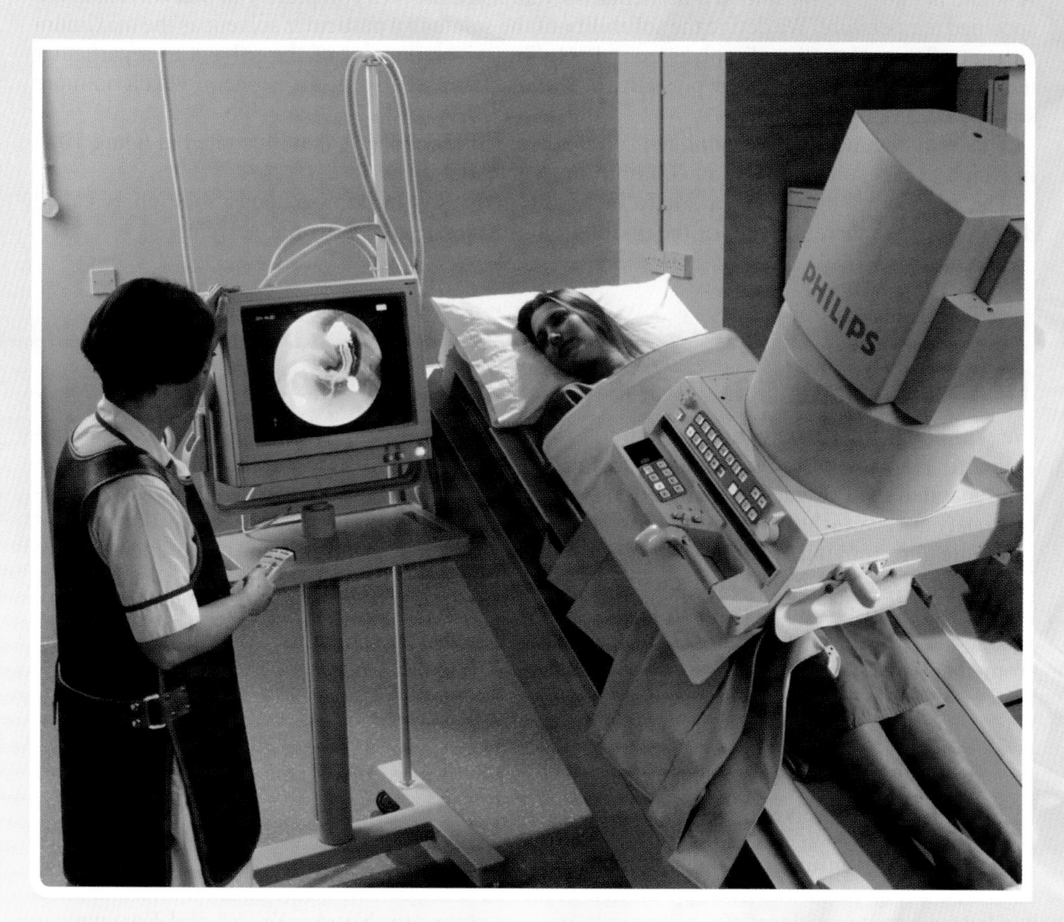

LEARNING OBJECTIVES

After studying this chapter, you should be able to:

10.1 define important terms related to solutions

10.2 describe why gases mix spontaneously

10.3 explain the energetics involved in the formation of liquid solutions

10.4 carry out calculations involving slightly soluble salts

10.5 quantify the effects that arise in solutions as a result of colligative properties.

10.1 Introduction to solutions and solubility

The ability of chemical substances to mix with each other is a very important aspect of chemistry. We are familiar with many examples from everyday life; for example, dissolving instant coffee in a cup of hot water, adding detergent to water and cleaning a used paintbrush with turpentine are all situations where we require two substances to mix completely. However, there are also substances that we cannot get to mix no matter how hard we try, the prime example being oil and water (so we don't use water to clean a paintbrush covered in oil-based paint). Many aspects of chemical synthesis, in particular the purification of products following a chemical reaction, rely on the fact that particular substances don't mix. An understanding of the principles involved in the formation of solutions is important in many areas of chemistry. In this chapter, we will focus on the energetics of formation of solutions and the quantification of physical properties of solutions.

FIGURE 10.1 A saturated solution. In a saturated solution, a dynamic equilibrium exists between the undissolved solute and the dissolved solute in the solution.

A **solution** can be defined as a homogeneous mixture of two or more pure substances ('homogeneous' means that all regions of the solution have exactly the same composition). Despite the fact that we often think of a solution as involving some type of liquid, we can have gaseous solutions comprising two or more gases (e.g. air) and even solid solutions comprising two or more solids (e.g. alloys such as brass and pewter) in addition to the 'usual' solutions containing a gas, liquid or solid dissolved in a liquid. In the case of liquid solutions, we call the liquid the **solvent**, while the dissolved substance is called the **solute**; the solute is present in smaller amounts than the solvent. A solute is said to be **soluble** in a particular solvent if it dissolves completely in that solvent at the specified temperature. We define the **solubility** of the solute in a particular solvent as the maximum amount of the solute that dissolves completely in a given amount of the solvent at a particular temperature, T, and a particular pressure, p, while a **saturated solution** is one in which no more solute will dissolve.

The most common type of saturated solution we will encounter is that illustrated in figure 10.1, in which excess solid solute is in equilibrium with its dissolved form.

The process of dissolving a solute in a solvent to give a homogeneous solution is called **dissolution**. We therefore talk of the dissolution of a solute in a particular solvent.

10.2 Gaseous solutions

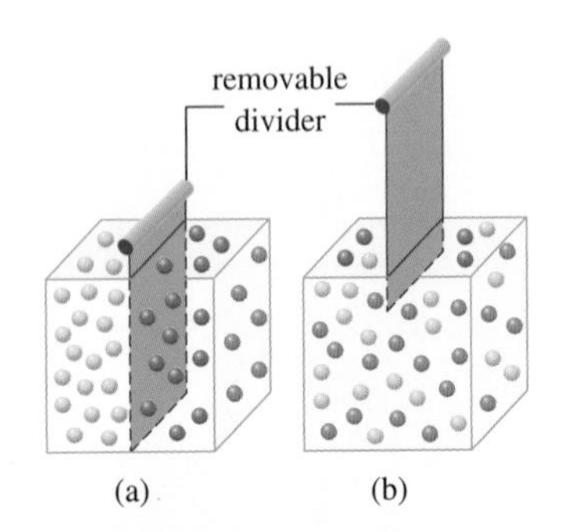

FIGURE 10.2 Mixing gases. When two gases, initially in separate compartments **(a)** suddenly find themselves in the same container **(b)**, they mix spontaneously.

We will begin our investigation of solutions by looking at mixing two gases to form a gaseous solution. We start with the situation depicted in figure 10.2, where two different pure gases are separated by a divider. When we remove the divider, we find that the two gases mix spontaneously to give a homogeneous gaseous solution. This is not surprising given that the gas molecules are moving randomly at hundreds of metres per second. Even if we cool the two gases, they will still mix, albeit more slowly. In fact, as we saw in chapter 9, all gases mix completely with all other gases in all proportions. The fact that the two gases in figure 10.2 mix spontaneously means that ΔG for the mixing process ($\Delta_{mix}G$) must be negative. Recalling from chapter 8 that:

$$\Delta G = \Delta H - T\Delta S$$

and given that the enthalpy change when two gases are mixed is usually small, it is the large positive $\Delta_{mix}S$ term that ensures that $\Delta_{mix}G$ is negative. The completely mixed state is far more statistically likely and, therefore, of higher entropy than the unmixed state; this leads to a large entropy increase on mixing, and a correspondingly large positive $\Delta_{mix}S$ value. In fact, we say that the mixing process is entropy driven. When we consider the gases at the molecular level (or indeed the atomic level for monatomic gases such as He, Ne, Ar, Kr, Xe and Rn), we can see that there is no impediment to mixing the gases, as the intermolecular forces between individual gas molecules are tiny, and the gas molecules are far apart. However, when we consider condensed-phase solutions (i.e. solutions involving liquids and/or solids), we will see that the magnitude of interatomic or intermolecular attractions in the condensed phase can determine whether or not two substances can mix. In these cases, the enthalpy of mixing becomes important.

10.3 Liquid solutions

As mentioned on the previous page, there are many possible types of solutions. In this section, we will discuss, in detail, the solutions that result from dissolution of a gas, liquid or solid in a liquid.

Gas–liquid solutions

As can be seen in table 10.1, gases display a wide range of solubilities in water. Molecules such as O_2 and N_2 have very limited solubility at room temperature, while the solubility of the polar molecule ammonia is a quite remarkable 51.8 g per 100 g of water at 20 °C. For a gas to dissolve in a liquid, the gas molecules must be able to disperse themselves evenly throughout the solvent. Such a process differs from that for formation of a gaseous solution in that the intramolecular forces between the solvent molecules are not negligible, and neither therefore is the value of $\Delta_{sol}H$, the enthalpy of solution.

TABLE 10.1 Solubilities of common gases in water.

Gas	Temperature		
	0 °C	20 °C	50 °C
nitrogen, N_2	0.0029	0.0019	0.0012
oxygen, O_2	0.0069	0.0043	0.0027
carbon dioxide, CO_2	0.335	0.169	0.076
sulfur dioxide, SO_2	22.8	10.6	4.3
ammonia, NH_3	89.9	51.8	28.4

Solubilities are in grams of solute per 100 g of water when the gaseous space over the liquid is saturated with the gas and the total pressure is 10^5 Pa.

The enthalpy change when a gas dissolves in a liquid has essentially two contributions, as shown in figure 10.3.

1. *Energy is required to open 'pockets' in the solvent that can hold gas molecules.* The solvent must be expanded slightly to accommodate the molecules of the gas. This is an endothermic process since attractions between solvent molecules must be overcome. Water is a special case — it already contains open holes in its network of loose hydrogen bonds around room temperature. For water, very little energy is required to create pockets that can hold gas molecules.
2. *Energy is released when gas molecules enter these pockets.* Intermolecular attractions between the dissolved gas molecules and the surrounding solvent molecules lower the total energy, and energy is released as heat. The stronger the attractions are, the more heat is released. Water can form hydrogen bonds with some gases, such as NH_3, whereas many organic solvents cannot. More heat is released when such a gas molecule is placed in a pocket in water than in organic solvents.

Enthalpies of solution for gases in organic solvents are often endothermic because the energy required to open up pockets is greater than the energy released by attractions formed between the gas and solvent molecules. Enthalpies of solution for gases in water are often exothermic because water already contains pockets to hold the gas molecules, and energy is released when water and gas molecules attract each other.

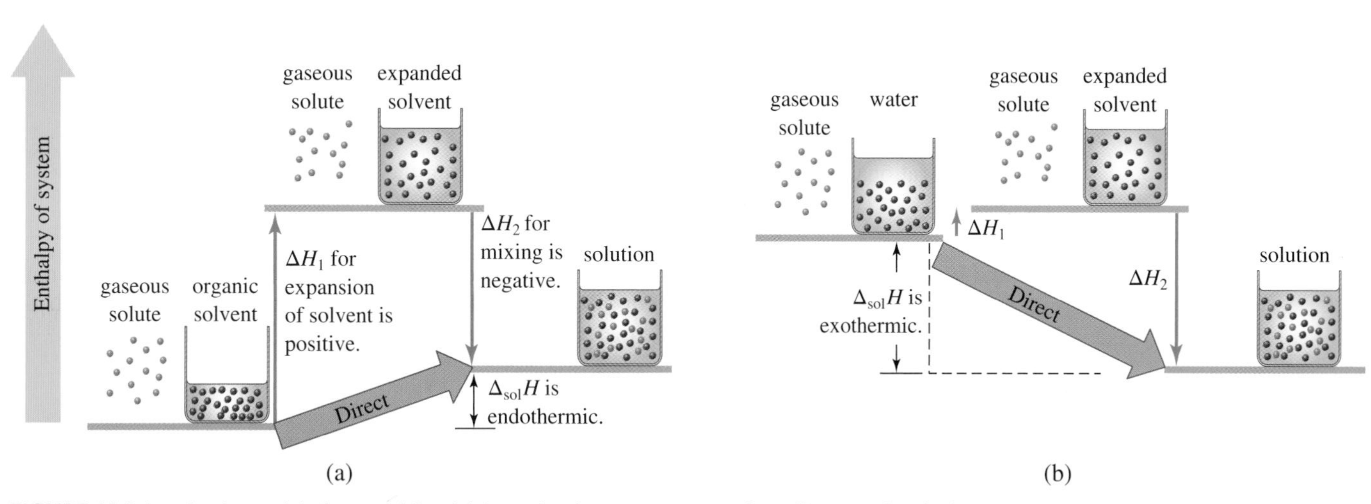

FIGURE 10.3 A molecular model of gas solubility: **(a)** A gas dissolves in an organic solvent. Energy is absorbed to open 'pockets' in the solvent that can hold the gas molecules. In the second step, energy is released when the gas molecules enter the pockets, where they are attracted to the solvent molecules. Here, the solution process is shown to be endothermic. **(b)** Around room temperature, water's loose network of hydrogen bonds already contains pockets that can accommodate gas molecules, so little energy is needed to prepare the solvent to accept the gas. In the second step, energy is released as the gas molecules take their place in the pockets where they experience attractions to the water molecules. In this case, the solution process is exothermic.

While we stated earlier that gaseous solutions form mainly because of the large entropy of the completely mixed state, the situation is not so straightforward with solutions of a gas in a liquid. Liquid water has a surprisingly ordered structure, and the addition of gas molecules does not necessarily disrupt this structure — in fact, there is experimental evidence that entropy *decreases* on formation of some gas–water solutions. Thus the solubility of a particular gas in water is a subtle balance of enthalpic and entropic effects, and it is often difficult to determine which factor predominates.

We can also see from table 10.1 that the solubility of gases in water varies significantly with temperature, with all the examples given becoming less soluble as the temperature increases. The decreasing solubility of O_2 with increasing temperature has obvious implications for aquatic life, given the gradual warming of the Earth's oceans.

If we write the dissolution process as an equilibrium between the undissolved gas and the dissolved gas:

$$\text{gas}_{\text{undissolved}} \rightleftharpoons \text{gas}_{\text{dissolved}}$$

we can see that the equilibrium constant for this process decreases with increasing temperature for the examples in table 10.1. Recalling the relationship between ΔH and K from chapter 9 (p. 362), we can conclude that the dissolution process for these gases is exothermic. However, not all gases become less soluble as the temperature increases; for example, the solubilities of H_2, N_2, CO, He and Ne in common organic solvents, such as toluene and acetone, actually increase with increasing temperature. You can see from these examples that predicting solubilities even in simple systems is fraught with difficulty!

The solubility of a gas in a liquid is affected not only by temperature but also by pressure. For example, the gases in air, chiefly nitrogen and oxygen, while not very soluble in water under ordinary pressures, become increasingly soluble at elevated pressures (see figure 10.4).

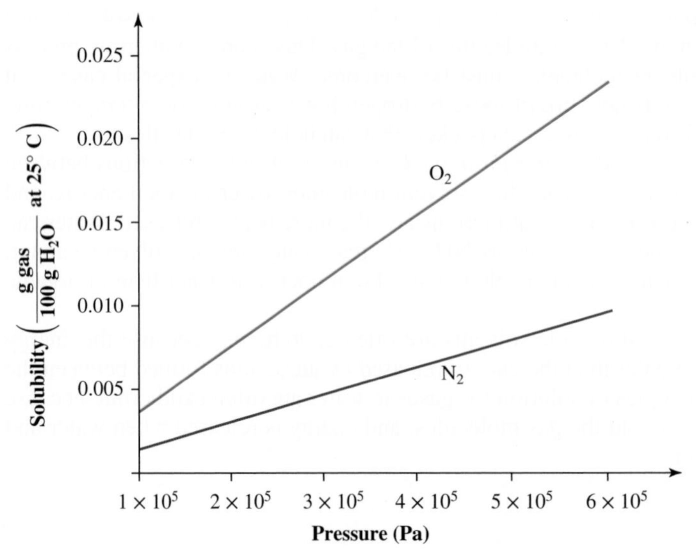

FIGURE 10.4 Solubility in water versus pressure for nitrogen and oxygen. The amount of gas that dissolves increases as the pressure is raised.

We can rationalise this observation by again considering the equilibrium:

$$\text{gas}_{\text{undissolved}} \rightleftharpoons \text{gas}_{\text{dissolved}}$$

and looking at the effect of increasing the pressure on the position of this equilibrium. The reaction quotient expression for this equilibrium can be written in a form that involves both the pressure of the undissolved gas and the concentration of the dissolved gas.

$$Q = \frac{[\text{gas}_{\text{dissolved}}]}{\left(\dfrac{p(\text{gas}_{\text{undissolved}})}{p^{\ominus}}\right)}$$

We can see from this that Q is inversely proportional to $p(\text{gas}_{\text{undissolved}})$. Thus, if we increase the pressure of the undissolved gas at constant temperature, we decrease Q so that $Q < K$. To restore equilibrium and make $Q = K$, the equilibrium position shifts towards the right-hand side to increase Q by increasing $[\text{gas}_{\text{dissolved}}]$. Of course, the converse is also true; if we decrease the pressure above a gas–liquid solution, we increase Q, and some of the dissolved gas leaves the solution in order to decrease Q and restore equilibrium. We see this most often whenever we open a bottle of anything fizzy; when the top is removed, bubbles of CO_2 emerge from the solution in response to the sudden lowering of pressure (figure 10.5).

To see how pressure affects gas solubility on a molecular level, imagine a closed container equipped with a piston, partly filled with a solution of a gas in a liquid. Figure 10.6a shows the system at equilibrium; gas molecules move at equal rates between the dissolved and undissolved states. Gas molecules enter the solution at a rate proportional to the frequency with which they collide with the surface of the solution. This frequency increases when we increase the pressure of the gas (figure 10.6b) because the gas molecules are squeezed closer together. Because liquids and liquid solutions are incompressible, the increased pressure alone has no effect on the frequency with which gas molecules can leave the solution. That means that increasing the pressure favours dissolution of the gas.

FIGURE 10.5 Bottled carbonated beverages fizz when the bottle is opened because the sudden drop in pressure causes a sudden drop in gas solubility.

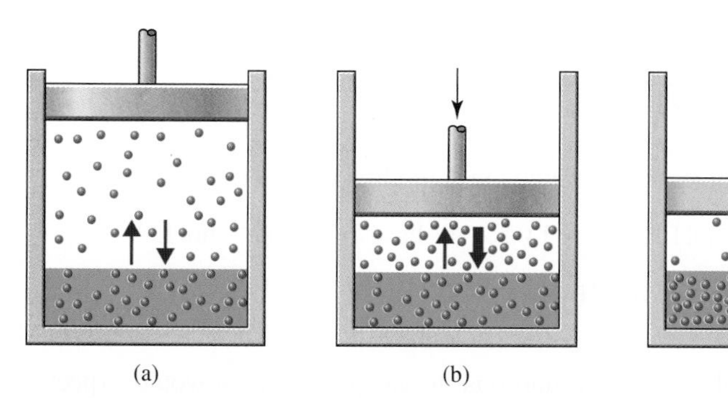

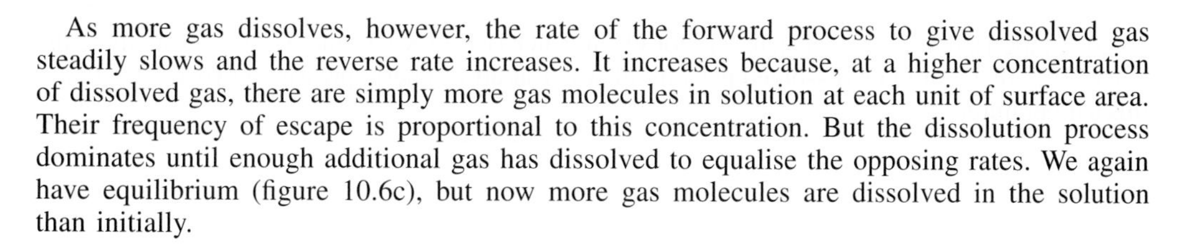

FIGURE 10.6 How pressure increases the solubility of a gas in a liquid: **(a)** At some specific pressure, equilibrium exists between the vapour phase and the solution. **(b)** An increase in pressure puts stress on the equilibrium. More gas molecules are dissolving than are leaving the solution. **(c)** More gas has dissolved and equilibrium is restored.

As more gas dissolves, however, the rate of the forward process to give dissolved gas steadily slows and the reverse rate increases. It increases because, at a higher concentration of dissolved gas, there are simply more gas molecules in solution at each unit of surface area. Their frequency of escape is proportional to this concentration. But the dissolution process dominates until enough additional gas has dissolved to equalise the opposing rates. We again have equilibrium (figure 10.6c), but now more gas molecules are dissolved in the solution than initially.

Henry's law

For gases *that do not react with the solvent*, Henry's law gives the relationship between gas pressure and gas solubility. **Henry's law** states that the concentration of a gas dissolved in a liquid at any given temperature is directly proportional to the partial pressure of the gas above the solution.

$$c_{gas} = k_H p_{gas} \text{ (at constant temperature)}$$

In the equation, c_{gas} is the concentration of the dissolved gas and p_{gas} is the partial pressure (pp. 231–2) of the gas above the solution. The proportionality constant, k_H, called the *Henry's law constant*, is unique to each gas. The equation is true only at low concentrations and pressures and, as we said, for gases that do not react with the solvent.

An alternative (and commonly used) expression of Henry's law is:

$$\frac{c_1}{p_1} = \frac{c_2}{p_2}$$

where c_1 and p_1 refer to initial conditions and c_2 and p_2 to final conditions.

WORKED EXAMPLE 10.1

Using Henry's law

At 20 °C, the solubility of N_2 in water, expressed in terms of mass of dissolved gas per volume of solution, is 0.0150 g L^{-1} when the partial pressure of nitrogen is 7.63×10^4 Pa. Use these data to calculate the Henry's Law constant for N_2 in water at this temperature. What is the solubility of N_2 in water at 20 °C when its partial pressure is 1.05×10^5 Pa?

Analysis

This problem deals with the effect of gas pressure on gas solubility, so Henry's law applies. We must first convert the solubility of N_2 in g L^{-1} to a solubility in mol L^{-1} to calculate the Henry's Law constant. We then use this to calculate the solubility at the higher pressure.

Solution

We use stoichiometry to determine the solubility of N_2 in mol L^{-1}. We know that 0.0150 g of N_2 will dissolve in 1.00 L of water. Therefore:

$$n_{N_2} = \frac{m}{M} = \frac{0.0150\,\text{g}}{28.02\,\text{g mol}^{-1}} = 5.35\times10^{-4}\,\text{mol}$$

This is the amount of N_2 that will dissolved in 1.00 L. Thus, the solubility of N_2 is 5.35×10^{-4} mol L^{-1}.

Gathering data, we now have $c_{gas} = 5.35 \times 10^{-4}$ mol L^{-1} and $p_{gas} = 7.63 \times 10^4$ Pa. We use these to calculate k_H by rearranging $c_{gas} = k_H p_{gas}$.

$$k_H = \frac{c_{gas}}{p_{gas}} = \frac{5.35\times10^{-4}\,\text{mol L}^{-1}}{7.63\times10^4\,\text{Pa}} = 7.01\times10^{-9}\,\text{mol L}^{-1}\,\text{Pa}^{-1}$$

We now use this calculated value of k_H to determine the solubility at higher pressure.

$$c_{gas} = k_H p_{gas} = 7.01 \times 10^{-9}\,\text{mol L}^{-1}\,\text{Pa}^{-1} \times 1.05 \times 10^5\,\text{Pa} = 7.36 \times 10^{-4}\,\text{mol L}^{-1}$$

Therefore, the solubility of N_2 at 1.05×10^5 Pa is 7.36×10^{-4} mol L^{-1}.

Is our answer reasonable?

As Henry's Law states that gas solubility is proportional to gas pressure, we would expect N_2 to be more soluble at higher pressure. Our answer shows this to be the case.

PRACTICE EXERCISE 10.1 At 20 °C, the Henry's Law constant, k_H, for O_2 in water is 1.34×10^{-8} mol L^{-1} Pa^{-1}. Calculate the mass of O_2 that would be dissolved per litre of water at atmospheric pressure (1.013×10^5 Pa) at this temperature.

Liquid–liquid solutions

The formation of a liquid–liquid solution requires that we overcome the attractive forces present between the molecules of the two pure liquids to get the liquids to mix completely. This is an even more complicated situation than formation of a gas–liquid solution, yet we can predict the ability of any two liquids to mix, with some confidence, simply by considering the polarities of the two liquids. For example, we find that ethanol and water mix completely in all proportions — we say they are **miscible** — while hexane and water are essentially **immiscible** and form two layers on addition of one liquid to the other (figure 10.7).

Both water and ethanol are polar molecules, while hexane is essentially nonpolar; these facts are central to explaining the behaviour of these liquids on mixing. The —OH functional group in the ethanol molecule leads to extensive hydrogen bonding interactions between ethanol molecules and between ethanol and water molecules (figure 10.8). Thus, the intermolecular attractive forces in the two pure liquids can be compensated for on mixing by the hydrogen bonding interactions between the ethanol and water molecules; the system, therefore, adopts the entropically favoured mixed state. Conversely, hexane molecules are nonpolar and so cannot interact strongly with water molecules; this means that the energy required to disrupt the hydrogen bonding network in water cannot be compensated for by water–hexane interactions on mixing. Even if we could disperse water molecules in hexane, each time two water molecules collided, they would tend to 'stick together' through hydrogen bonding interactions. This would continue to occur until all the water molecules had stuck together and the system would eventually become two phases.

We can use similar arguments to explain why nonpolar liquids tend to mix with each other. The intermolecular interactions in the pure liquids are low, so there is no energetic impediment to mixing.

In general, when two liquids are of similar polarity, they tend to be miscible, whereas liquids of widely differing polarities are often found to be immiscible. This **like-dissolves-like rule** works well in predicting solubilities, but, as always, there are exceptions. The presence of a polar —OH functional group does not guarantee that all alcohols are water soluble; while methanol, CH_3OH, ethanol, CH_3CH_2OH, and propan-1-ol, $CH_3CH_2CH_2OH$, are miscible with water in all proportions, butan-1-ol, $CH_3CH_2CH_2CH_2OH$, is only partially miscible and pentan-1-ol, $CH_3CH_2CH_2CH_2CH_2OH$, is immiscible.

```
    H   H                         H   H   H   H   H   H
    |   |   δ−  δ+                |   |   |   |   |   |
H — C — C — O — H             H — C — C — C — C — C — C — H
    |   |                         |   |   |   |   |   |
    H   H                         H   H   H   H   H   H
```

ethanol
(a polar —OH group)
$\mu = 5.63 \times 10^{-30}$ C m

hexane
(no polar group)
$\mu = 0$ C m

(a)

(b)

FIGURE 10.7 Like dissolves like. **(a)** Ethanol is soluble in water because the mixed ethanol–water attractions are not significantly weaker than the water–water or ethanol–ethanol attractions. **(b)** Hexane is not soluble in water because the mixed hexane–water attraction is much weaker than the water–water and hexane–hexane attractions. The stronger attractions in the pure liquids constitute an energy barrier to mixing.

hydrogen bond

C_2H_5—O ··· H—O—H; H···O(H)···H—O(C_2H_5)···H—O—H

FIGURE 10.8 Hydrogen bonds in aqueous ethanol. Ethanol molecules form hydrogen bonds to water molecules.

The differing miscibilities of the above alcohols with water cannot be explained in terms of differing polarities as they all have very similar dipole moments (~5.6 × 10^{-30} C m). We must therefore consider the relative energetic magnitudes of the hydrogen bonding interactions between the polar —OH functional groups and water, and the dispersion forces between the nonpolar hydrocarbon tails of neighbouring alcohol molecules. For methanol, ethanol and propan-1-ol, the hydrogen bonding interactions between the —OH groups and water dominate, leading to water miscibility. However, as the hydrocarbon tails on the alcohol molecules increase in length beyond three carbon atoms, the dispersion forces between these become significant. This causes the alcohol molecules to cluster together as a single phase, leading to immiscibility with water.

We can use similar arguments to explain the solubilities of alcohols in nonpolar solvents. For example, even though long-chain alcohols are appreciably polar molecules, they are miscible with hexane because of the dispersion forces between the hydrocarbon chain on the alcohol and the hexane molecules. These dispersion forces diminish as the chain length of the alcohol decreases, and we find that the shortest chain alcohol, methanol, is immiscible with hexane.

The long-chain alcohols discussed above are examples of molecules with both polar and nonpolar regions; we might therefore expect the different regions of such molecules to display different affinities towards both polar and nonpolar solvents. Indeed, we make use of this behaviour every day when we use soaps and detergents to make oils and water mix. Both detergents and soaps consist of molecules called surfactants (a contraction of the term 'surface active agents'), which comprise polar and nonpolar parts; the polar parts generally result from a full positive or negative charge. The structure of a common surfactant found in shampoos and toothpastes, sodium lauryl sulfate, is shown in figure 10.9. As the polar group is negatively charged, we call this an anionic surfactant. When added to a mixture of two immiscible liquids, the polar end of a surfactant molecule preferentially orients itself in the polar liquid, while the non-polar end is situated in the non-polar liquid. This helps bring the two immiscible phases together. This process will be discussed in more detail in chapter 23.

O, O, S, O, O^-, Na^+

FIGURE 10.9 The structure of sodium lauryl sulfate.

Liquid–solid solutions

In moving to solutions of solids in liquids, the basic principles remain the same. We will look first at what happens when sodium chloride, a crystalline salt, dissolves in water.

Figure 10.10 depicts a section of a crystal of NaCl in contact with water. The dipoles of water molecules orient themselves so that the negative ends of some point towards Na^+ ions and the positive ends of others point at Cl^- ions. In other words, *ion–dipole* attractions occur that tend to tug and pull ions from the crystal. At the corners and edges of the crystal, ions are held by fewer neighbours within the solid so they are more readily dislodged than those elsewhere on the crystal's surface. As water molecules dislodge these ions, new corners and edges are exposed and the crystal continues to dissolve.

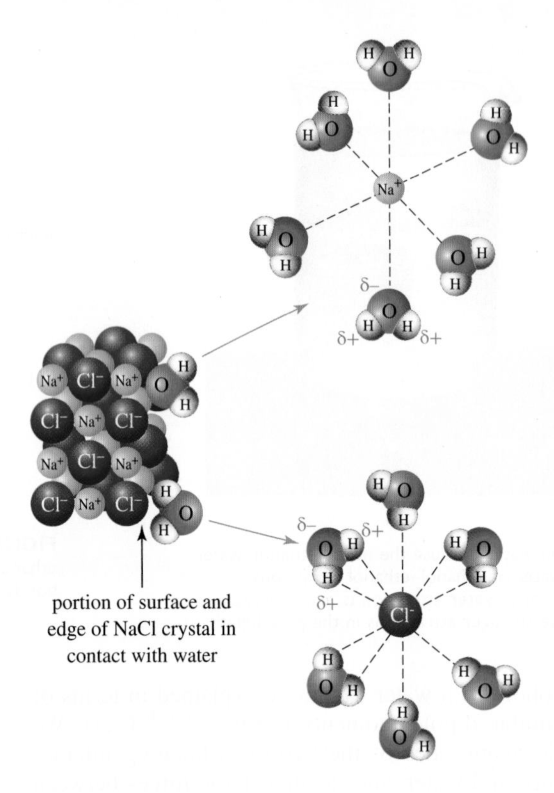

FIGURE 10.10 Hydration of ions involves a complex redirection of forces of attraction and repulsion. Before the solution forms, water molecules are attracted only to each other, and Na^+ and Cl^- ions are attracted only to each other in the crystal. In the solution, the ions have water molecules to take the places of their oppositely charged counterparts; in addition, water molecules are attracted to ions even more than they are to other water molecules.

As they are freed, the ions become completely surrounded by water molecules (see figure 10.10). This phenomenon is called the **hydration** of ions. The *general* term for the surrounding of a solute particle by solvent molecules is **solvation**, so hydration is just a special case of solvation.

Ionic compounds can dissolve in water when the attractions between water dipoles and ions overcome both the attractions of the ions for each other within the crystal and the attractive forces between water molecules in bulk water. Similar events explain why solids composed of polar molecules, such as those of sugar, dissolve in water (see figure 10.11). Attractions between the solvent and solute dipoles help to dislodge molecules from the crystal and bring them into solution. Again we see that 'like dissolves like'; a polar solute dissolves in a polar solvent.

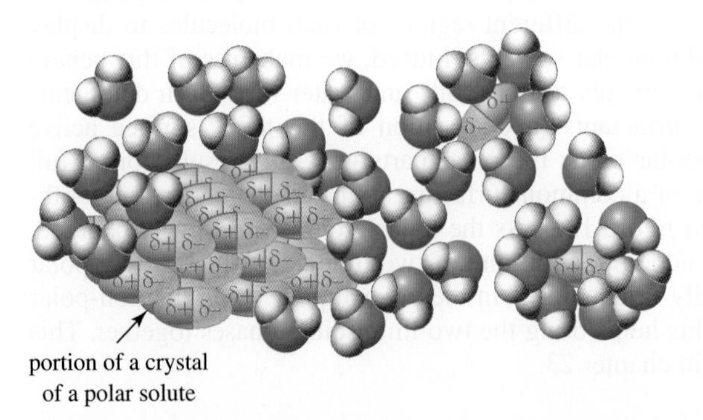

FIGURE 10.11 Hydration of a polar molecule. A polar solid dissolves in water because its molecules are attracted to the very polar water molecules, causing the solute molecules to become hydrated. Notice that the water molecules orient themselves so their positive ends are near the negative ends of the solute dipoles and their negative ends are near the positive ends of the solute dipoles.

The same reasoning explains why nonpolar solids, such as wax, are soluble in nonpolar solvents, such as benzene. Wax is a solid mixture of long-chain hydrocarbons held together by dispersion forces. The attractions between the molecules of the solvent (benzene) and the solute (wax) are also due to dispersion forces of comparable strength.

When intermolecular attractive forces within solute and solvent are sufficiently different, the two do not form a solution. For example, ionic solids or very polar molecular solids (such as sugar) are **insoluble** in nonpolar solvents such as benzene and hexane. The molecules of these hydrocarbon solvents are unable to attract ions or very polar molecules with enough force to overcome the much stronger attractions that the ions or polar molecules experience in the solid state.

We have already seen that, while the enthalpy changes on mixing in simple gaseous solutions are minimal, the same is not necessarily true in liquid–liquid solutions. The enthalpy changes on formation of a solution containing a dissolved ionic solid can be substantial and we will now consider these changes in some detail. Recalling from chapter 8 that enthalpy is a state function, and therefore does not depend on the way in which we get from the initial state to the final state, we can model the dissolution of an ionic solute in a solvent as a hypothetical two-step process, as shown in figure 10.12.

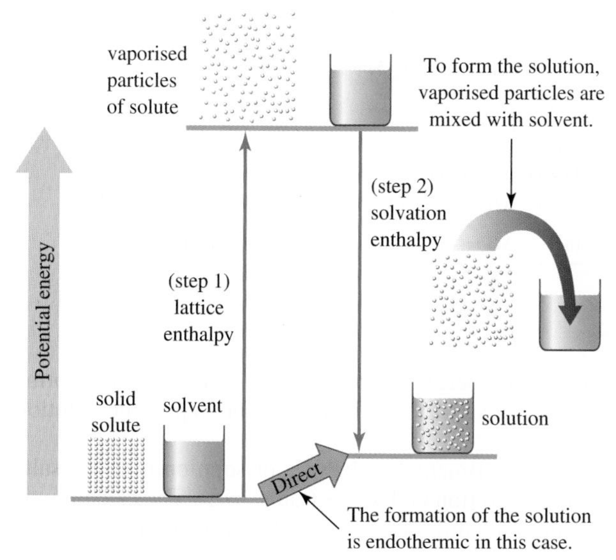

FIGURE 10.12 Enthalpy diagram for a solid dissolving in a liquid. In the real world, the solution is formed directly, as indicated by the purple arrow. We can analyse the enthalpy change by imagining the two separate steps, because enthalpy changes are state functions and so are independent of the path. The enthalpy change along the direct path is the algebraic sum of those for step 1 and step 2.

As you can see from figure 10.12, the first step is endothermic. The particles in the solid attract each other, and so energy must be supplied to separate them. The amount of energy required is the **lattice enthalpy**. Recall from chapter 8 that the lattice enthalpy is the enthalpy required to separate 1 mole of a crystalline compound into gaseous particles. For ionic compounds, the gaseous particles are ions; for molecular compounds, they are molecules. For example, the lattice enthalpy of potassium iodide, KI, is the enthalpy change for the process:

$$KI(s) \rightarrow K^+(g) + I^-(g) \qquad \Delta H = +632\ kJ\ mol^{-1}$$

In the second step, gaseous solute particles enter the solvent and are solvated. This step is exothermic. The enthalpy change when gaseous solute particles (obtained from 1 mole of solute) are dissolved in a solvent is called the **solvation enthalpy**. If the solvent is water, the solvation enthalpy can also be called the **hydration enthalpy**. For example, the hydration enthalpy of KI is the enthalpy change for the process:

$$K^+(g) + I^-(g) \rightarrow K^+(aq) + I^-(aq) \qquad \Delta H = -619\ kJ\ mol^{-1}$$

The **enthalpy of solution** ($\Delta_{sol}H$) is the enthalpy change when 1 mole of a crystalline substance is dissolved in a solvent; it is equal to the enthalpy difference between the endothermic step 1 and the exothermic step 2. For example, the enthalpy of solution of KI calculated from the data given on the previous page is the sum of the enthalpy changes for steps 1 and 2, namely:

$$KI(s) \rightarrow K^+(aq) + I^-(aq) \qquad \Delta_{sol}H = -619\ kJ\ mol^{-1} + 632\ kJ\ mol^{-1} = 13\ kJ\ mol^{-1}$$

This process is shown in detail in figure 10.13.

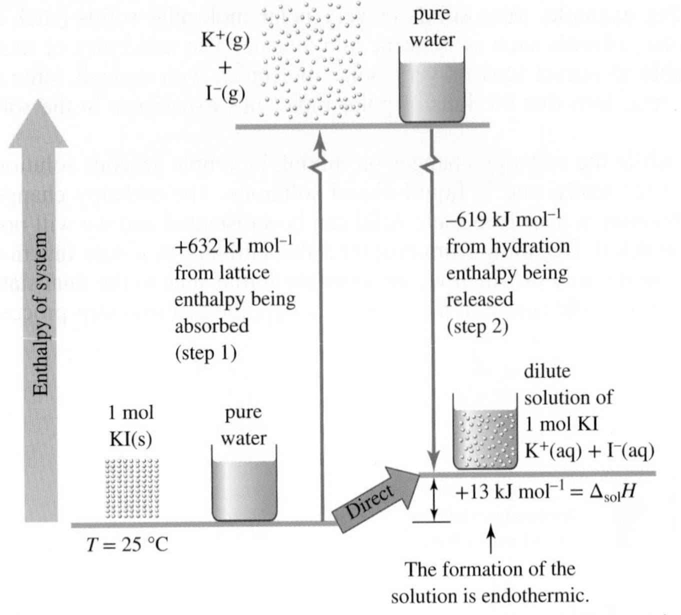

Lattice enthalpy:	$KI(s) \longrightarrow K^+(g) + I^-(g)$	$\Delta H = +632\ kJ\ mol^{-1}$
Hydration enthalpy:	$K^+(g) + I^-(g) \longrightarrow K^+(aq) + I^-(aq)$	$\Delta H = -619\ kJ\ mol^{-1}$
Total:	$KI(s) \longrightarrow K^+(aq) + I^-(aq)$	$\Delta_{sol}H = +13\ kJ\ mol^{-1}$

FIGURE 10.13 Enthalpy of solution — the formation of aqueous potassium iodide.

When the energy required for step 1 exceeds the energy released in step 2, the solution forms endothermically; when more energy is released in step 2 than is needed for step 1, the solution forms exothermically.

We can test this analysis by comparing the experimental enthalpies of solution of some salts in water with those calculated from lattice and hydration enthalpies (table 10.2). The calculations are described in figures 10.13 and 10.14 using enthalpy diagrams for the formation of aqueous solutions of two salts, KI and NaBr.

TABLE 10.2 Lattice enthalpies, hydration enthalpies and enthalpies of solution for some group I metal halides at 25 °C.

			$\Delta_{sol}H$[(a)]	
Compound	**Lattice enthalpy (kJ mol^{-1})**	**Hydration enthalpy (kJ mol^{-1})**	**Calculated $\Delta_{sol}H$ (kJ mol^{-1})[(b)]**	**Measured $\Delta_{sol}H$ (kJ mol^{-1})**
LiCl	+833	–883	–50	–37.0
NaCl	+766	–770	–4	+3.9
KCl	+690	–686	+4	+17.2
LiBr	+787	–854	–67	–49.0
NaBr	+728	–741	–13	–0.602
KBr	+665	–657	+8	+19.9
KI	+632	–619	+13	+20.33

(a) Enthalpies of solution refer to the formation of extremely dilute solutions.
(b) Calculated $\Delta_{sol}H$ = lattice enthalpy + hydration enthalpy.

The agreement between calculated and measured values in table 10.2 is not particularly close. This is partly because precise lattice and hydration enthalpies are not known and partly because the model used in our analysis is evidently too simple. Notice, however, that when 'theory' predicts relatively large enthalpies of solution, the experimental values are also relatively large and both values have the same sign (except for NaCl, where the values are close to 0 and fall either side of it). Notice also that the changes in values show the same trends when we compare the three chloride salts — LiCl, NaCl and KCl — and the three bromide salts — LiBr, NaBr and KBr.

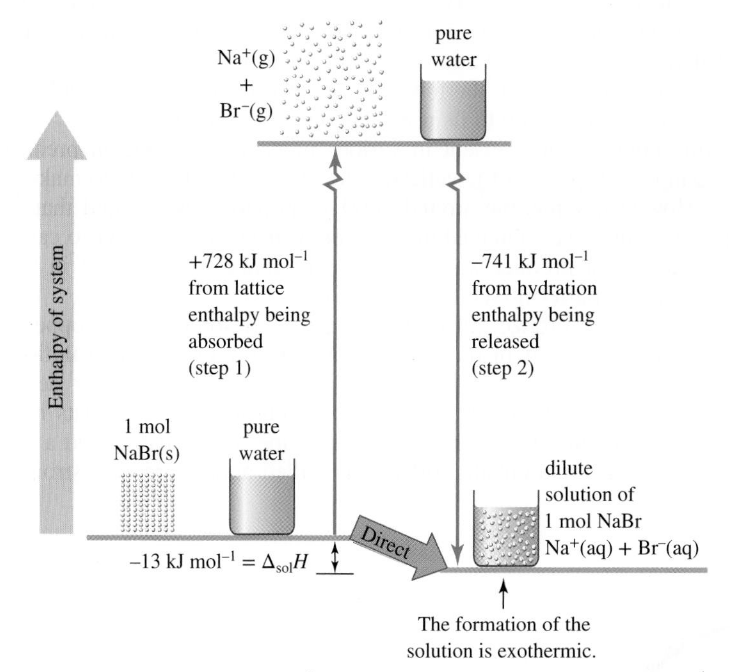

Lattice enthalpy:	$NaBr(s) \rightarrow Na^+(g) + Br^-(g)$	$\Delta H = +728\ kJ\ mol^{-1}$
Hydration enthalpy:	$Na^+(g) + Br^-(g) \rightarrow Na^+(aq) + Br^-(aq)$	$\Delta H = -741\ kJ\ mol^{-1}$
Total:	$NaBr(s) \rightarrow Na^+(aq) + Br^-(aq)$	$\Delta_{sol}H = -13\ kJ\ mol^{-1}$

FIGURE 10.14 Enthalpy of solution — the formation of aqueous sodium bromide.

Temperature can have a significant effect on the solubility of a solid solute in a liquid. Figure 10.15 shows a plot of solubility versus temperature for a number of ionic salts; it is obvious that all but one of these examples become more soluble as the temperature increases, which we might expect intuitively (there are few genuine examples of salts that become less soluble as the temperature increases).

However, not everything is as straightforward as we would wish. Many of you will have dissolved solid NaOH in water and noticed that the solution gets quite hot as the solid dissolves, meaning that the process is exothermic. From the van't Hoff equation (chapter 9, p. 370) we might therefore predict that the equilibrium constant for the dissolution of NaOH should decrease as the temperature increases; hence, NaOH should become less soluble with increasing temperature. As you can see in figure 10.15, the opposite trend is observed, and this can be explained by the nature of solid NaOH in a saturated solution. NaOH generally exists as an anhydrous solid — in other words, the crystalline solid contains no water molecules — and, as we have stated earlier, this dissolves exothermically. However, when we prepare a saturated solution of NaOH by adding excess NaOH to water, analysis of the solid phase shows it to consist not of anhydrous NaOH, but rather the solid *hydrate*, $NaOH{\cdot}H_2O$, in which one water molecule is associated with each NaOH unit. In fact, this species dissolves endothermically, so it is more soluble at higher temperatures. You should also appreciate that the van't Hoff equation uses $\Delta H^{\ominus}$ values, so $\Delta_{sol}H$ data measured under nonstandard conditions are of limited use in predicting the variation of solubilities with temperature.

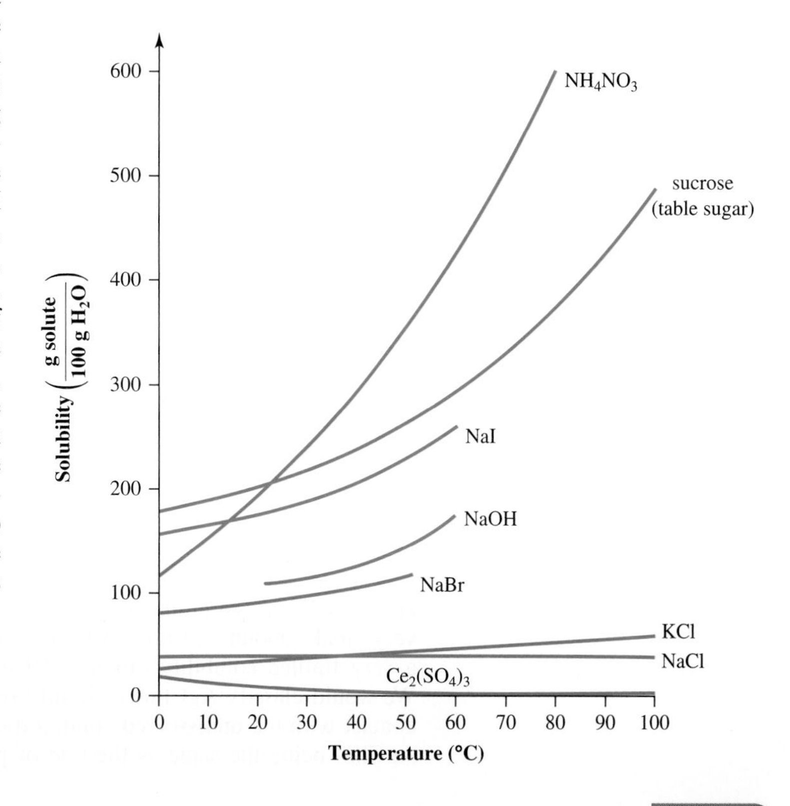

FIGURE 10.15 Solubility in water versus temperature for several substances. Most substances become more soluble when the temperature of the solution increases, but the amount of this increased solubility varies considerably.

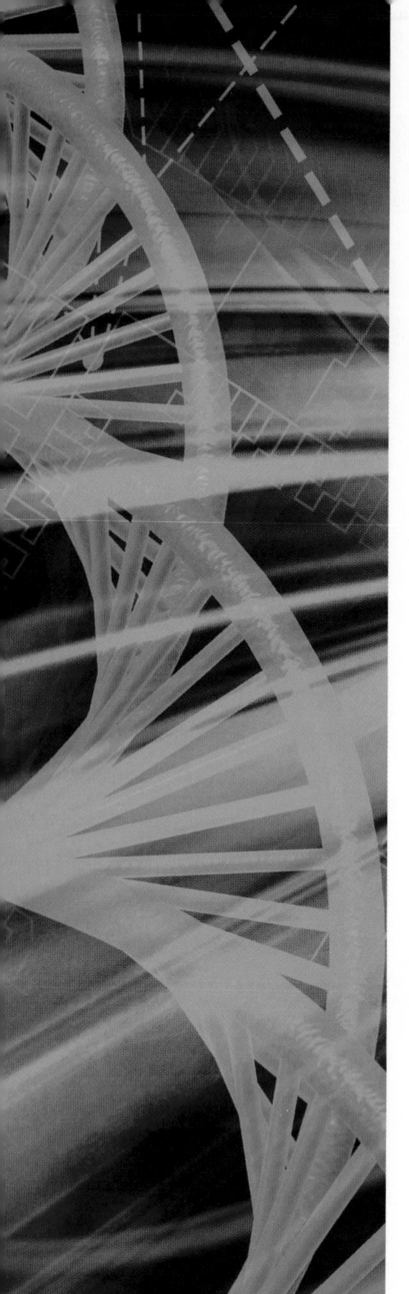

Chemical Connections

Solid solutions: alloys

Pure metals occur rarely in nature. The majority exist naturally as oxides through reaction with atmospheric oxygen, and require smelting in order to isolate the pure element. The best known of the so-called 'native' metals are gold and silver, both of which were primarily used by early humans for decoration or adornment. Of the few other metals found in their metallic state, copper is arguably the most important, owing to its use as a component of bronze.

Large deposits of metallic copper were present in various parts of the world in prehistoric times. While useful for making small pieces of jewellery, it was too soft and brittle to make large implements and weapons. However, sometime around 5000 years ago, it was found that when tin was heated with copper, a product was obtained that was both stronger and easier to cast than either of its component metals. That product was bronze, a substance so important in human history that an entire age was named after it (see figure 10.16).

There are many different types of bronze, all of which are solid solutions of copper with metals such as tin, zinc, lead, nickel and bismuth in various proportions. Solid solutions of metals are called alloys, and are the subject of enormous interest because of their desirable properties. For example, stainless steel is an alloy of iron with chromium that resists rusting, nitinol is an alloy of nickel and titanium that 'remembers' its shape when heated, and a family of iron alloys with aluminium, nickel and cobalt, collectively called alnico, are very strong permanent magnets.

FIGURE 10.16 Bronze is a solid solution of copper with a variety other metals. It has been used for thousands of years, beginning of course with the Bronze Age.

10.4 Quantification of solubility: the solubility product

Ionic salts are generally classified as being either soluble or insoluble in water. The word 'insoluble' suggests that salts classified as such do not dissolve at all in water, but this is not the case. If we mix aqueous solutions containing equal amounts of the soluble salts silver nitrate and sodium chloride (figure 10.17), we see immediate precipitation of the 'insoluble' salt AgCl, according to the equation:

$$AgNO_3(aq) + NaCl(aq) \rightarrow AgCl(s) + NaNO_3(aq)$$

However, if we carefully analyse the solution above the precipitate, we find that it contains very small amounts of both $Ag^+(aq)$ and $Cl^-(aq)$ ions; this means that AgCl does in fact have a very limited solubility (about 0.19 milligrams per 100 mL at 25.0 °C) in aqueous solution. We would classify AgCl as a slightly soluble salt. As is the case with any saturated solution in contact with the undissolved solid, a dynamic equilibrium exists, with the rate of dissolution of AgCl(s) being the same as the rate of precipitation of $Ag^+(aq)$ and $Cl^-(aq)$. We can, therefore, quantify the solubility of AgCl(s) with reference to the magnitude of the equilibrium constant for the dissolution equilibrium:

$$AgCl(s) \rightleftharpoons Ag^+(aq) + Cl^-(aq)$$

FIGURE 10.17 Precipitation of AgCl(s) on mixing aqueous solutions of $AgNO_3$ and NaCl.

The equilibrium constant for the dissolution of a slightly soluble salt is called the **solubility product** and is abbreviated K_{sp}. For the above equilibrium, we write K_{sp} as:

$$K_{sp} = [Ag^+][Cl^-]$$

Recall from chapter 9 that pure solids *never* appear in an equilibrium constant. Therefore, as we will see below, the expression for K_{sp} for any ionic solid always consists only of the product of the aqueous ions raised to their stoichiometric powers. We can derive an expression for the solubility product of any salt $M_aX_b(s)$ by considering the equilibrium constant for its dissolution in water.

$$M_aX_b(s) \rightleftharpoons aM^{b+}(aq) + bX^{a-}(aq)$$

$$K_{sp} = [M^{b+}]^a[X^{a-}]^b$$

For AgCl, $K_{sp} = 1.8 \times 10^{-10}$ at 25.0 °C, meaning that the solubility of AgCl in water is indeed small *but not zero*. Salts that we generally classify as slightly soluble have K_{sp} values of the order of 10^{-5} or less, and table 10.3 (overleaf) gives K_{sp} data for a variety of such salts (a more comprehensive list is given in appendix C). In general, the smaller the value of K_{sp}, the less soluble is the salt. However, you should appreciate that direct comparison of K_{sp} values can be made only for salts with the same ratio of anions to cations. For example, CaF_2 is more soluble on a molar basis than AgCl in water, even though K_{sp} for CaF_2 (3.9×10^{-11}) is smaller than that for AgCl. This is because the equilibrium constant expressions for the two salts do not have identical forms; the equation for dissolution of CaF_2 is:

$$CaF_2(s) \rightleftharpoons Ca^{2+}(aq) + 2F^-(aq)$$

and this leads to the equilibrium constant expression:

$$K_{sp} = [Ca^{2+}][F^-]^2$$

which obviously differs from that above for AgCl. We will explore this in more detail on the following pages.

The relationship between K_{sp} and solubility

One way to determine the value of K_{sp} for a slightly soluble salt is to measure its solubility, that is, how much of the salt is required to make a saturated solution of a specified volume. The molar concentration of a salt in its saturated solution is called the **molar solubility** (*s*); this equals the amount of salt dissolved in 1 litre of the saturated solution at the specified temperature. The molar solubility can be used to calculate K_{sp}, assuming that all of the salt that dissolves is 100% dissociated into its ions. This assumption works reasonably well for slightly soluble salts made of univalent ions, such as silver bromide. For simplicity, and to illustrate the nature of calculations involving solubility equilibria, we will work on the assumption that all salts behave as though they are 100% dissociated. This is not entirely true, especially for salts of polyvalent ions, so the accuracy of our calculations is limited.

TABLE 10.3 Solubility products for selected slightly soluble salts at 25 °C.

Salt		Dissolved ions	K_{sp}
halides			
$PbCl_2(s)$	$\rightleftharpoons$	$Pb^{2+}(aq) + 2Cl^-(aq)$	1.7×10^{-5}
$PbBr_2(s)$	$\rightleftharpoons$	$Pb^{2+}(aq) + 2Br^-(aq)$	2.1×10^{-6}
$PbF_2(s)$	$\rightleftharpoons$	$Pb^{2+}(aq) + 2F^-(aq)$	3.6×10^{-8}
$PbI_2(s)$	$\rightleftharpoons$	$Pb^{2+}(aq) + 2I^-(aq)$	7.9×10^{-9}
$AgCl(s)$	$\rightleftharpoons$	$Ag^+(aq) + Cl^-(aq)$	1.8×10^{-10}
$CaF_2(s)$	$\rightleftharpoons$	$Ca^{2+}(aq) + 2F^-(aq)$	3.9×10^{-11}
$AgBr(s)$	$\rightleftharpoons$	$Ag^+(aq) + Br^-(aq)$	5.0×10^{-13}
$AgI(s)$	$\rightleftharpoons$	$Ag^+(aq) + I^-(aq)$	8.3×10^{-17}
hydroxides			
$Ca(OH)_2(s)$	$\rightleftharpoons$	$Ca^{2+}(aq) + 2OH^-(aq)$	6.5×10^{-6}
$Mg(OH)_2(s)$	$\rightleftharpoons$	$Mg^{2+}(aq) + 2OH^-(aq)$	7.1×10^{-12}
$Fe(OH)_2(s)$	$\rightleftharpoons$	$Fe^{2+}(aq) + 2OH^-(aq)$	7.9×10^{-16}
$Zn(OH)_2(s)$	$\rightleftharpoons$	$Zn^{2+}(aq) + 2OH^-(aq)$	3.0×10^{-16}(a)
$Al(OH)_3(s)$	$\rightleftharpoons$	$Al^{3+}(aq) + 3OH^-(aq)$	3×10^{-34}(b)
$Fe(OH)_3(s)$	$\rightleftharpoons$	$Fe^{3+}(aq) + 3OH^-(aq)$	1.6×10^{-39}
carbonates			
$NiCO_3(s)$	$\rightleftharpoons$	$Ni^{2+}(aq) + CO_3^{2-}(aq)$	1.3×10^{-7}
$MgCO_3(s)$	$\rightleftharpoons$	$Mg^{2+}(aq) + CO_3^{2-}(aq)$	3.5×10^{-8}
$BaCO_3(s)$	$\rightleftharpoons$	$Ba^{2+}(aq) + CO_3^{2-}(aq)$	5.0×10^{-9}
$CaCO_3(s)$	$\rightleftharpoons$	$Ca^{2+}(aq) + CO_3^{2-}(aq)$	4.5×10^{-9}(c)
$SrCO_3(s)$	$\rightleftharpoons$	$Sr^{2+}(aq) + CO_3^{2-}(aq)$	9.3×10^{-10}
$CoCO_3(s)$	$\rightleftharpoons$	$Co^{2+}(aq) + CO_3^{2-}(aq)$	1.0×10^{-10}
$ZnCO_3(s)$	$\rightleftharpoons$	$Zn^{2+}(aq) + CO_3^{2-}(aq)$	1.0×10^{-10}
$Ag_2CO_3(s)$	$\rightleftharpoons$	$2Ag^+(aq) + CO_3^{2-}(aq)$	8.1×10^{-12}
chromates			
$Ag_2CrO_4(s)$	$\rightleftharpoons$	$2Ag^+(aq) + CrO_4^{2-}(aq)$	1.2×10^{-12}
$PbCrO_4(s)$	$\rightleftharpoons$	$Pb^{2+}(aq) + CrO_4^{2-}(aq)$	2.5×10^{-13}
sulfates			
$CaSO_4(s)$	$\rightleftharpoons$	$Ca^{2+}(aq) + SO_4^{2-}(aq)$	2.4×10^{-5}
$PbSO_4(s)$	$\rightleftharpoons$	$Pb^{2+}(aq) + SO_4^{2-}(aq)$	6.3×10^{-7}
$SrSO_4(s)$	$\rightleftharpoons$	$Sr^{2+}(aq) + SO_4^{2-}(aq)$	3.2×10^{-7}
$BaSO_4(s)$	$\rightleftharpoons$	$Ba^{2+}(aq) + SO_4^{2-}(aq)$	1.1×10^{-10}
oxalates			
$MgC_2O_4(s)$	$\rightleftharpoons$	$Mg^{2+}(aq) + C_2O_4^{2-}(aq)$	8.6×10^{-5}
$FeC_2O_4(s)$	$\rightleftharpoons$	$Fe^{2+}(aq) + C_2O_4^{2-}(aq)$	2.1×10^{-7}
$BaC_2O_4(s)$	$\rightleftharpoons$	$Ba^{2+}(aq) + C_2O_4^{2-}(aq)$	1.2×10^{-7}
$CaC_2O_4(s)$	$\rightleftharpoons$	$Ca^{2+}(aq) + C_2O_4^{2-}(aq)$	2.3×10^{-9}
$PbC_2O_4(s)$	$\rightleftharpoons$	$Pb^{2+}(aq) + C_2O_4^{2-}(aq)$	2.7×10^{-11}

A more comprehensive list is given in appendix C.

WORKED EXAMPLE 10.2

Calculating K_{sp} from solubility data

In the time before digital photography, silver bromide, AgBr, was the light-sensitive compound used in nearly all photographic film. The solubility of AgBr in water is 1.3×10^{-4} g L^{-1} at 25 °C. Calculate K_{sp} for AgBr at this temperature.

Analysis

We begin by writing the balanced chemical equation for the dissolution of AgBr and use this to determine the expression for the solubility product, K_{sp}.

$$AgBr(s) \rightleftharpoons Ag^+(aq) + Br^-(aq)$$
$$K_{sp} = [Ag^+][Br^-]$$

To calculate K_{sp}, we require the concentrations of Ag^+ and Br^- in the saturated solution expressed in mol L^{-1}, so we must convert the given value of 1.3×10^{-4} g L^{-1} to mol L^{-1}. We then insert these values into the above expression for K_{sp}.

Solution

We know that 1.3×10^{-4} g of AgBr is dissolved in every litre of the saturated solution, so we convert 1.3×10^{-4} g of AgBr to amount in moles.

$$n = \frac{m}{M} = \frac{1.3 \times 10^{-4}\,g}{187.77\,g\,mol^{-1}} = 6.9 \times 10^{-7}\,mol$$

From the balanced chemical equation for the dissolution reaction, we know that, when 6.9×10^{-7} mol of AgBr(s) dissolves in water, it produces 6.9×10^{-7} mol of $Ag^+(aq)$ ions and 6.9×10^{-7} mol of $Br^-(aq)$ ions. Hence, the molar concentrations of Ag^+ and Br^- are both 6.9×10^{-7} mol L^{-1}. Inserting these values into the expression for K_{sp} gives:

$$K_{sp} = [Ag^+][Br^-] = (6.9 \times 10^{-7}\,mol\,L^{-1})(6.9 \times 10^{-7}\,mol\,L^{-1}) = 4.8 \times 10^{-13} \text{ at } 25\,°C$$

As is usual with equilibrium constants, K_{sp} is dimensionless.

Is our answer reasonable?

We have divided roughly 1×10^{-4} by a number that is about 2×10^2, which gives a value of about 5×10^{-7}, so our molar solubility seems reasonable. (Also, we know AgBr has a very low solubility in water, so we expect the molar solubility to be very small.) The amount of each ion formed per litre must equal the amount of AgBr that dissolves. Finally, if we round 6.9×10^{-7} to 7×10^{-7} and square it, we obtain $49 \times 10^{-14} = 4.9 \times 10^{-13}$. Our answer of 4.8×10^{-13} therefore seems reasonable.

PRACTICE EXERCISE 10.2

The solubility of thallium(I) iodide, TlI, in water at 20 °C is 1.8×10^{-5} mol L^{-1}. Calculate K_{sp} for TlI assuming that it is 100% dissociated in the solution.

Obviously, if we can calculate K_{sp} from molar solubility data, we should also be able to calculate the molar solubility from K_{sp}. Worked example 10.3 shows how this is done.

WORKED EXAMPLE 10.3

Calculating molar solubility from K_{sp}

Calculate the molar solubility of lead iodide, PbI_2, given that $K_{sp}(PbI_2) = 7.9 \times 10^{-9}$ at 25 °C.

Analysis

We start, as usual, by writing the balanced chemical equation for the dissolution of $PbI_2(s)$ and write the K_{sp} expression from this.

$$PbI_2(s) \rightleftharpoons Pb^{2+}(aq) + 2I^-(aq)$$
$$K_{sp} = [Pb^{2+}][I^-]^2$$

In this case, we know the value of K_{sp} and must determine $[Pb^{2+}]$ and $[I^-]$. However, we have one equation containing two unknowns. To solve this, we must make use of the stoichiometric relationship between $[Pb^{2+}]$ and $[I^-]$ in the saturated solution.

Solution

As this is an equilibrium problem, we can solve it using a concentration table in a similar manner to the examples in chapter 9 (pp. 373–81). We set up the concentration table in the usual fashion, recognising that the initial concentrations of the ions will be 0 M. As the salt dissolves, s moles of PbI_2 will dissolve to increase the concentrations of Pb^{2+} and I^- by s and $2s$, respectively. Note that we use s rather than x in this concentration table, as s is defined as the molar solubility; in other words the amount of the salt dissolved in 1 litre of solution.

Here is the concentration table.

	$PbI_2(s)$	$\rightleftharpoons$	$Pb^{2+}(aq)$	+	$2I^-(aq)$
Initial concentration (M)			0		0
Change in concentration (M)			$+s$		$+2s$
Equilibrium concentration (M)			s		$2s$

We now substitute the equilibrium concentrations into the equilibrium constant expression, K_{sp}.

$$K_{sp} = [Pb^{2+}][I^-]^2$$
$$= (s)(2s)^2$$
$$= (s)(4s^2)$$
$$= 4s^3$$

We have now gone from one equation with two unknowns to one equation with a single unknown, s, which we can solve.

$$4s^3 = 7.9 \times 10^{-9}$$
$$s^3 = \frac{7.9 \times 10^{-9}}{4}$$
$$= 2.0 \times 10^{-9}$$
$$s = 1.3 \times 10^{-3}$$

Thus, the molar solubility of $PbI_2(s)$ in water at 25 °C is 1.3×10^{-3} mol L^{-1}.

Is our answer reasonable?

A molar solubility of 10^{-3} mol L^{-1} is low, implying that the salt is not very soluble and that our answer is at least reasonable. We can quickly check our answer by carrying out the reverse calculation, cubing s and multiplying the answer by 4. This should equal K_{sp} (allowing for rounding errors).

PRACTICE EXERCISE 10.3

Use the K_{sp} data in table 10.3 to calculate the molar solubility of (a) PbI_2 and (b) $Fe(OH)_3$ in water at 25°C.

We have seen in worked example 10.3 that solubility, s, can be related to K_{sp}. The exact nature of this relationship depends on the chemical formula of the salt. Above, we showed that, for PbI_2, $K_{sp} = 4s^3$, and this relationship will be the same for all 2 : 1 electrolytes (that is, those having the formula MX_2 or M_2X). Table 10.4 shows the possible forms of the relationship between solubility, s, and the solubility product K_{sp} for all common types of electrolytes. You should be able to derive these expressions using the method outlined in worked example 10.3.

TABLE 10.4 The relationship between solubility, s, and the solubility product K_{sp} for common electrolytes.

Type of electrolyte	Formula of electrolyte	K_{sp} expression in terms of concentration	Solubility, s, in terms of $[M^{n+}]$ and $[X^{n-}]$	K_{sp} expression in terms of solubility, s
1 : 1	MX	$K_{sp} = [M^{n+}][X^{n-}]$	$[M^{n+}] = [X^{n-}] = s$	$K_{sp} = s^2$
2 : 1	MX_2	$K_{sp} = [M^{2n+}][X^{n-}]^2$	$[M^{2n+}] = s, [X^{n-}] = 2s$	$K_{sp} = 4s^3$
	M_2X	$K_{sp} = [M^{n+}]^2[X^{2n-}]$	$[M^{n+}] = 2s, [X^{2n-}] = s$	
3 : 1	MX_3	$K_{sp} = [M^{3n+}][X^{n-}]^3$	$[M^{3n+}] = s, [X^{n-}] = 3s$	$K_{sp} = 27s^4$
	M_3X	$K_{sp} = [M^{n+}]^3[X^{3n-}]$	$[M^{n+}] = 3s, [X^{3n-}] = s$	
3 : 2	M_2X_3	$K_{sp} = [M^{3n+}]^2[X^{2n-}]^3$	$[M^{3n+}] = 2s, [X^{2n-}] = 3s$	$K_{sp} = 108s^5$
	M_3X_2	$K_{sp} = [M^{2n+}]^3[X^{3n-}]^2$	$[M^{2n+}] = 3s, [X^{3n-}] = 2s$	

The common ion effect

Until now, we have considered the solubility only of slightly soluble ionic salts in pure water. Does the solubility of such salts differ if the solvent is not pure water but a solution that already contains one of the ions in the slightly soluble salt? Likewise, what happens if we take a saturated solution of a slightly soluble salt and add a solution of another salt that has one ion in common with the first salt?

We can answer these questions with reference to a saturated solution of $PbCl_2(s)$ ($K_{sp} = 1.7 \times 10^{-5}$) at equilibrium. To this saturated solution we add a small volume of a concentrated solution of $Pb(NO_3)_2$, a soluble lead salt, and look at the effect on the equilibrium position by comparing Q_{sp} and K_{sp}. We begin by writing the balanced chemical equation for the dissolution of $PbCl_2$ and derive the expression for K_{sp} from this, recalling that the expressions for K_{sp} and Q_{sp} have the same form. (In chapter 9, we called Q the reaction quotient. In the case of Q_{sp}, that name is rather incongruous, given that the expression for Q_{sp} is never a fraction. In some texts, Q_{sp} is called the **ionic product**.)

$$PbCl_2(s) \rightleftharpoons Pb^{2+}(aq) + 2Cl^-(aq)$$

$$K_{sp} = [Pb^{2+}][Cl^-]^2$$

Adding a small volume of concentrated $Pb(NO_3)_2(aq)$ to the saturated solution instantaneously increases $[Pb^{2+}]$ and therefore the value of Q_{sp}. This means that $Q_{sp} > K_{sp}$ and therefore, to restore equilibrium, the value of Q_{sp} must be decreased. The system achieves this by precipitating $PbCl_2(s)$, which reduces both $[Pb^{2+}]$ and $[Cl^-]$, thereby decreasing the value of Q_{sp} to the point where the equilibrium condition $Q_{sp} = K_{sp}$ is again satisfied. If we were to analyse the saturated solution at this point, we would find $[Cl^-]$ to be less than it was in the original solution. As all the Cl^- in solution comes from dissolution of $PbCl_2$, this means that $PbCl_2(s)$ is *less soluble* in the presence of another source of Pb^{2+}. This is a manifestation of the **common ion effect**, which states that any ionic salt is less soluble in the presence of a **common ion**, an ion that is a component of the salt.

We have shown above that $PbCl_2$ is less soluble in the presence of Pb^{2+} than it is in pure water, and we can also show by the same reasoning that it is less soluble in the presence of another source of Cl^-. The common ion effect is not limited to solutions of slightly soluble salts; for example, even the relatively soluble salt NaCl can be precipitated from solution on addition of concentrated HCl, a source of the common ion Cl^-. The addition or presence of a common ion *always* lowers the solubility of a salt providing that no further chemical reaction is possible. An example of a situation where a chemical reaction occurs is provided by the slightly soluble salt AgCl(s), which can react with excess Cl^- to form the complex anion $[AgCl_2]^-$, which, in turn, forms soluble salts with a variety of cations.

WORKED EXAMPLE 10.4

Calculations involving the common ion effect

What is the molar solubility of PbI_2 in a 0.10 M NaI solution at 25 °C?

Analysis

As usual, we begin with the balanced chemical equation and the expression for K_{sp} derived from it.

$$PbI_2(s) \rightleftharpoons Pb^{2+}(aq) + 2I^-(aq)$$

$$K_{sp} = [Pb^{2+}][I^-]^2 = 7.9 \times 10^{-9} \text{ (from table 10.3)}$$

In this case, we construct a concentration table as we did in chapter 9 (worked example 9.10, pp. 373–5). The initial concentration of Pb^{2+} is 0; we assume that NaI dissociates completely so the initial concentration of I^- is 0.10 M. As in worked example 10.3, we represent the molar solubility of $PbI_2(s)$ in the NaI solution as s mol L^{-1}. Therefore, when the $PbI_2(s)$ dissolves, $[Pb^{2+}]$ changes by $+s$ mol L^{-1} and $[I^-]$ by $+2s$ mol L^{-1}. We can then calculate the equilibrium concentrations of Pb^{2+} and I^- by summing the initial concentrations and the changes. Here is the concentration table.

	$PbI_2(s)$	$\rightleftharpoons$	$Pb^{2+}(aq)$	+	$2I^-(aq)$
Initial concentration (M)			0		0.10
Change in concentration (M)			$+s$		$+2s$
Equilibrium concentration (M)			s		$0.10 + 2s$

Solution

Substituting equilibrium values into the K_{sp} expression gives:

$$K_{sp} = [Pb^{2+}][I^-]^2 = s(0.10 + 2s)^2 = 7.9 \times 10^{-9}$$

A brief inspection reveals that solving this expression for s will be difficult if we cannot simplify the maths. Fortunately, a simplification is possible, because the small value of K_{sp} for PbI_2 tells us that the salt has a very low solubility. We also know that the presence of the common ion I^- makes PbI_2 even less soluble. This means very little of the salt will dissolve, so s (or even $2s$) is quite small. Let's assume that $2s$ is much smaller than 0.10, in a similar fashion to simplifications we made in worked example 9.14 on pages 380–1. If this is so:

$$0.10 + 2s \approx 0.10 \text{ (assuming } 2s \text{ is negligible compared with 0.10)}$$

Substituting 0.10 M for the I^- concentration gives:

$$K_{sp} = s(0.10)^2 = 7.9 \times 10^{-9}$$

$$s = \frac{7.9 \times 10^{-9}}{(0.10)^2}$$

$$= 7.9 \times 10^{-7} \text{ M}$$

Thus, the molar solubility of PbI_2 in 0.10 M NaI solution is 7.9×10^{-7} M.

Is our answer reasonable?

Check the entries in the table. The initial concentrations come from the NaI solution, which contains no Pb^{2+} but does contain 0.10 M I^-. By letting s equal the molar solubility, the coefficients of s in the 'Change in concentration' row equal the coefficients in the equation for the equilibrium. We should also check to see if our simplifying assumption is valid. Notice that $2s$, which equals 1.6×10^{-6}, is indeed vastly smaller than 0.10, just as we anticipated. (If we add 1.6×10^{-6} to 0.10 and round correctly, we obtain 0.10.)

In worked example 10.3 we found that the molar solubility of PbI_2 in *pure* water is 1.3×10^{-3} M. In water that contains 0.10 M NaI (worked example 10.4), the solubility of PbI_2 is 7.9×10^{-7} M, well over 1000 times smaller. As we said, the common ion effect can cause huge reductions in the solubilities of slightly soluble compounds.

PRACTICE EXERCISE 10.4

Calculate the molar solubility of AgI in 0.20 M NaI solution at 25 °C. Compare the answer with the calculated molar solubility of AgI in pure water.

Will a precipitate form?

We can use K_{sp} values to predict whether a precipitate will form when solutions containing ionic salts are mixed. We calculate Q_{sp} by determining the molar concentration of the appropriate ions immediately after mixing and compare this value with K_{sp}. If $Q_{sp} > K_{sp}$ at the instant the solutions are mixed, the equilibrium position will move to the left to make $Q_{sp} = K_{sp}$. This will result in the formation of a precipitate. Conversely, if $Q_{sp} < K_{sp}$ on mixing the solutions, the equilibrium position will move to the right, the dissolved ions will be favoured, and no precipitate will form.

WORKED EXAMPLE 10.5

Predicting whether a precipitate will form

Will a precipitate of AgCl(s) form on mixing 50.0 mL of 1.0×10^{-4} M NaCl(aq) with 50.0 mL of 1.0×10^{-6} M $AgNO_3$(aq) at 25 °C?

Analysis

We have to calculate Q_{sp} for the solution immediately after mixing and compare its value with K_{sp} for AgCl. Thus, we require the molar concentrations of Ag^+(aq) and Cl^-(aq), remembering that there is a dilution factor on mixing the two solutions.

Solution

We begin by writing the balanced chemical equation for the dissolution of AgCl(s) and deriving the expression for K_{sp} from it.

$$AgCl(s) \rightleftharpoons Ag^+(aq) + Cl^-(aq)$$

$$K_{sp} = [Ag^+][Cl^-] = 1.8 \times 10^{-10} \text{ (from table 10.3)}$$

We then calculate the molar concentrations of Ag^+ and Cl^- in the final mixed solution by determining the amount of each and dividing it by the volume of the final solution.

$$n_{Ag^+} = cV = (1.0 \times 10^{-6}\,\text{mol L}^{-1}) \times (50.0 \times 10^{-3}\,\text{L}) = 5.0 \times 10^{-8}\,\text{mol}$$

$$n_{Cl^-} = cV = (1.0 \times 10^{-4}\,\text{mol L}^{-1}) \times (50.0 \times 10^{-3}\,\text{L}) = 5.0 \times 10^{-6}\,\text{mol}$$

$$[Ag^+]_{final} = \frac{5.0 \times 10^{-8}\,\text{mol}}{100 \times 10^{-3}\,\text{L}} = 5.0 \times 10^{-7}\,\text{mol L}^{-1}$$

$$[Cl^-]_{final} = \frac{5.0 \times 10^{-6}\,\text{mol}}{100 \times 10^{-3}\,\text{L}} = 5.0 \times 10^{-5}\,\text{mol L}^{-1}$$

We now substitute these values into the expression for Q_{sp}.

$$Q_{sp} = [Ag^+][Cl^-] = (5.0 \times 10^{-7}\,\text{mol L}^{-1}) \times (5.0 \times 10^{-5}\,\text{mol L}^{-1}) = 2.5 \times 10^{-11}$$

Thus, $Q_{sp} < K_{sp}$, so we predict that no precipitate of AgCl(s) will form when the solutions are mixed. Note that, like K_{sp}, Q_{sp} is dimensionless.

Is our answer reasonable?

As this is essentially a stoichiometric problem, it is hard to tell whether the answer makes sense. In cases like these, you should check your arithmetic and ensure you have transcribed the correct numbers into your calculations. The most common mistake to make in this type of problem is to forget to take account of the final volume of the mixed solution. We have used the correct numbers, and we have factored in the dilution, so our answer should be correct. This example clearly shows that AgCl(s) is not insoluble.

PRACTICE EXERCISE 10.5

Determine whether or not a precipitate of $NiCO_3$(s) will form if 100.0 mL of 1.0×10^{-3} M $Ni(NO_3)_2$(aq) is mixed with 100.0 mL of 2.0×10^{-3} M Na_2CO_3(aq) at 25 °C.

Chemical Connections

Kidney stones

The formation of kidney stones (figure 10.18) is one of the most obvious applications of solubility in the human body. These solids, which range in size from a grain of sand to a golf ball, can lead to the blockage of the ureter, which carries urine from the kidney to the bladder. The most common symptom of kidney stones is excruciating pain, and their removal can require invasive techniques.

Kidney stones generally contain slightly soluble calcium salts of either oxalate ($C_2O_4^{2-}$) or phosphate, with the most common components being calcium oxalate monohydrate ($CaC_2O_4{\cdot}H_2O$), hydroxyapatite ($Ca_5OH(PO_4)_3$) and brushite ($CaHPO_4{\cdot}2H_2O$). These compounds are thought to be formed when the concentrations of the Ca^{2+}, $C_2O_4^{2-}$ and PO_4^{3-}/HPO_4^{2-} ions in the urine are high. We can rationalise this behaviour on the basis of what we have learned about solubility in this chapter. We know that when $Q_{sp} > K_{sp}$, a precipitate is expected to form, so a high concentration of ions will favour the formation of the solid.

One of the recommended measures to prevent the formation of kidney stones is the daily consumption of significant volumes of water. This will lower the concentration of the ions in the urine, which will lower Q_{sp}, hopefully to the point where $Q_{sp} < K_{sp}$, thereby ensuring no precipitate will form. Another reported method of avoiding kidney stones is to drink orange juice; this is high in citric acid, which binds strongly to the Ca^{2+} ion, thereby lowering the concentration of free Ca^{2+} and consequently lowering Q_{sp}. Regardless of how you do it, if you want to avoid the pain of a kidney stone, keep your Q_{sp} low!

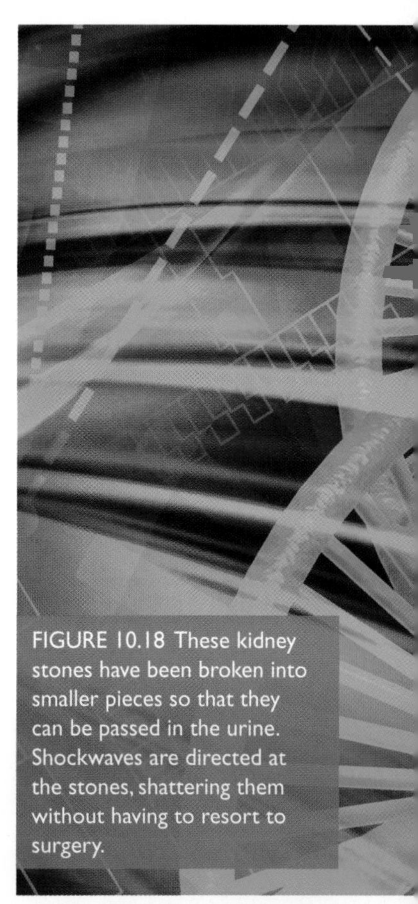

FIGURE 10.18 These kidney stones have been broken into smaller pieces so that they can be passed in the urine. Shockwaves are directed at the stones, shattering them without having to resort to surgery.

10.5 Colligative properties of solutions

When we prepare a solution by dissolving a small amount of a **nonvolatile** solute, in other words, a solute that does not have an appreciable vapour pressure, in a solvent, we find that particular properties of the resulting solution differ quite significantly from those of the pure solvent. For example, the boiling point of the solution is higher than that of the pure solvent and the freezing point is lower. Interestingly, the magnitudes of the boiling point elevation and the freezing point depression do not necessarily depend on the identity of the solute; if we prepared separate aqueous solutions of NaCl and NaBr by dissolving 1 mole of each salt in 1 kg of water, we would find that the boiling points (about 101 °C) and freezing points (about −3.4 °C) of both solutions would be the same. Properties such as vapour pressure, boiling point elevation, freezing point depression and osmotic pressure, which depend only on the *number* of solute particles in the solution, are called **colligative properties** and can be very useful in elucidating the exact nature of solute particles in a solution.

The word 'colligative' is derived from the Latin *colligare* meaning 'to bind together'. Such a derivation is appropriate as colligative properties do in fact arise from the attractive interactions between solute particles and solvent molecules.

Colligative properties are important in everyday situations. Adding antifreeze to a car's radiator lowers the freezing point of the coolant solution to well below 0 °C; this reduces the probability of damage to the engine caused by the expansion that occurs on formation of ice. Antifreeze also raises the boiling point of the radiator coolant. Reverse osmosis, a process that relies on colligative properties, is used extensively in the food industry as a method of concentrating heat-sensitive materials, such as fruit and vegetable juices, as it requires only the application of pressure. It is also used for desalination of sea water and to remove acetic acid from fermenting wine.

To be able to quantify the effects of colligative properties, we must first look at ways in which we can express the composition of solutions.

Molarity

The most common method of expressing the concentration of a solution, and the one we have used throughout the book thus far, is in $mol\,L^{-1}$; that is, the amount of substance in a particular volume of solution. This is more properly called the **molar concentration** or **molarity** of a solution and has the symbol c.

$$\text{molarity } (c) = \frac{\text{amount of solute (mol)}}{\text{volume of solution (L)}}$$

However, because solutions (usually) increase their volume when the temperature is raised, and vice versa, the molarity of a solution changes as the temperature changes. For example, a saline solution of known molarity prepared at 25 °C has a different molarity when it is warmed to body temperature; this makes molarity a less than ideal measure of concentration in situations involving colligative properties.

Molality

Molality is the preferred method of expressing solution composition in situations that involve colligative properties, as these properties are proportional to molality. We define the **molal concentration** or **molality** of a solution as the amount of solute per kilogram of solvent, and use the symbol b.

$$\text{molality } (b) = \frac{\text{amount of solute (mol)}}{\text{mass of solvent (kg)}}$$

The molality of a solution is temperature independent as mass does not depend on temperature. The molarity and molality of aqueous solutions are numerically different but, as aqueous solutions become more and more dilute, the numerical value of the molarity approaches that of the molality.

WORKED EXAMPLE 10.6

Calculation to prepare a solution of a given molality

What mass of NaCl would have to be dissolved in 500.0 g of water to prepare a solution of molality $0.150\,mol\,kg^{-1}$?

Analysis

Rearranging the equation for molality given above, we obtain:

$$\text{amount of solute (mol)} = \text{molality } (mol\,kg^{-1}) \times \text{mass of solvent (kg)}$$

Knowing the amount of solute, NaCl, we then relate amount to mass through the molar mass of NaCl.

Solution

$$\begin{aligned}\text{amount of NaCl} &= \text{molality} \times \text{mass of solvent}\\ &= 0.150\,\text{mol kg}^{-1} \times 0.5000\,\text{kg}\\ &= 0.0750\,\text{mol}\end{aligned}$$

$$m = nM = 0.0750\,\text{mol} \times 58.44\,\text{g mol}^{-1} = 4.38\,\text{g}$$

Therefore, 4.38 g of NaCl needs to be dissolved in 500.0 g of water to give a solution of molality 0.150 mol kg^{-1}.

Is our answer reasonable?

If we round the molar mass of NaCl to 60, 0.1 mol is 6 g and 0.2 mol is 12 g. We also notice that 0.150 mol kg^{-1} is halfway between 0.1 and 0.2 mol kg^{-1}. So, for 1 kilogram of water, we'd need halfway between 6 g (0.1 mol) and 12 g (0.2 mol) of NaCl. For half as much water (500 g), we cut these limits in two, so we need halfway between 3 and 6 g of NaCl. Our answer is in this range.

PRACTICE EXERCISE 10.6

Water freezes at a lower temperature when it contains solutes. To study the effect of methanol on the freezing point of water, we might begin by preparing a series of solutions of known molalities. Calculate the mass of methanol, CH_3OH, needed to prepare a 0.250 mol kg^{-1} solution using 2000 g of water.

Mole fraction

In chapter 6 we introduced the mole fraction as a method of expressing the composition of gas mixtures. We can also use it for liquid solutions. We define the **mole fraction (x)** of a particular component of a solution as the amount of that component divided by the total amount of material in the solution. For example, the mole fraction of A, x_A, in a solution containing substances A, B and C would be defined as:

$$x_A = \frac{n_A}{n_A + n_B + n_C}$$

Mole fraction is also temperature independent.

We will now see how these alternative measures of solution composition are used when quantifying colligative properties.

Raoult's law

As mentioned earlier, the boiling point of a solution containing a nonvolatile solute is higher than that of the pure solvent. The boiling point of a solvent is the temperature at which the vapour pressure of the solvent is equal to the atmospheric pressure; therefore, the observation of a higher boiling point must mean that the solution has a lower vapour pressure than the pure solvent, as it now takes greater energy input to raise the vapour pressure of the solution to equal the atmospheric pressure. Providing that the solution is sufficiently dilute (i.e. x_{solvent} is close to 1), **Raoult's law** describes the relationship between the vapour pressure of the solution (p_{solution}), the mole fraction of solvent in the solution (x_{solvent}) and the vapour pressure of the pure solvent (p^*_{solvent}) in a simple two-component system.

$$p_{\text{solution}} = x_{\text{solvent}}\,p^*_{\text{solvent}} \text{ (at constant temperature)}$$

This equation describes a straight line and so a plot of p_{solution} versus x_{solvent} is linear, as shown in figure 10.19. Because x_{solvent} and x_{solute} are related in a two-component system:

$$x_{\text{solvent}} = 1 - x_{\text{solute}}$$

we can also write Raoult's law in terms of x_{solute}.

$$\begin{aligned}p_{\text{solution}} &= (1 - x_{\text{solute}})p^*_{\text{solvent}}\\ &= p^*_{\text{solvent}} - x_{\text{solute}}\,p^*_{\text{solvent}}\end{aligned}$$

This expression can be rearranged to give the following relationship for the *pressure difference* (Δp) between the pure solvent and the solution.

$$\Delta p = x_{\text{solute}}\,p^*_{\text{solvent}}$$

where $\Delta p = p^*_{\text{solvent}} - p_{\text{solution}}$.

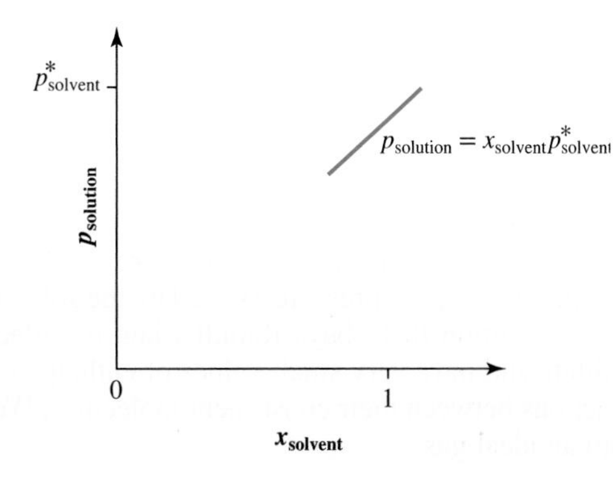

FIGURE 10.19 A Raoult's law plot. When the vapour pressure of a solution is plotted against the mole fraction of the solvent, a straight line results. It should be stressed that such behaviour is observed only in dilute solutions, that is, those in which x_{solvent} is close to 1.

WORKED EXAMPLE 10.7

Calculation using Raoult's law

What is the vapour pressure of a solution of 10.0 g of candle wax in 40.0 g of carbon tetrachloride, CCl_4, at 23 °C? The molecular formula of candle wax is $C_{22}H_{46}$ and the vapour pressure of pure CCl_4 at 23 °C is 1.32×10^4 Pa.

Analysis

We use Raoult's law so we must calculate the mole fraction of solvent, $x_{solvent}$, from the data given by calculating the amount of both $C_{22}H_{46}$ and CCl_4.

Solution

$$n_{CCl_4} = \frac{m}{M} = \frac{40.0\,\text{g}}{153.81\,\text{g mol}^{-1}} = 0.260\,\text{mol}$$

$$n_{C_{22}H_{46}} = \frac{m}{M} = \frac{10.0\,\text{g}}{310.59\,\text{g mol}^{-1}} = 0.0322\,\text{mol}$$

Therefore, $n_{total} = 0.260\,\text{mol} + 0.0322\,\text{mol} = 0.292\,\text{mol}$. We can now calculate the mole fraction of solvent.

$$x_{solvent} = \frac{n_{CCl_4}}{n_{total}} = \frac{0.260\,\text{mol}}{0.292\,\text{mol}} = 0.890$$

We use this value in Raoult's law.

$$p_{solution} = x_{solvent}\,p^*_{solvent} = 0.890 \times (1.32 \times 10^4\,\text{Pa}) = 1.17 \times 10^4\,\text{Pa}$$

Hence, dissolving 10.0 g of candle wax in 40.0 g of CCl_4 lowers the vapour pressure by about 11%.

Is our answer reasonable?

The pressure we obtained is less than the vapour pressure of the pure solvent, which is what we would expect from Raoult's law. Barring arithmetic errors, our answer looks sensible.

PRACTICE EXERCISE 10.7

Di(2-ethylhexyl)phthalate, $C_{24}H_{38}O_4$ (molar mass 390.544 g mol^{-1}), is a viscous liquid used in huge quantities as a plasticiser for PVC. Its vapour pressure is negligible at room temperature. The vapour pressure of pure octane at 20 °C is 1.38×10^3 Pa. What is the vapour pressure at 20 °C of a solution of 10.0 g of di(2-ethylhexyl)phthalate in 60.0 g of octane, C_8H_{18} (molar mass 114.224 g mol^{-1})?

While Raoult's law shows that a solution has a lower vapour pressure than the pure solvent, it does not explain why this is the case. If we compare a pure solvent with a solution containing a nonvolatile solute, we see that the surface of the solution comprises a mixture of solvent and solute molecules, while the surface of the solvent consists only of solvent molecules (figure 10.20).

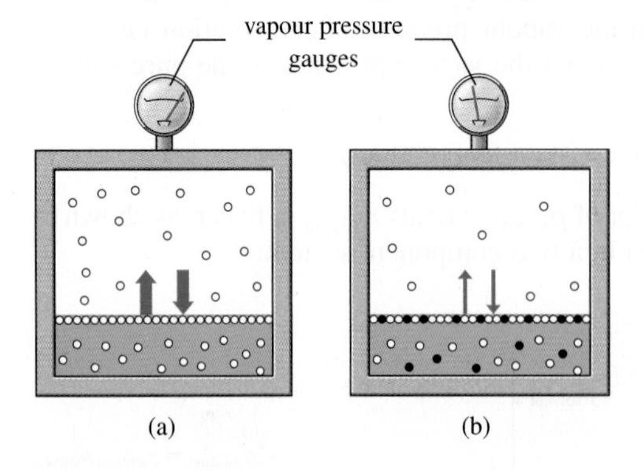

FIGURE 10.20 Effect of a nonvolatile solute on the vapour pressure of a solvent: **(a)** Equilibrium between a pure solvent and its vapour. With a high number of solvent molecules at the surface of the liquid, the rate of evaporation is relatively high. **(b)** In the solution, some of the solvent molecules at the surface have been replaced by solute molecules so there are fewer solvent molecules available to evaporate from the surface of the solution. This lowers the evaporation rate, so, when equilibrium is established, there are fewer molecules in the vapour. The vapour pressure of the solution is therefore less than that of the pure solvent.

Therefore, there are fewer molecules of solvent at the surface of a solution than of a pure solvent. This means that, in the solution, there are fewer solvent molecules available to evaporate, so we expect the vapour pressure exerted by the solvent molecules to be lower than in the pure solvent.

A solution that obeys Raoult's law is called an **ideal solution**. Such solutions are generally dilute and have very small values of enthalpy of mixing, implying that there are only small interactions between their constituent molecules. We can think of an ideal solution as being analogous to an ideal gas.

Solutions containing more than one volatile component

When two (or more) components of a liquid solution can evaporate, the vapour contains molecules of each. Each volatile component contributes its own partial pressure to the total pressure. By Raoult's law, the partial pressure of a particular component is directly proportional to the component's mole fraction in the solution, and the total vapour pressure is the sum of the partial pressures. To calculate these partial pressures, we use Raoult's law for each component. For component *A*:

$$p_A = x_A p_A^*$$

and for component *B*:

$$p_B = x_B p_B^*$$

The total pressure above a solution of liquids *A* and *B* is then the sum of p_A and p_B.

$$p_{\text{total}} = x_A p_A^* + x_B p_B^* \text{ (at constant temperature)}$$

Figure 10.21 shows how the vapour pressure of an ideal, two-component solution of volatile compounds changes with composition.

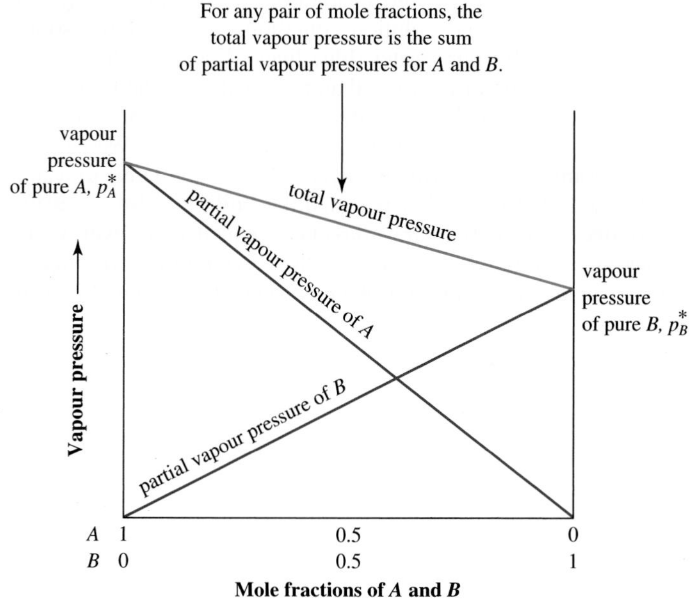

FIGURE 10.21 The vapour pressure of an ideal, two-component solution of volatile compounds.

WORKED EXAMPLE 10.8

Calculating the vapour pressure of a solution of two volatile liquids

At 20 °C, acetone has a vapour pressure of 2.13×10^4 Pa. The vapour pressure of water at 20 °C is 2.30×10^3 Pa. Assuming that the mixture obeys Raoult's law, calculate the vapour pressure of a solution of acetone and water containing 80.0 mol % of water.

Analysis

To find p_{total}, we need to calculate the individual partial pressures and then add them.

Solution

A concentration of 80.0 mol % water corresponds to a mole fraction of 0.800 for water and 0.200 for acetone, so:

$$p_{\text{acetone}} = 0.200 \times 2.13 \times 10^4 \text{ Pa} = 4.26 \times 10^3 \text{ Pa}$$

$$p_{\text{water}} = 0.800 \times 2.30 \times 10^3 \text{ Pa} = 1.84 \times 10^3 \text{ Pa}$$

$$p_{\text{total}} = 6.10 \times 10^3 \text{ Pa}$$

Is our answer reasonable?

The vapour pressure of the solution (6.10×10^3 Pa) has to be higher than that of pure water (2.30×10^3 Pa), because of the volatile acetone, but less than that of pure acetone (2.13×10^4 Pa), because of the high mole fraction of water. The answer seems to be reasonable.

PRACTICE EXERCISE 10.8 At 20 °C, the vapour pressure of cyclohexane (a hydrocarbon solvent) is 8.80×10^3 Pa and that of toluene (another solvent) is 2.78×10^3 Pa. What is the vapour pressure of a solution of the two at 20 °C when toluene is present at a mole fraction of 0.700?

Because solutes influence the vapour pressures of solutions, they also affect their boiling and freezing points. The boiling point of a solution of a nonvolatile solute, for example, is higher than that of the pure solvent, and the freezing point of the solution is lower than that of the solvent. We can explain this for aqueous solutions by considering the phase diagram for water (see figure 7.20, p. 270).

Boiling point elevation and freezing point depression

We have seen how the presence of a nonvolatile solute in a solution increases the boiling point of the solution relative to that of the pure solvent. Such solutions also have lower freezing points than those of the pure solvents, meaning that the attractive forces between solvent molecules in the solution must be smaller than those in the pure solvent at a given temperature. This is due to the presence of solute molecules and the corresponding solute–solvent interactions, which are not present in the pure solvent.

We can illustrate the changes in boiling and freezing points in solutions using a phase diagram. Figure 10.22 shows a phase diagram for an aqueous solution containing a nonvolatile solute (red lines) superimposed on the phase diagram for pure water (blue lines). We know that, at any temperature, the aqueous solution has a lower vapour pressure than pure water, so the liquid–gas equilibrium line lies below that for pure water. This has the effect of lowering the pressure and temperature at the triple point, so the ice–liquid equilibrium line is also displaced. We can see from the phase diagram that the boiling point of the solution is increased by ΔT_b, relative to water, and the freezing point is depressed below 0 °C by ΔT_f. ΔT_b is called the **boiling point elevation** and ΔT_f is called the **freezing point depression** and both are a direct consequence of lowering the vapour pressure in the solution. They are colligative properties as their magnitudes are directly proportional to the number of solute particles in the solution. We can calculate ΔT_b and ΔT_f from the equations:

$$\Delta T_b = K_b b$$

$$\Delta T_f = K_f b$$

where K_b and K_f are the **molal boiling point elevation constant** and the **molal freezing point depression constant**, respectively, and b is the molality of the solution in mol kg^{-1}. K_b and K_f are properties of the solvent only and are independent of the identity of the solute; table 10.5 gives values of K_b and K_f for a number of solvents. These values apply exactly to very dilute solutions only, but we can use them to estimate the freezing points and boiling points of fairly concentrated solutions with reasonable precision.

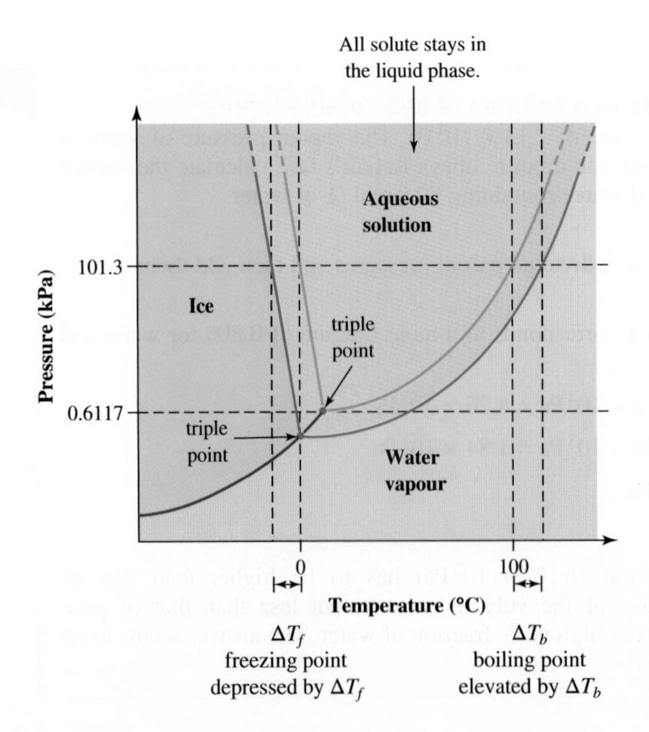

FIGURE 10.22 Phase diagram for an aqueous solution of a nonvolatile solute.

TABLE 10.5 Molal boiling point elevation and freezing point depression constants.

Solvent	K_b (K mol^{-1} kg)	K_f (K mol^{-1} kg)
water	0.51	1.86
acetic acid	3.07	3.57
benzene	2.53	5.07
chloroform	3.63	–
camphor	–	37.7
cyclohexane	2.69	20.0

WORKED EXAMPLE 10.9

Estimating a freezing point using a colligative property

Estimate the freezing point of a solution made from 10.00 g of urea, $CO(NH_2)_2$ (molar mass 60.06 g mol^{-1}), in 100.0 g of water.

Analysis

$\Delta T_f = K_f b$ relates concentration to freezing point depression. To use the equation, we must first calculate the molality of the solution.

Solution

$$n_{CO(NH_2)_2} = \frac{10.00\,\text{g}}{60.06\,\text{g mol}^{-1}} = 0.1665\,\text{mol}$$

This is the amount of $CO(NH_2)_2$ in 100.0 g or 0.1000 kg of water. Therefore:

$$b = \frac{0.1665\,\text{mol}}{0.1000\,\text{kg}} = 1.665\,\text{mol kg}^{-1}$$

For our estimate, we must next use $\Delta T_f = K_f b$. From table 10.5, K_f for water is 1.86 K mol^{-1} kg.

$$\Delta T_f = K_f b = (1.86\,\text{K mol}^{-1}\,\text{kg})(1.665\,\text{mol kg}^{-1}) = 3.10\,\text{K}$$

Remembering that a temperature difference of 3.10 K is equal to 3.10 °C, the solution should freeze at 3.10 °C below 0 °C, or at –3.10 °C.

Is our answer reasonable?

For a solution of molality 1 mol kg^{-1}, the freezing point depression is:

$$\Delta T_f = K_f b = (1.86\,\text{K mol}^{-1}\,\text{kg})(1\,\text{mol kg}^{-1}) = 1.86\,\text{K}$$

Therefore, for every unit of molality, the freezing point must be depressed by about 2 °C. The molality of this solution is between 1 mol kg^{-1} and 2 mol kg^{-1}, so we expect the freezing point depression to be between about 2 °C and 4 °C. It is.

PRACTICE EXERCISE 10.9

Glycerol ($C_3H_8O_3$) is a nonvolatile compound that has been used as an antifreeze in car engines. However, it is not always realised that an antifreeze also raises the boiling point of water in the cooling system as well. At what temperature does a 5.00% (by mass) aqueous solution of glycerol boil at 1.013×10^5 Pa?

We have described the properties of freezing point depression and boiling point elevation as *colligative*; that is, they depend on the relative *numbers* of particles, rather than their identities. Because the effects are proportional to molal concentrations, experimental values of ΔT_f or ΔT_b can be useful for calculating the molar masses of unknown solutes. We'll see how this works in worked example 10.10.

WORKED EXAMPLE 10.10

Calculating a molar mass from freezing point depression data

A solution made by dissolving 5.65 g of an unknown molecular compound in 110.0 g of benzene froze at 4.39 °C. Given that the normal melting point of benzene is 5.45 °C, calculate the molar mass of the solute.

Analysis

We can use $\Delta T_f = K_f b$ to calculate the amount of solute in the given solution. Then we can divide the mass of solute by the amount of solute to find the molar mass.

Solution

From table 10.5, the value of K_f for benzene is 5.07 K mol^{-1} kg. The magnitude of the freezing point depression is:

$$\Delta T_f = 5.45\ °C - 4.39\ °C = 1.06\ °C = 1.06\ K$$

We now use $\Delta T_f = K_f b$ to find the molality of the solution.

$$\Delta T_f = K_f b$$

$$b = \frac{\Delta T_f}{K_f}$$

$$= \frac{1.06\ K}{5.07\ K\ mol^{-1}\ kg}$$

$$= 0.209\ mol\ kg^{-1}$$

This means that there is 0.209 mol of solute for every kilogram of benzene in the solution. But we have only 110.0 g or 0.1100 kg of benzene. So the actual amount of solute in the given solution is:

$$0.1100\ kg \times 0.209\ mol\ kg^{-1} = 0.0230\ mol$$

We can now obtain the molar mass.

$$M = \frac{m}{n} = \frac{5.65\ g}{0.0230\ mol} = 246\ g\ mol^{-1}$$

The mass of 1 mole of the solute is 246 g.

Is our answer reasonable?

The easiest way to check our answer is to do the reverse calculation, starting with 5.65 g of a compound having a molar mass of 246 g mol^{-1}, and determining the resultant freezing point depression when dissolved in 110.0 g of benzene. If you do this (it's good practice!), you'll find that the answer is correct.

PRACTICE EXERCISE 10.10 A solution made by dissolving 3.46 g of an unknown compound in 85.0 g of benzene froze at 4.13 °C. What is the molar mass of the compound?

Osmosis and osmotic pressure

In living things, membranes of various kinds keep mixtures and solutions organised and separated. Yet some substances have to be able to pass through membranes so that nutrients and products of chemical reactions can be distributed correctly. In other words, these membranes must have a selective *permeability*. They must keep some substances from going through while allowing others to pass. Such membranes are said to be *semipermeable*.

The degree of permeability varies with the kind of membrane. Cellophane, for example, is permeable to water and small solute particles — ions or molecules — but impermeable to very large molecules, such as those of starch or proteins. Special membranes can even be prepared that are permeable only to water but not to any solutes.

Depending on the kind of membrane separating solutions of different concentration, two similar phenomena, *dialysis* and *osmosis*, can be observed. Both are functions of the relative populations of particles in the dissolved materials on either side of the membrane, so they are colligative properties.

When a membrane is capable of allowing both water and small solute particles to pass through, such as membranes found in living systems, the process is called **dialysis**, and the membrane is called a *dialysing membrane*. It does not permit huge molecules to pass through, such as those of proteins and starch. Artificial kidney machines use dialysing membranes to help filter and remove the smaller molecules of wastes from the blood while the larger protein molecules, which cannot pass through the membrane, are retained in the blood.

Osmosis is a net shift of only solvent through a membrane, in a direction from the side less concentrated in *solute* to the side more concentrated. The special membrane required is called an **osmotic membrane**. Such membranes are rare, but they can be made.

The *direction* in which osmosis occurs, namely, the net movement of solvent from the more dilute solution (or pure solvent) into the more concentrated solution, may be explained with the help of figure 10.23. Water molecules move freely across the membrane. The concentration of water in the aqueous solution (*B*) is lower than the concentration in pure water (*A*). More water molecules are available to cross the membrane from *A* to *B*, so the flow of water molecules *into* the solution is somewhat greater than the flow *out of* the solution. In figure 10.23b, the net flow of water into the solution has visibly increased the volume of *B*.

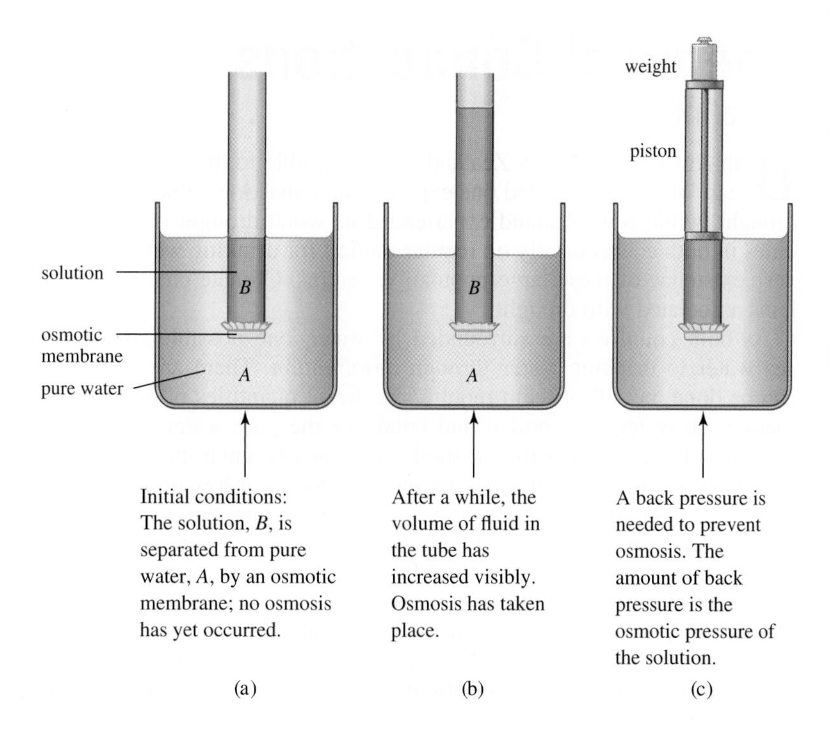

weight
piston
solution
osmotic membrane
pure water
B
A
B
A
Initial conditions: The solution, B, is separated from pure water, A, by an osmotic membrane; no osmosis has yet occurred.
After a while, the volume of fluid in the tube has increased visibly. Osmosis has taken place.
A back pressure is needed to prevent osmosis. The amount of back pressure is the osmotic pressure of the solution.
(a)
(b)
(c)

FIGURE 10.23 Osmosis and osmotic pressure.

If we push on the solution, we can force water back through the membrane (from B to A). The weight of the rising fluid column in figure 10.23b provides a push or opposing pressure that eventually stops osmosis. If we apply further pressure (figure 10.23c), we can force enough water back through the membrane to restore the system to its original condition. The exact opposing pressure needed to prevent any osmotic flow *when one of the liquids is pure solvent* is called the **osmotic pressure** of the solution.

By *exceeding* the osmotic pressure, osmosis can be reversed and we refer to this situation as reverse osmosis. Reverse osmosis allows pure water to be obtained from an aqueous salt solution simply by forcing it through a membrane under high pressure. It is therefore widely used to obtain drinking water from sea water. There are currently six reverse osmosis desalination plants operating in Australia.

To further illustrate the principles involved in osmosis, consider the situation in figure 10.24; we have a beaker of water in an enclosed space *already saturated in water vapour*, and we have just added a beaker of an aqueous solution of a nonvolatile solute. Water molecules tend to escape into the vapour phase from both beakers, but their rate of escape from the pure water (larger upward arrow) is greater than their rate of escape from the solution (smaller upward arrow). This must be so because the vapour pressure of pure water is greater than that of the solution, as predicted by Raoult's law. The rate at which water molecules return to the solution from the saturated vapour state is not diminished by the presence of the solute. At the solution–vapour interface, more water molecules return to the solution than leave it, and this takes water from the vapour state. To keep the vapour phase saturated, more of the pure water must evaporate. The net effect is that water moves from the beaker of pure water into the vapour space and then into the solution. In osmosis, an osmotic membrane instead of beakers separates the liquids, but the same principles are at work.

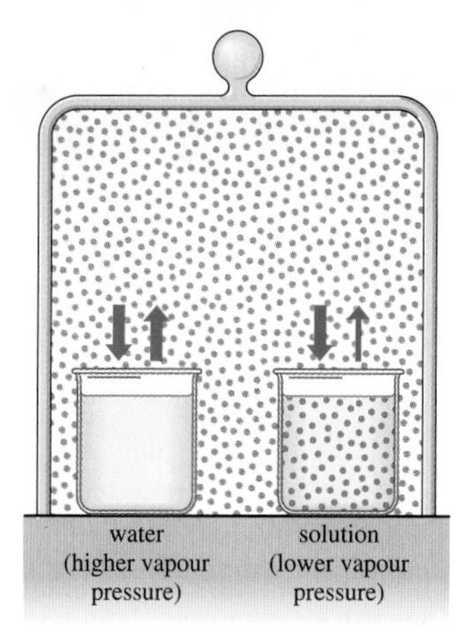

FIGURE 10.24 Principles at work in osmosis. Here, beakers, instead of a semipermeable membrane, separate the pure solvent (water) from an aqueous solution containing a nonvolatile solute. Because the vapour pressure of pure water is greater than the vapour pressure of the aqueous solution, molecules transfer from the beaker of pure water to the aqueous solution.

The higher the concentration of the *solution*, the more water is transferred to it (figure 10.24). In other words, the osmotic pressure of a solution is proportional to the relative concentrations of its solute and solvent.

The symbol used for osmotic pressure is Π, the capital Greek letter *pi*. In a dilute aqueous solution, Π (Pa) is proportional both to temperature, T (K), and molar concentration, c (mol m^{-3}).

$$\Pi \propto cT$$

The proportionality constant turns out to be the gas constant, R, so for a *dilute* aqueous solution we can write:

$$\Pi = cRT$$

As $c = \frac{n}{V}$:

$$\Pi V = nRT$$

This is the *van't Hoff equation for osmotic pressure* and is analogous to the ideal gas law. Note that we work in terms of concentration rather than molality here to emphasise the similarity of this equation to the ideal gas law. Recall from page 410 that, provided the solution is sufficiently dilute, the molality and concentration are numerically essentially equal.

Chemical Connections

Desalination

Both Australia and New Zealand are susceptible to drought. Record low rainfalls in the early years of this century led one expert to state that Australia was in the grip of a 1-in-1000-year drought, while New Zealand experienced its worst drought in 100 years in the 1990s. As major cities in both countries rely on regular rainfall for drinking water, drought conditions threaten the very existence of these large population centres. Climate change may also exacerbate the problems associated with drought.

As both countries are surrounded by water, one possible solution lies in the conversion of sea water to drinking water through *desalination*. There are a number of ways in which this can be done, but all of them require significant quantities of energy. The simplest way of desalinating sea water is to boil it and condense the pure water vapour that results. The high heat capacity of water makes this method economically, environmentally and energetically infeasible on a large scale. However, carrying out the process at low pressure means that the water boils at a lower temperature, so less energy is required to heat the water. Of course, the energy required to lower the pressure also needs to be taken into account in the overall economics of the process. Several desalination plants in the Middle East use a variant of this technology and some have capacities of over 250 million litres per day.

Currently, the most popular method of desalination is *reverse osmosis* (see figure 10.25). As we have seen in this chapter, pure water will flow spontaneously through a semipermeable membrane from a dilute electrolyte solution to a concentrated electrolyte solution, thereby exerting an osmotic pressure. If a pressure greater than this is exerted on the more concentrated solution, the direction of flow can be reversed, and pure water results. The Kurnell desalination plant in Sydney operates using this principle (figure 10.26). It was designed to provide a minimum of 15% of Sydney's water needs. The electricity required to generate the required high pressures comes from a wind farm, meaning that this plant runs entirely on renewable energy.

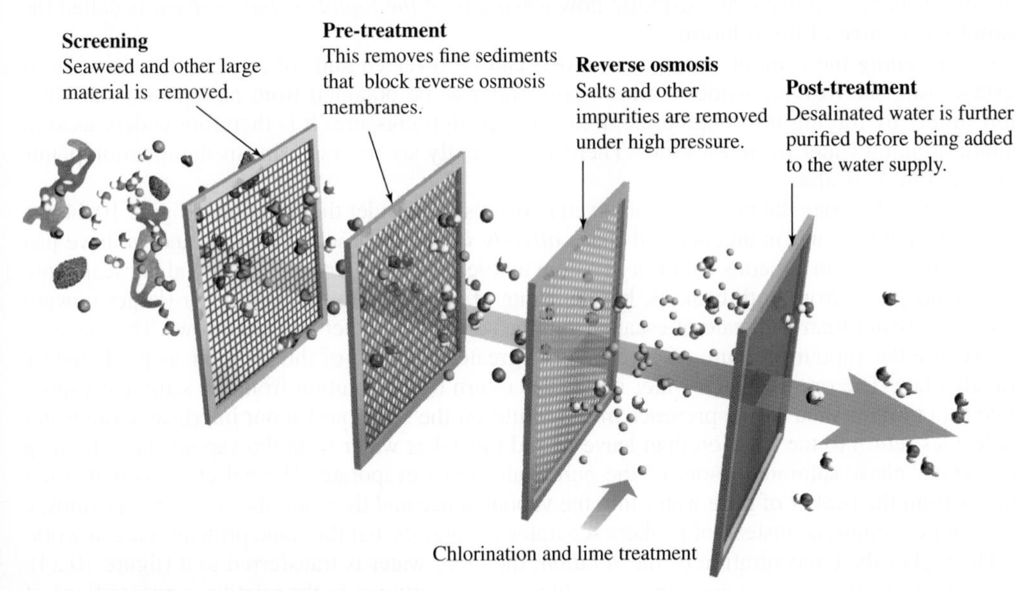

Screening
Seaweed and other large material is removed.
Pre-treatment
This removes fine sediments that block reverse osmosis membranes.
Reverse osmosis
Salts and other impurities are removed under high pressure.
Post-treatment
Desalinated water is further purified before being added to the water supply.
Chlorination and lime treatment

FIGURE 10.25 The four major steps in conversion of sea water to drinking water using reverse osmosis.

FIGURE 10.26 Kurnell desalination plant in Sydney, New South Wales.

One less-than-obvious problem with the use of reverse osmosis plants that must be overcome is the disposal of the concentrated salt solutions that are formed. Pumping these wastes back into the ocean at source can lead to high levels of salt in the local marine environment, especially in shallow and slow-moving estuarine waters, and this can have a harmful effect on marine life. A possible solution is to pump the waste into deep offshore waters where the sea currents disperse the salt; this however does have attendant energy costs. Another solution is the harvesting of salt from the concentrated residues, and this may indeed be economically viable.

Osmotic pressure is of tremendous importance in biology and medicine. Cells are surrounded by membranes that restrict the flow of salts but allow water to pass through freely. To maintain a constant amount of water, the osmotic pressure of solutions on either side of a cell membrane must be identical. For example, a solution that is 0.9% NaCl by mass has the same osmotic pressure as the contents of red blood cells, and red blood cells bathed in this solution can maintain their normal water content. The solution is said to be **isotonic** with red blood cells. Blood plasma is an isotonic solution.

If a cell is placed in a solution with a salt concentration higher than the concentration within the cell, osmosis causes water to flow out of the cell. Such a solution is said to be **hypertonic**. The cell shrinks and dehydrates, and eventually dies. This process kills freshwater fish and plants that are washed out to sea.

On the other hand, water flows into a cell if it is placed into a solution with an osmotic pressure that is much lower than the osmotic pressure of the cell's contents. Such a solution is called a **hypotonic** solution. A cell placed in distilled water, for example, swells and bursts. The effects of isotonic, hypertonic and hypotonic solutions on red blood cells are shown in figure 10.27.

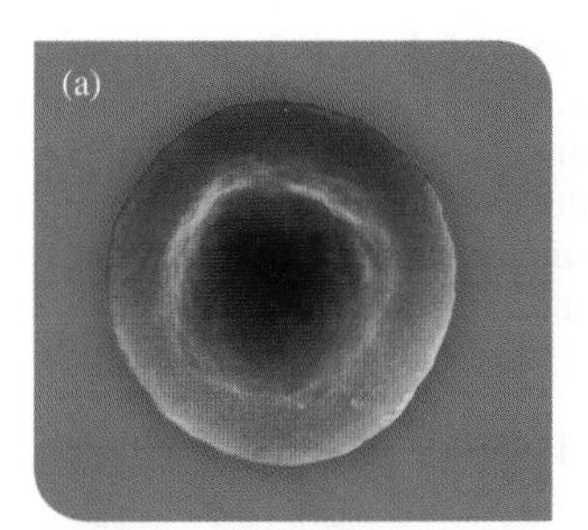

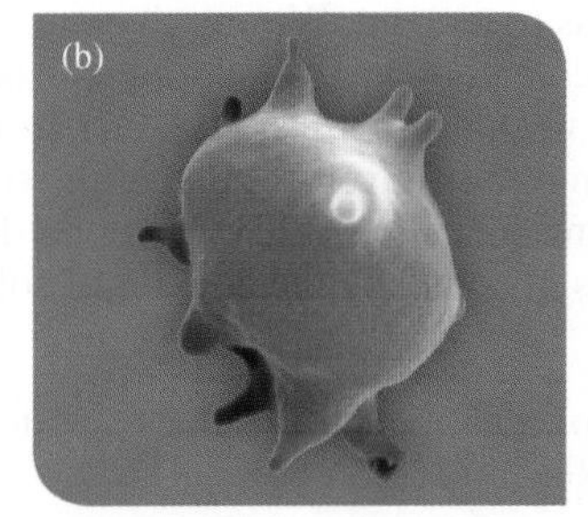

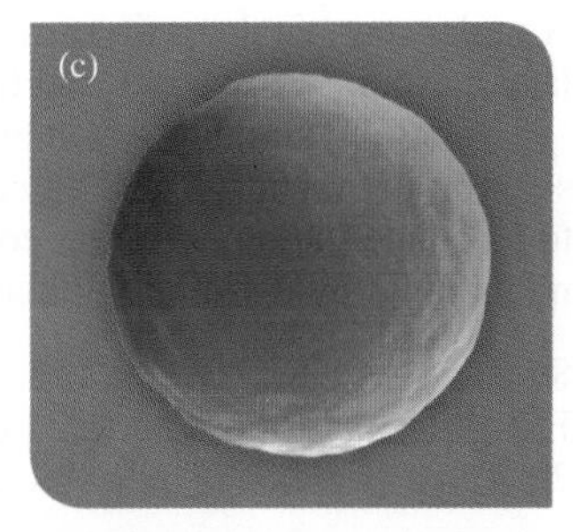

FIGURE 10.27 Effects on red blood cells that have been placed in isotonic, hypertonic and hypotonic solutions, all at the same magnification: **(a)** In an isotonic solution (0.9% NaCl by mass), solutions on either side of the cell membrane have the same osmotic pressure, so there is no net flow of water in one direction across the membrane. **(b)** In a hypertonic solution (5.0% NaCl by mass), water flows from an area of lower salt concentration (inside the cell) to higher concentration (the hypertonic solution), causing the cell to dehydrate. **(c)** In a hypotonic solution (0.1% NaCl by mass), water flows from an area of lower salt concentration (the hypotonic solution) to higher concentration (inside the cell). The cell swells and bursts.

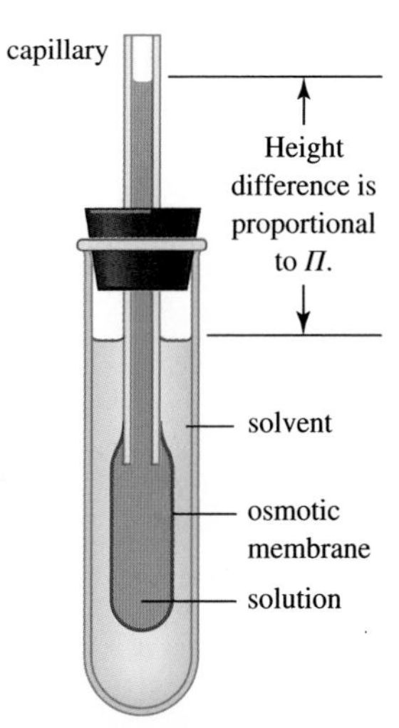

FIGURE 10.28 Simple osmometer. When solvent moves into the solution by osmosis, the level of the solution in the capillary rises. The height reached is related to the osmotic pressure of the solution.

Obviously, the measurement of osmotic pressure can be very important in preparing solutions used to culture tissues or to administer medicines intravenously. Osmotic pressures can be measured by an instrument called an *osmometer*, illustrated and explained in figure 10.28. Osmotic pressures can be very high, even in dilute solutions, as worked example 10.11 shows.

WORKED EXAMPLE 10.11

Calculating osmotic pressure

A solution of 1.00×10^{-3} M sugar in water is separated from pure water by an osmotic membrane. What osmotic pressure (in Pa) develops at 25 °C?

Analysis

We can use $\Pi = cRT$. Recall that $R = 8.314\,\text{J mol}^{-1}\,\text{K}^{-1}$. When we use this equation, the concentration must be expressed in mol m^{-3} to give an answer in Pa (1 Pa = 1 J m^{-3}, p. 31).

Solution

We multiply the concentration by 1000 to convert from mol L^{-1} to mol m^{-3}; thus, $c = 1.00\,\text{mol m}^{-3}$.

$$\Pi = (1.00\,\text{mol m}^{-3})(8.314\,\text{J mol}^{-1}\,\text{K}^{-1})(298\,\text{K})$$
$$= 2.48 \times 10^{3}\,\text{J m}^{-3} = 2.48 \times 10^{3}\,\text{Pa}$$

Thus, the osmotic pressure of 1.00×10^{-3} M sugar in water is 2.48×10^{3} Pa.

Is our answer reasonable?

The units are the greatest cause of error in calculations like this. Remember to express T in K and c in mol m^{-3}, not mol L^{-1}. This will give an answer in Pa.

PRACTICE EXERCISE 10.11

Determine the osmotic pressure of a 0.0156 M glucose solution at body temperature (37 °C).

In worked example 10.11, you saw that a 1.00×10^{-3} M solution of sugar has an osmotic pressure of 2.48×10^3 Pa. This pressure would support a column of the solution roughly 25 cm high. If the solution had been 100 times as concentrated (0.100 M sugar — still relatively dilute), the height of the column supported would be roughly 25 m.

An osmotic pressure measurement of a dilute solution can be used to determine the molar concentration and, therefore, the molar mass of a molecular solute. Determination of molar mass by osmotic pressure is much more sensitive than determination by freezing point depression or boiling point elevation. Worked example 10.12 shows how experimental measurements of osmotic pressure can be used to determine molar mass.

WORKED EXAMPLE 10.12

Calculating molar mass from osmotic pressure

An aqueous solution with a volume of 100 mL and containing 0.122 g of an unknown molecular compound has an osmotic pressure of 2.11×10^3 Pa at 20.0 °C. What is the molar mass of the solute?

Analysis

A molar mass is the ratio of mass to amount. We are given the mass of the solute; to calculate the ratio, we need to find the amount equivalent to this mass.

To find the amount of solute, we must first use $\Pi = cRT$ to find the concentration of the solution in mol m^{-3}. Then we can use the given volume of the solution and the concentration to calculate the amount that corresponds to 0.122 g of solute. Finally, we can calculate the molar mass by dividing the mass of solute by the amount of solute.

Solution

First, calculate the solution's concentration using the osmotic pressure. The temperature, 20.0 °C, corresponds to 293 K. Recall that 1 Pa = 1 J m^{-3}.

$$\Pi = cRT$$

$$c = \frac{\Pi}{RT} = \frac{2.11 \times 10^3\ \text{Pa}}{8.314\ \text{J mol}^{-1}\text{K}^{-1} \times 293\ \text{K}} = 0.866\ \text{mol m}^{-3} = 8.66 \times 10^{-4}\ \text{mol L}^{-1}$$

Therefore, in 100 mL or 0.100 L of solution, we have:

$$n = cV = 8.66 \times 10^{-4}\ \text{mol L}^{-1} \times 0.100\ \text{L} = 8.66 \times 10^{-5}\ \text{mol}$$

The molar mass of the solute is then obtained in the usual way.

$$M = \frac{m}{n} = \frac{0.122\ \text{g}}{8.66 \times 10^{-5}\ \text{mol}} = 1.41 \times 10^3\ \text{g mol}^{-1}$$

Is our answer reasonable?

All we can do is double-check the calculation of the amount, which turns out to be a little less than 10^{-4} mol. Dividing the mass, roughly 0.12 g, by 10^{-4} mol gives 0.12×10^4 g mol^{-1}, or 1.2×10^3 g mol^{-1}, which is very roughly what we found.

PRACTICE EXERCISE 10.12

A solution of a carbohydrate prepared by dissolving 72.4 mg in 100 mL of solution has an osmotic pressure of 3.29×10^3 Pa at 25.0 °C. What is the molar mass of the compound?

Measurement of solute dissociation

The molal freezing point depression constant for water is 1.86 K mol^{-1} kg, and therefore you might think that a 1.00 mol kg^{-1} solution of NaCl would freeze at −1.86 °C. Instead, it freezes at about −3.37 °C. This greater depression of the freezing point by the salt, which is almost twice 1.86 °C, is not hard to understand if we remember that colligative properties depend on the number of solute particles. NaCl(s) undergoes complete **dissociation** into its constituent ions in water according to the equation:

$$NaCl(s) \rightarrow Na^+(aq) + Cl^-(aq)$$

Therefore, if we dissolve 1 mole of NaCl(s) (58.5 g) in 1 kg of water, the resulting solution has a total molality of *dissolved solute particles* of 2 mol kg^{-1}: 1 mol kg^{-1} of Na^+ ions and 1 mol kg^{-1} of Cl^- ions. Theoretically, a 1.00 mol kg^{-1} NaCl solution should freeze at 2 × (−1.86 °C) or −3.72 °C. (Why it actually freezes a little higher than this, at −3.37 °C, will be discussed shortly.)

If we made up a 1.00 mol kg^{-1} solution of $(NH_4)_2SO_4$, we would have to consider the following dissociation.

$$(NH_4)_2SO_4(s) \rightarrow 2NH_4^+(aq) + SO_4^{2-}(aq)$$

So, 1 mole of $(NH_4)_2SO_4$ gives a total of 3 moles of ions (2 moles of NH_4^+ and 1 mole of SO_4^{2-}). We expect the freezing point of this solution of $(NH_4)_2SO_4$ to be 3 × (−1.86 °C) = −5.58 °C.

You should ensure you understand the distinction between *dissolution* and *dissociation*. A slightly soluble ionic solid such as AgCl dissolves only to a very small extent (i.e. it does not undergo complete dissolution), but the small amount of resulting Ag^+ and Cl^- ions in solution are essentially completely dissociated (i.e. they show little tendency to stick together).

When we want to estimate a colligative property of a solution of an electrolyte, we recalculate the solution's molality assuming that the salt dissociates completely. This is illustrated in worked example 10.13.

WORKED EXAMPLE 10.13

Estimating the freezing point of a salt solution

Estimate the freezing point of an aqueous solution of 0.106 mol kg^{-1} $MgCl_2$, assuming that it dissociates completely.

Analysis

The equation that relates a change in freezing temperature to molality is:

$$\Delta T_f = K_f b$$

From table 10.5, K_f for water is 1.86 K mol^{-1} kg. We cannot simply use 0.106 mol kg^{-1} for the molarity because $MgCl_2$ is an ionic compound that dissociates in water; we must use the total molality of ions in the equation.

Solution

When $MgCl_2$ dissolves in water, it dissociates as follows.

$$MgCl_2(s) \rightarrow Mg^{2+}(aq) + 2Cl^-(aq)$$

Because 1 mole of $MgCl_2$ gives 3 moles of ions, the effective (assumed) molality of what we have called 0.106 mol kg^{-1} $MgCl_2$ is three times as great.

$$\text{effective molality} = 3 \times 0.106 \text{ mol kg}^{-1} = 0.318 \text{ mol kg}^{-1}$$

Now we can use $\Delta T_f = K_f b$.

$$\Delta T_f = (1.86 \text{ K mol}^{-1} \text{ kg})(0.318 \text{ mol kg}^{-1})$$
$$= 0.591 \text{ K} = 0.591 \text{ °C}$$

The freezing point is depressed below 0.000 °C by 0.591 °C, so we calculate that this solution freezes at −0.591 °C.

Is our answer reasonable?

The molality, as recalculated, is roughly 0.3, so 0.3 of 1.86 (call it 2) is 0.6, which, after we add the unit °C and subtract from 0 °C, gives −0.6 °C, which is close to the answer.

PRACTICE EXERCISE 10.13

Calculate the freezing point of an aqueous solution of 0.172 mol kg^{-1} $Al_2(SO_4)_3$ on the assumption that it is 100% dissociated. Calculate its freezing point assuming 0% dissociation.

Experiments show that neither the 1.00 mol kg^{-1} $(NH_4)_2SO_4$ nor the 1.00 mol kg^{-1} NaCl solution described above and on the previous page freezes at as low a temperature as calculated. Our assumption that an electrolyte separates *completely* into its ions is incorrect. Some oppositely charged ions exist as very closely associated pairs, called **ion pairs**, which behave as single 'molecules'. Clusters larger than two ions probably exist also. The formation of ion pairs and clusters makes the actual *particle* concentration in a 1.00 mol kg^{-1} NaCl solution somewhat less than 2.00 mol kg^{-1}. As a result, the freezing point depression of 1.00 mol kg^{-1} NaCl is not quite as large as that calculated on the basis of 100% dissociation.

As solutions of electrolytes are made more *dilute*, the observed and calculated freezing points come closer together. At ever greater dilutions, the association of ions is less of a complication because the ions are further apart, so the solutes behave more and more as if they were 100% separated into their ions.

Chemists compare the degrees of dissociation of electrolytes at different dilutions by a quantity called the **van't Hoff factor** (i). This is the ratio of the observed freezing point depression to the value calculated on the assumption that the solute dissolves as a nonelectrolyte.

$$i = \frac{(\Delta T_f)_{\text{measured}}}{(\Delta T_f)_{\text{calculated as a nonelectrolyte}}}$$

The theoretical van't Hoff factor, i, is 2 for NaCl, KCl and $MgSO_4$, which break up into two ions on 100% dissociation. For K_2SO_4, the theoretical value of i is 3 because each K_2SO_4 unit gives three ions. The actual van't Hoff factors for several electrolytes at different dilutions are given in table 10.6. Notice that with decreasing concentration (i.e. at higher dilutions) the experimental van't Hoff factors agree better with their corresponding theoretical van't Hoff factors.

TABLE 10.6 Effect of concentration on van't Hoff factor.

	van't Hoff factor (i)			
	Molal concentration (mol salt/kg water)			If 100% dissociation occurred
Salt	0.1	0.01	0.001	
NaCl	1.87	1.94	1.97	2.00
KCl	1.85	1.94	1.98	2.00
K_2SO_4	2.32	2.70	2.84	3.00
$MgSO_4$	1.21	1.53	1.82	2.00

The increase in the percentage of complete dissociation that comes with greater dilution is not the same for all salts. In going from molalities of 0.1 to 0.001 mol kg^{-1}, the increase in % dissociation of KCl, as measured by the change in i, is only about 7%. For K_2SO_4, the increase for the same dilution is about 22%, a difference caused by the anion SO_4^{2-}. This anion has twice the charge of the anion in KCl, so the SO_4^{2-} ion attracts K^+ more strongly than can Cl^-. Hence, letting an ion of 2– charge and an ion of 1+ charge get farther apart by dilution has a greater effect on their acting independently than giving ions of 1– and 1+ charge more room. When *both* cation and anion are doubly charged, the improvement in % dissociation with dilution is even greater. We can see from table 10.6 that there is an almost 50% increase in the value of i for $MgSO_4$ as we go from a 0.1 to a 0.001 mol kg^{-1} solution.

There are many substances in which dissociation is far from complete. For example, in 1.00 mol kg^{-1} aqueous acetic acid (a weak acid), the following equilibrium exists.

$$CH_3COOH(aq) + H_2O(l) \rightleftharpoons H_3O^+(aq) + CH_3COO^-(aq)$$

The solution freezes at −1.90 °C. This is only a little lower than that expected (−1.86 °C) if no ionisation occurred, which is evidence for only a small amount of ionisation. We can estimate % ionisation by the following calculation.

We first use the data to calculate the *apparent molality* of the solution: i.e. the molality of all dissolved species, CH_3COOH, CH_3COO^- and H_3O^+. We again use $\Delta T_f = K_f b$, but now letting b stand for the apparent molality. K_f is 1.86 K mol^{-1} kg so:

$$b = \frac{\Delta T_f}{K_f} = \frac{1.90\,\text{K}}{1.86\,\text{K mol}^{-1}\,\text{kg}}$$
$$= 1.02\,\text{mol kg}^{-1}$$

If there is 1.02 mol of all solute species in 1 kg of solvent, we have 0.02 mol of solute species more than we started with, because we began with 1.00 mol of acetic acid. Since we get just one *extra* particle for each acetic acid molecule that ionises, the extra 0.02 mol of particles must have come from the ionisation of 0.02 mol of acetic acid. So:

$$\%\text{ ionisation} = \frac{\text{mol of acid ionised}}{\text{mol of acid available}} \times 100\%$$
$$= \frac{0.02}{1.00} \times 100\%$$
$$= 2\%$$

In other words, this procedure estimates the % ionisation in 1.00 mol kg^{-1} acetic acid to be 2%. (Other kinds of measurements offer better precision, and the % ionisation of acetic acid at this concentration is actually less than 1%.) We will discuss the ionisation of acids in greater detail in chapter 11.

Some molecular solutes produce smaller colligative effects than their molal concentrations would lead us to predict. This is often evidence that solute molecules are clustering or undergoing **association** in solution. For example, when dissolved in benzene, benzoic acid molecules associate as **dimers** held together by hydrogen bonds.

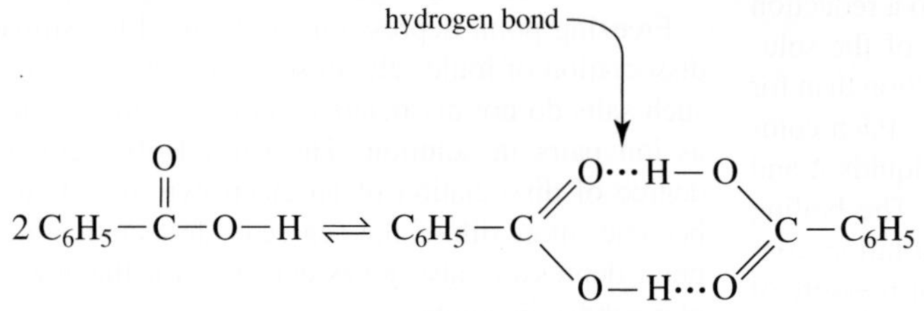

$$2\,C_6H_5-\overset{\displaystyle O}{\overset{\|}{C}}-O-H \rightleftharpoons C_6H_5-C\begin{matrix} \diagup\!\!\!\!\!\diagup O\cdots H-O \diagdown \\ \diagdown O-H\cdots O \diagup\!\!\!\!\!\diagup \end{matrix}C-C_6H_5$$

hydrogen bond

benzoic acid — benzoic acid dimer

Because of association, the depression of the freezing point of a 1.00 mol kg^{-1} solution of benzoic acid in benzene is only about half of the calculated value. By forming a dimer, benzoic acid has an effective molecular mass twice that normally calculated. The larger effective molar mass reduces the molal concentration of particles by a factor of two, and the effect on the freezing point depression is halved.

It is appropriate that we have finished this chapter with two examples involving acids. In the next chapter, we will look at the solution properties of acids and bases in detail, and we will see that the behaviour of these compounds in solution is governed by the principles we have outlined in this and the previous two chapters.

SUMMARY

LO 1 Introduction to solutions and solubility

A solution is a homogeneous mixture of two or more pure substances that can be gases, liquids or solids. The liquid in a liquid solution is called the solvent, and the dissolved substance is the solute. A saturated solution contains the maximum possible amount of dissolved solute at the given temperature and pressure.

LO 2 Gaseous solutions

All gases mix spontaneously in all proportions. As the enthalpy change for mixing is usually small, owing to the very low interatomic or intermolecular forces between the gas atoms or molecules, the mixing process is essentially entropy driven.

LO 3 Liquid solutions

In contrast to gaseous solutions, liquid solutions have significant intermolecular interactions involving solvent molecules so they generally have significant enthalpies of solution, $\Delta_{sol}H$. The two contributions to $\Delta_{sol}H$ when a gas dissolves in a liquid are the energy required to open 'pockets' in the solvent and the energy released when gas molecules occupy these pockets. $\Delta_{sol}H$ values are usually positive for gases in organic solvents and negative for gases in water. The solubilities of gases in water usually decrease with increasing temperature. Henry's law, $c_{gas} = k_H p_{gas}$, states that the concentration of a gas in a liquid at constant temperature is directly proportional to the partial pressure of the gas above the solution.

Two liquids that mix completely in all proportions are said to be miscible, while those that do not mix are immiscible. Liquids of similar polarities are often miscible; in other words, 'like dissolves like'. Dissolution of an ionic compound in water gives the hydrated free ions, a specific form of solvation. Polar molecules are generally soluble in polar solvents, and vice versa.

The molar enthalpy of solution of a solid is the enthalpy change when 1 mole of a crystalline substance is dissolved in a solvent. The solubility of a solid in a liquid is temperature dependent and generally increases with increasing temperature.

LO 4 Quantification of solubility: the solubility product

The solubility product, K_{sp}, is the equilibrium constant for the dissolution of a slightly soluble salt in a solvent (usually water). For the general equilibrium:

$$M_aX_b(s) \rightleftharpoons aM^{b+}(aq) + bX^{a-}(aq)$$

it is written as $K_{sp} = [M^{b+}]^a[X^{a-}]^b$. The molar solubility ($s$) of a salt at a specified temperature is the amount of the salt dissolved in 1 litre of a saturated solution at that temperature. The relationship between s and K_{sp} depends on the nature of the salt. For a 1 : 1 salt, $K_{sp} = s^2$, for a 1 : 2 salt, $K_{sp} = 4s^3$, for a 1 : 3 salt, $K_{sp} = 27s^4$ and, for a 2 : 3 salt, $K_{sp} = 108s^5$.

The effect of adding a common ion to a saturated solution of a slightly soluble salt can be calculated by comparing Q_{sp} (sometimes called the ionic product) with K_{sp}. We can use the same method to predict if a precipitate will form when two solutions are mixed. The common ion effect shows that salts are always less soluble in the presence of a common ion, providing no further chemical reaction is possible.

LO 5 Colligative properties of solutions

A solution containing a nonvolatile solute boils at a higher temperature and freezes at a lower temperature than the pure solvent. This can be explained in terms of the colligative properties of the solvent. In so doing, instead of expressing solution composition in terms of molar concentration or molarity, it is necessary to use molal concentration or molality (b).

We use mole fractions, x, when explaining the effect of a solute on boiling and freezing points. Raoult's law, $p_{\text{solution}} = x_{\text{solvent}}p^*_{\text{solvent}}$, shows that the pressure above a solution is proportional to the mole fraction of the solvent in a dilute solution. This is due to a reduction in the number of molecules of solvent at the surface of the solution, which leads to a lower vapour pressure for the solution than for the pure solvent. An ideal solution obeys Raoult's law. Each component of an ideal solution consisting of two volatile liquids A and B obeys Raoult's law, such that $p_{\text{total}} = x_A p^*_A + x_B p^*_B$. The boiling point elevation and freezing point depression of a solution containing a nonvolatile solute is due to the reduced vapour pressure of the solution relative to that of the pure solvent. The magnitudes of these effects can be determined using the equations $\Delta T_b = K_b b$ and $\Delta T_f = K_f b$, where K_b and K_f are the molal boiling point elevation constant and the molal freezing point depression constant of the solvent, respectively, and b is the molality of the solution (mol kg^{-1}).

Dialysis is the passage of water and small solute particles through a semipermeable membrane, while osmosis is the passage of solvent only. Osmosis occurs through an osmotic membrane and results in a net shift of solvent from the side less concentrated in solute to the side more concentrated. The osmotic pressure (Π) is that required to stop this net flow of solvent when one of the liquids is pure water. The van't Hoff equation for osmotic pressure is $\Pi V = nRT$ and is analogous to the ideal gas equation.

Freezing point depression can be used to estimate the degree of dissociation of ionic salts in solution. These results often show that such salts do not dissociate completely but exist to a certain extent as ion pairs in solution. The van't Hoff factor (i) measures the degree of dissociation of an electrolyte in solution. As a solution becomes more dilute, the degree of dissociation increases. Freezing point depression also gives evidence for the association of molecular solutes in solution.

KEY CONCEPTS AND EQUATIONS

Henry's law
(section 10.3)

This law:

$$c_{\text{gas}} = k_{\text{H}} p_{\text{gas}}$$

allows calculation of the solubility of a gas in a solvent at a given pressure from its solubility at another pressure.

Like-dissolves-like rule
(section 10.3)

This rule uses chemical composition and structure to predict whether two substances can form a solution.

Solubility product constant, K_{sp}
(section 10.4)

We use K_{sp} to calculate the molar solubility of a salt, either in pure water or in a solution that contains a common ion. Comparison of K_{sp} with Q_{sp} allows us to decide whether a precipitate will form on mixing ionic solutions.

Molar solubility
(section 10.4)

We can use molar solubility to calculate the value of K_{sp} for a slightly soluble salt.

Common ion effect
(section 10.4)

The presence of a common ion in a solution lowers the solubility of a salt containing that common ion in the solution.

Molal concentration (molality)
(section 10.5)

The law:

$$\text{molality} = \frac{\text{moles of solute}}{\text{kg of solvent}}$$

provides a temperature-independent concentration expression for use with colligative properties.

Raoult's law
(section 10.5)

This expression:

$$p_{\text{solution}} = x_{\text{solvent}} p^*_{\text{solvent}}$$

is used to calculate the effect of a solute on the vapour pressure of a solution.

Freezing point depression (ΔT_f) and boiling point elevation (ΔT_b)
(section 10.5)

The expressions:

$$\Delta T_f = K_f b$$

and:

$$\Delta T_b = K_b b$$

enable us to estimate freezing and boiling points, molar masses and % dissociation of a weak electrolyte.

Equation for osmotic pressure
(section 10.5)

We use this equation:

$$\Pi V = nRT$$

to estimate the osmotic pressure of a solution or to calculate molar masses from osmotic pressure and concentration data.

KEY TERMS

REVIEW QUESTIONS

LO 1 Introduction to solutions and solubility

10.1 Define the following terms.
(a) solution
(b) solvent
(c) solute
(d) solubility
(e) saturated solution
(f) dissolution

LO 2 Gaseous solutions

10.2 Why do two gases mix spontaneously when they are brought into contact?

LO 3 Liquid solutions

10.3 When substances form liquid solutions, what two factors determine the solubility of the solute in the solvent?

10.4 Methanol, CH_3OH, and water are miscible in all proportions. What does this mean? Explain how the —OH unit in methanol contributes to this.

10.5 Petrol and water are immiscible. What does this mean? Explain why they are immiscible in terms of structural features of their molecules and forces of attraction between them.

10.6 Iodine, I_2, is only very slightly soluble in water but dissolves readily in carbon tetrachloride, a nonpolar solvent. Why do you think this is?

10.7 Iodine dissolves far better in ethanol (to form 'tincture of iodine', an old antiseptic) than in water. What does this tell us about the alcohol molecules compared with water molecules?

10.8 The value of $\Delta_{sol}H$ for LiBr is $-67\ kJ\ mol^{-1}$. If an aqueous solution of LiBr is prepared in an insulated container, will the system's temperature increase or decrease as the LiBr dissolves?

10.9 The Na^+ and Ca^{2+} ions are about the same size. Which would be expected to have the larger hydration enthalpy? Why?

10.10 The value of $\Delta_{sol}H$ for the formation of an acetone–water solution is negative. Explain this in terms of intermolecular forces of attraction.

10.11 The value of $\Delta_{sol}H$ for the formation of an ethanol–hexane solution is positive. Explain this in terms of intermolecular forces of attraction.

10.12 If the value of $\Delta_{sol}H$ for the formation of a mixture of two liquids A and B is 0, what does this imply about the relative strengths of A—A, B—B and A—B intermolecular attractions?

10.13 If a saturated solution of NH_4NO_3 in 100 g of solvent at 70 °C is cooled to 10 °C, what mass of solute will precipitate? (Use data in figure 10.15.)

10.14 Use the data in figure 10.15, to determine which salt would precipitate first if a hot concentrated solution in which equal masses of NaBr and NaCl are dissolved is slowly cooled from 50 °C.

10.15 Fishermen know that, on hot summer days, the largest fish are found in deep sinks in lake bottoms, where the water is coolest. Use the temperature dependence of oxygen solubility in water to explain why.

10.16 What is Henry's law?

10.17 Mountain streams often contain fewer living things than equivalent streams at sea level. In terms of oxygen solubility at different pressures, give one reason why this might be true.

10.18 Why does a bottled carbonated beverage fizz when you take the cap off?

LO 4 Quantification of solubility: the solubility product

10.19 What is the difference between Q_{sp} and K_{sp}?

10.20 Illustrate the common ion effect by showing what happens to the value of Q_{sp} when a small volume of concentrated NaCl solution is added to a saturated solution of AgCl.

10.21 With respect to K_{sp}, what conditions must be met before a precipitate forms in a solution?

LO 5 Colligative properties of solutions

10.22 How does the molality of a solution vary with increasing temperature? How does the molarity of a solution vary with increasing temperature?

10.23 Suppose a $1.0\ mol\ kg^{-1}$ solution of a solute is made using a solvent with a density of $1.15\ g\ mL^{-1}$. Is the molarity of this solution numerically greater than or less than 1.0? Explain.

10.24 What specific fact about a physical property of a solution must be true to call it a colligative property?

10.25 Viewed at a molecular level, what causes a solution containing a nonvolatile solute to have a lower vapour pressure than the solvent at the same temperature?

10.26 What kinds of data are needed to find out if a solution of two miscible liquids is almost exactly an ideal solution?

10.27 When an aqueous solution of sodium chloride starts to freeze, why don't the ice crystals contain ions of the salt?

10.28 Explain why a nonvolatile solute dissolved in water gives the system:
(a) a higher boiling point than water
(b) a lower freezing point than water.

10.29 Why do we call dialysing and osmotic membranes 'semipermeable'? What is the opposite of 'permeable'?

10.30 What is the key difference between dialysing and osmotic membranes?

10.31 What is meant by 'dialysis'?

10.32 At a molecular level, explain why in osmosis there is a net migration of solvent from the side of the membrane less concentrated in solute to the side more concentrated in solute.

10.33 Two glucose solutions of unequal molarity are separated by an osmotic membrane. Which solution *loses* water, the one with the higher or the lower molarity?

10.34 What is the difference between a hypertonic solution and a hypotonic solution?

10.35 Why are colligative properties of solutions of ionic compounds usually more pronounced than those of solutions of molecular compounds of the same molalities?

10.36 What is the van't Hoff factor? What is its expected value for all nondissociating molecular solutes? If its measured value is slightly larger than 1.0, what does this suggest about the solute? What is suggested by a van't Hoff factor of approximately 0.5?

10.37 Which aqueous solution, if either, is likely to have the higher boiling point, 0.50 mol kg^{-1} NaI or 0.50 mol kg^{-1} Na_2CO_3?

REVIEW PROBLEMS

10.38 Consider the formation of a solution of aqueous lithium chloride. Write the thermochemical equations for the:
(a) conversion of solid LiCl into its gaseous ions
(b) subsequent formation of the solution by hydration of the ions. The lattice enthalpy of LiCl is 833 kJ mol^{-1}, and the hydration enthalpy of the ions is −883 kJ mol^{-1}.
(c) Calculate the enthalpy of solution of LiCl in kJ mol^{-1}. **LO3**

10.39 If the lattice enthalpy of a particular ionic compound is 720 kJ mol^{-1} and its enthalpy of solution in water is −22 kJ mol^{-1}, estimate the hydration enthalpy of the ions in the compound. **LO3**

10.40 If the enthalpy of solution of an ionic compound in water is −50 kJ mol^{-1}, and the hydration enthalpy is −890 kJ mol^{-1}, estimate the lattice enthalpy of the ionic compound. **LO3**

10.41 The solubility of methane, the chief component of natural gas, in water at 20 °C and 1×10^5 Pa pressure is 0.025 g L^{-1}. What is its solubility in water at 1.25×10^5 Pa and 20 °C? **LO2**

10.42 At 9.74×10^4 Pa and 20 °C, nitrogen has a solubility in water of 0.018 g L^{-1}. At 8.16×10^4 Pa and 20 °C, its solubility is 0.015 g L^{-1}. Show that nitrogen obeys Henry's law. **LO4**

10.43 A particular gas has a solubility in water of 0.011 g L^{-1} at 25 °C. Under these conditions, the partial pressure of the gas over the solution is 0.80×10^5 Pa. Use these data to determine the solubility of the gas at the same temperature but at double the pressure. **LO2**

10.44 Write the K_{sp} expressions for each of the following compounds. (a) AgCN (b) $Mg(OH)_2$ (c) SrF_2 (d) CuBr (e) $BaSO_3$ (f) $V(OH)_3$ (g) Ag_2SO_4 (h) $Ca_3(PO_4)_2$ (i) $AuCl_3$ **LO4**

10.45 As we saw on the front page of this chapter, barium sulfate, $BaSO_4$, can be safely ingested despite the fact that the Ba^{2+} ion is toxic. This is due to the very low solubility of $BaSO_4$, with only 0.002 45 g of $BaSO_4$ dissolving per litre of water at 25 °C. Calculate K_{sp} for $BaSO_4$. **LO4**

10.46 In an experiment to determine the solubility of silver acetate, $AgOOCCH_3$ in water, it was found that exactly 8.00 g of the salt would dissolve in 1.00 L of water at 25 °C. Using these data, calculate the molar solubility and K_{sp} for silver acetate at this temperature. **LO4**

10.47 When 269 mL of a saturated solution of CuCl was evaporated to dryness, 0.0116 g of solid CuCl was obtained. What is the value of K_{sp} for this salt? **LO4**

10.48 At 25 °C, the molar solubility of Ag_3PO_4 is 1.8×10^{-5} mol L^{-1}. Calculate K_{sp} for this salt. **LO4**

10.49 The molar solubility of $AuCl_3$ in water at 25 °C is 3.3×10^{-7} mol L^{-1}. What is the value of K_{sp} for this salt? **LO4**

10.50 Use the data in table 10.3 to determine the molar solubilities of the following compounds in water. (a) $PbBr_2$ (b) Ag_2CrO_4 (c) PbI_2 **LO4**

10.51 At 25 °C, K_{sp} for AgCl = 1.8×10^{-10} while that for Ag_2CO_3, 8.1×10^{-12}, is smaller. Which salt, AgCl or Ag_2CO_3, has the larger molar solubility in water? **LO4**

10.52 At 25 °C, the value of K_{sp} for AgCN is 2.2×10^{-16} and that for $Zn(CN)_2$ is 3×10^{-16}. In terms of grams per 100 mL of solution, which salt is the more soluble in water? **LO4**

10.53 A salt with a formula MX has $K_{sp} = 2.7 \times 10^{-7}$. What value of K_{sp} must another slightly soluble salt, MX_3, have if the molar solubilities of the two salts are to be identical? **LO4**

10.54 Calcium sulfate is found in plaster. At 25 °C, the value of K_{sp} for $CaSO_4$ is 2.4×10^{-5}. What is the calculated molar solubility of $CaSO_4$ in water? **LO4**

10.55 Silver chloride, AgCl, has $K_{sp} = 1.8 \times 10^{-10}$ at 25 °C. Calculate the molar solubility of AgCl in: (a) pure water, (b) 0.0100 M HCl(aq) solution, (c) 0.100 M HCl(aq) solution and (d) 0.100 M $MgCl_2$(aq) solution at 25 °C. Assume that AgCl(s) does not react with excess Cl^- to form $[AgCl_2]^-$. **LO4**

10.56 Gold(III) chloride, $AuCl_3$, has $K_{sp} = 3.2 \times 10^{-25}$ at 25 °C. Calculate the molar solubility of $AuCl_3$ in: (a) pure water, (b) 0.010 M HCl solution, (c) 0.010 M $MgCl_2$ solution and (d) 0.010 M $Au(NO_3)_3$ solution at 25 °C. Assume that $AuCl_3$ does not react with any of the compounds in (b), (c) or (d). **LO4**

10.57 Calculate the molar solubility of PbI_2 at 25 °C in: (a) a 0.200 M $Pb(NO_3)_2$(aq) solution and (b) a 0.200 M MgI_2(aq) solution. PbI_2 has $K_{sp} = 7.9 \times 10^{-9}$ at 25 °C. **LO4**

10.58 What is the molar solubility of $Mg(OH)_2$ in 0.20 M NaOH at 25 °C? (For $Mg(OH)_2$, $K_{sp} = 7.1 \times 10^{-12}$ at 25 °C.) **LO4**

10.59 Calculate the molar solubility of $CaCO_3$ in a 0.012 M $CaBr_2$(aq) solution at 25 °C. **LO4**

10.60 Calculate the molar solubility of AgCl in 0.050 M $AlCl_3$ at 25 °C. **LO4**

10.61 What mass of $Fe(OH)_2$ would be formed if 2.75 g of NaOH(s) was added to 200 mL of 0.10 M $FeCl_2$(aq) solution at 25 °C? What is the molar concentration of Fe^{2+} in the final solution? **LO4**

10.62 Suppose that 1.75 g of NaOH(s) is added to 250 mL of 0.10 M $NiCl_2$ solution at 25 °C. What mass, in grams, of $Ni(OH)_2$ is formed? **LO4**

10.63 Does a precipitate of $PbCl_2$ form when 0.0150 mol of $Pb(NO_3)_2$ and 0.0120 mol of NaCl is dissolved in 1.00 L of solution at 25 °C? **LO5**

10.64 Silver acetate, $AgOOCCH_3$, has $K_{sp} = 2.3 \times 10^{-3}$ at 25 °C. Determine if a precipitate will form when 0.028 mol of $AgNO_3$ and 0.31 mol of $Ca(OOCCH_3)_2$ are dissolved in a total volume of 1.00 L of solution at 25 °C. **LO4**

10.65 Does a precipitate of $PbBr_2$ form if 50.0 mL of 0.0100 M $Pb(NO_3)_2$ is mixed with (a) 50.0 mL of 0.0100 M KBr and (b) 50.0 mL of 0.100 M NaBr at 25 °C? **LO5**

10.66 Does a precipitate of barium sulfate, $BaSO_4$, form if 17.5 mL of 0.100 M $Ba(NO_3)_2(aq)$ is added to 42.7 mL of 0.0310 M $Na_2SO_4(aq)$ at 25 °C? ($BaSO_4$ has $K_{sp} = 1.1 \times 10^{-10}$ at 25 °C.) **LO4**

10.67 Determine the molality of a 3.000 mol L^{-1} aqueous solution of NaCl that has a density of 1.07 g mL^{-1}. **LO5**

10.68 What is the molality of a 0.104 mol L^{-1} solution of acetic acid, CH_3COOH, that has a density of 1.00 g mL^{-1}? **LO5**

10.69 What is the molal concentration of glucose, $C_6H_{12}O_6$, (a sugar found in many fruits) in a solution made by dissolving 24.0 g of glucose in 1.00 kg of water? What is the mole fraction of glucose in the solution? **LO5**

10.70 An aqueous solution of propan-2-ol, $CH_3CH(OH)CH_3$, has a mole fraction of alcohol equal to 0.194. What is the molality of the alcohol? **LO5**

10.71 An aqueous solution of $NaNO_3$ has a molality of 0.363 mol kg^{-1} and a density of 1.0185 g mL^{-1}. Calculate the molar concentration of $NaNO_3$ and mole fraction of $NaNO_3$ in the solution. **LO5**

10.72 The vapour pressure of water is 3.13×10^3 Pa at 25 °C. What is the vapour pressure of a solution prepared by dissolving 33.0 g of the nonvolatile solute $C_6H_{12}O_6$ in 225 g of water? Assume the solution is ideal. **LO5**

10.73 At 25 °C, the vapour pressures of benzene, C_6H_6, and toluene, C_7H_8, are 1.23×10^4 and 3.54×10^3 Pa, respectively. A solution is prepared by mixing 60.0 g of benzene and 40.0 g of toluene. At what pressure will this solution boil? **LO5**

10.74 Pentane (C_5H_{12}, density = 0.626 g mL^{-1}) and heptane (C_7H_{16}, density = 0.684 g mL^{-1}) are two liquid hydrocarbons present in petrol. At 20 °C, the vapour pressure of pentane is 5.53×10^4 Pa and the vapour pressure of heptane is 4.74×10^3 Pa. If equal volumes of the two liquids are mixed, what is the total vapour pressure of the resulting solution? **LO5**

10.75 The vapour pressure of pure methanol, CH_3OH, at 30 °C is 2.11×10^4 Pa. What mass of the nonvolatile solute glycerol, $HOCH_2CH(OH)CH_2OH$, must be added to 100 g of methanol to obtain a solution with a vapour pressure of 1.84×10^4 Pa? **LO5**

10.76 A solution containing 6.8 g of a nonvolatile, nondissociating substance dissolved in 1 mole of chloroform, $CHCl_3$, has a vapour pressure of 6.78×10^4 Pa. The vapour pressure of pure $CHCl_3$ at the same temperature is 6.92×10^4 Pa. Calculate: (a) the mole fraction of the solute, (b) the amount of solute in the solution and (c) the molar mass of the solute. **LO5**

10.77 At 21.0 °C, a solution of 18.26 g of a nonvolatile, nonpolar compound dissolved in 33.25 g of bromoethane, CH_3CH_2Br, has a vapour pressure of 4.42×10^4 Pa. The vapour pressure of pure bromoethane at this temperature is 5.26×10^4 Pa. What is the molar mass of the compound? **LO5**

10.78 The sugar in ice-cream lowers the freezing point of the mixture and can make it difficult to keep ice-cream frozen hard. Estimate the boiling and freezing points of a 2.50 mol kg^{-1} solution of sugar in water. **LO5**

10.79 A solution is made by dissolving 52.7 g of the nonvolatile liquid glycerol ($HOCH_2CH(OH)CH_2OH$) in 250 g of water. Calculate the approximate freezing point of the solution and the boiling point of the solution. **LO5**

10.80 What mass of urea, NH_2CONH_2, is needed to lower the freezing point of 100 g of water by 2.50 °C? **LO5**

10.81 What is the boiling point of the solution described in question 10.80? **LO5**

10.82 A solution containing 15.00 g of an unknown nondissociating compound dissolved in 235.0 g of benzene freezes at 2.93 °C. Calculate the molar mass of the unknown, given that the normal freezing point of benzene is 5.45 °C. **LO5**

10.83 A solution of 75.0 g of a nonvolatile, nondissociating compound in 1.00 kg of benzene boils at 81.7 °C. Calculate the molar mass of the unknown, given that the normal boiling point of benzene is 80.1 °C. **LO5**

10.84 When 7.76 g of a nondissociating molecular compound was dissolved in 1000 g of benzene, the resulting solution displayed a freezing point depression of 0.307 °C. If the empirical formula of the compound is C_4H_2N, what are its molar mass and molecular formula? **LO5**

10.85 The osmotic pressure of an aqueous solution of a compound having a concentration of 1.80 g L^{-1} at 25 °C was 2.94 Pa. Calculate the molar mass of the compound. **LO5**

10.86 A saturated solution of 0.400 g of a polypeptide in 1.00 L of an aqueous solution has an osmotic pressure of 4.92×10^2 Pa at 27 °C. What is the approximate molar mass of the polypeptide? **LO5**

10.87 The vapour pressure of water at 20 °C is 2.30×10^3 Pa. Calculate the vapour pressure at 20 °C of a solution made by dissolving 5.00 g of NaCl in 75.0 g of water. Assume that the solute dissociates completely to give an ideal solution. **LO5**

10.88 What mass of $AlCl_3$ would have to be dissolved in 150 mL of water to give a solution that has a vapour pressure of 5.09×10^3 Pa at 35 °C? (Assume complete dissociation of the solute and ideal solution behaviour. At 35 °C, the vapour pressure of pure water is 5.55×10^3 Pa.) **LO5**

10.89 Below are the concentrations of the most abundant ions in sea water.

Ion	Molality (mol kg^{-1})
chloride	0.566
sodium	0.486
magnesium	0.055
sulfate	0.029
calcium	0.011
potassium	0.011
hydrogen carbonate	0.002

Use these data to estimate the osmotic pressure of sea water at 25 °C. What is the minimum pressure needed to desalinate sea water by reverse osmosis? **LO5**

10.90 What is the expected freezing point of a 0.20 mol kg^{-1} solution of $CaCl_2$? (Assume complete dissociation.) **LO5**

10.91 A 0.050 mol kg^{-1} aqueous solution of mercury(I) nitrate has a freezing point of approximately −0.27 °C. Show how this suggests that the formula of the mercury(I) ion is Hg_2^{2+}. **LO5**

10.92 A 1.00 mol kg^{-1} aqueous solution of HF freezes at −1.91 °C. What is the percentage ionisation of HF in the solution? **LO5**

10.93 Calculate the percentage ionisation of a weak electrolyte HX if an aqueous solution of HX of molality 0.106 mol kg^{-1} has a freezing point of −0.212 °C. **LO5**

10.94 The van't Hoff factor for the solute in 0.100 mol kg^{-1} $NiSO_4(aq)$ is 1.19. What would this factor be if the solution behaved as if it were 100% dissociated? **LO5**

10.95 What is the expected van't Hoff factor for $Al_2(SO_4)_3$ in an aqueous solution, assuming 100% dissociation? **LO5**

10.96 A 0.118 mol kg^{-1} solution of LiCl has a freezing point of −0.415 °C. What is the van't Hoff factor for this solute at this concentration? **LO5**

10.97 Use the data in question 10.96 to calculate the approximate osmotic pressure of a 0.118 mol kg^{-1} solution of LiCl at 20 °C. **LO5**

ADDITIONAL EXERCISES

10.98 The 'bends' is a medical emergency caused when a scuba diver rises too quickly to the surface from a deep dive. The origin of the problem can be illustrated by the results of the calculations in this question. At 37 °C (normal body temperature), the solubility of N_2 in water is 0.015 g L^{-1} when its pressure over the solution is 1.0×10^5 Pa. Air is approximately 78 mol % N_2. **LO4**

(a) What amount of N_2 is dissolved per litre of blood (essentially an aqueous solution) when a diver inhales air at a pressure of 1.0×10^5 Pa?

(b) What amount of N_2 dissolves per litre of blood when the diver is submerged to a depth of approximately 30 m, where the total pressure of the air being breathed is 4.0×10^5 Pa?

(c) If the diver surfaces quickly, what volume of N_2 gas (in mL), in the form of tiny bubbles, is released into the bloodstream from each litre of blood (at 37 °C and 1.0×10^5 Pa)?

10.99 Deep-sea divers sometimes substitute helium for nitrogen in the air that they carry in their tanks to prevent the 'bends', because helium is much less soluble in blood than nitrogen. Use the simple molecular model of gas solubility discussed on page 393 to explain why helium is relatively insoluble in water. **LO2**

10.100 Use the simple molecular model of gas solubility discussed on pages 393–4 to explain why N_2 gas solubility in water decreases as the temperature of the solution rises from room temperature to about 70 °C, and then begins to increase. **LO2**

10.101 The vapour pressure of a mixture of 250 g of carbon tetrachloride, CCl_4, and 28.9 g of an unknown compound is 1.82×10^4 Pa at 30 °C. The vapour pressure of pure tetrachloromethane at 30 °C is 1.88×10^4 Pa, while that of the pure unknown is 1.39×10^4 Pa. What is the approximate molar mass of the unknown? **LO5**

10.102 Benzene reacts with hot concentrated nitric acid dissolved in sulfuric acid to give chiefly nitrobenzene, $C_6H_5NO_2$. A by-product is often obtained, which comprises 42.86% C, 2.40% H and 16.67% N (by mass). The boiling point of a solution of 5.5 g of the by-product in 45 g of benzene was 1.84 °C higher than that of benzene. **LO5**

(a) Calculate the empirical formula of the by-product.

(b) Calculate a molar mass of the by-product and determine its molecular formula.

10.103 Excess solid $Mn(OH)_2$ is added to a solution of 0.100 M $FeCl_2$. After reaction, what are the molar concentrations of Mn^{2+} and Fe^{2+} in the solution? (For $Mn(OH)_2$, $K_{sp} = 1.6 \times 10^{-13}$.) **LO4**

10.104 Suppose that 50.0 mL of 0.12 M $AgNO_3$ is added to 50.0 mL of 0.048 M NaCl solution. **LO4**

(a) What mass of AgCl forms?

(b) Calculate the final concentrations of all of the ions in the solution that is in contact with the precipitate.

(c) What percentage of the Ag^+ ions precipitates?

10.105 The high concentration of Ca^{2+} ions in hard water makes it difficult for soap and shampoos to lather properly. A sample of hard water was found to have 244 ppm (parts per million on a weight per volume ratio) Ca^{2+} ions. A 1.25 g sample of K_2CO_3 was dissolved in 1.00 L of this water. What is the new concentration of Ca^{2+} in ppm? (Assume that the addition of K_2CO_3 does not change the volume, and assume that the densities of the aqueous solutions involved are all 1.00 g mL^{-1}.) **LO5**

10.106 In question 10.38, we calculated the enthalpy of solution, $\Delta_{sol}H$, of LiCl(s). Suppose the lattice enthalpy of LiCl was just 2% greater than the value given in table 10.2. What then would be the calculated $\Delta_{sol}H$ of LiCl(s)? How would this new value of $\Delta_{sol}H$ compare with that given in the table? By what percentage is the new value different from the value in the table? (The aim of this calculation is to show that small percentage errors in two large numbers, such as lattice enthalpy and hydration enthalpy, can cause large percentage changes in the absolute difference between them.) **LO3**

10.107 The osmotic pressure of a dilute solution of a slightly soluble polymer in water was measured using the osmometer in figure 10.28. The difference in the heights of the liquid levels was 1.17 cm at 25 °C. Assume the solution has a density of 1.00 g mL^{-1}. **LO5**

(a) What is the osmotic pressure of the solution?

(b) What is the molarity of the solution?

(c) At what temperature would the solution freeze?

(d) Use your answers to (a)–(c) to explain why freezing point depression cannot be used to determine the molar masses of compounds composed of very large molecules.

10.108 Consider an aqueous 1.00 mol kg^{-1} solution of Na_3PO_4, a compound with useful detergent properties. Calculate the boiling point of the solution assuming that: **LO5**

(a) it does not ionise at all in solution

(b) the van't Hoff factor for Na_3PO_4 reflects 100% dissociation into its ions.

The 1.00 mol kg^{-1} solution boils at 100.83 °C at 1.0×10^5 Pa.

(c) Calculate the van't Hoff factor for the solute in this solution.

10.109 In an aqueous solution of KNO_3, the concentration is 0.9159 mol % of the salt. The solution's density is 1.0489 g mL^{-1}. Calculate the: (a) molal concentration of KNO_3, (b) percentage (w/w) of KNO_3 and (c) molarity of the KNO_3 in the solution. **LO5**

10.110 It was found that the molar solubility of $BaSO_3$ in 0.10 M $BaCl_2$ is 8.0×10^{-6} M. What is the value of K_{sp} for $BaSO_3$? **LO4**

10.111 Both AgCl and AgI are very slightly soluble salts, but the solubility of AgI is much less than that of AgCl, as can be seen by their K_{sp} values. Suppose that a solution contains both Cl^- and I^- with $[Cl^-] = 0.050$ M and $[I^-] = 0.050$ M at 25 °C. If solid $AgNO_3$ is added to 1.00 L of this mixture (so that no appreciable change in volume occurs), what is the value of $[I^-]$ when AgCl first begins to precipitate? **LO5**

11 Acids and bases

Of all the planets and moons in our solar system, the Earth is the only one that has been found to support life as we know it. This is due to a combination of factors, including the distance from the Sun, an atmosphere containing oxygen, and the presence of plentiful amounts of water that is neither too acidic nor too basic. For many years, it was wondered if life existed on Venus, our closest planetary neighbour, as it is about the same size as Earth and has an atmosphere. However, a number of probes have shown that, in addition to the enormous surface pressure (about 100 times that on Earth) and high surface temperature (~460 °C), the upper atmosphere contains significant amounts of sulfuric acid, a strong, corrosive acid, inhalation of which would be fatal to any pioneering astronauts. Attention has therefore turned to Mars as a target for human exploration.

However, there are some forms of life on Earth, called acidophiles, which can exist, and indeed thrive, in highly acidic environments with pH values approaching 0. These are often found in geothermal areas in which the concentration of sulfur-containing compounds is high. One such acidophile is the alga *Cyanidium caldarium*, whose green colour is often seen around New Zealand's thermal areas. The photograph shows the crater lake on New Zealand's White Island, with the green colour arising, in part, from the presence of this alga. *Cyanidium caldarium* has managed to adapt to the highly acidic conditions by being able to pump acid out of its cells and using acid-resistant polysaccharide molecules to protect the cells from the acidic environment.

In this chapter, we will discuss a number of aspects of acids and bases. We will investigate the various definitions of acids and bases, show that there is an enormous range of acid and base strengths and that these can be rationalised on the basis of molecular structure, quantify Brønsted–Lowry acid and base behaviour using the concepts of pH, K_a and K_b, and study titrations of acids and bases.

LEARNING OBJECTIVES

After studying this chapter, you should be able to:

11.1 define acids and bases

11.2 identify acid–base reactions in water

11.3 perform calculations involving strong acids and bases

11.4 perform calculations involving weak acids and bases

11.5 explain the relationship between structure and acidity

11.6 perform calculations involving buffer solutions

11.7 analyse acid–base titrations

11.8 identify Lewis acids and bases.

11.1 The Brønsted–Lowry definition of acids and bases

In chapter 10, we looked at the nature of solutes and showed how freezing point depression measurements give us information about the degree of dissociation of a particular solute. Similar information can also be obtained by measurement of the conductivity of a solution, and this technique is both simpler to carry out and more sensitive than freezing point depression. It is effective because only solutions that contain ions will conduct an electric current. The magnitude of the electric current depends on the concentration and nature of the ions in the solution. A high concentration of ions allows the passage of a larger current, and therefore gives a larger conductivity than a low concentration, while a pure solvent that contains no ions will have a conductivity of zero.

We tend to think of water as consisting solely of H_2O molecules. Each molecule comprises a central oxygen atom that is covalently bound to two hydrogen atoms, and we might therefore expect that liquid water would have no ionic character. However, measurements show that pure water does exhibit a small conductivity, consistent with the presence of ions. (This is the reason that, unlike the unfortunate French singer Claude François, writer of the song 'My Way', you should never try to change a light bulb while standing in the bath.) These ions arise from the reaction of water with itself, according to the following equation.

$$H_2O(l) + H_2O(l) \rightleftharpoons H_3O^+(aq) + OH^-(aq)$$

The net result of this reaction is the transfer of a proton (H^+, a hydrogen atom minus an electron) from one molecule of water to another, and formation of the ionic species H_3O^+ (the **hydronium ion**) and OH^- (the **hydroxide ion**). Reactions such as this, which involve the transfer of a single proton from one species to another are called **acid–base reactions**. The reaction of water with itself represents arguably the simplest example of such a reaction. In this case, water acts as both an acid and a base.

FIGURE 11.1 Svante Arrhenius proposed the first comprehensive theory of acids and bases.

The concept of acids and bases has been known for hundreds of years, but it is only relatively recently that definitions of the words 'acid' and 'base' have been agreed upon. The word acid derives from the Latin *acidus*, meaning sour; acids do indeed taste sour, but it is not a good idea to taste them in the laboratory. The word alkali, which is synonymous with base, comes from the Arabic *al-qaliy*, literally 'the ashes', referring to the ashes of the saltwort plant that grows in alkaline soils. The first comprehensive theory concerning acids and bases appeared in 1884 in the PhD thesis of the Swedish chemist, Svante Arrhenius (1859–1927, Nobel Prize in chemistry, 1903; figure 11.1), who was nearly failed for proposing that ions could exist in solution. He defined an acid as a substance that released H^+ ions when dissolved in water, and a base as a substance that gave rise to OH^- ions on dissolution in water. This definition is similar to the more general **Brønsted–Lowry** definition of acids and bases, which we will use from this point. This was proposed by Johannes Brønsted, a Danish chemist, and Thomas Lowry, an English chemist, independently of each other in 1923. They considered that acid–base reactions involved proton (H^+) transfer between an acid and a base, which they defined as follows.

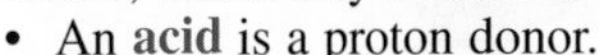

- An **acid** is a proton donor.
- A **base** is a proton acceptor.

In other words, a Brønsted–Lowry acid will donate a proton to a Brønsted–Lowry base, which will accept it. From now on, we will refer to Brønsted–Lowry acids and bases simply as acids and bases. To illustrate the Brønsted–Lowry concept, let's return to the example involving water that we introduced previously. We stated that water acted as both an acid and a base, and we can now explain this with reference to the Brønsted–Lowry definition.

One water molecule acts as a proton donor, and is therefore an acid. This water molecule becomes OH^- on the right-hand side of the equation. The other water molecule accepts the donated proton, and therefore acts as a base. This water molecule is converted to H_3O^+ on the right-hand side of the equation. It is instructive to consider this reaction in terms of the structures shown in figure 11.2.

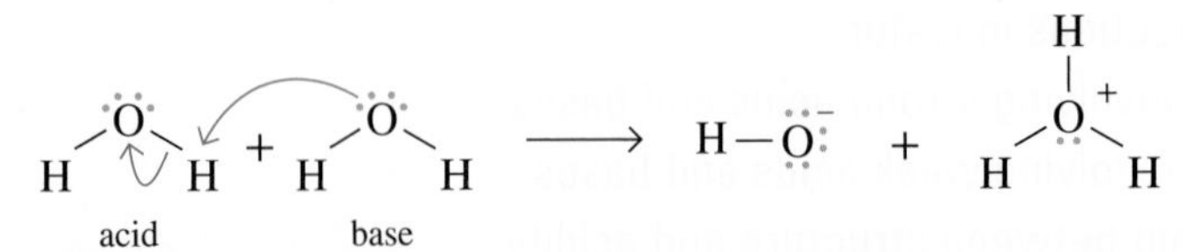

FIGURE 11.2 A mechanistic view of the acid–base reaction of one water molecule with another.

Obviously, in order for H_2O to act as an acid, it must, according to the Brønsted–Lowry definition, have a proton to donate. However, the mere presence of hydrogen atoms, which are potential protons, is not sufficient for a chemical species to act as an acid. For example, methane, CH_4, contains four hydrogen atoms but shows essentially no acidic properties. For a hydrogen atom to be measurably acidic, it must be bound to another atom via an appreciably polar bond, and therefore acids tend to contain hydrogen atoms bound to group 16 or 17 elements. Figure 11.2 also shows that, for H_2O to act as a base and accept the donated proton, it must have a lone pair of electrons to which the donated proton can covalently bond. The presence of one or more lone pairs is a prerequisite for a species to act as a base, but not all species containing lone pairs act as bases; for example, the chloride ion, Cl^-, contains four lone pairs, but displays negligible basic properties. Bases usually contain group 15 or 16 elements, the atoms of which (especially group 16) are often deprotonated and hence negatively charged. Table 11.1 lists a number of common acids and bases that would be found in any chemistry laboratory. You can see that the acids contain a very polar H—*X* bond, while the bases contain N and deprotonated O atoms with one and three lone pairs, respectively.

TABLE 11.1 Some common acids and bases.

Formula	Name	Structure
Acids		
HCl	hydrochloric acid[(a)]	H—Cl:
HNO_3	nitric acid	O=N(=O)—O—H
H_2SO_4	sulfuric acid	H—O—S(=O)(=O)—O—H
CH_3COOH	acetic acid	H—C(H)(H)—C(=O)—O—H
H_3PO_4	phosphoric acid	H—O—P(=O)(O—H)(O—H)
Bases		
NaOH	sodium hydroxide	Na^+ $^-$:O—H
NH_3	ammonia[(b)]	H—N(H)—H
C_5H_5N	pyridine	(pyridine ring, N:)
Na_2CO_3	sodium carbonate	Na^+ $^-$O—C(=O)—O$^-$ Na^+

(a) Pure HCl is a gas at room temperature, and is called hydrogen chloride. Hydrochloric acid refers to a solution of HCl in water.
(b) Pure NH_3 is a gas at room temperature. In the laboratory, it is often used as an aqueous solution.

It can be seen from the structures of the acids in table 11.1 that, in some cases, there is more than one very polar H—X bond, and these acids can donate more than one proton. Such acids are collectively called **polyprotic acids**. Hydrochloric acid, nitric acid and acetic acid, each of which can donate only one proton, are examples of **monoprotic acids**. Sulfuric acid can donate two protons and is called a **diprotic acid** whereas phosphoric acid, which can donate three protons, is a **triprotic acid**.

Similarly, some bases can accept more than one proton and are collectively called **polyprotic bases**. Hydroxide ion, ammonia and pyridine are examples of **monoprotic bases**, whereas carbonate ion is a **diprotic base**.

We now consider the reactions that occur when we place an acid or a base in water. We begin by using a chemical equation and a mechanistic diagram to see what happens when we dissolve gaseous hydrogen chloride in water.

$$H_2O(l) + HCl(g) \longrightarrow H_3O^+(aq) + Cl^-(aq)$$

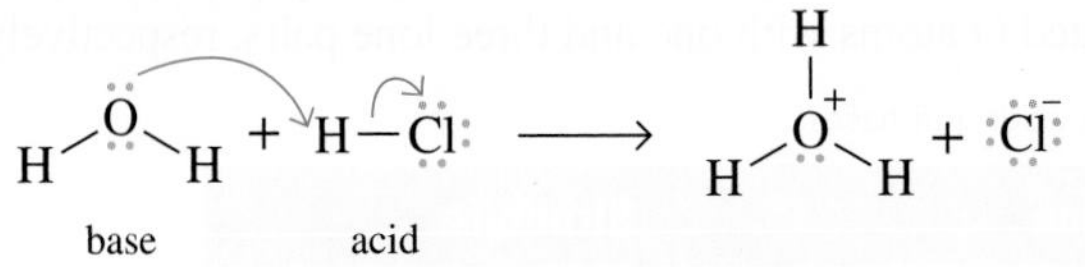

$$H{-}\ddot{O}{-}H + H{-}Cl \longrightarrow H_3O^+ + :Cl:^-$$

base acid

Here, HCl is the acid and H_2O is the base. HCl donates a proton to H_2O, thereby becoming Cl^-, while H_2O, on accepting the proton, becomes H_3O^+, the hydronium ion. This reaction goes essentially to completion and we say that HCl dissociates completely in aqueous solution.

Similarly, if we dissolve gaseous ammonia, NH_3, in water, the following reaction occurs.

$$NH_3(g) \quad + \quad H_2O(l) \longrightarrow NH_4^+(aq) \quad + \quad OH^-(aq)$$

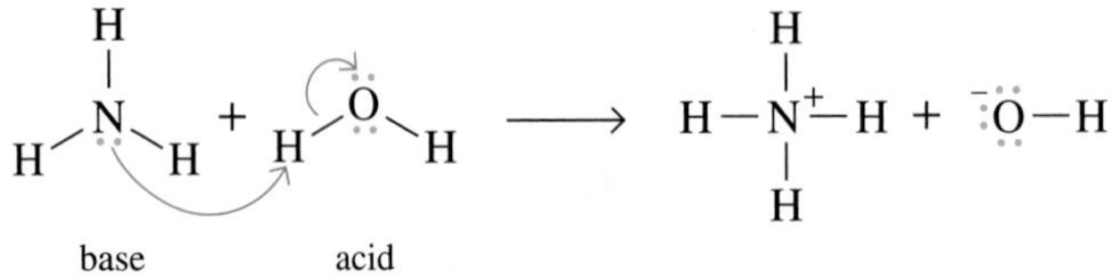

$$H_3\ddot{N} + H{-}\ddot{O}{-}H \longrightarrow H{-}\overset{+}{N}H_3 + {}^{-}:\ddot{O}{-}H$$

base acid

In this reaction, H_2O acts as an acid, donating a proton to NH_3 and thus becoming OH^-, while NH_3 acts as a base by accepting the proton from H_2O to give NH_4^+. However, this reaction differs from that for HCl in that it proceeds almost negligibly in the forward direction, and the reaction of ammonia with water is therefore far from complete. Note that water, the solvent in both reactions, acts as a base in the former reaction and an acid in the latter; compounds such as this that can act as either an acid or a base are called **amphiprotic**.

Conjugate acid–base pairs

It is informative to write the reverse equations for the reactions of HCl and NH_3 with water.

$$Cl^-(aq) + H_3O^+(aq) \rightarrow HCl(aq) + H_2O(l)$$
$$NH_4^+(aq) + OH^-(aq) \rightarrow NH_3(aq) + H_2O(l)$$

We can see that the reverse reactions also involve transfer of a proton, and are therefore themselves acid–base reactions. In the reaction:

$$Cl^-(aq) + H_3O^+(aq) \rightarrow HCl(aq) + H_2O(l)$$

H_3O^+ acts as an acid, donating a proton to Cl^-, which acts as a base by accepting it. However, this reaction proceeds almost imperceptibly in the forward direction; we will see why later in this chapter. Similarly, in the reaction:

$$NH_4^+(aq) + OH^-(aq) \rightarrow NH_3(aq) + H_2O(l)$$

NH_4^+ acts as an acid, donating a proton to the base OH^-. This illustrates a general feature of all Brønsted–Lowry acid–base reactions: *both the forward and reverse reactions are acid–base reactions*. Furthermore, there are always two sets of species on either side of the equation that differ only by a proton, and we call these **conjugate acid–base pairs**. In the reaction:

$$HCl(aq) + H_2O(l) \rightarrow H_3O^+(aq) + Cl^-(aq)$$

both HCl and Cl^-, and H_2O and H_3O^+ differ only by a proton. We showed that HCl acts as an acid in the forward reaction and Cl^- acts as a base in the reverse reaction. Therefore, Cl^- is the **conjugate base** of HCl, and HCl is the conjugate acid of Cl^-. Using the same reasoning, we can see that H_2O is the conjugate base of H_3O^+, and H_3O^+ is the conjugate acid of H_2O.

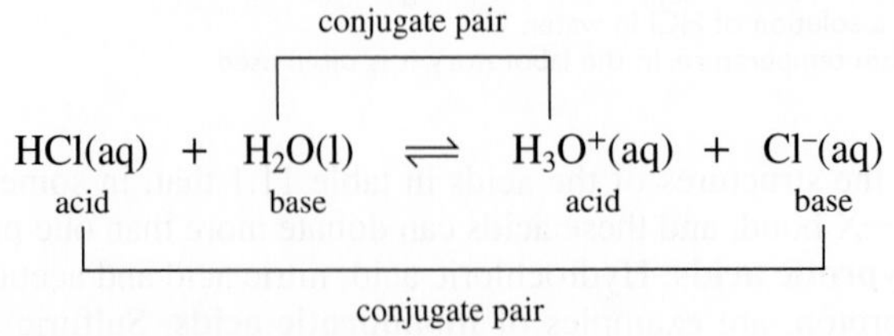

conjugate pair

$$\underset{\text{acid}}{HCl(aq)} + \underset{\text{base}}{H_2O(l)} \rightleftharpoons \underset{\text{acid}}{H_3O^+(aq)} + \underset{\text{base}}{Cl^-(aq)}$$

conjugate pair

As both the forward and reverse reactions are acid–base reactions, and occur to some extent, it is usual to write acid–base reactions as equilibria. We will look at how we quantify these equilibria later in this chapter.

WORKED EXAMPLE 11.1

Determining the conjugate bases of Brønsted–Lowry acids

What are the conjugate bases of nitric acid, HNO_3, and the hydrogen sulfate ion, HSO_4^-?

Analysis

The members of a conjugate acid–base pair differ by one proton, with the acid having the greater number of protons. To find the formula of the base, we remove one proton from the acid.

Solution

Removing one proton from HNO_3 leaves NO_3^-. The nitrate ion, NO_3^-, is thus the conjugate base of HNO_3. Removing a proton from HSO_4^- leaves its conjugate base, SO_4^{2-}.

Is our answer reasonable?

As a check, we can compare the two formulae in each pair.

$$HNO_3 \qquad NO_3^-$$
$$HSO_4^- \qquad SO_4^{2-}$$

In each case, the formula on the right has one fewer proton than the formula on the left, so it is the conjugate base. We have answered the question correctly.

PRACTICE EXERCISE 11.1

1. Write the formula of the conjugate base of each of the following Brønsted–Lowry acids.
   (a) H_2O (b) H_3PO_4 (c) $HClO_4$ (d) HF (e) H_2SO_3 (f) HIO_3
2. Write the formula of the conjugate acid of each of the following Brønsted–Lowry bases.
   (a) H_2O (b) NH_3 (c) F^- (d) OH^- (e) CO_3^{2-} (f) PO_4^{3-}

WORKED EXAMPLE 11.2

Identifying conjugate acid–base pairs in a Brønsted–Lowry acid–base reaction

The sodium hydrogen sulfate ion, HSO_4^-, reacts as follows with the phosphate ion, PO_4^{3-}.

$$HSO_4^-(aq) + PO_4^{3-}(aq) \rightleftharpoons SO_4^{2-}(aq) + HPO_4^{2-}(aq)$$

Identify the two conjugate acid–base pairs.

Analysis

There are two things that help us to identify the conjugate acid–base pairs in an equation. One is that the members of a conjugate pair differ by a proton. The second is that the members of each pair must be on opposite sides of the reaction arrow. In each pair, the acid has the larger number of protons.

Solution

Two of the formulae in the equation contain PO_4, so they must belong to the same conjugate pair. The one with the most protons, HPO_4^{2-}, is the acid, and the other, PO_4^{3-}, is the base. Therefore, one conjugate acid–base pair is HPO_4^{2-} and PO_4^{3-}. The other two ions, HSO_4^- and SO_4^{2-}, belong to the second conjugate acid–base pair; HSO_4^- is the conjugate acid and SO_4^{2-} is the conjugate base.

conjugate pair

$$\underset{\text{acid}}{HSO_4^-(aq)} + \underset{\text{base}}{PO_4^{3-}(aq)} \rightleftharpoons \underset{\text{base}}{SO_4^{2-}(aq)} + \underset{\text{acid}}{HPO_4^{2-}(aq)}$$

conjugate pair

Is our answer reasonable?

A check satisfies us that we have fulfilled the requirements that each conjugate pair has one member on each side of the arrow and that the members of each pair differ from each other by one (and only one) proton.

PRACTICE EXERCISE 11.2 One kind of baking powder contains sodium bicarbonate and calcium dihydrogen phosphate. When water is added, the following reaction occurs.

$$HCO_3^-(aq) + H_2PO_4^-(aq) \rightleftharpoons H_2CO_3(aq) + HPO_4^{2-}(aq)$$

Identify the two acids and the two bases in this reaction. (The H_2CO_3 decomposes to release CO_2, which causes the cake to rise.)

11.2 Acid–base reactions in water

In many of the examples we have considered so far, water has acted as both the solvent, and either the acid or the base. However, acid–base reactions can occur in any solvent, and can even occur in the absence of solvent, as shown in figure 11.3.

$$NH_3(g) + HCl(g) \longrightarrow NH_4Cl(s)$$

$$\underset{\text{base}}{H_2\ddot{N}H} + \underset{\text{acid}}{H{-}\ddot{\underset{..}{Cl}}:} \longrightarrow H{-}\overset{H}{\underset{H}{N^+}}{-}H \quad :\ddot{\underset{..}{Cl}}:^-$$

This is also a proton transfer reaction, in which gaseous HCl donates a proton to gaseous NH_3, as shown in the mechanistic diagram. Given that most acid–base reactions of interest occur in aqueous solution, we will concentrate on water as the solvent for the majority of our discussions.

FIGURE 11.3 The reaction of gaseous HCl with gaseous NH_3. As each gas escapes from its concentrated aqueous solution and mingles with the other, a cloud of tiny crystals of $NH_4Cl(s)$ forms above the bottles.

The autoprotolysis of water

We have shown that water can react with itself to generate hydronium and hydroxide ions according to the equation:

$$H_2O(l) + H_2O(l) \rightleftharpoons H_3O^+(aq) + OH^-(aq)$$

This reaction, in which a proton is transferred between identical molecules, is called **autoprotolysis** (sometimes called autoionisation). The extent of the autoprotolysis of water can be determined using the equilibrium constant for this process. Using the methods developed in chapter 9, and remembering that pure liquids do not appear in an equilibrium constant expression, we write the equilibrium constant for the equilibrium as:

$$K_w = [H_3O^+][OH^-] = 1.0 \times 10^{-14} \; (T = 25.0\,°C)$$

The extraordinary importance of this equilibrium means that the equilibrium constant is given the special symbol K_w and is called the **autoprotolysis constant of water**. Its very small value shows that the autoprotolysis of water proceeds only to a very small extent and that the equilibrium concentrations of H_3O^+ and OH^- in pure water are small. We can calculate just how small by using the value of K_w and realising from the stoichiometry of the reaction that the equilibrium concentrations of H_3O^+ and OH^- in pure water must be the same.

$$K_w = 1.0 \times 10^{-14} = [H_3O^+][OH^-]$$

If $[H_3O^+] = [OH^-]$, then:

$$[H_3O^+] = [OH^-] = \sqrt{1.0 \times 10^{-14}} = 1.0 \times 10^{-7}\ mol\,L^{-1}$$

Therefore, the $[H_3O^+]$ and $[OH^-]$ in pure water at 25.0 °C are both $1.0 \times 10^{-7}\ mol\,L^{-1}$. Like most equilibrium constants, the value of K_w is temperature dependent and increases as the temperature increases. For example, at 40.0 °C, $K_w = 3.0 \times 10^{-14}$ and pure water at this temperature has $[H_3O^+] = [OH^-] = 1.7 \times 10^{-7}\ mol\,L^{-1}$. Table 11.2 lists values of K_w at a number of temperatures.

TABLE 11.2 K_w at various temperatures.

Temperature (°C)	K_w
0	1.5×10^{-15}
10	3.0×10^{-15}
20	6.8×10^{-15}
25	1.0×10^{-14}
30	1.5×10^{-14}
40	3.0×10^{-14}
50	5.5×10^{-14}
60	9.5×10^{-14}

We say that pure water is **neutral**, because it contains equal concentrations of H_3O^+ and OH^-. Aqueous solutions in which $[H_3O^+] > [OH^-]$ are termed **acidic**, while **basic** (or **alkaline**) solutions have $[H_3O^+] < [OH^-]$. There is an inverse relationship between $[H_3O^+]$ and $[OH^-]$ in aqueous solution at constant temperature; as one increases, the other must decrease in order to keep K_w constant. If we know the value of only one of $[H_3O^+]$ or $[OH^-]$ in a solution, we can use K_w to calculate the other, as illustrated in worked example 11.3.

WORKED EXAMPLE 11.3

Finding $[H_3O^+]$ from $[OH^-]$ or $[OH^-]$ from $[H_3O^+]$

In a sample of blood at 25 °C, $[H_3O^+] = 4.6 \times 10^{-8}$ M. Find $[OH^-]$ and decide if the sample is acidic, basic or neutral.

Analysis

We know that the values of $[H_3O^+]$ and $[OH^-]$ are related to each other through K_w at 25 °C as follows.

$$[H_3O^+][OH^-] = 1.0 \times 10^{-14}$$

If we know one concentration, we can always find the other.

Solution

We substitute the given value of $[H_3O^+]$ into this equation and solve for $[OH^-]$.

$$1.0 \times 10^{-14} = (4.6 \times 10^{-8})[OH^-]$$

Solve for $[OH^-]$, remembering that its units are $mol\,L^{-1}$ or M.

$$[OH^-] = \frac{1.0 \times 10^{-14}}{4.6 \times 10^{-8}}\ M$$
$$= 2.2 \times 10^{-7}\ M$$

When we compare a $[H_3O^+]$ of 4.6×10^{-8} M with a $[OH^-]$ of 2.2×10^{-7} M, we see that $[OH^-] > [H_3O^+]$. Our answer, then, is that the blood is slightly basic.

Is our answer reasonable?

We know $K_w = 1.0 \times 10^{-14} = [H_3O^+][OH^-]$. So one check is to note that, if $[H_3O^+]$ is slightly *less* than 1×10^{-7}, $[OH^-]$ will be slightly *more* than 1×10^{-7}, as it is.

PRACTICE EXERCISE 11.3 A sample of rainwater was found to have $[OH^-] = 2.6 \times 10^{-9}$ M. What is $[H_3O^+]$? Is the solution acidic, basic or neutral?

The concept of pH

As we have seen previously, the concentrations of H_3O^+ and OH^- in pure water are very small, and the same is often true of $[H_3O^+]$ and $[OH^-]$ in aqueous solutions of acids and bases. To avoid working with such inconveniently small numbers, we commonly express $[H_3O^+]$ in terms of the **pH** of the solution. This approach was developed by the Danish chemist Søren Sørensen (1868–1939) in 1909. pH is defined as the negative logarithm of the concentration of H_3O^+ in a solution, or, in equation form:

$$\text{pH} = -\log[H_3O^+]$$

We can also define the **pOH** of a solution as:

$$\text{pOH} = -\log[OH^-]$$

In both cases, p is simply an abbreviation for '–log' (p is derived from the German word *Potenz* meaning 'power'). It is important to realise that log means $\log_{10}$, rather than natural logarithm (ln or $\log_e$). It is vital that you appreciate the difference between the two types of logarithms — for example, $\log(1 \times 10^2) = 2$, whereas $\ln(1 \times 10^2) \approx 4.605$. You should also ensure that you know under which circumstances the different types of logarithms are used; you generally use log only in calculations involving pH or (as we will see) pK_a.

Having shown that $[H_3O^+] = 1.0 \times 10^{-7}$ mol L^{-1} in pure water at 25.0 °C, we can now calculate the pH of pure water under these conditions.

$$\text{pH} = -\log[H_3O^+] = -\log(1.0 \times 10^{-7}) = 7.00$$

There are two things to note here. Firstly, you cannot take the logarithm of a quantity that has units, and therefore we must either implicitly drop the concentration units of $[H_3O^+]$ before taking the logarithm or treat the term $[H_3O^+]$ as being the dimensionless quantity $\frac{[H_3O^+]}{c^{\ominus}}$, as we did in similar situations in chapter 9. Secondly, the number of decimal places in a logarithm of any number is equal to the number of significant figures in that number. Thus, our pH is given to two decimal places, 7.00.

The fact that pure water has a pH of 7 at 25.0 °C is probably a familiar result to many of you. Not so familiar perhaps is the pOH of pure water at 25.0 °C. Again, we have shown that $[OH^-] = 1.0 \times 10^{-7}$ under these conditions and therefore:

$$\text{pOH} = -\log[OH^-] = -\log(1.0 \times 10^{-7}) = 7.00$$

In both cases, you can see that it is much more convenient to write 7.00 than 1.0×10^{-7} (or indeed 0.000 000 10).

We showed previously that solutions with $[H_3O^+] > [OH^-]$ are acidic, while those with $[H_3O^+] < [OH^-]$ are basic. This means that (at 25.0 °C) solutions with pH < 7 are acidic, and those with pH > 7 are basic (or alkaline).

It is important that you are proficient in dealing with logarithms on your calculator and are able to convert from $[H_3O^+]$ to pH and from pH to $[H_3O^+]$ without any trouble. The latter involves taking the antilogarithm of the *negative* of the pH, i.e. $[H_3O^+] = 10^{-\text{pH}}$.

A simple relationship between pH and pOH can be derived, starting from the K_w expression. To do this, we make use of the property of logarithms which says that the logarithm of the product of two numbers is equal to the sum of the logarithms of those numbers, that is, $\log(ab) = \log(a) + \log(b)$.

We know that $[H_3O^+][OH^-] = K_w$. If we now take the negative logarithm of both sides, we obtain:

$$-\log[H_3O^+] + -\log[OH^-] = -\log K_w$$

But $-\log[H_3O^+] = pH$, $-\log[OH^-] = pOH$ and $-\log K_w = pK_w$.
Therefore:

$$pH + pOH = pK_w$$

where **pK_w** is the negative logarithm of K_w. At 25.0 °C, $K_w = 1.0 \times 10^{-14}$ and therefore this becomes:

$$pH + pOH = 14$$

Note that this equation applies only at 25.0 °C and we will assume this temperature in calculations throughout this chapter. Because pH and pOH are related by such a simple expression, it is usual to talk almost exclusively in terms of the pH of a solution, and pOH is rarely used.

It should be noted that our definition of pH:

$$pH = -\log[H_3O^+]$$

is an approximation. The exact definition involves the use of activity (a), a concept we have met briefly in chapter 9 (p. 355), rather than concentration, and is written as:

$$pH = -\log a_{H_3O^+}$$

However, provided that we work under dilute conditions, concentration is an acceptable approximation to activity, and we will use the equation $pH = -\log[H_3O^+]$ throughout the rest of the book.

Worked examples 11.4, 11.5 and 11.6 illustrate the types of problems you may encounter concerning the manipulations of pH, pOH, $[H_3O^+]$ and $[OH^-]$.

WORKED EXAMPLE 11.4

Calculating pH and pOH from $[H_3O^+]$

The dune lakes on Fraser Island, off the Queensland coast, are appreciably acidic due to the presence of natural organic acids from plant material. The water in Lake McKenzie (figure 11.4), the most acidic lake, was found to have an H_3O^+ concentration of 7.1×10^{-5} M at 25 °C. What are the pH and pOH values of the lake's water?

FIGURE 11.4 Lake McKenzie, the most acidic lake on Fraser Island, Queensland.

Analysis

If we know the value of $[H_3O^+]$, we can use $pH = -\log[H_3O^+]$ to find the pH.

Solution

$$pH = -\log[H_3O^+]$$

We substitute $[H_3O^+] = 7.1 \times 10^{-5}$ M into this expression to give:

$$pH = -\log(7.1 \times 10^{-5})$$

We first take the logarithm of this number and then take the negative of the logarithm. Using a calculator to find the logarithm of 7.1×10^{-5} gives the value −4.15 (note the negative sign). To find the pH, we change its algebraic sign.

$$pH = -(-4.15) = 4.15$$

Once we know the pH, the pOH is easily found by using the equation $pH + pOH = 14.00$ at 25 °C. Therefore, the pOH of the lake water is:

$$pOH = 14.00 - 4.15$$
$$= 9.85$$

Is our answer reasonable?

The value of $[H_3O^+]$ is between 1×10^{-5} and 1×10^{-4}, so the pH has to be between 5 and 4 (the antilogarithms of 1×10^{-5} and 1×10^{-4}, respectively), which it is. Therefore, the pOH must be between 9 and 10, as it is.

PRACTICE EXERCISE 11.4

A soft drink has an H_3O^+ concentration of 3.67×10^{-4} mol L^{-1}. What are the pH and pOH values of this drink? Is it acidic, basic or neutral?

WORKED EXAMPLE 11.5

Calculating pH from $[OH^-]$

What is the pH of a sodium hydroxide solution at 25 °C in which the hydroxide ion concentration is 0.0026 M?

Analysis

There are two ways to solve this problem. We could use the K_w expression:

$$K_w = [H_3O^+][OH^-] = 1 \times 10^{-14}$$

and the given value of the hydroxide ion concentration to find $[H_3O^+]$ and then calculate the pH using $pH = -\log[H_3O^+]$.

The second way is to calculate the pOH from the hydroxide ion concentration ($pOH = -\log[OH^-]$) and then subtract the pOH from 14.00 to find the pH. While both methods are valid and give the same answer, the second path requires less effort, so let's proceed that way.

Solution

We use the equation:

$$pOH = -\log[OH^-]$$

and substitute $[OH^-] = 0.0026$ M into this to give:

$$pOH = -\log(0.0026)$$

Taking the logarithm gives −2.59. Therefore:

$$pOH = -(-2.59) = 2.59$$

Then we subtract this pOH value from 14.00 to find the pH.

$$pH = 14.00 - 2.59$$
$$= 11.41$$

Notice that the pH in this basic solution is well above 7.

Is our answer reasonable?

The molar concentration of the hydroxide ion is between 1×10^{-3} and 1×10^{-2}, so the pOH must be between 3 and 2 (the antilogarithms of 1×10^{-3} and 1×10^{-2}, respectively) which it is. Thus, the pH should be between 11 and 12.

PRACTICE EXERCISE 11.5

The ammonia often found in household cleaning products leads to these being basic. A sample of one brand was found to have $[OH^-] = 2.00 \times 10^{-3}$ mol L^{-1}. What is the pH of the solution?

WORKED EXAMPLE 11.6

Calculating $[H_3O^+]$ from pH

'Calcareous soil' is soil rich in calcium carbonate and is particularly abundant along the south coast of Australia. The pH of the moisture in such soil generally ranges from just over 7 to as high as 8.3. After one particular soil sample was soaked in water, the pH of the water was measured to be 8.14. What value of $[H_3O^+]$ corresponds to a pH of 8.14? Is the soil acidic or basic?

Analysis

The equation we use to calculate $[H_3O^+]$ from pH is $[H_3O^+] = 10^{-pH}$.

Solution

We substitute the pH value, 8.14, into the equation above, which gives:

$$[H_3O^+] = 10^{-8.14}$$

We therefore need to calculate 10 to the power of −8.14. The easiest way to do this on a calculator is to press the '10^x' key and then enter −8.14. (Note that, on some calculators, you may have to do these in the opposite order.) Because the pH has two digits following the decimal point, we obtain two significant figures in the H_3O^+ concentration. Therefore, the answer, correctly rounded, is:

$$[H_3O^+] = 7.2 \times 10^{-9}\ \text{M}$$

Because pH > 7 and $[H_3O^+] < 1 \times 10^{-7}$ M, the soil is basic.

Is our answer reasonable?

The given pH value of 8.14 is between 8 and 9, so the H_3O^+ concentration must lie between 10^{-8} and 10^{-9} mol L^{-1}, as we found.

PRACTICE EXERCISE 11.6

Find the values of $[H_3O^+]$ and $[OH^-]$ that correspond to each of the following pH values. State whether each solution is acidic or basic.

(a) 2.30 (the approximate pH of lemon juice)
(b) 3.85 (the approximate pH of sauerkraut)
(c) 10.81 (the pH of milk of magnesia, a laxative and an antacid)
(d) 3.50 (the pH of orange juice, on average)
(e) 11.61 (the pH of dilute, household ammonia)

The pH scale ranges from 0 to 14 for most practical applications, although negative values of pH can be obtained in very acidic solutions and pH values greater than 14 occur in concentrated aqueous bases. A 0.10 mol L^{-1} solution of HCl(aq) has a pH of 1.00, while a 0.10 mol L^{-1} solution of NaOH(aq) has a pH of 13.00. Figure 11.5 shows the pH values of some common aqueous solutions. Accurate pH values (±0.01 pH units) can be easily determined with a pH meter, which uses a glass electrode that is sensitive to H_3O^+ ions (see figure 11.6). Less accurate values can be obtained using pH paper (figure 11.7), while blue and red litmus paper is used only in a qualitative manner to determine whether a solution is acidic or basic.

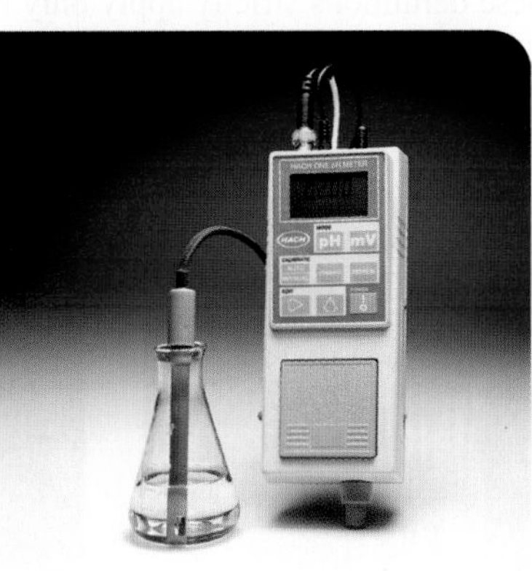

FIGURE 11.6 A pH meter has a special combination electrode that is sensitive to hydrogen ion concentration. After the instrument has been calibrated using solutions of known pH, the electrode is dipped into the solution to be tested and the pH is read from the meter. The pH meter detects the difference between the H_3O^+ concentrations in the tested solution and the solution inside the electrode as a voltage, which is then converted to a pH value.

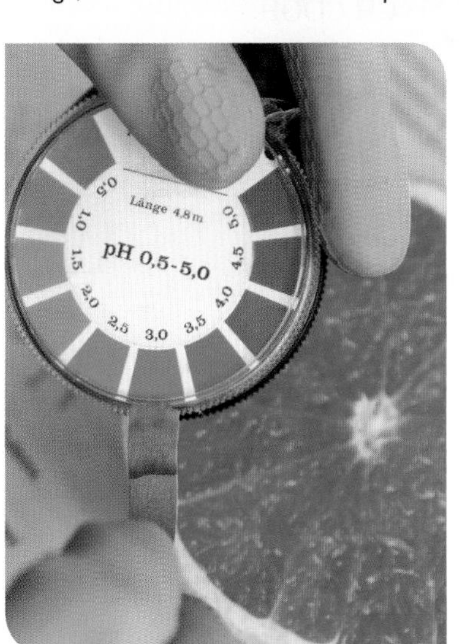

FIGURE 11.7 A pH test paper. The colour of this Hydrion® test strip changed to green when a drop of the orange juice was placed on it. According to the colour code, the pH of the juice is approximately 3.

most acidic

pH

- 0 — 1.0 M HCl
- 1 — 0.1 M HCl
- stomach contents (about 1–2)
- 0.3 M citric acid
- 2
- lemon juice
- 0.1 M acetic acid
- 3
- soft drinks (about 2–4)
- orange juice
- carbonic acid (saturated)
- 4
- beer
- 5
- 0.03 M boric acid
- 6
- milk
- saliva (about 6–7)
- 7 neutral — pure water
- blood
- 8
- sea water
- 9
- 10
- 11 — 0.1 M ammonia
- 0.05 M Na_2CO_3
- 12 — 0.03 M Na_3PO_4
- limewater
- 13 — oven cleaner
- 14 — 1.0 M NaOH

most basic

FIGURE 11.5 The pH scale.

The strength of acids and bases

We stated previously that the pH of a 0.10 M solution of HCl(aq) is 1.00. This means that $[H_3O^+]$ in this solution is 0.10 M and therefore implies that the reaction:

$$HCl(aq) + H_2O(l) \rightleftharpoons H_3O^+(aq) + Cl^-(aq)$$

proceeds to completion; in other words, HCl is completely dissociated in water. However, if we measure the pH of a 0.10 M aqueous solution of the seemingly similar acid HF, we find that it is not 1.00 but 2.10. You can show therefore that $[H_3O^+]$ in the HF solution is not 0.10 M, but 7.9×10^{-3} M, and this means that the reaction:

$$HF(aq) + H_2O(l) \rightleftharpoons H_3O^+(aq) + F^-(aq)$$

does not proceed to completion. In fact, at any particular time, only about 1 in 13 molecules of HF have reacted with water to form H_3O^+ and F^-, with the majority remaining undissociated. We find a similar situation with a 0.10 M aqueous solution of acetic acid, CH_3COOH. The measured pH of this solution is again not 1.00 but 2.88, which corresponds to $[H_3O^+] = 1.3 \times 10^{-3}$ M and again means that the equilibrium:

$$CH_3COOH(aq) + H_2O(l) \rightleftharpoons H_3O^+(aq) + CH_3COO^-(aq)$$

lies very much towards the reactants. Obviously there is a fundamental difference between HCl and both HF and CH_3COOH. HCl is able to transfer its proton completely to water, while HF and CH_3COOH are very poor proton donors towards water. This means they exist predominantly as molecules in aqueous solution, with the result being a low concentration of ions in solution. This can be illustrated using conductivity, the concept we used to open this chapter (p. 432). Figure 11.8 shows the different conductivities of aqueous solutions containing HCl and CH_3COOH. This behaviour shows the different strengths of the acids. We say that HCl is a **strong acid**, whereas both HF and CH_3COOH are **weak acids**. A general definition of these terms is as follows.

- A strong acid reacts completely with water to give quantitative formation of H_3O^+.
- A weak acid reacts incompletely with water to form less than stoichiometric amounts of H_3O^+.

There are analogous definitions for bases.

- A **strong base** reacts completely with water to give quantitative formation of OH^-.
- A **weak base** reacts incompletely with water to form less than stoichiometric amounts of OH^-.

Note that these definitions strictly apply only to Brønsted–Lowry acids in aqueous solution.

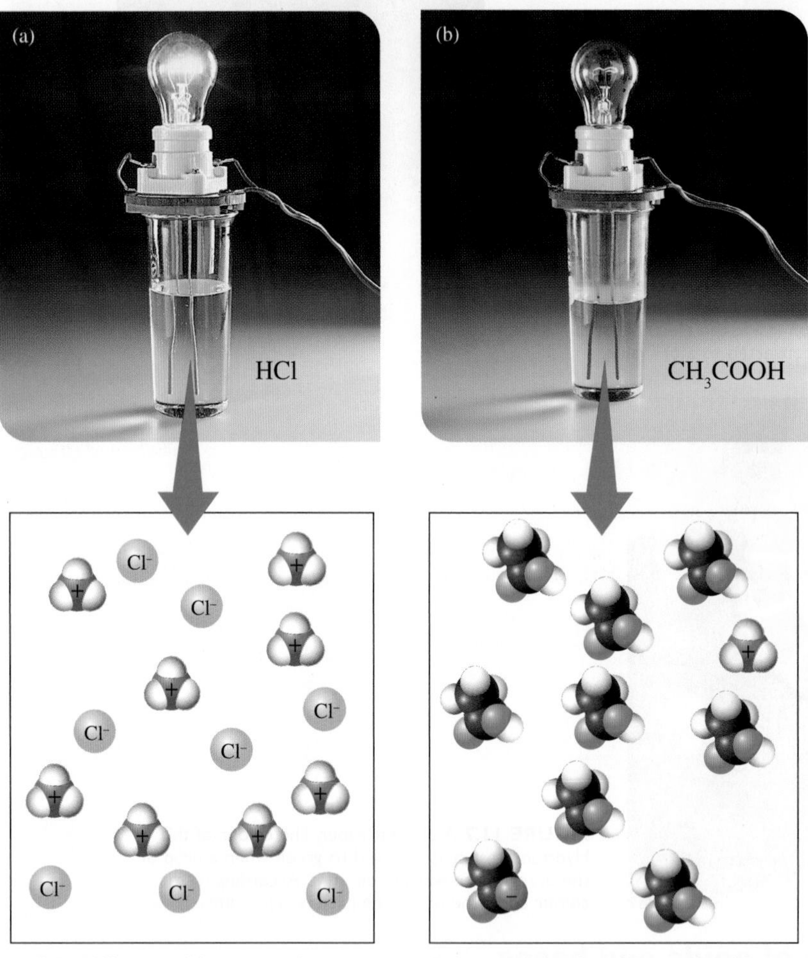

(a)
(b)
HCl
CH_3COOH
Cl⁻
All the HCl reacts with water, so there are many ions present.
Only a small fraction of the acetic acid reacts with water, so there are few ions to conduct electricity. Most of the acetic acid is present as neutral molecules of CH_3COOH.

FIGURE 11.8 Electrical conductivity of solutions of strong and weak acids at equal concentrations. The wires lead to a battery (not shown), which provides electrical power to light the bulb: **(a)** HCl reacts completely with water to give H_3O^+ and Cl^-, resulting in high conductivity, thus enabling the light to glow brightly. **(b)** CH_3COOH reacts with water to give only very small amounts of H_3O^+ and CH_3COO^-, so the light is dimmer.

The use of the words 'strong' and 'weak' is somewhat unfortunate, as it implies that there is a distinct cut-off point between strong and weak acids. This is not the case. Our definition of strong acids and bases implies that they are completely dissociated in water and should therefore have comparable strengths — this is true only if we restrict ourselves to relatively dilute aqueous solutions. However, weak acids and bases show an enormous range of strengths in aqueous solution; rather than relying on the qualitative concept 'weak' to describe their behaviour, it is more informative to quantify their ability to either donate or accept a proton by looking at the value of the equilibrium constant for their reaction with water, and we will do this in section 11.4 (p. 447). We will first look at the nature of strong acids and bases in aqueous solution.

Chemical Connections

Swimming pools and soil

Swimming pools and soil are just two examples of many situations where pH is important.

The water in swimming pools is more than just water; it is a dilute solution of compounds that prevent the growth of bacteria and stabilise the pool lining while allowing us to swim safely in the pool. The pH of a swimming pool should be between 7.2 and 7.6.

The pH can also affect the availability of substances that plants need for growth (figure 11.9). A pH of 6 to 7 is best for most plants; most nutrients are more soluble when the soil is slightly acidic than when it is neutral or slightly basic. If the soil pH is too high, metal ions that plants need, such as iron and manganese, will precipitate and not be available in the ground water. A very low pH is not good either; if the pH drops to 4–5, metal ions such as aluminium, that are toxic to many plants, are released as their compounds become more soluble. A low pH also inhibits the growth of beneficial bacteria that decompose organic matter in the soil to release nutrients, especially nitrogen.

FIGURE 11.9 Hydrangeas grown in acidic soil have blue flowers (left), while those grown in alkaline soils have pink/red flowers (right).

It is important to be able to measure the pH of these systems. Commercial pH test kits use acid–base indicators. Kits for testing the pH of pool water (figure 11.10) use phenol red indicator, which changes colour over the pH range of 6.4–8.2, while soil test kits (figure 11.11) use a 'universal indicator', which is a mixture of indicators that enable estimation of pH over a wide range.

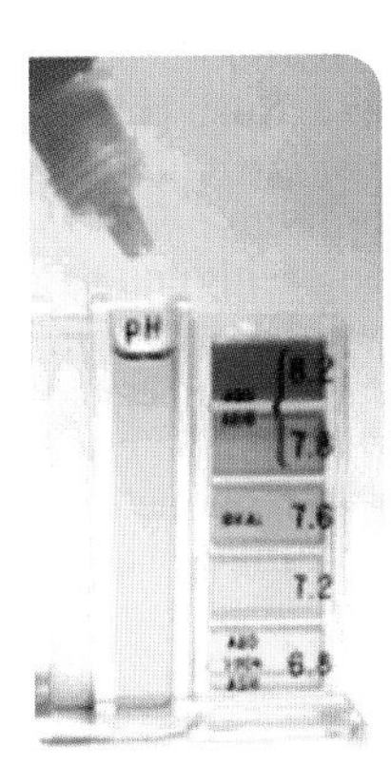

FIGURE 11.10 Swimming pool test kit. This apparatus is used to test both pH and chlorine concentration. The colour of the indicator tells us that the pH of the pool water is approximately 7.6, which is slightly basic.

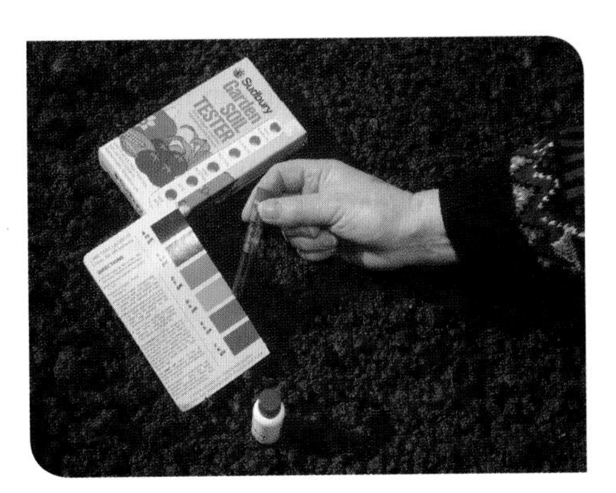

FIGURE 11.11 Testing soil pH. The indicator reveals that the soil sample being tested has a pH of approximately 6.5, which is slightly acidic.

11.3 Strong acids and bases

We have defined a strong acid as an acid that donates a proton completely to water. Using this definition, for any strong acid HA, the reaction:

$$HA(aq) + H_2O(l) \rightleftharpoons H_3O^+(aq) + A^-(aq)$$

proceeds to completion. Similarly, our definition of a strong base says that OH^- is formed quantitatively on reaction with water, and therefore the reaction of any strong base B with water:

$$B(aq) + H_2O(l) \rightleftharpoons BH^+(aq) + OH^-(aq)$$

should also proceed to completion.

A list of some strong acids and bases is given in table 11.3.

TABLE 11.3 Examples of strong acids and bases.

Strong acids[a]	Strong bases
$HClO_4$ ($HOClO_3$), perchloric acid	LiOH, lithium hydroxide
HCl, hydrochloric acid	NaOH, sodium hydroxide
HBr, hydrobromic acid	KOH, potassium hydroxide
HI, hydroiodic acid	$Ca(OH)_2$, calcium hydroxide
HNO_3 ($HONO_2$), nitric acid	RbOH, rubidium hydroxide
H_2SO_4 ($(HO)_2SO_2$), sulfuric acid	$NaOCH_3$, sodium methoxide
HCF_3SO_3 ($HOSO_2CF_3$), trifluoromethanesulfonic acid	CsOH, caesium hydroxide

(a) The molecular formulae of the acids in parentheses emphasise the actual structure of each acid; note that the acidic proton is attached to an O atom in all cases.

The fact that strong acids undergo complete dissociation tells us something about the basicity of their conjugate bases. For example, the equilibrium:

$$HCl(aq) + H_2O(l) \rightleftharpoons H_3O^+(aq) + Cl^-(aq)$$

lies almost completely to the right-hand side, and therefore the reverse reaction occurs to only a negligible extent. This means that Cl^-, the conjugate base of HCl, does not readily accept a proton from H_3O^+, and consequently Cl^- must be a very weak base. Such behaviour is common to all strong acids and we can make the generalisation that *the conjugate base of a strong acid is very weak*. We will investigate this concept further when we discuss weak acids and bases (section 11.4). When writing equations for the dissociation of strong acids and bases in water it is therefore usual to replace the equilibrium arrow with a single forward arrow. We should also note that H_3O^+ is the strongest acid and OH^- is the strongest base that can exist in aqueous solution. We have already seen that the strong acid HCl donates its proton almost completely to water to form H_3O^+, and the same is true of all acids stronger than H_3O^+; they react completely when dissolved in water to form H_3O^+ and none of the undissociated acid is present. Likewise, any base stronger than OH^- will, when dissolved in water, quantitatively deprotonate water to give OH^-. Hence, there are limits to the range of acidities and basicities that can be accessed in water and, if we want to go beyond these limits, we must use a different solvent.

pH calculations in solutions of strong acids and bases

As strong acids and bases react with water almost completely, the calculation of pH in solutions containing strong acids or bases is a relatively straightforward exercise in stoichiometry.

When the solute in an aqueous solution is a strong monoprotic acid, such as HCl or HNO_3, we expect to obtain 1 mole of H_3O^+ for every mole of the acid in the solution. Thus, a 1.0×10^{-2} M solution of HCl contains 1.0×10^{-2} mol L^{-1} of H_3O^+, and a 2.0×10^{-3} M solution of HNO_3 contains 2.0×10^{-3} mol L^{-1} of H_3O^+.

To calculate the pH of a solution of a strong monoprotic acid, we use $[H_3O^+]$ obtained from the stated molar concentration of the acid. Thus, the 1.0×10^{-2} M HCl solution mentioned above has $[H_3O^+] = 1.0 \times 10^{-2}$ M, and therefore the $pH = -\log[H_3O^+] = -\log(1.0 \times 10^{-2}) = 2.00$.

For strong bases, calculating the pH from the OH^- concentration is similarly straightforward. A 5.0×10^{-2} M solution of NaOH contains 5.0×10^{-2} mol L^{-1} of OH^- because the base is fully dissociated and each mole of NaOH releases 1 mole of OH^- when it dissociates. Therefore, $pOH = -\log(5.0 \times 10^{-2}) = 1.30$, and $pH = 14.00 - 1.30 = 12.70$ at 25 °C.

For bases such as $Ba(OH)_2$, we have to recognise that 2 moles of OH^- are released by each mole of the base.

$$Ba(OH)_2(s) \rightarrow Ba^{2+}(aq) + 2OH^-(aq)$$

Therefore, if a solution contained 1.0×10^{-2} mol $Ba(OH)_2$ per litre, the concentration of OH^- would be 2.0×10^{-2} M. Of course, once we know the OH^- concentration we can calculate pOH, from which we can calculate the pH. Worked example 11.7 illustrates the kinds of calculations we have just described.

WORKED EXAMPLE 11.7

Calculating pH, pOH, $[H_3O^+]$ and $[OH^-]$ for aqueous solutions of strong acids or bases

Calculate the pH, pOH, $[H_3O^+]$ and $[OH^-]$ of the following solutions at 25 °C: (a) 2.0×10^{-2} M HCl and (b) 3.5×10^{-4} M $Ba(OH)_2$. HCl is a strong acid and $Ba(OH)_2$ is a strong base.

Analysis

The solute in (a) is HCl, a strong acid. It gives quantitative formation of H_3O^+ and therefore the reaction:

$$HCl(aq) + H_2O(l) \rightleftharpoons H_3O^+(aq) + Cl^-(aq)$$

proceeds to completion. From each mole of HCl, we expect 1 mole of H_3O^+, so we can use the molar concentration of HCl to obtain $[H_3O^+]$, from which we can calculate the pH, pOH and $[OH^-]$.

The solute in (b) is $Ba(OH)_2$, a strong base. From each mole of $Ba(OH)_2$, 2 moles of OH^- are liberated, according to the equation:

$$Ba(OH)_2(s) \rightleftharpoons Ba^{2+}(aq) + 2OH^-(aq)$$

and this proceeds to completion. We use the molarity of $Ba(OH)_2$ to calculate the hydroxide ion concentration. From $[OH^-]$ we can calculate $[H_3O^+]$, pOH and pH.

Solution

(a) Because HCl is completely dissociated, in 2.0×10^{-2} M HCl, $[H_3O^+] = 2.0 \times 10^{-2}$ M. Therefore:

$$pH = -\log[H_3O^+] = -\log(2.0 \times 10^{-2}) = 1.70$$

Thus, in 2.0×10^{-2} M HCl, the pH is 1.70. The pOH is $(14.00 - 1.70) = 12.30$. To find $[OH^-]$, we can use this value of pOH.

$$[OH^-] = 10^{-pOH} = 10^{-12.30} = 5.0 \times 10^{-13}\text{ M}$$

Note how much smaller $[OH^-]$ is in this acidic solution than it is in pure water. This makes sense because the solution is quite acidic.

(b) As noted, $Ba(OH)_2$ is a strong base, and so for each mole of $Ba(OH)_2$ we obtain 2 moles of OH^-. Therefore:

$$[OH^-] = 2 \times (3.5 \times 10^{-4})\text{ M} = 7.0 \times 10^{-4}\text{ M}$$

$$pOH = -\log[OH^-] = -\log(7.0 \times 10^{-4}) = 3.15$$

Thus, the pOH of this solution is 3.15, and the $pH = (14.00 - 3.15) = 10.85$. We can then obtain $[H_3O^+]$ from the equation:

$$[H_3O^+] = 10^{-pH}$$
$$= 10^{-10.85} = 1.4 \times 10^{-11}\text{ M}$$

We would expect the concentration of H_3O^+ to be very low because the solution is quite basic.

Is our answer reasonable?

In (a), the concentration of H_3O^+ is between 10^{-2} M and 10^{-1} M and so the pH must be between 1 and 2, as we found. In (b), the molarity of OH^- is between 10^{-4} M and 10^{-3} M. Hence, the pOH must be between 3 and 4, which it is.

PRACTICE EXERCISE 11.7

Calculate $[H_3O^+]$ and the pH of an aqueous 7.4×10^{-2} M NaOH solution.

Suppression of the autoprotolysis of water

In the preceding calculations, we have assumed that all of the H_3O^+ or OH^- in the solutions of strong acid or base comes from the acid or the base, respectively. However, there is another potential source of both H_3O^+ and OH^- in an aqueous solution, namely water. We have seen that the autoprotolysis of water gives rise to low concentrations of H_3O^+ and OH^-, and we will now show the conditions under which we are justified in neglecting these contributions. We will consider only a solution of strong acid, but realise that analogous arguments apply for a solution of a strong base.

In an aqueous solution of any acid, there are two sources of H_3O^+; the acid itself, and water. Thus:

$$[H_3O^+]_{total} = [H_3O^+]_{from\ acid} + [H_3O^+]_{from\ H_2O}$$

Except in *very* dilute solutions of strong acids, the amount of H_3O^+ contributed by water ($[H_3O^+]_{from\ H_2O}$) is negligible compared with the amount of H_3O^+ from the acid ($[H_3O^+]_{from\ acid}$). For instance, in worked example 11.7 we saw that in 2.0×10^{-2} M HCl the $[OH^-]$ was 5.0×10^{-13} M. The only source of OH^- in this acidic solution is from the autoprotolysis of water, and the amounts of OH^- and H_3O^+ formed by the autoprotolysis of water must be equal. Therefore, $[H_3O^+]_{from\ H_2O}$ also equals 5.0×10^{-13} M. If we now look at the total $[H_3O^+]$ for this solution, we have:

$$[H_3O^+]_{total} = \underset{\text{(from HCl)}}{2.0 \times 10^{-2}\ \text{M}} + \underset{\text{(from } H_2O)}{5.0 \times 10^{-13}\ \text{M}}$$

$$= 2.0 \times 10^{-2}\ \text{M (rounded correctly)}$$

In any solution of an acid, the autoprotolysis of water is suppressed by the H_3O^+ produced by the acid. This is a demonstration of the common ion effect we first discussed in chapter 10 (p. 407); in this case, H_3O^+ is the common ion, and the large excess of H_3O^+ from the acid suppresses the autoprotolysis of water.

We can rationalise this by considering the autoprotolysis equilibrium of water.

$$H_2O(l) + H_2O(l) \rightleftharpoons H_3O^+(aq) + OH^-(aq)$$

If we add another source of H_3O^+ to water, we instantaneously increase the reaction quotient Q_w (where $Q_w = [H_3O^+][OH^-]$), so that $Q_w > K_w$. Therefore, the equilibrium shifts towards the reactants to decrease Q_w, resulting in lower amounts of both H_3O^+ and OH^- *from the autoprotolysis of water*. Similar reasoning also applies to the addition of another source of OH^- to water. Therefore, addition of either H_3O^+ or OH^- to water suppresses the autoprotolysis of water. The maximum possible contribution of $[H_3O^+]$ or $[OH^-]$ from the autoprotolysis of water is 1.0×10^{-7} M at 25.0 °C, and, in most cases, especially when strong acids or bases are involved, this can be safely ignored. However, as shown in worked example 11.8, there are exceptions.

WORKED EXAMPLE 11.8

Calculating the pH of *very* dilute strong acid solutions

Calculate the pH of a 1.0×10^{-10} M solution of HCl(aq).

Analysis

At first glance, this appears a simple matter of equating $[H_3O^+]$ to the concentration of the acid, as we have done in previous examples concerning strong acid solutions. However, if we do this, we obtain a pH of 10, which is obviously incorrect as it would mean that the solution of HCl is basic! In this case, we need to take *both* possible sources of H_3O^+ into account — HCl and H_2O — and see which one, if any, dominates.

Solution

We start by calculating the contribution of H_3O^+ from both water and HCl. We know that the HCl solution is 1.0×10^{-10} M, and therefore:

$$[H_3O^+]_{from\ HCl} = 1.0 \times 10^{-10}\ \text{M}$$

Ordinarily, the addition of acid to water would suppress the autoprotolysis of water through the common ion effect (H_3O^+ is the common ion), meaning that the contribution of H_3O^+ from this would be less than 1.0×10^{-7} M. However, in this case, the concentration of acid is very small compared with that resulting from the autoprotolysis of water, and we can therefore say, to a very good approximation:

$$[H_3O^+]_{from\ H_2O} = 1.0 \times 10^{-7}\ \text{M}$$

Hence:

$$[H_3O^+]_{total} = 1.0 \times 10^{-7}\ \text{M} + 1.0 \times 10^{-10}\ \text{M}$$

$$= 1.0 \times 10^{-7}\ \text{M (rounded correctly)}$$

and the calculated pH becomes:

$$pH = -\log[H_3O^+]$$
$$= -\log(1.0 \times 10^{-7}) = 7.00$$

Is our answer reasonable?

A pH of 7 is certainly much more reasonable than a pH of 10. The strong acid is so dilute that, in this case, water is the major contributor of H_3O^+ to the solution.

PRACTICE EXERCISE 11.8

Calculate the pH of a 1.0×10^{-11} M solution of NaOH.

From now on, we assume that, except for *very* dilute solutions (10^{-6} M or less), the autoprotolysis of water can be neglected when calculating the pH of strong acids or bases.

11.4 Weak acids and bases

Earlier in this chapter (p. 442), we stated that weak acids and weak bases react incompletely with water to give less than stoichiometric amounts of H_3O^+ and OH^-, respectively. We can quantify the extent to which these reactions occur by looking at the values of their equilibrium constants. If we consider a weak monoprotic acid HA, we can write its reaction with water as:

$$HA(aq) + H_2O(l) \rightleftharpoons H_3O^+(aq) + A^-(aq)$$

The equilibrium constant expression for this is therefore:

$$K_a = \frac{[H_3O^+][A^-]}{[HA]}$$

The equilibrium constant for the dissociation of an acid in aqueous solution is called the **acidity constant** and is given the symbol K_a. We can see that the value of K_a tells us how far the reaction has proceeded towards completion when equilibrium is established.

Similarly, we can write the equation for the reaction of a weak monoprotic base B with water as:

$$B(aq) + H_2O(l) \rightleftharpoons BH^+(aq) + OH^-(aq)$$

and the corresponding equilibrium constant expression is:

$$K_b = \frac{[BH^+][OH^-]}{[B]}$$

This is called the **basicity constant** and is given the symbol K_b.

When polyprotic acids or bases are considered, a separate equilibrium constant must be written for each proton loss or proton gain. For example, the diprotic acid H_2A will undergo two separate proton loss reactions in water, one involving H_2A and the other involving its conjugate base, HA^-. This gives rise to two separate K_a expressions.

$$H_2A + H_2O \rightleftharpoons HA^- + H_3O^+ \qquad K_{a_1} = \frac{[H_3O^+][HA^-]}{[H_2A]}$$

$$HA^- + H_2O \rightleftharpoons A^{2-} + H_3O^+ \qquad K_{a_2} = \frac{[H_3O^+][A^{2-}]}{[HA^-]}$$

Analogous K_b expressions can be written for a polyprotic base.

Because weak acids and weak bases react with water only to a very small extent, both K_a and K_b values are generally significantly less than 1. For example, K_a for acetic acid is 1.8×10^{-5}, while K_b for pyridine is 1.7×10^{-9}. To put these values in perspective, it is again instructive (as we did in chapter 9) to write these numbers in an alternative form — for example, knowing that $10^{-5} = \frac{1}{10^5}$, we can write K_a for acetic acid as:

$$K_a = \frac{[H_3O^+][CH_3COO^-]}{[CH_3COOH]} = 1.8 \times 10^{-5} = \frac{1.8}{10^5}$$

When written in this fashion, it is evident that the concentrations on the top line are *much* smaller than the concentration on the bottom. In other words, the concentration of undissociated acid at equilibrium is enormous compared with $[H_3O^+]$ and $[CH_3COO^-]$. This means that acetic acid is indeed a weak acid. A similar approach can be used to emphasise the weakly basic nature of pyridine.

Because K_a and K_b values are usually significantly less than 1, it is often more convenient to use **pK_a** and **pK_b** values, where:

$$pK_a = -\log K_a$$

$$pK_b = -\log K_b$$

Both pK_a and pK_b are analogous to pH, and the defining equations are therefore similar. We can also write each equation in terms of K_a and K_b respectively, and these then become:

$$K_a = 10^{-pK_a}$$

$$K_b = 10^{-pK_b}$$

Note that polyprotic acids and bases have more than one pK_a or pK_b value.

The strength of a weak acid is determined by its K_a value: the larger the K_a, the stronger the acid and the greater the degree of dissociation at equilibrium. Because of the negative sign in the defining equation for pK_a, the stronger the acid, the *smaller* is its pK_a value. The values of K_a and pK_a for some typical weak acids are given in table 11.4. A more complete list is located in appendix E. Similar arguments apply to K_b and pK_b values, some of which are given in table 11.5.

TABLE 11.4 K_a and pK_a values for weak monoprotic acids at 25 °C.

Name of acid	Formula	K_a	pK_a
chloroacetic acid	$ClCH_2COOH$	1.4×10^{-3}	2.85
nitrous acid	HNO_2	7.1×10^{-4}	3.15
hydrofluoric acid	HF	6.8×10^{-4}	3.17
cyanic acid	$HOCN$	3.5×10^{-4}	3.46
formic acid	$HCOOH$	1.8×10^{-4}	3.74
barbituric acid	$C_4H_4N_2O_3$	9.8×10^{-5}	4.01
acetic acid	CH_3COOH	1.8×10^{-5}	4.74
hydrazoic acid	HN_3	1.8×10^{-5}	4.74
butanoic acid	$CH_3CH_2CH_2COOH$	1.5×10^{-5}	4.82
propanoic acid	CH_3CH_2COOH	1.4×10^{-5}	4.89
hypochlorous acid	$HOCl$	3.0×10^{-8}	7.52
hydrogen cyanide (aq)	HCN	6.2×10^{-10}	9.21
phenol	C_6H_5OH	1.3×10^{-10}	9.89
hydrogen peroxide	H_2O_2	1.8×10^{-12}	11.74

TABLE 11.5 K_b and pK_b values for weak molecular bases at 25 °C.

Name of base	Formula	K_b	pK_b
butylamine	$C_4H_9NH_2$	5.9×10^{-4}	3.23
methylamine	CH_3NH_2	4.4×10^{-4}	3.36
ammonia	NH_3	1.8×10^{-5}	4.74
hydrazine	N_2H_4	1.7×10^{-6}	5.77
strychnine	$C_{21}H_{22}N_2O_2$	1.0×10^{-6}	6.00
morphine	$C_{17}H_{19}NO_3$	7.5×10^{-7}	6.13
hydroxylamine	$HONH_2$	6.6×10^{-9}	8.18
pyridine	C_5H_5N	1.7×10^{-9}	8.77
aniline	$C_6H_5NH_2$	4.1×10^{-10}	9.36

WORKED EXAMPLE 11.9

Interpreting pK_a and finding K_a

A certain acid was found to have a pK_a of 4.88. Is this acid stronger or weaker than acetic acid? What is the K_a for the acid?

Analysis

We can compare the acid strengths by comparing the pK_a values. The larger the pK_a, the weaker is the acid. Finding K_a from pK_a involves the same kind of calculation as finding $[H_3O^+]$ from pH, which you learned to do earlier in this chapter.

Solution

From table 11.4, the pK_a of acetic acid is 4.74. The acid referred to in the problem has a pK_a of 4.88. Because the acid has a larger pK_a than acetic acid, it is a weaker acid.

To find K_a from pK_a, we use the equation:

$$K_a = 10^{-pK_a}$$

Substituting pK_a = 4.88:

$$K_a = 10^{-4.88} = 1.3 \times 10^{-5}$$

Is our answer reasonable?

As a quick check, we can compare the K_a values. For acetic acid, $K_a = 1.8 \times 10^{-5}$. This is larger than 1.3×10^{-5}, which tells us that acetic acid is the stronger acid. This agrees with our conclusion based on the pK_a values.

PRACTICE EXERCISE 11.9

Three weak acids, HX, HY and HZ, have pK_a values of 4.13, 3.91 and 4.02, respectively. Rank these acids in order from strongest to weakest. What are the K_a values of these acids?

Some molecules contain both an acidic site and a basic site, and this leads to interesting structural consequences. Consider, for example, the amino acid glycine. We would usually write the condensed structural formula of this molecule as NH_2CH_2COOH. However, it actually exists in the solid state as a **zwitterion**, a structure which contains both a positive and negative charge but which is overall neutral.

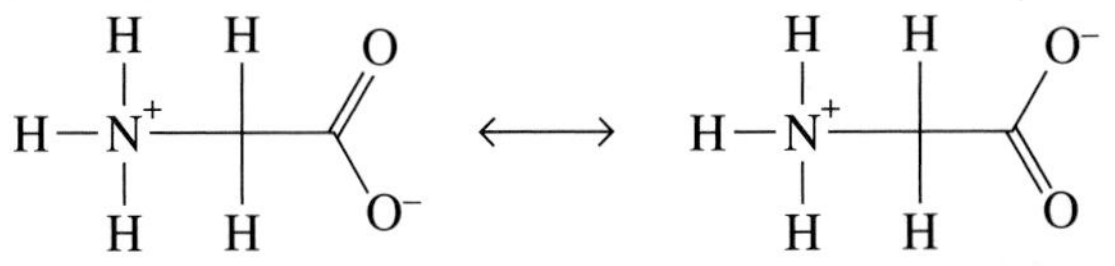

We will discuss such compounds in more detail in chapter 24.

We learned in chapter 9 how to manipulate equilibrium constant expressions and found that, when two chemical equations were added together, the equilibrium constant for the resulting equation was equal to the product of the equilibrium constants for the two reactions. This approach leads to an important result when applied to an acid and its conjugate base. Consider, for example, formic acid, HCOOH, a weak acid partly responsible for the sting of some ants (the word *formica* is Latin for 'ant'). The acid reacts with water according to the equation:

$$HCOOH(aq) + H_2O(l) \rightleftharpoons H_3O^+(aq) + HCOO^-(aq)$$

for which:

$$K_a = \frac{[H_3O^+][HCOO^-]}{[HCOOH]}$$

The conjugate base of formic acid, the formate ion, $HCOO^-$, reacts with water as follows.

$$HCOO^-(aq) + H_2O(l) \rightleftharpoons HCOOH(aq) + OH^-(aq)$$

Hence:

$$K_b = \frac{[HCOOH][OH^-]}{[HCOO^-]}$$

If we add these equations, we obtain the equation for the autoprotolysis of water.

$$\begin{array}{l} \cancel{HCOOH(aq)} + H_2O(l) \rightleftharpoons H_3O^+(aq) + \cancel{HCOO^-(aq)} \\ \cancel{HCOO^-(aq)} + H_2O(l) \rightleftharpoons \cancel{HCOOH(aq)} + OH^-(aq) \\ \hline H_2O(l) + H_2O(l) \rightleftharpoons H_3O^+(aq) + OH^-(aq) \end{array}$$

If we multiply the expressions for K_a and K_b together, we obtain:

$$\frac{[H_3O^+][HCOO^-]}{[HCOOH]} \times \frac{[HCOOH][OH^-]}{[HCOO^-]} = [H_3O^+][OH^-] = K_w$$

or

$$K_aK_b = K_w$$

This relationship holds for *any* acid–base conjugate pair, regardless of the strength of the acid or base.

Another useful relationship can be derived by taking the negative logarithm of each side of $K_aK_b = K_w$ and remembering that the logarithm of the product of two numbers is the sum of their logarithms.

$$pK_a + pK_b = pK_w = 14.00 \text{ (at 25 °C)}$$

There are some important consequences of the relationship $K_aK_b = K_w$. One is that it is not necessary to tabulate both K_a and K_b for the members of an acid–base pair; if one K is known, the other can be calculated. For example, K_a for HCOOH and K_b for NH_3 are found in most tables of acid–base equilibrium constants, but these tables usually do not contain K_b for $HCOO^-$ or K_a for NH_4^+. If we need them, we can calculate them using $K_aK_b = K_w$.

PRACTICE EXERCISE 11.10 The value of K_a for HCOOH is 1.8×10^{-4}. What is the K_b for the $HCOO^-$ ion?

Another interesting and useful observation is that *there is an inverse relationship between the strengths of the acid and base members of a conjugate pair.* This is illustrated graphically in figure 11.12. Because the product of K_a and K_b is a constant, the larger the value of K_a, the smaller the value of K_b. In other words, *the stronger the conjugate acid, the weaker its conjugate base.* Do not make the mistake of assuming that the conjugate base of a weak acid is strong. Inspection of figure 11.12 and use of the relationship $pK_a + pK_b = 14.00$ show that, in fact, the conjugate base of a weak acid is weak. For example, the pK_a of acetic acid, a weak acid, is 4.74. Its conjugate base, the acetate ion, must therefore have a pK_b of $(14.00 - 4.74) = 9.26$, a value indicative of a weak base. Similar arguments apply for the conjugate acid of a weak base.

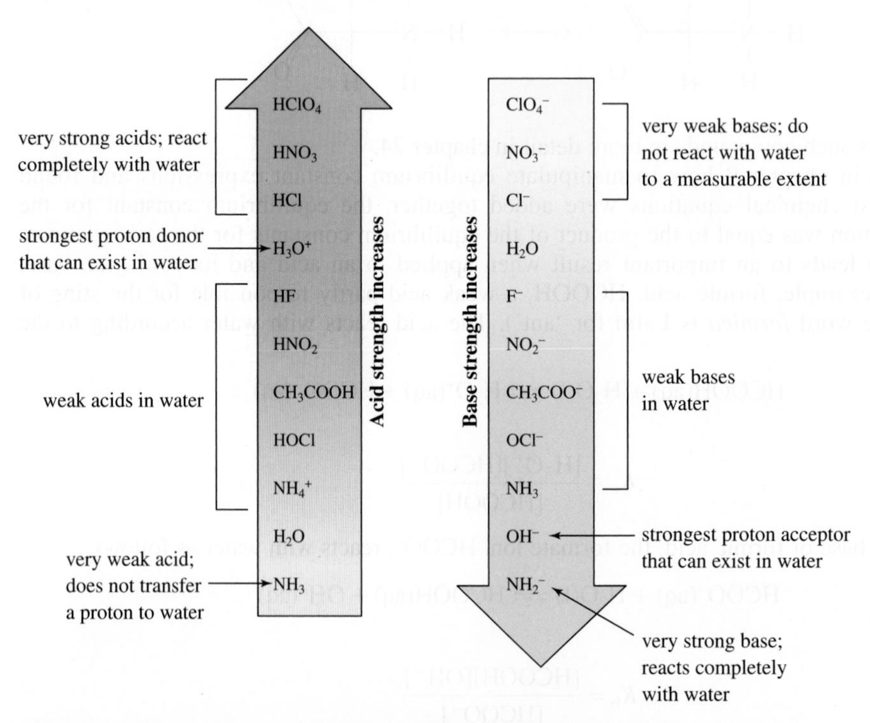

FIGURE 11.12 The relative strengths of conjugate acid–base pairs. The stronger the acid, the weaker its conjugate base. The weaker the acid, the stronger its conjugate base.

pH calculations in solutions of weak acids and bases

In section 11.3, we learned how to calculate the pH of solutions of strong acids and bases. The calculations relied on the fact that, because strong acids and bases dissociate completely in water, the concentration of H_3O^+ or OH^- was equal to the initial concentration of the acid or base.

However, because weak acids and bases react incompletely with water, this is no longer the case. For weak acids and bases, we must use the value of K_a or K_b to determine $[H_3O^+]$ or $[OH^-]$ and hence pH. We will illustrate this by calculating the pH of a 1.0 M solution of acetic acid, CH_3COOH.

We begin by writing the balanced chemical equation for the reaction of the acid with water, and obtain the expression for K_a from this.

$$CH_3COOH(aq) + H_2O(l) \rightleftharpoons H_3O^+(aq) + CH_3COO^-(aq)$$

$$K_a = \frac{[H_3O^+][CH_3COO^-]}{[CH_3COOH]} = 1.8 \times 10^{-5} \text{ (from table 11.4)}$$

We cannot solve this equation for $[H_3O^+]$ directly in this form as there are two unknowns: $[CH_3COO^-]$ and $[CH_3COOH]$. The fact that these unknowns are related (the sum of $[CH_3COO^-]$ and $[CH_3COOH]$ must be 1.0 M, the initial concentration of the acid) allows the equation to be solved using arithmetic manipulations, but this involves solving a quadratic equation. However, it is usual to use a concentration table for calculations of this sort. We set this up in the same way as we did in chapter 9. We assume that the initial concentration of CH_3COOH is 1.0 M and the initial concentrations of both H_3O^+ and CH_3COO^- are 0. (Recall that there is a small nonzero concentration of H_3O^+ arising from the autoprotolysis of water, but we can safely neglect this as it is a maximum of 1.0×10^{-7} M.) As the reaction proceeds to equilibrium, the concentration of CH_3COOH decreases by $-x$, and the concentrations of both H_3O^+ and CH_3COO^- increase by $+x$, owing to the stoichiometry of the reaction. At equilibrium, the concentrations of CH_3COOH, CH_3COO^- and H_3O^+ are $(1.0 - x)$, x and x, respectively. Hence, we obtain the following concentration table. (As in chapter 9, concentrations given in the question are highlighted.)

	H_2O + CH_3COOH	$\rightleftharpoons$ H_3O^+	+ CH_3COO^-
Initial concentration ($mol\ L^{-1}$)	1.0	0	0
Change in concentration ($mol\ L^{-1}$)	$-x$	$+x$	$+x$
Equilibrium concentration ($mol\ L^{-1}$)	$1.0 - x$	x	x

When the values in the last row of the table are substituted into the expression for K_a we obtain:

$$K_a = \frac{[H_3O^+][CH_3COO^-]}{[CH_3COOH]} = \frac{(x)(x)}{1.0 - x} = 1.8 \times 10^{-5}$$

This equation involves a term in x^2, and it could be solved exactly using the quadratic formula. However, this and many other similar calculations involving weak acids and bases can be simplified. Recall from chapter 9 (p. 381) that, if the concentration from which x is subtracted is at least 400 times as large as K, we can make simplifying assumptions. In this case, the initial concentration of 1.0 M is indeed at least 400 times as large as K_a, 1.8×10^{-5}. Therefore, we can make the approximation $(1.0 - x) \approx 1.0$. If we now replace $1.0 - x$ with 1.0 in the equation above, we obtain:

$$\frac{x^2}{1.0} = 1.8 \times 10^{-5}$$

We can now solve this directly:

$$x^2 = (1.0)(1.8 \times 10^{-5}) = 1.8 \times 10^{-5}$$

$$x = \sqrt{1.8 \times 10^{-5}} = 4.2 \times 10^{-3}\ \text{M}$$

We see that the value of x is indeed negligible compared with 1.0 M (i.e. if we subtract 4.2×10^{-3} M from 1.0 M and round correctly, we obtain 1.0 M). We also note that the same answer (to two significant figures) is obtained by solving the quadratic equation exactly, further justifying our assumption. This value corresponds to approximately 1 molecule in 240 of CH_3COOH being dissociated at equilibrium; CH_3COOH is indeed a weak acid. Having calculated x, and knowing that $x = [H_3O^+]$ at equilibrium, we can find the pH of the solution by taking the negative logarithm of this number.

$$\text{pH} = -\log(4.2 \times 10^{-3}) = 2.38$$

Notice that when we make the approximation above, the initial concentration of the acid is used as if it were the equilibrium concentration. This approximation is valid when the equilibrium constant is small and the concentrations of the solutes are reasonably high — conditions that apply to most situations you will encounter. On pages 458–60, we will discuss the conditions under which the approximation is not valid.

Recall also that we neglected any contribution of H_3O^+ from the autoprotolysis of water to the total $[H_3O^+]$ in the solution. We can show that this assumption is justified by looking at the magnitude of $[H_3O^+]$ due to the acid compared with $[H_3O^+]$ from water. The maximum possible $[H_3O^+]$ contributed by water is 1.0×10^{-7} M, and this will be smaller in reality owing to the common ion effect from the presence of the acid. Even if we compare the maximum possible $[H_3O^+]$ from water (1.0×10^{-7} M) with $[H_3O^+]$ from the acid (4.2×10^{-3} M), we see that it is negligibly small (4.2×10^{-3} M + 1.0×10^{-7} M = 4.2×10^{-3} M) and so our assumption is justified.

We will now look at some examples that illustrate typical weak acid and weak base equilibrium problems.

WORKED EXAMPLE 11.10

Calculating $[H_3O^+]$ and pH values for a solution of a weak acid from its K_a value

Calcium propanoate, a compound used as a preservative in baked goods (figure 11.13), is a salt of the weak acid propanoic acid, CH_3CH_2COOH. A student planned an experiment that would use propanoic acid at a concentration of 0.10 M. Calculate the values of $[H_3O^+]$ and pH for this solution at 25 °C. For propanoic acid, $K_a = 1.4 \times 10^{-5}$ at 25 °C.

$$CH_3-CH_2-\overset{\overset{\displaystyle O}{\|}}{C}-O-H$$

FIGURE 11.13 The calcium salt of propanoic acid, calcium propanoate, is used as a preservative in baked products.

Analysis

First, we write the balanced equation for the reaction of the acid with water and obtain the expression for K_a from this.

$$CH_3CH_2COOH(aq) + H_2O(l) \rightleftharpoons CH_3CH_2COO^-(aq) + H_3O^+(aq)$$

$$K_a = \frac{[H_3O^+][CH_3CH_2COO^-]}{[CH_3CH_2COOH]}$$

The initial concentration of the acid is 0.10 M, and the initial concentrations of both ions are 0 M. The concentration of CH_3CH_2COOH will change by $-x$ and those of $CH_3CH_2COO^-$ and H_3O^+ will increase by $+x$ as the reaction proceeds to equilibrium. We can now construct the concentration table.

Solution

	H_2O + CH_3CH_2COOH	$\rightleftharpoons$ H_3O^+	+ $CH_3CH_2COO^-$
Initial concentration (mol L^{-1})	0.10	0	0
Change in concentration (mol L^{-1})	$-x$	$+x$	$+x$
Equilibrium concentration (mol L^{-1})	$0.10 - x$	x	x

As propanoic acid is a weak acid, it dissociates only to a small extent and therefore x is very small. We can then make the simplifying approximation that $(0.10 - x) \approx 0.10$, so we take the equilibrium concentration of CH_3CH_2COOH to be 0.10 M. Substituting these quantities into the K_a expression gives:

$$K_a = \frac{[H_3O^+][CH_3CH_2COO^-]}{[CH_3CH_2COOH]} = \frac{(x)(x)}{0.10 - x} \approx \frac{(x)(x)}{0.10} = 1.4 \times 10^{-5}$$

We can now solve this for x.

$$x^2 = (0.10)(1.4 \times 10^{-5}) = 1.4 \times 10^{-6}$$

$$x = \sqrt{1.4 \times 10^{-6}} = 1.2 \times 10^{-3}$$

We know that $x = [H_3O^+]$ and therefore:

$$[H_3O^+] = 1.2 \times 10^{-3}\ \text{M}$$

Finally, we calculate the pH.

$$\begin{aligned} pH &= -\log[H_3O^+] = -\log(1.2 \times 10^{-3}) \\ &= 2.92 \end{aligned}$$

Is our answer reasonable?

First, we see that the calculated pH is less than 7. This tells us that the solution is acidic, which it should be for a solution of an acid. Also, the pH is higher than it would be if the acid were strong. (A 0.10 M solution of a strong acid would have $[H_3O^+] = 0.10$ M and pH = 1.0.) If we wish to further check the accuracy of the calculation, we can substitute the calculated equilibrium concentrations into the expression for K_a. If the calculated quantities are correct, we should obtain the numerical value of K_a.

$$\frac{[H_3O^+][CH_3CH_2COO^-]}{[CH_3CH_2COOH]} = \frac{(x)(x)}{0.10} = \frac{(1.2 \times 10^{-3})^2}{0.10} = 1.4 \times 10^{-5} = K_a$$

The check works, so we have done the calculation correctly.

PRACTICE EXERCISE 11.11

Pentanoic acid, $CH_3CH_2CH_2CH_2COOH$, is sometimes called valeric acid, as it can be extracted from the plant *valeriana officinalis*. It is a weak acid with $K_a = 1.4 \times 10^{-5}$. Calculate $[H_3O^+]$ and the pH of a 0.025 M solution of pentanoic acid.

WORKED EXAMPLE 11.11

Calculating the pH of a solution of a weak base

A solution of hydrazine, N_2H_4, a weak base, has a concentration of 0.25 M. What is the pH of the solution? Hydrazine has $K_b = 1.7 \times 10^{-6}$.

Analysis

This example is analogous to worked example 11.10, except that we are now dealing with a weak base and using K_b rather than K_a.

Solution

We begin with the balanced chemical equation and obtain the K_b expression.

$$N_2H_4(aq) + H_2O(l) \rightleftharpoons N_2H_5^+(aq) + OH^-(aq)$$

$$K_b = \frac{[N_2H_5^+][OH^-]}{[N_2H_4]}$$

Again, the concentration of N_2H_4 will change by $-x$ and those of $N_2H_5^+$ and OH^- will increase by $+x$ as the reaction proceeds to equilibrium. We construct the concentration table as follows.

	H_2O +	N_2H_4	$\rightleftharpoons$	$N_2H_5^+$ +	OH^-
Initial concentration (mol L^{-1})		0.25		0	0
Change in concentration (mol L^{-1})		$-x$		$+x$	$+x$
Equilibrium concentration (mol L^{-1})		$0.25 - x$		x	x

Because hydrazine is a weak base, it reacts with water only to a small extent. We can therefore make the same approximation that we made for a weak acid; that is, $(0.25 - x) \approx 0.25$. We substitute the values from the concentration table into the K_b expression.

$$K_b = \frac{[N_2H_5^+][OH^-]}{[N_2H_4]} = \frac{(x)(x)}{0.25 - x} \approx \frac{(x)(x)}{0.25} = 1.7 \times 10^{-6}$$

We solve this for x as follows.

$$x^2 = (0.25)(1.7 \times 10^{-6}) = 4.3 \times 10^{-7}$$

$$x = \sqrt{4.3 \times 10^{-7}} = 6.5 \times 10^{-4}$$

This value represents the hydroxide ion concentration, from which we can calculate the pOH.

$$\begin{aligned} pOH &= -\log[OH^-] \\ &= -\log(6.5 \times 10^{-4}) \\ &= 3.19 \end{aligned}$$

The pH of the solution can then be obtained from the relationship:

$$\begin{aligned} pH + pOH &= 14.00 \\ pH &= 14.00 - 3.19 \\ &= 10.81 \end{aligned}$$

Is our answer reasonable?

As stated in the question, hydrazine is a weak base, so we would expect a pH > 7. We can quickly check to see whether the value of x and our assumed equilibrium concentration of N_2H_4 (0.25 M) are correct by substituting them into the expression for K_b and comparing the result with the numerical value of K_b. Doing this gives 1.7×10^{-6}, which is the same as K_b, so the value of x is correct.

PRACTICE EXERCISE 11.12

Pyridine, C_5H_5N, is a toxic, foul-smelling liquid for which $K_b = 1.7 \times 10^{-9}$. What is the pH of a 0.010 M aqueous solution of pyridine?

In worked examples 11.10 and 11.11, we have calculated the pH given the value of K_a or K_b. We can also use measured values of pH to calculate K_a or K_b values, as shown in worked example 11.12.

WORKED EXAMPLE 11.12

Calculating K_a and pK_a from pH

Lactic acid, $CH_3CH(OH)COOH$, is a monoprotic acid that is present in sour milk and yoghurt. The pH of a 0.100 M solution of lactic acid is 2.44 at 25 °C. Calculate K_a and pK_a for lactic acid at this temperature.

Analysis

In this case we are given the pH of the solution, from which we can calculate $[H_3O^+]$. In essence, this is the reverse of the previous examples. We use the balanced chemical equation to construct a concentration table and then substitute the calculated equilibrium concentrations into the expression for K_a.

Solution

We begin by writing the balanced chemical equation and obtaining the expression for K_a.

$$CH_3CH(OH)COOH(aq) + H_2O(l) \rightleftharpoons H_3O^+(aq) + CH_3CH(OH)COO^-(aq)$$

$$K_a = \frac{[H_3O^+][CH_3CH(OH)COO^-]}{[CH_3CH(OH)COOH]}$$

In this problem, the only source of the ions H_3O^+ and $CH_3CH(OH)COO^-$ is the acid. As a result, the equilibrium concentration of H_3O^+ must be identical to that of $CH_3CH(OH)COO^-$ because the reaction gives these ions in a 1 : 1 ratio. However, we do not have the values of either $[H_3O^+]$ or $[CH_3CH(OH)COO^-]$; all we have is the pH of the 0.100 M solution. We start by finding $[H_3O^+]$ from the pH. This gives us the *equilibrium* value of $[H_3O^+]$.

$$\begin{aligned} [H_3O^+] &= 10^{-pH} = 10^{-2.44} \\ &= 3.6 \times 10^{-3}\,M \end{aligned}$$

Because $[CH_3CH(OH)COO^-] = [H_3O^+]$, we know that $[CH_3CH(OH)COO^-] = 3.6 \times 10^{-3}$ M at equilibrium. We now set up the concentration table, given that the initial concentration of lactic acid is 0.100 M.

$H_2O + CH_3CH(OH)COOH$	$\rightleftharpoons$	H_3O^+	$+ CH_3CH(OH)COO^-$
Initial concentration ($mol\ L^{-1}$)	0.100	0	0
Change in concentration ($mol\ L^{-1}$)	-3.6×10^{-3}	$+3.6 \times 10^{-3}$	$+3.6 \times 10^{-3}$
Equilibrium concentration ($mol\ L^{-1}$)	$0.100 - 3.6 \times 10^{-3} = 0.096$	3.6×10^{-3}	3.6×10^{-3}

This problem differs from the previous worked examples in that we know the value of x (3.6×10^{-3} M) but we do not know K_a. We proceed by substituting the equilibrium values into the K_a expression. We know that both $[H_3O^+]$ and $[CH_3CH(OH)COO^-] = 3.6 \times 10^{-3}$ M, and, from the concentration table, that $[CH_3CH(OH)COOH] = (0.100 - x) = (0.100 - 3.6 \times 10^{-3}) = 0.096$ M. (Note that, because we know the value of x in this case, we do not have to make the assumption that $(0.100 - x) \approx 0.100$.) Therefore:

$$K_a = \frac{[H_3O^+][CH_3CH(OH)COO^-]}{[CH_3CH(OH)COOH]} = \frac{(3.6\times10^{-3})(3.6\times10^{-3})}{0.096}$$
$$= 1.4 \times 10^{-4}$$

Thus the K_a for lactic acid is 1.4×10^{-4}. To find pK_a, we take the negative logarithm of K_a.

$$pK_a = -\log K_a$$
$$= -\log(1.4 \times 10^{-4})$$
$$= 3.85$$

Is our answer reasonable?

Weak acids have small acidity constants, so the value we obtained for K_a seems to be reasonable. We should also check the entries in the concentration table to be sure they are reasonable. For example, the 'changes' for the ions are both positive, meaning both of their concentrations are increasing. This is the way it should be. Also, we have the concentration of lactic acid decreasing, as it must do.

PRACTICE EXERCISE 11.13

Ethanolamine, $NH_2CH_2CH_2OH$, is a very water-soluble amine that is used to remove H_2S and CO_2 from industrial waste gas streams. Given that the pH of a 0.100 M aqueous solution of ethanolamine is 11.25, calculate its K_b and pK_b.

pH calculations in solutions of salts of weak acids and bases

We have seen that Cl^-, the conjugate base of the strong acid HCl, is a very weak base and shows essentially no tendency to react with H_3O^+ to form HCl. This means that Cl^- is even less likely to act as a base towards H_2O, and we can say that Cl^- has essentially no basic properties. Therefore, an aqueous solution of NaCl has a pH of 7.00 at 25 °C. This is true of the conjugate bases of all strong monoprotic acids. Thus, dilute aqueous solutions of salts such as NaBr, NaI, $NaNO_3$, and, indeed, any sodium salt of a strong monoprotic acid, have pH values of 7.00 at 25 °C. However, this is not the case for the salt of a weak acid, such as $NaOOCCH_3$, which has a pH > 7, or the salt of a weak base, such as NH_4Cl, which has a pH < 7 at this temperature. If we dissolve each of these in water, we get complete dissociation into the respective ions as follows.

$$NaOOCCH_3(s) \xrightarrow{H_2O} Na^+(aq) + CH_3COO^-(aq)$$

$$NH_4Cl(s) \xrightarrow{H_2O} NH_4^+(aq) + Cl^-(aq)$$

The CH_3COO^- and NH_4^+ ions are then involved in the following equilibria:

$$CH_3COO^-(aq) + H_2O(l) \rightleftharpoons CH_3COOH(aq) + OH^-(aq)$$
$$NH_4^+(aq) + H_2O(l) \rightleftharpoons NH_3(aq) + H_3O^+(aq)$$

and the two solutions are therefore slightly basic and slightly acidic, respectively.

As illustrated in figure 11.12 (p. 450), conjugate base strength increases as acid strength decreases, and therefore the salts of weak acids can be appreciably basic. Similarly, the salts of weak bases are generally slightly acidic. We can see that this is the case by looking at pK_a and pK_b values for conjugate pairs. For example, as we have seen, acetic acid, CH_3COOH, has a pK_a of 4.74, and this means that the acetate ion, CH_3COO^-, has a pK_b of 14.00 − 4.74 = 9.26. If we consider HCN, an even weaker acid (pK_a = 9.21), we can see that the cyanide ion, CN^-, has a pK_b of 14.00 − 9.21 = 4.79, making it almost as strong a base as ammonia, NH_3. We can use the pK_a and pK_b values to calculate the pH of aqueous solutions containing the salts of weak acids and bases, as shown in worked examples 11.13 and 11.14.

WORKED EXAMPLE 11.13

Calculating the pH of a salt solution

What is the pH of a 0.10 M solution of NaOCl at 25 °C? For HOCl, $K_a = 3.0 \times 10^{-8}$.

Analysis

OCl^- is the conjugate base of the weak acid HOCl, and we therefore expect the solution to be slightly basic. As we are given only K_a for HOCl and we are asked to find the pH, we use K_b, which we can obtain from K_a, to solve this problem. As usual, we start with a balanced chemical equation and obtain the expression for the appropriate equilibrium constant from this.

Solution

First, we write the chemical equation for the reaction of OCl^- with water and the corresponding K_b expression.

$$OCl^-(aq) + H_2O(l) \rightleftharpoons HOCl(aq) + OH^-(aq)$$

$$K_b = \frac{[HOCl][OH^-]}{[OCl^-]}$$

The data provided in the problem include the K_a for HOCl, but we can calculate K_b because $K_aK_b = K_w$ for a conjugate pair.

$$K_b = \frac{K_w}{K_a} = \frac{1.0 \times 10^{-14}}{3.0 \times 10^{-8}} = 3.3 \times 10^{-7}$$

Now let us set up the concentration table. The only source of HOCl and OH^- is the reaction of OCl^-, so their concentrations each increase by x and the concentration of OCl^- decreases by x as the reaction proceeds to equilibrium.

	H_2O + OCl^-	$\rightleftharpoons$	HOCl	+ OH^-
Initial concentration (mol L^{-1})	0.10		0	0
Change in concentration (mol L^{-1})	$-x$		$+x$	$+x$
Equilibrium concentration (mol L^{-1})	$(0.10 - x)$		x	x

At equilibrium, the concentrations of HOCl and OH^- are the same, x.

$$[HOCl] = [OH^-] = x$$

As OCl^- is a weak base, it reacts with water only to a small extent, and therefore x is small. We can then make the approximation that $(0.10 - x) \approx 0.10$ because $0.1 > 400 \times K_b$, and so $[OCl^-] = 0.10$ M at equilibrium. We substitute the values into the K_b expression to give:

$$K_b = \frac{[HOCl][OH^-]}{[OCl^-]} = \frac{(x)(x)}{0.10 - x} \approx \frac{(x)(x)}{0.10} = 3.3 \times 10^{-7}$$

We solve this for x as follows.

$$x^2 = (0.10)(3.3 \times 10^{-7}) = 3.3 \times 10^{-8}$$

$$x = \sqrt{3.3 \times 10^{-8}} = 1.8 \times 10^{-4}\ \text{M}$$

This value of x represents the OH^- concentration, from which we can calculate the pOH and then the pH.

$$\begin{aligned} pOH &= -\log[OH^-] \\ &= -\log(1.8 \times 10^{-4}) \\ &= 3.74 \\ pH &= 14.00 - pOH \\ &= 14.00 - 3.74 \\ &= 10.26 \end{aligned}$$

The pH of this solution is 10.26.

Is our answer reasonable?

From the nature of the salt, we expect the solution to be basic. The calculated pH corresponds to a basic solution, so the answer seems reasonable. We can also check the accuracy of the answer by substituting the calculated equilibrium concentrations into the expression for K_b.

$$\frac{[HOCl][OH^-]}{[OCl^-]} = \frac{(x)(x)}{0.10} = \frac{(1.8\times10^{-4})^2}{0.10} = 3.2\times10^{-7}$$

The result is acceptably close to the value of K_b, so our answers are correct.

What is the pH of a 0.10 M solution of $NaNO_2$? Use the pK_a of HNO_2 from table 11.4.

PRACTICE EXERCISE 11.14

WORKED EXAMPLE 11.14

Calculating the pH of a salt solution

What is the pH of a 0.20 M solution of hydrazinium chloride, N_2H_5Cl at 25 °C? Hydrazine, N_2H_4, is a weak base with $K_b = 1.7 \times 10^{-6}$.

Analysis

The hydrazinium ion, $N_2H_5^+$, is the conjugate acid of the weak base hydrazine, N_2H_4, and therefore we expect a solution containing this ion to be slightly acidic. We proceed in a similar fashion to worked example 11.13, starting, as ever, with the balanced chemical equation and the appropriate equilibrium constant expression. As we have seen, Cl^- is an extremely weak base; it reacts with water to a negligible extent and does not affect the pH of the solution. It can therefore be neglected in our calculations.

Solution

We will begin with the balanced chemical equation for the reaction of $N_2H_5^+$ with water and write the K_a expression.

$$N_2H_5^+(aq) + H_2O(l) \rightleftharpoons H_3O^+(aq) + N_2H_4(aq)$$

$$K_a = \frac{[H_3O^+][N_2H_4]}{[N_2H_5^+]}$$

The problem has given us K_b for N_2H_4, but we need K_a for $N_2H_5^+$. We obtain this by solving the equation $K_aK_b = K_w$ for K_a.

$$K_a = \frac{K_w}{K_b} = \frac{1.0\times10^{-14}}{1.7\times10^{-6}} = 5.9 \times 10^{-9}$$

Now we set up the concentration table. The initial concentrations of H_3O^+ and N_2H_4 are both set to zero. Next, we indicate that the concentration of $N_2H_5^+$ decreases by x and the concentrations of H_3O^+ and N_2H_4 both increase by x as the reaction proceeds to equilibrium.

	H_2O + $N_2H_5^+$	$\rightleftharpoons$	H_3O^+ +	N_2H_4
Initial concentration (mol L^{-1})	0.20		0	0
Change in concentration (mol L^{-1})	$-x$		$+x$	$+x$
Equilibrium concentration (mol L^{-1})	$0.20 - x$		x	x

At equilibrium, equal amounts of H_3O^+ and N_2H_4 are present, and their equilibrium concentrations are each equal to x.

$$[H_3O^+] = [N_2H_4] = x$$

As $N_2H_5^+$ is a weak acid, it reacts with water to a small extent. We can assume that $(0.20 - x) \approx 0.20$ because $0.2 > 400 \times K_a$, and therefore $[N_2H_5^+] = 0.20$ M at equilibrium. We then substitute quantities into the equilibrium constant expression.

$$K_a = \frac{[H_3O^+][N_2H_4]}{[N_2H_5^+]} = \frac{(x)(x)}{0.20-x} \approx \frac{(x)(x)}{0.20} = 5.9\times10^{-9}$$

We solve this for x as follows.

$$x^2 = (0.20)(5.9 \times 10^{-9}) = 1.2 \times 10^{-9}$$

$$x = \sqrt{1.2 \times 10^{-9}} = 3.4 \times 10^{-5}\text{ M}$$

Since $x = [H_3O^+]$, the pH of the solution is:

$$pH = -\log[H_3O^+] = -\log(3.4 \times 10^{-5})$$
$$= 4.47$$

Is our answer reasonable?

The active solute species in the solution is a weak acid and the calculated pH is less than 7, so the answer seems reasonable. Check the accuracy yourself by substituting equilibrium concentrations into the expression for K_a.

PRACTICE EXERCISE 11.15 What is the pH of a 0.0750 M solution of NH_4ClO_4? Use the pK_b of NH_3 from table 11.5.

Solutions that contain the salt of a weak acid and a weak base

We saw previously that both the NH_4^+ cation and CH_3COO^- anion affect the pH of an aqueous solution. We now address what happens when both the cation and anion in a single salt are able to affect the pH. Whether or not the salt has a net effect on the pH now depends on the relative strengths of its ions, one functioning as an acid and the other as a base. If they are matched in their respective strengths, the salt has no net effect on pH. Consider, for example, a solution of ammonium acetate, NH_4OOCCH_3, in which the ammonium ion is an acidic cation and the acetate ion is a basic anion. However, K_a of NH_4^+ is 5.6×10^{-10} and K_b of CH_3COO^- just happens to be the same, 5.6×10^{-10}. The cation tends to produce H_3O^+ ions to the same extent that the anion tends to produce OH^-. So, in aqueous ammonium acetate, $[H_3O^+] = [OH^-]$, and the solution has a pH of 7.

Consider, now, ammonium formate, NH_4OOCH. The formate ion, $HCOO^-$, is the conjugate base of the weak acid formic acid and has $K_b = 5.6 \times 10^{-11}$. Comparing this value with the (slightly larger) K_a of the ammonium ion, 5.6×10^{-10}, we see that NH_4^+ is slightly stronger as an acid than the formate ion is as a base. As a result, a solution of ammonium formate is slightly acidic. The exact calculation of pH in such solutions is difficult, and we are concerned here only in predicting if the solution is acidic, basic or neutral.

WORKED EXAMPLE 11.15

Predicting how a salt affects the pH of its solution

Is a 0.20 M aqueous solution of NH_4F acidic, basic or neutral?

Analysis

This is a salt in which the cation is a weak acid (it is the conjugate acid of a weak base, NH_3) and the anion is a weak base (it is the conjugate base of a weak acid, HF). The question is how do the two ions compare in their abilities to affect the pH of the solution? We have to compare their respective K_a and K_b values to compare their strengths.

Solution

K_a of NH_4^+ (calculated from the K_b of NH_3) is 5.6×10^{-10}. Similarly, K_b of F^- is 1.5×10^{-11} (calculated from the K_a of HF, 6.8×10^{-4}). Comparing the two equilibrium constants, we see that K_a for the acid, NH_4^+, is greater than K_b for the base, F^-. Therefore, the reaction of NH_4^+ with water to give H_3O^+ proceeds to a greater extent than the reaction of F^- with water to give OH^-. This means that, at equilibrium, $[H_3O^+] > [OH^-]$, and we expect the solution to be slightly acidic.

Is our answer reasonable?

There is not much to check here except to be sure that we have done the arithmetic correctly.

PRACTICE EXERCISE 11.16 Is an aqueous solution of ammonium cyanide, NH_4CN, acidic, basic or neutral?

Situations where simplifying assumptions do not work

On pages 450–4, we used initial concentrations of acids and bases as though they were equilibrium concentrations when we performed calculations. This is only an approximation, as we discussed on page 451, but it works most of the time. Unfortunately, it does not work in all cases, so we will now examine those conditions under which simplifying approximations do not work. We will also study how to solve problems when the approximations cannot be used.

When a weak acid HA reacts with water, its concentration is reduced as the ions form. If we let x represent the amount of acid that reacts per litre, the equilibrium concentration becomes:

$$[HA]_{equilibrium} = [HA]_{initial} - x$$

In previous problems, we have assumed that $[HA]_{equilibrium} \approx [HA]_{initial}$ because x is small compared with $[HA]_{initial}$. However, this is the case only if $[HA]_{initial}$ is greater than, or equal to, 400 times the value of K_a. As the value of K_a of the weak acid increases, our assumption becomes less valid because the extent of dissociation of the acid increases, and $[HA]_{equilibrium}$ is no longer approximately equal to $[HA]_{initial}$.

When our assumption is not valid, we must solve a quadratic equation as we did in chapter 9. We also assumed that we could safely neglect any contribution to $[H_3O^+]$ or $[OH^-]$ from the autoprotolysis of water. This is the case only if the value of $[H_3O^+]$ or $[OH^-]$ from the acid or base is greater than 1×10^{-5} (i.e. at least 100 times 1×10^{-7}, the maximum possible contribution from water). This will generally be the case, and we assume we can continue to neglect the autoprotolysis of water from here on. We illustrate a situation that requires use of the quadratic formula in worked example 11.16.

WORKED EXAMPLE 11.16

Using the quadratic formula in equilibrium products

Chloroacetic acid, $ClCH_2COOH$, is used as a herbicide and in the manufacture of dyes and other organic chemicals. It is a weak acid with $K_a = 1.4 \times 10^{-3}$. What is the pH of a 0.010 M solution of $ClCH_2COOH$ at 25 °C?

Analysis

Before we begin the solution, we check to see whether we can use our usual simplifying approximation. We do this by calculating $400 \times K_a$ and then comparing the result with the initial concentration of the acid.

$$400 \times K_a = 400(1.4 \times 10^{-3}) = 0.56$$

The initial concentration of $ClCH_2COOH$ is *less than* 0.56 M, so we know the simplification does not work. We therefore have to set up a concentration table and then use a quadratic equation to obtain the solution.

Solution

We begin by writing the balanced chemical equation and the K_a expression.

$$ClCH_2COOH(aq) + H_2O(l) \rightleftharpoons H_3O^+(aq) + ClCH_2COO^-(aq)$$

$$K_a = \frac{[H_3O^+][ClCH_2COO^-]}{[ClCH_2COOH]} = 1.4 \times 10^{-3}$$

The initial concentration of the acid will be reduced by an amount, $-x$, as it reacts with water to form the ions. From this, we build the concentration table, assuming, as always, that the initial concentrations of the ions are zero.

	$H_2O + ClCH_2COOH$	$\rightleftharpoons$	H_3O^+ +	$ClCH_2COO^-$
Initial concentration (mol L^{-1})	0.010		0	0
Change in concentration (mol L^{-1})	$-x$		$+x$	$+x$
Equilibrium concentration (mol L^{-1})	$0.010 - x$		x	x

Substituting equilibrium concentrations into the equilibrium expression gives:

$$K_a = \frac{[H_3O^+][ClCH_2COO^-]}{[ClCH_2COOH]} = \frac{(x)(x)}{0.010 - x} = 1.4 \times 10^{-3}$$

This time we cannot assume $(0.010 - x) = 0.010$. To solve the problem, we first rearrange the expression to obtain the familiar form of a quadratic equation. We do this by multiplying both sides by $(0.010 - x)$. This gives:

$$x^2 = (0.010 - x)(1.4 \times 10^{-3})$$
$$x^2 = (1.4 \times 10^{-5}) - (1.4 \times 10^{-3})x$$
$$x^2 + (1.4 \times 10^{-3})x - (1.4 \times 10^{-5}) = 0$$

This is a general quadratic equation of the form:

$$ax^2 + bx + c = 0$$

We now use the quadratic formula:

$$x = \frac{-b \pm \sqrt{b^2 - 4ac}}{2a}$$

and make the substitutions $a = 1$, $b = 1.4 \times 10^{-3}$ and $c = -1.4 \times 10^{-5}$. Entering these into the quadratic formula gives:

$$x = \frac{-1.4\times10^{-3} \pm \sqrt{(1.4\times10^{-3})^2 - 4(1)(-1.4\times10^{-5})}}{2(1)}$$

$$= 3.1 \times 10^{-3}\ \text{M and } x = -4.5 \times 10^{-3}\ \text{M}$$

We know that x cannot be negative because that would give negative concentrations for the ions, which is impossible, so we choose the first value as the correct one. This yields the following equilibrium concentrations.

$$[H_3O^+] = 3.1 \times 10^{-3}\ \text{M}$$

$$[ClCH_2COO^-] = 3.1 \times 10^{-3}\ \text{M}$$

$$[ClCH_2COOH] = 0.010 - x = 0.010 - 0.0031 = 0.007\ \text{M}$$

Finally, we calculate the pH of the solution.

$$\text{pH} = -\log(3.1 \times 10^{-3}) = 2.51$$

Notice that x is not negligible compared with the initial concentration, so the simplifying approximation would not have been valid. We can calculate the error that would be produced if we made the simplifying approximation in this case. We would obtain the same equation.

$$K_a = \frac{[H_3O^+][ClCH_2COO^-]}{[ClCH_2COOH]} = \frac{(x)(x)}{0.010 - x} = 1.4\times10^{-3}$$

and, making the assumption that $(0.010 - x) \approx 0.010$, we would then obtain:

$$K_a = \frac{[H_3O^+][ClCH_2COO^-]}{[ClCH_2COOH]} = \frac{(x)(x)}{0.010} = 1.4\times10^{-3}$$

Solving this for x would give:

$$x^2 = (0.010)(1.4 \times 10^{-3}) = 1.4 \times 10^{-5}$$

$$x = \sqrt{1.4\times10^{-5}} = 3.7 \times 10^{-3}$$

Therefore, making the assumption gives $[H_3O^+] = 3.7 \times 10^{-3}$ M, compared with the correct value of 3.1×10^{-3} M from solution of the quadratic equation, approximately a 20% error.

Is our answer reasonable?

As before, a quick check can be performed by substituting the calculated equilibrium concentrations into the equilibrium expression.

$$K_a = \frac{[H_3O^+][ClCH_2COO^-]}{[ClCH_2COOH]} = \frac{(x)(x)}{0.010 - x} = \frac{(3.1\times10^{-3})^2}{0.007} = 1.4\times10^{-3}$$

The value we obtain equals K_a, so the equilibrium concentrations are correct.

PRACTICE EXERCISE 11.17 The weak base trimethylamine, $(CH_3)_3N$, has $K_b = 6.3 \times 10^{-5}$. Calculate the pH of a 0.0010 M solution of trimethylamine.

11.5 The molecular basis of acid strength

Up until this point, we have labelled acids as either strong or weak, but have not explained what determines the strength of a particular acid. In this section, we will show that a variety of factors contribute to the overall strength of an acid.

Binary acids

A binary acid is defined as an acid containing H and only one other element, generally a nonmetal. The simplest binary acids are the monoprotic species HF, HCl, HBr and HI containing group 17 elements. Diprotic binary acids include H_2O and H_2S, whereas NH_3 and CH_4 are examples of, albeit extremely weak, triprotic and tetraprotic binary acids, respectively. The observed order of acid strengths of these acids is (from weakest to strongest).

$$CH_4 < NH_3 < H_2O < H_2S < HF \ll HCl < HBr < HI$$

In order for each of these compounds to act as an acid, a single H—X bond must be broken, so it might be expected there would be some correlation between the H—X bond enthalpy (which can be thought of as the ease with which the bond can be broken) and the acid strength. As can be seen from table 11.6, this is not the case.

TABLE 11.6 Bond enthalpies for cleavage of some H—X bonds.

Bond	Bond enthalpy for homolytic cleavage ($kJ\,mol^{-1}$)	Bond enthalpy for heterolytic cleavage ($kJ\,mol^{-1}$)
C—H	412	1744
N—H	388	1688
O—H	463	1633
S—H	338	1468
F—H	565	1554
Cl—H	431	1395
Br—H	366	1353
I—H	299	1314

This is perhaps not surprising when you consider the definition of bond enthalpy. Recall (pp. 318–19) that a bond enthalpy refers to the process:

$$H{-}X(g) \rightarrow H(g) + X(g)$$

in which the bond in the gas phase molecule undergoes **homolytic cleavage** (the bond breaks evenly so that one electron is given to each of the atoms involved in the bond) to give electrically neutral species; such a process is in fact very different from that involved in proton donation from H—X to water, which occurs via **heterolytic cleavage** (the bond breaks unevenly so that both electrons are given to one of the atoms involved in the bond) to form solvated ions, according to the equation:

$$H{-}X(aq) + H_2O(l) \rightarrow H_3O^+(aq) + X^-(aq)$$

A good correlation between the acidity and the heterolytic H—X bond energies in the gas phase does exist, as can be seen in table 11.6; these gas phase data, however, neglect solvent effects and do not therefore constitute an acceptable approximation of reality.

A more realistic approach that does include solvent effects uses the Hess's law cycle shown in figure 11.14.

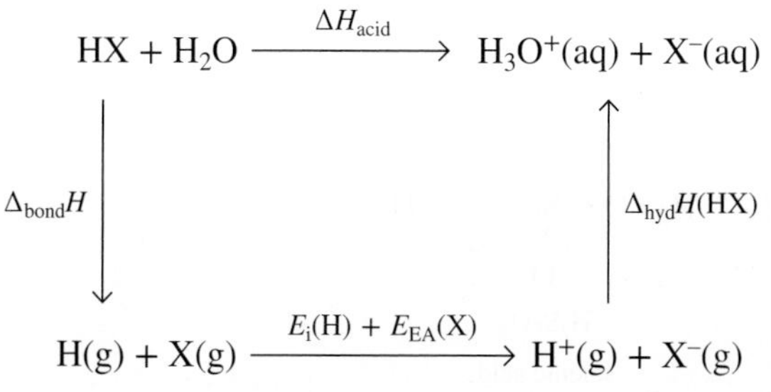

$HX + H_2O \xrightarrow{\Delta H_{acid}} H_3O^+(aq) + X^-(aq)$

$\Delta_{bond}H$ (down: $HX + H_2O \rightarrow H(g) + X(g)$)

$\Delta_{hyd}H(HX)$ (up: $H^+(g) + X^-(g) \rightarrow H_3O^+(aq) + X^-(aq)$)

$H(g) + X(g) \xrightarrow{E_i(H) + E_{EA}(X)} H^+(g) + X^-(g)$

FIGURE 11.14 A Hess's law cycle for the dissociation of an acid in aqueous solution.

The value of ΔH_{acid}, the enthalpy change for the dissociation of the acid in its standard state (gas for all acids except H_2O) in water, can be calculated for the binary acids listed in table 11.7 using available thermochemical data; $\Delta_{bond}H$ is the H—X bond energy (homolytic cleavage), E_i(H) is the ionisation energy of the gaseous hydrogen atom (1312 kJ mol^{-1}), E_{EA} is the electron affinity of X(g), and $\Delta_{hyd}H$ (HX) is the hydration enthalpy (p. 399) of gaseous HX.

TABLE 11.7 Enthalpy and energy changes involved in the dissociation of a variety of acids in aqueous solution at 25 °C.

Acid (HX)	$\Delta_{bond}H$ (kJ mol^{-1})	E_i(H) + E_{EA}(X) (kJ mol^{-1})	$\Delta_{hyd}H$(HX) (kJ mol^{-1})	ΔH_{acid} (kJ mol^{-1})
CH_4 (X = CH_3)	412	1303	−1583	132
NH_3 (X = NH_2)	388	1241	−1603	26
H_2O (X = OH)	463	1134	−1623	15[(a)]
H_2S (X = SH)	338	1090	−1443	−15
HF (X = F)	565	990	−1613	−58
HCl (X = Cl)	431	963	−1470	−76
HBr (X = Br)	366	987	−1439	−86
HI (X = I)	299	1017	−1394	−78

(a) To provide a meaningful comparison with the other acids, H_2O must be vaporised, a process which requires 41 kJ mol^{-1}. The final value of 15 kJ mol^{-1} includes this contribution.
(b) This approach is based on that of Fridgen, T. D., J. Chem. Educ., 2008, 85, 1220

It can be seen, from the data in table 11.7, that the trends in the ΔH_{acid} values correlate well with the acid strengths; as ΔH_{acid} becomes less positive (more negative), the acid strength increases. It can also be seen that no single contributing factor to ΔH_{acid} is primarily responsible for the observed trend in acid strengths. While it may be tempting to attribute the increasing acidity in the monoprotic acids from HF to HI to the significant decrease in $\Delta_{bond}H$, this is almost exactly opposed by the decrease in $\Delta_{hyd}H$ values. Similarly, an increase in electron affinity of X does appear to correlate well with an increase in acidity, but this breaks down for HBr and HI. It should also be noted that the above approach neglects entropic effects, which will undoubtedly contribute to the overall energetics of the acid dissociation process.

Having said this, two general observations concerning the acid strength of binary acids can be made.

1. The acid strength of binary acids increases going across a period (e.g. $CH_4 < NH_3 < H_2O < HF$).
2. The acid strength of binary acids increases going down a group (e.g. HF < HCl < HBr < HI).

The first observation has been attributed to the increasing electronegativity of X, rendering the H—X bond more polar, while the second has been explained by the increasing size of X, making the H—X bond longer and, therefore, weaker. As the data in table 11.7 show, however, such simple interpretations should be treated with caution, and are generally best avoided.

PRACTICE EXERCISE 11.18

Using only the periodic table, choose the stronger acid of each of the following pairs.
(a) H_2Se or HBr (b) H_2Se or H_2Te (c) CH_3OH or CH_3SH

Oxoacids

Acids composed of hydrogen, oxygen and some other element are called oxoacids (see table 11.8). Those that are strong acids in water are marked in the table by asterisks.

A feature common to the structures of all oxoacids is the presence of O—H groups bonded to some central atom. For example, the structures of two oxoacids of the group 16 elements are:

```
        O                   O
        ‖                   ‖
H—O—S—O—H           H—O—Se—O—H
        ‖                   ‖
        O                   O
```

H_2SO_4 sulfuric acid H_2SeO_4 selenic acid

TABLE 11.8 Some oxoacids of nonmetals and metalloids[a].

Group 14		**Group 17**	
H_2CO_3	carbonic acid[b]	HOF	hypofluorous acid[e]
Group 15		*$HClO_4$	perchloric acid
*HNO_3	nitric acid[a]	*$HClO_3$	chloric acid
HNO_2	nitrous acid	$HClO_2$	chlorous acid
H_3PO_4	phosphoric acid	HOCl	hypochlorous acid
H_3PO_3	phosphonic acid[c]	*$HBrO_4$	perbromic acid[f]
H_3AsO_4	arsenic acid	*$HBrO_3$	bromic acid
H_3AsO_3	arsenous acid	HOBr	hypobromous acid
Group 16		HIO_4 (H_5IO_6)[g]	periodic acid
*H_2SO_4	sulfuric acid	HOI	hypoiodous acid
H_2SO_3	sulfurous acid[d]	HIO_3	iodic acid
*H_2SeO_4	selenic acid		
H_2SeO_3	selenous acid		

(a) Strong acids are marked with asterisks.
(b) Predominantly $CO_2(aq)$.
(c) Phosphonic acid, despite its formula, is only a diprotic acid. It should more correctly be written as $HPO(OH)_2$.
(d) Hypothetical. An aqueous solution actually contains just dissolved sulfur dioxide, $SO_2(aq)$.
(e) Unstable at room temperature
(f) Pure perbromic acid is unstable; a dihydrate is known.
(g) H_5IO_6 is formed from $HIO_4 + 2H_2O$.

As with the binary acids, several general observations about the acid strengths of oxoacids can be made.

1. Within groups of oxoacids with the same number of oxygen atoms, the acid strength increases going up a group. This is best illustrated by oxoacids containing group 17 elements. The hypohalous acids, all of which are weak, increase in strength in the order HOI < HOBr < HOCl, with pK_a values of 10.64, 8.68 and 7.52, respectively. (Note that HOF is unstable, decomposing rapidly at room temperature.) For the strong group 17 oxoacids with the general formula HXO_4, the order of acid strength is $HIO_4 < HBrO_4 < HClO_4$. (Note that HFO_4 is unknown, as F cannot accommodate more than an octet of electrons.) Similarly, in the group 16 acids, we find that H_2SeO_4 is weaker than H_2SO_4 (the third member of this series, 'H_2TeO_4' actually exists as $Te(OH)_6$ and does not therefore provide a valid comparison); in group 15, H_3AsO_4 ($pK_a = 2.25$) is slightly weaker than H_3PO_4 ($pK_a = 2.15$).

PRACTICE EXERCISE 11.19

Which is the stronger acid, H_3AsO_3 or H_3PO_3?

2. Within groups of oxoacids with the same number of oxygen atoms, the acid strength increases across a period. This can be seen in oxoacids containing period 3 elements, where acid strength increases in the order $H_3PO_4 < H_2SO_4 < HClO_4$.
3. Acid strength increases as the number of lone oxygen atoms increases. (A lone oxygen atom is one that is bonded only to the central atom and not a hydrogen atom.) There are numerous examples that demonstrate this observation. For example, nitrous acid, HNO_2, which contains one lone oxygen atom, is a weak acid ($pK_a = 3.15$) while nitric acid, HNO_3, is a strong acid.

$$O{=}N{-}O{-}H \quad < \quad O_2N{-}O{-}H$$

nitrous acid nitric acid

A similar situation is seen with selenous acid, H_2SeO_3, and selenic acid, H_2SeO_4, where the latter, having two lone oxygen atoms, is a much stronger acid.

$$H{-}O{-}Se({=}O){-}O{-}H \quad < \quad H{-}O{-}Se({=}O)_2{-}O{-}H$$

H_2SeO_3 selenous acid

H_2SeO_4 selenic acid

Among oxoacids of the halogens, we find the same trend in acid strengths. For the oxoacids of chlorine, for instance, we find the trend:

$$HClO < HClO_2 < HClO_3 < HClO_4$$

Comparing their structures we have:

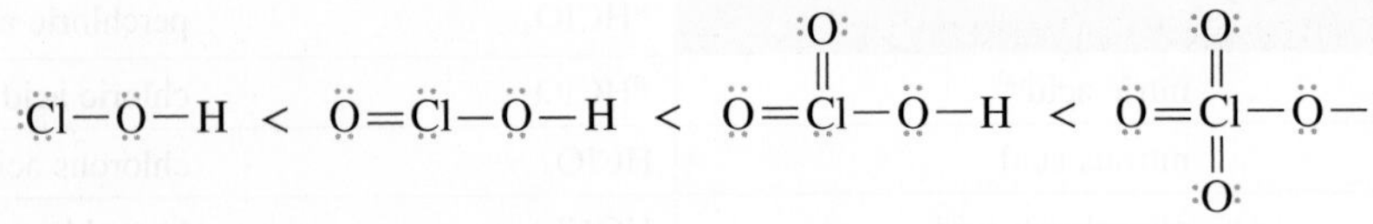

$$\text{:}\ddot{Cl}-\ddot{O}-H < \ddot{O}=\ddot{Cl}-\ddot{O}-H < \ddot{O}=\overset{\overset{:O:}{\|}}{\ddot{Cl}}-\ddot{O}-H < \ddot{O}=\underset{\underset{:O:}{\|}}{\overset{\overset{:O:}{\|}}{Cl}}-\ddot{O}-H$$

HClO $\quad$ $HClO_2$ $\quad$ $HClO_3$ $\quad$ $HClO_4$

The acid-strengthening effect of lone oxygen atoms is not limited to inorganic oxoacids. Consider, for example, ethanol and acetic acid.

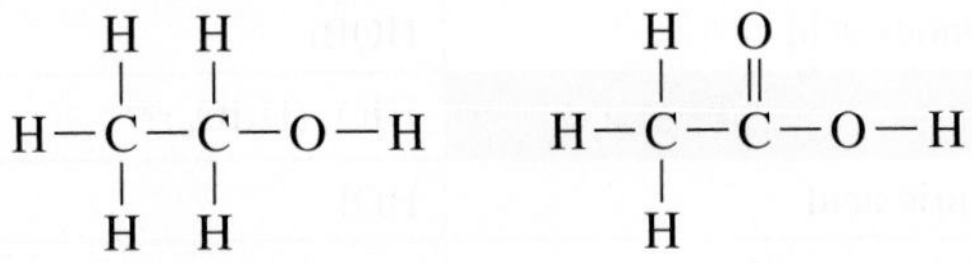

$$H-\underset{\underset{H}{|}}{\overset{\overset{H}{|}}{C}}-\underset{\underset{H}{|}}{\overset{\overset{H}{|}}{C}}-O-H \qquad H-\underset{\underset{H}{|}}{\overset{\overset{H}{|}}{C}}-\overset{\overset{O}{\|}}{C}-O-H$$

ethanol $\qquad$ acetic acid

Ethanol displays negligible acid properties in water ($pK_a = 15.9$). However, incorporation of a lone oxygen atom as a replacement for two protons yields acetic acid, which is eleven orders of magnitude more acidic ($pK_a = 4.74$). This enormous increase in acidity has been ascribed in part to the differing stabilities of the respective conjugate bases, which is governed by their ability to delocalise the negative charge arising from the loss of a proton.

The acetate ion can be drawn as two equivalent resonance structures (chapter 5), with the negative charge on either of the carboxylate O atoms. The actual structure of the acetate ion lies somewhere between these two forms, with the negative charge delocalised over the two O atoms and the carboxylate carbon atom.

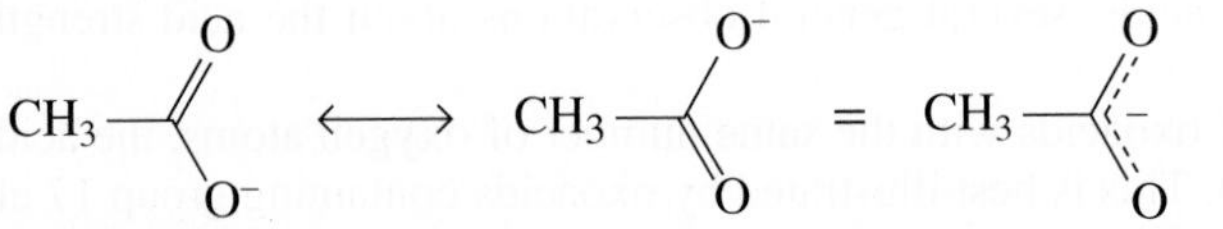

$$CH_3-C(=O)O^- \longleftrightarrow CH_3-C(O^-)=O \;=\; CH_3-C(\text{O}\cdots\text{O})^-$$

acetate ion

This 'spreading out' or **resonance stabilisation** of the negative charge results in a lowering in energy relative to the situation where the charge is localised on a single atom. The ethoxide ion cannot undergo such resonance stabilisation, and we therefore expect ethanol to be a weaker acid than acetic acid. An analogous situation occurs in the oxoacids of main-group elements. Consider the acids H_2SO_4 and H_3PO_4.

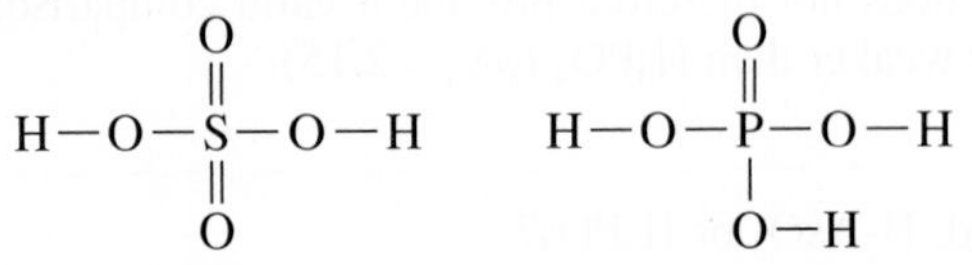

$$H-O-\underset{\underset{O}{\|}}{\overset{\overset{O}{\|}}{S}}-O-H \qquad H-O-\underset{\underset{O-H}{|}}{\overset{\overset{O}{\|}}{P}}-O-H$$

H_2SO_4 $\qquad$ H_3PO_4

sulfuric acid $\qquad$ phosphoric acid

As you can see in their structures, H_2SO_4 has two lone oxygen atoms and H_3PO_4 has only one. Loss of a proton from each gives the following species.

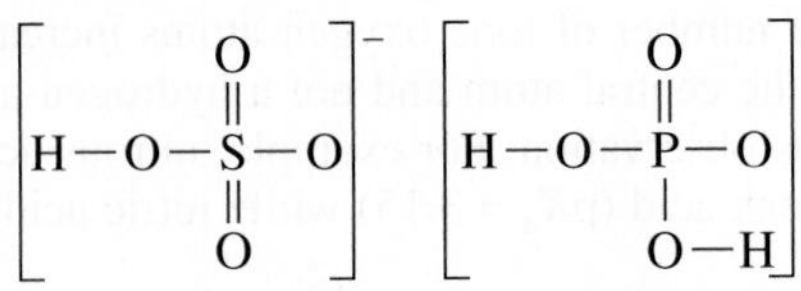

$$\left[H-O-\underset{\underset{O}{\|}}{\overset{\overset{O}{\|}}{S}}-O\right]^- \qquad \left[H-O-\underset{\underset{O-H}{|}}{\overset{\overset{O}{\|}}{P}}-O\right]^-$$

HSO_4^- $\qquad$ $H_2PO_4^-$

In oxoanions such as these, the lone oxygen atoms carry most of the negative charge, rather than the oxygen atoms bonded to hydrogen atoms. This charge actually spreads over the lone oxygen atoms. In HSO_4^-, therefore, each lone oxygen atom carries a charge of about $\frac{1}{3}-$. By the same reasoning, in $H_2PO_4^-$ each lone oxygen atom carries a charge of about $\frac{1}{2}-$. The smaller negative charge on the lone oxygen atoms in HSO_4^- makes this ion less able to attract H_3O^+ ions, so HSO_4^- is a weaker base than $H_2PO_4^-$. In other words, HSO_4^- cannot become H_2SO_4 again as readily as $H_2PO_4^-$ can become H_3PO_4. There are thus two factors that make H_2SO_4 a stronger acid than H_3PO_4 in water. One is the greater tendency of H_2SO_4 to donate a proton and the other is the lesser tendency of HSO_4^- to accept a proton.

PRACTICE EXERCISE 11.20

In each pair, select the stronger acid.
(a) HIO_3 or HIO_4 (b) H_3AsO_3 or H_3AsO_4

Chemical Connections

The strongest acid and the strongest base

In this chapter, we have discussed the concept of acid and base strength, and we have related this to the extent to which the acid or base donates a proton to, or accepts a proton from, water. The obvious question that then arises is: what is the strongest acid and what is the strongest base? In water, the answer to this question is, as we have seen (p. 444), H_3O^+ and OH^-. However, if we use different experimental conditions, then we find that much stronger acids and bases exist. Of these, both the strongest acid and strongest base have an Australasian connection.

In 2004, Professor Christopher Reed, an expat New Zealand Professor of Chemistry at the University of California at Riverside, in the USA, published a paper with a number of colleagues titled 'The strongest isolable acid'. The acid in question is a carborane, a molecule with a framework comprising carbon and boron atoms, and has the chemical formula $H(CHB_{11}Cl_{11})$. The structure of this acid, which is a solid at room temperature, is shown in figure 11.15. The extraordinary acid strength of $H(CHB_{11}Cl_{11})$ was quantified using NMR (nuclear magnetic resonance) and IR (infrared) spectroscopies (see chapter 20) and these techniques showed it to be at least 1 million times stronger than sulfuric acid.

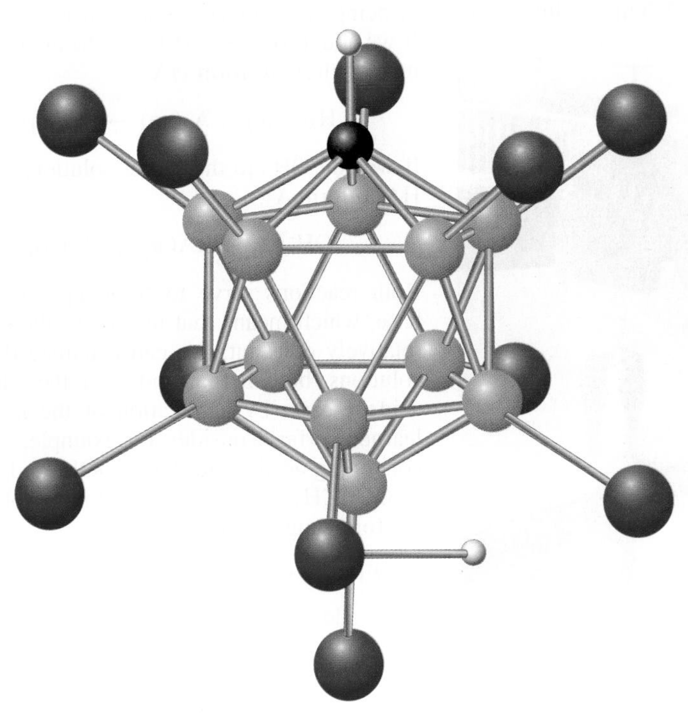

FIGURE 11.15 Representation of the structure of a carborane acid with the chemical formula, $H(CHB_{11}Cl_{11})$

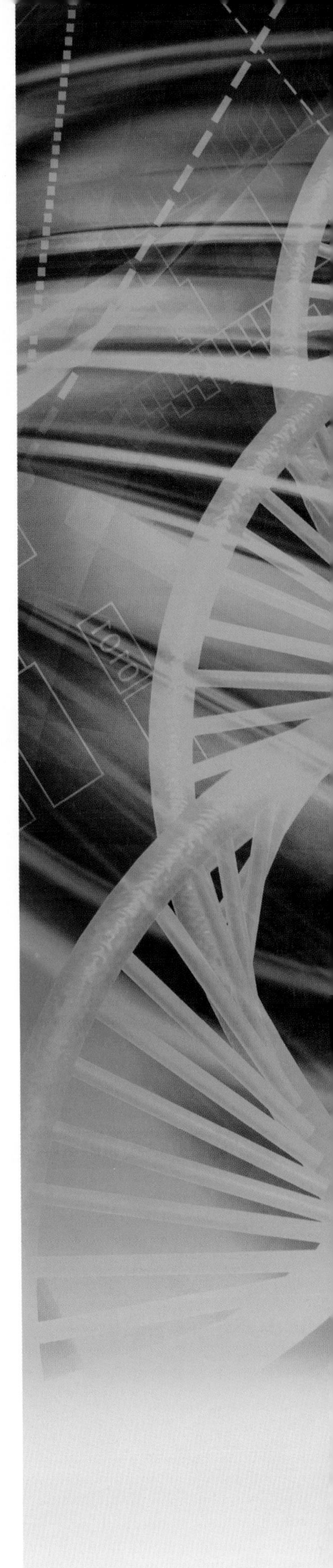

The strongest known base, by contrast, has the very simple chemical formula LiO^-. This was shown by Professor Leo Radom, from the University of Sydney, along with Australian and US colleagues, in 2008. Using a combination of mass spectrometry (see chapter 20) and sophisticated computational techniques, these workers demonstrated that, in the gas phase, LiO^- was a stronger base than CH_3^-, the previous strongest gas-phase base. They also showed that it will be difficult to find any stronger base than this. However, unlike the strongest acid, gas-phase LiO^- cannot be put in a bottle, so its extreme basicity cannot be exploited in the chemistry laboratory. It was, however, 'isolated' in a mass spectrometer, which makes it a real chemical species.

Extraordinarily strong acids and bases (superacids and superbases) are not merely scientific curiosities. Indeed, the Hungarian-born US chemist George Olah was awarded the Nobel Prize in chemistry in 1994 for his seminal work in the field of superacids, results of which found significant applications in the petrochemical industry.

11.6 Buffer solutions

The control and maintenance of pH is crucial in many chemical and biological systems. For example, if the pH of your blood, which is typically 7.35 to 7.42, were to change to either 7.00 or 8.00, you would die. Fortunately, nature has developed a sophisticated chemical system which maintains blood pH within the vital limits. Figure 11.16 shows another application of a buffer solution.

A **buffer solution** is one that contains appreciable amounts of both a weak acid and its conjugate base, or a weak base and its conjugate acid. As a result of this, a buffer solution resists change in pH after the addition of small amounts of either acid or base, and also on moderate dilution. The pH values over which a buffer is effective at maintaining the pH are determined by either pK_a or pK_b of the weak acid or weak base, respectively, and the ratio of the concentrations of conjugate pairs present in the solution.

pH calculations in buffer solutions

We will first consider a buffer solution containing equal amounts of a weak acid HA and its conjugate base A^-. A buffer solution works by being able to react to neutralise added H_3O^+ or OH^-. If we add a source of H_3O^+ to the above solution, it will react with A^- to form HA.

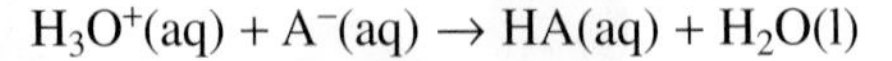

$$H_3O^+(aq) + A^-(aq) \rightarrow HA(aq) + H_2O(l)$$

If we add OH^- to the buffer solution, it will react with HA to form A^-.

$$OH^-(aq) + HA(aq) \rightarrow A^-(aq) + H_2O(l)$$

Both reactions serve to 'mop up' any added acid or base, which means that the pH of the solution remains relatively constant. We can calculate the pH of buffer solutions, both before and after the addition of either acid or base, by application of the methods we have learned so far. Consider, for example, a buffer solution containing acetic acid, CH_3COOH, and sodium acetate, $NaOOCCH_3$. Addition of acid to this solution will cause the following reaction to occur:

$$H_3O^+(aq) + CH_3COO^-(aq) \rightarrow CH_3COOH(aq) + H_2O(l)$$

while added base will be neutralised as follows:

$$OH^-(aq) + CH_3COOH(aq) \rightarrow CH_3COO^-(aq) + H_2O(l)$$

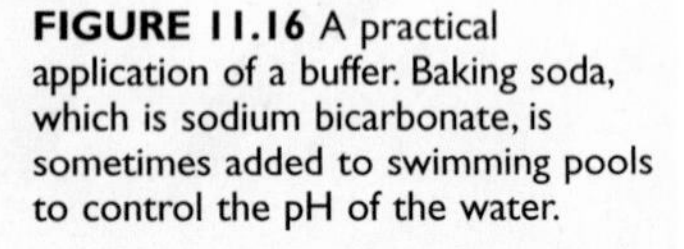

FIGURE 11.16 A practical application of a buffer. Baking soda, which is sodium bicarbonate, is sometimes added to swimming pools to control the pH of the water.

Worked example 11.17 shows how we carry out calculations involving the pH of buffer solutions.

WORKED EXAMPLE 11.17

Calculating the pH of a buffer

To study the effects of a weakly acidic medium on the rate of corrosion of a metal alloy, a student prepared a buffer solution in which $[NaOOCCH_3] = 0.11$ M and $[CH_3COOH] = 0.090$ M. What is the pH of this solution at 25 °C?

Analysis

The buffer solution contains both the weak acid CH_3COOH and its conjugate base CH_3COO^-, and we are given the concentrations of both. We can solve the problem by using either K_a for CH_3COOH or K_b for CH_3COO^-. We will show in detail how to use the former, and will look at the latter in practice exercise 11.24. We begin by looking up the value of K_a for CH_3COOH, then set up a concentration table and solve for $[H_3O^+]$ using the simplifying approximations developed earlier.

Solution

We begin with the balanced chemical equation and the expression for K_a.

$$CH_3COOH(aq) + H_2O(l) \rightleftharpoons CH_3COO^-(aq) + H_3O^+(aq)$$

$$K_a = \frac{[H_3O^+][CH_3COO^-]}{[CH_3COOH]} = 1.8 \times 10^{-5}$$

We set up the concentration table and take the initial concentrations of CH_3COOH and CH_3COO^- to be the values given in the problem. We assume the initial $[H_3O^+] = 0$ and therefore this must increase by $+x$ as the reaction proceeds to equilibrium. The other changes follow from that. The completed table is shown below.

	H_2O + CH_3COOH	$\rightleftharpoons$	H_3O^+	+ CH_3COO^-
Initial concentration (mol L^{-1})	0.090		0	0.11
Change in concentration (mol L^{-1})	$-x$		$+x$	$+x$
Equilibrium concentration (mol L^{-1})	$0.090 - x$		x	$0.11 + x$

For buffer solutions, the quantity x will be very small. This is because the conjugate base of the weak acid is already present in solution, so this suppresses the dissociation of the weak acid as a result of the common ion effect. We can therefore safely assume that $(0.090 - x) \approx 0.090$ and $(0.11 + x) \approx 0.11$. Substituting these quantities into the K_a expression gives:

$$K_a = \frac{[H_3O^+][CH_3COO^-]}{[CH_3COOH]} = \frac{(x)(0.11+x)}{0.090-x} \approx \frac{(x)(0.11)}{0.090} = 1.8 \times 10^{-5}$$

We solve for x as follows.

$$0.11x = (0.090)(1.8 \times 10^{-5})$$

$$x = \frac{(0.090)(1.8 \times 10^{-5})}{0.11}$$

$$= 1.5 \times 10^{-5}$$

Because x equals $[H_3O^+]$, we now have $[H_3O^+] = 1.5 \times 10^{-5}$ M. Then we calculate pH.

$$pH = -\log[H_3O^+] = -\log(1.5 \times 10^{-5})$$

$$= 4.82$$

Thus the pH of the buffer is 4.82.

Is our answer reasonable?

We can check the answer in the usual way by substituting our calculated equilibrium values into the expression for K_a.

$$K_a = \frac{[H_3O^+][CH_3COO^-]}{[CH_3COOH]} = \frac{(1.5 \times 10^{-5})(0.11)}{0.090} = 1.8 \times 10^{-5}$$

This equals K_a, so the values we have obtained are correct equilibrium concentrations.

PRACTICE EXERCISE 11.21

Calculate the pH of the buffer solution in the preceding example by using the K_b for CH_3COO^-. Be sure to write the chemical equation for the equilibrium as the reaction of CH_3COO^- with water. Then use the chemical equation as a guide in setting up the equilibrium expression for K_b. If you work through the problem correctly, you should obtain the same answer as above.

We can rearrange the expression for K_a into a form that is particularly useful for calculations involving buffer solutions, and which also removes the need to write a concentration table for each calculation.

We know that, for the equilibrium:

$$HA(aq) + H_2O(l) \rightleftharpoons H_3O^+(aq) + A^-(aq)$$

we can write K_a as:

$$K_a = \frac{[H_3O^+][A^-]}{[HA]}$$

If we multiply both sides of the equation by [HA] and divide both sides by $[A^-]$, we obtain:

$$[H_3O^+] = K_a \frac{[HA]}{[A^-]}$$

If we take the negative logarithm (p) of both sides, realising that the logarithm of a product of two numbers is the sum of the logarithms of those numbers, and that $-\log \frac{x}{y} = \log \frac{y}{x}$, we obtain:

$$\text{pH} = \text{p}K_a + \log \frac{[\text{A}^-]}{[\text{HA}]}$$

This rearrangement of the K_a expression is called the **Henderson–Hasselbalch equation**. This form of the K_a expression emphasises the fact that the pH of a buffer solution depends on both the $\text{p}K_a$ of the weak acid and the ratio of conjugate base to acid. Recalling that $c = \frac{n}{V}$, we can rewrite the equation as:

$$\text{pH} = \text{p}K_a + \log \left(\frac{\frac{n_{\text{A}^-}}{V}}{\frac{n_{\text{HA}}}{V}} \right)$$

and therefore:

$$\text{pH} = \text{p}K_a + \log \frac{n_{\text{A}^-}}{n_{\text{HA}}}$$

Therefore, we can express this relationship in terms of either a concentration ratio or a mole ratio. Note that, under the special condition where $[\text{HA}] = [\text{A}^-]$ (or $n_{\text{HA}} = n_{\text{A}^-}$), $\frac{[\text{A}^-]}{[\text{HA}]} = 1$, $\log(1) = 0$, and therefore $\text{pH} = \text{p}K_a$.

A further consequence of the relationship derived above is that the pH of a buffer should not change if the buffer undergoes moderate dilution. Dilution changes the volume of a solution but it does not change the amounts of the solutes, so their mole ratio remains constant and so does $[\text{H}_3\text{O}^+]$.

WORKED EXAMPLE 11.18

Calculating the pH of an ammonia/ammonium ion buffer

To study the influence of an alkaline medium on the rate of a reaction, a student prepared a buffer solution by dissolving 0.12 mol of NH_3 and 0.095 mol of NH_4Cl in 250 mL of water. What is the pH of the buffer at 25 °C?

Analysis

The pH of the buffer is determined by the mole ratio of the members of the acid–base pair; so to calculate the pH we use the amounts of NH_3 and NH_4^+ directly in the Henderson–Hasselbalch equation. We also require the $\text{p}K_a$ of NH_4^+.

Solution

Given that $\text{p}K_a(\text{NH}_4^+) = 9.26$, we can insert the values directly into the Henderson–Hasselbalch equation.

$$\text{pH} = \text{p}K_a + \log \frac{n_{\text{A}^-}}{n_{\text{HA}}} = 9.26 + \log \frac{0.12}{0.095} = 9.36$$

Note that we assumed that the initial concentrations of the acid and conjugate base are the equilibrium concentrations. We are usually justified in doing this.

Is our answer reasonable?

There is more base than acid in the buffer. Therefore, the ratio $\frac{[\text{A}^-]}{[\text{HA}]}$ is greater than 1, and the log of this ratio will be positive. We would therefore expect a pH greater than the $\text{p}K_a$, which we have obtained.

PRACTICE EXERCISE 11.22

Determine the pH of the buffer in worked example 11.18 when $[\text{NH}_3] = 0.08$ M and $[\text{NH}_4\text{Cl}] = 0.15$ M.

Buffer solutions function most efficiently when the ratio $\frac{[\text{A}^-]}{[\text{HA}]}$ is close to 1. When this is the case, the concentrations of both the weak acid and its conjugate base are essentially equal, and the buffer can therefore resist change in pH on the addition of either acid or base. As we showed above, when $\frac{[\text{A}^-]}{[\text{HA}]} = 1$, $\text{pH} = \text{p}K_a$, and therefore the $\text{p}K_a$ of the weak acid determines where, on the pH scale, a buffer can work best. Thus, to prepare a specific buffer for use at a particular pH, we first select a weak acid with a $\text{p}K_a$ near the desired pH. Then, by experimentally adjusting the ratio $\frac{[\text{A}^-]}{[\text{HA}]}$, we can make a final adjustment to obtain the desired pH.

If the ratio $\frac{[A^-]}{[HA]}$ is significantly different from 1, then the concentrations of the weak acid and its conjugate base in solution are no longer similar. If, for example, $[HA] > [A^-]$, the buffer will resist change in pH well on the addition of base, but it will be less effective at resisting change in pH on addition of acid, as there is less A^- present for the added acid to react with. Once all the A^- has reacted, the pH of the solution will change rapidly. As long as the ratio $\frac{[A^-]}{[HA]}$ in a buffer solution is between $\frac{1}{10}$ and $\frac{10}{1}$, the buffer will maintain pH effectively. If we substitute these values into the Henderson–Hasselbalch equation, we obtain $pH = pK_a - 1$ and $pH = pK_a + 1$, respectively. This means that the operational range of any buffer is 1 pH unit either side of the pK_a of the weak acid, i.e. $pH = pK_a \pm 1$.

WORKED EXAMPLE 11.19

Preparing a buffer solution with a predetermined pH

A solution buffered at a pH of 5.00 is needed in an experiment. Can we use acetic acid and sodium acetate to prepare the buffer? If so, what amount of $NaOOCCH_3$ must be added to 1.0 L of a solution that contains 1.0 mol $NaOOCCH_3$ at 25 °C?

Analysis

There are two parts to this problem. To answer the first, we check the pK_a of acetic acid to see if it is in the desired range of $pH = pK_a \pm 1$. If it is, we can calculate the necessary mole ratio. Once we have this, we can proceed to calculate the amount of CH_3COO^- needed and then the amount of $NaOOCCH_3$.

Solution

The pK_a of acetic acid (4.74) falls in the desired range and therefore acetic acid can be used with the acetate ion to make the buffer.

We find the required mole ratio of solutes by using the Henderson–Hasselbalch equation.

$$pH = pK_a + \log \frac{n_{A^-}}{n_{HA}}$$

$$5.00 = 4.74 + \log \frac{n_{CH_3COO^-}}{n_{CH_3COOH}}$$

Therefore:

$$\log \frac{n_{CH_3COO^-}}{n_{CH_3COOH}} = 0.26$$

and hence:

$$\frac{n_{CH_3COO^-}}{n_{CH_3COOH}} = 1.82$$

$$n_{CH_3COO^-} = 1.82 \times n_{CH_3COOH}$$

We know that $n_{CH_3COOH} = 1.0$ mol and so $n_{CH_3COO^-} = 1.82$ mol.

Because each mole of $NaOOCCH_3$ contains 1 mole of CH_3COO^-, we require 1.82 mol $NaOOCCH_3$.

Is our answer reasonable?

A 1 : 1 mole ratio of CH_3COO^- to CH_3COOH would give $pH = pK_a = 4.74$. The desired pH of 5.00 is slightly more basic than 4.74, so the amount of conjugate base should be larger than the amount of conjugate acid. Our answer of 1.82 mol of $NaOOCCH_3$ appears to be reasonable.

PRACTICE EXERCISE 11.23

As part of a study into the rate of a particular chemical reaction, a student requires an aqueous buffer with a pH of 9.40. Why would a mixture of ammonia (NH_3) and ammonium chloride (NH_4Cl) be suitable for buffering at this pH? What mole ratio of ammonia to ammonium chloride is needed? What mass of NH_4Cl would have to be added to a solution that contains 0.10 mol NH_3 at 25 °C?

Once we have prepared a buffer solution, we can calculate the pH change on addition of a specified amount of strong acid or base by assuming that the added strong acid or base reacts quantitatively with the weak acid or base component of the buffer solution. This is illustrated in worked example 11.20.

WORKED EXAMPLE 11.20

Calculating the pH change in a buffer solution

What pH change will occur after the addition of 0.10 L of 0.10 M HCl to 1.0 L of a buffer solution containing 0.10 mol CH_3COOH and 0.10 mol $NaOOCCH_3$ at 25 °C? How does this compare with the pH change that would occur after the addition of the same amount of HCl to 1.0 L of pure water?

Analysis

We are asked for a pH *change*, which means we have to calculate the pH of the initial buffer solution. We then calculate the amount of HCl added and assume that this reacts completely with CH_3COO^- to form CH_3COOH. We then use the new amounts of CH_3COO^- and CH_3COOH in the Henderson–Hasselbalch equation to calculate the final pH. The question involving pure water is a straightforward calculation of the pH of a strong acid.

Solution

The initial pH can be obtained from the initial $[CH_3COOH]$ and $[CH_3COO^-]$ values using the Henderson–Hasselbalch equation.

$$\begin{aligned} pH &= pK_a + \log\frac{[A^-]}{[HA]} \\ &= 4.74 + \log\frac{0.1}{0.1} \\ &= 4.74 \end{aligned}$$

We now carry out a stoichiometric calculation to determine the equilibrium n_{CH_3COOH} and $n_{CH_3COO^-}$ values. The reaction which will occur on addition of the strong acid is:

$$HCl(aq) + CH_3COO^-(aq) \rightarrow CH_3COOH(aq) + Cl^-(aq)$$

We assume this reaction goes to completion, so that:

$$\begin{aligned} n_{HCl\ (added)} &= n_{CH_3COO^-\ (used)} = n_{CH_3COOH\ (formed)} \\ n_{HCl\ (added)} &= c \times V \\ &= 0.10\ mol\ L^{-1} \times 0.10\ L \\ &= 0.010\ mol \end{aligned}$$

Therefore, 0.010 mol HCl(aq) reacts with 0.010 mol $CH_3COO^-(aq)$ to form 0.010 mol $CH_3COOH(aq)$. Initially we had 0.10 mol CH_3COOH and 0.10 mol CH_3COO^-, and so the final amounts are:

$$\begin{aligned} n_{CH_3COOH\ (final)} &= 0.10\ mol + 0.010\ mol = 0.11\ mol \\ n_{CH_3COO^-\ (final)} &= 0.10\ mol - 0.010\ mol = 0.090\ mol \end{aligned}$$

We now substitute these values into the Henderson–Hasselbalch equation.

$$\begin{aligned} pH &= 4.74 + \log\frac{0.090}{0.11} \\ &= 4.65 \end{aligned}$$

Therefore, the *change* in pH is –0.09 pH units, the negative sign showing that this is a decrease in pH.

When we add 0.10 L of 0.10 mol L^{-1} HCl to 1.0 L of water, the final solution contains 0.010 mol HCl in 1.1 L. As HCl is a strong acid, it reacts completely with water to form H_3O^+ and hence:

$$[H_3O^+] = \frac{0.010\ mol}{1.1\ L} = 0.0091\ mol\ L^{-1}$$

Thus, $pH = -\log[H_3O^+] = -\log(0.0091) = 2.04$.

We started with pure water, and so the initial pH was 7.00. Hence, the pH change is –4.96 pH units.

Is our answer reasonable?

We started with the buffer solution at pH 4.74 and added a small amount of strong acid to it. We expect that the final pH should be slightly less than this, which it is.

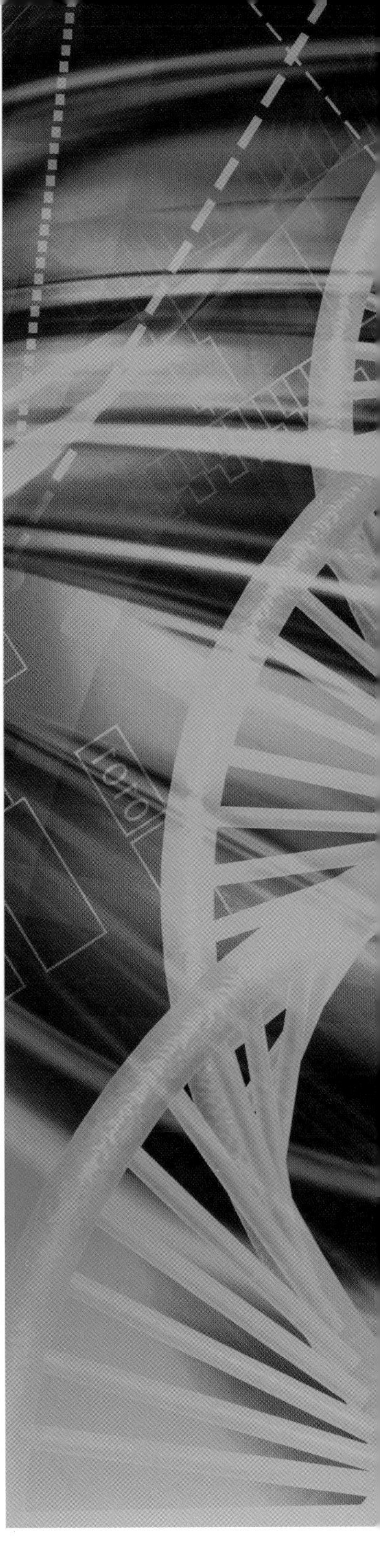

Chemical Connections

Buffers in biological systems

As we stated at the start of section 11.6, it is crucial that the pH of our blood is kept within certain limits. Indeed, this is true not only of our blood, but of the majority of our bodily fluids — the pH of the fluid inside our cells (the intracellular fluid) needs to be maintained at a nearly constant value while chemical reactions involving the consumption and release of H_3O^+ ions are occurring at rapid rates. So how does nature accomplish this? Not surprisingly, through the use of buffers.

The $H_2PO_4^-/HPO_4^{2-}$ buffer system is particularly important in the intracellular fluid, as well as in the urine. The $H_2PO_4^-$ anion has a pK_a of 7.20, meaning that the $H_2PO_4^-/HPO_4^{2-}$ system will act as an effective buffer around this pH, as both $H_2PO_4^-$ and HPO_4^{2-} will be present in significant amounts.

The H_2CO_3/HCO_3^- buffer system aids in regulating the pH of the blood and the extracellular fluid (the fluid outside our cells). Although we write the formula H_2CO_3 and call the molecule 'carbonic acid', as you will see in chapter 14 (p. 635), H_2CO_3 should be thought of as predominantly $CO_2(g)$ dissolved in water. The apparent pK_a of H_2CO_3 is 6.35, which means it should not be a good buffer at all for pH values around 7.4. However, the actual buffering process is rather complicated and the fact that the body can independently regulate both the partial pressure of CO_2 (and hence $[H_2CO_3]$) and $[HCO_3^-]$ renders this buffer system ideal for blood.

Other buffers in our body include the NH_4^+/NH_3 system, bone, in the form of $CaCO_3$, and a variety of proteins, of which one important example is haemoglobin. Proteins act as buffers around physiological pH because of the presence of the amino acid histidine (His). The sidechain of this amino acid contains an imidazole ring, which acts as a weak base through use of its non-protonated N atom to accept a proton from water. The pK_a of the conjugate acid of imidazole is 7.00, which makes this group perfect for buffering biological systems. Therefore, proteins with high numbers of histidine residues can be very good buffers around pH 7.

Histidine (His) — imidazole

$$\text{imidazolium (N–H)} + H_2O \xrightleftharpoons{pKa = 7.00} \text{imidazole} + H_3O^+$$

All of the buffers discussed above are necessary to keep what in reality are tiny concentrations of H_3O^+ constant, thereby keeping us alive.

PRACTICE EXERCISE 11.24

Calculate the pH change on the addition of 0.10 L of 0.10 M NaOH to 1.0 L of a buffer solution containing 0.08 mol $NH_3(aq)$ ($pK_b = 4.74$) and 0.12 mol $NH_4Cl(aq)$ at 25 °C.

Worked example 11.20 illustrates the effectiveness of a buffer solution. When the strong acid is added to water, the solution becomes 4.96 pH units more acidic, whereas the buffer solution is little affected, becoming only 0.09 pH units more acidic.

If enough strong acid is added to react with all of the base component of a buffer solution, the mixture is no longer able to neutralise any more strong acid. We can illustrate this with reference to worked example 11.20. Initially there is 0.10 mol CH_3COO^- in the buffer solution; if 0.10 mol HCl is added, this will react completely with the 0.10 mol CH_3COO^- to give 0.10 mol CH_3COOH. When this happens, the concentration of CH_3COO^- is very small (all of the CH_3COO^- in solution at this point will come from dissociation of CH_3COOH) and the solution no longer functions as a buffer towards the addition of more acid. Similarly, if we added 0.10 mol NaOH to the original buffer solution in the example, it would react completely with the 0.10 mol CH_3COOH present to form 0.10 mol CH_3COO^-, thereby leaving essentially no CH_3COOH in the solution to react with any further added NaOH. This means that there is a limit to the amount of added acid or base that a buffer solution can absorb. The **buffer capacity** of any buffer is a measure of the amount of H_3O^+ or OH^- that can be added without significant change in the pH. In the example, the addition of ~0.10 mol HCl or ~0.10 mol NaOH would exhaust the capacity of the buffer solution.

11.7 Acid–base titrations

We often encounter solutions of acid or bases with unknown concentrations in chemistry. One method of determining the concentration of an acid or a base is to carry out an **acid–base titration** (figure 11.17). This involves gradual addition of an acid or base of known concentration (the **titrant**) from a **burette** to an accurately known volume of the unknown solution. An **acid–base indicator** changes colour when the **endpoint** of the titration is reached, and, if we have chosen the indicator correctly, the endpoint will be a good approximation of the **equivalence point** of the titration. The equivalence point is the point at which the reaction stoichiometry is satisfied, or, in other words, where the amount of titrant added exactly equals the amount of acid or base initially present. The equivalence point is sometimes called the stoichiometric point. In all acid–base titrations, regardless of the strengths of the acid or base involved, the essential reaction that occurs is:

$$H_3O^+(aq) + OH^-(aq) \rightarrow 2H_2O(l)$$

FIGURE 11.17 An acid–base titration.

Knowledge of both the stoichiometry of the reaction and the volume of acid or base added allows us to calculate the concentration of the unknown solution. We will now examine acid–base titrations in a little more detail, with particular emphasis on the variations in pH that occur as the titrations proceed.

Strong acid – strong base and strong base – strong acid titrations

Figure 11.18 shows a plot of pH versus volume of titrant added for the titration of 25.00 mL of 0.200 M HCl(aq) with 0.200 M NaOH(aq). We call such a plot a **titration curve**, and it has the same general shape for the titration of any strong acid with any strong base. Likewise, the titration curve for the titration of 25.00 mL of 0.200 M NaOH with 0.200 M HCl is depicted in figure 11.19, and this too is general for the titration of any strong base with any strong acid.

We will consider the strong acid – strong base (note that the titrant comes last when describing a titration) titration curve in detail, realising that analogous arguments apply for the strong base – strong acid titration.

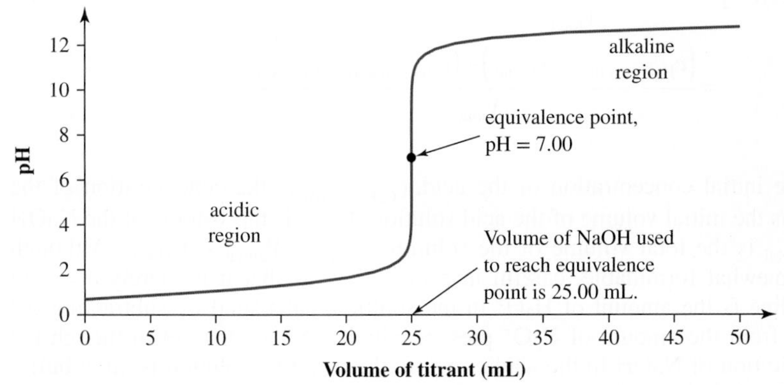

FIGURE 11.18 Titration curve for the titration of a strong acid with a strong base. Here we follow how the pH changes during the titration of 25.00 mL of 0.200 M HCl(aq) with 0.200 M NaOH(aq).

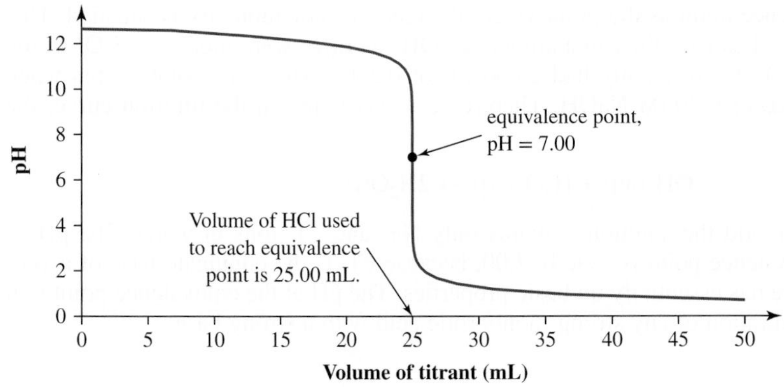

FIGURE 11.19 Titration curve for the titration of a strong base with a strong acid (25.00 mL 0.200 M NaOH(aq) with 0.200 M HCl(aq)).

There are four important points and regions to consider for a strong acid – strong base titration curve:

1. the initial pH
2. the acidic region
3. the equivalence point
4. the alkaline region.

The initial pH

The initial pH in a strong acid – strong base titration is the pH of the strong acid being titrated. Because the acid is strong, it will react completely with water to generate H_3O^+, and therefore we can calculate the pH by taking the negative logarithm of the acid concentration. For the titration curve in figure 11.18, the initial pH will be:

$$\begin{aligned} pH &= -\log[H_3O^+] \\ &= -\log(0.200) \\ &= 0.70 \end{aligned}$$

The acidic region

In the acidic region, we are adding OH^- to a solution of H_3O^+, and the essential reaction that occurs is:

$$OH^-(aq) + H_3O^+(aq) \rightarrow 2H_2O(l)$$

As you can see from figure 11.18, the pH initially changes very slowly on addition of NaOH(aq), and it is not until the equivalence point is approached that the pH starts to increase significantly. In the acidic region, the solution contains H_3O^+ and Cl^- ions from the acid, and Na^+ ions from the base, and the pH is governed by the amount of unreacted H_3O^+ in the solution. The concentration of H_3O^+ decreases as NaOH is added, and, at any point in this region, it can be calculated from the equation:

$$[H_3O^+] = \frac{n_{H_3O^+(\text{initial})} - n_{OH^-(\text{added})}}{V_{\text{total}}} = \frac{\left(c_{H_3O^+(\text{initial})} \times V_{\text{initial}}\right) - \left(c_{OH^-(\text{added})} \times V_{\text{added}}\right)}{V_{\text{total}}}$$

where $c_{H_3O^+(\text{initial})}$ is the initial concentration of the acid, $c_{OH^-(\text{added})}$ is the concentration of the NaOH solution, V_{initial} is the initial volume of the acid solution, V_{added} is the volume of the NaOH solution added and V_{total} is the total volume of the solution ($V_{\text{total}} = V_{\text{initial}} + V_{\text{added}}$). Although this equation looks somewhat formidable, careful inspection reveals that it is merely $c = \frac{n}{V}$ in another form; the top line is the amount of H_3O^+ in the solution, calculated by subtracting the amount of OH^- added from the amount of H_3O^+ present initially. Notice that, even though the pH changes little on addition of NaOH in the acidic region, the resulting solution is not a buffer solution — the pH of this solution will change significantly on moderate dilution, in contrast to a solution containing significant amounts of a weak acid and its conjugate base.

The equivalence point

We defined the equivalence point as the point where the reaction stoichiometry is satisfied. This means that we have added exactly the same amount of OH^- as there were moles of H_3O^+ in the initial solution of acid. As we originally had 25.00 mL of 0.200 M HCl, this point occurs when we have added 25.00 mL of 0.200 M NaOH. Therefore, at this point on the titration curve, the reaction:

$$OH^-(aq) + H_3O^+(aq) \rightarrow 2H_2O(l)$$

has gone to completion, and the solution contains only Na^+ and Cl^- ions in water. The pH of the solution at the equivalence point is exactly 7.00, because Cl^- is the conjugate base of a very strong acid and therefore has essentially no basic properties. The pH at the equivalence point will *always* be 7.00 for the titration of any strong monoprotic acid with a strong base.

The alkaline region

Beyond the equivalence point, we are adding excess NaOH(aq) to a solution containing Na^+(aq) and Cl^-(aq), so there is no chemical reaction occurring. The pH in this region is governed by the amount of excess NaOH(aq) added to the solution and will always be > 7. The concentration of OH^-, and hence the pH, can be calculated from the following equation:

$$[OH^-] = \frac{n_{OH^-(\text{total})} - n_{OH^-(\text{equiv.})}}{V_{\text{total}}} = \frac{\left(c_{OH^-(\text{initial})} \times V_{\text{added}}\right) - \left(c_{OH^-(\text{initial})} \times V_{\text{equiv.}}\right)}{V_{\text{total}}}$$

where $c_{OH^-(\text{initial})}$ is the concentration of the NaOH solution, V_{added} is the total volume of NaOH solution added, $V_{\text{equiv.}}$ is the volume of NaOH solution added at the equivalence point (in this case, 25.00 mL) and V_{total} is the total volume of the solution. The top line gives the amount of OH^- in excess of that required to reach the equivalence point; this is obtained by subtracting the amount of OH^- used to reach the equivalence point from the total amount of OH^- added.

Weak acid – strong base and weak base – strong acid titrations

Figure 11.20 shows the titration curve for the titration of 25.00 mL of 0.200 M CH_3COOH(aq) with 0.200 M NaOH(aq), while the titration curve for the titration of 25.00 mL of 0.200 M NH_3 with 0.200 M HCl is given in figure 11.21. We will focus on the weak acid – strong base curve, realising that analogous arguments can be made for the weak base – strong acid case. As you can see, the weak acid – strong base curve looks rather different from the strong acid – strong base curve (figure 11.18), especially in the acidic region. We will again consider in detail the four important points and regions on this curve.

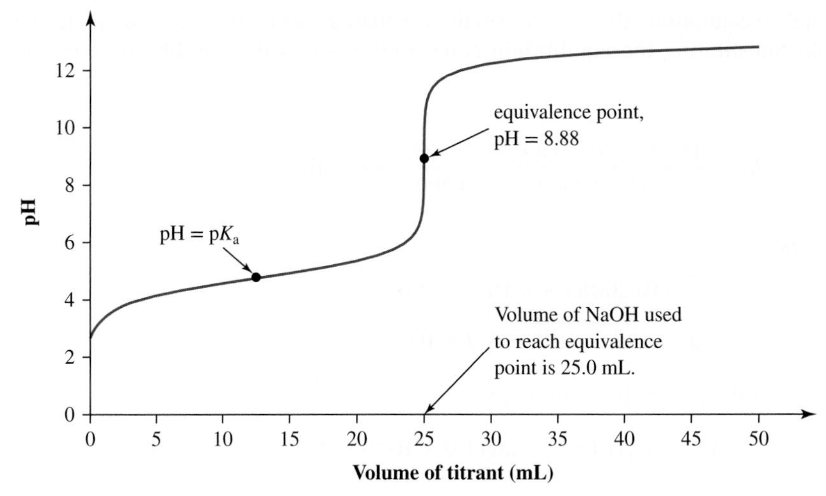

FIGURE 11.20 Titration curve for the titration of a weak acid with a strong base. In this titration, we follow the pH as 25.00 mL of 0.200 M $CH_3COOH(aq)$ is titrated with 0.200 M NaOH(aq).

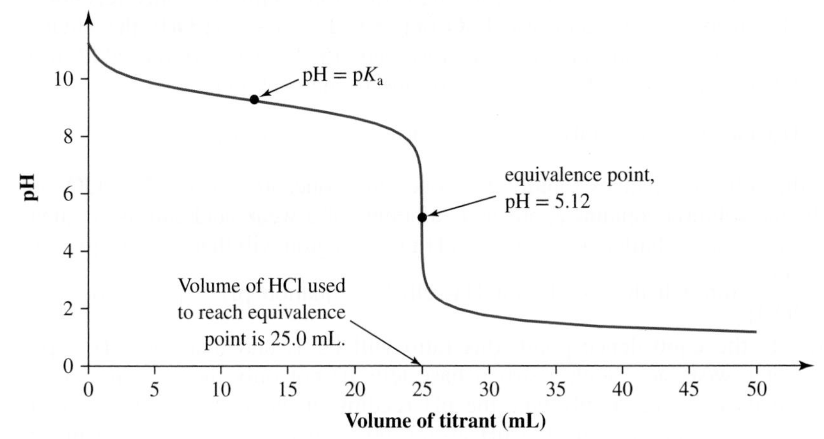

FIGURE 11.21 Titration curve for the titration of a weak base with a strong acid. Here we follow the pH as 25.00 mL of 0.200 M $NH_3(aq)$ is titrated with 0.200 M HCl(aq).

The initial pH

The initial pH of 2.72 in this titration is significantly higher than that for the strong acid – strong base titration. This is to be expected as CH_3COOH is a weak acid and dissociates only to a small extent. We can calculate the initial pH of the acid solution in a weak acid – strong base titration with the method that was used for the calculation of the pH of a solution of a weak acid. We therefore set up a concentration table for a 0.200 M CH_3COOH solution. The balanced chemical equation is:

$$CH_3COOH(aq) + H_2O(l) \rightleftharpoons CH_3COO^-(aq) + H_3O^+(aq)$$

and the equilibrium constant expression is:

$$K_a = \frac{[H_3O^+][CH_3COO^-]}{[CH_3COOH]} = 1.8 \times 10^{-5}$$

The concentration table can then be written as:

	H_2O + CH_3COOH	$\rightleftharpoons$	H_3O^+ +	CH_3COO^-
Initial concentration (mol L^{-1})	0.200		0	0
Change in concentration (mol L^{-1})	$-x$		$+x$	$+x$
Equilibrium concentration (mol L^{-1})	$0.200 - x$		x	x

We make the usual assumption that x is small compared with 0.200, and therefore $(0.200 - x) \approx 0.200$. Substituting the equilibrium concentrations into the equilibrium constant expression gives:

$$K_a = \frac{[H_3O^+][CH_3COO^-]}{[CH_3COOH]} = \frac{(x)(x)}{0.200} = 1.8 \times 10^{-5}$$

and solving for x gives:

$$x^2 = (0.200)(1.8 \times 10^{-5}) = 3.6 \times 10^{-6}$$

$$x = \sqrt{3.6 \times 10^{-6}} = 1.9 \times 10^{-3}$$

As $x = [H_3O^+]$, we obtain the pH of the solution by:

$$pH = -\log[H_3O^+] = -\log(1.9 \times 10^{-3}) = 2.72$$

The acidic region

The general shape of the titration curve in this region differs from that for the strong acid – strong base titration in that there is an initial rapid increase in pH. The curve also lies at higher pH values than the analogous points of the strong acid – strong base titration. While the only reaction of note in the strong acid – strong base titration was $H_3O^+(aq) + OH^-(aq) \rightarrow 2H_2O(l)$, the situation is more complicated when weak acids or bases are involved. In this case, initial addition of $NaOH(aq)$ to $CH_3COOH(aq)$ forms $CH_3COO^-(aq)$ according to the equation:

$$CH_3COOH(aq) + NaOH(aq) \rightarrow NaOOCCH_3(aq) + H_2O(l)$$

Initially, the pH of the solution changes rapidly, but, once the concentration of CH_3COO^- has increased sufficiently, the solution contains appreciable amounts of a weak acid and its conjugate base — in other words, we have a buffer solution. The pH in this region will therefore be governed by the ratio $\frac{[CH_3COO^-]}{[CH_3COOH]}$ through the Henderson–Hasselbalch equation $pH = pK_a + \log\frac{[A^-]}{[HA]}$. At the point halfway to the equivalence point, this ratio will be 1, and therefore $pH = pK_a$. Therefore, a titration of a weak acid with a strong base tells us not only the concentration of the weak acid but also its pK_a; we simply take the pH reading at the volume halfway to the equivalence point. While we can calculate the pH at any point in the acidic region using the Henderson–Hasselbalch equation, it can be more conveniently written in terms of concentrations and volumes of the solutions involved in the titration. The amount of CH_3COO^- is equal to the amount of NaOH added, and the amount of CH_3COOH in the solution at any point is equal to the initial amount of CH_3COOH minus the amount of NaOH added, so we can rewrite this equation as follows:

$$\begin{aligned} pH &= pK_a + \log\frac{[A^-]}{[HA]} \\ &= pK_a + \log\frac{n_{OH^-\,(added)}}{n_{HA\,(initial)} - n_{OH^-\,(added)}} \\ &= pK_a + \log\frac{\left(c_{OH^-\,(initial)} \times V_{OH^-\,(added)}\right)}{\left(c_{HA\,(initial)} \times V_{initial}\right) - \left(c_{OH^-\,(initial)} \times V_{OH^-\,(added)}\right)} \end{aligned}$$

where $c_{OH^-\,(initial)}$ is the initial concentration of the NaOH solution, $V_{OH^-\,(added)}$ is the volume of the NaOH solution added, $c_{HA\,(initial)}$ is the initial concentration of the acid solution and $V_{initial}$ is the initial volume of the acid solution.

Note that this equation is not exact, as it does not take into account the dissociation of the acid or the reaction of the conjugate base with water. However, as both of these reactions are suppressed by the common ion effect in a buffer solution, it gives a very good approximation, especially when the concentrations of the acid and conjugate base are similar. You should not attempt to memorise the previous equation, as it can be derived from the Henderson–Hasselbalch equation, which itself is derived from the K_a expression for a weak acid.

As we approach the equivalence point of the titration, the reaction:

$$CH_3COOH(aq) + NaOH(aq) \rightarrow NaOOCCH_3(aq) + H_2O(l)$$

proceeds further towards completion and the pH starts to increase rapidly. This is because we have exhausted the capacity of the buffer solution; nearly all the CH_3COOH has been used up, and there is no longer sufficient acid present to be able to react with any added OH^-.

The equivalence point

At the equivalence point, the reaction:

$$CH_3COOH(aq) + NaOH(aq) \rightarrow NaOOCCH_3(aq) + H_2O(l)$$

has gone to completion and the solution contains 0.100 M $Na^+(aq)$ and 0.100 M $CH_3COO^-(aq)$. In other words, at this point we have a 0.1 M aqueous solution of $NaOOCCH_3$. We have seen previously that CH_3COO^-, being the conjugate base of a weak acid, is appreciably basic, and therefore the following reaction will occur.

$$CH_3COO^-(aq) + H_2O(l) \rightleftharpoons CH_3COOH(aq) + OH^-(aq)$$

This means that, at the equivalence point, there is a small amount of OH^- in the solution, and therefore the pH is greater than 7.00 (in this case it is 8.88, as we show below). In the titration of any weak acid with any strong base, the pH at the equivalence point is always greater than 7, because the conjugate base of the weak acid is measurably basic. The exact pH depends on both the K_b and the concentration of the conjugate base; the weaker the acid, the stronger its conjugate base, and therefore the more basic the solution at the equivalence point. Calculation of the pH at the equivalence point follows the method we outlined previously for the determination of the pH of a salt of a weak acid. The balanced chemical equation we need to consider is:

$$CH_3COO^-(aq) + H_2O(l) \rightleftharpoons CH_3COOH(aq) + OH^-(aq)$$

We will use K_b to determine $[OH^-]$, and hence pH:

$$K_b = \frac{[CH_3COOH][OH^-]}{[CH_3COO^-]} = 5.6 \times 10^{-10}$$

The concentration table is as follows. Note that the initial concentration of CH_3COO^- is 0.100 M; this is because at the equivalence point the solution has twice its starting volume due to the added NaOH solution.

	$H_2O + CH_3COO^-$	$\rightleftharpoons$	CH_3COOH +	OH^-
Initial concentration (mol L^{-1})	0.100		0	0
Change in concentration (mol L^{-1})	$-x$		$+x$	$+x$
Equilibrium concentration (mol L^{-1})	$0.100 - x$		x	x

We make the assumption that x will be small compared with 0.100, because CH_3COO^- is a weak base. Therefore $(0.100 - x) \approx 0.100$. We then substitute values into the K_b expression:

$$K_b = \frac{[CH_3COOH][OH^-]}{[CH_3COO^-]} = \frac{(x)(x)}{0.100} = 5.6 \times 10^{-10}$$

Solving this for x gives:

$$x^2 = (0.100)(5.6 \times 10^{-10}) = 5.6 \times 10^{-11}$$

$$x = \sqrt{5.6 \times 10^{-11}} = 7.5 \times 10^{-6}$$

As $x = [OH^-]$, we obtain the pH of the solution as follows.

$$pOH = -\log[OH^-] = -\log(7.5 \times 10^{-6}) = 5.12$$

$$pH = 14.00 - pOH = 14.00 - 5.12 = 8.88$$

Analogous arguments apply for the titration of a weak base with a strong acid, and the pH at the equivalence point in such titrations is < 7. The exact pH can be calculated using the K_a for the conjugate acid of the weak base.

The alkaline region

Beyond the equivalence point, we are simply adding excess OH^- to a solution of a weak base, and the pH in this region is controlled by the excess amount of NaOH. The titration curve in this region is essentially identical to that for a strong acid – strong base titration, and the pH at any point in this region may therefore be calculated using the same method as described previously for such titrations.

Diprotic acids

When a weak diprotic acid such as ascorbic acid (vitamin C) is titrated with a strong base, there are two protons available to react with the base and there are therefore two equivalence points. Provided that the values of pK_{a1} and pK_{a2} differ by several powers of 10, the neutralisation takes place stepwise and the resulting titration curve shows two sharp increases in pH as shown in figure 11.22. Calculations involving acids containing two or more acidic protons can be difficult, especially if the values of pK_{a1} and pK_{a2} are similar, and lie outside the scope of this book.

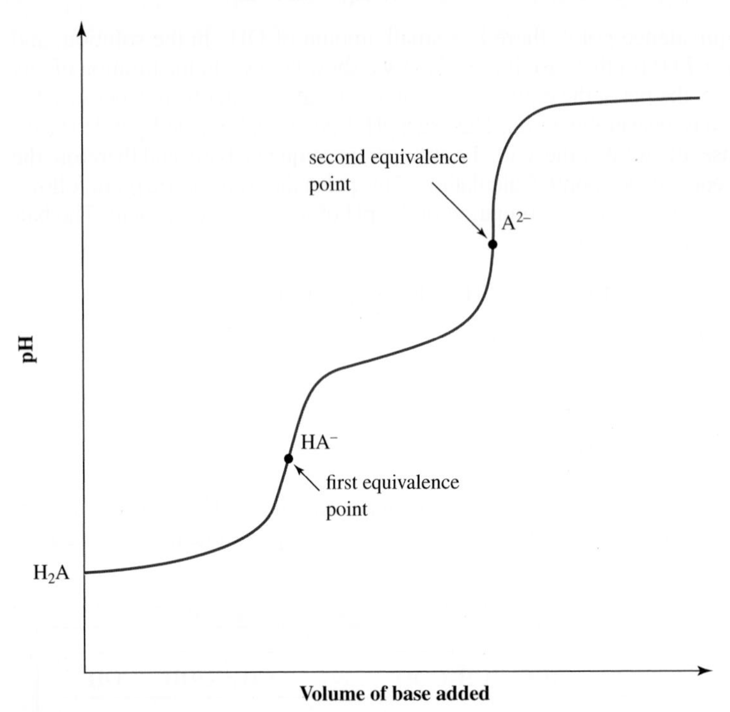

FIGURE 11.22 Titration of a diprotic acid, H_2A, by a strong base. As each equivalence point is reached, there is a sharp rise in the pH.

Speciation diagrams

It is often useful to know the relative amounts of a weak acid and its conjugate base that are present in an aqueous solution having a particular pH. We can do this with the aid of a speciation diagram, which is a plot of mole ratio versus pH. From the speciation diagram for acetic acid in figure 11.23, we can see at a glance that undissociated CH_3COOH predominates at pH values less than 4.74 (the pK_a of acetic acid). At the point where pH = pK_a, there are equal amounts of CH_3COOH and CH_3COO^-, and the two lines therefore intersect. At higher pH values, the conjugate base, CH_3COO^-, is present in greater amounts than the undissociated acid; beyond pH 8, there is essentially no CH_3COOH present in the solution.

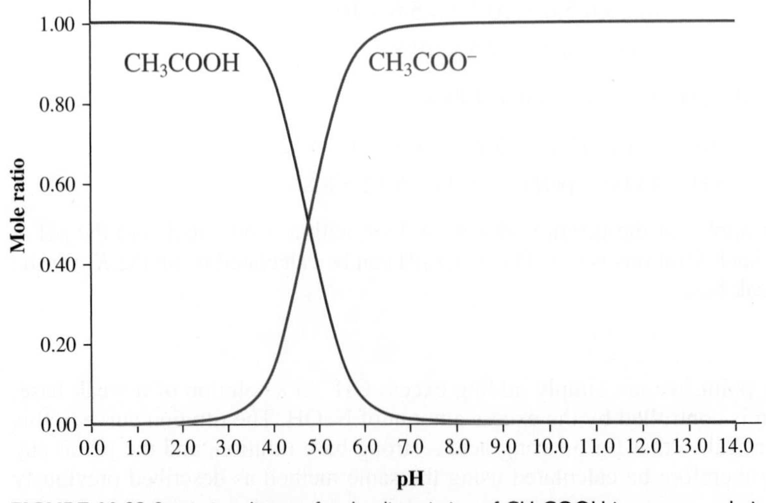

FIGURE 11.23 Speciation diagram for the dissociation of CH_3COOH in aqueous solution.

Speciation diagrams are relatively simple for monoprotic acids, but become somewhat more complicated, and consequently somewhat more useful, for polyprotic acids. Figure 11.24 shows the speciation diagram for H_3PO_4, a triprotic acid, which yields the anions $H_2PO_4^{2-}$, HPO_4^{2-} and PO_4^{3-} on successive deprotonations. The pK_a values are:

$$pK_a\,(H_3PO_4) = 2.15$$
$$pK_a\,(H_2PO_4^-) = 7.20$$
$$pK_a\,(HPO_4^{2-}) = 12.3$$

It can be seen that H_3PO_4 predominates in very acidic regions, but, by pH 4, it has mostly been deprotonated to form $H_2PO_4^-$. As we proceed past the pK_a of $H_2PO_4^-$ (7.20), its conjugate base, HPO_4^{2-}, is present in the largest amount until the pH increases past 12.3, the pK_a of HPO_4^{2-}. Under these conditions, PO_4^{3-} then becomes the dominant species in solution.

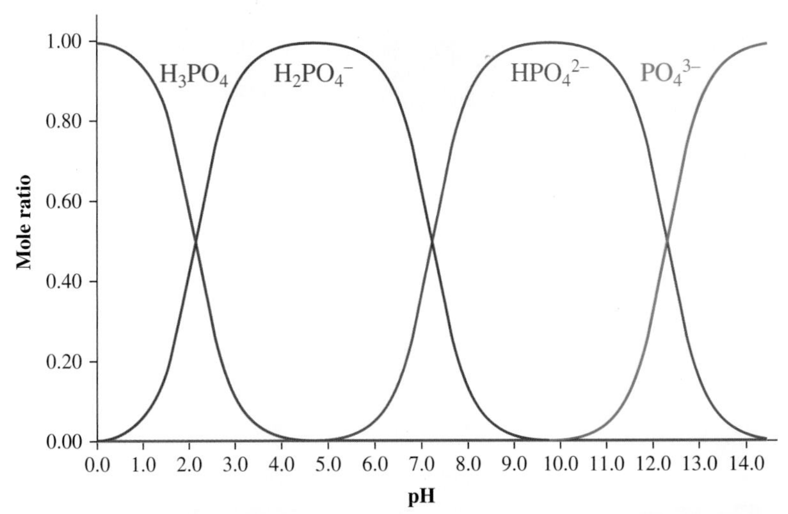

FIGURE 11.24 Speciation diagram for the dissociation of H_3PO_4 in aqueous solution.

Speciation diagrams are extremely useful not only in situations involving polyprotic acids, but also in systems where metal ions react with ligands (see chapter 13), which themselves contain acidic protons. The calculations involved can be rather complex, especially in systems involving a number of components, but can usually be carried out satisfactorily using spreadsheets.

Acid–base indicators

The indicators used in acid–base titrations are themselves weak acids, which we can represent by the formula H*In*. A requirement of an acid–base indicator is that its 'acid form' H*In* and conjugate 'base form' In^- have different colours. In solution, the indicator is involved in a typical acid–base equilibrium.

$$\underset{\substack{\text{acid form}\\\text{colour 1}}}{HIn(aq)} + H_2O(l) \rightleftharpoons H_3O^+(aq) + \underset{\substack{\text{base form}\\\text{colour 2}}}{In^-(aq)}$$

The corresponding acid ionisation constant K_{In} is given by:

$$K_{In} = \frac{[H_3O^+][In^-]}{[HIn]}$$

In a strongly acidic solution, when $[H_3O^+]$ is high, the equilibrium is shifted to the left, and most of the indicator exists in its acid form. Under these conditions, the colour we observe is that of H*In*. If base is added to the solution, the $[H_3O^+]$ drops and the equilibrium shifts to the right, towards In^-. The colour we then observe is that of the base form of the indicator (figure 11.25).

How acid–base indicators work

In a typical acid–base titration, as we pass the equivalence point, there is a sudden and large change in the pH. For example, in the titration of HCl with NaOH described earlier (figure 11.18, overleaf), the pH just before the equivalence point (when 24.97 mL of the base has been added) is 3.92. Just one drop later (when 25.03 mL of base has been added), we have passed the equivalence point and the pH has risen to 10.08. This large swing in pH causes a sudden shift in the position of equilibrium for the indicator, and we go from a condition where most of the indicator is in its acid form to a condition in which most is in the base form. As a result, the solution changes colour and we say that the endpoint has been reached.

FIGURE 11.25 Colours of the acid form and base form of some common acid–base indicators.

Selecting the best indicator for an acid–base titration

The explanation in the preceding paragraph assumes that the *midpoint* of the colour change range of the indicator corresponds to the pH at the equivalence point. At this midpoint, we expect that there are equal amounts of both forms of the indicator, which means that $[In^-] = [HIn]$. If this condition is true, then from the equation:

$$\text{pH} = \text{p}K_{In} + \log \frac{[In^-]}{[\text{H}In]}$$

we find that $\text{p}K_{In}$ = pH at the equivalence point.

This important result tells us that, once we know the pH of the solution at the equivalence point, we also know the $\text{p}K_{In}$ that the indicator should have to function most effectively. Thus, we want an indicator with a $\text{p}K_{In}$ equal to (or as close as possible to) the pH at the equivalence point. This will ensure that the observed endpoint will be as close as possible to the equivalence point. Phenolphthalein, for example, changes a solution from colourless to pink as the pH of the solution changes over a range of 8.2 to 10.0 (see table 11.9). Phenolphthalein is therefore a nearly perfect indicator for the titration of a weak acid by a strong base, where the pH at the equivalence point is on the basic side (figure 11.20).

Phenolphthalein also works very well for the titration of a strong acid with a strong base. As discussed earlier, just one drop of titrant can bring the pH from below 7 to nearly 10, which spans the pH range for phenolphthalein.

When we perform a titration, we want to use as little indicator as possible. The reason for this is that indicators are weak acids and also react with the titrant. If we were forced to use a lot of indicator, it would affect our measurements and make the titration less accurate. Therefore, the best indicators are those with the most intense colours. Then, even extremely small amounts can give striking colour changes without consuming too much of the titrant.

TABLE 11.9 Common acid–base indicators.

Indicator	Approximate pH range for colour change	Colour change (lower to higher pH)
methyl green	0.2–1.8	yellow to blue
thymol blue	1.2–2.8	red to yellow
methyl orange	3.2–4.4	red to yellow
ethyl red	4.0–5.8	colourless to red
methyl purple	4.8–5.4	purple to green
bromocresol purple	5.2–6.8	yellow to purple
bromothymol blue	6.0–7.6	yellow to blue
phenol red	6.4–8.2	yellow to red/violet
litmus	4.7–8.3	red to blue
cresol red	7.0–8.8	yellow to red
thymol blue	8.0–9.6	yellow to blue
phenolphthalein	8.2–10.0	colourless to pink
thymolphthalein	9.4–10.6	colourless to blue
alizarin yellow R	10.1–12.0	yellow to red
clayton yellow	12.2–13.2	yellow to amber

11.8 Lewis acids and bases

We stated earlier in this chapter that there was more than one definition of acids and bases, and, while we have used the Brønsted–Lowry definition throughout this chapter, there is a more general definition first proposed by the American chemist Gilbert Lewis.

- A **Lewis acid** is an electron-pair acceptor.
- A **Lewis base** is an electron-pair donor.

To illustrate these definitions, we will consider the reaction between BF_3 and NH_3.

$$BF_3(g) + NH_3(g) \rightarrow F_3B{-}NH_3(s)$$

In this reaction, NH_3 formally donates an electron pair to BF_3, resulting in formation of a B—N covalent bond, as shown in figure 11.26.

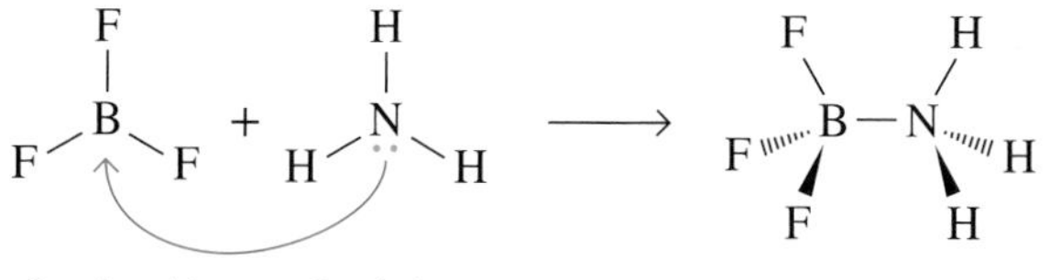

Lewis acid Lewis base

FIGURE 11.26 The Lewis acid–base reaction of BF_3 with NH_3. A covalent bond between B and N is formed, with both electrons in the bond formally deriving from the Lewis basic N atom.

Thus, BF_3 is a Lewis acid and accepts an electron pair from the Lewis base NH_3 in this reaction. Compounds such as F_3BNH_3, which are products of the reaction between a Lewis acid and a Lewis base, are sometimes called Lewis adducts, or simply **adducts**. As we will see in chapter 13, the chemistry of transition metal ions is dominated by Lewis adducts called coordination complexes. The formation of these involves the reaction of a Lewis acidic transition metal ion with one or more Lewis basic species called ligands, each containing lone pairs of electrons. For example, the coordination complex $[Ni(NH_3)_4]^{2+}$ is formed from reaction of Ni^{2+} with four NH_3 ligands, according to the equation:

$$\underset{\text{Lewis acid}}{Ni^{2+}(aq)} + \underset{\text{Lewis base}}{4NH_3(aq)} \rightarrow \underset{\text{Lewis adduct (coordination complex)}}{[Ni(NH_3)_4]^{2+}(aq)}$$

Organic chemistry, as we will see from chapters 16 to 26, is full of examples of Lewis acids and bases. Much of organic chemistry is concerned with the making and breaking of covalent bonds involving carbon, and such reactions can be considered from a Lewis acid – Lewis base perspective. For example, the reaction of a ketone with a primary amine gives an imine and involves the formation of a C=N bond (see chapter 21, p. 960). The first step of this reaction involves the reaction of the Lewis basic amine with the Lewis acidic ketone, as follows.

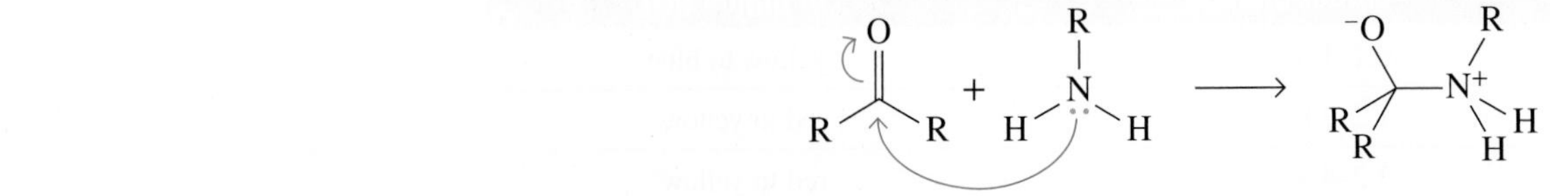

Lewis acid Lewis base adduct

The similarity to the BF_3/NH_3 reaction is obvious. The nomenclature of organic chemistry dictates that Lewis acids are called **electrophiles** and Lewis bases are called **nucleophiles**.

As the Lewis definition of acids and bases is the most general of the three we have met so far, it follows that all Brønsted–Lowry acids and bases must be Lewis acids and bases. However, the reverse does not necessarily hold; for example, the Lewis acid BF_3 has no protons to donate and therefore cannot be a Brønsted–Lowry acid.

The strengths of Lewis acids and bases are not as readily quantified as those of their Brønsted–Lowry counterparts. While the latter are measured relative to a single solvent (water), the large variety of solvents in which Lewis acids and bases are used makes a single scale of strengths difficult to achieve. However, it is known that a mixture of the Brønsted–Lowry acid HF and the Lewis acid antimony pentafluoride, SbF_5, is capable of protonating extremely weak bases, such as alkanes, and is in fact one of the strongest acids known. Such mixtures are called superacids; in fact, we have already seen an example of a superacid on page 465.

The concept of Lewis acids and bases is particularly useful in understanding the reactivities of *p*-block elements. As we will see in chapters 13 and 14, in nature, *p*-block elements such as oxygen, sulfur and the halogens tend to be found in combination with particular metals. For example, copper, lead and mercury are most often found as sulfide ores; sodium and potassium are found as their chloride salts; magnesium and calcium exist as carbonates; and aluminium, titanium and iron are all found as oxides. The underlying principle that determines which particular combination is favoured is the strength with which the atoms involved bind their valence electrons. This is related to the Lewis acid and base characteristics of the atoms and we can therefore use the Lewis acid–base model to describe the many different affinities that exist among elements. This notion not only explains the natural distribution of minerals, but also can be used to predict patterns of chemical reactivity.

Recognising Lewis acids and bases

A Lewis base must have valence electrons available for bond formation. Any molecule with a Lewis structure that shows nonbonding electron pairs can act as a Lewis base. Ammonia, phosphorus trichloride and dimethyl ether, each of which contains lone pairs, are Lewis bases. Anions can also act as Lewis bases.

For example, in the reaction:

$$\underset{\text{acid}}{SiF_4} + \underset{\text{base}}{:F^-} \longrightarrow \underset{\text{adduct}}{SiF_5^-}$$

the fluoride ion, with eight valence electrons in its 2*s* and 2*p* orbitals, acts as a Lewis base. A Lewis acid must be able to accept electrons to form a new bond. Because bond formation can occur in several ways, compounds with several different structural characteristics can act as Lewis acids. Nevertheless, most Lewis acids fall into the following categories.

1. *A molecule that has vacant valence orbitals.* A good example is BF_3, which uses a vacant 2*p* orbital on boron to form an adduct with ammonia. The elements in the *p* block beyond period 2 of the periodic table have *d* orbitals which can potentially accept an electron pair, thus allowing them to act as Lewis acids. The silicon atom in SiF_4 is an example.
2. *A molecule with delocalised π bonds involving oxygen.* Examples are CO_2, SO_2 and SO_3. Each of these molecules can form a σ bond between its central atom and a Lewis base, at the expense of a π bond. For example, the hydroxide anion, a good Lewis base, attacks the carbon atom of CO_2 to form hydrogen carbonate.

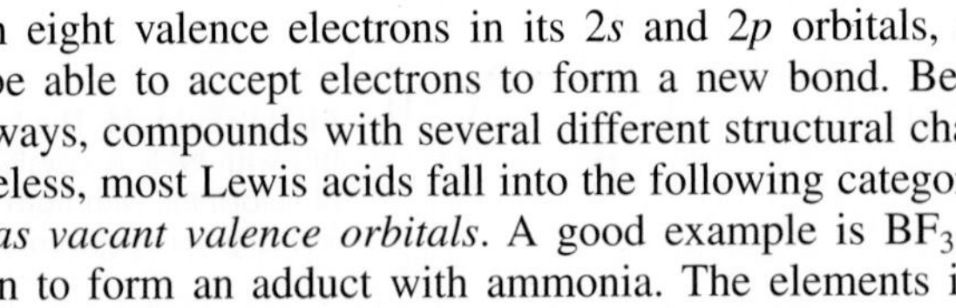

Lewis acid Lewis base hydrogen carbonate adduct

In this reaction, the oxygen atom of the hydroxide ion donates a pair of electrons to make a new C—O bond. Because all the valence orbitals of the carbon atom in CO_2 are involved in bonding to oxygen, one of the C—O π bonds must be broken to make an orbital available to overlap with the occupied orbital of the hydroxide anion.

3. *A metal cation.* Removing electrons from a metal atom always generates vacant valence orbitals. As described on the previous page, many transition metal cations form complexes with ligands in aqueous solution. In these complexes, the ligands act as Lewis bases, donating pairs of electrons to form metal–ligand bonds. The metal cation accepts these electrons, so it acts as a Lewis acid. Metal cations from the *p* block also act as Lewis acids. For example, $Pb^{2+}(aq)$ (group 14) forms a Lewis acid–base adduct with four CN^- anions, each of which donates a pair of electrons.

$$Pb^{2+}(aq) + 4CN^-(aq) \longrightarrow [Pb(CN)_4]^{2-}(aq)$$

Worked example 11.21 provides practice in recognising Lewis acids and bases.

WORKED EXAMPLE 11.21

Lewis acids and bases

Identify the Lewis acids and bases in each of the following reactions and draw the structures of the resulting adducts.

(a) $AlCl_3 + Cl^- \longrightarrow AlCl_4^-$

(b) $Co^{3+} + 6NH_3 \longrightarrow [Co(NH_3)_6]^{3+}$

(c) $SO_2 + OH^- \longrightarrow HSO_3^-$

Analysis

Every Lewis base has one or more lone pairs of valence electrons. A Lewis acid can have vacancies in its valence shell, or it can sacrifice a π bond to make a valence orbital available for adduct formation. To decide whether a molecule or ion acts as a Lewis acid or base, examine its Lewis structure for these features. Review pages 177–9 in chapter 5 for procedures used to determine Lewis structures.

Solution

(a) Both reactants contain chlorine atoms with lone pairs, so either might act as a Lewis base if a suitable Lewis acid is present. The aluminium atom of $AlCl_3$ has a vacant $3p$ orbital perpendicular to the molecular plane. The empty *p* orbital accepts a pair of electrons from the Cl^- anion to form the fourth Al—Cl bond. The Lewis acid is $AlCl_3$, and the Lewis base is Cl^-.

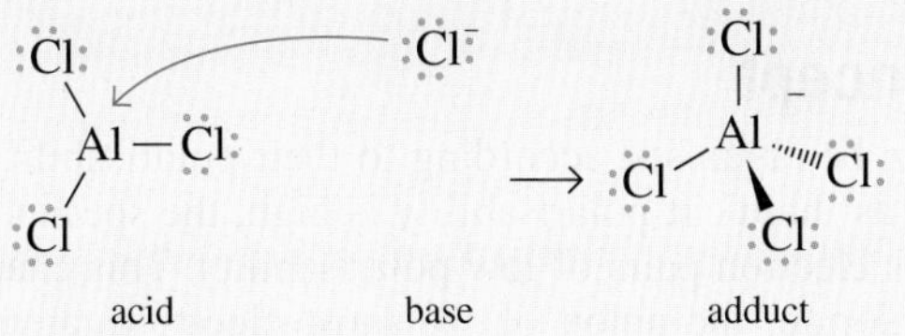

acid base adduct

(b) As already noted, ammonia is a Lewis base because it has a lone pair of electrons on the nitrogen atom. Like other transition metal cations, Co^{3+} is a Lewis acid and uses vacant $3d$ orbitals to form bonds to NH_3.

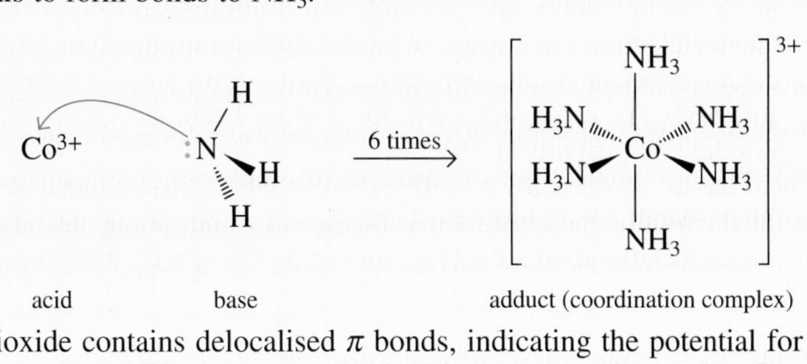

6 times

acid base adduct (coordination complex)

(c) Sulfur dioxide contains delocalised π bonds, indicating the potential for Lewis acidity. The sulfur atom of SO_2 has a set of $3d$ orbitals that can be used to form an adduct. In this case, the hydroxide ion acts as a Lewis base. The anion uses one lone pair of electrons to form a new bond to sulfur.

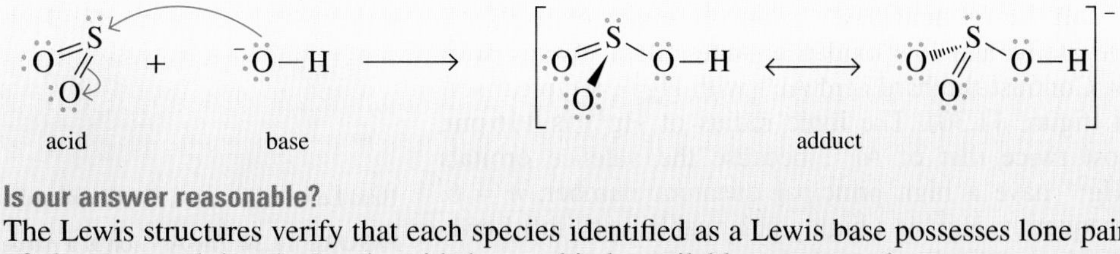

acid base adduct

Is our answer reasonable?

The Lewis structures verify that each species identified as a Lewis base possesses lone pairs of electrons and that the Lewis acids have orbitals available to accept electrons.

PRACTICE EXERCISE 11.25 The reaction of $SbCl_5$ with Cl^- gives the $SbCl_6^-$ ion. Identify the Lewis acid and Lewis base in this reaction, and draw a reaction scheme showing all electron transfers.

Polarisability

Polarisability, described in chapter 6, is a measure of the ease with which the electron cloud of an atom, ion or molecule can be distorted by an electrical charge. An electron cloud is polarised towards a positive charge and away from a negative charge, as shown in figure 11.27. Pushing the electron cloud to one side of an atom causes a polarisation of charge. The side with the concentrated electron density has a small negative charge; the protons in the nucleus give the opposite side a small positive charge.

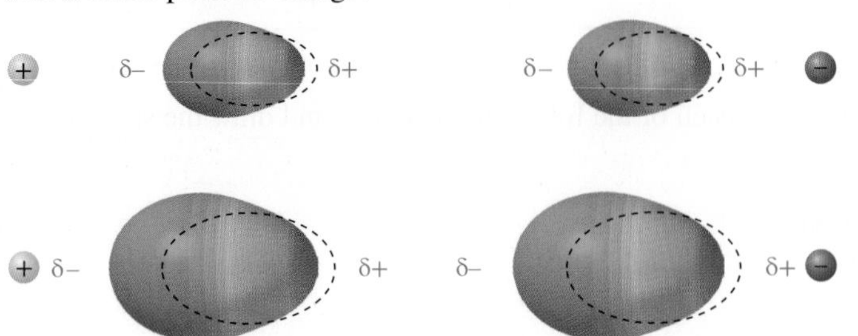

FIGURE 11.27 The electron cloud of an atom is polarised towards a positive charge and away from a negative charge. A smaller atom binds its valence electrons more tightly and is less polarisable than a larger atom, which binds its electrons loosely.

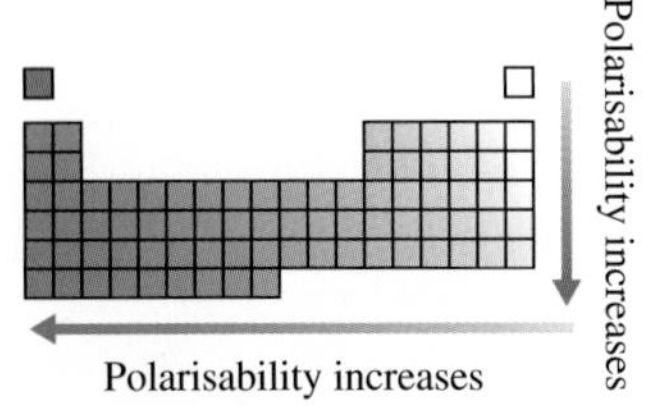

FIGURE 11.28 Polarisability decreases across a period and increases down a group.

Polarisability shows periodic variations that correlate with periodic trends in how tightly valence electrons are bound to the nucleus (figure 11.28).

1. Polarisability decreases from left to right in any period of the periodic table. As the effective nuclear charge (Z_{eff}) increases, the nucleus holds the valence electrons more tightly.
2. Polarisability increases from top to bottom in any group of the periodic table. As the principal quantum number (n) increases, the valence orbitals become larger. This reduces the net attraction between valence electrons and the nucleus.

The hard–soft concept

Lewis acids and bases can be organised according to their polarisability. If polarisability is low, the species is categorised as 'hard'. If polarisability is high, the species is 'soft'.

A **hard Lewis base** has electron pairs of low polarisability. This characteristic correlates with high electronegativity. Fluoride, the anion of the most electronegative element, is the hardest Lewis base because it contains a small, dense sphere of negative charge. Molecules and ions that contain oxygen or nitrogen atoms are usually also hard bases, although not as hard as fluorine. Examples include H_2O, CH_3OH, OH^-, NH_3 and H_2NCH_3.

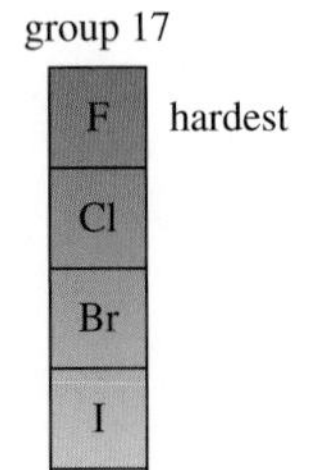

FIGURE 11.29 The donor atom becomes softer down a group.

A **soft Lewis base** has a large donor atom of high polarisability and low electronegativity. The iodide ion has its valence electrons in large $n = 5$ orbitals, making this anion highly polarisable and a very soft base. Other molecules and polyatomic anions with donor atoms from periods 3 to 6 are usually also soft bases. To summarise, the donor atom becomes softer from top to bottom of a group of the periodic table (figure 11.29).

A **hard Lewis acid** has an acceptor atom with low polarisability. Most metal atoms and ions are hard acids. In general, the smaller the ionic radius and the larger the charge, the harder the acid. The Al^{3+} ion, with an ionic radius of only 67 pm, is a prime example of a hard Lewis acid. The nucleus exerts a strong pull on the compact electron cloud, giving the ion very low polarisability.

The designation of hard acids is not restricted to metal cations. For example, in BF_3 the small boron atom in its +3 oxidation state is bonded to three highly electronegative fluorine atoms. All the B—F bonds are polarised away from a boron centre that is already electron deficient. Boron trifluoride is therefore a hard Lewis acid.

A **soft Lewis acid** has a relatively high polarisability. Large atoms and low oxidation states often convey softness. Contrast the hard acid Al^{3+} with Hg^{2+}, a typical soft acid (figure 11.30). The ionic radius of Hg^{2+} is 116 pm, almost twice that of Al^{3+}, because the valence orbitals of Hg^{2+} have a high principal quantum number, $n = 6$. Consequently, Hg^{2+} is a highly polarisable, very soft Lewis acid. The relatively few soft Lewis acid transition metal ions are located around gold in the periodic table.

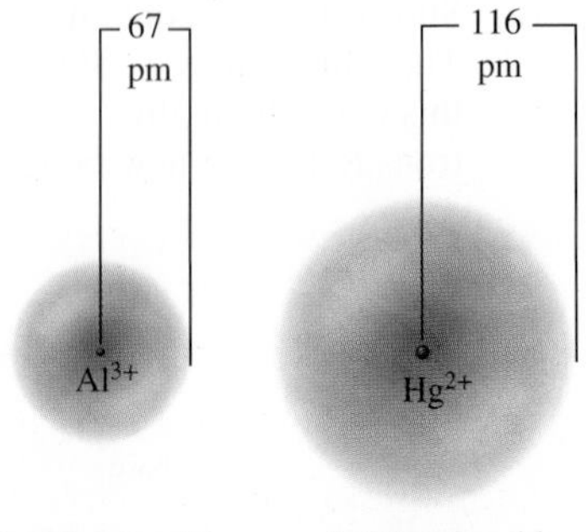

FIGURE 11.30 The Al^{3+} ion is a harder Lewis acid than Hg^{2+} because of its smaller ionic radius and higher charge.

The terms 'hard' and 'soft' are relative, so there is no sharp dividing line between the two, and many Lewis acids and bases are intermediate between hard and soft. Worked example 11.22 shows how to categorise Lewis acids and bases according to their hard–soft properties.

WORKED EXAMPLE 11.22

Ranking hardness and softness

Rank the following groups of Lewis acids and bases from softest to hardest. (a) H_2S, H_2O and H_2Se (b) Fe^0, Fe^{3+} and Fe^{2+} (c) BCl_3, $GaCl_3$ and $AlCl_3$

Analysis

The first task is to decide whether the members of a given group are Lewis acids or bases. Then evaluate the relative softness and hardness based on polarisability, taking into account correlations with electronegativity, size and charge. Refer to the periodic table to assess the trends.

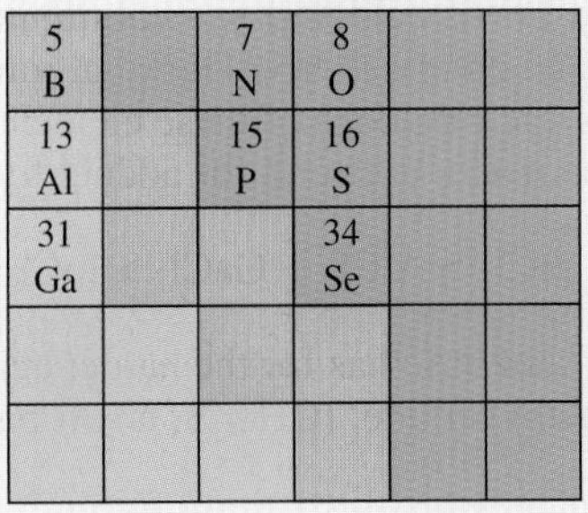

5 B		7 N	8 O		
13 Al		15 P	16 S		
31 Ga			34 Se		

Solution

(a) H–Se–H, H–S–H, H–O–H

These three molecules have lone pairs, so they are Lewis bases. The central atoms are in the same group of the periodic table, so their polarisability and the softness of the molecules increases moving down the group. Thus, H_2Se is softer than H_2S, which is softer than H_2O.

(b) Fe^0 Fe^{2+} Fe^{3+}

Metal atoms and cations are Lewis acids. As valence electrons are removed from a metal atom, the remaining electron cloud undergoes an ever larger pull from the nuclear charge. This decreases the size of the ion as well as its polarisability. Thus, Fe^0 is softer than Fe^{2+}, which is softer than Fe^{3+}.

(c) $GaCl_3$, $AlCl_3$, BCl_3 (trigonal planar structures, Cl atoms bonded to central Ga, Al, B)

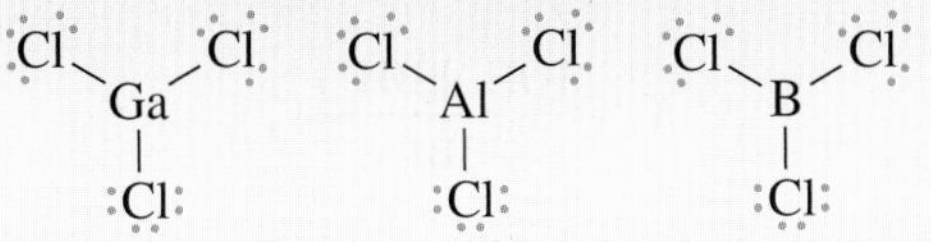

These three molecules have trigonal planar geometries with sp^2 hybridised central atoms. Each has a vacant valence *p* orbital perpendicular to the molecular plane, making the molecules Lewis acids. The size, polarisability and softness of the central acceptor atom increases going down the group. A gallium atom is larger and more polarisable than an aluminium atom. Thus, $GaCl_3$ is softer than $AlCl_3$, which is softer than BCl_3. We have already noted that the Al^{3+} ion is a hard Lewis acid, and is therefore likely to form compounds which are themselves hard Lewis acids. Thus, gallium trichloride is a soft Lewis acid, whereas $AlCl_3$ and BCl_3 are both hard.

Is our answer reasonable?

Hardness and softness are primarily determined by atomic size, and the order of hardness in each of these sets is consistent with the size trends among the species.

PRACTICE EXERCISE 11.26

Identify H_2O, NH_3, and PH_3 as Lewis acids or Lewis bases, and rank them in order of increasing hardness.

The hard–soft acid–base principle

The concept of hard and soft acids and bases can be used to interpret many trends in chemical reactivity. These trends are summarised in the hard–soft acid–base principle (HSAB principle), an empirical summary of results collected from many chemical reactions.

- Hard Lewis acids tend to combine with hard Lewis bases.
- Soft Lewis acids tend to combine with soft Lewis bases.

The geochemical distribution of metals conforms to the HSAB principle. Metals that form hard acid cations have strong affinities for hard bases such as oxide, fluoride and chloride. Most metal ions that are hard acids are found bonded to the oxygen atoms of various silicate anions. These elements are concentrated in the Earth's mantle. Hard acid metals also occur in combinations with other hard bases, including oxides or, less often, halides, sulfates and carbonates. Examples include rutile, TiO_2, limestone, $CaCO_3$, gypsum, $CaSO_4$, and sylvite, KCl.

Metals that are soft acids, such as gold and platinum, have low affinities for hard oxygen atoms, so they are not affected by O_2 in the atmosphere. Consequently, these metals, including Ru, Rh, Pd, Os, Ir, Pt and Au, are often found in the crust of the Earth in their elemental form. Other soft metals occur in nature as sulfides. The sulfide anion is a soft base, so this category includes some soft acids and many intermediate cases. Soft acids may also occur either in elemental form or as arsenide or telluride minerals. Examples include galena, PbS, cinnabar, HgS, chalcopyrite, $CuFeS_2$, argentite, Ag_2S, calaverite, $AuTe_2$, and sperrylite, $PtAs_2$.

The group 13 elements illustrate the trend in hardness and softness among Lewis acids. At the top of the group, boron is a hard Lewis acid and so is aluminium. Moving down the group, the valence orbitals increase in size and polarisability. Thus, gallium is a borderline acid, and indium is soft. The order of reactivity for the trihalides of these elements depends on the Lewis base. For a hard base such as ammonia, the reactivity trend for adduct formation is:

$$BCl_3 > AlCl_3 > GaCl_3 > InCl_3$$

reflecting the preference that the hard base has for the harder acid. The order is reversed when the trihalides form adducts with dimethylsulfide, $(CH_3)_2S$, a soft base.

SUMMARY

LO 1 The Brønsted–Lowry definition of acids and bases

A Brønsted–Lowry acid is a proton donor; a Brønsted–Lowry base is a proton acceptor. According to the Brønsted–Lowry approach, an acid–base reaction is a proton transfer event. In an equilibrium involving a Brønsted–Lowry acid and base, there are two conjugate acid–base pairs. The members of any given pair differ from each other by only one proton. A solvent that can be either an acid or a base, depending on the nature of the other reactant, is amphiprotic.

LO 2 Acid–base reactions in water

Water reacts with itself to produce small amounts of H_3O^+ and OH^- ions. The concentrations of these ions in both pure water and in dilute aqueous solutions are related by the expression:

$$[H_3O^+][OH^-] = K_w = 1.0 \times 10^{-14} \quad \text{(at 25 °C)}$$

K_w is the autoprotolysis constant of water. In pure water:

$$[H_3O^+] = [OH^-] = 1.0 \times 10^{-7}\,M \quad \text{(at 25 °C)}$$

The pH of a solution is a measure of acidity and is normally measured with a pH meter. As the pH decreases, the acidity increases. The defining equation for pH is $pH = -\log[H_3O^+]$. In exponential form, this relationship between $[H_3O^+]$ and pH is given by $[H_3O^+] = 10^{-pH}$.

Comparable expressions can be used to describe low OH^- ion concentrations in terms of pOH values: $pOH = -\log[OH^-]$ and $[OH^-] = 10^{-pOH}$. At 25 °C, pH + pOH = 14.00. A solution is acidic if its pH is less than 7.00 and basic if its pH is greater than 7.00. A neutral solution has a pH of 7.00.

LO 3 Strong acids and bases

In calculating the pH of solutions of strong acids and bases, we assume that they react completely with water. The autoprotolysis of water contributes negligibly to $[H_3O^+]$ in a solution of an acid. It also contributes negligibly to the $[OH^-]$ in a solution of a base.

LO 4 Weak acids and bases

An acid HA reacts with water according to the general equation:

$$HA + H_2O \rightleftharpoons H_3O^+ + A^-$$

The equilibrium constant is called the acidity constant, K_a.

$$K_a = \frac{[H_3O^+][A^-]}{[HA]}$$

A base B reacts by the general equation:

$$B + H_2O \rightleftharpoons BH^+ + OH^-$$

The equilibrium constant is called the basicity constant, K_b.

$$K_b = \frac{[OH^-][BH^+]}{[B]}$$

The smaller the value of K_a or K_b, the weaker the substances are as Brønsted–Lowry acids or bases.

Another way to compare the relative strengths of acids or bases is to use the negative logarithms of K_a and K_b, called pK_a and pK_b, respectively.

$$pK_a = -\log K_a$$
$$pK_b = -\log K_b$$

The *smaller* the pK_a or pK_b, the *stronger* the acid or base. For a conjugate acid–base pair:

$$K_a K_b = K_w$$

and:

$$pK_a + pK_b = 14.00 \quad \text{(at 25 °C)}$$

The values of K_a and K_b can be obtained from initial concentrations of the acid or base and the pH of the solution. The measured pH gives the equilibrium value for $[H_3O^+]$.

When the initial concentration of the acid (or base) is larger than 400 times the value of K_a (or K_b), it is safe to use initial concentrations of acid or base as though they were equilibrium values when calculating pH or pOH. When this approximation cannot be used, we use the quadratic formula.

The weaker an acid, the stronger its conjugate base, and vice versa. Anions of strong acids, such as Cl^- and NO_3^-, are such weak Brønsted–Lowry bases that they cannot affect the pH of a solution.

If a salt is derived from a weak acid *and* a weak base, its net effect on pH has to be established on a case-by-case basis by determining which of the two ions reacts with water to a greater extent.

LO 5 The molecular basis of acid strength

Binary acids contain only hydrogen and another nonmetal. Their strengths increase from top to bottom within a group and left to right across a period. Oxoacids, which contain oxygen atoms in addition to hydrogen and another element, increase in strength as the number of oxygen atoms on the same central atom increases. Oxoacids with the same number of oxygen atoms generally increase in strength as the central atom moves from bottom to top within a group and from left to right across a period.

LO 6 Buffer solutions

A solution that contains appreciable amounts of both a weak acid and a weak base (usually an acid–base conjugate pair) is called a buffer solution, because it is able to absorb H_3O^+ from a strong acid or OH^- from a strong base without suffering large changes in pH.

For the general acid–base pair, HA and A^-, the following reactions occur in a buffer solution:

$$H_3O^+ \text{ is added to the buffer:} \quad A^- + H_3O^+ \rightarrow HA + H_2O$$

$$OH^- \text{ is added to the buffer:} \quad HA + OH^- \rightarrow A^- + H_2O$$

The pH of a buffer is controlled by the ratio of weak acid to weak base, expressed either in terms of concentration or amount.

Because the $[H_3O^+]$ is determined by the mole ratio of HA to A^-, dilution does not change the pH of a buffer. The equation:

$$\text{pH} = \text{p}K_a + \log\frac{[A^-]}{[HA]}$$

can be used to calculate the pH directly from the $\text{p}K_a$ of the acid and the concentrations of the conjugate acid HA and base A^-.

In performing buffer calculations, the usually valid simplifications are:

$$[HA]_{\text{equilibrium}} = [HA]_{\text{initial}}$$
$$[A^-]_{\text{equilibrium}} = [A^-]_{\text{initial}}$$

Buffers are most effective when the $\text{p}K_a$ of the acid member of the buffer pair lies within ±1 unit of the desired pH.

Although the mole ratio of A^- to HA determines the pH at which a buffer works, the amounts of the acid and its conjugate base give the buffer its capacity for taking on strong acid or base with little change in pH.

LO 7 Acid–base titrations

The pH at the equivalence point depends on the ions of the salt formed during the titration. In the titration of a strong acid with a strong base, the pH at the equivalence point is 7 because neither of the ions produced in the reaction affects the pH of an aqueous solution.

For the titration of a weak acid with a strong base, the approach towards calculation of the pH varies depending on which stage of the titration is considered.

1. Before the titration begins, the only solute is the weak acid, and so K_a is used.
2. After starting, but before the equivalence point, the solution is a buffer. Either K_a or K_b can be used.
3. At the equivalence point, the active solute is the conjugate base of the weak acid. K_b for the conjugate base must be used.
4. After the equivalence point, excess strong base determines the pH of the mixture. No equilibrium calculations are required. At the equivalence point, the pH is above 7.

Similar calculations apply in the titration of a weak base by a strong acid. In such titrations, the pH at the equivalence point is below 7.

The indicator chosen for a titration should have the centre of its colour change range near the pH at the equivalence point. For the titration of a strong or weak acid with a strong base, phenolphthalein is the best indicator. When a weak base is titrated with a strong acid, several indicators, including methyl orange, work satisfactorily.

LO 8 Lewis acids and bases

A Lewis acid is an electron-pair acceptor, while a Lewis base is an electron-pair donor. Electrophiles are Lewis acids and nucleophiles are Lewis bases.

A Lewis base must have one or more pairs of valence electrons available for bond formation. A Lewis acid must be able to accept an electron pair to form a new bond. Most Lewis acids are one of the following: a molecule with vacant valence orbitals, a molecule with delocalised π bonds involving oxygen, or a metal cation.

A hard Lewis base has electron pairs of low polarisability and high electronegativity. A soft Lewis base has a large donor atom of high polarisability and low electronegativity. A hard Lewis acid has an acceptor atom with low polarisability. A soft Lewis acid has a relatively high polarisability. The hard–soft acid–base (HSAB) principle states that hard Lewis acids tend to combine with hard Lewis bases, while soft Lewis acids tend to combine with soft Lewis bases. This is reflected in nature, where hard metal ions tend to be found bonded to hard bases, and vice versa.

KEY CONCEPTS AND EQUATIONS

Autoprotolysis constant for water $[H_3O^+][OH^-] = K_w$
(section 11.2)

This equation is used to calculate $[H_3O^+]$ if $[OH^-]$ is known, and vice versa.

$$[H_3O^+][OH^-] = K_w$$

pH and pOH
(section 11.2)

These provide a convenient method for representing $[H_3O^+]$ and $[OH^-]$ in aqueous solution.

Relationship between pH and pOH: pH + pOH = 14.00 at 25 °C
(section 11.2)

This relationship can be used to calculate pH if pOH is known, and vice versa.

$$\text{pH} + \text{pOH} = 14.00 \text{ (at 25 °C)}$$

$K_aK_b = K_w$
(section 11.4)

This equation is used to calculate K_a given K_b, or vice versa.

Periodic trends in strengths of acids
(section 11.5)

Periodic trends can be used to predict the relative acidities of acids hydrids themselves and for molecules that contain *X*—H bonds. The principles involved also let you compare the acidities of compounds containing different electronegative elements.

$pH = pK_a + \log \frac{[A^-]}{[HA]}$ *(section 11.6)*	This equation can be used to calculate the pH of a buffer solution.
Hard–soft acid–base principle *(section 11.8)*	Hard Lewis acids tend to combine with hard Lewis bases. Soft Lewis acids tend to combine with soft Lewis bases.

KEY TERMS

acid *p. 432*
acid–base indicator *p. 472*
acid–base reaction *p. 432*
acid–base titration *p. 472*
acidic *p. 437*
acidity constant (K_a) *p. 447*
adduct *p. 481*
alkaline *p. 437*
amphiprotic *p. 434*
autoprotolysis *p. 436*
autoprotolysis constant of water (K_w) *p. 437*
base *p. 432*
basic *p. 437*
basicity constant (K_b) *p. 447*
Brønsted–Lowry *p. 432*
buffer capacity *p. 472*
buffer solution *p. 466*
burette *p. 472*
conjugate acid–base pair *p. 434*
conjugate base *p. 434*
diprotic acid *p. 434*
diprotic base *p. 434*
electrophile *p. 482*
endpoint *p. 472*
equivalence point *p. 472*
hard Lewis acid *p. 484*
hard Lewis base *p. 484*
Henderson–Hasselbalch equation *p. 468*
heterolytic cleavage *p. 461*
homolytic cleavage *p. 461*
hydronium ion (H_3O^+) *p. 432*
hydroxide ion (OH^-) *p. 432*
Lewis acid *p. 481*
Lewis base *p. 481*
monoprotic acid *p. 433*
monoprotic base *p. 434*
neutral *p. 437*
nucleophile *p. 482*
pH *p. 438*
pK_a *p. 448*
pK_b *p. 448*
pK_w *p. 439*
pOH *p. 438*
polyprotic acid *p. 433*
polyprotic base *p. 434*
resonance stabilisation *p. 464*
soft Lewis acid *p. 484*
soft Lewis base *p. 484*
strong acid *p. 442*
strong base *p. 442*
titrant *p. 472*
titration curve *p. 472*
triprotic acid *p. 434*
weak acid *p. 442*
weak base *p. 442*
zwitterion *p. 449*

REVIEW QUESTIONS

LO 1 The Brønsted–Lowry definition of acids and bases

11.1 Give the definition of both a Brønsted–Lowry acid and a Brønsted–Lowry base.

11.2 How are the formulae of the members of a conjugate acid–base pair related to each other? Within the pair, how can you tell which is the acid?

11.3 When asked to identify the conjugate acid of the HPO_4^{2-} ion, a student wrote H_3PO_4. Was the student correct? Explain your answer.

11.4 Define the term 'amphiprotic'.

11.5 Write the structural formulae of the conjugate acids of the following bases.

(a) $CH_3-CH_2-\ddot{N}H_2$

(b) $\ddot{N}H_2-\ddot{N}H_2$

(c) :N⁀NH (imidazole ring)

11.6 Write the structural formulae for the conjugate bases of the following structures.

(a) $C_6H_5-NH_3^+$ (benzene ring bearing NH_3^+)

(b) $[(CH_3)_3N-H]^+$ (CH_3-N-H with CH_3 above and below N)

(c) $[H_2N-NH_3]^+$ (H–N(H)–N(H)(H)–H with lone pair on the left N)

LO 2 Acid–base reactions in water

11.7 Explain why the concentration of H_3O^+ in pure water at 25 °C cannot be greater than 1×10^{-7} M.

11.8 At 40.0 °C, $K_w = 3.0 \times 10^{-14}$. Calculate the pH of pure water at this temperature. Comment on the statement 'pure water always has a pH of 7' with reference to your answer.

LO 3 Strong acids and bases

11.9 Give the definition of both a strong acid and a strong base.

11.10 Will the conjugate acid of a strong base be very strong, strong, weak or very weak? Give your reasoning.

11.11 The $CH_3CH_2O^-$ (ethoxide) ion is a stronger base than the OH^- ion. Why can the ethoxide ion not exist in aqueous solution? Use a chemical equation that shows the reaction that occurs when the ethoxide ion is placed in water to illustrate your answer.

LO 4 Weak acids and bases

11.12 Write the general equation for the reaction of a weak acid with water. Give the expression for K_a.

11.13 Write the chemical equation for the reaction of each of the following weak acids with water and write the appropriate K_a expression.
(a) HF (b) HSO_4^- (c) HOBr (d) $(CH_3)_2NH_2^+$

11.14 Write the general equation for the reaction of a weak base with water. Give the expression for K_b.

11.15 Write the chemical equation for the reaction of each of the following weak bases with water and write the appropriate K_b expression.
(a) $H_2PO_4^-$ (b) NH_3 (c) NO_2^- (d) $CH_3CH_2CH_2NH_2$

11.16 The pK_a of HCN is 9.21 and that of HF is 3.17. Which is the stronger Brønsted–Lowry base, CN^- or F^-?

11.17 Acetylsalicylic acid is more commonly known by its trademarked name, Aspirin. It is a monoprotic acid with a K_a value of 3.27×10^{-4}. Is a solution of the sodium salt of aspirin in water acidic, basic or neutral? Explain.

11.18 Consider the following compounds and suppose that a 0.1 M solution is prepared of each: NaI, KF, $(NH_4)_2SO_4$, KCN, $KOOCCH_3$, $CsNO_3$ and KBr. Write the formulae of those that have solutions that are: (a) acidic (b) basic (c) neutral.

11.19 Gardeners often apply ammonium nitrate (NH_4NO_3) to soil as a source of nitrogen for plant growth. Will this compound have any effect on the acidity of the moisture in the ground? Explain.

11.20 A solution of hydrazinium acetate is slightly acidic. Without looking at the tables of equilibrium constants, is K_a for acetic acid larger or smaller than K_b for hydrazine? Justify your answer.

11.21 Under what circumstances can we assume that the equilibrium concentration of a weak acid or base is equal to its initial concentration when we calculate the pH of a solution?

11.22 For which of the following are we permitted to use the initial concentration of the acid or base to calculate the pH of the solution specified?

(a) 0.020 M CH_3COOH
(b) 0.10 M CH_3NH_2
(c) 0.002 M N_2H_4
(d) 0.050 M HCOOH

LO 5 The molecular basis of acid strength

11.23 How do the strengths of the binary acids vary from left to right across a period of the periodic table? How do they vary from top to bottom within a group?

11.24 Astatine, At, atomic number 85, is radioactive and does not occur in appreciable amounts in nature. Based on what you have learned in this chapter, answer the following.

(a) How would the acidity of HAt compare with HI?
(b) How would the acidity of $HAtO_3$ compare with $HBrO_3$?

11.25 Explain why perchloric acid ($HClO_4$) is a stronger acid than chloric acid ($HClO_3$)

11.26 Explain why $HClO_4$ is a stronger acid than H_2SeO_4.

11.27 Explain why the OBr^- ion is a stronger base than the NO_2^- ion.

11.28 The position of equilibrium in the equation below lies far to the left. Identify the conjugate acid–base pairs. Which of the two acids is stronger?

$$HOCl(aq) + H_2O \rightleftharpoons H_3O^+(aq) + OCl^-(aq)$$

11.29 The phosphate ion PO_4^{3-} is a weaker base than hydroxide ion, and HPO_4^{2-} is a stronger acid than water. Use these data to determine if the position of the equilibrium:

$$PO_4^{3-} + H_2O \rightleftharpoons HPO_4^{2-} + OH^-$$

lies to the left or the right. Justify your answer.

11.30 Acetic acid, CH_3COOH, is a weaker acid than nitrous acid, HNO_2. How do the strengths of the bases CH_3COO^- and NO_2^- compare?

11.31 In water, both $HClO_4$ and HNO_3, appear to be acids of equal strength; they are both 100% ionised. However, by using solvents other than water, it can be shown that $HClO_4$ is a stronger acid than HNO_3. What solvent property would be necessary in order to distinguish between the acidities of these two acids?

11.32 Formic acid, HCOOH, and acetic acid, CH_3COOH, are classified as weak acids, but in water HCOOH is more dissociated than CH_3COOH. However, if we use liquid ammonia as a solvent for these acids, they both appear to be of equal strengths and 100% dissociated. Explain why this is so.

LO 6 Buffer solutions

11.33 Write ionic equations that show how the components of the following buffer solutions react on addition of either H_3O^+ or OH^-.

(a) NH_4Cl and NH_3
(b) H_2CO_3 and $NaHCO_3$ (the 'carbonate' buffer in blood)
(c) NaH_2PO_4 and Na_2HPO_4 (the 'phosphate' buffer inside body cells)

11.34 Buffer 1 is a solution containing 0.10 M NH_4Cl and 1.0 M NH_3. Buffer 2 is a solution containing 1.0 M NH_4Cl and 0.10 M NH_3. Which buffer is better able to maintain a steady pH on the addition of strong acid? Explain.

11.35 A student attempted to prepare a buffer solution by adjusting the pH of a 0.100 M aqueous solution of CH_3COOH to 7.00 with concentrated aqueous NaOH. Why was the final solution not a buffer solution?

LO 7 Acid–base titrations

11.36 When an aqueous solution containing 0.0100 mol of propanoic acid is titrated with sodium hydroxide, will the solution be acidic, neutral or basic when exactly 0.0100 mol of sodium hydroxide has been added?

11.37 When a solution of hydrazine is titrated by hydrochloric acid, will the solution be acidic, neutral or basic at the equivalence point?

11.38 Two terms that are often used in titrations are 'endpoint' and 'equivalence point'. What is the difference between them?

11.39 Qualitatively, describe how an acid–base indicator works. Why do we want to use a minimum amount of indicator in a titration?

11.40 Which indicator, phenolphthalein or methyl purple, should be used in the titration of dilute ammonia by dilute hydrochloric acid. Give your reasons.

11.41 What is a good indicator for titrating potassium hydroxide with hydrobromic acid? Explain.

11.42 A student chose to use thymol blue in a titration of CH_3COOH with NaOH. Was this the correct decision? Justify your answer.

LO 8 Lewis acids and bases

11.43 Define the terms 'Lewis acid' and 'Lewis base'.

11.44 The addition of a proton to a water molecule to give H_3O^+ is the archetypal example of a Brønsted–Lowry acid–base reaction. Explain why this is also a Lewis acid–base reaction.

11.45 Methylamine has the formula CH_3NH_2 and the structure:

```
    H   H
    |   |
H — C — N:
    |   |
    H   H
```

Use Lewis structures to illustrate the reaction of methylamine with boron trifluoride, BF_3.

11.46 The fluoride ion, F^-, is a Lewis base. Explain why it cannot function as a Lewis acid.

11.47 Draw the Lewis structures of each of the reactants in the following reactions.

(a) $SO_3 + OH^- \longrightarrow HSO_4^-$
(b) $SnCl_2 + Cl^- \longrightarrow [SnCl_3]^-$
(c) $AsF_3 + SbF_5 \longrightarrow [AsF_2]^+[SbF_6]^-$

11.48 Identify the Lewis acid and the Lewis base in each of the reactions that appears in question 11.47.

11.49 Draw mechanistic arrows that show the donation of electrons that takes place in each reaction in question 11.47 and draw the Lewis structures of the products.

11.50 What is the definition of a hard acid? What is the definition of a soft base?

11.51 Rank the three molecules or ions in each of the following groups from softest to hardest.
(a) NCl_3, NH_3 and NF_3
(b) ClO_4^-, ClO_2^- and ClO_3^-
(c) Pb^{2+}, Pb^{4+} and Zn^{2+}
(d) PCl_3, $SbCl_3$ and PF_3

11.52 Boric acid, H_3BO_3 has the structure $B(OH)_3$, with three OH groups bonded to a central B atom. However, the anion present in aqueous solutions of boric acid is $[B(OH)_4]^-$ rather than $OB(OH)_2^-$. This is because H_3BO_3 undergoes Lewis acid–base adduct formation with one water molecule, and the adduct then transfers a proton to a second water molecule to generate $[B(OH)_4]^-$ and a hydronium ion. Draw Lewis structural diagrams illustrating these two transfer reactions; include arrows that show the movement of electrons.

11.53 Iron is never found as the free metal in nature; it is often found combined with oxygen or sulfur in compounds. The other members of group 8 in the periodic table, ruthenium and osmium, occur in elemental form. State the hard–soft acid–base (HSAB) principle and use this to explain these observations.

11.54 Lead poisoning can be treated using the dianion of 2,3-dimercaptopropanol. Two ions bind one Pb^{2+} ion in a soluble tetrahedral complex that can be excreted from the body. Using hard–soft acid–base concepts, draw the expected structure of this complex.

2,3-dimercaptopropanol dianion

REVIEW PROBLEMS

11.55 Write the formula of the conjugate acid of each of the following. (a) OBr^- (b) O^{2-} (c) $H_2PO_4^-$ (d) ClO_4^- (e) NH_3 **LO1**

11.56 Write the formula for the conjugate base of each of the following. **LO1**
(a) NH_2OH
(b) HSO_3^-
(c) HCN
(d) H_5IO_6
(e) HNO_2

11.57 Identify the conjugate acid–base pairs in the following reactions. **LO1**
(a) $CH_3COOH + NH_3 \rightleftharpoons NH_4^+ + CH_3COO^-$
(b) $HSO_4^- + N_2H_4 \rightleftharpoons N_2H_5^+ + SO_4^{2-}$
(c) $HOCl + ClO_2^- \rightleftharpoons HClO_2 + OCl^-$
(d) $HCO_3^- + C_5H_5N \rightleftharpoons C_5H_5NH^+ + CO_3^{2-}$

11.58 Identify the conjugate acid–base pairs in the following reactions. **LO1**
(a) $HSO_4^- + SO_3^{2-} \rightleftharpoons HSO_3^- + SO_4^{2-}$
(b) $S^{2-} + H_2O \rightleftharpoons HS^- + OH^-$
(c) $CN^- + H_3O^+ \rightleftharpoons HCN + H_2O$
(d) $H_2Se + H_2O \rightleftharpoons HSe^- + H_3O^+$

11.59 Calculate $[H_3O^+]$ in each of the following aqueous solutions in which $[OH^-]$ is: **LO2**
(a) 4.7×10^{-2} M
(b) 1.1×10^{-5} M
(c) 2.8×10^{-8} M
(d) 9.4×10^{-11} M

11.60 Calculate $[OH^-]$ in each of the following aqueous solutions in which the H_3O^+ concentrations are: **LO2**
(a) 3.5×10^{-8} M
(b) 0.0065 M
(c) 2.5×10^{-13} M
(d) 7.5×10^{-5} M

11.61 Calculate the pH of each solution in question 11.59. **LO2**

11.62 Calculate the pH of each solution in question 11.60. **LO2**

11.63 Calculate $[H_3O^+]$ and $[OH^-]$ in solutions that have the following pH values. **LO2**
(a) 1.37 (b) 9.18 (c) 7.02 (d) 13.63 (e) 4.42

11.64 What is the $[H_3O^+]$ in 0.030 M HNO_3? What is the pH of the solution? What is the OH^- concentration in the solution? **LO2**

11.65 A sodium hydroxide solution is prepared by dissolving 2.75 g NaOH in 750 mL of water. Calculate $[OH^-]$ and $[H_3O^+]$ in the solution, and the pOH and the pH of the solution. **LO2**

11.66 An aqueous solution was made by dissolving 0.837 g $Ba(OH)_2$ in water to give a final volume of 100 mL. What is the molar concentration of OH^- in the solution? What are the pOH and the pH? What is the H_3O^+ concentration in the solution? **LO2**

11.67 What volume of 0.317 M KOH is needed to completely neutralise the HNO_3 in 300 mL of a nitric acid solution that has a pH of 1.77? **LO2**

11.68 The K_a for HF is 6.8×10^{-4}. What is K_b for F^-? **LO2**

11.69 The phenolate ion, $C_6H_5O^-$, has $K_b = 7.8 \times 10^{-5}$. What is the K_a of phenol? **LO2**

11.70 Hydrogen peroxide, H_2O_2, is a weak acid with $K_a = 1.8 \times 10^{-12}$. What is the value of K_b for the HO_2^- ion? **LO2**

11.71 Trimethylamine, $(CH_3)_3N$, has an odour reminiscent of rotting fish. Its K_b is 6.3×10^{-5}. Calculate the K_a of its conjugate acid. **LO2**

11.72 Lactic acid, $CH_3CH(OH)COOH$, is responsible for the sour taste of sour milk. At 25 °C, its $K_a = 1.4 \times 10^{-4}$. What is the K_b of its conjugate base, the lactate ion, $CH_3CH(OH)COO^-$? **LO2**

11.73 Chlorous acid, $HClO_2$, has a pK_a of 1.95.
(a) What is the K_b of its conjugate base, the chlorite ion?
(b) Is the chlorite ion a stronger or weaker base than the acetate ion? **LO2**

11.74 At the temperature of the human body, 37 °C, the value of K_w is 2.4×10^{-14}. Calculate the $[H_3O^+]$, $[OH^-]$, pH and pOH of pure water at this temperature. What is the relation between pH, pOH and K_w at this temperature? Is water neutral at this temperature? **LO2**

11.75 Hydrazine, NH_2NH_2, is a Brønsted–Lowry base. A 0.050 M solution has a pH of 10.46. What are K_b and pK_b for hydrazine? **LO2**

11.76 What are the concentrations of all the solute species in 0.150 M lactic acid, $CH_3CH(OH)COOH$? What is the pH of the solution? This acid has $K_a = 1.4 \times 10^{-4}$. **LO3**

11.77 Calculate the pH of a 0.085 M solution of cyanic acid, HOCN, given that $K_a = 3.5 \times 10^{-4}$ for this acid. **LO3**

11.78 Phenol, also known as carbolic acid, is sometimes used as a disinfectant. What are the concentrations of all of the solute species in a 0.050 M solution of phenol, C_6H_5OH? For this acid, $K_a = 1.3 \times 10^{-10}$. **LO4**

11.79 Codeine is a naturally occurring molecule that is used for the relief of pain. It is a weak base with a pK_b of 5.79. Calculate the pH of a 0.014 M solution of codeine. **LO4**

11.80 Pyridine, C_5H_5N, is a bad-smelling liquid that is a weak base in water. Its pK_b is 8.82. What is the pH of a 0.20 M aqueous solution of this compound? **LO4**

11.81 Calculate the amount of NH_3 that must be dissolved in water to give 750 mL of solution with a pH of 11.48? **LO4**

11.82 Calculate the pH of 0.20 M NaCN. What is the concentration of HCN in the solution? **LO4**

11.83 What is the pH of a 0.35 M aqueous solution of KF. What is the concentration of HF in the solution? **LO4**

11.84 Calculate the pH of 0.15 M methylammonium chloride, CH_3NH_3Cl. For methylamine, CH_3NH_2, $K_b = 4.4 \times 10^{-4}$. **LO4**

11.85 What is the pH of a 0.22 M aqueous solution of hydroxylammonium chloride, $HONH_3Cl$? Hydroxylamine, NH_2OH, is a weak base with $K_b = 6.6 \times 10^{-9}$. **LO4**

11.86 A 0.18 M solution of the sodium salt of nicotinic acid (also known as niacin) has a pH of 9.05. What is the value of K_a for nicotinic acid? **LO4**

11.87 Calculate the mass of NH_4Br that has to be dissolved in 1.00 L of water at 25 °C to give a solution with a pH of 5.16. **LO4**

11.88 If chlorine gas is bubbled through a basic aqueous solution, the hypochlorite ion, OCl^-, is formed. Chlorine bleach is generally a 5% aqueous solution of NaOCl by weight. Calculate the approximate pH of chlorine bleach solution, assuming that NaOCl is the only solute and that the bleach solution has a density of $1.0\,g\,mL^{-1}$. **LO4**

11.89 What is the pH of a 0.0050 M solution of sodium cyanide? The K_b of NaCN is 1.62×10^{-5}. **LO4**

11.90 Calculate the pH of a 0.015 M solution of chloroacetic acid. $K_a = 1.36 \times 10^{-3}$ for this acid. **LO4**

11.91 The compound *para*-aminobenzoic acid (PABA) is a powerful sunscreening agent whose salts were once used widely in suntan and sunscreen lotions. The parent acid is a weak acid with a pK_a of 4.92 (at 25 °C). What is the $[H_3O^+]$ and pH of a 0.030 M solution of this acid? **LO4**

11.92 Barbituric acid, $HC_4H_3N_2O_3$, was discovered by the Nobel Prize-awarded organic chemist Adolph von Baeyer in 1864. It is the parent compound of the barbiturates, a class of sedative drugs. Its pK_a is 4.01. What is $[H_3O^+]$ and the pH of a 0.012 M solution of barbituric acid? **LO4**

11.93 Choose the stronger acid in each of the following. (a) H_2S or H_2Se (b) H_2Te or HI (c) PH_3 or NH_3 **LO5**

11.94 Choose the stronger acid in each of the following. (a) HIO_3 or HIO_4 (b) H_3AsO_4 or H_3AsO_3 **LO5**

11.95 Choose the stronger acid in each of the following. (a) H_3AsO_4 or H_3PO_4 (b) H_2CO_3 or HNO_3 (c) H_2SeO_4 or $HClO_4$ **LO5**

11.96 What is the pH of a solution that contains 0.15 M CH_3COOH and 0.25 M CH_3COO^-? $K_a = 1.8 \times 10^{-5}$ for CH_3COOH. **LO5**

11.97 Calculate the pH of a buffer solution containing 0.15 M NH_3 and 0.28 M NH_4^+. **LO6**

11.98 Suppose 25.0 mL of 0.10 M HCl is added to 250 mL of a buffer composed of 0.25 M NH_3 and 0.20 M NH_4Cl. By how much will the concentrations of the NH_3 and NH_4^+ ions change after the addition of the strong acid? **LO6**

11.99 A student added 100 mL of 0.10 M NaOH to 325 mL of a buffer that contained 0.19 M CH_3COOH and 0.14 M CH_3COO^-. By how much did the concentrations of CH_3COOH and CH_3COO^- change after the addition of the strong base? **LO6**

11.100 What mole ratio of NH_4Cl to NH_3 would buffer a solution at pH 9.25? **LO6**

11.101 Determine the mass of sodium acetate, $NaOOCCH_3$, that needs to be added to 0.75 L of 0.18 M acetic acid (pK_a 4.74) to make a buffer solution having a pH of 5.00. **LO6**

11.102 What mass of sodium formate, NaOOCH, needs to be dissolved in 1.0 L of 0.12 M formic acid (pK_a 3.74) to make a buffer solution with pH 3.80? **LO6**

11.103 Determine the mass of ammonium chloride, NH_4Cl, that has to be dissolved in 450 mL of 0.15 M NH_3 to prepare a buffer solution having a pH of 10.00. **LO6**

11.104 What mass of ammonium chloride has to be dissolved in 125 mL of 0.10 M NH_3 to make it a buffer with a pH of 9.15? **LO6**

11.105 Calculate the pH of a buffer solution prepared by dissolving 0.115 mol of acetic acid and 0.108 mol of sodium acetate in water. If 20.00 mL of 0.138 M HCl is added to the buffer, what is the final pH value? What would the pH be if the same volume of the HCl solution was added to 125 mL of pure water? **LO6**

11.106 What volume of 0.15 M HCl has to be added to the buffer described in question 11.105 to make the pH decrease by 0.05 pH units? What volume of the same HCl solution would, if added to 100 mL of pure water, make the pH decrease by 0.05 pH units? **LO6**

11.107 In a titration of a 20.00 mL sample of lactic acid with sodium hydroxide, it was found that 19.38 mL of 0.126 M NaOH was required to reach the equivalence point. What amount of lactic acid was in the sample? **LO7**

11.108 Ascorbic acid (vitamin C) is a diprotic acid with the formula $H_2C_6H_6O_6$. A sample of a vitamin supplement was analysed by titrating a 0.1000 g sample dissolved in water with 0.0200 M NaOH. A volume of 15.20 mL of the base was required to react completely with the ascorbic acid. What was the percentage by mass of ascorbic acid in the sample? **LO7**

11.109 A sample of NaCl was found to be contaminated with Na_2CO_3. In order to determine the amount of contaminant, 50.00 mL of 0.225 M HCl (an excess of HCl) was added to 1.351 g of the mixture of NaCl and Na_2CO_3. The unreacted HCl was then titrated with 0.100 M NaOH. The titration required 20.74 mL of the NaOH solution. What was the percentage by mass of NaCl in the original mixture of NaCl and Na_2CO_3? **LO7**

11.110 Aspirin is the common name of a monoprotic acid called acetylsalicylic acid. Its formula is $HOOCC_6H_4OCOCH_3$. A certain pain reliever was analysed for aspirin by dissolving 0.250 g of it in water and titrating it with 0.0300 M KOH solution. The titration required 29.40 mL of base. What is the percentage by mass of aspirin in the drug? **LO7**

11.111 When 30.00 mL of 0.10 M hypochlorous acid, HOCl, is titrated with 0.10 M sodium hydroxide, what is the pH at the equivalence point? Select a suitable indicator for this titration from table 11.9. **LO7**

11.112 When 25 mL of 0.10 M aqueous ammonia is titrated with 0.10 M hydrobromic acid, what is the pH at the equivalence point? Select a good indicator for this titration from table 11.9. **LO7**

11.113 What is the pH of a solution prepared by mixing 23.0 mL of 0.146 M CH_3COOH with 30.0 mL of 0.235 M NaOH? **LO7**

11.114 What is the pH of a solution prepared by mixing 30.0 mL of 0.200 M CH_3COOH with 15.0 mL of 0.400 M KOH? **LO7**

11.115 For the titration of 20.00 mL of 0.1575 M acetic acid with 0.1575 M NaOH, calculate the pH: **LO7**
(a) before the addition of any NaOH solution
(b) after 7.50 mL of the base has been added
(c) after half of the acetic acid has been neutralised
(d) at the equivalence point.

11.116 For the titration of 25.00 mL of 0.1000 M ammonia with 0.1000 M HCl, calculate the pH: **LO7**
(a) before the addition of any HCl solution
(b) after 10.00 mL of the acid has been added
(c) after half of the NH_3 has been neutralised
(d) at the equivalence point.

11.117 Identify the Lewis acid and Lewis base in the reaction **LO8**

$$NH_2^- + H_3O^+ \rightarrow NH_3 + H_2O$$

11.118 Identify the Lewis acid and Lewis base in the reaction. **LO8**

$$BF_3 + F^- \rightarrow BF_4^-$$

11.119 Aluminium chloride, $AlCl_3$, forms molecules with the formula Al_2Cl_6. The structure of these molecules is:

Cl Cl Cl
Al Al
Cl Cl Cl

Use Lewis structures to show how the reaction $2AlCl_3 \rightarrow Al_2Cl_6$ is a Lewis acid–base reaction. **LO8**

11.120 In each of the following reactions, identify the Lewis acid and the Lewis base. **LO8**
(a) $Ni + 4CO \longrightarrow [Ni(CO)_4]$
(b) $SbCl_3 + 2Cl^- \longrightarrow [SbCl_5]^{2-}$
(c) $(CH_3)_3P + AlBr_3 \longrightarrow (CH_3)_3P{-}AlBr_3$
(d) $BF_3 + ClF_3 \longrightarrow [ClF_2]^+[BF_4]^-$

11.121 Identify the Lewis acid and the Lewis base in each of the following reactions. **LO8**
(a) $SnCl_4 + 2Cl^- \longrightarrow SnCl_6^{2-}$
(b) $Co^{3+} + 6NH_3 \longrightarrow [Co(NH_3)_6]^{3+}$
(c) $Br_2 + Br^- \longrightarrow Br_3^-$
(d) $MgO + CO_2 \longrightarrow MgCO_3$

11.122 Using liquid ammonia, NH_3, as a solvent, sodium amide, $NaNH_2$, reacts with ammonium chloride, NH_4Cl, in an acid–base neutralisation reaction. Assuming that these compounds are completely dissociated in liquid ammonia, write molecular, ionic and net ionic equations for the reaction. Which substance is the acid and which is the base? **LO8**

11.123 Rank the following ions in order of increasing polarisability, and explain your reasoning: Ag^+, Cu^+, Au^+. **LO8**

11.124 Rank the three Lewis acids in each of the following groups from hardest to softest, and explain your reasoning. **LO8**
(a) BCl_3, BF_3 and $AlCl_3$
(b) Al^{3+}, Tl^{3+} and Tl^+
(c) $AlCl_3$, AlI_3 and $AlBr_3$

11.125 Rank the three Lewis bases in each of the following groups from hardest to softest, and explain your rankings. **LO8**
(a) NH_3, SbH_3 and PH_3
(b) PO_4^{3-}, ClO_4^- and SO_4^{2-}
(c) O^{2-}, Se^{2-} and S^{2-}

11.126 Explain why SO_3 is a harder Lewis acid than SO_2. **LO8**

11.127 Explain why iodide is a soft base but chloride is a hard base. **LO8**

ADDITIONAL EXERCISES

11.128 Solutions of biological molecules often require buffering, and it is important that the buffer molecules are chemically inert and unreactive towards the biological molecules. With this in mind, a series of buffers called Good's Buffers were developed, which buffer over a wide pH range. One of the most these buffers is called HEPES [4-(2-hydroxyethyl)-1-piperazineethanesulfonic acid], and the nominal structure is given below. However, the molecule actually exists as a zwitterion. Given that there are two basic sites in the molecule, suggest, with reasons, which of the N atoms is protonated in the zwitterion. **LO6**

HO–CH₂CH₂–N(piperazine)N–CH₂CH₂–SO_3H

11.129 Hydrazine, N_2H_4, is a weaker Brønsted–Lowry base than ammonia. In the following reaction, would the position of equilibrium lie to the left or to the right? Justify your answer. **LO3**

$$N_2H_5^+ + NH_3 \rightleftharpoons N_2H_4 + NH_4^+$$

11.130 Using a spreadsheet program, and the equations on pages 474 and 476, plot, on the same graph, the titration curves for the acidic region of titrations of 25.00 mL of 0.200 M HCl against 0.200 M NaOH, and 25.00 mL of 0.200 M CH_3COOH against 0.200 M NaOH. When generating the data for the curves, add the NaOH 0.1 mL at a time and ignore the final point when 25.0 mL of NaOH has been added (why?). Comment on any obvious differences between the two curves. **LO7**

11.131 Calculate the pH of a solution that is 0.120 M in HCl and also 0.145 M in CH_3COOH. What is the concentration of acetate ion in this solution? **LO6**

11.132 A solution is prepared by mixing 300 mL of 0.500 M NH_3 and 100 mL of 0.500 M HCl. Assuming that the volumes are additive, what is the pH of the resulting mixture? **LO3**

11.133 The introduction of a nitro (NO_2) group into an organic molecule is called nitration, and often requires the generation of the NO_2^+ electrophile. This can be accomplished in a mixture of sulfuric and nitric acids. Given that sulfuric acid is a stronger acid than nitric acid, suggest a possible reaction scheme that could account for the formation of NO_2^+. **LO5**

11.134 Predict whether the pH of 0.120 M NH_4CN is greater than, less than or equal to 7.00. Give your reasons. **LO5**

11.135 When a water molecule is bonded to a positively charged transition metal ion, it becomes significantly more acidic. In fact, the pK_a of the $[Fe(OH_2)_6]^{3+}$ ion (2.17), which corresponds to the equilibrium:

$$[Fe(OH_2)_6]^{3+}(aq) + H_2O(l) \rightleftharpoons [Fe(OH_2)_5(OH)]^{2+}(aq) + H_3O^+(aq)$$

makes this an acid of comparable strength to phosphoric acid. Explain why this is the case. **LO5**

11.136 The fluorides BF_3, AlF_3, SiF_4 and PF_5 are Lewis acids. They all form very stable fluoroanions when treated with lithium fluoride. In contrast, three other fluorides, CF_4, NF_3 and SF_6, do not react with lithium fluoride. Use Lewis acid–base concepts to explain this behaviour. **LO8**

11.137 Complete each of the following reactions by considering the Lewis acid and base characteristics of the starting materials. **LO8**
(a) $AlCl_3 + LiCH_3 \longrightarrow ?$
(b) SO_3 + excess $H_2O \longrightarrow ?$
(c) $SbF_5 + LiF \longrightarrow ?$
(d) $SF_4 + AsCl_5 \longrightarrow ?$

11.138 Certain proteolytic enzymes react in alkaline solutions. One of these enzymes produces 1.8 micromoles of hydrogen ions per second in a 2.50 mL portion of a buffer composed of 0.25 M NH_3 and 0.20 M NH_4Cl. By how much will the concentrations of the NH_3 and NH_4^+ ions change after the reaction has run for 35 seconds? **LO6**

12 Oxidation and reduction

In the preceding chapters, we learned about some important reactions that take place in aqueous solutions. In this chapter, we will expand on that knowledge with a discussion of reactions that involve the transfer of one or more electrons from one chemical species to another. Chemists call these electron transfer processes oxidations and reductions or, simply, redox reactions.

Redox reactions are very common. For example, the photo on this page shows the rusting remains of the MV *Sygna*, a bulk carrier that was wrecked on Stockton Beach near Newcastle during a violent storm in 1974. Fortunately the 31 crew on board were rescued by RAAF helicopter. Multiple salvage attempts failed and 40 years later the Sygna still stands about 80 m offshore. It is believed the Sygna will rust down to the waterline within the next 10 years. Formation of rust is a redox process in which electrons are transferred from iron to oxygen.

In this chapter, we will learn how to recognise and analyse the changes that occur in these reactions. We will also learn that redox reactions can be used to generate electricity in galvanic cells (commonly called batteries) and that, by reversing this process, electricity can be used to cause nonspontaneous redox reactions. Because electricity plays a role in these systems, the processes involved are electrochemical changes. The study of such changes is called electrochemistry.

LEARNING OBJECTIVES

After studying this chapter, you should be able to:

12.1 describe the characteristics of a redox reaction

12.2 write balanced net ionic equations for redox reactions

12.3 describe the operation of a galvanic cell

12.4 use reduction potentials to determine whether a reaction will be spontaneous

12.5 describe the relationship between cell potential, concentration and Gibbs energy, and perform calculations using the Nernst equation

12.6 explain the process of corrosion

12.7 describe the processes in electrolysis and perform calculations using Faraday's law

12.8 describe the electrode processes during charge and discharge in batteries.

12.1 Oxidation and reduction

Among the first reactions studied by early scientists were those that involved oxygen. According to the original definition, *oxidation* is a reaction with oxygen. If a metal atom M reacts with molecular oxygen, O_2, according to the equation $M + \frac{1}{2}O_2 \rightarrow MO$, formation of the metal oxide MO occurs through transfer of electrons (e^-) from the metal atom:

$$M \rightarrow M^{2+} + 2e^-$$

to the oxygen atom:

$$\tfrac{1}{2}O_2 + 2e^- \rightarrow O^{2-}$$

Hence:

$$M + \tfrac{1}{2}O_2 \rightarrow M^{2+} + O^{2-}$$

By taking up two electrons, oxygen acquires a completely filled valence shell with eight electrons. On the other hand, the reverse reaction (removal of oxygen from the metal oxide to give the metal in its elemental form) was described as *reduction*.

Only much later did chemists realise that the reactions of oxygen constitute a special case of a more general phenomenon, one in which electrons are transferred from one substance to another. For example, the reaction between sodium and chlorine to give sodium chloride proceeds through transfer of electrons from sodium (*oxidation* of sodium) to chlorine (*reduction* of chlorine).

$$Na \rightarrow Na^+ + e^- \qquad \text{(oxidation)}$$
$$\tfrac{1}{2}Cl_2 + e^- \rightarrow Cl^- \qquad \text{(reduction)}$$

Notice that the electron appears as a 'product' if the process is oxidation and as a 'reactant' if the process is reduction.

Based on our understanding of this process, we define **oxidation** as the loss of electrons. An **oxidising agent** (or oxidant) is then a substance that has the ability to remove electrons from a substance. **Reduction** is the gain of electrons. A **reducing agent** (or reductant) is a substance with the ability to donate electrons to another substance. Thus, the oxidising agent is in fact reduced, and the reducing agent is oxidised. Collectively, *electron transfer* reactions are called **reduction–oxidation reactions** or **redox reactions.**

Redox reactions are summarised in the following equation.

$$\text{reducing agent} \underset{\text{reduction}}{\overset{\text{oxidation}}{\rightleftharpoons}} \text{oxidising agent} + \text{electron(s)}$$

To help remember, think of the mnemonic OIL RIG, which stands for 'oxidation is loss, reduction is gain'. Since positive and negative charges must always be balanced in a reaction, *oxidation and reduction always occur together.* No substance is ever oxidised unless something else is reduced, and the total number of electrons lost by one substance is always the same as the total number gained by the other. In the reaction of sodium with chlorine, for example, the overall reaction is:

$$2Na + Cl_2 \rightarrow 2NaCl$$

When two sodium atoms are oxidised, two electrons are lost, which is exactly the number of electrons gained when one Cl_2 molecule is reduced.

WORKED EXAMPLE 12.1

Identifying reduction–oxidation

The bright light produced by the reaction between magnesium and oxygen is often used in fireworks displays (figure 12.1). The product of the reaction is magnesium oxide, MgO.

$$2Mg + O_2 \rightarrow 2MgO$$

Which element is oxidised and which is reduced? What are the oxidising and reducing agents?

FIGURE 12.1 The reaction between magnesium and oxygen produces bright white light.

Analysis

Magnesium oxide is an ionic substance consisting of a metal and a nonmetal. The locations of the elements in the periodic table (groups 2 and 16) tell us that the ions in MgO are Mg^{2+} and O^{2-}.

Solution

When a magnesium atom becomes a magnesium ion, it must lose two electrons.

$$Mg \rightarrow Mg^{2+} + 2e^-$$

Because magnesium is oxidised, it must be the reducing agent. Oxygen can gain these two electrons to give an O^{2-} ion.

$$\tfrac{1}{2}O_2 + 2e^- \rightarrow O^{2-}$$

O_2 is reduced and so it must be the oxidising agent.

Is our answer reasonable?

There are two things we can do to check our answers. First, we can check to be sure that we have placed electrons on the correct sides of the equations. As with ionic equations, the number of atoms of each kind and net charge must be the same on both sides. We see that this is true for both equations. (If we had placed the electrons on the wrong side, the charges would not balance.) As oxidation reactions always have electrons on the right-hand side, and reduction reactions always have electrons on the left-hand side, by observing the locations of the electrons in the equations, we come to the same conclusions that Mg is oxidised and O_2 is reduced.

Another check is noting that we have identified one substance as being oxidised and the other as being reduced. If we had made a mistake, we might have found that both were oxidised, or both were reduced. This is impossible because, in every reaction in which there is oxidation, there must also be reduction, and vice versa.

PRACTICE EXERCISE 12.1

Identify the substances oxidised and reduced and the oxidising and reducing agents in the reaction of copper and chlorine to form $CuCl_2$.

Oxidation numbers

The reaction of magnesium with oxygen in worked example 12.1 is clearly a redox reaction. However, not all reactions with oxygen produce ionic products. For example, the reaction of sulfur with oxygen is a redox reaction, but the product sulfur dioxide, SO_2, is covalent. To help us follow electron transfers in such reactions, we use the convenient concept of the oxidation number. (Note that oxidation numbers in covalent compounds are used for convenience and should not be misinterpreted as ionic charges.) The **oxidation number** (or **oxidation state**) is the hypothetical charge that an individual atom or ion in a molecule *would* possess *if* the shared electron pairs in each covalent bond were assigned to the more electronegative element in the bond. Thus, the oxidation number is the charge that each atom would have if the compound were divided into monatomic ions. Oxidation numbers are always given for our individual atom or ion and not for groups of atoms or ions.

The oxidation number of an element in a particular compound is assigned according to the following basic rules.

1. The oxidation number of any free element (an element not combined chemically with a different element) is 0. For example, Ar, Fe, O in O_2, P in P_4 and S in S_8 all have oxidation numbers of 0.
2. The oxidation number for any simple, monatomic ion (e.g. Na^+ and Cl^-) is equal to the charge on the ion.
3. The sum of the oxidation numbers of all atoms in a neutral molecule must equal zero. The sum of all oxidation numbers in a polyatomic ion must equal the charge on the ion.
4. In all of its compounds, fluorine has an oxidation number of –1.
5. In most of its compounds, hydrogen has an oxidation number of +1.
6. In most of its compounds, oxygen has an oxidation number of –2.

In addition to these basic rules, some other chemical knowledge is required. As shown in chapter 4, the periodic table can be used to determine the charges on certain ions of the elements. For instance, all the metals in group 1 form ions with a 1+ charge, and all those in group 2 form ions with a 2+ charge. This means that when we find sodium in a compound, we can assign it an oxidation number of +1 because its simple ion, Na^+, has a charge of 1+. Similarly, calcium in a compound exists as Ca^{2+} and has an oxidation number of +2. In binary ionic compounds with metals, the nonmetals have oxidation numbers that are equal to the charges on their anions. For example, the compound Fe_2O_3 contains the oxide ion, O^{2-}, which is assigned an oxidation number of –2. Similarly, Mg_3P_2 contains the phosphide ion, P^{3-}, which has an oxidation number of –3.

It is important to realise that the oxidation number does not actually equal a charge on an atom. To differentiate oxidation numbers from actual electrical charges, the following convention is used; oxidation numbers are given with the sign *before* the number, whereas electrical charges specify the sign *after* the number. For example, the sodium ion has a charge of 1+ and an oxidation number of +1. In a molecular formula, oxidation numbers are placed as a small Arabic numeral above the corresponding element symbol.

$$H_2\overset{+6}{S}O_4 \quad K\overset{+7}{Mn}O_4 \quad Na\overset{+5}{N}O_3 \quad \overset{-3}{N}H_4Cl$$

Sometimes it is convenient to specify the oxidation number of an element when its name is written out. This is done by writing the oxidation number as a Roman numeral in parentheses after the name of the element. For example, 'iron(III)' means iron with an oxidation number of +3.

The numbered rules given on the previous page usually come into play when an element can have more than one oxidation number, such as transition metals. Iron can form Fe^{2+} and Fe^{3+} ions so, in an iron compound, the rules help to determine which one is present. Nonmetals can also exhibit more than one oxidation number, especially when they are combined with hydrogen and oxygen in compounds or polyatomic ions, and their oxidation numbers must be calculated using the rules.

WORKED EXAMPLE 12.2

Assigning oxidation numbers

Molybdenum disulfide, MoS_2, has a structure that allows it to be used as a dry lubricant, much like graphite. What are the oxidation numbers of the atoms in MoS_2?

Analysis

This is a binary compound of a metal and a nonmetal so, for the purposes of assigning oxidation numbers, we assume it to contain ions. Molybdenum is a transition metal, so there is no simple rule to tell what its ions are. However, sulfur is a nonmetal in group 16, and its ion is S^{2-}.

Solution

Because S^{2-} is a simple monatomic ion, its charge equals its oxidation number, so sulfur has an oxidation number of −2. Now we can use rule 3 (the summation rule) to determine the oxidation number of molybdenum, which we will represent by x.

S	(2 atoms) × (−2) = −4	(rule 2)
Mo	(1 atom) × (x) = x	
	Sum = 0	(rule 3)

The value of x must be +4 for the sum to be 0 (MoS_2 is a neutral molecule). Therefore, the oxidation numbers are Mo = +4 and S = −2.

PRACTICE EXERCISE 12.2

Assign oxidation numbers to each atom in: (a) $NiCl_2$, (b) Mg_2TiO_4, (c) $K_2Cr_2O_7$, (d) HPO_4^{2-}, (e) $(NH_4)_2Ce(NO_3)_6$, (f) $K_4[Fe(CN)_6]$ and (g) $Na_3[Fe(CN)_6]$.

Some classes of compounds have atoms in oxidation states that are exceptions to the rules on the previous page, but these are rare. Examples include the oxygen atoms in peroxides (for example, in hydrogen peroxide, H_2O_2, which is used for bleaching hair, the oxygen atoms have an oxidation number of −1, instead of −2) and the hydrogen atoms in hydrides (for example, in calcium hydride, CaH_2, the hydrogen atoms have an oxidation number of −1, instead of +1). Oxidation numbers calculated by the rules on the previous page can have fractional values (e.g. the nitrogen atoms in the ionic compound sodium azide, NaN_3 (see figure 12.2), have an oxidation number of $-\frac{1}{3}$).

FIGURE 12.2 The airbags used as safety devices in modern cars are inflated by the explosive decomposition of sodium azide, which gives elemental sodium and gaseous nitrogen through the reaction:

$$2NaN_3 \rightarrow 2Na + 3N_2$$

Sodium metal, which is highly reactive, is converted into a harmless silicate glass through reaction with additives, such as potassium nitrate and silicon dioxide. Because of the high toxicity of sodium azide, it is now sometimes replaced by less harmful nitrogen-rich compounds, such as tetrazoles (five-membered rings with one carbon and four nitrogen atoms).

Using the concept of oxidation numbers, we can now view a redox reaction as a chemical reaction in which changes in oxidation numbers occur. *Oxidation is an increase in oxidation number. Reduction is a decrease in oxidation number.*

WORKED EXAMPLE 12.3

Using oxidation numbers to analyse redox reactions

The reaction of concentrated hydrochloric acid, HCl, with potassium permanganate, $KMnO_4$, is a convenient method of preparing chlorine gas in the laboratory.

$$6HCl + 2KMnO_4 + 2H^+ \rightarrow 3Cl_2 + 2MnO_2 + 4H_2O + 2K^+$$

Identify the substances oxidised and reduced as well as the oxidising and reducing agents in this reaction.

Analysis

To identify the redox species, we need to consider what is happening with the oxidation numbers. This will tell us what is oxidised and reduced. Then we recall that the substance oxidised is the reducing agent, and the substance reduced is the oxidising agent.

Solution

To identify redox changes using oxidation numbers, we first assign an oxidation number to each atom on both sides of the equation.

$$\overset{+1\,-1}{6HCl} + \overset{+1\,+7\,-2}{2KMnO_4} + \overset{+1}{2H^+} \longrightarrow \overset{\pm 0}{3Cl_2} + \overset{+4\,-2}{2MnO_2} + \overset{+1\,-2}{4H_2O} + \overset{+1}{2K^+}$$

Next we look for changes, keeping in mind that an increase in oxidation number is oxidation and a decrease is reduction.

A change from –1 to 0 is an increase in oxidation number.

oxidation

A change from +7 to +4 is a decrease in oxidation number.

reduction

$$\overset{+1\,-1}{6HCl} + \overset{+1\,+7\,-2}{2KMnO_4} + \overset{+1}{2H^+} \longrightarrow \overset{\pm 0}{3Cl_2} + \overset{+4\,-2}{2MnO_2} + \overset{+1\,-2}{4H_2O} + \overset{+1}{2K^+}$$

Thus, the Cl in HCl is oxidised and the Mn in $KMnO_4$ is reduced. The reducing agent is HCl and the oxidising agent is $KMnO_4$. (Note that, in identifying the oxidising and reducing agents, we specify the entire formulae for the substances containing the atoms changing oxidation numbers.)

Is our answer reasonable?

There are lots of things we could check here. We have found changes that lead us to identify an oxidation and a reduction process; this gives us confidence that we have done the rest of the work correctly.

PRACTICE EXERCISE 12.3

Identify the substances oxidised and reduced, as well as the oxidising and reducing agents in the following reaction.

$$H_3AsO_4 + SnCl_2 + 2HCl \longrightarrow H_3AsO_3 + SnCl_4 + H_2O$$

12.2 Balancing net ionic equations for redox reactions

Many redox reactions take place in aqueous solution and many of these involve ions; they are ionic reactions. In studying redox reactions, it is often helpful to write ionic and net ionic equations. We do this by dividing the oxidation and reduction processes into individual equations called **half-equations**, which are then balanced separately. There, we combine the balanced half-equations to obtain the fully balanced net ionic equation.

To illustrate the method, we will balance the net ionic equation for the reaction of iron(III) chloride, $FeCl_3$, with tin(II) chloride, $SnCl_2$, in aqueous solution, which changes the Fe^{3+} to Fe^{2+} and the Sn^{2+} to Sn^{4+}. In the reaction, the chloride ion is unaffected — it is a 'spectator ion'.

We begin by writing an equation that shows only the species involved in the reaction. In this case, the reactants are Fe^{3+} and Sn^{2+}, and the products are Fe^{2+} and Sn^{4+}. The equation is therefore:

$$Fe^{3+} + Sn^{2+} \rightarrow Fe^{2+} + Sn^{4+}$$

Note that, while the equation is balanced in terms of mass, it is not yet balanced in terms of charge on either side of the arrow. To balance this equation in terms of both mass and charge, we must first identify the reactant and product of the two half-equations.

$$Sn^{2+} \rightarrow Sn^{4+}$$
$$Fe^{3+} \rightarrow Fe^{2+}$$

Next, we balance the half-equations so that each obeys both criteria for a balanced ionic equation; both atoms and charge have to balance. Obviously, the atoms are already balanced in each equation. The charge, however, is not. We add electrons as necessary to balance the charges. For the first half-equation, two electrons are added to the right, so the net charge on both sides will be 2+. In the second half-equation, one electron is added to the left, which makes the net charge on both sides equal to 2+.

$$Sn^{2+} \rightarrow Sn^{4+} + 2e^-$$
$$Fe^{3+} + e^- \rightarrow Fe^{2+}$$

Note that an oxidation half-equation will *always* have the electrons on the right-hand side and a reduction half-equation will *always* have the electrons on the left-hand side. Therefore, the first half-equation corresponds to an oxidation reaction and the second to a reduction reaction.

In any overall redox reaction, the number of electrons gained always equals the number lost. Given that two electrons are lost by Sn^{2+} in the oxidation process:

$$Sn^{2+} \rightarrow Sn^{4+} + 2e^-$$

two electrons must be used in the reduction process. Therefore, we need to multiply each of the coefficients in the reduction half-equation by 2.

$$2Fe^{3+} + 2e^- \rightarrow 2Fe^{2+}$$

We combine the balanced half-equations by adding them to give:

$$\begin{array}{r} Sn^{2+} \rightarrow Sn^{4+} + 2e^- \\ 2Fe^{3+} + 2e^- \rightarrow 2Fe^{2+} \\ \hline Sn^{2+} + 2Fe^{3+} + 2e^- \rightarrow Sn^{4+} + 2Fe^{2+} + 2e^- \end{array}$$

Finally, we note that two electrons appear on each side of the equation. We can cancel these to give the final balanced equation.

$$Sn^{2+} + 2Fe^{3+} \rightarrow Sn^{4+} + 2Fe^{2+}$$

Notice that both the charge *and* the number of each type of atom are now balanced.

PRACTICE EXERCISE 12.4

Balance the following redox equation.

$$Al + Cu^{2+} \rightarrow Al^{3+} + Cu$$

Redox reactions in acidic and basic solutions

In many redox reactions in aqueous solution, H_3O^+ or OH^- ions play an important role, as do water molecules. For example, when solutions of $K_2Cr_2O_7$ and $FeSO_4$ are mixed, the acidity of the mixture decreases as the reaction proceeds, as dichromate ions, $Cr_2O_7^{2-}$, oxidise Fe^{2+}. This is because the reaction uses H_3O^+ as a reactant and produces H_2O as a product. In other reactions, OH^- is consumed, while in still others H_2O is a reactant. Also, in many cases, the products (or even the reactants) of a redox reaction will depend on the acidity of the solution. For example, in an acidic solution, MnO_4^- is reduced to Mn^{2+} ions, but in a neutral or slightly basic solution the reduction product is insoluble MnO_2, with Mn having an oxidation number of +4. We will first learn how to balance acidic solutions. Balancing basic solutions uses the same concepts but requires an additional step at the end to ensure that the final equation does not show H^+.

Acidic solutions

$Cr_2O_7^{2-}$ (orange-red) reacts with Fe^{2+} (almost colourless) in an acidic solution to give Cr^{3+} (green) and Fe^{3+} (orange) as products (figure 12.3). Therefore, the equation we must balance is:

$$Cr_2O_7^{2-} + Fe^{2+} \rightarrow Cr^{3+} + Fe^{3+}$$

FIGURE 12.3 A solution of $K_2Cr_2O_7$ being added to an acidic solution containing Fe^{2+}. Note the green-orange swirl of products.

The oxidation number of Cr in $Cr_2O_7^{2-}$ is +6.

The balanced equation can be found through the following steps.

Step 1: Identify the reactant and product of each of the oxidation and reduction processes.

$$Cr_2O_7^{2-} \rightarrow Cr^{3+} \quad \text{(reduction)}$$
$$Fe^{2+} \rightarrow Fe^{3+} \quad \text{(oxidation)}$$

Step 2: Balance atoms other than H and O.
There are two Cr atoms on the left and only one on the right, so we place a coefficient of 2 in front of Cr^{3+}. The oxidation half-equation is already balanced in terms of atoms.

$$Cr_2O_7^{2-} \rightarrow 2Cr^{3+}$$
$$Fe^{2+} \rightarrow Fe^{3+}$$

Step 3: Balance oxygen by adding H_2O.
There are seven oxygen atoms on the left of the reduction half-equation and none on the right. Therefore, we add $7H_2O$ to the right side of the reduction half-equation.

$$Cr_2O_7^{2-} \rightarrow 2Cr^{3+} + 7H_2O$$
$$Fe^{2+} \rightarrow Fe^{3+}$$

Step 4: Balance hydrogen by adding H^+. Note that, for simplicity, we use H^+, rather than H_3O^+, when balancing redox equations.
After adding the water, we see that we have created an imbalance; the first half-equation has 14 hydrogen atoms on the right and none on the left. To balance these, we add $14H^+$ to the left side of the half-equation.

$$14H^+ + Cr_2O_7^{2-} \rightarrow 2Cr^{3+} + 7H_2O$$
$$Fe^{2+} \rightarrow Fe^{3+}$$

Now each half-equation is balanced in terms of number and type of atoms involved. Next we will balance the charge.

Step 5: Balance the charge by adding electrons.
First we calculate the net charge on each side. For the reduction half-equation, we have:

$$\underbrace{14H^+ + Cr_2O_7^{2-}}_{\text{net charge} = (14+)\ +\ (2-)\ =\ 12+} \quad \rightarrow \quad \underbrace{2Cr^{3+} + 7H_2O}_{\text{net charge} = 2(3+)\ +\ 0\ =\ 6+}$$

To balance the net charge, we need to add six electrons to the left-hand side of the half-equation. Recall that, in balanced reduction half-equations, electrons always appear on the left-hand side.

$$6e^- + 14H^+ + Cr_2O_7^{2-} \rightarrow 2Cr^{3+} + 7H_2O$$

To balance the oxidation half-equation, we need to add one electron to the right.

$$Fe^{2+} \rightarrow Fe^{3+} + e^-$$

Step 6: Make the number of electrons gained equal to the number lost.
At this point we have the two balanced half-equations:

$$6e^- + 14H^+ + Cr_2O_7^{2-} \rightarrow 2Cr^{3+} + 7H_2O$$
$$Fe^{2+} \rightarrow Fe^{3+} + e^-$$

Because six electrons are gained in the reduction process, but only one is lost in the oxidation process, we multiply all of the coefficients of the oxidation half-equation by 6.

$$6Fe^{2+} \rightarrow 6Fe^{3+} + 6e^-$$

Step 7: Add the balanced half-equations.

$$6e^- + 14H^+ + Cr_2O_7^{2-} \rightarrow 2Cr^{3+} + 7H_2O$$
$$6Fe^{2+} \rightarrow 6Fe^{3+} + 6e^-$$
$$6e^- + 14H^+ + Cr_2O_7^{2-} + 6Fe^{2+} \rightarrow 2Cr^{3+} + 7H_2O + 6Fe^{3+} + 6e^-$$

Step 8: Cancel any species that is the same on both sides.
Cancel six electrons from both sides to give the final balanced equation.

$$14H^+ + Cr_2O_7^{2-} + 6Fe^{2+} \rightarrow 2Cr^{3+} + 7H_2O + 6Fe^{3+}$$

Step 9: Check that the final equation is balanced in terms of number and type of atoms and charge. In step 4, we used H^+ instead of H_3O^+. If you want to express your equation with H_3O^+, take note of the number of H^+ in the balanced equations and add the same number of H_2O to *both* sides. Then combine H^+ and H_2O as H_3O^+.

Summary of the steps for balancing an equation for a redox reaction in an acidic solution

Step 1: Identify reactants and products for each half-equation.
Step 2: Balance atoms other than H and O.
Step 3: Balance oxygen by adding H_2O.
Step 4: Balance hydrogen by adding H^+.
Step 5: Balance net charge by adding e^-.
Step 6: Make e^- gain equal e^- loss.
Step 7: Add the balanced half-equations.
Step 8: Cancel any species that is the same on both sides.
Step 9: Check that there are the same number and type of atoms and the same charge on each side of the reaction.

WORKED EXAMPLE 12.4

Balancing redox equations in acidic solutions

Balance the following equation. The reaction occurs in acidic solution.

$$MnO_4^- + H_2SO_3 \rightarrow SO_4^{2-} + Mn^{2+}$$

Solution

We follow the steps just given. The oxidation number of Mn in MnO_4^- is +7, of S in H_2SO_3 is +4 and of S in SO_4^{2-} is +6.

Step 1: Identify the reactants and products of the reduction and oxidation processes.

$$MnO_4^- \rightarrow Mn^{2+} \quad \text{(reduction)}$$

$$H_2SO_3 \rightarrow SO_4^{2-} \quad \text{(oxidation)}$$

Step 2: There is nothing to do for this step. All the atoms except H and O are already balanced.

Step 3: Add H_2O to balance oxygen.

$$MnO_4^- \rightarrow Mn^{2+} + 4H_2O$$

$$H_2O + H_2SO_3 \rightarrow SO_4^{2-}$$

Step 4: Add H^+ to balance hydrogen.

$$8H^+ + MnO_4^- \rightarrow Mn^{2+} + 4H_2O$$

$$H_2O + H_2SO_3 \rightarrow SO_4^{2-} + 4H^+$$

Step 5: Balance the charge by adding electrons.

$$5e^- + 8H^+ + MnO_4^- \rightarrow Mn^{2+} + 4H_2O$$

$$H_2O + H_2SO_3 \rightarrow SO_4^{2-} + 4H^+ + 2e^-$$

Step 6: Make electron loss equal to electron gain by multiplying the first equation by 2 and the second equation by 5.

$$10e^- + 16H^+ + 2MnO_4^- \rightarrow 2Mn^{2+} + 8H_2O$$

$$5H_2O + 5H_2SO_3 \rightarrow 5SO_4^{2-} + 20H^+ + 10e^-$$

Step 7: Add the balanced half-equations.

$$10e^- + 16H^+ + 2MnO_4^- \rightarrow 2Mn^{2+} + 8H_2O$$

$$5H_2O + 5H_2SO_3 \rightarrow 5SO_4^{2-} + 20H^+ + 10e^-$$

$$10e^- + 16H^+ + 2MnO_4^- + 5H_2O + 5H_2SO_3 \rightarrow 2Mn^{2+} + 8H_2O + 5SO_4^{2-} + 20H^+ + 10e^-$$

Step 8: Cancel $10e^-$, $16H^+$ and $5H_2O$ from both sides. The final equation is:

$$2MnO_4^- + 5H_2SO_3 \rightarrow 2Mn^{2+} + 3H_2O + 5SO_4^{2-} + 4H^+$$

Is our answer reasonable?

This is step 9. The check is simple. Notice that each side of the equation has the same number of atoms of each element and the same net charge. This is what makes it a balanced equation and confirms that we have completed the problem correctly.

PRACTICE EXERCISE 12.5

The element technetium (atomic number 43) is radioactive; one of its isotopes, ^{99m}Tc, is used in medicine for diagnostic imaging. The isotope is usually obtained in the form of the pertechnetate anion, TcO_4^-, but its use sometimes requires the technetium to be in a lower oxidation state. Reduction can be carried out using Sn^{2+} in an acidic solution. The equation is:

$$TcO_4^- + Sn^{2+} \rightarrow Tc^{4+} + Sn^{4+} \quad \text{(acidic solution)}$$

Balance the equation.

PRACTICE EXERCISE 12.6

Balance the following equation.

$$CuS + NO_3^- \rightarrow Cu^{2+} + NO + HSO_4^- \quad \text{(acidic solution)}$$

Basic solutions

In basic solutions, the dominant species are H_2O and OH^-. Strictly speaking, these should be used to balance the half-equations. However, the simplest way to obtain a balanced equation for a basic solution is to first balance it as if it were in acidic solution. We balance the equation using the nine steps just described, and then we use the four-step procedure described below to convert the equation to the correct form for a basic solution. The conversion uses the fact that H^+ and OH^- react in a 1 : 1 ratio to give H_2O.

Additional steps for balancing an equation for a redox reaction in a basic solution

Step 10: Take note of the number of H^+ in the balanced equation and add the same number of OH^- to *each* side.

Step 11: Combine each pair of H^+ and OH^- to form one H_2O.

Step 12: Cancel any H_2O molecules that occur on both sides.

Step 13: Check that there are the same number and type of atoms and the same charge on each side of the reaction.

As an example, suppose we wanted to balance the following equation in a basic solution:

$$SO_3^{2-} + MnO_4^- \rightarrow SO_4^{2-} + MnO_2$$

Following steps 1 to 9 for acidic solutions gives:

$$2H^+ + 3SO_3^{2-} + 2MnO_4^- \rightarrow 3SO_4^{2-} + 2MnO_2 + H_2O$$

Conversion of this equation to one appropriate for a basic solution proceeds as follows.

Step 10: Take note of the number of H^+ in the balanced equation and add the same number of OH^- to *each* side.
The equation for an acidic solution has $2H^+$ on the left, so we add $2OH^-$ to *each* side.

$$2OH^- + 2H^+ + 3SO_3^{2-} + 2MnO_4^- \rightarrow 3SO_4^{2-} + 2MnO_2 + H_2O + 2OH^-$$

Step 11: Combine H^+ and OH^- to form H_2O.
The left side has $2OH^-$ and $2H^+$, which become $2H_2O$.

$$2OH^- + 2H^+ + 3SO_3^{2-} + 2MnO_4^- \longrightarrow 3SO_4^{2-} + 2MnO_2^- + H_2O + 2OH^-$$

$$\downarrow$$

$$2H_2O + 3SO_3^{2-} + 2MnO_4^- \longrightarrow 3SO_4^{2-} + 2MnO_2^- + H_2O + 2OH^-$$

Step 12: Cancel any H_2O molecules that occur on both sides.
In this equation, one H_2O can be eliminated from both sides. The final equation, balanced in a basic solution, is:

$$H_2O + 3SO_3^{2-} + 2MnO_4^- \rightarrow 3SO_4^{2-} + 2MnO_2 + 2OH^-$$

Step 13: Check that the equation is balanced in terms of number and type of atoms and charge.

WORKED EXAMPLE 12.5

Balance the following equation.

$$Ag + Zn^{2+} \rightarrow Ag_2O + Zn \quad \text{(basic solution)}$$

Solution

Step 1: Identify reactants and products for each half-equation.

$$\overset{\pm 0}{Ag} \rightarrow \overset{+1}{Ag_2}O \quad \text{(oxidation)}$$

$$Zn^{2+} \rightarrow \overset{\pm 0}{Zn} \quad \text{(reduction)}$$

Step 2: Balance atoms other than H and O.

$$2Ag \rightarrow Ag_2O$$

$$Zn^{2+} \rightarrow Zn$$

Step 3: Balance oxygen by adding H_2O.

$$H_2O + 2Ag \rightarrow Ag_2O$$

$$Zn^{2+} \rightarrow Zn$$

Step 4: Balance hydrogen by adding H^+.

$$H_2O + 2Ag \rightarrow Ag_2O + 2H^+$$

$$Zn^{2+} \rightarrow Zn$$

Step 5: Balance net charge by adding e^-.

$$H_2O + 2Ag \rightarrow Ag_2O + 2H^+ + 2e^-$$

$$Zn^{2+} + 2e^- \rightarrow Zn$$

Step 6: Make e^- gain equal e^- loss.

Step 7: Add the balanced half-equations.

$$H_2O + 2Ag + Zn^{2+} + \cancel{2e^-} \rightarrow Ag_2O + 2H^+ + \cancel{2e^-} + Zn$$

Step 8: Cancel any species that is the same on both sides.

$$\Rightarrow H_2O + 2Ag + Zn^{2+} \rightarrow Ag_2O + 2H^+ + Zn$$

Step 9: Check that there are the same number and type of atoms and the same charge on each side of the reaction.

Step 10: Take note of the number of H^+ in the balanced equation and add the same number of OH^- to *each* side.

$$2Ag + Zn^{2+} + H_2O + 2OH^- \rightarrow 2Ag_2O + \underbrace{2H_2O}_{2H^+ + 2\bar{O}H \text{ combined}} + Zn$$

Step 11: Combine each pair of H^+ and OH^- to form one H_2O.

Step 12: Cancel any H_2O molecules that occur on both sides.

$$2Ag + Zn^{2+} + 2OH^- \rightarrow Ag_2O + Zn + H_2O$$

Step 13: Check that there are the same number and type of atoms and the same charge on each side of the reaction.

PRACTICE EXERCISE 12.7

Balance the following equation in a basic solution.

$$MnO_4^- + C_2O_4^{2-} \rightarrow MnO_2 + CO_3^{2-}$$

12.3 Galvanic cells

The previous discussion has shown that in chemical processes an oxidation cannot occur without a simultaneous reduction, and vice versa. This is because one substance has to release electrons (the reducing agent, which becomes the oxidised species after the reaction), which have to be taken up by another substance (the oxidising agent, which becomes the reduced species after the reaction). A specific substance cannot donate electrons to or take up electrons from every other compound. Because of this, absolute oxidising or reducing agents do not exist. In fact, whether a substance acts as an oxidant or reductant in a reaction depends on the nature of the reaction partner to be oxidised or reduced, respectively. We will illustrate this with three examples.

Example 1

If a strip of metallic zinc is dipped into a solution of copper sulfate, a reddish-brown deposit of metallic copper forms on the zinc (see figure 12.4). Analysis of the solution reveals that it now contains zinc ions, as well as some remaining unreacted copper ions.

Zinc transfers electrons to copper ions, and the results of this experiment can be summarised by the equation:

$$Zn(s) + CuSO_4(aq) \rightarrow ZnSO_4(aq) + Cu(s)$$

The term '(aq)' expresses the fact that the molecules or ions are dissolved in water, where they become surrounded by water molecules. (This is called hydration.)

The redox changes become clear if both half-equations are analysed. Both copper sulfate and zinc sulfate are soluble salts, and they are completely dissociated. The sulfate ion is a spectator ion and not involved in the redox process.

$$\begin{array}{rl} Zn(s) \rightarrow Zn^{2+}(aq) + 2e^- & \text{(oxidation)} \\ Cu^{2+}(aq) + 2e^- \rightarrow Cu(s) & \text{(reduction)} \\ \hline Zn(s) + Cu^{2+}(aq) \rightarrow Zn^{2+}(aq) + Cu(s) & \text{(redox reaction)} \end{array}$$

FIGURE 12.4 The reaction of zinc with copper ions: **(a)** A piece of shiny zinc is held next to a beaker containing a copper sulfate solution. **(b)** When zinc is placed in the solution, copper ions are reduced to elemental copper while the zinc dissolves. **(c)** After a while, the zinc becomes coated with a red-brown layer of copper. The blue colour of the solution fades as Cu^{2+} is reduced.

An atomic-level view of the processes at the surface of zinc during the reaction is depicted in figure 12.5.

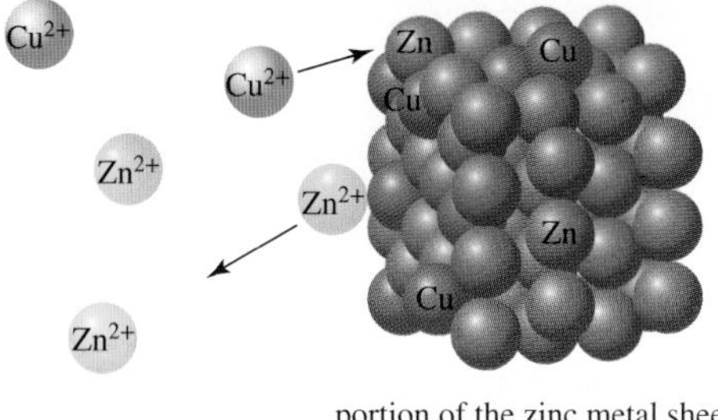

Two electrons are transferred from the zinc atom to the copper ion.

The result is a zinc ion and a copper atom.

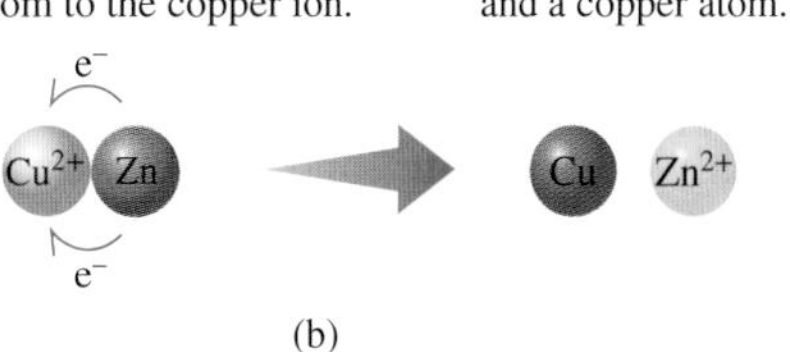

FIGURE 12.5 The reaction of copper ions with zinc, viewed at the atomic level: **(a)** Copper ions (pink) collide with the zinc surface where they pick up electrons from zinc atoms (grey). The zinc atoms become zinc ions (yellow) and enter the solution. The copper ions become copper atoms (brown) and stick to the surface of the zinc. (For clarity, the water molecules of the solution and the sulfate ions are not shown.) **(b)** A close-up view of the exchange of electrons that leads to the reaction.

Example 2

When a piece of copper is dipped into a solution of zinc sulfate (see figure 12.6), no reaction occurs. Copper cannot reduce the zinc ions to metallic zinc.

$$Cu(s) + ZnSO_4(aq) \rightarrow \text{no reaction}$$

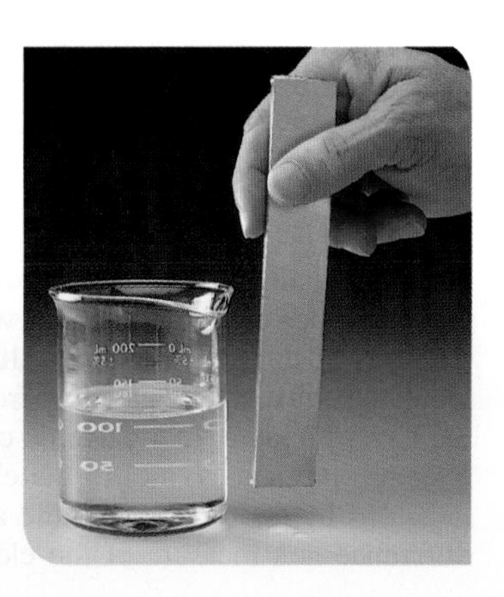

FIGURE 12.6 Copper cannot reduce zinc ions. Although metallic zinc will displace copper from a solution containing Cu^{2+} ions, metallic copper will not displace Zn^{2+} from its solutions. Here we see that the copper bar is unaffected by being dipped into a solution of zinc sulfate.

Example 3

When a coil of copper is dipped into a solution that contains silver ions, these are reduced to elemental silver and copper is oxidised (see figure 12.7). The redox processes are:

$$\begin{array}{rl} Cu(s) \rightarrow Cu^{2+}(aq) + 2e^- & \text{(oxidation)} \\ 2Ag^+(aq) + 2e^- \rightarrow 2Ag(s) & \text{(reduction)} \\ \hline Cu(s) + 2Ag^+(aq) \rightarrow Cu^{2+}(aq) + 2Ag(s) & \text{(redox reaction)} \end{array}$$

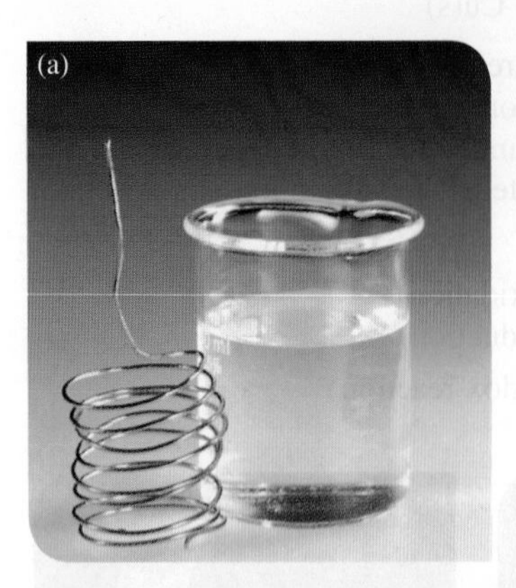
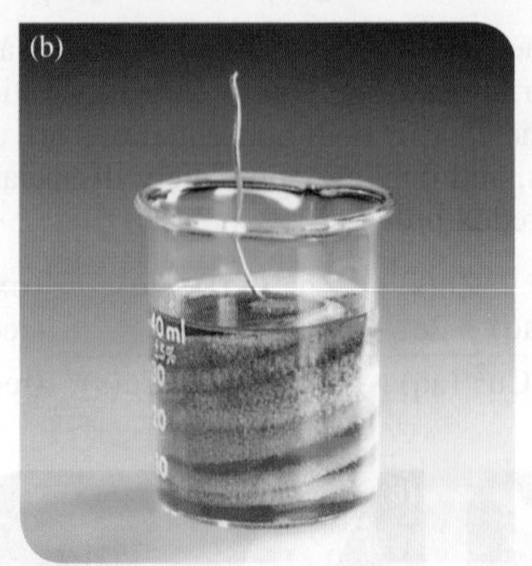

FIGURE 12.7 Reaction of copper with a solution of silver nitrate. **(a)** A coil of copper wire stands next to a beaker containing a silver nitrate solution. **(b)** When the copper wire is placed in the solution, copper dissolves, giving the solution its blue colour, and metallic silver deposits as glittering crystals on the wire. **(c)** After a while, much of the copper has dissolved and nearly all of the silver has deposited in its elemental form.

To gain insights into the different oxidising and reducing behaviours of these substances, the driving force of the electron transfer needs to be understood. The finding that, in example 1, an electrical current (moving electrons) flows between the zinc and copper systems means that a **potential difference** exists between the two systems. (Remember that a current, e.g. water, heat, gas and electric current, can flow only if a difference in level of height, temperature, pressure or potential exists.)

The potential difference is expressed in the electrical unit **volt (V)**, which is a measure of the amount of energy, in **joules (J)**, that can be delivered per SI unit of charge, **coulomb (C)**, as the current moves through a circuit. Thus, a current flowing under a potential difference of 1 volt can deliver 1 joule of energy per coulomb.

$$1\text{ V} = \frac{1\text{ J}}{1\text{ C}} = 1\text{ J C}^{-1}$$

Setting up a galvanic cell

The potential difference between zinc and copper cannot be determined experimentally simply by dipping zinc metal into a solution of copper ions, because the electron transfer takes place directly between Cu^{2+} and Zn, as shown in figure 12.7. Instead, if a zinc rod (electrode) is dipped into a solution of $ZnSO_4$ and a copper rod (electrode) is dipped into a solution of $CuSO_4$, and the solutions are connected with a salt bridge (its function will be described shortly), zinc can transfer its electrons to the copper ions only through an external circuit (see figure 12.8a). An arrangement where a metal, M, is dipped into a solution containing a salt of the respective metal ion, $M^{n+}X_n$, (if X is a monoanion) is called a **half-cell**. In a half-cell, oxidation and reduction can occur according to the equilibrium:

$$M \rightleftharpoons M^{n+} + ne^-$$

The chemical process is identical to that in example 1 (on the previous page). However, in this case, the existing potential difference can be measured using a highly sensitive current meter or by inserting a **voltmeter** with a very high resistance. Such a combination of two half-cells is called a **galvanic cell** (after Luigi Galvani, 1737–1798, an Italian anatomist who discovered that electricity can cause the contraction of muscles). The overall reaction that takes place in the galvanic cell is the **cell reaction**.

In the zinc–copper system, if the solutions have a concentration of 1.00 M Cu^{2+} and 1.00 M Zn^{2+}, the potential difference, or **electrochemical potential**, is 1.10 V. The potential difference between the electrodes can be compared with the pressure difference between two gas containers filled with gas at different pressures. If both containers were connected, the pressure in the system would be equalised by gas flow from the container with the higher pressure to the container with the lower pressure. In a galvanic cell, the 'electron gas' flows from the electrode with the higher 'electron pressure' (zinc) to the electrode with the lower 'electron pressure' (copper), which is indicated by the direction of the arrow above the voltmeter in figure 12.8a. Thus, the potential difference is a measure of the electron pressure difference between two electrodes. The redox reaction between Zn and Cu^{2+} (example 1) is finished when the electron pressures between zinc and copper are in equilibrium. In many textbooks, the potential difference in a galvanic cell is called electromotive force. This term is discouraged, since potential difference is not a 'force'. The unit of potential difference in a galvanic cell is the volt (V).

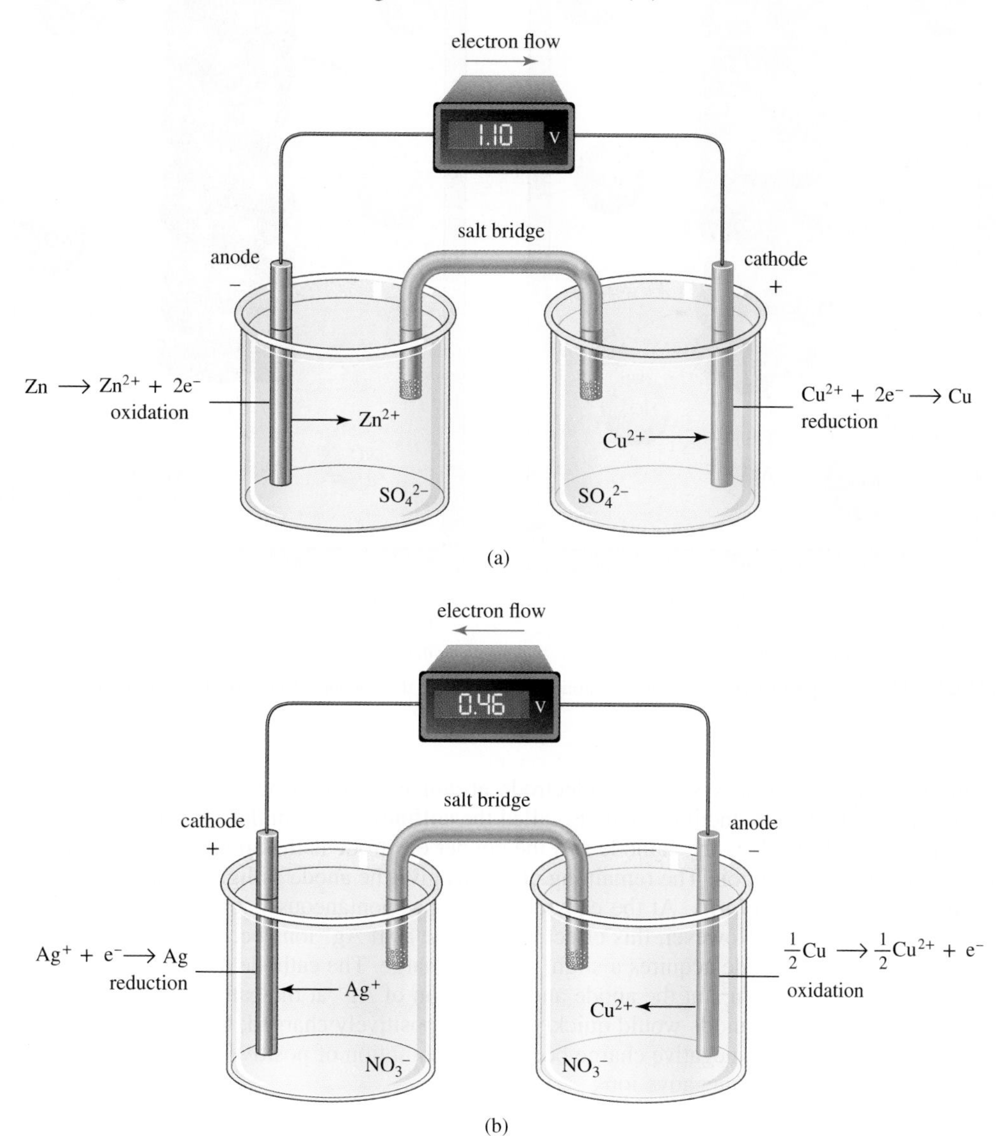

FIGURE 12.8 Galvanic cells: **(a)** the zinc–copper cell and **(b)** the silver–copper cell. At 25 °C under standard conditions (1×10^5 Pa and concentration of 1 M for all solutes), the potential is 1.10 V in the zinc–copper cell and 0.46 V in the silver–copper cell.

When a copper electrode in a solution of $Cu(NO_3)_2$ is connected to a silver electrode in a solution of $AgNO_3$, the current flows in the opposite direction (see figure 12.8b). When the concentrations of the $Cu(NO_3)_2$ and $AgNO_3$ solutions are 1.00 M, the potential difference is 0.46 V. Likewise, a zinc electrode in a solution of $ZnSO_4$ connected to a silver electrode in a solution of $AgNO_3$ will form a galvanic cell with a potential difference of 1.56 V (the sum of the potential differences of the two previously discussed galvanic cells, 1.10 V + 0.46 V = 1.56 V) and in which the electrons flow from the zinc electrode to the silver electrode. We will explain this finding later.

Processes in galvanic cells

As we have seen, a redox system consisting of two metals and solutions of their salts (two half-cells) can be combined to form a galvanic cell, which produces electricity. In other words, a galvanic cell consists of an oxidising agent in one compartment (the first half-cell) that pulls electrons through a wire from a reducing agent in the other compartment (the other half-cell).

We will now have a closer look at the processes occurring in galvanic cells. The processes involved are described as **electrochemical changes**, and the study of such changes is called **electrochemistry**. Figure 12.9 (overleaf) shows schematically the processes that take place in the two half-cells of the copper–silver galvanic cell. At the copper electrode, Cu^{2+} ions *enter* the solution when copper atoms are oxidised, leaving electrons behind on the electrode. The solution around the copper electrode becomes positively charged, unless Cu^{2+} ions move away from the electrode or NO_3^- ions move towards it. At the silver electrode, Ag^+ ions *leave* the solution and become silver atoms by acquiring electrons from the electrode surface. The solution around the silver electrode becomes negatively charged, unless more Ag^+ ions move towards the electrode or NO_3^- ions move away.

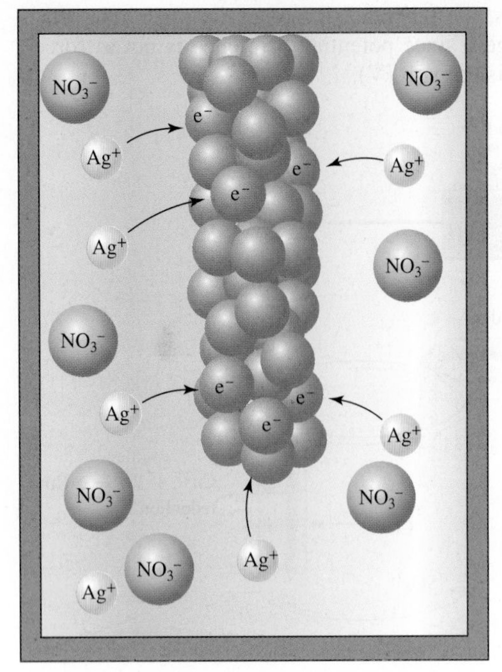

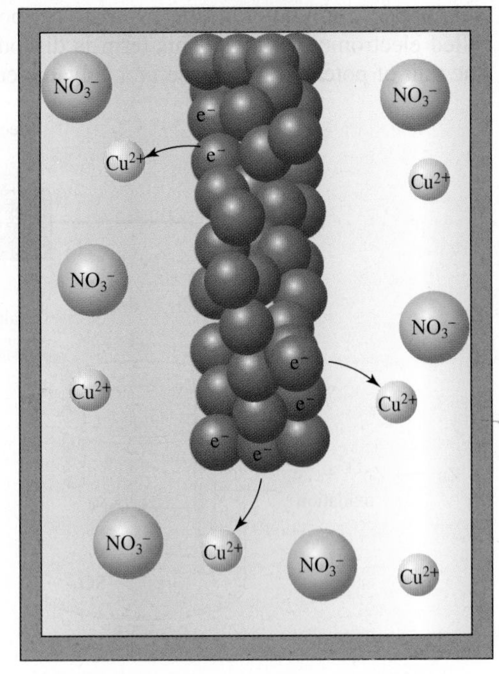

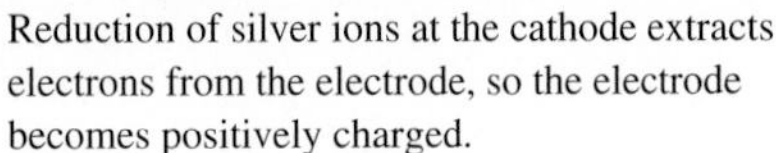
Reduction of silver ions at the cathode extracts electrons from the electrode, so the electrode becomes positively charged.

Oxidation of copper atoms at the anode leaves electrons behind on the electrode, which becomes negatively charged.

FIGURE 12.9 Changes that take place at the anode and cathode in the copper–silver galvanic cell (not drawn to scale).

In any electrochemical system, the electrode at which oxidation occurs is called the **anode**; the electrode at which reduction occurs is called the **cathode**. Thus, in the copper–silver galvanic cell, the silver electrode is the cathode and the copper electrode is the anode. At the anode, Cu^{2+} is released into the solution. The remaining electrons give the anode a slight negative charge; the anode has a *negative polarity*. At the cathode, electrons spontaneously join Ag^+ ions to become a part of the electrode. However, this effect is the same as if Ag^+ ions become a part of the electrode, so that the electrode acquires a slight positive charge. The cathode has a *positive polarity*.

If the production of Cu^{2+} at the anode and deposition of Ag^+ at the cathode were to continue, the solution around the anode would quickly become positively charged; the solution around the cathode would acquire a negative charge because consumption of positive ions from the solution would leave an excess of negative ions.

Nature does not permit large amounts of positive and negative charge to accumulate, so, for the reactions to continue and for electricity to continue to flow through the external circuit, there must be a means of balancing the charges in the solutions around the electrodes. To understand how this happens, we will examine how electric charge is conducted in the cell.

In the external circuit of the cell, electric charge is transported from one electrode to the other by the movement of electrons through the wires. This type of conduction is called **electronic conduction** and is how metals in general, and other materials, conduct electricity. In the cell, electrons always travel from the negatively charged anode, where they are left behind by the oxidation process, to the positively charged cathode, where they are picked up by the substance being reduced.

In electrochemical cells, there is another kind of electrical conduction that also takes place to balance charges in the two half-cells. In a solution that contains ions (or in a molten ionic compound), *electric charge is carried through the liquid by the movement of ions, not electrons*. The transport of electric charge by ions is called **electrolytic conduction**.

For a galvanic cell to work, the solutions in both half-cells must remain electrically neutral. This requires that ions be permitted to enter or leave the solutions. For example, when copper is oxidised, the solution surrounding the electrode becomes filled with Cu^{2+} ions, so negative ions are needed to balance their charge. Similarly, when Ag^+ ions are reduced, NO_3^- ions are left behind in the solution and positive ions are needed to maintain neutrality. The salt bridge shown in figure 12.8 allows the movement of ions required to keep the solutions neutral.

A **salt bridge** is a tube filled with a solution of a salt composed of relatively inert ions that are not involved in the cell reaction. Often KNO_3 or KCl are used. In the most simplified version, the tube is fitted with a porous plug at each end that prevents the solution from pouring out but, at the same time, enables the solution in the salt bridge to exchange ions with the solutions in the half-cells.

During operation of the cell, negative ions can diffuse from the salt bridge into the copper half-cell, or Cu^{2+} ions can leave the solution and enter the salt bridge. Both processes together keep the copper half-cell electrically neutral. In the silver half-cell, positive ions from the salt bridge can enter or negative NO_3^- ions can leave the half-cell to keep this half electrically neutral, too.

Without the salt bridge, electrical neutrality could not be maintained and no electric current could be produced by the cell. Therefore, *electrolytic contact* — contact by means of a solution containing ions — must be maintained for the cell to function. A closer look at the overall ion movement during the operation of the galvanic cell shows that negative ions (*anions*) move away from the cathode, where they are present in excess, *towards the anode*, where they are needed to balance the charge of the positive ions (*cations*) formed. Similarly, cations move away from the anode, where they are in excess, *towards the cathode*, where they balance the anions left in excess. In summary, in galvanic cells:

- the *cathode* is the electrode at which reduction (electron gain) occurs. Cations move in the general direction of the cathode
- the *anode* is the electrode at which oxidation (electron loss) occurs. Anions move in the general direction of the anode.

Figure 12.10 provides a summary of the components of, and processes in, a basic galvanic cell.

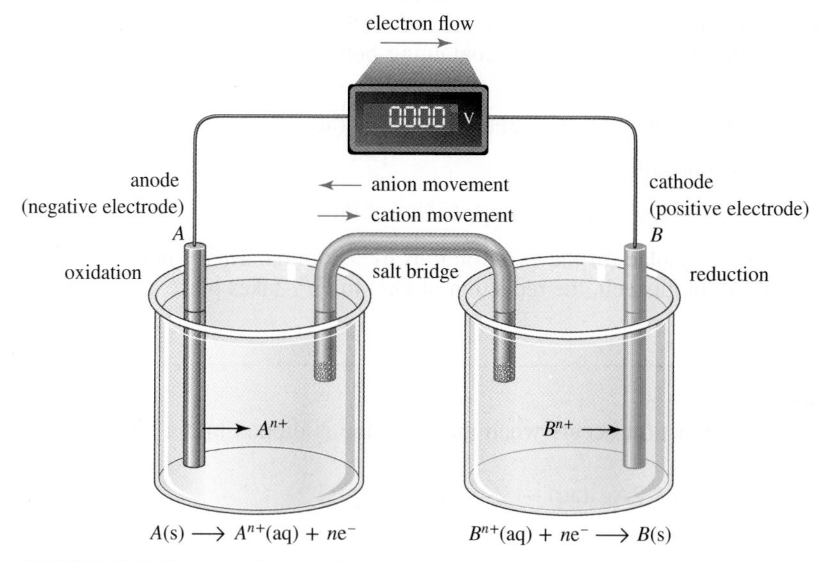

FIGURE 12.10 Schematic diagram of a galvanic cell. Electrons always flow from the negative electrode (anode) to the positive electrode (cathode). In this diagram, the anode is the left-hand electrode at which oxidation occurs, and the cathode is the right-hand electrode where reduction occurs.

After this theoretical description of galvanic cells, you may perhaps wonder where you would encounter galvanic cells in your daily life. Basically, a galvanic cell can be considered as a battery, and we will discuss batteries later in this chapter. (More precisely, a battery consists of several galvanic cells.) But you may have also created an unwanted galvanic cell in your mouth; if you have old fillings in your teeth, you may have experienced a strange and perhaps unpleasant sensation, like an electric shock, while accidentally biting on a piece of aluminium foil. Most old fillings are made of an amalgam of mercury and silver. When the aluminium foil touches the filling, a galvanic cell is created in the mouth, in which aluminium is the anode, the filling is the cathode and saliva is the electrolyte salt bridge. In essence, the contact short-circuits the galvanic cell and a small amount of current flows, which is sensed by the nerves in the teeth.

Notation of galvanic cells and cell reactions

As a matter of convenience, chemists use a **cell diagram** as a shorthand way of describing the make-up of a galvanic cell. The cell diagram, which is also known as **standard cell notation**, for the galvanic cell in figure 12.10 is:

$$A(s)\,|\,A^{n+}(aq)\,\|\,B^{n+}(aq)\,|\,B(s)$$

By convention, *the half-cell containing the anode (that is, where oxidation occurs) is specified on the left of a cell diagram*, with the electrode material of the anode given first. The single vertical line represents a *phase boundary* — here, between the solid electrode and the solution that surrounds it. The double vertical dashed lines represent the salt bridge, which connects the solutions in the two half-cells. On the right, the half-cell containing the cathode is described, with the material of the cathode given last. *The defined cell reaction for any cell diagram is obtained based on the fact that reduction occurs at the right-hand electrode (cathode).* Therefore, the defined cell reaction for the cell in figure 12.10 is:

$$A(s) + B^{n+}(aq) \rightarrow A^{n+}(aq) + B(s)$$

We will see in section 12.4 how to determine whether the defined cell reaction occurs spontaneously.

The copper–silver cell discussed on pages 507–8 is represented by the following cell diagram.

$$Cu(s) \mid Cu^{2+}(aq) \parallel Ag^{+}(aq) \mid Ag(s)$$

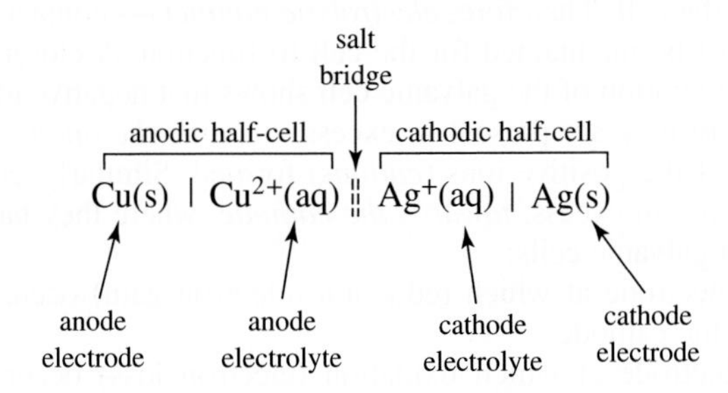

Sometimes, the oxidised and reduced forms of the reactants in a half-cell are both in solution. For example, a galvanic cell can be made using an anodic half-cell composed of a zinc electrode dipping into a solution containing Zn^{2+} and a cathodic half-cell composed of an inert (nonreacting) platinum electrode dipping into a solution containing both Fe^{2+} and Fe^{3+} (figure 12.11). The cell reaction is:

$$2Fe^{3+}(aq) + Zn(s) \rightarrow 2Fe^{2+}(aq) + Zn^{2+}(aq)$$

The cell diagram for this galvanic cell is:

$$Zn(s) \mid Zn^{2+}(aq) \parallel Fe^{3+}(aq), Fe^{2+}(aq) \mid Pt(s)$$

where we have separated the formulae for the two iron ions by a comma. This is done when the species are in the same phase. In this cell, the reduction of Fe^{3+} to Fe^{2+} takes place at the surface of the inert platinum electrode.

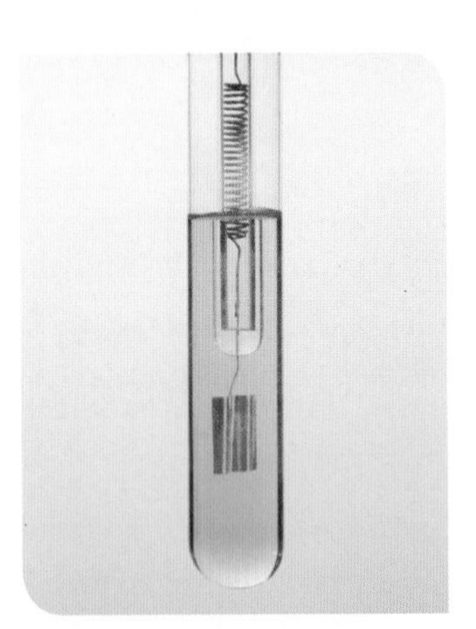

FIGURE 12.11 The cathodic half-cell for the galvanic cell represented by the cell diagram on the left, showing the platinum gauze cathode in a solution of Fe^{2+} and Fe^{3+}.

WORKED EXAMPLE 12.6

Describing galvanic cells

The following spontaneous reaction occurs when metallic zinc is dipped into a solution of silver nitrate.

$$Zn(s) + 2Ag^{+}(aq) \rightarrow Zn^{2+}(aq) + 2Ag(s)$$

Describe the galvanic cell for this reaction. What is the reaction in each half-cell? What is the cell diagram? Sketch the cell and label the cathode and anode, the charges on each electrode, the directions of the flow of ions and the direction of electron flow. Also indicate the positive and negative electrodes.

Analysis

Answering all these questions relies on identifying the anode and cathode from the equation for the cell reaction. By definition, the anode is the electrode at which oxidation occurs, and the cathode is where reduction occurs. The first step, therefore, is to determine which reactant is oxidised and which is reduced. One way to do this is to divide the cell reaction into half-reactions and balance them by adding electrons. Then, if electrons appear as a product, the half-reaction is oxidation; if the electrons appear as a reactant, the half-reaction is reduction.

Solution

The balanced half-equations are:

$$Zn(s) \rightarrow Zn^{2+}(aq) + 2e^{-}$$

$$2Ag^{+}(aq) + 2e^{-} \rightarrow 2Ag(s)$$

Zn(s) loses electrons and is oxidised, so it is the anode. The anodic half-cell is, therefore, a zinc electrode dipping into a solution that contains Zn^{2+} (e.g. from dissolved $Zn(NO_3)_2$ or $ZnSO_4$). Within the cell diagram, the anodic half-cell is written on the left-hand side with the electrode material at the left of the vertical line and the oxidation product at the right.

$$Zn(s) \mid Zn^{2+}(aq)$$

Silver ions gain electrons and are reduced to metallic silver, so the cathodic half-cell consists of a silver electrode dipping into a solution containing Ag^{+} (e.g. from dissolved $AgNO_3$). Within the cell diagram, the silver cathodic half-cell is written on the right-hand side, with the electrode material to the right of the vertical line and the substance reduced on the left.

$$Ag^{+}(aq) \mid Ag(s)$$

In the cell diagram, the zinc anodic half-cell (left) and the silver cathodic half-cell (right) are separated by double vertical dashed lines that represent the salt bridge.

$$\underset{\text{anode}}{Zn(s)\,|\,Zn^{2+}(aq)} \,\vdots\vdots\, \underset{\text{cathode}}{Ag^{+}(aq)\,|\,Ag(s)}$$

The cell may be sketched as shown on the right.

The anode always carries a negative charge in a galvanic cell, so the zinc electrode is negative and the silver electrode is positive. Electrons in the external circuit travel from the negative electrode to the positive electrode (i.e. from the Zn anode to the Ag cathode). The anions move towards the anode, and the cations move towards the cathode.

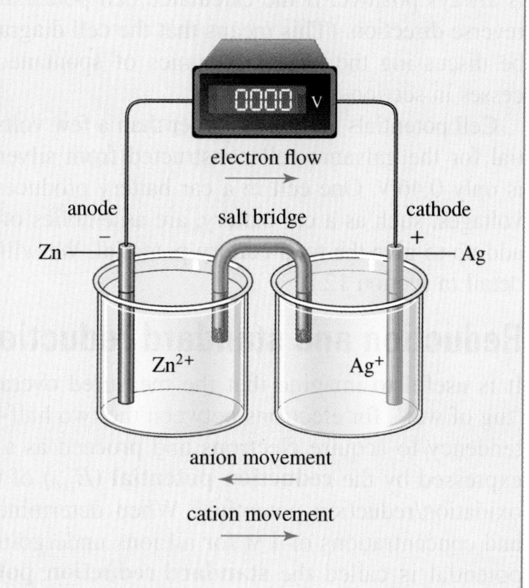

Is our answer reasonable?

All of the answers depend on determining which substance is oxidised and which is reduced, so that is what to check. Oxidation is electron loss, and Zn must lose electrons to become Zn^{2+}, so zinc is oxidised and must be the anode. If zinc is the anode, then silver must be the cathode. All the rest follows by reasoning.

PRACTICE EXERCISE 12.8

Sketch and label a galvanic cell for the following spontaneous redox reaction.

$$Fe(s) + Sn^{2+}(aq) \rightarrow Fe^{2+}(aq) + Sn(s)$$

Write the half-equations for the reactions that occur at the anode and cathode. Give the cell diagram. Sketch the cell and label the cathode and anode, the charges on each electrode, the directions of the flow of ions and the direction of electron flow.

12.4 Reduction potentials

Cell and standard cell potentials

The voltage or potential of a galvanic cell varies with the amount of current flowing through the circuit. The *maximum* potential that a given cell can generate is called its **cell potential** (E_{cell}), or electrochemical potential, an expression introduced on page 506. As mentioned previously, the convention in writing a cell diagram is to put the cathode (reduction process) on the right-hand side and the anode (oxidation process) on the left-hand side. The cell potential is then the potential difference between the cathodic and anodic half-cells, that is, the right (R) and the left (L) half-cells, respectively.

$$E_{cell} = E_R - E_L$$

E_{cell} depends on the composition of the electrodes and the concentrations of the ions in the half-cells, as well as the temperature and pressure. To compare the potentials for different cells, we use the **standard cell potential** ($E^{\ominus}_{cell}$). This is the potential of the cell when all of the ion concentrations are 1 M and any gases involved in the cell reaction are at a pressure of 1×10^5 Pa. The temperature is generally assumed to be 298 K (25 °C), but should always be specified. The IUPAC convention for determining $E^{\ominus}_{cell}$ is:

$$E^{\ominus}_{cell} = E^{\ominus}_R - E^{\ominus}_L$$

A positive $E^{\ominus}_{cell}$ means that reduction occurs at the right-hand electrode, whereas a negative $E^{\ominus}_{cell}$ tells us that reduction occurs at the left-hand electrode. The cell diagram for the cell in figure 12.10 (p. 509) can be written as:

$$A(s)\,|\,A^{n+}(aq) \,\vdots\vdots\, B^{n+}(aq)\,|\,B(s)$$

A positive $E^{\ominus}_{cell}$ for any cell means that the defined cell reaction is spontaneous or, equivalently, the polarity of the right-hand electrode (cathode) is positive. Therefore, the cell reaction for the defined cell:

$$A(s) + B^{n+}(aq) \rightarrow A^{n+}(aq) + B(s)$$

will be spontaneous and the B^{n+} ions in the right-hand half-cell will be reduced to B. On the other hand, a negative $E^{\ominus}_{cell}$ means that the defined cell reaction is not spontaneous, and the reverse reaction will be spontaneous. In a galvanic cell, the cell potential for the spontaneous reaction is always positive. If the calculated cell potential is negative, the reaction is spontaneous in the reverse direction. (This means that the cell diagram was written in the wrong direction.) We will be discussing the thermodynamics of spontaneous and non-spontaneous electrochemical processes in section 12.5.

Cell potentials are rarely larger than a few volts. As mentioned before, the standard cell potential for the galvanic cell constructed from silver and copper electrodes shown in figure 12.8b is only 0.46 V. One cell in a car battery produces only about 2 V. Batteries that generate higher voltages, such as a car battery, are assemblies of cells arranged in series so that their potentials add up to give the total battery potential. We will discuss the chemistry of important batteries in detail in section 12.8.

Reduction and standard reduction potentials

It is useful to imagine that the measured overall cell potential arises from a competition, or 'tug of war', for electrons between the two half-cells. Thus, each half-cell has a certain natural tendency to acquire electrons and proceed as a *reduction*. The magnitude of this tendency is expressed by the **reduction potential** (E_{red}) of the reaction (also known as redox potential or oxidation/reduction potential). When determined under standard conditions ($p = 1 \times 10^5$ Pa and concentrations of 1 M for all ions undergoing either oxidation or reduction), the reduction potential is called the **standard reduction potential**. These values are usually tabulated at 25 °C. To represent a standard reduction potential, we add a subscript to the symbol $E^{\ominus}$ that identifies the substance undergoing reduction. Thus, the standard reduction potential for the half-reaction:

$$Cu^{2+}(aq) + 2e^- \rightarrow Cu(s)$$

is specified as $E^{\ominus}_{Cu^{2+}/Cu}$.

When two half-cells are connected, the one with the larger reduction potential (the greater tendency to undergo reduction) acquires electrons from the half-cell with the lower reduction potential, which is therefore forced to undergo oxidation.

The measured cell potential represents the magnitude and sign of the *difference* between the reduction potential of one half-cell and the reduction potential of the other. As we have discussed before, when $E^{\ominus}_{cell}$ is positive, the cell reaction is spontaneous as written. We can see from $E^{\ominus}_{cell} = E^{\ominus}_R - E^{\ominus}_L$ that, if $E^{\ominus}_{cell}$ is positive, $E^{\ominus}_R$ is greater than $E^{\ominus}_L$. An example of this is the copper–silver cell discussed previously.

$$Cu(s) \mid Cu^{2+}(aq) \,\|\, Ag^+(aq) \mid Ag(s)$$

The two possible reduction processes in this cell are:

$$Ag^+(aq) + e^- \rightarrow Ag(s)$$
$$Cu^{2+}(aq) + 2e^- \rightarrow Cu(s)$$

As we have seen in figure 12.9, the spontaneous process in this cell is reduction of $Ag^+(aq)$ and oxidation of Cu(s) according to the equation:

$$2Ag^+(aq) + Cu(s) \rightarrow 2Ag(s) + Cu^{2+}(aq)$$

This means that the standard reduction potential of the Ag^+/Ag half-cell must be larger than the standard reduction potential of the Cu^{2+}/Cu half-cell. In other words, if we knew the values of $E^{\ominus}_{Ag^+/Ag}$ and $E^{\ominus}_{Cu^{2+}/Cu}$ and calculated $E^{\ominus}_{cell}$ by:

$$E^{\ominus}_{cell} = E^{\ominus}_{Ag^+/Ag} - E^{\ominus}_{Cu^{2+}/Cu}$$

we would find that $E^{\ominus}_{cell}$ is positive.

A common error is to multiply the standard reduction potentials by the factors used to balance the two half-reactions. In this case, the temptation is to multiply $E^{\ominus}_{Ag^+/Ag}$ by a factor of 2. However, cell potentials are given in volts, which is the energy per unit of charge. Thus, stoichiometric factors are not required and half-cell potentials are simply combined to give the cell potentials:

$$E^{\ominus}_{cell} = E^{\ominus}_R - E^{\ominus}_L$$

Determining standard reduction potentials

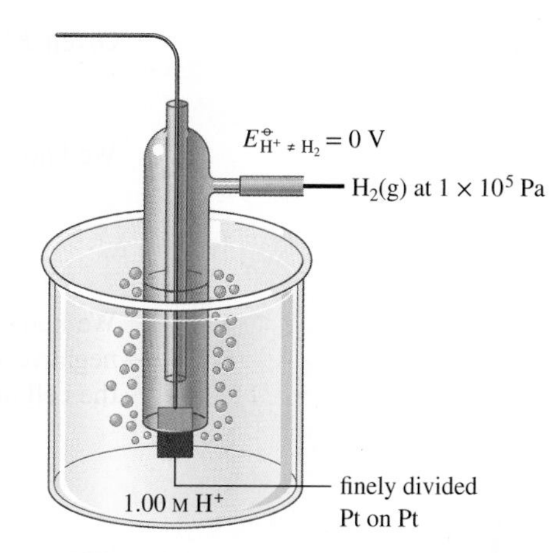

FIGURE 12.12 The standard hydrogen electrode (T = 298 K).

Unfortunately, we have no way to measure the absolute standard reduction potential of an isolated half-cell. All that can be measured are potential differences when two half-cells are connected. Therefore, to assign values to the various standard reduction potentials, a reference electrode has been arbitrarily chosen and its standard reduction potential has been assigned a value of *exactly* 0 V. This reference electrode is called the **standard hydrogen electrode**, SHE (see figure 12.12). Gaseous hydrogen at a pressure of 1×10^5 Pa is bubbled over a platinum electrode coated with very finely divided platinum, which provides a large catalytic surface area on which the electrode reaction can occur. This electrode is surrounded by a solution in which the concentration of hydrogen ions, H^+, is 1 M. The reaction at the platinum surface, written as a reduction, is:

$$2H^+(aq,\ 1\ M) + 2e^- \rightleftharpoons H_2(g,\ 1 \times 10^5\ Pa,\ 298\ K) \qquad E^\ominus_{H^+/H_2} = 0\ V \quad \text{(by definition)}$$

where the double arrows indicate only that the reaction is reversible, not that there is true equilibrium. Whether the half-reaction occurs as a reduction or an oxidation depends on the reduction potential of the half-cell with which it is paired.

Figure 12.13a illustrates the hydrogen electrode connected to a copper half-cell to form a galvanic cell. When calculating a standard reduction potential, *the convention is to put the hydrogen half-cell on the left-hand side of the cell diagram*. Therefore, the cell diagram for this cell is:

$$Pt(s)\,|\,H_2(g)\,|\,H^+(aq)\,||\,Cu^{2+}(aq)\,|\,Cu(s)$$

Given $E^\ominus_{cell} = E^\ominus_R - E^\ominus_L$:

$$E^\ominus_{cell} = E^\ominus_{Cu^{2+}/Cu} - E^\ominus_{H^+/H_2}$$

We know that $E^\ominus_{H^+/H_2} = 0$ V, so:

$$E^\ominus_{cell} = E^\ominus_{Cu^{2+}/Cu} - 0$$
$$= E^\ominus_{Cu^{2+}/Cu}$$

We know from our measurement in figure 12.13a that $E^\ominus_{cell} = +0.34$ V so $E^\ominus_{Cu^{2+}/Cu} = +0.34$ V. The positive value means that reduction is occurring at the right-hand electrode in the cell diagram so the spontaneous cell reaction is:

$$Cu^{2+}(aq) + H_2(g) \rightarrow Cu(s) + 2H^+(aq)$$

Now let's look at a galvanic cell set-up between a zinc electrode and a hydrogen electrode (see figure 12.13b). Remembering that the hydrogen electrode by definition appears on the left-hand side, the cell diagram is:

$$Pt(s)\,|\,H_2(g)\,|\,H^+(aq)\,||\,Zn^{2+}(aq)\,|\,Zn(s)$$

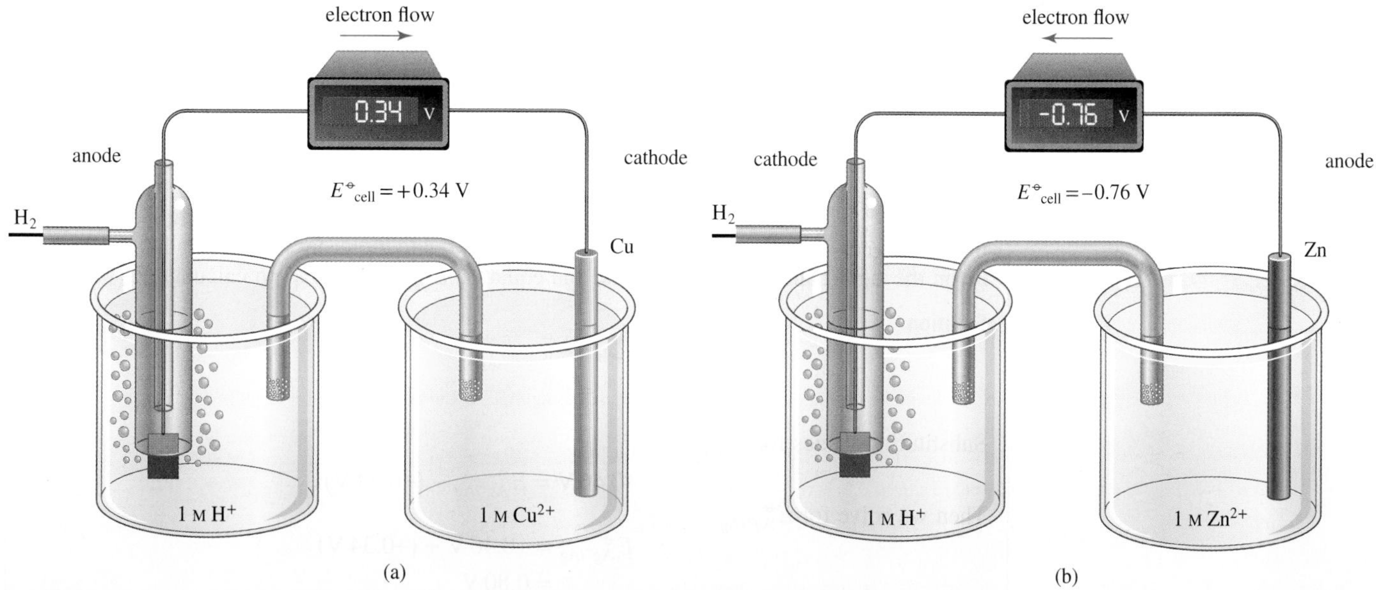

FIGURE 12.13 Galvanic cells composed of: **(a)** copper and hydrogen half-cells and **(b)** zinc and hydrogen half-cells.

Given $E^{\ominus}_{cell} = E^{\ominus}_{R} - E^{\ominus}_{L}$:

$$E^{\ominus}_{cell} = E^{\ominus}_{Zn^{2+}/Zn} - E^{\ominus}_{H^+/H_2}$$

We know that $E^{\ominus}_{H^+/H_2} = 0\ V$, so:

$$E^{\ominus}_{cell} = E^{\ominus}_{Zn^{2+}/Zn} - 0$$
$$= E^{\ominus}_{Zn^{2+}/Zn}$$

We know from our measurement in figure 12.13b that $E^{\ominus}_{cell} = -0.76\ V$ so $E^{\ominus}_{Zn^{2+}/Zn} = -0.76\ V$. The negative value in this case means that reduction is in fact occurring at the left-hand electrode of the cell diagram so the spontaneous cell reaction is:

$$2H^+(aq) + Zn(s) \rightarrow H_2(g) + Zn^{2+}(aq)$$

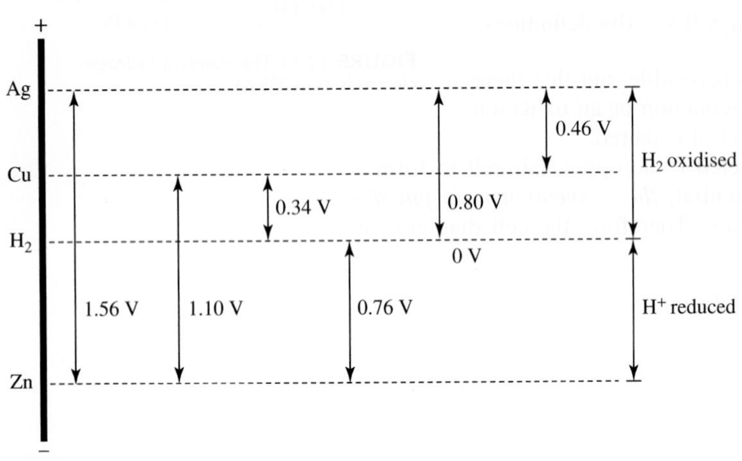

FIGURE 12.14 Choice of an arbitrary zero-point to determine standard reduction potentials $E^{\ominus}$ for a galvanic cell.

We can visualise these results in a scaled graph of the reduction potentials, where a lower positioned species can donate electrons to a higher one (figure 12.14). From this, we can directly calculate the electrochemical potential in any galvanic cell. You can see from this figure that the values of $E^{\ominus}_{cell}$ are additive. We know that $E^{\ominus}_{cell}$ of the zinc–copper galvanic cell = 1.10 V and $E^{\ominus}_{cell}$ of the copper–silver cell = 0.46 V, so $E^{\ominus}_{cell}$ of the zinc–silver cell is the sum of both (1.10 V + 0.46 V = 1.56 V).

However, this set-up enables us to measure only potential *differences*. The *absolute* potentials of the respective electrodes are not known. But we don't need to know these because we are interested only in the $E^{\ominus}_{cell}$ of the galvanic cell itself, so it is sufficient to choose an arbitrary zero-point. (This is similar to defining zero on the celsius scale as the temperature of melting ice, or to measuring elevation relative to sea level rather than from the centre of the Earth.)

The standard reduction potentials of many half-reactions can be compared with that of the standard hydrogen electrode as described on the previous page. Table 12.1 lists values obtained for some typical half-reactions. They are arranged in decreasing order — the half-reactions at the top have the greatest tendency to occur as a reduction, while those at the bottom have the highest tendency to occur as an oxidation. Table 12.1 shows the following.

- Substances located to the left of the double arrows are oxidising agents, because they become reduced when the reactions proceed in the forward direction.
- The strongest oxidising agents are those most easily reduced, and they are located at the top left of the table (e.g. F_2).
- Substances located to the right of the double arrows are reducing agents; they become oxidised when the reactions proceed from right to left.
- The strongest reducing agents are those found at the bottom right of the table (e.g. Li).

WORKED EXAMPLE 12.7

Calculating half-cell potentials

The standard cell potential $E^{\ominus}_{cell}$ of the silver–copper galvanic cell has a value of 0.46 V. The cell reaction is:

$$2Ag^+(aq) + Cu(s) \rightarrow 2Ag(s) + Cu^{2+}(aq)$$

and $E^{\ominus}_{Cu^{2+}/Cu} = +0.34\ V$. What is $E^{\ominus}_{Ag^+/Ag}$?

Analysis

Because $E^{\ominus}_{cell}$ is positive, we know that reduction must be occurring at the right-hand electrode. We know from the equation that Ag^+ is reduced and therefore the Ag^+/Ag half-cell must be on the right-hand side. We then use the equation $E^{\ominus}_{cell} = E^{\ominus}_{R} - E^{\ominus}_{L}$ to calculate $E^{\ominus}_{Ag^+/Ag}$.

Solution

$$E^{\ominus}_{cell} = E^{\ominus}_{R} - E^{\ominus}_{L}$$
$$= E^{\ominus}_{Ag^+/Ag} - E^{\ominus}_{Cu^{2+}/Cu}$$

Substituting values for $E^{\ominus}_{cell}$ and $E^{\ominus}_{Cu^{2+}/Cu}$:

$$+0.46\ V = E^{\ominus}_{Ag^+/Ag} - (+0.34\ V)$$

Then we solve for $E^{\ominus}_{Ag^+/Ag}$:

$$E^{\ominus}_{Ag^+/Ag} = +0.46\ V + (+0.34\ V)$$
$$= 0.80\ V$$

Is our answer reasonable?

The half-cell with the larger $E^\ominus$ value will always undergo reduction. $E^\ominus_{Ag^+/Ag}$ is larger than $E^\ominus_{Cu^{2+}/Cu}$ so we predict that Ag^+ will be reduced, as is observed. Our calculated reduction potential appears to be correct. Keep in mind that stoichiometric factors are not considered.

TABLE 12.1 Standard reduction potentials $E^\ominus$ at 298 K.

	Half reaction			$E^\ominus$ (volts)	
Strongest oxidant	$F_2(g) + 2e^-$	$\rightleftharpoons$	$2F^-(aq)$	+2.87	Weakest reductant
	$S_2O_8^{2-}(aq) + 2e^-$	$\rightleftharpoons$	$2SO_4^{2-}(aq)$	+2.01	
	$PbO_2(s) + HSO_4^-(aq) + 3H^+(aq) + 2e^-$	$\rightleftharpoons$	$PbSO_4(s) + 2H_2O(l)$	+1.69	
	$2HOCl(aq) + 2H^+(aq) + 2e^-$	$\rightleftharpoons$	$Cl_2(g) + 2H_2O(l)$	+1.63	
	$MnO_4^-(aq) + 8H^+(aq) + 5e^-$	$\rightleftharpoons$	$Mn^{2+}(aq) + 4H_2O(l)$	+1.51	
	$BrO_3^-(aq) + 6H^+(aq) + 6e^-$	$\rightleftharpoons$	$Br^-(aq) + 3H_2O(l)$	+1.47	
	$PbO_2(s) + 4H^+(aq) + 2e^-$	$\rightleftharpoons$	$Pb^{2+}(aq) + 2H_2O(l)$	+1.46	
	$ClO_3^-(aq) + 6H^+(aq) + 6e^-$	$\rightleftharpoons$	$Cl^-(aq) + 3H_2O(l)$	+1.45	
	$Au^{3+}(aq) + 3e^-$	$\rightleftharpoons$	$Au(s)$	+1.42	
	$ClO_4^-(aq) + 8H^+ + 8e^-$	$\rightleftharpoons$	$Cl^-(aq) + 4H_2O$	+1.38	
	$Cl_2(g) + 2e^-$	$\rightleftharpoons$	$2Cl^-(aq)$	+1.36	
	$O_2(g) + 4H^+(aq) + 4e^-$	$\rightleftharpoons$	$2H_2O(l)$	+1.23	
	$Br_2(aq) + 2e^-$	$\rightleftharpoons$	$2Br^-(aq)$	+1.07	
	$NO_3^-(aq) + 4H^+(aq) + 3e^-$	$\rightleftharpoons$	$NO(g) + 2H_2O(l)$	+0.96	
	$Ag^+(aq) + e^-$	$\rightleftharpoons$	$Ag(s)$	+0.80	
	$Fe^{3+}(aq) + e^-$	$\rightleftharpoons$	$Fe^{2+}(aq)$	+0.77	
	$I_2(s) + 2e^-$	$\rightleftharpoons$	$2I^-(aq)$	+0.54	
	$NiO_2(s) + 2H_2O(l) + 2e^-$	$\rightleftharpoons$	$Ni(OH)_2(s) + 2OH^-(aq)$	+0.49	
	$Cu^{2+}(aq) + 2e^-$	$\rightleftharpoons$	$Cu(s)$	+0.34	
	$Cl_2(aq) + 4OH^-(aq)$	$\rightleftharpoons$	$2OCl^-(aq) + 2H_2O(l) + 2e^-$	+0.32	
	$Hg_2Cl_2(s) + 2e^-$	$\rightleftharpoons$	$2Hg(l) + 2Cl^-(aq)$	+0.27	
	$AgCl(s) + e^-$	$\rightleftharpoons$	$Ag(s) + Cl^-(aq)$	+0.23	
	$SO_4^{2-}(aq) + 4H^+(aq) + 2e^-$	$\rightleftharpoons$	$H_2SO_3(aq) + H_2O(l)$	+0.17	
	$Sn^{4+}(aq) + 2e^-$	$\rightleftharpoons$	$Sn^{2+}(aq)$	+0.15	
	$AgBr(s) + e^-$	$\rightleftharpoons$	$Ag(s) + Br^-(aq)$	+0.07	
	$2H^+(aq) + 2e^-$	$\rightleftharpoons$	$H_2(g)$	0	
	$Sn^{2+}(aq) + 2e^-$	$\rightleftharpoons$	$Sn(s)$	−0.14	
	$Ni^{2+}(aq) + 2e^-$	$\rightleftharpoons$	$Ni(s)$	−0.25	
	$Co^{2+}(aq) + 2e^-$	$\rightleftharpoons$	$Co(s)$	−0.28	
	$PbSO_4(s) + H^+(aq) + 2e^-$	$\rightleftharpoons$	$Pb(s) + HSO_4^-(aq)$	−0.36	
	$Cd^{2+}(aq) + 2e^-$	$\rightleftharpoons$	$Cd(s)$	−0.40	
	$Fe^{2+}(aq) + 2e^-$	$\rightleftharpoons$	$Fe(s)$	−0.44	
	$Cr^{3+}(aq) + 3e^-$	$\rightleftharpoons$	$Cr(s)$	−0.74	
	$Zn^{2+}(aq) + 2e^-$	$\rightleftharpoons$	$Zn(s)$	−0.76	
	$2H_2O(l) + 2e^-$	$\rightleftharpoons$	$H_2(g) + 2OH^-(aq)$	−0.83	
	$Al^{3+}(aq) + 3e^-$	$\rightleftharpoons$	$Al(s)$	−1.66	
	$Mg^{2+}(aq) + 2e^-$	$\rightleftharpoons$	$Mg(s)$	−2.37	
	$Na^+(aq) + e^-$	$\rightleftharpoons$	$Na(s)$	−2.71	
	$Ca^{2+}(aq) + 2e^-$	$\rightleftharpoons$	$Ca(s)$	−2.76	
	$K^+(aq) + e^-$	$\rightleftharpoons$	$K(s)$	−2.92	
Weakest oxidant	$Li^+(aq) + e^-$	$\rightleftharpoons$	$Li(s)$	−3.05	Strongest reductant

PRACTICE EXERCISE 12.9 The galvanic cell described in practice exercise 12.8 has a standard cell potential of $E^{\ominus}_{cell} = 0.30$ V. Given that $E^{\ominus}_{Fe^{2+}/Fe} = -0.44$ V, calculate $E^{\ominus}_{Sn^{2+}/Sn}$. Check your answer by referring to table 12.1.

We can use the data in table 12.1 to show that we obtain the same spontaneous reaction regardless of how the cell diagram is written. For example, writing the silver–copper cell diagram with the silver half-cell on the right-hand side and the copper half-cell on the left:

$$Cu(s) | Cu^{2+}(aq) \,\vdots\vdots\, Ag^{+}(aq) | Ag(s)$$

gives:

$$\begin{aligned} E^{\ominus}_{cell} &= E^{\ominus}_{Ag^{+}/Ag} - E^{\ominus}_{Cu^{2+}/Cu} \\ &= +0.80 \text{ V} - (+0.34 \text{ V}) \\ &= +0.46 \text{ V} \end{aligned}$$

and the positive value predicts that $Ag^{+}(aq)$ will be reduced and Cu(s) will be oxidised. Conversely, if we write the cell diagram with the copper half-cell on the right-hand side and the silver half-cell on the left:

$$Ag(s) | Ag^{+}(aq) \,\vdots\vdots\, Cu^{2+}(aq) | Cu(s)$$

then:

$$\begin{aligned} E^{\ominus}_{cell} &= E^{\ominus}_{Cu^{2+}/Cu} - E^{\ominus}_{Ag^{+}/Ag} \\ &= +0.34 \text{ V} - (+0.80 \text{ V}) \\ &= -0.46 \text{ V} \end{aligned}$$

The negative value tells us that reduction does not occur at the right-hand electrode and that the reverse of the defined cell reaction is spontaneous. Therefore, Cu(s) is oxidised and $Ag^{+}(aq)$ is reduced.

PRACTICE EXERCISE 12.10 Using the data from table 12.1, calculate the standard cell potential, $E^{\ominus}_{cell}$, for the following galvanic cell (at 298 K).

$$Fe | 1.0 \text{ M } FeCl_2 \,\vdots\vdots\, 1.0 \text{ M } MgCl_2 | Mg$$

Is the reaction spontaneous?

The standard hydrogen electrode, SHE, is not the only reference electrode used. The **saturated calomel electrode** (SCE) is also common and consists of elemental mercury, mercury(I) chloride and Hg_2Cl_2 (calomel, Cl—Hg—Hg—Cl) in a saturated KCl solution. The reaction:

$$Hg_2Cl_2(s) + 2e^{-} \rightleftharpoons 2Hg(l) + 2Cl^{-}(aq)$$

has a reduction potential of 0.244 V at 298 K versus SHE. Another popular reference electrode is the **silver/silver chloride electrode**, which contains an Ag rod coated with silver chloride and bathed in a saturated KCl solution. This has a reduction potential of 0.199 V at 298 K versus SHE for the reaction:

$$AgCl(s) + e^{-} \rightleftharpoons Ag(s) + Cl^{-}(aq)$$

These electrodes are widely used because they are more stable and easier to use than the SHE.

Spontaneous and nonspontaneous reactions

One of the goals of chemistry is to predict reactions. This can be done for redox reactions — whether they occur in a galvanic cell or just in a container with all the chemicals combined in one reaction mixture — using the half-reactions and standard reduction potentials in table 12.1. We will illustrate this with some examples.

The reactants and products of *spontaneous* redox reactions are easy to spot when reduction potentials are listed in order of most positive to least positive (most negative), as in table 12.1. *For any pair of half-reactions, the one with the more positive reduction potential will occur as a reduction. The other half-reaction is reversed and occurs as an oxidation.* It is important to note that, when the direction of a reaction is reversed, the sign of the reduction potential is not changed when calculating $E^{\ominus}_{cell}$. The minus sign in the equation $E^{\ominus}_{cell} = E^{\ominus}_{R} - E^{\ominus}_{L}$ takes care of this, essentially reversing the reaction listed in table 12.1.

The IUPAC convention requires that you:

- *never change the sign of a reduction potential when calculating* E_{cell}
- *never multiply by stoichiometric coefficients when balancing a cell equation.*

Worked examples 12.8 and 12.9 demonstrate this.

WORKED EXAMPLE 12.8

Predicting the outcome of redox reactions

Predict the reaction that will occur when both Ni and Fe are added to a solution that contains both Ni^{2+} and Fe^{2+}, each at 1 M concentration.

Analysis

The first question would be, 'What *possible* reactions could occur?' The system involves a possible redox reaction that can be predicted using data in table 12.1. One way to do this is to note the relative positions of the half-reactions when arranged as they are in table 12.1.

$$Ni^{2+}(aq) + 2e^- \rightleftharpoons Ni(s) \qquad E^\ominus_{Ni^{2+}/Ni} = -0.25\ V$$

$$Fe^{2+}(aq) + 2e^- \rightleftharpoons Fe(s) \qquad E^\ominus_{Fe^{2+}/Fe} = -0.44\ V$$

In table 12.1, the half-reaction with the more positive (in this case, less negative) reduction potential will occur as a reduction. Therefore $Ni^{2+}(aq)$ will be reduced and Fe(s) will be oxidised. The products are the substances on the opposite sides of each half-reaction, namely Ni(s) and $Fe^{2+}(aq)$.

Solution

We have done nearly all the work in our analysis of the problem. All that is left is to write the equation. The reactants are Ni^{2+} and Fe; the products are Ni and Fe^{2+}.

$$Ni^{2+}(aq) + Fe(s) \rightarrow Ni(s) + Fe^{2+}(aq)$$

The equation is balanced in terms of both atoms and charges, so this is the spontaneous reaction that will occur in the system specified in the problem.

Is our answer reasonable?

If our prediction of the spontaneous reaction is correct, $E^\ominus_{cell}$ should be positive.

$$E^\ominus_{cell} = E^\ominus_R - E^\ominus_L$$

As Ni^{2+} is being reduced, Ni^{2+}/Ni will be the right-hand electrode (cathode).

$$E^\ominus_{cell} = E^\ominus_{Ni^{2+}/Ni} - E^\ominus_{Fe^{2+}/Fe}$$

From table 12.1:

$$E^\ominus_{cell} = -0.25\ V - (-0.44\ V)$$
$$= +0.19\ V$$

As $E^\ominus_{cell}$ is positive, our prediction is correct. Remember that, although the reaction $Fe^{2+} + 2e^- \rightarrow Fe$ is reversed, the sign of $E^\ominus_{Fe^{2+}/Fe}$ (−0.44 V) is not changed.

PRACTICE EXERCISE 12.11

What spontaneous reaction occurs if Cl_2 and Br_2 are added to a solution that contains both Cl^- and Br^-, each at 1 M concentration? (Use data from table 12.1.)

If you intend to use a particular spontaneous redox reaction in a galvanic cell, the reduction potentials can be used to predict what the standard cell potential will be, as illustrated in worked example 12.9.

WORKED EXAMPLE 12.9

Predicting the cell reaction and cell potential of a galvanic cell

A typical cell of a lead storage battery of the type used to start cars is constructed using electrodes made of lead and lead(IV) oxide, PbO_2, and with sulfuric acid as the electrolyte. The half-reactions and their reduction potentials in this system are:

$$PbO_2(s) + 3H^+(aq) + HSO_4^-(aq) + 2e^- \rightleftharpoons PbSO_4(s) + 2H_2O(l) \qquad E^\ominus_{PbO_2/PbSO_4} = +1.69\ V$$

$$PbSO_4(s) + H^+(aq) + 2e^- \rightleftharpoons Pb(s) + HSO_4^-(aq) \qquad E^\ominus_{PbSO_4/Pb} = -0.36\ V$$

What is the spontaneous cell reaction and what is the standard potential of this cell?

Analysis

In the spontaneous cell reaction, the half-reaction with the larger (more positive) reduction potential will take place as a reduction, while the other half-reaction will be reversed and occur as an oxidation. The cell potential is the difference between the two reduction potentials.

Solution

PbO_2 has a larger, more positive reduction potential than $PbSO_4$, so the first half-reaction will occur in the direction written. The second must be reversed to occur as an oxidation. In the cell, therefore, the half-reactions are:

$$PbO_2(s) + 3H^+(aq) + HSO_4^-(aq) + 2e^- \rightarrow PbSO_4(s) + 2H_2O(l) \qquad \text{(reduction)}$$
$$Pb(s) + HSO_4^-(aq) \rightarrow PbSO_4(s) + H^+(aq) + 2e^- \qquad \text{(oxidation)}$$
$$PbO_2(s) + Pb(s) + 2H^+(aq) + 2HSO_4^-(aq) \rightarrow 2PbSO_4(s) + 2H_2O(l) \qquad \text{(cell reaction)}$$

(*Hint:* Use changes in oxidation numbers to check when balancing complex redox reactions.)
The cell potential is obtained by:

$$E^\ominus_{cell} = E^\ominus_R - E^\ominus_L$$

Because the first half-reaction occurs as a reduction, this becomes the right-hand electrode. Therefore:

$$E^\ominus_{cell} = E^\ominus_{PbO_2/PbSO_4} - E^\ominus_{PbSO_4/Pb}$$
$$= (+1.69\text{ V}) - (-0.36\text{ V})$$
$$= +2.05\text{ V}$$

Again, even though we reversed the direction of the reaction:

$$PbSO_4(s) + H^+(aq) + 2e^- \rightleftharpoons Pb(s) + HSO_4^-(aq)$$

we did not change the sign of the reduction potential when calculating $E^\ominus_{cell}$. The convention $E^\ominus_{cell} = E^\ominus_R - E^\ominus_L$ took care of this.

PRACTICE EXERCISE 12.12 Determine the cell reaction and the standard cell potential of a galvanic cell employing the following half-reactions.

$$Al^{3+}(aq) + 3e^- \rightleftharpoons Al(s) \qquad E^\ominus_{Al^{3+}/Al} = -1.66\text{ V}$$
$$Cu^{2+}(aq) + 2e^- \rightleftharpoons Cu(s) \qquad E^\ominus_{Cu^{2+}/Cu} = +0.34\text{ V}$$

Which electrode (Al or Cu) would be the anode?

Because the spontaneous redox reaction that takes place among a mixture of reactants can be predicted from the standard reduction potentials, it should be possible to predict whether a particular reaction, *as written*, can occur spontaneously. We can do this by calculating the cell potential that corresponds to the reaction in question and seeing if the potential is *positive*.

Oxidising and nonoxidising acids

The standard reduction potentials in table 12.1 would predict that acids (represented by $H^+(aq)$ or H_3O^+) can oxidise certain metals, such as Fe, Zn, Sn and Mg (see figure 12.15), but not other metals, such as Ag, Cu and Au.

For example, when dilute aqueous sulfuric acid reacts with zinc, a zinc salt is formed and bubbles of gaseous hydrogen are released.

$$Zn(s) + H_2SO_4(aq) \rightarrow ZnSO_4(aq) + H_2(g)$$

FIGURE 12.15 Zinc reacts with aqueous sulfuric acid. Bubbles of hydrogen are formed when a solution of sulfuric acid comes in contact with metallic zinc.

Dilute aqueous sulfuric acid ionises in water to give positively charged hydrogen ions (protons) and negatively charged sulfate ions. The proton can act as an oxidising agent by being reduced to H_2. Although the sulfate ion can be reduced (to SO_3^{2-}, for example), protons are much more easily reduced under these dilute conditions. Therefore, when a metal such as zinc is added to dilute aqueous sulfuric acid, it is H^+, rather than SO_4^{2-}, that removes the electrons from zinc.

A similar situation can be found with hydrochloric acid, which contains H^+ and Cl^- ions. H^+ is the oxidising agent, as the Cl^- ion shows no tendency to gain an electron under these (or any other) conditions.

Compared with many other chemicals, the solvated proton, $H^+(aq)$, is a rather poor oxidising agent, so hydrochloric acid and dilute aqueous sulfuric acid have only poor oxidising abilities. For this reason, they are often called **nonoxidising acids**, even though their protons can oxidise certain metals. Actually, when we call something a nonoxidising acid, we are saying that the *anion* of the acid is a weaker oxidising agent than H^+ (which is equivalent to saying that the anion of the acid is more difficult to reduce than H^+). Examples of nonoxidising acids are listed in table 12.2.

TABLE 12.2 Nonoxidising and oxidising acids.

Nonoxidising acid	
$HCl(aq)$	
$H_2SO_4(aq)$[a]	
$H_3PO_4(aq)$	
most carboxylic acids, e.g. CH_3COOH (acetic acid), $HCOOH$ (formic acid)	
Oxidising acid	**Reduction reaction**
HNO_3	(conc.) $NO_3^-(aq) + 2H^+(aq) + e^- \rightarrow NO_2(g) + H_2O(l)$
	(dilute) $NO_3^-(aq) + 4H^+(aq) + 3e^- \rightarrow NO(g) + 2H_2O(l)$
	(very dilute, with strong reducing agent) $NO_3^-(aq) + 10H^+(aq) + 8e^- \rightarrow NH_4^+(aq) + 3H_2O(l)$
H_2SO_4	(hot, conc.) $SO_4^{2-}(aq) + 4H^+(aq) + 2e^- \rightarrow SO_2(g) + 2H_2O(l)$
	(hot, conc. with strong reducing agent) $SO_4^{2-}(aq) + 10H^+(aq) + 8e^- \rightarrow H_2S(g) + 4H_2O(l)$
$HClO_4$	$ClO_4^-(aq) + 8H^+(aq) + 8e^- \rightarrow Cl^-(aq) + 4H_2O$

(a) H_2SO_4 is a nonoxidising acid when cold and dilute.

Not all acids are like HCl and dilute aqueous H_2SO_4. **Oxidising acids** have anions that are stronger oxidising agents than H^+ (see table 12.2). An example is concentrated nitric acid, HNO_3. When dissolved in water, nitric acid ionises to give H^+ and NO_3^- ions. However, in this solution the nitrate ion is a more powerful oxidising agent than a proton. In the competition for electrons, therefore, the nitrate ion wins and, when nitric acid reacts with a metal, the nitrate ion is reduced.

Because the NO_3^- ion is a stronger oxidising agent than H^+, it can oxidise metals that H^+ cannot. For example, if a copper coin is dropped into concentrated nitric acid, it reacts violently (figure 12.16). The reddish-brown gas is the highly toxic nitrogen dioxide, NO_2, which is formed by reduction of the NO_3^- ion.

$$\overset{0}{Cu}(s) + 4H\overset{+5}{N}O_3(aq) \rightarrow \overset{+2}{Cu}(NO_3)_2(aq) + 2\overset{+4}{N}O_2(g) + 2H_2O(l)$$

Notice that the oxidation number of nitrogen decreases; we identify the structural unit that contains this nitrogen, the NO_3^- ion, as the oxidising agent.

FIGURE 12.16 A copper coin in concentrated nitric acid showing the violent reaction, formation of the blue Cu^{2+} ion and the evolution of reddish-brown $NO_2(g)$.

When oxidising acids react with metals, it is often difficult to predict the products. The particular reaction that occurs depends very much on the concentration of the acid and experimental conditions (e.g. whether the reaction was performed with heating). Reduction of the nitrate ion, for example, can produce all sorts of compounds with different oxidation states of nitrogen, depending on the reducing power of the metal and the concentration of the acid. When concentrated nitric acid reacts with a metal, nitrogen dioxide, NO_2, is often formed as the reduction product. When dilute nitric acid is used to dissolve a metal, the product is often nitrogen monoxide (also called nitric oxide), NO, instead. Copper, for example, can display either behaviour towards nitric acid. The net ionic equations for the reactions are as shown overleaf.

Concentrated HNO_3:

$$Cu(s) + 4H^+(aq) + 2NO_3^-(aq) \rightarrow Cu^{2+}(aq) + 2NO_2(g) + 2H_2O(l)$$

Dilute HNO_3:

$$3Cu(s) + 8H^+(aq) + 2NO_3^-(aq) \rightarrow 3Cu^{2+}(aq) + 2NO(g) + 4H_2O(l)$$

If very dilute nitric acid reacts with a metal that is a particularly strong reducing agent, such as zinc, the nitrogen can be reduced all the way down to the −3 oxidation state that it has in NH_4^+ (or NH_3). The net ionic equation is:

$$4Zn(s) + 10H^+(aq) + NO_3^-(aq) \rightarrow 4Zn^{2+}(aq) + NH_4^+(aq) + 3H_2O(l)$$

The nitrate ion in the presence of protons makes nitric acid a powerful oxidising acid. All metals except the very unreactive ones, such as platinum and gold, are attacked by it. Nitric acid also easily oxidises organic substances, so it is wise to be especially careful when working with this acid in the laboratory. Very serious accidents have occurred when concentrated nitric acid has been used around organic substances due to the violent reactious that could result in heat, gas or fire.

In a dilute solution, the sulfate ion of sulfuric acid has little tendency to serve as an oxidising agent. However, if the sulfuric acid is both concentrated and hot, it becomes a fairly potent oxidant. For example, copper does not react with cool, dilute H_2SO_4, but it is attacked by hot, concentrated H_2SO_4 according to the equation:

FIGURE 12.17 Aqua regia, obtained by mixing concentrated nitric acid with concentrated hydrochloric acid.

$$Cu(s) + 2H_2SO_4(\text{hot, conc.}) \rightarrow CuSO_4(aq) + SO_2(g) + 2H_2O(l)$$

Because of this oxidising ability, hot, concentrated sulfuric acid can be very dangerous. The liquid is viscous and can stick to the skin, causing severe burns.

The mixture obtained by mixing concentrated nitric acid and concentrated hydrochloric acid (usually in a volumetric ratio of 1 : 3) is called aqua regia (figure 12.17). The name comes from the fact that the strong oxidising power of this acid mixture can dissolve even the so-called 'royal' ('regal' or 'noble') metals, gold and platinum.

Aqua regia is used to produce chloroauric acid, $HAuCl_4$, which serves as the electrolyte in the Wohlwill process for the production of gold with 99.999% purity. We will discuss electrolytic processes later in this chapter.

12.5 Relationship between cell potential, concentration and Gibbs energy

The fact that cell potentials can be used to predict the spontaneity of redox reactions is no coincidence. There is a relationship between the cell potential and the Gibbs energy change for a reaction. In chapter 8, we saw that ΔG for a reaction is a measure of the maximum useful work that can be obtained from a chemical reaction at constant temperature and pressure. Specifically, the relationship is:

$$\Delta G = \text{maximum non-}p\text{V work}$$

The Gibbs energy change, ΔG

In an electrical system, work is supplied by the electric current that is pushed along by the potential of the cell. It can be calculated from the equation:

$$\text{maximum non-}p\text{V work} = -zFE_{\text{cell}}$$

where z is the amount of electrons transferred, F is a constant called the **Faraday constant**, which is equal to the number of coulombs of charge equivalent to 1 mole of electrons ($1\,F = 96\,485\ \text{C mol}^{-1}$), and E_{cell} is the potential of the cell in volts. Analysis of the units in the equation shows that the answer will be in joules, the unit of energy. Recall that 1 volt = 1 J C^{-1}, so:

$$\begin{aligned}\text{maximum non-}p\text{V work} &= [(\text{mol}) \times (\text{C mol}^{-1}) \times (\text{J C}^{-1})] = \text{joule}\\ &= -z \times F \times E_{\text{cell}}\end{aligned}$$

Combining our two equations gives:

$$\Delta G = -zFE_{\text{cell}}$$

What does this equation mean? It tells us that the change in Gibbs energy, ΔG, between the reactants and products of the reaction is directly related to the maximum potential of a cell, E_{cell}. When E_{cell} is positive, ΔG will be negative and the cell reaction will be spontaneous. On the other

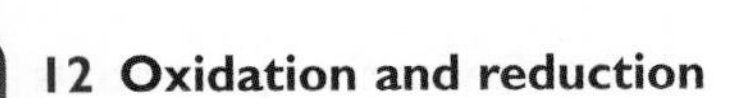

hand, when E_{cell} is negative, ΔG will be positive and the cell reaction will be spontaneous in the reverse direction. If the *standard* cell potential $E^{\ominus}_{cell}$ is used, the standard Gibbs energy change, $\Delta G^{\ominus}$, can be calculated:

$$\Delta G^{\ominus} = -zFE^{\ominus}_{cell}$$

WORKED EXAMPLE 12.10

Calculating the standard Gibbs energy change

Calculate $\Delta G^{\ominus}$ for the following reaction, given that its standard cell potential is 0.32 V at 25 °C.

$$NiO_2(s) + 2Cl^-(aq) + 4H^+(aq) \rightarrow Cl_2(g) + Ni^{2+}(aq) + 2H_2O(l)$$

Analysis

This is a straightforward application of $\Delta G^{\ominus} = -zFE^{\ominus}_{cell}$. Because 2 moles of Cl^- are oxidised to Cl_2 and 2 moles of electrons are therefore transferred, the amount z = 2 mol. We will also use the Faraday constant, $1\ F = 96\,485\ C\ mol^{-1}$.

Solution

$$\begin{aligned}\Delta G^{\ominus} &= -zFE^{\ominus}_{cell} \\ &= -(2\ mol) \times (96\,485\ C\ mol^{-1}) \times (0.32\ J\ C^{-1}) \\ &= -6.18 \times 10^4\ J \\ &= -61.8\ kJ\end{aligned}$$

PRACTICE EXERCISE 12.13

Calculate $\Delta G^{\ominus}$ for the reactions that take place in the galvanic cell described in practice exercise 12.12.

Equilibrium constant, *K*

One useful application of electrochemistry is the determination of equilibrium constants. In chapter 9 we saw that $\Delta G^{\ominus}$ is related to the equilibrium constant by the expression:

$$\Delta G^{\ominus} = -RT\ln K_c$$

Note that we use K_c for the equilibrium constant because the electrochemical reactions described here occur in solution.

Combining:

$$\Delta G^{\ominus} = -zFE^{\ominus}_{cell}$$

with:

$$\Delta G^{\ominus} = -RT\ln K_c$$

we get:

$$-zFE^{\ominus}_{cell} = -RT\ln K_c$$

so:

$$E^{\ominus}_{cell} = \frac{RT}{zF}\ln K_c$$

This equation shows the relationship between $E^{\ominus}_{cell}$ and the equilibrium constant.

In this equation, the value of $R = 8.314\ J\ mol^{-1}\ K^{-1}$, T is the temperature in kelvin, $F = 96\,485$ C per mole of e^-, and z is the amount of electrons transferred in the reaction.

For historical reasons:

$$E^{\ominus}_{cell} = \frac{RT}{zF}\ln K_c$$

is sometimes expressed using the common logarithm, log (logarithm with base 10). Natural and common logarithms are related by the equation:

$$\ln x = 2.303\log x$$

For reactions at 298 K (25 °C), all of the constants (R, T and F) can be combined with the factor 2.303 to give $0.0592\ J\ C^{-1}$. Because $J\ C^{-1} = V$, the equation reduces to:

$$E^{\ominus}_{cell} = \frac{0.0592\ V}{z}\log K_c$$

where z is the amount of electrons transferred in the cell reaction as it is written.

WORKED EXAMPLE 12.11

Calculating equilibrium constants from $E^{\ominus}_{cell}$

Calculate K_c for the reaction in worked example 12.10.

Analysis

The problem simply involves substituting values into:

$$E^{\ominus}_{cell} = \frac{RT}{zF}\ln K_c$$

and solving for the equilibrium constant.

Solution

The reaction in worked example 12.10 has $E^{\ominus}_{cell} = 0.32$ V and $z = 2$. The temperature is 25 °C (298 K). We can solve:

$$E^{\ominus}_{cell} = \frac{RT}{zF}\ln K_c$$

for $\ln K_c$ and then substitute values:

$$\ln K_c = \frac{E^{\ominus}_{cell}\, zF}{RT}$$

Substituting values (1 V = 1 J C^{-1}):

$$\ln K_c = \frac{0.32\text{ J C}^{-1} \times 2 \times 96\,485\text{ C mol}^{-1}}{8.314\text{ J mol}^{-1}\text{K}^{-1} \times 298\text{ K}}$$

$$= 24.9$$

Taking the antilogarithm:

$$K_c = e^{24.9} = 7 \times 10^{10}$$

Is our answer reasonable?

As a rough check, we can look at the magnitude of $E^{\ominus}_{cell}$ and apply some simple reasoning. When $E^{\ominus}_{cell}$ is positive, $\Delta G^{\ominus}$ is negative. In chapter 9 we learned that, when $\Delta G^{\ominus}$ is negative, the reaction proceeds far towards completion when equilibrium is reached. Therefore, we expect that K_c will be large, and that agrees with our answer.

PRACTICE EXERCISE 12.14

The calculated standard cell potential for the reaction:

$$Cu^{2+}(aq) + 2Ag(s) \rightleftharpoons Cu(s) + 2Ag^{+}(aq)$$

is $E^{\ominus}_{cell} = -0.46$ V. Calculate K_c for this reaction at 298 K.

The Nernst equation

When all of the ion concentrations in a cell are 1 M, and the partial pressures of any gases involved in the cell reaction are at 1×10^5 Pa, the cell potential is equal to the standard potential. When the concentrations or pressures change, however, so does the cell potential. For example, in an operating cell or battery, the potential gradually drops as the reactants are used up and as the cell reaction approaches its natural equilibrium status. At equilibrium, the potential has dropped to 0 — the battery is dead.

The effect of concentration on the cell potential can be obtained from thermodynamics. In chapter 9, we learned that the Gibbs energy change is related to the reaction quotient (Q, see section 9.2) by the equation:

$$\Delta G = \Delta G^{\ominus} + RT\ln Q$$

(Remember that the reaction quotient, Q, is used when the reactants are not in equilibrium; in other words, the equilibrium constant, K, is a special case of Q, used when the system is in equilibrium.)

Substituting for ΔG and $\Delta G^{\ominus}$ from:

$$\Delta G = -zFE_{cell}$$

and:

$$\Delta G^{\ominus} = -zFE^{\ominus}_{cell}$$

gives:

$$-zFE_{cell} = -zFE^{\ominus}_{cell} + RT\ln Q$$

Dividing both sides by $-zF$ gives:

$$E_{cell} = E^{\ominus}_{cell} - \frac{RT}{zF}\ln Q$$

This equation is commonly known as the **Nernst equation**, named after Walter Nernst, a German chemist and physicist who was awarded the Nobel Prize in chemistry in 1920 in recognition of his work in thermochemistry. Using common logarithms instead of natural logarithms and calculating the constants for 298 K (25 °C) gives another form of the Nernst equation that is sometimes used:

$$E_{cell} = E^{\ominus}_{cell} - \frac{0.0592 \text{ V}}{z}\log Q$$

In writing the Nernst equation for a galvanic cell, we will construct the reaction quotient (Q) using molar concentrations for ions and partial pressures in pascals for gases. Thus, for a cell using a hydrogen electrode (with the partial pressure of H_2 not necessarily equal to 1×10^5 Pa) for which the cell reaction is:

$$Cu^{2+}(aq) + H_2(g) \rightarrow Cu(s) + 2H^+(aq)$$

the Nernst equation is:

$$E_{cell} = E^{\ominus}_{cell} - \frac{RT}{zF}\ln\frac{[H^+]^2}{[Cu^{2+}]\frac{p_{H_2}}{p^{\ominus}}}$$

(Remember that we do not include concentrations of pure solids or pure liquids in the reaction quotient, Q.)

Because of interionic interactions, ions do not always behave as though their concentrations are equal to their molarities. Strictly speaking, effective concentrations (called activities, see section 9.2) should be used in the expression for Q. Effective concentrations are difficult to calculate, so for simplicity we will use molarities and accept the fact that our calculations are not entirely accurate.

Alternatively, we could add a nonreacting background electrolyte at a much higher concentration than the reactants or products and report this concentration with our results. Often 1 M KCl or 1 M $LiClO_4$ is used as the background electrolyte. The reaction is then studied under conditions of constant ionic strength since the overall changes of ion concentration as a result of the reaction that is actually studied, are small. Under such conditions, the reaction (and cell potentials) behaves simply according to molarities. In chapter 15, we will see that similar considerations are made when studying the kinetics of ionic reactions. For now, in worked examples, we will ignore ionic strength effects. In the laboratory, you should take care in your experimental set-ups to control these effects.

WORKED EXAMPLE 12.12

Calculating the effect of concentration on E_{cell}

Consider a galvanic cell involving the following half-reactions.

$$Ni^{2+}(aq) + 2e^- \rightleftharpoons Ni(s) \qquad E^{\ominus}_{Ni^{2+}/Ni} = -0.25 \text{ V}$$

$$Cr^{3+}(aq) + 3e^- \rightleftharpoons Cr(s) \qquad E^{\ominus}_{Cr^{3+}/Cr} = -0.74 \text{ V}$$

Calculate the cell potential when $[Ni^{2+}] = 1.0 \times 10^{-4}$ M and $[Cr^{3+}] = 2.0 \times 10^{-3}$ M.

Analysis

Because the concentrations are not 1 M, we must use the Nernst equation. First, we need the cell reaction to determine z, the number of electrons transferred, and the correct form of the reaction quotient, Q. We must also note that the reacting system is heterogeneous; solid metals and a liquid solution of their dissolved ions are involved, so we have to remember that Q does not contain concentration terms for solids, such as Ni(s) and Cr(s).

Solution

The nickel half-cell has the more positive reduction potential, so its half-reaction will occur as a reduction. This means that Cr(s) will be oxidised. Making electron gain equal to electron loss, the cell reaction is:

$$3[Ni^{2+}(aq) + 2e^- \rightarrow Ni(s)] \qquad \text{(reduction)}$$

$$2[Cr(s) \rightarrow Cr^{3+}(aq) + 3e^-] \qquad \text{(oxidation)}$$

$$3Ni^{2+}(aq) + 2Cr(s) \rightarrow 3Ni(s) + 2Cr^{3+}(aq) \qquad \text{(cell reaction)}$$

In this reaction, $6e^-$ are transferred ($z = 6$). The Nernst equation for the system is therefore:

$$E_{cell} = E^{\ominus}_{cell} - \frac{RT}{6F}\ln\frac{[Cr^{3+}]^2}{[Ni^{2+}]^3}$$

Note that we calculate the reaction quotient using the concentrations of the ions raised to powers equal to their coefficients in the net cell reaction, and that we have not included concentration terms for the two solids. This is the procedure we followed for heterogeneous equilibria in chapter 9 (pp. 358–9).

$E^{\ominus}_{cell}$ is determined according to:

$$E^{\ominus}_{cell} = E^{\ominus}_{Ni^{2+}/Ni} - E^{\ominus}_{Cr^{3+}/Cr}$$

$$= (-0.25 \text{ V}) - (-0.74 \text{ V})$$

$$= 0.49 \text{ V}$$

Now we can substitute this value for $E^{\ominus}_{cell}$ along with $R = 8.314\ J\ mol^{-1}\ K^{-1}$, $T = 298\ K$, $z = 6$, $F = 96\,485\ C\ mol^{-1}$, $[Ni^{2+}] = 1.0 \times 10^{-4}\ M$ and $[Cr^{3+}] = 2.0 \times 10^{-3}\ M$ into the Nernst equation. Remembering that $1\ V = 1\ J\ C^{-1}$, we obtain:

$$E_{cell} = 0.49\ V - \frac{8.314\ J\ mol^{-1}K^{-1} \times 298\ K}{6 \times 96\,485\ C\ mol^{-1}} \ln \frac{(2.0 \times 10^{-3})^2}{(1.0 \times 10^{-4})^3}$$

$$= 0.49\ V - (0.004\,28\ V)\ln(4.0 \times 10^6)$$

$$= 0.49\ V - 0.0651\ V$$

$$= 0.42\ V$$

The potential of the cell is expected to be 0.42 V.

Is our answer reasonable?

There is no simple way to check the answer. However, there are certain critical points to look over. First, check that the half-reactions have been combined correctly to give the balanced cell reaction (check the oxidation numbers), because we need the coefficients of the equation to obtain the correct superscripts in the Nernst equation. Then, check that the temperature in kelvin has been used, that $R = 8.314\ J\ mol^{-1}\ K^{-1}$ has been used, and that the other substitutions have been made correctly.

PRACTICE EXERCISE 12.15

In a certain zinc–copper cell:

$$Zn(s) + Cu^{2+}(aq) \rightarrow Zn^{2+}(aq) + Cu(s)$$

the concentrations of the ions are $[Cu^{2+}] = 0.01\ M$ and $[Zn^{2+}] = 1.0\ M$. Use the reduction potentials in table 12.1 to calculate the cell potential at 25 °C.

One of the principal uses of the relationship between concentration and cell potential is in the measurement of concentrations. Experimental determination of cell potentials, combined with modern developments in electronics, has provided a means of monitoring and analysing the concentrations of all sorts of substances in solution, even some that are not themselves ionic and that are not involved directly in electrochemical changes. Worked example 12.12 illustrates how the Nernst equation is applied in determining concentrations.

WORKED EXAMPLE 12.13

Using the Nernst equation to determine concentrations

In a large number of samples of water, in which the copper ion concentration is expected to be quite small, $[Cu^{2+}]$ was measured using an electrochemical cell. This consisted of a silver electrode dipping into a 1.00 M solution of $AgNO_3$, which was connected by a salt bridge to a second half-cell containing a copper electrode. The copper half-cell was then filled with one water sample after another, and the cell potential was measured for each sample. In the analysis of one particular sample, the cell potential at 25 °C measured was 0.62 V, with the copper electrode being the anode. What was the concentration of Cu^{2+} in this sample?

Analysis

In this problem, we have been given the cell potential, E_{cell}, and we can calculate $E^{\ominus}_{cell}$ from the reduction potentials in table 12.1. The unknown quantity is one of the concentration terms in the Nernst equation.

Solution

The first step is to write the defined cell reaction, because we need it to calculate $E^{\ominus}_{cell}$ and to write the expression for Q for use in the Nernst equation. Because copper is the anode, it is being oxidised. This also means that Ag^+ is being reduced. Therefore, the equation for the cell reaction is:

$$Cu(s) + 2Ag^+(aq) \rightarrow Cu^{2+}(aq) + 2Ag(s)$$

Because two electrons are transferred, $z = 2$ and the Nernst equation is therefore:

$$E_{cell} = E^{\ominus}_{cell} - \frac{RT}{2F} \ln \frac{[Cu^{2+}]}{[Ag^+]^2}$$

The value of $E^{\ominus}_{cell}$ can be obtained using the data in table 12.1. Following our usual procedure and recognising that silver ions are reduced:

$$E^{\ominus}_{cell} = E^{\ominus}_{Ag^+} - E^{\ominus}_{Cu^{2+}}$$

$$= +0.80\ V - (+0.34\ V)$$

$$= 0.46\ V$$

Substitution of the values into the Nernst equation (remembering that $1\,V = 1\,J\,C^{-1}$) gives:

$$0.62\,V = 0.46\,V - \frac{8.314\,J\,mol^{-1}\,K^{-1} \times 298\,K}{2 \times 96\,485\,C\,mol^{-1}} \ln \frac{[Cu^{2+}]}{[Ag^+]^2}$$

Solving for $\ln([Cu^{2+}]/[Ag^+]^2)$ gives:

$$\ln \frac{[Cu^{2+}]}{[Ag^+]^2} = -12$$

Taking the antilogarithm gives us the value of Q:

$$\frac{[Cu^{2+}]}{[Ag^+]^2} = 6 \times 10^{-6}$$

Because the concentration of Ag^+ is known (1.00 M):

$$\frac{[Cu^{2+}]}{(1.00)^2} = 6 \times 10^{-6}$$

$$[Cu^{2+}] = 6 \times 10^{-6}\,M$$

Is our answer reasonable?

We should first check that we have written the correct chemical equation, which the rest of the solution to the problem relies on. We can then insert our calculated value of $[Cu^{2+}]$ back into the Nernst equation and ensure that we obtain $E_{cell} = 0.62\,V$. This will show with certainty that our answer is correct.

As a final point, notice that the Cu^{2+} concentration is indeed very small and that it can be obtained very easily simply by measuring the potential generated by the electrochemical cell. Determining the concentrations in many samples is also very simple — just change the aqueous sample and measure the potential again.

PRACTICE EXERCISE 12.16

A galvanic cell was constructed by connecting a nickel electrode dipping into 1.20 M $NiSO_4$ solution to a chromium electrode dipping into a solution containing Cr^{3+} at an unknown concentration. The potential of the cell was measured to be 0.552 V, with the chromium serving as the anode. The standard cell potential for this system was determined to be 0.487 V. What was the concentration of Cr^{3+} in the solution of unknown concentration?

It should be noted that the ability of certain compounds to act as either oxidant or reductant can depend strongly on the pH of the reaction solution. For example, as can be seen from table 12.1, the standard reduction potential, $E^{\ominus}_{MnO_4^-, H^+/Mn^{2+}}$ is +1.51 V for the half-reaction:

$$MnO_4^-(aq) + 8H^+ + 5e^- \rightarrow Mn^{2+}(aq) + 4H_2O(l)$$

The Nernst equation for this system ($z = 5$ electrons exchanged) can also be applied to half-cell reactions so that we obtain:

$$E_{MnO_4^-, H^+/Mn^{2+}} = 1.51\,V - \frac{RT}{5F} \ln \frac{[Mn^{2+}]}{[MnO_4^-][H^+]^8}$$

Thus, the more acidic the solution becomes (i.e. the higher the $[H^+]$), the greater the reduction potential. In other words, MnO_4^- is a much stronger oxidant in acidic solutions than in neutral or even basic solutions.

Concentration cells

Thus far, we have considered only cells with different chemical species at the right-hand and left-hand electrodes, and we have shown that a current invariably results from such an arrangement because one of the half-cells has a larger $E^{\ominus}$ than the other. We can also show that current will flow in an electrochemical cell when the chemical species at both electrodes are the same, providing that they are present at different concentrations. Such cells are called **concentration cells**.

Consider the cell on the right at 298 K. We can write the cell diagram for this cell as:

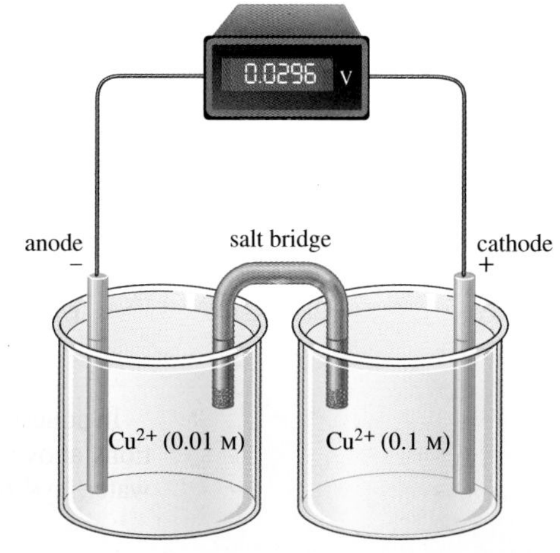

$$Cu(s)\,|\,Cu^{2+}(aq, 0.01\,M)\,\|\,Cu^{2+}(aq, 0.1\,M)\,|\,Cu(s)$$

The two half reactions are:

$$Cu(s) \rightarrow Cu^{2+}(aq, 0.01\,M) + 2e^-$$
$$Cu^{2+}(aq, 0.1\,M) + 2e^- \rightarrow Cu(s)$$

and the overall cell reaction is therefore:

$$Cu^{2+}(aq, 0.1\,M) \rightarrow Cu^{2+}(aq, 0.01\,M)$$

We can write the Nernst equation for this reaction as:

$$E = E^{\ominus} - \frac{RT}{zF}\ln Q$$

$$= E^{\ominus} - \frac{RT}{zF}\ln\frac{[0.01\,M]}{[0.1\,M]}$$

Since the redox processes in the left and right half-cells are identical, we have:

$$E^{\ominus}_{cell} = E^{\ominus}_{R} - E^{\ominus}_{L} = 0\,V$$

Hence:

$$E_{cell} = 0\,V - 0.0128\ln\frac{(0.01)}{(0.1)}$$

$$= 0.0296\,V$$

Therefore, there is a small, but measurable, potential difference between the electrodes that results solely from the different concentrations of the two Cu^{2+} solutions. As the reaction proceeds, the concentrations of the two solutions will eventually become the same and there will be no potential difference between the electrodes. This is similar to the spontaneous process of mixing that occurs when a 0.01 M solution of Cu^{2+} is poured into an equal volume of a 0.1 M solution of Cu^{2+} ions (and vice versa).

Interestingly, many biochemical processes are powered by concentration gradients. For example, adenosine-5′-triphosphate (ATP), which is the main energy-transfer molecule in cells, is synthesised from adenosine diphosphate (ADP) and inorganic phosphate in mitochondria; this process is powered by a proton gradient (a build-up of positive charge on one side of a membrane).

12.6 Corrosion

The corrosion of iron and other metals is one of the most commonly encountered redox processes in our daily life, and it has plagued humanity ever since these metals were first discovered. At first glance, the rusting of a car or ship (see p. 494) does not seem to be a complicated process, since iron reacts with oxygen to give iron(II) oxide.

$$2Fe \rightarrow 2Fe^{2+} + 4e^- \quad \text{(oxidation)}$$

$$O_2 + 4e^- \rightarrow 2O^{2-} \quad \text{(reduction)}$$

$$2Fe + O_2 \rightarrow 2FeO \quad \text{(net reaction)}$$

The electrons released when the iron is oxidised travel through the metal to some other place where the iron is exposed to air. This is where reduction occurs. Is this process really that easy? Why does a car rust faster in moist air? Why doesn't it rust in pure water that is oxygen free? The corrosion process is apparently electrochemical in nature with the iron acting as the anode, as shown in figure 12.18.

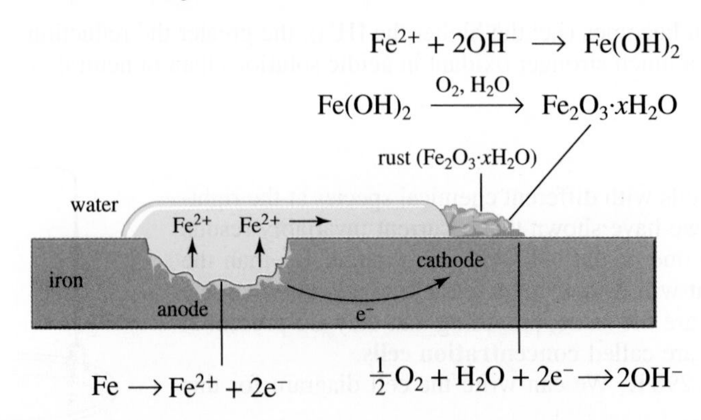

FIGURE 12.18 Corrosion of iron. Iron dissolves in anodic regions to give Fe^{2+}. Electrons travel through the metal to cathodic sites where oxygen is reduced, forming OH^-. The combination of the Fe^{2+} and OH^-, followed by oxidation in air, gives rust.

In aqueous solution, the reduction of oxygen proceeds differently from that shown in the equations above. Oxide anions, O^{2-}, are not very stable in aqueous solution. When they react with water, hydroxide anions are produced. The reduction process of oxygen in water is therefore:

$$O_2 + 4e^- + 2H_2O \rightarrow 4OH^-$$

The rusting of iron can now be written as:

$$2Fe \rightarrow 2Fe^{2+} + 4e^- \quad \text{(oxidation)}$$
$$O_2 + 4e^- + 2H_2O \rightarrow 4OH^- \quad \text{(reduction)}$$
$$2Fe + O_2 + 2H_2O \rightarrow 2Fe(OH)_2 \quad \text{(net reaction)}$$

However, we are not finished yet. Rust contains not only iron(II) ions but also iron(III) ions. Thus, the iron(II) ions formed in the anodic regions gradually diffuse through the water and eventually contact the hydroxide ions. This causes a precipitate of iron(II) hydroxide, $Fe(OH)_2$, to form, which is further oxidised by oxygen to iron(III) hydroxide, $Fe(OH)_3$.

$$4Fe(OH)_2 \rightarrow 4Fe(OH)_2^+ + 4e^- \quad \text{(oxidation)}$$
$$O_2 + 4e^- + 2H_2O \rightarrow 4OH^- \quad \text{(reduction)}$$
$$4Fe(OH)_2 + O_2 + 2H_2O \rightarrow 4Fe(OH)_3 \quad \text{(net reaction)}$$

We are still not finished. Actually, rust does not consist of iron(III) hydroxide, but is iron oxide hydroxide with the formula FeO(OH). FeO(OH) is formed by dehydration of $Fe(OH)_3$.

$$Fe(OH)_3 \rightarrow FeO(OH) + H_2O$$

This mechanism of rusting explains one of the more interesting aspects of this damaging process. Perhaps you have noticed that, when rusting occurs on the body of a car, the rust appears at and around a break or a scratch in the surface of the paint, but the damage extends under the surface for some distance. Apparently, the Fe^{2+} ions formed at the anode sites can diffuse rather long distances to the hole in the paint, where they finally react with air and water to form the rust.

Corrosion can be inhibited by several techniques. In a process called **galvanisation**, iron objects are coated with zinc. Since the reduction potential of Zn^{2+} is more negative than that of Fe^{2+}:

$$Zn^{2+} + 2e^- \rightarrow Zn \quad E^\ominus_{Zn^{2+}/Zn} = -0.76\ V$$
$$Fe^{2+} + 2e^- \rightarrow Fe \quad E^\ominus_{Fe^{2+}/Fe} = -0.44\ V$$

the oxidation of zinc is favoured. This leads to formation of a protective layer of a zinc oxide which lasts until all the zinc is corroded away. Tin plating as in 'tin' cans, in contrast, leads to a very rapid corrosion of the iron once its surface is scratched and the iron is exposed, because the Sn^{2+}/Sn reduction potential:

$$Sn^{2+} + 2e^- \rightarrow Sn \quad E^\ominus_{Sn^{2+}/Sn} = -0.14\ V$$

is greater than that of Fe^{2+}/Fe, and therefore Sn^{2+} oxidises Fe. Some oxides are stable in a sense because they adhere to the metal surface and form an impermeable layer over a fairly wide pH range. This is why aluminium reacts only very slowly in air even though its reduction potential is strongly negative ($Al(s) \rightarrow Al^{3+}(aq) + 3e^-$, $E^\ominus_{Al^{3+}/Al} = -1.66$ V). Such a phenomenon is called **passivation**.

Another protecting method is to change the potential of the subject by pumping in electrons, which can be used to satisfy the demands of oxygen without involving oxidation of the metal. **Cathodic protection** is a concept used to protect large objects, such as ships, pipelines and buildings, from rusting (figure 12.19). The object is connected to a metal with a more negative electrode potential, such as magnesium ($Mg^{2+}(aq) + 2e^- \rightarrow Mg(s)$, $E^\ominus_{Mg^{2+}/Mg} = -2.37$ V). Magnesium acts as a *sacrificial anode* by supplying its electrons to the iron, while it is oxidised to Mg^{2+}. You can see from this that the occasional replacement of a piece of magnesium is much cheaper than replacing an entire ship.

FIGURE 12.19 Before launching, a shiny new zinc anode disk was attached to the bronze rudder of this boat to provide cathodic protection. Over time, the zinc has corroded, instead of the less reactive bronze.

12.7 Electrolysis

The preceding sections have shown how spontaneous redox reactions can be used to generate electrical energy. We now turn our attention to the opposite process: the use of electrical energy to force nonspontaneous redox reactions to occur. In fact, these are precisely the kinds of reactions that take place when recharging batteries.

What is electrolysis?

When electricity is passed through a molten (melted) ionic compound or through a solution of an electrolyte, a chemical reaction called **electrolysis** can occur. An example of an electrolysis apparatus, called an **electrolysis cell** or **electrolytic cell**, is shown in figure 12.20. This particular cell contains molten sodium chloride. (A substance undergoing electrolysis must be molten or in solution so its ions can move freely and conduction can occur.) Inert electrodes — electrodes that will not react with the molten NaCl — are dipped into the cell and then connected to a source of direct current (DC) electricity.

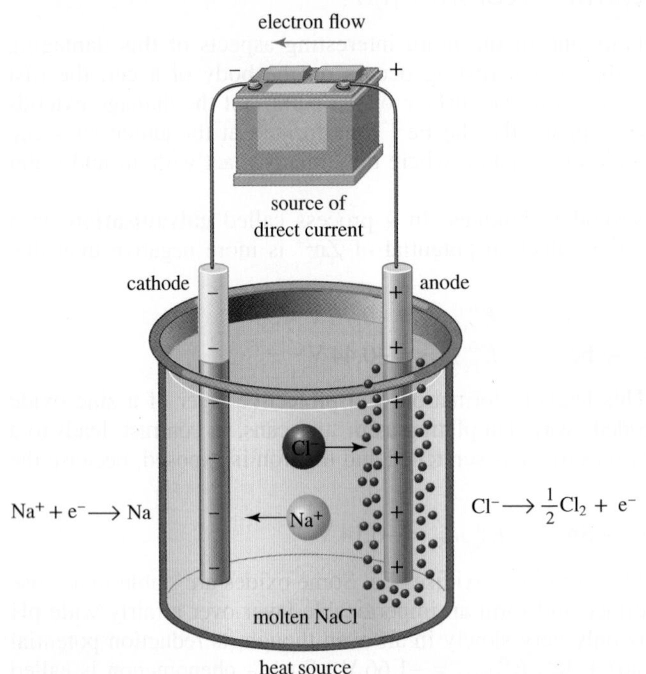

FIGURE 12.20 Electrolysis of molten sodium chloride. The passage of an electric current decomposes molten sodium chloride into metallic sodium and gaseous chlorine.

The DC source serves as an 'electron pump', pulling electrons away from one electrode and pushing them through the external wiring onto the other electrode. The electrode from which electrons are removed becomes positively charged, while the other electrode becomes negatively charged. In this example, the following occurs.

- At the positive electrode, oxidation occurs as electrons are pulled away from negatively charged chloride ions. Because of the nature of the chemical change in an electrolysis cell, the *positive electrode becomes the anode, to which the anions move*.
- The DC source pumps the electrons through the external electric circuit to the negative electrode. Here, reduction takes place as the electrons are forced onto positively charged sodium ions, so the *negative electrode is the cathode, to which the cations move*.

The chemical changes that occur at the electrodes can be described by the following equations.

$$2Na^+(l) + 2e^- \rightarrow 2Na(l) \qquad \text{(cathode)}$$
$$2Cl^-(l) \rightarrow Cl_2(g) + 2e^- \qquad \text{(anode)}$$
$$\overline{2Na^+(l) + 2Cl^-(l) \rightarrow 2Na(l) + Cl_2(g)} \qquad \text{(cell reaction)}$$

(At the melting point of NaCl, 801 °C, metallic sodium is a liquid.)

Comparison of electrolytic and galvanic cells

- In a *galvanic cell*, the spontaneous cell reaction deposits electrons on the anode and removes them from the cathode. As a result, the anode carries a slight negative charge and the cathode a slight positive charge.

- In an *electrolytic cell*, the situation is reversed. Here, oxidation at the anode must be forced to occur, which requires that the anode is positive so it can remove electrons from the reactant at that electrode. On the other hand, the cathode must be made negative so it can force the reactant at the electrode to accept electrons.

By agreement among scientists, the names *anode* and *cathode* are always assigned according to the nature of the reaction taking place at the electrode.
- If the reaction is *oxidation*, the electrode is called the *anode*.
- If the reaction is *reduction*, the electrode is called the *cathode*.

It is important to remember:
- in an *electrolytic cell*, the cathode is negative (reduction) and the anode is positive (oxidation)
- in a *galvanic cell*, the cathode is positive (reduction) and the anode is negative (oxidation).

Electrolysis in aqueous solutions

When electrolysis is carried out in an aqueous solution, the electrode reactions can be more complicated; we must consider oxidation and reduction of the solute as well as oxidation and reduction of water. (It is important to note that the nature of the electrode material itself can also strongly influence the outcome of the electrolysis. We will not discuss this further here.) For example, electrolysis of a solution of potassium sulfate (figure 12.21) gives hydrogen and oxygen. At the cathode, water is reduced, not K^+.

$$2H_2O(l) + 2e^- \rightarrow H_2(g) + 2OH^-(aq) \quad \text{(cathode)}$$

At the anode, water is oxidised, not the sulfate ion.

$$2H_2O(l) \rightarrow O_2(g) + 4H^+(aq) + 4e^- \quad \text{(anode)}$$

We can understand why these redox reactions occur if we examine reduction potential data from table 12.1. For example, at the cathode we have the following competing reactions.

$$K^+(aq) + e^- \rightarrow K(s) \qquad E^{\ominus}_{K^+/K} = -2.92\text{ V}$$
$$2H_2O(l) + 2e^- \rightarrow H_2(g) + 2OH^-(aq) \qquad E^{\ominus}_{H_2O/H_2} = -0.83\text{ V}$$

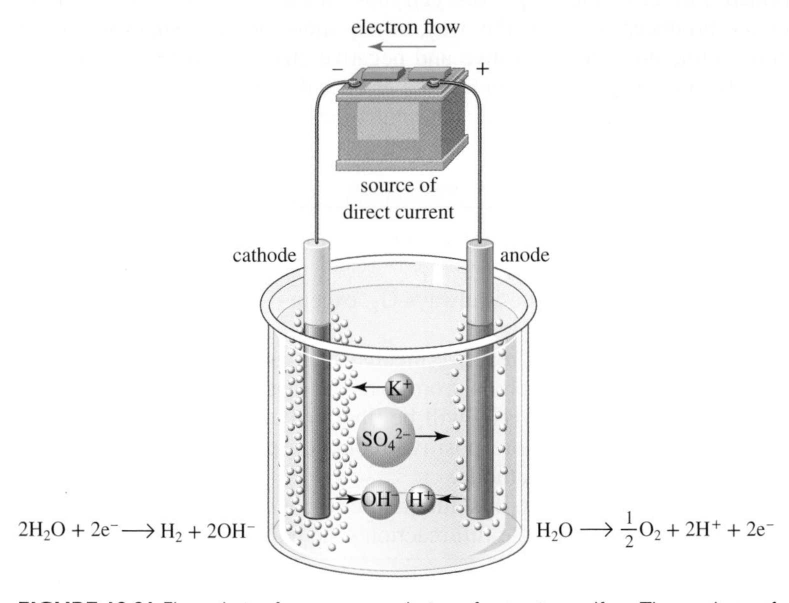

FIGURE 12.21 Electrolysis of an aqueous solution of potassium sulfate. The products of the electrolysis are the gases hydrogen and oxygen.

Water has a much less negative reduction potential than K^+, which means H_2O is much easier to reduce than K^+. During electrolysis, the more easily reduced substance is reduced and H_2 is formed at the cathode.

At the anode, the possible oxidation half-reactions are:

$$2SO_4^{2-}(aq) \rightarrow S_2O_8^{2-}(aq) + 2e^-$$
$$2H_2O(l) \rightarrow 4H^+(aq) + O_2(g) + 4e^-$$

In table 12.1, we find them written in the opposite direction.

$$S_2O_8^{2-}(aq) + 2e^- \rightarrow 2SO_4^{2-}(aq) \qquad E^{\ominus}_{S_2O_8^{2-}/SO_4^{2-}} = +2.01\text{ V}$$
$$O_2(g) + 4H^+(aq) + 4e^- \rightarrow 2H_2O(l) \qquad E^{\ominus}_{O_2,\,H^+/H_2O} = +1.23\text{ V}$$

The $E^{\ominus}$ values tell us that $S_2O_8^{2-}$ is more easily reduced than O_2. But if $S_2O_8^{2-}$ is *more easily reduced*, then the product, SO_4^{2-}, must be *less easily oxidised*. Stated another way, *the half-reaction with the smaller (less positive) reduction potential occurs more easily as an oxidation*. As a result, during electrolysis, water is oxidised instead of SO_4^{2-}, and O_2 is formed at the anode.

The overall cell reaction for the electrolysis of the K_2SO_4 solution is:

$$4H_2O(l) + 4e^- \rightleftharpoons 2H_2(g) + 4OH^-(aq) \quad \text{(cathode)}$$
$$2H_2O(l) \rightleftharpoons O_2(g) + 4H^+(aq) + 4e^- \quad \text{(anode)}$$
$$6H_2O(l) \rightarrow 2H_2(g) + O_2(g) + \underbrace{4H^+(aq) + 4OH^-(aq)}_{4H_2O(l)} \quad \text{(cell reaction)}$$

The net change then is:

$$2H_2O(l) \xrightarrow{\text{electrolysis}} 2H_2(g) + O_2(g)$$

The word 'electrolysis' above the arrow in the equation shows that electricity is the driving force for this otherwise nonspontaneous reaction.

What is the role of potassium sulfate in this electrolysis, given that neither K^+ nor SO_4^{2-} ions are changed by the reaction? If the electrolysis is attempted with pure distilled water, nothing happens. There is no current flow, and no H_2 or O_2 forms. Apparently, the potassium sulfate must have some purpose.

The function of K_2SO_4 (or any other inert electrolyte) is to maintain electrical neutrality in the vicinity of the electrodes. If K_2SO_4 were not present and the electrolysis were to occur anyway, the solution around the anode would become positively charged. It would become filled with H^+ ions, with no negative ions to balance their charge. Similarly, the solution surrounding the cathode would become negatively charged as it is filled with OH^- ions, with no nearby positive ions. The formation of positively or negatively charged solutions requires too much energy so, in the absence of an electrolyte, the electrode reactions cannot take place.

When K_2SO_4 is in the solution, K^+ ions can move towards the cathode and mingle with the OH^- ions as they are formed. Similarly, the SO_4^{2-} ions can move towards the anode and mingle with the H^+ ions as they are produced there. In this way, at any moment, each small region of the solution can contain the same number of positive and negative charges and thereby remain neutral; in other words, the K^+ and SO_4^{2-} ions complete the electrical circuit.

Below is a set of rules that can be used to predict the outcomes of electrolysis reactions in aqueous solution (using inert electrodes).

1. Identify all cations and anions in the system. Remember that in aqueous solutions we have also H^+ and OH^-.
2. The product of the electrolysis at the anode is always O_2, except if the electrolyte contains high concentrations of Cl^-, Br^- or I^- ions.
3. During electrolysis of aqueous solutions, reactive metals, such as Na or K, are never produced at the cathode. If the cations come from a metal that has a less positive reduction potential than water, hydrogen (H_2) will be liberated at the cathode. If the cations come from a metal that has a more positive reduction potential than water, then the metal will be deposited at the cathode.
4. Identify the cations and anions that remain in solution after the electrolysis.
5. Summarise all reactions in 1–4 to obtain the full reaction scheme.

Reduction potentials can be used to anticipate the products of an electrolysis. This is illustrated by worked example 12.14.

WORKED EXAMPLE 12.14

Predicting the products in an electrolysis reaction

Electrolysis is planned for an aqueous solution that contains a mixture of 0.50 M $ZnSO_4$ and 0.50 M $NiSO_4$. On the basis of reduction potentials, what products are expected to be observed at the electrodes? What is the expected net cell reaction?

Analysis

We need to consider the competing reactions at the cathode and the anode. At the cathode, the half-reaction with the *most positive* reduction potential will be the one expected to occur as a reduction. At the anode, the half-reaction with the *least positive* reduction potential is the one that should occur as an oxidation.

Solution

At the cathode, the competing reduction reactions involve the two cations and water. The reactions and their reduction potentials are:

$$Ni^{2+}(aq) + 2e^- \rightleftharpoons Ni(s) \qquad E^{\ominus}_{Ni^{2+}/Ni} = -0.25\text{ V}$$
$$Zn^{2+}(aq) + 2e^- \rightleftharpoons Zn(s) \qquad E^{\ominus}_{Zn^{2+}/Zn} = -0.76\text{ V}$$
$$2H_2O(l) + 2e^- \rightleftharpoons H_2(g) + 2OH^-(aq) \qquad E^{\ominus}_{H_2O/H_2,\,OH^-} = -0.83\text{ V}$$

The most positive reduction potential is that of Ni^{2+}, so we expect this ion to be reduced at the cathode and solid nickel to be formed.

At the anode, the competing oxidation reactions are for water and SO_4^{2-} ions. In table 12.1, oxidised substances are found on the right-hand side of the half-reactions. The two half-reactions with these as products are:

$$S_2O_8^{2-}(aq) + 2e^- \rightleftharpoons 2SO_4^{2-}(aq) \qquad E^{\ominus}_{S_2O_8^{2-}/SO_4^{2-}} = +2.01\text{ V}$$
$$O_2(g) + 4H^+(aq) + 4e^- \rightleftharpoons 2H_2O(l) \qquad E^{\ominus}_{O_2,H^+/H_2O} = +1.23\text{ V}$$

The half-reaction with the less positive $E^{\ominus}$ occurs more easily as an oxidation, so we expect the oxidation half-reaction to be:

$$2H_2O(l) \rightleftharpoons O_2(g) + 4H^+(aq) + 4e^-$$

At the anode, we expect O_2 to be formed.

The predicted net cell reaction is obtained by combining the two expected electrode half-reactions, making the electron loss equal to the electron gain:

$$2H_2O(l) \rightarrow O_2(g) + 4H^+(aq) + 4e^- \qquad \text{(anode)}$$
$$2Ni^{2+}(aq) + 4e^- \rightarrow 2Ni(s) \qquad \text{(cathode)}$$
$$2H_2O(l) + 2Ni^{2+}(aq) \rightarrow O_2(g) + 4H^+(aq) + 2Ni(s) \qquad \text{(cell reaction)}$$

Is our answer reasonable?

We can check the locations of the half-reactions in table 12.1 to confirm our conclusions. For the reduction step, the more positive the reduction potential, the greater its tendency to occur as a reduction. Among the competing half-reactions at the cathode, the reduction potential for Ni^{2+} is highest, so we expect that Ni^{2+} is the easiest to reduce and Ni(s) should be formed at the cathode.

For the oxidation step, the more negative the reduction potential, the easier it is to occur as an oxidation. On this basis, the oxidation of water is easier than the oxidation of SO_4^{2-}, so we expect H_2O to be oxidised and O_2 to be formed at the anode.

PRACTICE EXERCISE 12.17

In the electrolysis of an aqueous solution containing both Cd^{2+} and Cr^{3+}, what products do we expect at the electrodes?

Stoichiometry of electrochemical reactions

In about 1833, British scientist Michael Faraday discovered that the amount of chemical change that occurs during electrolysis is directly proportional to the amount of electric charge that is passed through an electrolytic cell. For example, the reduction of copper ions at a cathode is given by the equation:

$$Cu^{2+}(aq) + 2e^- \rightarrow Cu(s)$$

Deposition of 1 mol of metallic copper requires 2 mol of electrons. Therefore, to deposit 2 mol of copper requires 4 mol of electrons, and that takes twice as much electricity. The half-reaction for an oxidation or reduction, therefore, relates the amount of chemical substance consumed or produced to the amount of electrons that the electric current must supply.

The SI unit of electric current is the **ampere (A)** and of charge is the coulomb (C). A coulomb is the amount of charge that passes by a given point in a wire when an electric current of 1 ampere flows for 1 second.

$$1\text{ coulomb} = 1\text{ ampere} \times 1\text{ second}$$
$$1\text{ C} = 1\text{ A} \times 1\text{ s}$$

For example, if a current of 4 A flows through a wire for 10 s, 40 C passes by a given point in the wire.

As we noted earlier, 1 mol of electrons carries a charge of 96 485 C. With this, laboratory measurements can be related to the amount of chemical change that occurs during an electrolysis. The amount of product formed can be calculated using **Faraday's law**:

$$\text{amount of product} = \frac{It}{zF}$$

where I = current (A), t = time (s), z = number of electrons transferred in the balanced equation and F = Faraday constant ($C\ mol^{-1}$). Worked example 12.15 demonstrates the use of this equation.

WORKED EXAMPLE 12.15

Calculations related to electrolysis

What mass of copper is deposited on the cathode of an electrolytic cell if an electric current of 2.00 A is run through a solution of $CuSO_4$ for a period of 20.0 min?

Analysis

We obtain the amount of copper by using Faraday's law. To determine the mass of copper, we multiply the amount of copper by the molar mass of copper.

Solution

First we convert minutes to seconds: 20.0 min = 1.20×10^3 s. We then use Faraday's law to determine the amount of copper (remember that $F = 96\,485\ C\ mol^{-1}$ or $9.6485 \times 10^4\ A\ s\ mol^{-1}$).

$$\text{amount of product} = \frac{It}{zF}$$

$$= \frac{2.00\ A \times 1.20 \times 10^3\ s}{2 \times 9.6485 \times 10^4\ A\ s\ mol^{-1}}$$

$$= 1.24 \times 10^{-2}\ mol$$

We obtain the mass by multiplying by the molar mass of copper.

$$\text{mass of copper} = 1.24 \times 10^{-2}\ mol \times 63.55\ g\ mol^{-1}$$

The electrolysis will deposit 0.79 g of copper on the cathode.

Is our answer reasonable?

Check that you have inserted the correct numbers into the equation. It is important to remember to convert minutes to seconds in this type of problem.

PRACTICE EXERCISE 12.18

Electrolysis provides a useful way to deposit a thin metallic coating on an electrically conducting surface. The technique is called *electroplating*. How much time would it take in minutes to deposit 0.50 g of metallic nickel on a metal object using a current of 3.00 A? The nickel is reduced from the +2 oxidation state.

12.8 Batteries

As mentioned earlier, the potential produced by one galvanic cell is not enough to power electronic instruments such as toys, torches, electronic calculators, laptop computers, heart pacemakers, video cameras or mobile phones, let alone a car. A **battery** (figure 12.22) contains a group of galvanic cells usually connected in series so that the potentials of the individual cells add up. These devices are classified as being either **primary cells** (cells not designed to be recharged; they are discarded after their energy is depleted) or **secondary cells** (cells designed for repeated use; they can be recharged). In this section, we will focus on the most important types of batteries, such as car batteries, dry cells and rechargeable batteries, as well as fuel cells.

FIGURE 12.22 The oldest known electric battery in existence (dating from about 2200 years ago), discovered in 1938 in Baghdad, Iraq, consists of a copper tube surrounding an iron rod. If filled with an acidic liquid such as vinegar, the cell could produce a small electric current.

The lead storage battery

The common **lead storage battery** (figure 12.23) used to start a car is composed of a number of secondary cells, each having a potential of about 2 V, that are connected in series so that their voltages are additive (see figure 12.24). Most car batteries contain six such cells and give about 12 V, but 6, 24 and 32 V batteries are also available.

The anode of each cell in a typical lead storage battery is composed of a set of lead plates, the cathode consists of another set of plates that hold a coating of PbO_2, and the electrolyte is sulfuric acid (see worked example 12.9, p. 517). When the battery is discharging, the electrode reactions are:

$$PbO_2(s) + 3H^+(aq) + HSO_4^-(aq) + 2e^- \rightarrow PbSO_4(s) + 2H_2O(l) \qquad \text{(cathode)}$$

$$Pb(s) + HSO_4^-(aq) \rightarrow PbSO_4(s) + H^+(aq) + 2e^- \qquad \text{(anode)}$$

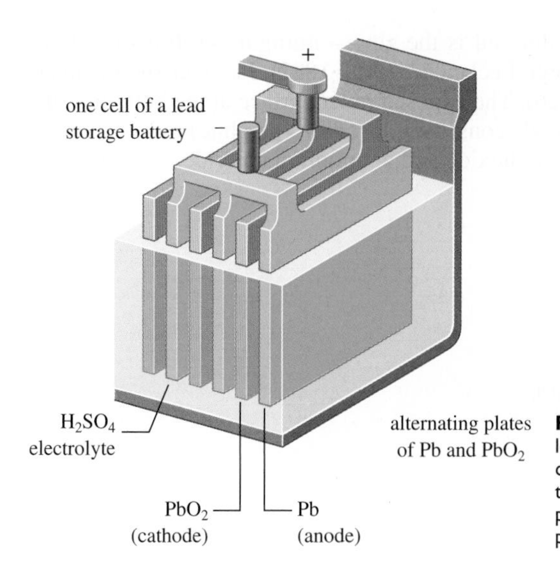

FIGURE 12.23 Lead storage battery. A 12-volt lead storage battery, such as those used in most cars, consists of six cells like the one shown here. Notice that the anode and cathode each consists of several plates connected together. This allows the cell to produce the large currents necessary to start a car.

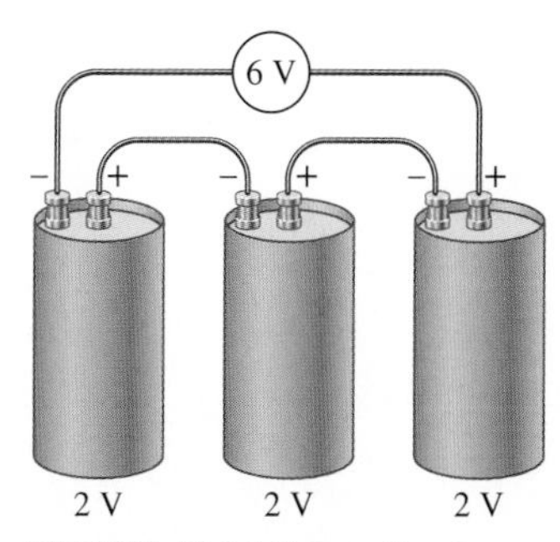

FIGURE 12.24 If three 2-volt cells are connected in series, their voltages are additive to provide a total of 6 volts. Cars generally use 12-volt batteries containing six cells.

The net cell reaction is therefore:

$$PbO_2(s) + Pb(s) + \underbrace{2H^+(aq) + 2HSO_4^-(aq)}_{2H_2SO_4(aq)} \rightarrow 2PbSO_4(s) + 2H_2O(l)$$

The spontaneous reaction occurring in a lead battery is an example of a **symproportionation** (or comproportionation) reaction, where a chemical reaction occurs between two reactants that contain the same element in different oxidation states. The product has an oxidation number intermediate to the two reactants. In this specific reaction, PbO_2 (oxidation state of Pb = +4) undergoes symproportionation with elemental Pb (oxidation state = 0) to give $PbSO_4$ (oxidation state of Pb = +2).

The reverse of a symproportionation reaction is a **disproportionation** reaction, where a species is simultaneously oxidised and reduced to form two different products. Such occurs, when the battery is recharged by electrolysis through the application of a voltage from an external source. The reaction for battery recharge is:

$$2PbSO_4(s) + 2H_2O(l) \xrightarrow{\text{electrolysis}} PbO_2(s) + Pb(s) + 2H^+(aq) + 2HSO_4^-(aq)$$

As a lead battery discharges, the sulfuric acid concentration decreases, which provides a convenient means of checking the state of the battery. Because the density of a sulfuric acid solution decreases as its concentration drops, the concentration can be determined very simply by measuring the density with a **hydrometer**, which consists of a rubber bulb that is used to draw the battery fluid into a glass tube containing a float (see figure 12.25). The depth to which the float sinks is inversely proportional to the density of the liquid — the deeper the float sinks, the lower is the density of the acid and the weaker is the charge of the battery. The narrow neck of the float is usually marked to indicate the state of charge of the battery.

Disadvantages of the lead storage battery are that it is very heavy and that its corrosive sulfuric acid can spill. The most modern lead storage batteries use a lead–calcium alloy as the anode. This reduces the need to vent individual cells so the battery can be sealed, preventing spillage of the electrolyte.

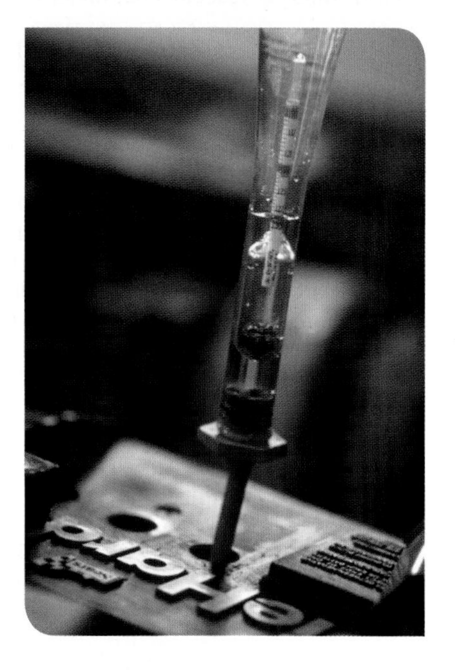

FIGURE 12.25 A battery hydrometer. Battery acid is drawn into the glass tube. The depth to which the float sinks is inversely proportional to the concentration of the acid and, therefore, to the state of charge of the battery.

Dry cell batteries

Electronic household instruments, such as remote controls, wristwatches, DVD and MP3 players, torches and radios, are powered by small, highly efficient **dry cell batteries**. They are manufactured in different sizes, named by the A-classification, where AAA and AA are the most common sizes used in remote controls, digital cameras and DVD players. The larger C size is most often used in portable stereos and many electrical toys, whereas size D is the standard battery used in baton-shaped torches.

The ordinary, relatively inexpensive 1.5 V dry cell is the **zinc – manganese dioxide cell**, or **Leclanché cell** (named after its inventor, George Leclanché, 1839–1882). Its outer shell is made of zinc, which serves as the anode (figure 12.26). The exposed outer surface at the bottom of the cell is the negative end of the battery. The cathode consists of a carbon (graphite) rod surrounded by a moist paste of graphite powder, manganese dioxide and ammonium chloride.

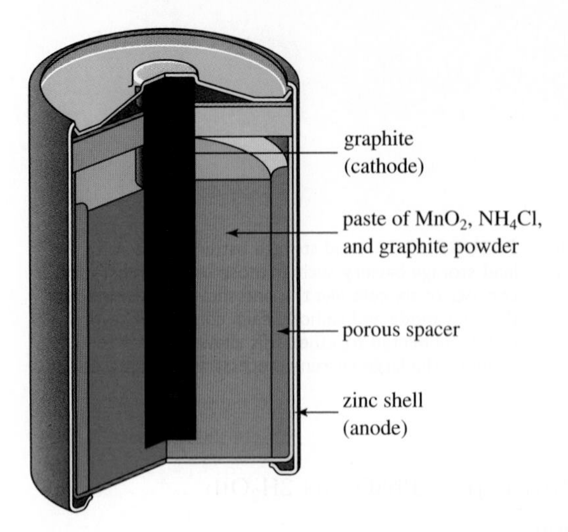

FIGURE 12.26 A cut-away view of a zinc – manganese dioxide dry cell (Leclanché cell).

The anode reaction is simply the oxidation of zinc.

$$Zn(s) \rightarrow Zn^{2+}(aq) + 2e^- \qquad \text{(anode)}$$

The cathode reaction is complex, and no simple overall cell reaction can be written. One of the major reactions is:

$$2MnO_2(s) + 2NH_4^+(aq) + 2e^- \rightarrow Mn_2O_3(s) + 2NH_3(aq) + H_2O(l) \qquad \text{(cathode)}$$

The ammonia that forms at the cathode reacts with some of the Zn^{2+} produced from the anode to form a complex ion, $Zn(NH_3)_4^{2+}$.

A more popular version of the Leclanché battery uses a basic, or *alkaline*, electrolyte (generally potassium hydroxide) and is called an **alkaline battery** or **alkaline dry cell**. It, too, uses Zn and MnO_2 as reactants but under basic conditions (figure 12.27). The half-cell reactions are:

$$Zn(s) + 2OH^-(aq) \rightarrow ZnO(s) + H_2O(l) + 2e^- \qquad \text{(anode)}$$

$$2MnO_2(s) + H_2O(l) + 2e^- \rightarrow Mn_2O_3(s) + 2OH^-(aq) \qquad \text{(cathode)}$$

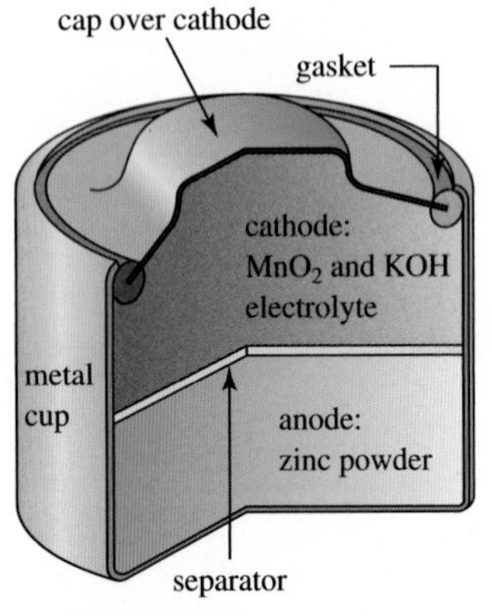

FIGURE 12.27 A simplified diagram of an alkaline zinc – manganese dioxide dry cell.

and the voltage is about 1.54 V. The alkaline dry cell has a longer shelf life and can deliver higher currents for longer than the less expensive Leclanché cell shown in figure 12.26. Over time, however, alkaline batteries are prone to corrosion and leaking. The released corrosive potassium hydroxide, which can also emerge from seams around the battery, forms a feathery crystalline structure on the outside of the battery (figure 12.28). The damage can spread further through the metal electrodes to circuit boards leading to oxidation of copper traces and other components, which can result in permanent damage to the circuitry and destruction of the equipment.

FIGURE 12.28 Alkaline dry cells are prone to leakage and corrosion.

The **nickel–cadmium storage cell,** or **nicad battery**, (commonly abbreviated to NiCd or nicad) is a secondary cell that produces a potential of about 1.3–1.4 V. The cathode in the NiCd cell is NiO(OH), a compound of nickel in the +3 oxidation state, and the electrolyte is a solution of KOH. The electrode reactions in the cell during discharge are:

$$Cd(s) + 2OH^-(aq) \rightarrow Cd(OH)_2(s) + 2e^- \quad \text{(anode)}$$

$$2NiO(OH)(s) + 2H_2O + 2e^- \rightarrow 2Ni(OH)_2(s) + 2OH^-(aq) \quad \text{(cathode)}$$

$$Cd(s) + 2NiO(OH)(s) + 2H_2O \rightarrow 2Ni(OH)_2(s) + Cd(OH)_2(s) \quad E^{\ominus}_{cell} = 1.3\text{–}1.4\text{ V} \quad \text{(cell reaction)}$$

The nickel–cadmium battery can be recharged by reversing the anode and cathode reactions above to regenerate the reactants. The battery can be sealed to prevent leakage, which is particularly important in electronic devices.

Nickel–cadmium batteries work especially well in applications such as portable power tools, DVD players and even electric cars. They have a high *energy density* (available energy per unit volume), release energy quickly, and can be recharged rapidly.

Modern high-performance batteries

Nickel – metal hydride battery

Nickel – metal hydride (NiMH) batteries are secondary cells and have been used extensively in recent years to power devices such as mobile phones, video cameras and even electric vehicles. They are similar in many ways to the alkaline nickel–cadmium cells discussed earlier, except for the anode reactant, which is hydrogen. At first, this seems odd, because hydrogen is a gas at room temperature and atmospheric pressure. However, in the late 1960s, it was discovered that some metal alloys (such as $LaNi_5$, an alloy of lanthanum and nickel, and Mg_2Ni, an alloy of magnesium and nickel) can absorb and hold substantial amounts of hydrogen in its atomic form and that the hydrogen could be made to participate in reversible electrochemical reactions. (*Note:* The term 'metal hydride' has come to be used to describe the hydrogen-holding alloy in which hydrogen is contained in its atomic form. These should not be mixed up with compounds of hydrogen with metals such as sodium that actually contain the *hydride ion*, H^-. The metal hydrides described here are not of that type.)

As in the NiCd cell, the cathode in NiMH batteries is NiO(OH) and the electrolyte is a solution of KOH. A diagram of a cylindrical cell is shown in figure 12.29. Using the symbol MH to stand for the metal hydride, the reactions in the cell during discharge are:

$$MH(s) + OH^-(aq) \rightarrow M(s) + H_2O(l) + e^- \quad \text{(anode)}$$

$$NiO(OH)(s) + H_2O(l) + e^- \rightarrow Ni(OH)_2(s) + OH^-(aq) \quad \text{(cathode)}$$

$$MH(s) + NiO(OH)(s) \rightarrow Ni(OH)_2(s) + M(s) \quad E^{\ominus}_{cell} = 1.35\text{ V} \quad \text{(cell reaction)}$$

When the cell is recharged, these reactions are reversed. The principal advantage of the NiMH cell over the NiCd cell is that it can store about 50% more energy in the same volume.

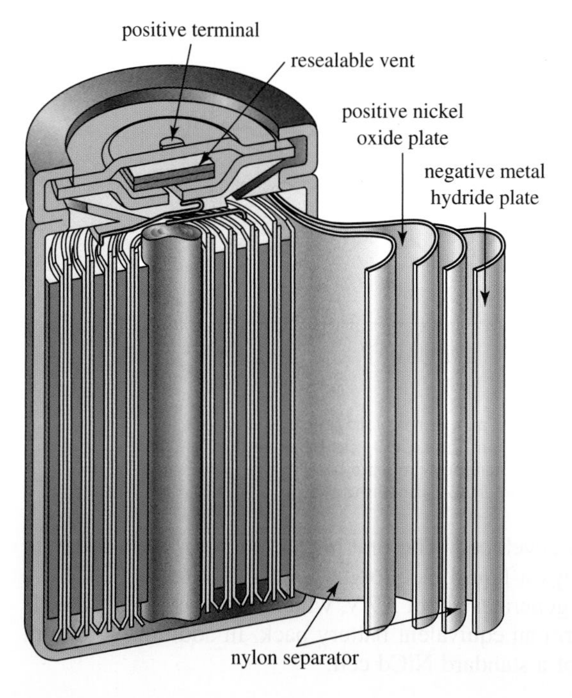

FIGURE 12.29 Cutaway view of a nickel – metal hydride cell. The electrode sandwich is rolled up; this yields a large effective electrode area and enables the cell to deliver large amounts of energy quickly.

Lithium ion cells

If you look at the location of lithium in the table of reduction potentials (table 12.1), you will see that it has the most negative reduction potential of any metal. This means that lithium is very easily oxidised electrochemically, and its large negative reduction potential suggests it has a lot of appeal as an anode material. Furthermore, lithium is a very lightweight metal, so a cell employing lithium as a reactant would also be lightweight.

Attempts to make rechargeable lithium batteries containing lithium metal electrodes have been plagued with safety problems. Success came with the development of the **lithium ion cell**, which uses lithium ions rather than metallic lithium. The transport of Li^+ ions through the electrolyte from one electrode to the other is accompanied by the transport of electrons through the external circuit to maintain charge balance. The following paragraphs describe how lithium ions are used in a rechargeable cell.

It was discovered that, because they are small, Li^+ ions can slip between layers of atoms in certain crystalline substances (a process called **intercalation**). Graphite is such a substance. As we saw in chapter 7, graphite consists of layers of fused hexagonal rings of carbon atoms. The other most commonly used compound able to intercalate Li^+ ions is $LiCoO_2$. These are the materials used to make the electrodes.

When the cell is constructed, it is in its 'uncharged' state with no Li^+ ions between the layers of carbon atoms in the graphite. When the cell is being charged (figure 12.30), Li^+ ions leave $LiCoO_2$ (with x being the amount of transferred Li^+) and travel through the electrolyte to the graphite.

$$LiCoO_2 + \text{graphite} \rightarrow Li_{1-x}CoO_2 + Li_x\text{graphite} \qquad \text{(initial charging)}$$

When the cell spontaneously discharges to provide electrical power (figure 12.30), Li^+ ions move back through the electrolyte to the cobalt oxide, while electrons move through the external circuit from the graphite electrode (anode) to the cobalt oxide electrode (cathode). If we represent the amount of Li^+ transferring by y, the discharge 'reaction' is:

$$Li_{1-x}CoO_2 + Li_x\text{graphite} \rightarrow Li_{1-x+y}CoO_2 + Li_{x-y}\text{graphite} \qquad \text{(discharging)}$$

Thus, the charging and discharging cycles simply sweep Li^+ ions back and forth between the two electrodes, with electrons flowing through the external circuit to keep the charge in balance.

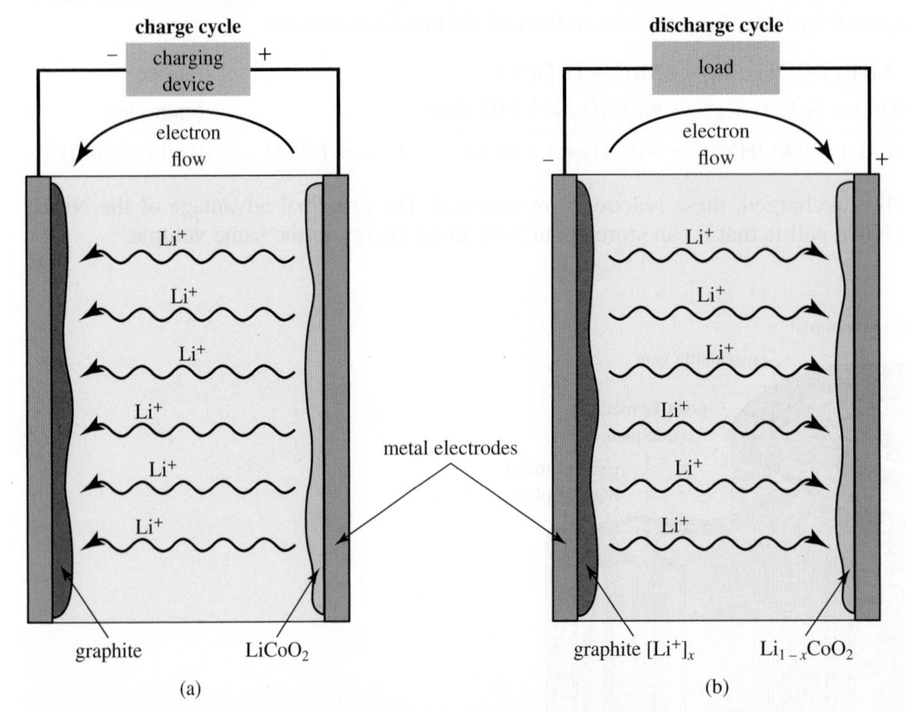

FIGURE 12.30 Lithium ion cell. **(a)** During the charging cycle, an external voltage forces electrons through the external circuit and causes lithium ions to travel from the $LiCoO_2$ electrode to the graphite electrode. **(b)** During discharge, the lithium ions spontaneously migrate back to the $LiCoO_2$ electrode, and electrons flow through the external circuit to balance the charge.

Two types of lithium ion cells have been developed. The one most commonly used today in mobile phones and laptop computers employs a liquid electrolyte (usually containing $LiPF_6$, a compound with Li^+ and PF_6^- ions). The cell generates about 3.7 V, which means three NiCd cells connected in series would be required to form an equivalent battery pack. In addition, a lithium ion cell has about twice the energy density of a standard NiCd cell.

Fuel cells

The galvanic cells discussed so far can produce power only for a limited time because the electrode reactants are eventually degraded. Fuel cells are different; they are electrochemical cells in which the electrode reactants are supplied continuously and can operate without a theoretical limit as long as the supply of reactants is maintained. This makes fuel cells an attractive source of power where long-term generation of electrical energy is required.

Figure 12.31 illustrates an early design of a hydrogen–oxygen **fuel cell**. The electrolyte, a hot (~200 °C), concentrated solution of potassium hydroxide in the centre compartment, is in contact with two porous electrodes that contain a catalyst (usually platinum) to facilitate the electrode reactions. Gaseous hydrogen and oxygen under pressure are circulated so as to come in contact with the electrodes. At the cathode, oxygen is reduced.

$$O_2(g) + 2H_2O(l) + 4e^- \rightarrow 4OH^-(aq) \qquad \text{(cathode)}$$

At the anode, hydrogen is oxidised to water.

$$H_2(g) + 2OH^-(aq) \rightarrow 2H_2O(l) + 2e^- \qquad \text{(anode)}$$

Some of the water formed at the anode leaves as steam mixed with the circulating hydrogen gas. The net cell reaction, after making electron loss equal to electron gain, is:

$$2H_2(g) + O_2(g) \rightarrow 2H_2O(l) \qquad \text{(cell reaction)}$$

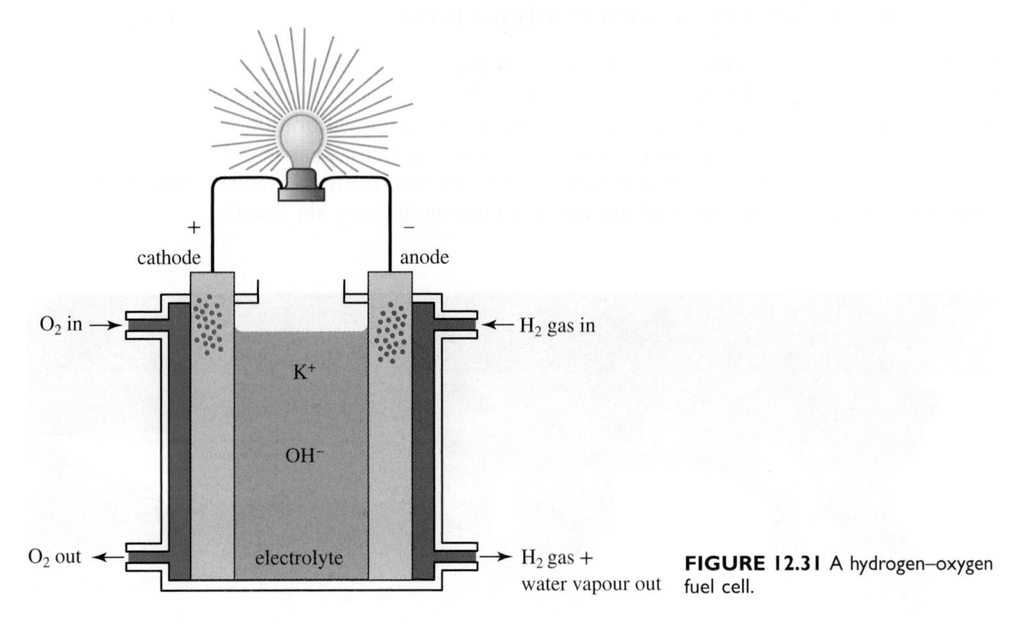

FIGURE 12.31 A hydrogen–oxygen fuel cell.

A major advantage of the fuel cell is that there is no electrode material to be replaced, as there is in an ordinary battery. The fuel can be fed in continuously to produce power. In fact, hydrogen–oxygen fuel cells were used in the Gemini and Apollo missions in the second half of the twentieth century and other space programs for just this reason.

One reason fuel cells are so appealing is their thermodynamic efficiency. In a fuel cell, the net reaction is equivalent to combustion. The production of usable energy by the direct combustion of fuels is an extremely inefficient process. Largely because of the constraints set by fundamental thermodynamic principles, modern electrical power plants are unable to harness more than about 35% to 40% of the potential energy in oil, coal or natural gas. A petrol or diesel engine has an efficiency of only about 25% to 30%. The rest of the energy is lost to the surroundings as heat, which is why a vehicle must have an effective cooling system.

Fuel cells 'burn' fuel under conditions that are much more thermodynamically reversible than simple combustion. They therefore achieve much greater efficiencies — 75% is quite feasible. In addition, hydrogen–oxygen fuel cells are essentially pollution free. The only product formed by the cell is water. However, hydrogen fuel is not necessarily produced by pollution-free processes.

Advances in fuel cell development have led to lower operating temperatures and the ability to use methanol, a liquid at room temperature, as a source of hydrogen. The hydrogen is generated from methanol vapour, $CH_3OH(g)$, by a catalytic process. The net reaction is:

$$CH_3OH(g) + H_2O(g) \rightarrow CO_2(g) + 3H_2(g)$$

DaimlerChrysler originally demonstrated the feasibility of such a system by driving its NECAR5 vehicle, powered by five fuel cells, 5000 kilometres across the United States in 16 days. An onboard fuel reformer extracted hydrogen from methanol to feed the fuel cells.

The development of efficient fuel cells is an active research area, especially for the transportation market.

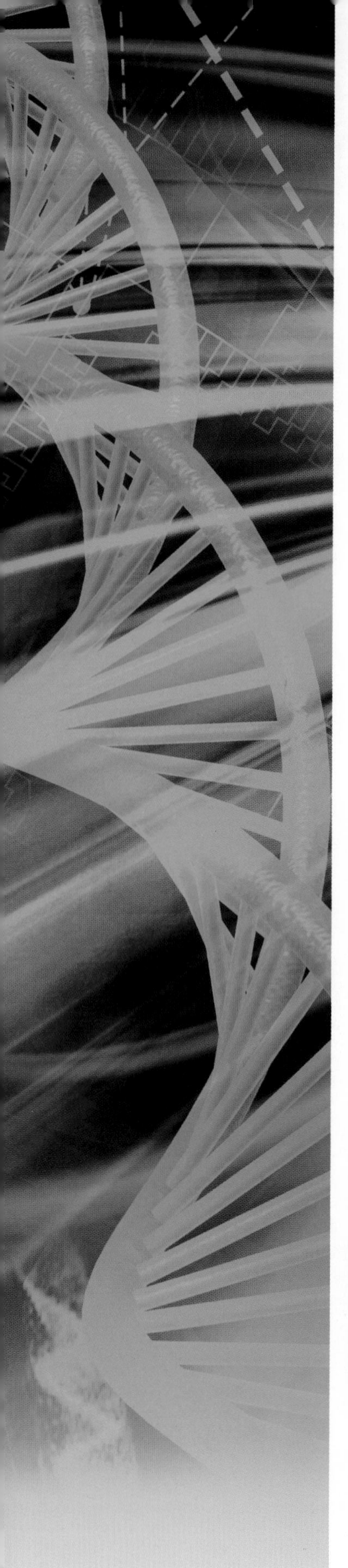

Chemical Connections
Breathalysers

A breathalyser is a device to estimate the alcohol content in blood from a breath sample. The first practical roadside breathalyser was developed in 1931 where the breath was pumped through a solution of acidified potassium permanganate. Ethanol in the breath leads to a colour change that results from the reduction of the dark purple Mn^{7+} to the pale pink Mn^{2+} by ethanol, which is oxidised to acetic acid, according to:

$$4KMnO_4^- + 5CH_3CH_2OH + 12H^+ \rightarrow 4Mn^{2+} + 5CH_3COOH + 11H_2O$$

Modern handheld breathalysers (figure 12.32) are fuel cells consisting of two platinum electrodes that are separated by a porous acid-electrolyte material. The exhaled air flows past one side of the fuel cell, where ethanol is oxidised at the anode and atmospheric oxygen is reduced at the cathode.

$$CH_3CH_2OH(g) + H_2O(l) \rightarrow CH_3CO_2H(l) + 4H^+(aq) + 4e^- \qquad \text{(anode)}$$

$$O_2(g) + 4H^+(aq) + 4e^- \rightarrow 2H_2O(l) \qquad \text{(cathode)}$$

$$CH_3CH_2OH(l) + O_2(g) \rightarrow CH_3COOH(l) + H_2O(l) \qquad \text{(cell reaction)}$$

The device measures the electrical current that is produced by this reaction. The more ethanol is in the breath the higher the current, which is then displayed as an approximation of the overall blood alcohol content. These fuel cell breath analysers require recalibration usually once a year.

Breath analysers sold to consumers often use a semiconductor sensor (silicon oxide), which is considerably more prone to contamination and interference from substances other than breath alcohol. These analysers require recalibration or replacement every six months.

FIGURE 12.32 Modern breathalysers are fuel cells that pass a higher current when there is ethanol in the breath.

SUMMARY

LO 1 Oxidation and reduction

Oxidation is the loss of electrons or an increase in oxidation number; reduction is the gain of electrons or a decrease in oxidation number. Both always occur together in reduction–oxidation (or redox) reactions. The substance oxidised is the reducing agent; the substance reduced is the oxidising agent. Oxidation numbers are a bookkeeping device that we use to follow changes in redox reactions. The term 'oxidation state' is equivalent to oxidation number.

LO 2 Balancing net ionic equations for redox reactions

In a balanced redox equation, the number of electrons gained by one substance is always equal to the number lost by another substance. To balance a redox equation, the skeleton net ionic equation is divided into two half-equations; for the oxidation and reduction process which are balanced separately before being recombined to give the final balanced net ionic equation. For reactions in basic solution, the equation is first balanced as if it occurred in an acidic solution, and then the balanced equation is converted to its proper form for a basic solution by adding an appropriate number of OH^- ions.

LO 3 Galvanic cells

A galvanic cell is composed of two half-cells, each containing an electrode in contact with an electrolyte reactant. A spontaneous redox reaction is thus divided into separate oxidation and reduction half-reactions, with the electron transfer occurring through an external electric circuit. Reduction occurs at the cathode; oxidation occurs at the anode. In a galvanic cell, the cathode is positively charged and the anode is negatively charged. The half-cells are usually connected by a salt bridge to complete the electric circuit; this permits electrical neutrality to be maintained by allowing cations to move towards the cathode and anions towards the anode.

LO 4 Reduction potentials

The potential (expressed in volts) produced by a cell is equal to the standard cell potential when all ion concentrations are 1 M and the partial pressure of any gas involved equals 1×10^5 Pa. The cell potential can be considered to be the difference between the reduction potentials of the half-cells. In the spontaneous reaction, the half-cell with the more positive reduction potential undergoes reduction and forces the other to undergo oxidation. The reduction potentials of isolated half-cells cannot be measured, but values are assigned by choosing the standard hydrogen electrode as a reference electrode; its reduction potential is assigned a value of exactly 0 V. Species more easily reduced than H^+ have positive reduction potentials; those less easily reduced have negative reduction potentials. Reduction potentials can be used to predict the cell reaction and to calculate $E^{\ominus}_{cell}$, as well as to predict whether a given reaction is spontaneous.

In solutions containing nonoxidising acids, the strongest oxidising agent is H^+. The reaction of a metal with a nonoxidising acid gives hydrogen gas and a salt of the acid. Only metals with more negative reduction potentials than hydrogen react this way. Oxidising acids, such as concentrated HNO_3 or concentrated H_2SO_4, contain an anion that is a stronger oxidising agent than H^+, and they can oxidise many metals that nonoxidising acids cannot.

LO 5 Relationship between cell potential, concentration and Gibbs energy

The values of $\Delta G^{\ominus}$ and K_c for a reaction can be calculated from $E^{\ominus}_{cell}$. They all involve the Faraday constant, F, which equals the number of coulombs (C) of charge per mole of electrons ($1\,F = 96\,485$ C mol^{-1}). The Nernst equation relates the cell potential to the standard cell potential and the reaction quotient. It allows the cell potential to be calculated for ion concentrations other than 1 M.

LO 6 Corrosion

Metals corrode because they oxidise easily. The corrosion of iron (formation of rust) is an electrochemical reaction that involves both oxygen and water. Some metals form insoluble oxides, which prevent them from further oxidation (passivation). Corrosion can be combatted by processes such as galvanisation and cathodic protection.

LO 7 Electrolysis

In an electrolytic cell, a flow of electricity causes an otherwise nonspontaneous reaction to occur. A negatively charged cathode causes reduction of one reactant and a positively charged anode causes oxidation of another. The electrode reactions are determined by which species is most easily reduced and which is most easily oxidised; however, in aqueous solutions, complex surface effects at the electrodes can alter the natural order. In the electrolysis of water, an electrolyte must be present to maintain electrical neutrality at the electrodes. The relationship between Faraday constant, F, current and time and the half-reactions that occur at the anode or cathode allow us to calculate the amount of chemical change.

LO 8 Batteries

The zinc – manganese dioxide cell (the Leclanché cell or common dry cell) and the common alkaline battery (which uses essentially the same reactions as the less expensive dry cell) are primary cells and are not rechargeable. The lead storage battery and the nickel–cadmium (nicad) battery are secondary cells and are rechargeable. The state of charge of a lead storage battery can be tested with a hydrometer, which measures the density of the sulfuric acid electrolyte. The rechargeable nickel – metal hydride battery (NiMH) uses hydrogen contained in a metal alloy as its anode reactant and has a higher energy density than the nicad battery. The rechargeable lithium ion cell produces a large cell potential and has a very high energy density. Lithium ion cells store and release energy by transferring lithium ions between electrodes where the Li^+ ions are intercalated between layers of atoms in the electrode materials. Fuel cells, which have high thermodynamic efficiencies, can provide continuous power because they consume fuel that can be fed continuously.

KEY CONCEPTS AND EQUATIONS

Rules for assigning oxidation numbers *(section 12.1)*

Use these rules to assign oxidation numbers to the atoms in a chemical formula. Use changes in oxidation numbers to identify oxidation and reduction.

1. The oxidation number of any free element (an element not combined chemically with a different element) is 0. For example, Ar, Fe, O in O_2, P in P_4 and S in S_8 all have oxidation numbers of 0.
2. The oxidation number of any simple, monatomic ion (e.g. Na^+ and Cl^-) is equal to the charge on the ion.
3. The sum of all the oxidation numbers of the atoms in a neutral molecule must equal 0. The sum of all the oxidation numbers in a polyatomic ion must equal the charge on the ion.

4. In all of its compounds, fluorine has an oxidation number of –1.
5. In most of its compounds, hydrogen has an oxidation number of +1.
6. In most of its compounds, oxygen has an oxidation number of –2.

Balancing redox equations *(section 12.2)*

The following steps are used to balance an equation for a redox reaction in an acidic solution.
1. Identify reactants and products for each half-equation.
2. Balance atoms other than H and O.
3. Balance oxygen by adding H_2O.
4. Balance hydrogen by adding H^+.
5. Balance net charge by adding e^-.
6. Make e^- gain equal e^- loss.
7. Add the balanced half-equations.
8. Cancel any species that is the same on both sides.
9. Check that there are the same number and type of atoms and the same charge on each side of the reaction.

The following additional steps are used to balance an equation for a redox reaction in a basic solution.

10. Take note of the number of H^+ in the balanced equation and add the same number of OH^- to *each* side.
11. Combine each pair of H^+ and OH^- to form one H_2O.
12. Cancel any H_2O that occur on both sides.
13. Check that there are the same number and type of atoms and the same charge on each side of the reaction.

Standard reduction potentials *(section 12.4)*

In a galvanic cell, the difference between two standard reduction potentials equals the standard cell potential. Comparing reduction potentials lets us predict the electrode reactions in electrolysis.

Standard cell potentials *(section 12.4)*

By calculating $E^{\ominus}_{cell}$, we can predict the spontaneity of a redox reaction. We can use $E^{\ominus}_{cell}$ to calculate $\Delta G^{\ominus}$ and equilibrium constants. They are also needed in the Nernst equation to relate concentrations of species in galvanic cells to the cell potential.

$\Delta G^{\ominus} = -zFE^{\ominus}_{cell}$ *(section 12.5)*

This equation is used to calculate standard Gibbs energy changes from standard cell potentials, and vice versa.

$E^{\ominus}_{cell} = \frac{RT}{zF}\ln K_c$ *(section 12.5)*

This equation is used to calculate equilibrium constants from standard cell potentials.

Nernst equation $E_{cell} = E^{\ominus}_{cell} - \frac{RT}{zF}\ln Q$ *(section 12.5)*

This equation is used to calculate the cell potential from $E^{\ominus}_{cell}$ and concentration data; we can also calculate the concentration of a species in solution from $E^{\ominus}_{cell}$ and a measured value of E_{cell}.

Faraday constant, $F = 96\,485$ C mol^{-1} *(section 12.5)*

Besides being a constant in the equations above, the Faraday constant relates coulombs (obtained from the product of current and time) to moles of chemical change in electrochemical reactions.

Faraday's law *(section 12.7)*

$$\text{amount of product} = \frac{It}{zF}$$

KEY TERMS

REVIEW QUESTIONS

LO 1 Oxidation and reduction

12.1 Define 'oxidation' and 'reduction' in terms of (a) electron transfer and (b) oxidation numbers.

12.2 In the reaction $2Zn + O_2 \rightarrow 2ZnO$, which substance is the oxidising agent and which is the reducing agent? Which substance is oxidised and which is reduced?

12.3 Why must both oxidation and reduction occur simultaneously during a redox reaction? What is an oxidising agent? What happens to it in a redox reaction? What is a reducing agent? What happens to it in a redox reaction?

12.4 Are the following reactions redox reactions? Explain.

(a) $NO_2 + NO \rightarrow N_2O_3$ (b) $2CrO_4^{2-} + 2H^+ \rightarrow Cr_2O_7^{2-} + H_2O$

12.5 If the oxidation number of manganese in a certain molecule changes from +7 to +4 during a reaction, is manganese oxidised or reduced? How many electrons are gained (or lost) by each manganese atom?

12.6 Bromine gas is used to make fire-retardant materials used on children's pyjamas. However, to dispose of excess bromine, it is passed through a solution of sodium hydroxide. The products are NaBr and NaOBr. What is being reduced and what is being oxidised? Write a balanced equation for the reaction.

LO 2 Balancing net ionic equations for redox reactions

12.7 The following equations are not balanced. Why? Balance them.
(a) $Na + Fe_2 \rightarrow Na^+ + Fe$
(b) $Ag^+ + Cr \rightarrow Ag + Cr^{3+}$
(c) $Ag + Au^{3+} \rightarrow Ag^+ + Au$
(d) $Ca + Al^{3+} \rightarrow Ca^{2+} + Al$

12.8 What are the net charges on the left and right sides of the following equations? Add electrons as necessary to make each of them a balanced half-reaction.
(a) $NO_3^- + 10H^+ \rightarrow NH_4^+ + 3H_2O$
(b) $Cl_2 + 4H_2O \rightarrow 2ClO_2^- + 8H^+$

12.9 In question 12.8, which half-reaction represents oxidation? Which represents reduction?

12.10 At first glance, the following equation may appear to be balanced.

$$MnO_4^- + Sn^{2+} \rightarrow SnO_2 + MnO_2$$

What is wrong with it?

LO 3 Galvanic cells

12.11 What is a galvanic cell? What is a half-cell?

12.12 What is the function of a salt bridge?

12.13 In the copper–silver cell, why must the Cu^{2+} and Ag^+ solutions be kept in separate containers?

12.14 Which redox processes take place at the anode and cathode in a galvanic cell? What electric charges do the anode and cathode carry in a galvanic cell?

12.15 How do electrolytic conduction and metallic conduction differ?

12.16 Why is the movement of the ions relative to the electrodes the same in both galvanic and electrolytic cells? Explain.

12.17 When iron metal is placed into a solution of silver nitrate, the iron dissolves to give Fe^{2+}, and silver metal is formed. Write a net ionic equation for this reaction. Describe how you could use the reaction in a galvanic cell. Which metal, silver or iron, is the cathode?

12.18 Aluminium will displace tin from solution according to the equation:

$$2Al(s) + 3Sn^{2+}(aq) \rightarrow 2Al^{3+}(aq) + 3Sn(s)$$

What would be the individual half-cell reactions if this were the cell reaction in a galvanic cell? Which metal would be the anode and which the cathode?

12.19 Sketch a galvanic cell for which the cell diagram is:

$$Sn(s)\,|\,Sn^{2+}(aq)\,\|\,Ag^+(aq)\,|\,Ag(s)$$

(a) Label the anode and the cathode.
(b) Indicate the charge on each electrode.
(c) Indicate the direction of electron flow in the external circuit.
(d) Write the equation for the net cell reaction.

12.20 Sketch a galvanic cell in which inert platinum electrodes are used in the half-cells for the system:

$$Pt(s)\,|\,Fe^{2+}(aq), Fe^{3+}(aq)\,\|\,Br_2(aq), Br^-(aq)\,|\,Pt(s)$$

Label the diagram to indicate the composition of the electrolytes in the two cell compartments. Show the signs of the electrodes and label the anode and cathode. Write the equation for the net cell reaction.

LO 4 Reduction potentials

12.21 For a galvanic cell, what is the meaning of the term 'potential'? What are its units?

12.22 What is the difference between a cell potential and a standard cell potential?

12.23 How are standard reduction potentials combined to give the standard cell potential for a spontaneous reaction?

12.24 Is it possible to measure the potential of an isolated half-cell? Explain your answer.

12.25 Describe the hydrogen electrode. What is the value of its standard reduction potential?

12.26 What do the positive and negative signs of reduction potentials tell us?

12.27 If $E^{\ominus}_{Cd^{2+}/Cd}$ had been chosen as the standard reference electrode and had been assigned a potential of 0 V, what would the reduction potential of the hydrogen electrode be relative to it?

12.28 If you set up a galvanic cell using metals not found in table 12.1, what experimental information will tell you which is the anode and which is the cathode in the cell?

12.29 What is a nonoxidising acid? Give two examples. What is the oxidising agent in a nonoxidising acid?

12.30 What is the strongest oxidising agent in a dilute aqueous solution of sulfuric acid?

12.31 If a metal can react with a solution of HCl, is its reduction potential more negative or more positive than hydrogen?

12.32 Where in table 12.1 do we find the best reducing agents? Where do we find the best oxidising agents?

12.33 Which metals in table 12.1 will react with water? Write chemical equations for each of these reactions.

LO 5 Relationship between cell potential, concentration and Gibbs energy

12.34 Write the equation that relates the standard cell potential to the standard Gibbs energy change for a reaction.

12.35 What is the equation that relates the equilibrium constant to the standard cell potential?

12.36 Show how the equation that relates the equilibrium constant to the standard cell potential can be derived from the Nernst equation.

12.37 You have learned that the principles of thermodynamics allow the following equation to be derived:

$$\Delta G = \Delta G^{\ominus} + RT\ln Q$$

where Q is the reaction quotient. Without referring to the text, use this equation and the relationship between ΔG and the cell potential to derive the Nernst equation.

12.38 The cell reaction during the discharge of a lead storage battery is:

$$Pb^{2+}(aq) + Cd(s) \rightarrow Cd^{2+}(aq) + Pb(s)$$

The standard cell potential, $E^{\ominus}_{cell}$, is 0.277 V. What is the correct form of the Nernst equation for this reaction at 25 °C?

LO 6 Corrosion

12.39 Explain the chemical processes involved in the rusting of iron.

12.40 Explain how galvanisation works.

12.41 Define 'passivation'.

12.42 What is cathodic protection?

LO 7 Electrolysis

12.43 What electric charges do the anode and the cathode carry in an electrolytic cell? What does the term 'inert electrode' mean?

12.44 In a galvanic cell, do electrons travel from anode to cathode, or from cathode to anode? Explain.

12.45 Why must NaCl be melted before it is electrolysed to give Na and Cl_2? Write the anode, cathode and overall cell reactions for the electrolysis of molten NaCl.

12.46 Write half-reactions for the oxidation and the reduction of water.

12.47 What happens to the pH of the aqueous solution near the cathode and anode during the electrolysis of K_2SO_4? What function does K_2SO_4 serve in the electrolysis of a K_2SO_4 solution?

12.48 What is described by the Faraday constant, F? What relationships relate F to current and time measurements?

12.49 Using the same current, which will require more time: depositing 0.10 mol Cu from a Cu^{2+} solution or depositing 0.10 mol of Au from a Au^{3+} solution? Explain your reasoning.

12.50 An electric current is passed through two electrolysis cells connected in series (so the same amount of current passes through each of them). One cell contains Cu^{2+} and the other contains Fe^{2+}. In which cell will the greater mass of metal be deposited? Explain your answer.

LO 8 Batteries

12.51 What are the anode and cathode reactions during discharge of a lead storage battery? How can a battery produce a potential of 12 V if the cell reaction has a standard potential of only 2 V?

12.52 What is a symproportionation reaction?

12.53 What are the anode and cathode reactions during the charging of a lead storage battery?

12.54 What reactions occur at the electrodes in the ordinary dry cell?

12.55 What chemical reactions take place at the electrodes in an alkaline dry cell?

12.56 Give the half-cell reactions and the cell reaction that take place in a nicad battery during discharge. What are the reactions that take place during charging of the cell?

12.57 How is hydrogen held as a reactant in a nickel – metal hydride battery? Write the chemical formula for a typical alloy used in this battery. What is the electrolyte?

12.58 What are the anode, cathode and net cell reactions that take place in a nickel – metal hydride battery during discharge? What are the reactions while the battery is being charged?

12.59 What are the electrode materials in a typical lithium ion cell? Explain what happens when the cell is charged. Explain what happens when the cell is discharged.

12.60 Write the cathode, anode and net cell reaction in a hydrogen–oxygen fuel cell.

REVIEW PROBLEMS

12.61 For the following reactions, identify the substance oxidised, the substance reduced, the oxidising agent and the reducing agent. **LO1**

(a) $2HNO_3 + 3H_3AsO_3 \rightarrow 2NO + 3H_3AsO_4 + H_2O$

(b) $NaI + 3HOCl \rightarrow NaIO_3 + 3HCl$

(c) $2KMnO_4 + 5H_2C_2O_4 + 3H_2SO_4 \rightarrow 10CO_2 + K_2SO_4 + 2MnSO_4 + 8H_2O$

(d) $3CuS + 8HNO_3 \rightarrow 3CuSO_4 + 8NO + 4H_2O$

(e) $6H_2SO_4 + 2Fe \rightarrow 3Fe_2(SO_4)_3 + 6H_2O + 3SO_2$

(f) $2Na + 2H_2O \rightarrow 2NaOH + H_2$

(g) $2K_2Cr_2O_7 + 3CH_3CH_2OH + 8H_2SO_4 \rightarrow 2Cr_2(SO_4)_3 + 3CH_3COOH + 11H_2O$

(h) $I_2 + 10HNO_3 \rightarrow 2HIO_3 + 10NO_2 + 4H_2O$

(i) $HCl + NaCl + 2NaClO_4 \rightarrow 2ClO_2 + 2NaCl + NaOH$

12.62 Assign oxidation numbers to all of the atoms in the following compounds and ions. **LO1**

(a) H_2O
(b) N_2O_5
(c) NOCl
(d) $HClO_3$
(e) ClO_2
(f) ClO_2^-
(g) ClO_3^-
(h) ClO_4^-
(i) $Ca(VO_3)_2$
(j) $SnCl_4$
(k) MnO_4^{2-}

12.63 When chlorine is added to drinking water to kill bacteria, some of the chlorine is converted to chloride ions by the following equilibrium. **LO1**

$$Cl_2(aq) + H_2O(l) \rightleftharpoons H^+(aq) + Cl^-(aq) + HOCl(aq)$$

In the forward reaction (the reaction going from left to right), which substance is oxidised and which is reduced? In the reverse reaction, which is the oxidising agent and which is the reducing agent?

12.64 Nitrogen dioxide, NO_2, is a pollutant in smog. The gas has is responsible for the red-brown colour associated with this type of air pollution. Nitrogen dioxide is also a contributor to acid rain because, when rain passes through air contaminated with NO_2, it dissolves and undergoes the following reaction. **LO1**

$$3NO_2(g) + H_2O(l) \rightarrow NO(g) + 2H^+(aq) + 2NO_3^-(aq)$$

In this reaction, which element is reduced and which is oxidised?

12.65 Chlorine dioxide, ClO_2, is used to kill bacteria in the dairy, meat and other food and beverage industries. It is unstable, but can be made by the following reaction. **LO1**

$$Cl_2 + 2NaClO_2 \rightarrow 2ClO_2 + 2NaCl$$

Label the oxidation state of each element and identify the substances oxidised and reduced as well as the oxidising and reducing agents in the reaction.

12.66 Balance the following equations for reactions occurring in an acidic aqueous solution. **LO2**

(a) $S_2O_3^{2-} + OCl^- \rightarrow Cl^- + S_4O_6^{2-}$
(b) $NO_3^- + Cu \rightarrow NO_2 + Cu^{2+}$
(c) $IO_3^- + AsO_3^{3-} \rightarrow I^- + AsO_4^{3-}$
(d) $SO_4^{2-} + Zn \rightarrow Zn^{2+} + SO_2$
(e) $NO_3^- + Zn \rightarrow NH_4^+ + Zn^{2+}$
(f) $Cr^{3+} + BiO_3^- \rightarrow Cr_2O_7^{2-} + Bi^{3+}$
(g) $I_2 + OCl^- \rightarrow IO_3^- + Cl^-$
(h) $Mn^{2+} + BiO_3^- \rightarrow MnO_4^- + Bi^{3+}$
(i) $H_3AsO_3 + Cr_2O_7^{2-} \rightarrow H_3AsO_4 + Cr^{3+}$
(j) $I^- + HSO_4^- \rightarrow I_2 + SO_2$
(k) $Sn + NO_3^- \rightarrow SnO_2 + NO$
(l) $PbO_2 + Cl^- \rightarrow PbCl_2 + Cl_2$
(m) $Ag + NO_3^- \rightarrow NO_2 + Ag^+$
(n) $Fe^{3+} + NH_3OH^+ \rightarrow Fe^{2+} + N_2O$
(o) $HNO_2 + I^- \rightarrow I_2 + NO$
(p) $C_2O_4^{2-} + HNO_2 \rightarrow CO_2 + NO$
(q) $HNO_2 + MnO_4^- \rightarrow Mn^{2+} + NO_3^-$
(r) $H_3PO_2 + Cr_2O_7^{2-} \rightarrow H_3PO_4 + Cr^{3+}$
(s) $VO_2^+ + Sn^{2+} \rightarrow VO^{2+} + Sn^{4+}$
(t) $XeF_2 + Cl^- \rightarrow Xe + F^- + Cl_2$
(u) $OCl^- + S_2O_3^{2-} \rightarrow Cl^- + SO_4^{2-}$

12.67 Balance equations for the following reactions occurring in a basic aqueous solution. **LO2**

(a) $CrO_4^{2-} + S^{2-} \rightarrow S + CrO_2^-$
(b) $MnO_4^- + C_2O_4^{2-} \rightarrow CO_2 + MnO_2$
(c) $ClO_3^- + N_2H_4 \rightarrow NO + Cl^-$
(d) $NiO_2 + Mn(OH)_2 \rightarrow Mn_2O_3 + Ni(OH)_2$
(e) $SO_3^{2-} + MnO_4^- \rightarrow SO_4^{2-} + MnO_2$
(f) $CrO_2^- + S_2O_8^{2-} \rightarrow CrO_4^{2-} + SO_4^{2-}$
(g) $SO_3^{2-} + CrO_4^{2-} \rightarrow SO_4^{2-} + CrO_2$
(h) $O_2 + N_2H_4 \rightarrow H_2O_2 + N_2$
(i) $Fe(OH)_2 + O_2 \rightarrow Fe(OH)_3 + OH^-$
(j) $Au + CN^- + O_2 \rightarrow [Au(CN)_4]^- + OH^-$

12.68 Ozone, O_3, is a very powerful oxidising agent, and can be used to treat water to kill bacteria and make it safe to drink. One of the problems with this method of purifying water is that any bromide ions (Br^-) in the water are oxidised to bromate ions, BrO_3^-, which have shown evidence of causing cancer in test animals. Assuming that ozone is reduced to water, write a balanced chemical equation for the reaction acidic solution. **LO2**

12.69 On the basis of the discussions in this chapter, suggest chemical equations for the oxidation of metallic silver to Ag^+ ions with: (a) dilute HNO_3 and (b) concentrated HNO_3. **LO2**

12.70 Hot and concentrated, sulfuric acid is a fairly strong oxidising agent. Write a balanced chemical equation for the oxidation of metallic copper to copper(II) ion by hot, concentrated H_2SO_4, in which the sulfur is reduced to SO_2. **LO2**

12.71 Use table 12.1 to predict the outcome of the following reactions in aqueous solution. If no reaction occurs, write NR. If a reaction occurs, write a balanced chemical equation for it. **LO2**

(a) $Li^+ + Cr \rightarrow$
(b) $Cr + Pb^{2+} \rightarrow$
(c) $Ag^+ + Fe \rightarrow$
(d) $Ag + Au^{3+} \rightarrow$
(e) $Mn + Fe^{2+} \rightarrow$
(f) $Cl^- + Co \rightarrow$
(g) $Mg + Co^{2+} \rightarrow$
(h) $Cr + Sn^{2+} \rightarrow$
(i) $Ni^{2+} + Na \rightarrow$
(j) $Cr + H^+ \rightarrow$
(k) $Pb + Cd^{2+} \rightarrow$
(l) $Mn + Pb^{2+} \rightarrow$
(m) $Zn + Co^{2+} \rightarrow$
(n) $K + H_2O \rightarrow$

12.72 Write the half-equations and the balanced cell reaction for each of the following galvanic cells. **LO3**

(a) $Cd(s) | Cd^{2+}(aq) || Au^{3+}(aq) | Au(s)$
(b) $Pb(s), PbSO_4(s) | HSO_4^-(aq) || H^+(aq), HSO_4^-(aq) | PbO_2(s), PbSO_4(s)$
(c) $Cr(s) | Cr^{3+}(aq) || Cu^{2+}(aq) | Cu(s)$
(d) $Mg(s) | Mg^{2+}(aq) || Al^{3+}(aq) | Al(s)$
(e) $Fe(s) | Fe^{2+}(aq) || Br_2(aq), Br^-(aq) | Pt(s)$
(f) $C(s) | Cr^{2+}(aq), Cr^{3+}(aq) || Ti^{3+}(aq), Ti^+(aq) | C(s)$

12.73 Write the cell diagram for a cell with a standard hydrogen electrode as anode and a cathode consisting of platinum wire dipped into an electrolyte solution of 0.5 M Cu^{2+} and 0.1 M Cu^+. The two half-cells are connected by a salt bridge.

$$Pt(s) | H_2(1 \times 10^5\ Pa) | H^+(1\ M) || Cu^{2+}(0.5\ M), Cu^+(0.1\ M) | Pt(s)$$

12.74 Write the cell diagrams for the following galvanic cells. For half-reactions in which all the reactants are in solution or are gases, assume the use of inert platinum electrodes. **LO3**

(a) $Cd^{2+}(aq) + Fe(s) \rightarrow Cd(s) + Fe^{2+}(aq)$
(b) $Cl_2(g) + 2Br^-(aq) \rightarrow Br_2(aq) + 2Cl^-(aq)$
(c) $Au^{3+}(aq) + 3Ag(s) \rightarrow Au(s) + 3Ag^+(aq)$
(d) $NO_3^-(aq) + 4H^+(aq) + 3Fe^{2+}(aq) \rightarrow 3Fe^{3+}(aq) + NO(g) + 2H_2O(l)$
(e) $NiO_2(s) + 4H^+(aq) + 2Ag(s) \rightarrow Ni^{2+}(aq) + 2H_2O(l) + 2Ag^+(aq)$
(f) $Mg(s) + Cd^{2+}(aq) \rightarrow Mg^{2+}(aq) + Cd(s)$

12.75 Use the data in table 12.1 to determine which of the following reactions should occur spontaneously under standard conditions in aqueous solution at 25 °C. **LO3**

(a) $2Au^{3+} + 6I^- \rightarrow 3I_2 + 2Au$
(b) $3Fe^{2+} + 2NO + 4H_2O \rightarrow 3Fe + 2NO_3^- + 8H^+$
(c) $3Ca + 2Cr^{3+} \rightarrow 2Cr + 3Ca^{2+}$
(d) $Br_2 + 2Cl^- \rightarrow Cl_2 + 2Br^-$
(e) $Ni^{2+} + Fe \rightarrow Fe^{2+} + Ni$
(f) $H_2SO_3 + H_2O + Br_2 \rightarrow 4H^+ + SO_4^{2-} + 2Br^-$
(g) $SO_4^{2-} + 4H^+ + 2I^- \rightarrow H_2SO_3 + I_2 + H_2O$

12.76 Write cell diagrams using the following pairs of half-reactions. For half-reactions in which all the reactants are in solution, assume the use of inert platinum electrodes. Calculate their standard cell potentials. For each pair, identify the anode and cathode. **LO4**

(a) $BrO_3^-(aq) + 6H^+(aq) + 6e^- \rightleftharpoons Br^-(aq) + 3H_2O(l)$
$Cu^{2+}(aq) + 2e^- \rightleftharpoons Cu(s)$
(b) $Fe^{3+}(aq) + e^- \rightleftharpoons Fe^{2+}(aq)$
$Ag^+(aq) + e^- \rightleftharpoons Ag(s)$
(c) $NO_3^-(aq) + 4H^+(aq) + 3e^- \rightleftharpoons NO(g) + 2H_2O(l)$
$MnO_4^-(aq) + 8H^+(aq) + 5e^- \rightleftharpoons Mn^{2+}(aq) + 4H_2O(l)$

12.77 Determine the spontaneous reaction between H_2SO_3, $S_2O_3^{2-}$, HOCl and Cl_2. The half-equations involved are: **LO4**

$$2H_2SO_3 + 2H^+ + 4e^- \rightleftharpoons S_2O_3^{2-} + 3H_2O \quad E^{\ominus}_{H_2SO_3/S_2O_3} = +0.40\ V$$

$$2HOCl + 2H^+ + 2e^- \rightleftharpoons Cl_2 + 2H_2O \quad E^{\ominus}_{HOCl/Cl_2} = +1.63\ V$$

12.78 Calculate $\Delta G^{\ominus}$ for the reaction: **LO5**

$$2Al(s) + 3Br_2(l) \rightarrow 2Al^{3+} + 6Br^-(aq)$$

for which $E^{\ominus}_{cell} = 2.741\ V$.

12.79 Given the following half-equations and their standard reduction potentials: **LO5**

$$2ClO_3^- + 12H^+ + 10e^- \rightleftharpoons Cl_2 + 6H_2O \quad E^{\ominus}_{ClO_3^-/Cl_2} = +1.47\,V$$
$$S_2O_8^{2-} + 2e^- \rightleftharpoons 2SO_4^{2-} \quad E^{\ominus}_{S_2O_8^{2-}/SO_4^{2-}} = +2.01\,V$$

calculate (a) $E^{\ominus}_{cell}$, (b) $\Delta G^{\ominus}$ for the cell reaction and (c) K_c for the cell reaction.

12.80 Calculate K_c for the following reactions, using the data in table 12.1. Assume $T = 298$ K. **LO5**

(a) $Ni^{2+} + Co \rightleftharpoons Ni + Co^{2+}$

(b) $2H_2O + 2Cl_2 \rightleftharpoons 4H^+ + 4Cl^- + O_2$

12.81 The cell reaction: **LO5**

$$NiO_2(s) + 4H^+(aq) + 2Ag(s) \rightarrow Ni^{2+}(aq) + 2H_2O(l) + 2Ag^+(aq)$$

has $E^{\ominus}_{cell} = +2.48$ V. What will the cell potential be at a pH of 3.5 when the concentration of $Ni^{2+} = 0.015$ M and of $Ag^+ = 0.1$ M?

12.82 A cell having the following cell-reaction was set up. **LO5**

$$Mg(s) + Cd^{2+}(aq) \rightarrow Mg^{2+}(aq) + Cd(s) \quad E^{\ominus}_{cell} = +1.97\,V$$

The magnesium electrode was dipped into a 1.00 M solution of $MgSO_4$ and the cadmium electrode was dipped into a solution of unknown Cd^{2+} concentration. The potential of the cell was measured to be +1.54 V. What was the unknown Cd^{2+} concentration?

12.83 A silver wire coated with AgCl is sensitive to the presence of chloride ions because of the half-cell reaction. **LO3**

$$AgCl(s) + e^- \rightleftharpoons Ag(s) + Cl^-$$

A student, wishing to measure the chloride ion concentration in a number of water samples, constructed a galvanic cell using the AgCl electrode as one half-cell and a copper wire dipped into a 1.00 M $CuSO_4$ solution as the other half-cell. In one analysis, the potential of the cell was measured to be +0.0925 V, with the copper half-cell serving as the cathode. What was the chloride ion concentration in the water, assuming a temperature of 25 °C? Use the data in table 12.1.

12.84 Suppose a galvanic cell was constructed at 25 °C using a Cu^{2+}/Cu half-cell (in which the molar concentration of Cu^{2+}/Cu was 1.00 M) and a hydrogen electrode having a partial pressure of H_2 equal to 1.0×10^5 Pa. The hydrogen electrode dipped into a solution of unknown hydrogen ion concentration, and the two half-cells were connected by a salt bridge. Use the data in table 12.1. **LO3**

(a) Derive an equation for the pH of the solution with the unknown hydrogen ion concentration, expressed in terms of E_{cell} and $E^{\ominus}_{cell}$. You will need to use the equation:

$$\ln x = 2.303 \log x$$

to answer this question.

(b) If the pH of the solution was 5.15, what would be the observed potential of the cell?

(c) If the potential of the cell was 0.645 V, what would be the pH of the solution?

12.85 How many coulombs are passed through an electrolysis cell by: (a) a current of 4.0 A for 600 s, (b) a current of 10.0 A for 20 min and (c) a current of 1.5 A for 6 h? What amount of electrons corresponds to the answers to each part of this question? **LO3**

12.86 What amount of Cr^{3+} would be reduced to Cr by the same amount of electricity that produces 12.0 g Ag from a solution of $AgNO_3$? If a current of 4.0 A was used, how many minutes would the electrolysis take? **LO7**

12.87 What mass of $Fe(OH)_2$ is produced at an iron anode when a basic solution undergoes electrolysis at a current of 8.0 A for 12.0 min? **LO7**

12.88 How long must a current of 3.0 A be applied to a solution of Ag^+ to produce 10.5 g of silver metal? **LO7**

12.89 How long would it take to produce 75.0 g of metallic chromium by the electrolytic reduction of Cr^{3+} with a current of 2.25 A? **LO7**

12.90 How many grams of Cd are consumed in a nickel-cadmium battery if it operates at a constant current of 0.25 A for 15 s? **LO7**

12.91 How many amperes would be needed to produce 60.0 g of magnesium during the electrolysis of molten $MgCl_2$ in 2.0 hours? **LO7**

12.92 What volume of dry gaseous H_2, measured at 20 °C and 1.0×10^5 Pa, would be produced at the cathode in the electrolysis of dilute H_2SO_4 with a current of 0.75 A for 15.0 min? **LO7**

12.93 Write the anode reaction for the electrolysis of an aqueous solution that contains: (a) SO_4^{2-}, (b) Br^- and (c) SO_4^{2-} and Br^-. **LO7**

12.94 What products would we expect at the electrodes if a solution containing: (a) both KBr and $CuSO_4$ and (b) both $BaCl_2$ and CuI_2 were electrolysed? Write the equations for the respective net cell reactions. **LO7**

ADDITIONAL EXERCISES

12.95 The following chemical reactions are observed to occur in aqueous solution. **LO1**

$$2Al + 3Cu^{2+} \rightarrow 2Al^{3+} + 3Cu$$
$$2Al + 3Fe^{2+} \rightarrow 3Fe + 2Al^{3+}$$
$$Pb^{2+} + Fe \rightarrow Pb + Fe^{2+}$$
$$Fe + Cu^{2+} \rightarrow Fe^{2+} + Cu$$
$$2Al + 3Pb^{2+} \rightarrow 3Pb + 2Al^{3+}$$
$$Pb + Cu^{2+} \rightarrow Pb^{2+} + Cu$$

Arrange the metals Al, Pb, Fe and Cu in order of increasing ease of oxidation. Were all the experiments described actually necessary to establish the order?

12.96 In each pair below, choose the metal that would most likely react more rapidly with a nonoxidising acid such as HCl: (a) aluminium or iron, (b) zinc or cobalt, (c) cadmium or magnesium. **LO1**

12.97 Lead(IV) oxide reacts with hydrochloric acid to give chlorine. The equation for the reaction is: **LO2**

$$PbO_2 + 4Cl^- + 4H^+ \rightarrow PbCl_2 + 2H_2O + Cl_2$$

What mass of PbO_2 must react to give 15.0 g of Cl_2?

12.98 Biodiesel is formed from the reaction of oils with methanol or ethanol. One of the products is methyl octanoate, $C_9H_{18}O_2$,

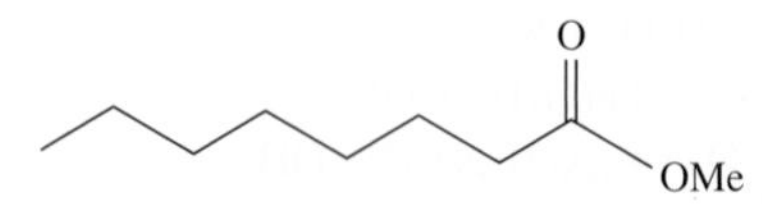

which burns completely in a diesel engine. Give a balanced equation for this reaction. If the density of methyl octanoate is 0.877 g mL^{-1}, what mass of CO_2 will be formed from the burning of 1 litre of methyl octanoate? **LO5**

12.99 Suppose that a galvanic cell was set up having the net cell reaction: **LO3**

$$Zn(s) + 2Ag^+(aq) \rightarrow Zn^{2+}(aq) + 2Ag(s)$$

The Ag^+ and Zn^{2+} concentrations in their respective half-cells are initially 1.00 M, and each half-cell contains 100 mL of electrolyte solution. If this cell delivers current at a constant rate of 0.1 A, what will the cell potential be after 15.0 h?

12.100 To determine the reduction potential of the Pt^{2+} ion, a galvanic cell was set up in which one half-cell consisted of a

Pt electrode dipped into a 0.01 M solution of $Pt(NO_3)_2$ and the other of a silver wire coated with AgCl dipped into a 0.10 M solution of HCl. The potential of the cell was measured to be 0.778 V, and it was found that the Pt electrode carried a positive charge. Given the following half-reaction and its reduction potential: **LO 3**

$$AgCl(s) + e^- \rightleftharpoons Ag(s) + Cl^-(aq) \qquad E^{\ominus}_{AgCl/Ag} = +0.23\ V$$

calculate the standard reduction potential for the half-reaction:

$$Pt^{2+}(aq) + 2e^- \rightleftharpoons Pt(s)$$

12.101 A student set up an electrolysis apparatus and passed a current of 1.22 A through a 3 M H_2SO_4 solution for 30.0 min. The H_2 formed at the cathode was collected and found to have a volume, over water at 27 °C, of 288 mL at a total pressure of 1.0×10^5 Pa. Use these data to calculate the charge on the electron, expressed in coulombs. **LO 7**

12.102 A hydrogen electrode is immersed in a 0.10 M solution of acetic acid at 25 °C. This electrode is connected to another consisting of an iron nail dipped into 0.10 M $FeCl_2$. What will be the measured potential of this cell? Assume $p_{H_2} = 1.0 \times 10^5$ Pa. **LO 4**

12.103 Consider the following half-reactions at 25 °C. **LO 4**

$$MnO_4^- + 8H^+ + 5e^- \rightarrow Mn^{2+} + 4H_2O$$
$$ClO_3^- + 6H^+ + 6e^- \rightarrow Cl^- + 3H_2O$$

(a) Use the data in table 12.1 to determine the value of:
  (i) $E^{\ominus}_{cell}$
  (ii) $\Delta G^{\ominus}$ for this reaction
  (iii) K_c for this reaction at 25 °C.
(b) Write the Nernst equation for this reaction.
(c) What is the potential of the cell when $[MnO_4^-] = 0.20$ M, $[Mn^{2+}] = 0.050$ M, $[Cl^-] = 0.0030$ M, $[ClO_3^-] = 0.110$ M and the pH of the solution is 4.25?

12.104 A solution of NaCl is neutral, with a pH of 7 at 25 °C. If electrolysis is carried out on 500 mL of an NaCl solution with a current of 0.5 A, how long would it take for the pH of the solution to rise to a value of 9? **LO 4**

12.105 Consider the following galvanic cell. **LO 3**

$$Ag(s)\,|\,Ag^+(3.0 \times 10^{-4}\ M)\,||\,Fe^{3+}(1.1 \times 10^{-3}\ M),\ Fe^{2+}(0.040\ M)\,|\,Pt(s)$$

Calculate the cell potential. Determine the sign of the electrodes in the cell. Write the equation for the spontaneous cell reaction.

12.106 Consider a galvanic cell involving chromium(III) and gold(III) and the corresponding Cr and Au electrodes. (Use the data in table 12.1.) **LO 3**
(a) Sketch this cell.
(b) Identify the anode and the cathode.
(c) Write the balanced overall reaction.
(d) Which electrode will lose mass?
(e) Indicate the direction of the electron flow.
(f) Indicate the flow of the anions and cations.
(g) What is the potential of this cell under standard conditions?
(h) Write the shorthand notation.
(i) Calculate $\Delta G^{\ominus}$ for this reaction.

12.107 The silvery look of 'white gold' results from plating with rhodium (Rh). How many coulombs of electricity must be pumped through an rhodium(III) solution to plate 1 gram of solid rhodium? **LO 7**

12.108 Calculate the quantity of electricity (coulombs) necessary to deposit 100.0 g of copper from a $CuSO_4$ solution. **LO 7**

12.109 How many minutes will it take to plate out 30.0 g of Ni from a solution of $NiSO_4$ using a current of 3.45 A? **LO 7**

12.110 A constant electric current deposits 0.365 g of silver metal in 12 960 seconds from a solution of silver nitrate. Calculate the current. What is the half-equation for the deposition of silver? **LO 7**

13 Transition metal chemistry

Australia is one of the world's main sources of rubies and sapphires. These precious gemstones are prized for their beautiful colours; rubies are always red, whereas sapphires can exhibit a range of colours. Remarkably, both rubies and sapphires are composed primarily of colourless aluminium oxide, Al_2O_3, and their colours are due to the presence of small amounts of transition metal ions — chromium in rubies, and titanium, nickel and iron in sapphires. The colours that these transition metals impart result from both the energies of the *d* orbitals of the transition metal ions and how the *d* electrons are arranged in these orbitals.

In this chapter, we will investigate what makes transition metals unique among the elements of the periodic table with respect to the colours and magnetic behaviours of their compounds.

LEARNING OBJECTIVES

After studying this chapter, you should be able to:

13.1 outline the properties of metals

13.2 describe the special features of transition metals

13.3 detail the structures and properties of common ligands

13.4 describe the structural and bonding characteristics of transition metal complexes

13.5 explain how transition metals are used in biological systems

13.6 detail the processes used in the isolation and purification of transition metals

13.7 outline some applications of transition metals.

13.1 Metals in the periodic table

In chapter 1, we learned that the chemical elements can be divided into three groups: metals, metalloids and nonmetals. Of the 118 known elements, 91 or 92 are metals (polonium is variously described as either a metal or metalloid). Metals can be classified into different types on the periodic table, generally according to the nature of their valence orbitals (figure 13.1).

Periodic table of the elements

main-group metals; transition metals; lanthanoids; metalloids; nonmetals; actinoids

Key: atomic number / Symbol / standard atomic mass

Group labels: 1 alkali metals; 2 alkaline earth metals; 15 pnictogens; 16 chalcogens; 17 halogens; 18 noble gases

	1	2	3	4	5	6	7	8	9	10	11	12	13	14	15	16	17	18
1	1 H [1.008]																	2 He 4.003
2	3 Li [6.94]	4 Be 9.012											5 B [10.81]	6 C [12.01]	7 N [14.01]	8 O [16.00]	9 F 19.00	10 Ne 20.18
3	11 Na 22.99	12 Mg [24.31]											13 Al 26.98	14 Si [28.09]	15 P 30.97	16 S [32.06]	17 Cl [35.45]	18 Ar 39.95
4	19 K 39.10	20 Ca 40.08	21 Sc 44.96	22 Ti 47.87	23 V 50.94	24 Cr 52.00	25 Mn 54.94	26 Fe 55.85	27 Co 58.93	28 Ni 58.69	29 Cu 63.55	30 Zn 65.38(2)	31 Ga 69.72	32 Ge 72.63	33 As 74.92	34 Se 78.97	35 Br [79.90]	36 Kr 83.80
5	37 Rb 85.47	38 Sr 87.62	39 Y 88.91	40 Zr 91.22	41 Nb 92.91	42 Mo 95.95	43 Tc 97.91	44 Ru 101.1	45 Rh 102.9	46 Pd 106.4	47 Ag 107.9	48 Cd 112.4	49 In 114.8	50 Sn 118.7	51 Sb 121.8	52 Te 127.6	53 I 126.9	54 Xe 131.3
6	55 Cs 132.9	56 Ba 137.3	57–71	72 Hf 178.5	73 Ta 180.9	74 W 183.8	75 Re 186.2	76 Os 190.2	77 Ir 192.2	78 Pt 195.1	79 Au 197.0	80 Hg 200.6	81 Tl [204.4]	82 Pb 207.2	83 Bi 209.0	84 Po (209.0)	85 At (210.0)	86 Rn (222.0)
7	87 Fr (223.0)	88 Ra (226.0)	89–103	104 Rf (265.1)	105 Db (268.1)	106 Sg (271.1)	107 Bh (270.1)	108 Hs (277.2)	109 Mt (276.2)	110 Ds (281.2)	111 Rg (280.2)	112 Cn (285.2)	113 Uut (284.2)	114 Fl (289.2)	115 Uup (288.2)	116 Lv (293.2)	117 Uus (294.2)	118 Uuo (294.2)

lanthanoid series	57 La 138.9	58 Ce 140.1	59 Pr 140.9	60 Nd 144.2	61 Pm 144.9	62 Sm 150.4	63 Eu 152.0	64 Gd 157.3	65 Tb 158.9	66 Dy 162.5	67 Ho 164.9	68 Er 167.3	69 Tm 168.9	70 Yb 173.1	71 Lu 175.0
actinoid series	89 Ac 227.0	90 Th 232.0	91 Pa 231.0	92 U 238.0	93 Np 237.0	94 Pu 244.1	95 Am 243.1	96 Cm 247.1	97 Bk 247.1	98 Cf 251.1	99 Es 252.1	100 Fm 257.1	101 Md 258.1	102 No 259.1	103 Lr 262.1

FIGURE 13.1 The metals in the periodic table of the elements. The transition metals are shown in yellow and occupy groups 3 to 12 of the periodic table.

The metals in groups 1 and 2 have *s* valence orbitals, whereas those in groups 13 to 16 have *p* valence orbitals. (The former are sometimes called pre-transition metals and the latter post-transition metals.) Collectively, these metals are known as **main-group metals**. The lanthanoid and actinoid metals are characterised by partially filled *f* orbitals. The lanthanoids, with the exception of promethium, Pm, occur naturally; the majority of the actinoid elements must be prepared by nuclear reactions.

The metals in groups 3 to 12 of the periodic table are collectively called the **transition metals**, and their chemistry contrasts significantly with other metals in the periodic table. For example, if we dissolve any group 1 or 2 metal (carefully!) in nitric acid, we obtain a colourless solution in all cases. However, repeating the experiment using metallic cobalt, nickel or copper produces purple, green and blue solutions, respectively. In fact, we obtain coloured solutions for many of the transition metals; there are some exceptions (silver, zinc, cadmium and mercury, for example), and gold does not dissolve in nitric acid. If we reduce the solutions containing cobalt, nickel or copper to dryness, we find that the resulting solids are **paramagnetic**, meaning that they are attracted into a magnetic field, and, indeed, this is the case for many (but not all) of the transition metals. The corresponding solids obtained from the group 1 and 2 metals exhibit **diamagnetic** behaviour, being very slightly repelled from a magnetic field.

Colour and magnetic behaviour are two of the defining characteristics of compounds containing transition metals. As we saw in chapter 1, transition metal elements are sometimes called the *d*-block elements, because the valence orbitals of these elements are *d* orbitals, and we will see that many of the spectral and magnetic properties of transition metal compounds can be related to the energies of the five *d* orbitals of the transition metal ion.

Transition metals are distinguished by their ability to form **complexes**. In these chemical species, transition metal ions (usually positively charged) are surrounded by one or more ions (usually negatively charged) or neutral molecules, which are called **ligands**. When writing the formula of a transition metal complex, we enclose the metal and the bound ligands in square brackets (the metal ion first, followed by the ligands in alphabetical order), with the overall charge (if any) superscripted outside the brackets.

The $[Co(OH_2)_6]^{2+}$ ion is a simple example of a transition metal complex, in which a single cobalt ion is bonded to six water molecules. The structure of this pink complex, which results from the dissolution of metallic cobalt in nitric acid as described on the previous page, is shown in figure 13.2a. In this complex, the cobalt ion has an oxidation state of +2, and is surrounded by six H_2O ligands to give an octahedral geometry about the metal ion.

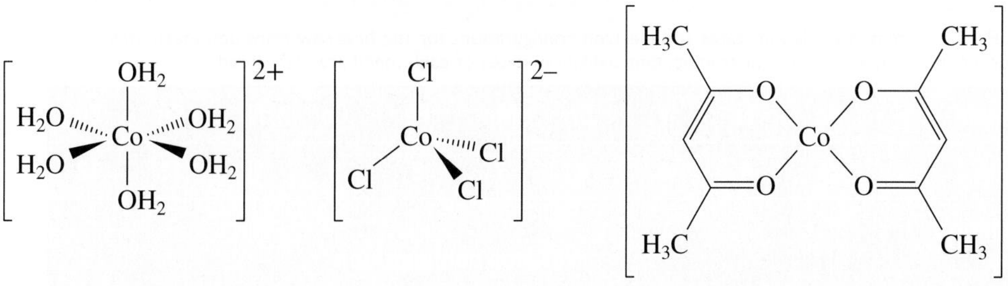

FIGURE 13.2 The structures of three transition metal complexes in which the cobalt ion has an oxidation state of +2: **(a)** the $[Co(OH_2)_6]^{2+}$ complex cation, **(b)** the $[CoCl_4]^{2-}$ complex anion and **(c)** the neutral complex $[Co(acac)_2]$.

If we dissolve metallic cobalt in hydrochloric acid, rather than nitric acid, we obtain a deep blue solution, due to the formation of the $[CoCl_4]^{2-}$ complex anion. The cobalt ion again has an oxidation state of +2, but in this case is surrounded by four Cl^- ligands in a tetrahedral arrangement (figure 13.2b).

Finally, if we react a suitable cobalt salt with the acetylacetonate (acac) ligand, we obtain the neutral complex $[Co(acac)_2]$. As for $[Co(OH_2)_6]^{2+}$ and $[CoCl_4]^{2-}$, the cobalt ion has an oxidation state of +2, but here it is bonded to four oxygen atoms of two acac ligands (figure 13.2c). The geometry around the cobalt ion is square planar, and the complex is green.

The examples above give some idea of the wide range of structures, charges and colours of transition metal complexes. Most importantly, you can see that:

- transition metal complexes can be positively or negatively charged, or neutral
- the number of ligands around the central metal ion can vary
- there are a number of different possible geometries of the ligands around the central metal ion.

In addition, as we will see, there is usually a variety of oxidation states available to the transition metal ion, and transition metal complexes can often contain more than one metal ion.

Transition metal complexes are essential to many biological processes; for example, oxygen is transported around our bodies in a complex containing iron, a zinc complex ensures that we can process potentially toxic CO_2, and a manganese complex in green plants facilitates the conversion of water to oxygen. Transition metal complexes find extensive use in industrial processes, often being used as catalysts for the formation of plastics, bulk chemicals and pharmaceuticals. They are also used increasingly as drugs in chemotherapy and photodynamic therapy.

As we first saw in section 11.8, transition metal complexes are formed by a Lewis acid – Lewis base interaction between a transition metal cation and one or more ligands, with each ligand formally donating one or more electron pairs to the metal ion. Therefore, we will investigate the two component parts of a complex, the transition metal and the ligand, before we consider transition metal complexes in detail. Note that transition metals are not unique in their ability to form complexes, and other Lewis acidic metal ions and main-group elements in the periodic table can also be surrounded by one or more ligands to form complexes. However, we concentrate on transition metal complexes in this chapter because of their extremely unusual properties mentioned previously. Any use of the term 'complex' in this chapter therefore refers to a transition metal complex.

13.2 Transition metals

As we saw in chapter 4, the transition metals are characterised by d valence orbitals, so the neutral atoms generally have valence electron configurations of $(n + 1)s^2nd^{(x-2)}$, where x is the group number of the metal in the periodic table, and n is the principal quantum number. The group 4 element titanium, for example, has the valence configuration $4s^23d^2$. Recall from chapter 4, however, that there are exceptions to this general rule, which initially appear in the first-row ($3d$)

elements Cr($4s^13d^5$) and Cu($4s^13d^{10}$). More important for transition metal complexes are the electron configurations of the ions. In transition metal complexes, the nd orbitals are always of lower energy than the $(n + 1)s$ orbitals; these complexes therefore always have vacant $(n + 1)s$ orbitals. Knowledge of the electron configuration of transition metal ions is important as another of the defining characteristics of transition metals is their ability, in contrast to most other metals in the periodic table, to exist in a variety of oxidation states, the values of which, for transition metal ions, we denote using Roman numerals. For example, the oxidation state of iron in the majority of its complexes is either Fe(II) or Fe(III), while complexes containing manganese in every oxidation state from Mn(0) to Mn(VII) are known. Table 13.1 shows the common oxidation states encountered for the first-row transition metal ions, along with their corresponding electron configurations.

TABLE 13.1 Common oxidation states and electron configurations for the first-row transition metal ions in transition metal complexes. The most important oxidation states of each metal are highlighted.

Element	Sc	Ti	V	Cr	Mn	Fe	Co	Ni	Cu	Zn
Group	3	4	5	6	7	8	9	10	11	12
Oxidation state	Valence configuration									
I			d^4	d^5	d^6	d^7	d^8	d^9	d^{10}	
II		d^2	d^3	$\boxed{d^4}$	$\boxed{d^5}$	$\boxed{d^6}$	$\boxed{d^7}$	$\boxed{d^8}$	$\boxed{d^9}$	$\boxed{d^{10}}$
III	$\boxed{d^0}$	d^1	d^2	$\boxed{d^3}$	$\boxed{d^4}$	$\boxed{d^5}$	$\boxed{d^6}$	d^7	d^8	
IV		$\boxed{d^0}$	d^1	d^2	$\boxed{d^3}$	d^4	d^5	d^6		
V			$\boxed{d^0}$	d^1	d^2	d^3	d^4			
VI				$\boxed{d^0}$	d^1	d^2				
VII					$\boxed{d^0}$					

WORKED EXAMPLE 13.1

Determining *d*-electron configurations

What are the *d*-electron configurations of the transition metal ions Mn^{2+}, Cu^{2+}, Co^{3+}, Ti^{3+} and Cr^{2+} in transition metal complexes?

Analysis

Although the required *d*-electron configurations can be obtained from table 13.1, you should determine the answers by first calculating the number of electrons in each ion and then using these to fill the orbitals in each ion.

Solution

Each transition metal is in the first row, so they share a common [Ar] core electron configuration for their first 18 electrons. Recall also that the 4*s* orbitals are of higher energy than the 3*d* orbitals in transition metal complexes, so we start filling the 3*d* orbitals first. The numbers of electrons in the neutral atoms are obtained from the atomic numbers, and we remove the appropriate number depending on the charge on the ion. Given that the first 18 electrons are core electrons, any electrons in excess of 18 go directly into the 3*d* orbitals.

- Mn: $Z = 25$, so there are 25 electrons in the neutral atom. Hence, Mn^{2+} has 23 electrons. Mn^{2+} has five *d* electrons (23 – 18), so its electron configuration is [Ar]$3d^5$.
- Cu: $Z = 29$, so there are 29 electrons in the neutral atom. Hence, Cu^{2+} has 27 electrons. Cu^{2+} has nine *d* electrons (27 – 18), so its electron configuration is [Ar]$3d^9$.
- Co: $Z = 27$, so there are 27 electrons in the neutral atom. Hence, Co^{3+} has 24 electrons. Co^{3+} has six *d* electrons (24 – 18), so its electron configuration is [Ar]$3d^6$.
- Ti: $Z = 22$, so there are 22 electrons in the neutral atom. Hence, Ti^{3+} has 19 electrons. Ti^{3+} has one *d* electron (19 – 18), so its electron configuration is [Ar]$3d^1$.
- Cr: $Z = 24$, so there are 24 electrons in the neutral atom. Hence, Cr^{2+} has 22 electrons. Cr^{2+} has four *d* electrons (22 – 18), so its electron configuration is [Ar]$3d^4$.

Is our answer reasonable?

All the answers indicate that the 3*d* orbitals are partially filled, as we would expect for a transition metal complex. Our answers are therefore likely to be correct, barring arithmetical errors.

One quick way of determining *d* electron configurations is to subtract the charge on the ion from the group number of the transition metal. For example, Co is in group 9 of the periodic table, so the Co^{3+} ion has (9 – 3) = 6 *d* electrons in a transition metal complex. Using this method, we can confirm that our answers are correct. This method is particularly useful in determining the *d* electron configurations of third row transition metal ions, as the lanthanoid elements lie between the main-group metals and these transition metals, thereby complicating the core electron count. You should also be aware that the *d* electron configuration of transition metal ions belonging to the same group is the same. For example, Co^{3+}, Rh^{3+} and Ir^{3+} all have a d^6 electron configuration in transition metal complexes.

PRACTICE EXERCISE 13.1

What are the *d*-electron configurations of the transition metal ions V^{2+}, Fe^{2+}, Zn^{2+}, Au^{3+} and W^{3+} in transition metal complexes?

13.3 Ligands

Transition metal ions are Lewis acids (see chapter 11), and therefore act as electron pair acceptors towards one or more ligands when forming metal complexes. The word 'ligand' is derived from the Latin *ligare* meaning 'to bind'. Ligands are Lewis bases and can donate an electron pair to a Lewis acidic transition metal ion. The electron pair is usually a lone pair located on an atom or ion. The atom on which the lone pair is located is called the **donor atom**. Ligands can have one or more donor atoms. The necessity for a lone pair of electrons means that relatively few elements in the periodic table can act as donor atoms; the most common are F, Cl, Br, I, O, S, N and P.

As stated earlier, ligands can be either ions or molecules, and they can be negatively charged, neutral or, in very rare cases, positively charged. Anions that serve as ligands include many simple monatomic ions, such as the halide ions (F^-, Cl^-, Br^-, I^-) and the sulfide ion (S^{2-}), each of which contains four lone pairs of electrons. Common polyatomic anions that are ligands are the nitrite ion (NO_2^-), cyanide ion (CN^-), hydroxide ion (OH^-), thiocyanate ion (SCN^-) and acetate ion (CH_3COO^-) (see table 13.2). The most common neutral molecule that serves as a ligand is water, which contains two lone pairs of electrons on the O atom.

Most of the reactions of transition metal ions in aqueous solutions are actually reactions of complex ions in which the metal is attached to a number of water molecules, depending on the identity of the metal ion. Therefore, the notation $M^{n+}(aq)$ actually refers to a complex ion with the formula $[M(OH_2)_x]^{n+}$. $Cu^{2+}(aq)$, for example, is believed to exist as the complex ion $[Cu(OH_2)_5]^{2+}$ in aqueous solution, but $Co^{2+}(aq)$ forms $[Co(OH_2)_6]^{2+}$ under the same conditions. Another common neutral ligand is ammonia, NH_3, which has one lone pair of electrons on the nitrogen atom. If ammonia is added to an aqueous solution containing the $[Ni(OH_2)_6]^{2+}$ ion, for example, the colour changes almost instantaneously from green to blue as ammonia molecules bind preferentially to the metal ion, displacing water molecules (figure 13.3) according to the equation:

$$[Ni(OH_2)_6]^{2+}(aq) + 6NH_3(aq) \rightarrow [Ni(NH_3)_6]^{2+}(aq) + 6H_2O(l)$$

FIGURE 13.3 Complex ions of nickel. Addition of ammonia to a solution of green $[Ni(OH_2)_6]^{2+}$ ions (left) leads to the formation of blue $[Ni(NH_3)_6]^{2+}$ ions (right).

Ligands that use only one atom to bond to a metal ion are called **monodentate** (literally 'one-toothed') ligands. There are also many ligands that have two or more donor atoms. Collectively they are referred to as **polydentate** ligands. Ligands with two donor atoms are called **bidentate** ligands, and, when they form complexes, both donor atoms can become bound to the same metal ion. The best known example of a bidentate ligand is ethylenediamine, $NH_2CH_2CH_2NH_2$, which is usually abbreviated en (this is a historical name but is used almost exclusively in preference to its IUPAC name, ethane-1,2-diamine). Oxalate ion, $C_2O_4^{2-}$ (ox), is another common bidentate ligand, and the structures of both are shown in figure 13.4.

When bidentate ligands bind to a metal ion, they form **chelate rings** (see figure 13.4). Complexes containing chelate rings are often called **chelate complexes** (the word 'chelate' is derived from the Greek word for 'claw'), while ligands that can form chelate rings are often called chelating ligands. Ligands containing three or more donor atoms can potentially form more than one chelate ring with the same metal ion and such ligands often have particularly high affinities for transition metal ions. An excellent example of this is the polydentate ligand ethylenediaminetetraacetic acid, abbreviated H_4EDTA (figure 13.5, overleaf).

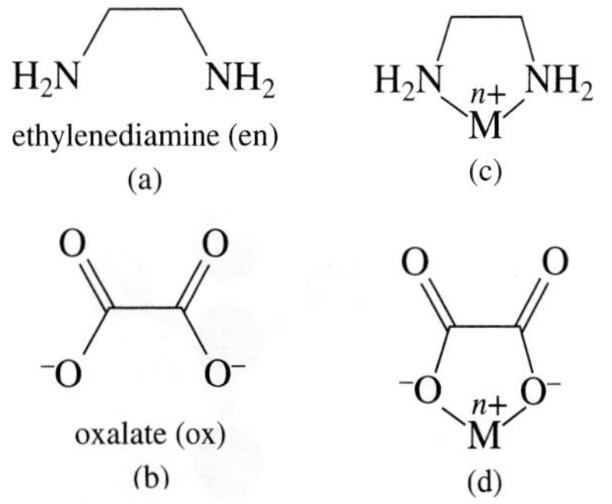

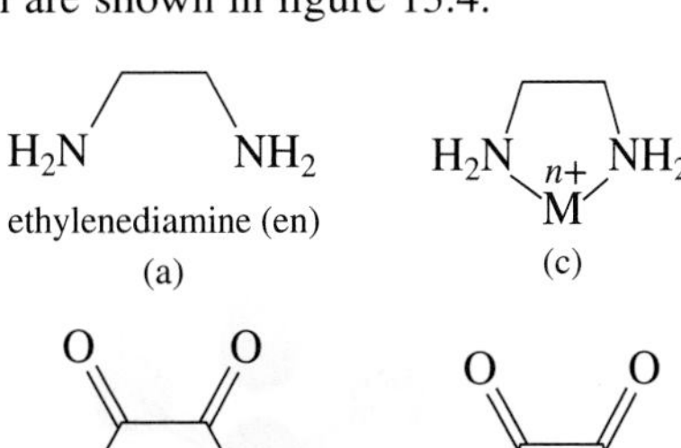

FIGURE 13.4 The structures of the bidentate ligands **(a)** ethylenediamine (en) and **(b)** oxalate (ox), and **(c)** and **(d)** the five-membered chelate rings formed on binding to a metal ion, M.

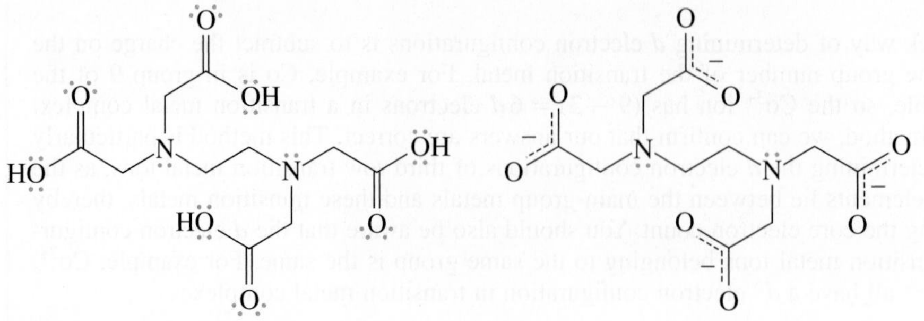

H_4EDTA

(a)

$EDTA^{4-}$

(b)

FIGURE 13.5 Representations of the structures of: **(a)** H_4EDTA and **(b)** the tetraanion $EDTA^{4-}$.

The neutral H_4EDTA molecule loses four protons to become the $EDTA^{4-}$ tetraanion, which is a hexadentate ligand that binds strongly through four O atoms and the two N atoms to a large number of metal ions. $EDTA^{4-}$ is a particularly useful and important ligand. It is relatively non-toxic, which allows it to be used in small amounts in foods to retard spoilage. Many shampoos contain the tetrasodium salt, Na_4EDTA, to soften water; the $EDTA^{4-}$ binds to Ca^{2+}, Mg^{2+} and Fe^{3+} ions, which removes them from the water and prevents them from interfering with the action of soaps in the shampoo. Sometimes, $EDTA^{4-}$ is added in small amounts to whole blood to prevent clotting; it works by binding calcium ions, which the clotting process requires. $EDTA^{4-}$ has even been used as a treatment in cases of poisoning because it can help remove poisonous heavy metal ions such as Pb^{2+} from the body when they have been accidentally ingested. Figure 13.6 shows how the $EDTA^{4-}$ ligand wraps around the metal in a transition metal complex.

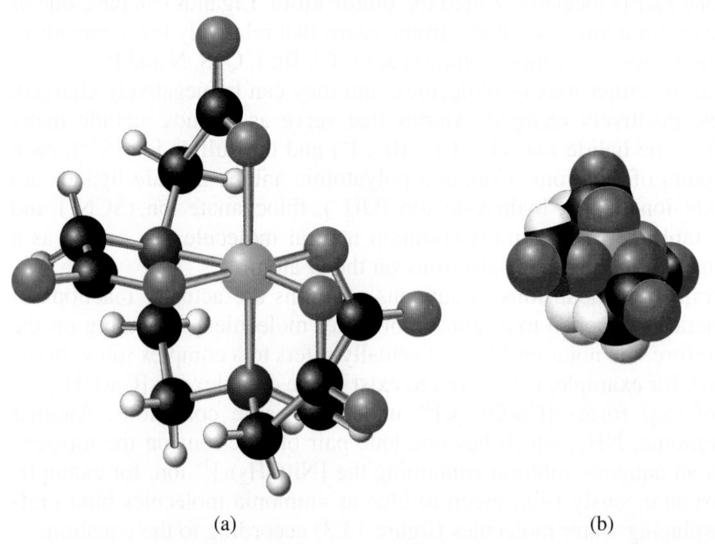

FIGURE 13.6 Two representations of the transition metal complex $[Co(EDTA)]^-$: **(a)** ball-and-stick model and **(b)** space-filling model (the Co(III) ion is orange). Note the almost complete encapsulation of the metal ion in the space-filling model.

Ligands containing two or more donor atoms can potentially act as bridges between two or more metal ions, provided the donor atoms are oriented correctly. An example of this is provided by the copper acetate complex, $[Cu_2(OH_2)_2(OOCCH_3)_4]$, in which the four acetate ligands bridge the two Cu(II) ions, as shown in figure 13.7.

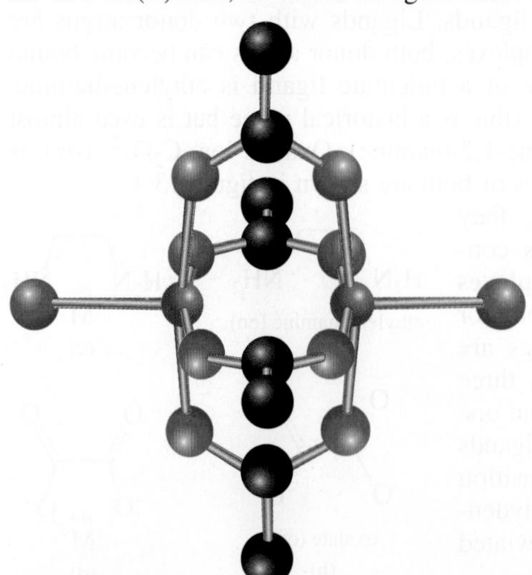

FIGURE 13.7 Ball-and-stick model of $[Cu_2(OH_2)_2(OOCCH_3)_4]$ (the Cu ions are brown). Hydrogen atoms have been omitted for clarity.

There are a vast number of known ligands, and table 13.2 summarises some of the more common and important of those used in the synthesis of transition metal complexes.

TABLE 13.2 Common ligands found in transition metal complexes. Rules for naming ligands are found in section 13.4.

Type of ligand	Ligand name[a]	Formula or abbreviation	Structure	Donor atom(s)
monodentate	bromido	Br^-	$:\ddot{Br}:^-$	Br
	chlorido	Cl^-	$:\ddot{Cl}:^-$	Cl
	fluorido	F^-	$:\ddot{F}:^-$	F
	iodido	I^-	$:\ddot{I}:^-$	I
	cyanido	CN^-	$:N{\equiv}C:^-$	C
	carbonyl	CO	$:O{\equiv}C:$	C
	ammine[b]	NH_3	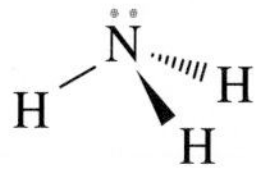 $\ddot{N}H_3$ (N with H, H, H)	N
	nitrito-*κN*	NO_2^-	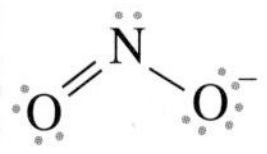 $O{=}\ddot{N}{-}O^-$	N
	pyridine	C_5H_5N, py	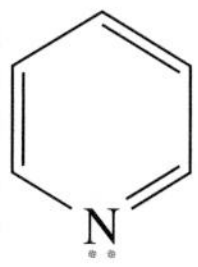 (pyridine ring, N)	N
	thiocyanato-*κN*	NCS^-	$:\ddot{S}{=}C{=}\ddot{N}:^-$	N
	thiocyanato-*κS*	NCS^-	$:\ddot{S}{=}C{=}\ddot{N}:^-$	S
	aqua	H_2O	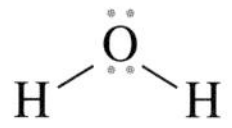 $H{-}\ddot{O}{-}H$	O
	oxido	O^{2-}	$:\ddot{O}:^{2-}$	O
	hydroxido	OH^-	$H{-}\ddot{O}:^-$	O
	nitrito–*κO*	NO_2^-	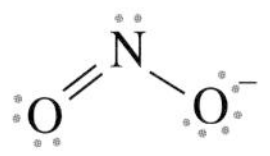 $O{=}\ddot{N}{-}O^-$	O
	hydrido	H^-	$H:^-$	H
	trimethylphosphane[c]	$P(CH_3)_3$, PMe_3	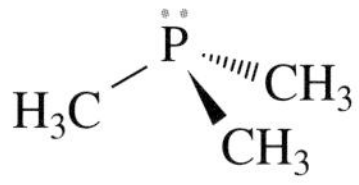 $H_3C{-}\ddot{P}(CH_3)CH_3$	P
bidentate	ethylenediamine	en	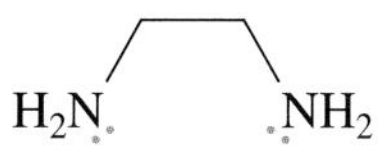 $H_2\ddot{N}{-}CH_2CH_2{-}\ddot{N}H_2$	N, N
	bipyridine	bipy	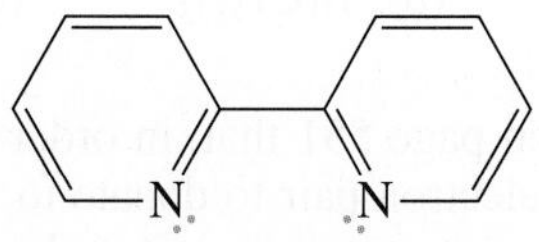 (two linked pyridine rings, N N)	N, N

(continued)

TABLE 13.2 *(continued)*

Type of ligand	Ligand name	Formula or abbreviation	Structure	Donor atom(s)
bidentate	glycinato(d)	gly	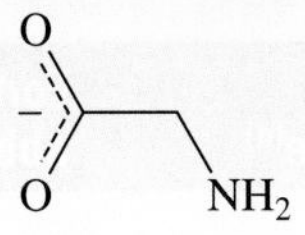	N, O
	oxalato	ox	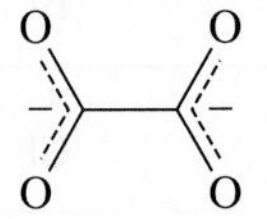	O, O
	acetato	OAc	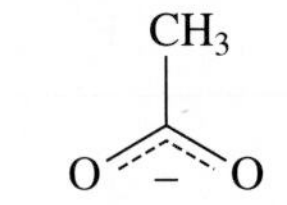	O, O
	acetylacetonato	acac	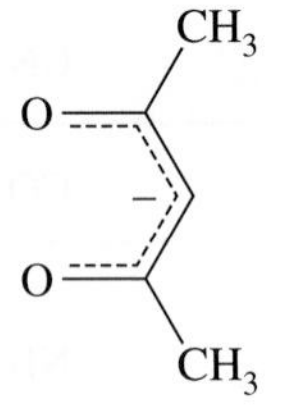	O, O
tetradentate	porphyrin-21, 23-diido		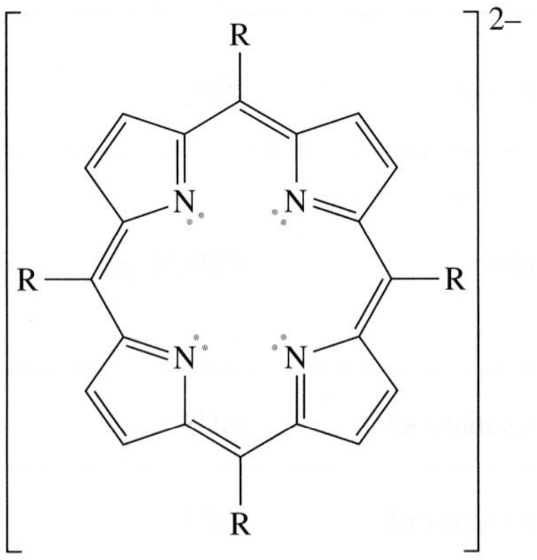	N, N, N, N
hexadentate	ethylenediaminetetraacetato	EDTA	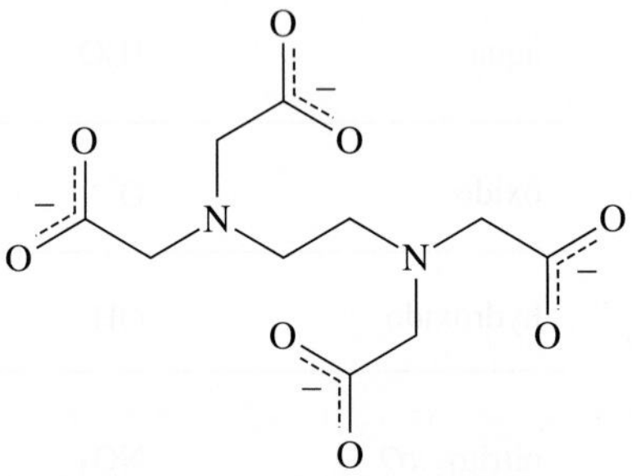	N, N, O, O, O, O

(a) Note that ligands are named according to the rules given on pp. 562–3.

(b) Amines (derivatives of NH_3), such as methylamine, CH_3NH_2, also are common ligands.

(c) This is the simplest of a group of ligands called *phosphanes*, which have three organic groups bonded to phosphorus.

(d) The anionic forms of all common amino acids act as bidentate ligands.

WORKED EXAMPLE 13.2

Identifying ligands

Which of the following species would be unlikely to function as a ligand in a transition metal complex?

(a) NH_4^+ (b) $B(CH_3)_3$ (c) HS^-

Analysis

We learned on page 551 that, in order to function as a ligand, a molecule or ion must have an available electron pair to donate to a transition metal ion. Therefore, we will proceed by drawing the Lewis structures of (a), (b) and (c) and looking for the presence of electron pairs.

Solution

The Lewis structures of (a), (b) and (c) are as follows.

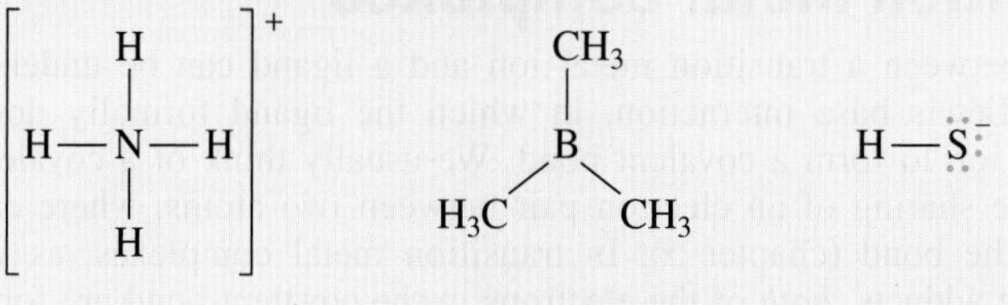

(a) (b) (c)

From these, it can be seen that only HS^- has electron pairs available for donation to a transition metal ion and could thus potentially act as a ligand.

PRACTICE EXERCISE 13.2

By drawing Lewis structures, determine which of the following species could act as a ligand towards a transition metal ion by donating a lone pair of electrons: BH_3, BH_4^-, CH_4, CH_3^-.

Many Australian and New Zealand chemists have been instrumental in the development of novel ligands, and two in particular deserve special mention. New Zealand chemist Neil Curtis and coworkers at Victoria University in the early 1960s reported the first metal complex containing a macrocyclic ligand, shown in figure 13.8. Macrocyclic ligands are simply large cyclic ligands that contain three or more donor atoms. The work of Curtis has led to the development of macrocyclic chemistry as an important subdiscipline of inorganic chemistry, and literally thousands of complexes containing macrocyclic ligands have been prepared in the years since his pioneering research.

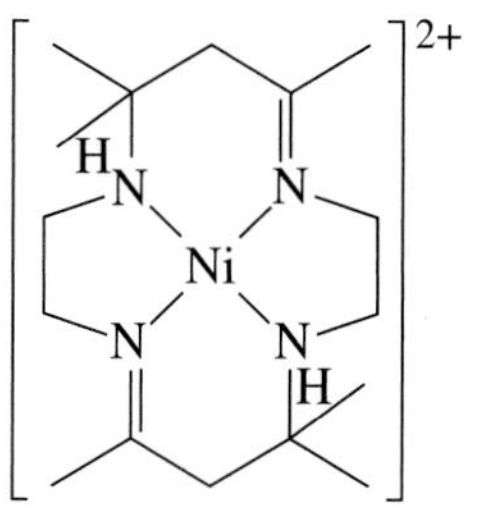

FIGURE 13.8 The first transition metal complex containing a macrocyclic ligand, prepared by New Zealand chemists in the early 1960s.

In 1977, Australian chemist Alan Sargeson and coworkers at the Australian National University prepared the first examples of groups of ligands called sepulchrates (from 'sepulchre' meaning 'tomb') and sarcophagines (from 'sarcophagus' meaning 'coffin'), examples of which are shown in figure 13.9.

These hexadentate ligands, which are trivially called 'cage' ligands, completely encapsulate a transition metal ion, and, once inserted inside the ligand cavity, the metal ion is extremely difficult to remove. Complexes of cage ligands show extremely interesting properties, particularly in biological systems. Such complexes have effective antiviral action against viruses such as hepatitis and herpes, and work by inhibiting their replication. Cage compounds can also kill tapeworms, both in vitro (in a test tube) and in vivo (in a live organism), and destroy giardia in vitro at micromolar concentrations of the complex.

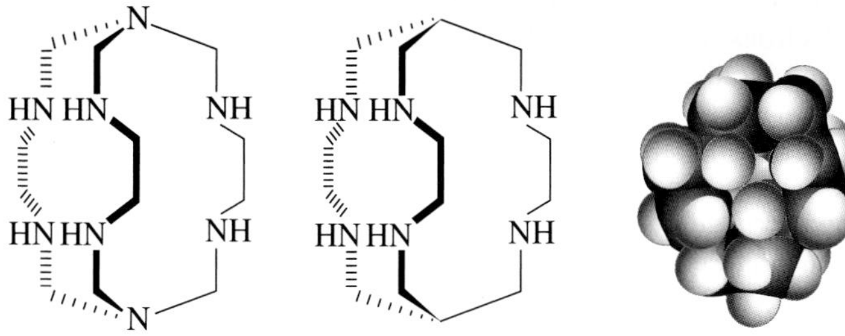

(a) (b) (c)

FIGURE 13.9 (a) A sepulchrate ligand, **(b)** a sarcophagine ligand and **(c)** a space-filling model of a Co(III) complex of a sarcophagine ligand. Note that the orange Co(III) ion is completely enclosed inside the ligand.

All of the ligands described thus far contain a donor atom with at least one lone pair of electrons. There is also a variety of organic molecules that can function as ligands in transition metal complexes, in which one or more carbon atoms are bound to the transition metal ion. Neutral unsaturated organic molecules, including alkenes, alkynes and aromatic species such as benzene, can bind to metal ions using electrons from their π orbitals (chapter 16); alkenes and alkynes generally donate two electrons, while benzene usually donates six. Negatively charged organic ligands can be generated by deprotonation, thus allowing coordination of, for example, the anions of alkanes, while the cyclopentadienide anion, $C_5H_5^-$, has also found extensive use as a ligand. Compounds which contain organic ligands bound to a metal ion through one or more carbon atoms are called **organometallic compounds** (figure 13.10).

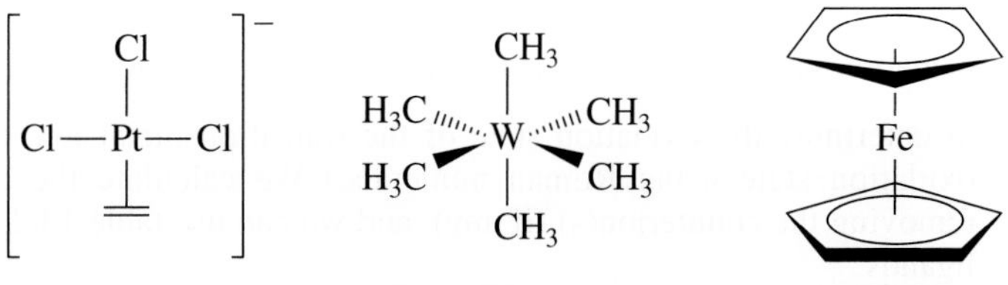

(a) (b) (c)

FIGURE 13.10 Representations of the structures of some organometallic complexes: **(a)** $[Pt(C_2H_4)Cl_3]^-$, the anion of Zeise's salt, contains a molecule of ethene bound to the Pt ion. **(b)** $[W(CH_3)_6]$ contains six CH_3^- ligands, which are formally derived from deprotonation of CH_4. **(c)** In $[Fe(C_5H_5)_2]$, commonly known as ferrocene, each $C_5H_5^-$ ligand donates six electrons to the Fe(II) ion.

13.4 Transition metal complexes

The interaction between a transition metal ion and a ligand can be understood in terms of a Lewis acid – Lewis base interaction, in which the ligand formally donates an electron pair to the metal ion to form a covalent bond. We usually think of a covalent single bond as resulting from the sharing of an electron pair between two atoms, where each atom donates one electron to the bond (chapter 5). In transition metal complexes, as well as all Lewis acid – Lewis base adducts, *both* of the electrons in the covalent bond are formally contributed by the ligand (Lewis base). Although metal–ligand covalent bonds in metal complexes and single covalent bonds in organic compounds both involve sharing a pair of electrons, the former are sometimes called **coordinate bonds**, **donor covalent bonds** or **dative bonds**, and we often say that a ligand is 'coordinated to' rather than 'bound to' a transition metal ion. Transition metal complexes in general are often called **coordination compounds**, while the discipline of the study of transition metal complexes is called **coordination chemistry**. Coordinate bonds are sometimes drawn with an arrow on one end to show the direction of electron donation, but it is more common to depict such bonds in the same way as all other covalent bonds (figure 13.11).

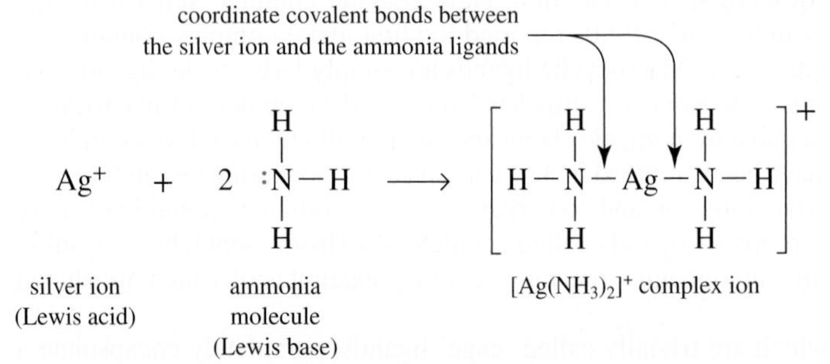

coordinate covalent bonds between the silver ion and the ammonia ligands

$$Ag^+ + 2\ :NH_3 \longrightarrow [H_3N{-}Ag{-}NH_3]^+$$

silver ion (Lewis acid) | ammonia molecule (Lewis base) | $[Ag(NH_3)_2]^+$ complex ion

FIGURE 13.11 Formation of a coordinate bond between Ag^+ and NH_3. Note that the overall charge on the complex is the same as the sum of the charges on the metal ion and the ligands.

When assigning formal charges to the coordinated ligand atoms in a transition metal complex, we assume that both electrons in the donated electron pair 'belong' to the ligand donor atom. This is illustrated in the example above, where, in $[Ag(NH_3)_2]^+$, the N atoms are regarded as having five electrons, with both electrons in each Ag—N bond assigned to the N atom. Therefore, both N atoms are formally neutral, rather than positively charged, and are coordinated to an Ag(I) ion.

When writing the formula of a transition metal complex, we always enclose the metal-containing section in square brackets, with the ligands written in alphabetical order, and with the overall charge on the complex (if any) superscripted outside the brackets. For example, we would write the formula of the complex containing an Fe(III) ion surrounded by six Cl^- ligands as $[FeCl_6]^{3-}$. The sodium salt of this complex anion would be written as $Na_3[FeCl_6]$ with the charges omitted, as is usual for ionic compounds. Ions outside the square brackets are termed **counterions** and may be either cations or anions, depending on the charge on the complex ion. It is important that you can determine the oxidation state of the metal ion in a transition metal complex from the formula of the complex, and this can usually be done using the equation:

oxidation state of transition metal ion = charge on the complex – sum of charges on the ligands

WORKED EXAMPLE 13.3

Determining the oxidation state of a transition metal ion in a complex

What is the oxidation state of the transition metal ion in each of the following complexes?

(a) $Na_3[FeCl_6]$ (b) $[CoCl(NH_3)_5]Cl_2$ (c) $[Re(CO)_5Cl]$

Analysis

We will use the equation:

oxidation state of transition metal ion = charge on the complex – sum of charges on the ligands

to determine the oxidation state of the transition metal ion. (Remember that we denote oxidation state using Roman numerals.) We calculate the charge on the complex by removing the counterion(s) (if any), and we can use table 13.2 to obtain the charges of the ligands.

Solution

(a) Removing three Na^+ ions means that the charge on the complex is 3−; that is, $[FeCl_6]^{3-}$. Each of the Cl^- ligands has a 1− charge, so the total charge on the ligands is 6−. Therefore:

$$\text{oxidation state of the transition metal ion} = (3-) - (6-) = \text{III}$$

(b) Removing the two Cl^- counterions leaves a charge on the complex of 2+; that is, $[CoCl(NH_3)_5]^{2+}$. There are five NH_3 ligands, which are neutral, and one Cl^- ligand with a 1− charge. This gives a total charge on the ligands of 1−. Therefore:

$$\text{oxidation state of the transition metal ion} = (2+) - (1-) = \text{III}$$

(c) There are no counterions to remove, so the charge on the complex must be 0. There are five CO ligands, which are neutral, and one Cl^- ligand with a 1− charge. This gives a total charge on the ligands of 1−. Therefore:

$$\text{oxidation state of the transition metal ion} = 0 - (1-) = \text{I}$$

PRACTICE EXERCISE 13.3

Determine the oxidation states of the transition metal ions in the following complexes: $[CoCl_3(NH_3)_3]$, $K[MnO_4]$, $[Mo(CO)_6]$, $K_2[ReH_9]$ and $[CrO_2Cl_2]$.

When writing the formulae of transition metal complexes, it is usual to write ligands with their donor atom(s) first. For example, we would write $[Fe(OH_2)_6]^{2+}$ instead of $[Fe(H_2O)_6]^{2+}$. In the case of potentially bidentate ligands such as CO_3^{2-} and $C_2O_4^{2-}$, we write the two donor atoms first, followed by the rest of the atoms. Therefore, the formulae $[Co(O_2CO)_3]^{3-}$ and $[Fe(O_2C_2O_2)_3]^{3-}$ imply that the carbonate and oxalate ligands are chelated to the Co(III) and Fe(III) centres, respectively. Such notation allows us to easily distinguish between complexes in which the ligands act as chelates and, for example, the monodentate carbonate ligand found in $[Co(NH_3)_5OCO_2]^+$.

Structures of transition metal complexes

Transition metal complexes exhibit a huge variety of structures, as is to be expected when the central metal ion can be surrounded by anything from one to nine monodentate ligands. Central to the structure adopted by any transition metal complex is the **coordination number** of the metal ion. This is defined as the number of donor atoms directly attached to the metal ion. For example, the complex $[CuCl_4]^{2-}$ has four Cl^- ligands bound to a Cu(II) ion, so the coordination number of the metal ion in this complex is 4; we also say that the metal ion is 4-coordinate. Similarly, the complex $[Co(NH_3)_6]^{3+}$ contains a 6-coordinate Co(III) ion bonded to the N atoms of the six surrounding NH_3 ligands. The coordination number of a metal ion may not be so obvious when the metal complex contains polydentate ligands. For example, the Ni(II) ion in $[Ni(en)_3]^{2+}$ (figure 13.12) is 6-coordinate, not 3-coordinate, as each en ligand contains two donor atoms, meaning that the Ni(II) ion is bound to six N atoms. A similar argument can be used to show that the Co(III) ion in $[Co(EDTA)]^-$ (see figure 13.6) is also 6-coordinate. The most common coordination number among transition metal complexes is 6, while 4-coordinate complexes are the next most common. Certain geometries are associated with each coordination number, and these are outlined on the following pages.

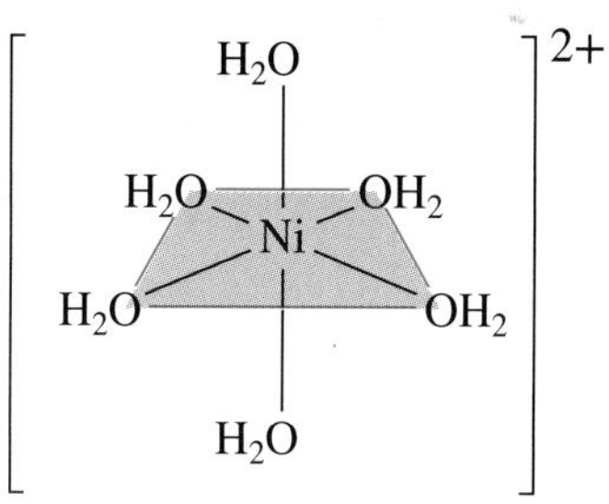

$[Ni(OH_2)_6]^{2+}$ complex with monodentate ligands

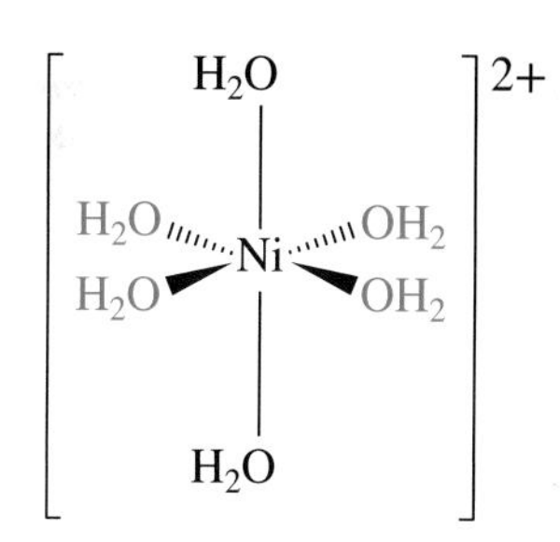

The aqua ligands in red are equatorial ligands, while those in black are axial ligands.

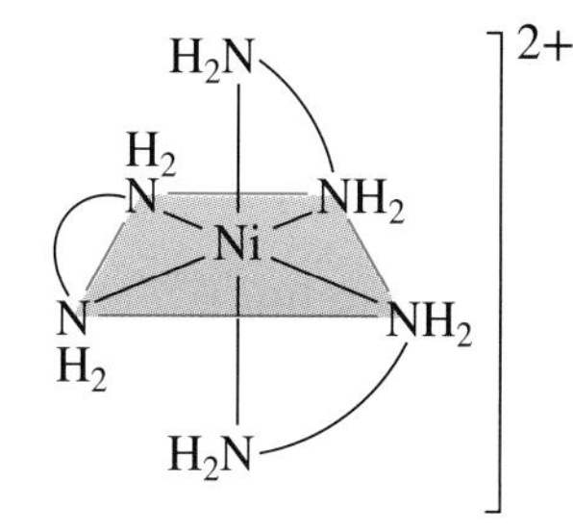

$[Ni(en)_3]^{2+}$ complex with bidentate ligands

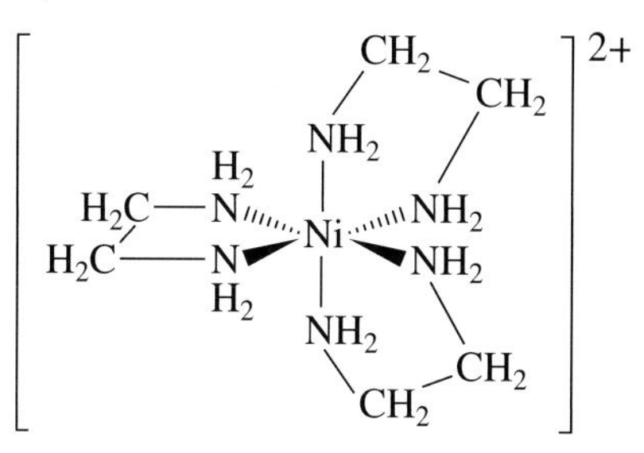

FIGURE 13.12 Octahedral complexes. Complexes with this geometry can be formed with either monodentate ligands, such as water, or polydentate ligands, such as ethylenediamine (en). To simplify drawing the ethylenediamine complex, the two methylene groups joining the donor nitrogen atoms in the ligand, $—CH_2—CH_2—$, are represented as the curved line between the N atoms. Also notice that the nitrogen atoms of the bidentate ligand span adjacent positions within the octahedron, as the methylene chain is not long enough to allow them to span opposite positions.

Coordination number 6

Most 6-coordinate complexes adopt an **octahedral** geometry, in which the six ligand donor atoms are positioned at the vertices of an octahedron (figure 13.13, overleaf). The octahedron is so named because it contains eight faces. The two ligands in the plane of the page are called the **axial** ligands, while the remaining four are called the **equatorial** ligands.

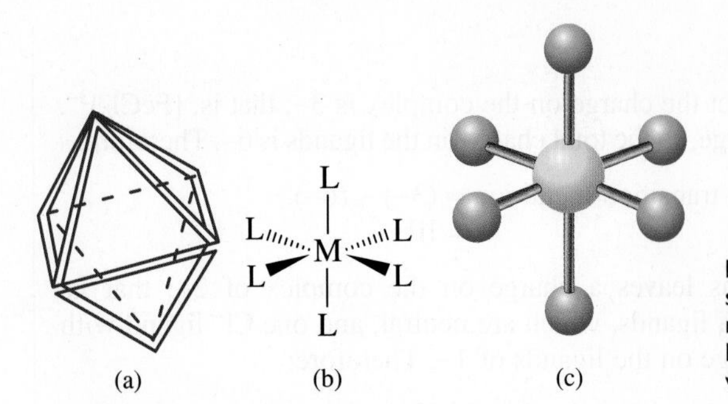

FIGURE 13.13 **(a)** An octahedron, **(b)** a structural diagram and **(c)** a ball-and-stick representation of an octahedral ML_6 complex. Note that each ligand is situated at one of the six vertices of the octahedron.

Such an arrangement minimises ligand–ligand repulsions by placing the ligands as far away as possible from each other, and this geometry is in fact that predicted by VSEPR theory (chapter 5, pp. 182–8). The octahedral geometry can accommodate both monodentate and polydentate ligands, as shown in figure 13.12.

Coordination number 4

VSEPR predicts that a tetrahedral geometry is favoured for a coordination number of 4, but both **tetrahedral** and **square planar** geometries are found for 4-coordinate complexes, and these are illustrated in figures 13.14 and 13.15.

Tetrahedral geometries are often observed for complexes of d^{10} metal ions such as Cu(I) and Zn(II), while square planar complexes are often formed by Ni(II), Pt(II) and Au(III) metal ions with d^8 electron configurations. In fact, a perfect tetrahedral or square planar geometry is rarely observed, and many 4-coordinate complexes adopt a geometry intermediate between the two.

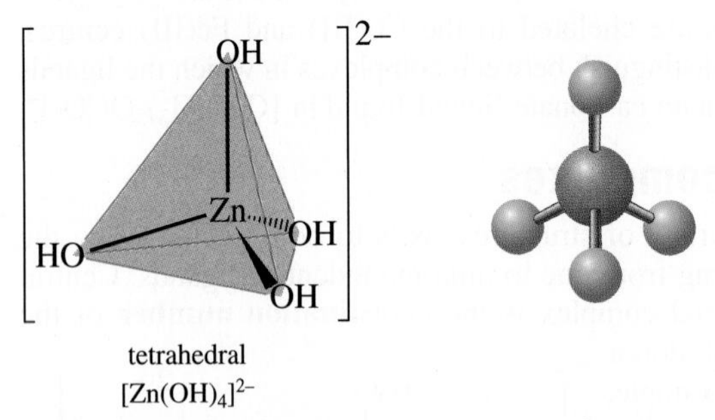

FIGURE 13.14 Tetrahedral geometry. $[Zn(OH)_4]^{2-}$ adopts a tetrahedral structure.

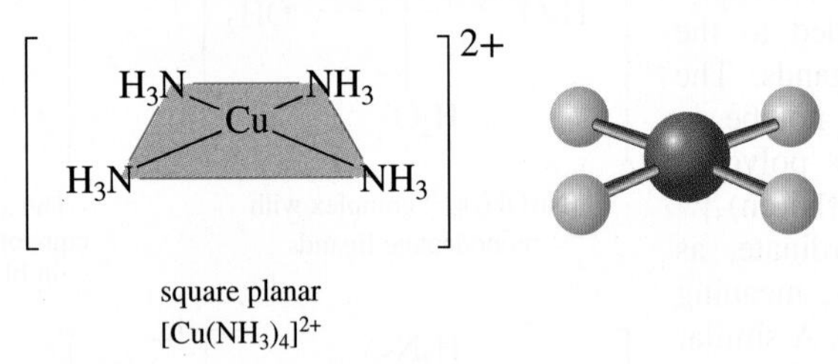

FIGURE 13.15 Square planar geometry. $[Cu(NH_3)_4]^{2+}$ adopts a square planar structure.

Coordination number 5

This is probably the most important coordination number following 6 and 4. There are two possible geometries for 5-coordinate complexes: **trigonal bipyramidal** and **square pyramidal**, which are illustrated in figure 13.16.

Trigonal bipyramidal geometry is predicted on the basis of VSEPR, but, as we found for 4-coordinate complexes, VSEPR is of limited use when applied to transition metal complexes, especially where the metal ion does not have a d^0 or d^{10} configuration. For example, the $[CuCl_5]^{3-}$ anion is trigonal bipyramidal, while the $[Ni(CN)_5]^{3-}$ anion in the complex $[Cr(en)_3][Ni(CN)_5]$ simultaneously displays both square pyramidal and trigonal bipyramidal geometries in the same compound.

It is common to refer to the ligands pointing above and below the trigonal plane in a trigonal bipyramidal complex as axial, while the remaining three ligands are called equatorial.

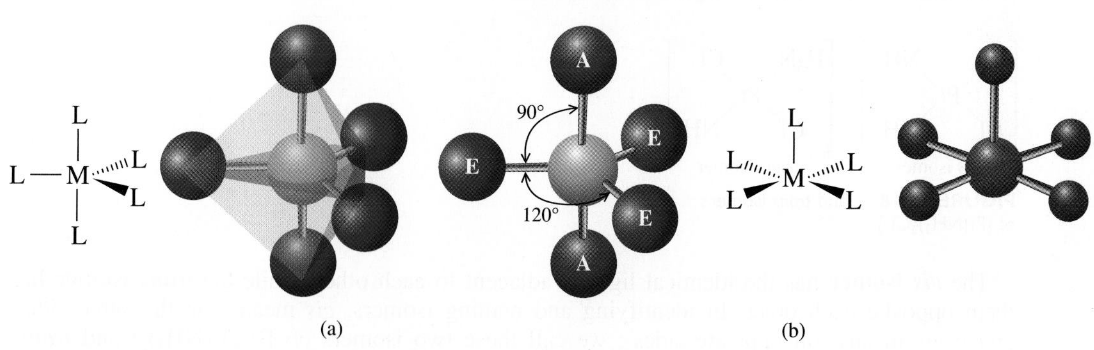

FIGURE 13.16 (a) Trigonal bipyramidal and **(b)** square pyramidal geometries found in 5-coordinate complexes.

Isomerism in transition metal complexes

We were first introduced to the concept of isomerism in chapter 2, where we discussed constitutional isomers of alkanes. There are a number of different types of isomerism in transition metal complexes, some of which are detailed below.

Structural isomerism

Structural isomers are molecules with the same molecular formula but with the constituent atoms joined together in different ways. In coordination chemistry, we will consider four types of structural isomerism.

Ionisation isomerism

The purple complex $[CoBr(NH_3)_5]SO_4$ and the red complex $[Co(NH_3)OSO_3]Br$ (figure 13.17) are examples of **ionisation isomers**. The former has a bromido ligand and sulfate as the counterion, while the roles are reversed in the latter, with a sulfato ligand coordinated to the metal ion and bromide acting as the counterion.

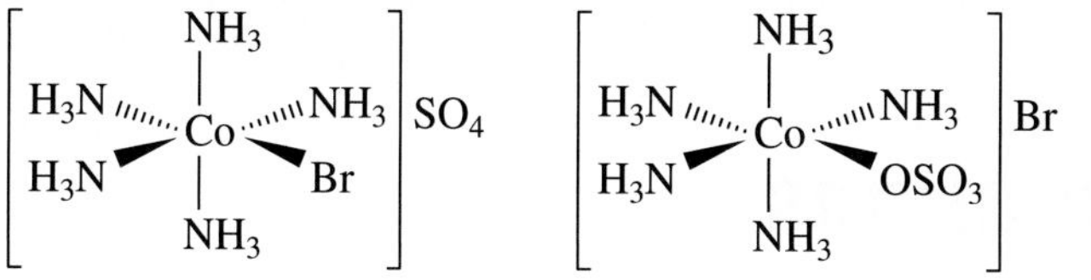

$[CoBr(NH_3)_5]SO_4$: Co bonded to NH_3, H_3N, NH_3, H_3N, Br, NH_3; counterion SO_4

$[Co(NH_3)_5OSO_3]Br$: Co bonded to NH_3, H_3N, NH_3, H_3N, OSO_3, NH_3; counterion Br

FIGURE 13.17 The ionisation isomers $[CoBr(NH_3)_5]SO_4$ and $[Co(NH_3)OSO_3]Br$.

Hydrate isomerism

This is similar to ionisation isomerism, and results from the different numbers of water molecules that can be coordinated to a metal ion. For example, there are three complexes with the empirical formula $CrCl_3{\cdot}6H_2O$:

- $[Cr(OH_2)_6]Cl_3$ (purple)
- $[CrCl(OH_2)_5]Cl_2{\cdot}H_2O$ (blue–green)
- $[CrCl_2(OH_2)_4]Cl{\cdot}2H_2O$ (green).

These complexes are **hydrate isomers**.

Coordination isomerism

Coordination isomers result when ligands are exchanged between a complex cation and a complex anion of the same coordination compound. For example, $[Co(NH_3)_6][Cr(CN)_6]$ and $[Cr(NH_3)_6][Co(CN)_6]$ are coordination isomers.

Linkage isomerism

Ligands containing more than one donor atom can potentially form **linkage isomers**, which result from the different possible ways in which the ligand can bind to the metal ion. For example, the thiocyanato ligand, NCS^-, can bind using a lone pair on either the N atom, to give the brick-red $[CoNCS(NH_3)_5]^{2+}$ ion, or on the S atom, to give the purple complex $[Co(NH_3)_5SCN]^{2+}$. These complexes are called linkage isomers. The NO_2^- ligand can also form linkage isomers by bonding to a transition metal ion either through the N atom or one of the O atoms. Ligands, such as NCS^- and NO_2^-, with two or more different potential donor atoms are called **ambidentate**.

Stereoisomerism

Stereoisomers are isomers having the same connections between the constituent atoms but different arrangements of those atoms in space. We met an example of stereoisomerism in chapter 2 (p. 46) where we found that the two methyl groups in 1,2-dimethylcyclopentane could either be on the same side or different sides of the five-membered ring; we designated the two possible isomers as *cis* and *trans,* respectively. A similar type of isomerism can exist in transition metal complexes and is typified by the square planar complex $[PtCl_2(NH_3)_2]$.

The two possible ***cis–trans* isomers** of the square planar complex with the empirical formula $[PtCl_2(NH_3)_2]$ are shown in figure 13.18.

cis isomer *trans* isomer

FIGURE 13.18 *cis* and *trans* isomers of $[Pt(NH_3)_2Cl_2]$.

The ***cis* isomer** has the identical ligands adjacent to each other, while the ***trans* isomer** has them opposite each other. In identifying and naming isomers, *cis* means 'on the same side', and *trans* means 'on opposite sides'; we call these two isomers *cis*-$[PtCl_2(NH_3)_2]$ and *trans*-$[PtCl_2(NH_3)_2]$. The two isomers have distinctly different chemical properties, most interesting of which is the fact that the *cis* isomer (common name cisplatin) is a particularly effective chemotherapy agent against certain types of cancer, while the *trans* isomer shows little activity (see the Chemical Connections feature below). It should be noted that a tetrahedral arrangement of the ligands around the Pt ion would not give rise to stereoisomers, and this can be used to differentiate between square planar and tetrahedral geometries.

Octahedral complexes can also show *cis–trans* isomerism. While there are no such possible isomers of complexes of the type $[ML_6]$ and $[ML_5X]$, *cis* and *trans* isomers are possible for $[ML_4X_2]$ species, with the classic example being the complexes *cis*-$[CoCl_2(NH_3)_4]^+$ (violet) and *trans*-$[CoCl_2(NH_3)_4]^+$ (green). The structures of these complexes are given in figure 13.20.

FIGURE 13.20 The structures of the isomers *cis*-$[CoCl_2(NH_3)_4]^+$ (left) and *trans*-$[CoCl_2(NH_3)_4]^+$ (right).

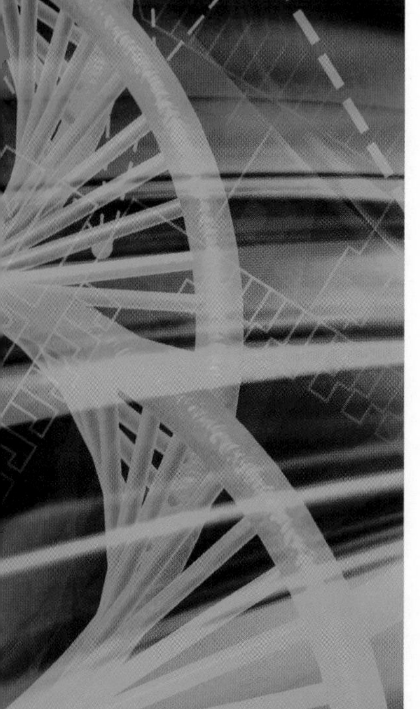

Chemical Connections

Cisplatin: an accidental discovery

At the beginning of the 1970s, 70% of those who developed testicular cancer died within one year, but by the end of that decade this proportion was down to 10% and it is now effectively zero. This is largely due to the *cis*-diamminedichloridoplatinum(II) complex *cis*-$[PtCl_2(NH_3)_2]$, commonly known as cisplatin (figure 13.19), an extremely effective chemotherapy agent with a deceptively simple chemical structure. Interestingly, the discovery of its anticancer properties, like many discoveries in science, was completely accidental.

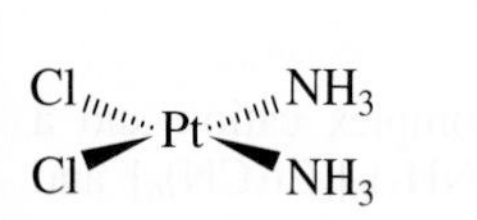

FIGURE 13.19 Models of cisplatin.

In 1961, Barnett (Barney) Rosenberg (1926–2009) was a newly appointed associate professor of biophysics at Michigan State University in the USA when he began studying the effect of electric fields on cell division. Using platinum electrodes, he passed a current through cells of the bacterium *E. coli* in an ammonia buffer, and found that their shape changed enormously, owing to the fact that cell division in the bacterium had been inhibited. However, painstaking investigations over several years showed that this was not a result of the electric field, but rather that minute quantities of platinum from the electrodes had reacted with ammonia in the buffer and chloride in the solution to form *cis*-$[PtCl_2(NH_3)_2]$, a complex that had been known since 1844, and it was this complex that was inhibiting cell division. It was then but a short step to investigating the anti-cancer properties of this complex, and cisplatin was eventually approved by the Federal Drug Administration for use against testicular and ovarian cancers in 1978.

Rosenberg's discovery initiated an entirely new field of research, transition metal complexes as anti-cancer agents, and showed that, in contrast to drugs like Taxol® (see chapter 17), compounds with very simple molecular structures can be very effective in the treatment of cancer.

Note that these are the only possible isomers of this complex; any other arrangement of the ligands is identical to one of the two structures in figure 13.19.

Isomers are also possible for octahedral complexes with the formula $[ML_3X_3]$ and these are depicted in figure 13.21 for the complex $[CoCl_3(NH_3)_3]$. These two isomers are designated ***mer* (meridional)** and ***fac* (facial)**. The *mer* isomer has the two sets of three identical ligands lying in perpendicular planes around the meridian of the complex, while the *fac* isomer has each set of three ligands occupying a face of the octahedron, so that the two sets of ligands lie in parallel planes. Again, these are the only possible geometric isomers of these complexes, and there is no other arrangement of ligands that is different from those in figure 13.21.

We will see in chapter 17 that another type of stereoisomerism exists in transition metal complexes containing polydentate ligands. These complexes, with mirror images that are not superimposable on each other, are said to be chiral.

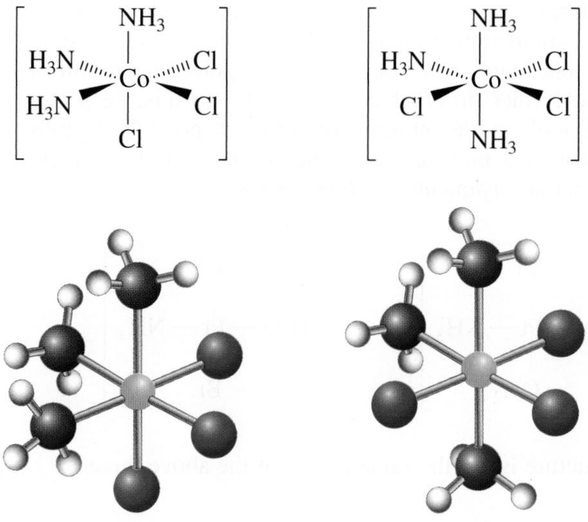

FIGURE 13.21 The *fac* isomer (left) and *mer* isomer (right) of $[CoCl_3(NH_3)_3]$.

WORKED EXAMPLE 13.4

Identifying and drawing stereoisomers

Draw all the possible stereoisomers of the following transition metal complexes.

(a) $[Co(NH_3)_5OH_2]^{3+}$ (The complex is octahedral.)
(b) $[RhBr_2(NH_3)_4]^+$ (The complex is octahedral.)
(c) $[PtBrCl(NH_3)(OH_2)]$ (The complex is square planar.)

Analysis

We need to see if changing the relative positioning of the ligands around the central metal ion gives rise to different complexes. If so, then stereoisomers are present.

Solution

(a) In this complex, we have five NH_3 ligands and a single OH_2 ligand arranged in an octahedron around the metal ion. As all the positions in an octahedron are equivalent, we can place the unique aqua ligand anywhere, and its position relative to the five NH_3 ligands is the same; there are always four *cis* neighbouring NH_3 ligands and one *trans* NH_3 ligand opposite. The three depictions of the cation below are therefore identical and hence there are no stereoisomers of this complex.

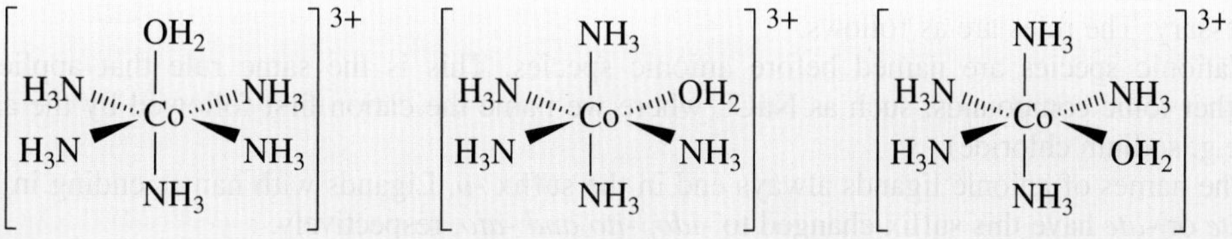

(b) In this complex, we have a set of four NH_3 ligands and a set of two Br^- ligands. In an octahedron, we can place the two Br^- ligands so that they are either adjacent to each other, or on opposite sides from each other. This gives rise to two different complexes, and therefore these are stereoisomers. An inspection of the depictions below should convince you that these are the only stereoisomers possible.

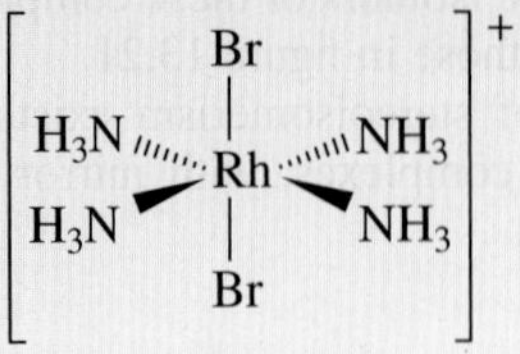

Br; H_3N, NH_3, Rh, H_3N, NH_3; Br — $[\ ]^+$ *trans*

Br; H_3N, NH_3, Rh, H_3N, Br; NH_3 — $[\ ]^+$ *cis*

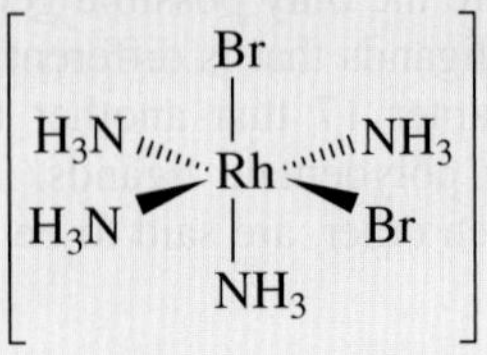

(c) In this complex, we have four different ligands arranged in a square plane around the metal ion. Therefore, you can anticipate that there should be a number of possible stereoisomers. Begin by drawing one, and then systematically change the positions of the ligands until you obtain no further different complexes. To do this, keep one ligand in the same position and look at the number of different possible ligands that can be placed *trans* to this. You will find that, no matter which ligand you pick, there are only three possible different arrangements, as shown below.

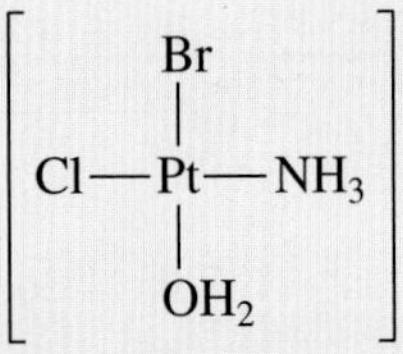

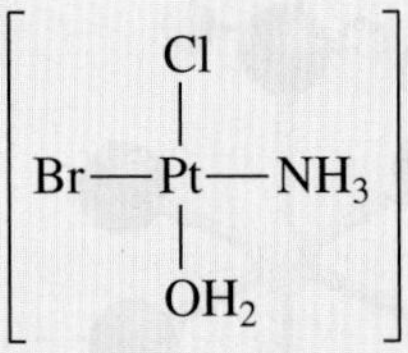

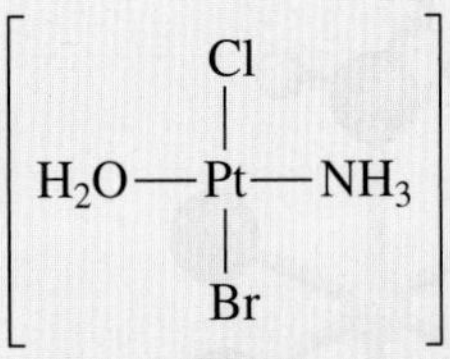

[Br; Cl—Pt—NH_3; OH_2]   [Cl; Br—Pt—NH_3; OH_2]   [Cl; H_2O—Pt—NH_3; Br]

You might think that the following structure is not the same as any of the above three.

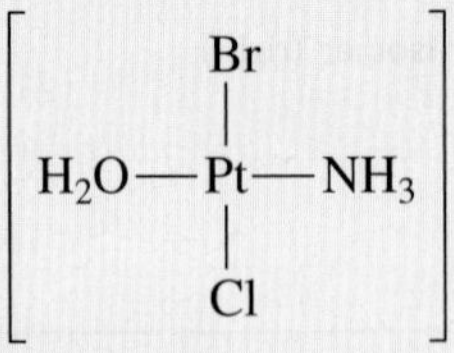

[Br; H_2O—Pt—NH_3; Cl]

However, if you rotate it 180° around the H_2O—Pt—NH_3 axis, you will obtain the third structure above.

PRACTICE EXERCISE 13.4 Determine the number of possible stereoisomers of the following complexes, and sketch their structures.

(a) $[PtBrCl_3]^{2-}$ (The complex is square planar.)

(b) $[CoCl_2(PPh_3)_2]$ (The complex is tetrahedral; note that PPh_3 is shorthand for $P(C_6H_5)_3$, a monodentate phosphane ligand.)

(c) $[CoCl(NH_3)_3(OH_2)_2]^{2+}$ (The complex is octahedral.)

The nomenclature of transition metal complexes

Chapter 2 introduced the IUPAC rules for naming simple organic compounds and we found that such compounds were named on the basis of the longest carbon chain in the molecule. We cannot use these rules for naming coordination compounds so a separate nomenclature system is necessary. The rules are as follows.

1. Cationic species are named before anionic species. This is the same rule that applies to other ionic compounds, such as NaCl, where we name the cation first followed by the anion (e.g. sodium chloride).
2. The names of anionic ligands always end in the suffix *-o*. Ligands with names ending in *-ide*, *-ite* or *-ate* have this suffix changed to *-ido, -ito and -ato*, respectively.

Anion		Ligand name
chloride	Cl^-	chlorido
bromide	Br^-	bromido
cyanide	CN^-	cyanido
oxide	O^{2-}	oxido
carbonate	CO_3^{2-}	carbonato
thiosulfate	$S_2O_3^{2-}$	thiosulfato
thiocyanate	SCN^-	thiocyanato-κ*S* (when bonded through sulfur) thiocyanato-κ*N* (when bonded through nitrogen)
oxalate	$C_2O_4^{2-}$	oxalato
nitrite	NO_2^-	nitrito-κ*O* (when bonded through oxygen) nitrito-κ*N* (when bonded through nitrogen)

When a choice of donor atom exists, the coordinated atom is italicised and preceded by the Greek letter *kappa*, κ.

3. A neutral ligand is usually given the same name as the neutral molecule. Thus the molecule ethylenediamine, when serving as a ligand, is called ethylenediamine in the name of the complex. Three very important exceptions to this, however, are water, ammonia and carbon monoxide. These are named as follows when they serve as ligands:
   - H_2O aqua
   - NH_3 ammine (note the double m)
   - CO carbonyl.
4. When there is more than one of a particular ligand, their number is specified by the prefixes *di-* = 2, *tri-* = 3, *tetra-* = 4, *penta-* = 5, *hexa-* = 6 etc. If the name of the ligand already incorporates one of these prefixes (such as ethylenediamine), enclose the ligand in parentheses and use the following prefixes instead: *bis-* = 2, *tris-* = 3, *tetrakis-* = 4. Following this rule, the presence of two Cl^- ligands in a complex would be indicated as *dichlorido-*. If two ethylenediamine ligands are present, we would write bis(ethylenediamine).
5. In the name of the complex, the ligands are named first, in alphabetical order without regard to charge, followed by the name of the metal. For example, the ligand set in the neutral complex $[Co(CN)Cl_2(NH_3)_3]$ would be named as *triamminedichloridocyanido-*: *triammine-* for the three NH_3 ligands, *dichlorido-* for the two Cl^- ligands and *cyanido-* for the CN^- ligand. Notice that, in alphabetising the names of the ligands, we ignore the prefixes *tri-* and *di-*. Thus *triammine-* is written before *dichlorido-* because *ammine-* precedes *chlorido-* alphabetically. For the same reason, *dichlorido-* is written before *cyanido-*.
6. Negative (anionic) complex ions always end in the suffix *-ate*. This suffix is appended to the English name of the metal atom in most cases. However, if the name of the metal ends in *-ium* or *-ese*, the ending is dropped and replaced by *-ate*.

Metal	Metal as named in an anionic complex
chromium	chromate
manganese	manganate
nickel	nickelate
cobalt	cobaltate
zinc	zincate
platinum	platinate
vanadium	vanadate

For metals with symbols derived from their Latin names, the suffix *-ate* is appended to the Latin stem.

Metal	Stem	Metal as named in an anionic complex
iron	ferr-	ferrate
copper	cupr-	cuprate
silver	argent-	argentate
gold	aur-	aurate

Mercury, antimony and tungsten are the exceptions; in an anion, they are named mercurate, antimonate and tungstate, respectively. For neutral or positively charged complexes, the metal is always specified with the English name for the element, without any suffix.

7. The oxidation state of the metal in a complex may be written in Roman numerals within parentheses following the name of the metal, or the charge on the coordination entity may be specified in Arabic numbers followed by the charge sign, all enclosed in parentheses. For example:
   - $[Co(NH_3)_6]^{3+}$ is the hexaamminecobalt(III) ion or the hexaamminecobalt(3+) ion
   - $[CuCl_4]^{2-}$ is the tetrachloridocuprate(II) ion or the tetrachloridocuprate(2–) ion.

   Notice that there are no spaces between the name of the ligands and the name of the metal, and there is no space between the name of the metal and the parentheses that enclose the oxidation state expressed in either Roman or Arabic numerals.
8. The number of counterions need not be specified, as it can be inferred from the charge on the complex ion. For example, $K_4[NiCl_6]$ would be named potassium hexachloridonickelate(II), not tetrapotassium hexachloridonickelate(II). Note also that there is a space between the names of the cation and anion.
9. Include any stereochemical descriptors (e.g. *cis, trans, mer, fac*) at the start of the name, italicised and hyphenated.

The following are some additional examples. Study them carefully to see how the rules given above apply, then try worked examples 13.5 and 13.6 and practice exercises 13.5 and 13.6.

$[Ni(CN)_4]^{2-}$	tetracyanidonickelate(II) ion
$K_3[CoCl_6]$	potassium hexachloridocobaltate(III)
$[CoCl_2(NH_3)_4]^+$	tetraamminedichloridocobalt(III) ion
$[Ag(NH_3)_2]^+$	diamminesilver(I) ion
$[Ag(S_2O_3)_2]^{3-}$	dithiosulfatoargentate(I) ion
$[Mn(en)_3]Cl_2$	tris(ethylenediamine)manganese(II) chloride
$[PtCl_2(NH_3)_2]$	diamminedichloridoplatinum(II)

WORKED EXAMPLE 13.5

Naming coordination complexes

What is the IUPAC name for each of the following coordination compounds?
(a) $[Ni(OH_2)_6]SO_4$ (b) $[CrCl_2(en)_2]Cl$ (c) $K_2[CoCl_4]$

Analysis

We simply follow the guidelines given above for naming coordination compounds.

Solution

(a) The nickel ion is surrounded by six ligands, so we use the prefix *hexa-*. The ligands are water molecules, which have the special name *aqua*, so the ligand set is named *hexaaqua-*. The complex is cationic, and the 2– charge on sulfate means that the cation has a 2+ charge. As the aqua ligands are neutral, the oxidation state of Ni is Ni(II). Therefore, the name of the complex is hexaaquanickel(II) sulfate.

(b) The chromium ion is coordinated to two en ligands and two Cl^- ligands. Naming the ligands alphabetically puts the Cl^- ligands first, followed by the en ligands. The two Cl^- ligands are named *dichlorido-*, while, as we have seen in the guidelines, two en ligands are named bis(ethylenediamine). Hence, the ligand set is named dichloridobis(ethylenediamine). The presence of a single Cl^- counterion means that the charge on the complex cation is 1+; the oxidation state of the metal ion must be Cr(III), as it has two Cl^- ions coordinated to it. Therefore, the name of the complex is dichloridobis(ethylenediamine) chromium(III) chloride.

(c) The cation, potassium, is named first. The complex anion contains a cobalt ion surrounded by four Cl^- ligands, so the ligand set is named *tetrachlorido-*. The complex is an anion with a 2– charge, so the oxidation state of the metal ion is Co(II). Since the complex is an anion, the metal is given the suffix *-ate*. Therefore, the name of the complex is potassium tetrachloridocobaltate(II).

Is our answer reasonable?

Verify that the numerical prefixes give the correct total number of ligands, that the ligands are named alphabetically and that the overall species is charge neutral. This should ensure that your answer is correct.

PRACTICE EXERCISE 13.5

Name the following coordination compounds.

(a) $[CoCl(NH_3)_5]Cl_2$

(b) $[W(CO)_3(P(CH_3)_3)_3]$

(c) $K_3[Fe(ox)_3]$

(d) $[Cr(en)_3]Cl_3$

(e) $K_3[Ni(CN)_5]$

Having learned how to name coordination compounds, you should be able to determine the formula from the name.

WORKED EXAMPLE 13.6

Writing a formula from the name of a coordination compound

Determine the formulae of the following coordination compounds.

(a) *fac*-triamminetriiodidoruthenium(III)

(b) sodium hexacyanidoferrate(II)

Analysis

Construct the formula by breaking the name down, one piece at a time. Begin by determining the number and type of ligands around the metal ion. From these, and the oxidation state of the metal ion, you can then determine the correct number of counterions (if any).

Solution

(a) The prefix *triammine-* corresponds to three NH_3 ligands and *triiodido-* to three I^- ligands. Therefore, there are six ligands surrounding the Ru(III) ion. The *fac-* prefix means that each set of three ligands is arranged such that they occupy a face of an octahedron. The formula of the complex is *fac*-$[RuI_3(NH_3)_3]$.

(b) The prefix *hexacyanido-* means that there are six CN^- ligands around the Fe(II) ion, and this gives the anion a charge of 4–. Sodium is the cation, so four of these are required to balance the charge on the complex anion. The formula of the complex is $Na_4[Fe(CN)_6]$.

Is our answer reasonable?

Conversion from the name to the formula is usually fairly straightforward, and our answers do correspond to the names. As shown in (b), make sure you include the correct number of cations and anions to ensure charge neutrality.

PRACTICE EXERCISE 13.6

Determine the formulae of the following coordination compounds.

(a) tris(ethylenediamine)ruthenium(II) chloride

(b) potassium hexachloridoplatinate(IV)

(c) ammonium diaquatetracyanidoferrate(II)

(d) *trans*-dichloridobis(ethylenediamine)cobalt(III) chloride

The chelate effect

In chapter 11, we quantified the strength of Brønsted–Lowry acids using K_a, the equilibrium constant for donation of a proton to water. As the formation of transition metal complexes from their constituent metal ions and ligands is also an equilibrium process, we can use an analogous method to quantify the extent of such reactions. Consider, for example, the formation of a hypothetical complex $[ML_n]^{x+}$ from the metal ion, M^{x+}, and the ligands, nL, in aqueous solution according to the equation:

$$M^{x+}(aq) + nL(aq) \rightleftharpoons [ML_n]^{x+}(aq)$$

(Note that, for simplicity, we have assumed that the ligand is neutral.) The equilibrium constant for this process, which we can write in the same way as we have done in preceding chapters, is called the **cumulative formation constant** (β_n). Therefore, for the above equilibrium:

$$\beta_n = \frac{[{ML_n}^{x+}(aq)]}{[M^{x+}(aq)][L(aq)]^n}$$

Obviously, the larger the value of β_n, the further to the right-hand side the equilibrium position lies, and therefore the more complete the formation of the complex. Table 13.3 gives values of β_n for a number of transition metal complexes, while a more comprehensive list may be found in appendix D. It can be seen that, in all cases, the equilibrium for the formation reaction lies very much towards products. However, there is a range of nearly 44 orders of magnitude in the values of β_n, showing that the formation reactions of some complexes proceed much further towards completion than others.

TABLE 13.3 β_n values for a variety of transition metal complexes containing monodentate and polydentate ligands at 25 °C.

Ligand	Metal	Equilibrium	n	β_n
NH_3	Co(II)	$Co^{2+}(aq) + 6NH_3(aq) \rightleftharpoons [Co(NH_3)_6]^{2+}(aq)$	6	5.0×10^4
	Co(III)	$Co^{3+}(aq) + 6NH_3(aq) \rightleftharpoons [Co(NH_3)_6]^{3+}(aq)$	6	4.6×10^{33}
	Ni(II)	$Ni^{2+}(aq) + 6NH_3(aq) \rightleftharpoons [Ni(NH_3)_6]^{2+}(aq)$	6	2.0×10^8
	Cu(II)	$Cu^{2+}(aq) + 4NH_3(aq) \rightleftharpoons [Cu(NH_3)_4]^{2+}(aq)$	4	1.1×10^{13}
en	Co(II)	$Co^{2+}(aq) + 3en(aq) \rightleftharpoons [Co(en)_3]^{2+}(aq)$	3	1.0×10^{14}
	Co(III)	$Co^{3+}(aq) + 3en(aq) \rightleftharpoons [Co(en)_3]^{3+}(aq)$	3	5.0×10^{48}
	Ni(II)	$Ni^{2+}(aq) + 3en(aq) \rightleftharpoons [Ni(en)_3]^{2+}(aq)$	3	4.1×10^{17}
	Cu(II)	$Cu^{2+}(aq) + 2en(aq) \rightleftharpoons [Cu(en)_2]^{2+}(aq)$	2	4.0×10^{19}
$EDTA^{4-}$	Co(II)	$Co^{2+}(aq) + EDTA^{4-}(aq) \rightleftharpoons [Co(EDTA)]^{2-}(aq)$	1	2.8×10^{16}
	Co(III)	$Co^{3+}(aq) + EDTA^{4-}(aq) \rightleftharpoons [Co(EDTA)]^{-}(aq)$	1	2.5×10^{41}

Inspection of the data in table 13.3 reveals some interesting trends. For example, coordination of the same ligand to different metal ions gives very different β_n values, showing that a given ligand binds more strongly to some metal ions than to others. Ligands can also show different affinities for various oxidation states of the same metal ion. For example, notice the very different values for the Co(II) and Co(III) complexes listed, with Co(III) favoured in each case by a factor of up to 10^{34}. Of particular interest are the values for complexes of the monodentate ligand NH_3 and the bidentate ligand en. For example, $[Ni(NH_3)_6]^{2+}$ and $[Ni(en)_3]^{2+}$ have very similar structures, with Ni bonded to six N atoms in both cases, but β_3 for $[Ni(en)_3]^{2+}$ is larger than β_6 for $[Ni(NH_3)_6]^{2+}$ by a factor of 10^9. This trend is repeated in all of the complexes of NH_3 and ethylenediamine listed in table 13.3 and is found to be fairly general for complexes containing these ligands.

The larger values of β_n for complexes containing polydentate ligands than those for analogous complexes containing monodentate ligands is called the **chelate effect**. We will explain this with reference to the formation reactions of the two Ni complexes discussed above.

$$[Ni(OH_2)_6]^{2+}(aq) + 6NH_3(aq) \rightleftharpoons [Ni(NH_3)_6]^{2+}(aq) + 6H_2O(l) \quad \beta_6 = \frac{[Ni(NH_3)_6{}^{2+}]}{[Ni(OH_2)_6{}^{2+}][NH_3]^6}$$

$$[Ni(OH_2)_6]^{2+}(aq) + 3en(aq) \rightleftharpoons [Ni(en)_3]^{2+}(aq) + 6H_2O(l) \quad \beta_2 = \frac{[Ni(en)_3{}^{2+}]}{[Ni(OH_2)_6{}^{2+}][en]^3}$$

Recall from chapter 9 (p. 364) that ΔG° for a reaction is related to the equilibrium constant for that reaction by the equation $\Delta G^\circ = -RT\ln K$. Therefore, a large negative value of ΔG° leads to a large equilibrium constant (or, in the case of transition metal complexes, a large cumulative formation constant) for the reaction, meaning that the forward reaction will proceed essentially to completion. We also learned in chapter 8 (p. 331) that $\Delta G^\circ = \Delta H^\circ - T\Delta S^\circ$, and that the relative magnitudes and signs of both ΔH° and ΔS° are crucial in determining the value and sign of ΔG°, and hence the value of K. If we examine the two formation reactions above, we can see that both involve making six Ni—N bonds and this should result in similar negative values of ΔH° for the two reactions. However, the two reactions differ in their values of ΔS°; while formation of $[Ni(NH_3)_6]^{2+}$ gives no net change in the number of species in solution, and would thus be expected to have a ΔS° value around zero, formation of $[Ni(en)_3]^{2+}$ gives a net increase in the number of species in solution, and hence has a positive ΔS° value. Remembering that the ΔH° values are similar, $[Ni(en)_3]^{2+}$ will have a larger $T\Delta S^\circ$ term than $[Ni(NH_3)_6]^{2+}$, and so, from the equation $\Delta G^\circ = \Delta H^\circ - T\Delta S^\circ$ we therefore expect ΔG° for formation of $[Ni(en)_3]^{2+}$ to be more negative than that for $[Ni(NH_3)_6]^{2+}$. This means that β_3 for $[Ni(en)_3]^{2+}$ should be greater than β_6 for $[Ni(NH_3)_6]^{2+}$, as is indeed found (4.1×10^{17} versus 2.0×10^8).

We can also rationalise the chelate effect by considering the mechanisms by which monodentate and bidentate ligands dissociate from a transition metal ion in a complex.

The initial mechanistic process that occurs in both cases is cleavage of a single Ni—N bond; in the case of $[Ni(NH_3)_6]^{2+}$, this leads to the loss of one NH_3 ligand, which is immediately replaced by a water molecule.

$$[Ni(NH_3)_6]^{2+}(aq) + H_2O(l) \rightleftharpoons [Ni(NH_3)_5(OH_2)]^{2+}(aq) + NH_3(aq)$$

However, cleavage of an Ni—N bond in $[Ni(en)_3]^{2+}$ leads to a situation in which one end of one of the en ligands becomes detached from the complex but remains close to the metal ion because the other donor atom of the en ligand is still bonded to the metal ion. Therefore, there is a significant probability that the 'dangling' donor atom can reattach to the metal ion before the other end of the ligand detaches (figure 13.22); from this, it is apparent that it should be more difficult to lose a bidentate ligand than a monodentate ligand.

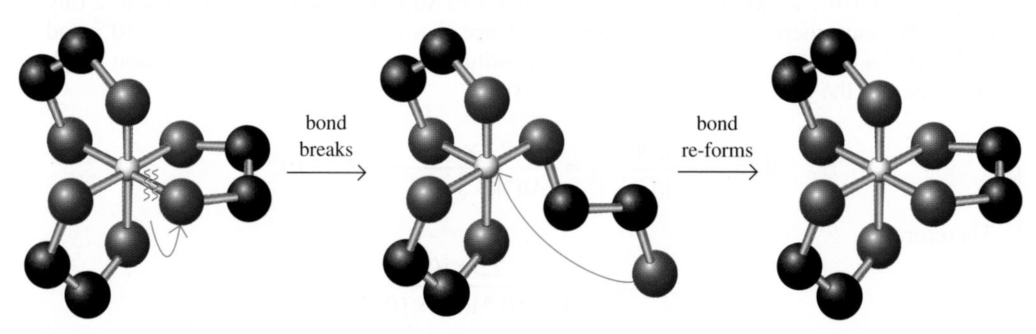

FIGURE 13.22 The chelate effect. Breaking one Ni—N bond in $[Ni(en)_3]^{2+}$ leaves the en ligand dangling but 'tethered' to the metal by the second nitrogen atom (hydrogen atoms have been omitted for clarity).

The loss of a ligand becomes progressively more difficult as the number of donor atoms increases. Therefore, in general, the more donor atoms a ligand contains, the larger is the value of β_n for its complexes.

We can obtain some appreciation of the magnitudes of β_n values by carrying out calculations similar to those we have performed for other equilibrium systems in earlier chapters. For example, worked example 13.7 looks at a reaction used to extract gold from ores.

WORKED EXAMPLE 13.7

Calculations involving β_n

The small amounts of gold in low-grade ores (figure 13.23) can be extracted using a combination of oxidation and complexation. Elemental gold is oxidised to Au^+, which forms a complex with cyanide anions according to the equation:

$$Au^+(aq) + 2CN^-(aq) \rightleftharpoons [Au(CN)_2]^-(aq) \qquad \beta_2 = 2 \times 10^{38}$$

Suppose that a sample of ore containing 2.5×10^{-3} mol of gold is extracted with 1.0 L of 4.0×10^{-2} M aqueous KCN solution under oxidising conditions. Calculate the concentrations of the three species involved in the complexation equilibrium.

FIGURE 13.23 Gold in low-grade ores can be extracted using oxidation and complexation.

Analysis

We treat this in a similar fashion to problems we have seen in earlier chapters by using a concentration table. However, in this case, we assume complete initial formation of $[Au(CN)_2]^-$. We are justified in this assumption because of the enormous value of β_2. We then realise that a very small amount of Au^+ and CN^- will be formed by dissociation of $[Au(CN)_2]^-$ and, using the concentration table and the value of β_2, we can determine the equilibrium concentrations.

Solution

We assume all of the gold is initially converted to $[Au(CN)_2]^-$, so our initial concentrations for the concentration table are:

$$[Au(CN)_2]^- = 2.5 \times 10^{-3}\,\text{M}$$
$$[Au^+] = 0\,\text{M}$$

We obtain the starting $[CN^-]$ from $[CN^-]_{added} - [CN^-]_{converted\ to\ [Ag(CN)_2]^-}$, so:

$$[CN^-] = [4 \times 10^{-2} - (2 \times 2.5 \times 10^{-3})]\,\text{M}$$
$$= 3.5 \times 10^{-2}\,\text{M}$$

The system then comes to equilibrium through dissociation of $[Au(CN)_2]^-$ to form both Au^+ and CN^-. The concentration table is then:

	$Au^+(aq)$	+	$2CN^-(aq)$	$\rightleftharpoons$	$[Au(CN)_2]^-$
Initial concentration (M)	0		3.5×10^{-2}		2.5×10^{-3}
Change in concentration (M)	$+x$		$+2x$		$-x$
Equilibrium concentration (M)	x		$3.5 \times 10^{-2} + 2x$		$2.5 \times 10^{-3} - x$

As $\beta_2 = 2 \times 10^{38}$, we know that the dissociation of $[Au(CN)_2]^-$ will proceed only to a tiny extent. We can, therefore, assume with confidence that $(2.5 \times 10^{-3} - x) \approx 2.5 \times 10^{-3}$ and $(3.5 \times 10^{-2} + 2x) \approx 3.5 \times 10^{-2}$. We can now substitute these values into the expression for β_2 and solve for the only remaining unknown, $[Au^+]$.

$$\beta_2 = \frac{[Au(CN)_2^-]}{[Au^+][CN^-]^2} = \frac{2.5\times10^{-3}}{[Au^+](3.5\times10^{-2})^2} = 2\times10^{38}$$

Therefore:

$$[Au^+] = \frac{2.5\times10^{-3}}{(2\times10^{38})(3.5\times10^{-2})^2}$$

$$= 1.0 \times 10^{-38}\,\text{M}$$

$$[CN^-] = 3.5 \times 10^{-2}\,\text{M}$$

$$[Au(CN)_2]^- = 2.5 \times 10^{-3}\,\text{M}$$

The tiny value of $[Au^+]$ shows the extraordinarily high affinity of CN^- for Au^+ in aqueous solution.

Is our answer reasonable?

Owing to the large value of β_2, we expect an extremely low concentration of Au^+ in solution, which is what we have found. Our assumptions are therefore reasonable and formation of $[Au(CN)_2]^-$ is essentially complete.

PRACTICE EXERCISE 13.7

Wilson's disease is a condition caused by excess copper in both the liver and the brain. One of the early drugs used to remove Cu^{2+} in the treatment of this disease was triethylenetetramine, $NH_2CH_2CH_2NHCH_2CH_2NHCH_2CH_2NH_2$ (trien), a tetradentate ligand. This forms the 1:1 complex $[Cu(trien)]^{2+}$ with $Cu^{2+}(aq)$, for which $\beta_1 = 2.5 \times 10^{20}$. If 0.500 g of $[Cu(OH_2)_6]Cl_2$ is dissolved in 0.200 L of 0.100 M aqueous trien, calculate the equilibrium concentrations of $Cu^{2+}(aq)$ and $[Cu(trien)]^{2+}$ in the final solution.

Inert and labile transition metal complexes

The magnitude of β_n for a complex is a measure of how far the formation reaction for that complex proceeds towards completion. For example, if β_n is small, the complex shows little tendency to form from the free metal ion and ligand(s); if β_n is large, the complex forms almost completely.

We might be tempted to say that the larger the value of β_n, the more *stable* the complex; this is true from a thermodynamic viewpoint, because the larger the value of β_n, the larger and more negative the value of ΔG° for the formation of a particular complex. However, we find that many complexes with large β_n values are, in fact, quite *unstable* with respect to exchange of ligands. This is illustrated by the complex $[Ni(CN)_4]^{2-}$, for which $\beta_4 = 3.2 \times 10^{30}$. This extremely large value means that addition of CN^- to a solution containing Ni(II) ions results in essentially complete formation of $[Ni(CN)_4]^{2-}$. However, addition of $^{14}CN^-$ (cyanide containing a radioactive ^{14}C label) to a solution of $[Ni(CN)_4]^{2-}$ results in very rapid incorporation of $^{14}CN^-$ into the complex, as shown in the following equation.

$$[Ni(CN)_4]^{2-} + {}^{14}CN \xrightarrow{\text{rapid}} [Ni({}^{14}CN)(CN)_3]^{2-} + CN^-$$

This means that the cyanide ligands in $[Ni(CN)_4]^{2-}$ are not bonded strongly to the Ni(II) ion, so they exchange very rapidly with added CN^-. We say that the $[Ni(CN)_4]^{2-}$ ion is **labile**, meaning that it undergoes rapid ligand exchange. $[Ni(CN)_4]^{2-}$ is, therefore, an example of a thermodynamically stable, but kinetically labile, transition metal complex. The opposite situation is observed for the Co(III) complex $[Co(NH_3)_6]^{3+}$. The equilibrium constant for the reaction:

$$[Co(NH_3)_6]^{3+} + 6H_3O^+ \rightleftharpoons [Co(OH_2)_6]^{3+} + 6NH_4^+$$

is of the order of 10^{25} at 25 °C, corresponding to a $\Delta G^\ominus$ value of $-143\ \text{kJ mol}^{-1}$. This means that the $[Co(NH_3)_6]^{3+}$ ion is thermodynamically unstable with respect to formation of $[Co(OH_2)_6]^{3+}$ in acidic solution. Despite this, $[Co(NH_3)_6]^{3+}$ can be recovered unchanged after several days in 1.0 M H_3O^+, meaning that the NH_3 ligands cannot easily exchange with H_2O. We say that $[Co(NH_3)_6]^{3+}$ is **inert** to ligand substitution and, therefore, is an example of a thermodynamically unstable, but kinetically inert, transition metal complex. Most first-row transition metal complexes are labile, but complexes containing metals with a d^3 or d^6 electron configuration are often inert, with Cr(III) and Co(III) being the classic examples of inert metal centres. It is important, therefore, that you are careful when using the word 'stable', and that you realise that thermodynamic stability and kinetic stability are entirely separate concepts.

Electrochemical aspects of transition metal complexes

Electrochemistry provides another convenient method of quantifying the thermodynamic stability of transition metal complexes. As we saw in chapter 12 (table 12.1), the reduction potentials of simple aqueous transition metal ions span a wide range of values and, therefore, electrochemical behaviour. Thus, the $Ag^+(aq)$ ion is moderately oxidising ($E^\ominus_{Ag^+/Ag} = +0.80$ V) while the $Cr^{3+}(aq)$ ion shows negligible oxidising ability ($E^\ominus_{Cr^{3+}/Cr} = -0.74$ V). However, coordination of ligands to aqueous transition metal ions can result in the formation of complexes having very different electrochemical behaviours from those of the parent aqueous transition metal ions. The classic example of this is provided by the $Co^{3+}(aq)$ (or $[Co(OH_2)_6]^{3+}$) ion. Under standard conditions in aqueous solution, this ion is extremely strongly oxidising ($E^\ominus_{Co^{3+}/Co^{2+}} = +1.9$ V), and as a result is generally stable only under ice-cold acidic conditions. However, replacement of the six H_2O ligands with six NH_3 or three ethylenediamine (en) ligands results in enormous stabilisation of the resulting $[Co(NH_3)_6]^{3+}$ ($E^\ominus_{[Co(NH_3)_6]^{3+}/[Co(NH_3)_6]^{2+}} = 0.1$ V) and $[Co(en)_3]^{3+}$ ($E^\ominus_{[Co(en)_3]^{3+}/[Co(en)_3]^{2+}} = -0.25$ V) ions. This observation is general for the majority of N-donor ligands and, as a result, Co(III) complexes in which the metal ion is bonded to six oxygen atoms are rather rare.

When a transition metal ion is surrounded by anionic ligands, the resulting complex generally becomes more difficult to reduce, as addition of an electron to a negatively charged complex is unfavourable on electrostatic grounds. An example of this is the anionic Fe(III) complex $[Fe(CN)_6]^{3-}$, for which $E^\ominus_{[Fe(CN)_6]^{3-}/[Fe(CN)_6]^{4-}} = +0.28$ V, significantly less than that for the cationic $Fe^{3+}(aq)$ ion ($E^\ominus_{Fe^{3+}/Fe^{2+}} = +0.77$ V).

The correct choice of ligand can assist in stabilising transition metal ions that are unstable with respect to **disproportionation**. This is a redox reaction in which the metal ion is simultaneously oxidised and reduced. The most celebrated example is the $Cu^+(aq)$ ion, for which the following $E^\ominus$ values have been obtained.

$$E^\ominus_{Cu^{2+}/Cu^+} = 0.16\ \text{V}$$

$$E^\ominus_{Cu^+/Cu} = 0.52\ \text{V}$$

The disproportionation of Cu^+ proceeds according to the equation:

$$2Cu^+(aq) \rightarrow Cu^{2+}(aq) + Cu(s)$$

for which $E^\ominus = E_R - E_L = 0.52\ \text{V} - 0.16\ \text{V} = +0.36\ \text{V}$ (assuming that Cu^+ is reduced at the right-hand electrode). The positive value of $E^\ominus$ shows that this reaction is spontaneous under standard conditions, and aqueous solutions of Cu^+ are often found to be unstable. However, use of strong-field ligands (see p. 576) such as CN^- or various phosphanes gives Cu(I) complexes which show significant stability. Note that disproportionation is not limited to transition metal ions, and species such as O_2^- (superoxide) and HOCl (hypochlorous acid) can disproportionate.

Bonding in transition metal complexes

Earlier in this chapter, we stated that two of the defining characteristics of transition metal complexes are that they are usually coloured and that they are often paramagnetic. It is not unusual for a given metal ion to form complexes with different ligands to give a rainbow of colours, as illustrated in figure 13.24 for a series of Co(III) complexes.

FIGURE 13.24 Aqueous solutions of the Co(III) complexes (from left to right): $[Co(O_2CO)(en)_2]^+$, *trans*-$[CoCl_2(en)_2]^+$, $[Co(en)_3]^{3+}$, $[CoCl(NH_3)_5]^{2+}$ and $[Co(NH_3)_5OH_2]^{3+}$. As the metal ion is the same in all cases, the variety of colours arises from the different ligands surrounding the Co(III) ion.

Also, because transition metal ions often have incompletely filled *d* orbitals, we expect to find many of them with unpaired *d* electrons, and such compounds should therefore be paramagnetic. However, for a given metal ion in a particular oxidation state, the number of unpaired electrons is not necessarily the same from one complex to another. Consider, for example, the Fe(II) complexes $[Fe(OH_2)_6]^{2+}$ and $[Fe(CN)_6]^{4-}$. As Fe^{2+} has the electron configuration $[Ar]3d^6$, both contain six *d* electrons, but four of these are unpaired in the $[Fe(OH_2)_6]^{2+}$ ion, while all of them are paired in the $[Fe(CN)_6]^{4-}$ ion. As a result, the $[Fe(OH_2)_6]^{2+}$ ion is paramagnetic and the $[Fe(CN)_6]^{4-}$ ion is diamagnetic. We can explain both the colours and magnetic properties of transition metal complexes using **crystal field theory**, a theory of bonding that looks at how the energies of the *d* orbitals on the metal ion are influenced by the surrounding ligands.

Crystal field theory of bonding in octahedral coordination complexes

Crystal field theory is based on the premise that the ligands surrounding the metal ion in a transition metal complex exert an electric field that affects the energies of the *d* orbitals on the metal ion to different extents. It assumes that the interaction between the metal and the ligands is purely electrostatic, an unrealistic situation, given that all metal–ligand bonds exhibit some degree of covalency. Nevertheless, crystal field theory can explain many of the unusual features of transition metal complexes and is conceptually straightforward.

We begin by looking at the electron distributions (figure 13.25) and energies of the five *d* orbitals in a free transition metal ion.

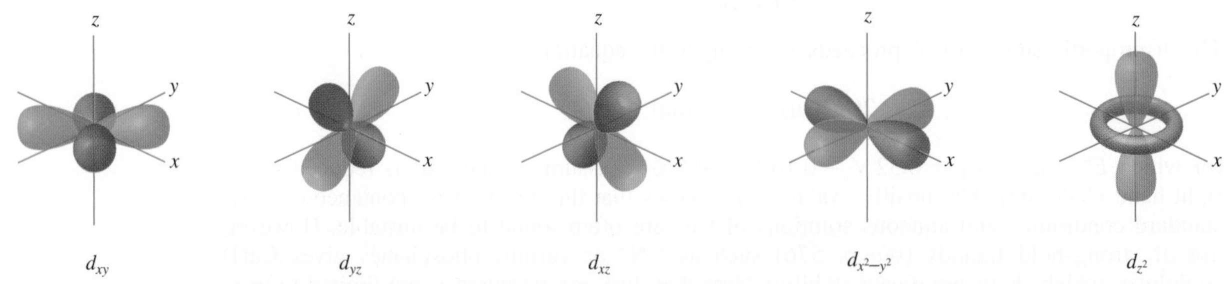

FIGURE 13.25 The directional properties of the five *d* orbitals in a free transition metal ion.

The d_{xy}, d_{xz} and d_{yz} orbitals all have lobes pointing between the *x*-, *y*- and *z*-axes, while the lobes of the $d_{x^2-y^2}$ and d_{z^2} orbitals point along the axes. In the absence of any ligands, the five *d* orbitals are **degenerate**, meaning they are all of the same energy. However, this is not the case in the presence of ligands, where we find that the degeneracy is removed. We illustrate this by considering the formation of an octahedral transition metal complex by the introduction of six ligands at bonding distance along the *x*-, *y*- and *z*-axes, as shown in figure 13.26 on the next page, and looking at the effect of this on the energies of the five *d* orbitals.

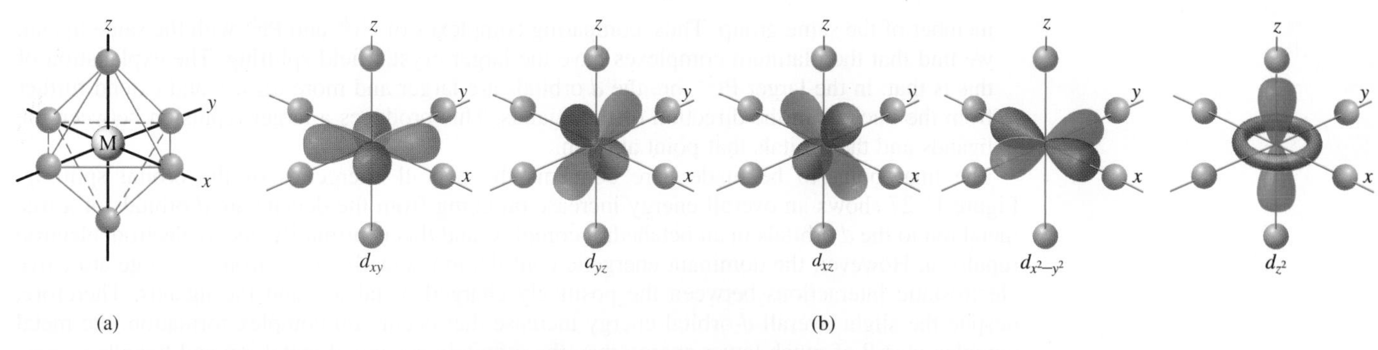

FIGURE 13.26 (a) An octahedral ML_6 complex and **(b)** the orientation of the metal orbitals with respect to the ligands in an octahedral ML_6 complex.

In an isolated atom or ion, all the *d* orbitals of a given *d* subshell are degenerate. Therefore, an electron has the same energy regardless of which *d* orbital it occupies. In an octahedral complex, however, this is no longer the case. As the lobes of the $d_{x^2-y^2}$ and d_{z^2} orbitals point directly towards the ligands, an electron in either of these orbitals is nearer the electron pairs of the ligands than if it was in a d_{xy}, d_{xz} or d_{yz} orbital, with lobes pointing between the ligands. Since the electron itself is negatively charged and is repelled by the ligand electrons, an electron in the $d_{x^2-y^2}$ or d_{z^2} orbital is of higher energy than one in a d_{xy}, d_{xz} or d_{yz} orbital. Therefore, we can see that introduction of the ligands removes the degeneracy of the five *d* orbitals, splitting them into two sets of degenerate orbitals; the lower energy set consists of the d_{xy}, d_{xz} and d_{yz} orbitals, and the higher energy set comprises the $d_{x^2-y^2}$ and d_{z^2} orbitals (figure 13.27). The three lower energy orbitals are collectively called the t_{2g} orbitals while the two higher energy orbitals are called the e_g orbitals. These labels refer to the symmetry of the sets of orbitals, and are obtained using a branch of mathematics called group theory, which is described in advanced chemistry courses.

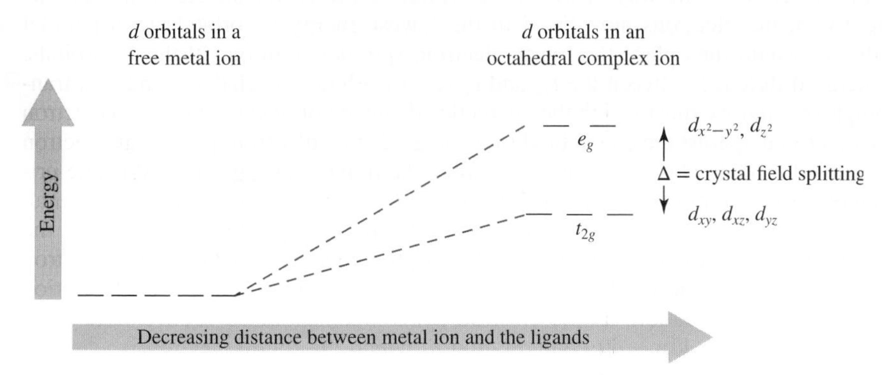

FIGURE 13.27 The changes in the energies of the *d* orbitals of a metal ion as an octahedral complex is formed. As the ligands approach the metal ion, the initially degenerate *d* orbitals split into two new sets of degenerate orbitals.

Figure 13.27 shows that, regardless of which orbital the electron occupies, its energy increases because it is repelled by the electrons of the approaching ligands. However, the electron is repelled more (and has a higher energy) if it is in an orbital that points directly at the ligands than if it occupies an orbital that points between them. In an octahedral complex, the energy difference between the two sets of *d*-orbital energy levels is called the **crystal field splitting energy** (CFSE). It is usually given the symbol Δ_o in octahedral complexes (delta oh, the 'o' standing for octahedral), and its magnitude depends on the following factors.

- *The nature of the ligand.* Some ligands produce a larger splitting of the energies of the *d* orbitals than others. For example, for a given metal ion, CN^- nearly always gives a large value of Δ_o, and F^- always gives a small value. This is due to the more extensive orbital interactions between the metal and ligand that occur in the case of CN^-. We will discuss this further later in this section.
- *The oxidation state of the metal.* For a given metal and ligand, the size of Δ_o increases with an increase in the oxidation state of the metal. As electrons are removed from the metal, the charge on the ion becomes more positive and the ion becomes smaller. This means that the ligands are attracted to the metal more strongly and they can approach the metal ion more closely. As a result, they also approach the *d* orbitals along the *x*-, *y*- and *z*-axes more closely, causing greater repulsion. This produces a greater splitting of the t_{2g} and e_g orbitals and a correspondingly larger value of Δ_o.
- *The position of the transition metal in the periodic table.* For a given ligand and oxidation state, the magnitude of Δ_o increases going down a group. In other words, an ion of an element in the first row of transition elements has a smaller value of Δ_o than the ion of a heavier

member of the same group. Thus, comparing complexes of Ni^{2+} and Pt^{2+} with the same ligand, we find that the platinum complexes have the larger crystal field splitting. The explanation of this is that, in the larger Pt^{2+} ion, the *d* orbitals are larger and more diffuse and extend further from the nucleus in the direction of the ligands. This produces a larger repulsion between the ligands and the orbitals that point at them.

The final point to be made here concerns the overall energetics of the orbital splitting. Figure 13.27 shows an overall energy increase on going from the degenerate *d* orbitals of a free metal ion to the *d* orbitals in an octahedral complex, and this is primarily due to electron–electron repulsion. However, the dominant energetic contribution actually arises from the large attractive electrostatic interactions between the positively charged metal ion and the ligands. Therefore, despite the slight overall *d*-orbital energy increase that occurs on complex formation, the metal complex is still of much lower energy than the infinitely separated metal ion and ligands.

Electron configurations in octahedral transition metal complexes

Having shown that the *d* orbitals in a transition metal complex are not degenerate, we now look at how these orbitals are populated with *d* electrons. We learned in chapter 4 that orbitals are filled in order of increasing energy (the Aufbau principle); a set of degenerate orbitals has a single electron placed in each of the orbitals before any pairing of electrons takes place (Hund's rule); and no two electrons of the same spin may occupy the same orbital (the Pauli exclusion principle). These rules give us the ground-state electron configuration of any chemical species. We will, therefore, consider the situations outlined in figure 13.28 for octahedral complexes having d^1, d^2 and d^3 electron configurations.

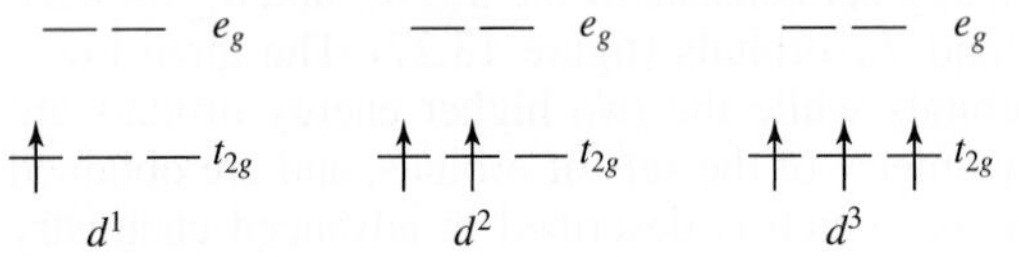

FIGURE 13.28 Orbital occupancies for d^1, d^2 and d^3 electron configurations.

There is no ambiguity about the way in which the orbitals are filled for the electron configurations in figure 13.28; the electrons are placed in the lowest energy t_{2g} orbitals with parallel spins. Ordinarily, we would then place the fourth electron, spin down, in one of the t_{2g} orbitals. However, the energy difference between the t_{2g} and e_g sets of orbitals is relatively small in transition metal complexes, and we must weigh the energetic advantage of placing the fourth electron in a low-energy t_{2g} orbital against the energetic disadvantage that results from placing an electron in an already occupied orbital. This latter energy is called the **pairing energy (*P*)** and is due primarily to the interelectronic repulsions that result from having two electrons in the same orbital. The magnitude of *P*, compared with the crystal field splitting energy, Δ_o, determines which of the two possible electron configurations is favoured. If $P > \Delta_o$, the energetically favourable electron configuration is that in which the fourth electron occupies an e_g orbital. Such a configuration is called the **high-spin** configuration, as the electron spins are maximised. Conversely, when $P < \Delta_o$, the complex adopts the **low-spin** configuration in which the fourth electron is added to one of the lower energy orbitals and is paired with one of the t_{2g} electrons. In this case, the electron spins are minimised and the lowest number of unpaired electrons results. These possibilities are shown in figure 13.29 for the d^4 Cr(II) complexes $[Cr(OH_2)_6]^{2+}$ and $[Cr(CN)_6]^{4-}$.

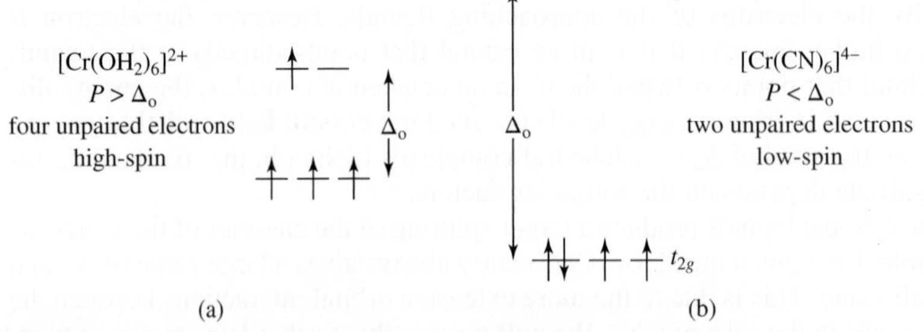

FIGURE 13.29 The two possible electron configurations for a d^4 complex:
(a) When Δ_o is small, as for $[Cr(OH_2)_6]^{2+}$, the high-spin configuration is favoured.
(b) A large value of Δ_o favours the low-spin configuration, as shown for $[Cr(CN)_6]^{4-}$.

Because of its relatively small value of Δ_o, $[Cr(OH_2)_6]^{2+}$ adopts the high-spin configuration, while $[Cr(CN)_6]^{4-}$, which has a significantly larger value of Δ_o, is low-spin. In octahedral complexes, high-spin and low-spin possibilities exist only for d^4, d^5, d^6 and d^7 electron configurations (figure 13.30). There is only one possible configuration for each of the d^8, d^9 and d^{10} configurations (figure 13.31).

Crystal field theory applied to 4-coordinate complexes

As mentioned earlier, 4 is the second most common coordination number for metal ions in coordination compounds. We can apply crystal field theory to the two possible 4-coordinate geometries, square planar and tetrahedral, using the same principles we learned earlier.

Square planar complexes

We can form a square planar complex from an octahedral complex simply by removing the two axial ligands that lie along the *z*-axis. This lowers the energy of any of the *d* orbitals with a *z* component (d_{z^2}, d_{xz} and d_{yz}) as electronic repulsion between the electrons in these orbitals and the ligand electrons is reduced. Removing the axial ligands along the *z*-axis allows closer approach to the metal by the ligands in the *xy*-plane; this means that the energies of the orbitals in this plane ($d_{x^2-y^2}$ and d_{xy}) are raised slightly, with the energy of the $d_{x^2-y^2}$ orbital being more affected as its lobes point directly towards the ligands. Figure 13.32 shows these energy changes and the resulting *d*-orbital splitting diagram for a square planar complex.

Tetrahedral complexes

The derivation of a *d*-orbital splitting diagram for a tetrahedral complex is not as conceptually straightforward as that for an octahedral or square planar complex, as none of the four ligands lies directly along the *x*-, *y*- or *z*-axes (figure 13.33).

Notice that the order of the orbitals is exactly opposite to that for an octahedral complex (see figure 13.28). In a tetrahedral complex, the crystal field splitting energy is termed Δ_t, and it can be shown that $\Delta_t \approx \frac{4}{9}\Delta_o$ for the same metal ion with the same ligands. This small value of Δ_t is usually less than the pairing energy, *P*, so tetrahedral complexes almost always adopt high-spin electron configurations. Note that high-spin and low-spin configurations are theoretically possible for d^3, d^4, d^5 and d^6 tetrahedral complexes.

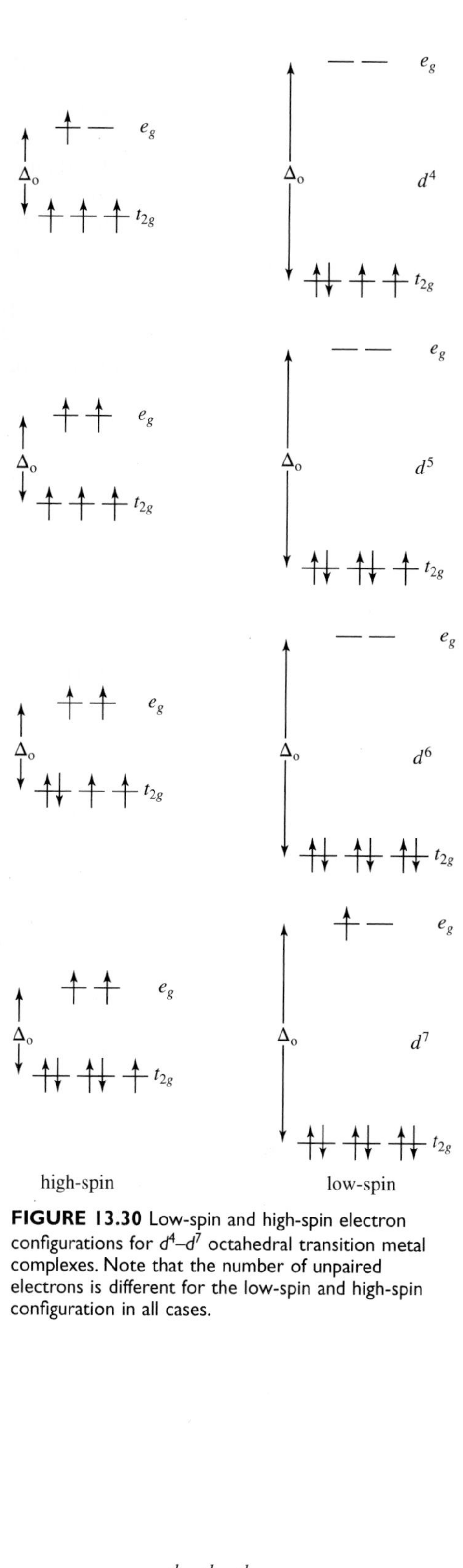

FIGURE 13.30 Low-spin and high-spin electron configurations for d^4–d^7 octahedral transition metal complexes. Note that the number of unpaired electrons is different for the low-spin and high-spin configuration in all cases.

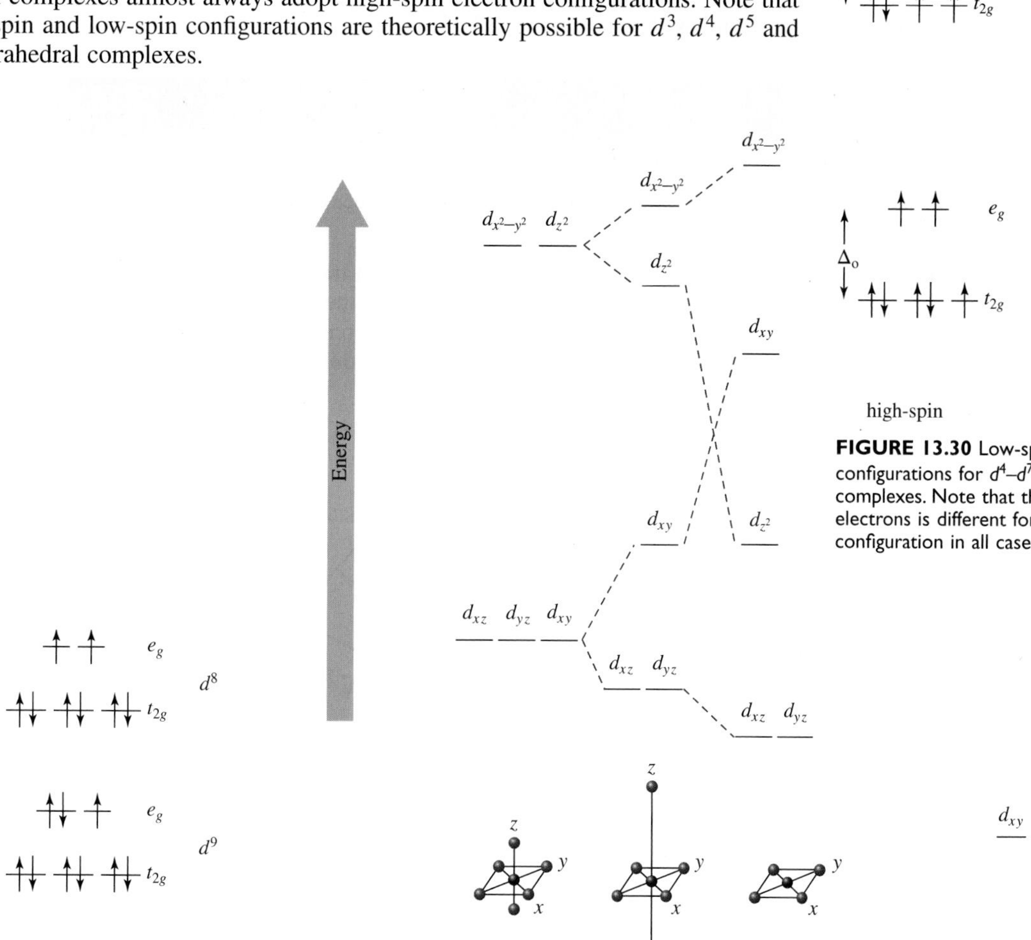

FIGURE 13.31 Electron configurations for d^8, d^9 and d^{10} octahedral transition metal complexes.

FIGURE 13.32 The derivation of a *d*-orbital splitting diagram (not to scale) for a square planar complex by removal of axial ligands from an octahedral complex.

FIGURE 13.33 The *d*-orbital splitting diagram for a tetrahedral complex.

The colours of transition metal complexes

We saw in chapter 4 that excited-state atoms emit photons of particular energies only. A similar situation occurs when atoms, ions or molecules absorb light; we find that these species absorb photons of certain energies only, rather than absorbing all photons irrespective of their energy. Absorption of a photon by an octahedral transition metal complex can occur if its energy exactly matches Δ_o, the energy difference between the t_{2g} and e_g sets of orbitals. This then leads to an **electronic transition** from a t_{2g} orbital to a higher energy e_g orbital. Such electronic transitions, which involve electrons in the t_{2g} and e_g sets of orbitals only, are called ***d–d* transitions**. In many octahedral transition metal complexes, the energy difference, Δ_o, corresponds to photons of visible light, as shown in figure 13.34 for the d^1 complex $[Ti(OH_2)_6]^{3+}$. This is the reason that many transition metal complexes appear coloured.

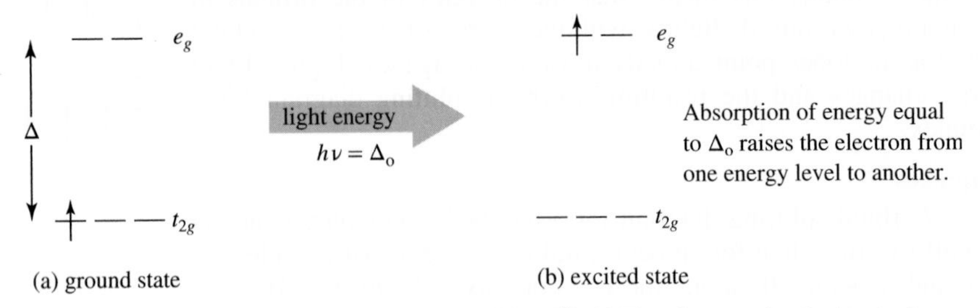

FIGURE 13.34 Absorption of a photon by $[Ti(OH_2)_6]^{3+}$. **(a)** The electron distribution in the ground state of the $[Ti(OH_2)_6]^{3+}$ ion. **(b)** Absorption of a photon results in an electronic transition from the t_{2g} set of *d* orbitals to the higher energy e_g set.

As you know, white light contains photons with energies that correspond to all of the colours in the visible spectrum, as shown in figure 13.35.

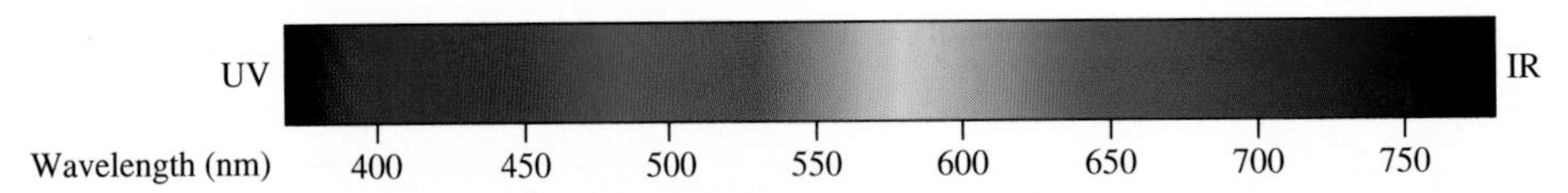

FIGURE 13.35 The visible spectrum, showing the colours as a function of wavelength.

If we shine white light through a solution of a coloured transition metal complex, the light that emerges after passing through the solution contains all the colours except those that have been absorbed, and it is these transmitted colours that we perceive when we look at the solution. If we know what colour the solution appears, we can determine which colour is being absorbed by using a colour wheel, like that shown in figure 13.36.

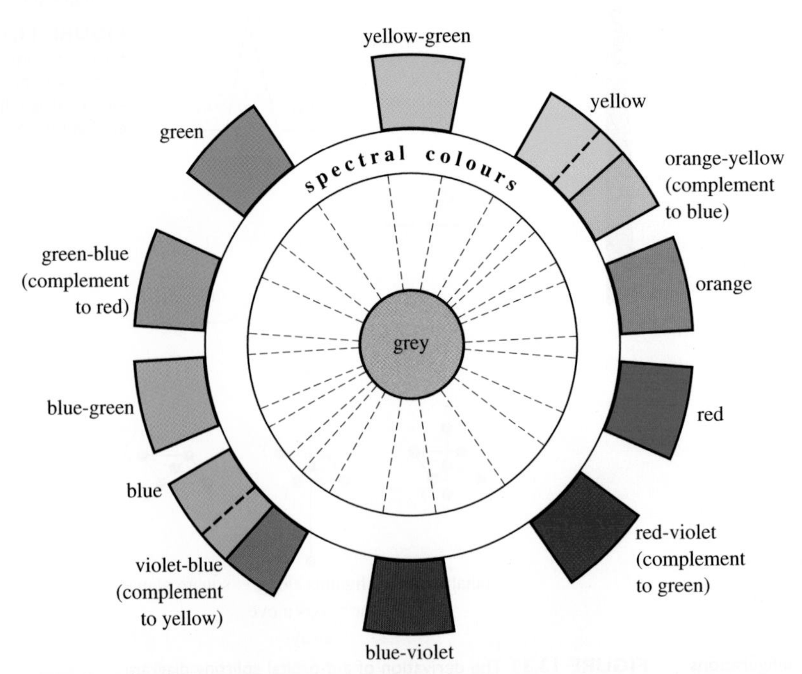

FIGURE 13.36 A colour wheel. Colours opposite each other are called complementary colours. When a substance absorbs light of a particular colour, the light that is reflected or transmitted has the colour of its complement. Thus, something that absorbs red light appears green-blue, and vice versa.

The colour lying opposite any colour on the wheel is called its **complementary colour**; for example, red-violet is the complementary colour of green, while green-blue is the complementary colour of red. If a substance absorbs a particular colour when illuminated with white light, the perceived colour of the reflected or transmitted light is the complementary colour. In the case of the $[Ti(OH_2)_6]^{3+}$ ion, the electronic transition from the t_{2g} set to the e_g set occurs on absorption of green light and this is why a solution of this ion appears red-violet. We can quantify the absorption of visible light using a UV/visible spectrophotometer, a device that determines both the wavelength and intensity of light absorbed by chemical species. Spectrophotometers output a spectrum, a plot of **absorbance (A)** versus wavelength, where absorbance is defined as:

$$A = \log\frac{I_o}{I}$$

with I_o being the intensity of the incident light and I the intensity of the transmitted light. Note that absorbance is a dimensionless quantity.

Figure 13.37 shows the UV/visible spectrum of the red-violet $[Ti(OH_2)_6]^{3+}$ ion, and we can see that this displays a single peak, with a maximum absorbance at 514 nm.

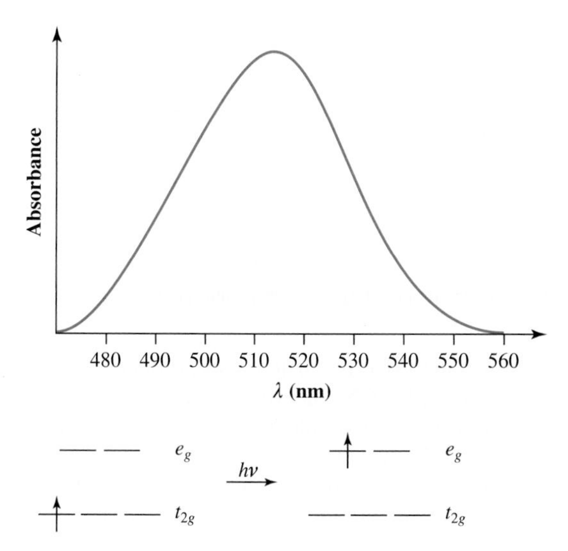

FIGURE 13.37 The UV/visible spectrum of the $[Ti(OH_2)_6]^{3+}$ ion. The peak corresponds to an electronic transition of the single d electron from the t_{2g} to the e_g orbital set.

As Ti(III) is a d^1 transition metal ion, we know that this absorbance band is due to the promotion of the single d electron from the t_{2g} set of orbitals to the e_g set of orbitals, and we can use the highest point on the peak to calculate the energy difference between these sets of orbitals — in other words, we can calculate the value of Δ_o for $[Ti(OH_2)_6]^{3+}$ from this spectrum. Recall from chapter 4 that there is an inverse relationship between wavelength (λ) and energy (E), given by the equation:

$$E = \frac{hc}{\lambda}$$

where $h = 6.626 \times 10^{-34}$ J s (Planck's constant) and $c = 2.998 \times 10^8$ m s^{-1} (the speed of light). We can now find the energy corresponding to a photon of wavelength 514 nm by using the equation:

$$\begin{aligned} E &= \frac{(6.626 \times 10^{-34}\,\text{J s}) \times (2.998 \times 10^8\,\text{m s}^{-1})}{514 \times 10^{-9}\,\text{m}} \\ &= 3.86 \times 10^{-19}\,\text{J} \end{aligned}$$

This value tells us that the t_{2g} and e_g orbitals are separated by 3.86×10^{-19} J in a single $[Ti(OH_2)_6]^{3+}$ ion. It is usual to report the value of Δ_o for a complex on a molar basis, rather than for an individual ion, and we do this by multiplying E by the Avogadro constant (6.022×10^{23} mol^{-1}). Therefore:

$$\begin{aligned} E &= (3.86 \times 10^{-19}\,\text{J}) \times (6.022 \times 10^{23}\,\text{mol}^{-1}) \\ &= 232 \times 10^3\,\text{J mol}^{-1} \\ &= 232\,\text{kJ mol}^{-1} \end{aligned}$$

Therefore, Δ_o for $[Ti(OH_2)_6]^{3+}$ = 232 kJ mol^{-1}. It should be noted that the *exact* value of Δ_o can be determined in this way only for a small number of metal complexes. Electron–electron repulsions in complexes with more than one *d* electron mean that the exact determination of Δ_o for such complexes is difficult, and, indeed, beyond the scope of this book. However, the previous treatment does give a fair approximation of Δ_o in these cases and we will continue to use it throughout this chapter.

We have already seen in a series of Co(III) complexes that the nature of the ligands is critical in determining the colour of transition metal complexes (see figure 13.24) and must therefore influence the value of Δ_o. The greater the value of Δ_o, the further apart in energy are the t_{2g} and e_g orbitals and, therefore, the shorter the wavelength of light required to promote an electron from the t_{2g} set to the e_g set of orbitals. This can be illustrated by looking at the Ti(III) complexes $[TiF_6]^{3-}$ and $[Ti(NCS)_6]^{3-}$, which have maximum absorbance wavelengths (λ_{max}) at 526 nm and 543 nm, respectively. These are at a longer wavelength than that of $[Ti(OH_2)_6]^{3+}$, which has a maximum absorbance wavelength of 514 nm. The Δ_o values corresponding to these absorption maxima are 232 kJ mol^{-1} for $[Ti(OH_2)_6]^{3+}$, 227 kJ mol^{-1} for $[TiF_6]^{3-}$ and 220 kJ mol^{-1} for $[Ti(NCS)_6]^{3-}$, meaning that the separation of the t_{2g} and e_g orbitals varies significantly with the identities of the ligands for the same metal in the same oxidation state (figure 13.38).

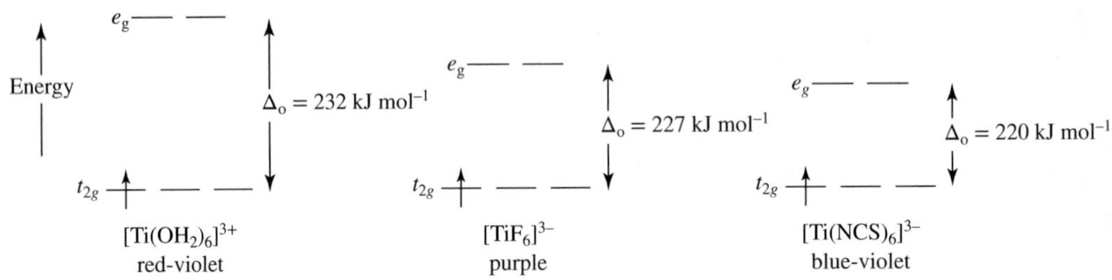

FIGURE 13.38 Energy separations between their t_{2g} and e_g orbitals for $[Ti(OH_2)_6]^{3+}$, $[TiF_6]^{3-}$ and $[Ti(NCS)_6]^{3-}$.

A ligand that produces a large crystal field splitting with one metal ion also produces a large value of Δ_o in complexes with other metals. For example, CN^- nearly always gives a very large value of Δ_o regardless of the metal to which it is bound. Complexes containing NH_3 have lower values of Δ_o than complexes containing CN^-, while Δ_o values for complexes containing H_2O are smaller still. Thus, ligands can be arranged in order of the magnitude of their crystal field splitting energies. This sequence is called the **spectrochemical series**. Such a series containing some common ligands arranged in order of decreasing Δ_o values is:

$$CO \approx CN^- > NO_2^- > en > NH_3 > H_2O > C_2O_4^{2-} > OH^- > F^- > Cl^- > Br^- > I^-$$

For a given metal ion, the carbonyl ligand produces the largest value of Δ_o, and iodide produces the smallest. Ligands such as CO, CN^- and NO_2^-, which give large values of Δ_o, are called **strong-field** ligands, while the halide ions and others that induce small *d*-orbital splittings are called **weak-field** ligands. The *d*-orbital splittings induced by strong-field ligands are often so large that the energy gap between the t_{2g} and e_g sets of orbitals is greater than the energy of visible photons. Therefore, electronic transitions between the t_{2g} and e_g orbitals in octahedral complexes containing only CO or CN^- ligands are often caused only by UV photons, so such complexes are usually very pale yellow or colourless.

Octahedral complexes of Zn(II) and high-spin Mn(II) are also often colourless, or nearly so, but for a very different reason; Zn(II) is a d^{10} metal ion, meaning that the t_{2g} and e_g sets of orbitals are full (figure 13.39).

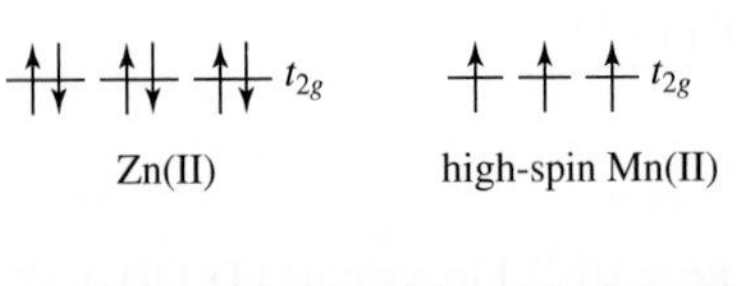

e_g t_{2g} Zn(II) e_g t_{2g} high-spin Mn(II)

FIGURE 13.39 The *d*-electron configurations for complexes of Zn(II) (left) and high-spin Mn(II) (right).

No electronic transition between the t_{2g} and e_g orbitals is possible, as the e_g orbitals are unable to accommodate an extra electron. High-spin Mn(II) has a d^5 electron configuration as shown in figure 13.39; to promote an electron from a t_{2g} orbital to an e_g orbital, we would require the electron involved to change its spin during the transition, otherwise a situation would result in which two electrons with the same spin occupy a single orbital, in violation of the Pauli exclusion principle. We say, therefore, that the $t_{2g} \rightarrow e_g$ electronic transition in such complexes is spin forbidden; the probability of such a transition occurring is very low, so these complexes are usually very pale in colour.

We might expect complexes containing transition metal ions with a d^0 electron configuration to be colourless, as obviously there are no possible *d–d* transitions between the t_{2g} and e_g orbitals, and this is borne out in many cases. However, ions such as $[MnO_4]^-$ (purple) and $[Cr_2O_7]^{2-}$ (orange) are very intensely coloured, despite the metal ion having a d^0 configuration in each case. The colour in such compounds is due to **ligand-to-metal charge transfer (LMCT) transitions**, which involve the formal transfer of an electron on the ligand to an orbital on the metal ion. While such transitions usually occur in the ultraviolet part of the spectrum and are, therefore, invisible to the eye, some low-energy LMCT transitions are observed, such as those in $[MnO_4]^-$ and $[Cr_2O_7]^{2-}$. LMCT transitions have a much higher probability of occurring than transitions between t_{2g} and e_g orbitals, and this explains the extremely intense colour of, for example, the $[MnO_4]^-$ ion. In some complexes where the metal ion contains occupied *d* orbitals, an electron can be formally transferred in the opposite direction, i.e. from the metal to the ligand. Such processes are called **metal-to-ligand charge transfer (MLCT) transitions** and often occur in complexes having aromatic, pyridine-based ligands. These transitions also have high probabilities and lead to very intensely coloured complexes. One of the best known examples of a complex that displays an MLCT transition is $[Ru(bipy)_3]^{2+}$, which contains three bidentate 2,2'-bipyridine ligands.

It should be noted that it is not the nature of the ligands alone that ultimately determines the colour of a transition metal complex. For example, the isomeric complexes *cis*-$[CoCl_2(en)_2]^+$ and *trans*-$[CoCl_2(en)_2]^+$, which contain identical ligands coordinated to a Co(III) centre, are violet and green, respectively. Other factors such as the arrangement of the ligands and the metal–ligand bond lengths can also be important.

WORKED EXAMPLE 13.8

Colours of transition metal complexes

A student prepared five Co(III) complexes in an undergraduate chemistry laboratory and stored them in sample tubes. However, the student forgot to label the tubes and, at the start of the next laboratory session, could not remember which complex was which. To identify them, the student recorded the UV/visible spectra of the complexes and found that the five complexes exhibited maximum absorbance values in the visible region at 506 nm, 490 nm, 534 nm, 441 nm and 471 nm. The five complexes prepared by the student were $[CoCl(NH_3)_5]^{2+}$, $[Co(NH_3)_6]^{3+}$, $[Co(NH_3)_5OH]^{2+}$, $[CoCN(NH_3)_5]^{2+}$ and $[Co(NH_3)_5OH_2]^{3+}$. Match the UV/visible data to the complexes.

Analysis

We have five very similar complexes, all of which contain a $[Co(NH_3)_5]^{3+}$ moiety and differ only in the nature of the sixth ligand. Therefore, we should be able to use the spectrochemical series to decide the order of the UV/visible absorption maxima across the five complexes; the strongest field ligand should have the largest energy gap between the t_{2g} and e_g orbitals, and should therefore have the shortest wavelength absorption maximum.

Solution

We begin by ordering the five ligands according to their positions in the spectrochemical series, from strongest field to weakest field. The order is therefore:

$$CN^- > NH_3 > OH_2 > OH^- > Cl^-$$

We would expect the complex containing CN^- to have the largest energy gap between the t_{2g} and e_g sets of orbitals, as CN^- is the strongest field ligand, and therefore this complex will exhibit the shortest wavelength (highest energy) absorption maximum. The complex containing Cl^- should have the smallest energy gap between the t_{2g} and e_g sets of orbitals, as Cl^- is the weakest field ligand, and therefore this complex will exhibit the longest wavelength (lowest energy) absorption maximum. The remaining three complexes will be ordered according to the positions of the ligands in the spectrochemical series above. Therefore, the identities of the complexes are:

$[CoCN(NH_3)_5]^{2+}$: $\lambda_{max} = 441$ nm

$[Co(NH_3)_6]^{3+}$: $\lambda_{max} = 471$ nm

$[Co(NH_3)_5OH_2]^{3+}$: $\lambda_{max} = 490$ nm

$[Co(NH_3)_5OH]^{2+}$: $\lambda_{max} = 506$ nm

$[CoCl(NH_3)_5]^{2+}$: $\lambda_{max} = 534$ nm

PRACTICE EXERCISE 13.8

The octahedral complex $[CoCl_6]^{4-}$ and the tetrahedral complex $[CoCl_4]^{2-}$ both contain Co(II). Which is larger: Δ_o for $[CoCl_6]^{4-}$ or Δ_t for $[CoCl_4]^{2-}$? Explain your answer.

The absorbance of a solution of a transition metal complex (and indeed, of any solution) is related to the concentration of the absorbing species by the **Beer–Lambert law** (also often called **Beer's law**), which states that:

$$A = \varepsilon c l$$

where A = absorbance (dimensionless), ε = molar absorption coefficient ($mol^{-1}\,L\,cm^{-1}$), c = concentration of absorbing species ($mol\,L^{-1}$) and l = pathlength of cell (cm). The somewhat unusual units of ε arise from the fact that UV/visible spectra are generally measured using cells having a pathlength of 1 cm.

The ε values for transition metal complexes depend on both the geometry of the metal ion and the type of electronic transition occurring and, can be useful in characterising these species. For example, *d–d* transitions for octahedral complexes generally have ε values $<100\,mol^{-1}\,L\,cm^{-1}$ and, as a result, such complexes are usually weakly coloured. Tetrahedral complexes are often more intensely coloured, with ε values for *d–d* transitions in these lying in the range 200–250 $mol^{-1}\,L\,cm^{-1}$. More intense still, as discussed on the previous page, are LMCT and MLCT transitions, which can have ε values up to $10^4\,mol^{-1}\,L\,cm^{-1}$.

You will notice that the UV/visible spectra arising from *d–d* transitions in transition metal complexes look very different from the atomic spectra we saw in chapter 4 (p. 128). While the latter display sharp lines, indicative of electronic transitions between well-defined, quantised energy levels, the former show broad bands, an observation seemingly at odds with the definite energies of the t_{2g} and e_g sets of orbitals between which these electronic transitions occur. The broad bands can be rationalised by considering vibrations of the complexes. Consider for example the octahedral $[Ti(OH_2)_6]^{3+}$ ion, the UV/visible spectrum of which is given in figure 13.37; if all the $[Ti(OH_2)_6]^{3+}$ ions in the sample exhibited identical octahedral geometry with all Ti—O bond distances the same, then the *d–d* absorbance band would indeed be much narrower than that observed. However, vibrations within the cations at room temperature ensure that there is a variety of Ti—O bond distances and, therefore, geometries at any given time. As the energies of the t_{2g} and e_g sets of orbitals depend on the Ti—O bond distances, this therefore leads to a range of energy differences between the t_{2g} and e_g sets, which in turn results in broad absorption bands. Single atoms cannot undergo vibrations in the same way as molecules or polyatomic ions, and so their spectra consist of sharp lines. This is illustrated in figure 13.40.

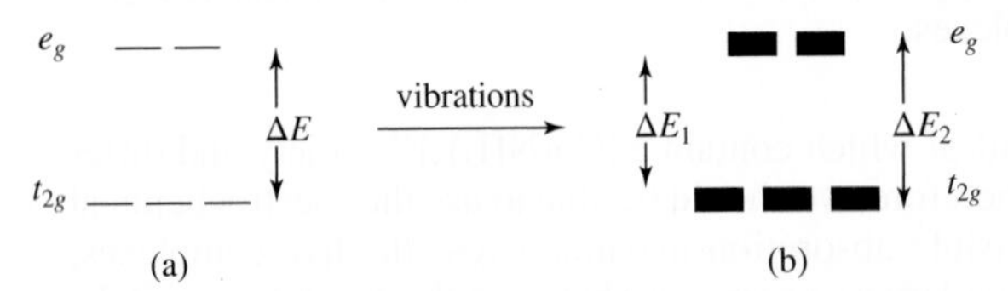

FIGURE 13.40 (a) In the absence of vibrations, the energies of the t_{2g} and e_g sets of orbitals are precisely defined, and an electronic transition between these therefore results in a narrow spectral line. **(b)** When vibrations occur, they result in a range of energies for each of the t_{2g} and e_g sets of orbitals. Electronic transitions between these therefore span the range of energies between ΔE_1 and ΔE_2, thereby leading to broad bands.

The magnetic properties of transition metal complexes

As we have seen both in chapter 4 and earlier in this chapter, a paramagnetic substance is one that is attracted into a magnetic field. For a substance to be paramagnetic, it must contain unpaired electrons, the number of which determines the magnitude of its response to a magnetic field. This response can be measured by using a very accurate balance to determine the mass difference of a sample of the complex in the presence and absence of a magnetic field. This allows us to calculate the **magnetic moment** (μ) of the complex and, hence, to determine the number of unpaired electrons it contains. The magnetic moment of a complex is measured in Bohr magnetons (μ_B) and is defined as:

$$\mu = \sqrt{n(n+2)}\ \mu_B$$

where n is the number of unpaired electrons in the complex. Therefore, a complex containing one unpaired electron should have $\mu = \sqrt{3} = 1.73\ \mu_B$ while the value of μ for a complex containing three unpaired electrons would be $\sqrt{15} = 3.87\ \mu_B$. Note that this equation applies exactly only to complexes in which the orbital motion of the unpaired electrons between degenerate d orbitals is negligible. We will assume that this is the case in the examples presented.

Having introduced the concept of high-spin and low-spin electron configurations, we can now explain the unusual magnetic behaviour of the $[Fe(OH_2)_6]^{2+}$ and $[Fe(CN)_6]^{4-}$ ions discussed on page 570. Both complexes contain d^6 Fe(II) metal ions, but $[Fe(OH_2)_6]^{2+}$ is paramagnetic ($\mu \approx 4.9\ \mu_B$) while $[Fe(CN)_6]^{4-}$ is diamagnetic ($\mu \approx 0\ \mu_B$). We know from the spectrochemical series that H_2O induces a much smaller value of Δ_o than does CN^-, meaning that $[Fe(OH_2)_6]^{2+}$ is high-spin while $[Fe(CN)_6]^{4-}$ is low-spin, as shown in figure 13.41.

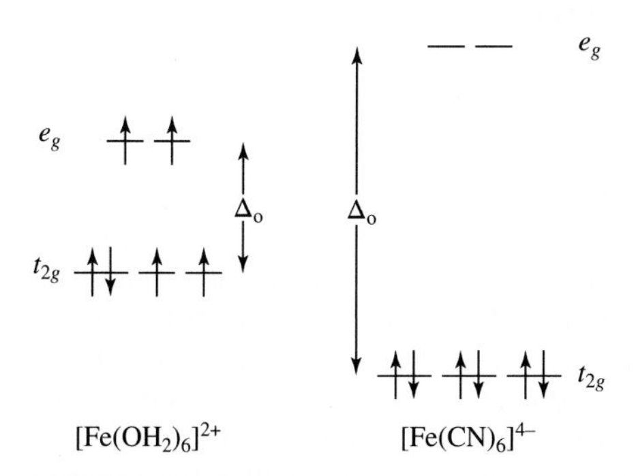

FIGURE 13.41 The *d* electron configurations of $[Fe(OH_2)_6]^{2+}$ and $[Fe(CN)_6]^{4-}$. The magnitude of Δ_o for the cyanide complex is much larger than that for the complex containing water. This produces a maximum pairing of electrons in the t_{2g} orbitals in $[Fe(CN)_6]^{4-}$, and a corresponding low-spin configuration for this complex.

Therefore, $[Fe(OH_2)_6]^{2+}$ contains four unpaired electrons and is paramagnetic, while $[Fe(CN)_6]^{4-}$ has no unpaired electrons and is diamagnetic.

Magnetic measurements allow us to distinguish between high-spin and low-spin possibilities in transition metal complexes. For example, the magnetic moments of the d^4 Cr(II) complexes $[Cr(OH_2)_6]^{2+}$ and $[Cr(CN)_6]^{4-}$ are 4.9 μ_B and 2.8 μ_B, respectively. The value of 4.9 μ_B for $[Cr(OH_2)_6]^{2+}$ is consistent with the presence of four unpaired electrons and, therefore, a high-spin $(t_{2g})^3(e_g)^1$ configuration. The value of 2.8 μ_B for $[Cr(CN)_6]^{4-}$ arises from two unpaired electrons, which suggests a low-spin $(t_{2g})^4(e_g)^0$ configuration.

WORKED EXAMPLE 13.9

The magnetic behaviour of octahedral transition metal complexes

Predict whether the following octahedral transition metal complex ions are paramagnetic or diamagnetic: $[Co(CN)_6]^{3-}$, $[TiCl_6]^{3-}$, $[V(OH_2)_6]^{3+}$, $[CoF_6]^{3-}$. If they are paramagnetic, estimate their magnetic moments.

Analysis

We can calculate the *d*-electron configurations of the complexes by first determining the oxidation states and then numbers of *d* electrons for each metal ion. Where high-spin and low-spin possibilities occur, we use the position of the ligand in the spectrochemical series to decide between the two configurations.

Solution

The oxidation states are Co(III), Ti(III), V(III) and Co(III), which correspond to d^6, d^1, d^2 and d^6 electron configurations, respectively. High-spin and low-spin possibilities exist for both $[Co(CN)_6]^{3-}$ and $[CoF_6]^{3-}$. We know from the spectrochemical series on page 576 that CN^- is a strong-field ligand, while F^- is a weak-field ligand. We therefore predict that $[Co(CN)_6]^{3-}$ is d^6 low-spin and diamagnetic, while $[CoF_6]^{3-}$ is d^6 high-spin, with four unpaired electrons, and paramagnetic. $[TiCl_6]^{3-}$ must be paramagnetic, as it contains an odd number of *d* electrons, while $[V(OH_2)_6]^{3+}$ is also paramagnetic, as the two *d* electrons occupy two of the t_{2g} orbitals with parallel spins. Therefore, the electron configurations and magnetic moments are as shown on the right.

$[Co(CN)_6]^{3-}$ — e_g, t_{2g} — $\mu = 0\ \mu_B$

$[TiCl_6]^{3-}$ — e_g, t_{2g} — $\mu = \sqrt{3} = 1.73\ \mu_B$

$[V(OH_2)_6]^{3+}$ — e_g, t_{2g} — $\mu = \sqrt{8} = 2.83\ \mu_B$

$[CoF_6]^{3-}$ — e_g, t_{2g} — $\mu = \sqrt{24} = 4.90\ \mu_B$

Is our answer reasonable?

Provided we have counted the *d* electrons correctly and filled the orbitals in the appropriate fashion, our answers should be correct.

PRACTICE EXERCISE 13.9 Determine how many unpaired electrons are contained in the following octahedral complex ions.

$$[Ni(NH_3)_6]^{2+}, [Cu(en)_3]^{2+}, [ZnCl_6]^{4-}.$$

Estimate the magnetic moments of these complexes.

It is important to realise that the simple paramagnetic complexes we have discussed, while attracted into a magnetic field, do not themselves act as permanent magnets. The presence of unpaired electrons is not sufficient in itself for a substance to act as a permanent magnet; for this, a substance must have a nonzero magnetic field that persists over time. In the absence of an external magnetic field, the unpaired electron spins in a paramagnetic material are aligned randomly; therefore, the tiny magnetic fields resulting from each electron spin cancel each other and the net magnetic moment is zero. Application of an external magnetic field results in the electron spins aligning, and this ordering of the spins results in a net magnetic moment within the material. However, when the external magnetic field is removed, the electron spins again align randomly and the individual magnetic fields again cancel each other. In order to act as a permanent magnet, these unpaired spins must behave cooperatively to give a nonzero magnetic moment in the absence of an external magnetic field. Of all the metals in the periodic table, only five, iron, cobalt, nickel, dysprosium and gadolinium, have this ability and can therefore act as permanent magnets. They display a property called **ferromagnetism**, which means that all of the electron spins are oriented in the same direction in the absence of a magnetic field.

Figure 13.42 shows the difference between paramagnetic and ferromagnetic materials. Ferromagnetic materials display the alignment of electron spins within small areas called domains in the absence of a magnetic field. Again, because the ordering of the domains is randomised, the individual magnetic fields due to each domain cancel out and there is no net magnetic moment. When an external magnetic field is applied, each domain aligns with all the others, and this alignment remains when the magnetic field is switched off, resulting in the material having a nonzero magnetic moment and behaving as a permanent magnet. Many of you will have observed this behaviour when you have magnetised a needle; stroking a needle with a permanent magnet causes the individual domains within the needle (made of steel, a ferromagnetic material) to align, and this alignment remains when the permanent magnet is removed.

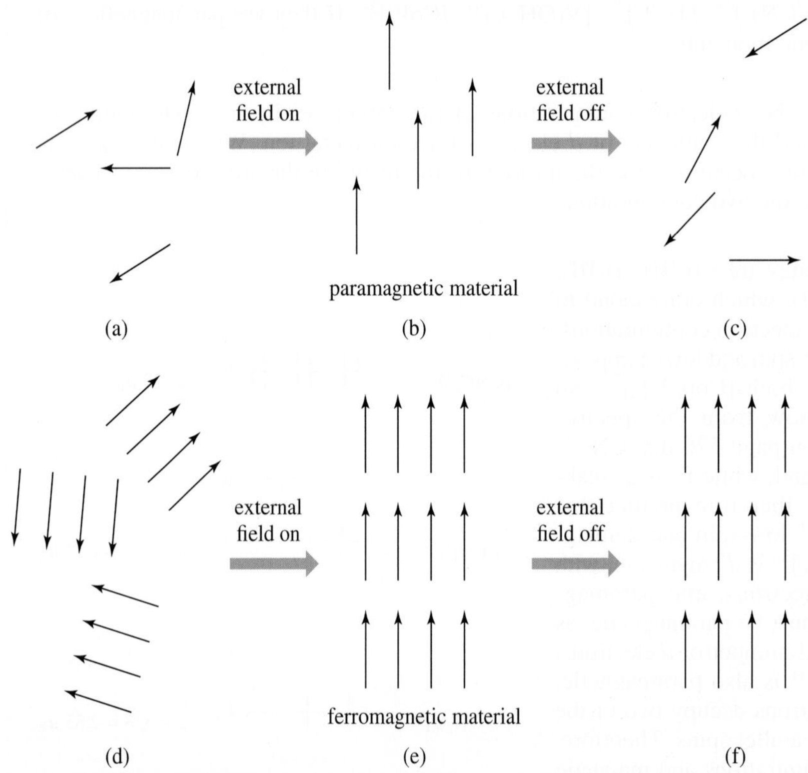

FIGURE 13.42 The differing behaviours of paramagnetic and ferromagnetic materials. Each arrow represents a single electron spin. In a paramagnetic material: **(a)** In the absence of an external magnetic field, the directions of the electron spins are randomised. **(b)** Applying an external magnetic field causes the electron spins to align. **(c)** When the external magnetic field is switched off, the electron spins again randomise. In a ferromagnetic material: **(d)** In the absence of an external magnetic field, there is short-range ordering of the electron spins in individual domains within the material. **(e)** Applying an external magnetic field causes all the electron spins within all the domains to align. **(f)** When the external magnetic field is switched off, the electron spins remain aligned to give a permanent magnet.

Chemical Connections

The most important chemical reaction of all?

Chances are, if you posed the question 'What is the most important chemical reaction' to 100 chemists, you would get 100 different answers. But without doubt, the formation of oxygen from water, which occurs in photosynthesis, is one of the most important, given that life on Earth depends on it.

Photosynthesis is an extremely complicated chemical process, the net result of which is the reduction of CO_2 to carbohydrates and the oxidation of water to O_2. The latter reaction is a four-electron process, and is mediated by a transition metal complex, called the oxygen evolving complex (OEC). However, at present, we do not know the exact structure of this complex. What we do know is that it contains 4 manganese ions and a calcium ion, with 3 of the manganese ions and the calcium ion forming a cube, in which they are bridged by oxide (O^{2-}) ions.

The oxidation of water is thermodynamically uphill, with $E^{\ominus} = 1.23$ V for the reaction:

$$O_2 + 4H^+ + 4e^- \rightarrow 2H_2O$$

under standard conditions, or, more realistically $E = 0.82$ V at pH 7. Therefore, a strong oxidising agent is required that will not indiscriminately oxidise biological molecules. Nature undoubtedly chose manganese as the metal ion in the OEC for two reasons.

1. It can exist in a variety of oxidation states in its complexes (Mn(II), Mn(III) and Mn(IV) are relatively common) and can therefore cycle between these on gaining or losing electrons.
2. Both Mn(III) and Mn(IV) are strongly oxidising, a necessary requirement for the oxidation of water.

Recognition of the importance of manganese in the OEC led to an extraordinary resurgence in studies of the coordination chemistry of manganese in the 1980s that has continued until the present day. It is in no small part thanks to these studies that we are coming ever closer to an understanding of how the remarkable transition metal complex that is the OEC works.

While magnetic materials are widely used in data storage (the magnetic strips on the back of your ATM and credit cards, for instance), nanoscale magnetic materials offer potentially enormous possibilities as replacements for current silicon-based technology in computers. Central to this is the development of both magnetic nanoparticles, which are well-defined magnetic particles with a diameter of less than 100 nm and generally comprising collections of metal atoms or metal oxides, and single-molecule magnets, in which individual molecules containing unpaired electrons with aligned spins act as magnets. Both areas are currently the focus of extensive worldwide research.

13.5 Transition metal ions in biological systems

Almost 90% of the atoms that make up the human body are either hydrogen or oxygen. Most of these are in water, which constitutes around 70% of our bodies. The organic molecules that make up the body, as well as the molecules involved in biosynthesis and energy production, consist almost entirely of C, H, N and O. These four elements account for 99% of all the atoms present in a human. Another seven elements — Na, K, Ca, Mg, P, S and Cl — are also essential for all known life-forms. These seven add another 0.9% to the total atom count of a human being. The remaining 0.1%, the so-called trace elements, are required by most organisms.

Although trace elements are present in only minute amounts, they are essential for healthy function. Of these trace elements, nine are transition metals, namely the elements V, Cr, Mn, Fe, Co, Ni, Cu and Zn from the first row of the transition metals and Mo (molybdenum) from the second row. Many of these are natural constituents of proteins, biological macromolecules made of long chains of amino acids that are described in chapter 24. These metalloproteins play three essential roles in biochemistry. Some act as transport and storage agents, moving small molecules from place to place within an organism. Others are enzymes, catalysts for a diverse group of biochemical reactions. Both transport and catalysis depend on the ability of transition metals to bind and release ligands. The third role of metalloproteins is to serve as redox reagents, adding or removing electrons in many different reactions. Transition metals are ideal for this purpose because of their ability to cycle between two or more oxidation states.

Transport and storage metalloproteins

Organisms extract energy from food by using molecular oxygen to oxidise fats and carbohydrates. For most animals, the movement and storage of O_2 is accomplished by the iron-containing proteins haemoglobin and myoglobin. The iron atom in haemoglobin binds O_2 and transports this vital molecule from the lungs to various parts of the body where oxidation takes place. Haemoglobin makes O_2 about 70 times more soluble in blood than in water. Whereas haemoglobin transports O_2, myoglobin stores O_2 in tissues, such as muscle, that require large amounts of oxygen. Haemoglobin and myoglobin have closely related structures and contain one (myoglobin) or four (haemoglobin) haem groups (figure 13.43). These are essentially iron–porphyrin complexes, as shown in figure 13.44 on the next page.

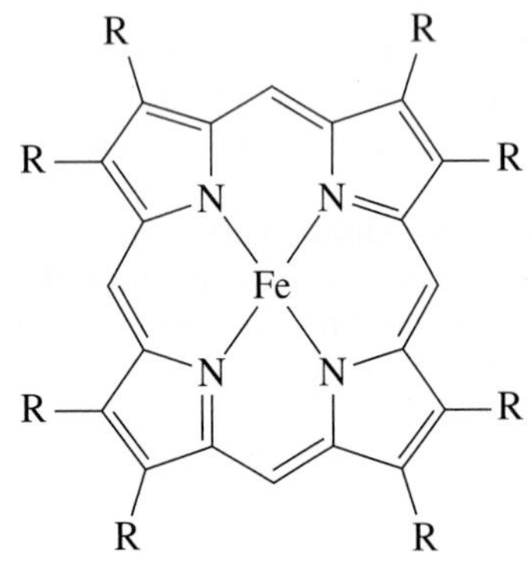

FIGURE 13.43 The haem group, as found in haemoglobin and myoglobin.

In myoglobin, the haem group is bound to a polypeptide chain of 153 amino acids arranged in helical arrays, and the ribbon structure of myoglobin is shown in figure 13.44. The polypeptide chain folds to create a 'pocket' in the protein for a haem group.

Haemoglobin is made up of four polypeptide chains, each of which is similar in shape and structure to a myoglobin molecule. Each haem unit in both myoglobin and haemoglobin contains one Fe^{2+} ion bound to four nitrogen donor atoms in a square planar arrangement. This leaves the metal with two axial coordination sites to bind other ligands. One of these sites is occupied by an N-donor ligand from a protein side chain that holds the haem in the pocket of the protein, while the sixth coordination site is where reversible binding of O_2 takes place. Ordinarily, interaction of an Fe(II) ion with molecular O_2 would result in irreversible oxidation to Fe(III) ($E^{\ominus}_{O_2/H_2O} = 0.82$ V at pH = 7, $E^{\ominus}_{Fe^{3+}/Fe^{2+}} = 0.77$ V) but, remarkably, as a result of the protein surrounding the haem, both haemoglobin and myoglobin are able to bind O_2 without being irreversibly oxidised. (Oxidation of haemoglobin gives a compound called methaemoglobin, which cannot bind O_2.)

In the lungs, haemoglobin 'loads' four oxygen molecules and then moves through the bloodstream. In tissues, the concentration of O_2 is very low, but there is plenty of CO_2, the end-product of metabolism. The concentration of CO_2 has an important effect on haemoglobin–oxygen binding. Like oxygen, CO_2 can bind to haemoglobin. However, CO_2 binds to specific amino acid side chains of the protein, rather than to the haem group. Binding CO_2 to the protein causes the shape of the haemoglobin molecule to change in ways that reduce the equilibrium constant for O_2 binding. The reduced binding constant allows haemoglobin to release its O_2 molecules in oxygen-deficient, CO_2-rich tissue. The bloodstream carries this deoxygenated haemoglobin back to the lungs, where it releases CO_2 and binds four more molecules of O_2. This CO_2 effect does not operate in myoglobin, which binds and stores the oxygen released by haemoglobin.

Carbon monoxide seriously impedes transport of O_2. The deadly effect of inhaled CO results from its reaction with haemoglobin. A CO molecule is almost the same size and shape as O_2, so it fits into the binding pocket of the haemoglobin molecule. In addition, the carbon atom of CO forms a stronger bond with Fe(II) than does O_2. Under typical conditions in the lungs, haemoglobin binds CO over 200 times more strongly than it binds O_2. Haemoglobin complexed to CO cannot transport oxygen so, when a significant fraction of haemoglobin contains CO, oxygen 'starvation' occurs at the cellular level, leading to loss of consciousness and death.

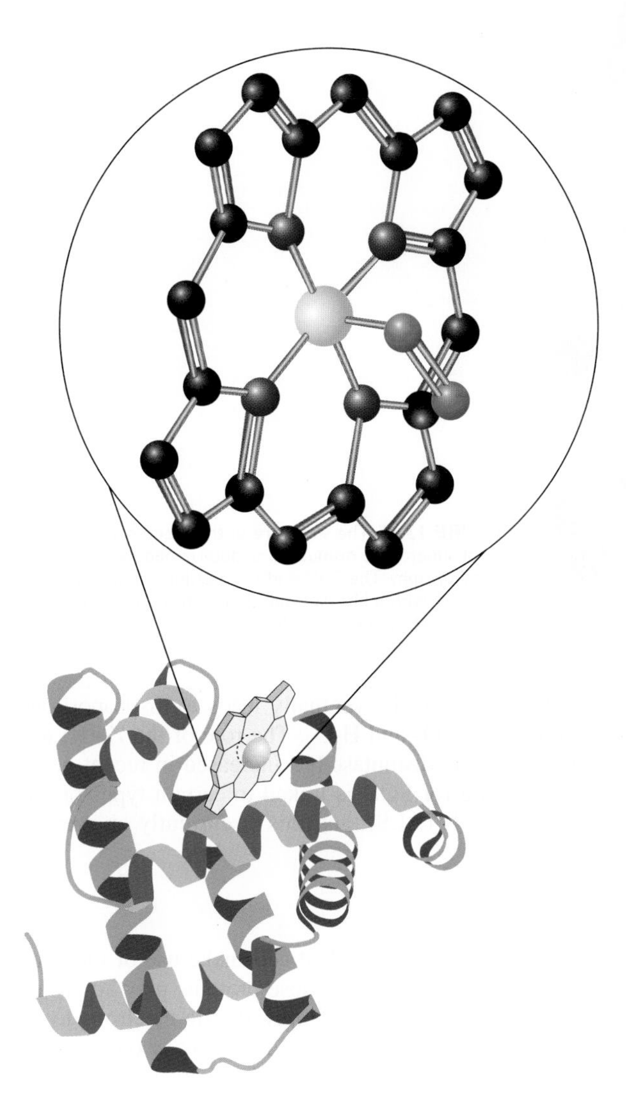

FIGURE 13.44 A ribbon depiction of the structure of myoglobin, determined by X-ray crystallography, showing the haem unit embedded in a pocket created by the folding of the protein chain. (Fe is shown as pink and O as red.)

An adult human contains only about 4 g of iron, most of it in the form of haem-containing proteins. Yet the daily requirement of iron in the diet is only about 7–18 mg, indicating that the body recycles iron. The recycling of iron requires a transport system and a storage mechanism. Iron is transported by a protein called transferrin, which collects iron in the spleen and liver, where haemoglobin is degraded, and carries it to the bone marrow where fresh red blood cells are synthesised. The protein ferritin stores iron in the body. A ferritin molecule consists of a protein coat and an iron-containing core. The outer coat is made up of 24 polypeptide chains, each comprising about 175 amino acids, which pack together to form a sphere. The sphere is hollow, and channels through the protein coat allow movement of iron in and out of the molecule. The core of the protein contains hydrated iron(III) oxide, $Fe_2O_3{\cdot}H_2O$. The protein retains its shape whether or not iron is stored on the inside. When filled to capacity, one ferritin molecule holds as many as 4500 iron atoms, but the core is only partially filled under normal conditions. In this way, the protein can provide iron as needed for haemoglobin synthesis or to store iron if an excess is absorbed by the body.

Metalloenzymes

As we have seen in this section, molecular oxygen is essential to life. However, the superoxide ion, O_2^-, which is the product of one-electron reduction of O_2, can damage cells and is thought to play a role in the ageing process. The enzyme superoxide dismutase (SOD) is abundant in virtually every type of aerobic organism; its role in cells is to destroy superoxide ions. The active site of the SOD enzyme, the region of the enzyme where the catalytic reaction occurs, contains one copper ion and one zinc ion, each of which displays a tetrahedral geometry. As figure 13.45 (overleaf) shows, the two metals are held close together by a histidine ligand (histidine is an amino acid, chapter 24) that forms a bridge between the Cu(II) and Zn(II) ions.

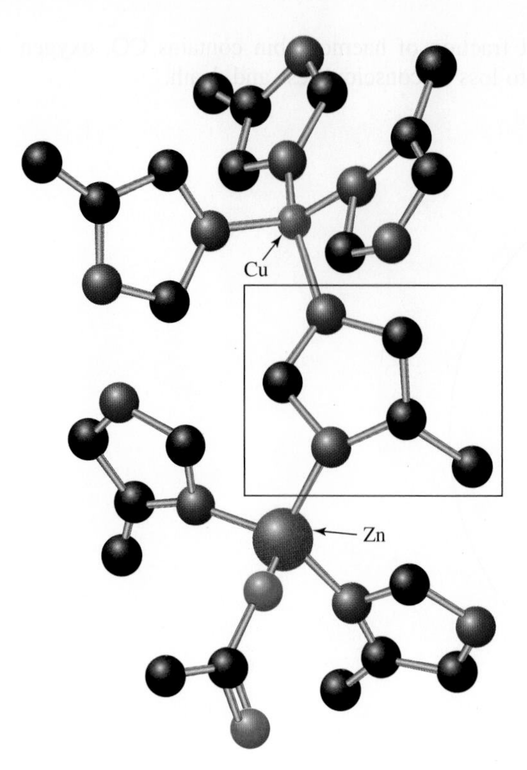

FIGURE 13.45 The structure of the metal-containing site of superoxide dismutase, as determined by X-ray crystallography. The Zn^{2+} and Cu^{2+} cations lie in close proximity, with a histidine side chain (in the box) acting as a bridge between the metals.

It is thought that O_2^- initially binds to the Cu centre, which eventually leads to disproportionation (also called dismutation) of O_2^- and formation of O_2 and H_2O_2. The role of the Zn(II) ion is thought to be purely structural. Interest in superoxide dismutase has increased in recent years with the discovery that a mutation in the gene coding for SOD is linked to certain types of the neurodegenerative disease amyotrophic lateral sclerosis (ALS), and this is currently an area of intense research.

Electron transfer proteins

The many redox reactions that take place within an animal or a plant cell make use of metalloproteins with a wide range of electron transfer potentials. Some of these proteins play key roles in respiration, photosynthesis and nitrogen fixation, while others simply shuttle electrons to or from enzymes that require electron transfer as part of their catalytic activity. In many other cases, a complex enzyme may incorporate its own electron transfer centres. There are three general categories of transition metal redox centres: cytochromes, blue copper proteins and iron–sulfur proteins. Cytochromes are iron-containing proteins that incorporate haem groups and that facilitate electron transfer by cycling the iron between the (II) and (III) states. A typical blue copper redox protein contains a single copper atom in a distorted tetrahedral environment. The copper centre can cycle between the Cu(I) and Cu(II) oxidation states, thereby aiding electron transfer. Iron–sulfur proteins display a variety of nuclearities (one, two or four iron centres) and geometries; the two available oxidation states of iron, Fe(II) and Fe(III), allow electron transfer to take place.

13.6 Isolation and purification of transition metals

The production and purification of metals from their ores is called **metallurgy**. This technology has an ancient history and may represent one of the earliest useful applications of chemistry. Metallurgical advances have had profound influences on the course of human civilisation, so much so that historians speak of the Bronze Age (about 3000 to 5000 years ago) and the Iron Age (starting about 3000 years ago). Metallurgy has also had an enormous impact on the histories of both New Zealand and Australia, with gold rushes in the nineteenth century contributing significantly to the economy and social fabric of both countries. Australia is one of the most important mineral producers in the world, and the mining industry is a major source of overseas earnings. In this section, we discuss some of the techniques of metallurgy as applied to transition metals, and we also look at some of the areas in which transition metals are used.

Nearly all transition metals are readily oxidised, so most ores are compounds in which the transition metal has a positive oxidation state. Examples include oxides (TiO_2, rutile; Fe_2O_3, haematite; Cu_2O, cuprite), sulfides (ZnS, sphalerite; MoS_2, molybdenite) and carbonates ($FeCO_3$, siderite). Other minerals contain oxoanions ($MnWO_4$, wolframite) and those with even more complex structures such as carnotite, $K_2(UO_2)_2(VO_4)_2 \cdot 3H_2O$. Figure 13.46 shows in schematic fashion some of the alternative paths leading from ores to pure metals, all of which involve reduction as the essential chemical process.

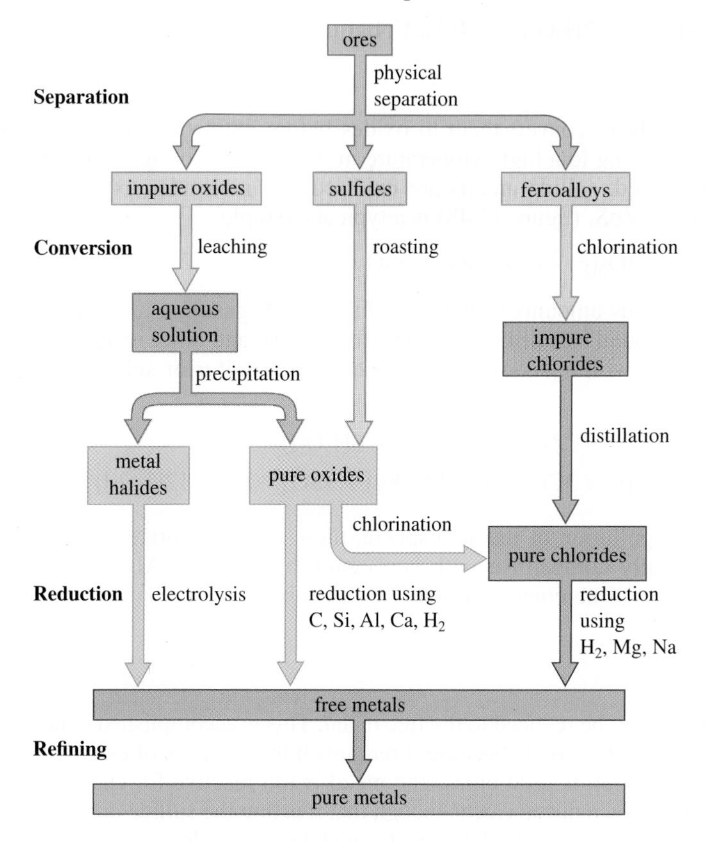

FIGURE 13.46 Metallurgy includes separation, conversion, reduction and refining steps. The starting material is an impure ore, and the end-product is pure metal.

The four main steps in the conversion of an ore to a pure metal are *separation*, *conversion*, *reduction* and *refining*. Separation involves the removal of a particular compound of some desired metal from other contaminants. The conversion step entails chemical treatment of the separated material to convert it into a form that can be easily reduced. Chemical or electrolytic reduction then gives the impure metal, which is finally refined to the pure metal.

Separation

Ore obtained from a mining operation contains a desired mineral contaminated with other components, which may include sand, clay and organic matter. This economically valueless portion of the ore, which is called gangue (pronounced 'gang'), must be removed before the metal can be extracted and refined. Ores can be separated into components by physical or chemical methods.

Flotation, a common physical separation process, was discovered and developed in Australia, and was first used on a large scale at Broken Hill in the early years of the twentieth century. During flotation, the ore is crushed and mixed with water to form a thick slurry. As shown in figure 13.47, the slurry is transferred to a flotation vessel and mixed with oil and a surfactant (in this case, a detergent).

The polar head groups of the detergent coat the surface of the mineral particles, but the nonpolar tails point outwards, making the detergent-coated mineral particles hydrophobic. Air is blown vigorously through the mixture, carrying the oil and the coated mineral to the surface, where they become trapped in the froth. Because the gangue has a much lower affinity for the detergent, it absorbs water and sinks to the bottom of the flotation vessel. The froth is removed at the top, and the gangue is removed at the bottom.

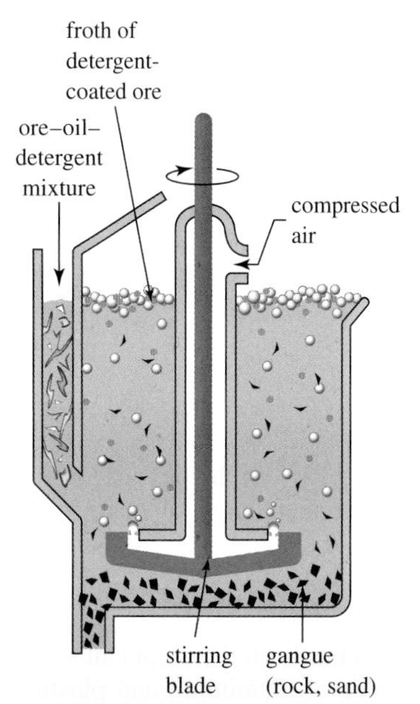

FIGURE 13.47 In the flotation process, detergent-coated mineral particles float to the surface of the mixture, where they become trapped in the froth. The gangue settles to the bottom.

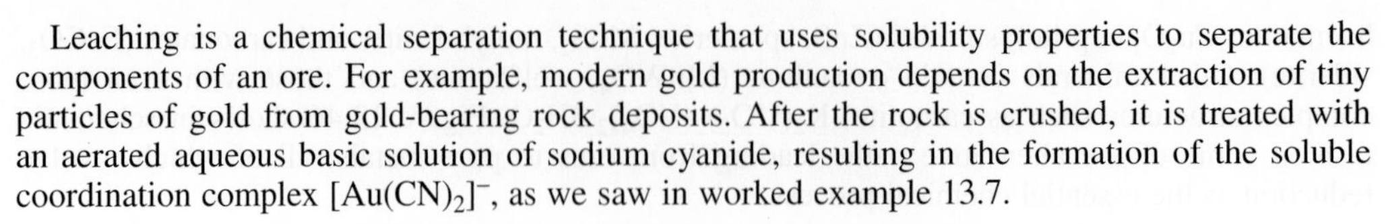

Leaching is a chemical separation technique that uses solubility properties to separate the components of an ore. For example, modern gold production depends on the extraction of tiny particles of gold from gold-bearing rock deposits. After the rock is crushed, it is treated with an aerated aqueous basic solution of sodium cyanide, resulting in the formation of the soluble coordination complex $[Au(CN)_2]^-$, as we saw in worked example 13.7.

$$4Au(s) + 8CN^-(aq) + O_2(g) + 2H_2O(l) \rightarrow 4[Au(CN)_2]^-(aq) + 4OH^-(aq)$$

Conversion

FIGURE 13.48 Sphalerite.

Sulfide ores may be treated chemically to convert them to oxides before extraction can occur. The process of oxidising an ore by heating to a high temperature in the presence of air is known as roasting. When sulfide ore is roasted, sulfide anions are oxidised and molecular oxygen is reduced. The conversion of sphalerite, ZnS, (figure 13.48) is a typical example.

$$2ZnS(s) + 3O_2(g) \rightarrow 2ZnO(s) + 2SO_2(g)$$

Unfortunately, roasting produces copious amounts of highly polluting SO_2 gas that, in the past, seriously damaged the environment around sulfide ore smelters. Today, zinc and other metals can be extracted from sulfides by reaction with aqueous acid to generate free sulfur or sulfate ions rather than SO_2. Here are two examples.

$$2ZnS(s) + 4H_3O^+(aq) + O_2(g) \rightarrow 2Zn^{2+}(aq) + 2S(s) + 6H_2O(l)$$

$$3CuS(s) + 8NO_3^-(aq) + 8H_3O^+(aq) \rightarrow 3Cu^{2+}(aq) + 8NO(g) + 3SO_4^{2-}(aq) + 12H_2O(l)$$

Taking into account the cost of SO_2 pollution of the atmosphere, these more elaborate aqueous separation procedures are economically competitive with conversion by roasting. Also, in the case of CuS, the liberated $SO_4^{2-}(aq)$ can be converted to H_2SO_4 for use in the manufacture of fertiliser (see chapter 14, p. 633).

Reduction

Once an ore is in suitably pure form, it can be reduced to the free metal. This is accomplished either chemically or electrolytically. Electrolysis is costly because it requires huge amounts of electrical energy. For this reason, chemical reduction is used unless the metal is too reactive for chemical reducing agents to be effective. Mercury is reduced easily enough that roasting the sulfide ore frees the metal. The sulfide ion is the reducing agent, and both O_2 and the metal ion gain electrons.

$$HgS(s) + O_2(g) \rightarrow Hg(l) + SO_2(g)$$

However, this reduction produces SO_2, which must be removed from the exhaust gases. One of the most common chemical reducing agents for metallurgy is coke, a form of carbon made by heating coal at high temperature in the absence of O_2 until all of the volatile impurities have been removed. Metals such as Co, Ni, Fe and Zn that have cations with a moderately negative reduction potential can be reduced by coke. For example, direct reaction with coke in a furnace frees nickel from its oxide.

$$NiO(s) + C(s) \rightarrow Ni(l) + CO(g)$$

Refining

Chemical reduction of an ore usually gives metal that is not pure enough for its intended use. Further refining of the metal removes undesirable impurities. Several important metals, including Cu, Ni, Zn and Cr, are refined by electrolysis, either from an aqueous solution of the metal salt or from anodes prepared from the impure metal. To give one example, Zn(II) ions, obtained by dissolving ZnS or ZnO in acidic solution, can be reduced while water is oxidised.

$$Zn^{2+}(aq) + 2e^- \rightarrow Zn(s)$$

$$6H_2O(l) \rightarrow O_2(g) + 4H_3O^+(aq) + 4e^-$$

$$2Zn^{2+}(aq) + 6H_2O(l) \rightarrow 2Zn(s) + O_2(g) + 4H_3O^+(aq)$$

Table 13.4 on the next page provides a summary of the chemical species and processes involved in the metallurgy of many transition metals. The following survey of several metals provides further examples of the four phases of metallurgy.

Iron and steel

Iron has been the dominant structural material of modern times and, despite the growth in importance of aluminium and plastics, it still ranks first in total use. Worldwide production of steel (iron strengthened by additives) is in the order of 1.5 billion tonnes per year. Australia produces about 18% of the world's iron ore, second behind China, with most coming from mines in Western Australia (figure 13.49). Significant deposits of ironsands are found on the west coast of the North Island of New Zealand.

TABLE 13.4 Metallurgy of transition metals.

Metal	Ore	Separation method	Intermediate	Reducing agent
Ti	TiO_2	chlorination	$TiCl_4$	Mg
Cr	$FeCr_2O_4$	roasting	Cr_2O_3	Al
Mn	MnO_2		Mn_2O_3	Al
Fe	Fe_3O_4	slag formation	(a)	C
Co	CoAsS	roasting	CoO	C
Ni	Ni_9S_8	complexation	$Ni(CO)_4$	H_2
Cu	$CuFeS_2$	leaching	Cu^{2+} (b)	
Zn	ZnS	roasting	ZnO	C
Mo	MoS_2	roasting	MoO_3	H_2
W	$CaWO_4$	leaching	WO_4^{2-}, WO_3	H_2
Au	Au	leaching	$[Au(CN_2)]^-$	Zn
Hg	HgS	roasting	(c)	S^{2-}

(a) $CaSiO_3$ is formed as an impurity.
(b) SO_4^{2-} is formed as an impurity.
(c) SO_2 is formed as an impurity.

FIGURE 13.49 Iron ore stockpiles in the Pilbara region of Western Australia.

The most important iron ores are two oxides, haematite, Fe_2O_3, and magnetite, Fe_3O_4. The production of iron from its ores involves several chemical processes that take place in a blast furnace. As shown in figure 13.50 (overleaf), this is an enormous chemical reactor in which heating, reduction and purification occur together.

The raw materials placed in a blast furnace include the ore (usually haematite) and coke, which serves as the reducing agent. The ores always contain various amounts of silicon dioxide, SiO_2, which is removed chemically by reaction with limestone, $CaCO_3$. To begin the conversion process, pellets of ore, coke and limestone are mixed and fed into the top of the furnace, and a blast of hot air is blown in at the bottom. As the starting materials fall through the furnace, the burning coke generates intense heat.

$$2C(s) + O_2(g) \rightarrow 2CO(g) \qquad \Delta H^{\circ} = -221 \text{ kJ mol}^{-1}$$

The result is a temperature gradient ranging from about 800 °C at the top of the furnace to 1900 °C at the bottom.

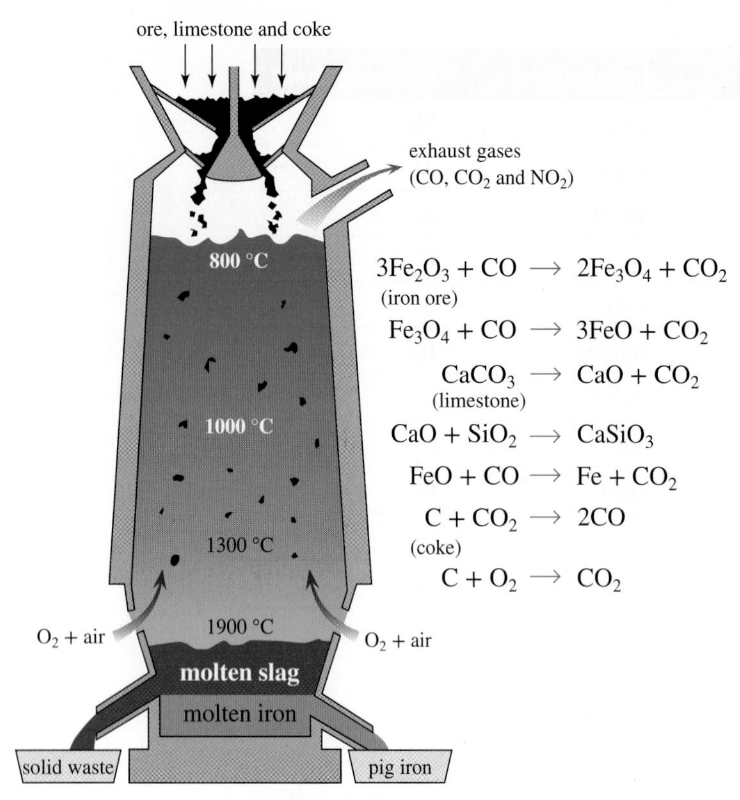

FIGURE 13.50 A diagrammatic view showing the chemical reactions occurring within the pictured blast furnace.

The reduction of iron oxide takes place in several stages in different temperature zones within the furnace. The reducing agent is CO produced from burning coke. The key reactions are:

$$3Fe_2O_3(s) + CO(g) \rightarrow 2Fe_3O_4(s) + CO_2(g)$$
$$Fe_3O_4(s) + CO(g) \rightarrow 3FeO(s) + CO_2(g)$$
$$FeO(s) + CO(g) \rightarrow Fe(l) + CO_2(g)$$

Once liberated from its oxides, the iron melts when the temperature reaches 1500 °C and collects in a pool at the bottom of the furnace.

While heating and reduction occur, limestone decomposes into calcium oxide and CO_2. The CaO then reacts with SiO_2 impurities in the ore to generate calcium silicate.

$$CaCO_3(s) \xrightarrow{\text{heat}} CaO(s) + CO_2(g)$$
$$CaO(s) + SiO_2(s) \xrightarrow{\text{heat}} CaSiO_3(l)$$

At blast furnace temperatures, calcium silicate is a liquid, called slag. Being less dense than iron, slag pools on the surface of the molten metal. Both products are drained periodically through openings in the bottom of the furnace. Although this chemistry is complex, the basic process is reduction of iron oxide by carbon in an atmosphere depleted of oxygen. Archaeologists have found ancient smelters in Tanzania that exploited this chemistry to produce iron around 2000 years ago. Early African peoples used a hole lined with termite residue as a fuel before adding iron ore. Charred reeds and charcoal provided the reducing substance. Finally, a chimney of mud was added. When this furnace was 'fired', a pool of iron collected in the bottom.

The iron formed in a blast furnace, called pig iron, contains impurities that make the metal brittle. These include phosphorus and silicon from silicate and phosphate minerals that contaminate the original ore, as well as carbon and sulfur from the coke. This iron is refined in a converter furnace. Here, a stream of O_2 gas blows through molten impure iron. Oxygen reacts with the nonmetal impurities, converting them to oxides. As in the blast furnace, CaO is added to convert SiO_2 into liquid calcium silicate, in which the other oxides dissolve. The molten iron is analysed at intervals until its impurities have been reduced to satisfactory levels. Then the liquid metal, now in a form called steel, is poured from the converter and allowed to solidify. Most steels contain various amounts of other elements that are added deliberately to give the metal particular properties. These additives may be introduced during the converter process or when the molten metal is poured off. One of the most important additives is manganese, which adds strength and hardness to steel. Manganese is added to nearly every form of steel in amounts ranging from less than 1% to more than 10%; in fact, more than 80% of all the manganese produced is incorporated into steel.

Titanium

The metallurgy of titanium illustrates the purification of one metal by another. The major titanium ores are rutile, TiO_2, and ilmenite, $FeTiO_3$, and Australia provides a significant proportion of the world's supply of these (about 15% of ilmenite and 58% of rutile). These ores are converted to titanium(IV) chloride by a redox reaction with chlorine gas and coke. For rutile:

$$TiO_2(s) + C(s) + 2Cl_2(g) \xrightarrow{500\,°C} TiCl_4(g) + CO_2(g)$$

In this reaction, carbon is oxidised and chlorine is reduced. When the hot gas cools, titanium tetrachloride (bp = 140 °C) condenses to a liquid that is purified by distillation. Titanium metal is obtained by reduction of $TiCl_4$ with molten magnesium metal at high temperature. The reaction gives solid titanium metal (mp = 1660 °C) and liquid magnesium chloride (mp = 714 °C).

$$TiCl_4(g) + 2Mg(l) \xrightarrow{850\,°C} Ti(s) + 2MgCl_2(l)$$

Copper

Copper is found mainly in the sulfide ore chalcopyrite, $FeCuS_2$ (figure 13.51), but chalcocite, Cu_2S, cuprite, Cu_2O, and malachite, $Cu_2CO_3(OH)_2$, are also important, while elemental copper nuggets can also be found in parts of the world. Copper ores often have concentrations of copper less than 1% by mass, so achieving economic viability requires mining operations on a huge scale. The extraction and purification of copper from chalcopyrite is complicated by the need to remove the iron. The first step in the process is flotation, which concentrates the ore to around 15% Cu by mass. In the next step, the concentrated ore is roasted to convert $FeCuS_2$ to CuS and FeO.

FIGURE 13.51 Chalcopyrite.

$$2FeCuS_2(s) + 3O_2(g) \rightarrow 2CuS(s) + 2FeO(s) + 2SO_2(g)$$

Copper(II) sulfide is unaffected if the temperature is kept below 800 °C. Heating the mixture of CuS and FeO to 1400 °C in the presence of silica, SiO_2, causes the material to melt and separate into two layers. The top layer is molten $FeSiO_3$ formed from the reaction of SiO_2 with FeO. As this takes place, the copper in the bottom layer is reduced from CuS to Cu_2S. This bottom layer consists of molten Cu_2S contaminated with FeS. Reduction of the Cu_2S takes place in a converter furnace following the same principle that converts impure iron into steel. Silica is added, and oxygen gas is blown through the molten mixture. Iron impurities are converted first to FeO and then to $FeSiO_3$, which is a liquid that floats to the surface. At the same time, Cu_2S is converted to Cu_2O, which reacts with more Cu_2S to give copper metal and SO_2.

$$2Cu_2S(l) + 3O_2(g) \rightarrow 2Cu_2O(l) + 2SO_2(g)$$
$$2Cu_2O(l) + Cu_2S(l) \rightarrow 6Cu(l) + SO_2(g)$$

Copper metal obtained from the converter furnace must be refined to better than 99.95% purity before it can be used to make electrical wiring. This is accomplished by electrolysis, as illustrated in figure 13.52.

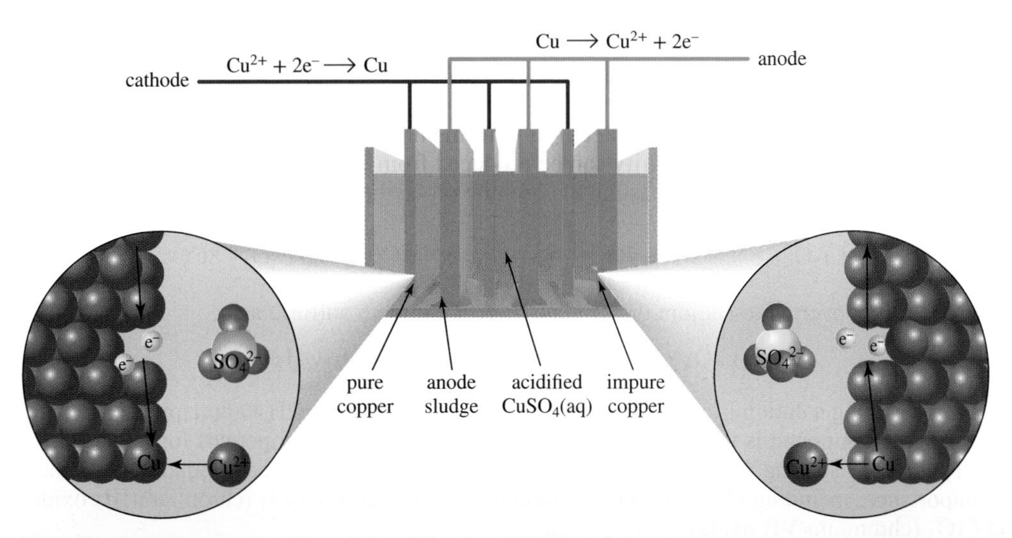

FIGURE 13.52 Diagram of an electrolytic cell for the purification of copper.

The impure copper is formed into slabs that serve as anodes in electrolytic cells. The cathodes are constructed from thin sheets of very pure copper. These electrodes are immersed in a solution of $CuSO_4$ dissolved in dilute sulfuric acid. Application of a controlled voltage causes oxidation in which copper, along with iron and nickel impurities, is oxidised to its cations. Less reactive metal contaminants, including silver, gold and platinum, are not oxidised. As the electrolysis proceeds and the anode dissolves, these and other insoluble impurities fall to the bottom of the cell. The metal cations released from the anode migrate through the solution to the cathode. Because Cu(II) is easier to reduce than Fe(II) and Ni(II), careful control of the applied voltage makes it possible to reduce Cu(II) to Cu metal, leaving Fe(II) and Ni(II) dissolved in solution.

13.7 Applications of transition metals

A complete discussion of all the transition metals is beyond the scope of a first-year chemistry course. Instead, we provide a brief survey of several metals that highlights the diversity and utility of this group of elements.

Titanium

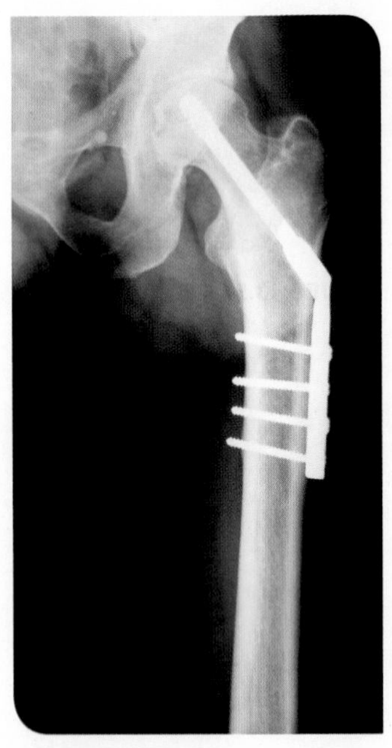

FIGURE 13.53 Titanium plates and pins are used to hold this broken leg together.

Titanium, the ninth most abundant element in the Earth's crust, is characterised by its high strength, its low density (–57% that of steel) and its stability at very high temperatures. When alloyed with small amounts of aluminium or tin, titanium has the highest strength-to-weight ratio of all the engineering metals. Its major use is in the construction of aircraft frames and jet engines. Because titanium is also highly resistant to corrosion, it is used in the construction of pipes, pumps and vessels for the chemical industry and for surgical implants (figure 13.53). Because it is difficult to purify and fabricate, titanium is an expensive metal. For example, although titanium bicycles are highly prized by avid cyclists, they are quite expensive, with high-end models costing several thousand dollars for just the frame. The most important compound of titanium is titanium(IV) oxide, TiO_2. More than 7 million tonnes of TiO_2 are produced worldwide every year, the majority by the controlled combustion of $TiCl_4$.

$$TiCl_4(g) + O_2(g) \xrightarrow{1200\,°C} TiO_2(s) + 2Cl_2(g)$$

The Cl_2 produced in the combustion reaction is recycled to produce more $TiCl_4$ from rutile ore. Titanium dioxide is brilliant white, owing to the d^0 electron configuration of the Ti(IV) ions, highly opaque, chemically inert and nontoxic. Consequently, it finds wide use as a pigment in paints and other coatings, paper, sunscreens, cosmetics and toothpaste. TiO_2 is also finding increasing use in solar cells and self-cleaning windows because of its electrochemical properties.

Chromium

Chromium makes up just 0.012% of the Earth's crust, yet it is an important industrial metal. The main use of chromium is in metal alloys. Stainless steel, for example, contains as much as 20% chromium. Nichrome, a 60:40 alloy of nickel and chromium, is used to make heat-radiating wires in electrical devices such as toasters and hair dryers. Another important application of chromium metal is as a protective and decorative coating for the surface of metal objects. The only important ore of chromium is chromite, $FeCr_2O_4$. Reduction of chromite with coke gives ferrochrome, an iron–chromium compound.

$$FeCr_2O_4(s) + 4C(s) \rightarrow FeCr_2(s) + 4CO(g)$$

Ferrochrome is mixed directly with molten iron to form chromium-containing stainless steel.

Chromium compounds of high purity can be produced from chromite ore without reduction to the free metal. The first step is the roasting of chromite ore in the presence of sodium carbonate.

$$4FeCr_2O_4 + 8Na_2CO_3 + 7O_2 \xrightarrow{1100\,°C} 8Na_2CrO_4 + 2Fe_2O_3 + 8CO_2$$

The product is converted to sodium dichromate by reaction with sulfuric acid.

$$2Na_2CrO_4 + H_2SO_4 \rightarrow Na_2Cr_2O_7 + Na_2SO_4 + H_2O$$

When the resulting solution is concentrated by evaporation, $Na_2Cr_2O_7{\cdot}2H_2O$ precipitates from the solution. This compound is the most important source of chromium compounds for the chemical industry. It is the starting material for most other chromium-containing compounds of commercial importance, including $(NH_4)_2Cr_2O_7$ (ammonium dichromate), Cr_2O_3 (chromium(III) oxide) and CrO_3 (chromium(VI) oxide).

Pure chromium metal is made by a two-step reduction sequence. First, sodium dichromate is reduced to chromium(III) oxide by heating in the presence of charcoal.

$$Na_2Cr_2O_7 + 2C \xrightarrow{heat} Cr_2O_3 + Na_2CO_3 + CO$$

Dissolving Cr_2O_3 in sulfuric acid gives an aqueous solution of Cr(III) cations.

$$Cr_2O_3(s) + 6H_3O^+(aq) \rightarrow 2Cr^{3+}(aq) + 9H_2O(l)$$

Electrolysis of this solution reduces the cations to pure Cr metal, which forms a hard, durable film on the surface of the object serving as the cathode. The name 'chromium' is derived from the Greek *chroma* meaning 'colour', because compounds of this metal display a wide variety of colours (figure 13.54).

FIGURE 13.54 Chromium compounds display a striking range of colours. Shown here are Na_2CrO_4 (yellow), $K_2Cr_2O_7$ (orange), $CrCl_3$ (green) and CrO_3 (dark purple).

Chromium compounds have been used for many years as pigments in paints and other coatings; for example, $Na_2Cr_2O_7$ is bright orange, Cr_2O_3 is green and the salts of CrO_4^{2-} are bright yellow. However, in recent years, the use of chromium pigments has diminished because chromium in the VI oxidation state is carcinogenic.

Chromium is also important in converting animal hides into leather. In the tanning process, hides are treated with basic solutions of Cr(III) salts, which cause cross-linking of collagen proteins. The hides toughen and become pliable and resistant to biological decay.

Copper, silver and gold

The first three pure metals known to humanity were probably copper, silver and gold, collectively known as the coinage metals because they found early use as coins. All three can be found in nature in their pure elemental form, and all have been highly valued throughout civilisation. The oldest known gold coins were used in Egypt around 5400 years ago. At about the same time, copper was obtained in the Middle East from charcoal reduction of its ores. The first metallurgy of silver was developed in Asia Minor (Turkey) about 500 years later. All three of these metals are excellent electrical conductors and are highly resistant to corrosion. These properties, coupled with its relatively low cost, make copper one of the most useful metals in modern society. About half of all copper produced is for electrical wiring, and the metal is also widely used for plumbing pipes.

Copper is used to make several important alloys, the most important of which are bronze and brass. Both alloys contain copper mixed with smaller amounts of tin and zinc in various proportions. The amount of tin in bronze exceeds that of zinc, whereas the opposite is true for brass. The discovery of bronze around 5000 years ago launched the advance of civilisation known today as the Bronze Age. Because bronze is harder and stronger than other metals known in antiquity, it became a mainstay of the civilisations of India and the Mediterranean, being used for tools, cookware, weapons, coins and objects of art. Today, the principal use of bronze is for bearings, fittings and machine parts.

Copper is resistant to oxidation, but over time the metal acquires a coating of green corrosion called patina (figure 13.55). The green compound is a mixed salt of Cu(II), hydroxide, sulfate and carbonate ions that is formed by air oxidation in the presence of carbon dioxide and small amounts of sulfur dioxide.

$$3Cu + 2H_2O + SO_2 + 2O_2 \rightarrow Cu_3(OH)_4SO_4$$

$$2Cu + H_2O + CO_2 + O_2 \rightarrow Cu_2(OH)_2CO_3$$

Although trace amounts of copper are essential for all forms of life, the metal is toxic in large amounts. Thus, copper(II) salts, particularly $CuSO_4{\cdot}5H_2O$, are used as pesticides and wood preservatives. Wood soaked in solutions of Cu(II) or coated with paints containing Cu(II) resist degradation by bacteria, algae and fungi.

Silver is usually found as a minor component of ores of more abundant metals such as copper and zinc. Most commercial silver is produced as a by-product of the production of these common metals. For example, electrolytic refining of copper generates a solid anodic residue that is rich in silver and other precious metals. The silver from this residue can be isolated by oxidising the metal into nitrate-containing solutions, silver nitrate being one of the few soluble silver salts. The pure metal is then deposited electrolytically. Silver is used for tableware in the form of sterling silver, an alloy containing small amounts of copper to make the metal harder. Silver is also used

FIGURE 13.55 Over time, copper metal develops a green coating called patina.

in jewellery, mirrors and batteries, and, prior to the advent of digital cameras, one of its main uses was in photography. Silver does not form a simple oxide by direct oxidation in air, but the metal does form a black tarnish with oxygen and trace amounts of hydrogen sulfide in the atmosphere.

$$4Ag + 2H_2S + O_2 \rightarrow \underset{\text{black}}{2Ag_2S} + 2H_2O$$

The extraction of gold by leaching has been described on page 586. Gold is used extensively in the manufacture of jewellery. Interestingly, Au(I) compounds are very effective in the treatment of rheumatoid arthritis, and certain gold-containing compounds appear to have anticancer properties.

Zinc and mercury

Zinc and mercury are found in the Earth's crust as sulfide ores, the most common of which are sphalerite, ZnS, and cinnabar, HgS; the isolation and purification of the metals from these is described on page 586. Most of the world's zinc output is used to prevent the corrosion of steel. Zinc is easier to oxidise than iron, as shown by the more negative reduction potential of Zn(II).

$$Fe^{2+} + 2e^- \rightarrow Fe \qquad E^\ominus = -0.447\text{ V}$$

$$Zn^{2+} + 2e^- \rightarrow Zn \qquad E^\ominus = -0.7618\text{ V}$$

Consequently, a zinc coating oxidises preferentially and protects steel from corrosion. Zinc coatings are applied in several ways: by immersion in molten zinc, by paint containing powdered zinc and by electroplating. Zinc is also combined with copper and tin to make brass and bronze, and large amounts of zinc are used to make several types of batteries. Zinc oxide is the most important zinc compound. Its principal industrial use is as a catalyst to shorten the time of vulcanisation in the production of rubber. The compound is also used as a white pigment in paints, cosmetics and photocopy paper. In everyday life, ZnO is used as a common sunscreen (figure 13.56).

FIGURE 13.56 ZnO is a common sunscreen.

The use of mercury for extracting silver and gold from their ores has been known for many centuries. Gold and silver form amalgams with liquid mercury, which can then be removed by distillation to leave the pure precious metal. The Romans mined the mineral cinnabar, HgS, from deposits in Spain 2000 years ago, and in the sixteenth century the Spanish shipped mercury obtained from the same ore deposits to the Americas for the extraction of silver. Mercury is an important component of street lamps and fluorescent lights. It is used in thermometers and barometers and in gas-pressure regulators, electrical switches and electrodes.

The platinum metals

Six of the transition metals — Ru, Os, Rh, Ir, Pd and Pt — are known collectively as the platinum metals. The group is named for the most familiar and most abundant of the six. These elements are usually found mingled together in ore deposits, and they share many common features. Although they are rare (total annual production is only about 400 tonnes), the platinum metals play important roles in modern society. They are valuable by-products of the extraction of common metals such as copper and nickel, and the anodic residue that results from copper refining (see p. 589) is a particularly important source. The chemistry involved in their purification is too complicated to describe here, except to note that the final reduction step involves reaction of molecular hydrogen with metal halide complexes.

By far the most important use of the platinum metals is for catalysis. The largest single use is in catalytic converters in the exhaust systems of cars. Platinum is the principal catalyst, but catalytic converters also contain rhodium and palladium. These elements catalyse a wide variety of reactions in the chemical and petroleum industries. For example, platinum metal is the catalyst for ammonia oxidation in the production of nitric acid.

$$4NH_3 + 5O_2 \xrightarrow{\text{Pt gauze, ~900 °C}} 4NO + 6H_2O$$

Palladium is used as a catalyst for hydrogenation reactions in the food industry, and a rhodium catalyst is used in the production of acetic acid.

$$CH_3OH + CO \xrightarrow{\text{Rh catalyst, 175 °C}} CH_3COOH$$

Osmium and iridium hold the distinction of being the two most dense elements in the periodic table, with a tennis ball-sized sphere of either metal weighing a quite astonishing 3.4 kg.

LO 1 Metals in the periodic table

The majority of known elements are metals, and they occur in distinct sections of the periodic table. Metals in groups 1, 2 and 13–16 are called main-group metals, while those in groups 3–12 are called transition metals. Compounds of the former are usually diamagnetic and colourless while those of the latter are often paramagnetic and coloured. Transition metal ions form complexes in which the metal ion is coordinated to one or more ligands. Complexes can be positively or negatively charged or neutral.

LO 2 Transition metals

Transition metals are characterised by *d* valence orbitals, and can exist in more than one oxidation state. The *nd* orbitals in transition metal complexes are always of lower energy than the $(n + 1)s$ orbitals so are filled with electrons first. Transition metal ions are Lewis acids.

LO 3 Ligands

Ligands are Lewis bases that contain one or more electron pairs and can be negatively charged, neutral or, very rarely, positively charged. The atom on which a lone pair is located is called a donor atom; common donor atoms are F, Cl, Br, I, O, S, N and P, although organic compounds with C donor atoms are also known. Ligands that bind to a metal ion using only one donor atom are called monodentate, those using two are called bidentate and those using two or more are collectively called polydentate. When polydentate ligands bind to a single transition metal ion, they form one or more chelate rings, and complexes containing these are called chelates.

LO 4 Transition metal complexes

The study of transition metal complexes is called coordination chemistry. Transition metal complexes form as a result of a Lewis acid – Lewis base interaction between a transition metal ion and a ligand. The bonds between the metal ion and the ligand are sometimes called coordinate, donor covalent or dative bonds, while the complexes are often called coordination compounds. Complexes are written enclosed in square brackets, while the counterions (if any) that balance the charge are written outside the square brackets.

The most common coordination number of the central metal ion in transition metal complexes is 6, followed by 4 and 5. Most 6-coordinate complexes display an octahedral geometry. The 4-coordinate complexes adopt either a square planar or tetrahedral geometry, while 5-coordinate complexes are usually trigonal bipyramidal or square pyramidal.

There are numerous isomeric possibilities in transition metal complexes. Structural isomers have the same molecular formula, but different orders of attachments of the constituent atoms. Four important types in coordination chemistry are ionisation isomers, coordination isomers, linkage isomers and hydrate isomers. Linkage isomers are formed with ambidentate ligands, ligands with two or more different types of donor atom. Stereoisomers have the same connections between the constituent atoms, but different arrangements of the atoms in space. An example of such isomerism are *cis–trans* isomers; a *cis* isomer has two groups on the same side of some reference plane and a *trans* isomer has the groups on different sides. Meridional (*mer*) and facial (*fac*) isomers are also possible in complexes of the form $[ML_3X_3]$. Transition metal complexes are named according to IUPAC rules.

The formation of transition metal complexes can be quantified by the cumulative formation constant (β_n), which is the equilibrium constant for the process:

$$M^{x+}(aq) + nL(aq) \rightleftharpoons [ML_n]^{x+}(aq)$$

where $\beta_n = \dfrac{[ML_n{}^{x+}(aq)]}{[M^{x+}(aq)][L(aq)]^n}$

β_n values for most transition metal complexes are extremely large, with those for complexes containing bidentate or polydentate ligands being significantly greater than those for the corresponding monodentate ligands, a manifestation of the chelate effect. Complexes that readily exchange their ligands are said to be labile, while those with slow rates of ligand exchange are called inert.

Crystal field theory can be used to describe the bonding in transition metal complexes, and it predicts that the *d* orbitals, which are degenerate in a free transition metal ion, are split by interactions with the ligand electrons. In an octahedral complex, they are split into a lower energy t_{2g} set of three orbitals and a higher energy e_g set of two orbitals. The energy difference between these sets of orbitals is called the crystal field splitting energy (CFSE), which is given the symbol Δ_o. The way in which these orbitals are filled with electrons depends on the comparative magnitudes of Δ_o and the pairing energy, *P*. If $P > \Delta_o$, a high-spin configuration results; if $P < \Delta_o$, a low-spin configuration results. In octahedral complexes, these possibilities arise only for d^4, d^5, d^6 and d^7 configurations. Different *d*-orbital splitting diagrams are obtained for the 4-coordinate square planar and tetrahedral geometries.

Transition metal complexes are often coloured because of electronic transitions between the nondegenerate *d* orbitals; the colour of the complex is complementary to that absorbed. We quantify the absorbed light by measuring the absorbance with a spectrophotometer, and we can calculate the value of Δ_o from the wavelength of the absorbance peak in the UV/visible spectrum. The position of this peak is found to depend on the nature of the ligands, and we can arrange the ligands in order of the magnitude of their crystal field splitting energies to give the spectrochemical series. Ligands giving a large Δ_o are called strong-field while those that induce small splittings are called weak-field. In addition to electronic transitions between the *d* orbitals, very intense ligand-to-metal charge transfer transitions can also occur in transition metal complexes.

LO 5 Transition metal ions in biological systems

The transition metals V, Cr, Mn, Fe, Co, Ni, Cu, Zn and Mo are essential for healthy biological function. Transition metal ions are important constituents of proteins and enzymes involved in transport and storage (Fe in haemoglobin and ferritin), catalysis (Cu and Zn in superoxide dismutase) and electron transfer (Fe in cytochromes and iron–sulfur proteins and Cu in blue copper proteins). The ability of the transition metal ion to adopt more than one oxidation state and coordination number is crucial to the operation of these biological systems.

LO 6 Isolation and purification of transition metals

Metallurgy is the study of the production and purification of metals from their ores. As ores contain metals in positive oxidation states, reduction is required to liberate the free metals. The four steps involved in isolating a pure metal from its ore are separation, conversion, reduction and refinement. The reduction step is carried out either chemically or electrolytically, depending on the ore. Iron is the most important metal and is obtained from its ore using a blast furnace, with coke as the reductant. Titanium is isolated from TiO_2 by conversion to $TiCl_4$ and reduction with molten Mg, while copper is obtained from its sulfide and oxide ores and purified by electrolysis.

LO 7 Applications of transition metals

Transition metals have a vast number of uses. Titanium has high strength and low density, making it useful for aircraft and bike frames and jet engines. Most chromium is used to make steel and other alloys. Copper, silver and gold, the coinage metals, are quite resistant to direct oxidation; copper is mainly used in electrical wiring, while some gold complexes are useful for treating arthritis. Zinc is used to galvanise iron, while mercury is used in thermometers and fluorescent lights. Ru, Os, Rh, Ir, Pd and Pt constitute the platinum metals and are chemically important as solid catalysts in industrial processes as well as catalytic converters in cars.

KEY CONCEPTS AND EQUATIONS

Determining *d*-electron configurations *(section 13.2)*

This allows us to calculate the number of *d* electrons in any transition metal ion. Neutral atoms have valence electron configurations of $(n+1)s^2nd^{x-2}$, where x is the group number of the metal in the periodic table and n is the principal quantum number. The nd orbitals are filled before the $(n+1)s$ orbitals in transition metal complexes.

Cumulative formation constant (β_n) *(section 13.4)*

This quantifies the extent to which the formation reaction of a transition metal complex occurs. For the equilibrium:

$$M^{x+}(aq) + nL(aq) \rightleftharpoons [ML_n]^{x+}(aq)$$

$$\beta_n = \frac{[ML_n{}^{x+}(aq)]}{[M^{x+}(aq)][L(aq)]^n}$$

Naming transition metal complexes *(section 13.4)*

The IUPAC rules allow us to determine the correct name of any transition metal complex.

1. Cationic species are named before anionic species.
2. The names of anionic ligands always end in the suffix *-o*. Ligands with names ending in *-ide*, *-ite* and *-ate* have this suffix changed to *-ido*, *-ito* and *-ato*, respectively.
3. A neutral ligand is usually given the same name as the neutral molecule, except aqua for H_2O, ammine for NH_3 and carbonyl for CO.
4. The number of a particular ligand is specified by the prefixes *di-* = 2, *tri-* = 3, *tetra-* = 4, *penta-* = 5, *hexa-* = 6 etc.
5. Ligands are named first, in alphabetical order without regard to charge, followed by the name of the metal.
6. Negative (anionic) complex ions always end in the suffix *-ate*. For neutral or positively charged complexes, the metal is always specified with the English name for the element, without any suffix.
7. The oxidation state of the metal in the complex is written in Roman numerals within parentheses following the name of the metal.
8. The number of counterions need not be specified.
9. Include any stereochemical descriptors (e.g. *cis, trans, mer, fac*) at the start of the name, italicised and hyphenated.

Crystal field theory *(section 13.4)*

This allows us to determine the *d*-orbital splitting diagrams for transition metal complexes of any geometry, and hence the *d*-electron configuration in these complexes. It is based on the crystal field splitting energy, which is the energy difference (Δ_o) between the d_{xy}, d_{xz} and d_{yz} orbital set and the $d_{x^2-y^2}$ and d_{z^2} orbital set.

The spectrochemical series *(section 13.4)*

The position of a ligand in the spectrochemical series tells us the magnitude of the *d*–orbital splitting it can induce. In order of decreasing Δ_o values, the series is:

$$CO \approx CN^- > NO_2{}^- > en > NH_3 > H_2O > C_2O_4{}^{2-} > OH^- > F^- > Cl^- > Br^- > I^-$$

KEY TERMS

absorbance (*A*) *p. 575*
ambidentate *p. 559*
axial *p. 557*
Beer–Lambert law *p. 578*
Beer's law *p. 578*
bidentate *p. 551*
chelate complex *p. 551*
chelate effect *p. 566*
chelate ring *p. 551*
cis isomer *p. 559*
cis–trans isomers *p. 560*
complementary colour *p. 575*
complex *p. 549*
coordinate bond *p. 556*
coordination chemistry *p. 556*
coordination compound *p. 556*
coordination isomers *p. 559*
coordination number *p. 557*
counterion *p. 556*
crystal field splitting energy (Δ_o) *p. 571*
crystal field theory *p. 570*
cumulative formation constant (β_n) *p. 565*
d–d transition *p. 574*
dative bond *p. 556*
degenerate *p. 570*
diamagnetic *p. 548*
disproportionation *p. 569*
donor atom *p. 551*
donor covalent bond *p. 556*
electronic transition *p. 574*
equatorial *p. 557*
facial (*fac*) *p. 561*
ferromagnetism *p. 580*
high-spin *p. 572*
hydrate isomers *p. 559*
inert *p. 569*
ionisation isomers *p. 559*
labile *p. 568*
ligand *p. 549*
ligand-to-metal charge transfer (LMCT) transition *p. 577*
linkage isomers *p. 559*
low-spin *p. 572*
magnetic moment (μ) *p. 578*
main-group metals *p. 548*
meridional (*mer*) *p. 561*
metal-to-ligand charge transfer (MLCT) transition *p. 577*
metallurgy *p. 584*
monodentate *p. 551*
octahedral *p. 557*
organometallic compound *p. 555*
pairing energy (*P*) *p. 572*
paramagnetic *p. 548*
polydentate *p. 551*
spectrochemical series *p. 576*
square planar *p. 558*
square pyramidal *p. 558*
stereoisomers *p. 559*
strong-field *p. 576*
structural isomers *p. 559*
tetrahedral *p. 558*
trans isomer *p. 559*
transition metals *p. 548*
trigonal bipyrimidal *p. 558*
weak-field *p. 576*

LO 1 LO 2 Metals in the periodic table and transition metals

13.1 Starting materials for the synthesis of transition metal complexes, such as those below, can be purchased from chemical supply companies. Determine the oxidation state of the transition metal in. (a) MnI_2, (b) $NiCl_2{\cdot}6H_2O$, (c) $MoOCl_4$, (d) $AuCl_3{\cdot}3H_2O$, (e) $TiCl_4$, (f) V_2O_5, (g) K_2PtCl_6, (h) $NaNbO_3$, (i) Re_2O_7, (j) $(NH_4)_2[Fe(OH_2)_6](SO_4)_2$.

13.2 Give the names and symbols for the elements that have the following valence configurations. (a) $4s^13d^5$ (b) $5s^24d^{10}$ (c) $4s^13d^{10}$ (d) $5s^14d^8$ (e) $4s^23d^2$ (f) $5s^14d^4$

13.3 Write the d electron configurations of the following transition metal complexes. (a) Co^{2+} (b) Os^{3+} (c) Cu^+ (d) Pd^{2+} (e) V^{2+} (f) Au^{3+} (g) Ta^{5+} (h) Re^{e+}

LO 3 Ligands

13.4 To be a ligand, a substance should also generally be a Lewis base. Explain.

13.5 Give the names and chemical formulae of two electrically neutral, monodentate ligands.

13.6 Give the formulae of four ions that have 1− charges and are monatomic, monodentate ligands.

13.7 What is the maximum number of bidentate ligands that can be present in a 6-coordinate metal complex?

13.8 What is a chelating ligand? Use Lewis structures to illustrate how the oxalate ion, $C_2O_4^{2-}$, can function as a chelating ligand.

13.9 The $EDTA^{4-}$ ligand has 10 potential donor atoms. Why do you think that generally only 6 of them bind to a transition metal ion?

LO 4 Transition metal complexes

13.10 The formation of the complex ion $[Cu(OH_2)_5]^{2+}$ is described as a Lewis acid–base reaction.
(a) Explain this statement.
(b) What are the formulae of the Lewis acid and the Lewis base in this reaction?
(c) What is the formula of the ligand?
(d) What is the name of the species that provides the donor atom?
(e) What atom is the donor atom, and why is it so designated?
(f) What is the name of the species that is the Lewis acid?

13.11 Give three synonyms for the covalent bonds between the metal ion and the ligand(s) in a transition metal complex.

13.12 Use Lewis structures to illustrate the formation of $[Cu(NH_3)_4]^{2+}$ and $[CuCl_4]^{2-}$ ions from their respective components.

13.13 How does the process for determining the formal charges of atoms in a transition metal complex differ from that outlined in chapter 5?

13.14 The cobalt(III) ion, Co^{3+}, forms a 1 : 1 complex with $EDTA^{4-}$. What is the net charge, if any, on this complex, and what would be a suitable formula for it (using the symbol EDTA)?

13.15 The chelate effect is used to explain the fact that the $[Cr(en)_3]^{3+}$ cation has a higher β_n value than the $[Cr(NH_3)_6]^{3+}$ cation. Outline the importance of entropy in this explanation.

13.16 What is a coordination number? What structures are generally observed for complexes in which the central metal ion has a coordination number of 4?

13.17 The two common geometries observed for 5-coordinate complexes are trigonal bipyramidal and square pyramidal. Why does VSEPR favour the former over the latter?

13.18 Sketch the structure of an octahedral complex that contains only identical monodentate ligands.

13.19 Outline the four types of structural isomers commonly found in coordination complexes.

13.20 Define 'stereoisomerism'.

13.21 Why are *cis* and *trans* isomers possible in square planar complexes but not in tetrahedral complexes?

13.22 What is cisplatin? Draw its structure.

13.23 Sketch and label the five d orbitals, being sure to show how these are oriented relative to the x, y and z axes.

13.24 Which d orbitals point between the x-, y- and z-axes and which point along the axes?

13.25 In an octahedral complex, the d orbitals are no longer degenerate as they are in the free metal ion, and are split into two sets. Show how this arises from a consideration of the repulsions between the electrons on the metal ion and those on the ligand(s).

13.26 Sketch the d-orbital energy level diagram for a typical octahedral complex.

13.27 What is the process that gives rise to the colours of transition metal complexes?

13.28 If a complex appears red, what colour light does it absorb? What colour light is absorbed if the complex appears yellow?

13.29 What value is used to order the ligands in the spectrochemical series?

13.30 What do the terms 'low-spin complex' and 'high-spin complex' mean?

13.31 Only particular d electron configurations can give rise to high-spin and low-spin octahedral and tetrahedral complexes. Give these d electron configurations for each geometry.

13.32 Sketch what happens to the d-orbital electron configuration of the $[Fe(CN)_6]^{4-}$ ion when the complex absorbs a photon of visible light.

13.33 The octahedral $[CoF_6]^{3-}$ anion, in contrast to the majority of 6-coordinate Co(III) complexes, is paramagnetic. Sketch the d-orbital energy level diagram of this complex and indicate the electron occupancy of the orbitals.

13.34 Consider the complex $[MCl_4(OH_2)_2]^-$ illustrated in the centre below. Suppose the structure of this complex is distorted to give the structure on the right, where the water molecules along the z-axis have moved away from the metal somewhat and the four chloride ions along the x- and y-axes have moved closer. What effect will this distortion have on the splitting pattern of the d orbitals? Use a sketch of the splitting pattern to illustrate your answer.

before distortion

after distortion

LO 5 Transition metal ions in biological systems

13.35 What transition metals are essential for humans?

13.36 What features do myoglobin and the cytochromes have in common?

LO 6 Isolation and purification of transition metals

13.37 Why do the majority of transition metals in nature not exist as free metals?

13.38 What is gangue?

13.39 What are the four main steps used in metallurgy to convert ores to pure metals?

13.40 Describe the flotation process.

13.41 Why is the roasting of sulfide ores in air potentially damaging to the environment?

13.42 Why is reduction, rather than oxidation, necessary to extract metals from their compounds?

13.43 Why is chemical reduction generally preferred to electrolysis in the reduction of metal ores to the free metals?

13.44 Describe the chemical reactions involved in the reduction of Fe_2O_3 that take place in a blast furnace. What is the active reducing agent in the blast furnace?

13.45 Magnetite, Fe_3O_4, used as the production of iron and steel, is an example of a mixed valence compound, in which the iron ions display different oxidation states. What are the oxidation states of iron in Fe_3O_4?

13.46 What is the difference between pig iron and steel?

LO 7 Applications of transition metals

13.47 Identify the coinage metals and describe some of their applications.

13.48 Identify the platinum metals and describe some of their applications.

13.49 What features of titanium account for its use as an engineering metal?

13.50 Explain why titanium(IV) oxide is used extensively as a white pigment.

REVIEW PROBLEMS

13.51 The manganese(III) ion forms a complex with five chloride ions, called the pentachloridomanganate(III) ion. What is the net charge on this complex ion, and what is its formula? **LO 4**

13.52 Write the formula, including the correct charge, for a complex that contains Co(III), two Cl^- ligands and two ethylenediamine ligands. **LO 4**

13.53 A complex contains Rh(III), four ammine ligands and two Cl^- ligands. Write the formula, being sure to include the correct charge. Are there possible isomers of this complex? **LO 4**

13.54 How would the following molecules or ions be named as ligands when writing the name of a complex ion? **LO 3**

(a) $C_2O_4^{2-}$
(b) S^{2-}
(c) Cl^-
(d) $(CH_3)_2NH$ (dimethylamine)
(e) NH_3
(f) N^{3-}
(g) SO_4^{2-}
(h) CH_3COO^-

13.55 Give IUPAC names for each of the following. **LO 2**

(a) $[Co(NH_3)_6]^{3+}$
(b) *trans*-$[CrCl_2(NH_3)_4]^+$
(c) $[CuCl_5]^{3-}$
(d) $[Mo(CN)_8]^{4-}$
(e) $[Fe(OH_2)_6]^{2+}$
(f) $[AgCl_4]^{3-}$
(g) *cis*-$[CoBr_2(en)_2]Br$
(h) $[Cr(NH_3)_5OH]SO_4$
(i) $K_3[Co(O_2CO)_3]$

13.56 Write chemical formulae for each of the following. **LO 3**

(a) tetraaquadicyanidoiron(III) ion
(b) tetraammineoxalatonickel(II)
(c) diaquatetracyanidoferrate(III) ion
(d) potassium hexathiocyanatomanganate(III)
(e) tetrachloridocuprate(II) ion
(f) tetrachloridoaurate(III) ion
(g) bis(ethylenediamine)dinitroiron(III) ion
(h) tetraamminedicarbonatocobalt(III) nitrate
(i) ethylenediaminetetraacetatoferrate(II) ion
(j) diamminedichloridoplatinum(II)

13.57 What are the coordination number and the oxidation state of the ruthenium ion in the complex $K_2[RuCl_5(OH_2)]$. **LO 4**

13.58 What is the coordination number of nickel in $[Ni(NO_2)_2(O_2C_2O_2)_2]^{4-}$? **LO 3**

13.59 Multidentate ligands containing more than one pyridine ring are widely used in coordination chemistry. One such ligand, shown above right, is commonly called terpyridine (not its IUPAC name) and is abbreviated 'terpy'. **LO 4**

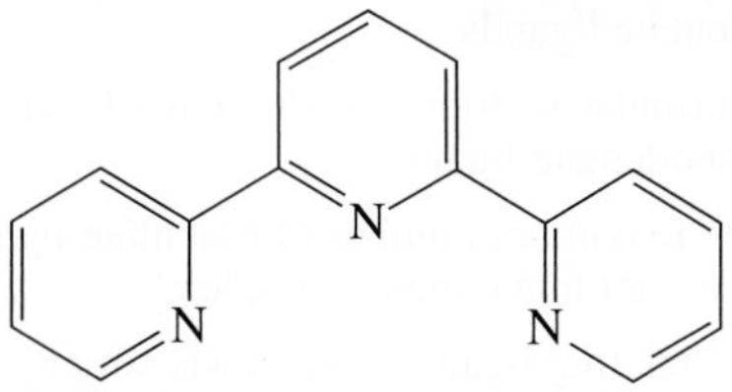

It uses three donor atoms when it bonds to a transition metal ion.

(a) Identify the donor atoms.
(b) What is the coordination number of the ruthenium ion in the complex $[Ru(terpy)_2]^{2+}$?
(c) Which complex would be expected to have the larger β_n value, $[Ru(NH_3)_6]^{2+}$ or $[Ru(terpy)_2]^{2+}$? **LO 4**

13.60 The following are two structures for a complex. Are they different isomers, or are they identical? Explain your answer. **LO 4**

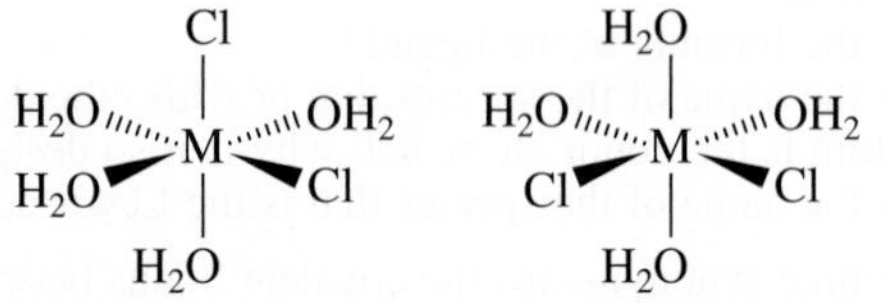

structure 1 structure 2

13.61 The related square planar complexes $[PtBrCl(NH_3)_2]$ and $[PtBrCl(en)]$ have different numbers of possible isomers. Draw all the isomers for each complex and account for this difference. **LO 4**

13.62 The complex $[CoCl_3(NH_3)_3]$ can exist in two isomeric forms. Sketch them. **LO 4**

13.63 Which complex in the following pairs should exhibit the larger value of Δ_o? Give your reasoning in each case. **LO 4**

(a) $[Cr(OH_2)_6]^{2+}$ or $[Cr(OH_2)_6]^{3+}$
(b) $[Co(en)_3]^{3+}$ or $[CoF_6]^{3-}$

13.64 Arrange the following complexes according to the wavelengths of their absorption maxima, from lowest to highest: $[Cr(OH_2)_6]^{3+}$, $[CrCl_6]^{3-}$, $[Cr(en)_3]^{3+}$, $[Cr(CN)_6]^{3-}$, $[Cr(NO_2)_6]^{3-}$, $[CrF_6]^{3-}$, $[Cr(NH_3)_6]^{3+}$. **LO 4**

13.65 Which of the following complexes is expected to exhibit its maximum absorbance at the longest wavelength: $[Co(OH_2)_6]^{3+}$, $[Co(en)_3]^{3+}$ or $[Co(CN)_6]^{3-}$? Give your reasoning. **LO 4**

13.66 In each of the following pairs, which complex should absorb light of the longer wavelength? **LO 4**

(a) $[Fe(OH_2)_6]^{2+}$ or $[Fe(CN)_6]^{4-}$
(b) $[Mn(CN)_6]^{3-}$ or $[Mn(CN)_6]^{4-}$

13.67 Which complex in each of the following pairs is expected to exhibit its maximum absorbance at the shortest wavelength? Give your reasoning in each case. **LO 4**
(a) $[RuCl(NH_3)_5]^{2+}$ or $[FeCl(NH_3)_5]^{2+}$
(b) $[Ru(NH_3)_6]^{2+}$ or $[Ru(NH_3)_6]^{3+}$

13.68 An octahedral complex $[CoA_6]^{3+}$ is red. The complex $[CoB_6]^{3+}$ is green. Which ligand, *A* or *B*, produces the larger crystal field splitting energy, Δ_o? Explain your answer. **LO 4**

13.69 Which of the two complexes described in question 13.68 is expected to be more easily reduced, $[CoA_6]^{3+}$ or $[CoB_6]^{3+}$? Explain your answer. **LO 6**

13.70 Referring to the complexes in questions 13.68 and 13.69, is the colour of $[CoA_6]^{2+}$ more likely to be red or blue? **LO 4**

13.71 Is the complex $[Co(CN)_6]^{4-}$ more likely to be low-spin or high-spin? Why could it not be diamagnetic? **LO 4**

13.72 Sketch the *d*-orbital energy levels for $[Fe(OH_2)_6]^{3+}$ and $[Fe(CN)_6]^{3-}$ and predict the number of unpaired electrons in each. **LO 4**

13.73 Determine the oxidation state and *d* electron configuration of the metal ion in each of the following coordination complexes. **LO 2,4**
(a) $[Co(en)_3]Cl_3$
(b) *trans*-$[CrI_2(NH_3)_4]I$
(c) $[Mn(acac)_3]$
(d) $[Ru(bipy)_3]Cl_2$
(e) $[W(CO)_6]$
(f) $[RhCl(NH_3)_5]Cl_2$
(g) $K_3[Fe(ox)_3]$
(h) $K_3[Ni(CN)_5]$
(i) $[Fe(CO)_3(P(CH_3)_3)_2]$
(j) *mer*-$[IrBr_3(NH_3)_3]$

13.74 Name the compounds in question 13.73. **LO 4**

13.75 Draw the probable structures of the metal complexes in question 13.73. **LO 4**

13.76 Write the formulae of the following complexes. **LO 3,4**
(a) *cis*-tetraamminechloridonitrocobalt(III)
(b) amminetrichloridoplatinate(II)
(c) *trans*-diaquabis(ethylenediamine)copper(II)
(d) tetrachloridoferrate(III).
(e) potassium tetrachloridoplatinate(II)
(f) pentaammineaquachromium(III) iodide
(g) tris(ethylenediamine)manganese(II) chloride
(h) pentaammineiodidocobalt(III) nitrate.

13.77 Draw the probable structure of each complex ion in question 13.76. **LO 4**

13.78 For an octahedral complex of each of the following metal ions, draw a crystal field energy diagram that shows the electron occupancy of the various *d* orbitals: (a) Ti(II), (b) Cr(III), (c) Mn(II), (d) Fe(III), (e) Zn(II), (f) Cr(II), (g) Co(II), (h) Rh(III). Where appropriate, show both the high-spin and low-spin configurations. **LO 4**

13.79 Predict whether each of the following complexes is diamagnetic or paramagnetic. If the complex is paramagnetic, state the number of unpaired electrons it has and estimate its magnetic moment. **LO 4**
(a) $[Co(OH_2)_6]^{2+}$
(b) $[Cr(en)_3]^{3+}$
(c) $[CoCl_4]^{2-}$ (tetrahedral)
(d) $[Pd(P(CH_3)_3)_4]$ (square planar)
(e) $[PtCl_2(en)]$ (square planar)
(f) $[Mn(CN)_6]^{3-}$
(g) $[ZnCl_4]^{2-}$
(h) $[FeF_6]^{3-}$

13.80 Compounds of Zr(II) are dark purple, but most Zr(IV) compounds are colourless. Explain. **LO 4**

13.81 The Co(III) complexes $[Co(OH_2)_6]^{3+}$ and $[Co(NH_3)_6]^{3+}$ have very different colours – one is blue and the other is orange. Decide which is which, and explain your reasoning. **LO 4**

13.82 The complex $[Fe(en)_3]Cl_3$ is low-spin. State: **LO 4,7**
(a) the coordination number of the metal
(b) the oxidation number and number of *d* electrons of the metal
(c) the geometry of the complex
(d) whether the complex is diamagnetic or paramagnetic
(e) the number of unpaired electrons in the complex
(f) the magnetic moment of the complex.

13.83 Consider the complex $[Ni(OH_2)_6]SO_4$. Determine: **LO 4**
(a) the coordination number of the metal ion
(b) the oxidation number and *d* electron configuration of the metal ion
(c) the geometry of the complex
(d) whether the complex is diamagnetic or paramagnetic
(e) the number of unpaired electrons in the complex
(f) the magnetic moment of the complex.

13.84 A solution containing Fe(II)(aq) is paramagnetic. Addition of sodium cyanide makes the solution diamagnetic. Use crystal field energy diagrams to explain why the magnetic properties of the solutions are different. **LO 6**

13.85 The $[Co(NH_3)_6]^{3+}$ cation has its maximum absorbance at 475 nm. Calculate the crystal field splitting energy for the compound and predict its colour. **LO 4**

13.86 The complex $[Fe(OH_2)_6]^{3+}$ has its maximum absorbance at 724 nm. Calculate the crystal field splitting energy for the compound and predict its colour. **LO 4**

13.87 Draw all possible isomers of the following compounds. **LO 4**
(a) $[Co(en)I_2(NH_3)_2]$ (b) $[FeCl_2(CO)_4]$
(c) $[PtCl_2(P(CH_3)_3)_2]$ (square planar) (d) $[RhCl_3(NH_3)_3]$

13.88 Draw the structures of all possible isomers of each of the following coordination compounds. **LO 4**
(a) tetraamminedibromidocobalt(III) bromide
(b) triamminetrichloridochromium(III)
(c) dicarbonylbis(trimethylphosphane)platinum(0) (square planar)

13.89 The nitrato ligand can bind to a single transition metal ion in either a monodentate or a bidentate fashion. Draw each binding mode of this ligand. **LO 4**

13.90 Write electron configurations for the following. (a) Cr Cr(II) and Cr(III) (b) V, V(II), V(III), V(IV) and V(V) (c) Ti, Ti(II) and Ti(IV) (d) Au, Au(I) and Au(II) (e) Ni, Ni(II) and Ni(III) (f) Mn, Mn(II), Mn(IV) and Mn(VII) **LO 2**

13.91 Name the following coordination compounds. **LO 4**

(a) $[Cr(OH_2)_3Cl_2(OH_2)]Cl$ — octahedral Cr with axial OH_2 and Cl; equatorial Cl, H_2O, OH_2, OH_2; counter-ion Cl
(b) Mn with axial Br and CO; equatorial OC, OC, CO, CO
(c) Pt with Cl, Cl, H_3N, NH_3
(d) $[Zn(H_3N)_2(NH_3)_2]$ with H_3N, H_3N, NH_3, NH_3; counter-ion SO_4
(e) K_3[Fe] with axial CN and CN; equatorial NC, NC, CN, CN
(f) [Co] with axial NH_3 and NH_3; equatorial H_3N, Cl, Cl, NH_3; counter-ion Cl

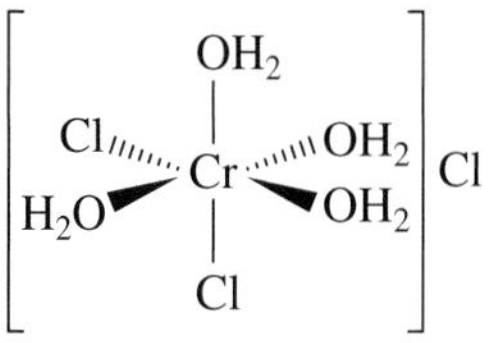

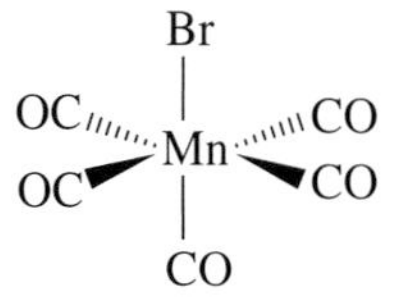

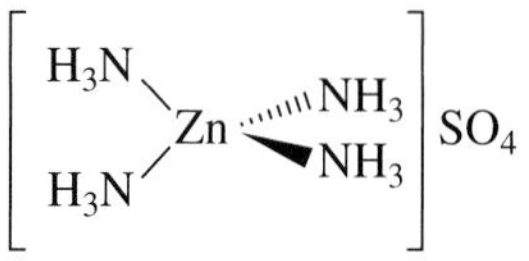

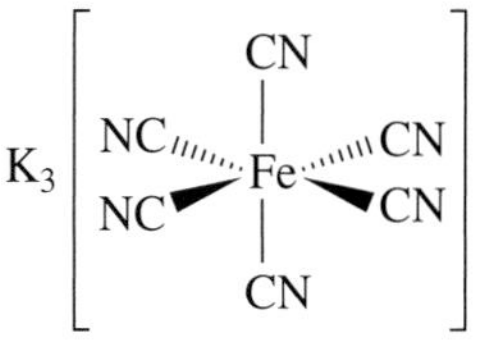

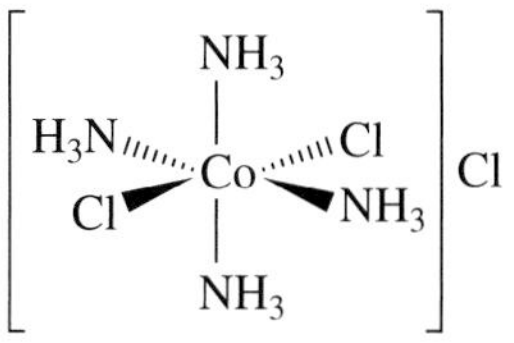

13.92 The Cu(II) ion forms tetrahedral complexes with some anionic ligands. When $CuSO_4{\cdot}5H_2O$ dissolves in water, a blue solution results. The addition of aqueous KF solution results in a green precipitate, but the addition of aqueous KCl results in a bright green solution. Identify each green species and write chemical reactions for these processes. **LO 4**

13.93 The hydrated Ni(II) ion, $[Ni(OH_2)_6]^{2+}$, is green. When ammonia is added to an aqueous solution of this ion, the colour changes to deep blue. Addition of ethylenediamine to the deep blue solution again results in a colour change, this time to purple. Identify the complexes giving rise to the deep blue and purple colours. **LO 4**

13.94 Draw a ball-and-stick model of the *mer* isomer of $[NiCl_3F_3]^{4-}$, oriented so that the fluoride ions are in the top, bottom and left forward positions. **LO 4**

13.95 Draw a ball-and-stick model of the *fac* isomer of $[NiCl_3F_3]^{4-}$, oriented so that the fluoride ions are in the top and two rear positions. **LO 4**

13.96 As ligands, chloride and cyanide are at opposite ends of the spectrochemical series. Nevertheless, experiments show that $[CrCl_6]^{3-}$ and $[Cr(CN)_6]^{3-}$ display similar magnetic behaviour. Explain how this can be so. **LO 8**

13.97 Octahedral complexes containing the Ni(II) ion usually display relatively constant magnetic moments around 2.5 to 3.0 μ_B, while those containing Fe(II) can exhibit variable magnetic moments as low as 0 μ_B or as high as 4.90 μ_B. Why? **LO 4**

13.98 The $d_{x^2-y^2}$ and d_{z^2} orbitals are degenerate in an octahedral complex. However, in a square planar complex, the $d_{x^2-y^2}$ orbital is of higher energy than the d_{z^2} orbital. Use orbital sketches to explain the difference. **LO 4**

13.99 Despite appearing extremely similar, the d_{xy} and $d_{x^2-y^2}$ orbitals have different energies in octahedral coordination complexes. Use sketches of the orbital interactions in such complexes to illustrate the reason for this. **LO 4**

13.100 Determine the Lewis structure of the dichromate anion, $[Cr_2O_7]^{2-}$, which contains one bridging oxygen atom, and draw a model showing its geometry. **LO 4**

13.101 Superoxide dismutase catalyses the disproportionation of the superoxide ion, O_2^-. Write a balanced chemical equation for this process. **LO 5**

13.102 Some researchers use the term 'the brass enzyme' to describe superoxide dismutase. Can you suggest a reason for this nickname? **LO 7**

13.103 Blue copper proteins are only blue when the copper ion is in the Cu(II) oxidation state — when the metal is reduced to Cu(I), they become colourless. The extreme intensity of the blue colour in Cu(II) blue copper proteins comes not from a *d-d* transition, but from a ligand-to-metal charge transfer transition involving an electron being transferred from a sulfur lone pair on a cysteine ligand to the copper centre. Why does this charge transfer interaction occur for Cu(II) but not Cu(I)? **LO 5**

13.104 Solutions containing Cu^{2+}(aq) ions are blue, whereas solutions containing Zn^{2+}(aq) ions are colourless. Explain this observation. **LO 4**

13.105 Explain why UV/vis spectroscopy might not be the method of choice for the study of a metalloenzyme containing Mn(II) ions. **LO 5**

13.106 Explain how liquid mercury can be used to purify metals such as gold and silver. **LO 6**

13.107 Write balanced chemical equations for each of the following metallurgical processes. (a) formation of $TiCl_4$ from TiO_2 using coke and chlorine gas (b) reduction of HgS by reaction with O_2 (show, using oxidation numbers what is reduced and what is oxidised in this reaction) (c) roasting of ZnS **LO 7**

13.108 A copper ore contains 2.37% Cu_2S by mass. If 5.60×10^4 kg of this ore is heated in air, calculate the mass of copper metal that is obtained and the volume of SO_2 gas produced in ambient conditions (9.93×10^4 Pa and 23.5 °C). **LO 6**

13.109 Chromite, $FeCr_2O_4$, is reduced to ferrochrome, $FeCr_2$, using coke. What mass of coke is required for each kilogram of chromite, assuming the reaction proceeds to completion? **LO 7**

13.110 Both vanadium and silver are lustrous silvery metals. Suggest why silver is widely used for jewellery but vanadium is not. **LO 6**

13.111 Titanium is the second most abundant transition metal in the Earth's crust, after iron. However, unlike iron and many other transition metals, titanium has essentially no biological role in humans. Why you think this is so? **LO 7**

13.112 What is the chemical name and the formula of the black tarnish that accumulates on objects made of silver? Write a balanced equation to show how this black tarnish forms. **LO 6,7**

13.113 The black tarnish on silver may be removed by placing the tarnished object on a sheet of aluminium foil in water containing a salt. Write a balanced chemical equation for this process, specifying which substances are oxidised and reduced. **LO 6**

ADDITIONAL EXERCISES

13.114 The complex $[PtCl_2(NH_3)_2]$ can be obtained as two distinct isomeric forms. Show that this means the complex cannot be tetrahedral. **LO 4**

13.115 A green, octahedral Co(III) complex, prepared from $[Co(OH_2)_6]Cl_2$ and ethylenediamine was found to have the following elemental analysis. **LO 4,6**

C: 16.83 %; H: 5.65 %; N: 19.63 %; Cl: 37.25 %.

Treatment of an aqueous solution containing 1.00 g of the complex with excess $AgNO_3$ resulted in immediate precipitation of 0.502 g of AgCl.

(a) Use the elemental analysis to determine the empirical formula of the green complex.
(b) Use the mass of AgCl obtained to determine the actual formula of the green complex. (Note that ionic chloride is precipitated immediately as AgCl on adding $AgNO_3$, whereas chloride bound to Co(III) is precipitated only slowly).
(c) Heating the green compound in water gives a purple compound which has the same elemental analysis and reacts identically towards $AgNO_3$. Explain these observations.
(d) Given that the purple complex can be resolved into enantiomers (see chapter 17) while the green complex cannot, draw the structures of both the green and purple complexes.

13.116 The octahedral chromium complexes $[Cr(NH_3)_5L]^{n+}$, where L is Cl^-, H_2O or NH_3, have absorption maxima at 515 nm, 480 nm and 465 nm, respectively. **LO 4**
(a) What is the value of n for each of the complexes?
(b) Predict the colours of the three complexes.
(c) Calculate the crystal field splitting energy in kJ mol^{-1} for each of the complexes and explain the trend in values.

13.117 Two-coordinate complexes are relatively rare. One example is the $[Ag(NH_3)_2]^+$ cation, which has a linear geometry. Predict the crystal field splitting diagram for this complex, placing the ligands along the z-axis. **LO 4**

13.118 Tetracarbonylnickel(0) is $[Ni(CO)_4]$, and tetracyanidozinc(II) is $[Zn(CN)_4]^{2-}$. Predict the geometry and colour of each complex. **LO 4**

13.119 The group 6 transition metals Cr, Mo and W all form colourless complexes having the general formula $[M(CO)_6]$. Determine the oxidation state of the metal ion in these complexes. Hence, obtain the *d*-orbital splitting diagrams for these complexes, and explain why they are colourless.

13.120 The complex $[Ni(CN)_4]^{2-}$ is diamagnetic, but $[NiCl_4]^{2-}$ is paramagnetic. Propose structures for the two complexes and explain why they have different magnetic properties. **LO 4**

13.121 In addition to high-spin and low-spin possibilities, so-called 'intermediate spin' complexes can occasionally be isolated. These have a number of unpaired electrons between those of the high-spin and low-spin cases. For example, six-coordinate d^6 intermediate-spin complexes contain two unpaired electrons. Such complexes generally result from very strong bonding in the equatorial (*xy*) plane and very weak bonding in the *z* direction. Sketch a *d*-orbital splitting diagram that might be expected for these complexes. **LO 4**

13.122 A portion of the absorption spectrum of the octahedral complex ion $[CrCl_2(OH_2)_4]^+$ is shown below. **LO 4**

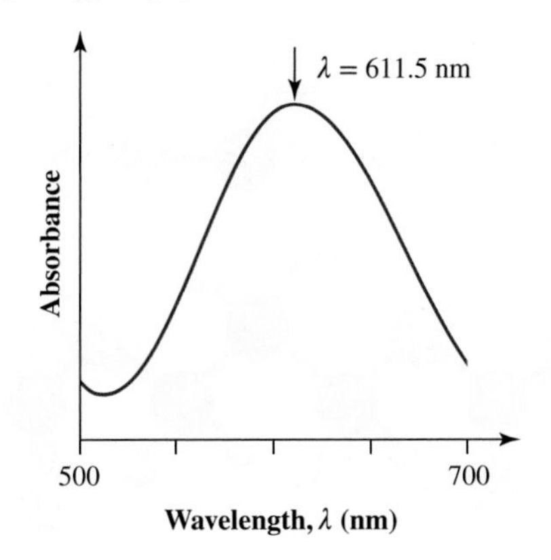

(a) Estimate the crystal field splitting energy, Δ_o, in kJ mol^{-1}.
(b) What colour is the complex?
(c) Name the complex cation.
(d) Draw all possible isomers of the complex.
(e) Draw the *d*-orbital splitting diagram and show the electronic transition that gives the complex its colour.

13.123 In the 1890s, the Swiss chemist Alfred Werner prepared several platinum complexes that contained both ammonia and chlorine. He determined the formulae of these species by precipitating the chloride ions with Ag^+. The following table shows the empirical formulae and number of chloride ions that precipitate per formula unit.

Empirical formula	Number of Cl^- ions
$PtCl_4{\cdot}2NH_3$	0
$PtCl_4{\cdot}3NH_3$	1
$PtCl_4{\cdot}4NH_3$	2
$PtCl_4{\cdot}5NH_3$	3
$PtCl_4{\cdot}6NH_3$	4

Determine the molecular formulae of these platinum complexes, name them and draw their structures. **LO 4,6**

13.124 Silverplating processes often use $Ag(CN)_2^-$ as the source of Ag^+ in solution. To make a solution of this ion, a chemist used 4.0 L of 3.00 mol L^{-1} NaCN and 50 L of 0.2 mol L^{-1} $AgNO_3$. What is the concentration of free Ag^+ ions in this solution? Why is AgCl not used in this process? β_2, = 2×10^{38} for $Ag(CN)_2^-$ **LO 7**

14 The *p*-block elements

What do dynamite and angina (chest pain) have in common? The answer is nitroglycerin, $C_3H_5N_3O_9$, (sometimes known as glyceryl trinitrate) the structure of which is shown below.

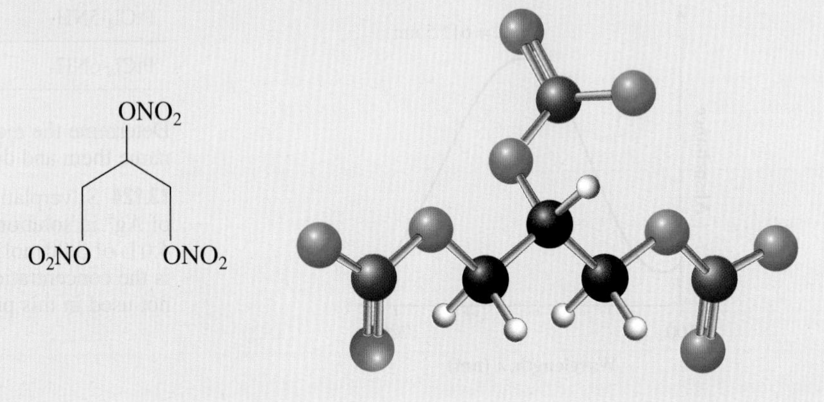

Nitroglycerin is a violently explosive colourless liquid that was first prepared by the Italian chemist Ascanio Sobrero in 1847. Its explosive properties attracted immediate interest, with one Alfred Nobel being the first to overcome its notoriously unpredictable nature by mixing it into a paste with diatomaceous earth to give dynamite. Remarkably, nitroglycerin was also found to have medical uses; it has been prescribed for the treatment of angina since 1878 as its use results in dilation of blood vessels, leading to decreasing blood pressure and increasing bloodflow. However, the mechanism by which this occurs was discovered only recently; to general astonishment from the medical and scientific community, the simple but very reactive molecule nitric oxide (NO), formed from nitroglycerine, was found to be the active agent. Since this time, NO has been implicated in a number of important biological processes, particularly in the brain. There is evidence that the neural processes leading to memory include the production and diffusion of NO, with an enzyme in the brain producing NO from the amino acid arginine.

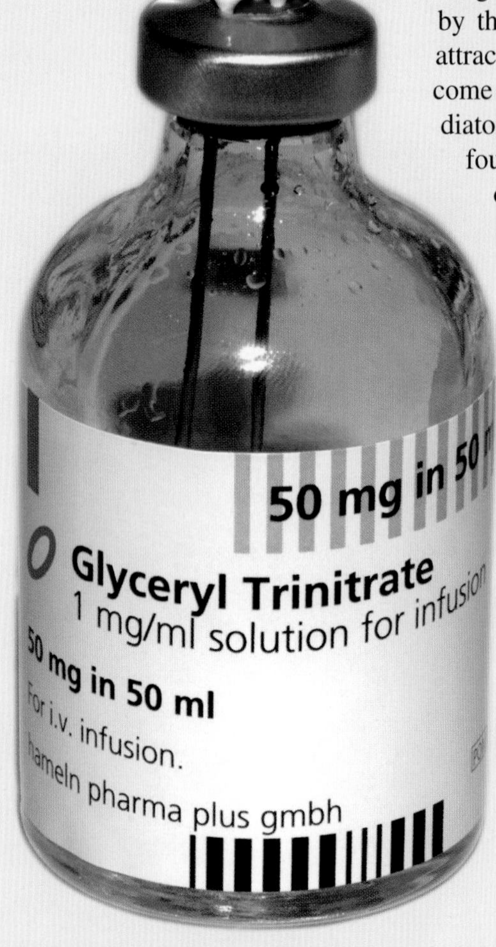

The components of the NO molecule, nitrogen and oxygen, are both *p*-block elements, a name given to elements belonging to groups 13 to 18 of the periodic table and characterised by the gradual filling of the *p* orbitals. In this chapter, we describe selected features of the diverse chemistry of the *p*-block elements.

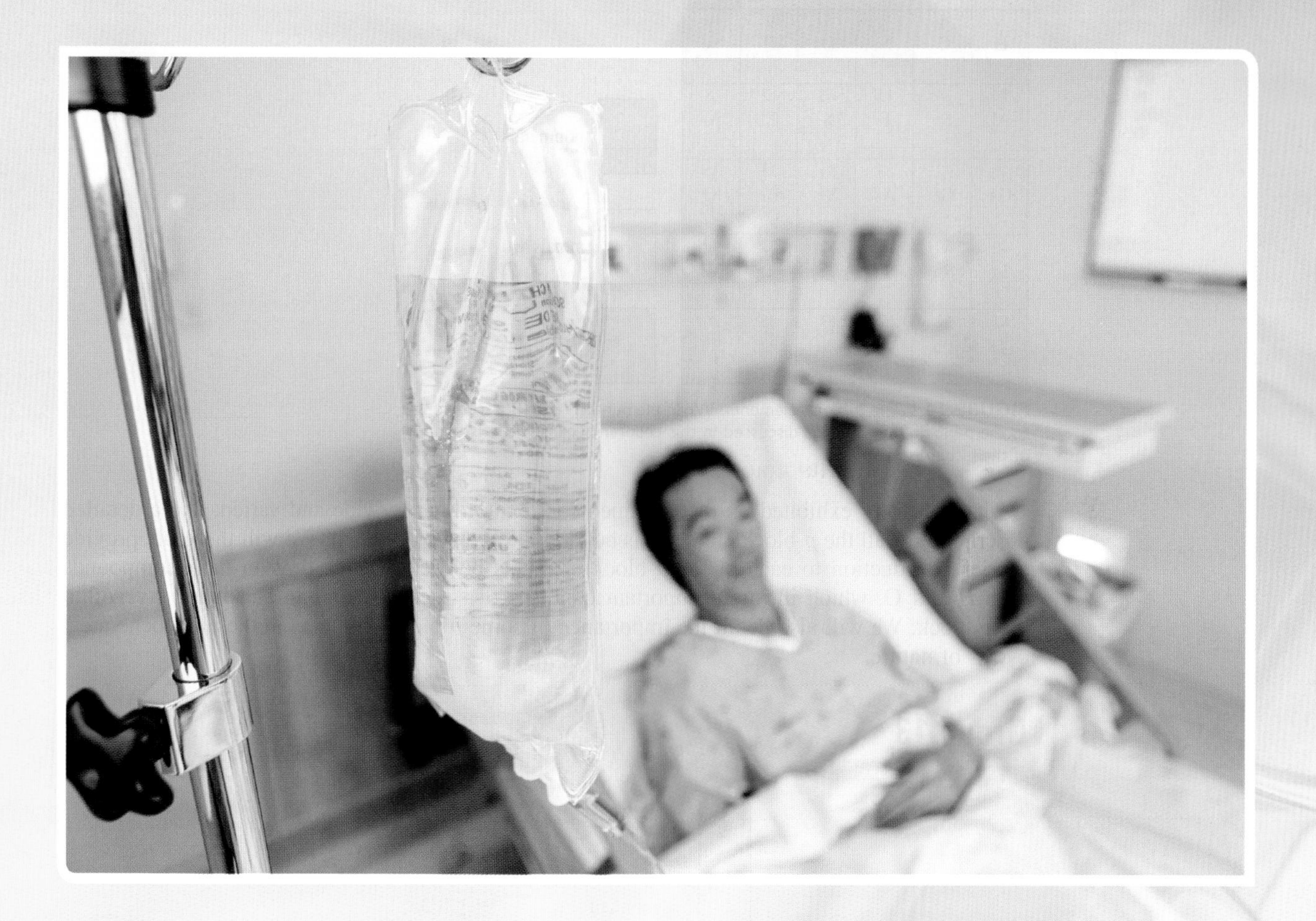

LEARNING OBJECTIVES

After studying this chapter, you should be able to:

14.1 describe the general behaviour of the *p*-block elements

14.2 rationalise the chemistry of compounds of the *p*-block elements with H and O

14.3 describe the involvement of *p*-block elements in biogeochemical cycles.

14.1 The *p*-block elements

In chapter 1 (figure 1.16, p. 17), we showed that all elements in the periodic table could be classified as one of three types: metals, nonmetals and metalloids. We also showed that the different groups of the periodic table (1 to 18) are defined by the valence electron configuration; for example, groups 1 and 2 are called *s*-block elements, owing to their ns^1 or ns^2 configurations, while groups 3 to 12 are called *d*-block elements, with valence electron configurations of $(n + 1)s^2nd^1$ to $(n + 1)s^2nd^{10}$. While the elements of groups 1 to 12, and the lanthanoid and actinoid elements are, with the exception of hydrogen, exclusively metals, groups 13 to 18 contain elements which are predominantly nonmetals or metalloids. The elements of groups 13 to 18 are collectively called ***p*-block elements** and they are characterised by the gradual filling of the *p* orbitals across each period. As can be seen in figure 14.1, the valence-shell electron configuration of these elements ranges from ns^2np^1 (group 13) to ns^2np^6 (group 18).

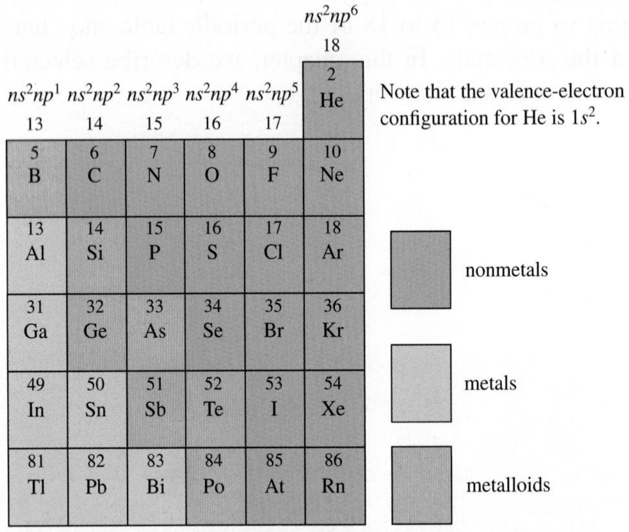

ns^2np^1 13	ns^2np^2 14	ns^2np^3 15	ns^2np^4 16	ns^2np^5 17	ns^2np^6 18
					2 He
5 B	6 C	7 N	8 O	9 F	10 Ne
13 Al	14 Si	15 P	16 S	17 Cl	18 Ar
31 Ga	32 Ge	33 As	34 Se	35 Br	36 Kr
49 In	50 Sn	51 Sb	52 Te	53 I	54 Xe
81 Tl	82 Pb	83 Bi	84 Po	85 At	86 Rn

Note that the valence-electron configuration for He is $1s^2$.

nonmetals

metals

metalloids

FIGURE 14.1 The *p*-block elements. While elements 113–118 also belong to the *p*-block, they are not discussed in this chapter because little is known of their chemistry.

The chemistry exhibited by these elements is extraordinarily wide and varied, and a detailed description of all the *p*-block elements is beyond the scope of this text. We will therefore provide a brief introduction to each of the *p*-block elements before concentrating on their compounds with *H* and O, which illustrate important concepts of structure, bonding and reactivity within the *p* block. We will also detail the importance of some *p*-block elements in the biogeochemical cycles of nature.

Group 13

13		
5 B	boron	[He] $2s^22p^1$
13 Al	aluminium	[Ne] $3s^23p^1$
31 Ga	gallium	[Ar] $4s^23d^{10}4p^1$
49 In	indium	[Kr] $5s^24d^{10}5p^1$
81 Tl	thallium	[Xe] $6s^25d^{10}4f^{14}6p^1$

The elements in group 13, with the exception of boron, are metals, and therefore display the expected characteristics outlined on page 17 in chapter 1. The elements have the valence-electron configuration ns^2np^1, and tend to exhibit a formal oxidation number of +3 in their compounds. However, as we will see, both indium and thallium have a significant chemistry in the +1 oxidation state owing to the fact that the valence ns^2 electrons experience a relatively high Z_{eff}, and are therefore held relatively tightly, due to the poor shielding characteristics of the inner *d* and *f* electrons.

Boron is a metalloid and, as befits its small size and consequent hard Lewis acidity, is found in combination with oxygen in nature, in a variety of borate minerals. The element itself is a shiny black solid, with a very high melting point of 2077 °C (in fact, among the *p*-block elements the melting point of boron is exceeded only by that of carbon) which is attributable to its unusual solid state structure. The valence-electron configuration of boron, $2s^22p^1$, means that it is unable to complete an octet (p. 178) through the formation of three single bonds, and elemental boron therefore adopts a rather complex molecular structure that cannot be described by simple Lewis structures. Boron exists in several different crystalline forms (allotropes), each of which is characterised by clusters of 12 boron atoms located at the vertices of an icosahedron (a 20-sided geometric figure), as shown in figure 14.2.

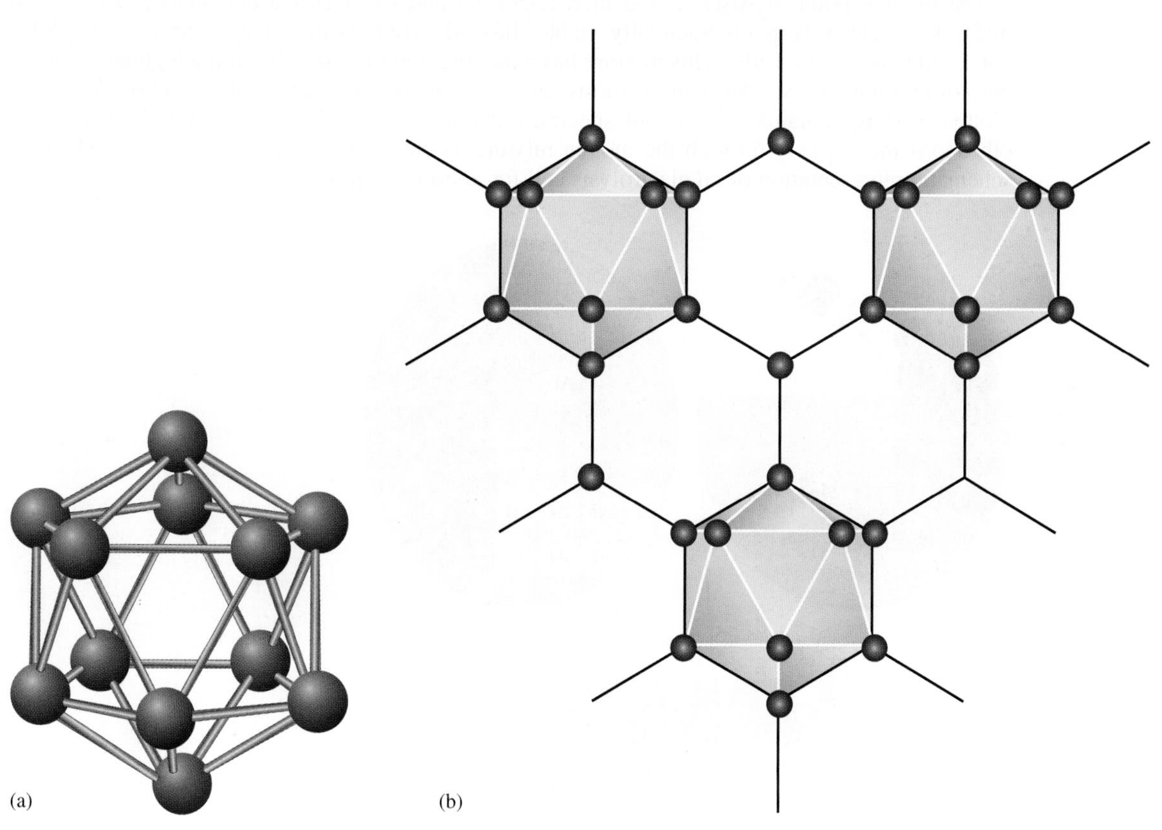

FIGURE 14.2 The structure of an allotrope of elemental boron. **(a)** The arrangement of 12 boron atoms in a B_{12} cluster. **(b)** The element boron consists of an interconnecting network of B_{12} clusters that produces a very hard and high-melting solid.

Each boron atom within a given cluster is equidistant from five others and, in the solid, each of these is also joined to yet another boron atom outside the cluster (figure 14.2b). The electrons available for bonding are therefore delocalised to a large extent over many boron atoms, and the simple Lewis idea of pairs of electrons in a covalent bond does not apply in this case. The linking together of B_{12} units produces a large, three-dimensional covalent solid that is very difficult to break down. As a result, boron is very hard (it is the second hardest element) and has a very high melting point. Boron atoms share with carbon atoms (albeit to a much lesser extent) the ability to undergo catenation (p. 42), bonding to themselves to form moderately sized molecules. However, in contrast to carbon, such molecules generally adopt three-dimensional polyhedral structures, rather than chains or rings.

Aluminium is the only one of the four group 13 metals not to contain filled *d* orbitals; it therefore has a small atomic radius which renders it, like boron, a hard Lewis acid. It is found in nature as bauxite, Al(O)OH, a compound of which Australia is the world's largest producer, and in a variety of compounds called aluminosilicates which are composed of aluminium, silicon and oxygen; in all these compounds, aluminium exists as the 3+ cation. Reduction of bauxite, or indeed any aluminium compound, to metallic aluminium is difficult and, as a result, the pure element was first isolated only in 1825 from the reduction of $AlCl_3$ with potassium. Its rarity for a time following this meant that aluminium was more expensive than gold or platinum, and Emperor Napoleon III of France reputedly had aluminium dinnerware which was used for important guests. Nowadays, aluminium is produced from purified Al_2O_3 by electrolysis, using the Hall–Héroult process which was developed in the 1880s (see Chemical Connections, p. 605).

Bauxite is first purified of insoluble impurities through conversion to the soluble complex ion $[Al(OH)_4]^-$ by dissolution in a strong base.

$$Al(O)OH(s) + OH^-(aq) + H_2O \longrightarrow [Al(OH)_4]^-(aq)$$

Dilution of this solution with water causes the precipitation of $Al(OH)_3$.

$$[Al(OH)_4]^-(aq) \xrightarrow{\text{added water}} Al(OH)_3(s) + OH^-(aq)$$

Dehydration of this gives pure Al_2O_3.

$$2Al(OH)_3(s) \xrightarrow{125\,°C} Al_2O_3(s) + 3H_2O(g)$$

The melting point of Al_2O_3 is too high (2015 °C) and its electrical conductivity too low to make direct electrolysis commercially viable. Instead, Al_2O_3 is mixed with cryolite, Na_3AlF_6, containing about 10% CaF_2. This mixture has a melting point of 1000 °C, still a high temperature but not prohibitively so. Aluminium forms several complex ions with fluoride and oxide, so the molten mixture contains a variety of species, including AlF_4^-, AlF_6^{3-} and $AlOF_3^{2-}$. These and other ions move freely through the molten mixture as electrolysis occurs. Figure 14.3 shows a schematic representation of an electrolysis cell for aluminium production.

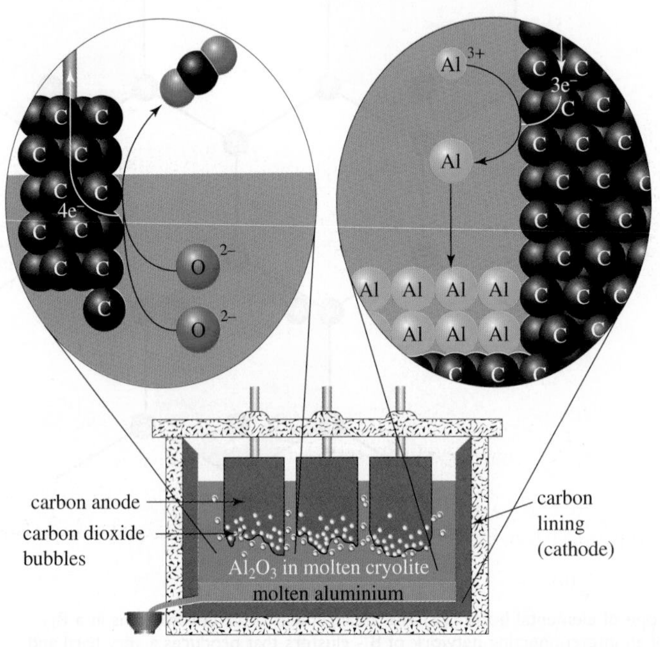

FIGURE 14.3 Aluminium metal is produced by electrolysis of aluminium oxide dissolved in molten cryolite. Al^{3+} is reduced to Al at the cathode, and C is oxidised to CO_2 at the anode.

An external electrical potential drives electrons into a graphite cathode, where Al^{3+} ions are reduced to Al metal.

$$Al^{3+}(melt) + 3e^- \longrightarrow Al(l) \qquad \text{(cathode)}$$

The anode, which is also made of graphite, is oxidised during electrolysis. Carbon from the anode combines with oxide ions to form CO_2 gas.

$$2O^{2-}(melt) + C(s) \longrightarrow CO_2(g) + 4e^- \qquad \text{(anode)}$$

Because of the variety of ionic species present in the melt, the reactions that take place at the electrodes are considerably more complex than the simple representations given here. Nevertheless, the overall reaction is the one given by the sum of these simplified reactions.

$$4Al^{3+}(melt) + 6O^{2-}(melt) + 3C(s) \longrightarrow 3CO_2(g) + 4Al(l)$$

The electrolysis apparatus operates well above the melting point of aluminium (660 °C) and, as liquid aluminium has a higher density than the molten salt mixture, pure liquid metal settles to the bottom of the reactor. The pure metal is drained through a plug and cast into ingots (figure 14.4). Aluminium refining consumes huge amounts of electricity. In fact, the aluminium smelter at Tiwai Point in New Zealand consumes 15% of the country's entire energy output.

FIGURE 14.4 Aluminium refining uses very large amounts of electricity.

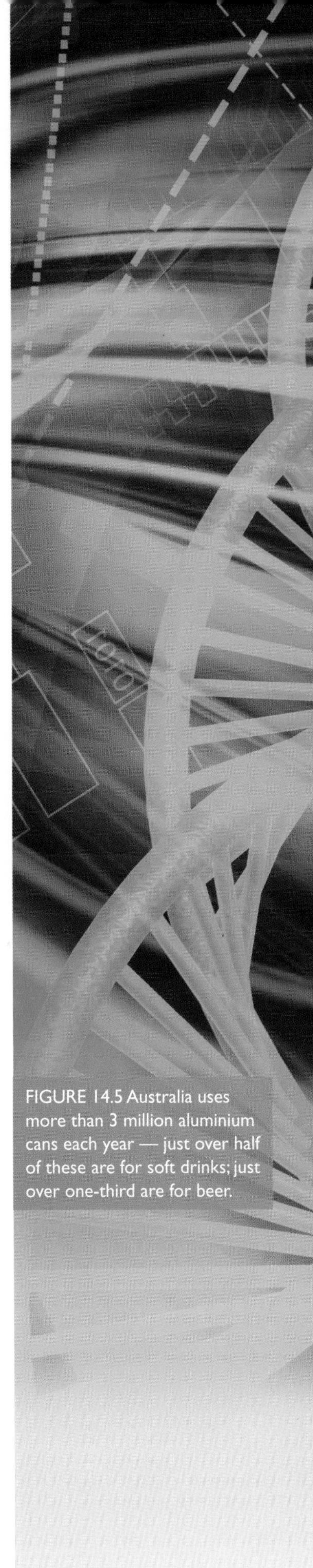

Chemical Connections

The Hall–Héroult process

Apart from its obvious uses in cans (figure 14.5), foil, and windows and doors, aluminium's strength, lightness, high electrical and thermal conductivity, and resistance to corrosion make it useful in kitchen utensils, car, aircraft and rocket construction, electrical transmission lines and in coatings for high-precision mirrors such as those in telescopes.

Aluminium is the third most abundant element in the Earth's crust and the most abundant metal. Nevertheless, aluminium was not isolated until 1825 and was still a precious rarity 60 years later. The reason for this elusiveness is the high stability of Al^{3+}. The reduction of aluminium compounds to the free metal requires stronger reducing power than common chemical reducing agents can provide. The extraction of aluminium had to await the birth of electrochemistry and the development of electrolysis (see chapter 12).

Two 22-year-old scientists, the American chemist Charles Hall and the French metallurgist Paul Héroult, developed the same process independently in 1886, both becoming famous as founders of the aluminium industry, Hall in the United States and Héroult in Europe.

Hall was inspired by his chemistry professor, who observed that whoever perfected an inexpensive way of producing aluminium would become rich and famous. After his graduation, Hall set to work in his home laboratory trying to electrolyse various compounds of aluminium. Working with his sister Julia, who had also studied chemistry, Hall successfully produced globules of the metal within 8 months. Meanwhile, Héroult was developing the identical process in France. Hall founded a company for the manufacture of aluminium. That company became immensely successful, eventually growing into Alcoa.

FIGURE 14.5 Australia uses more than 3 million aluminium cans each year — just over half of these are for soft drinks; just over one-third are for beer.

Successful electrolysis of aluminium requires a liquid medium other than water that can conduct electricity. The key to the Hall–Héroult process is the use of molten cryolite, Na_3AlF_6, as a solvent. Cryolite melts at an accessible temperature, it dissolves Al_2O_3, and it is available in the necessary purity. A second important feature is the choice of graphite to serve as the anode. Graphite provides an easy oxidation process, the oxidation of carbon to CO_2.

The reasoning that led Hall and Héroult to the identical process was probably similar. Electrolysis was recognised as a powerful reducing method. All attempts to reduce aqueous aluminium cations failed, making it clear that some molten salt would have to be used. Experimenting with various salts, no doubt guided by the principle that 'like dissolves like', the two young men eventually tried Na_3AlF_6, a mineral whose constituent elements should not interfere with the reduction of aluminium. Graphite electrodes were already in use, so experimenting with them would have been a natural choice. All the ingredients for this invention may have been in place, but that does not detract from its brilliance.

In more than 100 years of growth in the aluminium industry, the only significant change to the Hall–Héroult process has been the addition of CaF_2 to the melt to lower the operating temperature.

Gallium is a silvery white metal that is obtained primarily as a by-product of the purification of bauxite, discussed on page 604. Removal of $Al(OH)_3$ leads to concentration of gallium compounds in the remaining alkaline solution and electrolysis eventually yields the pure metal. Gallium was one of the 'missing' elements predicted by Mendeleev in his original periodic table (see figure 1.15, p. 16). Gallium is of interest, among other things, for its very low melting point (around 30 °C) and the extremely large temperature range over which it exists as a liquid (it boils around 2200 °C). Compounds of gallium find extensive use in light-emitting diodes (LEDs).

Indium is not found in nature in its elemental state. Its relatively large size means it behaves as a soft Lewis acid, and this is reflected in the fact that it is generally found in sulfide ore deposits along with zinc, copper and tin. The pure metal is obtained as a by-product of the electrolytic refining of zinc and is known for the fact that it emits a high-pitched 'cry' when it is bent. The major use of indium is in thin films of indium tin oxide (ITO) which are used in liquid crystal displays.

Thallium is a soft metal which exhibits a metallic lustre when freshly cut. However, it quickly develops a grey oxide layer on exposure to air and generally resembles lead in appearance. It is found in sulfide and selenium ore deposits, reflecting its soft Lewis acidic nature. Thallium and its compounds are highly toxic and were once poisons of choice for serial killers. In fact, Sydney woman Caroline Grills was sentenced to death in 1953 (later commuted to life in prison) for the attempted murder of two of her in-laws using thallium, and over the previous year there were some 46 reported cases of thallium poisoning, with 10 deaths, in Sydney. The toxicity of the element, which limits its usefulness, is thought to arise from both its similarity in size to potassium and its affinity for sulfur-containing amino acids. Despite this, there has been some interest in thallium-containing materials as high-temperature superconductors.

Group 14

14		
6 C	carbon	[He] $2s^22p^2$
14 Si	silicon	[Ne] $3s^23p^2$
32 Ge	germanium	[Ar] $4s^23d^{10}4p^2$
50 Sn	tin	[Kr] $5s^24d^{10}5p^2$
82 Pb	lead	[Xe] $6s^25d^{10}4f^{14}6p^2$

Group 14 comprises a nonmetal, two metalloids and two metals. The elements have valence electron configurations ns^2np^2 and the predominant oxidation number of group 14 elements in their compounds is +4. However, as we also saw in group 13, the heavier elements can display an oxidation state two lower than that usually found in the group. Thus both tin and lead have significant chemistry in the +2 oxidation state, with compounds of the latter in the +4 oxidation state often being unstable with respect to reduction.

We have already briefly discussed the different allotropes of elemental carbon, and chapter 16 is devoted to the chemistry of carbon, so we will not concern ourselves further with this element here.

Silicon behaves as a hard Lewis acid, and occurs in nature as SiO_2 (silica) and various silicate anions. The element may be obtained by the high-temperature reduction of silica using coke in an electric arc furnace, according to the equation.

$$SiO_2 + 2C \longrightarrow Si + 2CO$$

but this gives silicon with a purity of only 98% — not high enough for its most important use in silicon chips. High-purity silicon is obtained by reduction of $SiCl_4$, itself prepared by reaction of SiO_2 and coke in the presence of Cl_2 at high temperature.

$$SiO_2 + 2C + 2Cl_2 \longrightarrow SiCl_4 + 2CO$$

The $SiCl_4$ is distilled and then reduced with magnesium to give silicon and $MgCl_2$, which is removed by washing with water.

$$SiCl_4 + 2Mg \longrightarrow Si + 2MgCl_2$$

The silicon is purified further by zone refining, a process depicted in figure 14.6. A heating coil is passed slowly along a rod of impure silicon; this causes the silicon to melt in the region being heated, and then resolidify as the heating coil moves past. During this process, impurities remain preferentially in the liquid phase, which floats to the top, yielding solid silicon in the lower part of the rod containing less than 1 part per billion of impurities.

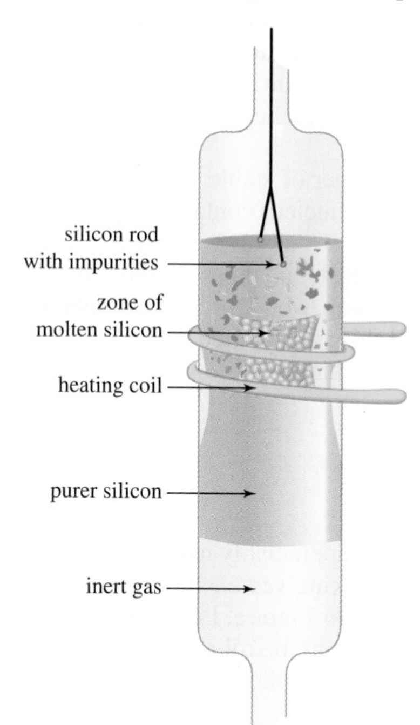

FIGURE 14.6 In zone refining, a heating coil passes along a rod of impure silicon. The impurities concentrate in the molten zone, leaving behind solid material that is of higher purity. Repeated passage of the coil moves the impurities to one end of the rod.

Germanium is found in some sulfide minerals and zinc ores, and in coal. Like silicon, germanium finds extensive use in semiconductors, and the ultrapure element is also prepared by zone refining. Germanium crystallises as grey-white crystals with the same structure as diamond, but these are brittle and nowhere near as hard. The main uses of germanium and its compounds are in fibre optics, infrared optics and solar cells. On reduction by sodium in liquid ammonia, germanium, like both tin and lead below it in the periodic table, forms interesting polyatomic anions; examples of these include Ge_4^{2-}, Ge_9^{4-} and Ge_9^{2-}.

Tin is found in nature as SnO_2 in the mineral cassiterite and this can be reduced to the metal using charcoal at high temperature. Historically, tin was used widely as a constituent of metal alloys, of which bronze, solder and pewter are common examples. Bronze (figure 14.7) is an alloy of copper containing approximately 20% tin and smaller amounts of zinc. Pewter (figure 14.8) is another Cu–Sn alloy that contains tin as the major component (≈85%), with roughly equal portions of copper, bismuth and antimony. Solder consists of 67% lead and 33% tin.

FIGURE 14.7 Bronze.

FIGURE 14.8 Pewter.

Because of its relatively low melting point (232 °C) and good resistance to oxidation, tin is used to provide protective coatings on metals such as iron that oxidise more readily. 'Tin cans' are actually iron cans dipped in molten tin to provide a thin surface film of tin. Metals that are soft Lewis acids, such as cadmium, mercury and lead, are extremely hazardous to living organisms. Tin, in contrast, is not. One reason is that tin oxide is highly insoluble, so tin is seldom found at measurable levels in aqueous solution. Perhaps more importantly, the toxic metals generally act by binding to sulfur in essential enzymes. Tin is a harder Lewis acid than the other heavy metals, so it has a lower affinity for sulfur, a relatively soft Lewis base.

As we saw in chapter 1, tin holds the record for the greatest number of stable isotopes of any element (10), and this is probably not unrelated to the fact that the tin nucleus contains 50 protons; as we will see in e-chapter 27, 50 is a magic number.

Lead is one of the oldest known metals and was used extensively in pipes in the Roman Empire; in fact, our modern word 'plumbing' derives from the Latin word for lead, *plumbum*. It is found in nature primarily as PbS in the mineral galena, and this is converted to the element first by roasting in air to form PbO, and subsequent reduction with charcoal.

$$2PbS + 3O_2 \xrightarrow{\text{heat}} 2PbO + 2SO_2$$

$$PbO + C \xrightarrow{\text{heat}} Pb + CO$$

Lead is a soft blue–grey metal, and both it and its compounds are appreciably toxic to humans. Indeed, the extensive use of lead by the Romans in cooking and drinking vessels has led some to speculate that lead poisoning was a factor in the decline of the Roman Empire. Lead compounds were used in huge quantities in the twentieth century as additives in both petrol and paint; thankfully, the incredible folly of this has finally been appreciated and such applications are now rare. The major use of lead currently is in car batteries.

Group 15

15		
7 N	nitrogen	[He] $2s^22p^3$
15 P	phosphorus	[Ne] $3s^23p^3$
33 As	arsenic	[Ar] $4s^23d^{10}4p^3$
51 Sb	antimony	[Kr] $5s^24d^{10}5p^3$
83 Bi	bismuth	[Xe] $6s^25d^{10}4f^{14}6p^3$

FIGURE 14.9 Liquid nitrogen boils at −196 °C.

Group 15 comprises two nonmetals, two metalloids and a metal. Elemental nitrogen makes up 78% (by volume) of the air around us. It exists as the diatomic molecule N_2, in which the two nitrogen atoms are held together by a triple bond of considerable strength (942 kJ mol^{-1}, see worked example 5.10, pp. 210–11). This makes the molecule unreactive under ambient conditions, a fact for which humankind can be extraordinarily grateful, given that this means it does not instantly react with the 21% by volume of oxygen in the atmosphere. However, its lack of reactivity also means that, in order for it to be converted into compounds that can be utilised by living systems, nature has had to develop an extremely complex enzyme called nitrogenase, generally found in legumes, to catalyse the conversion of N_2 to ammonia. Despite the best efforts of some of the finest scientists over many years, we have yet to be able to replicate this process usefully in the laboratory, and in order to prepare nitrogen-based fertilisers, we use the rather less elegant Haber–Bosch process, in which N_2 and H_2 are heated together at high pressure over an iron catalyst. The importance of this process is shown by the fact that it consumes an estimated 1–2% of the world's energy supply each year. Elemental nitrogen finds extensive use in its liquid form as a coolant; liquid nitrogen (figure 14.9) boils at −196 °C at atmospheric pressure.

Phosphorus was isolated from the distillation of urine by Hennig Brand in 1669, making it the element having the earliest authenticated date of discovery. It exists as a number of different-coloured allotropes, the most common being white, red and black. It is, like nitrogen, essential to life, being a constituent of the biologically important molecules DNA, RNA, and adenosine mono-, di- and triphosphate. Perhaps surprisingly then, pure white phosphorus is highly toxic, with a dose of only 50 mg considered lethal. Modern production of elemental phosphorus uses a technique similar to the metallurgical processes described in chapter 13. Apatite is mixed with silica and coke and then heated strongly in the absence of oxygen. Under these conditions, coke reduces phosphate to elemental phosphorus, the silica forms liquid calcium silicate, and the fluoride ions in apatite dissolve in the liquid calcium silicate. The reactions are not fully understood, but the stoichiometry for the calcium phosphate part of apatite is as follows.

$$2Ca_3(PO_4)_2 + 6SiO_2 + 10C \xrightarrow{1450\,°C} P_4 + 6CaSiO_3 + 10CO$$

Elemental phosphorus is a gas at the temperature of the reaction, so the product distils off along with carbon monoxide. When cooled to room temperature, P_4 condenses as a waxy white solid.

White phosphorus consists of individual P_4 molecules with the four atoms at the corners of a tetrahedron. Each atom bonds to three others and has one lone pair of electrons, giving a total of four electron pairs around each P atom (chapter 5). However, the triangular geometry of the faces of the tetrahedron constrains the bond angles in the P_4 tetrahedron to 60°, far from the optimal 4-coordinate geometry of 109.5°. As a result, P_4 is highly reactive. Samples of white phosphorus are stored under water because P_4 burns spontaneously in the presence of oxygen (figure 14.10). The glow that emanates from P_4 as it burns in the dark led to its name (from the Greek *phos*, meaning 'light', and *phoros*, meaning 'bringing').

FIGURE 14.10 The strained 60° bond angles in tetrahedral P_4 make white phosphorus highly reactive. This form of the element must be kept out of contact with air, in which it burns spontaneously.

Heating white phosphorus in the absence of oxygen causes the discrete P_4 units to link together, giving a chemically distinct elemental form, red phosphorus. As figure 14.11 shows, one P—P bond of each tetrahedron breaks to allow formation of the bonds that link the P_4 fragments. The bond angles that involve the links are much closer to 109.5°, making red phosphorus less strained and less reactive than white phosphorus. Red phosphorus undergoes the same chemical reactions as the white form, but higher temperatures are required. In addition, red phosphorus is essentially nontoxic, so it is easier and safer to handle.

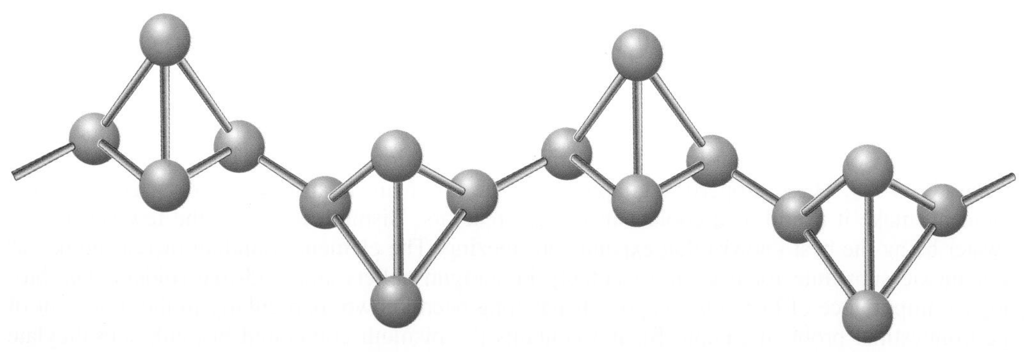

FIGURE 14.11 Red phosphorus consists of chains of P_4 tetrahedra linked through P—P bonds.

Another stable form of the element is black phosphorus, which can be prepared by heating red phosphorus under high pressure. The black form contains chains of P_4 units cross-linked by P—P bonds, making this form even more polymerised and less strained than red phosphorus. Worked example 14.1 explores another difference between the elemental forms of phosphorus.

WORKED EXAMPLE 14.1

Melting points of phosphorus

The melting point of white phosphorus is 44.1 °C. In contrast, red phosphorus remains a solid up to 400 °C. Account for this large difference in melting point.

Analysis

As described in chapter 7, melting points of solids may depend on both covalent bonds and intermolecular forces, so we must explain the melting point difference with reference to the bonding differences between the two forms.

Solution

White phosphorus consists of individual P_4 molecules. Because there are no polar bonds, the molecules are held in the solid state by dispersion forces only, and it would be expected that white phosphorus would have a relatively low melting point. Red phosphorus is made of long chains of P_4 groups, each of which is held together by covalent P—P bonds, as shown in figure 14.11 on the previous page. Thus, red phosphorus consists of macromolecules held together by dispersion forces. To melt, red phosphorus must break some chemical bonds. Bond breaking requires much more energy than the energy needed to overcome dispersion forces, giving red phosphorus the higher melting temperature.

Is our answer reasonable?

The molecular models in figures 14.10 and 14.11 show that white phosphorus is a molecular solid, whereas red phosphorus is intermediate between molecular and network. This correlates well with the melting points.

PRACTICE EXERCISE 14.1

Predict whether black phosphorus melts at a higher or lower temperature than the white and red forms, and rank the three forms of elemental phosphorus in order of increasing density, giving a reason for your ranking.

Arsenic is possibly best known among the general public for being a poison. In addition to its many human victims, it also has been recently found to have accounted for the early demise of Phar Lap, the famous New Zealand-born and Australia-raised racehorse; analysis of hairs from Phar Lap's mane using sophisticated X-ray based spectroscopic techniques showed the ingestion of arsenic while the horse was still alive. Elemental arsenic exists as the grey allotropic form at room temperature and pressure, while at least two other allotropes are known. It is often found in combination with sulfur in a variety of minerals, with FeSAs (arsenopyrite) being the most common; heating this in the absence of air gives the pure element. Arguably the most important use of arsenic is in the semiconductor gallium arsenide (GaAs), which shows properties similar, and in some cases superior, to silicon in diodes and solar cells.

Antimony compounds have been known for over 2000 years, with Sb_2S_3 (stibnite) being used as both a medicine and a black eye makeup in ancient Egypt. The element is a grey metallic solid at ambient temperatures and at least four other allotropes are known. Stibnite is the most important ore of antimony, and the pure element is obtained from either roasting in air and subsequent reduction of the resulting Sb_2O_3 with coke, or direct reduction by Fe. Antimony is used primarily to harden the lead used in car batteries, while some antimony compounds have been found to be effective against parasite-borne tropical diseases.

Bismuth is primarily obtained as a by-product of the smelting of lead and copper. The element is a soft grey metal having a relatively low melting point (271 °C) and a high boiling point (1564 °C). These properties, coupled with the low neutron absorption cross-section of the element, make it useful as a coolant in nuclear reactors. Bismuth is one of the few substances (water being the best known) that expands on freezing. The element is finding increasing use as a nontoxic substitute for lead, most notably in shotgun pellets and lead-free solders. The biological importance of bismuth compounds has long been known, particularly in the treatment of gastrointestinal problems; Pepto-Bismol contains the bismuth compound bismuth subsalicylate (figure 14.12) as its active ingredient.

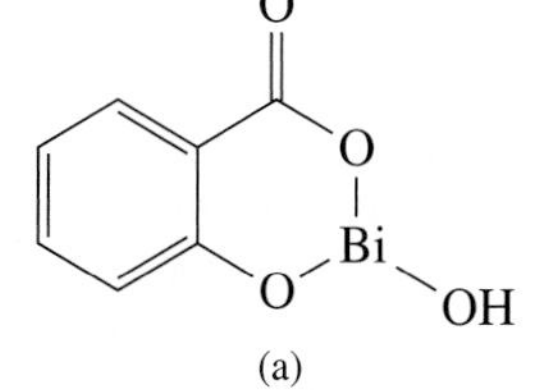

FIGURE 14.12 (a) The structure of bismuth subsalicylate, the active ingredient in **(b)** Pepto-Bismol.

Group 16

16		
8 O	oxygen	[He] $2s^2 2p^4$
16 S	sulfur	[Ne] $3s^2 3p^4$
34 Se	selenium	[Ar] $4s^2 3d^{10} 4p^4$
52 Te	tellurium	[Kr] $5s^2 4d^{10} 5p^4$
84 Po	polonium	[Xe] $6s^2 5d^{10} 4f^{14} 6p^4$

The ability of photosynthetic bacteria over two billion years ago to produce elemental oxygen is responsible for life on Earth evolving as we now know it. Oxygen and its compounds are ubiquitous throughout the Earth's atmosphere, crust and biosphere, and it is the third most abundant element in the universe, after hydrogen and helium. It forms stable compounds with the majority of elements in the periodic table, the only exceptions being some of the elements in group 18. The pure gaseous element is colourless, odourless and tasteless, while the liquid form exhibits a pale blue colour. The first person to isolate and study oxygen gas was the Swedish pharmacist Carl Wilhelm Scheele over the period 1771–73, but the English part-time scientist Joseph Priestley, who isolated oxygen in 1774, is usually credited as the discoverer of the element. This is because Scheele did not publish his work until 1777 and was thus beaten into print by Priestley by some two years. Antoine Lavoisier (chapter 1, p. 5) was also instrumental in early work on oxygen, especially in recognising its importance in combustion reactions. Ozone, O_3, is an allotrope of oxygen; it occurs in the atmosphere, and a layer in the stratosphere acts as a shield for potentially damaging ultraviolet radiation from the Sun (pp. 678–9). It is also an excellent oxidising agent and is increasingly being used in place of chlorine-based oxidants in the pulp and paper industry, as well as for the treatment of drinking water.

Sulfur is encountered most often in sulfide minerals such as pyrite, FeS_2, molybdenite, MoS_2, chalcocite, Cu_2S, cinnabar, HgS, and galena, PbS. It is also present in huge amounts as H_2S in natural gas and in sulfur-containing organic compounds in crude oil and coal. In addition, sulfur is found in abundance as the pure element, particularly around hot springs and volcanoes and in capping layers over natural salt deposits. References to sulfur occur throughout recorded history, dating back more than 3500 years. Under ambient conditions elemental sulfur is a yellow crystalline solid that consists of individual S_8 molecules. As shown in figure 14.13, the eight atoms in each molecule form a ring that puckers in such a way that four atoms lie in one plane and the other four atoms lie in a second plane. There are a large number of sulfur allotropes having cyclic structures, ranging from S_6 to S_{20}, while some catenated chain forms are also known.

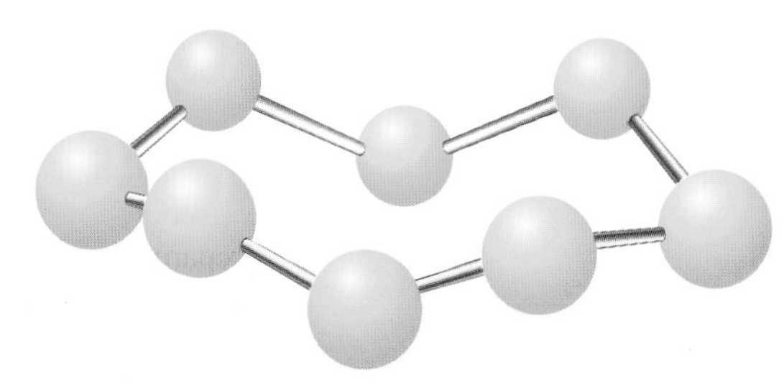

FIGURE 14.13 Under normal conditions, sulfur forms yellow crystals. The crystals consist of individual S_8 molecules, with the eight sulfur atoms of each molecule arranged in a puckered ring.

The major present-day source of the element is from hydrogen sulfide produced as a by-product of oil and gas refining. Many petroleum and natural gas supplies contain some sulfur — up to 25% in some cases. Besides being undesirable in the final products, sulfur poisons many of the catalysts used in oil refining; hence, it must be removed from crude petroleum as a first step in the refining process. About 90% of world output of sulfur is converted to sulfuric acid, the most produced industrial chemical worldwide, and more than 60% of that sulfuric acid is used to extract phosphoric acid from phosphate minerals to prepare fertilisers. This major use exploits the fact that sulfuric acid is the least expensive strong Brønsted–Lowry acid. In addition to protonating the phosphate anions, sulfuric acid sequesters Ca^{2+} as $CaSO_4$, a waste by-product of the process.

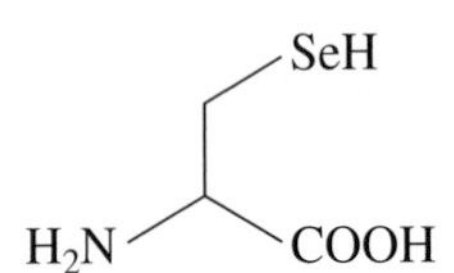

FIGURE 14.14 Structure of the amino acid selenocysteine.

Selenium is a rare element, and mining it directly is not economic. It is therefore obtained as a by-product from the electrolytic purification of copper (p. 589). At least six allotropes are known, four of which are red, one metallic grey (the stable form under ambient conditions), and one black. Three of the red allotropes consist of Se_8 rings, thus mirroring the structure of elemental sulfur. Selenium is a trace element but is toxic in high doses. It is incorporated into proteins via selenocysteine (figure 14.14), a modified amino acid in which the sulfur atom of cysteine (p. 1064) has been replaced by a selenium atom. The resulting proteins act, among other things, to reduce reactive oxygen species and to regulate thyroid hormones. The glass industry is the major user of selenium; it is used to remove the green tint in glass, caused by iron, and also to absorb solar radiation in plate glass. Selenium also finds uses in photocopiers; being a semiconductor, it can be used as a photoreceptor to obtain the latent image, which is then developed by the toner.

Tellurium, like selenium, is obtained predominantly from the electrolytic purification of copper. The pure element is silver-white with a metallic lustre and is rather brittle. In contrast to the two elements directly above it in the periodic table, the stable form under ambient conditions does not consist of eight-atom rings; rather, it comprises helical chains. The major use of tellurium is as an additive to a number of metals to improve their machining characteristics. It is also starting to find extensive use in CdTe (cadmium telluride) solar cells, and also in high-speed rewriteable phase-change memory (PCM) chips. Compounds of tellurium are considered moderately toxic, and ingestion of even tiny amounts leads to garlic-smelling 'tellurium breath' from the formation of dimethyltellurium (Me_2Te) in the body. The element forms a remarkable variety of polyatomic cations and anions; examples include Te_n^{2-} (n = 2–5, 7, 8), Te_6^{3-}, Te_4^{2+}, Te_6^{2+}, Te_6^{4+}, Te_8^{4+} and Te_{10}^{2+}.

Polonium was discovered by the Curies in 1898 and was named after Marie's home country of Poland. All isotopes of polonium are radioactive and this, coupled with its very low abundance (estimated at 0.1 mg per tonne of pitchblende, an ore of uranium), therefore limits the usefulness of the element. In addition to its radioactivity, polonium is toxic in the same way that other heavy metals such as lead and mercury are toxic. In 2006, the Russian journalist Alexander Litvinenko was murdered using what was thought to be as little as 1 mg of the isotope ^{210}Po. The intense energy of the radiation emitted by polonium can be appreciated by the observation that a sample of around 0.5 g of the pure metal will heat itself to over 500 °C.

Group 17

17		
9 F	fluorine	[He] $2s^2 2p^5$
17 Cl	chlorine	[Ne] $3s^2 3p^5$
35 Br	bromine	[Ar] $4s^2 3d^{10} 4p^5$
53 I	iodine	[Kr] $5s^2 4d^{10} 5p^5$
85 At	astatine	[Xe] $6s^2 5d^{10} 4f^{14} 6p^5$

Elemental fluorine, F_2, is a pale green gas of extraordinary reactivity. It reacts directly with all elements except O_2, N_2, He, Ne, Kr and Ar. Compounds of fluorine with all elements except He and Ne are known. It does not exist in nature as the gaseous element, but as the fluoride ion, predominantly in minerals such as fluorspar, CaF_2, and cryolite, Na_3AlF_6. Because F_2 is an extremely powerful oxidising agent ($E^{\ominus}$ = +2.87 V, table 12.1, p. 515), the element cannot be prepared by chemical oxidation of F^- and is therefore obtained by electrolysis of KF·2HF, as depicted in figure 14.15; here, KF acts as an electrolyte.

$$2F^- \longrightarrow F_2 + 2e^- \quad \text{(anode)}$$

$$2HF + 2e^- \longrightarrow H_2 + 2F^- \quad \text{(cathode)}$$

$$2HF \longrightarrow F_2 + H_2 \quad \text{(cell reaction)}$$

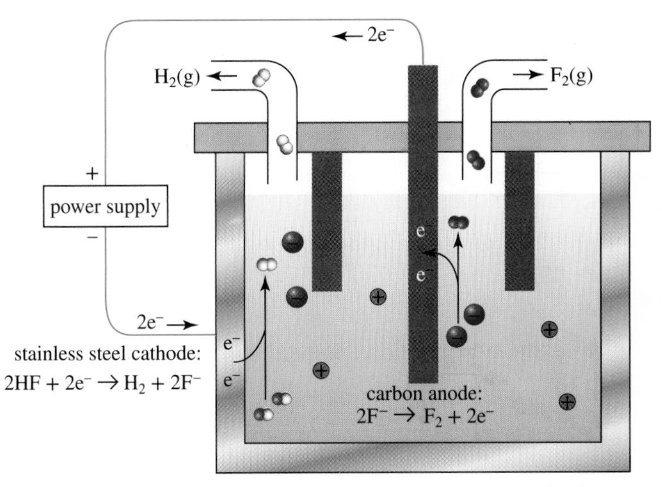

FIGURE 14.15 Schematic view of an electrolytic cell used for the production of molecular fluorine, showing the molecular species involved in the redox reactions.

Over half of the F_2 produced annually is used to prepare uranium hexafluoride (UF_6), a compound used in the separation of the fissile isotope ^{235}U (p. 1177) from the predominant ^{238}U isotope for use in nuclear power stations and nuclear weapons. The introduction of fluorine atoms into organic molecules gives them interesting and unusual properties, the most notable example of this being Teflon, the non-stick polymer derived from tetrafluoroethene, and such molecules represent an important use of fluorine.

The stable form of chlorine that exists under ambient conditions is the diatomic, pale green gas Cl_2, but, like the other members of group 17, it is found in nature predominantly as the halide ion, in this case Cl^-. Chlorine is an important industrial chemical, and some industrial applications of chlorine derivatives are outlined in table 14.1.

TABLE 14.1 Major uses of chlorine.

Reactant	Intermediate	Final products
Organic reagents		
benzene	chlorobenzenes	plastics, dyestuffs
butadiene	chloroprene	neoprene (chlorinated rubber)
ethene	chloroethenes	plastics (PVC), solvents
methane	chloromethanes	silicones
Inorganic reagents		
CO	$COCl_2$	plastics (polycarbonate)
NaOH	NaOCl	bleaches, disinfectants
phosphorus	PCl_3, $POCl_3$	pesticides, flame retardants
rutile (TiO_2 ore)	$TiCl_4$	paint pigment (pure TiO_2)

The starting material for all industrial chlorine chemistry is sodium chloride, obtained primarily by evaporation of sea water. Like fluorine, chemical oxidation of this is not feasible because of the strongly oxidising nature of the parent halogen, and so chloride must be oxidised electrolytically to produce chlorine gas. This is carried out on an industrial scale using the chlor–alkali process, which is shown schematically in figure 14.16 (overleaf). As with all electrolytic processes, the energy costs are very high, but the process is economically feasible because it generates three commercially valuable products: H_2 gas, aqueous NaOH and Cl_2 gas.

The major use of chlorine is in the production of chlorine-containing organic compounds such as 1,2-dichloroethane and chloroethene (vinyl chloride) for the production of PVC (polyvinylchloride). It is also widely used in the paper industry, as an oxidising agent to bleach both pulp and paper, and also as a bleaching agent for textiles. Molecular chlorine is used extensively as a purifying agent for water supplies because it destroys harmful bacteria, producing Cl^- in the process. Dissolution of Cl_2 in water results in formation of hypochlorous acid, HOCl, according to the equation:

$$Cl_2(aq) + H_2O(l) \longrightarrow HOCl(aq) + Cl^-(aq) + H^+(aq)$$

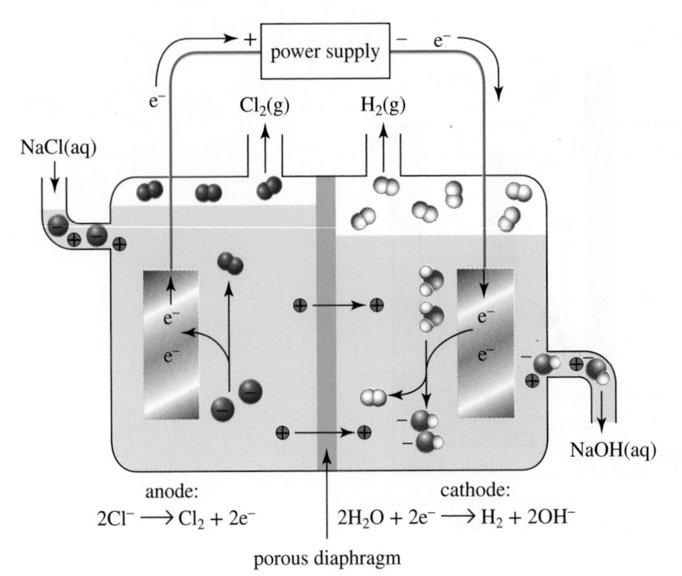

FIGURE 14.16 Schematic view of the electrolytic chlor–alkali process showing the molecular species involved in the redox reactions.

The hypochlorite ion, OCl^-, the conjugate base of hypochlorous acid, is the active ingredient in most bleaches. This ion oxidises many organic materials that are coloured, breaking them into smaller, colourless substances that are easily removed by detergents. Like all the other halogens, molecular chlorine is toxic, and it holds the dubious distinction of being the first chemical warfare agent to be used on a large scale. During World War I, nearly 6000 cylinders of chlorine were used at the Second Battle of Ypres, on 22 April 1915.

WORKED EXAMPLE 14.2

Disproportionation

Show, by assigning oxidation numbers to all chlorine-containing species in the equation:

$$Cl_2(aq) + H_2O(l) \rightarrow HOCl(aq) + Cl^-(aq) + H^+(aq)$$

that this is an example of a disproportionation reaction. Use electrochemical data from chapter 12 to determine $E^{\ominus}$ for this process, and comment on its spontaneity under standard conditions in both acidic ($[H^+] = 1.0$ M) and basic ($[OH^-] = 1.0$ M) solution.

Analysis

As we saw in chapter 13, disproportionation is the simultaneous oxidation and reduction of a single chemical species. In this case, molecular chlorine, Cl_2, is the species undergoing disproportionation. Assign oxidation numbers using the rules from chapter 12 (p. 497), and calculate $E^{\ominus}$ using $E^{\ominus} = E_R - E_L$, as outlined in chapter 12.

Solution

The oxidation number of chlorine in Cl_2 is 0, as is the case for any free element.

The oxidation number of chlorine in HOCl is +1. The compound is neutral, H has an oxidation number of +1, and O has an oxidation number of –2. Therefore, Cl must have an oxidation number of +1 so that the sum of all oxidation numbers in the molecule is zero.

The oxidation number of chlorine in Cl^- is –1, the same as the charge on the ion.

Thus, we can see that Cl_2, in which the oxidation number of chlorine is zero, is being converted to species with a lower (Cl^-) and higher (HOCl) oxidation number for chlorine; this therefore is an example of disproportionation.

If we assume that reduction occurs at the right-hand electrode of a galvanic cell, then we calculate $E^{\ominus}$ as follows (data from table 12.1, p. 515):

$$Cl_2(aq) + 2e^- \rightarrow 2Cl^-(aq) \qquad \text{(reduction)}$$

$$Cl_2(aq) + 2H_2O(l) \rightarrow 2HOCl(aq) + 2H^+(aq) + 2e^- \qquad \text{(oxidation)}$$

$$2Cl_2(aq) + 2H_2O(l) \rightarrow 2HOCl(aq) + 2Cl^-(aq) + 2H^+(aq)$$

Dividing this by 2 gives:

$$Cl_2(aq) + H_2O(l) \rightarrow HOCl(aq) + Cl^-(aq) + H^+(aq)$$

$$E^\circ = E_R - E_L = +1.36\text{ V} - (+1.63\text{ V}) = -0.27\text{ V}$$

Thus, under standard conditions in acidic solution, this reaction is not spontaneous.
In basic solution, the reactions are as follows.

$$Cl_2(aq) + 2e^- \rightarrow 2Cl^-(aq) \qquad \text{(reduction)}$$

$$Cl_2(aq) + 4OH^-(aq) \rightarrow 2OCl^-(aq) + 2H_2O(l) + 2e^- \qquad \text{(oxidation)}$$

$$2Cl_2(aq) + 4OH^-(aq) \rightarrow 2OCl^-(aq) + 2Cl^-(aq) + 2H_2O(l)$$

Dividing this by 2 gives:

$$Cl_2(aq) + 2OH^-(aq) \rightarrow OCl^-(aq) + Cl^-(aq) + H_2O(l)$$

Knowing that $E^\circ_{OCl^-/Cl_2} = +0.32$ V, we can then calculate E° as:

$$E^\circ = E_R - E_L = +1.36\text{ V} - (+0.32\text{ V}) = +1.04\text{ V}$$

Thus, under standard conditions in basic solution, this reaction is spontaneous.
Recall from our discussion on pages 361–2 that a positive value of ΔG° (which corresponds to a negative value of E°) does not mean the forward reaction does not occur at all; it merely means that it never proceeds to such an extent that all products will be present at c°. Thus, dissolution of Cl_2 in 1.0 M H_3O^+ will result in the formation of small amounts of HOCl and Cl^-. However, the forward reaction is much more favoured in basic solution, as evidenced by its large positive value of E°.

Bromine is a dense deep red liquid, one of only two elements on the periodic table which exist as a liquid under ambient conditions (the other is mercury). As Br_2 is a weaker oxidising agent than Cl_2, chemical oxidation of Br^- is feasible and, indeed, Br_2 is prepared on an industrial scale by the oxidation of Br^- with Cl_2. Br^- is found primarily in sea water, salt lakes and brine deposits. The most well-known source is probably the Dead Sea, on the border of Israel and Jordan. Bromine finds its major use in flame retardants, accounting for over 50% of the world's annual production. Like chlorine, it is also used in water purification. Before the advent of digital cameras, silver bromide was used extensively as the light-sensitive material in photographic film. Bromomethane (methyl bromide) is used in large quantities as a pesticide; however, its use is decreasing as it is recognised to be an ozone-depleting chemical.

Iodine, like bromine, is prepared by oxidation of the halide ion by Cl_2. It is a deep purple solid with a significant vapour pressure, and a sample of I_2 in a closed container will be seen to have a purple vapour above it (see figure 14.17). There are no major industrial uses of iodine, but it is increasingly being used as an alternative to chlorine in water treatment, especially in Australia. Iodine is a trace element, being a constituent of thyroxine (p. 1065), a hormone which regulates the rate of cellular use of oxygen. Iodine deficiency was historically manifested by a goitre, a swollen thyroid gland, but the compulsory addition of small amounts of NaI to table salt (in 1924, New Zealand was one of the first countries in the world to do this) has all but eliminated this in developed countries. However, it is also now known that iodine deficiency can lead to developmental disabilities and ironically the current fad for using 'natural' salt containing no iodine appears to have contributed to iodine deficiency again being a problem in both Australia and New Zealand. For this reason, bakers in both countries have been required to use iodised salt in their products since 2009.

FIGURE 14.17 Iodine sublimes at atmospheric pressure, and purple iodine vapour can be seen above a sample of solid iodine.

Astatine occurs in tiny amounts on Earth owing to its presence in three radioactive disintegration series and the extremely short lifetime (8.3 h) of its longest lived isotope. All of its isotopes are radioactive. It competes with francium for the distinction of being the rarest element in the Earth's crust.

Group 18

18		
2 He	helium	$1s^2$
10 Ne	neon	[He] $2s^22p^6$
18 Ar	argon	[Ne] $3s^23p^6$
36 Kr	krypton	[Ar] $4s^23d^{10}4p^6$
54 Xe	xenon	[Kr] $5s^24d^{10}5p^6$
86 Rn	radon	[Xe] $6s^25d^{10}4f^{14}6p^6$

There was no place in Mendeleev's original periodic table (chapter 1, p. 16) for the elements of group 18, simply because none of them had been discovered at that time. They were originally called the noble gases, as they were thought to be chemically nonreactive, but the isolation of the first noble gas compound in 1962 (see Chemical Connections, p. 618) showed them to have significant reactivity under the correct conditions. They all exist as monatomic gases under ambient conditions.

Although helium is the second most abundant element in the universe, it is present on Earth in very low concentrations, owing to the fact that it is light enough to escape the Earth's gravitational pull. Helium is obtained primarily from natural gas deposits, having been formed as a result of radioactive decay of ^{238}U, and is used extensively in cryogenic (very low temperature) applications. Helium has the lowest boiling point of any element; it condenses to a liquid at 4.2 K. Several materials become superconducting at very low temperatures and liquid helium is used to obtain these conditions. Of particular interest to chemists is the use of liquid helium to cool the superconducting magnets used in NMR spectrometers (p. 896). Helium is unusual in that it cannot be solidified at atmospheric pressure. It also exhibits a phenomenon called superfluidity when it is cooled below 2.17 K; this means the liquid has zero viscosity and will coat the entire surface of any container in which it is placed. The superfluid also shows extraordinarily high thermal conductivity, greater than that of copper.

Neon is found in the atmosphere at concentrations of 1 part in 65 000, and it is usually obtained from the fractional distillation of liquid air. The passage of an electric discharge through the gas causes a beautiful orange-red glow (figure 14.18), due to a large number of electronic transitions that give rise to many atomic emission lines around 600 nm (figure 4.16, p. 128). This effect was used by Georges Claude in the construction of the first neon light in 1910. The emission properties of neon are also used in helium–neon (HeNe) lasers; these emit red light of wavelength 632.8 nm.

FIGURE 14.18 Emission of photons from excited neon atoms gives a brilliant red light.

Chemical Connections

Compounds of group 18 elements

Prior to 1962, most chemists thought that the group 18 elements were chemically nonreactive and would not form compounds with any element on the periodic table. However, the work of Neil Bartlett, a young British chemist working at the University of British Columbia in Vancouver in the early 1960s, changed this misconception forever.

Bartlett's original interest was not in the group 18 elements but, rather, in compounds formed between fluorine and platinum. As a graduate student in England, he had isolated an unusual red solid with the formula PtF_6O_2 while studying the reactions of both metallic platinum and some platinum compounds with elemental fluorine in the presence of small amounts of oxygen. When he moved to Vancouver he set about trying to characterise this further. He eventually showed, using a variety of techniques, that the correct formula of this solid was $O_2^+PtF_6^-$. This was the first example of a compound containing an oxidised O_2 molecule; we usually think of O_2 as being an excellent oxidising agent, and therefore it is easily reduced. Oxidising O_2 should be extremely difficult and would require an oxidising agent of enormous strength. As one of the ways in which he had prepared the solid involved the reaction of the gas PtF_6 with O_2, Bartlett realised that PtF_6 must be an exceptionally good oxidising agent. Isolation of $O_2^+PtF_6^-$ from this reaction was an achievement in itself, but even more remarkable results lay in store.

While preparing for an undergraduate lecture, Bartlett happened to look at a table of ionisation potentials and found that the energy required to remove an electron from Xe(g) (1170 kJ mol^{-1}) was essentially equal to that required for O_2(g) (1177 kJ mol^{-1}). To put these numbers in perspective, recall from chapter 5 that the N≡N triple bond, one of the strongest known covalent bonds, has a bond energy of 942 kJ mol^{-1}; removing an electron from either O_2 or Xe is *really* difficult! Bartlett therefore reasoned that if PtF_6 was a strong enough oxidising agent to oxidise O_2, it should also be able to oxidise Xe and possibly form a compound with this group 18 element. Such reasoning went against all perceived chemical wisdom of the time but, nevertheless, Bartlett set up a simple apparatus in which deep red gaseous PtF_6 and colourless Xe at room temperature were separated by a seal; when the seal was broken, a yellow solid formed immediately, and chemical history was made (figure 14.19).

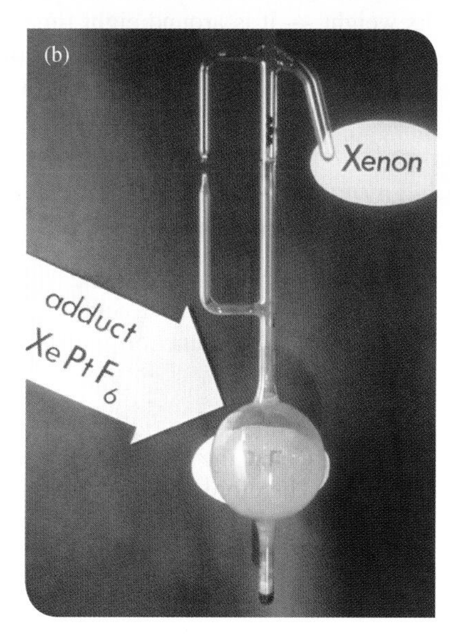

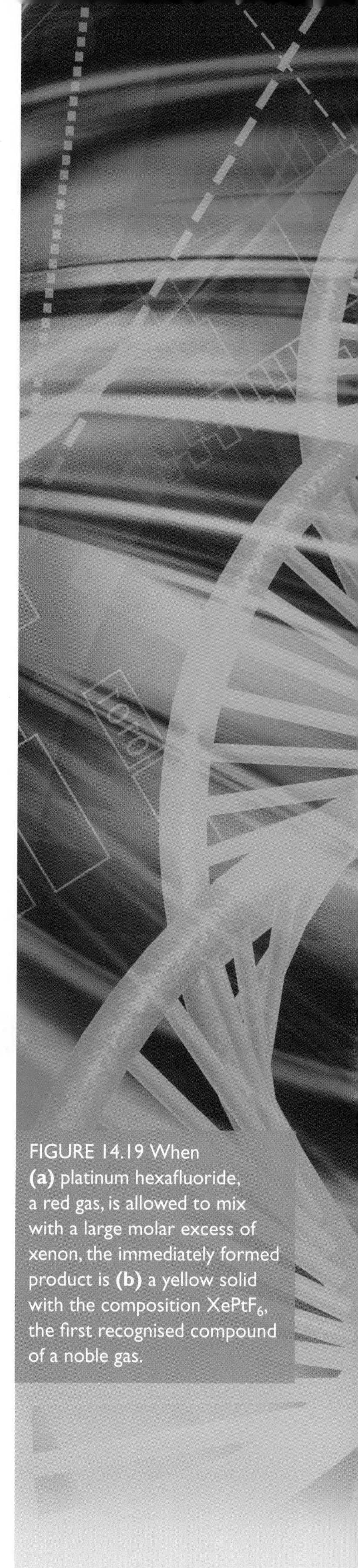

FIGURE 14.19 When **(a)** platinum hexafluoride, a red gas, is allowed to mix with a large molar excess of xenon, the immediately formed product is **(b)** a yellow solid with the composition $XePtF_6$, the first recognised compound of a noble gas.

Even today, the exact identity of the yellow solid is not certain. Originally, it was thought to be $Xe^+PtF_6^-$, by analogy with the O_2^+ compound. However, it now appears that the compound is better described as $[XeF]^+[PtF_5]^-$. Regardless, Bartlett's experiment inspired chemists around the world to study the reactivity of the group 18 elements with strong oxidising agents. The first krypton compound was reported in 1963, while the first xenon oxide, XeO_3, was also prepared that year. Bartlett was possibly the first person to experience the highly unstable nature of xenon oxides when what was thought to be a sample of XeO_2 exploded in a quartz tube while he was not wearing eye protection; 27 years later, a piece of glass was removed from his eye.

There are now several hundred known compounds containing group 18 elements, and research into their synthesis and properties continues to this day. Bartlett's extraordinary discovery changed chemistry overnight and forced chemists to fundamentally alter their ideas concerning reactivity; we now no longer think of group 18 elements as being chemically inert. Surprisingly, Bartlett, who died in 2008, was not awarded the Nobel Prize for his achievement; however, his name is immortalised in every inorganic chemistry textbook as the man who showed that the noble gases are not quite so noble.

Argon is the most abundant of the group 18 elements. It is the third most abundant component of air (0.934% by volume) after nitrogen and oxygen, and is obtained from this source. Its relative profusion and lack of chemical reactivity make it the inert gas of choice in applications which require exclusion of oxygen that are carried out under conditions where nitrogen might react. As we will see in e-chapter 27, the argon isotope ^{40}Ar is important in the dating of rocks. Although argon is considered chemically inert, the compound HArF has been recently reported; however, it is stable only in a matrix of frozen argon below 27 K.

Krypton has an abundance in air around one-fifth that of helium. Its emission properties make it ideal for use in high-speed photographic flash bulbs, as it gives a high-intensity white light, while it also finds use in krypton–fluoride (KrF) lasers, which give ultraviolet light of wavelength 248 nm. Krypton is the first of the group 18 elements that show appreciable chemical reactivity; the compound KrF_2 was prepared in 1963 by low temperature (20 K) photolysis of a mixture of krypton and fluorine, and it is stable up to 250 K. Since this time, over 30 Kr compounds, all containing Kr in the +2 oxidation state and all of which have krypton bonded to fluorine, have been prepared.

Xenon is the least abundant of the group 18 elements found in the atmosphere (around 0.09 parts per million) but the most chemically interesting. The first xenon compounds were prepared in 1962, thereby demonstrating the potential reactivity of the so-called 'noble' gases. The fluorides XeF_2, XeF_4 and XeF_6 are all known, as are the oxides XeO_3 and XeO_4; the last two are both excellent oxidising agents as evidenced by the following reduction potentials.

$$XeO_3 + 6H^+ + 6e^- \rightarrow Xe + 3H_2O \qquad E^\ominus = +2.10\text{ V}$$

$$H_4XeO_6 + 8H^+ + 8e^- \rightarrow Xe + 6H_2O \qquad E^\ominus = +2.18\text{ V}$$

However, both are dangerously explosive. Several mixed oxofluorides have been prepared, while compounds containing Xe bonded to C, N, Cl and some transition metals are also known. The element itself is used in xenon lamps, lasers and as an anaesthetic.

Radon is formed from the radioactive decay of radium, which is in turn formed from both uranium and thorium. All isotopes of radon are radioactive, and the longest lived has a half-life of less than four days. This means that radon occurs in miniscule amounts on Earth. However, houses constructed of stone, or built in areas containing high levels of uranium or thorium minerals, can accumulate potentially dangerous concentrations of radon in their basements owing to its weight — it is around eight times the density of air — and kits which test radon concentrations in air are commercially available. Although some controversy exists, recent research has proposed Ernest Rutherford as the original discoverer of radon.

WORKED EXAMPLE 14.3

Lewis structure of a group 18 compound

Draw the Lewis structure of XeO_3, and deduce the most probable geometry of this molecule.

Analysis

We obtain the Lewis structure of the molecules using the rules in chapter 5, and we then use these, in conjunction with VSEPR theory (chapter 5) to propose the geometry of the molecules.

Solution

Step 1: Count the valence electrons; Xe has 8 valence electrons and each O atom has 6. Therefore, there is a total of 26 valence electrons.

Step 2: Draw the bonding framework using single bonds.

```
      O
      |
  O—Xe—O
```

Electrons used = 6. Therefore, there are 20 electrons still to be allocated.

Step 3: Place three nonbonding pairs on each outer atom.

```
      :Ö:
       |
  :Ö—Xe—Ö:
   ¨      ¨
```

Electrons used = 18. Therefore, there are 2 electrons still to be allocated.

Step 4: Assign the remaining electrons to the inner atom.

```
      :Ö:
       |
  :Ö—Xe—Ö:
   ¨  ¨   ¨
```

Electrons used = 2. All electrons are now assigned.

Step 5: Minimise formal charges on all atoms.
The formal charges on each atom are as shown below. Moving electron pairs to make three Xe=O double bonds will result in each atom having a formal charge of zero.

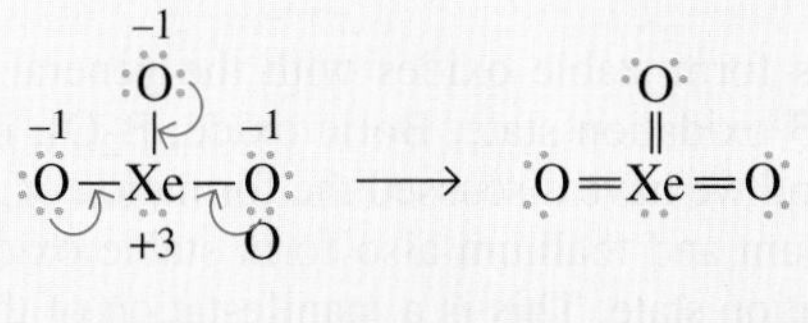

$$:\ddot{O}-Xe-\ddot{O}: \;(\text{formal charges } -1, +3, -1, -1) \longrightarrow O=Xe=O \text{ (with } =O\text{)}$$

There are four sets of electron pairs around the central Xe atom, so the structure of XeO_3 will be based on a tetrahedron. As only three of the sets are equivalent, the molecule will adopt a trigonal pyramidal geometry.

PRACTICE EXERCISE 14.2

Draw the Lewis structures of XeO_4 and KrF_2 (Xe and Kr are the central atoms, respectively), and determine their most probable geometries.

14.2 Reactivity of the *p*-block elements

In chapter 1, we saw that the elements in the periodic table could be divided into the *s*-block, *p*-block, *d*-block and *f*-block elements, as well as into metals, metalloids and nonmetals. The *s*-, *d*- and *f*-block elements are all metals, and it is only within the *p*-block elements that we find examples of metalloids and nonmetals. This observation has important implications for the reactivity of the *p*-block elements. In this section we will look at some representative examples of the reactivity of the *p*-block elements, with an emphasis on their compounds with hydrogen and oxygen. Note that oxoacids of these elements, some of which are discussed below, have been introduced previously in section 11.5.

Group 13 compounds

Boron is a metalloid and exhibits vastly different reactivity to the other members of group 13, which are all metals. From its position in group 13, with a valence electron configuration of $2s^22p^1$, it might be expected that the formula of the simplest compound of boron and hydrogen would be BH_3. In fact, this compound does not exist, and the simplest compound of boron and hydrogen is diborane, B_2H_6. This molecule has the structure outlined in figure 14.20.

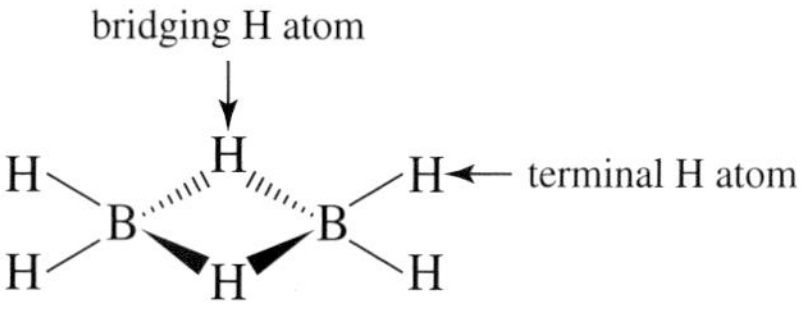

FIGURE 14.20 The structure of B_2H_6. The B—H bonds involving the four terminal H atoms are two-centre, two-electron bonds, while the B—H—B bonds, involving the bridging H atoms are two-centre, three electron bonds. The geometry about each B atom is approximately tetrahedral.

If we attempt to draw the Lewis structure corresponding to the above molecule, it becomes obvious that there are too few electrons for each B—H bond to contain two electrons. There are a total of 12 valence electrons (3 from each boron atom and 1 from each hydrogen atom) which must be used to construct 8 B—H bonds; clearly, we are several electrons short. Experiments show that the terminal B—H bonds are shorter, and therefore stronger than those involving the bridging H atoms. We therefore assign the four terminal B—H bonds as being single covalent bonds involving two electrons. We then use the remaining 4 electrons to construct two three-centre, two electron bonds; each of these involves two electrons being delocalised over each set of central B—H—B atoms. This leads to weaker, and therefore longer B—H bonds in the bridge. B_2H_6 provides an example of a molecule in which the bonding cannot be adequately described using the Lewis electron pair approach we outlined in chapter 5. In fact, B_2H_6 is the simplest of a family of molecules and anions comprising only boron and hydrogen, called the boranes. These molecules adopt structures in which the boron atoms are located at the vertices of polyhedra, somewhat similar to the structure of elemental boron discussed on page 603. The bonding in these molecules is of particular interest, since all of these molecules are, like B_2H_6, formally electron deficient. This results in significant reactivity towards Lewis bases; many of the boranes react rapidly and violently with atmospheric oxygen, and it was this property that led to boranes being studied extensively in the early years of the cold war as potential rocket fuels.

The hydrides of the remaining elements in group 13 do not display the extraordinary structural variety of those of boron, although both Al_2H_6 and Ga_2H_6, which have structures analogous to that of B_2H_6, are known. The hydrides decrease in stability down the group, with those of both indium and thallium being stable only at extremely low temperatures.

Two important group 13 hydride anions are BH_4^- and AlH_4^-. The salts $NaBH_4$ and $LiAlH_4$ are used extensively in organic chemistry as reducing agents. Examples of their use can be found in chapters 21 and 23.

All the group 13 elements form stable oxides with the general formula X_2O_3, in which the element is formally in the +3 oxidation state. Boric oxide, B_2O_3, is important as a precursor to other compounds of boron, and we have discussed the formation of Al_2O_3 from bauxite in section 14.1. However, gallium, indium and thallium also form stable oxides of the formula X_2O, with the metal ion in the +1 oxidation state. This is a manifestation of the **inert pair effect**, in which the oxidation state two units lower than that normally exhibited by elements in a group becomes more stable proceeding down the group. This can be explained in terms of the increasing stability of the valence *s* orbital going down a group, thereby making loss of the valence *s* electrons more difficult. Therefore, of the valence ns^2np^1 electrons, only the *p* electron is lost, leading to formation of the +1 oxidation state. This phenomenon is not unique to group 13, being observed also in groups 14 to 16.

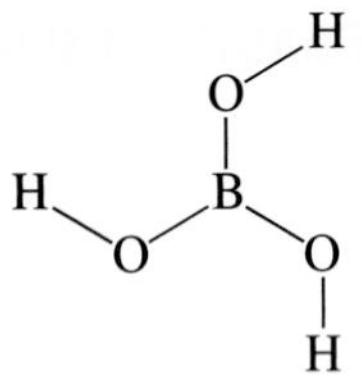

The acid–base characteristics of the oxides change markedly on going down the group. B_2O_3 is Lewis acidic, reacting with water to give boric acid, $B(OH)_3$, according to the equation:

$$B_2O_3 + 3H_2O \rightarrow 2B(OH)_3$$

FIGURE 14.21 The structure of $B(OH)_3$.

Boric acid is a solid, in which individual $B(OH)_3$ molecules are hydrogen-bonded to each other in sheets. The structure of the $B(OH)_3$ molecule is shown in figure 14.21.

$B(OH)_3$ acts as a Lewis acid, not a Brønsted–Lowry acid. For example, the reaction of $B(OH)_3$ with water involves attack of water at the boron atom and subsequent proton loss from this — not proton transfer from a B—O—H unit — as shown in figure 14.22.

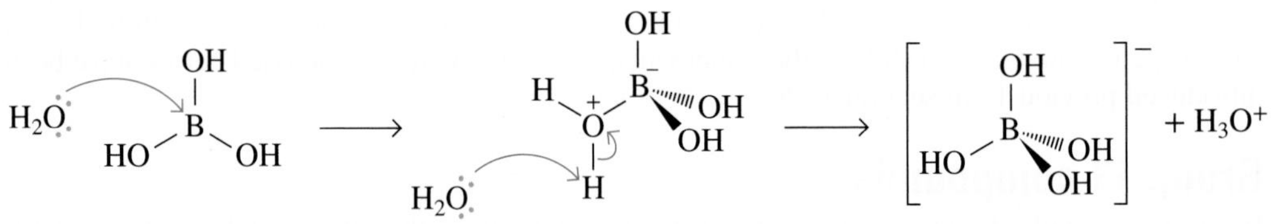

FIGURE 14.22 Mechanism of the reaction of $B(OH)_3$ with water, with the overall reaction being: $B(OH)_3 + 2H_2O \rightarrow B(OH)_4^- + H_3O^+$

This gives rise to an apparent $pK_a = 9.24$. The Lewis acidity of boric acid can be explained by considering the bonding in the molecule. The boron atom is sp^2-hybridised, with O—B—O bond angles of 120°, and therefore has a vacant *p* orbital, which is oriented perpendicular to the plane of the molecule. This is analogous to the bonding situation in BF_3, which we discussed on pages 200–1. This vacant *p* orbital can accept a pair of electrons, thereby accounting for the Lewis acidity of $B(OH)_3$.

Both Al_2O_3 and Ga_2O_3 are amphoteric, reacting with both acids and bases, while In_2O_3 and Tl_2O_3 are basic.

Group 14 compounds

The simplest compound of carbon and hydrogen is methane, CH_4. However, in contrast to the behaviour of every other element on the periodic table, there are literally millions of compounds containing only C and H — we saw on page 62 that 36 797 588 constitutional isomers are possible for the formula $C_{25}H_{52}$ alone! Some of the compounds containing only carbon and hydrogen will be discussed in chapter 16.

Silane, SiH_4, is the simplest of a small number of compounds containing only silicon and hydrogen. Others include Si_2H_6, Si_3H_8 and Si_4H_{10}. In contrast to their carbon analogues, all of these are extremely reactive in air, and must be kept under an inert atmosphere. These molecules are structurally similar to short-chain alkanes (see chapter 16). Germanium also forms a number of compounds of general formula Ge_nH_{2n+2}, while both SnH_4 and PbH_4 are unstable at room temperature.

Carbon forms two common oxides, CO and CO_2, both of which are gases at room temperature. These are formed from the combustion of any carbon-containing material, the former generally when the supply of oxygen is limited. Dissolution of CO_2 in water gives carbonic acid, which will be discussed in section 14.3. The unusual molecule carbon suboxide, C_3O_2, is also known. It is a gas at room temperature and can be formed by passing an electric current through CO, or by the dehydration of malonic acid, $HOOCCH_2COOH$, with P_2O_5. The structure of this molecule is given in figure 14.23; the C atoms are all formally *sp*-hybridised and the molecule is essentially linear.

$$:\ddot{O}=C=C=C=\ddot{O}:$$

FIGURE 14.23 The Lewis structure of carbon suboxide, C_3O_2.

The chemistry of silicon and oxygen is extensive, with silicate minerals comprising a significant proportion of the Earth's crust. The simplest compound of silicon and oxygen, silicon dioxide, has the empirical formula SiO_2, but is not a discrete molecule. In contrast to CO_2, SiO_2, generally called silica, exists in a number of 3-dimensional network solid forms at room temperature and pressure, in all of which each Si atom is bonded to four oxygen atoms and has a tetrahedral geometry. The most common of these forms is quartz (figure 14.24), of which sand is an impure form. One of the most important uses of silica is in glass; details of the compositions of various glasses are given in section 7.6. Silicate minerals are primarily composed of tetrahedral SiO_4^{4-} units which form chains, rings or sheets through the sharing of oxygen atoms between units. Substitution of Si with Al in silicates leads to aluminosilicates, which include feldspars (the major component of igneous rocks) and zeolites (useful as absorbents and ion-exchange materials).

FIGURE 14.24 Quartz is the most common form of silica.

The oxides GeO_2, SnO_2 and PbO_2, in which the group 14 elements are in the +4 oxidation state, are all known. However, the oxides GeO, SnO and PbO are also known. These oxides, in which the group 14 elements are in the +2 oxidation state, reflect the increasing stability of the +2 oxidation state going down the group. Again, this is a manifestation of the inert pair effect discussed above. In fact, PbO_2 is an extremely good oxidising agent, as shown by the following reduction potentials.

$$PbO_2(s) + 4H^+(aq) + 2e^- \rightarrow Pb^{2+}(aq) + 2H_2O(l) \qquad E^\ominus = 1.46\text{ V}$$

$$PbO_2(s) + 4H^+(aq) + SO_4^{2-}(aq) + 2e^- \rightarrow PbSO_4(s) + 2H_2O(l) \qquad E^\ominus = 1.69\text{ V}$$

The latter is the half equation for the reduction reaction that occurs in lead batteries, with sulfate coming from the sulfuric acid electrolyte. The corresponding oxidation process is the conversion of lead to lead sulfate:

$$Pb(s) + SO_4^{2-}(aq) \rightarrow PbSO_4(s) + 2e^-$$

with $E^\ominus_{PbSO_4/Pb} = -0.36$ V, which gives an equation for the overall reaction of

$$Pb(s) + PbO_2(s) + 4H^+(aq) + 2SO_4^{2-}(aq) \rightarrow 2PbSO_4(s) + 2H_2O(l)$$

and $E^\ominus_{cell} = E^\ominus_R - E^\ominus_L = 1.69\text{ V} - (-0.36\text{ V}) = 2.05\text{ V}$.

A typical 12 V lead battery (figure 14.25) contains 6 of these cells arranged in series, with $PbSO_4(s)$ being formed as the battery discharges.

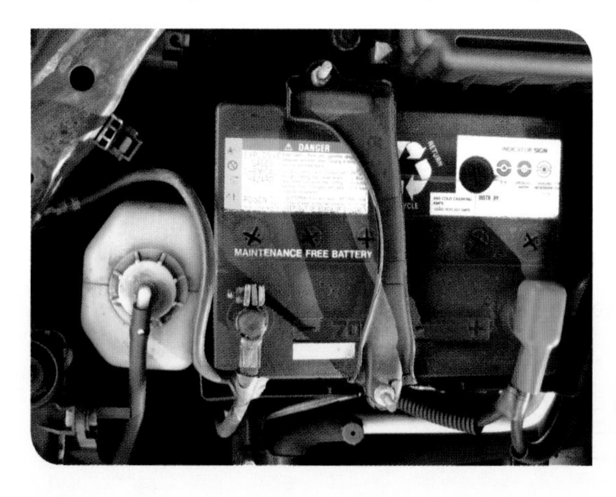

FIGURE 14.25 A typical 12 V lead battery, like those used in most cars, has six $Pb(s) + PbO_2(s) + 4H^+(aq) + 2SO_4^{2-}(aq) \rightarrow 2PbSO_4(s) + 2H_2O(l)$ cells arranged in series.

A third oxide of lead, Pb_3O_4, is a **mixed-valence** compound. Using the rules for assigning oxidation numbers from section 12.1, the 8– charge from the four oxygen atoms must be balanced by an 8+ charge from the three Pb ions. This can be achieved by assigning two of the three Pb ions a +2 charge, and the remaining Pb ion a +4 charge. We can designate this using the notation $Pb^{II}Pb^{II}Pb^{IV}O_4$, which makes apparent the origin of the term 'mixed valence'.

None of the group 14 oxoacids are particularly well characterised. The predominant species in a solution of carbonic acid is, as we shall see, dissolved CO_2, while silicic acid, H_4SiO_4, polymerises rapidly in anything but the most dilute of solutions. The aqueous chemistry of the remaining three elements in the group is similarly uncertain.

Group 15 compounds

The three compounds of nitrogen with hydrogen, NH_3, N_2H_4 and N_2H_2 (ammonia, hydrazine and diazene, respectively) are discussed in section 14.3. Phosphane, PH_3, adopts a pyramidal structure similar to that of ammonia, but with somewhat smaller bond angles (≈94°). It is a highly toxic gas which, owing to the presence of small amounts of impurities, is spontaneously flammable in air. In contrast to ammonia, phosphane does not act as a Brønsted–Lowry base in aqueous solution; it also does not exhibit the extraordinary solubility of ammonia in water (≈0.3 g L^{-1} versus ≈518 g L^{-1}), reflecting the importance of hydrogen-bonding in aqueous ammonia. Arsane, AsH_3, and stibane, SbH_3, both have similar properties to those of phosphane, while bismuthane, BiH_3, is unstable, decomposing to its elements below 0 °C.

Nitrogen forms eight neutral compounds with oxygen, which are detailed in the next section. Phosphorus forms two main oxides, one each in the +3 and +5 oxidation state. While the empirical (and commonly quoted) formulae of these are P_2O_3 and P_2O_5, their respective molecular formulae and systematic names are P_4O_6 (tetraphosphorus hexaoxide) and P_4O_{10} (tetraphosphorus decaoxide). The structure of the latter is shown in figure 14.31. Arsenic is similar, in that both As_4O_6 and 'As_2O_5' are known, although the latter has a polymeric structure. The decreasing stability of the +5 oxidation state going down the group is reflected in the fact that the major oxides of both antimony and bismuth are the +3 species Sb_4O_6 and Bi_2O_3, with the +5 oxides being very unstable in both cases.

The oxoacids of both nitrogen and phosphorus are extremely important industrial compounds. HNO_3, nitric acid, is a strong acid of which over 50 million tonnes are produced annually by the Ostwald process (named after Wilhelm Ostwald, Nobel Prize in chemistry 1909). This starts with the high-temperature air-oxidation of ammonia over a platinum/rhodium catalyst to give nitric oxide, NO, according to the equation:

$$4NH_3(g) + 5O_2(g) \xrightarrow{Pt/Rh} 4NO(g) + 6H_2O(g)$$

The NO is then air oxidised to NO_2:

$$2NO(g) + O_2(g) \rightarrow 2NO_2(g)$$

and in the final step, the NO_2 dissolved in water at high temperature undergoes disproportionation to give HNO_3 and NO; the latter is recirculated and air-oxidised to form more NO_2.

$$3NO_2(aq) + H_2O(l) \rightarrow 2HNO_3(aq) + NO(g)$$

The aqueous solution of HNO_3 formed by this method has a concentration around 60% by weight, and subsequent distillation gives 'concentrated' nitric acid, which is 70% by weight, and has a concentration of 16.0 mol L^{-1}. The concentrated acid is a very good oxidising agent, and dissolves all metals with the exception of Re, Ir, Au and Pt. Its primary use is in the manufacture of nitrate fertilisers.

HNO_2, nitrous acid, is a weak acid ($K_a = 7.1 \times 10^{-4}$) again formed by dissolving NO_2 in water. The exact details of this reaction are complicated, but it again involves a disproportionation to give HNO_2 and HNO_3.

$$2NO_2(g) + H_2O(l) \rightarrow HNO_2(aq) + HNO_3(aq)$$

In contrast to HNO_3, HNO_2 cannot be obtained as a pure liquid. It undergoes disproportionation in all but very dilute aqueous solution. The following standard reduction potentials are relevant in this respect.

$$HNO_2 + H^+ + e^- \rightarrow NO + H_2O \qquad E^\ominus = 0.983\ V$$

$$NO_3^- + 3H^+ + 2e^- \rightarrow HNO_2 + H_2O \qquad E^\ominus = 0.934\ V$$

Multiplying the first reaction by 2 and reversing the second equation gives:

$$2HNO_2 + 2H^+ + 2e^- \rightarrow 2NO + 2H_2O$$

$$HNO_2 + H_2O \rightarrow NO_3^- + 3H^+ + 2e^-$$

$$\overline{3HNO_2 \rightarrow 2NO + NO_3^- + H^+ + H_2O}$$

with $E^{\ominus}_{\text{cell}} = E^{\ominus}_{\text{R}} - E^{\ominus}_{\text{L}} = 0.983\text{ V} - (0.934\text{ V}) = 0.049\text{ V}$. The overall disproportionation process is therefore spontaneous under standard conditions.

Phosphoric acid, H_3PO_4, is a triprotic acid containing three ionisable P—O—H protons, as can be seen from its structure in figure 14.26. It is discussed in more detail in section 14.3.

```
       O                   O                O
       ||                  ||               ||
H—O—P—O—H         H—O—P—O—H         H—P—O—H
       |                   |                |
       O                   H                H
       |
       H
```

H_3PO_4	H_3PO_3	H_3PO_2
phosphoric acid	phosphonic acid	phosphinic acid
$pKa_1 = 2.15$	$pKa_1 = 0.97$	$pKa = 0.80$
$pKa_2 = 7.20$	$pKa_2 = 5.78$	
$pKa_3 = 12.38$		

FIGURE 14.26 The structures of, and acidity constant data for, H_3PO_4, H_3PO_3 and H_3PO_2.

Two other important phosphorus oxoacids are H_3PO_3 (phosphonic acid) and H_3PO_2 (phosphinic acid), which are diprotic and monoprotic, respectively. The structures of these acids, represented in figure 14.26 show that not all of the hydrogen atoms are ionisable. As we learned in chapter 11, in order for a hydrogen atom to be appreciably acidic, it must be bonded to a significantly electronegative element, generally O or a halogen. Only the hydrogen atoms bonded to oxygen in H_3PO_3 and H_3PO_2 are acidic, and the chemical formulae of these acids should therefore be written $HPO(OH)_2$ and $H_2PO(OH)$, respectively, to better reflect this.

H_3AsO_4, arsenic acid, is similar in both structure and acidity to phosphoric acid, with $pK_{a1} = 2.24$, $pK_{a2} = 6.96$ and $pK_{a3} = 11.50$. However, H_3AsO_3, arsenous acid, in contrast to H_3PO_3, behaves as a monoprotic acid with $pK_a = 9.29$; this implies a structure of $As(OH)_3$, similar to that of boric acid. The oxoacids — if any — of antimony and bismuth are poorly defined.

Group 16 compounds

Oxygen forms two major compounds with hydrogen, namely H_2O (water) and H_2O_2 (hydrogen peroxide). The former is ubiquitous not only on Earth, but also throughout the universe; Jupiter's satellite Europa is thought to have an ocean of liquid water beneath its ice-covered surface (figure 14.27). The Hubble Space Telescope has provided evidence for water being present in the atmospheres of five planets beyond our solar system, and, in 2011, astronomers detected a massive (140 trillion times the amount of the water in the Earth's oceans) amount of water surrounding a black hole 12 billion light years from Earth.

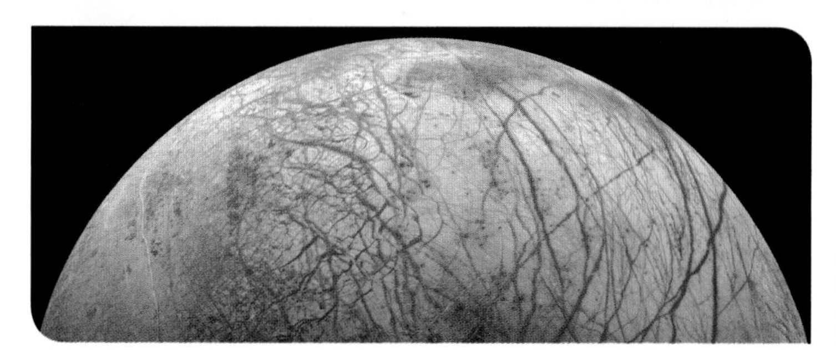

FIGURE 14.27 Jupiter's moon Europa is thought to have an ocean of liquid water under its icy shell.

We have already discussed some properties of water, most notably its acid–base properties (chapter 11) and its propensity to hydrogen-bond (chapter 6). Pure hydrogen peroxide is a colourless, viscous liquid which can explode when heated or mixed with organic compounds. It is generally provided as a 30% by weight aqueous solution for laboratory work and is sold in pharmacies as a 3% aqueous solution as a disinfectant. It is an excellent oxidising agent in acidic solution:

$$H_2O_2 + 2H^+ + 2e^- \rightarrow 2H_2O \qquad E^{\ominus} = 1.76\text{ V}$$

while it can also be oxidised relatively easily:

$$O_2 + 2H^+ + 2e^- \rightarrow H_2O_2 \qquad E^{\ominus} = 0.70\text{ V}$$

The above reduction potentials show that H_2O_2 is unstable with respect to disproportionation.

$$2H_2O_2 \rightarrow 2H_2O + O_2 \qquad E^{\ominus} = (1.76\text{ V} - 0.70\text{ V}) = 1.06\text{ V}$$

This normally slow reaction can be catalysed by trace amounts of metal ions. The fact that the product of H_2O_2 reduction is water makes this an extremely attractive replacement for chlorine-based compounds as an industrial oxidising agent. H_2O_2 is very weakly acidic, with $pK_a = 11.74$ for the reaction:

$$H_2O_2(aq) + H_2O(l) \rightarrow HO_2^-(aq) + H_3O^+(aq)$$

while it is significantly less basic than water, and can only be protonated by extremely strong acids.

The only important hydrides of the remaining elements in group 16 have the formula H_2X. H_2S, hydrogen sulfide (or sulfane, to give it its IUPAC name) is sometimes known as 'rotten egg gas' because of its distinctive smell. Indeed, H_2S is formed from the decomposition of sulfur-containing proteins, and its familiar odour is often evident around thermal areas. It is highly toxic, and deaths have occurred through exposure to H_2S around hot pools. Despite this, H_2S has been recently found to be an important signalling molecule in the human body, and this is currently the subject of extensive research. Like all of the group 16 H_2X molecules, H_2S exhibits a bent geometry, with a bond angle of ≈92°. It is significantly more acidic than water, with $pK_{a1} = 7.02$ and $pK_{a2} = 13.9$; the reasons for this increase in acidity have been discussed in section 11.5. Both H_2Se (common name hydrogen selenide, IUPAC name selane) and H_2Te (common name hydrogen telluride, IUPAC name tellane) exhibit the same odoriferous qualities as H_2S, but are even more disagreeable. They are both stronger acids than H_2S, with $pK_{a1} = 3.89$ (H_2Se); 2.6 (H_2Te) and $pK_{a2} = 11.0$ (H_2Se); 11 (H_2Te). H_2Po (IUPAC name polane) is poorly characterised and appears to be unstable at room temperature.

Ozone, O_3, is best considered as an allotrope of oxygen, rather than a compound of oxygen with oxygen, and is discussed in the next section. The chemistry of sulfur with oxygen is extensive, with a number of sulfur oxides known, containing sulfur in oxidation states ranging from +0.2 ($S_{10}O$) to +6 (SO_3). The most important sulfur oxides, SO_2 and SO_3, are discussed in the next section. Selenium, tellurium and polonium all form oxides having the formula XO_2, with the group 16 element in the +4 oxidation state, in preference to the +6 compounds XO_3. Both SeO_3 and TeO_3 can be prepared, but are highly oxidising, again reflecting the increasing importance of lower oxidation states towards the bottom of groups in the *p*-block elements.

The most important sulfur oxoacid, and indeed one of the most important of all oxoacids is sulfuric acid, H_2SO_4. This is one of the most produced industrial chemicals on Earth, with production in 2014 believed to exceed 284 million tonnes. It is predominantly manufactured using the contact process; this involves initial combustion of sulfur to give $SO_2(g)$:

$$S(s) + O_2(g) \rightarrow SO_2(g)$$

which is then oxidised to SO_3 over a solid V_2O_5 catalyst:

$$SO_2(g) + 1/2O_2(g) \xrightarrow{V_2O_5(s)} SO_3(g)$$

The gaseous SO_3 is then bubbled through concentrated H_2SO_4 to form oleum ($H_2S_2O_7$), also sometimes known as fuming sulfuric acid.

$$SO_3(g) + H_2SO_4(l) \rightarrow H_2S_2O_7(l)$$

This is then diluted with water to give concentrated sulfuric acid.

$$H_2S_2O_7(l) + H_2O(l) \rightarrow 2H_2SO_4(l)$$

Theoretically, dissolution of SO_3 in water could be used to form H_2SO_4 directly, according to the equation:

$$SO_3(g) + H_2O(l) \rightarrow H_2SO_4(l)$$

However, on an industrial scale, this very exothermic reaction leads to the formation of a fine fog of H_2SO_4, and it is for this reason that H_2SO_4 must be used in the production of H_2SO_4! Concentrated H_2SO_4 is a viscous liquid, with a concentration of 18 mol L^{-1}. It is a strong diprotic acid, with $pK_{a2} = 1.99$. Its reaction with water is extremely exothermic, and for this reason, whenever concentrated sulfuric acid requires dilution with water, the acid should always be added slowly, with stirring (and cooling, if necessary), to the water. Bubbling $SO_2(g)$ through water gives sulfurous acid, H_2SO_3. This is somewhat analogous to H_2CO_3 (see p. 636) in that this process does not gives rise to H_2SO_3 itself, but rather, hydrated sulfur dioxide, $SO_2(aq)$. The acidic solution obtained gives an apparent $pK_{a1} = 1.86$. A number of other sulfur oxoacids are known. Replacement of one O atom of H_2SO_4 with an *S* atom gives thiosulfuric acid, $H_2S_2O_3$, which is extremely unstable and cannot be isolated from aqueous solution. As a result, there is some uncertainty about its structure, with calculations suggesting a proton is attached to an S, rather than O atom; the two possible structures are shown in figure 14.28. Both H_2SO_5 (peroxysulfuric acid) and $H_2S_2O_8$ (peroxydisulfuric acid), contain an —O—O— peroxo unit, as can be seen from their structures in figure 14.28.

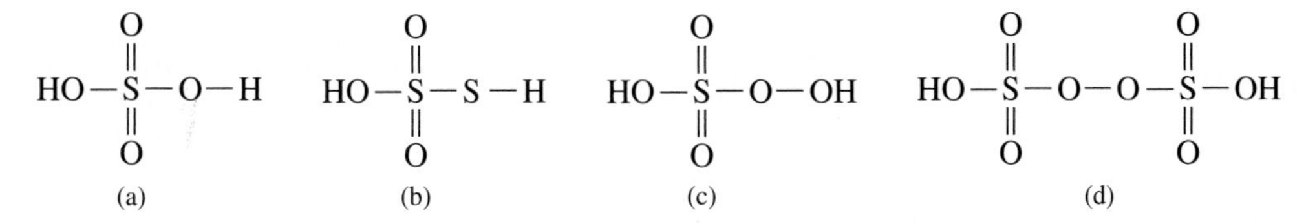

FIGURE 14.28 The structures of: **(a)** $H_2S_2O_3$ with both protons on O atoms, **(b)** $H_2S_2O_3$ with a proton on an S atom, and **(c)** H_2SO_5, and (d) $H_2S_2O_8$.

The latter, in particular, is an excellent oxidising agent, with $E^{\ominus}_{S_2O_8^{2-}/SO_4^{2-}} = 2.01$ V. Both selenic acid, H_2SeO_4 and selenous acid, H_2SeO_3 are known; the former is similar to H_2SO_4 in structure and acid strength, but is more strongly oxidising, while the latter is a colourless solid with the structure $(HO)_2SeO$. The structure of H_2TeO_3, tellurous acid, is not known with certainty, while the Te(VI) oxoacid has the formula $Te(OH)_6$; this displays an octahedral geometry about Te, and is a weak acid, with $pK_{a1} = 7.68$. Oxoacids of polonium do not appear to have been unequivocally characterised.

Group 17 compounds

The hydrides of the group 17 elements, HF (hydrogen fluoride), HCl (hydrogen chloride), HBr (hydrogen bromide) and HI (hydrogen iodide) are all gases at room temperature and pressure. Both HF(g) and HCl(g) can be prepared by reacting concentrated sulfuric acid with a solid fluoride or chloride salt, respectively. However, this method fails when a bromide or iodide salt is used, as concentrated sulfuric acid is a sufficiently strong oxidising agent to oxidise both bromide and iodide to the respective halogens. HBr(g) and HI(g) are therefore generally prepared by reaction of PBr_3 or PI_3 with water. When dissolved in water, the group 17 hydrides form an important series of acids, all of which, with the exception of HF ($pK_a = 3.17$), are strong. The trends in the strengths of these acids have been discussed in section 11.5. General data concerning these acids are given in table 14.2.

TABLE 14.2 Compositions of the concentrated hydrohalic acids.

Acid name	Weight %	Density/g mL^{-1}	Concentration/mol L^{-1}
Hydrofluoric acid (HF(aq))	46	1.15	32.1
Hydrochloric acid (HCl(aq))	37	1.18	11.6
Hydrobromic acid (HBr(aq))	48	1.50	8.89
Hydroiodic acid (HI(aq))	57	1.70	7.57

While hydrochloric acid, hydrobromic acid, and, to a lesser extent, hydroiodic acid are all commonly used reagents in chemistry laboratories, the use of hydrofluoric acid is limited by its physical properties. Despite HF(aq) being a weak acid, it is extremely corrosive towards many substances; for example, solutions of HF(aq) are generally handled in Teflon containers, as, in contrast to the other hydrohalic acids, hydrofluoric acid etches glass. It must also be handled with enormous care, as it causes severe burns; it readily penetrates the skin causing tissue necrosis and, in severe cases, decalcification of bone. This can last for several days. Should skin contact occur, the affected area must be treated immediately with a solution of calcium gluconate; this precipitates fluoride ion as CaF_2.

A variety of compounds of the group 17 elements with oxygen are known. In these, the halogens formally exhibit positive oxidation states, an observation somewhat at odds with the position of these elements in the periodic table, which would suggest preferential formation of the −1 oxidation state. As a result, the majority of the halogen oxides are excellent oxidising agents. Fluorine forms two oxides, OF_2 (oxygen difluoride) and O_2F_2 (dioxygen difluoride); the former, a gas which is stable at room temperature, has a bent geometry similar to that of water while the structure of the latter (unstable above 223 K) resembles that of hydrogen peroxide. The most important oxides of chlorine are Cl_2O (dichlorine monoxide, Cl +1 oxidation state), ClO_2 (chlorine dioxide, Cl +4 oxidation state), and Cl_2O_7 (dichlorine heptaoxide, Cl +7 oxidation state), all of which are highly reactive and can undergo explosive decomposition when pure. The two former compounds are gases at room temperature while the latter is a liquid. ClO_2 is an important industrial compound; it is used as an aqueous solution in both the bleaching of paper and in water treatment. The unstable bromine oxides Br_2O_3 and Br_2O_5 have been characterised in the solid state, but little is known about them, while I_2O_5, a stable white solid, is the most important oxide of iodine.

The oxoacids again contain the group 17 elements in formal positive oxidation states. The general formulae of these acids are commonly written as HOX (hypohalous acids, formal X oxidation state +1), HXO_2 (halous acids, formal X oxidation state +3), HXO_3 (halic acids, formal X oxidation state +5) and HXO_4 (perhalic acids, formal X oxidation state +7). However, in all cases the proton is attached to an oxygen atom, and better representations of the latter three acids are OX(OH), $O_2X(OH)$ and $O_3X(OH)$. The structures of the acids are given in figure 14.29 (overleaf).

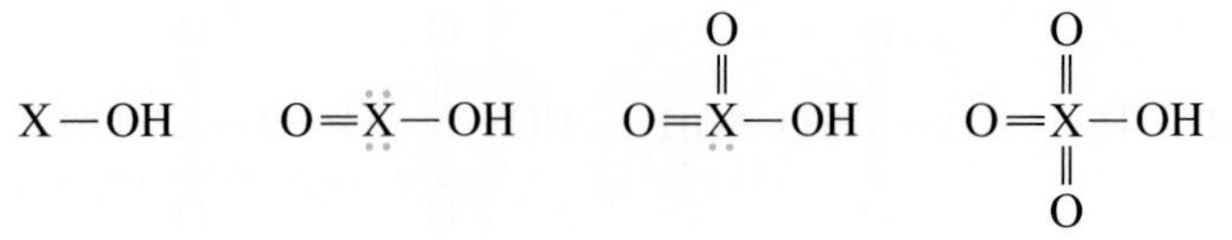

X—OH   O=X—OH   O=X(=O)—OH   O=X(=O)(=O)—OH

FIGURE 14.29 The structures of the group 17 oxoacids. Halous acids exhibit a bent geometry about the X atom, halic acids are pyramidal, and perhalic acids are tetrahedral.

The hypohalous acids are all prepared by reaction of the parent halogen with water (or in the case of HOF, with ice at very low temperature), according to the equation:

$$X_2 + 2H_2O \rightarrow HOX + H_3O^+ + X^-$$

HOF is the only hypohalous acid that can be prepared in a pure form (a gas at room temperature and pressure), while the others are used in aqueous solution. They are all weak acids, with $pK_a = 7.53$ (HOCl), 8.63 (HOBr) and 10.64 (HOI) (the pK_a of HOF is not known, as the pure acid reacts rapidly with water). HOF is the only known oxoacid of fluorine, which may be rationalised by realising that the most electronegative element on the periodic table would be unlikely to be stable in a significantly positive oxidation state.

Chlorous acid, $HClO_2$, is the only one of the halous acids to have been unequivocally characterised, although it cannot be isolated in its pure form. While still a weak acid, it is much stronger than HOCl, with $pK_a = 1.95$. Chloric acid, $HClO_3$ and bromic acid, $HBrO_3$, are both moderately strong acids which exist only as aqueous solutions, while iodic acid, HIO_3, is also moderately strong and can be isolated as a white solid. $HClO_4$, perchloric acid, is an extremely strong, oxidising acid which is commonly used in the digestion of organic material. In its pure form, it is dangerously explosive and it is usually supplied as either a 60% or 70% aqueous solution; even at these concentrations (especially the latter), explosions can occur if the acid comes in contact with organic material. Perbromic acid, $HBrO_4$, was unknown until 1968. It is a strong acid which cannot be prepared in pure form, although aqueous solutions are stable. HIO_4 is a solid at room temperature. Dissolution in water does not result in simple dissociation of the acid to give H_3O^+ and IO_3^-; instead, a number of iodine-containing species are formed in a series of complex equilibria.

14.3 The biogeochemical cycles of nature

Having introduced the *p*-block elements and some of their compounds, we will now concentrate on what are arguably the most important of these: oxygen, sulfur, nitrogen, phosphorus and carbon, which all lie in the second and third periods of the periodic table. We choose these elements because of their importance in the Earth's biochemical and geochemical systems.

Bonding in the *p*-block elements

As we have already seen, the elements of the *p* block adopt a wide variety of structures, and these can be related to the valence-electron configuration of each element and the bonding which results as a consequence of this. We first introduced the concept of an octet in chapter 5 (p. 178). From what we have learned about the lack of reactivity of the group 18 elements in this chapter and their valence-electron configurations (ns^2np^6), we can see that the presence of eight electrons in the valence shell appears to confer some stability to an atom. The group 18 elements are reluctant to undergo reaction with any but the most reactive chemical reagents and exist as monatomic gases. We also often find that an octet of electrons about an atom in a molecule appears to be a particularly stable energetic situation, particularly for elements in the second period, and that the structures adopted by elements can be rationalised on the basis of this. For example, if we start at Ne (group 18) and go across the periodic table to C (group 14), we find the number of valence electrons decreases by one for each group we traverse. Therefore, in order to obtain an octet, atoms must share one or more electrons. This is illustrated in table 14.3.

TABLE 14.3 Single and multiple covalent bonding in second period *p*-block elements.

Element	Valence-electron configuration	Electrons required for octet	Structure
neon	$2s^22p^6$	0	:Ne:
fluorine	$2s^22p^5$	1	:F—F:
oxygen	$2s^22p^4$	2	:O=O:
nitrogen	$2s^22p^3$	3	:N≡N:
carbon	$2s^22p^2$	4	—C—C— C=C —C≡C—

A fluorine atom can obtain the single electron required for an octet by sharing a single electron with another fluorine atom; this gives both atoms a share of an octet, and so elemental fluorine exists as F_2, with a single two-electron bond between the atoms. An oxygen atom requires two electrons for an octet and will therefore share two electrons with another oxygen atom. This results in O_2 molecules, which contain a double bond between the atoms. The triple bond in elemental nitrogen results from each atom sharing three electrons with the other, to give N_2. If this trend were to be continued, we might expect carbon to exist as C_2 molecules with a quadruple bond between the atoms. However, this is not energetically favourable, and a carbon atom instead obtains an octet by sharing its four valence electrons with four, three or two neighbouring atoms to give alkanes, alkenes and alkynes, respectively. Inspection of the final column of table 14.3 in fact shows the relationship between the structures of F_2, O_2 and N_2, and C_2H_6, C_2H_4 and C_2H_2, respectively. We say that F_2 and C_2H_6 are isoelectronic (they contain the same number of electrons), as are O_2 and C_2H_4, and N_2 and C_2H_2. The multiple bonds that hold the latter molecules together are formed from overlap of *p* orbitals in a π-type fashion (chapter 5, pp. 203–4); this is favourable for elements from the second period owing to their relatively small sizes, which allow close approach of the atoms and correspondingly efficient orbital overlap. However, this is not the case for elements from the third period as their atomic radii are larger, and π bonding between these is therefore significantly weaker. Therefore, these elements tend to adopt structures based on σ bonding only between neighbouring atoms. Thus, the structure of elemental silicon is the same as that of diamond (pp. 264–5), with each Si atom connected to four other Si atoms; phosphorus exists as σ-bonded P_4 tetrahedra with each P atom bonded to three others (p. 609), and sulfur is composed of S_8 rings with single S—S bonds (p. 611).

While there are significant differences between the structures of the second and third period elements, the compounds that these elements form with hydrogen unanimously reflect the group to which they belong. As can be seen below, the number of H atoms in each compound corresponds to the number of electrons required by the *p*-block element to obtain an octet.

Group 14	Group 15	Group 16	Group 17	Group 18
CH_4	NH_3	OH_2 (H_2O)	FH (HF)	Ne
SiH_4	PH_3	SH_2 (H_2S)	ClH (HCl)	Ar

All of these compounds contain 4 sets of electron pairs and therefore, according to VSEPR theory (p. 184), their structures are based on the tetrahedron, with 0, 1, 2 or 3 lone pairs.

The group 16 cycles

The oxygen cycle

We have seen earlier in this chapter that there are two allotropes of elemental oxygen: the O_2 molecule and the O_3 (ozone) molecule. Both are vital to life on Earth. There are also three reduced forms of oxygen, O_2^- (superoxide), O_2^{2-} (peroxide) and O^{2-}(oxide), all of which are important in living systems. These species are related as shown in the oxygen cycle below.

$$O_3 \rightleftarrows O_2 \xrightarrow{+e^-} O_2^- \quad (HO_2)$$

$$O_2 \underset{}{\overset{+4e^-,\ +4H^+}{\rightleftarrows}} 2H_2O \qquad O_2^- \xrightarrow{+e^-} O_2^{2-}$$

$$2H_2O \xleftarrow[+4H^+]{+2e^-} O_2^{2-} \quad (H_2O_2)$$

The ozone molecule can be written as two resonance forms (p. 180):

$$O{=}O{-}O \longleftrightarrow O{-}O{=}O$$

with the true structure being the average of these. This is reflected by the fact that the oxygen–oxygen bond lengths in O_3 are identical (128 pm) and lie between the lengths of the O—O single bond in H_2O_2 (148 pm) and the O=O double bond in O_2 (121 pm), thereby attesting to their partial double bond character. The presence of a lone pair on the central O atom results in the molecule adopting a bent geometry, with an O—O—O angle of 117°. Ozone is prepared in the laboratory by the passage of an electric discharge through O_2, and the pungent odour of ozone can often be detected in the vicinity of high-voltage electrical equipment. In nature, it is formed in the upper atmosphere by the action of ultraviolet radiation from the Sun on O_2. It is also formed at lower altitudes in photochemical smog, owing to the presence of nitrogen oxides (NO_x) originating from vehicle and industrial emissions. These are converted to NO_2, which releases an oxygen atom on photolysis, and this then reacts with O_2 to give O_3. The ozone molecule is highly reactive, as might be expected from its positive standard Gibbs energy of formation:

$$\tfrac{3}{2}O_2(g) \longrightarrow O_3(g) \qquad \Delta_f G^\circ = +163\,\text{kJ mol}^{-1}$$

and its large positive standard reduction potential:

$$O_3(g) + 2H^+(aq) + 2e^- \longrightarrow O_2(g) + H_2O(l) \qquad E^\circ = +2.08\,\text{V}$$

Oxygen, O_2, is involved in what are arguably the two most important chemical reactions in the biosphere: photosynthesis and respiration. In photosynthesis, green plants use the Sun's energy to convert CO_2 and H_2O to carbohydrates and O_2. The oxidation half-equation in this redox process involves the oxidation of H_2O to O_2:

$$2H_2O(l) \longrightarrow O_2(g) + 4H^+(aq) + 4e^-$$

and the electrons released are then used to reduce CO_2 to (in this case) glucose:

$$6CO_2(g) + 24H^+(aq) + 24e^- \longrightarrow C_6H_{12}O_6(s) + 6H_2O(l)$$

The overall reaction then becomes:

$$6CO_2(g) + 6H_2O(l) \rightarrow C_6H_{12}O_6(s) + 6O_2(g)$$

Given that $\Delta_f G^\circ$ for $C_6H_{12}O_6(s)$ is $-911\,\text{kJ mol}^{-1}$, we can use the data in table 8.7 (p. 335) to show that, for the overall reaction, $\Delta G^\circ = +2879\,\text{kJ mol}^{-1}$. This corresponds to $K = 2 \times 10^{-505}$ at 298 K and, in the absence of energy from sunlight, we would expect this reaction not to proceed at all.

Respiration is essentially the opposite of photosynthesis. In this, glucose reacts with O_2 to give CO_2 and H_2O, plus a large amount of energy which can be used by the organism. The overall reaction involves the four-electron reduction of O_2 to H_2O, which does not occur as a single process but in a series of steps involving stepwise reduction to O_2^- and O_2^{2-}. Of these, superoxide, O_2^-, is toxic to cells, and nature uses the enzyme superoxide dismutase to control the amount of this potentially dangerous ion within an organism.

WORKED EXAMPLE 14.4

Electronic structures of reduced oxygen species

What are the bond orders in the superoxide, O_2^-, and peroxide, O_2^{2-}, anions? Order the three species O_2^-, O_2^{2-} and O_2 in order of decreasing O—O bond length.

Analysis

We will use the MO diagram we derived for the O_2 molecule in chapter 5 (figure 5.50, p. 209), add the appropriate number of electrons, and determine the bond order using the equation:

$$\text{bond order} = \tfrac{1}{2}(\text{number of electrons in bonding molecular orbitals} - \text{number of electrons in antibonding molecular orbitals})$$

introduced on page 206.

Solution

Using figure 5.50 (p. 209), we find that the electronic configuration of the formally double-bonded O_2 molecule is $(\sigma_s)^2 (\sigma_s^*)^2 (\sigma_p)^2 (\pi_x)^2 (\pi_y)^2 (\pi_x^*)^1 (\pi_y^*)^1$. One electron reduction to give O_2^- requires the added electron to be placed in either of the degenerate antibonding π_x^* or π_y^* orbitals. This gives the electronic configuration $(\sigma_s)^2 (\sigma_s^*)^2 (\sigma_p)^2 (\pi_x)^2 (\pi_y)^2 (\pi_x^*)^2 (\pi_y^*)^1$. The bond order in O_2^- is therefore $\tfrac{1}{2}(8 - 5) = 1.5$.

Addition of a further electron to give O_2^{2-} results in the electronic configuration $(\sigma_s)^2 (\sigma_s^*)^2 (\sigma_p)^2 (\pi_x)^2 (\pi_y)^2 (\pi_x^*)^2 (\pi_y^*)^2$, in which the π^* orbitals are full. The bond order of O_2^{2-} is therefore $\tfrac{1}{2}(8 - 6) = 1$.

On the basis of bond orders, we would predict O_2^{2-} to have the longest O—O bond, followed by O_2^- and O_2. The respective O—O bond lengths of 149 pm, 128 pm and 121 pm are consistent with this molecular orbital electronic description.

The sulfur cycle

Sulfur is a nutrient that is essential for life. It is found in the amino acids cysteine (Cys) and methionine (Met) (table 24.1, p. 1064) in which the sulfur-containing functional group is a thiol and thioether, respectively. In these compounds, sulfur exhibits its lowest possible formal oxidation number of −2. Sulfur as S^{2-} is also found coordinated to metal ions in a variety of metalloproteins and metalloenzymes. However, the presence of O_2 in Earth's atmosphere and its thermodynamically favourable reactions with many sulfur-containing species mean that oxidation is generally a facile process, resulting ultimately in compounds such as SO_2, SO_3 and H_2SO_4. The thermodynamic stability of these molecules is reflected in their values of $\Delta_f G^\circ$; these are $-300.4\,\text{kJ mol}^{-1}$, $-370.4\,\text{kJ mol}^{-1}$ and $-689.9\,\text{kJ mol}^{-1}$ for $SO_2(g)$, $SO_3(g)$ and $H_2SO_4(l)$, respectively, at 25 °C (appendix A). Investigation of the Lewis structures of these molecules shows that an octet of electrons about the central sulfur atom is not necessarily the most energetically favourable arrangement, as is often found with elements from the third period. The SO_2 molecule adopts a bent geometry, with two S═O double bonds and a lone pair on the sulfur atom.

$$:\ddot{O}{=}\ddot{S}{=}\ddot{O}:$$

The S═O double bonds arise from the necessity to minimise the formal charges on the oxygen and sulfur atoms, and doing this therefore results in 10, rather than 8, electrons surrounding the central sulfur atom. A similar situation arises for SO_3, where the Lewis structure in which the formal charges are minimised has 12 electrons surrounding the S atom:

$$O{=}S({=}O){=}O$$

while the optimal Lewis structure for H_2SO_4 also places 12 electrons around the sulfur atom:

$$H{-}O{-}S({=}O)_2{-}O{-}H$$

The ability of third period elements and below to be surrounded by more than eight electrons (sometimes called expanding the octet) is attributed by some to participation of *d* orbitals on the central atom in bonding. However, it should be noted that this interpretation is controversial and other explanations have been advanced.

The biological sulfur cycle is shown below. As can be seen, those sulfur compounds important to life tend to contain the element in the thermodynamically disfavourable (in the presence of O_2) −2 oxidation state, and nature therefore requires a method by which both sulfur and sulfur–oxygen compounds can be reduced. This is achieved by a variety of micro-organisms under anaerobic (absence of oxygen) conditions. When mammals die, their reduced sulfur species are oxidised to sulfate by biological oxidants, and this is either incorporated in sulfate minerals or dissolved in aqueous solution. The same result can be achieved chemically through the oxidation of sulfur species in the atmosphere; the decomposition of plankton results in the formation of volatile dimethyl sulfide, Me_2S, which escapes into the atmosphere and is photochemically oxidised to SO_2. This then reacts further with O_2 to give SO_3, and reaction of this with water gives sulfuric acid, H_2SO_4, the most important component of acid rain. While this process occurs naturally in the atmosphere, industrial production of SO_2 over the last century has increased the amount of H_2SO_4 in the atmosphere to the point where acid rain has led to significant environmental damage, mainly in the Northern Hemisphere. The sulfur impurities in both coal and fossil fuels initially form SO_2 on combustion, and this is then converted into H_2SO_4 in the atmosphere via the above route. The reduction of sulfate to sulfite and eventually sulfide takes place in plants and bacteria, which use a number of enzymes to effect this conversion. The resulting sulfide is then incorporated into cysteine and methionine, and the cycle starts again. It should also be noted that sulfide itself occurs as a ligand in the M—S (M = a metal ion) centres of a number of metalloproteins, most importantly iron–sulfur proteins, and appears to derive from breakdown of organic sulfur species, rather than being directly incorporated as S^{2-}.

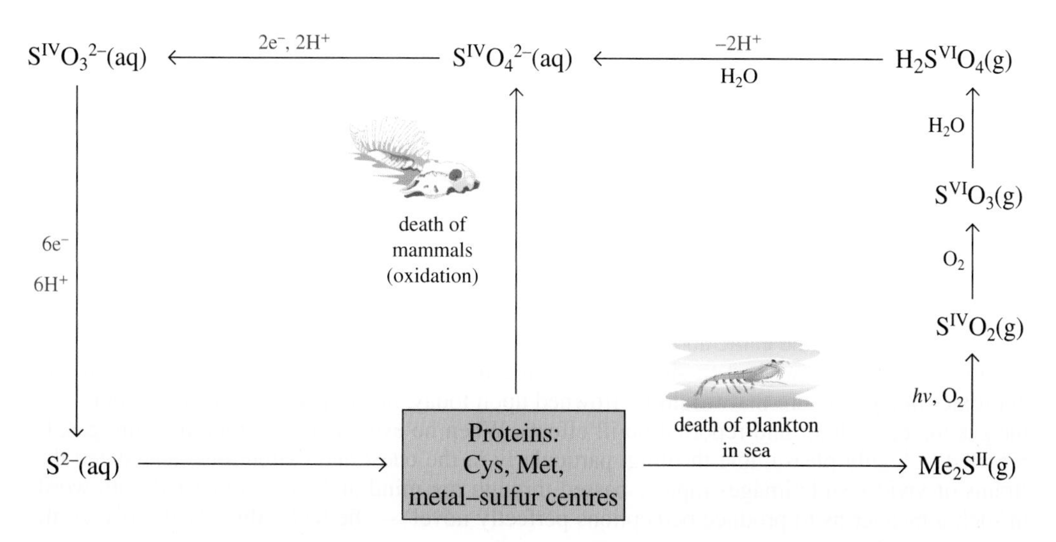

The group 15 cycles

The nitrogen cycle

Earlier in this chapter, we discussed the extraordinary stability of the N_2 molecule, a consequence of the strong N≡N triple bond, and how nature uses the nitrogenase enzyme to allow conversion of N_2 to NH_3. As with sulfur, the reduced states of the element are important in living systems, and incorporation of nitrogen into biological molecules such as amino acids and the nucleic acid bases generally occurs through NH_3.

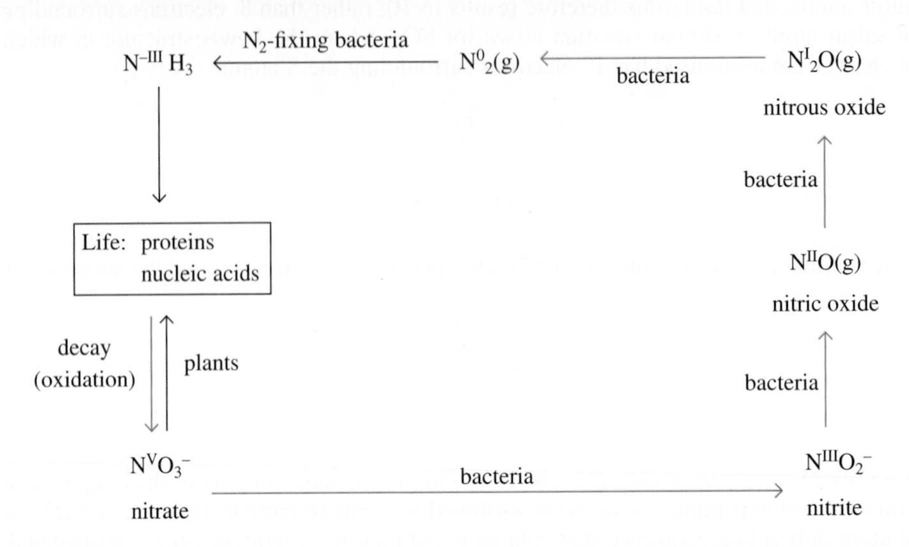

Decomposition of living matter ultimately results in oxidation of N-containing molecules to NO_3^- in which N displays its highest oxidation state of +5. Remembering that N is a second period element, it prefers to be surrounded by an octet of electrons, and this is shown in the three possible resonance structures of the NO_3^- ion in which the formal charges on the atoms are minimised.

Nitrate can be reduced by anaerobic bacteria to various nitrogen oxides, and ultimately to molecular N_2. Such bacteria essentially 'respire' using these N-containing species and are actually poisoned by O_2! The initial reduction step involves formation of nitrite ion, NO_2^-; this is often used as a preservative in bacon, ham and corned beef, although there have been some recent health concerns regarding its use. The ion adopts a bent geometry, and two resonance structures can be drawn.

Subsequent reduction gives nitric oxide, NO(g), a molecule of considerable biological importance as discussed in the introduction to this chapter. This is one of eight known neutral compounds containing only N and O (the others are N_2O, NO_2, NO_3, N_2O_2, N_2O_3, N_2O_4 and N_2O_5). The Lewis structure of NO is interesting, as, owing to the odd number of electrons in a nitrogen atom ($1s^22s^22p^3$) and the even number of electrons in an oxygen atom ($1s^22s^22p^4$), the molecule contains an odd number of electrons. The structure can be drawn such that there is a formal double bond between the nitrogen and oxygen atoms, and an unpaired electron on the N atom:

$$\dot{N}=O$$

although it appears that, in reality, the unpaired electron is involved in N—O bonding to give an N—O bond order of 2.5. Reduction of NO yields nitrous oxide, N_2O, commonly referred to as laughing gas. This name came about as a result of an experiment in 1799 by the English chemist Humphry Davy, who wished to show that the recently prepared (1793) molecule was, in fact, not toxic. In a procedure that would be frowned upon today, he simply inhaled small amounts of the gas mixed with air and reported no ill effects. When he eventually breathed the pure gas, he reported 'a highly pleasurable thrilling, particularly in the chest and extremities' and stated that 'trains of vivid visible images rapidly passed through my mind and were connected with words in such a manner as to produce perceptions perfectly novel' — he had indirectly discovered the anaesthetic properties of nitrous oxide, although it was not used as such until 1844. The molecule is linear and, as we saw on pages 181–2, three resonance structures can be drawn, two of which are the major contributors to the actual structure.

$$:\ddot{N}=N=\ddot{O}: \longleftrightarrow :N\equiv N-\ddot{O}:$$

Figure 14.30 shows the electron distributions in NO_3^-, NO_2^-, NO and N_2O.

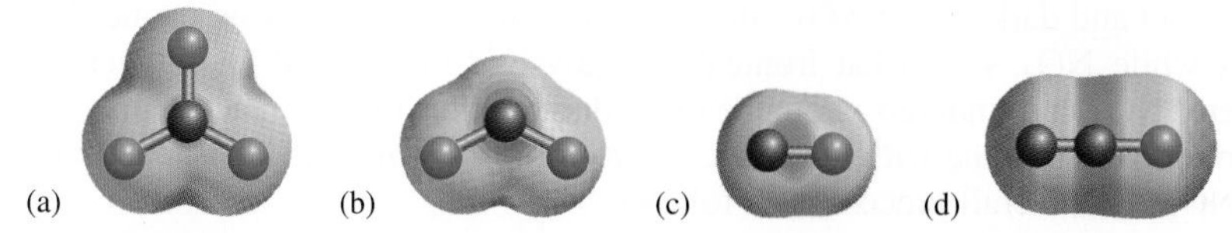

FIGURE 14.30 The electron density distributions for: **(a)** NO_3^-, **(b)** NO_2^-, **(c)** NO and **(d)** N_2O. Red indicates the highest electron density and blue indicates the lowest.

Reduction of N_2O gives $N_2(g)$, which is then used as a feedstock in N_2-fixing bacteria to form ammonia. The overall reaction for this process is:

$$N_2 + 8H^+ + 16MgATP + 8e^- \longrightarrow 2NH_3 + H_2 + 16MgADP + 16P_i$$

where P_i is inorganic phosphate, and MgATP and MgADP are the magnesium salts of adenosine triphosphate (p. 634) and adenosine diphosphate, respectively. This reaction is catalysed by a variety of nitrogenase enzymes, which are metalloenzymes containing Fe, Fe and Mo, or Fe and V, at the active sites, and which generally operate under anaerobic conditions. Although the exact mechanism of the reduction process has yet to be determined, it appears plausible that the reaction proceeds via two intermediate compounds of nitrogen and hydrogen, diazene, N_2H_2, and hydrazine, N_2H_4.

$$\text{H-N=N-H (cis)} \qquad \text{H-N=N-H (trans)}$$

cis-and *trans*-diazene

$$H_2N-NH_2$$

hydrazine

Diazene is unstable, with a lifetime of only minutes at room temperature, and can exist as *cis* and *trans* isomers owing to the relative orientations of the hydrogen atoms. Pure hydrazine, although much more stable than diazene at room temperature, can undergo violent decomposition at higher temperatures and this property has led to its use as a rocket fuel. In the chemical laboratory (where it is usually handled as an aqueous solution!), it is an excellent reducing agent, especially in basic aqueous solution.

$$NH_3 + \tfrac{1}{2}N_2 + H_2O + e^- \longrightarrow N_2H_4 + OH^- \qquad E^\ominus = -2.42\ \text{V}$$

Table 14.4 shows how the N—O and N—N bond lengths in the nitrogen compounds discussed thus far correlate with the structures we have drawn. Both NO_3^- and NO_2^- have N—O bond orders greater than 1 and less than 2, and this is reflected in their N—O bond lengths of 124 pm, significantly shorter than the N—O single bond in gaseous HNO_3 (141 pm). The neutral N_2O molecule has a slightly shorter N—O bond length (119 pm), again consistent with a bond order between 1 and 2, whereas the 115 pm N—O bond length in NO reflects a bond order of at least 2.

TABLE 14.4 N—O and N—N bond lengths in selected nitrogen compounds.

Compound	N—O length (pm)	N—N length (pm)
NO_3^-	124	–
NO_2^-	124	–
NO	115	–
N_2O	119	113
N_2	–	110
N_2H_2	–	125
N_2H_4	–	145

Similar trends are seen in the N—N bond lengths of the compounds in table 14.4. N_2H_4, with a formal N—N single bond, displays the longest N—N distance (145 pm), which shortens to 125 pm in double-bonded diazene. The resonance structures of N_2O predict an N—N bond order between 2 and 3, which is consistent with the still shorter bond length (113 pm), while triple-bonded N_2 displays the shortest N—N distance (110 pm).

Nitrogen oxides (NO_x) resulting from the combination of nitrogen and oxygen at high temperature in internal combustion engines have been implicated in atmospheric pollution. Of these species, both colourless NO and dark brown NO_2 cause particular problems involving ozone; NO destroys O_3 molecules while NO_2, somewhat ironically, catalyses formation of O_3 molecules. In the lower atmosphere, NO_2 can undergo photochemical dissociation to give reactive oxygen atoms and NO; the former can combine with O_2 molecules to form O_3 while the latter react with oxygen to regenerate NO_2. The overall process is as follows.

$$\begin{aligned} NO_2 &\xrightarrow{h\nu} NO + O \\ O + O_2 &\longrightarrow O_3 \\ NO + \tfrac{1}{2}O_2 &\longrightarrow NO_2 \\ \hline \tfrac{3}{2}O_2 &\longrightarrow O_3 \end{aligned}$$

NO_2 is a catalyst in this process, and hence a single molecule of NO_2 can potentially generate an enormous number of O_3 molecules. While O_3 in the upper atmosphere is beneficial in that it absorbs damaging ultraviolet radiation from the Sun, at lower altitudes O_3 reacts with hydrocarbons in the atmosphere to form toxic peroxy acids, inhalation of which can lead to respiratory problems.

$$RCH_3 + O_3 \longrightarrow R{-}C(=O){-}O{-}O{-}H$$

This combination of atmospheric pollutants is called photochemical smog.

In the upper atmosphere, the NO formed from photolysis of NO_2 reacts with O_3, causing its depletion. Again, this is a catalytic process, as NO is regenerated from the reaction of NO_2 with an oxygen atom.

$$\begin{aligned} NO + O_3 &\longrightarrow NO_2 + O_2 \\ O + NO_2 &\longrightarrow NO + O_2 \\ \hline O + O_3 &\longrightarrow 2O_2 \end{aligned}$$

Catalytic converters in the exhaust systems of cars have been used since the mid 1970s to minimise the amount of NO_x species emitted; these generally use catalysts containing precious metals such as platinum and rhodium to convert NO_x species to N_2 and O_2.

The phosphorus cycle

The phosphorus cycle differs from those we have investigated so far, in that there is essentially no redox chemistry involved. In contrast to both sulfur and nitrogen, where nature uses the reduced states of the elements, phosphorus in nature exists nearly exclusively in the +5 oxidation state as phosphate, or a phosphate derivative. (It should be noted that the term 'phosphate' technically refers only to the PO_4^{3-} ion, but it is often used to denote any phosphorus-containing species in which there are four phosphorus–oxygen bonds. In order to distinguish PO_4^{3-}, and any of its protonated derivatives, from other phosphates, it is often referred to as 'inorganic' phosphate, abbreviated P_i.) Being a third period element, phosphorus is able to accommodate more than eight electrons around it and, as we saw on page 181, there are four resonance structures that can be drawn for the PO_4^{3-} ion.

$$[\,\text{P with three P–O and one P=O (bottom)}\,]^{3-} \longleftrightarrow [\,\text{P=O (right)}\,]^{3-} \longleftrightarrow [\,\text{P=O (top)}\,]^{3-} \longleftrightarrow [\,\text{P=O (left)}\,]^{3-}$$

The PO_4^{3-} ion can be thought of as deriving from complete deprotonation of phosphoric acid, H_3PO_4. This is a rare example of a triprotic acid, with pK_a values of 2.15, 7.20 and 12.38 corresponding to the three deprotonations.

$$\mathrm{HO{-}P(OH)({=}O){-}OH} + H_2O \rightleftharpoons \left[\mathrm{O{-}P(OH)({=}O){-}OH}\right]^{-} + H_3O^+ \quad pK_a = 2.15$$

$$\left[\mathrm{O{-}P(OH)({=}O){-}OH}\right]^{-} + H_2O \rightleftharpoons \left[\mathrm{O{-}P(O)({=}O){-}OH}\right]^{2-} + H_3O^+ \quad pK_a = 7.20$$

$$\left[\mathrm{O{-}P(O)({=}O){-}OH}\right]^{2-} + H_2O \rightleftharpoons \left[\mathrm{O{-}P(O)({=}O){-}O}\right]^{3-} + H_3O^+ \quad pK_a = 12.38$$

The first pK_a of 2.15 shows phosphoric acid to be of significant strength, but it is far from being a strong acid, as is often incorrectly stated.

Phosphoric acid consistently ranks among the top 10 industrial chemicals produced, owing to its subsequent conversion to fertilisers. Almost all phosphoric acid is produced directly from fluoroapatite. The ore is partially purified, crushed and then slurried with aqueous sulfuric acid.

$$Ca_5(PO_4)_3F(s) + 5H_2SO_4(aq) \longrightarrow 3H_3PO_4(aq) + 5CaSO_4(s) + HF(aq)$$

The dilute phosphoric acid obtained from this process is concentrated by evaporation and is often highly coloured because of the presence of many metal ion impurities in the phosphate rock. However, this impure acid is suitable for the manufacture of phosphate fertilisers, which accounts for almost 90% of phosphoric acid production. High-purity H_3PO_4 is obtained using a more expensive redox process that starts from the pure element. Controlled combustion of white phosphorus gives phosphorus(V) oxide, P_4O_{10} (figure 14.31), and addition of water to this generates high-purity phosphoric acid.

$$P_4O_{10}(s) + 6H_2O(l) \longrightarrow 4H_3PO_4(aq)$$

This highly pure product, which constitutes about 10% of the total industrial output of phosphoric acid, is the starting material for making food additives, pharmaceuticals and detergents.

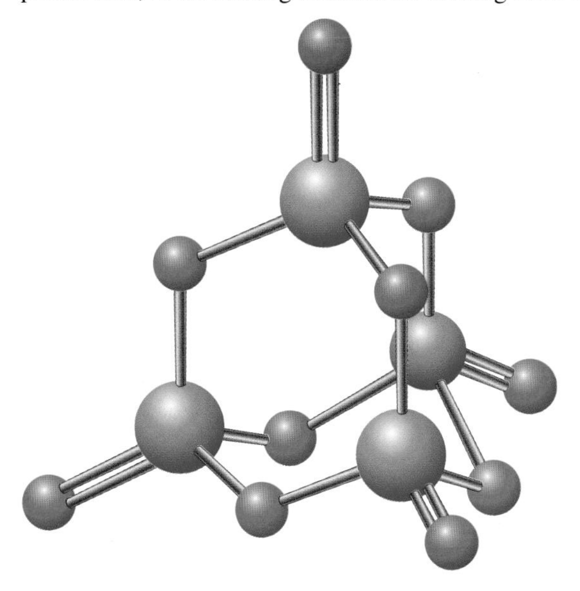

FIGURE 14.31 Phosphorus(V) oxide, P_4O_{10}, contains the same tetrahedral arrangement of phosphorus atoms as in P_4 (see figure 14.11), but an oxygen atom is inserted into each P—P bond. An additional terminal O atom is double-bonded to each P atom.

Plants typically contain about 0.2% phosphorus by weight, but the element is easily depleted from soils. For this reason, the major commercial application of phosphorus is in fertilisers. There are several common phosphorus fertilisers, but the most important one is ammonium hydrogen phosphate, $(NH_4)_2HPO_4$. This phosphate compound is particularly valuable because it is highly soluble and provides both phosphorus and nitrogen. The compound can be made by treating phosphoric acid with ammonia.

$$H_3PO_4 + 2NH_3 \longrightarrow (NH_4)_2HPO_4$$

This is the predominant industrial route to $(NH_4)_2HPO_4$, accounting for the single largest use of phosphoric acid.

An important reaction of phosphoric acid (and its derivatives) is phosphate condensation, a process that is similar to the condensation reactions of other species that contain O—H bonds.

$$(HO)_2P(O)O{-}H + H{-}OP(O)(OH)_2 \longrightarrow (HO)_2P(O){-}O{-}P(O)(OH)_2 + H_2O$$

The product, $H_4P_2O_7$, is pyrophosphoric acid, a tetraprotic acid. A second condensation leads to triphosphoric acid, which can lose five protons.

$$H_4P_2O_7 + H_3PO_4 \longrightarrow H_5P_3O_{10} + H_2O$$

The sodium salts of these acids, sodium pyrophosphate, $Na_4P_2O_7$, and sodium triphosphate, $Na_5P_3O_{10}$, have been used widely in detergents. The polyphosphate anions are good additives for cleaning agents because they form complexes with metal ions, including those that make water 'hard' (Ca^{2+}, Mg^{2+}) and those that can cause colour stains (Fe^{3+}, Mn^{2+}). Moreover, polyphosphates are nontoxic, nonflammable and noncorrosive; they do not attack dyes or fabrics, and they are readily decomposed during wastewater treatment. These advantages appear to make polyphosphates ideal for use in cleaning agents. Unfortunately, adding phosphates to water leads to an imbalance in aquatic biosystems, particularly in lakes. Inorganic phosphate is rapidly incorporated as a nutrient by algae, leading to excess algal growth, which can overwhelm a lake with decaying organic matter. Organic decay consumes oxygen, depleting the lake of this life-sustaining substance and leading to the death of fish and other aquatic life forms. This contributes more organic decay with a corresponding increase in oxygen depletion, until eventually the lake supports no animal life.

Phosphate condensation reactions play an essential role in metabolism. The conversion of adenosine diphosphate (ADP) to adenosine triphosphate (ATP) requires an input of Gibbs energy.

$$ADP + H_3PO_4 \longrightarrow ATP + H_2O \qquad \Delta G^{\ominus} = +30.6\,kJ\,mol^{-1}$$

ATP serves as a major biochemical energy source, releasing energy in the reverse, hydrolysis, reaction. The ease of interchanging O—H and O—P bonds probably accounts for the fact that nature chose a phosphate condensation–hydrolysis reaction for energy storage and transport.

phosphoric acid + ADP (diphosphate, adenosine, adenine) ⟶ ATP (triphosphate) $+ H_2O$

phosphoric acid ADP ATP

In ADP and ATP, one end of the polyphosphate chain links to adenosine through an O—C bond. These bonds form in condensation reactions between hydrogen phosphate and alcohols. Other biochemical substances contain P—O—C linkages, and these are generally referred to as phosphate esters, by analogy with 'organic' esters (chapter 23). However, while it is possible to replace only a single proton in a carboxylic acid with an alkyl group, replacement of one, two or three protons in phosphoric acid by an alkyl group gives phosphate mono-, di- and triesters, respectively.

$(HO)_2P(O)OR$ phosphate monoester $HO{-}P(O)(OR)_2$ phosphate diester $P(O)(OR)_3$ phosphate triester

Both ADP and ATP are phosphate monoesters, while the most important biological phosphate diesters are DNA and RNA (see figure 25.5).

The major repository for phosphorus on Earth is in phosphate minerals, the most important of which are apatites. These have the general formula $Ca_5(PO_4)_3X$, where X = F (fluoroapatite), Cl (chloroapatite) and OH (hydroxyapatite), all of which are sparingly soluble under neutral conditions. Tooth enamel is essentially pure hydroxyapatite, and this is also an important component of bones. Use of fluoridated toothpaste is thought to result in conversion of hydroxyapatite to the less soluble fluoroapatite, thereby slowing the rate of tooth decay. Important phosphate deposits are found across the world, perhaps most interestingly on the tiny (21 km^2) Pacific island of Nauru. At one time in the twentieth century, Nauru had one of the highest per capita incomes in the world solely because of its phosphate exports. Today, however, extensive strip mining has exhausted the phosphate deposits and, with these providing the only source of foreign income, the island's future is uncertain.

The phosphorus cycle (figure 14.32) essentially involves the interconversion between inorganic and organic phosphate (here, organic phosphate refers to biological molecules such as DNA, RNA, ADP and ATP). Inorganic phosphate, from the weathering of rocks, is assimilated by plants and converted to organic phosphate. The plants are consumed by animals and, when both die, bacterial action converts the organic phosphate back to inorganic phosphate.

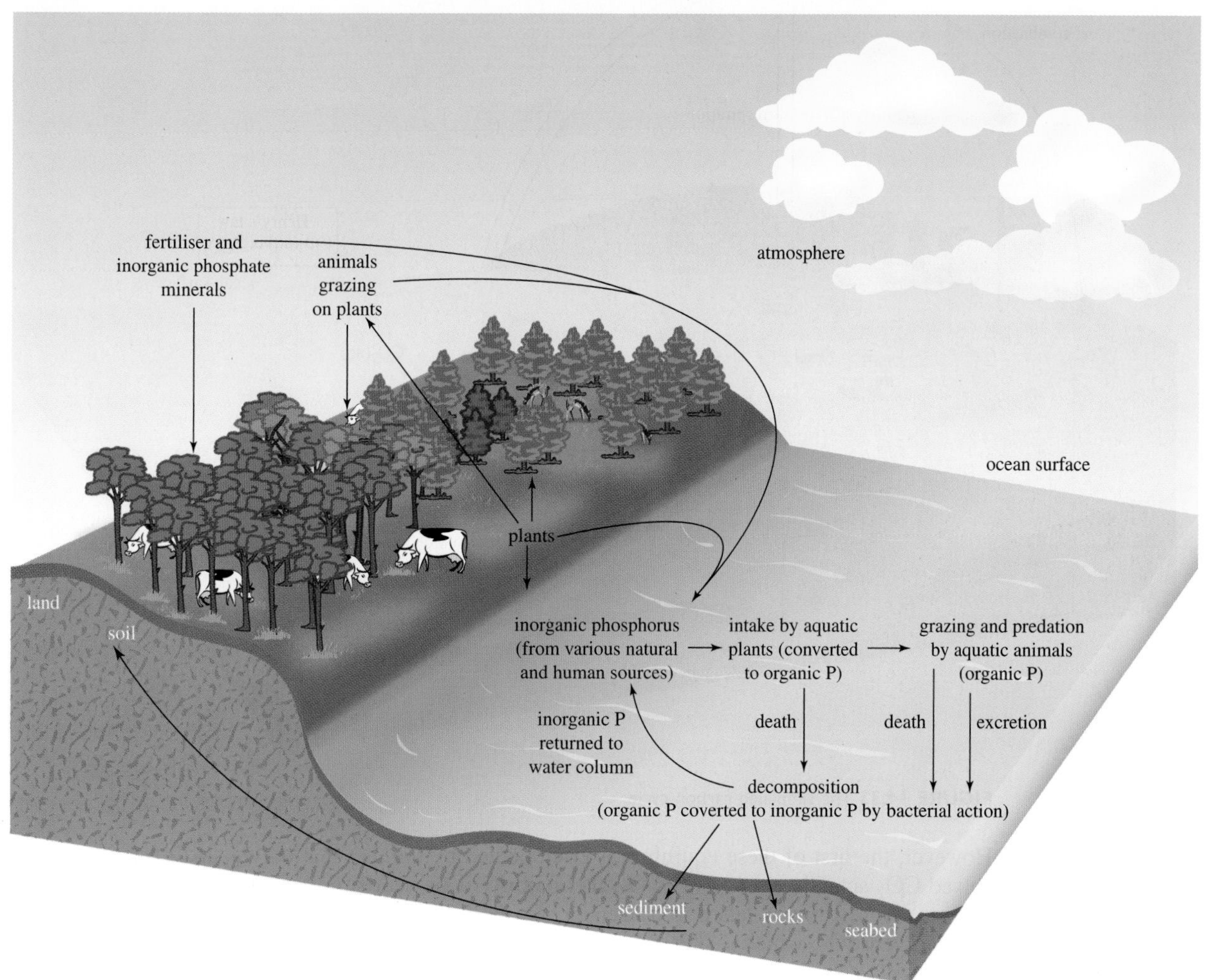

FIGURE 14.32 The phosphorus cycle.

The carbon cycle

Carbon accounts for only 0.08% of the mass of Earth, yet all life on this planet is based on carbon. While we will discuss the organic chemistry of carbon in detail over the second half of this book, we have as yet only touched upon the inorganic chemistry of carbon in detailing how CO_2 can act as a greenhouse gas. Carbon appears in nature in oxidation states ranging from –4 to +4, and its cycling between compounds exhibiting these oxidation states is extremely complex. We will therefore restrict ourselves to the way in which CO_2 travels between the atmosphere, aquatic systems and the Earth's crust. This simplified cycle (figure 14.33, overleaf) does not therefore involve any redox chemistry.

Carbon dioxide is the ultimate product of the combustion of any carbon-containing compound. It is a linear molecule, with formal double bonds between the C and O atoms.

$$\ddot{O}=C=\ddot{O}$$

CO_2 currently comprises nearly 400 parts per million of the Earth's atmosphere and this amount is steadily increasing owing to burning of fossil fuels. It is slightly soluble in water, with a solubility of 0.145 g L^{-1} at 25 °C. The resulting solution is measurably acidic and this is ascribed to the formation of carbonic acid, H_2CO_3.

$$H_2O(l) + CO_2(g) \longrightarrow H_2CO_3(aq)$$

Given the importance of this molecule, it is surprising that conclusive evidence for its existence was obtained only in 1987, on the basis of mass spectrometric measurements.

H_2CO_3 is a diprotic acid which can undergo two dissociations:

$$H_2CO_3 + H_2O \rightleftharpoons HCO_3^- + H_3O^+ \qquad pK_a = 6.35$$

$$HCO_3^- + H_2O \rightleftharpoons CO_3^{2-} + H_3O^+ \qquad pK_a = 10.33$$

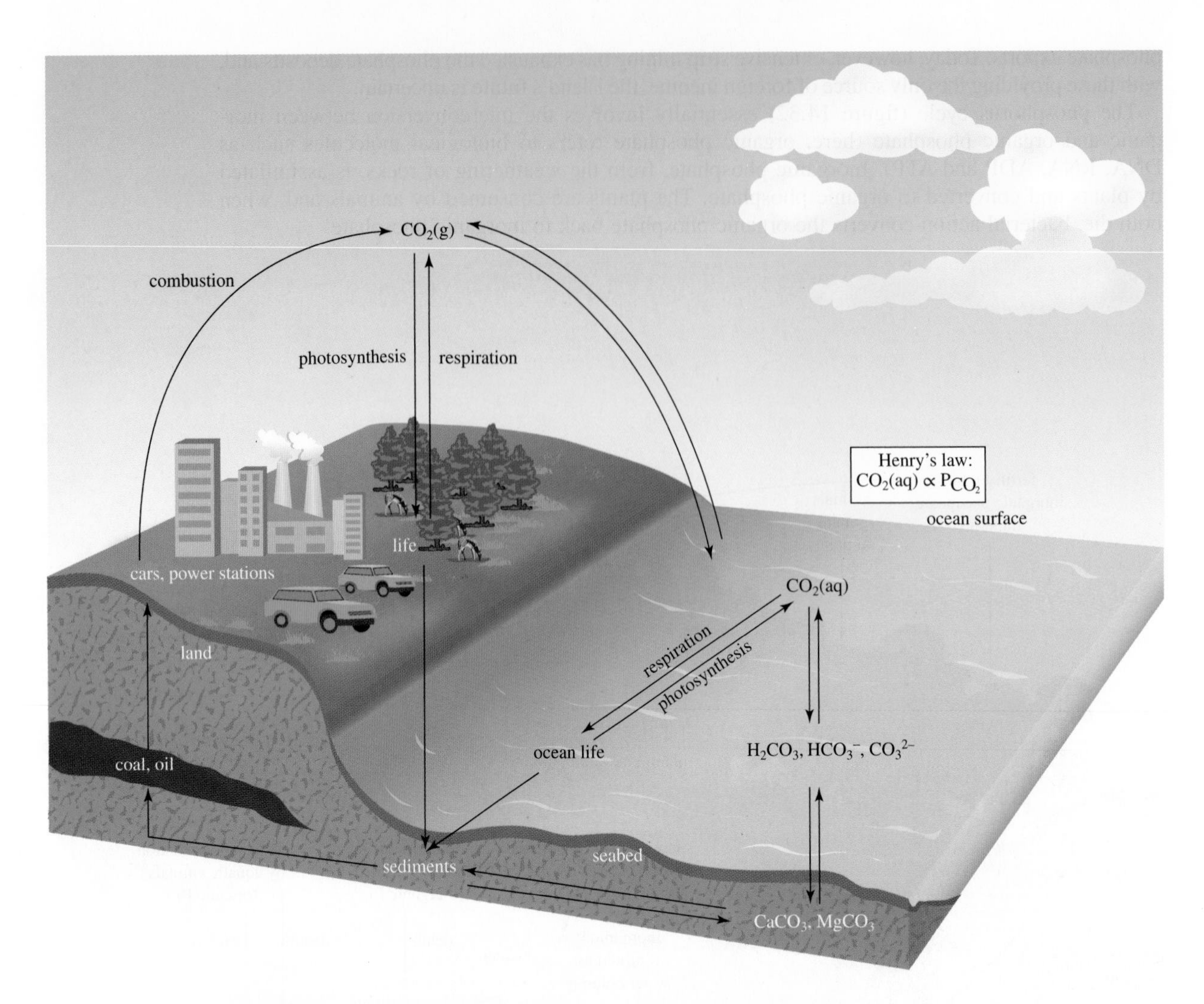

FIGURE 14.33 The simplified carbon cycle.

However, the first of these equilibria is complicated by the fact that only a small fraction of dissolved CO_2 actually exists as H_2CO_3; the majority is present simply as hydrated carbon dioxide, $CO_2(aq)$. When this is taken into account, the first pK_a drops to around 3.7. It is the presence of dissolved CO_2, which makes natural fresh water slightly acidic, and this can easily be shown (on a small scale) by measuring the pH of tap water, bottled water or water in the laboratory — provided it is not freshly distilled, it will have a pH close to 5.

Inspection of the two equilibria above shows that they provide a mechanism for the formation of both hydrogencarbonate (HCO_3^-, also called bicarbonate) and carbonate (CO_3^{2-}) ions which can then be incorporated into minerals in the Earth's crust by precipitation as metal salts. The two most important carbonate salts in this respect are $MgCO_3$ and $CaCO_3$, both of which are only slightly soluble (K_{sp} values of 2×10^{-5} and 3.3×10^{-9}, respectively) and precipitate readily.

Nature has provided a method by which carbon dioxide can be used as a chemical feedstock by plants. The process of photosynthesis takes carbon dioxide and water as reactants and turns them into carbohydrates and oxygen, as we outlined in the oxygen cycle earlier. Therefore, carbon in the atmosphere is found primarily as CO_2; in the biosphere it is found mainly as carbohydrates, and in the lithosphere it is found as carbonate minerals and sediments. Until recently, the distribution of carbon between these terrestrial spheres was stable. However, with the advent of the industrial revolution and burning of fossil fuels, the atmospheric concentration of CO_2 has steadily increased over the last century. Increasing CO_2 in the atmosphere may contribute to global warming that could have catastrophic consequences. In a warmer climate, for example, much of the polar ice caps would melt, and the oceans would expand as their temperatures increased. Sea levels could rise by several metres and inundate many of the world's most populous regions, such as Bangladesh, the Netherlands, and the East and Gulf Coasts of the United States. In addition, even small temperature fluctuations may alter global weather patterns and perhaps lead to serious droughts.

Our planet is such a complicated, dynamic set of interconnected systems that scientists do not know with certainty the result of increasing the concentration of atmospheric CO_2. They are studying the carbon cycle in hopes of learning what lies ahead for the planet. There is some urgency to such studies because recent global weather patterns show some of the characteristics predicted from computer models of global warming: average temperatures higher than those in previous years and an increased incidence of extreme weather such as cyclones and heavy rainfall.

SUMMARY

LO 1 The *p*-block elements

Groups 13 to 18 of the periodic table contain elements which are predominantly nonmetals or metalloids. The elements of groups 13 to 18 are collectively called *p*-block elements and are characterised by the gradual filling of the *p* orbitals across each period. The valence-shell electron configuration of these elements ranges from ns^2np^1 (group 13) to ns^2np^6 (group 18).

Group 13: boron (B), aluminium (Al), gallium (Ga), indium (In), thallium (Tl)
Group 14: carbon (C), silicon (Si), germanium (Ge), tin (Sn), lead (Pb)
Group 15: nitrogen (N), phosphorus (P), arsenic (As), antimony (Sb), bismuth (Bi)
Group 16: oxygen (O), sulfur (S), selenium (Se), tellurium (Te), polonium (Po)
Group 17: fluorine (F), chlorine (Cl), bromine (Br), iodine (I), astatine (At)
Group 18: helium (He), neon (Ne), argon (Ar), krypton (Kr), xenon (Xe), radon (Rn)

LO 2 Reactivity of the *p*-block elements

Group 13 elements: The hydrides decrease in stability going down the group. Boron forms a large number of hydrides, all of which are electron-deficient. Oxides of the elements are known with the elements in both the +3 and +1 oxidation state, with the latter observed going down the group. The +3 oxides go from being Lewis acidic to basic going down the group. $B(OH)_3$ is a Lewis, rather than Brønsted, acid.

Group 14 elements: Carbon forms a vast number of compounds containing only C and H. Silicon forms some hydrides of similar structure to short-chain alkanes. The hydrides decrease in stability going down the group, as do the oxides having the group 14 element in the +4 oxidation state. This is illustrated by the fact that PbO_2 is a strong oxidant. Lead also forms the mixed-valence oxide Pb_3O_4, which formally contains lead in both +2 and +4 oxidation states. The group 14 oxoacids are not well characterised.

Group 15 elements: The hydrides having general formula XH_3 decrease in stability going down the group. Numerous well-characterised oxides of the elements exist, with those having the group 15 element in the +3 oxidation state becoming more important going down the group. The oxoacids of nitrogen and phosphorus are important industrial compounds, and exhibit a wide range of strengths. H_3AsO_3 exhibits a single, anomalous pK_a of 9.29, suggesting a structure similar to that of $B(OH)_3$.

Group 16 elements: All of the hydrides having the general formula H_2X are weakly acidic, with pK_{a1} decreasing down the group. Oxides having the formulae XO_2 and XO_3 are known for all elements (with the exception of oxygen), with the former becoming more stable going down the group. Numerous sulfur oxoacids are known, with H_2SO_4 being one of the most produced industrial chemicals. The oxoacids become more poorly characterised going down the group.

Group 17 elements: The gaseous hydrides, HX, all form important monoprotic acids on dissolution in water, with the acid strength increasing down the group. The oxides all feature the group 17 elements in formally positive oxidation states, and are therefore generally highly reactive. The oxoacids are mostly well-characterised, with compounds of general formula HOX, HXO_2, HXO_3 and HXO_4 known.

LO 3 The biogeochemical cycles of nature

The oxygen cycle involves O_2, O_3, O_2^-, O_2^{2-} and H_2O, which interconvert through redox processes. Photosynthesis involves the formation of O_2 from H_2O, while respiration is essentially the reverse reaction, the formation of H_2O from O_2 via reaction with glucose. Ozone is formed in the upper atmosphere through the action of ultraviolet light on oxygen, while in the lower atmosphere it is produced from photochemical smog.

The sulfur cycle involves S^{2-}, Me_2S, SO_2, SO_3, H_2SO_4, SO_4^{2-}, SO_3^{2-} and sulfur-containing amino acids and M–S centres. Formation of compounds containing sulfur in a high oxidation state (+4 or +6) is thermodynamically favourable. To be useful to living organisms, these are reduced by micro-organisms under anaerobic conditions. Acid rain results from the burning of fossil fuels containing sulfur impurities.

The nitrogen cycle involves N_2, NH_3, NO_3^-, NO_2^-, NO, N_2O and amino and nucleic acids. The N_2 molecule, while usually chemically nonreactive, is converted to ammonia by the nitrogenase enzyme, and this then acts as the source of nitrogen in amino acids and nucleic acids. NO_3^- is formed as the end result of decomposition of living matter, and this is reduced to the other nitrogen oxides by anaerobic bacteria. Various nitrogen oxides are involved in atmospheric pollution.

The phosphorus cycle involves the interconversion of organic and inorganic phosphate, and differs from the other cycles of importance in that no redox chemistry is involved. Inorganic phosphate includes various metal salts of phosphate in varying states of protonation, along with phosphate minerals such as apatites. Molecules such as ATP, ADP, DNA and RNA constitute organic phosphate; these molecules are important in energy storage and the transfer of genetic information.

CO_2 travels between the Earth's atmosphere, oceans and crust. Dissolution of CO_2 in water gives small amounts of carbonic acid, H_2CO_3, with the majority of CO_2 existing simply as hydrated CO_2. Deprotonation reactions give both HCO_3^- (hydrogen carbonate) and CO_3^{2-} (carbonate) which are incorporated into the Earth's crust by precipitation as slightly soluble metal salts such as $MgCO_3$ and $CaCO_3$.

KEY CONCEPTS AND EQUATIONS

The biogeochemical cycles of nature ***(section 14.3)***

The *p*-block elements oxygen, sulfur, nitrogen, phosphorus and carbon are important in nature and, with the exception of phosphorus, form compounds containing the elements in a variety of oxidation states. These compounds are involved in chemical cycles in nature.

KEY TERMS

p-block elements *p. 602*

inert pair effect *p. 620*

mixed valence *p. 622*

REVIEW QUESTIONS

LO3 The biogeochemical cycles of nature

14.1 Figure 14.31 highlights the triangular arrangement of phosphorus atoms in P_4O_{10}. As every phosphorus atom bonds to four oxygen atoms, however, the geometry around each phosphorus atom is tetrahedral, just as in the phosphate anion. Redraw figure 14.31 in a way that highlights one of the tetrahedral PO_4 units.

14.2 Shown below is a ball-and-stick model of the active ingredient (glyphosate) in Roundup®, a commercial herbicide. Draw the Lewis structure of Roundup and describe its functional groups.

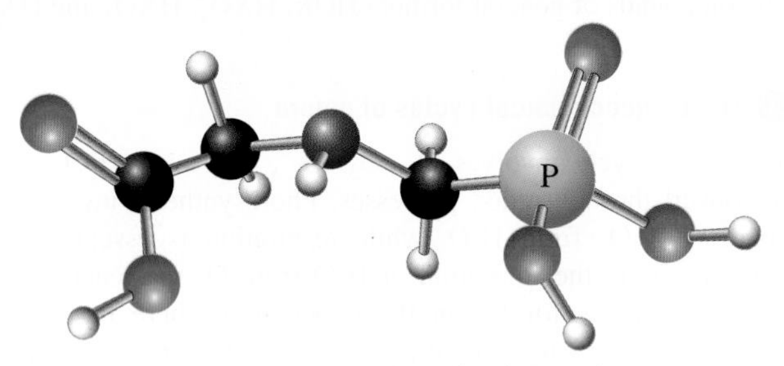

glyphosate, the active ingredient in Roundup

14.3 Determine the changes in oxidation states for the sulfur atoms in the following reactions that convert H_2S into S_8.

$$2H_2S + 3O_2 \longrightarrow 2SO_2 + 2H_2O$$

$$16H_2S + 8SO_2 \xrightarrow{Fe_2O_3} 3S_8 + 16H_2O$$

14.4 Determine the Lewis structures and describe the geometry of: (a) SF_4, (S is the central atom) and (b) BrF_5 (Br is the central atom).

14.5 Which of the following statements about species involved in the biogeochemical sulfur cycle and the chemical transformations in this cycle is *incorrect*?

A. The transformation of $(CH_3)_2S$ to SO_2 is an oxidation reaction.
B. SO_2 is best described as being V-shaped.
C. H_2SO_4 is a stronger acid than H_2SO_3.
D. The sulfite ion is trigonal planar in shape.
E. The reaction of SO_3 with water to form H_2SO_4 is a nonredox process.

14.6 Which one of the following statements about species present in the biogeochemical carbon cycle and the chemical transformations in this cycle is *incorrect*?

A. The transformation of coal into CO_2 is an oxidation reaction.
B. The shape of the CO_3^{2-} ion is best described as trigonal planar.
C. HCO_3^- is a weaker acid than H_2CO_3.
D. CO_2 reacts completely with water to form H_2CO_3.
E. The reaction of CO_2 with water is a nonredox process.

14.7 Which one of the following statements about species involved in the biogeochemical oxygen cycle is *correct*?

A. Photosynthesis leads to production of carbon dioxide.
B. Anaerobic organisms use dioxygen to eliminate electrons produced by oxidation of food molecules.
C. O_2 is evolved as a by-product of respiration.
D. H_2O_2 is the neutral form of the peroxide anion.
E. Two electrons are required to reduce dioxygen to the superoxide anion.

REVIEW PROBLEMS

14.8 Write Lewis structures for the following three species found in the solution that is electrolysed to form aluminium metal: $[AlF_4]^-$, $[AlF_6]^{3-}$ and $[AlOF_3]^{2-}$. Al is the central atom in each case. **LO1**

14.9 Although Cl_2 can be manufactured by electrolysis of aqueous NaCl, the analogous reaction cannot be used to make F_2. Explain why anhydrous HF must be used for production of F_2. **LO1**

14.10 Pyrophosphate, $P_2O_7^{4-}$, and triphosphate, $P_3O_{10}^{5-}$, are the two smallest polyphosphate anions. What is the chemical formula of the next largest polyphosphate anion? Draw the structure of this anion. **LO1**

14.11 Three phosphate anions can condense to form a ring with the chemical formula $P_3O_9^{3-}$. Determine the Lewis structure and draw the structure of this anion. **LO1**

14.12 Write the reaction that converts, fluoroapatite to phosphoric acid. Identify any Brønsted–Lowry acid–base or redox processes. **LO1**

14.13 Give the reaction that converts calcium phosphate to phosphoric acid. Identify any Brønsted–Lowry acid–base or redox processes. **LO1**

14.14 Write structural formulae showing the conversion of white phosphorus into red phosphorus. Use curved arrows to show how electrons move during this conversion. **LO1**

14.15 Write structural formulae showing the reaction of ATP with water to form ADP. What is the other product? **LO1**

14.16 Identify the oxidising agent, reducing agent and changes of oxidation state that occur in the reaction forming $SiCl_4$ from SiO_2. **LO1**

14.17 Each of the following molecules or ions plays a crucial role in the biogeochemical cycle for sulfur. For each of these species draw a Lewis structure, and predict the electron pair geometry, molecular shape and bond angles. Show formal charges where required. **LO3**

(a) S_8
(b) SO_3^{2-}
(c) SO_4^{2-}
(d) Me_2S

Which of these species has a non-zero dipole moment?

14.18 Denitrifying bacteria respire on the different oxides of nitrogen and carry out reductive chemical changes such as: **LO3**

$$NO_3^- \longrightarrow NO_2^- \quad NO_2^- \longrightarrow NO \quad NO \longrightarrow N_2O \quad N_2O \longrightarrow N_2$$

Write balanced half-equations for each of these steps in acidic solution.

14.19 Nitrogen-fixing bacteria carry out the oxidative change: **LO3**

$$NH_4^+ \longrightarrow NO_3^-$$

Write a balanced half-equation for this reaction in acidic solution.

14.20 Carbonate buffers are important in regulating the pH of blood at 7.40. What is the concentration ratio of $CO_2(aq)$ (usually written H_2CO_3) to $HCO_3^-(aq)$ in blood at pH = 7.40? **LO3**

$$CO_2(aq) + 2H_2O(l) \rightleftharpoons H_2CO_3(aq) + H_2O(l)$$

$$H_2CO_3(aq) + H_2O(l) \rightleftharpoons HCO_3^-(aq) + H_3O^+(aq) \quad pK_a = 6.35$$

14.21 Phosphate buffers are important in regulating the pH of intracellular fluids at pH values generally between 7.1 and 7.2. What is the concentration ratio of $H_2PO_4^-$ to HPO_4^{2-} in intracellular fluid at pH = 7.15? **LO3**

$$H_2PO_4^-(aq) + H_2O(l) \rightleftharpoons HPO_4^{2-}(aq) + H_3O^+(aq) \quad pK_a = 7.20$$

14.22 Why is a buffer composed of H_3PO_4 and $H_2PO_4^-$ ineffective in buffering the pH of intracellular fluid? **LO1**

$$H_3PO_4(aq) + H_2O(l) \rightleftharpoons H_2PO_4^-(aq) + H_3O^+(aq) \quad pK_a = 2.15$$

14.23 As stated in section 14.3, nitrogen forms eight neutral compounds with oxygen, namely NO, N_2O, NO_2, NO_3, N_2O_2, N_2O_3, N_2O_4 and N_2O_5. Determine the oxidation state of nitrogen in each of these compounds. **LO2**

14.24 White phosphorus consists of P_4 tetrahedra. Draw the Lewis structure of P_4, showing the location of all lone pairs. **LO2**

14.25 The atoms in the Br_2O_3 molecule are connected in the following manner. (Note that all bonds in the molecule are not necessarily single.) **LO2**

```
O
 \
  Br—O—Br
 /
O
```

Draw the Lewis structure of this molecule, showing the location of all lone pairs.

14.26 The atoms in the Br_2O_5 molecule are connected in the following manner. (Note that all the bonds in the molecule are not necessarily single.) **LO2**

```
O             O
 \           /
  Br—O—Br
 /           \
O             O
```

Draw the Lewis structure of this molecule, showing the location of all lone pairs, and give the oxidation states of the bromine atoms.

14.27 Show, by writing half-equations for the oxidation and reduction processes, that the reaction that occurs when NO_2 is dissolved in water: **LO2**

$$2NO_2 + H_2O \rightarrow HNO_2 + HNO_3$$

is a disproportionation.

14.28 Draw the biological nitrogen cycle, giving the formula of each of the molecules and ions present in the cycle. **LO3**

14.29 Write equations to demonstrate that the presence of nitrogen dioxide in the lower atmosphere leads to the production of ozone in the lower atmosphere. **LO3**

14.30 Write equations to demonstrate that diffusion of nitrogen dioxide to the upper atmosphere leads to the loss of ozone in the upper atmosphere. **LO3**

ADDITIONAL EXERCISES

14.31 Water in thermal hot springs is often unpalatable due to dissolved H_2S. Treatment with Cl_2 oxidises the sulfur to S_8, which precipitates (the other product is HCl). Balance this reaction and calculate the mass of Cl_2 required to purify 6.0×10^3 L of water containing 25 parts per million (by mass) of dissolved H_2S. **LO3**

14.32 Calcium dihydrogen phosphate is a common phosphorus fertiliser that is made by treating fluoroapatite with phosphoric acid. Hydrogen fluoride is a by-product of the synthesis. Write a balanced equation for the production of this fertiliser and calculate the mass percentage of phosphorus in the fertiliser. **LO3**

14.33 The black tarnish that forms on pure silver metal is the sulfide Ag_2S, which is formed by reaction with H_2S in the atmosphere. **LO1**

$$4Ag + 2H_2S + O_2 \longrightarrow 2Ag_2S + 2H_2O$$

The sulfide forms in preference to Ag_2O, even though the atmosphere is 20% O_2 with just a trace of H_2S. Use Lewis acid–base arguments to explain this behaviour.

14.34 Thionyl chloride, $SOCl_2$, is used to remove water of hydration from metal halide hydrates. Besides the anhydrous metal halide, the products are SO_2 and HCl. **LO1**

(a) Draw the Lewis structure of $SOCl_2$. S is the central atom.
(b) Write the balanced equation for the dehydration reaction of iron (III) chloride hexahydrate with $SOCl_2$.

14.35 Ammonium dihydrogen phosphate and ammonium hydrogen phosphate are common fertilisers that provide both nitrogen and phosphorus to growing plants. In contrast, ammonium phosphate is rarely used as a fertiliser because this compound has a high vapour pressure of toxic ammonia gas. Write balanced equations showing how each of these solid phosphates could generate ammonia gas. Why does ammonium phosphate have the highest vapour pressure of NH_3 of these three compounds? **LO3**

14.36 Boron trichloride is a gas, boron tribromide is a liquid, and boron triiodide is a solid. Explain this trend in terms of intermolecular forces and polarisability. **LO1**

14.37 Aluminium refining requires large amounts of electricity. Calculate the masses of Al and Na that are produced per mole of charge by electrolytic refining of Al_2O_3 and NaCl. **LO1**

14.38 Trisodium phosphate forms strongly basic solutions that are used as cleaning agents. Write balanced equations that show why Na_3PO_4 solutions are strongly basic. Include pK_a values to support your equations. **LO1**

14.39 The first commercially successful method for the production of aluminium metal was developed in 1854 by Henri Deville. The process relied on earlier work by the Danish scientist Hans Oersted, who discovered that aluminium chloride is produced when chlorine gas is passed over hot aluminium oxide. Deville found that aluminium chloride reacts with sodium metal to give aluminium metal. Write balanced equations for these two reactions. **LO1**

14.40 Metal oxides and sulfide ores are usually contaminated with silica, SiO_2. This impurity must be removed when the ore is reduced to the pure element. Silica can be removed by adding calcium oxide to the reactor. Silica reacts with CaO to give $CaSiO_3$. Write a balanced equation for this reaction, and describe the reaction in terms of Lewis acids and bases. **LO1**

14.41 Phosphorus(V) oxide has a very strong affinity for water; hence, it is often used as a drying agent in laboratory desiccators. One mole of P_4O_{10} reacts with 6 moles of water. Based on this stoichiometry, identify the product of the reaction and balance the equation. **LO1**

14.42 An old method by which the concentration of H_2O_2 was expressed was in terms of 'volume'. For example, '100 volume H_2O_2', a solution of H_2O_2 in water, meant that 100 volumes of $O_2(g)$ were released from 1 volume of $H_2O_2(l)$ at 273 K and 1.013×10^5 Pa, according to the following equation. **LO2**

$$2H_2O_2(l) \rightarrow 2H_2O(l) + O_2(g)$$

Calculate the concentration of '100 volume H_2O_2' in mol L^{-1}.

14.43 In section 14.1, we found that, while the action of $H_2SO_4(l)$ on NaF or NaCl can be used to prepare HF or HCl, the analogous reaction with NaBr or NaI cannot be used to generate pure HBr or HI, as Br^- and I^- are oxidised by H_2SO_4. Use the following reduction potentials:

$$Br_2(aq) + 2e^- \rightarrow 2Br^-(aq) \qquad E^\ominus = +1.07\text{ V}$$
$$I_2(aq) + 2e^- \rightarrow 2I^-(aq) \qquad E^\ominus = +0.54\text{ V}$$
$$H_2SO_4(aq) + 2H^+(aq) + 2e^- \rightarrow SO_2(aq) + 2H_2O(l) \qquad E^\ominus = +0.16\text{ V}$$

to show that $E^\ominus$ for the oxidation of both Br^- and I^- by $H_2SO_4(aq)$ is, in fact, negative. Given this result, why do you think $H_2SO_4(l)$ can actually oxidise Br^- and I^-? **LO1**

15 Reaction kinetics

Chemical reactions are essentially the movement of electrons to make and break bonds. Some reactions are very slow and others are very fast. An example of a slow reaction is the tarnishing of silver. The inflation of an airbag in a car during an accident, on the other hand, is a very fast reaction. Understanding why chemical reactions are occurring with different rates and determining the step-by-step process of these reactions at a molecular level (the so-called reaction mechanism) is a very important field in chemistry. In fact, the Nobel Prize in chemistry was awarded in 1999 to Ahmed H Zewail for developing ultra-fast laser techniques, which enable chemists to observe the motion of atoms in a molecule during a chemical reaction and to determine the mechanism of reactions.

The area of chemistry that deals with reaction rates and mechanisms is called *chemical kinetics*. Reaction rates can be influenced by many variables, for example, temperature, pressure and concentration of reactants.

We have all heard of spontaneous combustions that have occurred in mines. In particular, open-cut coal mines are surrounded by spoil piles that often contain waste coal and other carbonaceous materials with small particle sizes that react with oxygen in the atmosphere to produce heat.

If the rate at which heat is generated is greater than the rate at which heat can be dissipated, the temperature of the spoil pile rises. If this heating is not controlled, spontaneous and very violent combustions can occur, which cause significant safety, environmental and economic problems.

LEARNING OBJECTIVES

After studying this chapter, you should be able to:

15.1 explain what is meant by 'rate of reaction'
15.2 describe the five factors that affect reaction rates
15.3 explain what a 'rate law' represents
15.4 use the different rate laws to calculate rate constants, reaction times, concentrations and reaction orders from experimental data
15.5 explain the general concept of collision theory
15.6 outline the concept of a reaction mechanism
15.7 describe the role of catalysts in a reaction.

15.1 Reaction rates

In chapter 8 we learned about the definition of a spontaneous reaction. While the value and sign of ΔG indicate the spontaneity of a chemical reaction, they do not tell us anything about how fast the reaction might be. For example, the thermodynamic data for the conversion of diamond to graphite (figure 15.1) tell us that this is a spontaneous process under standard conditions ($\Delta G^{\ominus} = -2.9\,\text{kJ mol}^{-1}$), but everybody — not just those lucky enough to own precious diamond jewellery — knows that this reaction is very, very slow! The formation of ammonia from gaseous hydrogen and nitrogen through the process:

$$3H_2(g) + N_2(g) \rightarrow 2NH_3(g)$$

also has a thermodynamic tendency to occur, but no products are observed to form at room temperature and atmospheric pressure. There are many other examples of spontaneous reactions that occur so slowly that we cannot observe any change over long periods of time (in some cases, years).

FIGURE 15.1 (a) Diamond and **(b)** graphite are two different allotropes of carbon. Although $\Delta G^{\ominus}$ for the conversion of diamond into graphite is negative, a diamond will not convert into graphite at a measurable rate under normal conditions.

Chemical kinetics can help us understand why reactions proceed in a particular way. It is the study of the rates by which the concentrations of reactants and products change in a chemical reaction. Figure 12.7 on page 506 shows a copper wire being dipped into a solution of silver nitrate. With increasing reaction time, the blue colour of the aqueous solution intensifies, indicating formation of Cu^{2+} ions, whereas elemental silver continuously deposits on the wire. The rates of chemical reactions are studied by monitoring the concentrations of reactants and/or products over time. There are many different ways to observe concentration changes as the reaction proceeds. Figure 15.2 shows a *concentration–time* profile for the hypothetical reaction $A \rightarrow 2B$.

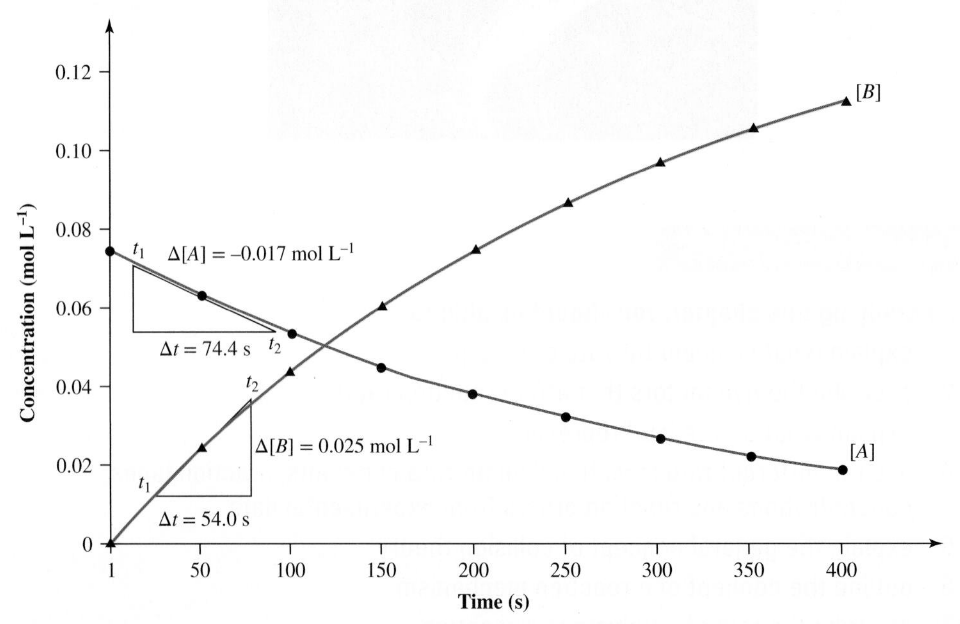

FIGURE 15.2 The progress (concentration–time profile) of the reaction $A \rightarrow 2B$, showing the concentrations of A and B over time.

Figure 15.2 shows that the concentration of the reactant, A, decreases with time, while the concentration of the product, B, increases with time, and we usually talk of the rate of consumption of a reactant and the rate of formation of a product. We can determine the rate of consumption of A and the rate of formation of B by measuring the concentrations of A and B at two different times, t_1 (early in the reaction) and t_2 (later in the reaction).

$$\text{rate of consumption of } A = \frac{\text{concentration of } A \text{ at time } t_2 - \text{concentration of } A \text{ at time } t_1}{t_2 - t_1} = \frac{\Delta[A]}{\Delta t}$$

The square brackets indicate concentration in mol L^{-1}, and the symbol Δ indicates a change in a given quantity (final value of quantity – initial value of quantity). Similarly, the rate of formation of B is:

$$\text{rate of formation of } B = \frac{\text{concentration of } B \text{ at time } t_2 - \text{concentration of } B \text{ at time } t_1}{t_2 - t_1} = \frac{\Delta[B]}{\Delta t}$$

By convention, rates of formation and consumption are always reported as a positive value, irrespective of whether something increases or decreases in concentration. (This can be compared with the speed of a car; it may drive forwards or in reverse, but the speed is always a positive number.)

The second and third columns of table 15.1 show measurements of the concentrations of A and B, respectively, over time as our hypothetical reaction proceeds.

TABLE 15.1 Concentration of A and B and average rates as a function of time for the reaction $A \rightarrow 2B$, as shown in figure 15.2.

Time (s)	$[A]$ (mol L^{-1})	$[B]$ (mol L^{-1})	Time period (s)	$-\frac{\Delta[A]}{\Delta t}$ (mol L^{-1} s^{-1})
0	0.0750	0.0		
50	0.0629	0.0242	$0 \rightarrow 50$	2.4×10^{-4}
100	0.0529	0.0442	$50 \rightarrow 100$	2.0×10^{-4}
150	0.0444	0.0612	$100 \rightarrow 150$	1.8×10^{-4}
200	0.0372	0.0756	$150 \rightarrow 200$	1.4×10^{-4}
250	0.0313	0.0874	$200 \rightarrow 250$	1.2×10^{-4}
300	0.0262	0.0976	$250 \rightarrow 300$	1.0×10^{-4}
350	0.0220	0.106	$300 \rightarrow 350$	8.0×10^{-5}
400	0.0185	0.113	$350 \rightarrow 400$	7.0×10^{-5}

We can use the data in table 15.1 to calculate the **average rate** of consumption of A in the first 50 seconds of the reaction.

$$\begin{aligned}\text{rate of consumption of } A &= \frac{\Delta[A]}{\Delta t} \\ &= \frac{[A]_{t=50} - [A]_{t=0}}{50\,\text{s} - 0\,\text{s}} \\ &= \frac{0.0629\,\text{mol L}^{-1} - 0.0750\,\text{mol L}^{-1}}{50\,\text{s}} \\ &= -2.4 \times 10^{-4}\,\text{mol L}^{-1}\,\text{s}^{-1}\end{aligned}$$

As you can see, $\Delta[A]$ is a negative quantity, because the concentration of A decreases with time. Since rates of change of concentration are always positive numbers, the rate of consumption of a reactant is defined as:

$$\text{rate of consumption} = -\frac{\Delta[\text{reactant}]}{\Delta t}$$

The average rate of consumption of A in the first 50 seconds of the reaction is therefore:

$$\begin{aligned}\text{rate of consumption of } A &= -\frac{\Delta[A]}{\Delta t} \\ &= -(-2.4 \times 10^{-4}\,\text{mol L}^{-1}\,\text{s}^{-1}) \\ &= 2.4 \times 10^{-4}\,\text{mol L}^{-1}\,\text{s}^{-1}\end{aligned}$$

The last column of table 15.1 gives average rates of consumption of A for this reaction during some other 50-second time intervals. The data show that average rates of consumption of A decrease with time as A is used up. This is because these rates usually depend on the concentration of A, which changes as the reaction proceeds. Therefore, to minimise errors it is better to determine reaction rates over short periods of time.

The rate of change of concentration at a particular time is called the **instantaneous rate of change of concentration**. Looking back to figure 15.2, we see that the instantaneous rate of consumption of A or formation of B can be determined from the slope of the tangent to the curve. For example, in figure 15.2 a tangent is drawn at $t = 50$ seconds on the A curve. The slope of this line, as calculated from the values of $\Delta[A]$ and Δt given in figure 15.2, gives the instantaneous rate of consumption of A at $t = 50$ seconds. (Remember that we are looking at the consumption of a reaction and that the slope will therefore be negative.)

$$\text{rate of consumption of } A = -(\text{slope of tangent}) = -\frac{d[A]}{dt}$$

$$= \frac{-0.017 \text{ mol L}^{-1}}{74.4 \text{ s}}$$

$$= 2.3 \times 10^{-4} \text{ mol L}^{-1} \text{ s}^{-1}$$

In this equation, 'd' indicates an infinitesimally small change. Remember that the rate of a reaction in solution has the unit *concentration per unit time* or (concentration) (time)$^{-1}$ and this is usually mol L^{-1} s^{-1}.

Now, look at the rate of formation of the product, B. For this, the coefficients in the balanced equation for the reaction have to be taken into account, because the stoichiometry determines the relative rates at which reactants are consumed and products formed. In the case of our hypothetical reaction, $A \rightarrow 2B$, B is produced twice as fast as A is consumed, which can be verified from figure 15.2. The slope of the tangent to the B curve at $t = 50$ seconds gives the instantaneous rate of formation of B at 50 seconds, which is:

$$\text{rate of formation of } B = (\text{slope of tangent}) = \frac{d[B]}{dt}$$

$$= \frac{0.025 \text{ mol L}^{-1}}{54.0 \text{ s}}$$

$$= 4.6 \times 10^{-4} \text{ mol L}^{-1} \text{ s}^{-1}$$

This shows that the rate of formation of B is indeed twice that of consumption of A. Note that the slope of the tangent is positive.

As you can see, the rates of consumption of reactants and rates of formation of products in a particular reaction are not necessarily the same. It is much more meaningful for us to define the **rate of reaction** whose value is the same regardless of whether we are monitoring rates of consumption of reactants or rates of formation of products. For the general reaction:

$$aA + bB \rightarrow cC + dD$$

where a to d are the stoichiometric coefficients of the reactants A and B and the products C and D respectively, we define the rate of reaction as:

$$\text{rate of reaction} = -\frac{1}{a}\frac{d[A]}{dt} = -\frac{1}{b}\frac{d[B]}{dt} = \frac{1}{c}\frac{d[C]}{dt} = \frac{1}{d}\frac{d[D]}{dt}$$

Therefore, for our reaction $A \rightarrow 2B$, we would write:

$$\text{rate of reaction} = -\frac{d[A]}{dt} = \frac{1}{2}\frac{d[B]}{dt}$$

$$= 2.3 \times 10^{-4} \text{ mol L}^{-1} \text{s}^{-1}$$

The rate of a reaction has units of (concentration) (time)$^{-1}$. It is important that you appreciate the difference between the rate of reaction and the rates of consumption of reactants and rates of formation of products. The rate of reaction at any instant in time has only one value for any particular reaction and is independent of the reaction stoichiometry, whereas the rates of reactant consumption and product formation are different if the stoichiometric coefficients differ.

WORKED EXAMPLE 15.1

Estimating the initial rate of a reaction

The following data have been measured at 508 °C for the reaction $2HI(g) \rightarrow H_2(g) + I_2(g)$.

Time (s)	[HI] (mol L^{-1})	Time (s)	[HI] (mol L^{-1})
0	0.100	200	0.0387
50	0.0716	250	0.0336
100	0.0558	300	0.0296
150	0.0457	350	0.0265

What is the initial rate of consumption of HI, and what is the initial rate of reaction at this temperature?

Analysis

The initial rate of consumption of HI is the instantaneous rate of consumption of HI at time zero. Once we plot the data, we can draw a tangent to the curve showing [HI] as a function of time at time zero. The instantaneous rate of consumption of HI is the slope of the tangent. The slope of the line can be calculated from the coordinates of any two points (x_1, y_1) and (x_2, y_2) using the equation:

$$\text{slope} = \frac{y_2 - y_1}{x_2 - x_1} = \frac{\text{d[HI]}}{\text{d}t}$$

The initial rate of reaction can be calculated using the stoichiometry of the reaction.

$$\text{rate of reaction} = -\frac{1}{2}\frac{\text{d[HI]}}{\text{d}t}$$

Solution

Because we want to study the rates at which concentrations change, we have to use differential calculus. Thus, since a tangent to a curve touches the curve at only one point, we can draw the line as shown in the diagram below.

To determine the slope of the tangent precisely, we should choose two points that are as far apart as possible. We could use, for example, the point on the curve (0 s, 0.10 mol L^{-1}) and the intersection of the tangent with the time axis (150 s, 0 mol L^{-1}), which are widely separated.

$$\text{slope} = \frac{\text{d[HI]}}{\text{d}t} = \frac{0.00\,\text{mol L}^{-1} - 0.10\,\text{mol L}^{-1}}{150\,\text{s} - 0\,\text{s}}$$
$$= -6.7 \times 10^{-4}\,\text{mol L}^{-1}\,\text{s}^{-1}$$

(There is no requirement to restrict the extrapolation to concentration = 0 to find the slope.)

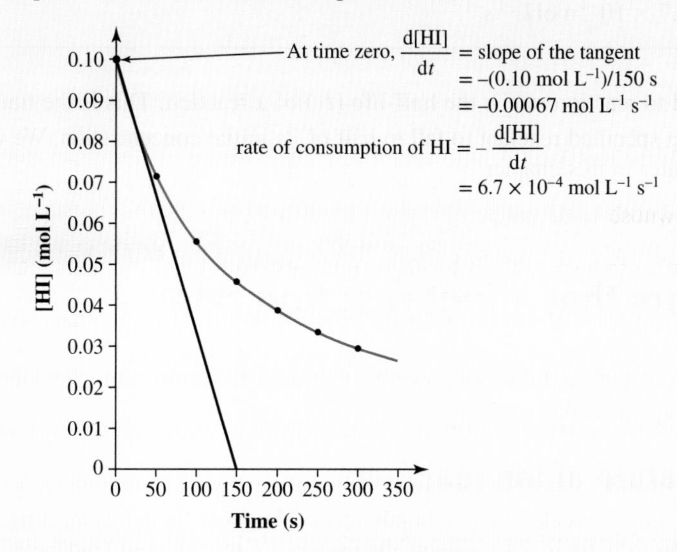

The slope is negative because the concentration of HI is decreasing as time increases. Reaction rates are positive quantities, so we report the initial rate of consumption of HI as 6.7×10^{-4} mol L^{-1} s^{-1}. To do this calculation, we rely on accurately drawing by hand a tangent to the curve at $t = 0$. Because of this, there is a degree of uncertainty in the accuracy of our answer. The initial rate of reaction is defined by the equation:

$$\text{rate of reaction} = -\frac{1}{2}\frac{\text{d[HI]}}{\text{d}t}$$

We have already calculated the initial $\frac{\text{d[HI]}}{\text{d}t}$ at time zero as being -6.7×10^{-4} mol L^{-1} s^{-1}, so the initial rate of reaction is:

$$-\tfrac{1}{2}(-6.7\times10^{-4}) = 3.4\times10^{-4}\,\text{mol L}^{-1}\,\text{s}^{-1}$$

Is our answer reasonable?

As noted previously, drawing a tangent to a curve by eye is not precise. There may be considerable variation in the determined time difference, which affects the accuracy of our calculated rate. One way of checking our answer is to realise by inspection of the graph that the instantaneous rate of consumption of HI at time zero should be close to, but slightly greater than, the average rate of consumption of HI between 0 s and 50 s.

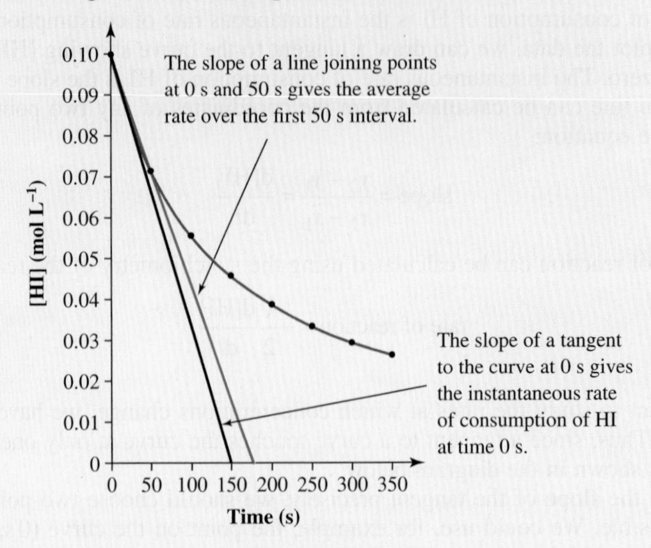

We can calculate the average rate of consumption of HI directly from the data in the table on page 644, which have been accurately measured.

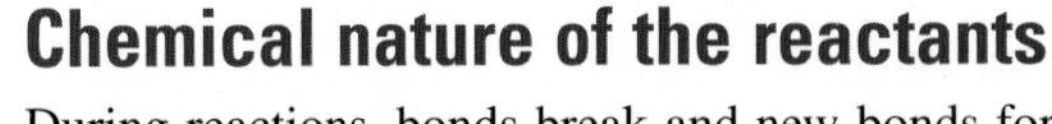

$$\text{slope} = \frac{0.0716\,\text{mol L}^{-1} - 0.100\,\text{mol L}^{-1}}{50\,\text{s} - 0\,\text{s}} = -5.7 \times 10^{-4}\,\text{mol L}^{-1}\,\text{s}^{-1}$$

So the average rate of consumption of HI from 0 s to 50 s is $5.7 \times 10^{-4}\,\text{mol L}^{-1}\,\text{s}^{-1}$. As expected, this is close to but slightly less than the instantaneous rate of consumption of HI at time zero, $6.7 \times 10^{-4}\,\text{mol L}^{-1}\,\text{s}^{-1}$.

A concept related to reaction rate is the **half-life** ($t_{\frac{1}{2}}$) of a reaction. This is the time taken for the concentration of a specified reactant to fall to half of its initial concentration. We will learn more about half-lives later in this chapter.

15.2 Factors that affect reaction rates

Five principal factors influence reaction rates: the chemical nature of the reactants, the physical nature of the reactants, the concentrations of the reactants, the temperature, and the availability of rate-accelerating agents called catalysts.

Chemical nature of the reactants

During reactions, bonds break and new bonds form. The most fundamental differences between reaction rates, therefore, lie in the reactants themselves, in the inherent tendencies of their atoms, molecules or ions to undergo bond-breaking and/or bond-making processes. Some reactions are fast by nature and others are slow. As you have seen in chapter 4 (figure 4.45, p. 157), alkali metals can be easily oxidised, which leads to their high reactivity. Thus, a freshly exposed surface of metallic sodium tarnishes almost instantly when exposed to air and moisture. Under identical conditions, potassium also reacts with air and moisture, but the reaction is much faster because the ionisation energy of potassium is even lower than that of sodium (figure 15.3). Compared with these fast reactions, the reaction of silver with water is very slow, which you can check by putting a silver ring into a glass of water and observing that nothing happens.

FIGURE 15.3 The chemical nature of the reactants affects reaction rates. **(a)** Sodium is easily oxidised, so it reacts quickly with water. **(b)** Potassium is even more easily oxidised than sodium, so its reaction with water is explosively fast.

Physical nature of the reactants

Most reactions involve two or more reactants whose constituent atoms, ions or molecules must collide for the reaction to occur. This is why reactions are most commonly carried out in liquid solutions or in the gas phase where the reactants are able to intermingle and their constituent atoms, molecules or ions can collide with each other easily.

Consider, for example, that petrol vapour explodes when mixed with air in the right proportions and ignited. (An explosion is an extremely rapid reaction that generates hot expanding gases.) The combustion of vaporised petrol illustrates a **homogeneous reaction**, because all of the reactants are in the same phase (in this case in the gas phase).

When the reactants are present in different phases, for example, if one is a gas and the other is a liquid or a solid, the reaction is called a **heterogeneous reaction**. In a heterogeneous reaction, the reactants are able to meet only at the interface between the phases, *so the area of contact between the phases affects the reaction rate*. This area is controlled by the sizes of the particles of the reactant constituents. By pulverising a solid, the total surface area can be significantly increased (figure 15.4). This maximises contact between the atoms, ions or molecules in the solid state with those in a different phase.

Although heterogeneous reactions are obviously important, they are very complex and difficult to analyse. In this chapter, therefore, we will focus mostly on homogeneous systems.

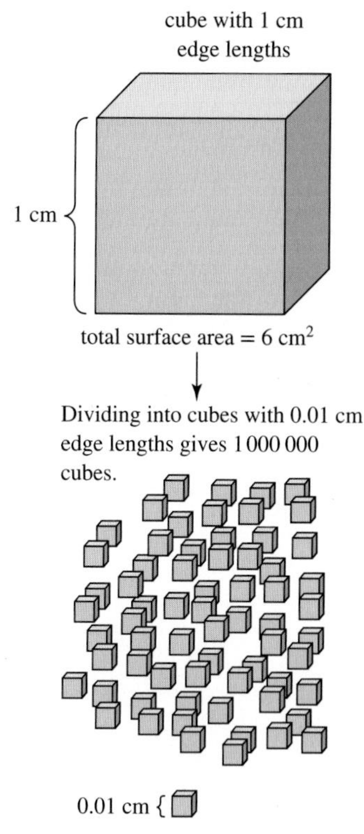

FIGURE 15.4 Effect of crushing a solid. When a single solid is subdivided into much smaller pieces, the total surface area of all of the pieces becomes very large.

Concentrations of the reactants

The rates of both homogeneous and heterogeneous reactions are affected by the concentrations of the reactants. For example, wood burns relatively quickly in air but extremely rapidly in pure oxygen. Even red-hot steel wool, which only sputters and glows in air, bursts into flame when thrust into pure oxygen. Studies have shown that the Earth's atmosphere has had an oxygen content as high as 35% in the past (nowadays, it is only 21%). This higher oxygen concentration had some negative consequences, such as fires burning hotter and faster.

Temperature of the system

Almost all chemical reactions occur faster at higher temperatures. Milk doesn't spoil as quickly when it is cold. It takes longer to cook an egg at 80 °C than at 100 °C. You may also have noticed that insects move more slowly when the air is cool. Insects are cold-blooded creatures, which means that their body temperature is determined by the temperature of their surroundings. As the air cools, insects cool, and so the rates of their chemical metabolism slow down, making them sluggish.

Presence of catalysts

Catalysts are substances that increase the rate of chemical reactions without being used up. Catalysts affect every moment of our lives because the enzymes that direct our body chemistry are all catalysts. Many processes used by the chemical industry to make petrol, plastics, fertilisers and other everyday products are only enabled by catalysts. We will discuss how catalysts affect reaction rates in section 15.7.

15.3 Overview of rate laws

In worked example 15.1, we discussed the decomposition of hydrogen iodide.

$$2HI(g) \rightarrow H_2(g) + I_2(g) \qquad \text{(forward reaction)}$$

We have seen in chapter 9 that chemical reactions are reversible. As the decomposition proceeds, H_2 and I_2 are formed and can themselves react to form HI (see figure 15.5).

$$H_2(g) + I_2(g) \rightarrow 2HI(g) \qquad \text{(reverse reaction)}$$

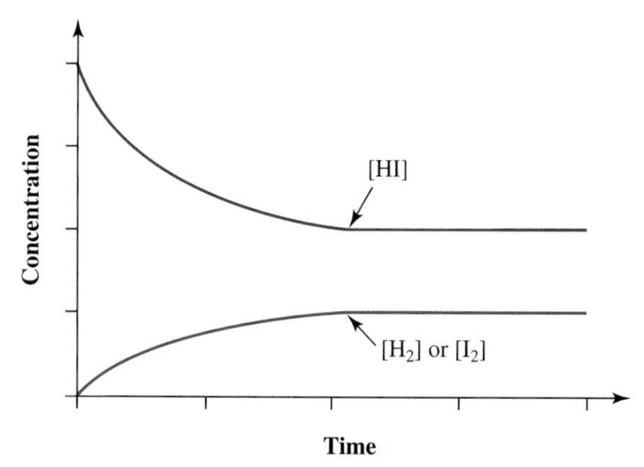

FIGURE 15.5 Concentration–time profile for the decomposition of hydrogen iodide.

When hydrogen iodide gas is first placed in an empty container, the forward reaction is dominant and the rate of change of concentration of HI depends only on the rate of the forward reaction. When sufficient products have formed, the rate of the reverse reaction becomes important too, and the rate of change of concentration of HI then depends on the difference in the rates of the forward and reverse reactions.

Generally, we want to determine the rate of reaction under conditions in which the reverse reaction can be ignored, that is, when the rate of the reaction depends only on the concentration of the reactants. This can be done if we determine the rate of reaction immediately after placing the reactants into the reaction vessel, when the reaction time has been too short to build up significant amounts of products. This is called the **initial rate** of a reaction.

rate constant/coefficient ↘ ↙ order

$$\text{rate of reaction} = k[\text{HI}]^n$$

Such an equation in which the rate of reaction is given as a function of reactant concentrations (in this case [HI]) is called a **rate law**. Note that, because the reaction has been studied under conditions in which the reverse reaction does not contribute to the overall rate, the products do not appear in the rate law. The rate law contains a proportionality constant k, which is called the **rate constant** or **rate coefficient** for the reaction, and an exponent n, called the **order** with respect to the reactant. The terms 'rate constant' and 'rate coefficient' are often used interchangeably. In this book we will use the term 'rate constant'. The value of the rate constant depends on the particular reaction being studied, as well as on the conditions, such as temperature and pressure, under which the reaction occurs. It is not dependent on the concentrations of reactants. The order can be positive, negative or zero, an integer or a fraction, but *cannot* be deduced from the balanced equation except in very special circumstances. This is a very important point, which we will examine in more detail soon.

As we saw earlier, the rate of a reaction does not depend on whether we monitor the consumption of reactants or the formation of products. In fact, which reactant or product concentration we choose to monitor often depends on which data are the easiest to obtain. For example, to study the decomposition of hydrogen iodide into H_2 and I_2, it is easiest to monitor the I_2 concentration, because it is the only coloured substance in the reaction. As the reaction proceeds, purple iodine vapour forms, which can be measured with instruments that relate the intensity of the colour to the iodine concentration.

Thus, if we define the rate of reaction in terms of formation of I_2, we get:

$$\text{rate of reaction} = \frac{\mathrm{d}[I_2]}{\mathrm{d}t} = k[\text{HI}]^n$$

As we have discussed before, in this equation, 'd' stands for an extremely small change. We could also define the reaction rate in terms of consumption of HI.

$$\text{rate of reaction} = -\frac{1}{2}\frac{\mathrm{d}[\text{HI}]}{\mathrm{d}t} = k[\text{HI}]^n$$

The rate of consumption of HI is twice the rate of formation of I_2. Once we know the rate of formation of iodine, we also know the rate of formation of hydrogen, because the stoichiometric coefficients of I_2 and H_2 are the same.

WORKED EXAMPLE 15.2

Relationships of rates within a reaction

Butane, C_4H_{10}, burns in oxygen to give CO_2 and H_2O according to the equation:

$$2C_4H_{10}(g) + 13O_2(g) \rightarrow 8CO_2(g) + 10H_2O(g)$$

If the butane concentration is decreasing at a rate of $0.20\,\text{mol}\,\text{L}^{-1}\,\text{s}^{-1}$, what is the rate at which the oxygen concentration is decreasing, and what are the rates at which the product concentrations are increasing?

Analysis

We need to relate the rate of consumption of oxygen and the rates of formation of the products to the given rate of consumption of butane. The chemical equation links amounts of these substances to the amount of butane. The magnitudes of the rates relative to each other are in the same relationship as the coefficients in the balanced equation.

Solution

From our definition of rate of reaction, we can write:

$$\text{rate of reaction} = -\frac{1}{2}\frac{\mathrm{d}[C_4H_{10}]}{\mathrm{d}t} = -\frac{1}{13}\frac{\mathrm{d}[O_2]}{\mathrm{d}t} = \frac{1}{8}\frac{\mathrm{d}[CO_2]}{\mathrm{d}t} = \frac{1}{10}\frac{\mathrm{d}[H_2O]}{\mathrm{d}t}$$

We are told that:

$$-\frac{\mathrm{d}[C_4H_{10}]}{\mathrm{d}t} = 0.20\,\text{mol}\,\text{L}^{-1}\,\text{s}^{-1}$$

and we can therefore use this, together with our expression for the rate of reaction, to determine $-\frac{d[O_2]}{dt}$.

$$\frac{1}{2}\frac{d[C_4H_{10}]}{dt} = -\frac{1}{13}\frac{d[O_2]}{dt}$$

Thus:

$$\begin{aligned}\frac{d[O_2]}{dt} &= \frac{13}{2}\frac{d[C_4H_{10}]}{dt} \\ &= \frac{13}{2} \times 0.20\ \text{mol L}^{-1}\,\text{s}^{-1} \\ &= 1.30\ \text{mol L}^{-1}\,\text{s}^{-1}\end{aligned}$$

We can carry out similar calculations for both H_2O and CO_2. For CO_2:

$$-\frac{1}{2}\frac{d[C_4H_{10}]}{dt} = \frac{1}{8}\frac{d[CO_2]}{dt}$$

Therefore:

$$\begin{aligned}\frac{d[CO_2]}{dt} &= -\frac{8}{2}\frac{d[C_4H_{10}]}{dt} = -4\frac{d[C_4H_{10}]}{dt} \\ &= 4 \times 0.20\ \text{mol L}^{-1}\,\text{s}^{-1}\end{aligned}$$

(Note that the negative sign in the calculation disappears since rates are defined as positive.)

$$= 0.80\ \text{mol L}^{-1}\,\text{s}^{-1}$$

For H_2O:

$$-\frac{1}{2}\frac{d[C_4H_{10}]}{dt} = \frac{1}{10}\frac{d[H_2O]}{dt}$$

Therefore:

$$\begin{aligned}\frac{d[H_2O]}{dt} &= -\frac{10}{2}\frac{d[C_4H_{10}]}{dt} = -5\frac{d[C_4H_{10}]}{dt} \\ &= 5 \times 0.20\ \text{mol L}^{-1}\,\text{s}^{-1}\end{aligned}$$

(Again, we take the positive value as the rate of formation.)

$$= 1.00\ \text{mol L}^{-1}\,\text{s}^{-1}$$

Is our answer reasonable?

If what we have calculated is correct, then the ratio of the numerical values of the rates of formation of CO_2 and H_2O, namely 0.80 to 1.00, should be the same as the ratio of the corresponding coefficients in the chemical equation, namely 8 to 10 (the same as 0.80 to 1.00). The ratios match, so we can be confident in our answers.

PRACTICE EXERCISE 15.1

Hydrogen sulfide burns in oxygen to form sulfur dioxide and water.

$$2H_2S(g) + 3O_2(g) \rightarrow 2SO_2(g) + 2H_2O(g)$$

If sulfur dioxide is being formed at a rate of 0.30 mol L^{-1} s^{-1}, what are the rates of consumption of hydrogen sulfide and oxygen?

15.4 Types of rate laws: differential and integrated

In our discussion, we defined the rate law for the decomposition of HI as:

$$\text{rate of reaction} = -\frac{1}{2}\frac{d[HI]}{dt} = k[HI]^n$$

This expresses the *rate of reaction as a function of concentration* and is known as a **differential rate law** (often abbreviated to rate law). Once the order, n, is known, we can determine exactly how the rate of reaction depends on the concentration of HI. A second important type of rate law expresses the *concentration as a function of time*. This is called an **integrated rate law**.

We can determine the differential rate law by measuring rate changes when the concentrations are changed. We can determine the integrated rate law by measuring changes in concentrations over time. The two types of rate law are related and, if we know one rate law for a reaction, we can, in most cases, determine the other.

Rate laws help us to identify the steps by which a chemical reaction occurs and thus understand the reaction. Most reactions involve a series of sequential steps. The sum of the individual reaction steps is called the **reaction mechanism**. The formation of ammonia from nitrogen and hydrogen, which we encountered at the beginning of the chapter, is extremely slow at room temperature under atmospheric pressure. This is because the strong H—H single bond and the N≡N triple bond must be broken. To speed this reaction up and make it synthetically useful by using catalysts, the steps that are involved in the reaction must be known.

The differential rate law

To understand the mechanism of a chemical reaction, we must first determine the form of the rate law. Let us look at another hypothetical reaction:

$$C \rightarrow 2D$$

Table 15.2 and figure 15.6 present the data for this reaction. In this reaction we assume that the reverse reaction $2D \rightarrow C$ is much slower than the forward reaction, so we are not concerned about the effect of the reverse reaction on the rate of reaction at any time.

TABLE 15.2 Concentration and time data for the reaction $C \rightarrow 2D$.

Time (s)	$[C]$ (mol L^{-1})	Time (s)	$[C]$ (mol L^{-1})	Time (s)	$[C]$ (mol L^{-1})
0	0.70	0.75	0.40	1.50	0.23
0.25	0.58	1.00	0.33	1.75	0.19
0.50	0.48	1.25	0.27	2.00	0.16

From figure 15.6, we can see that the slopes of the tangents to the curve at $[C] = 0.60$ M and 0.30 M give the rates of reaction, 0.42 mol L^{-1} s^{-1} and 0.21 mol L^{-1} s^{-1}, respectively.

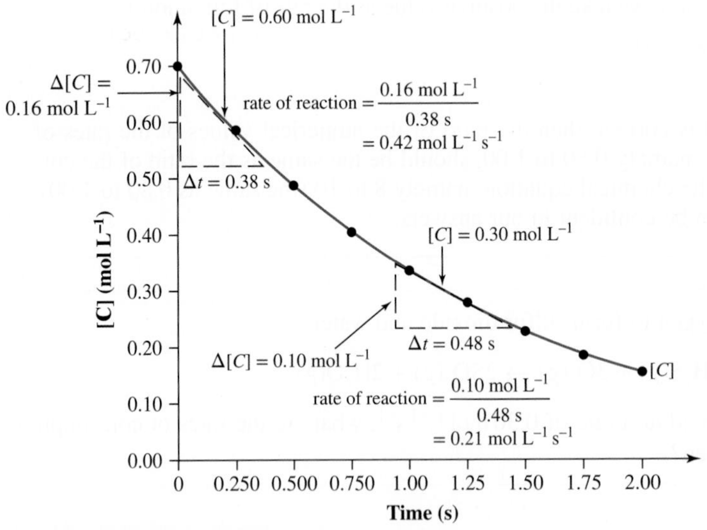

FIGURE 15.6 Determination of the order of the reaction $C \rightarrow 2D$.

When the concentration of C is halved, the rate of reaction is also halved. The differential rate law for this reaction is therefore:

$$\text{rate of reaction} = -\frac{d[C]}{dt} = k[C]^1 = k[C]$$

That is, the rate of reaction depends on the concentration of C to the first power and we say that the reaction is 'first order in C'. (Remember that the order cannot be determined from the stoichiometric coefficient of C — in this example it is just coincidence that they are the same. The order can be determined only from experiments.)

WORKED EXAMPLE 15.3

Calculating a rate constant from the rate data

Use the data in figure 15.6 to calculate the rate constant for the reaction $C \rightarrow 2D$.

Analysis

We know that the rate law for the reaction $C \rightarrow 2D$ is:

$$\text{rate of reaction} = k[\text{C}]$$

Therefore:

$$k = \frac{\text{rate of reaction}}{[C]}$$

We substitute data from figure 15.6 into this equation and solve for k.

Solution

For $[C] = 0.60\ \text{mol L}^{-1}$, we have found that rate of reaction = $0.42\ \text{mol L}^{-1}\ \text{s}^{-1}$. Inserting this into the equation gives:

$$k = \frac{0.42\ \text{mol L}^{-1}\ \text{s}^{-1}}{0.60\ \text{mol L}^{-1}} = 0.70\ \text{s}^{-1}$$

Is our answer reasonable?

We can check whether our calculation is correct in 2 ways.

1. The calculated unit of the rate constant is s^{-1}, or $(\text{time})^{-1}$. Since the rate of reaction always has the unit (concentration) $(\text{time})^{-1}$, we can check whether the unit for k is correct by inserting then into the rate law:

$$\underbrace{(\text{concentration})\,(\text{time})^{-1}}_{\text{rate of reaction}} = \underbrace{(\text{time})^{-1}}_{k}\underbrace{(\text{concentration})}_{[C]}$$

   The unit of k of a first order reaction is $(\text{time})^{-1}$.

2. We can also calculate the rate constant for the other value of the reaction rate shown in figure 15.6. It should be the same as that calculated above, as the rate constant for a reaction is constant at constant temperature.

   For $[C] = 0.30\ \text{mol L}^{-1}$, rate of reaction = $0.21\ \text{mol}^{-1}\ \text{L}^{-1}\ \text{s}^{-1}$; therefore:

$$k = \frac{0.21\ \text{mol L}^{-1}\text{s}^{-1}}{0.30\ \text{mol L}^{-1}} = 0.70\ \text{s}^{-1}$$

   As the rate constant is the same under both concentration conditions, we can be confident that our answer is correct.

Let's go a step further. Consider the following hypothetical reaction:

$$A + B \rightarrow C$$

Suppose that the data in table 15.3 have been obtained in a series of five experiments at the same temperature. The form of the rate law for the reaction is:

$$\text{rate of reaction} = k[A]^n[B]^m$$

The values of n and m are found mathematically, and can sometimes be solved by looking for patterns in the rate data. One of the easiest ways to reveal patterns in data is to form ratios of results using different sets of conditions. This procedure is also called the 'method of initial rates'. Because this technique is quite generally useful, let's look in some detail at how it is applied to the problem of finding the rate law exponents.

TABLE 15.3 Concentration and rate data for the hypothetical reaction $A + B \rightarrow C$ determined at the same temperature.

Experiment	Initial concentrations A (mol L^{-1})	B (mol L^{-1})	Initial rate of formation of C (mol L^{-1} s^{-1})
1	0.10	0.10	0.20
2	0.20	0.10	0.40
3	0.30	0.10	0.60
4	0.30	0.20	2.40
5	0.30	0.30	5.40

For experiments 1, 2 and 3 in table 15.3, the concentration of B was held constant at 0.10 M. Any change in the rate of reaction for these first three experiments must be due to the change in $[A]$. The rate law tells us that, when $[B]$ is held constant, the rate of reaction must be proportional to $[A]^n$ so, if we take the ratio of rate laws for experiments 1 and 2, (since $[B]^m$ cancels out) we obtain:

$$\frac{\text{rate of reaction}_{\text{experiment 2}}}{\text{rate of reaction}_{\text{experiment 1}}} = \left(\frac{[A]_{\text{experiment 2}}}{[A]_{\text{experiment 1}}}\right)^n$$

Let's look at the left-hand side of this equation first.

$$\frac{\text{rate of reaction}_{\text{experiment 2}}}{\text{rate of reaction}_{\text{experiment 1}}} = \frac{0.40 \text{ mol L}^{-1} \text{ s}^{-1}}{0.20 \text{ mol L}^{-1} \text{ s}^{-1}} = 2.0$$

For the right-hand side of the equation we have:

$$\left(\frac{[A]_{\text{experiment 2}}}{[A]_{\text{experiment 1}}}\right)^n = \left(\frac{0.20 \text{ mol L}^{-1}}{0.10 \text{ mol L}^{-1}}\right)^n = 2.0^n$$

So doubling $[A]$ from experiment 1 to experiment 2 doubles the rate of reaction. If we now combine the left- and right-hand sides of the equation, we can write:

$$2.0 = 2.0^n$$

for experiments 1 and 2.

For the other possible combinations of experiments, we have:

$$3.0 = 3.0^n \text{ (for experiments 1 and 3)}$$
$$1.5 = 1.5^n \text{ (for experiments 2 and 3)}$$

The only value of n that makes all of these equations true is $n = 1$. The reaction must be first order with respect to A.

In the final three experiments, the concentration of B changes while the concentration of A is held constant. This time it is $[B]$ that affects the rate. Taking the ratio of rate laws for experiments 3 and 4, we have:

$$\frac{\text{rate of reaction}_{\text{experiment 4}}}{\text{rate of reaction}_{\text{experiment 3}}} = \left(\frac{[B]_{\text{experiment 4}}}{[B]_{\text{experiment 3}}}\right)^m$$

For each unique combination of experiments 3, 4 and 5, we have:

$$4.0 = 2.0^m \text{ (for experiments 3 and 4)}$$
$$9.0 = 3.0^m \text{ (for experiments 3 and 5)}$$
$$2.25 = 1.5^m \text{ (for experiments 4 and 5)}$$

The only value of m that makes all of these equations true is $m = 2$. The reaction must be second order with respect to B.

Having determined the exponents for the concentration terms, we now know that the rate law for the reaction is:

$$\text{rate of reaction} = k[A]^1[B]^2$$

The reaction is first order with respect to A and second order with respect to B. *The overall order of a reaction is defined as the sum of the orders for each reactant in the rate law.* Thus, our hypothetical reaction $A + B \rightarrow C$ has an overall order of 3.

PRACTICE EXERCISE 15.2

The following reaction:

$$BrO_3^- + 3SO_3^{2-} \rightarrow Br^- + 3SO_4^{2-}$$

has the rate law:

$$\text{rate of reaction} = k[BrO_3^-][SO_3^{2-}].$$

What is the order of the reaction with respect to each reactant? What is the overall order of the reaction?

WORKED EXAMPLE 15.4

Determining the exponents of a rate law

Sulfuryl chloride, SO_2Cl_2, is a dense liquid with a pungent odour and is corrosive to the skin and lungs. It is used to manufacture the antiseptic chlorophenol. The following data were collected on the decomposition of SO_2Cl_2 at a certain temperature.

$$SO_2Cl_2(g) \rightarrow SO_2(g) + Cl_2(g)$$

Initial concentration of SO_2Cl_2 (mol L^{-1})	Initial rate of formation of SO_2 (mol L^{-1} s^{-1})
0.100	2.2×10^{-6}
0.200	4.4×10^{-6}
0.300	6.6×10^{-6}

Determine the rate law and the value of the rate constant for this reaction.

Analysis

The first step is to write the general form of the expected rate law so we can see which exponents have to be determined. Then we study the data to see how the rate of reaction changes when the concentration is changed by a certain factor.

Solution

We expect the rate law to have the form:

$$\text{rate of reaction} = k[SO_2Cl_2]^n$$

Let's examine the data from the first two experiments. Notice that, when the concentration doubles from 0.100 mol L^{-1} to 0.200 mol L^{-1}, the initial rate of reaction also doubles (from 2.2×10^{-6} mol L^{-1} s^{-1} to 4.4×10^{-6} mol L^{-1} s^{-1}). If we look at the first and third experiments, we see that, when the concentration triples (from 0.100 M to 0.300 M), the rate of reaction also triples (from 2.2×10^{-6} mol L^{-1} s^{-1} to 6.6×10^{-6} mol L^{-1} s^{-1}). This behaviour tells us that the reaction must be first order in the SO_2Cl_2 concentration. The rate law is therefore:

$$\text{rate of reaction} = k[SO_2Cl_2]^1 = k[SO_2Cl_2]$$

To evaluate k, we can use any of the three sets of data. Choosing the first set:

$$k = \frac{\text{rate of reaction}}{[SO_2Cl_2]} = \frac{2.2\times10^{-6}\ \text{mol L}^{-1}\text{s}^{-1}}{0.100\ \text{mol L}^{-1}} = 2.2\times10^{-5}\ \text{s}^{-1}$$

Is our answer reasonable?

We should obtain the same value of k by picking any other pair of values. With the last pair of data, at an initial $[SO_2Cl_2]$ of 0.300 mol L^{-1} and an initial rate of reaction of 6.6×10^{-6} mol L^{-1} s^{-1}, we calculate k again to be 2.2×10^{-5} s^{-1}. Keep in mind that the unit of a first-order rate constant is $(\text{time})^{-1}$ (in this example, s^{-1}).

Note that we could also use experiments 2 and 3 to determine the order of the reaction. From the second to the third experiment, the rate of reaction increases by the same factor (1.5) as the concentration, so, using these data too, the reaction must be first order.

To calculate the value of k for the $A + B \rightarrow$ products reaction, we need only substitute rate and concentration data into the rate law.

$$k = \frac{\text{rate}}{[A]^1[B]^2}$$

Using the data from experiment 1 in table 15.3:

$$k = \frac{0.20\ \text{mol L}^{-1}\text{s}^{-1}}{(0.10\ \text{mol L}^{-1})(0.10\ \text{mol L}^{-1})^2} = \frac{0.20\ \text{mol L}^{-1}\text{s}^{-1}}{0.0010\ \text{mol}^3\ \text{L}^{-3}}$$

After cancelling units, the value of k with the net units is:

$$k = 2.0 \times 10^2\ \text{mol}^{-2}\ \text{L}^2\ \text{s}^{-1}$$

Note that the unit of a third-order rate constant is $(\text{concentration})^{-2}$ $(\text{time})^{-1}$.

As mentioned before, the rate of a reaction always has the unit (concentration) (time)$^{-1}$. However, it is important to understand that the units of the rate constant depend on the order of the reaction.

For example, for a first-order reaction, the unit is (time)$^{-1}$, and for a second-order reaction the unit is (concentration)$^{-1}$ (time)$^{-1}$. This can also be rationalised from:

$$\text{rate of reaction} = k[A]^n$$

- If $n = 1$, the unit for $[A]^n$ is concentration, so the unit for k is (time)$^{-1}$.
- If $n = 2$, the unit for $[A]^n$ is (concentration)^2, so the unit for k is (concentration)$^{-1}$ (time)$^{-1}$.

PRACTICE EXERCISE 15.3

Use the data from the other four experiments (table 15.3) to calculate k for this reaction. What do you notice about the values of k?

Table 15.4 summarises the reasoning used to determine the order for each reactant from experimental data.

TABLE 15.4 Relationship between the order of a reaction and changes in concentration and rate.

Factor by which the concentration is changed	Factor by which the rate of reaction changes	Exponent of the concentration term in the rate law
2	rate of reaction is unchanged	0
3	rate of reaction is unchanged	0
4	rate of reaction is unchanged	0
2	$2 = 2^1$	1
3	$3 = 3^1$	1
4	$4 = 4^1$	1
2	$4 = 2^2$	2
3	$9 = 3^2$	2
4	$16 = 4^2$	2
2	$8 = 2^3$	3
3	$27 = 3^3$	3
4	$64 = 4^3$	3

PRACTICE EXERCISE 15.4

The following data were measured at the same temperature for the reduction of nitric oxide with hydrogen.

$$2NO(g) + 2H_2(g) \rightarrow N_2(g) + 2H_2O(g)$$

Initial concentration		Initial rate of formation of H_2O (mol L^{-1} s^{-1})
[NO] (mol L^{-1})	[H_2] (mol L^{-1})	
0.10	0.10	1.23×10^{-3}
0.10	0.20	2.46×10^{-3}
0.20	0.10	4.92×10^{-3}

What is the rate law for the reaction?

WORKED EXAMPLE 15.5

Calculating reaction rates from the rate law

In the stratosphere, molecular oxygen, O_2, can be broken down into two oxygen atoms by ultraviolet radiation from the sun. When one of these oxygen atoms collides with an ozone molecule, O_3, in the stratosphere, the ozone molecule is destroyed and two oxygen molecules are formed.

$$O(g) + O_3(g) \rightarrow 2O_2(g)$$

This reaction is part of the natural cycle of ozone destruction and formation in the stratosphere. The experimentally determined rate law for the reaction is:

$$\text{rate of reaction} = k[O_3][O]$$

with a rate constant of $k = 4.15 \times 10^5\ \text{mol}^{-1}\ \text{L s}^{-1}$ at the temperature of the experiment. The reactant concentrations at an altitude of 25 km are: $[O_3] = 1.2 \times 10^{-8}\ \text{mol L}^{-1}$ and $[O] = 1.7 \times 10^{-14}\ \text{mol L}^{-1}$. What is the rate of ozone destruction for this reaction at an altitude of 25 km?

Analysis

Because we already know the rate law and that the reaction is first order for both [O] and $[O_3]$, the answer to this question is obtained by substituting the given molar concentrations into the rate law.

Solution

Substituting the given values into the rate law gives:

$$\text{rate of reaction} = (4.15\times10^5\ \text{mol}^{-1}\ \text{L s}^{-1})(1.2\times10^{-8}\ \text{mol L}^{-1})(1.7\times10^{-14}\ \text{mol L}^{-1})$$

$$= 8.5\times10^{-17}\ \text{mol L}^{-1}\ \text{s}^{-1}$$

Note that the units of rate of reaction are $\text{mol L}^{-1}\ \text{s}^{-1}$, which is (concentration) $(\text{time})^{-1}$, consistent with what we learned earlier (see p. 643).

Is our answer reasonable?

There is obviously no simple check. Multiplying the powers of 10 for the rate constant and the concentrations together reassures us that the rate of reaction is of the correct order of magnitude, and we can see that the answer has the correct units. It is important to realise that the units of the rate constant k are *always* such that the calculated rate of reaction has the units (concentration) $(\text{time})^{-1}$, $\text{mol L}^{-1}\ \text{s}^{-1}$ in this example. This means that the unit of a rate constant is different for a reaction of overall zero, first, second or third order etc. We show this in other examples later in this chapter.

PRACTICE EXERCISE 15.5

Azomethane gas ($CH_3N{=}NCH_3$) undergoes decomposition to produce ethane gas (C_2H_6) and nitrogen gas (N_2).

$$CH_3N{=}NCH_3(g) \rightarrow C_2H_6(g) + N_2(g)$$

The rate of azomethane decomposition is directly proportional to its concentration. The rate constant for this decomposition at a particular temperature has been measured as $1.2 \times 10^{-2}\ \text{min}^{-1}$. Calculate the rate of the reaction at this particular temperature, when $[CH_3N{=}NCH_3] = 0.25\ \text{mol L}^{-1}$.

The integrated rate law

So far, we have looked at differential rate laws, that is, rate laws that express the rate of reaction as a function of concentration. Integrated rate laws are also useful. Recall that integrated rate laws express concentration as a function of time. Consider the hypothetical reaction:

$$A \rightarrow \text{products}$$

The differential rate law has the form:

$$\text{rate of reaction} = -\frac{d[A]}{dt} = k[A]^n$$

The integrated rate laws for first order ($n = 1$), second order ($n = 2$) and zero order ($n = 0$) reactions are developed individually.

First-order rate laws

Let's assume that our hypothetical reaction $A \rightarrow$ products is first order in A. The differential rate law is then:

$$\text{rate of reaction} = -\frac{d[A]}{dt} = k[A]$$

Rearrangement gives:

$$-\frac{d[A]}{[A]} = k dt$$

We now want to look at the development of the concentration from the beginning of the reaction (at $t = 0$) until a certain reaction time t. This means we have to integrate this rate equation.

$$-\int_{[A]_0}^{[A]_t} \frac{d[A]}{[A]} = k\int_{t=0}^{t} dt$$

Since $\int \frac{d[A]}{[A]} = \ln[A]$ (ignoring the constant of integration), we obtain:

$$-(\ln[A]_t - \ln[A]_0) = kt - kt_0$$
$$-\ln[A]_t + \ln[A]_0 = kt$$
$$\ln[A]_t = -kt + \ln[A]_0$$

in which $[A]_t$ is the concentration at time t and $[A]_0$ is the initial concentration. The *integrated first-order rate law* is $\ln[A]_t = -kt + \ln[A]_0$, which shows how the concentration of A changes with time. Thus, if the initial concentration and the rate constant are known, $[A]$ at any time can be calculated. The integrated first-order rate law can also be expressed in terms of the ratio of $[A]_t$ and $[A]_0$.

$$\ln\frac{[A]_0}{[A]_t} = kt$$

$$[A]_t = [A]_0 e^{-kt}$$

The last equation shows that, for first-order reactions, the concentration of A decays exponentially with time.

WORKED EXAMPLE 15.6

Concentration–time calculations for first-order reactions

Dinitrogen pentoxide, N_2O_5, is not very stable. In the gas phase or dissolved in a nonpolar solvent, such as tetrachloromethane, it decomposes by a first-order reaction into N_2O_4 and O_2.

$$2N_2O_5 \rightarrow 2N_2O_4 + O_2$$

The experimentally determined rate law is:

$$\text{rate of reaction} = k[N_2O_5]$$

At 45 °C, the rate constant for the reaction in tetrachloromethane is $6.22 \times 10^{-4}\,\text{s}^{-1}$. If the initial concentration of N_2O_5 in the solution is 0.100 M, how long will it take for the concentration to drop to 0.0100 M?

Analysis

To solve the problem, we substitute quantities into:

$$\ln\frac{[N_2O_5]_0}{[N_2O_5]_t} = kt$$

and solve for t.

Solution

First, we assemble the data.

$$[N_2O_5]_0 = 0.100\,\text{M} \qquad [N_2O_5]_t = 0.0100\,\text{M}$$

$$k = 6.22 \times 10^{-4}\,\text{s}^{-1} \qquad t = ?$$

Substituting this into the equation:

$$\ln\frac{0.100\,\text{M}}{0.0100\,\text{M}} = (6.22\times10^{-4}\,\text{s}^{-1})t$$

Since the unit for concentration (M) cancels out, we have:

$$\ln 10.0 = (6.22\times10^{-4}\,\text{s}^{-1})t$$

$$2.303 = (6.22\times10^{-4}\,\text{s}^{-1})t$$

As we have seen in earlier chapters, when taking a logarithm of a quantity, the number of digits written *after the decimal point* equals the number of significant figures in the quantity. Here, 10.0 has three significant figures, so the natural logarithm of 10.0, which is 2.303, has three digits after the decimal.

Solving for t:

$$t = \frac{2.303}{6.22\times10^{-4}\,\text{s}^{-1}}$$

$$= 3.70 \times 10^3\,\text{s}$$

This is the time in seconds.

Is our answer reasonable?

We can substitute the time we obtained, along with the given initial concentration, into our equation and see whether we get back the given concentration at time t.

$$[N_2O_5]_t = [N_2O_5]_0 e^{-kt} = 0.100\,\text{M} \times e^{(-6.22\times10^{-4}\,\text{s}^{-1}\times 3.70\times10^3\,\text{s})}$$

$$= 0.01\,\text{M}$$

which is exactly what we started with.

PRACTICE EXERCISE 15.6

If the initial concentration of N_2O_5 in a tetrachloromethane solution at 45 °C is 0.500 M (see worked example 15.6), what will its concentration be after exactly 1 hour?

The equation $\ln[A]_t = -kt + \ln[A]_0$ is of the form $y = mx + c$, where a plot of y versus x gives a straight line with the slope 'm' and the intercept 'c'.

$$\begin{array}{ccccc} \ln[A]_t & = & -k\,t & + & \ln[A]_0 \\ \updownarrow & & \updownarrow\updownarrow & & \updownarrow \\ y & = & \mathrm{m}\,x & + & \mathrm{c} \end{array}$$

For first-order reactions a plot of the natural logarithm of concentration versus reaction time always gives a straight line. This fact is often used to determine the order of a reaction. For the reaction $A \rightarrow$ products, the plot of $\ln[A]_t$ versus t is a straight line, if the reaction is first order. Conversely, if the plot does not give a straight line, then the reaction is not first order. Worked example 15.7 illustrates the procedure.

WORKED EXAMPLE 15.7

Determining the order of a reaction

The following data are for the reaction $N_2O_5(g) \rightarrow N_2O_4(g) + \frac{1}{2}O_2(g)$.

Time (s)	$[N_2O_5]$ (mol L^{-1})	Time (s)	$[N_2O_5]$ (mol L^{-1})
0	2.000×10^{-1}	200	5.000×10^{-2}
50	1.414×10^{-1}	300	2.500×10^{-2}
100	1.000×10^{-1}	400	1.250×10^{-2}

(a) Show that the rate law is first order in $[N_2O_5]$.
(b) Calculate the value of the rate constant.

Analysis

(a) If a plot of $\ln[N_2O_5]$ versus time, according to $\ln[N_2O_5]_t = -kt + \ln[N_2O_5]_0$, gives a straight line, the reaction must be first order in $[N_2O_5]$.
(b) The slope of the plot in (a), if it is a straight line, is equivalent to the negative of the rate constant.

Solution

First we calculate the natural logarithms of the $[N_2O_5]$ values, which gives the following data.

Time (s)	$\ln[N_2O_5]$	Time (s)	$\ln[N_2O_5]$
0	−1.609	200	−2.996
50	−1.956	300	−3.689
100	−2.303	400	−4.382

We then plot $\ln[N_2O_5]$ versus time.

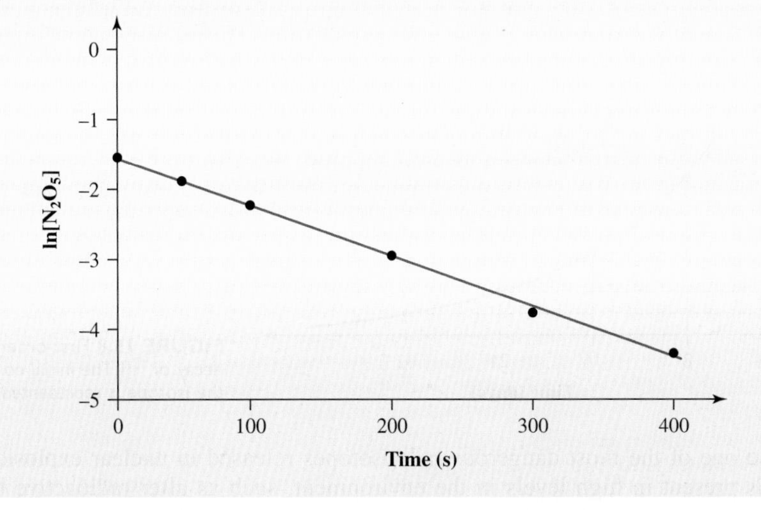

The plot is a straight line, which confirms that the reaction is first order in $[N_2O_5]$ (we could also use a calculator to do this analysis). In this case, the slope of the line equals $-k$. This gives us:

$$k = -\text{slope} = 6.930 \times 10^{-3}\ \text{s}^{-1}$$

Is our answer reasonable?

We can substitute this value back into the original equation $\ln[N_2O_5]_t = -kt + \ln[N_2O_5]_0$ and ensure we obtain the same $[N_2O_5]$ given in the table of data. Doing this shows that our answer is correct.

PRACTICE EXERCISE 15.7 Using the data given in worked example 15.7, calculate $[N_2O_5]$ at $t = 150$ s.

Half-life of first-order reactions

For a first-order reaction, the half-life of the reactant can be obtained from:

$$\ln\frac{[A]_0}{[A]_t} = kt$$

by setting $[A]_t$ equal to half of the initial concentration $[A]_0$.

$$[A]_t = \tfrac{1}{2}[A]_0$$

Substituting $\frac{1}{2}[A]_0$ for $[A]_t$ and $t_{\frac{1}{2}}$ for t in the original equation gives:

$$\ln\frac{[A]_0}{\frac{1}{2}[A]_0} = kt_{\frac{1}{2}}$$

Note that the left-hand side of the equation simplifies to ln2, and so, to solve the equation for $t_{\frac{1}{2}}$, we have:

$$t_{\frac{1}{2}} = \frac{\ln 2}{k} = \frac{0.693}{k}$$

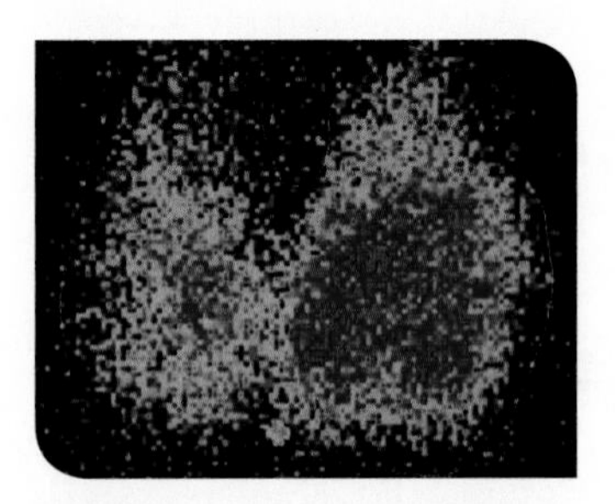

FIGURE 15.7 Scintigram of thyroid glands after the patient was injected with radioactive iodine. This shows the activity of the gland from blue (most active) to yellow (least active).

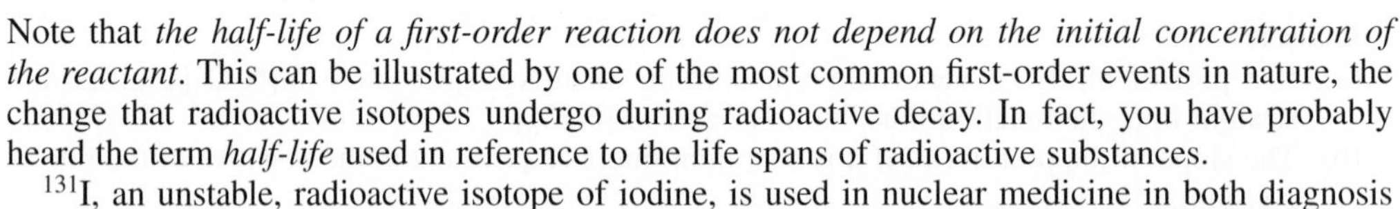

Note that *the half-life of a first-order reaction does not depend on the initial concentration of the reactant*. This can be illustrated by one of the most common first-order events in nature, the change that radioactive isotopes undergo during radioactive decay. In fact, you have probably heard the term *half-life* used in reference to the life spans of radioactive substances.

^{131}I, an unstable, radioactive isotope of iodine, is used in nuclear medicine in both diagnosis and therapy of thyroid disorders, such as thyroid cancer and thyrotoxicosis. The thyroid gland is a small organ located just below the 'Adam's apple' and astride the windpipe. It uses iodide ions to make a hormone, so, when a patient is given a dose of $^{131}I^-$ mixed with nonradioactive I^-, both ions are taken up by the thyroid gland (see figure 15.7). The temporary change in radioactivity of the gland is a measure of thyroid activity. ^{131}I undergoes a nuclear reaction where it emits beta radiation and changes into a stable isotope of xenon. The intensity of the radiation decreases, or decays, with time (see figure 15.8). Notice that the time it takes for the first half of the ^{131}I to disappear is 8 days. Then, during the next 8 days, half of the remaining ^{131}I disappears, and so on. Regardless of the initial amount, it takes 8 days for half of that amount of ^{131}I to disappear, which means that the half-life of ^{131}I is a constant.

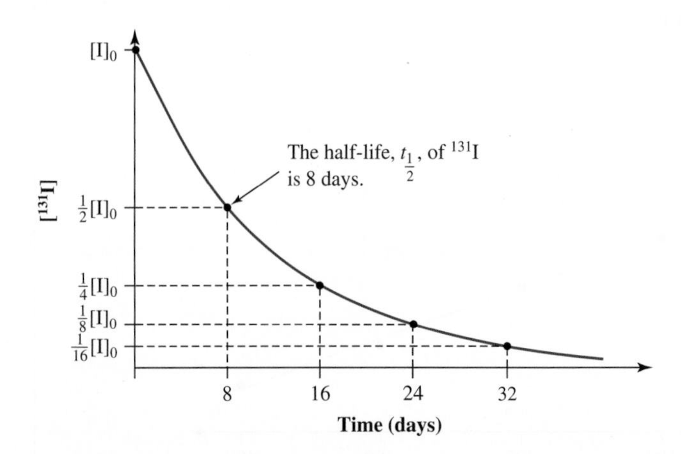

FIGURE 15.8 First-order radioactive decay of ^{131}I. The initial concentration of the isotope is represented by $[I]_0$.

^{131}I is also one of the most dangerous radioisotopes released in nuclear explosions or accidents. If it is present in high levels in the environment, such as after radioactive fallout, it is absorbed by the body after eating contaminated food and concentrates in the thyroid gland. When ^{131}I decays, it can damage the thyroid gland and may lead to thyroid cancer. One possible treatment is an iodine supplement, which increases the total amount of iodine in the body. This, therefore, reduces the relative proportion of ^{131}I and also lowers the uptake and retention of this radioactive isotope in the body. Such iodine supplements were distributed to people living near the Fukushima nuclear power plant in Japan after it exploded in 2011.

WORKED EXAMPLE 15.8

Half-life calculations

The half-life of radioactive ^{131}I is 8.0 days. What fraction of the initial ^{131}I would be present in a patient after 32.0 days if none of it were eliminated through natural body processes?

Analysis

Having learned that ^{131}I is a radioactive isotope and so decays by a first-order process, we know that the half-life is constant and does not depend on the ^{131}I concentration.

Solution

We can solve the problem using the integrated first-order rate law. We'll need the first-order rate constant k, which we can obtain from the half-life by rearranging:

$$t_{\frac{1}{2}} = \frac{\ln 2}{k} = \frac{0.693}{k}$$

$$k = \frac{0.693}{t_{\frac{1}{2}}} = \frac{0.693}{8.0\text{ day}} = 0.0866\text{ day}^{-1}$$

Then we can use $\ln\frac{[^{131}\text{I}]_0}{[^{131}\text{I}]_t} = kt$ to calculate the fraction $\frac{[^{131}\text{I}]_0}{[^{131}\text{I}]_t}$.

$$\ln\frac{[^{131}\text{I}]_0}{[^{131}\text{I}]_t} = kt = (0.0866\text{ day}^{-1})(32.0\text{ day}) = 2.77$$

Taking the antilogarithm of both sides, we have:

$$\frac{[^{131}\text{I}]_0}{[^{131}\text{I}]_t} = e^{2.77} = 16.0$$

Is our answer reasonable?

A period of 32.0 days is exactly four half-lives. If we take the fraction initially present to be 1, we can set up a table.

Half-life	0	1	2	3	4
Fraction	1	$\frac{1}{2}$	$\frac{1}{4}$	$\frac{1}{8}$	$\frac{1}{16}$

Half of the ^{131}I is lost in the first half-life, half of that disappears in the second half-life, and so on. Therefore, the fraction remaining after four half-lives is $\frac{1}{16}$.

The initial concentration, $[^{131}\text{I}]_0$, is 16.0 times as large as the concentration after 32.0 days. The fraction remaining after 32.0 days is $\frac{1}{16}$, which is exactly what we obtained using the mathematical approach.

Second-order rate laws

For a general reaction involving one reactant, which is second order in $[A]$:

$$A \rightarrow \text{products}$$

the second-order differential rate law is:

$$\text{rate of reaction} = -\frac{d[A]}{dt} = k[A]^2$$

Again, to look at how the concentration of B changes over time, we have to rearrange this equation to give $-\frac{d[A]}{[A]^2} = k\text{d}t$, and then integrate this between the start of the reaction ($t = 0$) and a certain reaction time (t).

$$-\int_{[A]_0}^{[A]_t} \frac{d[A]}{[A]^2} = k\int_{t=0}^{t} dt$$

As $\int \frac{d[A]}{[A]^2} = -\frac{1}{[A]}$, and ignoring the constant of integration, we obtain:

$$-\left(\frac{-1}{[A]_t} - \frac{-1}{[A]_0}\right) = kt - kt_0$$

$$-\left(\frac{-1}{[A]_t} + \frac{1}{[A]_0}\right) = kt$$

$$\frac{1}{[A]_t} - \frac{1}{[A]_0} = kt$$

$$\frac{1}{[A]_t} = kt + \frac{1}{[A]_0}$$

This equation is the *integrated second-order rate law*. It shows that for second-order reactions the reciprocal of the concentration is related linearly to time.

WORKED EXAMPLE 15.9

Concentration–time calculations for second-order reactions

Nitrosyl chloride, NOCl, decomposes slowly to NO and Cl_2.

$$2NOCl \rightarrow 2NO + Cl_2$$

The rate law shows that the rate of reaction is second order in [NOCl].

$$\text{rate of reaction} = k[NOCl]^2$$

The rate constant k equals 0.020 $mol^{-1}\,L\,s^{-1}$ at a certain temperature. If the initial concentration of NOCl in a closed reaction vessel is 0.050 M, what is the concentration after 30 minutes?

Analysis

We are given a rate law and can see that it is for a second-order reaction. We must calculate $[NOCl]_t$, the molar concentration of NOCl, after 30 min (1800 s) using $\frac{1}{[NOCl]_t} = kt + \frac{1}{[NOCl]_0}$.

Solution

We know that:

$[NOCl]_0 = 0.050\,M\ (= 0.050\,mol\,L^{-1})$ $k = 0.020\,mol^{-1}\,L\,s^{-1}$

$[NOCl]_t = ?\,M$ $t = 1800\,s$

The general integrated second-order rate law is:

$$\frac{1}{[NOCl]_t} = kt + \frac{1}{[NOCl]_0}$$

Making the substitution gives:

$$\frac{1}{[NOCl]_t} = (0.020\,mol^{-1}\,L\,s^{-1} \times 1800\,s) + \frac{1}{0.050\,mol\,L^{-1}}$$

Solving for $\frac{1}{[NOCl]_t}$ gives:

$$\frac{1}{[NOCl]_t} = 36\,mol^{-1}\,L + 20\,mol^{-1}\,L$$

$$\frac{1}{[NOCl]_t} = 56\,mol^{-1}\,L$$

Taking the reciprocals of both sides gives us the value of [NOCl]:

$$[NOCl]_t = \frac{1}{56\,mol^{-1}\,L}$$

$$= 0.018\,mol\,L^{-1}$$

$$= 0.018\,M$$

The molar concentration of NOCl has decreased from 0.050 M to 0.018 M after 30 minutes.

Is our answer reasonable?

The concentration of NOCl has decreased, as it must. We can check the calculation by substituting this value into the original equation.

PRACTICE EXERCISE 15.8

For the reaction in worked example 15.9, determine how long it would take for the NOCl concentration to drop from 0.040 M to 0.010 M.

The rate constant for a second-order reaction with a rate law following $\frac{1}{[A]_t} = kt + \frac{1}{[A]_0}$ can be determined graphically by a method similar to that used for a first-order reaction. The equation corresponds to an equation for a straight line:

$$\frac{1}{[A]_t} = kt + \frac{1}{[A]_0}$$

$$y = mx + c$$

When a reaction is second order, a plot of $\frac{1}{[A]_t}$ versus t should yield a straight line with a slope k. This is illustrated in figure 15.9b for the decomposition of HI, according to $2HI(g) \rightarrow H_2(g) + I_2(g)$, using data from worked example 15.1. Figure 15.9a shows a plot of ln[HI] versus time; the fact that this is a curve confirms the reaction is not first order in [HI].

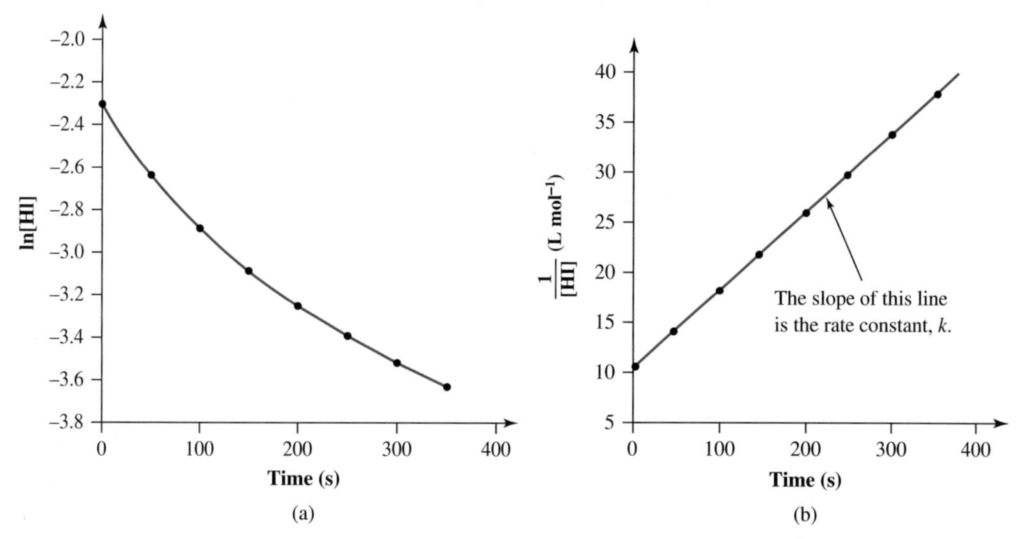

FIGURE 15.9 Second-order kinetics: **(a)** a graph of ln[HI] versus time and **(b)** a graph of $\frac{1}{[HI]}$ versus time for the data in worked example 15.1.

Half-life of second-order reactions

After one half-life of a second-order reaction has elapsed, we have:

$$[A]_t = \frac{[A]_0}{2}$$

Inserting this into $\frac{1}{[A]_t} = kt + \frac{1}{[A]_0}$ gives:

$$\frac{1}{\frac{[A]_0}{2}} = kt_{\frac{1}{2}} + \frac{1}{[A]_0}$$

$$\frac{1}{[A]_0} = kt_{\frac{1}{2}}$$

$$t_{\frac{1}{2}} = \frac{1}{k[A]_0}$$

Thus, in contrast to first-order reactions, the half-life of a second-order reaction is inversely proportional to the initial concentration of the reactant.

WORKED EXAMPLE 15.10

Half-life calculations

The reaction $2HI(g) \rightarrow H_2(g) + I_2(g)$ has the rate law, rate $= k[HI]^2$, with $k = 0.079\ mol^{-1}\ L\ s^{-1}$ at 508 °C. What is the half-life for this reaction at this temperature when the initial HI concentration is 0.375 M?

Analysis

The rate law tells us that the reaction is second order. To calculate the half-life, we need to use $t_{\frac{1}{2}} = \frac{1}{k[HI]_0}$.

Solution

The initial concentration is 0.375 mol L^{-1} and $k = 0.079\ mol^{-1}\ L\ s^{-1}$. Substituting these values into $t_{\frac{1}{2}} = \frac{1}{k[HI]_0}$ gives:

$$t_{\frac{1}{2}} = \frac{1}{(0.079\ mol^{-1}\ L\ s^{-1})(0.375\ mol\ L^{-1})}$$
$$= 33.7\ s$$

Is our answer reasonable?

Substituting this value back into the equation above shows that this answer is correct.

PRACTICE EXERCISE 15.9 Suppose that the value of $t_{\frac{1}{2}}$ for a certain reaction was found to be independent of the initial concentration of the reactants. What could you say about the order of the reaction?

PRACTICE EXERCISE 15.10 Nitrogen dioxide gas (NO_2) decomposes to produce nitrogen oxide gas (NO) and oxygen gas (O_2).

$$NO_2(g) \rightarrow NO(g) + O_2(g)$$

At 100 °C, the rate of this decomposition is directly proportional to the square of the NO_2 concentration. The rate constant for the decomposition at this temperature has been determined as 10.1 L $mol^{-1}\ s^{-1}$. For $[NO_2]_0 = 3.75 \times 10^{-3}\ mol\ L^{-1}$, calculate:

(a) the rate of the reaction at 100 °C

(b) the half-life of this decomposition.

Zero-order rate laws

While most reactions that involve only one reactant are first- or second-order reactions, there are reactions for which the rate of the reaction is constant; that is, it does not vary with concentration. An example is the thermal decomposition of bromoethane on a zinc surface, which acts as a catalyst.

$$C_2H_5Br(g) \xrightarrow{Zn} CH_2{=}CH_2(g) + HBr(g)$$

The reaction occurs at the surface of the zinc metal, which is saturated with bromoethane. Therefore, the reaction rate is not dependent on the concentration of bromoethane. Such reactions are called *zero-order reactions*. For the general zero-order reaction $A \rightarrow$ products, we can write the rate law as:

$$\text{rate of reaction} = -\frac{d[A]}{dt} = k[A]_0 = k$$

How does the concentration of A change over time? As before, we arrange the rate law to give $-d[A] = kdt$ and, by integration between the beginning of the reaction ($t = 0$) and a certain reaction time (t), the *integrated rate law for a zero-order reaction* can be obtained.

$$-\int_{[A]_0}^{[A]_t} d[A] = k\int_{t=0}^{t} dt$$
$$-([A]_t - [A]_0) = kt - kt_0$$
$$[A]_0 - [A]_t = kt$$
$$[A]_t = -kt + [A]_0$$

This equation corresponds to an equation for a straight line.

$$\begin{array}{ccccc} [A]_t & = & -k\,t & + & [A]_0 \\ \updownarrow & & \updownarrow\updownarrow & & \updownarrow \\ y & = & m\,x & + & c \end{array}$$

A plot of $[A]$ versus time gives a straight line of slope $-k$, as shown in figure 15.10.

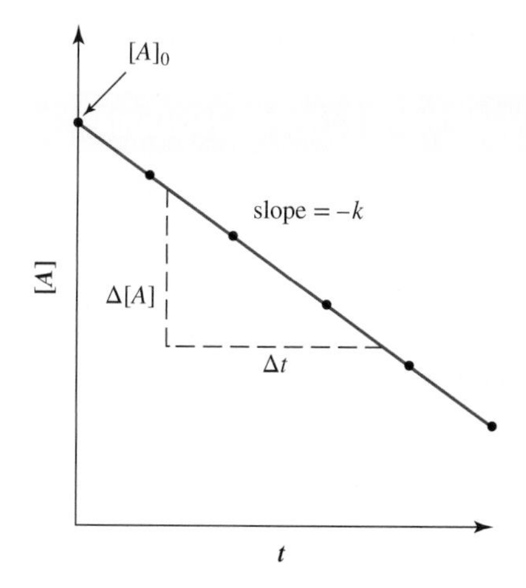

FIGURE 15.10 For a zero-order reaction, [*A*] versus *t* is a straight line with slope –*k*.

The half-life of a zero-order reaction is $[A]_t = \frac{[A]_0}{2}$ when $t = t_{\frac{1}{2}}$. Inserting this into the equation $[A]_t = -kt + [A]_0$ gives:

$$\frac{[A]_0}{2} = -kt_{\frac{1}{2}} + [A]_0$$

$$kt_{\frac{1}{2}} = \frac{[A]_0}{2}$$

$$t_{\frac{1}{2}} = \frac{[A]_0}{2k}$$

For a zero-order reaction, the half-life is proportional to the concentration of the reactant.

Many reactions which require surfaces of metals show zero-order kinetics. The example shown in figure 15.11 is the decomposition of dinitrogen oxide, N_2O, into nitrogen and oxygen on a hot platinum surface.

$$2N_2O(g) \xrightarrow{Pt} 2N_2(g) + O_2(g)$$

When the platinum surface is completely covered with N_2O molecules, an increase in $[N_2O]$ does not change the rate of the reaction. N_2O is known as 'laughing gas' and is used as an anaesthetic in medicine and dentistry.

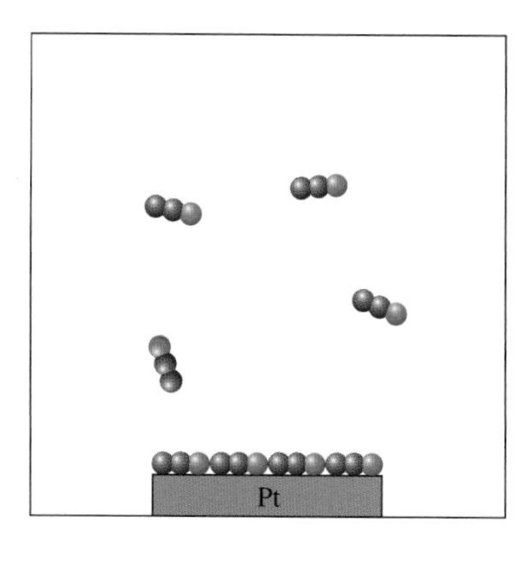

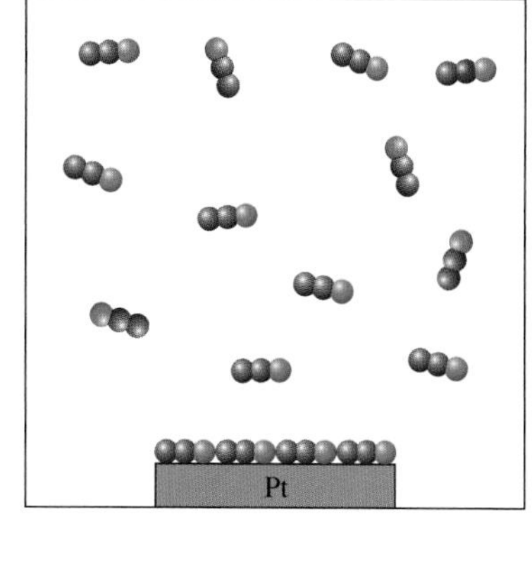

N_2O

(a) (b)

FIGURE 15.11 The decomposition reaction $2N_2O(g) \rightarrow 2N_2(g) + O_2(g)$ takes place on a platinum surface. The rate of decomposition of N_2O is the same in **(a)** and **(b)**, regardless of the concentration of N_2O, because the platinum surface can accommodate only a certain number of molecules. As a result, this reaction is zero order in $[N_2O]$.

Also, many thermal decomposition reactions follow zero-order kinetics, where the reaction rate is independent of the concentration of the reactant. Ethanol metabolism in the body is catalysed by the enzyme alcohol dehydrogenase. Because only a limited amount of ethanol can be removed in this way, if there is a high concentration of ethanol in the blood, metabolism can follow zero-order kinetics (see also pp. 682–3).

Before moving on to look at the theory of chemical kinetics, study table 15.5, which summarises the kinetics for reactions of different orders.

TABLE 15.5 Summary of the kinetics for reactions of the type $A \rightarrow$ products that are zero, first or second order in $[A]$.

	Zero order	First order	Second order
Rate law	rate = k	rate = $k[A]$	rate = $k[A]^2$
Integrated rate law	$[A] = -kt + [A]_0$	$\ln[A] = -kt + \ln[A]_0$	$\frac{1}{[A]} = kt + \frac{1}{[A]_0}$
Linear graph	$[A]$ versus t	$\ln[A]$ versus t	$\frac{1}{[A]}$ versus t
Slope of linear graph	$-k$	$-k$	k
Units of rate constant	mol L^{-1} s^{-1}	s^{-1}	mol^{-1} L s^{-1}
Half-life ($t_{\frac{1}{2}}$)	$\frac{[A]_0}{2k}$	$\frac{\ln 2}{k}$	$\frac{1}{k[A]_0}$

15.5 Theory of chemical kinetics

In section 15.2 we mentioned that nearly all reactions proceed faster at higher temperatures. Generally, the rate of reaction increases by a factor of about two or three for each 10 °C increase in temperature, although the exact amount of increase differs from one reaction to another. Temperature evidently has a significant effect on reaction rate. To understand why, we need to develop some theoretical models that explain our observations. One of the simplest models is called **collision theory**.

Collision theory

The basic principle of collision theory is that the rate of a reaction is proportional to the number of effective collisions per second among the reactant molecules. An *effective collision* is one that actually results in the formation of product molecules. Anything that can increase the frequency of effective collisions should, therefore, increase the rate.

One of the several factors that influence the number of effective collisions per second is *concentration*. As reactant concentrations increase, the number of collisions per second of all types, including effective collisions, also increases.

Not every collision between reactant molecules actually results in a chemical change. We know this because the reactant atoms or molecules in a gas or a liquid undergo an enormous number of collisions per second with each other. If every collision were effective, all reactions would be over in an instant. *Only a very small fraction of all the collisions can really lead to a net change.* We will now explore the reason for this behaviour.

Molecular orientation

In most reactions, when two reactant molecules collide they must be oriented correctly for a reaction to occur. For example, the reaction represented by the following equation:

$$2NO_2Cl \rightarrow 2NO_2 + Cl_2$$

appears to proceed by a two-step mechanism, which we will discuss in section 15.6. At this point, it is sufficient to know that one step involves the collision of an NO_2Cl molecule with a chlorine atom, Cl.

$$NO_2Cl + Cl \rightarrow NO_2 + Cl_2$$

The orientation of the NO_2Cl molecule when hit by the Cl atom is important (see figure 15.12). The orientation shown in figure 15.12a cannot result in the formation of Cl_2 because the two Cl atoms are not being brought close enough together for a new Cl—Cl bond to form as an N—Cl bond breaks. Figure 15.12b shows the necessary orientation of NO_2Cl and Cl that leads to successful product formation.

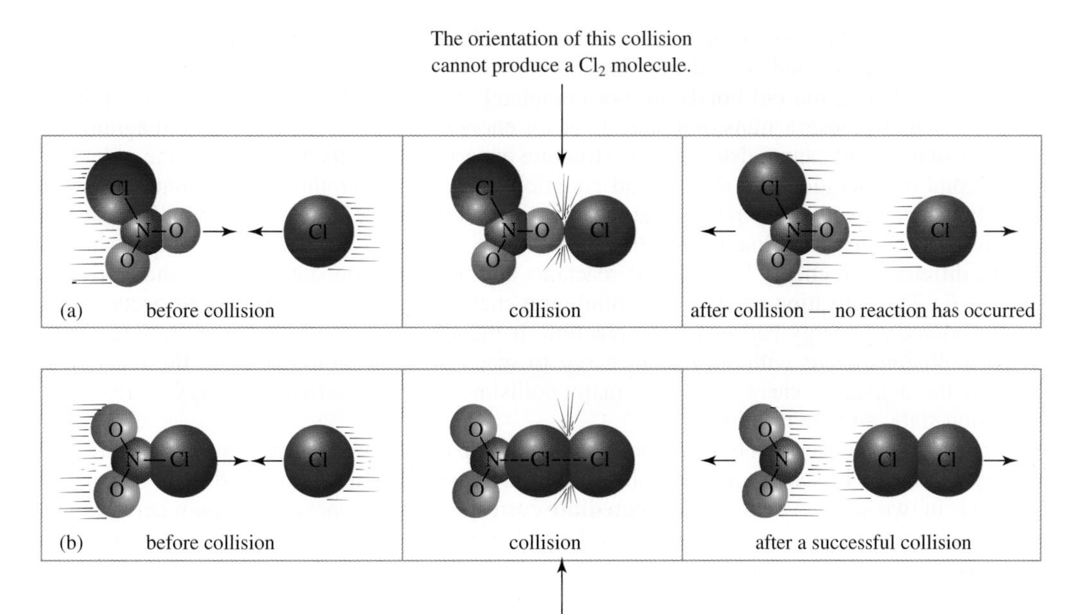

FIGURE 15.12 The importance of molecular orientation during a collision in a reaction. The key step in the decomposition of NO_2Cl to NO_2 and Cl_2 is the collision of Cl with a NO_2Cl molecule.
(a) An ineffectively oriented collision.
(b) An effectively oriented collision.

Activation energy

Not all collisions, even those correctly oriented, are energetic enough to result in the formation of products; this is the major reason why only a small percentage of all collisions actually lead to a chemical change.

Why is this so? This question was first addressed by Svante Arrhenius (Nobel Prize in chemistry, 1903) in the 1880s. He proposed the existence of a *threshold* energy, the **activation energy (E_a)**, which must be overcome for a reaction to occur. This can be rationalised by considering the previous reaction of NO_2Cl with Cl. In this reaction, an N—Cl bond must break and a Cl—Cl bond must form. Breaking an N—Cl bond requires a considerable energy of 188 kJ mol^{-1}, which must come from somewhere. The collision theory postulates that the energy required to break the bond comes from the *kinetic energy* (KE) of the molecules before the collision. This kinetic energy is converted into *potential energy* (PE) as the molecules are distorted during a collision; old bonds in the reactant molecules are broken, and new bonds are formed in the product molecules. For most chemical reactions, the activation energy is quite large, and only a small fraction of all well-oriented, colliding molecules have it.

The reaction progress can be plotted as shown in figure 15.13, which is called a *potential energy diagram*.

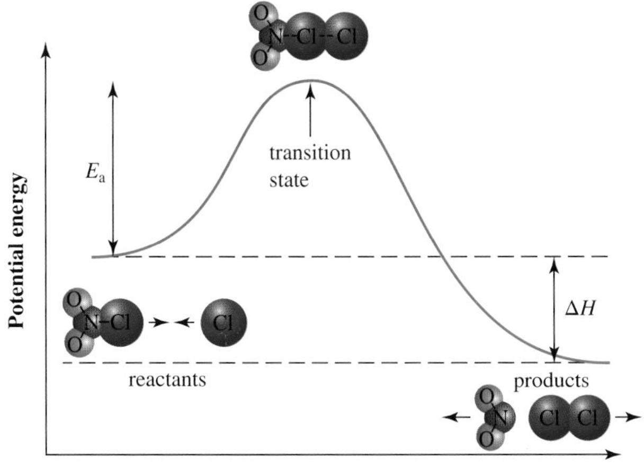

FIGURE 15.13 Potential energy diagram for the reaction:

$$NO_2Cl + Cl \rightarrow NO_2 + Cl_2$$

The activation energy, E_a, is the combined kinetic energy that the colliding particles must have for the reaction to occur. ΔH is the enthalpy difference between products and reactants. In an exothermic reaction, which is shown here, ΔH is negative ($\Delta H < 0$).

The vertical axis represents potential energy, which changes during a reaction as the kinetic energy of the colliding particles is converted into potential energy. The horizontal axis is called the **reaction coordinate**, and it represents the extent to which the reactants have changed to the products. The arrangement found on the top of the potential energy 'hill' (or barrier) is called the **transition state**. At the transition state, sufficient energy is concentrated to allow the bonds in the reactants to break. As they break, energy is redistributed and new bonds form, giving products. Once the transition state is reached, the reaction can either proceed to give products or, in the reverse direction, to give reactants. A transition state has a definite geometry, a definite arrangement of bonding and nonbonding electrons, and a definite distribution of electron density and charge; the old bonds are not completely broken and the new bonds have not yet been formed. Because a transition state is at an energy maximum on an energy diagram, we cannot isolate it and cannot determine its structure experimentally. Its lifetime is of the order of a picosecond (the duration of a single bond vibration). However, even though we cannot observe a transition state directly by experimental means, we can often infer a great deal about its probable structure from other experimental observations.

The difference in energy between the reactants and the transition state is called the activation energy, E_a. The activation energy is the minimum energy required for a reaction to occur; it can be considered an energy barrier for the reaction. If the activation energy is large, very few molecular collisions occur with sufficient energy to reach the transition state, and the reaction is slow. If the activation energy is small, many collisions generate sufficient energy to reach the transition state, and the reaction is fast.

In a reaction that occurs in two or more steps, each step has its own transition state and activation energy. Figure 15.14 shows an energy diagram for a hypothetical reaction $A + B \rightarrow C + D$, which proceeds in two steps. A **reaction intermediate** corresponds to an energy minimum between two transition states, in this case transition states 1 and 2. Note that, because the energies of the reaction intermediates are higher than the energies of either the reactants or the products, these intermediates are highly reactive and can only rarely be isolated. However, because intermediates have a certain lifetime (in contrast to transition states), searching for the presence of a reaction intermediate with optical or other fast recording methods may provide experimental support for a reaction mechanism. It is important that you become familiar with the definition of reaction intermediates and transition states in order to avoid mixing them up.

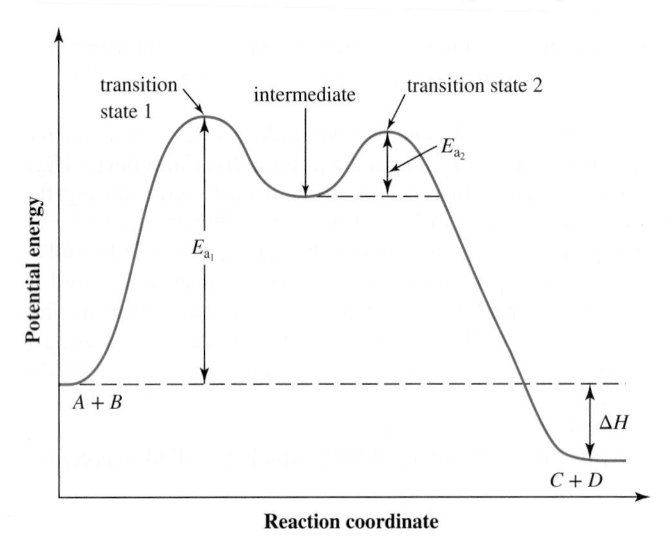

FIGURE 15.14 Energy diagram for a two-step reaction involving the formation of an intermediate. The enthalpy of the reactants is higher than that of the products, and ΔH for the conversion of $A + B$ to $C + D$ is therefore negative.

Both the reaction of NO_2Cl with Cl to give NO_2 and Cl_2 shown in figure 15.13 and our hypothetical reaction $A + B \rightarrow C + D$ in figure 15.14 are *exothermic*, as indicated by the fact that the products have a lower potential energy than the reactants (energy difference, or enthalpy of reaction $\Delta H < 0$). A potential energy diagram for an *endothermic* reaction, where the potential energy of the products is higher than that of the reactants ($\Delta H > 0$), is shown in figure 15.15.

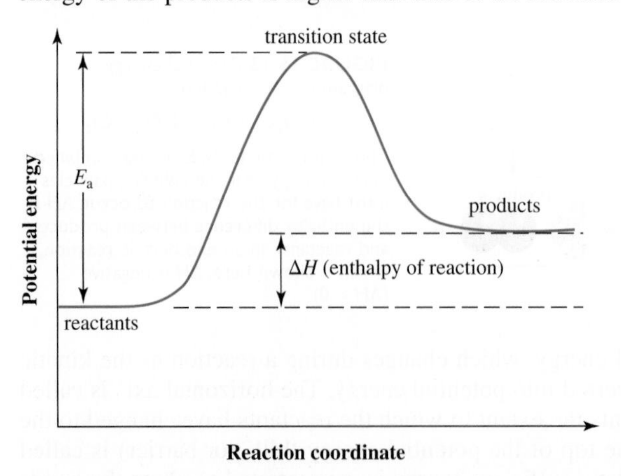

FIGURE 15.15 A potential energy diagram for an endothermic reaction.

It is important to understand that ΔH has no influence on the rate of the reaction. *The rate of a reaction is solely determined by the height of the activation barrier, E_a.*

Let's come back to our reaction of NO_2Cl with Cl atoms in figure 15.13. A certain minimum energy, E_a, is required for an NO_2Cl and a Cl atom species to get over the hill and form products. This is energy provided by the collision between these species. A collision between an NO_2Cl molecule and a Cl atom with small kinetic energies does not provide enough energy to get over the barrier, and no reaction occurs — even if their orientation is ideal. At a given temperature, only a certain fraction of the collisions possess enough energy to be effective and form products.

The concept that a reaction involving two molecules can occur only after successful a collision between these species seems quite logical. However, many reactions involve only one molecule: for example, the decomposition of N_2O_5 to N_2O_4 and O_2 shown in worked example 15.7. How do molecules in such reactions acquire the energy required to surmount the energy barrier associated with breaking the bonds? This question was addressed for reactions in the gas phase by the English scientists Sir Cyril Hinshelwood (Nobel Prize in chemistry, 1956; figure 15.16) and Fredrick Alexander Lindemann (figure 15.17) who suggested that the required activation energy is accumulated through collisions between these molecules, which produce energised ('activated') molecules. Qualitatively, the rate of the unimolecular reaction is therefore proportional to the concentration of activated molecules, which in turn is proportional to the concentration of unenergised molecules.

FIGURE 15.16 Sir Cyril Hinshelwood (Nobel Prize in chemistry, 1956).

FIGURE 15.17 Frederick Lindemann, Chief Scientific Advisor to Winston Churchill during World War II.

Temperature effects: the Arrhenius equation

With the concept of activation energy, we can explain why the rate of a reaction increases so much with increasing temperature. The two plots in figure 15.18 correspond to different temperatures for the same mixture of reactants. Each curve is a plot of the different fractions of all collisions (vertical axis) that have particular values of kinetic energy of collision (horizontal axis). The total area under a curve then represents the total number of collisions, because all of the fractions must add up to this total. Notice what happens to the plots when the temperature is increased; the maximum point shifts to the right and the curve flattens somewhat.

The shaded areas under the curves in figure 15.18 represent the sum of all the fractions of the total collisions that equal or exceed the activation energy. This sum — we could call it the *reacting fraction* — is much greater at the higher temperature than at the lower temperature because a significant fraction of the curve shifts beyond the activation energy with even a modest increase in temperature. In other words, at the higher temperature, a much greater fraction of the collisions result in a chemical change, so the reactants react faster at the higher temperature.

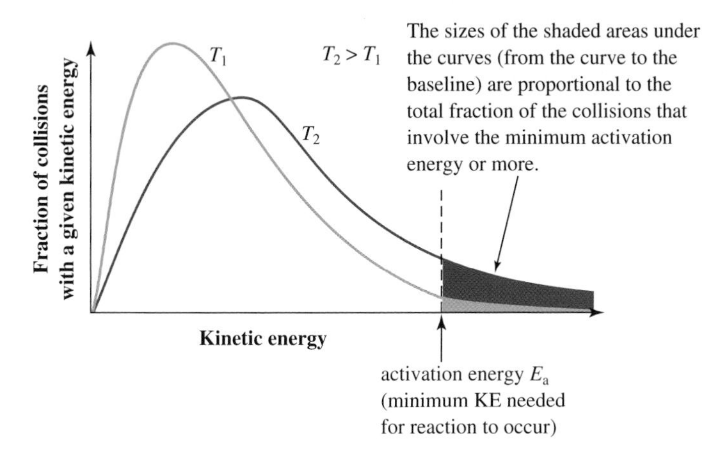

FIGURE 15.18 Kinetic energy distributions for a reaction mixture at two different temperatures T_1 and T_2.

In fact, the fractions of effective collisions increase *exponentially* with temperature. Arrhenius postulated that the number of collisions having an equal or greater energy than the activation energy is given by the expression:

$$\frac{N}{N_0} = e^{-E_a/RT}$$

where: N = number of collisions with at least E_a
N_0 = total number of collisions
E_a = activation energy (J mol^{-1})
R = gas constant (8.314 J mol^{-1} K^{-1})
T = temperature (K)

The term $e^{-E_a/RT}$ is the probability that any given collision will result in a reaction.

The shapes of the curves in figure 15.18 are given by the so-called Maxwell–Boltzmann distributions. All molecules of a particular chemical species have the same mass so their kinetic energies depend only on the velocities of the particles (kinetic energy = $\frac{1}{2}mv^2$). In any mixture of moving particles, some particles have a low speed (low energy), some have a high speed (high energy), but most have a speed very close to the average. The Maxwell–Boltzmann distribution shows how the speeds (energies) of moving particles in a mixture change with temperature. Figure 15.18 illustrates some important aspects of Maxwell–Boltzmann distributions, for example, there are no molecules at zero energy and few molecules at high energy. Most importantly, there is no maximum energy value (the curve does not touch the x-axis).

Thus, summarising our discussion so far, two requirements must be satisfied for a successful reaction.

1. The collision energy must equal or exceed the activation energy.
2. The relative orientation of the reactants must allow formation of any new bonds that are necessary to produce the products of the reaction.

With these two requirements, the rate constant k can now be expressed as:

$$k = zpe^{-E_a/RT}$$

In this equation, z represents the collision frequency (the total number of collisions per second). The factor p is called the **steric factor**, and represents the fraction of collisions with correct orientations so that the actual chemical reaction can occur. The factor $e^{-E_a/RT}$ reflects the fraction of collisions with sufficient energy to surmount the activation barrier. This equation is most commonly written as:

$$k = Ae^{-E_a/RT}$$

This is called the **Arrhenius equation**. The factors zp are replaced by A, which is called the **pre-exponential** or **frequency factor** for this reaction. A has the same units as the rate constant and this depends on the order of the reaction. In principle, A is dependent on the temperature, since the frequency of collisions increases with increasing temperature. However, this temperature dependence is much smaller than the exponential term. As an approximation, A is therefore usually considered as a constant.

How can the activation energy be determined experimentally? Taking the natural logarithm of each side of the Arrhenius equation gives:

$$\ln k = -\frac{E_a}{R}\left(\frac{1}{T}\right) + \ln A$$

This is a linear equation of the type $y = mx + c$,

$$\ln k = -\frac{E_a}{R}\left(\frac{1}{T}\right) + \ln A$$

$$\updownarrow \quad \updownarrow \quad \updownarrow \quad \updownarrow$$

$$y = \quad m \quad x \quad + \ c$$

Thus, by measuring the rate constant at different temperatures for a reaction, a plot of $\ln k$ versus $\frac{1}{T}$ gives a straight line (if the reaction obeys the Arrhenius equation). From the slope and the intercept, E_a and the pre-exponential factor A can be determined. Indeed, most reactions obey the Arrhenius equation to a good approximation, which indicates that the collision theory is a physically reasonable model.

WORKED EXAMPLE 15.11

Determination of E_a and the pre-exponential factor A

Consider the second-order decomposition of NO_2 into NO and O_2. The equation is:

$$2NO_2(g) \rightarrow 2NO(g) + O_2(g)$$

The following data were collected for the reaction.

k ($mol^{-1}\ L\ s^{-1}$)	Temperature (°C)
7.8	400
10	410
14	420
18	430
24	440

Determine the activation energy in kilojoules per mole and the pre-exponential factor A for this reaction.

Analysis

The Arrhenius equation, $\ln k = -\frac{E_a}{R}\left(\frac{1}{T}\right) + \ln A$, applies, but the use of rate data to determine the activation energy graphically requires that we plot $\ln k$, not k, versus the reciprocal of the temperature in kelvin, so we have to convert the given data into $\ln k$ and $\frac{1}{T}$ before we can plot the graph. From the slope of the straight line, E_a can be obtained. The factor A can be determined either graphically, from the intercept, or algebraically, by substitution of the data into the Arrhenius equation. Note that A must have the same units as the rate constant k.

Solution

To illustrate, using the first set of data, the conversions are:

$$\ln k = \ln(7.8) = 2.05$$

$$\frac{1}{T} = \frac{1}{(400 + 273)\,\text{K}} = \frac{1}{673\,\text{K}} = 1.486 \times 10^{-3}\,\text{K}^{-1}$$

We are carrying extra 'significant figures' for the purpose of graphing the data. The remaining conversions give the following table.

ln*k*	$\frac{1}{T}$**(K^{-1})**
2.05	1.486×10^{-3}
2.30	1.464×10^{-3}
2.64	1.443×10^{-3}
2.89	1.422×10^{-3}
3.18	1.403×10^{-3}

Then we plot ln$k$ versus $\frac{1}{T}$ as shown in the figure below.

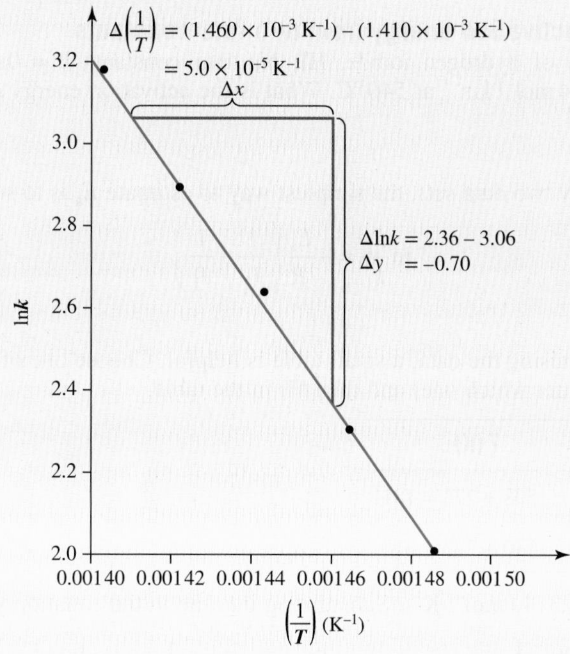

Alternatively, you can use a computer or scientific calculator to calculate the slope of the straight line that best fits the data (linear regression). This will indicate the quality of the linear fit.

(a) To determine E_a, calculate the slope of the straight line as follows:

$$\text{slope} = \frac{\Delta(\ln k)}{\Delta(\frac{1}{T})} = \frac{-0.70}{5.0 \times 10^{-5}\,\text{K}^{-1}}$$

$$= -1.4 \times 10^{4}\,\text{K} = -\frac{E_a}{R}$$

After changing signs and solving for E_a we have:

$$E_a = (8.314\,\text{J mol}^{-1}\,\text{K}^{-1})(1.4 \times 10^{4}\,\text{K})$$

$$= 1.2 \times 10^{5}\,\text{J mol}^{-1}$$

$$= 1.2 \times 10^{2}\,\text{kJ mol}^{-1}$$

(b) To determine A, we must find the y-intercept of the graph. (If you used a computer or calculator to determine the slope, the linear regression also gives you the intercept.) We do this by using the equation $y = mx + c$, the slope calculated in part (a), and the ln$k$ versus $\frac{1}{T}$ data given above. Putting the data into the equation gives five different values for c (22.82, 22.80, 22.85, 22.84 and 22.80), the average of which is 22.82. This corresponds to ln$A$, so taking the exponential of this gives $A = 8.1 \times 10^9 \text{ mol}^{-1}\text{ L s}^{-1}$. The units of A are those for a second-order reaction.

Is our answer reasonable?

The activation energy and pre-exponential factor are positive. You can check these values by substituting them back into the original equation.

The activation energy can also be obtained from two rate constants measured at different temperatures. At temperature T_1, the rate constant k_1 is expressed as:

$$\ln k_1 = \frac{E_a}{RT_1} + \ln A$$

At temperature T_2, we then have:

$$\ln k_2 = \frac{E_a}{RT_2} + \ln A$$

Since the pre-exponential factor A is approximately independent of the temperature, subtraction of the first equation from the second gives:

$$\ln\frac{k_2}{k_1} = -\frac{E_a}{R}\left(\frac{1}{T_2} - \frac{1}{T_1}\right)$$

Worked example 15.12 illustrates the use of this equation.

WORKED EXAMPLE 15.12

Calculating the activation energy from two rate constants

The decomposition of hydrogen iodide, HI, has rate constants $k = 0.079 \text{ mol}^{-1}\text{ L s}^{-1}$ at 508 °C and $k = 0.24 \text{ mol}^{-1}\text{ L s}^{-1}$ at 540 °C. What is the activation energy of this reaction in kJ mol^{-1}?

Analysis

When there are only two data sets, the simplest way to estimate E_a is to solve the equation:

$$\ln\frac{k_2}{k_1} = -\frac{E_a}{R}\left(\frac{1}{T_2} - \frac{1}{T_1}\right)$$

Solution

Let's begin by organising the data; a small table is helpful. Choose one of the rate constants as k_1 (it doesn't matter which one) and then fill in the table.

k ($\text{mol}^{-1}\text{ L s}^{-1}$)	T (K)
0.079	508 + 273 = 781
0.24	540 + 273 = 813

We also have $R = 8.314 \text{ J mol}^{-1}\text{ K}^{-1}$. Substituting into the initial equation yields:

$$\ln\frac{0.24 \text{ mol}^{-1}\text{L s}^{-1}}{0.079 \text{ mol}^{-1}\text{L s}^{-1}} = -\frac{E_a}{8.314 \text{ J mol}^{-1}\text{K}^{-1}}\left(\frac{1}{813\text{ K}} - \frac{1}{781\text{ K}}\right)$$

$$\ln 3.0 = -\frac{E_a}{8.314 \text{ J mol}^{-1}\text{K}^{-1}}(0.00123\text{ K}^{-1} - 0.00128\text{ K}^{-1})$$

$$1.10 = -\frac{E_a}{8.314 \text{ J mol}^{-1}\text{K}^{-1}}(0.000\,050)$$

Multiplying both sides by $8.314 \text{ J mol}^{-1}\text{ K}^{-1}$ gives:

$$9.15 \text{ J mol}^{-1}\text{ K}^{-1} = E_a(0.000\,050\text{ K}^{-1})$$

Solving for E_a, we have:

$$E_a = 1.8 \times 10^5 \text{ J mol}^{-1}$$
$$= 1.8 \times 10^2 \text{ kJ mol}^{-1}$$

Note that this is of the same order of magnitude as the energy of a comparably weak single bond.

PRACTICE EXERCISE 15.11

The reaction $2NO_2 \rightarrow 2NO + O_2$ has an activation energy of 111 kJ mol^{-1}. At 400 °C, $k = 7.8 \text{ mol}^{-1}\text{ L s}^{-1}$. What is the value of k at 430 °C?

Recall from the start of the chapter that the conversion of diamond into graphite is very slow, even though graphite is thermodynamically stable under standard conditions at room temperature. The reason for this is that the activation energy for the process, which involves breaking strong carbon–carbon bonds, is high (it requires temperatures higher than 1500 °C under exclusion of oxygen). Just because a product is thermodynamically stable does not mean that its formation will be rapid. A thermodynamically unstable product can exist under standard conditions if a large activation energy is required for its reaction. Because of this, diamond is called *metastable* at atmospheric pressure and room temperature.

15.6 Reaction mechanisms

At the beginning of this chapter we mentioned that most reactions do not occur in a single step. Instead, the net overall reaction is the result of a series of simple reactions, each of which is called an elementary reaction. An **elementary reaction** is a reaction in which one or more of the chemical species reacts directly to form products in a single reaction step and with a single transition state. In these cases, the rate law can be written from its chemical equation, using its coefficients as exponents for the concentration terms without requiring experiments to determine the exponents. The entire series of elementary reactions is called the reaction mechanism. For most reactions, the individual elementary reactions cannot actually be observed; instead, we see only the net reaction. Therefore, the mechanism a chemist writes is really a *theory* about what occurs step-by-step as the reactants are changed to the products.

What makes an elementary reaction 'elementary'? Remember that the exponents of the concentration terms for the rate law of an *overall* reaction (the order) bear no necessary relationship to the coefficients in the overall balanced equation. As we keep emphasising, the exponents in the overall rate law must be determined experimentally. So what makes an elementary reaction 'elementary' is that a simple relationship between coefficients and exponents *does exist*.

Because the individual steps in a mechanism sometimes cannot be observed directly, devising a mechanism for a reaction requires some ingenuity. However, we can immediately tell whether a proposed mechanism is feasible. The *overall rate law derived from the mechanism must agree with the observed rate law for the overall reaction.*

Let's see how the exponents of the concentration terms in a rate law for an elementary reaction can be devised. Consider the following example of an elementary reaction that involves collisions between two identical molecules leading directly to the products.

$$2NO_2 \rightarrow NO_3 + NO$$
$$\text{rate of reaction} = k[NO_2]^n$$

How can we predict the value of the exponent n? If the NO_2 concentration was doubled, there would be *twice* as many individual NO_2 molecules and *each* would have *twice* as many neighbours with which to collide. The number of NO_2-to-NO_2 collisions per second therefore would increase by a factor of 4, resulting in an increase in the rate by a factor of 4, which is 2^2. Earlier we saw that, when doubling a concentration leads to a fourfold increase in the rate, the concentration of that reactant is raised to the second power in the rate law. Thus, if $2NO_2 \rightarrow NO_3 + NO$ represents an elementary reaction, its rate law should be:

$$\text{rate of reaction} = k[NO_2]^2$$

By definition, the exponent in the rate law for an elementary reaction is the same as the coefficient in the chemical equation. If we found, experimentally, that the exponent in this rate law was not equal to 2, the reaction is not really an elementary reaction. Another way to define an elementary reaction is by its molecularity. **Molecularity** is the number of species that must collide to produce the reaction in an elementary reaction. The order of an elementary reaction is the same as the molecularity of that reaction.

- An elementary reaction involving one molecule is *unimolecular* and has a first-order rate law.
- An elementary reaction involving two species is *bimolecular* and has a second-order rate law.
- An elementary reaction involving three species is *termolecular* and has a third-order rate law. (There is low probability that three molecules will collide with the correct orientation and sufficient energy to surmount the activation barrier, so termolecular reactions are extremely rare.)

Remember that this rule applies only to *elementary reactions*. If all we know is the balanced equation for the overall reaction, the only way we can find the exponents of the rate equation is by doing experiments.

The rate-determining step

The rate law of an elementary reaction helps to determine possible reaction mechanisms. How does this work? Let us consider the gaseous reaction:

$$2NO_2Cl \rightarrow 2NO_2 + Cl_2$$

Experimentally, the rate is first order in NO_2Cl, so the rate law is:

$$\text{rate of reaction} = k[NO_2Cl]$$

The first question we might ask is: Could the overall reaction ($2NO_2Cl \rightarrow 2NO_2 + Cl_2$) occur in a single step by the collision of two NO_2Cl molecules? The answer is no, because then it would be an elementary reaction and the predicted rate law would include a squared term, $[NO_2Cl]^2$. But the experimental rate law is first order in NO_2Cl. So the predicted and experimental rate laws don't agree, and we must look further to find the mechanism of the reaction.

On the basis of chemical intuition and other information, it is now believed that the actual mechanism of the reaction is the following two-step sequence of elementary reactions.

$$NO_2Cl \xrightarrow{k_1} NO_2 + Cl \qquad \text{(slow)}$$

$$NO_2Cl + Cl \xrightarrow{k_2} NO_2 + Cl_2 \qquad \text{(fast)}$$

The Cl atom formed here in the first step is an intermediate. In this reaction, we never actually observe the Cl atom because, once it is formed, it quickly reacts in the second step. Recall from section 15.5 that an intermediate is higher in energy and has a more limited lifetime than the reactants and products.

Notice that when the two reactions are added, the intermediate, Cl, drops out and we obtain the net overall reaction $2NO_2Cl \rightarrow 2NO_2 + Cl_2$. *Being able to add the elementary reactions and obtain the overall reaction is another major test of a mechanism.*

In any multistep mechanism, one step is usually much slower than the others. In this mechanism, for example, it is believed that the first step is slow, and that, once a Cl atom forms, it reacts very rapidly with another NO_2Cl molecule to give the final products.

The final products of a multistep reaction cannot appear more rapidly than do the products of the slow step, so the slow step in a mechanism is called the **rate-determining step** or rate-limiting step. In the two-step mechanism for this particular reaction, then, the first reaction is the rate-determining step because the final products cannot be formed faster than the rate at which Cl atoms form.

The rate-determining step is similar to a slow worker on an assembly line. The production rate depends on how quickly the slow worker works, regardless of how fast the other workers are. The factors that control the speed of the rate-determining step therefore also control the overall rate of the reaction. This means that *the rate law for the rate-determining step is directly related to the rate law for the overall reaction.* Note that the rate-determining step can be any of the steps in a sequence of elementary reactions (and not necessarily only the first step, as in our example).

Because the rate-determining step is an elementary reaction, we can predict its rate law from the coefficients of its reactants. The coefficient of NO_2Cl in its relatively slow breakdown to NO_2 and Cl is 1. Therefore, the rate law predicted for the first step is:

$$\text{rate} = k_1[NO_2Cl]$$

Notice that the predicted rate law derived for the two-step mechanism agrees with the experimentally measured rate law. Although scientists may never actually *prove* the correctness of a proposed mechanism, at least considerable support for it can be provided. From the standpoint of kinetics, therefore, the mechanism is reasonable.

WORKED EXAMPLE 15.13

Drawing an energy diagram

Draw an energy diagram for a two-step exothermic reaction in which the second step is rate determining.

Analysis

A two-step reaction involves the formation of an intermediate. If the reaction is exothermic, the products must be lower in energy than the reactants. In order for the second step to be rate determining, it must have a higher energy barrier.

Solution

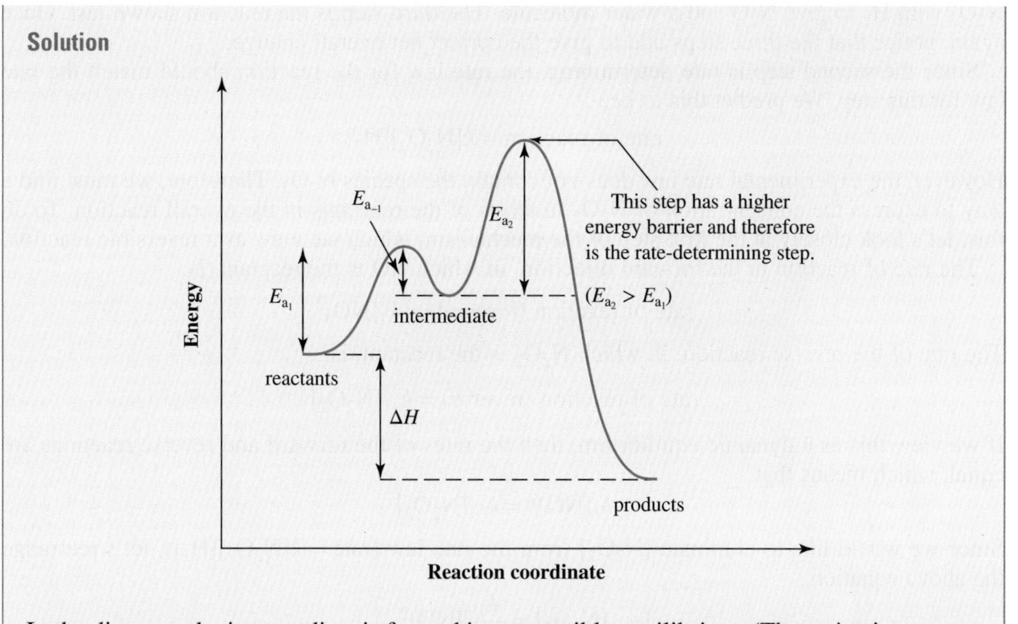

In the diagram, the intermediate is formed in a reversible equilibrium. (The activation energy required to re-form the reactants is $E_{a_{-1}}$.) The effective energy required for the reaction is the difference between the energies of the second transition state and the reactants.

PRACTICE EXERCISE 15.12

In what way would the energy diagram drawn in worked example 15.13 change if: (a) the reaction were endothermic, or (b) the first step was rate determining?

Mechanisms with fast, reversible steps

We will now study the following gas phase reaction.

$$2NO + 2H_2 \rightarrow N_2 + 2H_2O$$

The experimentally determined rate law is:

$$\text{rate of reaction} = k[NO]^2[H_2] \qquad \text{(experimental)}$$

We can quickly tell from this rate law that $2NO + 2H_2 \rightarrow N_2 + 2H_2O$ is not itself an elementary reaction. If it were, the exponent for $[H_2]$ would be 2. Obviously, a mechanism involving two or more steps must be involved.

A chemically reasonable mechanism that yields the correct form for the rate law consists of the following two steps.

$$2NO + H_2 \rightarrow N_2O + H_2O \qquad \text{(slow)}$$
$$N_2O + H_2 \rightarrow N_2 + H_2O \qquad \text{(fast)}$$

One test of the mechanism, as we said, is that the two equations must add to give the correct overall equation, and they do. Further, the chemistry of the second step has actually been observed in separate experiments, where it was shown that N_2O reacts with H_2 to give N_2 and H_2O. Another test of the mechanism involves the coefficients of NO and H_2 in the predicted rate law for the first step, the supposed rate-determining step.

$$\text{rate of reaction} = k[NO]^2[H_2] \qquad \text{(predicted)}$$

This rate equation does match the experimental rate law, but there is still a serious flaw in the proposed mechanism. If the postulated slow step actually describes an elementary reaction, it would involve the simultaneous collision between three molecules, two NO molecules and one H_2 molecule. As mentioned previously, termolecular processes are very rare. Thus, it is believed that the reaction proceeds by the following three-step sequence of bimolecular elementary reactions.

$$2NO \underset{k_{-1}}{\overset{k_1}{\rightleftharpoons}} N_2O_2 \qquad \text{(fast)}$$
$$N_2O_2 + H_2 \xrightarrow{k_2} N_2O + H_2O \qquad \text{(slow)}$$
$$N_2O + H_2 \xrightarrow{k_3} N_2 + H_2O \qquad \text{(fast)}$$

In this mechanism the first step is proposed to be a rapidly established equilibrium in which the unstable intermediate N_2O_2 forms by dimerisation of NO in the forward reaction and then quickly decomposes into NO by the reverse reaction. The rate-determining step is the reaction of

N_2O_2 with H_2 to give N_2O and a water molecule. The third step is the reaction shown last. Once again, notice that the three steps add to give the correct net overall change.

Since the second step is rate determining, the rate law for the reaction should match the rate law for this step. We predict this to be:

$$\text{rate of reaction} = k[N_2O_2][H_2]$$

However, the experimental rate law does not contain the species N_2O_2. Therefore, we must find a way to express the concentration of N_2O_2 in terms of the reactants in the overall reaction. To do this, let's look closely at the first step of the mechanism, which we view as a reversible reaction.

The rate of reaction in the forward direction, in which NO is the reactant, is:

$$\text{rate of reaction (forward)} = k_1[NO]^2$$

The rate of the reverse reaction, in which N_2O_2 is the reactant, is:

$$\text{rate of reaction (reverse)} = k_{-1}[N_2O_2]$$

If we view this as a dynamic equilibrium, then the rates of the forward and reverse reactions are equal, which means that:

$$k_1[NO]^2 = k_{-1}[N_2O_2]$$

Since we would like to eliminate $[N_2O_2]$ from the rate law (rate = $k[N_2O_2][H_2]$), let's rearrange the above equation.

$$[N_2O_2] = \frac{k_1}{k_{-1}}[NO]^2$$

Substituting into the rate law:

$$\text{rate of reaction} = k[N_2O_2][H_2]$$

yields:

$$\text{rate of reaction} = k\left(\frac{k_1}{k_{-1}}\right)[NO]^2[H_2]$$

We can now combine all the constants into one constant (k') and get:

$$\text{rate of reaction} = k'[NO]^2[H_2]$$

Now the rate law derived from the mechanism matches the rate law obtained experimentally. The three-step mechanism does appear to be reasonable on the basis of kinetics.

The procedure we have worked through here applies to many reactions that proceed by mechanisms involving sequential steps. Steps that precede the rate-determining step are considered to be rapidly established equilibria involving unstable intermediates.

PRACTICE EXERCISE 15.13 Ozone, O_3, reacts with nitric oxide, NO, to form nitrogen dioxide and oxygen.

$$NO + O_3 \rightarrow NO_2 + O_2$$

This is one of the reactions involved in the formation of photochemical smog. If this reaction occurs in a single step, what is the expected rate law for the reaction?

PRACTICE EXERCISE 15.14 For the reaction of $Mo(CO)_6$:

$$Mo(CO)_6 + P(CH_3)_3 \rightarrow Mo(CO)_5P(CH_3)_3 + CO$$

the following mechanism has been proposed:

$$Mo(CO)_6 \rightarrow Mo(CO)_5 + CO \qquad \text{(step 1)}$$
$$Mo(CO)_5 + P(CH_3)_3 \rightarrow Mo(CO)_5P(CH_3)_3 \qquad \text{(step 2)}$$

(a) Is the proposed mechanism consistent with the equation for the overall reaction?
(b) Identify the reaction intermediate or intermediates.
(c) Write the rate law if the first step would be rate determining.

The steady-state approximation

Often, in simple cases of reaction mechanisms, one step is the rate-determining step. However, it is not unusual that, in complex reaction mechanisms, the rate-determining step may vary when the reaction conditions are changed. In such cases, where the rate-determining step cannot easily be chosen, the **steady-state approximation** can be used to analyse the reaction. The basic idea of this method is the assumption that the concentration of any intermediate remains constant as the reaction proceeds. (Remember that an intermediate is neither a reactant nor a product, but it is produced and consumed during the reaction.) This will generally be the case only when the concentration of the intermediate remains low over the duration of the reaction.

Let's consider the following hypothetical reaction.

$$2AB + C_2 \rightarrow A_2B + C_2B$$

This reaction may proceed via the following mechanism.

$$2AB \underset{k_{-1}}{\overset{k_1}{\rightleftharpoons}} A_2B_2$$

$$A_2B_2 + C_2 \xrightarrow{k_2} A_2B + C_2B$$

The intermediate in this mechanism is A_2B_2. If the concentration of A_2B_2 remains constant throughout the reaction, or is assumed to remain constant, that is:

$$\frac{d[A_2B_2]}{dt} = 0$$

the steady-state approximation can be applied.

Next, all steps that are producing and consuming intermediate A_2B_2 need to be identified, and the rate law for each has to be written. Then the steady-state approximation is applied by setting the total rate of formation of A_2B_2 equal to its total rate of consumption.

$$\text{rate of } A_2B_2 \text{ formation} = \text{rate of } A_2B_2 \text{ consumption, therefore } \frac{d[A_2B_2]}{dt} = 0$$

The general procedure is demonstrated below.

1. Rate of formation of A_2B_2
   In this mechanism, A_2B_2 is produced only in the forward reaction of the first elementary step. The rate law for this step is:

$$\frac{d[A_2B_2]}{dt} = k_1[AB]^2$$

2. Rate of consumption of A_2B_2
   A_2B_2 is consumed in the reverse reaction of the first step and in the second step. The rate laws for these steps are:

$$-\frac{d[A_2B_2]}{dt} = k_{-1}[A_2B_2]$$

   and

$$-\frac{d[A_2B_2]}{dt} = k_2[A_2B_2][C_2]$$

3. Application of the steady-state approximation for intermediate A_2B_2:
   rate of formation = rate of consumption

$$k_1[AB]^2 = k_{-1}[A_2B_2] + k_2[A_2B_2][C_2] = [A_2B_2](k_{-1} + k_2[C_2])$$

   If we solve for the concentration of our intermediate $[A_2B_2]$, we get:

$$[A_2B_2] = \frac{k_1[AB]^2}{k_{-1} + k_2[C_2]}$$

4. The rate law for the overall reaction
   Now, the rate law for the overall reaction $2AB + C_2 \rightarrow A_2B + C_2B$ is written. This can be done in various ways, depending on whether we chose a reactant or a product to represent the rate of reaction. In this example we use the consumption of C_2 to define the rate of reaction.

$$\text{rate of reaction} = -\frac{d[C_2]}{dt}$$

   C_2 is consumed only in the second step of the mechanism, therefore:

$$-\frac{d[C_2]}{dt} = k_2[A_2B_2][C_2]$$

The equation $[A_2B_2] = \frac{k_1[AB]^2}{k_{-1}+k_2[C_2]}$ provides us with an expression for the concentration of our intermediate A_2B_2. Substitution of this expression into the rate law gives:

$$\text{rate of reaction} = -\frac{d[C_2]}{dt} = k_2[C_2]\left(\frac{k_1[AB]^2}{k_{-1} + k_2[C_2]}\right)$$

$$= \frac{k_1k_2[C_2][AB]^2}{k_{-1} + k_2[C_2]}$$

This is the overall rate law for the proposed mechanism based on the steady-state approximation. This rate law is quite complicated (this is common for rate laws determined using the steady-state analysis). The usual way to test the validity of the rate law involves choosing concentration conditions that result in a simplification of the rate law.

For example, if the reaction between AB and C_2 is studied under conditions where the concentration of C_2 is large, the reverse reaction of step 1 can be neglected, since all A_2B_2 formed in the first reaction rapidly reacts with C_2 to produce $A_2B + C_2B$. Then we have:

$$k_2[C_2] >> k_{-1}$$

This reduces the full rate law to:

$$\text{rate of reaction} = \frac{k_1k_2[C_2][AB]^2}{k_2[C_2]} = k_1[AB]^2$$

Thus, if the suggested mechanism is valid, at sufficiently high concentrations of C_2, the reaction should show a second-order dependence on the concentration of AB.

On the other hand, if the reaction is performed under conditions of low C_2 concentrations, the intermediate A_2B_2 will preferentially decompose into AB through the reverse reaction of step 1 rather than reacting with C_2. Then, we have:

$$k_{-1} >> k_2[C_2]$$

This leads to a rate law of:

$$\text{rate of reaction} = \frac{k_1k_2}{k_{-1}}[C_2][AB]^2 = k'[C_2][AB]^2$$

Studies under these conditions should show first-order dependence on $[C_2]$ and second-order dependence on $[AB]$, if the assumed mechanism is correct.

WORKED EXAMPLE 15.14

Using the steady-state approximation for constructing a rate law

The experimentally determined rate law for the decomposition of ozone ($2O_3 \rightarrow 3O_2$) is:

$$-\frac{d[O_3]}{dt} = k\frac{[O_3]^2}{[O_2]}$$

The following mechanism has been proposed to account for the rate law.

$$O_3 \underset{k_{-1}}{\overset{k_1}{\rightleftharpoons}} O_2 + O \qquad \text{(step 1)}$$

$$O + O_3 \xrightarrow{k_2} 2O_2 \qquad \text{(step 2)}$$

Derive the rate law for the decomposition of ozone by applying the steady-state approximation to the concentration of atomic oxygen.

Analysis

The procedure to solve such a problem consists of several steps.

(a) Construct the rate law in terms of consumption of ozone.

(b) Construct the steady-state expression for the intermediate O by applying the criterion $\frac{d[O]}{dt} = 0$, which means that rate of formation of O = rate of consumption of O. We have to identify each step that produces or consumes O and write the appropriate rate law for each. The sum of the rate laws that produce O are then set equal to the sum of the rate laws that consume O.

(c) The resulting steady-state approximation is then solved for [O].

(d) The expression from step (b) for [O] is used to substitute for the concentration of the intermediate found in the rate law of step (a). By doing this we obtain an overall rate law that contains only reactant and/or product concentrations.

Solution

(a) The overall reaction is the sum of the two reaction steps. Ozone is consumed in the forward reaction of step 1 and step 2, and is produced in the reverse reaction of step 1. The rate law for the consumption of ozone can then be written as:

$$-\frac{d[O_3]}{dt} = k_1[O_3] + k_2[O_3][O] - k_{-1}[O_2][O]$$

$$= k_1[O_3] + [O](k_2[O_3] - k_{-1}[O_2]) \qquad \text{(equation 1)}$$

(b) O is produced only in the forward reaction of step 1. Thus, rate of formation of $O = k_1[O_3]$. O is consumed in the reverse reaction of step 1 and in step 2. Therefore, the rate of consumption of $O = k_{-1}[O_2][O] + k_2[O_3][O]$.
Applying the steady-state condition gives:

$$k_1[O_3] = k_{-1}[O_2][O] + k_2[O_3][O]$$

(c) Solving for the concentration of O:

$$k_1[O_3] = [O](k_{-1}[O_2] + k_2[O_3])$$

$$[O] = \frac{k_1[O_3]}{k_{-1}[O_2]+k_2[O_3]}$$

(d) Substitute for [O] in equation 1:

$$-\frac{d[O_3]}{dt} = k_1[O_3] + \frac{k_1[O_3]k_2[O_3] - k_{-1}[O_2]}{k_{-1}[O_2]+k_2[O_3]}$$

$$= k_1[O_3] + \frac{k_1k_2[O_3]^2 - k_1k_{-1}[O_3][O_2]}{k_{-1}[O_2]+k_2[O_3]}$$

Multiply the first term on the right-hand side by:

$$\frac{k_{-1}[O_2]+k_2[O_3]}{k_{-1}[O_2]+k_2[O_3]}$$

to obtain:

$$-\frac{d[O_3]}{dt} = \frac{k_1[O_3](k_{-1}[O_2]+k_2[O_3]) + k_1k_2[O_3]^2 - k_1k_{-1}[O_2][O_3]}{k_{-1}[O_2]+k_2[O_3]}$$

$$= \frac{k_1k_{-1}[O_3][O_2] + k_1k_2[O_3]^2 + k_1k_2[O_3]^2 - k_1k_{-1}[O_3][O_2]}{k_{-1}[O_2]+k_2[O_3]}$$

$k_1k_{-1}[O_2][O_3]$ cancels out giving:

$$-\frac{d[O_3]}{dt} = \frac{2k_1k_2[O_3]^2}{k_{-1}[O_2]+k_2[O_3]} \quad \text{(equation 2)}$$

Now when $[O_2]$ is large ($k_{-1}[O_2] >> k_2[O_3]$) the denominator reduces to $k_{-1}[O_2]$, which means that equation 2 reduces to:

$$-\frac{d[O_3]}{dt} = \frac{2k_1k_2[O_3]^2}{k_{-1}[O_2]}$$

which is effectively:

$$-\frac{d[O_3]}{dt} = \frac{k_{obs}[O_3]^2}{[O_2]}$$

the same as the experimentally determined rate law.

Is our answer reasonable?

The rate law obtained using the steady-state approximation is the same as the observed rate law. This is what we wanted.

PRACTICE EXERCISE 15.15

A possible mechanism for the decomposition of NO_2Cl is:

$$NO_2Cl \rightarrow NO_2 + Cl$$

$$NO_2Cl + Cl \rightarrow NO_2 + Cl_2$$

What would the predicted rate law be if the second step in the mechanism were the rate-determining step?

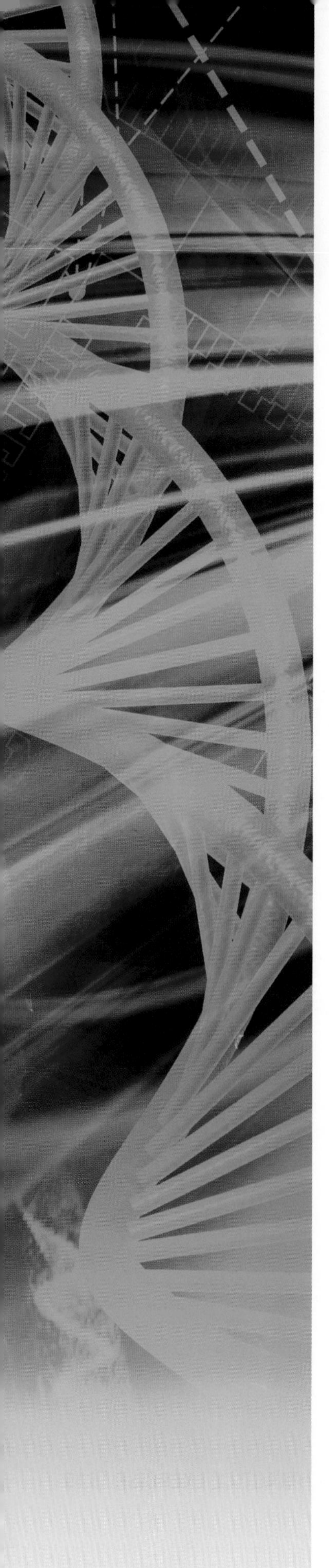

Chemical Connections

The Antarctic ozone hole

The ozone layer in the stratosphere (>12 km high) shields the Earth from damaging UV radiation from the Sun. There is measurable clear evidence has been building that human activities have caused a dangerous reduction in the stratospheric ozone layer. When the ozone hole was detected over Antarctica, its creation was soon linked to an increase in halogen-containing compounds produced from chlorofluorocarbons (CFCs), chlorinated solvents, halons (bromine-containing compounds) and bromomethane released into the atmosphere. These chemicals were once widely used as refrigerants, aerosols, cleaning solvents, fire-fighting chemicals and fumigants. Figure 15.19 shows the development of the ozone hole in the Antarctic spring from 1979 to 2014. The amount of ozone is measured in Dobson units (DU), where 1 DU is defined as the amount of ozone that would constitute a layer 0.01 mm thick at 0 °C and 1.013×10^5 Pa.

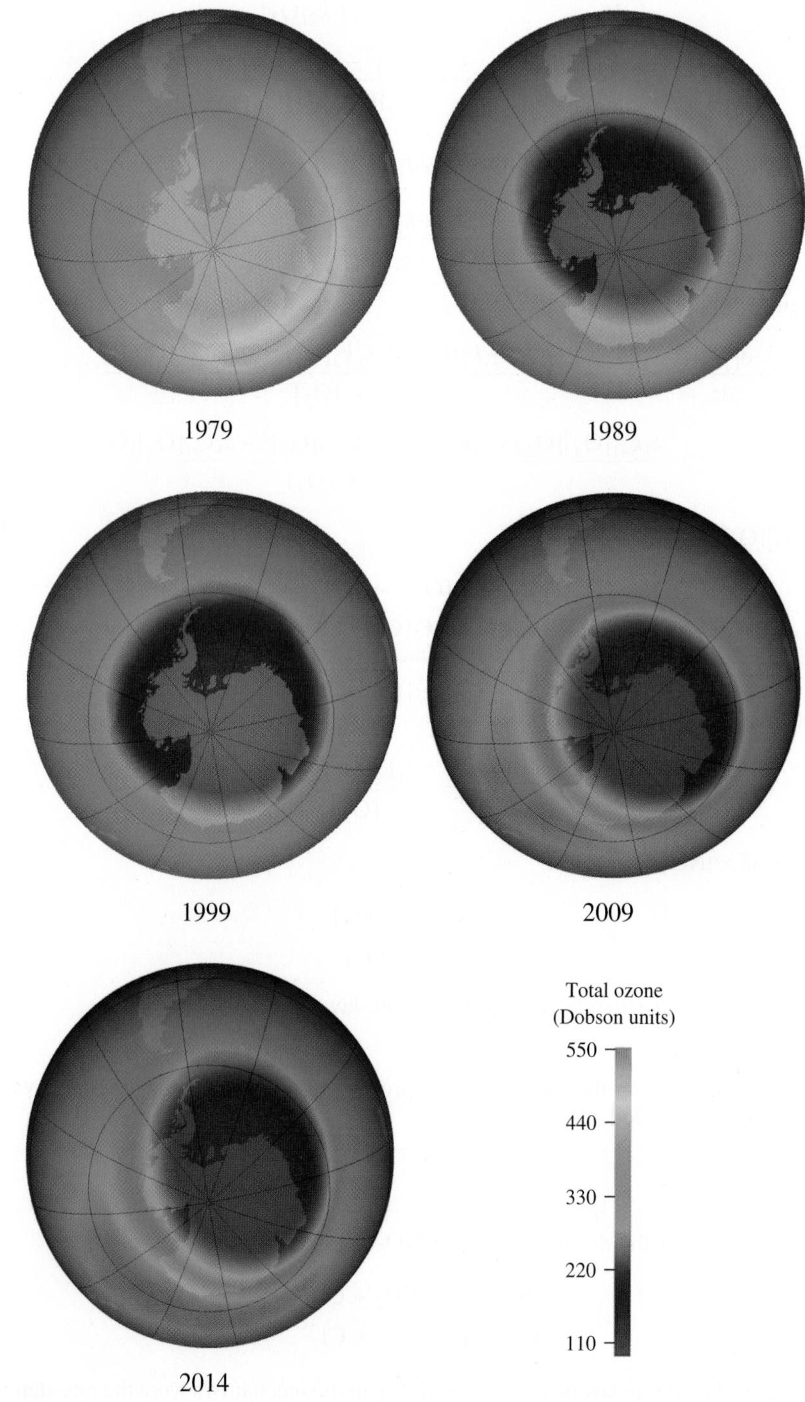

FIGURE 15.19 Average ozone level (Dobson units) for October over the Southern Hemisphere in 1979, 1989, 1999, 2009 and 2014.

How does the ozone loss occur? The most commonly found breakdown products (often called *reservoirs*) of CFCs are hydrochloric acid, HCl, and chlorine nitrate, $ClONO_2$. In addition, dinitrogen pentoxide, N_2O_5, is a reservoir of nitrogen oxides and plays an important role in the chemistry of ozone depletion. HCl and $ClONO_2$ (and their bromine counterparts) are converted into more active forms of chlorine (or bromine) on the surface of *polar stratospheric clouds* (PSCs). These PSCs are formed during winter polar nights, when sunlight does not reach the South Pole. A strong circumpolar wind (polar vortex) develops in the middle of the stratosphere; this isolates the air over the polar region from the remaining atmosphere. Because of the lack of sunlight, air within the vortex can get very cold; PSCs form when the temperature falls below −80 °C. These clouds consist of nitric acid trihydrate, $HNO_3{\cdot}3H_2O$, but, as the temperature decreases even further, larger droplets of water ice can form with nitric acid dissolved in them. These PSCs are the key to ozone loss over Antarctica.

The most important reactions leading eventually to the destruction of ozone on PSCs are those that form HNO_3, H_2O and Cl_2 from $ClONO_2$, HCl and N_2O_5 through the intermediate HOCl.

$$HCl + ClONO_2 \rightarrow HNO_3 + Cl_2$$
$$ClONO_2 + H_2O \rightarrow HNO_3 + HOCl$$
$$HCl + HOCl \rightarrow H_2O + Cl_2$$
$$N_2O_5 + HCl \rightarrow HNO_3 + ClONO_2$$
$$N_2O_5 + H_2O \rightarrow 2HNO_3$$

Once sunlight returns to the polar region in the Southern Hemisphere spring, the Cl_2 formed in the reactions above is photolysed into chlorine atoms.

$$Cl_2 + h\nu \rightarrow 2Cl$$

Chlorine atoms have an unpaired electron. Such species are called free radicals. Radicals are highly reactive because of the tendency of electrons to become paired by forming ions or covalent bonds.

Although there are high concentrations of active forms of chlorine above the South Pole, there are many more ozone molecules than chlorine species. So, how can the active chlorine destroy nearly all of the ozone? The answer lies in what is known as a *catalytic cycle*. Here, one molecule significantly changes or enables a reaction cycle without itself being altered by the cycle. The following catalytic cycle is believed to be responsible for 70% of the ozone loss over Antarctica.

$$2Cl + 2O_3 \rightarrow 2ClO + 2O_2$$
$$ClO + ClO + M \rightarrow Cl_2O_2 + M$$
$$Cl_2O_2 + h\nu \rightarrow Cl + ClO_2$$
$$ClO_2 + M \rightarrow Cl + O_2 + M$$
$$\text{Net:} \quad 2O_3 \rightarrow 3O_2$$

In these equations, *M* stands for a *collision partner*, which assists in providing the energy required to surmount the activation barrier for the reaction, but which is not involved in the reaction itself. In this cycle, two chlorine atoms catalyse the destruction of two ozone molecules to give three oxygen molecules.

A considerable depletion of the ozone layer over the Antarctic region was monitored during the 2004–05 winter, where unusually low stratospheric temperatures led to abundant PSCs. In general, stratospheric temperatures in the polar regions are not sufficiently low to enable formation of large quantities of PSCs.

As a consequence of the ozone hole, Australians and New Zealanders suffer the highest rates of skin cancer in the world (around 1400 Australians and more than 300 New Zealanders die from skin cancer each year). An international agreement called the Montreal Protocol, originally signed in 1987, limits the production and use of ozone-depleting substances. Since that time, a reduction in the rate of ozone loss has been measured, and the concentration of CFCs in the atmosphere is levelling off. In fact, in the last few years, it appears that the ozone holes have been less severe. It is estimated that, if all countries achieve the targets set by the Montreal Protocol, the ozone in the stratosphere should eventually recover. However, because of a long lag time in chemicals reaching the stratosphere, that recovery is likely to take 50 years.

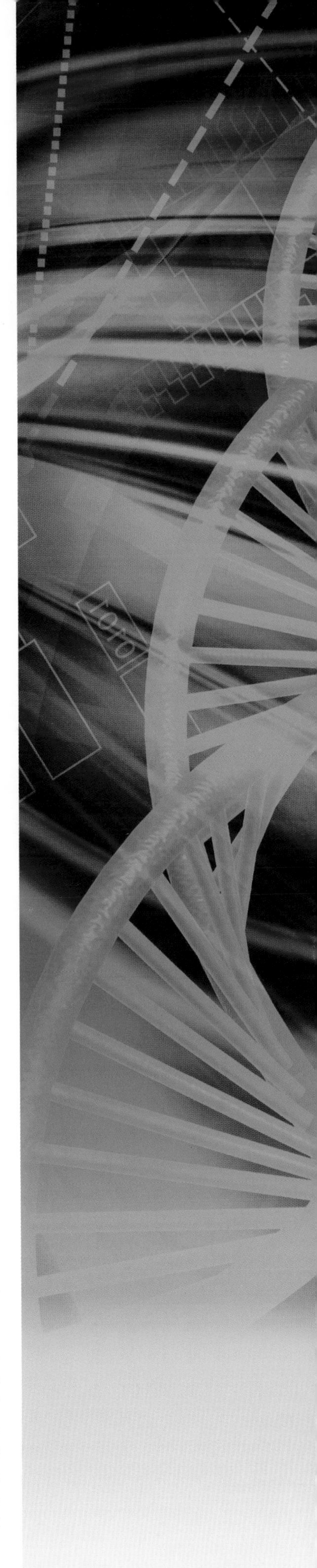

15.7 Catalysts

A **catalyst** is a substance that changes the rate of a chemical reaction without being used up. In other words, all of the catalyst added at the start of a reaction is present chemically unchanged after the reaction has gone to completion. The action caused by a catalyst is called **catalysis**. Broadly speaking, there are two kinds of catalysts; *positive catalysts* speed up reactions, and *negative catalysts,* usually called *inhibitors,* slow reactions down. In future, when we use the term 'catalyst' we mean a positive catalyst.

Although the catalyst is not part of the overall reaction, it does participate by changing the mechanism of the reaction. The catalyst provides a path to the products that has a rate-determining step with a lower activation energy than that of the uncatalysed reaction (see figure 15.20). Because the activation energy along this new route is lower, a greater fraction of the collisions of the reactant molecules have the minimum energy needed to react, so the reaction proceeds faster. Note that a catalyst cannot change ΔH for a reaction. An endothermic reaction can not become exothermic by using a catalyst.

Catalysts can be divided into two groups: **homogeneous catalysts**, which exist in the same phase as the reactants, and **heterogeneous catalysts**, which exist in a separate phase.

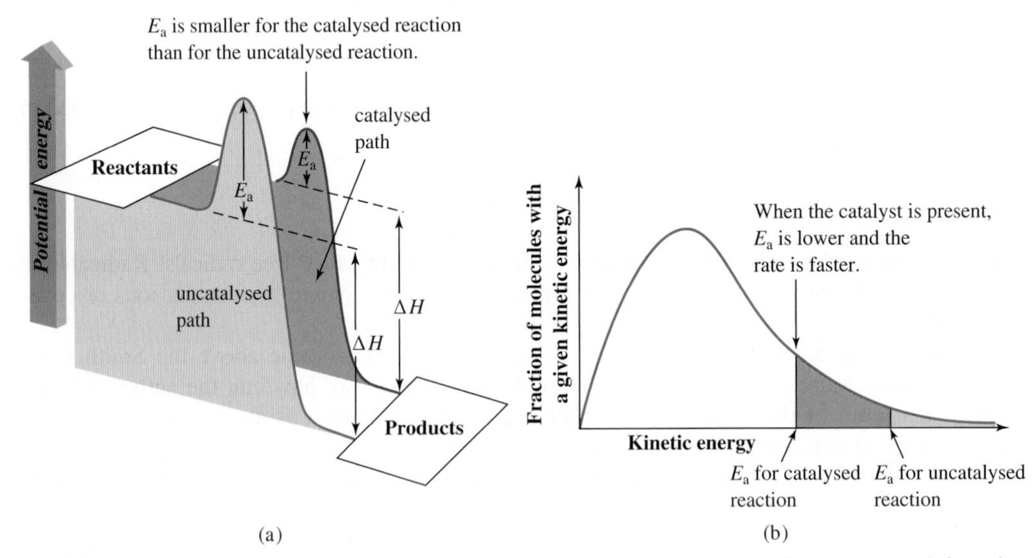

FIGURE 15.20 Effect of a catalyst on a reaction: **(a)** The catalyst provides an alternative, lower energy path from the reactants to the products, which proceeds through a different pathway involving different intermediates. **(b)** A larger fraction of molecules have sufficient energy to react when the catalysed path is available. Although not shown, the purple shading continues to the end of the graph.

Homogeneous catalysts

Homogeneous catalysts are in the same phase as the reactants. An example of homogeneous catalysis is found in the *lead chamber process* for manufacturing sulfuric acid. Sulfuric acid is one of the most important chemicals in industry. It is also the acid used in lead car batteries (see section 12.8).To make sulfuric acid by this process, sulfur is burned to give SO_2, which is then oxidised to SO_3. The SO_3 is dissolved in water to give H_2SO_4.

$$S + O_2 \rightarrow SO_2$$
$$SO_2 + \tfrac{1}{2}O_2 \rightarrow SO_3$$
$$SO_3 + H_2O \rightarrow H_2SO_4$$

Unassisted, the second reaction, oxidation of SO_2 to SO_3, occurs slowly. In the lead chamber process, the SO_2 is combined with a mixture of NO, NO_2, air and steam in large lead-lined reaction chambers. NO_2 readily oxidises the SO_2 to give NO and SO_3. NO is then reoxidised to NO_2 by oxygen.

$$NO_2 + SO_2 \rightarrow NO + SO_3$$
$$NO + \tfrac{1}{2}O_2 \rightarrow NO_2$$

In this process, NO_2 serves as a catalyst by being an oxygen carrier and by providing a lower energy path for the oxidation of SO_2 to SO_3. Notice, as must be true for any catalyst, the NO_2 is regenerated during the reaction; it has not been permanently changed.

Heterogeneous catalysts

Heterogeneous catalysts are in a separate phase from the reactants. A heterogeneous catalyst is commonly a solid and usually functions by promoting a reaction on its surface. One or more of the reactant molecules are adsorbed onto the surface of the catalyst, where an interaction with the surface increases their reactivity. An example is the industrial synthesis of ammonia from hydrogen and nitrogen by the Haber–Bosch process.

$$3H_2 + N_2 \rightarrow 2NH_3$$

The Haber–Bosch (figure 15.21) process is one of the most important processes in industry, as it harnesses the abundance of nitrogen in the atmosphere to generate ammonia (nitrogen fixation). Ammonia is the precursor to nitrate fertilisers and many other products.

As we have mentioned earlier, without a catalyst, this reaction is extremely slow. In the Haber–Bosch process, the reaction takes place on the surface of an iron catalyst that contains traces of aluminium and potassium oxides. The ammonia synthesis is usually performed under high pressure, because for every 2 moles of product 4 moles of reactants are required. (This is an example for the application of Le Châtelier's principle; see section 9.4.) The detailed mechanism of the Haber–Bosch process has been determined by Gerhard Ertl (Nobel Prize in chemistry, 2007). Hydrogen and nitrogen molecules dissociate into atoms while being held on the catalytic surface. The hydrogen atoms then combine with the nitrogen atoms to form ammonia. Finally, the completed ammonia molecule breaks away, freeing the surface of the catalyst for further reaction. This sequence of steps is illustrated in figure 15.22.

FIGURE 15.21 **(a)** Fritz Haber (Nobel Prize in chemistry, 1918) and **(b)** Carl Bosch (Nobel Prize in chemistry, 1931), who developed the Haber–Bosch process of ammonia synthesis.

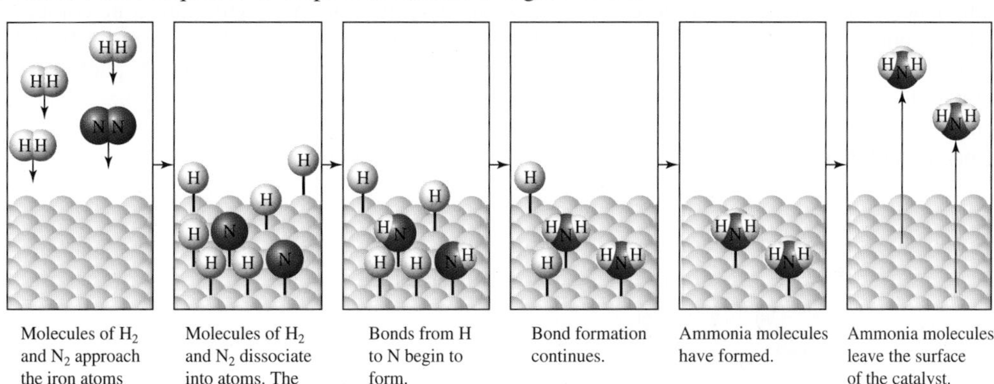

Molecules of H_2 and N_2 approach the iron atoms (bottom) on the surface of the catalyst.
Molecules of H_2 and N_2 dissociate into atoms. The dissociation of N_2 into N atoms is the rate-determining step.
Bonds from H to N begin to form.
Bond formation continues.
Ammonia molecules have formed.
Ammonia molecules leave the surface of the catalyst.

FIGURE 15.22 The Haber–Bosch process. Catalytic formation of ammonia molecules from hydrogen and nitrogen on the surface of a catalyst.

Heterogeneous catalysts are used in many important commercial processes. The petroleum industry uses heterogeneous catalysts to crack hydrocarbons into smaller fragments and then re-form them, also assisted by catalysts, into useful commodities for industrial synthetic processes that produce petrol, polymers, pharmaceuticals and so on.

Modern vehicles are equipped with a catalytic converter (figure 15.23) designed to lower the concentrations of exhaust pollutants, such as carbon monoxide, unburned hydrocarbons and nitrogen oxides. Air is introduced into the exhaust stream that then passes over a catalyst that adsorbs CO, NO and O_2. The NO dissociates into N and O atoms, and the O_2 also dissociates into atoms. Pairing of nitrogen atoms then produces N_2, and oxidation of CO by oxygen atoms produces CO_2. Unburned or incompletely burned hydrocarbons are also oxidised to CO_2 and H_2O. The catalysts in catalytic converters are deactivated or 'poisoned' by lead-based octane boosters such as tetraethyl lead, $Pb(C_2H_5)_4$, and 'leaded' petrol cannot be legally used in vehicles anymore.

FIGURE 15.23 A modern catalytic converter of the type used in most new cars.

Enzyme kinetics

Some of the most powerful homogeneous catalysts are enzymes. Enzymes (often called *biocatalysts*) consist of proteins and contain a specially shaped area called an 'active site' that lowers the energy of the transition state of the reaction being catalysed. They are very specific and usually have a dramatic effect on the reaction they control. For example, the activation energy for the acidic hydrolysis of sucrose into glucose and fructose is 107 kJ mol^{-1}. The enzyme saccharase (enzyme names have the suffix *-ase*) reduces the activation barrier of this reaction to 36 kJ mol^{-1}, which accelerates the rate by a factor of 10^{12} at blood temperature (310 K). Enzyme catalysis can be represented by a series of elementary reactions (see p. 671). Qualitatively, the starting materials, or *substrates*, *S*, are reversibly bound in the *active sites* of the enzyme, *E*, where they form an *enzyme–substrate complex*, *E·S*. The enzyme enables the conversion of the substrate into the product, *P*, which is then released from the enzyme complex. Like all catalysts, the enzyme remains unchanged after the reaction.

There are two different hypotheses for how the substrate is bound into the enzyme, which we very briefly discuss here. The *lock-and-key* hypothesis assumes that the substrate simply 'fits' into the active site to form the enzyme–substrate complex. This is schematically shown in figure 15.24.

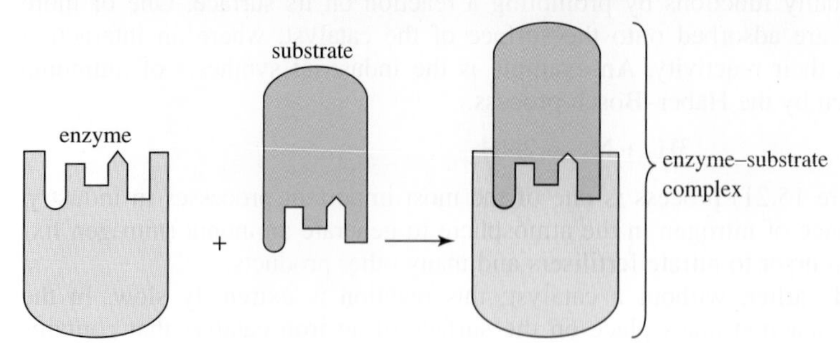

FIGURE 15.24 Schematic presentation of the lock-and-key model in substrate binding by an enzyme.

The *induced fit* hypothesis assumes that the enzyme molecule changes shape as the substrate molecule comes close — the incoming substrate *induces* the structural change. This more sophisticated model is based on the fact that the free rotation along single bonds makes molecules flexible.

Unprotected exposure to intense sunlight can lead to the formation of genotoxic products in DNA. The importance of efficient mechanisms for the repair of such UV lesions is demonstrated by hereditary diseases such as xeroderma pigmentosum; patients suffering this disease lack enzymes that repair UV-induced DNA damage, which leads to carcinomas and other skin malignancies at a young age. Figure 15.25 shows the structure of an enzyme–substrate complex of DNA photolyase and a DNA duplex containing a damaged site, which was caused by UV irradiation. In prokaryotes, plants and many animals (such as marsupials), DNA photolyases are mainly responsible for repairing UV-light-induced lesions in DNA. In the enzyme–substrate complex, the lesion is 'flipped' out of the duplex DNA into the active site of the enzyme and 'flipped' back into the DNA helix after the repair process is complete.

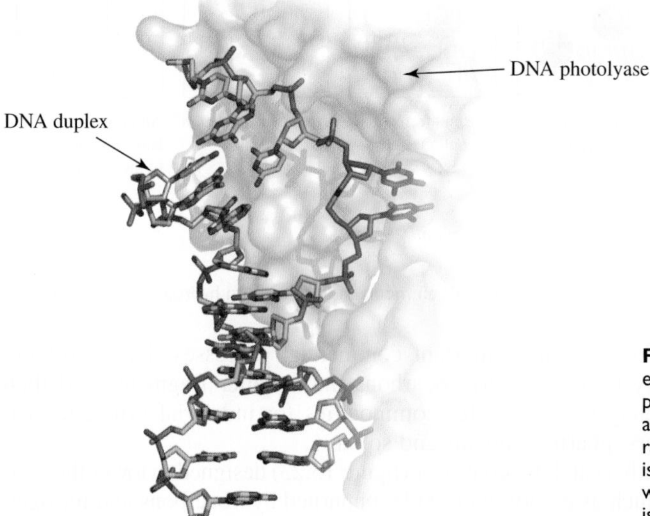

FIGURE 15.25 Illustration of an enzyme–substrate complex of DNA photolyase and a DNA duplex containing a site damaged by exposure to UV radiation. The surface of the enzyme is represented by van der Waals radii, whereas the structure of the DNA duplex is shown as a tube model.

The action of an enzyme can be described by the **Michaelis–Menten mechanism**. The rate of an enzyme-catalysed reaction in which the substrate S is converted into the product P is found to be dependent on the concentration of the enzyme E, even though the enzyme does not undergo a net change. Thus, the mechanism of the reaction can be written as:

$$E + S \underset{k_{-1}}{\overset{k_1}{\rightleftharpoons}} E{\cdot}S$$

$$E{\cdot}S \xrightarrow{k_2} P + E$$

Since the product P is formed only in the second reaction and not in the first reaction, the rate law for formation of P is:

$$\frac{d[P]}{dt} = k_2[E{\cdot}S]$$

$E{\cdot}S$ is the reactive intermediate, and its concentration can be described using the steady-state approximation, where rate of formation equals rate of consumption, that is, $\frac{d[E\cdot S]}{dt} = 0$.

Therefore, we have:

$$\frac{d[E{\cdot}S]}{dt} = \underbrace{k_1[E][S]}_{\text{formation of } E\cdot S} - \underbrace{(k_{-1}[E{\cdot}S] + k_2[E{\cdot}S])}_{\text{consumption of } E\cdot S} = 0$$

This gives:

$$k_1[E][S] = k_{-1}[E{\cdot}S] + k_2[E{\cdot}S]$$

$$[E{\cdot}S] = \frac{k_1[E][S]}{k_2+k_{-1}}$$

$[E]$ and $[S]$ are the concentrations of the free enzyme and free substrate, respectively. If we consider $[E]_0$ as the total concentration of the enzyme, we have:

$$[E]_0 = [E{\cdot}S] + [E] = \text{constant}$$

(because E is a catalyst, which does not undergo net change) or after rearrangement:

$$[E] = [E]_0 - [E{\cdot}S]$$

Since the concentration of S is much larger than that of the enzyme, the free substrate concentration is almost the same as the total substrate concentration $[S]_{total}$. As an approximation, we can therefore assume that $[S] = [S]_{total}$. We now have:

$$[E{\cdot}S] = \frac{k_1([E]_0-[E{\cdot}S])[S]}{k_2+k_{-1}}$$

$$[E{\cdot}S] = \frac{k_1[E]_0[S]-k_1[E{\cdot}S][S]}{k_2+k_{-1}}$$

$$[E{\cdot}S](k_2+k_{-1}) = k_1[E]_0[S] - k_1[E{\cdot}S][S]$$

$$[E{\cdot}S](k_2+k_{-1}) + k_1[E{\cdot}S][S] = k_1[E]_0[S]$$

$$[E{\cdot}S](k_2+k_{-1}+k_1[S]) = k_1[E]_0[S]$$

$$[E{\cdot}S] = \frac{k_1[E_0][S]}{k_2+k_{-1}+k_1[S]}$$

With this expression for $[E{\cdot}S]$, we can now rewrite the rate law for the formation of P as shown below.

$$\frac{d[P]}{dt} = k_2[E{\cdot}S] = \frac{k_2k_1[E]_0[S]}{k_2+k_{-1}+k_1[S]} = \frac{k_2[E]_0[S]}{K_M+[S]} \qquad \text{(equation 1)}$$

This is called the **Michaelis–Menten equation** and includes the **Michaelis constant**:

$$K_M = \frac{k_2+k_{-1}}{k_1}$$

According to equation 1, the rate of enzyme catalysis (enzymolysis) depends *linearly* on the enzyme concentration $[E]_0$, but in a more complicated way on the substrate concentration $[S]$. When $[S] >> K_M$, the rate law in equation 1 reduces to:

$$\frac{d[P]}{dt} = k_2[E]_0 = \text{rate}_{max}$$

showing that the rate is zero order in S. Under these conditions the enzyme is saturated, the rate of the enzyme reaction is constant and at its maximum, rate_{max} (see pp. 662–3). The Michaelis–Menten equation can then be written as:

$$\text{rate of reaction} = \frac{d[P]}{dt} = \frac{\text{rate}_{max}[S]}{K_M+[S]} \qquad \text{(equation 2)}$$

On the other hand, when the substrate concentration is small and $[S] << K_M$, the rate law in equation 1 reduces to:

$$\frac{d[P]}{dt} = \frac{k_2}{K_M}[E]_0[S]$$

which shows that the rate of product formation depends both on the concentration of enzyme and substrate. This behaviour is shown in figure 15.26 (overleaf).

The Michaelis–Menten equation is the basic equation of enzyme kinetics, in which K_M has a simple operational definition. At the substrate concentration where $[S] = K_M$, equation 2 becomes:

$$\text{rate of reaction} = \frac{\text{rate}_{max}}{2}$$

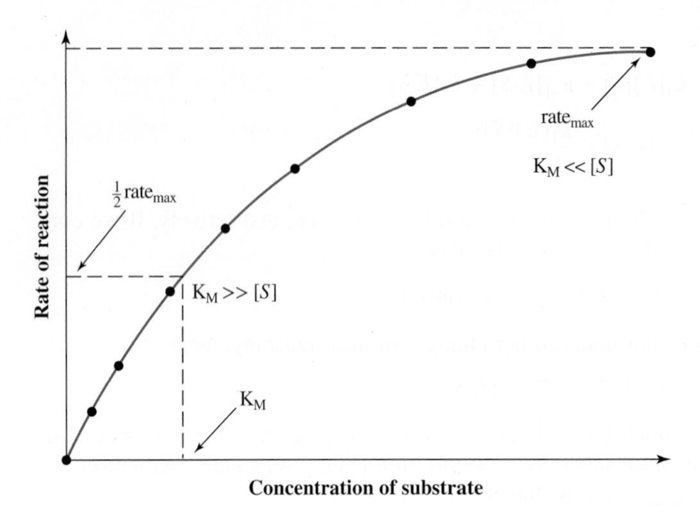

FIGURE 15.26 Saturation curve for an enzyme reaction showing the relationship between the substrate concentration and reaction rate.

Here, K_M is the substrate concentration at which the rate of reaction is half of the maximum. Thus, if an enzyme has a small K_M (K_M varies widely with the identity of the enzyme and the nature of the substrate), it achieves maximal catalytic efficiency at low substrate concentrations.

The activity of enzymes can be reduced or even eliminated by enzyme inhibitors, which can bind to the enzyme or the enzyme–substrate complex (reversible inhibitors). Irreversible inhibitors inactivate enzymes by forming covalent bonds which modify the active site of the enzyme.

Enzymes are not used only in biological systems. Their unique properties as highly specific catalysts have inspired scientists to use them in the chemical industry and other applications. Because of the limited number of available enzymes and their lack of stability in organic solvents and at higher temperatures, protein engineering is a highly active research field, where new enzymes with novel properties are designed. Examples of enzymes in our daily life include: contact lens cleaners, which contain enzymes called proteases that remove proteins on contact lenses to prevent infections; biological detergents such as laundry soap, which contain proteases to assist in the removal of protein stains from clothes; machine dishwashing detergents, which contain amylases to remove resistant starch residues, and lipases to help in the removal of fatty and oily stains; and cellulases used in biological fabric conditioners.

In this chapter, we have learned the fundamentals of chemical kinetics. We will apply some of the concepts we have learned here in chapter 18 when we look at the reactions of haloalkanes.

SUMMARY

LO 1 Reaction rates

Chemical kinetics is the study of the rate of a chemical reaction by determining the changes in concentrations of reactants and products. The rate of change in concentration (for reactions in solution) or pressure (for gas phase reactions) at a particular time is called the instantaneous rate of change of concentration.

The time required for half of a reactant to disappear is the half-life, $t_{\frac{1}{2}}$.

LO 2 Factors that affect reaction rates

Reaction rates are controlled by five factors: (1) the chemical nature of the reactants, (2) the physical nature of the reactants, (3) the concentrations of the reactants, (4) the temperature and (5) the presence of catalysts. The rates of heterogeneous reactions are determined largely by the area of contact between the phases; the rates of homogeneous reactions are determined by the concentrations of the reactants. The rate of a reaction is measured by monitoring the change in reactant or product concentrations with time and, for the general reaction $aA + bB \rightarrow cC + dD$, is defined as:

$$\text{rate of reaction} = -\frac{1}{a}\frac{\text{d}[A]}{\text{d}t} = -\frac{1}{b}\frac{\text{d}[B]}{\text{d}t} = \frac{1}{c}\frac{\text{d}[C]}{\text{d}t} = \frac{1}{d}\frac{\text{d}[D]}{\text{d}t}$$

LO 3 Overview of rate laws

The rate law for a reaction relates the rate of the reaction to the molar concentrations of the reactants. The rate of a reaction is proportional to the product of the molar concentrations of the reactants, each raised to an appropriate power (order of the reactant). These exponents must be determined by experiments in which the concentrations are varied and the effects on the rate of the reaction are measured. The proportionality constant, k, is called the rate constant. Its value depends on temperature but not on the concentrations of the reactants. The sum of the exponents in the rate law is defined as the order (or overall order) of the reaction.

$$\text{rate of reaction} = k[\text{reactants}]^n$$

LO 4 Types of rate laws: differential and integrated

The differential rate law shows how the rate of a reaction depends on concentrations. The integrated rate law shows how the concentrations depend on time.

Equations exist that relate the concentration of a reactant at a given time t to the initial concentration and the rate constant. For a first-order reaction, the half-life is a constant that depends only on the rate constant for the reaction; it is independent of the initial concentration. The half-life of a second-order reaction is inversely proportional to both the initial concentration of the reactant and to the rate constant. The half-life of a zero-order reaction is proportional to the initial concentration of the reactant but is inversely proportional to the rate constant.

LO 5 Theory of chemical kinetics

According to the collision theory, the rate of a reaction depends on the number of effective collisions per second between the reactant molecules, which is an extremely small fraction of the total number of collisions per second. This fraction is small partly because collisions usually do not result in products unless the reactant molecules are suitably oriented. The major reason, however, is that the colliding molecules must jointly possess a minimum molecular kinetic energy, called the activation energy, E_a. As the temperature increases, a larger fraction of the collisions has this necessary energy, making more collisions effective each second and the reaction faster.

The enthalpy of a reaction is the net potential energy difference between the reactants and the products. In reversible reactions, the values of E_a for both the forward and reverse reactions can be identified on a potential energy diagram. The species at the maximum on an energy diagram is called the transition state. The transition state has a lifetime of one vibration and cannot be isolated. In a multi-step reaction, a reaction intermediate is characterised as an energy minimum between two transition states. The lifetime of the intermediate depends on the magnitude of E_a associated with the transition states between which the intermediate exists. Intermediates have higher energies than reactants and products.

The Arrhenius equation shows how changes in temperature affect a rate constant. The Arrhenius equation also enables us to determine E_a, either graphically or by a calculation using the appropriate form of the Arrhenius equation. The calculation requires two rate constants determined at two different temperatures. The activation energy and the rate constant at one temperature can be used to calculate the rate constant at another temperature.

LO 6 Reaction mechanisms

The detailed sequence of elementary reactions that lead to the net chemical change is the mechanism of the reaction. Support for a mechanism comes from matching the predicted rate law for the mechanism with the rate law obtained from experimental data. For the rate-determining step, or for any elementary reaction, the corresponding rate law has exponents equal to the coefficients in the balanced equation. In more complicated reaction mechanisms, where the rate-determining step cannot be easily identified, the steady-state approximation can be used to derive the rate law.

LO 7 Catalysts

Catalysts are substances that change a reaction rate but are not consumed by the reaction. Negative catalysts inhibit reactions. Positive catalysts provide alternative paths for reactions for which at least one step has a smaller activation energy than the uncatalysed reaction. Homogeneous catalysts are in the same phase as the reactants. Heterogeneous catalysts provide a path of lower activation energy by having a surface on which the reactants are adsorbed and react. Catalysts in living systems are called enzymes. Enzyme kinetics can be described by the Michaelis–Menten equation, which contains the Michaelis constant K_M; this represents the ratio of the rate constants for formation of the enzyme–substrate complex and the enzyme reaction to form products.

KEY CONCEPTS AND EQUATIONS

Rate law of a reaction
(section 15.4)

This shows the dependence of the rate of a reaction on the concentration of one or more reactants (and occasionally products), each of which is raised to a particular exponent. The overall order of a reaction is given by the sum of these exponents. The rate constant for a reaction can be calculated from this, given the appropriate rate data.

Integrated first-order rate law
(section 15.4)

$$\ln[A]_t = -kt + \ln[A]_0$$

For a first-order reaction with known k, this equation is used to calculate the concentration of a reactant at some specified time after the start of the reaction, or to calculate the time required for the concentration to drop to a specified value.

Integrated second-order rate law
(section 15.4)

$$\frac{1}{[A]_t} = kt + \frac{1}{[A]_0}$$

For a second-order reaction with known k, this equation is used to calculate the concentration of a reactant at a specified time after the start of the reaction, or to calculate the time required for the concentration to drop to a specified value.

Integrated zero-order rate law
(section 15.4)

$$[A]_t = -kt + [A]_0$$

For a zero-order reaction with known k, this equation is used to calculate the concentration of a reactant at some specified time after the start of the reaction, or to calculate the time required for the concentration to drop to a specified value.

Half-life
(section 15.4)

Zero order:

$$t_{\frac{1}{2}} = \frac{[A]_0}{2k}$$

First order:

$$t_{\frac{1}{2}} = \frac{\ln 2}{k}$$

Second order:

$$t_{\frac{1}{2}} = \frac{1}{k[A]_0}$$

These equations are used to calculate a half-life from experimental data and reaction order, or to calculate the rate of disappearance of a reactant.

Arrhenius equation
(section 15.5)

$$k = Ae^{-E_a/RT}$$

This equation allows graphical determination of an activation energy and relates rate constants, activation energy and temperature.

Michaelis–Menten equation
(section 15.7)

$$\text{rate} = \frac{\text{rate}_{max}[S]}{K_M + [S]}$$

This is the basic equation of enzyme kinetics.

KEY TERMS

activation energy (E_a) *p. 665*
Arrhenius equation *p. 668*
average rate *p. 643*
catalysis *p. 680*
catalyst *p. 680*
chemical kinetics *p. 642*
collision theory *p. 664*
differential rate law *p. 649*
elementary reaction *p. 671*
frequency factor *p. 668*
half-life ($t_{\frac{1}{2}}$) *p. 646*
heterogeneous catalyst *p. 680*
heterogeneous reaction *p. 647*
homogeneous catalyst *p. 680*
homogeneous reaction *p. 647*
initial rate *p. 648*
instantaneous rate of change of concentration *p. 644*
integrated rate law *p. 649*
Michaelis constant *p. 683*
Michaelis–Menten equation *p. 683*
Michaelis–Menten mechanism *p. 682*
molecularity *p. 671*
order *p. 648*
pre-exponential factor *p. 668*
rate coefficient *p. 648*
rate constant *p. 648*
rate-determining step *p. 672*
rate law *p. 648*
rate of reaction *p. 644*
reaction coordinate *p. 665*
reaction intermediate *p. 666*
reaction mechanism *p. 650*
steady-state approximation *p. 674*
steric factor *p. 668*
transition state *p. 665*

REVIEW QUESTIONS

LO 1 Reaction rates

15.1 What does 'rate of reaction' mean in qualitative terms?

15.2 In terms of rates of reaction, what is an explosion?

15.3 What is meant by the term 'half-life'?

LO 2 Factors that affect reaction rates

15.4 Suppose we compared two reactions, one requiring the simultaneous collision of three molecules and the other requiring a collision between two molecules. From the standpoint of statistics, and all other factors being equal, which reaction should be faster? Explain your answer.

15.5 List the five factors that affect the rates of chemical reactions.

15.6 How does an instantaneous rate of reaction differ from an average rate of reaction?

15.7 Explain how the initial instantaneous rate of reaction can be determined from experimental concentration versus time data.

15.8 What is a 'homogeneous reaction'? Give an example.

15.9 What is a 'heterogeneous reaction'? Give an example.

15.10 Why are chemical reactions usually carried out in solution?

15.11 How does particle size affect the rate of a heterogeneous reaction? Why?

15.12 The rate of hardening of epoxy glue depends on the amount of hardener that is mixed into the glue. What factor affecting rates of reaction does this illustrate?

15.13 What is a 'catalyst'?

15.14 On the basis of what you learned in chapters 6 and 7, why do foods cook faster in a pressure cooker than in an open pot of boiling water?

15.15 People who have been submerged in very cold water and who are believed to have drowned can sometimes be revived. On the other hand, people who have been submerged in warmer water for the same length of time have died. Explain this in terms of factors that affect the rates of chemical reactions.

LO 3 Overview of rate laws

15.16 What are the units of rate of reaction?

15.17 What is a 'rate law'? What is the proportionality constant called?

15.18 What is meant by the 'order' of a reaction?

15.19 Can the exponents in a rate law be determined from the balanced equation? Explain.

LO 4 Types of rate laws: differential and integrated

15.20 How does the dependence of rate of reaction on concentration differ between a zero-order and a first-order reaction?

15.21 If the concentration of a reactant is doubled and the rate of reaction is unchanged, what is the order of the reaction with respect to that reactant?

15.22 If the concentration of a reactant is doubled and the rate of the reaction doubles, what is the order of the reaction with respect to that reactant?

15.23 If the concentration of a reactant is doubled, by what factor will the rate of reaction increase if the reaction is second order with respect to that reactant?

15.24 In an experiment, the concentration of a reactant was tripled. The rate of reaction increased by a factor of 27. What is the order of the reaction with respect to that reactant?

15.25 Biological reactions usually involve the interaction of an enzyme with a substrate, the substance that actually undergoes the chemical change. In many cases, the rate of reaction depends on the concentration of the enzyme, but is independent of the substrate concentration. What is the order of the reaction with respect to the substrate in such instances?

15.26 A reaction has the following rate law.

$$\text{rate of reaction} = k[A]^2[B][C]$$

What are the units of the rate constant, k?

15.27 Give the equations that relate concentration to time for: (a) a first-order reaction and (b) a second-order reaction.

15.28 How is the half-life of a first-order reaction affected by the initial concentration of the reactant?

15.29 How is the half-life of a second-order reaction affected by the initial reactant concentration?

15.30 Derive the equations for $t_{\frac{1}{2}}$ for first- and second-order reactions.

15.31 The integrated rate law for a zero-order reaction is:

$$[A]_t - [A]_0 = kt$$

Derive an equation for the half-life of a zero-order reaction.

15.32 The rate law for a certain enzymatic reaction is zero order with respect to the substrate. The rate constant for the reaction is 6.4×10^2 mol L^{-1} s^{-1}. If the initial concentration of the substrate is 0.275 mol L^{-1}, what is the initial rate of the reaction?

LO 5 Theory of chemical kinetics

15.33 What is the basic postulate of the collision theory?

15.34 What two factors influence the effectiveness of molecular collisions in producing chemical change?

15.35 In terms of the kinetic theory, why does an increase in temperature increase the rate of reaction?

15.36 Explain, in terms of the law of conservation of energy, why an endothermic reaction leads to a cooling of the reaction mixture (provided heat cannot enter from outside the system).

15.37 Define the term 'transition state'.

15.38 Draw the potential energy diagrams for an exothermic and an endothermic reaction. Indicate for both diagrams the transition state, activation barrier for forward and reverse reaction, and enthalpy of reaction.

15.39 Suppose a certain slow reaction is found to have a very small activation energy. What does this suggest about the importance of molecular orientation in the formation of the transition state?

15.40 State the Arrhenius equation (which relates the rate constant to temperature and activation energy), and define the symbols.

LO 6 Reaction mechanisms

15.41 What is the definition of an 'elementary reaction'? How are elementary reactions related to the mechanism of a reaction?

15.42 What is a 'rate-determining step'?

15.43 In what way is the rate law for a reaction related to the rate-determining step?

15.44 The following mechanism has been proposed for a particular reaction.

$$2NO \rightarrow N_2O_2$$
$$N_2O_2 + H_2 \rightarrow N_2O + H_2O$$
$$N_2O + H_2 \rightarrow N_2 + H_2O$$

What is the net overall change that occurs in this reaction?

15.45 If the experimental rate law for the mechanism in question 15.44 is second order in [NO] and first order in [H_2], which of the reaction steps is the rate-determining step?

15.46 If the reaction $NO_2 + CO \rightarrow NO + CO_2$ occurs by a one-step collision process, what would be the expected rate law for the reaction? The actual rate law is rate = $k[NO_2]^2$. Could the reaction actually occur by a one-step collision between NO_2 and CO? Explain.

15.47 The experimental rate law for the reaction $NO_2 + CO \rightarrow CO_2 + NO$ is:

$$\text{rate of reaction} = k[NO_2]^2$$

If the mechanism is:

$$2NO_2 \rightarrow NO_3 + NO \quad \text{(slow)}$$
$$NO_3 + CO \rightarrow NO_2 + CO_2 \quad \text{(fast)}$$

show that the predicted rate law is the same as the experimental rate law.

15.48 The following mechanism has been proposed for the reduction of NO_3^- by $MoCl_6^{2-}$.

$$MoCl_6^{2-} \underset{k_{-1}}{\overset{k_1}{\rightleftharpoons}} MoCl_5^- + Cl^-$$
$$NO_3^- + MoCl_5^- \xrightarrow{k_2} MoOCl_5 + NO_2^-$$

What is the intermediate?
Derive an expression for the rate law, where:

$$\text{rate of reaction} = \frac{d[NO_2^-]}{dt}$$

for the overall reaction, using the steady-state approximation.

15.49 Consider the hypothetical reaction:

$$Y \rightarrow A + Z$$

which is assumed to occur by the following mechanism:

$$2Y \underset{k_{-1}}{\overset{k_1}{\rightleftharpoons}} Y^* + Y$$
$$Y^* \xrightarrow{k_2} A + Z$$

(Y^* represents an *activated* molecule, which has enough energy to surmount the activation barrier).

(a) Derive a rate law for the production of A using the steady-state approximation.
(b) Assuming that this reaction is known to be first order, under what conditions does your derived rate law agree with this observation?
(c) Explain how a chemical reaction can be first order when, even in a simple case (for example the present reaction), molecules must collide to build up enough energy to get over the energy barrier? Why aren't all reactions at least second order?

LO 7 Catalysts

15.50 Enzymes are biological catalysts. One proposed mechanism for enzyme of the reaction $A \rightarrow B$ is:

$$E + A \rightarrow E{\cdot}A \quad \text{(step 1)}$$
$$E{\cdot}A \rightarrow E + B \quad \text{(step 2)}$$

where E is the enzyme. What is the overall rate law for this reaction if step 1 is the slower step? What is the overall rate law if step 2 is the slower step?

15.51 How does a catalyst increase the rate of a chemical reaction?

15.52 What is a 'homogeneous catalyst'? How does it function?

15.53 What is a 'heterogeneous catalyst'? How does it function?

REVIEW PROBLEMS

15.54 The following data were collected at a certain temperature for the decomposition of sulfuryl chloride, SO_2Cl_2, a chemical used in a variety of organic syntheses. **LO 1**

$$SO_2Cl_2 \rightarrow SO_2 + Cl_2$$

Time (min)	$[SO_2Cl_2]$ (mol L^{-1})
0	0.1000
100	0.0876
200	0.0768
300	0.0673
400	0.0590
500	0.0517
600	0.0453
700	0.0397
800	0.0348
900	0.0305
1000	0.0267
1100	0.0234

Draw a graph of concentration versus time and determine the rate of formation of SO_2 at $t = 200$ min and $t = 600$ min.

15.55 If the following reaction were to occur in one step, what would be the rate law for this reaction? **LO 3**

$$AB + C \rightarrow AC + B$$

15.56 In the catalysed formation of ammonia according to the reaction $3H_2 + N_2 \rightarrow 2NH_3$, how does the rate of consumption of hydrogen compare with the rate of consumption of nitrogen? How does the rate of formation of NH_3 compare with the rate of consumption of nitrogen? **LO 1,7**

15.57 The rate constant for the first-order decomposition of N_2O_5 to NO_2 and O_2 has a value of $4.8 \times 10^{-4}\,s^{-1}$. **LO 4**
(a) What is the half-life of the reaction (in seconds)?
(b) The initial pressure is 66.7 kPa. What will be the pressure:
(i) 10 s and (ii) 10 min after initiation of the reaction?

15.58 If it takes 75 min for the concentration of a reactant to drop to 25% of its initial value through a first order reaction, what is the rate constant for the reaction in units min^{-1}? **LO 4**

15.59 The concentration of a drug in a body is commonly expressed in units of milligrams per kilograms of body weight. Consider an animal with an initial dose of 25.0 mg/kg body weight. After 2 hours the concentration has dropped to 15 mg/kg body weight. The drug is metabolically eliminated through a first-order process. What is the rate constant in units min^{-1}? **LO 4**

15.60 At a certain moment in the reaction: **LO 1**

$$2N_2O_5 \rightarrow 4NO_2 + O_2$$

N_2O_5 was found to be decomposing at a rate of $2.5 \times 10^{-6}\,mol\,L^{-1}\,s^{-1}$. What were the rates of formation of NO_2 and O_2?

15.61 Calculate the rate of the reaction: **LO 1,4**

$$H_2SeO_3 + 6I^- + 4H^+ \rightarrow Se + 2I_3^- + 3H_2O$$

given that the rate law for the reaction (at 0 °C) is:

$$\text{rate of reaction} = (5.0 \times 10^5\,L^5\,mol^{-5}\,s^{-1})[H_2SeO_3][I^-]^3[H^+]^2$$

and the reactant concentrations are $[H_2SeO_3] = 2.0 \times 10^{-2}$ M, $[I^-] = 2.0 \times 10^{-3}$ M, and $[H^+] = 1.0 \times 10^{-3}$ M.

15.62 The rate law for the reaction $2H_2S + 3O_2 \rightarrow 2SO_2 + 2H_2O$ was found to be: rate of reaction = $k[H_2S]^2$. If the reaction rate for the consumption of O_2 is $2.6 \times 10^{-4}\,mol\,L^{-1}\,s^{-1}$ at a particular temperature, what is the reaction rate for production of H_2O? **LO 4**

15.63 The reaction $2NO_2 \rightarrow NO_3 + NO$ was found to follow the rate law: **LO 4**

$$\text{rate of reaction} = k[NO_2]^2$$

Given that the concentration of NO_2 is 0.75 M and the rate constant is $1.25 \times 10^2\,mol^{-1}\,L\,s^{-1}$ at a particular temperature, calculate the rate of the reaction.

15.64 The oxidation of NO (released in small amounts in the exhaust of cars) produces the brownish-red gas NO_2, which is a component of urban air pollution. **LO 1,4**

$$2NO + O_2 \rightarrow 2NO_2$$

The rate law for the reaction is rate of reaction = $k[NO]^2[O_2]$. At 25 °C, $k = 7.1 \times 10^9\,L^2\,mol^{-2}\,s^{-1}$. What would the rate of the reaction be if $[NO] = 0.0010\,mol\,L^{-1}$ and $[O_2] = 0.034\,mol\,L^{-1}$?

15.65 The rate law for the decomposition of N_2O_5 is: **LO 1,4**

$$\text{rate of reaction} = k[N_2O_5]$$

If $k = 1.0 \times 10^{-5}\,s^{-1}$, what is the rate of reaction when the N_2O_5 concentration is $0.0010\,mol\,L^{-1}$?

15.66 The reaction $4NH_3 + 5O_2 \rightarrow 4NO + 6H_2O$ was found to follow the rate law: **LO 4**

$$\text{rate of reaction} = k[NH_3]^2[O_2]^2$$

If the concentrations of NH_3 and O_2 are 0.19 M and 0.79 M respectively and the reaction rate is $1.43 \times 10^{-4}\,mol\,L^{-1}\,s^{-1}$, what is the rate constant for this reaction?

15.67 For the reaction: **LO 4**

$$2HCrO_4^- + 3HSO_3^- + 5H^+ \rightarrow 2Cr^{3+} + 3SO_4^{2-} + 5H_2O$$

the rate law is:

$$\text{rate of reaction} = k[HCrO_4^-][HSO_3^-]^2[H^+]$$

(a) What is the order of the reaction with respect to each reactant?
(b) What is the overall order of the reaction?

15.68 A key elementary reaction in the destruction of stratospheric ozone from nitrogen oxides in jet exhaust of high-flying aircraft is the reaction: **LO 4**

$$NO + O_3 \rightarrow NO_2 + O_2$$

The rate law is:

$$\text{rate of reaction} = k[NO][O_3]$$

(a) What is the order with respect to each reactant?
(b) What is the overall order of the reaction?

15.69 The reaction $Cl_2 + 3F_2 \rightarrow 2ClF_3$ has a rate law proportional to the square of the fluorine concentration. Using a rate constant of $1.4 \times 10^{-3}\,mol^{-1}\,L\,s^{-1}$ and a rate of $9.3 \times 10^{-5}\,mol\,L^{-1}\,s^{-1}$, calculate the fluorine concentration. **LO 4**

15.70 The following data were obtained for the reaction: $2HgCl_2 + C_2O_4^{2-} \rightarrow 2Cl^- + 2CO_2 + Hg_2Cl_2$. **LO 4**

Experiment	$[HgCl_2]_0$ (mol L^{-1})	$[C_2O_4^{2-}]_0$ (mol L^{-1})	Initial rate (mol L^{-1} min^{-1})
1	0.105	0.15	1.8×10^{-5}
2	0.105	0.30	7.1×10^{-5}
3	0.052	0.30	3.5×10^{-5}

(a) Determine the rate law for the reaction.
(b) Calculate the value of the rate constant.

15.71 The major pollutants NO(g), CO(g), NO_2(g) and CO_2(g), which are emitted by cars, can react according to the following equation. **LO4**

$$NO_2(g) + CO(g) \rightarrow NO(g) + CO_2(g)$$

The following rate data were collected at 225 °C.

Experiment	$[NO_2]_0$ (mol L^{-1})	$[CO]_0$ (mol L^{-1})	Initial rate $d[NO_2]/dt$ (mol L^{-1} s^{-1})
1	0.263	0.826	1.44×10^{-5}
2	0.263	0.413	1.44×10^{-5}
3	0.526	0.413	5.76×10^{-5}

(a) Determine the rate law for the reaction.
(b) Calculate the value of the rate constant at 225 °C.
(c) Calculate the rate of appearance of CO_2 when $[NO_2] = [CO] =$ 0.500 M.

15.72 Cyclopropane, C_3H_6, is a gas used as a general anaesthetic. It undergoes a slow rearrangement to propene. **LO4**

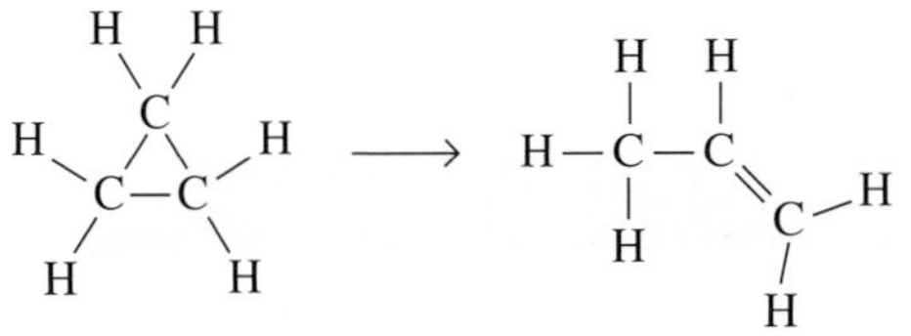

cyclopropane → propene

At a certain temperature, the following rate data were obtained.

Initial concentration of C_3H_6 (mol L^{-1})	Rate of formation of propene (mol L^{-1} s^{-1})
0.050	2.95×10^{-5}
0.100	5.90×10^{-5}
0.150	8.85×10^{-5}

What is the rate law for the reaction? What is the value of the rate constant? Include the correct units.

15.73 The initial rate of formation of a substance Q in the gas phase depends on the concentration. **LO4**

Experiment	$[Q]_0$ (molecules cm^{-3})	Initial rate (molecules cm^{-3} s^{-1})
1	5.0×10^3	3.6×10^{-4}
2	8.2×10^3	9.6×10^{-4}
3	17×10^3	41×10^{-4}
4	30×10^3	130×10^{-4}

(a) Find the order of the reaction.
(b) Calculate the value of the rate constant.

15.74 The formation of small amounts of nitric oxide, NO, in car engines is the first step in the formation of smog. Nitric oxide is readily oxidised to nitrogen dioxide by the reaction: **LO4**

$$2NO(g) + O_2(g) \rightarrow 2NO_2(g)$$

The following data were collected in a study of the rate of this reaction.

Initial concentration		Rate of formation of NO_2 (mol L^{-1} s^{-1})
$[O_2]$ (mol L^{-1})	[NO] (mol L^{-1})	
0.0010	0.0010	7.10
0.0040	0.0010	28.4
0.0040	0.0030	255.6

What is the rate law for the reaction? What is the rate constant with its correct units?

15.75 The following data were obtained for the reaction of $(CH_3)_3CBr$ with hydroxide ions at 55 °C. **LO4**

$$(CH_3)_3CBr + OH^- \rightarrow (CH_3)_3COH + Br^-$$

Initial concentration		Rate of formation of $(CH_3)_3COH$ (mol L^{-1} s^{-1})
$[(CH_3)_3CBr]$ (mol L^{-1})	$[OH^-]$ (mol L^{-1})	
0.10	0.10	1.0×10^{-3}
0.20	0.10	2.0×10^{-3}
0.30	0.10	3.0×10^{-3}
0.10	0.20	1.0×10^{-3}
0.10	0.30	1.0×10^{-3}

What is the rate law for the reaction? What is the value of the rate constant (with correct units) at this temperature?

15.76 Data for the decomposition of SO_2Cl_2 according to the equation $SO_2Cl_2(g) \rightarrow SO_2(g) + Cl_2(g)$ were given in question 15.54. Decide graphically whether the reaction is first or second order. Graphically determine the rate constant for this reaction. **LO3**

15.77 The decomposition of SO_2Cl_2 described in question 15.54 has a first-order rate constant $k = 2.2 \times 10^{-5}$ s^{-1} at 320 °C. If the initial SO_2Cl_2 concentration in a container is 4×10^{-3} M, what is its concentration: (a) after 1 hour and (b) after 1 day?

15.78 If it takes 75 min for the concentration of a reactant to drop to 20% of its initial value in a first-order reaction, what is the rate constant for the reaction in the units min^{-1}? **LO4**

15.79 The decomposition of hydrogen iodide follows the equation $2HI(g) \rightarrow H_2(g) + I_2(g)$. The reaction is second order and has a rate constant of 1.6×10^{-3} mol^{-1} L s^{-1} at 700 °C. If the initial concentration of HI in a container is 3.4×10^{-2} M, how long will it take for the concentration to be reduced to 8.0×10^{-4} M? **LO4**

15.80 The following table shows the results of the alkaline hydrolysis of ethyl nitrobenzoate. **LO4**

t (s)	0	100	200	300	400	500	600	700	800
[ethyl nitrobenzoate] (10^{-2} mol L^{-1})	5	3.55	2.75	2.25	1.85	1.6	1.48	1.4	1.38

Determine the order of the reaction and the rate constant.

15.81 One of the hazards of nuclear explosions is the generation of ^{90}Sr and its subsequent incorporation in the bones in place of calcium. ^{90}Sr has a half-life of 28.1 years. A newly born child was exposed to nuclear fall-out and absorbed 1 μg of ^{90}Sr. How much will remain after (a) 18 years and (b) 70 years, if none is lost metabolically? **LO4**

15.82 The progress of the liquid phase reaction $2A \rightarrow C$ was monitored as a function of time by a spectroscopic method to give the following data. **LO4**

t (min)	0	10	20	30	40	N
[C] (mol L^{-1})	0	0.089	0.153	0.200	0.230	0.312

(a) What is the order of this reaction?
(b) What is the rate constant?

15.83 The progress of the gas phase reaction mixture $2B \rightarrow D$ was monitored by measuring the total pressure as a function of time. Use the data in the table below to determine: **LO4**
(a) the order of the reaction
(b) the rate constant.

t (s)	0	100	200	300	400
p (Torr)	400	322	288	268	256

(*Note:* Torr is a unit for pressure: -> 1 Torr = 133.322 Pa.)

15.84 The second-order rate constant for the decomposition of HI at 700 °C was given in question 15.79. At 2.5×10^3 min after a particular experiment had begun, the HI concentration was equal to 4.5×10^{-4} mol L^{-1}. What was the initial molar concentration of HI in the reaction vessel? **LO4**

15.85 A reaction has the rate constant of 0.0117 s^{-1} at 400 K and 0.689 s^{-1} at 450 K. **LO4**
(a) Determine the activation barrier.
(b) What is the value of the rate constant at 425 K?

15.86 ^{220}Rn decays into ^{216}Po through a first-order process with a rate constant of 0.0125 s^{-1}. When the concentration of ^{220}Rn is 2.3×10^{-9} mol L^{-1}, what is the rate of the reaction? **LO4**

15.87 The half-life for the radioactive decay of ^{14}C is 5730 years. An archaeological sample contains wood that has only 72% of the ^{14}C found in living trees. What's the age of the sample? **LO4**

15.88 What ratio ^{14}C/^{12}C ratio would a wooden door lintel from an excavated site in Mexico be expected to have if the lintel is 9000 years old? **LO4**

15.89 ^{90}Sr has a half-life of 28 years. How long will it take for all of the ^{90}Sr presently on Earth to be reduced to $\frac{1}{32}$ of its present amount? **LO4**

15.90 Using the graph from question 15.54, determine the time required for the SO_2Cl_2 concentration to drop from 0.100 mol L^{-1} to 0.050 mol L^{-1}. How long does it take for the concentration to drop from 0.050 mol L^{-1} to 0.025 mol L^{-1}? What is the order of this reaction? **LO4**

15.91 The second-order rate constant for the reaction between ethyl acetate and sodium hydroxide, for example, EtOAc (aq) + NaOH (aq) $\rightarrow$ AcONa (aq) + EtOH (aq), is 0.11 mol^{-1} L s^{-1}. What is the concentration of ethyl acetate after: (a) 10 s and (b) 10 min, when the initial concentrations are [EtOAc] = 0.100 M and [NaOH] = 0.05 M? **LO4**

15.92 The decomposition of NOCl, the compound that gives a yellow-orange colour to aqua regia (a mixture of concentrated HCl and HNO_3 that is able to dissolve gold and platinum) follows the reaction $2NOCl \rightarrow 2NO + Cl_2$. It is a second-order reaction with $k = 6.7 \times 10^{-4}$ mol^{-1} L s^{-1} at 400 K. What is the half-life of this reaction if the initial concentration of NOCl is 0.20 mol L^{-1}? **LO4**

15.93 Rate constants were measured at various temperatures for the reaction: **LO4,5**

$$HI(g) + CH_3I(g) \rightarrow CH_4(g) + I_2(g)$$

The following data were obtained.

Rate constant (mol^{-1} L s^{-1})	Temperature (°C)
1.91×10^{-2}	205
2.74×10^{-2}	210
3.90×10^{-2}	215
5.51×10^{-2}	220
7.73×10^{-2}	225
1.08×10^{-1}	230

Determine the activation energy in kJ mol^{-1} both graphically and by calculation using $\ln\frac{k_2}{k_1} = -\frac{E_a}{R}\left(\frac{1}{T_2} - \frac{1}{T_1}\right)$. For the calculation of E_a, use the first and last rows of data in the table above.

15.94 The following table shows the rate constants for the second-order decomposition of acetaldehyde (ethanal, CH_3CHO) over a temperature range 300–500 K. **LO4,6**

T (K)	300	350	400	450	500
k (mol^{-1} L s^{-1})	7.9×10^6	3.0×10^7	7.9×10^7	1.7×10^8	3.2×10^8

Find the activation energy and the pre-exponential factor.

15.95 The reaction $2NOCl \rightarrow 2NO + Cl_2$ has $k = 9.3 \times 10^{-5}$ mol^{-1} L s^{-1} at 100 °C and $k = 1.0 \times 10^{-3}$ mol^{-1} L s^{-1} at 130 °C. What is E_a for this reaction in kJ mol^{-1}? Use the data at 100 °C to calculate the pre-exponential factor *A*. **LO5**

15.96 Calculate the activation energy for a second order reaction, using the following data. **LO6**

$k_1 = 4.0$ mol^{-1} L s^{-1} at 37 °C

$k_2 = 8.0$ mol^{-1} L s^{-1} at 87 °C

15.97 Calculate the activation energy for the reaction $ClO_3^- + H_2O \rightarrow ClO_4^- + H_2$, using the data in the table below. **LO6**

Rate constant *k* (s^{-1})	Temperature (C^0)
2.0×10^{-3}	25
4.0×10^{-3}	35
8.0×10^{-4}	45
1.6×10^{-2}	55

15.98 The rate constants for the reaction of benzene with oxygen atoms are 1.44×10^7 at 300.3 K, 3.03×10^7 at 341.2 K and 6.90×10^7 at 392.2 K, all in mol^{-1} L s^{-1}. **LO6**
Determine:
(a) the activation energy
(b) the pre-exponential factor A.

15.99 The conversion of cyclopropane, an anaesthetic, to propene (see question 15.72) has a rate constant $k = 1.3 \times 10^{-6}$ s^{-1} at 400 °C and $k = 1.1 \times 10^{-5}$ s^{-1} at 430 °C. **LO6**
(a) What is the activation energy in kJ mol^{-1}?
(b) What is the value of the pre-exponential factor, *A*, for this reaction?
(c) What is the rate constant for the reaction at 350 °C?

15.100 The reaction of CO_2 with water to form carbonic acid, $CO_2(aq) + H_2O(l) \rightarrow H_2CO_3(aq)$, has $k = 3.75 \times 10^{-2}$ s^{-1} at 25 °C and $k = 2.1 \times 10^{-3}$ s^{-1} at 0 °C. What is the activation energy for this reaction in kJ mol^{-1}? **LO6**

15.101 If a reaction has $k = 3.0 \times 10^{-4}$ s^{-1} at 25 °C and an activation energy of 100 kJ mol^{-1}, what is the value of *k* at 50 °C? **LO6**

15.102 The rate constant of a polymerisation reaction (see chapter 14) was established as a function of temperature. **LO5,6**
(a) How can it be shown that the kinetics of this reaction follow Arrhenius behaviour?
(b) If the reaction follows Arrhenius behaviour, how can the activation energy for the reaction and the pre-exponential factor A be derived?

15.103 The decomposition of N_2O_5 has an activation energy of 103 kJ mol^{-1} and a pre-exponential factor of 4.3×10^{13} s^{-1}. What is the rate constant for this decomposition at: (a) 20 °C and (b) 100 °C? **LO6**

15.104 At 35 °C, the rate constant for the reaction: **LO6**

$$\underset{\text{sucrose}}{C_{12}H_{22}O_{11}} + H_2O \rightarrow \underset{\text{glucose}}{C_6H_{12}O_6} + \underset{\text{fructose}}{C_6H_{12}O_6}$$

is $k = 6.2 \times 10^{-5}\ s^{-1}$. The activation energy for the reaction is 108 kJ mol^{-1}. What is the rate constant for the reaction at 45 °C?

15.105 A sucrase enzyme breaks down 0.15 mol L^{-1} lactose every 37 minutes. What is the reaction rate in: **LO6**
(a) mol L^{-1} min^{-1}
(b) mmol L^{-1} min^{-1}
(c) mol min^{-1} in a 500 mL flask?

15.106 The following mechanism has been suggested for the reaction $H_2(g) + I_2(g) \rightarrow 2HI(g)$. **LO4**

$$I_2(g) \xrightarrow{k_1} 2I\ (g)$$
$$2I\ (g) \xrightarrow{k_2} I_2(g)$$
$$H_2(g) + 2I\ (g) \xrightarrow{k_3} 2HI(g)$$

Derive the rate law using the steady-state approximation.

15.107 A possible mechanism for the reaction $2N_2O_5 \rightarrow 4NO_2 + O_2$ is: **LO4**

$$N_2O_5 \underset{k_{-1}}{\overset{k_1}{\rightleftharpoons}} NO_2 + NO_3$$
$$NO_2 + NO_3 \xrightarrow{k_2} NO + NO_2 + O_2$$
$$NO + NO_3 \xrightarrow{k_3} 2NO_2$$

What is the rate law derived using the steady-state approximation?

15.108 The decomposition of many substances on the surface of a heterogeneous catalyst shows the following relationship between reaction rate and substrate concentration. **LO4**

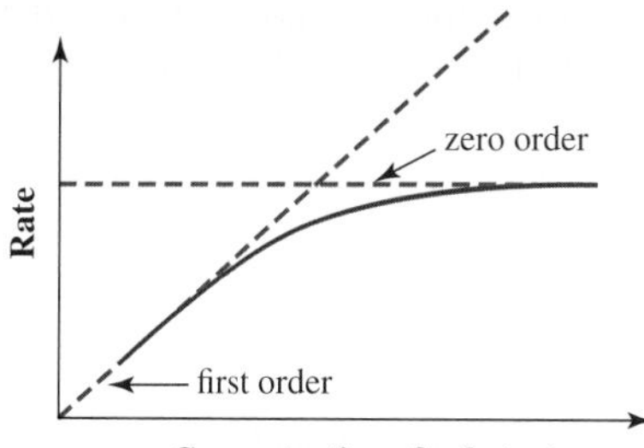

Why does the rate law change from first order to zero order in the concentration of the substrate?

ADDITIONAL EXERCISES

15.109 The disproportionation of hydrogen peroxide into oxygen and water has an enthalpy of reaction of −98.2 kJ mol^{-1} and an activation barrier of 75 kJ mol^{-1}. Iodide ions act as a catalyst for this reaction, with an activation barrier of 56 kJ mol^{-1}. The enzyme, catalase, is also a catalyst for this reaction, and this pathway has an activation barrier of 23 kJ mol^{-1}. **LO6**
(a) Draw a labelled potential energy diagram for this process both without and with each of the catalysts.
(b) Calculate the factor by which the reaction speeds up due to the presence of each of these two catalysts at a temperature of 37 °C. Assume that the pre-exponential Arrhenius factor remains constant.

15.110 The rate law for a certain enzymatic reaction is zero order with respect to the substrate. The rate constant for the reaction is 6.4×10^2 mol L^{-1} s^{-1}. If the initial concentration of the substrate is 0.275 mol L^{-1}, what is the initial rate of the reaction? **LO3**

15.111 The initial rate of oxygen production by an enzyme was measured for a range of substrate concentrations. **LO7**

[substrate] (mol L^{-1})	0.050	0.017	0.010	0.005	0.002
Rate (mm^3 min^{-1})	16.6	12.4	10.1	6.6	3.3

What is the value for the Michaelis constant for the reaction?

15.112 ^{14}C dating can be used to estimate the age of formerly living materials because the uptake of ^{14}C from carbon dioxide in the atmosphere stops once the organism dies. If tissue samples from a mummy contain about 81% of the ^{14}C expected in living tissue, how old is the mummy? The half-life for decay of ^{14}C is 5730 years. **LO4**

15.113 One of the reactions that occurs in polluted air in urban areas is $2NO_2(g) + O_3(g) \rightarrow N_2O_5(g) + O_2(g)$. It is believed that a species with the formula NO_3 is involved in the mechanism, and the observed rate law for the overall reaction is: **LO6**

$$\text{rate of reaction} = k[NO_2][O_3]$$

Propose a mechanism for this reaction that includes the species NO_3 and is consistent with the observed rate law.

15.114 The oxidation of NO to NO_2 in the troposphere involves carbon monoxide, CO. A possible mechanism is: **LO7**

$$CO + OH \rightarrow CO_2 + H$$
$$H + O_2 \rightarrow HO_2$$
$$HO_2 + NO \rightarrow HO + NO_2$$

(a) Identify the intermediates in this mechanism.
(b) Write the net chemical equation for this mechanism.

15.115 Suppose a reaction occurs with the mechanism: **LO4**

$$2A \rightleftharpoons A_2 \quad \text{(step 1, fast)}$$
$$A_2 + E \longrightarrow B + C \quad \text{(step 2, slow)}$$

in which step 1 is a very rapid reversible reaction that can be considered to be essentially an equilibrium (forward and reverse reactions occurring at the same rate) and step 2 is slow.
(a) What is the chemical equation for the net reaction that occurs in this chemical change?
(b) Write the rate law for the forward reaction in step 1.
(c) Write the rate law for the reverse reaction in step 1.
(d) Write the rate law for the rate-determining step.
(e) Use the results from parts (b) and (c) to rewrite the rate law of the rate-determining step in terms of the concentrations of the reactants in the overall balanced equation for the reaction.

15.116 The decomposition of urea, $(NH_2)_2CO$, in 0.10 M HCl follows the equation: **LO4**

$$(NH_2)_2CO(aq) + H_3O^+(aq) \rightarrow 2NH_4^+(aq) + CO_2(g) + H_2O(l)$$

At 60 °C, $k = 5.84 \times 10^{-6}\ min^{-1}$ and, at 70 °C, $k = 2.25 \times 10^{-5}\ min^{-1}$. If this reaction is run at 80 °C starting with a urea concentration of 2×10^{-3} M, how long will it take for the urea concentration to drop to 1.2×10^{-3} M?

15.117 It was mentioned that the rates of many reactions approximately double for each 10 °C rise in temperature. Assuming a starting temperature of 25 °C, what would the activation energy be, in kJ mol^{-1}, if the rate of a reaction were to be twice as large at 35 °C? **LO2,4**

15.118 The cooking of an egg involves the denaturation of a protein called albumen. The time required to achieve a particular degree of denaturation is inversely proportional to the rate constant for the process. This reaction has a high activation energy, $E_a = 418\ kJ\ mol^{-1}$. Calculate how long it would take to cook a traditional 3-minute egg on top of a cold mountain on a day when the atmospheric pressure there is 47 kPa. **LO3**

15.119 The enthalpies of formation of Al_2O_3 and B_2O_3 are −1676 and −1274 kJ mol^{-1}, respectively. Do these data indicate the rate of oxidation of aluminium and boron? Explain your answer. **LO4**

15.120 The rate constant for a reaction is $2.90 \times 10^{-9}\ s^{-1}$ at 454 K with an activation energy of 178.5 kJ mol^{-1}. What is the value of the rate constant at 395 K? **LO4**

15.121 The rate constant for a reaction is $1.72 \times 10^{-4}\ s^{-1}$ at 298 K and $2.75 \times 10^{-3}\ s^{-1}$ at 301 K. What is the activation energy in kJ mol^{-1}? **LO6**

15.122 The activation energy of a unimolecular reaction is 190.2 kJ mol^{-1} at 295 K. Assuming a frequency factor of $10^{13}\ s^{-1}$, what is the value of the rate constant? **LO6**

15.123 Calculate the activation energy for a second-order reaction using the following kinetic data: $k = 4\ mol^{-1}\ L\ s^{-1}$ at 37 °C and $k = 8.0\ mol^{-1}\ L\ s^{-1}$ at 87 °C. **LO6**

15.124 (a) Calculate the average rate of a chemical reaction if the concentration of reactant A is 0.45 M after 3 min and 0.20 M after 8 min.
(b) Why do we refer to the average rate when we are calculating a reaction time interval? **LO6**

15.125 A reaction is believed to take place in the following three steps. **LO6**

$$HBr(g) + O_2(g) \rightarrow HOOBr(g) \quad \text{(step 1, slow)}$$

$$HOOBr(g) + HBr(g) \rightarrow 2HOBr(g) \quad \text{(step 2, fast)}$$

$$2HOBr(g) + 2HBr(g) \rightarrow 2H_2O(g) + 2Br_2(g) \quad \text{(step 3, fast)}$$

(a) Add the three steps to show that it gives the correct overall reaction:

$$4HBr(g) + O_2(g) \rightarrow 2H_2O(g) + 2Br_2(g)$$

(b) Identify any reaction intermediates or catalysts.
(c) Which step is the rate-determining step? Explain.

15.126 Which of the five factors that affect the rate of a reaction is illustrated in each of the following. **LO2**
(a) Food is sometimes frozen before it is used.
(b) Small sticks of wood are often used to start a fire.
(c) In hospitals, the speed of the healing process is often increased in an oxygen tent.

15.127 Why would you expect the rate of the reaction: **LO2**

$$Ag^+(aq) + Br^-(aq) \rightarrow AgBr(s)$$

at room temperature to be much faster than the rate of the reaction:

$$CH_4(g) + 2O_2(g) \rightarrow CO_2(g) + 2H_2O(l)$$

at room temperature?

15.128 Consider the reaction $Zn(s) + 2HCl(aq) \rightarrow ZnCl_2(aq) + H_2(g)$. What would be the effect on the rate of reaction if: **LO6**
(a) powdered zinc was used instead of a solid piece of zinc metal
(b) the concentration of HCl(aq) was doubled
(c) the temperature was increased?

MATHS FOR CHEMISTRY

In chemistry, we often study particular aspects of a chemical reaction over a period of time, or as we change the temperature. For example, we might be interested in how the concentration of a reactant in a chemical reaction changes with time, or how the rate of a chemical reaction changes as the temperature is increased. One of the best ways of displaying trends in these data is to plot them on a graph. Generally, we do this in two dimensions, by plotting a graph of y (called the dependent variable) versus x (called the independent variable).

Consider the following set of (x,y) data.

(0,2), (1,4), (2,6), (3,8), (4,10), (5,12), (6,14), (7,16), (8,18), (9,20), (10,22)

Plotting these on an (x,y) graph gives the following.

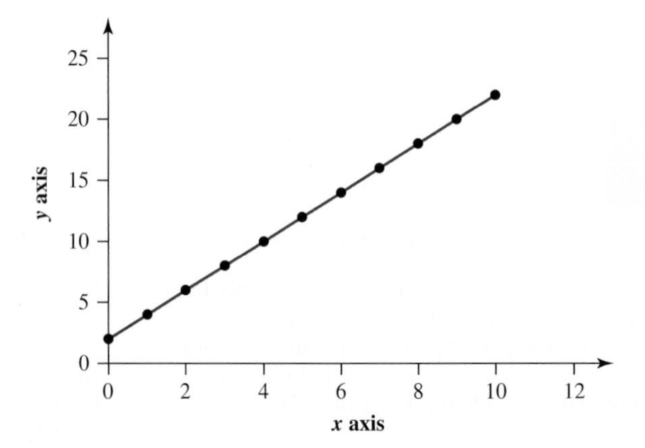

It is immediately obvious that a straight line can be drawn through the data points, and we say that there is therefore a linear relationship between x and y. Whenever there is a straight line relationship between the x and y variables we can say that the x and y variables are directly proportional. The straight line on such graphs can be described by the general equation:

$$y = mx + c$$

where x and y are the two variables, m is the slope (sometimes called the gradient) of the graph, and c is a constant, the value of which corresponds to the point where the line intercepts the y axis.

The slope of the graph is defined by the equation:

$$\text{slope} = \frac{\Delta y}{\Delta x}$$

where Δ (the Greek letter, capital delta) means 'the change in', and is generally defined as the final value of x or y minus the initial value of x or y. Therefore, to obtain the slope of the above graph, we choose two (x,y) points, as follows.

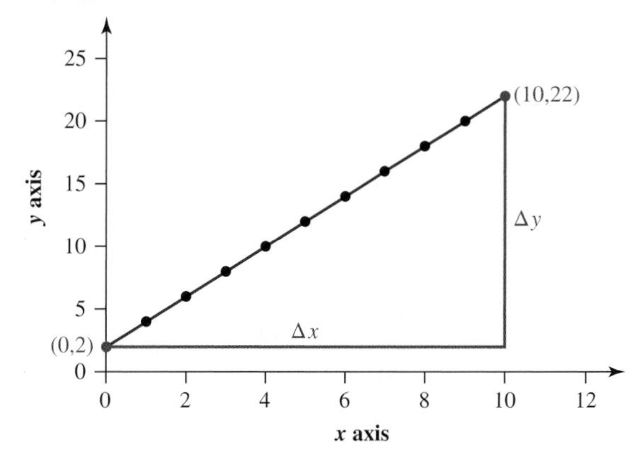

The slope of this graph can then be determined.

$$\text{slope} = \frac{\Delta y}{\Delta x} = \frac{(22-2)}{10-0} = \frac{20}{10} = 2$$

Notice that it does not matter which two (x,y) points we choose in order to obtain the slope — we will still obtain the same value. For example, if we chose the (x,y) points (18,8) and (3,8), the equation would then become:

$$\text{slope} = \frac{\Delta y}{\Delta x} = \frac{(18-8)}{(8-3)} = \frac{10}{5} = 2$$

Notice also that the order in which we carry out the subtractions does not matter. For the above example, if we reversed the two (x,y) points, we would still obtain the same slope:

$$\text{slope} = \frac{\Delta y}{\Delta x} = \frac{(8-18)}{(3-8)} = \frac{-10}{-5} = 2$$

Having determined that the slope of the straight line (m) is 2, and that the straight line intercepts the y axis at $y = 2$, then inserting these values into the general equation $y = mx + c$, gives the equation of this straight line as:

$$y = 2x + 2$$

This equation now allows us to find the value of y that is associated with any value of x.

A little thought about the consequences of the above sentences should convince you that a straight line on a graph has a constant slope along its entire length. But what happens when we do not have a straight line relationship between the variables?

Consider the following (x,y) points.

(0,2), (1,3), (2,6), (3,11), (4,18), (5,27), (6,38), (7,51), (8,66), (9,83), (10,102)

Plotting these on an (x,y) graph gives the following.

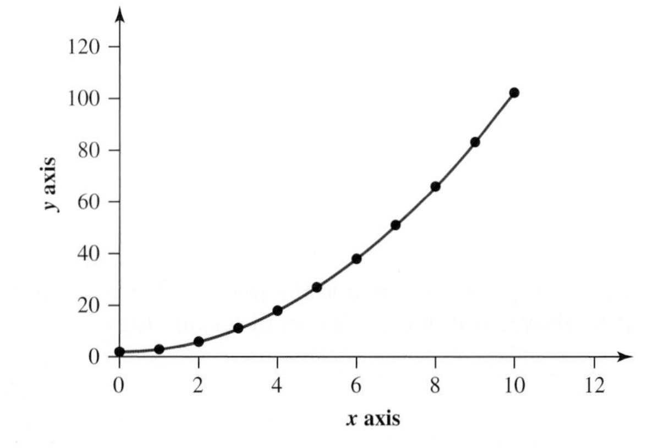

Now we do not have a straight-line relationship between the variables x and y and so they are not directly proportional. The line joining the points is now a curve, and you should hopefully be able to see from the above graph that the slope of this curve, which is still defined exactly as above, changes along its length; near the point (0,2) the curve is nearly flat, while as x increases the curve becomes steeper and steeper. Because of this, instead of talking about the slope of the curve, we have to specify the slope of the curve *at a particular point* on the curve. For example, what is the slope of the curve at the point (5,27)? To determine this, let's proceed as we did above, by picking two (x,y) points, one of which is (5,27) and calculating the slope of the line joining those two points from the equation:

$$\text{slope} = \frac{\Delta y}{\Delta x}$$

If we choose (2,6) and (5,27), we obtain:

$$\text{slope} = \frac{\Delta y}{\Delta x} = \frac{(27-6)}{(5-2)} = \frac{16}{3} = 7$$

If we choose (3,11) and (5,27), we obtain:

$$\text{slope} = \frac{\Delta y}{\Delta x} = \frac{(27-11)}{(5-3)} = \frac{16}{2} = 8$$

If we choose (4,18) and (5,27), we obtain

$$\text{slope} = \frac{\Delta y}{\Delta x} = \frac{(27-18)}{(5-4)} = \frac{9}{1} = 9$$

As you can see, the slope of the curve at the point (5,27) changes, depending on which other point we choose in the calculation. The slope steadily increases as we get closer and closer to the point (5,27).

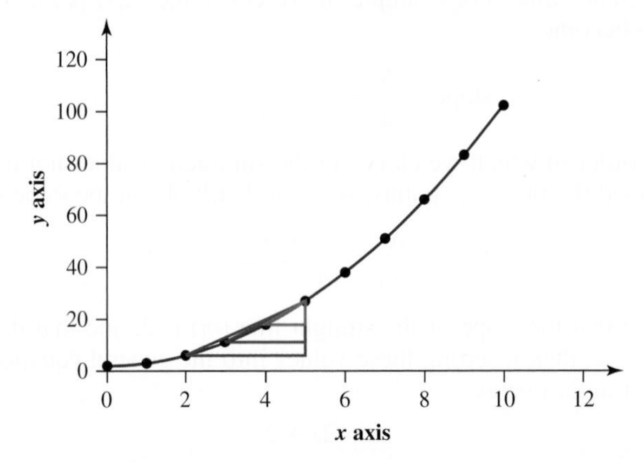

What we need to know, in order to determine the slope of the graph at the point (5,27) is the slope of the tangent to the curve at this point. A tangent is a straight line that touches a curve only at a single point. From the above plot, you can see that what we are actually calculating when we use the equation:

$$\text{slope} = \frac{\Delta y}{\Delta x}$$

is the slope of the longest pink, green and red lines, which are all approximations to the tangent to the curve at the point (5,27). As Δy and Δx get smaller, the slope of this line increases, and gets closer to the actual slope of the tangent. If we could make Δy and Δx infinitesimally small, the slope that we obtain from the equation:

$$\text{slope} = \frac{\Delta y}{\Delta x}$$

would be exactly that of the tangent.

A branch of mathematics called calculus provides methods by which we can do this, and we can use this to show that the slope of the curve at the point (5,27) is 10. It uses exactly the same principles as are described above, with the exception of terminology.

Instead of the equation:

$$\text{slope} = \frac{\Delta y}{\Delta x}$$

calculus calculates the slope using the equation:

$$\text{slope} = \frac{dy}{dx}$$

where the Δ symbols have been replaced by d. The d simply means 'an infinitesimal change in' — in other words, dy simply means a really, really small change in the value of y, or a really really small value of Δy.

Calculus is often used to determine how the value of a variable changes over time — in other words, the rate of change of a variable. In chemistry, this is often used in chemical kinetics to determine the rate of a chemical reaction — we monitor the concentration of a reactant or product over time. This then leads to an equation of the type:

$$\text{rate of reaction} = \frac{d[A]}{dt}$$

This means that the rate of the reaction can be obtained simply by determining the slope of a graph of $[A]$ (y axis) versus time (x axis).

So whenever you see the terminology $\frac{dy}{dx}$, don't be afraid! It is simply the slope of a graph.

16 The chemistry of carbon

The Argyle diamond mine (pictured opposite) is located in the remote Kimberley region of Western Australia. It has produced more than 800 million carats of rough diamonds since it started operations in 1983. A 1 carat diamond weighs only 200 mg, so the estimated 20 million carats currently produced each year by Argyle represent only 4 tonnes of pure carbon. Diamond is one of the naturally occurring forms of the element carbon. Of course this form of carbon is extremely valuable. In November 2013 the 59.60 carat (less than 12 g) 'Pink Star' diamond sold for the highest auction price ever paid for a gemstone (A$94 million or more than NZ$100 million). Large lumps of this type of carbon are worth more than $7 500 000 per gram. There are some rational reasons for this value. Diamond has a remarkable ability to disperse light of differing wavelengths, which gives it visual beauty. Diamond also has the highest hardness and thermal conductivity of any natural material, and diamond cutting and polishing tools are essential for many industrial applications.

Thermodynamically, however, diamond is less stable than the 'lead' found in pencils. In fact, the 'lead' of lead pencils is really clay mixed with a soft, dark grey form of pure carbon called graphite. Along with coal, graphite represents the most common form of elemental carbon found on Earth. All three materials represent pure forms of carbon, yet they all have different characteristics, with graphite used as a lubricant and coal used as a fuel. Although we might think of diamonds as being indestructible, almost 250 years ago Antoine Lavoisier used a large lens to focus the Sun's rays and proved that diamond burns in an atmosphere of oxygen to give carbon dioxide, just like coal.

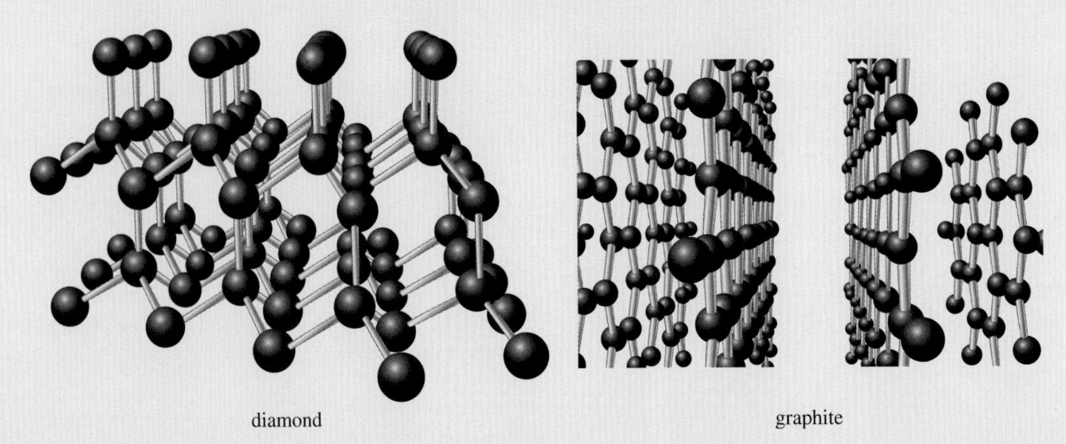

Carbon atoms can combine covalently with each other and with other elements to form an enormous variety of molecules. There are many millions of compounds that contain carbon including colourful dyes, drugs, perfumes, synthetic and natural fibres, plastics, flavourings, foods and fuels. The list covers almost everything we encounter in our normal lives and also includes the carbohydrates, proteins, vitamins, fats, cell membranes and enzymes that make up our own bodies.

LEARNING OBJECTIVES

After studying this chapter, you should be able to:

16.1 distinguish between the four classes of hydrocarbon

16.2 describe the properties of alkanes using standard nomenclature and representations

16.3 describe the properties of alkenes and alkynes

16.4 describe the limited reactivity of alkanes

16.5 write mechanisms for each of the four main types of addition reactions of alkenes

16.6 correlate the reactivity of alkynes with alkenes and describe how alkynes may be converted to alkenes

16.7 describe how aromatic hydrocarbons differ from alkenes and alkynes through the presence of remarkably stable cyclic π-bonds

16.8 write mechanisms that describe each of the five major types of substitution reactions of aromatic hydrocarbons.

16.1 Introduction to hydrocarbons

A **hydrocarbon** is a compound composed of only carbon and hydrogen. Figure 16.1 shows the four classes of hydrocarbons, along with the characteristic type of bonding between carbon atoms in each class.

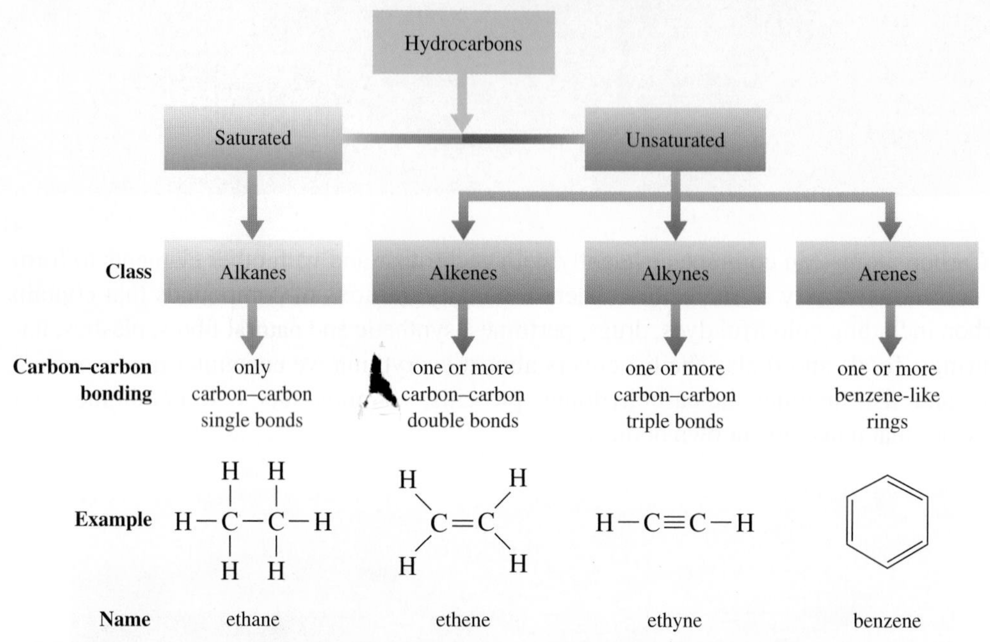

FIGURE 16.1 The four classes of hydrocarbons.

Alkanes are hydrocarbons that contain only carbon–carbon single bonds. Such hydrocarbons are said to be **saturated hydrocarbons**, meaning that each carbon atom has the maximum possible number of atoms (four) bonded to it. We often refer to alkanes as **aliphatic hydrocarbons**, because the physical properties of the higher molecular weight members of this class resemble those of the long-carbon-chain molecules we find in animal fats and plant oils ('aliphatic' is derived from the Greek word *aleiphar* meaning 'fat' or 'oil').

Alkenes are hydrocarbons that contain one or more carbon–carbon double bonds. **Alkynes** are hydrocarbons that contain one or more carbon–carbon triple bonds. **Arenes** are cyclic structures containing carbon–carbon bonds that impart special stability. Alkenes, alkynes and arenes are said to be **unsaturated hydrocarbons**.

16.2 Alkanes

Recall that alkanes are saturated hydrocarbons; they contain only single bonds and each carbon has the maximum number of atoms (four) bonded to it. Methane, CH_4, and ethane, C_2H_6, are the first two members of the alkane family. Figure 16.2 shows Lewis structures and ball-and-stick models for these molecules. The Lewis structures show the atom connectivity but do not reflect the three-dimensional shapes of the molecules. We learned in section 5.4 that methane is tetrahedral, and all H—C—H bond angles are 109.5°. Each carbon atom in ethane is also tetrahedral, and all bond angles are about 109.5°. Although the three-dimensional shapes of larger alkanes are more complex than those of methane and ethane, the four bonds around each carbon atom are still arranged in a tetrahedral manner, and all bond angles are still approximately 109.5°.

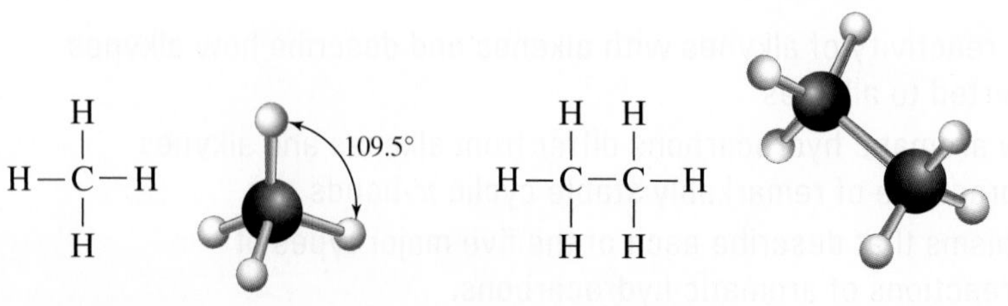

109.5°

methane, CH_4

ethane, C_2H_6

FIGURE 16.2 Structures of methane and ethane.

In chapter 2 we learned various ways to depict chemical structures. In figure 16.3, the next members of the alkane family — propane, butane and pentane — are drawn, first as condensed structural formulae that show all carbons and hydrogens, then as line structures, and then as ball-and-stick models. Recall that, in this type of representation, a line represents a carbon–carbon bond and an angle represents a carbon atom. A line ending represents a $—CH_3$ group. Although hydrogen atoms are not shown in line structures, they are assumed to be there in sufficient numbers to give each carbon atom four bonds.

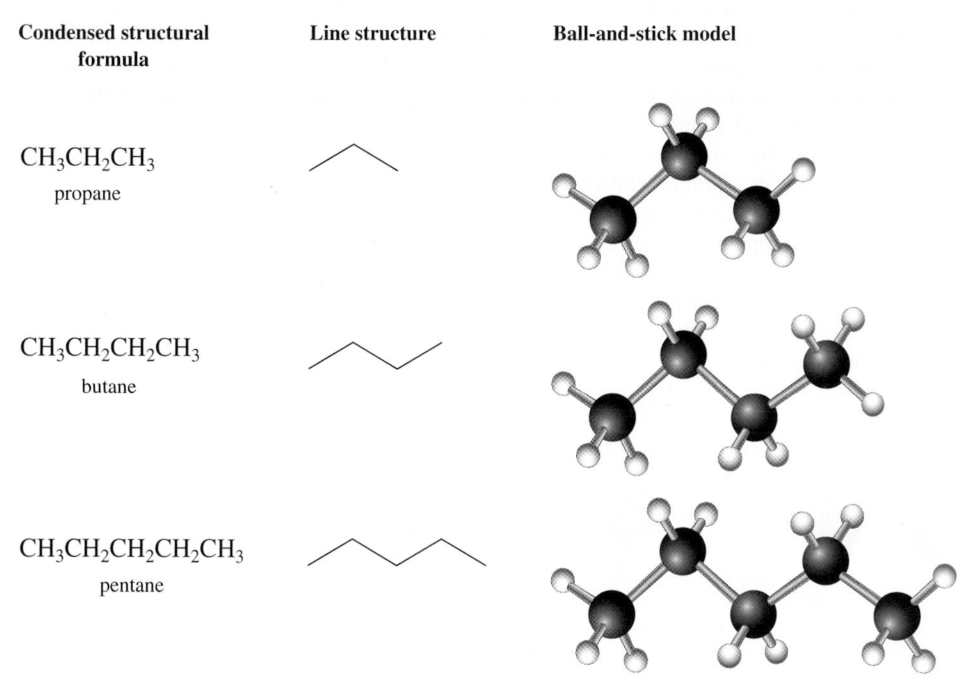

FIGURE 16.3 Structures of propane, butane and pentane.

We can write condensed structural formulae for alkanes in still another abbreviated form. For example, the condensed structural formula of pentane, $CH_3CH_2CH_2CH_2CH_3$, contains three CH_2 (methylene) groups in the middle of the chain. We can collect these groups together and write the condensed structural formula as $CH_3(CH_2)_3CH_3$. Table 16.1 gives the IUPAC names and molecular formulae of the first 10 alkanes. Note that the names of all alkanes end in *-ane*.

Alkanes have the general molecular formula C_nH_{2n+2} (with the exception of cycloalkanes, which we will deal with later in the chapter). Thus, given the number of carbon atoms in an alkane, it is easy to determine the number of hydrogen atoms in the molecule and also its molecular formula. For example, decane, with 10 carbon atoms, must have $(2 \times 10) + 2 = 22$ hydrogen atoms and a molecular formula of $C_{10}H_{22}$.

TABLE 16.1 Names, molecular formulae and condensed structural formulae for the first 10 straight-chain alkanes.

Name	Molecular formula	Condensed structural formula	Melting point (°C)	Boiling point (°C)	Density of liquid (g mL^{-1} at 0 °C)[(a)]
methane	CH_4	CH_4	−182	−164	(a gas)
ethane	C_2H_6	CH_3CH_3	−183	−88	(a gas)
propane	C_3H_8	$CH_3CH_2CH_3$	−190	−42	(a gas)
butane	C_4H_{10}	$CH_3(CH_2)_2CH_3$	−138	0	(a gas)
pentane	C_5H_{12}	$CH_3(CH_2)_3CH_3$	−130	36	0.626
hexane	C_6H_{14}	$CH_3(CH_2)_4CH_3$	−95	69	0.659
heptane	C_7H_{16}	$CH_3(CH_2)_5CH_3$	−90	98	0.684
octane	C_8H_{18}	$CH_3(CH_2)_6CH_3$	−57	126	0.703
nonane	C_9H_{20}	$CH_3(CH_2)_7CH_3$	−51	151	0.718
decane	$C_{10}H_{22}$	$CH_3(CH_2)_8CH_3$	−30	174	0.730

(a) For comparison, the density of H_2O is 1 g mL^{-1} at 4 °C.

Conformation of alkanes

Even though structural and condensed structural formulae are useful for showing the order of attachment of atoms, they do not show three-dimensional shapes. To recognise the relationships between the structure and properties of molecules, it is crucial to understand the three-dimensional shapes of molecules. Molecules are three-dimensional objects, and it is essential that you become comfortable in dealing with them as such.

Alkanes of two or more carbon atoms can be twisted into a number of different three-dimensional arrangements by rotating around one or more carbon–carbon bonds. Any three-dimensional arrangement of atoms that results from rotation around a single bond is called a **conformer**; the representation of the position of the atoms in the molecule is called its **conformation**. Figure 16.4a shows a ball-and-stick model of a *staggered conformation* of ethane. In this conformation, the three C—H bonds on one carbon atom are as far away as possible from the three C—H bonds on the adjacent carbon atom. Figure 16.4b shows a **Newman projection**, which is a shorthand way of representing the staggered conformation of ethane. In a Newman projection, we view a molecule along the axis of a C—C bond. The three atoms or groups of atoms nearer your eye appear on lines extending from the centre of the circle at angles of 120°. The three atoms or groups of atoms on the carbon atom further from your eye appear on lines extending from the circumference of the circle at angles of 120°. Remember that bond angles around each carbon atom in ethane are approximately 109.5°, and not 120° as this Newman projection might suggest.

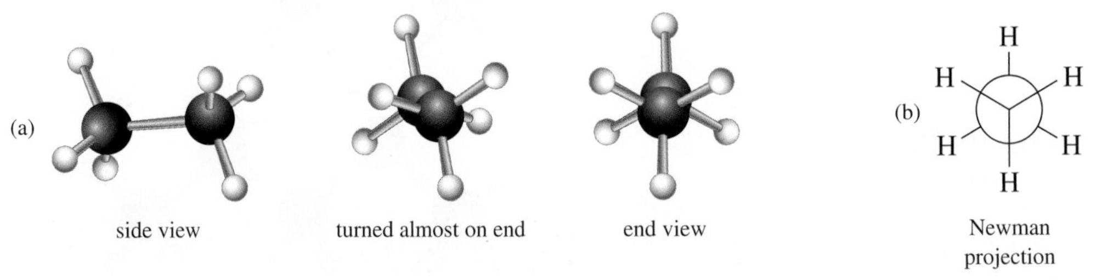

FIGURE 16.4 A staggered conformation of ethane: **(a)** ball-and-stick models and **(b)** Newman projection.

Figure 16.5 shows a ball-and-stick model and a Newman projection of an *eclipsed conformation* of ethane. In this conformation, the three C—H bonds on one carbon atom are as close as possible to the three C—H bonds on the adjacent carbon atom. In other words, hydrogen atoms on the back carbon atom are eclipsed by the hydrogen atoms on the front carbon atom. (Note that in the Newman projection the bonds have been offset a little for clarity.)

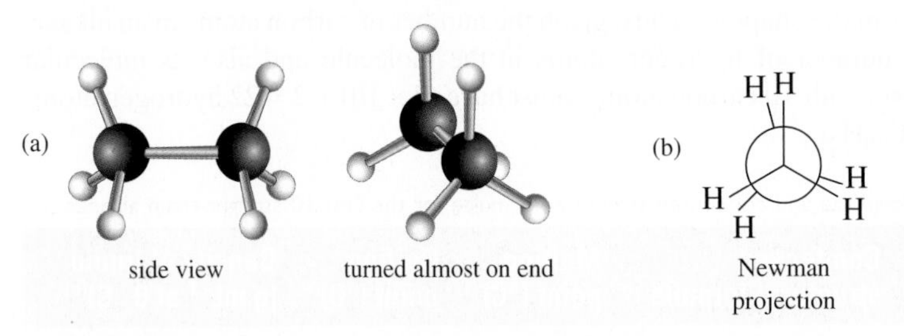

FIGURE 16.5 An eclipsed conformation of ethane: **(a)** ball-and-stick models and **(b)** Newman projection, with the bonds offset slightly for clarity.

The difference in potential energy between these two conformations is approximately 12.6 kJ mol^{-1} (see figure 16.6), which means that, at room temperature, the ratio of ethane molecules in a staggered conformation to those in an eclipsed conformation is approximately 100 to 1.

It is important to realise that molecules are not static like the pictures you see on this page and in figure 16.7 on the next page. At room temperature the bonds can be stretching, shaking, bending and vibrating up to 10^{13} times per second! The groups rotate more slowly, but typical C—C bond rotation is still more than a *billion* revolutions per second. You will learn more about how chemists detect some of these atom movements when we deal with infrared spectroscopy in section 20.3. The key point here is that, at room temperature, movement between the various conformers is constant and occurs *extremely* rapidly, even with the energy barriers that arise from eclipsing interactions. Thus, while you may be able to draw many conformational forms for a molecule, in practical terms, for a typical noncyclic alkane, they do not exist long enough to display the differences in properties that constitutional isomers possess.

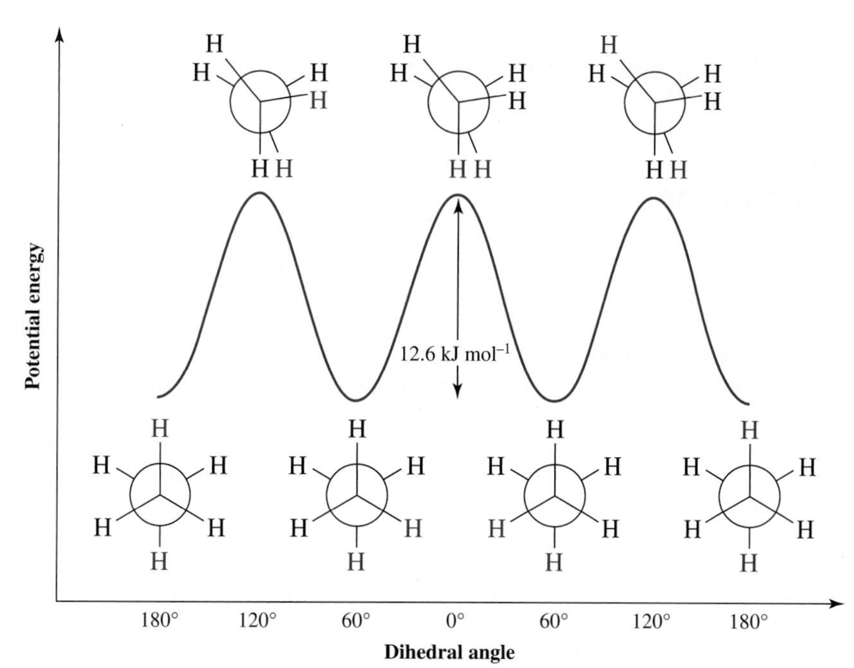

FIGURE 16.6 An energy diagram showing the conformational analysis of ethane.

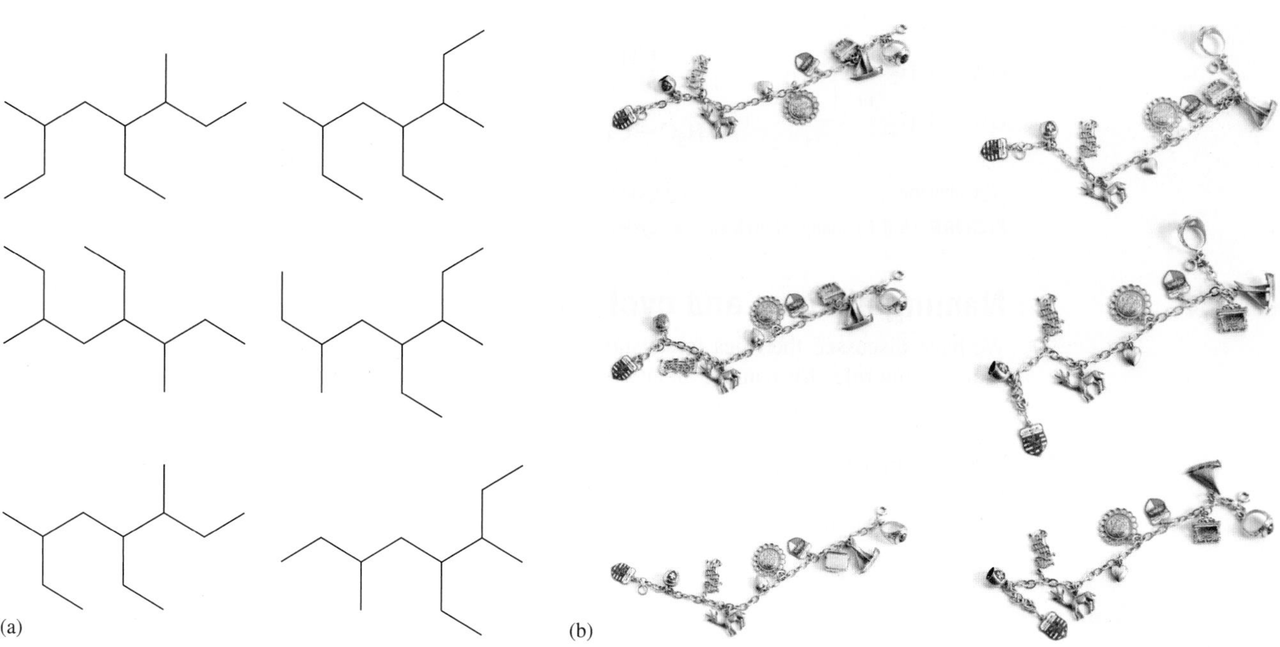

FIGURE 16.7 **(a)** None of these structures represent different isomers. They are all 4-ethyl-3,6-dimethyloctane drawn in different orientations or conformations. **(b)** These photos do not represent different pieces of jewellery. They are all of the same charm bracelet shown in different orientations or conformations.

WORKED EXAMPLE 16.1

Drawing Newman projections

Draw Newman projections for one staggered conformation and one eclipsed conformation of propane.

Solution

The following are Newman projections and ball-and-stick models of these conformations. Note that we slightly offset the ball-and-stick models for clarity.

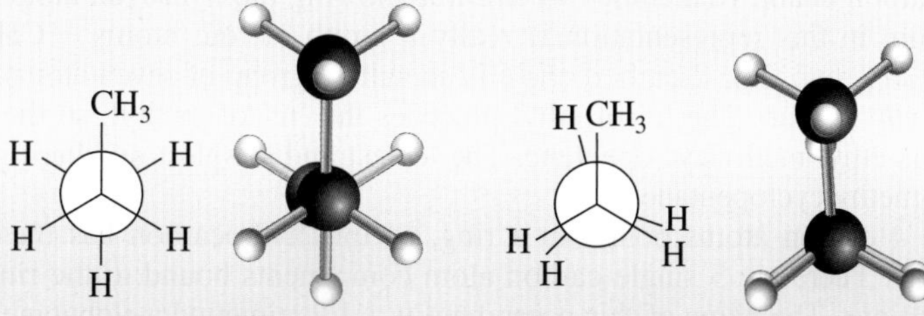

Viewed end on along the central carbon–carbon bond, we can see the atoms are distributed as far apart as possible in the staggered conformation whereas, in the eclipsed conformation, they align.

PRACTICE EXERCISE 16.1 Draw Newman projections for two staggered and two eclipsed conformations of butane as viewed along the central C—C bond.

Cycloalkanes

A hydrocarbon that contains carbon atoms joined to form a ring is called a cyclic hydrocarbon. When all carbon atoms of the ring are saturated, we call the hydrocarbon a **cycloalkane**. Cycloalkanes with ring sizes of three to over 30 abound in nature, but, in principle, there is no limit to ring size. Five-membered (cyclopentane) and six-membered (cyclohexane) rings are especially abundant in nature.

Cycloalkanes contain two fewer hydrogen atoms than an alkane with the same number of carbon atoms. For instance, compare the molecular formula of cyclohexane, C_6H_{12}, with that of hexane, C_6H_{14}. The general formula of a cycloalkane is C_nH_{2n}. Note, however, that cycloalkanes have essentially the same chemical reactivity as linear alkanes, so this degree of 'unsaturation' does not provide any chemical functionality.

Figure 16.8 shows the structural formulae of cyclobutane, cyclopentane and cyclohexane. When writing structural formulae for cycloalkanes, chemists rarely show all carbon and hydrogen atoms. Rather, they use line structures to represent cycloalkane rings. Each ring is represented by a regular polygon having the same number of sides as there are carbon atoms in the ring. For example, chemists represent cyclobutane by a square, cyclopentane by a pentagon and cyclohexane by a hexagon.

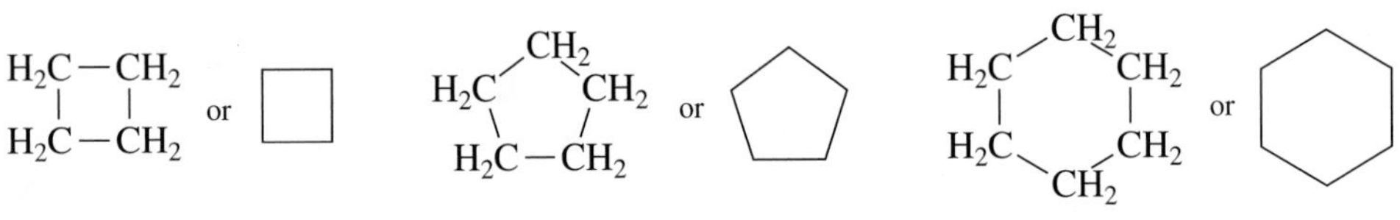

cyclobutane cyclopentane cyclohexane

FIGURE 16.8 Examples of cycloalkanes: cyclobutane, cyclopentane and cyclohexane.

Naming alkanes and cycloalkanes

We have discussed the rules for designating correct names to alkanes in chapter 2 (pp. 58–60). Many of the rules for naming alkanes carry over in the naming of cycloalkanes.

To name a cycloalkane, prefix the name of the corresponding open-chain hydrocarbon with *cyclo-*, and name each substituent on the ring. If there is only one substituent, there is no need to give it a number. If there are two substituents, number the ring beginning with the substituent of lower alphabetical order. If there are three or more substituents, number the ring to give them the lowest set of numbers, and then list the substituents in alphabetical order.

WORKED EXAMPLE 16.2

Naming cycloalkanes

Write the molecular formula and IUPAC name for each of the following cycloalkanes.

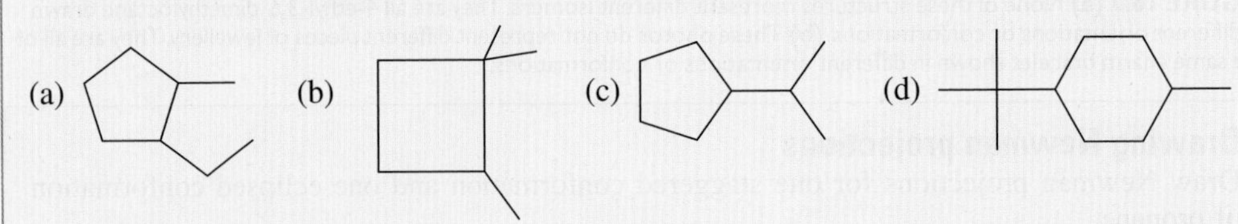

Analysis and solution

(a) There are 5 carbon atoms are bound together to form a ring. The parent name is therefore cyclopentane. Attached to this ring of 5 carbon atoms are a single carbon atom and a 2-carbon chain. Remember we are not showing the hydrogen atoms, so a single carbon atom in this representation is really a group of four atoms $—CH_3$, which we call a *methyl* group. The 2-carbon chain is actually a group of seven atoms, $—CH_2CH_3$, called an *ethyl* group. The 'e' of ethyl precedes the 'm' of methyl, so the name of the molecule is ethyl methylcyclopentane. The lowest and simplest numbering sequence is 1-ethyl-2-methylcyclopentane.

(b) There are 4 carbon atoms bound in a ring, so this hydrocarbon has the parent name cyclobutane. There are 3 single carbon atom components bound to the ring at two different positions. The name of this compound is 1,1,2-trimethylcyclobutane.

(c) The molecular formula of this cycloalkane is C_8H_{16}. Because there is only one substituent on the ring, there is no need to number the atoms of the ring. IUPAC accepts two names for this compound: isopropylcyclopentane and 1-methylethylcyclopentane.

(d) Number the atoms of the cyclohexane ring beginning with *tert*-butyl, the substituent of lower alphabetical order. The compound's name is 1-*tert*-butyl-4-methylcyclohexane and its molecular formula is $C_{11}H_{22}$. IUPAC also accepts 1-(1,1-dimethylethyl)-4-methylcyclohexane.

PRACTICE EXERCISE 16.2

Write the molecular formula and IUPAC name for each of the following cycloalkanes.

(a) 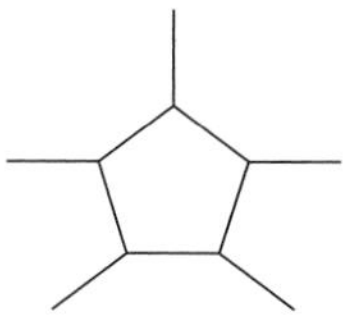 (c) 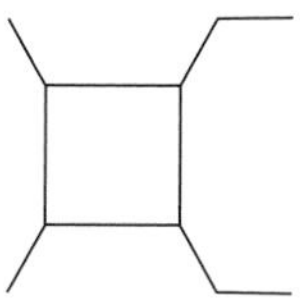 (e)

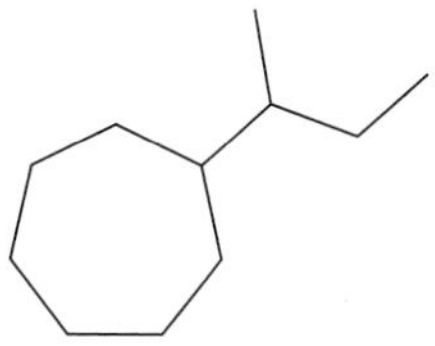

(b) 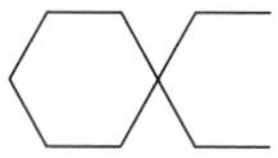 (d) 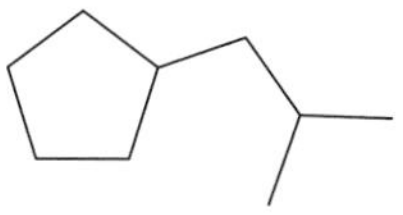 (f)

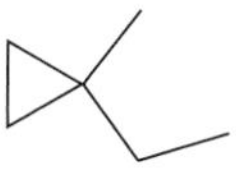

Conformations of cycloalkanes

When using line structure polygons to represent cycloalkane rings, it is important to realise that cycloalkanes are not flat structures. A pentagon possesses angles of 108°, which is close to the ideal tetrahedral angle of 109.5° found in methane (see section 5.4). This implies that the **angle strain** in a flat pentagon would be small. Angle strain results when a bond angle in a molecule differs from the optimal tetrahedral angle. Yet cyclopentane is not flat but in fact forms an 'open envelope' structure.

This structure arises to reduce the eclipsing interactions that would occur if the molecule was a flat structure. Recall from the discussions on the conformation of alkanes (p. 700) that the lowest energy for a molecule arises when bonds are staggered and are not eclipsed in the Newman projection (figure 16.4). A flat pentagon structure for pentane would give rise to 10 eclipsed C—H bonds. This unfavoured interaction produces **torsional strain**. Also called eclipsed interaction strain, torsional strain arises when nonbonded atoms separated by three bonds are forced from a staggered conformation to an eclipsed conformation. For a flat pentagon structure this torsional strain would equate to about 42 kJ mol^{-1}. To relieve some of this strain, the atoms twist into the 'envelope' conformation (figure 16.9) where four carbon atoms are in a plane and the fifth carbon lies above the plane, somewhat like an envelope with the flap folded outward.

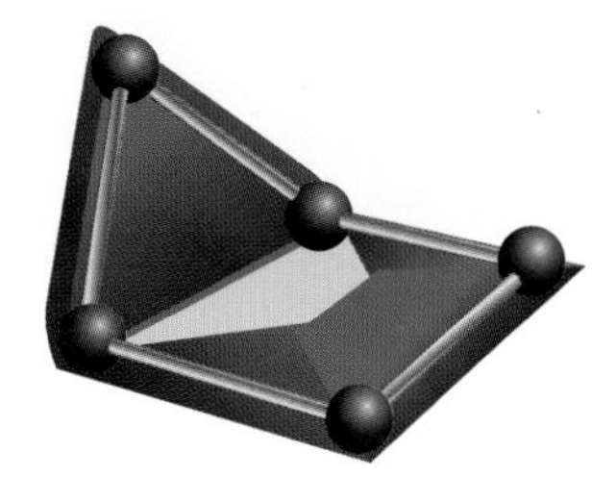

FIGURE 16.9 'Envelope' conformation.

In this envelope conformation the number of eclipsing interactions is reduced, but to attain this shape the C—C—C bond angles are also reduced to 105°, thus indicating an increase in angle strain. The total strain energy in cyclopentane in the envelope conformation is about 23.4 kJ mol^{-1}, which is substantially less than if the molecule was completely flat. It is important to remember, however, that molecules exist in a dynamic situation and these cyclopentane molecules are not rigid and unmoving. Although the envelope conformation represents a preferred lower energy state, at normal temperatures these molecules vibrate and wiggle as bonds stretch and contract. The out-of-plane carbon atom can move to the other side of the plane, and indeed other carbon atoms in the ring can assume the out-of-plane position. This property is particularly important with other rings, especially cyclohexane.

Cyclohexane can readily adopt a number of puckered conformations, the most stable of which is called the **chair conformation**. In the chair conformation, four carbon atoms lie in a plane with one carbon atom above the plane and another carbon atom on the opposite side of the ring taking a position below the plane. The shape of the chair conformation is somewhat like the low reclining chairs that you might find beside a swimming pool (figure 16.10). The chair conformation for cyclohexane (figure 16.11, overleaf) is the most stable arrangement because the C—C—C bond angles are all very close to the ideal 109.5° (minimising angle strain) and the C—H bonds can all form a low-energy staggered orientation (minimising torsional strain). The consequence of this is that in the chair conformation there are two sets of C—H bonds, depending on their orientation in space. Six C—H bonds are oriented essentially in the same plane as the four carbon atoms that make up the 'seat' of the chair. The other six C—H bonds point either up or down in the approximate direction of the chair 'back' or 'legs'. Arranging the bonds in this conformation minimises angle and torsional strain, and so cyclohexane in the chair conformation has almost no strain energy; this explains why these six-membered rings of carbon are found so commonly in nature.

FIGURE 16.10 'Chair' conformation.

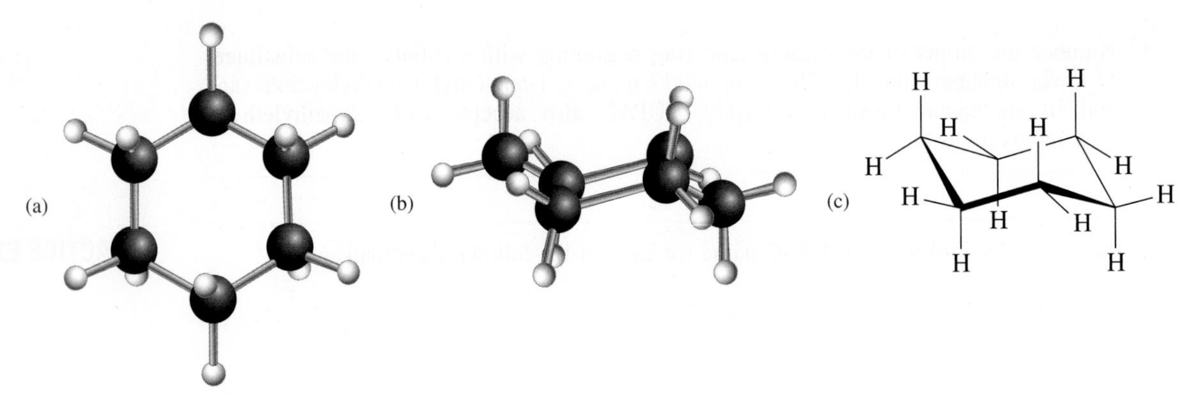

FIGURE 16.11 Cyclohexane. The most stable conformation is the 'chair' conformation: **(a)** ball-and-stick model viewed from above, **(b)** ball-and-stick model viewed from the side, **(c)** skeletal model, **(d)** skeletal model viewed from the front of the 'chair' with some H removed for clarity, and **(e)** Newman projection.

Equatorial bonds are those that are oriented more or less in the plane of the seat of the chair (figure 16.12b) and the hydrogen atoms therefore are called equatorial hydrogen atoms. The other bonds are called **axial bonds** (figure 16.12c) and the hydrogen atoms in this orientation are called axial hydrogen atoms. Three axial bonds are directed upwards and three downwards. Equatorial bonds are aligned approximately in the plane of the seat of the imaginary chair, but close inspection shows that these also involve three bonds that point slightly up and three that point slightly down. Notice also that the orientation of either the axial or the equatorial bonds alternates, first up and then down, as you move from one carbon atom in the ring to the next. If the axial bond on one carbon atom points up then the equatorial bond on that carbon atom points slightly downwards. Conversely, if the axial bond on a particular carbon atom points downwards, then the equatorial bond on that atom points slightly upwards.

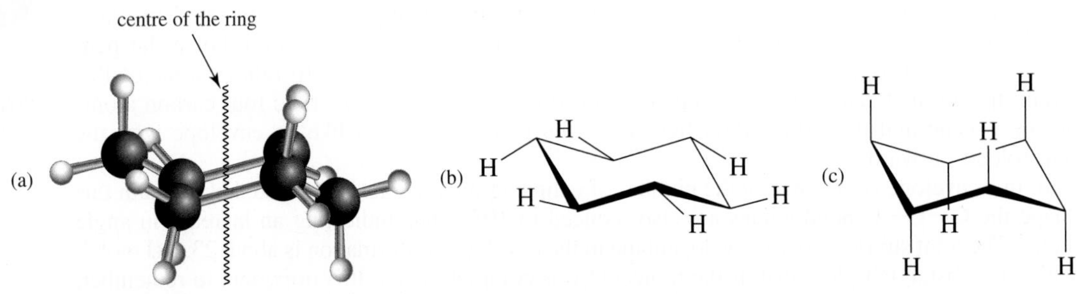

FIGURE 16.12 Chair conformation of cyclohexane showing axial and equatorial C—H bonds: **(a)** ball-and-stick model showing all 12 hydrogen atoms, **(b)** the six equatorial C—H bonds and **(c)** the six axial C—H bonds.

Remember that, like cyclopentane, cyclohexane exists in a dynamic state and there are many other nonplanar conformations that the ring easily attains. Apart from other chair conformations, there is also a 'boat' conformation. The **boat conformation** occurs when both of the out-of-plane carbon atoms are on the same side of the plane made from the other four carbon atoms. This conformation is readily formed from the chair conformation by bending the ring and swinging the carbon atom on the chair 'leg' up so that it reaches the same position as the carbon atom on the chair 'back' (figure 16.13).

The boat conformation is less stable than the chair conformation. Torsional strain is present in the boat conformation as four sets of hydrogen atoms become eclipsed. Another type of strain, called **steric strain**, is also generated. Steric strain is also called nonbonded interaction strain, and it involves strain which arises in the molecule when two parts try to occupy the same space. In the case of the boat conformation of cyclohexane, the two axial hydrogen atoms and their attendant electron clouds interact across the ring, and this unfavourable action leads to strain in the molecule. The difference in energy between the chair and the boat conformations is about 27 kJ mol^{-1}. This means that, at room temperature, fewer than 0.001% of the molecules might be found in the boat conformation at any time. In fact, to lower the energy slightly, the boat conformation twists into a structure called the twist boat, which is a few kJ mol^{-1} lower in energy. Nevertheless, the boat conformation is important, as it is through this higher energy orientation that one chair conformation is converted into another. Two equivalent-energy chair conformations interconvert rapidly at room temperature by first twisting into the higher energy boat conformation (figure 16.13) and then relaxing into the lower energy chair conformation. When one chair is converted into another, a change occurs in relative orientations in space of the hydrogen atoms bound to the carbon atoms of the ring. All equatorial hydrogen atoms in one chair become axial hydrogen atoms in the other and vice versa (figure 16.14).

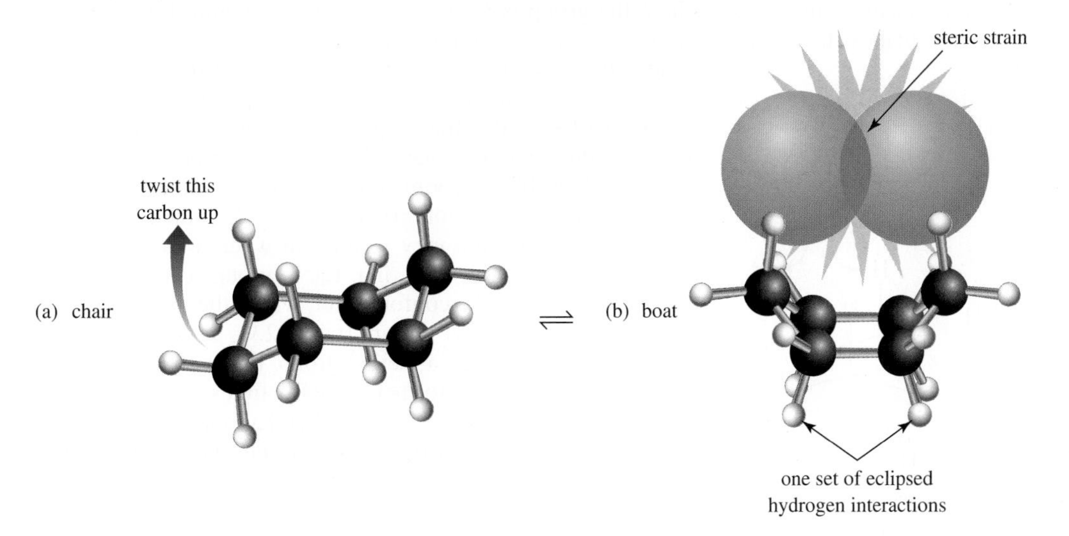

FIGURE 16.13 Conversion of **(a)** a chair conformation to **(b)** a boat conformation. In the boat conformation, there is both torsional strain, due to the four sets of eclipsed hydrogen interactions, and steric strain. A chair conformation is more stable than a boat conformation.

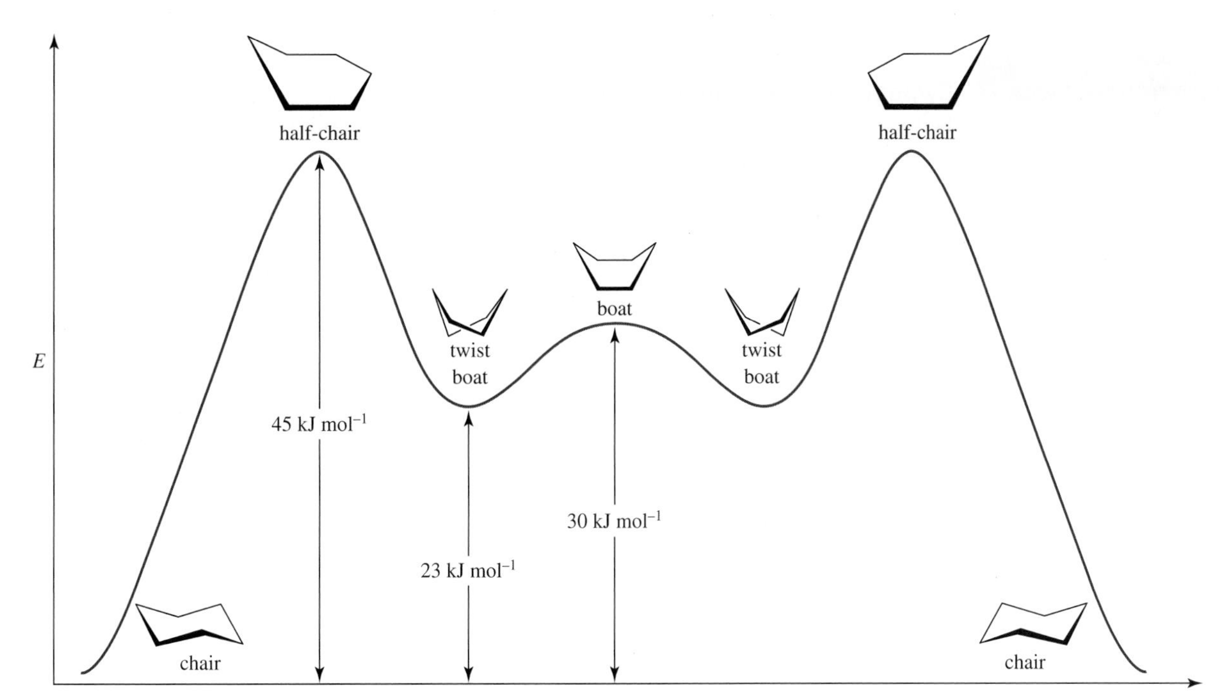

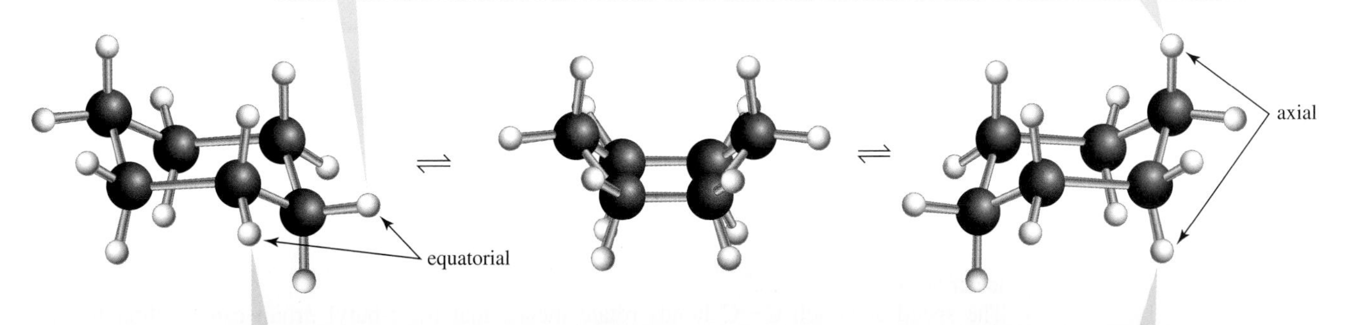

FIGURE 16.14 Interconversion of chair cyclohexanes via a boat conformation. All C—H bonds that are equatorial in one chair are axial in the other chair, and vice versa.

This process becomes much more significant when one of the hydrogen atoms is replaced by another group such as a methyl or other alkyl group. Interconversion of one chair to another now has the consequence of converting the substituted group from an equatorial to an axial orientation (or vice versa) and this now has energy implications. A large group can impose considerable steric strain across the ring when it is in an axial orientation. This type of strain is called an axial–axial (or diaxial) interaction. The presence of such a group on the ring favours the chair conformation where the group is in the equatorial orientation. For a methyl group, this makes the axial orientation about 7.28 kJ mol^{-1} higher in energy and so, at room temperature and at equilibrium, about 95% of these molecules would have the methyl group in the equatorial orientation (figure 16.15). Like the higher energy boat conformation, however, this does not mean that such a conformation is not readily achievable at room temperature but rather that such molecules prefer one conformation over the other. However, this effect can become very significant when a large group is present on the cyclohexane ring. Large groups can essentially 'lock' the ring into the conformation where the group is equatorial. For a *tert*-butyl group (table 2.9, p. 59), which is considerably larger than a methyl group, this preference means that the equatorial conformer is 4000 times more abundant than the conformer where the *tert*-butyl group is in the axial orientation. In effect, the ring is locked into one conformation, and chemists actually call such molecules conformationally locked isomers.

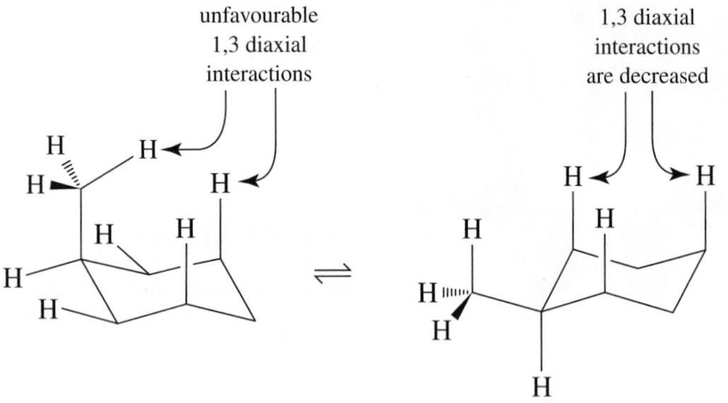

unfavourable 1,3 diaxial interactions

1,3 diaxial interactions are decreased

FIGURE 16.15 1,3 diaxial interaction leads to steric strain.

WORKED EXAMPLE 16.3

Cycloalkane conformational isomers

Is it possible to draw a lower energy conformation for the following molecule?

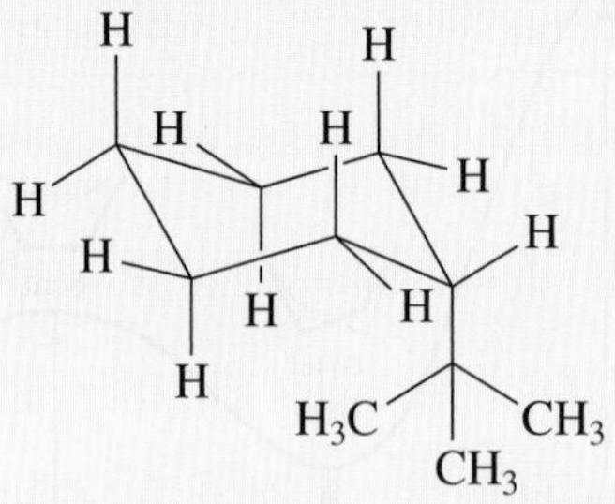

Analysis

In a lower energy conformation, the large *t*-butyl group would not be in an axial position as it interacts with the axial hydrogen atoms on carbons 3 and 5. We know that cyclohexanes can easily flip their orientation from one chair conformer to the other, so we should see what the structure looks like after such a transformation.

Solution

Bringing the left-hand side of the ring downwards and lifting the right-hand side upwards swaps the positions of the current axial groups with the equatorial position.

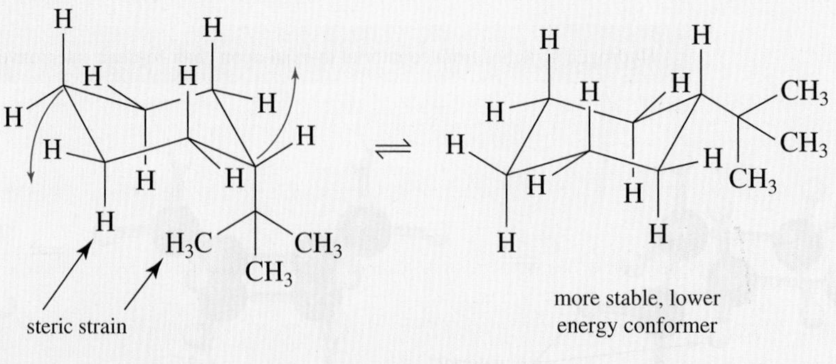

steric strain

more stable, lower energy conformer

Is our answer reasonable?

The speed at which C—C bonds rotate means that the *t*-butyl group can be thought of as a high-speed fan with the methyl group 'blades', striking the axial hydrogen atoms on carbons 3 and 5. The large *t*-butyl group is now placed further away from the other parts of the cyclohexane, with which it was previously interacting.

PRACTICE EXERCISE 16.3

For each of the following pairs, which is the lower energy conformation? Explain your answers.

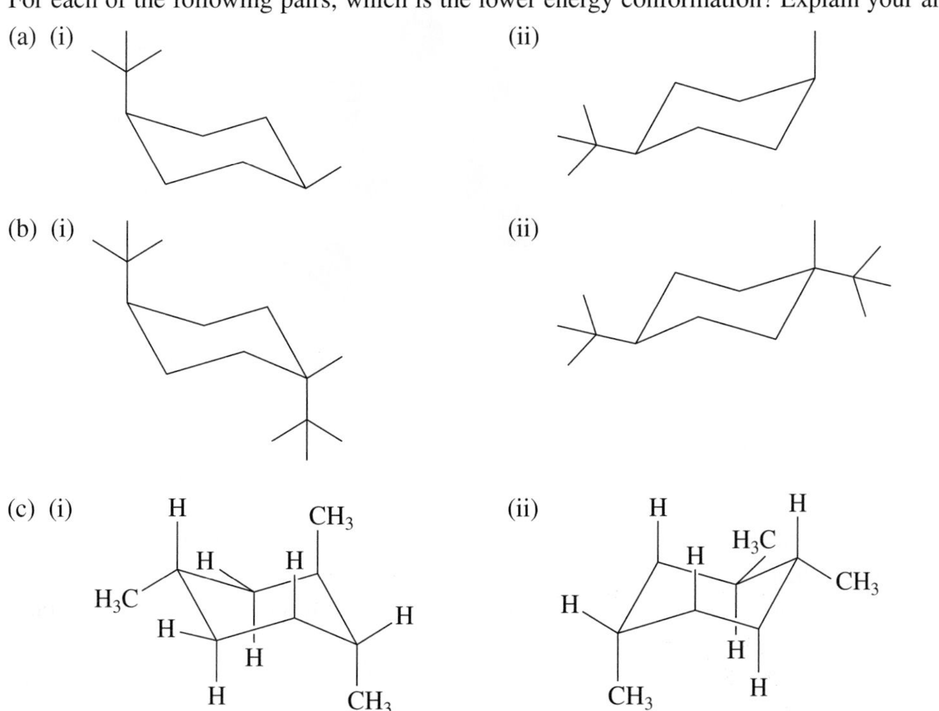

cis–trans isomerism in cycloalkanes

Cycloalkanes with substituents on two or more carbon atoms of the ring are called ***cis–trans* isomers**. All *cis–trans* isomers have (1) the same molecular formula, (2) the same order of attachment of atoms and (3) an arrangement of atoms that cannot be interchanged by rotation around σ bonds under ordinary conditions (see chapter 5). In *cis* isomers, the groups are 'on the same side'; in *trans* isomers, the groups are 'across from each other'.

In chapter 2, we illustrated *cis–trans* isomerism in cycloalkanes using 1,2-dimethylcyclopentane.

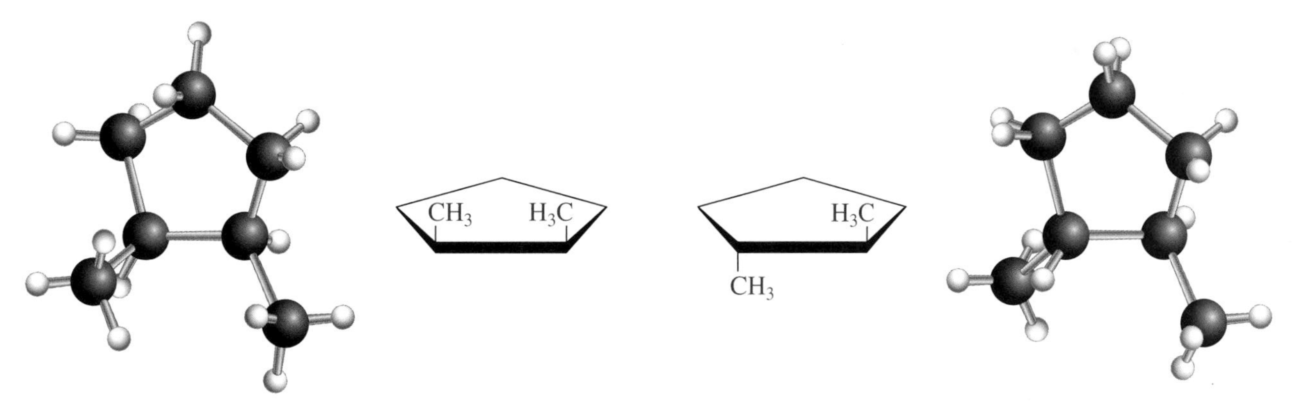

cis-1,2-dimethylcyclopentane *trans*-1,2-dimethylcyclopentane

For simplicity, some hydrogen atoms have been omitted and, to add clarity, the cyclopentane ring has been drawn as a planar pentagon viewed edge on. (We know that cyclopentane exists primarily in an 'envelope' conformation shown in the ball-and-stick models above. However, to analyse the structural relationship between the substituents, we can represent this as a pentagonal line drawing.) Carbon–carbon bonds of the ring that project forwards are shown above as heavy lines. When viewed from this perspective, substituents bonded to the cyclopentane ring project above and below the plane of the ring. In the isomer of 1,2-dimethylcyclopentane shown above left, the methyl groups are on the same side of the ring (either both above or both below the plane of the ring) and we call this arrangement *cis*; in the isomer shown on the right, the methyl groups are on opposite sides of the ring (one above and one below the plane of the ring) and we call this arrangement *trans*.

Alternatively, the cyclopentane ring can be viewed from above, with the ring in the plane of the paper. Substituents on the ring then either project towards you (that is, they project above the plane of the page) and are shown by solid wedges, or project away from you (they project below the plane of the page) and are shown by hashed wedges. In the following structural formulae, only the two methyl groups are shown (hydrogen atoms of the ring are not shown).

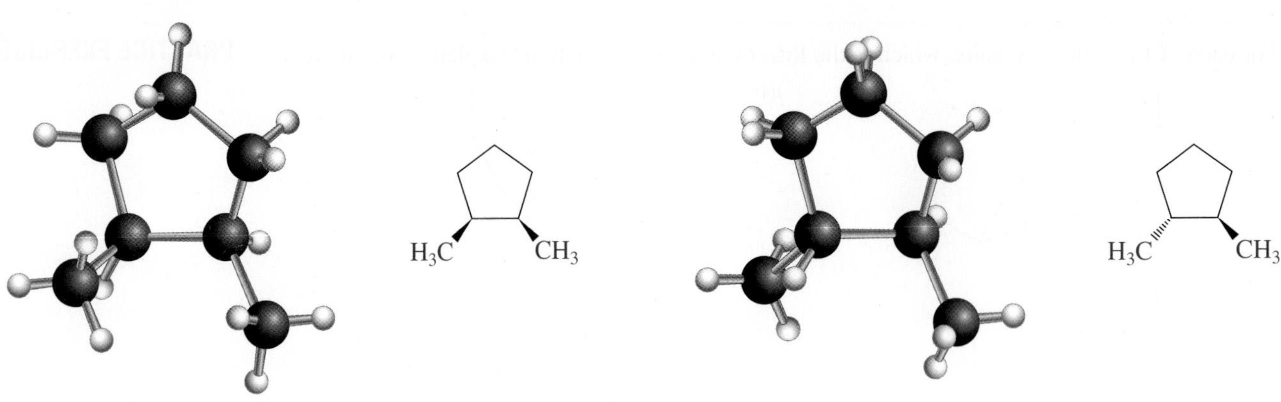

cis-1,2-dimethylcyclopentane

trans-1,2-dimethylcyclopentane

WORKED EXAMPLE 16.4

Drawing *cis–trans* isomers

Which of the following cycloalkanes show *cis–trans* isomerism? For each that does, draw both forms.

(a) methylcyclopentane (b) 1,1-dimethylcyclobutane (c) 1,3-dimethylcyclobutane

Analysis and solution

(a) Methylcyclopentane does not show cis–trans isomerism. It has only one substituent on the ring.

(b) Only one arrangement is possible for the two methyl groups on the ring, which must be *trans*, so 1,1-dimethylcyclobutane does not show *cis–trans* isomerism.

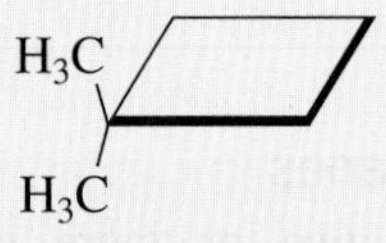

1,1-dimethylcyclobutane

(c) The following diagram shows the *cis–trans* isomerism of 1,3-dimethylcyclobutane. Note that, in these structural formulae, we show only the hydrogen atoms on carbon atoms bearing the methyl groups.

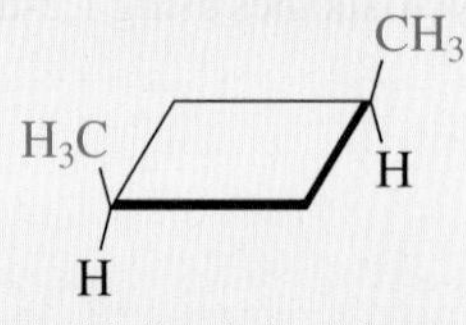

cis-1,3-dimethylcyclobutane

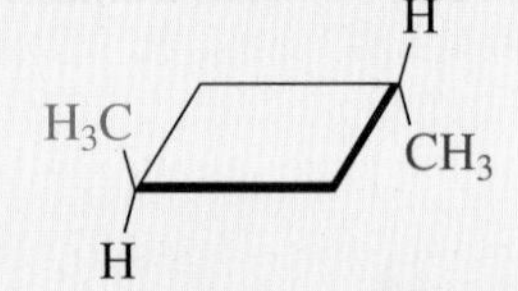

trans-1,3-dimethylcyclobutane

PRACTICE EXERCISE 16.4

Which of the following cycloalkanes show *cis–trans* isomerism? For each that does, draw both isomers.

(a) 1,3-dimethylcyclopentane (b) ethylcyclopentane (c) 1-ethyl-2-methylcyclobutane

Physical properties of alkanes

The most important property of alkanes is their almost complete lack of polarity. This is illustrated in figure 16.16, which shows the uniform distribution of the outer bonding electrons in pentane. As we saw in chapter 5, the difference in electronegativity between carbon and hydrogen is 2.5 – 2.1 = 0.4 on the Pauling scale and, given this small difference, we classify a C—H bond as a nonpolar covalent bond. Therefore, alkanes are nonpolar compounds, and there are only weak interactions between molecules.

Boiling points

As we saw in chapter 6 (p. 246), interactions between alkane molecules consist of only weak dispersion forces. Because of this, the boiling points of alkanes are lower than those of almost any other type of compound of the same molar mass. As the number of atoms and electrons and, therefore, the molar mass of an alkane

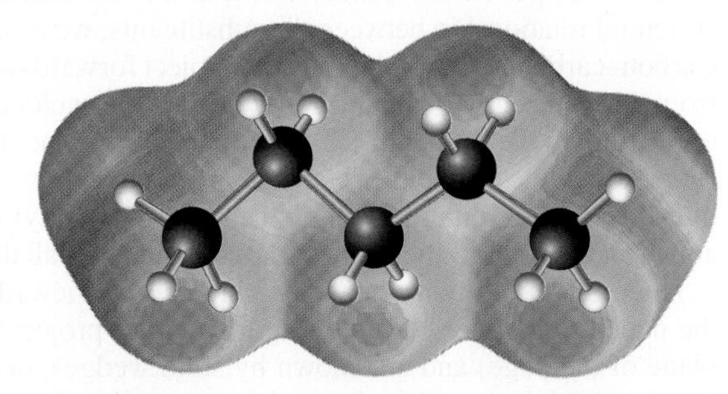

FIGURE 16.16 The electron density model of pentane (like all alkanes) shows no evidence of polarity.

increase, the number, and hence total strength, of dispersion forces between alkane molecules increases, thereby increasing the boiling point (see figure 6.34).

Alkanes containing one to four carbon atoms are gases at room temperature and atmospheric pressure, and those containing five to 17 carbon atoms are colourless liquids. High-molar-mass alkanes (those with 18 or more carbon atoms) are white, waxy solids. Several plant waxes are high-molar-mass alkanes. The wax found naturally in apple skins, for example, is an unbranched alkane with the molecular formula $C_{27}H_{56}$. Paraffin wax, a mixture of high-molar-mass alkanes, is used for wax candles, in lubricants and to seal home-made jams and other preserves. Petrolatum, so named because it is derived from petroleum refining, is a liquid mixture of high-molar-mass alkanes. Sold as mineral oil and Vaseline®, petrolatum is used as an ointment base in pharmaceuticals and cosmetics and as a lubricant and rust preventative.

Melting point and density

The melting points of alkanes increase with increasing molar mass. The increase, however, is not as regular as that observed for boiling points, because the ability of molecules to pack into ordered patterns of solids changes as the molecular size and shape change.

The average density of the alkanes listed in table 16.1 is about $0.7\,g\,mL^{-1}$ whereas that of higher molar mass alkanes is about $0.8\,g\,mL^{-1}$. All liquid and solid alkanes are less dense than water ($1.0\,g\,mL^{-1}$) and so float on water.

Isomeric alkanes

Recall that isomers are molecules with the same molecular formula but different structures, and constitutional isomers are compounds with the same molecular formula but different structures because of different sequences of atom connectivity. Alkanes exhibit constitutional isomerism arising from side chains in the carbon sequence. As we learned in chapter 2 (pp. 58–62), each isomer has a unique name according to a set of rules in which numbers are used to indicate the positions of the side chains.

The ability of carbon atoms to form strong stable bonds with other carbon atoms results in a staggering number of constitutional isomers. As table 16.2 shows, there are three constitutional isomers with the molecular formula C_5H_{12}, 75 constitutional isomers with the molecular formula $C_{10}H_{22}$ and more than 36 million constitutional isomers with the molecular formula $C_{25}H_{52}$.

TABLE 16.2 Constitutional isomers of various C_nH_{2n+2} hydrocarbons.

Molecular formula	No. of constitutional isomers
CH_4	0
C_5H_{12}	3
$C_{10}H_{22}$	75
$C_{15}H_{32}$	4347
$C_{25}H_{52}$	36 797 588

Thus, for even a small number of carbon and hydrogen atoms, a very large number of constitutional isomers is possible. In fact, the potential for structural and functional group individuality among organic molecules made from just the basic building blocks of carbon, hydrogen, nitrogen and oxygen is practically limitless.

WORKED EXAMPLE 16.5

Hydrocarbon isomers

How many types of different hydrocarbon molecules are there with the formulae C_4H_{10}, C_5H_{12}, C_6H_{14}, $C_{11}H_{24}$ and $C_{30}H_{62}$?

Analysis

The question seems straightforward and, at least initially, it is. The compounds listed are ordinary hydrocarbons. There are no other elements present so all structural variety arises from the different sequences by which carbon atoms and hydrogen atoms can be assembled. There is no simple shortcut to the answer to this question and, in fact, we are better served by drawing out the possible structures to better understand the nature of the molecules involved.

Solution

Starting with the simplest, C_4H_{10}, there are two possible structures. The most obvious answer is a simple chain of carbon atoms. There is another possible sequence, however, involving a branch.

$$H_3C-CH_2-CH_2-CH_3 \qquad H_3C-\underset{}{\overset{\displaystyle CH_3}{\overset{|}{C}H}}-CH_3$$

With C_5H_{12}, three different molecules are possible.

$$H_3C-CH_2-CH_2-CH_2-CH_3 \qquad H_3C-CH(CH_3)-CH_2-CH_3 \qquad H_3C-C(CH_3)_2-CH_3$$

For C_6H_{14}, there are five different ways to assemble the carbon and hydrogen atoms.

$$H_3C-CH_2-CH_2-CH_2-CH_2-CH_3 \qquad H_3C-CH(CH_3)-CH_2-CH_2-CH_3$$

$$H_3C-C(CH_3)_2-CH_2-CH_3 \qquad H_3C-CH_2-CH(CH_3)-CH_2-CH_3 \qquad H_3C-CH(CH_3)-CH(CH_3)-CH_3$$

Here you will notice that, for clarity, we have not used line structures. However, even if we did, space precludes drawing all the possible structures for $C_{11}H_{24}$; there are 159!

You will not be surprised to know that there are a lot of isomers of $C_{30}H_{62}$, but you will still no doubt be shocked to know just how many. There are more than 4 billion ways of assembling these 92 atoms. Most organic compounds contain many more than this number of atoms and almost all involve other elements and more complicated bonding than is present in simple hydrocarbons. This is why organic chemistry is such a challenging field of science and why we need to find groupings of similar properties and categories to simplify the topic.

Is our answer reasonable?

To confirm that the structural formulae drawn represent constitutional isomers, write the molecular formula of each and check that they are the same, and check each structure to ensure that simple rotation around any bond does not reproduce one of the other structures.

As we learned in chapter 2, constitutional isomers have different physical properties. Table 16.3 lists the boiling points, melting points and densities of the five compounds with the molecular formula C_6H_{14}. The boiling point of each of the branched-chain molecules is lower than that of hexane itself, and the more branching there is the lower the boiling point is. These differences in boiling points are related to molecular shape. The only forces of attraction between alkane molecules are dispersion forces. As branching increases, the shape of an alkane molecule becomes more compact, and its surface area decreases. As the surface area decreases, so too does the area of contact between molecules. This decrease leads to weaker dispersion forces, so boiling points also decrease (see figure 16.17). Thus, for any group of constitutional isomers, it is usually observed that the least branched isomer has the highest boiling point and the most branched isomer has the lowest boiling point. The trend in melting points is less obvious, but, as previously mentioned, it correlates with a molecule's ability to pack into ordered arrays of solids.

TABLE 16.3 Physical properties of the isomeric alkanes with the molecular formula C_6H_{14}.

Name	Boiling point (°C)	Melting point (°C)	Density (g mL^{-1} at 0 °C)
hexane	69	–95	0.659
3-methylpentane	63	–118	0.664
2-methylpentane	60	–153	0.653
2,3-dimethylbutane	58	–128	0.662
2,2-dimethylbutane	50	–100	0.649

larger surface area, an increase in dispersion forces and a higher boiling point

hexane

smaller surface area, a decrease in dispersion forces and a lower boiling point

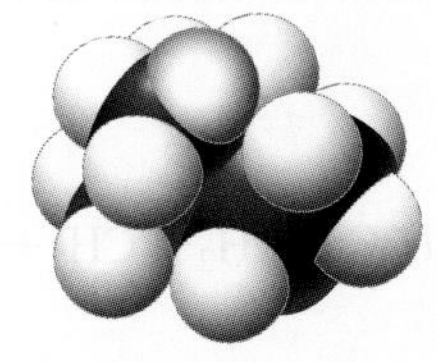

2,2-dimethylbutane

FIGURE 16.17 As branching increases, the shape of an alkane molecule becomes more compact, and its surface area decreases. As the surface area decreases, the strength of the dispersion forces decreases, and the boiling point also decreases.

WORKED EXAMPLE 16.6

Physical properties of alkanes

Arrange the alkanes in each of the following sets in order of increasing boiling point.

(a) butane, decane and hexane

(b) 2-methylheptane, octane and 2,2,4-trimethylpentane

Analysis and solution

(a) All of the compounds are unbranched alkanes. As the number of carbon atoms in the chain increases, the dispersion forces between molecules increase, and the boiling point increases. Decane has the highest boiling point and butane the lowest.

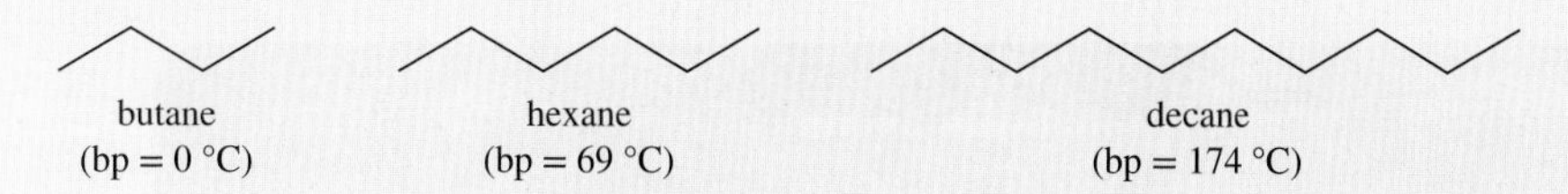

(b) These three alkanes are constitutional isomers with the molecular formula C_8H_{18}. Their relative boiling points depend on the degree of branching. The most highly branched isomer, 2,2,4-trimethylpentane, has the smallest surface area and the lowest boiling point. Octane, the unbranched isomer, has the largest surface area and the highest boiling point.

2,2,4-trimethylpentane (bp = 99 °C) 2-methylheptane (bp = 118 °C) octane (bp = 126 °C)

PRACTICE EXERCISE 16.5

Arrange the alkanes in each of the following sets in order of increasing boiling point.

(a) 2-methylbutane, 2,2-dimethylpropane and pentane

(b) 3,3-dimethylheptane, 2,2,4-trimethylhexane and nonane

16.3 Alkenes and alkynes

Some hydrocarbons have double bonds or triple bonds between carbon atoms and, hence, fewer hydrogen atoms than the corresponding alkanes. These are called unsaturated hydrocarbons. There are three classes of unsaturated hydrocarbons: alkenes, alkynes and arenes. Alkenes contain one or more carbon–carbon double bonds, and alkynes contain one or more carbon–carbon triple bonds. Alkenes have the general formula C_nH_{2n}. Alkynes have the general formula C_nH_{2n-2}. Ethene is the simplest alkene, and ethyne is the simplest alkyne.

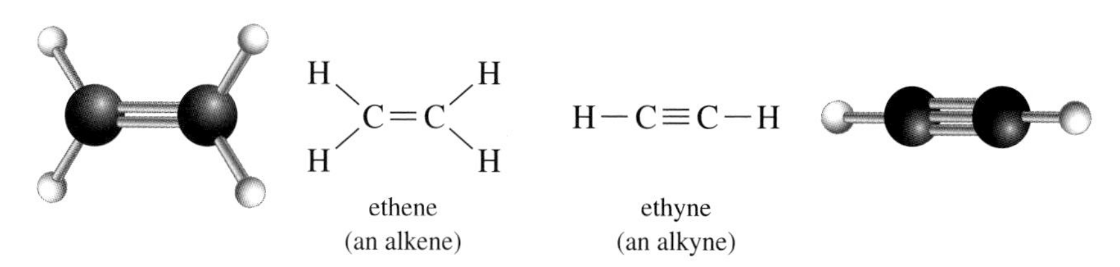

Arenes are the third class of unsaturated hydrocarbons. The simplest arene is benzene.

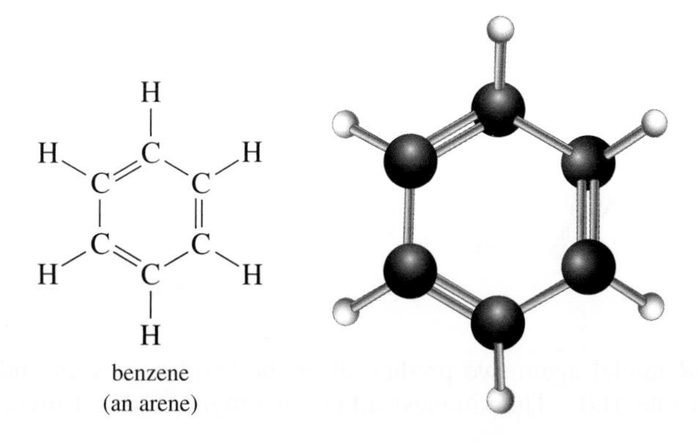

Arenes are compounds that contain one or more 'benzene' rings. The chemistry of benzene and its derivatives is quite different from that of alkenes and alkynes and will be described later in the chapter. All we need to remember at this point is that a benzene ring is not chemically reactive under any of the conditions we describe for simple alkenes and alkynes.

Compounds containing carbon–carbon double bonds are especially widespread in nature. Many organic compounds found in nature are derived from an alkene called isoprene. Isoprene is a major contributor to the haze and scent that are characteristic of Australian eucalyptus forests (figure 16.18). Furthermore, several low-molar-mass alkenes, including ethene and propene, have enormous commercial importance in our modern, industrialised society. The organic chemical industry produces more ethene worldwide than any other chemical.

FIGURE 16.18 The blue haze that gives the Blue Mountains their name arises from isoprene and other hydrocarbons released from eucalyptus trees on exposure to sunlight and high temperatures.

Ethene is unusual among the alkenes in that it occurs only in trace amounts in nature (where it plays an essential role in the process by which fruit ripens). The enormous amounts of it required to meet the needs of the chemical industry are generated from the refining of crude oil into petrol or the conversion of ethane extracted from natural gas as shown below.

$$\underset{\text{ethane}}{CH_3CH_3} \xrightarrow[\text{(thermal cracking)}]{800\text{–}900\ °C} \underset{\text{ethene}}{CH_2{=}CH_2} + H_2$$

The crucial point to recognise is that ethene and all of the commercial and industrial products made from it (such as plastic shopping bags) are derived from either natural gas or crude oil — both nonrenewable natural resources!

Shapes of alkenes and alkynes

Using the valence-shell-electron-pair repulsion (VSEPR) model (see chapter 5), we predict a value of 120° for the bond angles around each carbon atom in a double bond. The observed H—C—C bond angle in ethene is 121.7°, a value close to that predicted by this model. In other alkenes, deviations from the predicted angle of 120° may be somewhat larger as a result of strain between groups bonded to one or both carbon atoms of the double bond. The C—C—C bond angle in propene, for example, is 124.7°.

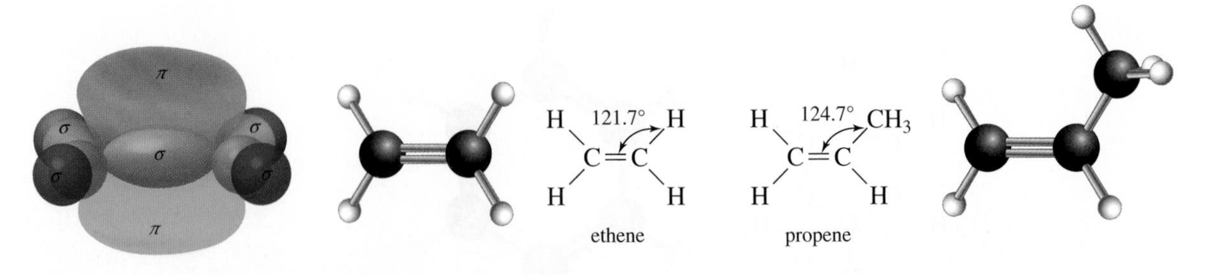

Using the VSEPR model again, we predict all of the bond angles around each carbon atom in the triple bond to be 180°. The simplest alkyne is ethyne, C_2H_2. Ethyne is indeed a linear molecule; all of its bond angles are 180°. (*Note:* All orbitals, except pink π orbitals, are the same phase. Colours are used for clarity only.)

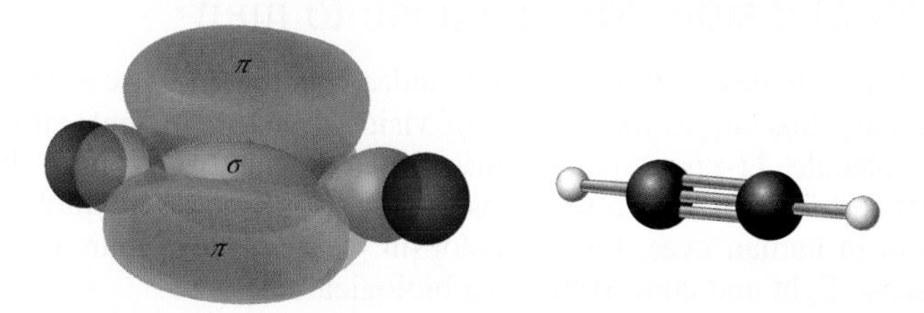

As we saw in chapter 5, a C≡C triple bond is described in terms of the overlap of *sp* hybrid orbitals of adjacent carbon atoms to form a σ bond, the overlap of parallel $2p_y$ orbitals to form one π bond, and the overlap of parallel $2p_z$ orbitals to form a second π bond. In ethyne, each carbon atom bonds to a hydrogen atom by the overlap of an *sp* hybrid orbital of carbon with a 1*s* atomic orbital of hydrogen.

cis–trans isomerism in alkenes

In chapter 5, we described the formation of a carbon–carbon double bond in terms of the overlap of atomic orbitals. A carbon–carbon double bond consists of one σ bond and one π bond. Each carbon atom of the double bond uses its three sp^2 hybrid orbitals to form σ bonds with three atoms. The unhybridised 2*p* atomic orbitals, which lie perpendicular to the plane created by the axes of the three sp^2 hybrid orbitals, combine to form the π bond of the carbon–carbon double bond.

Whereas rotation around a single bond is relatively free (the energy barrier in ethane is approximately 12.5 kJ mol^{-1}), it takes considerably more energy to rotate a double bond. To break the π bond in ethene (i.e. to rotate one carbon atom by 90° with respect to the other so that no overlap occurs between 2*p* orbitals on adjacent carbon atoms) requires approximately 264 kJ mol^{-1} (figure 16.19). This energy is considerably greater than the thermal energy available at room temperature, so rotation around a carbon–carbon double bond is severely restricted.

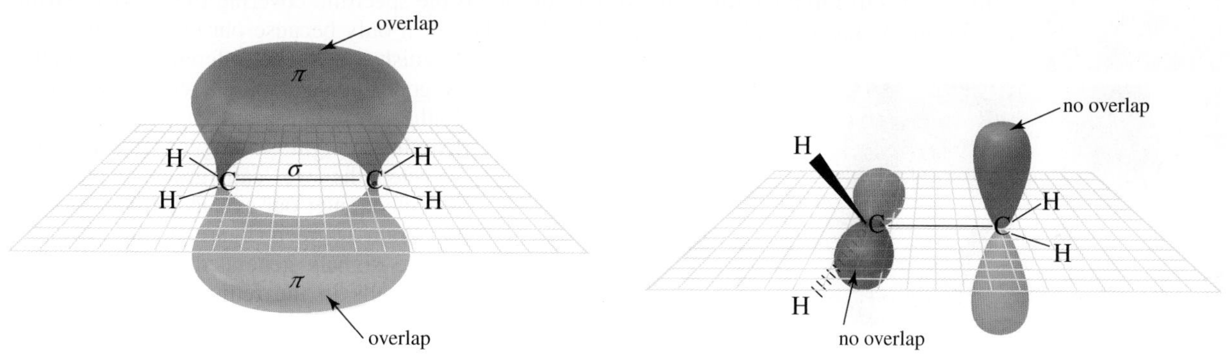

FIGURE 16.19 Restricted rotation around the carbon–carbon double bond in ethene: **(a)** orbital overlap model showing the π bond; **(b)** the π bond is broken by rotating the plane of one H—C—H group by 90° with respect to the plane of the other H—C—H group.

Because of restricted rotation around a carbon–carbon double bond, an alkene in which each carbon atom of the double bond has two different groups bonded to it shows *cis–trans* isomerism. Consider, for example, but-2-ene. In *cis*-but-2-ene, the two methyl groups are on the same side of the double bond; in *trans*-but-2-ene, the two methyl groups are on opposite sides of the double bond.

H H
C=C
H_3C CH_3

cis-but-2-ene
(mp = −139 °C, bp = 4 °C)

H CH_3
C=C
H_3C H

trans-but-2-ene
(mp = −106 °C, bp = 1 °C)

These two compounds cannot be converted into one another at room temperature because of the restricted rotation around the double bond; they are different compounds with different physical and chemical properties and are called **configurational isomers**.

Because of repulsion between alkyl substituents on the same side of the double bond in the *cis* isomer, *cis* alkenes are less stable than their *trans* isomers. This can be seen from the forcing together of the methyl hydrogen atoms in the space-filling model of *cis*-but-2-ene model above.

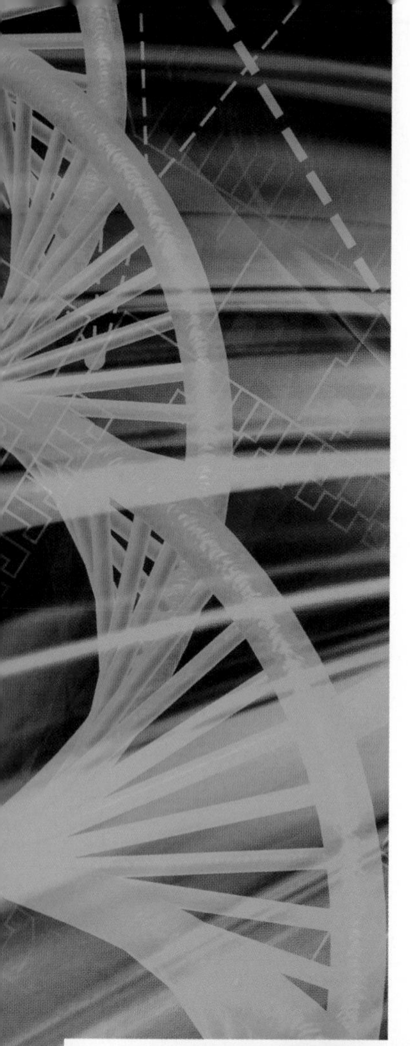

Chemical Connections

The chemistry of vision: from mollusc to man

Remarkably, all vertebrates, arthropods and, indeed, some molluscs share aspects of the amazing chemistry that drives the process of vision. To some extent, all of these creatures rely on the same molecule, 11-*cis*-retinal, an unsaturated alkene that strongly absorbs blue-green light (between 450 and 550 nm with a peak absorption at 498 nm). This molecule, when bound to the protein opsin in human eyes, forms rhodopsin — a purple-coloured material that is the basis of how we sense light and convert this to a biological response.

1 6 11 12 5 15 H C O + H_2N—opsin ⟶ H C N—opsin

11-*cis*-retinal

rhodopsin
visual purple

Our eyes are sophisticated organs with two forms of light-sensing cells that, because of their shape, are called rods and cones. About 100 million rod cells are located primarily on the periphery of the retina, and these allow vision in low light. However, this sensitivity does not extend to colour recognition, so the view is monochromatic. To detect colour, there are three types of cone cells, each of which uses retinal bound to proteins, slightly different from opsin, called iodopsins. Variation in a few key amino acids in the iodopsin chain distorts retinal's π bond system to change the position of the peak of maximum light absorption.

In the human eye, there are only about 3 million cone cells, and the photoreceptors in these cells are sensitive to three different portions of the visible spectrum covering from about 380 nm to 750 nm. A 'green' apple is perceived by us to be green only because our eye is able to distinguish between different wavelengths. A green apple does not emit green light. Rather, it simply absorbs all the wavelengths of light shining on it except the wavelengths we call green, which are reflected, enter our eye and are detected by our brain as green (figure 16.20).

FIGURE 16.20 A green apple absorbs all wavelengths of light except green, which it reflects. The reflected light enters our eyes and the brain detects it as green.

The signals generated by the rod and cone cells in the retina are converted by our brain to a visual image of the colour and brightness of an object. This process relies on a remarkable fusion of physics, chemistry and biology: the physics of light interacting with matter, the chemistry of how a molecule's shape is controlled by the nature of its bonds and the biology of the way a molecule's shape governs a cellular membrane to create a flow of ions and a nerve response. To interact with light, a molecule must have bonds that allow absorption of the precise energy of the light we see as 'visible'. Most organic molecules do not absorb any light in this region and so, naturally, cannot form the basis of a chemical light detector. However, when 11-*cis*-retinal is bound to opsin, it absorbs the energy of photons of wavelength around 500 nm — right in the middle of our visible region. This is only the first step in a complicated process, however (see figure 16.21). Absorption of light allows 11-*cis*-retinal to relax from a bent shape (arising from the *cis* double bond) to a lower energy, all-*trans* arrangement. The straighter all-*trans*-retinal can no longer fit in the receptor designed for 11-*cis*-retinal and is ejected. This change leads to differences in cell membrane potential and, as ions are pumped into the cell, this response turns into a nerve impulse that travels to the brain. That is not the end of the matter; the all-*trans*-retinal generated by this process is reconverted to 11-*cis*-retinal by specific enzymes, which can re-bind into the receptor, ready to interact with another photon of light.

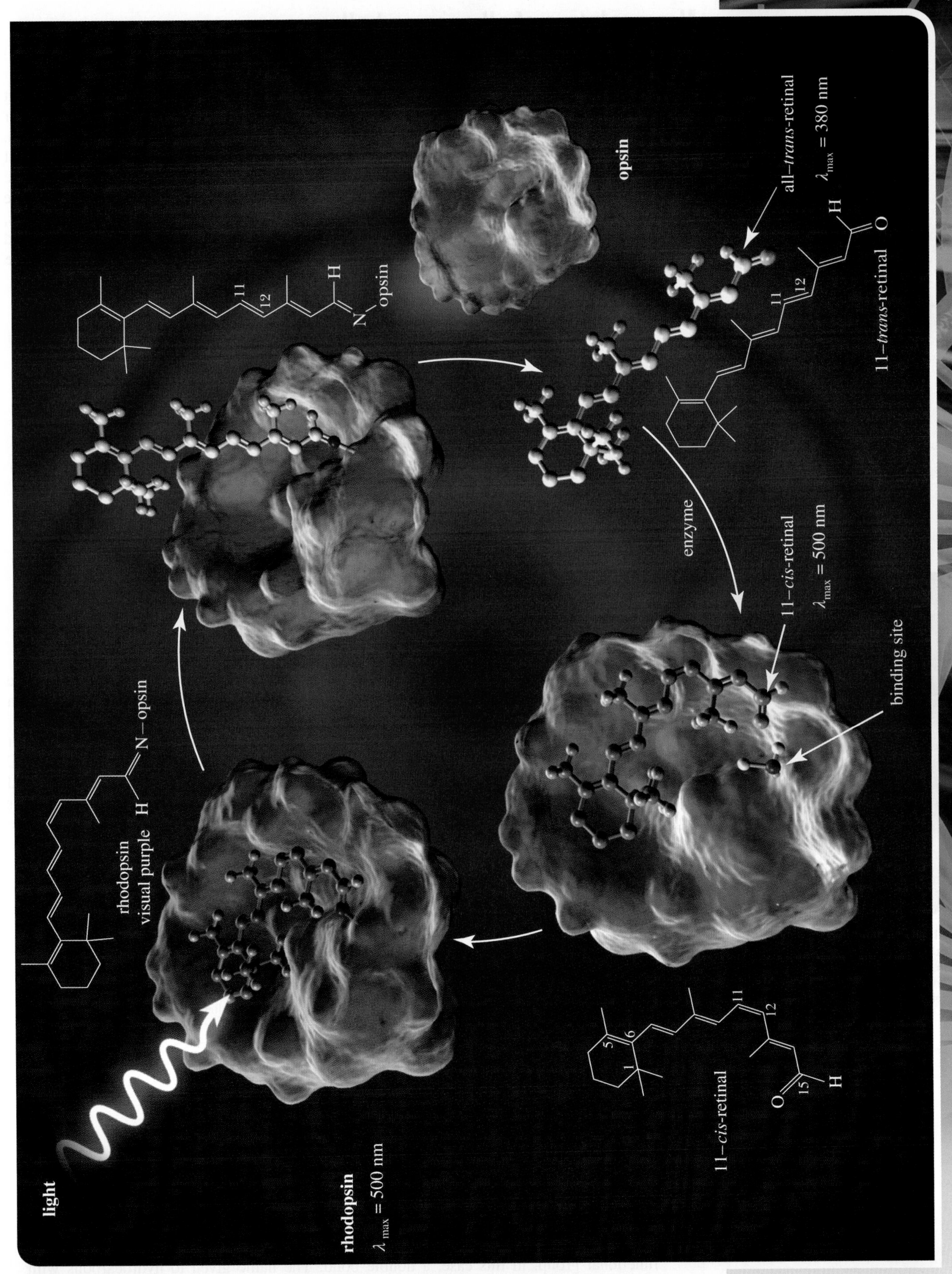

FIGURE 16.21 The chemistry of colour detection in the eye following absorption of a particular wavelength of light by rhodopsin.

Nomenclature of alkenes and alkynes

Like alkanes, alkenes are named using the IUPAC system, but, as we shall see, some are still referred to by their common names.

We form IUPAC names of alkenes by changing the *-an-* component of the name of the parent alkane to *-en-*. Hence, $CH_2{=}CH_2$ is named ethene, and $CH_3CH{=}CH_2$ is named propene. In higher alkenes where the location of the double bond differs between isomers, we use a numbering system. We identify the longest carbon chain that contains the double bond and number it in the direction that gives the carbon atoms of the double bond the lower set of numbers. We then use the number of the first carbon atom of the double bond to show its location. As seen below, hex-1-ene has the double bond at the end of the chain, the double bond in hex-2-ene involves the 2nd and 3rd carbon atoms, and the double bond in hex-3-ene is in the middle of the chain. We name branched or substituted alkenes in a manner similar to the way we name alkanes. We number the carbon atoms, locate the double bond, locate and name substituent groups, and name the main chain.

$\overset{6}{C}H_3\overset{5}{C}H_2\overset{4}{C}H_2\overset{3}{C}H_2\overset{2}{C}H{=}\overset{1}{C}H_2$ — hex-1-ene

$CH_3CH_2CH_2CH{=}CHCH_3$ — hex-2-ene

$CH_3CH_2CH{=}CHCH_2CH_3$ — hex-3-ene

$CH_3CH_2CH(CH_3)CH_2CH{=}CH_2$ — 4-methylhex-1-ene

$CH_3CH_2CH(CH_3)C(CH_2CH_3){=}CH_2$ — 2-ethyl-3-methylpent-1-ene

Note that there is a six-carbon chain in 2-ethyl-3-methylpent-1-ene. However, because the longest chain that contains the double bond has only five carbon atoms, the parent hydrocarbon is pentane, and we name the molecule as a disubstituted pent-1-ene.

We form IUPAC names of alkynes by changing the *-an-* component of the name of the parent alkane to *-yn-*. Thus, $HC{\equiv}CH$ is named ethyne, and $CH_3C{\equiv}CH$ is named propyne. (The IUPAC system retains the name *acetylene*, so there are two acceptable names for ethyne: ethyne and acetylene. Of these two names, acetylene is used much more frequently (figure 16.22).) For larger molecules, such as those shown below, we number the longest carbon chain that contains the triple bond from the end that gives the triple-bonded carbon atoms the lower set of numbers. We indicate the location of the triple bond by the number of the first carbon atom of the triple bond.

FIGURE 16.22 The combustion of ethyne (commonly called acetylene) in an oxyacetylene torch yields energy that produces very high temperatures.

$\overset{4}{C}H_3\overset{3}{C}H(CH_3)\overset{2}{C}{\equiv}\overset{1}{C}H$ — 3-methylbut-1-yne

$\overset{1}{C}H_3\overset{2}{C}H_2\overset{3}{C}{\equiv}\overset{4}{C}\overset{5}{C}H_2\overset{6}{C}(CH_3)_2\overset{7}{C}H_3$ — 6,6-dimethylhept-3-yne

WORKED EXAMPLE 16.7

Naming alkenes and alkynes

Write the IUPAC name of each of the following unsaturated hydrocarbons.

(a) $CH_2{=}CH(CH_2)_5CH_3$ (b) $(H_3C)_2C{=}C(CH_3)H$ (c) $CH_3(CH_2)_2C{\equiv}CCH_3$

Analysis and solution

(a) As with all nomenclature, the first task is to find the longest chain involving the functional group: in this case, a double bond. The name given to hydrocarbons with double bonds is 'alkene'. The eight carbon atoms in this chain give the parent name, octene. We then number the chain to give the carbon atoms involved in the double bond the smallest numbers: in this case, 1 and 2. Hence, the name is oct-1-ene.

(b) Use the same process as for (a). In this case, the functional group is again a double bond and the longest chain is four carbon atoms, so the parent name is butene. The double bond is between carbon atoms 2 and 3, so we have but-2-ene. There is a methyl substituent, also on carbon atom 2. Hence, the name is 2-methylbut-2-ene.

(c) In this case, the functional group is a triple bond, which makes the hydrocarbon an alkyne, and the longest chain is six carbon atoms. The triple bond is between carbon atoms 2 and 3, so the name is hex-2-yne.

PRACTICE EXERCISE 16.6

Write the IUPAC name of each of the following unsaturated hydrocarbons.

(a) 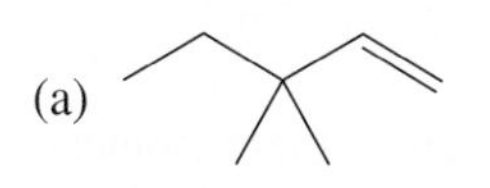 (b) (c)

Despite the precision and universal acceptance of IUPAC nomenclature, some alkenes and alkynes — particularly those of low molar mass — are known almost exclusively by their common names, as illustrated below.

	$CH_2{=}CH_2$	$CH_3CH{=}CH_2$	$CH_3C(CH_3){=}CH_2$	$H{-}C{\equiv}C{-}H$
IUPAC name	ethene	propene	2-methylpropene	ethyne
Common name	ethylene	propylene	isobutylene	acetylene

Designating configuration in alkenes

As we saw previously (p. 713), the physical properties of but-2-ene depend on the orientation of the substituents (or bonds to the substituents) in relation to the double bond. These different orientations represent different molecules (*cis–trans* isomers). We need a precise way of describing the different isomers that can arise when a double bond is present in a hydrocarbon. There are currently two ways of designating the orientation of groups attached to carbon atoms in a double bond: *cis–trans* isomers and *E,Z* isomers.

The *cis–trans* system

The most common method for specifying the configuration of a disubstituted alkene uses the prefixes *cis* and *trans*. In this system, the orientation of the atoms of the parent chain determines whether the alkene is *cis* or *trans*. The following is the structural formula for the *cis* isomer of 4-methylpent-2-ene.

cis-4-methylpent-2-ene

In this example, carbon atoms of the main chain (carbons 1 and 4) are on the same side of the double bond so the configuration of this alkene is *cis*.

WORKED EXAMPLE 16.8

Describing *cis–trans* isomers in alkenes

Name each of the following alkenes, and, using the *cis–trans* system, specify the configuration around each double bond.

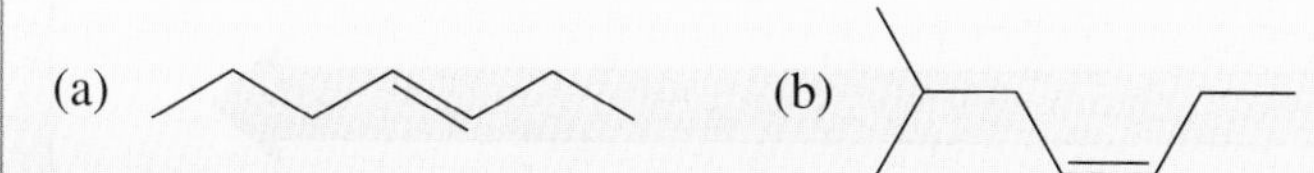

Analysis and solution

(a) The chain contains seven carbon atoms and is numbered from the end that gives the lower number to the first carbon atom of the double bond. The carbon atoms of the parent chain are on opposite sides of the double bond. The compound's name is *trans*-hept-3-ene.

(b) The longest chain contains seven carbon atoms and is numbered from the right, so that the first carbon atom of the double bond is C(3) of the chain. The carbon atoms of the parent chain are on the same side of the double bond. The compound's name is *cis*-6-methylhept-3-ene.

PRACTICE EXERCISE 16.7 Name each of the following alkenes, and, using the *cis–trans* system, specify its configuration.

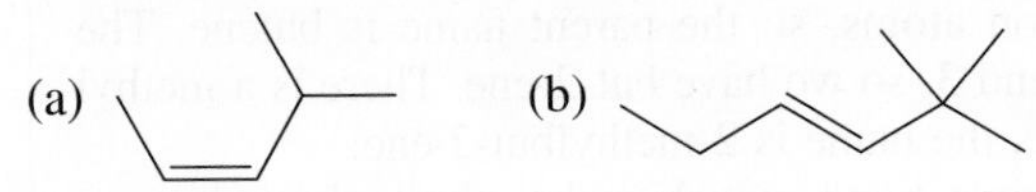

The *E,Z* system

Consider the two possible structures of the molecule 1-bromo-2-chloro-1-fluoroethene.

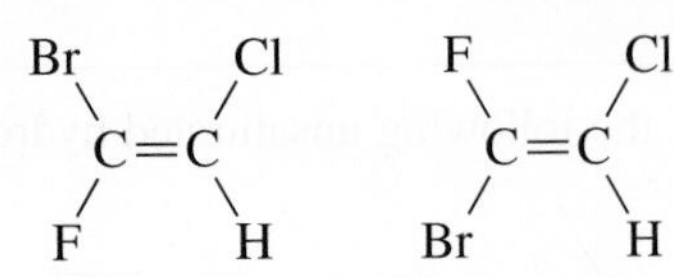

The *cis–trans* convention fails when trying to give a definitive name to each of these isomers. The ***E,Z* system** must be used for tri- and tetrasubstituted alkenes. This system uses a set of rules to assign priorities to the substituents on each carbon atom of a double bond. If the groups of higher priority are on the same side of the double bond, the configuration of the alkene is *Z* (from the German word *zusammen* meaning 'together'). If the groups of higher priority are on opposite sides of the double bond, the configuration is *E* (from the German word *entgegen* meaning 'opposite').

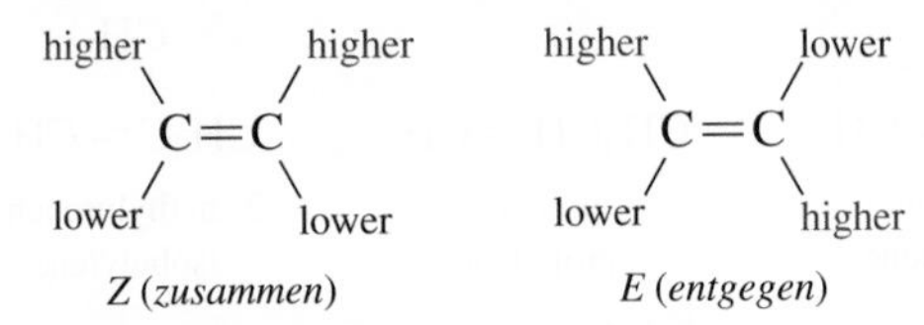

The most important step in determining an *E* or a *Z* configuration is to correctly assign a priority to each of the groups bonded to each carbon atom. The priority rules are shown in figure 16.23. (The priority rules are often called the Cahn–Ingold–Prelog (CIP) rules after the chemists who developed the system.)

1. Priority is based on atomic number. The higher the atomic number, the higher the priority. The following are several substituents arranged in order of increasing priority. (The atomic number of the atom determining priority is shown in parentheses.)

$$\overset{(1)}{-H},\ \overset{(6)}{-CH_3},\ \overset{(7)}{-NH_2},\ \overset{(8)}{-OH},\ \overset{(16)}{-SH},\ \overset{(17)}{-Cl},\ \overset{(35)}{-Br},\ \overset{(53)}{-I}$$

Increasing priority →

2. If priority cannot be assigned on the basis of the atoms bonded directly to the double bond, look at the next set of atoms, and continue until a priority can be assigned. Priority is assigned at the first point of difference. If, for example, you need to assign the priority of a methyl group compared to an ethyl group, the first atom bound to the alkene in both cases is a carbon atom. As these are equivalent, you then assess the atoms that are bound to the carbon in each case. For the methyl group, only hydrogen atoms are bound to the carbon atom. However, with an ethyl group there is another carbon atom attached and this therefore gives the ethyl group higher priority than the methyl group. The following is a series of groups, arranged in order of increasing priority. (Again, numbers in parentheses give the atomic number of the atom on which the assignment of priority is based.)

$$-CH_2-\overset{(1)}{H},\ -CH_2-\overset{(6)}{CH_3},\ -CH_2-\overset{(7)}{NH_2},\ -CH_2-\overset{(8)}{OH},\ -CH_2-\overset{(17)}{Cl}$$

Increasing priority →

3. We treat atoms participating in a double or triple bond as if they were bonded to an equivalent number of similar atoms by single bonds; that is, atoms of a double bond are replicated. Accordingly:

$-CH{=}CH_2$ is treated as $-\underset{}{\overset{C}{\overset{|}{C}}}H-\overset{C}{\overset{|}{C}}H_2$ and $-\overset{O}{\overset{\|}{C}}H$ is treated as $-\underset{\underset{H}{|}}{\overset{O}{\overset{|}{C}}}-O$

FIGURE 16.23 Priority rules for the *E,Z* system.

WORKED EXAMPLE 16.9

Assigning priorities in the *E*,*Z* system

Assign priorities to the groups in each of the following sets.

(a) $-\overset{\overset{\displaystyle O}{\|}}{C}OH$ and $-\overset{\overset{\displaystyle O}{\|}}{C}H$ (b) $-CH_2-NH_2$ and $-CH_2-OH$

Analysis and solution

(a) The first point of difference is the O of the —OH in the carboxyl group, compared with the —H in the aldehyde group. The carboxyl group has higher priority.

$-\overset{\overset{\displaystyle O}{\|}}{C}-O-H$ carboxyl group (higher priority) $-\overset{\overset{\displaystyle O}{\|}}{C}-H$ aldehyde group (lower priority)

(b) Oxygen has higher priority (higher atomic number) than nitrogen. Therefore, the carboxyl group has higher priority than the primary amino group.

$-CH_2-NH_2$ lower priority $-CH_2-OH$ higher priority

WORKED EXAMPLE 16.10

Specifying configurations of alkenes using the *E*,*Z* system

Name each of the following alkenes and specify its configuration using the *E*,*Z* system.

(a) 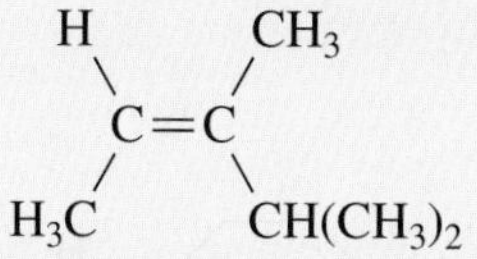 H and CH_3 above, H_3C and $CH(CH_3)_2$ below, C=C (b) 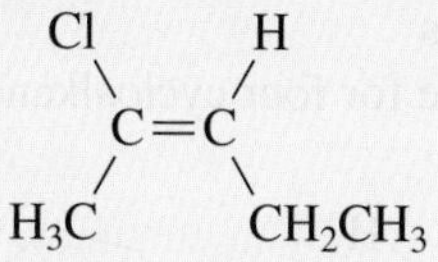 Cl and H above, H_3C and CH_2CH_3 below, C=C

Analysis and solution

(a) The group of higher priority on C(2) is methyl, $—CH_3$; that of higher priority on C(3) is isopropyl, $—CH(CH_3)_2$. Because the groups of higher priority are on the same side of the carbon–carbon double bond, the alkene has the *Z* configuration. Its name is (*Z*)-3,4-dimethylpent-2-ene.

(b) The groups of higher priority on C(2) and C(3) are —Cl and $—CH_2CH_3$. Because these groups are on opposite sides of the double bond, the configuration of this alkene is *E*, and its name is (*E*)-2-chloropent-2-ene.

PRACTICE EXERCISE 16.8

Name each of the following alkenes and specify its configuration using the *E*,*Z* system.

(a) 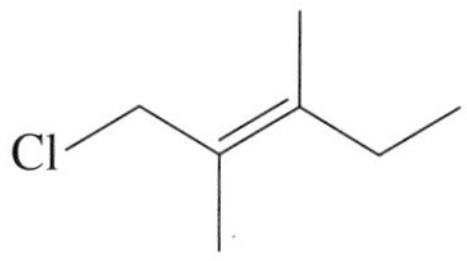 Cl (b) 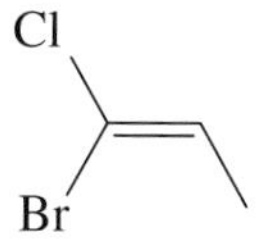 Cl Br (c)

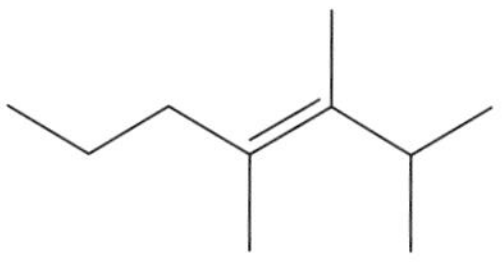

Cycloalkenes

Double bonds can be present in cyclic hydrocarbons and these molecules are called **cycloalkenes**. The rules for naming cycloalkenes follow the same logic we have seen previously. In naming cycloalkenes, we number the carbon atoms of the ring double bond 1 and 2 in the direction that gives the substituents the smallest possible numbers. We name and locate substituents and list them in alphabetical order, as in the following compounds.

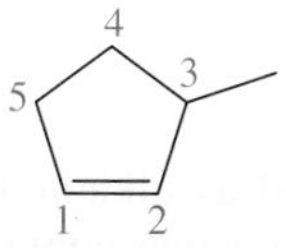

4 3 5 1 2

3-methylcyclopentene
(not 5-methylcyclopentene)

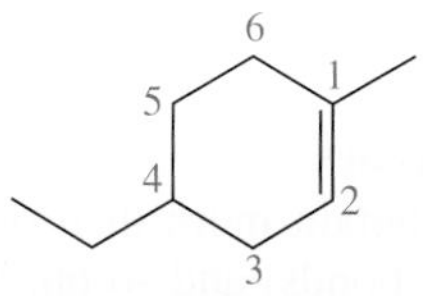

6 1 5 4 2 3

4-ethyl-1-methylcyclohexene
(not 5-ethyl-2-methylcyclohexene)

WORKED EXAMPLE 16.11

Naming cycloalkenes

Write the IUPAC names for the following cycloalkenes.

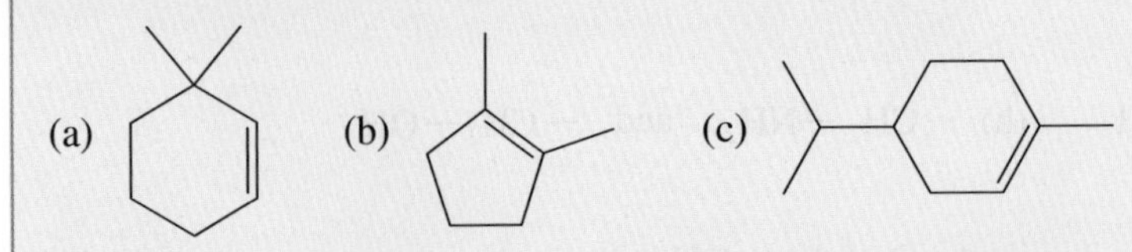

Analysis

We number the carbon atoms of the double bond 1 and 2 in the direction that gives the substituents the smallest possible numbers. We name and locate substituents and list them in alphabetical order.

Solution

(a) 3,3-dimethylcyclohexene
(b) 1,2-dimethylcyclopentene
(c) 4-isopropyl-1-methylcyclohexene

PRACTICE EXERCISE 16.9

Write the IUPAC names for the following cycloalkenes.

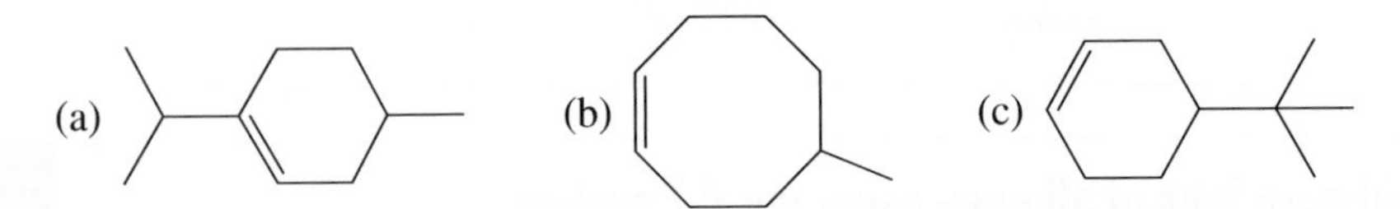

cis–trans isomerism in cycloalkenes

The following are structural formulae for four cycloalkenes.

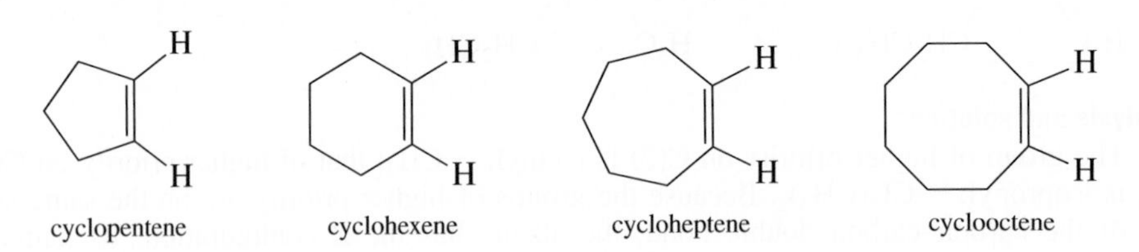

In these representations, the configuration around each double bond is *cis*. Because of angle strain, it is not possible to have a *trans* configuration in cycloalkenes of seven or fewer carbon atoms. To date, *trans*-cyclooctene (shown below) is the smallest *trans*-cycloalkene that has been prepared in pure form and that is stable at room temperature. Yet, even in this *trans*-cycloalkene, there is considerable intramolecular strain. The isomer *cis*-cyclooctene is more stable than its *trans* isomer by 38 kJ mol^{-1}.

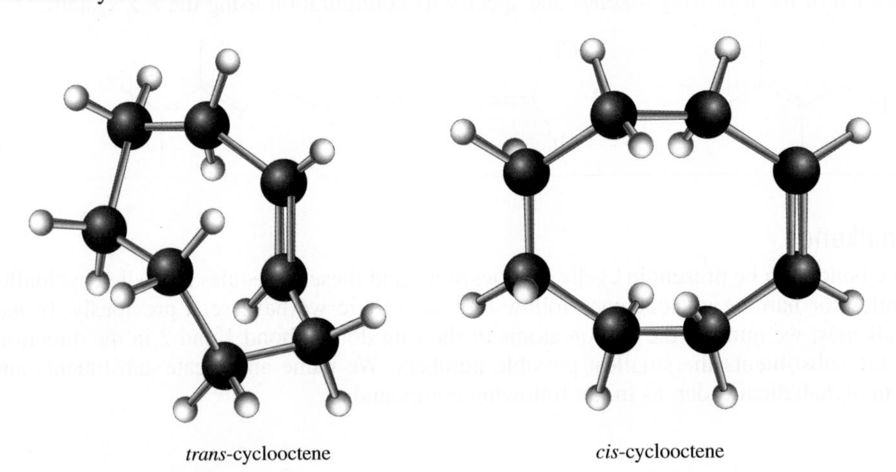

trans-cyclooctene *cis*-cyclooctene

Dienes, trienes and polyenes

We name alkenes that contain more than one double bond as alkadienes (two double bonds), alkatrienes (three double bonds) and so on. We refer to those that contain several double bonds more generally as polyenes (from the Greek word *poly* meaning 'many'). Three examples of dienes are shown on the next page.

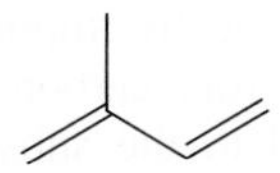

$CH_2{=}CHCH_2CH{=}CH_2$
penta-1,4-diene

$CH_2{=}C(CH_3){-}CH{=}CH_2$
2-methylbuta-1,3-diene
(isoprene)

cyclopenta-1,3-diene

Thus far, we have considered *cis–trans* isomerism in alkenes containing only one carbon–carbon double bond. For an alkene with one carbon–carbon double bond that can show *cis–trans* isomerism, there are two isomers: one *cis* and one *trans*. For an alkene with *n* carbon–carbon double bonds, each of which can show *cis–trans* isomerism, a maximum of 2^n *cis–trans* isomers are possible.

WORKED EXAMPLE 16.12

cis–trans isomerism in polyenes

(a) How many *cis–trans* isomers are possible for hepta-2,4-diene?
(b) Name these isomers using the *E*,*Z* nomenclature.

Analysis and solution

(a) This molecule has two carbon–carbon double bonds, each of which exhibits *cis–trans* isomerism. As the following table shows, $2^2 = 4$ *cis–trans* isomers are possible (to the right of the table are line structures for two of these isomers).

Double bond	
$C_2{=}C_3$	$C_4{=}C_5$
trans	*trans*
trans	*cis*
cis	*trans*
cis	*cis*

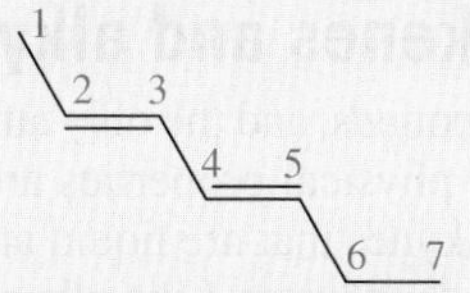

trans,trans-hepta-2,4-diene

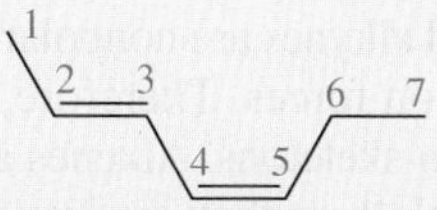

trans,cis-hepta-2,4-diene

(b) Because all of these isomers are disubstituted alkenes it is acceptable to use the terms *cis* and *trans*. They can also be defined using the more encompassing IUPAC terms *Z* and *E*. Therefore *cis* can be replaced with *Z* and *trans* can be replaced with *E*. So, for instance, the names of the two isomers shown become (*E*,*E*)-hepta-2,4-diene and (*E*,*Z*)-hepta-2,4-diene, respectively.

PRACTICE EXERCISE 16.10

Draw structural formulae for the two *cis–trans* isomers of hepta-2,4-diene not shown in worked example 16.12.

WORKED EXAMPLE 16.13

Drawing *cis–trans* isomers of polyenes

Draw all possible *cis–trans* isomers for the following unsaturated alcohol.

$CH_3C(CH_3){=}CHCH_2CH_2C(CH_3){=}CHCH_2OH$

Analysis and solution

It is possible to have *cis–trans* isomerism only around the double bond between C(2) and C(3) of the chain. It is not possible for the other double bond, because C(7) has two identical groups on it. Thus, $2^1 = 2$ *cis–trans* isomers are possible. The *trans* isomer of this alcohol, named geraniol, is a major component of the oils of rose, citronella and lemongrass.

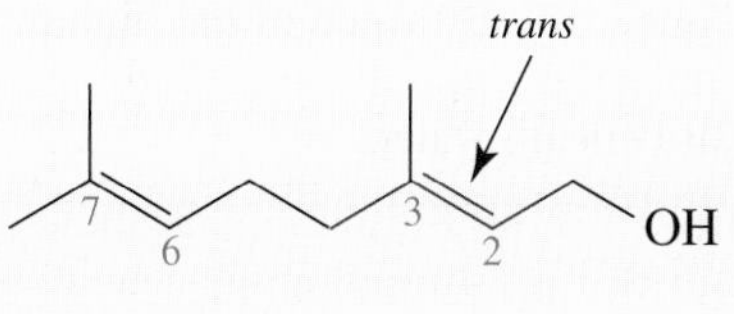

trans — OH — the *trans* isomer

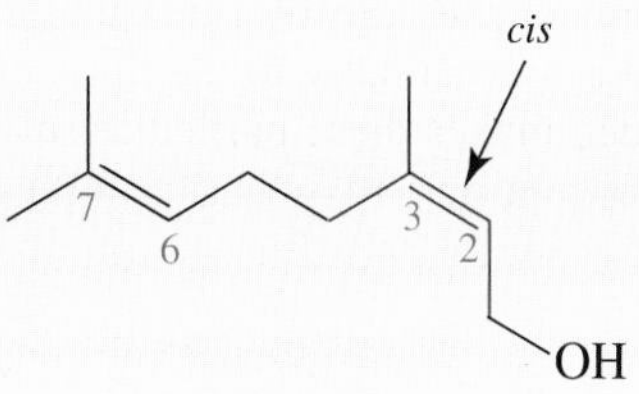

cis — OH — the *cis* isomer

PRACTICE EXERCISE 16.11

How many *cis–trans* isomers are possible for the following unsaturated alcohol?

$CH_3C(CH_3){=}CHCH_2CH_2C(CH_3){=}CHCH_2CH_2C(CH_3){=}CHCH_2OH$

Vitamin A is an example of a biologically important compound for which a number of *cis–trans* isomers are possible. There are four carbon–carbon double bonds (highlighted in blue) in the chain of carbon atoms bonded to the substituted cyclohexene ring, and each has the potential for *cis–trans* isomerism. Thus, $2^4 = 16$ *cis–trans* isomers are possible for a molecule with this structural formula. Vitamin A is the all-*trans* isomer. The enzyme-catalysed oxidation of vitamin A converts the primary hydroxyl group (highlighted in red) to an aldehyde group (red) to give retinal, the biologically active form of the vitamin.

OH
enzyme-catalysed oxidation →

vitamin A (retinol)

H
O

vitamin A aldehyde (retinal)

Physical properties of alkenes and alkynes

Alkenes and alkynes are nonpolar compounds, and the only attractive forces between their molecules are dispersion forces. Therefore, their physical properties are similar to those of alkanes with the same carbon skeletons. Alkenes and alkynes that are liquid at room temperature have densities less than $1.0\,g\,mL^{-1}$, so they are less dense than water. Like alkanes, alkenes and alkynes are soluble in each other. Because of their contrasting polarity with water, they do not dissolve. Instead, they form a separate layer when mixed with water or another polar organic liquid such as methanol.

WORKED EXAMPLE 16.14

Physical properties of alkenes and alkynes

Non-1-ene is a liquid at room temperature. Describe what happens when non-1-ene is mixed with water. What happens when non-1-ene is mixed with 8-methylnon-1-yne?

Analysis

Alkenes and alkynes are soluble in each other, but not soluble in polar organic liquids. We need to determine whether each substance is polar or nonpolar.

Solution

Non-1-ene is an alkene. Since all alkenes are nonpolar, non-1-ene is nonpolar. Water, H_2O, is polar. Since a nonpolar substance will not dissolve in a polar substance, we know that water and non-1-ene will form two layers. Water has a higher density than non-1-ene and will therefore be the bottom layer, and non-1-ene will be the upper layer.

8-methylnon-1-yne is an alkyne. Since all alkynes are nonpolar, 8-methylnon-1-yne is nonpolar. Since non-1-ene and 8-methylnon-1-yne are both nonpolar, they will — like all alkenes and alkynes — dissolve in one another to form a homogeneous solution.

Is our answer reasonable?

There is little to check with this example. The key is simply to remember that alkenes and alkynes are soluble in each other, but insoluble in polar organic liquids. Remember that almost all nonpolar organic liquids have a lower density than water. Think of oil floating on water. The few exceptions to this 'rule' are some of the halogenated solvents such as dichloromethane and chloroform (trichloromethane). The presence of the high atomic mass halogens adds substantially to the mass of the molecule without significantly increasing the density, which makes these particular solvents sit beneath water.

16.4 Reactions of alkanes

For completeness, we should start with the reactions of alkanes, yet alkanes actually have very little reactivity. It is for this reason that chemists describe alkanes as unfunctionalised. They are quite unreactive with most reagents, a behaviour consistent with the fact that they are nonpolar compounds containing only strong σ bonds.

Under certain conditions, however, alkanes and cycloalkanes react with oxygen, O_2. By far their most important reaction with oxygen is oxidation (combustion) to form carbon dioxide and water. The oxidation of saturated hydrocarbons is the basis for their use as energy sources for heat (natural gas, liquefied petroleum gas (LPG) and fuel oil) and power (petrol, diesel fuel and aviation fuel). The following are balanced equations for the complete combustion of methane, the major component of natural gas, and for propane, the major component of LPG.

$$\underset{\text{methane}}{CH_4} + 2O_2 \rightarrow CO_2 + 2H_2O \qquad \Delta_c H^\ominus = -886\,\text{kJ mol}^{-1}$$

$$\underset{\text{propane}}{CH_3CH_2CH_3} + 5O_2 \rightarrow 3CO_2 + 4H_2O \qquad \Delta_c H^\ominus = -2220\,\text{kJ mol}^{-1}$$

Another reaction of alkanes involves a substitution reaction where halogens replace hydrogen atoms, and this will be discussed in chapter 18.

16.5 Reactions of alkenes

The most characteristic reaction of alkenes is **addition to the carbon–carbon double bond** so that the π bond is broken and, in its place, σ bonds are formed to two new atoms or groups of atoms. Several examples of reactions at the carbon–carbon double bond are shown in table 16.4, along with the descriptive name(s) associated with each.

TABLE 16.4 Characteristic addition reactions of alkenes.

Reaction	Descriptive name(s)
$>C{=}C< + HCl \longrightarrow -\underset{H}{C}-\underset{Cl}{C}-$	hydrochlorination (an example of hydrohalogenation)
$>C{=}C< + H_2O \longrightarrow -\underset{H}{C}-\underset{OH}{C}-$	hydration
$>C{=}C< + Br_2 \longrightarrow -\underset{Br}{C}-\underset{Br}{C}-$	bromination (an example of halogenation)
$>C{=}C< + H_2 \longrightarrow -\underset{H}{C}-\underset{H}{C}-$	hydrogenation (reduction)

From the perspective of the chemical industry, the most important reaction of low-molar-mass alkenes is the production of polymers (e.g. polyethylene and polystyrene). Polymers arise from the sequential addition of many low-molar-mass molecules to create very large molecules of high molar mass, as illustrated here by the formation of polyethylene from ethene (ethylene).

$$nCH_2{=}CH_2 \xrightarrow{\text{initiator}} \left(CH_2CH_2\right)_n$$

To achieve such an addition, alkenes first react with a specific reagent called an initiator and then with each other to form a steadily growing chain. In alkene-derived polymers of industrial and commercial importance, n is a large number, typically several thousand. We discuss the formation of polymers from alkenes in more detail in e-chapter 26.

Another reaction of alkenes is reduction to alkanes, which is essentially the addition of H_2 across the double bond. We will discuss this after we look at the addition reactions.

Electrophilic addition reactions

The basis of reactivity is the attraction between positive and negative species. The double bond in alkenes is an electron-rich (i.e. negative) target for positive species. These positive species are called **electrophiles** (which literally means 'attracted to electrons'). Alkenes undergo addition reactions with electrophiles to produce saturated compounds.

In this section we examine the three most important types of electrophilic addition reactions: the addition of hydrogen halides (HCl, HBr, and HI), water (H_2O) and halogens (Br_2, Cl_2). We first study some of the experimental observations about each addition reaction and then its mechanism. By examining these particular reactions, we develop a general understanding of how alkenes undergo addition reactions.

Addition of hydrogen halides

The hydrogen halides HCl, HBr and HI (commonly abbreviated HX) add to alkenes to give haloalkanes (alkyl halides). These additions may be carried out either with the pure reagents or in the presence of a polar solvent such as acetic acid. The addition of HCl to ethene gives chloroethane.

$$\underset{\text{ethene}}{CH_2{=}CH_2} + HCl \longrightarrow \underset{\text{chloroethane}}{\overset{H\quad\ Cl}{CH_2{-}CH_2}}$$

The addition of HCl to propene gives 2-chloropropane; hydrogen adds to C(1) of propene, and chlorine adds to C(2). If the orientation of addition were reversed, 1-chloropropane would be formed.

The observed result is that 2-chloropropane is formed to the virtual exclusion of 1-chloropropane.

$$\underset{\text{propene}}{CH_3CH{=}CH_2} + HCl \longrightarrow \underset{\text{2-chloropropane}}{\overset{Cl\qquad H}{CH_3CH{-}CH_2}} + \underset{\text{1-chloropropane (trace amounts)}}{\overset{H\qquad Cl}{CH_3CH{-}CH_2}}$$

We say that the addition of HCl to propene is highly regioselective and that 2-chloropropane is the major (indeed almost exclusive) product of the reaction. A **regioselective reaction** is one in which one direction of bond forming or breaking occurs in preference to all other directions.

Nineteenth century Russian chemist Vladimir Markovnikov observed this regioselectivity and made the generalisation, known as **Markovnikov's rule**, that, in the addition of HX to an alkene, hydrogen adds to the double-bonded carbon atom with the greater number of hydrogen atoms already bonded to it.

WORKED EXAMPLE 16.15

Alkene addition reactions

Name and draw a structural formula for the major product of each of the following alkene addition reactions.

(a) $CH_3\overset{CH_3}{C}{=}CH_2 + HI \longrightarrow$ (b) 1-methylcyclopentene + HCl $\longrightarrow$

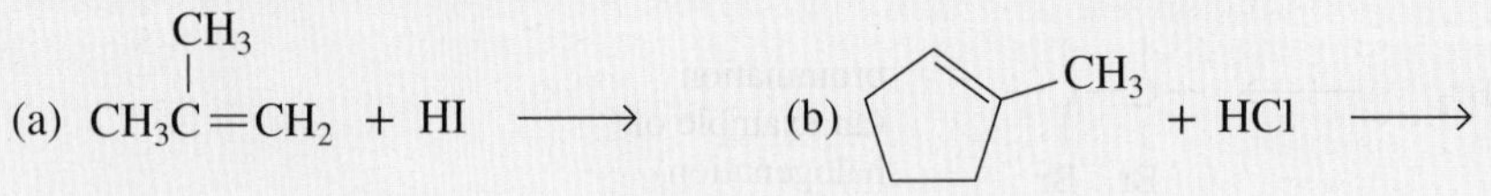

Analysis and solution

Using Markovnikov's rule, we predict that 2-iodo-2-methylpropane is the product in (a) and 1-chloro-1-methylcyclopentane is the product in (b).

(a) $CH_3\overset{CH_3}{\underset{I}{C}}CH_3$ 2-iodo-2-methylpropane

(b) cyclopentane ring bearing Cl and CH_3 on the same carbon: 1-chloro-1-methylcyclopentane

PRACTICE EXERCISE 16.12

Name and draw the structural formula of the major product of each of the following alkene addition reactions.

(a) $CH_3CH{=}CH_2 + HI \longrightarrow$ (b) cyclohexane ring$=CH_2 + HI \longrightarrow$

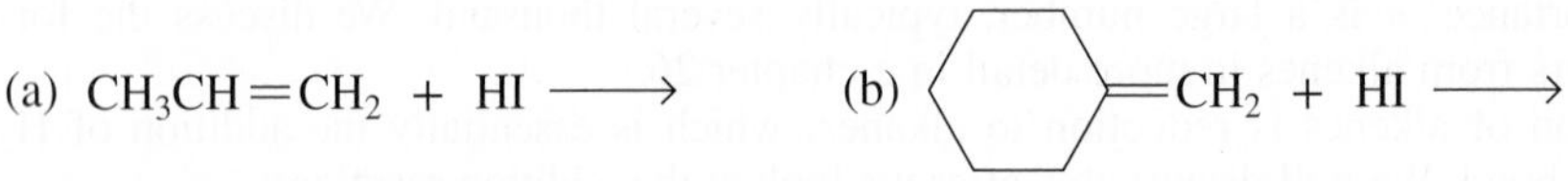

Although Markovnikov's rule provides a way to predict the product of many alkene addition reactions, it does not explain why one product predominates over other possible products. To do this, we need to understand the mechanism of the process. As we saw in chapter 15, mechanisms provide a complete description of the bond-breaking and bond-forming processes that occur in the transformation of starting materials to intermediate species and, then, to products. We represent the movement of the two electrons involved in these bond-breaking and bond-forming steps with a curved arrow. For example, we can represent the breaking of the single bond in the hypothetical *AB* molecule by using curved arrows.

$$A{-}B \longrightarrow A^+ + B^-$$

Note that the arrow begins at the bond that is being broken and the head of the arrow shows the destination of the pair of electrons. We use arrows to show the movement of electrons in each bond-forming and bond-breaking step in a mechanism. Once we have a complete mechanistic description for a particular reaction, it is possible to make generalisations and then predict how other similar reactions might occur.

Chemists account for the addition of HX to an alkene by a two-step mechanism, which we illustrate by the reaction of but-2-ene with hydrogen chloride to give 2-chlorobutane. Let us first look at this two-step mechanism in general and then go back and study each step in detail.

Step 1: The reaction begins with the transfer of a proton from HCl to but-2-ene, as shown by the two curved arrows on the left side of the following equation.

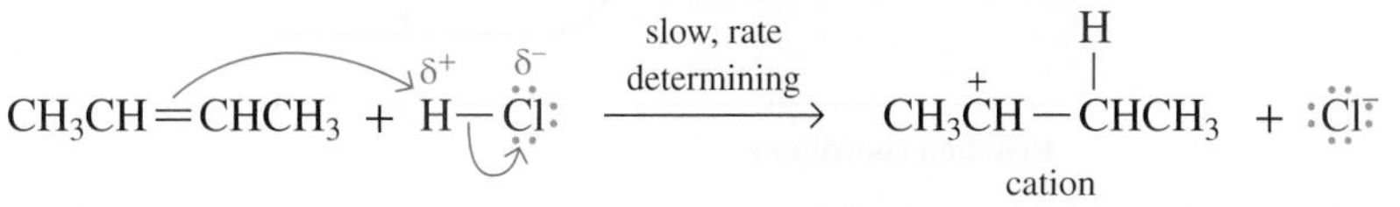

$$CH_3CH{=}CHCH_3 + \overset{\delta+}{H}{-}\overset{\delta-}{\ddot{Cl}}: \xrightarrow[\text{determining}]{\text{slow, rate}} \underset{\text{cation}}{CH_3\overset{+}{C}H{-}\overset{H}{C}HCH_3} + :\ddot{Cl}:^-$$

The first curved arrow shows the breaking of the π bond of the alkene and its electron pair now forming a new covalent bond with the hydrogen atom of HCl. The second curved arrow shows the breaking of the polar covalent bond in HCl and this electron pair being given entirely to chlorine, forming a chloride ion. Step 1 in this mechanism results in the formation of an organic cation and a chloride ion. It is important to remember that, even though we described this process as a proton transfer, a mechanistic arrow always shows the movement of electrons, not protons. A mechanistic arrow should *never* start from a hydrogen atom in an organic molecule.

Step 2: The reaction of the cation (a Lewis acid) with a chloride ion (a Lewis base) completes the valence shell of carbon and gives 2-chlorobutane.

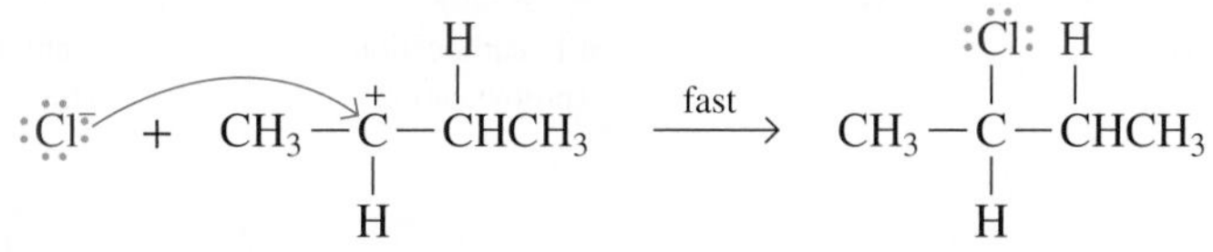

$$\underset{\substack{\text{chloride ion}\\\text{(a Lewis base)}}}{:\ddot{Cl}:^-} + \underset{\substack{\text{cation}\\\text{(a Lewis acid)}}}{CH_3{-}\overset{+}{C}H{-}CH_2CH_3} \xrightarrow{\text{fast}} \underset{\text{2-chlorobutane}}{CH_3{-}CHCl{-}CH_2CH_3}$$

Now let us look at the individual steps in more detail. There is a great deal of important organic chemistry embedded in these two steps, and it is crucial that you understand it now.

Step 1 results in the formation of an organic cation. One carbon atom in this cation has only six electrons in its valence shell and carries a charge of +1. A species containing a positively charged carbon atom is called a **carbocation** (*carbon* + *cation*). Carbocations are classified as primary (1°), secondary (2°) or tertiary (3°), depending on the number of carbon atoms bonded to the carbon atom bearing the positive charge. The cation in the reaction given above is a 2° carbocation. All carbocations are Lewis acids (chapter 11). They are also electrophiles.

In a carbocation, the carbon atom bearing the positive charge is bonded to three other atoms, and, as predicted by the valence-shell-electron-pair repulsion (VSEPR) model, the three bonds around that carbon atom are coplanar and form bond angles of approximately 120°. According to the valence bond theory, the electron-deficient carbon atom of a carbocation uses its sp^2 hybrid orbitals to form σ bonds to the three attached groups. The unhybridised $2p$ orbital lies perpendicular to the σ bond framework and contains no electrons. A Lewis structure and an orbital overlap diagram for a common tertiary carbocation ($C_4H_9^+$) are shown in figure 16.24. This carbocation is given the common name *tertiary* butyl cation or *tert*-butyl cation.

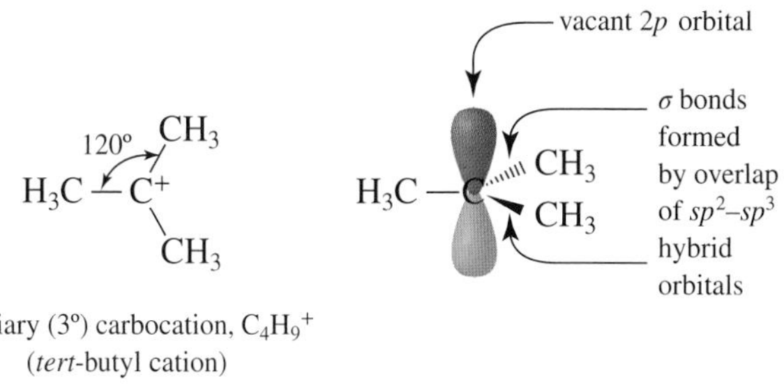

FIGURE 16.24 Representations of the structure of a tertiary (3°) carbocation: **(a)** Lewis structure and **(b)** an orbital picture.

Recall from chapter 15 that the process of a reaction can also be depicted by the energy changes from starting material to intermediate species through to products. Figure 16.25 (overleaf) shows such an energy diagram for the two-step reaction of but-2-ene with HCl. The slower, rate-determining step (that has to pass over the higher energy barrier) is step 1 (i.e. E_{a1} is greater than E_{a2}), which leads to the formation of the 2° carbocation intermediate. This intermediate lies in an energy minimum between the transition states for steps 1 and 2. As soon as the carbocation intermediate (a Lewis acid) forms, it reacts with a chloride ion (a Lewis base) in a Lewis acid–base reaction to give 2-chlorobutane. Note that the energy of 2-chlorobutane (the product) is less than the energy of but-2-ene and HCl (the reactants). Thus, in this alkene addition reaction, heat is released so the reaction is exothermic.

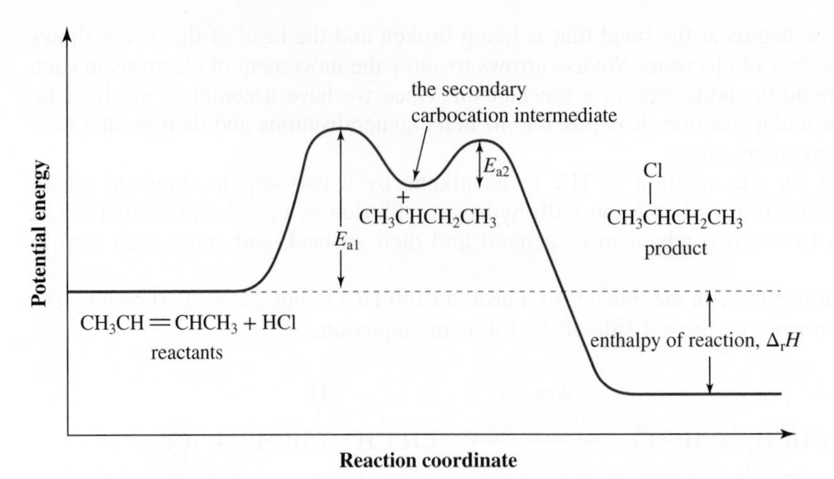

FIGURE 16.25 Energy diagram for the two-step addition of HCl to but-2-ene. The reaction is exothermic.

Given that the rate-determining step of the reaction in figure 16.25 involves both but-2-ene and HCl as reactants, we would expect the rate law for this reaction to have the form:

$$\text{rate of reaction} = k[\text{but-2-ene}][\text{HCl}]$$

and the reaction should display second-order kinetics (see chapter 15).

Relative stabilities of carbocations: regioselectivity and Markovnikov's rule

The reaction of HX with an asymmetrical alkene can, at least in principle, give two different carbocation intermediates, depending on which of the double-bonded carbon atoms has H^+ added to it, as illustrated by the reaction of HCl with propene.

$$\underset{\text{propene}}{CH_3CH{=}CH_2} + H{-}\ddot{\underset{..}{Cl}}{:} \longrightarrow \underset{\substack{\text{a 1° carbocation}\\ \text{(propyl cation)}}}{CH_3CH_2\overset{+}{C}H_2} \xrightarrow{:\ddot{Cl}:^-} \underset{\substack{\text{1-chloropropane}\\ \text{(trace amounts)}}}{CH_3CH_2CH_2Cl}$$

$$\underset{\text{propene}}{CH_3CH{=}CH_2} + H{-}\ddot{\underset{..}{Cl}}{:} \longrightarrow \underset{\substack{\text{a 2° carbocation}\\ \text{(isopropyl cation)}}}{CH_3\overset{+}{C}HCH_3} \xrightarrow{:\ddot{Cl}:^-} \underset{\substack{\text{2-chloropropane}\\ \text{(major product formed)}}}{CH_3CH(Cl)CH_3}$$

The major product formed is 2-chloropropane, and 1-chloropropane is formed in only trace amounts. Because carbocations react very quickly with chloride ions, the virtual absence of 1-chloropropane as a product tells us that the 2° carbocation is formed in preference to the 1° carbocation.

Similarly, in the reaction of HCl with 2-methylpropene, the transfer of a proton to the carbon–carbon double bond might form either a 1° carbocation (isobutyl cation) or a 3° carbocation (*tert*-butyl cation).

$$\underset{\text{2-methylpropene}}{(CH_3)_2C{=}CH_2} + H{-}\ddot{\underset{..}{Cl}}{:} \longrightarrow \underset{\substack{\text{a 1° carbocation}\\ \text{(isobutyl cation)}}}{(CH_3)_2CH{-}\overset{+}{C}H_2} \xrightarrow{:\ddot{Cl}:^-} \underset{\substack{\text{1-chloro-}\\ \text{2-methylpropane}\\ \text{(trace amounts)}}}{(CH_3)_2CH{-}CH_2Cl}$$

$$\underset{\text{2-methylpropene}}{(CH_3)_2C{=}CH_2} + H{-}\ddot{\underset{..}{Cl}}{:} \longrightarrow \underset{\substack{\text{a 3° carbocation}\\ \text{(}\textit{tert}\text{-butyl cation)}}}{CH_3\overset{+}{C}(CH_3){-}CH_2H} \xrightarrow{:\ddot{Cl}:^-} \underset{\substack{\text{2-chloro-}\\ \text{2-methylpropane}\\ \text{(major product formed)}}}{CH_3C(CH_3)(Cl){-}CH_2H}$$

In this reaction, the observed product is 2-chloro-2-methylpropane, indicating that the 3° carbocation forms in preference to the 1° carbocation.

From such experiments and a great amount of other experimental evidence, we learn that a 3° carbocation is more stable and has a lower activation energy for its formation than a 2° carbocation. A 2° carbocation, in turn, is more stable and has a lower activation energy for its formation than a 1° carbocation. It follows that a more stable carbocation intermediate forms more rapidly than a less stable carbocation intermediate. Figure 16.26 shows the order of stability of four types of alkyl carbocations.

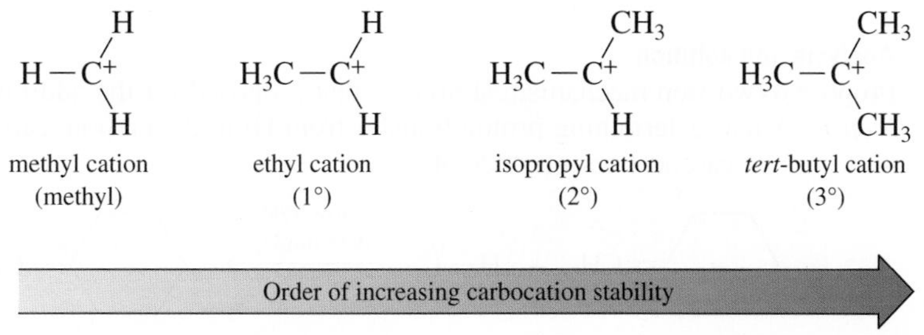

FIGURE 16.26 The order of stability of four types of alkyl carbocations.

Although the concept of the relative stabilities of carbocations had not been developed in Markovnikov's time, it is the underlying basis for his rule; that is, the proton of H—X adds to the less substituted carbon atom of a double bond, because this mode of addition produces the more stable carbocation intermediate.

Now that we know the order of stability of carbocations, how do we account for it? The principles of physics teach us that a system bearing a charge (either positive or negative) is more stable if the charge is distributed over the system rather than localised at a particular point within the system. If we assume that alkyl groups bonded to a positively charged carbon atom release electrons towards the cationic carbon atom, thereby distributing the charge of the cation over the structure, this explains the order of stability of carbocations. The electron-releasing ability of alkyl groups bonded to a cationic carbon atom is accounted for by the **inductive effect**.

The inductive effect operates in the following way: the electron deficiency of the carbon atom bearing a positive charge exerts an electron-withdrawing inductive effect that polarises electrons from adjacent σ bonds towards it. Thus, the positive charge of the cation is not localised on the trivalent carbon atom, but rather is partially distributed over nearby atoms. As the number of alkyl groups bonded to the cationic carbon atom increases, the positive charge can be distributed over more atoms and the stability of the cation increases.

The inductive effect is not the only factor that influences the stability of carbocations. In future studies in chemistry, you may learn about the principle of hyperconjugation, which also affects carbocation stability. Hyperconjugation involves that interaction of adjacent *p* orbitals of alkyl groups, helping to stabilise the carbocation. Again, as with the inductive effect, the more alkyl groups bound to the carbocation, the greater is its stability.

WORKED EXAMPLE 16.16

Relative stabilities of carbocations

Arrange the following carbocations in order of increasing stability.

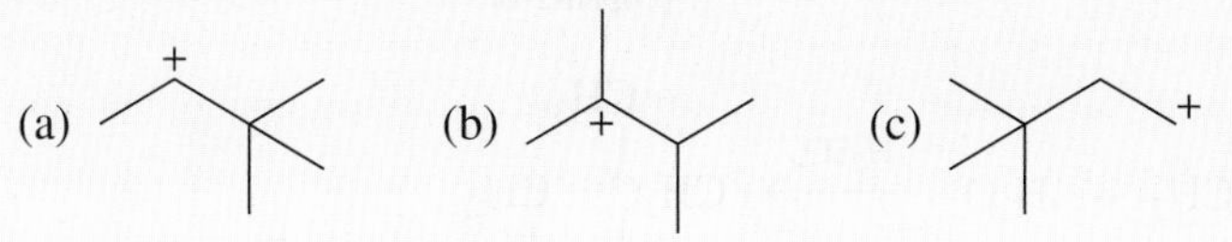

Solution

Carbocation (a) is secondary, (b) is tertiary and (c) is primary. In order of increasing stability, they are (c), (a) and (b).

PRACTICE EXERCISE 16.13

Arrange the following carbocations in order of increasing stability.

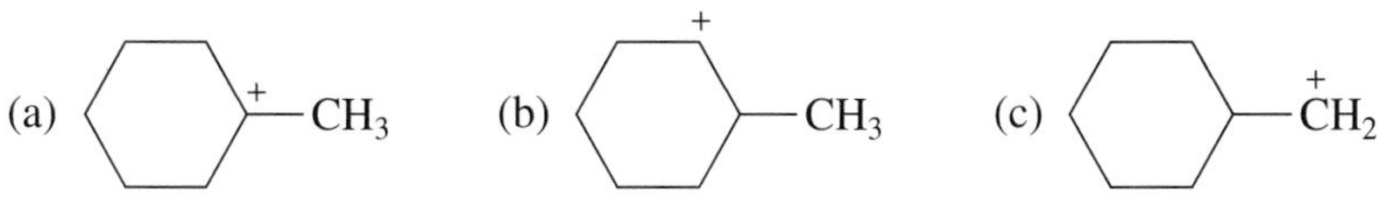

WORKED EXAMPLE 16.17

The mechanism of addition

Propose a mechanism for the addition of HI to methylenecyclohexane, which gives 1-iodo-1-methylcyclohexane.

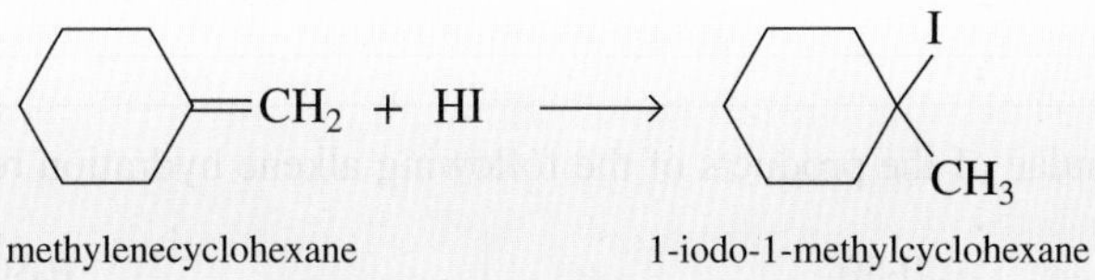

Which step in your mechanism is rate determining?

Analysis and solution

Propose a two-step mechanism similar to that proposed for the addition of HCl to propene.

Step 1: A rate-determining proton transfer from HI to the carbon–carbon double bond gives a 3° carbocation intermediate.

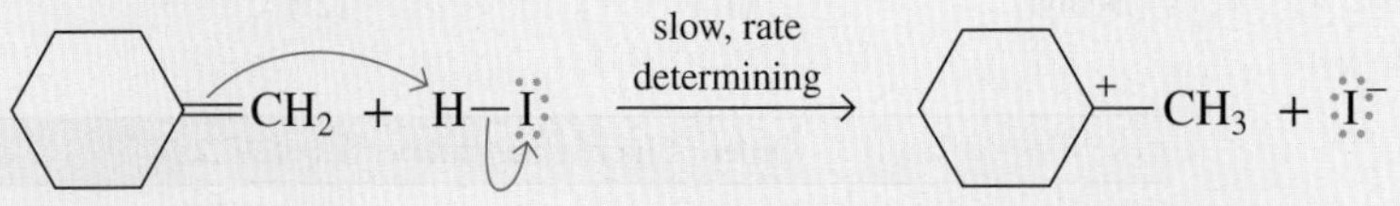

$$\text{methylenecyclohexane} + H{-}\ddot{I}: \xrightarrow{\text{slow, rate determining}} \text{a 3° carbocation intermediate} + :\ddot{I}:^-$$

Step 2: Reaction of the 3° carbocation intermediate (a Lewis acid) with an iodide ion (a Lewis base) completes the valence shell of carbon and gives the product.

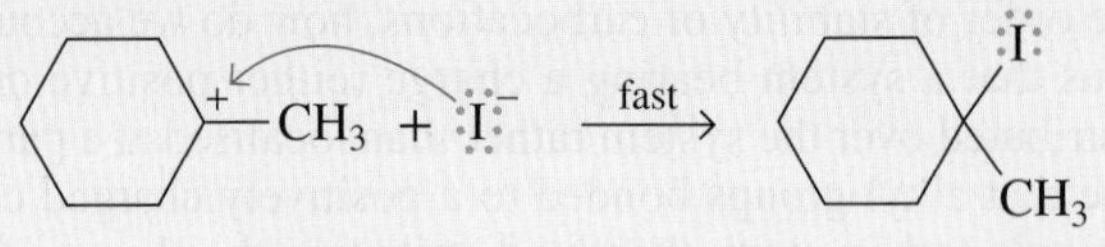

$$\text{carbocation} + :\ddot{I}:^- \xrightarrow{\text{fast}} \text{1-iodo-1-methylcyclohexane}$$

PRACTICE EXERCISE 16.14 Propose a mechanism for the addition of HI to 1-methylcyclohexene, which gives 1-iodo-1-methylcyclohexane. Which step in your mechanism is rate determining?

Addition of water: acid-catalysed hydration

In the presence of an acid catalyst — most commonly, sulfuric acid — water adds to the carbon–carbon double bond of an alkene to give an alcohol. The addition of water is called **hydration**. The precise mechanism of this hydration is given below but, in effect, for simple alkenes, H is added to the carbon atom of the double bond with the greater number of hydrogen atoms, and OH is added to the carbon atom with the lower number of hydrogen atoms. Thus, H—OH adds to alkenes in accordance with Markovnikov's rule.

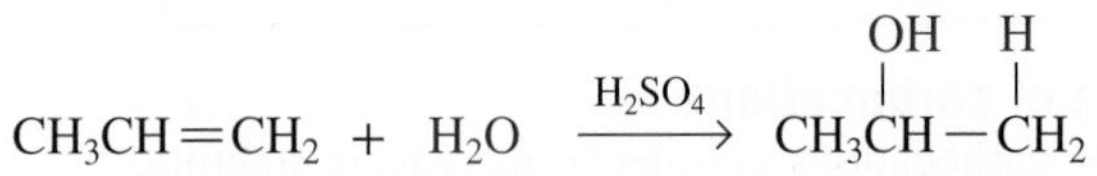

$$CH_3CH{=}CH_2 + H_2O \xrightarrow{H_2SO_4} CH_3CH(OH){-}CH_2(H)$$

propene → propan-2-ol

$$CH_3C(CH_3){=}CH_2 + H_2O \xrightarrow{H_2SO_4} CH_3C(CH_3)(HO){-}CH_2(H)$$

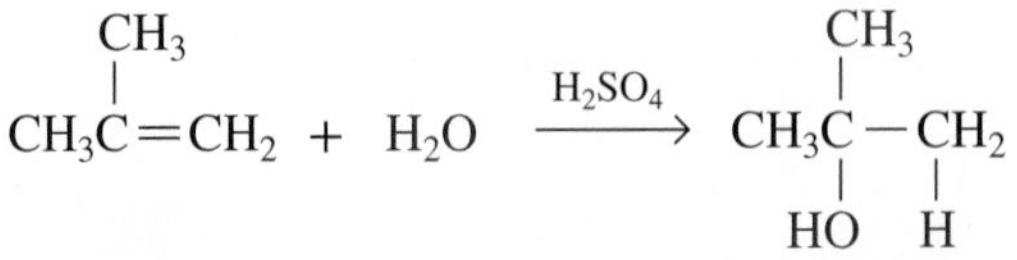

2-methylpropene → 2-methylpropan-2-ol

WORKED EXAMPLE 16.18

Acid-catalysed hydration

Draw the structural formula of the product of the acid-catalysed hydration of 1-methylcyclohexene.

Analysis

We know that a hydrogen atom from water will add to the carbon atom of the double bond bearing more attached hydrogen atoms. This then means that the —OH will add to the carbon atom bearing the methyl group. We can therefore write the structural formula of the product.

Solution

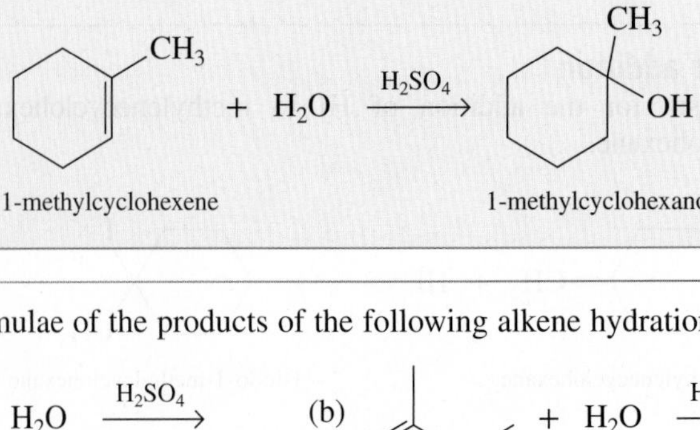

$$\text{1-methylcyclohexene} + H_2O \xrightarrow{H_2SO_4} \text{1-methylcyclohexanol}$$

PRACTICE EXERCISE 16.15 Draw structural formulae of the products of the following alkene hydration reactions.

(a) 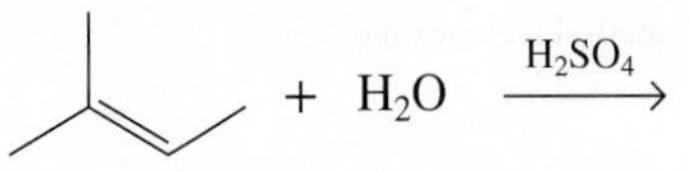 2-methylbut-2-ene $+ H_2O \xrightarrow{H_2SO_4}$ (b) 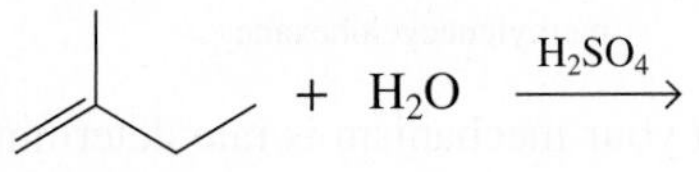 2-methylbut-1-ene $+ H_2O \xrightarrow{H_2SO_4}$

The mechanism of the acid-catalysed hydration of alkenes is similar to the mechanism we have described for the addition of HCl, HBr and HI to alkenes and is illustrated by the hydration of propene to propan-2-ol. This mechanism is consistent with the fact that acid is a catalyst. For every H_3O^+ consumed in step 1, another is generated in step 3.

Step 1: Proton transfer from the acid catalyst to propene gives a 2° carbocation intermediate (a Lewis acid).

$$CH_3CH{=}CH_2 + H{-}\ddot{O}^+{-}H \ (\text{with } H) \xrightarrow[\text{determining}]{\text{slow, rate}} CH_3\overset{+}{C}H{-}CH_3 + H{-}\ddot{O}{-}H$$

a 2° carbocation intermediate

Step 2: Reaction of the carbocation intermediate (a Lewis acid) with water (a Lewis base) completes the valence shell of carbon and gives an **oxonium ion**.

$$CH_3\overset{+}{C}H{-}CH_3 + H{-}\ddot{O}{-}H \xrightarrow{\text{fast}} CH_3CH(\overset{+}{O}H_2){-}CH_3$$

an oxonium ion

Step 3: Proton transfer from the oxonium ion to water gives the alcohol and generates a new molecule of the catalyst.

$$CH_3CH(\overset{+}{O}H_2){-}CH_3 + H{-}\ddot{O}{-}H \xrightarrow{\text{fast}} CH_3CH(\ddot{O}H){-}CH_3 + H{-}\ddot{O}^+{-}H \ (\text{with } H)$$

WORKED EXAMPLE 16.19

The mechanism of acid-catalysed hydration

Propose a mechanism for the acid-catalysed hydration of methylenecyclohexane to give 1-methylcyclohexanol. Which step in your mechanism is rate determining?

Solution

The mechanism involves three steps, similar to those for the acid-catalysed hydration of propene. The formation of the 3° carbocation intermediate in step 1 is rate determining.

Step 1: Proton transfer from the acid catalyst to the alkene gives a 3° carbocation intermediate (a Lewis acid).

$$C_6H_{10}{=}CH_2 + H{-}\ddot{O}^+{-}H \ (\text{with } H) \xrightarrow[\text{determining}]{\text{slow, rate}} C_6H_{10}^{+}{-}CH_3 + H{-}\ddot{O}{-}H$$

a 3° carbocation intermediate

Step 2: Reaction of the carbocation intermediate (a Lewis acid) with water (a Lewis base) completes the valence shell of carbon and gives an oxonium ion.

$$C_6H_{10}^{+}{-}CH_3 + H{-}\ddot{O}{-}H \xrightarrow{\text{fast}} C_6H_{10}(CH_3)(\overset{+}{O}H_2)$$

an oxonium ion

Step 3: Proton transfer from the oxonium ion to water gives the alcohol and generates a new molecule of the catalyst.

$$C_6H_{10}(CH_3)(\overset{+}{O}H_2) + H{-}\ddot{O}{-}H \xrightarrow{\text{fast}} C_6H_{10}(CH_3)(\ddot{O}H) + H{-}\ddot{O}^+{-}H \ (\text{with } H)$$

PRACTICE EXERCISE 16.16 Propose a mechanism for the acid-catalysed hydration of 1-methylcyclohexene to give 1-methylcyclohexanol. Which step in your mechanism is rate determining?

Addition of bromine and chlorine

Chlorine, Cl_2, and bromine, Br_2, react with alkenes at room temperature by the addition of halogen atoms to the two carbon atoms of the double bond, forming two new carbon–halogen bonds.

$$CH_3CH{=}CHCH_3 + Br_2 \xrightarrow[CH_2Cl_2]{} CH_3CH(Br){-}CH(Br)CH_3$$

but-2-ene → 2,3-dibromobutane

Fluorine, F_2, also adds to alkenes, but, because its reactions are very fast and difficult to control, addition of fluorine is not a useful laboratory reaction. Iodine, I_2, also adds to alkenes, but the products of the reaction can in some cases be unstable and subsequently degrade, generating other products.

The addition of bromine and chlorine to a cycloalkene gives a *trans*-dihalocycloalkane. For example, the addition of bromine to cyclohexene gives *trans*-1,2-dibromocyclohexane; the *cis* isomer is not formed. Thus, the addition of a halogen to a cycloalkene is stereospecific. A **stereospecific reaction** is a reaction in which one stereoisomer is formed or destroyed in preference to all others that might be formed or destroyed. Addition of bromine to a cycloalkene is highly stereospecific; the halogen atoms always add *trans* to each other.

cyclohexene + Br_2 $\xrightarrow[CH_2Cl_2]{}$ *trans*-1,2-dibromocyclohexane

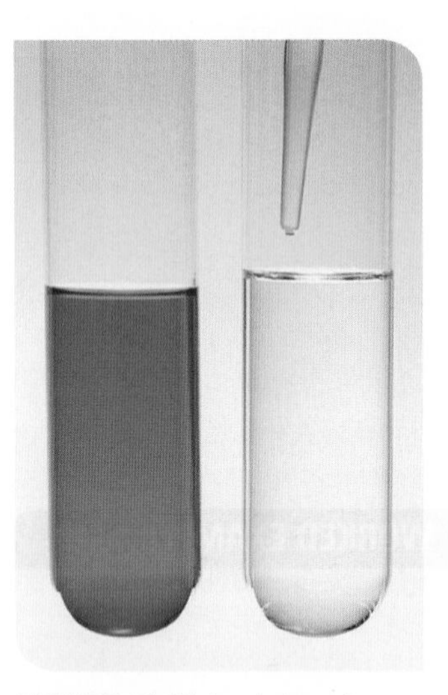

FIGURE 16.27 A solution of bromine in dichloromethane, CH_2Cl_2, is red (left). Add a few drops of an alkene and the red colour disappears (right).

The reaction of bromine with an alkene is a particularly useful qualitative test for the presence of a carbon–carbon double bond. If we dissolve bromine in dichloromethane, CH_2Cl_2, the solution becomes red due to the presence of the red-coloured bromine (figure 16.27). Both alkenes and dibromoalkanes are colourless. If we now mix a few drops of the bromine solution with an alkene, the bromine is consumed, a dibromoalkane is formed and, with the removal of the bromine, the solution becomes colourless.

WORKED EXAMPLE 16.20

Addition of bromine and chlorine

Complete the following reactions, showing the relative orientations of the substituents in the products.

(a) cyclopentene + Br_2 $\xrightarrow[CH_2Cl_2]{}$

(b) 1-methylcyclohexene (CH_3) + Cl_2 $\xrightarrow[CH_2Cl_2]{}$

Solution

The halogen atoms are *trans* to each other in each product.

(a) cyclopentene + Br_2 $\xrightarrow[CH_2Cl_2]{}$ *trans*-1,2-dibromocyclopentane (Br, Br)

(b) 1-methylcyclohexene (CH_3) + Cl_2 $\xrightarrow[CH_2Cl_2]{}$ *trans*-1,2-dichloro-1-methylcyclohexane (CH_3, Cl, Cl)

Complete the following reactions. **PRACTICE EXERCISE 16.17**

(a) $CH_3C(CH_3)_2CH{=}CH_2$ + Br_2 $\xrightarrow{CH_2Cl_2}$ (b) methylenecyclohexane ($=CH_2$) + Cl_2 $\xrightarrow{CH_2Cl_2}$

Bridged halonium ion intermediates and *anti* selectivity

We explain the addition of bromine and chlorine to cycloalkenes, as well as their selectivity (they always add *trans* to each other), by a two-step mechanism that involves a halogen bearing a positive charge, called a **halonium ion**. The cyclic structure of which this ion is a part is called a **bridged halonium ion**. The bridged bromonium ion shown in step 1 of the mechanism below may look odd to you, but it is an acceptable Lewis structure. A calculation of formal charge places a positive charge on bromine. Then, in step 2, a bromide ion reacts with the bridged intermediate from the side opposite that occupied by the bromine atom, giving the dibromoalkane. Thus, bromine atoms add from opposite faces of the carbon–carbon double bond. We say that this addition occurs with ***anti* selectivity**. Alternatively, we say that the addition of halogens is stereospecific involving ***anti* addition** of the halogen atoms.

Step 1: Reaction of the π electrons of the carbon–carbon double bond with bromine forms a bridged bromonium ion intermediate in which bromine bears a positive formal charge.

a bridged bromonium ion intermediate

Step 2: A bromide ion (a Lewis base) attacks a carbon atom of the three-membered ring (a Lewis acid) from the side opposite the bridged bromonium ion, opening the three-membered ring.

anti orientation of added bromine atoms

a Newman projection of the product

As we can see from the Newman projection above, these bromine atoms are *trans* to each other but, in open-chain alkanes, this relative position is rapidly scrambled by normal bond rotation around the carbon–carbon bonds. On the other hand, such rotation is not possible in a cycloalkene so the bromine atoms remain on opposite sides of the ring.

cyclopentene + Br_2 ⟶ *trans*-1,2-dibromocyclopentane

Reduction of alkenes: formation of alkanes

Most alkenes react quantitatively with molecular hydrogen, H_2, in the presence of a transition metal catalyst to give alkanes. Commonly used transition metal catalysts include platinum, palladium, ruthenium and nickel. Yields are usually quantitative or nearly so. Because the conversion of an alkene to an alkane involves reduction by hydrogen in the presence of a catalyst, the process is called **catalytic reduction** or **catalytic hydrogenation**.

cyclohexene + H_2 $\xrightarrow[25\ °C,\ 3 \times 10^5\ Pa]{Pd}$ cyclohexane

The metal catalyst is used as a finely powdered solid, which may be supported on some inert material such as powdered charcoal or alumina. The reaction is carried out by dissolving the alkene in ethanol or another nonreacting organic solvent, adding the solid catalyst, and exposing the mixture to hydrogen gas at pressures ranging from 1×10^5 to 100×10^5 Pa (i.e. up to 100 times normal atmospheric pressure). Alternatively, the metal may be bound to certain organic molecules and used in the form of a soluble complex.

Catalytic reduction is stereospecific, the most common pattern being the ***syn* addition** of hydrogen atoms to the carbon–carbon double bond (meaning the hydrogen atoms are added to the same side of the carbon–carbon double bond). The catalytic reduction of 1,2-dimethylcyclohexene, for example, yields *cis*-1,2-dimethylcyclohexane.

$$\text{1,2-dimethylcyclohexene} + H_2 \xrightarrow{Pt} \textit{cis}\text{-1,2-dimethylcyclohexane}$$

The transition metals used in catalytic reduction can adsorb large quantities of hydrogen onto their surfaces, probably by forming metal–hydrogen σ bonds. Similarly, these transition metals adsorb alkenes on their surfaces, forming carbon–metal bonds (figure 16.28a). Hydrogen atoms are added to the alkene in more than one step.

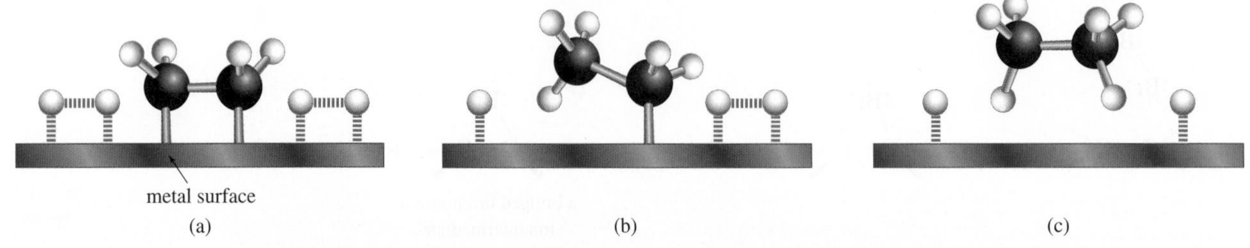

FIGURE 16.28 *syn* addition of hydrogen to an alkene involving a transition metal catalyst. **(a)** Hydrogen and the alkene are adsorbed on the metal surface, and **(b)** one hydrogen atom is transferred to the alkene, forming a new C—H bond. The other carbon atom remains adsorbed on the metal surface. **(c)** A second C—H bond forms and the alkane is desorbed.

Enthalpies of hydrogenation and the relative stabilities of alkenes

The **enthalpy of hydrogenation** of an alkene is defined as its enthalpy of reaction, $\Delta_r H^\ominus$, with hydrogen to form an alkane. Table 16.5 lists the enthalpies of hydrogenation of several alkenes.

TABLE 16.5 Enthalpies of hydrogenation of several alkenes.

Name	Structural formula	$\Delta_r H^\ominus$ (kJ mol^{-1})
ethene	$CH_2{=}CH_2$	–137
propene	$CH_3CH{=}CH_2$	–126
but-1-ene	$CH_3CH_2CH{=}CH_2$	–127
cis-but-2-ene	H_3C CH_3 C=C H H	–120
trans-but-2-ene	H_3C H C=C H CH_3	–116
2-methylbut-2-ene	H_3C CH_3 C=C H_3C H	–113
2,3-dimethylbut-2-ene	H_3C CH_3 C=C H_3C CH_3	–111

Three important points follow from the information in table 16.5.

1. The reduction of an alkene to an alkane is an exothermic process. This observation is consistent with the fact that, during hydrogenation, there is net conversion from weaker π bonding to stronger σ bonding; that is, one σ bond (H—H) and one π bond (C=C) are broken, and two new σ bonds (C—H) are formed.
2. The enthalpies of hydrogenation depend on the degree of substitution of the carbon–carbon double bond: the greater the substitution, the lower the enthalpy of hydrogenation. Compare, for example, the enthalpies of hydrogenation of ethene (no substituents), propene (one substituent), but-1-ene (one substituent) and the *cis* and *trans* isomers of but-2-ene (two substituents each).
3. The enthalpy of hydrogenation of a *trans* alkene is less than that of the isomeric *cis* alkene. Compare, for example, the enthalpies of hydrogenation of *cis*-but-2-ene and *trans*-but-2-ene. Because the reduction of both alkenes gives butane, any difference in their enthalpies of hydrogenation must be due to a difference in relative energy between the two alkenes (figure 16.29). The alkene with the lower (less negative) value of $\Delta_r H^\ominus$ is the more stable alkene. We explain the greater stability of *trans* alkenes relative to *cis* alkenes in terms of nonbonded interaction strain. In *cis*-but-2-ene, the two —CH_3 groups are sufficiently close to each other that there is repulsion between their electron clouds. This repulsion is reflected in the larger enthalpy of hydrogenation (decreased stability) of *cis*-but-2-ene compared with that of *trans*-but-2-ene.

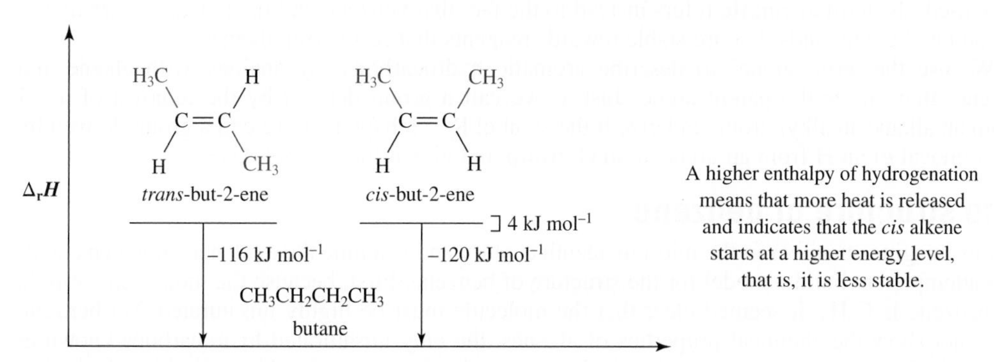

FIGURE 16.29 Enthalpies of hydrogenation of *cis*-but-2-ene and *trans*-but-2-ene; *trans*-but-2-ene is more stable than *cis*-but-2-ene by 4 kJ mol^{-1}.

We have looked at molecules with one double bond. Molecules with more than one carbon–carbon double bond undergo the same addition reactions. However, there is a class of unsaturated molecules that do not undergo any of these addition reactions. These molecules are called aromatic compounds and are covered in section 16.7.

16.6 Reactions of alkynes

Much of the chemistry of alkynes mirrors the chemistry of alkenes; they undergo the same reduction reactions, as well as hydrogen halide addition and halogen addition reactions. Alkynes also undergo hydration, but unlike alkenes the synthetic outcome is a ketone and this is covered in chapter 21.

Reduction of alkynes is particularly important in the synthesis of complicated molecules used in making pharmaceuticals. Alkynes are easily reduced to alkanes by addition of hydrogen gas using a metal catalyst. The significant differences from the reduction of alkenes are that this reduction occurs in stages and the choice of catalyst can control the synthetic outcome. Complete reduction of the alkyne occurs when palladium coated onto carbon is used as the catalyst. Another catalyst involving deactivated palladium, called Lindlar catalyst, produces the *cis* alkene from the triple bond. The *trans* alkene can be generated using sodium or lithium dissolved in liquid ammonia.

$CH_3-CH_2-C\equiv C-CH_3$

$\xrightarrow[\text{Pd/C}]{2H_2}$ $CH_3-CH_2-CH_2-CH_2-CH_3$ pentane

$\xrightarrow[\text{Lindlar catalyst}]{H_2}$ *cis*-pent-2-ene (H and H on the same side of C=C; CH_3-CH_2 and CH_3 on the same side)

$\xrightarrow[NH_3]{Li}$ *trans*-pent-2-ene (CH_3-CH_2 and H on one carbon, H and CH_3 on the other, on opposite sides)

16.7 Aromatic compounds

The simplest example of an aromatic compound is benzene, C_6H_6. Benzene, a colourless liquid, was first isolated by Michael Faraday in 1825 from the oily residue that collected in the gas pipes of London. Benzene's molecular formula, C_6H_6, suggests a high degree of unsaturation. Remember, C_nH_{2n+2} is the formula for alkanes, so an alkane with six carbon atoms has a molecular formula of C_6H_{14} and a cycloalkane with six carbon atoms has a molecular formula of C_6H_{12}. Considering benzene's high degree of *un*saturation, it might be expected to show many of the reactions characteristic of alkenes. Yet, benzene is remarkably unreactive! It does not undergo the addition, oxidation and reduction reactions characteristic of alkenes. For example, benzene does not undergo addition reactions with bromine, hydrogen chloride or other reagents that usually add to carbon–carbon double bonds. When benzene reacts, it does so by substitution (see chapter 18) in which a hydrogen atom is replaced by another atom or a group of atoms.

The term 'aromatic' was originally used to classify benzene and its derivatives because many of them have distinctive odours. It became clear, however, that a better classification for these compounds would be one based on structure and chemical reactivity, rather than aroma. As it is now used, the term **aromatic** refers instead to the fact that benzene and its derivatives are highly unsaturated compounds that are stable towards reagents that react with alkenes.

We use the term 'arene' to describe aromatic hydrocarbons, by analogy with alkane and alkene. Benzene is the parent arene. Just as we call a group derived by the removal of an H from an alkane an alkyl group and give it the symbol R— (chapter 2), we call a group derived by the removal of an H from an arene an **aryl group** and give it the symbol Ar—.

The structure of benzene

Let us imagine ourselves in the mid-nineteenth century and examine the evidence on which chemists attempted to build a model for the structure of benzene. First, because the molecular formula of benzene is C_6H_6, it seemed clear that the molecule must be highly unsaturated. Yet benzene does not show the chemical properties of alkenes, the only unsaturated hydrocarbons known at that time. Benzene does undergo chemical reactions, but its characteristic reaction is substitution rather than addition. When benzene is treated with bromine in the presence of iron(III) chloride as a catalyst, for example, only one compound, with molecular formula C_6H_5Br, forms.

$$\underset{\text{benzene}}{C_6H_6} + Br_2 \xrightarrow{FeCl_3} \underset{\text{bromobenzene}}{C_6H_5Br} + HBr$$

Chemists concluded, therefore, that all six carbon atoms and all six hydrogen atoms of benzene must be equivalent. When bromobenzene is treated with bromine in the presence of iron(III) chloride, three isomeric dibromobenzenes are formed.

$$\underset{\text{bromobenzene}}{C_6H_5Br} + Br_2 \xrightarrow{FeCl_3} \underset{\text{dibromobenzene}}{C_6H_4Br_2} + HBr$$

(formed as a mixture of three constitutional isomers)

For chemists in the mid-nineteenth century, the problem was to incorporate these observations, along with the accepted tetravalence of carbon, into a structural formula for benzene. Before we examine their proposals, we should note that the problem of the structure of benzene and other aromatic hydrocarbons occupied the efforts of chemists for over a century. It was not until the 1930s that chemists developed a general understanding of the unique chemical properties of benzene and its derivatives.

Kekulé's model of benzene

The structure for benzene proposed by August Kekulé in 1872 consisted of a six-membered ring with alternating single and double bonds and with one hydrogen atom bonded to each carbon atom. Kekulé further proposed that the ring contains three double bonds that shift back and forth so rapidly that the two forms cannot be separated. Each structure has become known as a **Kekulé structure**.

rapid interconversion

a Kekulé structure, showing all atoms

Kekulé structures as line structures

Because all of the carbon atoms and hydrogen atoms of Kekulé's structure are equivalent, substituting bromine for any one of the hydrogen atoms gives the same compound. Thus, Kekulé's proposed structure was consistent with the fact that treating benzene with bromine in the presence of iron(III) chloride gives only one compound, with molecular formula C_6H_5Br.

His proposal also accounted for the fact that the bromination of bromobenzene gives three (and only three) isomeric dibromobenzenes.

$$C_6H_5Br + Br_2 \xrightarrow{FeCl_3} C_6H_4Br_2 + C_6H_4Br_2 + C_6H_4Br_2 + HBr$$

the three isomeric dibromobenzenes

Although Kekulé's proposal was consistent with many experimental observations, it was contested for years. The major objection was that it did not account for the unusual chemical behaviour of benzene. If benzene contains three double bonds, why, his critics asked, doesn't it show the reactions typical of alkenes? Why doesn't it add 3 moles of bromine to form 1,2,3,4,5,6-hexabromocyclohexane? Why, instead, does benzene react by substitution rather than addition?

The valence bond model of benzene

The concepts of the **hybridisation of atomic orbitals** and the **theory of resonance**, both components of valence bond theory (see chapter 5) developed by Linus Pauling in the 1930s, provided the first adequate description of the structure of benzene. The carbon skeleton of benzene forms a regular hexagon with C—C—C and H—C—C bond angles of 120°. For this type of bonding, carbon uses sp^2 hybrid orbitals. Each carbon atom forms σ bonds to two adjacent carbon atoms by the overlap of sp^2–sp^2 hybrid orbitals and one σ bond to hydrogen by the overlap of sp^2–$1s$ orbitals. As determined experimentally, all carbon–carbon bonds in benzene are the same length, 1.39 Å, a value between the length of a single bond between sp^3 hybridised carbons (1.54 Å) and that of a double bond between sp^2 hybridised carbons (1.30 Å).

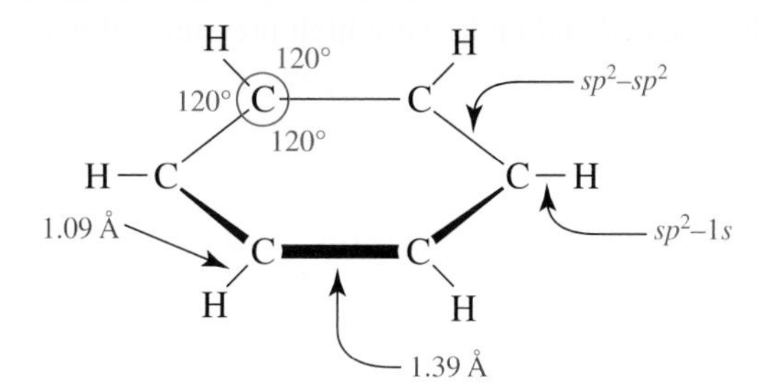

Each carbon atom also has a single unhybridised $2p$ orbital that contains one electron. These six $2p$ orbitals lie perpendicular to the plane of the ring and overlap to form a continuous π cloud encompassing all six carbon atoms. The electron density of the π system of a benzene ring lies in one torus (a doughnut-shaped region) above the plane of the ring and in a second torus below the plane (figure 16.30).

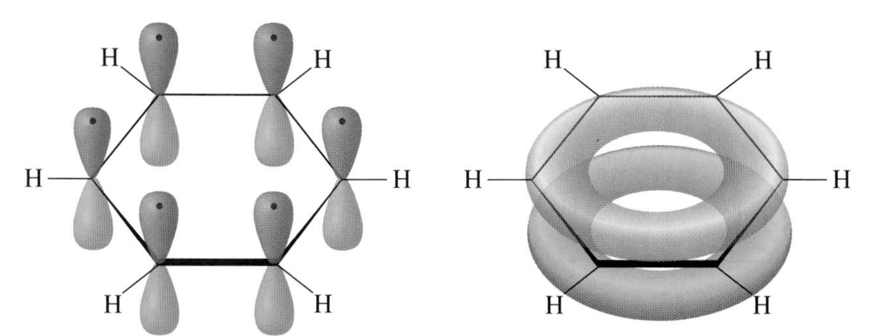

FIGURE 16.30 Orbital overlap model of bonding in benzene: **(a)** The carbon–hydrogen framework. The six $2p$ orbitals, each with one electron, are shown uncombined. **(b)** The overlap of parallel $2p$ orbitals forms a continuous π cloud, shown by one torus above the plane of the ring and a second below the plane of the ring.

The resonance model of benzene

The concept of resonance introduced in chapter 5 tells us that, if we can represent a molecule or ion by two or more contributing structures, that molecule cannot be adequately represented by any single contributing structure. We represent benzene as a hybrid of two equivalent contributing structures, often referred to as Kekulé structures and we indicate the relationship between the two by the use of a double-headed arrow.

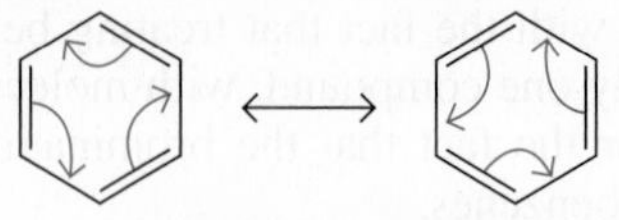

benzene as a hybrid of two equivalent contributing structures

Each Kekulé structure makes an equal contribution to the hybrid; thus, the C—C bonds are neither single nor double bonds, but something intermediate. We recognise that neither of these contributing structures exists (they are merely alternative ways to pair $2p$ orbitals, with no reason to prefer one over the other) and that the actual structure is something in between. An alternative means of representing the benzene ring that highlights the special nature of the aromatic structure is to show a circle within the six-membered ring of carbon atoms.

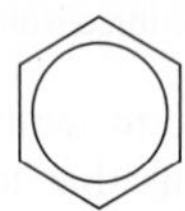

Nevertheless, we continue to use a single contributing structure, also called a resonance structure, to represent this molecule because it serves to remind us of the tetravalence of the carbon atoms involved and allows us to more easily represent the movement of electrons in reaction mechanisms.

The resonance energy of benzene

Resonance energy is the difference in energy between a resonance hybrid and its most stable hypothetical contributing structure. One way to estimate the resonance energy of benzene is to compare the enthalpies of hydrogenation of cyclohexene and benzene. In the presence of a transition metal catalyst, hydrogen readily reduces cyclohexene to cyclohexane (p. 731).

$$\text{cyclohexene} + H_2 \xrightarrow[1\text{–}2 \times 10^5\ \text{Pa}]{\text{Ni}} \text{cyclohexane} \qquad \Delta_r H^\ominus = -120\ \text{kJ mol}^{-1}$$

By contrast, benzene is reduced only very slowly to cyclohexane under these conditions. It is reduced more rapidly when heated and under very high pressures of hydrogen.

$$\text{benzene} + 3H_2 \xrightarrow[200\text{–}300 \times 10^5\ \text{Pa}]{\text{Ni}} \text{cyclohexane} \qquad \Delta_r H^\ominus = -208\ \text{kJ mol}^{-1}$$

The catalytic reduction of an alkene is an exothermic reaction (p. 308). The enthalpy of hydrogenation per double bond varies somewhat with the degree of substitution of the double bond; for cyclohexene, $\Delta_r H^\ominus = -120\ \text{kJ mol}^{-1}$. If we consider benzene to be 1,3,5-cyclohexatriene, a hypothetical compound with alternating single and double bonds, we might expect its enthalpy of hydrogenation to be $3 \times -120 = -360\ \text{kJ mol}^{-1}$. Instead, the enthalpy of hydrogenation of benzene is only $-208\ \text{kJ mol}^{-1}$. The difference of $152\ \text{kJ mol}^{-1}$ between the expected value and the experimentally observed value is the resonance energy of benzene. Figure 16.31 shows these experimental results in the form of a graph.

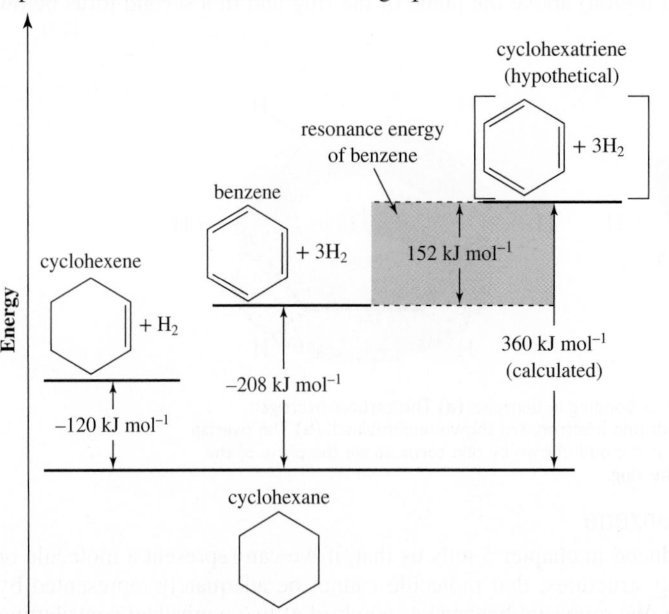

FIGURE 16.31 The resonance energy of benzene, as determined by a comparison of the enthalpies of hydrogenation of cyclohexene, benzene and the hypothetical 1,3,5-cyclohexatriene.

For comparison, the strength of a carbon–carbon single bond is about 333–418 kJ mol^{-1}, and that of hydrogen bonding in water and low-molar-mass alcohols is about 8.4–21 kJ mol^{-1}. Thus, although the resonance energy of benzene is less than the strength of a carbon–carbon single bond, it is considerably greater than the strength of hydrogen bonding in water and alcohols. In section 19.1, we will see that hydrogen bonding has a greater effect on the physical properties of alcohols and little impact on those of alkanes. Figure 16.31 shows that the resonance energy of benzene and other aromatic hydrocarbons has a dramatic effect on their chemical reactivity.

The following diagram shows resonance energies for benzene and several other aromatic hydrocarbons.

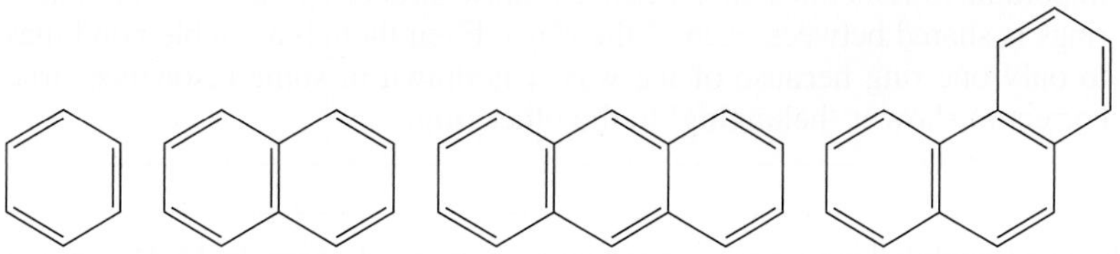

Resonance energy (kJ mol^{-1})	benzene	naphthalene	anthracene	phenanthrene
	152	255	347	381

WORKED EXAMPLE 16.21

Drawing resonance structures for aromatic compounds

Naphthalene, an aromatic compound, can be represented as three equivalent resonance structures. One of these resonance structures is shown below. Draw the remaining resonance structures and show how each one is related to the next using curved arrows.

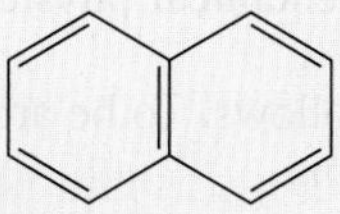

Analysis

Resonance structures show the movement of π electrons between one structure and the next. This means that all we have to do is move these electrons around in going from one resonance structure to the next. When moving π electrons, we need to make sure that the carbons we move them to and from all end up with an octet (i.e. four bonds) in the final structure.

Solution

The first new resonance structure we can draw is one in which all of the double bonds in the ring on the left are moved around by one position, as was shown for the Kekulé structures of benzene.

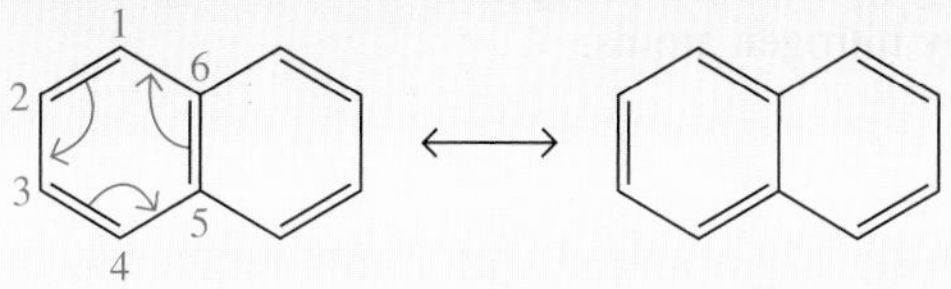

1 2 3 4 5 6 ⟷

Following the electrons around this ring, you can see that the π electrons between C(5) and C(6) are moved to end up between C(1) and C(6), and so on around the ring, until the π electrons between C(3) and C(4) are moved to end up between C(4) and C(5). This completes the series of movements and ensures that the electrons we moved away from C(5) initially are restored by the final electron movement step, so that once again all carbons in the ring have four bonds.

The second new resonance structure (and the third and final one overall for naphthalene) can be drawn by once again beginning with the original provided structure, but this time looking at the ring on the right. As before, all of the double bonds in this ring can be moved around by one position to give a new resonance structure.

⟷

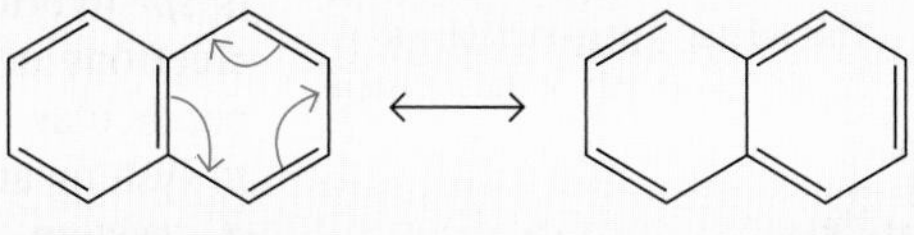

Combining the structures:

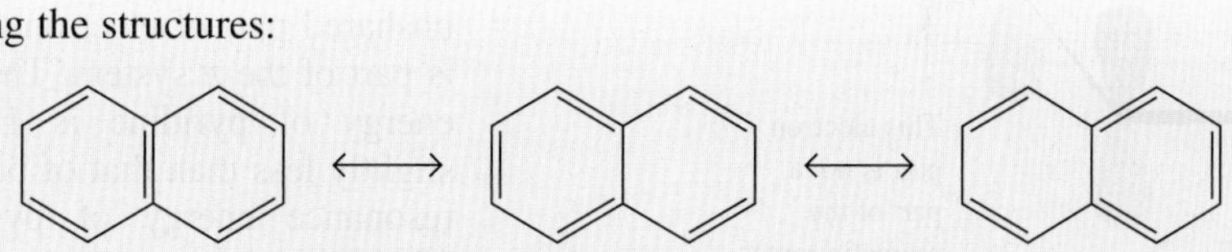

⟷ ⟷

Is our answer reasonable?

Note that all of the resonance structures above differ only in where we have decided to draw the double bonds formed by their π electrons. In reality the molecule does not have alternating single and double bonds and there is actually a continuous π-bond system shaped like a 'figure 8' that stretches uniformly over the two rings (and another that stretches below the two rings). Drawing resonance structures like this, however, aids our understanding of the bonding involved and helps us remember the valency of carbon (four) so that we do not make mistakes in determining the exact bonds and the numbers of atoms present. It is also important to remember that when we draw structures like this the bond that joins the two rings is shared between each of the rings. Even though a double bond may appear to belong to only one ring because of the way it is drawn in some resonance structures, it is just as correct to show it 'belonging' to the other ring.

PRACTICE EXERCISE 16.18 The third naphthalene resonance structure in worked example 16.21 can also be derived directly from the resonance structure derived in step 1. Using curved arrows, show the electron movement required for this.

The concept of aromaticity

Many types of molecules other than benzene and its derivatives show aromatic character; that is, they contain high degrees of unsaturation, yet fail to undergo characteristic alkene addition and oxidation–reduction reactions. What chemists had long sought to understand were the principles underlying aromatic character. German chemical physicist Erich Hückel solved this problem in the 1930s.

Hückel's criteria are summarised as follows. To be aromatic, a ring must:

1. have one $2p$ orbital on each of its atoms
2. be planar or nearly planar, so that there is continuous overlap or nearly continuous overlap of all $2p$ orbitals of the ring
3. have 2, 6, 10, 14, 18, and so on, π electrons in the cyclic arrangement of $2p$ orbitals. (Note that these numbers are solutions to the equation $4n + 2$, where n is equal to 0, 1, 2, 3, 4 and so on.)

Benzene meets these criteria. It is cyclic, planar, has one $2p$ orbital on each carbon atom of the ring, and has six π electrons (an aromatic sextet) in the cyclic arrangement of its $2p$ orbitals. If structures with more than one ring meet these criteria, they are also aromatic and some examples of these appear above.

The unsaturated rings may also contain atoms other than carbon, and these molecules are called **heterocyclic compounds**. For example, pyridine and pyrimidine are analogues of benzene. In pyridine, one CH group of benzene is replaced by a nitrogen atom and, in pyrimidine, two CH groups are replaced by nitrogen atoms.

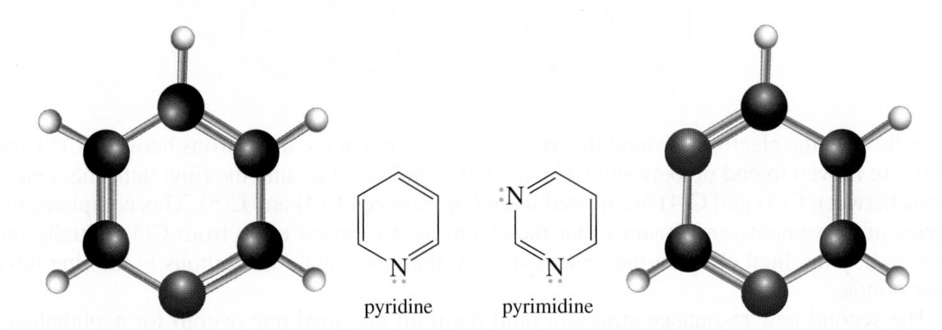
N
pyridine
N
N
pyrimidine

Both the pyridine and pyrimidine molecules meet the Hückel criteria for aromaticity. They are both cyclic and planar, have one $2p$ orbital on each atom of the ring, and have six electrons in the π system. In pyridine, nitrogen is sp^2 hybridised; its unshared pair of electrons occupies an sp^2 orbital perpendicular to the $2p$ orbitals of the π system and so is not a part of the π system. In pyrimidine, neither unshared pair of electrons of nitrogen is part of the π system. The resonance energy of pyridine is $134\,kJ\,mol^{-1}$, slightly less than that of benzene. The resonance energy of pyrimidine is $109\,kJ\,mol^{-1}$.

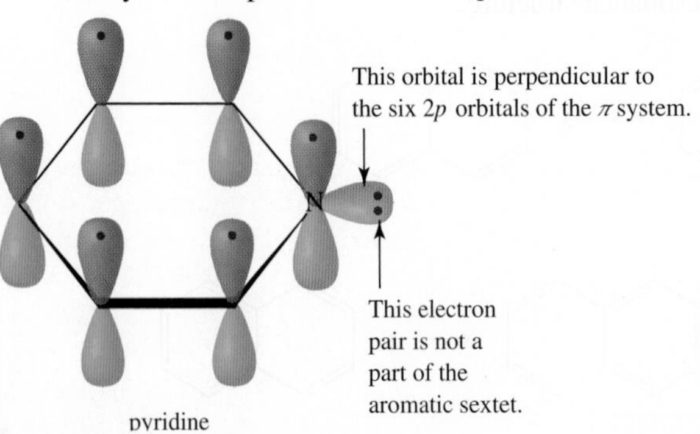
This orbital is perpendicular to the six $2p$ orbitals of the π system.
N
This electron pair is not a part of the aromatic sextet.
pyridine

The five-membered-ring compounds furan, pyrrole and imidazole are also aromatic.

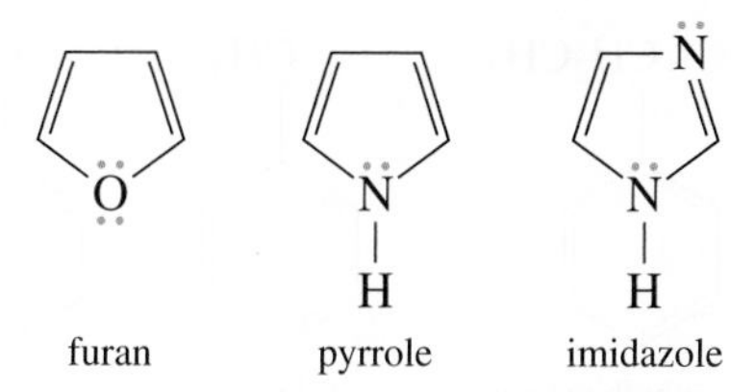

furan
pyrrole
imidazole

In these planar compounds, each heteroatom (non-carbon atom) is sp^2 hybridised, and its unhybridised $2p$ orbital is part of a continuous cycle of five $2p$ orbitals. In furan, one unshared pair of electrons of the heteroatom lies in the unhybridised $2p$ orbital and is a part of the π system (figure 16.32). The other unshared pair of electrons lies in an sp^2 hybrid orbital, perpendicular to the $2p$ orbitals, and is not a part of the π system. In pyrrole, the unshared pair of electrons on nitrogen is part of the aromatic sextet. In imidazole, the unshared pair of electrons on one nitrogen is part of the aromatic sextet; the unshared pair on the other nitrogen is not.

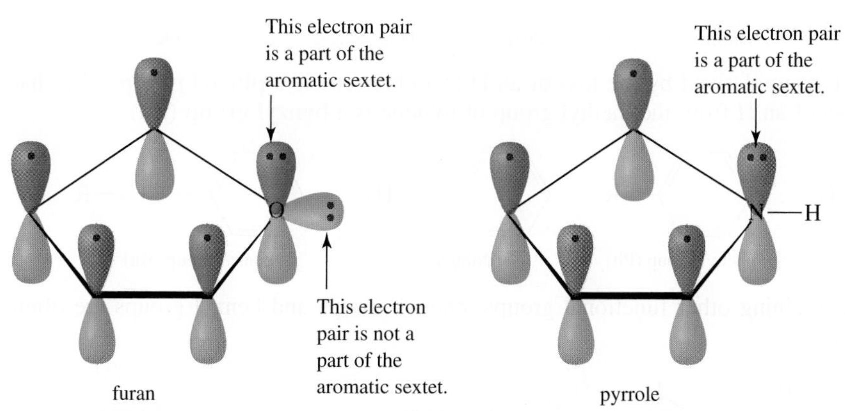

This electron pair is a part of the aromatic sextet.
This electron pair is not a part of the aromatic sextet.
furan
This electron pair is a part of the aromatic sextet.
pyrrole

FIGURE 16.32 Origin of the six π electrons (the aromatic sextet) in furan and pyrrole. The resonance energy of furan is 67 kJ mol^{-1} and that of pyrrole is 88 kJ mol^{-1}.

Nature abounds with compounds having a heterocyclic ring fused to one or more other rings. Two such compounds especially important in the biological world are indole and purine.

indole   serotonin (a neurotransmitter)   purine   adenine

Indole contains a pyrrole ring fused to a benzene ring. Compounds derived from indole include the amino acid L-tryptophan (section 24.1) and the neurotransmitter serotonin. Purine contains a six-membered pyrimidine ring fused to a five-membered imidazole ring. Adenine is one of the building blocks of deoxyribonucleic acid (DNA) and ribonucleic acid (RNA), as described in chapter 25. It is also a component of the biological oxidising agent nicotinamide adenine dinucleotide, abbreviated NAD^+.

Nomenclature

The simplest aromatic hydrocarbon is benzene, which contains only hydrogen atoms bound to the aromatic ring. Of course, there are many aromatic hydrocarbons with numerous different substituents in various positions around the ring. As always, we need to have a clear means to describe each of these individual molecules.

Monosubstituted benzenes

Monosubstituted alkylbenzenes are named as derivatives of benzene; an example is ethylbenzene. The IUPAC system retains certain common names for several of the simpler monosubstituted alkylbenzenes. Examples are **toluene** (rather than methylbenzene) and **styrene** (rather than phenylethene).

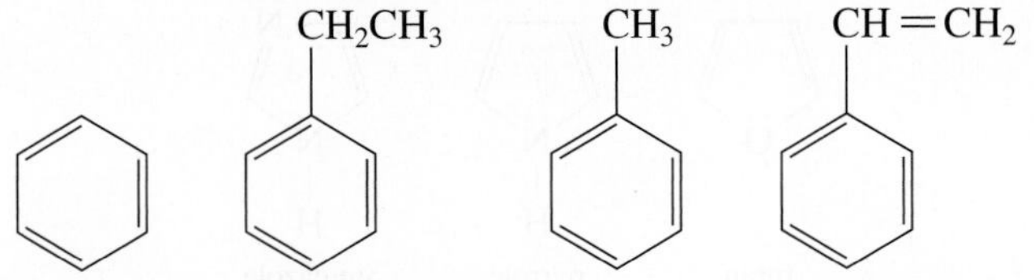

CH_2CH_3 CH_3 $CH{=}CH_2$

benzene ethylbenzene toluene styrene

The common names **phenol, aniline, benzaldehyde, benzoic acid** and **anisole** are also retained by the IUPAC system.

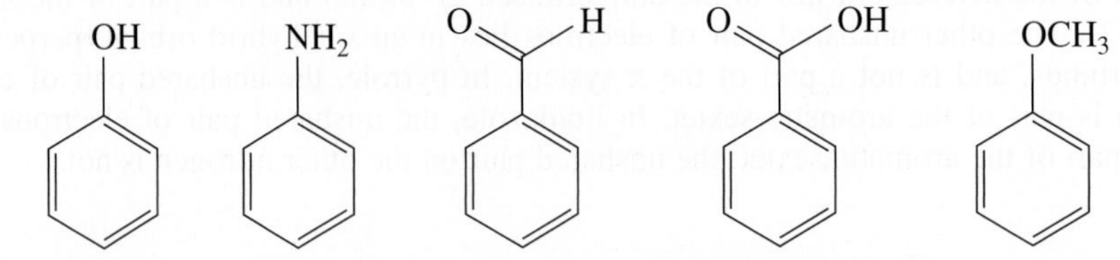

OH NH_2 O H O OH OCH_3

phenol aniline benzaldehyde benzoic acid anisole

The substituent group derived by the loss of an H from benzene is a **phenyl group** (Ph); that derived by the loss of an H from the methyl group of toluene is a **benzyl group** (Bn).

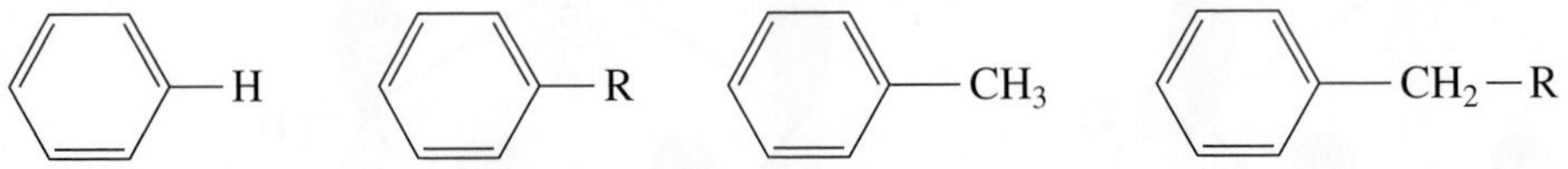

H R CH_3 $CH_2{-}R$

benzene phenyl group (Ph) toluene benzyl group (Bn)

In molecules containing other functional groups, phenyl groups and benzyl groups are often named as substituents.

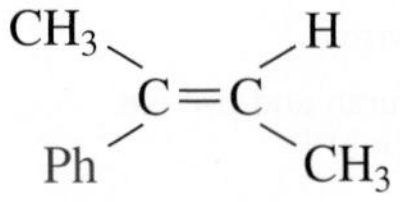

CH_3 H C=C Ph CH_3 $PhCH_2CH_2OH$ $PhCH_2Cl$

(*Z*)-2-phenylbut-2-ene 2-phenylethanol benzyl chloride

Disubstituted benzenes

When two substituents occur on a benzene ring, three constitutional isomers are possible. We locate substituents either by numbering the atoms of the ring or by using the locators ***ortho***, ***meta*** and ***para***, which are abbreviated to *o*, *m* and *p* respectively. The numbers 1,2- are equivalent to *ortho* (Greek word meaning 'straight'), 1,3- to *meta* (Greek word meaning 'after') and 1,4- to *para* (Greek word meaning 'beyond').

When one of the two substituents on the ring imparts a special name to the compound, such as toluene, phenol and aniline, we name the compound as a derivative of that parent molecule. In this case, the special substituent occupies ring position number 1. The IUPAC system retains the common name **xylene** for the three isomeric dimethylbenzenes. When neither group imparts a special name, we locate the two substituents and list them in alphabetical order before the word 'benzene'. The carbon atom of the benzene ring with the substituent that comes first in alphabetical order is numbered C(1).

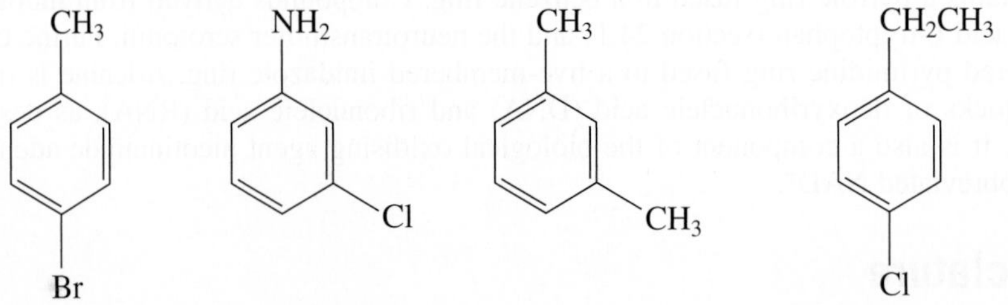

CH_3 NH_2 CH_3 CH_2CH_3

Cl CH_3

Br Cl

4-bromotoluene (*p*-bromotoluene) 3-chloroaniline (*m*-chloroaniline) 1,3-dimethylbenzene (*m*-xylene) 1-chloro-4-ethylbenzene (*p*-chloroethylbenzene)

Polysubstituted benzenes

When three or more substituents are present on a ring, we specify their locations by numbers. If one of the substituents imparts a special name, the molecule is named as a derivative of that parent molecule. If none of the substituents imparts a special name, we number them to give the smallest set of numbers and list them in alphabetical order before the word 'benzene'. In the following examples, the first compound is a derivative of toluene, and the second is a derivative of phenol. Because there is no special name for the third compound, we number the carbon atoms using the smallest possible set of numbers, then list its three substituents in alphabetical order, followed by the word 'benzene'.

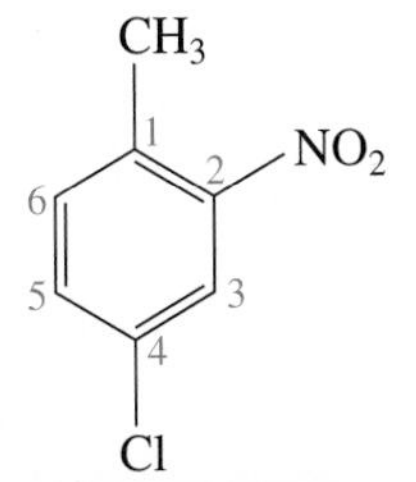

4-chloro-2-nitrotoluene

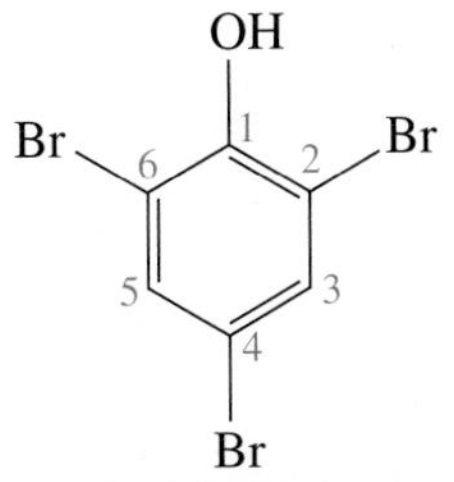

2,4,6-tribromophenol

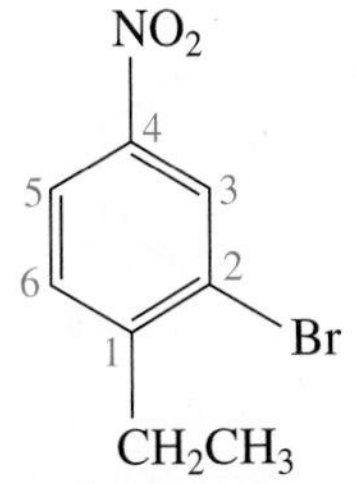

2-bromo-1-ethyl-4-nitrobenzene

WORKED EXAMPLE 16.22

Nomenclature of substituted benzenes

Write names for the following compounds.

(a) 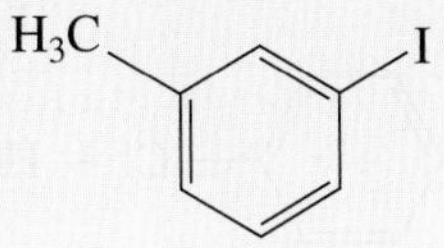 (b) 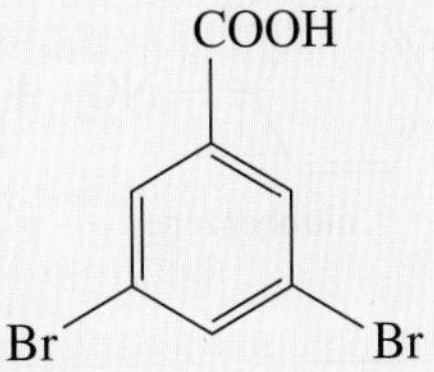 (c) (d)

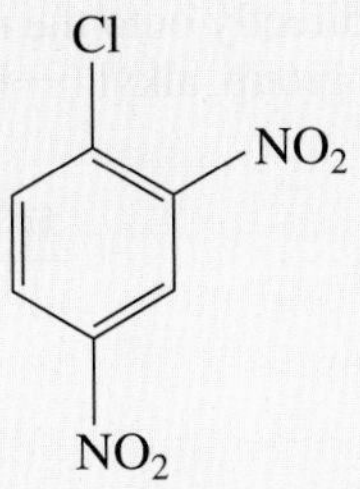

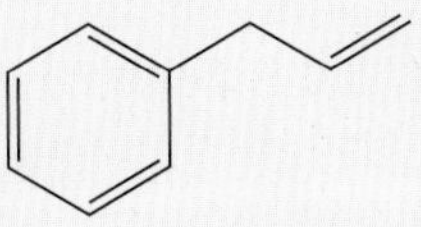

Analysis

We look for the important functional groups that may govern the parent name. For example, the presence of a methyl group bound to benzene generates the parent name toluene; a carboxyl group, benzoic acid. When the substituent becomes complicated we name the benzene ring as the substituent *phenyl*.

Solution

(a) 3-iodotoluene or *m*-iodotoluene
(b) 3,5-dibromobenzoic acid
(c) 1-chloro-2,4-dinitrobenzene
(d) 3-phenylpropene

PRACTICE EXERCISE 16.19

Write names for the following compounds.

(a) 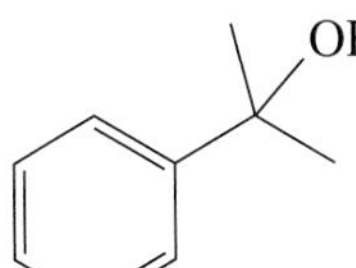 (b) 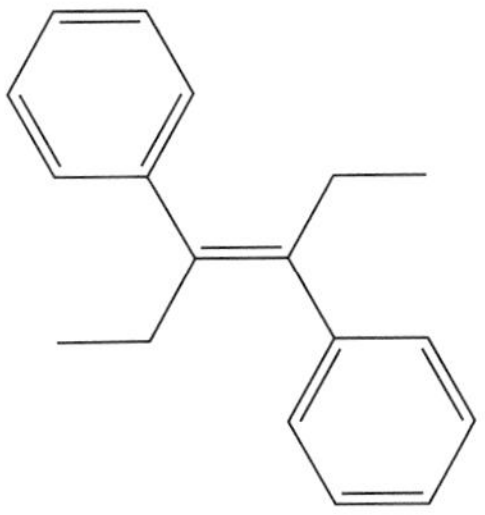 (c)

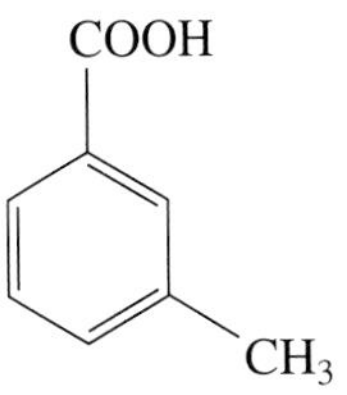

Polycyclic aromatic hydrocarbons (PAHs) contain two or more aromatic rings, and each pair of rings shares two carbon atoms. Naphthalene, anthracene and phenanthrene, the most common PAHs, and substances derived from them are found in coal tar and high-boiling-point petroleum residues. At one time, naphthalene was used as a moth repellent and insecticide in preserving woollens, but its use has decreased due to the introduction of chlorinated hydrocarbons such as *p*-dichlorobenzene. Also found in coal tar are lesser amounts of a compound called benzo[*a*]pyrene. This compound is also found in the exhausts of petrol-powered internal combustion engines (such as car engines) and in cigarette smoke. Benzo[*a*]pyrene is a very potent carcinogen and mutagen.

naphthalene anthracene phenanthrene benzo[*a*]pyrene

16.8 Reactions of aromatic compounds: electrophilic aromatic substitution

By far the most characteristic reaction of aromatic compounds is substitution at a ring carbon atom. Some groups that can be introduced directly onto the ring are the halogens (—X), the nitro (—NO_2) group, the sulfonic acid (—SO_3H) group, alkyl (—R) groups and acyl (RCO—) groups.

Halogenation $C_6H_5\text{—H} + Cl_2 \xrightarrow{FeCl_3} C_6H_5\text{—Cl} + HCl$ (chlorobenzene)

Nitration $C_6H_5\text{—H} + HNO_3 \xrightarrow{H_2SO_4} C_6H_5\text{—}NO_2 + H_2O$ (nitrobenzene)

Sulfonation $C_6H_5\text{—H} + H_2SO_4 \longrightarrow C_6H_5\text{—}SO_3H + H_2O$ (benzenesulfonic acid)

Alkylation $C_6H_5\text{—H} + RX \xrightarrow{AlCl_3} C_6H_5\text{—R} + HX$ (an alkylbenzene)

Acylation $C_6H_5\text{—H} + R\text{—}\overset{O}{\overset{\|}{C}}\text{—X} \xrightarrow{AlCl_3} C_6H_5\text{—}\overset{O}{\overset{\|}{C}}R + HX$

an acyl halide an acylbenzene

There are several types of **electrophilic aromatic substitution** reactions; that is, reactions in which a hydrogen of an aromatic ring is replaced by an electrophile, E^+. The mechanisms of these reactions are actually very similar. In fact, they can be broken down into three common steps.

Step 1: Formation of the electrophile

$$\text{reagent(s)} \rightarrow E^+$$

Step 2: Reaction of the electrophile with the aromatic ring to give a resonance-stabilised cation intermediate

H + E^+ $\xrightarrow{\text{slow}}$ H E + ⟷ H E + ⟷ H + E

resonance-stabilised cation intermediate

Step 3: Proton transfer to a base to regenerate the aromatic ring

$$\text{(cation intermediate)-E} + \text{:base} \xrightarrow{\text{fast}} \text{C}_6\text{H}_5\text{E} + \text{base—H}$$

The reactions we are about to study differ only in the way the electrophile is generated and in the base that removes the proton to re-form the aromatic ring. You should keep this principle in mind as we explore the details of each reaction.

Halogenation

Chlorine alone does not react with benzene, in contrast to its instantaneous addition to cyclohexene (section 16.5). However, in the presence of a Lewis acid catalyst, such as iron(III) chloride or aluminium chloride, chlorine reacts to give chlorobenzene and HCl. Chemists account for this type of electrophilic aromatic substitution by the following three-step mechanism.

Step 1: Formation of the electrophile by reaction between chlorine (a Lewis base) and $FeCl_3$ (a Lewis acid), giving an ion pair containing a chloronium ion (an electrophile)

$$\text{:Cl—Cl:} + \text{FeCl}_3 \rightleftharpoons \text{:Cl—}\overset{+}{\text{Cl}}\text{—}\overset{-}{\text{Fe}}\text{Cl}_3 \rightleftharpoons \text{:Cl}^{+}\ \text{:Cl—}\overset{-}{\text{Fe}}\text{Cl}_3$$

chlorine (a Lewis base); iron(III) chloride (a Lewis acid); a molecular complex with a positive charge on chlorine and a negative charge on iron; an ion pair containing a chloronium ion

Step 2: Reaction of the Cl_2–$FeCl_3$ ion pair with the π electron cloud of the aromatic ring, forming a resonance-stabilised cation intermediate, represented here as a hybrid of three contributing structures

$$\text{C}_6\text{H}_6 + \text{:Cl}^{+}\ \text{:Cl—}\overset{-}{\text{Fe}}\text{Cl}_3 \xrightarrow{\text{slow, rate determining}}$$

$$\text{(H, Cl)cation} \longleftrightarrow \text{(H, Cl)cation} \longleftrightarrow \text{(H, Cl)cation}$$

resonance-stabilised cation intermediate

Step 3: Proton transfer from the cation intermediate to $FeCl_4^-$, forming HCl, regenerating the Lewis acid catalyst and giving chlorobenzene

$$\text{cation intermediate} + \text{:Cl—}\overset{-}{\text{Fe}}\text{Cl}_3 \xrightarrow{\text{fast}} \text{C}_6\text{H}_5\text{—Cl:} + \text{H—Cl:} + \text{FeCl}_3$$

cation intermediate; chlorobenzene

Treatment of benzene with bromine in the presence of a Lewis acid gives bromobenzene and HBr. The mechanism for this reaction is the same as that for the chlorination of benzene.

The major difference between the addition of a halogen to an alkene and substitution by a halogen on an aromatic ring is the fate of the cation intermediate formed in the first step of each reaction (figure 16.33). Recall that the addition of chlorine (or bromine) to an alkene is a two-step process; the first and slower step is the formation of a bridged chloronium ion intermediate. This intermediate then reacts with a chloride ion to complete the addition. With aromatic compounds, the cation intermediate loses H^+ to regenerate the aromatic ring and regain its large resonance stabilisation. There is no such resonance stabilisation to be regained in the case of an alkene.

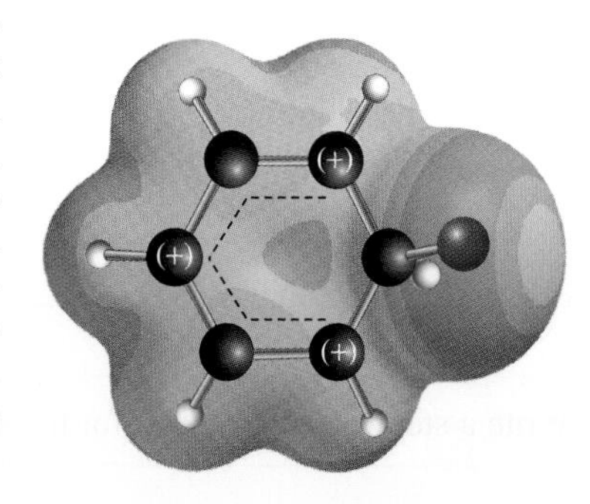

FIGURE 16.33 The positive charge on the resonance-stabilised intermediate is distributed approximately equally on carbon atoms 2, 4 and 6 of the ring relative to the point of substitution.

Nitration and sulfonation

The sequence of steps for the nitration and sulfonation of benzene is similar to that for chlorination and bromination. For *nitration*, the electrophile is the **nitronium ion**, NO_2^+, generated by the reaction of nitric acid with sulfuric acid. In the following equations nitric acid is written $HONO_2$ to show more clearly the origin of the nitronium ion.

Step 1: Formation of the electrophile by (a) proton transfer from sulfuric acid to the —OH group of nitric acid, giving the conjugate acid of nitric acid:

$$\text{H—\ddot{O}—NO}_2 + \text{H—\ddot{O}—SO}_3\text{H} \rightleftharpoons \text{H—}\overset{+}{\text{O}}(\text{H})\text{—NO}_2 + \text{HSO}_4^-$$

nitric acid — conjugate acid of nitric acid

and (b) loss of water from this conjugate acid, giving the nitronium ion, NO_2^+:

$$\text{H—}\overset{+}{\text{O}}(\text{H})\text{—NO}_2 \rightleftharpoons \text{H—\ddot{O}(H):} + \text{O=}\overset{+}{\text{N}}\text{=O}$$

the nitronium ion

Steps 2 and 3 of the nitration are given in worked example 16.23.

The *sulfonation* of benzene is carried out using hot, concentrated sulfuric acid. The electrophile under these conditions is either SO_3 or HSO_3^+, depending on the experimental conditions. The HSO_3^+ electrophile is formed from sulfuric acid in the following way.

Step 1: Formation of the electrophile by acid-catalysed dehydration

$$\text{HO—S(=O)}_2\text{—\ddot{O}H} + \text{H}^+ \rightleftharpoons \text{HO—S(=O)}_2\text{—}\overset{+}{\text{O}}\text{H}_2 \rightleftharpoons \text{HO—}\overset{+}{\text{S}}\text{(=O)}_2 + \text{H}_2\text{\ddot{O}:}$$

sulfuric acid — the electrophile

Steps 2 and 3 of sulfonation follow the same mechanism as for nitration.

WORKED EXAMPLE 16.23

Electrophilic aromatic substitution of benzene

Write a stepwise mechanism for the nitration of benzene.

Analysis and solution

Step 1: Formation of the electrophile, as given above

Step 2: Reaction of the nitronium ion (an electrophile) with the benzene ring (a nucleophile), giving a resonance-stabilised cation intermediate

$$\text{C}_6\text{H}_6 + NO_2^+ \longrightarrow [\text{C}_6\text{H}_6\text{NO}_2]^+ \longleftrightarrow [\text{C}_6\text{H}_6\text{NO}_2]^+ \longleftrightarrow [\text{C}_6\text{H}_6\text{NO}_2]^+$$

Step 3: Proton transfer from this intermediate to H_2O regenerating the aromatic ring and giving nitrobenzene

$$\text{H}_2\ddot{\text{O}}\text{:} + [\text{C}_6\text{H}_6\text{NO}_2]^+ \longrightarrow \text{C}_6\text{H}_5\text{NO}_2 + \text{H}_3\text{O}^+$$

nitrobenzene

PRACTICE EXERCISE 16.20 Write a stepwise mechanism for the sulfonation of benzene. Use HSO_3^+ as the electrophile.

Alkylation

Alkylation is a very important synthetic outcome as the new carbon–carbon bond between benzene and an alkyl group builds up the complexity of the molecule. This is illustrated by the reaction of benzene with 2-chloropropane in the presence of aluminium chloride.

$$\text{benzene} + \text{2-chloropropane (isopropyl chloride)} \xrightarrow{AlCl_3} \text{isopropylbenzene (cumene)} + HCl$$

This form of alkylation, often called **Friedel–Crafts alkylation** after the two chemists who discovered it in 1877, is among the most important methods for adding new carbon–carbon bonds to aromatic rings. The reaction involves the following steps.

Step 1: Formation of the electrophile by reaction of a haloalkane (a Lewis base) with aluminium chloride (a Lewis acid), giving a molecular complex in which aluminium has a negative formal charge and the halogen of the haloalkane has a positive formal charge. Redistribution of electrons in this complex then gives an alkyl carbocation as part of an ion pair.

$$R-\ddot{Cl}: + AlCl_3 \rightleftharpoons R-\overset{+}{Cl}-\overset{-}{Al}Cl_3 \rightleftharpoons R^+ \; :\ddot{Cl}-\overset{-}{Al}Cl_3$$

a molecular complex with a positive charge on chlorine and a negative charge on aluminium

an ion pair containing a carbocation

the electrophile

Step 2: Reaction of the alkyl carbocation with the π electrons of the aromatic ring, giving a resonance-stabilised cation intermediate

The positive charge is delocalised onto three atoms of the ring.

Step 3: Proton transfer regenerating the aromatic character of the ring and the Lewis acid catalyst

$$+ :\ddot{Cl}-\overset{-}{Al}Cl_3 \longrightarrow C_6H_5-R + AlCl_3 + H-\ddot{Cl}:$$

There are two major limitations on Friedel–Crafts alkylations. The first is that it is practical only with stable carbocations, such as 3° and 2° carbocations. You might explore the reasons for this in subsequent chemistry courses.

The second limitation on Friedel–Crafts alkylation is that it fails altogether on benzene rings bearing one or more strongly electron-withdrawing groups (indicated by the letter Y).

$$C_6H_5Y + RX \xrightarrow{AlCl_3} \text{no reaction}$$

When any of the following groups is present on the aromatic ring, the benzene ring does not undergo Friedel–Crafts alkylation.

$$-\overset{O}{\overset{\|}{C}}H \quad -\overset{O}{\overset{\|}{C}}R \quad -\overset{O}{\overset{\|}{C}}OH \quad -\overset{O}{\overset{\|}{C}}OR \quad -\overset{O}{\overset{\|}{C}}NH_2$$

$$-SO_3H \quad -C\equiv N \quad -NO_2 \quad -NR_3^+$$

$$-CF_3 \quad -CCl_3$$

A common characteristic of these groups is that each has either a full or partial positive charge on the atom bonded to the benzene ring (see examples on the next page). For carbonyl-containing compounds, this partial positive charge arises because of the difference in electronegativity between the carbonyl oxygen and carbon atoms. For the $—CF_3$ and $—CCl_3$ groups, the partial positive charge on carbon arises because of the difference in electronegativity between the carbon atom and the halogens bonded to it. In both the nitro group, $—NO_2$, and the trialkylammonium group, such as $—N(CH_3)_3^+$ below, there is a positive charge on nitrogen.

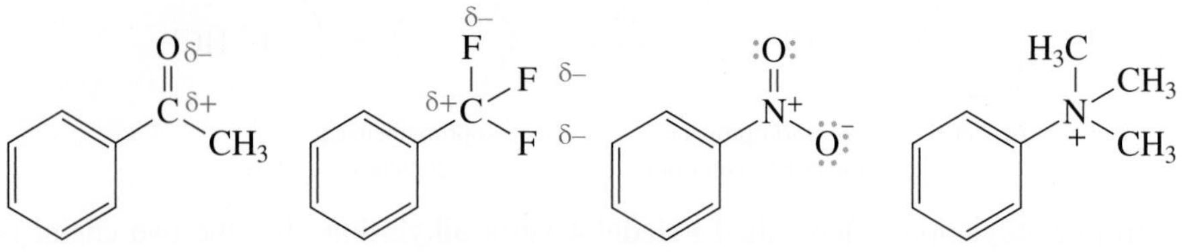

the carbonyl group of a ketone

a trifluoromethyl group

a nitro group

a trimethylammonium group

Acylation

Treating an aromatic hydrocarbon with an acyl halide in the presence of aluminium chloride forms a carbon–carbon bond and gives a ketone in a process known as **acylation**. An acyl halide is a derivative of a carboxylic acid in which the —OH of the carboxyl group is replaced by a halogen, most commonly chlorine (see chapter 23). Acyl halides are also referred to as acid halides. An RCO— group is known as an acyl group. The reaction of an acyl halide with an aromatic hydrocarbon is illustrated below by the reaction of benzene and acetyl chloride in the presence of aluminium chloride to give acetophenone.

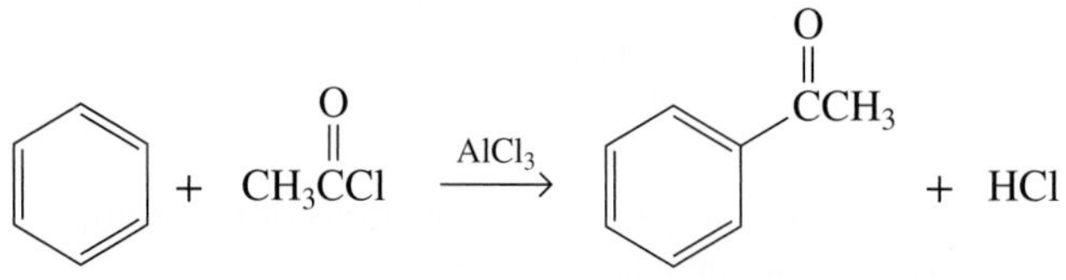

$$C_6H_6 + CH_3COCl \xrightarrow{AlCl_3} C_6H_5COCH_3 + HCl$$

benzene; acetyl chloride (an acyl halide); acetophenone (a ketone)

This reaction is related to the Friedel–Crafts alkylation in that aluminium chloride is used to generate the electrophile, which in this case is called an acylium ion. This reaction was also discovered by Friedel and Crafts so is often called the **Friedel–Crafts acylation**. The mechanism is shown below.

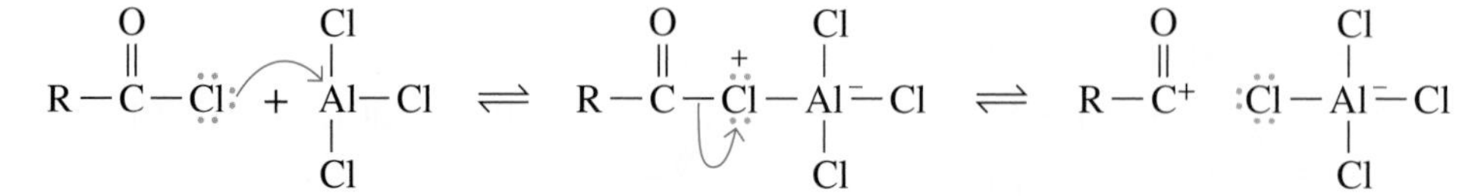

$$R{-}CO{-}Cl + AlCl_3 \rightleftharpoons R{-}CO{-}Cl^+{-}AlCl_3^- \rightleftharpoons R{-}C^+{=}O \; Cl{-}AlCl_3^-$$

an acyl chloride (a Lewis base); aluminium chloride (a Lewis acid); a molecular complex with a positive charge on chlorine and a negative charge on aluminium; an ion pair containing an acylium ion

WORKED EXAMPLE 16.24

Forming carbon–carbon bonds to aromatic rings

Write a structural formula for the product formed by Friedel–Crafts alkylation or acylation of benzene with each of the following.

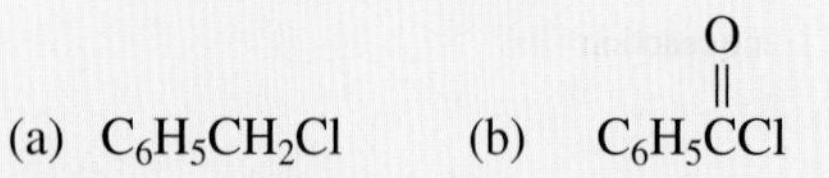

(a) $C_6H_5CH_2Cl$ benzyl chloride (b) C_6H_5COCl benzoyl chloride

Analysis and solution

Benzene is unsubstituted so can undergo both alkylations and acylations.

(a) Treatment of benzyl chloride with aluminium chloride gives the resonance-stabilised benzyl cation. Reaction of this cation with benzene, followed by loss of H^+, gives diphenylmethane

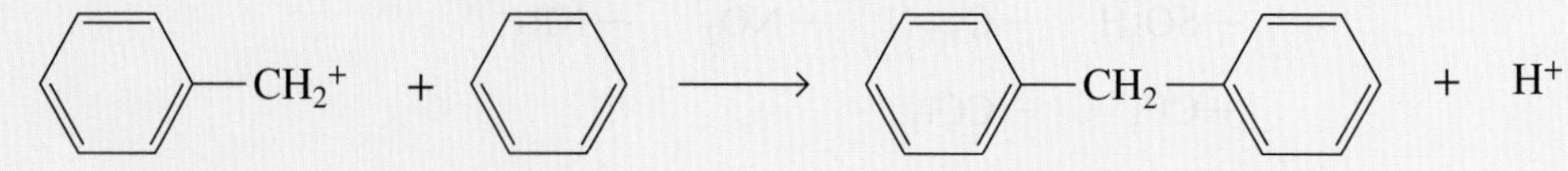

$$C_6H_5CH_2^+ + C_6H_6 \longrightarrow C_6H_5CH_2C_6H_5 + H^+$$

benzyl cation; diphenylmethane

(b) Treatment of benzoyl chloride with aluminium chloride gives an acyl cation. Reaction of this cation with benzene, followed by loss of H^+, gives benzophenone.

$$C_6H_5C^+{=}O \text{ (benzoyl cation)} + C_6H_6 \longrightarrow C_6H_5C(=O)C_6H_5 \text{ (benzophenone)} + H^+$$

SUMMARY

LO 1 Introduction to hydrocarbons

A hydrocarbon is a compound that contains only carbon and hydrogen atoms. The four classes of hydrocarbons are alkanes, alkenes, alkynes and arenes.

LO 2 Alkanes

A saturated hydrocarbon contains only single bonds. Alkanes have the general formula C_nH_{2n+2}. Constitutional isomers have the same molecular formula but a different connectivity (a different order of attachment of their atoms). Alkanes are named according to a set of rules developed by IUPAC and described in chapter 2.

An alkane that contains carbon atoms bonded to form a ring is called a cycloalkane. To name a cycloalkane, prefix the name of the open-chain hydrocarbon with '*cyclo*-'. Compounds with the same molecular formula and the same order of attachment of atoms, but arrangements of atoms in space that cannot be interconverted by rotation around single bonds, are called *cis–trans* isomers. The term *cis* means that substituents are on the same side of the ring; *trans* means that they are on opposite sides of the ring. Most cycloalkanes with substituents on two or more carbon atoms of the ring show *cis–trans* isomerism.

Alkanes are nonpolar compounds, and the only forces of attraction between their molecules are dispersion forces, which are weak electrostatic interactions between temporary partial positive and negative charges on atoms or molecules.

LO 3 Alkenes and alkynes

An alkene is an unsaturated hydrocarbon that contains a carbon–carbon double bond. Alkenes have the general formula C_nH_{2n}. An alkyne is an unsaturated hydrocarbon that contains a carbon–carbon triple bond. Alkynes have the general formula C_nH_{2n-2}. We show the presence of a carbon–carbon double bond by changing the naming component of the parent hydrocarbon from *-an-* to *-en-*. We show the presence of a carbon–carbon triple bond by changing the naming component of the parent alkane from *-an-* to *-yn-*.

The structural feature that makes *cis–trans* isomerism possible in alkenes is restricted rotation around the two carbon atoms of the double bond. If atoms of the parent chain are on the same side of the double bond, the configuration of the alkene is *cis*; if they are on opposite sides, the configuration is *trans*. Using a set of priority rules, we can also specify the configuration of a carbon–carbon double bond by the *E,Z* system. If the two groups of higher priority are on the same side of the double bond, the configuration of the alkene is *Z*; if they are on opposite sides, it is *E*.

LO 4 Reactions of alkanes

Alkanes have very limited reactivity. The most important reaction of alkanes is combustion with oxygen, which is the basis of their use as energy sources.

LO 5 Reactions of alkenes

A characteristic reaction of alkenes is addition, during which a π bond is broken and σ bonds to two new atoms or groups of atoms are formed. This process is described as *syn* addition or *anti* addition depending on whether the atoms appear on the same or opposite sides of the original carbon–carbon bond.

An electrophile is any molecule or ion that can accept a pair of electrons to form a new covalent bond. All electrophiles are Lewis acids. The rate-determining step in electrophilic addition to an alkene is the reaction of an electrophile with a carbon–carbon double bond to form a carbocation — an ion that contains a carbon atom with only six electrons in its valence shell and that has a positive charge. Carbocations are planar, with bond angles of 120° around the positive carbon atom. The order of stability of carbocations is 3° > 2° > 1° > methyl.

LO 6 Reactions of alkynes

Alkynes can undergo similar reactions to alkenes: addition, reduction and oxidation. Of particular importance is reduction using hydrogen gas where the choice of catalyst determines whether an alkene or an alkane is produced.

LO 7 Aromatic compounds

Benzene and its alkyl derivatives are classified as aromatic hydrocarbons, or arenes. The concepts of hybridisation of atomic orbitals and the theory of resonance are used to describe the structure of benzene. The resonance energy of benzene is about $152\,kJ\,mol^{-1}$.

According to the Hückel criteria for aromaticity, a ring is aromatic if it (1) has one *p* orbital on each atom of the ring, (2) is planar, so that overlap of all *p* orbitals of the ring is continuous or nearly so, and (3) has 2, 6, 10, 14, 18, and so on (i.e. $4n + 2$), π electrons in the overlapping system of *p* orbitals. A heterocyclic aromatic compound contains one or more atoms other than carbon in an aromatic ring.

Aromatic compounds are named by the IUPAC system. The common names toluene, xylene, styrene, phenol, aniline, benzoic acid and benzaldehyde are retained. The C_6H_5— group is named phenyl, and the $C_6H_5CH_2$— group is named benzyl. To locate two substituents on a benzene ring, either number the atoms of the ring or use the locators *ortho* (*o*), *meta* (*m*) and *para* (*p*).

Polycyclic aromatic hydrocarbons contain two or more fused benzene rings. Particularly abundant are naphthalene, anthracene and phenanthrene, and their derivatives.

LO 8 Reactions of aromatic compounds: electrophilic aromatic substitution

Aromatic compounds react with many forms of positively charged electrophiles to give substituted aromatic systems.

KEY CONCEPTS AND EQUATIONS

Oxidation of alkanes
(section 16.4)

Alkanes display virtually no reactivity, except under harsh conditions where they may be completely oxidised to carbon dioxide and water. This process liberates energy and provides the basis for their use as sources of heat and power:

$$CH_3CH_2CH_3 + 5O_2 \rightarrow 3CO_2 + 4H_2O + \text{energy}$$

Addition of H—X to alkenes
(section 16.5)

Addition of H—X is regioselective and follows Markovnikov's rule. Reaction occurs in two steps and involves the formation of a carbocation intermediate.

$$\text{1-methylcyclopentene} + HCl \longrightarrow \text{1-chloro-1-methylcyclopentane}$$

Acid-catalysed hydration of alkenes
(section 16.5)

Hydration is regioselective and follows Markovnikov's rule. Reaction involves two steps and forms a carbocation intermediate.

$$CH_3C(CH_3){=}CH_2 + H_2O \xrightarrow{H_2SO_4} CH_3C(CH_3)(OH)CH_3$$

Addition of bromine and chlorine to alkenes
(section 16.5)

Addition occurs in two steps and involves *anti* addition by way of a bridged bromonium or chloronium ion intermediate.

$$\text{cyclohexene} + Br_2 \xrightarrow{CH_2Cl_2} \textit{trans}\text{-1,2-dibromocyclohexane}$$

Reduction of alkenes: formation of alkanes
(section 16.5)

Catalytic reduction involves, predominantly, the *syn* addition of hydrogen.

$$\text{1,2-dimethylcyclohexene} + H_2 \xrightarrow[\text{metal catalyst}]{\text{transition}} \textit{cis}\text{-1,2-dimethylcyclohexane}$$

Reduction of alkynes to alkenes or alkanes
(section 16.6)

The choice of catalyst controls the synthetic outcome of the reduction of alkynes.

$$CH_3{-}CH_2{-}C{\equiv}C{-}CH_3 \xrightarrow[Pd/C]{2H_2} CH_3{-}CH_2{-}CH_2{-}CH_2{-}CH_3 \ \text{(pentane)}$$

$$CH_3{-}CH_2{-}C{\equiv}C{-}CH_3 \xrightarrow[\text{Lindlar catalyst}]{H_2} \textit{cis}\text{-pent-2-ene}$$

$$CH_3{-}CH_2{-}C{\equiv}C{-}CH_3 \xrightarrow[NH_3]{Li} \textit{trans}\text{-pent-2-ene}$$

Chlorination and bromination of aromatic compounds
(section 16.8)

The electrophile is a halonium ion, Cl^+ or Br^+, formed by treating Cl_2 or Br_2 with $AlCl_3$ or $FeCl_3$.

$$\text{benzene} + Cl_2 \xrightarrow{AlCl_3} C_6H_5{-}Cl + HCl$$

Nitration of aromatic compounds
(section 16.8)

The electrophile is the nitronium ion, NO_2^+, formed by treating nitric acid with sulfuric acid.

$$\text{benzene} + HNO_3 \xrightarrow{H_2SO_4} C_6H_5{-}NO_2$$

Sulfonation of aromatic compounds *(section 16.8)*	The electrophile is HSO_3^+. benzene + H_2SO_4 ⟶ benzene—SO_3H + H_2O
Alkylation of aromatic compounds *(section 16.8)*	The electrophile is an alkyl carbocation formed by treating an alkyl halide with a Lewis acid. benzene + $(CH_3)_2CHCl$ $\xrightarrow{AlCl_3}$ benzene—$CH(CH_3)_2$ + HCl
Acylation of aromatic compounds *(section 16.8)*	The electrophile is an acyl cation formed by treating an acyl halide with a Lewis acid. benzene + $CH_3C(=O)Cl$ $\xrightarrow{AlCl_3}$ benzene—$C(=O)CH_3$ + HCl

KEY TERMS

acylation *p. 746*
addition to the carbon–carbon double bond *p. 723*
aliphatic hydrocarbon *p. 698*
alkane *p. 698*
alkene *p. 698*
alkylation *p. 745*
alkyne *p. 698*
angle strain *p. 703*
aniline *p. 740*
anisole *p. 740*
anti addition *p. 731*
anti selectivity *p. 731*
arene *p. 698*
aromatic *p. 734*
aryl group *p. 734*
axial bonds *p. 704*
benzaldehyde *p. 740*
benzoic acid *p. 740*
benzyl group *p. 740*
boat conformation *p. 704*
bridged halonium ion *p. 731*
carbocation *p. 725*
catalytic hydrogenation *p. 731*
catalytic reduction *p. 731*
chair conformation *p. 703*
cis–trans isomers *p. 707*
configurational isomers *p. 713*
conformation *p. 700*
conformer *p. 700*
cycloalkane *p. 702*
cycloalkene *p. 719*
electrophile *p. 723*
electrophilic aromatic substitution *p. 742*
enthalpy of hydrogenation *p. 732*
equatorial bonds *p. 704*
E,Z system *p. 718*
Friedel–Crafts acylation *p. 746*
Friedel–Crafts alkylation *p. 745*
halonium ion *p. 731*
heterocyclic compound *p. 738*
hybridisation of atomic orbitals *p. 735*
hydration *p. 728*
hydrocarbon *p. 698*
inductive effect *p. 727*
Kekulé structure *p. 734*
Markovnikov's rule *p. 724*
meta *p. 740*
Newman projection *p. 700*
nitronium ion *p. 744*
ortho *p. 740*
oxonium ion *p. 729*
para *p. 740*
phenol *p. 740*
phenyl group *p. 740*
polycyclic aromatic hydrocarbon (PAH) *p. 741*
regioselective reaction *p. 724*
resonance energy *p. 736*
saturated hydrocarbon *p. 698*
stereospecific reaction *p. 730*
steric strain *p. 704*
styrene *p. 740*
syn addition *p. 732*
theory of resonance *p. 735*
toluene *p. 740*
torsional strain *p. 703*
unsaturated hydrocarbon *p. 698*
xylene *p. 740*

REVIEW QUESTIONS

LO 1 Introduction to hydrocarbons

16.1 List the four types of hydrocarbon.

16.2 How do unsaturated hydrocarbons differ from alkanes?

16.3 (a) What is the molecular formula for a non-cyclic saturated hydrocarbon that has five carbons?
(b) How does this differ from the molecular formula for a cyclic hydrocarbon that has five carbons?
(c) What would the molecular formula be for a cyclic hydrocarbon that is also an alkene?

LO 2 Alkanes

16.4 Which of these statements are true about constitutional isomers?
(a) They have the same molecular formula.
(b) They have the same molar mass.
(c) They have the same order of attachment of atoms.
(d) They have the same physical properties.

16.5 Write a line structure for each of these structural formulae.
(a) $CH_3C(CH_3)_3$
(b) $(CH_3)_2CHCH(CH_3)_2$
(c) $CH_3CH_2CCH_2CH_3$ with CH_2CH_3 above and CH_2CH_3 below the central C
(d) $(CH_3)_4C$
(e) $CH_3CH_2CHCHCH_2CH_2CHCH_3$ with $CH_2(CH_3)_2$ above the first CH, $CH_2C(CH_3)_3$ below the second CH, and CH_2CH_3 above the last CH
(f) $(CH_3)_3CCHCH_2CH_3$ with CH_2CH_3 below the CH

16.6 Write a condensed structural formula and the molecular formula for each of the following alkanes.

(a) 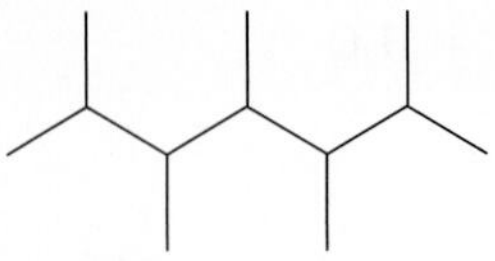

(b) 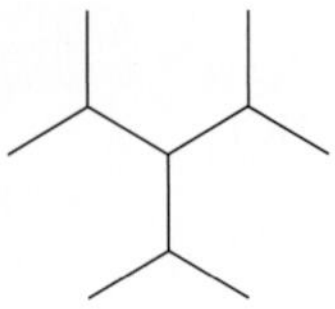

(c) 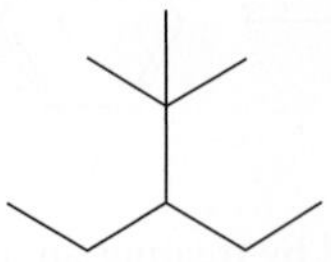

16.7 For each of the following condensed structural formulae, provide an even more abbreviated formula, using parentheses and subscripts.

(a) $CH_3CH_2CH_2CH_2CH_2\underset{}{\overset{\overset{CH_3}{|}}{C}}HCH_3$

(b) $H\overset{\overset{CH_2CH_2CH_3}{|}}{\underset{\underset{CH_2CH_2CH_3}{|}}{C}}CH_2CH_2CH_3$

(c) $CH_3\overset{\overset{CH_2CH_2CH_3}{|}}{\underset{\underset{CH_2CH_2CH_3}{|}}{C}}CH_2CH_2CH_2CH_2CH_3$

16.8 Name and draw line structures for the nine constitutional isomers with the molecular formula C_7H_{16}.

16.9 For each of the following, state whether the two structural formulae shown represent constitutional isomers.

(a)  and 

(b) 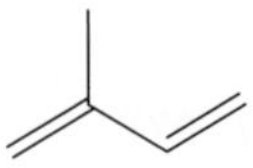 and 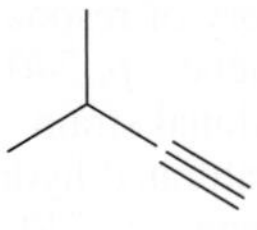

(c) 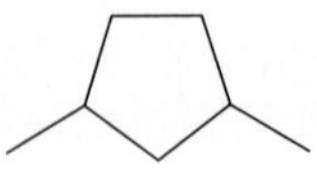 and 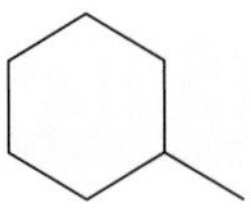

(d) 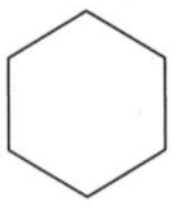 and 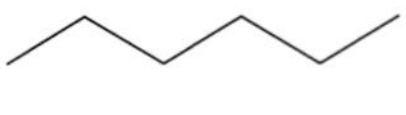

(e) 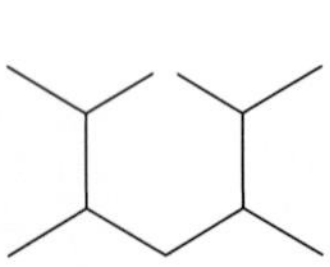 and 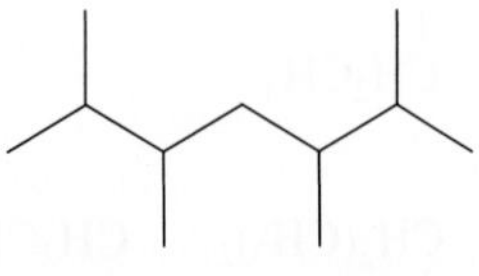

(f) 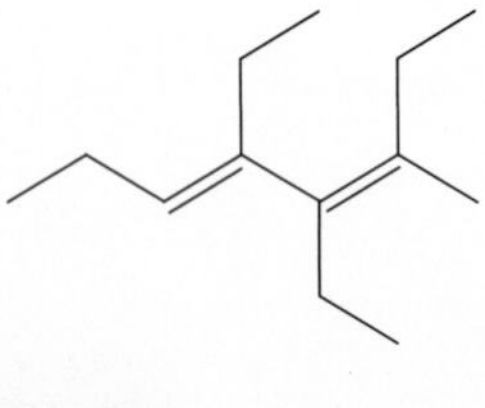 and 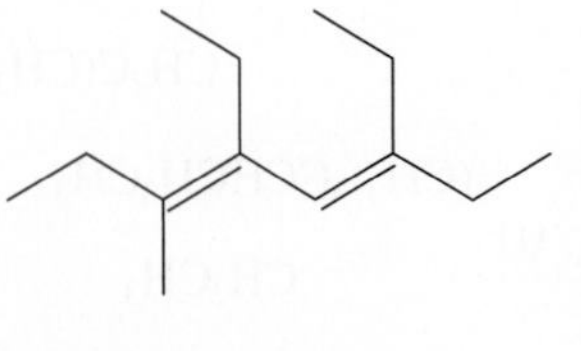

16.10 In each of the following, are the two compounds constitutional isomers?

(a) CH_3CH_2OH and CH_3OCH_3

(b) $CH_3\overset{\overset{O}{\|}}{C}CH_3$ and $CH_3CH_2\overset{\overset{O}{\|}}{C}H$

(c) $CH_3\overset{\overset{O}{\|}}{C}OCH_3$ and $CH_3CH_2\overset{\overset{O}{\|}}{C}OH$

(d) $CH_3\overset{\overset{OH}{|}}{C}HCH_2CH_3$ and $CH_3\overset{\overset{O}{\|}}{C}CH_2CH_3$

(e) 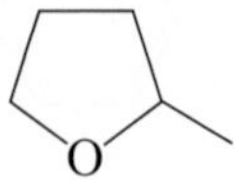 and 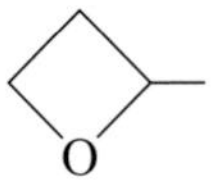

(f) 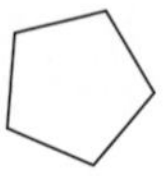 and $CH_2{=}CHCH_2CH_3$

16.11 Write IUPAC names for each of the following alkanes and cycloalkanes.

(a) $CH_3\underset{\underset{CH_3}{|}}{C}HCH_2CH_2CH_3$

(b) $CH_3\underset{\underset{CH_3}{|}}{C}HCH_2CH_2\underset{\underset{CH_3}{|}}{C}HCH_3$

(c) $CH_3(CH_2)_4\underset{\underset{CH_2CH_3}{|}}{C}HCH_2CH_3$

(d) 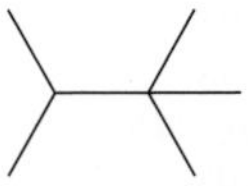

(e) 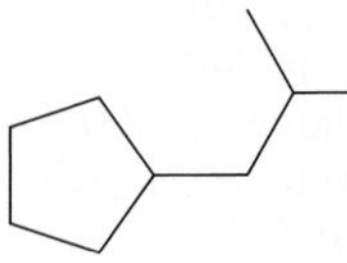

(f) 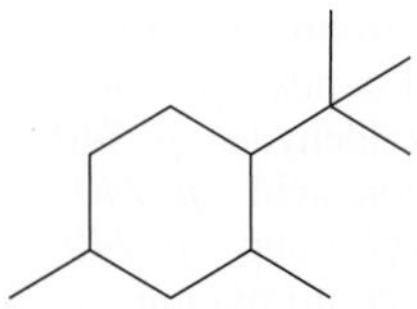

16.12 Write line structures for these alkanes.
(a) 2,2,4-trimethylhexane
(b) 2,2-dimethylpropane
(c) 3-ethyl-2,4,5-trimethyloctane
(d) 5-butyl-2,2-dimethylnonane
(e) 4-isopropyloctane
(f) 3,3-dimethylpentane
(g) *trans*-1,3-dimethylcyclopentane
(h) *cis*-1,2-diethylcyclobutane

16.13 Explain why each of the following names is an incorrect IUPAC name, and write the correct IUPAC name for the intended compound.
(a) 1,3-dimethylbutane
(b) 4-methylpentane
(c) 2,2-diethylbutane
(d) 2-ethyl-3-methylpentane
(e) 2-propylpentane
(f) 2,2-diethylheptane
(g) 2,2-dimethylcyclopropane
(h) 1-ethyl-5-methylcyclohexane

16.14 What structural feature of cycloalkanes makes it possible for them to exhibit *cis–trans* isomerism?

16.15 Is *cis–trans* isomerism possible in alkanes?

16.16 Name and draw structural formulae for the *cis* and *trans* isomers of 1,2-dimethylcyclopentare.

16.17 Name and draw structural formulae for all cycloalkanes with the molecular formula C_5H_{10}. Be certain to include *cis–trans* isomers, as well as constitutional isomers.

16.18 Using a planar pentagon representation for the cyclobutane ring, draw structural formulae for the *cis* and *trans* isomers of the following.
(a) 1,2-dimethylcyclobutane
(b) 1,3-dimethylcyclobutane

LO 3 Alkenes and alkynes

16.19 Draw a structural formula for each of these compounds.
(a) *trans*-3-methylpent-2-ene
(b) 3-methylhex-1-yne
(c) 3-methylbut-1-ene
(d) 3-ethyl-3-methylpent-1-yne
(e) 2,3-dimethylbut-2-ene
(f) *cis*-pent-2-ene
(g) (*Z*)-1-bromopropene
(h) 1-ethylcyclopentene

16.20 Draw a structural formula for each of these compounds.
(a) 1-isopropyl-4-methylcyclohexene
(b) (6*E*)-2,6-dimethylocta-2,6-diene
(c) *trans*-1,2-diisopropylcyclopropane
(d) 2-methylhex-3-yne
(e) 2-chloropropene
(f) tetrachloroethene

16.21 Write the IUPAC name for each of these compounds.

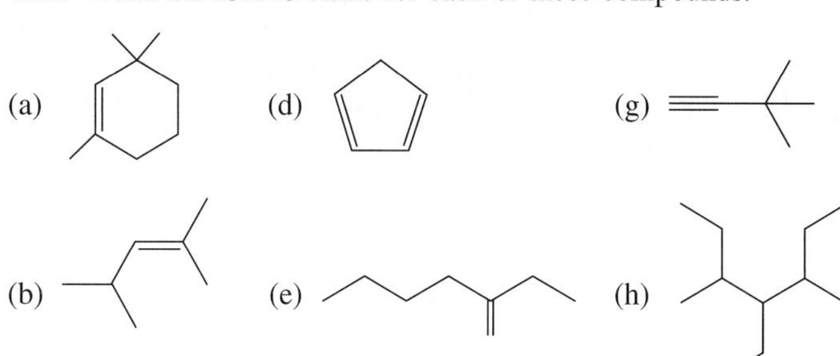

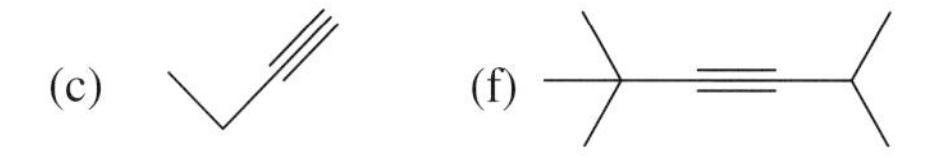

16.22 Write the structural formulae and give the IUPAC names for all possible alkynes with each of the following molecular formulae.
(a) C_5H_8 (b) C_6H_{10} (c) C_8H_{12}

16.23 Give the IUPAC names for the following.

(a) $CH_3CH_2-C\equiv CC(CH_3)_2CH_3$ (with CH_3 above and below the fourth carbon)

(b) $CH_3C\equiv CCH_2C\equiv CCH_2CH_3$

(c) $H_3CHC=C-C\equiv CCHCH_3$ (with CH_3 on the second carbon and on the CH of $CHCH_3$)

(d) $HC\equiv C-CH_2-C\equiv CCH_2CH_3$

16.24 Explain why the following names are incorrect. Give the correct name and draw the structural formula.
(a) but-3-yne
(b) pent-3-yne-1-ene
(c) but-3-en-1-yne

16.25 Explain why these names are incorrect, and write a correct name for the intended compound.
(a) 1-methylpropene
(b) pent-3-ene
(c) 2-methylcyclohexene
(d) 3,3-dimethylpentene
(e) hex-4-yne
(f) 2-isopropylbut-2-ene

16.26 Explain why these names are incorrect, and write a correct name for the intended compound.
(a) 2-ethylprop-1-ene
(b) 5-isopropylcyclohexene
(c) 4-methylhex-4-ene
(d) 2-*sec*-butylbut-1-ene
(e) 6,6-dimethylcyclohexene
(f) 2-ethylhex-2-ene

16.27 Which of these alkenes show *cis–trans* isomerism? For each that does, draw structural formulae for both isomers.
(a) hex-1-ene
(b) hex-2-ene
(c) hex-3-ene
(d) 2-methylhex-2-ene
(e) 3-methylhex-2-ene
(f) 2,3-dimethylhex-2-ene

16.28 Which of these alkenes show *cis–trans* isomerism? For each that does, draw structural formulae for both isomers.
(a) pent-1-ene
(b) pent-2-ene
(c) 3-ethylpent-2-ene
(d) 2,3-dimethylpent-2-ene
(e) 2-methylpent-2-ene
(f) 2,4-dimethylpent-2-ene

16.29 Which of these alkenes can exist as pairs of *cis-trans* isomers? For each alkene that can, draw the *trans* isomer.
(a) $CH_2{=}CHBr$
(b) $CH_3CH{=}CHBr$
(c) $(CH_3)_2C{=}CHCH_3$
(d) $(CH_3)_2CHCH{=}CHCH_3$

16.30 Arrange the groups in each of these sets in order of increasing priority.
(a) $-CH_3$, $-Br$, $-CH_2CH_3$
(b) $-OCH_3$, $-CH(CH_3)_2$, $-CH_2CH_2NH_2$
(c) $-CH_2OH$, $-COOH$, $-OH$
(d) $-CH{=}CH_2$, $-CH{=}O$, $-CH(CH_3)_2$

16.31 For each of the following molecules that possess *cis–trans* isomers, draw the *cis* isomer.

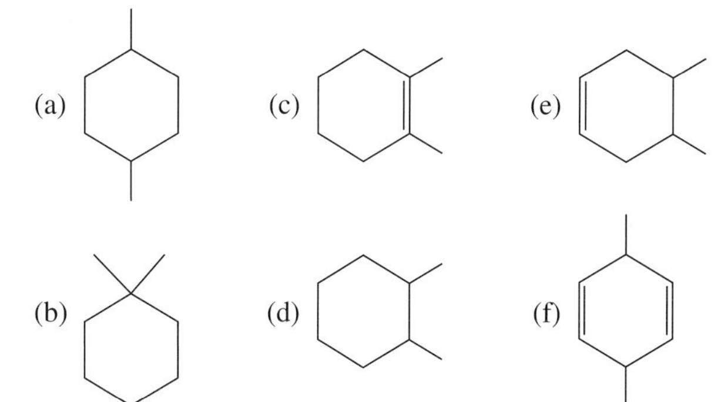

16.32 Explain why each of these names is incorrect or incomplete, and then write a correct name.
(a) (*Z*)-2-methylpent-1-ene
(b) (*E*)-3,4-diethylhex-3-ene
(c) *trans*-2,3-dimethylhex-2-ene
(d) (1*Z*,3*Z*)-2,3-dimethylbuta-1,3-diene

16.33 Draw structural formulae for all compounds with the molecular formula C_5H_{10} that are:
(a) alkenes that do not show *cis–trans* isomerism
(b) alkenes that do show *cis–trans* isomerism
(c) cycloalkanes that do not show *cis–trans* isomerism
(d) cycloalkanes that do show *cis–trans* isomerism.

LO 4 Reactions of alkanes

16.34 Write balanced equations for the combustion of each of the following hydrocarbons. Assume that each is converted completely to carbon dioxide and water.
(a) hexane
(b) cyclohexane
(c) 2-methylpentane

LO 5 Reactions of alkenes

16.35 Select the more stable carbocation from each of these pairs.

(a) $CH_3CH_2\overset{+}{C}H_2$ or $CH_3\overset{+}{C}HCH_3$

(b) $CH_3CH(CH_3)\overset{+}{C}HCH_3$ or $CH_3\overset{+}{C}(CH_3)CH_2CH_3$

16.36 Select the more stable carbocation from each of these pairs.

(a) 1-methylcyclohexyl cation (ring carbon with + and CH_3) or 2-methylcyclohexyl cation

(b) methylcyclohexyl cation (ring +, CH_3) or cyclohexyl–$\overset{+}{C}H_2$

16.37 Write a balanced equation for the combustion of 2-methylpropene in air to give carbon dioxide and water. The oxidising agent is O_2, which makes up approximately 20% of air.

16.38 Draw the product formed by treating each of these alkenes with H_2/Ni.

(a) (b) (c) (d)

16.39 Show how to convert ethene into these compounds.
(a) ethane
(b) ethanol
(c) bromoethane
(d) 1,2-dibromoethane
(e) chloroethane

LO 6 Reactions of alkynes

16.40 Using any alkyne and the required catalyst, describe how you would prepare the following alkenes.
(a) *cis*-oct-2-ene
(b) *trans*-hept-2-ene
(c) 3-methylhex-1-ene

16.41 Predict the products of the reaction of hex-2-yne with the following.
(a) H_2 and Pd/C
(b) H_2 and Na/NH_3
(c) H_2 using Lindlar catalyst
(d) H_2 using Lindlar catalyst followed by the addition of Br_2

16.42 Write equations for the following reactions of alkynes.
(a) pent-2-yne (1 mole) + H_2 (1 mole) + Lindlar catalyst
(b) hex-3-yne (1 mole) + Br_2 (2 moles)
(c) but-2-yne (1 mole) + Li (1 mole) in NH_3(l)
(d) but-1-yne (1 mole) + H_2 (2 moles) + Pd/C catalyst

16.43 What is the product of the following reaction?

C≡CH $\xrightarrow{H_2/Pd}$

LO 7 Aromatic compounds

16.44 Name these compounds.

(a) NO_2, Br
(e) COOH, NO_2, Cl
(b) CH_3, Cl
(f) Br, C_6H_5
(c) OH
(g) C_6H_5, C_6H_5

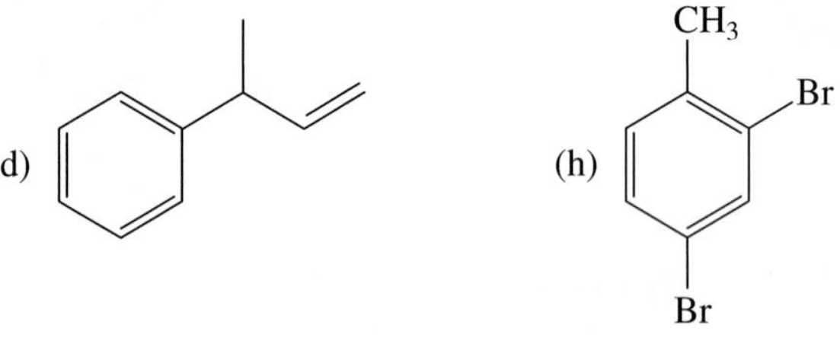

16.45 Draw structural formulae for these compounds.
(a) 1-bromo-2-chloro-4-ethylbenzene
(b) 4-iodo-1,2-dimethylbenzene
(c) 2,4,6-trinitrotoluene
(d) 4-phenylpentan-2-ol
(e) *p*-cresol (*p*-methylphenol)
(f) 2,4-dichlorophenol
(g) 1-phenylcyclopropanol
(h) styrene (phenylethene)
(i) *m*-bromophenol
(j) 2,4-dibromoaniline
(k) isobutylbenzene
(l) *m*-xylene

LO 8 Reactions of aromatic compounds: electrophilic aromatic substitution

16.46 Draw a structural formula for the compound formed by treating benzene with each of the following combinations of reagents.
(a) $CH_3CH_2Cl/AlCl_3$
(b) $CH_2{=}CH_2/H_2SO_4$
(c) CH_3CH_2OH/H_2SO_4

16.47 Show three different combinations of reagents you might use to convert benzene to isopropylbenzene.

16.48 (a) Saturated, non-cyclic hydrocarbons have the general molecular formula C_nH_{2n+2}. If a double bond is introduced this changes to C_nH_{2n}. Can you draw a molecule that has no double bonds and has the molecular formula C_8H_{14}?
(b) How many structures meet this requirement? **LO1**

16.49 Following on from question 16.48, can you draw a saturated hydrocarbon that has the molecular formula C_6H_6? **LO1**

16.50 In question 16.8, you drew structural formulae for all constitutional isomers with molecular formula C_7H_{16}. Predict which of those isomers has the lowest boiling point and which has the highest. **LO2**

16.51 What generalisations can you make about the densities of alkanes relative to that of water? **LO2**

16.52 Which unbranched alkane has about the same boiling point as water (see table 16.1)? Calculate the molar mass of this alkane, and compare it with that of water. **LO2**

16.53 Table 16.1 shows that each $—CH_2$ group added to the carbon chain of an alkane increases the boiling point of the alkane. The increase is greater going from CH_4 to C_2H_6 and from C_2H_6 to C_3H_8 than it is from C_8H_{18} to C_9H_{20} or from C_9H_{20} to $C_{10}H_{22}$. What do you think is the reason for this trend? **LO2**

16.54 As stated in section 16.2, the wax found naturally in apple skins is an unbranched alkane with the molecular formula $C_{27}H_{56}$. Explain how the presence of this alkane prevents the loss of moisture from within an apple. **LO2**

16.55 Predict answers to the following questions about dodecane, $C_{12}H_{26}$, an unbranched alkane. **LO2**
(a) Will it dissolve in water?
(b) Will it dissolve in hexane?
(c) Will it burn when ignited?
(d) Is it a liquid, solid or gas at room temperature and atmospheric pressure?
(e) Is it more or less dense than water?

16.56 The following table shows enthalpies of combustion for methane and propane. **LO4**

Hydrocarbon	Component of	$\Delta_c H^\circ$ (kJ mol^{-1})
CH_4	natural gas	–886
$CH_3CH_2CH_3$	LPG	–2220

On a gram-for-gram basis, which of these hydrocarbons is the better source of heat energy?

16.57 There are three compounds with the molecular formula $C_2H_2Br_2$. Two of these compounds have a dipole moment greater than zero, and one has no dipole moment (see chapter 5, p. 189). Draw structural formulae for the three compounds, and explain why two have dipole moments but the third one does not. **LO2**

16.58 Name and draw structural formulae for all alkenes with the molecular formula C_5H_{10}. As you draw these alkenes, remember that *cis* and *trans* isomers are different compounds and must be counted separately. **LO3**

16.59 Name and draw structural formulae for all alkenes with the molecular formula C_6H_{12} that have the following carbon skeletons (count *cis* and *trans* isomers separately). **LO3**

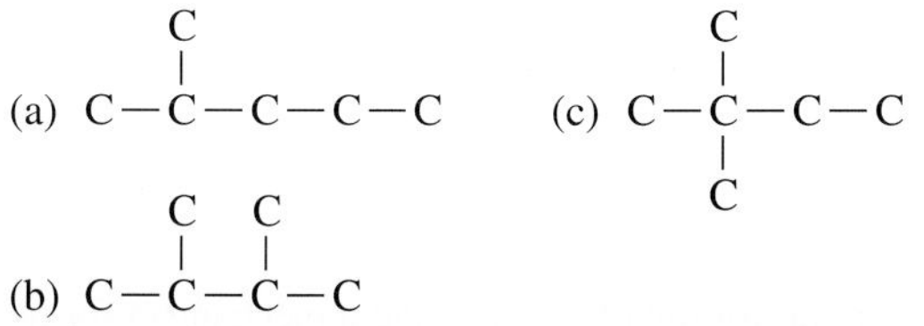

(a) C—C(C)—C—C—C

(b) C—C(C)—C(C)—C

(c) C—C(C)(C)—C—C

16.60 Draw the structural formula for at least one bromoalkene with the molecular formula C_5H_9Br that: **LO3**
(a) shows *E,Z* isomerism (b) does not show *E,Z* isomerism.

16.61 A triene found in the fragrance of cotton blossoms and several essential oils has the IUPAC name (3*Z*)-3,7-dimethylocta-1,3,6-triene and the common name *β*-ocimene. Draw its structural formula. **LO3**

16.62 Determine whether the structures in each of the following sets represent the same molecule, *cis–trans* isomers or constitutional isomers. If they are the same molecule, determine whether they are in the same or different conformations. **LO3**

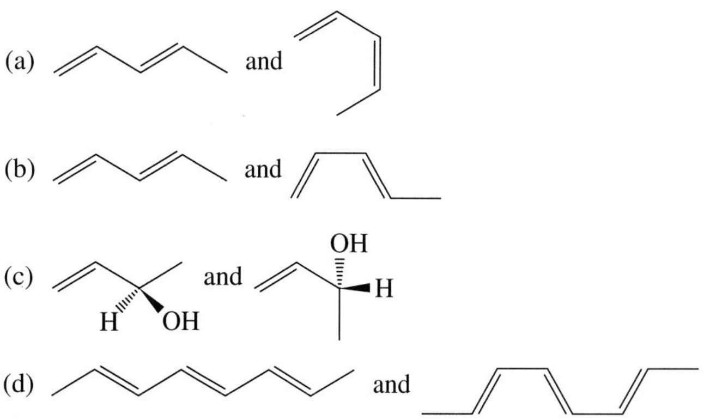

(a) and

(b) and

(c) H OH and OH H

(d) and

16.63 How many moles of CO_2 would be produced when 2 moles of the following are completely combusted? **LO4**
(a) octane
(b) cycloheptane
(c) 1-ethyl-2,3-dimethylcyclohexane

16.64 (a) Draw three of the possible compounds that would be formed if one of the hydrogen atoms of pentane was substituted with a bromine atom.
(b) How many possible compounds can be formed if one of the hydrogen atoms of cyclobutane is substituted with a chlorine atom? **LO4**

16.65 Which alkyne and which reagent could be used to give each of the following? **LO6**
(a) 2,2-dichlorobutane
(b) butan-2-ol
(c) 2,2,3,3-tetrabromobutane

16.66 How would you carry out the following transformations? **LO6**

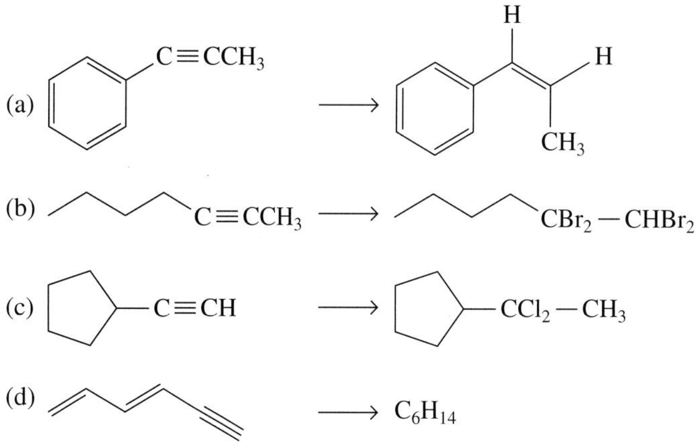

(a) $C{\equiv}CCH_3$ ⟶ H, H, CH_3

(b) $C{\equiv}CCH_3$ ⟶ $CBr_2—CHBr_2$

(c) $C{\equiv}CH$ ⟶ $CCl_2—CH_3$

(d) ⟶ C_6H_{14}

16.67 Draw structural formulae for the isomeric carbocation intermediates formed by the reaction of each of the following alkenes with HCl. Label each carbocation as primary, secondary or tertiary, and state which, if either, of the isomeric carbocations is formed more readily. **LO5**

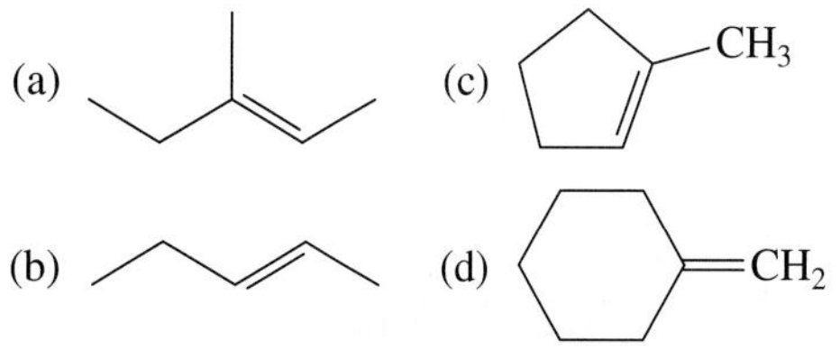

(a)

(b)

(c) CH_3

(d) $=CH_2$

16.68 From each of these pairs of compounds, select the one that reacts more rapidly with HI, draw the structural formula of the major product formed in each case, and explain the basis for your ranking. **LO5**

(a) 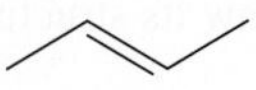 and

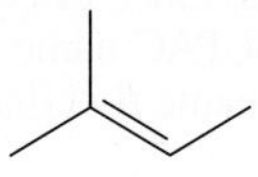

(b) 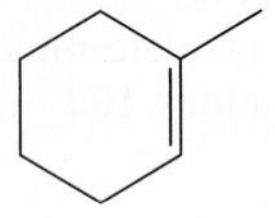 and

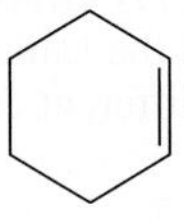

16.69 Complete these equations by predicting the major product formed in each reaction. **LO5**

(a) 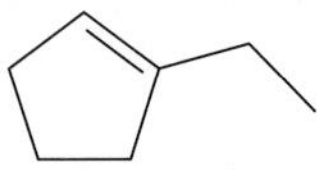 + HCl ⟶

(b) 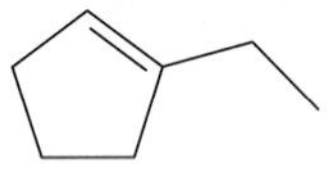 + H_2O $\xrightarrow{H_2SO_4}$

(c) 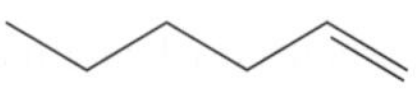 + HI ⟶

(d) 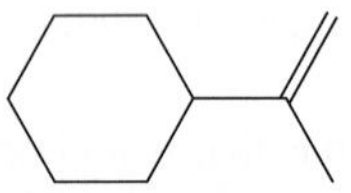 + HCl ⟶

(e) 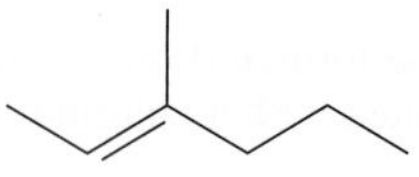 + H_2O $\xrightarrow{H_2SO_4}$

(f)  + H_2O $\xrightarrow{H_2SO_4}$

16.70 The reaction of 2-methylpent-2-ene with each of the following reagents is regioselective. Draw the structural formula for the product of each reaction, and account for the observed regioselectivity. **LO5**
(a) HI
(b) H_2O in the presence of H_2SO_4

16.71 The addition of bromine and chlorine to cycloalkenes is stereospecific. Predict the stereochemistry of the product formed in each of these reactions. **LO5**
(a) 1-methylcyclohexene + Br_2
(b) 1,2-dimethylcyclopentene + Cl_2

16.72 In each of the following reactions, draw the structural formula of an alkene with the indicated molecular formula that gives the compound shown as the major product. Note that more than one alkene may give the same compound as the major product. **LO5**

(a) C_5H_{10} + H_2O $\xrightarrow{H_2SO_4}$ OH

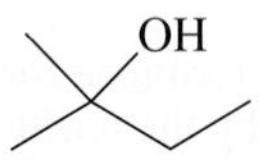

(b) C_5H_{10} + Br_2 ⟶ Br Br

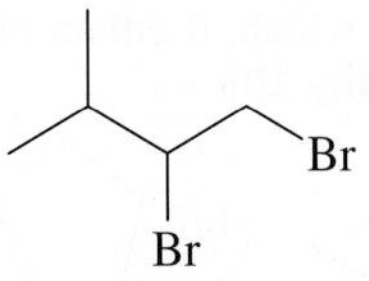

(c) C_7H_{12} + HCl ⟶ Cl

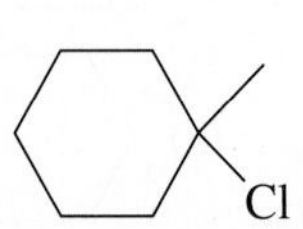

16.73 Draw the structural formula of an alkene with the molecular formula C_5H_{10} that reacts with Br_2 to give each of these roducts. **LO5**

(a) 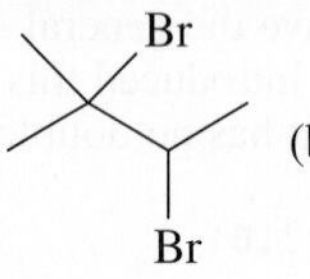 Br Br (b) 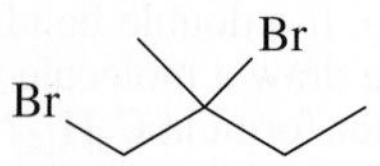 Br Br (c) 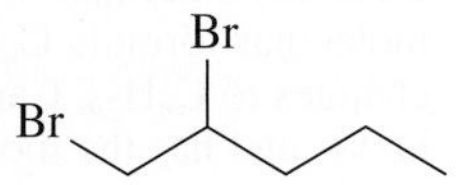 Br Br

16.74 Draw the structural formula of a cycloalkene with the molecular formula C_6H_{10} that reacts with Cl_2 to give each of these compounds. **LO5**

(a) 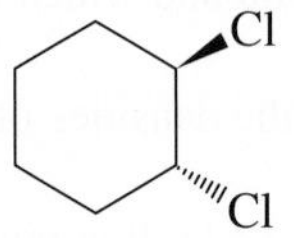 Cl Cl (c) 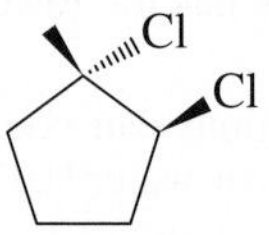 Cl Cl

(b) 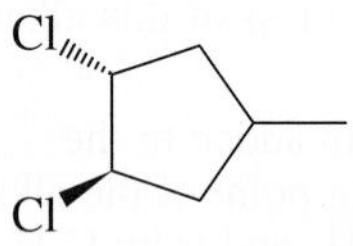 Cl Cl (d) 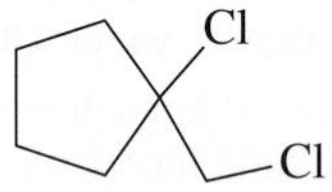 Cl Cl

16.75 Draw the structural formula of an alkene with the molecular formula C_5H_{10} that reacts with HCl to give each of the following chloroalkanes as the major product. **LO5**

(a) 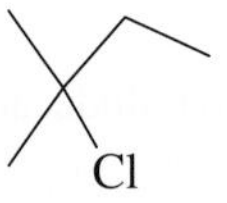 Cl (b) 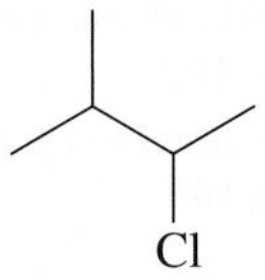 Cl (c) 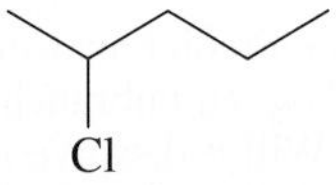 Cl

16.76 The compounds *cis*-hex-3-ene and *trans*-hex-3-ene are different and have different physical and chemical properties. Yet, when treated with H_2O/H_2SO_4, each gives the same alcohol. What is the alcohol, and how do you account for the fact that each alkene gives the same one? **LO5**

16.77 Draw the structural formula of an alkene that undergoes acid-catalysed hydration to give each of the following alcohols as the major product. Note that more than one alkene may give the same compound as the major product. **LO5**
(a) hexan-3-ol
(b) 1-methylcyclobutanol
(c) 2-methylbutan-2-ol
(d) propan-2-ol

16.78 Draw the structural formula of an alkene that undergoes acid-catalysed hydration to give each of the following alcohols as the major product. Note that more than one alkene may give the same compound as the major product. **LO5**
(a) cyclohexanol
(b) 1,2-dimethylcyclopentanol
(c) 1-methylcyclohexanol
(d) 1-isopropyl-4-methylcyclohexanol

16.79 Terpin is prepared commercially by the acid-catalysed hydration of limonene. **LO5**

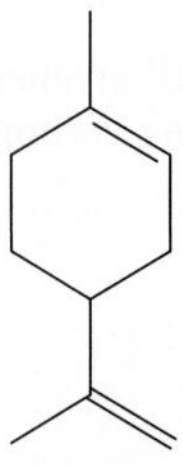

limonene + $2H_2O$ $\xrightarrow{H_2SO_4}$ $C_{10}H_{20}O_2$
terpin

(a) Propose a structural formula for terpin and a mechanism for its formation.
(b) How many *cis–trans* isomers are possible for the structural formula you proposed in (a)?

16.80 The treatment of 2-methylpropene with methanol in the presence of a sulfuric acid catalyst gives *tert*-butyl methyl ether. **LO5**

$$CH_3C(CH_3){=}CH_2 + CH_3OH \xrightarrow{H_2SO_4} CH_3C(CH_3)_2{-}OCH_3$$

2-methylpropene methanol *tert*-butyl methyl ether

Propose a thermodynamically sound mechanism for the formation of this ether.

16.81 The treatment of 1-methylcyclohexene with methanol in the presence of a sulfuric acid catalyst gives a compound with the molecular formula $C_8H_{16}O$. **LO5**

1-methylcyclohexene $+ CH_3OH$ (methanol) $\xrightarrow{H_2SO_4} C_8H_{16}O$

Propose a structural formula for this compound and a mechanism for its formation.

16.82 Hydrocarbon *A*, C_5H_8, reacts with 2 moles of Br_2 to give 1,2,3,4-tetrabromo-2-methylbutane. What is the structure of hydrocarbon *A*? **LO5**

16.83 Alkenes *A* and *B* both have the formula C_5H_{10}. Both react with H_2/Pt and with HBr to give identical products. What are the structures of *A* and *B*? **LO5**

16.84 Show how to convert cyclopentene into these compounds. **LO5**

(a) Br, Br (b) OH (c) Br (d)

16.85 Show how to convert but-1-ene into these compounds. **LO5**

(a) butane (c) 2-bromobutane
(b) butan-2-ol (d) 1,2-dibromobutane

16.86 Show how the following compounds can be synthesised in good yields from an alkene. **LO5**

(a) Br (b) OH (c) Br, Br

16.87 Show that pyridine can be represented as a hybrid of two equivalent contributing structures. **LO6**

16.88 Show that naphthalene (pp. 741–2) can be represented as a hybrid of three contributing structures. Show also, by the use of curved arrows, how one contributing structure is related to the next. **LO6**

16.89 Draw four contributing structures for anthracene (p. 741). **LO6**

16.90 Which of the following compounds are aromatic? **LO6**

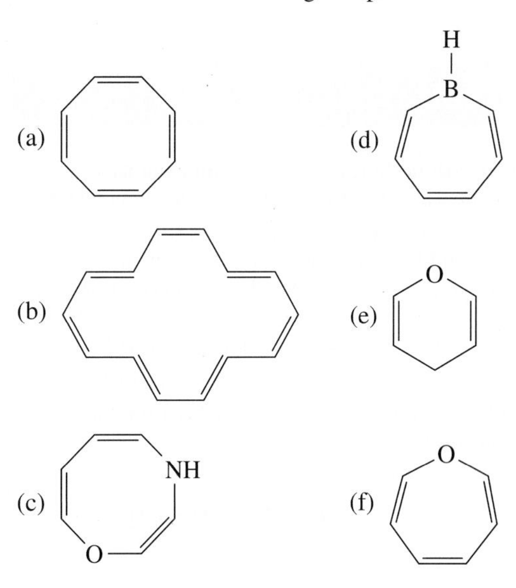

(c) NH, O (f) O

16.91 Explain why cyclopentadiene ($pK_a = 16$) is many orders of magnitude more acidic than cyclopentane ($pK_a > 50$) (*Hint:* Draw the structural formula of the anion formed by removing one of the protons on the $—CH_2—$ group, and then apply the Hückel criteria for aromaticity.) **LO6**

cyclopentadiene cyclopentane

16.92 How many monochlorination products are possible when naphthalene is treated with $Cl_2/AlCl_3$? **LO6**

16.93 Write a stepwise mechanism for the following reaction, using curved arrows to show the flow of electrons in each step. **LO6**

benzene + *tert*-butyl chloride (Cl) $\xrightarrow{AlCl_3}$ *tert*-butylbenzene + HCl

16.94 Write a stepwise mechanism for the preparation of diphenylmethane by treating benzene with dichloromethane in the presence of an aluminium chloride catalyst. **LO6**

16.95 Using styrene, $C_6H_5CH{=}CH_2$, as the only aromatic starting material, show how to synthesise the following compounds. In addition to styrene, use any other necessary organic or inorganic chemicals. Any compound synthesised in one part of your answer may be used to make any other compound in the answer. **LO6**

(a) C_6H_5–CH(Br)CH$_3$ (c) C_6H_5–CH$_2$CH$_3$

(b) C_6H_5–CH(OH)CH$_3$

16.96 When naphthalene is treated with bromine in the presence of a catalyst, how many possible dibromo naphthalenes could be formed? **LO7**

16.97 Name the following aromatic compounds. **LO7**

16.98 Give the product of the following electrophilic aromatic substitution reactions. **LO8**

16.99 An unknown compound is found to have the molecular formula C_7H_8. This implies a considerable degree of unsaturation in the molecule, or a number of cyclic structures are present. When treated with bromine no reaction takes place (see p. 730), which indicates no double bonds are present. Draw two possible structures for compounds that would meet these criteria.

When reacted with bromine in the presence of ferric bromide however the unknown compound readily reacts to give a new species with the molecular formula C_7H_7Br.

What is the structure and name of the unknown compound? **LO8**

ADDITIONAL EXERCISES

16.100 Explain why 1,2-dimethylcyclohexane can exist as *cis–trans* isomers, while 1,2-dimethylcyclododecane cannot. **LO3**

16.101 Each carbon atom in ethane and in ethene is surrounded by eight valence electrons and has four bonds to it. Explain how the valence-shell-electron-pair repulsion (VSEPR) model (section 5.4) predicts a bond angle of 109.5° around each carbon atom in ethane, but an angle of 120° around each carbon atom in ethene. **LO1**

16.102 Use the valence-shell-electron-pair repulsion (VSEPR) model to predict all bond angles around each of the following highlighted carbon atoms. **LO3**

(a) (c) $HC{\equiv}C{-}CH{=}CH_2$

(b) $-CH_2OH$ (d)

16.103 For each highlighted carbon atom in question 16.102, identify which orbitals are used to form each π bond and which are used to form each π bond. **LO3**

16.104 Predict all bond angles around each of the following highlighted carbon atoms. **LO3**

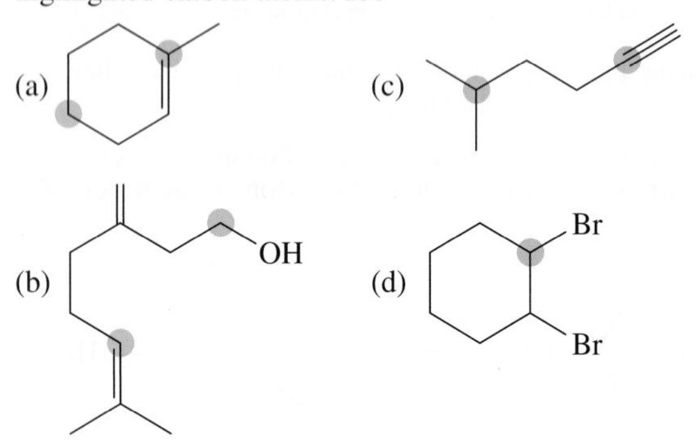

16.105 For each highlighted carbon atom in question 16.104, identify which orbitals are used to form each σ bond and which are used to form each π bond. **LO3**

16.106 The structure of propa-1,2-diene (allene) is shown below. The plane created by H—C—H of C(1) is perpendicular to that created by H—C—H of C(3). **LO3**

propa-1,2-diene (allene) ball-and-stick model

(a) State the orbital hybridisation of each carbon atom in allene.
(b) Account for the molecular geometry of allene in terms of the orbital overlap model. Specifically, explain why all four hydrogen atoms are not in the same plane.

16.107 Each of the following secondary carbocations is more stable than the tertiary carbocation shown. **LO5**

a tertiary carbocation

(b) (c)

Explain why each of the carbocations might be more stable than the tertiary carbocation.

16.108 Oleic acid and elaidic acid are, respectively, the *cis* and *trans* isomers of octadec-9-enoic acid. One of these fatty acids, a colourless liquid that solidifies at 4 °C, is a major component of butterfat. The other, a white solid with a melting point of 44–45 °C, is a major component of partially hydrogenated vegetable oils. Which of these two fatty acids is the *cis* isomer and which is the *trans* isomer? **LO3**

16.109 Recall that an alkene possesses a π cloud of electrons above and below the plane of the C=C bond. Any reagent can, therefore, react with either face of the double bond. Determine whether the reaction of each of the following reagents with the top face of *cis*-but-2-ene will produce the same product as the reaction of the same reagent with the bottom face. (*Hint:* Build molecular models of the products and compare them.) **LO5**

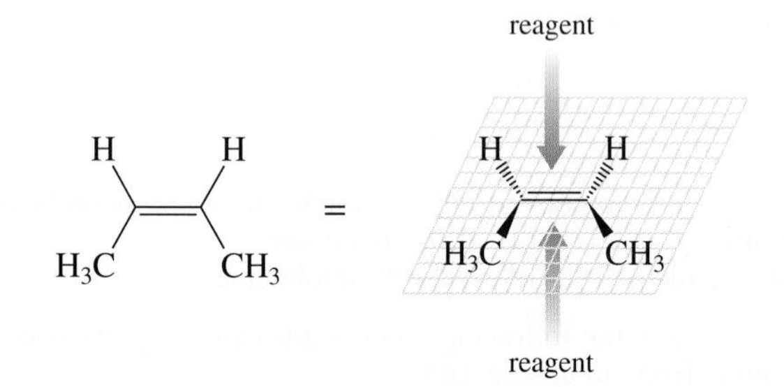

(a) H_2/Pt
(b) Br_2/CH_2Cl_2

16.110 The bombardier beetle generates *p*-quinone, an irritating chemical, by the enzyme-catalysed oxidation of hydroquinone, using hydrogen peroxide as the oxidising agent. Heat generated in this oxidation produces superheated steam, which is ejected, along with *p*-quinone, with explosive force. **LO8**

hydroquinone $+ H_2O_2 \xrightarrow{\text{enzyme catalyst}}$ *p*-quinone $+ H_2O$ + heat

(a) Balance the equation.
(b) Is this an oxidation reaction (see chapter 12)?

17 Chirality

More than 175 years ago, a German chemist named Justus von Liebig developed a method to coat glass with metallic silver by the chemical reduction of silver nitrate. The invention of silvered-glass mirrors allowed mirrors to be manufactured on a large scale, and for the first time in history, ordinary people could afford to buy a mirror and look at their reflections. Today mirrors are common and inexpensive items of furniture and decoration in our homes and businesses. We have become familiar with looking at our images in mirrors and used to following our movements when shaving, applying make-up and brushing our hair, even though the images we see are actually reversed versions of our true selves.

Reflected images are surprisingly common in nature. About one-quarter of identical twins are 'mirror-image twins' for instance, with skin features such as moles appearing on the opposite sides of their bodies and with teeth arrangements and hair patterns swapped from one side to the other. It is not unusual for one mirror-image twin to be left-handed and the other right-handed. This mirrored symmetry even carries over to the genetic level, with several examples recorded of such twins developing diseases such as cancers at similar times, but in the opposite sides of their bodies.

Molecules can also be found as mirror-image twins. Chemists call these forms of isomers 'enantiomers'. Molecules that are enantiomers involve the same atoms and the same sequence of atom connection, but the atoms are positioned differently in three-dimensional space. The differences in such isomers can be subtle and hard to identify, but can lead to substantially different properties, especially in the chemistry of our bodies. For example, the drug thalidomide was developed in the early 1960s to treat morning sickness in pregnant women. Tragically, thalidomide exists as two enantiomers, one of which is an effective treatment for morning sickness, but the other causes deformities in the developing foetus. The Australian obstetrician Dr William McBride first alerted the world to the link between thalidomide and birth defects in a letter to the medical journal *The Lancet* in December 1961.

It took until 2010, however, before scientists were able to discover the exact cause behind the problem: one enantiomer of thalidomide inhibits production of a crucial protein called cereblon that creates the key enzymes needed for limb development. Today, pharmaceutical companies are required to prove that both forms of any such drugs are safe for their intended use.

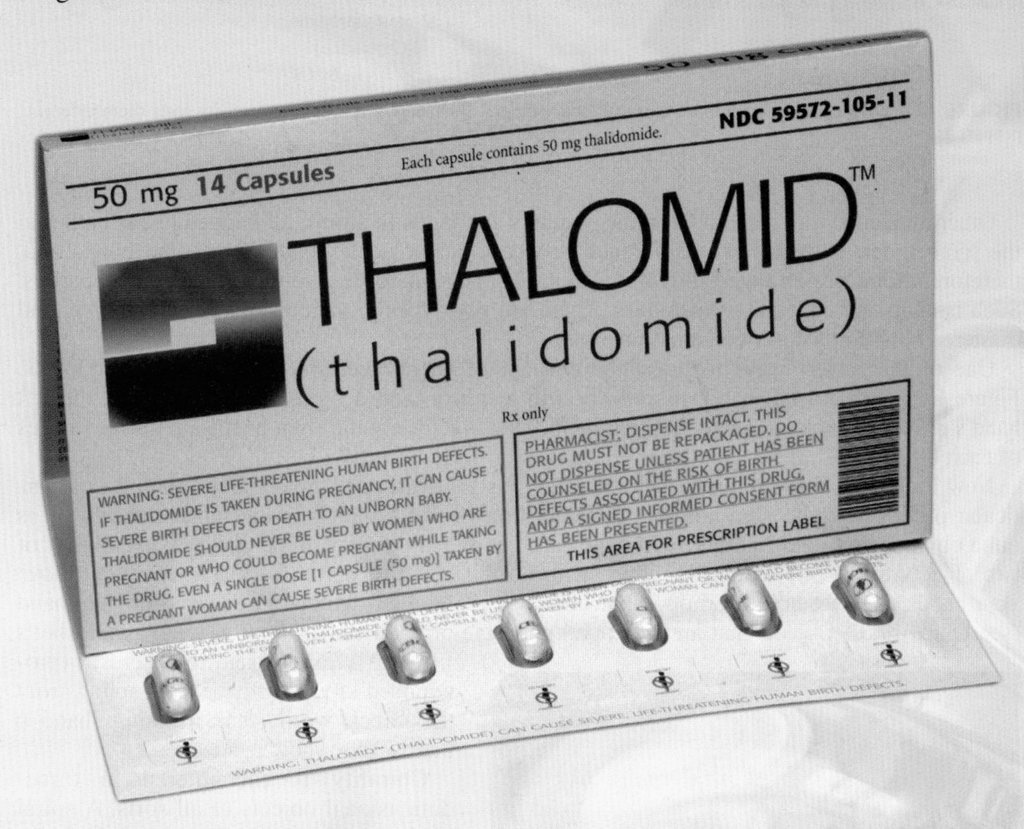

LEARNING OBJECTIVES

After studying this chapter, you should be able to:

17.1 define stereoisomers and recognise stereocentres

17.2 recognise enantiomers as non-equivalent mirror images

17.3 apply the *R*,*S* system to name enantiomers

17.4 differentiate isomers with more than one stereocentre

17.5 define and measure optical activity

17.6 describe how enantiomers are important in biology

17.7 describe the production of single enantiomer drugs.

17.1 Stereoisomers

In chapter 2 we learned that **isomers** are molecules with the same molecular formula (the same numbers of each type of atom) but different structures. We described how different sequences of atom connectivity lead to molecules with different properties and we call these **constitutional isomers**. For example, butane and 2-methylpropane (isobutane) have the same numbers of carbon atoms (four) and hydrogen atoms (ten), but their carbon atoms are connected in a different sequence. Butane has a linear chain and isobutane has a branch, as shown in figure 17.1. These different sequences can give rise to differences in properties, such as boiling point.

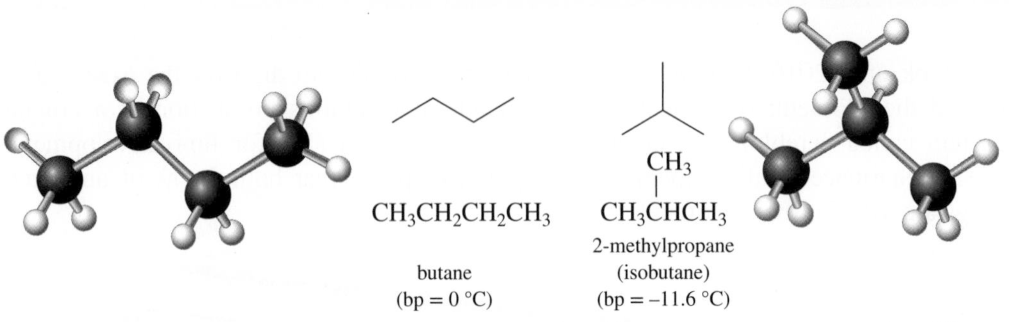

FIGURE 17.1 Butane and isobutane are constitutional isomers. Their different atom connectivity gives them different properties.

mirror image of left hand

left hand

right hand

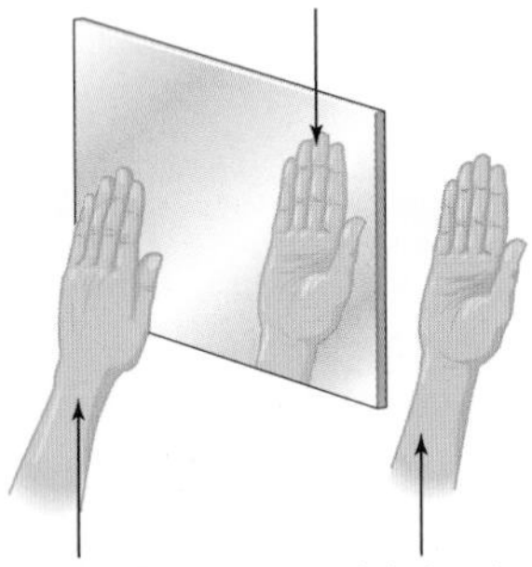

FIGURE 17.2 The left and right hands are nonsuperimposable mirror images of each other. Stereoisomers that are nonsuperimposable mirror images of each other are called enantiomers.

Other molecules can contain the same numbers and kinds of atoms, all bonded to each other in the same order, with the only difference between the molecules being that some bonds and, therefore, atoms are arranged differently in space. This can also give them different properties. Such isomers are called **stereoisomers**. There are two types of stereoisomers: enantiomers and diastereomers. Let's describe enantiomers first.

One of the easiest ways to grasp the concept of enantiomers is by looking at our hands. Figure 17.2 shows a left hand, its reflection in a mirror, and a right hand. Notice that the left hand's reflection looks just like the right hand. That is, our left and right hands are mirror images of each other.

Now look at your own hands. The left hand has four different fingers and a thumb in a particular order, the right hand has four different fingers and a thumb in the same order, but it is quite obvious that your left and right hands are not the same. Superimposability is the ability of two objects to fit 'one within the other' with no mismatch of parts. No matter how you move your hands, they are not superimposable. Just try putting a left-handed glove on your right hand (see figure 17.3). Objects that are nonsuperimposable on their mirror images, such as the shells in figure 17.5, are said to be **chiral** (pronounced ki-ral to rhyme with spiral, from the Greek word *cheir* meaning 'hand'); that is, they show handedness.

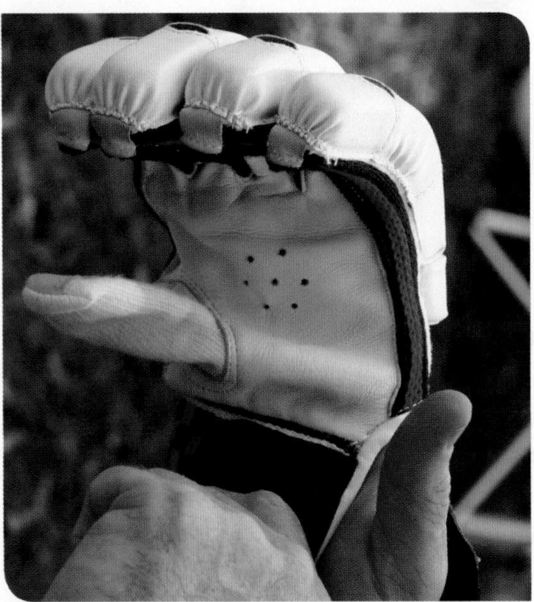

FIGURE 17.3 A left-handed glove does not fit; that is, it is not superimposable on the right hand.

Chirality is encountered in three-dimensional objects of all sorts. A spiral binding on a notebook is chiral. A clock-face is also chiral. As you examine the world around you, you will see that many objects are chiral (e.g. a ceiling fan, the thread of a screw, a computer keyboard). Chirality even underpins the workings of things that you cannot see. Modern liquid crystal displays that are found in HD televisions, smartphones (figure 17.4) and computer screens work because the structures of the materials that are present in these displays are arranged in chiral orientations. *Even individual molecules can be chiral.* Such molecules, stereoisomers that cannot be superimposed on their mirror images, are called called **enantiomers**.

FIGURE 17.4 LCD screens rely on materials that are arranged in chiral orientations.

FIGURE 17.5 The next time you are at the beach, keep an eye out for a coiled shell. You will see that the coiling forms a spiral. Most coiled shells spiral in the same direction; with the shell apex pointing upwards, the shell's opening is to the right, as in these examples. While this is the case for most coiled shells, it is not true for all species; some occur in both left-handed and right-handed forms.

The two types of molecules present in the drug thalidomide are enantiomers. A simpler example of enantiomers is glucose (β-D-glucopyranose) and its mirror image, as can be seen in figure 17.6.

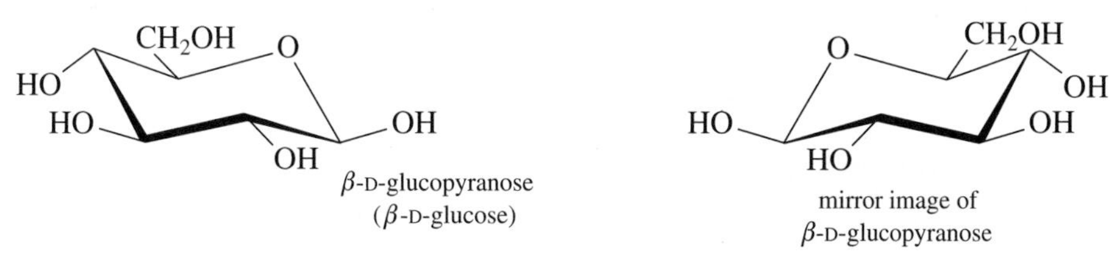

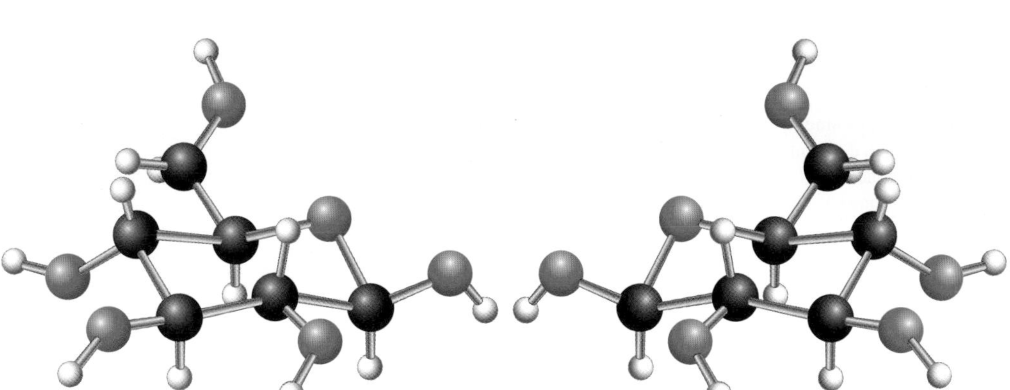

FIGURE 17.6 A pair of enantiomers: β-D-glucopyranose and its nonsuperimposable mirror image.

If our hands were superimposable, we would have two left hands or two right hands. If the mirror images of molecules were superimposable, they would be the same molecule. Objects that have superimposable mirror images are said to be **achiral**. That is, they do not exhibit handedness.

Imagine for a moment that you can make the little finger of your right hand stand at 90° to the other fingers on your right hand (perhaps you can). Hold it next to your left hand. There are still four fingers and a thumb on each hand, connected in the same sequence, but, as shown in figure 17.7, the hands no longer look like mirror images. That is, if you held your left hand to a mirror, its reflection would not look like the right hand in this position.

Stereoisomers that are not mirror images of each other are called **diastereomers**. An example of diastereomers that we have already encountered (p. 707) is shown in figure 17.8 (overleaf): *cis–trans* isomers in cycloalkanes. (Remember that the solid wedge represents a bond pointing above the plane of the page and a hashed wedge represents a bond

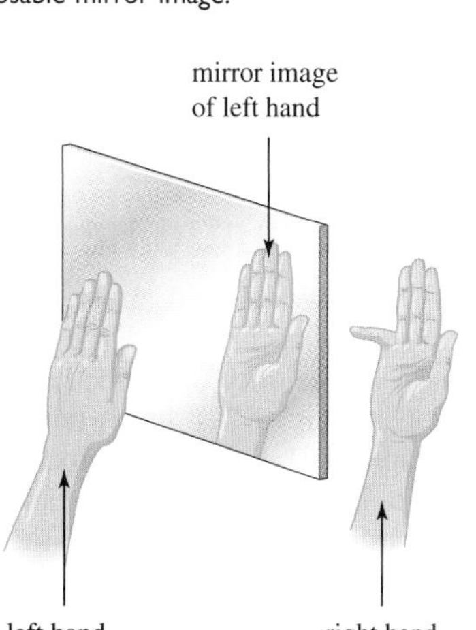

FIGURE 17.7 Provided that you don't move the little finger on your right hand from this position, the left and right hands are no longer mirror images of each other. Stereoisomers that are not mirror images of each other are called diastereomers.

pointing below the plane of the page.) In *cis*-1,2-dimethylcyclopentane, both —CH_3 groups are above the plane of the page. In *trans*-1,2-dimethylcyclopentane, one —CH_3 group is below the plane of the page and one is above the plane of the page.

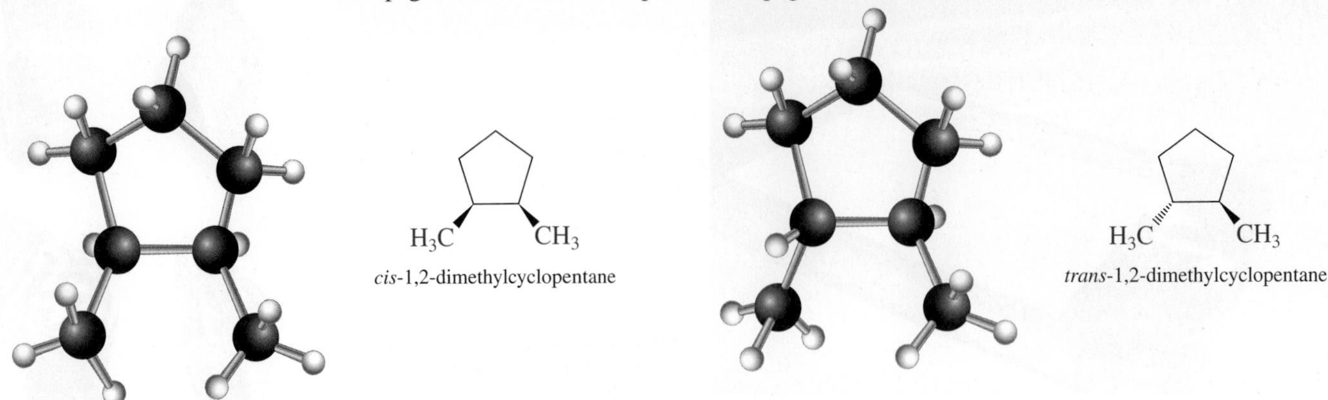

FIGURE 17.8 The *cis–trans* isomers of 1,2-dimethylcyclopentane are a pair of diastereomers. They are stereoisomers, but they are not mirror images and so they are not enantiomers.

The different types of isomers are summarised in figure 17.9. This chapter will focus on enantiomers and, to a lesser extent, on diastereomers. The significance of diastereomers will become clear when carbohydrates are discussed in chapter 22.

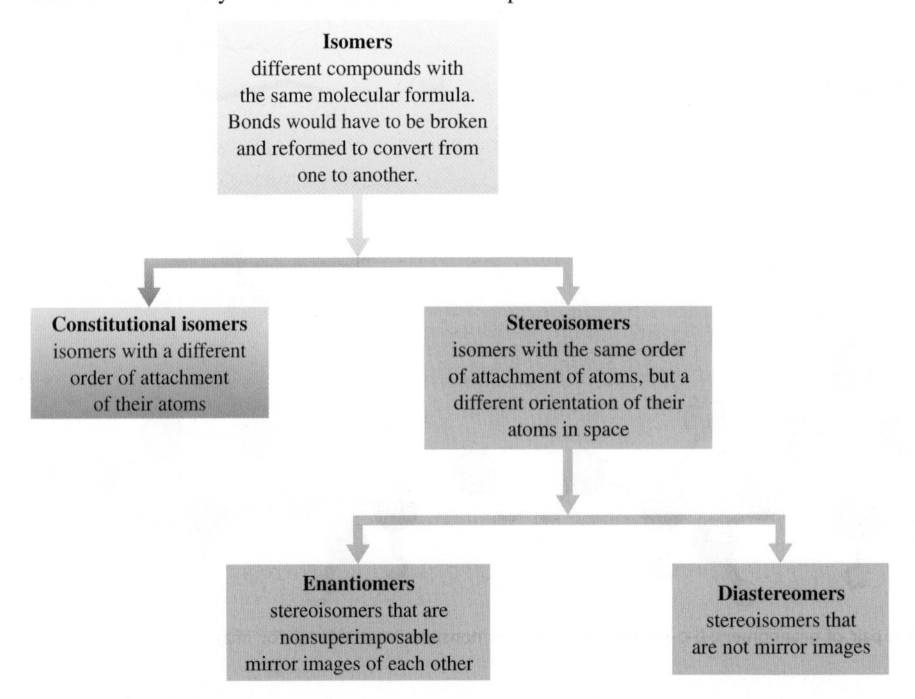

FIGURE 17.9 Relationships among isomers. (*Note:* This description excludes conformers (see p. 700), which can interconvert without breaking bonds.)

17.2 Enantiomerism

As we have learned above, enantiomers are stereoisomers that are nonsuperimposable mirror images of each other. Except for inorganic compounds and a few simple organic compounds, the vast majority of molecules in the biological world show enantiomerism, including carbohydrates (chapter 22), lipids (chapter 23), amino acids and proteins (chapter 24), and nucleic acids (chapter 25). Further, approximately half of all pharmaceuticals show enantiomerism.

To understand the significance of enantiomerism, recall that enantiomers have some different properties. While they have the same boiling points, melting points and solubilities, each of a pair of enantiomers reacts differently towards other chiral molecules. This is especially important in biology. For example, one form of thalidomide acts in the body to produce a sedative/hypnotic effect that controls the symptoms of morning sickness, whereas the other form acts to produce birth defects.

As the structure of thalidomide is complicated, we will start by looking at a simpler example. In figure 17.10, different atoms are presented as different shapes and colours to help us visualise and mentally manipulate the molecules in three dimensions. In figure 17.10, (a) is an atom with four different groups attached to it. We can use a mirror to produce (b), its mirror image. Imagine that we can lift (b) out of the mirror. No matter how we try to move (b), we cannot superimpose the two molecules. Because they are mirror images and nonsuperimposable, they are enantiomers.

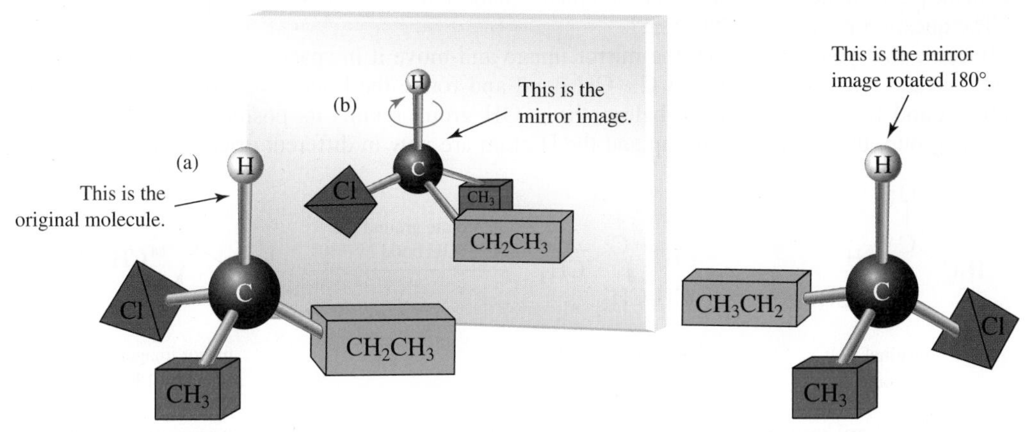

FIGURE 17.10 Enantiomers are mirror images of each other that cannot be superimposed, no matter how we rotate the original molecule and its mirror image.

Now let's look at a real example. The molecule butan-2-ol, a natural by-product of fermentation used as a flavouring agent and as a solvent for paints, has four different groups bonded to a central carbon atom.

$$CH_3CH(OH)CH_2CH$$

butan-2-ol

The structural formula above does not show the shape of butan-2-ol or the orientation of its atoms in space. Let's look at the molecule in three dimensions:

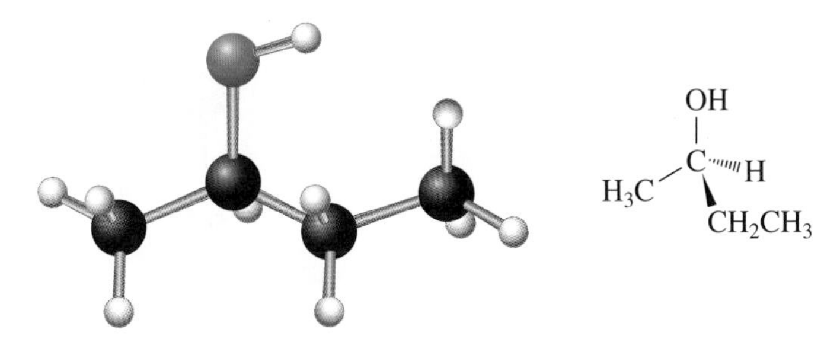

On the left is a ball-and-stick model and on the right is how we draw the three-dimensional line structure.

Now let's look at the mirror image of this structure. Imagine if you could somehow look into both the 'mirror world' and the real world. Then you could see both the original molecule and its mirror image as three-dimensional molecules.

OH, H_3C, C, H, CH_2CH_3 — original | HO, H, C, CH_3, CH_3CH_2 — mirror image → OH, HC_3, C, H, CH_2CH_3 — real world | HO, H, C, CH_3, H_3CCH_2 — 'mirror world' → OH, H_3C, C, H, CH_2CH_3 — original | HO, H, C, CH_3, CH_3CH_2 — mirror image

On the left at the bottom of the previous page is what we will call the original molecule. On the right is the mirror image. At first glance, they may look the same: The —OH and —CH_3 groups on C(2) are in the plane of the paper, the —H is behind the plane and the —CH_2CH_3 group is in front of it.

The question is: are they the same?

Imagine that you can pick up the mirror image and move it in space in any way you wish. If you hold the mirror image by the C—OH bond and rotate the bottom part of the molecule by 180° around this bond as shown below, the —OH group retains its position in space, but the —CH_3 group, the —CH_2CH_3 group and the H atom are now in different orientations.

original molecule | mirror image of original molecule —(rotate around the C—OH bond by 180°)→ mirror image rotated by 180°

No matter how you move the mirror image in space, you cannot fit (superimpose) it on the original so that all bonds and atoms match.

mirror image rotated by 180° → H pointing towards you; CH_2CH_3 pointing away

original molecule → H pointing away; CH_2CH_3 pointing towards you

No matter which way you orient the molecule, as long as no bonds are broken or rearranged, only two of the four groups bonded to the central carbon atom of the mirror image can be made to coincide with those on the original. The molecules are mirror images *and* nonsuperimposable. Recall that we describe such objects as *chiral* and such molecules as *enantiomers*. Like gloves, each enantiomer always has a partner; enantiomers occur in pairs with one being the reflection of the other.

Not all mirror images of molecules are different. If we can move the mirror image in space and find that it fits over the original so that every bond, atom and detail of the mirror image exactly matches the bonds, atoms and details of the original, then the two are *superimposable*. This is shown in figure 17.11: (a) is an atom with four groups attached to it, and we can use a mirror to produce (b), its mirror image. Imagine that we can lift (b) out of the mirror. We can see that we can indeed superimpose the two molecules. In this case, the mirror image and the original represent the same molecule.

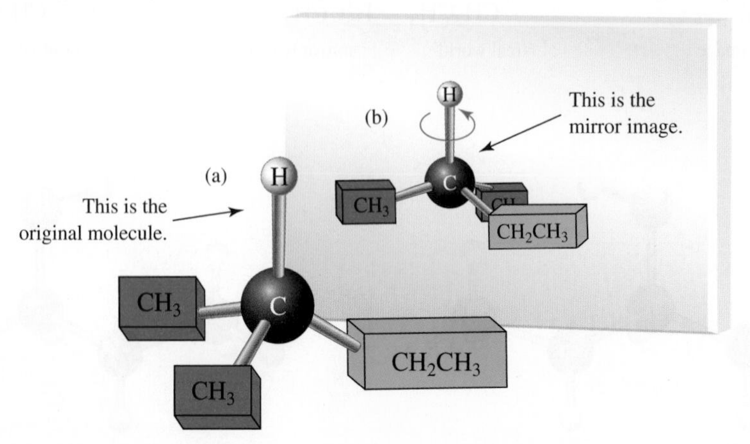

FIGURE 17.11 Superimposable mirror images are exactly the same molecule.

Let's consider a molecule such as propan-2-ol, which has a central carbon atom bonded to four groups, two of which are the same, a $—CH_3$ group.

The following diagram shows a three-dimensional representation of propan-2-ol on the left and its mirror image on the right.

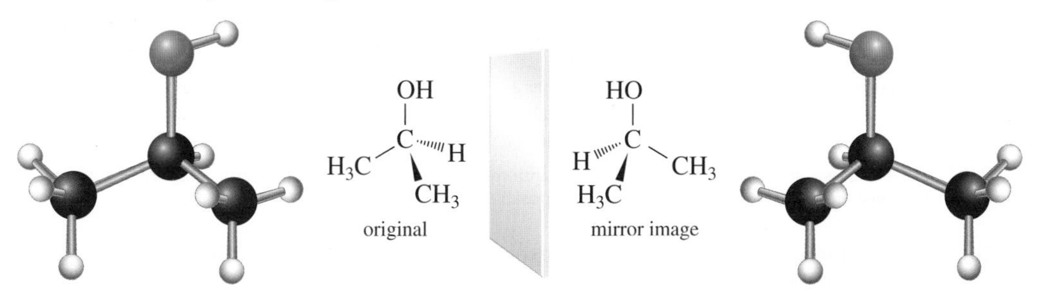

As we can see below, the mirror image of this molecule is superimposable on the original.

original

mirror image

60°

rotate around the C — OH bond by 60°

mirror image rotated by 60°

If a molecule and its mirror image are superimposable, then the molecule and its mirror image are identical, and there is no possibility of enantiomerism. Recall that we describe such a molecule as *achiral* (without chirality). It does not have left-hand and right-hand versions.

Any achiral object generally has at least one **plane of symmetry**. A plane of symmetry (also called a *mirror plane*) is an imaginary plane passing through an object and dividing it so that one half of the object is the reflection of the other half. The beaker shown in figure 17.12 has a single plane of symmetry, whereas a cube (figure 17.12b) has several planes of symmetry. An example of a molecule with a single plane of symmetry is propan-2-ol (figure 17.12c).

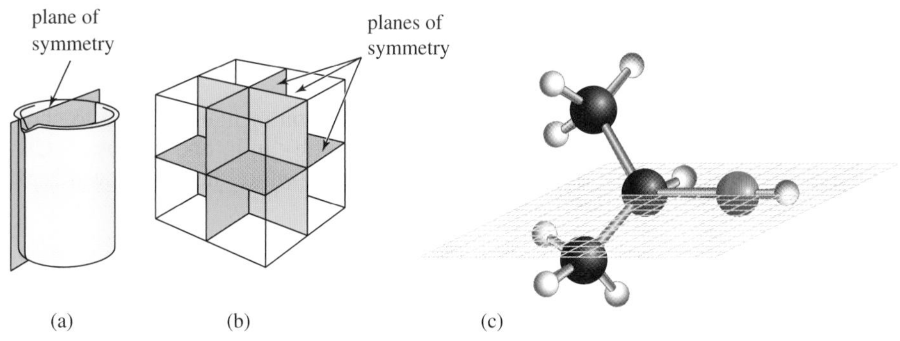

FIGURE 17.12 Planes of symmetry in: **(a)** a beaker, **(b)** a cube and **(c)** propan-2-ol. The beaker and propan-2-ol each has one plane of symmetry; the cube has several planes of symmetry, only three of which are shown in the figure (the diagonals form the other planes of symmetry).

Stereocentres

The most common basis for enantiomerism in organic molecules is the presence of a carbon atom bonded to four different groups. Such a carbon atom is an example of a **stereocentre**. A stereocentre is the part of a molecule that can be assembled in two different ways to form stereoisomers.

Inorganic molecules can also exhibit enantiomerism. Whereas the majority of examples of enantiomers in organic chemistry are based on a carbon atom bonded to four different groups, a metal atom is often the stereocentre in inorganic enantiomers. Inorganic enantiomers exhibit a wide variety of geometries. A classic example of a coordination complex (see chapter 13) that can exist as enantiomers is the golden-yellow $[Co(en)_3]^{3+}$ ion (figure 17.13).

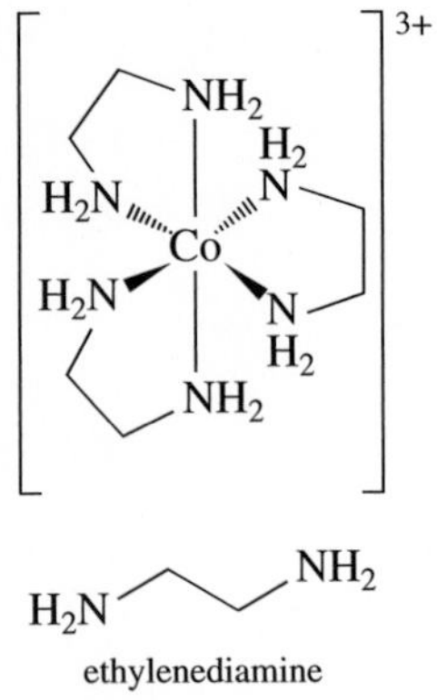

$[Co(en)_3]^{3+}$ comprises a central 6-coordinate Co^{3+} ion bound to three bidentate ethylenediamine ligands. Ethylenediamine, which is also known more correctly as ethane-1,2-diamine, is given the abbreviation 'en'. Here the stereocentre is the central cobalt atom, with the enantiomers arising from the two different ways in which the three ligands are arranged around it (figure 17.14). (Notice that the bidentate ethylenediamine ligand itself is achiral.)

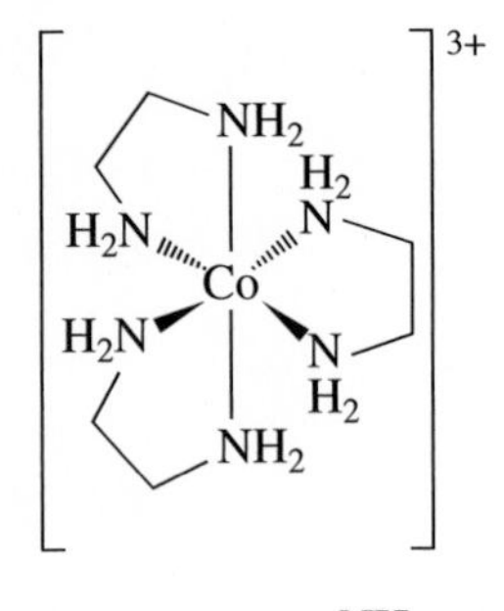

H_2N–CH_2CH_2–NH_2

ethylenediamine

FIGURE 17.13 The structures of $[Co(en)_3]^{3+}$ and ethylenediamine, or ethane-1,2-diamine (en).

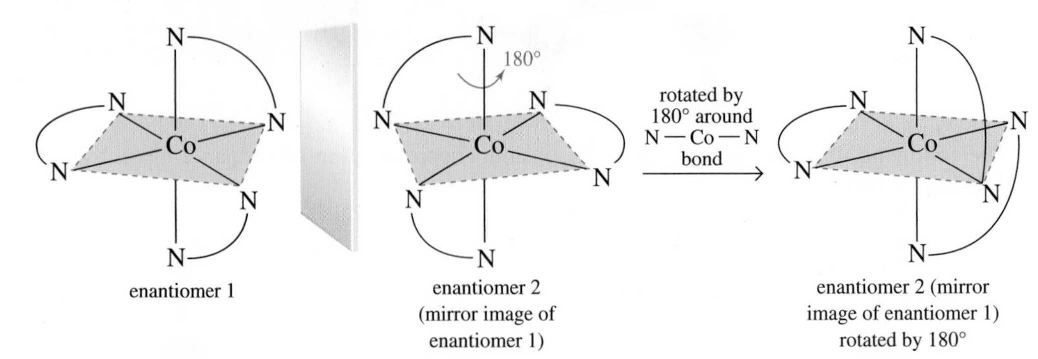

FIGURE 17.14 The two enantiomers of $[Co(en)_3]^{3+}$. Enantiomer 2 is constructed as the mirror image of enantiomer 1. No matter how enantiomer 2 is turned, it is not superimposable on enantiomer 1.

Similarly, enantiomers can exist for complexes that contain only two en ligands, such as *cis*-$[Co(en)_2Cl_2]^+$ (figure 17.15).

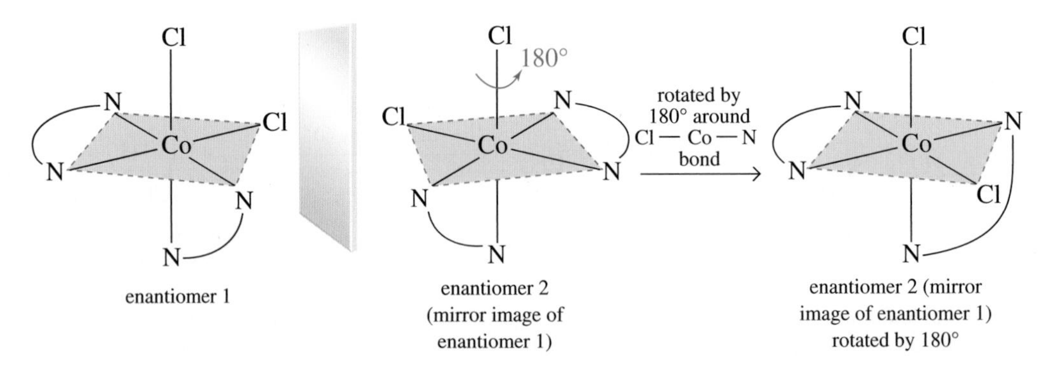

FIGURE 17.15 Isomers of the $[Co(en)Cl_2]^+$ ion. The mirror image of the *trans* isomer (not shown) can be superimposed exactly on the original, so the *trans* isomer is not chiral. However, the *cis* isomer (enantiomer 1) is chiral because its mirror image (enantiomer 2) cannot be superimposed on the original.

Enantiomers can also exist for complexes containing one en ligand; an example of this is *cis*-$[Co(en)(NH_3)_2Cl_2]^+$.

Representing enantiomers of complicated organic molecules

It is important that we can clearly represent the three-dimensional structures of enantiomers on a two-dimensional page. This is relatively straightforward for simple molecules such as those we have already encountered in this chapter, but it becomes more challenging for complicated molecules.

In our initial discussions of butan-2-ol, we used the representation in figure 17.16a to show the tetrahedral geometry of the central carbon atom (the stereocentre); in it, two groups (—CH_3 and —OH) are in the plane of the paper, one (—CH_2CH_3) is coming out of the plane towards us, and one (—H) is behind the plane, away from us.

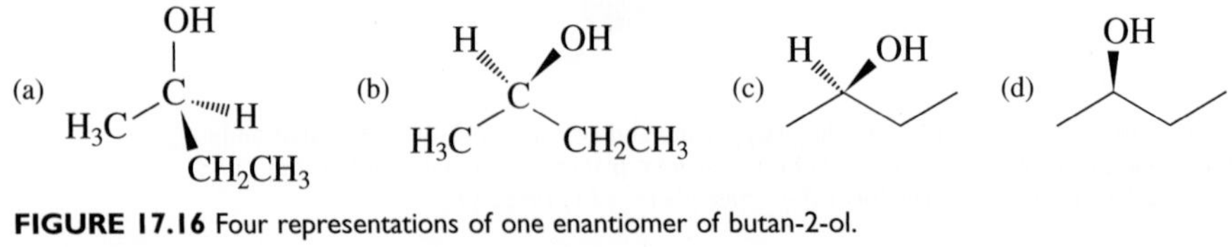

(a) OH, H_3C, C, H, CH_2CH_3 (b) H, OH, C, H_3C, CH_2CH_3 (c) H, OH (d) OH

FIGURE 17.16 Four representations of one enantiomer of butan-2-ol.

In figure 17.16, we can turn (a) slightly in space and tip it a bit to place the carbon framework in the plane of the paper. Doing so gives us representation (b), in which we still have two groups in the plane of the paper (—CH_3 and —CH_2CH_3), one coming towards us (—OH) and one going away from us (—H). For an even more abbreviated representation of this enantiomer of butan-2-ol, we can turn (b) into the line structure (c). Although we don't normally show hydrogen atoms in a line structure, we do in (c) just to remind ourselves that the fourth group on this stereocentre is really there and that it is —H. Finally, we can carry the abbreviation a step further and write butan-2-ol as (d). Here, we omit the —H on the stereocentre, but we know that it must be there (carbon needs four bonds), and we know that it must be behind the plane of the paper. Clearly, the abbreviated representations (c) and (d) are the easiest to draw, especially for very complicated structures, and we will tend to rely on these abbreviated representations throughout the remainder of the text.

When you have to write three-dimensional representations of stereocentres, try to keep the carbon framework in the plane of the paper and the other two atoms or groups of atoms on the stereocentre towards and away from you, respectively. Using representation (d) as a model, we get the following two different representations of its enantiomer.

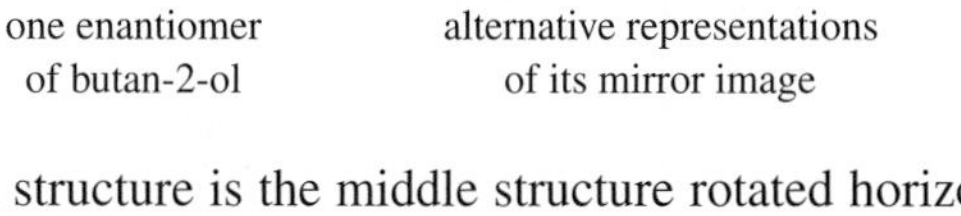

OH OH OH

one enantiomer of butan-2-ol

alternative representations of its mirror image

Notice that the right-hand structure is the middle structure rotated horizontally by 180°.

When drawing the enantiomeric pairs of more complicated structures, you may find it useful to 'map' a molecule's image across an imaginary mirror, which we can represent by a vertical line. If the mirror 'reflection' is not superimposable, it is a different molecule (an enantiomer). Figure 17.17 shows this technique for 4-methylcyclopent-2-enone.

Rotating the mirror image by 180° brings the double bond in the ring into alignment with the molecule on the left. However, notice that, in the molecule on the left of the mirror, the methyl group sits above the plane of the page whereas it lies beneath in the rotated molecule on the right of the mirror. Similarly, the hydrogen atom, which is below the plane of the page in the molecule on the left of the mirror is above the plane in the rotated mirror image.

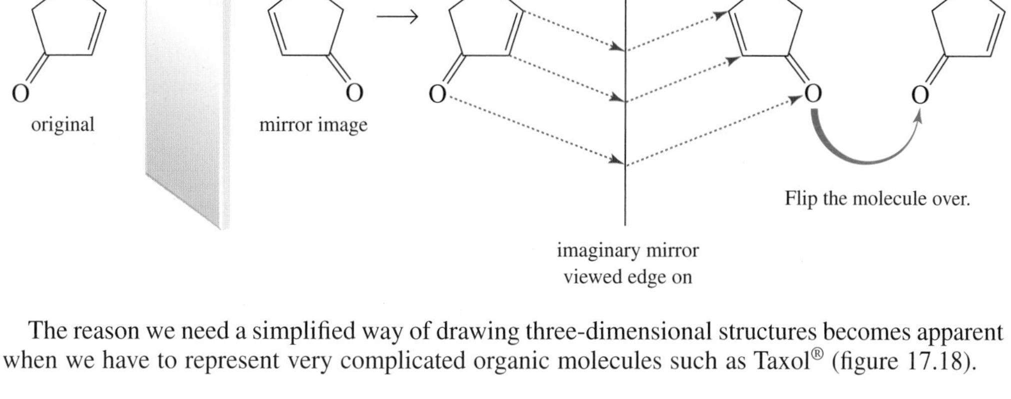

H CH_3 O original

H_3C H O mirror image

H CH_3 O → H_3C H O

H_3C H O

Flip the molecule over.

imaginary mirror viewed edge on

FIGURE 17.17 An imaginary mirror can be used to map a molecule's mirror image: in this case, 4-methylcyclopent-2-enone.

The reason we need a simplified way of drawing three-dimensional structures becomes apparent when we have to represent very complicated organic molecules such as Taxol® (figure 17.18).

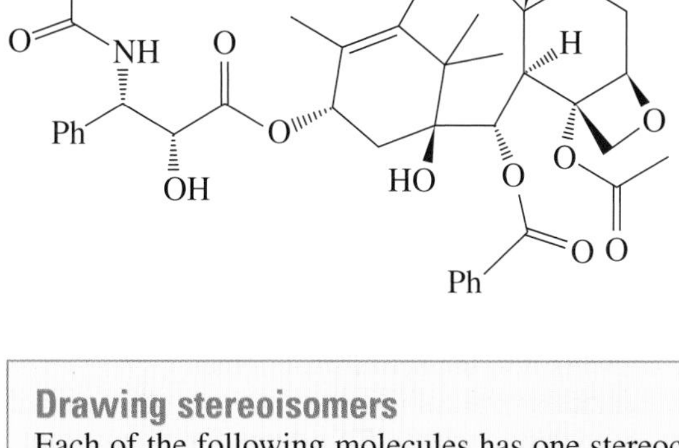

O O O OH Ph O NH O H Ph O O OH HO O O O Ph O

FIGURE 17.18 Structure of the important anti-cancer drug Taxol® showing the many stereocentres present in this complicated molecule.

WORKED EXAMPLE 17.1

Drawing stereoisomers

Each of the following molecules has one stereocentre.

(a) Cl
$CH_3CHCH_2CH_3$

(b) 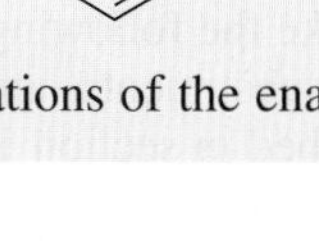 Cl (cyclohexene ring)

Draw three-dimensional representations of the enantiomers of each of these molecules.

Analysis

You will find it helpful to study the models of each pair of enantiomers and to view them from different perspectives. As you work with these models, notice that each enantiomer has a carbon atom bonded to four different groups, which makes the molecule chiral.

You may find it helpful to use the mapping process to represent the enantiomers involved. For (b), this process gives:

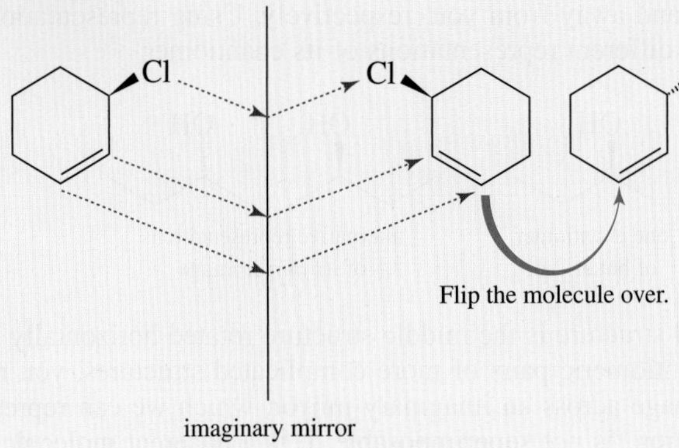

Flip the molecule over.

imaginary mirror

Solution

The hydrogen atom at the stereocentre is shown in (a), but not in (b).

(a) (b)

PRACTICE EXERCISE 17.1 Each of the following molecules has one stereocentre.

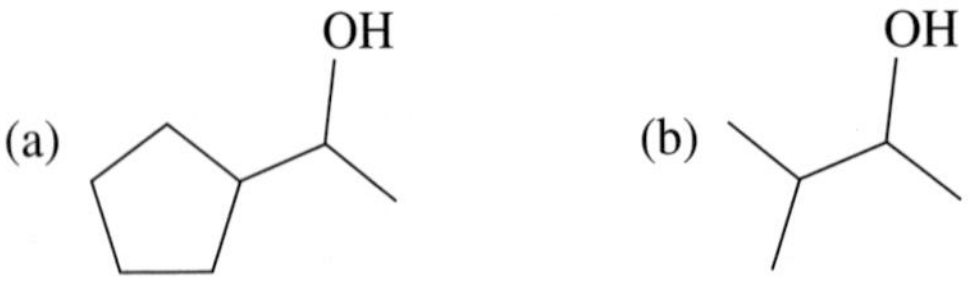

(a) (b)

Draw three-dimensional representations of the enantiomers of each of these molecules.

17.3 Naming stereocentres: the *R,S* system

Because enantiomers are different compounds, each must have a different name. The over-the-counter drug ibuprofen (an analgesic sold under various names including Nurofen®), for example, shows enantiomerism and can exist as the following pair of enantiomers.

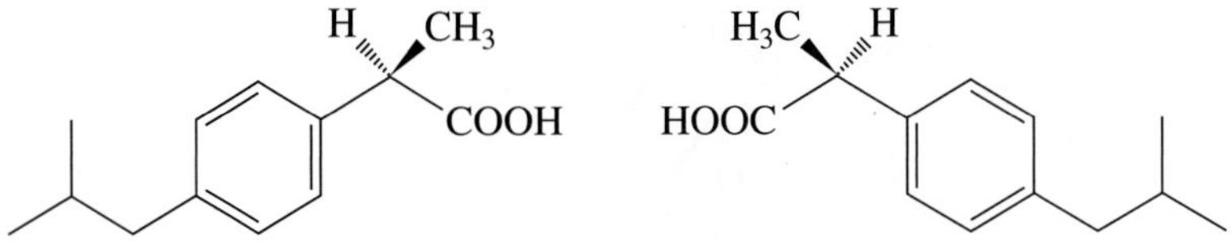

the inactive enantiomer of ibuprofen the active enantiomer of ibuprofen

Only one enantiomer of ibuprofen has a therapeutic effect. This enantiomer reaches effective concentrations in the human body in approximately 12 minutes. However, in this case, the inactive enantiomer is not wasted. The body slowly converts it to the active enantiomer.

We need to be able to assign a unique name to each enantiomer of ibuprofen (or any other pair of enantiomers, for that matter). To do this, chemists have developed the *R,S* system for organic molecules. The ***R,S* system** is a set of rules for specifying the configuration around a stereocentre. This can be an *R* configuration or an *S* configuration. The first step in assigning an *R* or *S* configuration is to arrange the groups bonded to the stereocentre in *order of priority.* For this, we use the same set of priority rules we used in section 16.3 to assign an *E,Z* configuration to an alkene.

To assign an *R* or *S* configuration, undertake the following steps.

1. Locate the stereocentre, identify its four substituents, and assign a priority from 1 (highest) to 4 (lowest) to each substituent (as described in section 16.3).

2. Orient the molecule in space so that the group of lowest priority (4) is directed away from you, as the steering column of a car would be. The three groups of higher priority (1–3) then project towards you, as the spokes of a steering wheel would.
3. Read the three groups projecting towards you in order, from highest priority (1) to lowest priority (3).
4. If reading the groups proceeds in a clockwise direction, the configuration is designated ***R*** (from the Latin word *rectus* meaning 'straight' or 'correct'); if reading proceeds in an anticlockwise direction, the configuration is ***S*** (from the Latin word *sinister* meaning 'left'). You can also visualise this situation as follows: turning the steering wheel to the right equals *R*, and turning it to the left equals *S*.

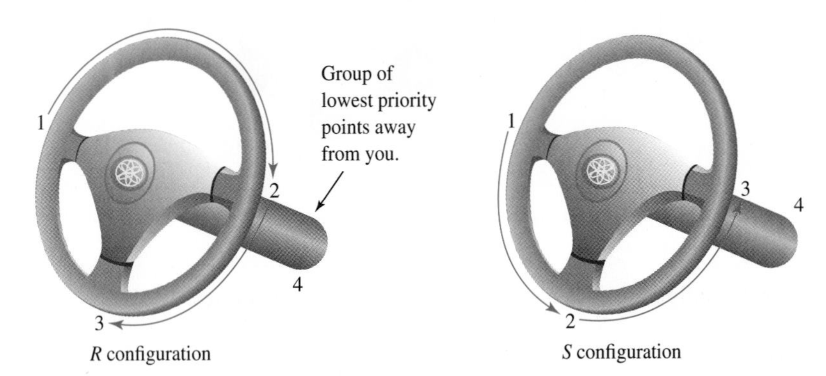

Now let us assign *R* and *S* configurations to our three-dimensional drawings of the enantiomers of ibuprofen. In order of decreasing priority, the groups bonded to the stereocentre are —COOH > —C_6H_4 > —CH_3 > —H. In the enantiomer below on the left, the sequence of groups on the stereocentre in order of priority occurs clockwise. Therefore, this enantiomer is (*R*)-ibuprofen, and its mirror image is (*S*)-ibuprofen.

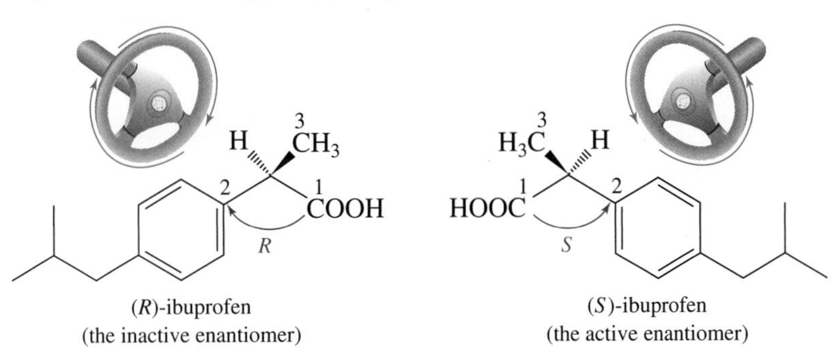

(*R*)-ibuprofen
(the inactive enantiomer)

(*S*)-ibuprofen
(the active enantiomer)

WORKED EXAMPLE 17.2

Using the *R,S* system

Assign an *R* or *S* configuration to the stereocentres in the following molecules.

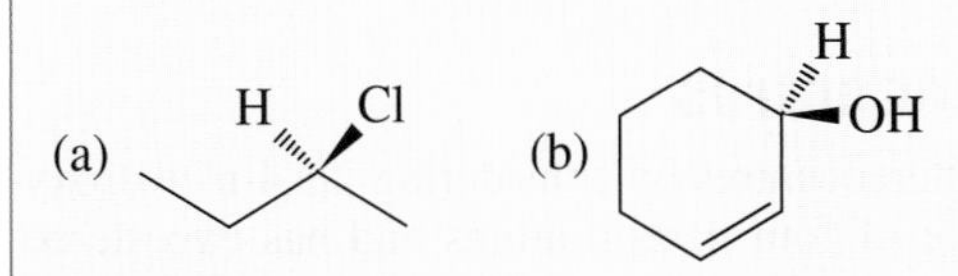

Analysis and solution

View each molecule through the stereocentre and along the bond from the stereocentre towards the group of lowest priority.

(a) The order of priority is —Cl > —CH_2CH_3 > —CH_3 > —H. The group of lowest priority, H, points away from you. Reading the groups in the order 1, 2, 3 occurs in an anticlockwise direction, so the configuration is *S*.

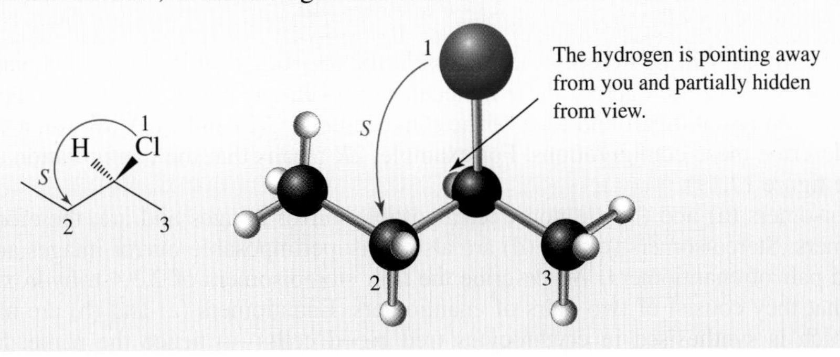

(b) The order of priority is $—OH > —CH{=}CH > —CH_2—CH_2 > —H$. With hydrogen, the group of lowest priority, pointing away from you, reading the groups in the order 1, 2, 3 occurs in a clockwise direction, so the configuration is *R*.

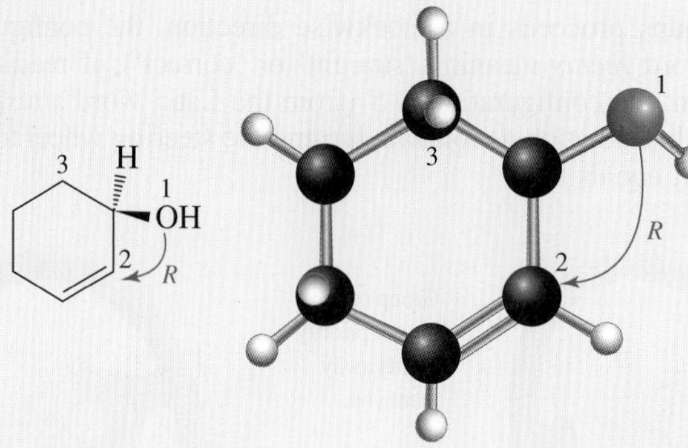

PRACTICE EXERCISE 17.2 Assign an *R* or *S* configuration to each of the following stereocentres.

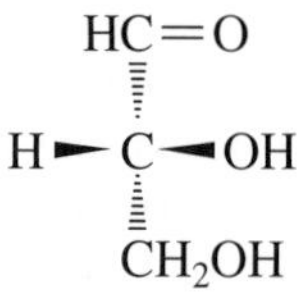

17.4 Molecules with more than one stereocentre

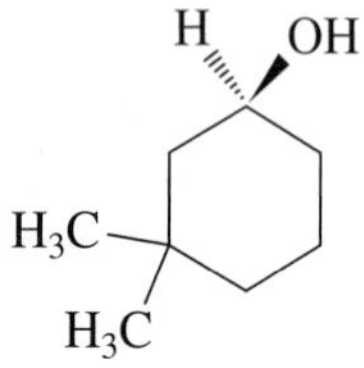

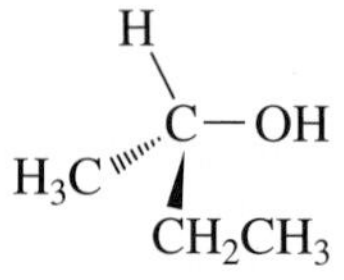

We have now seen several examples of molecules with one stereocentre and verified that, for each, two stereoisomers (one pair of enantiomers) are possible. Now let us consider molecules with multiple stereocentres. To generalise, for a molecule with *n* stereocentres, the maximum possible number of stereoisomers is 2^n. We know that, for a molecule with one stereocentre, $2^1 = 2$ stereoisomers are possible. For a molecule with two stereocentres, $2^2 = 4$ stereoisomers are possible; for a molecule with three stereocentres, $2^3 = 8$ stereoisomers are possible, and so on. However, it should be noted that sometimes the maximum number of stereoisomers is not possible, as we will see on page 772 for *meso* compounds.

Acyclic molecules with two stereocentres

We begin our study of acyclic molecules with two stereocentres by considering 2,3,4-trihydroxybutanal. This aldehyde comprises a linear sequence of four carbon atoms and has two stereocentres: that is, two carbon atoms with four different groups attached. The two stereocentres, at C(2) and C(3), are marked with asterisks.

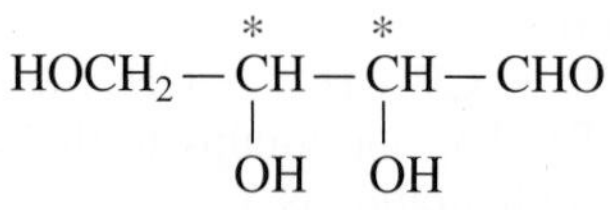

$$HOCH_2-\overset{*}{C}H(OH)-\overset{*}{C}H(OH)-CHO$$

2,3,4-trihydroxybutanal

The maximum number of stereoisomers possible for this molecule is $2^n = 2^2 = 4$, each of which is drawn in figure 17.19. Each of these molecules has a unique name, and we use the *R,S* system to assign a configuration around each of the stereocentres, C(2) and C(3). We use a special notation to describe these configurations. For example, 2*R* means that the configuration around C(2) is *R* (see figure 17.19).

Stereoisomers (a) and (b) are nonsuperimposable mirror images and are, therefore, a pair of enantiomers. Stereoisomers (c) and (d) are also nonsuperimposable mirror images and represent a second pair of enantiomers. We describe the four stereoisomers of 2,3,4-trihydroxybutanal by saying that they consist of two pairs of enantiomers. Enantiomers (a) and (b) are named erythrose, which is synthesised in erythrocytes (red blood cells) — hence the name. Enantiomers

(c) and (d) are named threose. Erythrose and threose belong to the class of compounds called carbohydrates.

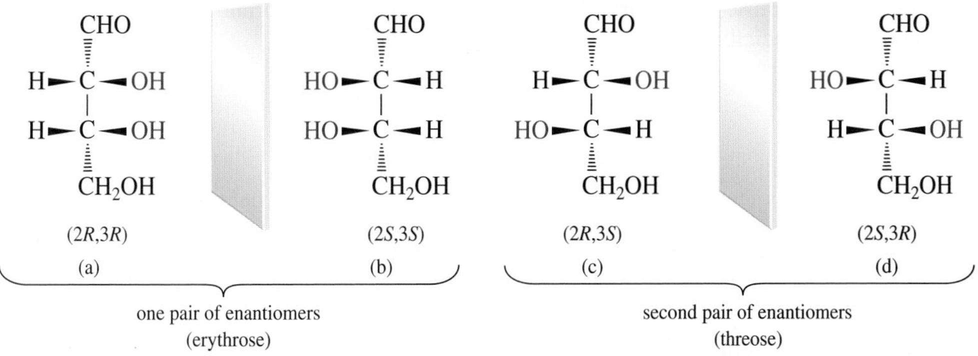

FIGURE 17.19 The four stereoisomers of 2,3,4-trihydroxybutanal, a compound with two stereocentres.

We also need to define the relationship between (a) and (c). They are stereoisomers but are not mirror images of each other. As you will recall from page 761, we call stereoisomers of this type diastereomers. Similarly, (a) and (d), (b) and (c), and (b) and (d) are pairs of diastereomers. *Diastereomers are stereoisomers that are not enantiomers; that is, they are stereoisomers that are not mirror images of each other.* Like enantiomers, diastereomers have the same sequence of connection. However, while enantiomers have mostly the same physical properties (melting point, boiling point, density etc.), diastereomers can have completely different properties. This is very evident with carbohydrates, which we will talk about in chapter 22.

WORKED EXAMPLE 17.3

Enantiomers and diastereomers

The following are stereorepresentations of the four stereoisomers of butane-1,2,3-triol.

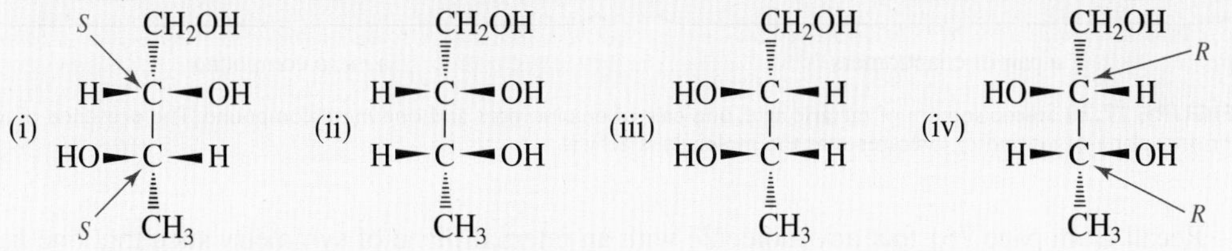

(i)	(ii)	(iii)	(iv)
CH_2OH	CH_2OH	CH_2OH	CH_2OH
H—C—OH (S)	H—C—OH	HO—C—H	HO—C—H (R)
HO—C—H (S)	H—C—OH	HO—C—H	H—C—OH (R)
CH_3	CH_3	CH_3	CH_3

Configurations are given for the stereocentres in (i) and (iv).

(a) Which pairs of compounds are enantiomers?

(b) Which pairs of compounds are diastereomers?

Analysis and solution

(a) Enantiomers are stereoisomers that are nonsuperimposable mirror images. Compounds (i) and (iv) are one pair of enantiomers, and compounds (ii) and (iii) are a second pair of enantiomers. Note that the configurations of the stereocentres in (i) are the opposite of those in (iv), its enantiomer.

(b) Diastereomers are stereoisomers that are not mirror images. Compounds (i) and (ii), (i) and (iii), (ii) and (iv), and (iii) and (iv) are diastereomers.

PRACTICE EXERCISE 17.3

The following are stereorepresentations of the four stereoisomers of 3-chlorobutan-2-ol.

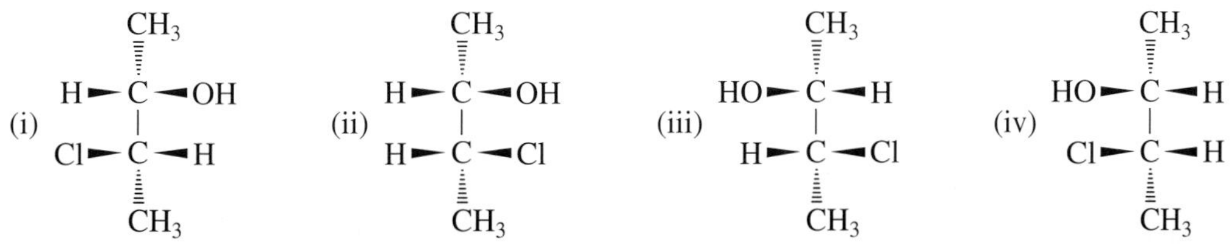

(i)	(ii)	(iii)	(iv)
CH_3	CH_3	CH_3	CH_3
H—C—OH	H—C—OH	HO—C—H	HO—C—H
Cl—C—H	H—C—Cl	H—C—Cl	Cl—C—H
CH_3	CH_3	CH_3	CH_3

(a) Which pairs of compounds are enantiomers?

(b) Which pairs of compounds are diastereomers?

Meso compounds

Certain molecules with two or more stereocentres do not have as many stereoisomers as you might expect from the 2^n rule. One such molecule is 2,3-dihydroxybutanedioic acid, more commonly named tartaric acid. Tartaric acid is a colourless, crystalline compound occurring largely in plants, especially in grapes. The structure of tartaric acid is:

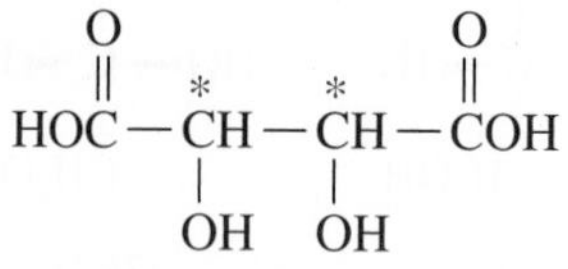

$$\mathrm{HO\overset{O}{\overset{\|}{C}}-\overset{*}{C}H(OH)-\overset{*}{C}H(OH)-\overset{O}{\overset{\|}{C}}OH}$$

2,3-dihydroxybutanedioic acid
(tartaric acid)

C(2) and C(3) of tartaric acid are stereocentres, and, from the 2^n rule, the maximum number of stereoisomers you might expect is $2^2 = 4$. Figure 17.20 shows the two pairs of mirror images of this compound. Structures (a) and (b) are nonsuperimposable mirror images and, therefore, they are a pair of enantiomers. Structures (c) and (d) are also mirror images, but they are superimposable if they are rotated to the appropriate orientation. Therefore, (c) and (d) are *not* different molecules; they are the same molecule, just oriented differently. Because (c) and its mirror image (d) are superimposable, they are the same molecule and achiral. This means that they are optically inactive and do not interact with plane-polarised light (see section 17.5 for more details).

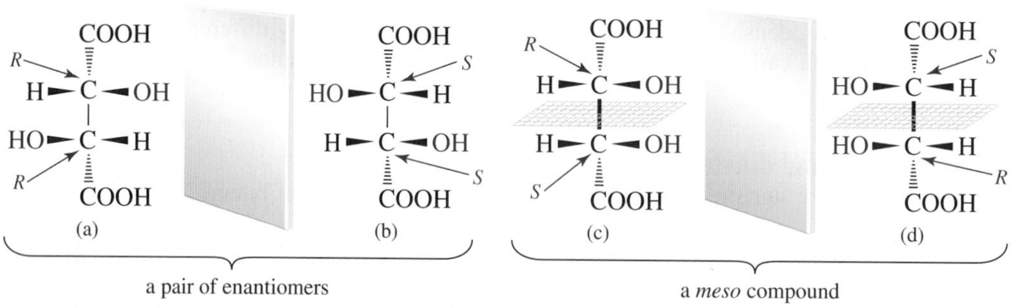

FIGURE 17.20 Stereoisomers of tartaric acid, one pair of enantiomers and one *meso* compound. The presence of an internal plane of symmetry indicates that the molecule is achiral.

Recall from page 761 that any molecule with an internal plane of symmetry such that one half is the reflection of the other half is achiral and does not have enantiomers. Thus, even though (c) has two stereocentres, it is achiral. Its plane of symmetry is shown in figure 17.20. The stereoisomer of tartaric acid represented by (c) or (d) is called a ***meso* compound**, defined as an achiral compound with two or more stereocentres.

We can now return to the original question: how many stereoisomers are there of tartaric acid? The answer is three: one *meso* compound and one pair of enantiomers. Note that the *meso* compound is a diastereomer of each of the other stereoisomers.

WORKED EXAMPLE 17.4

Enantiomers and *meso* compounds

The following are stereorepresentations of the three stereoisomers of butane-2,3-diol.

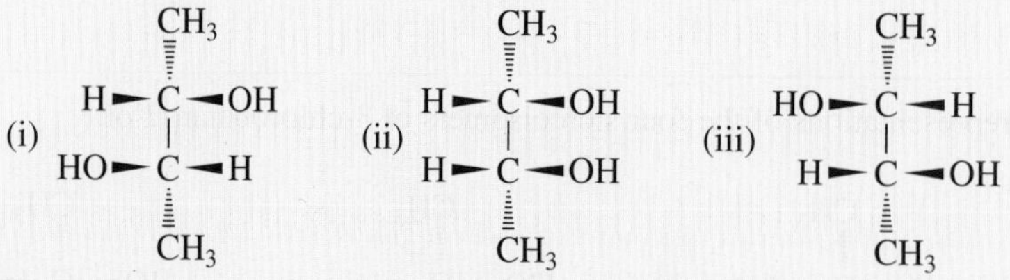

(a) Which pair of molecules are enantiomers?
(b) Which is the *meso* compound?

Solution

(a) Compounds (i) and (iii) are enantiomers.
(b) Compound (ii) is a *meso* compound.

The following are four Newman projection formulae (see section 16.2) of tartaric acid.

(i) (ii) (iii) (iv)

(a) Which formulae represent the same compound?
(b) Which formulae represent enantiomers?
(c) Which formula(e) represent(s) *meso*-tartaric acid?

Cyclic molecules with two stereocentres

In this section, we concentrate on derivatives of cyclopentane and cyclohexane containing two stereocentres. We can analyse chirality in these cyclic compounds in the same way we did for acyclic compounds.

Disubstituted derivatives of cyclopentane

Let us start with 2-methylcyclopentanol, a compound with two stereocentres. Using the 2^n rule, we predict a maximum of $2^2 = 4$ stereoisomers. Both the *cis* isomer and the *trans* isomer (see section 16.2) are chiral, with the *cis* isomer existing as one pair of enantiomers and the *trans* isomer existing as a second pair:

cis-2-methylcyclopentanol
(a pair of enantiomers)

trans-2-methylcyclopentanol
(a pair of enantiomers)

Because 1,2-cyclopentanediol also has two stereocentres, the 2^n rule predicts a maximum of $2^2 = 4$ stereoisomers. However, as seen in the following diagram, only three stereoisomers exist for this compound.

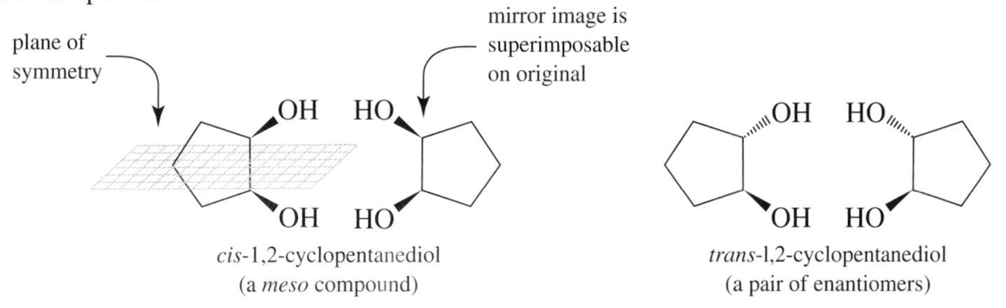

cis-1,2-cyclopentanediol
(a *meso* compound)

trans-1,2-cyclopentanediol
(a pair of enantiomers)

The *cis* isomer is achiral (*meso*) because it is superimposable on its mirror image. To put it another way, the *cis* isomer is achiral because it possesses a plane of symmetry that bisects the molecule into two mirror-image halves. The *trans* isomer is chiral and exists as a pair of enantiomers.

WORKED EXAMPLE 17.5

Stereoisomers for 3-methylcyclopentanol

How many stereoisomers are possible for 3-methylcyclopentanol?

Analysis and solution

There are four stereoisomers of 3-methylcyclopentanol, with the *cis* isomer existing as one pair of enantiomers and the *trans* isomer as a second pair.

cis-1,3-methylcyclopentanol
(a pair of enantiomers)

trans-3-methylcyclopentanol
(a pair of enantiomers)

PRACTICE EXERCISE 17.5 How many stereoisomers are possible for 1,3-cyclopentanediol?

Disubstituted derivatives of cyclohexane

As an example of a disubstituted cyclohexane, let us consider the methylcyclohexanols. Firstly, 4-methylcyclohexanol can exist as two stereoisomers — a pair of *cis–trans* isomers. Both the *cis* and *trans* isomers are *meso* compounds and are achiral. In each, a plane of symmetry runs through the —CH_3 and —OH groups and the two attached carbon atoms.

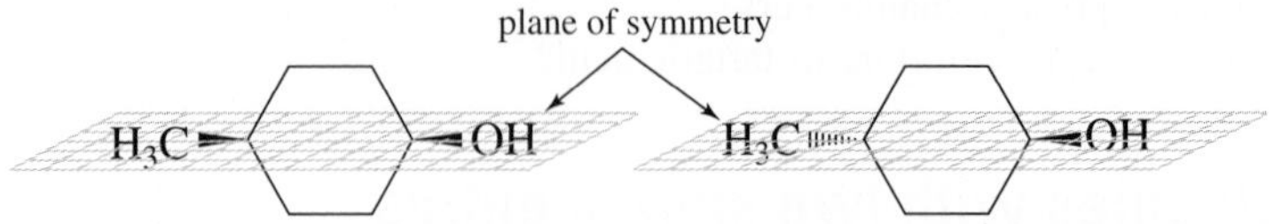

plane of symmetry

H_3C OH H_3C OH

cis-4-methylcyclohexanol *trans*-4-methylcyclohexanol

Similarly, 3-methylcyclohexanol contains two stereocentres and exists as $2^2 = 4$ stereoisomers, with the *cis* isomer as one pair of enantiomers and the *trans* isomer as a second pair.

OH HO OH HO

CH_3 H_3C CH_3 H_3C

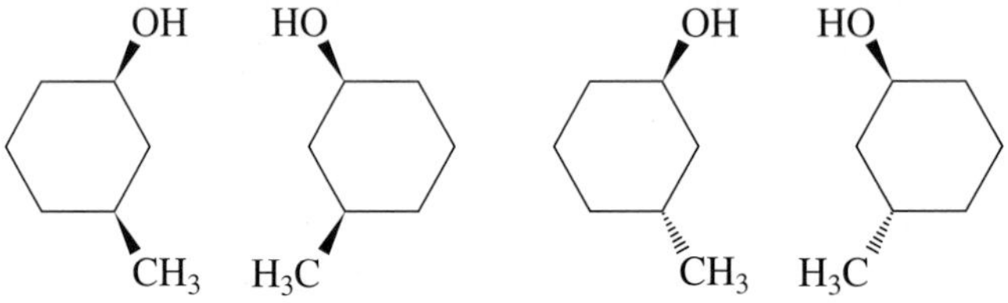

cis-3-methylcyclohexanol
(a pair of enantiomers)

trans-3-methylcyclohexanol
(a pair of enantiomers)

Again, 2-methylcyclohexanol has two stereocentres and exists as $2^2 = 4$ stereoisomers, with the *cis* isomer as one pair of enantiomers and the *trans* isomer as a second pair.

OH HO OH HO

CH_3 H_3C CH_3 H_3C

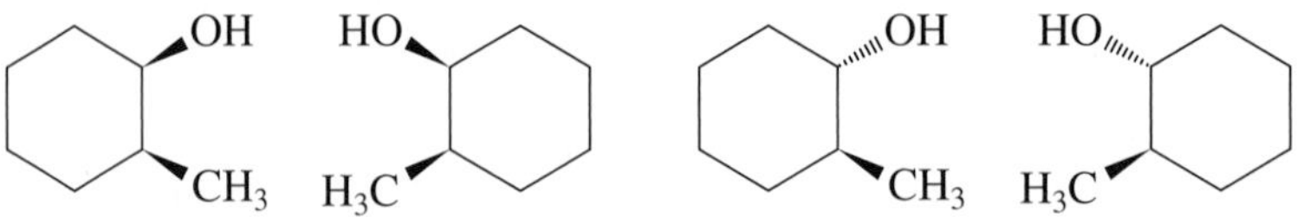

cis-2-methylcyclohexanol
(a pair of enantiomers)

trans-2-methylcyclohexanol
(a pair of enantiomers)

WORKED EXAMPLE 17.6

Stereoisomers for cyclohexane-1,3-diol

How many stereoisomers exist for cyclohexane-1,3-diol?

Analysis and solution

According to the 2^n rule, cyclohexane-1,3-diol has two stereocentres, so it has a maximum of $2^2 = 4$ stereoisomers. The *trans* isomer of this compound exists as a pair of enantiomers. The *cis* isomer has a plane of symmetry so is a *meso* compound.

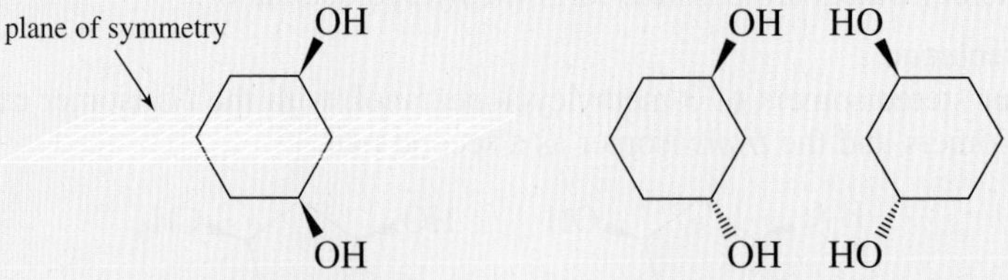

plane of symmetry

OH OH HO

OH OH HO

cis-cyclohexane-1,3-diol
(a *meso* compound)

trans-cyclohexane-1,3-diol
(a pair of enantiomers)

Therefore, although the 2^n rule predicts a maximum of four stereoisomers for cyclohexane-1,3-diol, only three exist: one pair of enantiomers and one *meso* compound.

PRACTICE EXERCISE 17.6 How many stereoisomers exist for cyclohexane-1,4-diol?

Molecules with three or more stereocentres

The 2^n rule applies equally to molecules with three or more stereocentres. Here is a disubstituted cyclohexanol with three stereocentres, each marked with an asterisk.

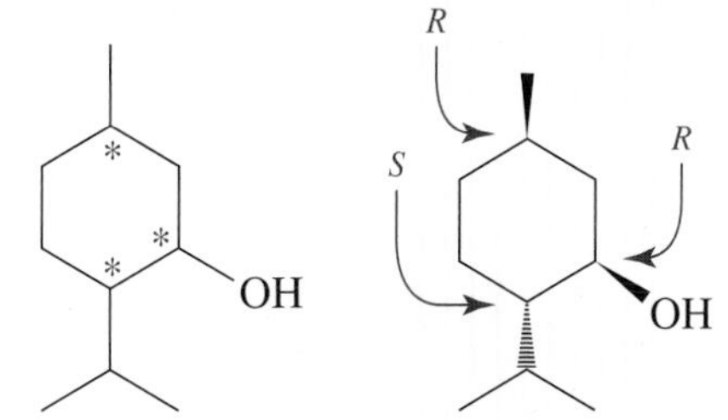

2-isopropyl-5-methyl-cyclohexanol

(–)-menthol (1*R*,2*S*,5*R*)-2-isopropyl-5-methyl-cyclohexanol

There is a maximum of $2^3 = 8$ stereoisomers possible for this molecule. Menthol, one of the eight, has the configuration shown above on the right. The configuration at each stereocentre is indicated. Menthol is the major component present in peppermint and other mint oils that gives these their characteristic taste.

Cholesterol, a more complicated molecule, has eight stereocentres.

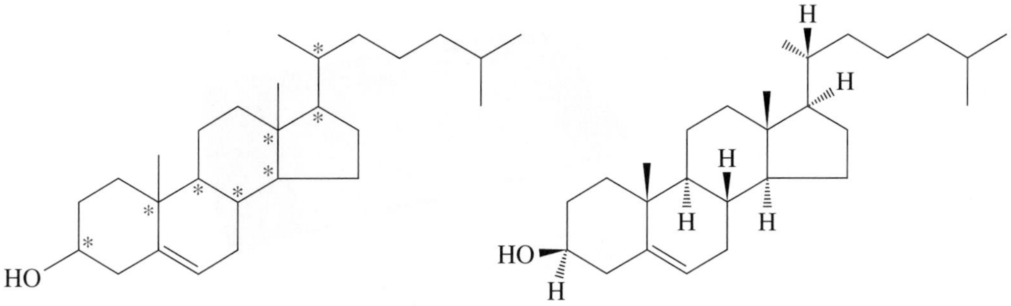

Cholesterol has eight stereocentres; 256 stereoisomers are possible.

This is the stereoisomer found in human metabolism.

To identify the stereocentres, remember to add an appropriate number of hydrogen atoms to complete the tetravalence of each carbon atom you think might be a stereocentre.

17.5 Optical activity: detecting chirality in the laboratory

We indicated in the previous section that one diastereomer (menthol) of 2-isopropyl-5-methyl-cyclohexanol is primarily responsible for the flavour of peppermint. Diastereomers have markedly different properties. Enantiomers also have different properties, but the differences are more subtle. In general, enantiomers have identical physical and chemical properties. For instance, they have the same melting point, the same boiling point and the same solubilities in water and other common solvents. They do differ, however, in their **optical activity** (the ability to rotate a plane of polarised light). This property is particularly useful to chemists as it allows us to detect enantiomers in the laboratory.

Each member of a pair of enantiomers rotates a plane of polarised light in the opposite direction. To understand how optical activity is detected in the laboratory, we must first understand plane-polarised light and a polarimeter, the instrument used to detect optical activity.

Plane-polarised light

Ordinary light consists of waves oscillating in all planes perpendicular to its direction of propagation (shown on the left side of figure 17.21, overleaf). Certain materials, such as calcite and Polaroid™ sheet (a plastic film with oriented crystals of an organic substance embedded in it), selectively transmit light waves vibrating in parallel planes (shown on the right side of figure 17.21, overleaf). Electromagnetic radiation vibrating only in parallel planes is said to be **plane polarised**.

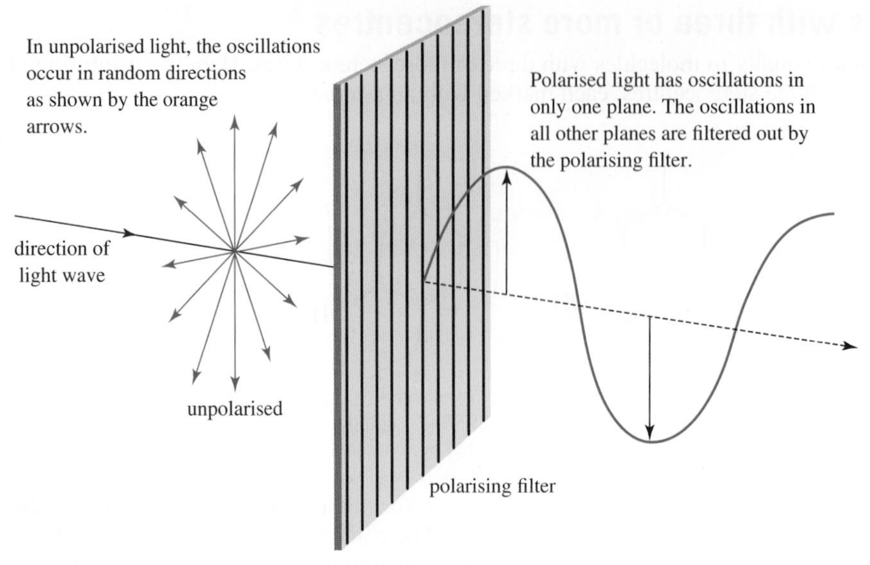
In unpolarised light, the oscillations occur in random directions as shown by the orange arrows.
Polarised light has oscillations in only one plane. The oscillations in all other planes are filtered out by the polarising filter.
direction of light wave
unpolarised
polarising filter

FIGURE 17.21 The effect of a polarising filter on light.

Polarimeters

FIGURE 17.22 A chemical polarimeter.

A **polarimeter** consists of a light source, a polarising filter and an analysing filter (each made of calcite or Polaroid film), and a sample tube (figures 17.22 and 17.23). If the sample tube is empty, the intensity of light reaching the detector (in this case, your eye) is at its maximum when the polarising axes of the two filters are parallel. If the analysing filter is turned either clockwise or anticlockwise, less light is transmitted. When the axis of the analysing filter is at right angles to the axis of the polarising filter, the field of view is dark. This position of the analysing filter is taken to be 0° on the optical scale.

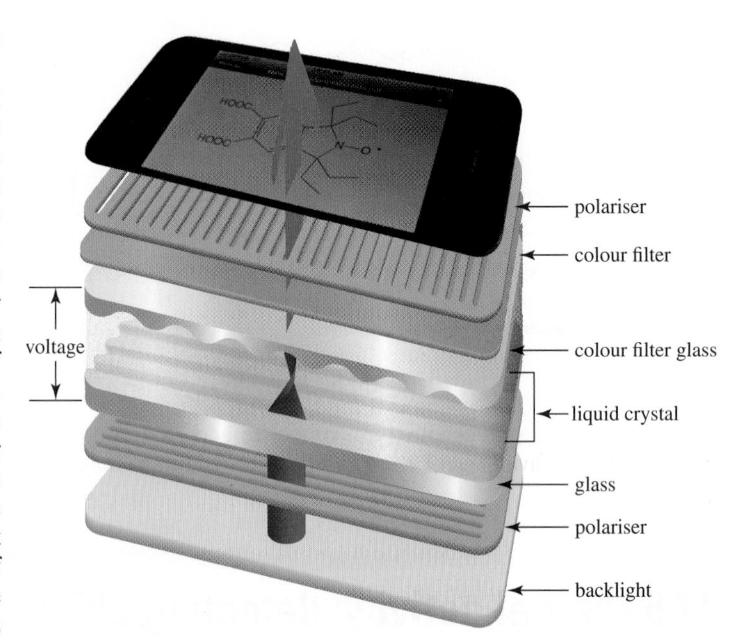
voltage
polariser
colour filter
colour filter glass
liquid crystal
glass
polariser
backlight

FIGURE 17.23 Polarising filters are used in liquid crystal displays (LCD).

This same principle forms the basis for how images are produced on the LCD screens of smart-phones and laptops (see figure 17.23).

Measuring the rotation of plane-polarised light

The ability of molecules to rotate a plane of polarised light can be observed with the use of a polarimeter in the following way (see figure 17.24). First, a sample tube filled with solvent is placed in the polarimeter, and the analysing filter is adjusted so that no light passes through to the observer; that is, the filter is set to 0°. Then we place a solution of an optically active compound in the sample tube. When we do so, we find that a certain amount of light now passes through the analysing filter. We also find that the plane of polarised light from the polarising filter has been rotated so that it is no longer at an angle of 90° to the analysing filter. We then rotate the analysing filter to restore darkness in the field of view. The number of degrees, α, through which we must rotate the analysing filter to restore darkness to the field of view is called the **observed rotation**. If we must turn the analysing filter to the right (clockwise) to restore the dark field, we say that the compound is **dextrorotatory** (from the Latin word *dexter* meaning 'on the right side'); if we must turn it to the left (anticlockwise), we say that the compound is **levorotatory** (from the Latin word *laevus* meaning 'on the left side').

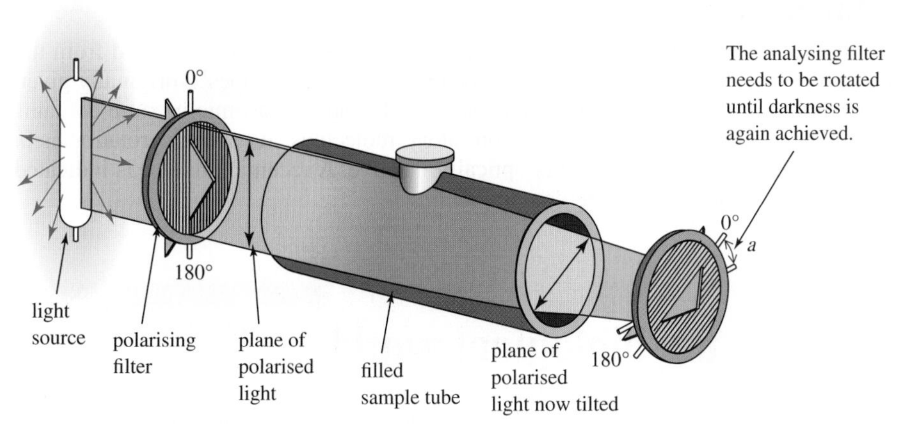

FIGURE 17.24 Schematic diagram of a polarimeter with its sample tube containing a solution of an optically active compound. The analysing filter would need to be turned clockwise by α degrees to restore the dark field.

The magnitude of the observed rotation for a particular compound depends on its concentration, the length of the sample tube, the temperature, the solvent and the wavelength of the light used. The **specific rotation** ($[\alpha]$) is defined as the observed rotation at a specific cell length and sample concentration:

$$\text{specific rotation} = [\alpha]_{\lambda}^{T} = \frac{\text{observed rotation (degrees)}}{\text{length (dm)} \times \text{concentration (g/mL)}}$$

The standard cell length is 1 decimetre (1 dm = 0.1 m, see figure 17.25), although cells of different lengths can be used. The concentration of a sample dissolved in a solvent is expressed as grams per millilitre of solution. The temperature (T, in degrees Celsius) and wavelength (λ, in nanometres) of light are designated, respectively, as superscript and subscript. The light source most commonly used in polarimetry is the so-called 'sodium D line' (λ = 589 nm), first named by German physicist Joseph von Fraunhofer when he observed the dark bands in the optical spectrum of the Sun. The same line is responsible for the yellow colour of sodium-vapour lamps seen in many street lights (see worked example 4.3 on pp. 126–7).

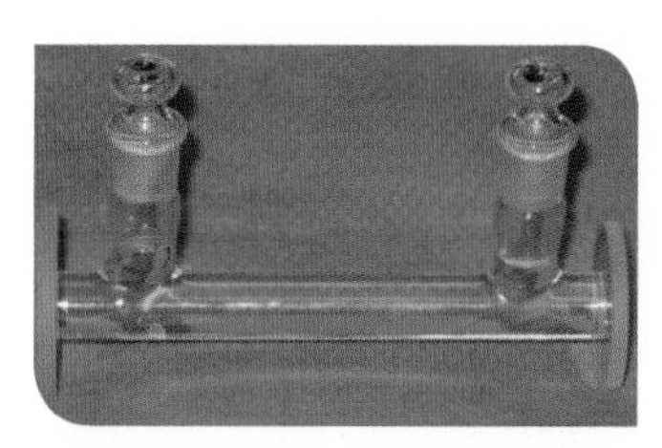

FIGURE 17.25 A 1 dm polarimeter cell.

In reporting either observed or specific rotation, it is common to indicate a dextrorotatory compound with a plus sign in parentheses, (+), and a levorotatory compound with a minus sign in parentheses, (–). For any pair of enantiomers, one enantiomer is dextrorotatory and the other is levorotatory. For each pair, the values of the specific rotation are exactly the same, but the signs are opposite. The following are the specific rotations of the enantiomers of butan-2-ol at 25 °C, observed with the D line of sodium.

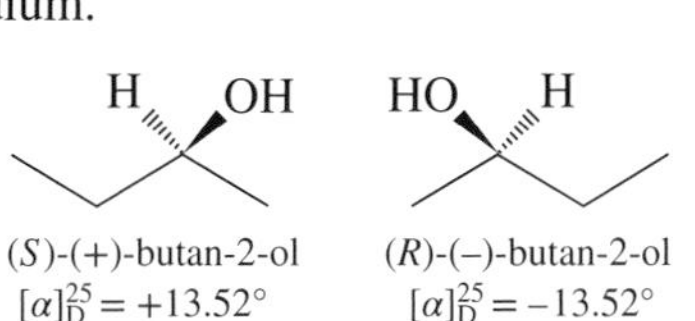

You will notice in the specific rotations for butan-2-ol that S is the dextrorotatory enantiomer and R is the levorotatory enantiomer. The R and S designation is not related to the optical activity; that is, S enantiomers may be either levorotatory or dextrorotatory, and vice versa.

WORKED EXAMPLE 17.7

Calculating specific rotation

A solution is prepared by dissolving 4.00 g of testosterone, a male sex hormone, in 100 mL of ethanol and placing it in a sample tube 1.00 dm in length. The observed rotation of this sample at 25 °C, using the D line of sodium, is +4.36°. Calculate the specific rotation of testosterone.

Analysis and solution

The concentration of testosterone is 4.00 g/100 mL = 0.0400 g/mL. The length of the sample tube is 1.00 dm. Inserting these values into the equation for calculating specific rotation gives:

$$\text{specific rotation} = \frac{\text{observed rotation}}{\text{length} \times \text{concentration}} = \frac{+4.36°}{1.00 \times 0.0400} = +109°$$

PRACTICE EXERCISE 17.7

The specific rotation of progesterone, a female sex hormone, is +172°, measured at 20 °C. Calculate the observed rotation for a solution prepared by dissolving 4.00 g of progesterone in 100 mL of dioxane and placing it in a sample tube 1.00 dm long.

Racemic mixtures

An equimolar mixture of two enantiomers is called a **racemic mixture**, a term derived from the name 'racemic acid' (from the Latin word *racemus* meaning 'cluster of grapes'), originally given to an equimolar mixture of the enantiomers of tartaric acid. Because a racemic mixture contains equal numbers of the dextrorotatory and the levorotatory molecules, its specific rotation is 0. Alternatively, we say that a racemic mixture is optically inactive. A racemic mixture is indicated by adding the prefix (±) to the name of the compound.

17.6 Chirality in the biological world

Almost all the molecules in living systems, both plant and animal, are chiral. Although these chiral molecules can exist as a number of stereoisomers, almost invariably only one stereoisomer is found in nature. Of course, instances do occur in which more than one stereoisomer is found, but these rarely exist together in the same biological system.

Perhaps the most conspicuous examples of chirality among biological molecules are the enzymes, all of which have many stereocentres. An example is chymotrypsin, an enzyme found in the intestines of animals that catalyses the digestion of proteins. Chymotrypsin has 251 stereocentres. The maximum possible number of stereoisomers is thus staggeringly large, almost beyond comprehension. Fortunately, nature does not squander its precious energy and resources unnecessarily; only one of these stereoisomers is produced and used by any given organism. Because enzymes are chiral substances, most produce or react only with substances that match their stereochemical requirements.

How an enzyme distinguishes between enantiomers

For an enzyme to catalyse a biological reaction, the molecule involved must first attach to a chiral binding site on the enzyme's surface. An enzyme with binding sites specific for three of the four groups on a stereocentre can distinguish between a molecule and its enantiomer or one of its diastereomers. Let's look at figure 17.26, which is a representation of an enzyme that catalyses a reaction of glyceraldehyde. The enzyme has three sites arranged on its surface: a binding site specific for —H, a second specific site for —OH and a third specific site for —CHO. The enzyme can distinguish (*R*)-(+)-glyceraldehyde (the natural, or biologically active, form) from its enantiomer because the natural enantiomer can be bound, with three groups interacting with their appropriate binding sites; for the *S* enantiomer, only two groups at most can interact with these binding sites.

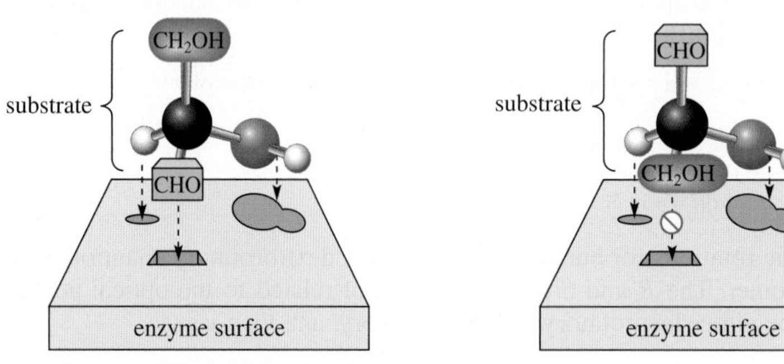

This enantiomer of glyceraldehyde fits the three specific binding sites on the enzyme surface.

This enantiomer of glyceraldehyde does not fit the same binding sites.

FIGURE 17.26 A schematic diagram of an enzyme surface capable of interacting with (*R*)-(+)-glyceraldehyde at three binding sites, but with (*S*)-(−)-glyceraldehyde at only two of these sites.

Because interactions between molecules in living systems take place in a chiral environment, we should expect that a molecule and its enantiomer or one of its diastereomers elicit different physiological responses. As we have already seen, (*S*)-ibuprofen is active as a pain and fever reliever, whereas its *R* enantiomer is inactive. The *S* enantiomer of the closely related analgesic naproxen is also the active pain-relieving form of this compound, but its *R* enantiomer is a liver toxin!

H_3C H
HOOC

(*S*)-ibuprofen

H_3C H
HOOC
OCH_3

(*S*)-naproxen

Chemical Connections

Chemistry and forensics

Some of the most popular shows on TV today involve scientists solving crimes using chemistry skills. Crime stories have always been popular; however, today's forensic scientists have access to tools and techniques that allow modern scriptwriters to develop complex and wide-ranging story-lines for these shows, even if they are often viewed as unrealistic by actual forensic scientists. However, even 80 years ago, crime writers were using their knowledge of chemistry to produce original stories. One such example is the 1930 crime story *The Documents in the Case* co-written by Dorothy L Sayers and Robert Euston.

In possibly the first example of the involvement of realistic forensic science to solve a whodunit, a murderer was brought to justice through an understanding of chirality in organic chemistry. The 1930 story by Sayers and Euston concerned George Harrison, found dead in the kitchen of his small cottage in the woods called *The Shack*. Beside Harrison's body were some wild mushrooms, which it was known he was fond of collecting from the woods. In the subsequent inquest, the Home Analyst, Sir James Lubbock, testified that the cause of death was poisoning by muscarine, the active principle from the mushroom *Amanita muscaria* (figure 17.27), a large amount of which was found in the stomach and vomitus of the victim. Sir James had no hesitation in declaring to the coroner that the death arose through accidental ingestion of the poison during the preparation of a meal involving the mushrooms. The jury agreed and brought in a verdict of 'Accidental Death due to poisoning by *Amanita muscaria*'.

FIGURE 17.27 *Amanita muscaria.*

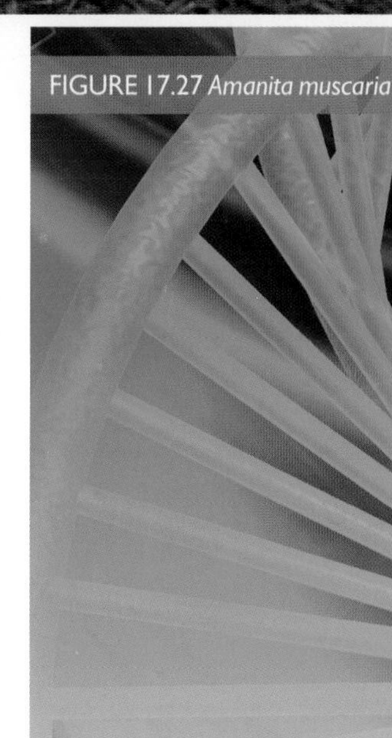

And there the matter would have ended except for the efforts of John Bunting, a friend of the deceased, who could not believe that Harrison would have mistaken the mushroom for an edible variety. Bunting's initial investigations were all fruitless until he talked to an organic chemist. In doing so, he discovered that 'life has a kind of bias — a lopsidedness so to speak', and it is sometimes possible to distinguish between synthetic substances and those derived from living organisms. Synthesis generally produces the racemate, the combined chiral forms of a molecule, whereas the natural products are usually of one form only, single enantiomers. The organic chemist told Bunting that a simple polarimeter measurement would confirm that the poison came from the mushroom and, after some effort, Bunting was able to convince the Home Analyst to undertake the analysis. When Lubbock's measurement showed no optical activity, it was clear that the muscarine (figure 17.28) was a racemic mixture and therefore of synthetic origin. Harrison's death must have been due to foul play. Subsequently, the source of the racemic mixture of muscarine was traced to Mrs Harrison's lover who paid the full penalty under the law — death by hanging.

FIGURE 17.28 (a) Muscarine activates the nervous system by mimicking acetylcholine. **(b)** Acetylcholine — the messenger molecule involved in nerve impulse transmission.

Chirality is much more involved with saving lives than ending them, as many modern medicines are sold as a single enantiomer. The chiral drug industry now represents close to one-third of all drug sales worldwide. Single enantiomer drugs and intermediates represent more than 150 billion dollars in sales each year. AstraZeneca for instance sold over 5.7 billion dollars worth of one drug alone (esomeprazole, used to treat ulcers) in one year, making it the third biggest selling pharmaceutical drug that year. The industry's growth is rooted, in part, in the chemistry by which molecules interact with biological systems. Biological messenger molecules and cell surface receptors that medicinal chemists try to affect or mimic are chiral, so drug molecules must match this asymmetry. The chiral drugs synthesised by the pharmaceutical industry today are generated by synthetic reactions that lead predominantly to one enantiomer, or by special separation techniques used to isolate the enantiomers.

17.7 Synthesising chiral drugs

Many of today's medicines involve chiral compounds. Some of these, such as ibuprofen (see pp. 768–9), are sold as a mixture of both enantiomers. Others, however, must be enantiomerically pure because only one enantiomer has the desired therapeutic effect. Pharmaceutical companies approach the production of enantiomerically pure medicines in two ways; they can synthesise a racemic mixture of the drug and then separate the two enantiomers (resolution) or they can control the synthetic conditions so that only one enantiomer is formed (asymmetric synthesis).

Resolution

Resolution is the separation of a mixture into its individual components. Resolution of enantiomers is, in general, difficult, but several laboratory methods can be used. In this section, we illustrate just one: the use of enzymes as chiral catalysts. The principle involved in this method is that a particular enzyme will catalyse a reaction of a chiral molecule, but not of its enantiomer (see figure 17.26, p. 778).

The esterases have received particular attention in this regard. This class of enzyme catalyses the hydrolysis of esters to give an alcohol and a carboxylic acid. Figure 17.29 illustrates the resolution of (*R*,*S*)-naproxen. The ethyl esters of both (*R*)- and (*S*)-naproxen are solids with very low solubilities in water. Chemists use an esterase in alkaline solution to selectively hydrolyse the (*S*)-ester, which goes into aqueous solution as the sodium salt of the (*S*)-carboxylic acid. The (*R*)-ester is unaffected by these conditions. Filtering the alkaline solution recovers the crystals of the (*R*)-ester. After the crystals are removed, the alkaline solution is acidified to precipitate pure (*S*)-naproxen. The recovered (*R*)-ester can be racemised (converted to an *R*,*S*-mixture) and treated again with the esterase. By recycling the (*R*)-ester, the racemic ester is converted to (*S*)-naproxen.

H, CH_3, OCH_2CH_3, C, O, H_3CO — ethyl ester of (*S*)-naproxen

+

H_3C, H, H_3CCH_2O, C, O, OCH_3 — ethyl ester of (*R*)-naproxen (not affected by the esterase)

1. esterase, NaOH, H_2O
2. HCl, H_2O

H, CH_3, OH, C, O, H_3CO — (*S*)-naproxen

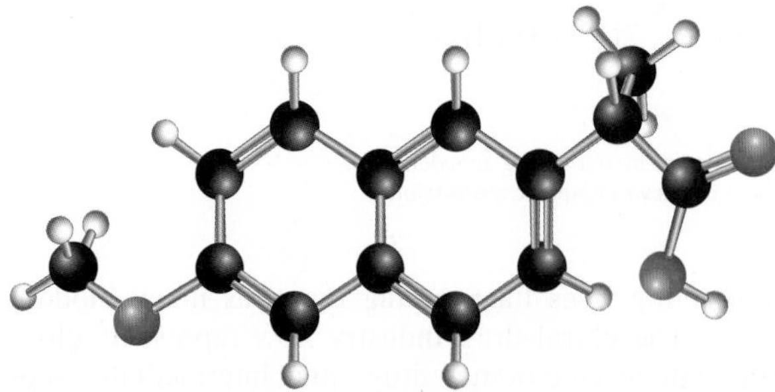

FIGURE 17.29 The resolution of (*R*,*S*)-naproxen.

The sodium salt of (*S*)-naproxen is the active ingredient in many over-the-counter non-steroidal anti-inflammatory drugs (NSAIDs). The (*S*)-enantiomer is much more effective than the (*R*)-enantiomer, and more importantly the (*R*)-enantiomer has significant liver toxicity so must not be present in the drug.

Removing the (*R*)-enantiomer of naproxen to provide enantiomerically pure (*S*)-naproxen makes this a useful and effective drug, but this is not the case for all single enantiomers. For example, resolution of the enantiomers of thalidomide would still result in a drug with unwanted side-effects because the body is able to convert one enantiomer to the other.

Asymmetric synthesis

Inorganic stereoisomers are essential for the industrial synthesis of some important chiral drugs. In 2001 the Nobel Prize in chemistry was awarded to chemists William Knowles, Ryoji Noyori and Barry Sharpless who developed chiral inorganic catalysts that, for the first time, enabled the practical synthesis of chirally pure compounds used to make medicines, such as L-DOPA, for treatment of Parkinson's disease, and naproxen, the structures of which are shown below.

Asymmetric synthesis is the name chemists give to procedures that give rise to predominantly one enantiomer. Efficient asymmetric synthesis is crucial, particularly for companies that make pharmaceuticals. The first commercialised catalytic asymmetric synthesis using a chiral transition-metal complex as the catalyst was the production of L-DOPA from a nonchiral aromatic alkene. This reaction has been in commercial use since 1974 and is now known as the *Monsanto process* (figure 17.30) after the name of the company that developed it. The commercial success of the synthesis of L-DOPA stimulated the subsequent development and application of other catalytic asymmetric reactions.

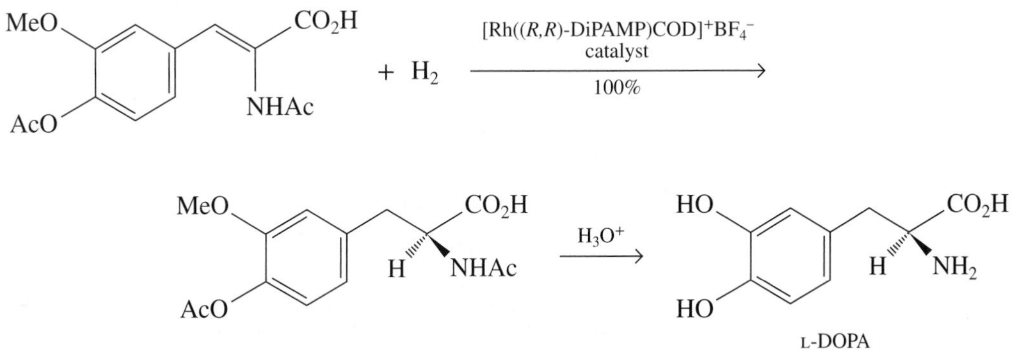

MeO, AcO, CO_2H, NHAc + H_2 → $[Rh((R,R)\text{-DiPAMP})COD]^+BF_4^-$ catalyst, 100%

MeO, AcO, CO_2H, H, NHAc → H_3O^+ → HO, HO, CO_2H, H, NH_2

L-DOPA

FIGURE 17.30 The Monsanto process for the production of L-DOPA, which is used to treat Parkinson's disease.

As well as being produced by resolution as discussed previously, (*S*)-naproxen can also be synthesised industrially using special catalysts with chiral ligands. This asymmetric synthesis (figure 17.31) uses the chiral BINAP complex. This clean, simple and economical approach may be applied on any scale from <100 mg to >100 kg. BINAP-catalysed asymmetric syntheses are applied to the industrial production of medicinal compounds in quantities of well over 100 tonnes a year. This approach has been used to develop a range of pharmaceuticals, agrochemicals, flavours and fragrances.

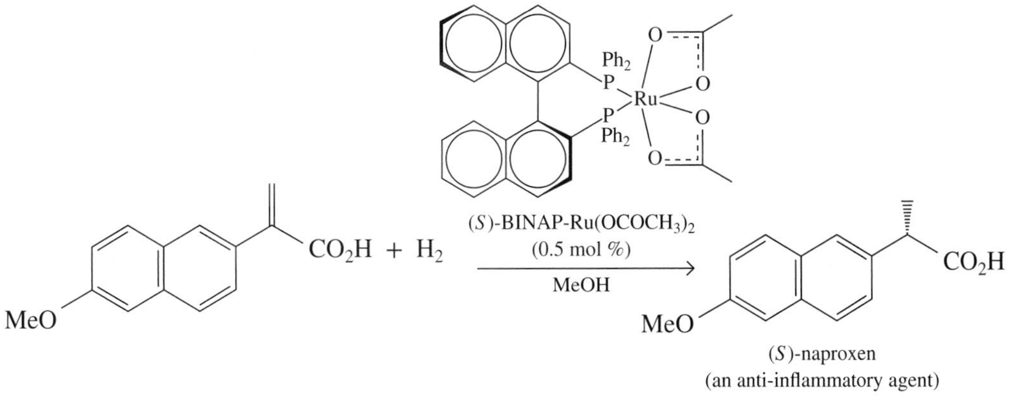

Ph_2 P, Ru, O, O, O, O, P Ph_2

MeO, CO_2H + H_2 → (*S*)-BINAP-Ru$(OCOCH_3)_2$ (0.5 mol %), MeOH → MeO, CO_2H

(*S*)-naproxen
(an anti-inflammatory agent)

FIGURE 17.31 The commercial asymmetric synthesis of the anti-inflammatory drug naproxen (sold as Naprogesic™ in Australia and New Zealand).

Because asymmetric synthesis is difficult, even for relatively small molecules, the complete synthesis of complex organic molecules such as the important cancer-fighting drug Taxol (see p. 767) provides a tremendous challenge. Chemists produce Taxol by starting with a simpler compound found in nature and then making a few, less demanding, asymmetric transformations to achieve the desired material.

Asymmetric synthesis allows us to create new medicines and compounds with specific biological effects. Achieving stereochemical control over synthetic reactions, however, is one of the greatest challenges facing chemists today. To master such challenging methodology, we must first learn about the key functional groups, their properties and how they react; this will be covered in the following chapters.

SUMMARY

LO 1 Stereoisomers

Isomers are molecules with the same molecular formula but different structures. Constitutional isomers are isomers with different sequences of atom connectivity. Stereoisomers have the same order of attachment of atoms, but a different three-dimensional orientation of their atoms in space. A chiral object is one that is not superimposable on its mirror image. Enantiomers are stereoisomers that are chiral. An achiral object is the same as its mirror image. Diastereomers are stereoisomers that are not superimposable and are not mirror images.

LO 2 Enantiomerism

Most biological molecules are enantiomeric. Enantiomers have the same physical and chemical properties but each of a pair reacts differently with other chiral molecules. A plane of symmetry is an imaginary plane passing through an object, dividing it such that one half is the reflection of the other half. Objects with planes of symmetry are achiral. A stereocentre is the part of a molecule that can be assembled in two different ways to generate stereoisomers. The most common type of stereocentre among organic compounds is a tetrahedral carbon atom with four different groups bonded to it. Inorganic molecules can also have stereocentres and, therefore, enantiomers and diastereomers.

LO 3 Naming stereocentres: the *R,S* system

The configuration at a stereocentre can be specified by the *R,S* system. To apply this naming system: (1) each atom or group of atoms bonded to the stereocentre is assigned a priority and is numbered from highest priority to lowest priority, (2) the molecule is oriented in space so that the group of lowest priority is directed away from the observer, (3) the remaining three groups are read in order, from highest priority to lowest priority, and (4) if the sequence of the groups is clockwise, the configuration is *R*, if anticlockwise, the configuration is *S*.

LO 4 Molecules with more than one stereocentre

For a molecule with n stereocentres, the maximum possible number of stereoisomers is 2^n. A plane of symmetry in a molecule reduces the number of stereoisomers to fewer than that predicted by the 2^n rule. A *meso* compound contains two or more stereocentres assembled so that its molecules are achiral.

LO 5 Optical activity: detecting chirality in the laboratory

Enantiomers differ in their optical activity — that is, their ability to rotate a plane of polarised light. Plane-polarised light oscillates only in parallel planes. A polarimeter is an instrument used to detect and measure the magnitude of optical activity. Observed rotation is the number of degrees a plane of polarised light is rotated. Specific rotation is the observed rotation measured with a cell 1 dm long. Levorotatory compounds rotate a plane of polarised light anticlockwise. Dextrorotatory compounds rotate a plane of polarised light clockwise. A racemic mixture is a mixture of equal amounts of two enantiomers and has a specific rotation of 0.

LO 6 Chirality in the biological world

Chirality is very important in biology. Almost all the molecules in living systems, both plant and animal, are chiral. Enzymes catalyse biological reactions for one specific enantiomer. Each enantiomer of a drug can have different biological activity.

LO 7 Synthesising chiral drugs

Enantiomerically pure drugs can be produced via two approaches: a racemic mixture of the drug can be separated in a process called resolution, or the synthetic conditions can be controlled so that only one enantiomer is formed. One means of resolution is to treat the racemic mixture with an enzyme that catalyses a specific reaction of one enantiomer but not the other. Chiral inorganic catalysts can be useful tools in asymmetric synthesis to control reactions so that only one enantiomer is formed.

KEY CONCEPTS AND EQUATIONS

The *R,S* system
(section 17.3)

Orienting molecules with the lowest priority bond away from the point of view and prioritising the three remaining groups gives a clockwise (*R*) or an anticlockwise (*S*) sequence.

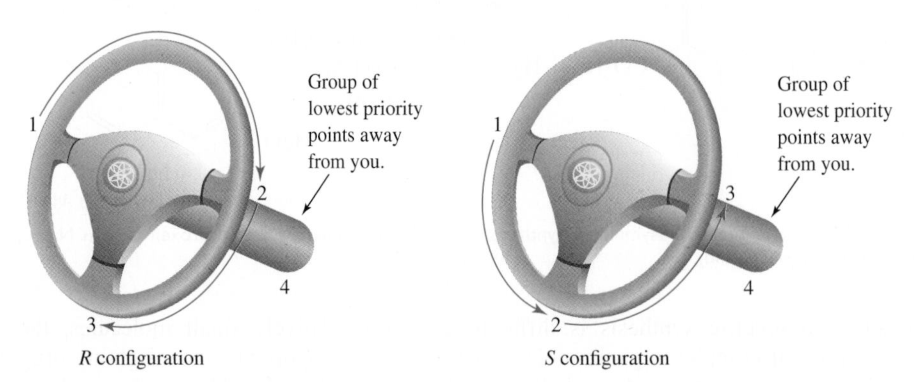

1
2
3
4
Group of lowest priority points away from you.
1
3
2
4
Group of lowest priority points away from you.

R configuration

S configuration

The 2^n rule
(section 17.4)

For a molecule with n stereocentres, there will be at most $2n$ stereoisomers.

Specific rotation formula for enantiomers
(section 17.5)

$$\text{specific rotation} = [\alpha]_\lambda^T = \frac{\text{observed rotation (degrees)}}{\text{length (dm)} \times \text{concentration (g/mL)}}$$

KEY TERMS

achiral *p. 761*
asymmetric synthesis *p. 781*
chiral *p. 760*
constitutional isomers *p. 760*
dextrorotatory *p. 776*
diastereomers *p. 761*
enantiomers *p. 760*
isomers *p. 760*
levorotatory *p. 776*
meso compound *p. 772*
observed rotation *p. 776*
optical activity *p. 775*
plane of symmetry *p. 765*
plane polarised *p. 775*
polarimeter *p. 776*
R *p. 769*
R,S system *p. 768*
racemic mixture *p. 778*
resolution *p. 780*
S *p. 769*
specific rotation ([α]) *p. 777*
stereocentre *p. 765*
stereoisomers *p. 760*

REVIEW QUESTIONS

LO 1 Stereoisomers

17.1 Define the term 'stereoisomer'.

17.2 How are constitutional isomers different from stereoisomers? How are they the same?

17.3 What is meant by the term 'chiral'?

17.4 Which of these objects are chiral? (Assume there are no labels or identifying marks present.)

(a) cricket ball

(b) iPhone

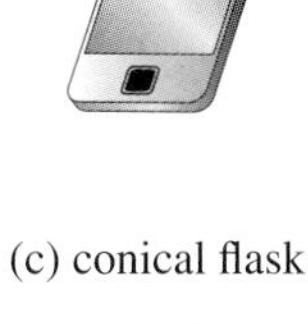

(c) conical flask

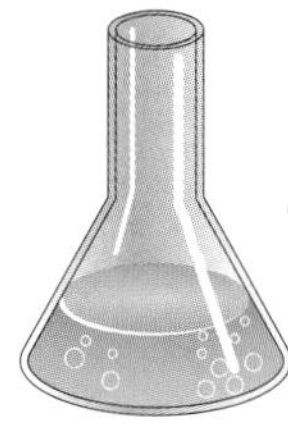

(d) paper clip

(e) bolt

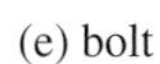

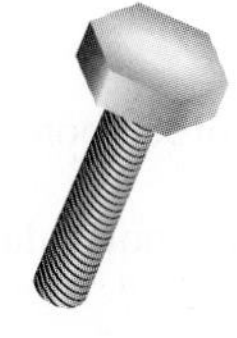

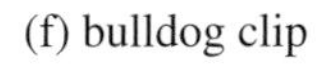

(f) bulldog clip

(g) signature

(h) the word

(i) the word

17.5 Next time you have the opportunity to view a collection of coneshells or other seashells with a helical twist, study the chirality of their twists. Do you find an equal number of left-handed and right-handed shells or, for example, do they all have the same handedness?

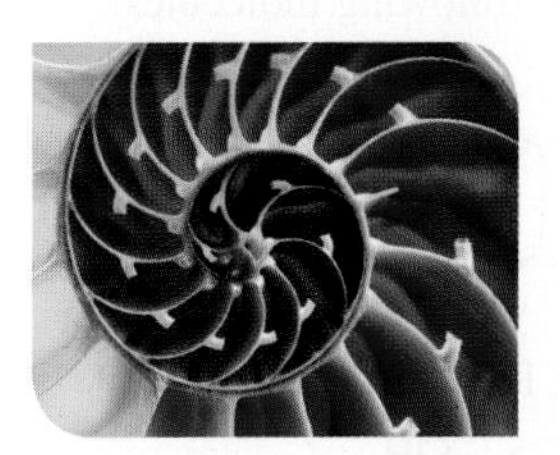

Median cross-section through the shell of a chambered nautilus found in the deep waters of the Pacific Ocean. The shell shows handedness; this cross-section is a right-handed spiral.

17.6 When you next have an opportunity to examine any of the seemingly endless varieties of spiral pasta, look how they are twisted to form a spiral. Do pieces of any one kind of pasta all twist in the same direction? (That is, do they all have a right-handed twist or a left-handed twist, or are they a mixture, possibly even a racemic mixture?)

17.7 Which of the following objects have mirror images that are not the same as the original object?

(a)

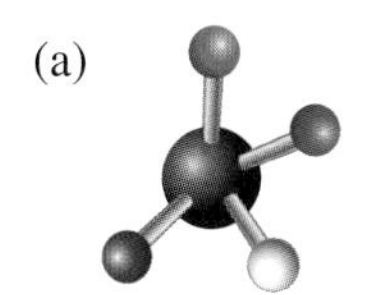

(b)

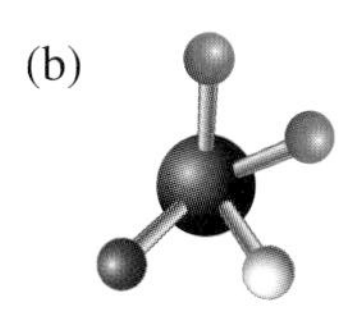

(c)

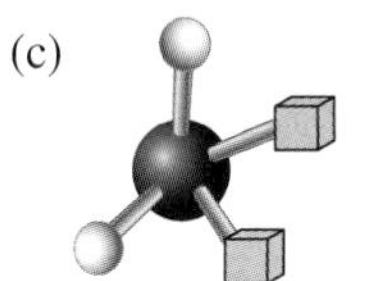

(d)

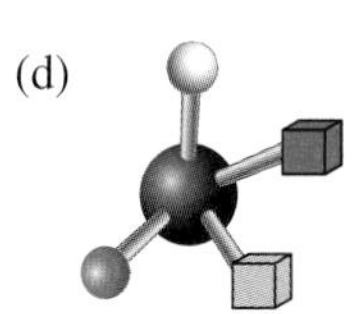

LO 2 Enantiomerism

17.8 State whether each of the following statements is true.

(a) All enantiomers are chiral.
(b) A diastereomer of a chiral molecule must also be chiral.
(c) A molecule with an internal plane of symmetry can never be chiral.
(d) All achiral molecules have enantiomers.
(e) All achiral molecules have diastereomers.
(f) All chiral molecules have enantiomers.
(g) All chiral molecules have diastereomers.

17.9 Which of the following compounds contain stereocentres?

(a) 2-bromobutane
(b) 3-bromopentane
(c) 3-bromohex-1-ene
(d) 1,2-dibromopropane
(e) 2,2-dibromopropane

17.10 Which of the following hydrocarbons have enantiomers? Indicate the stereocentre that makes the molecule chiral.

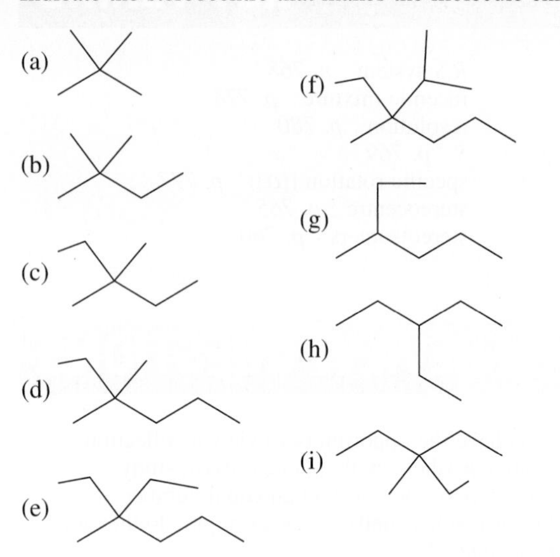

17.11 Draw the enantiomer of each of the following molecules.

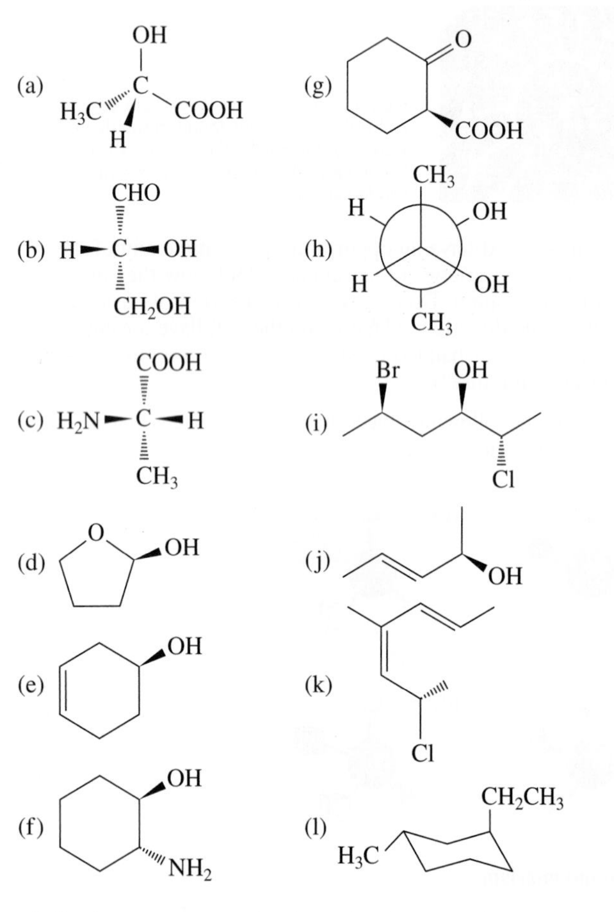

17.12 Closely examine the following four metal complexes. Which of these complexes are chiral, that is, has an enantiomer? Draw these enantiomers.

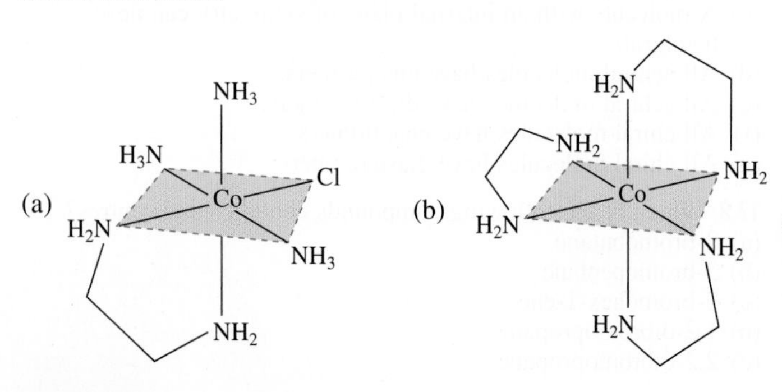

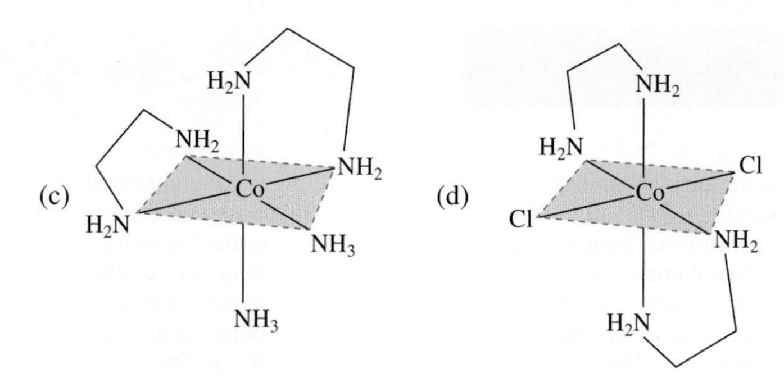

17.13 Draw the two enantiomers of the complex $[Co(C_2O_4)_3]^{3-}$, which has three bidentate oxalate ligands (see chapter 13, p. 551).

17.14 Mark each stereocentre in the following molecules with an asterisk. (*Note:* Not all contain stereocentres.)

(a) OH, CH_3, N, CH_3 (c) O, O^-, $^+NH_3$

(b) O, OH (d) O

17.15 Mark each stereocentre in the following molecules with an asterisk. (*Note:* Not all contain stereocentres.)

(a) HO, OH, OH (c) OH

(b) HO, OH (d) OH

LO 3 Naming stereocentres: the *R,S* system

17.16 Which of the following molecules have an *R* configuration?

(a) CH_3, Br, C, H, CH_2OH (c) CH_2OH, H_3C, C, Br, H

(b) H, $HOCH_2$, C, CH_3, Br (d) Br, H, C, CH_2OH, CH_3

17.17 Assign priorities to the groups in each of the following sets.

(a) —H, —CH_3, —OH, —CH_2OH
(b) —$CH_2CH{=}CH_2$, —$CH{=}CH_2$—CH_3, —CH_2COOH
(c) —CH_3, —H, —COO^-, —$^+NH_3$
(d) —CH_3, —CH_2SH, —$^+NH_3$, —COO^-

LO 4 Molecules with more than one stereocentre

17.18 (a) What is the smallest noncyclic saturated hydrocarbon (C_nH_{2n+2}) that can contain a stereocentre?
(b) Write the structural formula of a molecule with the molecular formula $C_6H_{14}Br$ that contains two stereocentres.

17.19 Assign *R* or *S* configuration to the following.

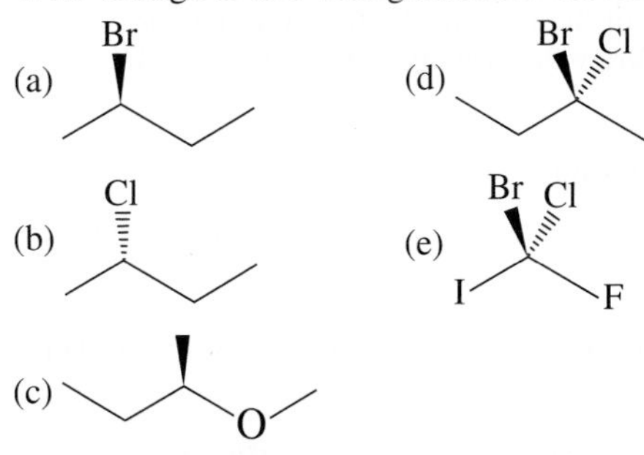

(g) (h)

How many stereoisomers are possible for each molecule?

17.20 For centuries, Chinese herbal medicine has used extracts of *Ephedra sinica* to treat asthma. Investigation of this plant resulted in the isolation of ephedrine, a potent dilator of the air passages of the lungs. The naturally occurring stereoisomer is levorotatory and has the following structure.

ephedrine

Assign an *R* or *S* configuration to each stereocentre.

17.21 Atorvastatin is sold under the trade name Lipitor and is used for lowering cholesterol. Annual global sales of this compound exceed $13 billion. Assign a configuration to each chirality centre in atorvastatin.

atorvastatin

17.22 Label each stereocentre in the following molecules with an asterisk.

(a) $CH_3CH(OH)CH(OH)COOH$ (d)

(b) $CH_2—COOH$ / $CH—COOH$ / $HO—CH—COOH$ (e)

(c) (f)

17.23 Identify the number of stereoisomers expected for each of the following.

(a) (d)

(b) (e)

(c) (f)

17.24 Label with asterisks the four stereocentres in amoxicillin, which belongs to the family of semisynthetic penicillins.

amoxicillin

17.25 In the more stable chair conformation of glucose, all groups on the 6-membered ring are equatorial.

(a) Identify all stereocentres in this molecule.
(b) How many stereoisomers are possible?
(c) How many pairs of enantiomers are possible?
(d) What is the configuration (*R* or *S*) at C(1) and C(5) in the stereoisomer shown?

LO 5 Optical activity: detecting chirality in the laboratory

17.26 What is a racemic mixture? Is a racemic mixture optically active (i.e. will it rotate a plane of polarised light)?

17.27 The specific rotation of naturally occurring ephedrine, shown in question 17.20, is –41°. What is the specific rotation of its enantiomer?

LO 6 Chirality in the biological world

17.28 Insects use particular organic molecules called pheromones as signals to attract mates. As an example, the pine sawfly species *Neodiprion* and *Diprion* use an ester derivative of 3,7-dimethylpentadecan-2-ol as a sex pheromone.

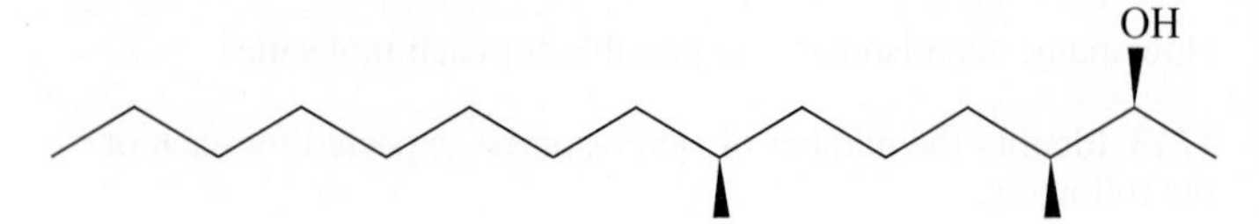

(a) How many different stereoisomers exist for this molecule?
(b) The stereoisomer shown produces much of the pheromone activity of this molecule. Using the *R*,*S* naming system, give the correct name for this molecule.
(c) Interestingly, the stereoisomer shown must be mixed with precisely 0.1% of the (2*S*,3*R*,7*S*) diastereomer for maximum activity. Draw the structure of this key (2*S*,3*R*,7*S*) isomer.

LO 7 Synthesising chiral drugs

17.29 What are the two approaches to achieving enantiomerically pure compounds?

17.30 Why is it necessary for some drugs to be resolved into their separate enantiomers?

17.31 Ibuprofen is sold as a (±) racemic mixture. Why is it not necessary to resolve this mixture into the pure enantiomers?

REVIEW PROBLEMS

17.32 One reason we can be sure that sp^3 hybridised carbon atoms are tetrahedral is the number of stereoisomers that can exist for different organic compounds. **LO1**
(a) How many stereoisomers are possible for $CHCl_3$, CH_2Cl_2 and CHBrClF if the four atoms bonded to the carbon atom have a tetrahedral arrangement?
(b) How many stereoisomers are possible for each of the compounds if the four atoms bonded to the carbon atom have a square planar arrangement?

17.33 Using only C, H and O, write structural formulae for the lowest molar mass chiral molecule of each of the following classes of compounds. **LO1**
(a) alkane
(b) alcohol
(c) aldehyde
(d) ketone
(e) carboxylic acid

17.34 Which alcohols with the molecular formula $C_5H_{12}O$ are chiral? **LO1**

17.35 Which carboxylic acids with the molecular formula $C_6H_{12}O_2$ are chiral? **LO1**

17.36 Draw all possible stereoisomers of hex-1-en-2-ol. **LO1**

17.37 Mark each stereocentre in the following molecules with an asterisk. (*Note:* Not all contain stereocentres.) **LO2**

(a) $CH_3C(CH_3)(OH)CH{=}CH_2$

(b) $HOOC{-}CH(OH){-}CH_3$

(c) $CH_3CH(CH_3)CH(NH_2)COOH$

(d) $CH_3C(=O)CH_2CH_3$

(e) $HOCH_2{-}CH(OH){-}CH_2OH$

(f) $CH_3CH_2CH(OH)CH{=}CH_2$

(g) $HOC(CH_2COOH)_2COOH$

17.38 The following are eight stereorepresentations of lactic acid. **LO2**

(a) C with COOH (up), OH, H (hashed), H_3C (wedge)

(b) C with CH_3 (up), H, HO (hashed), HOOC (wedge)

(c) C with COOH (up), CH_3, HO (hashed), H (wedge)

(d) C with CH_3 (up), COOH, H (hashed), HO (wedge)

(e) H—C—OH with COOH (top), CH_3 (bottom)

(f) H—C—OH with CH_3 (top), COOH (bottom)

(g) H_3C—C—COOH with OH (top), H (bottom)

(h) H—C—COOH with CH_3 (top), OH (bottom)

Take (a) as the reference structure. Which stereorepresentations are identical to (a) and which are mirror images of (a)?

17.39 The following are structural formulae for the enantiomers of carvone. **LO3**

(−)-carvone (in spearmint oil)

(+)-carvone (in caraway and dillseed oil)

Each enantiomer has a distinctive odour characteristic of its source. Assign an *R* or *S* configuration to the stereocentre in each. How can they have such different properties when they are so similar in structure?

17.40 The following is a staggered conformation of one of the stereoisomers of butan-2-ol. **LO3**

(H_3C, H, H on one carbon; H, OH, CH_3 on the other)

(a) Is this (*R*)-butan-2-ol or (*S*)-butan-2-ol?
(b) Draw a Newman projection for this staggered conformation, viewed along the bond between C(2) and C(3).
(c) Draw a Newman projection for one more staggered conformation of this molecule. Which of your conformations is the more stable? Assume that the —OH and —CH_3 groups are comparable in size.

17.41 Label all stereocentres in loratadine (Claritin® in New Zealand or Claratyne® in Australia) and fexofenadine (Telfast®), now the top-selling antihistamines in Australia and New Zealand. **LO2**

(a) 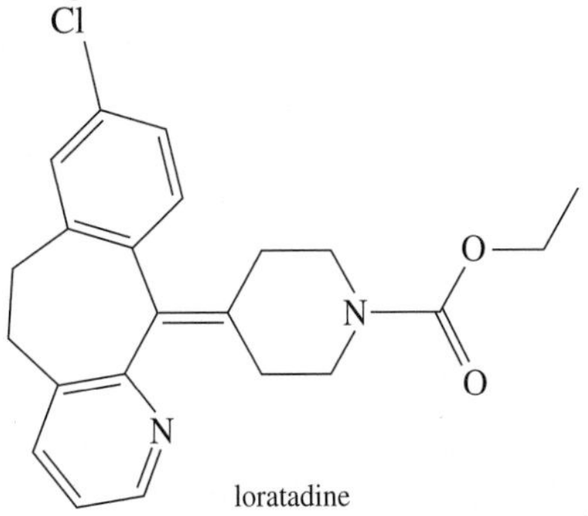 loratadine (Claritin® or Claratyne®)

(b) 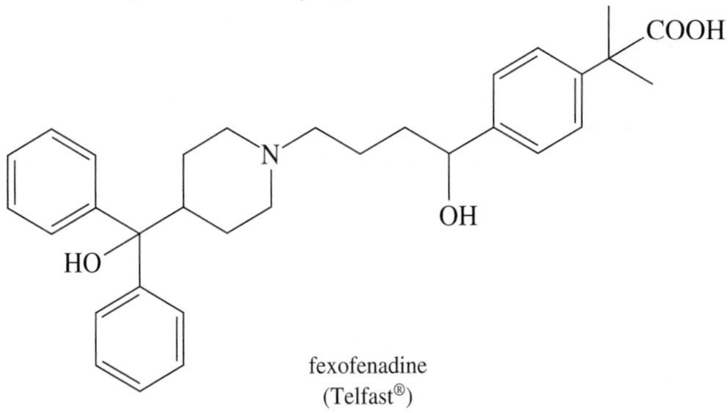 fexofenadine (Telfast®)

How many stereoisomers are possible for each compound?

17.42 The following are structural formulae for three of the most widely prescribed drugs used to treat depression. **LO2**

(a) 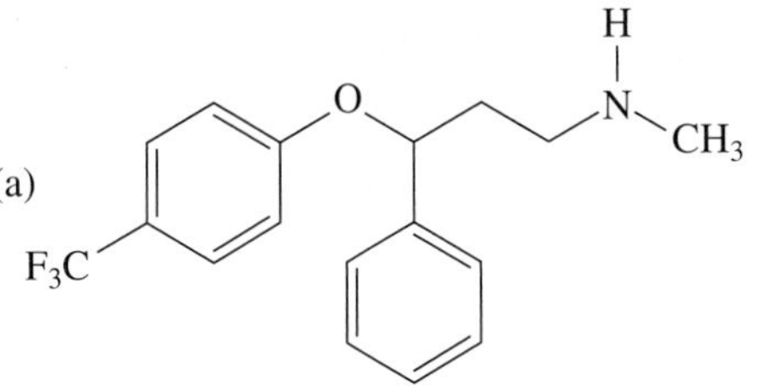 fluoxetine (Prozac®)

(b) 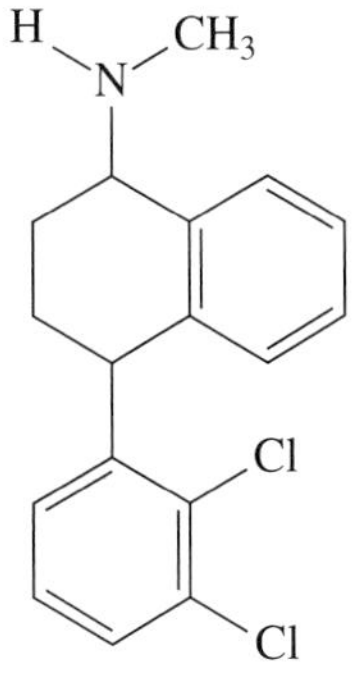 Sertraline (Zoloft®)

(c) 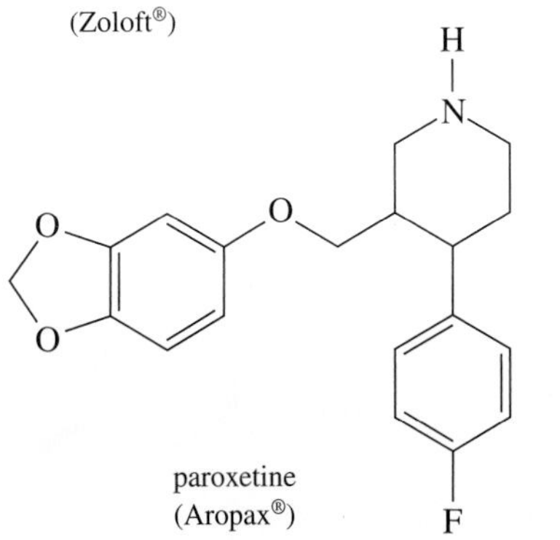 paroxetine (Aropax®)

Label all stereocentres in each compound, and state the number of stereoisomers possible for each.

17.43 Triamcinolone acetonide, the active ingredient in Azmacort® inhalers, is a steroid used to treat bronchial asthma. **LO4**

triamcinolone acetonide

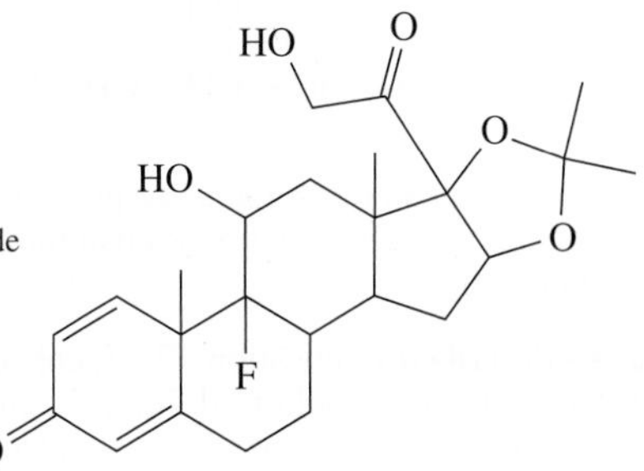

(a) Label the eight stereocentres in this molecule.
(b) How many stereoisomers are possible for the molecule? (Only one stereoisomer is the active ingredient in Azmacort.)

17.44 How many stereoisomers are possible for 1,2-cyclobutanediol? **LO4**

17.45 How many stereoisomers are possible for 1,3-cyclobutanediol? **LO4**

17.46 How many stereoisomers are possible for 1,2-cyclopentanediol? **LO4**

17.47 How many stereoisomers are possible for 1-ethyl-2-methylcyclobutane? **LO4**

17.48 How many stereoisomers are possible for 1-ethyl-3-methylcyclobutane? **LO4**

17.49 How many stereoisomers are possible for each of the following compounds? **LO4**

(a) 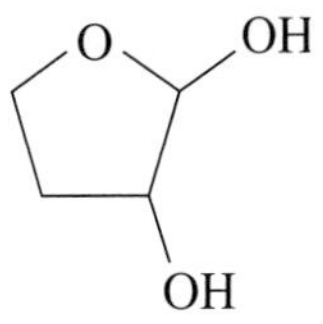 (b) 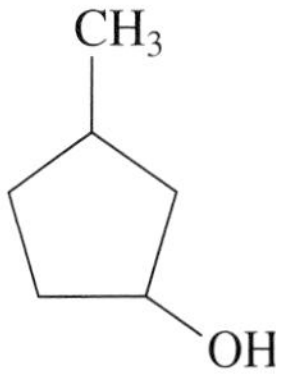 (c) (d)

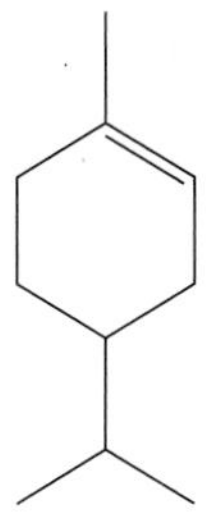

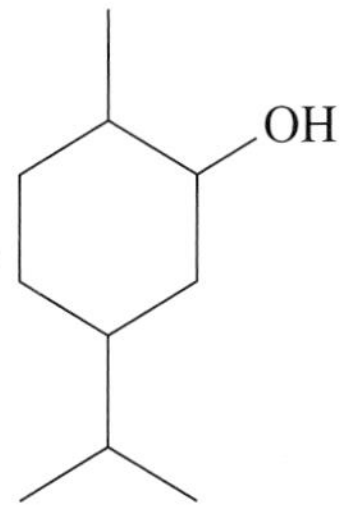

17.50 Which of the following structural formulae represent *meso* compounds? **LO4**

(a) 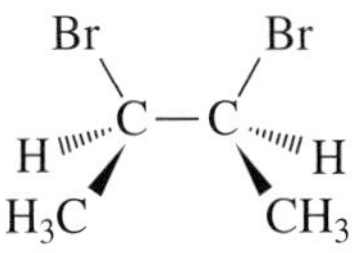 (b) 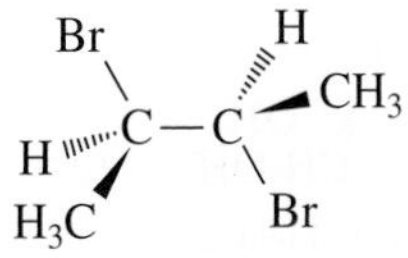 (c) (d) 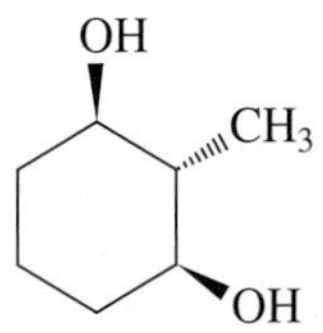 (e) (f)

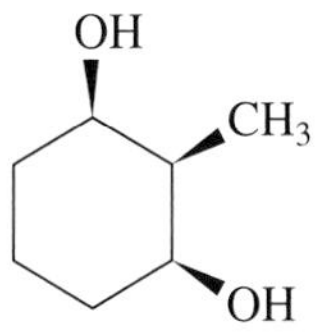

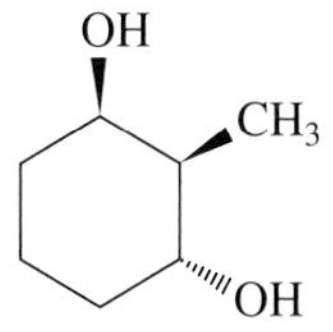

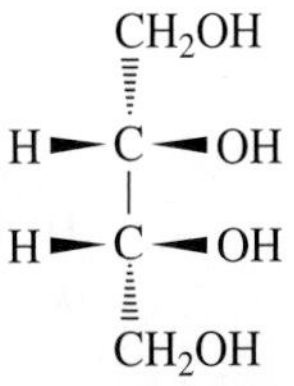

17.51 Draw a Newman projection, viewed along the bond between C(2) and C(3), for both the most stable and the least stable conformations of *meso*-tartaric acid. **LO4**

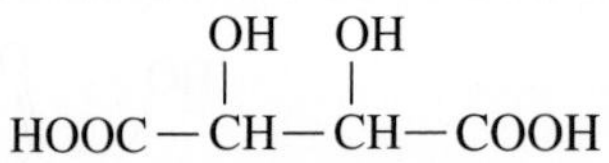

$$HOOC-\underset{OH}{CH}-\underset{OH}{CH}-COOH$$

17.52 How many stereoisomers are possible for 1,3-dimethylcyclopentane? Which are pairs of enantiomers? Which are *meso* compounds? **LO4**

17.53 As described in question 17.27, ephedrine is optically active and rotates a plane of polarised light 41° anticlockwise. Which of the following compounds would rotate plane-polarised light 41° in a clockwise direction? **LO5**

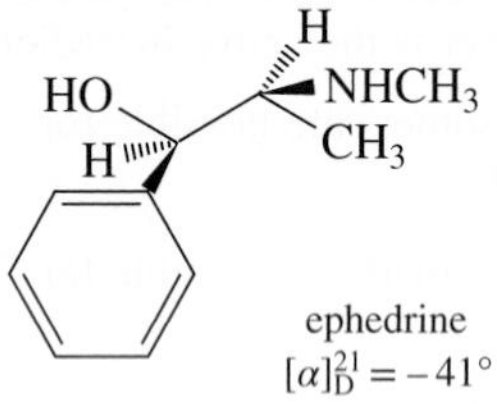

ephedrine
$[\alpha]_D^{21} = -41°$

(a) 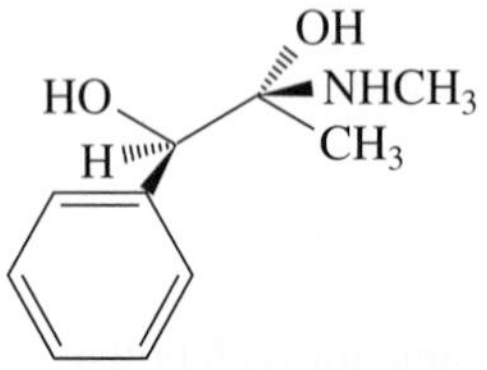 (b) 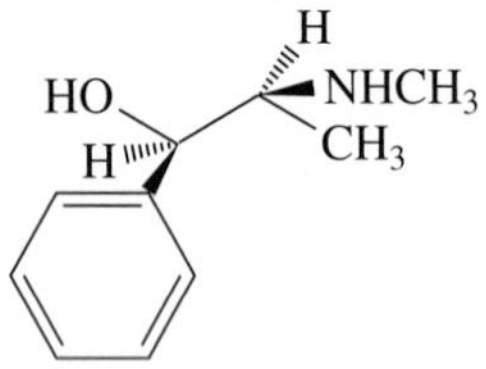 (c) 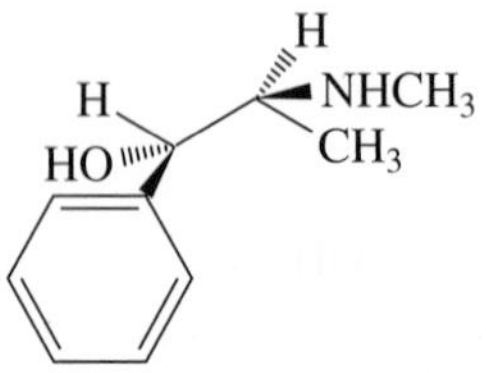 (d) 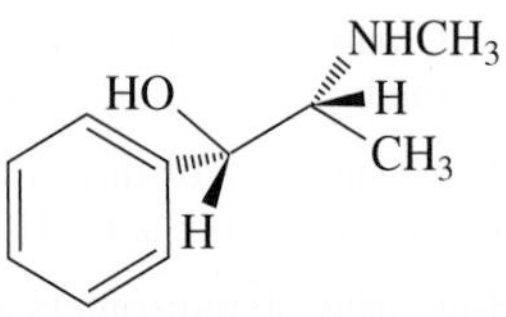 (e)

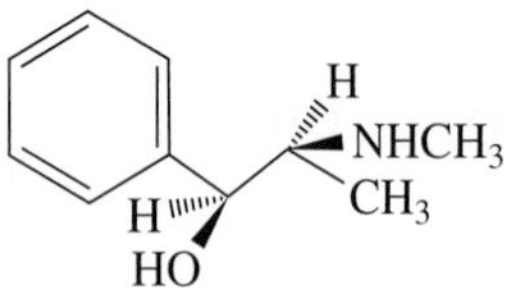

17.54 The natural product (*R*)-glyceraldehyde shown below has a specific rotation, $[\alpha]_D$ at 25 °C, of +13.5° when measured in aqueous solution. When a sample from the laboratory was measured, the observed rotation was +9.4°. **LO5**

(a) Use the formula for calculating the optical activity of enantiomers to determine the concentration of (*R*)-glyceraldehyde used in this measurement.
(b) What would be the observed optical activity of this compound if the concentration used in the measurement was halved?
(c) What would be the value measured if the mixture contained half the original concentration of (*R*)-glyceraldehyde and the same amount of (*S*)-glyceraldehyde?

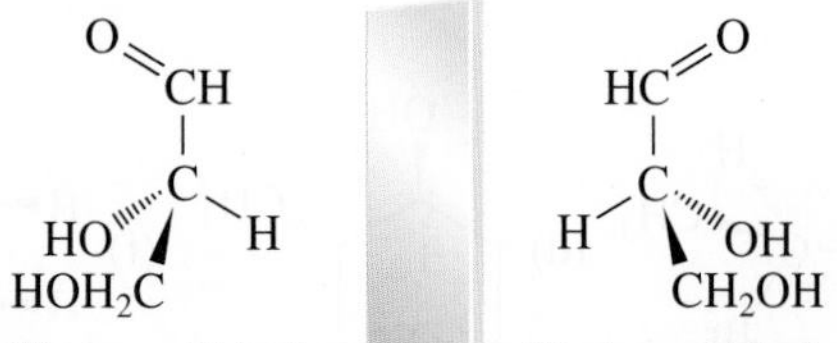

(*R*)-glyceraldehyde (*S*)-glyceraldehyde

17.55 The (*S*)-enantiomer of the following molecule is (+)-carvone, which is extracted from the oil of caraway seeds. **LO5**

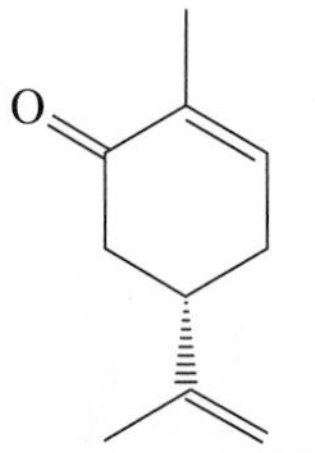

The (*R*)-enantiomer is (–)-carvone, and it is isolated from the oil of spearmint. Draw the correct molecular structures for these molecules. If the specific rotation of a pure sample of carvone was measured as +62.5° at 20 °C, which enantiomer of carvone was measured in the polarimeter?

17.56 Atropine, extracted from the plant *Atropa belladonna*, has been used in the treatment of bradycardia (low heart rate) and cardiac arrest. Draw the enantiomer of atropine. **LO6**

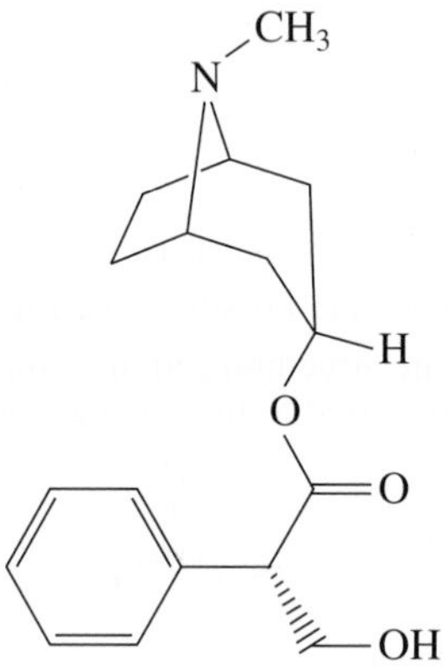

atropine

17.57 (*R*)-limonene is found in many citrus fruits, including oranges and lemons. **LO6**

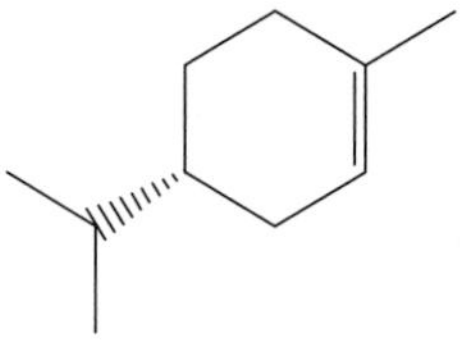

(*R*)-limonene

For each of the following compounds, identify whether it is (*R*)-limonene or its enantiomer, (*S*)-limonene.

(a) 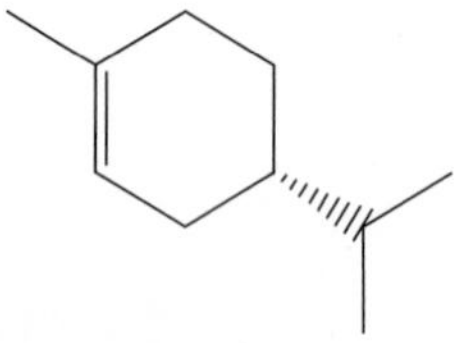 (b) 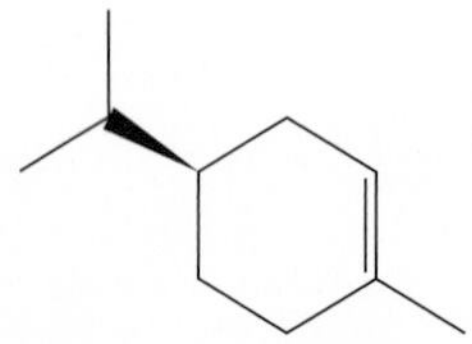 (c) 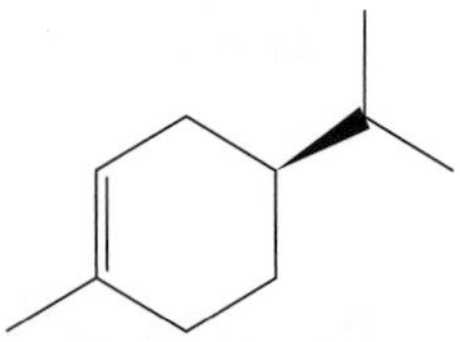 (d)

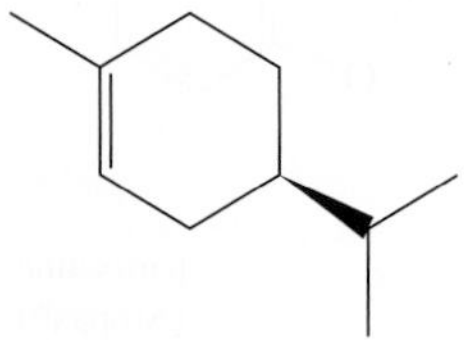

17.58 State whether each of the following statements is true for molecules (i) and (ii). **LO1**

(i) (ii)

(a) Molecules (i) and (ii) are the same.
(b) Molecules (i) and (ii) are diastereomers of each other.
(c) Gentle heating of molecule (i) converts it to molecule (ii), so these compounds are conformational isomers.
(d) Both molecules (i) and (ii) have cyclic imide functional groups so cannot have enantiomeric isomers.
(e) Molecule (i) would rotate a plane of polarised light in the opposite direction to molecule (ii).

17.59 Predict the product(s) of the following reactions (where more than one stereoisomer is possible, show each stereoisomer). **LO7**

(a) $\xrightarrow[CH_2Cl_2]{Br_2}$ (b) $\xrightarrow[H_2SO_4]{H_2O}$

17.60 Which alkene, (a) or (b), when treated with H_2/Pd will ensure a high yield of the following stereoisomer, *cis*-decalin. **LO7**

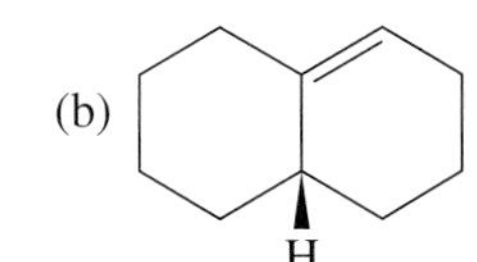

cis-decalin

(a) (b)

17.61 Which of the following reactions will yield a racemic mixture of products? **LO7**

(a) $\xrightarrow{HCl}$ (c) $\xrightarrow[CH_2Cl_2]{Br_2}$

(b) $\xrightarrow[Pt]{H_2}$ (d) $\xrightarrow[CH_2Cl_2]{Br_2}$

17.62 Draw all the stereoisomers that can be formed in the following reaction. Comment on how useful this particular reaction would be as a synthetic method. **LO7**

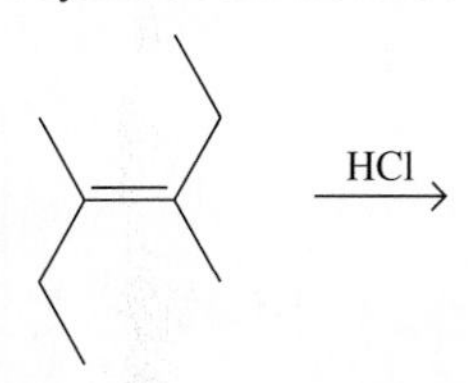

17.63 Explain why the product of the following reaction does not rotate plane-polarised light. **LO7**

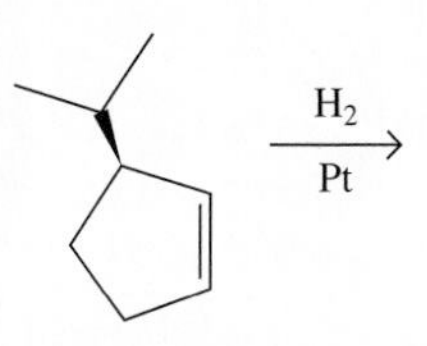

17.64 Using only ethylenediamine (en = $H_2NCH_2CH_2NH_2$) and bromide anions as ligands, construct a cationic octahedral complex of cobalt(III); your complex cation should have a charge of +1 and it should be chiral. Draw a three-dimensional structure for this coordination complex. Then draw the structure of a diastereomer of this complex. **LO2**

17.65 Draw all possible stereoisomers of $[CoCl_2(en)(NH_3)_2]^+$. Which of these stereoisomers are chiral? (en = $H_2NCH_2CH_2NH_2$) **LO2**

17.66 It is possible for a compound to be chiral even though it lacks a carbon atom with four different groups. For example, consider the structure of the following compound, which belongs to the class of compounds called allenes. This allene is chiral. Draw its enantiomer, and explain why this compound is chiral. **LO2**

18 Haloalkanes

Capitalising on the limitless energy provided by the sun is one of the most attractive ways to tackle our society's clean energy needs. As old buildings are demolished and replaced with modern, energy-efficient designs, it is not unusual to see the roofs and even walls of new buildings decked-out with extensive arrays of solar panels. This well-established photovoltaic technology is based on silicon and, although modern crystalline silicon devices are approaching the theoretical efficiency limit of ≈30%, these devices are expensive and require rare inorganic materials. While these panels are suited to rooftop installations on buildings and increasingly on our homes, there is a large scientific research effort underway to develop new kinds of devices to generate electricity from the sun based on carbon-based materials that are flexible and lightweight. Scientists are seeking to exploit types of molecules called organic photovoltaics in softer, polymer-based devices such as backpacks that may serve to charge our phones and tablets when we are on the move.

The synthesis of many of these complex molecules relies on linking together complex aromatic rings using reactions that exploit the chemistry of aryl carbon to halogen bonds. Typically molecules with two bromine atoms can be efficiently reacted to form larger, complex molecules with extended conjugation. The chemistry of organic photovoltaics can also be modified through the presence of other halogens, usually fluorine, that have a strong electronegative influence on the physical properties. Such atoms also can impart lower reactivity towards the degradation that can occur with temperature and environmental exposure. One recent example is a fluorine-containing polymeric aromatic system which, in conjunction with a substituted fullerene, gives solar efficiency of over 6%.

n [thiophene: F, $COO(CH_2)_{19}CH_3$, Br, S, Br] $+ \; n$ Me_3Sn–[benzodithiophene: $O(CH_2)_7CH_3$, S, S, $O(CH_2)_7CH_3$]–$SnMe_3$ $\longrightarrow$ [F, $COO(CH_2)_{19}CH_3$, S, S, S, S, $O(CH_2)_7CH_3$, $O(CH_2)_7CH_3$]$_n$

The carbon–halogen bond can be very strong, especially for fluorine, and the properties that halogen atoms impart on the molecule can be very useful. Haloalkanes are used as flame retardants, in fire extinguishers and as refrigerants and solvents. While these applications arise from the low reactivity of haloalkanes, chemists understand that, under the right reaction conditions, haloalkanes can be converted to alcohols, ethers, thiols, amines and alkenes. This makes them very versatile synthetic building blocks that are often used to generate important compounds used in medicine, materials and agriculture. Understanding how haloalkanes are transformed allows great control over the transformations of simple molecules into more complex and valuable products. In this chapter, we study two characteristic reactions of haloalkanes: nucleophilic substitution and β-elimination.

LEARNING OBJECTIVES

After studying this chapter, you should be able to:

18.1 describe the chemical properties of haloalkanes using correct chemical terminology

18.2 describe the sequence of bond transformations involved in the nucleophilic substitution of haloalkanes

18.3 describe the sequence of bond transformations involved in the β-elimination reactions of haloalkanes

18.4 explain how the reaction conditions and reagents influence the reaction processes of haloalkanes.

18.1 Haloalkanes

Haloalkanes are compounds containing a halogen atom covalently bonded to an sp^3 hybridised carbon atom. The general symbol for a haloalkane is R—X, where X may be F, Cl, Br or I.

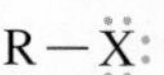

a haloalkane
(an alkyl halide)

FIGURE 18.1 In 1989, Australia banned the use of CFCs in aerosol cans. New Zealand stopped their use in 1990.

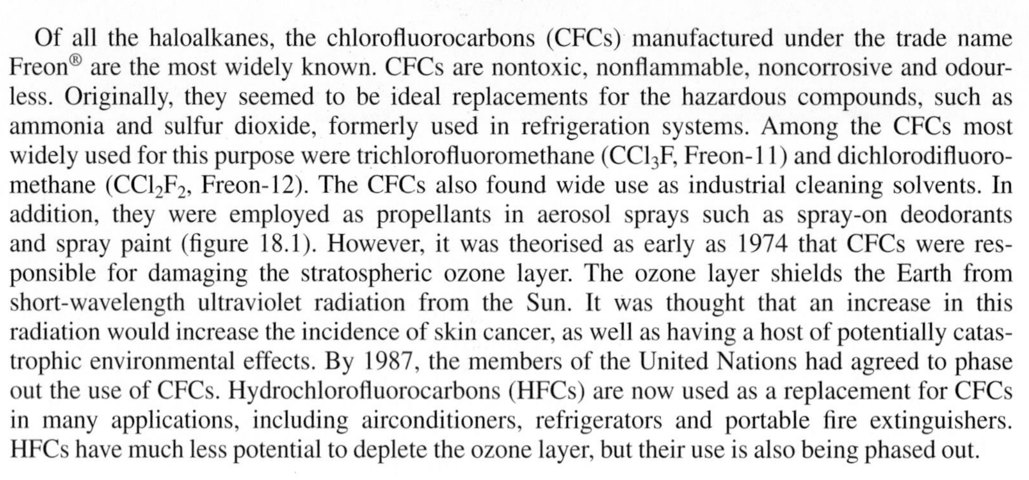

Of all the haloalkanes, the chlorofluorocarbons (CFCs) manufactured under the trade name Freon® are the most widely known. CFCs are nontoxic, nonflammable, noncorrosive and odourless. Originally, they seemed to be ideal replacements for the hazardous compounds, such as ammonia and sulfur dioxide, formerly used in refrigeration systems. Among the CFCs most widely used for this purpose were trichlorofluoromethane (CCl_3F, Freon-11) and dichlorodifluoromethane (CCl_2F_2, Freon-12). The CFCs also found wide use as industrial cleaning solvents. In addition, they were employed as propellants in aerosol sprays such as spray-on deodorants and spray paint (figure 18.1). However, it was theorised as early as 1974 that CFCs were responsible for damaging the stratospheric ozone layer. The ozone layer shields the Earth from short-wavelength ultraviolet radiation from the Sun. It was thought that an increase in this radiation would increase the incidence of skin cancer, as well as having a host of potentially catastrophic environmental effects. By 1987, the members of the United Nations had agreed to phase out the use of CFCs. Hydrochlorofluorocarbons (HFCs) are now used as a replacement for CFCs in many applications, including airconditioners, refrigerators and portable fire extinguishers. HFCs have much less potential to deplete the ozone layer, but their use is also being phased out.

Nomenclature

IUPAC names for haloalkanes are derived by naming the parent alkane according to the rules given in section 2.3.

- Locate and number the parent chain from the direction that gives the substituent encountered first the lower number.
- Show halogen substituents by the prefixes *fluoro-*, *chloro-*, *bromo-* and *iodo-*, and list them in alphabetical order, along with other substituents.
- Use a number preceding the name of the halogen to locate each halogen on the parent chain.

3-bromo-2-methylpentane

In molecules containing functional groups designated by a suffix (such as *-ol* for alcohols, *-al* for aldehydes, *-one* for ketones and *-oic acid* for carboxylic acids), the functional groups take precedence in the nomenclature. The location of the functional group indicated by the suffix then determines the numbering, and the halogen is simply classed as a substituent akin to an alkyl side-chain.

trans-2-chlorocyclohexanol

In haloalk*e*nes, the location of the double bond determines the numbering of the parent hydrocarbon; again, the halogen is classed as a substituent.

4-bromocyclohexene

Common names of haloalkanes consist of the common name of the alkyl group, followed by the name of the halide as a separate word. Hence, the term *alkyl halide* is a common name for this class of compounds. In the following examples, the IUPAC name of the compound is given first, followed by its common name in parentheses.

$$\underset{\text{2-fluorobutane (\textit{sec}-butyl fluoride)}}{CH_3\overset{\overset{F}{|}}{C}HCH_2CH_3} \qquad \underset{\text{chloroethene (vinyl chloride)}}{CH_2{=}CHCl}$$

FIGURE 18.2 Methyl chloroform and trichloroethene are solvents used for commercial dry-cleaning.

Methane may have one or more hydrogen atoms replaced by halogen atoms. Halogenated methanes containing more than one halogen atom are useful solvents and are generally referred to by their common names. Dichloromethane (methylene chloride) is the most widely used haloalkane solvent (see figure 18.2). Compounds of the type CHX_3 are called **haloforms**. The common name for $CHCl_3$, for example, is *chloroform*. The common name for CH_3CCl_3 is *methyl chloroform.*

CH_2Cl_2	$CHCl_3$	CH_3CCl_3	$CCl_2{=}CHCl$
dichloromethane (methylene chloride)	trichloromethane (chloroform)	1,1,1-trichloroethane (methyl chloroform)	trichloroethene (trichloroethylene or trichlor)

WORKED EXAMPLE 18.1

Naming haloalkanes

Write the IUPAC name for each of the following compounds.

(a) Br (b) Br (c) H Br

Analysis and solution

(a) The longest continuous carbon chain has three carbon atoms. There are no functional groups apart from the halogen, so the correct solution involves numbering these three carbon atoms to give the carbon atom bonded to the bromine atom the lowest number: in this case, 1. Consequently, the methyl substituent is on the second carbon atom of the chain. The IUPAC name is 1-bromo-2-methylpropane. Its common name is isobutyl bromide.

(b) There is an alkene functional group present, and this is indicated by using the suffix *–ene.* The first step involves finding the longest carbon chain that includes the double bond. Then we number this chain so that the first carbon atom of the double bond has the lowest possible number.

1 2 3 4 Br 5

We then prioritise the groups present on the alkene. Of the groups bound to C(3), bromoethyl has priority over methyl (see the CIP rules in chapter 16 on p. 718). On C(2), the methyl group has priority over the hydrogen atom, which is present but not shown. The two priority groups are on opposite sides of the double bond, and this is designated *E*. Therefore, the IUPAC name is (*E*)-4-bromo-3-methylpent-2-ene. Note that C(4) is a stereocentre but, as the configuration is not shown, we cannot designate it as either *R* or *S*.

(c) There are no functional groups apart from the halogen; we number the six carbon atoms present so that the bromine atom is on C(2). In this case, C(2) is a stereocentre and there are two possible three-dimensional structures. Recall from chapter 17 that we have to prioritise the groups present on this chiral carbon atom. The bromine atom, which has the highest atomic number, has the highest priority, followed by C(3), C(1) and the hydrogen atom. Viewed with this hydrogen atom directed away from the eye, the priority groups are arranged in an anticlockwise sequence and the molecule is given the designation *S*. Therefore, the IUPAC name is (*S*)-2-bromohexane.

Is our answer reasonable?

Remember that, when an alkene is present, there can be two ways of constructing a molecule, so the name must include a descriptor to specify which isomer is involved. There are two ways of representing the groups on a stereocentre and the name must describe the specific isomer present.

PRACTICE EXERCISE 18.1 Write the IUPAC name for each of the following compounds.

(a) (b) Br (c) CH_3CHCH_2Cl with Cl on C-2 (d) Cl

Synthesis of haloalkanes

We learned in chapter 16 that alkanes are described as being unfunctionalised because of their very unreactive nature. Alkanes possess only strong σ bonds. Furthermore, as the electronegativities of carbon and hydrogen are so similar, the electrons that are shared in these σ bonds are evenly distributed. This means that there are no regions of increased partial charge in alkanes; without such dipoles, neither nucleophiles nor electrophiles are attracted. This lack of reactivity led early chemists to classify alkanes as *paraffins* after the Latin term *parum affinis*, which means 'little affinity'. Alkanes can react under extreme conditions; in chapter 16 we described how complete oxidation of alkanes generates heat, CO_2 and water. In this chapter we will learn how, again under extreme reaction conditions, alkanes can be pushed to undergo a reaction called halogenation. **Halogenation** describes a reaction where some of the hydrogen atoms in alkanes are substituted by halogen atoms, most commonly chlorine or bromine.

Chlorination and bromination

If chlorine, Cl_2, and methane gas, CH_4, are mixed together at room temperature in the absence of strong light, no reaction occurs. However, if the two gases are heated to a high temperature, or if they are exposed to an intense light source, a reaction begins and heat is liberated. As well as unreacted chlorine and methane gas, two new substances can be detected: chloromethane and hydrogen chloride. If the reaction is allowed to progress further with more chlorine gas being added, chloromethane itself begins to react and a mixture of dichloromethane, CH_2Cl_2, trichloromethane, $CHCl_3$ (also known as chloroform), and tetrachloromethane, CCl_4 (also known as carbon tetrachloride), is generated.

$$\underset{\text{methane}}{CH_4} + \underset{\text{chlorine}}{Cl_2} \xrightarrow[\text{heat}]{\text{light or}} \underset{\text{chloromethane}}{CH_3Cl} + HCl$$

$$\underset{\text{chloromethane}}{CH_3Cl} + \underset{\text{chlorine}}{Cl_2} \xrightarrow[\text{heat}]{\text{light or}} \underset{\text{dichloromethane}}{CH_2Cl_2} + HCl$$

$$\underset{\text{dichloromethane}}{CH_2Cl_2} + \underset{\text{chlorine}}{Cl_2} \xrightarrow[\text{heat}]{\text{light or}} \underset{\text{chloroform}}{CHCl_3} + HCl$$

$$\underset{\text{chloroform}}{CHCl_3} + \underset{\text{chlorine}}{Cl_2} \xrightarrow[\text{heat}]{\text{light or}} \underset{\text{carbon tetrachloride}}{CCl_4} + HCl$$

The strong conditions required for the reaction make it difficult to control, and only by using a very large excess of alkane or halogen can the reaction be limited to a single substitution product.

The mechanism by which the substitution reaction occurs is well understood and it occurs in three stages: initiation, propagation and termination. We will illustrate this process for the monochlorination of methane mentioned above.

$$\underset{\text{methane}}{CH_4} + \underset{\text{chlorine}}{Cl_2} \xrightarrow[\text{heat}]{\text{light or}} \underset{\text{chloromethane}}{CH_3Cl} + HCl$$

Initiation

The first stage involves breaking a bond in a chlorine molecule as a result of high temperature or absorption of light. This bond cleavage occurs evenly (homolytically) to give two chlorine atoms, 2Cl•, each with an unpaired electron in its outer shell.

$$Cl{-}Cl \xrightarrow[\text{heat}]{\text{light or}} 2Cl\bullet$$

Such odd-numbered electron species are called **radicals** (or free radicals) and the overall reaction to give the chlorinated alkane is defined as a **radical substitution** reaction.

Propagation

The second stage involves formation of the product and regeneration of the radical. The initially formed chlorine radicals are extraordinarily reactive and can remove (or abstract) hydrogen atoms from other molecules to re-acquire the stable, filled, outer electron shell. In this case, the chlorine atom radical reacts with methane to give rise to HCl. (We use single-headed arrows rather than double-headed arrows to indicate that the movement of only one electron is involved.) The other product of the reaction is also a radical, in this case, a methyl radical, $\cdot CH_3$, and it is reactive enough to abstract a chlorine atom from Cl_2. This, in turn, produces another radical, and so on. We call this sequence of events, where the product of one step is the reactant in the next step, a chain reaction.

$$Cl\cdot + H-CH_3 \longrightarrow Cl-H + \cdot CH_3$$

$$H_3C\cdot + Cl-Cl \longrightarrow H_3C-Cl + \cdot Cl$$

In this manner, the generation of only a few radicals in the initial homolytic stage gives rise to a substantial number of reactions with the alkane, CH_4, and, consequently, conversion to the haloalkane, CH_3Cl.

Termination

The reaction is stopped when two radicals encounter each other to produce nonradical species. Note that, because the concentration of radicals generated at any one time is so low, the chances of two radicals encountering each other are much lower than that of a radical encountering an alkane molecule.

$$Cl\cdot + \cdot Cl \longrightarrow Cl-Cl$$

$$H_3C\cdot + \cdot Cl \longrightarrow H_3C-Cl$$

$$H_3C\cdot + \cdot CH_3 \longrightarrow H_3C-CH_3$$

Bromine can also undergo this radical substitution reaction. For example, bromine can react with ethane to generate bromoethane and, if the reaction is not controlled, dibromoethane and tribromoethane.

$$\underset{\text{ethane}}{CH_3CH_3} + \underset{\text{bromine}}{Br_2} \xrightarrow[\text{heat}]{\text{light or}} \underset{\text{bromoethane}}{CH_3CH_2Br} + HBr$$

While this substitution reaction can occur for the other halogens as well, it is generally used only with chlorine or bromine for a number of purely practical reasons. Fluorine, F_2, reacts with alkanes under these conditions, but the reactions are highly exothermic (reflecting the strength of the C—F bond) and so are very difficult to control. Iodine, I_2, on the other hand is seldom used because the reaction is endothermic and too much energy is required to initiate the reaction, making it difficult to control.

When an alkane with different types of C—H bonds is reacted with a halogen under these conditions, all of the possible substitution products may be detected. For instance, reacting 2-methylbutane with chlorine at 300 °C gives the following mixture of four products.

$$CH_3CH(CH_3)CH_2CH_3 \xrightarrow[300\ °C]{Cl_2} \underset{33.5\%}{Cl-CH_2CH(CH_3)CH_2CH_3} + \underset{22\%}{CH_3CCl(CH_3)CH_2CH_3} + \underset{28\%}{CH_3CH(CH_3)CHClCH_3} + \underset{16.5\%}{CH_3CH(CH_3)CH_2CH_2-Cl}$$

This is not a statistical distribution. As there are nine C—H bonds that are part of $-CH_3$ groups, two C—H bonds from $-CH_2-$ groups and only one C—H bond on the tertiary carbon atom, the statistical distribution would be 75%, 16.7% and 8.3%. Clearly some types of C—H bonds are more susceptible to reaction than others. The products of the reaction are determined both by the numbers of C—H bonds and by the stability of the radicals that would be formed by the abstraction. For this chlorination reaction, a 3° hydrogen atom is about 4 times more likely to be substituted than a 1° C—H, and a 2° C—H is about 2.5 times more likely to be substituted than a 1° C—H. This arises from the different stability of the radicals generated by the abstraction. The halogen atom radicals Cl• and Br• have different reactivities, but the different stabilities of the carbon-centred radicals that may be formed also influence the reaction outcome. The more stable the radical, the more easily it is formed. This is because the stability of the radical is reflected in the stability of the transition state leading to its formation. Therefore, it is easier to remove a hydrogen atom from a 2° carbon atom to generate a 2° radical than it is to remove a hydrogen atom from a 1° carbon atom to produce a 1° radical. Tertiary radicals are more stable than 2° radicals. Other radicals, especially some that involve heteroatoms such as those derived from phenols, are even more stable and are important antioxidants used to interrupt the damage that more reactive radicals can generate.

WORKED EXAMPLE 18.2

Determining the products of radical substitution reactions

Give the structures and IUPAC names of the products of the following radical substitution reactions (ignoring stereochemistry and assuming monosubstitution).

(a) $H_3C-C(CH_3)_2-H + Br_2 \xrightarrow{\text{high temperature}}$

(b) $\text{cyclohexane} + Cl_2 \xrightarrow{\text{light}}$

(c) $H_3C-CH_2-C(CH_3)_2-CH_3 + Br_2 \xrightarrow{\text{light}}$

Analysis and solution

(a) The compound to be reacted with bromine is called 2-methylpropane. There are only two types of C—H bonds present in this compound. There is a single 3° C—H present, and the rest of the hydrogen atoms are part of methyl groups. This means that only two possible monosubstituted haloalkanes can arise.

$CH_3-C(CH_3)_2-Br$ — 2-bromo-2-methylpropane

$CH_3-CH(CH_3)-CH_2-Br$ — 1-bromo-2-methylpropane

(b) The compound to be reacted with chlorine is called cyclohexane. As the molecule is flexible (only one monosubstitution product is possible, and no stereochemistry issues are involved), we consider that there is only one type of C—H bond present, so the product of monosubstitution with chlorine would be called chlorocyclohexane.

(c) The compound to be reacted with bromine is called 2,2-dimethylbutane, and there are three positions where hydrogen can be substituted by bromine.

$H_3C-CH_2-C(CH_3)_2-CH_2-Br$ — 1-bromo-2,2-dimethylbutane

$CH_3-CH(Br)-C(CH_3)_2-CH_3$ — 3-bromo-2,2-dimethylbutane

$Br-CH_2-CH_2-C(CH_3)_2-CH_3$ — 1-bromo-3,3-dimethylbutane

Is our answer reasonable?

Remember that single bonds are all freely rotatable, so all three CH_3 groups bound to a single carbon atom are equivalent. Under the strong conditions used for halogenation of alkanes, all possible hydrogen atoms can be substituted. The presence of halogens in an alkane is indicated by adding the prefix *chloro-* or *bromo-* to the parent alkane name. The position numbers should be chosen to give the smallest total.

PRACTICE EXERCISE 18.2

Give the structures and write the IUPAC names for all of the possible products of the following substitution reactions (ignoring stereochemistry and assuming monosubstitution).

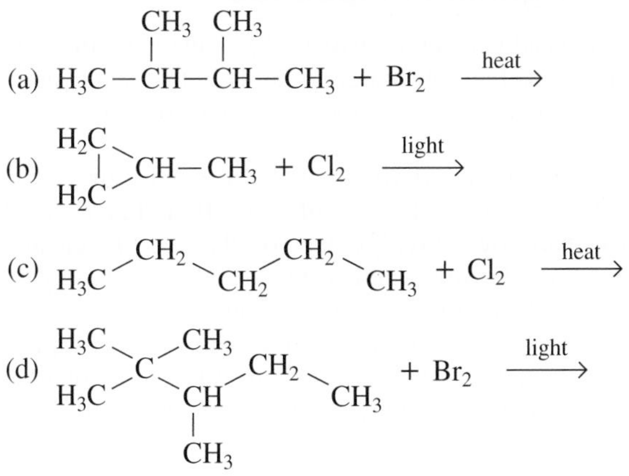

(a) $H_3C-CH(CH_3)-CH(CH_3)-CH_3 + Br_2 \xrightarrow{\text{heat}}$

(b) $(H_2C)_2CH-CH_3 + Cl_2 \xrightarrow{\text{light}}$ (cyclopropane ring: H_2C, H_2C, CH)

(c) $H_3C-CH_2-CH_2-CH_2-CH_3 + Cl_2 \xrightarrow{\text{heat}}$

(d) $(H_3C)_3C-CH(CH_3)-CH_2-CH_3 + Br_2 \xrightarrow{\text{light}}$

Principal reactions of haloalkanes

Haloalkanes are useful in the synthesis of more complicated organic molecules. This is because the polarised carbon–halogen bond is readily attacked by species with a negative character. Such species are called **nucleophiles** (which literally means 'attracted to the nucleus'). A nucleophile is, in fact, any reagent with an unshared pair of electrons that can be donated to another atom or ion to form a new covalent bond — 'nucleophile' is another name for a Lewis base. **Nucleophilic substitution** is any reaction in which one nucleophile is substituted for another. In the following general equation, $Nu{:}^-$ is the nucleophile, X is the leaving group, and substitution takes place on an sp^3 hybridised carbon atom.

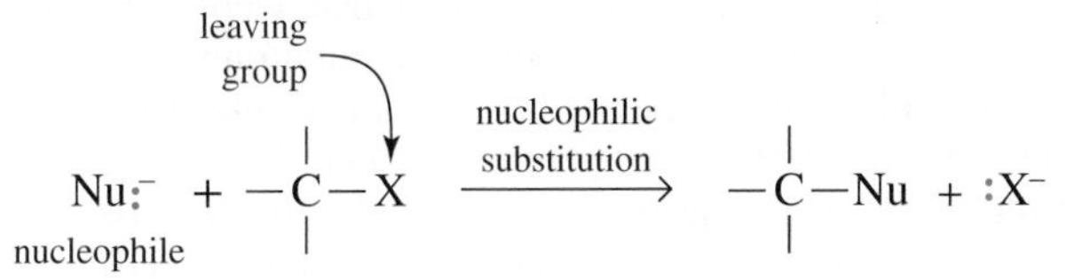

$$\underset{\text{nucleophile}}{Nu{:}^-} + -\overset{|}{\underset{|}{C}}-\overset{\text{leaving group}}{X} \xrightarrow{\text{nucleophilic substitution}} -\overset{|}{\underset{|}{C}}-Nu + {:}X^-$$

Halide ions, with a filled outer electron shell equivalent to the noble gases, make excellent leaving groups, which explains the value of haloalkanes for the synthesis of other molecules.

The process of nucleophilic substitution is evident from the name; an atom or group is replaced (substituted) by another. This is not so clear with **β-elimination**, where atoms or groups are removed from two adjacent carbon atoms. For example, H and X could be removed from a haloalkane, or H and OH from an alcohol, to give a carbon–carbon double bond in both cases.

Because all nucleophiles are also bases, nucleophilic substitution and base-catalysed β-elimination are competing reactions. (The degree to which each occurs is governed by subtle variations in structure and reaction conditions; you may encounter more on this topic in later courses.) The ethoxide ion, $CH_3CH_2O^-$, for example, is both a nucleophile and a base. With bromocyclohexane, it reacts as a nucleophile (pathway shown in red) to give ethoxycyclohexane (cyclohexyl ethyl ether) and as a base (pathway shown in blue) to give cyclohexene and ethanol.

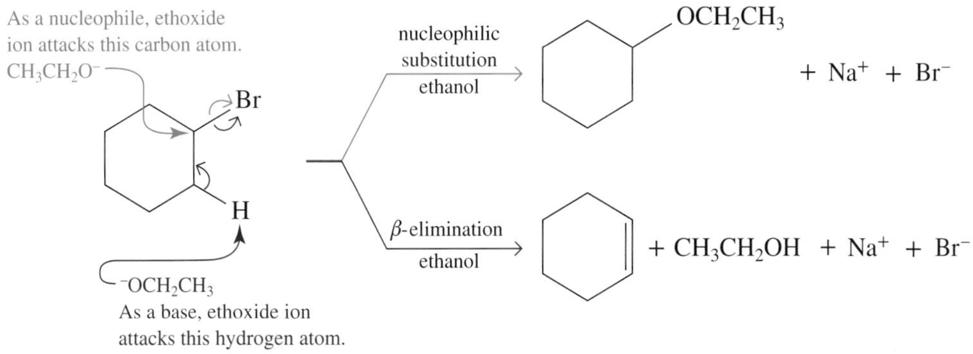

As a nucleophile, ethoxide ion attacks this carbon atom. $CH_3CH_2O^-$

Br

H

$^-OCH_2CH_3$

As a base, ethoxide ion attacks this hydrogen atom.

nucleophilic substitution, ethanol → (cyclohexyl)OCH_2CH_3 $+ Na^+ + Br^-$

β-elimination, ethanol → cyclohexene $+ CH_3CH_2OH + Na^+ + Br^-$

In this chapter, we study both of these organic reactions. Using them, we can convert haloalkanes to compounds with other functional groups including alcohols, ethers, thiols, sulfides, amines, nitriles, alkenes and alkynes. Molecules with these functional groups are important as medicines, industrial chemicals and modern materials. An understanding of the nucleophilic substitution and β-elimination reactions of haloalkanes enables chemists to create some of the complex organic molecules we use in our modern society.

Chemical Connections

Battling bugs with halogenated organic compounds

How is your scalp? *Itchy*? Have you ever had head lice? Don't worry; the chances are that you are not alone. Even if you personally haven't had an infestation, it is likely that someone in your family or one of your friends has suffered pediculosis (the medical name for louse infestations; figure 18.3).

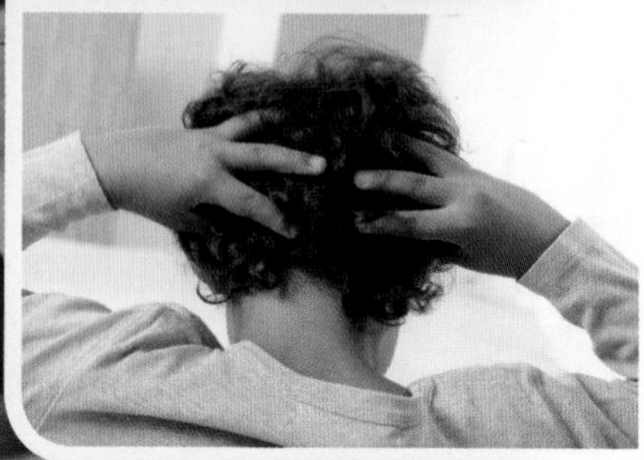

Head lice are small, wingless, blood sucking insects (figure 18.4) that only survive on humans. They are most prevalent in children between the ages of 3 and 10 and their direct family members. Girls are more at risk than boys, possibly because the lice like cleaner hair. If removed from the head, the lice die relatively quickly and so infestations largely occur through head to head contact or through sharing of combs or hats.

Head lice have been around probably as long as we have. In the 15th century, topical mercury treatment was used to treat head lice. Today we recognise mercury as a potential poison, though absorption of elemental mercury through the skin does not occur very readily.

In 1943 the US Army issued DDT to its troops to kill lice. It was very effective and so DDT grew to be a popular treatment for head lice from the 1950s through to the 1970s. DDT is probably the most well-known example of an organochlorine insecticide. Other examples include chlordane and lindane. Organochlorines were used widely in Australia and across the world as powerful insecticides; however, concerns about their persistence in the environment and potential health effects have led to many of them now being banned.

FIGURE 18.3 Head lice infestations, or pediculosis, are very common. The characteristic 'symptom' is an itchy head.

In 2009, an international ban on the use of organochlorines including DDT and lindane was implemented under the Stockholm Convention on Persistent Organic Pollutants. However, a short-term exemption in this ban was allowed (until 2014) specifically for treatments of head lice. Despite this international ban now being in force, lindane — a compound thought by some to be too dangerous to be present in the environment even in trace amounts — remains in use, in a much more concentrated form, in some US anti-lice shampoos.

Even before DDT was banned, DDT-resistant head lice were already becoming common. Treatments then turned to permethrin, another chlorine-containing compound that is a chemical relative of the insecticidal substances generated by certain plants. However, by the 1990s head lice that were resistant even to the effects of permethrin were beginning to be detected.

Despite the use of these powerful insecticides, the number of diagnosed cases of human louse infestations has increased worldwide since the mid-1960s and today reaches hundreds of millions annually.

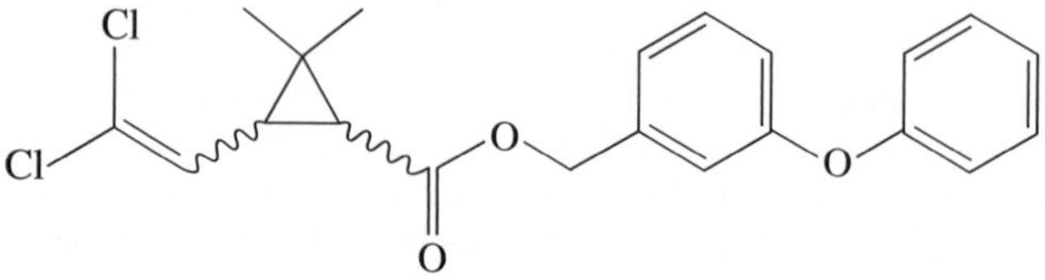

FIGURE 18.4 A close-up looking at the human head louse.

permethrin

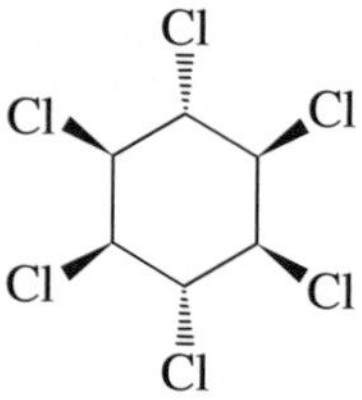

lindane

Why do these insecticides all contain halogen atoms? Usually this is not because the halogen atoms form the reactive part of the molecule. Chemists introduce halogens into these molecules not because of their reactivity, but because the electronegativity of the halogen atoms imparts a difference in reactivity in other parts of the molecule. It is not only in agricultural and insecticidal chemicals that halogens are present. Many pharmaceutical drugs contain halogens, especially fluorine. The C—F bond is very strong, so it does not easily react, degrade or get converted metabolically. The fluorine atom is not too different in size to a hydrogen atom, but it has a very strong influence on the electronegativity within the structure. Increased polarity generates stronger dipoles that can make the drug bind more strongly to the chemical structures in the receptors and active sites in cells that give rise to the therapeutic or biological effect.

Whether it is a new medicine or an insecticide, chemists are constantly seeking to make new active molecules that may provide a greater effect at lower exposures and thus help us to keep ahead in the ongoing battles against bugs and disease, while decreasing the 'collateral damage' — the side-effects and risks associated with contact with any active compound.

18.2 Nucleophilic substitution

Nucleophilic substitution is one of the most important reactions of haloalkanes and can lead to a wide variety of new functional groups, several of which are illustrated in table 18.1.

TABLE 18.1 Some nucleophilic substitution reactions.

Reaction: $Nu:^- + CH_3X \longrightarrow CH_3Nu + :X^-$			
Nucleophile		**Product**	**Class of compound formed**
$HO:^-$	$\longrightarrow$	CH_3OH	alcohol
$RO:^-$	$\longrightarrow$	CH_3OR	ether
$HS:^-$	$\longrightarrow$	CH_3SH	thiol (mercaptan)
$RS:^-$	$\longrightarrow$	CH_3SR	sulfide (thioether)
$:I:^-$	$\longrightarrow$	$CH_3I:$	alkyl iodide
$:NH_3$	$\longrightarrow$	$CH_3NH_3^+$	alkylammonium ion
HOH	$\longrightarrow$	$CH_3O^+(H)—H$	alcohol (after proton transfer)
CH_3OH	$\longrightarrow$	$CH_3O^+(H)—CH_3$	ether (after proton transfer)

Note the following points from table 18.1.

1. While the symbol $Nu:^-$ is used to represent any nucleophile, not all nucleophiles are negatively charged.
2. If the nucleophile is negatively charged, as for OH^- and RS^-, the atom donating the pair of electrons in the substitution reaction becomes neutral in the product.
3. If the nucleophile is uncharged, as for NH_3 and CH_3OH, the atom donating the pair of electrons in the substitution reaction becomes positively charged in the product. Often, the product then undergoes a second step involving proton transfer to yield a neutral substitution product.

WORKED EXAMPLE 18.3

Determining the products of nucleophilic substitution

Complete the following nucleophilic substitution reactions.

(a) (1-bromobutane, Br) + Na^+OH^- $\longrightarrow$

(b) (1-chlorobutane, Cl) + NH_3 $\longrightarrow$

Analysis and solution

The nucleophile attacks the haloalkane and a halide ion, X^-, is ejected.

(a) Hydroxide ion is the nucleophile and bromine is the leaving group.

(Br) + Na^+OH^- $\longrightarrow$ (OH) + Na^+Br^-

1-bromobutane; sodium hydroxide; butan-1-ol; sodium bromide

(b) Ammonia is the nucleophile and chlorine is the leaving group.

1-chlorobutane + NH_3 (ammonia) ⟶ butylammonium chloride ($NH_3^+Cl^-$)

Is our answer reasonable?

Check the structures to be sure each atom has the correct valency; that is, they are bound to the appropriate number of groups and have charges as required.

PRACTICE EXERCISE 18.3 Complete the following nucleophilic substitution reactions.

(a) cyclopentyl—Br + $CH_3CH_2S^-Na^+$ ⟶

(b) cyclopentyl—Br + $CH_3C(=O)O^-Na^+$ ⟶

Mechanisms of nucleophilic substitution

On the basis of decades of experimental observations, chemists have proposed two controlling mechanisms for nucleophilic substitutions. A fundamental difference between them is the timing of bond breaking between carbon and the leaving group and of bond forming between carbon and the nucleophile.

S_N2 mechanism

At one extreme, bond breaking and bond forming occur simultaneously. Thus, the departure of the leaving group is assisted by the incoming nucleophile. This mechanism is designated S_N2, where S stands for substitution, N for nucleophilic and 2 for a bimolecular reaction. A **bimolecular reaction** is one in which two species are involved in the transition state of the rate-determining step. This type of substitution reaction is classified as bimolecular because both the haloalkane and the nucleophile are involved in the rate-determining step and contribute to the rate law of the reaction:

$$\text{rate of reaction} = k[\text{haloalkane}][\text{nucleophile}]$$

Figure 18.5 shows the S_N2 mechanism for the reaction of a hydroxide ion and bromomethane to form methanol and a bromide ion. The nucleophile attacks the reactive centre from the side opposite the leaving group, as shown in the figure below. You can see that, for a chiral compound, an S_N2 reaction at a stereocentre gives rise to a chiral product. This occurs with inversion of the configuration around the stereocentre.

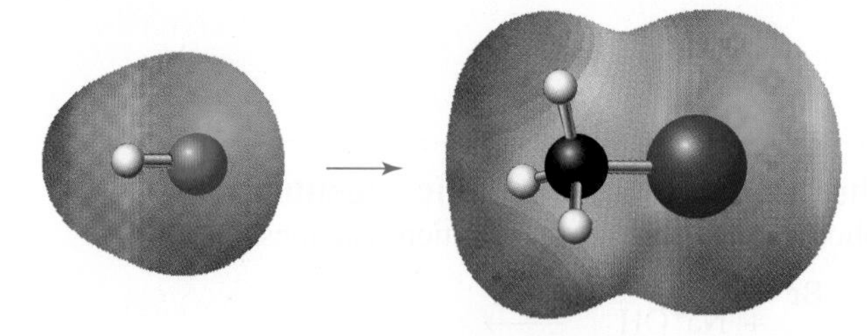

nucleophilic attack from the side opposite the leaving group

HO^- + CH_3Br (reactants) → $[HO^{\delta-}\cdots CH_3 \cdots Br^{\delta-}]^-$ (transition state with simultaneous bond breaking and bond forming) → $HO-CH_3$ + Br^- (products)

FIGURE 18.5 The S_N2 reaction is driven by the attraction between the negative charge of the nucleophile (in this case, the negatively charged oxygen atom of the hydroxide ion) and the centre of positive charge of the electrophile (in this case, the partial positive charge on the carbon atom bearing the bromine leaving group).

Figure 18.6 shows an energy diagram for an S_N2 reaction. There is a single transition state and no reactive intermediate.

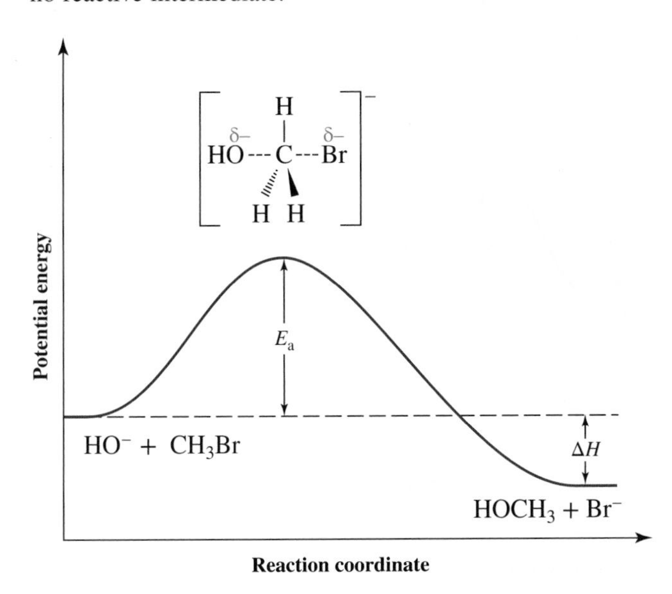

FIGURE 18.6 An energy diagram for an S_N2 reaction. There is one transition state and no reactive intermediate.

S_N1 mechanism

In the other limiting mechanism, called S_N1, bond breaking between carbon and the leaving group is completed before bond forming with the nucleophile begins. In the designation S_N1, S stands for substitution, N for nucleophilic and 1 for a unimolecular reaction. A **unimolecular reaction** is one in which only one species is involved in the transition state of the rate-determining step. This type of substitution is unimolecular because only the haloalkane is involved in the rate-determining step and contributes to the rate law:

$$\text{rate of reaction} = k[\text{haloalkane}]$$

The S_N1 mechanism is illustrated by the reaction of 2-bromo-2-methylpropane (*tert*-butyl bromide) in methanol, CH_3OH, to form 2-methoxy-2-methylpropane (*tert*-butyl methyl ether). This is an example of a **solvolysis** reaction — a reaction in which the solvent plays the role of the nucleophile in the substitution reaction.

Step 1: The ionisation of a C—X bond forms a 3° carbocation intermediate.

$(CH_3)_3C{-}Br \xrightarrow{\text{slow, rate determining}} (CH_3)_3C^+ + Br^-$

a carbocation intermediate; carbon is trigonal planar

Step 2: Reaction of methanol at either face of the planar carbocation intermediate gives an oxonium ion.

either CH_3OH + $(CH_3)_3C^+$ or CH_3OH $\xrightarrow{\text{fast}}$ $(CH_3)_3C{-}O^+(H)CH_3$ + $(CH_3)_3C{-}O^+(H)CH_3$

Step 3: Proton transfer from the oxonium ion to methanol (the solvent) completes the reaction and gives *tert*-butyl methyl ether.

$(CH_3)_3C{-}O^+(H)CH_3 + H{-}O{-}CH_3 \overset{\text{fast}}{\rightleftharpoons} (CH_3)_3C{-}O{-}CH_3 + H{-}O^+(H){-}CH_3$

Figure 18.7 (overleaf) shows an energy diagram for the S_N1 reaction of 2-bromo-2-methylpropane and methanol. There is one transition state leading to formation of the carbocation intermediate in step 1 and a second transition state for the reaction of the carbocation intermediate with methanol in step 2 to give the oxonium ion. The reaction leading to formation of the carbocation intermediate has the higher energy barrier (E_{a1}), so it is the rate-determining step. Recall that you have previously encountered this type of energy diagram in chapter 15 (p. 666).

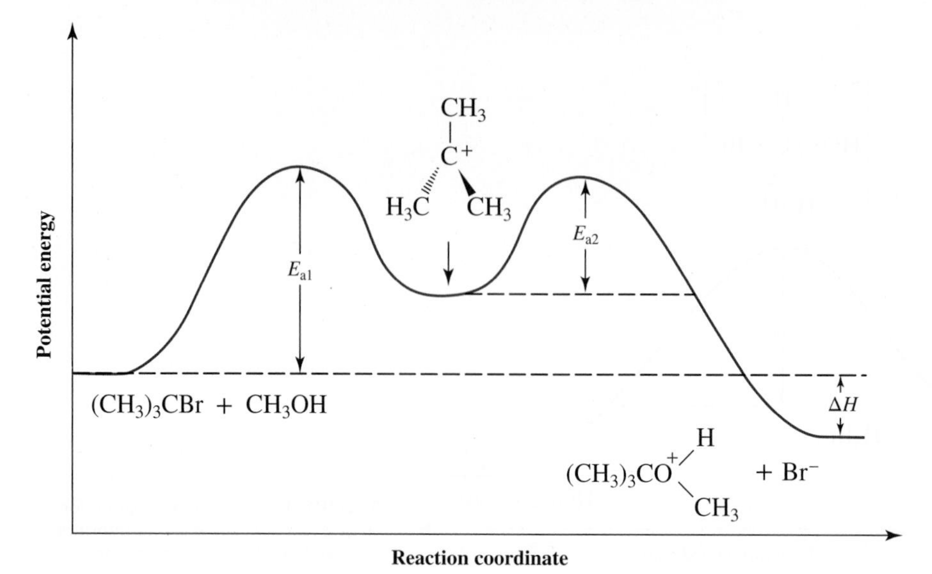

FIGURE 18.7 An energy diagram for the S_N1 reaction of 2-bromo-2-methylpropane and methanol. There is one transition state leading to formation of the carbocation intermediate in step 1 and a second transition state for the reaction of the carbocation intermediate with methanol in step 2. Step 1 crosses the higher energy barrier and so is rate determining.

If an S_N1 reaction is carried out at a tetrahedral stereocentre, the product is a racemic mixture (see chapter 17). We can illustrate this result with the following example. Upon ionisation, the *R* enantiomer forms an achiral carbocation intermediate. Attack by the nucleophile from the left face of the carbocation intermediate gives the *S* enantiomer; attack from the right face gives the *R* enantiomer. Because attack by the nucleophile occurs with equal probability from either face of the planar carbocation intermediate, the *R* and *S* enantiomers are formed in equal amounts, and the product is a racemic mixture.

R enantiomer —ionisation, $-Cl^-$→ planar carbocation (achiral) —1. CH_3OH 2. proton transfer ($-H^+$)→ *S* enantiomer + *R* enantiomer (a racemic mixture)

Experimental evidence for S_N1 and S_N2 mechanisms

Let us now examine some of the experimental evidence on which these two contrasting mechanisms are based. As we do, we consider the following questions.

1. What effect does the nature of the nucleophile have on the rate of reaction?
2. What effect does the structure of the haloalkane have on the rate of reaction?
3. What effect does the structure of the leaving group have on the rate of reaction?
4. What is the role of the solvent?

Nature of the nucleophile

Nucleophilicity is a kinetic property, which we measure by relative rates of reaction. We can establish the relative nucleophilicities for a series of nucleophiles by measuring the rate at which each displaces a leaving group from a haloalkane under defined conditions. For example, we could measure the rate at which various nucleophiles displace bromide from bromoethane in ethanol at 25 °C. The following equation shows this reaction with ammonia, NH_3, as the nucleophile.

$$CH_3CH_2Br + NH_3 \xrightarrow[25\,^\circ C]{\text{ethanol}} CH_3CH_2NH_3^+ + Br^-$$

Comparing the rate of reaction of bromoethane under these defined conditions with the nucleophiles listed in table 18.2 allows us to define their relative nucleophilicity. The more effective nucleophiles are those that react more rapidly. The nucleophiles in table 18.2 are those we deal with most commonly in this text.

TABLE 18.2 Examples of common nucleophiles and their relative effectiveness.

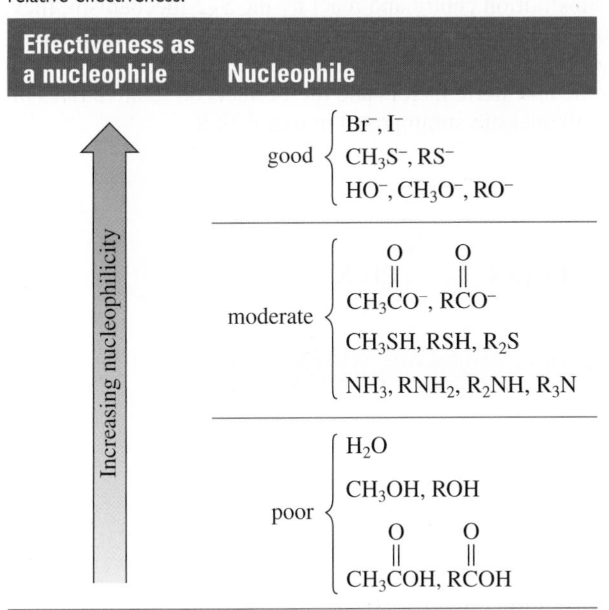

Effectiveness as a nucleophile	Nucleophile
good	Br^-, I^- CH_3S^-, RS^- HO^-, CH_3O^-, RO^-
moderate	CH_3COO^-, $RCOO^-$ (C=O) CH_3SH, RSH, R_2S NH_3, RNH_2, R_2NH, R_3N
poor	H_2O CH_3OH, ROH CH_3COOH, $RCOOH$ (C=O)

Increasing nucleophilicity ↑

Because the nucleophile participates in the rate-determining step in an S_N2 reaction, the better the nucleophile, the more likely it is that the reaction will occur by that mechanism. The nucleophile does not participate in the rate-determining step in an S_N1 reaction. Thus, an S_N1 reaction can, in principle, occur at approximately the same rate with any of the common nucleophiles, regardless of their relative nucleophilicities.

Structure of the haloalkane

S_N1 reactions are governed mainly by the relative stabilities of carbocation intermediates. S_N2 reactions, by contrast, are governed mainly by factors involving the size and bulkiness of the molecules involved. We describe these factors as **steric hindrance**, which refers to the ability of groups, because of their size and shape, to hinder access to a reaction site within a molecule. S_N2 transition states are particularly sensitive to crowding around the site of reaction. The distinction is as follows.

1. *Relative stabilities of carbocations:* As we learned in section 16.5 (figure 16.26), 3° carbocations are the most stable carbocations, requiring the lowest activation energy for their formation; 1° carbocations are the least stable, requiring the highest activation energy for their formation. In fact, 1° carbocations are so unstable that they have never been observed in solution. Therefore, 3° haloalkanes are the most likely to react by carbocation formation; 2° haloalkanes are less likely to react in this manner; and methyl and 1° haloalkanes never react in this manner.
2. *Steric hindrance:* In an S_N2 reaction, the nucleophile begins to form a new covalent bond to the substitution centre by approaching at 180° to the leaving group. If we compare the ease of approach by the nucleophile to the substitution centre of a 1° haloalkane with that of a 3° haloalkane, we see that this approach is considerably easier in the case of the 1° haloalkane. Two hydrogen atoms and one alkyl group shield the substitution centre of a 1° haloalkane. In contrast, three alkyl groups shield the substitution centre of a 3° haloalkane. This centre in bromoethane is easily accessed by a nucleophile, while there is extreme crowding around it in 2-bromo-2-methylpropane.

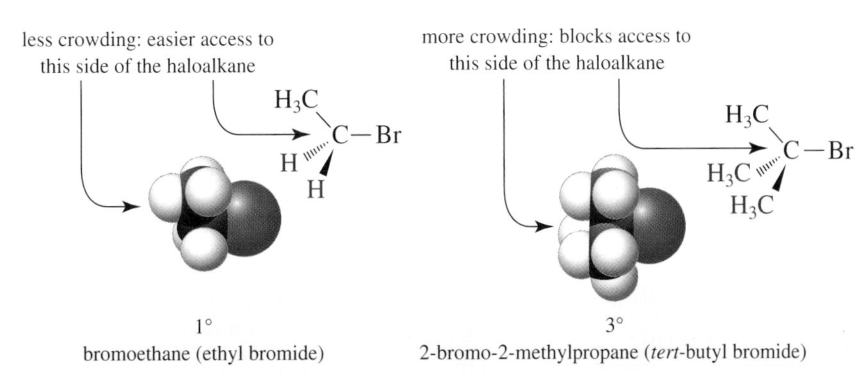

Given the competition between electronic and steric factors, we find that 3° haloalkanes react by the S_N1 mechanism because 3° carbocation intermediates are particularly stable and because the approach of a nucleophile to the substitution centre in a 3° haloalkane is hindered by the three groups surrounding it; 3° haloalkanes never react by the S_N2 mechanism. Halomethanes and 1° haloalkanes have little crowding around the substitution centre and react by the S_N2 mechanism; they never react by the S_N1 mechanism, because methyl and 1° carbocations are so unstable. Secondary haloalkanes may react by either the S_N1 or the S_N2 mechanism, depending on the nucleophile and solvent. The competition between electronic and steric factors and their effects on relative rates of nucleophilic substitution reactions of haloalkanes are summarised in figure 18.8.

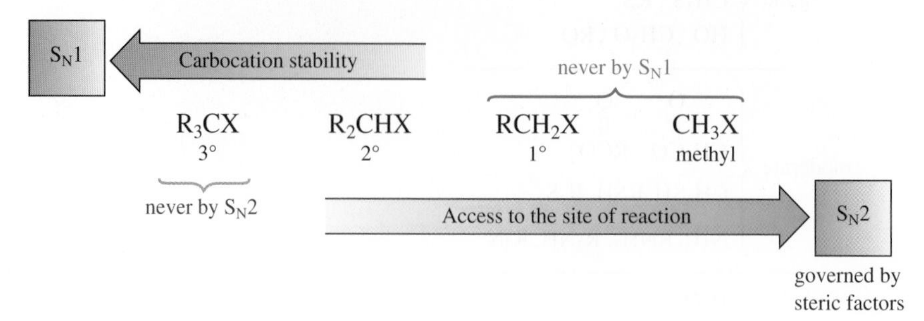

FIGURE 18.8 Effect of electronic and steric factors in competition between S_N1 and S_N2 reactions of haloalkanes.

The leaving group

In the transition state for nucleophilic substitution on a haloalkane, the leaving group develops a partial negative charge in both S_N1 and S_N2 reactions; therefore, the ability of a group to function as a leaving group is related to how stable it is as an anion. The most stable anions and the best leaving groups are the conjugate bases of strong acids. Thus, we can use information on the relative strengths of organic and inorganic acids in appendix E to determine which anions are the best leaving groups.

Greater ability to act as a leaving group ←

rarely act as leaving groups in nucleophilic substitution and β-elimination reactions: $HO^- > CH_3O^- > NH_2^-$

$$I^- > Br^- > Cl^- \gg F^- > CH_3\overset{\overset{\large O}{\|}}{C}O^- > HO^- > CH_3O^- > NH_2^-$$

← Greater stability of anion; greater strength of conjugate acid

The best leaving groups in this series are the halide ions I^-, Br^- and Cl^-. Hydroxide ions, OH^-, methoxide ions, CH_3O^-, and amide ions, NH_2^-, are such poor leaving groups that they rarely, if ever, are displaced in nucleophilic substitution reactions.

The solvent

Solvents provide the medium in which reactants are dissolved and in which nucleophilic substitution reactions take place. Common solvents for these reactions are divided into two groups: protic and aprotic.

Protic solvents contain —OH groups and are hydrogen-bond donors. Common protic solvents for nucleophilic substitution reactions are water, low-molar-mass alcohols and low-molar-mass carboxylic acids (table 18.3). Each can solvate (see chapter 10, p. 399) both the anionic and cationic components of ionic compounds by electrostatic interaction between its partially negatively charged oxygen atom(s) and the cation and between its partially positively charged hydrogen atom and the anion. These same properties aid in the ionisation of C—X bonds to give an X^- anion and a carbocation; thus, protic solvents are good solvents in which to carry out S_N1 reactions.

TABLE 18.3 Common protic solvents.

Protic solvent	Structure	Polarity of solvent	Notes
water	H_2O	↑ Increasing	These solvents favour S_N1 reactions. The greater the polarity of the solvent, the easier it is to form carbocations in it, because both the carbocation and the negatively charged leaving group can be solvated.
formic acid	$HCOOH$		
methanol	CH_3OH		
ethanol	CH_3CH_2OH		
acetic acid	CH_3COOH		

Aprotic solvents do not contain —OH groups and cannot function as hydrogen-bond donors. Table 18.4 lists the aprotic solvents most commonly used for nucleophilic substitution reactions. Dimethyl sulfoxide and acetone are polar aprotic solvents; dichloromethane and diethyl ether are relatively nonpolar aprotic solvents. The aprotic solvents listed in table 18.4 are particularly good for S_N2 reactions. Polar aprotic solvents can solvate cations only; they cannot solvate anions (i.e. the anions are not closely surrounded by solvent molecules), so anions in aprotic solvents are much more reactive as nucleophiles.

TABLE 18.4 Common aprotic solvents.

Aprotic solvent	Structure	Polarity of solvent	Notes
dimethyl sulfoxide (DMSO)	$CH_3S(=O)CH_3$	Increasing (↑)	These solvents favour S_N2 reactions. Although solvents at the top of this list are polar, the formation of carbocations in them is far more difficult than in protic solvents, because the anionic leaving group cannot be solvated by these solvents.
acetone	$CH_3C(=O)CH_3$		
dichloromethane	CH_2Cl_2		
diethyl ether	$(CH_3CH_2)_2O$		

Table 18.5 summarises the factors favouring S_N1 and S_N2 reactions; it also shows the change in configuration when nucleophilic substitution takes place at a stereocentre.

TABLE 18.5 Comparison between S_N1 and S_N2 reactions of haloalkanes.

Type of haloalkane	S_N2	S_N1
methyl, CH_3X	S_N2 is favoured.	S_N1 does not occur. The methyl cation is so unstable that it is never observed in solution.
primary, RCH_2X	S_N2 is favoured.	S_N1 does not occur. Primary carbocations are so unstable that they are never observed in solution.
secondary, R_2CHX	S_N2 is favoured in aprotic solvents with good nucleophiles.	S_N1 is favoured in protic solvents with poor nucleophiles.
tertiary, R_3CX	S_N2 does not occur because of steric hindrance around the substitution centre.	S_N1 is favoured because of the ease of formation of 3° carbocations.
substitution at a stereocentre	Inversion of configuration — the nucleophile attacks the stereocentre from the side opposite the leaving group.	Racemisation — the carbocation intermediate is planar, and an attack by the nucleophile occurs with equal probability from either side.

Analysis of several nucleophilic substitution reactions

Predictions about the mechanism of a particular nucleophilic substitution reaction must be based on considerations of the structure of the haloalkane, the nucleophile, the leaving group and the solvent. The following are analyses of three such reactions.

Example 1

Methanol is a polar protic solvent and a good one in which to form carbocations. For example, 2-chlorobutane ionises in methanol to form a 2° carbocation intermediate. Methanol is a weak nucleophile. From this analysis, we predict that reaction occurs by the S_N1 mechanism. The 2° carbocation intermediate (an electrophile) then reacts with methanol (a nucleophile) followed by proton transfer to give the observed product. The product is formed as a 50:50 mixture of *R* and *S* configurations; that is, it is formed as a racemic mixture.

$$2\ \text{(Cl)} + 2CH_3OH \longrightarrow \text{(OCH}_3\text{)} + \text{(OCH}_3\text{)} + 2H^+Cl^-$$

R enantiomer

Example 2
The equation below shows a 1° bromoalkane in the presence of an iodide ion, a good nucleophile. Because 1° carbocations are so unstable, they never form in solution, and an S_N1 reaction is not possible. Dimethyl sulfoxide (DMSO), a polar aprotic solvent, is a good solvent in which to carry out S_N2 reactions. From this analysis, we predict that reaction occurs by the S_N2 mechanism.

$$\text{(CH}_3)_2\text{CHCH}_2\text{Br} + Na^+I^- \xrightarrow{\text{DMSO}} \text{(CH}_3)_2\text{CHCH}_2\text{I} + Na^+Br^-$$

Example 3
Bromide is a good leaving group on a 2° carbon atom. The methylsulfide ion is a good nucleophile. Acetone, a polar aprotic solvent, is a good medium in which to carry out S_N2 reactions, but a poor medium in which to carry out S_N1 reactions. Therefore, we predict that the reaction below occurs by the S_N2 mechanism, and that the product formed has the *R* configuration.

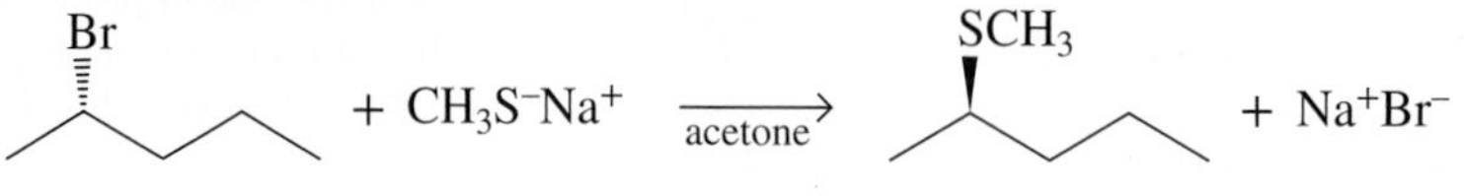

$$\text{Br (S enantiomer)} + CH_3S^-Na^+ \xrightarrow{\text{acetone}} SCH_3 \text{ (R enantiomer)} + Na^+Br^-$$

S enantiomer — *R* enantiomer

WORKED EXAMPLE 18.4

Predicting reaction products and mechanisms

Write the expected product of each of the following nucleophilic substitution reactions and predict the mechanism by which the product is formed.

(a) 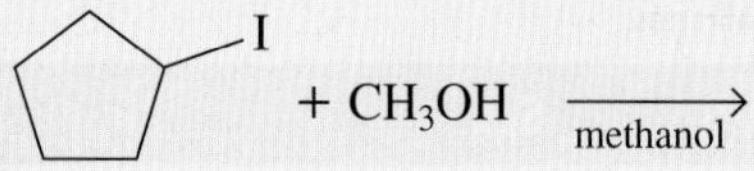 iodocyclopentane $+ CH_3OH \xrightarrow{\text{methanol}}$

(b) 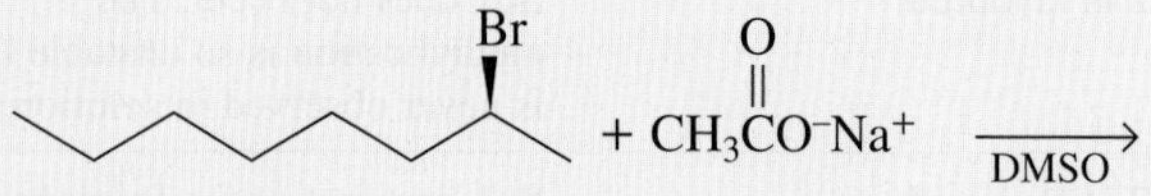 2-bromoheptane (Br) $+ CH_3\overset{O}{\overset{\|}{C}}O^-Na^+ \xrightarrow{\text{DMSO}}$

Analysis and solution

With our knowledge of the factors that control the reaction mechanisms of haloalkanes, we can predict the products and the manner in which they are formed.

(a) Methanol is a poor nucleophile. It is also a polar protic solvent that can solvate carbocations. Ionisation of the carbon–iodine bond forms a 2° carbocation intermediate. We predict that the reaction occurs by the S_N1 mechanism.

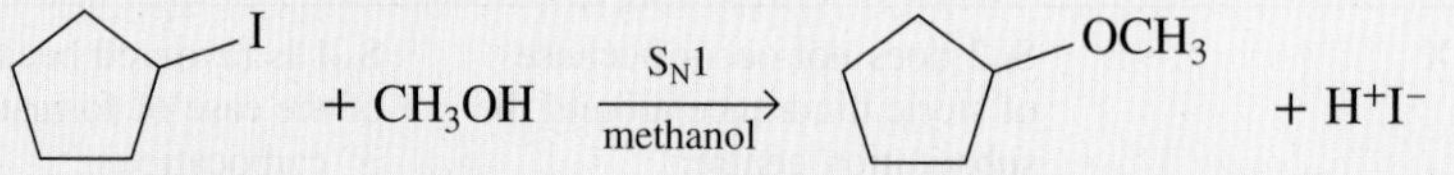

$$\text{C}_5\text{H}_9\text{I} + CH_3OH \xrightarrow[\text{methanol}]{S_N1} \text{C}_5\text{H}_9\text{OCH}_3 + H^+I^-$$

(b) Bromide is a good leaving group on a 2° carbon atom. The acetate ion is a moderate nucleophile. DMSO is a particularly good solvent for S_N2 reactions. We predict substitution by the S_N2 mechanism with inversion of configuration at the stereocentre.

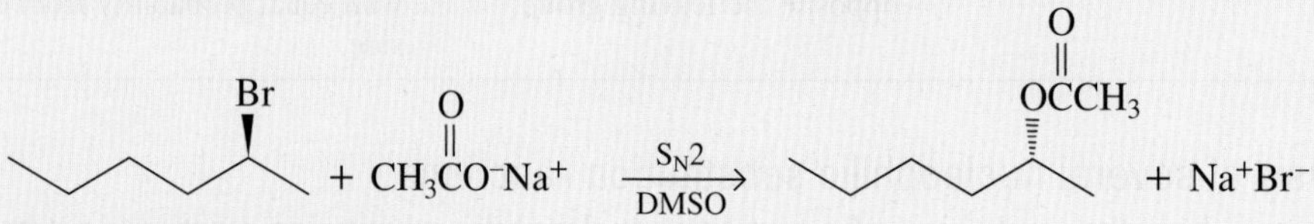

$$\text{Br} + CH_3\overset{O}{\overset{\|}{C}}O^-Na^+ \xrightarrow[\text{DMSO}]{S_N2} \overset{O}{\overset{\|}{OCCH_3}} + Na^+Br^-$$

Is our answer reasonable?

Check your answer by answering these two questions: Are the products sensible in terms of the groups and bonding? Do the products have poorer leaving groups than the starting materials?

PRACTICE EXERCISE 18.4

Write the expected product of each of the following nucleophilic substitution reactions and predict the mechanism by which the product is formed.

(a) 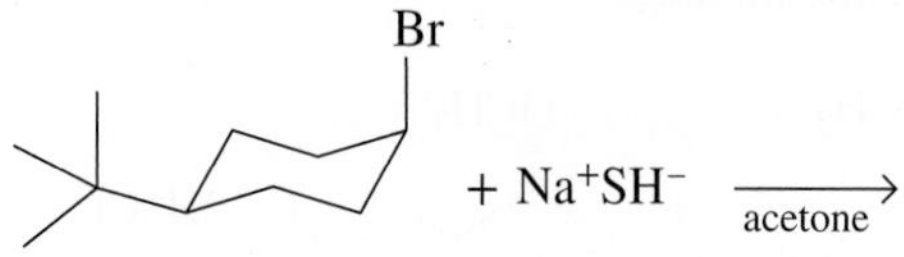 Br (tert-butylcyclohexyl bromide) $+ Na^+SH^- \xrightarrow{\text{acetone}}$

(b) 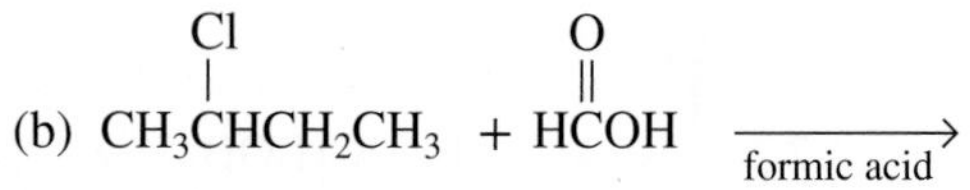 $CH_3\overset{Cl}{\overset{|}{C}}HCH_2CH_3 + H\overset{O}{\overset{\|}{C}}OH \xrightarrow{\text{formic acid}}$

18.3 β-elimination

Earlier in this chapter (p. 797), we described one of the reactions of haloalkanes as elimination to produce alkenes. These are called β-eliminations and, in this section, we study a type of β-elimination called **dehydrohalogenation.** In this process, in the presence of a strong base such as hydroxide ions or ethoxide ions, hydrogen can be removed from a carbon atom adjacent to the halogen, followed by removal of the halogen, to form a carbon–carbon double bond.

$$\underset{\text{a haloalkane}}{-\overset{|\beta}{\underset{H}{\underset{|}{C}}}-\overset{|\alpha}{\underset{X}{\underset{|}{C}}}-} + \underset{\text{base}}{CH_3CH_2O^-Na^+} \xrightarrow{CH_3CH_2OH} \underset{\text{an alkene}}{>C=C<} + CH_3CH_2OH + Na^+X^-$$

As the equation shows, we call the carbon atom bearing the halogen the α-carbon atom and the adjacent carbon atom the β-carbon atom.

Because most nucleophiles can also act as bases and vice versa, it is important to keep in mind that β-elimination and nucleophilic substitution are competing reactions. At this level, it is not necessary to explore which is more favoured, although we will deal with some of the principles that govern the outcome in section 18.4. At this stage, note that both processes may occur in the reaction of some molecules.

Common strong bases used for β-elimination are OH^-, OR^- and NH_2^-. The following are three examples of base-promoted reactions:

1-bromooctane (β, α, Br) + t-BuO^-K^+ (potassium *tert*-butoxide) ⟶ oct-1-ene + t-BuOH + Na^+Br^-

2-bromo-2-methylbutane $\xrightarrow[CH_3CH_2OH]{CH_3CH_2O^-Na^+}$ 2-methylbut-2-ene (major product) + 2-methylbut-1-ene

1-bromo-1-methylcyclopentane $\xrightarrow[CH_3OH]{CH_3O^-Na^+}$ 1-methylcyclopentene (major product) + methylenecyclopentane (CH_2)

In the first example, the base is shown as a reactant on the left-hand side of the arrow and the reaction equation is balanced. In the second and third examples, although the base is still a reactant, it is shown over the reaction arrow and the equation is not balanced. In organic chemistry, we focus primarily on changes to the organic structure and we do not necessarily show the presence of reactants and spectator ions. The second and third examples are more complicated than the first; there are nonequivalent β-carbon atoms, each bearing a hydrogen atom. Therefore, two alkenes are possible from each β-elimination reaction. In each case, the major product of these and most other β-elimination reactions is the more substituted alkene (it is also the more stable alkene). This is known as **Zaitsev's rule** (or Zaitsev elimination) in honour of the chemist who first made this generalisation.

WORKED EXAMPLE 18.5

Predicting β-elimination products

Predict the β-elimination product(s) formed when each of the following bromoalkanes is treated with sodium ethoxide in ethanol (if two products might be formed, predict the major product).

(a) Br (b) Br

Analysis and solution

When strong bases can remove a hydrogen atom to form a relatively stable carbocation, β-elimination becomes possible.

(a) There are two nonequivalent β-carbon atoms in this bromoalkane, so two alkenes are possible. The major product is 2-methylbut-2-ene, the more substituted alkene.

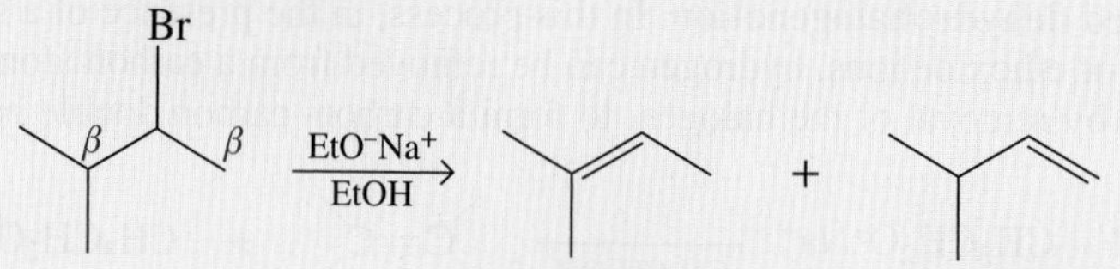

Br β β $\xrightarrow[\text{EtOH}]{\text{EtO}^-\text{Na}^+}$ +

2-methylbut-2-ene (major product) 3-methylbut-1-ene

(b) There is only one β-carbon atom in this bromoalkane, so only one alkene is possible.

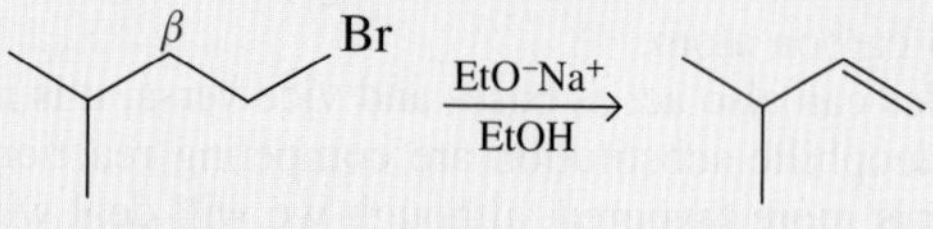

β Br $\xrightarrow[\text{EtOH}]{\text{EtO}^-\text{Na}^+}$

3-methylbut-1-ene

Is our answer reasonable?

You should check your answer to ensure that the newly formed double bond involves the carbon atom that was bonded to the halogen. Also, ensure that the valencies of all the atoms in the products are correct. Remember that a neutral carbon atom must always have four bonds.

PRACTICE EXERCISE 18.5 Predict the β-elimination products formed when each of the following chloroalkanes is treated with sodium ethoxide in ethanol (if two products might be formed, predict which is the major product).

(a) 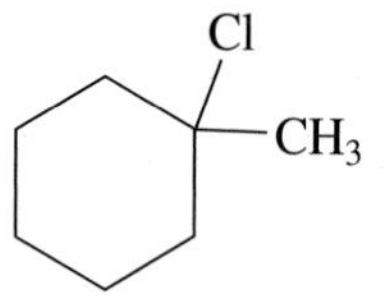 Cl, CH_3 (b) 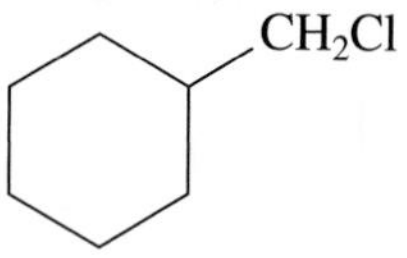 CH_2Cl (c) 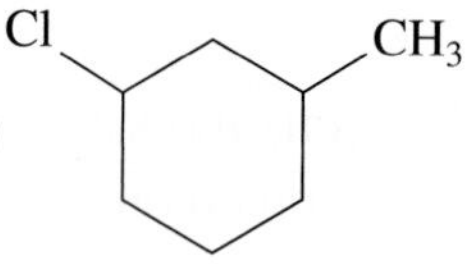 Cl, CH_3

Mechanisms of β-elimination

There are two types of mechanisms for β-elimination reactions. A fundamental difference between them is the timing of the bond-breaking and bond-forming steps. Recall that we made this same statement about the two limiting mechanisms for nucleophilic substitution reactions in section 18.2.

E1 mechanism

When the carbon–halogen bond is completely broken before the base removes a hydrogen atom to give the carbon–carbon double bond, the mechanism is designated **E1**. E stands for elimination and 1 for a unimolecular reaction; only one species, in this case the haloalkane, is involved in the rate-determining step. The rate law for an E1 reaction has the same form as that for an S_N1 reaction.

$$\text{rate of reaction} = k[\text{haloalkane}]$$

The mechanism for an E1 reaction is illustrated by the reaction of 2-bromo-2-methylpropane to form 2-methylpropene. In this two-step mechanism, the rate-determining step is the breaking of the carbon–halogen bond to form a carbocation intermediate (just as it is in the S_N1 mechanism).

Step 1: Rate-determining breaking of the C—Br bond gives a carbocation intermediate.

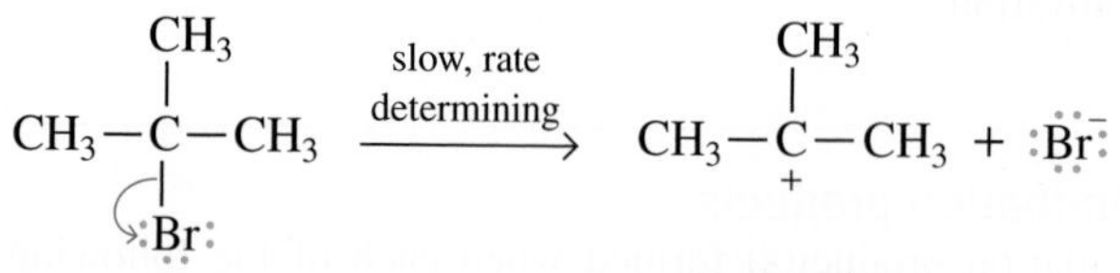

$$CH_3-\underset{Br}{\overset{CH_3}{C}}-CH_3 \xrightarrow[\text{determining}]{\text{slow, rate}} CH_3-\underset{+}{\overset{CH_3}{C}}-CH_3 + :\ddot{Br}:^-$$

a carbocation intermediate

Step 2: Proton transfer from the carbocation intermediate to methanol (which in this instance is both the solvent and a reactant) gives the alkene.

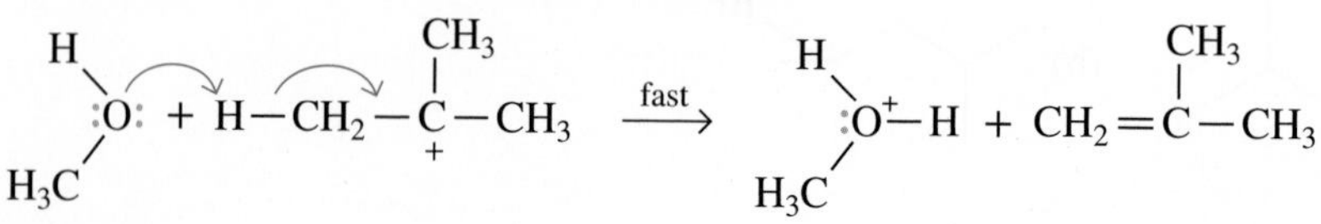

$$H_3C(H)\ddot{O}: + H-CH_2-\underset{+}{\overset{CH_3}{C}}-CH_3 \xrightarrow{\text{fast}} H_3C(H)\overset{+}{O}-H + CH_2=\overset{CH_3}{C}-CH_3$$

E2 mechanism

When the base removes a β-hydrogen atom at the same time as the carbon–halogen bond is broken, it is designated an **E2** reaction. E stands for elimination and 2 for a bimolecular reaction. The rate law for the rate-determining step depends on both the haloalkane and the base.

$$\text{rate of reaction} = k[\text{haloalkane}][\text{base}]$$

The stronger the base, the more likely it is that the E2 mechanism will dominate. We illustrate the E2 mechanism by the reaction of 1-bromopropane with sodium ethoxide. In this mechanism, proton transfer to the base, formation of the carbon–carbon double bond and ejection of the bromide ion occur simultaneously; that is, all bond-forming and bond-breaking steps occur at the same time.

$$CH_3CH_2\ddot{\underset{\cdot\cdot}{O}}{:}^- + H{-}\underset{}{\overset{CH_3}{\overset{|}{C}H}}{-}CH_2{-}\ddot{\underset{\cdot\cdot}{Br}}{:} \longrightarrow CH_3CH_2\ddot{\underset{\cdot\cdot}{O}}{-}H + CH_3CH{=}CH_2 + {:}\ddot{\underset{\cdot\cdot}{Br}}{:}^-$$

For both E1 and E2 reactions, the major product is that formed in accordance with Zaitsev's rule as illustrated by this E2 reaction.

$$\text{2-bromohexane} \xrightarrow[CH_3OH]{CH_3O^-Na^+} \text{hex-2-ene} + \text{hex-1-ene}$$

(products formed in the ratio 7 hex-2-ene : 3 hex-1-ene)

Table 18.6 summarises the generalisations made on this and the previous pages about reactions of haloalkanes.

TABLE 18.6 Comparison between E1 and E2 reactions of haloalkanes.

Haloalkane	E1	E2
primary, RCH_2X	E1 does not occur. Primary carbocations are so unstable that they are never observed in solution.	E2 is favoured.
secondary, R_2CHX	E1 is the main reaction with weak bases such as H_2O and ROH.	E2 is the main reaction with strong bases such as OH^- and OR^-.
tertiary, R_3CX	E1 is the main reaction with weak bases such as H_2O and ROH.	E2 is the main reaction with strong bases such as OH^- and OR^-.

WORKED EXAMPLE 18.6

Predicting β-elimination reaction mechanisms

Predict whether each of the following β-elimination reactions proceeds predominantly by the E1 or E2 mechanism, and write a structural formula for the major organic product.

(a) $CH_3\underset{Cl}{\overset{CH_3}{C}}CH_2CH_3 + NaOH \xrightarrow[H_2O]{80\,°C}$ (b) $CH_3\underset{Cl}{\overset{CH_3}{C}}CH_2CH_3 \xrightarrow{CH_3COOH}$

Analysis and solution

Based on the structure of the haloalkane and the strength of the acid, the likely reaction mechanism can be determined.

(a) A 3° chloroalkane is heated with a strong base. Elimination by an E2 reaction predominates, giving 2-methylbut-2-ene as the major product.

$$CH_3\underset{Cl}{\overset{CH_3}{C}}CH_2CH_3 + NaOH \xrightarrow[H_2O]{80\,°C} CH_3\overset{CH_3}{C}{=}CHCH_3 + NaCl + H_2O$$

(b) A 3° chloroalkane is dissolved in acetic acid, a solvent that promotes the formation of carbocations, to form a 3° carbocation that then loses a proton to give 2-methylbut-2-ene as the major elimination product. The reaction occurs by the E1 mechanism.

$$CH_3C(CH_3)(Cl)CH_2CH_3 \xrightarrow{CH_3COOH} CH_3C(CH_3){=}CHCH_3 + HCl$$

Is our answer reasonable?

Check your answer by asking these two questions: Do the structures make sense? Are the conditions and base strength sufficient for the reaction described?

PRACTICE EXERCISE 18.6

Predict whether each of the following β-elimination reactions proceeds predominantly by the E1 or E2 mechanism, and write a structural formula for the major organic product.

(a) (2-bromopentane, skeletal structure with Br) + $NaOCH_3 \xrightarrow{\text{methanol}}$

(b) (cyclohexane with Cl and methyl substituents) + $NaOCH_2CH_3 \xrightarrow{\text{ethanol}}$

18.4 Substitution versus elimination

Thus far, we have considered two types of reactions of haloalkanes: nucleophilic substitution and β-elimination. Many of the nucleophiles we have examined, such as hydroxide ions and alkoxide ions, are also strong bases. Accordingly, nucleophilic substitution and β-elimination often compete with each other, and the ratio of products formed by these reactions depends on the relative rates of the two reactions.

$$H{-}C{-}C{-}X + :Nu^- \xrightarrow{\text{nucleophilic substitution}} H{-}C{-}C{-}Nu + :X^-$$

$$H{-}C{-}C{-}X + :Nu^- \xrightarrow{\beta\text{-elimination}} C{=}C + H{-}Nu + :X^-$$

S_N1 versus E1 reactions

Reactions of 2° and 3° haloalkanes in polar protic solvents give mixtures of substitution and elimination products. In both reactions, step 1 is the formation of a carbocation intermediate. This step is then followed by either (1) the loss of a hydrogen atom to give an alkene (E1) or (2) reaction with the solvent to give a substitution product (S_N1). In polar protic solvents, the products formed depend only on the structure of the particular carbocation. For example, *tert*-butyl chloride and *tert*-butyl iodide in 80% aqueous ethanol both react with the solvent, giving the same mixture of substitution and elimination products.

$$(CH_3)_3C{-}I \xrightarrow{-I^-} (CH_3)_3C^+ \qquad (CH_3)_3C{-}Cl \xrightarrow{-Cl^-} (CH_3)_3C^+$$

$$(CH_3)_3C^+ \xrightarrow{E1} CH_2{=}C(CH_3)_2 + H^+$$

$$(CH_3)_3C^+ \xrightarrow[H_2O]{S_N1} (CH_3)_3C{-}OH + H^+$$

$$(CH_3)_3C^+ \xrightarrow[CH_3CH_2OH]{S_N1} (CH_3)_3C{-}OCH_2CH_3 + H^+$$

Because the iodide ion is a better leaving group than the chloride ion, *tert*-butyl iodide reacts over 100 times faster than *tert*-butyl chloride. However, the ratio of products is the same.

S_N2 versus E2 reactions

It is considerably easier to predict the ratio of substitution to elimination products for reactions of haloalkanes with reagents that act as both nucleophiles and bases. The guiding principles are as follows.

1. Branching at the α-carbon or β-carbon(s) increases steric hindrance around the α-carbon and significantly retards S_N2 reactions. By contrast, branching at the α-carbon or β-carbon(s) increases the rate of E2 reactions because of the increased stability of the alkene product.
2. The greater the nucleophilicity of the attacking reagent, the greater is the S_N2:E2 ratio. Conversely, the greater the basicity of the attacking reagent, the lower is the S_N2:E2 ratio.

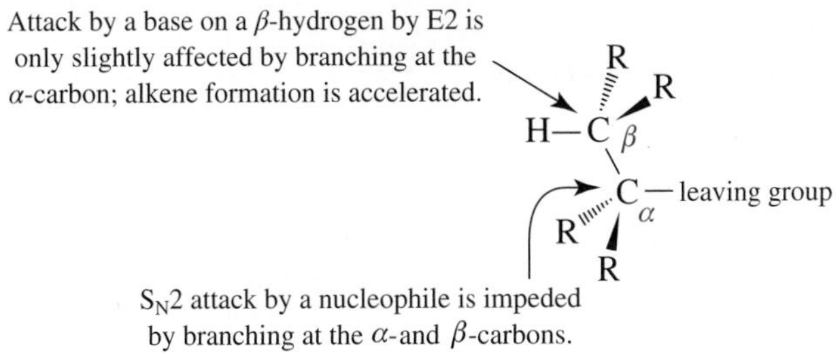

Attack by a base on a β-hydrogen by E2 is only slightly affected by branching at the α-carbon; alkene formation is accelerated.

R R H—C β C — leaving group α R R

S_N2 attack by a nucleophile is impeded by branching at the α-and β-carbons.

Primary haloalkanes react with bases/nucleophiles to give predominantly substitution products. With strong bases, such as hydroxide ions and ethoxide ions, a percentage of the product is formed by an E2 reaction, but it is generally small compared with that formed by an S_N2 reaction. With strong, bulky bases, such as the *tert*-butoxide ion, the E2 product becomes the major product. Tertiary haloalkanes react with all strong bases/good nucleophiles to give only elimination products.

Secondary haloalkanes are borderline, and substitution or elimination may be favoured, depending on the particular base/nucleophile, solvent and temperature at which the reaction is carried out. Elimination is favoured with strong bases/good nucleophiles, such as hydroxide ions and ethoxide ions. Substitution is favoured with weak bases/poor nucleophiles, such as acetate ions. Table 18.7 summarises this information about substitution versus elimination reactions of haloalkanes.

TABLE 18.7 Summary of substitution versus elimination reactions of haloalkanes.

Haloalkane	Mechanism	Comments
methyl, CH_3X	S_N2	S_N2 is the only substitution reaction observed.
	~~S_N1~~	S_N1 reactions of methyl halides are never observed. The methyl cation is so unstable that it is never formed in solution.
primary, RCH_2X	S_N2	S_N2 is the main reaction with strong bases, such as OH^- and EtO^-, and with good nucleophiles/weak bases, such as I^- and CH_3COO^-.
	E2	E2 is the main reaction with strong, bulky bases, such as potassium *tert*-butoxide.
	~~S_N1/E1~~	Primary carbocations are never formed in solution so S_N1 and E1 reactions of 1° haloalkanes are never observed.
secondary, R_2CHX	S_N2	S_N2 is the main reaction with weak bases/good nucleophiles, such as I^- and CH_3COO^-.
	E2	E2 is the main reaction with strong bases/good nucleophiles, such as OH^- and $CH_3CH_2O^-$.
	S_N1/E1	S_N1 and E1 are common in reactions with weak nucleophiles in polar protic solvents, such as water, methanol and ethanol.
tertiary, R_3CX	~~S_N2~~	S_N2 reactions of 3° haloalkanes are never observed because of the extreme crowding around the 3° carbon atom.
	E2	E2 is the main reaction with strong bases, such as HO^- and RO^-.
	S_N1/E1	S_N1 and E1 are the main reactions with poor nucleophiles/weak bases.

WORKED EXAMPLE 18.7

Substitution versus elimination

Predict whether each of the following reactions proceeds predominantly by substitution (S_N1 or S_N2) or elimination (E1 or E2) or whether the two compete. Write structural formulae for the major organic product(s).

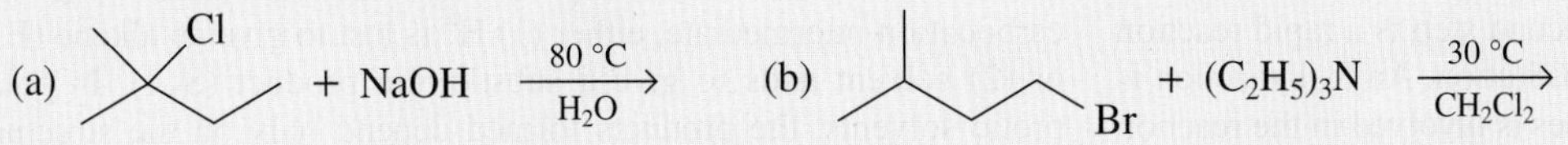

(a) Cl + NaOH $\xrightarrow[H_2O]{80\ °C}$ (b) Br + $(C_2H_5)_3N$ $\xrightarrow[CH_2Cl_2]{30\ °C}$

Analysis and solution

(a) A 3° chloroalkane is heated with a strong base/good nucleophile. Elimination by an E2 reaction predominates to give 2-methylbut-2-ene as the major product.

$$\text{2-chloro-2-methylbutane} + NaOH \xrightarrow[H_2O]{80\ °C} \text{2-methylbut-2-ene} + NaCl + H_2O$$

(b) Reaction of a 1° bromoalkane with triethylamine, a moderate nucleophile/weak base, gives substitution by an S_N2 reaction.

$$\text{1-bromo-3-methylbutane} + (C_2H_5)_3N \xrightarrow[CH_2Cl_2]{30\ °C} \text{3-methylbutyl}\text{-}\overset{+}{N}(C_2H_5)_3Br^-$$

PRACTICE EXERCISE 18.7 Predict whether each of the following reactions proceeds predominantly by substitution (S_N1 or S_N2) or elimination (E1 or E2) or whether the two compete. Write structural formulae for the major organic product(s).

(a) 2-bromobutane (Br) $+ CH_3O^-Na^+ \xrightarrow{\text{methanol}}$

(b) 1-chloro-4-methylcyclohexane (Cl) $+ Na^+I^- \xrightarrow{\text{acetone}}$

SUMMARY

LO 1 Haloalkanes

Haloalkanes contain a halogen atom covalently bonded to an sp^3 hybridised carbon atom. In the IUPAC system, halogen atoms are named as *fluoro-*, *chloro-*, *bromo-* or *iodo-* substituents and are listed with other substituents in alphabetical order. Haloalkanes are often called alkyl halides. Common names are derived by naming the alkyl group, followed by the name of the halide as a separate word. Compounds of the type CHX_3 are called haloforms.

Haloalkanes can be generated in several ways but one of the most important is direct halogenation of an alkane via a free radical substitution reaction. The reaction is not very selective and generally all possible monosubstitution products are formed. If the reaction is not controlled and an excess of alkane is not present, multiple substitution can occur.

A nucleophile is any reagent with an unshared pair of electrons that can be donated to another atom or ion to form a new covalent bond. Nucleophilic substitution is any reaction in which one nucleophile is substituted for another. β-elimination is any reaction in which atoms or groups are removed from two adjacent carbon atoms. Because all nucleophiles are also bases, nucleophilic substitution and base-promoted β-elimination are competing reactions.

LO 2 Nucleophilic substitution

In a nucleophilic substitution reaction, negatively charged nucleophiles become neutral after the reaction. Uncharged nucleophiles become positively charged and must undergo a second step to lose the proton.

An S_N2 reaction occurs in one step. The loss of one group is assisted by the incoming nucleophile, and both are involved in the transition state of the rate-determining step. As two species are involved, it is called a bimolecular reaction.

An S_N1 reaction occurs in two steps. The first step involves loss of a halogen to form a carbocation intermediate; this is slow so it governs the rate of the reaction. The second step is a rapid reaction with a nucleophile to complete the substitution. An S_N1 reaction is a unimolecular reaction; only one species is involved in the reaction leading to the transition state of the rate-determining step.

Solvolysis describes a reaction where the solvent plays the role of the nucleophile in the substitution reaction. For S_N1 reactions taking place at a stereocentre, the reaction occurs with racemisation.

The nucleophilicity of a reagent is measured by the rate of its reaction in a reference nucleophilic substitution. S_N1 reactions are governed by electronic factors, namely, the relative stabilities of carbocation intermediates. S_N2 reactions are governed by steric hindrance, the ability of larger groups to hinder access to the site of substitution. The ability of a group to function as a leaving group is related to its stability as an anion. The most stable anions and the best leaving groups are the conjugate bases of strong acids. Protic solvents contain —OH groups. They interact strongly with polar molecules and ions and are good solvents in which to form carbocations. Protic solvents favour S_N1 reactions. Aprotic solvents do not contain —OH groups. They do not interact as strongly with polar molecules and ions, and carbocations are less likely to form in them. Aprotic solvents favour S_N2 reactions.

LO 3 β-elimination

Dehydrohalogenation, a type of β-elimination reaction, is the removal of H and X from adjacent carbon atoms. A β-elimination that gives the most highly substituted, and therefore most stable, alkene is said to follow Zaitsev's rule.

An E1 reaction occurs in two steps: breaking the C—X bond to form a carbocation intermediate, followed by the loss of H^+ to form an alkene. An E2 reaction occurs in one step: reaction with a base to remove H^+, formation of the alkene and departure of the leaving group, all occurring simultaneously.

LO 4 Substitution versus elimination

When a nucleophile is also a strong base, nucleophilic substitution and β-elimination often compete with each other. Reactions of 2° and 3° haloalkanes in polar protic solvents give mixtures of substitution and elimination products. After the formation of the carbocation intermediate, either (1) H^+ is lost to give an alkene (E1) or (2) solvent adds to give a substitution product (S_N1). In polar protic solvents, the products formed depend only on the structure of the particular carbocation. For reactions of haloalkanes with reagents that act as both nucleophiles and bases, steric hindrance significantly retards S_N2 reactions, while branching at the α-carbon or β-carbon(s) increases the rate of E2 reactions to give the alkene product. The greater the nucleophilicity of the attacking reagent, the greater is the S_N2:E2 ratio. Conversely, the greater the basicity of the attacking reagent, the smaller is the S_N2:E2 ratio.

KEY CONCEPTS AND EQUATIONS

Halogenation: a radical substitution reaction
(section 18.1)

Substitution of hydrogens atoms on an alkane can be achieved using high temperatures or intense light to induce a radical chain reaction. Only bromine and chlorine are practical for this halogenation reaction.

$$CH_3CH(CH_3)CH_2CH_3 \xrightarrow[300\ °C]{Cl_2} Cl{-}CH_2CH(CH_3)CH_2CH_3\ (33.5\%) + CH_3CCl(CH_3)CH_2CH_3\ (22\%) + CH_3CH(CH_3)CHClCH_3\ (28\%) + CH_3CH(CH_3)CH_2CH_2{-}Cl\ (16.5\%)$$

Nucleophilic substitution: S_N2
(section 18.2)

S_N2 reactions occur in one step, and both the nucleophile and the leaving group are involved in the transition state of the rate-determining step. The nucleophile may be negatively charged or neutral. S_N2 reactions result in an inversion of configuration at the reaction centre. They are accelerated in polar aprotic solvents compared with polar protic solvents. S_N2 reactions are governed by steric factors, namely, the degree of crowding around the site of reaction.

$$I^- + CH_3CH_2C(H)(H_3C){-}Cl \longrightarrow I{-}C(CH_2CH_3)(H)(CH_3) + Cl^-$$

$$(CH_3)_3N + CH_3CH_2C(H)(H_3C){-}Cl \longrightarrow (CH_3)_3\overset{+}{N}{-}C(CH_2CH_3)(H)(CH_3) + Cl^-$$

Nucleophilic substitution: S_N1
(section 18.2)

An S_N1 reaction occurs in two steps. Step 1 is the slow, rate-determining cleavage of the C—X bond to form a carbocation intermediate, followed in step 2 by the intermediate's rapid reaction with a nucleophile to complete the substitution. Reaction at a stereocentre gives a racemic product. S_N1 reactions are governed by electronic factors, namely, the relative stabilities of carbocation intermediates.

$$\text{1-chloro-1-methylcyclohexane (CH}_3\text{, Cl)} + CH_3CH_2OH \longrightarrow \text{1-methylcyclohexyl (CH}_3\text{)}{-}OCH_2CH_3 + HCl$$

β-elimination: E1
(section 18.3)

E1 reactions involve the elimination of atoms or groups of atoms from adjacent carbon atoms. Reactions occur in two steps and involve the formation of a carbocation intermediate.

$$\text{(2-chloro-3-methylbutane)} \xrightarrow{CH_3COOH} \text{(2-methylbut-2-ene)} + HCl$$

β-elimination: E2
(section 18.3)

An E2 reaction occurs in one step: reaction with base to remove a hydrogen atom, formation of the alkene and departure of the leaving group, all occurring simultaneously.

$$\text{(2-bromohexane)} \xrightarrow[CH_3OH]{CH_3O^-Na^+} \text{hex-2-ene} + \text{hex-1-ene}$$

(products formed in the ratio 7 hex-2-ene : 3 hex-1-ene)

KEY TERMS

aprotic solvents *p. 805*
β-elimination *p. 797*
bimolecular reaction *p. 800*
dehydrohalogenation *p. 807*
E1 *p. 808*
E2 *p. 809*
haloalkanes *p. 792*
haloform *p. 793*
halogenation *p. 794*
nucleophile *p. 797*
nucleophilic substitution *p. 797*
nucleophilicity *p. 802*
protic solvents *p. 804*
radical *p. 794*
radical substitution *p. 794*
S_N1 *p. 801*
S_N2 *p. 800*
solvolysis *p. 801*
steric hindrance *p. 803*
unimolecular reaction *p. 801*
Zaitsev's rule *p. 807*

REVIEW QUESTIONS

LO1 Haloalkanes

18.1 Write the IUPAC name of each of the following compounds.

(a) $Br_2C{=}CH_2$

(b)

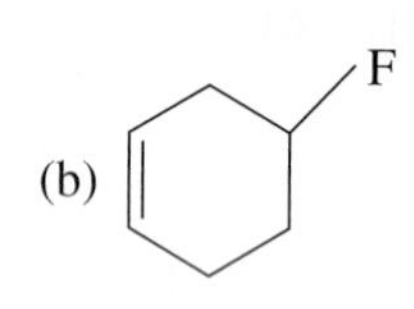

(c) $Cl(CH_2)_5Br$

(d)

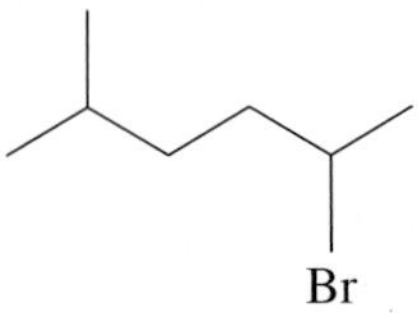

Br

(e) $ClFBr_2$

(f)

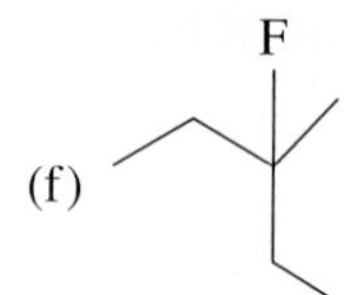

18.2 Write the IUPAC name of each of the following compounds (be certain to include a designation of stereochemical configuration, where appropriate, in your answer).

(a) 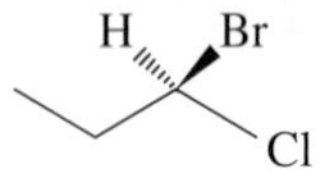 H, Br, Cl

(b) 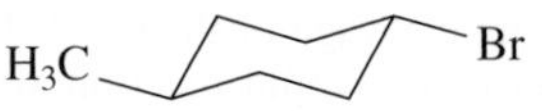 H_3C, Br

(c) 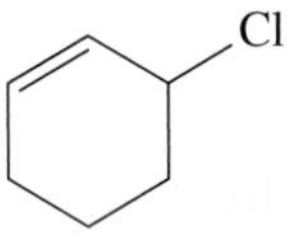 Cl

(d) 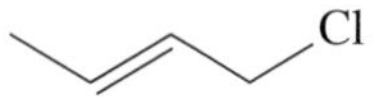 Cl

(e) 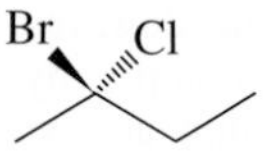 Br, Cl

(f) 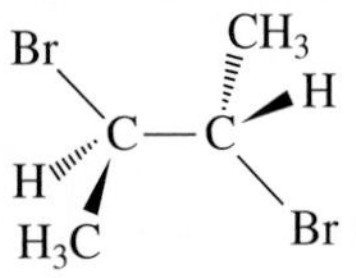 Br, CH_3, H, C—C, H, H_3C, Br

18.3 Draw a structural formula for each of the following compounds with the IUPAC name given.
(a) 3-bromopropene
(b) (*R*)-2-chloropentane
(c) *meso*-3,4-dibromohexane
(d) *trans*-1-bromo-3-isopropylcyclohexane
(e) 1,2-dichloroethane
(f) bromocyclobutane

18.4 Draw a structural formula for each of the following compounds with the common name given.
(a) isopropyl chloride
(b) *sec*-butyl bromide
(c) allyl iodide
(d) methylene chloride
(e) chloroform
(f) *tert*-butyl chloride
(g) isobutyl chloride

18.5 Which of the following compounds are 2° haloalkanes?
(a) isobutyl chloride
(b) 2-iodooctane
(c) *trans*-1-chloro-4-methylcyclohexane

18.6 Ignoring stereochemistry, write the structural formulae and give the IUPAC names for the products of the following halogenation reactions.

(a) $H_3C-CH_2-CH_2-CH_3 + Cl_2 \xrightarrow{\text{heat}}$

(b) 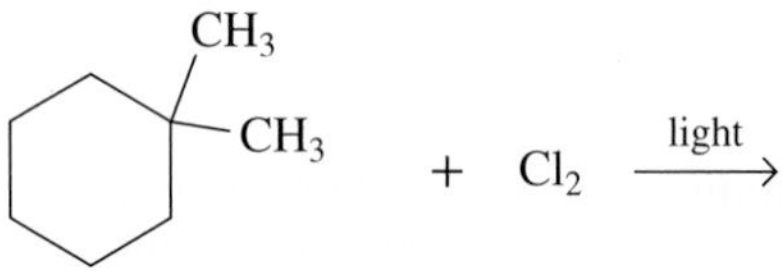 1,1-dimethylcyclohexane (CH_3, CH_3) $+ Cl_2 \xrightarrow{\text{light}}$

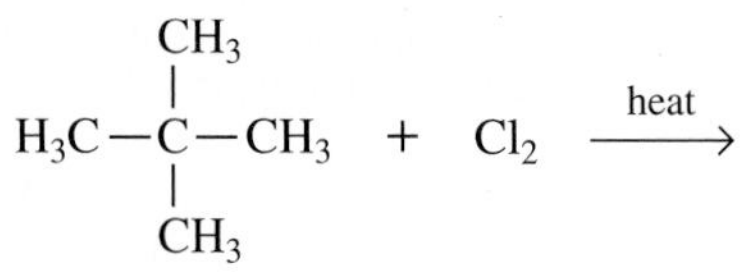

(c) $H_3C-C(CH_3)_2-CH_3 + Cl_2 \xrightarrow{\text{heat}}$

(d) $(H_3C)_2C(CH_3)$–cyclohexane $+ Br_2 \xrightarrow{\text{light}}$

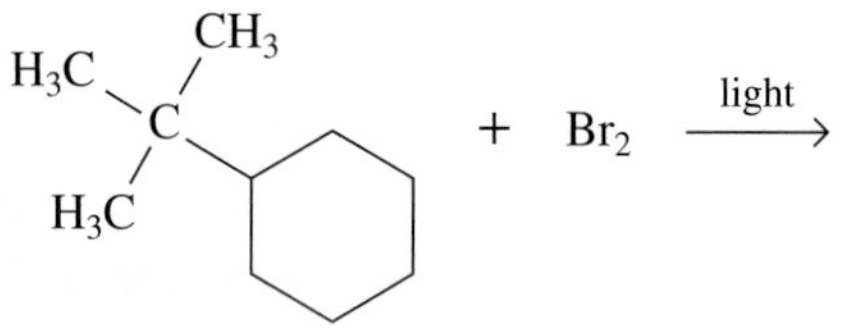

LO2 Nucleophilic substitution

18.7 Write structural formulae for the following organic solvents commonly used in substitution reactions to influence the reaction mechanism to be either S_N1 or S_N2. Indicate whether the solvents are protic or aprotic, polar or non-polar.
(a) dichloromethane
(b) acetone
(c) ethanol
(d) diethyl ether
(e) dimethyl sulfoxide

18.8 Arrange these protic solvents in order of increasing polarity: H_2O, CH_3CH_2OH, CH_3OH.

18.9 Arrange these aprotic solvents in order of increasing polarity: acetone, pentane, diethyl ether.

18.10 From each of the following pairs, select the better nucleophile.
(a) H_2O or OH^-
(b) CH_3COO^- or OH^-
(c) CH_3SH or CH_3S^-

18.11 Which statements are true for S_N2 reactions of haloalkanes?
(a) Both the haloalkane and the nucleophile are involved in the transition state of the rate-determining step.
(b) The reaction proceeds with inversion of configuration at the substitution centre.
(c) The reaction proceeds with retention of optical activity.
(d) The order of reactivity is 3° > 2° > 1° > methyl.
(e) The nucleophile must have an unshared pair of electrons and bear a negative charge.
(f) The greater the nucleophilicity of the nucleophile, the greater is the rate of reaction.

LO 3 *β*-elimination

18.12 Draw structural formulae for the alkene(s) formed by treating each of the following haloalkanes with sodium ethoxide in ethanol. Assume that elimination occurs by the E2 mechanism. Where two alkenes are possible, use Zaitsev's rule to predict which alkene is the major product.

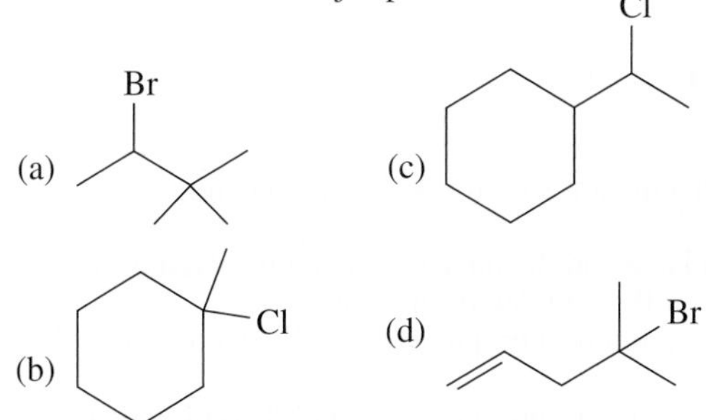

18.13 Which of the following haloalkanes undergo dehydrohalogenation to give alkenes that do not show *cis–trans* isomerism?

(a) 2-chloropentane
(b) 2-chlorobutane
(c) chlorocyclohexane
(d) isobutyl chloride

18.14 What haloalkane might you use as starting material to produce each of the following alkenes in high yield and uncontaminated by isomeric alkenes?

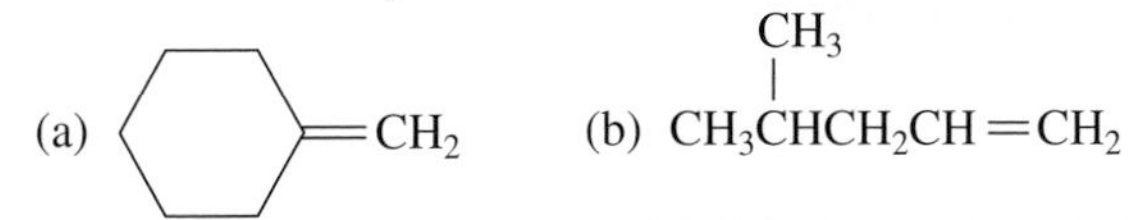

LO 4 Substitution versus elimination

18.15 The reaction of methyl iodide with sodium hydroxide to give methanol always involves the S_N2 mechanism. Why is it that S_N1 cannot play a role in this conversion?

18.16 What is the product of the reaction of 2-bromoethane with potassium *tert*-butoxide? Why is it that little substitution occurs in this reaction?

18.17 What is the major product of the reaction of 2-iodopropane with sodium ethoxide?

18.18 The reaction of 2-bromo-2-methylpropane and sodium hydroxide occurs primarily by which mechanism: S_N1, S_N2, E1 or E2?

18.19 Consider the following substitution reaction.

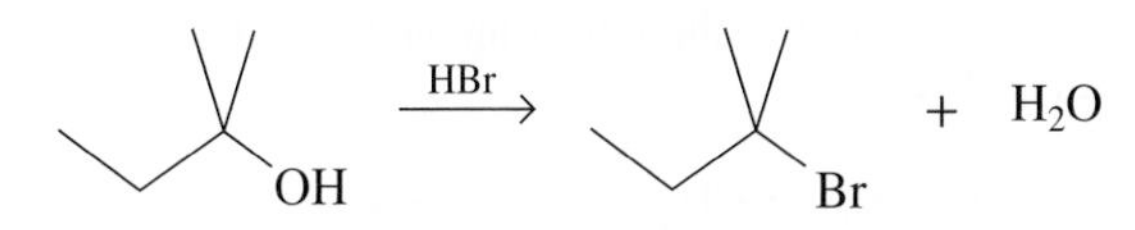

(a) Determine whether this reaction proceeds via an S_N1 or S_N2 process.
(b) Draw the mechanism for this reaction.
(c) What is the rate equation for this reaction?
(d) Would the reaction occur at a faster rate if sodium bromide were added to the reaction mixture?
(e) Draw an energy diagram for this reaction.

18.20 Consider the following substitution reaction.

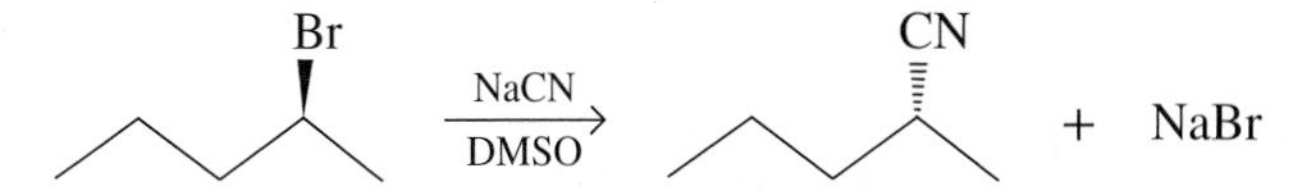

(a) Determine whether this reaction proceeds via an S_N1 or S_N2 process.
(b) Draw the mechanism for this reaction.
(c) What is the rate equation for this reaction?
(d) Would the reaction occur at a faster rate if the concentration of cyanide were doubled?
(e) Draw an energy diagram for this reaction.

REVIEW PROBLEMS

18.21 (a) Draw the structures for all of the haloalkanes that have a molecular formula of $C_5H_{11}F$. (*Hint:* There are more than eight.)
(b) Are any of these isomers chiral?
(c) Can you give the IUPAC name for these molecules, including any chiral isomers? **LO 1**

18.22 Draw the structures of the following molecules. **LO 1**
(a) 1-ethyl-1-fluorcyclopentane
(b) 1-fluoro-1-methylcyclopentane
(c) 1-bromo-1-fluoro-2,6-dimethylcyclohexane

18.23 Which of the following molecules has the most polarised carbon–halogen bond? **LO 1**

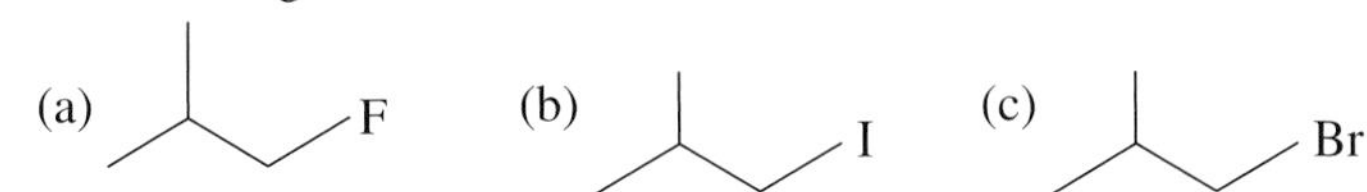

18.24 Give the structures and write the IUPAC names for all of the possible products of the following free radical substitution reactions. (Assume mono substitution only, and ignore possible stereoisomers.) **LO 1**

(a) cyclopentane + $Br_2 \xrightarrow{\text{heat}}$

(b) methylcyclohexane + $Cl_2 \xrightarrow{\text{light}}$

(c) 2,2,3,3-tetramethylbutane + $Cl_2 \xrightarrow{\text{light}}$

18.25 Which of the following molecules will most easily form a carbocation and why? **LO 2**

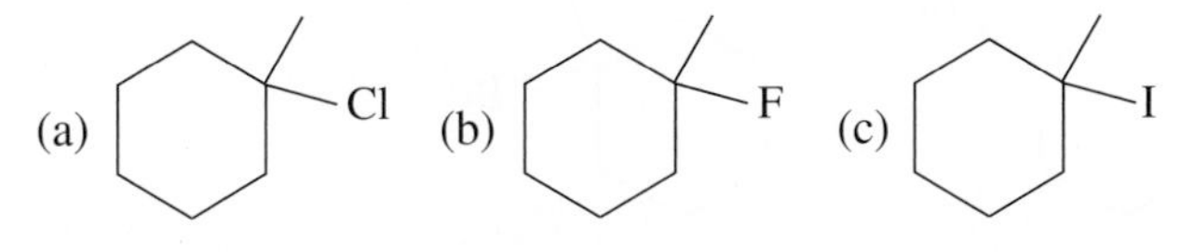

18.26 Could 1-bromo-methylcyclopentane easily undergo an S_N2 reaction with a nucleophile? Why or why not? **LO 2**

18.27 Is the following proposed reaction likely to produce the product described? **LO 2**

OH $+ Na^+ + :\ddot{Br}:^- \longrightarrow$ Br $+ :\ddot{O}H^- + Na^+$

18.28 (a) Is the reaction of 5-bromohexa-1,3-diene with sodium hydroxide likely to proceed via a substitution or an elimination process?
(b) The exclusive product of this reaction has the molecular formula C_6H_6. Can you explain why this is the only product generated and why the reaction proceeds by this mechanism? **LO 3**

18.29 Give a bromoalkane starting material that could react to give the following elimination reactions where the structure shown is the *major* product. **LO 3**

A. $\xrightarrow[\text{methanol, heat}]{CH_3O^-Na^+}$

B. $\xrightarrow[\text{methanol, heat}]{CH_3O^-Na^+}$

C. $\xrightarrow[\text{methanol, heat}]{CH_3O^-Na^+}$

18.30 Ignoring stereochemistry, write the structural formulae and give the IUPAC names for the products of the following halogenation reactions. **LO 1,4**

(a) 1,4-dimethylcyclohexane + Cl_2 $\xrightarrow{\text{heat}}$

(b) 1,1,3,3-tetramethylcyclopentane + Br_2 $\xrightarrow{\text{light}}$

(c) 3,4-diethylhexane + Br_2 $\xrightarrow{\text{heat}}$

(d) 3-ethylpentane + Br_2 $\xrightarrow{\text{light}}$

18.31 Give the product of the following nucleophilic substitutions. **LO 2**

(a) Na^+I^- + $CH_3CH_2CH_2Cl$ $\xrightarrow[\text{acetone}]{}$

(b) 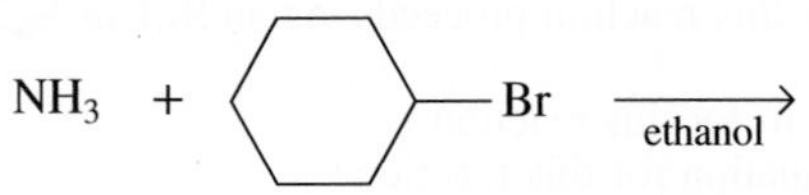 NH_3 + (cyclohexyl)—Br $\xrightarrow[\text{ethanol}]{}$

(c) $CH_3CH_2O^-Na^+$ + $CH_2{=}CHCH_2Cl$ $\xrightarrow[\text{ethanol}]{}$

18.32 Give the product of the following substituton reactions, all of which proceed by an S_N2 process. **LO 4**

(a) 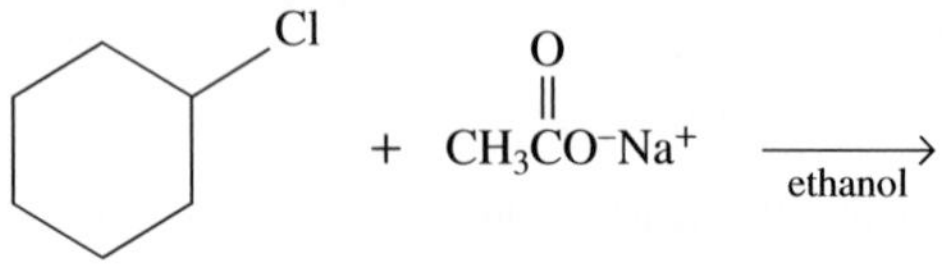 (chloromethyl)cyclohexane (Cl) + $CH_3\overset{O}{\overset{\|}{C}}O^-Na^+$ $\xrightarrow[\text{ethanol}]{}$

(b) CH_3—$\overset{I}{\overset{|}{C}}H$—$CH_2CH_3$ + CH_3CH_2—S^- Na^+ $\xrightarrow[\text{acetone}]{}$

(c) $CH_3\overset{CH_3}{\overset{|}{C}}HCH_2CH_2Br$ + Na^+I^- $\xrightarrow[\text{acetone}]{}$

(d) $(CH_3)_3N$ + CH_3I $\xrightarrow[\text{acetone}]{}$

(e) 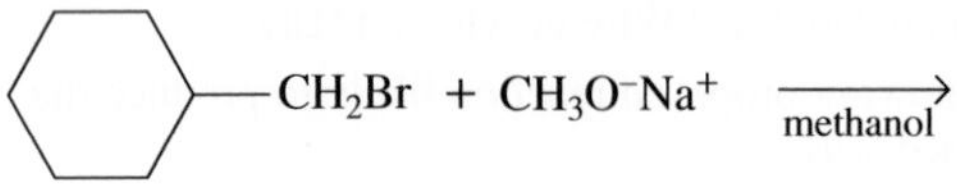 (cyclohexyl)—CH_2Br + $CH_3O^-Na^+$ $\xrightarrow[\text{methanol}]{}$

(f) 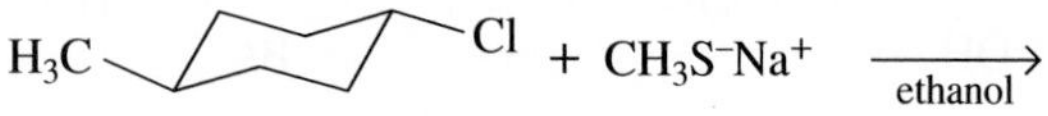 H_3C—(cyclohexane chair)—Cl + $CH_3S^-Na^+$ $\xrightarrow[\text{ethanol}]{}$

(g) 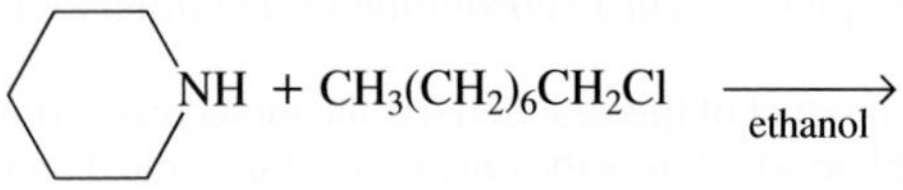 (piperidine) NH + $CH_3(CH_2)_6CH_2Cl$ $\xrightarrow[\text{ethanol}]{}$

(h) (cyclopentyl)—CH_2Cl + NH_3 $\xrightarrow[\text{ethanol}]{}$

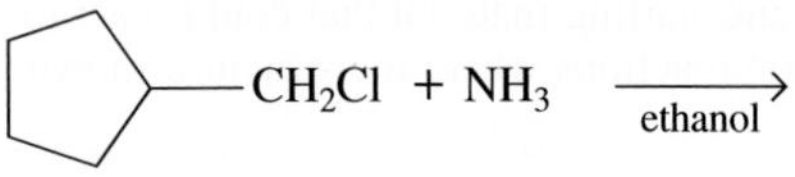

18.33 You were told that each reaction in question 18.32 proceeds by the S_N2 mechanism. Suppose you were not told the mechanism. Describe how you could conclude, from the structures of the haloalkane, the nucleophile and the solvent, that each reaction is in fact an S_N2 reaction. **LO 4**

18.34 In the following reactions, a haloalkane is treated with a compound that has two nucleophilic sites. Select the more nucleophilic site in each part, and show the product of each substitution reaction. **LO 2**

(a) $HOCH_2CH_2NH_2$ + CH_3I $\xrightarrow[\text{ethanol}]{}$

(b) 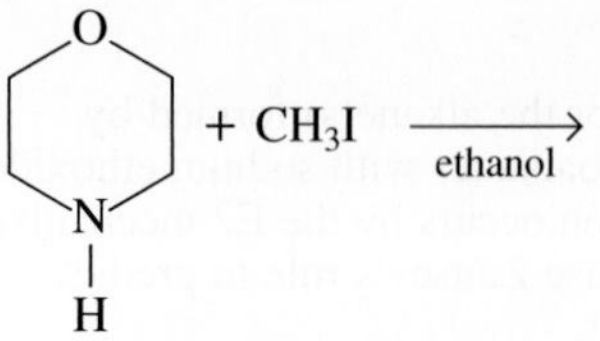 (morpholine: O, N—H) + CH_3I $\xrightarrow[\text{ethanol}]{}$

(c) $HOCH_2CH_2SH$ + CH_3I $\xrightarrow[\text{ethanol}]{}$

18.35 Which statements are true for S_N1 reactions of haloalkanes? **LO 2**

(a) Both the haloalkane and the nucleophile are involved in the transition state of the rate-determining step.
(b) The reaction at a stereocentre proceeds with retention of configuration.
(c) The reaction at a stereocentre proceeds with loss of optical activity.
(d) The order of reactivity is 3° > 2° > 1° > methyl.
(e) The greater the steric crowding around the reactive centre, the lower is the rate of reaction.
(f) The rate of reaction is greater with good nucleophiles than with poor nucleophiles.

18.36 Draw a structural formula for the product of each of the following S_N1 reactions. **LO 2**

(a) $CH_3\overset{Cl}{\overset{|}{C}}HCH_2CH_3$ + CH_3CH_2OH $\xrightarrow[\text{ethanol}]{}$
S enantiomer

(b) 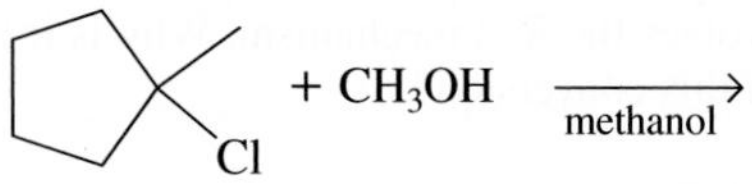 (1-chloro-1-methylcyclopentane) + CH_3OH $\xrightarrow[\text{methanol}]{}$

(c) $CH_3\overset{CH_3}{\underset{CH_3}{C}}Cl$ + $CH_3\overset{O}{\overset{\|}{C}}OH$ $\xrightarrow[\text{acetic acid}]{}$

(d) 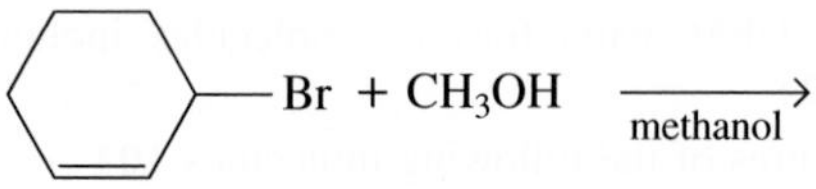 (cyclohexenyl)—Br + CH_3OH $\xrightarrow[\text{methanol}]{}$

18.37 You were told that each substitution reaction in question 18.36 proceeds by the S_N1 mechanism. Suppose that you were not told the mechanism. Describe how you could conclude, from the structures of the haloalkane, the nucleophile and the solvent, that each reaction is in fact an S_N1 reaction. **LO 2,4**

18.38 Select the member of each of the following pairs that undergoes nucleophilic substitution in aqueous ethanol more rapidly. **LO 4**

(a) 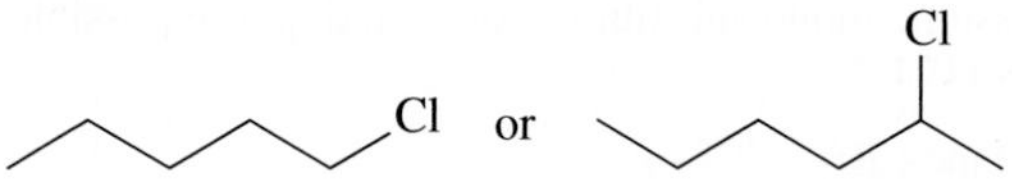 (1-chloropentane) Cl or (secondary chloride) Cl

(b) 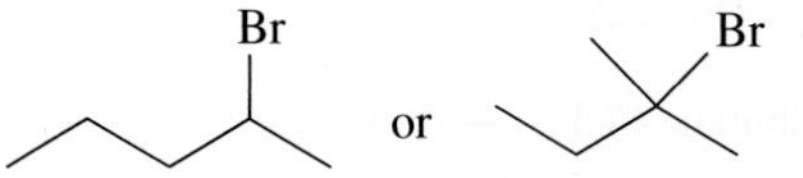 (2-bromobutane) Br or (tertiary bromide) Br

(c) 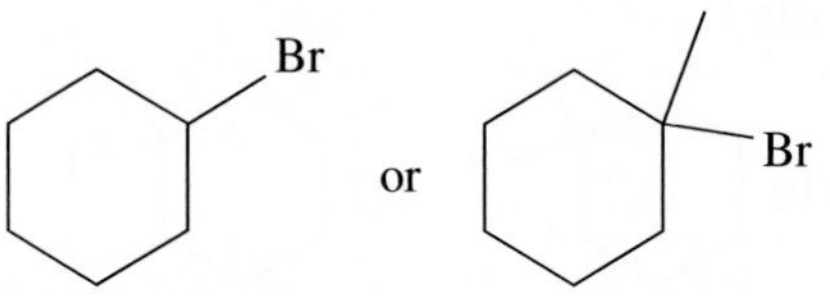 (bromomethylcyclohexane) Br or (tertiary bromide on cyclohexane) Br

18.39 Identify the reagent you would use to accomplish each of the following transformations. **LO4**

(a) cyclobutanol ⟶ bromocyclobutane
(b) *tert*-butanol ⟶ *tert*-butyl chloride
(c) ethyl chloride ⟶ ethanol

18.40 Propose a mechanism for the formation of the products (but not their relative percentages) in this reaction. **LO2,3**

$$(CH_3)_3CCl \xrightarrow[25\ °C]{20\%\ H_2O,\ 80\%\ CH_3CH_2OH} \underbrace{(CH_3)_3COCH_2CH_3 + (CH_3)_3COH}_{85\%} + \underbrace{CH_3C(CH_3){=}CH_2}_{15\%} + HCl$$

18.41 The rate of reaction in question 18.40 increases by 140 times when carried out in 80% water/20% ethanol compared with 40% water/60% ethanol. Account for this difference. **LO2,3,4**

18.42 What hybridisation best describes the reacting carbon atom in the S_N2 transition state? **LO2**

18.43 Haloalkenes such as vinyl bromide, $CH_2{=}CHBr$, undergo neither S_N1 nor S_N2 reactions. What factors account for this lack of reactivity? **LO4**

18.44 Show how you might synthesise the following compounds from a haloalkane and a nucleophile. **LO4**

(a) CN (b) CN (c) O, O (d) SH (e) OCH_3 (f) O (g) SH

18.45 Show how you might synthesise the following compounds from a haloalkane and a nucleophile. **LO4**

(a) NH_2 (b) CH_2NH_2 (c) $OCCH_3$ (O) (d) S (e) O, O (f) $(CH_3CH_2CH_2CH_2)_2O$

18.46 For each of the following alkenes, draw structural formulae of all chloroalkanes that undergo dehydrohalogenation when treated with KOH to give that alkene as the major product. (For some alkenes, only one chloroalkane gives the desired alkene as the major product; for other alkenes, two chloroalkanes may work.) **LO3**

(a) (b) CH_2

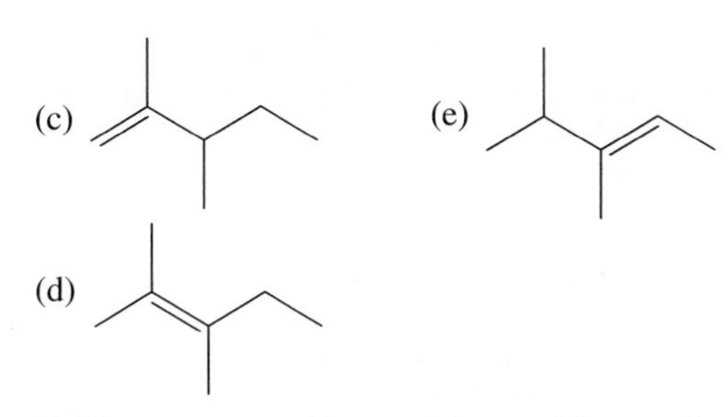

(c) (e)

(d)

18.47 When *cis*-4-chlorocyclohexanol is treated with sodium hydroxide in ethanol, it gives only one substitution product, *trans*-1,4-cyclohexanediol (1). Under the same experimental conditions, *trans*-4-chlorocyclohexanol gives cyclohex-3-enol (2) and the bicyclic ether (3). **LO3,4**

cis-4-chloro-cyclohexanol $\xrightarrow[CH_3CH_2OH]{NaOH}$ (1)

trans-4-chloro-cyclohexanol $\xrightarrow[CH_3CH_2OH]{NaOH}$ (2) + (3)

(a) Propose a mechanism for the formation of product (1), and account for its configuration.
(b) Propose a mechanism for the formation of product (2).
(c) Account for the fact that the bicyclic ether (3) is formed from the *trans* isomer, but not from the *cis* isomer.

18.48 In each of the following reactions, show how to convert the given starting material into the desired product. (Note that some syntheses require only one step, whereas others require two or more steps.) **LO2,3,4**

(a) Cl ⟶
(b) ⟶ Br
(c) Cl ⟶ OH
(d) Br ⟶
(e) Br ⟶ OH, OH
(f) Br ⟶ Br, Br

18.49 How many isomers, including *cis–trans* isomers, are possible for the major product of dehydrohalogenation of each of the following haloalkanes? **LO 3**
(a) 3-chloro-3-methylhexane
(b) 3-bromohexane

18.50 Give the products of the reaction of 1-iodopropane with each of the following. **LO 4**

(a) NaOH
(b) $NaNH_2$
(c) NaCN
(d) $NaOOCCH_3$
(e) NaI
(f) $NaOC(CH_3)_3$

18.51 Below are two potential methods for preparing the same ether, but only one of them is successful. Identify the successful approach and explain your choice. **LO 4**

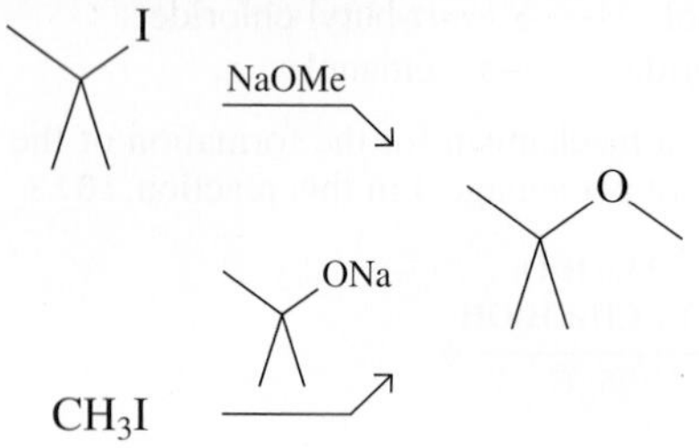

NaOMe
ONa
CH_3I

ADDITIONAL EXERCISES

18.52 The Williamson ether synthesis involves treating a haloalkane with a metal alkoxide. The following two reactions are intended to give benzyl *tert*-butyl ether. One reaction gives the ether in good yield, but the other does not. Which reaction gives the ether? What is the product of the other reaction, and how do you account for its formation? **LO 2**

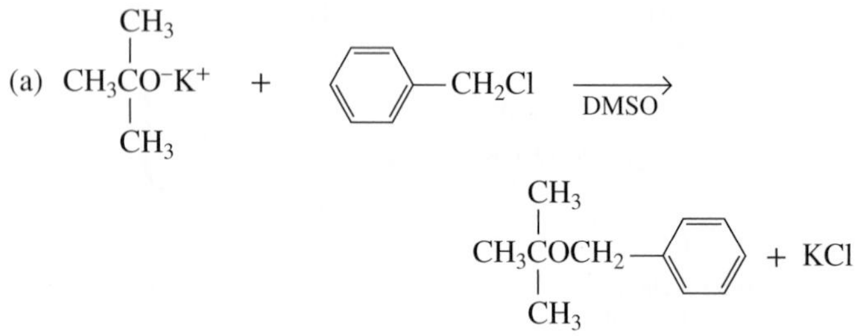

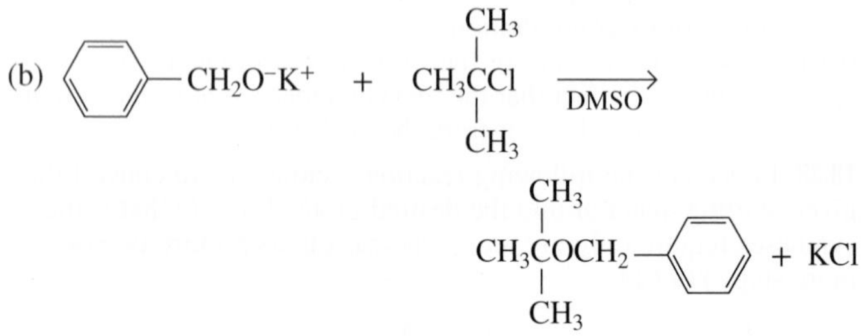

(a) $(CH_3)_3CO^-K^+$ + $C_6H_5CH_2Cl$ $\xrightarrow{DMSO}$ $(CH_3)_3COCH_2C_6H_5$ + KCl

(b) $C_6H_5CH_2O^-K^+$ + $(CH_3)_3CCl$ $\xrightarrow{DMSO}$ $(CH_3)_3COCH_2C_6H_5$ + KCl

18.53 Propose a mechanism for the following reaction. **LO 2,4**

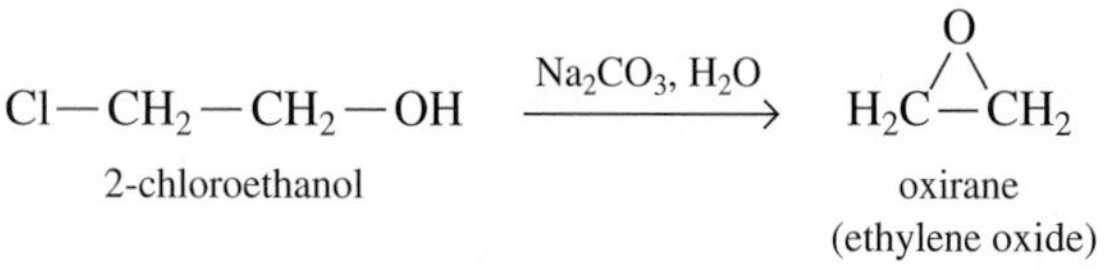

$$Cl-CH_2-CH_2-OH \xrightarrow{Na_2CO_3,\ H_2O} H_2C-CH_2 \text{ (with bridging O)}$$

2-chloroethanol → oxirane (ethylene oxide)

18.54 An —OH group is a poor leaving group, yet substitution occurs readily in the following reaction. Propose a mechanism for this reaction that shows how OH overcomes its limitation of being a poor leaving group. **LO 4**

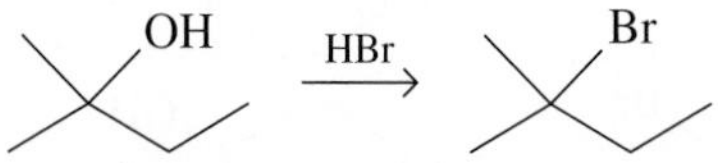

OH $\xrightarrow{HBr}$ Br

18.55 Explain why (*S*)-2-bromobutane becomes optically inactive when treated with sodium bromide in DMSO. **LO 2**

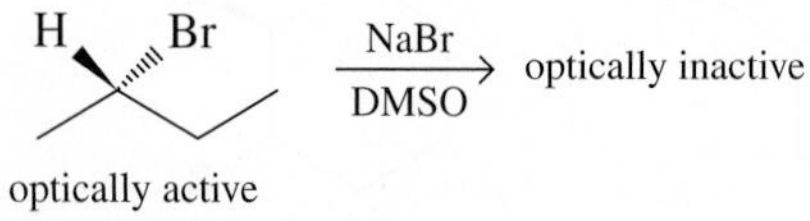

H, Br $\xrightarrow[DMSO]{NaBr}$ optically inactive

optically active

18.56 Explain why phenoxide is a much poorer nucleophile than cyclohexoxide. **LO 4**

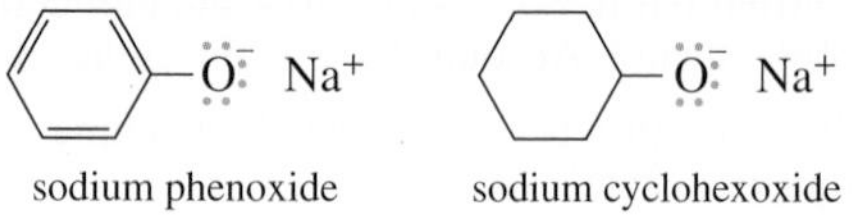

O^- Na^+ sodium phenoxide

O^- Na^+ sodium cyclohexoxide

18.57 In ethers, each side of the oxygen is essentially an —OR group and so is a poor leaving group. Epoxides are three-membered ring ethers (oxirane in question 18.53 is an example of an epoxide). Explain why an epoxide reacts readily with a nucleophile despite being an ether. **LO 4**

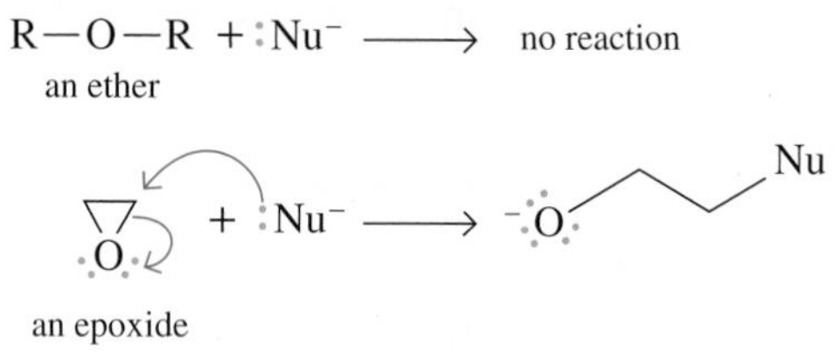

$R-O-R + :Nu^- \longrightarrow$ no reaction
an ether

an epoxide $+ :Nu^- \longrightarrow$ ^-O ... Nu

18.58 What alkene(s) and reaction conditions give each of the following haloalkanes in good yield? (*Hint:* Review chapter 16.) **LO 4**

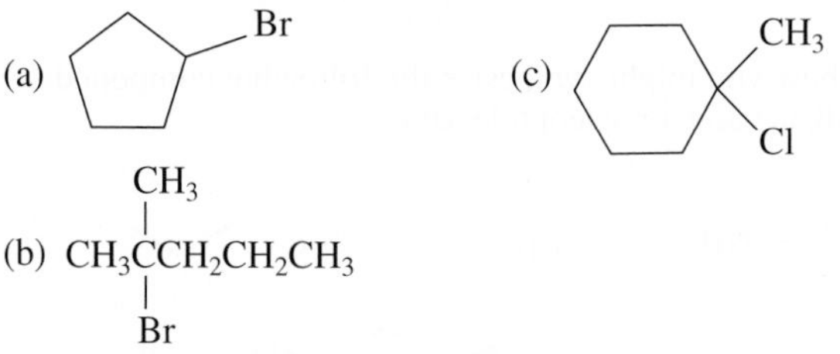

(a) bromocyclopentane (Br)

(b) $CH_3CBr(CH_3)CH_2CH_2CH_3$

(c) 1-chloro-1-methylcyclohexane (CH_3, Cl)

18.59 Show reagents and conditions that bring about the following conversions. **LO 4**

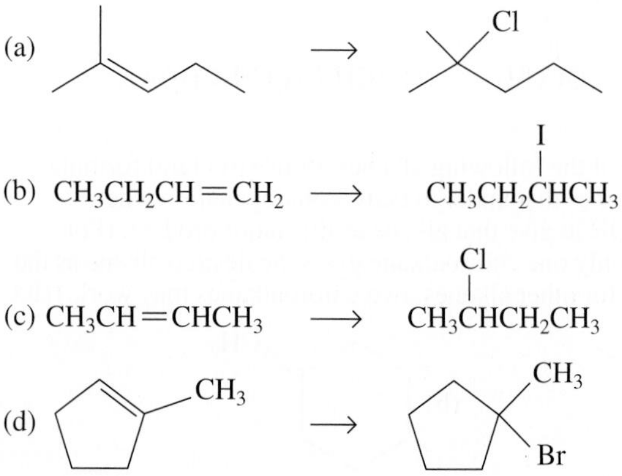

(a) 2-methylpent-2-ene ⟶ 2-chloro-2-methylpentane (Cl)

(b) $CH_3CH_2CH{=}CH_2 \longrightarrow CH_3CH_2CHICH_3$

(c) $CH_3CH{=}CHCH_3 \longrightarrow CH_3CHClCH_2CH_3$

(d) 1-methylcyclopentene (CH_3) ⟶ 1-bromo-1-methylcyclopentane (CH_3, Br)

19 Alcohols, amines and related compounds

On a per capita basis, Australia and New Zealand have the highest incidence of asthma in the world. The inhaler in the picture delivers puffs of salbutamol (also known as albuterol or, more commonly, Ventolin®), which is a potent synthetic bronchodilator used in the treatment of asthma. The structure of salbutamol (shown opposite) is patterned after that of epinephrine (adrenaline) and, like many drugs, salbutamol is a polyfunctional compound (i.e. it contains more than one type of functional group). The functional groups present in salbutamol include a primary and secondary alcohol, a phenol and a secondary amine. These functional groups are important in the binding of salbutamol to the receptors (active site) in the lungs, which causes smooth muscle relaxation. The binding occurs through hydrogen bonding between the hydroxyl groups, and ion–ion attractions between the protonated amine and a carboxylate group in the active site. The bulky tertiary butyl group helps to reduce side-effects by preventing binding at similar shaped receptors that cause smooth muscle contraction.

Alcohols, phenols, ethers and amines are extremely common in nature and these functional groups play important roles in many biological, pharmaceutical and industrial applications. Thiols, sulfur-containing analogues of alcohols, play important roles in biological systems including flavour and odour chemistry.

In this chapter, we study the physical and chemical properties of these functional groups. We will learn how they can be prepared from, and transformed into, other functional groups.

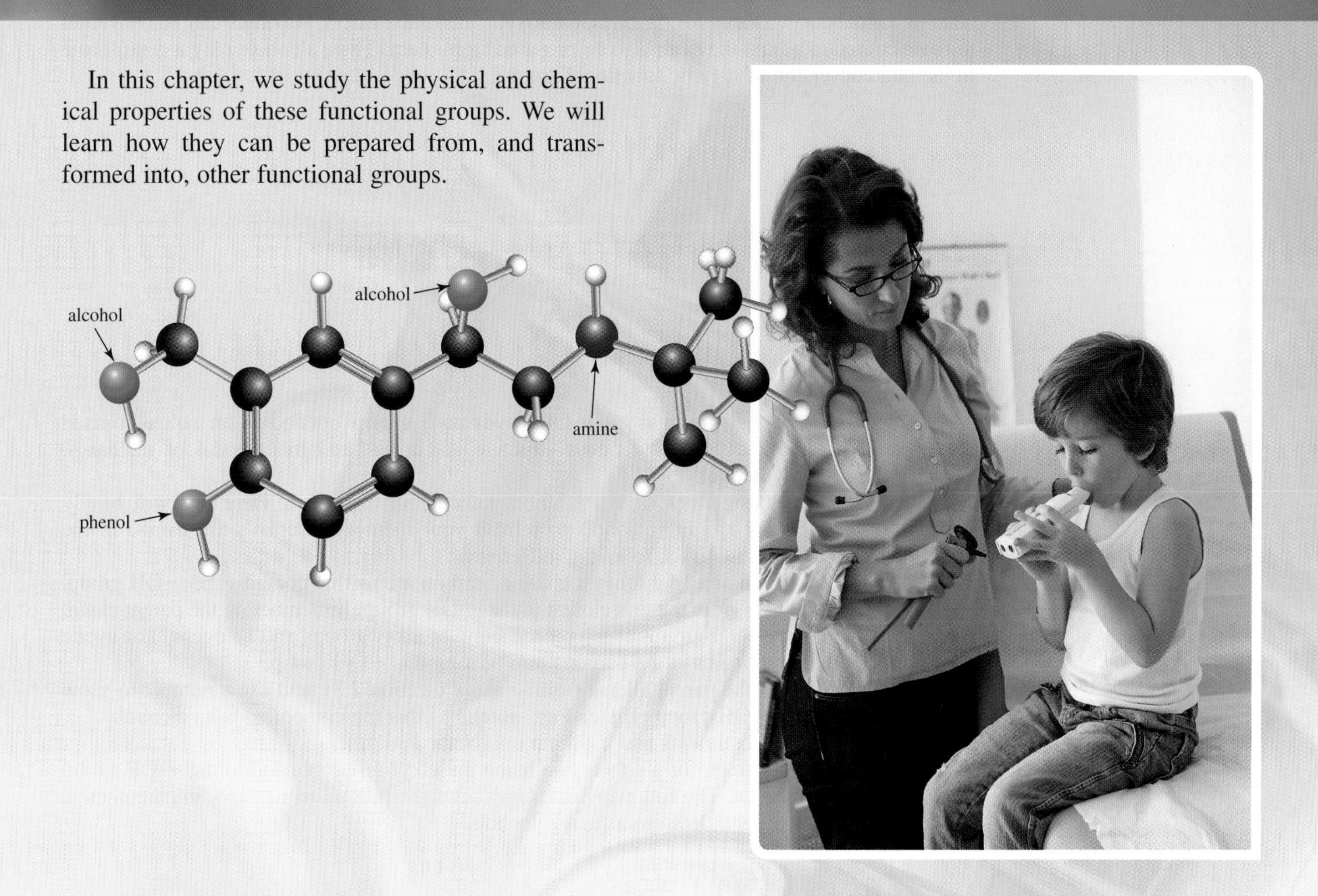

LEARNING OBJECTIVES

After studying this chapter, you should be able to:

19.1 describe the functional group, nomenclature, physical properties and preparation of alcohols

19.2 explain, using examples, the reactions of alcohols

19.3 describe the functional group, nomenclature and reactions of phenols

19.4 describe the functional group, nomenclature, physical properties and reactions of ethers

19.5 describe the functional group, nomenclature, physical properties and reactions of thiols

19.6 describe the functional group, nomenclature, physical properties and preparation of amines

19.7 explain, using examples, the reactions of amines.

19.1 Alcohols

Alcohols, phenols and ethers can be thought of as three classes of compounds that are related to water, while thiols are sulfur-containing analogues of alcohols.

H–O–H (water) R–O–H (an alcohol) Ar–O–H (a phenol) R–O–R (an ether) R–S–H (a thiol)

Alcohols are particularly important in both laboratory reactions and biochemical transformations of organic compounds. They can be converted into other types of compounds, such as alkenes, haloalkanes, aldehydes, ketones, carboxylic acids and esters. Alcohols can be converted to these compounds, and they can also be prepared from them. Thus, alcohols play a central role in the interconversion of organic functional groups.

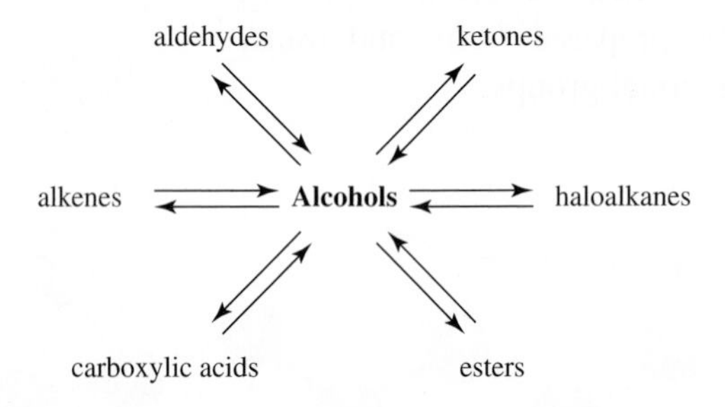

(a)

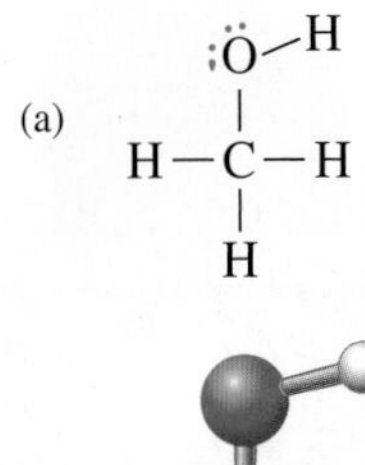

(b)

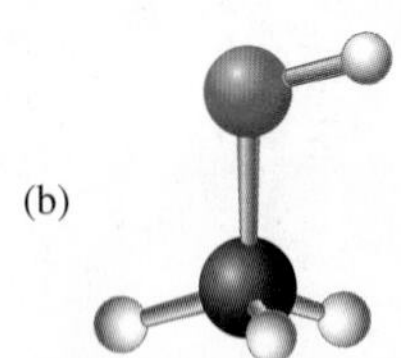

FIGURE 19.1 Methanol, CH_3OH: **(a)** Lewis structure and **(b)** ball-and-stick model. The bond angles in methanol are very close to the tetrahedral angle.

The **alcohol** functional group is an —OH **(hydroxyl) group** bonded to an sp^3 hybridised carbon atom. Figure 19.1 shows a Lewis structure and a ball-and-stick model of methanol, CH_3OH, the simplest alcohol.

We explained the basic rules of IUPAC nomenclature in chapter 2. Here, we will briefly describe the process for naming alcohols to refresh your memory. Alcohols are named in the same manner as alkanes, with the following differences.

1. Select, as the parent alkane, the longest chain of carbon atoms that contains the —OH group, and number that chain from the end closest to the —OH group. In numbering the parent chain, the location of the —OH group takes precedence over alkyl groups and halogens. For cyclic alcohols, numbering begins at the carbon atom bearing the —OH group.
2. Change the suffix of the parent alkane from *-e* to *-ol* (section 2.3), and use a number to show the location of the —OH group. The ending *-ol* tells us that the compound is an alcohol.
3. Name and number substituents and list them in alphabetical order.

To derive common names for alcohols, we name the alkyl group bonded to the —OH group and add the word *alcohol*. The following diagram shows the IUPAC names and, in parentheses, the common names of eight low-molar-mass alcohols.

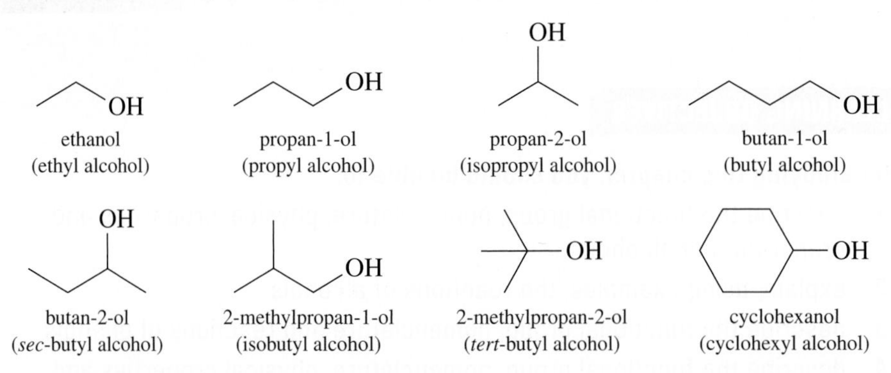

WORKED EXAMPLE 19.1

Naming alcohols

Write the IUPAC name for each of the following alcohols.

(a) $CH_3(CH_2)_6CH_2OH$ (b) OH (c) OH

Analysis

A general strategy for nomenclature questions is to draw the structure (if it is not already in a suitable form), identify the key functional group (in this case the —OH group), and then number the carbon chain starting from the end closest to the functional group. Finally, identify all other substituents. As the compounds in this example are alcohols, the names will all end in *-ol*. Compound (a) is derived from an unbranched alkane so we need to identify only the alkane and the position of the —OH group. Compound (b) includes a branch on the second last carbon atom, and the —OH group is on the second carbon atom. Compound (c) is a little more complex. It is a cyclic alcohol, so we start numbering from the —OH group. There is also a methyl group on the second carbon atom which is *trans-* to the —OH group. Thus, at a minimum, we need to indicate the relative stereochemistry, the position of the substituent and that it is a cyclic alcohol. To be more precise, we should indicate the absolute stereochemistry of the two stereocentres using the (*R*,*S*) system from section 17.3.

Solution

(a) octan-1-ol (b) 4-methylpentan-2-ol

(c) *trans*-2-methylcyclohexanol or (1*R*,2*R*)-2-methylcyclohexanol

Is our answer reasonable?

The best way to check if your answer is reasonable is to draw a structure from your answer and determine if the structure represents the same molecule as the question.

PRACTICE EXERCISE 19.1

Write the IUPAC name for each of the following alcohols.

(a) OH (b) OH (c) OH

In the IUPAC system, a compound containing two hydroxyl groups is named as a **diol**, one containing three hydroxyl groups is named as a **triol** and so on. In IUPAC names for diols, triols and so on, the final *-e* of the parent alkane name is retained, for example, ethane-1,2-diol.

As with many organic compounds, common names for certain diols and triols have persisted. Compounds containing two hydroxyl groups on adjacent carbons are often referred to as **glycols**. Ethylene glycol (see figure 19.2) and propylene glycol are synthesised from ethylene (ethene) and propylene (propene), respectively; hence, their common names, as shown below.

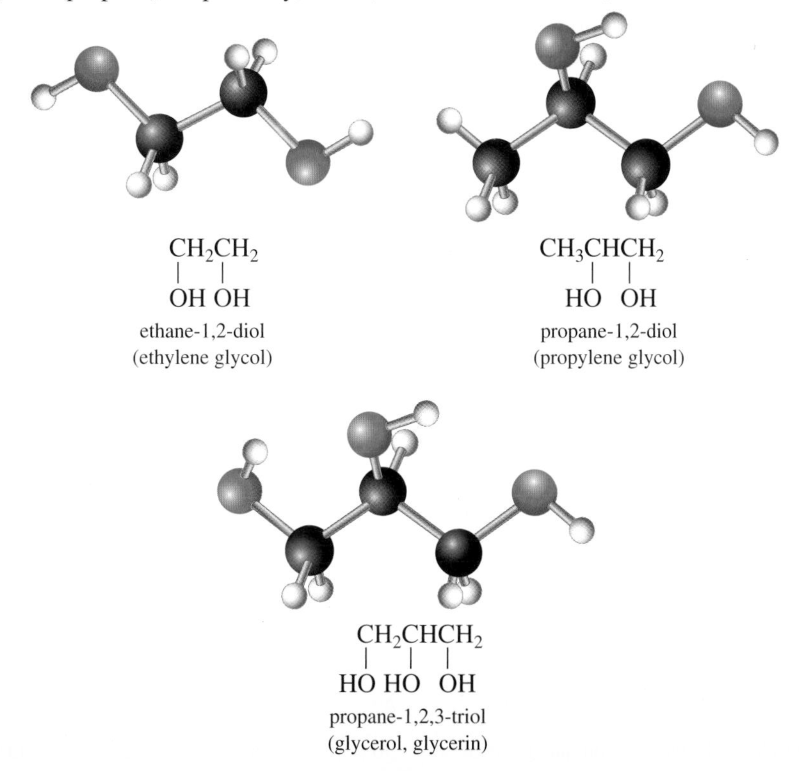

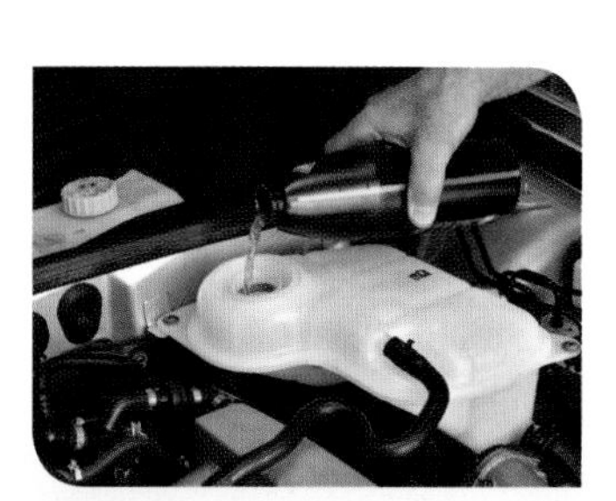

FIGURE 19.2 Ethylene glycol is a polar molecule and dissolves readily in water, a polar solvent. One of its most common uses is in antifreeze. (*Note:* Ethylene glycol itself is colourless.)

We classify alcohols as primary (1°), secondary (2°) or tertiary (3°), depending on whether the —OH group is on a primary, secondary or tertiary carbon atom.

WORKED EXAMPLE 19.2

Primary, secondary and tertiary alcohols

Classify each of the following alcohols as primary, secondary or tertiary.

(a) OH (b) CH_3 | CH_3COH | CH_3 (c) OH

Analysis

To determine whether an alcohol is primary, secondary or tertiary, identify the carbon atom bonded to the —OH group. Now count the number of alkyl groups bonded to this carbon atom. If there is one, the alcohol is primary, if two, secondary, and if three, tertiary. Alternatively, you may wish to consider the number of hydrogen atoms on the carbon atom with the —OH group; two hydrogen atoms = primary, one hydrogen atom = secondary, and no hydrogen atoms = tertiary.

Solution

(a) secondary (2°) (b) tertiary (3°) (c) primary (1°)

Is our answer reasonable?

As there are two 'counting' methods (number of alkyl groups or number of hydrogen atoms), you should double check your answer by using the alternative method.

PRACTICE EXERCISE 19.2 Classify each of the following alcohols as primary, secondary or tertiary.

(a) OH (b) OH (c) $CH_2{=}CHCH_2OH$ (d) OH

Physical properties

The most important physical property of alcohols arises from the polarity of their —OH groups. The large difference in electronegativity (figure 5.5, p. 174) between oxygen and carbon (3.5 − 2.5 = 1.0) and between oxygen and hydrogen (3.5 − 2.1 = 1.4), means both the C—O and O—H bonds are polar covalent, and alcohols are polar molecules, as shown for methanol in figure 19.3.

Table 19.1 lists the boiling points and water solubility for five groups of alcohols and alkanes with the same number of electrons and similar molar mass. Notice that, of the compounds compared in each group, the alcohol has the higher boiling point and is more soluble in water.

(a)

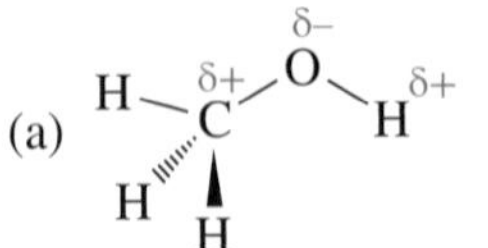

(b)

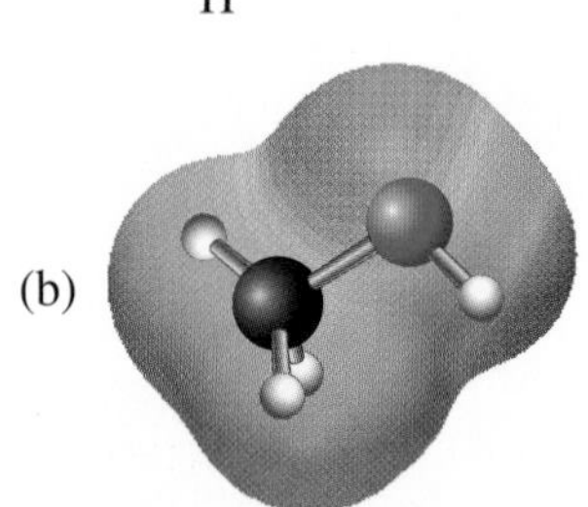

FIGURE 19.3 Polarity of the C—O—H bond in methanol: **(a)** There are partial positive charges on carbon and hydrogen and a partial negative charge on oxygen. **(b)** An electron density map showing the partial negative charge (in red) around oxygen and a partial positive charge (in blue) around hydrogen of the —OH group.

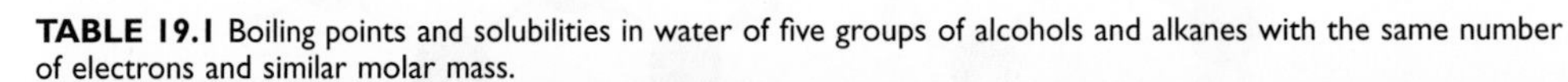

TABLE 19.1 Boiling points and solubilities in water of five groups of alcohols and alkanes with the same number of electrons and similar molar mass.

Structural formula	Name	Molar mass	Total electrons	Boiling point (°C)	Solubility in water
CH_3OH	methanol	32	18	65	infinite
CH_3CH_3	ethane	30	18	−89	insoluble
CH_3CH_2OH	ethanol	46	26	78	infinite
$CH_3CH_2CH_3$	propane	44	26	−42	insoluble
$CH_3CH_2CH_2OH$	propan-1-ol	60	34	97	infinite
$CH_3CH_2CH_2CH_3$	butane	58	34	0	insoluble
$CH_3CH_2CH_2CH_2OH$	butan-1-ol	74	42	117	8 g/100 g
$CH_3CH_2CH_2CH_2CH_3$	pentane	72	42	36	insoluble
$CH_3CH_2CH_2CH_2CH_2OH$	pentan-1-ol	88	50	138	2.3 g/100 g
$HOCH_2CH_2CH_2CH_2OH$	butane-1,4-diol	90	50	230	infinite
$CH_3CH_2CH_2CH_2CH_2CH_3$	hexane	86	50	69	insoluble

The boiling points of all types of compounds, including alcohols, increase with increasing number of electrons because of increased dispersion forces between larger molecules (see chapter 6, pp. 246–7). Compare, for example, the boiling points of ethanol, propan-1-ol, butan-1-ol and pentan-1-ol in table 19.1.

Alcohols have higher boiling points than alkanes with the same or a similar number of electrons, because alcohols are polar and can also associate in the liquid state by **hydrogen bonding** as depicted for ethanol in figure 19.4. We first encountered this type of intermolecular attraction in section 6.8.

FIGURE 19.4 A representation of the association of ethanol molecules in the liquid state. Each —OH group can participate in up to three hydrogen bonds (one through hydrogen and two through oxygen). Only two of these three possible hydrogen bonds per molecule are shown in the figure. (Other intermolecular forces such as dispersion and dipole/induced dipole are not shown here.)

Extra energy is required to separate each alcohol molecule from its neighbours because of hydrogen bonding between alcohol molecules in the liquid state. This results in the relatively high boiling points of alcohols compared with alkanes. Additional hydroxyl groups in a molecule further increase the extent of hydrogen bonding, as can be seen by comparing the boiling points of pentan-1-ol (138 °C) and butane-1,4-diol (230 °C), which have the same number of electrons (and hence the same dispersion forces).

Alcohols are much more soluble in water than alkanes, alkenes and alkynes with a comparable number of electrons. The hydroxyl group in alcohols readily forms hydrogen bonds with water molecules, resulting in the higher solubility of alcohols in water. Methanol, ethanol and propan-1-ol are soluble in water in all proportions. As the number of electrons (and size) increases, the physical properties of alcohols become more like those of hydrocarbons with a similar number of electrons. Alcohols of higher molar mass are much less soluble in water because of the increased size of the hydrocarbon parts of the molecules.

Preparation of alcohols

Alcohols can be prepared from many functional groups using many different reactions. Some of these reactions are discussed briefly here and in more detail elsewhere in this text. As ethanol is arguably the most important alcohol, we will discuss its manufacture separately before going on to the general synthetic methods available for the preparation of alcohols.

FIGURE 19.5 Sugar cane can be used for ethanol production.

In Australia, ethanol is produced by the fermentation of sugars. There are two main sources of these sugars: sugar cane (figure 19.5) and grains. Sugars from cane and molasses (a by-product of the sugar industry) can be converted directly into ethanol. Grains, however, produce starches, which must first be hydrolysed (broken down) to fermentable sugars, before being used to produce ethanol.

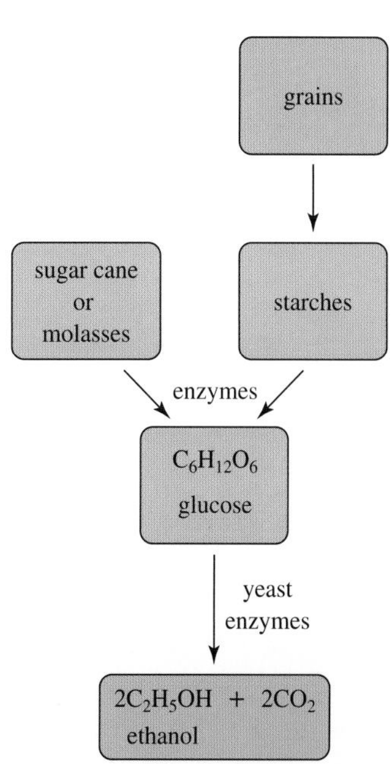

Ethanol is also produced industrially by the vapour phase hydration of ethylene (ethene) in the presence of an acid catalyst.

$$H_2C{=}CH_2 + H_2O \xrightarrow[\text{catalyst}]{\text{acid}} CH_3CH_2OH$$

Preparation from alkenes

Alcohols can be prepared by the acid-catalysed hydration of alkenes (see section 16.5). When the alkene is asymmetrically substituted, the addition obeys Markovnikov's rule (i.e. the hydrogen atom adds to the carbon atom of the double bond with more hydrogen atoms, thus producing the more stable carbocation intermediate).

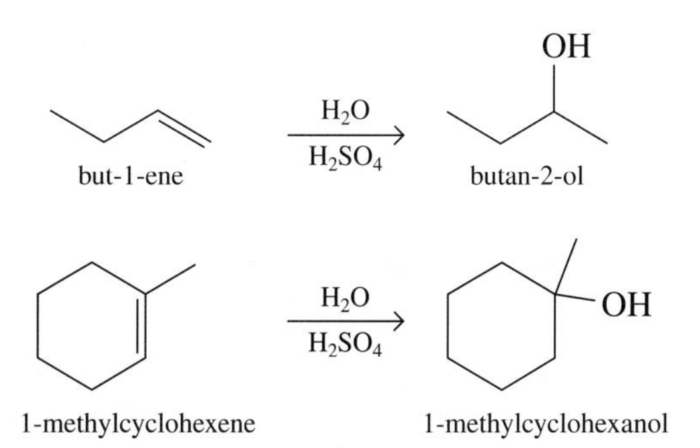

Preparation from haloalkanes

Alcohols can be prepared from haloalkanes by nucleophilic substitution with hydroxide ions or water (see section 18.2). Tertiary haloalkanes are readily converted to alcohols by water.

$$(CH_3)_3C{-}Br + H_2O \rightleftharpoons (CH_3)_3C{-}OH + HBr$$

2-bromo-2-methylpropane (*tert*-butyl bromide) → 2-methylpropan-2-ol (*tert*-butyl alcohol)

Secondary and primary alcohols are better prepared by reaction with hydroxide ions.

1-bromo-2-methylpropane —(dilute NaOH, 25 °C)→ 2-methylpropan-1-ol

Reduction of carbonyl compounds

Alcohols can be prepared by the reduction of carbonyl compounds. Aldehydes are reduced to primary alcohols, and ketones are reduced to secondary alcohols (discussed further in section 21.5). This is most commonly carried out using metal hydride reducing agents such as sodium borohydride, $NaBH_4$, or lithium aluminium hydride, $LiAlH_4$.

butanal —(1. $NaBH_4$ 2. H_3O^+)→ butan-1-ol (85%)

cyclohex-2-enone —(1. $LiAlH_4$ 2. H_3O^+)→ cyclohex-2-enol (94%)

Carboxylic acids and esters can be reduced to form primary alcohols (discussed further in section 23.5). This reduction requires the use of the more powerful reducing agent, $LiAlH_4$.

$$CH_3(CH_2)_7CH{=}CH(CH_2)_7COOH \xrightarrow[2.\ H_3O^+]{1.\ LiAlH_4} CH_3(CH_2)_7CH{=}CH(CH_2)_7CH_2OH$$

oleic acid → octadec-9-en-1-ol (87%)

methylpent-2-enoate (OCH_3) —(1. $LiAlH_4$ 2. H_3O^+)→ pent-2-en-1-ol (91%) + CH_3OH

Addition of Grignard reagents to carbonyl compounds

Alcohols can be prepared by the addition of Grignard reagents to carbonyl compounds. Aldehydes react with Grignard reagents to give secondary alcohols (except formaldehyde (methanal), which gives primary alcohols), while ketones give tertiary alcohols (discussed in greater detail in section 21.5).

3-methylbutanal + phenylmagnesium bromide (MgBr) —(then H_3O^+)→ 3-methyl-1-phenylbutan-1-ol (73%) (a 2° alcohol)

cyclohexanone + CH_3CH_2MgBr (ethylmagnesium bromide) —(then H_3O^+)→ 1-ethylcyclohexanol (89%) (a 3° alcohol)

Esters can also be converted into tertiary alcohols by reaction with Grignard reagents (discussed further in section 21.5). In this case, 2 equivalents of Grignard reagent add to the ester to give the tertiary alcohol.

$$\text{methyl butanoate} + 2CH_3MgI \xrightarrow[\text{2. } H_3O^+]{\text{1. heat}} \text{2-methylpentan-2-ol} + CH_3OH$$

methyl butanoate — methylmagnesium iodide — 2-methylpentan-2-ol

19.2 Reactions of alcohols

In this section, we study the acidity and basicity of alcohols, their dehydration to alkenes, their conversion to haloalkanes, and their oxidation to aldehydes, ketones or carboxylic acids.

Acidity of alcohols

Alcohols have similar acid ionisation constants (pK_a) to water (15.7), which means that aqueous solutions of alcohols have a pH close to that of pure water. The pK_a of ethanol, for example, is 15.9. The reaction below is a simple acid–base reaction in which ethanol is acting as an acid and water is acting as a base.

$$CH_3CH_2O{-}H + H{-}O{-}H \rightleftharpoons CH_3CH_2O^- + H{-}O^+(H){-}H \qquad K_a = \frac{[CH_3CH_2O^-][H_3O^+]}{[CH_3CH_2OH]} = 1.3 \times 10^{-16}$$

$$pK_a = 15.9$$

Table 19.2 gives the acid ionisation constants for several low-molar-mass alcohols. Methanol and ethanol are about as acidic as water. Secondary and tertiary water-soluble alcohols are slightly weaker acids than water (pH of aqueous solutions = 7). (At this point, it would be worthwhile to review the position of equilibrium in acid–base reactions discussed in chapter 11.) Note that, although acetic acid is a 'weak acid' compared with acids such as HCl, its K_a is still 10^{10} times greater than that of alcohols.

TABLE 19.2 pK_a values for selected alcohols in dilute aqueous solution[(a)].

Compound	Structural formula	pK_a	
hydrogen chloride	HCl	−7	stronger acid
acetic acid	CH_3COOH	4.74	↑
methanol	CH_3OH	15.5	
water	H_2O	15.7	
ethanol	CH_3CH_2OH	15.9	
propan-2-ol	$(CH_3)_2CHOH$	17	
2-methylpropan-2-ol	$(CH_3)_3COH$	18	weaker acid

(a) Also given for comparison are pK_a values for water, acetic acid and hydrogen chloride.

Basicity of alcohols

In the presence of a strong acid, the oxygen atom of an alcohol is a weak base and reacts with the acid by proton transfer to form an oxonium ion. The reaction below is a simple acid–base reaction in which ethanol is acting as a base and the hydronium ion is acting as an acid.

$$CH_3CH_2{-}O{-}H + H{-}O^+(H){-}H \rightleftharpoons CH_3CH_2{-}^+O(H){-}H + H{-}O{-}H$$

ethanol — hydronium ion (pK_a = −1.7) — ethyloxonium ion (pK_a = −2.4)

Thus, depending on conditions, alcohols can function as either weak acids or weak bases.

FIGURE 19.6 Methanol reacts with sodium metal to produce hydrogen gas.

Reaction with active metals

Like water, alcohols react with Li, Na, K, Mg and other active metals to liberate hydrogen gas. However, in the case of alcohols, metal alkoxides rather than hydroxides are formed. In the following oxidation–reduction reaction, Na is oxidised to Na^+ and H^+ is reduced to H_2 (figure 19.6).

$$2CH_3OH + 2Na \longrightarrow \underset{\text{sodium methoxide}}{2CH_3O^- Na^+} + H_2$$

To name a metal alkoxide, write the name of the cation followed by the name of the anion. The name of an alkoxide ion is derived from a prefix showing the number of carbon atoms and their arrangement (e.g. *meth-*, *eth-*, *isoprop-*, *tert-but-*) followed by the suffix *-oxide*. Alkoxides can also be named as alkanolates by adding the suffix *-ate* to the alcohol name (e.g. methanolate, ethanolate).

Alkoxide ions are generally stronger bases than the hydroxide ion. In addition to sodium methoxide, the following metal salts of alcohols are commonly used in organic reactions that require a strong base in a nonaqueous solvent: sodium ethoxide in ethanol, and potassium *tert*-butoxide in 2-methylpropan-2-ol (*tert*-butyl alcohol).

$$\underset{\text{sodium ethoxide}}{CH_3CH_2O^- Na^+} \qquad \underset{\text{potassium } tert\text{-butoxide}}{CH_3\overset{\overset{\large CH_3}{|}}{\underset{\underset{\large CH_3}{|}}{C}}O^- K^+}$$

As we saw in chapter 18, alkoxide ions can also be used as nucleophiles in substitution reactions.

Conversion to haloalkanes

The conversion of an alcohol to a haloalkane involves substituting a halogen for the —OH group at a saturated carbon atom. The most common reagents for this conversion are the hydrohalic acids (halogen acids), thionyl chloride and phosphorus halides.

Reaction with hydrohalic acids (HCl, HBr and HI)

Water-soluble tertiary alcohols react very rapidly with HCl, HBr and HI. Mixing a tertiary alcohol with concentrated hydrochloric acid for a few minutes at room temperature converts the alcohol to a water-insoluble chloroalkane that separates from the aqueous layer.

$$\underset{\text{2-methylpropan-2-ol}}{CH_3\overset{\overset{\large CH_3}{|}}{\underset{\underset{\large CH_3}{|}}{C}}OH} + HCl \xrightarrow{25\,°C} \underset{\text{2-chloro-2-methylpropane}}{CH_3\overset{\overset{\large CH_3}{|}}{\underset{\underset{\large CH_3}{|}}{C}}Cl} + H_2O$$

Low-molar-mass, water-soluble primary and secondary alcohols do not react under these conditions.

Water-insoluble tertiary alcohols are converted to tertiary halides by bubbling gaseous HX through a solution of the alcohol dissolved in diethyl ether or tetrahydrofuran (THF).

$$\text{1-methylcyclohexanol} + HCl \xrightarrow[\text{ether}]{0\,°C} \text{1-chloro-1-methylcyclohexane} + H_2O$$

Water-insoluble primary and secondary alcohols react only slowly under these conditions.

Primary and secondary alcohols are converted to bromoalkanes and iodoalkanes by treatment with concentrated hydrobromic and hydroiodic acids. The conditions required to convert primary alcohols to bromoalkanes are very harsh and cannot be used in the presence of most other functional groups. For example, boiling butan-1-ol with concentrated HBr gives 1-bromobutane.

$$\underset{\text{butan-1-ol}}{\text{CH}_3\text{CH}_2\text{CH}_2\text{CH}_2\text{OH}} + HBr \longrightarrow \underset{\substack{\text{1-bromobutane} \\ \text{(butyl bromide)}}}{\text{CH}_3\text{CH}_2\text{CH}_2\text{CH}_2\text{Br}} + H_2O$$

From observations of the relative ease of reaction of alcohols with HX (3° > 2° > 1°), it was proposed that the conversion of tertiary and secondary alcohols to haloalkanes by concentrated HX occurs by an S_N1 mechanism (see section 18.2 for further discussion of nucleophilic substitution mechanisms) and involves the formation of carbocation intermediates. This was later confirmed by appropriate kinetics experiments. The reaction is believed to occur through the three steps shown below.

Step 1: Rapid and reversible proton transfer from the acid to the —OH group gives an oxonium ion (a simple acid–base reaction). This proton transfer converts the poor leaving group, OH^-, into a better leaving group, H_2O.

$$(CH_3)_3C{-}\ddot{O}{-}H + H{-}\overset{+}{O}H_2 \underset{}{\overset{\text{rapid and reversible}}{\rightleftharpoons}} (CH_3)_3C{-}\overset{+}{O}H_2 + H_2\ddot{O}$$

2-methylpropan-2-ol (*tert*-butyl alcohol) — an oxonium ion

Step 2: Loss of water from the oxonium ion gives a 3° carbocation intermediate (or a 2° carbocation intermediate if we started with a 2° alcohol).

$$(CH_3)_2(H_3C)C{-}\overset{+}{O}H_2 \overset{S_N1\text{ slow, rate determining}}{\rightleftharpoons} H_3C{-}\overset{+}{C}(CH_3)_2 + H_2\ddot{O}$$

oxonium ion — 3° carbocation

When we say that a reaction is slow, we are *not* suggesting that individual bonds are broken/formed slowly. The *number* of molecules that undergo the described transformation in a given period of time determines the rate of the reaction (see section 15.1 for more details).

Step 3: Reaction of the 3° carbocation intermediate (an electrophile) with a chloride ion (a nucleophile) gives the haloalkane product.

$$H_3C{-}\overset{+}{C}(CH_3)_2 + :\ddot{Cl}:^- \overset{\text{fast}}{\rightleftharpoons} H_3C{-}C(CH_3)_2{-}\ddot{Cl}:$$

2-chloro-2-methylpropane (*tert*-butyl chloride)

Primary alcohols react with HX by an S_N2 mechanism. In the rate-determining step, the halide ion displaces H_2O from the carbon atom bearing the oxonium ion. The displacement of H_2O and the formation of the C—X bond occur simultaneously.

Step 1: Rapid and reversible proton transfer to the —OH group (a simple acid–base reaction) converts the leaving group from OH^- to H_2O, which is a better leaving group.

$$CH_3CH_2CH_2CH_2{-}\ddot{O}H + H{-}\overset{+}{O}H_2 \overset{\text{rapid and reversible}}{\rightleftharpoons} CH_3CH_2CH_2CH_2{-}\overset{+}{O}H_2 + H_2\ddot{O}$$

an oxonium ion

Step 2: The nucleophilic displacement of H_2O by Br^- then gives the bromoalkane.

$$:\ddot{Br}:^- + CH_3CH_2CH_2CH_2{-}\overset{+}{O}H_2 \xrightarrow{S_N2\text{ slow, rate determining}} CH_3CH_2CH_2CH_2{-}\ddot{Br}: + H_2\ddot{O}$$

(Theoretically this step is reversible, but in practice the reverse reaction is so slow that we generally consider it as nonreversible.)

Why do tertiary alcohols react with HX by formation of carbocation intermediates (S_N1), whereas primary alcohols react by direct displacement of the $—OH_2^+$ group (the protonated —OH group) (S_N2)? The answer is a combination of the same two factors involved in nucleophilic substitution reactions of haloalkanes (section 18.2).

1. *Electronic factors.* Tertiary carbocations are the most stable and most readily formed, whereas primary carbocations are the least stable and hardest to form. Therefore, tertiary alcohols are most likely to react by carbocation formation, secondary alcohols are intermediate and primary alcohols rarely, if ever, react by carbocation formation.
2. *Steric factors.* To form a new carbon–halogen bond by an S_N2 mechanism, the halide ion must approach the carbon atom bearing the leaving group from the side directly opposite that group and begin to form a new covalent bond. If we compare the carbon atom attached to the oxygen

atom of an oxonium ion generated from a primary alcohol with one generated from a tertiary alcohol, we see that the approach is considerably easier in the case of the primary alcohol. For a primary alcohol, the approach from the side opposite the carbon atom bearing the oxonium ion is screened by two hydrogen atoms and one alkyl group, whereas, in tertiary alcohols, this type of approach is screened by three alkyl groups.

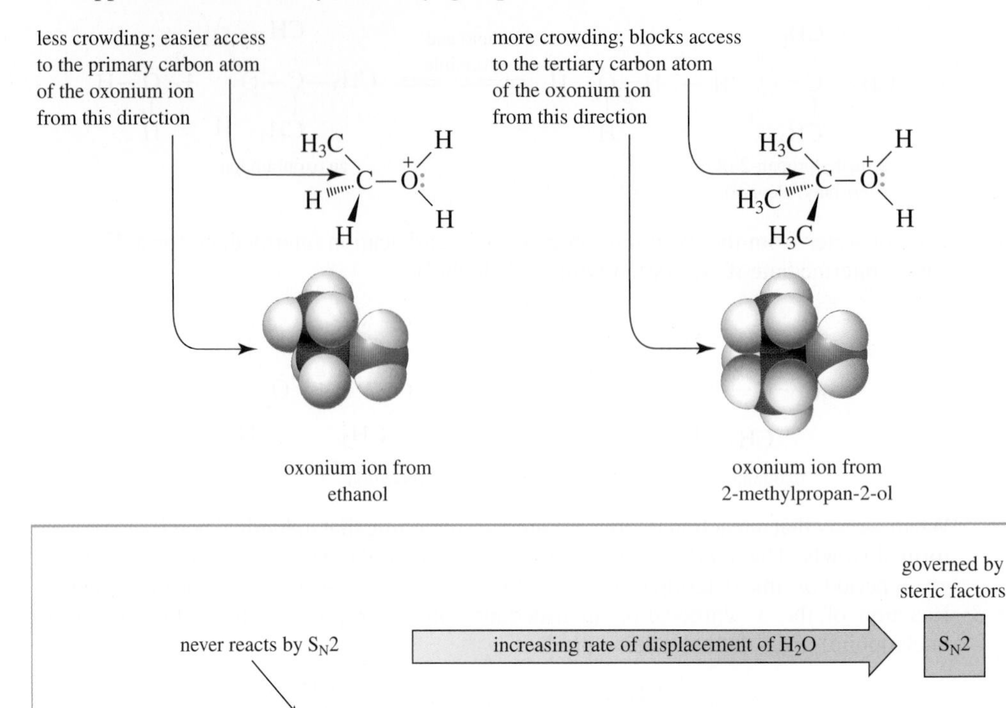

oxonium ion from ethanol

oxonium ion from 2-methylpropan-2-ol

governed by steric factors

never reacts by S_N2 — increasing rate of displacement of H_2O → S_N2

3° alcohol **2° alcohol** **1° alcohol**

S_N1 ← increasing rate of carbocation formation — never reacts by S_N1

governed by electronic factors

Reaction with thionyl chloride, $SOCl_2$

The most widely used reagent for the conversion of primary and secondary alcohols to alkyl chlorides is thionyl chloride, $SOCl_2$. The by-products of this nucleophilic substitution reaction are HCl and SO_2, both given off as gases. Often, an organic base such as pyridine is added to neutralise the HCl by-product.

heptan-1-ol + $SOCl_2$ (thionyl chloride) $\longrightarrow$ 1-chloroheptane + SO_2 + HCl

HCl (acid) + pyridine (base) $\longrightarrow$ pyridinium chloride (salt)

Reaction with phosphorus halides

Alcohols can also be converted into haloalkanes by reaction with phosphorus halides (PX_3 or PX_5, where X = Cl or Br). These are particularly useful for the formation of alkyl bromides. The reaction by-products are water soluble so they are easily separated from the haloalkane as haloalkanes are not normally soluble in water. Note that 3 equivalents of alcohol can react with 1 equivalent of phosphorus tribromide.

3 pentan-1-ol + PBr_3 $\longrightarrow$ 3 1-bromopentane + H_3PO_3

hexan-2-ol + PBr_5 $\longrightarrow$ 2-bromohexane + $POBr_3$ + HBr

Acid-catalysed dehydration to alkenes

An alcohol can be converted to an alkene by **dehydration**, that is, by the elimination of a molecule of water from adjacent carbon atoms. In section 16.5, we discussed the acid-catalysed hydration of alkenes to give alcohols. In this section, we will discuss the acid-catalysed dehydration of alcohols to give alkenes. In fact, hydration–dehydration reactions are reversible. Alkene hydration and alcohol dehydration are competing reactions, and the following equilibrium exists:

$$\underset{\text{an alkene}}{>C{=}C<} + H_2O \underset{}{\overset{\text{acid catalyst}}{\rightleftharpoons}} \underset{\text{an alcohol}}{-\underset{H}{C}-\underset{OH}{C}-}$$

How, then, do we control which product will predominate? Recall that Le Châtelier's principle (section 9.4, pp. 366–7) states that a system in equilibrium will respond to a stress in the equilibrium by counteracting that stress. This response allows us to control these two reactions to give the desired product. Large amounts of water (achieved with the use of dilute aqueous acid) favour alcohol formation, whereas a scarcity of water (achieved using concentrated acid) or experimental conditions that remove water (e.g. heating the reaction mixture above 100 °C) avour alkene formation. Thus, depending on the experimental conditions, it is possible to use the hydration–dehydration equilibrium to selectively prepare either alcohols or alkenes in high yields.

In the laboratory, the dehydration of an alcohol is most often carried out by heating it with either 85% phosphoric acid or concentrated sulfuric acid (typically 98%). Primary alcohols are the most difficult to dehydrate and generally require heating in concentrated sulfuric acid at temperatures as high as 180 °C. Secondary alcohols undergo acid-catalysed dehydration at somewhat lower temperatures. The acid-catalysed dehydration of tertiary alcohols often requires temperatures only slightly above room temperature.

$$CH_3CH_2OH \xrightarrow[180\,°C]{H_2SO_4} CH_2{=}CH_2 + H_2O$$

$$\underset{\text{cyclohexanol}}{C_6H_{11}OH} \xrightarrow[100\,°C]{H_2SO_4} \underset{\text{cyclohexene}}{C_6H_{10}} + H_2O$$

$$\underset{\substack{\text{2-methylpropan-2-ol}\\(\textit{tert}\text{-butyl alcohol})}}{(CH_3)_3C{-}OH} \xrightarrow[50\,°C]{H_2SO_4} \underset{\text{2-methylpropene}}{H_2C{=}C(CH_3)_2} + H_2O$$

Thus, the ease of acid-catalysed dehydration of alcohols occurs in the order:

1° alcohol < 2° alcohol < 3° alcohol

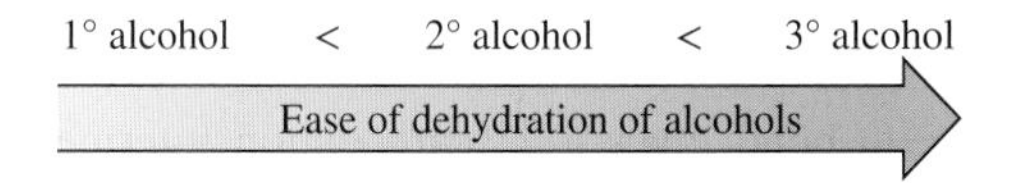

When isomeric alkenes are obtained in the acid-catalysed dehydration of an alcohol, the more stable alkene (the one with the greater number of substituents on the double bond generally predominates; that is, the acid-catalysed dehydration of alcohols follows Zaitsev's rule (section 18.3).

$$\underset{\text{butan-2-ol}}{CH_3CH_2\overset{OH}{C}HCH_3} \xrightarrow[\text{heat}]{85\%\ H_3PO_4} \underset{\substack{\text{but-2-ene}\\(80\%)}}{CH_3CH{=}CHCH_3} + \underset{\substack{\text{but-1-ene}\\(20\%)}}{CH_3CH_2CH{=}CH_2}$$

WORKED EXAMPLE 19.3

Acid-catalysed dehydration

Draw structural formulae for the alkenes that form upon acid-catalysed dehydration of the alcohol on the right, and predict which alkene is the major product.

$$\text{(alcohol, OH)} \xrightarrow[\text{heat}]{H_2SO_4}$$

Analysis

Identify the carbon atom bearing the —OH group, which is C(2) in this example. In a dehydration reaction, a hydrogen ion is removed from one of the carbon atoms neighbouring this carbon atom, C(1) or C(3) in this case. The double bond is formed between the carbon atom with the —OH group and the carbon atom that loses the hydrogen ion, in this example, between C(2) and C(3), or C(2) and C(1). To predict the major product, use Zaitsev's rule.

Solution

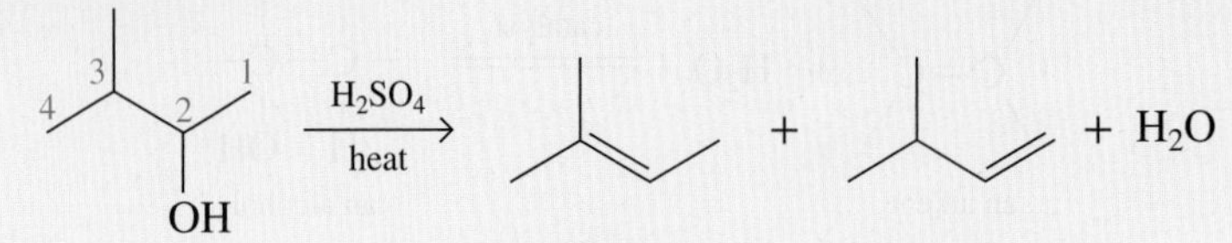

3-methylbutan-2-ol — 2-methylbut-2-ene (major product) — 3-methylbut-1-ene

With three alkyl groups (three methyl groups) on the double bond, 2-methylbut-2-ene is the major product and is formed by elimination between C(2) and C(3). With only one alkyl group (an isopropyl group) on the double bond, 3-methylbut-1-ene is the minor product and is formed by elimination between C(2) and C(1).

Is our answer reasonable?

You should check that, in each of the alkenes, the carbon atom that was bonded to the —OH group is now one of the carbon atoms in the double bond.

PRACTICE EXERCISE 19.3 Draw structural formulae for the alkenes that form upon acid-catalysed dehydration of the following alcohols, and predict which alkene is the major product.

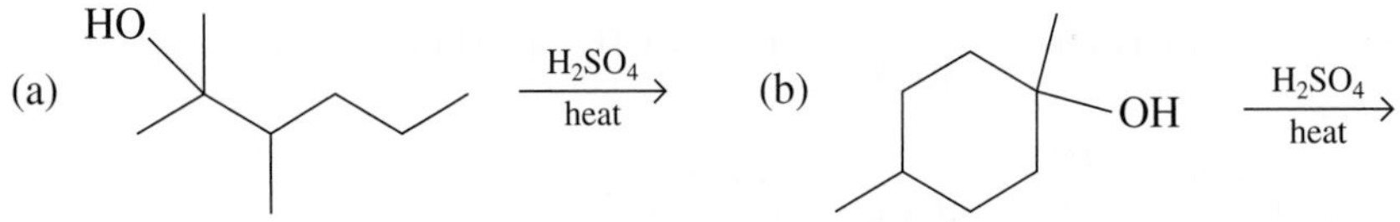

From the relative ease of dehydration of alcohols (3° > 2° > 1°), chemists proposed a three-step mechanism for the acid-catalysed dehydration of secondary and tertiary alcohols. This mechanism involves the formation of a carbocation intermediate in the rate-determining step and is therefore an E1 mechanism.

Step 1: Proton transfer from H_3O^+ to the —OH group of the alcohol gives an oxonium ion (a simple acid–base reaction). This converts OH^-, a poor leaving group, into H_2O, a better leaving group.

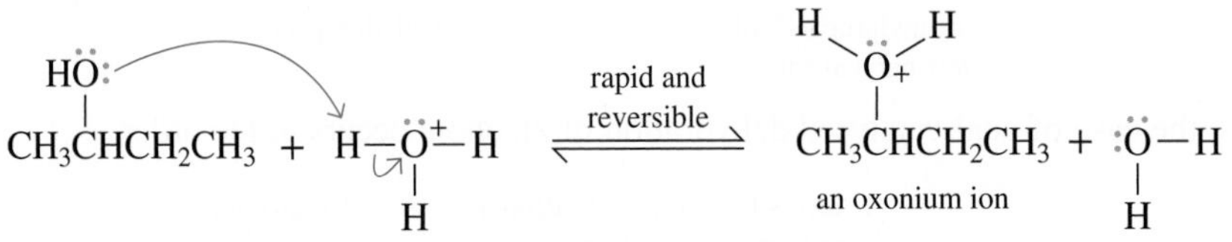

Step 2: Breaking of the C—O bond gives a 2° carbocation intermediate (or a 3° carbocation intermediate if we started with a 3° alcohol) and H_2O.

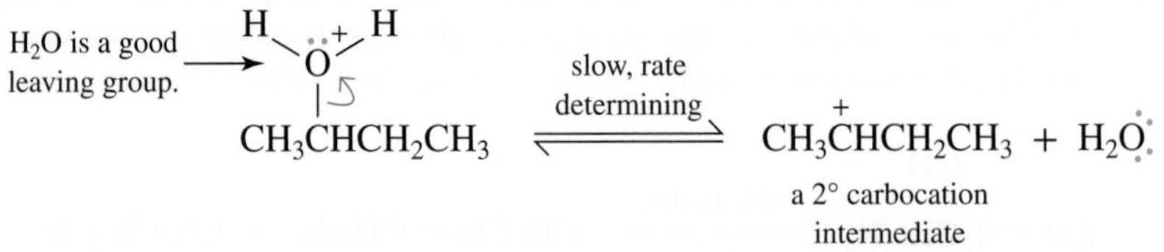

You should note that the two steps above are the same as those for the acid-catalysed conversion of secondary and tertiary alcohols into haloalkanes (p. 826).

Step 3: Proton transfer to H_2O from the carbon atom adjacent to the positively charged carbon atom gives the alkene and regenerates the catalyst. The σ electrons of a C—H bond become the π electrons of the carbon–carbon double bond.

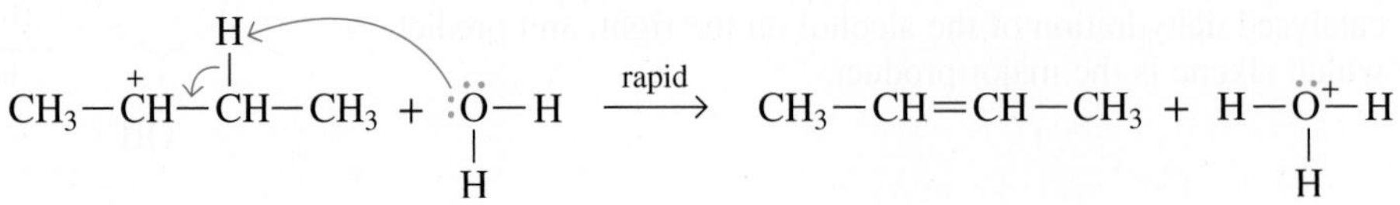

Note that but-2-ene, the major product as predicted by Zaitsev's rule, is obtained as a mixture of *cis* and *trans* isomers. Removal of a proton from the methyl group adjacent to the positively charged carbon atom will produce but-1-ene.

Because the rate-determining step in the acid-catalysed dehydration of secondary and tertiary alcohols is the formation of a carbocation intermediate, the relative ease of dehydration of these alcohols parallels the ease of formation of carbocations.

Primary alcohols do not readily dehydrate, but, when they do, it is believed that they react by the following two-step E2 mechanism, in which step 2 is the rate-determining step.

Step 1: Proton transfer from H_3O^+ to the —OH group of the alcohol gives an oxonium ion (a simple acid–base reaction).

$$CH_3CH_2-\ddot{O}-H + H-\overset{+}{O}(H)-H \xrightleftharpoons[]{\text{rapid and reversible}} CH_3CH_2-\overset{+}{O}(H)-H + :\ddot{O}(H)-H$$

Step 2: Simultaneous proton transfer to the solvent and loss of H_2O gives the alkene.

$$H-\ddot{O}(H): + H-C(H)(H)-CH_2-\overset{+}{O}(H)H \xrightarrow[\text{E2}]{\text{slow, rate determining}} H-\overset{+}{O}(H)-H + H_2C=CH_2 + :\ddot{O}(H)-H$$

Oxidation of primary and secondary alcohols

The oxidation of alcohols is another example of oxidation–reduction (redox) chemistry, as discussed in chapter 12 (see sections 12.1 and 12.2). Oxidation of primary alcohols gives aldehydes or carboxylic acids, depending on the experimental conditions. Secondary alcohols are oxidised to ketones. Tertiary alcohols are not easily oxidised. The following is a series of transformations in which primary alcohols are oxidised first to aldehydes and then to carboxylic acids. The fact that each transformation involves oxidation is indicated by the symbol [O] over the reaction arrow.

$$\underset{\text{primary alcohol}}{CH_3-CH_2-OH} \xrightarrow{[O]} \underset{\text{aldehyde}}{CH_3-\overset{O}{\overset{\|}{C}}-H} \xrightarrow{[O]} \underset{\text{carboxylic acid}}{CH_3-\overset{O}{\overset{\|}{C}}-OH}$$

There are many oxidising agents available for the oxidation of alcohols. A small selection of these reagents includes potassium permanganate, chromium(VI) compounds, sodium hypochlorite (bleach) and nitric acid. However, the reagents most commonly used in the laboratory for the oxidation of primary alcohols to carboxylic acids, and secondary alcohols to ketones, are potassium permanganate, $KMnO_4$, and chromic acid, H_2CrO_4. Chromic acid is a prepared by dissolving either chromium(VI) oxide or potassium dichromate in aqueous sulfuric acid.

$$\underset{\text{chromium(VI) oxide}}{CrO_3} + H_2O \xrightarrow{H_2SO_4} \underset{\text{chromic acid}}{H_2CrO_4}$$

$$\underset{\text{potassium dichromate}}{K_2Cr_2O_7} \xrightarrow{H_2SO_4} H_2Cr_2O_7 \xrightarrow{H_2O} \underset{\text{chromic acid}}{2H_2CrO_4}$$

The oxidation of octan-1-ol by chromic acid in aqueous sulfuric acid gives octanoic acid in high yield. These experimental conditions are more than sufficient to oxidise the intermediate aldehyde to a carboxylic acid.

$$\underset{\text{octan-1-ol}}{CH_3(CH_2)_6CH_2OH} \xrightarrow[H_2SO_4,\ H_2O]{CrO_3} \underset{\text{octanal (not isolated)}}{\left[CH_3(CH_2)_6\overset{O}{\overset{\|}{C}}H\right]} \longrightarrow \underset{\text{octanoic acid}}{CH_3(CH_2)_6\overset{O}{\overset{\|}{C}}OH}$$

Let us consider the two half-equations for the reaction above (for revision, see section 12.2) and the balanced equation. This allows us to determine the number of equivalents of oxidising agent required. For convenience, we will use the molecular formulae for octan-1-ol and octanoic acid.

$$C_8H_{18}O + H_2O \rightarrow C_8H_{16}O_2 + 4H^+ + 4e^- \quad \text{(oxidation)}$$

$$H_2CrO_4 + 6H^+ + 3e^- \rightarrow Cr^{3+} + 4H_2O \quad \text{(reduction)}$$

$$3C_8H_{18}O + 4H_2CrO_4 + 12H^+ \rightarrow 3C_8H_{16}O_2 + 4Cr^{3+} + 13H_2O \quad \text{(overall reaction)}$$

To prepare aldehydes by oxidising primary alcohols, mild conditions are required. These can include using dilute solutions and low temperatures or mild oxidising agents such as, **pyridinium chlorochromate (PCC)**. As shown below, this is prepared by dissolving CrO_3 in aqueous HCl and adding pyridine to precipitate the PCC as an orange solid. (*Note:* The oxidation state of chromium is still +6.) PCC oxidations are carried out in aprotic solvents, most commonly dichloromethane, CH_2Cl_2.

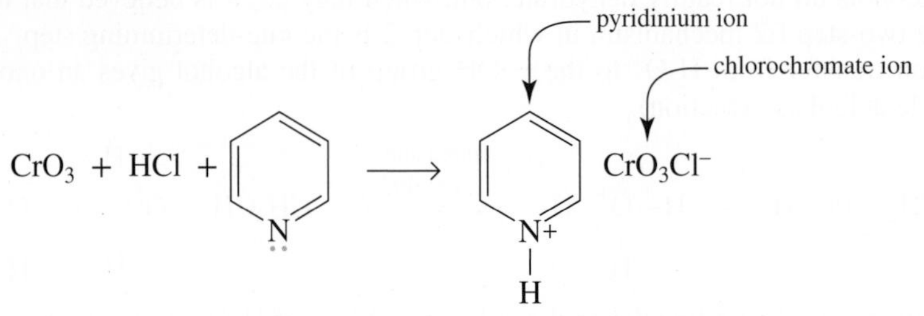

$$CrO_3 + HCl + \text{pyridine} \longrightarrow \text{pyridinium ion} \; CrO_3Cl^-$$

pyridinium ion

chlorochromate ion

pyridine

pyridinium chlorochromate (PCC)

This reagent is selective for the oxidation of primary alcohols to aldehydes and also has little effect on carbon–carbon double bonds or other easily oxidised functional groups. In the following example, geraniol, a molecule produced by honey bees, is oxidised to geranial, a lemon-scented molecule used in perfumes, without affecting either carbon–carbon double bond.

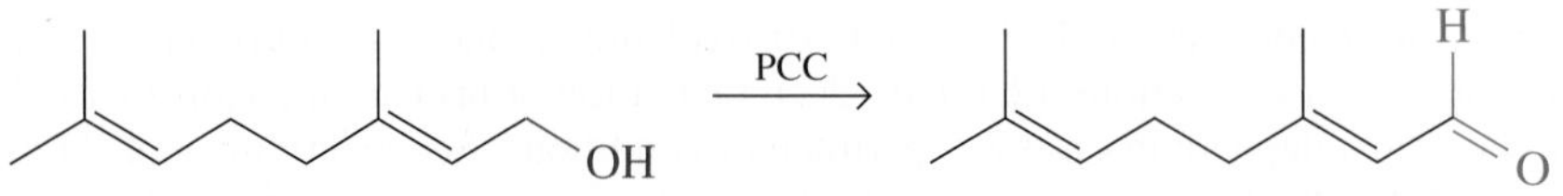

geraniol $\xrightarrow{\text{PCC}}$ geranial

Secondary alcohols are oxidised to ketones by both chromic acid and PCC.

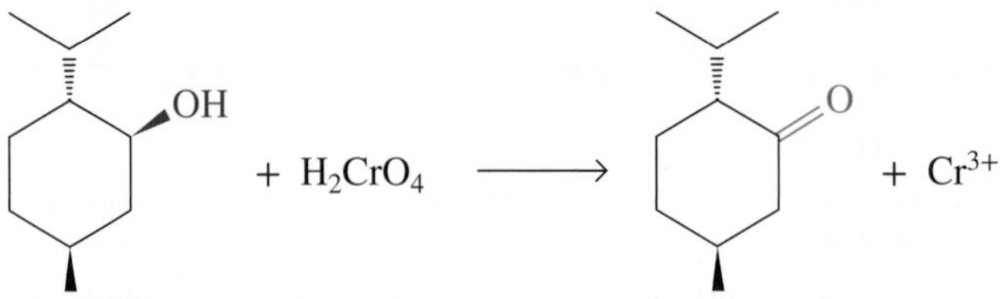

$$\text{2-isopropyl-5-methylcyclohexanol (menthol)} + H_2CrO_4 \longrightarrow \text{2-isopropyl-5-methylcyclohexanone (menthone)} + Cr^{3+}$$

Tertiary alcohols are resistant to oxidation because the carbon atom bearing the —OH group is bonded to three carbon atoms and therefore cannot form a carbon–oxygen double bond.

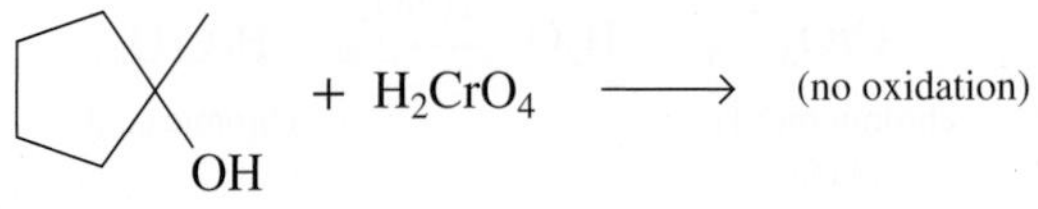

$$\text{1-methylcyclopentanol} + H_2CrO_4 \longrightarrow \text{(no oxidation)}$$

Note that the essential feature of the oxidation of an alcohol is the presence of at least one hydrogen atom on the carbon atom bearing the —OH group. Tertiary alcohols lack such a hydrogen, so they are not oxidised.

WORKED EXAMPLE 19.4

Pyridinium chlorochromate (PCC) oxidations

Draw the product of the treatment of each of the following alcohols with PCC.

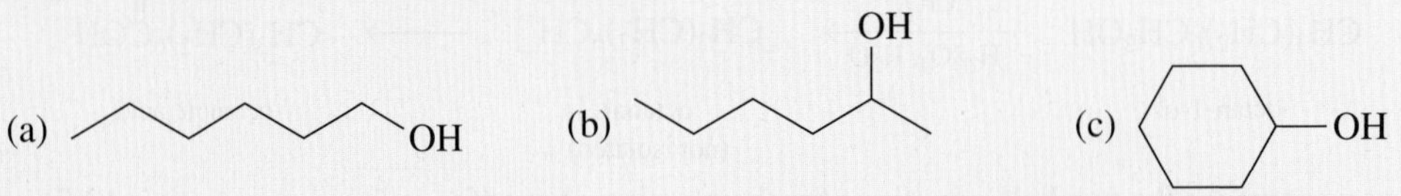

(a) hexan-1-ol (b) hexan-2-ol (c) cyclohexanol

Analysis

As the oxidising agent is PCC, primary alcohols will be oxidised to aldehydes and secondary alcohols to ketones. Identify the carbon atom bearing the —OH group; the carbonyl group is formed at this position.

Solution

The primary alcohol hexan-1-ol is oxidised to hexanal. The secondary alcohol hexan-2-ol is oxidised to hexan-2-one. Cyclohexanol, a secondary alcohol, is oxidised to cyclohexanone.

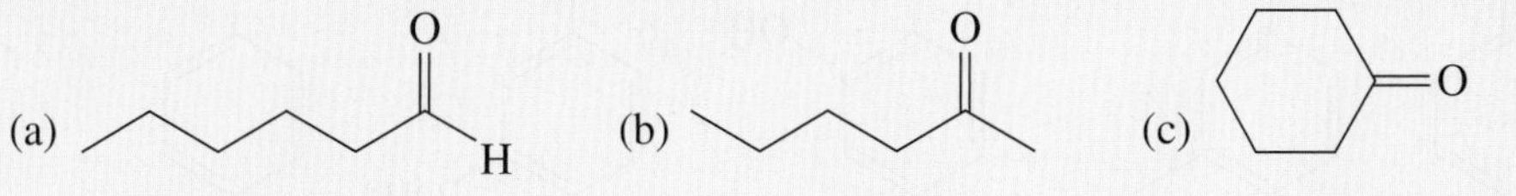

(a) hexanal (b) hexan-2-one (c) cyclohexanone

Is our answer reasonable?

Check that an aldehyde has been formed from a primary alcohol, and a ketone has been formed from a secondary alcohol. Check also that the position of the carbonyl group is correct (particularly important for ketones).

PRACTICE EXERCISE 19.4

What is the product obtained on treating each of the alcohols in worked example 19.4 with excess potassium permanganate?

Ester formation

An important reaction of alcohols is the formation of esters by condensation with carboxylic acids, acid chlorides or anhydrides. The chemistry will be introduced briefly here and discussed in greater detail in chapter 23.

$$CH_3CH_2CH_2\overset{O}{\overset{\|}{C}}OH + HOCH_2CH_3 \underset{\text{heat}}{\overset{\text{acid catalyst}}{\rightleftharpoons}} CH_3CH_2CH_2\overset{O}{\overset{\|}{C}}OCH_2CH_3 + H_2O$$

butanoic acid (bp = 163 °C); ethanol (bp = 78 °C); ethyl butanoate (bp = 120 °C) (fragrance of pineapple)

Alcohols can also form esters with inorganic acids such as nitric acid, phosphoric acid, sulfuric and sulfonic acids (RSO_3H). Nitroglycerin is an example of an ester formed between glycerol and three mole equivalents of nitric acid. DNA and RNA (see chapter 25) are polymers in which the backbone consists of alternating units of phosphoric acid and a monosaccharide bonded together through ester linkages between the alcohol groups of the monosaccharide and the phosphoric acid molecules. Figure 19.7 shows some examples of esters formed between alcohols and inorganic acids.

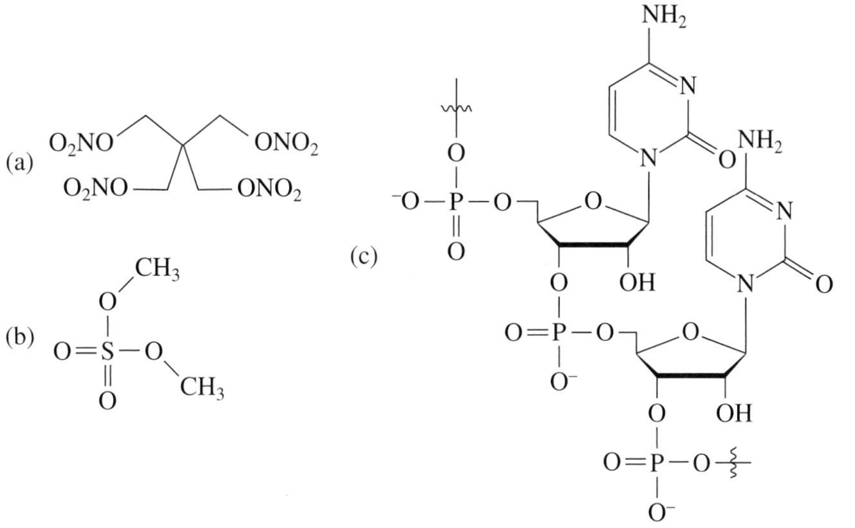

(a) O_2NO, ONO_2, O_2NO, ONO_2

(b) CH_3, O, O=S—O, O, CH_3

(c) NH_2, N, N, O, NH_2, N, N, O, ^-O—P—O, O, O, OH, O=P—O, O^-, O, OH, O=P—O, O^-

FIGURE 19.7 Examples of esters of inorganic acids: **(a)** pentaerythritol tetranitrate (an explosive), **(b)** dimethyl sulfate (an industrial methylating agent) and **(c)** a fragment of RNA.

19.3 Phenols

Phenols are compounds with one or more hydroxyl groups bonded directly to an aromatic ring. In the previous section we discussed the chemistry of alcohols (ROH) and here we will introduce the chemistry of phenols (ArOH). Although these functional groups are similar (they both form esters and ethers), they have some important differences, particularly their acidity.

We name substituted phenols either as derivatives of phenol or by common names as shown below.

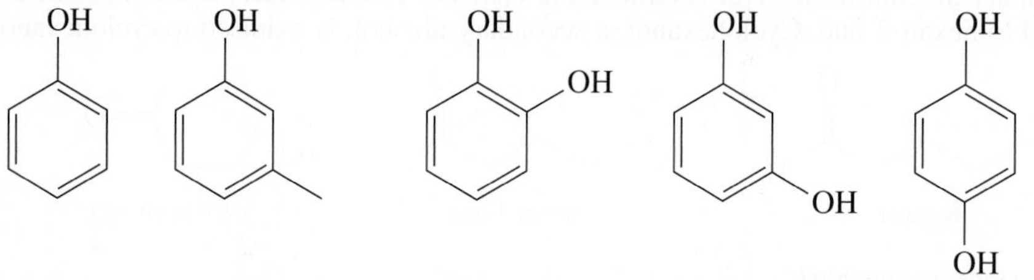

OH OH OH OH OH

phenol 3-methylphenol (m-cresol) benzene-1,2-diol (catechol) benzene-1,3-diol (resorcinol) benzene-1,4-diol (hydroquinone)

Note that benzyl alcohol (phenylmethanol) is *not* a phenol because the hydroxyl group is not connected directly to the aromatic ring. It is the direct connection of the hydroxyl group to an aromatic ring that gives rise to the different chemical properties of phenols; these will be discussed below.

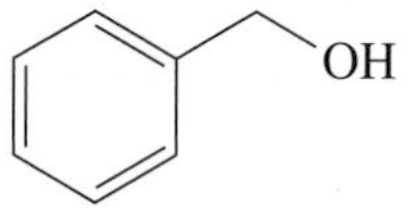

OH

benzyl alcohol

Phenols are widely distributed in nature. Phenol itself and the isomeric cresols (o-, m- and p-cresol) are found in coal tar (the liquid distilled from coal in the production of coke). The phenol derivatives thymol and vanillin are important constituents of thyme and vanilla beans, respectively.

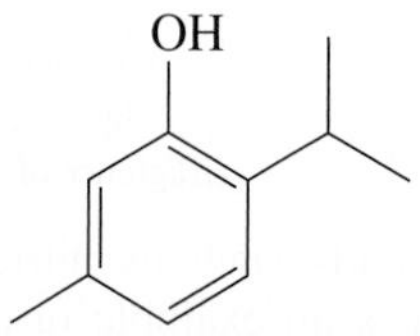

OH CHO OCH_3 OH

2-isopropyl-5-methylphenol (thymol) 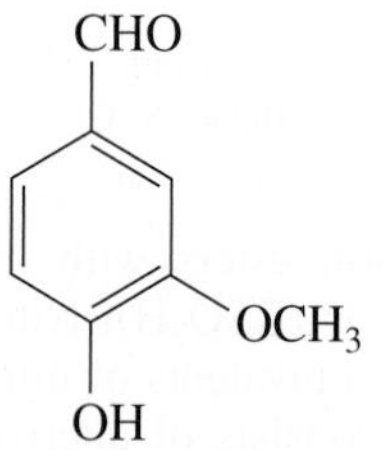 4-hydroxy-3-methoxy-benzaldehyde (vanillin)

Phenol, or carbolic acid, as it was once called, is a solid with a low melting point and is only slightly soluble in water. In sufficiently high concentrations, it is corrosive to all kinds of cells, because it has a high permeability across the cell membrane and disrupts biochemical processes within the cell. In dilute solutions, phenol has some antiseptic properties and was introduced into the practice of surgery by Joseph Lister, who demonstrated his technique of aseptic surgery at the University of Glasgow School of Medicine in 1865. Nowadays, phenol has been replaced by antiseptics that are more powerful and have fewer undesirable side-effects. Among these is hexylresorcinol, which is widely used in nonprescription preparations as a mild antiseptic and disinfectant. Eugenol, which can be isolated from the flower buds of *Eugenia aromatica* (clove), is used as a dental antiseptic and analgesic.

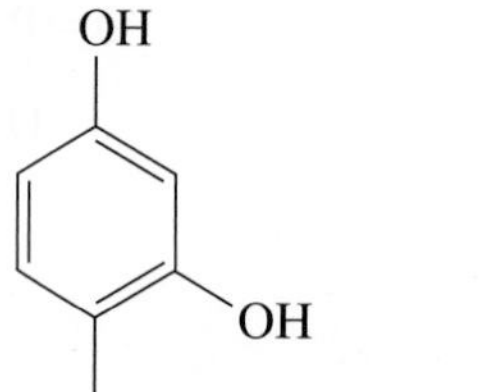

OH OH OH OCH_3

hexylresorcinol 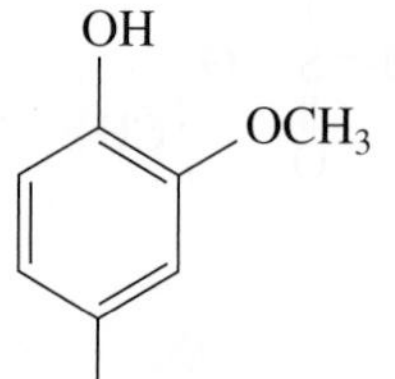 eugenol

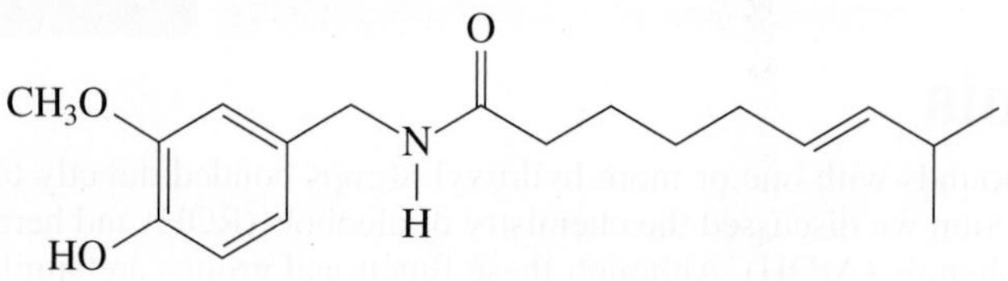

O CH_3O N H HO

capsaicin

Chemical Connections

Some like it hot!

Capsaicin is the active ingredient in chillies that makes food containing chilli exciting to eat. It is classified as an irritant as it causes a burning sensation when it comes into contact with mucus membranes and sensitive skin. It is this burning sensation that has made the use of chillies in food very popular in some cuisines. There are many chilli varieties with different levels of 'hottness'. The hotness is measured by the Scoville scale. Some of the hottest chillies have a Scoville score over 1 million. Pure capsaicin scores around 15 million. Capsaicin is also the active ingredient in oleoresin capsicum spray (commonly known as pepper spray, figure 19.8).

O

CH_3O N

H

HO

capsaicin

FIGURE 19.8 Pepper spray contains the phenol derivative capsaicin.

Capsaicin, belongs to a family of compounds known as the vanilloids as it contains the vanillyl group. The most important member of the vanilloid family is vanillin (see p. 836), which is the principal component of vanilla bean extract. Vanillin is widely used in flavourings (figure 19.9) and is also a principal base note in perfumes.

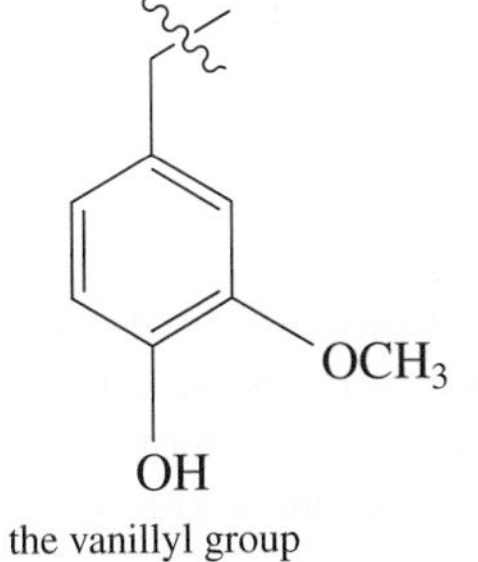

the vanillyl group

FIGURE 19.9 Capsaicin is a member of the vanilloid family of compounds, all of which contain the vanillyl group. Vanillin is an important consistuent of vanilla beans, which are used to flavour a variety of foods, including vanilla ice-cream.

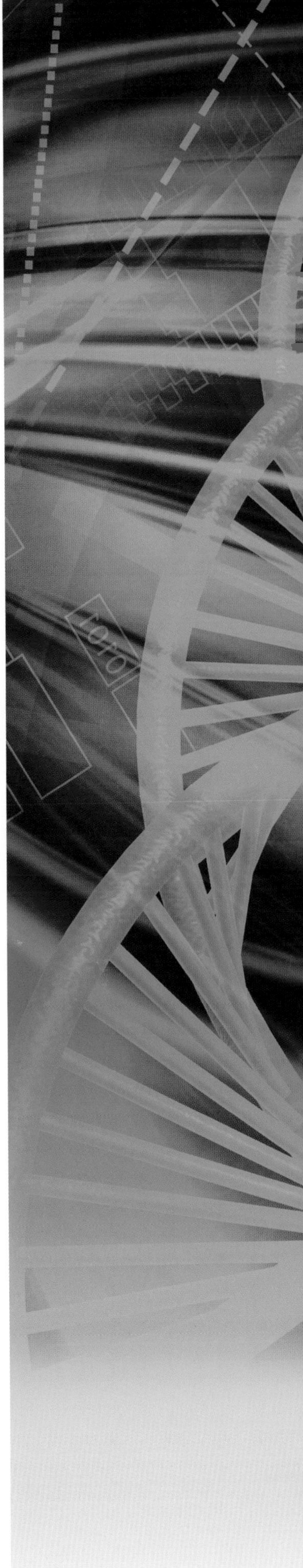

Acidity of phenols

Phenols and alcohols both contain an —OH group. We group phenols as a separate class of compounds, however, because their chemical properties are quite different from those of alcohols. One of the most important differences is that phenols are significantly more acidic than alcohols. Indeed, the acidity constant for phenol is 10^6 times larger than that of ethanol.

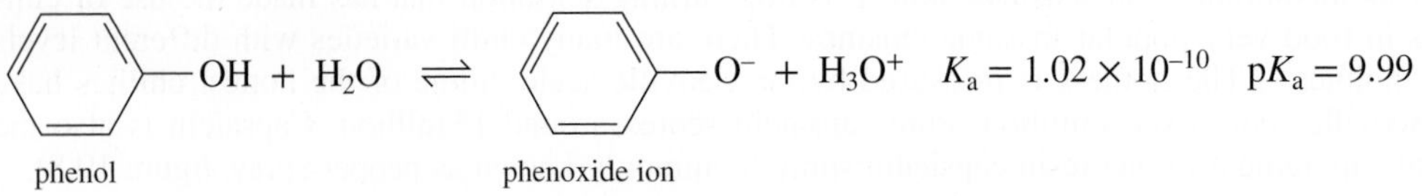

$$C_6H_5OH + H_2O \rightleftharpoons C_6H_5O^- + H_3O^+ \quad K_a = 1.02 \times 10^{-10} \quad pK_a = 9.99$$

phenol — phenoxide ion

$$CH_3CH_2OH + H_2O \rightleftharpoons CH_3CH_2O^- + H_3O^+ \quad K_a = 1.3 \times 10^{-16} \quad pK_a = 15.9$$

ethanol — ethoxide ion

Another way to compare the relative acid strengths of ethanol and phenol is to look at the hydronium ion concentration and pH of a 0.1 M aqueous solution of each (table 19.3). For comparison, the $[H_3O^+]$ and pH of 0.1 M HCl are also included.

TABLE 19.3 Relative acidities of 0.1 M solutions of ethanol, phenol and HCl.

Acid ionisation equation	$[H_3O^+]$	pH
$CH_3CH_2OH + H_2O \rightleftharpoons CH_3CH_2O^- + H_3O^+$	1.0×10^{-7}	7.0
$C_6H_5OH + H_2O \rightleftharpoons C_6H_5O^- + H_3O^+$	3.3×10^{-6}	5.4
$HCl + H_2O \rightleftharpoons Cl^- + H_3O^+$	0.1	1.0

In aqueous solution, alcohols are neutral substances, and the hydronium ion concentration of 0.1 M ethanol is the same as that of pure water. A 0.1 M solution of phenol is slightly acidic and has a pH of 5.4. By contrast, 0.1 M HCl, a strong acid (completely ionised in aqueous solution), has a pH of 1.0.

The greater acidity of phenols compared with alcohols results from the greater stability of the phenoxide ion compared with an alkoxide ion. The negative charge on the phenoxide ion is delocalised by resonance. The two contributing structures below left for the phenoxide ion place the negative charge on oxygen, while the three on the right place the negative charge on the *ortho* and *para* positions of the ring. Thus, in the resonance hybrid, the negative charge of the phenoxide ion is delocalised over four atoms, which stabilises the phenoxide ion relative to an alkoxide ion, for which no delocalisation is possible.

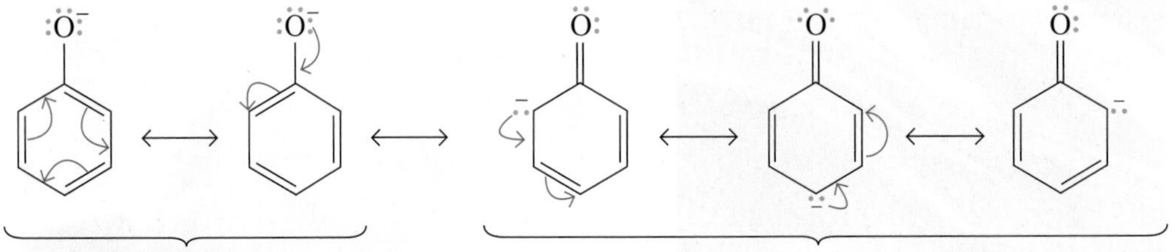

These two Kekulé structures are equivalent.

These three contributing structures delocalise the negative charge onto carbon atoms of the ring.

Note that, although the resonance model helps us understand why phenol is a stronger acid than ethanol, it does not provide any quantitative means of predicting just how much stronger an acid it might be. To find out how much stronger one acid is than another, we must determine their pK_a values experimentally and compare them.

Ring substituents have marked effects on the acidities of phenols through a combination of inductive and resonance effects. Electron-withdrawing groups such as chloro, cyano and nitro withdraw electron density from the aromatic ring, weaken the O—H bond and stabilise the phenoxide ion. The diagram at the top of the next page shows the resonance forms for the 4-nitrophenoxide ion. You will notice that there is an extra resonance form (only one of the Kekulé forms has been shown) for this ion where the charge has been delocalised on an oxygen atom of the nitro group. This delocalisation on the oxygen atom increases the stability of the phenoxide ion and makes 4-nitrophenol a stronger acid than phenol.

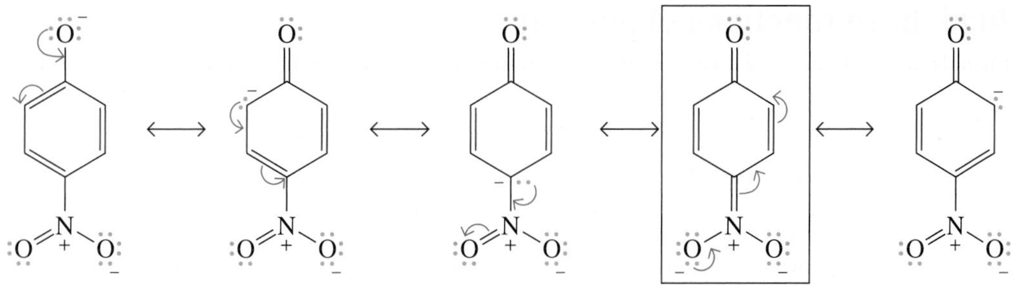

charge delocalised on the nitro group

Conversely, electron-donating groups like amino, alkoxy and alkyl groups increase the electron density of the aromatic ring and destabilise the phenoxide ion. Table 19.4 lists the pK_a values for some phenols.

TABLE 19.4 The pK_a values for some phenols. The pK_{a1} value for acetic acid is given for comparison.

Name	pK_a
2,4,6-trinitrophenol	0.42
acetic acid	4.76
4-nitrophenol	7.15
4-chlorophenol	9.41
phenol	9.99
4-methoxyphenol	10.21
4-methylphenol	10.26

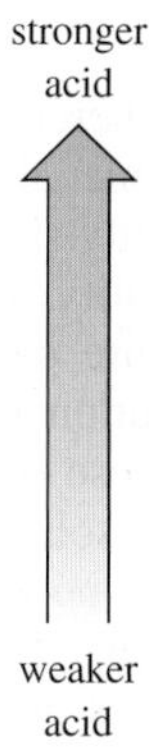

stronger acid

weaker acid

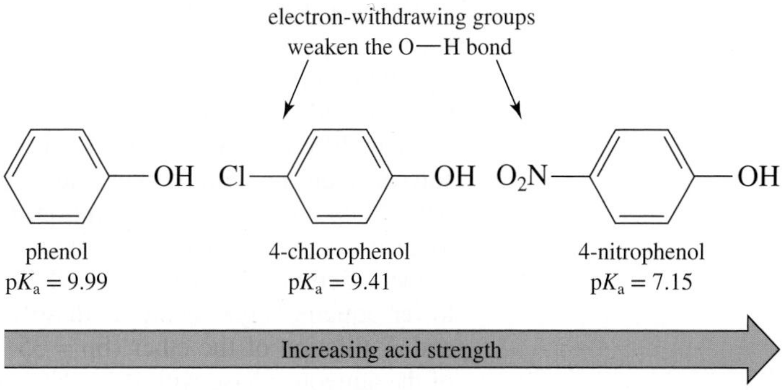

electron-withdrawing groups weaken the O—H bond

phenol pK_a = 9.99

4-chlorophenol pK_a = 9.41

4-nitrophenol pK_a = 7.15

Increasing acid strength

WORKED EXAMPLE 19.5

Acidity

Arrange these compounds in order of increasing acidity: 2,4-dinitrophenol, phenol and benzyl alcohol.

Analysis

As alcohols are significantly less acidic than phenols, these should be listed first. Of the phenols, determine whether the substituents are electron withdrawing or electron donating. Electron-withdrawing groups increase the acidity of the phenol, while electron-donating groups decrease the acidity.

Solution

Benzyl alcohol, a primary alcohol, has a pK_a of approximately 16. The pK_a of phenol is 9.99. Nitro groups are electron withdrawing and increase the acidity of the phenolic —OH group. In order of increasing acidity, these compounds are:

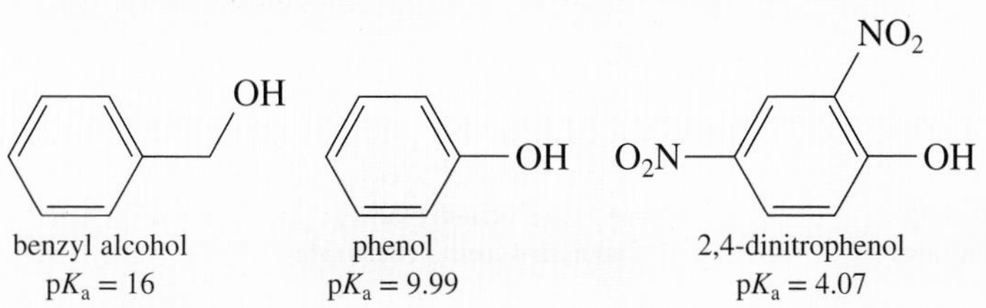

benzyl alcohol pK_a = 16

phenol pK_a = 9.99

2,4-dinitrophenol pK_a = 4.07

Is our answer reasonable?

Check that your answer has the compounds in the order of:

alcohols < phenols with electron-donating groups < phenol < phenols with electron-withdrawing groups.

PRACTICE EXERCISE 19.5

Arrange these compounds in order of increasing acidity: 2,4-dichlorophenol, 4-chlorophenol and 2,4-dimethoxyphenol.

Acid–base reactions of phenols

Phenols are weak acids and react with strong bases, such as NaOH, to form water-soluble salts.

$$\text{C}_6\text{H}_5\text{—OH} + \text{NaOH} \longrightarrow \text{C}_6\text{H}_5\text{—O}^-\text{Na}^+ + \text{H}_2\text{O}$$

phenol	sodium hydroxide	sodium phenoxide	water
$pK_a = 9.99$			$pK_a = 15.7$
(stronger acid)	(stronger base)	(weaker base)	(weaker acid)

Most phenols do not react with weaker bases, such as sodium bicarbonate, and do not dissolve in aqueous sodium bicarbonate. Carbonic acid is a stronger acid than most phenols and, consequently, the equilibrium for the reaction of phenols with bicarbonate ions lies far to the left (see chapter 11).

$$\text{C}_6\text{H}_5\text{—OH} + \text{NaHCO}_3 \rightleftharpoons \text{C}_6\text{H}_5\text{—O}^-\text{Na}^+ + \text{H}_2\text{CO}_3$$

phenol	sodium bicarbonate	sodium phenoxide	carbonic acid
$pK_a = 9.99$			$pK_a = 6.36$
(weaker acid)	(weaker base)	(stronger base)	(stronger acid)

The fact that phenols are weakly acidic, whereas alcohols are neutral, provides a convenient way to separate phenols from water-insoluble alcohols. Suppose that we want to separate 4-methylphenol from cyclohexanol. Each is only slightly soluble in water; therefore, they cannot be separated on the basis of their water solubility. They can be separated, however, on the basis of their difference in acidity. First, the mixture of the two is dissolved in a water-immiscible solvent such as diethyl ether. Next, this solution is placed in a separating funnel and shaken with dilute aqueous NaOH solution. Under these conditions, 4-methylphenol reacts with NaOH to give sodium 4-methylphenoxide, a water-soluble salt. The upper layer in the separating funnel, diethyl ether (density = $0.74\ \text{g mL}^{-1}$), now contains only dissolved cyclohexanol. The lower aqueous layer contains dissolved sodium 4-methylphenoxide. The layers are separated, and distillation of the ether (bp = 35 °C) leaves pure cyclohexanol (bp = 161 °C). Acidification of the aqueous phase with 0.1 M HCl or another strong acid converts sodium 4-methylphenoxide to 4-methylphenol, which is sparingly soluble in water and can be extracted with ether and recovered in pure form. The flowchart below summarises these experimental steps.

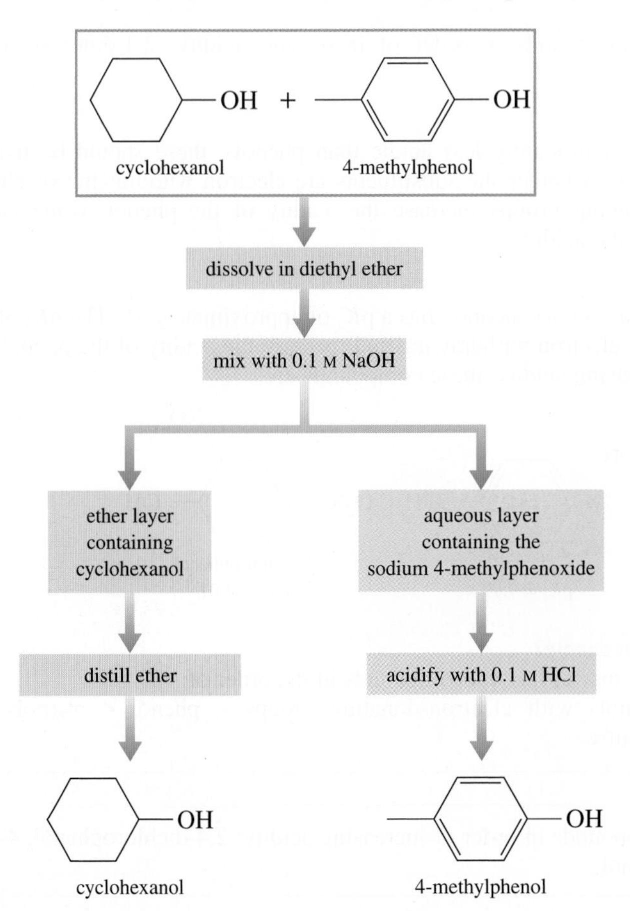

Oxidation of phenols

Phenols are readily oxidised to give quinones. Quinones are intensely coloured compounds ranging from red through to violet. The pink tinge found in old bottles of phenol is due to trace amounts of quinones. The pH indicator alizarin is a quinone. It has two acidic hydrogens, so it can be used for two different pH regions. It is yellow at pH < 5.5, red at pH 6.8–11 and purple at pH > 12.

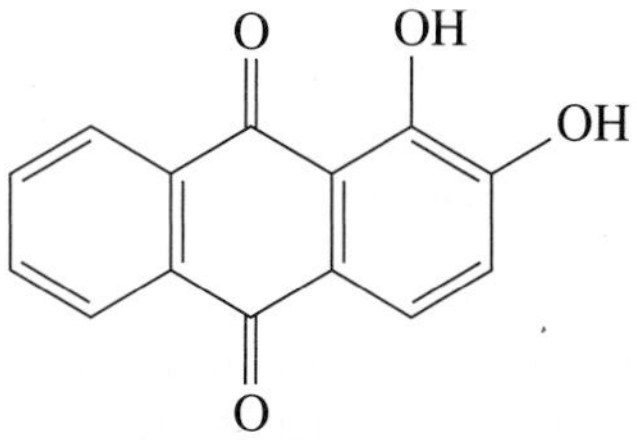

alizarin
acidic form
(yellow)

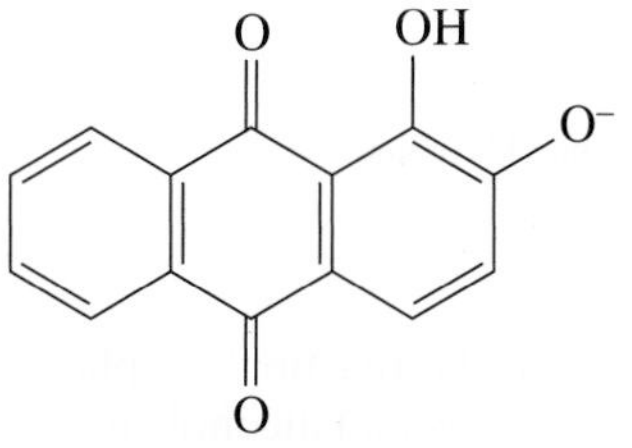

alizarin
mono-basic form
(red)

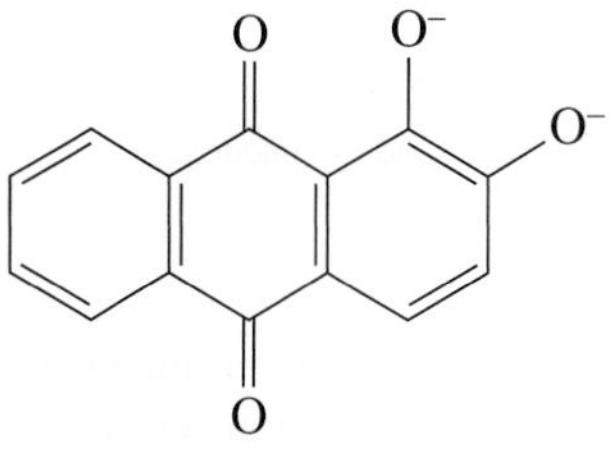

alizarin
di-basic form
(purple)

The oxidation of phenols is quite different from the oxidation of alcohols (see section 19.2) because they do not have a hydrogen atom on the carbon atom bearing the hydroxyl group. The parent compound, phenol, is easily oxidised and gives rise to cyclohexa-2,5-dien-1,4-dione (*p*-benzoquinone).

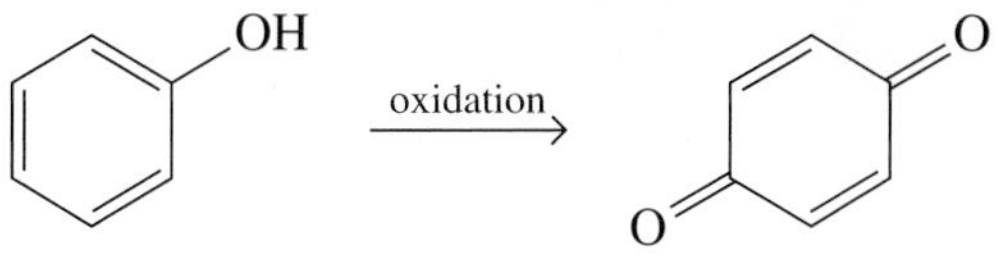

phenol —oxidation→ *p*-benzoquinone (a quinone with a brilliant yellow colour)

The oxidation of 1,2- and 1,4-dihydroxybenzene also occurs readily and gives the corresponding *o*- and *p*-benzoquinone. The benzoquinone can also be easily reduced to the dihydroxybenzene by treatment with a mild reducing agent.

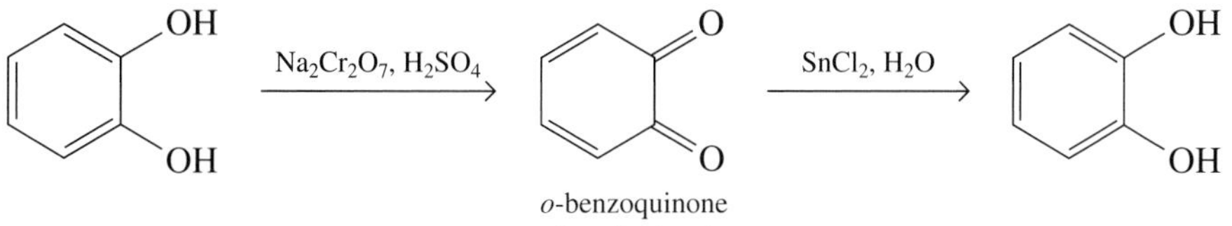

$\xrightarrow{Na_2Cr_2O_7,\ H_2SO_4}$ *o*-benzoquinone (a red colour) $\xrightarrow{SnCl_2,\ H_2O}$

Ester and ether formation

Phenols can form esters by reaction with acid chlorides or acid anhydrides. However, unlike the alcohols (see section 19.2), esters cannot be prepared by treatment of phenols with carboxylic acids as phenols are generally too unreactive. This lack of reactivity is due in part to the delocalisation of the electrons on the oxygen atom of the phenol, making it less nucleophilic.

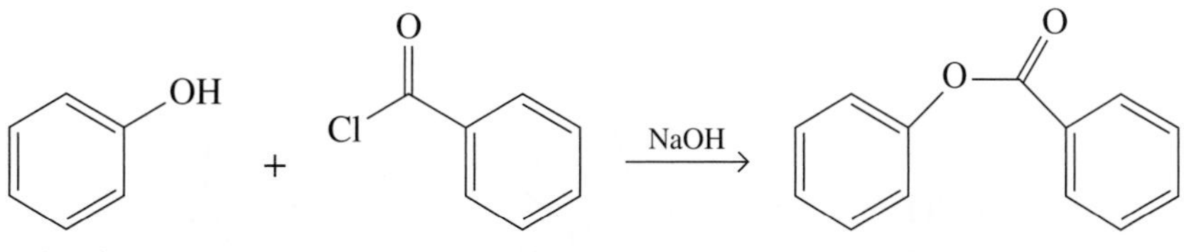

phenol + benzoyl chloride —NaOH→ phenyl benzoate (76%)

Since phenols are quite acidic, they are readily deprotonated to give phenoxide ions. These ions are good nucleophiles and will displace halogens from haloalkanes (see section 18.2) to produce ethers.

sodium phenoxide + butyl bromide ⟶ butyl phenyl ether (80%)

You may have noticed that many of the reactions of phenols are similar to those of alcohols. Careful inspection will reveal that phenols and alcohols react in the same way when the reaction involves breaking the O—H bond (reactions with bases, oxidation, and ether and ester formation). Phenols do not readily undergo reactions that require breaking the C—O bond (reaction with HX to form aryl halides or dehydration).

19.4 Ethers

In the previous sections we studied compounds related to water in which a hydrogen atom was replaced by an alkyl group (alcohols) or an aryl group (phenols). In this section we will investigate 'relatives' of water where both hydrogen atoms have been replaced by either alkyl or aryl groups — ethers. The functional group of an **ether** is an atom of oxygen bonded to two carbon atoms that are not bonded to **heteroatoms** (e.g. O, N, S, halogen). Figure 19.10 shows a Lewis structure and ball-and-stick model of dimethyl ether, CH_3OCH_3, the simplest ether.

(a) (b)

FIGURE 19.10 Dimethyl ether, CH_3OCH_3: **(a)** Lewis structure and **(b)** ball-and-stick model.

Ethers are named in the IUPAC system by selecting the longest carbon chain as the parent alkane and naming the —OR group bonded to it as an **alkoxy** (*alkyl + oxygen*) group. Common names are derived by listing the alkyl groups bonded to oxygen in alphabetical order and adding the word *ether*.

$CH_3CH_2OCH_2CH_3$ — ethoxyethane (diethyl ether)

$CH_3OC(CH_3)_2CH_3$ — 2-methoxy-2-methylpropane (methyl *tert*-butyl ether, MTBE)

trans-2-ethoxycyclohexanol

Chemists almost invariably use common names for low-molar-mass ethers. For example, although ethoxyethane is the IUPAC name for $CH_3CH_2OCH_2CH_3$, it is rarely called that, but instead is called diethyl ether, ethyl ether, or, even more commonly, simply ether.

Cyclic ethers are heterocyclic compounds in which the ether oxygen atom is one of the atoms in a ring. These ethers are generally known by their common names.

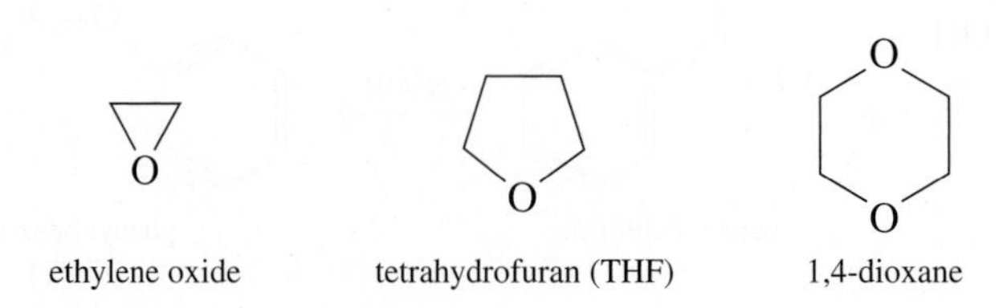

WORKED EXAMPLE 19.6

Naming ethers

Write the IUPAC and common names for each of the following ethers.

(a) $CH_3C(CH_3)_2OCH_2CH_3$ (b) cyclohexyl–O–cyclohexyl

Analysis

To name ethers, you should first identify the ether oxygen atom and then identify the two alkyl groups attached to this oxygen atom. In the IUPAC system, the larger of the two groups (if they are different) is the parent alkane and the other is the alkoxy group. Using the common nomenclature, the alkyl groups are listed alphabetically followed by the word 'ether'.

Solution

(a) 2-ethoxy-2-methylpropane. Its common name is *tert*-butyl ethyl ether.
(b) cyclohexoxycyclohexane. Its common name is dicyclohexyl ether.

Is our answer reasonable?

To check that your names are sensible, draw structures that correspond to your answers and compare these with the structures provided in the question.

PRACTICE EXERCISE 19.6

Write the IUPAC and common names for each of the following ethers.

(a) (isopropyl–O–isopropyl skeletal structure) (b) (cyclohexyl–O–ethyl skeletal structure)

Physical properties

Ethers are moderately polar compounds in which the oxygen atom has a partial negative charge, and each carbon atom bonded to it has a partial positive charge (figure 19.11). Because of steric hindrance, however, only weak forces of attraction exist between ether molecules in the pure liquid. Consequently, boiling points of ethers are much lower than those of alcohols with a similar number of electrons (table 19.5) and are closer to hydrocarbons of comparable number of electrons (table 19.1).

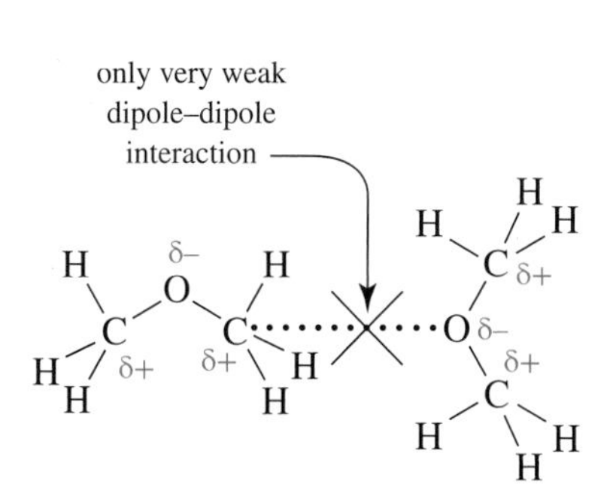

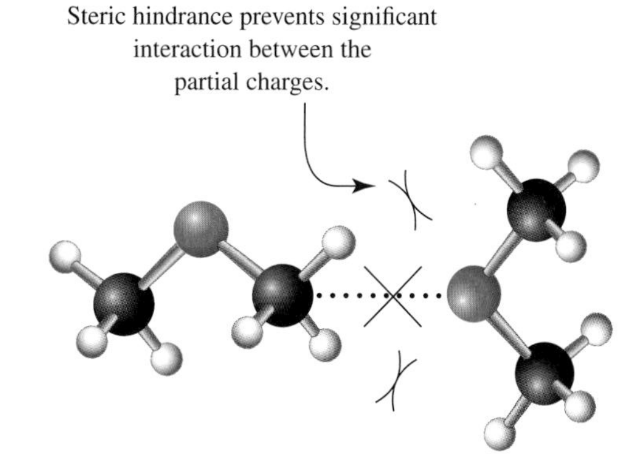

FIGURE 19.11 Ethers are polar molecules, but, because of steric hindrance, only weak attractive interactions exist between their molecules in the pure liquid.

TABLE 19.5 Boiling points and solubilities in water of some alcohols and ethers of comparable number of electrons.

Structural formula	Name	Total electrons	Boiling point (°C)	Solubility in water
CH_3CH_2OH	ethanol	26	78	infinite
CH_3OCH_3	dimethyl ether	26	−24	7.8 g/100 g
$CH_3CH_2CH_2CH_2OH$	butan-1-ol	42	117	7.4 g/100 g
$CH_3CH_2OCH_2CH_3$	diethyl ether	42	35	8 g/100 g
$CH_3CH_2CH_2CH_2CH_2OH$	pentan-1-ol	50	138	2.3 g/100 g
$CH_3CH_2CH_2CH_2OCH_3$	butyl methyl ether	50	71	slight
$CH_3OCH_2CH_2OCH_3$	1,2-dimethoxyethane	50	84	infinite

Because the oxygen atom of an ether carries a partial negative charge, ethers form hydrogen bonds with water molecules (figure 19.12) and are more soluble in water than hydrocarbons with a similar number of electrons and shape (compare data in tables 19.1 and 19.5).

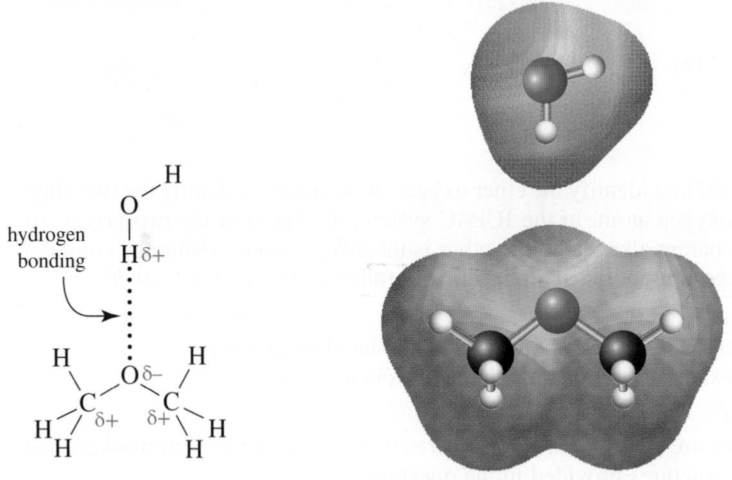

FIGURE 19.12 Dimethyl ether in water. The partially negative oxygen atom of the ether is the hydrogen bond acceptor, and a partially positive hydrogen atom of a water molecule is the hydrogen bond donor. Ethers are hydrogen bond acceptors only; they are not hydrogen bond donors.

The effect of hydrogen bonding is illustrated dramatically by comparing the boiling points of ethanol (78 °C) and its constitutional isomer dimethyl ether (−24 °C). The difference in boiling points between these two compounds is due to the polar O—H group in the alcohol, which is capable of forming intermolecular hydrogen bonds. This hydrogen bonding increases the attractive force between molecules of ethanol; thus, ethanol has a higher boiling point (and water solubility) than dimethyl ether.

CH_3CH_2OH	CH_3OCH_3
ethanol	dimethyl ether
bp = 78 °C	bp = −24 °C

WORKED EXAMPLE 19.7

Solubility

Arrange the following compounds in order of increasing solubility in water.

$CH_3OCH_2CH_2OCH_3$	$CH_3CH_2OCH_2CH_3$	$CH_3CH_2CH_2CH_2CH_2CH_3$
1,2-dimethoxyethane	diethyl ether	hexane

Analysis

As water is a polar solvent we expect polar compounds to have a greater solubility in water than nonpolar compounds. Both diethyl ether and 1,2-dimethoxyethane are polar compounds due to the presence of polar C—O—C groups, and each interacts with water as a hydrogen bond acceptor. Hexane is a nonpolar hydrocarbon and therefore we expect it to have the lowest solubility in water. Generally, compounds with more polar groups are more soluble in water.

Solution

$CH_3CH_2CH_2CH_2CH_2CH_3$	$CH_3CH_2OCH_2CH_3$	$CH_3OCH_2CH_2OCH_3$
insoluble	8 g/100 g water	soluble in all proportions

Is our answer reasonable?

Check the number of polar groups in each compound. Are your compounds listed in order of lowest number of polar groups to highest number of polar groups?

PRACTICE EXERCISE 19.7 Arrange the following compounds in order of increasing boiling point.

(a) O O (b) HO OH (c) O OH

Reactions of ethers

Ethers, R—O—R, resemble hydrocarbons in their resistance to chemical reactions. They do not react readily with oxidising agents (e.g. potassium dichromate or potassium permanganate) or reducing agents (e.g. sodium borohydride or lithium aluminium hydride). They are not affected by most acids or bases at moderate temperatures. However, they can be cleaved with concentrated hydrobromic or hydroiodic acid to give the corresponding haloalkane and alcohol. If an excess of HX is used, the alcohol can react further to give a haloalkane and water (see section 19.2).

dipropyl ether + HBr ⟶ 1-bromopropane + propan-1-ol

dipropyl ether + 2HBr ⟶ 2 1-bromopropane + H_2O

This reaction is another example of a nucleophilic substitution and is similar to the reaction of alcohols with HX. The first step involves the protonation of the ether oxygen to form an oxonium ion (R_2OH^+).

Because of their good solvent properties and general inertness to chemical reaction, ethers are excellent solvents in which to carry out many organic reactions.

19.5 Thiols

The sulfur analogue of an alcohol is called a **thiol** (thi- from the Greek: *theion*, sulfur) or, in the older literature, a **mercaptan**, which literally means 'mercury capturing'. Since sulfur and oxygen belong to the same group of the periodic table, they share similar chemical properties. Thiols form thioesters, thioethers and thioacetals in the same way alcohols form esters, ethers and acetals (see chapter 21). The functional group of a thiol is an —SH (sulfhydryl) group.

Figure 19.13 shows a Lewis structure and a ball-and-stick model of methanethiol, CH_3SH, the simplest thiol. The electronegativities of carbon and sulfur in methanethiol are virtually identical (2.55 and 2.58 respectively), while sulfur is slightly more electronegative than hydrogen (2.58 versus 2.20). The electron density model shows some slight partial positive charge on hydrogen atom of the S—H group and some slight partial negative charge on the sulfur atom.

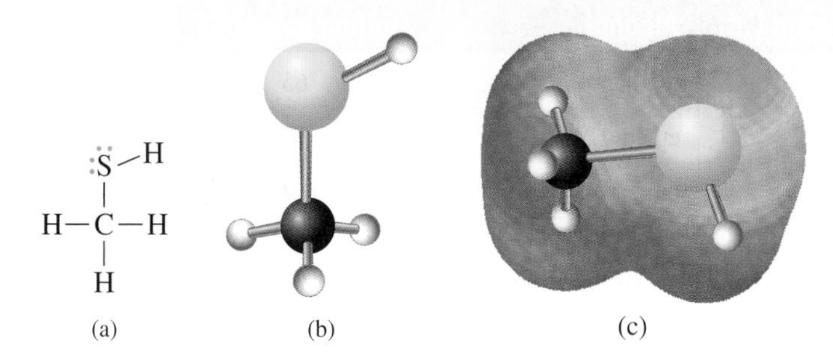

FIGURE 19.13 Methanethiol, CH_3SH: **(a)** Lewis structure, **(b)** ball-and-stick model and **(c)** electron density model.

In the IUPAC system, thiols are named in the same way as alcohols, except that the ending *-thiol* is added to the name of the parent alkane. Common names for simple thiols are derived by naming the alkyl group bonded to the —SH group and adding the word *mercaptan*. In compounds containing other functional groups, the presence of an —SH group is indicated by the prefix *mercapto-*.

CH_3CH_2SH	$CH_3CH(CH_3)CH_2SH$	$HSCH_2CH_2OH$
ethanethiol (ethyl mercaptan)	2-methylpropane-1-thiol (isobutyl mercaptan)	2-mercaptoethanol

Sulfur analogues of ethers are named by using the word *sulfide* to show the presence of the —S— group. The following are common names of two sulfides.

CH_3SCH_3	$CH_3CH_2SCH(CH_3)CH_3$
dimethyl sulfide	ethyl isopropyl sulfide

The most outstanding property of low-molar-mass thiols is their strong, often unpleasant odour, such as those from rotten eggs and sewage. However, not all thiols have an unpleasant odour; for example, grapefruits, mushrooms, onions, garlic and coffee (figure 19.14) all contain sulfur compounds.

FIGURE 19.14 Garlic is one of several common foods that contain sulfur compounds.

The aroma of grapefruit is primarily due to a thiol, commonly known as grapefruit mercaptan. The structural formula and a ball-and-stick model of grapefruit mercaptan are shown below. Only the *R*-enantiomer has the characteristic odour.

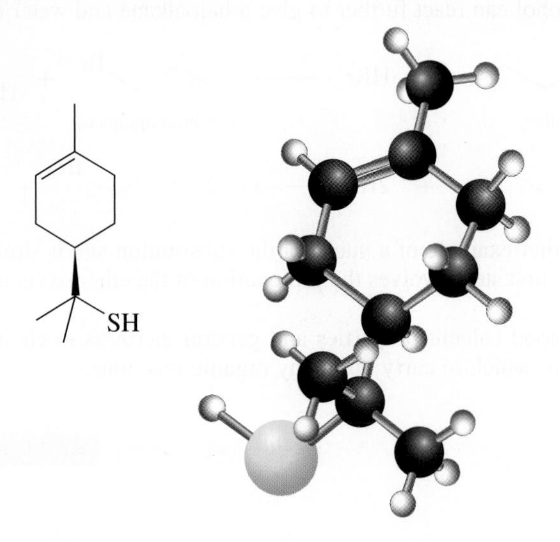

Physical properties

Because of the small difference in electronegativity between sulfur and hydrogen (2.5 − 2.1 = 0.4), we classify the S—H bond as nonpolar covalent. Because of this lack of polarity, thiols show little intermolecular association by hydrogen bonding. Consequently, they have lower boiling points and are less soluble in water and other polar solvents than the corresponding alcohol. Table 19.6 gives the boiling points of three low-molar-mass thiols. For comparison, the table also gives the boiling points of alcohols with the same number of carbon atoms.

TABLE 19.6 Boiling points of three thiols and three alcohols with the same number of carbon atoms.

Thiol	Boiling point (°C)	Alcohol	Boiling point (°C)
methanethiol	6	methanol	65
ethanethiol	35	ethanol	78
butane-1-thiol	98	butan-1-ol	117

Earlier, we illustrated the importance of hydrogen bonding in alcohols by comparing the boiling points of ethanol (78 °C) and its constitutional isomer dimethyl ether (–24 °C). By comparison, the boiling point of ethanethiol is 35 °C, and that of its constitutional isomer dimethyl sulfide is 37 °C.

CH_3CH_2SH	CH_3SCH_3
ethanethiol	dimethyl sulfide
bp = 35 °C	bp = 37 °C

The fact that the boiling points of these constitutional isomers are almost identical indicates that little or no hydrogen bonding occurs between thiol molecules.

Reactions of thiols

In this section, we discuss the acidity of thiols, and their reaction with strong bases and with molecular oxygen.

Acidity

Hydrogen sulfide is a stronger acid than water.

$$H_2O + H_2O \rightleftharpoons HO^- + H_3O^+ \quad pK_a = 15.7$$

$$H_2S + H_2O \rightleftharpoons HS^- + H_3O^+ \quad pK_a = 7.0$$

Similarly, thiols are stronger acids than alcohols. Compare, for example, the pK_a of ethanol and the pK_a of ethanethiol in dilute aqueous solution.

$$CH_3CH_2OH + H_2O \rightleftharpoons CH_3CH_2O^- + H_3O^+ \quad pK_a = 15.9$$

$$CH_3CH_2SH + H_2O \rightleftharpoons CH_3CH_2S^- + H_3O^+ \quad pK_a = 8.5$$

Thiols are sufficiently strong acids that, when dissolved in aqueous sodium hydroxide, they are converted completely to alkylsulfide salts.

$$\underset{\substack{pK_a = 8.5 \\ \text{(stronger acid)}}}{CH_3CH_2SH} + \underset{\text{(stronger base)}}{Na^+OH^-} \longrightarrow \underset{\text{(weaker base)}}{CH_3CH_2S^-Na^+} + \underset{\substack{pK_a = 15.7 \\ \text{(weaker acid)}}}{H_2O}$$

Oxidation to disulfides

Many of the chemical properties of thiols occur because the sulfur atom of a thiol is easily oxidised to several higher oxidation states. The most common reaction of thiols in biological systems is their oxidation to disulfides, the functional group of which is a **disulfide** (—S—S—) bond. Thiols are readily oxidised to disulfides by molecular oxygen. In fact, they are so susceptible to oxidation that they must be protected from contact with air during storage. Disulfides, in turn, are easily reduced to thiols by several reagents. This easy interconversion between thiols and disulfides is very important in protein chemistry, as we will see in chapter 24.

$$\underset{\text{thiol}}{2HOCH_2CH_2SH} \underset{\text{reduction}}{\overset{\text{oxidation}}{\rightleftharpoons}} \underset{\text{disulfide}}{HOCH_2CH_2S{-}SCH_2CH_2OH}$$

We derive common names of simple disulfides by listing the names of the groups bonded to sulfur and adding the word *disulfide,* for example, $CH_3S{-}SCH_3$ is named dimethyldisulfide.

19.6 Amines

Amines are derivatives of ammonia in which one or more hydrogen atoms are replaced by alkyl or aryl groups. Amines are classified as primary (1°), secondary (2°) or tertiary (3°), depending on the number of hydrogen atoms of ammonia that are replaced. Note that 1°, 2° and 3° refer to the degree of substitution on the *nitrogen* atom, not the carbon atom bearing the nitrogen atom (unlike haloalkanes and alcohols — see section 19.1).

ammonia

methylamine (a 1° amine)

dimethylamine (a 2° amine)

trimethylamine (a 3° amine)

Amines are further divided into aliphatic amines and aromatic amines. In an **aliphatic amine**, all the carbon atoms bonded directly to the nitrogen atom are derived from alkyl groups; in an **aromatic amine**, one or more of the groups bonded directly to the nitrogen atom are aryl groups. Note that the third example below (benzyldimethylamine) is not classed as an aromatic amine because the nitrogen atom is not bonded directly onto the aromatic ring.

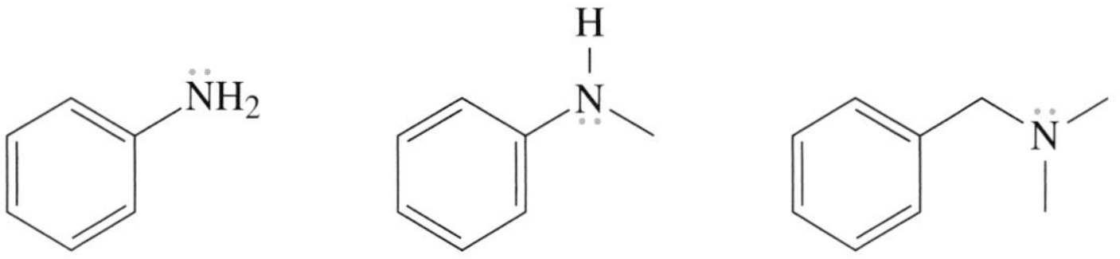

aniline (a 1° aromatic amine)

N-methylaniline (a 2° aromatic amine)

benzyldimethylamine (a 3° aliphatic amine)

An amine in which the nitrogen atom is part of a ring is classified as a **heterocyclic amine**. When the nitrogen atom is part of an aromatic ring (section 16.7, p. 738), the amine is classified as a **heterocyclic aromatic amine**. The following are structural formulae for two heterocyclic aliphatic amines and two heterocyclic aromatic amines.

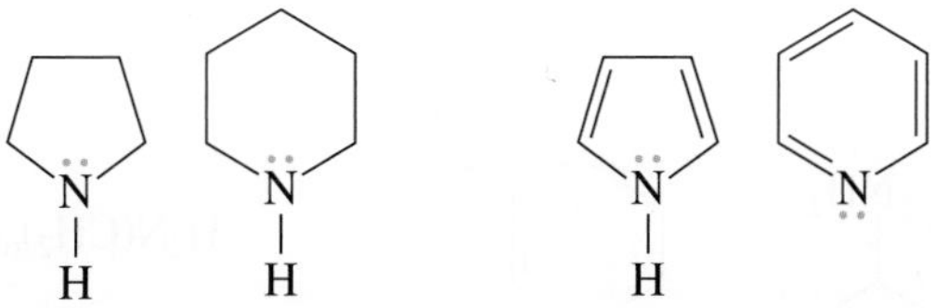

pyrrolidine piperidine (heterocyclic aliphatic amines)

pyrrole pyridine (heterocyclic aromatic amines)

WORKED EXAMPLE 19.8

Classifying amines

Alkaloids are basic nitrogen-containing compounds found in plants. Many alkaloids have physiological activity when administered to humans.

Examples of alkaloids that are physiologically active in humans include scopolamine, nicotine and cocaine. Scopolamine (also known as hyoscine) is obtained from the leaves of the Australian *Duboisia* tree and is classified as an anticholinergic (it blocks the action of acetylcholine). The hydrobromide salt is used in the prevention of motion sickness. In small doses, nicotine is an addictive stimulant. In larger doses, it causes depression, nausea and vomiting. In still larger doses, it is a deadly poison. Solutions of nicotine in water are used as insecticides. Cocaine is a central nervous system stimulant obtained from the leaves of the coca plant.

Classify each amino group in the following alkaloids according to type (primary, secondary, tertiary, heterocyclic, aliphatic or aromatic).

(a) 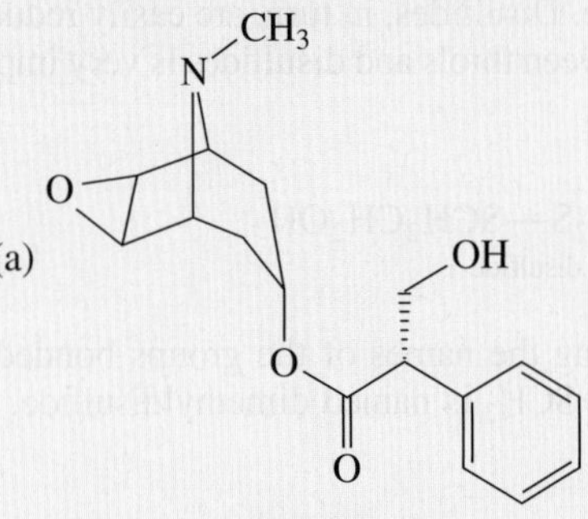 scopolamine (b) 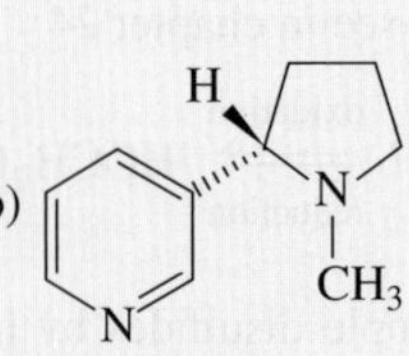 (S)-nicotine (c) 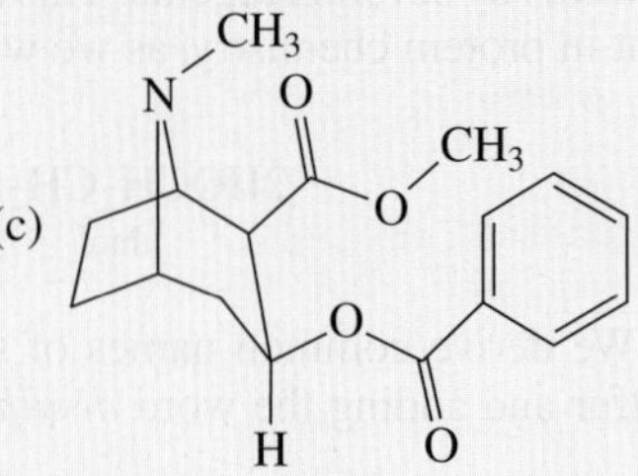 cocaine

Analysis

First identify the nitrogen atom in the compound. Now, check the following.

(i) Ensure that the nitrogen atom is part of an amine. (There are other functional groups where the nitrogen atom is not bound to alkyl or aryl groups. These are called amide, nitrile and nitro groups.)

(ii) If the nitrogen atom is part of a ring, the compound is heterocyclic.

(iii) If the nitrogen atom is bonded directly to an aromatic ring, or part of an aromatic ring, then the compound is an aromatic amine.

Then count the number of groups bonded to the nitrogen atom to determine if it is primary, secondary or tertiary.

Solution

(a) a tertiary heterocyclic aliphatic amine

(b) one tertiary heterocyclic aliphatic amine and one heterocyclic aromatic amine

(c) a tertiary heterocyclic aliphatic amine

PRACTICE EXERCISE 19.8

Classify each amino group in the following amines according to type (primary, secondary, tertiary, heterocyclic, aliphatic or aromatic).

(a) 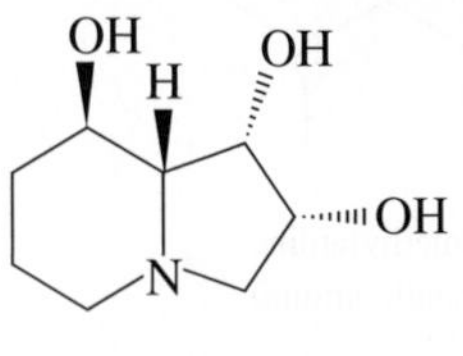 (b) 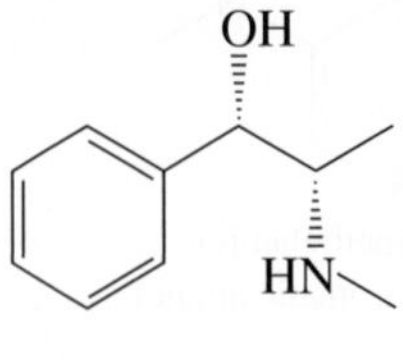 (c)

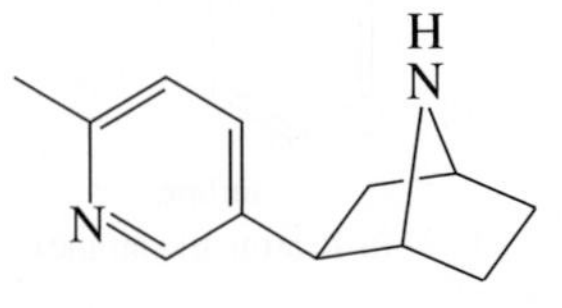

Systematic names for aliphatic amines are derived just as they are for alcohols. The suffix *-e* of the parent alkane is dropped and is replaced by *-amine*; that is, they are named alkanamines.

$H_2N(CH_2)_6NH_2$

butan-2-amine (S)-1-phenylethanamine hexane-1,6-diamine

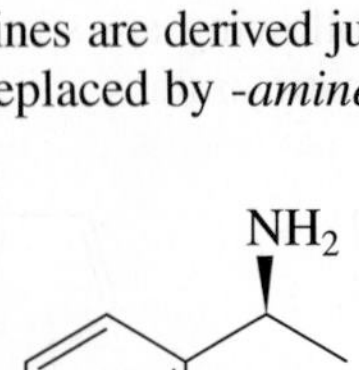

Common names for most aliphatic amines are derived by listing the alkyl groups bonded to nitrogen in alphabetical order in one word ending in the suffix *-amine*; that is, they are named as **alkylamines.**

CH_3NH_2 — methylamine
tert-butylamine
dicyclopentylamine
triethylamine

WORKED EXAMPLE 19.9

Naming amines

Write the IUPAC name for each of the following amines.

(a) (b) (c)

Analysis

First, you need to identify the longest chain as this will be the parent amine. The suffix *-e* of the parent alkane is then dropped and replaced by *-amine* to give an alkanamine. If there is more than one amino group, then use *-diamine*, *-triamine* and so on, remembering to indicate the position of the amino groups. Next you should identify any substituents. These are listed in alphabetical order, with numbers indicating the position of the substituent.

Solution

(a) hexan-1-amine
(b) butane-1,4-diamine (This is also commonly known as putrescine because it smells putrid.)
(c) The systematic name of this compound is (*R*)-1-phenylpropan-2-amine. Its common name is amphetamine. The dextrorotatory isomer of amphetamine (shown here) is a central nervous system stimulant and is manufactured and sold under several trade names. The salt with sulfuric acid is marketed as Dexedrine® sulfate.

Is our answer reasonable?

To check that your names are correct, draw structures that correspond to your answers and compare these with the structures provided in the question.

PRACTICE EXERCISE 19.9

Write a structural formula for each of the following amines.
(a) 3,3-dimethylbutan-1-amine
(b) *trans*-4-methylcyclohexanamine
(c) (*S*)-pentan-2-amine

IUPAC nomenclature retains the common name **aniline** for $C_6H_5NH_2$, the simplest aromatic amine. Its simple derivatives are named with the prefixes *o*-, *m*- and *p*-, or numbers to locate substituents. Several derivatives of aniline have common names that are still widely used. Among these are toluidine, for a methyl-substituted aniline, and anisidine, for a methoxy-substituted aniline.

aniline

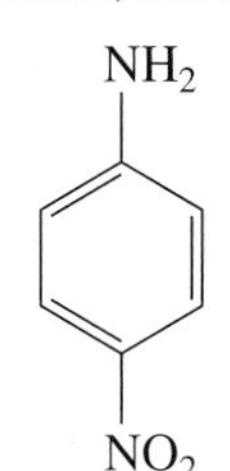

4-nitroaniline (*p*-nitroaniline)

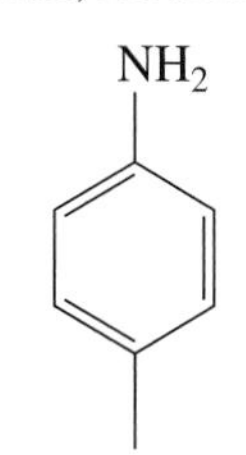

4-methylaniline (*p*-toluidine)

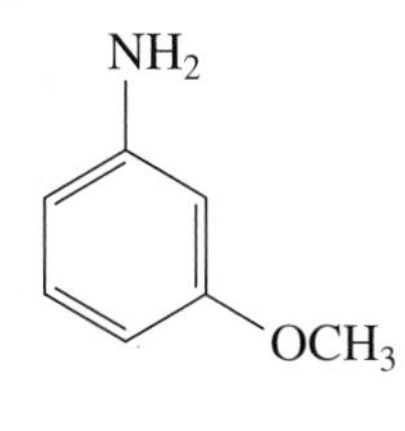

3-methoxyaniline (*m*-anisidine)

Secondary and tertiary amines are commonly named as *N*-substituted primary amines. For asymmetrical amines, the largest group is taken as the parent amine, then the smaller group or groups bonded to nitrogen are named, and their location is indicated by the prefix *N* (specifying that they are attached to nitrogen).

N-methylaniline
N,*N*-dimethylcyclopentanamine

Among the various functional groups discussed in this text, the $—NH_2$ group has one of the lowest nomenclature priorities (the priority list is provided in table 21.1, p. 947). The following compounds each contains a functional group of higher precedence than the amino group, and, accordingly, the $—NH_2$ group is indicated by the prefix *amino-*.

2-aminoethanol (ethanolamine)

2-aminobenzoic acid (anthranilic acid)

Chemical Connections

The opioid analgesics

Morphine and codeine (a monomethyl ether of morphine) are opioid analgesics obtained from unripe seed pods of the opium poppy, *Papaver somniferum* (figure 19.15), and have been known for centuries. Heroin is not naturally occurring but is synthesised by treating morphine with 2 mole equivalents of acetic anhydride.

FIGURE 19.15 The opium poppy, *Papaver somniferum*, is the source of morphine and codeine.

morphine codeine heroin

Even though morphine is one of modern medicine's most effective painkillers, it has two serious side-effects; it is addictive, and it depresses the respiratory control centre of the central nervous system. Large doses of morphine (or heroin) can lead to death by respiratory failure. For these reasons, chemists have sought to produce painkillers without these serious disadvantages.

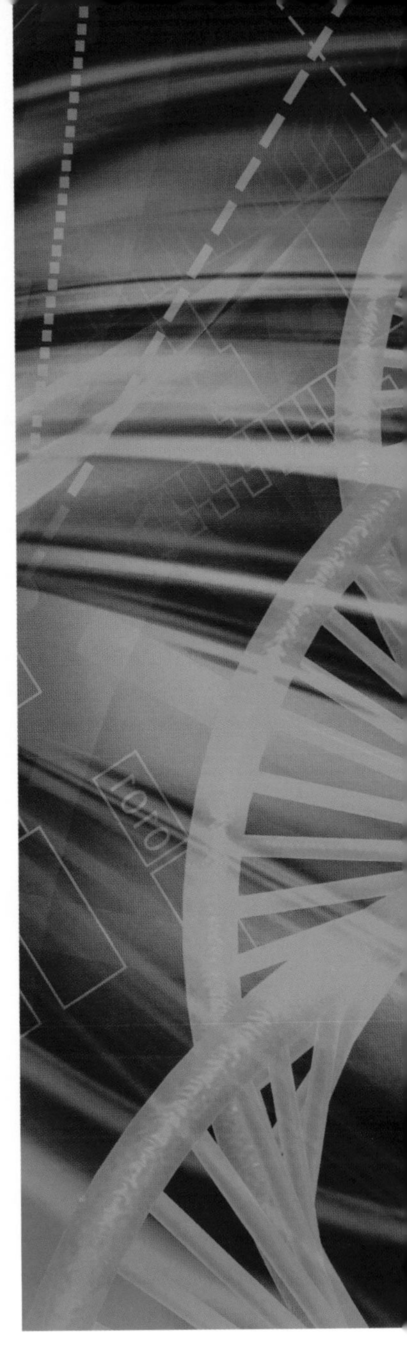

One strategy in this ongoing research has been to synthesise compounds related in structure to morphine in the hope that they would be equally effective analgesics, but with diminished side-effects. The following are structural formulae for two such compounds that have proven clinically useful.

(L)-enantiomer = levomethorphan
(D)-enantiomer = dextromethorphan

naltrexone

Levomethorphan is a potent analgesic. Interestingly, its dextrorotatory enantiomer, dextromethorphan, has no analgesic activity. It does, however, show approximately the same cough-suppressing activity as morphine and is used extensively in cough remedies. Although naltrexone has a similar structure to that of morphine, it counters morphine's (and heroin's) effect and has been used to combat heroin addiction.

aromatic ring
quaternary carbon
two carbon spacer
tertiary amine
redraw

the requirements for the 'morphine rule'

meperidine

It has been discovered that there can be even further simplification in the structure of morphine-like analgesics; the 'morphine rule' describes the minimum requirements for opiate-like effects. The rule states that, for a compound to exhibit opiate-like effects, it should contain (1) an aromatic ring, (2) a quaternary carbon atom attached to the ring and (3) a tertiary amine two carbon atoms from the quaternary carbon atom, and (4) one of the groups on the tertiary amine must be small (e.g. a methyl group). One such simplification is represented by meperidine, the hydrochloride salt of which is the widely used analgesic Demerol®. It was hoped that meperidine and related synthetic drugs would be free of many of the morphine-like undesirable side-effects. It is now clear, however, that they are not.

Physical properties

Amines are polar compounds, and both primary and secondary amines form intermolecular hydrogen bonds (figure 19.16).

An N—H···N hydrogen bond is weaker than an O—H···O hydrogen bond, because the difference in electronegativity between nitrogen and hydrogen (3.0 − 2.1 = 0.9) is less than that between oxygen and hydrogen (3.5 − 2.1 = 1.4). We can illustrate the effect of intermolecular hydrogen bonding by comparing the boiling points of methylamine and methanol as shown in the table below.

	CH_3NH_2	CH_3OH
number of electrons	18	18
boiling point (°C)	−6.3	65.0

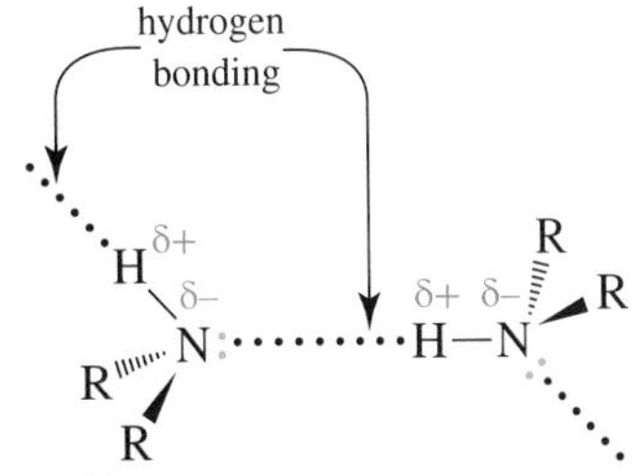

FIGURE 19.16 Intermolecular association of 1° and 2° amines by hydrogen bonding. Nitrogen is approximately tetrahedral in shape, with the axis of the hydrogen bond along the fourth axis of the tetrahedron.

Both compounds are polar molecules and interact in the pure liquid by hydrogen bonding. Methanol has the higher boiling point because hydrogen bonding between its molecules is stronger than that between molecules of methylamine.

All classes of amines form hydrogen bonds with water and are more soluble in water than hydrocarbons of a comparable number of electrons. Most low-molar-mass amines are completely soluble in water (table 19.7). Higher- molar-mass amines are only moderately soluble or insoluble.

TABLE 19.7 Physical properties of selected amines.

Name	Structural formula	Melting point (°C)	Boiling point (°C)	Solubility in water
ammonia	NH_3	–78	–33	very soluble
Primary amines				
methylamine	CH_3NH_2	–95	–6	very soluble
ethylamine	$CH_3CH_2NH_2$	–81	17	very soluble
propylamine	$CH_3CH_2CH_2NH_2$	–83	48	very soluble
cyclohexylamine	$C_6H_{11}NH_2$	–17	135	slightly soluble
Secondary amines				
diethylamine	$(CH_3CH_2)_2NH$	–48	56	very soluble
Tertiary amines				
triethylamine	$(CH_3CH_2)_3N$	–114	89	slightly soluble
Aromatic amines				
aniline	$C_6H_5NH_2$	–6	184	slightly soluble
Heterocyclic aromatic amines				
pyridine	C_5H_5N	–42	116	very soluble

Preparation of amines

Amines can be prepared by a variety of methods, from a range of starting materials. Some of the more common methods will be discussed here.

Preparation from haloalkanes

Amines can be prepared from haloalkanes by nucleophilic substitution reactions (see section 18.2). Given that ammonia is a good nucleophile, it can be used to prepare amines from haloalkanes.

$$\underset{\text{1-bromopropane}}{CH_3CH_2CH_2Br} + :NH_3 \longrightarrow \underset{\text{propylammonium}}{CH_3CH_2CH_2\overset{+}{N}H_3} + \underset{\text{bromide}}{Br^-}$$

Note that the charge is balanced (both sides have zero net charge). The alkyl ammonium salt is readily converted to the amine by reaction with a stronger base than the amine (e.g. NaOH).

$$CH_3CH_2CH_2\overset{+}{N}H_3 + {}^-OH \longrightarrow \underset{\text{propylamine}}{CH_3CH_2CH_2NH_2} + H_2O$$

A problem with using ammonia to prepare primary amines is that these amines are also nucleophilic species and can react in the same way as ammonia to produce secondary amines. Similarly, secondary amines can react to give tertiary amines. Consequently, mixtures of 1°, 2° and 3° amines are usually obtained.

$$CH_3CH_2CH_2Br + CH_3CH_2CH_2NH_2 \longrightarrow \underset{\text{dipropylammonium}}{(CH_3CH_2CH_2)_2\overset{+}{N}H_2} + \underset{\text{bromide}}{Br^-}$$

Primary amines can be prepared, in a two-step process, by using the azide ion (N_3^-) or phthalimide ion (use of the latter is known as the Gabriel synthesis) as nucleophiles. Using these processes, secondary and tertiary amines are not produced.

$$CH_3CH_2CH_2Br + {:}N_3^- \longrightarrow CH_3CH_2CH_2N_3 + Br^-$$

propyl azide

$$CH_3CH_2CH_2N_3 \xrightarrow[Pd/C]{H_2} CH_3CH_2CH_2NH_2 + N_2$$

PhCH₂Br + phthalimide ion ⟶ N-benzylphthalimide

benzyl bromide phthalimide ion *N*-benzylphthalimide

N-benzylphthalimide $\xrightarrow[2.\ NaOH]{1.\ N_2H_4/HCl}$ $PhCH_2NH_2$ + phthalate (COO^-, COO^-)

benzylamine (81%)

Synthesis of arylamines: reduction of the —NO_2 group

As we have already seen (section 16.8), the nitration of an aromatic ring introduces a —NO_2 group. A particular value of nitration is the fact that the resulting nitro group can be reduced to a primary amino group, —NH_2, by hydrogenation in the presence of a transition metal catalyst such as nickel, palladium or platinum.

3-nitrobenzoic acid (COOH, NO_2) + $3H_2$ $\xrightarrow{Ni}$ 3-aminobenzoic acid (COOH, NH_2) + $2H_2O$

3-nitrobenzoic acid 3-aminobenzoic acid

This method has the potential disadvantage that other susceptible groups, such as a carbon–carbon double bond, and the carbonyl group of an aldehyde or ketone, may also be reduced. Note that neither the —COOH nor the aromatic ring is reduced under these conditions.

Alternatively, a nitro group can be reduced to a primary amino group by a metal in acid.

2,4-dinitrotoluene (NO_2, NO_2) $\xrightarrow[C_2H_5OH,\ H_2O]{Fe,\ HCl}$ ($NH_3^+Cl^-$, $NH_3^+Cl^-$) $\xrightarrow{NaOH,\ H_2O}$ 2,4-diaminotoluene (NH_2, NH_2)

2,4-dinitrotoluene 2,4-diaminotoluene

The most commonly used metal reducing agents are iron, zinc or tin in dilute HCl. When reduced by this method, the amine is obtained as a salt, which is then treated with a strong base to liberate the free amine.

Other methods for the preparation of amines

There are many other methods available for the preparation of amines. Some of these, discussed in more detail later, include the reductive amination of aldehydes and ketones (section 21.5), and the reduction or hydrolysis of amides (section 23.5).

19.7 Reactions of amines

The reactions of amines are governed by the unshared pair of electrons on the nitrogen atom. Because of these unshared electrons, amines are both basic and nucleophilic. Consequently, amines react readily with acids to form salts and also react with electrophilic species like haloalkanes, acyl halides and many others.

Basicity of amines

Like ammonia, all amines are weak bases, and aqueous solutions of amines are basic. The following acid–base reaction between an amine and water is written using curved arrows to emphasise that, in this proton-transfer reaction, the unshared pair of electrons on nitrogen forms a new covalent bond with hydrogen and displaces the hydroxide ion.

$$CH_3-NH_2: + H-\ddot{O}-H \rightleftharpoons CH_3-NH_3^+ + :\ddot{O}-H^-$$

methylamine | methylammonium ion | hydroxide ion

The value of K_b, the base ionisation constant, for methylamine is 4.37×10^{-4} ($pK_b = 3.36$).

$$K_b = \frac{[CH_3NH_3^+][OH^-]}{[CH_3NH_2]} = 4.37 \times 10^{-4}$$

It is also common to discuss the basicity of amines by referring to the acidity constant of the corresponding conjugate acid, as illustrated for the ionisation of the methylammonium ion below.

$$CH_3NH_3^+ + H_2O \rightleftharpoons CH_3NH_2 + H_3O^+$$

$$K_a = \frac{[CH_3NH_2][H_3O^+]}{[CH_3NH_3^+]} = 2.29 \times 10^{-11}$$

$$pK_a = 10.64$$

Values of pK_a and pK_b for any acid–conjugate base pair are related by the equation:

$$pK_a + pK_b = 14.00 \qquad \text{(at 25 °C)}$$

Values of pK_a and pK_b for selected amines are given in table 19.8 on the next page. You may wish to review chapter 11 for further discussion of acid–base equilibria.

WORKED EXAMPLE 19.10

Predicting equilibrium

Predict the position of equilibrium for the acid–base reaction below.

$$CH_3NH_2 + CH_3COOH \rightleftharpoons CH_3NH_3^+ + CH_3COO^-$$

Analysis

Use the approach we developed earlier to predict the position of equilibrium in acid–base reactions. Equilibrium favours the reaction between the stronger acid and stronger base to form the weaker acid and weaker base.

Solution

Equilibrium favours the formation of methylammonium ions and acetate ions.

$$CH_3NH_2 + CH_3COOH \rightleftharpoons CH_3NH_3^+ + CH_3COO^-$$

CH_3NH_2	CH_3COOH	$CH_3NH_3^+$	CH_3COO^-
	$pK_a = 4.76$	$pK_a = 10.64$	
(stronger base)	(stronger acid)	(weaker acid)	(weaker base)

PRACTICE EXERCISE 19.10

Predict the position of equilibrium for the following acid–base reaction.

$$PhNH_3^+ + H_2O \rightleftharpoons PhNH_2 + H_3O^+$$

TABLE 19.8 Base strengths (pK_b) of selected amines and acid strengths (pK_a) of their conjugate acids at 25 °C[a].

Amine	Structure	pK_b	pK_a
ammonia	NH_3	4.74	9.26
Primary amines			
ethylamine	$CH_3CH_2NH_2$	3.19	10.81
cyclohexylamine	$C_6H_{11}NH_2$	3.34	10.66
Secondary amines			
diethylamine	$(CH_3CH_2)_2NH$	3.02	10.98
Tertiary amines			
triethylamine	$(CH_3CH_2)_3N$	3.25	10.75
Aromatic amines			
aniline	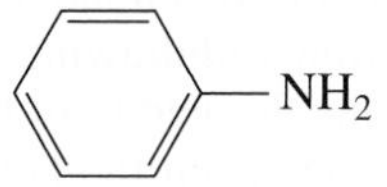 NH_2	9.36	4.64
4-methylaniline	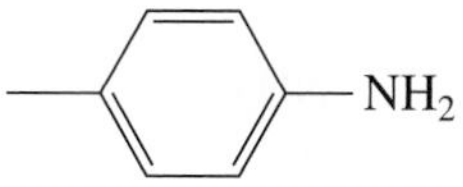 NH_2	8.92	5.08
4-nitroaniline	 O_2N — NH_2	13.0	1.0
Heterocyclic aromatic amines			
pyridine	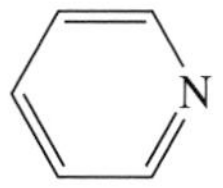 N	8.82	5.18

(a) For each amine, $pK_a + pK_b = 14.00$ at 25 °C.

From the information in table 19.8, we can make the following generalisations about the acid–base properties of the various classes of amines.

1. All aliphatic amines have about the same base strength, with pK_b values from 3.0 to 4.0, and are slightly stronger bases than ammonia.
2. Aromatic amines are considerably weaker bases than aliphatic amines. Compare, for example, the values of pK_b for aniline and cyclohexylamine shown below.

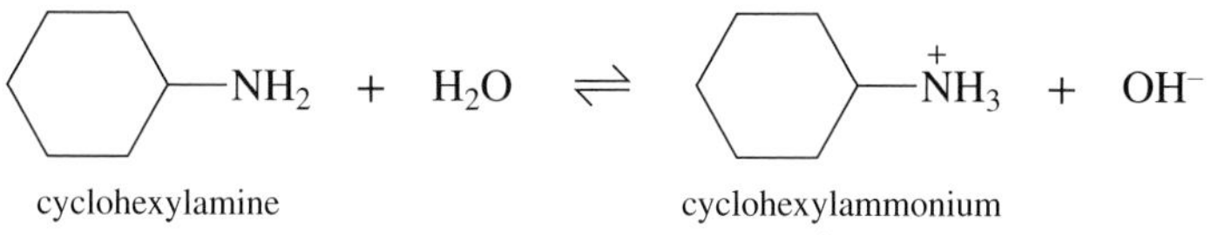

NH_2 + H_2O ⇌ $\overset{+}{N}H_3$ + OH^- $\quad pK_b = 3.34$, $K_b = 4.5 \times 10^{-4}$

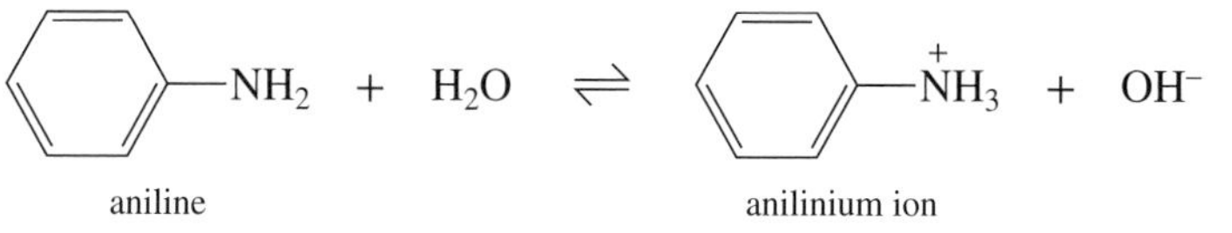

cyclohexylamine — cyclohexylammonium ion

NH_2 + H_2O ⇌ $\overset{+}{N}H_3$ + OH^- $\quad pK_b = 9.36$, $K_b = 4.5 \times 10^{-10}$

aniline — anilinium ion

The basicity constant for aniline is approximately one million times smaller (the larger the value of pK_b, the weaker is the base) than that for cyclohexylamine.

Aromatic amines are weaker bases than aliphatic amines because of the resonance interaction of the unshared pair of electrons on nitrogen with the π electron system of the aromatic ring. This resonance interaction is also the reason that phenols are more acidic than alcohols (compare the resonance diagram below with that for phenoxide ion on p. 838). This resonance reduces the electron density on the nitrogen atom (or oxygen atom in the case of phenols) making it a poorer proton acceptor (or a better proton donor in the case of phenols) than the non-aromatic analogues. Because no resonance interaction is possible for an alkylamine, the electron pair on its nitrogen atom is more available for reaction with an acid.

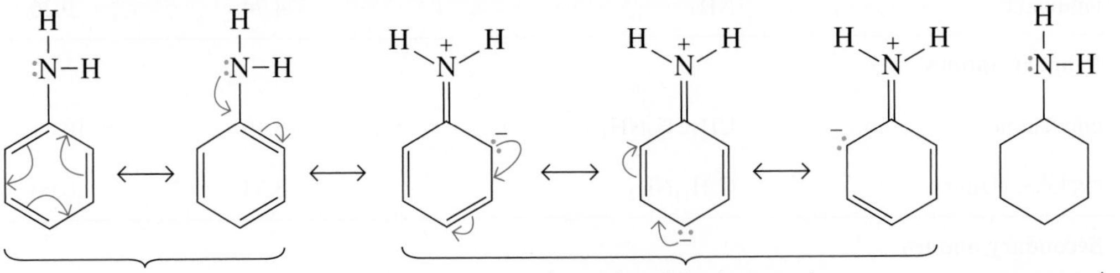

two Kekulé structures

interaction of the electron pair on nitrogen with the π system of the aromatic ring

no resonance is possible with alkylamines

3. Electron-withdrawing groups such as halogens, and nitro and carbonyl groups decrease the basicity of substituted aromatic amines by further decreasing the availability of the electron pair on nitrogen. The diagram below shows the resonance forms for 4-nitroaniline showing how the nitro group reduces the basicity. You will notice that there are four resonance forms in which the nitrogen atom of the amino group does not have a nonbonding pair of electrons (in the case above for aniline there are three). This electron-withdrawing effect of the nitro group has been discussed previously with respect to increasing the acidity of phenols (see section 19.3). In both cases, the nitro group is withdrawing electrons from the heteroatom (nitrogen in anilines, or oxygen in phenols).

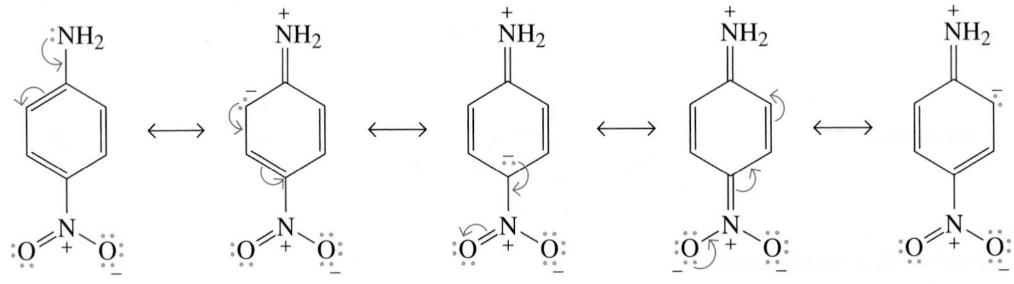

WORKED EXAMPLE 19.11

Strength of bases

Select the stronger base in each of the following pairs of amines.

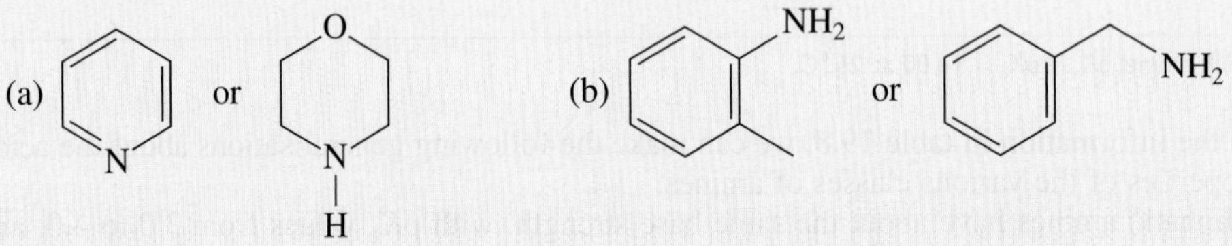

(a) A or B (b) C or D

Analysis

In worked example 19.8, we classified amines as primary, secondary, tertiary, heterocyclic, aliphatic or aromatic. From table 19.8 we deduce that aliphatic amines are stronger bases than aromatic amines. Therefore, once you have classified the amines you should be able to determine which is the stronger base.

Solution

(a) Morpholine (B) is the stronger base ($pK_b = 5.79$). It has a basicity comparable to that of secondary aliphatic amines. Pyridine (A), a heterocyclic aromatic amine ($pK_b = 8.82$), is considerably less basic than aliphatic amines.

(b) Benzylamine (D), a primary aliphatic amine, is the stronger base ($pK_b = 3–4$). The aromatic amine, *o*-toluidine (C), is the weaker base ($pK_b = 9–10$).

PRACTICE EXERCISE 19.11

Select the stronger acid from each of the following pairs of ions.

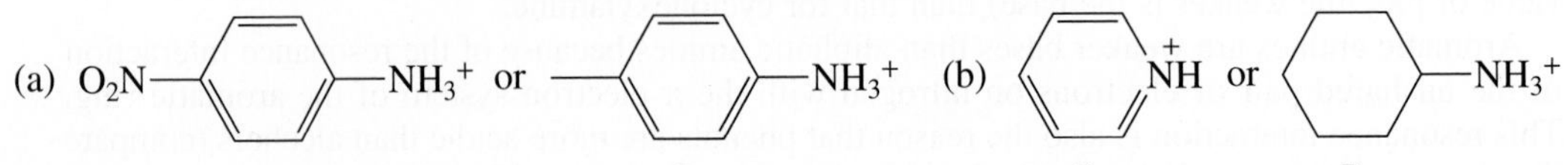

(a) O_2N–C₆H₄–NH_3^+ (A) or CH_3–C₆H₄–NH_3^+ (B) (b) pyridinium $\overset{+}{N}H$ (C) or cyclohexyl–NH_3^+ (D)

Guanidine, with a pK_b value of 0.4, is one of the strongest bases among neutral compounds.

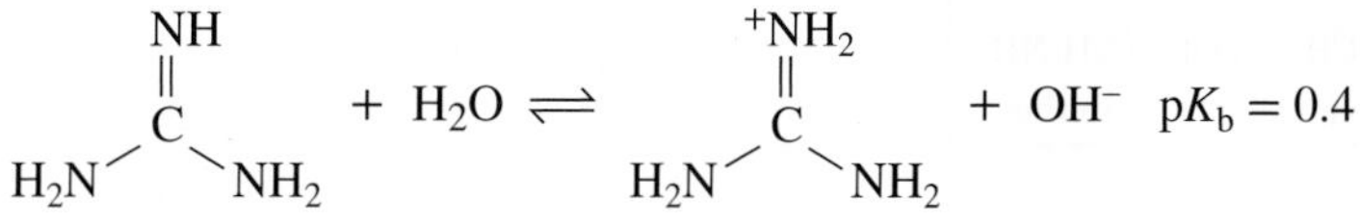

$$H_2N{-}C({=}NH){-}NH_2 + H_2O \rightleftharpoons H_2N{-}C({=}NH_2^+){-}NH_2 + OH^- \quad pK_b = 0.4$$

guanidine guanidinium ion

The remarkable basicity of guanidine is attributed to the fact that the positive charge on the guanidinium ion is delocalised equally over the three nitrogen atoms, as shown by the three equivalent contributing structures below.

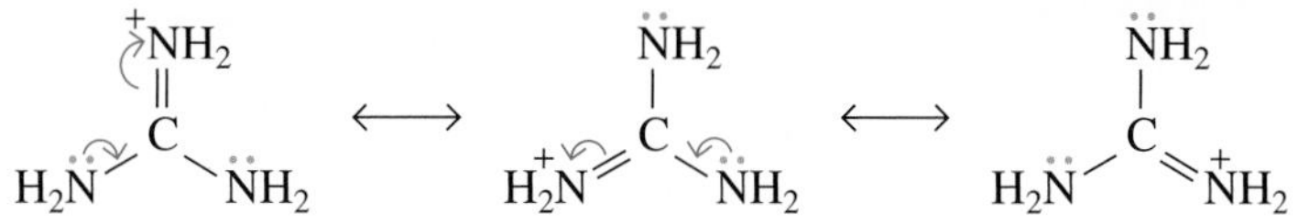

three equivalent contributing structures

Hence, the guanidinium ion is a highly stable cation. The presence of a guanidine group in the amino acid arginine accounts for the basicity of its side chain (section 24.2).

Reaction with acids

Amines, whether soluble or insoluble in water, react quantitatively with strong acids to form water-soluble salts, as illustrated below by the reaction of (*R*)-norepinephrine (noradrenaline) with aqueous HCl to form a hydrochloride salt.

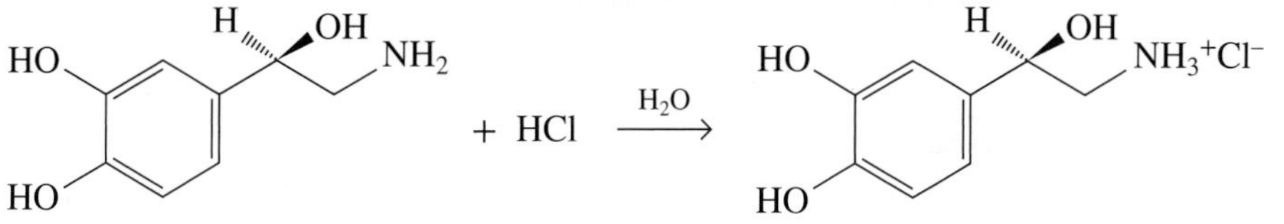

$$+ \ HCl \xrightarrow{H_2O}$$

(*R*)-norepinephrine
(only slightly soluble in water)

(*R*)-norepinephrine hydrochloride
(a water-soluble salt)

Norepinephrine is a neurotransmitter and a stress hormone secreted by the medulla of the adrenal gland. It plays an important part in attention and focus. People with ADHD (attention deficit hyperactivity disorder) are prescribed drugs that help increase the levels of norepinephrine in the brain.

Drug companies have taken advantage of the reaction of amines with strong acids to improve the solubility of many drugs, including codeine and salbutamol. In the treatment of asthma, salbutamol is sold as the sulfate salt for increased water solubility in the lungs.

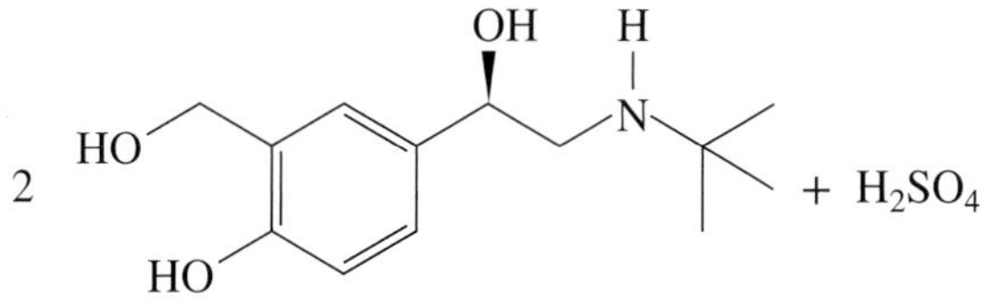

$$2 \ (R\text{-salbutamol}) + H_2SO_4$$

(*R*)-salbutamol
(slightly soluble in water)

$$SO_4^{2-}$$

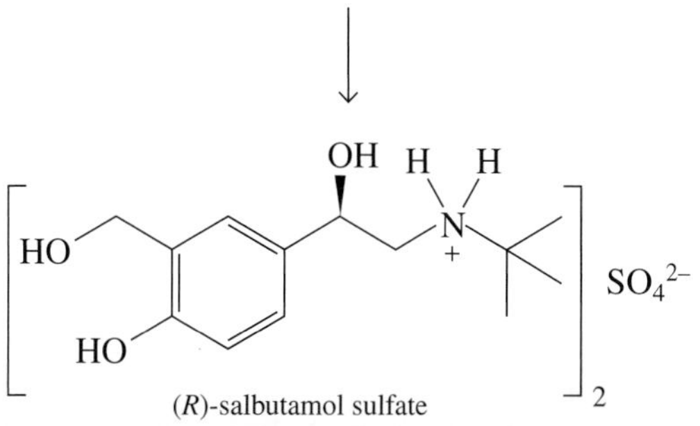

(*R*)-salbutamol sulfate
(soluble in water)

The basicity of amines and the solubility of amine salts in water can be used to separate amines from water-insoluble, nonbasic compounds. Figure 19.17 (overleaf) is a flowchart for the separation of aniline from anisole. Note that aniline is recovered from its salt by treatment with NaOH.

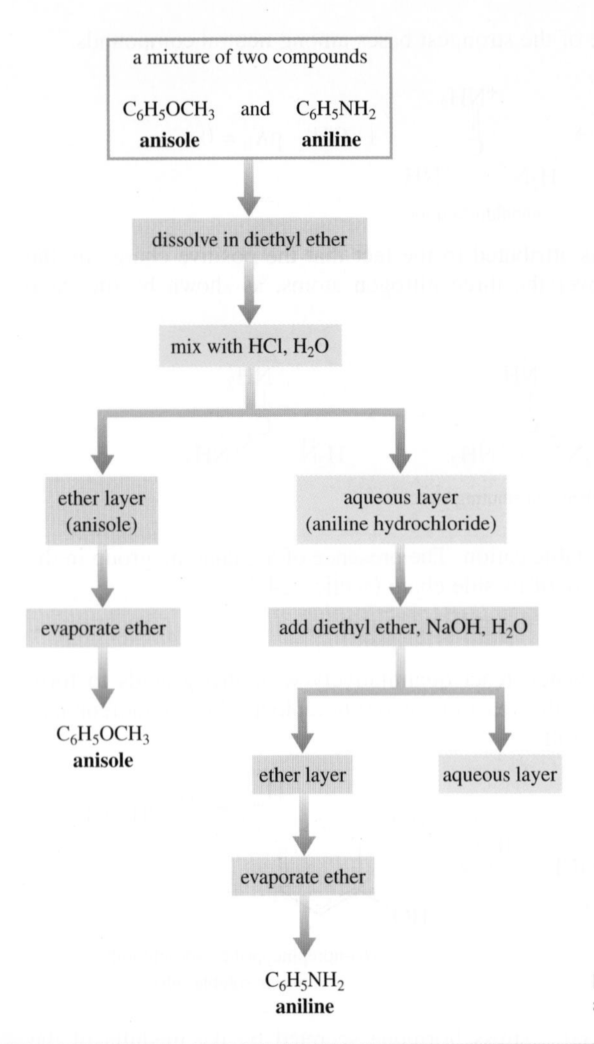

FIGURE 19.17 Separation and purification of an amine and a neutral compound.

WORKED EXAMPLE 19.12

Amino acid — alanine

The following are two structural formulae for alanine, 2-aminopropanoic acid, one of the building blocks of proteins (chapter 24).

$$\underset{\text{A}}{CH_3CH(NH_2)C(=O)OH} \quad \text{or} \quad \underset{\text{B}}{CH_3CH(NH_3^+)C(=O)O^-}$$

Is alanine better represented by structural formula A or B?

Analysis

This question is similar to worked example 19.10 except that the amine and the carboxylic acid are on the same molecule. Structural formula A contains both an amino group (a base) and a carboxyl group (an acid), while structure B contains the ammonium group (the conjugate acid) and the carboxylate group (the conjugate base).

Solution

Proton transfer from the stronger acid (—COOH) to the stronger base (—NH_2) gives an internal salt; therefore, B is the better representation of alanine. Within the field of amino acid chemistry, the internal salt represented by B is called a **zwitterion** (chapter 24).

PRACTICE EXERCISE 19.12

As shown in worked example 19.12, alanine is better represented as a zwitterion. Suppose that the zwitterion is dissolved in water.

(a) In what way would you expect the structure of alanine in aqueous solution to change if concentrated HCl were added to adjust the pH of the solution to 2.0?

(b) In what way would you expect the structure of alanine in aqueous solution to change if concentrated NaOH were added to bring the pH of the solution to 12.0?

Chemical Connections

Drug bioavailability

Many drugs contain at least one nitrogen atom and you will notice that the nitrogen atoms in the medicinal compounds mentioned previously (e.g. salbutamol and those in worked example 19.8) are all present as amino groups. Why is the amino group important for pharmacological activity? Because amines are weak bases, at physiological pH (~7.4) they are present as an equilibrium mixture of the protonated (ionic) and nonprotonated (neutral) forms; at this pH, the ionic form predominates. The equilibrium concentrations of the two forms depend on the pKa value of the ionic form and the pH of the environment. The ionic form is soluble in aqueous solvents (such as blood) and this allows the transport of the medication through the circulatory system to the appropriate site of action. Furthermore it is the ionic form that binds to the active site; the positively charged ammonium ion binds to a negatively charged carboxylate ion. However, the ionic form cannot cross the nonpolar environment of the cell membranes to bind to the active site. So how does the drug reach the target?

The answer lies in the equilibrium between the ionic and neutral forms of the drug. The neutral form crosses the membrane and, once on the other side, re-establishes equilibrium, producing some of the ionic form that binds to the receptor. On the other side of the membrane, the ionic form that could not cross the membrane also re-establishes equilibrium by forming more of the neutral form that can cross the membrane. In this way, the drug can cross the membrane in the neutral form and interact with the active site as the ionic form. The rate at which this process occurs depends on the equilibrium concentrations of the two forms.

$$\underset{\text{ionic form: soluble and readily transported in blood fluid}}{HNR_3^+} \underset{}{\overset{\text{deprotonation}}{\rightleftharpoons}} \underset{\text{neutral form: can cross the cell membrane}}{:NR_3} \overset{\text{reprotonation}}{\rightleftharpoons} \underset{\text{ionic form: binds to the active site}}{HNR_3^+}$$

Reaction of primary aromatic amines with nitrous acid

Nitrous acid, HNO_2, is an unstable compound that is prepared by adding sulfuric or hydrochloric acid to an aqueous solution of sodium nitrite, $NaNO_2$. Nitrous acid is a weak acid and ionises according to the following equation.

$$\underset{\text{nitrous acid}}{HNO_2} + H_2O \rightleftharpoons H_3O^+ + NO_2^- \quad K_a = 4.26 \times 10^{-4}$$
$$pK_a = 3.57$$

Nitrous acid reacts with amines in different ways, depending on whether the amine is primary, secondary or tertiary, and whether it is aliphatic or aromatic. We concentrate on the reaction of nitrous acid with primary aromatic amines, because this reaction is useful in organic synthesis.

Treatment of a primary aromatic amine, for example aniline, with nitrous acid gives a relatively stable diazonium salt.

$$\underset{\substack{\text{aniline}\\\text{(a 1° aromatic amine)}}}{C_6H_5\text{—}NH_2} + \underset{\substack{\text{sodium}\\\text{nitrite}}}{NaNO_2} + HCl \xrightarrow[0\ °C]{H_2O} \underset{\substack{\text{benzenediazonium}\\\text{chloride}}}{C_6H_5\text{—}\overset{+}{N}{\equiv}N\ Cl^-} + H_2O$$

We can also write the equation for this reaction in the following more abbreviated form.

$$C_6H_5\text{—}NH_2 \xrightarrow[0\ °C]{NaNO_2,\ HCl} \underset{\substack{\text{benzenediazonium}\\\text{chloride}}}{C_6H_5\text{—}N_2^+Cl^-}$$

Primary aliphatic amines also give diazonium salts, but these are unstable and decompose immediately to give a complex mixture of products. When we warm an aqueous solution of an aromatic diazonium salt, the —N_2^+ group is replaced by an —OH group. This reaction is one of the few methods we have for the synthesis of phenols. It enables us to convert an aromatic amine to a phenol by first forming the aromatic diazonium salt and then heating the solution. In this manner, we can convert 2-bromo-4-methylaniline to 2-bromo-4-methylphenol.

2-bromo-4-methylaniline $\xrightarrow[\text{2. Warm the solution.}]{\text{1. } NaNO_2,\ HCl,\ H_2O,\ 0\ °C}$ 2-bromo-4-methylphenol

WORKED EXAMPLE 19.13

Conversion of toluene to 4-hydroxybenzoic acid

Show the reagents that will bring about each step in this conversion of toluene to 4-hydroxybenzoic acid.

toluene $\xrightarrow{(1)}$ (4-nitrotoluene, NO_2) $\xrightarrow{(2)}$ (COOH, NO_2) $\xrightarrow{(3)}$ (COOH, NH_2) $\xrightarrow{(4)}$ (COOH, OH) 4-hydroxy-benzoic acid

Analysis

In this question you need to identify, in each step, the change in going from the starting material to the product.

Step 1: This is an electrophilic aromatic substitution reaction (section 16.8), specifically a nitration reaction.

Step 2: This is an oxidation of the benzylic carbon atom (section 23.4).

Step 3: This is a reduction of the nitro group (section 19.6).

Step 4: This is a conversion of a primary aromatic amine to a phenol (section 19.7).

Solution

Step 1: Treatment with nitric acid/sulfuric acid, followed by separation of the *ortho* and *para* isomers.

Step 2: Treatment with chromic acid or potassium permanganate.

Step 3: Treatment with H_2 in the presence of a transition metal catalyst *or* using Fe, Sn or Zn in the presence of aqueous HCl, followed by neutralisation of the hydrochloride salt.

Step 4: Treatment with $NaNO_2$/HCl to form the diazonium ion salt and then warming the solution.

PRACTICE EXERCISE 19.13

Show how you can use a similar set of steps to those in worked example 19.13 to convert chlorobenzene into 4-chlorophenol.

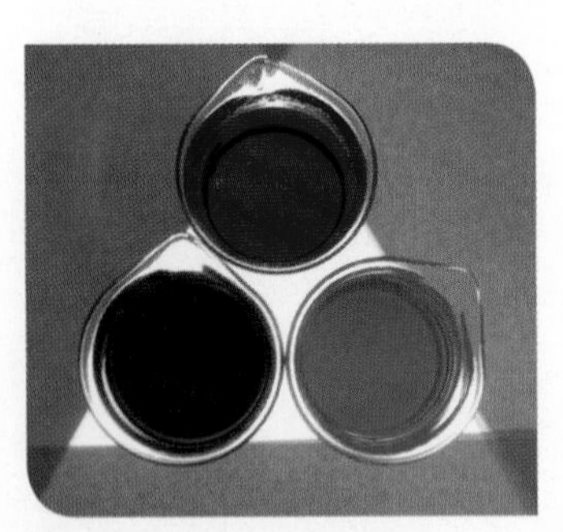

FIGURE 19.18 Azo dyes produced by the addition of 4-nitrobenzenediazonium chloride to phenol (red), 1-naphthol (blue) and 2-naphthol (purple).

Diazonium salts can also undergo coupling reactions with phenols and anilines to give **azo compounds** (Ar—N=N—Ar). These compounds are brightly coloured and are commonly used as dyes. This coupling is another example of an electrophilic aromatic substitution reaction (see section 16.8), where the electrophile is the diazonium salt. The addition of the diazonium salt derived from 4-nitroaniline to a basic solution of phenol instantly gives a brilliant red dye (figure 19.18).

4-O_2N-C_6H_4-$N_2^+Cl^-$ + $C_6H_5O^-$ $\xrightarrow{NaOH}$ ^-O-C_6H_4-N=N-C_6H_4-NO_2

a brilliant red dye

The reaction of secondary amines with nitrous acid produces *N*-nitrosamines.

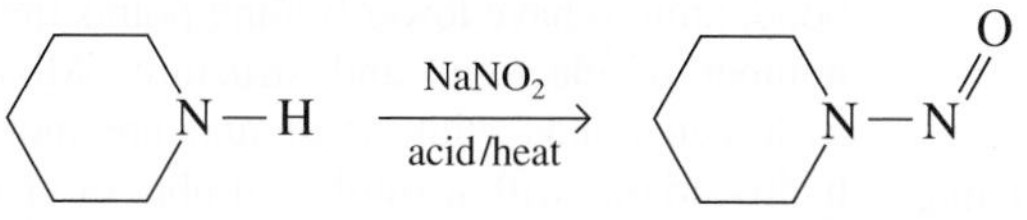

$$\text{piperidine} \xrightarrow[\text{acid/heat}]{NaNO_2} N\text{-nitrosopiperidine}$$

piperidine

N-nitrosopiperidine

Nitrosamines are important because many are known to be carcinogens in animals, and they have been found in cooked foods that have been preserved with sodium nitrite. Sodium nitrite is added to meats to inhibit the growth of *Clostridium botulinum*, which causes botulism, a deadly form of food poisoning.

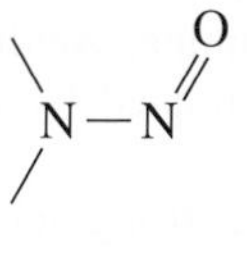

N-nitrosodimethylamine (found in tobacco smoke, beer and hot dogs)

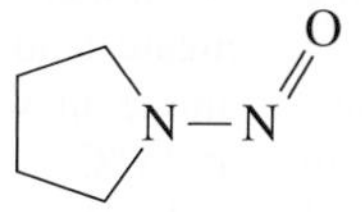

N-nitrosopyrrolidine (formed when bacon cured with sodium nitrite is fried)

Amide formation

An important reaction of amines is the formation of amides by condensation with acid chlorides or carboxylic anhydrides. The chemistry will be discussed in greater detail in chapter 23 (carboxylic acids) and chapter 24 (amino acids and proteins).

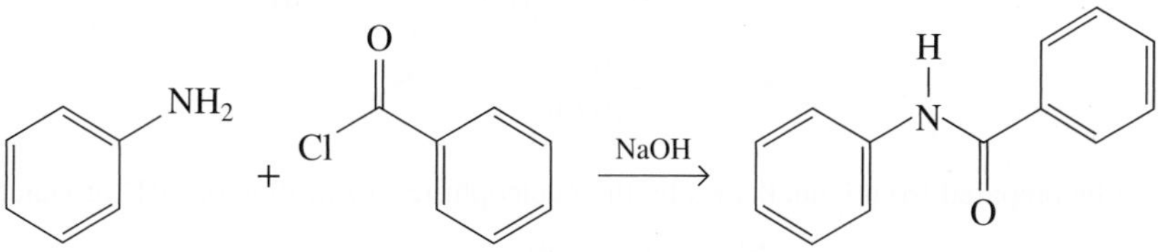

$$\text{aniline} + \text{benzoyl chloride} \xrightarrow{NaOH} N\text{-phenylbenzamide}$$

aniline

benzoyl chloride

N-phenylbenzamide (benzanilide) (75%)

SUMMARY

LO 1 Alcohols

The alcohol functional group is an —OH (hydroxyl) group bonded to an sp^3 hybridised carbon atom. Alcohols are classified as primary, secondary or tertiary, depending on whether the —OH group is bonded to a primary, secondary or tertiary carbon atom. IUPAC names for alcohols are formed by changing the suffix of the parent alkane from *-e* to *-ol*. Common names for alcohols are derived by naming the alkyl group bonded to the —OH group and adding the word 'alcohol'.

Alcohols are polar compounds with the oxygen atom bearing a partial negative charge and both the carbon atom and hydrogen atom bonded to it bearing partial positive charges. Because of intermolecular hydrogen bonding, the boiling points of alcohols are higher than those of hydrocarbons with a similar number of electrons. Because of increased dispersion forces, the boiling points of alcohols increase with increasing number of electrons (molar mass). Alcohols interact with water by hydrogen bonding and therefore are more soluble in water than hydrocarbons with a similar number of electrons.

LO 2 Reactions of alcohols

Since alcohols have a reactivity similar to water, they can be readily deprotonated to form alkoxides, or protonated to form oxonium ions. They are readily converted into alkenes, haloalkanes, aldehydes, ketones, carboxylic acids and esters. They can also be prepared from these functional groups.

LO 3 Phenols

The phenol functional group is an —OH group bonded directly to a benzene ring. Phenol and its derivatives are weak acids, with pK_a values of approximately 10.0, but are considerably stronger acids than alcohols, which have pK_a values of 16–18. They react quantitatively with strong bases to give water-soluble phenoxide salts.

LO 4 Ethers

The ether functional group is an oxygen atom bonded to two carbon atoms that are not also bonded to a heteroatom. In the IUPAC name of an ether, the parent alkane is named, and then the —OR group is named as an alkoxy substituent. Common names are derived by naming the two groups bonded to oxygen, followed by the word 'ether'. Ethers are moderately polar compounds. Their boiling points are close to those of hydrocarbons with a similar number of electrons. Because ethers are hydrogen bond acceptors, they are more soluble in water than are hydrocarbons with a similar number of electrons. Ethers are resistant to chemical reactions and do not react readily with oxidising or reducing agents or with acids or bases under moderate conditions.

LO 5 Thiols

A thiol is the sulfur analogue of an alcohol; it contains an —SH (sulfhydryl) group in place of an —OH group. Thiols are named in the same manner as alcohols, but the suffix *-e* is replaced by *-thiol*. Common names for thiols are derived by naming the alkylgroup bonded to —SH and adding the word 'mercaptan'. In compounds containing functional groups of higher precedence, the presence of a thiol is indicated by the prefix *mercapto-*. For thioethers, name the two groups bonded to sulfur, followed by the word 'sulfide'. The S—H bond is nonpolar, and the physical properties of thiols are similar to those of hydrocarbons with a similar number of electrons.

LO 6 Amines

Amines are classified as primary, secondary or tertiary, depending on the number of hydrogen atoms of ammonia replaced by alkyl or aryl groups. In an aliphatic amine, all carbon atoms bonded to nitrogen are derived from alkyl groups. In an aromatic amine, one or more of the groups bonded to nitrogen are aryl groups. A heterocyclic amine is an amine in which the nitrogen atom is part of a ring. A heterocyclic aromatic amine is an amine in which the nitrogen atom is part of an aromatic ring.

In IUPAC nomenclature, aliphatic amines are named alkanamines. In the common system of nomenclature, aliphatic amines are named alkylamines, with the alkyl groups listed in alphabetical order in one word ending in the suffix *-amine*.

Amines are polar compounds, and primary and secondary amines associate by intermolecular hydrogen bonding. Because an N—H···N hydrogen bond is weaker than an O—H···O hydrogen bond, amines have lower boiling points than alcohols with a similar number of electrons and structure. All classes of amines form hydrogen bonds with water and are more soluble in water than hydrocarbons with a similar number of electrons.

LO 7 Reactions of amines

Amines are weak bases, and aqueous solutions of amines are basic. The basicity constant for an amine in water is given the symbol K_b. It is also common to discuss the acid–base properties of amines by reference to the acidity constant, K_a, for the conjugate acid of the amine. Acidity and basicity constants for an amine in water are related by the equation $pK_a + pK_b = 14.00$ at 25 °C.

Amines react with strong acids to produce water-soluble salts, a strategy used by drug companies to enhance the solubility of drugs. Amines can also react as nucleophiles in substitution reactions.

KEY CONCEPTS AND EQUATIONS

Preparation of alcohols from alkenes
(section 19.1)

Alcohols can be prepared by the acid-catalysed hydration of alkenes.

$$\text{1-pentene} \xrightarrow[H_2SO_4]{H_2O} \text{2-pentanol (OH)}$$

Preparation of alcohols from haloalkanes
(section 19.1)

Alcohols can be prepared from haloalkanes by the nucleophilic substitution of OH^- or water.

$$\text{1-bromopentane (Br)} \xrightarrow[25\ °C]{\text{dilute NaOH}} \text{1-pentanol (OH)}$$

Preparation of alcohols by reduction of carbonyl compounds
(section 19.1)

Alcohols can be prepared by the reduction of carbonyl compounds. Aldehydes, acids and esters are reduced to primary alcohols, and ketones are reduced to secondary alcohols.

$$\text{cyclohexanone (=O)} \xrightarrow[2.\ H_3O^+]{1.\ LiAlH_4} \text{cyclohexanol (—OH)}$$

Preparation of alcohols from Grignard reagents
(section 19.1)

Alcohols can be prepared by the addition of Grignard reagents to carbonyl compounds. Aldehydes give secondary alcohols, while ketones and esters give tertiary alcohols.

$$\text{butanal (O, H)} + CH_3CH_2MgBr \xrightarrow[H_3O^+]{\text{then}} \text{3-hexanol (OH)}$$

Acidity of alcohols
(section 19.2)

In dilute aqueous solution, methanol and ethanol are comparable in acidity to water. Secondary and tertiary alcohols are weaker acids than water.

$$CH_3OH + H_2O \rightleftharpoons CH_3O^- + H_3O^+ \quad pK_a = 15.5$$

Reaction of alcohols with active metals
(section 19.2)

Alcohols react with Li, Na, K and other active metals to form metal alkoxides, which are generally stronger bases than NaOH and KOH.

$$2CH_3CH_2OH + 2Na \rightarrow 2CH_3CH_2O^-\,Na^+ + H_2$$

Reaction of alcohols with HCl, HBr and HI
(section 19.2)

Primary alcohols react with HBr and HI to form haloalkanes by an S_N2 mechanism.

$$CH_3CH_2CH_2CH_2OH + HBr \rightarrow CH_3CH_2CH_2CH_2Br + H_2O$$

Tertiary alcohols react with HCl, HBr and HI by an S_N1 mechanism, with the formation of a carbocation intermediate.

$$(CH_3)_3COH + HCl \xrightarrow{25\ °C} (CH_3)_3CCl + H_2O$$

Secondary alcohols may react with HCl, HBr and HI to form haloalkanes by an S_N2 or an S_N1 mechanism, depending on the alcohol and experimental conditions.

Reaction of alcohols with $SOCl_2$
(section 19.2)

The reaction of alcohols with thionyl chloride is often the method of choice for converting alcohols into a chloroalkanes.

$$CH_3(CH_2)_5OH + SOCl_2 \rightarrow CH_3(CH_2)_5Cl + SO_2 + HCl$$

Reaction of alcohols with phosphorus halides
(section 19.2)

The reaction of alcohols with phosphorus halides is often used for the preparation of bromoalkanes.

OH $\xrightarrow{PBr_3}$ Br

Acid-catalysed dehydration of alcohols
(section 19.2)

When isomeric alkenes are possible, the major product of acid-catalysed dehydration of alcohols is generally the more substituted alkene (Zaitsev's rule).

$$CH_3CH_2CH(OH)CH_3 \xrightarrow[heat]{H_3PO_4} CH_3CH{=}CHCH_3 + CH_3CH_2CH{=}CH_2 + H_2O$$

major product

Oxidation of primary alcohols to aldehydes
(section 19.2)

The oxidation of primary alcohols to a aldehydes is most conveniently carried out using pyridinium chlorochromate (PCC).

OH $\xrightarrow[CH_2Cl_2]{PCC}$ H, O

Oxidation of primary alcohols to carboxylic acids
(section 19.2)

Primary alcohols are oxidised to carboxylic acids by strong oxidising agents such as chromic acid and potassium permanganate.

$$CH_3(CH_2)_4CH_2OH + H_2CrO_4 \xrightarrow[acetone]{H_2O} CH_3(CH_2)_4COH + Cr^{3+}$$

(with O double-bonded to C in the product)

Oxidation of secondary alcohols to ketones
(section 19.2)

Secondary alcohols are oxidised to ketones by strong oxidising agents and by PCC.

$$CH_3(CH_2)_4CH(OH)CH_3 + H_2CrO_4 \longrightarrow CH_3(CH_2)_4C(=O)CH_3 + Cr^{3+}$$

Ester formation
(section 19.2)

Alcohols react readily with carboxylic acids under acid catalysis to produce esters (this reaction will be discussed in more detail in chapter 23).

OH + HO, O $\xrightleftharpoons{\text{acid catalyst}}$ O, O + H_2O

Acidity of phenols
(section 19.3)

Phenols are weak acids.

$$\text{C}_6\text{H}_5\text{–OH} + H_2O \rightleftharpoons \text{C}_6\text{H}_5\text{–O}^- + H_3O^+ \quad K_a = 1.02 \times 10^{-10} \quad pK_a = 9.99$$

phenol — phenoxide ion

Substitution by electron-withdrawing groups, such as halogens or the nitro group, increases the acidity of phenols.

Reaction of phenols with strong bases
(section 19.3)

Water-insoluble phenols react quantitatively with strong bases to form water-soluble salts.

$$\text{C}_6\text{H}_5\text{–OH} + NaOH \longrightarrow \text{C}_6\text{H}_5\text{–O}^-Na^+ + H_2O$$

phenol	sodium hydroxide	sodium phenoxide	water
$pK_a = 9.99$			$pK_a = 15.7$
(stronger acid)	(stronger base)	(weaker base)	(weaker base)

Oxidation of phenols *(section 19.3)*

Phenols can be easily oxidised to their corresponding quinones.

$$\text{phenol (C}_6\text{H}_5\text{OH)} \xrightarrow{\text{oxidation}} \text{1,4-benzoquinone}$$

Ester formation *(section 19.3)*

Esters of phenols can be prepared by reaction with acid chlorides or anhydrides.

$$C_6H_5OH + CH_3COCl \xrightarrow{\text{pyridine}} C_6H_5OCOCH_3$$

This reaction is often carried out in the presence of a base to neutralise the HCl as it is formed.

Ether formation *(section 19.3)*

Due to the acidity of phenols, they react with haloalkanes to produce ethers under mildly basic conditions.

$$C_6H_5OH + CH_2{=}CHCH_2Br \xrightarrow{K_2CO_3} C_6H_5OCH_2CH{=}CH_2$$

Acidity of thiols *(section 19.5)*

Thiols are weak acids, pK_a = 8–9, but are considerably stronger acids than alcohols, pK_a = 16–18.

$$CH_3CH_2SH + H_2O \rightleftharpoons CH_3CH_2S^- + H_3O^+ \quad pK_a = 8.5$$

Oxidation to disulfides *(section 19.5)*

Oxidation of a thiol by O_2 gives a disulfide.

$$2RSH + \tfrac{1}{2}O_2 \longrightarrow RS{-}SR + H_2O$$

Preparation of amines from haloalkanes *(section 19.6)*

Amines can be prepared from haloalkanes by nucleophilic substitution with an appropriate nucleophile.

$$(CH_3)_2CHBr + NH_3 \longrightarrow (CH_3)_2CHNH_2$$

Reduction of aromatic NO_2 groups *(section 19.6)*

A nitro groups, —NO_2, on an aromatic ring can be reduced to an amino group by catalytic hydrogenation, or by treatment with a metal and hydrochloric acid, followed by a strong base to liberate the free amine.

$$C_6H_5NO_2 + 3H_2 \xrightarrow[(3 \times 10^5\ \text{Pa})]{\text{Ni}} C_6H_5NH_2$$

$$C_6H_4(NO_2)_2 \xrightarrow[C_2H_5OH,\ H_2O]{\text{Fe, HCl}} C_6H_4(NH_3^+Cl^-)_2 \xrightarrow{\text{NaOH, } H_2O} C_6H_4(NH_2)_2$$

Basicity of aliphatic amines *(section 19.7)*

Most aliphatic amines have comparable basicities (pK_b = 3.0–4.0) and are slightly stronger bases than ammonia.

$$CH_3NH_2 + H_2O \rightleftharpoons CH_3NH_3^+ + OH^- \quad pK_b = 3.36$$

Basicity of aromatic amines *(section 19.7)*

Aromatic amines (pK_b = 9.0–10.0) are considerably weaker bases than aliphatic amines. Resonance stabilisation from interaction of the unshared electron pair on nitrogen with the π system of the aromatic ring decreases the availability of that electron pair for reaction with an acid. Substitution on the ring by electron-withdrawing groups further decreases the basicity of the —NH_2 group.

$$C_6H_5{-}NH_2 + H_2O \rightleftharpoons C_6H_5{-}NH_3^+ + OH^- \quad pK_b = 9.36$$

Reaction of amines with strong acids
(section 19.7)

All amines react quantitatively with strong acids to form water-soluble salts.

insoluble in water — a water-soluble salt

Conversion of a primary aromatic amine to a phenol
(section 19.7)

Treatment of a primary aromatic amine with nitrous acid gives an arenediazonium salt. Heating the aqueous solution of this salt produces N_2 gas and a phenol.

Formation of azo dyes
(section 19.7)

Arenediazonium salts also couple with phenols and anilines to give azo dyes.

Formation of *N*-nitrosamines
(section 19.7)

Secondary amines react with nitrites under acidic conditions to produce *N*-nitrosamines.

Formation of amides
(section 19.7)

Amines can react with acid chlorides and anhydrides to produce amides (this reaction will be discussed in more detail in chapter 23).

KEY TERMS

alcohol *p. 822*
aliphatic amine *p. 847*
alkoxy *p. 842*
alkylamines *p. 849*
aniline *p. 849*
aromatic amine *p. 847*
azo compounds *p. 860*
cyclic ether *p. 842*
dehydration *p. 831*
diol *p. 823*
disulfide *p. 847*
ether *p. 842*
glycol *p. 823*
heteroatom *p. 842*
heterocyclic amine *p. 847*
heterocyclic aromatic amine *p. 847*
hydrogen bonding *p. 825*
hydroxyl group *p. 822*
mercaptan *p. 845*
phenol *p. 835*
pyridinium chlorochromate (PCC) *p. 834*
thiol *p. 845*
triol *p. 823*
zwitterion *p. 858*

REVIEW QUESTIONS

LO 1 Alcohols

19.1 Classify the following alcohols as primary, secondary or tertiary.

(a) (b)

19.2 Classify the following alcohols as primary, secondary or tertiary.

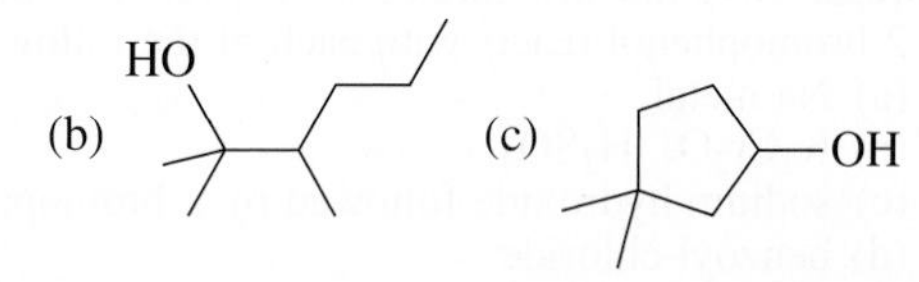

19.3 Name the following compounds.

(a) (b) (c) (d)

19.4 Name the following compounds.

(a) (c)

(b) (d)

19.5 '3-butanol' is not a proper name, but a structure could still be written for it. What is this structure, and what IUPAC name should be used?

19.6 '1-bromohexan-4-ol' is not a proper name, but a structure could still be written for it. What is this structure, and what IUPAC name should be used?

19.7 Explain why $CH_3CH_2CH_2OH$ is more soluble in water than $CH_3CH_2CH_2CH_3$.

19.8 Examine the structures of the following compounds.

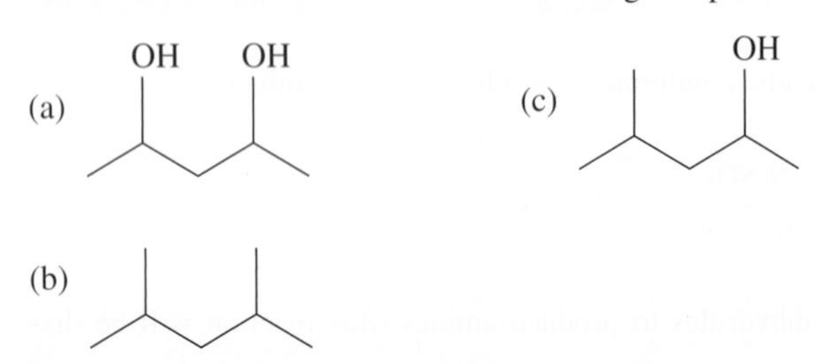

(a) Which has the highest solubility in water? Explain.
(b) Which has the lowest solubility in water? Explain.

19.9 Arrange the compounds in question 19.8 in order of increasing boiling point.

19.10 Arrange these compounds in order of increasing boiling point (values in °C are –42, 78, 117 and 198).
(a) $CH_3CH_2CH_2CH_2OH$
(b) CH_3CH_2OH
(c) $HOCH_2CH_2OH$
(d) $CH_3CH_2CH_3$

19.11 Give the structural formula of an alkene or alkenes from which each of the following alcohols can be prepared.
(a) butan-2-ol
(b) 1-methylcyclohexanol
(c) 2-methylpentan-2-ol

19.12 Give the structural formula of an alkene or alkenes from which each of the following alcohols can be prepared.
(a) octan-4-ol
(b) 4-methylhexan-2-ol
(c) cycloheptanol

19.13 Give the structural formula of the alcohol(s) formed on the reduction of the following carbonyl compounds.

(a) (b)

19.14 Give the structural formula of the alcohol(s) formed on the reduction of the following carbonyl compounds.

(a) (b)

LO 2 Reactions of alcohols

19.15 Show how to distinguish between cyclohexanol and cyclohexene using a simple chemical test.

19.16 Show how to distinguish between pentan-2-ol and 2-methylpentan-2-ol using a simple chemical test.

19.17 Write equations for the reaction of 3-methylpentan-1-ol with the following reagents.
(a) Na metal
(b) HBr, heat
(c) $K_2Cr_2O_7$, H_2SO_4, heat
(d) $SOCl_2$
(e) pyridinium chlorochromate (PCC)
(f) PBr_3

19.18 Write equations for the reaction of (i) 3-methylpentan-2-ol, and (ii) 1-methylcyclohexanol, with the following reagents.
(a) Na metal
(b) H_2SO_4, heat
(c) HBr, heat
(d) $K_2Cr_2O_7$, H_2SO_4, heat
(e) $SOCl_2$
(f) pyridinium chlorochromate (PCC)

19.19 Write the equation for the equilibrium that is present in a solution of butanoic acid and ethanol with a trace of strong acid.

19.20 When the reaction described in 19.19 is carried out in the laboratory, ethanol is used in vast excess, usually as the solvent, even though only one mole equivalent is required. What is the advantage of using such an excess of ethanol?

LO 3 Phenols

19.21 A monofunctional organic oxygen compound dissolves in aqueous base but not in aqueous acid. To which family of organic compounds that we studied in this chapter does this compound belong? Explain.

19.22 Why do water-insoluble carboxylic acids ($pK_a = 4–5$) dissolve in 10% sodium bicarbonate with the evolution of a gas, but water-insoluble phenols ($pK_a = 9.5–10.5$) do not show this chemical behaviour?

19.23 Use resonance theory to explain why phenol ($pK_a = 9.99$) is a stronger acid than cyclohexanol ($pK_a = 18$).

19.24 Arrange the compounds in each of the following sets in order of increasing acidity (from least acidic to most acidic).

(a) C_6H_5–OH, C_6H_{11}–OH, CH_3COOH

(b) C_6H_5–OH, $NaHCO_3$, H_2O

(c) O_2N–C_6H_4–OH, C_6H_5–OH, C_6H_5–CH_2OH

19.25 Give the structure of the expected product formed when 4-methoxyphenol reacts with each of the following reagents.
(a) sodium hydroxide
(b) chromic acid
(c) acetyl chloride
(d) sodium hydroxide followed by methyl iodide
(e) benzenediazonium chloride (section 19.7)

19.26 Give the structure of the expected product formed when 2-bromophenol reacts with each of the following reagents.
(a) Na metal
(b) $K_2Cr_2O_7$, H_2SO_4
(c) sodium hydroxide followed by 2-bromopropane
(d) benzoyl chloride

LO 4 Ethers

19.27 Write names for the following ethers.

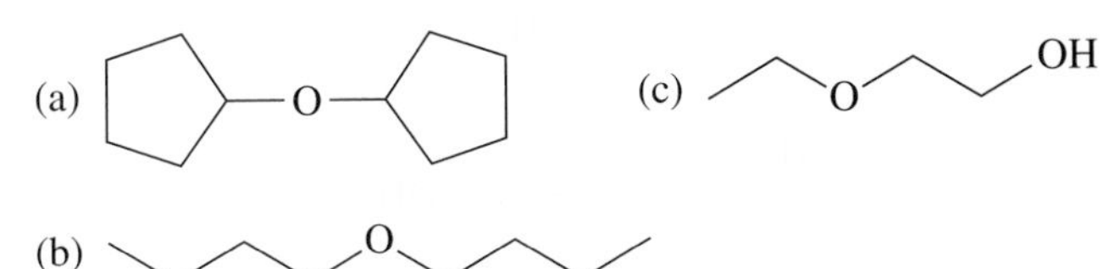

19.28 Draw a structural formula for each of the following ethers.
(a) 2-methyl-1-ethoxypropane
(b) 1,2,3-trimethoxybenzene
(c) 3-chloro-l-ethoxybutane

19.29 Which of the following compounds is more soluble in water? Explain.

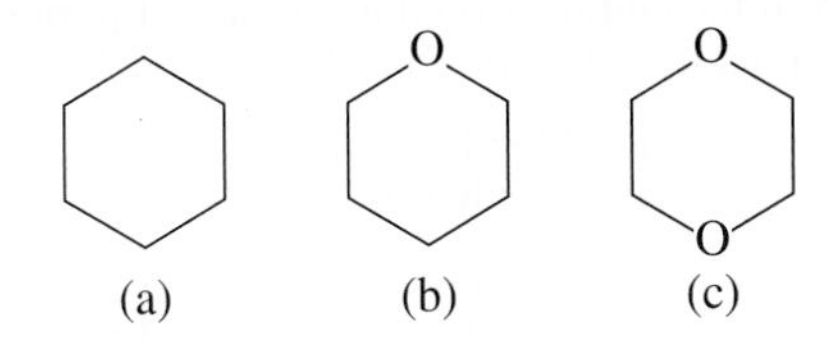

19.30 Arrange the following compounds in order of increasing boiling point (values in °C are –42, –24, 78 and 118).
A CH_3CH_2OH
B CH_3OCH_3
C $CH_3CH_2CH_3$
D CH_3COOH

19.31 Each of the following compounds is a common organic solvent. From each pair of compounds, select the solvent with the greater solubility in water.
(a) $CH_3CH_2OCH_2CH_3$ or CH_2Cl_2
(b) $CH_3CH_2OCH_2CH_3$ or CH_3CH_2OH

19.32 Each of the following compounds is a common organic solvent. From each pair of compounds, select the solvent with the greater solubility in water.
(a) $CH_3CH_2OCH_2CH_3$ or $CH_3(CH_2)_3CH_3$
(b) $CH_3CH_2OCH_2CH_3$ or $CH_3OCH_2CH_2OCH_3$
(c) $CH_3CH_2OCH_2CH_3$ or $(CH_3)_3COCH_2CH_3$

LO 5 Thiols

19.33 Draw structures corresponding to the following IUPAC names.
(a) 3-methylpentane-1-thiol
(b) diisopropyl sulfide
(c) 3-mercaptopropanol
(d) dimethyl disulfide

18.34 Name the following compounds.

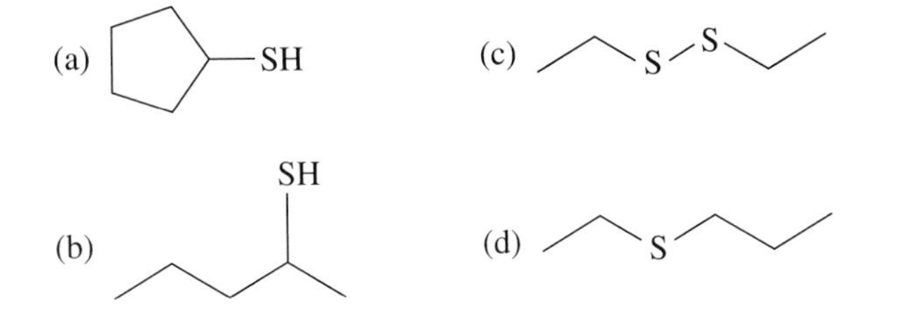

19.35 The following are structural formulae for butan-1-ol and butane-1-thiol. One of these compounds has a boiling point of 98.5 °C, the other has a boiling point of 117 °C. Which compound has which boiling point? Explain.

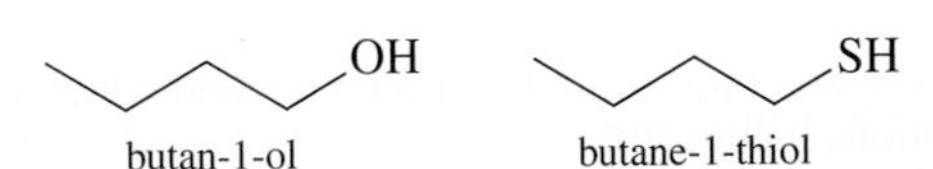

19.36 Which of the two compounds in question 19.35 would you expect to have greater solubility in water? Which would be more soluble in dilute sodium hydroxide solution? Explain.

19.37 From each of the following pairs, select the stronger acid, and write a structural formula for its conjugate base.
(a) H_2O or H_2S
(b) CH_3OH or CH_3SH
(c) CH_3COOH or CH_3CH_2SH

19.38 From each of the following pairs, select the stronger base, and write the structural formula of its conjugate acid.
(a) $CH_3CH_2CH_2CH_2O^-$ or $CH_3CH_2CH_2CH_2S^-$
(b) $CH_3CH_2CH_2S^-$ or OH^-

LO 6 Amines

19.39 Classify each amino group in the following compounds as primary, secondary or tertiary, and as aliphatic or aromatic.

(a)

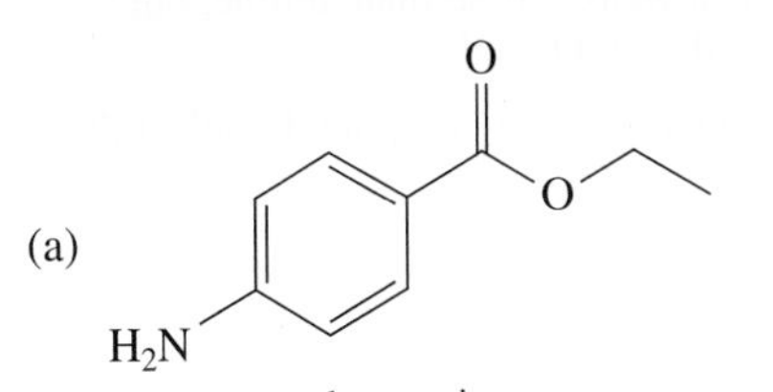

benzocaine
(a topical anaesthetic)

(b)

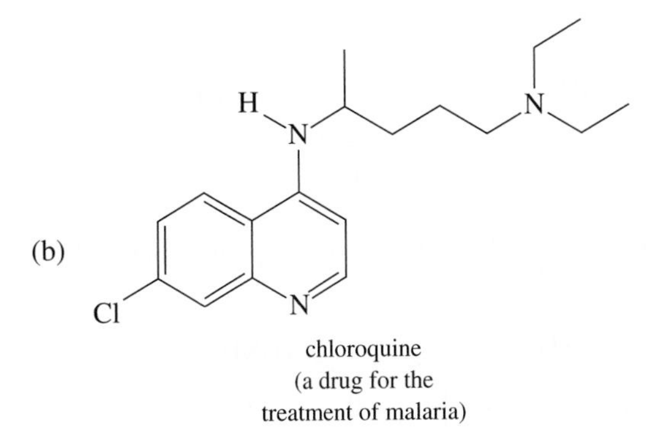

chloroquine
(a drug for the
treatment of malaria)

19.40 Draw examples of 1°, 2° and 3° amines that contain at least five carbon atoms. Using the same criterion, provide examples of 1°, 2° and 3° alcohols. How does the classification system differ between the two functional groups?

19.41 Draw a structural formula for each of the following amines.
(a) (*R*)-butan-2-amine
(b) octan-1-amine
(c) 2,2-dimethylpropan-1-amine
(d) 2-bromoaniline

19.42 Draw a structural formula for each of the following amines.
(a) hexanane-1-4 diamine
(b) tripropylamine
(c) *N,N*-dimethylaniline
(d) benzylamine

19.43 Why does butan-1-amine have a lower boiling point than butan-1-ol?

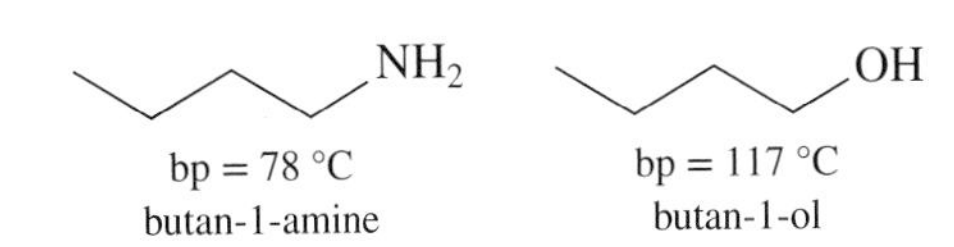

19.44 Propylamine, ethylmethylamine, and trimethylamine are constitutional isomers with molecular formula C_3H_9N.

$CH_3CH_2CH_2NH_2$	$CH_3CH_2NHCH_3$	$(CH_3)_3N$
bp = 48 °C	bp = 37 °C	bp = 3 °C
propylamine	ethylmethylamine	trimethylamine

Explain why trimethylamine has the lowest boiling point of the three, and propylamine has the highest.

LO7 Reactions of amines

19.45 A monofunctional organic nitrogen compound dissolves in aqueous hydrochloric acid but not in aqueous sodium hydroxide. What kind of organic compound is it?

19.46 A crystalline monofunctional organic nitrogen compound readily dissolves in water, but produces a precipitate when this solution is made alkaline with sodium hydroxide. What kind of organic compound is it?

19.47 Use resonance theory to explain why aniline ($pK_b = 9.36$) is a weaker base than cyclohexylamine ($pK_b = 3.34$).

19.48 Why does substitution of a nitro group make an aromatic amine a weaker base, but it makes a phenol a stronger acid? For example, 4-nitroaniline is a weaker base than aniline, but 4-nitrophenol is a stronger acid than phenol.

19.49 From each of the following pairs of compounds, select the stronger base.

(a) 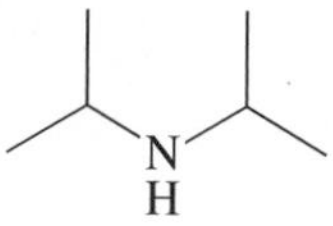 A (diisopropylamine, N–H) or B (pyridine)

(b) A (N-methylaniline, NH) or B (4-methylcyclohexylamine, NH_2)

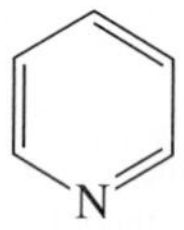

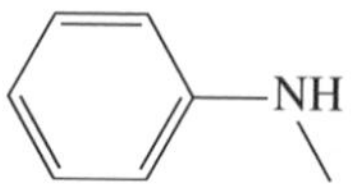

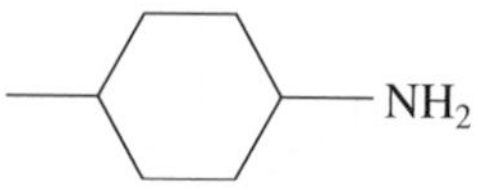

19.50 From each of the following pairs of compounds, select the stronger base.

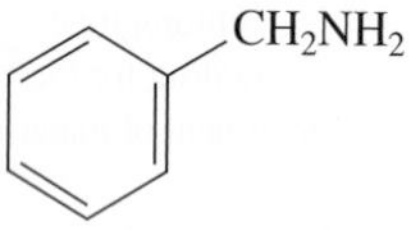

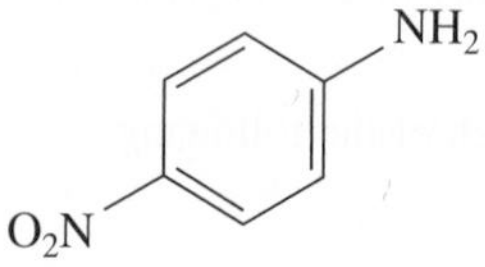

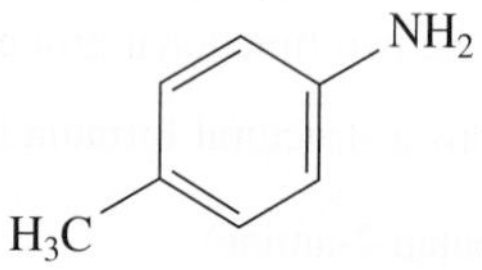

(a) 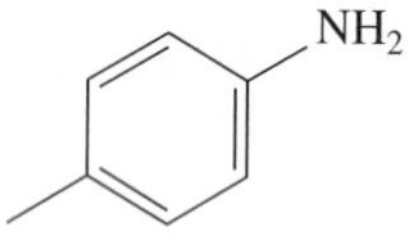 A (4-methylaniline, NH_2) or B (benzylamine, CH_2NH_2)

(b) A (4-nitroaniline, O_2N, NH_2) or B (4-methylaniline, H_3C, NH_2)

19.51 The following is a structural formula of pyridoxamine, one form of vitamin B_6.

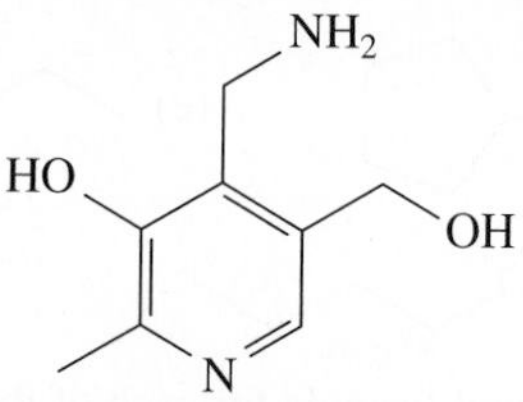

pyridoxamine (vitamin B_6)

(a) Which nitrogen atom of pyridoxamine is the stronger base?
(b) Draw the structural formula of the hydrochloride salt formed when pyridoxamine is treated with 1 mole of HCl.
(c) Is pyridoxamine chiral? Would a solution of pyridoxamine in water be optically active or optically inactive?

19.52 Procaine was one of the first local anaesthetics used for infiltration and regional anaesthesia.

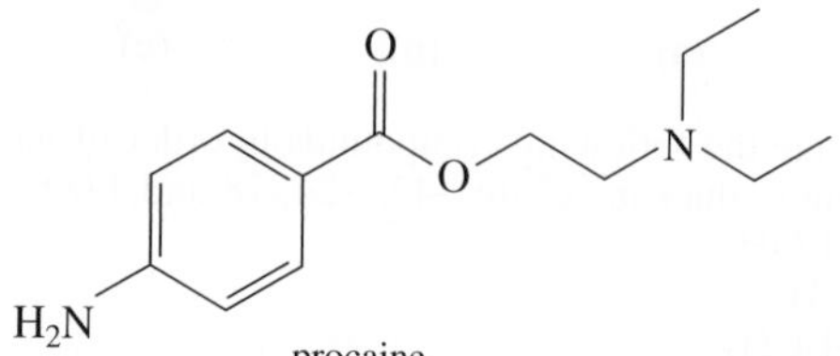

procaine

The hydrochloride salt of procaine is marketed as Novocaine®.
(a) Which nitrogen atom of procaine is the stronger base?
(b) Draw the formula of the salt formed by treating procaine with 1 mole of HCl.
(c) Is procaine chiral? Would a solution of Novocaine in water be optically active or optically inactive?

19.53 Give the structure of the expected product formed when benzylamine reacts with each of the following reagents.
(a) HBr
(b) iodomethane
(c) acetyl chloride
(d) acetic acid
(e) 1-bromobutane

19.54 Give the structure of the expected product formed when 4-methylaniline reacts with each of the following reagents.
(a) benzoyl chloride
(b) 2-bromopropane
(c) dilute hydrochloric acid solution
(d) nitrous acid
(e) propanoic acid

REVIEW PROBLEMS

19.55 Draw a structural formula for each of the following compounds. **LO1**
(a) isopropyl alcohol
(b) propylene glycol
(c) (*R*)-5-methyl-2-hexanol
(d) 4-aminobutanoic acid
(e) 2-aminoethanol (ethanolamine)
(f) 2-aminobenzoic acid

19.56 Draw a structural formula for each of the following compounds. **LO1**
(a) 2-methyl-2-propyl-1,3-propanediol
(b) 2,2-dimethyl-1-propanol
(c) 2-hydroxybutanoic acid
(d) (*S*)-2-aminopropanoic acid (alanine)
(e) 4-aminobutanal
(f) 4-amino-2-butanone

19.57 Name and draw structural formulae for the eight isomeric alcohols with molecular formula $C_5H_{12}O$. Which of these are chiral? **LO1**

19.58 There are eight constitutional isomers with molecular formula $C_4H_{11}N$. Name and draw a structural formula for each. Classify each amine as primary, secondary or tertiary. **LO6**

19.59 Draw a structural formula for each of the following compounds with the given molecular formula. **LO6**
(a) 2° arylamine, C_7H_9N
(b) 3° arylamine, $C_8H_{11}N$
(c) 1° aliphatic amine, C_7H_9N

19.60 Draw a structural formula for a compound which matches the descriptions below. **LO6**
(a) a chiral 1° amine with a molecular formula $C_5H_{13}N$
(b) a 3° heterocyclic amine with a molecular formula $C_6H_{13}N$
(c) a trisubstituted 1° aromatic amine with a molecular formula $C_{10}H_{16}N$

19.61 Epinephrine is a hormone secreted by the adrenal medulla. One of epinephrine's actions is as a bronchodilator. Albuterol, sold under several trade names including Ventolin® and Salbutamol®, is one of the most effective and widely prescribed anti-asthma drugs. The *R* enantiomer of albuterol is 68 times more effective in the treatment of asthma than the *S* enantiomer. **LO 1,6**

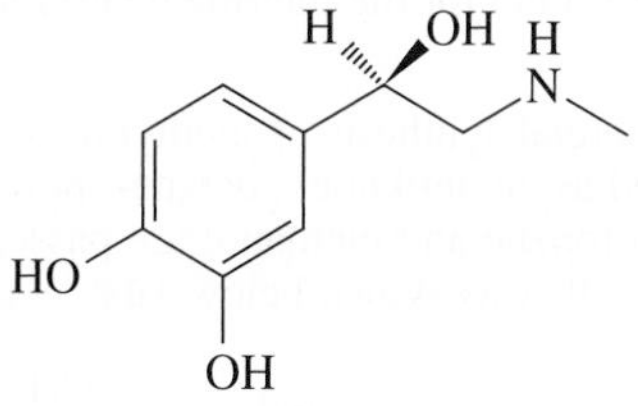

(*R*)-epinephrine
(adrenaline)

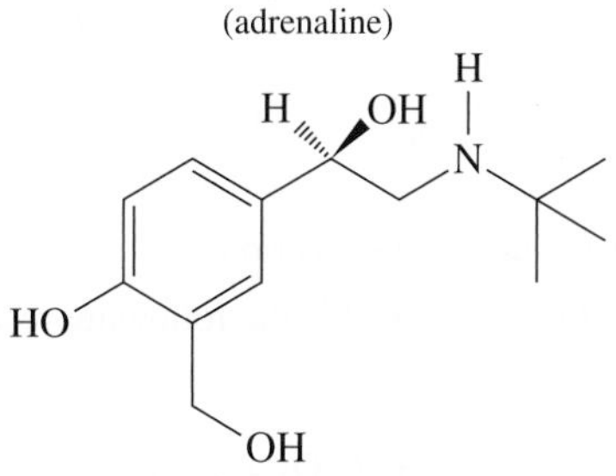

(*R*)-albuterol

(a) Label and name each functional group present.
(b) Classify each alcohol and amino group as primary, secondary or tertiary.
(c) List the similarities and differences between the structural formulae of these compounds.

19.62 Naltrexone is used to help recovering narcotic addicts stay drug free. **LO 1,6,7**

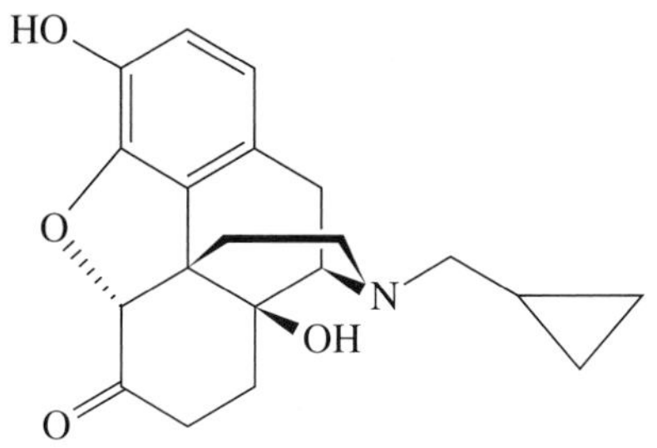

naltrexone

(a) Label and name all the functional groups. Where relevant, indicate whether the group is primary, secondary or tertiary.
(b) Upon addition of dilute hydrochloric acid solution, naltrexone forms a water-soluble salt. Draw the structure of this salt.

19.63 Swainsonine was first isolated from the Australian plant *Swainsona canescens*. Many plants containing swainsone were known as locoweeds because swainsonine causes a variety of neurological disorders in grazing livestock. **LO 1,6,7**

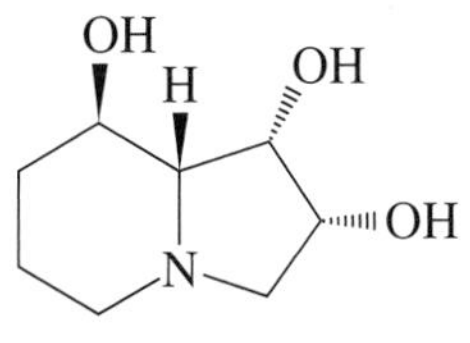

swainsonine

(a) Label and name all the functional groups. Where relevant, indicate whether the group is primary, secondary or tertiary.
(b) Upon addition of dilute hydrochloric acid solution, swainsonine forms a salt. Draw the structure of this salt.

19.64 Explain why putrescine, a foul-smelling compound produced by rotting flesh, ceases to smell upon treatment with two equivalents of HCl. **LO 6,7**

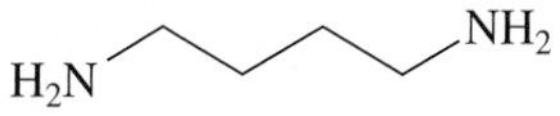

H_2N–$(CH_2)_4$–NH_2

butane-1,4-diamine
(putrescine)

19.65 Describe a procedure for separating a mixture of octan-1-ol and 4-bromophenol and recovering each in pure form. Both are insoluble in water but are soluble in diethyl ether. **LO 3**

19.66 Aniline is prepared by the catalytic reduction of nitrobenzene. **LO 7**

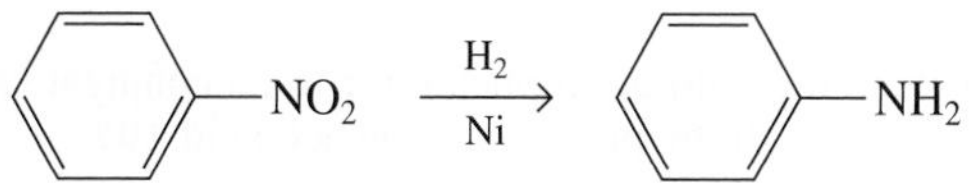

$$C_6H_5NO_2 \xrightarrow[Ni]{H_2} C_6H_5NH_2$$

Devise a chemical procedure based on the basicity of aniline to separate it from any unreacted nitrobenzene.

19.67 Suppose that you have a mixture of the following three compounds. **LO 3,7**

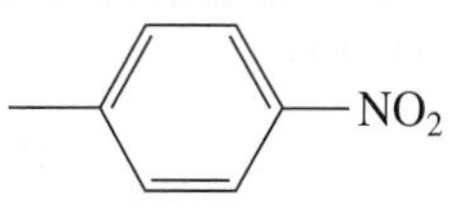

4-nitrotoluene
(*p*-nitrotoluene)

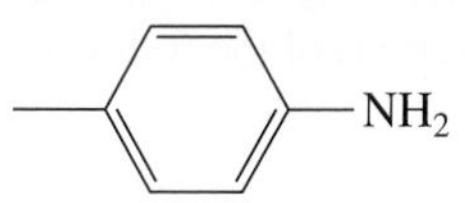

4-methylaniline
(*p*-toluidine)

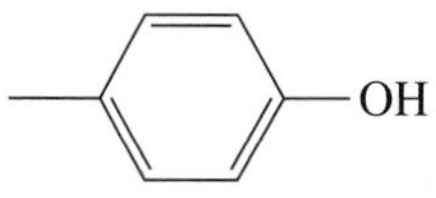

4-methylphenol
(*p*-cresol)

Devise a chemical procedure based on their relative acidity or basicity to separate and isolate each in pure form.

19.68 From each of the following pairs, select the stronger base. **LO 3**

(a) 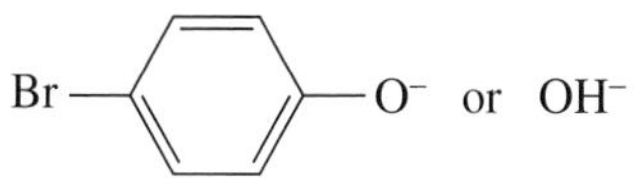 Br–C_6H_4–O^- or OH^-

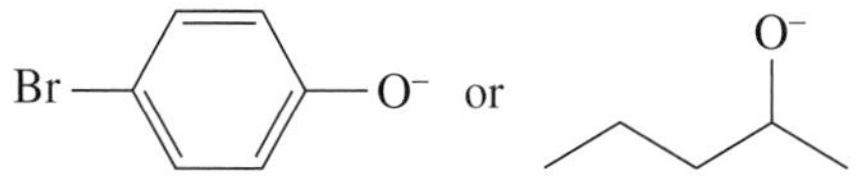

(b) Br–C_6H_4–O^- or butan-2-olate ($CH_3CH_2CH(O^-)CH_3$)

(c) Br–C_6H_4–O^- or HCO_3^-

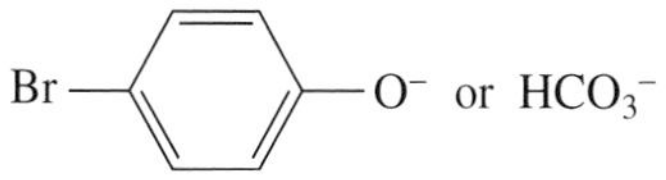

(d) 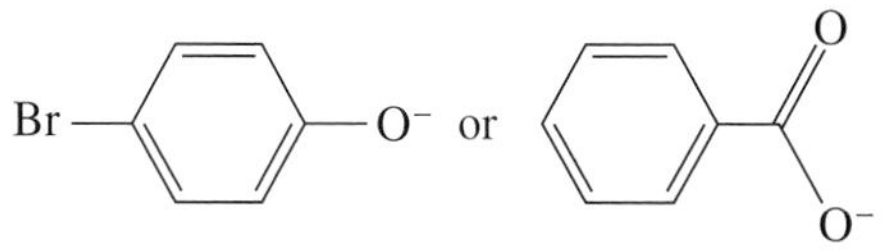 Br–C_6H_4–O^- or benzoate ($C_6H_5CO_2^-$)

19.69 Predict the position of equilibrium for each of the following acid–base reactions; that is, does each lie considerably to the left, does each lie considerably to the right, or are the concentrations evenly balanced? **LO 2,5**

(a) $CH_3CH_2OH + Na^+OH^- \rightleftharpoons CH_3CH_2O^-Na^+ + H_2O$
(b) $CH_3CH_2SH + Na^+OH^- \rightleftharpoons CH_3CH_2S^-Na^+ + H_2O$
(c) $CH_3CH_2OH + CH_3CH_2S^-Na^+ \rightleftharpoons CH_3CH_2O^-Na^+ + CH_3CH_2SH$
(d) $CH_3CH_2S^-Na^+ + CH_3COOH \rightleftharpoons CH_3CH_2SH + CH_3COO^-Na^+$

19.70 The pK_a value of the morpholinium ion is 8.33. **LO7**

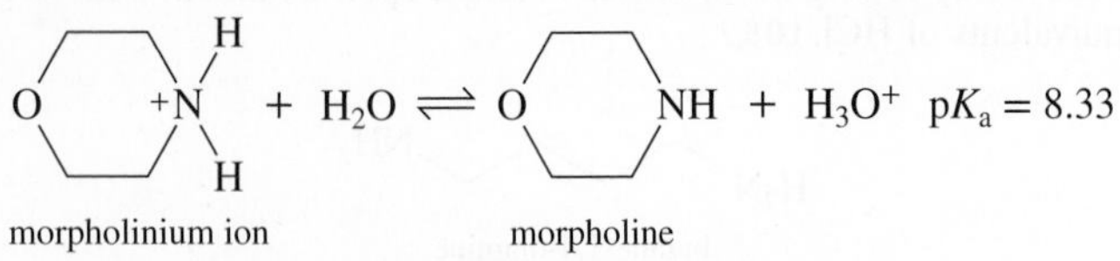

morpholinium ion + H_2O ⇌ morpholine + H_3O^+ pK_a = 8.33

(a) Calculate the ratio of morpholine to morpholinium ion in aqueous solution at pH 7.0.
(b) At what pH are the concentrations of morpholine and morpholinium ion equal?

19.71 The pK_b of amphetamine is approximately 3.2. Calculate the ratio of amphetamine to its conjugate acid at pH 7.4, the pH of blood plasma. **LO7**

19.72 Calculate the ratio of amphetamine to its conjugate acid at pH 1.0, such as might be present in stomach acid. **LO7**

19.73 Arrange the compounds in each of the following sets in order of decreasing solubility in water. **LO1,4**
(a) propanol, butane, diethyl ether
(b) hexan-1-ol, hexane-1,2-diol, dipropyl ether

19.74 Write the structures of the products of the acid-catalysed dehydrations of the following compounds. **LO2**

(a) 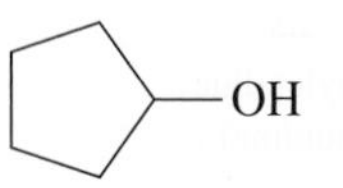 OH (c) 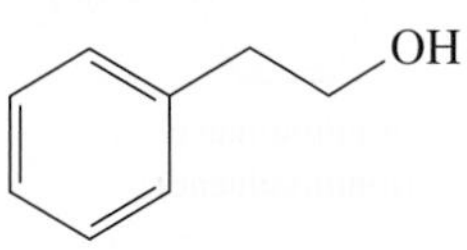 OH

(b) 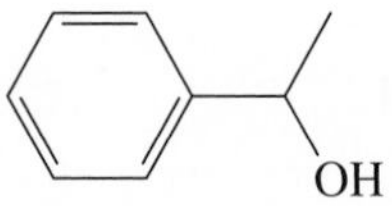 OH

19.75 Write the structures of the products of the acid-catalysed dehydrations of the following compounds. If more than one alkene is possible show them all, and indicate which isomer would predominate. **LO2**

(a) 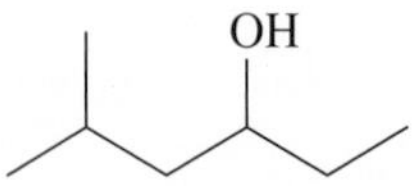 OH (c) 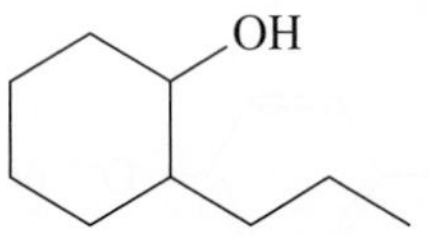 OH

(b) 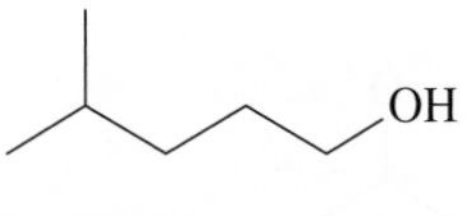 OH

19.76 Write the structures of the substitution products that form when the alcohols in question 19.74 are heated with thionyl chloride, $SOCl_2$. **LO2**

19.77 Write the structures of the substitution products that form when the alcohols in question 19.75 are heated with concentrated hydroiodic acid, HI. **LO2**

19.78 Write the structures of the products that can be prepared by the oxidation of the compounds in question 19.74 using chromic acid, H_2CrO_4. **LO2**

19.79 Write the structures of the products that can be prepared by the oxidation of the compounds in question 19.75 using pyridinium chlorochromate (PCC). **LO2**

19.80 Write the structure of the product of the acid-catalysed dehydration of pentan-3-ol. Write the mechanism of the reaction. **LO2**

19.81 When (*R*)-butan-2-ol is left in aqueous acid, it slowly loses its optical activity. When the organic material is recovered from the aqueous solution, only butan-2-ol is found. Account for the observed loss of optical activity. **LO2**

19.82 What is the most likely mechanism of the following reaction? **LO2**

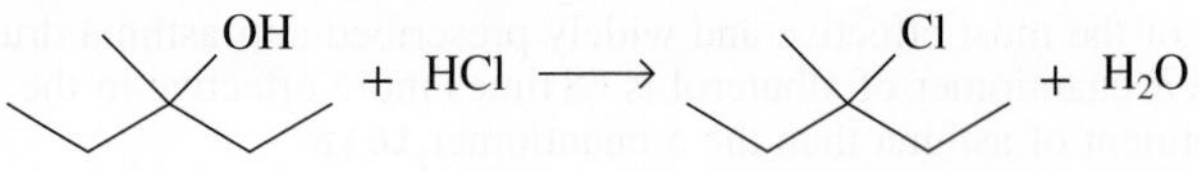

OH + HCl ⟶ Cl + H_2O

Draw a structural formula for the intermediate(s) formed during the reaction.

19.83 In the commercial synthesis of methyl *tert*-butyl ether (MTBE), once used as an antiknock, octane-improving petrol additive, 2-methylpropene and methanol are passed over an acid catalyst to give the ether as shown below. **LO4**

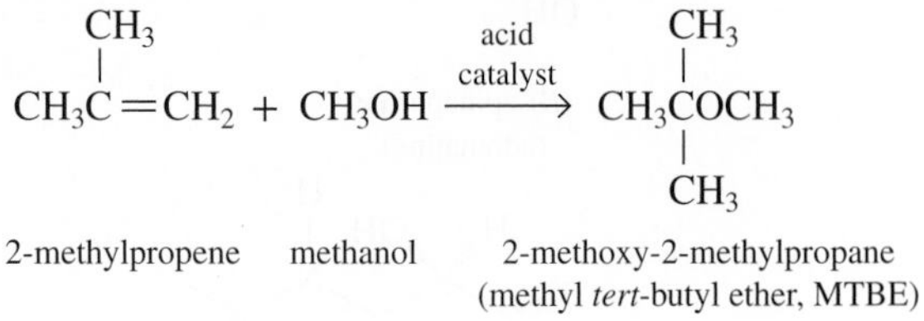

$$CH_3C(CH_3){=}CH_2 + CH_3OH \xrightarrow{\text{acid catalyst}} CH_3C(CH_3)_2OCH_3$$

2-methylpropene methanol 2-methoxy-2-methylpropane (methyl *tert*-butyl ether, MTBE)

Propose a mechanism for this reaction.

19.84 Complete the equations for the following reactions. **LO2**

(a) 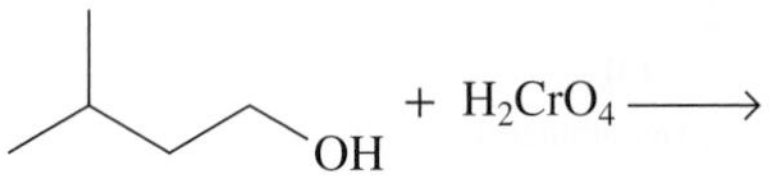 OH + H_2CrO_4 ⟶

(b) 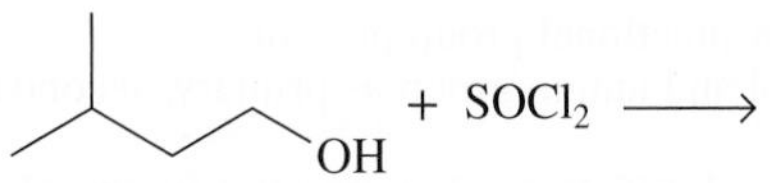 OH + $SOCl_2$ ⟶

(c) 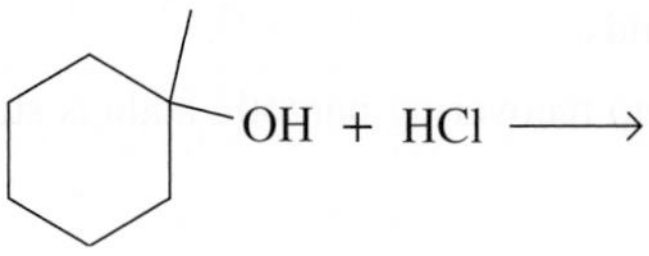 OH + HCl ⟶

19.85 Complete the equations for the following reactions. **LO2**

(a) 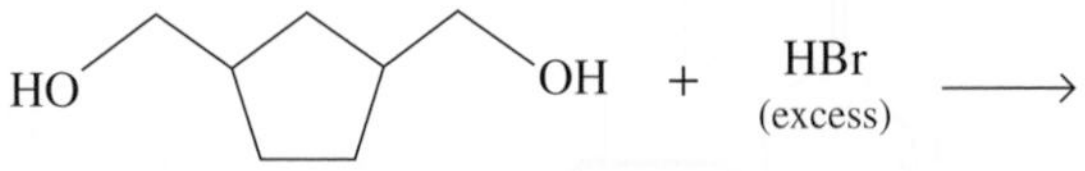 HO OH + HBr (excess) ⟶

(b)  OH + H_2CrO_2 ⟶

19.86 Write equations showing the reaction between the following pairs of substances. **LO2,3**

(a) 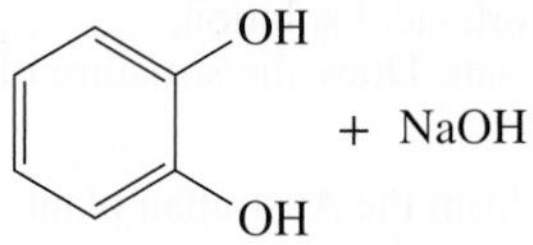 OH OH + NaOH

(b) 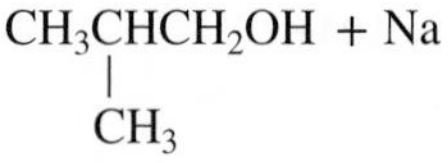 $CH_3CH(CH_3)CH_2OH$ + Na

(c) 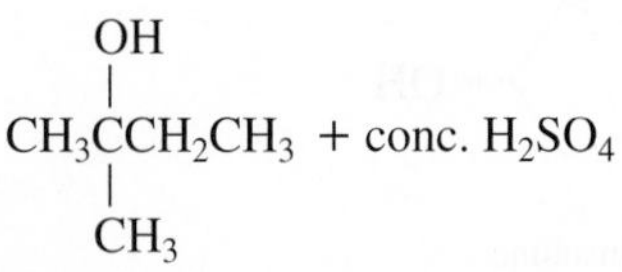 $CH_3C(OH)(CH_3)CH_2CH_3$ + conc. H_2SO_4

(d) 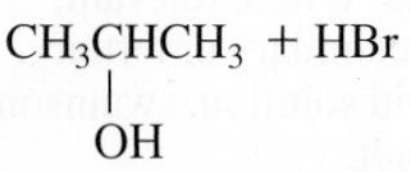 $CH_3CH(OH)CH_3$ + HBr

19.87 Write equations showing the reaction between the following pairs of substances. **LO 2,3**

(a) 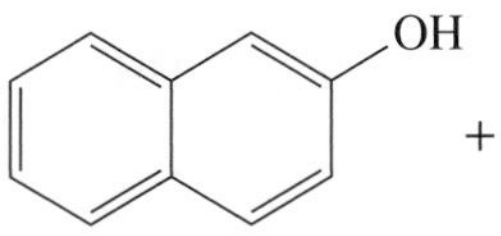 [2-naphthol structure, OH] + NaOH

(b) $CH_3C(OH)(CH_3)CH_3$ + HI

(c) $CH_3CH_2CH_2OH + SOCl_2$

(d) $CH_3CH(OH)CH_3 + K_2Cr_2O_7$

(e) $CH_3CH(OH)CH_3 + PBr_3$

19.88 Explain how you could use simple chemical tests to distinguish between the following sets of compounds. Explain your observations and give equations where necessary. **LO 2,3**

(a)  [4-tert-butylphenol, OH] [4-tert-butylcyclohexanol, OH]

(b) 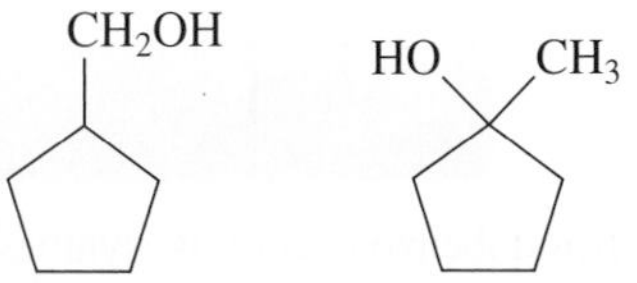 [cyclopentylmethanol, CH_2OH] [1-methylcyclopentanol, HO, CH_3]

19.89 Explain how you could use simple chemical tests to distinguish between the following set of compounds. Explain your observations and give equations where necessary. **LO 2,4**

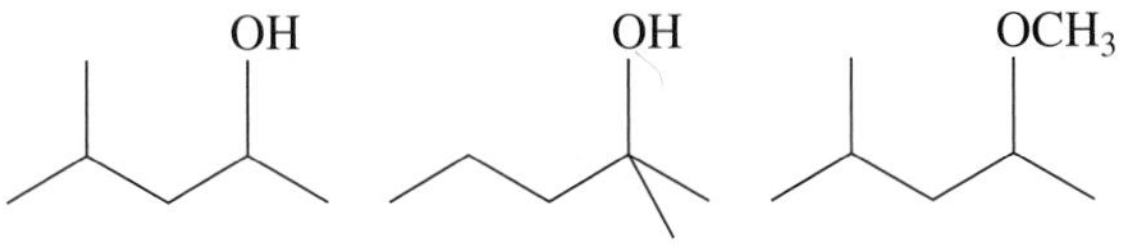

[structures: OH, OH, OCH_3]

19.90 Show how to convert cyclohexanol to the following compounds. **LO 1**
(a) cyclohexene
(b) cyclohexane
(c) cyclohexanone

19.91 Show how to convert: **LO 1,2**
(a) butan-1-ol to butan-2-ol in two steps
(b) hex-1-ene to hexan-2-one in two steps
(c) propene to acetone (propanone) in two steps.

19.92 Show how to prepare each of the following compounds from 2-methylpropan-1-ol. **LO 1,2**

(a) $CH_3C(CH_3){=}CH_2$ (b) $CH_3C(CH_3)(OH)CH_3$ (c) $CH_3CH(CH_3)COOH$

For any preparation involving more than one step, show each intermediate compound formed.

19.93 Show how to prepare each of the following compounds from 2,4-dimethylhexan-1-ol. **LO 1,2**

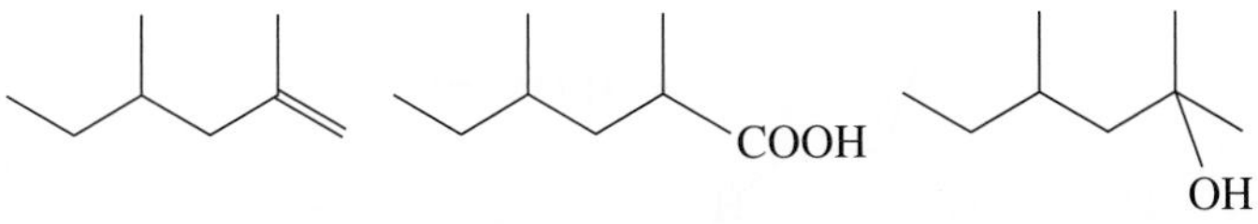

[structures: alkene; COOH; OH]

For any preparation involving more than one step, show each intermediate compound formed.

19.94 Show how to convert the alcohol on the left to compounds (a), (b) and (c). **LO 2**

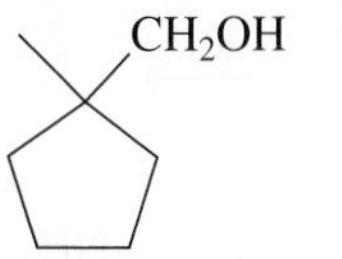

[structures: 1-methylcyclopentyl CH_2OH; (a) CH_2Cl; (b) CHO (O, H, H); (c) COOH (O, OH)]

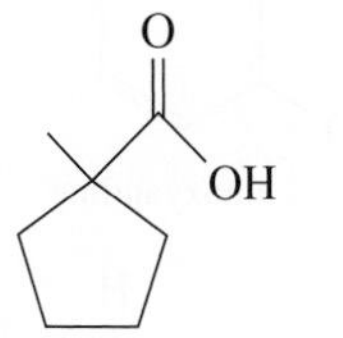

19.95 Show reagents and experimental conditions that can be used to synthesise the following compounds from propan-1-ol (any derivative of propan-1-ol prepared in an earlier part of this question may be used for a later synthesis). **LO 1,2**
(a) propanal
(b) propanoic acid
(c) propene
(d) propan-2-ol
(e) 2-bromopropane
(f) 1-chloropropane
(g) acetone

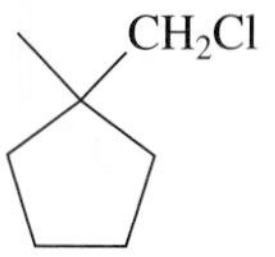

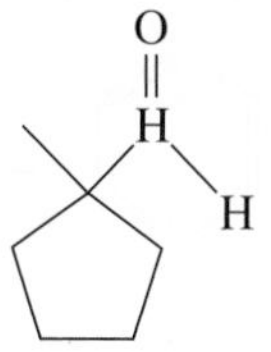

19.96 The compound 4-aminophenol is a building block in the synthesis of the analgesic acetaminophen. Show how this building block can be used in the synthesis of acetaminophen in three steps from phenol. **LO 6,7**

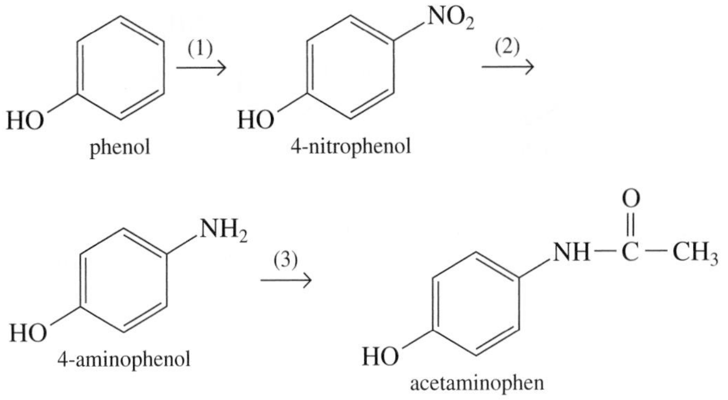

phenol (HO) —(1)→ 4-nitrophenol (HO, NO_2) —(2)→ 4-aminophenol (HO, NH_2) —(3)→ acetaminophen (HO, $NH{-}C(=O){-}CH_3$)

19.97 The topical anaesthetic benzocaine can be synthesised in four steps from toluene. Show how this can be achieved. **LO 2,6**

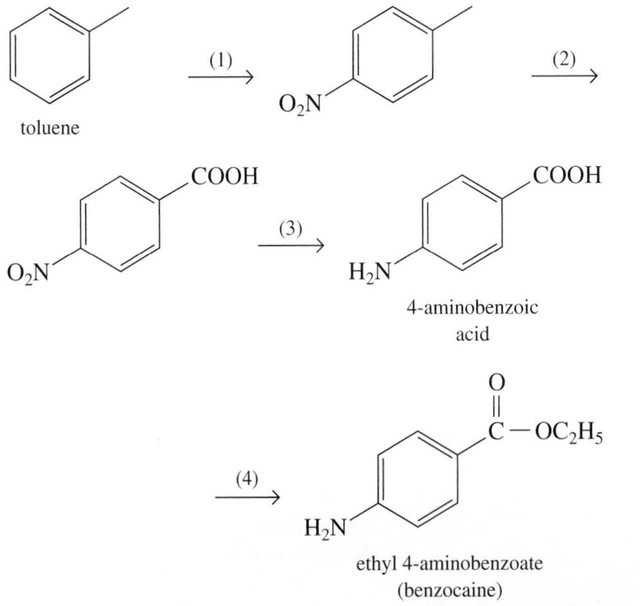

toluene —(1)→ (O_2N) —(2)→ (O_2N, COOH) —(3)→ 4-aminobenzoic acid (H_2N, COOH) —(4)→ ethyl 4-aminobenzoate (benzocaine) (H_2N, $C(=O){-}OC_2H_5$)

19.98 The over-the-counter analgesic phenacetin is synthesised in four steps from phenol. **LO 3,4,6**

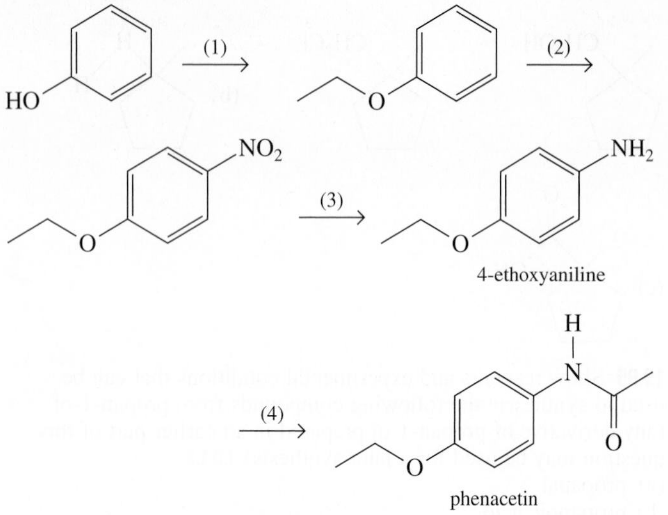

(1) → (2) → (3) → 4-ethoxyaniline (4) → phenacetin

Show reagents for each step of the synthesis of phenacetin.

19.99 The intravenous anaesthetic propofol is synthesised from phenol in four steps. Show the reagents required for steps (1) to (3). **LO 3,6,7**

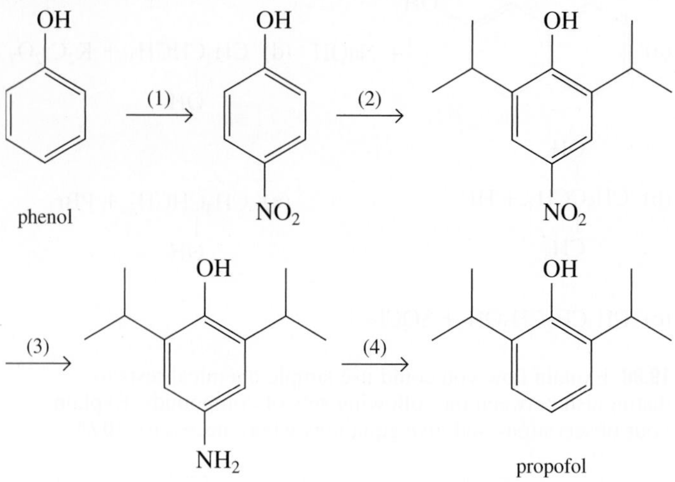

phenol (1) → (2) → (3) → (4) → propofol

ADDITIONAL EXERCISES

19.100 Which of the following compounds is a better nucleophile? Give reasons for your answer. **LO 7**

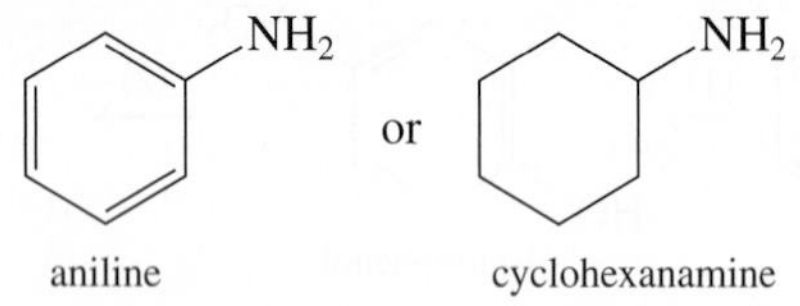

aniline or cyclohexanamine

19.101 Radiopaque imaging agents are substances, administered either orally or intravenously, that absorb X-rays more strongly than body material does. One of the best known of these agents is barium sulfate, the key ingredient in the 'barium cocktail' used for imaging the gastrointestinal tract. Among other X-ray imaging agents are the triiodoaromatics. You can get some idea of the kinds of imaging they are used for from the following selection of trade names: Angiografin®, Gastrografin, Cardiografin, Cholografin, Renografin, and Urografin®. The most common of the triiodoaromatics are derivatives of the following three triiodobenzenecarboxylic acids. **LO 6**

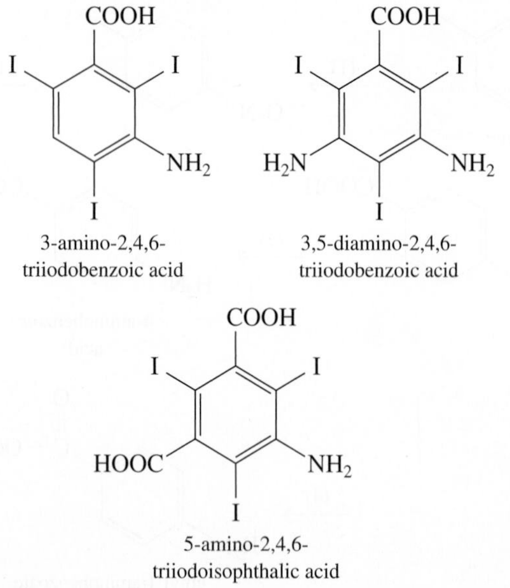

3-amino-2,4,6-triiodobenzoic acid

3,5-diamino-2,4,6-triiodobenzoic acid

5-amino-2,4,6-triiodoisophthalic acid

The compound 3-amino-2,4,6-triiodobenzoic acid is synthesised from benzoic acid in three steps.

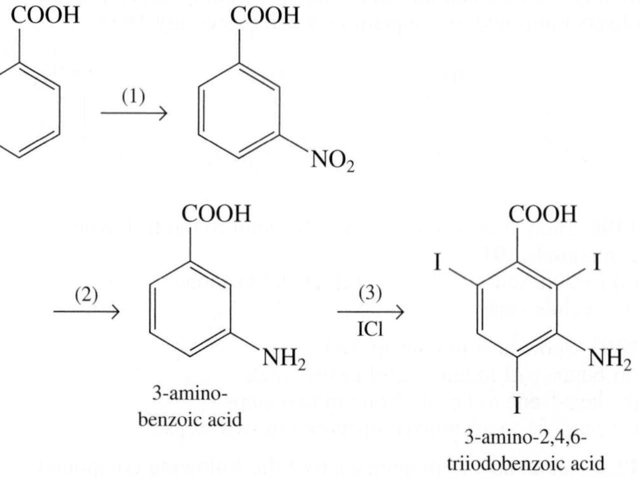

(1) → (2) → 3-aminobenzoic acid (3) ICl → 3-amino-2,4,6-triiodobenzoic acid

(a) Show the reagents required for steps (1) and (2).
(b) Iodine monochloride, ICl, a black crystalline solid with a melting point of 27.2 °C and a boiling point of 97 °C, is prepared by mixing equimolar amounts of I_2 and Cl_2. Propose a mechanism for the iodination of 3-aminobenzoic acid using this reagent.
(c) Show how to prepare 3,5-diamino-2,4,6-triiodobenzoic acid from benzoic acid.

19.102 Predict the product of the following acid–base reaction.

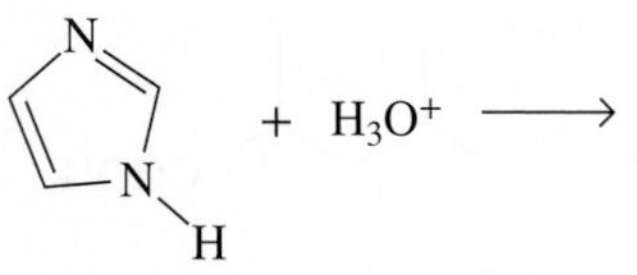

$+ H_3O^+ \longrightarrow$

19.103 Amines can act as nucleophiles. For each of the following molecules, circle the atom that would most likely be attacked by the nitrogen atom of an amine. **LO7**

(a) 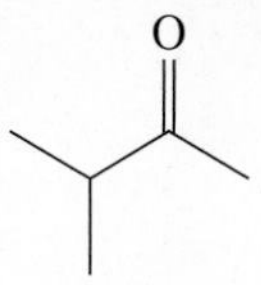 O

(b) 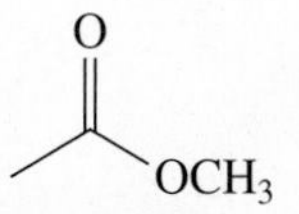 O OCH$_3$

(c) Cl Br

19.104 How can the following transformation be accomplished? (*Hint:* This is a two-step process involving an alcohol intermediate.) **LO4**

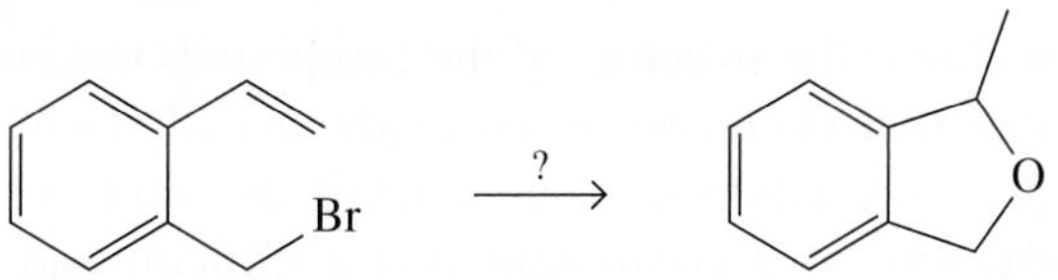

Br ? O

19.105 Give at least two chemically reasonable syntheses of ethyl 2-methylpropyl ether (ethyl isobutyl ether) using ethanol and methylpropan-1-ol as your starting materials. **LO4**

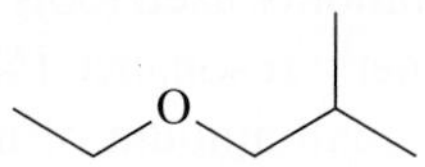

O

ethyl 2-methylpropyl ether

19.106 A pure liquid, *X*, reacts with sodium to produce hydrogen. It also reacts with phosphorus pentachloride to yield compound *Y*, which on treatment with alcoholic potassium hydroxide solution gives pent-1-ene only. What are the structures of *X* and *Y* and the reactions involved? **LO2**

19.107 When neopentyl alcohol (2,2-dimethylpropan-1-ol), $(CH_3)_3CCH_2OH$, is heated with acid, it is slowly converted into an 85 : 15 mixture of two alkenes with the formula C_5H_{10}. What are these alkenes and how are they formed? What do you think is the major product and why? **LO2**

19.108 Compound *A*, $C_5H_{10}O$, absorbs 1 mole of hydrogen on catalytic hydrogenation to give *B*. Both *A* and *B* react with sodium metal to produce hydrogen gas. *B*, when treated with phosphorus tribromide, gives *C*, $C_5H_{11}Br$, which when treated with lithium aluminium hydride gives *D*, C_5H_{12}. Compound *A* contains no methyl groups. Draw a cyclic and an acyclic structure for *A*, and explain all reactions involved. **LO2**

19.109 An unknown compound *X*, C_7H_9N, is only sparingly soluble in water, but dissolves readily in dilute hydrochloric acid. The resulting solution, on treatment with sodium nitrite, does not form an azo dye when added to an alkaline solution of 2-naphthol. However, *X* gives a positive test for a primary amine. Deduce the structure of *X*. **LO7**

19.110 Acid-catalysed hydration of 1-methylcyclohexene yields two alcohols. The major product does not undergo oxidation, while the minor product does undergo oxidation. Explain. **LO1,2**

20 Spectroscopy

Have you ever wondered how chemists know the structure of the compounds used or produced in various reactions? Remember that chemistry is an experimental science and evidence must be provided to support any claimed results. Until the mid-twentieth century, the determination of the structure of a compound was a difficult and time-consuming exercise. With the advent of modern instrumental techniques, this process has been greatly simplified (but it is *not* as easy as depicted on forensic science TV shows).

In this chapter we will discuss the three most commonly used tools: mass spectrometry (MS), infrared (IR) spectroscopy and nuclear magnetic resonance (NMR) spectroscopy. These techniques have the advantages that only a small amount of material is required, complex structures can be readily analysed and the process is relatively quick. We will introduce some of the features and applications of these techniques that aid in the determination of the structure of molecules.

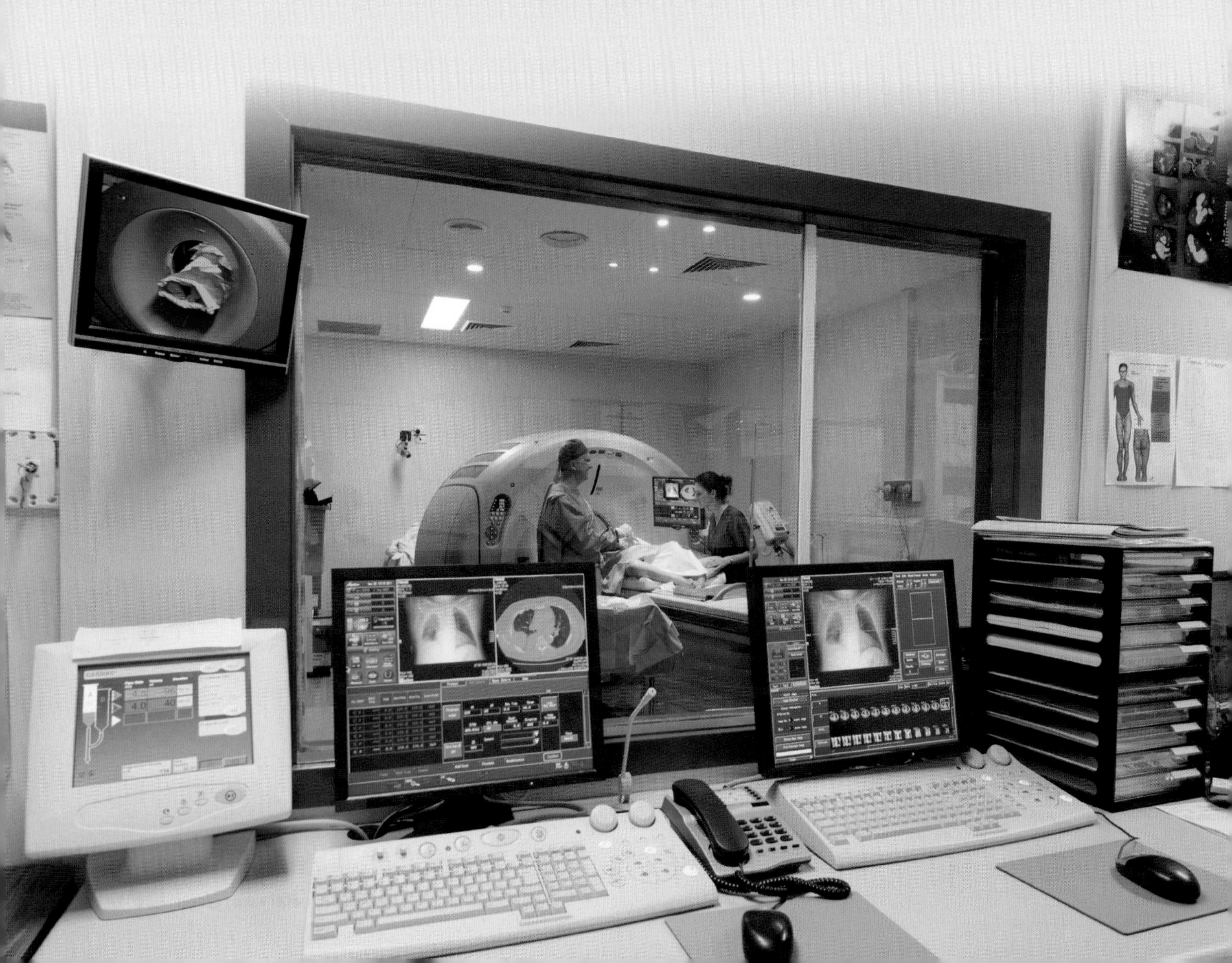

One of the most powerful diagnostic tools in modern medicine is magnetic resonance imaging (MRI), a technique founded on the principles of NMR spectroscopy. MRI differentiates the different tissues in the body by detecting the hydrogen nuclei of water molecules in the differing environments within those tissues. Analogously, the different environments that nuclei such as hydrogen, carbon or phosphorus reside in give rise to different NMR signals.

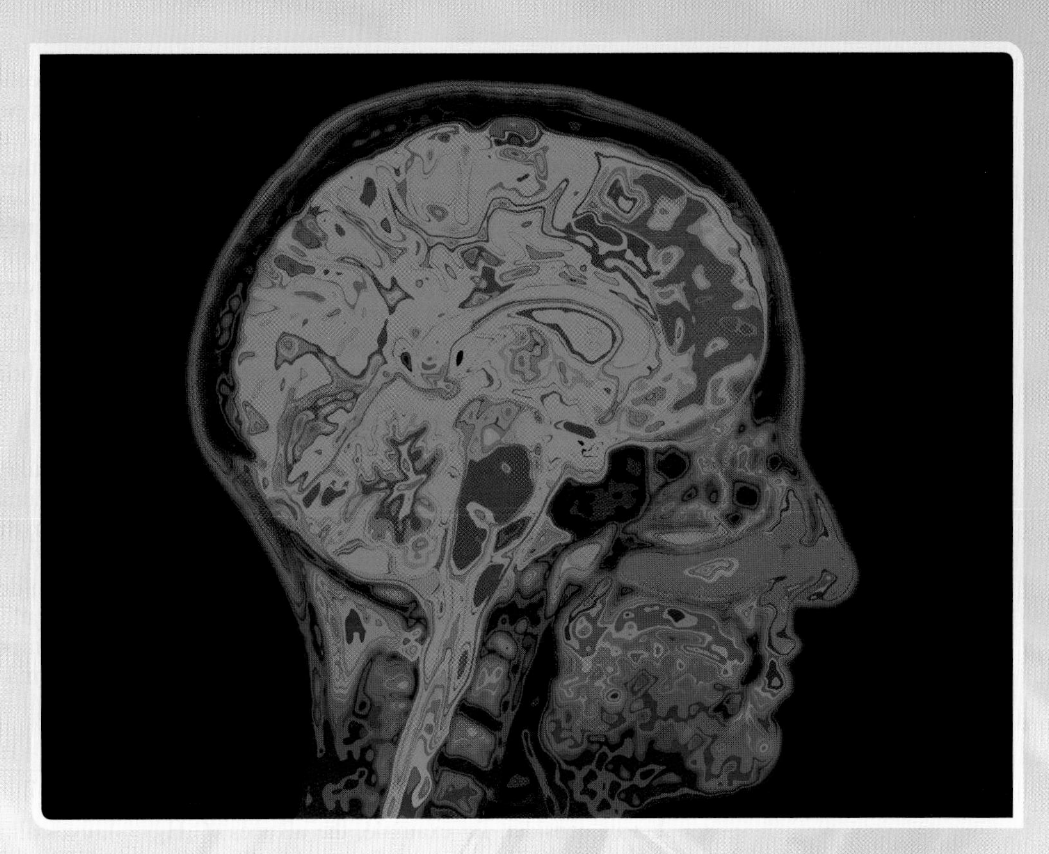

LEARNING OBJECTIVES

After studying this chapter, you should be able to:

20.1 determine the index of hydrogen deficiency (IHD) from a given molecular formula

20.2 apply mass spectral data to determine the molecular formula of an unknown compound

20.3 identify/predict key vibrational frequencies in infrared (IR) spectra of common functional groups

20.4 interpret and deduce structures from IR spectra

20.5 apply the basic principles of ^{1}H and ^{13}C NMR spectroscopy to analyse NMR spectra

20.6 interpret and analyse NMR spectra to determine the structure of compounds

20.7 describe the application of UV/visible spectroscopy and X-ray crystallography in chemistry.

20.1 Tools for determining structure

Let us consider two reactions we have studied previously: the hydration of 3-methylbut-1-ene and the dehydration of 3-methylbutan-2-ol.

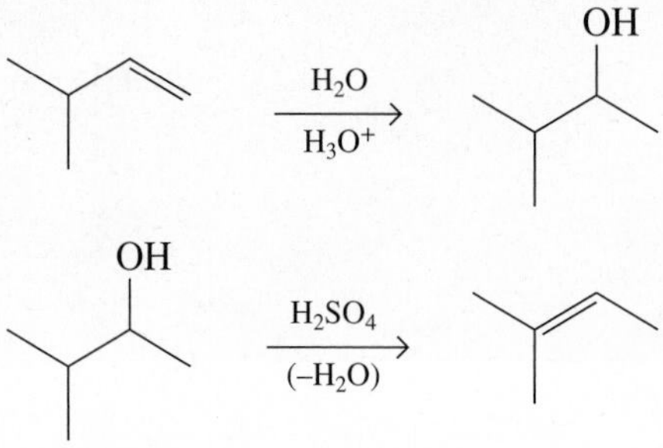

H_2O / H_3O^+

OH

H_2SO_4 / $(-H_2O)$

How do chemists determine that the product formed is an alcohol in the first reaction and an alkene in the second? Furthermore, they need to determine the position of the —OH group and the alkene group. How can this be done? Answering the first question by classical methods (those methods available to the early chemists) is not so hard; there are numerous test-tube tests available to identify an alcohol or alkene group. The second question can also be answered by classical techniques, but this is not so simple. Today, chemists rely almost exclusively on instrumental methods of analysis for structure determination. Although in this chapter we will focus on mass spectrometry (MS), infrared (IR) spectroscopy and nuclear magnetic resonance (NMR) spectroscopy, these are by no means the only tools chemists use. Some other tools used by chemists include X-ray crystallography, UV/visible spectroscopy and electron spin resonance. The general application of these other techniques will be described briefly in section 20.7.

The index of hydrogen deficiency

We can obtain valuable information about the structural formula of an unknown compound by inspecting its molecular formula. In addition to learning the number of each type of atom in a molecule of the compound, we can determine its **index of hydrogen deficiency (IHD)** (also known as **double bond equivalents** or degrees of unsaturation), which is the sum of the number of rings and π bonds in a molecule. For a hydrocarbon, we can determine this quantity by comparing the number of hydrogen atoms in the molecular formula of a compound of unknown structure with the number of hydrogen atoms in a **reference compound** with the same number of carbon atoms and with no rings or π bonds. The molecular formula of the reference hydrocarbon is C_nH_{2n+2}.

$$\text{index of hydrogen deficiency} = \frac{(\text{H}_{\text{reference}} - \text{H}_{\text{molecule}})}{2}$$

Let us consider, for example, the alkanes (C_nH_{2n+2}), cycloalkanes (C_nH_{2n}), alkenes (C_nH_{2n}) and alkynes (C_nH_{2n-2}) (see chapter 16). These have an IHD of 0, 1, 1 and 2 respectively. The IHD (sometimes called the 'unsaturation index' or the 'double bond equivalents') can be very useful in determining the structure of a molecule. If an unknown compound has an IHD of 0, we know immediately it does not contain any rings or π bonds, while an IHD of 1 would indicate the presence of either a ring or a π bond. Since benzene has an IHD of 4 (one ring and the equivalent of three double bonds), compounds containing a benzene ring must have an IHD ≥ 4; an IHD < 4 indicates that such a ring is not present in the molecule.

WORKED EXAMPLE 20.1

Calculating the IHD of hydrocarbons

Calculate the index of hydrogen deficiency for isopropylbenzene (cumene) with the molecular formula C_9H_{12} and account for this deficiency by reference to the structural formula of the compound.

Analysis

The molecular formula of the reference hydrocarbon with nine carbon atoms is C_9H_{20}.

Solution

The index of hydrogen deficiency of cumene is $(20 - 12)/2 = 4$ and is accounted for by the three 'double bonds' and one ring in cumene.

Is our answer reasonable?

Draw the structure of the compound and count the number of 'double bonds' and rings. Benzenoid compounds (compounds based on benzene) must have an IHD of at least 4.

PRACTICE EXERCISE 20.1

Calculate the index of hydrogen deficiency of cyclohepta-1,3,5-triene, C_7H_8, and account for this deficiency by reference to the structural formula of the compound.

To determine the molecular formula of a reference compound containing elements other than carbon and hydrogen, write the formula of the reference hydrocarbon, add to it other elements contained in the unknown compound, and make the following adjustments to the number of hydrogen atoms.

1. For each atom of a monovalent group 17 element (F, Cl, Br, I) added to the reference hydrocarbon, subtract one hydrogen atom; halogen substitutes for hydrogen and reduces the number of hydrogen atoms by one per halogen atom. The general formula of an acyclic monochloroalkane, for example, is $C_nH_{2n+1}Cl$.
2. No correction is necessary for the addition of atoms of group 16 elements (O, S, Se) to the reference hydrocarbon. Inserting a divalent group 16 element into a reference hydrocarbon does not change the number of hydrogen atoms.
3. For each atom of a trivalent group 15 element (N and P) added to the formula of the reference hydrocarbon, add one hydrogen atom. Inserting a trivalent group 15 element adds one hydrogen atom to the molecular formula of the reference compound. The general molecular formula for an acyclic alkylamine, for example, is $C_nH_{2n+3}N$.

WORKED EXAMPLE 20.2

Calculating the IHD

Isopentyl acetate (3-methylbutyl acetate), a compound with a banana-like odour, is a component of the alarm pheromone of honey bees. Its molecular formula is $C_7H_{14}O_2$. Calculate the index of hydrogen deficiency of this compound.

Analysis

The molecular formula of the reference hydrocarbon with seven carbon atoms is C_7H_{16}. Adding oxygen atoms to this formula does not change the number of hydrogen atoms.

Solution

The molecular formula of the reference compound is $C_7H_{16}O_2$, and the index of hydrogen deficiency is $(16 - 14)/2 = 1$, indicating either one ring or one π bond. As can be seen from the structural formula of isopentyl acetate, it contains one π bond, the carbon–oxygen double bond.

O

O

isopentyl acetate

Is our answer reasonable?

Draw the structure of the compound and count the number of double bonds and rings.

PRACTICE EXERCISE 20.2

The index of hydrogen deficiency of paracetamol is 5. Account for this value by reference to the structural formula of paracetamol.

H
N
O
HO

paracetamol

20.2 Mass spectrometry

In chapter 10 you learned that the colligative properties of solutions can be used to determine the molar mass of the dissolved solutes. An alternative and more widely used method is **mass spectrometry**. This technique allows us to determine the mass of molecules and fragments of molecules. The three key steps in mass spectrometry are (1) ionisation of the sample, (2) separation of the resulting ions and (3) detection of the ions. The fundamental principle is that, when a molecule is ionised, it can be accelerated through an electric or magnetic field where ions of different mass are deflected to different extents. The ions are then counted by the detector and their abundance is plotted on a graph against their mass to charge ratios (*m*/*z*). A schematic representation of a mass spectrometer is given in figure 20.1.

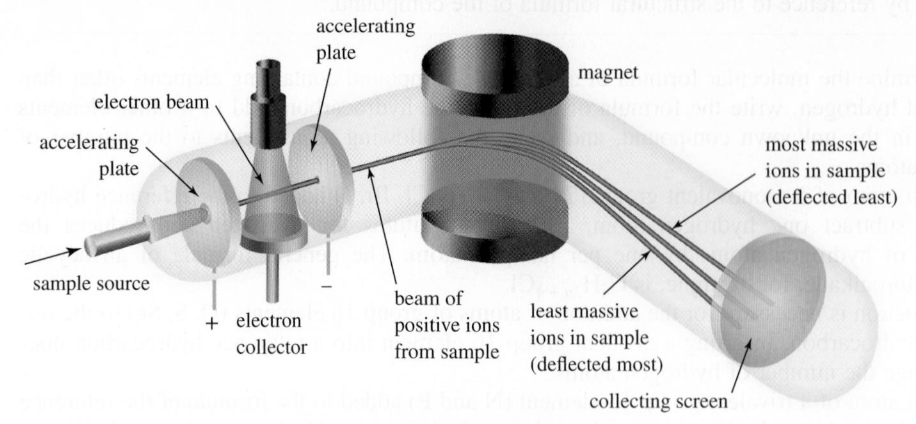

FIGURE 20.1 Schematic representation of one type of mass spectrometer. In the sample source, an electron beam ionises gas molecules into positively charged ions. The ions are accelerated and then deflected by a magnet. Each fragment follows a trajectory that depends on its mass to charge ratio (*m*/*z*).

When a sample is introduced into a mass spectrometer, it is bombarded with a stream of high-energy electrons, which causes some of the molecules to be ionised by the ejection of an electron. This produces ions that are **radical cations** ('radical' because there is an unpaired electron and 'cation' because of the positive charge). These ions are termed **molecular ions** (or parent ions) and have essentially the same mass as the original molecules because the mass of the lost electron is negligible.

$$\underset{\text{molecule}}{M} + \underset{\text{high energy electron}}{e^-} \rightarrow \underset{\text{molecular ion}}{(M)^{+\bullet}} + 2e^-$$

The peak due to the molecular ion (usually the one with the highest *m*/*z* value) can be used to quickly differentiate between molecules of similar mass. Consider, for example, the problem of differentiating between samples of hexane and cyclohexane. Mass spectrometry will allow identification because they have different molecular masses (86 and 84, respectively) as shown in figure 20.2. In chapter 1, we defined atomic mass as the mass of one atom, measured in amu. We can do the same for a single molecule. Thus, molecular mass is the mass of one molecule, measured in amu.

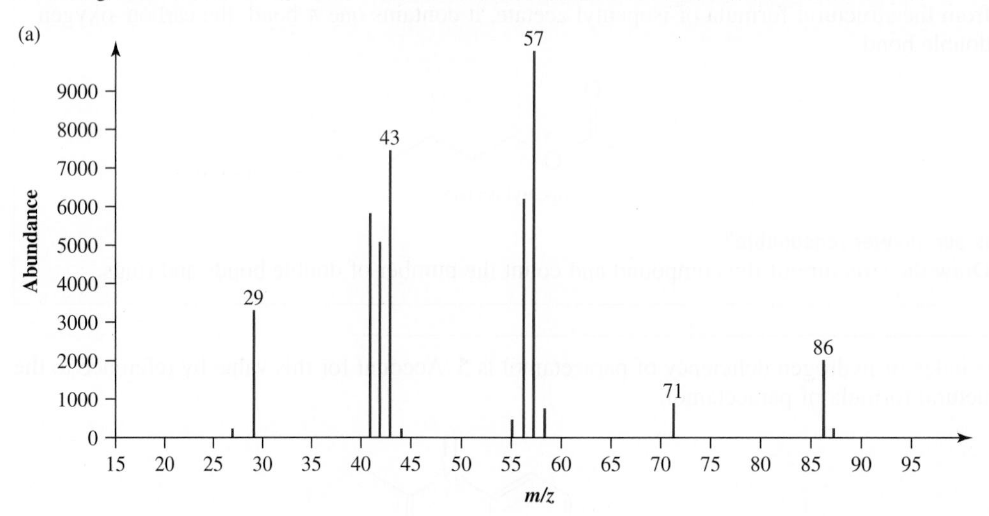

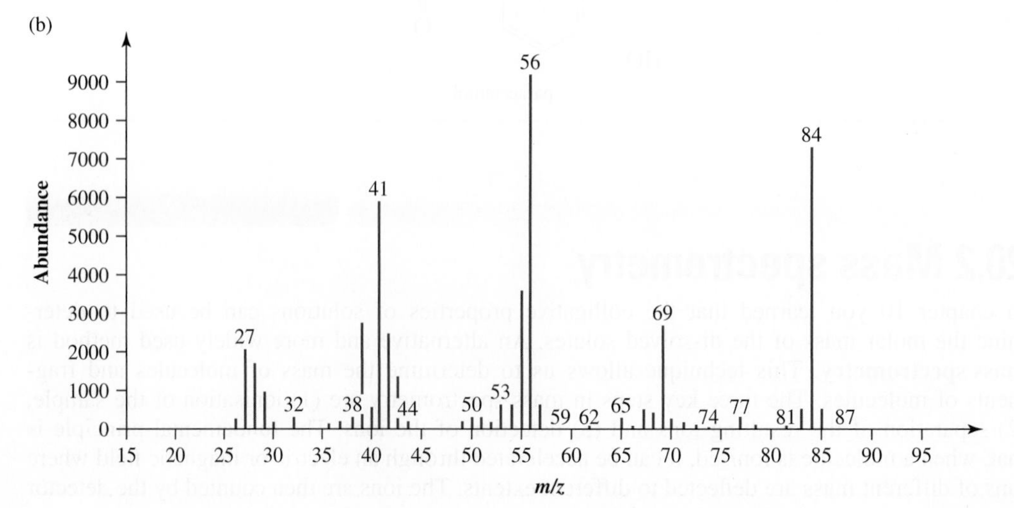

FIGURE 20.2 The mass spectra of hexane and cyclohexane: **(a)** The molecular ion signal at *m*/*z* 86 indicates that this spectrum is of hexane, **(b)** while *m*/*z* 84 indicates cyclohexane. The other peaks in the spectra are due to fragmentation products (see the next page).

WORKED EXAMPLE 20.3

Determining molecular formulae from mass spectra

The mass spectrum of an unknown alkane showed a molecular ion peak at m/z 128. What is the molecular formula of this compound?

Analysis

As this is an alkane we know it contains only C and H and it has the general formula C_nH_{2n+2}.

Solution

Substituting our known values into C_nH_{2n+2} gives the equation $12 \times n + [1 \times (2n + 2)] = 128$. Solving for n gives $n = 9$. Therefore, the molecular formula of the compound is C_9H_{20}.

Is our answer reasonable?

You should check that the molecular mass of C_9H_{20} is 128.

PRACTICE EXERCISE 20.3

(a) The mass spectrum of an unknown alkane showed a molecular ion peak at m/z 184. What is the molecular formula of this compound?
(b) The mass spectrum of an unknown cycloalkane showed a molecular ion peak at m/z 154. What is the molecular formula of this compound?

When ionising the sample in a mass spectrometer, many of the resulting molecular ions fragment to give cations of lower mass (sometimes called daughter ions) and neutral radicals (which are not detected as they are not charged). These fragment cations are sorted according to their mass to charge ratios (m/z). As the charge is almost always +1 when molecules are ionised in this way, m/z is equivalent to the mass of the ion. Analysis of these fragmentation patterns can be used to help determine the structure of a compound; however, details of this process will not be covered here. Instead, we will focus on the use of mass spectrometry to determine molar masses and molecular formulae.

Isotopes in mass spectrometry

Upon inspection of the spectra in figure 20.2, you will notice that the m/z values for all the peaks are given as whole numbers. This is sufficient for most purposes. However, it does not allow us to differentiate between molecules with the same mass (when quoted in whole numbers) but different formulae. The accurate mass of the molecular ion, and consequently the molecular formula, of a compound can be determined by using a technique called **high resolution mass spectrometry**. In this process, the mass to charge ratio of ions (including the molecular ion) is measured with a high degree of precision (usually to four decimal places). For example, consider the four compounds listed in table 20.1, all of which have different molecular formulae but the same nominal molecular mass. You will also notice that the molecular mass and the accurate mass are not the same. This is because, in determining the accurate mass, we use the accurate mass of the most abundant isotope of each element, rather than the atomic mass of the element. (Recall from section 1.3 that atomic mass is defined as the average mass per atom of a naturally occurring sample of atoms of the element.) Table 20.2 (overleaf) lists some accurate masses for the elements commonly encountered in organic chemistry, along with their natural abundances.

TABLE 20.1 Comparison of the masses of four compounds with molar mass of 88.

Name	Structure	Molecular formula	Average molecular mass (amu)	Accurate mass (amu)
butanoic acid	O, ‖, C, OH	$C_4H_8O_2$	88.11	88.0524
butane-2,3-diamine	H_2N, NH_2	$C_4H_{12}N_2$	88.15	88.1001
pentan-3-ol	OH	$C_5H_{12}O$	88.15	88.0888
N,N′-dimethylurea	O, ‖, C, H_3CHN, $NHCH_3$	$C_3H_8N_2O$	88.11	88.0637

TABLE 20.2 Isotope abundance and accurate mass of selected elements that are commonly encountered in organic chemistry.

Element	Molar mass ($g\ mol^{-1}$)	Isotope	% Natural abundance	Accurate mass (amu)
hydrogen	1.008	^{1}H	99.985	1.0078
		^{2}H	0.015	2.0140
carbon	12.011	^{12}C	98.89	12.0000
		^{13}C	1.11	13.0034
nitrogen	14.007	^{14}N	99.64	14.0031
		^{15}N	0.36	15.0001
oxygen	15.999	^{16}O	99.76	15.9949
		^{17}O	0.04	16.9991
		^{18}O	0.20	17.9992
phosphorus	30.9738	^{31}P	100.00	30.9738
sulfur	32.06	^{32}S	95.00	31.9721
		^{33}S	0.76	32.9715
		^{34}S	4.22	33.9679
		^{36}S	0.02	35.9671
fluorine	18.9984	^{19}F	100.00	18.9984
chlorine	35.45	^{35}Cl	75.77	34.9688
		^{37}Cl	24.23	36.9659
bromine	79.904	^{79}Br	50.69	78.9183
		^{81}Br	49.31	80.9163
iodine	126.9045	^{127}I	100.00	126.9044

WORKED EXAMPLE 20.4

Determining molecular formulae from high resolution mass spectra

Which of the molecular formulae C_7H_{16}, $C_6H_{14}N$, $C_5H_{12}N_2$, $C_6H_{12}O$ and $C_5H_8O_2$ corresponds to an accurate mass of *m*/*z* 100.0888?

Analysis

As a starting point, you will notice all of these molecular formulae have a nominal molecular mass of 100. To solve this problem, simply determine the accurate mass corresponding to each formula using the accurate mass value for the *most abundant isotope* of each element (not the accurate molar mass of the atoms).

Solution

$C_6H_{12}O$

Is our answer reasonable?

Double-check your calculations for the accurate mass. It is possible to work backwards from an accurate mass to determine the molecular formula, but this requires a computer program (or extensive trial and error).

PRACTICE EXERCISE 20.4

Which of the molecular formulae C_8H_{18}, $C_7H_{14}O$, $C_7H_{16}N$, $C_6H_{14}N_2$, $C_6H_{12}NO$ and $C_6H_{10}O_2$ corresponds to an accurate mass of 114.1158?

You may have noticed that, of the elements listed in table 20.2, many had one isotope with a natural abundance of more than 99%. In the other cases, we can use the isotope abundance to help determine the presence of particular elements. Bromine is particularly easy to recognise by the presence of two signals of equal intensity and a mass difference of 2, due to the ^{79}Br (M^+ ion) and ^{81}Br [$(M + 2)^+$ ion] isotopes, which are nearly equally abundant. The presence of a chlorine atom in a molecule will give rise to an $(M + 2)^+$ signal one-third the height of the M^+ signal (the ratio of ^{35}Cl to ^{37}Cl is 3 : 1). Isotope distribution has also been used to help identify many metallic and organometallic materials in the environment. For example, the pollutants lead, mercury and tin have four, seven and ten natural isotopes, respectively.

20.3 Infrared spectroscopy

Spectroscopic techniques, including infrared (IR) spectroscopy and nuclear magnetic resonance (NMR) spectroscopy, involve the interaction of molecules with electromagnetic radiation. Thus, to understand the fundamentals of spectroscopy, we must first review some of the fundamentals of electromagnetic radiation.

Electromagnetic radiation

In section 4.2, we studied the characteristics of light, a form of electromagnetic radiation, and we will provide only a summary here. Gamma rays, X-rays, ultraviolet light, visible light, **infrared radiation**, microwaves and radio waves are all examples of **electromagnetic radiation** and part of the electromagnetic spectrum. Because electromagnetic radiation behaves as a wave travelling at the speed of light, it is described in terms of its wavelength and frequency. Figure 20.3 summarises the wavelengths and frequencies of some regions of the electromagnetic spectrum.

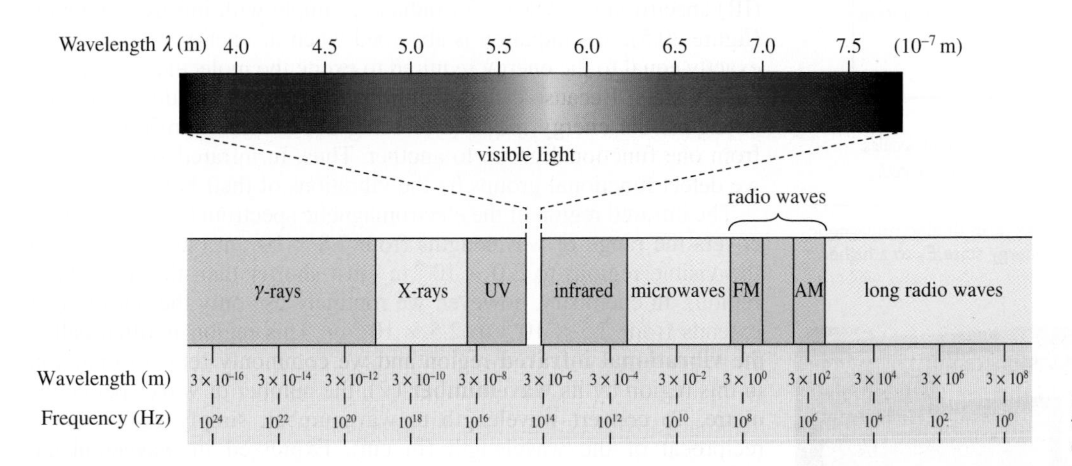

FIGURE 20.3 Wavelength and frequency of some regions of the electromagnetic spectrum.

Wavelength is the distance between any two consecutive equivalent points on the wave (e.g. crest to crest). Wavelength is given the symbol λ (Greek lowercase *lambda*) and is usually expressed in the SI base unit of metres. Other derived units commonly used to express wavelength are given in table 20.3.

TABLE 20.3 Common units used to express wavelength (λ).

Unit	Relation to metre
metre (m)	
millimetre (mm)	$1 \text{ mm} = 10^{-3} \text{ m}$
micrometre (μm)	$1 \mu\text{m} = 10^{-6} \text{ m}$
nanometre (nm)	$1 \text{ nm} = 10^{-9} \text{ m}$

λ (wavelength)

The **frequency** of a wave is the number of full cycles of the wave that pass a given point in a second. Frequency is given the symbol ν (Greek lowercase *nu*) and is reported in s^{-1}, also called **hertz (Hz)**. Wavelength and frequency are inversely proportional, and we can calculate one from the other using the relationship:

$$c = \nu\lambda$$

where ν is frequency in s^{-1}, c is the velocity of light ($2.998 \times 10^8 \text{ m s}^{-1}$), and λ is the wavelength in metres.

An alternative way to describe electromagnetic radiation is in terms of particles. We call these particles photons. The energy of a photon and the frequency of radiation are related by the equation:

$$E = h\nu = h\frac{c}{\lambda}$$

where E is the energy in kJ and h is Planck's constant, 6.626×10^{-34} J s. This equation tells us that high-energy radiation corresponds to short wavelengths, and vice versa. Thus, ultraviolet light (higher energy) has a shorter wavelength (approximately 10^{-7} m) than infrared radiation (lower energy), which has a wavelength of approximately 10^{-5} m.

The vibrational infrared spectrum

As we discussed in chapter 5, molecules are flexible structures with atoms and groups of atoms able to rotate around covalent single bonds. Covalent bonds stretch and bend just as if the atoms were joined by flexible springs. We know from experimental observations and from theories of molecular structure that all energy changes within a molecule are quantised; that is, they are subdivided into small, but well-defined, increments. For example, vibrations of bonds within molecules can undergo transitions only between allowed vibrational energy levels.

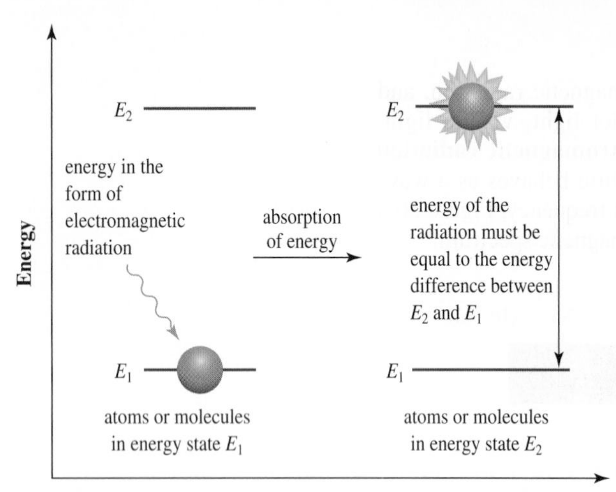

FIGURE 20.4 Absorption of energy in the form of electromagnetic radiation excites atoms or molecules in energy state E_1 to a higher energy state E_2.

We can cause atoms or molecules to undergo a transition from energy state E_1 to a higher energy state E_2 by irradiating them with electromagnetic radiation corresponding to the energy difference between states E_1 and E_2 as illustrated schematically in figure 20.4. When these atoms or molecules return to the ground state, an equivalent amount of energy is emitted. When a sample is irradiated with electromagnetic radiation of various wavelengths, it absorbs energy of particular wavelengths, while the wavelengths not absorbed simply pass through or are reflected from the sample unchanged. In infrared (IR) spectroscopy, when we irradiate a sample with infrared radiation (figure 20.5), this radiation is absorbed when its energy (frequency) is exactly equal to the energy required to excite the molecules to a higher energy state. Because different functional groups have different bond strengths, the energy required to bring about these transitions varies from one functional group to another. Thus, in infrared spectroscopy, we detect functional groups by the vibrations of their bonds.

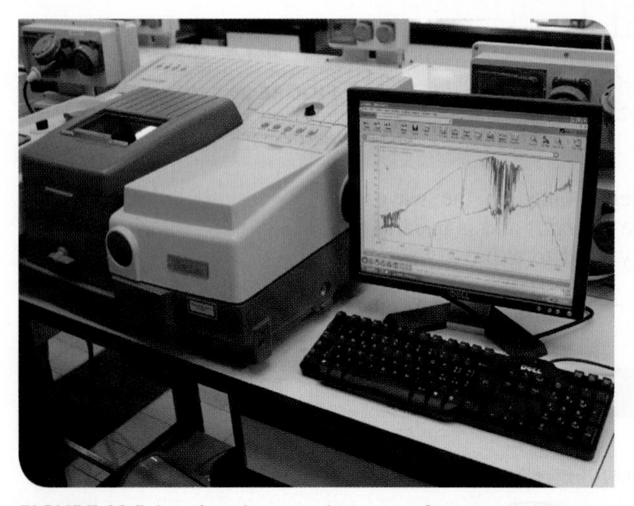

FIGURE 20.5 An infrared spectrophotometer. Spectra are shown on the monitor.

The infrared region of the electromagnetic spectrum (see figure 20.3) covers the range of wavelengths from 7.8×10^{-7} m (just longer than the visible region) to 2.0×10^{-3} m (just shorter than the microwave region). In chemistry, however, we routinely use only the portion that extends from 2.5×10^{-6} to 2.5×10^{-5} m. This region is often called the **vibrational infrared** region and we commonly refer to radiation in this region by its **wavenumber** ($\tilde{\nu}$), the number of waves per centimetre. To convert wavelength to wavenumber, simply calculate the reciprocal of the wavelength (in cm). Expressed in wavenumbers, the vibrational region of the infrared spectrum extends from 4000 to 400 cm^{-1} (the unit cm^{-1} is read 'reciprocal centimetre'). An advantage of using wavenumbers is that they are directly proportional to energy; the higher the wavenumber, the higher the energy of radiation.

Figure 20.6 shows an infrared spectrum of aspirin. The horizontal axis is calibrated in wavenumbers (cm^{-1}). The wavenumber scale is often divided into two or more linear regions. For all spectra reproduced in this text, it is divided into three linear regions: 4000–2200 cm^{-1}, 2200–1000 cm^{-1} and 1000–450 cm^{-1}. The vertical axis measures transmittance, with 100% transmittance at the top and 0% transmittance at the bottom. Thus, the baseline for an infrared spectrum (100% transmittance of radiation through the sample = 0% absorption) is at the top of the chart, and the absorption of radiation corresponds to a trough or valley. Strange as it may seem, we commonly refer to infrared absorptions as peaks, even though they are actually troughs. They are also called 'bands' or signals. If absorption of light is used on the vertical axis instead of transmittance, the signals are actually peaks.

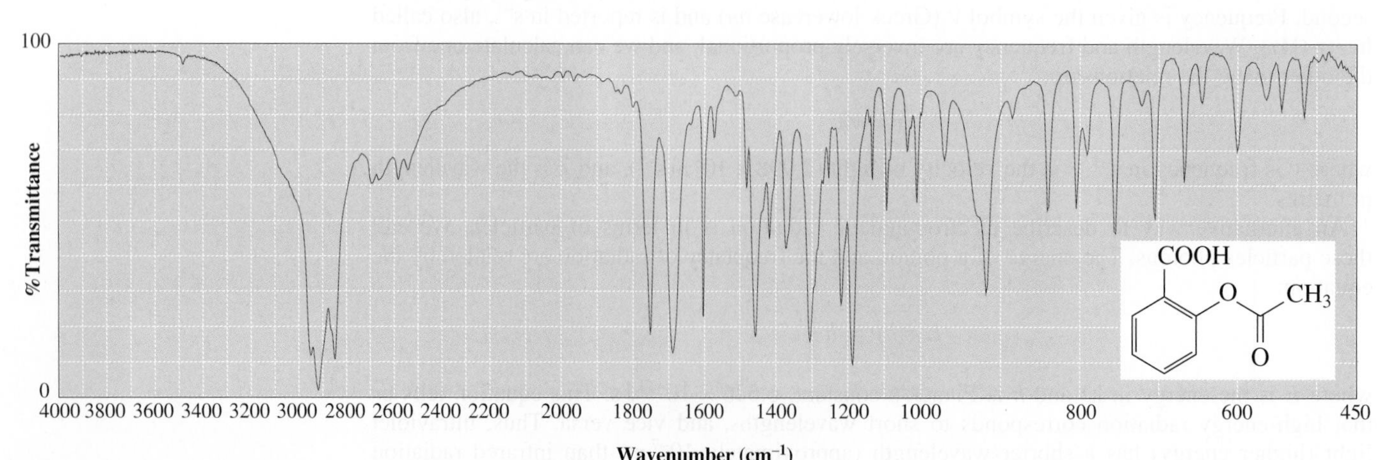

FIGURE 20.6 Infrared spectrum of aspirin.

Molecular vibrations

Atoms joined by covalent bonds are not permanently fixed in one position but, instead, undergo continuous vibrations relative to each other. The energies associated with transitions between vibrational energy levels in most covalent molecules range from 8 to 40 kJ mol^{-1}. Such transitions can be induced by the absorption of radiation in the infrared region of the electromagnetic spectrum.

For molecules to absorb infrared radiation, the bonds undergoing vibration must be polar, and their vibration must cause a periodic change in the bond dipole; the greater the polarity of the bonds, the more intense is the absorption. Any vibration that meets these criteria is said to be **infrared active**. Covalent bonds in homonuclear diatomic molecules, such as H_2 and Br_2, and some carbon–carbon double or triple bonds in symmetrical alkenes and alkynes do not absorb infrared radiation because they are not polar bonds. The multiple bonds in the following two compounds, for example, do not have a dipole moment so are not infrared active.

Neither the double nor the triple bond in these compounds is infrared active.

$(H_3C)_2C{=}C(CH_3)_2$ — 2,3-dimethylbut-2-ene

$H_3C-C{\equiv}C-CH_3$ — but-2-yne

For nonlinear molecules containing n atoms, $3n - 6$ allowed fundamental vibrations exist. For a compound as simple as ethanol, CH_3CH_2OH, there are 21 fundamental vibrations, and for hexanoic acid, $CH_3(CH_2)_4COOH$, there are 54. Thus, even for relatively simple molecules, a large number of vibrational energy levels exist, and the patterns of energy absorption for these and larger molecules are quite complex.

The simplest vibrational motions in molecules giving rise to the absorption of infrared radiation are **stretching** and **bending** motions. The fundamental stretching and bending vibrations for a methylene group are illustrated in figure 20.7.

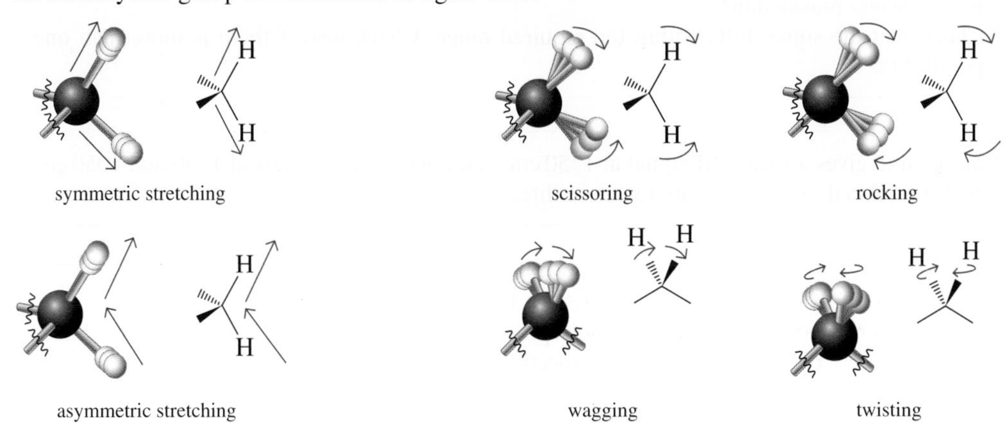

FIGURE 20.7 Fundamental modes of vibration for a methylene group.

To someone skilled in the interpretation of infrared spectra, absorption patterns can yield an enormous amount of information about chemical structure. The value of infrared spectra for us, however, is that we can use them to determine the presence or absence of particular functional groups. Carbonyl groups, for example, typically show strong absorption at approximately 1630–1800 cm^{-1}. The position of absorption for a particular carbonyl group depends principally on the functional group to which it belongs (aldehyde, ketone, carboxylic acid, anhydride, acid chloride, ester or amide). Other factors influencing the frequency of absorption include ring strain (if the carbonyl group is part of a ring) and conjugation (having an alkene or aromatic ring immediately adjacent to the carbonyl group).

Correlation tables

Data on absorption patterns of selected functional groups are collected in tables called **correlation tables**. Table 20.4 (overleaf) gives the characteristic infrared vibrational frequencies for the types of bonds and functional groups we deal with most often. Appendix H contains a cumulative correlation table. In these tables, we refer to the intensity of a particular absorption as strong (s), medium (m) or weak (w).

In general, we will pay most attention to the region from 3650 to 1500 cm^{-1}, where the characteristic stretching vibrations for most functional groups are found. Vibrations in the region from 1500 to 400 cm^{-1} can arise from phenomena such as combinations of two or more bands or harmonics of fundamental absorption bands. Such vibrations are much more complex and far more difficult to analyse. Because even slight variations in molecular structure lead to differences in absorption patterns in this region, it is often called the **fingerprint region**. If two compounds have even slightly different structures, the differences in their infrared spectra are most clearly discernible in the fingerprint region.

TABLE 20.4 Characteristic IR vibrational frequencies of selected functional groups.

Bond	Frequency (cm^{-1})	Intensity
O—H	3200–3500	strong and broad
N—H	3100–3500	medium
C—H	2850–3100	medium to strong
C≡C	2100–2260	weak
C=O	1630–1800	strong
C=C	1600–1680	weak
C—O	1050–1250	strong

WORKED EXAMPLE 20.5

Identifying functional groups using IR spectroscopy

Determine the functional group that is most likely present if a compound gives an IR signal at:
(a) 1705 cm^{-1} (b) 2150 cm^{-1}.

Analysis

Check the correlation tables for signals in the required regions. In this example, table 20.4 provides sufficient information.

Solution

(a) A carbonyl C=O group (signals between 1630 and 1800 cm^{-1})
(b) An alkyne C≡C group (signals between 2100 and 2260 cm^{-1})

Is our answer reasonable?

Check that the signal falls within the required range. Check also if there is more than one possibility.

PRACTICE EXERCISE 20.5

A compound gives a strong IR signal at 1750 cm^{-1} and two strong signals at 1250 and 1050 cm^{-1}. Which functional group accounts for these three signals?

WORKED EXAMPLE 20.6

Distinguishing between isomers using IR spectroscopy

Acetone (propanone) and allyl alcohol (prop-2-en-1-ol) are constitutional isomers. Show how these may be distinguished by IR spectroscopy.

$$CH_3-\overset{\overset{\displaystyle O}{\|}}{C}-CH_3 \qquad CH_2=CH-CH_2-OH$$

acetone (propanone) | allyl alcohol (prop-2-en-1-ol)

Analysis

Firstly, you need to identify the key functional group differences: in this case, a ketone (C=O) in acetone and an alcohol (OH) and alkene (C=C) in allyl alcohol. Now consult table 20.4.

Solution

Only acetone gives a strong signal in the C=O stretching region, 1630–1800 cm^{-1}. Alternatively, only allyl alcohol gives a strong signal in the O—H stretching region, 3200–3500 cm^{-1}.

Is our answer reasonable?

Check that you have chosen the appropriate strong signals. (Alkenes often display weak signals and are, therefore, not always a good choice).

PRACTICE EXERCISE 20.6

Pentan-3-one and cyclopentanol are constitutional isomers. Show how these may be distinguished by IR spectroscopy.

O OH

pentan-3-one cyclopentanol

Chemical Connections

The spectroscopy of fingerprints

Fingermarks are the most widely used method of personal identification in law enforcement. Their presence at a crime scene or on an object is instrumental in connecting individuals to a case. The most common form of fingerprint is the latent (hidden) fingermark. On deposition, the fingermark is a mixture of the natural secretions present on the skin, an emulsion of waxes, oils and aqueous components, and whatever surface contaminants may be present. Detection methods target differences between the latent fingermark and the surface upon which it is laid and are based either on physical attraction or a chemical reaction.

The successful development of latent fingermarks relies heavily upon the chemistry of the latent fingermark residue. With time, the chemical nature of the latent deposit changes due to evaporation of volatile components, bacterial action and oxidation. The rate of change depends on the initial chemical composition of the residue and the environmental conditions. This ageing process can have a significant effect on the successful development of a latent fingermark. Despite these issues, most fingermark detection techniques have been developed on the basis of knowledge of the components of human skin secretions, without regard for the potential for ageing of the print. In addition, there is currently no reliable technique for estimating the age of a latent fingermark.

The ability to make future improvements to current fingermark detection methods and the development of new approaches to the visualisation of latent fingermarks depends critically upon a greater understanding of the chemistry of the fingermark residue.

Researchers at Curtin University, in collaboration with scientists at the infra-red (IR) beamline at the Australian Synchrotron, have been using synchrotron IR microscopy to investigate the chemistry of latent fingermarks and how this changes with time. Fingermarks from donors of different age and gender were deposited on gold-plated glass (figure 20.8) and subjected to IR analysis at regular intervals over a period of 12 months. The synchrotron source is much brighter than conventional IR spectrometers, and this significantly enhances the signal-to-noise ratios and thereby reduces the time required to produce the spectra (figure 20.9). In addition, the synchrotron also provides much higher spatial resolution; this enables skin cell debris to be avoided by selecting very small areas for analysis. This approach has been very useful for monitoring the variation in lipid content of deposited latent fingermarks.

Future research will include investigation of the potential of mass spectral imaging in combination with IR data to obtain a better understanding of the fundamental chemistry of latent fingermark chemistry. This will provide a platform for the development of the next generation of latent fingermark detection methods. Observed changes in chemical composition over time and under different environmental conditions will also provide a valuable insight into the fingermark ageing process.

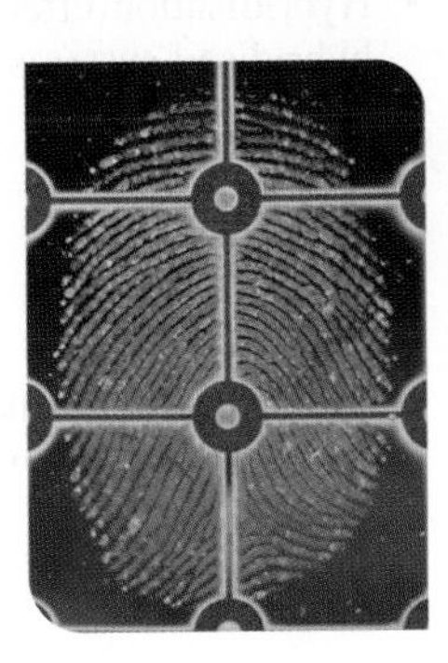

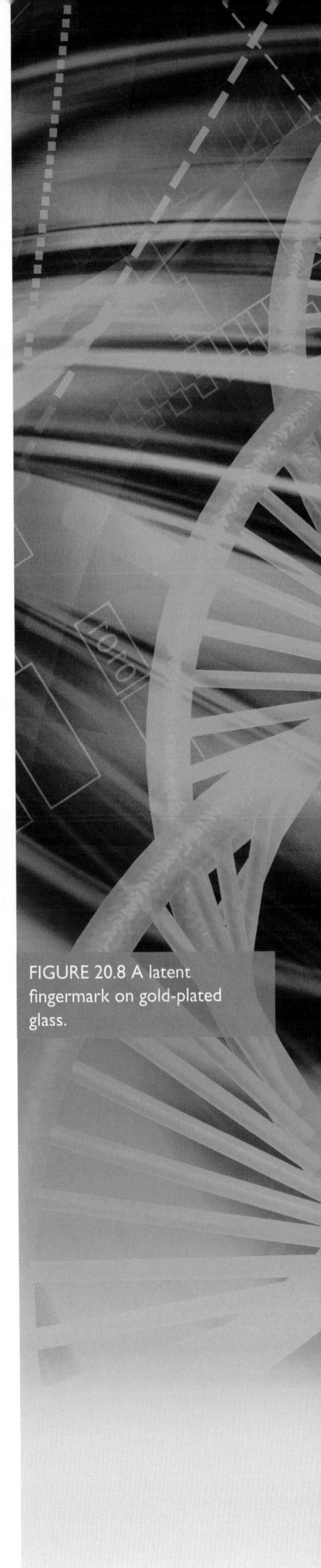

FIGURE 20.8 A latent fingermark on gold-plated glass.

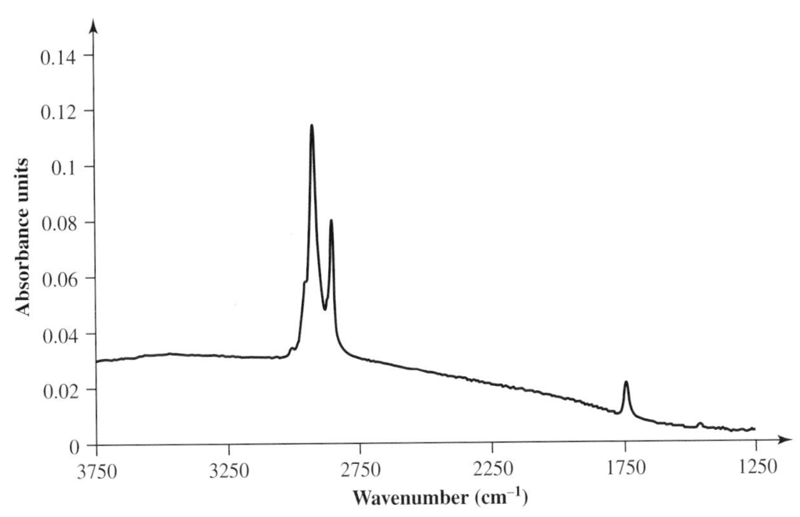

FIGURE 20.9 Typical synchrotron IR spectrum of a latent fingermark.

20.4 Interpreting infrared spectra

Interpreting spectroscopic data is a skill that is easy to acquire through practice and exposure to examples. An IR spectrum will reveal not only the functional groups that are present in a sample, but also those that can be excluded from consideration. Often, we can determine the structure of a compound solely from the data in the spectrum of the compound. At other times, we may need additional information, such as the molecular formula of the compound or knowledge of the reactions used to synthesise the molecule. In this section, we will see specific examples of IR spectra for characteristic functional groups. Familiarising yourself with them will help you to master the technique of spectral interpretation.

General rules for interpretation of IR spectra

Below is a set of general rules and guidelines for the analysis and interpretation of IR spectra.

- The stronger the bond, the higher is the vibrational frequency.

C≡C	C=C	C−C
2100–2260 cm^{-1}	1600–1680 cm^{-1}	800–1200 cm^{-1}

- The C—Y stretching frequency decreases with an increase in mass of Y.

C−H	C−O	C−Cl
2850–3000 cm^{-1}	1050–1250 cm^{-1}	600–800 cm^{-1}

- Bending usually occurs at a lower frequency than stretching.

C−H stretching	C−H bending
2850–3000 cm^{-1}	1340–1460 cm^{-1}

- Hybridisation effects cause the stretching frequency to increase as we move from sp^3 to sp hybridised systems.

sp	sp^2	sp^3
≡C−H	=C−H	−C−H
3280–3320 cm^{-1}	3000–3100 cm^{-1}	2850–3000 cm^{-1}

Alkanes

Infrared spectra of alkanes are usually simple, with few signals, the most common of which are given in table 20.5.

TABLE 20.5 Characteristic IR signals of alkanes, alkenes and alkynes.

Hydrocarbon	Vibration	Frequency (cm^{-1})	Intensity
Alkane			
C—H	stretching	2850–3000	strong
CH_2	bending	1450	medium
CH_3	bending	1375 and 1450	weak to medium
Alkene			
C—H	stretching	3000–3100	weak to medium
C=C	stretching	1600–1680	weak to medium
Alkyne			
C—H	stretching	3300	medium to strong
C≡C	stretching	2100–2260	weak to medium

Figure 20.10 shows an infrared spectrum of decane. The strong signal with multiple splittings between 2850 and 3000 cm^{-1} is characteristic of alkane C—H stretching. The C—H signal is strong in this spectrum because there are so many C—H bonds and no other functional groups. The other prominent signals in the spectrum are those due to methylene bending at 1465 cm^{-1} and methyl bending at 1380 cm^{-1}. Because alkane CH, CH_2 and CH_3 groups are present in many organic compounds, these peaks are among the most commonly encountered in infrared spectroscopy, and consequently they are usually not very useful as a diagnostic tool.

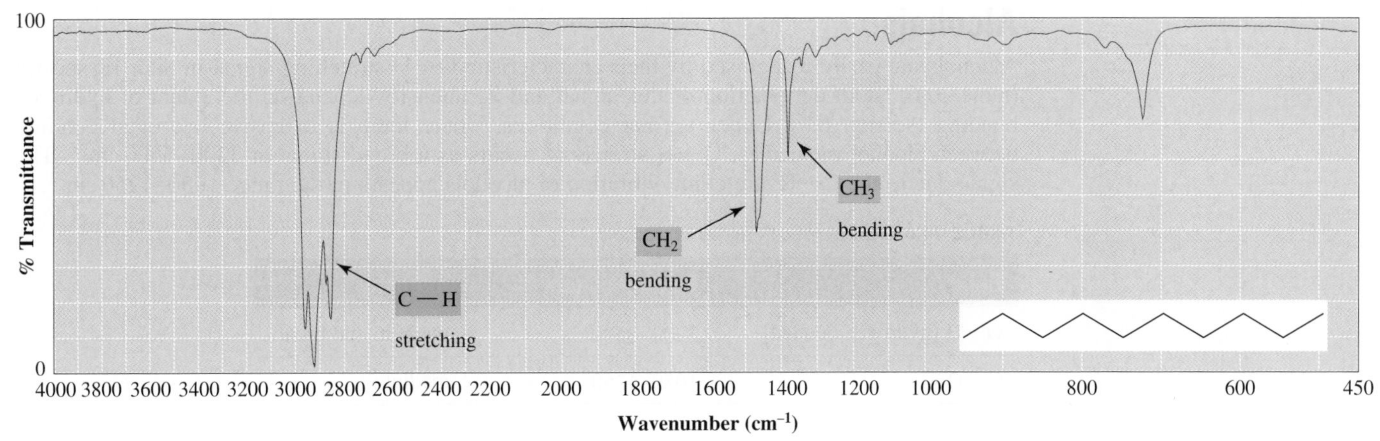

FIGURE 20.10 Infrared spectrum of decane.

Alkenes

An easily recognised alkene signal is the stretching band for the vinylic C—H in C=C—H at a slightly higher frequency (at 3000–3100 cm^{-1}) than the alkane C—H frequency (vinylic hydrogen atoms are those on a carbon atom of a carbon–carbon double bond). Also characteristic of alkenes is the C=C stretching band at 1600–1680 cm^{-1}. This vibration, however, is often weak and difficult to observe. Signals for both C—H stretching in vinylic C=C—H and C=C stretching can be seen in the infrared spectrum of cyclopentene (figure 20.11). Also visible are the signals for aliphatic C—H stretching near 2900 cm^{-1} and the methylene bending near 1440 cm^{-1}.

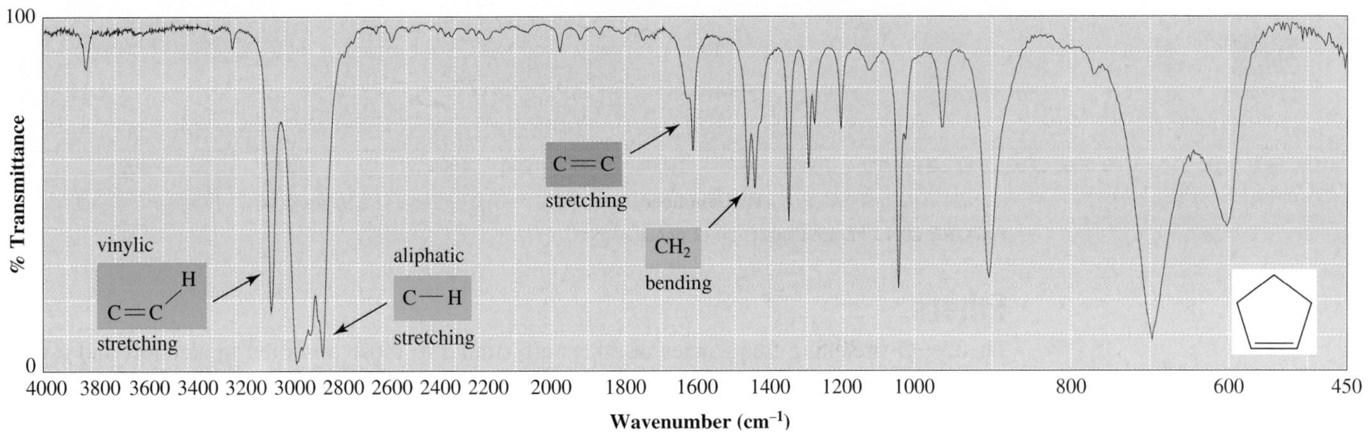

FIGURE 20.11 Infrared spectrum of cyclopentene.

Alkynes

The IR spectra of terminal alkynes exhibit a strong signal at 3300 cm^{-1} due to C—H stretching in C≡C—H. This absorption band is absent in internal alkynes because the triple bond is not bonded to a proton. (Terminal alkynes have the C≡C group at the end of the chain; in internal alkynes, the C≡C group is not at the end of the chain.) Alkynes give a weak signal between 2100–2260 cm^{-1}, due to C≡C stretching. The signal for this stretching is shown clearly in the spectrum of oct-1-yne (figure 20.12) but is absent in oct-4-yne.

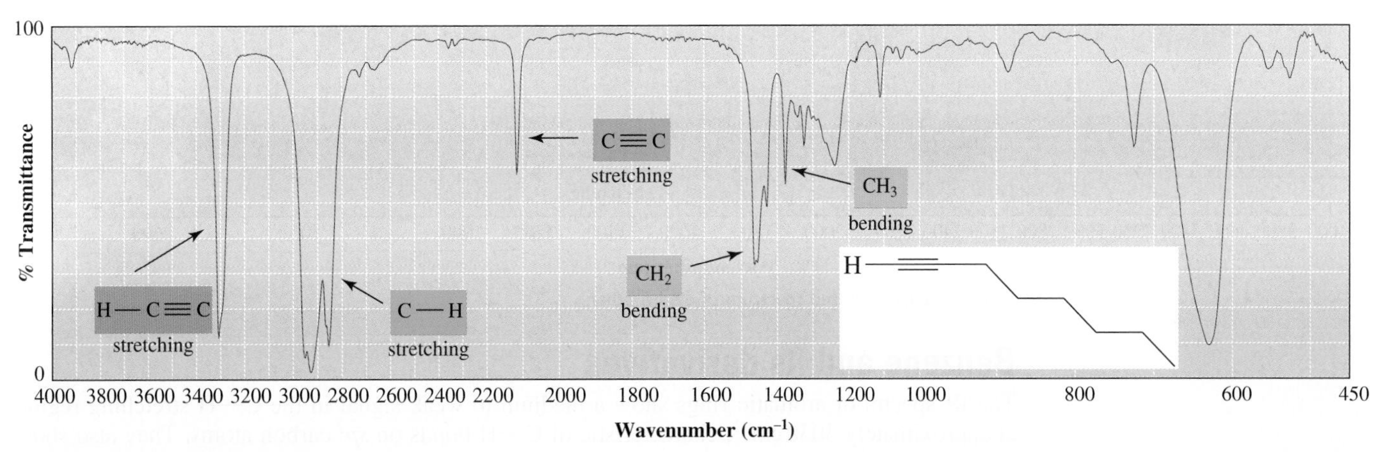

FIGURE 20.12 Infrared spectrum of oct-1-yne.

Alcohols

Alcohols are easily recognised by their characteristic O—H stretching signal in their IR spectra (table 20.6). Both the position of this signal and its intensity depend on the extent of hydrogen bonding (section 6.8). Under normal conditions, where there is extensive hydrogen bonding between alcohol molecules, O—H stretching occurs as a broad signal at 3200–3500 cm^{-1}. The signal due to the C—O stretching vibration of alcohols appears in the range 1050–1250 cm^{-1}.

TABLE 20.6 Characteristic IR signals of alcohols.

Bond	Frequency (cm^{-1})	Intensity
O—H (hydrogen bonded)	3200–3500	medium and broad
C—O	1050–1250	medium

Figure 20.13 shows an infrared spectrum of pentan-1-ol. The hydrogen-bonded O—H stretching appears as a strong, broad signal centred at 3340 cm^{-1}. The signal for C—O stretching appears near 1050 cm^{-1}, a value characteristic of primary alcohols.

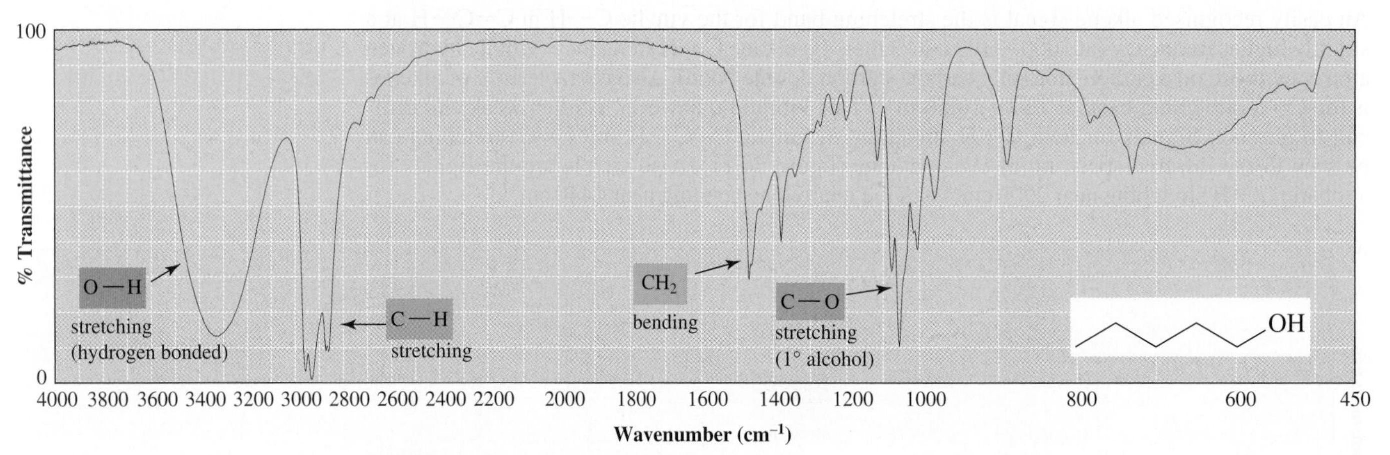

FIGURE 20.13 Infrared spectrum of pentan-1-ol.

Ethers

The C—O stretching frequencies of ethers are similar to those observed in alcohols and esters. Dialkyl ethers typically show a single signal in this region between 1070 and 1150 cm^{-1}. The presence or absence of a signal for O—H stretching at 3200–3500 cm^{-1} for a hydrogen-bonded O—H can be used to distinguish between an ether and an alcohol. The C—O stretching vibration is also present in esters. In this case, we can use the presence or absence of a signal for C=O stretching (1735 to 1750 cm^{-1}) to distinguish between an ether and an ester. Figure 20.14 shows an infrared spectrum of diethyl ether. Notice the absence of a signal for O—H stretching.

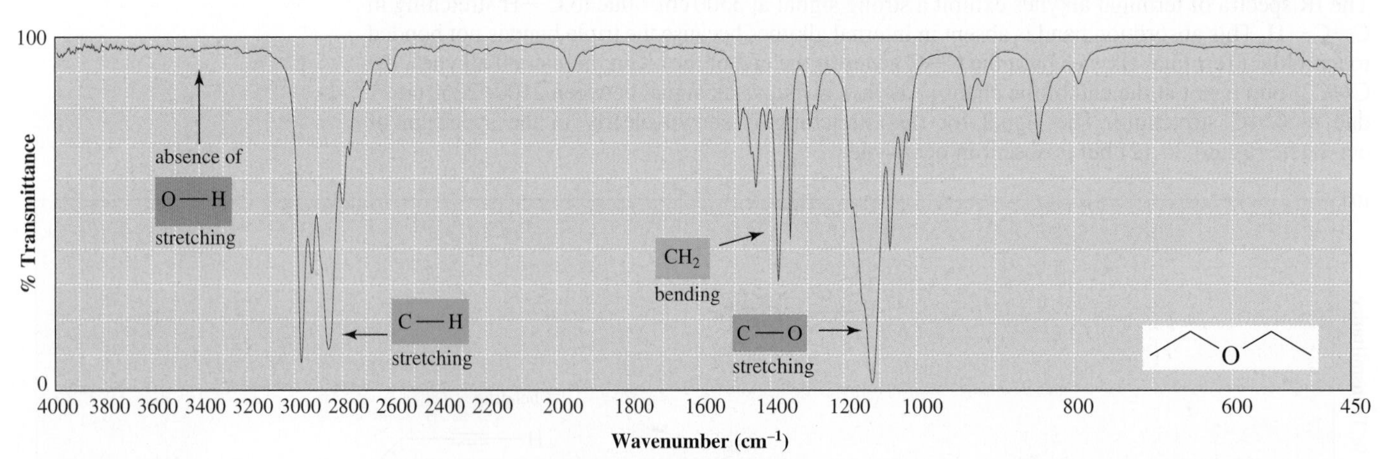

FIGURE 20.14 Infrared spectrum of diethyl ether.

Benzene and its derivatives

The IR spectra of aromatic rings show a medium to weak signal in the C—H stretching region at approximately 3030 cm^{-1}, characteristic of C—H bonds on sp^2 carbon atoms. They also show several signals due to C=C stretching between 1450 and 1600 cm^{-1}. In addition, they show strong signals in the region from 690 to 900 cm^{-1} due to C—H bending of the aromatic rings (table 20.7). Finally, the presence of weak, broad bands between 1700 and 2000 cm^{-1} can be an indicator of a benzene ring (often these are so weak that they are not observed).

TABLE 20.7 Characteristic IR signals of aromatic hydrocarbons.

Bond	Vibration	Frequency (cm^{-1})	Intensity
C—H	stretching	3030	medium to weak
C—H	bending	690–900	strong
C=C	stretching	1475 and 1600	strong to medium

The C—H and C=C signal patterns characteristic of aromatic rings can be seen in the infrared spectrum of toluene (figure 20.15).

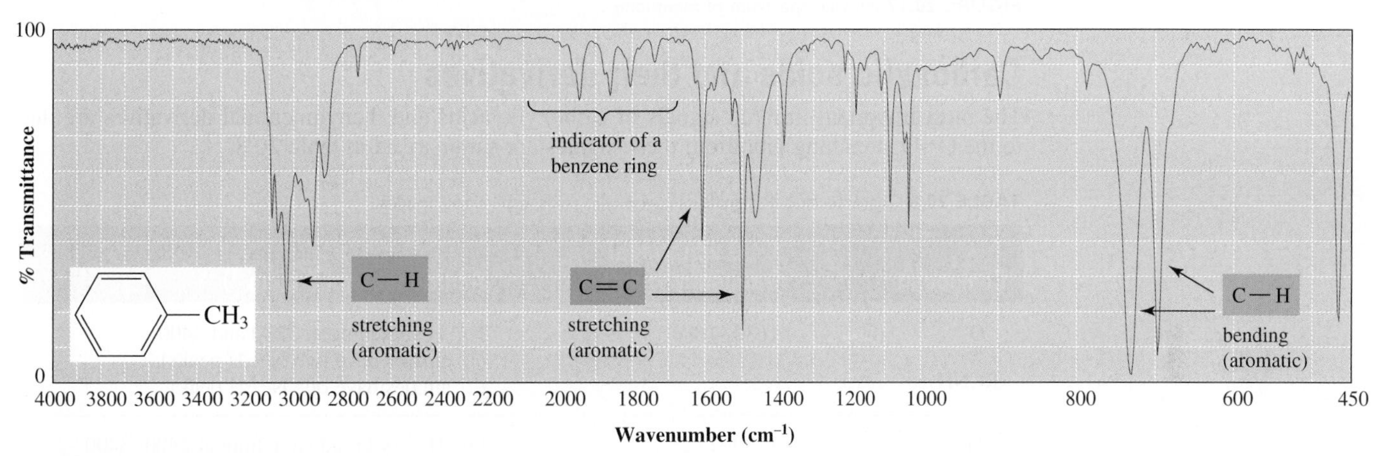

FIGURE 20.15 Infrared spectrum of toluene.

Amines

The most important and readily observed infrared signals of primary and secondary amines are due to N—H stretching vibrations, which appear in the region from 3100 to 3500 cm^{-1}. Primary amines have two signals in this region, one caused by symmetric stretching vibration and the other by asymmetric stretching. The two N—H stretching bands characteristic of a primary amine can be seen in the IR spectrum of butylamine (figure 20.16). Secondary amines give only one band in this region. Tertiary amines have no N—H and so are transparent in this region of the infrared spectrum.

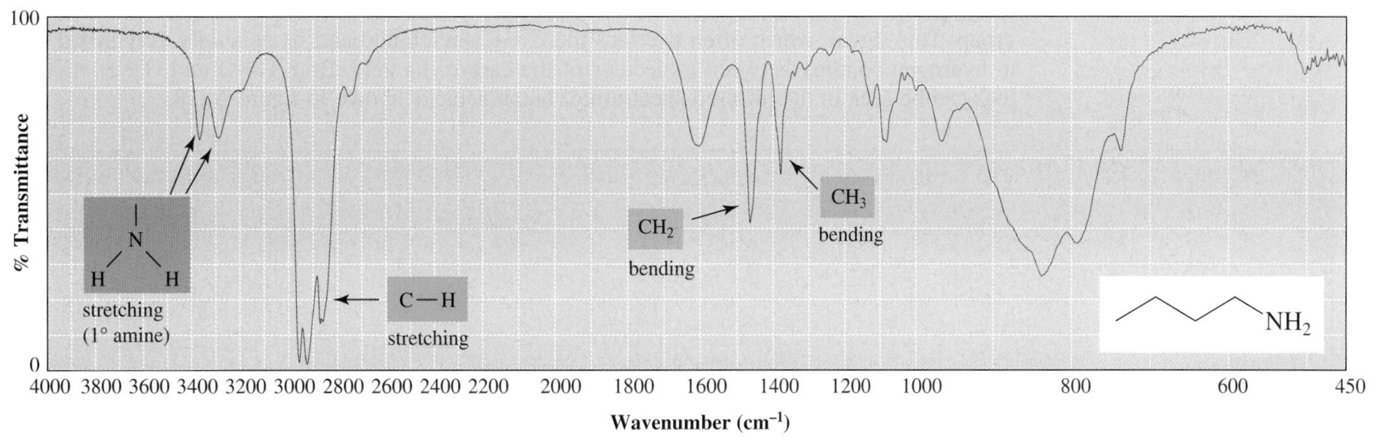

FIGURE 20.16 Infrared spectrum of butylamine, a primary amine.

Aldehydes and ketones

Aldehydes and ketones show characteristic strong infrared signals between 1705 and 1780 cm^{-1} associated with the stretching vibration of the carbon–oxygen double bond. The stretching vibration for the carbonyl group of menthone occurs at 1705 cm^{-1} (figure 20.17, overleaf).

Because several different functional groups contain a carbonyl group, it is often not possible to tell from signals in this region alone whether the carbonyl-containing compound is an aldehyde, a ketone, a carboxylic acid or an ester.

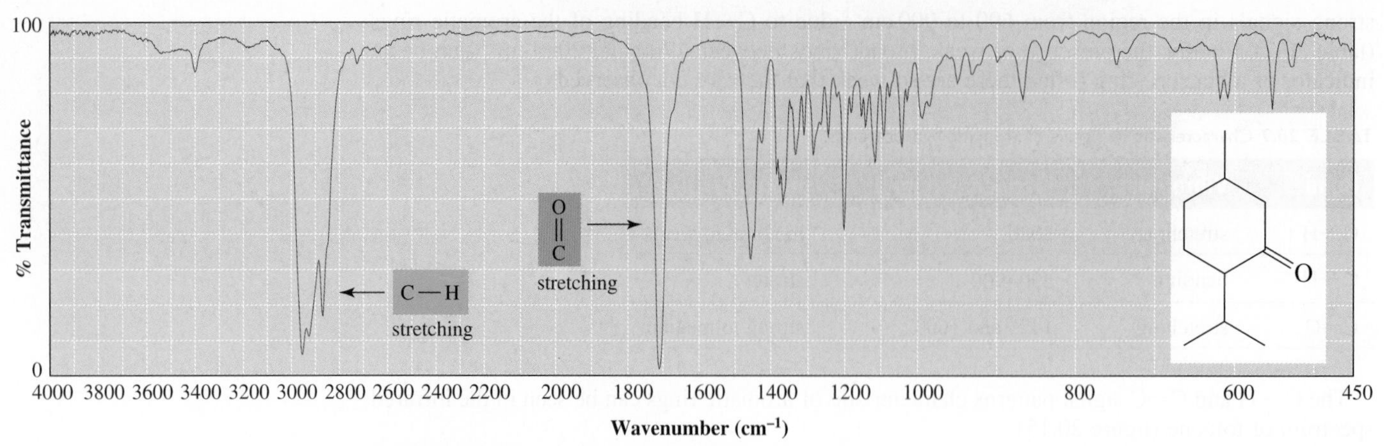

FIGURE 20.17 Infrared spectrum of menthone.

Carboxylic acids and their derivatives

The most important infrared signals of carboxylic acids and their functional derivatives are due to the C═O stretching vibration; these signals are summarised in table 20.8.

TABLE 20.8 Characteristic IR signals of carboxylic acids, esters and amides.

Compound	C═O vibrational frequency (cm^{-1})	Additional signals (cm^{-1})
$RC(=O)NH_2$	1630–1680	N—H stretching at 3200 and 3400 (1° amides have two N—H peaks) (2° amides have one N—H peak)
$RC(=O)OH$	1700–1725	O—H very broad stretching at 2400–3400 C—O stretching at 1210–1320
$RC(=O)OR$	1735–1750	C—O stretching at 1000–1100 and 1200–1250

The carboxyl group of a carboxylic acid gives rise to two characteristic signals in the infrared spectrum. One of these occurs in the region from 1700 to 1725 cm^{-1} and is associated with the stretching vibration of the carbonyl group. This region is essentially the same as that observed for the carbonyl groups of aldehydes and ketones. The other infrared signal characteristic of a carboxyl group is a band between 2400 and 3400 cm^{-1} due to the stretching vibration of the O—H group. This signal, which often overlaps the C—H stretching band, is generally very broad due to hydrogen bonding between molecules of the carboxylic acid. Both C═O and O—H stretchings can be seen in the infrared spectrum of butanoic acid, shown in figure 20.18.

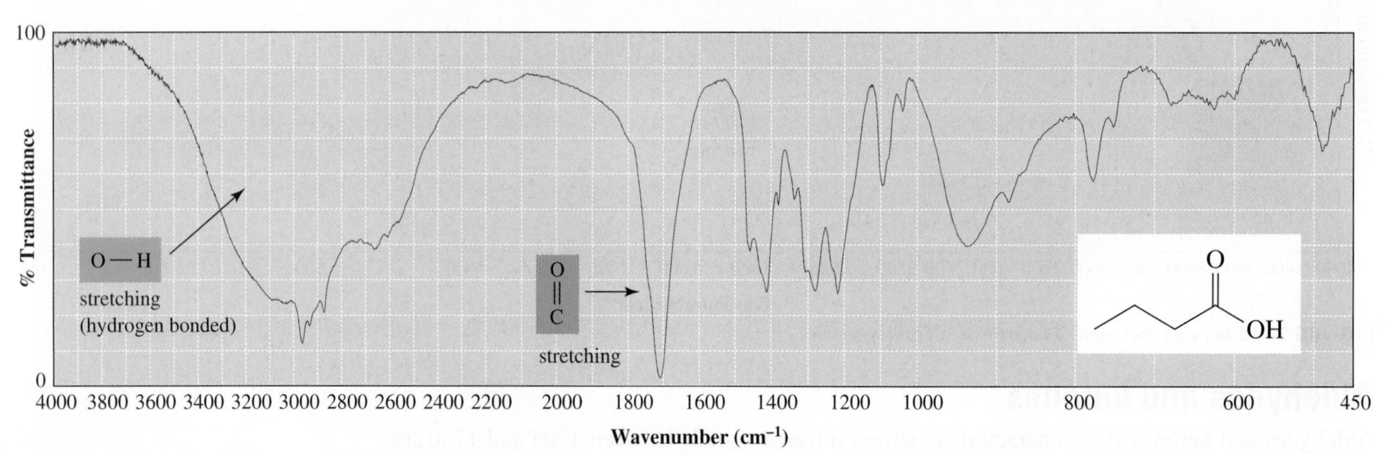

FIGURE 20.18 Infrared spectrum of butanoic acid.

The IR spectra of esters display strong signals for C═O stretching in the region between 1735 and 1750 cm^{-1}. In addition, they display strong signals for C—O stretching in the region from 1000 to 1250 cm^{-1} (figure 20.19).

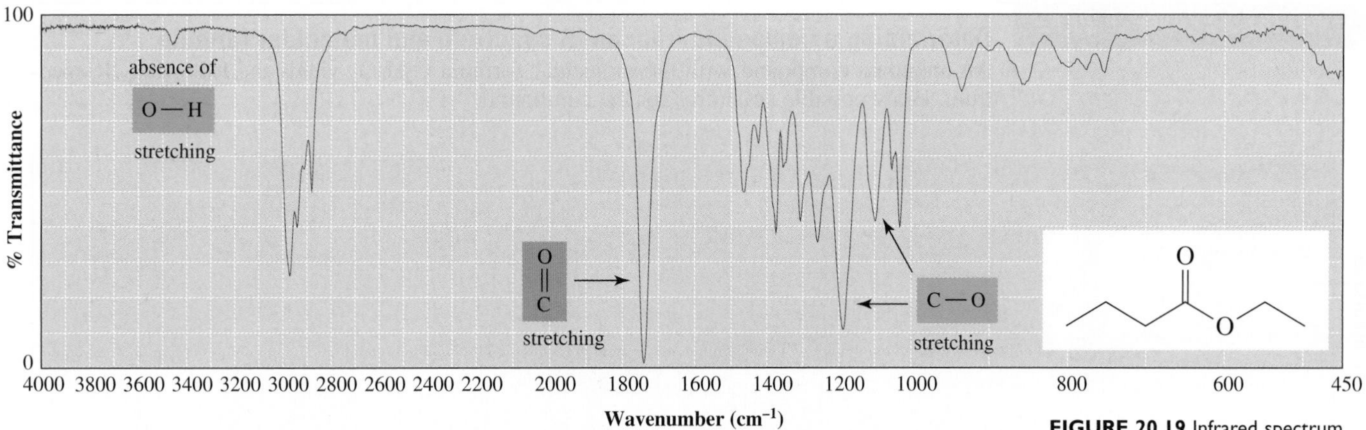

FIGURE 20.19 Infrared spectrum of ethyl butanoate.

The signal due to carbonyl stretching of amides occurs at lower wavenumbers (1630–1680 cm^{-1}) than other carbonyl compounds. Like the amines, primary and secondary amides show N—H stretching in the region from 3200 to 3400 cm^{-1}; primary amides ($RCONH_2$) show two N—H signals, whereas secondary amides (RCONHR) show only a single N—H signals. Tertiary amides, of course, do not show N—H stretching absorptions. Compare the three spectra in figure 20.20.

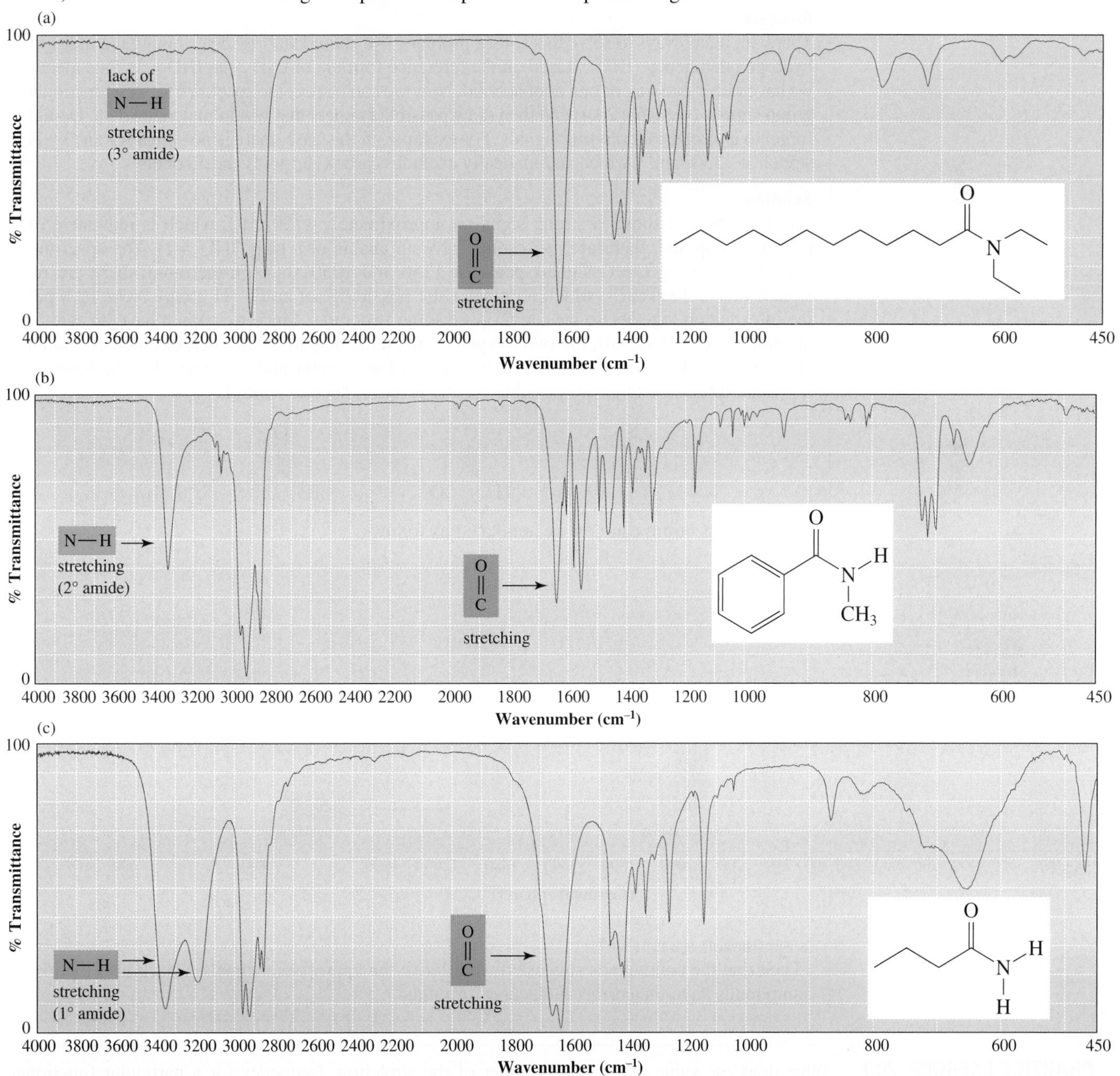

FIGURE 20.20 Infrared spectra of: **(a)** *N*,*N*-diethyldodecanamide, a tertiary amide, **(b)** *N*-methylbenzamide, a secondary amide, and **(c)** butanamide, a primary amide.

WORKED EXAMPLE 20.7

Determining a compound from an IR spectrum and molecular formula

An unknown compound with the molecular formula $C_3H_6O_2$ yields the following IR spectrum. Draw possible structures for the compound.

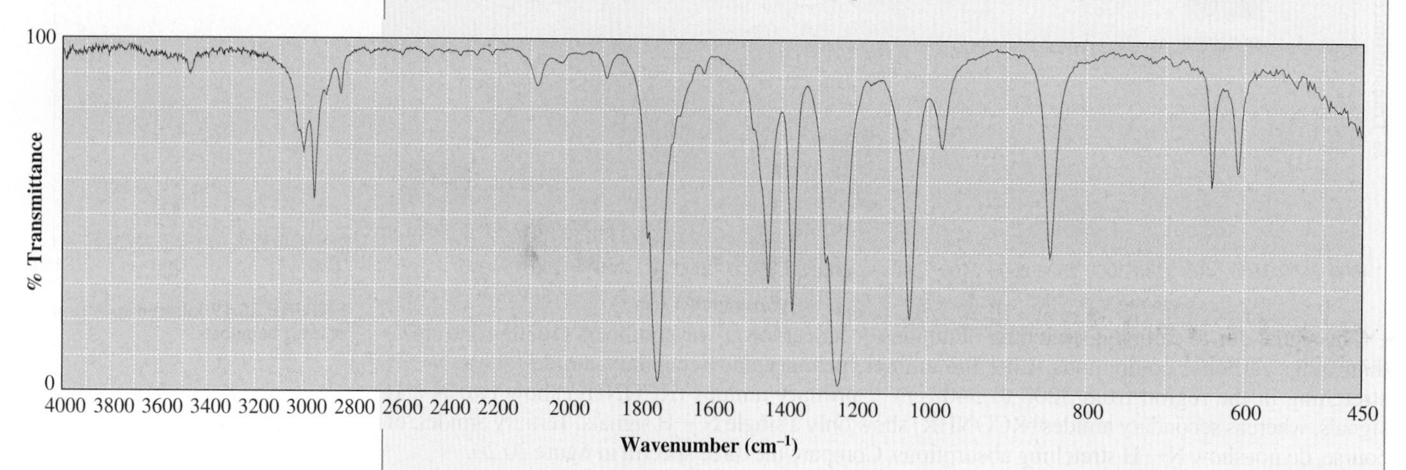

Analysis

There are many ways to solve this sort of problem. We should start by determining the index of hydrogen deficiency (IHD). Based on the reference formula of C_3H_8, the IHD = 1. This means we have either one ring or one double bond. Now we can inspect the spectrum for key functional group signals (see table 20.4) and then draw possible isomers that would match the data. Do not forget to also consider missing signals. For example, in this case, there is no signal for an O—H stretch at ~3300 cm^{-1}, which automatically excludes carboxylic acids and alcohols.

Solution

The IR spectrum shows a strong signal at approximately 1750 cm^{-1}, which is indicative of a C=O group and, therefore, we cannot have an alkene or a ring (IHD = 1). The spectrum also shows strong C—O signals peaks at 1250 and 1050 cm^{-1}. Furthermore, there are no signals above 3000 cm^{-1}, which eliminates the possibility of an O—H group. Now, let's put the information together so we can propose some structures. Since we cannot have an acid or alcohol (no O—H signal), the two oxygen atoms must be part of an ester, or an ether and an aldehyde (we do not have enough carbon atoms for an ether and a ketone). On the basis of this data, three structures are possible for the given molecular formula.

The given spectrum can now be annotated as follows.

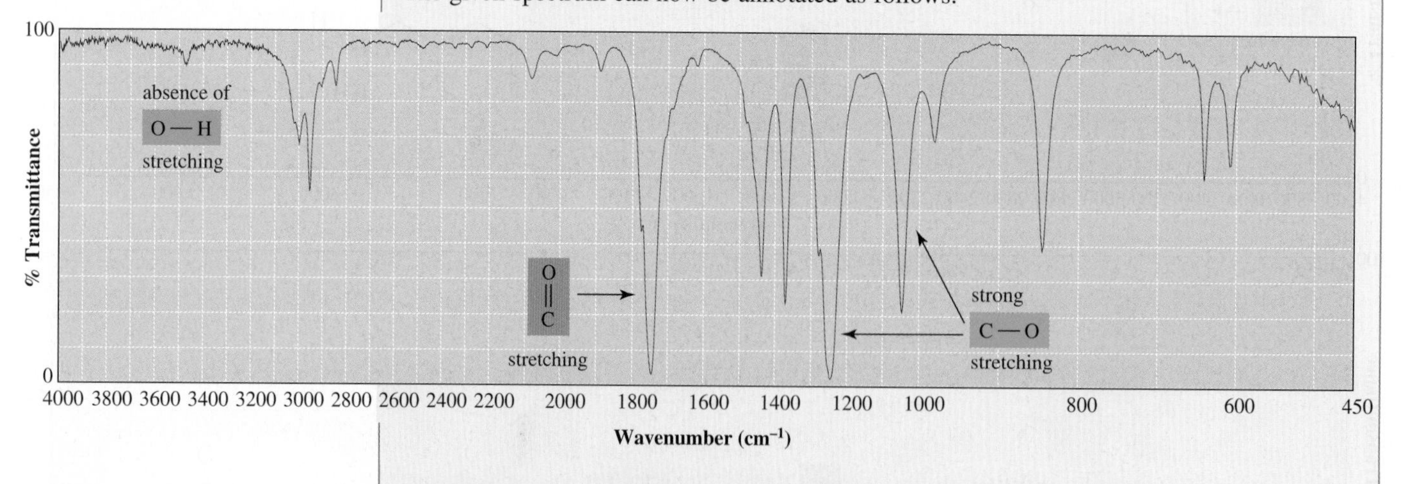

Is our answer reasonable?

Check that the compounds suggested would give the appropriate signals. Check also that the compounds have the correct molecular formula.

PRACTICE EXERCISE 20.7 What does the value of the wavenumber of the stretching frequency for a particular functional group indicate about the relative strength of the bond in that functional group?

Interpreting IR spectra

Determine possible structures for a compound that yields the following IR spectrum and has a molecular formula of C_7H_8O.

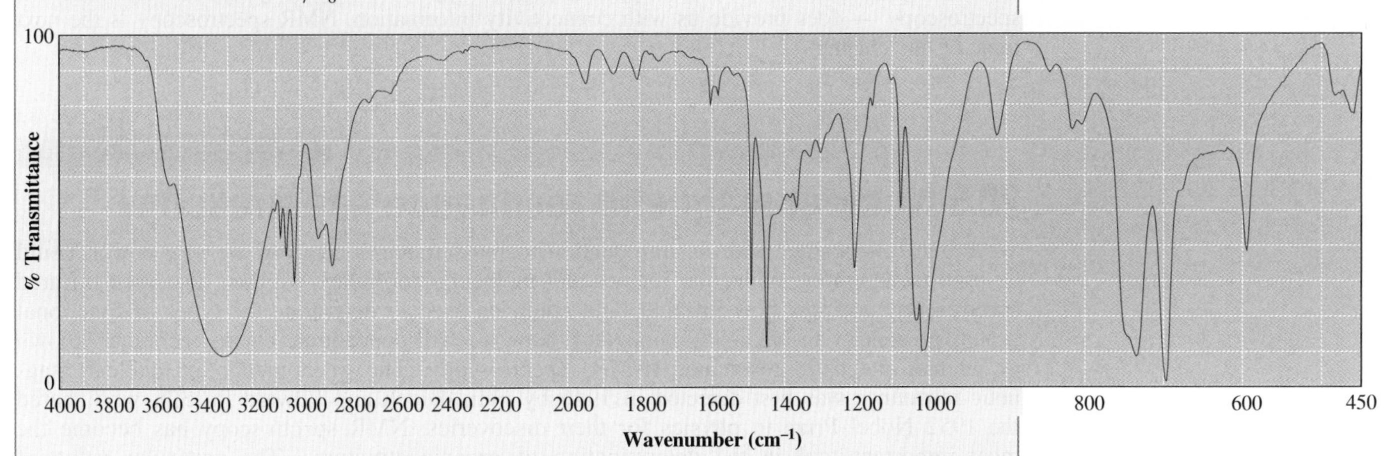

Analysis

Given the molecular formula of C_7H_8O, you should determine the index of hydrogen deficiency to be 4, based on the reference formula of C_7H_{16}. While accounting for this value may seem daunting at first (consider the possible combinations of four rings or π bonds), recall that the benzene ring has an index of hydrogen deficiency of 4. Look for characteristic bands for aromatic compounds in the IR spectrum; if they are present, you know that there are no other double bonds or rings in the molecule. The oxygen atom may be present as an alcohol/phenol, ether, aldehyde or ketone (the last two would not be possible if there was an aromatic ring present).

Solution

A quick inspection of the spectrum indicates a strong, broad signal at approximately 3310 cm^{-1} due to O—H stretching and the absence of a signal due to a carbonyl group between 1700 and 1800 cm^{-1}. Because we must have an —OH group, we cannot propose any structures with an ether group (or an aldehyde or a ketone). Closer inspection of the spectrum shows the characteristic aromatic C—H bending bands at 690 and 740 cm^{-1} and the weak, broad bands between 1700 and 2000 cm^{-1}, which support the existence of a benzene ring. Also, a signal for sp^2 C—H stretching is present just above 3000 cm^{-1}. Aromatic C=C stretching absorption bands at 1450 and 1490 cm^{-1} also indicate a benzene ring. Now, given that the molecular formula is C_7H_8O and we have a benzene ring and an —OH group, we have to account for only one more carbon atom. This last carbon atom must be bonded to the benzene ring. There are only four possible places we can attach an —OH group to toluene (methylbenzene) and these structures are given below.

OH OH OH HO

The given spectrum can now be annotated as follows.

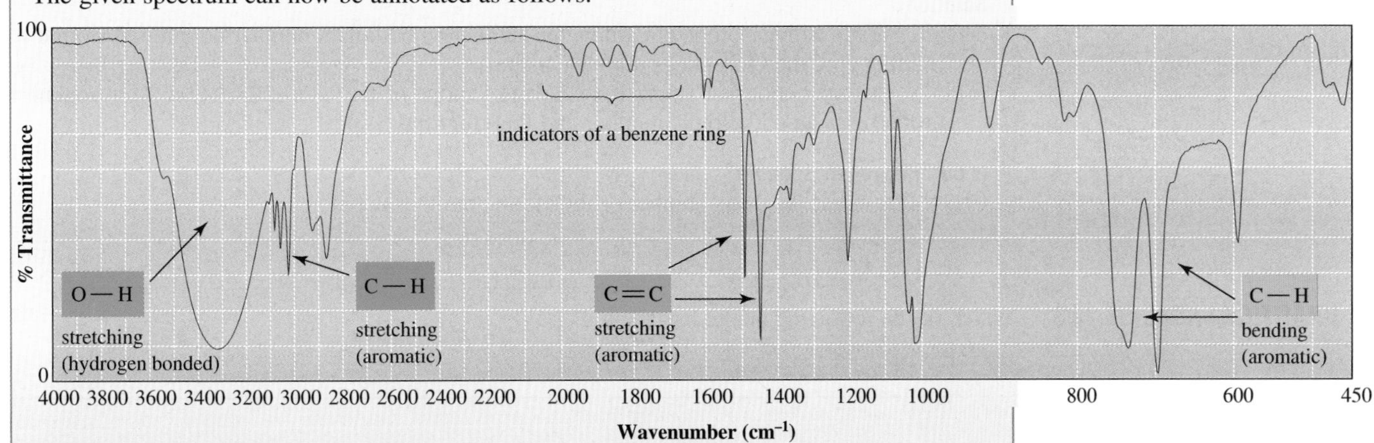

Is our answer reasonable?

You should check that all of the proposed compounds have the correct IHD and that the key signals for each compound are present in the spectrum.

The preceding examples illustrate the power and limitations of IR spectroscopy. The power lies in its ability to provide us with information about the functional groups in a molecule. IR spectroscopy does not, however, provide us with information on how those functional groups are connected. Fortunately, another type of spectroscopy — nuclear magnetic resonance (NMR) spectroscopy — does provide us with connectivity information. NMR spectroscopy is the next topic of this chapter.

20.5 Nuclear magnetic resonance spectroscopy

We began this chapter with an introduction to mass spectrometry and we saw how it could be used to determine molecular masses and molecular formulae. We then discussed infrared spectroscopy and saw how infrared light could be used to determine the types of functional groups present in an unknown compound. Now we will concentrate on the technique known as nuclear magnetic resonance (NMR) spectroscopy. The phenomenon of nuclear magnetic resonance was first detected in 1946 by Felix Bloch and Edward Purcell, who shared the 1952 Nobel Prize in physics for their discoveries. NMR spectroscopy has become the most important tool in the determination of organic structures. The particular value of **nuclear magnetic resonance (NMR) spectroscopy** is that it gives us information about the number and types of atoms in a molecule and can also provide information on how they are connected. **^{1}H-NMR spectroscopy** gives information about the number and types of hydrogen atoms within a molecule while **^{13}C-NMR spectroscopy** tells us about the number and types of carbon atoms. Although we will consider only hydrogen and carbon NMR, it is also possible to obtain NMR spectra of many other elements.

The origin of nuclear magnetic resonance

From our study in chapter 4, we are already familiar with the concept that an electron has a spin and that a spinning charge creates an associated magnetic field. In effect, an electron behaves as if it is a tiny bar magnet. An atomic nucleus that has an odd mass or an odd atomic number also has a spin (a nuclear spin) and also behaves as if it is a tiny bar magnet. Recall that, when designating isotopes, a superscript represents the mass of the element. Thus, the nuclei of ^{1}H and ^{13}C, isotopes of the two elements most common in organic compounds, also have a nuclear spin, whereas the nuclei of ^{12}C and ^{16}O do not have a nuclear spin. Accordingly, in this sense, nuclei of ^{1}H and ^{13}C are quite different from nuclei of ^{12}C and ^{16}O.

WORKED EXAMPLE 20.9

Determining whether an atomic nucleus has a spin

Which of the following nuclei has a spin?

(a) $^{14}_{6}C$ (b) $^{14}_{7}N$

Analysis

Recall that to have a nuclear spin an isotope must have either an odd atomic number or an odd mass.

Solution

(a) ^{14}C, a radioactive isotope of carbon, has neither an odd mass number nor an odd atomic number and, therefore, does not have a spin.

(b) ^{14}N, the most common naturally occurring isotope of nitrogen (99.63% of all nitrogen atoms), has an odd atomic number and therefore has a spin.

Is our answer reasonable?

Check that either the mass number or the atomic number is odd.

PRACTICE EXERCISE 20.8 Which of the following nuclei has a spin?

(a) $^{31}_{15}P$ (b) $^{195}_{78}Pt$

Within a collection of ^{1}H and ^{13}C atoms, the spins of their nuclei are completely random in orientation. When we place them in a powerful magnetic field, however, interactions between their nuclear spins and the applied magnetic field are quantised, and only two orientations are allowed (figure 20.21).

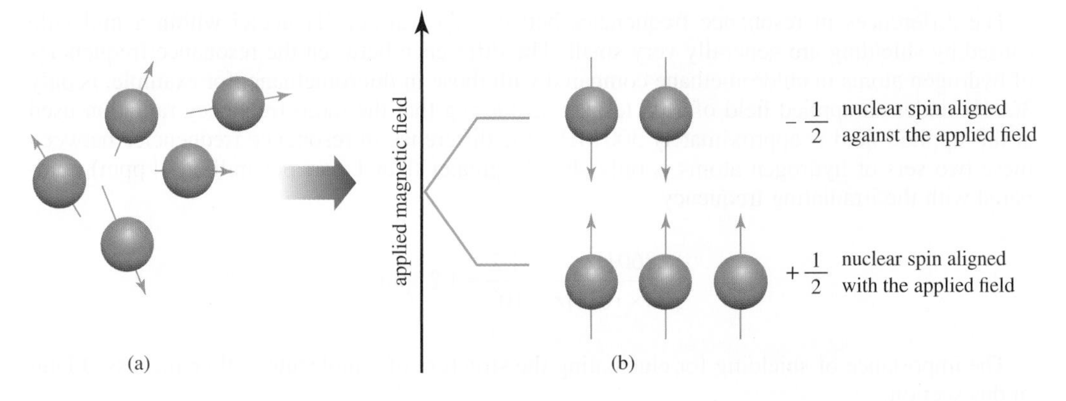

FIGURE 20.21 ^{1}H and ^{13}C nuclei **(a)** in the absence of an applied magnetic field and **(b)** in the presence of an applied field. ^{1}H and ^{13}C nuclei with spin $+\frac{1}{2}$ are aligned with the applied magnetic field and are in the lower spin energy state; those with spin $-\frac{1}{2}$ are aligned against the applied magnetic field and are in the higher spin energy state.

When hydrogen nuclei are placed in an applied magnetic field, a small majority of their nuclear spins align with the applied field in the lower energy state. Irradiation of nuclei with **radio-frequency radiation** of the appropriate energy causes some of those in the lower energy spin state to absorb energy and results in their nuclear spins flipping from the lower energy state to the higher energy state, as illustrated in figure 20.22. In this context, **resonance** is defined as the absorption (or emission when the nuclei return to equilibrium) of electromagnetic radiation by a spinning nucleus and the resulting flip of its nuclear spin state. (*Note:* Resonance used in this context is completely different from the resonance theory of chemical structures discussed earlier.)

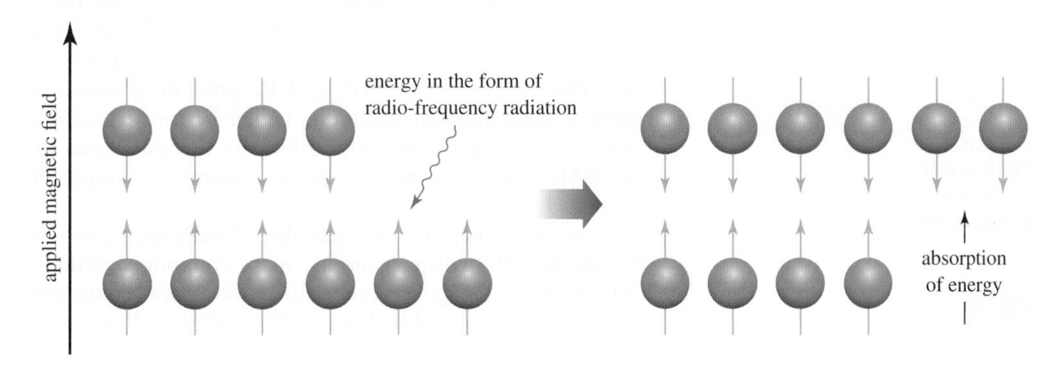

FIGURE 20.22 An example of resonance for nuclei of spin $+\frac{1}{2}$ and $-\frac{1}{2}$.

If a strong enough magnetic field is used, then the energy required to cause this flipping corresponds to that of radiowaves. For example, at an applied field strength of 7.05 tesla (T), which is readily available with present-day superconducting electromagnets, ^{1}H nuclei resonate over a range of frequencies of approximately 300 MHz (300 000 000 Hz). At the same magnetic field strength, ^{13}C nuclei resonate at approximately 75 MHz. Thus, we can use electromagnetic radiation in the radio-frequency range to detect changes in nuclear spin states for ^{1}H and ^{13}C. In the next several sections, we describe how these measurements are made for nuclear spin states of these two isotopes and then how this information can be correlated with molecular structure.

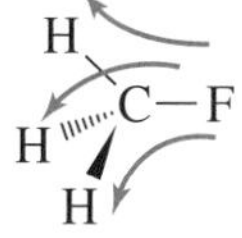

FIGURE 20.23 Fluorine's greater electronegativity produces a larger inductive effect and therefore reduces the electron density around each hydrogen atom. We say that these hydrogen atoms are deshielded.

Shielding

As atoms in molecules are surrounded by electrons and by other atoms they resonate at slightly different frequencies. The electrons that surround a nucleus also have spin so they create **local magnetic fields** that oppose the applied field. Consequently, the magnetic field experienced by a hydrogen atom in a molecule is slightly less than that of the applied magnetic field of the instrument. Although these local magnetic fields created by electrons are orders of magnitude weaker than the applied magnetic fields used in NMR spectroscopy, they are nonetheless significant at the molecular level. The result of these local magnetic fields is that hydrogen atoms are **shielded** from the applied field. The greater the shielding of a particular hydrogen atom by local magnetic fields, the greater is the energy (frequency of radiowaves) required to bring that hydrogen atom into resonance.

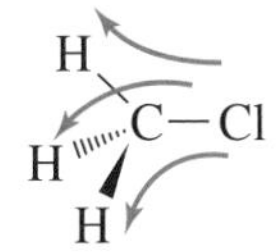

As we learned in previous chapters, the electron density around a nucleus can be influenced by the atoms that surround the nucleus. For example, the electron density around the hydrogen atoms in fluoromethane (figure 20.23) is less than that around the hydrogen atoms in chloromethane (figure 20.24), due to the greater electronegativity of fluorine relative to chlorine. Thus, we can say that the hydrogen atoms in chloromethane are *more shielded* than the hydrogen atoms in fluoromethane. Conversely, we can say the hydrogen atoms in fluoromethane are more **deshielded** (reduced electron density around each hydrogen atom) than those in chloromethane.

FIGURE 20.24 Chlorine is less electronegative than fluorine, resulting in a smaller inductive effect and therefore a greater electron density around each hydrogen atom. We say that the hydrogen atoms in chloromethane are more shielded (by their local environment) than those in fluoromethane.

The differences in resonance frequencies between the various ^{1}H nuclei within a molecule caused by shielding are generally very small. The difference between the resonance frequencies of hydrogen atoms in chloromethane compared with those in fluoromethane, for example, is only 360 Hz under an applied field of 7.05 tesla. Considering that the radio-frequency radiation used at this applied field is approximately 300 MHz, the difference in resonance frequencies between these two sets of hydrogen atoms is only slightly greater than 1 part per million (1 ppm) compared with the irradiating frequency.

$$\frac{360\,\text{Hz}}{300 \times 10^{6}\,\text{Hz}} = \frac{1.2}{10^{6}} = 1.2\,\text{ppm}$$

The importance of shielding for elucidating the structure of a molecule will be discussed later in this section.

An NMR spectrometer

Modern NMR spectrometers (figure 20.25) work by irradiating a sample with a short pulse of radiation that covers the entire range of relevant radiowave frequencies. All hydrogen atoms are simultaneously excited and then begin to return (or decay) to their original spin state. As each atom decays, it releases energy of a particular frequency that is detected as an electrical impulse in the receiver coils. The resulting combination of electrical impulses generated from all decay frequencies is called a free induction decay (FID). Typically an FID can be recorded every 1–2 seconds, and many are collected and then averaged to generate a spectrum using a mathematical technique called Fourier transformation. This signal averaging allows us to easily measure the NMR spectrum of very small amounts of sample (as little as 1 mg).

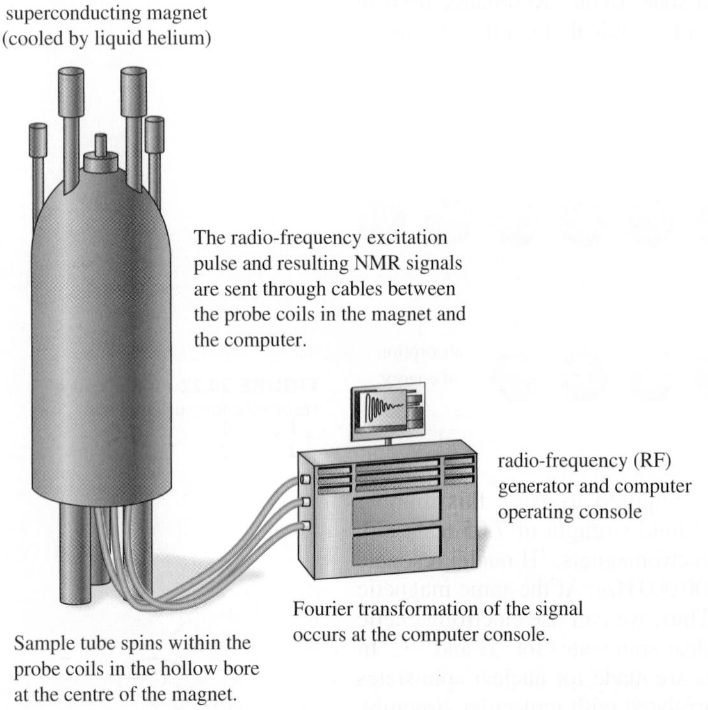

To record a spectrum, the sample is dissolved in a solvent having no ^{1}H hydrogen atoms, most commonly deuterochloroform, $CDCl_3$, or deuterium oxide, D_2O (deuterium, D, is the ^{2}H isotope of hydrogen). The sample cell is a small glass tube suspended in the hollow bore at the centre of the magnet and set spinning on its long axis to ensure that all parts of the sample experience a homogeneous applied field.

It is customary to measure the resonance frequencies of individual nuclei relative to the resonance frequency of the same nuclei in a reference compound. The reference compound now universally accepted for ^{1}H-NMR and ^{13}C-NMR spectroscopy in organic solvents is **tetramethylsilane (TMS)**.

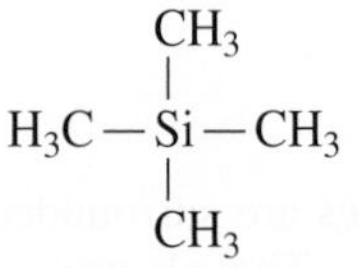

superconducting magnet (cooled by liquid helium)

The radio-frequency excitation pulse and resulting NMR signals are sent through cables between the probe coils in the magnet and the computer.

radio-frequency (RF) generator and computer operating console

Fourier transformation of the signal occurs at the computer console.

Sample tube spins within the probe coils in the hollow bore at the centre of the magnet.

FIGURE 20.25 Schematic diagram of a nuclear magnetic resonance spectrometer.

$$\begin{array}{c} CH_3 \\ | \\ H_3C-Si-CH_3 \\ | \\ CH_3 \end{array}$$

tetramethylsilane (TMS)

When we measure a ^{1}H-NMR (or ^{13}C-NMR) spectrum of a compound, we report how far the resonance signals of its hydrogen (or carbon) atoms are shifted from the resonance signal of the hydrogen (or carbon) atoms in TMS.

To standardise reporting of NMR data, scientists have adopted a quantity called the **chemical shift** (δ) expressed in parts per million.

$$\delta = \frac{\text{shift in frequency of a signal from TMS (Hz)}}{\text{operating frequency of the spectrometer (MHz)}}$$

In a typical ^{1}H-NMR spectrum, the horizontal axis represents the δ (delta) scale, with values from 0 on the right to 10 on the left, but values can fall outside this range (for example, see figures 20.30 and 20.46 and table 20.9). The vertical axis represents the intensity of the resonance signal. Figure 20.26 shows a ^{1}H-NMR spectrum of methyl acetate, a compound used in the manufacture of artificial leather. The small signal at δ 0 in this spectrum represents the hydrogen atoms of the reference compound, TMS. The remainder of the spectrum consists of two signals: one for the hydrogen atoms of the $—OCH_3$ group and one for the hydrogen atoms of the methyl attached to the carbonyl group. It is not our purpose at the moment to determine which hydrogen atoms give rise to which signal but only to recognise the form in which we record an NMR spectrum and to understand the meaning of the calibration marks.

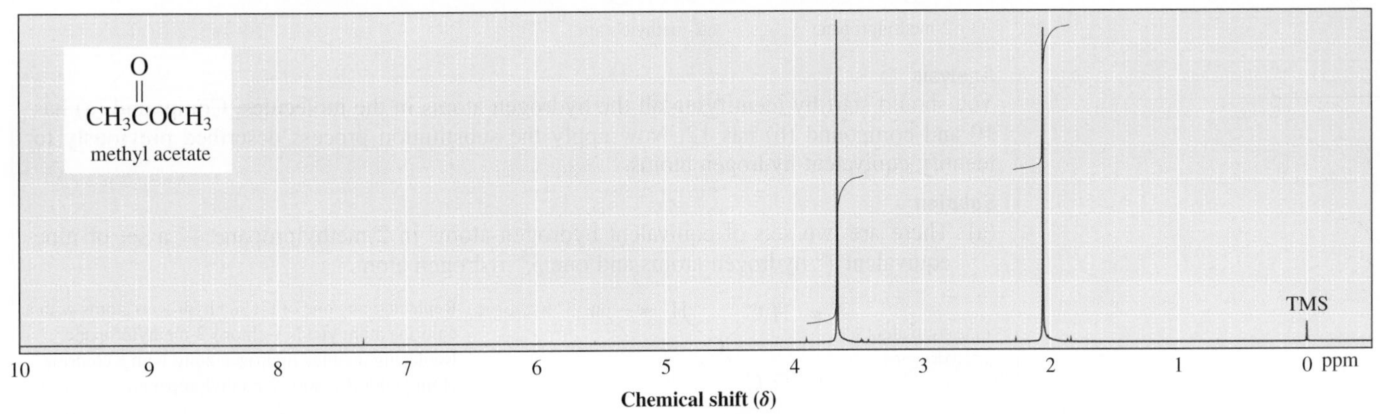

FIGURE 20.26 ^{1}H-NMR spectrum of methyl acetate.

A note on terminology: If a signal is shifted towards the left of the spectrum, we say that the nuclei giving rise to that signal are deshielded (also often referred to as shifted **downfield**). Conversely, if a signal is shifted towards the right of the spectrum, we say that the nuclei giving rise to that signal are shielded (also often referred to as shifted **upfield**). Figure 20.27 summarises some of the common NMR terms.

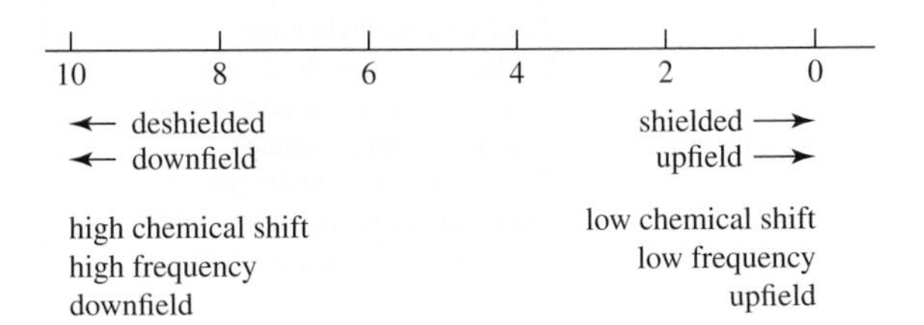

FIGURE 20.27 A summary of some common NMR terms.

Equivalent hydrogen atoms

Given the structural formula of a compound, how do we know how many signals to expect? The answer is that **equivalent hydrogen atoms** give the same ^{1}H-NMR signal; conversely, nonequivalent hydrogen atoms give different ^{1}H-NMR signals. A direct way to determine which hydrogen atoms in a molecule are equivalent is to replace each in turn by a test atom, such as a halogen atom. If replacement of two hydrogen atoms in turn gives the same compound, the two hydrogen atoms are equivalent. If replacement gives different compounds, the two hydrogen atoms are nonequivalent.

Using this substitution test, we can show that propane contains two sets of equivalent hydrogen atoms: a set of six equivalent 1° hydrogen atoms (hydrogen atoms attached to a 1° carbon atom) and a set of two equivalent 2° hydrogen atoms (hydrogen atoms attached to a 2° carbon atom). Thus we would expect to see two signals: one for the six equivalent methyl hydrogen atoms and one for the two equivalent methylene ($—CH_2—$) hydrogen atoms.

$CH_3—CH_2—CH_3$ (propane) → $CH_3—CH_2—CH_2Cl$ and $ClCH_2—CH_2—CH_3$

Replacement of any of the red hydrogen atoms by a chlorine atom gives 1-chloropropane; thus, all of the red hydrogen atoms in propane are equivalent.

$CH_3—CH_2—CH_3$ (propane) $\longrightarrow$ $CH_3—CHCl—CH_3$

Replacement of either of the blue hydrogen atoms by a chlorine atom gives 2-chloropropane; thus, both of the blue hydrogen atoms in propane are equivalent.

WORKED EXAMPLE 20.10

Equivalent hydrogen atoms

Determine the number of sets of equivalent hydrogen atoms in each of the following compounds and the number of hydrogen atoms in each set.

(a) 2-methylpropane (b) 2-methylbutane

Analysis

You should start by identifying all the hydrogen atoms in the molecules. Compound (a) has 10 and compound (b) has 12. Now apply the substitution process described previously to identify equivalent hydrogen atoms.

Solution

(a) There are two sets of equivalent hydrogen atoms in 2-methylpropane — a set of nine equivalent 1° hydrogen atoms and one 3° hydrogen atom.

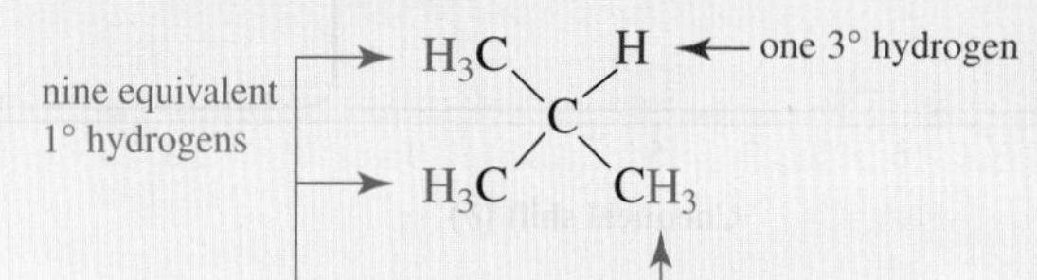

Replacing any one of the red hydrogen atoms with a chlorine atom yields 1-chloro-2-methylpropane. Replacing the blue hydrogen atom with a chlorine atom yields 2-chloro-2-methylpropane.

(b) There are four sets of equivalent hydrogen atoms in 2-methylbutane — two different sets of 1° hydrogen atoms, one set of 2° hydrogen atoms and one 3° hydrogen atom.

six equivalent 1° hydrogens → H_3C, H_3C; H ← one 3° hydrogen; CH_3 ← three equivalent 1° hydrogens; CH_2 ← two equivalent 2° hydrogens

Replacing any of the red hydrogen atoms with a chlorine atom yields 1-chloro-2-methylbutane. Replacing the blue hydrogen atom with a chlorine atom yields 2-chloro-2-methylbutane. Replacing a purple hydrogen atom with a chlorine atom yields 2-chloro-3-methylbutane. Replacing a green hydrogen atom with a chlorine atom yields 1-chloro-3-methylbutane.

Is our answer reasonable?

Check that you have identified all hydrogen atoms and that the different groups are actually different.

PRACTICE EXERCISE 20.9

Determine the number of sets of equivalent hydrogen atoms in each of the following compounds and the number of hydrogen atoms in each set.

(a) methylcyclopentane (b) 2,2,4-trimethylpentane

Here are four organic compounds, each of which has one set of equivalent hydrogen atoms and gives one signal in its ^{1}H-NMR spectrum.

$CH_3C(=O)CH_3$ acetone (propanone); $ClCH_2CH_2Cl$ 1,2-dichloroethane; cyclopentane; $(H_3C)_2C{=}C(CH_3)_2$ 2,3-dimethylbut-2-ene

The structures below show that molecules with two or more sets of equivalent hydrogen atoms give rise to a different resonance signal for each set. For example, 1,1-dichloroethane has (a) three equivalent 1° hydrogen atoms and (b) one 3° hydrogen atom, so there are two resonance signals in its ^{1}H-NMR spectrum.

CH_3CHCl_2 (a, b) 1,1-dichloroethane (two signals); cyclopentanone (a, b) (two signals); (Z)-1-chloropropene (a, b, c) (three signals); cyclohexene (a, b, c) (three signals)

You should see immediately that valuable information about a compound's molecular structure can be obtained simply by counting the number of signals in a ^{1}H-NMR spectrum of that compound. Consider, for example, the two constitutional isomers with molecular formula $C_2H_4Cl_2$, namely 1,2-dichloroethane and 1,1-dichloroethane. In 1,2-dichloroethane, all four hydrogen atoms are equivalent and therefore only one signal is seen in its ^{1}H-NMR spectrum. Its isomer 1,1-dichloroethane has two sets of equivalent hydrogen atoms and therefore shows two signals in its ^{1}H-NMR spectrum. Thus, simply counting signals can allow you to distinguish between these constitutional isomers. For example, in worked example 20.7 we narrowed down three possible candidates with a molecular formula of $C_3H_6O_2$ that could have produced the IR spectrum. Now, if we consider the number of signals in the ^{1}H-NMR spectrum of these compounds, we see that they give rise to two, three and three signals, respectively.

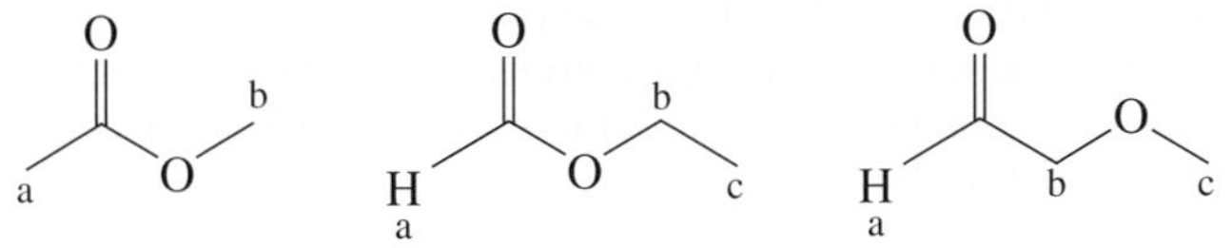

methyl acetate (methyl ethanoate) | ethyl formate (ethyl methanoate) | methoxyacetaldehyde (2-methoxyethanal)

Later we will see how we can distinguish between ethyl formate and methoxyacetaldehyde.

WORKED EXAMPLE 20.11

Structural formulae from number of signals in ^{1}H-NMR spectra

Each of the following compounds gives only one signal in its ^{1}H-NMR spectrum. Propose a structural formula for each.

(a) C_2H_6O (b) $C_3H_6Cl_2$ (c) C_6H_{12}

Analysis

We should start by determining the IHD as this will tell you if you need to consider rings or double bonds. With this information, you then need to consider how you can put the atoms in the formulae together so that all hydrogen atoms are equivalent (drawing possible isomers and checking the number of different hydrogen atoms is probably the easiest way).

Solution

Given that the IHD for the three compounds are 0, 0 and 1, respectively, we know that we need to consider a ring or alkene only for (c). The following are structural formulae for each of the given compounds. Notice that, for each structure, the replacement of any hydrogen atom with a chlorine atom will yield the same compound regardless of the hydrogen being replaced.

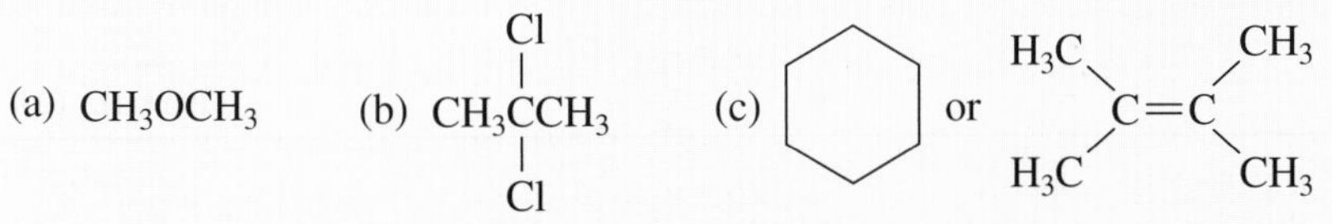

(a) CH_3OCH_3 (b) $CH_3CCl_2CH_3$ (c) cyclohexane or $(H_3C)_2C{=}C(CH_3)_2$

Is our answer reasonable?

Check that, for each structure, the replacement of any hydrogen atom with a chlorine atom will yield the same compound regardless of the hydrogen atom being replaced.

PRACTICE EXERCISE 20.10

Each of the following compounds gives only one signal in its ^{1}H-NMR spectrum. Propose a structural formula for each compound.

(a) $C_4H_6O_2$
(b) C_5H_{10}
(c) C_4H_9Br
(d) $C_4H_6Cl_4$

Signal areas

We have just seen that the number of signals in a ^{1}H-NMR spectrum gives us information about the number of sets of equivalent hydrogen atoms. Signal areas in a ^{1}H-NMR spectrum can be measured by a mathematical technique called **integration**. In most of the spectra shown in this text, this information is displayed in the form of a **line of integration** superimposed on the original spectrum. The vertical rise of the line of integration over each signal is proportional to the area under that signal, which, in turn, is proportional to the number of hydrogen atoms giving rise to the signal.

Figure 20.28 shows an integrated ^{1}H-NMR spectrum of the petrol additive *tert*-butyl acetate, $C_6H_{12}O_2$. The spectrum shows signals at δ 1.4 and 2.0. The integrated area of the more shielded signal (to the right: 67 chart divisions) is nearly three times that of the less shielded signal (to the left: 23 chart divisions). We have used the horizontal lines on the chart as a unit of measure (10 chart divisions are equivalent to the distance between two consecutive solid horizontal lines) — alternatively, we could measure the line separation with a ruler. This relationship corresponds to a ratio of 3 : 1. We know from the molecular formula that there is a total of 12 hydrogen atoms in the molecule. The ratios obtained from the integration lines are consistent with the presence of one set of nine equivalent hydrogen atoms and one set of three equivalent hydrogen atoms. We will often make use of shorthand notation in referring to an NMR spectrum of a molecule. The notation lists the chemical shift of each signal, beginning with the most deshielded signal and followed by the number of hydrogen atoms that give rise to each signal (based on the integration). The shorthand notation describing the spectrum of *tert*-butyl acetate (figure 20.28) would be δ 2.0 (3H) and δ 1.4 (9H).

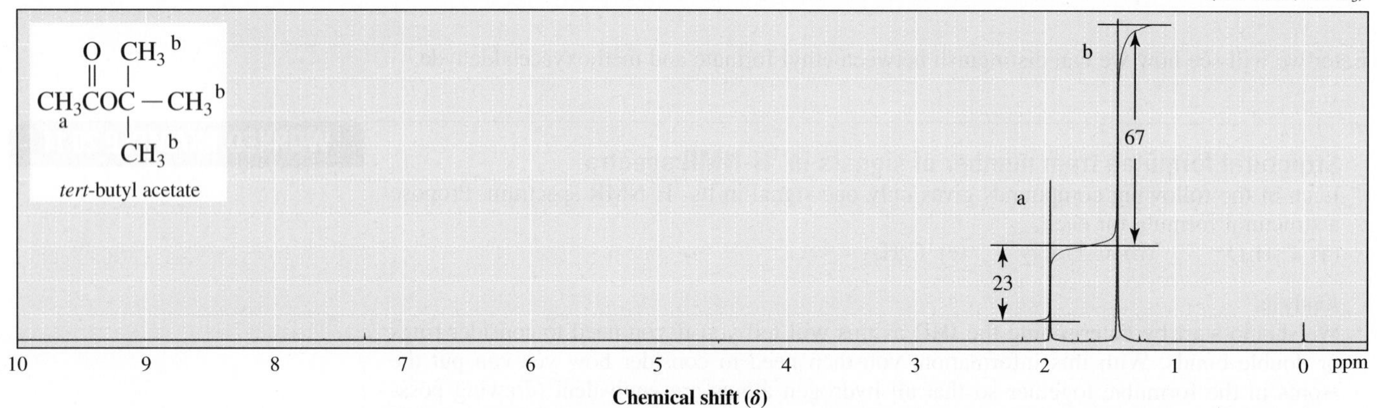

FIGURE 20.28 ^{1}H-NMR spectrum of *tert*-butyl acetate, $C_6H_{12}O_2$, showing a line of integration. The ratio of signal areas for the two peaks is 3 : 1, which, for a molecule possessing 12 hydrogen atoms, corresponds to nine equivalent hydrogen atoms of one set and three equivalent hydrogen atoms of another set.

An alternative way of representing integration values is to have the integration values printed below the horizontal scale axis. It is quite normal for the numerical values provided not to exactly match the ratio, but this is not a problem as we know we must have integer values for the number of hydrogen atoms. As an example, the spectrum of benzyl alcohol is given in figure 20.29.

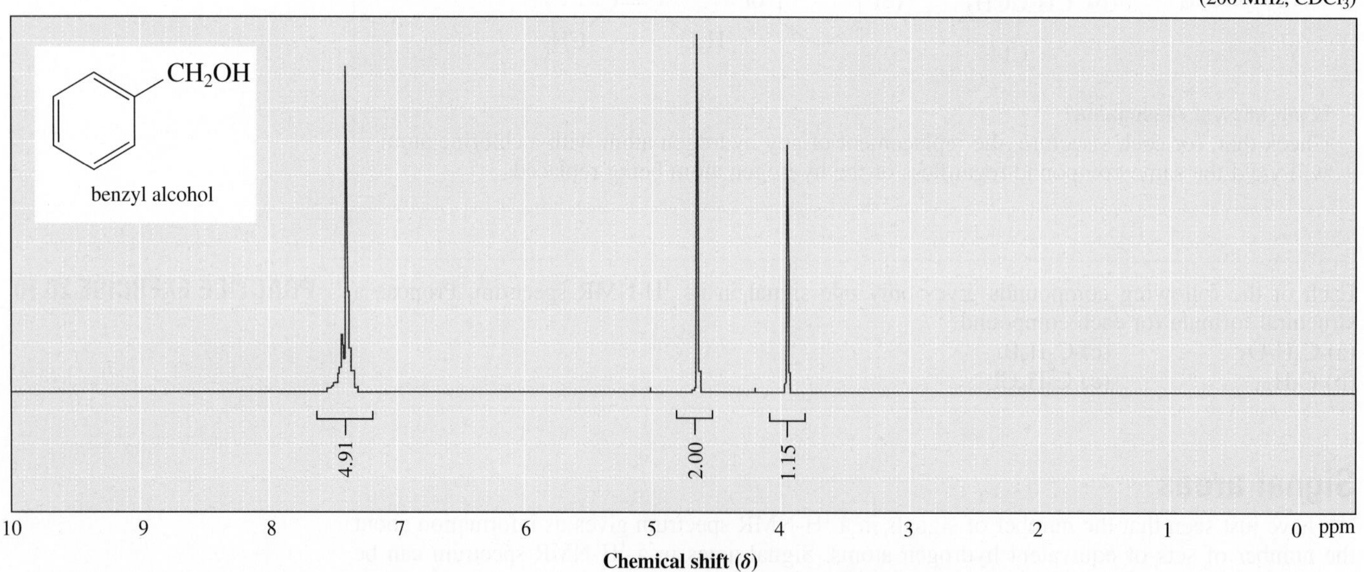

FIGURE 20.29 ^{1}H-NMR spectrum of benzyl alcohol, C_7H_8O, showing the integration values. The ratio of signal area for the three peaks is 5 : 2 : 1 (from left to right). Although the numerical values provided do not exactly match the ratio, this is not a problem as the number of hydrogen atoms must be integers.

WORKED EXAMPLE 20.12

Determining integration values

The following is a ^{1}H-NMR spectrum of a compound with the molecular formula $C_9H_{10}O_2$. From an analysis of the integration data, calculate the number of hydrogen atoms giving rise to each signal.

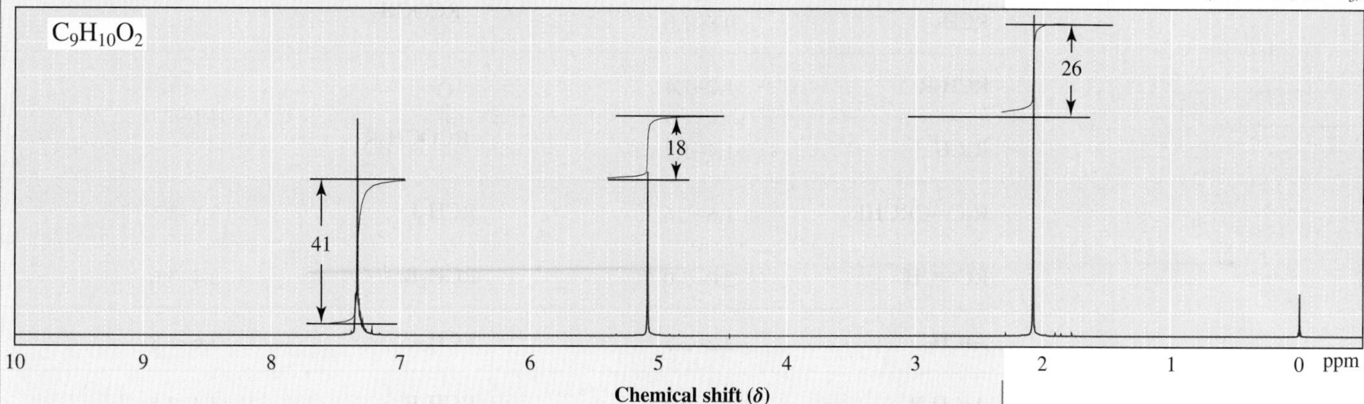

Analysis

Determine the relative ratio of the three signals, bearing in mind that there is a total of 10 hydrogen atoms in the molecule. Remember also that each signal must represent a whole number of hydrogen atoms (e.g. you cannot have 0.2 hydrogen atoms).

Solution

The ratio of the relative signal heights (obtained from the number of horizontal chart divisions) is 5 : 2 : 3. The molecular formula indicates that there are ten hydrogen atoms. Thus, the signal at δ 7.3 represents five hydrogen atoms, the signal at δ 5.1 represents two hydrogen atoms, and the signal at δ 2.1 represents three hydrogen atoms. Consequently, the signals and the number of hydrogen atoms each signal represents are δ 7.3 (5H), δ 5.1 (2H) and δ 2.1 (3H).

Is our answer reasonable?

Check that the sum of the number of hydrogen atoms for each signal equals the total number of hydrogen atoms in the molecule.

PRACTICE EXERCISE 20.11

The integration values of the two signals in the ^{1}H-NMR spectrum of a ketone with molecular formula $C_7H_{14}O$ are given as a ratio of 6.2 to 1.0. Calculate the number of hydrogen atoms giving rise to each signal, and propose a structural formula for this ketone.

Chemical shift

The position of a signal along the *x*-axis of an NMR spectrum is known as the chemical shift of that signal. The chemical shift of a signal in a ^{1}H-NMR spectrum can give us valuable information about the type of hydrogen atoms giving rise to that absorption. Hydrogen atoms on methyl groups bonded to sp^3 hybridised carbons, for example, typically give a signal near δ 0.8–1.0 (see figure 20.32 for an example). Hydrogen atoms on methyl groups bonded to a carbonyl carbon give signals near δ 2.0–2.3 (see figures 20.26 and 20.28), and hydrogen atoms on methyl groups bonded to oxygen give signals near δ 3.7–3.9 (see figure 20.26). Table 20.9 lists the average chemical shift for most of the types of hydrogen atoms we deal with in this text. Notice that most of the values shown fall within a rather narrow range from 0 to 12 δ units (ppm). In fact, although the table shows a variety of functional groups and hydrogen atoms bonded to them, we can use the rules of thumb in figure 20.30 to remember the chemical shifts of most types of hydrogen atom.

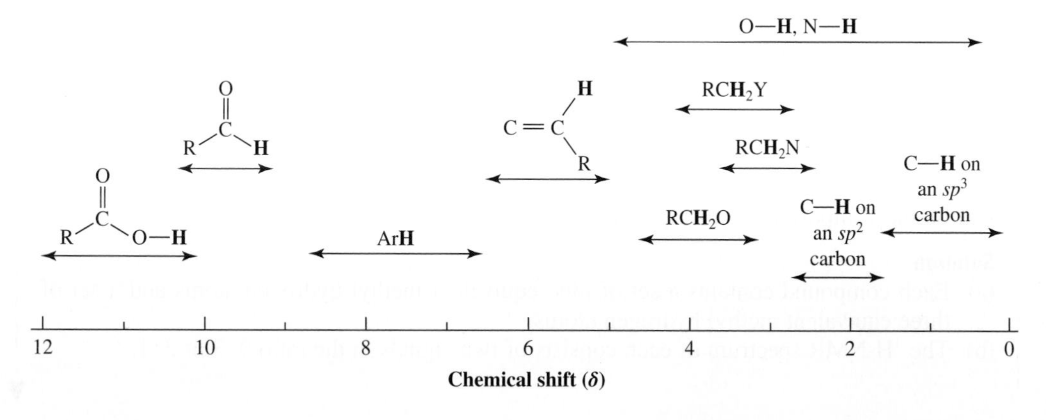

FIGURE 20.30 A simple correlation chart for ^{1}H-NMR chemical shifts. Where the range is labelled RCH$_2$Y (Y = halogen, N or O), shifts for methyl and methine hydrogen atoms, R$_2$CHY, also fall within this range (with the methyl being at slightly lower chemical shift and the methine at slightly higher chemical shift).

TABLE 20.9 Average values of chemical shifts of representative types of hydrogen atoms.

Type of hydrogen[(a)]	Chemical shift (δ)[(b)]	Type of hydrogen[(a)]	Chemical shift (δ)[(b)]
$(CH_3)_4Si$	0 (by definition)	$RC(=O)OCH_3$	3.7–3.9
RCH_3	0.8–1.0		
RCH_2R	1.2–1.4	$RC(=O)OCH_2R$	4.1–4.7
R_3CH	1.4–1.7		
$R_2C{=}CRCHR_2$	1.6–2.6	RCH_2I	3.1–3.3
$RC{\equiv}CH$	2.0–3.0	RCH_2Br	3.4–3.6
$ArCH_3$	2.2–2.5	RCH_2Cl	3.6–3.8
$ArCH_2R$	2.3–2.8	RCH_2F	4.4–4.5
ROH	0.5–6.0	ArOH	4.5–10
RCH_2OH	3.4–4.0	$R_2C{=}CH_2$	4.6–5.0
RCH_2OR	3.3–4.0	$R_2C{=}CHR$	5.0–5.7
R_2NH	0.5–5.0	ArH	6.5–8.5
$RC(=O)CH_3$	2.0–2.3	$RC(=O)H$	9.5–10.1
$RC(=O)CH_2R$	2.2–2.6	$RC(=O)OH$	10–12

(a) R = alkyl group, Ar = aryl group.
(b) Values are approximate. Other atoms within the molecule may cause the signal to appear outside these ranges.

WORKED EXAMPLE 20.13

Identifying isomers using chemical shift data

The following are two constitutional isomers with the molecular formula $C_6H_{12}O_2$.

$$H_3C{-}C({=}O){-}O{-}C(CH_3)_2{-}CH_3 \qquad H_3C{-}O{-}C({=}O){-}C(CH_3)_2{-}CH_3$$

isomer 1 isomer 2

(a) Predict the number of signals in the ¹H-NMR spectrum of each isomer.
(b) Predict the ratio of areas of the signals in each spectrum.
(c) Show how these isomers may be distinguished on the basis of chemical shift.

Analysis

First identify the types of different hydrogen atoms and then the numbers of each type. Now carefully examine the two structures and identify the key differences. If there are any hydrogen atoms in very different environments, then the signals due to these hydrogen atoms will be most important in distinguishing between the two structures.

Solution

(a) Each compound contains a set of nine equivalent methyl hydrogen atoms and a set of three equivalent methyl hydrogen atoms.
(b) The ¹H-NMR spectrum of each consists of two signals in the ratio 9 : 3 or 3 : 1.

(c) The two constitutional isomers can be distinguished by the chemical shift of the single —CH_3 group (shown in blue for each compound). Using our rules of thumb in figure 20.30, we find that the hydrogen atoms of CH_3O are less shielded (appear further downfield) than the hydrogen atoms of $CH_3C{=}O$. Table 20.9 gives approximate values for each chemical shift. Experimental values are as follows.

isomer 1: $H_3C-C(=O)-O-C(CH_3)_2-CH_3$ (δ 2.0 for $H_3C-C=O$; δ 1.4 for $C(CH_3)_3$)

isomer 2: $H_3C-O-C(=O)-C(CH_3)_2-CH_3$ (δ 3.7 for H_3C-O; δ 1.2 for $C(CH_3)_3$)

Is our answer reasonable?

Check the correlation tables to ensure that the chemical shifts are correct.

PRACTICE EXERCISE 22.12

(a) The following are two constitutional isomers with molecular formula $C_6H_{12}O_2$.

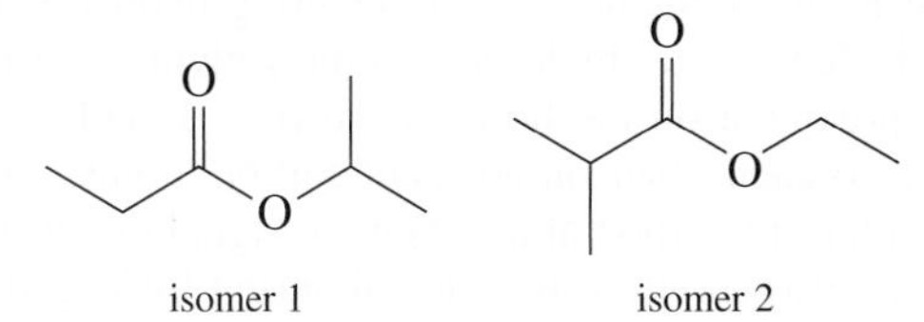

(i) Predict the number of signals in the ^{1}H-NMR spectrum of each isomer.
(ii) Predict the ratio of areas of the signals in each spectrum.
(iii) Show how these isomers may be distinguished on the basis of chemical shift.

(b) Consider the compounds in worked example 20.7. We have already determined the number of signals as shown below.

methyl acetate (signals a, b) — ethyl formate (signals a, b, c) — methoxyacetaldehyde (signals a, b, c)

(i) Predict the ratio of areas of the signals in each spectrum.
(ii) Show how these isomers may be distinguished on the basis of chemical shift.

Signal splitting and the $(n+1)$ rule

We have now seen three kinds of information that can be derived from an examination of a ^{1}H-NMR spectrum.

1. From the number of signals, we can determine the minimum number of sets of equivalent hydrogen atoms.
2. By integrating over signal areas, we can determine the relative numbers of hydrogen atoms giving rise to each signal.
3. From the chemical shift of each signal, we can derive information about the types of hydrogen atoms in each set.

We can derive a fourth kind of information from the **splitting pattern** of each signal. Consider the ^{1}H-NMR spectrum of 1,1,2-trichloroethane (figure 20.31, overleaf), a solvent for waxes and natural resins. This molecule contains two 2° hydrogen atoms and one 3° hydrogen atom, and, according to what we have learned so far, we predict two signals with relative areas 2 : 1, corresponding to the two hydrogen atoms of the —CH_2Cl group and the one hydrogen atom of the —$CHCl_2$ group. You see from the spectrum, however, that there are in fact five peaks. How can this be when we predict only two signals? The answer is that a hydrogen atom's resonance frequency can be affected by the tiny magnetic fields of other hydrogen atoms close by. Those fields cause the signal to be split into numerous peaks. A signal may be split into two peaks (called a **doublet**, d), three peaks (a **triplet**, t), four peaks (a **quartet**, q) and so on. A signal that has not been split is referred to as a **singlet** (s) and those with **complex splitting patterns** are termed **multiplets** (m). In general, signals are split if there are hydrogen atoms on adjacent carbon atoms. Signal splitting when there are more than three bonds between the hydrogen atoms, as in H—C—C—C—H, are uncommon, except when π bonds are involved (for example, in aromatic rings).

From the integration values and the chemical shift, we know that the signal at δ 4.0 in the ^{1}H-NMR spectrum of 1,1,2-trichloroethane is due to the hydrogen atoms of the —CH_2Cl group, and the signal at δ 5.8 is due to the single hydrogen of the —$CHCl_2$ group.

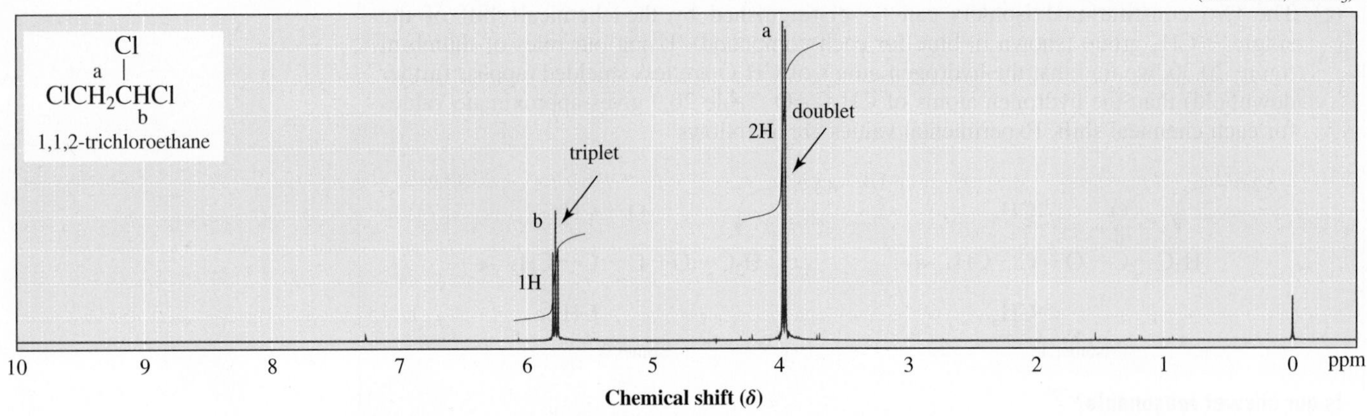

FIGURE 20.31 ^{1}H-NMR spectrum of 1,1,2-trichloroethane.

We say that the CH_2 signal at δ 4.0 is split into a doublet and that the CH signal at δ 5.8 is split into a triplet. In this phenomenon, called **signal splitting**, the ^{1}H-NMR signal from one set of hydrogen atoms is split by the influence of neighbouring hydrogen atoms that are equivalent to each other but not equivalent to the hydrogen atom(s) giving rise to the signal. The degree of signal splitting can be predicted on the basis of the **$(n + 1)$ rule**, where n is the number of hydrogen atoms equivalent to each other on adjacent carbon atoms. Note that, if the signal due to H_a is split by H_b on an adjacent carbon atom, then the signal due to H_b is also split by H_a. No signal splitting is observed between equivalent neighbouring hydrogen atoms. For example, the spectrum of 1,2-dichloroethane has only one signal (a singlet) at δ 3.7.

Pascal's triangle can be used to describe the splitting pattern (also called the **multiplicity**) and the peak intensities within split signals, as shown in table 20.10.

TABLE 20.10 Multiplicity and relative intensities of peaks in split signals.

Number of equivalent hydrogen atoms causing splitting	Multiplicity	Relative peak intensities
1	doublet	1 : 1
2	triplet	1 : 2 : 1
3	quartet	1 : 3 : 3 : 1
4	quintet	1 : 4 : 6 : 4 : 1
5	sextet	1 : 5 : 10 : 10 : 5 : 1
6	septet	1 : 6 : 15 : 20 : 15 : 6 : 1

Let us apply the $(n + 1)$ rule to the analysis of the spectrum of 1,1,2-trichloroethane (figure 20.32). The two hydrogen atoms of the —CH_2Cl group have one neighbouring hydrogen ($n = 1$), so their signal is split into a doublet ($1 + 1 = 2$). The single hydrogen atom of the —$CHCl_2$ group has a set of two equivalent neighbouring hydrogen atoms ($n = 2$), so its signal is split into a triplet ($2 + 1 = 3$).

For these hydrogen atoms, $n = 1$; their signal is split into $(1 + 1) = 2$ peaks — a doublet.

For this hydrogen atom, $n = 2$; its signal is split into $(2 + 1) = 3$ peaks — a triplet.

$$Cl-CH_2-\underset{\large Cl}{\underset{|}{C}}H-Cl$$

Let us now consider the splitting pattern in 1-chloropropane (figure 20.32). The two hydrogen atoms on C(1) have two neighbouring equivalent hydrogen atoms on the adjacent carbon atom and so the signal (δ 3.5) for these hydrogen atoms is split into a triplet $(2 + 1)$. Similarly, the signal due to the three hydrogen atoms on C(3) (δ 1.1) is split into a triplet. The two hydrogen atoms on C(2) are flanked on one side by a set of two hydrogen atoms on C(1), and on the other side by a set of three hydrogen atoms on C(3). Because the sets of hydrogen atoms on C(1) and C(3) are not equivalent to each other, they can cause the signal for the CH_2 group on C(2) to be split into a complex pattern, which we will refer to simply as a multiplet. If the magnitude of the splitting between the hydrogen atoms of C(1) and C(2) is equal to that between C(2) and C(3), then we can describe the signal as a sextet. The $(n + 1)$ rule can be used in cases where the coupling between the observed hydrogen atom(s) and adjacent nonequivalent hydrogen atoms is equal.

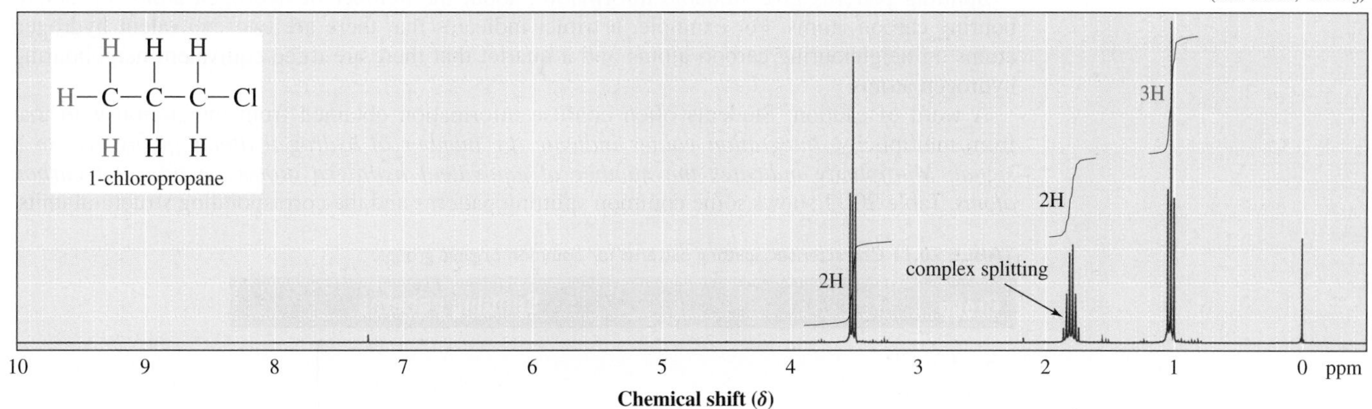

FIGURE 20.32 ^{1}H-NMR spectrum of 1-chloropropane.

WORKED EXAMPLE 24.14

Splitting patterns

Predict the number of signals and the splitting pattern of each signal in the ^{1}H-NMR spectrum of each of the following compounds.

(a) $CH_3C(=O)CH_2CH_3$ (b) $CH_3CH_2C(=O)CH_2CH_3$ (c) $CH_3C(=O)CH(CH_3)_2$

Analysis

First identify the different types of hydrogen atoms. Now consider the number of hydrogen atoms on the neighbouring carbon atom. Applying the $(n + 1)$ rule will give the splitting pattern.

Solution

The sets of equivalent hydrogen atoms in each molecule are colour coded below. In molecule (a), the signal for the red methyl group is unsplit (a singlet) because the group is too far (> 3 bonds) from any other hydrogen atoms. The blue $—CH_2—$ group has three equivalent neighbouring hydrogen atoms ($n = 3$) so it shows a signal split into a quartet ($3 + 1 = 4$). The green methyl group has two equivalent neighbouring hydrogen atoms ($n = 2$) so its signal is split into a triplet. The integration ratios for these signals would be 3:2:3.

Molecules (b) and (c) can be analysed in the same way. Thus, molecule (b) shows a triplet and a quartet in a 3:2 ratio. (Remember that the two CH_2 groups are equivalent, as are the two CH_3 groups.) Molecule (c) shows a singlet, a septet ($6 + 1 = 7$) and a doublet in the ratio 3:1:6.

(a) $CH_3—C(=O)—CH_2—CH_3$ (singlet, quartet, triplet) (b) $CH_3—CH_2—C(=O)—CH_2—CH_3$ (triplet, quartet, quartet, triplet) (c) $CH_3—C(=O)—CH(CH_3)_2$ (singlet, septet, doublet)

Is our answer reasonable?

Check that you have considered only the hydrogen atoms on neighbouring atoms.

PRACTICE EXERCISE 20.13

(a) The following are pairs of constitutional isomers. Predict the number of signals and the splitting pattern of each signal in the ^{1}H-NMR spectrum of each isomer.

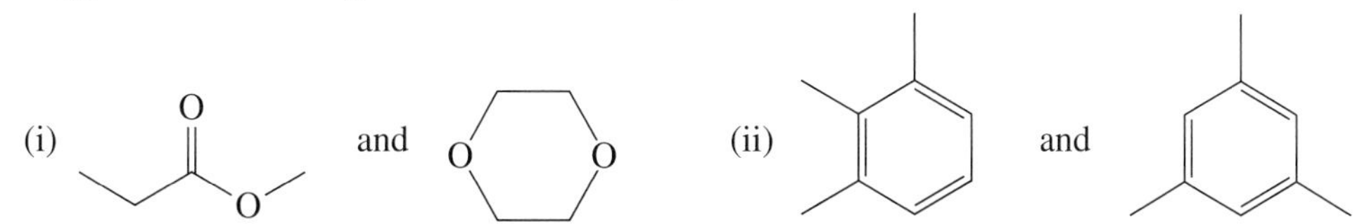

(b) Determine the multiplicities for each type of hydrogen atom in the esters from worked example 20.7 (p. 892). Although these compounds can be distinguished without considering multiplicity, it is a useful check that your initial analysis is correct.

methyl acetate (a, b) ethyl formate (a, b, c) methoxyacetaldehyde (a, b, c)

Splitting patterns give valuable information about the number of hydrogen atoms on neighbouring carbon atoms. For example, a triplet indicates that there are two equivalent hydrogen atoms on neighbouring carbon atoms and a quartet that there are three equivalent neighbouring hydrogen atoms.

A word of caution: Students often confuse information obtained from integration with that from multiplicity. *Integration values indicate the number of hydrogen atoms giving rise to a signal. Multiplicity indicates the number of equivalent hydrogen atoms on adjacent carbon atoms.* Table 20.11 shows some common splitting patterns and the corresponding structural units.

TABLE 20.11 Characteristic splitting patterns for common organic groups.

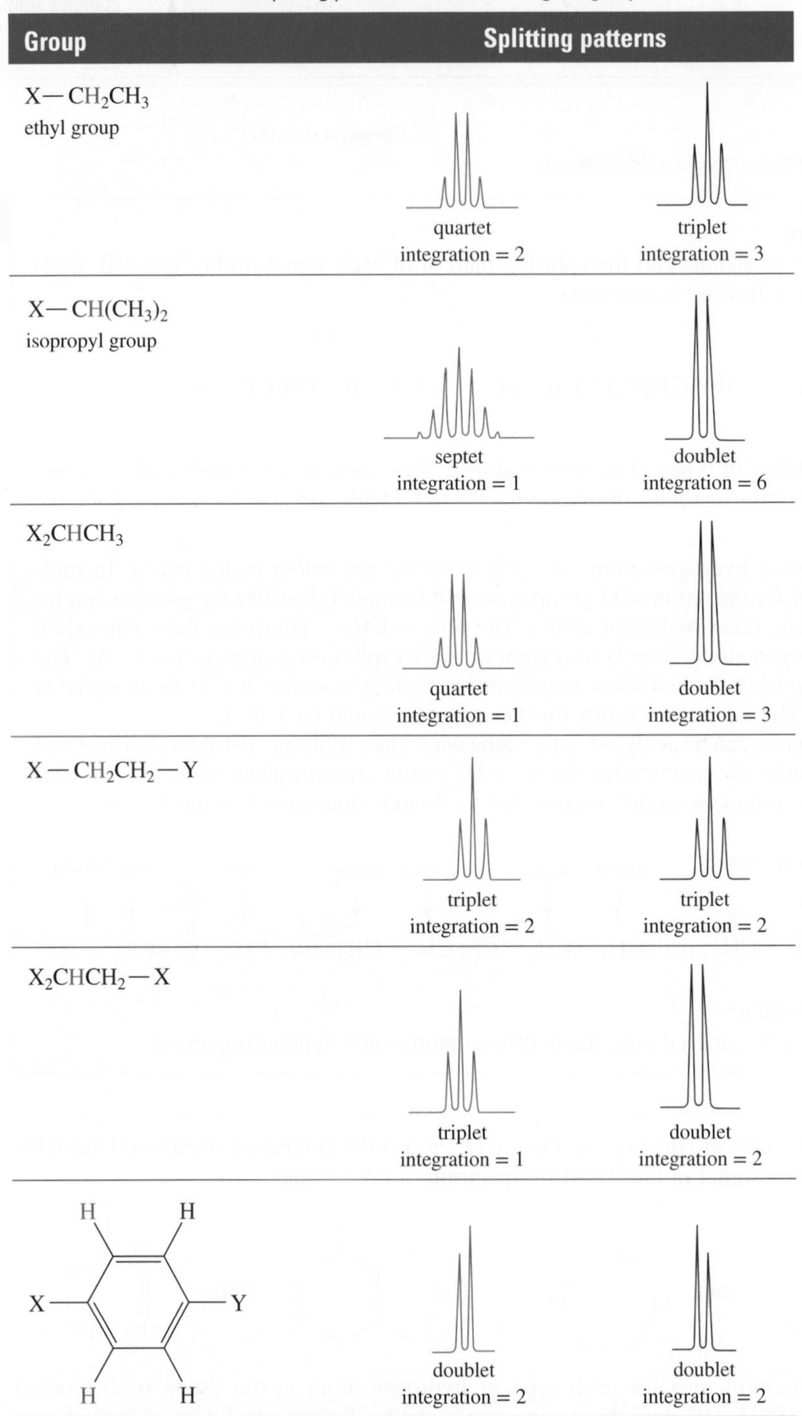

Group	Splitting patterns	
$X—CH_2CH_3$ ethyl group	quartet integration = 2	triplet integration = 3
$X—CH(CH_3)_2$ isopropyl group	septet integration = 1	doublet integration = 6
X_2CHCH_3	quartet integration = 1	doublet integration = 3
$X—CH_2CH_2—Y$	triplet integration = 2	triplet integration = 2
$X_2CHCH_2—X$	triplet integration = 1	doublet integration = 2
para-disubstituted benzene ring: X and Y on opposite carbons, with H, H, H, H on the other four	doublet integration = 2	doublet integration = 2

X is different from Y. It is assumed that the hydrogen atoms in the groups shown are not coupled to other hydrogen atoms.

^{13}C-NMR spectroscopy

Nuclei of ^{12}C, the most abundant (98.89%) natural isotope of carbon, do not have a nuclear spin and so are not detected by NMR spectroscopy. Nuclei of ^{13}C (natural abundance 1.11%), however, do have nuclear spin and so are detected by NMR spectroscopy in the same manner as hydrogen atoms are detected.

In the most common mode for recording a ^{13}C spectrum, a proton decoupled spectrum, all ^{13}C signals appear as singlets. The ^{13}C-NMR spectrum of citric acid (figure 20.33), a compound used to increase the solubility in water of many pharmaceutical drugs, consists of four singlets. Notice that, as in ^{1}H-NMR, equivalent carbon atoms generate only one signal.

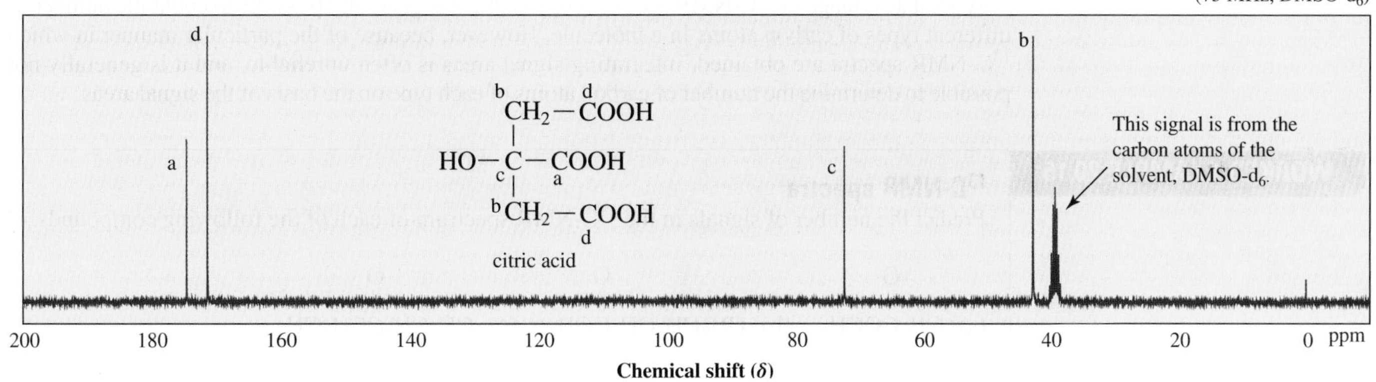

FIGURE 20.33 The ^{13}C-NMR spectrum of citric acid.

Figure 20.34 is a simple correlation chart for remembering the ^{13}C-NMR chemical shifts of various carbon atoms. If you compare this with figure 20.30, a correlation chart for ^{1}H-NMR chemical shifts, you will notice that signals for the C atoms fall in approximately the same relative region of the spectrum as their corresponding H atoms in the various functional groups. Table 20.12 shows approximate chemical shifts in ^{13}C-NMR spectroscopy.

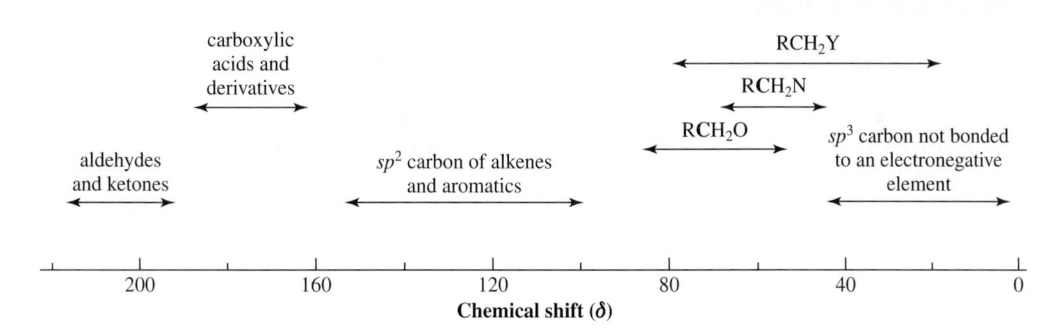

FIGURE 20.34 Carbon chemical shifts for common functional groups.

TABLE 20.12 ^{13}C-NMR chemical shifts.

Type of carbon	Chemical shift (δ)[a]	Type of carbon	Chemical shift (δ)[a]
RCH_3	0–40	C—R	110–160
RCH_2R	15–55		
R_3CH	20–60	$\mathrm{R\overset{O}{\overset{\|}{C}}OR}$	160–180
RCH_2I	0–40		
RCH_2Br	25–65	$\mathrm{R\overset{O}{\overset{\|}{C}}NR_2}$	165–180
RCH_2Cl	35–80		
R_3COH	40–80	$\mathrm{R\overset{O}{\overset{\|}{C}}OH}$	175–185
R_3COR	40–80		
$RC{\equiv}CR$	65–85	$\mathrm{R\overset{O}{\overset{\|}{C}}H}$, $\mathrm{R\overset{O}{\overset{\|}{C}}R}$	190–210
$R_2C{=}CR_2$	100–150		

(a) Values are approximate. Other atoms within the molecule may cause the signal to appear outside these ranges.

Notice how much broader the range of chemical shifts is for ^{13}C-NMR spectroscopy than for ^{1}H-NMR spectroscopy. Whereas most chemical shifts for ^{1}H-NMR spectroscopy fall within a rather narrow range of 0–12 ppm, those for ^{13}C-NMR spectroscopy cover 0–210 ppm. Because of this expanded scale, it is very unusual to find any two nonequivalent carbon atoms in the same molecule with identical chemical shifts. Most commonly, each different type of carbon atom within a molecule has a distinct signal that is clearly resolved from all other signals. Notice further that the chemical shift of carbonyl carbon atoms is quite distinct from the chemical shifts of sp^3 hybridised carbon atoms and other types of sp^2 hybridised carbon atoms. The presence or absence of a carbonyl carbon atom is quite easy to recognise in a ^{13}C-NMR spectrum.

A great advantage of ^{13}C-NMR spectroscopy is that it is generally possible to count the number of different types of carbon atoms in a molecule. However, because of the particular manner in which ^{13}C-NMR spectra are obtained, integrating signal areas is often unreliable, and it is generally not possible to determine the number of carbon atoms of each type on the basis of the signal areas.

WORKED EXAMPLE 20.15

^{13}C-NMR spectra

Predict the number of signals in the ^{13}C-NMR spectrum of each of the following compounds.

(a) $CH_3\overset{O}{\overset{\|}{C}}OCH_3$ (b) $CH_3CH_2CH_2\overset{O}{\overset{\|}{C}}CH_3$ (c) $CH_3CH_2\overset{O}{\overset{\|}{C}}CH_2CH_3$

Analysis

Check for symmetry. If there is no symmetry, the number of signals will equal the number of carbon atoms. If there is symmetry, the number of signals will be less than the number of carbon atoms. Count the number of nonequivalent carbon atoms.

Solution

The diagram below shows the number of signals in each spectrum, along with the chemical shift of each, colour coded to the carbon atom responsible for that signal. The chemical shifts of the carbonyl carbon atoms are quite distinctive (table 20.12) and are δ 171.4, 208.9, and 212.0 in these examples.

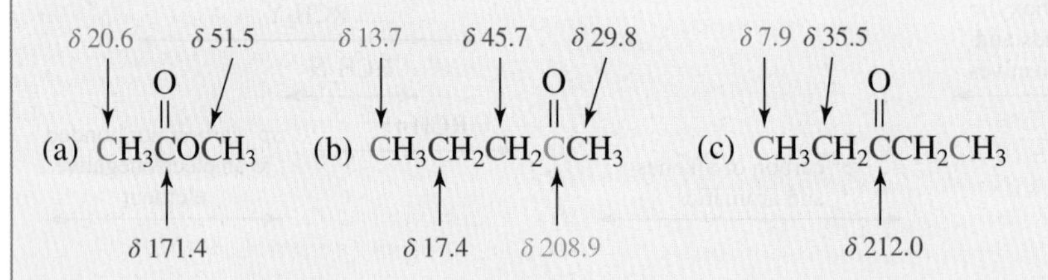

Is our answer reasonable?

Have you identified every carbon atom (this is particularly important when condensed structures or line structures are used)? Check also that you have not overlooked any symmetry.

PRACTICE EXERCISE 20.14

Explain how to distinguish between the members of each of the following pairs of constitutional isomers on the basis of the number of signals in the ^{13}C-NMR spectrum of each isomer.

(a) and (b) and

20.6 Interpreting NMR spectra

Interpreting NMR spectroscopic data is a skill that is best acquired through practice and exposure to examples. In this section, we will see specific examples of NMR spectra for characteristic functional groups. Familiarising yourself with them will help you to master the technique of spectral interpretation.

Alkanes

Because all hydrogen atoms in alkanes are in similar chemical environments, ^{1}H-NMR chemical shifts of their hydrogen atoms fall within a narrow range of δ 0.8–1.7, so signals often overlap.

Chemical shifts for alkane carbon atoms in ^{13}C-NMR spectroscopy fall within the wider range of δ 0–60. Notice how it is relatively easy to distinguish all the signals in the ^{13}C-NMR spectrum of 2,2,4-trimethylpentane (common name isooctane, figure 20.35), a major component in petrol, compared with the signals in the ^{1}H-NMR spectrum of 2,2,4-trimethylpentane (figure 20.36). In the ^{13}C-NMR spectrum, we can see all five signals; in the ^{1}H-NMR spectrum, we expect to see four signals but, in fact, we see only three. The reason is that nonequivalent hydrogen atoms often have similar chemical shifts, which lead to an overlap of signals. Note that, in the ^{13}C-NMR spectrum below, the three closely spaced peaks at δ 77 are due to the solvent $CDCl_3$. In fact, a signal due to the solvent is observed in all ^{13}C-NMR spectra, except those run in D_2O.

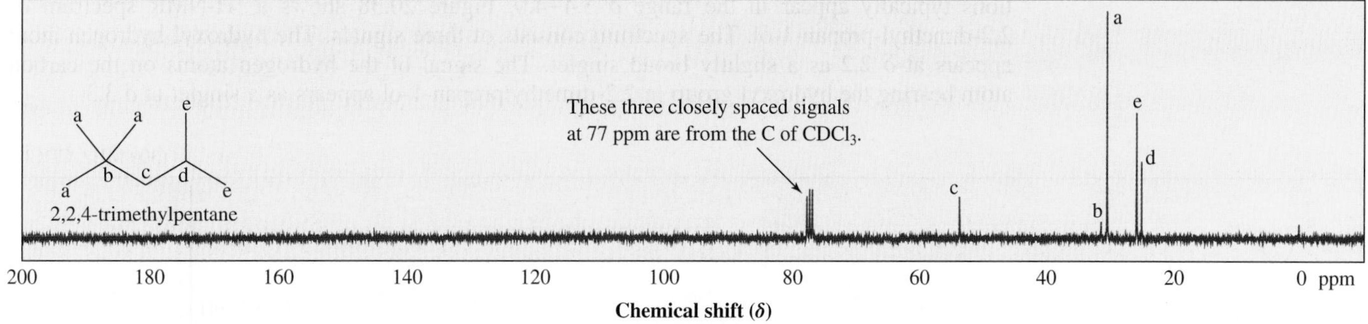

FIGURE 20.35 ^{13}C-NMR spectrum of 2,2,4-trimethylpentane, showing all five signals.

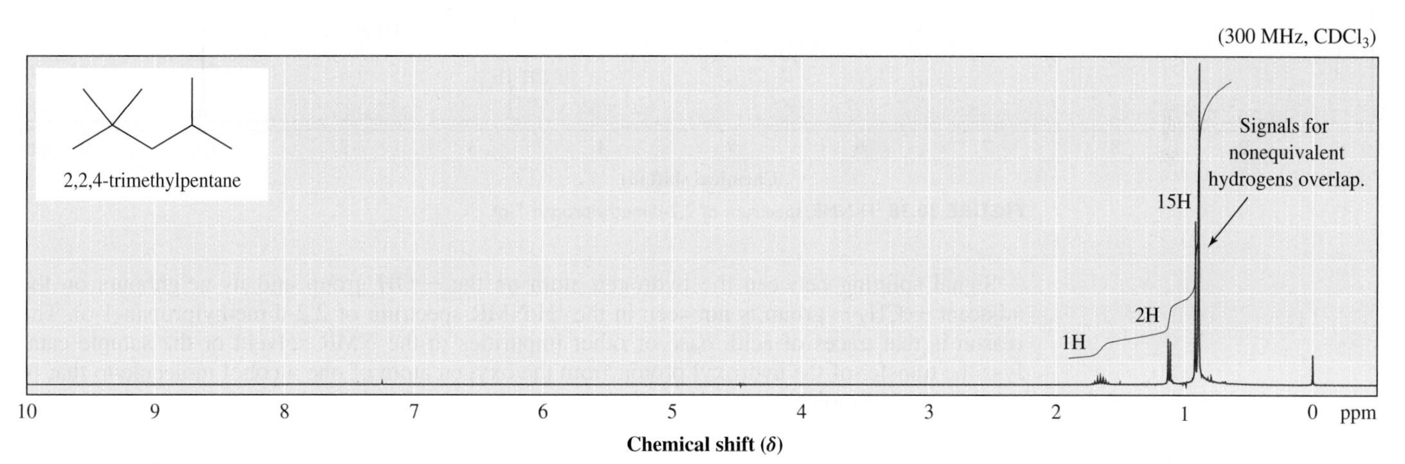

FIGURE 20.36 ^{1}H-NMR spectrum of 2,2,4-trimethylpentane, showing only three signals. Two sets of nonequivalent hydrogen atoms produce signals at δ 0.9, giving an integration ratio of 1 : 2 : 15 for the peaks at δ 1.7, 1.1 and 0.9, respectively.

Alkenes

The ^{1}H-NMR chemical shifts of vinylic hydrogen atoms (hydrogen atoms on a carbon atom of a carbon–carbon double bond) are larger than those of alkane hydrogen atoms and typically fall into the range δ 4.6–5.7. Figure 20.37 shows a ^{1}H-NMR spectrum of 1-methylcyclohexene. The signal for the one vinylic hydrogen atom appears at δ 5.4 split into a triplet by the two hydrogen atoms of the neighbouring —CH_2— group of the ring.

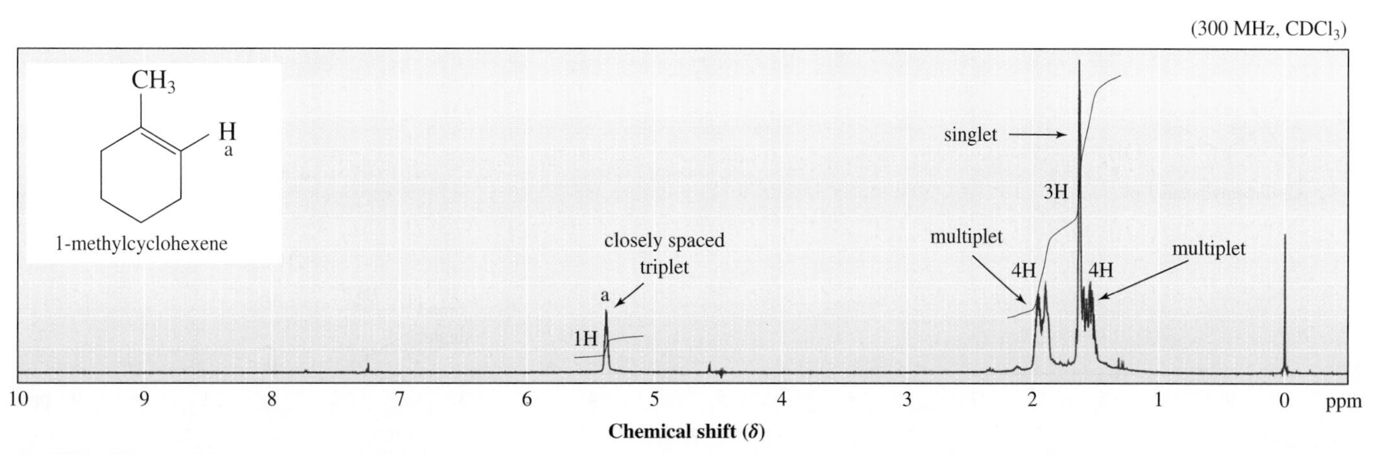

FIGURE 20.37 ^{1}H-NMR spectrum of 1-methylcyclohexene.

The sp^2 hybridised carbons of alkenes come into resonance in ^{13}C-NMR spectroscopy in the range δ 100–150 (see table 20.12), which is considerably downfield from resonances of sp^3 hybridised carbons.

Alcohols

The chemical shift of a hydroxyl hydrogen atom in a ^{1}H-NMR spectrum is variable and depends on the purity of the sample, the solvent, the concentration and the temperature. Normally, the shift appears in the range δ 3.0–4.5, but, depending on experimental conditions, it may have a chemical shift as low as δ 0.5. Hydrogen atoms on the carbon atom bearing the —OH group are deshielded by the electron-withdrawing inductive effect of the oxygen atom, and their absorptions typically appear in the range δ 3.4–4.0. Figure 20.38 shows a ^{1}H-NMR spectrum of 2,2-dimethyl-propan-1-ol. The spectrum consists of three signals. The hydroxyl hydrogen atom appears at δ 2.2 as a slightly broad singlet. The signal of the hydrogen atoms on the carbon atom bearing the hydroxyl group in 2,2-dimethylpropan-1-ol appears as a singlet at δ 3.3.

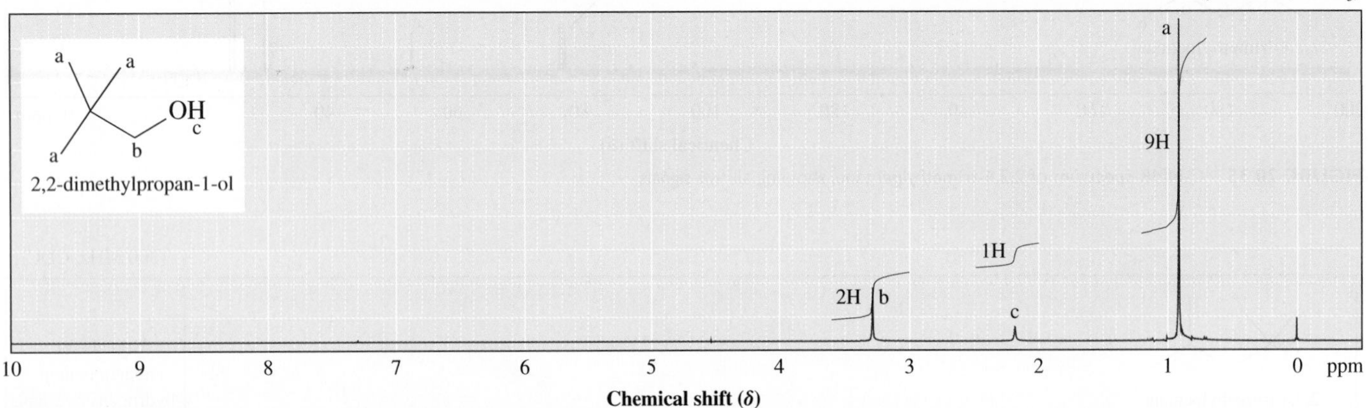

FIGURE 20.38 ^{1}H-NMR spectrum of 2,2-dimethylpropan-1-ol.

Signal splitting between the hydrogen atom on the —OH group and its neighbours on the adjacent —CH_2— group is not seen in the ^{1}H-NMR spectrum of 2,2-dimethylpropan-1-ol. The reason is that traces of acid, base or other impurities in the NMR solvent or the sample catalyse the transfer of the hydroxyl proton from the oxygen atom of one alcohol molecule to that of another alcohol molecule.

fast

hydrogen bonding

exchanged protons

Signals from alcohol hydrogen atoms typically appear as broad singlets, as shown in the ^{1}H-NMR spectrum of 3,3-dimethylbutan-1-ol (figure 20.39).

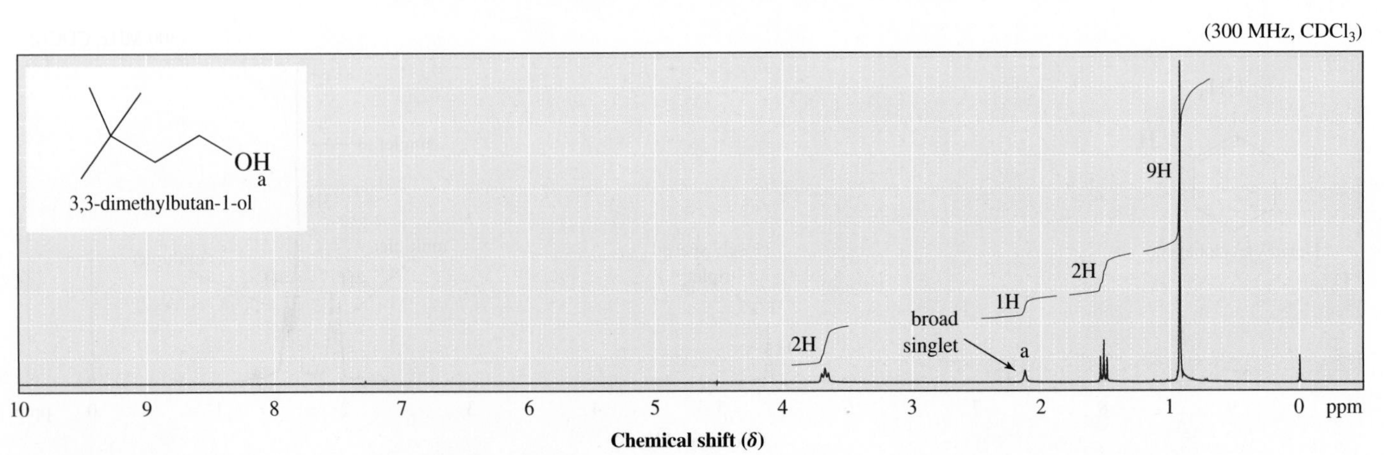

FIGURE 20.39 ^{1}H-NMR spectrum of 3,3-dimethylbutan-1-ol, showing the broad hydroxyl signal at δ 2.1.

Benzene and its derivatives

All six hydrogen atoms of benzene are equivalent, and their signal appears in its ^{1}H-NMR spectrum as a sharp singlet at δ 7.3. Hydrogen atoms bonded to a substituted benzene ring appear in the region δ 6.5–8.5. Few other types of hydrogen atoms give signals in this region; thus, aromatic hydrogen atoms are quite easily identifiable by their distinctive chemical shifts.

Recall that vinylic hydrogen atoms appear at δ 4.6–5.7. Hence, aromatic hydrogen atoms absorb radiation of even higher frequency than vinylic hydrogen atoms.

The ^{1}H-NMR spectrum of toluene (figure 20.40) shows a singlet at δ 2.3 for the three hydrogen atoms of the methyl group and a closely spaced multiplet at δ 7.3 for the five hydrogen atoms of the aromatic ring. The hydrogen atoms of an aromatic ring can produce quite complex signals as seen in the spectrum of ethyl benzoate (figure 20.41) with a multiplet at δ 8.1 (integrating for two hydrogen atoms) and a multiplet at δ 7.5 (integrating for three hydrogen atoms). The spectrum also shows the typical triplet (δ 1.4) and quartet (δ 4.4) pair of signals characteristic of an ethoxy group.

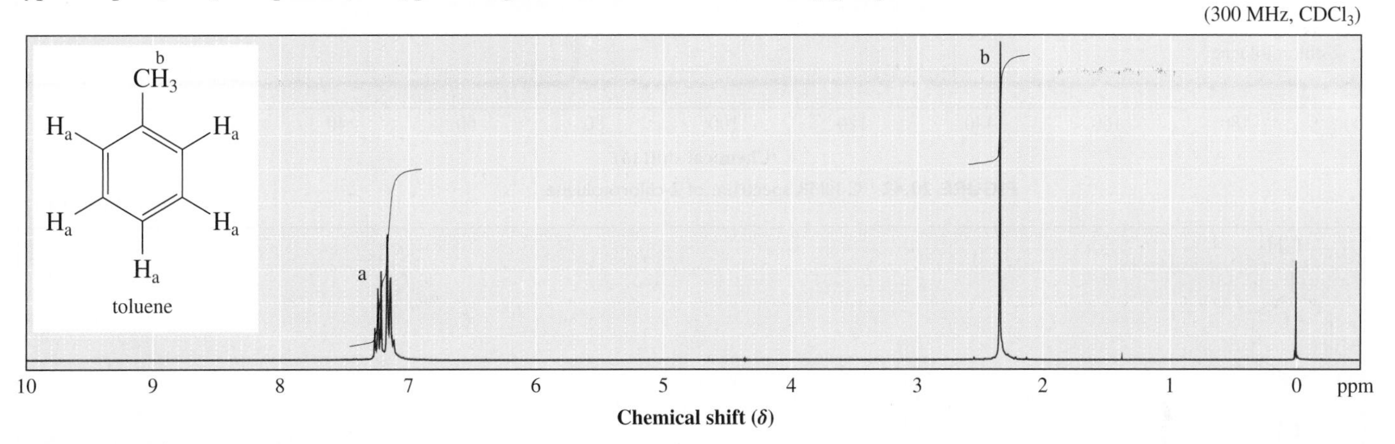

FIGURE 20.40 ^{1}H-NMR spectrum of toluene.

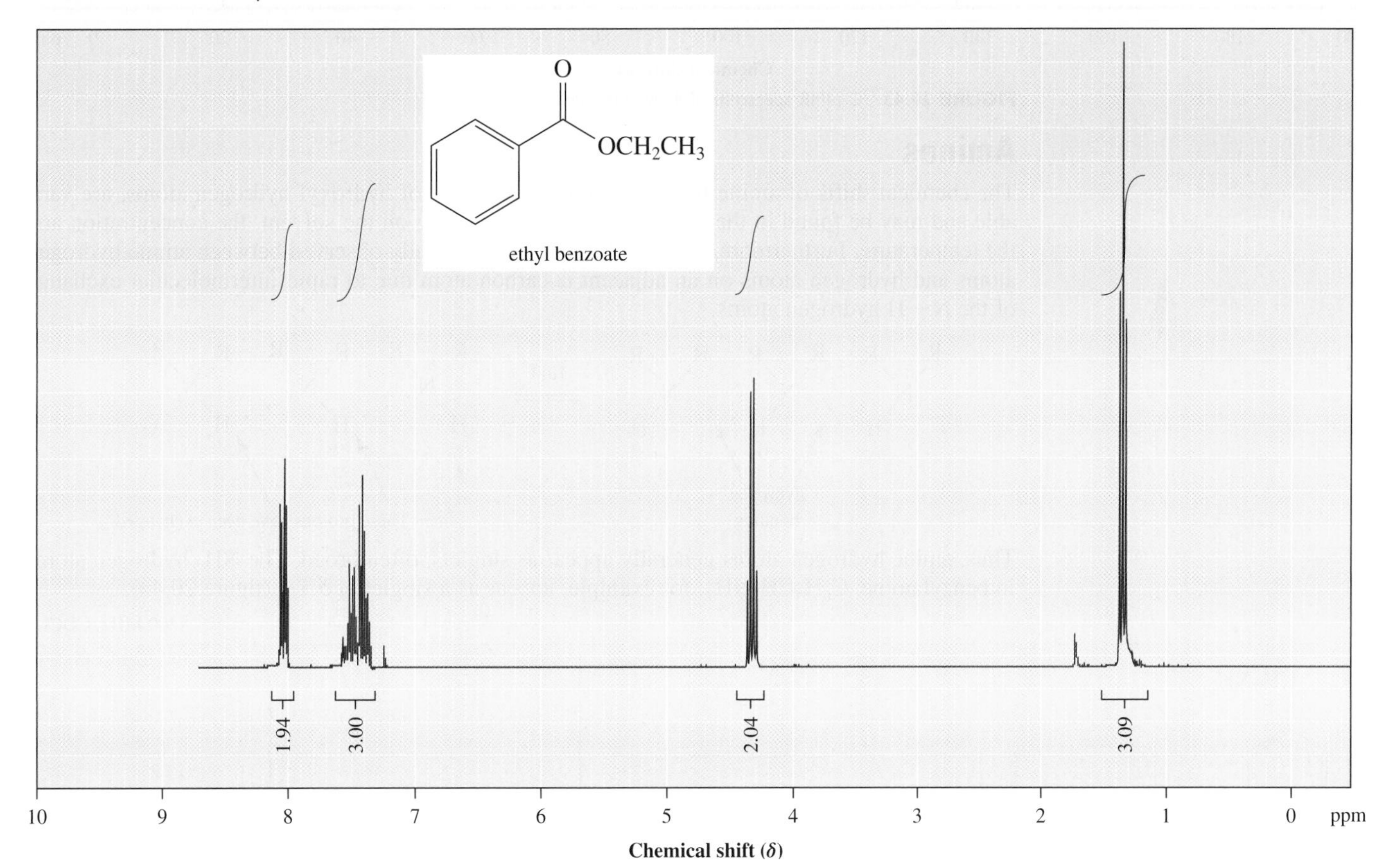

FIGURE 20.41 ^{1}H-NMR spectrum of ethyl benzoate.

In ^{13}C-NMR spectroscopy, carbon atoms of aromatic rings appear in the range δ 110–160. Benzene, for example, shows a single signal at δ 128. Because ^{13}C signals for alkene carbon atoms appear in the same range, it is generally not possible to establish the presence of an aromatic ring by ^{13}C-NMR spectroscopy alone. However, ^{13}C-NMR spectroscopy is particularly useful in establishing substitution patterns of aromatic rings. The ^{13}C-NMR spectrum of

2-chlorotoluene (figure 20.42) shows six signals in the aromatic region; the compound's more symmetric isomer 4-chlorotoluene (figure 20.43) shows only four signals in the aromatic region. Thus, to distinguish between these constitutional isomers, just count the signals. Notice in the ^{13}C-NMR spectra below that the signal intensities are not proportional to the number of equivalent carbon atoms giving rise to that signal, so integrating signal areas is often unreliable. This is a consequence of the way in which ^{13}C nuclei return to their lower energy states. Typically, quaternary (4°) carbon atoms (those not bonded to a hydrogen atom) give signals of low intensity.

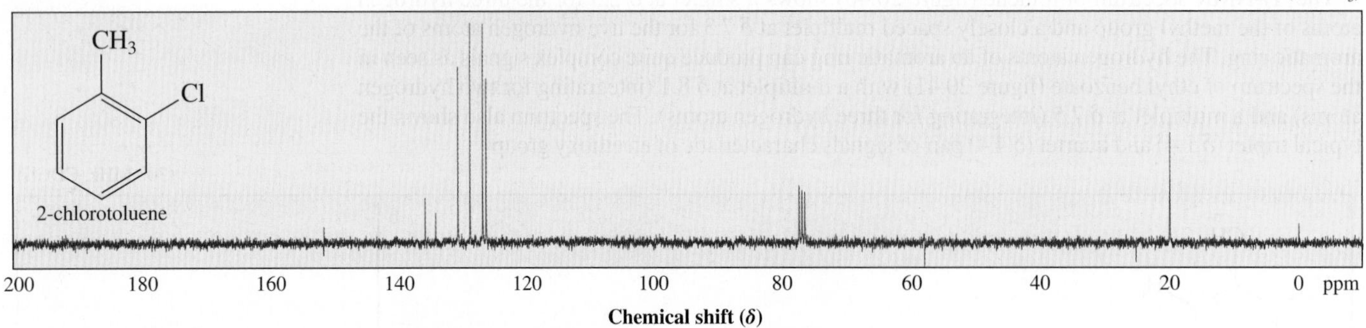

FIGURE 20.42 ^{13}C-NMR spectrum of 2-chlorotoluene.

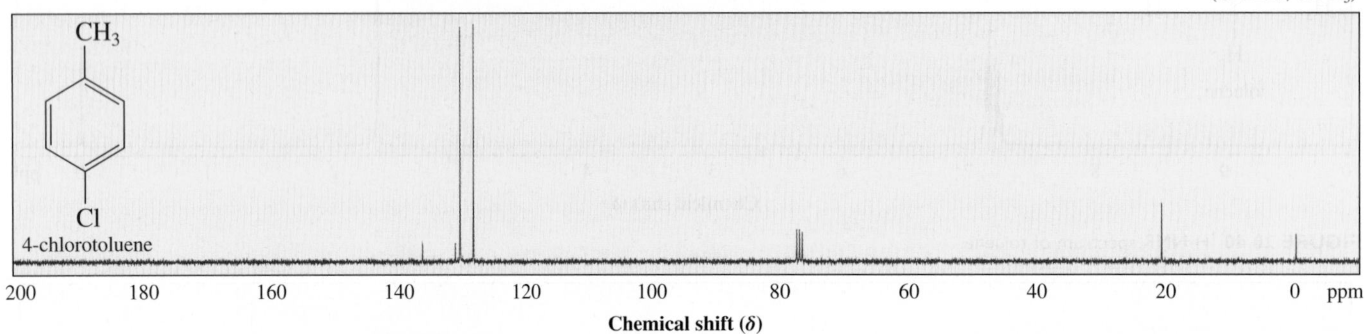

FIGURE 20.43 ^{13}C-NMR spectrum of 4-chlorotoluene.

Amines

The chemical shifts of amine hydrogen atoms, like those of hydroxyl hydrogen atoms, are variable and may be found in the region δ 0.5–5.0, depending on the solvent, the concentration and the temperature. Furthermore, signal splitting is not generally observed between amine hydrogen atoms and hydrogen atoms on an adjacent α-carbon atom due to rapid intermolecular exchange of the N—H hydrogen atoms.

fast

hydrogen bonding

These protons have been exchanged.

Thus, amine hydrogen atoms generally appear as singlets, often broad. The NH_2 hydrogen atoms in benzylamine, $C_6H_5CH_2NH_2$, for example, appear as a singlet at δ 1.4 (figure 20.44).

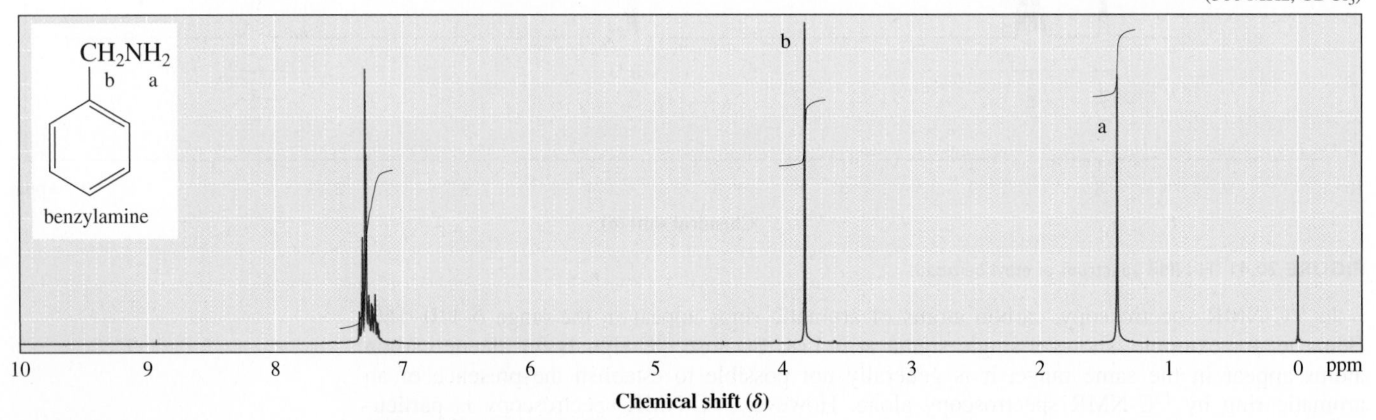

FIGURE 20.44 ^{1}H-NMR spectrum of benzylamine.

Aldehydes and ketones

^{1}H-NMR spectroscopy is an important tool for identifying aldehydes and for distinguishing between aldehydes and other carbonyl-containing compounds. The signal of an aldehyde hydrogen atom is typically in the region δ 9.5–10.1. Signal splitting between this hydrogen atom and those on the adjacent α-carbon atom is slight; consequently, the aldehyde hydrogen signal often appears as a closely spaced doublet or triplet. In the spectrum of butanal, for example, the triplet signal for the aldehyde hydrogen atom at δ 9.8 is so closely spaced that it looks almost like a singlet (figure 20.45).

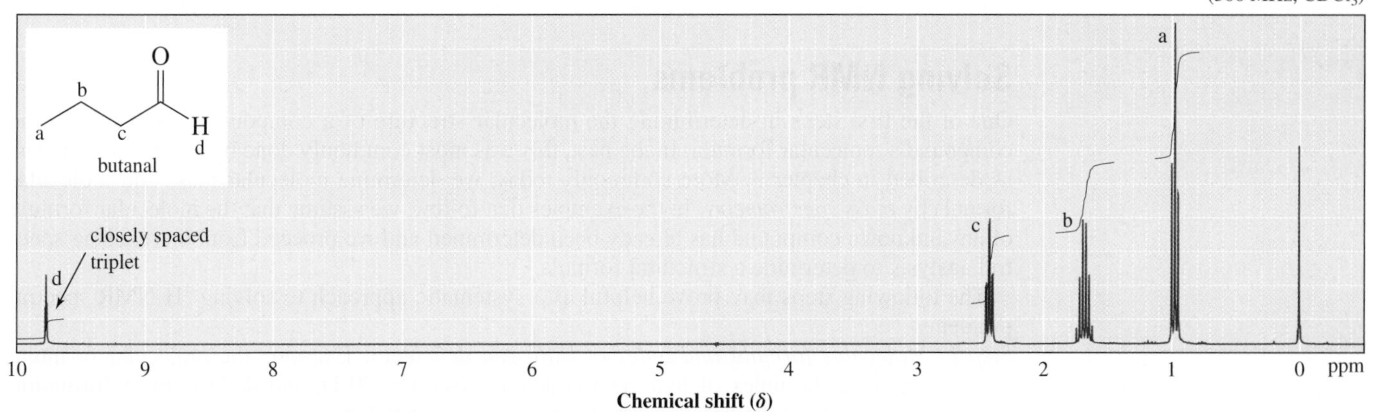

FIGURE 20.45 ^{1}H-NMR spectrum of butanal.

Just as the signal for an aldehyde hydrogen atom is only weakly split by the adjacent nonequivalent α-hydrogen atoms, the α-hydrogen atoms are weakly split by the aldehyde hydrogen atom. Hydrogen atoms on an α-carbon atom of an aldehyde or ketone typically appear around δ 2.1–2.6. The carbonyl carbon atoms of aldehydes and ketones are readily identifiable in ^{13}C-NMR spectroscopy by the position of their signal between δ 190 and 210.

Carboxylic acids

The hydrogen atom of the carboxyl group gives a signal in the range δ 10–12. The chemical shift of a carboxyl hydrogen atom is so large — even larger than the chemical shift of an aldehyde hydrogen atom (δ 9.5–10.1) — that it serves to distinguish carboxyl hydrogen atoms from most other types of hydrogen atoms. The signal of the carboxyl hydrogen of 2-methylpropanoic acid, labelled 'c' on the ^{1}H-NMR spectrum in figure 20.46, has been offset by δ 2.4. (Add 2.4 to the position at which the signal appears on the spectrum.) The chemical shift of this hydrogen is δ 12.0.

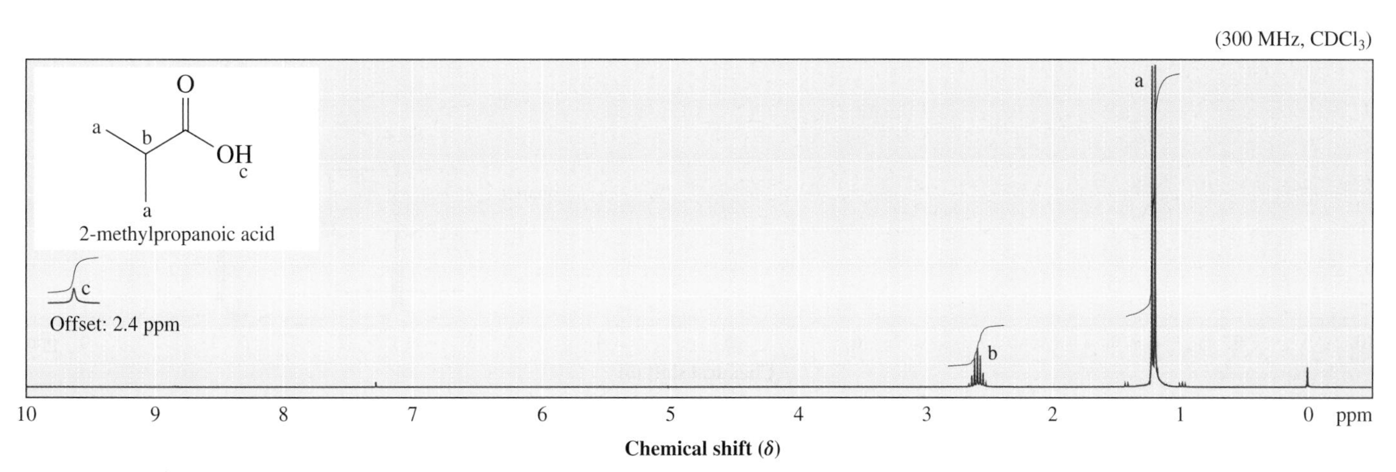

FIGURE 20.46 ^{1}H-NMR spectrum of 2-methylpropanoic acid (isobutyric acid).

Esters

Hydrogen atoms on the α-carbon atom of the carbonyl group of an ester are slightly deshielded and give signals at δ 2.1–2.6. Hydrogen atoms on the carbon atom bonded to the ester oxygen atom are more strongly deshielded and give signals at δ 3.7–4.7. It is thus possible to distinguish between ethyl acetate and its constitutional isomer, methyl propanoate, by the chemical shifts of either the singlet (s) —CH_3 absorption (compare δ 2.0 with 3.7) or the quartet (q) —CH_2— absorption (compare δ 4.1 with 2.3):

δ 2.0(s) δ 4.1(q) δ 2.3(q) δ 3.7(s)

$CH_3-C(=O)-O-CH_2-CH_3$ $CH_3-CH_2-C(=O)-O-CH_3$

ethyl acetate methyl propanoate

Solving NMR problems

One of the first steps in determining the molecular structure of a compound is to establish the compound's molecular formula. In the past, this was most commonly done by elemental analysis, as described in chapter 3. More commonly today, we determine molecular mass and molecular formula by *mass spectrometry*. In the examples that follow, we assume that the molecular formula of any unknown compound has already been determined and we proceed from there, using spectral analysis to determine a structural formula.

The following steps may prove helpful as a systematic approach to solving ^{1}H-NMR spectral problems.

Step 1: Molecular formula and index of hydrogen deficiency. Examine the molecular formula, calculate the index of hydrogen deficiency (section 20.1), and deduce any information you can about the presence or absence of rings or π bonds.

Step 2: Number of signals. Count the number of signals to determine the minimum number of sets of equivalent hydrogen atoms in the compound.

Step 3: Integration. Use signal integration and the molecular formula to determine the number of hydrogen atoms in each set.

Step 4: Pattern of chemical shifts. Examine the NMR spectrum for signals characteristic of the most common types of equivalent hydrogen atoms. (See the general rules of thumb for ^{1}H-NMR chemical shifts in section 20.5, p. 903.) Keep in mind that the ranges are broad and that hydrogen atoms of each type may be shifted outside the range, depending on details of the molecular structure in question.

Step 5: Splitting pattern. Examine splitting patterns for information about the number of nonequivalent neighbouring hydrogen atoms.

Step 6: Structural formula. Using the information gathered in steps 4 and 5, draw the structure of the fragments deduced and then piece them together to write a structural formula. Finally check that it is consistent with the information learned in steps 1 to 5.

Spectral problem 1: The compound is a colourless liquid with molecular formula $C_5H_{10}O$.

(300 MHz, $CDCl_3$)

$C_5H_{10}O$

δ 1.1

48

δ 2.4

quartet

triplet

33

10 9 8 7 6 5 4 3 2 1 0 ppm

Chemical shift (δ)

Analysis of spectral problem 1

Step 1: Molecular formula and index of hydrogen deficiency. The reference compound is $C_5H_{12}O$ so the index of hydrogen deficiency is 1 and the molecule contains either one ring or one π bond.

Step 2: Number of signals. There are two signals (a triplet and a quartet) so there are two sets of equivalent hydrogen atoms.

Step 3: Integration. By signal integration, we calculate that the number of hydrogen atoms giving rise to each signal is in the ratio 3:2. Because there are 10 hydrogen atoms, we conclude that the signal assignments are δ 1.1 (6H) and δ 2.4 (4H).

Step 4: *Pattern of chemical shifts.* The signal at δ 1.1 is in the alkyl region and, based on its chemical shift, most probably represents a methyl group (two methyl groups based on the integration data). No signal occurs at δ 4.6 to 5.7, so there are no vinylic hydrogen atoms. (If there is a carbon–carbon double bond in the molecule, there are no hydrogen atoms on it; that is, it is tetrasubstituted.) The signal at δ 2.4 is in the region for hydrogen atoms on the α-carbon atom of the carbonyl group (δ 2.1–2.6).

Step 5: *Splitting pattern.* The methyl signal at δ 1.1 is split into a triplet (t), so it must have two equivalent neighbouring hydrogen atoms, indicating —CH_2CH_3. The signal at δ 2.4 is split into a quartet (q), so it must have three equivalent neighbouring hydrogen atoms, which is also consistent with —CH_2CH_3. Consequently, an ethyl group accounts for these two signals. No other signals occur in the spectrum; therefore, there are no other types of hydrogen atoms in the molecule.

Step 6: *Structural formula.* So far we have learned that there are two signals with an integration of 6H and 4H (steps 2 and 3) and that these belong to two equivalent ethyl groups (steps 4 and 5). This accounts for all the hydrogen atoms in the formula and four of the carbon atoms; this leaves one carbon atom and one oxygen atom. Step 1 gave IHD = 1, so the remaining carbon and oxygen atoms must be a carbonyl group. Putting this information together we arrive at the structural formula shown at right. Note that the chemical shift of the methylene group (—CH_2—) at δ 2.4 is consistent with an alkyl group adjacent to a carbonyl group.

δ 2.4(q) δ 1.1(t)

$CH_3-CH_2-C(=O)-CH_2-CH_3$

pentan-3-one

Spectral problem 2: The compound is a colourless liquid with molecular formula $C_7H_{14}O$.

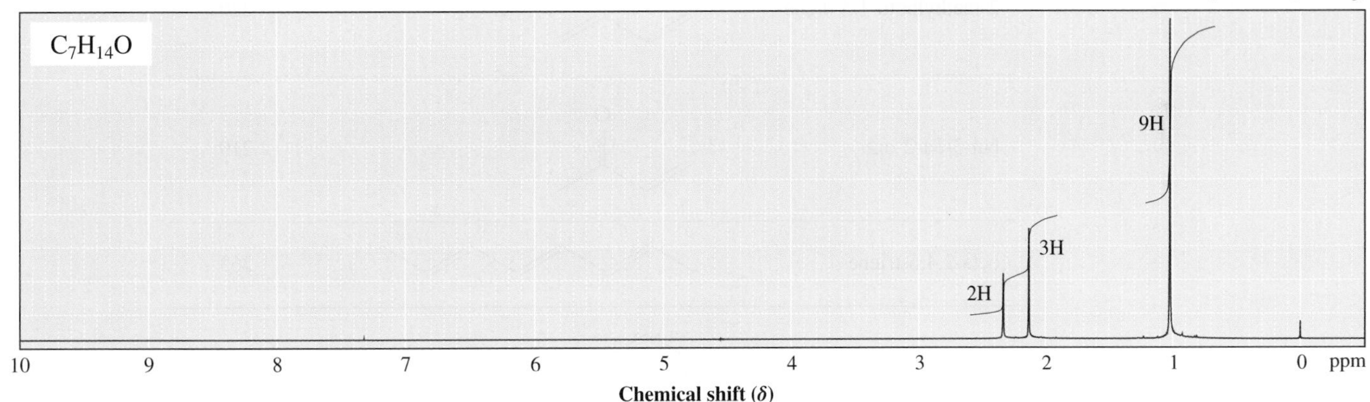

Analysis of spectral problem 2

Step 1: *Molecular formula and index of hydrogen deficiency.* The index of hydrogen deficiency is 1, so the compound contains one ring or one π bond.

Step 2: *Number of signals.* There are three signals and therefore three sets of equivalent hydrogen atoms.

Step 3: *Integration.* By signal integration, we calculate that the number of hydrogen atoms giving rise to each signal is in the ratio 2 : 3 : 9, reading from left to right.

Step 4: *Pattern of chemical shifts.* The singlet at δ 1.0 is characteristic of a methyl group adjacent to an sp^3 hybridised carbon atom (three methyl groups based on the integration). The singlets at δ 2.1 and 2.3 are characteristic of alkyl groups adjacent to a carbonyl group.

Step 5: *Splitting pattern.* All signals are singlets (s), which means that none of the hydrogen atoms are within three bonds of each other.

Step 6: *Structural formula.* Summarising the data above, we have three equivalent methyl groups (δ 1.0): another methyl group (δ 2.1) adjacent to a carbonyl group, and a methylene group (δ 2.3) also adjacent to a carbonyl group. We also have a carbonyl group as deduced from the IHD, molecular formula and the chemical shift of the methyl and methylene groups. This accounts for six carbon atoms, 14 hydrogen atoms and one oxygen atom. By comparison with the molecular formula, we know that we must also have a carbon atom not bonded to any hydrogen atoms. Now, to put this information together, let's start by considering the carbonyl group and the two attached groups, giving us —$CH_2C(O)CH_3$. The only possible arrangement for the remaining quaternary carbon atom and three methyl groups is —$C(CH_3)_3$. Therefore, we get the compound 4,4-dimethylpentan-2-one.

δ 1.0(s) δ 2.3(s) δ 2.1(s

$CH_3-C(CH_3)_2-CH_2-C(=O)-CH_3$

4,4-dimethylpentan-2-one

20.7 Other tools for determining structure

This chapter has focused on the tools most commonly used by chemists. However, there are other useful techniques including UV/visible spectroscopy, X-ray crystallography and electron spin resonance.

UV/visible spectroscopy is used in the study of compounds that absorb light in the ultraviolet–visible region of the spectrum (λ = 200–800 nm; see section 4.2, p. 122). The technique is applicable to organic compounds with conjugated π bonds (a sequence of four or more sp^2 hybridised atoms) and to coordination compounds of transition metals with partially filled *d* orbitals. UV/visible spectroscopy is a useful tool for studying the coordination chemistry of metals, as changes in coordination geometry and ligand donor atom can have marked effects on the spectrum. The colours and the theory of UV/visible spectroscopy of coordination compounds have been described in detail in section 13.4. As a general rule, for organic substances, the wavelength of maximum absorption (λ_{max}) increases with an increase in the extent of conjugation (see table 20.13 for some examples).

TABLE 20.13 Wavelength of maximum ultraviolet absorption of selected conjugated molecules.

Molecule	Structure	λ_{max} (nm)
2-methylbuta-1,3-diene		220
but-3-en-2-one	O	219
octa-2,4,6-triene		263
benzene		254

The spectra obtained have few absorption bands and generally lack fine detail. Consequently this technique is now rarely used to determine the structure of organic compounds. However, UV/visible spectroscopy is used widely in detectors for the analysis of liquid and gas samples as it can detect very low concentrations of analyte and provides quantitative data. The quantification is possible by applying the **Beer–Lambert law** (commonly known as Beer's law), which relates the absorbance to the concentration of the analyte and the sample cell pathlength:

$$A = \varepsilon bc$$

where ε is the molar absorptivity (L mol^{-1} cm^{-1}), b is the pathlength of the cell (cm), and c is the concentration of the sample (mol L^{-1}).

WORKED EXAMPLE 20.16

Application of the Beer–Lambert law

The molar absorptivity of vanillin at a wavelength of 231 nm is 14 200 L mol^{-1} cm^{-1}. A sample of vanillin was found to have an absorbance of 0.764 L mol^{-1} cm^{-1} at a wavelength of 231 nm when passing through a cell with a pathlength of 1.00 cm. What is the concentration of vanillin in this sample?

Analysis and solution

Before we start, it is good to check the data you are provided with and whether it is all required to solve the problem. Note that the wavelength of light is not part of the Beer–Lambert law, so this is not required to determine the concentration. It is provided because the value of the molar absorptivity is dependent on the wavelength of light absorbed. As the concentration is required, the Beer–Lambert law can be rewritten in terms of concentration.

$$c = \frac{A}{\varepsilon b}$$

Now substituting into the equation we get:

$$c = \frac{0.764}{14\,200} \times 1.00 = 5.38 \times 10^{-5}\,\text{M}$$

Is your answer reasonable?

A simple way to check your answer is to determine the absorbance using the value of the concentration you have determined.

PRACTICE EXERCISE 20.15

Colchicine is a natural product used in the treatment of gout. It has a molar absorptivity of 29 200 L mol^{-1} cm^{-1} at a wavelength of 231 nm. What is the concentration of a colchicine sample that has an absorbance of 0.845 L mol^{-1} cm^{-1} at a wavelength of 231 nm and cell pathlength of 5.00 cm?

An important application of UV-absorbing compounds is in sunscreens to protect against skin cancer (figure 20.47). Octyl methoxycinnamate and 4-methylbenzylidine camphor, which both absorb UVB radiation (290–320 nm), are the UV absorbers most commonly used in sunscreens.

X-ray crystallography is the study of crystal structures through X-ray diffraction techniques and is a powerful way of obtaining detailed information on structure and bonding within a crystal. When a beam of X-rays passes through a crystal, the X-rays are diffracted in a manner that is dependent on the size and position of the atoms within the crystal (see chapter 7, p. 284). Analysis of the diffraction pattern allows the determination of the three-dimensional structure of the compound, including bond lengths and bond angles. Each year, X-ray crystallography is used to determine the structure of thousands of inorganic, organic, organometallic and biological compounds. It is the most important technique available for the determination of the 3-D structure of proteins and other large biomolecules. However, this technique is limited to crystalline samples of suitable quality and size. Growing appropriate crystals, particularly of biomolecules, can be a significant challenge in itself.

Electron spin resonance (ESR) also known as **electron paramagnetic resonance (EPR)** is a type of magnetic resonance that is useful for the study of free radicals, paramagnetic species (see section 13.1, p. 548) and other substances containing unpaired electrons. It is arguably the most important technique for studying structure in free radicals and plays an important part in understanding reaction mechanisms.

FIGURE 20.47 UV-absorbing compounds are used in sunscreens.

SUMMARY

LO 1 Tools for determining structure

The index of hydrogen deficiency is the sum of the number of rings and π bonds in a molecule. It can be determined by comparing the number of hydrogen atoms in the molecular formula of a compound of unknown structure with the number of hydrogen atoms in a reference compound with the same number of carbon atoms and with no rings or π bonds.

LO 2 Mass spectrometry

Mass spectrometry is a technique that allows us to determine the mass of molecules and fragments of molecules. In mass spectrometry, a sample is ionised by bombardment with a stream of high-energy electrons to produce a molecular ion (an ion generated by removal of an electron from the molecule) and fragment ions. The molecular ion is a radical cation, because it has one unpaired electron and is positively charged, and has the same mass as the parent molecule. High resolution mass spectrometry is a process that measures the mass to charge ratio of ions (including the molecular ion) with a high degree of precision (usually to four decimal places), allowing the determination of a molecular formula. Relative abundances of isotopes may be used to identify elements present in a molecule.

LO 3 Infrared spectroscopy

Electromagnetic radiation can be described in terms of its wavelength (λ) and frequency (ν). Frequency is reported in s^{-1}. An alternative way to describe electromagnetic radiation is in terms of its energy, where $E = h\nu$. Molecular spectroscopy is the experimental process of measuring the frequencies of radiation that are absorbed or emitted by a substance and correlating these patterns with details of molecular structure. Interactions of molecules with infrared radiation excite covalent bonds to higher vibrational energy levels.

The vibrational infrared spectrum extends from 4000 to 400 cm^{-1}. Radiation in this region is referred to by its wavenumber ($\tilde{\nu}$) in reciprocal centimetres (cm^{-1}). To be infrared active, a bond must be polar; the more polar it is, the stronger is its absorption of IR radiation. There are $3n - 6$ fundamental vibrations for a nonlinear molecule containing n atoms. The simplest vibrations that give rise to the absorption of infrared radiation are stretching and bending vibrations. Stretching may be symmetrical or asymmetrical.

LO 4 Interpreting infrared spectra

Functional groups within molecules have characteristic vibrational frequencies, and these are often listed on correlation tables. The intensity of a signal is referred to as strong (s), medium (m) or weak (w). Stretching vibrations for most functional groups appear in the region from 3650 to 1500 cm^{-1}. The region from 1500 to 400 cm^{-1} is called the fingerprint region.

LO 5 Nuclear magnetic resonance spectroscopy

The interaction of molecules with radio-frequency radiation gives us information about nuclear spin energy levels. Nuclei of ^{1}H and ^{13}C, isotopes of the two elements most common to organic compounds, have a spin and behave like tiny bar magnets. When placed in a powerful magnetic field, the nuclear spins of these elements become aligned either with the applied field or against it. Nuclear spins aligned with the applied field are in the lower energy state; those aligned against the applied field are in the higher energy state. Resonance is the absorption of electromagnetic radiation by a nucleus and the resulting flip of its nuclear spin from a lower energy spin state to a higher energy spin state. An NMR spectrometer records such a resonance as a signal.

LO 6 Interpreting NMR spectra

In a ^{1}H-NMR spectrum, a resonance signal is reported by how far it is shifted from the resonance signal of the hydrogen atoms in tetramethylsilane (TMS). A resonance signal in a ^{13}C-NMR spectrum is reported by how far it is shifted from the resonance signal of the carbon atoms in TMS. A chemical shift (δ) is the frequency shift from TMS, divided by the operating frequency of the spectrometer.

Equivalent hydrogen atoms within a molecule have identical chemical shifts. The area of a ^{1}H-NMR signal is proportional to the number of equivalent hydrogen atoms giving rise to that signal. In signal splitting, the ^{1}H-NMR signal from one hydrogen atom or set of equivalent hydrogen atoms is split by the influence of hydrogen atoms on adjacent carbon atoms that are equivalent to each other but not to the atoms giving rise to the signal. According to the $(n + 1)$ rule, if a hydrogen atom has n hydrogen atoms that are not equivalent to it, but are equivalent to each other, on adjacent carbon atom(s), its ^{1}H-NMR signal is split into $(n + 1)$ peaks. Complex splitting patterns occur when a hydrogen atom is flanked by two or more sets of hydrogen atoms and those sets are nonequivalent. Splitting patterns are commonly referred to as singlets (s), doublets (d), triplets (t), quartets (q), quintets and multiplets (m).

In ^{13}C-NMR spectra, all the ^{13}C signals appear as singlets. The number of singlets almost always corresponds to the number of nonequivalent carbon atoms within a molecule. The area of a ^{13}C-NMR signal is *not* proportional to the number of equivalent carbon atoms giving rise to that signal.

LO 7 Other tools for determining structure

UV/visible spectroscopy is used in the study of compounds that absorb light in the ultraviolet–visible region of the spectrum (λ = 200–800 nm). UV/visible spectroscopy is commonly used in analytical chemistry to quantify very low levels of analytes. X-ray crystallography is the study of crystal structures through X-ray diffraction techniques. Electron spin resonance (ESR), also known as electron paramagnetic resonance (EPR), is a type of magnetic resonance that is useful for the study of free radicals, paramagnetic species and other substances containing unpaired electrons.

KEY CONCEPTS AND EQUATIONS

Index of hydrogen deficiency *(section 20.1)*
The sum of the number of rings and π bonds in a molecule.

$(n + 1)$ rule *(section 20.5)*
According to this rule for determining splitting patterns in NMR spectroscopy, if a hydrogen atom has n hydrogen atoms that are not equivalent to it, but equivalent to each other, on the same or adjacent atom(s), then the ^{1}H-NMR signal of the hydrogen atom is split into $(n + 1)$ peaks.

Beer–Lambert law *(section 20.7)*

Commonly known as Beer's law, this relates the absorbance of UV/visible light to the concentration of the analyte and the sample cell pathlength:

$$A = \varepsilon bc$$

where:
ε is the molar absorptivity (L mol^{-1} cm^{-1})
b is the pathlength of the cell (cm)
c is the concentration of the sample (mol L^{-1})

KEY TERMS

Beer–Lambert law *p. 916*
bending *p. 883*
chemical shift (δ) *p. 896*
^{13}C-NMR spectroscopy *p. 894*
complex splitting pattern *p. 903*
correlation table *p. 883*
deshielded *p. 895*
double bond equivalent *p. 876*
doublet *p. 903*
downfield *p. 897*
electromagnetic radiation *p. 881*
electron paramagnetic resonance (EPR) *p. 917*
electron spin resonance (ESR) *p. 917*
equivalent hydrogen atoms *p. 897*
fingerprint region *p. 883*
frequency (ν) *p. 881*
hertz (Hz) *p. 881*
high resolution mass spectrometry *p. 879*
^{1}H-NMR spectroscopy *p. 894*
index of hydrogen deficiency (IHD) *p. 876*
infrared active *p. 883*
infrared radiation *p. 881*
integration *p. 899*
line of integration *p. 899*
local magnetic field *p. 895*
mass spectrometry *p. 877*
molecular ion *p. 878*
multiplet *p. 903*
multiplicity *p. 904*
(n + 1) rule *p. 904*
nuclear magnetic resonance (NMR) spectroscopy *p. 894*
quartet *p. 903*
radical cation *p. 878*
radio-frequency radiation *p. 895*
reference compound *p. 876*
resonance *p. 895*
shielded *p. 895*
signal splitting *p. 904*
singlet *p. 903*
splitting pattern *p. 903*
stretching *p. 883*
tetramethylsilane (TMS) *p. 896*
triplet *p. 903*
upfield *p. 897*
UV/visible spectroscopy *p. 916*
vibrational infrared *p. 882*
wavelength (λ) *p. 881*
wavenumber ($\tilde{\nu}$) *p. 882*
X-ray crystallography *p. 917*

REVIEW QUESTIONS

LO 1 Tools for determining structure

20.1 Complete the following table.

Class of compound	Molecular formula	Index of hydrogen deficiency	Reason for hydrogen deficiency
alkane	C_nH_{2n+2}	0	(reference hydrocarbon)
alkene	C_nH_{2n}	1	one π bond
alkyne			
alkadiene			
cycloalkane			
cycloalkene			

20.2 Calculate the index of hydrogen deficiency of each of the following compounds.
(a) ibuprofen, $C_8H_{18}O_2$
(b) retinol (vitamin A), $C_{20}H_{30}O$
(c) nicotine, $C_{10}H_{14}N_2$
(d) carvone, $C_{10}H_{14}O$
(e) limonene, $C_{10}H_{16}$
(f) fluoroacetic acid (1080), $C_2H_3FO_2$

LO 2 Mass spectrometry

20.3 What information can we gain from mass spectrometry?

20.4 What is a molecular ion?

20.5 What does m/z stand for?

20.6 In the mass spectrum of decane, the peak due to the molecular ion (M)$^+$ is at m/z = 142. This, however, is not the peak with the highest m/z value. There is a peak at m/z = 143 with an intensity of approximately 10% that of the (M)$^+$ peak. This peak is referred to as the (M + 1)$^+$ peak. What is the source of this peak?

20.7 In the mass spectrum of bromobenzene, there is a peak at m/z = 156 (M)$^+$ and another at m/z = 158 (M + 2)$^+$. Explain the occurrence of these two peaks.

20.8 You are provided with the mass spectrum of a compound that contains one of the following halogens: fluorine, chlorine or bromine. Using your knowledge of isotopes, explain how you could identify the halogen present in the compound.

20.9 What mass spectrometry technique can be used to determine the molecular formula of a compound?

20.10 Determine the accurate mass of each of the following compounds.

(a) butan-2-ol
(b) 4-aminohexanal
(c) chlorobenzene
(d) trichloroacetic acid
(e) nicotine
(f) ibuprofen

20.11 Match each of the following compounds with the appropriate spectrum.

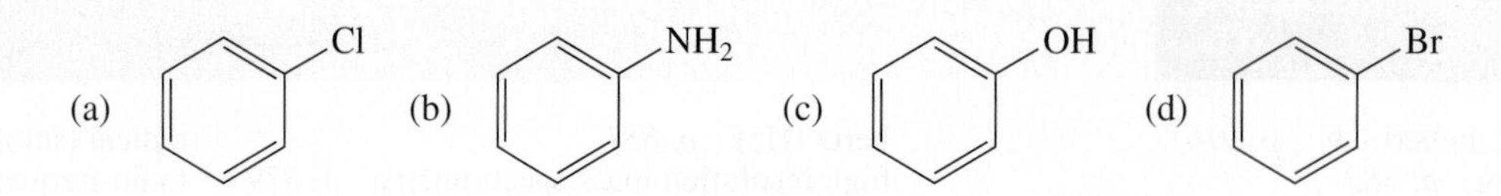

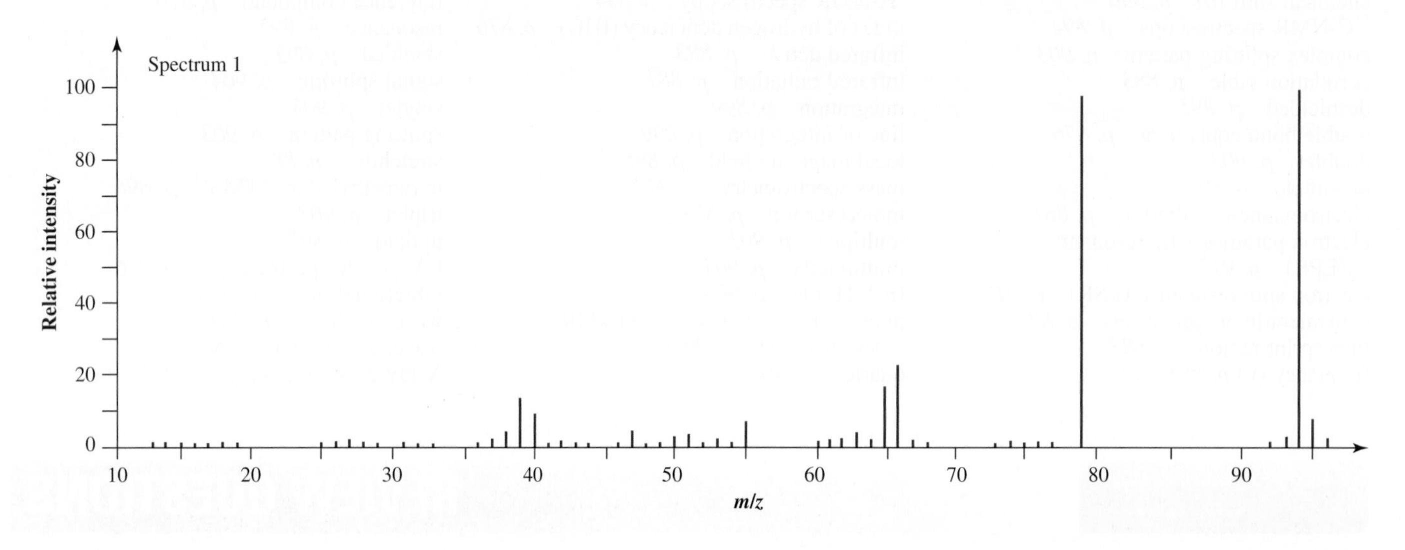

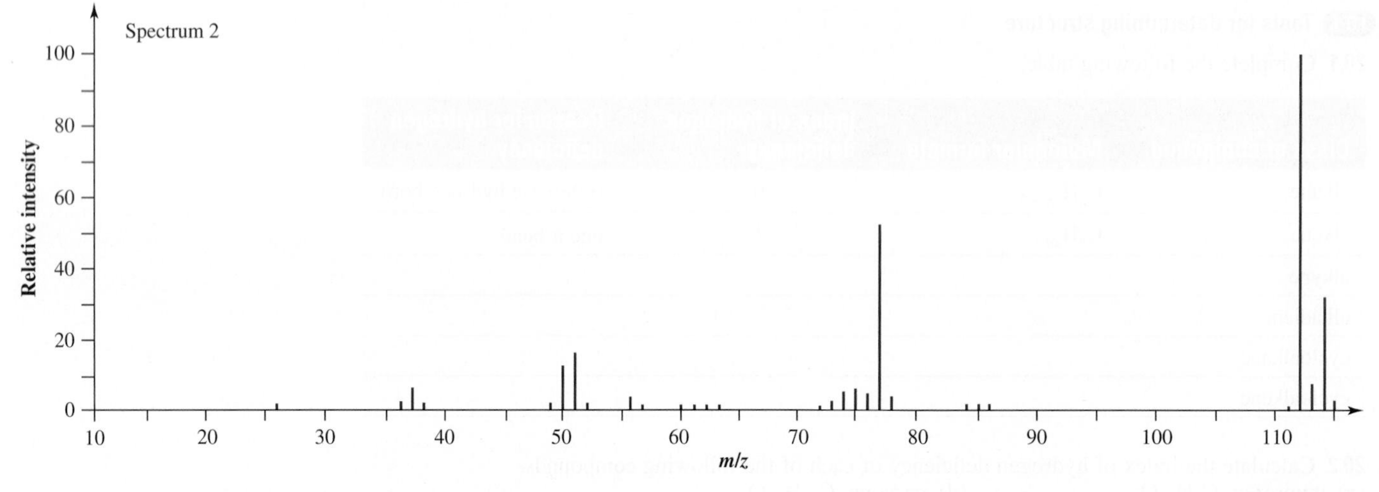

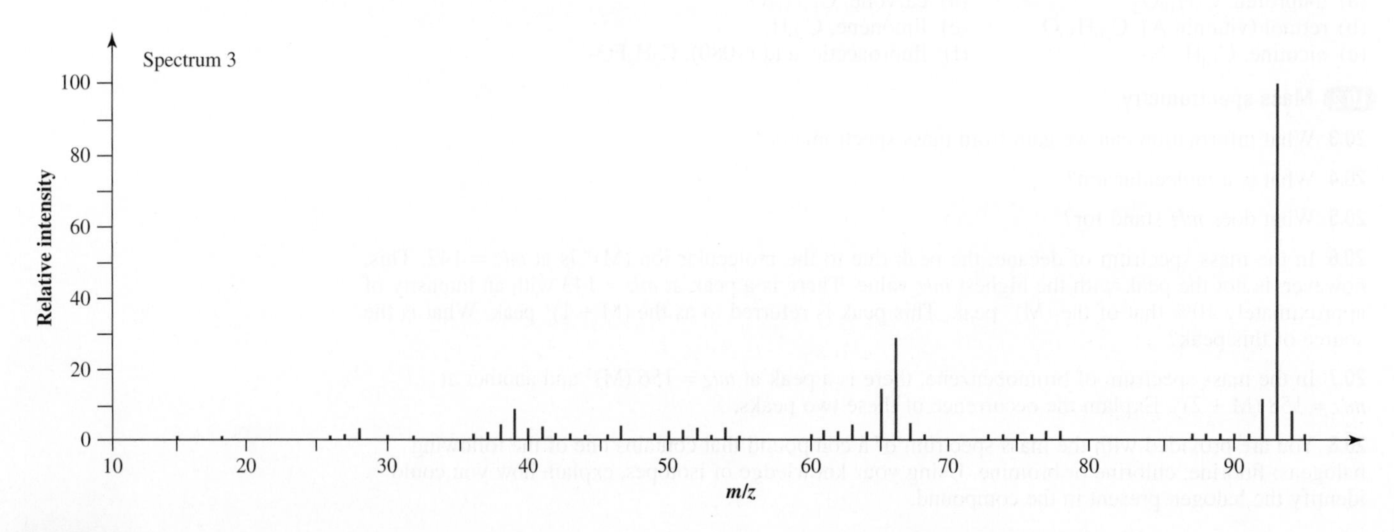

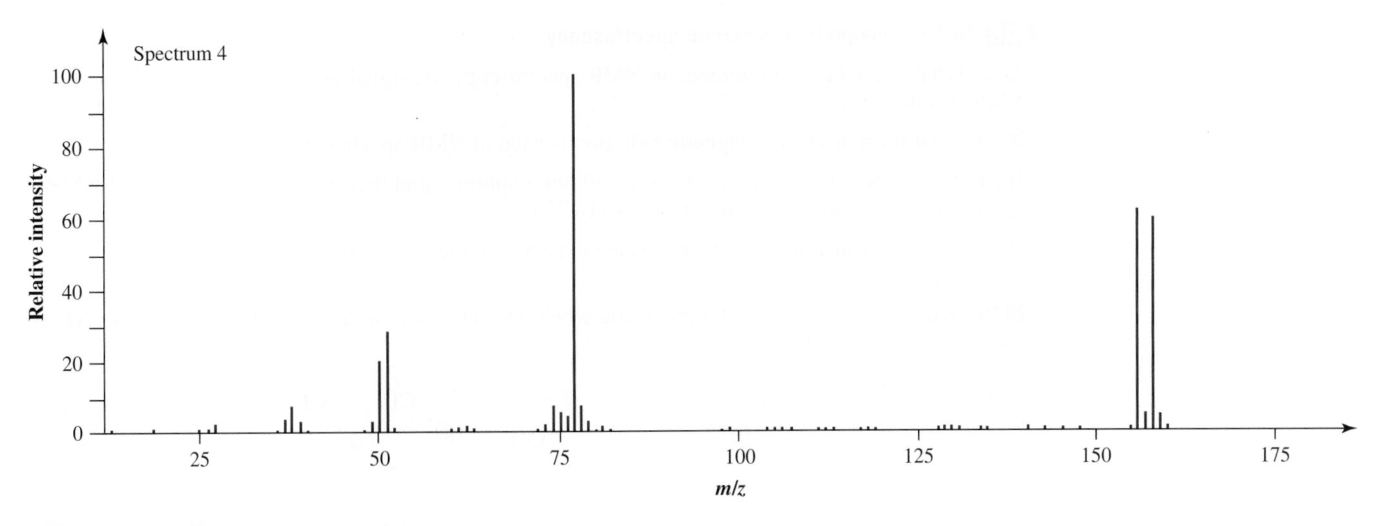

LO 3 Infrared spectroscopy

20.12 What information can we gain from infrared spectroscopy?

20.13 Which region of the electromagnetic spectrum, IR or UV, contains photons of higher energy?

20.14 Which has a lower characteristic stretching frequency, a C=C or a C=O bond? Explain briefly.

20.15 Which has a lower characteristic stretching frequency, a C=O bond or a C—O bond? Explain briefly.

20.16 How could IR spectroscopy be used to distinguish between the compounds $(CH_3)_3N$ and $CH_3NHCH_2CH_3$?

20.17 Predict the position of the C=O stretching absorption in acetate ion, $CH_3CO_2^-$, relative to that in acetic acid, CH_3CO_2H.

20.18 A chemist carried out a reaction in which hexan-3-one was reduced to hexan-3-ol. How could IR spectroscopy be used to determine:
(a) that an alcohol had been produced
(b) there was no ketone remaining in the product?

20.19 1-methylcyclohexanol was dehydrated to give 1-methylcyclohexene. How could IR spectroscopy be used to determine:
(a) that an alkene had been produced
(b) there was no alcohol remaining in the product?

LO 4 Interpreting infrared spectra

20.20 The following is the IR spectrum of L-tryptophan, a naturally occurring amino acid that is abundant in foods such as turkey.

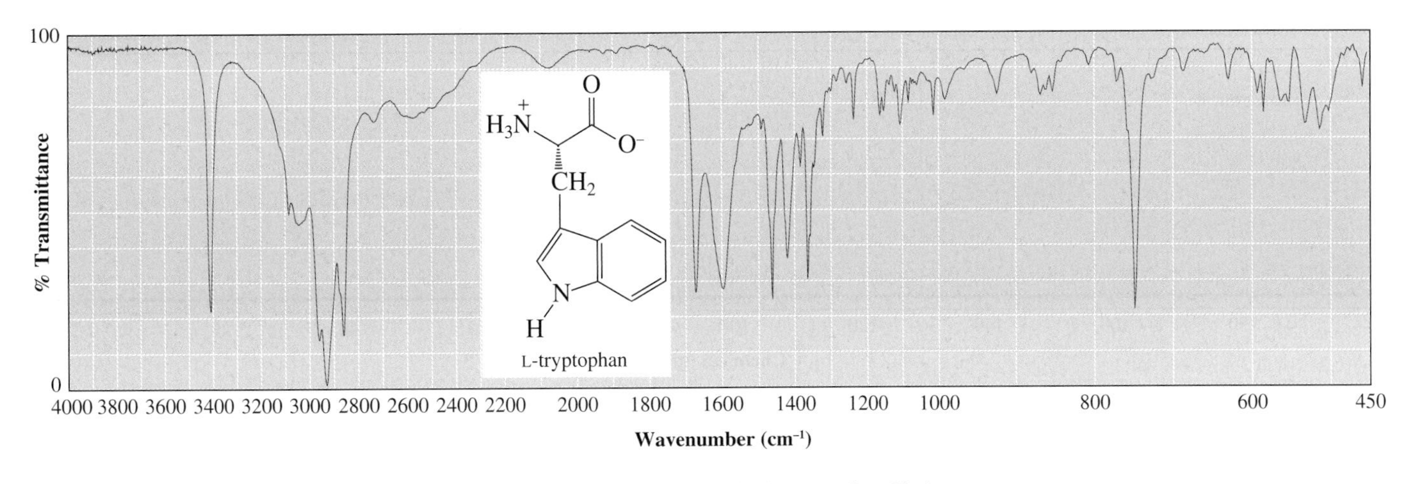

For many years, the L-tryptophan in turkey was believed to make people drowsy after Christmas dinner. Scientists now know that consuming L-tryptophan makes you drowsy only if the compound is taken on an empty stomach. Therefore, it is unlikely that your Christmas turkey is the cause of drowsiness. Notice that L-tryptophan contains one stereocentre. Its enantiomer, D-tryptophan, does not occur in nature, but can be synthesised in the laboratory. What would the IR spectrum of D-tryptophan look like?

20.21 The sex attractant of the codling moth has the molecular formula $C_{13}H_{24}O$. It gives an IR spectrum with a broad signal between 3200 and 3600 cm^{-1} and two signals between 1600 and 1700 cm^{-1}.
(a) Determine the index of hydrogen deficiency for this compound.
(b) What functional groups are present in this compound?

LO 5 Nuclear magnetic resonance spectroscopy

20.22 What is used as the reference in NMR spectroscopy; its signal is assigned $\delta = 0$ in ^{1}H and ^{13}C NMR spectroscopy?

20.23 What form of electromagnetic radiation is used in NMR spectroscopy?

20.24 Predict the number of signals expected, their splitting and their relative areas in the ^{1}H-NMR spectrum of 1,3-dibromopropane, $BrCH_2CH_2CH_2Br$.

20.25 How many distinct carbon signals are expected in the ^{13}C-NMR spectrum of 3-methylpentan-3-ol?

20.26 Determine the number of signals you would expect to see in the ^{1}H-NMR spectrum of each of the following compounds.

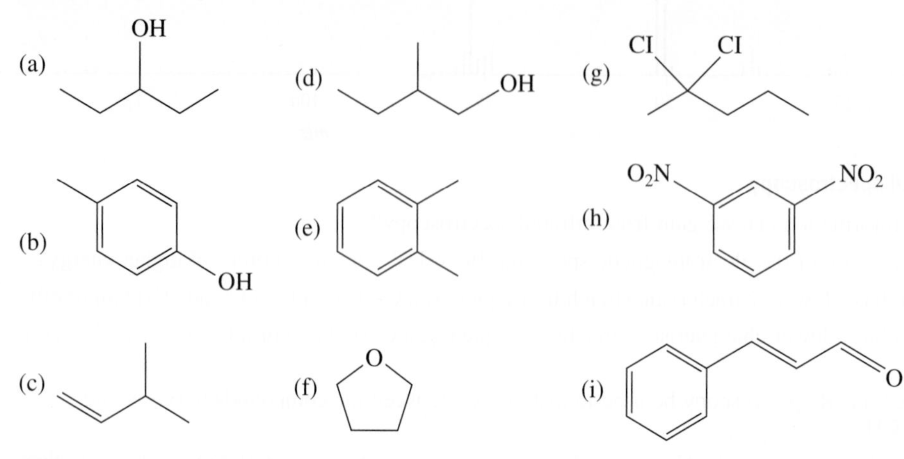

LO 6 Interpreting NMR spectra

20.27 Determine the number of signals you would expect to see in the ^{13}C-NMR spectrum of each of the compounds in question 20.26.

20.28 The following are structural formulae for the constitutional isomers of dimethylbenzene (*o*-xylene, *p*-xylene and *m*-xylene) and three sets of ^{13}C-NMR spectra. Assign each constitutional isomer to its correct spectrum.

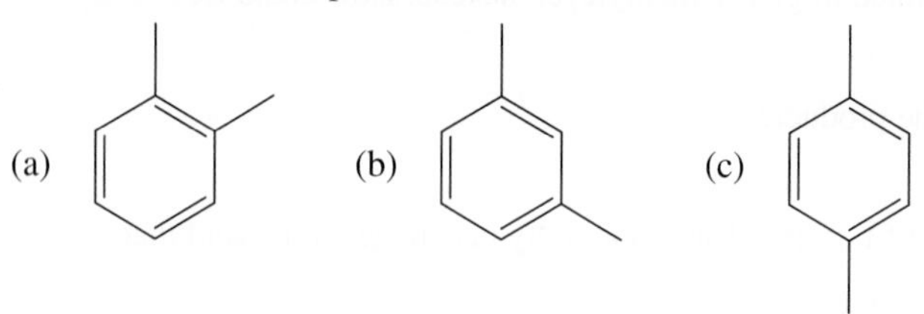

(75 MHz, $CDCl_3$)

Spectrum 1

200 180 160 140 120 100 80 60 40 20 0 ppm

Chemical shift (δ)

(75 MHz, $CDCl_3$)

Spectrum 2

200 180 160 140 120 100 80 60 40 20 0 ppm

Chemical shift (δ)

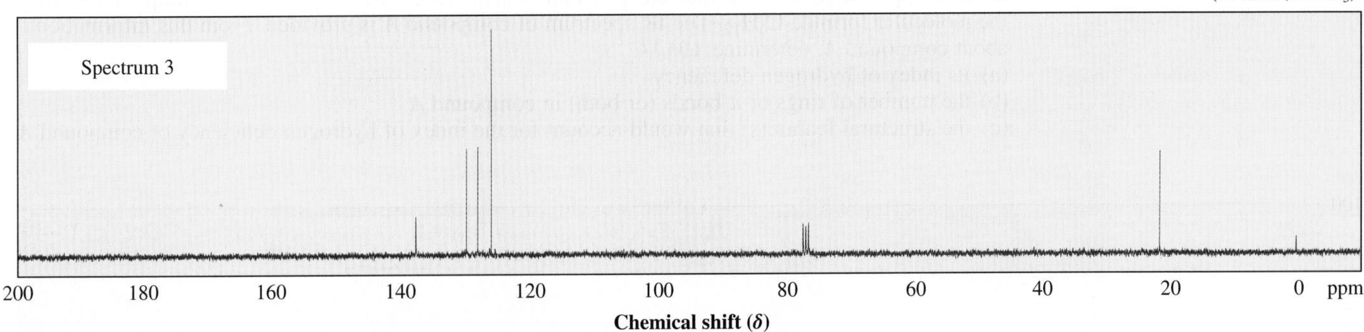

LO 7 Other tools for determining structure

20.29 What is the most important analytical technique for determining the three-dimensional structure of proteins?

20.30 What structural feature is required for organic molecules to absorb UV light?

20.31 Determine the molar absorptivity of a compound that has an absorbance of 0.644 L mol^{-1} cm^{-1} (pathlength = 2 cm) for a solution with a concentration of 4.3 × 10^{-5} M.

20.32 Codeine phosphate has a molar absorptivity of 1570 L mol^{-1} cm^{-1} in water at 284 nm. Determine the concentration of a solution of codeine phosphate that has an absorbance of 0.64.

REVIEW PROBLEMS

20.33 The mass spectrum below is that of an unbranched alkane. Which alkane produced this spectrum? **LO 2**

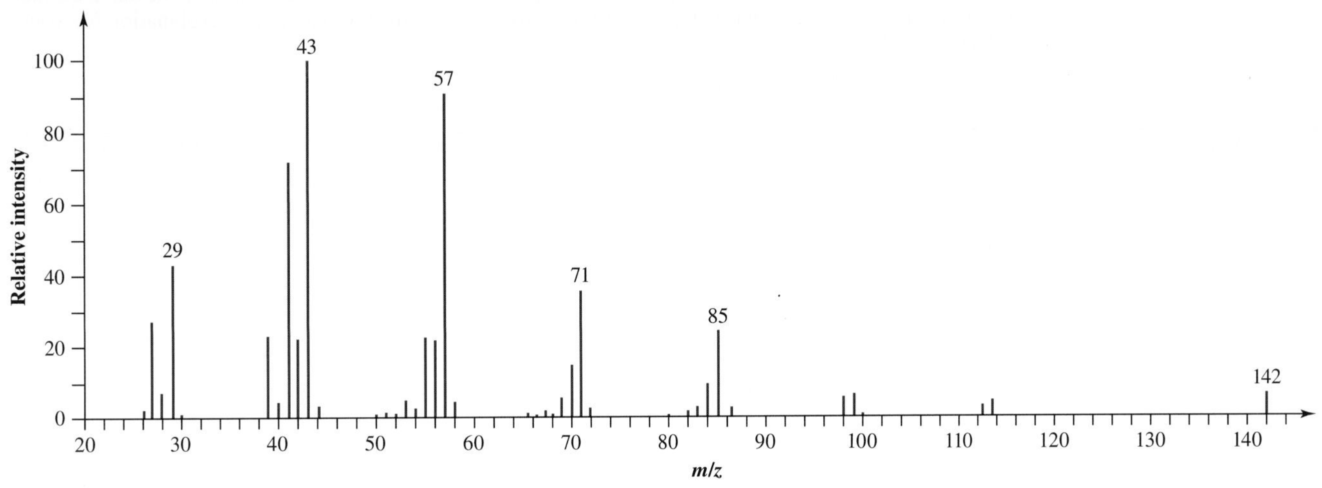

20.34 The mass spectrum below is that of a non-cyclic alkene. What is the molecular formula for the alkene that produced this spectrum? **LO 2**

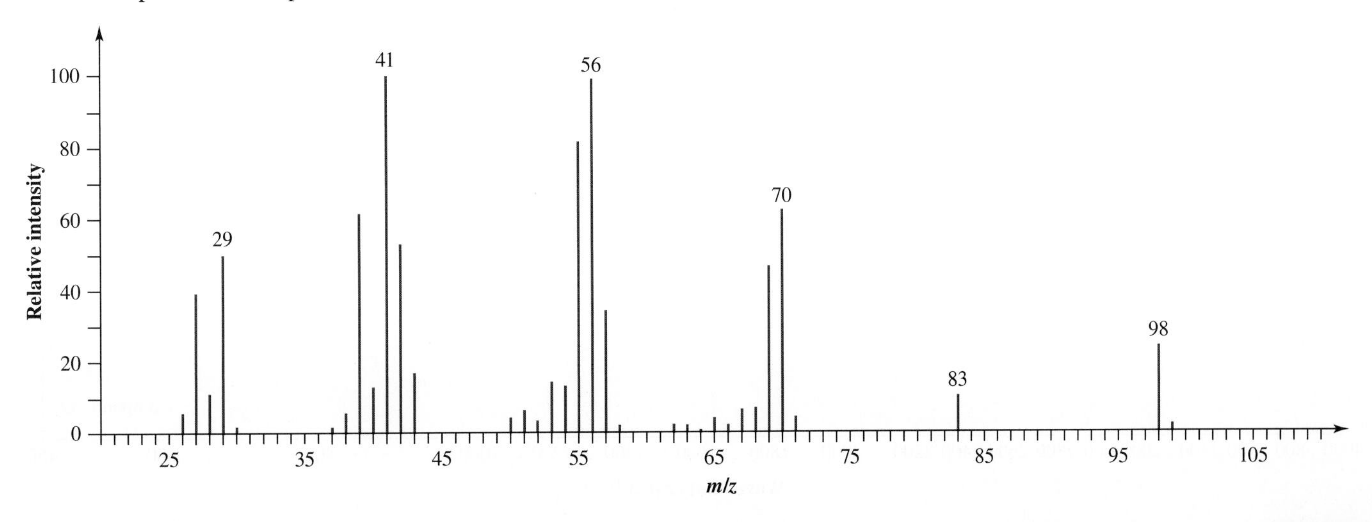

20.35 Compound *A*, with the molecular formula C_6H_{10}, reacts with H_2/Ni to give compound *B*, with the molecular formula C_6H_{12}. The IR spectrum of compound *A* is provided. From this information about compound *A*, determine: **LO 1,3,4**

(a) its index of hydrogen deficiency

(b) the number of rings or π bonds (or both) in compound *A*

(c) the structural feature(s) that would account for the index of hydrogen deficiency of compound *A*.

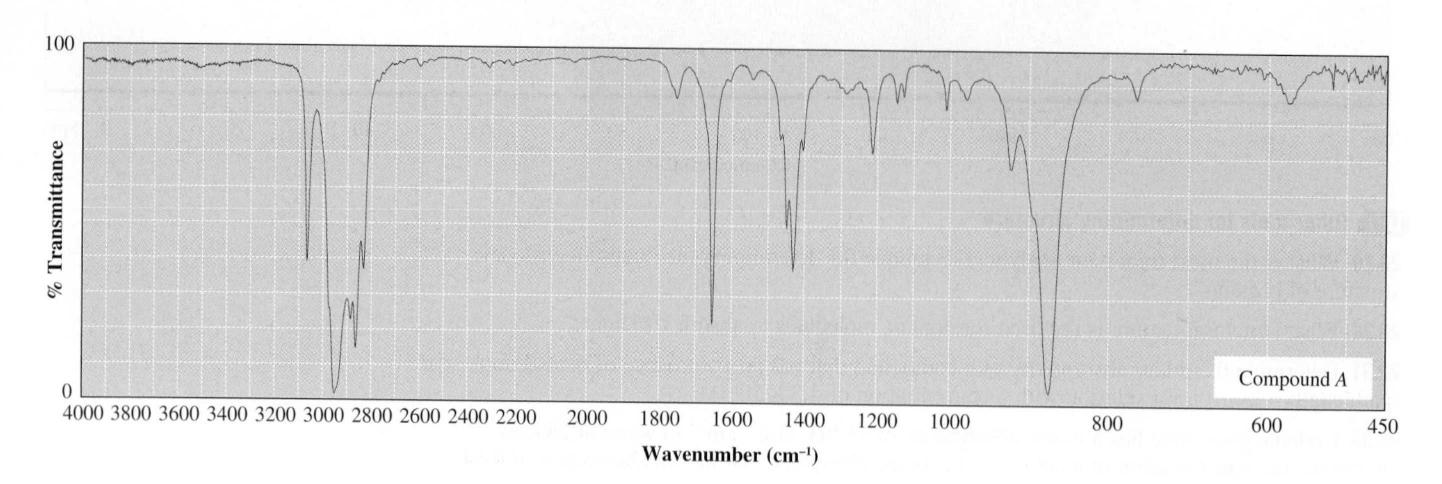

20.36 Benzyl bromide, $C_6H_5CH_2Br$, can be prepared by treating benzyl alcohol, $C_6H_5CH_2OH$, with phosphorus tribromide, PBr_3. **LO 2,3,4**

(a) How can mass spectrometry be used to confirm that benzyl bromide has been formed?

(b) What key signal in the IR spectrum of benzyl alcohol will be absent in the IR spectrum of benzyl bromide?

20.37 The following are infrared spectra of compounds *C* and *D*. One spectrum is of hexan-1-ol, the other of nonane. Assign each compound to its correct spectrum. Provide a brief explanation for your selection. **LO 3,4**

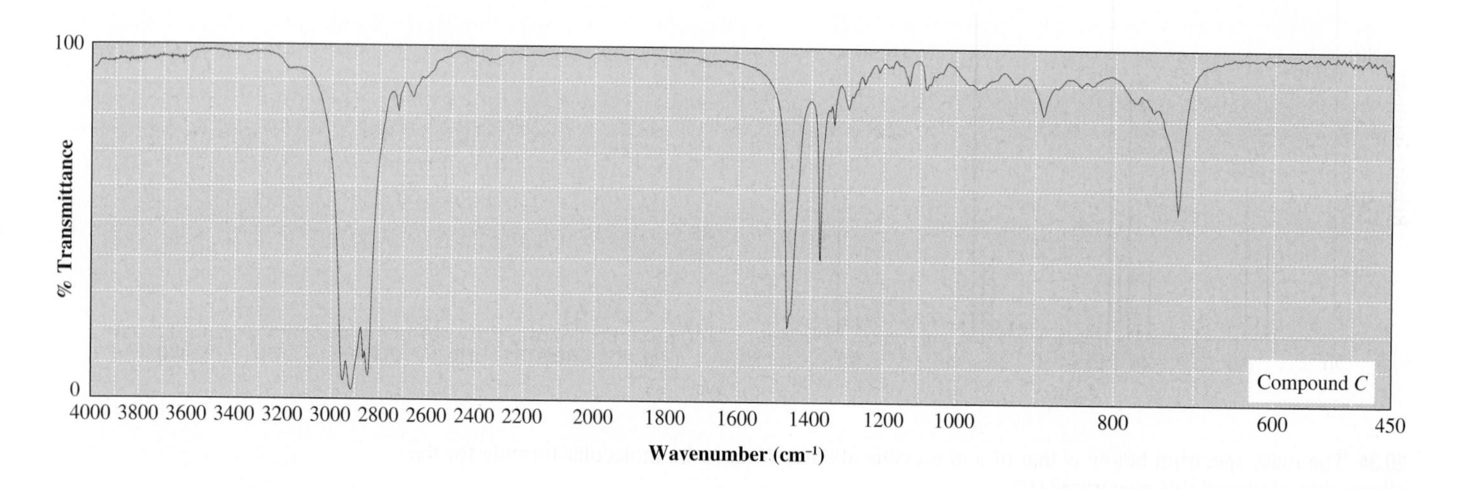

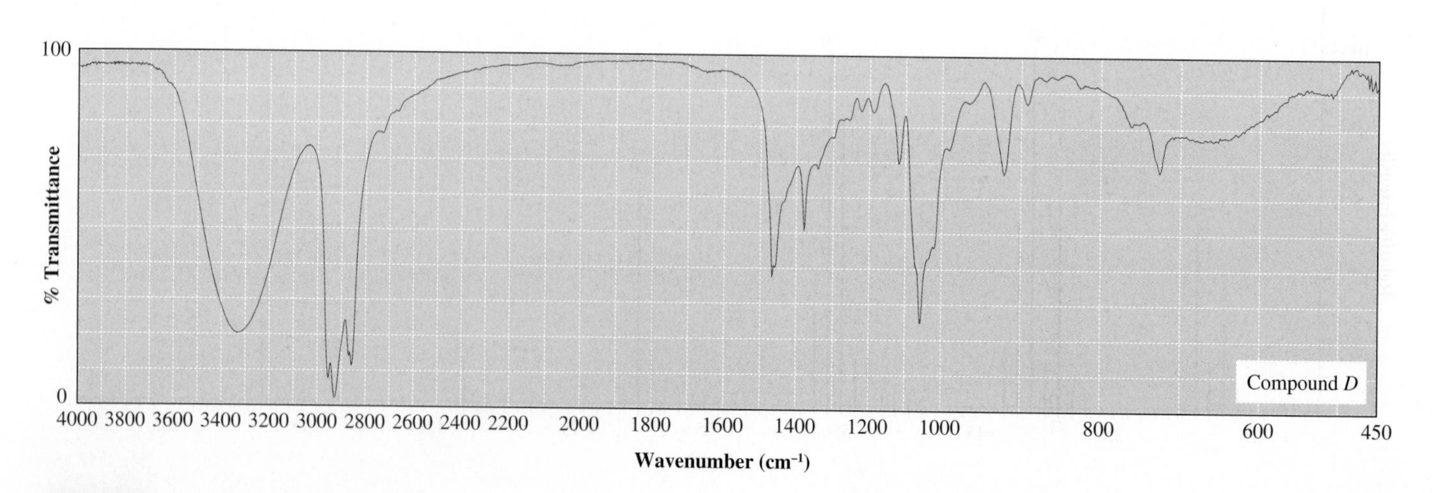

20.38 Compound *E*, with the molecular formula C_6H_{12}, reacts with H_2/Ni to give compound *F*, with the molecular formula C_6H_{14}. The IR spectrum of compound *E* is provided. From this information about compound *E*, determine: **LO 1,3,4**

(a) its index of hydrogen deficiency

(b) the number of rings or π bonds (or both) in compound *E*

(c) the structural feature(s) that would account for the index of hydrogen deficiency of compound *E*.

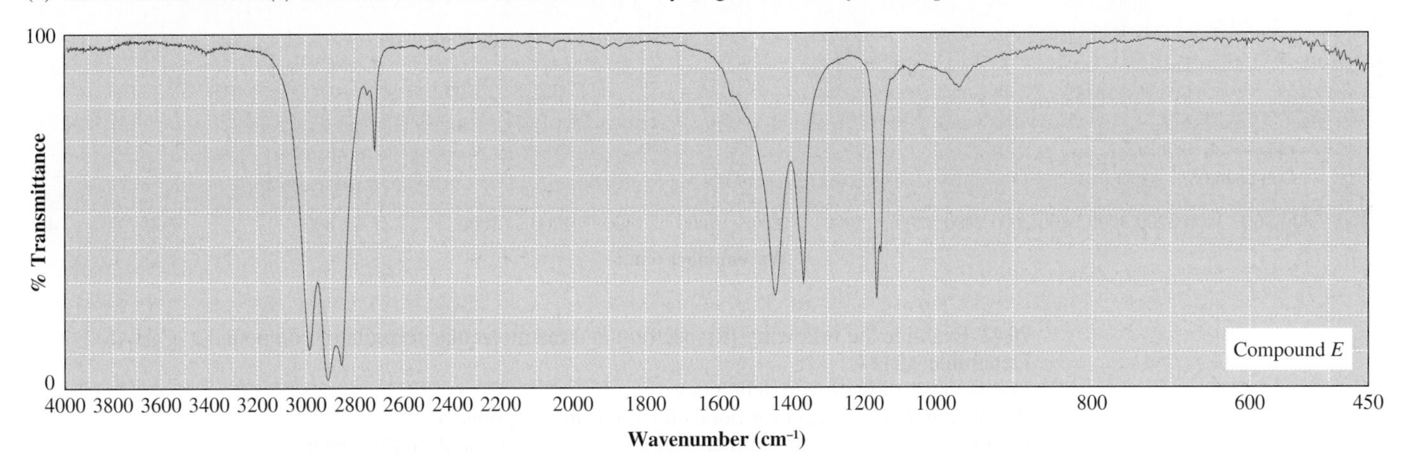

20.39 The constitutional isomers 2-methylbutan-1-ol and *tert*-butyl methyl ether have the molecular formula $C_5H_{12}O$. Assign each compound to its correct infrared spectrum, *G* or *H*. Provide a brief explanation for your selection. **LO 3,4**

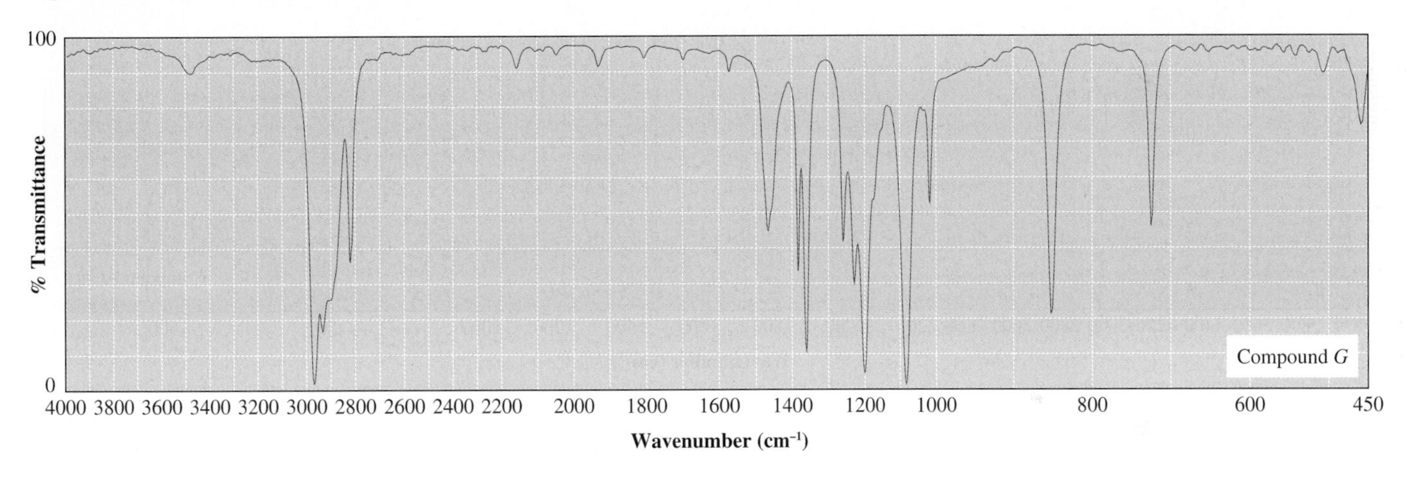

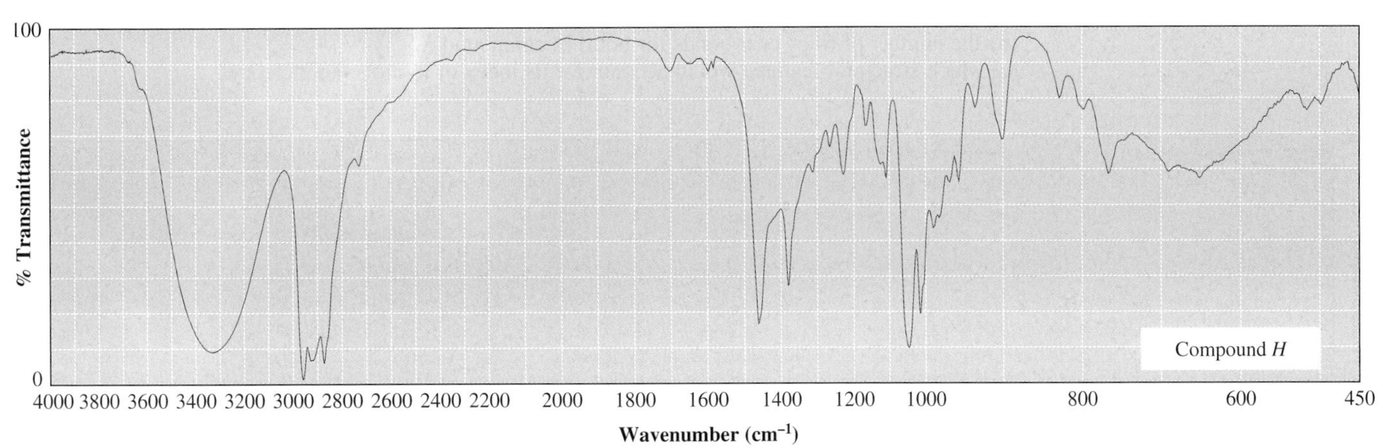

20.40 The oxidation of hexan-1-ol can produce hexanal and hexanoic acid. How can IR spectroscopy be used to distinguish between hexan-1-ol, hexanal and hexanoic acid? Give the approximate wavenumber values for the key diagnostic signals. **LO 3,4**

20.41 Examine the following IR spectrum and the molecular formula of compound *I*, $C_9H_{12}O$, and determine: **LO 1,3,4**

(a) its index of hydrogen deficiency

(b) the number of rings or π bonds (or both) in compound *I*

(c) which structural feature would account for its index of hydrogen deficiency

(d) which oxygen-containing functional group compound *I* contains.

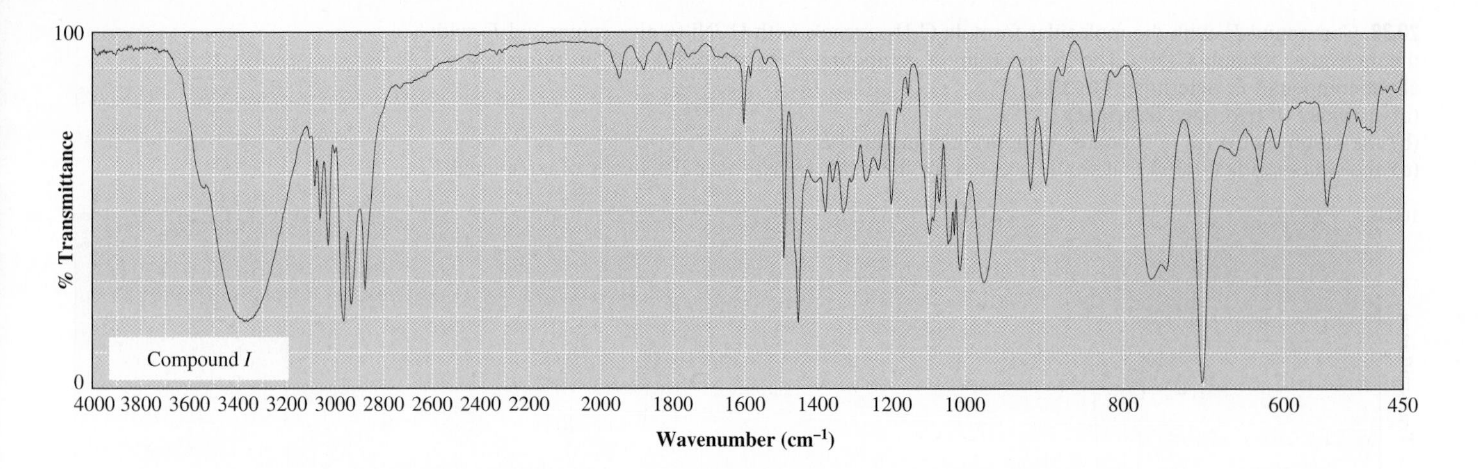

20.42 Examine the following IR spectrum and the molecular formula of compound *J*, $C_5H_{13}N$. Determine: **LO 1,3,4**

(a) its index of hydrogen deficiency
(b) the number of rings or π bonds (or both) in compound *J*
(c) the nitrogen-containing functional group(s) compound *J* might contain.

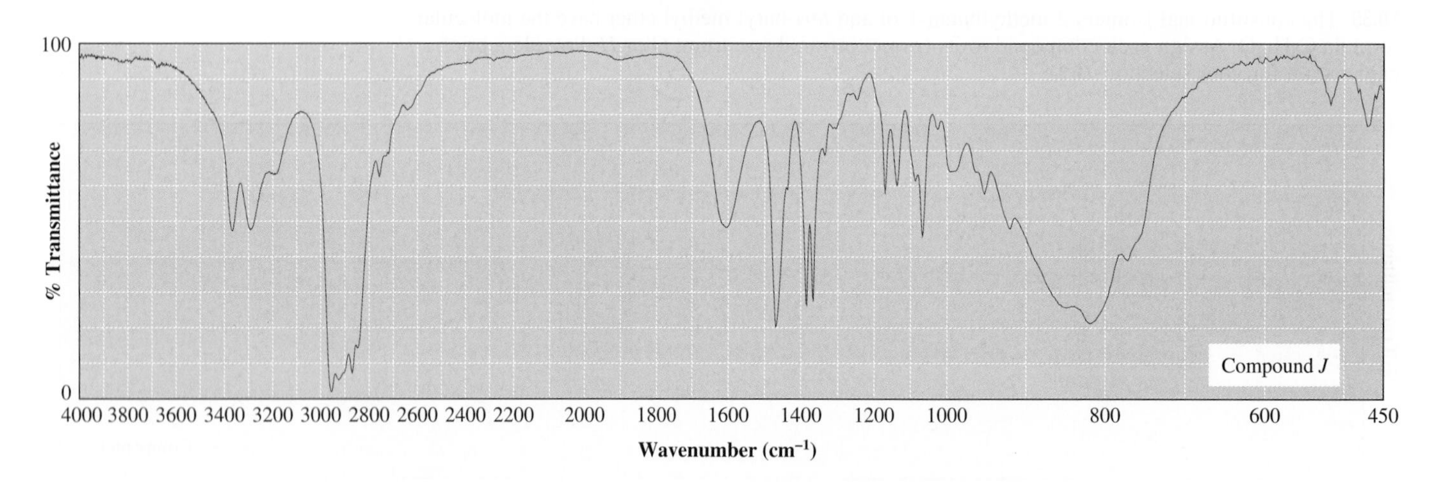

20.43 Examine the following IR spectrum and the molecular formula of compound *K*, $C_6H_{12}O$. Determine: **LO 1,3,4**

(a) its index of hydrogen deficiency
(b) the number of rings or π bonds (or both) in compound *K*
(c) which structural features would account for its index of hydrogen deficiency.

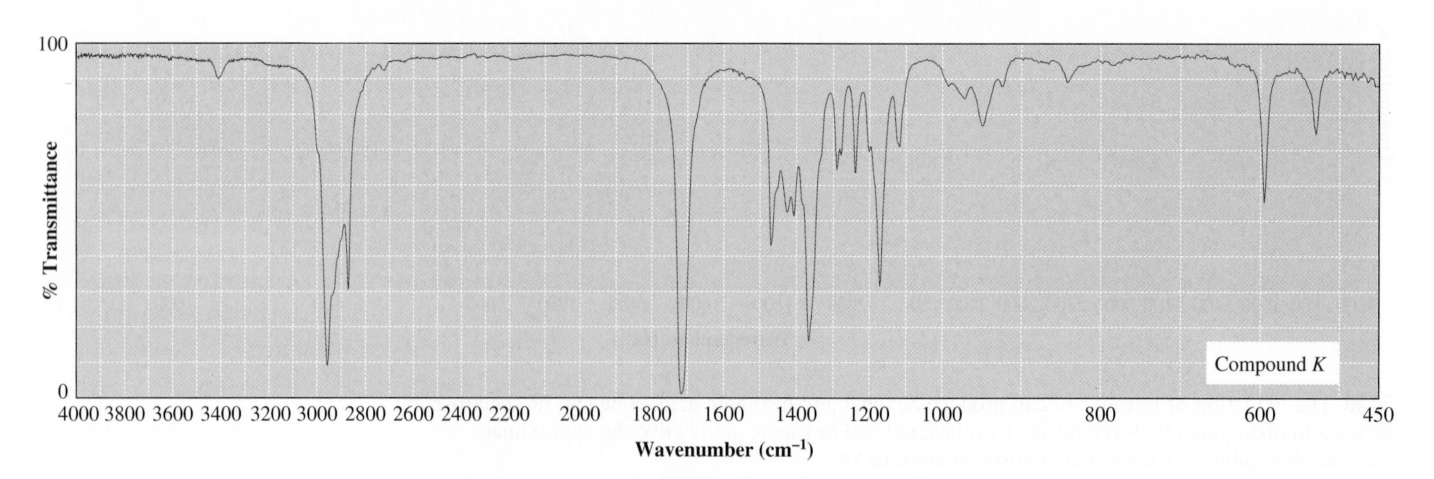

20.44 Examine the following IR spectrum and the molecular formula of compound *L*, $C_6H_{12}O_2$. Determine: **LO 1,3,4**

(a) its index of hydrogen deficiency
(b) the number of rings or π bonds (or both) in compound *L*
(c) the oxygen-containing functional group(s) compound *L* might contain.

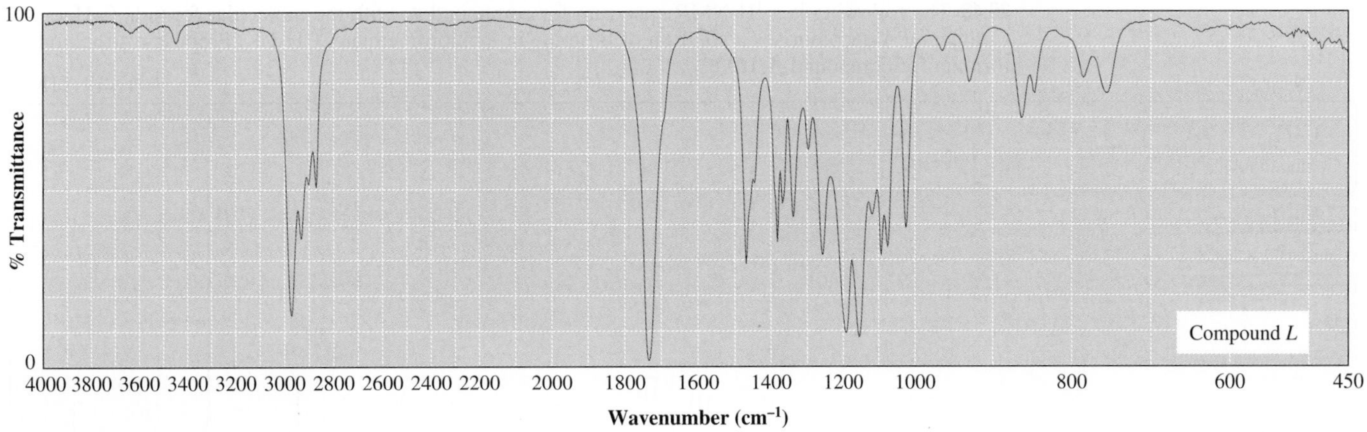

20.45 Examine the following IR spectrum and the molecular formula of compound *M*, C_3H_7NO. Determine: **LO 1,3,4**

(a) its index of hydrogen deficiency
(b) the number of rings or π bonds (or both) in compound *M*
(c) the oxygen- and nitrogen-containing functional group(s) in compound *M*.

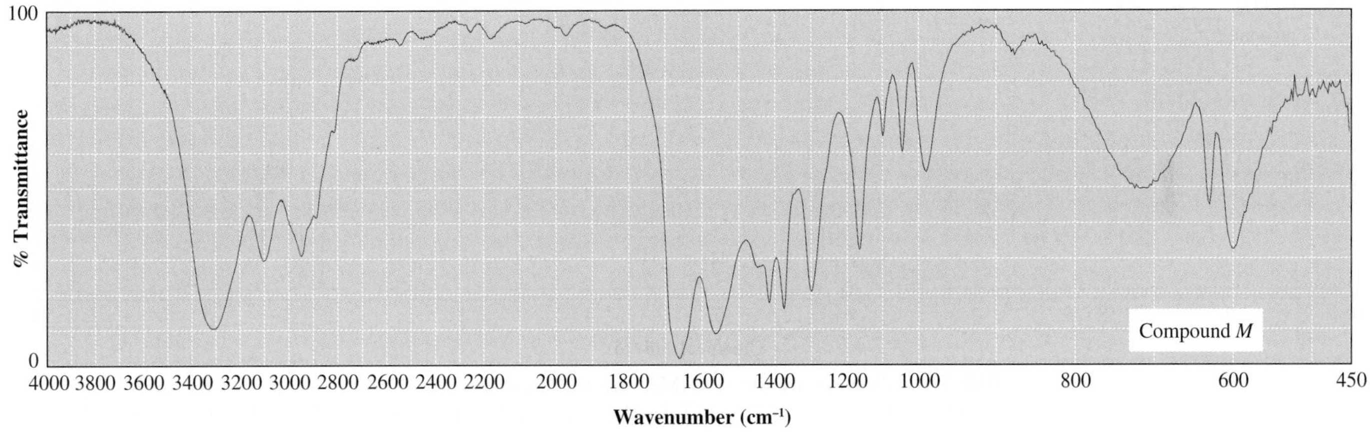

20.46 Show how IR spectroscopy can be used to distinguish between the following compounds. **LO 3,4**

(a) pentan-3-ol and pentan-3-one
(b) benzoic acid and phenol
(c) but-1-yne and but-2-yne
(d) hex-1-ene and hexanal
(e) propylamine and dipropylamine
(f) cyclohexane and cyclohexene

20.47 For each of the following pairs of compounds, list one major feature that appears in the IR spectrum of one compound but not the other. In your answer, state what type of bond vibration is responsible for the spectral feature you list, and give its approximate position in the IR spectrum. **LO 3,4**

(a) 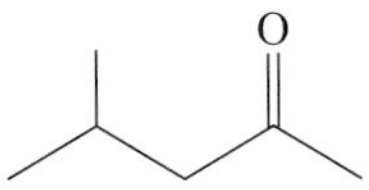 and 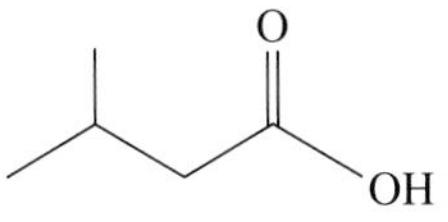

(b) 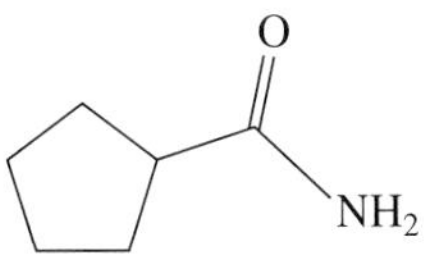 and 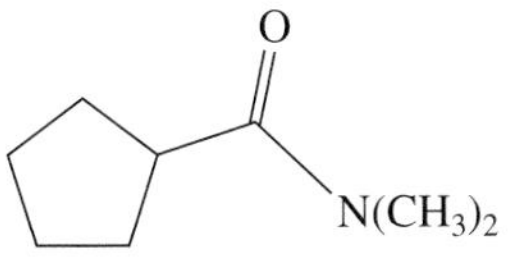

(c) 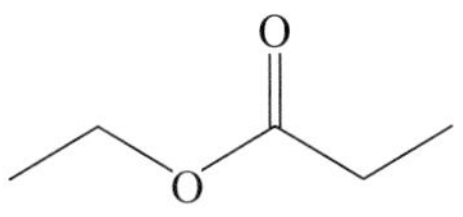 and 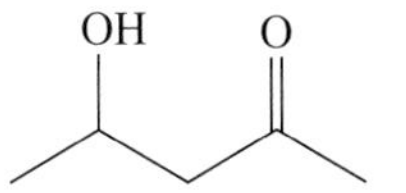

(d) 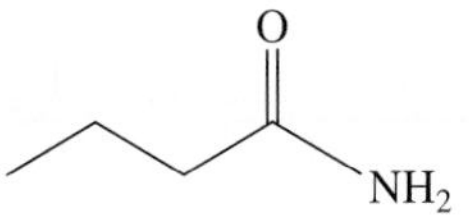 and 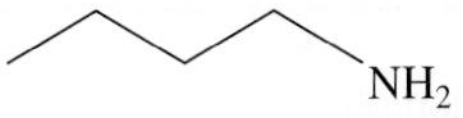

20.48 The following is a ^{1}H-NMR spectrum for compound *A*, with the molecular formula C_7H_{14}. Compound *A* decolourises a solution of bromine in dichloromethane, CH_2Cl_2. Propose a structural formula for compound *A*. **LO 5,6**

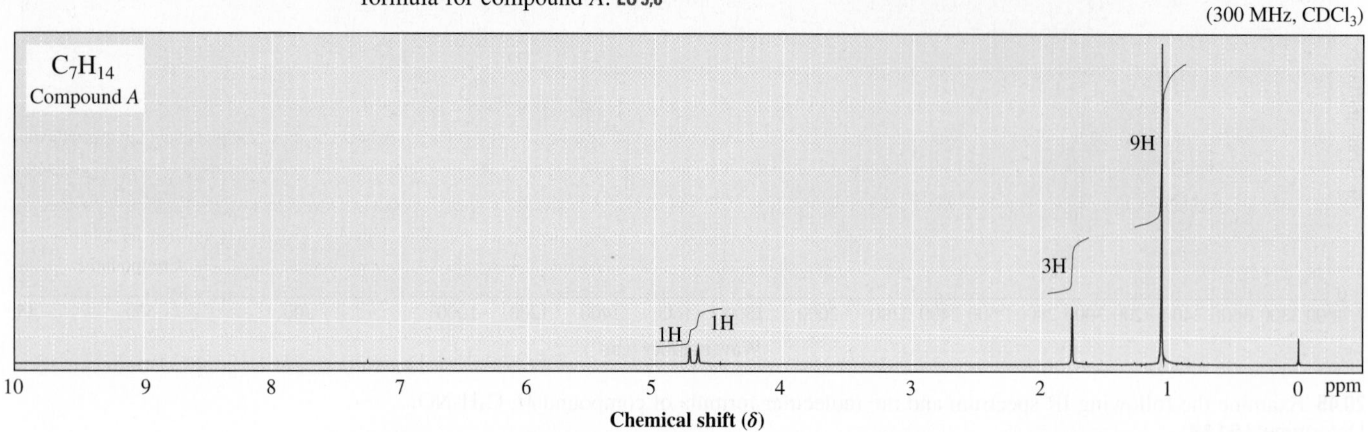

20.49 The following is a ^{1}H-NMR spectrum for compound *B*, with the molecular formula C_8H_{16}. Compound *B* decolourises a solution of Br_2 in CH_2Cl_2. Propose a structural formula for compound *B*. **LO 5,6**

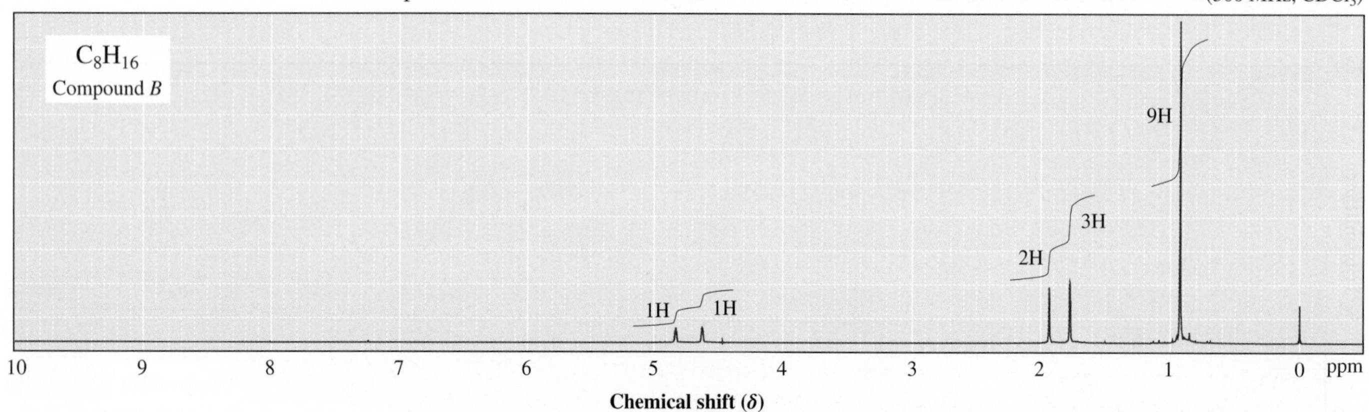

20.50 The following are the ^{1}H-NMR spectra of compounds *C* and *D*, both with the molecular formula C_4H_7Cl. Each compound decolourises a solution of Br_2 in CH_2Cl_2. Propose structural formulae for compounds *C* and *D*. **LO 5,6**

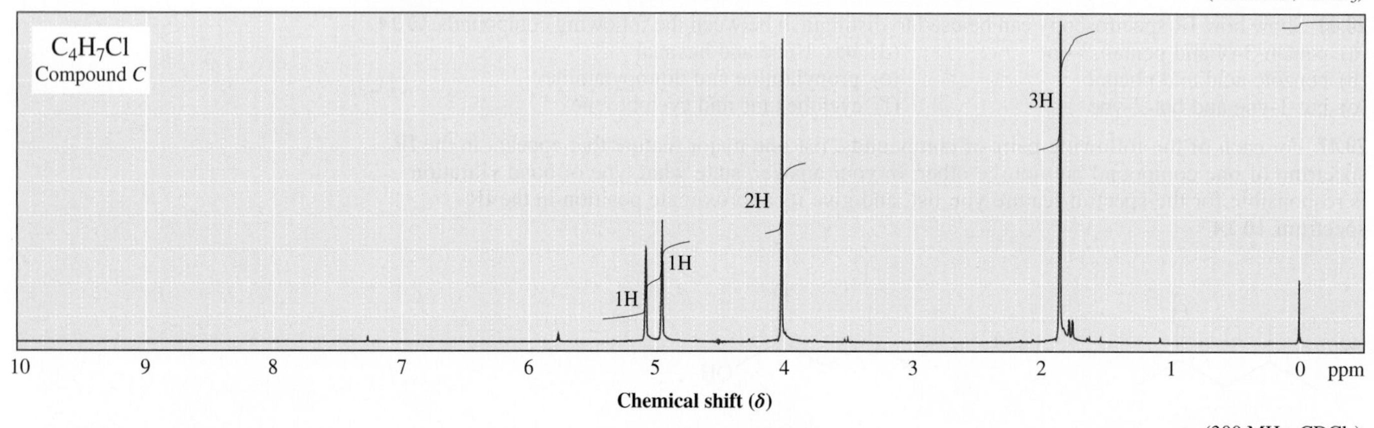

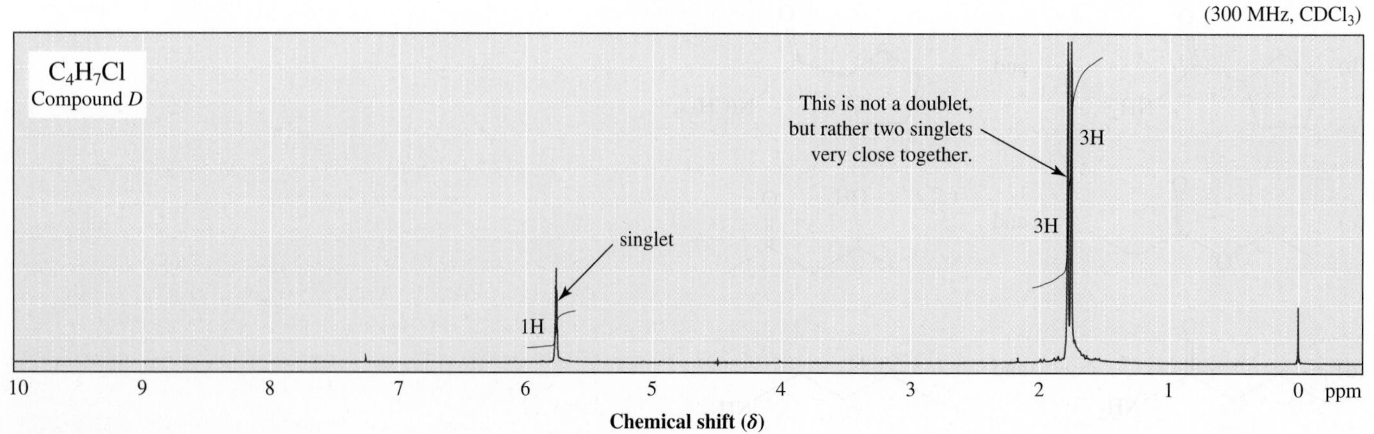

20.51 The following are the structural formulae of three alcohols with the molecular formula $C_7H_{16}O$ and three sets of ^{13}C-NMR spectral data. Assign each constitutional isomer to its correct spectral data. Provide a brief explanation for your selection. **LO 5,6**

(a) OH (c) OH

(b) OH

Spectrum 1	Spectrum 2	Spectrum 3
$\delta = 74.7$	$\delta = 71.0$	$\delta = 62.9$
30.5	43.7	32.8
7.7	29.2	31.9
	26.6	29.1
	23.3	25.8
	14.1	22.6
		14.1

20.52 Alcohol *E*, with the molecular formula $C_6H_{14}O$, undergoes acid-catalysed dehydration when it is warmed with phosphoric acid, giving compound *F*, with the molecular formula C_6H_{12}, as the major product. A ^1H-NMR spectrum of compound *E* shows peaks at δ 0.9 (t, 6H), 1.1 (s, 3H), 1.4 (s, 1H) and 1.5 (q, 4H). The ^{13}C-NMR spectrum of compound *E* shows peaks at δ 73.0, 33.7, 25.9 and 8.2. Propose structural formulae for compounds *E* and *F*. **LO 5,6**

20.53 Compound *G*, $C_6H_{14}O$, does not react with sodium metal and does not discharge the colour of Br_2 in CCl_4. The ^1H-NMR spectrum of compound *G* consists of only two signals: a 12H doublet at δ 1.1 and a 2H septet at δ 3.6. Propose a structural formula for compound *G*. **LO 5,6**

20.54 Propose a structural formula for each of the following haloalkanes. **LO 5,6**
(a) $C_2H_4Br_2$; δ 2.5 (d, 3H) and 5.9 (q, 1H)
(b) $C_4H_8Cl_2$; δ 1.7 (d, 6H) and 2.2 (q, 4H)
(c) $C_5H_8Br_4$; δ 3.6 (s, 8H)
(d) C_4H_9Br; δ 1.1 (d, 6H), 1.9 (m, 1H), and 3.4 (d, 2H)
(e) $C_5H_{11}Br$; δ 1.1 (s, 9H) and 3.2 (s, 2H)
(f) $C_7H_{15}Cl$; δ 1.1 (s, 9H) and 1.6 (s, 6H)

20.55 The following are structural formulae for esters (a), (b) and (c) and three ^1H-NMR spectra. Assign each compound to its correct spectrum and assign all signals to their corresponding hydrogen atoms. Provide a brief explanation for your selection. **LO 5,6**

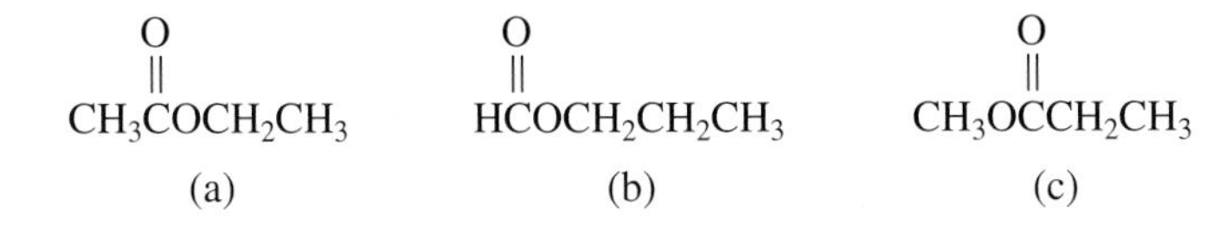

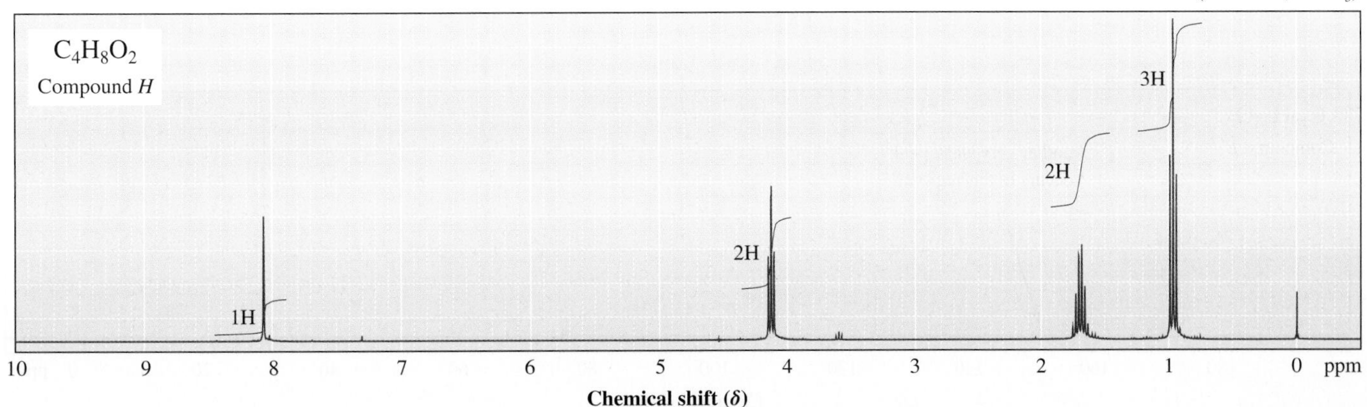

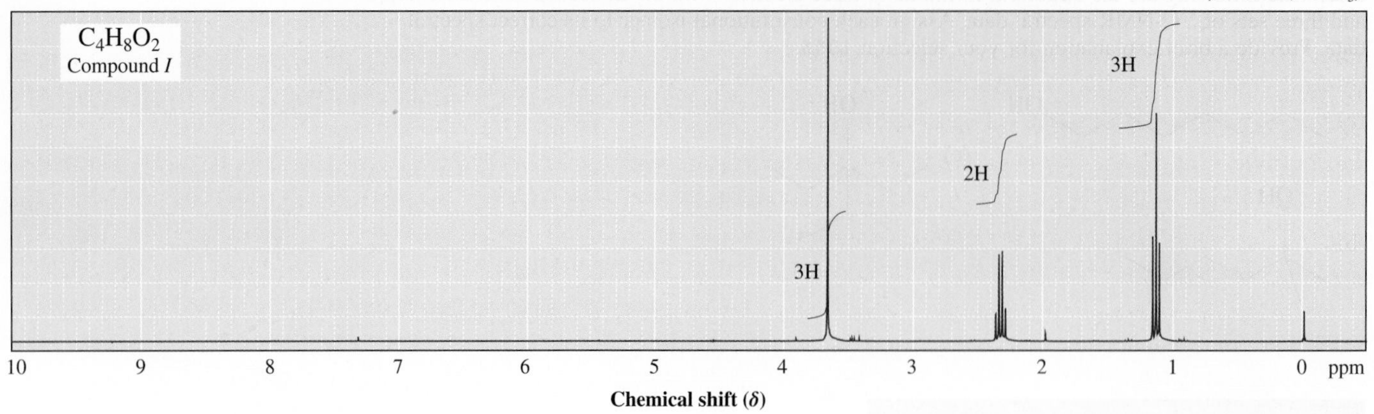

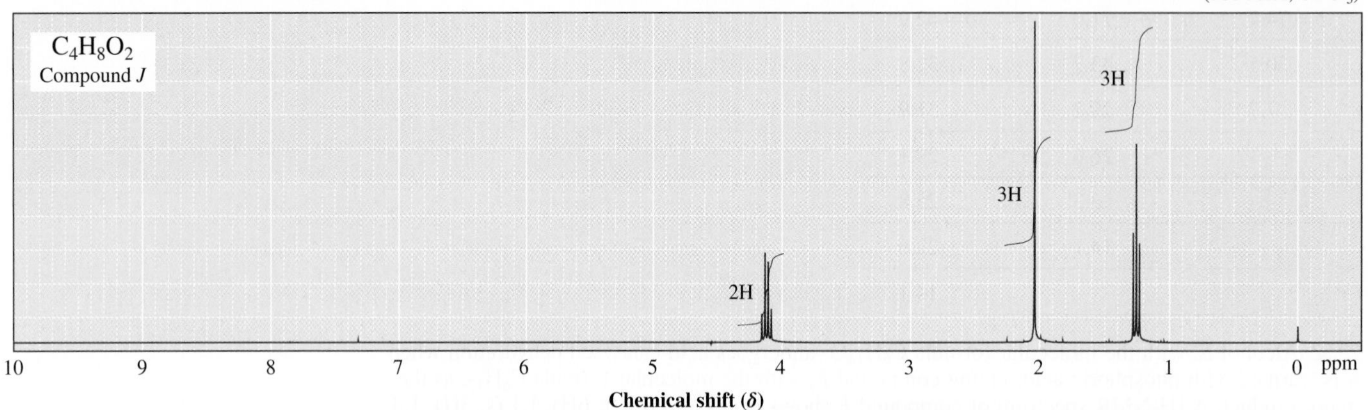

20.56 Compound *K*, $C_{10}H_{10}O_2$, is not soluble in water, 10% NaOH or 10% HCl. A ^{1}H-NMR spectrum of compound *K* shows signals at δ 2.6 (s, 6H) and 8.0 (s, 4H). A ^{13}C-NMR spectrum of compound *K* shows four signals. From this information, propose a structural formula for *K*. **LO 5,6**

20.57 Compound *L*, $C_{15}H_{24}O$, is used as an antioxidant in many commercial food products, synthetic rubbers and petroleum products. Propose a structural formula for compound *L* based on the following ^{1}H-NMR and ^{13}C-NMR spectra. **LO 5,6**

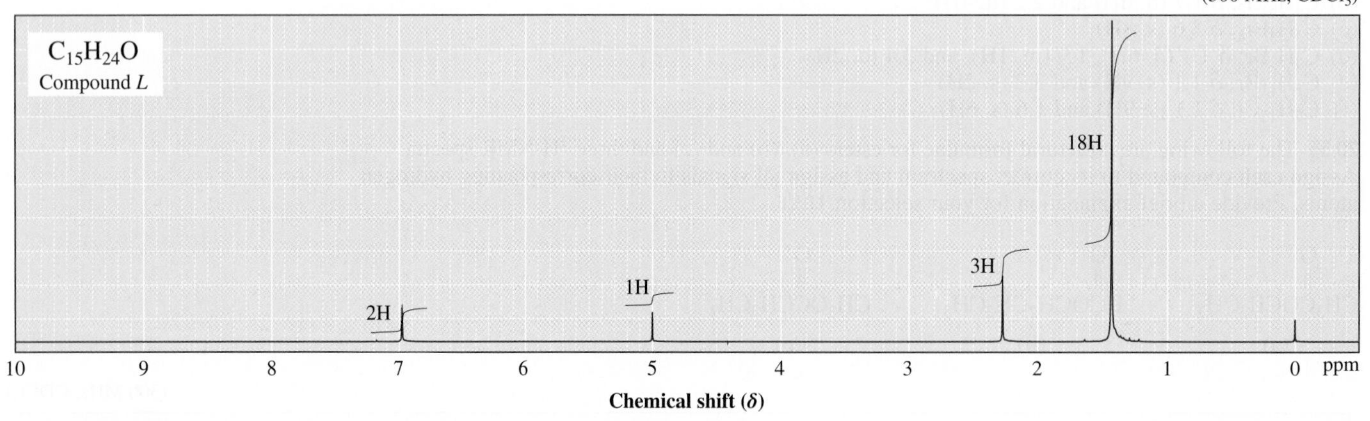

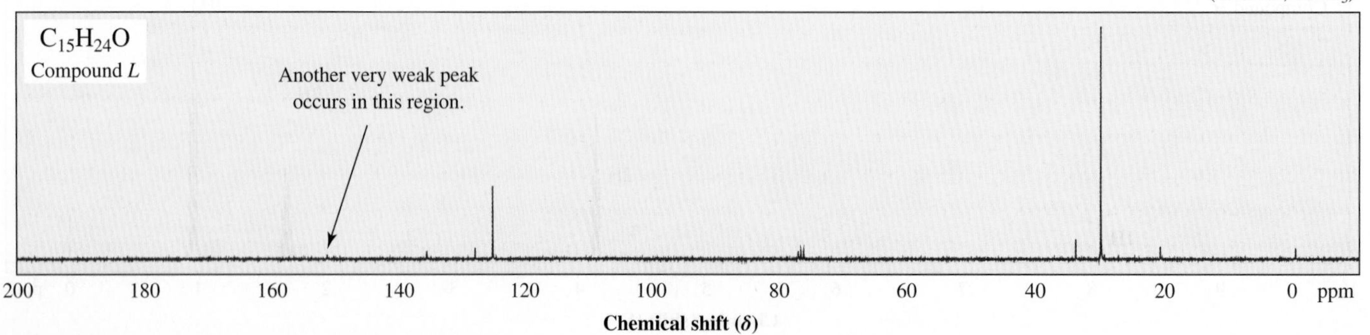

20.58 A compound with molecular formula $C_5H_8Cl_4$ exhibits a ^{1}H-NMR spectrum with only one signal and a ^{13}C-NMR spectrum with two signals. Propose a structure for this compound. **LO 5,6**

20.59 A compound with molecular formula $C_{12}H_{18}$ exhibits a ^{1}H-NMR spectrum with only one signal and a ^{13}C-NMR spectrum with two signals. Deduce the structure of this compound. **LO 5,6**

20.60 A compound with molecular formula $C_8H_{10}O$ produces six signals in its ^{13}C-NMR spectrum and exhibits the following ^{1}H-NMR spectrum. Deduce the structure of the compound. **LO 5,6**

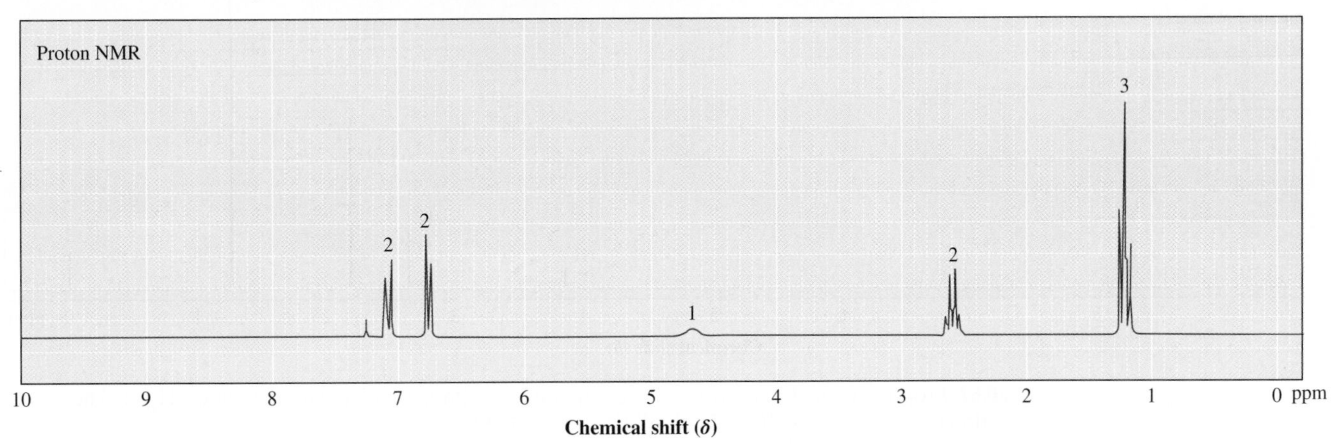

20.61 Deduce the structure of a compound with molecular formula C_9H_{12} that produces the following ^{1}H-NMR spectrum. **LO 5,6**

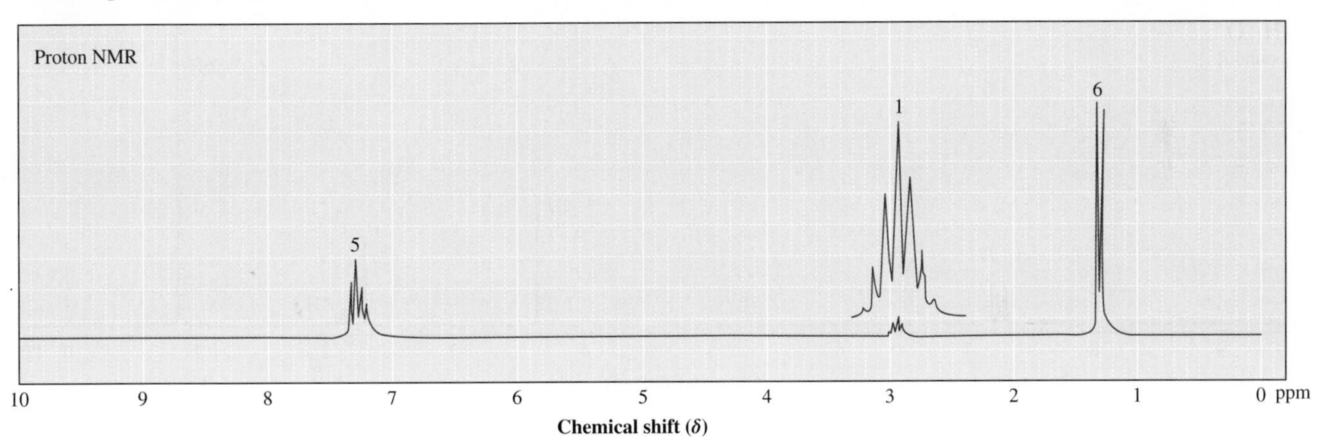

20.62 Propose a structural formula for each of the following compounds, which all contain an aromatic ring. **LO 5,6**

(a) $C_9H_{10}O$; δ 1.2 (t, 3H), 3.0 (q, 2H) and 7.4–8.0 (m, 5H)
(b) $C_{10}H_{12}O_2$; δ 2.2 (s, 3H), 2.9 (t, 2H), 4.3 (t, 2H) and 7.3 (s, 5H)
(c) $C_{10}H_{14}$; δ 1.2 (d, 6H), 2.3 (s, 3H), 2.9 (septet, 1H) and 7.0 (s, 4H)
(d) C_8H_9Br; δ 1.8 (d, 3H), 5.0 (q, 1H) and 7.3 (s, 5H)

20.63 Compound *M*, with the molecular formula $C_9H_{12}O$, readily undergoes acid-catalysed dehydration to give compound *N*, with the molecular formula C_9H_{10}. A ^{1}H-NMR spectrum of compound *M* shows signals at δ 0.9 (t, 3H), 1.8 (m, 2H), 2.3 (s, 1H), 4.6 (t, 1H) and 7.5 (m, 5H). From this information, propose structural formulae for compounds *M* and *N*. **LO 5,6**

20.64 Propose a structural formula for each of the following ketones. **LO 5,6**

(a) C_4H_8O; δ 1.0 (t, 3H), 2.1 (s 3H) and 2.4 (q, 2H)
(b) $C_7H_{14}O$; δ 0.9 (t, 6H), 1.6 (sextet, 4H) and 2.4 (t, 4H)

20.65 Propose a structural formula for compound *O*, a ketone with the molecular formula $C_{10}H_{12}O$ and the ^{1}H-NMR spectrum shown below. **LO 5,6**

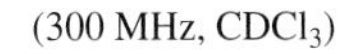

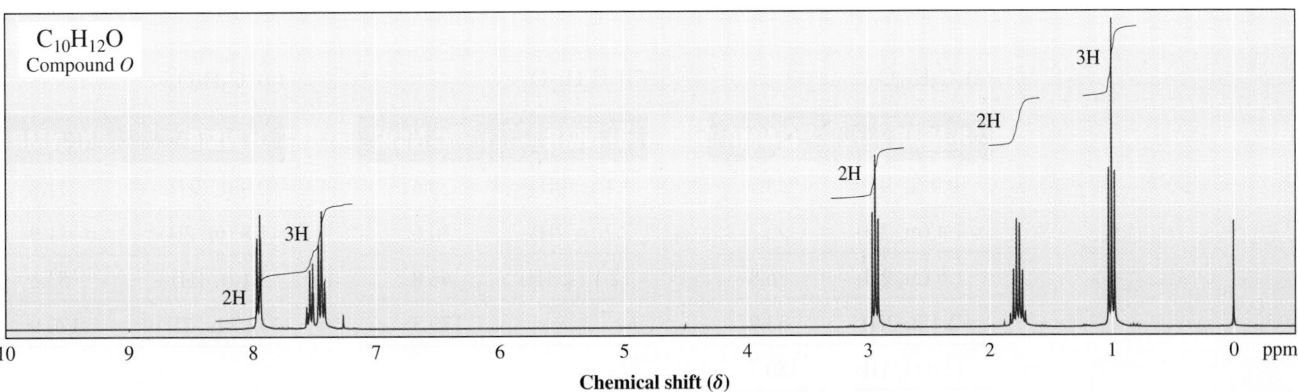

20.66 The following is a ^{1}H-NMR spectrum of compound *P*, with the molecular formula $C_6H_{12}O_2$. Compound *P* undergoes acid-catalysed dehydration to give compound *Q*, $C_6H_{10}O$. Propose structural formulae for compounds *P* and *Q*. **LO 5,6**

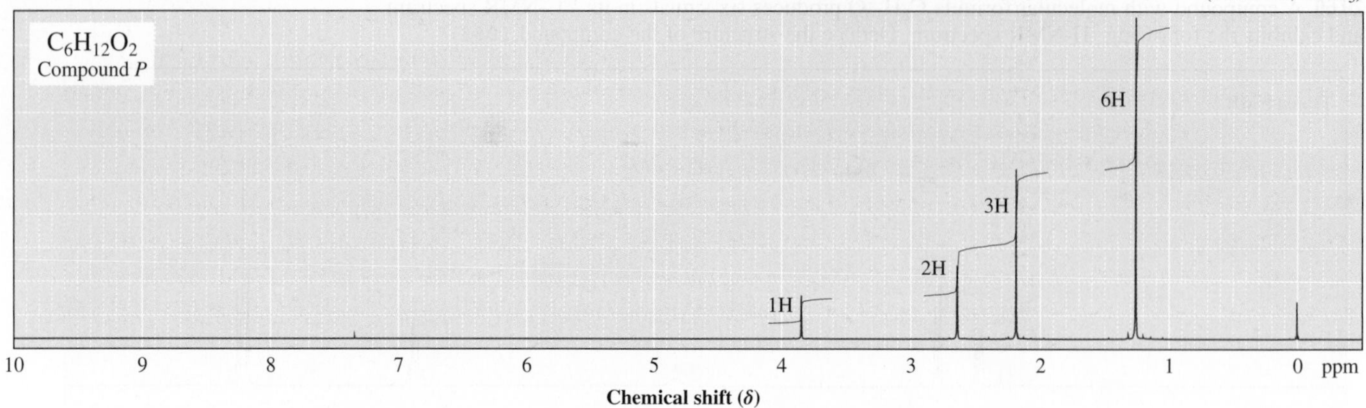

20.67 Propose a structural formula for compound *R*, with the molecular formula $C_{12}H_{16}O$. The following are its ^{1}H-NMR and ^{13}C-NMR spectra. **LO 5,6**

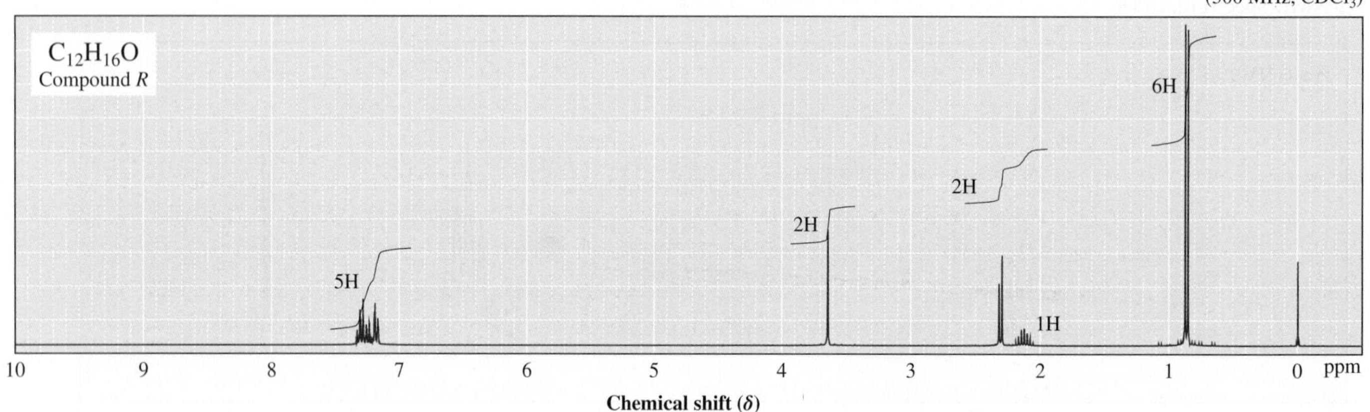

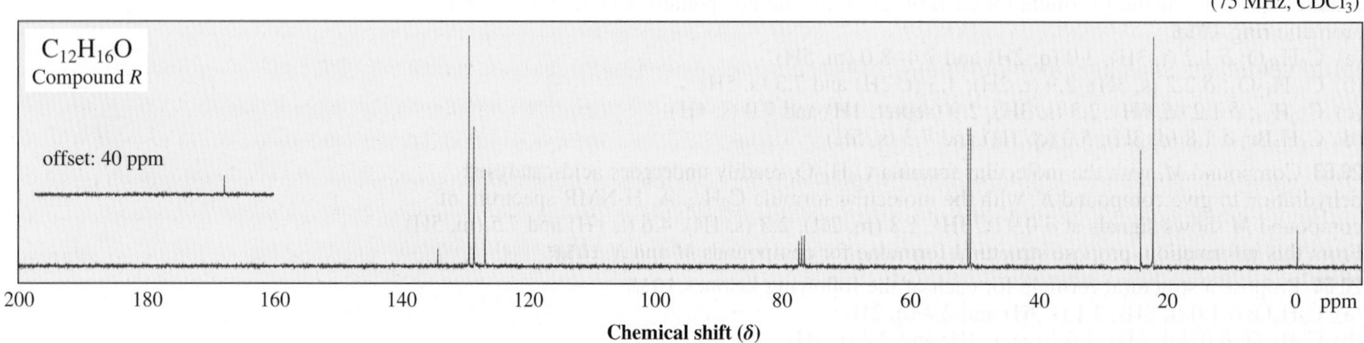

20.68 Propose a structural formula for each of the following carboxylic acids with the NMR signals given in the tables. **LO 5,6**

(a) $C_5H_{10}O_2$

^{1}H-NMR	^{13}C-NMR
0.9 (t, 3H)	13.7
1.4 (m, 2H)	22.2
1.6 (m, 2H)	26.8
2.4 (t, 2H)	33.9
12.0 (s, 1H)	180.7

(b) $C_6H_{12}O_2$

^{1}H-NMR	^{13}C-NMR
1.1 (s, 9H)	29.6
2.2 (s, 2H)	30.6
12.1 (s, 1H)	46.8
	179.3

(c) $C_5H_8O_4$

^{1}H-NMR	^{13}C-NMR
0.9 (t, 3H)	11.8
1.8 (m, 2H)	21.9
3.1 (t, 1H)	53.3
12.7 (s, 2H)	170.9

20.69 The following are ^{1}H-NMR and ^{13}C-NMR spectra of compound *S*, with the molecular formula $C_7H_{14}O_2$. Propose a structural formula for compound *S*. **LO 5,6**

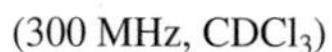

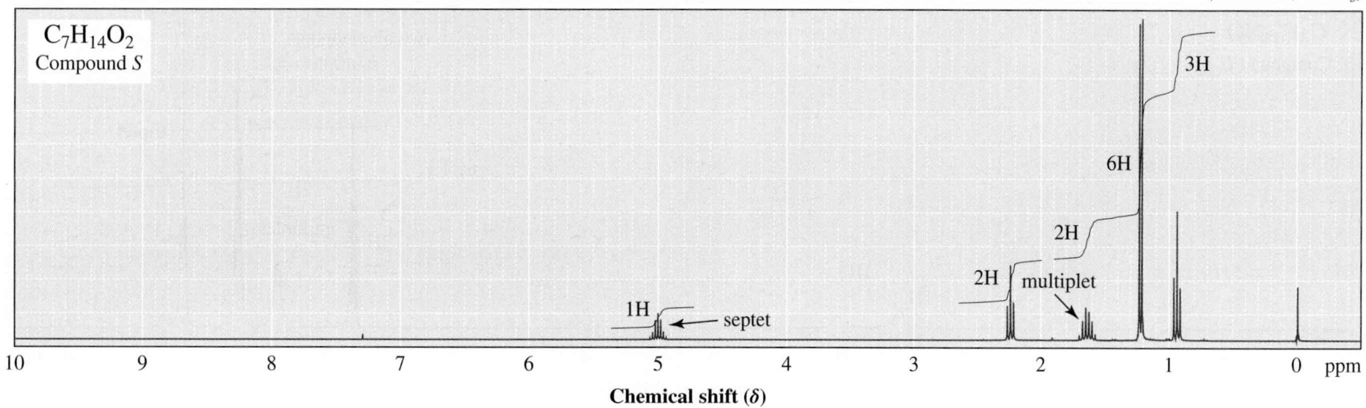

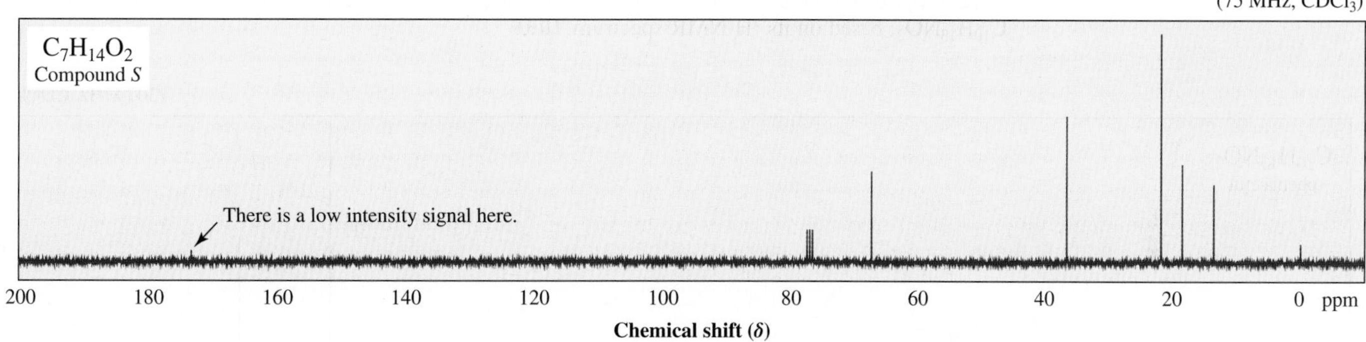

20.70 The following are ^{1}H-NMR and ^{13}C-NMR spectra of compound *T*, with the molecular formula $C_{10}H_{15}NO$. Propose a structural formula for this compound. **LO 5,6**

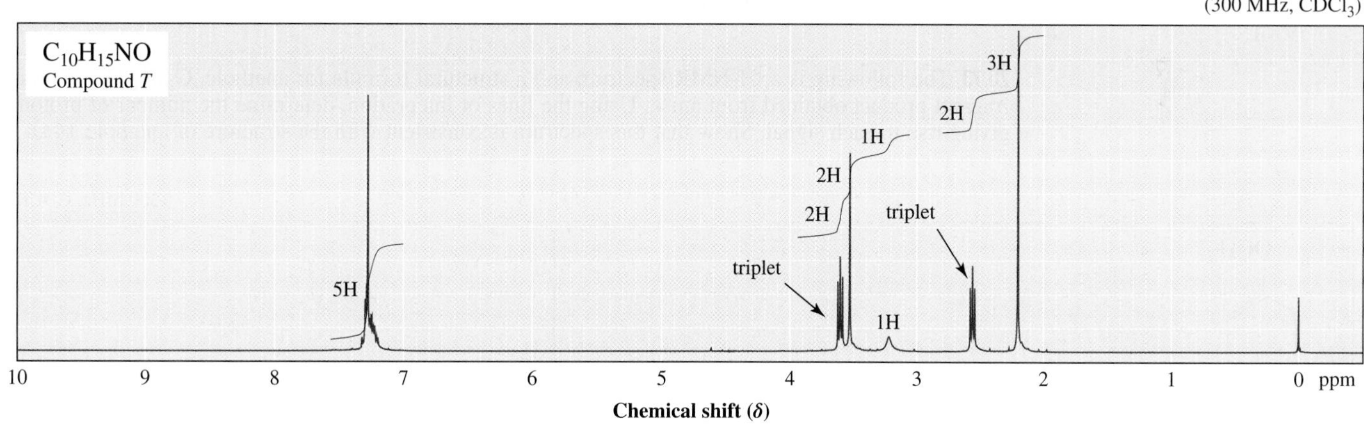

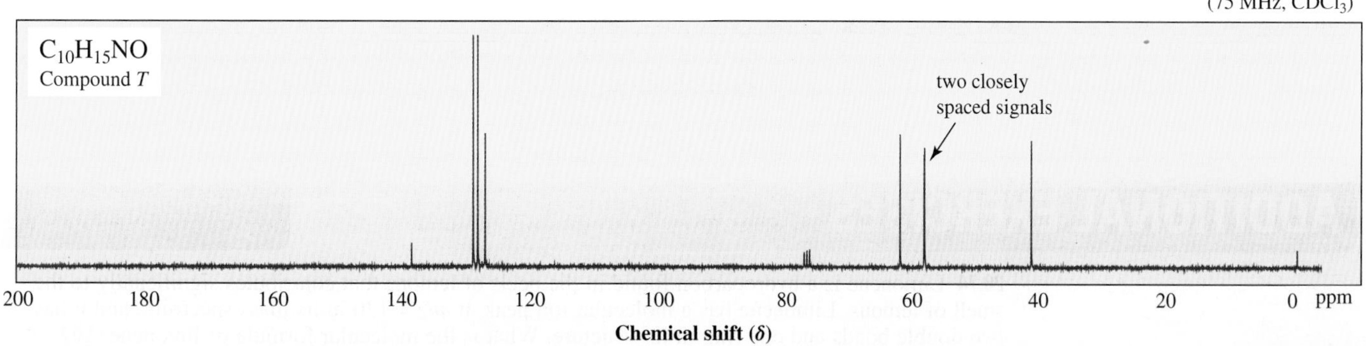

20.71 Propose a structural formula for amide *U*, with the molecular formula $C_6H_{13}NO$ and the ^{1}H-NMR spectrum shown below. **LO 5,6**

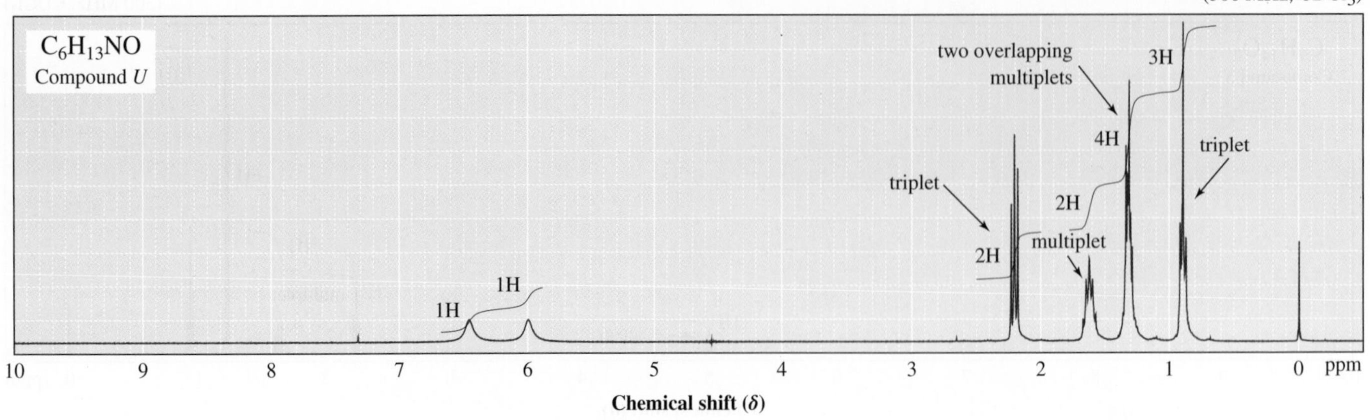

20.72 Propose a structural formula for the analgesic phenacetin, with the molecular formula $C_{10}H_{13}NO_2$, based on its ^{1}H-NMR spectrum. **LO 5,6**

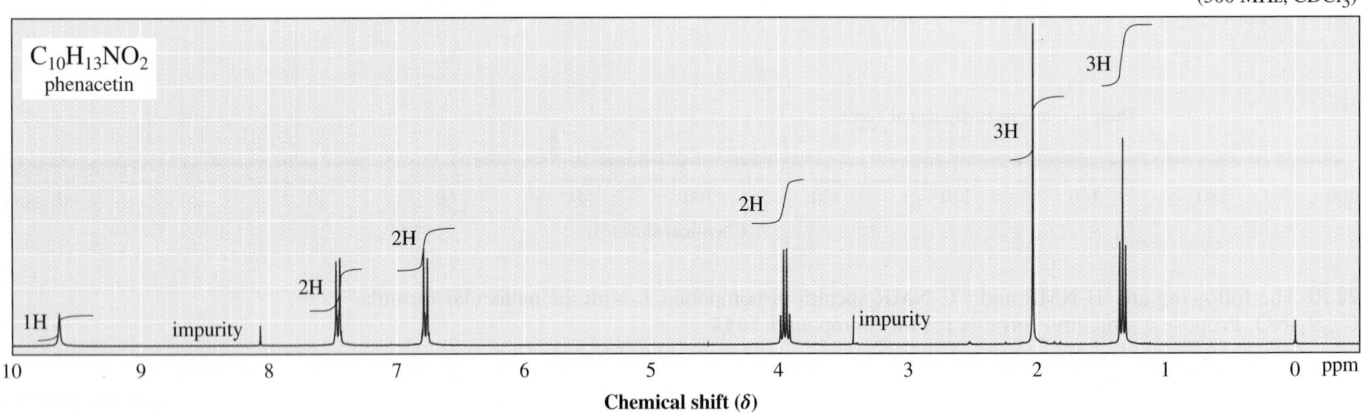

20.73 The following is a ^{1}H-NMR spectrum and a structural formula for anethole, $C_{10}H_{12}O$, a natural fragrant product obtained from anise. Using the lines of integration, determine the number of protons giving rise to each signal. Show that this spectrum is consistent with the structure of anethole. **LO 5,6**

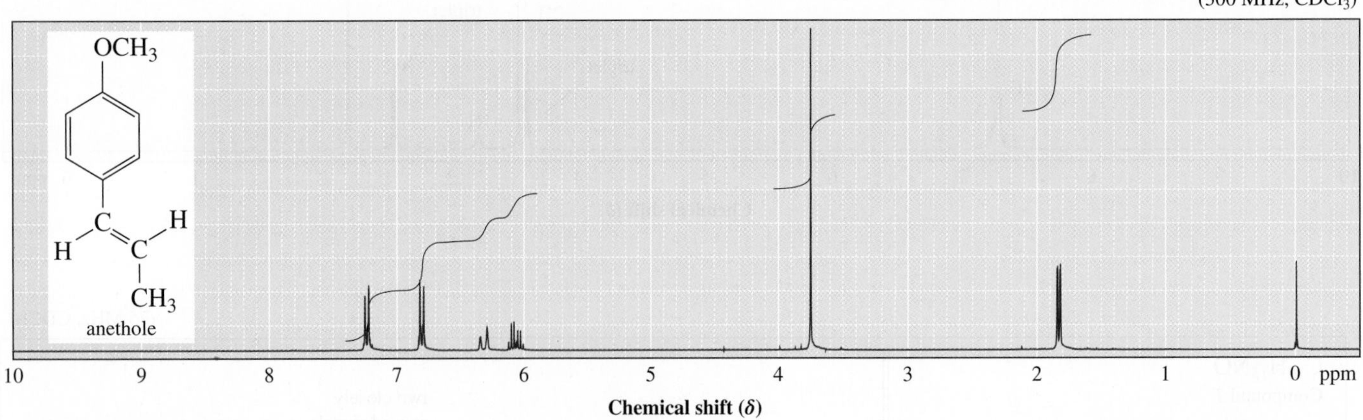

ADDITIONAL EXERCISES

20.74 Limonene is a hydrocarbon found in the peels of lemons that contributes significantly to the smell of lemons. Limonene has a molecular ion peak at $m/z = 136$ in its mass spectrum, and it has two double bonds and one ring in its structure. What is the molecular formula of limonene? **LO 2**

20.75 Following are the IR spectrum and mass spectrum of an unknown compound. Propose at least two possible structures for the unknown compound. **LO 2,3,4**

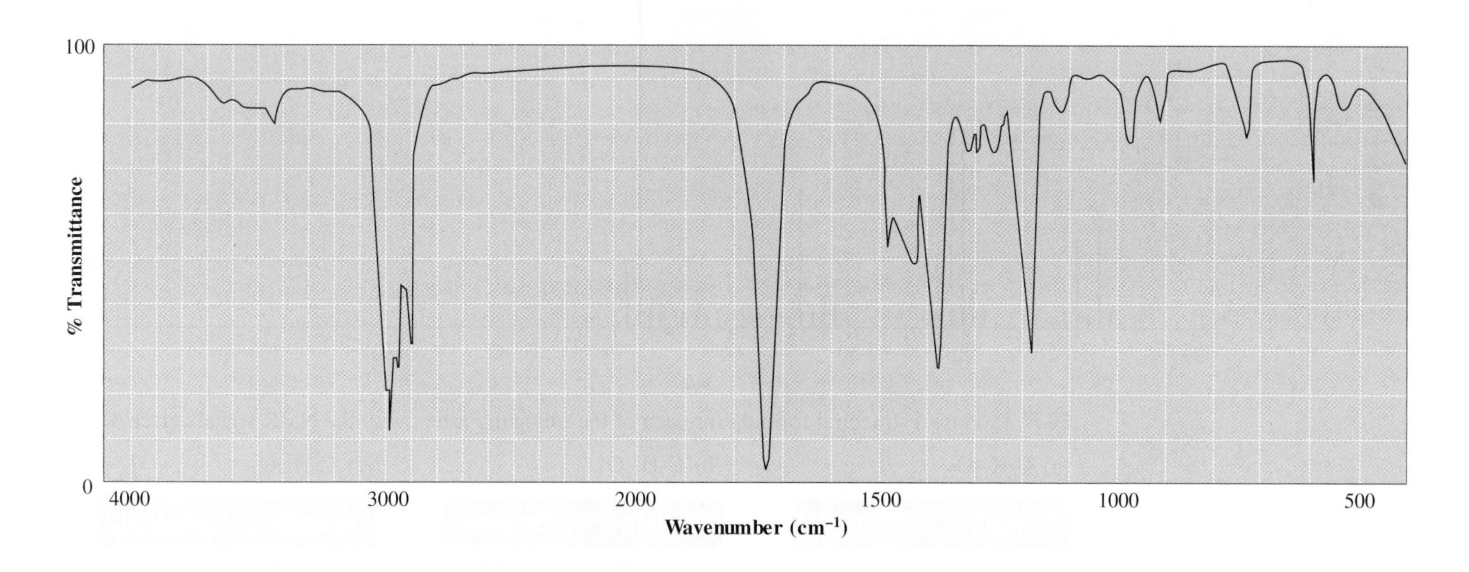

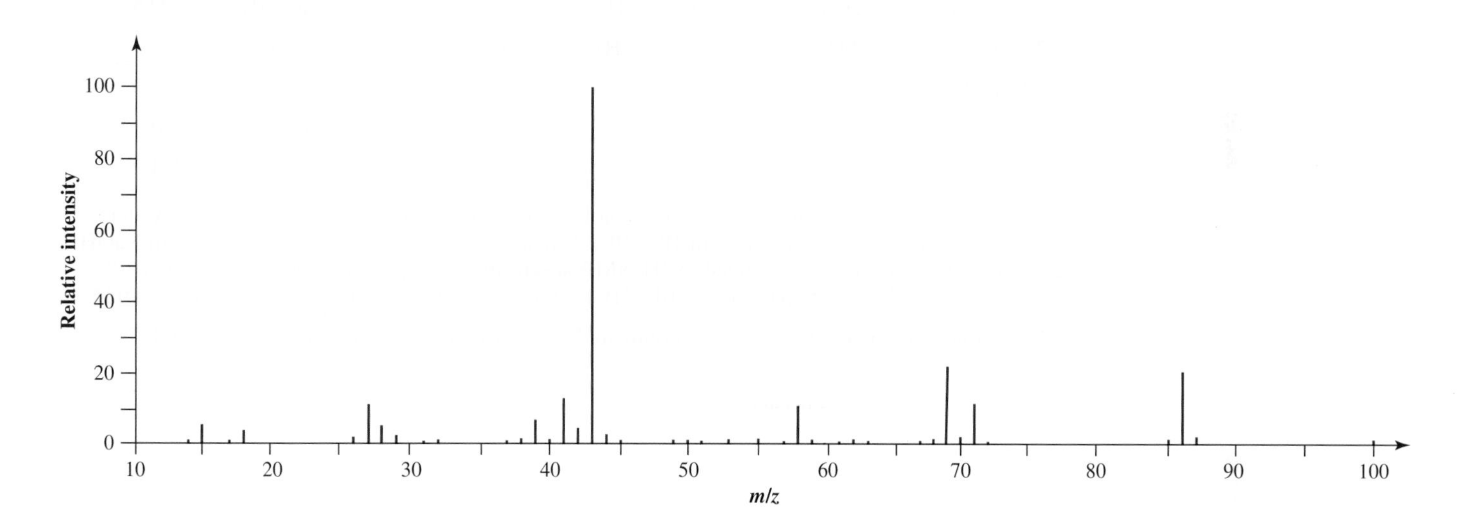

20.76 Following are the IR spectrum and mass spectrum of an unknown compound. Propose at least two possible structures for the unknown compound. **LO 2,3,4**

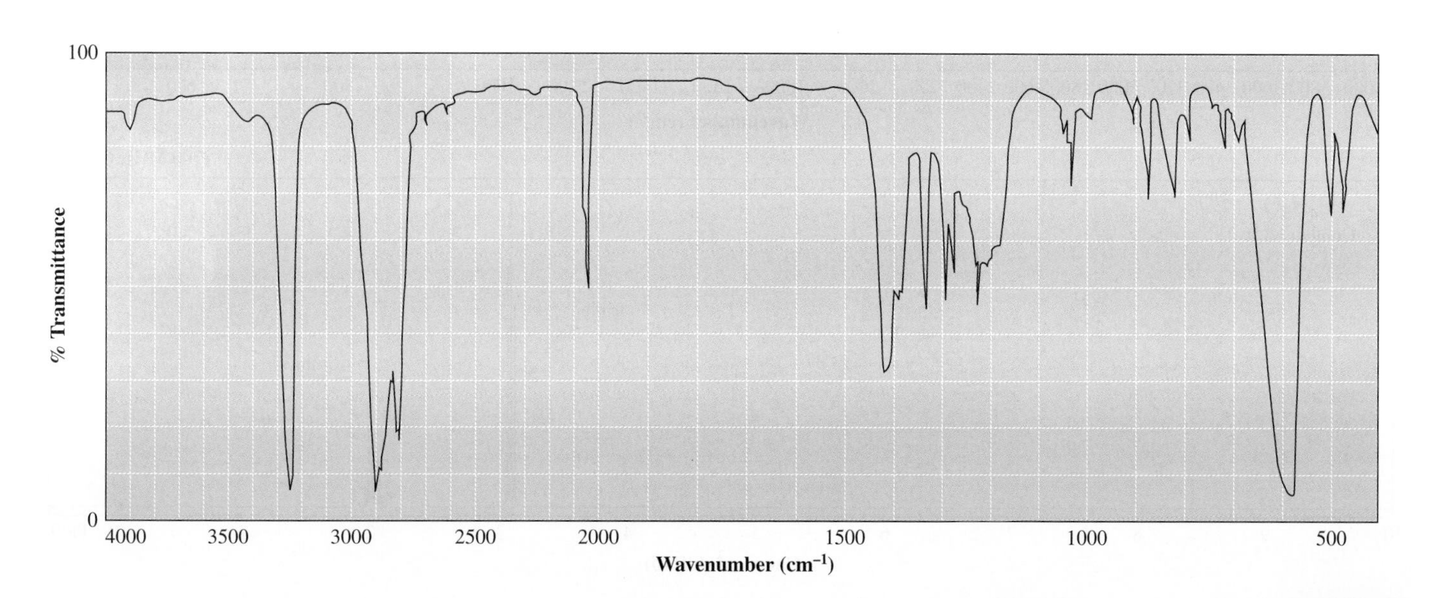

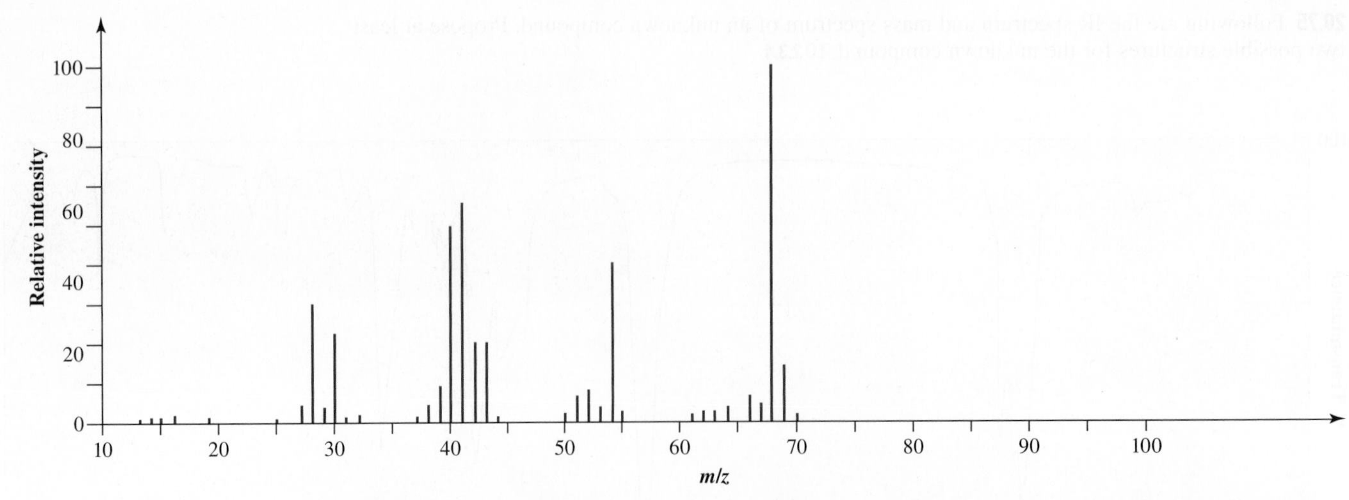

20.77 Propose a structural formula for each of the following esters with the NMR signals given. **LO 5,6**

(a) $C_6H_{12}O_2$

^{1}H-NMR	^{13}C-NMR
1.2 (d, 6H)	14.3
1.3 (t, 3H)	20.0
2.5 (m, 1H)	34.0
4.1 (q, 2H)	60.2
	177.2

(b) $C_7H_{12}O_4$

^{1}H-NMR	^{13}C-NMR
1.3 (t, 6H)	14.1
3.4 (s, 2H)	41.7
4.2 (q, 4H)	61.4
	166.5

(c) $C_7H_{14}O_2$

^{1}H-NMR	^{13}C-NMR
0.9 (d, 6H)	21.1
1.5 (m, 2H)	22.5
1.7 (m, 1H)	25.1
2.1 (s, 3H)	37.3
4.1 (t, 2H)	63.1
	171.2

20.78 Compound *V* is an oily liquid with the molecular formula $C_8H_9NO_2$. It is insoluble in water and aqueous NaOH, but it dissolves in 10% HCl. When its solution in HCl is neutralised with NaOH, compound *V* is recovered unchanged. A ^{1}H-NMR spectrum of compound *V* shows signals at δ 3.8 (s, 3H), 4.2 (s, 2H), 7.6 (d, 2H) and 8.7 (d, 2H). Propose a structural formula for compound *V*. **LO 5,6**

20.79 Propose a structural formula for compound *W*, C_4H_6O, based on the following IR and ^{1}H-NMR spectra. **LO 5,6**

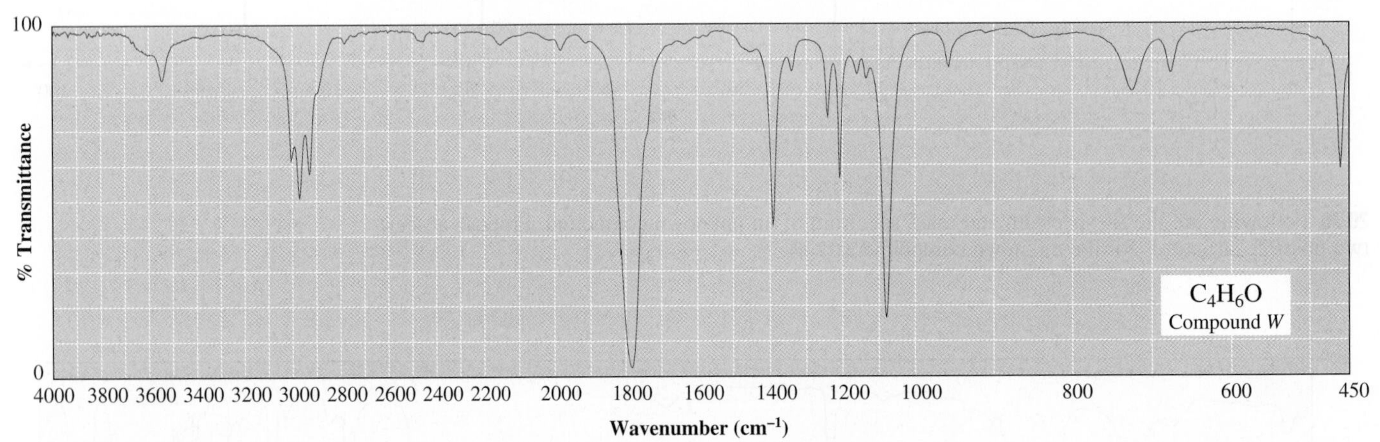

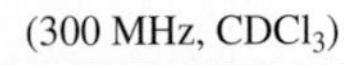

(300 MHz, $CDCl_3$)

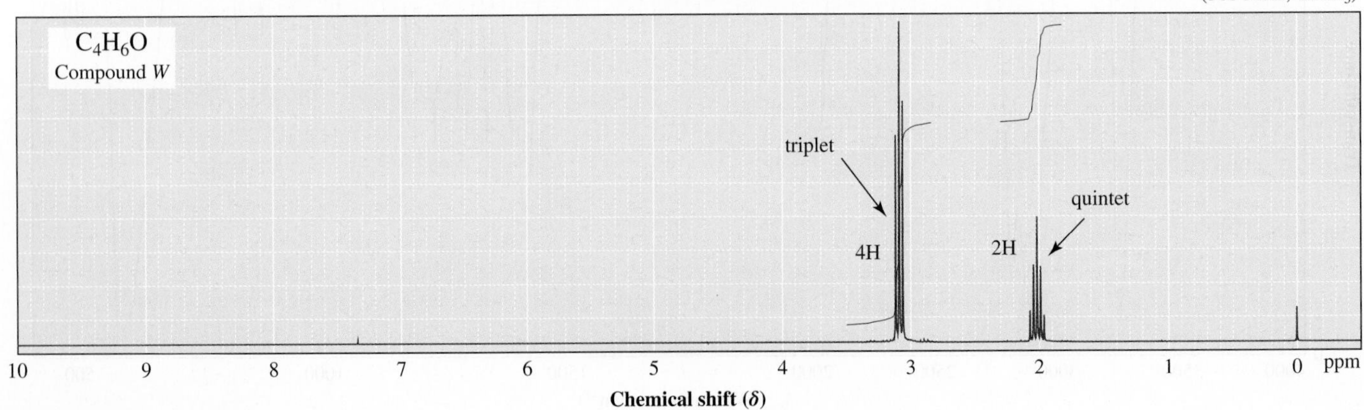

20.80 Propose a structural formula for compound *X*, $C_5H_{10}O_2$, based on the following IR and ^{1}H-NMR spectra.

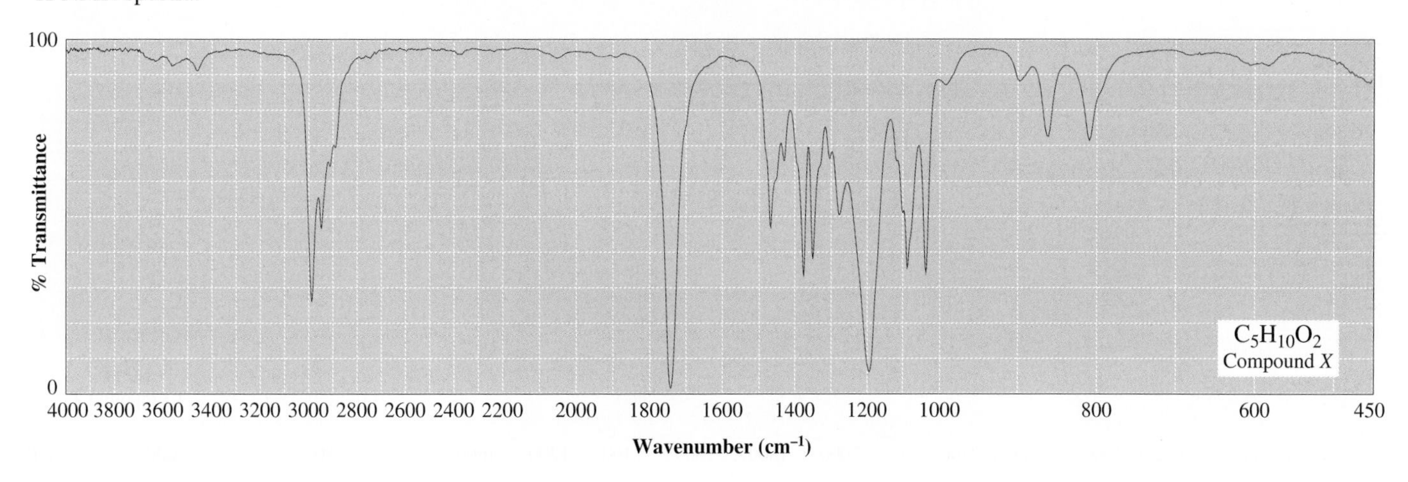

(300 MHz, $CDCl_3$)

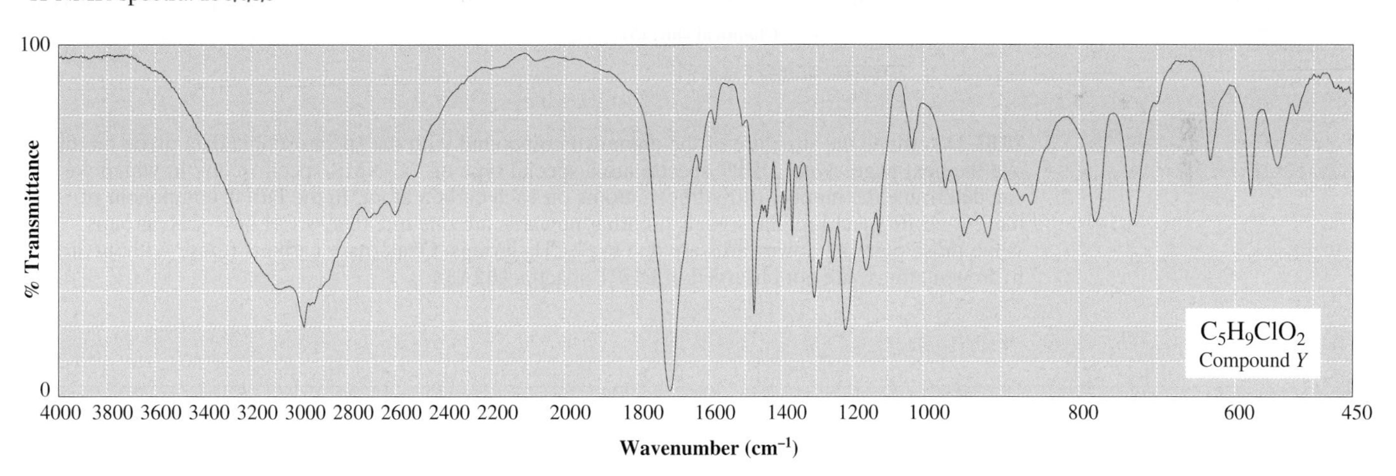

20.81 Propose a structural formula for compound *Y*, $C_5H_9ClO_2$, based on the following IR and ^{1}H-NMR spectra. **LO 3,4,5,6**

(300 MHz, $CDCl_3$)

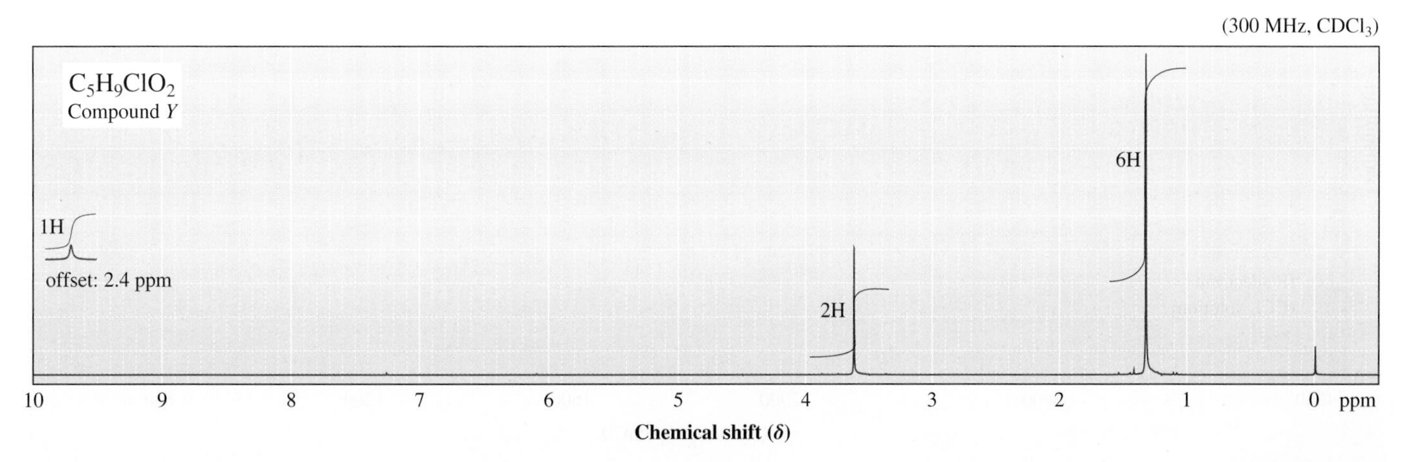

20.82 Propose a structural formula for compound *Z*, $C_6H_{14}O$, based on the following IR and ^{1}H-NMR spectra. **LO 3,4,5,6**

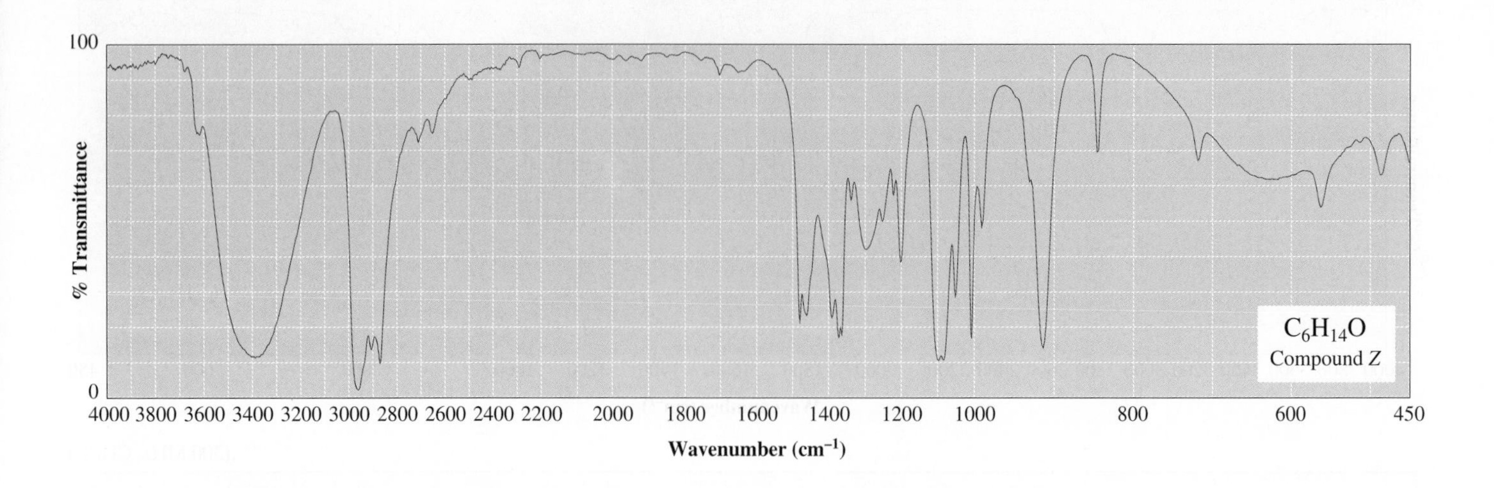

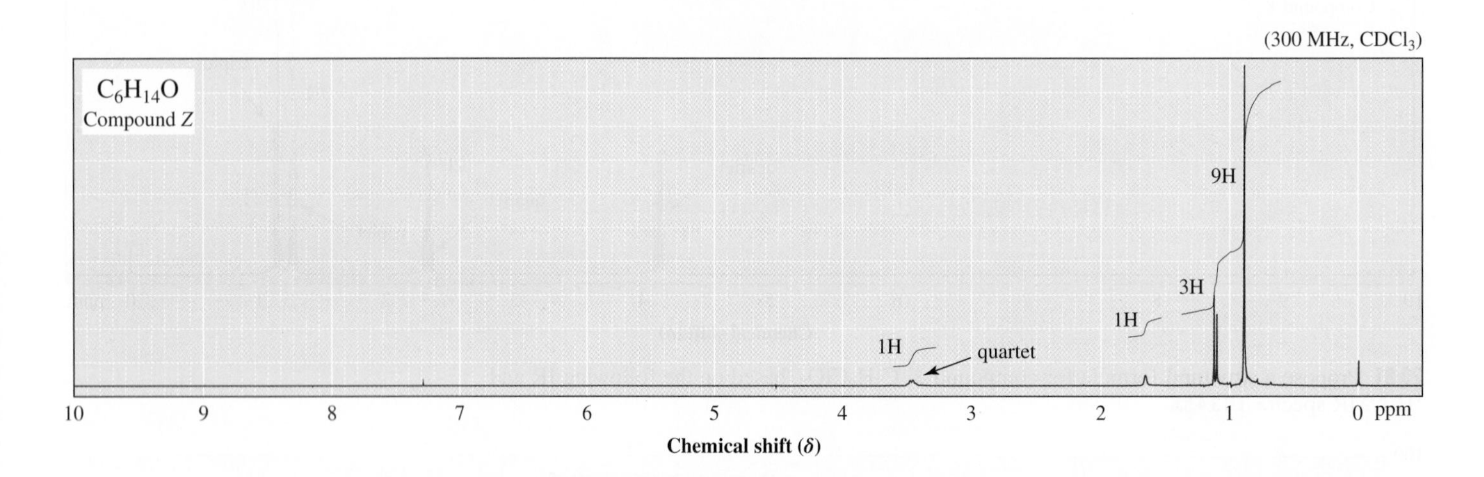

20.83 Determine the structure of the unknown compound with the spectroscopic data given on this and the next page. *Note:* DEPT spectra are a special type of ^{13}C-NMR spectroscopy in which we can determine the number of hydrogen atoms on each carbon atom. In the DEPT component of the ^{13}C-NMR spectrum, the signals pointing upwards are due to either —CH or —CH_3 groups, while those pointing downwards are due to —CH_2 groups. Quaternary carbons (those without any hydrogen atoms) are not observed in DEPT spectra. **LO 3,4,5,6**

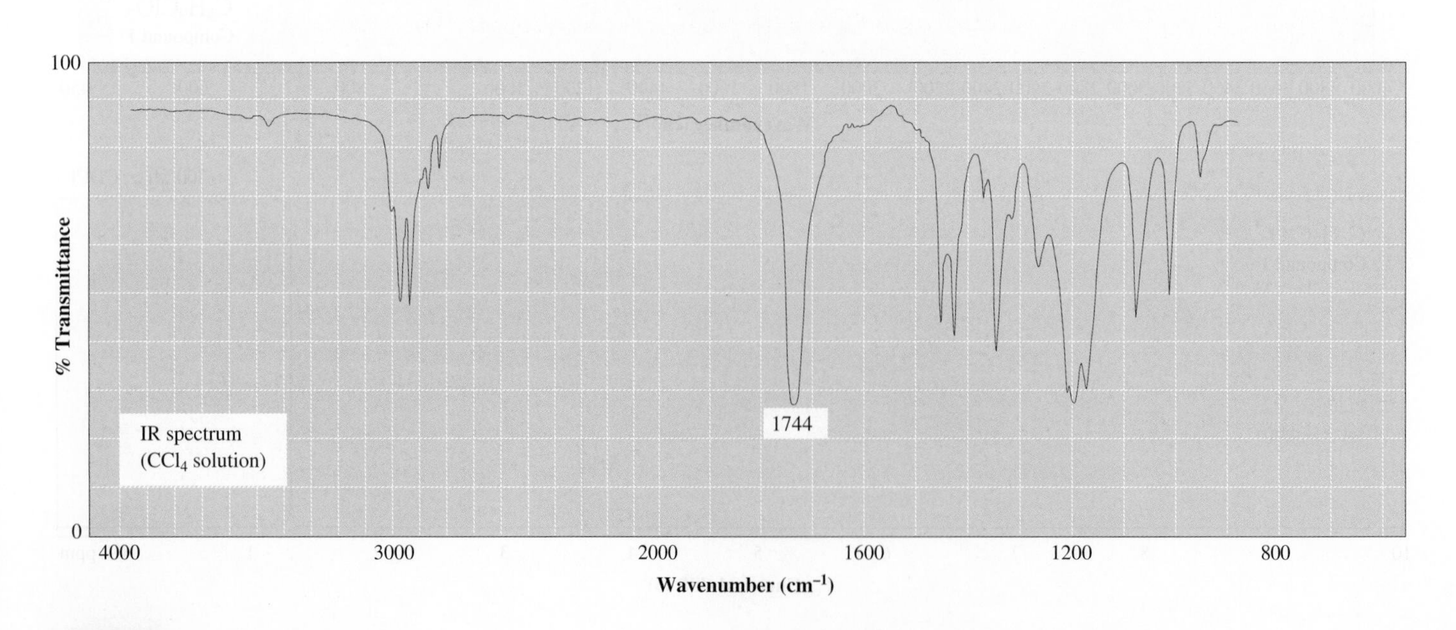

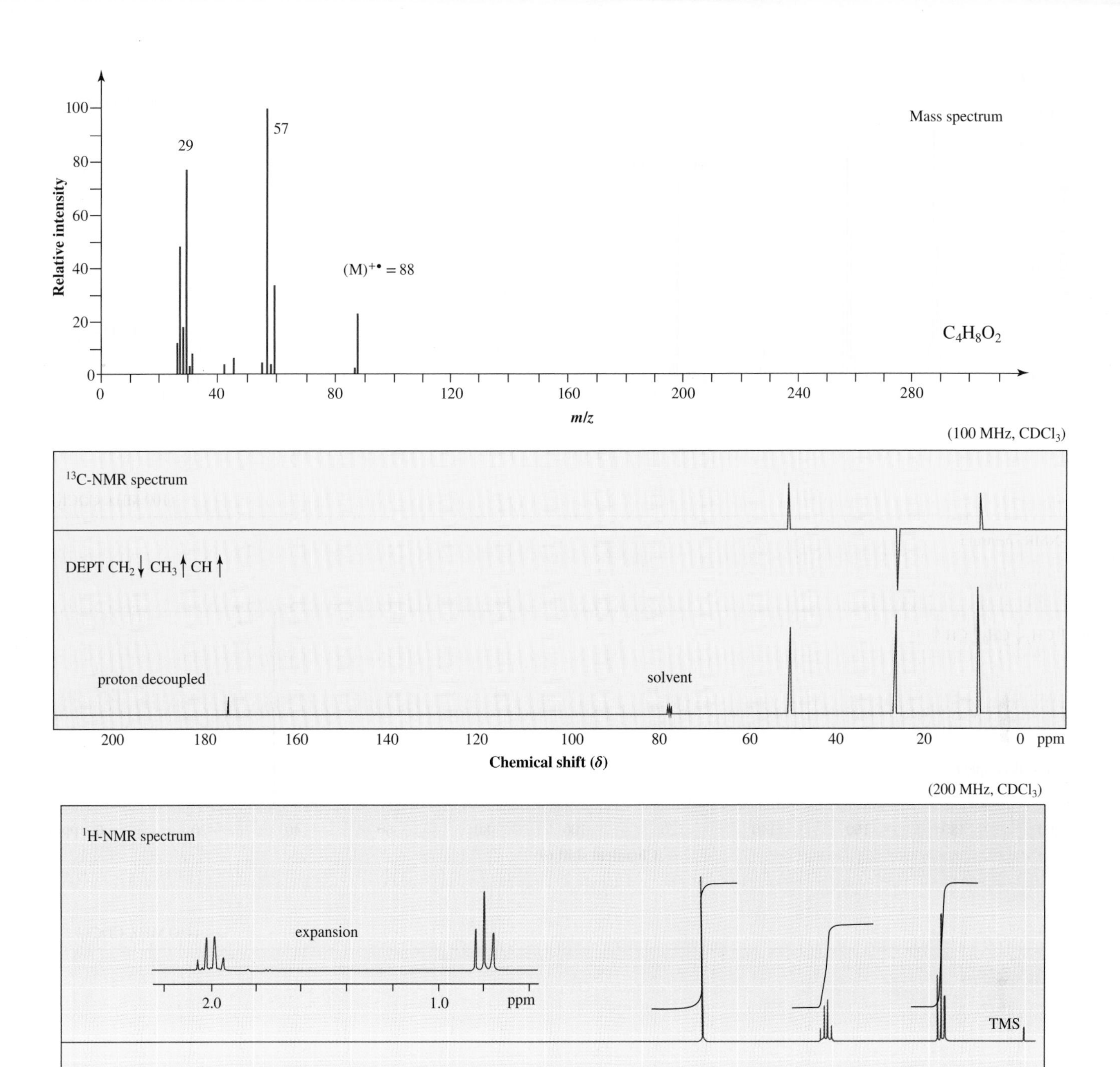

20.84 Determine the structure of the unknown compound with the spectroscopic data on this and the next page. **LO 3,4,5,6**

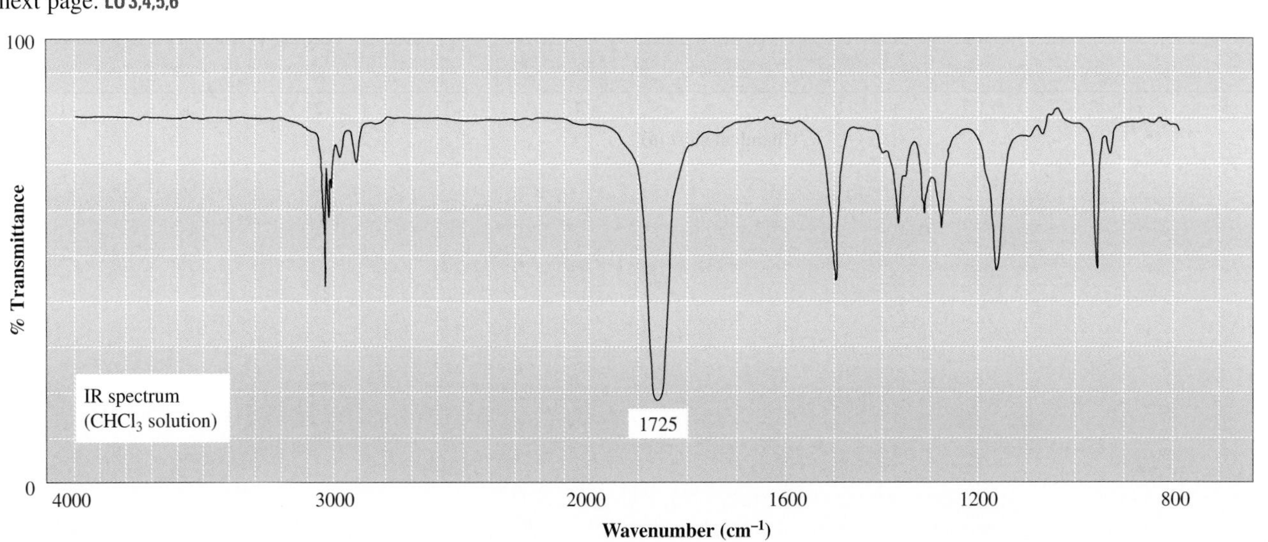

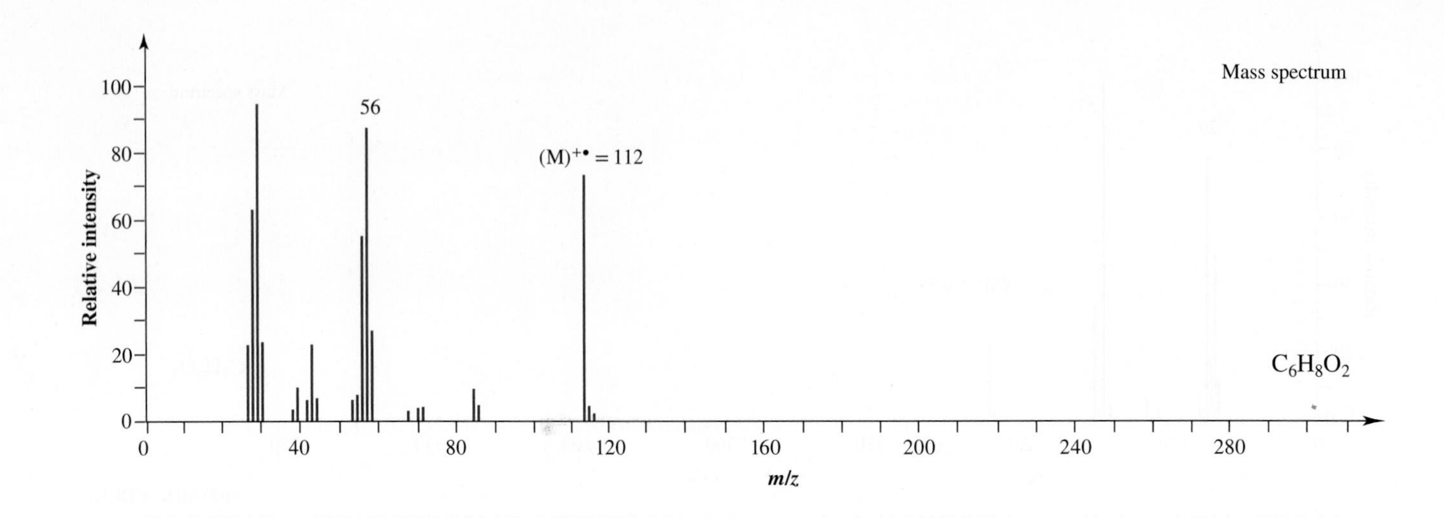

Mass spectrum
Relative intensity
100
80
60
40
20
0
56
(M)$^{+\bullet}$ = 112
$C_6H_8O_2$
0
40
80
120
160
200
240
280
m/z

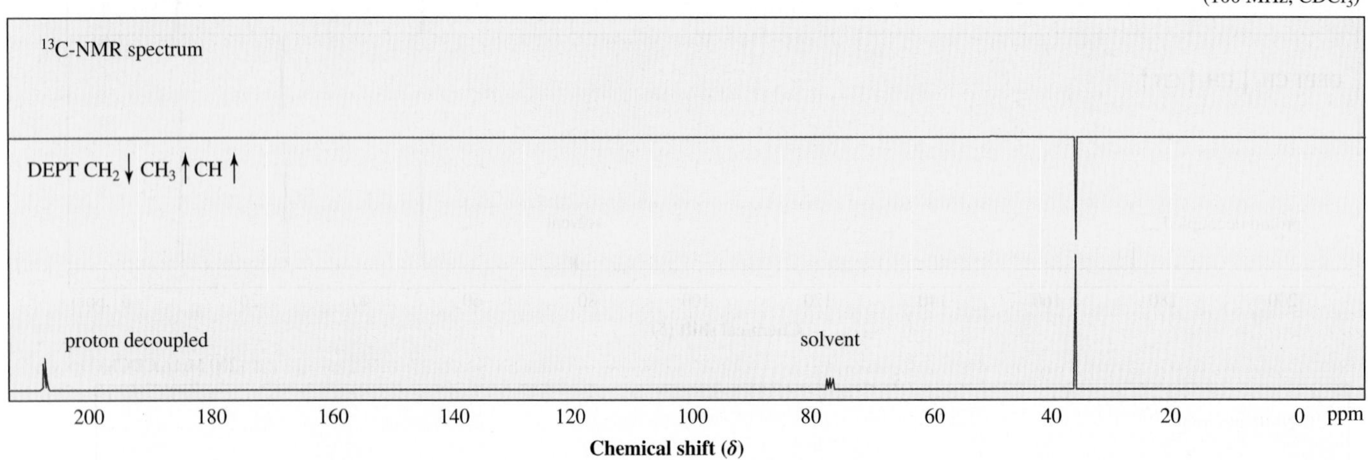

(100 MHz, $CDCl_3$)
^{13}C-NMR spectrum
DEPT CH_2 ↓ CH_3 ↑ CH ↑
proton decoupled
solvent
200
180
160
140
120
100
80
60
40
20
0 ppm
Chemical shift (δ)

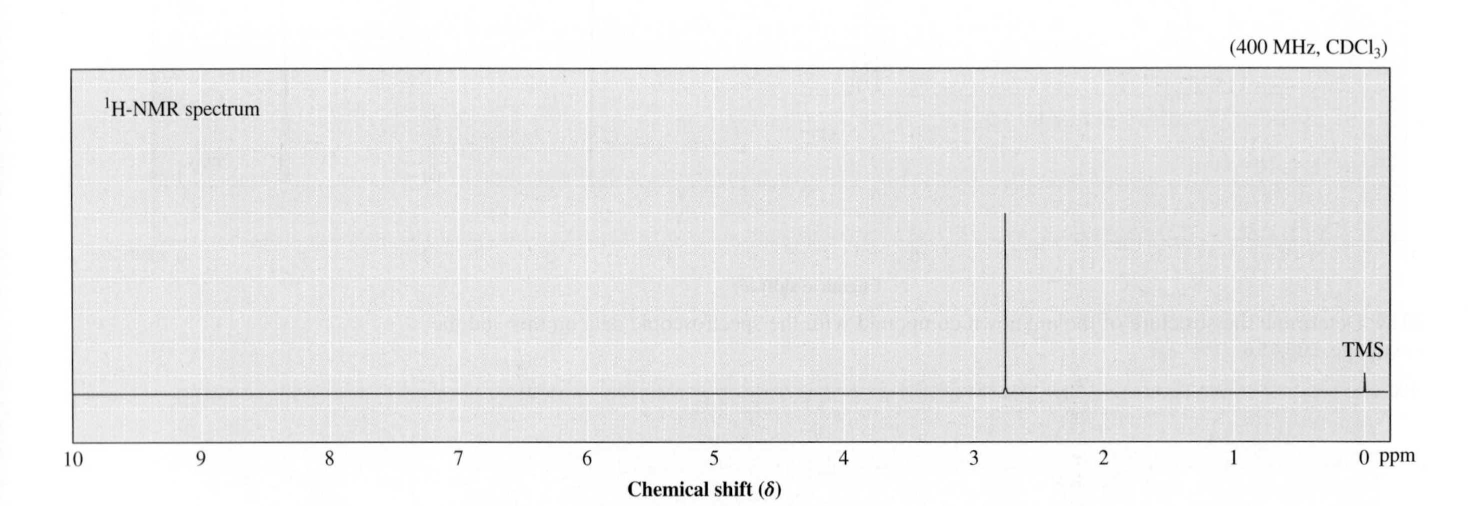

(400 MHz, $CDCl_3$)
^{1}H-NMR spectrum
TMS
10
9
8
7
6
5
4
3
2
1
0 ppm
Chemical shift (δ)

21 Aldehydes and ketones

Evening barbecues and pool parties are common summer leisure activities in Australia and New Zealand. However, also common are the dreaded mosquitoes and sandflies. Citronella torches are an increasingly popular weapon in the battle to prevent mosquito bites. Citronella (structure shown below) is a principal component of citronella oil and belongs to the family of natural products known as terpenes. It has a pleasant lemony scent and is widely used as a fragrance in soaps. Of the two functional groups in citronellal, an alkene and an aldehyde, it is the aldehyde group that is of most interest here.

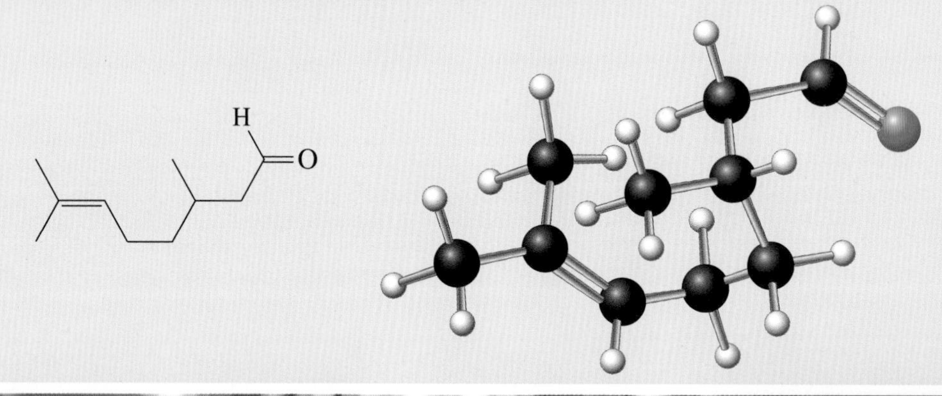

In this and several of the following chapters, we study the physical and chemical properties of compounds containing the carbonyl group, C═O. Since the carbonyl group is present in aldehydes, ketones, and carboxylic acids and their derivatives, it is one of the most important functional groups in organic chemistry. The chemical properties of the carbonyl group are straightforward, and an understanding of its characteristic reactions leads to an understanding of a wide variety of organic reactions.

LEARNING OBJECTIVES

After studying this chapter, you should be able to:

21.1 describe the structure and bonding in aldehydes and ketones

21.2 determine the IUPAC name of simple aldehydes and ketones given a structural formula (and vice versa)

21.3 explain how the intermolecular forces present between molecules of aldehydes and ketones affect the physical properties of these compounds

21.4 propose methods for the synthesis of aldehydes and ketones

21.5 predict the products of common reactions at the carbonyl group of aldehydes and ketones

21.6 describe the basic chemistry of keto–enol tautomerism and predict the products of halogenation reactions.

21.1 Structure and bonding

When a carbonyl group is bonded to a hydrogen atom and a hydrocarbon group (or a second hydrogen atom), the compound is an **aldehyde**. The simplest aldehyde, formaldehyde (methanal), has the carbonyl group bonded to two hydrogen atoms. In all other aldehydes, the carbonyl group is bonded to one hydrogen atom and one carbon atom. When the carbon atom of the C=O group is bonded to two hydrocarbon groups, the compound is a **ketone**. The following are Lewis structures for the aldehydes formaldehyde (methanal) and acetaldehyde (ethanal), and for the simplest ketones, acetone (propanone) and methyl ethyl ketone (butanone). The IUPAC name of each compound is in parentheses.

a carbonyl group

an aldehyde group | formaldehyde (methanal) | acetaldehyde (ethanal)

a ketone group | acetone (propanone) | methyl ethyl ketone (butanone)

The aldehyde group is often written —CHO, the double bond of the carbonyl group being implied. The abbreviation COH is *not* appropriate because it suggests an alcohol functional group. The ketone group is sometimes condensed to CO.

The carbonyl group is a planar group comprising a carbon atom bonded to an oxygen atom and two other atoms. The bond angles around the carbonyl carbon atom are approximately 120°. According to the valence bond theory (see section 5.6), a carbon–oxygen double bond consists of one σ bond formed by the overlap of the sp^2 hybrid orbitals of carbon and oxygen, and one π bond formed by the overlap of two parallel $2p$ orbitals. The two nonbonding pairs of electrons on the oxygen atom lie in the two remaining sp^2 hybrid orbitals (figure 21.1).

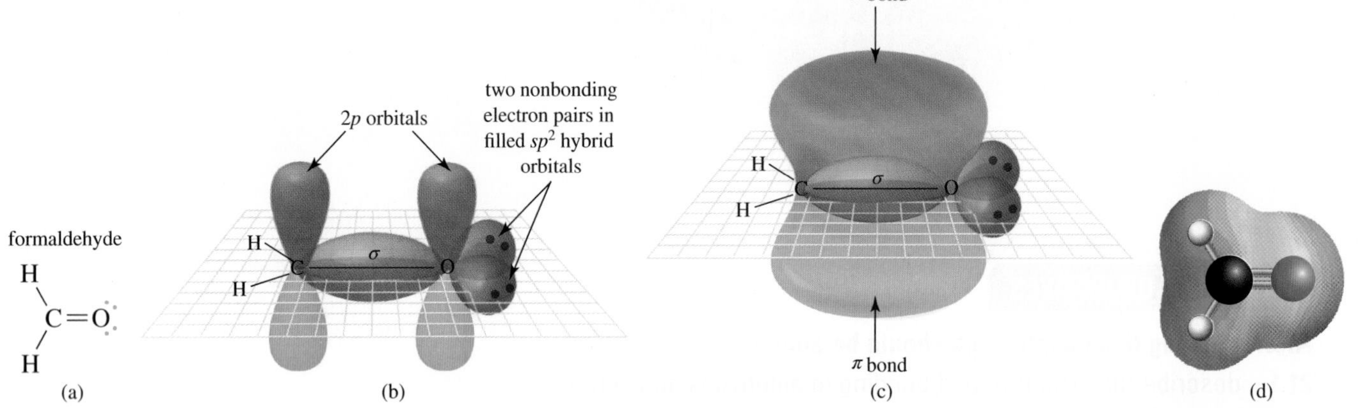

FIGURE 21.1 Representations of a carbon–oxygen double bond: **(a)** Lewis structure of formaldehyde, CH_2O, **(b)** the σ-bond framework and nonoverlapping parallel $2p$ atomic orbitals, **(c)** overlap of parallel $2p$ atomic orbitals to form a π bond and **(d)** an electron density model showing the uneven distribution of electrons in the carbonyl group.

21.2 Nomenclature

The IUPAC system of nomenclature for aldehydes and ketones follows the pattern of first selecting the longest chain of carbon atoms that contains the functional group as the parent alkane. We show the presence of the aldehyde group by changing the suffix *-e* of the parent alkane to *-al*, as in methanal (section 2.3). As the carbonyl group of an aldehyde can appear only at the end of a parent chain, its position is unambiguous — there is no need to use a number to locate it.

For cyclic molecules in which the —CHO group is bonded directly to the ring, we name the molecule by adding the suffix *-carbaldehyde* to the name of the ring. We number the atom of the ring bearing the aldehyde group as 1.

cyclopentane-carbaldehyde

trans-4-hydroxycyclo-hexanecarbaldehyde

The common name for an aldehyde is derived from the common name of the corresponding carboxylic acid by dropping the word 'acid' and changing the suffix *-ic* or *-oic* to *-aldehyde*. We can illustrate this using two common names of carboxylic acids with which you are familiar. The name formaldehyde is derived from formic acid, and the name acetaldehyde from acetic acid.

HCH (C=O)	HCOH (C=O)	CH_3CH (C=O)	CH_3COH (C=O)
formaldehyde	formic acid	acetaldehyde	acetic acid

Among the aldehydes for which the IUPAC system retains common names are: formaldehyde, acetaldehyde, benzaldehyde and cinnamaldehyde.

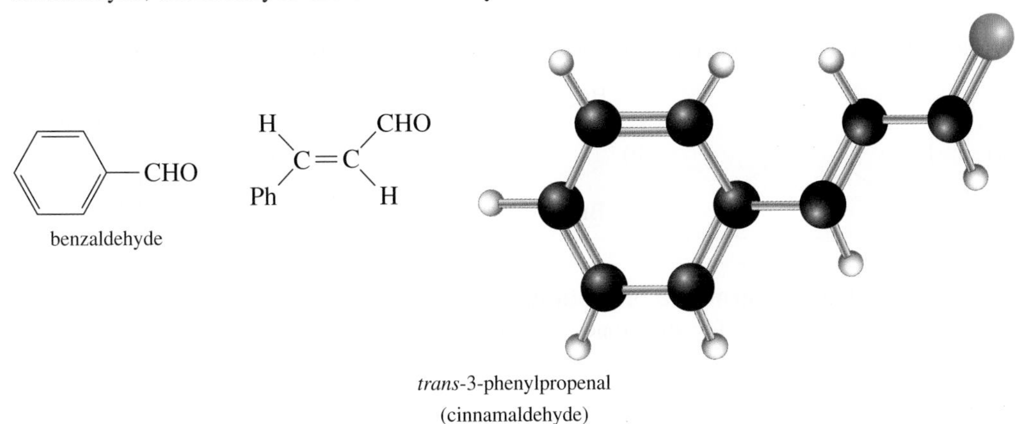

benzaldehyde

trans-3-phenylpropenal
(cinnamaldehyde)

Note here the alternative ways of writing the phenyl group (C_6H_5—). In benzaldehyde, it is written as a line-structure, while in cinnamaldehyde it is abbreviated to Ph.

In the IUPAC system, we name ketones by selecting the longest chain that contains the carbonyl group and making it the parent alkane. The presence of the ketone is indicated by changing the suffix from *-e* to *-one* (section 2.3). We number the parent chain so that the carbonyl carbon atom has the smallest number. The IUPAC system retains the common names acetone, acetophenone and benzophenone.

5-methylhexan-3-one

2-methyl-cyclohexanone

acetophenone

benzophenone

Common names for ketones are derived by naming each alkyl or aryl group bonded to the carbonyl group as a separate word, followed by the word 'ketone'. Groups are generally listed in order of increasing atomic weight. For example, methyl ethyl ketone, MEK, is a common solvent for varnishes and lacquers.

methyl ethyl ketone (MEK)
(butanone)

diethyl ketone
(pentan-3-one)

dicyclohexyl ketone

WORKED EXAMPLE 21.1

Naming aldehydes and ketones

Write the IUPAC name for each of the following compounds.

(a) 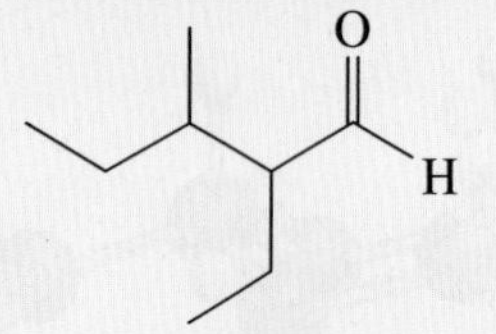 (b) 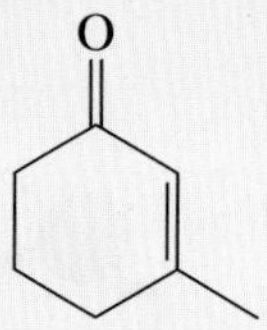 (c)

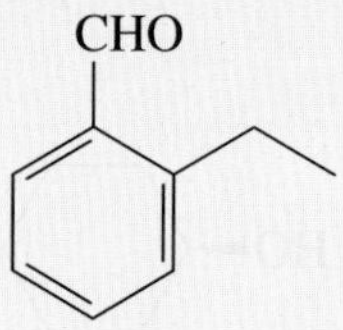

Analysis

(a) Identify the longest chain containing the aldehyde group. This will be the parent chain. Number the chain starting from the aldehyde group.
(b) Number the six-membered ring beginning with the carbonyl carbon atom.
(c) This molecule is derived from benzaldehyde, so it should be named as a substituted benzaldehyde.

Solution

(a) The IUPAC name of this compound is 2-ethyl-3-methylpentanal.
(b) The IUPAC name of this compound is 3-methylcyclohex-2-enone.
(c) The IUPAC name of this compound is 2-ethylbenzaldehyde.

Is our answer reasonable?

Check that the answer leads unambiguously to the given structure. Check that the structure is numbered from the correct end and that the substituents are numbered appropriately.

PRACTICE EXERCISE 21.1

Write the IUPAC name for each of the following compounds, and specify the configuration of (c).

(a) 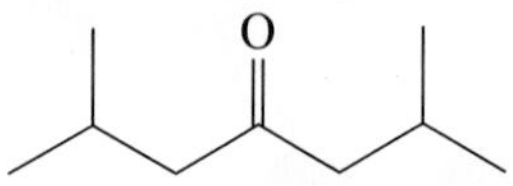 (b) 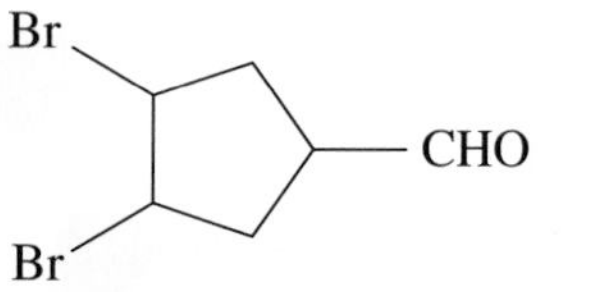 (c)

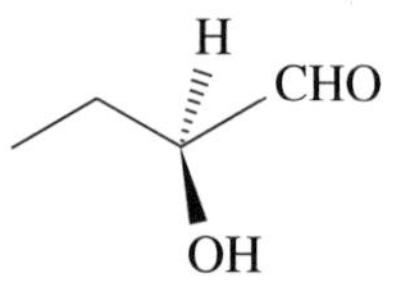

WORKED EXAMPLE 21.2

Drawing structural formulae for ketones

Write structural formulae for all ketones with the molecular formula $C_6H_{12}O$, and give each its IUPAC name. Which of these ketones are chiral?

Analysis

Firstly, identify the possible skeletal arrangements with six carbon atoms that could give a ketone. The possible ketones are derived from hexane, 2- and 3-methylpentane, and 2,2-dimethylbutane. A chiral ketone has at least one stereocentre (bonded to four different groups).

Solution

The following are line structures and IUPAC names for the six ketones with the molecular formula $C_6H_{12}O$.

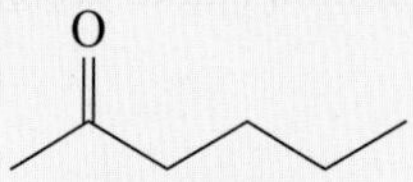

hexan-2-one 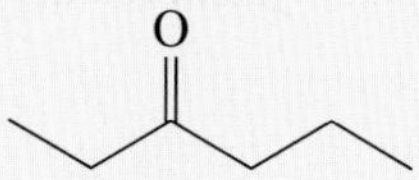 hexan-3-one 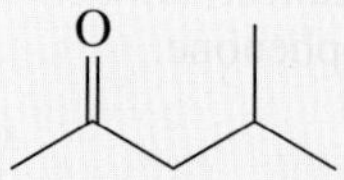 4-methylpentan-2-one

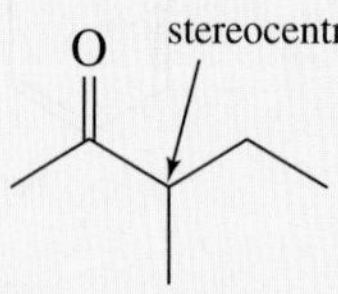

stereocentre

3-methylpentan-2-one 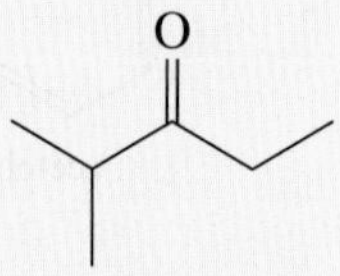 2-methylpentan-3-one 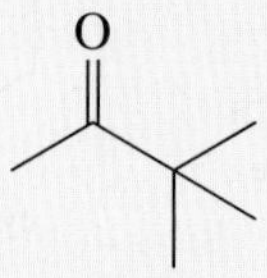 3,3-dimethylbutanone

Only 3-methylpentan-2-one has a stereocentre and is chiral.

Is our answer reasonable?

The most common error is to duplicate or miss some structures. Check that each structure is different (to avoid duplication). The best way to ensure that you have included all the possible structures is to check that the basic skeletal frameworks are covered (as given in the analysis) and then systematically move the carbonyl group to each possible position.

PRACTICE EXERCISE 21.2

Write structural formulae for all aldehydes with the molecular formula $C_6H_{12}O$, and give each its IUPAC name. Which of these aldehydes are chiral?

IUPAC names for compounds with more than one functional group

In naming compounds that contain more than one functional group, IUPAC has established an **order of precedence of functional groups**. That is, a system for prioritising functional groups for the purposes of IUPAC nomenclature. Table 21.1 gives the order of precedence for six functional groups.

TABLE 21.1 Decreasing order of precedence of six functional groups.

Functional group	Suffix	Prefix	Example of when the functional group is not highest priority and is used as a prefix
carboxyl group	-oic acid	—	
aldehyde group	-al	oxo-	3-oxopropanoic acid 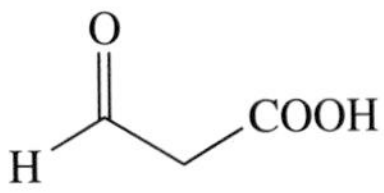
ketone group	-one	oxo-	3-oxobutanoic acid 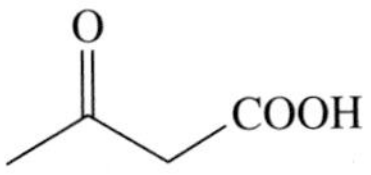
alcohol group	-ol	hydroxy-	4-hydroxybutanal 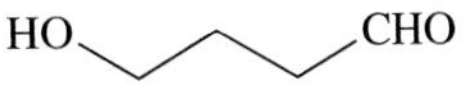
amino group	-amine	amino-	3-aminobutanal 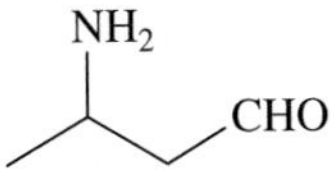
sulfhydryl	-thiol	mercapto-	2-mercaptoethanol 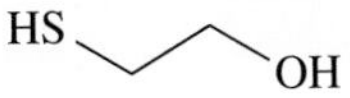

WORKED EXAMPLE 21.3

Naming complex aldehydes and ketones

Write the IUPAC name for each of the following compounds.

(a) 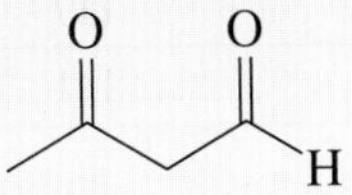 (b) 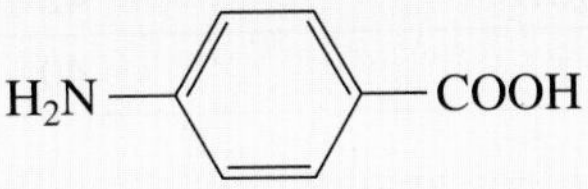 H_2N—◯—COOH (c)

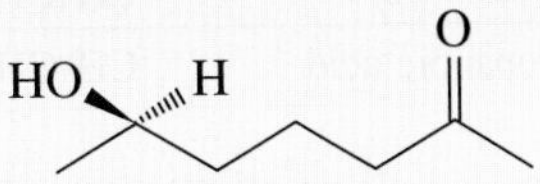

Analysis

(a) An aldehyde has higher precedence than a ketone, so we indicate the presence of the carbonyl group of the ketone by the prefix *oxo-*.

(b) The carboxyl group has higher precedence than the amino group, so we indicate the presence of the amino group by the prefix *amino-*.

(c) The carbonyl group has higher precedence than the hydroxyl group, so we indicate the hydroxyl group by the prefix *hydroxy-*. To determine the *R*,*S* configuration, follow the procedure in section 17.3.

Solution

(a) The IUPAC name of this compound is 3-oxobutanal.

(b) The IUPAC name is 4-aminobenzoic acid. Alternatively, the compound may be named *p*-aminobenzoic acid, abbreviated PABA. PABA, a growth factor of micro-organisms, is required for the synthesis of folic acid.

(c) The IUPAC name of this compound is (*R*)-6-hydroxyheptan-2-one.

Is our answer reasonable?

To check each answer, draw a structure from the name that has been determined. It should be the same as the compound from the question.

PRACTICE EXERCISE 21.3

Write IUPAC names for the following compounds.

(a) 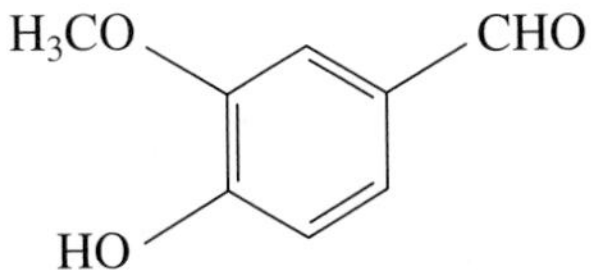 (b) 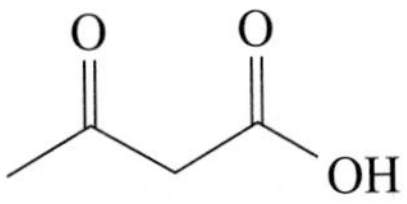 (c)

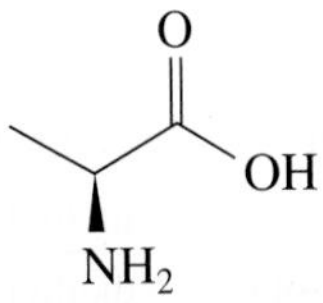

21.3 Physical properties

The carbon–oxygen double bond of the carbonyl group is very polar, due principally to the electronegativity difference between oxygen and carbon (χ = 3.5 compared with 2.5; see chapter 5, p. 189). Resonance effects also contribute to the polarity of the carbonyl groups.

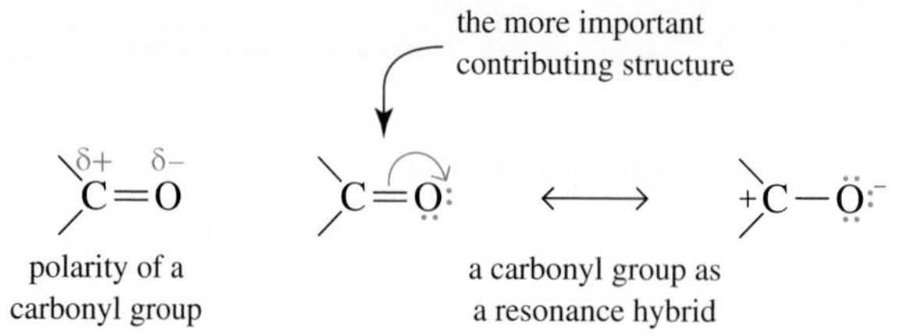

$$\overset{\delta+}{C}=\overset{\delta-}{O}$$

polarity of a carbonyl group

the more important contributing structure

$$C=\ddot{O}: \longleftrightarrow \overset{+}{C}-\ddot{O}:^-$$

a carbonyl group as a resonance hybrid

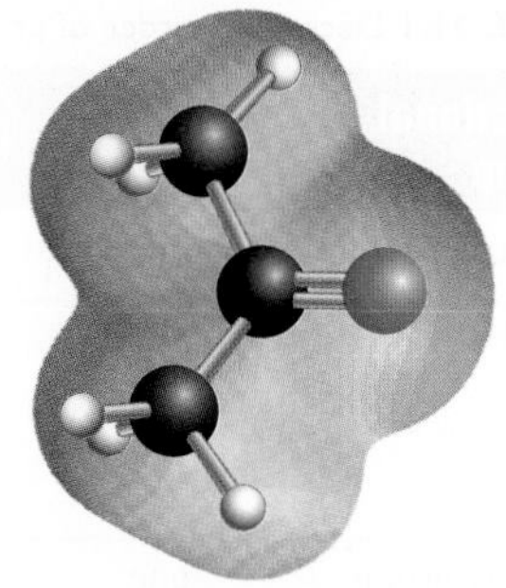

The electron density model (above) shows that the partial charge on an acetone molecule is distributed both on the carbonyl carbon atom and on the two attached groups. Because of the polarity of the carbonyl group, aldehydes and ketones are polar compounds and interact in the liquid state through dipole–dipole interactions. As a result, aldehydes and ketones have higher boiling points than nonpolar compounds with a comparable number of electrons. Table 21.2 lists the boiling points of six compounds with a similar number of electrons.

TABLE 21.2 Boiling points of six compounds of comparable molar mass and similar numbers of electrons.

Name	Structural formula	Number of electrons	Boiling point (°C)
diethyl ether	$CH_3CH_2OCH_2CH_3$	42	34
pentane	$CH_3CH_2CH_2CH_2CH_3$	42	36
butanal	$CH_3CH_2CH_2CHO$	40	76
butanone	$CH_3CH_2COCH_3$	40	80
butan-1-ol	$CH_3CH_2CH_2CH_2OH$	42	117
propanoic acid	CH_3CH_2COOH	40	141

Pentane and diethyl ether have the lowest boiling points of these six compounds. Both butanal and butanone are polar compounds, and have higher boiling points because of the intermolecular attraction between carbonyl groups. Alcohols (section 19.1) and carboxylic acids (section 23.1) are also polar compounds, and because their molecules associate by hydrogen bonding their boiling points are higher than those of butanal and butanone — compounds that cannot form intermolecular hydrogen bonds between like molecules.

Low-molar-mass aldehydes and ketones are more soluble in water than nonpolar compounds of similar molar mass because carbonyl groups interact with water molecules by hydrogen bonding. The carbonyl group is a hydrogen bond acceptor (the oxygen atom provides the electrons to form the hydrogen bond with water), as shown on the right.

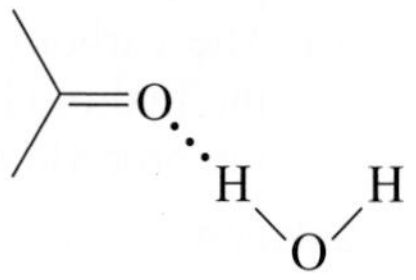

Table 21.3 lists the boiling points and solubilities in water of several low-molar-mass aldehydes and ketones.

TABLE 21.3 Physical properties of selected aldehydes and ketones.

IUPAC name	Common name	Structural formula	Boiling point (°C)	Solubility (g/100 g water)
methanal	formaldehyde	HCHO	–21	infinite
ethanal	acetaldehyde	CH_3CHO	20	infinite
propanal	propionaldehyde	CH_3CH_2CHO	49	16
butanal	butyraldehyde	$CH_3CH_2CH_2CHO$	76	7
hexanal	caproaldehyde	$CH_3(CH_2)_4CHO$	129	slight
propanone	acetone	CH_3COCH_3	56	infinite
butanone	methyl ethyl ketone	$CH_3COCH_2CH_3$	80	26
pentan-3-one	diethyl ketone	$CH_3CH_2COCH_2CH_3$	101	5

21.4 Preparation of aldehydes and ketones

Aldehydes and ketones can be prepared from many functional groups using various reactions. Some of these will be discussed briefly here and in more detail elsewhere in this text. However, before describing specific examples of the synthesis of aldehydes and ketones, we will look briefly at some of the more industrially important compounds in this class.

Industrially important aldehydes and ketones

Formaldehyde is probably the most important aldehyde produced on an industrial scale, with over 17 million tonnes produced annually, principally by the oxidation of methanol. It is used mainly in the manufacture of resins for the construction industry, particularly in the production of wood-based products. The next most abundantly produced aldehydes are the butanals (butanal and methylpropanal). These are produced by the hydroformylation (addition of carbon monoxide and hydrogen) of propene by a process known as the oxo-process (see the equation below). Approximately 6 million tonnes of butanals are produced annually, mainly for the manufacture of plasticisers. The aliphatic aldehydes with 6 to 13 carbon atoms are used extensively in the fragrance industry.

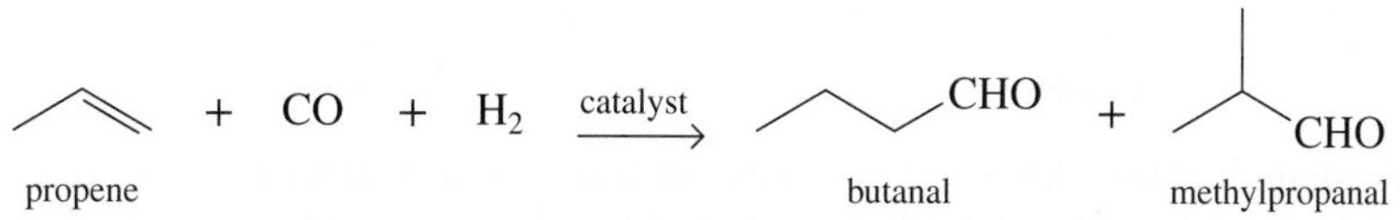

propene + CO + H_2 $\xrightarrow{\text{catalyst}}$ butanal (CHO) + methylpropanal (CHO)

Industrially important aromatic aldehydes include phenylacetaldehyde (used in the flavour and fragrance industry) and salicylaldehyde (2-hydroxybenzaldehyde), which is used in the synthesis of coumarin for the electroplating industry and also for the synthesis of agrochemicals, pharmaceuticals and fragrances.

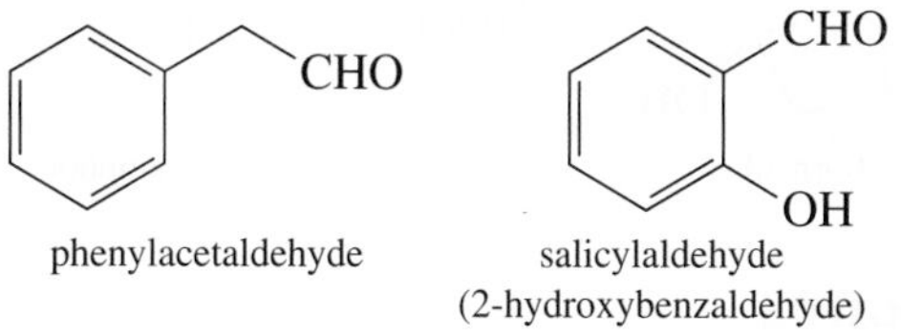

CHO — phenylacetaldehyde

CHO, OH — salicylaldehyde (2-hydroxybenzaldehyde)

Acetone is the simplest and most important ketone and finds ubiquitous use as a solvent. Higher members of the aliphatic methyl ketone series are also industrially significant solvents. These include methyl ethyl ketone (butanone), methyl isobutyl ketone (4-methylpentan-2-one) and methyl amyl ketone (heptan-2-one). Acetone has been produced industrially for over a century by a variety of methods. However, today, most of the world's supply is obtained as a co-product in the preparation of phenol by the hydroperoxidation of cumene. Acetone is used as a solvent in the preparation of methyl methacrylate and bisphenol A (both used in the production of polymers). The other 'methyl ketones', including methyl ethyl ketone (MEK, butanone) and methyl isobutyl ketone (MIBK, 4-methylpentanone) are also important industrial solvents, particularly for coatings and inks.

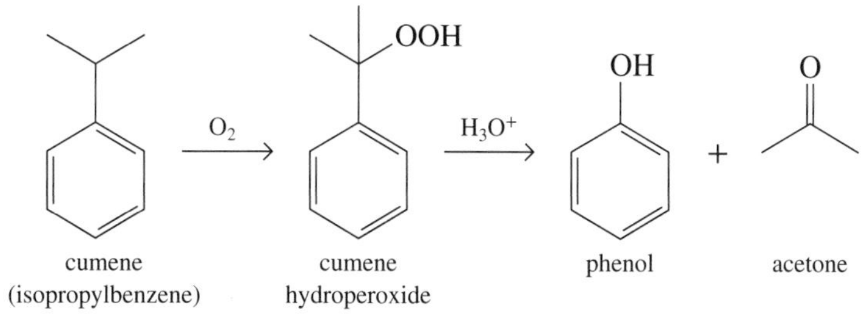

cumene (isopropylbenzene) $\xrightarrow{O_2}$ cumene hydroperoxide (OOH) $\xrightarrow{H_3O^+}$ phenol (OH) + acetone (O)

Important aromatic ketones include, acetophenone, propiophenone (ethyl phenyl ketone) (used in the manufacture of ephedrine and as a fragrance enhancer) and benzophenone (used as a photoinitiator in UV-curable inks and coatings).

Friedel–Crafts acylation

Ketones can be prepared by Friedel–Crafts acylation of aromatic compounds (see section 16.8) as illustrated by the reaction of benzene and acetyl chloride in the presence of aluminium chloride to give acetophenone.

Acetophenone is produced industrially by this process and is used in the preparation of paints and resins, and in pharmaceuticals, perfumes and pesticides.

benzene + acetyl chloride (an acyl halide) $\xrightarrow{AlCl_3}$ acetophenone (a ketone) + HCl

Oxidation of alcohols

We have already seen in chapter 19 that primary alcohols can be oxidised under mild conditions to give aldehydes (for more examples of this reaction see section 19.2).

citronellol $\xrightarrow{PCC}$ citronellal

PCC (pyridinium chlorochromate) is a mild oxidising agent that will oxidise primary alcohols to aldehydes but will not further oxidise aldehydes to carboxylic acids. Ketones can be obtained by the oxidation of secondary alcohols (for more examples of this reaction see section 19.2).

borneol $\xrightarrow[H_2SO_4]{Na_2Cr_2O_7}$ camphor

Ozonolysis of alkenes

Alkenes are readily cleaved by reaction with ozone followed by reductive hydrolysis to give aldehydes and/or ketones. Ozone is bubbled through a solution of the alkene in an inert solvent to produce an unstable ozonide. This ozonide is not isolated but treated with a mild reducing agent (typically zinc/water/acid or dimethyl sulfide) to give carbonyl compounds as shown below (along with inorganic products, which are not shown).

2-methylpent-2-ene $\xrightarrow{O_3}$ an ozonide $\xrightarrow{Zn/H_2O}$ acetone (a ketone) + propanal (an aldehyde)

Although the reaction may look complex, the final outcome is simply the cleavage of a C═C bond and its replacement with two C═O groups. When the alkene is part of a ring, a single compound with two carbonyl groups is formed.

1-methylcyclopentene $\xrightarrow[2.\ CH_3SCH_3]{1.\ O_3}$ 5-oxohexanal

Hydration of alkynes

In section 16.5, the electrophilic addition of water to alkenes (hydration) to give alcohols was discussed. Water can also be added to alkynes under acidic conditions in the presence of a mercuric sulfate catalyst. The initial product of this reaction is an enol (an alkene with an —OH group) which undergoes tautomerism (see section 21.6) to give a ketone. The addition obeys Markovnikov's rule (section 16.5), and terminal alkynes (those with the alkyne group at the end of the chain) give methyl ketones.

$$CH_3CH_2C{\equiv}C{-}H + H_2O \xrightarrow[HgSO_4]{H_2SO_4} \underset{\text{an enol}}{(CH_3CH_2)(HO)C{=}CH_2} \xrightleftharpoons{\text{tautomerism}} \underset{\text{butanone (ethyl methyl ketone)}}{CH_3CH_2COCH_3}$$

Internal alkynes give a mixture of two ketones (the intermediate enol is not shown).

$$CH_3CH_2C{\equiv}CCH_3 + H_2O \xrightarrow[HgSO_4]{H_2SO_4} \underset{\text{pentan-2-one}}{CH_3CH_2CH_2COCH_3} + \underset{\text{pentan-3-one}}{CH_3CH_2COCH_2CH_3}$$

21.5 Reactions

Aldehydes and ketones are very reactive due to the polarity and structure of the carbonyl group (figure 21.2). They are electron-rich groups due to the π bond and the two nonbonding pairs of electrons on the oxygen atom. The polarity of the C=O bond enhances the attraction of nucleophiles (to the δ+ carbon atom) and electrophiles (to the δ– oxygen atom), while the flat structure allows easy access to these reactive centres. Aldehydes are usually more reactive than ketones because access to the carbonyl carbon atom is easier as they have only one alkyl group (which is bulkier than a hydrogen atom) attached.

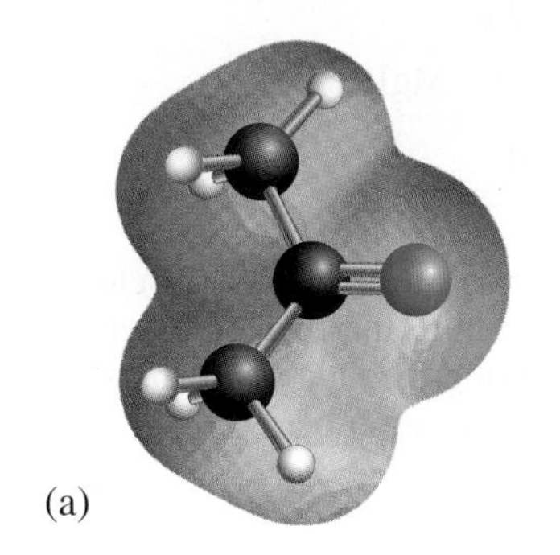

(a)

(b) The δ+ carbonyl carbon atom reacts with nucleophiles and bases. $R_2C^{\delta+}{=}O^{\delta-}$ The δ– carbonyl oxygen atom reacts with electrophiles and acids.

FIGURE 21.2 **(a)** An electron density model highlighting the polarity of the carbonyl group. **(b)** The polarity of the C=O bond enhances the attraction of nucleophiles and electrophiles.

The most common reaction theme of the carbonyl group is the addition of nucleophiles to form **tetrahedral carbonyl addition intermediates**. In the following general reaction, the nucleophilic reagent is written as Nu:$^-$ to emphasise the presence of an unshared pair of electrons.

$$Nu{:}^- + R_2C{=}\ddot{O}{:} \longrightarrow \underset{\text{tetrahedral carbonyl addition intermediate}}{Nu{-}CR_2{-}\ddot{O}{:}^-}$$

Addition of Grignard reagents

The addition of carbon nucleophiles is among the most important types of nucleophilic addition to carbonyl groups, because these reactions form new carbon–carbon bonds. In this section, we describe the preparation and reactions of Grignard reagents, and their reaction with aldehydes and ketones.

Formation and structure of organomagnesium compounds

Alkyl, aryl and vinylic halides react with group 1, group 2 and certain other metals to form **organometallic compounds**. An organometallic compound is a compound containing a carbon–metal bond. Organomagnesium compounds are among the most readily available, easily prepared and easily handled organometallic compounds. Organomagnesium compounds of the type RMgX or ArMgX are commonly named **Grignard reagents**, after Victor Grignard, who was awarded the 1912 Nobel Prize in chemistry for their discovery and their application to organic synthesis.

Grignard reagents are typically prepared by the slow addition of alkyl, aryl or vinyl halides to a stirred suspension of magnesium metal in an ether solvent. Organoiodides and bromides generally react rapidly under these conditions, whereas chlorides react more slowly. Butylmagnesium bromide, for example, is prepared by adding 1-bromobutane to magnesium metal. Aryl Grignards, such as phenylmagnesium bromide, are prepared in a similar manner.

1-bromobutane (Br) + Mg ⟶ butylmagnesium bromide (MgBr)

bromobenzene (Br) + Mg ⟶ phenylmagnesium bromide (MgBr)

Given that the difference in electronegativity between carbon and magnesium is 1.3 (2.5 – 1.2), the carbon–magnesium bond is best described as polar covalent, with the carbon atom bearing a partial negative charge and the magnesium atom bearing a partial positive charge. In the structural formula below right, the carbon–magnesium bond is shown as ionic to emphasise its nucleophilic character. Note that, although we can write a Grignard reagent as a **carbanion** (an anion in which the carbon atom has an unshared pair of electrons and bears a negative charge), a more accurate representation shows it as a polar covalent compound (below left) and, in fact, each of these represents a resonance form of the actual C—Mg bond.

$$CH_3(CH_2)_2\overset{\delta-}{C}H_2-\overset{\delta+}{MgBr} \longleftrightarrow CH_3(CH_2)_2CH_2{:}^- \; \overset{+}{MgBr}$$

The carbon atom is a nucleophile.

In chapter 18, you learned that the carbon atoms bonded to halogens are electrophilic (they are δ+ and undergo reactions with nucleophiles). The feature that makes Grignard reagents so valuable in organic synthesis is that the carbon atom bearing the halogen is transformed from an electrophile (δ+) into a nucleophile (δ–), as shown in the equation below.

1-bromobutane (δ+ C—Br δ–) + Mg ⟶ butylmagnesium bromide (δ– C—MgBr δ+)

Reaction with protic acids

Grignard reagents are very strong bases and will readily remove acidic protons from a wide variety compounds to form alkanes. Ethylmagnesium bromide, for example, reacts instantly with water to give ethane and magnesium salts. This reaction is an example of a very strong base completely deprotonating water.

$$CH_3\overset{\delta-}{CH_2}-\overset{\delta+}{MgBr} + H-OH \longrightarrow CH_3CH_2-H + Mg^{2+} + OH^- + Br^-$$

Any compound containing an O—H, N—H or S—H bond will react with a Grignard reagent by proton transfer. The following are examples of compounds containing those functional groups.

HOH	ROH	ArOH	RCOOH	RNH_2	RSH
water	alcohols	phenols	carboxylic acids	amines	thiols

Because Grignard reagents react so rapidly with these protic acids, they cannot be made from any halogen-containing compounds that also contain protic acids.

WORKED EXAMPLE 21.4

Reactions of Grignard reagents

Write an equation for the acid–base reaction between ethylmagnesium iodide and an alcohol. Use curved arrows to show the bond-breaking and bond-forming processes in this reaction. In addition, show that the reaction is an example of a Brønsted–Lowry acid–base reaction.

Analysis

First, identify the polarity of the species (δ+/δ–). The electrons will come from the δ– portion to form a bond with the δ+ portion. In this acid–base reaction the alcohol group provides the proton.

Solution

The alcohol is a weak acid and the ethyl carbanion is a very strong base.

$$\overset{\delta-}{CH_3CH_2}-\overset{\delta+}{MgI} + \overset{\delta+}{H}-\overset{\delta-}{\ddot{O}R} \longrightarrow CH_3CH_2-H + R\ddot{O}\!:^- [MgI]^+$$

ethylmagnesium iodide (stronger base); an alcohol (stronger acid); ethane (weaker acid); a magnesium alkoxide (weaker base)

Is our answer reasonable?

Check that the curved arrows start at an electron-rich region (atom or bond) and end at an electron-poor atom. Check that the δ+ atoms are paired with the δ– atoms.

PRACTICE EXERCISE 21.4

Explain why Grignard reagents cannot be formed from the following organohalides. What would be produced when magnesium metal is treated with: (a) 4-iodobenzoic acid and (b) 3-bromopropan-1-ol?

4-iodobenzoic acid; 3-bromopropan-1-ol

Addition of Grignard reagents to aldehydes and ketones

The special value of Grignard reagents is that they provide excellent ways to form new carbon–carbon bonds. In their reactions, Grignard reagents behave as though they were carbanions. A carbanion is a good nucleophile and adds to the carbonyl group of an aldehyde or ketone to form a tetrahedral carbonyl addition compound (see equation below). The reason for these reactions occurring so readily is the attraction between the partial negative charge on the carbon atom of the organometallic compound and the partial positive charge of the carbonyl carbon atom. In the examples that follow, the magnesium–oxygen bond is written as $—O^- [MgBr]^+$ to emphasise its ionic character. The alkoxide ions formed in Grignard reactions are strong bases and form alcohols when treated with an aqueous acid such as HCl or aqueous NH_4Cl during separation of the reaction products, by-products and unreacted starting material. A general scheme for the reaction of Grignard reagents with carbonyl compounds showing the polarity of the reacting species is given below.

$$\overset{\delta-}{R'}-\overset{\delta+}{MgX} + R_2\overset{\delta+}{C}=\overset{\delta-}{O} \longrightarrow R'-CR_2-O^-[MgX]^+$$

An easy way to determine the product is to identify the polarity of the species (δ+/δ–). The product is the result of the δ– part of the Grignard reagent forming a bond with the δ+ carbon atom of the carbonyl group (and between the δ+ part of the Grignard reagent and the δ– oxygen atom of the carbonyl group).

Addition to formaldehyde (methanal) gives 1° alcohols

Treatment of Grignard reagents with formaldehyde (methanal), followed by hydrolysis in aqueous acid, gives primary alcohols.

$$CH_3CH_2-MgBr + H-\underset{}{C}(=\ddot{O})-H \longrightarrow CH_3CH_2-CH_2-\ddot{O}\!:^- [MgBr]^+ \xrightarrow[H_2O]{HCl} CH_3CH_2-CH_2-\ddot{O}H + Mg^{2+}$$

formaldehyde; a magnesium alkoxide; propan-1-ol (a 1° alcohol)

Addition to an aldehyde (except formaldehyde) gives 2° alcohols

The treatment of Grignard reagents with any aldehydes other than formaldehyde, followed by hydrolysis in aqueous acid, gives secondary alcohols. The steps required for this reaction are shown below.

$$C_6H_{11}MgBr + CH_3CHO \longrightarrow C_6H_{11}CH(O^-[MgBr]^+)CH_3 \xrightarrow[H_2O]{HCl} C_6H_{11}CH(OH)CH_3 + Mg^{2+}$$

acetaldehyde — a magnesium alkoxide — 1-cylohexylethanol (a 2° alcohol)

Addition to ketones gives 3° alcohols

Treatment of Grignard reagents with ketones, followed by hydrolysis in aqueous acid, gives tertiary alcohols.

$$C_6H_5MgBr + CH_3COCH_3 \longrightarrow C_6H_5C(O^-[MgBr]^+)(CH_3)_2 \xrightarrow[H_2O]{HCl} C_6H_5C(OH)(CH_3)_2 + Mg^{2+}$$

acetone — a magnesium alkoxide — 2-phenylpropan-2-ol (a 3° alcohol)

Addition to carbon dioxide gives carboxylic acids

Grignard reagents react with carbon dioxide to give magnesium carboxylates, which are converted into carboxylic acids by acidification. This reaction is applicable to alkyl and aryl halides, and produces carboxylic acids with one carbon atom more than the original organohalide, as shown in the scheme below.

$$C_5H_9MgBr + O{=}C{=}O \longrightarrow C_5H_9CO_2^-[MgBr]^+ \xrightarrow[H_2O]{HCl} C_5H_9COOH$$

carbon dioxide — a magnesium carboxylate — cyclopentanecarboxylic acid

WORKED EXAMPLE 21.5

Synthesis of 3° alcohols using Grignard reagents

The tertiary alcohol 2-phenylbutan-2-ol can be synthesised by three different combinations of a Grignard reagent and a carbonyl compound. Show each combination.

Analysis

First, draw the structure of 2-phenylbutan-2-ol. As a tertiary alcohol, it can be produced by the reaction of a ketone with a Grignard reagent. To determine which ketones are suitable, identify the carbon atom bonded to the —OH group; this was the carbonyl carbon atom. Two of the groups bonded to this carbon atom form the ketone and the third forms the Grignard reagent. As there are three different groups we can have three different Grignard reagents and, by a process of elimination, three different ketones.

Solution

Curved arrows in each solution show the formation of the new carbon–carbon bond and the alkoxide ion, and labels on the final product show which set of reagents forms each bond.

(a) $C_6H_5MgBr + CH_3C(=O)CH_2CH_3$

(b) $C_6H_5C(=O)CH_3 + CH_3CH_2MgBr \longrightarrow C_6H_5C(O^-[MgBr]^+)(CH_3)CH_2CH_3 \xrightarrow[H_2O]{HCl} C_6H_5C(OH)(CH_3)CH_2CH_3$

(c) $C_6H_5C(=O)CH_2CH_3 + CH_3MgBr$

Is our answer reasonable?

A simple way to check if your answers make sense is to count the carbon atoms in the starting materials and compare this number with the product (10 carbon atoms). If these agree, then check that each case has a methyl, an ethyl and a phenyl group.

PRACTICE EXERCISE 21.5

Show how the following three compounds can be synthesised from the same Grignard reagent.

(a) OH (b) OH (c) OH

Addition of other carbon nucleophiles

Treatment of aldehydes and some ketones with sodium cyanide and dilute acid produces **cyanohydrins**, compounds with an —OH and a —CN group bonded to the same carbon atom. Ketones with bulky groups do not undergo this reaction. In this reaction, the negatively charged carbon atom of the cyanide ion adds to the δ+ carbonyl carbon atom, and then the proton from the acid adds to the resulting negative oxygen atom.

$$R_2C{=}O \xrightarrow[\text{2. HCl}]{\text{1. NaCN}} R_2C(OH)CN$$

Cyanohydrins are useful compounds in organic chemistry because they can be converted to α-hydroxycarboxylic acids (e.g. lactic acid) or α,β-unsaturated acids (e.g. methyl methacrylate), depending on conditions used. They can also be converted to β-amino alcohols (e.g. 1-aminopropan-2-ol).

$$H_3C{-}CHO \xrightarrow[\text{2. } H_3O^+]{\text{1. NaCN}} H_3C{-}CH(OH){-}CN \xrightarrow[H_3O^+/\text{heat}]{H_2O} H_3C{-}CH(OH){-}COOH$$

acetaldehyde — lactic acid

$$H_3C{-}CO{-}CH_3 \xrightarrow[\text{2. } H_3O^+]{\text{1. NaCN}} (H_3C)_2C(OH){-}CN \xrightarrow[CH_3OH/\text{heat}]{\text{conc. } H_2SO_4} H_2C{=}C(CH_3){-}COOCH_3$$

acetone — methyl methacrylate

$$H_3C{-}CHO \xrightarrow[\text{2. } H_3O^+]{\text{1. NaCN}} H_3C{-}CH(OH){-}CN \xrightarrow[\text{2. } H_2O]{\text{1. } LiAlH_4} H_3C{-}CH(OH){-}CH_2{-}NH$$

acetaldehyde — 1-aminopropan-2-ol

Note that we have now seen three convenient methods of adding one carbon atom to a carbon chain: the addition of cyanide ions to carbonyl compounds, and reactions of Grignard reagents with formaldehyde and with carbon dioxide.

Addition of alcohols

Treatment of aldehydes and ketones with alcohols in the presence of an acid catalyst produces **acetals**. The intermediate **hemiacetals** (half-acetals) are generally unstable and are only minor components of an equilibrium mixture, except in one important type of molecule (as discussed later).

$$H_3C{-}CO{-}CH_3 \underset{\text{acid}}{\overset{CH_3CH_2OH}{\rightleftharpoons}} (H_3C)_2C(OH){-}OCH_2CH_3 \underset{\text{acid}}{\overset{CH_3CH_2OH}{\rightleftharpoons}} (H_3C)_2C(OCH_2CH_3){-}OCH_2CH_3$$

acetone — a hemiacetal — an acetal

The hemiacetal group is a carbon atom bonded to an —OH group and an —OR or —OAr group. A hemiacetal derived from a ketone is sometimes referred to as a hemiketal.

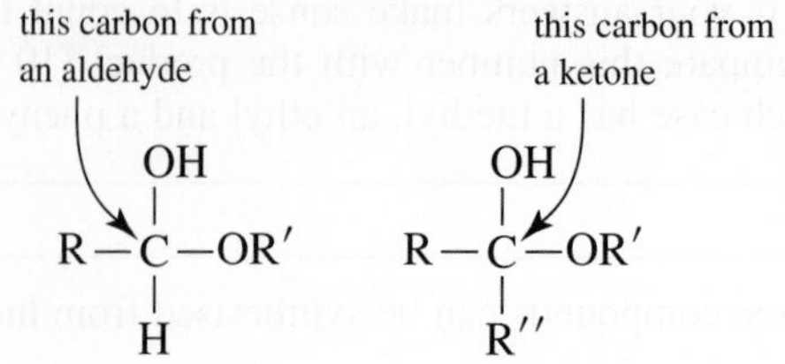

this carbon from an aldehyde

this carbon from a ketone

The acetal functional group is a carbon atom bonded to two — OR or — OAr groups. Acetals derived from ketones are sometimes referred to as **ketals**.

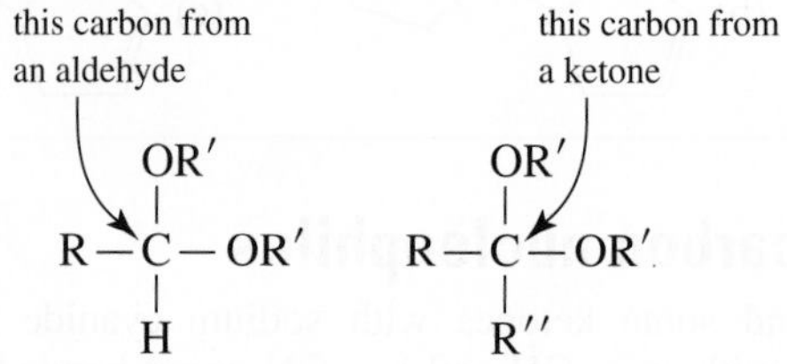

this carbon from an aldehyde

this carbon from a ketone

Hemiacetals are stable when a hydroxyl group is part of the same molecule that contains the carbonyl group, and five- or six-membered rings can form. In these cases, the compound exists almost entirely in a cyclic hemiacetal form.

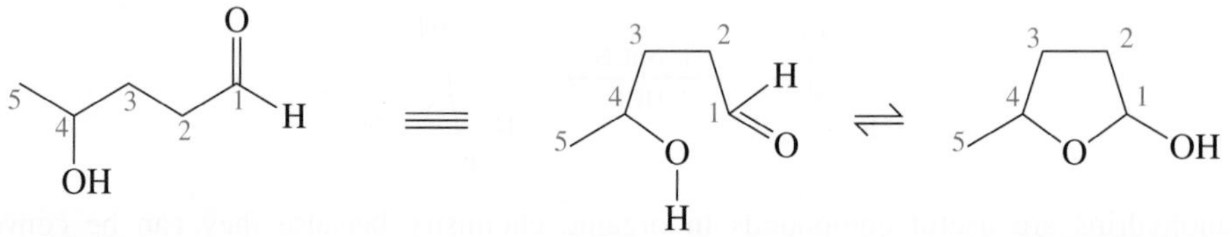

4-hydroxypentanal

4-hydroxypentanal

a cyclic hemiacetal (major form present at equilibrium)

We examine cyclic hemiacetals in more detail when we consider the chemistry of carbohydrates in chapter 22.

For clarity, we will study the reaction of alcohols with aldehydes and ketones in two stages: (i) the formation of hemiacetals and (ii) the subsequent transformation of hemiacetals into acetals.

The reaction of alcohols with aldehydes and ketones is an acid-catalysed example of addition to a carbonyl compound to produce a tetrahedral carbonyl addition intermediate (in this case, the hemiacetal). In this process, H^+ adds to the δ^- carbonyl oxygen atom, and the $\delta-$ oxygen atom of the alcohol adds to the $\delta+$ carbonyl carbon atom. This is another example of '$\delta-$ goes with $\delta+$'.

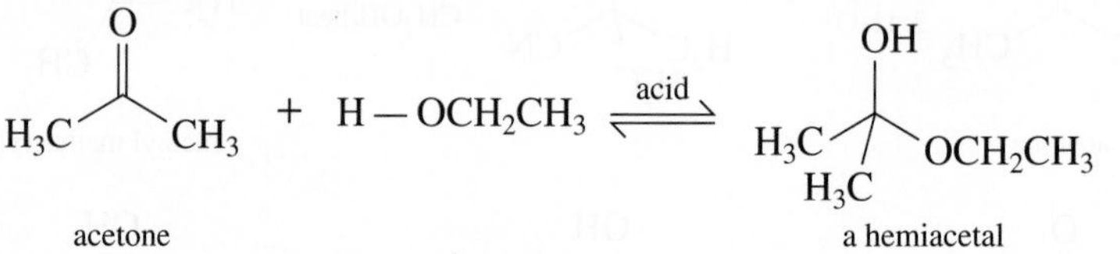

acetone

a hemiacetal

A mechanism for the formation of hemiacetals can be divided into three steps. Note that the acid H—A is a true catalyst in this reaction; it is used in step 1 and is regenerated in step 3.

Step 1: Proton transfer from the acid, H—A, to the carbonyl oxygen atom gives a resonance stabilised cation.

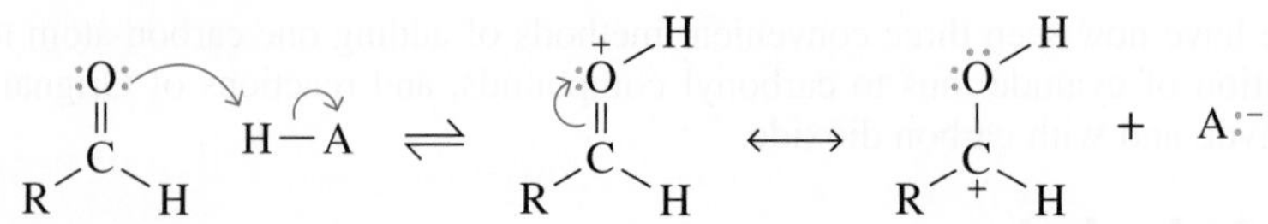

an aldehyde

resonance stabilised cation

Step 2: Reaction of the resonance-stabilised cation (an electrophile) with methanol (a nucleophile) gives the conjugate acid of a hemiacetal.

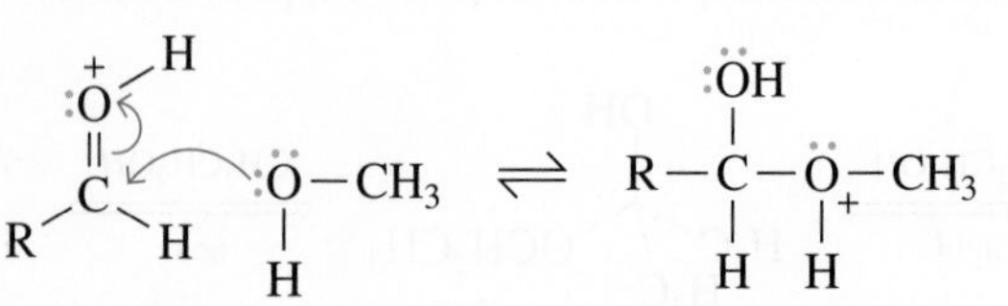

a protonated hemiacetal

Step 3: Proton transfer from the protonated hemiacetal to A^- gives the hemiacetal and regenerates the acid catalyst, H—A.

$$\text{R}-\underset{\text{H}}{\overset{\text{OH}}{\text{C}}}-\overset{+}{\underset{\text{H}}{\text{O}}}-\text{CH}_3 \; + \; :\text{A}^- \rightleftharpoons \text{R}-\underset{\text{H}}{\overset{\text{OH}}{\text{C}}}-\text{O}-\text{CH}_3 + \text{H}-\text{A}$$

a protonated hemiacetal   a hemiacetal

Hemiacetals can react further with alcohols to form acetals plus a mole equivalent of water.

$$(\text{H}_3\text{C})_2\text{C}(\text{OH})\text{OCH}_2\text{CH}_3 + \text{H}-\text{OCH}_2\text{CH}_3 \overset{\text{acid}}{\rightleftharpoons} (\text{H}_3\text{C})_2\text{C}(\text{OCH}_2\text{CH}_3)_2$$

a hemiacetal   a diethyl acetal

A mechanism for the acid-catalysed conversion of hemiacetals to acetals can be divided into four steps. Note that the acid, H—A, is also a catalyst in this reaction; it is used in step 1 and is regenerated in step 4.

Step 1: Proton transfer from the acid, H—A, to the hemiacetal —OH group gives an oxonium ion.

$$\text{R}-\underset{\text{H}}{\overset{\text{HO}}{\text{C}}}-\text{OCH}_3 + \text{H}-\text{A} \rightleftharpoons \text{R}-\underset{\text{H}}{\overset{\text{H}_2\overset{+}{\text{O}}}{\text{C}}}-\text{OCH}_3 + \text{A}:^-$$

an oxonium ion

Step 2: Loss of water from the oxonium ion gives a resonance-stabilised cation.

$$\text{R}-\underset{\text{H}}{\overset{\text{H}_2\overset{+}{\text{O}}}{\text{C}}}-\text{OCH}_3 \rightleftharpoons \text{R}-\underset{\text{H}}{\text{C}}=\overset{+}{\text{O}}\text{CH}_3 \longleftrightarrow \text{R}-\underset{\text{H}}{\overset{+}{\text{C}}}-\text{OCH}_3 + \text{H}_2\text{O}:$$

a resonance-stabilised cation

Step 3: Reaction of the resonance-stabilised cation (an electrophile) with methanol (a nucleophile) gives the conjugate acid of the acetal.

$$\text{CH}_3-\overset{\text{H}}{\text{O}}: + \text{R}-\underset{\text{H}}{\text{C}}=\overset{+}{\text{O}}\text{CH}_3 \rightleftharpoons \text{R}-\underset{\text{H}}{\overset{\text{H}\overset{+}{\text{O}}\text{CH}_3}{\text{C}}}-\text{OCH}_3$$

a protonated acetal

Step 4: Proton transfer from the protonated acetal to A^- gives the acetal and generates a new molecule of H—A.

$$\text{A}:^- + \text{R}-\underset{\text{H}}{\overset{\text{H}\overset{+}{\text{O}}\text{CH}_3}{\text{C}}}-\text{OCH}_3 \rightleftharpoons \text{HA} + \text{R}-\underset{\text{H}}{\overset{\text{OCH}_3}{\text{C}}}-\text{OCH}_3$$

a protonated acetal   an acetal

Formation of acetals is often carried out using the alcohol as a solvent and dissolving a dry protic or Lewis acid in the alcohol. Commonly used acids include hydrogen chloride gas, concentrated sulfuric acid, sulfonic acids (RSO_3H), zinc chloride and iron(III) chloride. Because the alcohol is both a reactant and the solvent, it is present in a large molar excess, which drives the reaction to the right and favours acetal formation. Alternatively, the reaction may be driven to the right by the removal of water as it is formed.

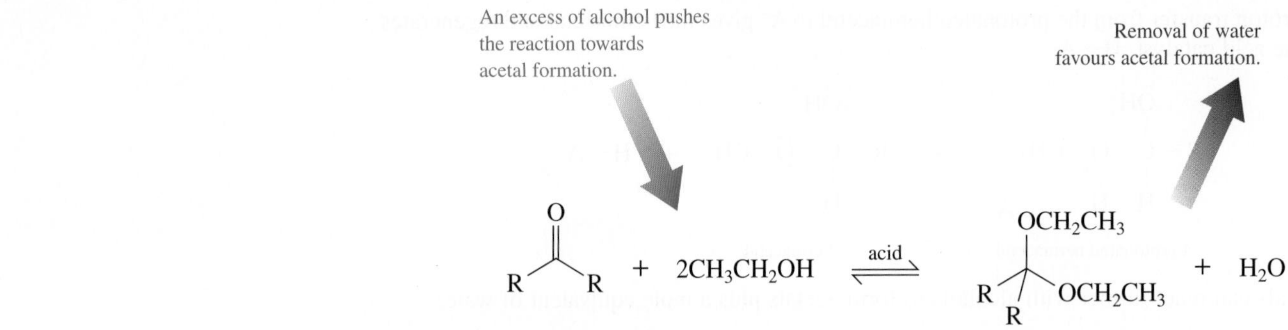

An excess of alcohol pushes the reaction towards acetal formation.

Removal of water favours acetal formation.

$$R_2C{=}O + 2CH_3CH_2OH \underset{}{\overset{\text{acid}}{\rightleftharpoons}} R_2C(OCH_2CH_3)_2 + H_2O$$

Note that, as water is a product in this reaction, the reagents (the carbonyl compound and the alcohol) must be anhydrous. Any excess water in the reaction mixture limits the amount of product formed. An important property of acetals is that they can be readily converted back to their parent carbonyl compound by the addition of excess water.

WORKED EXAMPLE 21.6

Formation of acetals

Show the reaction of the carbonyl group of each of the following ketones with one mole equivalent of alcohol to form a hemiacetal. Then show the reaction with a second mole equivalent of alcohol to form an acetal. (Note that, in part (b), ethane-1,2-diol (ethylene glycol) is a diol and provides both —OH groups.)

(a) 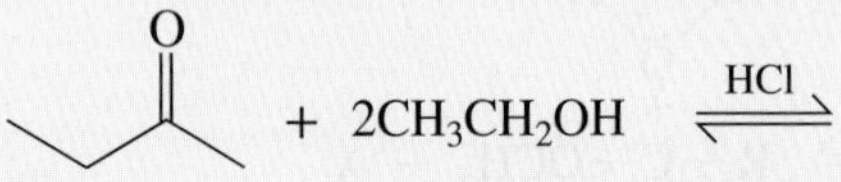 butan-2-one + $2CH_3CH_2OH$ $\overset{\text{HCl}}{\rightleftharpoons}$

(b) 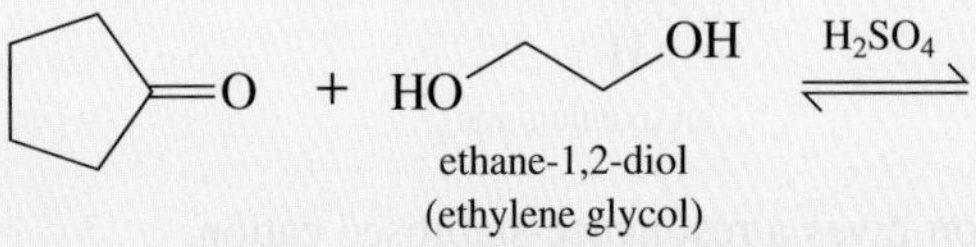 cyclopentanone + HO–CH₂CH₂–OH $\overset{H_2SO_4}{\rightleftharpoons}$

ethane-1,2-diol (ethylene glycol)

Analysis

The carbonyl carbon atom will become the hemiacetal, and then the acetal, so highlight this atom. The hemiacetal is formed by first forming an O—H bond between the carbonyl oxygen atom and the acid catalyst, and then a C—OR bond between the carbonyl carbon atom and the alcohol oxygen atom. The acetal is formed by replacing the —OH group with an —OR group. In example (b), the second OH group of the ethylene glycol in the hemiacetal is available to react to form the cyclic acetal.

Solution

Here are structural formulae of the hemiacetal and then the acetal.

(a) 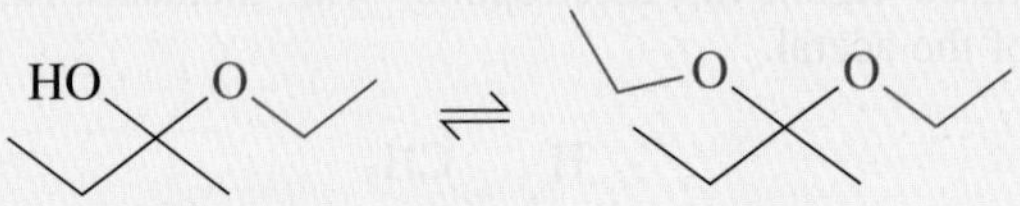 HO, O ⇌ O, O

(b) 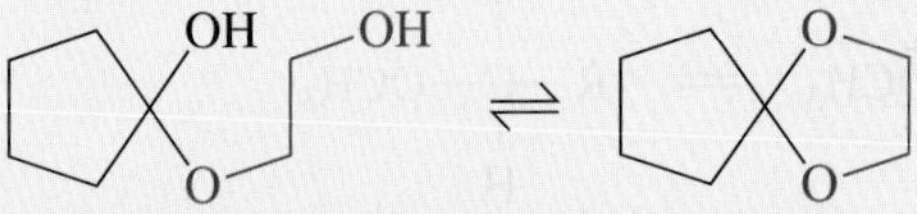 OH, OH, O ⇌ O, O

Is our answer reasonable?

A simple way to check if your structures make sense is to check that the number of carbon atoms on the left-hand side of the reaction is the same as the right-hand side. If these agree, then check that each hemiacetal has a carbon atom with an —OH and an —OR group and, for the acetal, this carbon atom now has two —OR groups.

PRACTICE EXERCISE 21.6

The hydrolysis of an acetal forms an aldehyde or a ketone and two equivalents of alcohol. The following are structural formulae for three acetals.

(a) 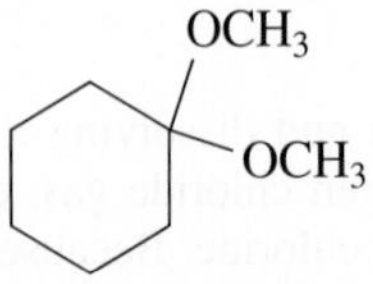 OCH_3, OCH_3 (b) 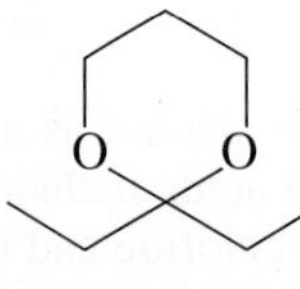 O, O (c) 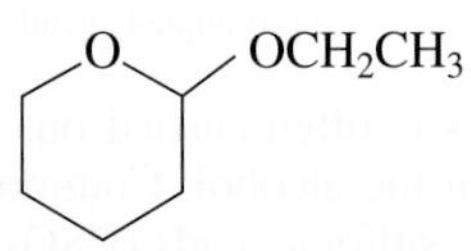 O, OCH_2CH_3

Draw the structural formulae for the products of the hydrolysis of each acetal in aqueous acid.

Like ethers, acetals are unreactive to bases, reducing agents, Grignard reagents and oxidising agents (except, of course, those which involve aqueous acid). Because of their lack of reactivity with these reagents, acetals are often used to protect the carbonyl groups of aldehydes and ketones while reactions are carried out on functional groups in other parts of the molecule. The acetal group can then be readily removed in high yield by mild acid hydrolysis.

Acetals as carbonyl-protecting groups

The use of acetals as carbonyl-protecting groups is illustrated by the synthesis of 5-hydroxy-5-phenylpentanal from benzaldehyde and 4-bromobutanal.

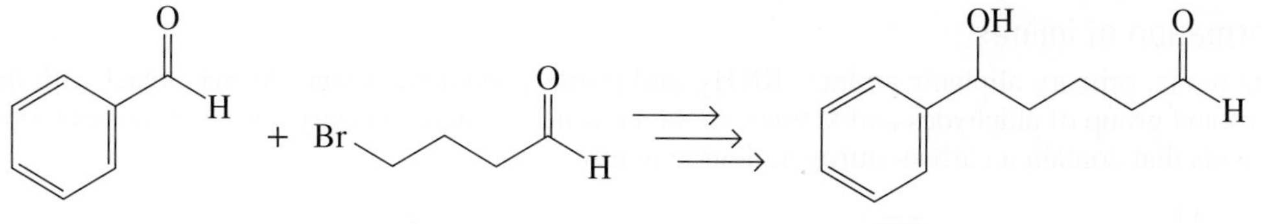

benzaldehyde + 4-bromobutanal ⟹ 5-hydroxy-5-phenylpentanal

One obvious way to form a new carbon–carbon bond between these two molecules is to treat benzaldehyde with the Grignard reagent formed from 4-bromobutanal. This Grignard reagent, however, would react immediately with the carbonyl group within the same molecule or in another molecule of 4-bromobutanal, causing it to self-destruct during preparation. A way to avoid this problem is to protect the carbonyl group of 4-bromobutanal by converting it to an acetal. Cyclic acetals are often used because they are particularly easy to prepare and are more stable to hydrolysis. This is similar to the chelate effect observed in the coordination of bidentate ligands to metal ions (see section 13.4).

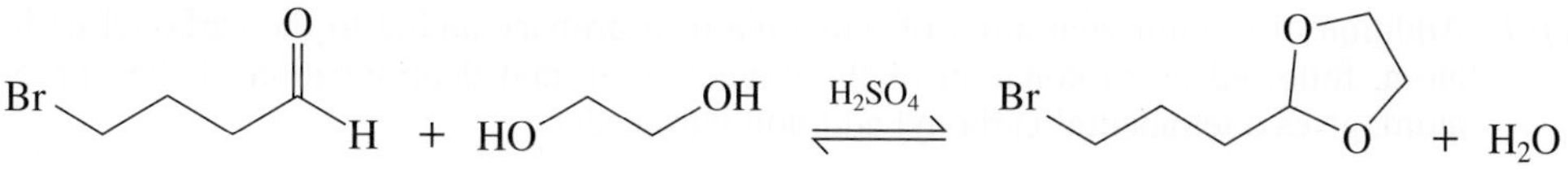

4-bromobutanal + ethylene glycol $\xrightleftharpoons{H_2SO_4}$ a cyclic acetal + H_2O

Treatment of the protected bromoaldehyde with magnesium in diethyl ether, followed by the addition of benzaldehyde, gives a magnesium alkoxide.

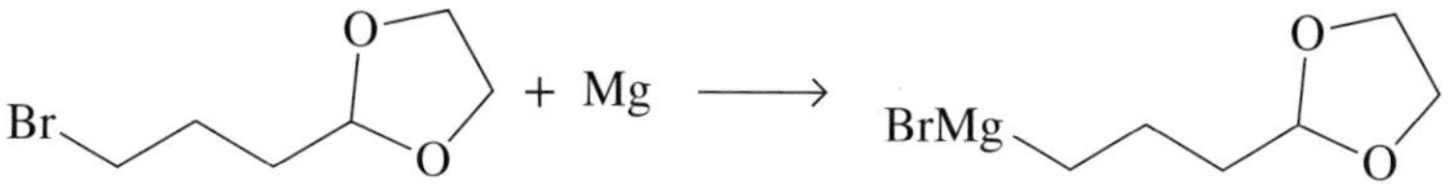

a cyclic acetal + Mg ⟶ a Grignard reagent

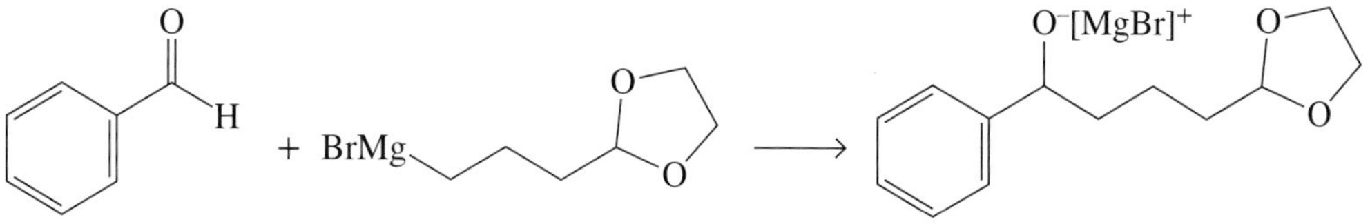

benzaldehyde + BrMg(CH₂)₃-acetal ⟶ a magnesium alkoxide ($O^-[MgBr]^+$)

Treatment of the magnesium alkoxide with aqueous acid accomplishes two things. First, protonation of the alkoxide anion gives the desired hydroxyl group, and then, hydrolysis of the cyclic acetal regenerates the aldehyde group.

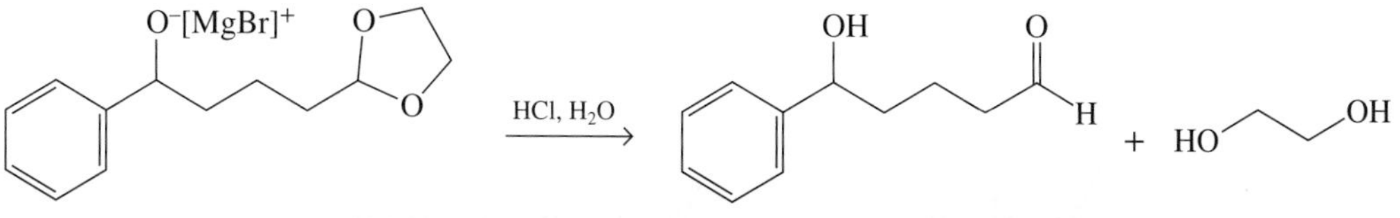

$O^-[MgBr]^+$ alkoxide $\xrightarrow{HCl,\ H_2O}$ 5-hydroxy-5-phenylpentanal + HO–CH₂CH₂–OH

Addition of ammonia, amines and related compounds

The chemistry in this section can be summarised by the general reaction shown at the top of the next page where G can be a range of groups including —H, —R, —Ar, —NH_2, —NHAr and

—OH but *not* an acyl group, —C(=O)R. Ultimately the carbonyl oxygen atom is replaced by a substituted nitrogen atom (N—G) with the formation of water as a by-product.

$$R_2C{=}O + H_2N{-}G \longrightarrow R_2C{=}N{-}G + H_2O$$

$$G = -H,\ -R,\ -Ar,\ -NH_2,\ -NHAr,\ -OH \quad G \neq -\overset{O}{\overset{\|}{C}}R$$

Formation of imines

Ammonia, primary aliphatic amines, RNH_2, and primary aromatic amine, $ArNH_2$, react with the carbonyl group of aldehydes and ketones in the presence of an acid catalyst to give imines (compounds that contain a carbon–nitrogen double bond).

$$CH_3\overset{O}{\overset{\|}{C}}H + H_2N{-}C_6H_5 \xrightleftharpoons{H_3O^+} CH_3CH{=}N{-}C_6H_5 + H_2O$$

acetaldehyde aniline an imine

$$\text{cyclohexanone}{=}O + NH_3 \xrightleftharpoons{H_3O^+} \text{cyclohexane}{=}NH + H_2O$$

cyclohexanone ammonia an imine

A mechanism is described in the following steps.

Step 1: Addition of the nitrogen atom of ammonia or a primary amine to the carbonyl carbon atom, followed by protonation of the oxygen atom and deprotonation of the nitrogen atom, gives a tetrahedral carbonyl addition intermediate.

$$\overset{\ddot{O}:}{\overset{\|}{C}} + H_2N{-}R \rightleftharpoons -\overset{:\ddot{O}:^-}{\overset{|}{C}}-\underset{H}{\overset{H}{N^+}}-R \rightleftharpoons -\overset{:\ddot{O}H}{\overset{|}{C}}-\underset{H}{\ddot{N}}-R$$

This step is analogous to the formation of hemiacetals, as both are examples of carbonyl addition reactions. The intermediate formed in step 1 is unstable, and step 2 occurs spontaneously.

Step 2: Protonation of the —OH group, followed by loss of water and proton transfer to the solvent gives the imine. Notice that the loss of water and the proton transfer have the characteristics of an E2 reaction (see section 18.3). Two things happen simultaneously in this dehydration: the carbon–nitrogen double bond forms, and the leaving group (in this case, a water molecule) departs.

$$-\overset{:\ddot{O}H}{\overset{|}{C}}-\underset{H}{\ddot{N}}-R \rightleftharpoons -\overset{H_2O^+}{\overset{|}{C}}-\underset{H}{\ddot{N}}-R \rightleftharpoons \;C{=}\underset{H}{\overset{+}{N}}-R \leftrightarrow \overset{+}{C}-\underset{H}{\ddot{N}}-R + 2H_2O$$

resonance-stabilised cation

Step 3: Deprotonation of the nitrogen atom in the resonance-stabilised cation gives the imine.

$$C{=}\underset{H}{\overset{+}{N}}-R + H_2\ddot{O} \rightleftharpoons C{=}\ddot{N}-R + H_3O^+$$

an imine

One example of the importance of imines in biological systems is the active form of vitamin A aldehyde (retinal), which is bound to the protein opsin in the human retina in the form of an imine called rhodopsin or visual purple (see chapter 16, pp. 714–5). The amino acid lysine (table 24.1, p. 1064) provides the primary amino group for this reaction.

11-*cis*-retinal + H_2N—opsin ⟶ rhodopsin visual purple

WORKED EXAMPLE 21.7

Structural formulae for imines

Write a structural formula for the imine formed in each of the following reactions.

(a) cyclopentanone + NH_2 (2-butylamine) $\xrightarrow[-H_2O]{HCl}$

(b) acetone + H_2N—C_6H_4—OCH_3 $\xrightarrow[-H_2O]{HCl}$

Analysis

The carbonyl carbon atom will be bonded to the nitrogen atom by a C=N double bond. The oxygen atom of the carbonyl group and the two hydrogen atoms of the —NH_2 group combine to form water.

Solution

Below is a structural formula for each imine.

(a) cyclopentylidene=N—(sec-butyl)

(b) $(CH_3)_2C$=N—C_6H_4—OCH_3

Is our answer reasonable?

Check that you get the starting materials if you cleave the C=N double bond and add an oxygen atom to the carbon atom and two hydrogen atoms to the nitrogen atom.

PRACTICE EXERCISE 21.7

Acid-catalysed hydrolysis of an imine gives an amine and an aldehyde or a ketone. When one equivalent of acid is used, the amine is converted to its ammonium salt. For each of the following imines, write a structural formula for the products of hydrolysis, using one equivalent of HCl.

(a) cyclopentyl—$CH_2N=C(CH_3)_2$ + H_2O $\xrightarrow{HCl}$

(b) phenyl—CH=N—cyclopentyl + H_2O $\xrightarrow{HCl}$

Formation of oximes and hydrazones

When aldehydes and ketones react with hydroxylamine, H_2NOH, they form **oximes** (compounds containing a C═NOH group).

$$\text{cyclohexanone}{=}O + H_2N{-}OH \longrightarrow \text{cyclohexyl}{=}N{-}OH + H_2O$$

an oxime

The LIX® reagents are examples of oximes and are produced industrially for the solvent extraction of copper and nickel from leach solutions. These reagents can be made to coordinate selectively to Cu^{2+} and Ni^{2+} ions by careful manipulation of the pH. The large alkyl group *para* to the phenol ensures the reagents have low solubility in aqueous solutions and remain in the organic solvent (usually kerosene).

OH NOH A R

$A = H$ or CH_3

$R = C_9H_{19}$ or $C_{12}H_{25}$

an LIX® reagent

A convenient spot test for aldehydes and ketones involves the addition of a solution of 2,4-dinitrophenylhydrazine (Brady's reagent), also known as the 2,4-DNP test. In the presence of aldehydes or ketones this reagent forms bright orange/red hydrazones.

$$(CH_3CH_2)_2C{=}O + H_2N{-}NH{-}C_6H_3(NO_2)_2 \longrightarrow (CH_3CH_2)_2C{=}N{-}NH{-}C_6H_3(NO_2)_2 + H_2O$$

2,4-dinitrophenylhydrazine

a hydrazone (a bright orange/red precipitate)

Reductive amination of aldehydes and ketones

One of the chief values of imines is that the carbon–nitrogen double bond can be reduced to a carbon–nitrogen single bond by hydrogen in the presence of a catalyst (typically nickel, palladium or platinum). By this two-step reaction, called **reductive amination**, a primary amine is converted into a secondary amine through an imine intermediate, as illustrated below by the conversion of cyclohexylamine to dicyclohexylamine.

$$\text{cyclohexanone} + \text{cyclohexylamine (a 1° amine)} \xrightarrow[-H_2O]{H_3O^+} [\text{(an imine)}]$$

$$[\text{(an imine)}] \xrightarrow{H_2/Ni} \text{dicyclohexylamine (a 2° amine)}$$

Conversion of an aldehyde or a ketone to an amine is generally carried out in one laboratory operation by mixing together the carbonyl-containing compound, the amine or ammonia, hydrogen, and the transition metal catalyst. The imine intermediate is not isolated.

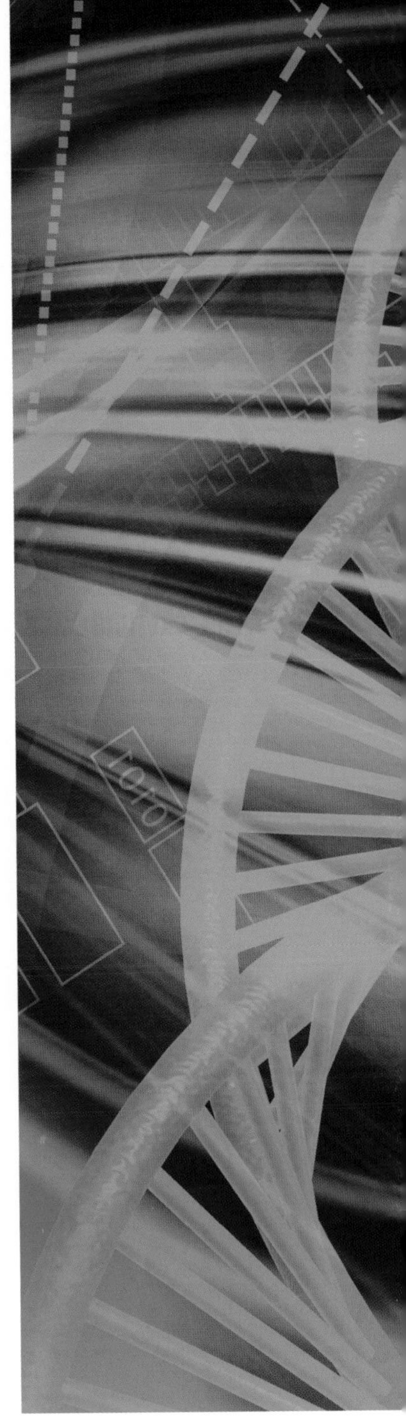

Chemical Connections

Drug delivery by hydrolysis

Formaldehyde has antiseptic properties and can be used to treat urinary tract infections due to its ability to react with nucleophiles present in urine. However, formaldehyde can be toxic when exposed to other regions of the body. Therefore, the use of formaldehyde as an antiseptic agent requires a method for selective delivery to the urinary tract. This can be accomplished by using a prodrug called methenamine.

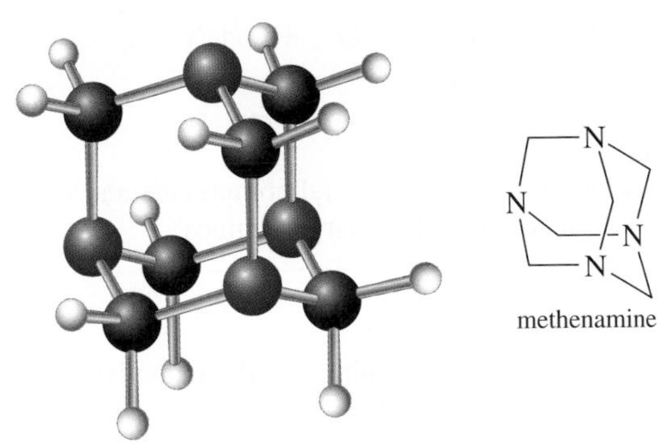

methenamine

This compound is a nitrogen analogue of an acetal. That is, each carbon atom is connected to two nitrogen atoms, very much like an acetal in which a carbon atom is connected to two oxygen atoms. A carbon atom that is connected to two heteroatoms (O or N) can undergo acid-catalysed hydrolysis.

$$R_2C(Z)_2 \xrightarrow{H_3O^+} R_2C{=}O$$

Z = O or N

Each of the carbon atoms in methenamine can be hydrolysed, releasing formaldehyde.

$$\text{methenamine} \xrightarrow{H_3O^+} 6\ H_2C{=}O + 4\overset{+}{N}H_4$$

methenamine — formaldehyde

Methenamine is placed in special tablets that do not dissolve as they travel through the acidic environment of the stomach, but do dissolve once they reach the basic environment of the intestinal tract. Methenamine is thereby released into the intestinal tract, where it is stable under basic conditions. Once it reaches the acidic environment of the urinary tract, methenamine is hydrolysed, releasing formaldehyde, as shown above. In this way, methenamine is used as a prodrug that enables delivery of formaldehyde specifically to the urinary tract. This method prevents the systemic release of formaldehyde to other organs of the body where it would be toxic.

WORKED EXAMPLE 21.8

Synthesis of amines

Show how to synthesise each of the following amines by reductive amination.

(a) 1-phenylethanamine (NH_2)　(b) diisopropylamine (H–N)

Analysis

In reductive amination, an imine is reduced to an amine by addition of hydrogen across the C═N double bond. Working backwards to determine the structure of the imine, simply 'draw in' a double bond between the C and N atoms (and remove a hydrogen atom from the nitrogen atom). In (a) there is only one C—N bond, so the choice is easy; in (b) there are two, but the groups are the same, so again the choice is easy. To determine the structure of the carbonyl compound, replace the C=NR group with a C=O group. The identity of the amine is determined by adding two hydrogen atoms to the nitrogen atom of the =NR group.

Solution

The carbonyl compound is treated with ammonia or an amine in the presence of H_2/Ni.

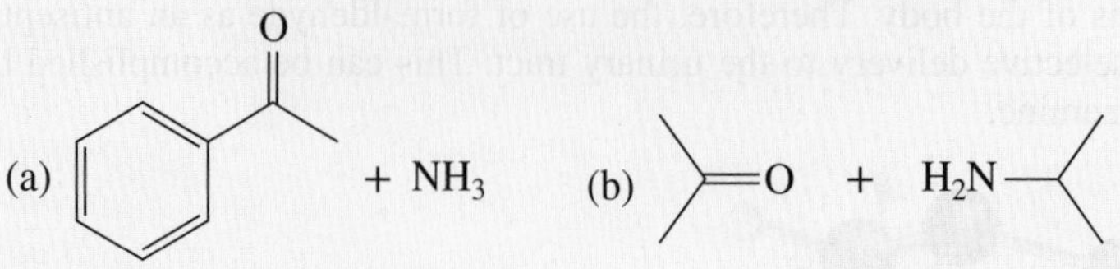

(a) [acetophenone] + NH_3 (b) [acetone] =O + H_2N—[isopropyl]

Is our answer reasonable?

Check that the skeletal structures of the carbonyl compounds and the amines are consistent with those of the imine intermediate and the target amines.

PRACTICE EXERCISE 21.8 Show how to prepare each of the following amines by the reductive amination of an appropriate aldehyde or ketone.

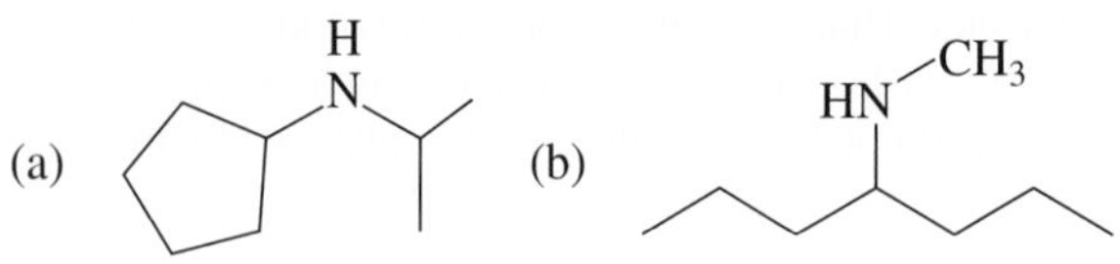

(a) [N-isopropylcyclopentylamine] (b) [HN–CH_3 on 3-hexyl chain]

Reduction

Aldehydes are reduced to primary alcohols, and ketones are reduced to secondary alcohols.

$$\underset{\text{an aldehyde}}{\overset{O}{\overset{\|}{RCH}}} \xrightarrow{\text{reduction}} \underset{\text{a primary alcohol}}{RCH_2OH} \qquad \underset{\text{a ketone}}{\overset{O}{\overset{\|}{RCR'}}} \xrightarrow{\text{reduction}} \underset{\text{a secondary alcohol}}{\overset{OH}{\overset{|}{RCHR'}}}$$

Metal hydride reductions

By far the most common laboratory reagents used to reduce the carbonyl group of an aldehyde or ketone to a hydroxyl group are sodium borohydride and lithium aluminium hydride. Each of these compounds behaves as a source of the **hydride ion**, H^-, which is a very good nucleophile. A hydride ion is a hydrogen atom with two electrons in its valence shell, $H:^-$. The structural formulae drawn below show formal negative charges on boron and aluminium.

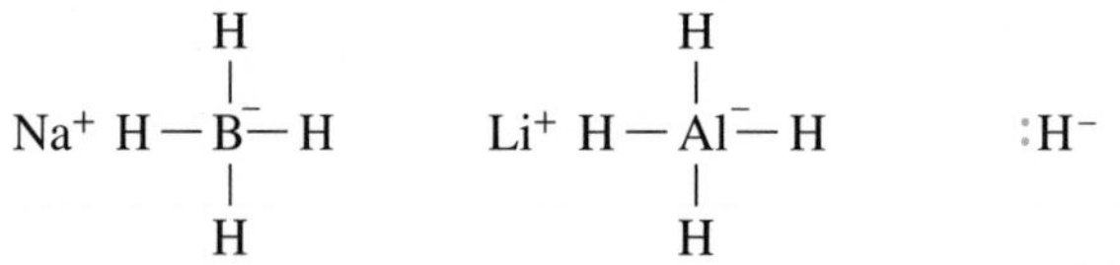

Na^+ H—$\bar{B}$—H (with H above and below) — sodium borohydride

Li^+ H—$\bar{Al}$—H (with H above and below) — lithium aluminium hydride

$:H^-$ — hydride ion

In fact, hydrogen is more electronegative than either boron or aluminium (H = 2.2, Al = 1.6 and B = 2.0) and the formal negative charge in the two reagents resides more on hydrogen atoms than on the metal.

Lithium aluminium hydride is a very powerful reducing agent; it rapidly reduces not only the carbonyl groups of aldehydes and ketones, but also those of carboxylic acids and their functional derivatives (section 23.5). Sodium borohydride is a much more selective reagent, reducing only aldehydes and ketones rapidly. Reductions using sodium borohydride are most commonly carried out in aqueous methanol, pure methanol or ethanol.

The key step in the metal hydride reduction of an aldehyde or ketone is the transfer of a hydride ion from the reducing agent to the carbonyl carbon atom to form a tetrahedral carbonyl addition compound. In this reaction, only the hydrogen atom attached to the carbon atom comes from the hydride-reducing agent; the hydrogen atom bonded to the oxygen atom comes from the water added to hydrolyse the metal alkoxide salt.

$$Na^+ H{-}\overset{H}{\underset{H}{\overset{|}{\underset{|}{B^-}}}}{-}H + R{-}\overset{\ddot{O}:}{\overset{\|}{C}}{-}R' \longrightarrow R{-}\overset{:\ddot{O}-\bar{B}H_3Na^+}{\overset{|}{\underset{\underset{H}{|}}{C}}}{-}R' \xrightarrow{H_2O} R{-}\overset{O-H}{\overset{|}{\underset{\underset{H}{|}}{C}}}{-}R'$$

This H comes from water during hydrolysis.

This H comes from the hydride-reducing agent.

Catalytic reduction

The carbonyl group of an aldehyde or a ketone is reduced to a hydroxyl group by hydrogen in the presence of a transition metal catalyst, most commonly finely divided palladium, platinum, nickel or rhodium. Reductions are generally carried out at temperatures from 25 °C to 100 °C and at pressures of hydrogen from 100 to 500 kPa (1–5 times atmospheric pressure). Under such conditions, cyclohexanone is reduced to cyclohexanol as shown below.

$$\text{cyclohexanone} + H_2 \xrightarrow[25\ °C,\ 2 \times 10^5\ Pa]{Pt} \text{cyclohexanol}$$

The catalytic reduction of aldehydes and ketones proceeds with generally very high yields, and isolation of the final product is very easy. However, a disadvantage is that some other functional groups are more reactive than aldehydes and ketones towards catalytic reduction (for example, carbon–carbon double bonds) and therefore are also reduced under these conditions.

$$\textit{trans}\text{-but-2-enal} \xrightarrow[Ni]{2H_2} \text{butan-1-ol}$$

The next two equations illustrate the selective reduction of a carbonyl group in the presence of a carbon–carbon double bond and, alternatively, the selective reduction of a carbon–carbon double bond in the presence of a carbonyl group.

Selective reduction of a carbonyl group can be carried out using sodium borohydride.

$$RCH{=}CH\overset{O}{\overset{\|}{C}}R' \xrightarrow[2.\ H_2O]{1.\ NaBH_4} RCH{=}CH\overset{OH}{\overset{|}{C}}HR'$$

Selective reduction of a carbon–carbon double bond can be achieved using hydrogen in the presence of a transition metal catalyst. Note that this reaction is possible because alkenes are easier to reduce by catalytic reduction than ketones. By carefully monitoring the reaction and stopping it after one mole equivalent of hydrogen has been consumed, we can readily obtain the carbonyl compound.

$$RCH{=}CH\overset{O}{\overset{\|}{C}}R' + H_2 \xrightarrow{Rh} RCH_2CH_2\overset{O}{\overset{\|}{C}}R'$$

WORKED EXAMPLE 21.9

Reduction of aldehydes and ketones

Complete the following reductions.

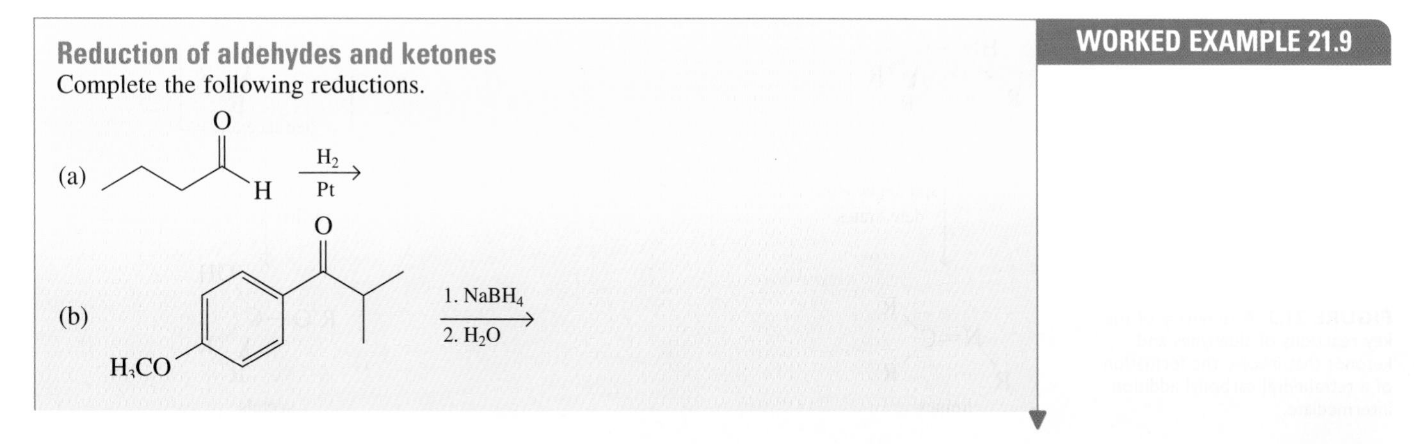

Analysis

The key here is to recognise that the carbon atom of the carbonyl group will become the alcohol carbon atom. To determine the structure of the reduction product, simply add a hydrogen atom to both the carbon and the oxygen atoms.

Solution

The carbonyl group of the aldehyde in (a) is reduced to a primary alcohol, and that of the ketone in (b) is reduced to a secondary alcohol.

(a) OH (b) H_3CO OH

Is our answer reasonable?

The reduction of aldehydes gives primary alcohols, and reduction of ketones gives secondary alcohols. Check that your proposed structures agree with this rule.

PRACTICE EXERCISE 21.9 What aldehyde or ketone gives each of the following alcohols upon reduction by $NaBH_4$?

(a) OH (b) O OH (c) OH

Figure 21.3 gives a summary of the key examples of reactions of aldehydes and ketones that involve the formation of a tetrahedral carbonyl addition intermediate.

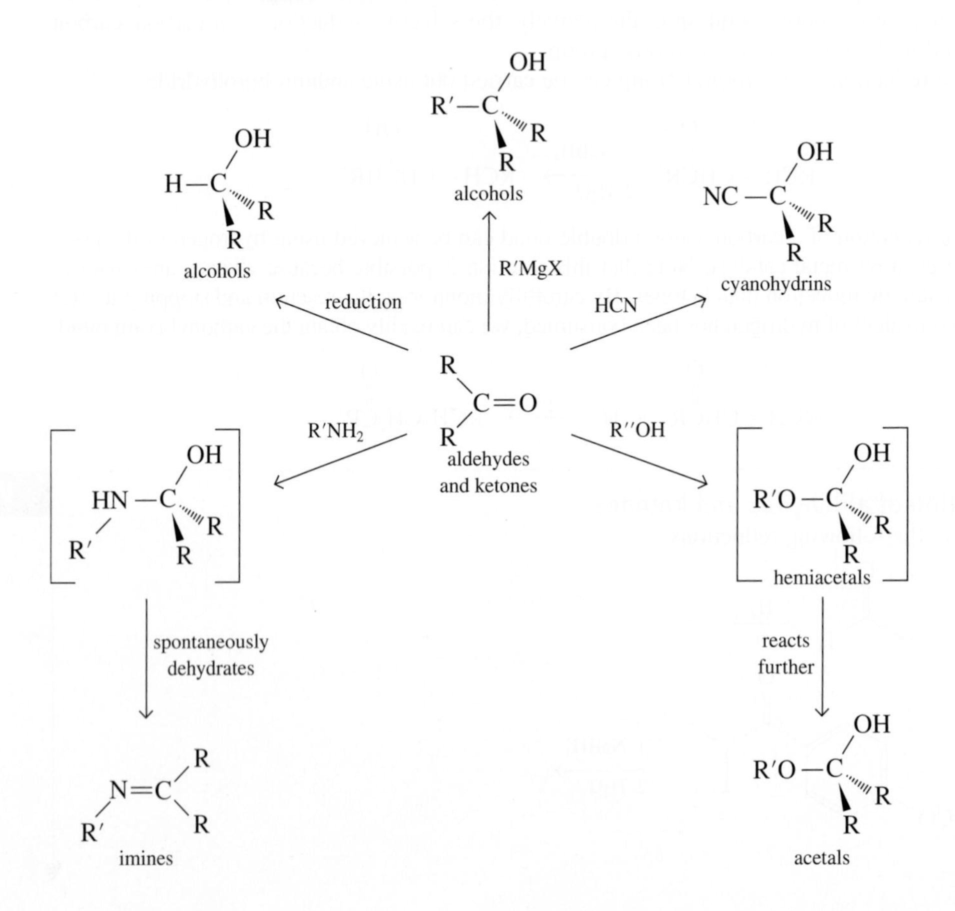

FIGURE 21.3 A summary of the key reactions of aldehydes and ketones that involve the formation of a tetrahedral carbonyl addition intermediate.

Oxidation of aldehydes to carboxylic acids

Aldehydes are oxidised to carboxylic acids by a variety of common oxidising agents, including chromic acid and molecular oxygen. In fact, aldehydes are one of the most easily oxidised functional groups. Oxidation by chromic acid is illustrated below by the conversion of hexanal to hexanoic acid.

$$\text{hexanal} \xrightarrow{H_2CrO_4} \text{hexanoic acid}$$

Aldehydes are also oxidised to carboxylic acids by silver(I) ions. One laboratory procedure is to shake a solution of the aldehyde dissolved in aqueous ethanol or tetrahydrofuran (THF) with a slurry of Ag_2O.

$$\text{vanillin (from vanilla)} + Ag_2O \xrightarrow[NaOH]{THF,\ H_2O} \xrightarrow[H_2O]{HCl} \text{vanillic acid} + Ag$$

Tollens' reagent, another form of silver(I) ion, is prepared by dissolving $AgNO_3$ in aqueous ammonia to give the silver–ammonia complex ion diamminesilver(I) nitrate.

$$Ag^+NO_3^- + 2NH_3 \xrightarrow{NH_3/H_2O} [Ag(NH_3)_2]^+NO_3^-$$

When Tollens' reagent is added to an aldehyde, the aldehyde is oxidised to a carboxylate anion, and silver ions are reduced to metallic silver. If this reaction is carried out properly, silver precipitates as a smooth, mirror-like deposit, hence the name **silver-mirror test** (figure 21.4).

$$RCH{=}O + 2[Ag(NH_3)_2]^+ \xrightarrow{NH_3/H_2O} RCO_2^- + 2Ag + 4NH_3$$

(2Ag: precipitates as silver mirror)

Nowadays, silver ions are rarely used for the oxidation of aldehydes because of the high cost of silver and because other, more convenient oxidation methods exist. The reaction, however, is still commonly used for silvering mirrors. In this process, formaldehyde or glucose is used as the aldehyde to reduce the silver(I) ions.

Aldehydes are also oxidised to carboxylic acids by molecular oxygen and by hydrogen peroxide.

$$2\ C_6H_5CHO + O_2 \longrightarrow 2\ C_6H_5COOH$$

(benzaldehyde → benzoic acid)

FIGURE 21.4 A silver mirror has been deposited on the inside of this flask by the reaction of an aldehyde with Tollens' reagent.

Molecular oxygen is the least expensive and most readily available of all oxidising agents, and, on an industrial scale, air oxidation of organic molecules, including aldehydes, is common. Air oxidation of aldehydes can also be a problem; aldehydes that are liquid at room temperature are so sensitive to oxidation by molecular oxygen that they must be protected from contact with air during storage. Often, this is done by sealing the aldehyde in a container under an atmosphere of nitrogen.

WORKED EXAMPLE 21.10

Oxidation of aldehydes

Draw a structural formula of the product formed by treating each of the following compounds with Tollens' reagent, followed by acidification with aqueous HCl.

(a) pentanal (b) cyclopentanecarbaldehyde

Analysis

Only the aldehyde group in each case is oxidised. This oxidation gives a carboxylic acid group.

Solution

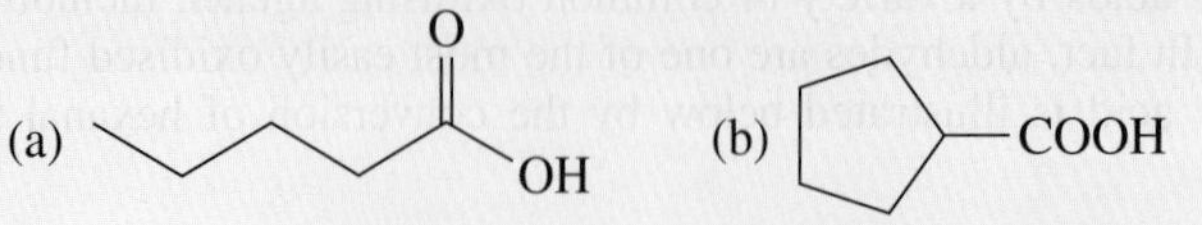

(a) pentanoic acid (b) cyclopentanecarboxylic acid

Is our answer reasonable?

Check that the products are carboxylic acids and that the number of carbon atoms in the starting material is the same as the product.

PRACTICE EXERCISE 21.10 Complete the following oxidations.

(a) 2-hydroxybenzaldehyde + O_2 → (b) phenylacetaldehyde + Tollen's reagent →

Oxidation of ketones to carboxylic acids

Ketones are much more resistant to oxidation than aldehydes. For example, ketones are not normally oxidised by chromic acid or potassium permanganate at room temperature. In fact, these reagents are used routinely to oxidise secondary alcohols to ketones in good yield (section 19.2).

Ketones undergo oxidative cleavage, via their enol form (see below), using potassium dichromate and potassium permanganate at higher temperatures, as well as by higher concentrations of nitric acid, HNO_3. The carbon–carbon double bond of the enol is cleaved to form two carboxyl or ketone groups, depending on the substitution pattern of the original ketone. An important industrial application of this reaction is the oxidation of cyclohexanone to hexanedioic acid (adipic acid), one of the two monomers required for the synthesis of the polymer nylon 6,6.

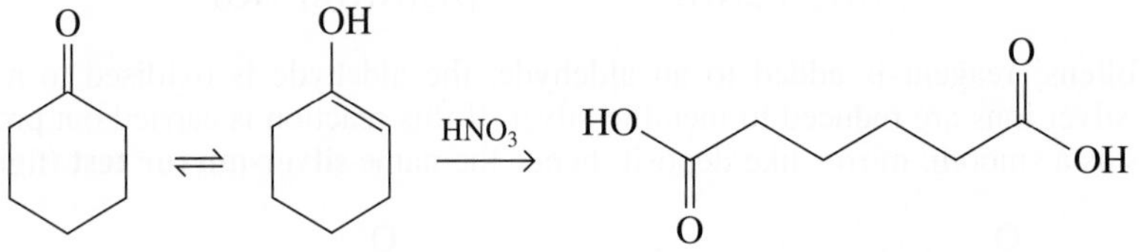

cyclohexanone (keto form) ⇌ cyclohexanone (enol form) $\xrightarrow{HNO_3}$ hexanedioic acid (adipic acid)

21.6 Keto–enol tautomerism

A carbon atom adjacent to a carbonyl group is called an **α-carbon**, and any hydrogen atoms bonded to it are called **α-hydrogens**.

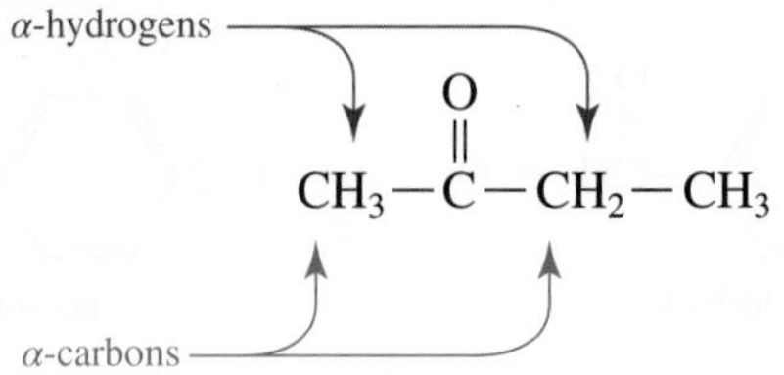

α-hydrogens

$CH_3-C(=O)-CH_2-CH_3$

α-carbons

An aldehyde or ketone that has at least one α-hydrogen atom is in equilibrium with a constitutional isomer called an **enol**. The name 'enol' is derived from the IUPAC designation of it as both an alkene (*-en-*) and an alcohol (*-ol*).

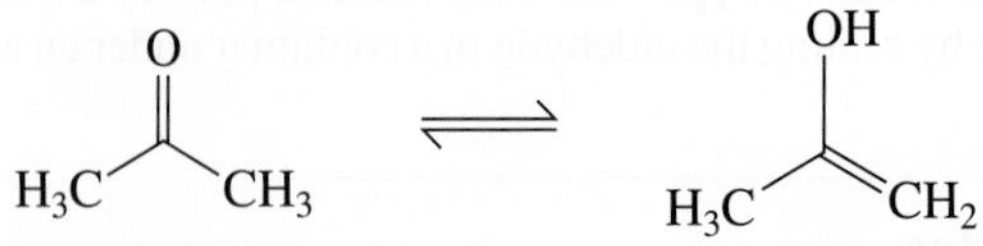

acetone (keto form) ⇌ acetone (enol form)

Keto and enol forms are examples of **tautomers**, which are constitutional isomers in equilibrium with each other. Tautomers differ in the location of a hydrogen atom and a double bond relative to a heteroatom, most commonly O, S or N. This type of isomerism is called **tautomerism**.

For most simple aldehydes and ketones, the position of the equilibrium in keto–enol tautomerism lies far on the side of the keto form (table 21.4), because a carbon–oxygen double bond is stronger than a carbon–carbon double bond.

TABLE 21.4 The position of keto–enol equilibrium for four aldehydes and ketones.

Keto form		Enol form	% Enol at equilibrium
$CH_3CH{=}O$	⇌	$CH_2{=}CH{-}OH$	6×10^{-5}
$CH_3C(=O)CH_3$	⇌	$CH_3C(OH){=}CH_2$	6×10^{-7}
cyclopentanone	⇌	cyclopent-1-en-1-ol	1×10^{-6}
cyclohexanone	⇌	cyclohex-1-en-1-ol	1×10^{-5}

The equilibration of keto and enol forms is catalysed by acid, as shown in the following two-step mechanism. (Note that a molecule of H—A is consumed in step 1, but another is generated in step 2.)

Step 1: Proton transfer from the acid catalyst, H—A, to the carbonyl oxygen atom forms the conjugate acid of the aldehyde or ketone.

$$CH_3{-}C(=\ddot{O}{:}){-}CH_3 + H{-}A \overset{\text{fast}}{\rightleftharpoons} CH_3{-}C(=\overset{+}{O}{-}H){-}CH_3 + {:}A^-$$

keto form — the conjugate acid of the ketone

Step 2: Proton transfer from the α-carbon atom to the base, A^-, gives the enol and generates a new molecule of the acid catalyst, H—A.

$$CH_3{-}C(=\overset{+}{O}{-}H){-}CH_2{-}H + {:}A^- \overset{\text{slow}}{\rightleftharpoons} CH_3{-}C(\ddot{O}H){=}CH_2 + H{-}A$$

enol form

This tautomerism can also be base catalysed. The process is the reverse of the acid-catalysed process in that the base removes a proton from the α-carbon atom first. The resulting alkoxide then removes a proton from the conjugate acid (formed when the base removes the α-hydrogen atom).

$$H_3C{-}C(=\ddot{O}{:}){-}CH_2{-}H \;\; {:}\ddot{O}H^- \rightleftharpoons H_3C{-}C(=\ddot{O}{:}){-}CH_2^- \longleftrightarrow H_3C{-}C({:}\ddot{O}{:}^-){=}CH_2$$

keto form — the conjugate base of the ketone (a resonance-stabilised anion)

$$H_3C{-}C({:}\ddot{O}{:}^-){=}CH_2 \;\; H{-}\ddot{O}H \rightleftharpoons H_3C{-}C(\ddot{O}H){=}CH_2 + {:}\ddot{O}H^-$$

WORKED EXAMPLE 21.11

Enol forms

Write two enol forms for each of the following compounds, and state which enol predominates at equilibrium.

(a) 2-methylcyclohexanone (b) 2-hexanone

Analysis

The carbon atom of the carbonyl group will remain sp^2 hybridised (it will be part of the alkene in the enol). The two possible enols are obtained by removal of an α-hydrogen atom from the two α-carbon atoms. Zaitsev's rule (section 18.3) will help predict the major enol form.

Solution

In each case, the major enol form has the more substituted (the more stable) carbon–carbon double bond.

(a) 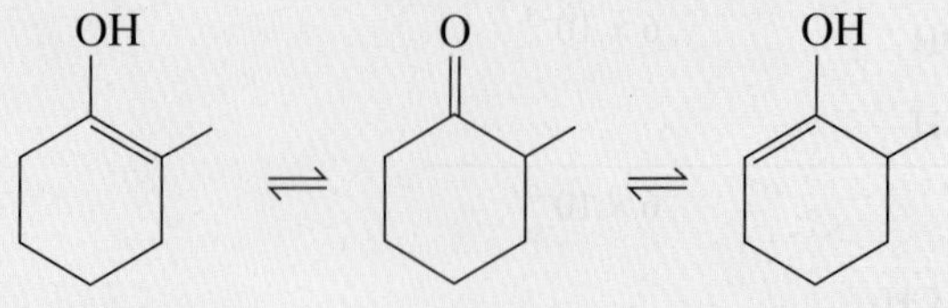 major enol ⇌ ketone ⇌ minor enol

(b) 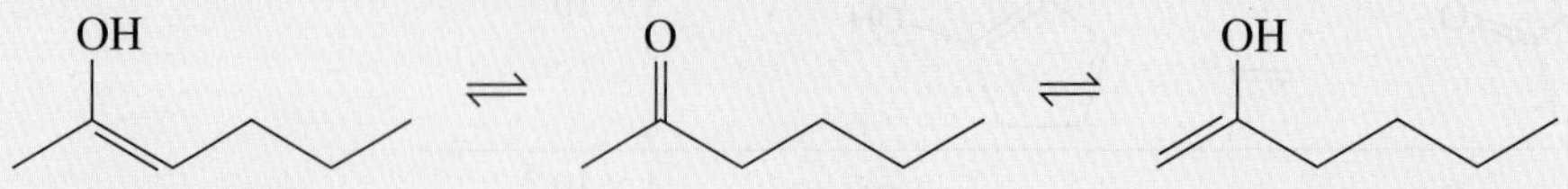 major enol ⇌ ketone ⇌ minor enol

Is our answer reasonable?

Check that one of the atoms in the alkene group was the carbonyl carbon atom.

PRACTICE EXERCISE 21.11 Draw the structural formula for the keto form of each of the following enols.

(a) 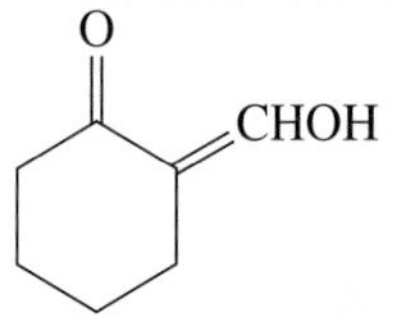 (b) 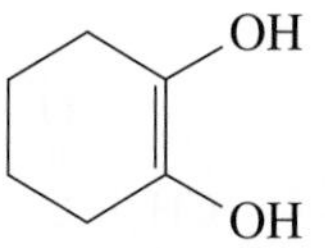 (c)

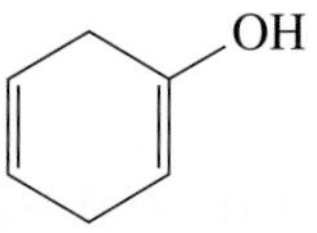

Racemisation at an α-carbon atom

When enantiomerically pure (either *R* or *S*) 3-phenylbutanone is dissolved in ethanol, no change occurs in the optical activity of the solution over time. If, however, a trace of acid is added, the optical activity of the solution begins to decrease and gradually drops to zero. When 3-phenylbutanone is isolated from this solution, it is found to be a racemic mixture (section 17.5). This observation can be explained by the acid-catalysed formation of an achiral enol intermediate. In the tautomerism of the achiral enol to the chiral keto form, the hydrogen ion can add to carbon-3 of the enol ether from above or below to generate the *R* and *S* enantiomers with equal probability.

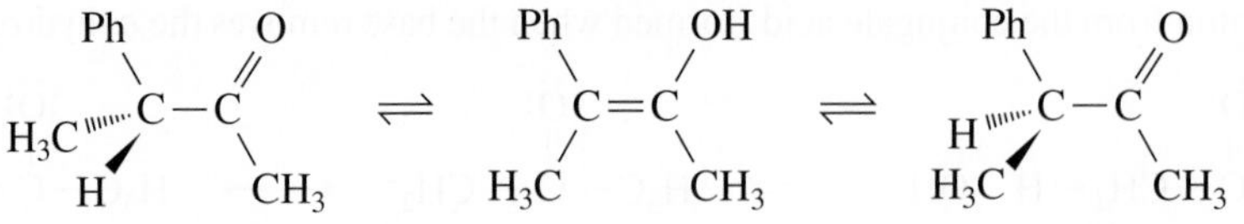

(*R*)-3-phenylbutanone ⇌ an achiral enol ⇌ (*S*)-3-phenylbutanone

Racemisation by this mechanism occurs only at α-carbon stereocentres with at least one α-hydrogen atom.

α-halogenation

Aldehydes and ketones with at least one α-hydrogen atom react with bromine and chlorine at the α-carbon atom to give an α-haloaldehyde or α-haloketone. Acetophenone, for example, reacts with bromine in acetic acid to give α-bromoacetophenone, an α-bromoketone.

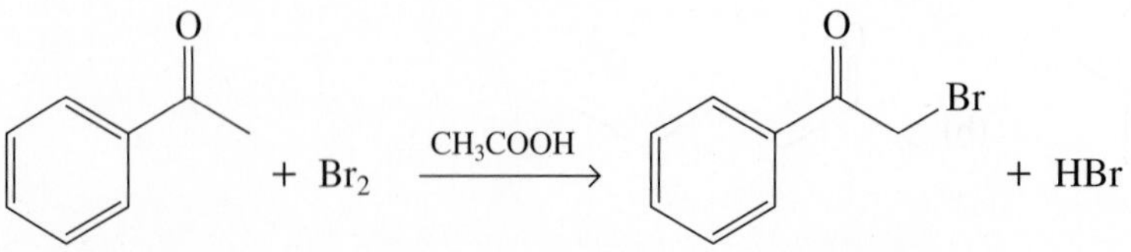

$$\text{acetophenone} + Br_2 \xrightarrow{CH_3COOH} \alpha\text{-bromoacetophenone} + HBr$$

Acidic and basic conditions both catalyse α-halogenation. For acid-catalysed halogenation, the HBr or HCl generated catalyses further reaction.

Step 1: Acid-catalysed keto–enol tautomerism gives the enol.

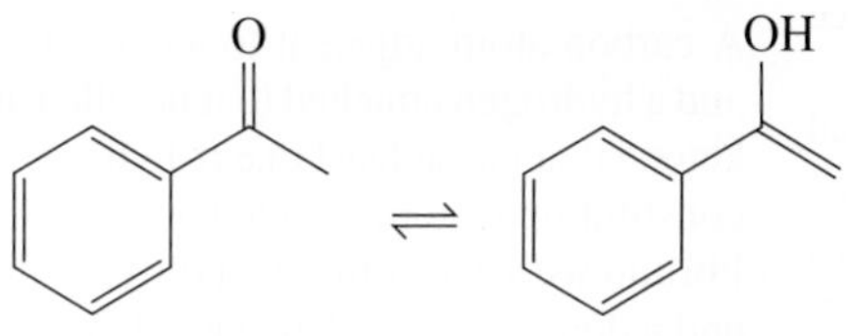

keto form ⇌ enol form

Step 2: Nucleophilic attack of the enol on the halogen molecule gives the α-haloketone.

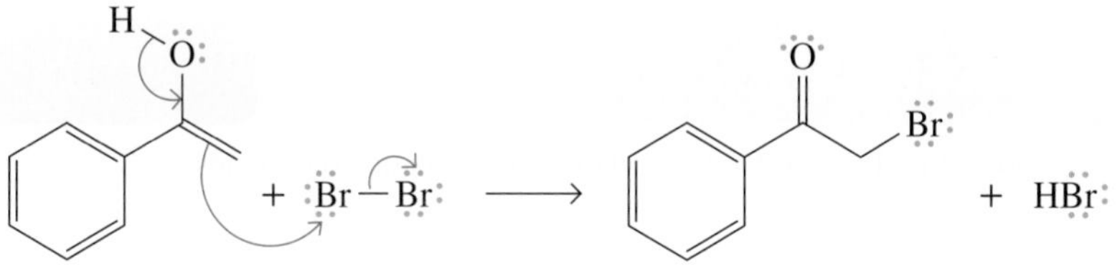

$$+ :Br-Br: \longrightarrow \quad + HBr:$$

The value of α-halogenation is that it converts an α-carbon atom into a centre that has a good leaving group and is therefore susceptible to substitution by a variety of good nucleophiles. In the following example, diethylamine (a nucleophile) reacts with the α-bromoketone to give an α-diethylaminoketone.

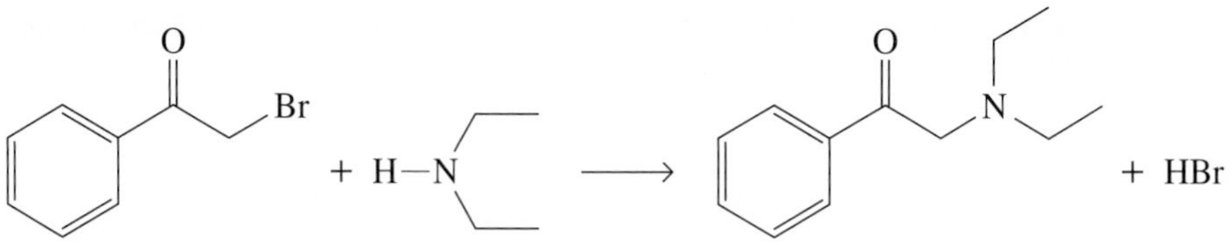

an α-bromoketone + H—N(Et)₂ ⟶ an α-diethylaminoketone + HBr

In practice, this type of nucleophilic substitution is generally carried out in the presence of a weak base, such as potassium carbonate, to neutralise the HX as it is formed.

SUMMARY

LO 1 Structure and bonding

An aldehyde contains a carbonyl group bonded to a hydrogen atom and a carbon atom. A ketone contains a carbonyl group bonded to two carbon atoms.

LO 2 Nomenclature

An aldehyde is named by changing the *-e* suffix of the parent alkane to *-al*. A —CHO group bonded to a ring is indicated by the suffix *-carbaldehyde*. A ketone is named by changing the *-e* suffix of the parent alkane to *-one* and using a number to locate the carbonyl group. In naming compounds that contain more than one functional group, the IUPAC system has established an order of precedence of functional groups. If the carbonyl group of an aldehyde or a ketone is lower in precedence than other functional groups in the molecule, it is indicated by the infix *-oxo-*.

LO 3 Physical properties

Aldehydes and ketones are polar compounds and interact in the pure state by dipole–dipole interactions; they have higher boiling points and are more soluble in water than nonpolar compounds of similar molar mass and similar number of electrons.

LO 4 Preparation of aldehydes and ketones

Aldehydes and ketones can be prepared from various functional groups using many different reactions. Friedel–Crafts acylation of aromatic compounds yields ketones (aldehydes can not be obtained by this method). Aldehydes and ketones can be prepared by the oxidation of alcohols; primary alcohols give aldehydes, and secondary alcohols give ketones. Ozonolysis of alkenes cleaves the carbon–carbon double bond to produce two carbonyl groups. Ketones can also be prepared by the hydration of alkynes. This process proceeds via an enol, which undergoes tautomerism to give the ketone.

LO 5 Reactions

Aldehydes and ketones are very reactive groups due to the polarity and structure of the carbonyl group. Aldehydes are usually more reactive than ketones because access to the carbonyl group is easier. The most common reaction of the carbonyl group is the addition of a nucleophile to form a tetrahedral carbonyl addition intermediate.

Alkyl, aryl and vinylic halides react with certain metals to form organometallic compounds. Organomagnesium compounds are commonly called Grignard reagents. The carbon–metal bond in Grignard reagents has a partial ionic character. Grignard reagents behave as carbanions and are both strong bases and good nucleophiles. Aldehydes and ketones react readily with Grignard reagents to form alcohols.

Treatment of aldehydes and some ketones with sodium cyanide and dilute acid produces cyanohydrins, compounds with an —OH and a —CN bonded to the same carbon atom.

The addition of an alcohol molecule to the carbonyl group of aldehydes or ketones forms hemiacetals. The hemiacetal functional group is a carbon atom bonded to an —OH group and an —OR or —OAr group. Hemiacetals can react further with alcohols to form acetals plus a molecule of water. The acetal functional group is a carbon atom bonded to two —OR or —OAr groups.

Ammonia, primary aliphatic amines, RNH_2, and primary aromatic amines, $ArNH_2$, react with the carbonyl group of aldehydes and ketones in the presence of an acid catalyst to give imines (compounds that contain a carbon–nitrogen double bond). When aldehydes and ketones react with hydroxylamine, H_2NOH, they form oximes (compounds containing a C═NOH group). A primary amine is converted to a secondary amine by way of an imine in a two-step reaction called reductive amination.

Aldehydes are reduced to primary alcohols. Ketones are reduced to secondary alcohols. Sodium borohydride and lithium aluminium hydride are the most common laboratory reagents used to reduce the carbonyl group of an aldehyde or ketone. Each acts as a source of a hydride ion, which is a very good nucleophile.

Aldehydes are one of the most easily oxidised of all functional groups. They oxidise to carboxylic acids. The addition of Tollens' reagent oxidises an aldehyde to a carboxylate anion and silver ions are reduced to metallic silver. Silver from this reaction precipitates as a smooth, mirror-like deposit, and the reaction is known as the silver-mirror test.

LO 6 Keto–enol tautomerism

A carbon atom adjacent to a carbonyl group is called an α-carbon, and a hydrogen attached to it is called an α-hydrogen. An aldehyde or ketone that has at least one α-hydrogen atom is in equilibrium with a constitutional isomer called an enol. Constitutional isomers in equilibrium with each other that differ in the location of a hydrogen atom and a double bond relative to a heteroatom are called tautomers, and this type of isomerism is called tautomerism.

KEY CONCEPTS AND EQUATIONS

Preparation of ketones by Freidel–Crafts acylation
(section 21.4)

Ketones can be prepared by Friedel–Crafts acylation of aromatic compounds.

$$\text{benzene} + CH_3CH_2COCl \xrightarrow{AlCl_3} C_6H_5COCH_2CH_3 + HCl$$

Preparation of aldehydes and ketones by oxidation of alcohols
(section 21.4)

Primary alcohols can be oxidised under mild conditions to give aldehydes, and secondary alcohols can be oxidised to give ketones.

$$CH_3(CH_2)_4OH \xrightarrow{PCC} CH_3(CH_2)_3CHO$$

$$CH_3(CH_2)_3CH(OH)CH_3 \xrightarrow[H_3O^+]{CrO_3} CH_3(CH_2)_3COCH_3$$

Preparation of aldehydes and ketones by ozonolysis of alkenes
(section 21.4)

Alkenes are readily cleaved by ozone, which when followed by reductive hydrolysis gives aldehydes and/or ketones.

$$\text{isopropylidenecyclohexane} \xrightarrow[2.\ Zn/H_2O]{1.\ O_3} \text{cyclohexanone} + O{=}C(CH_3)_2$$

Preparation of ketones by hydration of alkynes
(section 21.4)

Hydration of alkynes yields ketones.

$$CH_3CH_2C{\equiv}CCH_2CH_3 + H_2O \xrightarrow[HgSO_4]{H_2SO_4} CH_3CH_2CH_2COCH_2CH_3$$

Reaction with Grignard reagents
(section 21.5)

Treatment of formaldehyde with Grignard reagents, followed by hydrolysis in aqueous acid, gives primary alcohols. Similar treatment of other aldehydes gives secondary alcohols.

$$CH_3CHO \xrightarrow[2.\ HCl,\ H_2O]{1.\ C_6H_5MgBr} C_6H_5CH(OH)CH_3$$

Treatment of ketones with Grignard reagents gives tertiary alcohols.

$$CH_3COCH_3 \xrightarrow[2.\ HCl,\ H_2O]{1.\ C_6H_5MgBr} C_6H_5C(OH)(CH_3)_2$$

Addition of other carbon nucleophiles
(section 21.5)

Aldehydes and ketones will react with cyanide ions to form cyanohydrins.

$$H_3C{-}CO{-}CH_3 \xrightarrow[2.\ H_3O^+]{1.\ NaCN} H_3C{-}C(OH)(CN){-}CH_3$$

Addition of alcohols to form hemiacetals
(section 21.5)

Hemiacetals are only minor components of an equilibrium mixture of aldehydes or ketones, and alcohols, except where the —OH and C═O groups are parts of the same molecule and a five- or six-membered ring can form.

$$CH_3CH(OH)CH_2CH_2CHO \rightleftharpoons \text{2-methyl-5-hydroxytetrahydrofuran}$$

4-hydroxypentanal; a cyclic hemiacetal

Addition of alcohols to form acetals
(section 21.5)

The formation of acetals is catalysed by acid.

$$\text{cyclopentanone} + HOCH_2CH_2OH \underset{}{\overset{H_2SO_4}{\rightleftharpoons}} \text{(cyclic acetal)} + H_2O$$

Addition of ammonia and amines
(section 21.5)

The addition of ammonia or primary amines to the carbonyl group of aldehydes or ketones forms tetrahedral carbonyl addition intermediates. Loss of water from these intermediates gives imines.

$$\text{cyclopentanone} + H_2NCH_3 \overset{H_3O^+}{\rightleftharpoons} \text{cyclopentylidene}{=}NCH_3 + H_2O$$

Formation of oximes
(section 21.5)

The condensation of aldehydes and ketones with hydroxylamine gives oximes.

$$C_6H_5C(=O)CH_3 \xrightarrow{H_2NOH} C_6H_5C(=NOH)CH_3$$

Reductive amination to amines
(section 21.5)

The carbon–nitrogen double bond of imines can be reduced by hydrogen in the presence of a transition metal catalyst to a carbon–nitrogen single bond.

$$\text{cyclohexanone} + H_2N{-}C_6H_{11} \xrightarrow[-H_2O]{} [C_6H_{10}{=}N{-}C_6H_{11}]$$

$$[C_6H_{10}{=}N{-}C_6H_{11}] \xrightarrow{H_2/Ni} C_6H_{11}{-}NH{-}C_6H_{11}$$

Metal hydride reduction
(section 21.5)

Both $LiAlH_4$ and $NaBH_4$ reduce the carbonyl group of aldehydes or ketones to hydroxyl groups. They are selective in that neither reduces isolated carbon–carbon double bonds.

$$\text{cyclohex-2-enone} \xrightarrow[2.\ H_2O]{1.\ NaBH_4} \text{cyclohex-2-enol (OH)}$$

Catalytic reduction
(section 21.5)

Catalytic reduction of the carbonyl group of aldehydes or ketones to hydroxyl groups can be achieved in high yields. However, this occurs much more slowly than the reaction of alkenes and requires more vigorous conditions.

$$\text{cyclohexanone} + H_2 \xrightarrow[25\ °C,\ 2 \times 10^5\ Pa]{Pt} \text{cyclohexanol (OH)}$$

Oxidation of an aldehyde to a carboxylic acid
(section 21.5)

The aldehyde group is among the most easily oxidised functional groups. Oxidising agents include H_2CrO_4, silver(I) ions including Tollens' reagent, and O_2.

$$o\text{-}HOC_6H_4CH(=O) + Ag_2O \xrightarrow[2.\ H_2O,\ HCl]{1.\ THF,\ H_2O,\ NaOH} o\text{-}HOC_6H_4C(=O)OH + Ag$$

Keto–enol tautomerism
(section 21.6)

The keto form of an aldehyde or a ketone generally predominates over the enol form at equilibrium.

$$\underset{\substack{\text{keto form} \\ \text{(approx. 99.9\%)}}}{CH_3\overset{O}{\overset{\|}{C}}CH_3} \rightleftharpoons \underset{\text{enol form}}{CH_3\overset{OH}{\overset{|}{C}}{=}CH_2}$$

KEY TERMS

α-carbon *p. 968*
acetal *p. 955*
α-hydrogen *p. 968*
aldehyde *p. 944*
carbanion *p. 952*
cyanohydrins *p. 955*
enol *p. 968*
Grignard reagent *p. 951*
hemiacetal *p. 955*
hydride ion *p. 964*
ketal *p. 956*
ketone *p. 944*
order of precedence of functional groups *p. 947*
organometallic compounds *p. 951*
oximes *p. 962*
reductive amination *p. 962*
silver-mirror test *p. 967*
tautomerism *p. 968*
tautomers *p. 968*
tetrahedral carbonyl addition intermediate *p. 951*
Tollens' reagent *p. 967*

REVIEW QUESTIONS

LO 1 Structure and bonding

21.1 Draw structural formulae for the one ketone and two aldehydes with the molecular formula C_4H_8O.

21.2 Draw structural formulae for the four aldehydes with the molecular formula $C_6H_{12}O$.

21.3 Use the $\delta+$ and $\delta-$ notation to show the polarity of the following carbonyl compounds.

(a) CH_3O, CHO, HO
vanillin
4-hydroxy-3-methoxybenzaldehyde

(b) O
menthone
2-isopropyl-5-methylcyclohexanone

21.4 Compare and contrast the bonding in a carbonyl group (C=O) and an alkene group (C=C). What are the similarities and what are the differences?

LO 2 Nomenclature

21.5 Draw structural formulae corresponding to the following compound names.
(a) bromoacetone
(b) methylbutanone
(c) 3,5-dinitrobenzaldehyde
(d) 3,5-dimethylcyclohexanone
(e) tetramethylpentan-3-one
(f) butandial
(g) 3-hydroxybutanone
(h) 3-phenylpropenal

21.6 Draw structural formulae corresponding to the following compound names.
(a) 3-bromobutanone
(b) 3-hydroxypentanal
(c) 1-hydroxypentan-2-one
(d) 3-methyl-4-phenylbutanal
(e) (*S*)-4-bromohexan-2-one
(f) 3-methylbut-3-enone
(g) 4-oxoheptanal
(h) 3,4-dibromocyclopentanecarbaldehyde
(i) 5-oxopentanoic acid

21.7 Name the following compounds using the IUPAC system of nomenclature.

(a) $CH_3(CH_2)_8\overset{O}{\overset{\|}{C}}H$

(b) $CH_3(CH_2)_3\overset{O}{\overset{\|}{C}}CH_2CH_3$

(c) $(CH_3)_2C{=}CHCH_2\overset{O}{\overset{\|}{C}}H$

(d) $(CH_3)_3CCH_2\overset{O}{\overset{\|}{C}}H$

(e) $(CH_3)_3C\overset{Cl}{\overset{|}{C}}HCH_2\overset{O}{\overset{\|}{C}}CH_2CH_3$

(f) $CH_3\overset{CH_3}{\overset{|}{C}}HCH_2\overset{O}{\overset{\|}{C}}CH_2CH_3$

(g) O_2N–(benzene ring)–$\overset{O}{\overset{\|}{C}}H$

(h) H_3C–(cyclohexane ring)=O

21.8 Name the following compounds using the IUPAC system of nomenclature.

(a) O

(b) O

(c) H_3CO, O, H

(d) O

(e) O, OH

(f) CHO

(g) O, H, COOH

(h) CHO, HO

21.9 Provide a systematic (IUPAC) name for each of the following compounds.

(a)
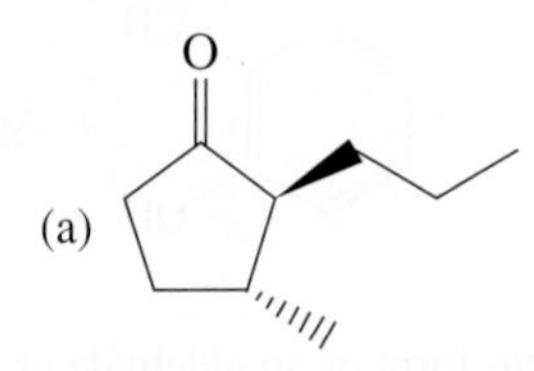

(b)
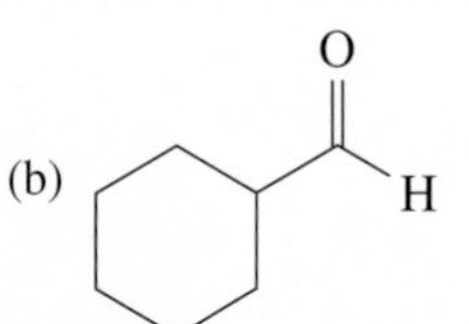

(c) O, H

(d)
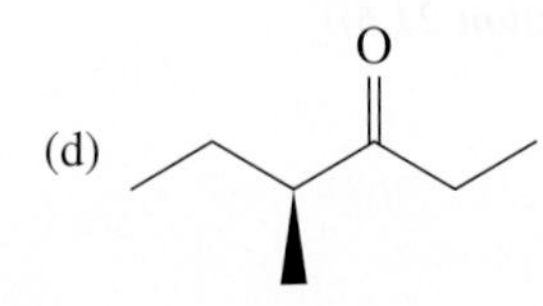

21.10 Name the family to which each of the following compounds belongs.

(a) 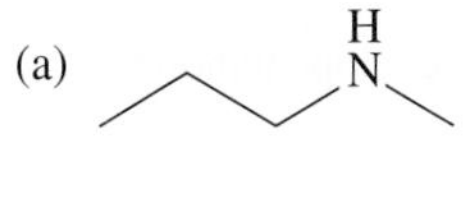

(d) 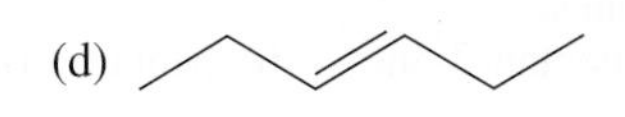

(b) 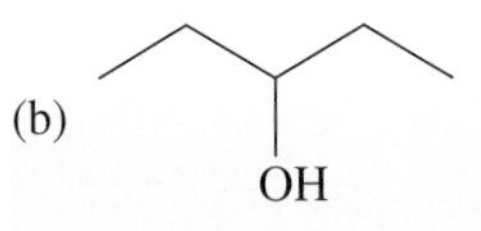

(e) O O

(c) COOH

(f) O

21.11 To which organic family does each of the following compounds belong?

(a) $CH_3C{\equiv}CH$

(b) 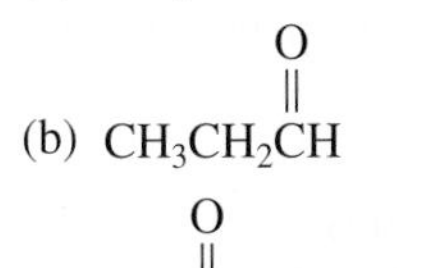 $CH_3CH_2\overset{O}{\overset{\|}{C}}H$

(c) $CH_3\overset{O}{\overset{\|}{C}}CH_2CH_2CH_3$

(d) $CH_3{-}O{-}CH_2CH_3$

(e) $CH_3CH_2NH_2$

(f) 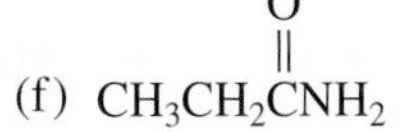 $CH_3CH_2\overset{O}{\overset{\|}{C}}NH_2$

LO 3 Physical properties

21.12 Why do aldehydes and ketones have boiling points that are lower than those of their corresponding alcohols?

21.13 Explain why aldehydes and ketones have higher boiling points than alkanes of similar molar mass and similar number of electrons.

21.14 Which would you expect to be more soluble in water, propanal or hexanal? Explain your answer.

21.15 Which would you expect to be more soluble in water, acetone (propanone) or hexan-3-one? Explain your answer.

21.16 With the aid of a sketch, show the hydrogen bonding between water and propanal.

21.17 Use a sketch to show the hydrogen bonding between methanol and butanone.

LO 4 Preparation of aldehydes and ketones

21.18 Complete the following reactions.

(a) + O Cl $\xrightarrow{AlCl_3}$

(b) $\xrightarrow[2.\ Zn/H_2O]{1.\ O_3}$

(c) $HC{\equiv}CCH_2CH_3 + H_2O \xrightarrow[HgSO_4]{H_2SO_4}$

(d) OH $\xrightarrow{Na_2Cr_2O_7}$

21.19 Give the structural formulae of the aldehyde or ketone obtained by the PCC oxidation of the following alcohols.
(a) 1-phenylethanol
(b) 2-phenylethanol

21.20 Give the structural formulae of the ketone(s) obtained by the hydration of the following alkynes.
(a) oct-1-yne
(b) oct-2-yne
(c) oct-3-yne

21.21 Give the structural formulae of the carbonyl compounds obtained by ozonolysis and reductive hydrolysis of the following alkenes.

(a)

(b)

(c)

(d)

LO 5 Reactions

21.22 Write the structures of the isomeric alcohols with the formula $C_4H_{10}O$ that could be oxidised to aldehydes. Write the structure of the isomer that could be oxidised to a ketone.

21.23 Which isomer of butan-1-ol cannot be oxidised by dichromate ions? Write its structure and IUPAC name.

21.24 Write the structure of the principal organic product of each of the following reactions.
(a) magnesium with iodoethane
(b) the product of (a) with formaldehyde, followed by dilute acid
(c) the product of (a) with pentan-3-one, followed by dilute acid
(d) the product of (a) with 4-methylbenzaldehyde, followed by dilute acid.

21.25 Write an equation for the acid–base reaction between phenylmagnesium iodide and a carboxylic acid. Use curved arrows to show the bond-breaking and bond-forming steps in this reaction. In addition, show that the reaction is an example of a stronger acid and stronger base reacting to form a weaker acid and weaker base.

21.26 Draw structural formulae for the product formed by treating each of the following compounds with butylmagnesium bromide, followed by hydrolysis in aqueous acid.

(a) CHO

(b) O

(c) O

21.27 Draw structural formulae for the product formed by treating each of the following compounds with propylmagnesium bromide, followed by hydrolysis in aqueous acid.

(a) O CHO

(b) O H_3CO

21.28 Predict the product of the reaction of 3-methylpentanal with each of the following reagents.
(a) lithium aluminium hydride, followed by treatment with water
(b) sodium borohydride in methanol
(c) hydrogen with a nickel catalyst
(d) methylmagnesium iodide, followed by dilute acid
(e) aniline ($PhNH_2$)
(f) 4-nitrophenylhydrazine
(g) sodium cyanide with addition of sulfuric acid
(h) silver–ammonia complex (Tollens' reagent)
(i) acidified potassium dichromate solution

21.29 Repeat question 21.28 with cyclopentanone as the reactant.

LO 6 Keto–enol tautomerism

21.30 Write the structural formulae for all possible enols of the following.
(a) hexanal (b) hexan-3-one (c) 3-methylcyclohexanone

21.31 Write the structural formulae of all possible products resulting from the α-bromination of the following ketones.
(a) pentan-2-one (b) pentan-3-one (c) cyclopentanone

REVIEW PROBLEMS

21.32 Complete the following reactions. **LO 4**

(a) benzene + acetic anhydride $\xrightarrow{AlCl_3}$

(b) 4-methylpentan-2-ol $\xrightarrow[H_2SO_4]{K_2Cr_2O_7}$

(c) 3-methylbutan-1-ol $\xrightarrow[H_2SO_4]{K_2Cr_2O_7}$

(d) 3-methylbutan-1-ol $\xrightarrow{PCC}$

21.33 For each of the following pairs, identify which compound would be expected to react more rapidly with a nucleophile. **LO 4**

(a)
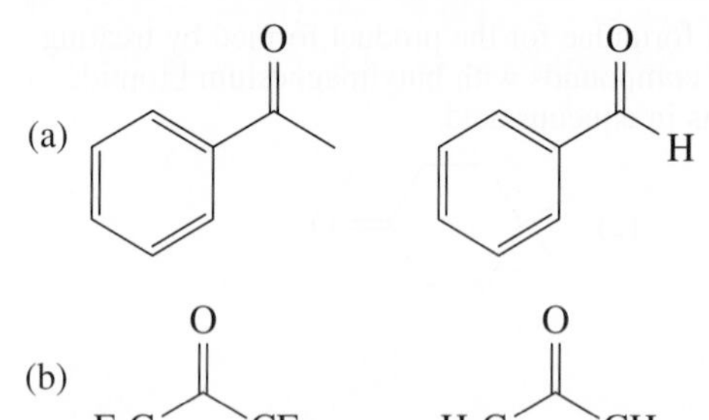

(b) $F_3C-CO-CF_3$ $H_3C-CO-CH_3$

21.34 How can the following conversions be carried out? **LO 4**
(a) hex-1-ene to hexan-2-one
(b) hexan-1-ol to hexanoic acid
(c) hexan-1-ol to hexanal
(d) oct-1-yne to octan-2-one
(e) octan-2-ol to octan-2-one
(f) cyclopentene to pentandial
(g) bromocycloheptane to cycloheptanone
(h) cycloheptene to cycloheptanone

21.35 Suggest a synthesis for each of the following alcohols, starting from an aldehyde or ketone and an appropriate Grignard reagent. (The number of combinations of Grignard reagent and aldehyde or ketone that might be used is shown in parentheses below each target molecule.) **LO 5**

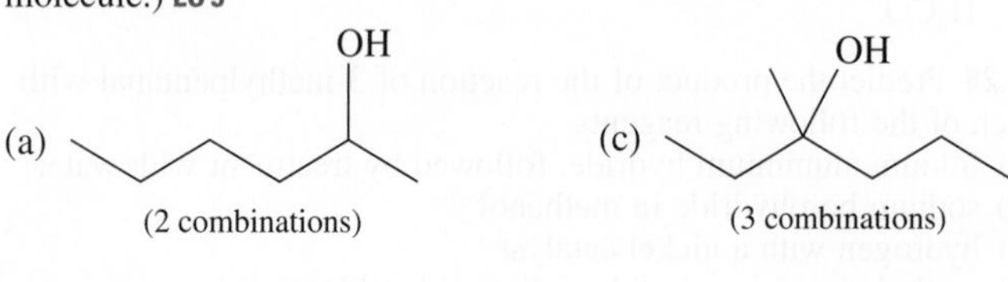

(b) (3-methoxyphenyl)(phenyl)methanol
(2 combinations)

21.36 Choose a Grignard reagent and a ketone that can be used to produce each of the following compounds. **LO 5**
(a) 1-phenylcyclopentanol
(b) 3-ethylpentan-3-ol
(c) 2,3-dimethylpentan-2-ol
(d) 2,3-dimethylpentan-3-ol

21.37 Complete the following reactions to show the major organic product. **LO 5**

(a) CH_3CH_2CHO + C_6H_5MgBr $\xrightarrow[H_3O^+]{H_2O}$

(b) $C_6H_5CH_2COCH_3$ + CH_3MgBr $\xrightarrow[H_3O^+]{H_2O}$

(c) $H_3CC{\equiv}CNa$ + CH_3COCH_3 $\xrightarrow[H_3O^+]{H_2O}$

(d) $CH_3COCH_2CH_3$ + C_6H_5MgBr $\xrightarrow[H_3O^+]{H_2O}$

21.38 Draw structural formulae for the hemiacetal and then the acetal formed from each of the following pairs of reactants in the presence of an acid catalyst. **LO 5**

(a) 1-(cyclopent-2-en-1-yl)ethanone + CH_3OH

(b) 4-tert-butylcyclohexanone + CH_3CH_2OH

(c) cyclopentane-1,2-diol (OH, OH) + 2-methylpropanal (CHO)

21.39 Draw structural formulae for the products of hydrolysis of each of the following acetals and ketals in aqueous acid. **LO 5**

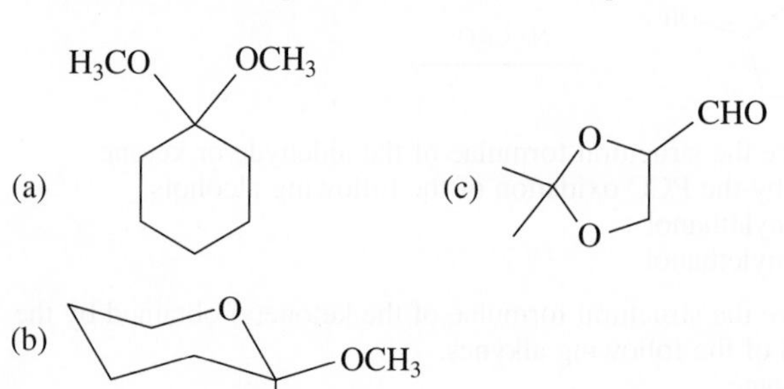

(b) 2-methoxytetrahydropyran (O, OCH_3, H)

21.40 The following compound is a component of the fragrance of jasmine. From what carbonyl-containing compound and alcohol is it derived? **LO5**

21.41 Propose a mechanism for the formation of the following cyclic acetal by treating acetone with ethylene glycol in the presence of an acid catalyst. Make sure that your mechanism is consistent with the fact that the oxygen atom of the water molecule is derived from the carbonyl oxygen atom of acetone. **LO4**

acetone + ethylene glycol $\xrightarrow{\text{acid}}$ cyclic acetal + H_2O

21.42 Propose a mechanism for the formation of a cyclic acetal from 4-hydroxypentanal and one equivalent of methanol. If the carbonyl oxygen atom of 4-hydroxypentanal is enriched with oxygen-18, does your mechanism predict that the oxygen label appears in the cyclic acetal or in the water? Explain. **LO4**

4-hydroxypentanal (^{18}O) + CH_3OH $\xrightarrow{\text{acid}}$ cyclic acetal (OCH_3) + H_2O

21.43 5-hydroxyhexanal forms a six-membered cyclic hemiacetal that predominates at equilibrium in acidic aqueous solution. **LO4**

5-hydroxyhexanal $\xrightleftharpoons{\text{acid}}$ a cyclic hemiacetal

(a) Draw a structural formula for this cyclic hemiacetal.
(b) How many stereoisomers are possible for 5-hydroxyhexanal?
(c) How many stereoisomers are possible for the cyclic hemiacetal?

21.44 Show how the secondary amine below can be prepared by two successive reductive aminations. **LO4**

ketone $\xrightarrow{(1)}$ amine (NH_2) $\xrightarrow{(2)}$ secondary amine (NH–CH_2Ph)

21.45 Show how to convert cyclohexanone to each of the following amines. **LO5**

(a) cyclohexyl–NH_2

(b) cyclohexyl–$NHCH(CH_3)_2$

(c) cyclohexyl–NH–phenyl

21.46 Rimantadine is effective in preventing infections caused by the influenza A virus and in treating the established illness. The drug is thought to exert its antiviral effect by blocking a late stage in the assembly of the virus. The following is the final step in the synthesis of rimantadine. **LO5**

adamantyl methyl ketone $\longrightarrow$ rimantadine (NH_2)

rimantadine
(an antiviral agent)

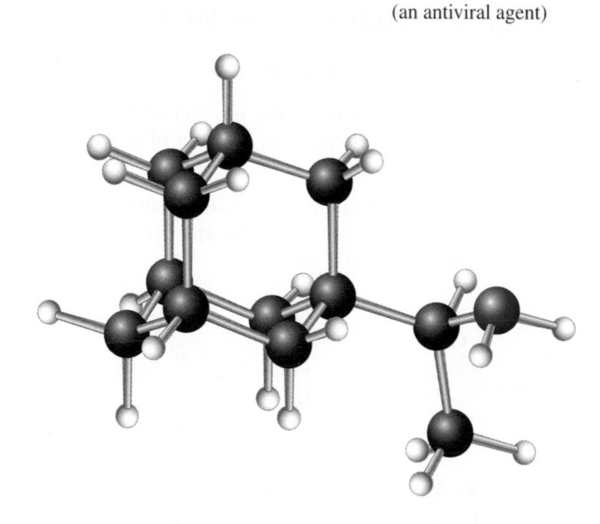

(a) Describe the experimental conditions which bring about this conversion.
(b) Is rimantadine chiral?

21.47 Methenamine, a product of the reaction of formaldehyde and ammonia, is a *prodrug* — a compound that is inactive by itself but is converted to an active drug in the body by a biochemical transformation. The strategy behind the use of methenamine as a prodrug is that nearly all bacteria are sensitive to formaldehyde at concentrations of 20 mg mL^{-1} or higher. Formaldehyde cannot be used directly in medicine because an effective concentration in plasma cannot be achieved with safe doses. Methenamine is stable at pH 7.4 (the pH of blood plasma) but undergoes acid-catalysed hydrolysis to formaldehyde and ammonium ions under the acidic conditions of the kidneys and the urinary tract. **LO5**

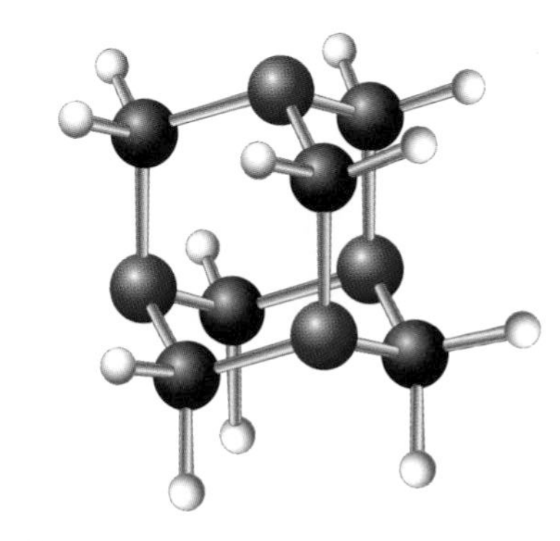

methenamine + H_2O $\xrightarrow{H^+}$ CH_2O + NH_4^+

methenamine

Thus, methenamine can be used as a site-specific drug to treat urinary infections.
(a) Balance the equation for the hydrolysis of methenamine to formaldehyde and ammonium ions.
(b) Does the pH of an aqueous solution of methenamine increase, remain the same or decrease as a result of the hydrolysis of the compound? Explain.

(c) Explain the meaning of the following statement: 'The functional group in methenamine is the nitrogen analogue of an acetal'.
(d) Account for the observation that methenamine is stable in blood plasma but undergoes hydrolysis in the urinary tract.

21.48 Draw a structural formula for the product formed by treating cyclopentanecarbaldehyde with each of the following sets of reagents. **LO4**
(a) $LiAlH_4$ followed by H_2O
(b) $NaBH_4$ in CH_3OH/H_2O
(c) H_2/Pt
(d) $[Ag(NH_3)_2]^+$ in NH_3/H_2O and then HCl/H_2O
(e) H_2CrO_4
(f) $C_6H_5NH_2$ in the presence of H_2/Ni

21.49 Draw a structural formula for the product of the reaction of *p*-bromoacetophenone with each set of reagents in question 21.48. **LO5**

21.50 Show how you would chemically distinguish between the compounds in the following sets. Describe any observations and give equations where applicable. **LO5**

(a) 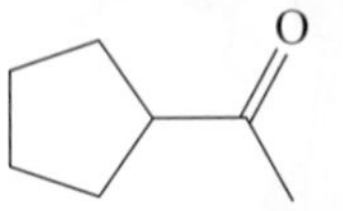 [cyclopentyl methyl ketone] and [3-methylcyclopentanecarbaldehyde]

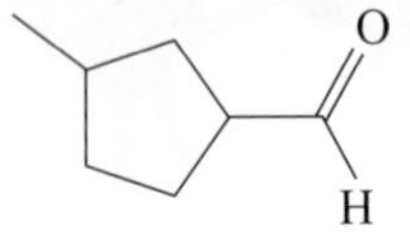

(b) 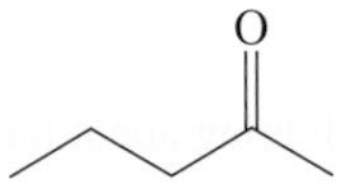 [butan-2-one] and [butan-2-ol]

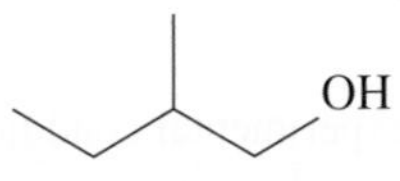

21.51 Show how you would chemically distinguish between the following compounds. Describe any observations and give equations where applicable. **LO5**

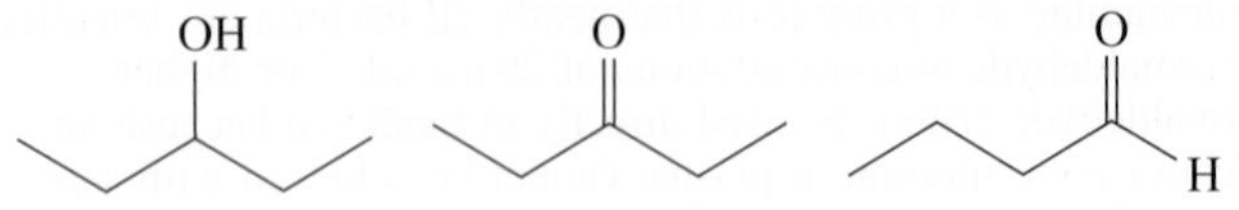

A (pentan-3-ol), B (pentan-3-one), C (butanal)

21.52 The following molecule belongs to a class of compounds called enediols. Each carbon atom of the double bond in an enediol carries an —OH group. **LO5**

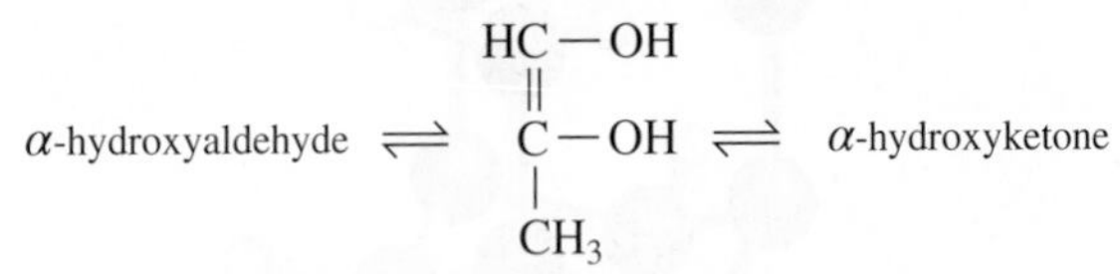

$$\alpha\text{-hydroxyaldehyde} \rightleftharpoons \begin{matrix} HC{-}OH \\ \| \\ C{-}OH \\ | \\ CH_3 \end{matrix} \rightleftharpoons \alpha\text{-hydroxyketone}$$

an enediol

Draw structural formulae for the α-hydroxyketone and the α-hydroxyaldehyde with which this enediol is in equilibrium.

21.53 In dilute aqueous acid, (*R*)-glyceraldehyde is converted into an equilibrium mixture of (*R,S*)-glyceraldehyde and dihydroxyacetone. **LO6**

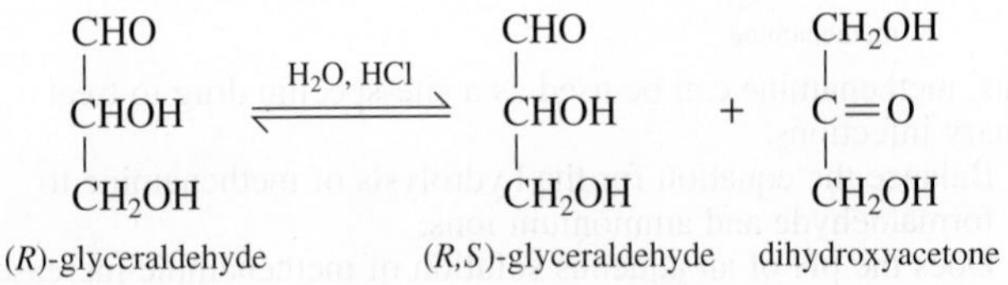

$$\underset{(R)\text{-glyceraldehyde}}{\begin{matrix} CHO \\ | \\ CHOH \\ | \\ CH_2OH \end{matrix}} \xrightleftharpoons{H_2O,\ HCl} \underset{(R,S)\text{-glyceraldehyde}}{\begin{matrix} CHO \\ | \\ CHOH \\ | \\ CH_2OH \end{matrix}} + \underset{\text{dihydroxyacetone}}{\begin{matrix} CH_2OH \\ | \\ C{=}O \\ | \\ CH_2OH \end{matrix}}$$

Propose a mechanism for this isomerisation.

21.54 Show the reagents and conditions that will cause the conversion of hexan-3-ol to 4-ethylheptan-3-one. **LO5**

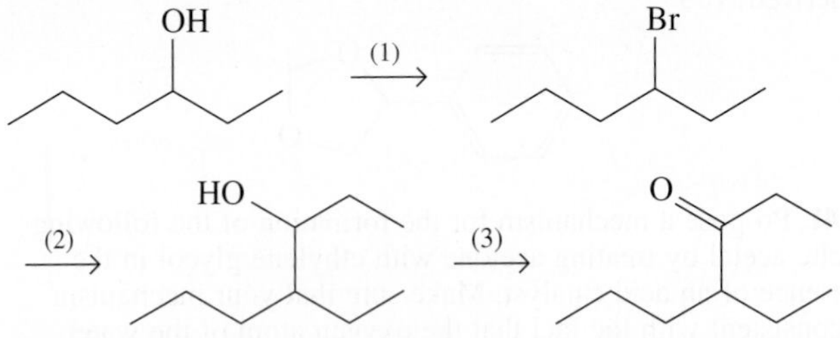

hexan-3-ol (OH) $\xrightarrow{(1)}$ 3-bromohexane (Br) $\xrightarrow{(2)}$ 4-ethylheptan-3-ol (HO) $\xrightarrow{(3)}$ 4-ethylheptan-3-one (O)

21.55 Starting with cyclohexanone, show how to prepare the following compounds. (In addition to the given starting material, use any other organic or inorganic reagents, as necessary.) **LO5**
(a) cyclohexanol
(b) cyclohexene
(c) 1-methylcyclohexanol

21.56 Starting with pentan-3-one, show how to prepare the following compounds. (In addition to the given starting material, use any other organic or inorganic reagents, as necessary.) **LO5**
(a) 3-phenylpentan-3-ol
(b) 3-phenylpent-2-ene
(c) 3-ethylpent-2-ene

21.57 Glutaraldehyde is a germicidal agent that is sometimes used to sterilise medical equipment that is too sensitive to be heated in an autoclave. In midly acidic conditions, glutaraldehyde exists in a cyclic form (below right). Draw a plausible mechanism for this transformation. **LO5**

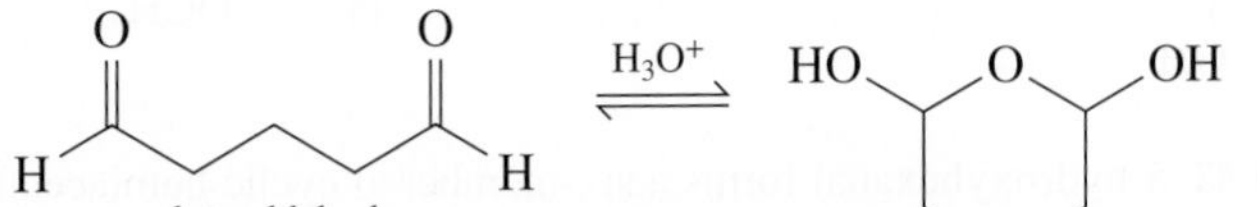

glutaraldehyde (H, O, O, H) $\xrightleftharpoons{H_3O^+}$ HO, O, OH (cyclic form)

21.58 Show how to bring about the following conversions. (In addition to the given starting material, use any other organic or inorganic reagents, as necessary.) **LO5**

(a) 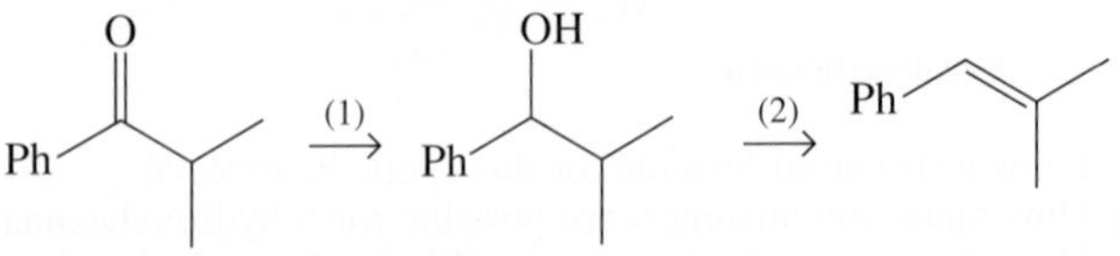 Ph ketone (O) $\xrightarrow{(1)}$ Ph alcohol (OH) $\xrightarrow{(2)}$ Ph alkene

(b) 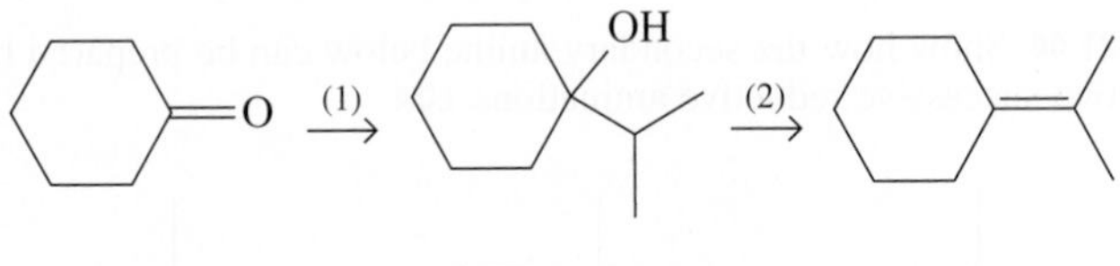 cyclohexanone (=O) $\xrightarrow{(1)}$ tertiary alcohol (OH) $\xrightarrow{(2)}$ alkene

(c) 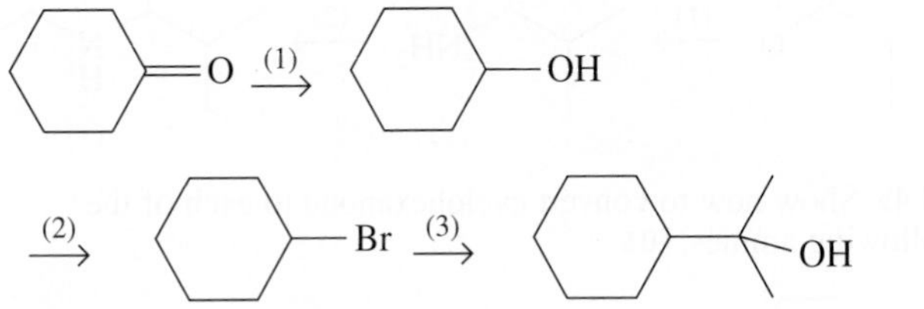 cyclohexanone (=O) $\xrightarrow{(1)}$ cyclohexanol (OH) $\xrightarrow{(2)}$ bromocyclohexane (Br) $\xrightarrow{(3)}$ alcohol (OH)

21.59 Using a Grignard reaction, show how you could prepare each of the following alcohols. **LO5**

(a) 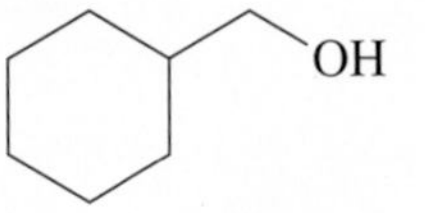 cyclohexylmethanol (OH)
(b) 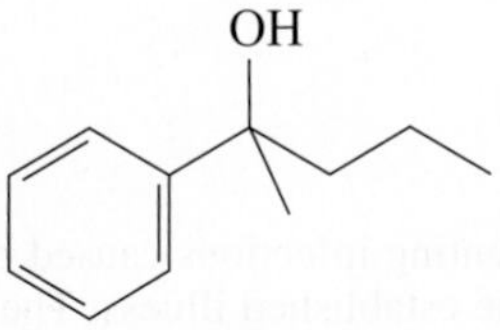 phenyl alcohol (OH)
(c) 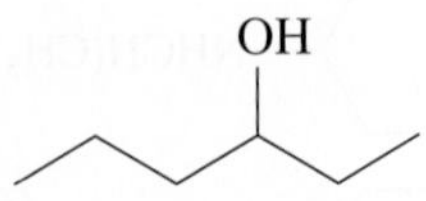 pentan-3-ol (OH)
(d) 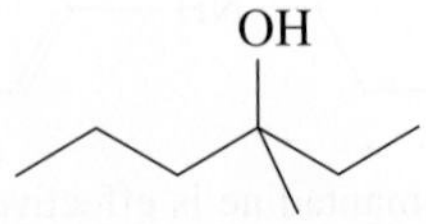 3-methylhexan-3-ol (OH)

21.60 What reagents would you use to perform each of the following transformations? **LO5**

(a)

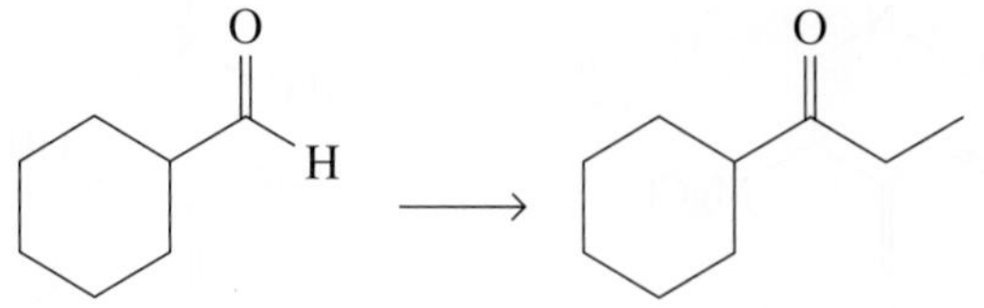

(b)

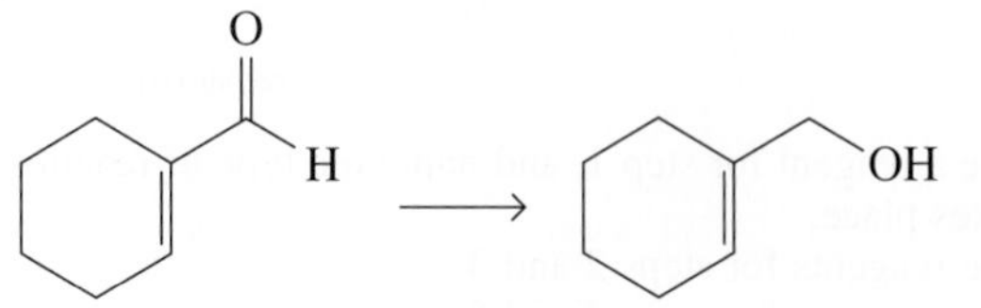

(c)

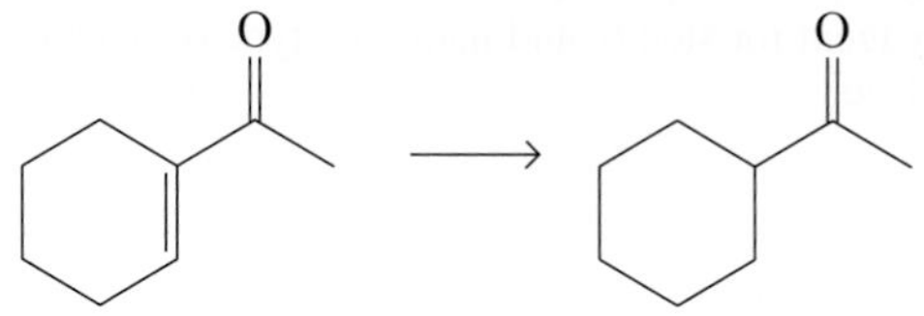

21.61 Many tumours of the breast are oestrogen dependent. Drugs that interfere with oestrogen binding have antitumour activity and may even help prevent the occurrence of tumours. A widely used antioestrogen drug is tamoxifen. **LO5**

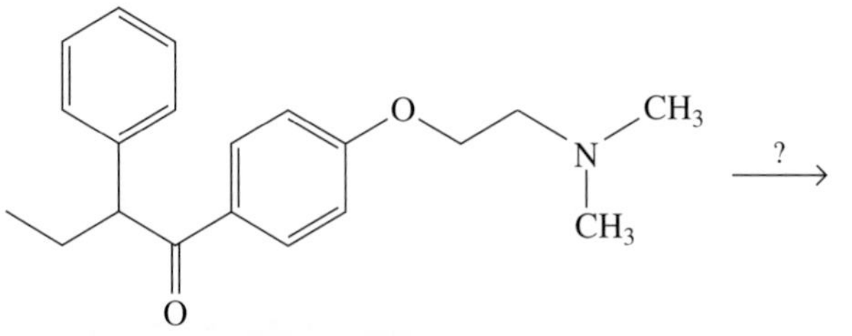

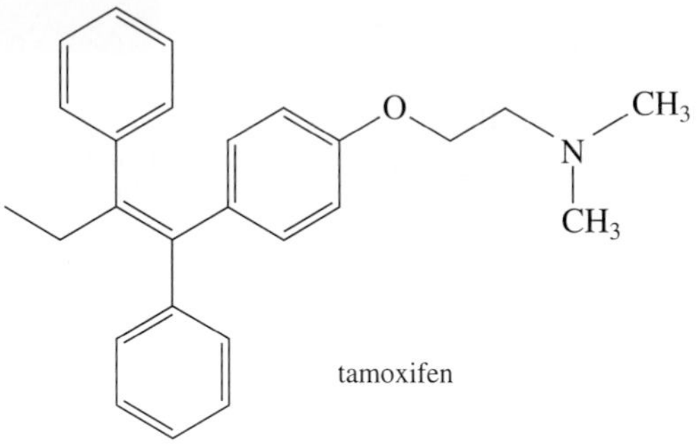

tamoxifen

(a) How many stereoisomers are possible for tamoxifen?
(b) Specify the configuration of the stereoisomer shown here.
(c) Show how tamoxifen can be synthesised from the given ketone using a Grignard reaction, followed by dehydration.

21.62 The following is a possible synthesis of the antidepressant bupropion (Wellbutrin®). **LO5**

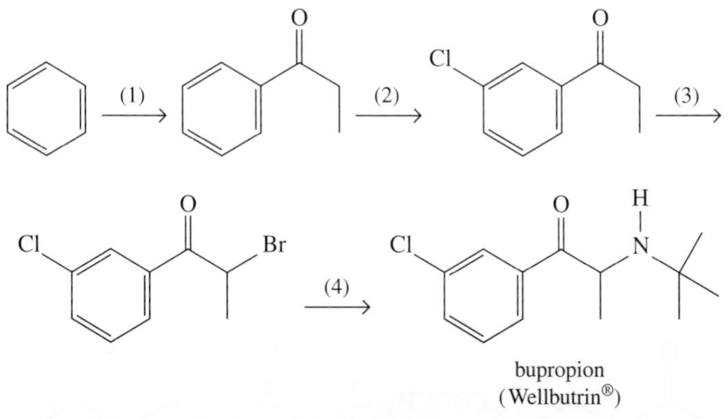

bupropion
(Wellbutrin®)

Show the reagents that will bring about each step in this synthesis.

21.63 An unknown compound *A* gave a positive silver-mirror test. Reaction of *A* with ethylmagnesium bromide followed by treatment with a dilute acid gave compound *B*, $C_6H_{14}O$. Compound *B*, upon treatment with concentrated sulfuric acid, gave compound *C*, C_6H_{12}. Reaction of *C* with ozone followed by zinc in water gave two products, propanal and acetone. Identify each of the compounds *A*, *B* and *C*, based on the chemical information given. Write an equation for each reaction described in the question. **LO5**

21.64 An unknown compound A ($C_6H_{12}O$) did not react with dilute potassium permanganate solution. Reaction of A with ethylmagnesium bromide followed by treatment with a dilute acid gave compound B, $C_8H_{18}O$. Compound B, upon treatment with concentrated sulfuric acid, gave compound C, C_8H_{16}. Reaction of C with ozone followed by zinc in water gave two products, pentan-3-one and acetone. Identify each of the compounds A, B and C, based on the chemical information given. Write an equation for each reaction described in the question. **LO5**

21.65 The following is a synthesis for diphenhydramine. **LO5**

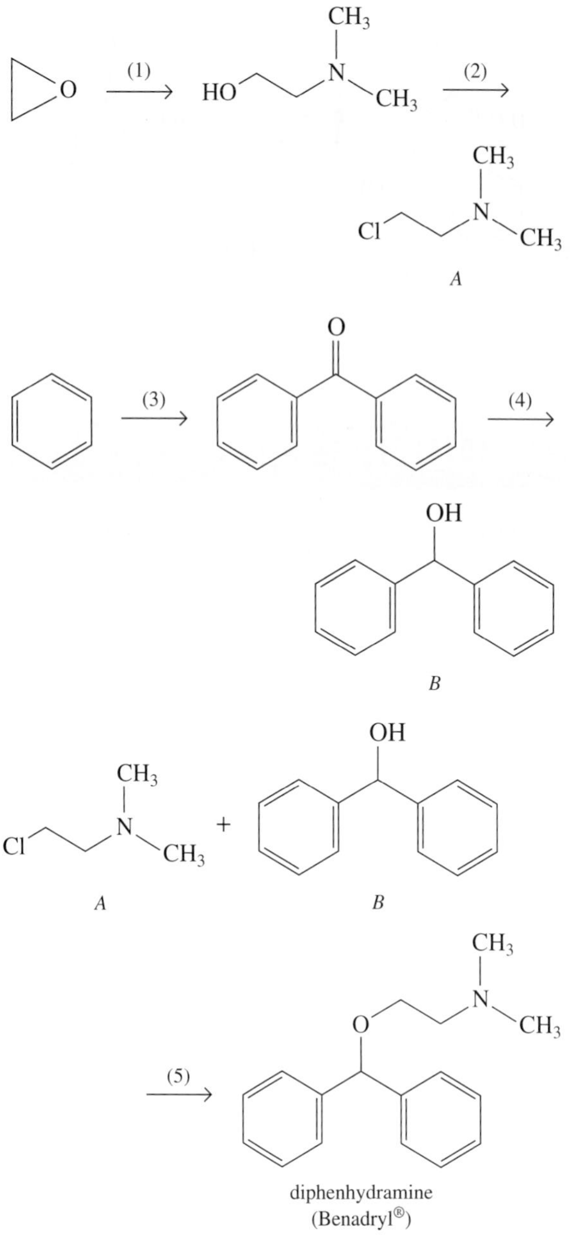

diphenhydramine
(Benadryl®)

The hydrochloride salt of this compound, best known by its trade name, Benadryl®, is an antihistamine.
(a) Propose reagents for steps 1 and 2.
(b) Propose reagents for steps 3 and 4.
(c) Show that step 5 is an example of nucleophilic aliphatic substitution. What type of mechanism, S_N1 or S_N2, is more likely for this reaction? Explain.

21.66 The following is a synthesis for the antidepressant venlafaxine. **LO5**

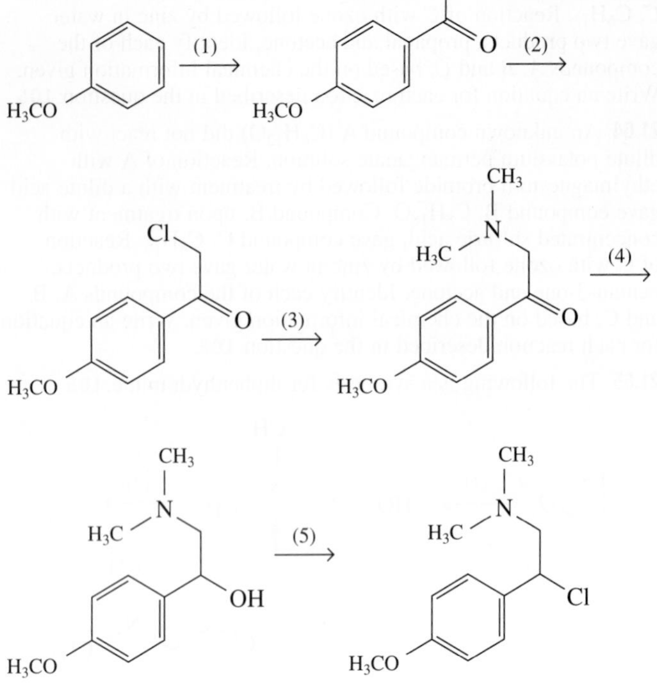

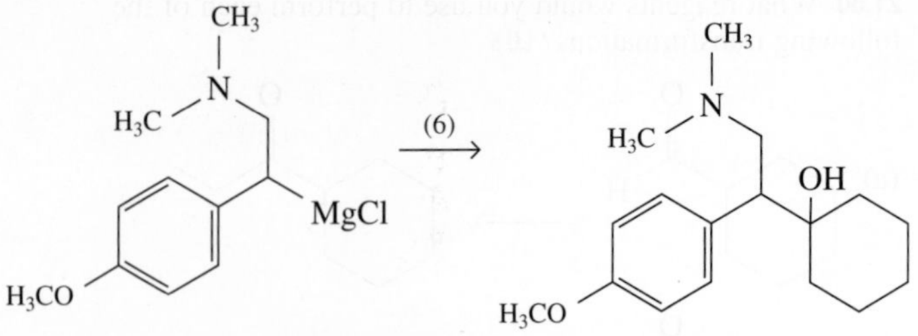

(1) → (2) → (3) → (4) → (5) → (6)

H_3CO ... CH_3 N H_3C ... Cl ... O ... OH ... MgCl

venlafaxine

(a) Propose a reagent for step 1, and name the type of reaction that takes place.
(b) Propose reagents for steps 2 and 3.
(c) Propose reagents for steps 4 and 5.
(d) Propose a reagent for step 6, and name the type of reaction that takes place.

ADDITIONAL EXERCISES

21.67 Diethyl ether is prepared on an industrial scale by the acid catalysed dehydration of ethanol. **LO5**

$$2CH_3CH_2OH \xrightarrow[180\ °C]{H_2SO_4} CH_3CH_2OCH_2CH_3 + H_2O$$

Explain why diethyl ether used in the preparation of Grignard reagents must be carefully purified to remove all traces of ethanol and water.

21.68 Identify the structures of compounds *A* to *E* below. **LO5**

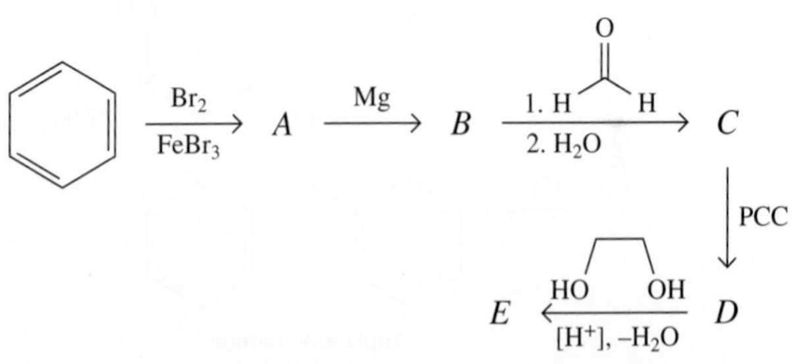

$$\text{benzene} \xrightarrow[FeBr_3]{Br_2} A \xrightarrow{Mg} B \xrightarrow[2.\ H_2O]{1.\ HCHO} C \xrightarrow{PCC} D \xrightarrow[[H^+],\ -H_2O]{HOCH_2CH_2OH} E$$

21.69 Reaction of a Grignard reagent with carbon dioxide, followed by treatment with aqueous HCl, gives a carboxylic acid. Propose a structural formula for the bracketed intermediate formed by the reaction of phenylmagnesium bromide with CO_2, and propose a mechanism for the formation of this intermediate. **LO5**

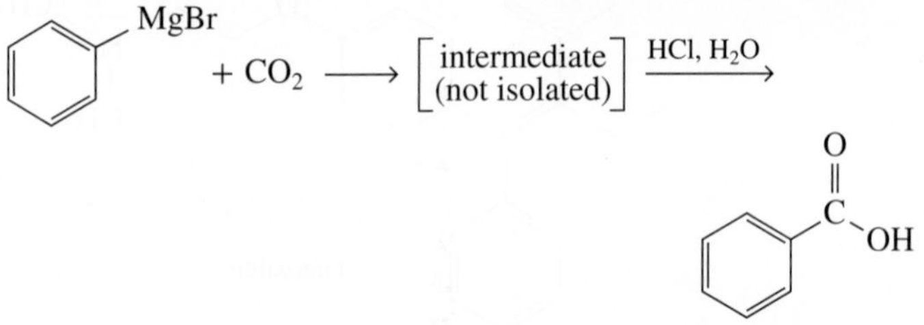

$$C_6H_5MgBr + CO_2 \longrightarrow \begin{bmatrix}\text{intermediate}\\ \text{(not isolated)}\end{bmatrix} \xrightarrow{HCl,\ H_2O} C_6H_5COOH$$

21.70 Provide the enol form of the ketone below and predict the position of equilibrium. **LO6**

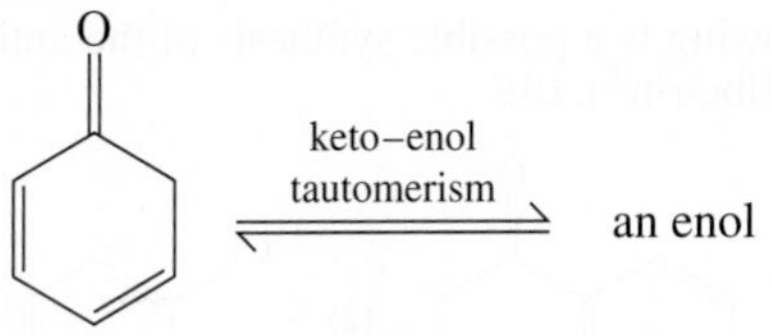

keto–enol tautomerism ⇌ an enol

21.71 How would you prepare 6-hydroxyhexan-2-one from methyl 5-oxohexanoate? Esters are readily reduced to primary alcohols with lithium aluminium hydride. More than one step is required. **LO5**

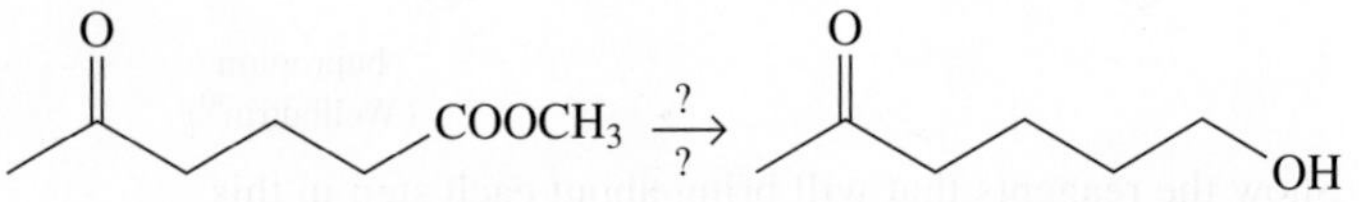

$$CH_3CO(CH_2)_3COOCH_3 \xrightarrow[?]{?} CH_3CO(CH_2)_3CH_2OH$$

methyl 5-oxohexanoate → 6-hydroxyhexan-2-one

21.72 Show how the following transformations can be carried out. Give the reagents required and the structure of any intermediate compounds. More than one step is required in all cases and more than one method may be possible. **LO5**

(a) OH → NHCH₃

(b)

(c)

(d) OH → OH

21.73 Propose a mechanism for the acid-catalysed reaction of the following hemiacetal, with an amine acting as a nucleophile. **LO5**

$\xrightarrow[H_2NCH_2CH_3]{H_3O^+}$ + H_2O

21.74 Identify the structures of compounds *A* to *C* below. Compound *C* can be converted into cyclohexane by heating with concentrated KOH solution. Identify the reagents that can be used to convert cyclohexene into cyclohexane in just one step. **LO5**

$\xrightarrow{H_3O^+}$ *A* $\xrightarrow{H_2CrO_4}$ *B* $\xrightarrow[(-H_2O)]{[H^+]\ NH_2NH_2}$ *C*

21.75 An aldehyde with the molecular formula C_4H_6O exhibits an IR signal at 1715 cm^{-1}. **LO5**

(a) Propose two possible structures that are consistent with this information.

(b) Describe how you could use ^{13}C-NMR spectroscopy to determine which of the two possible structures is correct.

21.76 The following are ^1H-NMR and IR spectra (see chapter 20) of compound *B*, $C_6H_{12}O_2$. Propose a structural formula for compound *B*. **LO1**

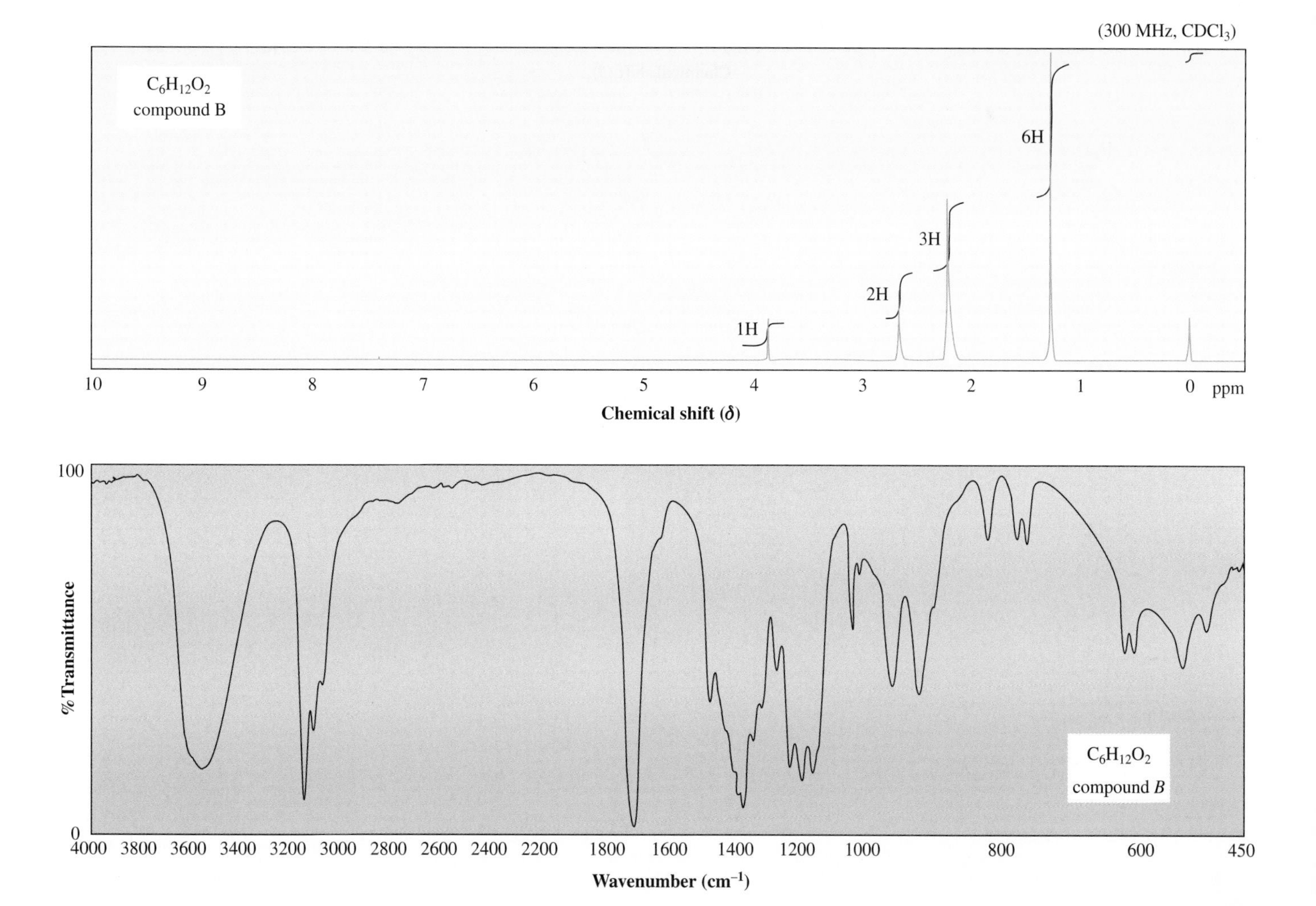

21.77 A compound with the molecular formula $C_9H_{10}O$ exhibits a strong signal at 1687 cm^{-1} in its IR spectrum. The ^1H- and ^{13}C-NMR spectra of the compound are shown below. Identify the structure of this compound. **LO 1**

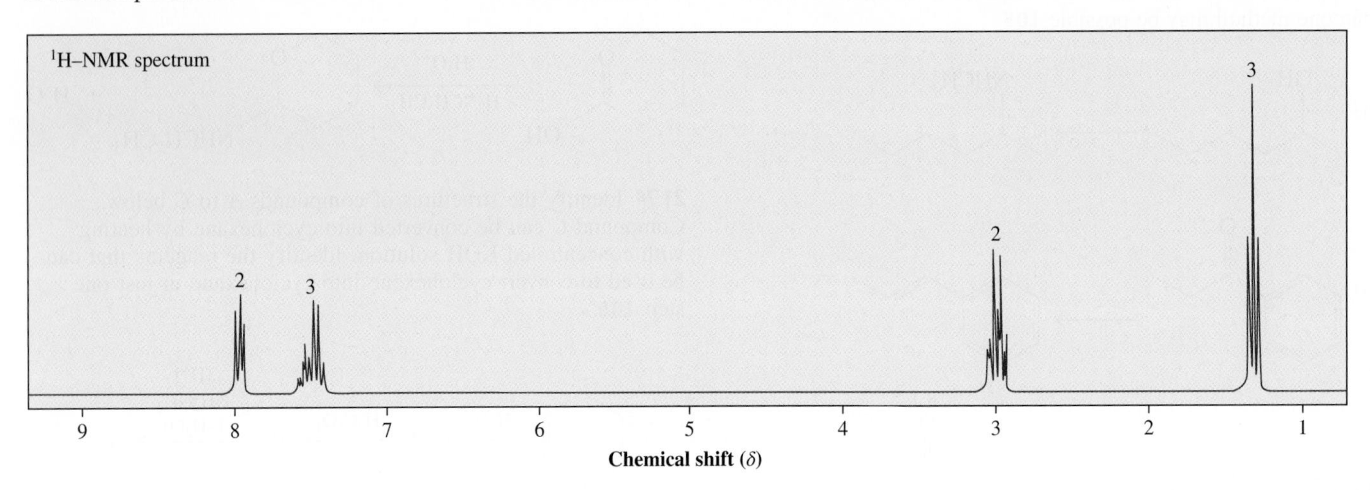

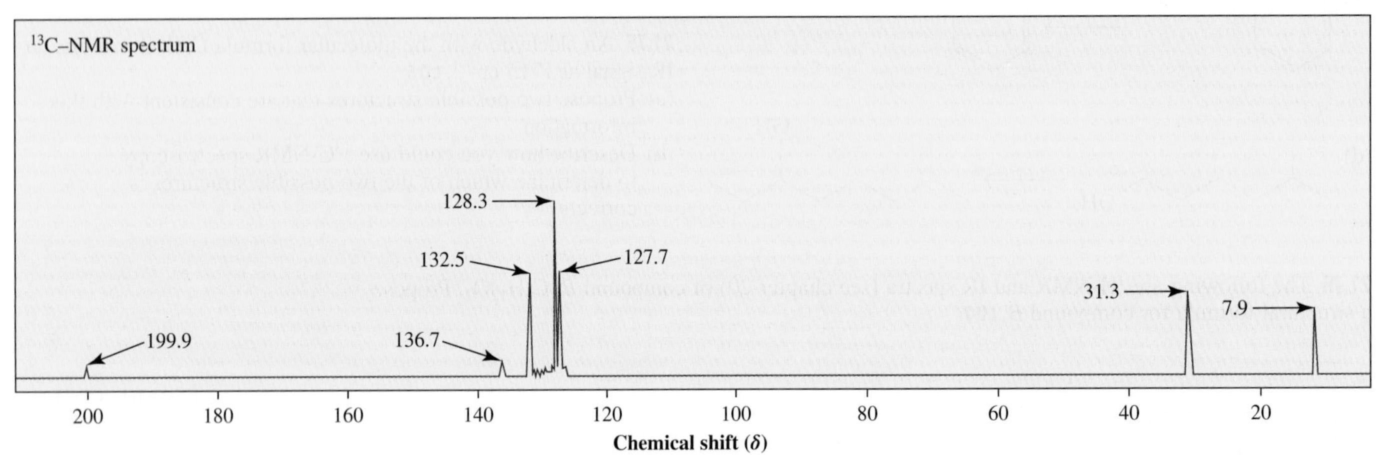

22 Carbohydrates

Carbohydrates is probably the chemical term that is most widely used by the general public. Commonly referred to as 'carbs', it seems everyone has an idea of how much, or how little, or what type we should be consuming in our diets.

Carbohydrates are in fact a major class of organic molecules that are important not only in food, but more broadly in biochemistry, medicines, agriculture and even as structural materials. Carbohydrates act as storehouses of chemical energy (glucose, starch, glycogen) and are components of supportive structures in plants (cellulose), crustacean shells (chitin) and connective tissues in animals (polysaccharides). Carbohydrates are also essential components in the nucleic acids RNA (D-ribose) and DNA (2-deoxy-D-ribose) and they play crucial roles in cell surface and membrane recognition that are necessary for cell function. Small carbohydrate molecules, such as glucose, are readily soluble in water and so can be transported through the vascular system to meet a plant's or animal's energy requirements.

Chemists are also becoming increasingly interested in carbohydrates as a potential solution for many of the world's energy problems that are generated by the burning of fossil fuels. Increasing research efforts are being focused on so called 'biofuels', largely ethanol, derived from cellulose. Cellulose accounts for approximately three-quarters of the dry weight of the plant, where it is used to provide plant cell walls with strength and rigidity. Presently much of the ethanol that is currently added to petrol is sourced from simpler carbohydrates (fructose and sucrose) derived from corn or sugar cane. This represents an important step to reduce fossil fuel consumption. However, the total energy provided by current biofuels, compared to the amount of fossil fuel required to produce them, is not high enough to provide sufficient impact on greenhouse gas emissions.

There is also a debate about whether directing food supplies to fuel production is sustainable if the world population continues to increase. As an alternative, efforts are underway to make biofuels from the non-edible cellulose found in the stems, stalks and leaves of plants grown for this sole purpose (like switchgrass). Such plants can be bred to generate higher yields and survive in harsher environments than current crops. The chemistry of converting cellulose to ethanol is difficult, but such processes have the potential to generate biofuels that provide many times the amounts of energy that are needed to generate them.

In this chapter we will look at a number of different types of carbohydrates and discuss some of the structural features and chemical reactions of this important class of organic compound.

LEARNING OBJECTIVES

After studying this chapter, you should be able to:

22.1 define carbohydrates

22.2 describe monosaccharides using aldose/ketose terminology

22.3 understand and describe the cyclic structure of monosaccharides

22.4 describe the chemical reactions of monosaccharides

22.5 explain disaccharides and oligosaccharides

22.6 define polysaccharides and describe starch, glycogen and cellulose.

22.1 Introduction to carbohydrates

Carbohydrates possess many functional groups, which leads to some interesting properties. They are also of fundamental importance as biochemicals. In chapters 19 and 21 we dealt with the individual functional groups found in carbohydrates (alcohols, aldehydes and ketones), so many of the properties, reactivities and concepts will be familiar. The range and variety of carbohydrate chemistry are immense and we will cover only the simplest forms. However, it is crucial to remember that all carbohydrates, no matter how complex, are governed by the properties of the individual chemical groups from which they are constructed.

The word 'carbohydrate' means 'hydrate of carbon' and derives from the formula $C_n(H_2O)_m$. Two examples of carbohydrates with molecular formulae that can be written as hydrates of carbon are:

- glucose (blood sugar), $C_6H_{12}O_6$, which can be written as $C_6(H_2O)_6$
- sucrose (table sugar), $C_{12}H_{22}O_{11}$, which can be written as $C_{12}(H_2O)_{11}$.

However, not all carbohydrates have this general formula. Some contain too few oxygen atoms, and some contain too many. Some also contain nitrogen. But the term 'carbohydrate' has become firmly rooted in chemical nomenclature and, although not completely accurate, it persists as the name for this class of compounds.

At the molecular level, most **carbohydrates** are polyhydroxyaldehydes, polyhydroxyketones or compounds that yield them after hydrolysis. Therefore, the chemistry of carbohydrates is essentially the chemistry of hydroxyl and carbonyl groups, and of the acetal bonds (section 21.5) formed between these two functional groups.

Almost all carbohydrates are chiral. This means that they interact with plane-polarised light (see chapter 17). Chirality is significant for their biological activity as many cellular interactions are controlled by specific carbohydrate stereoisomers. This is also why blood transfusions require specific matching blood types, the different blood types arise from different carbohydrates present on the surface of cells. Another important factor governed by the different structures of carbohydrates is their perceived sweetness. One carbohydrate may be very sweet while another with only a minor structural difference may have little apparent sweetness. Notably, all of the various carbohydrates — from the simplest through to the most complex stereoisomers important in biological activity — have ultimately arisen from the simple nonchiral CO_2 molecule via plant photosynthesis.

22.2 Monosaccharides

The word 'saccharide' comes from the ancient Latin word for 'sweet' and monosaccharides are the simplest forms of carbohydrate, unable to be hydrolysed to anything smaller.

Monosaccharides have the general formula $C_nH_{2n}O_n$, with one of the carbon atoms being part of a carbonyl group of either an aldehyde or a ketone. The most common monosaccharides contain from three to nine carbon atoms. There is a common naming system where the suffix *-ose* indicates that a molecule is a carbohydrate, and the prefixes *tri-*, *tetr-*, *pent-*, and so on, indicate the number of carbon atoms in the chain. Monosaccharides containing an aldehyde group are classified as **aldoses**; those containing a ketone group are classified as **ketoses**.

There are only two trioses — glyceraldehyde, which is an aldotriose, and dihydroxyacetone, which is a ketotriose.

glyceraldehyde (an aldotriose)	dihydroxyacetone (a ketotriose)
CHO	CH_2OH
|	|
CHOH	C=O
|	|
CH_2OH	CH_2OH

Often the designations *aldo-* and *keto-* are omitted, and these molecules are referred to simply as trioses, tetroses etc. Although these names do not describe the nature of the carbonyl group, they do indicate that the monosaccharide contains three and four carbon atoms, respectively.

Remember that these molecules can also be named using the IUPAC nomenclature system (see chapter 2). For example, the correct IUPAC name for glyceraldehyde is 2,3-dihydroxypropanal, and dihydroxyacetone is more correctly called 1,3-dihydroxypropanone.

Stereoisomerism

Glyceraldehyde contains one stereocentre and exists as a pair of enantiomers (see chapter 17). The stereoisomer shown at the top left on the next page has the *R* configuration and is named (*R*)-glyceraldehyde, while its enantiomer, shown at the top right on the next page, is named (*S*)-glyceraldehyde.

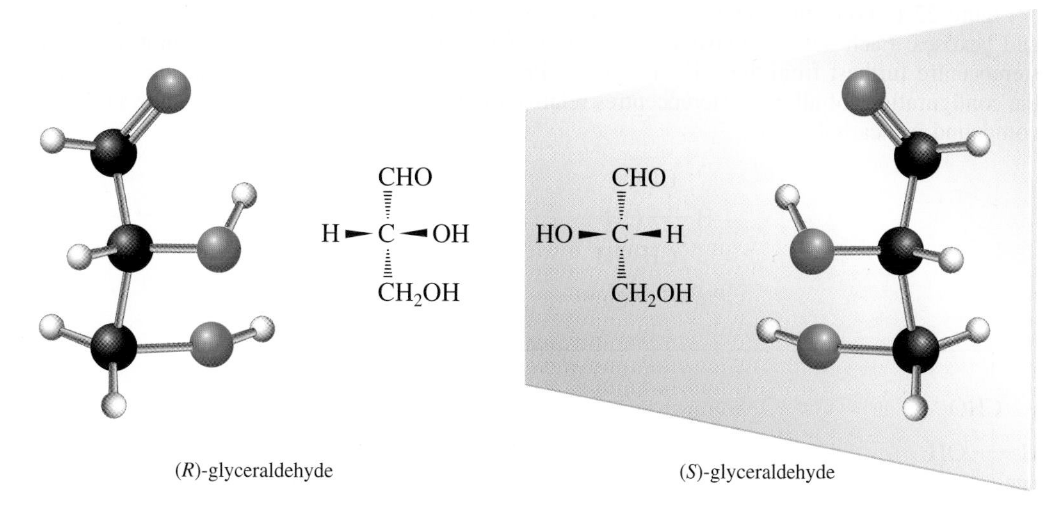

(*R*)-glyceraldehyde (*S*)-glyceraldehyde

Fischer projections

Chemists commonly use two-dimensional representations called **Fischer projections** to show the three-dimensional configuration of carbohydrates. To draw a Fischer projection, draw a three-dimensional representation with the most oxidised carbon atom towards the top and the molecule oriented so that the vertical bonds from each stereocentre are directed away from you and the horizontal bonds are directed towards you. Then write the molecule as a two-dimensional figure with each stereocentre indicated by the point at which the bonds cross. You now have a Fischer projection.

H, H, HO — C — C — CHO, OH, H

CHO
H — C — OH
CH_2OH

(*R*)-glyceraldehyde (three-dimensional representation)

convert to a Fischer projection →

CHO
H —+— OH
CH_2OH

(*R*)-glyceraldehyde (Fischer projection)

The two horizontal segments of this Fischer projection represent bonds directed towards you, and the two vertical segments represent bonds directed away from you. The only atom in the plane of the paper is the stereocentre. Note that you cannot treat a Fischer projection as a 3-D object as you can with the structure on the left. The horizontal and vertical lines in a Fischer projection have structural implications. If you rotate a Fischer projection by 90° then you create a new 3-D structure.

D- and L-monosaccharides

Even though the *R,S* system is widely accepted as a standard for designating the configuration of stereocentres, we still commonly indicate the configuration of carbohydrates using the D,L system proposed by Emil Fischer in 1891. He assigned the dextrorotatory and levorotatory enantiomers of glyceraldehyde the following configurations and named them D-glyceraldehyde and L-glyceraldehyde, respectively.

CHO
H —+— OH
CH_2OH

D-glyceraldehyde
$[\alpha]_D^{25} = +13.5°$

CHO
HO —+— H
CH_2OH

L-glyceraldehyde
$[\alpha]_D^{25} = -13.5°$

The monosaccharides D-glyceraldehyde and L-glyceraldehyde serve as reference points for the assignment of relative configurations for all other aldoses and ketoses. The reference point is the stereocentre furthest from the carbonyl group. Because this stereocentre is the next-to-the-last carbon atom on the chain, it is called the **penultimate carbon**. A **D-monosaccharide** is a monosaccharide that has the same configuration at its penultimate carbon atom as D-glyceraldehyde (its —OH is on the right in a Fischer projection). (Note that the D used in $[\alpha]_D^{25}$ refers to the sodium D line (see chapter 17) and is unrelated to the D in D-glyceraldehyde.) An **L-monosaccharide** has the same configuration at its penultimate carbon atom as L-glyceraldehyde (its —OH is on the left in a Fischer projection). Almost all monosaccharides in the biological world belong to the D series, and the majority of them are either hexoses or pentoses.

Figure 22.1 shows the names and Fischer projections for all D-aldotrioses, tetroses, pentoses and hexoses. Each name consists of three parts. The letter D specifies the configuration at the stereocentre furthest from the carbonyl group. Prefixes, such as *rib-*, *arabin-* and *gluc-*, specify the configurations of all other stereocentres relative to one another. The suffix *-ose* shows that the compound is a carbohydrate.

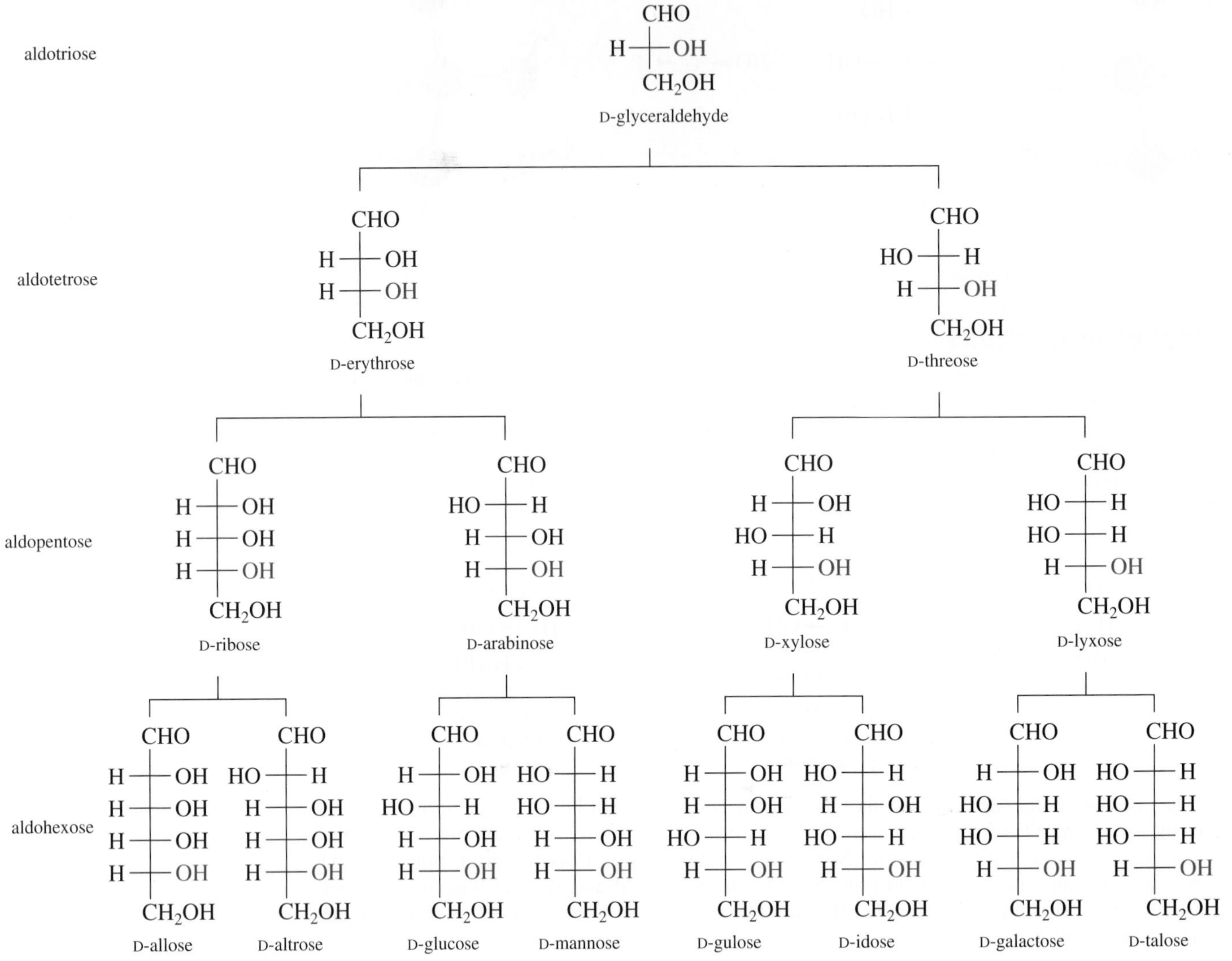

FIGURE 22.1 Configurational relationships among the isomeric D-aldotetroses, D-aldopentoses and D-aldohexoses. The configuration of the reference —OH on the penultimate carbon atom is shown in blue.

The three most abundant hexoses in the biological world are D-glucose, D-galactose and D-fructose. The first two (see figure 22.1) are D-aldohexoses while the third, fructose (at right), is a D-2-ketohexose. Glucose, by far the most abundant of the three, is also known as dextrose because it is dextrorotatory. Other names for this monosaccharide include *grape sugar* and *blood sugar*. Human blood normally contains 65–110 mg of glucose/100 mL of blood. D-fructose is one of the two monosaccharide building blocks of sucrose (table sugar, section 22.5).

CH_2OH
C=O
HO—H
H—OH
H—OH
CH_2OH
D-fructose

WORKED EXAMPLE 22.1

Naming and drawing simple carbohydrates

(a) Draw Fischer projections for the four aldotetroses.
(b) Which of the four aldotetroses are D-monosaccharides, which are L-monosaccharides and which are enantiomers?
(c) Refer to figure 22.1, and name each aldotetrose you have drawn.

Analysis

One way of approaching this question is to build chemical models of the aldotetroses and compare the relationships of the atoms and bonds in three-dimensional space. However, if you do not have a chemical model kit, you can draw and name the structures in order to properly visualise the relationships between the atoms and bonds involved in these molecules.

Solution

The Fischer projections for the four aldotetroses are:

D-erythrose	L-erythrose	D-threose	L-threose
CHO	CHO	CHO	CHO
H–C–OH	HO–C–H	HO–C–H	H–C–OH
H–C(3)–OH	HO–C(3)–H	H–C(3)–OH	HO–C(3)–H
CH_2OH	CH_2OH	CH_2OH	CH_2OH

(one pair of enantiomers) (a second pair of enantiomers)

Note that in the Fischer projection of a D-aldotetrose the —OH on C(3) is on the right and in an L-aldotetrose it is on the left.

Is our answer reasonable?

The most common mistake with Fischer projections is incorrectly counting the number of carbon atoms in the chain. Remember to put the most oxidised carbon atom at the top of the chain and put the (nonchiral) terminal CH_2OH at the bottom. Every intersection of vertical and horizontal lines represents a carbon atom.

PRACTICE EXERCISE 22.1

(a) Draw Fischer projections for all 2-ketopentoses.
(b) Which of the 2-ketopentoses are D-ketopentoses, which are L-ketopentoses, and which are enantiomers?

Amino sugars

Amino sugars contain an $—NH_2$ group in place of an —OH group. Only three amino sugars are common in nature: D-glucosamine, D-mannosamine and D-galactosamine. *N*-acetyl-D-glucosamine, a derivative of D-glucosamine, is a component of many polysaccharides, including connective tissue such as cartilage. It is also a component of chitin, the hard shell-like exoskeleton of lobsters, crabs, prawns and other shellfish. Several other amino sugars are components of naturally occurring antibiotics.

D-glucosamine	D-mannosamine (C(2) stereoisomer of D-glucosamine)	D-galactosamine (C(4) stereoisomer of D-glucosamine)	*N*-acetyl-D-glucosamine
CHO	CHO	CHO	CHO
H–C–NH_2	H_2N–C(2)–H	H–C–NH_2	H–C–$NHC(=O)CH_3$
HO–C–H	HO–C–H	HO–C–H	HO–C–H
H–C–OH	H–C–OH	HO–C(4)–H	H–C–OH
H–C–OH	H–C–OH	H–C–OH	H–C–OH
CH_2OH	CH_2OH	CH_2OH	CH_2OH

Physical properties

Monosaccharides are colourless, crystalline solids. Because hydrogen bonding is possible between their polar —OH groups and water, all monosaccharides are very soluble in water. They are only slightly soluble in ethanol and are insoluble in nonpolar solvents such as diethyl ether, dichloromethane and benzene.

22.3 The cyclic structure of monosaccharides

In section 21.5, we saw that aldehydes and ketones react with alcohols to form **hemiacetals**, sometimes referred to as hemiketals when derived from a ketone. We also saw that cyclic hemiacetals form readily when hydroxyl and carbonyl groups are parts of the same molecule and their interaction can form a five- or six-membered ring. For example, 4-hydroxypentanal forms a five-membered cyclic hemiacetal as shown on the next page. Note that 4-hydroxypentanal contains one stereocentre and that a second stereocentre is generated at C(1) by hemiacetal formation.

4-hydroxypentanal

a cyclic hemiacetal

Monosaccharides contain hydroxyl and carbonyl groups in the same molecule, and they exist almost exclusively as five- and six-membered cyclic hemiacetals.

Haworth projections

A common way of representing the cyclic structure of monosaccharides is the **Haworth projection**, which is named after the English chemist Sir Walter Norman Haworth (Nobel Prize in chemistry, 1937). In a Haworth projection, a five- or six-membered cyclic hemiacetal is represented as a planar pentagon or hexagon lying roughly perpendicular to the plane of the paper. Groups bonded to the carbon atoms of the ring then lie either above or below the plane of the ring. The new stereocentre created in forming the cyclic structure is called the **anomeric carbon**. Stereoisomers that differ in configuration only at the anomeric carbon atom are called **anomers**. The anomeric carbon atom of an aldose is C(1); in D-fructose, the most common ketose, it is C(2).

Typically, Haworth projections are written with the anomeric carbon atom at the right and the hemiacetal oxygen atom at the back right (figure 22.2).

D-glucose

β-D-glucopyranose (β-D-glucose)

α-D-glucopyranose (α-D-glucose)

FIGURE 22.2 Haworth projections for β-D-glucopyranose and α-D-glucopyranose.

As you study the open-chain and cyclic hemiacetal forms of D-glucose, note that in converting from a Fischer projection to a Haworth structure:

- groups on the right in the Fischer projection point down in the Haworth projection
- groups on the left in the Fischer projection point up in the Haworth projection
- for a D-monosaccharide, the terminal —CH_2OH points up in the Haworth projection
- the configuration of the anomeric —OH group is relative to the terminal —CH_2OH group: if the anomeric —OH group is on the same side as the terminal —CH_2OH, its configuration is β; if the anomeric —OH group is on the opposite side, it is α.

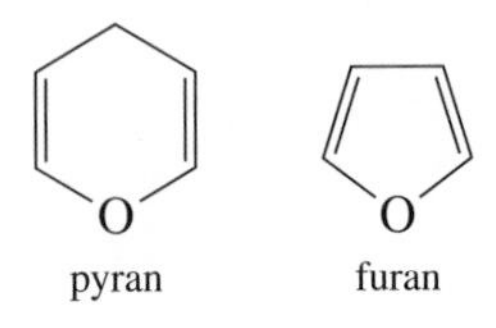

A six-membered hemiacetal ring is named with the component *-pyran-*, and a five-membered hemiacetal ring is named with the component *-furan-*. The terms **furanose** and **pyranose** are used because monosaccharide five- and six-membered rings correspond to the heterocyclic compounds pyran and furan.

A furanose is a five-membered cyclic hemiacetal form of a monosaccharide. A pyranose is a six-membered cyclic hemiacetal form of a monosaccharide.

Because the α and β forms of glucose are six-membered cyclic hemiacetals, they are named α-D-glucopyranose and β-D-glucopyranose, respectively. However, the designations *-furan-* and *-pyran-* are not always used in the names of monosaccharides. Thus, the glucopyranoses are often named simply α-D-glucose and β-D-glucose.

Aldopentoses also form cyclic hemiacetals. The most prevalent forms of D-ribose and other pentoses in the biological world are furanoses. Haworth projections for α-D-ribofuranose (α-D-ribose) and β-2-deoxy-D-ribofuranose (β-2-deoxy-D-ribose) are shown on the next page.

α-D-ribofuranose
(α-D-ribose)

β-2-deoxy-D-ribofuranose
(β-2-deoxy-D-ribose)

The prefix '2-deoxy' indicates the absence of oxygen at C(2). Units of D-ribose and 2-deoxy-D-ribose in nucleic acids and most other biological molecules are found almost exclusively in the β-configuration.

Fructose also forms five-membered cyclic hemiacetals. β-D-fructofuranose, for example, is found in the carbohydrate sucrose (see section 22.5).

α-D-fructofuranose
(α-D-fructose)

D-fructose

β-D-fructofuranose
(β-D-fructose)

anomeric carbon

Conformation representations

A five-membered ring is so close to being planar that Haworth projections are adequate to represent furanoses. For pyranoses, however, the six-membered ring is more accurately represented by what is known as a **chair conformation**, so-called because the molecule looks like a chair. The chair conformation allows all of the carbon atoms to have bond angles that are very close to the optimal tetrahedral bond angle and the strain in the molecule is therefore minimal. Figure 22.3 shows structural formulae for α-D-glucopyranose and β-D-glucopyranose, both drawn as chair conformations. The figure also shows the open-chain, or free, aldehyde form with which the cyclic hemiacetal forms are in equilibrium in aqueous solution. Notice that each group, including the anomeric —OH, in the chair conformation of β-D-glucopyranose is equatorial. Notice also that the —OH group on the anomeric carbon atom in α-D-glucopyranose is axial. Because of the equatorial orientation of the —OH on its anomeric carbon atom, β-D-glucopyranose is more stable and predominates in aqueous solution.

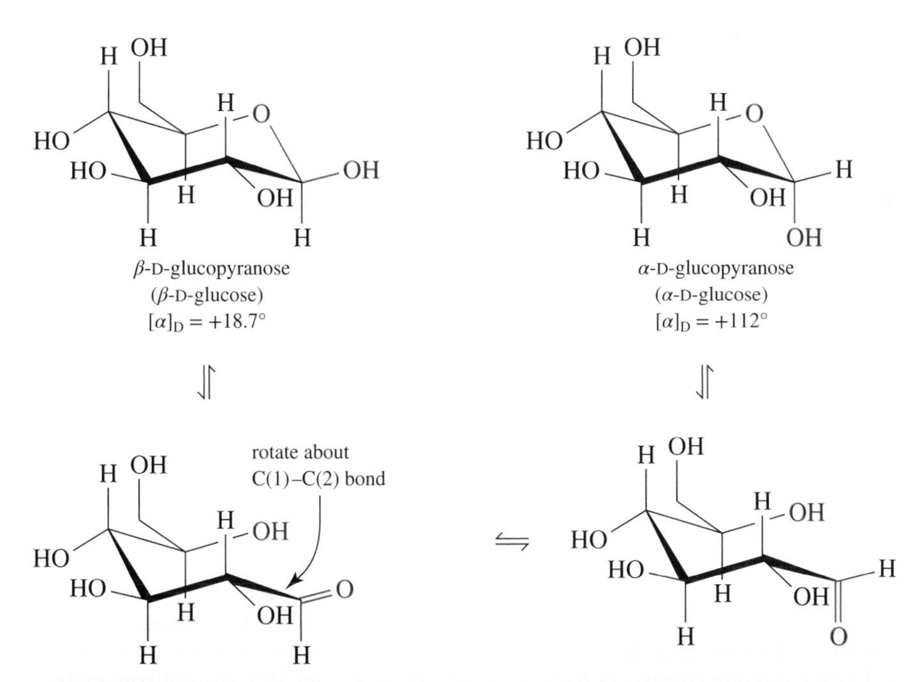

FIGURE 22.3 Chair conformations of α-D-glucopyranose and β-D-glucopyranose. Because α-D-glucose and β-D-glucose are different compounds (they are anomers), they have different specific rotations.

At this point, you should compare the relative orientations of groups on the D-glucopyranose ring in the Haworth projection and chair conformation.

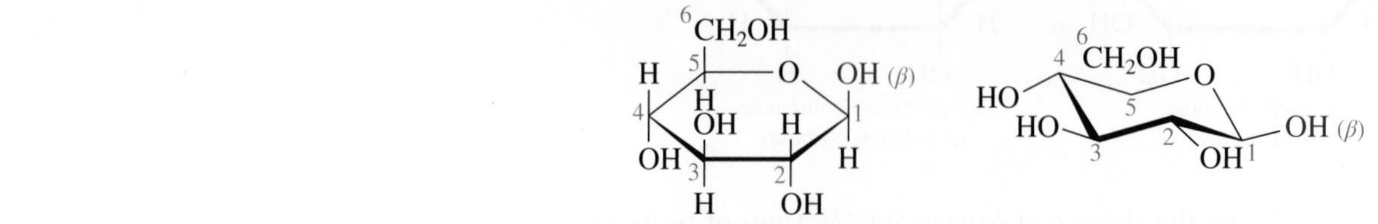

β-D-glucopyranose (Haworth projection)

β-D-glucopyranose (chair conformation)

Notice that the orientations of the groups on C(1) to C(5) in the Haworth projection of β-D-glucopyranose are up, down, up, down and up, respectively. The same is the case in the chair conformation, which has been drawn without showing the hydrogen atoms, for clarity.

WORKED EXAMPLE 22.2

Representing carbohydrates in the chair conformation

Draw chair conformations for α-D-galactopyranose and β-D-galactopyranose. Label the anomeric carbon atom in each cyclic hemiacetal.

Analysis and solution

D-galactose differs in configuration from D-glucose only at C(4). Therefore, to represent D-galactose, simply draw the α and β forms of D-glucopyranose and then interchange the positions of the —OH and —H groups on C(4). The specific rotations of each anomer are shown and you will note the impact on the optical rotation of the inversion of one stereocentre.

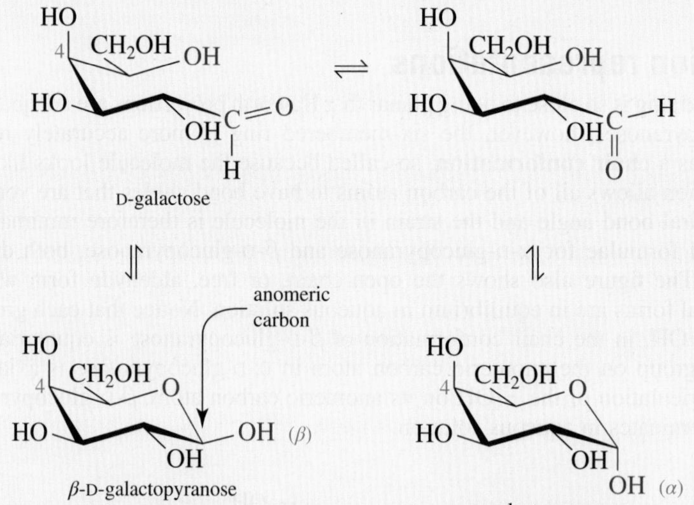

D-galactose

anomeric carbon

β-D-galactopyranose (β-D-galactose) $[\alpha]_D = +52.8°$

α-D-galactopyranose (α-D-galactose) $[\alpha]_D = +150.7°$

Is our answer reasonable?

Remember that the α and β anomers differ only in the orientation of the groups on the carbon atom of the hemiacetal group. Therefore, the structures of the α and β forms of the molecule should be the same in all aspects except for the stereochemistry of this carbon atom.

PRACTICE EXERCISE 22.2

Draw chair conformations for α-D-mannopyranose and β-D-mannopyranose. Label the anomeric carbon atom in each.

Mutarotation

Mutarotation is the change in specific rotation that accompanies the interconversion of α and β anomers in aqueous solution. As an example, a solution prepared by dissolving crystalline α-D-glucopyranose in water shows an initial rotation of +112° (figure 22.4), which gradually decreases to an equilibrium value of +52.7° as α-D-glucopyranose reaches an equilibrium with β-D-glucopyranose. A solution of β-D-glucopyranose also undergoes mutarotation, during which the specific rotation changes from an initial value of +18.7° to the same equilibrium value of +52.7°. The equilibrium mixture consists of 64% β-D-glucopyranose and 36% α-D-glucopyranose and only traces (0.003%) of the open-chain form. Mutarotation occurs in all carbohydrates that exist in hemiacetal forms.

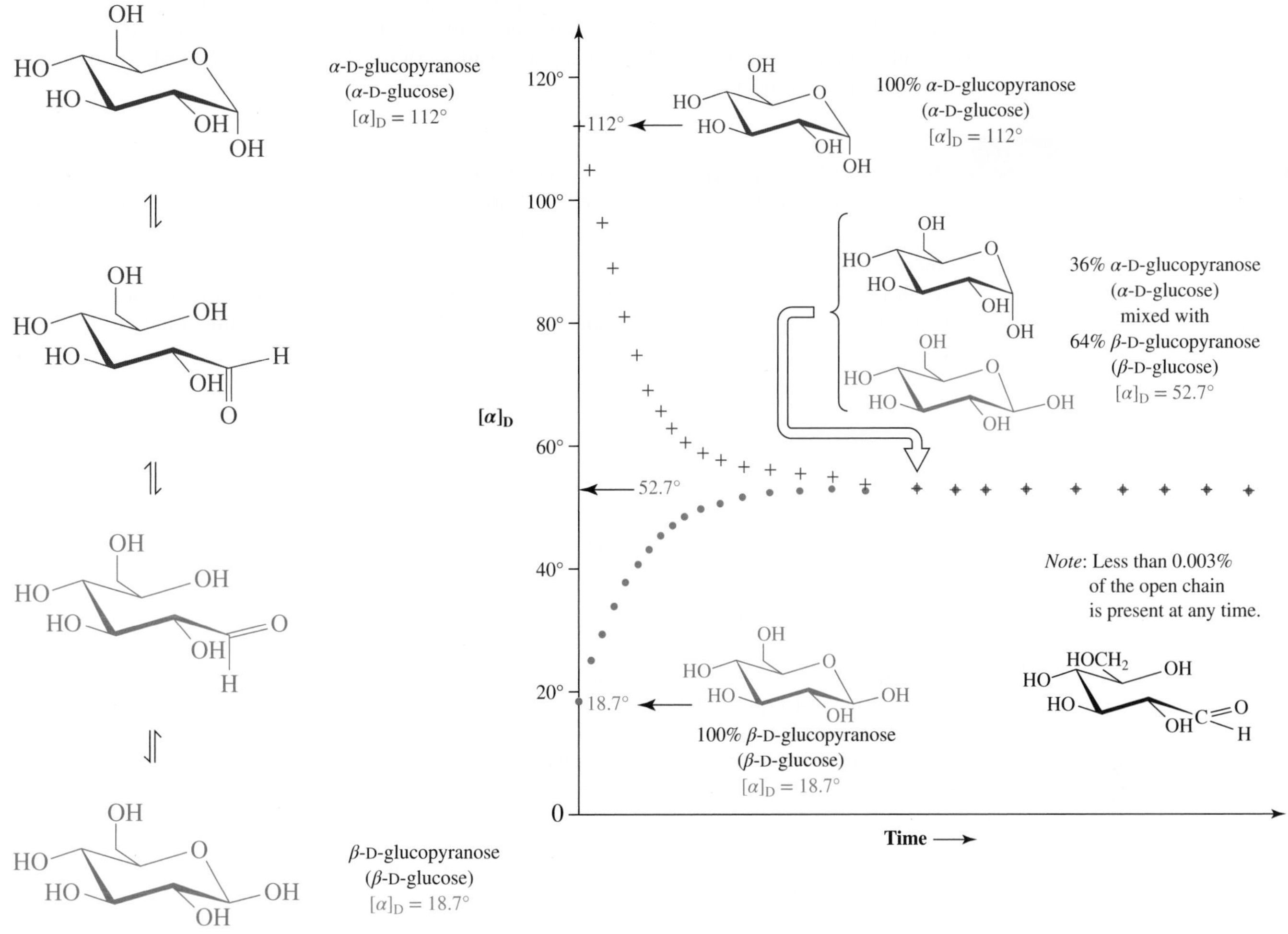

FIGURE 22.4 Specific rotations provided by each anomer provide evidence that interconversion arises.

22.4 Reactions of monosaccharides

In this section, we discuss the reactions of monosaccharides with alcohols, reducing agents and oxidising agents. We also examine how these reactions are useful in our everyday lives.

Formation of glycosides (acetals)

As we saw in section 21.5, treating an aldehyde or ketone with one molecule of an alcohol yields a hemiacetal, and treating the hemiacetal with a molecule of an alcohol yields an acetal. Treating a monosaccharide, all of which exist as cyclic hemiacetals, with an alcohol gives an acetal, as illustrated by the reaction of methanol with β-D-glucopyranose (β-D-glucose) drawn in the Haworth projection to emphasise the differences in the structures of the two products.

anomeric carbon

$+ CH_3OH \xrightarrow[-H_2O]{H^+}$

glycoside bond

glycoside bond

β-D-glucopyranose
(β-D-glucose)

methyl-β-D-glucopyranoside
(methyl-β-D-glucoside)

methyl-α-D-glucopyranoside
(methyl-α-D-glucoside)

A cyclic acetal derived from a monosaccharide is called a **glycoside**, and the bond from the anomeric carbon to the —OR group is called a **glycosidic bond**. Unlike a hemiacetal, an acetal is no longer in equilibrium with the open-chain carbonyl-containing compound in neutral or alkaline solutions. Like other acetals (section 21.5), glycosides are stable in water and aqueous base, but undergo hydrolysis in aqueous acid to form an alcohol and a monosaccharide.

We name glycosides by listing the alkyl or aryl group bonded to oxygen, followed by the name of the carbohydrate with the ending *-e* replaced with *-ide*. For example, glycosides derived from β-D-glucopyranose are named β-D-glucopyranosides; those derived from β-D-ribofuranose are named β-D-ribofuranosides.

WORKED EXAMPLE 22.3

Understanding the structures of glycosides

Draw a structural formula for methyl-β-D-ribofuranoside (methyl-β-D-riboside). Label the anomeric carbon atom and the glycosidic bond.

Analysis

By drawing a precise representation of the structure of a glycoside from its name you will gain an appreciation of how the atoms are arranged to give rise to this functional group.

Solution

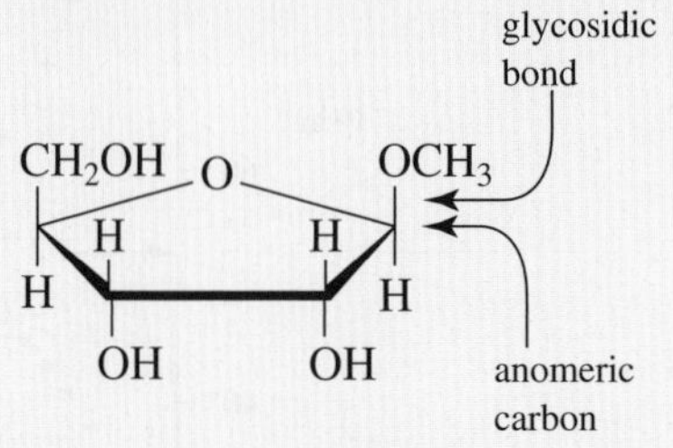

glycosidic bond

CH_2OH O OCH_3 H H H H OH OH

anomeric carbon

Is our answer reasonable?

Glycosides are acetals, so, unlike hemiacetals, they are always locked into a cyclic form. Consequently, you cannot use a Fischer projection — only a Haworth projection or perhaps a full structural representation. For simple glycosides, a Haworth projection clearly shows the relative arrangement of the —OH groups, which is crucial for determining which of the many monosaccharides is involved (figure 22.1).

PRACTICE EXERCISE 22.3

Draw a structural formula for the chair conformation of methyl-α-D-mannopyranoside (methyl-α-D-mannoside). Label the anomeric carbon atom and the glycosidic bond.

Just as the anomeric carbon atom of a cyclic hemiacetal reacts with the —OH group of an alcohol to form a glycoside, it can also react with the —NH group of an amine to form an *N*-glycoside. Especially important in the biological world are the *N*-glycosides formed between D-ribose and 2-deoxy-D-ribose (each as a furanose), and the heterocyclic aromatic amines uracil, cytosine, thymine, adenine and guanine (figure 22.5). *N*-glycosides of these compounds are the structural units of nucleic acids (chapter 25).

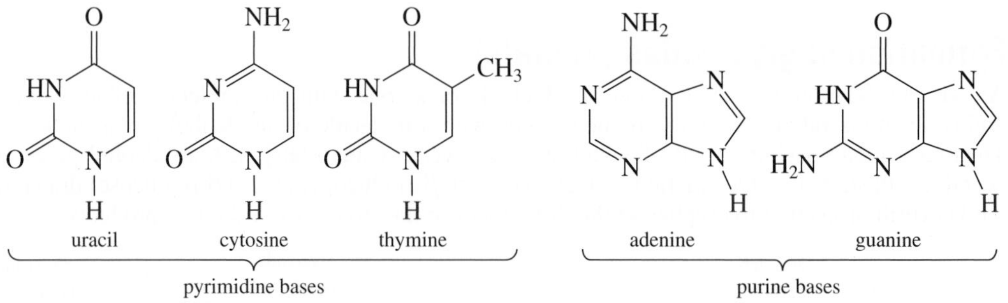

uracil cytosine thymine adenine guanine

pyrimidine bases purine bases

FIGURE 22.5 Structural formulae of the five most important purine and pyrimidine bases found in DNA and RNA. The hydrogen atom shown in colour is lost in the formation of an *N*-glycoside.

WORKED EXAMPLE 22.4

Understanding the structures of more complicated, biologically relevant glycosides

Draw a structural formula for the β-*N*-glycoside formed between D-ribofuranose and cytosine. Label the anomeric carbon atom and the *N*-glycosidic bond.

Analysis

Putting together the structural representations we have made for simple functional groups allows us to begin to build up representations of even the most complicated biological molecules, such as DNA. An *N*-glycoside contains an N atom bonded to the anomeric carbon atom of the carbohydrate. The β designation means that the N atom will lie on the same side of the ring as the CH_2OH group.

Solution

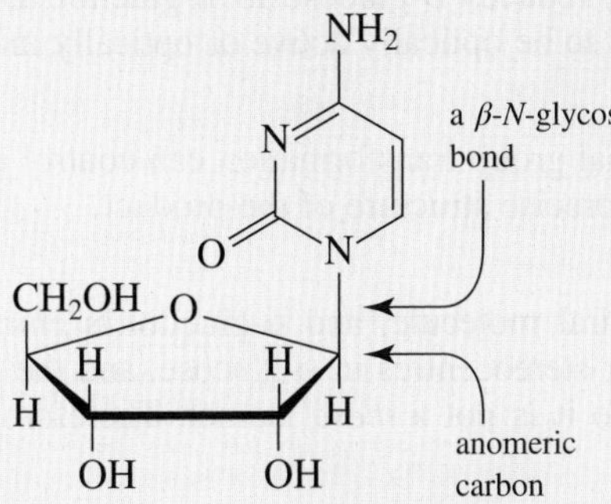

NH$_2$ ... a β-N-glycosidic bond ... anomeric carbon

(Structure: cytosine N linked to C-1 of β-D-ribofuranose; ring labels CH$_2$OH, O, H, H, H, H, OH, OH)

Is our answer reasonable?

The basis to the structure has to be the Haworth projection, so first check to ensure that it is correctly drawn. The acetal oxygen that forms the ring should be at the rear of the structure and the terminal —CH$_2$OH should also be to the rear, pointing up. Then check that the bond between D-ribofuranose and cytosine is correct. The bond is drawn upwards from the anomeric carbon atom to the nitrogen atom of cytosine and is therefore a β-N-glycosidic bond.

PRACTICE EXERCISE 22.4

Draw a structural formula for the β-N-glycoside formed between β-D-ribofuranose and adenine.

Reduction to alditols

The carbonyl group of a monosaccharide can be reduced to a hydroxyl group by a variety of reducing agents, including NaBH$_4$ (section 21.5, p. 965). The reduction products are known as **alditols**. Reduction of D-glucose gives D-glucitol, more commonly known as D-sorbitol. D-glucose is shown below in the open-chain form, only a small amount of which is present in solution. However, as it is reduced, the equilibrium between the cyclic hemiacetal form and the open-chain form shifts to replace the D-glucose. Note that D-glucose has also been represented in a chair conformation, which more realistically represents the reality of its shape.

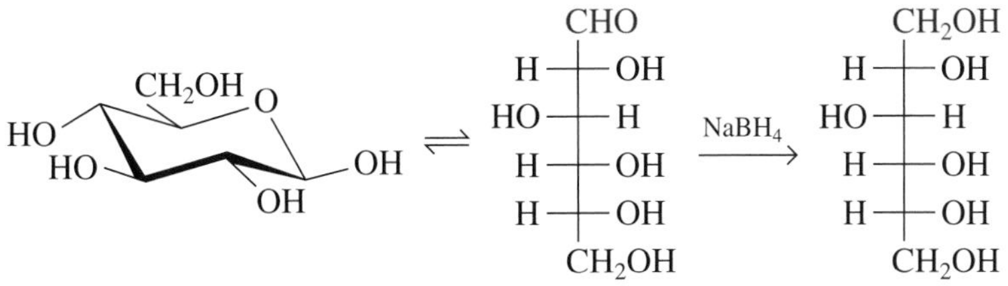

β-D-glucopyranose (β-D-glucose) ⇌ D-glucose $\xrightarrow{NaBH_4}$ D-glucitol (D-sorbitol)

Alditols are named by replacing the *-ose* in the name of the monosaccharide with *-itol*. Sorbitol is found in the plant world in many berries and in cherries, plums, pears, apples, seaweed and algae. It is about 60% as sweet as sucrose (table sugar) and is used in the manufacture of lollies and as a sugar substitute for diabetics. Among other alditols common in the biological world are erythritol, D-mannitol and xylitol, the last of which is used as a sweetening agent in 'sugarless' gum, confectionery and sweet cereals.

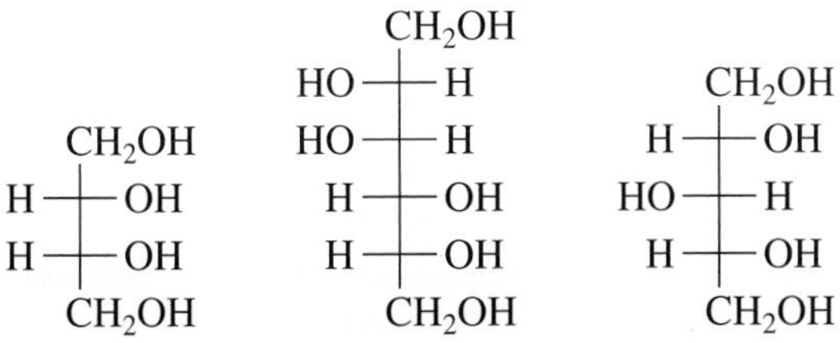

erythritol D-mannitol xylitol

WORKED EXAMPLE 22.5

Determining the structure of monosaccharide reaction products

Sodium borohydride, $NaBH_4$, reduces D-glucose to D-glucitol. Do you expect the alditol formed under these conditions to be optically active or optically inactive? Explain.

Analysis

Understanding how a functional group transformation can control a molecule's properties is best achieved by drawing the precise structure of the product.

Solution

Optical activity requires a chiral molecule, and D-glucitol is chiral. Reduction by $NaBH_4$ does not affect any of the four stereocentres in D-glucose, and the product does not have an internal plane of symmetry, so it is not a *meso* isomer. Therefore, we can predict that the product is optically active.

Is our answer reasonable?

Reduction using $NaBH_4$ removes the aldehyde functional group and therefore prevents hemiacetal formation. The generation of a hemiacetal creates a new chiral carbon atom, so this possibility is prevented. However, the other chiral carbon atoms are unaffected, so the product molecule still retains optical activity.

PRACTICE EXERCISE 22.5

Sodium borohydride, $NaBH_4$, reduces D-erythrose to erythritol. Do you expect the alditol formed under these conditions to be optically active or optically inactive? Explain.

Oxidation to aldonic acids (reducing sugars)

We saw in section 21.5 that several agents, including O_2, oxidise aldehydes (RCHO) to carboxylic acids (RCOOH). The ease of oxidation of aldehydes leads to a specific test for this functional group called the Tollens' test or silver-mirror test (chapter 21, p. 967), which is also relevant for monosaccharides. Similarly, under basic conditions, the aldehyde group of an aldose can be oxidised to a carboxylate group. Under these conditions, the cyclic form of the aldose is in equilibrium with the open-chain form, which is then oxidised by a mild oxidising agent. D-glucose, for example, is oxidised to D-gluconate (the anion of D-gluconic acid).

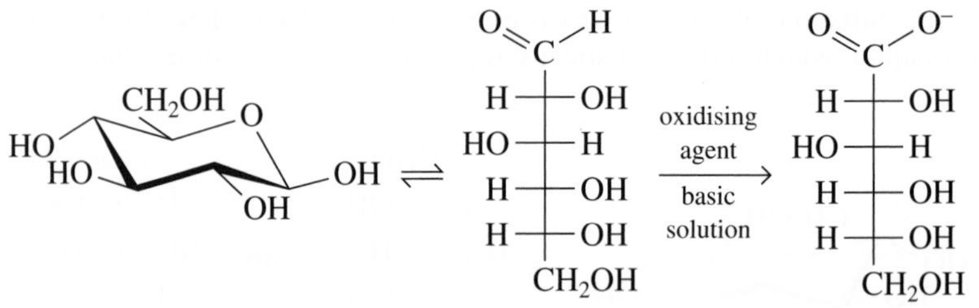

oxidising agent, basic solution

β-D-glucopyranose (β-D-glucose) ⇌ D-glucose → D-gluconate

Any carbohydrate that reacts with an oxidising agent to form an aldonic acid is classified as a **reducing sugar**. (It reduces the oxidising agent.)

Oxidation to uronic acids

Enzyme-catalysed oxidation of the primary alcohol at C(6) of a hexose yields a uronic acid. Enzyme-catalysed oxidation of D-glucose, for example, yields D-glucuronic acid, shown here in both its open-chain and cyclic hemiacetal forms (only the β form is shown).

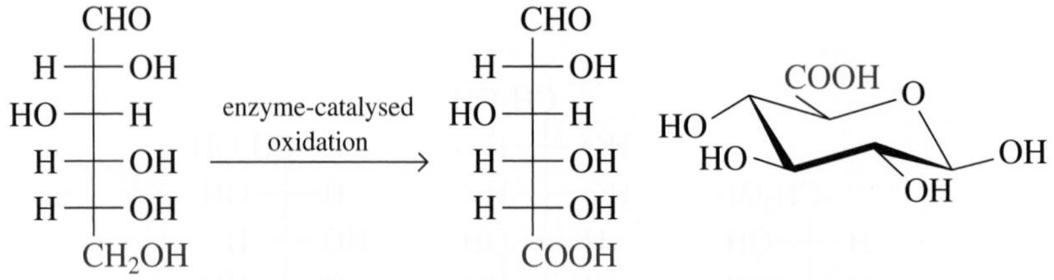

enzyme-catalysed oxidation

D-glucose → D-glucuronic acid (a uronic acid); chair conformation

D-glucuronic acid is widely distributed in both the plant and animal worlds. In humans, it is an important component of the acidic polysaccharides of connective tissues. The body also uses it to detoxify foreign phenols and alcohols. In the liver, these compounds are converted to glycosides of glucuronic acid (glucuronides), to be excreted in urine. The intravenous anaesthetic propofol, for example, is converted to the following water-soluble glucuronide and then excreted in urine.

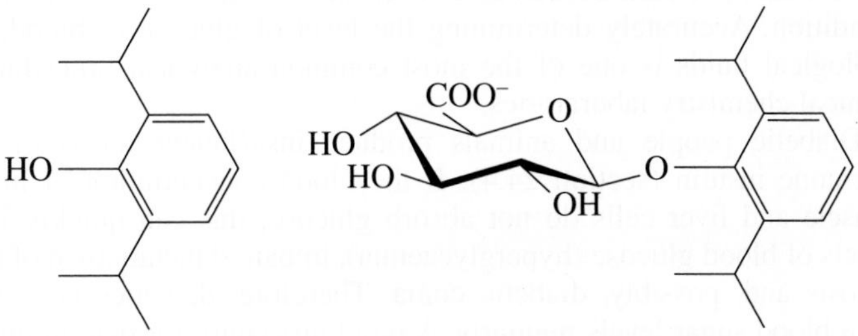

propofol   urine-soluble glucuronide

L-ascorbic acid (vitamin C)

One of the most important reactions of glucose is its conversion to L-ascorbic acid (vitamin C). This vitamin is synthesised biochemically by plants and some animals from D-glucose. Humans, however, do not have the enzyme systems required for this synthesis and so for us L-ascorbic acid is a vitamin, that is, an essential component of our diet. Captain Cook insisted that his crew consume sauerkraut and lime juice on the long voyages from England when he mapped the east coast of Australia and circumnavigated New Zealand. Limes are rich in vitamin C and so his crew did not suffer from scurvy, a vitamin C deficiency that was common among sailors of the time. Today, vitamin C is one of the most commonly used dietary supplements as there is a perception that it is an important antioxidant which controls free radicals and ageing (figure 22.6).

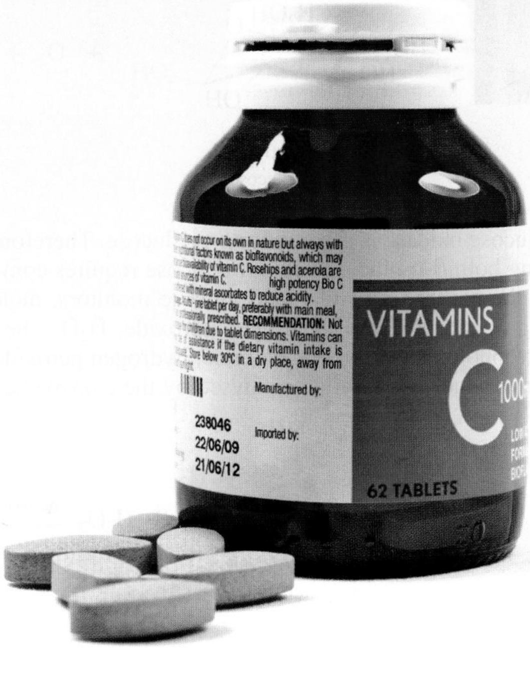

FIGURE 22.6 Vitamin C in an orange is identical to its synthetic tablet form.

The structural formula of vitamin C (L-ascorbic acid) resembles that of a monosaccharide. L-ascorbic acid is very easily oxidised to L-dehydroascorbic acid, a diketone.

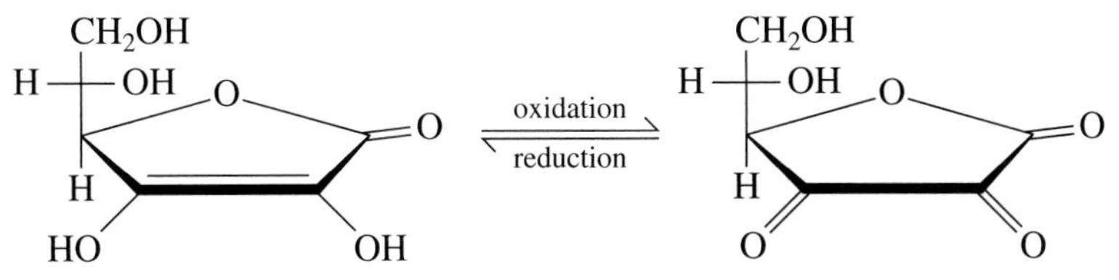

L-ascorbic acid (vitamin C)   L-dehydroascorbic acid

It is this ease of oxidation that makes vitamin C a crucial antioxidant in the body's biological defence mechanisms. Both L-ascorbic acid and L-dehydroascorbic acid are physiologically active and are found together in most body fluids.

Chemical Connections

Measuring blood sugar (glucose) to manage diabetes

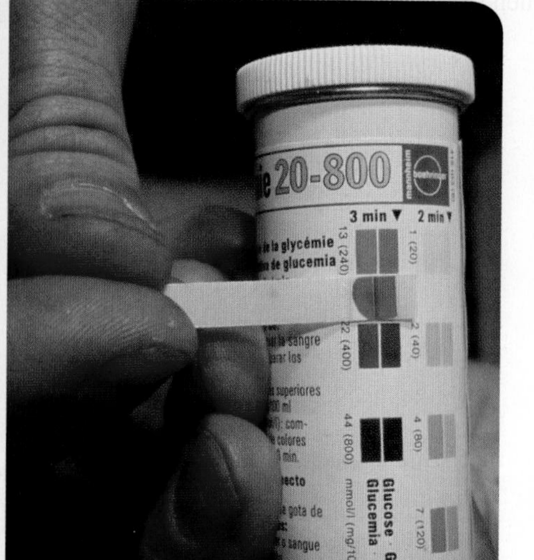

FIGURE 22.7 A blood test strip, which uses a colour intensity chart to measure glucose concentration.

Almost two million people in Australia and New Zealand suffer from diabetes, and more than 100 000 are diagnosed each year. These people need to closely monitor their blood sugar levels and administer medication to manage their condition. Accurately determining the level of glucose in blood, urine and other biological fluids is one of the most common analytical procedures performed in clinical chemistry laboratories.

Diabetic people and animals produce insufficient levels of the polypeptide hormone insulin (section 24.4). If the blood concentration of insulin is too low, muscle and liver cells do not absorb glucose; this can quickly lead to increased levels of blood glucose (hyperglycaemia), impaired metabolism of fats and proteins, ketosis and, possibly, diabetic coma. Therefore, diabetics need to measure their own blood sugar levels regularly. A rapid procedure for determining blood glucose levels is critical for early diagnosis and effective management of this disease. In addition to being rapid, a test must give an accurate measure of D-glucose; that is, it must be specific for D-glucose and not be affected by other substances present in blood. Modern glucose monitors use technology based on the enzyme glucose oxidase. This enzyme catalyses the oxidation of β-D-glucose to D-gluconic acid.

$$\beta\text{-D-glucopyranose } (\beta\text{-D-glucose}) + O_2 + H_2O \xrightarrow{\text{glucose oxidase}} \text{D-gluconic acid} + H_2O_2 \text{ (hydrogen peroxide)}$$

Glucose oxidase is specific for β-D-glucose. Therefore, complete oxidation of any sample containing both β-D-glucose and α-D-glucose requires conversion of the α form to the β form.

In the earlier forms of blood glucose monitors, molecular oxygen, O_2, was used as the oxidising agent, producing hydrogen peroxide, H_2O_2, the concentration of which can be detected spectrophotometrically. For example, hydrogen peroxide converts colourless *o*-toluidine to a coloured product in a reaction catalysed by the enzyme peroxidise.

$$\text{2-methylaniline } (o\text{-toluidine}) + H_2O_2 \xrightarrow{\text{peroxidase}} \text{coloured product}$$

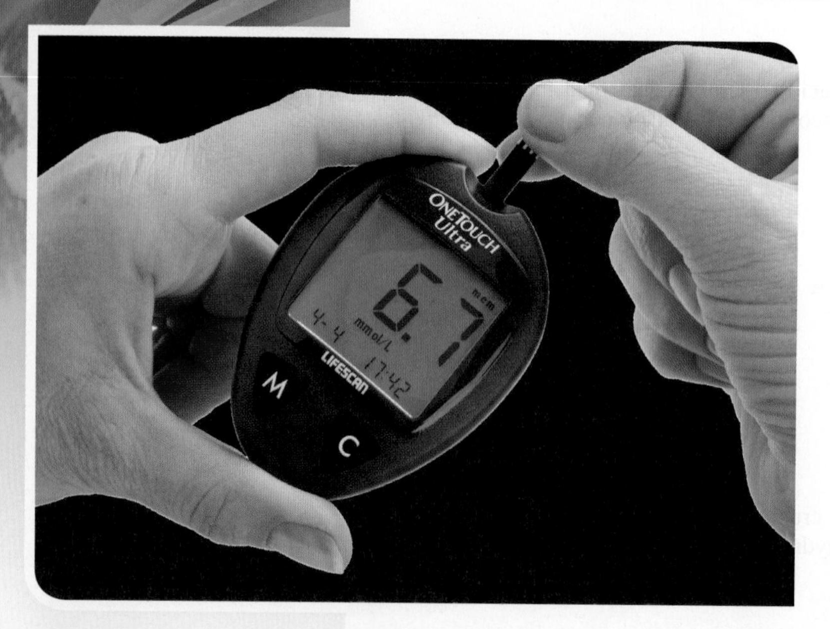

The original test strips used to measure blood sugar levels (see figure 22.7) relied on a colour intensity chart to determine the extent of the reaction and hence the concentration of glucose in the test solution. The first electronic testers used a spectrophotometric approach to determine the concentration of the coloured oxidation product.

Newer devices that do not rely on colour have come onto the market. Instead, an enzyme-linked electrode containing glucose oxidase is exposed to a precise amount of blood. The extent of oxidation of D-glucose is detected by the electrode and a glucose reading is given based on the current generated (figure 22.8).

FIGURE 22.8 A home blood sugar monitor.

22.5 Disaccharides and oligosaccharides

Most carbohydrates in nature contain more than one monosaccharide unit. Those which contain two units are called **disaccharides**; those which contain three units are called **trisaccharides**, and so on. The more general term, **oligosaccharide**, is often used for carbohydrates that contain from six to ten monosaccharide units. Carbohydrates containing larger numbers of monosaccharide units are called **polysaccharides**.

In a disaccharide, two monosaccharide units are joined by a glycosidic bond between the anomeric carbon atom of one unit and an —OH of the other. Sucrose, lactose and maltose are important disaccharides.

Sucrose

Sucrose (table sugar) is the most abundant disaccharide in the biological world. It is obtained principally from the juice of sugar cane and sugar beets. In sucrose, C(1) of α-D-glucopyranose bonds to C(2) of β-D-fructofuranose through an α-1,2-glycosidic bond.

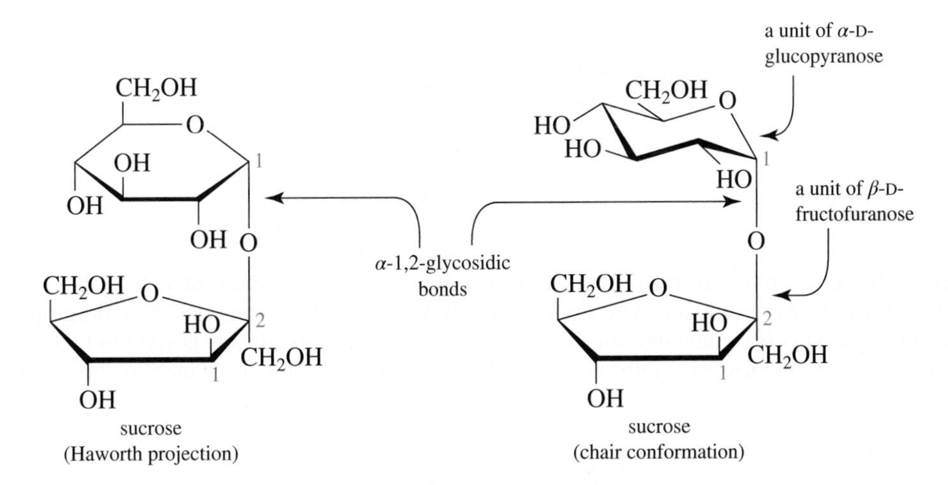

Because the anomeric carbon atoms of both the glucopyranose and fructofuranose units are involved in formation of the glycosidic bond, neither monosaccharide unit is in equilibrium with its open-chain form. Thus, sucrose is a nonreducing sugar.

Lactose

Lactose, the principal sugar in milk, accounts for 5% to 8% of human milk and 4% to 6% of cow's milk. This disaccharide consists of D-galactopyranose, bonded by a β-1,4-glycosidic bond to C(4) of D-glucopyranose.

β-1,4-glycosidic bond

lactose
(Haworth projection)

β-1,4-glycosidic bond

lactose
(chair conformation)

Lactose is a reducing sugar, because the cyclic hemiacetal of the D-glucopyranose unit is in equilibrium with its open-chain form and can be oxidised to a carboxyl group.

It has been estimated that 1 in 10 Australians and New Zealanders experience some degree of lactose intolerance. Lactose intolerance is an especially important health issue for Indigenous Australians and Maori among whom the incidence is much higher. People with lactose intolerance may suffer from calcium deficiencies unless their diet includes non-dairy sources of calcium.

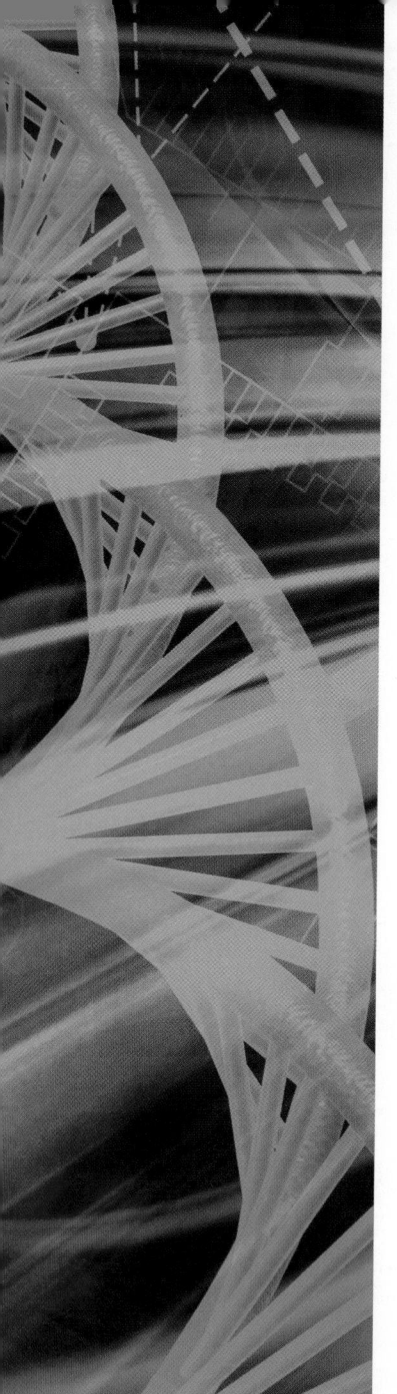

Chemical Connections

Honey is a drug!

Carbohydrates play critical roles in normal cell function, as well as in major diseases. Chemists are therefore constantly searching for new forms of carbohydrates and carbohydrate derivatives to use as drugs to fight disease. Due to their structural complexity, carbohydrates have generally not received as much attention as therapeutic drugs based on nucleic acids or proteins. This is changing, however, as chemists develop new synthetic methods and more powerful means of structural analysis. Experimental drugs with code names such as PAZ320 (for diabetes mellitus), GSC100 and GM-CT-01 (cancer) and GM-MD-02 (fibrosis) are being actively pursued by pharmaceutical companies across the world. In Australia the sulfonated carbohydrate derivatives PI-88 and PG545 (figure 22.9) are in development to fight cancer by blocking blood vessel growth in tumours.

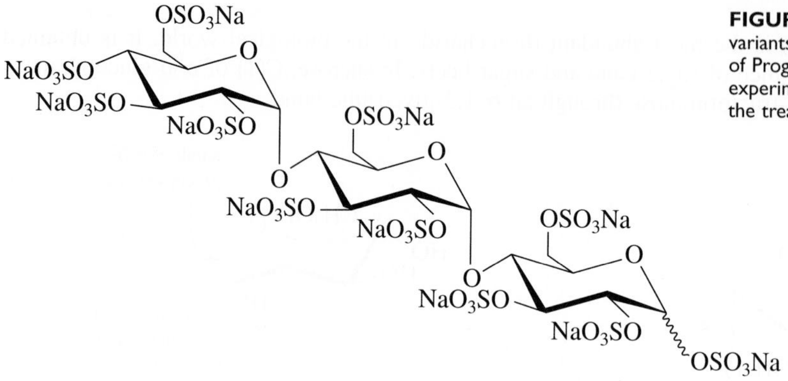

FIGURE 22.9 One of the variants of the structure of Progen Pharmaceuticals' experimental drug PG545 for the treatment of cancer.

The pathway from discovery to the release of a new drug is long and costly. It is also littered with failures. Many promising new drugs fail safety and efficacy tests, so drug development costs are high. This is a challenge for our society, as new drugs are always needed to better fight disease. New antibiotics especially are needed, as bacteria develop resistance to their activity (figure 22.10).

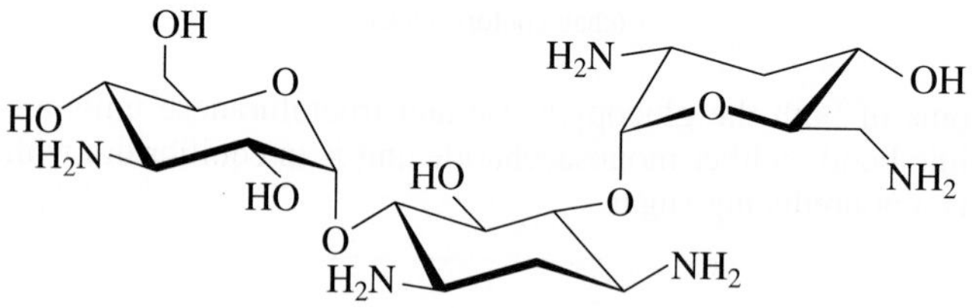

FIGURE 22.10 Tobramycin is a powerful aminotrisaccharide derivative used to fight infections. However, resistance to drugs such as Tobramycin can mean that infections in wounds or in diseases such as cystic fibrosis cannot be controlled.

The problem of antibacterial drug resistance is just as important in the developing world. Disease does not recognise borders, but expensive new treatments from research laboratories cannot help people without the resources to access them. In this context chemists are also investigating simpler solutions to fight disease. Honey (figure 22.11), for instance, has long been known to have antibacterial properties. This has been attributed to traces of hydrogen peroxide present in the complex mixture of carbohydrates that makes up this natural material. Recently a small tree called Mānuka (from Māori) found in New Zealand has been recognised as making the best honey for stopping bacterial infections in wounds. Mānuka trees also grow in southern Australia where they are called 'Tea trees', possibly because Captain Cook used the leaves to make a 'tea' drink. The extra efficacy of Mānuka honey over normal varieties is thought to arise from the presence of small amounts of methylglyoxal (2-oxopropanal) in the honey. However, the difference in antibiotic activity between Mānuka and other honeys cannot be explained by the different levels of methylglyoxal alone. Chemists believe it may also relate to a number of other components that make up the complex mixture that we know as honey. Honey is primarily a mixture of carbohydrates, mainly fructose (≈40%) and glucose (≈30%), but also some proportion of disaccharides including sucrose and maltose, as well as other more complex polysaccharides.

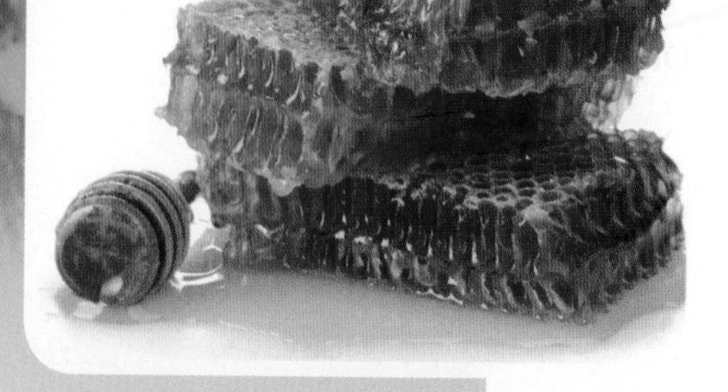

FIGURE 22.11 Honey has long been known to have antibacterial properties.

Honey has been used as a natural wound healer and infection fighter for thousands of years. More recently, the US Food and Drug Administration approved Mānuka-honey-based wound dressings for certain wounds. Mānuka honey has a remarkably broad spectrum of action, with efficacy towards bacteria, fungi and protozoa. Hopefully discoveries about the therapeutic activity of simple, readily available products such as honey can provide new and cost-effective treatments for diseases across the world.

Maltose

Maltose derives its name from its presence in malt, the juice from sprouted barley and other cereal grains. Maltose consists of two units of D-glucopyranose, joined by a glycosidic bond between C(1) (the anomeric carbon atom) of one unit and C(4) of the other unit. Because the oxygen atom on the anomeric carbon atom of the first glucopyranose unit is α, the bond joining the two units is called an α-1,4-glycosidic bond.

You might like to consider how you could best represent the structure of the carbohydrate rings and the α-disaccharide linkage that is present in maltose. Shown below are Haworth projections and a chair conformation for β-maltose, so named because the —OH group on the anomeric carbon atom of the glucose unit on the right is β. Remember that the protocol for Haworth projections is that the bonds coming from the rings are drawn vertically to the plane of the ring. This would mean that two Haworth projections linked together would require that the rings are oriented at an angle to each other. Perhaps a better representation keeps the rings in the horizontal plane orientation but shows the glycoside bond as not being vertical.

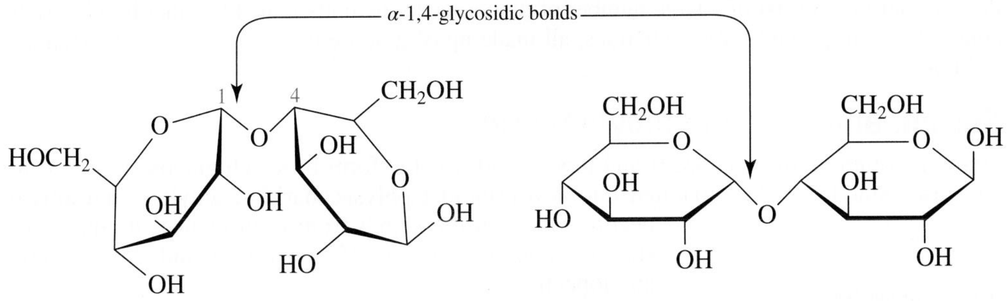

α-1,4-glycosidic bonds

two ways of showing the glycoside link between Haworth projections of glucose present in β-maltose

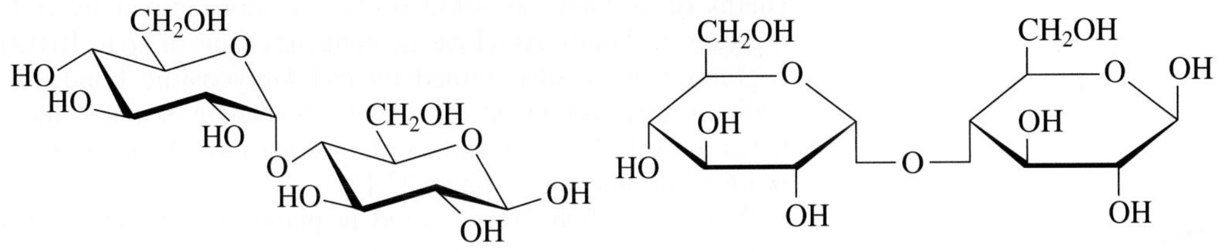

β-maltose
preferred structural representation

β-maltose
non-preferred structural representation

In reality the structure of disaccharides is best represented by the actual chair conformation, which more clearly shows the real bond angles and distances, as discussed in chapter 16 (pp. 703–4). In some texts you might see the glycoside linkage drawn as a U-shaped connection. This is not a very good representation, as the method for drawing structures outlined in chapter 16 (p. 699) suggests that two lines meeting at an angle indicate the presence of a carbon atom, and this is clearly not the case with a glycosidic bond.

WORKED EXAMPLE 22.6

Understanding the 3-D structure of disaccharides

Draw a chair conformation for the β anomer of a disaccharide in which two units of D-glucopyranose are joined by an α-1,6-glycosidic bond.

Analysis and solution

First draw a chair conformation of α-D-glucopyranose. Then connect the anomeric carbon atom of this monosaccharide to C(6) of a second D-glucopyranose unit by an α-glycosidic bond. The resulting molecule is either α or β depending on the orientation of the —OH group on the reducing end of the disaccharide. The disaccharide shown here is β.

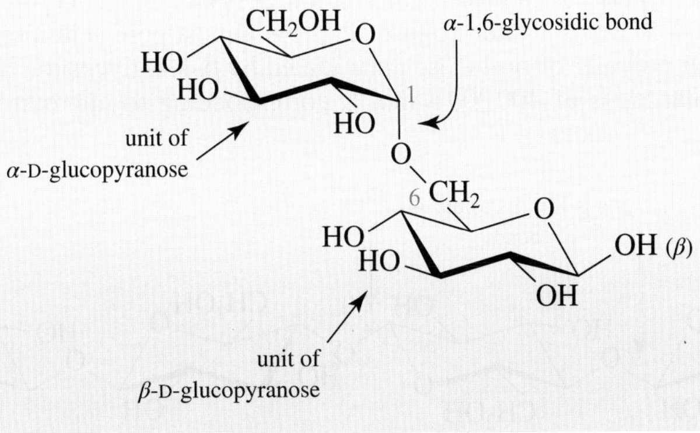

α-1,6-glycosidic bond

unit of α-D-glucopyranose

unit of β-D-glucopyranose

Is our answer reasonable?

Look closely at the drawing of the molecule. For a 1,6-connection, the first carbon atom of one ring, C(1), connects to the last carbon atom of the second ring, C(6). An α connection is oriented downwards from the first ring.

PRACTICE EXERCISE 22.6 Draw a Haworth projection and a chair conformation for the α form of a disaccharide in which two units of D-glucopyranose are joined by a β-1,3-glycosidic bond.

22.6 Polysaccharides

Polysaccharides consist of a large number of monosaccharide units joined together by glycosidic bonds. Three important polysaccharides, all made up of glucose units, are starch, glycogen and cellulose.

Starch: amylose and amylopectin

Starch is found in all plant seeds and tubers, and it is the form in which glucose is stored for later use. Starch can be separated into two principal polysaccharides: amylose and amylopectin. Although the starch from each plant is unique, most starches contain 20% to 25% amylose and 75% to 80% amylopectin.

Complete hydrolysis of both amylose and amylopectin yields only D-glucose. Amylose is composed of continuous, unbranched chains of as many as 4000 D-glucose units joined by α-1,4-glycosidic bonds. Amylopectin contains chains of up to 10 000 D-glucose units, also joined by α-1,4-glycosidic bonds. In addition, amylopectin has considerable branching from this linear network. At branch points, new chains of 24 to 30 units start by α-1,6-glycosidic bonds (figure 22.12).

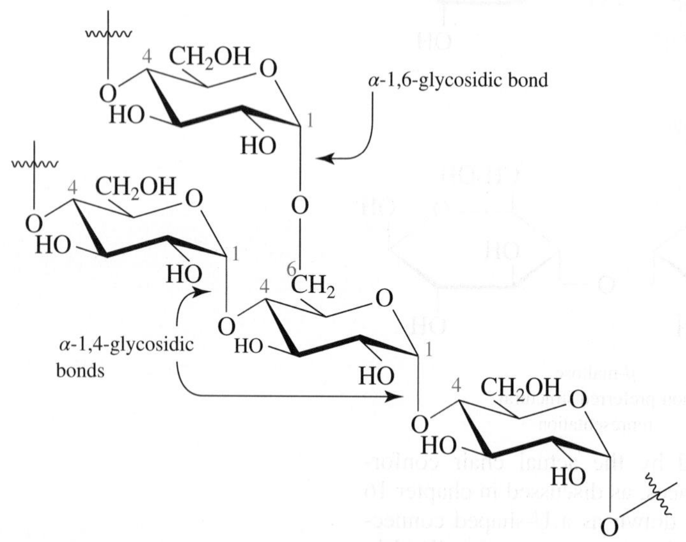

FIGURE 22.12 Amylopectin is a highly-branched polymer of D-glucose. Chains consist of 24 to 30 units of D-glucose, joined by α-1,4-glycosidic bonds, and branches created by α-1,6-glycosidic bonds.

Why are carbohydrates stored in plants as polysaccharides rather than monosaccharides, a more directly usable source of energy? The answer has to do with osmotic pressure, which is proportional to the molar concentration, not the molar mass, of a solute. If 1000 molecules of glucose are assembled into one starch macromolecule, a solution containing 1 g of starch per 10 mL will have only one-thousandth the osmotic pressure relative to a solution of 1 g of glucose in the same volume. This feat of packaging is a tremendous advantage because it reduces the strain on various membranes enclosing solutions of such macromolecules.

Glycogen

Glycogen is the reserve carbohydrate for animals. Like amylopectin, glycogen is a branched polymer of D-glucose. Each branch contains approximately 12–14 glucose units, joined by α-1,4- and α-1,6-glycosidic bonds. The total amount of glycogen in the body of a well-nourished adult human is about 350 g, divided almost equally between liver and muscle.

Cellulose

Cellulose, the most widely distributed plant skeletal polysaccharide, constitutes almost half of the cell-wall material of wood. Cotton (figure 22.14) is almost pure cellulose.

Cellulose, a linear polymer of D-glucose units joined by β-1,4-glycosidic bonds (figure 22.13), has an average molar mass of 400 000 g mol^{-1}, corresponding to approximately 2800 glucose units per molecule.

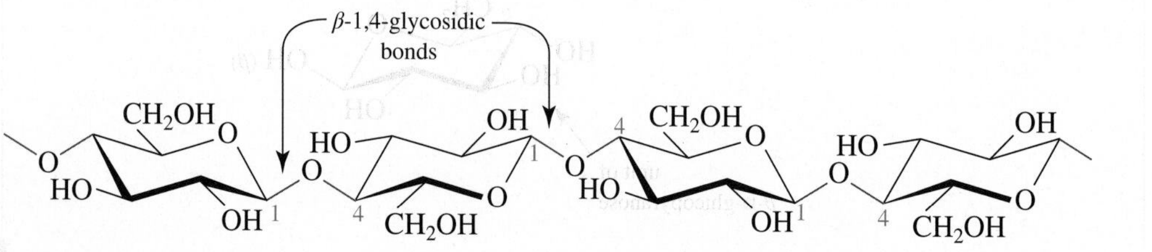

FIGURE 22.13 Cellulose is a linear polymer of D-glucose, joined by β-1,4-glycosidic bonds.

Cellulose molecules act like stiff rods by aligning themselves side-by-side into well-organised, water-insoluble fibres in which the —OH groups form numerous intermolecular hydrogen bonds. This arrangement of bundles of parallel chains gives cellulose fibres their high mechanical strength and explains why cellulose is insoluble in water. When a piece of cellulose-containing material is placed in water, there are not enough —OH groups on the surface of the fibre to pull individual cellulose molecules away from the strongly hydrogen-bonded fibre. At the molecular level, the chains of carbohydrates bind together through hydrogen bonding to form bundles (figure 22.15). These bundles of fibres are called microfibrils. They are glued together by smaller polysaccharide chains, called hemicelluloses, and act as the structural units that give plant walls their strength and rigidity.

FIGURE 22.14 Cotton ready to be harvested.

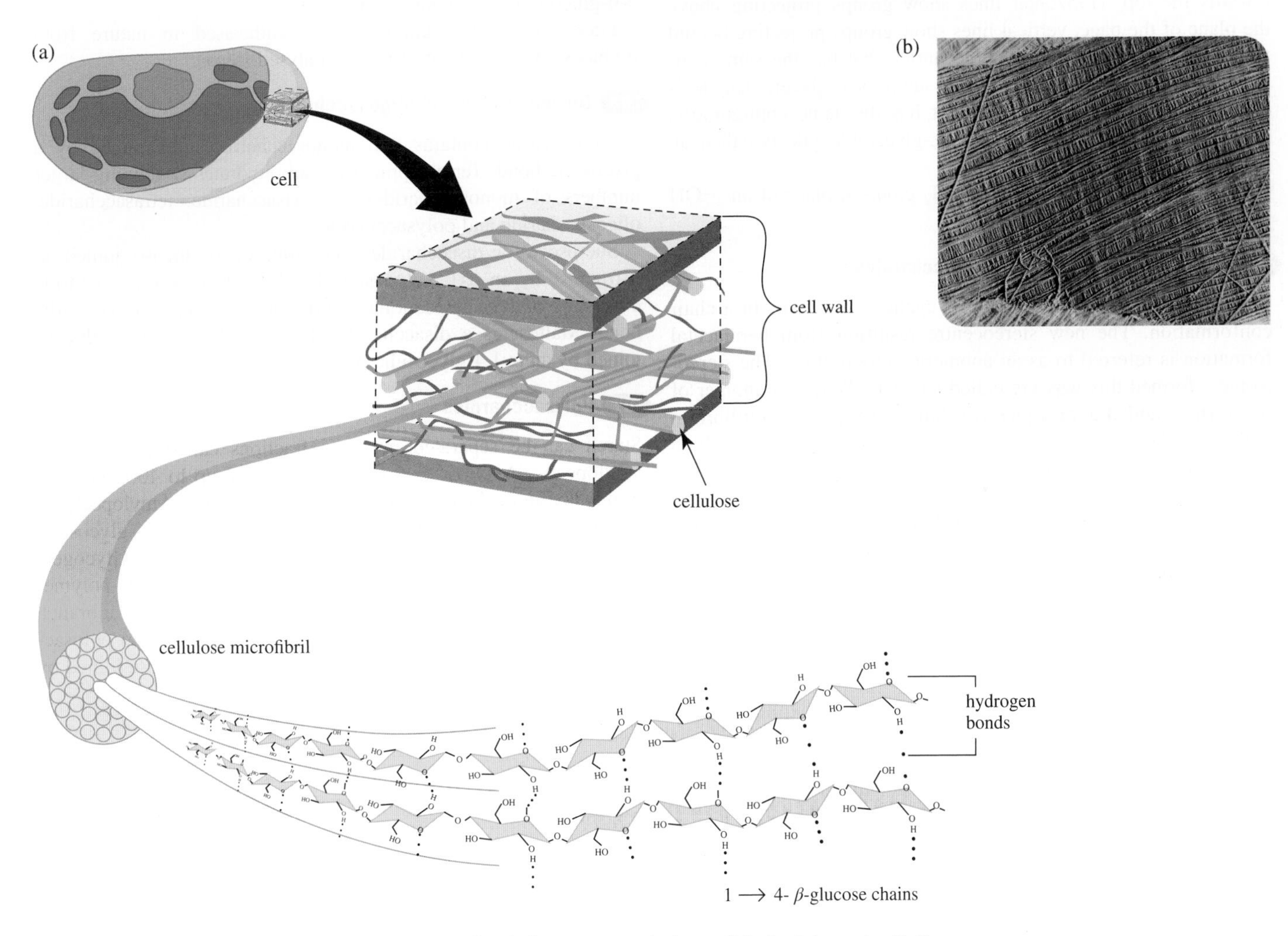

FIGURE 22.15 (a) Principal structures of plant cell walls, which are composed of cross-linked cellulose microfibrils constructed from linked β-glucose chains; and **(b)** electron micrograph of cell wall showing cross-linked microfibrils.

Humans cannot use cellulose as food because our digestive systems do not contain β-glucosidases, enzymes that catalyse the hydrolysis of β-glycosidic bonds. Instead, we have only α-glucosidases; hence, the polysaccharides we use as sources of glucose are starch and glycogen. By contrast, many bacteria and micro-organisms do contain β-glucosidases and can digest cellulose. Termites are fortunate (much to our regret) to have such bacteria in their intestines and can use wood as their principal food. Ruminants (cud-chewing animals) and horses can also digest grasses and hay, because β-glucosidase-containing micro-organisms are present within their alimentary systems.

SUMMARY

LO 1 Introduction to carbohydrates

Carbohydrates are polyhydroxyaldehydes, polyhydroxyketones or substances that yield these compounds after hydrolysis. Carbohydrates, no matter how complex, are governed by the properties of the individual chemical groups from which they are constructed.

LO 2 Monosaccharides

Monosaccharides are polyhydroxyaldehydes or polyhydroxyketones. The most common have the general formula $C_nH_{2n}O_n$, where n varies from 3 to 9. Their names contain the suffix *-ose*. The prefixes *tri-*, *tetr-*, *pent-* and so on show the number of carbon atoms in the chain. The prefix *aldo-* indicates an aldehyde, the prefix *keto-* designates a ketone. Monosaccharides containing an aldehyde group are classified as aldoses, and those containing a ketone group are ketoses.

In a Fischer projection of a carbohydrate, the carbon chain is written vertically, with the most highly oxidised carbon atom towards the top. Horizontal lines show groups projecting above the plane of the page, vertical lines show groups projecting behind the plane of the page. A monosaccharide that has the same configuration at the penultimate carbon atom as D-glyceraldehyde is called a D-monosaccharide; one that has the same configuration at the penultimate carbon atom as L-glyceraldehyde is called an L-monosaccharide.

An amino sugar contains an $—NH_2$ group in place of an —OH group.

LO 3 The cyclic structure of monosaccharides

Monosaccharides exist primarily as cyclic hemiacetals in a chair conformation. The new stereocentre resulting from hemiacetal formation is referred to as an anomeric carbon atom. The stereoisomers formed this way are called anomers. When an anomer of one form is added to an aqueous solution, it forms an equilibrium with the other anomer in a process called mutarotation, which is detected by the change in the optical rotation of the initial solution.

A six-membered cyclic hemiacetal is called a pyranose, a five-membered cyclic hemiacetal is a furanose. The symbol β indicates that the —OH on the anomeric carbon atom is on the same side of the ring as the terminal $—CH_2OH$. The symbol α indicates that —OH on the anomeric carbon atom is on the opposite side of the ring from the terminal $—CH_2OH$. Furanoses and pyranoses can be drawn as Haworth projections.

LO 4 Reactions of monosaccharides

A glycoside is an acetal derived from a monosaccharide. The bond from the anomeric carbon atom to the —OR group is called a glycosidic bond. The name of the glycoside is composed of the name of the alkyl or aryl group bonded to the acetal oxygen atom, followed by the name of the monosaccharide in which the terminal *-e* has been replaced by *-ide*.

An alditol is a polyhydroxy compound formed by the reduction of the carbonyl group of a monosaccharide. An aldonic acid is a carboxylic acid formed by oxidation of the aldehyde group of an aldose. The aldose reduces the oxidising agent and is therefore called a reducing sugar. Enzyme-catalysed oxidation of the terminal $—CH_2OH$ to a —COOH gives a uronic acid.

The determination of blood glucose levels is an important clinical test for the diagnosis and management of diabetes. The test uses the enzyme glucose oxidase, which catalyses the oxidation of β-D-glucose to D-gluconic acid.

L-ascorbic acid (vitamin C) is synthesised in nature from D-glucose by a series of enzyme-catalysed steps.

LO 5 Disaccharides and oligosaccharides

A disaccharide contains two monosaccharide units joined by a glycosidic bond. Terms applied to carbohydrates containing larger numbers of monosaccharides are: trisaccharide, tetrasaccharide, oligosaccharide and polysaccharide.

Sucrose is a disaccharide consisting of D-glucose joined to D-fructose by an α-1,2-glycosidic bond. Lactose is a disaccharide consisting of D-galactose joined to D-glucose by a β-1,4-glycosidic bond. Maltose is a disaccharide of two molecules of D-glucose joined by an α-1,4-glycosidic bond.

LO 6 Polysaccharides

Starch can be separated into two fractions called amylose and amylopectin. Amylose is a linear polymer of up to 4000 units of D-glucopyranose joined by α-1,4-glycosidic bonds. Amylopectin is a highly-branched polymer of D-glucose joined by α-1,4-glycosidic bonds and, at branch points, by α-1,6-glycosidic bonds. Glycogen, the reserve carbohydrate of animals, is a highly-branched polymer of D-glucopyranose joined by α-1,4-glycosidic bonds and, at branch points, by α-1,6-glycosidic bonds. Cellulose, the skeletal polysaccharide of plants, is a linear polymer of D-glucopyranose joined by β-1,4-glycosidic bonds.

KEY CONCEPTS AND EQUATIONS

Formation of cyclic hemiacetals *(section 22.3)*

A monosaccharide existing as a five-membered ring is a furanose; one existing as a six-membered ring is a pyranose. A pyranose is most commonly drawn as a Haworth projection or a chair conformation.

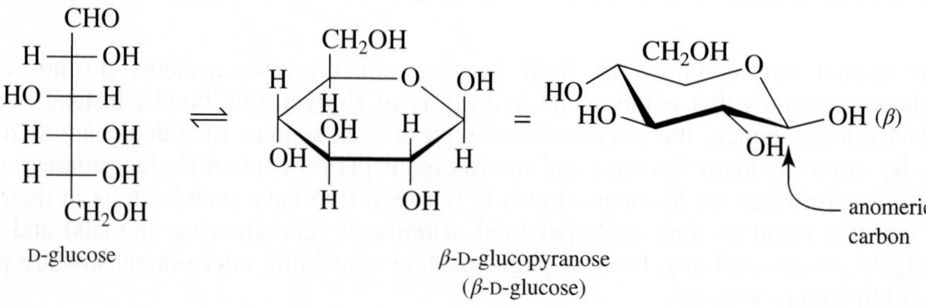

CHO, H—OH, HO—H, H—OH, H—OH, CH_2OH ⇌ CH_2OH, O, OH, H, H, OH, H, OH, H, H, OH = CH_2OH, O, HO, HO, OH (β), OH, anomeric carbon

D-glucose

β-D-glucopyranose (β-D-glucose)

Formation of glycosides *(section 22.4)*	Treatment of a monosaccharide with an alcohol in the presence of an acid catalyst forms a cyclic acetal called a glycoside. The bond to the new —OCH_3 group is called a glycosidic bond.
Reduction to alditols *(section 22.4)*	Reduction of the carbonyl group of an aldose or a ketose to a hydroxyl group yields a polyhydroxy compound called an alditol.
Oxidation to an aldonic acid *(section 22.4)*	A mild oxidising agent oxidises the aldehyde group of an aldose to a carboxyl group to give a polyhydroxycarboxylic acid called an aldonic acid.

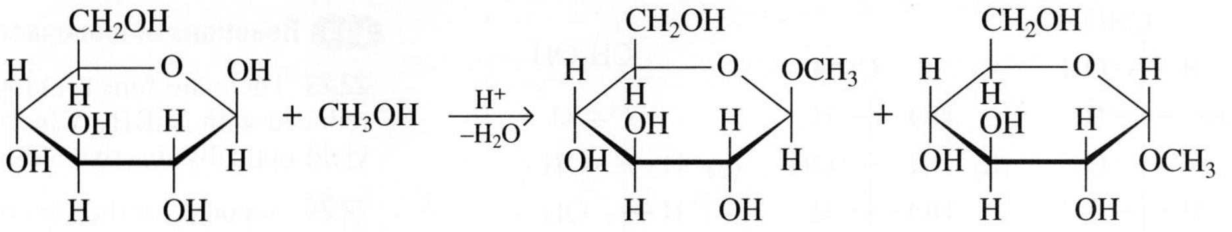

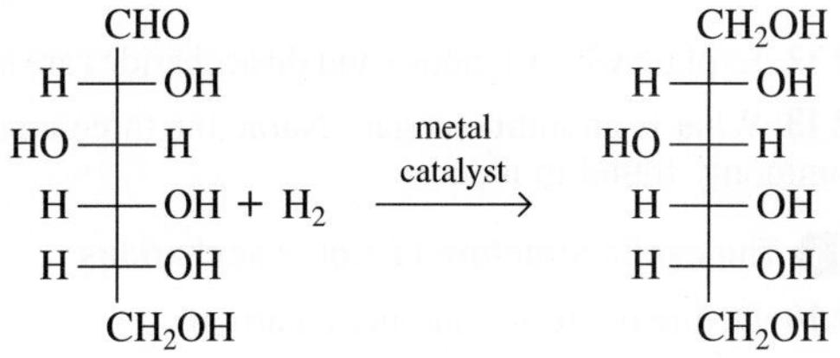

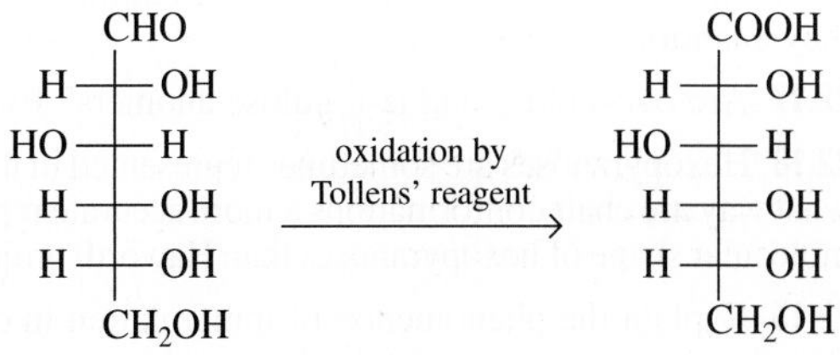

Glycoside formation: (glucopyranose, Haworth projection: CH_2OH, H, O, OH, H, OH, H, OH, H, H, OH) + CH_3OH $\xrightarrow[-H_2O]{H^+}$ (glycoside with OCH_3) + (glycoside with OCH_3)

Reduction: D-glucose (CHO; H–OH; HO–H; H–OH; H–OH; CH_2OH) + H_2 $\xrightarrow{\text{metal catalyst}}$ D-glucitol (D-sorbitol) (CH_2OH; H–OH; HO–H; H–OH; H–OH; CH_2OH)

Oxidation: D-glucose (CHO; H–OH; HO–H; H–OH; H–OH; CH_2OH) $\xrightarrow{\text{oxidation by Tollens' reagent}}$ D-gluconic acid (COOH; H–OH; HO–H; H–OH; H–OH; CH_2OH)

KEY TERMS

alditol *p. 995*
aldoses *p. 986*
anomeric carbon *p. 990*
anomers *p. 990*
carbohydrates *p. 986*
chair conformation *p. 991*
disaccharides *p. 999*
D-monosaccharide *p. 987*
Fischer projections *p. 987*
furanose *p. 990*
glycoside *p. 993*
glycosidic bond *p. 993*
Haworth projection *p. 990*
hemiacetal *p. 989*
ketoses *p. 986*
L-monosaccharide *p. 987*
monosaccharides *p. 986*
mutarotation *p. 992*
oligosaccharides *p. 999*
penultimate carbon *p. 987*
polysaccharides *p. 999*
pyranose *p. 990*
reducing sugar *p. 996*
trisaccharides *p. 999*

REVIEW QUESTIONS

LO 2 Monosaccharides

22.1 What is the difference in structure between: (a) an aldose and a ketose and (b) an aldopentose and a ketopentose?

22.2 Give the aldose/ketose designation for the following.

(a) $HOCH_2CHOHCH{=}O$

(b)  $HOCH_2CHOHC(=O)CH_2OH$

(c) 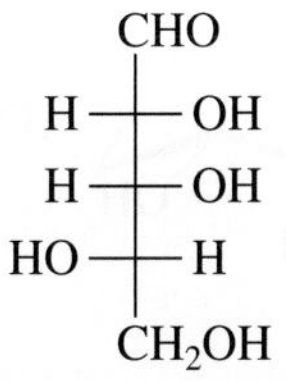 CHO; H–OH; H–OH; HO–H; CH_2OH

(d) 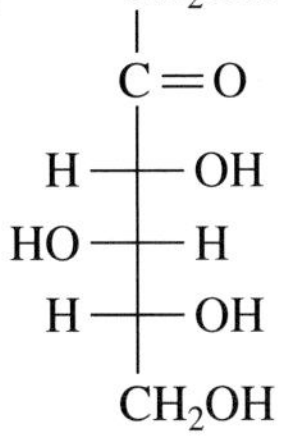 CH_2OH; C=O; H–OH; HO–H; H–OH; CH_2OH

(e) 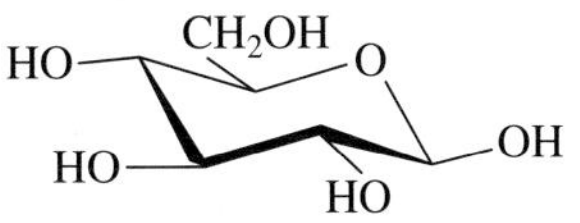 (pyranose chair: CH_2OH, HO, O, OH, HO, HO)

(f) 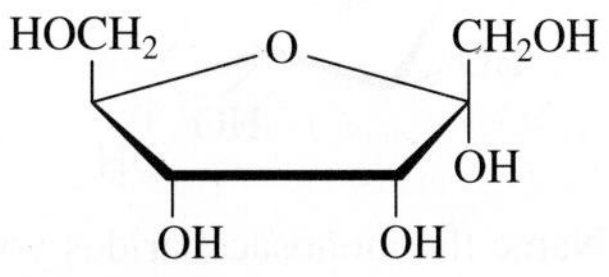 $HOCH_2$, O, CH_2OH, OH, OH, OH (furanose)

22.3 Which hexose is also known as dextrose?

22.4 How many different carbohydrates maybe defined as aldotrioses? Give skeletal line drawings for each possibility.

22.5 How many different carbohydrates exist that may be designated as ketopentoses. (Assume all reasonable structures are possible.)

22.6 Does a ketodiose exist? Why or why not? How about an aldodiose? Can you draw a skeletal line structure for such carbohydrates?

22.7 Draw all of the D-ketopentoses using Fischer projections.

22.8 D- and L-glyceraldehyde are enantiomers. Explain what this means.

22.9 Explain the meaning of the designations D and L as used to specify the configuration of carbohydrates.

22.10 How many stereocentres are present in D-glucose? How many stereocentres are present in D-ribose? How many stereoisomers are possible for each monosaccharide?

22.11 Which of the following compounds are D-monosaccharides and which are L-monosaccharides?

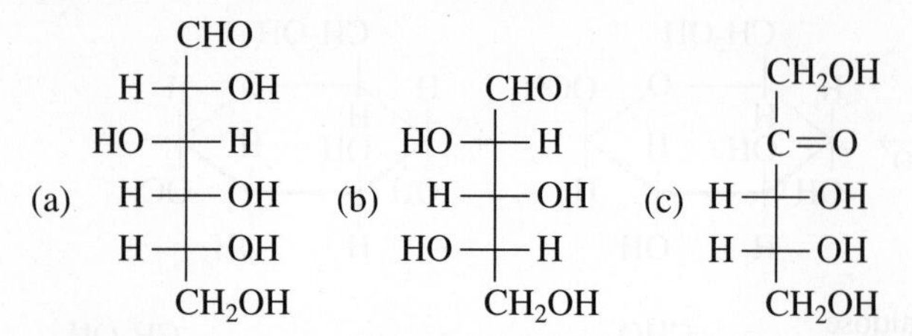

22.12 Explain why all mono- and disaccharides are soluble in water.

22.13 What is an amino sugar? Name the three amino sugars most commonly found in nature.

LO 3 The cyclic structure of monosaccharides

22.14 Define the term 'anomeric carbon'.

22.15 Explain the conventions for using α and β to designate the configurations of cyclic forms of monosaccharides.

22.16 Are α-D-glucose and β-D-glucose anomers? Explain. Are they enantiomers? Explain.

22.17 Are α-D-gulose and α-L-gulose anomers? Explain.

22.18 Hexopyranoses are sometimes represented in the chair form. In what way are chair conformations a more accurate representation of molecular shape of hexopyranoses than Haworth projections?

22.19 Explain the phenomenon of mutarotation in carbohydrates. By what means is mutarotation detected?

22.20 α-D-galactopyranose has an $[\alpha]_D = +150.7°$. When it is dissolved in water, however, the specific rotation of the solution after some time is measured as +80.2°. What would be the specific rotation of a solution of β-D-galactopyranose measured under the same conditions?

22.21 The specific rotation of a freshly prepared solution of pure α-D-glucose, $C_6(OH_2)_6$ changes over time to reach a value that is considerably lower than the initial reading. Would you expect this process to also potentially occur for the monosaccharide $C_3(OH_2)_3$?

22.22 (a) There are two D-ketopentoses that can form low-strain 5-membered hemiketal rings. Can you draw the skeletal line structure for the carbohydrates in the open chain and cyclic form?

(b) Pure ketopentoses and aldopentoses undergo mutarotation. Which smaller ketoses and aldoses also can undergo mutarotation?

LO 4 Reactions of monosaccharides

22.23 There are four D-aldopentoses (figure 22.1). If each is reduced with $NaBH_4$, which yield optically active alditols? Which yield optically inactive alditols?

22.24 Account for the observation that the reduction of D-glucose with $NaBH_4$ gives an optically active alditol, whereas the reduction of D-galactose with $NaBH_4$ gives an optically inactive alditol.

22.25 Which two D-aldohexoses give optically inactive (*meso*) alditols on reduction with $NaBH_4$? (*Hint:* See chapter 17.)

22.26 Name the two alditols formed by $NaBH_4$ reduction of D-fructose.

22.27 Is ascorbic acid a biological oxidising agent or a biological reducing agent? Explain.

LO 5 Disaccharides and oligosaccharides

22.28 Define the term 'glycosidic bond'.

22.29 What is the difference in meaning between the terms 'glycosidic bond' and 'glucosidic bond'?

22.30 Do glycosides undergo mutarotation?

22.31 In making lollies or syrup from sugar, sucrose is boiled in water with a little acid, such as lemon juice. Why does the product mixture taste sweeter than the starting sucrose solution?

22.32 Which of the following disaccharides are reduced by $NaBH_4$?
(a) sucrose (b) lactose (c) maltose

LO 6 Polysaccharides

22.33 What is the difference in structure between oligosaccharides and polysaccharides?

22.34 Name three polysaccharides that are composed of units of D-glucose. In which of the three polysaccharides are the glucose units joined by α-glycosidic bonds? In which are they joined by β-glycosidic bonds?

22.35 Starch can be separated into two principal polysaccharides: amylose and amylopectin. What is the major difference in structure between the two?

REVIEW PROBLEMS

22.36 2,6-dideoxy-D-altrose, known alternatively as D-digitoxose, is a monosaccharide obtained from the hydrolysis of digitoxin, a natural product extracted from purple foxglove (*Digitalis purpurea*). Digitoxin has found wide use in cardiology because it reduces the pulse rate, regularises heart rhythm and strengthens the heartbeat. Draw the structural formula of 2,6-dideoxy-D-altrose. **LO2**

The foxglove plant produces the important cardiac medication digitalis.

22.37 Draw Fischer projections for the monosaccharides L-ribose and L-arabinose. **LO2**

22.38 Draw α-D-glucopyranose (α-D-glucose) as a Haworth projection. Now, using only the following information, draw Haworth projections for the following monosaccharides. **LO3**

(a) α-D-mannopyranose (α-D-mannose). The configuration of D-mannose differs from that of D-glucose only at C(2).

(b) α-D-gulopyranose (α-D-gulose). The configuration of D-gulose differs from that of D-glucose at C(3) and C(4).

22.39 Convert each of the following Haworth projections to an open-chain form and then to a Fischer projection. **LO2,3**

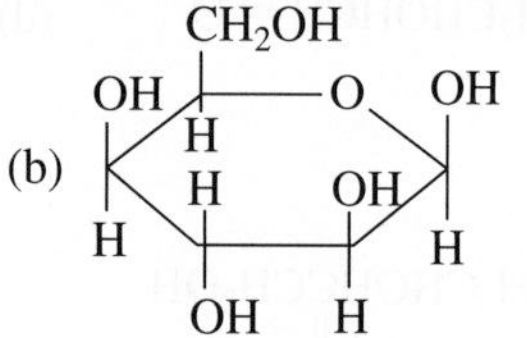

Name the monosaccharides you have drawn.

22.40 Convert each of the following chair conformations to an open-chain form and then to a Fischer projection. **LO2**

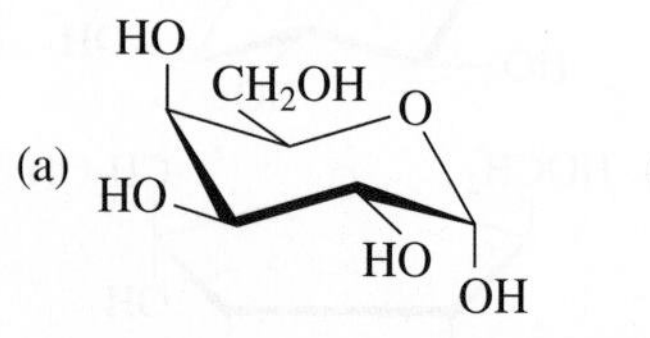

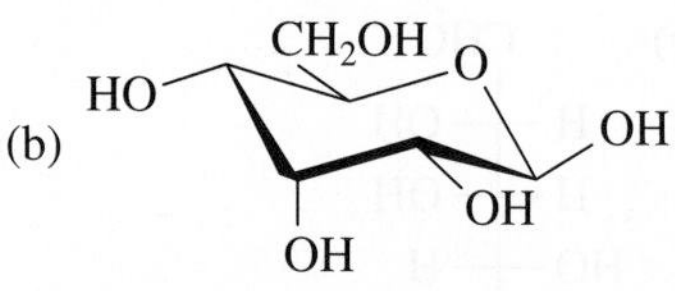

Name the monosaccharides you have drawn.

22.41 Use Haworth projections to represent the following molecules in an α-pyranose form. **LO2**

(a)

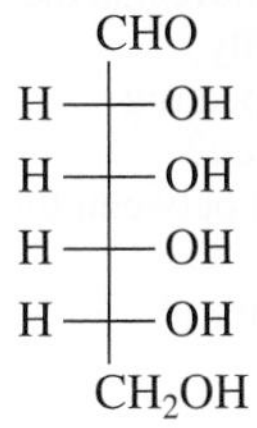

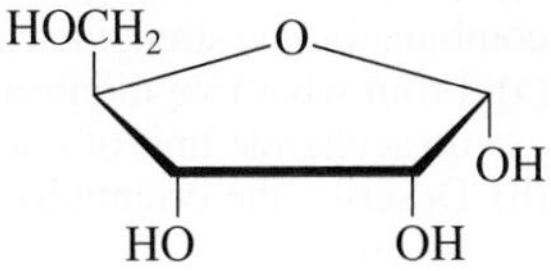

(b)

(c)

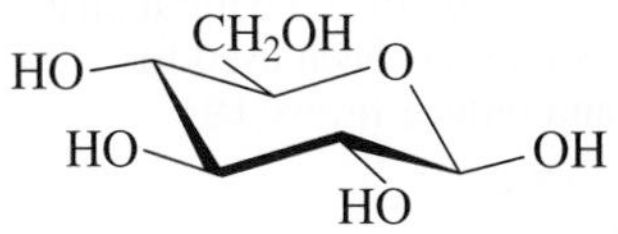

(d)

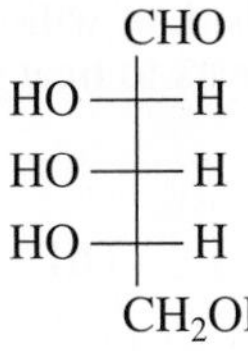

22.42 Using Haworth projections, convert the structures shown in problem 22.41 into β-furanose forms. **LO3**

22.43 The configuration of D-arabinose differs from the configuration of D-ribose only at C(2). Using this information, draw a Haworth projection for α-D-arabinofuranose (α-D-arabinose). **LO3**

22.44 Draw Fischer projections for the product(s) formed by the reaction of D-galactose with the following compounds, and state whether each product is optically active or optically inactive. **LO3**

(a) $NaBH_4$ in H_2O

(b) $AgNO_3$ in NH_3, H_2O

22.45 Repeat question 22.44, but use D-ribose in place of D-galactose. **LO3**

22.46 The reduction of D-fructose by $NaBH_4$ gives two alditols, one of which is D-sorbitol. Name and draw a structural formula for the other alditol. **LO4**

22.47 L-fucose, one of several monosaccharides commonly found in the surface polysaccharides of animal cells, is synthesised biochemically from D-mannose in the following eight-step sequence. **LO4,6**

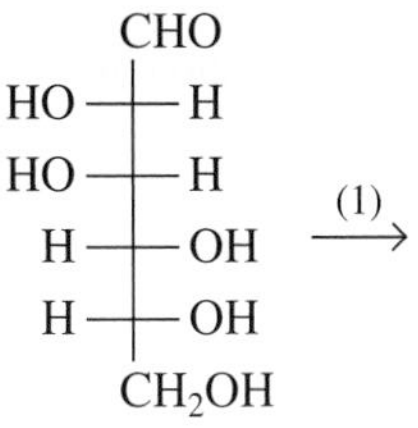

D-mannose

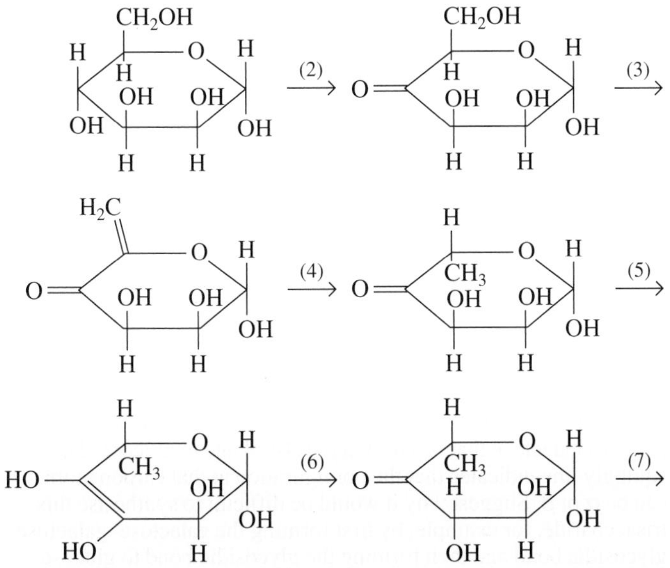

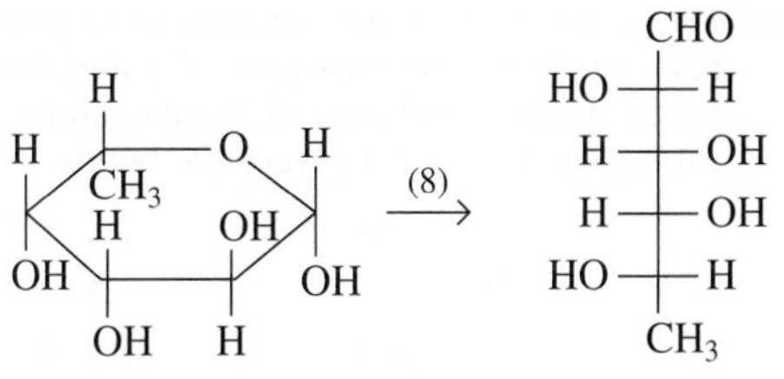

L-fucose

(a) Describe the type of reaction (oxidation, reduction, hydration, dehydration etc.) involved in each step.

(b) Explain why this monosaccharide, which is derived from D-mannose, now belongs to the L series.

22.48 Ascorbic acid is a diprotic acid with the following acid ionisation constants: $pK_{a1} = 4.17$ and $pK_{a2} = 11.57$. The two acidic hydrogen atoms are those connected with the enediol part of the molecule. Which hydrogen atom has which ionisation constant? (*Hint:* Draw separately the anion derived by the loss of one of these hydrogen atoms and that formed by the loss of the other hydrogen atom. Which anion has the greater degree of resonance stabilisation?) **LO4**

22.49 Trehalose is found in young mushrooms and is the chief carbohydrate in the blood of certain insects. Trehalose is a disaccharide consisting of two D-monosaccharide units, joined by an α-1,1-glycosidic bond. **LO4,5**

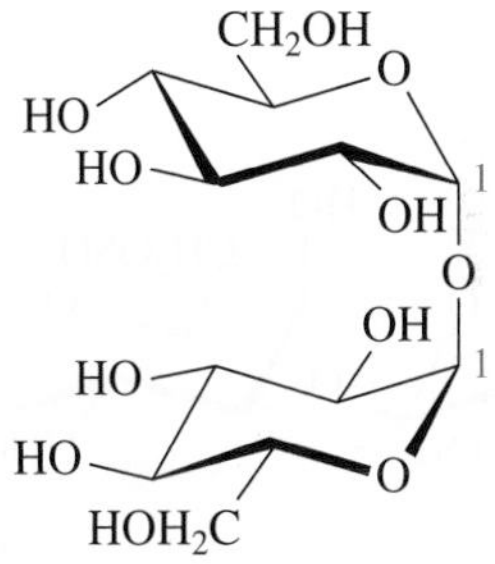

trehalose

(a) Is trehalose a reducing sugar?

(b) Does trehalose undergo mutarotation?

(c) Name the two monosaccharide units of which trehalose is composed.

22.50 Hot-water extracts of ground willow bark are an effective pain reliever. Unfortunately, the liquid is so bitter that most people refuse it. The pain reliever in these infusions is salicin. **LO2**

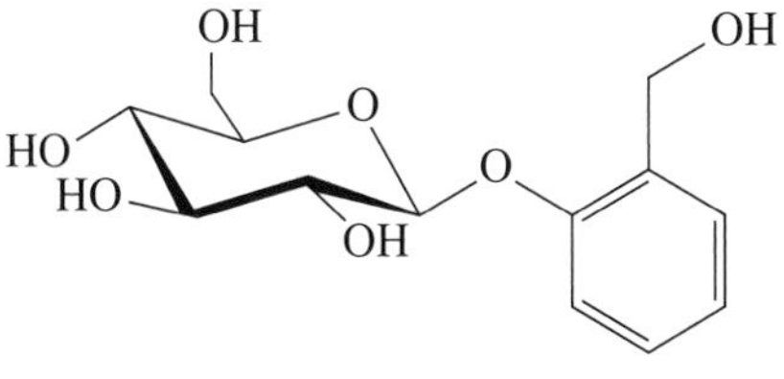

salicin

Name the monosaccharide unit in salicin.

22.51 A Fischer projection of *N*-acetyl-D-glucosamine is given on page 989. **LO2,3**

(a) Draw a Haworth projection and a chair conformation for the α- and β-pyranose forms of this monosaccharide.

(b) Draw a Haworth projection and a chair conformation for the disaccharide formed by joining two units of the pyranose form of *N*-acetyl-D-glucosamine by a β-1,4-glucosidic bond. If your drawing is correct, you have the structural formula for the repeating dimer of chitin, the structural polysaccharide component of the shell of lobsters and other crustaceans.

22.52 Propose structural formulae for the repeating disaccharide unit in the following polysaccharides. **LO3**

(a) Alginic acid, isolated from seaweed, is used as a thickening agent in ice-cream and other foods. Alginic acid is a polymer of D-mannuronic acid in the pyranose form, joined by β-1,4-glycosidic bonds.

(b) Pectic acid is the main component of pectin, which is responsible for the formation of jellies from fruits and berries. Pectic acid is a polymer of D-galacturonic acid in the pyranose form joined by α-1,4-glycosidic bonds.

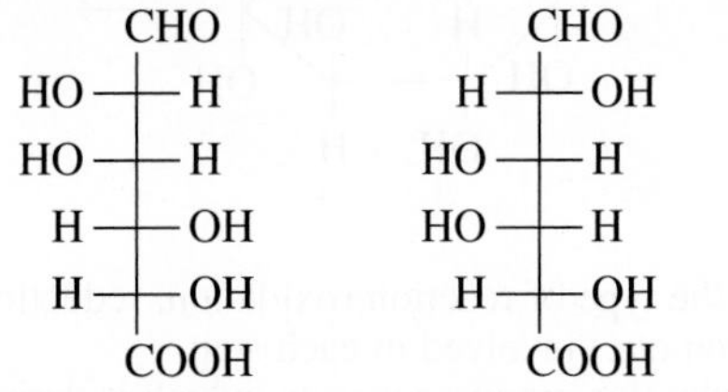

D-mannuronic acid D-galacturonic acid

22.53 The following are a Haworth projection and a chair conformation for the repeating disaccharide unit in chondroitin 6-sulfate. **LO3**

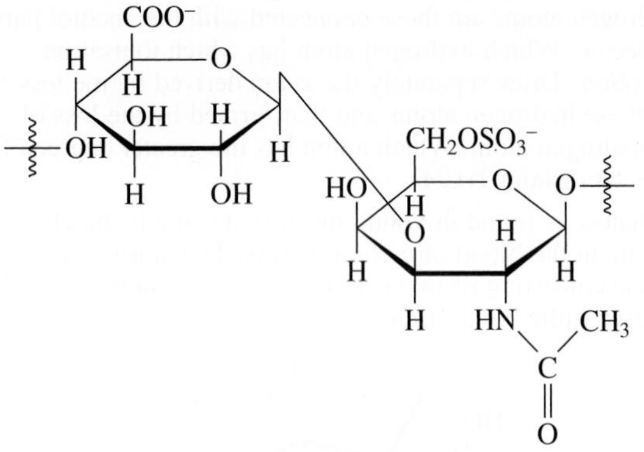

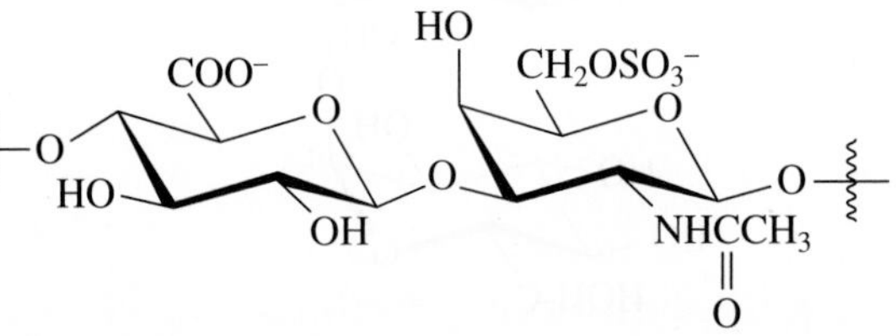

This biopolymer acts as a flexible connecting matrix between the tough protein filaments in cartilage and is available as a dietary supplement, often combined with D-glucosamine sulfate. Some believe that the combination can strengthen and improve joint flexibility.

(a) From what two monosaccharide units is the repeating disaccharide unit of chondroitin 6-sulfate derived?
(b) Describe the orientation of the glycosidic bond between the two units.
(c) Which is the better representation of the structure of chondroitin 6-sulfate?

22.54 As indicated in problem 22.50, salicin is a natural analgesic present in the bark of willow trees that has been used for thousands of years to treat pain and reduce fevers. **LO4**

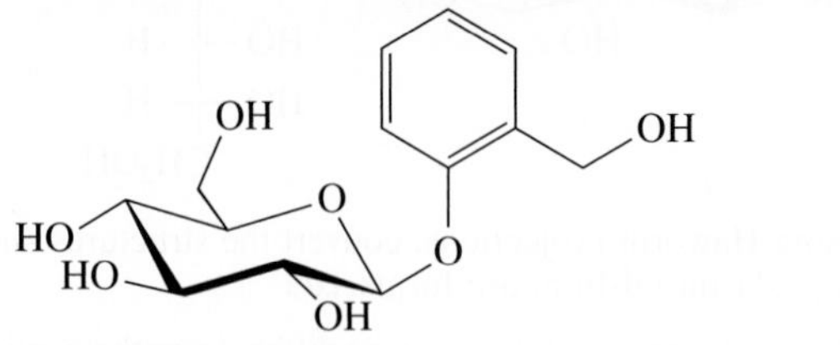

salicin

(a) Is salicin a reducing sugar?
(b) Identify the products obtained when salicin is hydrolysed in the presence of an acid.
(c) Is salicin an α-glycoside or a β-glycoside?
(d) Draw the major product expected when salicin is treated with excess acetic anhydride in the presence of pyridine.
(e) Would you expect salicin to exhibit mutarotation when dissolved in neutral water?

22.55 (a) Use skeletal line drawings, with the six-membered rings shown in a chair form to represent three glucose molecules linked by β-1,4-glycosidic bonds. **LO4**
(b) Draw the same trisaccharide, but this time use α-1,4-glycosidic linkages.
(c) Draw the same trisaccharide linked by α-1,6-glycosidic bonds.

ADDITIONAL EXERCISES

22.56 One step in glycolysis, the pathway that converts glucose to pyruvate, involves an enzyme-catalysed conversion of dihydroxyacetone phosphate to D-glyceraldehyde 3-phosphate. **LO4**

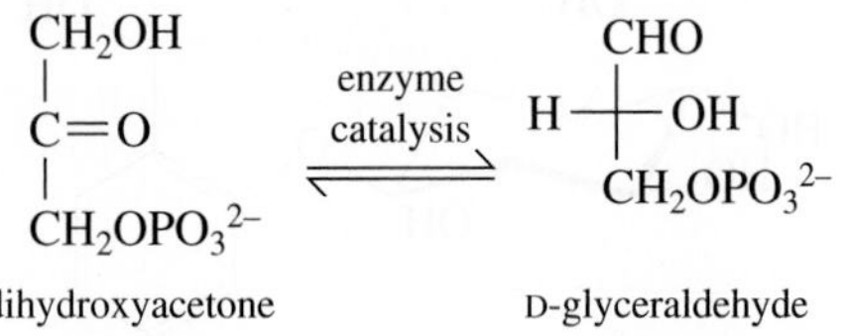

dihydroxyacetone phosphate $\rightleftharpoons$ (enzyme catalysis) D-glyceraldehyde 3-phosphate

Show that this transformation can be regarded as two enzyme-catalysed keto–enol tautomerisations (section 21.6).

22.57 One pathway for the metabolism of glucose 6-phosphate is its enzyme-catalysed conversion to fructose 6-phosphate. **LO4,6**

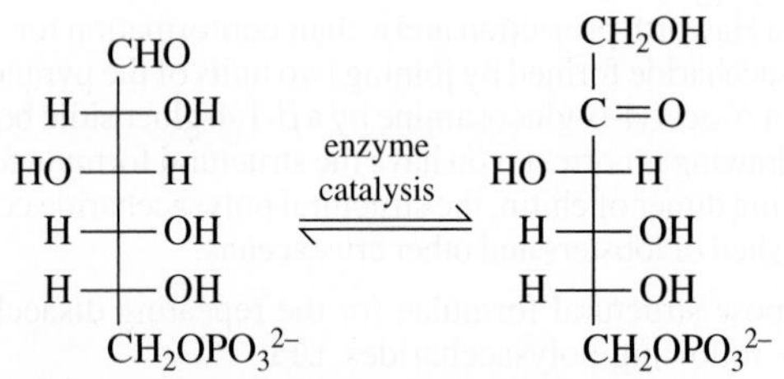

D-glucose 6-phosphate $\rightleftharpoons$ (enzyme catalysis) D-fructose 6-phosphate

Show that this transformation can be regarded as two enzyme-catalysed keto–enol tautomerisations.

22.58 Epimers are carbohydrates that differ in configuration at only one stereocentre. **LO2**

(a) Which of the aldohexoses are epimers of each other?
(b) Are all anomer pairs also epimers of each other? Explain. Are all epimers also anomers? Explain.

22.59 Oligosaccharides are very valuable therapeutically, but are especially difficult to synthesise, even though the starting materials are readily available. The structure of globotriose is shown below. This molecule is a receptor for a series of toxins synthesised by some strains of *E. coli*. **LO4,5**

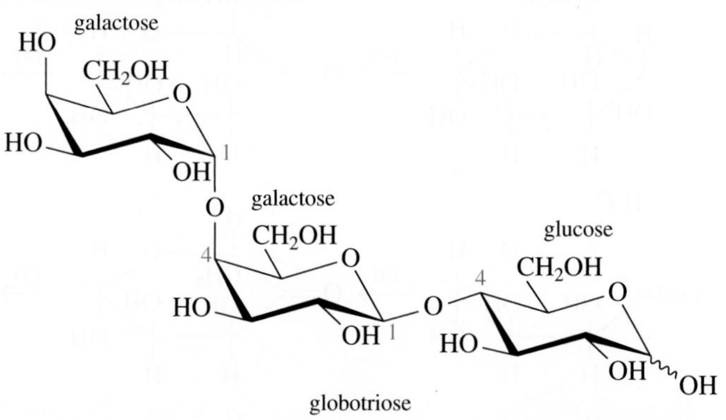

globotriose

From left to right, globotriose consists of an α-1,4-linkage of galactose to a galactose that is part of a β-1,4-linkage to glucose. The squiggly line indicates that the configuration at that carbon atom can be α or β. Suggest why it would be difficult to synthesise this trisaccharide, for example, by first forming the galactose–galactose glycosidic bond and then forming the glycosidic bond to glucose.

23 Carboxylic acids and their derivatives

Carboxylic acids and their derivatives are important compounds in our lives because they play a significant role in foods and many biological processes. They also find industrial applications in the form of polyesters, polyamides and more recently as biodiesel fuel. Carboxylic acids are weak acids and are responsible for the sour taste in many foods.

Carboxylic acids and their esters are important flavour components of fruits. Citric acid (opposite) is abundant in oranges, lemons, limes, grapefruits and mandarins. Malic acid was so named because it was isolated from apples (apple trees belong to the genus *Malus*). While tartaric acid is found in many fruits, it is an important acid in grapes and a by-product of the wine industry. Carboxylic acids also inhibit or retard microbial growth. This accounts for the use of vinegar (which contains about 5% acetic acid) to pickle (preserve) meats and vegetables. Benzoic acid (additive 210; 211–213 are its sodium, potassium and calcium salts) is commonly used as a food preservative, and citric acid (food acid 330; 331 is its sodium salt) is a common additive in soft drinks. The odour and flavour of fruits are due to a complex mixture of compounds, not just individual esters. However, individual esters can be important contributors to the odour and flavour. For example, the esters of acetic acid are important contributors to the odours of banana (isopentyl acetate, below right, and butyl acetate), pear (pentyl acetate) and orange (octyl acetate). Similarly, the esters of butanoic acid (a very unpleasant smelling acid) are common in apples and pineapples (methyl and ethyl butanoate), pears (pentyl butanoate), and strawberries and apricots (isopentyl butanoate).

In this chapter we will study the chemistry of carboxylic acids and four classes of organic compounds derived from the carboxyl group: acid halides, acid anhydrides, esters and amides.

LEARNING OBJECTIVES

After studying this chapter, you should be able to:

23.1 describe the structure and bonding in carboxylic acids and their derivatives

23.2 determine the IUPAC name of simple carboxylic acids and their derivatives given a structural formula (and vice versa)

23.3 explain how the intermolecular forces present between molecules of carboxylic acids and their derivatives affect their physical properties

23.4 propose methods for the synthesis of carboxylic acids

23.5 predict the products of common reactions of carboxylic acids and their derivatives

23.6 describe the basic chemistry of the triglycerides.

23.1 Structure and bonding

Carboxylic acids and their derivatives have similar structural and bonding features that result in similar reactions. However, their subtle structural differences also lead to quite different reactivities.

Carboxylic acids

The **carboxylic acid group** comprises the **carboxyl group**, so named because it is made up of a *carbonyl* group and a *hydroxyl* group. The following is a Lewis structure of the carboxyl group, as well as two alternative representations of it.

$$-C(=\ddot{O}:)-\ddot{O}-H \qquad -COOH \qquad -CO_2H$$

The general formula of an aliphatic carboxylic acid is RCOOH; that of an aromatic carboxylic acid is ArCOOH (Ar = aromatic ring). The general formula for each of the four carboxylic acid derived functional groups is given below, along with a drawing to help you see how the group is formally related to a carboxyl group. In all of these groups, the carbonyl carbon atom is bonded to a hetero atom (an atom other than carbon or hydrogen).

$RC(=O)Cl$ — an acid chloride $\xrightarrow[-HCl]{+H_2O}$ $RC(=O)OH$

$RC(=O)OC(=O)R$ — an acid anhydride $\xrightarrow[-RCOOH]{+H_2O}$ $RC(=O)OH$

$RC(=O)NH_2$ — an amide $\xrightarrow[-NH_3]{+H_2O}$ $RC(=O)OH$

$RC(=O)OR'$ — an ester $\xrightarrow[-R'OH]{+H_2O}$ $RC(=O)OH$

Acid halides

The **acid halide** (acyl halide) functional group comprises an **acyl group** (RCO—) bonded to a halogen atom. These are sometimes written in abbreviated form as RCOX or ArCOX (where X = halogen). The most common acid halides are the acid chlorides.

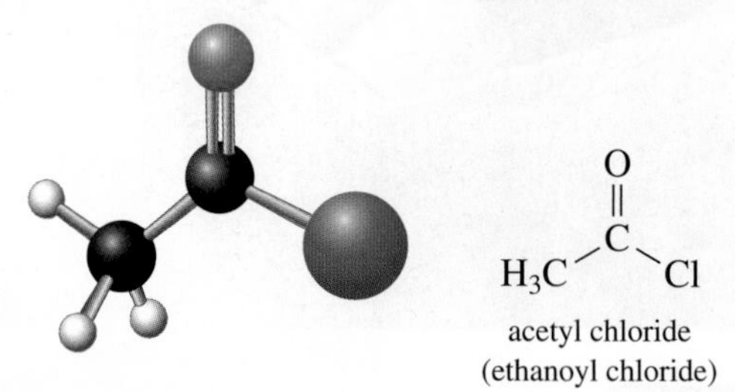

$H_3CC(=O)Cl$

acetyl chloride
(ethanoyl chloride)

Acid anhydrides

The **acid anhydride** functional group (commonly referred to simply as an anhydride) is two acyl groups bonded to an oxygen atom. The anhydride may be symmetrical (with two identical acyl groups), or it may be mixed (with two different acyl groups).

$H_3CC(=O)OC(=O)CH_3$

acetic anhydride
(ethanoic anhydride)

Esters of carboxylic acids

The **ester** functional group is an acyl group bonded to an —OR or an —OAr group.

$CH_3C(=O)OCH_2CH_3$

ethyl acetate
(ethyl ethanoate)

When esters are formed between an alcohol and a carboxylic acid group from within the same molecule, they are cyclic and are termed **lactones**.

Amides of carboxylic acids

The **amide** functional group is an acyl group bonded to a trivalent nitrogen atom. Amides can be classified as primary (N bonded to one carbon atom), secondary (N bonded to two carbon atoms) or tertiary (N bonded to three carbon atoms).

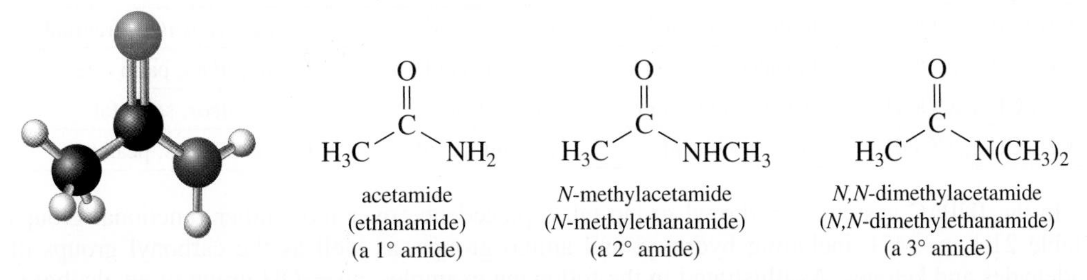

acetamide
(ethanamide)
(a 1° amide)

N-methylacetamide
(*N*-methylethanamide)
(a 2° amide)

N,*N*-dimethylacetamide
(*N*,*N*-dimethylethanamide)
(a 3° amide)

When amides are formed between an amino group and a carboxylic acid group from within the same molecule, they are cyclic and are termed **lactams**. Amide bonds are the key structural feature that joins amino acids together to form polypeptides and proteins (chapter 24).

23.2 Nomenclature

We explained the basic rules of IUPAC nomenclature in chapter 2. Here, we will briefly describe the process for carboxylic acids and their derivatives. We will commence with the carboxylic acids and work progressively through the acid halides, anhydrides, esters and finally the amides.

Carboxylic acids

We derive the IUPAC name of a carboxylic acid from that of the longest carbon chain that contains the carboxyl group by dropping the final *e* from the name of the parent alkane and replacing it with *-oic acid*. We number the chain beginning with the carbon atom of the carboxyl group; since it is understood to be C(1), there is no need to give it a number. In the following examples, the common name of each acid is given in parentheses.

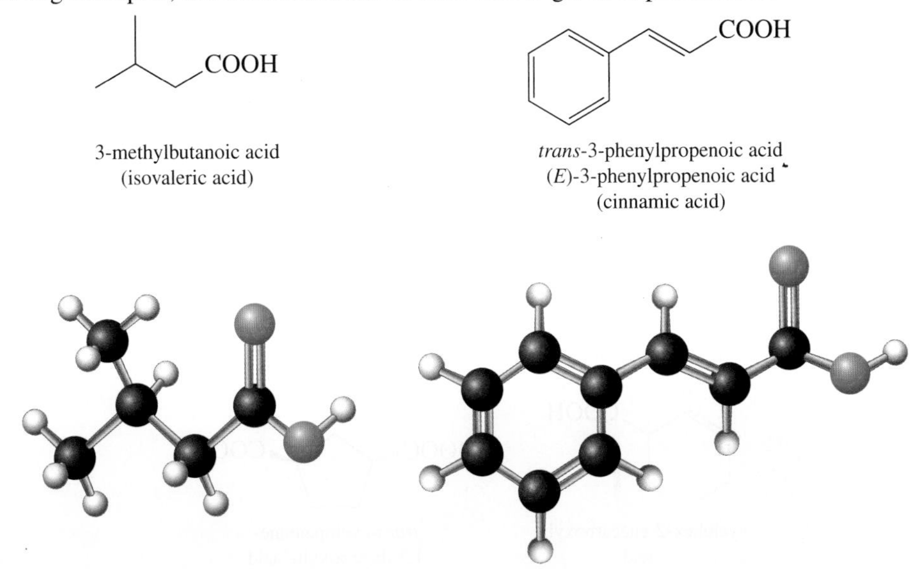

3-methylbutanoic acid
(isovaleric acid)

trans-3-phenylpropenoic acid
(*E*)-3-phenylpropenoic acid
(cinnamic acid)

FIGURE 23.1 Formic acid was first obtained in 1670 from the destructive distillation of ants belonging to the genus *Formica*. It is one of the components of their venom.

Many aliphatic carboxylic acids were known long before the development of structural theory and IUPAC nomenclature and were named according to their source or for some characteristic property. Table 23.1 lists several of the unbranched aliphatic carboxylic acids found in the biological world (for example, figure 23.1), along with the common name of each. Those with 16, 18 and 20 carbon atoms are particularly abundant in fats and oils (section 23.6).

TABLE 23.1 Several aliphatic carboxylic acids and their common names.

Structure	IUPAC name	Common name	Derivation
$HCOOH$	methanoic acid	formic acid	Latin: *formica*, ant
CH_3COOH	ethanoic acid	acetic acid	Latin: *acetum*, vinegar
CH_3CH_2COOH	propanoic acid	propionic acid	Greek: *propion*, first fat
$CH_3(CH_2)_2COOH$	butanoic acid	butyric acid	Latin: *butyrum*, butter
$CH_3(CH_2)_3COOH$	pentanoic acid	valeric acid	Latin: *valere*, to be strong
$CH_3(CH_2)_4COOH$	hexanoic acid	caproic acid	Latin: *caper*, goat
$CH_3(CH_2)_6COOH$	octanoic acid	caprylic acid	Latin: *caper*, goat
$CH_3(CH_2)_8COOH$	decanoic acid	capric acid	Latin: *caper*, goat
$CH_3(CH_2)_{10}COOH$	dodecanoic acid	lauric acid	Latin: *laurus*, laurel
$CH_3(CH_2)_{12}COOH$	tetradecanoic acid	myristic acid	Greek: *myristikos*, fragrant
$CH_3(CH_2)_{14}COOH$	hexadecanoic acid	palmitic acid	Latin: *palma*, palm tree
$CH_3(CH_2)_{16}COOH$	octadecanoic acid	stearic acid	Greek: *stear*, solid fat
$CH_3(CH_2)_{18}COOH$	eicosanoic acid	arachidic acid	Greek: *arachis*, peanut

In the IUPAC system, a carboxyl group takes precedence over most other functional groups (table 21.1, p. 947), including hydroxyl and amino groups, as well as the carbonyl groups of aldehydes and ketones. As illustrated in the following examples, an —OH group of an alcohol is indicated by the prefix *hydroxy-*, an $—NH_2$ group of an amine by *amino-*, and an =O group of an aldehyde or ketone by *oxo-*.

OH COOH — 5-hydroxyhexanoic acid

H_2N COOH — 4-aminobutanoic acid

O COOH — 5-oxohexanoic acid

Dicarboxylic acids are named by adding the suffix *-dioic*, followed by the word *acid*, to the name of the carbon chain that contains both carboxyl groups. Because the two carboxyl groups can be only at the ends of the parent chain, there is no need to number them. The following are IUPAC names and common names for several important aliphatic dicarboxylic acids.

HOOC—COOH — ethanedioic acid (oxalic acid)

HOOC COOH — propanedioic acid (malonic acid)

HOOC COOH — butanedioic acid (succinic acid)

HOOC COOH — pentanedioic acid (glutaric acid)

HOOC COOH — hexanedioic acid (adipic acid)

The name *oxalic acid* is derived from one of its sources in the biological world, namely, plants of the genus *Oxalis*, one of which is the soursob (sourgrass) plant, *O. pes-caprae*. Oxalic acid also occurs in human and animal urine, and calcium oxalate (the calcium salt of oxalic acid) is a major component of kidney stones. Adipic acid is one of the two monomers required for the synthesis of the polymer nylon 6,6.

A carboxylic acid containing a carboxyl group bonded to a cycloalkane ring is named by giving the name of the ring and adding the suffix *-carboxylic acid*. The atoms of the ring are numbered beginning with the carbon atom bearing the —COOH group.

3 2 1 COOH — cyclohex-2-enecarboxylic acid

HOOC COOH — *trans*-cyclopentane-1,3-dicarboxylic acid

The simplest aromatic carboxylic acid is benzoic acid. Derivatives are named by using numbers and prefixes to show the presence and location of substituents relative to the carboxyl group. Certain aromatic carboxylic acids have common names by which they are more usually known. For example, 2-hydroxybenzoic acid is more often called salicylic acid, named because it was first obtained from the bark of the willow, a tree of the genus *Salix*. Aromatic dicarboxylic acids are named by adding the words *dicarboxylic acid* to *benzene*. Examples are benzene-1,2-dicarboxylic acid and benzene-1,4-dicarboxylic acid; these are more usually known by their common names, phthalic acid and terephthalic acid, respectively. Terephthalic acid is one of the two organic components required for the synthesis of PET (polyethylene terephthalate) plastic bottles.

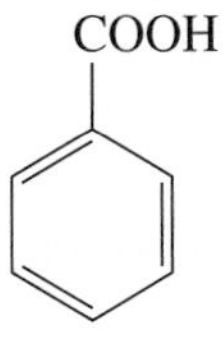

COOH — benzoic acid

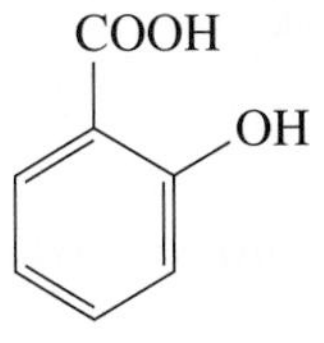

COOH, OH — 2-hydroxybenzoic acid (salicylic acid)

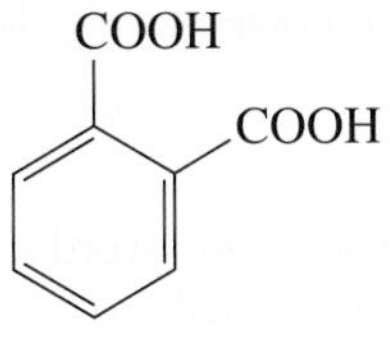

COOH, COOH — benzene-1,2-dicarboxylic acid (phthalic acid)

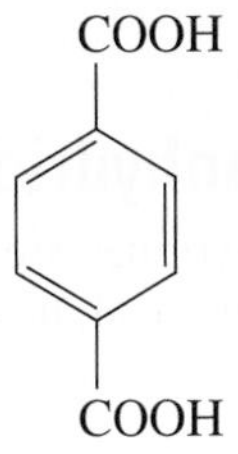

COOH, COOH — benzene-1,4-dicarboxylic acid (terephthalic acid)

WORKED EXAMPLE 23.1

Naming carboxylic acids

Write the IUPAC name for each of the following carboxylic acids.

(a) $CH_3(CH_2)_7$, $(CH_2)_7COOH$, C=C, H, H

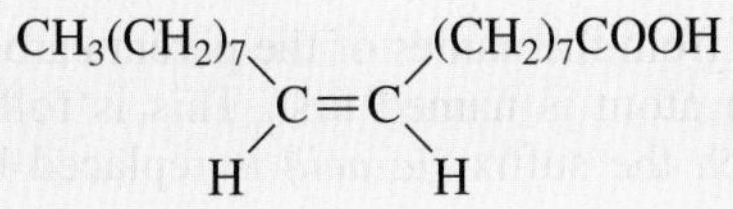

(c) OH, C, H, COOH, CH_3

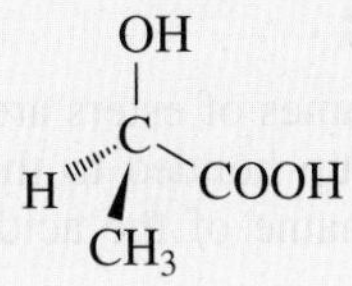

(b) COOH, OH

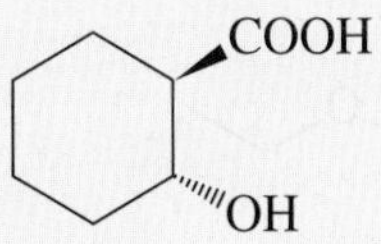

Analysis

(a) First determine the number of carbon atoms in the longest chain and then the position of any substituent or alkene (start numbering from the acid group). Where necessary, indicate the stereochemistry.
(b) This should be named as a cycloalkanecarboxylic acid.
(c) Determine the stereochemistry of the chiral centre (see section 17.3).

Solution

Common names are given in parentheses.

(a) *Z*-octadec-9-enoic acid (oleic acid, *cis*-octadec-9-enoic acid)
(b) (*R, R*)-2-hydroxycyclohexanecarboxylic acid
(c) (*R*)-2-hydroxypropanoic acid ((*R*)-lactic acid or D-lactic acid)

Is our answer reasonable?

To check your answer, draw a structure from the name and see if it matches the structure given. Is this the only structure possible from your answer?

PRACTICE EXERCISE 23.1

Each of the following compounds has a well-recognised common name. A derivative of glyceric acid is an intermediate in glycolysis. Maleic acid is an intermediate in the tricarboxylic acid (TCA) cycle (also known as the citric acid or Krebs cycle). Mevalonic acid is an intermediate in the biosynthesis of steroids.

(a) HO, COOH, HO, H — glyceric acid

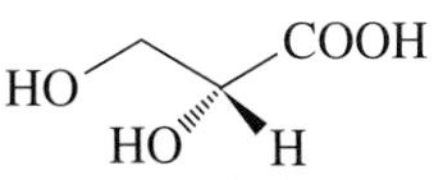

(b) COOH, COOH — maleic acid

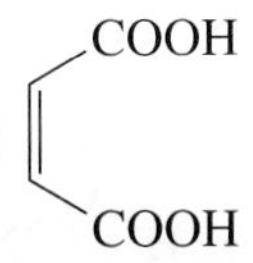

(c) HO, HO, CH_3, COOH — mevalonic acid

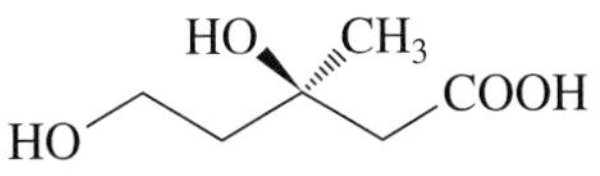

Write the IUPAC name for each of these compounds. Be certain to show the configuration of each.

Acid halides

Acid halides are named by changing the suffix *-ic acid* in the name of the parent carboxylic acid to *-yl halide*.

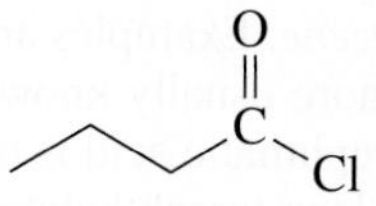

butanoyl chloride 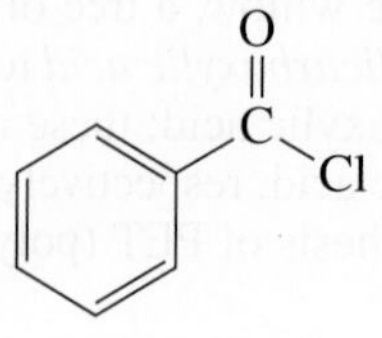   benzoyl chloride

Acid anhydrides

Acid anhydrides are named by replacing the word *acid* with *anhydride*. When the anhydride is not symmetrical, both acid groups are named.

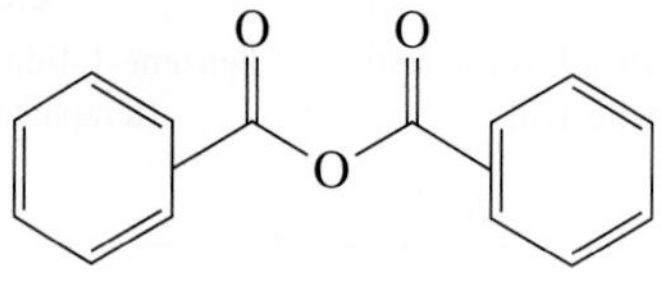

benzoic anhydride 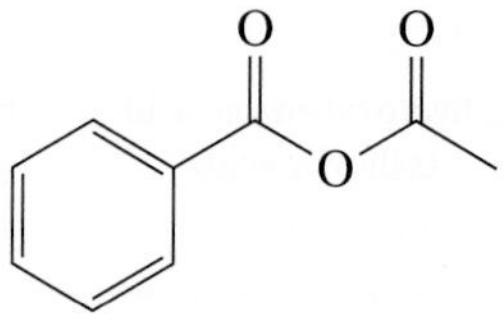   acetic benzoic anhydride

Esters and lactones

Both IUPAC and common names of esters are derived from the names of the parent carboxylic acids. The alkyl or aryl group bonded to the oxygen atom is named first. This is followed, as a separate word, by the name of the acid, in which the suffix *-ic acid* is replaced by the suffix *-ate*.

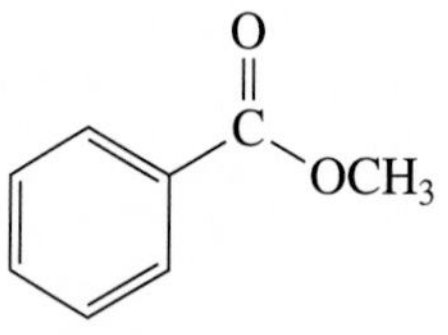

methyl benzoate 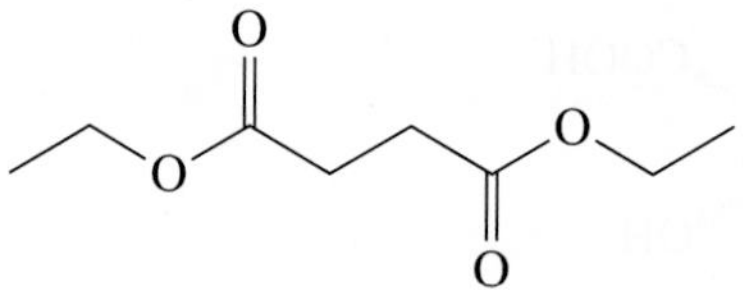   diethyl butanedioate (diethyl succinate)

The IUPAC name of a lactone is formed by dropping the suffix *-oic acid* from the name of the parent carboxylic acid and adding the suffix *-olactone*. The common name is derived similarly. The location of the oxygen atom in the ring is indicated by a number if the IUPAC name of the acid is used and by a Greek letter (α, β, γ, δ, ε etc.) if the common name of the acid is used.

4-butanolactone (a γ-lactone)

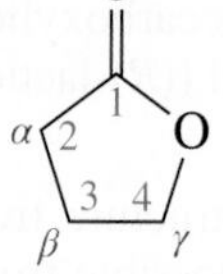

Amides and lactams

Amides are named by dropping the suffix *-oic acid* from the IUPAC name of the parent acid, or *-ic acid* from its common name, and adding *-amide*. If the nitrogen atom of an amide is bonded to an alkyl or aryl group, the group is named and its location on the nitrogen atom is indicated by *N*-. Two alkyl or aryl groups on the nitrogen atom are indicated by *N*,*N*-di- if the groups are identical or by -*N*-alkyl-*N*-alkyl if they are different.

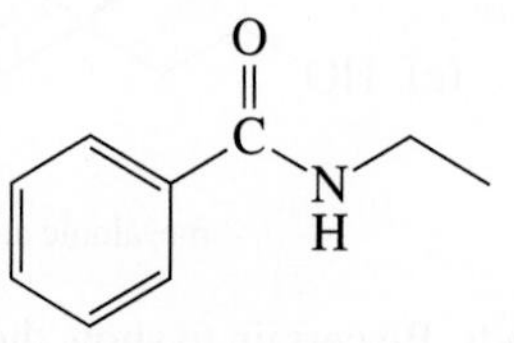

N-ethylbenzamide 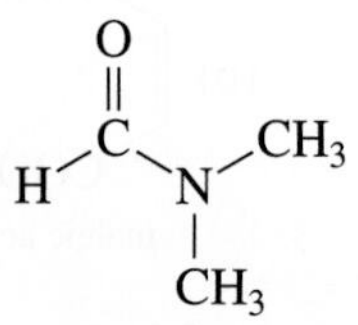   *N*,*N*-dimethylformamide

Cyclic amides are given the special name lactam. Their common names are derived in the same way as lactones except that the suffix is *-olactam*.

3-butanolactam
(a β-lactam)

6-hexanolactam
(an ε-lactam)

The compound 6-hexanolactam (caprolactam) is a key intermediate in the synthesis of nylon 6 (section 26.3, p. 1127). *Note:* The 'hex' in hexanolactam refers to the number of carbon atoms, rather than the ring size.

WORKED EXAMPLE 23.2

Nomenclature of carboxylic acid derivatives

Write the IUPAC name for each of the following compounds.

(a) (b) (c) (d)

Analysis

First identify the functional group present and then the acid from which it is derived. The ketone group in (b) must be named with the prefix *oxo*. Compound (d) has been derived from a substituted acetic acid.

Solution

Common names and derivations are given in parentheses.

(a) methyl 3-methylbutanoate (methyl isovalerate, from isovaleric acid)
(b) ethyl 3-oxobutanoate (ethyl acetoacetate, from acetoacetic acid; ethyl β-ketobutyrate from β-ketobutyric acid)
(c) hexanediamide (adipamide, from adipic acid)
(d) phenylethanoic anhydride (phenylacetic anhydride, from phenylacetic acid)

Is our answer reasonable?

To check your answer, draw a structure from the name and see if it matches the structure given.

PRACTICE EXERCISE 23.2

Draw a structural formula for each of the following compounds.

(a) butanoic anhydride
(b) pentyl propanoate
(c) dipropyl oxalate
(d) phenyl acetate
(e) *N,N*-diethylbenzamide
(f) 5-pentanolactam

23.3 Physical properties

In the liquid and solid states, carboxylic acids are associated by intermolecular hydrogen bonding into dimers, as shown below for acetic acid.

hydrogen bonding in the dimer

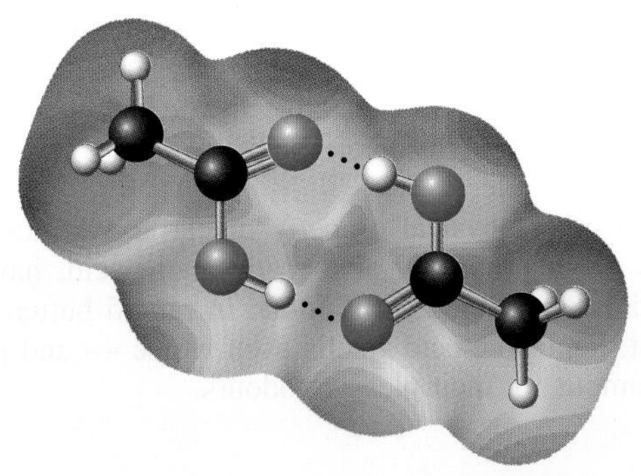

Carboxylic acids have significantly higher boiling points than other types of organic compounds with similar numbers of electrons (and, therefore, comparable molar mass), such as alcohols, aldehydes and ketones. For example, butanoic acid (table 23.2) has a higher boiling point than either pentan-1-ol or pentanal. This is because carboxylic acids are polar and form very strong intermolecular hydrogen bonds.

TABLE 23.2 Boiling points and solubilities in water of selected carboxylic acids, alcohols and aldehydes of comparable molar mass.

Structure	Name	Molar mass	Number of electrons	Boiling point (°C)	Solubility (g/100 mL H_2O)
CH_3COOH	acetic acid	60.1	32	118	infinite
$CH_3CH_2CH_2OH$	propan-1-ol	60.1	34	97	infinite
CH_3CH_2CHO	propanal	58.1	32	48	16
$CH_3(CH_2)_2COOH$	butanoic acid	88.1	48	163	infinite
$CH_3(CH_2)_3CH_2OH$	pentan-1-ol	88.1	50	137	2.7
$CH_3(CH_2)_3CHO$	pentanal	86.1	48	103	slight
$CH_3(CH_2)_4COOH$	hexanoic acid	116.2	64	205	1.0
$CH_3(CH_2)_5CH_2OH$	heptan-1-ol	116.2	66	176	0.2
$CH_3(CH_2)_5CHO$	heptanal	114.1	64	153	0.1

Carboxylic acids also interact with water molecules by hydrogen bonding through both their carbonyl and hydroxyl groups. Because of these hydrogen bonds, carboxylic acids are more soluble in water than are alcohols, ethers, aldehydes and ketones of comparable molar mass. The solubility of a carboxylic acid in water generally decreases as its molar mass increases (table 23.2). We account for this trend in the following way. A carboxylic acid has two regions of different polarity — a polar hydrophilic carboxyl group and, except for formic acid, a nonpolar hydrophobic hydrocarbon chain. The **hydrophilic** (water-loving) carboxyl group increases water solubility; the **hydrophobic** (water-hating) hydrocarbon chain decreases water solubility.

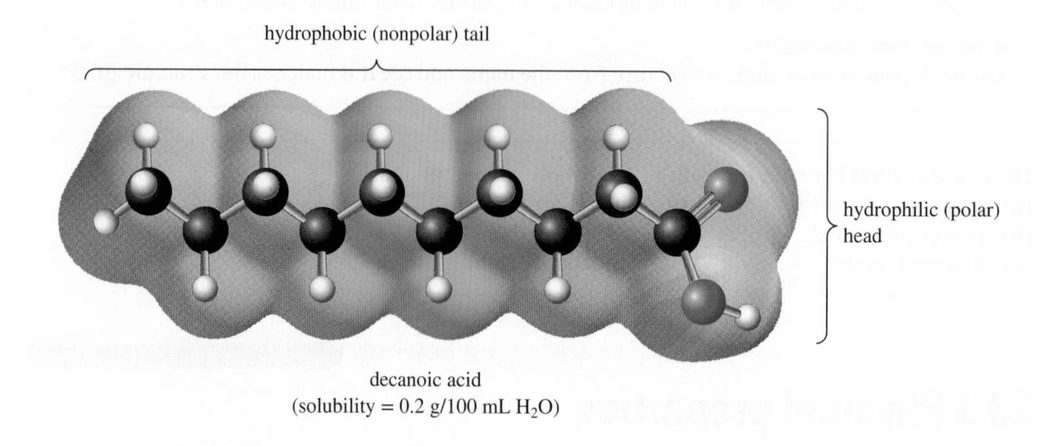

decanoic acid
(solubility = 0.2 g/100 mL H_2O)

The first four aliphatic carboxylic acids (formic, acetic, propanoic and butanoic acids) are infinitely soluble in water because the hydrophilic character of the carboxyl group more than compensates for the hydrophobic character of the hydrocarbon chain. As the size of the hydrocarbon chain increases relative to the size of the carboxyl group, water solubility decreases. The solubility of hexanoic acid in water is 1.0 g/100 g water; that of decanoic acid is only 0.2 g/100 g water.

One other physical property of carboxylic acids must be mentioned. The liquid carboxylic acids, from propanoic acid to decanoic acid, have foul odours, although not as bad as those of thiols. Butanoic acid is found in rancid butter, stale parmesan cheese, stale perspiration and vomit. Pentanoic acid smells even worse — and goats, which secrete C_6, C_8 and C_{10} acids, are not famous for their pleasant odours.

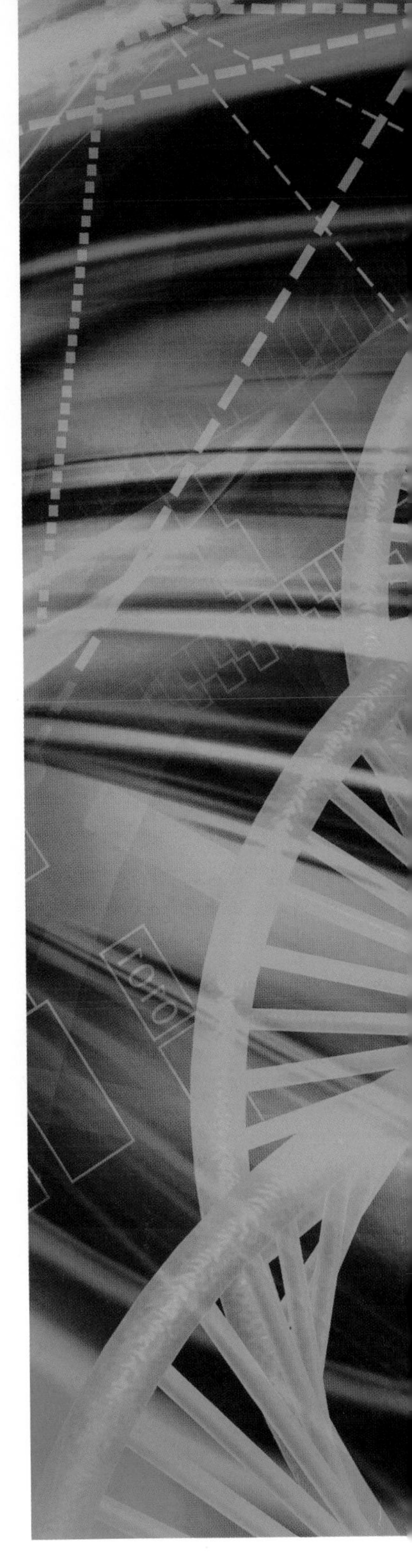

Chemical Connections

Carboxylic acids and derivatives in medicine

In 1933, a disgruntled farmer delivered a pail of unclotted blood to the laboratory of Dr Karl Link at the University of Wisconsin and told tales of cows bleeding to death from minor cuts. Over the next couple of years, Link and his collaborators discovered that, when cows are fed mouldy clover, their blood clotting is inhibited.

From the mouldy clover, Link isolated the anticoagulant dicoumarol, a substance that delays or prevents blood from clotting by interfering with vitamin K activity. Within a few years after its discovery, dicoumarol became widely used to treat victims of heart attack and others at risk of developing blood clots.

Coumarin is a lactone (cyclic ester) that gives clove its pleasant smell. It does not interfere with blood clotting and has been used as a flavouring agent in the food industry. Coumarin is converted to dicoumarol as sweet clover becomes mouldy.

as sweet clover becomes mouldy →

coumarin
(from sweet clover)

dicoumarol
(an anticoagulant)

In a search for even more potent anticoagulants, Link developed warfarin, now used primarily as a rat poison. When rats consume warfarin, their blood fails to clot and they bleed to death. Sold under the brand name Coumadin®, warfarin is also used as a blood thinner in humans. The commercial product is a racemic mixture, but the *S*-enantiomer is more active than the *R*-enantiomer.

(*S*) | (*R*)

warfarin
(a synthetic anticoagulant)

23.4 Preparation of carboxylic acids

There are many methods available for the preparation of carboxylic acids. Some of these will be discussed briefly here and in more detail elsewhere in this text.

Oxidation of primary alcohols and aldehydes

Primary alcohols and aldehydes are readily oxidised to carboxylic acids by a variety of common oxidising agents (discussed in greater detail in sections 19.2 and 21.5).

$$\underset{\text{hexan-1-ol}}{CH_3(CH_2)_4CH_2OH} \xrightarrow[H_2SO_4/H_2O]{CrO_3} \left[\underset{\substack{\text{hexanal}\\\text{(not isolated)}}}{CH_3(CH_2)_4\overset{O}{\overset{\|}{C}}H}\right] \longrightarrow \underset{\text{hexanoic acid}}{CH_3(CH_2)_4COOH}$$

$$\text{cyclopentanecarbaldehyde} \xrightarrow{KMnO_4} \text{cyclopentanecarboxylic acid}$$

cyclopentanecarbaldehyde   cyclopentanecarboxylic acid

Oxidation of alkylbenzenes

Although benzene is unaffected by strong oxidising agents, such as $KMnO_4$ and H_2CrO_4, alkylbenzenes are oxidised at the **benzylic carbon** (the carbon atom attached to the benzene ring) to give carboxylic acids. When we treat toluene with these oxidising agents under vigorous conditions, the side-chain methyl group is oxidised to a carboxyl group to give benzoic acid.

$$\underset{\text{toluene}}{C_6H_5CH_3} + H_2CrO_4 \longrightarrow \underset{\text{benzoic acid}}{C_6H_5COOH} + Cr^{3+}$$

Ethylbenzene and isopropylbenzene are also oxidised to benzoic acid under these conditions. The side chain of *tert*-butylbenzene, which has no **benzylic hydrogen** (a hydrogen atom bonded to a benzylic carbon atom), is not affected by these oxidising conditions. From these observations, we can generalise that:

- if a benzylic hydrogen atom exists, the benzylic carbon atom is oxidised to a carboxyl group and all other carbon atoms of the side chain are removed
- if no benzylic hydrogen atom exists, as in *tert*-butylbenzene, the side chain is not oxidised
- if more than one alkyl chain exists, each is oxidised to a carboxylic acid group.

Oxidation of 1,3-diethylbenzene gives benzene-1,3-dicarboxylic acid, more commonly named isophthalic acid. Note the product in the example below has two carbon atoms less than the starting material; under the reaction conditions used, these carbon atoms are converted to carbon dioxide.

$$\underset{\text{1,3-diethylbenzene}}{C_6H_4(CH_2CH_3)_2} \xrightarrow{H_2CrO_4} \underset{\substack{\text{benzene-1,3-dicarboxylic acid}\\\text{(isophthalic acid)}}}{C_6H_4(COOH)_2}$$

Carbonation of Grignard reagents

Grignard reagents (section 21.5) react with carbon dioxide to give a magnesium carboxylate that is converted into the carboxylic acid by acidification. This reaction is applicable to alkyl and aryl halides and produces carboxylic acids with one carbon atom more than the original organohalide. A general reaction is given on the next page, followed by two specific examples.

$$R{-}X + Mg \longrightarrow RMgX \xrightarrow{CO_2} RCO_2MgX \xrightarrow{H_3O^+} RCO_2H$$

$$\text{1-chlorobutane} \xrightarrow{Mg} \text{butylmagnesium chloride} \xrightarrow[2.\ H_3O^+]{1.\ CO_2} \text{pentanoic acid (80\% overall)}$$

$$\text{bromobenzene} \xrightarrow{Mg} \text{phenylmagnesium bromide} \xrightarrow[2.\ H_3O^+]{1.\ CO_2} \text{benzoic acid (85\% overall)}$$

This reaction follows the same process as that in the reaction of Grignard reagents with aldehydes and ketones (section 21.5).

$$\overset{\delta-}{R}{-}\overset{\delta+}{MgX} + \overset{\delta-}{O}{=}\overset{\delta+}{C}{=}\overset{\delta-}{O} \longrightarrow RC(=O)O^-[MgX]^+$$

Formation and hydrolysis of nitriles

Another method of preparing carboxylic acids from haloalkanes is by the formation and subsequent hydrolysis of **nitriles** (compounds containing a cyano group, —C≡N). Like the carbonation reaction described above, this process produces carboxylic acids with one carbon atom more than the original haloalkane. The nitrile is formed by a nucleophilic substitution reaction (section 18.2) by cyanide ions on a haloalkane. The resulting nitrile can then be hydrolysed by reaction with either hot aqueous acid or base.

$$\text{benzyl bromide} \xrightarrow{NaCN} \text{phenylacetonitrile} \xrightarrow[H_3O^+/\text{heat}]{H_2O} \text{phenylacetic acid} + NH_4^+$$

Hydrolysis of carboxylic acid derivatives

Carboxylic acids can also be obtained by the hydrolysis of carboxylic acid derivatives, particularly esters and amides — for example:

$$\text{ethyl benzoate} \xrightarrow[H_3O^+]{H_2O} \text{benzoic acid} + CH_3CH_2OH \text{ (ethanol)}$$

The chemistry of this reaction will be discussed in detail in section 23.5.

WORKED EXAMPLE 23.3

Formation of carboxylic acids

Show how each of the following compounds can be converted into 3-methylbutanoic acid.
(a) 3-methylbutanal (b) 1-bromo-2-methylpropane

Analysis

First, draw the structures of the product and the starting material. Now, compare the two compounds and determine what changes are required to go from the starting material to the product. The product has one more oxygen atom than (a), suggesting that oxidation is required. The product has one more carbon atom than (b), which means that we need to introduce another carbon atom; this might be done via a Grignard reaction or by the hydrolysis of a nitrile.

Solution

(a) 3-methylbutanal (CHO) $\xrightarrow{\text{oxidation}}$ 3-methylbutanoic acid (CO_2H)

A range of oxidising agents can be used, including $KMnO_4$ solution or chromic acid, H_2CrO_4. PCC (pyridinium chlorochromate) is not suitable.

(b) 1-bromo-2-methylpropane (Br) $\xrightarrow{\text{1. Mg; 2. } CO_2\text{; 3. } H_3O^+}$ 3-methylbutanoic acid (CO_2H)

or

1-bromo-2-methylpropane (Br) $\xrightarrow{\text{1. NaCN; 2. } H_3O^+\text{/heat}}$ 3-methylbutanoic acid (CO_2H)

Is our answer reasonable?

In each case, check that the carbon skeleton is consistent with the number of carbon atoms. You should also check that the reagents chosen are appropriate for the proposed transformations.

PRACTICE EXERCISE 23.3 Show how each of the following compounds can be converted into 3-methylpentanoic acid.

(a) 1-bromo-2-methylbutane (b) 3-methylpentanal

WORKED EXAMPLE 23.4

Oxidation at benzylic carbon atoms

Draw the structural formula for the product of vigorous oxidation by H_2CrO_4 of each of the following compounds.

(a) 1,4-dimethylbenzene (*p*-xylene) (b) 1-*tert*-butyl-4-ethylbenzene

Analysis

Start by checking whether the alkyl groups have at least one benzylic hydrogen atom. If they do, then chromic acid oxidises them to carboxylic acid groups, —COOH. If not, then it is not oxidised.

Solution

(a) $H_3C-C_6H_4-CH_3 \xrightarrow{H_2CrO_4} HOOC-C_6H_4-COOH$

1,4-dimethylbenzene (*p*-xylene) → benzene-1,4-dicarboxylic acid (terephthalic acid)

(b) 1-*tert*-butyl-4-ethylbenzene $\xrightarrow{H_2CrO_4}$ 4-*tert*-butylbenzoic acid (COOH)

Is our answer reasonable?

Check that the carbon skeleton is consistent with the number of carbon atoms. Check also that the oxidation has occurred at the benzylic carbon atom.

PRACTICE EXERCISE 23.4 Predict the products resulting from vigorous oxidation of each of the following compounds by $KMnO_4$.

(a) 1,4-dipropylbenzene structure (b) tetralin structure

23.5 Reactions of carboxylic acids and derivatives

The reactions of carboxylic acids can be classified into two general types: (a) acid–base and (b) nucleophilic acyl substitution.

$$R-\overset{\overset{\large O}{\|}}{C}-O-H$$

When this bond breaks, we have acid–base chemistry.

When this bond breaks, we have nucleophilic acyl substitution.

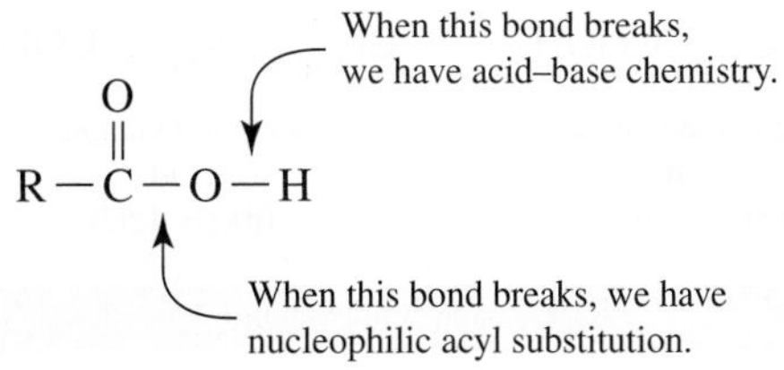

We will commence this section with the acid–base chemistry of carboxylic acids. We will then study the nucleophilic acyl substitution of the carboxylic acids and their derivatives.

Acidity

Although carboxylic acids are weak acids, relative to the mineral acids, they are significantly more acidic than alcohols and phenols (sections 19.2 and 19.3). At this point, it may be worthwhile to review chapter 11 and the discussion of the position of equilibrium in acid–base reactions.

Acid ionisation constants

Values of K_a for most unsubstituted aliphatic and aromatic carboxylic acids are between 10^{-4} and 10^{-5}. The value of K_a for acetic acid, for example, is 1.8×10^{-5}, so the pK_a of acetic acid is 4.74.

$$CH_3COOH + H_2O \rightleftharpoons CH_3COO^- + H_3O^+$$

$$K_a = \frac{[CH_3COO^-][H_3O^+]}{[CH_3COOH]} = 1.8 \times 10^{-5}$$

$$pK_a = 4.74$$

Carboxylic acids (pK_a = 4–5) are stronger acids than alcohols (pK_a = 16–18) because resonance stabilises the carboxylate anion by delocalising its negative charge. No comparable resonance stabilisation exists in alkoxide ions.

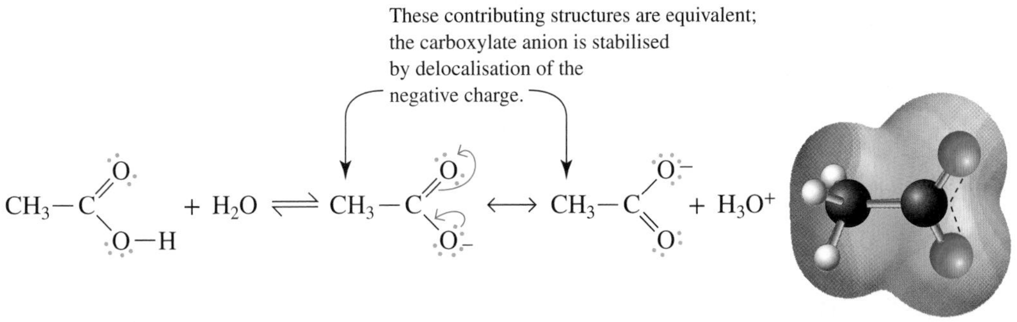

These contributing structures are equivalent; the carboxylate anion is stabilised by delocalisation of the negative charge.

$$CH_3-C(=O)-O-H + H_2O \rightleftharpoons CH_3-C(=O)-O^- \longleftrightarrow CH_3-C(-O^-)=O + H_3O^+$$

The electron density map shows the negative charge equally delocalised on the two carboxylate oxygen atoms.

Substitution at the α-carbon with an atom or a group of atoms of higher electronegativity than carbon increases the acidity of carboxylic acids, often by several orders of magnitude. Compare, for example, the acidities of acetic acid (pK_a = 4.74) and chloroacetic acid (pK_a = 2.85). A single chlorine substituent on the α-carbon atom increases the acidity constant by nearly 100 times! Both dichloroacetic acid and trichloroacetic acid are stronger acids than phosphoric acid (pK_a = 2.1).

Formula:	CH_3COOH	$ClCH_2COOH$	$Cl_2CHCOOH$	Cl_3CCOOH
Name:	acetic acid	chloroacetic acid	dichloroacetic acid	trichloroacetic acid
pK_a:	4.74	2.85	1.48	0.70

Increasing acid strength →

The acid-strengthening effect of halogen substitution falls off rather rapidly with increasing distance from the carboxyl group. Although the acidity constant of 2-chlorobutanoic acid ($pK_a = 2.83$) is 100 times that of butanoic acid, the acidity constant of 4-chlorobutanoic acid ($pK_a = 4.52$) is only about twice that of butanoic acid.

2-chlorobutanoic acid ($pK_a = 2.83$)

3-chlorobutanoic acid ($pK_a = 3.98$)

4-chlorobutanoic acid ($pK_a = 4.52$)

butanoic acid ($pK_a = 4.82$)

Decreasing acid strength →

Ring substituents have a moderate effect on the acidities of aromatic carboxylic acids through a combination of inductive and resonance effects (a similar effect was described for phenols in section 19.3). Electron-withdrawing groups, such as chloro, cyano and nitro, withdraw electron density from the aromatic ring, weaken the O—H bond and stabilise the carboxylate ion. Conversely, electron-donating groups, such as alkoxy and alkyl, increase the electron density of the aromatic ring and destabilise the carboxylate ion. Table 23.3 lists the pK_a values of some aromatic acids.

TABLE 23.3 The pK_a values of some aromatic acids.

Name	pK_a
2,4,6-trinitrobenzoic acid	0.65
4-nitrobenzoic acid	3.43
4-chlorobenzoic acid	4.00
benzoic acid	4.20
4-methylbenzoic acid	4.37
4-methoxybenzoic acid	4.50

Increasing acid strength ↑

WORKED EXAMPLE 23.5

Acidity

Which acid in each of the following sets is the stronger?

(a) propanoic acid or 2-hydroxypropanoic acid (lactic acid)

(b) 2-hydroxypropanoic acid (lactic acid) or 2-oxopropanoic acid (pyruvic acid)

Analysis

First identify the difference between the two acids. In both cases it is the substituent on the α-carbon atom. The compound with the better electron-withdrawing group will be the stronger acid.

Solution

(a) The stronger acid is 2-hydroxypropanoic acid because of the electron-withdrawing inductive effect of the hydroxyl oxygen atom.

(b) The stronger acid is 2-oxopropanoic acid because the electron-withdrawing inductive effect of the carbonyl oxygen atom is greater than that of the hydroxyl oxygen atom.

Is our answer reasonable?

Check the pK_a values of the acids; a lower pK_a value means a stronger acid — propanoic acid ($pK_a = 4.89$), 2-hydroxypropanoic acid ($pK_a = 3.85$) and 2-oxopropanoic acid ($pK_a = 2.39$).

PRACTICE EXERCISE 23.5

Match each of the following compounds with its appropriate pK_a value: 5.03, 2.86 or 0.22.

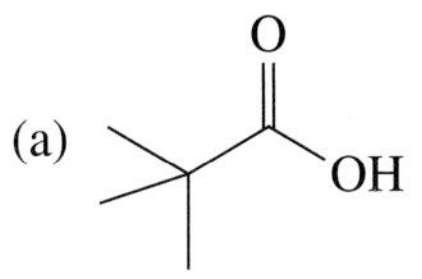

(a) 2,2-dimethylpropanoic acid

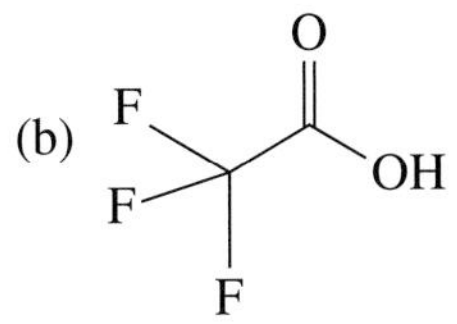

(b) trifluoroacetic acid

(c) bromoacetic acid

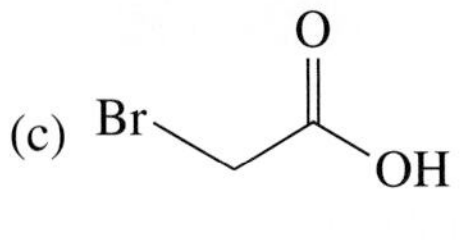

Reaction with bases

All carboxylic acids, whether soluble or insoluble in water, react with NaOH, KOH and other strong bases to form water-soluble salts.

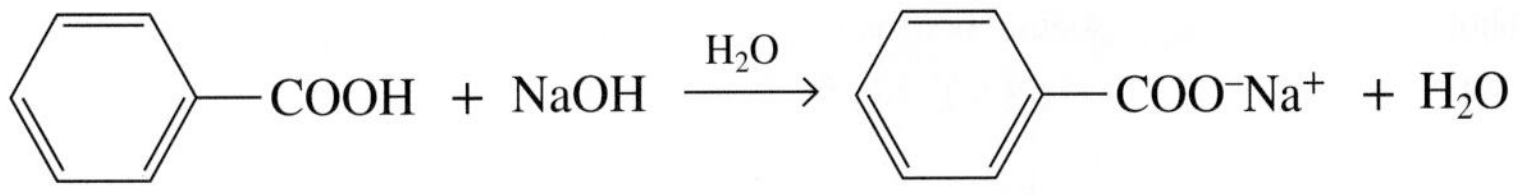

$$C_6H_5\text{—}COOH + NaOH \xrightarrow{H_2O} C_6H_5\text{—}COO^-Na^+ + H_2O$$

benzoic acid (slightly soluble in water) — sodium benzoate (solubility = 60 g/100 mL H_2O)

Sodium benzoate, a fungal growth inhibitor, is often added to baked goods 'to retard spoilage'. Calcium propanoate is used for the same purpose.

Carboxylic acids also form water-soluble salts with ammonia and amines.

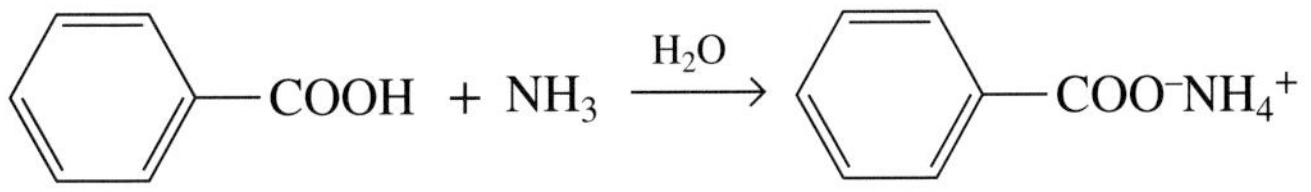

$$C_6H_5\text{—}COOH + NH_3 \xrightarrow{H_2O} C_6H_5\text{—}COO^-NH_4^+$$

benzoic acid (slightly soluble in water) — ammonium benzoate (solubility = 20 g/100 mL H_2O)

Carboxylic acids react with sodium hydrogencarbonate and sodium carbonate to form water-soluble sodium salts and carbonic acid (a relatively weak acid). Carbonic acid, in turn, decomposes to give water and carbon dioxide, which evolves as a gas.

$$CH_3COOH + Na^+HCO_3^- \xrightarrow{H_2O} CH_3COO^-Na^+ + H_2CO_3$$

$$H_2CO_3 \longrightarrow CO_2 + H_2O$$

$$CH_3COOH + Na^+HCO_3^- \longrightarrow CH_3COO^-Na^+ + CO_2 + H_2O$$

Salts of carboxylic acids are named in the same way as salts of inorganic acids: name the cation first and then the anion. Derive the name of the anion from the name of the carboxylic acid by dropping the suffix *-ic acid* and adding the suffix *-ate*. For example, the name of $CH_3CH_2COO^-Na^+$ is sodium propanoate, and that of $CH_3(CH_2)_{14}COO^-Na^+$ is sodium hexadecanoate (sodium palmitate).

Because carboxylic acid salts are water soluble, we can convert water-insoluble carboxylic acids to water-soluble alkali metal or ammonium salts and then extract them into aqueous solution. In turn, we can transform the salt into the free carboxylic acid by adding a strong acid, such as HCl or H_2SO_4. These reactions allow us to separate water-insoluble carboxylic acids from water-insoluble neutral compounds.

Figure 23.2 (overleaf) shows a flowchart for the separation of benzoic acid, a water-insoluble carboxylic acid, from benzyl alcohol, a water-insoluble nonacidic compound. First, we dissolve the mixture of benzoic acid and benzyl alcohol in diethyl ether (a solvent immiscible in water). Next, we shake the ether solution with aqueous NaOH to convert benzoic acid to its water-soluble sodium salt. Then we separate the ether from the aqueous phase. Distillation of the ether solution yields first diethyl ether (bp = 35 °C) and then benzyl alcohol (bp = 205 °C). When we acidify the aqueous solution with HCl, benzoic acid precipitates as a water-insoluble solid (mp = 122 °C) and is recovered by filtration. Similar separation schemes were described previously for phenols (section 19.3) and amines (section 19.7).

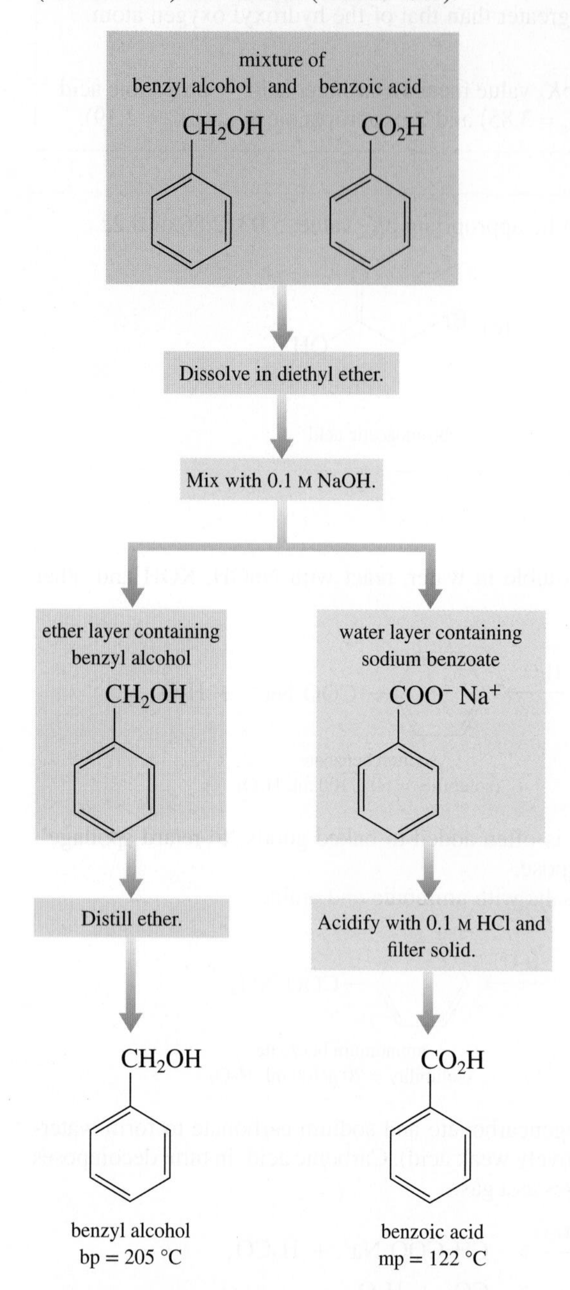

FIGURE 23.2 Flowchart for separating benzoic acid from benzyl alcohol.

Nucleophilic acyl substitution

The most common reaction of the carboxylic acid family (carboxylic acids, acid halides, anhydrides, esters and amides) is **nucleophilic acyl substitution**. The key step in this reaction is the addition of a nucleophile to the carbonyl carbon to form a tetrahedral carbonyl addition intermediate. In this respect, the reactions of these functional groups are similar to nucleophilic addition to the carbonyl groups in aldehydes and ketones (section 21.5). In the reactions of aldehydes and ketones, the tetrahedral carbonyl addition intermediate formed is then protonated. The result of this reaction is nucleophilic addition to a carbonyl group of an aldehyde or a ketone.

Nucleophilic acyl addition:

$$R_2C{=}\ddot{O}: \; + \; :Nu^- \longrightarrow \left[R_2C(Nu){-}\ddot{O}:^- \right] \xrightarrow{H_3O^+} R_2C(Nu){-}\ddot{O}H$$

an aldehyde or a ketone — tetrahedral carbonyl addition intermediate — addition product

For functional derivatives of carboxylic acids, the fate of the tetrahedral carbonyl addition intermediate is quite different from that of aldehydes and ketones. This intermediate collapses to expel the leaving group and regenerate the carbonyl group. The result of this addition–elimination sequence is nucleophilic acyl substitution.

Nucleophilic acyl substitution: $RC(=O)Y + :Nu^- \longrightarrow$ [tetrahedral carbonyl addition intermediate] $\longrightarrow RC(=O)Nu + :Y^-$

tetrahedral carbonyl addition intermediate — substitution product — leaving group

The major difference between these two types of carbonyl addition reactions is that aldehydes and ketones do not have a leaving group (group Y), that can leave as a stable anion. They undergo only nucleophilic acyl addition. The carboxylic acid derivatives we study in this chapter do have a leaving group (group Y), that can leave as a stable anion, so they can undergo nucleophilic acyl substitution.

In the general reaction equation above, we show the nucleophile and the leaving group as anions. That need not be the case, however. Neutral molecules, such as water, alcohols, ammonia and amines, may also serve as nucleophiles in the acid-catalysed version of the reaction. On the right, we show the leaving groups as anions to illustrate an important point: the weaker the base, the better is the leaving group (section 18.2).

$^-NR_2$ $\quad ^-OR \quad ^-OC(=O)R \quad X^-$

Increasing leaving ability →

← Increasing basicity

The weakest bases in the series on the right, and hence the best leaving groups, are the halide ions Cl^-, Br^- and I^-, and therefore acid halides are the most reactive towards nucleophilic acyl substitution. The strongest base, and hence the poorest leaving group, is the amide ion, and therefore amides are the least reactive towards nucleophilic acyl substitution. Acid halides and acid anhydrides are so reactive that they are not found in nature. Esters and amides, however, are extremely common.

$RC(=O)NH_2$ amide — $RC(=O)OR'$ ester — $RC(=O)OC(=O)R$ anhydride — $RC(=O)X$ acid halide

Increasing reactivity towards nucleophilic acyl substitution →

The figure below gives a summary of key examples of nucleophilic acyl substitution reactions that the carboxylic acid family can undergo. The amides ($Y = NH_2$), however, follow this general scheme only for the hydrolysis reaction (addition of water). All these will be discussed in more detail later in this chapter.

acid derivatives $RC(=O)Y$:
- H_2O → acids $RC(=O)OH$
- ROH → esters $RC(=O)OR$
- NH_3 → amides $RC(=O)NH_2$
- Grignard reagents → [ketones $RC(=O)R'$] → reacts further → tertiary alcohols
- reduction → [aldehydes $RC(=O)H$] → reacts further → primary alcohols

Acid halide formation

Among the acid halides, acid chlorides are the most frequently used in the laboratory and in industrial organic chemistry. Recall from section 19.2 that alcohols can be converted into chloroalkanes by treatment with thionyl chloride, $SOCl_2$, or with phosphorus chlorides, PCl_3 or PCl_5. These reagents also convert the —OH group of a carboxylic acid into a chloride, another example of a nucleophilic acyl substitution reaction. The most common way to prepare an acid chloride is to treat a carboxylic acid with thionyl chloride.

$$CH_3CH_2CH_2COOH + SOCl_2 \longrightarrow CH_3CH_2CH_2COCl + SO_2 + HCl$$

butanoic acid | thionyl chloride | butanoyl chloride

WORKED EXAMPLE 23.6

Formation of acid chlorides

Complete each of the following equations.

(a) $CH_3(CH_2)_4COOH + SOCl_2 \longrightarrow$

(b) $CH_3CH{=}CHCOOH + PCl_5 \longrightarrow$

Analysis

In this type of question, you should try to classify the reagent according to its purpose. In these examples, the reagents convert —OH groups to chlorides.

Solution

(a) $CH_3(CH_2)_4COCl + SO_2 + HCl$ (b) $CH_3CH{=}CHCOCl + POCl_3$

Is our answer reasonable?

The two principal reactions of carboxylic acids are breaking the O—H bond (acid dissociation) and breaking the C—OH bond (nucleophilic acyl substitution). In this case, it is the latter, so the products should show this; that is, has the —OH group been replaced with Cl?

PRACTICE EXERCISE 23.6

Complete each of the following equations.

(a) $Br\text{-}C_6H_4\text{-}COOH + SOCl_2 \longrightarrow$

(b) $HOOCCH_2COOH + 2PCl_5 \longrightarrow$

Reactions with alcohols

The reaction of carboxylic acids and their derivatives with alcohols is important for the preparation of esters. In the following pages, we will discuss the esterification reaction starting with the reaction of carboxylic acids with alcohols. We will then investigate the reactions of acid chlorides, acid anhydrides, esters and amides with alcohols.

Carboxylic acids — the Fischer esterification

Treatment of a carboxylic acid with an alcohol in the presence of an acid catalyst (most commonly, concentrated sulfuric acid) gives an ester. This method of forming an ester is given the special name **Fischer esterification** after the German chemist Emil Fischer (1852–1919, Nobel Prize in chemistry, 1902). As an example of a Fischer esterification, treating acetic acid with ethanol in the presence of concentrated sulfuric acid gives ethyl acetate, a common solvent (figure 23.3), and water.

$$CH_3\overset{\overset{O}{\|}}{C}OH + CH_3CH_2OH \overset{H_2SO_4}{\rightleftharpoons} CH_3\overset{\overset{O}{\|}}{C}OCH_2CH_3 + H_2O$$

acetic acid — ethanol — ethyl acetate (ethyl ethanoate)

FIGURE 23.3 Ethyl acetate is the solvent used in nail polish remover.

Acid-catalysed esterification is reversible, and generally, at equilibrium, significant quantities of carboxylic acid and alcohol remain. By controlling the experimental conditions, however, we can use Fischer esterification to prepare esters in high yields. If the alcohol is cheaper than the carboxylic acid, we can use a large excess of the alcohol to drive the reaction to the right and achieve a high conversion of carboxylic acid to its ester.

WORKED EXAMPLE 23.7

Fischer esterification

Complete the following Fischer esterification reactions.

(a) benzoic acid (C₆H₅COOH) + CH_3OH (excess) $\overset{H_2SO_4}{\rightleftharpoons}$

(b) HOOC–CH₂CH₂–COOH + CH_3CH_2OH (excess) $\overset{H_2SO_4}{\rightleftharpoons}$

Analysis

Since these are examples of nucleophilic acyl substitution, we simply need to identify the nucleophile and the group to be substituted. The nucleophile is the oxygen atom of the alcohol.

Solution

(a) $C_6H_5COOCH_3$ — methyl benzoate

(b) $CH_3CH_2OOCCH_2CH_2COOCH_2CH_3$ — diethyl butanedioate (diethyl succinate)

Is our answer reasonable?

Check that you have substituted all the —OH groups of the acids with —OR groups.

PRACTICE EXERCISE 23.7

Complete the following Fischer esterification reactions.

(a) $HOCH_2CH_2CH_2CH_2COOH$ $\overset{\text{acid catalyst}}{\rightleftharpoons}$ a cyclic ester

(b) $(CH_3)_2CHCH_2CH_2OH$ + $CH_3CH_2CH_2COOH$ $\overset{\text{acid catalyst}}{\rightleftharpoons}$ an ester found in strawberries

Note that, although we show the acid catalyst as H_2SO_4 when we write Fischer esterification reactions, the actual proton-transfer acid that initiates the reaction is the oxonium ion formed by the transfer of a proton from H_2SO_4 (the stronger acid) to the alcohol (the stronger base) used in the esterification reaction.

$$CH_3{-}\ddot{O}{-}H + H{-}O{-}SO_2{-}O{-}H \rightleftharpoons CH_3{-}\overset{+}{O}H_2 + {}^{-}O{-}SO_2{-}O{-}H$$

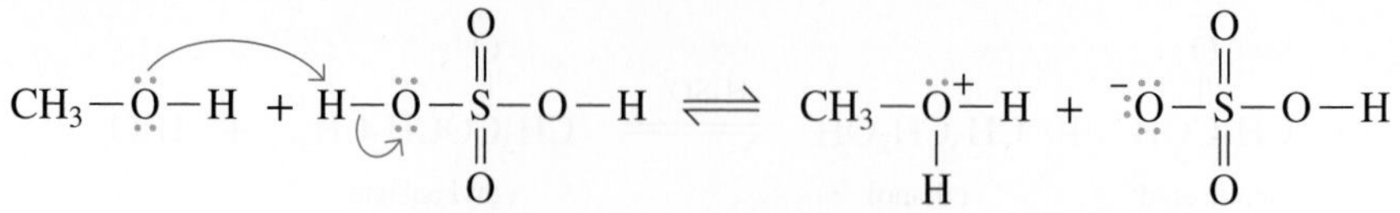

The reaction above is similar to the familiar reaction of water with sulfuric acid.

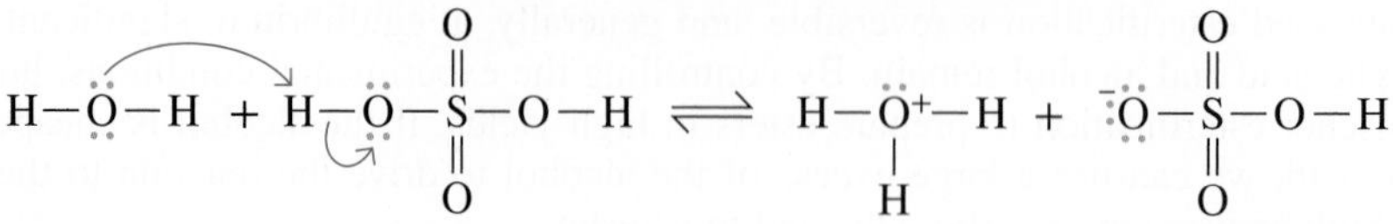

$$H{-}\ddot{O}{-}H + H{-}O{-}SO_2{-}O{-}H \rightleftharpoons H{-}\overset{+}{O}H{-}H + {}^{-}O{-}SO_2{-}O{-}H$$

A mechanism for the Fischer esterification reaction is shown below and we urge you to study it carefully. Do not be overwhelmed by the number of steps; it is still basically only a nucleophilic acyl substitution.

Step 1: Proton transfer from the acid catalyst to the carbonyl oxygen atom increases the electrophilicity of the carbonyl carbon atom.

Step 2: The nucleophilic oxygen atom of an alcohol adds to the electrophilic carbonyl carbon atom to form an oxonium ion. A C—O bond is formed in this step.

Step 3: Proton transfer from the oxonium ion to a second molecule of alcohol gives a tetrahedral carbonyl addition intermediate (TCAI).

Step 4: Proton transfer to one of the —OH groups of the TCAI gives a new oxonium ion.

Step 5: Water is lost from this oxonium ion. A C—O bond is broken in this step.

Step 6: Proton transfer to a molecule of alcohol gives the ester and water and regenerates the acid catalyst.

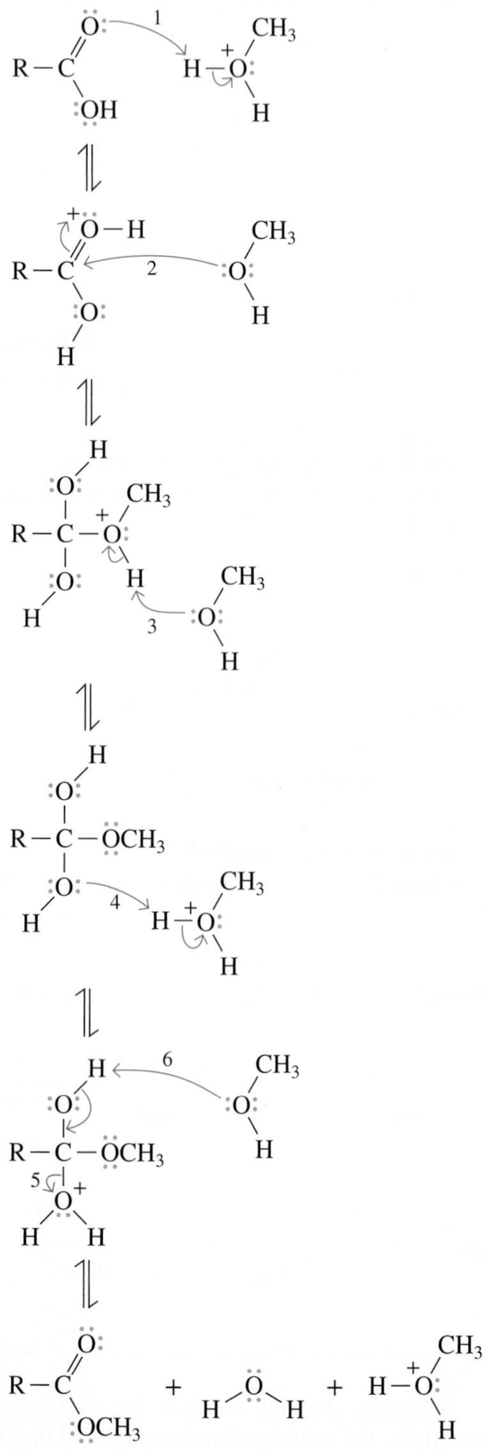

Note that, in the mechanism on the previous page, steps 1, 3, 4 and 6 are simply proton transfer processes. Step 2 is the formation of a C—O bond (the addition part of nucleophilic acyl substitution) and step 5 is the breaking of a C—O bond (the elimination part of nucleophilic acyl substitution). Note also that the oxygen atom of the alcohol is retained in the ester, while the —OH group of the carboxylic acid is lost (as water). This has been confirmed experimentally by using an alcohol labelled with a heavy isotope of oxygen (^{18}O). Upon separation of the components, it was found that the ester contained ^{18}O, while the water did not.

Acid chlorides

Acid chlorides react with alcohols via a nucleophilic acyl substitution reaction to give an ester and HCl.

butanoyl chloride + cyclohexanol ⟶ cyclohexyl butanoate + HCl

Because acid chlorides are so reactive towards even weak nucleophiles such as alcohols, no catalyst is necessary for these reactions. Phenol and substituted phenols also react with acid chlorides to give esters.

Acid anhydrides

Acid anhydrides react with alcohols to give 1 mole equivalent of ester and 1 mole equivalent of a carboxylic acid.

acetic anhydride + ethanol ⟶ ethyl acetate + acetic acid

Aspirin is synthesised on an industrial scale by reacting acetic anhydride with salicylic acid.

salicylic acid + acetic anhydride ⟶ acetylsalicylic acid (aspirin) + acetic acid

Note that, in the use of anhydrides for the preparation of esters (and amides), 1 mole of carboxylic acid is produced as a by-product. This is an inefficient use of the carboxylic acid and, consequently, anhydrides are used only when they are particularly cheap (e.g. acetic and phthalic anhydrides).

Esters

When treated with an alcohol in the presence of an acid or base catalyst, esters undergo an exchange reaction called **transesterification**. In this reaction, the original —OR group of the ester is exchanged for a new —OR group. In the following example, the transesterification can be driven to completion by heating the reaction at a temperature above the boiling point of methanol (65 °C) so that methanol distills from the reaction mixture.

2 methyl benzoate + ethane-1,2-diol (ethylene glycol) $\xrightleftharpoons{H_2SO_4}$ a diester of ethylene glycol + $2CH_3OH$

Amides

Amides do not react with alcohols under any experimental conditions. Alcohols are not strong enough nucleophiles to attack the carbonyl group of an amide.

The reactions of carboxylic acids, acid chlorides, anhydrides, esters and amides with alcohols are summarised below. Note that there are large differences in the rates and experimental conditions under which these functional groups undergo reactions with alcohols. At one extreme are acid chlorides and anhydrides, which react rapidly; at the other extreme are amides, which do not react at all.

Summary of the reactions of carboxylic acids, acid chlorides, anhydrides, esters and amides with alcohols

$$R-\overset{O}{\overset{\|}{C}}-OH + R'OH \underset{}{\overset{H_2SO_4}{\rightleftharpoons}} R-\overset{O}{\overset{\|}{C}}-OR' + H_2O$$

$$R-\overset{O}{\overset{\|}{C}}-Cl + R'OH \longrightarrow R-\overset{O}{\overset{\|}{C}}-OR' + HCl$$

$$R-\overset{O}{\overset{\|}{C}}-O-\overset{O}{\overset{\|}{C}}-R + R'OH \longrightarrow R-\overset{O}{\overset{\|}{C}}-OR' + R-\overset{O}{\overset{\|}{C}}-OH$$

$$R-\overset{O}{\overset{\|}{C}}-OR'' + R'OH \underset{}{\overset{H_2SO_4}{\rightleftharpoons}} R-\overset{O}{\overset{\|}{C}}-OR' + R''OH$$

$$R-\overset{O}{\overset{\|}{C}}-NH_2 + R'OH \longrightarrow \text{no reaction}$$

Reaction with water: hydrolysis

The hydrolysis of carboxylic acid derivatives is yet another example of a nucleophilic acyl substitution reaction. Essentially, carboxylic acid derivatives react with water to give carboxylic acids and H—Y (where Y is a leaving group). For some derivatives this hydrolysis requires acidic or basic conditions and heat.

$$R-\overset{O}{\overset{\|}{C}}-Y + H_2O \longrightarrow R-\overset{O}{\overset{\|}{C}}-OH + H-Y$$

Acid chlorides

Low-molar-mass acid chlorides react very rapidly with water to form carboxylic acids and HCl.

$$CH_3\overset{O}{\overset{\|}{C}}Cl + H_2O \longrightarrow CH_3\overset{O}{\overset{\|}{C}}OH + HCl$$

High-molar-mass acid chlorides are less water soluble and so react less rapidly with water.

Acid anhydrides

Acid anhydrides are generally less reactive than acid chlorides. Low-molar-mass anhydrides, however, react readily with water to form two mole equivalents of carboxylic acid. Both acid chlorides and acid anhydrides will be slowly hydrolysed by moisture in the atmosphere if they are not stored in air-tight containers.

$$CH_3\overset{O}{\overset{\|}{C}}O\overset{O}{\overset{\|}{C}}CH_3 + H_2O \longrightarrow CH_3\overset{O}{\overset{\|}{C}}OH + HO\overset{O}{\overset{\|}{C}}CH_3$$

Esters

Esters are hydrolysed only very slowly, even in boiling water. Hydrolysis becomes considerably more rapid, however, when esters are heated in aqueous acid or base. When we discussed acid-catalysed (Fischer) esterification earlier (pp. 1028–9), we pointed out that esterification is an equilibrium reaction. Hydrolysis of esters in aqueous acid is the same equilibrium reaction as the Fischer esterification, except in reverse. The role of the acid catalyst is to protonate the carbonyl oxygen atom, thereby increasing the electrophilic character of the carbonyl carbon atom and so facilitating addition by water to form a tetrahedral carbonyl addition intermediate. Collapse of this intermediate gives the carboxylic acid and an alcohol.

$$\mathrm{RC(=O)OCH_3} + \mathrm{H_2O} \overset{H^+}{\rightleftharpoons} \left[\mathrm{R{-}C(OH)_2(OCH_3)}\right] \overset{H^+}{\rightleftharpoons} \mathrm{RC(=O)OH} + \mathrm{CH_3OH}$$

tetrahedral carbonyl addition intermediate

In this reaction, the acid is a catalyst as it is consumed in the first step and is regenerated at the end of the reaction.

Hydrolysis of esters can also be carried out with hot aqueous base, such as aqueous NaOH. Hydrolysis of esters in aqueous base is often called **saponification**, a reference to the use of this reaction in the manufacture of soaps (section 23.6). Each mole of ester hydrolysed requires 1 mole of base, as shown in the following balanced equation.

$$\mathrm{R\overset{O}{\overset{\|}{C}}OCH_3} + \mathrm{NaOH} \xrightarrow{H_2O} \mathrm{R\overset{O}{\overset{\|}{C}}O^-Na^+} + \mathrm{CH_3OH}$$

The following shows a mechanism for hydrolysis of an ester in aqueous base.

Step 1: Addition of a hydroxide ion to the carbonyl carbon atoms of the ester gives a tetrahedral carbonyl addition intermediate.

$$\mathrm{R{-}\overset{O}{\overset{\|}{C}}{-}OCH_3} + {}^-\mathrm{OH} \rightleftharpoons \mathrm{R{-}\overset{O^-}{\overset{|}{\underset{OH}{\underset{|}{C}}}}{-}OCH_3}$$

Step 2: Collapse of this intermediate gives a carboxylic acid and an alkoxide ion.

$$\mathrm{R{-}\overset{O^-}{\overset{|}{\underset{OH}{\underset{|}{C}}}}{-}OCH_3} \rightleftharpoons \mathrm{R{-}\overset{O}{\overset{\|}{C}}{-}OH} + {}^-\mathrm{OCH_3}$$

Step 3: Proton transfer from the carboxyl group (an acid) to the alkoxide ion (a base) gives the carboxylate anion. This step is irreversible because the alcohol is not a strong enough nucleophile to react with carboxylate anion.

$$\mathrm{R{-}\overset{O}{\overset{\|}{C}}{-}O{-}H} + {}^-\mathrm{OCH_3} \longrightarrow \mathrm{R{-}\overset{O}{\overset{\|}{C}}{-}O^-} + \mathrm{H{-}OCH_3}$$

Note that steps 1 and 2 describe a typical nucleophilic acyl substitution reaction; step 3 is simply an acid–base reaction.

There are two major differences between the hydrolysis of esters in aqueous acid and that in aqueous base.

1. For hydrolysis in aqueous acid, acid is required in only catalytic amounts. For hydrolysis in aqueous base, base is required in equimolar amounts because it is a reactant, not just a catalyst.
2. Hydrolysis of an ester in aqueous acid is reversible. Hydrolysis in aqueous base is irreversible, because carboxylate anions do not react with alcohols.

WORKED EXAMPLE 23.8

Hydrolysis of esters

Complete and balance the following equations for the hydrolysis of each ester in aqueous sodium hydroxide, showing all products as they are ionised in aqueous NaOH.

(a) $\mathrm{C_6H_5COOCH(CH_3)_2} + \mathrm{NaOH} \xrightarrow{H_2O}$

(b) $\mathrm{CH_3COOCH_2CH_2OCOCH_3} + \mathrm{NaOH} \xrightarrow{H_2O}$

Analysis

First identify the carbonyl carbon atom(s) of the esters and the leaving groups. Substitution of the —OR group (the leaving group) by an —OH group occurs at these carbon atoms. Finally, the acid reacts with the strongly basic alkoxide ion (RO^-).

Solution

(a) $C_6H_5CO^-Na^+$ + $HOCH(CH_3)_2$

sodium benzoate; propan-2-ol (isopropyl alcohol)

(b) $2CH_3CO^-Na^+$ + $HOCH_2CH_2OH$

sodium acetate; ethane-1,2-diol (ethylene glycol)

Is our answer reasonable?

The products of hydrolysis are a carboxylic acid and an alcohol. In aqueous NaOH, the carboxylic acid is converted to its sodium salt. Therefore, 1 mole of NaOH is required for the hydrolysis of 1 mole of ester. Compound (b) is a diester, so 2 moles of NaOH are required.

PRACTICE EXERCISE 23.8 Complete and balance the following equations for the hydrolysis of each ester in aqueous solution, showing each product as it is ionised under the given experimental conditions.

(a) [cyclopentane-1,3-diyl diacetate] + NaOH (excess) $\xrightarrow{H_2O}$

(b) [diisopropyl oxalate] + H_2O $\overset{HCl}{\rightleftharpoons}$

Amides

Amides require considerably more vigorous conditions for hydrolysis in both acid and base than do esters. Amides undergo hydrolysis in hot aqueous acid to give carboxylic acids and ammonia (or amines). Hydrolysis is driven to completion by the acid–base reaction between ammonia or amines and acid to form ammonium salts. One mole of acid is required per mole of amide.

$$CH_3CH_2CH(Ph)CONH_2 + H_2O + HCl \xrightarrow{heat} CH_3CH_2CH(Ph)COOH + NH_4^+Cl^-$$

2-phenylbutanamide → 2-phenylbutanoic acid

In aqueous base, the products of amide hydrolysis are carboxylic acids and ammonia or amines. Base-catalysed hydrolysis is driven to completion by the acid–base reaction between carboxylic acids and base to form carboxylate salts. One mole of base is required per mole of amide.

$$CH_3CONH\text{—}C_6H_5 + NaOH \xrightarrow[heat]{H_2O} CH_3CO^-Na^+ + H_2N\text{—}C_6H_5$$

N-phenylethanamide (acetanilide); sodium acetate; aniline

The reactions of acid chlorides, anhydrides, esters and amides with water are summarised below. Remember that, although all four functional groups react with water, there are large differences in the rates and experimental conditions under which they are hydrolysed.

Summary of the reactions of acid chlorides, anhydrides, esters and amides with water

$$R-\overset{O}{\overset{\|}{C}}-Cl + H_2O \longrightarrow R-\overset{O}{\overset{\|}{C}}-OH + HCl$$

$$R-\overset{O}{\overset{\|}{C}}-O-\overset{O}{\overset{\|}{C}}-R + H_2O \longrightarrow R-\overset{O}{\overset{\|}{C}}-OH + HO-\overset{O}{\overset{\|}{C}}-R$$

$$R-\overset{O}{\overset{\|}{C}}-OR' + H_2O \xrightarrow{NaOH} R-\overset{O}{\overset{\|}{C}}-O^-Na^+ + R'OH$$

$$R-\overset{O}{\overset{\|}{C}}-OR' + H_2O \xrightarrow{H_2SO_4} R-\overset{O}{\overset{\|}{C}}-OH + R'OH$$

$$R-\overset{O}{\overset{\|}{C}}-NH_2 + H_2O \xrightarrow{NaOH} R-\overset{O}{\overset{\|}{C}}-O^-Na^+ + NH_3$$

$$R-\overset{O}{\overset{\|}{C}}-NH_2 + H_2O \xrightarrow{HCl} R-\overset{O}{\overset{\|}{C}}-OH + NH_4^+Cl^-$$

WORKED EXAMPLE 23.9

Hydrolysis of amides

Write an equation for the hydrolysis of each of the following amides in concentrated aqueous HCl, showing all products as they exist in aqueous HCl and showing the amount of HCl required for the hydrolysis of each amide.

(a) $CH_3\overset{O}{\overset{\|}{C}}N(CH_3)_2$ (b) [six-membered ring lactam: C=O adjacent to NH]

Analysis

First identify the carbonyl carbon atom of the amide. Substitution of the —NR_2 or —NHR group by an —OH group (the leaving group) will occur at this carbon atom. Remember that the hydrolysis of an amide gives a carboxylic acid and an amine (or ammonia). In (b), the carboxylic acid and the amine are part of the same molecule, so only one product will be formed. Finally consider the conditions of the reaction to determine the ionised form of the product(s).

Solution

(a) Hydrolysis of *N,N*-dimethylacetamide gives acetic acid and dimethylamine. Dimethylamine, a base, is protonated by HCl to form a dimethylammonium ion and is shown in the balanced equation as dimethylammonium chloride. Complete hydrolysis of this amide requires 1 mole of HCl for each mole of the amide.

$$CH_3\overset{O}{\overset{\|}{C}}N(CH_3)_2 + H_2O + HCl \xrightarrow{heat} CH_3\overset{O}{\overset{\|}{C}}OH + (CH_3)_2NH_2^+Cl^-$$

(b) Hydrolysis of this δ-lactam gives the protonated form of 5-aminopentanoic acid. One mole of acid is required per mole of lactam.

$$\text{[lactam ring, C=O, NH]} + H_2O + HCl \xrightarrow{heat} HO\overset{O}{\overset{\|}{C}}CH_2CH_2CH_2CH_2NH_3^+Cl^-$$

Is our answer reasonable?

Check that the carbonyl carbon atom is now part of a carboxylic acid group and that the N part of the amide is now an amino group. Now check that the state of ionisation is correct for the reaction conditions used (acid conditions give ammonium ions; basic conditions give carboxylate ions). Since the reaction is carried out under acidic conditions, the amine product will react further to produce an ammonium ion.

PRACTICE EXERCISE 23.9 Complete equations for the hydrolysis of the amides in worked example 23.9 in concentrated aqueous NaOH. Show all products as they exist in aqueous NaOH, and show the amount of NaOH required for the hydrolysis of each amide.

Reactions with ammonia and amines

Carboxylic acids react with ammonia and amines in an acid–base neutralisation reaction to give ammonium carboxylates (see reactions with bases, p. 1025).

Acid chlorides

Acid chlorides react readily with ammonia and with 1° and 2° amines to form amides. Complete conversion of acid chlorides to amides requires 2 mole equivalents of ammonia or amine, 1 mole equivalent to form the amide and 1 mole equivalent to neutralise the hydrogen chloride formed.

$$\text{hexanoyl chloride} + 2NH_3 \longrightarrow \text{hexanamide} + NH_4^+Cl^-$$

hexanoyl chloride — ammonia — hexanamide — ammonium chloride

Acid anhydrides

Acid anhydrides react with ammonia and with 1° and 2° amines to form amides. As with acid chlorides, 2 mole equivalents of ammonia or amine are required, 1 mole equivalent to form the amide and 1 mole equivalent to neutralise the carboxylic acid by-product. To help you see what happens, this reaction equation is broken into two steps, which, when added together, give the net reaction for the reaction of an anhydride with ammonia.

$$CH_3\overset{O}{\overset{\|}{C}}O\overset{O}{\overset{\|}{C}}CH_3 + NH_3 \longrightarrow CH_3\overset{O}{\overset{\|}{C}}NH_2 + CH_3\overset{O}{\overset{\|}{C}}OH$$

$$CH_3\overset{O}{\overset{\|}{C}}OH + NH_3 \longrightarrow CH_3\overset{O}{\overset{\|}{C}}O^-NH_4^+$$

$$CH_3\overset{O}{\overset{\|}{C}}O\overset{O}{\overset{\|}{C}}CH_3 + 2NH_3 \longrightarrow CH_3\overset{O}{\overset{\|}{C}}NH_2 + CH_3\overset{O}{\overset{\|}{C}}O^-NH_4^+$$

Esters

Esters react with ammonia and with 1° and 2° amines to form amides.

$$PhCH_2COOCH_2CH_3 + NH_3 \longrightarrow PhCH_2CONH_2 + HOCH_2CH_3$$

ethyl phenylacetate — phenylacetamide — ethanol

Because alkoxide anions are poorer leaving groups than halides or carboxylate ions, esters are less reactive towards ammonia, 1° amines and 2° amines than are acid chlorides or acid anhydrides.

Amides

Amides do not react with ammonia or amines.

The reactions of carboxylic acids, acid chlorides, anhydrides, esters and amides with ammonia and amines are summarised below.

Summary of the reactions of acid chlorides, anhydrides, esters and amides with ammonia and amines

$$R-\overset{\overset{\Large O}{\|}}{C}-OH + NH_3 \longrightarrow R-\overset{\overset{\Large O}{\|}}{C}-O^- + NH_4^+$$

$$R-\overset{\overset{\Large O}{\|}}{C}-Cl + 2NH_3 \longrightarrow R-\overset{\overset{\Large O}{\|}}{C}-NH_2 + NH_4^+\,Cl^-$$

$$R-\overset{\overset{\Large O}{\|}}{C}-O-\overset{\overset{\Large O}{\|}}{C}-R + 2NH_3 \longrightarrow R-\overset{\overset{\Large O}{\|}}{C}-NH_2 + R-\overset{\overset{\Large O}{\|}}{C}-O^-NH_4^+$$

$$R-\overset{\overset{\Large O}{\|}}{C}-OR' + NH_3 \longrightarrow R-\overset{\overset{\Large O}{\|}}{C}-NH_2 + R'OH$$

$$R-\overset{\overset{\Large O}{\|}}{C}-NH_2 \longrightarrow \text{no reaction with ammonia or amines}$$

WORKED EXAMPLE 23.10

Formation of amides from esters

Complete the following equations (the stoichiometry of each is given in the equation).

(a) ethyl butanoate $+ NH_3 \longrightarrow$

(b) diethyl carbonate $+ 2NH_3 \longrightarrow$

Analysis

First identify the carbonyl carbon atom of the ester. Nucleophilic substitution occurs at this carbon atom. Substitute the —OR group (the leaving group) with the nucleophile and then do the appropriate proton transfer reaction. In (b), the process occurs twice.

Solution

(a) butanamide $+ CH_3CH_2OH$

(b) $H_2N-\overset{\overset{\Large O}{\|}}{C}-NH_2 + 2CH_3CH_2OH$

urea

Is our answer reasonable?

Check that the substitution has occurred at the carbon atom of the carbonyl group. Check also that the alcohol formed corresponds to the —OR group of the starting esters.

PRACTICE EXERCISE 23.10 Complete the following equations (the stoichiometry of each is given in the equation).

(a) [cyclopentenyl acetate] + NH_3 ⟶ (b) [γ-butyrolactone] + NH_3 ⟶

Reduction

The carboxyl group is one of the organic functional groups that is most resistant to reduction. It is not affected by catalytic reduction under conditions that easily reduce aldehydes and ketones to alcohols, or alkenes to alkanes.

Most reductions of carbonyl compounds, including aldehydes and ketones, are now accomplished by transferring hydride ions from boron or aluminium hydrides. We have already seen the use of sodium borohydride and lithium aluminium hydride to reduce the carbonyl groups of aldehydes and ketones to hydroxyl groups (section 21.5). The most common reagent for the reduction of carboxylic acids to primary alcohols is the very powerful reducing agent lithium aluminium hydride; sodium borohydride is not a sufficiently powerful reducing agent to reduce the carboxylic acid derivatives.

Carboxylic acids

Lithium aluminium hydride, $LiAlH_4$, reduces carboxyl groups to primary alcohols in excellent yield. Reduction is most commonly carried out in diethyl ether or tetrahydrofuran (THF). An aluminium alkoxide is produced initially and is then treated with water to give primary alcohols, and lithium and aluminium hydroxides. These hydroxides are insoluble in diethyl ether and THF so can be removed by filtration. Evaporation of the solvent yields the primary alcohol.

$$\text{C}_5\text{H}_7\text{COOH} \xrightarrow[2.\ H_2O]{1.\ LiAlH_4} \text{C}_5\text{H}_7\text{CH}_2\text{OH} + LiOH + Al(OH)_3$$

cyclopent-3-ene-carboxylic acid — 4-hydroxymethyl-cyclopentene

Alkenes are generally not affected by metal hydride reducing reagents. These reagents function as hydride ion donors (nucleophiles) and alkenes do not normally react with nucleophiles.

Esters

Esters are reduced by lithium aluminium hydride to two alcohols. The alcohol derived from the acyl group is primary.

$$\text{PhCH(CH}_3\text{)COOCH}_3 \xrightarrow[2.\ H_2O,\ HCl]{1.\ LiAlH_4} \text{PhCH(CH}_3\text{)CH}_2\text{OH} + CH_3OH$$

methyl 2-phenylpropanoate — 2-phenylpropan-l-ol (a 1° alcohol) — methanol

Sodium borohydride is not normally used to reduce esters, because the reaction is very slow. Therefore, it is possible to use sodium borohydride to reduce the carbonyl group of aldehydes or ketones to hydroxyl groups without reducing ester or carboxyl groups in the same molecule.

$$\text{CH}_3\text{COCH}_2\text{COOCH}_2\text{CH}_3 \xrightarrow{NaBH_4} \text{CH}_3\text{CH(OH)CH}_2\text{COOCH}_2\text{CH}_3$$

Although it may not be immediately obvious, the reduction of esters involves a nucleophilic acyl substitution mechanism followed by reduction of the resulting aldehyde to give, ultimately, the primary alcohol. The metal hydride reduction of aldehydes has been discussed in section 21.5.

$$\text{RC(=O)OR}' + \text{H–AlH}_3^- \longrightarrow \text{R–C(O}^-\text{)(H)–OR}'' \longrightarrow \text{RC(=O)H} + \text{R}'\text{–O}^-$$

tetrahedral carbonyl addition intermediate — aldehyde, which reacts immediately to give the 1° alcohol

Amides

The reduction of amides by lithium aluminium hydride can be used to prepare 1°, 2° or 3° amines, depending on the degree of substitution of the amide.

$$\text{octanamide} \xrightarrow[\text{2. } H_2O]{\text{1. } LiAlH_4} \text{octan-1-amine (a 1° amine)}$$

$$N,N\text{-dimethylbenzamide} \xrightarrow[\text{2. } H_2O]{\text{1. } LiAlH_4} N,N\text{-dimethylbenzylamine (a 3° amine)}$$

WORKED EXAMPLE 23.11

Synthesis of amines from carboxylic acids

Show how to bring about each of the following conversions.

(a) $C_6H_5COOH \longrightarrow C_6H_5CH_2{-}N(\text{pyrrolidine ring})$

(b) $\text{cyclohexyl}{-}COOH \longrightarrow \text{cyclohexyl}{-}CH_2NHCH_3$

Analysis

The key step is to convert a carboxylic acid into an amide that can then be reduced with lithium aluminium hydride to give the required amine. We have discussed a couple of two-step methods of converting carboxylic acids to amides: via the acid chloride and via the ester.

Solution

The amide can be prepared by treating the carboxylic acid with $SOCl_2$ to form the acid chloride and then treating the acid chloride with an amine. Alternatively, the carboxylic acid can be converted to an ester by Fischer esterification and the ester treated with an amine to give the amide. To illustrate both methods, we have shown the acid chloride route for conversion (a) and the ester route for conversion (b).

(a) $C_6H_5COOH \xrightarrow{SOCl_2} C_6H_5COCl \xrightarrow{\text{HN(pyrrolidine)}} C_6H_5C(=O){-}N(\text{pyrrolidine}) \xrightarrow[\text{2. } H_2O]{\text{1. } LiAlH_4} C_6H_5CH_2{-}N(\text{pyrrolidine})$

(b) $\text{cyclohexyl}{-}COOH \xrightarrow{CH_3CH_2OH,\ H^+} \text{cyclohexyl}{-}COOCH_2CH_3 \xrightarrow{CH_3NH_2} \text{cyclohexyl}{-}CONHCH_3 \xrightarrow[\text{2. } H_2O]{\text{1. } LiAlH_4} \text{cyclohexyl}{-}CH_2NHCH_3$

Is our answer reasonable?

Check that the products of each reaction step have the appropriate number of carbon and hydrogen atoms.

PRACTICE EXERCISE 23.11 Show how to bring about each of the following conversions.

(a) HO(C=O)CH₂CH₂CH₂(C=O)OH ⟶ H_2N–(CH₂)₅–NH_2

(b) $(CH_3)_2CHCH_2COOH$ ⟶ $(CH_3)_2CHCH_2CH_2NHCH(CH_3)_2$

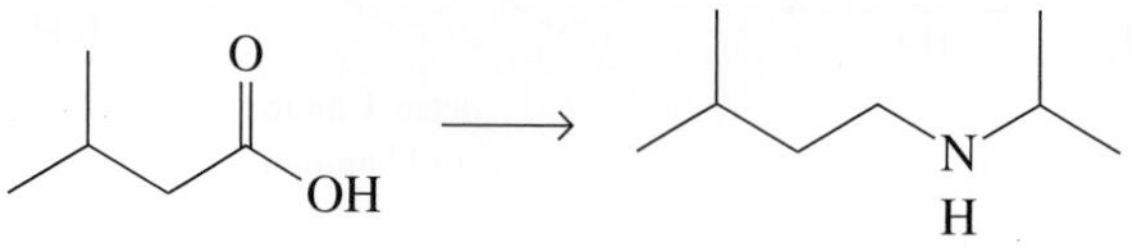

Selective reduction of other functional groups

Catalytic hydrogenation does not reduce carboxyl groups, but does reduce alkenes to alkanes. Therefore, we can use H_2 and a metal catalyst to reduce this functional group selectively in the presence of a carboxyl group.

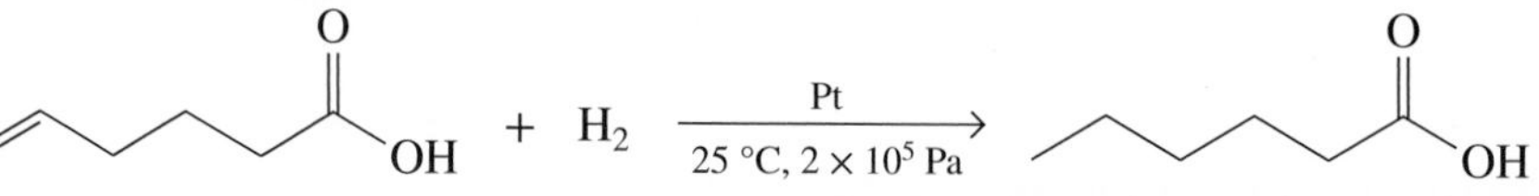

$$\text{hex-5-enoic acid} + H_2 \xrightarrow[25\ °C,\ 2 \times 10^5\ Pa]{Pt} \text{hexanoic acid}$$

We saw in section 21.5 that aldehydes and ketones are reduced to alcohols by both $LiAlH_4$ and $NaBH_4$. Only $LiAlH_4$, however, reduces carboxyl groups. Thus, it is possible to reduce an aldehyde or a ketone carbonyl group selectively in the presence of a carboxyl group by using the less reactive $NaBH_4$ as the reducing agent.

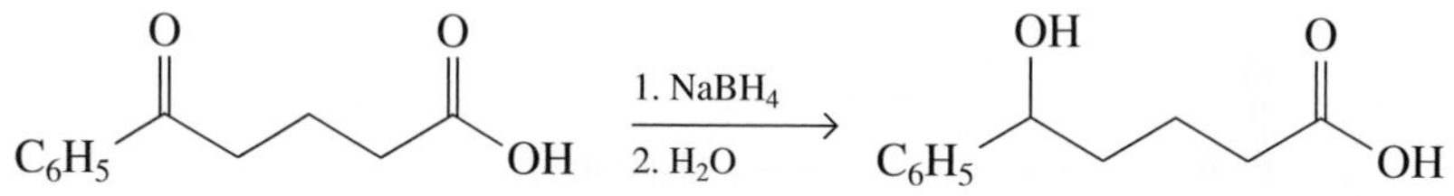

$$C_6H_5COCH_2CH_2CH_2COOH \xrightarrow[2.\ H_2O]{1.\ NaBH_4} C_6H_5CH(OH)CH_2CH_2CH_2COOH$$

5-oxo-5-phenylpentanoic acid → 5-hydroxy-5-phenylpentanoic acid

Esters with Grignard reagents

Treating esters of formic acid with 2 mole equivalents of a Grignard reagent, followed by protonation of the alkoxide salts, gives 2° alcohols; treating esters of any other carboxylic acid with Grignard reagents gives 3° alcohols in which two of the groups bonded to the carbon atom bearing the —OH group are the same.

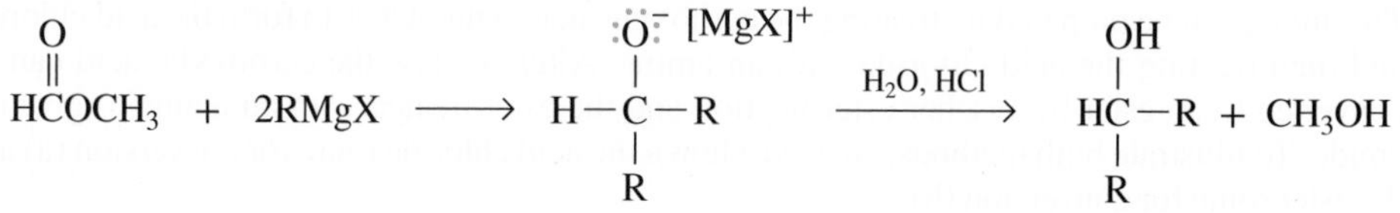

$$\underset{\text{an ester of formic acid}}{HCOCH_3} + 2RMgX \longrightarrow \underset{\text{magnesium alkoxide salt}}{H{-}C(R)(R){-}O^-\ [MgX]^+} \xrightarrow{H_2O,\ HCl} \underset{\text{a 2° alcohol}}{HC(OH)(R){-}R} + CH_3OH$$

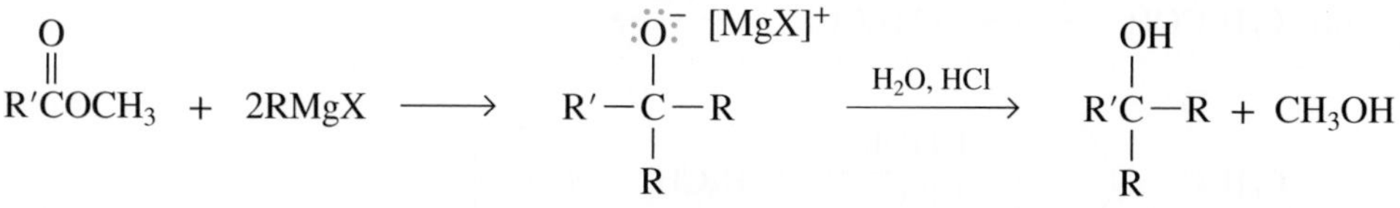

$$\underset{\text{an ester other than a formate ester}}{R'COCH_3} + 2RMgX \longrightarrow \underset{\text{magnesium alkoxide salt}}{R'{-}C(R)(R){-}O^-\ [MgX]^+} \xrightarrow{H_2O,\ HCl} \underset{\text{a 3° alcohol}}{R'C(OH)(R){-}R} + CH_3OH$$

The reaction of esters with Grignard reagents (shown above) involves the formation of two successive tetrahedral carbonyl addition intermediates. The first collapses to give a new carbonyl compound — an aldehyde from a formate ester, a ketone from all other esters. The second intermediate is stable and, when protonated, gives the final alcohol. It is important to realise that it is not possible to use RMgX and esters to prepare aldehydes or ketones; the intermediate aldehydes or ketones are more reactive than the esters and react immediately with the Grignard reagent to give alcohols.

Steps 1 and 2: These two steps describe the nucleophilic acyl substitution reaction. The reaction begins in step 1 with the addition of 1 mole equivalent of Grignard reagent to the carbonyl carbon to form a tetrahedral carbonyl addition intermediate. This intermediate then collapses in step 2 to give a new carbonyl-containing compound and a magnesium alkoxide salt.

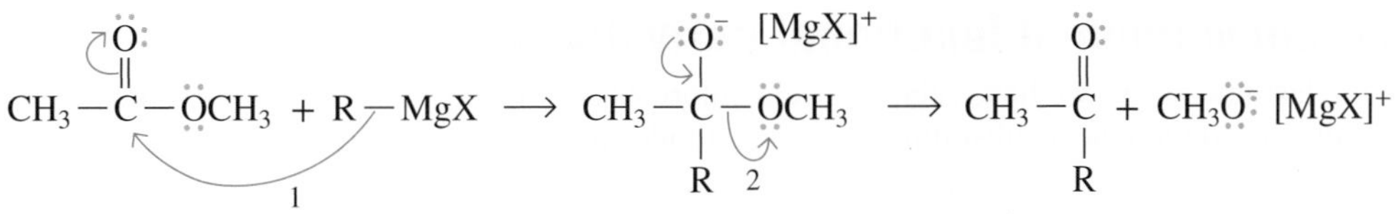

$$CH_3-\overset{\overset{\large O}{\|}}{C}-OCH_3 + R-MgX \xrightarrow{1} CH_3-\underset{R}{\overset{O^-\ [MgX]^+}{C}}-OCH_3 \xrightarrow{2} CH_3-\underset{R}{\overset{\overset{\large O}{\|}}{C}} + CH_3O^-\ [MgX]^+$$

a magnesium salt (a tetrahedral carbonyl addition intermediate) — a ketone

Steps 3 and 4: These two steps describe the typical carbonyl addition reaction of aldehydes and ketones (see section 21.5). The new carbonyl-containing compound reacts in step 3 with a second mole equivalent of Grignard reagent to form a second tetrahedral carbonyl addition compound, which, after hydrolysis in aqueous acid (step 4), gives 3° alcohols (or 2° alcohols if the starting ester was a formate).

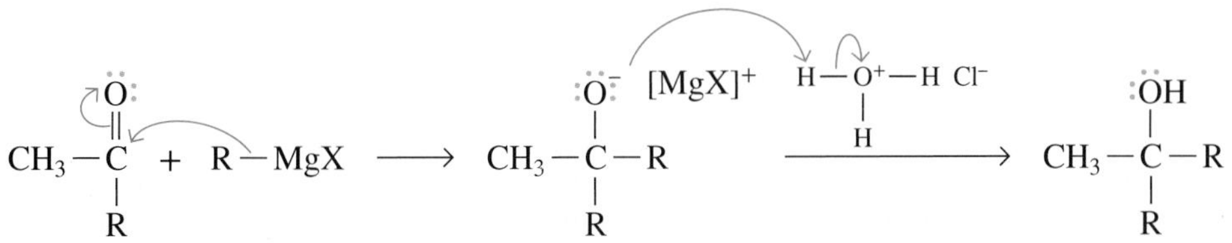

$$CH_3-\underset{R}{\overset{\overset{\large O}{\|}}{C}} + R-MgX \longrightarrow CH_3-\underset{R}{\overset{O^-\ [MgX]^+}{C}}-R \xrightarrow{H-\underset{H}{O^+}-H\ Cl^-} CH_3-\underset{R}{\overset{OH}{C}}-R$$

a ketone — magnesium salt — a 3° alcohol

Although it may not be obvious from the products obtained in the reaction of esters with Grignard reagents or reducing agents, the reaction proceeds via a nucleophilic acyl substitution. However, in both cases, the products of this initial reaction react readily with the reagents present to ultimately produce alcohols.

WORKED EXAMPLE 23.12

Reaction of esters with Grignard reagents

Complete each of the following Grignard equations.

(a) 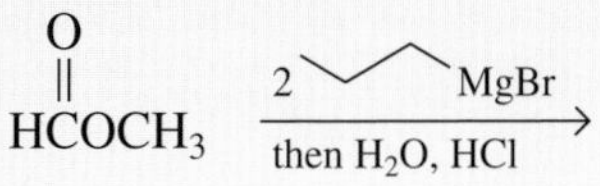 $HCOCH_3$ (with C=O) $\xrightarrow[\text{then } H_2O,\ HCl]{2\ \text{(propyl)MgBr}}$

(b) 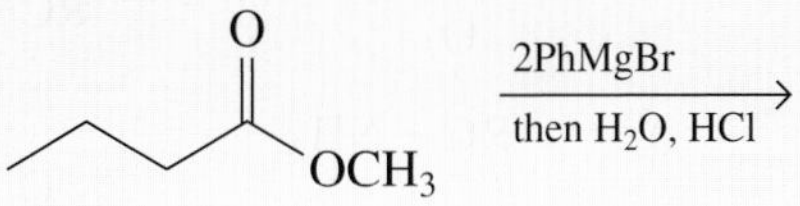 methyl butanoate (OCH_3 ester) $\xrightarrow[\text{then } H_2O,\ HCl]{2PhMgBr}$

Analysis

In both cases, identify the carbonyl carbon atom; this will be the carbon atom bearing the —OH group in the product. The first mole equivalent of Grignard reagent will substitute the alkoxy group (in these cases, the $—OCH_3$ group) to give an aldehyde or ketone. These then react further, in the usual fashion (see section 21.5), to produce alcohols.

Solution

(a) 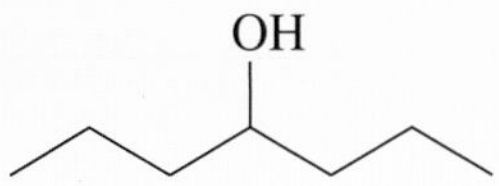 OH on central carbon of a chain with two propyl groups (4-heptanol)

(b) 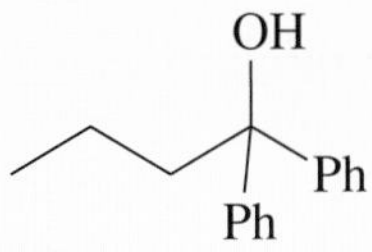 OH, Ph, Ph on carbon attached to a propyl chain

Is our answer reasonable?

Since (a) is an ester of formic acid, the final product should be a 2° alcohol. Ester (b) should produce a 3° alcohol. Both products should also have two equivalent —R groups.

PRACTICE EXERCISE 23.12

Show how to prepare each of the following alcohols by treating an ester with a Grignard reagent.

(a) 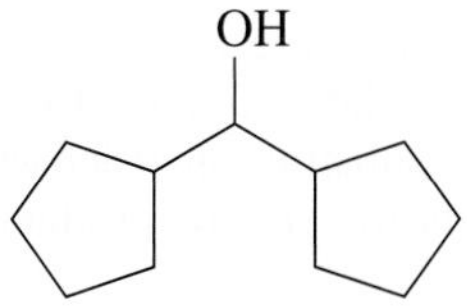 OH on carbon bearing two cyclopentyl groups

(b) 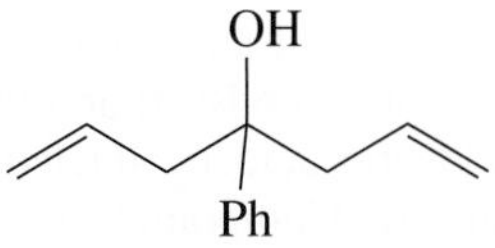 OH and Ph on carbon bearing two allyl-containing chains

Interconversion of functional derivatives

On the last few pages, we have seen that acid chlorides are the most reactive carboxyl derivatives towards nucleophilic acyl substitution and that amides are the least reactive.

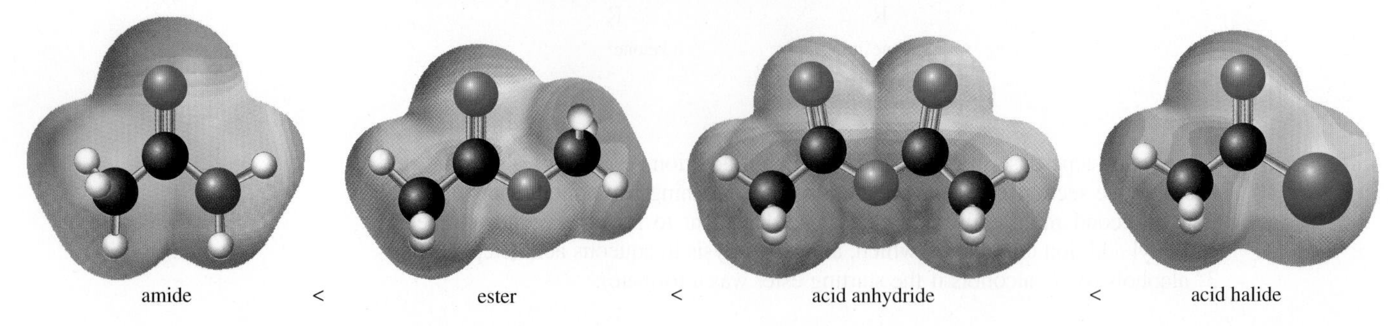

Increasing reactivity towards nucleophilic acyl substitution

Another useful way to think about the relative reactivities of these four functional derivatives of carboxylic acids is summarised in figure 23.4. Any functional group in this figure can be prepared from any functional group above it by treatment with an appropriate oxygen or nitrogen nucleophile. An acid chloride, for example, can be converted to an acid anhydride, an ester, an amide or a carboxylic acid. However, acid anhydrides, esters and amides do not react with chloride ions to give acid chlorides.

Notice that all carboxylic acid derivatives can be converted to carboxylic acids, which in turn can be converted to acid chlorides. Thus, any acid derivative can be used to synthesise another, either directly or via a carboxylic acid.

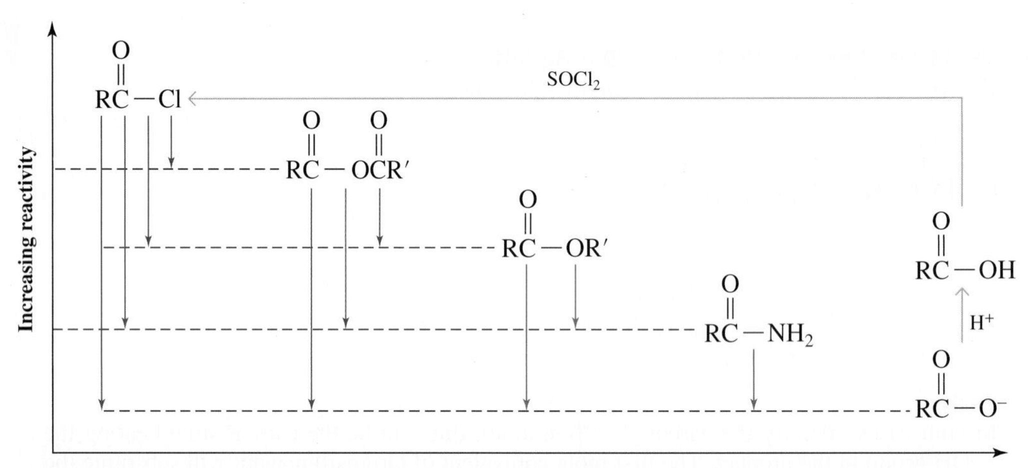

FIGURE 23.4 Relative reactivities of carboxylic acid derivatives towards nucleophilic acyl substitution. A more reactive derivative may be converted to a less reactive derivative by treatment with an appropriate reagent. Treatment of a carboxylic acid with thionyl chloride converts it to the more reactive acid chloride. Carboxylic acids are about as reactive as esters under acidic conditions but are converted to the unreactive carboxylate anions under basic conditions.

WORKED EXAMPLE 23.13

Interconversion of functional groups

Show how to convert phenylacetic acid into each of the following compounds.

(a) $PhCH_2C(=O)OCH_3$ (b) $PhCH_2C(=O)NH_2$

(c) $PhCH_2CH_2NH_2$ (d) $PhCH_2CH_2OH$

Analysis

First, draw the structure of phenylacetic acid and then compare this structure (the starting material) with those of the products. Compounds (a) and (b) are carboxylic acid derivatives, which can be obtained from phenylacetic acid by nucleophilic acyl substitution reactions (more than one step may be required). Compounds (c) and (d), however, are products obtained from the reduction of carboxylic acid derivatives.

Solution

Prepare the methyl ester (a) by Fischer esterification (pp. 1028–9) of phenylacetic acid with methanol. Then treat this ester with ammonia to prepare the amide (b). Alternatively, treat phenylacetic acid with thionyl chloride (figure 23.5) to give an acid chloride, and then treat the acid chloride with two equivalents of ammonia to give amide (b). Reduction of amide (b) by $LiAlH_4$ gives the 1° amine (c). Similar reduction of either phenylacetic acid or ester (a) gives the 1° alcohol (d).

Ph–CH₂–C(=O)–Cl

$SOCl_2$ — CH_3OH — NH_3 (2 eq)

phenylacetic acid —(CH_3OH, H_2SO_4, Fischer esterification)→ Ph–CH₂–C(=O)–OCH_3 (a) —(NH_3)→ Ph–CH₂–C(=O)–NH_2 (b)

phenylacetic acid —(1. $LiAlH_4$ 2. H_2O)→ Ph–CH₂–CH₂–OH (d)

(a) —(1. $LiAlH_4$ 2. H_2O)→ (d)

(b) —(1. $LiAlH_4$ 2. H_2O)→ Ph–CH₂–CH₂–NH_2 (c)

Is our answer reasonable?

Check that the carbon skeleton of phenylacetic acid is retained in each step; that is, the $PhCH_2$—C should be consistent throughout. In (a) and (b), the —C is part of a carbonyl group while in (c) and (d) it is a CH_2 group.

PRACTICE EXERCISE 23.13

Show how to convert (*R*)-3-phenylbutanoic acid to each of the following compounds.

(a) Ph–CH(CH₃)–CH₂–CH₂–OH
(*R*)-3-phenylbutan-l-ol

(b) Ph–CH(CH₃)–CH₂–CH₂–NH_2
(*R*)-3-phenylbutan-l-amine

23.6 Triglycerides

The triglycerides belong to a group of naturally occurring organic compounds called **lipids** that are classified together on the basis of their common solubility properties. Lipids are insoluble in water but soluble in relatively nonpolar, aprotic organic solvents, including diethyl ether and dichloromethane. (*Note:* This classification is based on a *physical property*, not a structural feature.)

Triglycerides or **triacylglycerols** are triesters of glycerol and long-chain carboxylic acids and they play a major role in human and animal biology. They are storage depots of chemical energy in the form of fats; humans store energy in the form of fat globules in adipose tissue.

Animal fats and vegetable oils, the most abundant naturally occurring lipids, are also called triglycerides or triacylglycerols. Hydrolysis of triglycerides in aqueous base, followed by acidification, gives glycerol and three fatty acids (acids with long carbon chains).

$$\underset{\text{a triglyceride}}{\begin{matrix} & CH_2OC(=O)R \\ R'C(=O)OCH & \\ & CH_2OC(=O)R'' \end{matrix}} \xrightarrow[\text{2. HCl, } H_2O]{\text{1. NaOH, } H_2O} \underset{\text{propane-1,2,3-triol (glycerol, glycerin)}}{\begin{matrix} CH_2OH \\ HOCH \\ CH_2OH \end{matrix}} + \underset{\text{fatty acids}}{\begin{matrix} RCOOH \\ R'COOH \\ R''COOH \end{matrix}}$$

Fatty acids

More than 500 different **fatty acids** have been isolated from various cells and tissues. Table 23.4 gives common names and structural formulae for the most abundant fatty acids. The number of carbons in a fatty acid and the number of carbon–carbon double bonds in its hydrocarbon chain are shown by two numbers separated by a colon. In this notation, linoleic acid, for example, is designated as an 18:2 fatty acid; its 18-carbon chain contains two carbon–carbon double bonds. The three most abundant fatty acids in nature are palmitic acid (16:0), stearic acid (18:0) and oleic acid (18:1).

TABLE 23.4 The most abundant fatty acids in animal fats, vegetable oils and biological membranes.

Carbon atoms: double bonds	Structure	Common name	Melting point (°C)
Saturated fatty acids			
12:0[a]	$CH_3(CH_2)_{10}COOH$	lauric acid	44
14:0	$CH_3(CH_2)_{12}COOH$	myristic acid	58
16:0	$CH_3(CH_2)_{14}COOH$	palmitic acid	63
18:0	$CH_3(CH_2)_{16}COOH$	stearic acid	70
20:0	$CH_3(CH_2)_{18}COOH$	arachidic acid	77
Unsaturated fatty acids			
16:1	$CH_3(CH_2)_5CH{=}CH(CH_2)_7COOH$	palmitoleic acid	1
18:1	$CH_3(CH_2)_7CH{=}CH(CH_2)_7COOH$	oleic acid	16
18:2	$CH_3(CH_2)_4(CH{=}CHCH_2)_2(CH_2)_6COOH$	linoleic acid	–5
18:3	$CH_3CH_2(CH{=}CHCH_2)_3(CH_2)_6COOH$	linolenic acid	–11
20:4	$CH_3(CH_2)_4(CH{=}CHCH_2)_4(CH_2)_2COOH$	arachidonic acid	–49

(a) The first number is the number of carbon atoms in the fatty acid; the second is the number of carbon–carbon double bonds in its hydrocarbon chain.

The most abundant fatty acids found in higher plants and animals have several important characteristics.

1. Nearly all fatty acids have an even number of carbon atoms, most between 12 and 20, in an unbranched chain.
2. In most unsaturated fatty acids, the *cis* isomer predominates; the *trans* isomer is rare.
3. Unsaturated fatty acids have lower melting points than their saturated counterparts. The greater the degree of unsaturation, the lower the melting point. Compare, for example, the melting points of these four 18-carbon fatty acids.

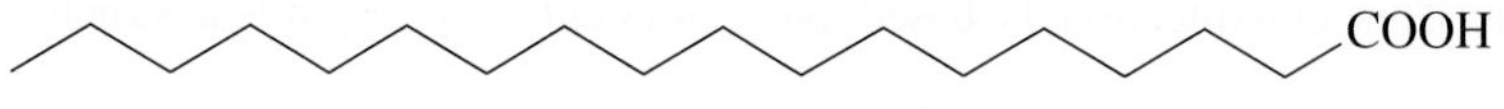

COOH stearic acid (18:0) (mp = 70 °C)

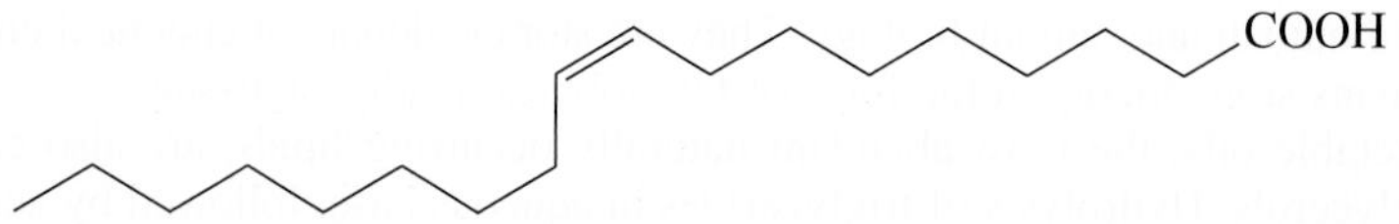

COOH oleic acid (18:1) (mp = 16 °C)

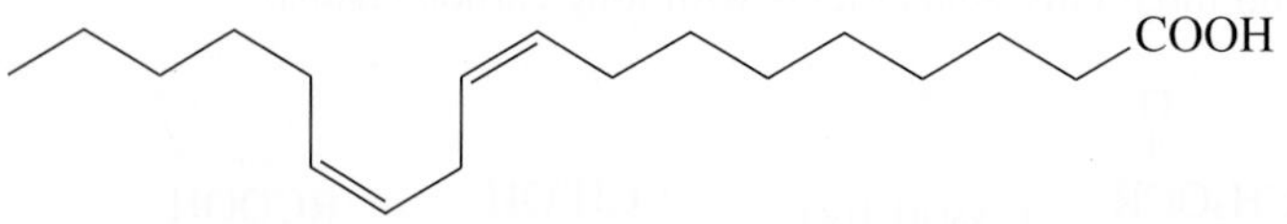

COOH linoleic acid (18:2) (mp = –5 °C)

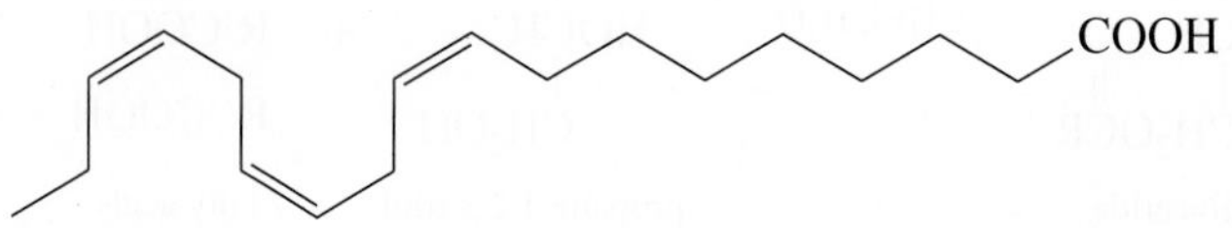

COOH linolenic acid (18:3) (mp = –11 °C)

WORKED EXAMPLE 23.14

Triglyceride structure

Draw the structural formula of a triglyceride derived from one molecule each of palmitic acid, oleic acid and stearic acid, the three most abundant fatty acids in the biological world.

Analysis

The triglyceride is a triester of glycerol (a triol) so each of the three carboxylic acids will form an ester with an —OH group of glycerol. There is more than one possible triglyceride using three different fatty acids.

Solution

In this structure, palmitic acid is esterified at C(1) of glycerol, oleic acid at C(2), and stearic acid at C(3).

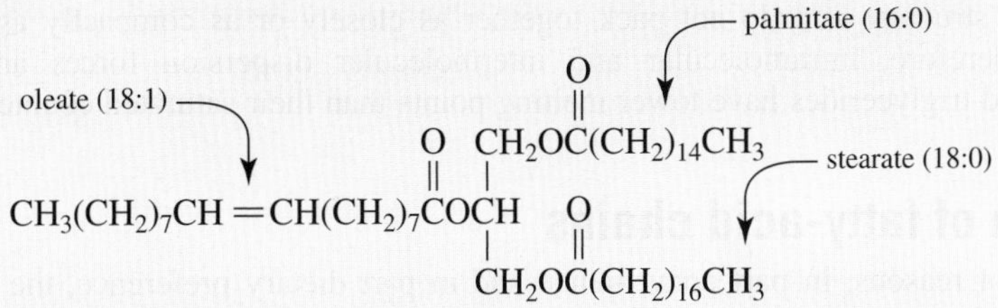

palmitate (16:0)

oleate (18:1)

$CH_2OC(=O)(CH_2)_{14}CH_3$

$CH_3(CH_2)_7CH{=}CH(CH_2)_7C(=O)OCH$

$CH_2OC(=O)(CH_2)_{16}CH_3$

stearate (18:0)

a triglyceride

Is our answer reasonable?

Check that the ester group is drawn the correct way around. That is, is the alkoxy group (—OR) derived from glycerol? Check also that the number of carbon atoms is correct.

PRACTICE EXERCISE 23.14

(a) How many constitutional isomers are possible for a triglyceride containing one molecule each of palmitic acid, oleic acid and stearic acid?
(b) Which of the constitutional isomers that you found in (a) are chiral?

Physical properties

The physical properties of a triglyceride depend on its fatty acid components. In general, the melting point of a triglyceride increases as the number of carbon atoms in its hydrocarbon chains increases and as the number of carbon–carbon double bonds decreases. Triglycerides rich in oleic acid, linoleic acid and other unsaturated fatty acids are generally liquids at room temperature and are called **oils** (e.g. corn oil and olive oil). Triglycerides rich in palmitic, stearic and other saturated fatty acids are generally semisolids or solids at room temperature and are called **fats** (e.g. human fat and butter fat). Fats of land animals typically contain approximately 40% to 50% saturated fatty acids by weight (table 23.5). Most plant oils, on the other hand, contain 20% or less saturated fatty acids and 80% or more unsaturated fatty acids. The notable exception to this generalisation about plant oils are the **tropical oils** (e.g. coconut and palm oils), which are considerably richer in low-molar-mass saturated fatty acids.

TABLE 23.5 Grams of fatty acid per 100 g of triglyceride of several fats and oils.

	Saturated fatty acids			Unsaturated fatty acids	
Fat or oil	Lauric (12:0)	Palmitic (16:0)	Stearic (18:0)	Oleic (18:1)	Linoleic (18:2)
human fat	–	24.0	8.4	46.9	10.2
beef fat	–	27.4	14.1	49.6	2.5
butter fat	2.5	29.0	9.2	26.7	3.6
canola oil	–	3.9	1.9	64.1	18.7
coconut oil	45.4	10.5	2.3	7.5	trace
corn oil	–	10.2	3.0	49.6	34.3
olive oil	–	6.9	2.3	84.4	4.6
palm oil	–	40.1	5.5	42.7	10.3
peanut oil	–	8.3	3.1	56.0	26.0
soybean oil	0.2	9.8	2.4	28.9	50.7
sunflower oil	0.5	6.8	4.7	18.6	68.7

Note: Only the most abundant fatty acids are given; other fatty acids are present in lesser amounts.

The lower melting points of triglycerides that are rich in unsaturated fatty acids are related to differences in three-dimensional shape between the hydrocarbon chains of their unsaturated and saturated fatty acid components. In a saturated triglyceride, the hydrocarbon chains are very flexible and can orient themselves to lie parallel to each other, giving the molecule an ordered, compact shape. Because of this compact three-dimensional shape and the resulting strength of the dispersion forces (section 6.8) between hydrocarbon chains of adjacent molecules, triglycerides that are rich in saturated fatty acids have melting points above room temperature.

The three-dimensional shape of an unsaturated fatty acid, particularly those containing *cis*-(*Z*)- alkenes, is quite different from that of a saturated fatty acid. Recall that unsaturated fatty acids of higher organisms are predominantly of the *cis*-(*Z*)- configuration; *trans*-(*E*)- configurations are rare.

Polyunsaturated triglycerides (triglycerides with several carbon–carbon double bonds) have a less ordered structure and do not pack together as closely or as compactly as saturated triglycerides. Therefore, intramolecular and intermolecular dispersion forces are weaker, so polyunsaturated triglycerides have lower melting points than their saturated counterparts.

Reduction of fatty-acid chains

For a variety of reasons, in part convenience and in part dietary preference, the conversion of oils to fats has become a major industry. The process is called **hardening** of oils and involves the catalytic reduction of some or all of an oil's carbon–carbon double bonds. In practice, the degree of hardening is carefully controlled to produce fats of a desired consistency. The resulting fats are sold for kitchen use. Margarine and other butter substitutes are produced by partial hydrogenation of polyunsaturated oils derived from canola, sunflower, olive and soybean oils (see figure 23.5). To the hardened oil are added β-carotene (to give it a yellow colour and make it look like butter), salt and about 15% milk by volume to form the final emulsion. Vitamins A and D may be added as well. Finally, because the product at this stage is tasteless, acetoin and diacetyl, two compounds that mimic the characteristic flavour of butter, are often added.

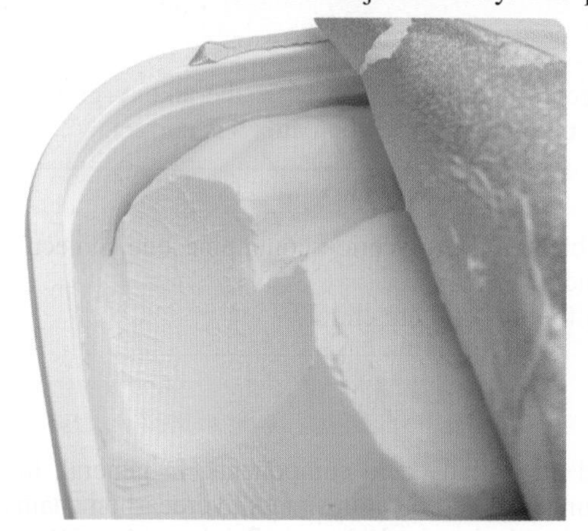

FIGURE 23.5 Margarine is produced by partial hydrogenation of polyunsaturated vegetable oils.

$$CH_3-CH(OH)-C(=O)-CH_3$$

3-hydroxybutanone (acetoin)

$$CH_3-C(=O)-C(=O)-CH_3$$

butanedione (diacetyl)

An unintended consequence of partial hydrogenation of the alkene groups in polyunsaturated oils is the isomerisation of a small proportion of the *cis* alkenes to *trans* alkenes. These are known as *trans*-fats and are of concern as they are believed to increase the risk of coronary heart disease. These *trans*-fats do occur naturally, although only to a small extent. Canola oil naturally has a relatively high level of *trans*-fats, but most other natural oils have very little *trans*-fat. Lamb and mutton also naturally contain moderate levels of *trans*-fats.

Rancidification of fats and oils

Unsaturated fats and oils, when exposed to air, can undergo a process of rancidification. In this process the alkene units are oxidatively cleaved to give low-molar-mass aldehydes and carboxylic acids. These compounds have unpleasant odours and flavours, thus making the food unpalatable.

$$\underset{\text{unsaturated fatty acid}}{RHC=CHR'} \longrightarrow \underset{\text{low-molar-mass aldehydes}}{RHC=O + O=CHR'} \longrightarrow \underset{\text{low-molar-mass acids}}{R(HO)C=O + O=C(OH)R'}$$

Often antioxidants are added to unsaturated fats and oils to retard this oxidation process.

Soaps and detergents

Natural **soaps** are prepared most commonly from a blend of tallow and coconut oil. In the preparation of tallow, the solid fats of cattle are melted with steam, and the tallow layer that forms on the top is removed. The preparation of soaps begins by boiling these triglycerides with sodium hydroxide. The reaction that takes place is called *saponification* (from the Latin *saponem* meaning 'soap').

$$\underset{\text{a triglyceride}}{\begin{array}{c} \mathrm{CH_2OCR} \\ | \\ \mathrm{RCOCH} \\ | \\ \mathrm{CH_2OCR} \end{array}} + 3\mathrm{NaOH} \xrightarrow{\text{saponification}} \underset{\substack{\text{propane-1,2,3-triol} \\ \text{(glycerol, glycerin)}}}{\begin{array}{c} \mathrm{CH_2OH} \\ | \\ \mathrm{CHOH} \\ | \\ \mathrm{CH_2OH} \end{array}} + \underset{\text{sodium soaps}}{3\mathrm{RCO^-Na^+}}$$

At the molecular level, saponification corresponds to base-promoted hydrolysis of the ester groups in triglycerides (section 23.5). The resulting soaps contain mainly the sodium salts of palmitic, stearic and oleic acids from tallow and the sodium salts of lauric and myristic acids from coconut oil.

After hydrolysis is complete, sodium chloride is added to precipitate the soap as thick curds. The water layer is then drawn off, and glycerol is recovered by vacuum distillation. The crude soap contains sodium chloride, sodium hydroxide and other impurities that are removed by boiling the curd in water and precipitating again with more sodium chloride. After several purifications, the soap can be used as an inexpensive industrial soap without further processing. Other treatments transform the crude soap into pH-controlled cosmetic soaps, medicated soaps and the like.

How soap cleans

Soap owes its remarkable cleaning properties to its ability to act as an emulsifying agent. Because the long hydrocarbon chains of natural soaps are insoluble in water, they tend to cluster to minimise their contact with surrounding water molecules. The polar carboxylate groups, by contrast, tend to remain in contact with the surrounding water molecules. Thus, in water, soap molecules spontaneously cluster into **micelles** (figure 23.6). A micelle is a spherical arrangement of organic molecules in water solution clustered so that their hydrophobic parts are buried inside the sphere and their hydrophilic parts are on the surface of the sphere and in contact with water.

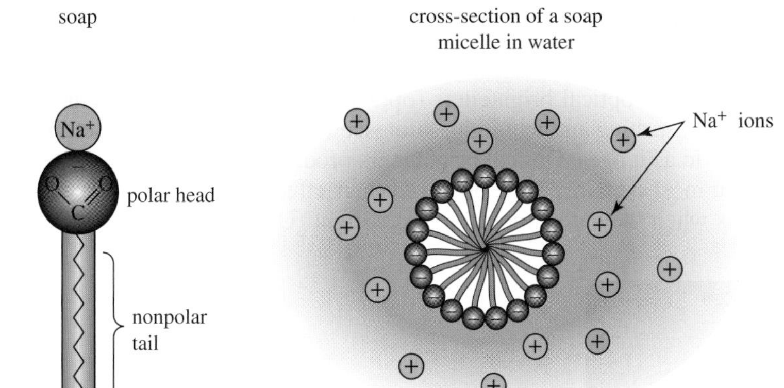

FIGURE 23.6 Soap micelles. Nonpolar (hydrophobic) hydrocarbon chains are clustered in the interior of the micelle, and polar (hydrophilic) carboxylate groups are on the surface of the micelle. Soap micelles repel each other because of their negative surface charges.

Most of the things we commonly think of as dirt (such as grease, oil and fat stains) are nonpolar and insoluble in water. When soap and this type of dirt are mixed together, as in a washing machine, the nonpolar hydrocarbon inner parts of the soap micelles 'dissolve' the nonpolar dirt molecules. In effect, new soap micelles are formed, this time with nonpolar dirt molecules in the centre (figure 23.7). In this way, nonpolar organic grease, oil and fat are 'dissolved' and washed away in the polar wash water.

Soaps, however, have their disadvantages, the main one being that they form water-insoluble salts when used in 'hard' water, which contains Ca(II), Mg(II) or Fe(III) ions.

$$\underset{\substack{\text{a sodium soap} \\ \text{(soluble in water as micelles)}}}{2\mathrm{CH_3(CH_2)_{14}COO^-Na^+}} + \mathrm{Ca^{2+}} \longrightarrow \underset{\substack{\text{calcium salt of a fatty acid} \\ \text{(insoluble in water)}}}{[\mathrm{CH_3(CH_2)_{14}COO^-}]_2\mathrm{Ca^{2+}}} + 2\mathrm{Na^+}$$

These calcium, magnesium and iron salts of fatty acids create problems, including rings around the bathtub, films that spoil the lustre of hair, and greyness and roughness that build up on textiles after repeated washings.

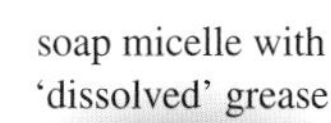

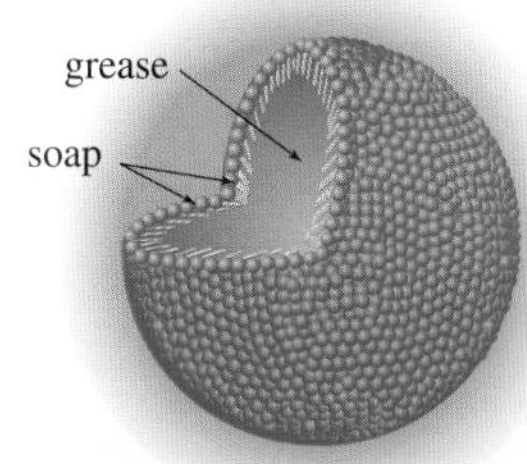

FIGURE 23.7 A soap micelle with a 'dissolved' oil or grease droplet.

Synthetic detergents

After the cleaning action of soaps was understood, chemists were in a position to design synthetic detergents. Molecules of a good detergent, they reasoned, must have a long hydrocarbon chain — preferably 12 to 20 carbon atoms long — and a polar group at one end of the molecule that does not form insoluble salts with the Ca(II), Mg(II) or Fe(III) ions in hard water. These essential characteristics of a soap, they recognised, could be produced in a molecule containing a sulfonate ($—SO_3^-$) group instead of a carboxylate ($—COO^-$) group. Calcium, magnesium and iron salts of monoalkylsulfuric and sulfonic acids are much more soluble in water than comparable salts of fatty acids.

The most widely used synthetic detergents today are the linear alkylbenzenesulfonates (LAS). One of the most common of these is sodium 4-dodecylbenzenesulfonate. To prepare this type of detergent, a linear alkylbenzene is treated with sulfuric acid to form an alkylbenzenesulfonic acid, followed by neutralisation of sulfonic acid with NaOH.

$$CH_3(CH_2)_{10}CH_2—C_6H_5 \xrightarrow[2.\ NaOH]{1.\ H_2SO_4} CH_3(CH_2)_{10}CH_2—C_6H_4—SO_3^-Na^+$$

dodecylbenzene

sodium 4-dodecylbenzenesulfonate (an anionic detergent)

The product is mixed with building agents and spray dried to give a smooth, flowing powder. The most common builder is sodium silicate. Alkylbenzenesulfonate detergents were introduced in the late 1950s, and today they command close to 90% of the market once held by natural soaps.

Among the most common additives to detergent preparations are foam stabilisers (to encourage longer lasting bubbles), bleaches and optical brighteners. The amide prepared from dodecanoic acid (lauric acid) and 2-aminoethanol (ethanolamine) is a common foam stabiliser added to liquid soaps, but not laundry detergents for obvious reasons — imagine a top-loading washing machine with foam spewing out of the lid! The most common bleach used in washing powders is sodium perborate, which decomposes at temperatures above 50 °C to give hydrogen peroxide, the actual bleaching agent.

$$CH_3(CH_2)_{10}\overset{O}{\overset{\|}{C}}NHCH_2CH_2OH$$

N-(2-hydroxyethyl)dodecanamide (a foam stabiliser)

$$O{=}B{-}O{-}O^-Na^+\cdot 4H_2O$$

sodium perborate tetrahydrate (a bleach)

Also added to laundry detergents are optical brighteners (optical bleaches). These substances are absorbed into fabrics and, after absorbing ambient light, fluoresce with a blue colour, offsetting the yellow colour acquired by fabric as it ages. Optical brighteners produce a 'whiter-than-white' appearance (see figure 23.8). You most certainly have observed their effects if you have seen the glow of white T-shirts or blouses when they are exposed to black light (UV radiation).

FIGURE 23.8 Optical brighteners convert UV light to brighter fluorescent colours.

SUMMARY

LO 1 Structure and bonding

The carboxylic acid functional group is the carboxyl group, —COOH. The acid halide functional group is an acyl group bonded to a halogen atom. The most common and widely used of the acid halides are the acid chlorides. The carboxylic anhydride functional group is two acyl groups bonded to an oxygen atom. The ester functional group is an acyl group bonded to an —OR or —OAr group. A cyclic ester is given the name lactone. The amide functional group is an acyl group bonded to a trivalent nitrogen atom. A cyclic amide is given the name lactam.

LO 2 Nomenclature

IUPAC names of carboxylic acids are derived from the parent alkane by dropping the suffix *-e* and adding *-oic acid*. Dicarboxylic acids are named as *-dioic acids*. Acid halides are named by changing the suffix *-ic acid* in the name of the parent carboxylic acid to *-yl halide*. Acid anhydrides are named by replacing the word *acid* with *anhydride*. When the anhydride is not symmetrical, both acid groups are named. In esters, the alkyl or aryl group bonded to oxygen is named first, followed as a separate word by the name of the acid in which the suffix *-ic acid* is replaced by the suffix *-ate*. Amides are named by dropping the suffix *-oic acid* from the IUPAC name of the parent acid, or *-ic acid* from its common name, and adding *-amide*.

LO 3 Physical properties

Carboxylic acids are polar compounds that associate by hydrogen bonding into dimers in the liquid and solid states. Carboxylic acids have higher boiling points and are more soluble in water than alcohols, aldehydes, ketones and ethers of comparable molar mass. A carboxylic acid consists of two regions of different polarity: a polar, hydrophilic carboxyl group, which increases solubility in water, and a nonpolar, hydrophobic hydrocarbon chain, which decreases solubility in water. The first four aliphatic carboxylic acids are infinitely soluble in water, because the hydrophilic carboxyl group more than offsets the hydrophobic hydrocarbon chain. As the size of the carbon chain increases, however, the hydrophobic group becomes dominant and solubility in water decreases.

LO 4 Preparation of carboxylic acids

Carboxylic acids can be prepared by the oxidation of primary alcohols, aldehydes and alkylbenzene derivatives. They can be prepared by the reaction of Grignard reagents with carbon dioxide and by the hydrolysis of nitriles and carboxylic acid derivatives.

LO 5 Reactions of carboxylic acids and derivatives

Values of pK_a for aliphatic carboxylic acids range from 4.0 to 5.0. Electron-withdrawing substituents near the carboxyl group increase acidity in both aliphatic and aromatic carboxylic acids.

A common reaction theme of functional derivatives of carboxylic acids is nucleophilic acyl addition to the carbonyl carbon atom to form a tetrahedral carbonyl addition intermediate, which then collapses to regenerate the carbonyl group. The result is nucleophilic acyl substitution. Listed in order of increasing reactivity towards nucleophilic acyl substitution, these functional derivatives are:

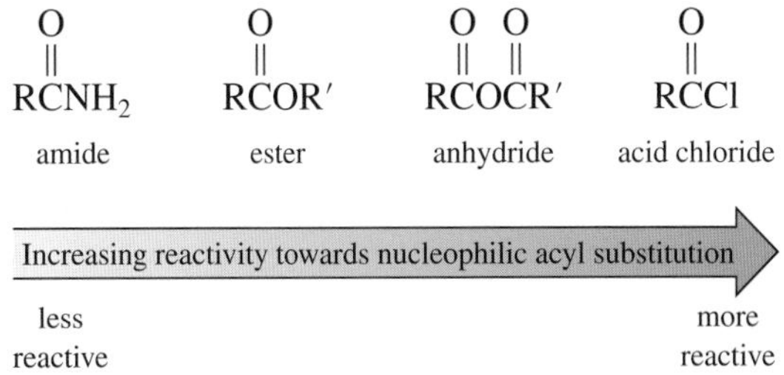

$RC(=O)NH_2$	$RC(=O)OR'$	$RC(=O)OC(=O)R'$	$RC(=O)Cl$
amide	ester	anhydride	acid chloride

Increasing reactivity towards nucleophilic acyl substitution →
less reactive … more reactive

Any more reactive functional derivative can be directly converted to any less reactive functional derivative by reaction with an appropriate oxygen or nitrogen nucleophile.

LO 6 Triglycerides

Lipids are a heterogeneous class of compounds grouped together on the basis of their solubility properties; they are insoluble in water but soluble in diethyl ether and dichloromethane. Carbohydrates, amino acids and proteins are largely insoluble in these organic solvents.

Triglycerides (triacylglycerols), the most abundant lipids, are triesters of glycerol and fatty acids. Fatty acids are long-chain carboxylic acids derived from the hydrolysis of fats, oils and the phospholipids of biological membranes. The melting point of a triglyceride increases as (1) the length of its hydrocarbon chains increases and (2) its degree of saturation increases. Triglycerides rich in saturated fatty acids are generally solids at room temperature; those rich in unsaturated fatty acids are generally oils at room temperature.

Soaps are sodium or potassium salts of fatty acids. In water, soaps form micelles, which 'dissolve' nonpolar organic grease and oil. Natural soaps precipitate as water-insoluble salts with Mg^{2+}, Ca^{2+} and Fe^{3+} ions in hard water. The most common and most widely used synthetic detergents are linear alkylbenzenesulfonates.

KEY CONCEPTS AND EQUATIONS

Preparation of carboxylic acids by oxidation of primary alcohols and aldehydes
(section 23.4)

Primary alcohols and aldehydes are readily oxidised to carboxylic acids with a variety of oxidising agents.

$$(CH_3)_2CHCH_2CH_2OH \xrightarrow{\text{oxidation}} (CH_3)_2CHCH_2COOH$$

Preparation of carboxylic acids by oxidation of alkylbenzenes
(section 23.4)

Alkylbenzenes are oxidised to benzoic acids under vigorous oxidising conditions.

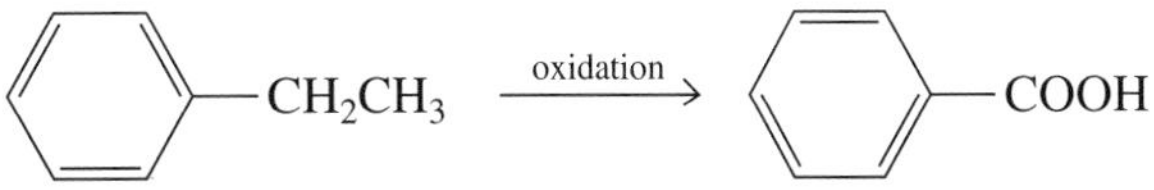

$$C_6H_5CH_2CH_3 \xrightarrow{\text{oxidation}} C_6H_5COOH$$

Preparation of carboxylic acids by carbonation of Grignard reagents
(section 23.4)

Carboxylic acids can be prepared by treatment of Grignard reagents with CO_2.

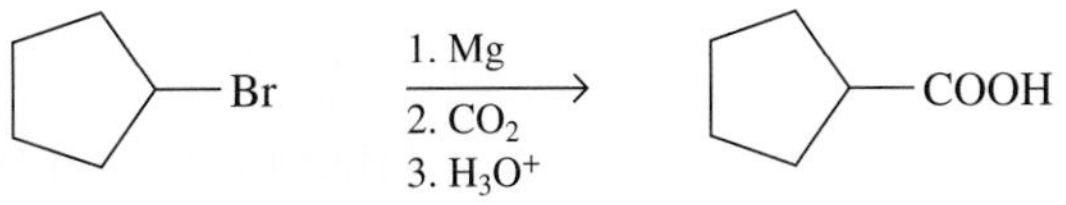

$$C_5H_9Br \xrightarrow[\text{2. } CO_2 \text{ 3. } H_3O^+]{\text{1. Mg}} C_5H_9COOH$$

Preparation of carboxylic acids by formation and hydrolysis of nitriles
(section 23.4)

Carboxylic acids can be prepared from alkyl halides by formation and subsequent hydrolysis of nitriles.

$$CH_3CH_2CH_2CH_2Br \xrightarrow[2.\ H_2O/H_3O^+]{1.\ NaCN} CH_3CH_2CH_2CH_2COOH$$

Acidity of carboxylic acids
(section 23.5)

Values of pK_a for most unsubstituted aliphatic and aromatic carboxylic acids range from 4 to 5.

$$CH_3COH + H_2O \rightleftharpoons CH_3CO^- + H_3O^+ \quad pK_a = 4.74$$

Substitution by electron-withdrawing groups decreases the pK_a value (increases acidity).

Reaction of carboxylic acids with bases
(section 23.5)

Carboxylic acids form water-soluble salts with alkali metal hydroxides, carbonates and bicarbonates, as well as with ammonia and amines.

$$C_6H_5COOH + NaOH \xrightarrow{H_2O} C_6H_5COO^-Na^+ + H_2O$$

Conversion to acid halides
(section 23.5)

Acid chlorides, the most common and widely used of the acid halides, are prepared by treating carboxylic acids with thionyl chloride.

$$CH_3CH_2CH_2COOH + SOCl_2 \longrightarrow CH_3CH_2CH_2COCl + SO_2 + HCl$$

Fischer esterification
(section 23.5)

Fischer esterification is reversible.

$$CH_3COOH + HOCH_2CH_2CH_3 \underset{}{\overset{H_2SO_4}{\rightleftharpoons}} CH_3COOCH_2CH_2CH_3 + H_2O$$

To force the reaction to the right, use excess alcohol (and/or remove water).

Reaction of acid chlorides with alcohols
(section 23.5)

Treatment of an acid chloride with an alcohol gives an ester and HCl.

$$CH_3CH_2CH_2COCl + CH_3OH \longrightarrow CH_3CH_2CH_2COOCH_3 + HCl$$

Reaction of acid anhydrides with alcohols
(section 23.5)

Treatment of an acid anhydride with an alcohol gives an ester and a carboxylic acid.

$$CH_3COCCH_3 + CH_3CH_2OH \longrightarrow CH_3COCH_2CH_3 + CH_3COH$$

Reaction of esters with alcohols
(section 23.5)

Treatment of an ester with an alcohol in the presence of an acid catalyst results in transesterification: that is, the replacement of one —OR group by a different —OR group.

$$C_6H_{11}COOCH_3 + HOCH_2CH_2CH(CH_3)_2 \overset{H_2SO_4}{\rightleftharpoons} C_6H_{11}COOCH_2CH_2CH(CH_3)_2 + CH_3OH$$

Hydrolysis of acid chlorides
(section 23.5)

Low-molar-mass acid chlorides react vigorously with water; high-molar-mass acid chlorides react less rapidly.

$$CH_3CCl + H_2O \longrightarrow CH_3COH + HCl$$

Hydrolysis of acid anhydrides
(section 23.5)

Low-molar-mass acid anhydrides react readily with water; high-molar-mass acid anhydrides react less rapidly.

$$CH_3COCCH_3 + H_2O \longrightarrow CH_3COH + HOCCH_3$$

Hydrolysis of esters *(section 23.5)*

Esters are hydrolysed only in the presence of base or acid; base is required in an equimolar amount; acid is a catalyst.

$$CH_3\overset{O}{\overset{\|}{C}}O-C_6H_{11} + NaOH \xrightarrow{H_2O} CH_3\overset{O}{\overset{\|}{C}}O^-Na^+ + HO-C_6H_{11}$$

$$CH_3\overset{O}{\overset{\|}{C}}O-C_6H_{11} + H_2O \underset{}{\overset{HCl}{\rightleftharpoons}} CH_3\overset{O}{\overset{\|}{C}}OH + HO-C_6H_{11}$$

Hydrolysis of amides *(section 23.5)*

Either acid or base is required in an amount equivalent to that of the amide.

$$CH_3CH_2CH_2\overset{O}{\overset{\|}{C}}NH_2 + H_2O + HCl \xrightarrow[heat]{H_2O} CH_3CH_2CH_2\overset{O}{\overset{\|}{C}}OH + NH_4^+Cl^-$$

$$CH_3\overset{O}{\overset{\|}{C}}NH-C_6H_5 + NaOH \xrightarrow[heat]{H_2O} CH_3\overset{O}{\overset{\|}{C}}O^-Na^+ + H_2N-C_6H_5$$

Reaction of acid chlorides with ammonia or amines *(section 23.5)*

This reaction requires 2 mole equivalents of ammonia or amine — 1 mole equivalent to form the amide and 1 mole equivalent to neutralise the HCl by-product.

$$CH_3\overset{O}{\overset{\|}{C}}Cl + 2NH_3 \longrightarrow CH_3\overset{O}{\overset{\|}{C}}NH_2 + NH_4^+Cl^-$$

Reaction of acid anhydrides with ammonia or amines *(section 23.5)*

This reaction requires 2 mole equivalents of ammonia or amine — 1 mole equivalent to form the amide and 1 mole equivalent to neutralise the carboxylic acid by-product.

$$CH_3\overset{O}{\overset{\|}{C}}O\overset{O}{\overset{\|}{C}}CH_3 + 2NH_3 \longrightarrow CH_3\overset{O}{\overset{\|}{C}}NH_2 + CH_3\overset{O}{\overset{\|}{C}}O^-NH_4^+$$

Reaction of esters with ammonia or amines *(section 23.5)*

Treatment of an ester with ammonia, a 1° amine or a 2° amine gives an amide.

$$PhCH_2COOCH_2CH_3 + NH_3 \longrightarrow PhCH_2CONH_2 + HOCH_2CH_3$$

ethyl phenylacetate → phenylacetamide + ethanol

Reduction by lithium aluminium hydride *(section 23.5)*

Lithium aluminium hydride reduces a carboxyl group to a primary alcohol.

$$C_5H_7-\overset{O}{\overset{\|}{C}}OH \xrightarrow[2.\ H_2O]{1.\ LiAlH_4} C_5H_7-CH_2OH$$

Reduction of esters *(section 23.5)*

Reduction of esters by lithium aluminium hydride gives two alcohols.

$$PhCH(CH_3)COOCH_3 \xrightarrow[2.\ H_2O,\ HCl]{1.\ LiAlH_4,\ ether} PhCH(CH_3)CH_2OH + CH_3OH$$

methyl 2-phenylpropanoate → 2-phenylpropan-1-ol + methanol

Reduction of amides *(section 23.5)*

Reduction of an amide by lithium aluminium hydride gives an amine.

$$CH_3(CH_2)_6CONH_2 \xrightarrow[2.\ H_2O]{1.\ LiAlH_4} CH_3(CH_2)_6CH_2NH_2$$

octanamide → octan-1-amine

Reaction of esters with Grignard reagents *(section 23.5)*

Treatment of esters with Grignard reagents, followed by hydrolysis, gives 3° alcohols, except for formate esters which give 2° alcohols.

$$C_6H_{11}COOCH_3 \xrightarrow[2.\ H_2O,\ HCl]{1.\ 2CH_3CH_2MgBr} C_6H_{11}C(OH)(CH_2CH_3)_2$$

KEY TERMS

acid anhydride *p. 1012*
acid halide *p. 1012*
acyl group *p. 1012*
amide *p. 1013*
benzylic carbon *p. 1020*
benzylic hydrogen *p. 1020*
carboxylic acid group *p. 1012*
carboxyl group *p. 1012*
ester *p. 1013*
fat *p. 1045*
fatty acid *p. 1044*
Fischer esterification *p. 1028*
hardening *p. 1046*
hydrophilic *p. 1018*
hydrophobic *p. 1018*
lactam *p. 1013*
lactone *p. 1013*
lipid *p. 1043*
micelle *p. 1047*
nitrile *p. 1021*
nucleophilic acyl substitution *p. 1026*
oil *p. 1045*
saponification *p. 1033*
soap *p. 1047*
transesterification *p. 1031*
triacylglycerol *p. 1043*
triglyceride *p. 1043*
tropical oil *p. 1045*

REVIEW QUESTIONS

LO 1 Structure and bonding

23.1 Draw structural formulae for the four carboxylic acids with the molecular formula $C_5H_{10}O_2$. Which of these carboxylic acids is chiral?

23.2 Draw the structural formulae for the seven carboxylic acids with the molecular formula $C_6H_{12}O_2$. Which of these is chiral?

23.3 Draw structural formulae for: (a) an ester, (b) an acid chloride, (c) an anhydride and (d) an amide.

23.4 On the structures you have drawn in question 23.3, show the polarity of the atoms in the functional groups using the δ+/δ− notation.

23.5 Identify the functional groups in each of the following compounds.

(a) 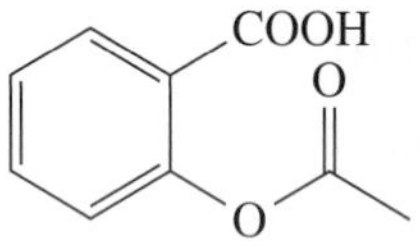 COOH O O

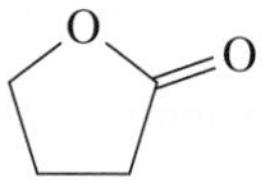

(b) 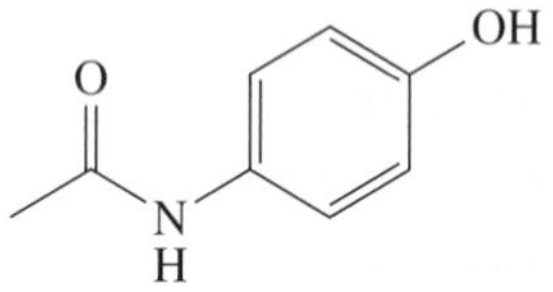 O OH N H

(c) O O

23.6 Identify the functional groups in each of the following compounds.

(a) 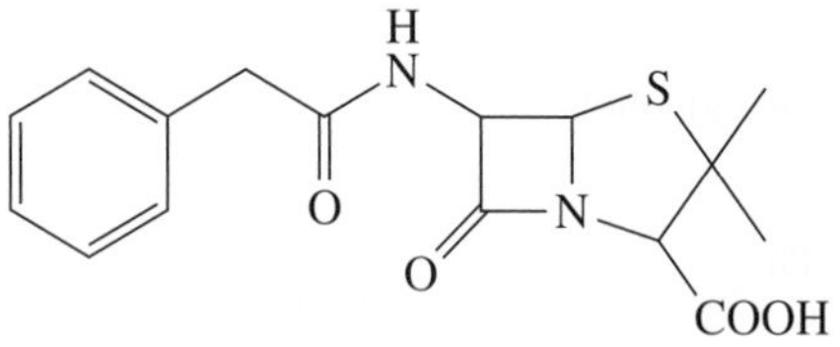 H N S O O N COOH

(b) 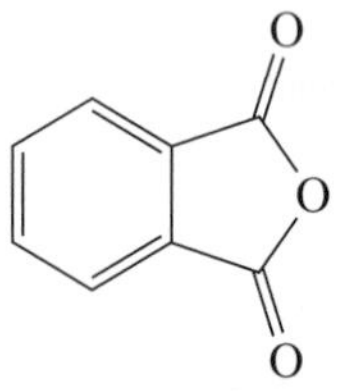 O O O

LO 2 Nomenclature

23.7 Write the IUPAC name for each of the following compounds.

(a) 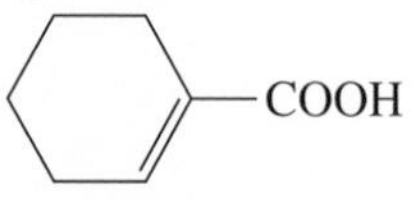 COOH

(b) 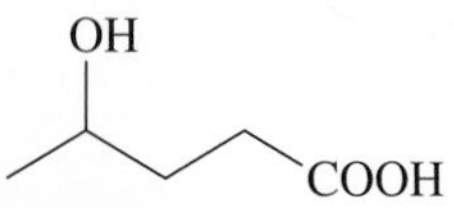 OH COOH

(c) 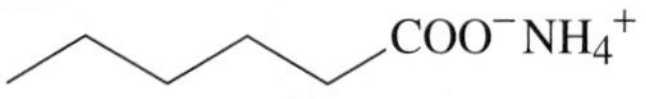 $COO^-NH_4^+$

(d) 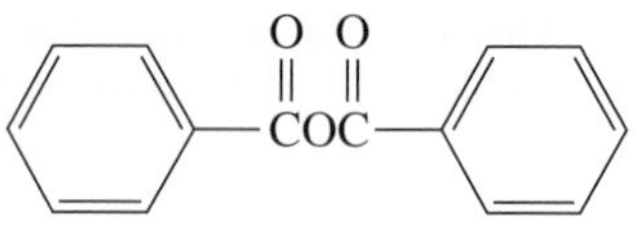 O O ‖ ‖ COC

(e) $CH_3(CH_2)_4\overset{O}{\overset{\|}{C}}NHCH_3$

(f) $CH_2(COOCH_2CH_3)_2$

23.8 Write the IUPAC name for each of the following compounds.

(a) 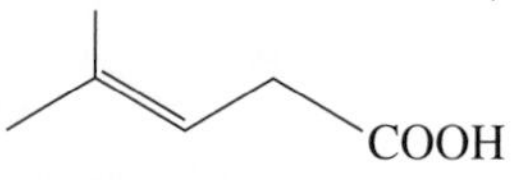 COOH

(b) 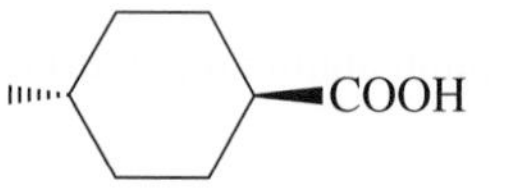 COOH

(c) 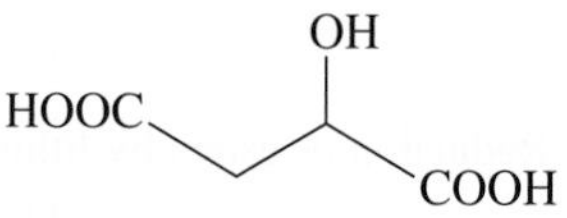 HOOC OH COOH

(d) 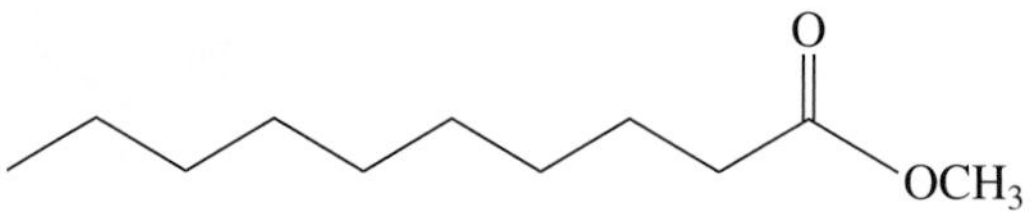 O OCH_3

(e) 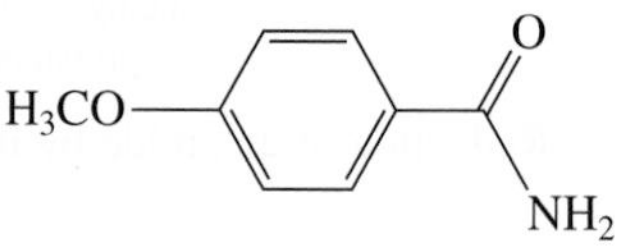 H_3CO O NH_2

(f) 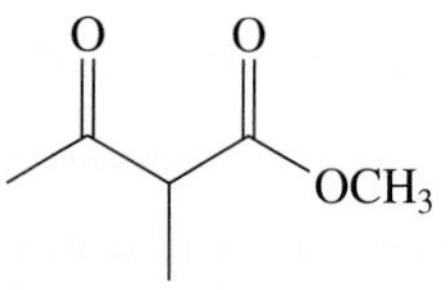 O O OCH_3

23.9 Draw the structural formula for each of the following carboxylic acids.

(a) 4-nitrophenylacetic acid
(b) 4-aminopentanoic acid
(c) 3-chloro-4-phenylbutanoic acid
(d) *cis*-hex-3-enedioic acid

23.10 Draw the structural formula for each of the following carboxylic acids.
(a) 2,2-diethylpentanoic acid
(b) 6-oxohexanoic acid
(c) 2,3-dichlorobutanoic
(d) pent-3-ynedioic acid

23.11 Draw the structural formula for each of the following carboxylic acid derivatives.
(a) dimethyl carbonate
(b) *p*-nitrobenzamide
(c) octanoyl chloride
(d) diethyl oxalate (diethyl ethanedioate)

23.12 Draw the structural formula for each of the following carboxylic acid derivatives.
(a) propyl butanoate
(b) benzoic anhydride
(c) butyl 3-bromopropanoate
(d) *N,N*-dimethylformamide

23.13 Draw the structural formula for methyl linoleate. Be certain to show the correct configuration of groups around each carbon–carbon double bond.

23.14 Draw the structural formula for each of the following carboxylic acid salts.
(a) potassium 4-methoxybenzoate
(b) sodium butanoate
(c) diammonium succinate
(d) calcium oxalate
(e) magnesium acetate
(f) potassium hydrogen phthalate (1,2-benzenedicarboxylic acid, mono potassium salt)

LO 3 Physical properties

23.15 Acetic acid and methyl formate are constitutional isomers. Both are liquids at room temperature, one with a boiling point of 32 °C, the other with a boiling point of 118 °C. Which of the two has the higher boiling point?

23.16 Butanoic acid has a boiling point of 164 °C, whereas its ethyl ester has a boiling point of 121 °C. Account for the fact that the boiling point of butanoic acid acid is higher than that of its ethyl ester, even though butanoic acid has a lower molar mass.

23.17 Arrange the compounds in each of the following sets in order of increasing boiling point.
(a) $CH_3(CH_2)_5COOH$, $CH_3(CH_2)_6CHO$, $CH_3(CH_2)_6CH_2OH$
(b) CH_3CH_2COOH, $CH_3CH_2CH_2CH_2OH$, $CH_3CH_2OCH_2CH_3$

23.18 Arrange the compounds in each of the following sets in order of increasing solubility in water.
(a) $CH_3(CH_2)_6OH$, $CH_3(CH_2)_4COOH$, $CH_3(CH_2)_5CHO$
(b) $CH_3(CH_2)_2COOH$, $CH_3(CH_2)_5OH$, $CH_3(CH_2)_2O(CH_2)_2CH_3$

23.19 Decanoic acid is not very soluble in water. Would it be more soluble under acidic or basic conditions? Explain your answer.

23.20 Characterise the structural features necessary to make a good synthetic detergent.

LO 4 Preparation of carboxylic acids

23.21 Draw a structural formula for the product formed by treating each of the following compounds with warm chromic acid, H_2CrO_4.

(a) $CH_3(CH_2)_4CH_2OH$

(b) 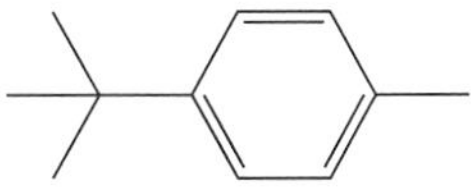 [structure: 1-*tert*-butyl-4-methylbenzene]

(c) 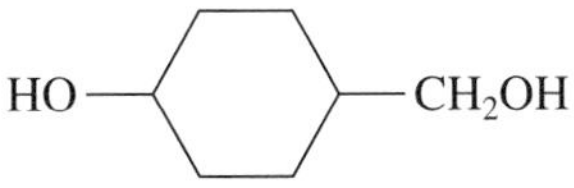 HO–(cyclohexane)–CH_2OH

23.22 Give the structure of the carboxylic acids formed in each of the following reactions.

(a) 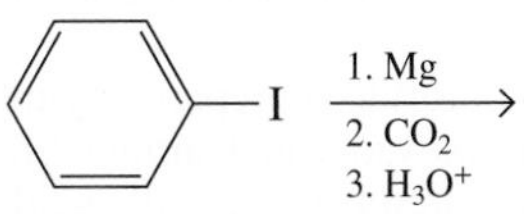 [iodobenzene] $\xrightarrow{\text{1. Mg; 2. } CO_2\text{; 3. } H_3O^+}$

(b) 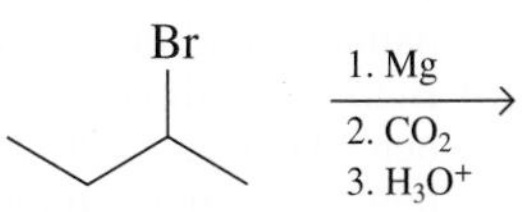 [2-bromobutane] $\xrightarrow{\text{1. Mg; 2. } CO_2\text{; 3. } H_3O^+}$

LO 5 Reactions of carboxylic acids and derivatives

23.23 Which is the stronger acid in each of the following pairs?
(a) phenol ($pK_a = 9.89$) or benzoic acid ($pK_a = 4.20$)
(b) lactic acid ($K_a = 1.4 \times 10^{-4}$) or ascorbic acid ($K_a = 7.9 \times 10^{-5}$)

23.24 Arrange these compounds in order of increasing acidity: phenylacetic acid, 2-phenylethanol, 2-ethylphenol

23.25 Assign the appropriate pK_a to each of the acids in the following sets.

(a) 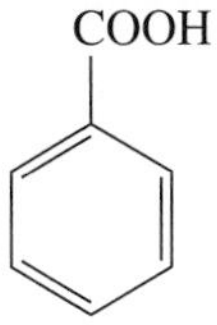 [benzoic acid: COOH on benzene] and [COOH on benzene with para NO_2] (pK_a = 4.20 and 3.43)

(b) 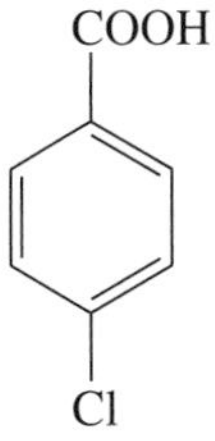 [COOH on benzene with para Cl] and [COOH on benzene with para OCH_3] (pK_a = 4.00 and 4.50)

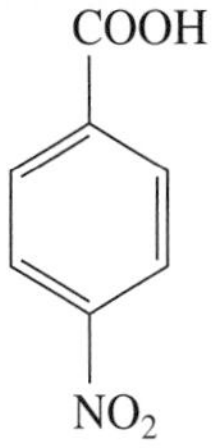

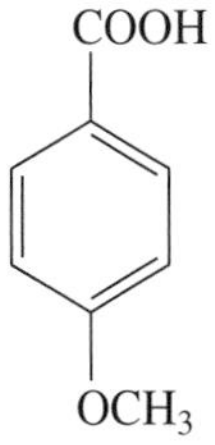

23.26 Assign the appropriate pK_a to each of the acids in the following sets.

(a) $CH_3C(=O)CH_2COOH$ and $CH_3C(=O)COOH$ (pK_a = 3.58 and 2.39)

(b) $CH_3CH(OH)COOH$ and CH_3CH_2COOH (pK_a = 4.78 and 3.85)

23.27 Rank the following in order of increasing acidity: butanoic acid, 2-chlorobutanoic acid, 3-chlorobutanoic acid, 4-chlorobutanoic acid.

23.28 Rank the following in order of increasing acidity: 2-bromopentanoic acid, 2-chloropentanoic acid, 2-fluoropentanoic acid, 2-iodopentanoic acid.

23.29 At room temperature, nicotinic acid has a moderate solubility in water of approximately 1 gram per 60 mL. How would its solubility change if the pH of the water was (a) decreased to pH = 2 and (b) increased to pH = 10? Use equations to explain your answer.

[structure: pyridine ring with COOH]

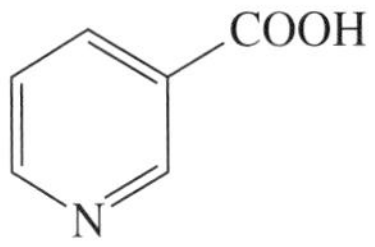

nicotinic acid
(niacin)

23.30 Hippuric acid is only slightly soluble (4 g/L) in water at room temperature. How would its solubility change if the pH of the water was (a) decreased to pH = 2 and (b) increased to pH = 10? Use equations to explain your answer.

[structure: $C_6H_5C(=O)NHCH_2COOH$]

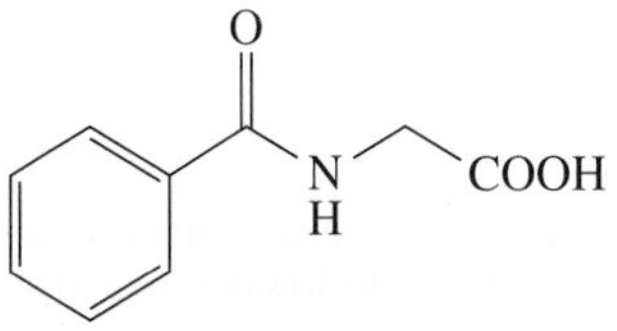

hippuric acid

23.31 Give the expected organic products formed (if any) when phenylacetic acid, $PhCH_2COOH$, is treated with each of the following reagents.
(a) $SOCl_2$
(b) $NaHCO_3$, H_2O
(c) NH_3, H_2O
(d) $LiAlH_4$ then H_2O
(e) CH_3OH, H_2SO_4 (catalyst)

23.32 Give the expected organic products formed (if any) when 4-methoxybenzoic acid is treated with each of the following reagents.
(a) PCl_3
(b) $NaOH$, H_2O
(c) $N(CH_2CH_3)_3$
(d) $NaBH_4$ then H_2O
(e) $CH_3CH_2CH_2OH$, H_2SO_4 (catalyst)

23.33 Arrange the following compounds in order of increasing reactivity towards nucleophilic acyl substitution.

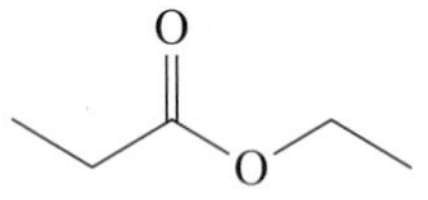

A 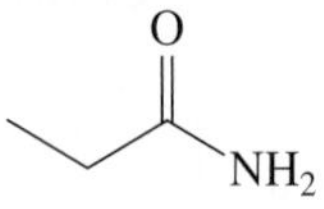 C 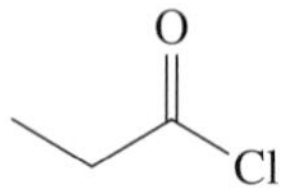 B 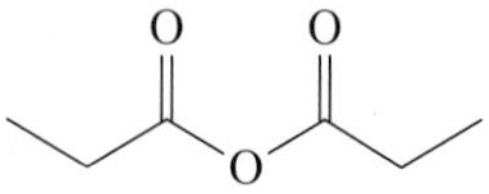 D

O O NH2 Cl

23.34 Complete the following Fischer esterification equations. (Assume excess carboxylic acid.)

(a) + $\xrightleftharpoons{H_2SO_4}$

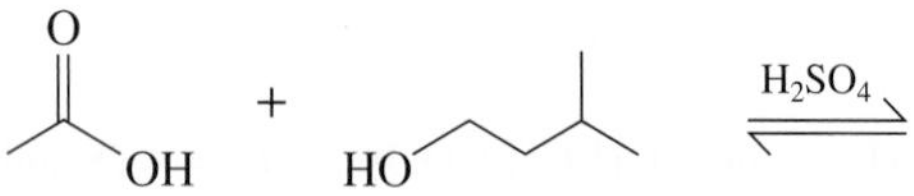

(b) HO OH OH + OH $\xrightleftharpoons{H_2SO_4}$

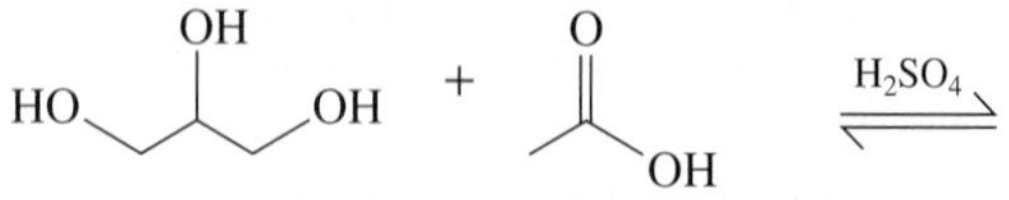

(c) HO OH + OH $\xrightleftharpoons{H_2SO_4}$

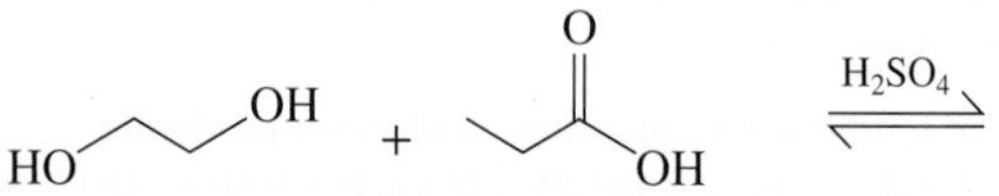

23.35 Identify the carboxylic acid and the alcohol that are necessary in order to make each of the following compounds via a Fischer esterification.

(a)

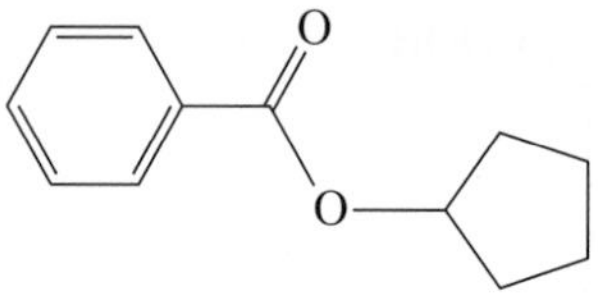

(b)

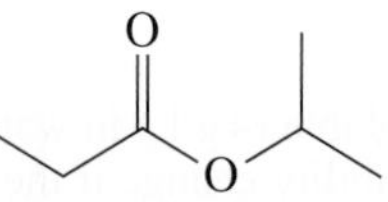

(c) $CH_3CH_2CO_2C(CH_3)_3$

23.36 Write the product(s) of the treatment of benzoyl chloride with an excess of each of the following reagents.
(a) propan-2-ol
(b) methylamine

23.37 Write the product(s) of the treatment of propanoic anhydride with an excess of each of the following reagents.
(a) butan-2-ol
(b) dimethylamine

23.38 What product(s) is(are) formed when methyl-4-methylbenzoate is treated with the following reagents?
(a) H_2O, NaOH, heat
(b) $LiAlH_4$ then H_2O
(c) H_2O, H_2SO_4, heat
(d) $CH_3CH_2CH_2CH_2NH_2$
(e) C_6H_5MgBr (2 moles) then H_2O, HCl

23.39 What product(s) is(are) formed when benzamide is treated with the following reagents?
(a) H_2O, HCl, heat
(b) NaOH, H_2O, heat
(c) $LiAlH_4$ then H_2O

23.40 Show how to convert pentanoic acid into each of the following compounds. (More than one step may be required.)
(a) methyl pentanoate
(b) 2-methylhexan-2-ol
(c) N-methylpentanamide
(d) ammonium butanoate
(e) pentanoyl chloride
(f) pentanal

23.41 Show how to convert ethyl benzoate into each of the following compounds. (More than one step may be required.)
(a) benzyl alcohol
(b) benzoic acid
(c) benzoyl chloride
(d) benzamide
(e) 3-phenylpentan-3-ol

LO 6 Triglycerides

23.42 Define the term 'hydrophobic'.

23.43 What is meant by the term 'hardening' as applied to vegetable oils?

23.44 Explain why the melting points of unsaturated fatty acids are lower than those of saturated fatty acids.

23.45 Which would you expect to have the higher melting point, glyceryl trioleate or glyceryl trilinoleate?

23.46 What amount of H_2 is used in the catalytic hydrogenation of 1 mole of a triglyceride derived from glycerol, stearic acid, linoleic acid and linolenic acid?

23.47 What amount of hydrogen is used in the catalytic hydrogenation of 1 mole of a triglyceride derived from glycerol, stearic acid, linoleic acid and arachidonic acid?

23.48 Draw the structure of a triglyceride derived from glycerol, palmitic acid, and two equivalents of oleic acid that is chiral. On this structure identify the hydrophobic and hydrophylic regions.

23.49 Draw the structure of a triglyceride derived from glycerol, stearic acid, and two equivalents of oleic acid that is chiral. If this was hydrogenated, would the product be chiral? Explain.

23.50 Show how to convert palmitic acid (hexadecanoic acid) into each of the following.
(a) isopropyl palmitate
(b) palmitoyl chloride
(c) hexadecan-1-ol (cetyl alcohol)
(d) hexadecan-1-amine
(e) hexadecanamide

23.51 Palmitic acid (hexadecanoic acid, 16:0) is the source of the hexadecyl (cetyl) group in the following compounds.

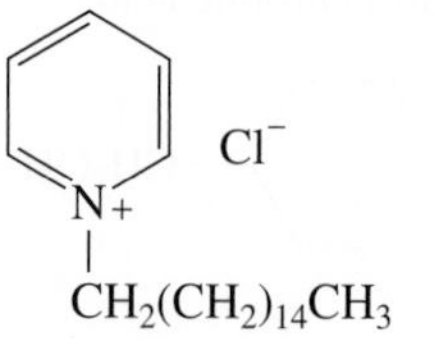

Cl^- N^+ $CH_2(CH_2)_{14}CH_3$

CH_3 Cl^- $CH_2\overset{+}{N}CH_2(CH_2)_{14}CH_3$ CH_3

cetylpyridinium chloride 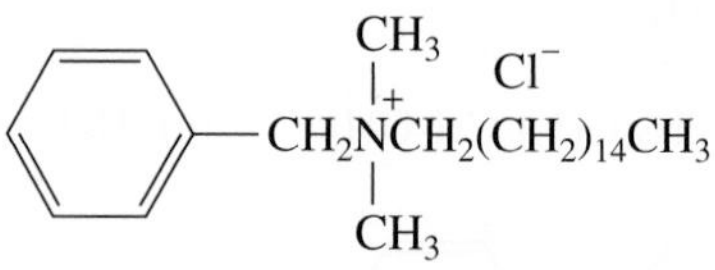 benzylcetyldimethylammonium chloride

Each compound is a mild, surface-acting germicide and fungicide and is used as a topical antiseptic and disinfectant.
(a) Cetylpyridinium chloride is prepared by treating pyridine with 1-chlorohexadecane (cetyl chloride). Show how to convert palmitic acid to cetyl chloride.
(b) Benzylcetyldimethylammonium chloride is prepared by treating benzyl chloride with *N*,*N*-dimethylhexadecan-1-amine. Show how this tertiary amine can be prepared from palmitic acid.

23.52 Megatomoic acid, the sex attractant of the female black carpet beetle, has the structure: **LO 1,2**

$$CH_3(CH_2)_7CH{=}CHCH{=}CHCH_2COOH$$

megatomoic acid

(a) What is the IUPAC name for megatomoic acid?
(b) How many stereoisomers are possible for this compound?

23.53 The IUPAC name for ibuprofen is 2-[4-(2-methylpropyl)phenyl]propanoic acid. Draw the structural formula for ibuprofen. On the structure, identify the 2-methylpropyl and the propanoic acid components. **LO 1,2**

23.54 The monopotassium salt of oxalic acid is present in certain leafy vegetables, including rhubarb. Both oxalic acid and its salts are poisonous in high concentrations. Draw the structural formula for monopotassium oxalate. **LO 1,2**

23.55 Potassium sorbate is added as a preservative to certain foods to prevent spoilage by bacteria and moulds and to extend the shelf life. The IUPAC name for potassium sorbate is potassium (2*E*,4*E*)-hexa-2,4-dienoate. Draw the structural formula for potassium sorbate. **LO 1,2**

23.56 Draw the structural formula of a compound with each of the following molecular formulae that, on oxidation by chromic acid, gives the carboxylic acid or dicarboxylic acid shown. **LO 1,4**

(a) $C_6H_{14}O$ $\xrightarrow{\text{oxidation}}$ COOH

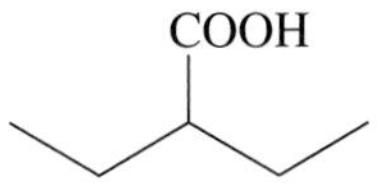

(b) C_8H_{10} $\xrightarrow{\text{oxidation}}$ COOH COOH

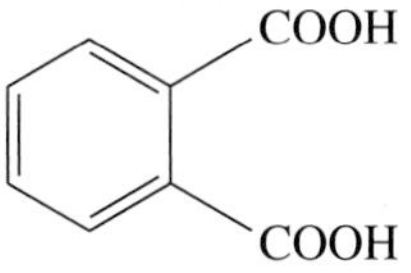

(c) $C_4H_6O_2$ $\xrightarrow{\text{oxidation}}$ HOOC COOH

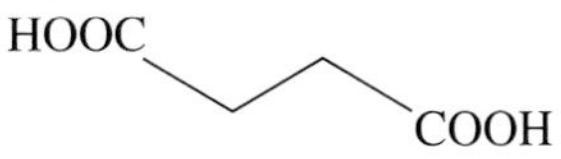

23.57 Complete the following acid–base reactions. **LO 5**

(a) 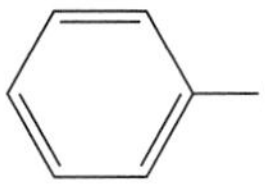 C_6H_5—$CH_2COOH + NaOH \longrightarrow$

(b) $CH_3CH{=}CHCH_2COOH + NaHCO_3 \longrightarrow$

(c) 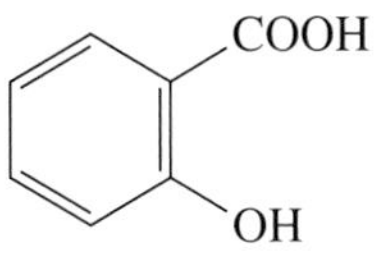 COOH OH + $NaHCO_3 \longrightarrow$

(d) $CH_3CH(OH)COOH + H_2NCH_2CH_2OH \longrightarrow$

(e) $CH_3CH{=}CHCH_2COO^-Na^+ + HCl \longrightarrow$

23.58 The normal pH range for blood plasma is 7.35–7.45. Under these conditions, would you expect the carboxyl group of lactic acid ($pK_a = 3.85$) to exist primarily as a carboxyl group or as a carboxylate anion? Explain. **LO 3,4,5**

23.59 The pK_a of aspirin is 3.49. Would you expect aspirin dissolved in blood plasma (pH = 7.35–7.45) to exist primarily in the carboxylic acid form or in the anionic carboxylate form? Explain. **LO 3,5**

23.60 Excess ascorbic acid ($pK_a = 4.10$) is excreted in the urine, the pH of which is normally 4.8–8.4. What form of ascorbic acid — ascorbic acid itself or ascorbate anions — would you expect to be present in urine with pH = 6.1?

23.61 The pH of human gastric juice is normally 1.0–3.0. What form of lactic acid ($pK_a = 3.85$) — lactic acid itself or its anion — would you expect to be present in the stomach? **LO 3,5**

23.62 In chapter 24, we discuss a class of compounds called amino acids, so named because they contain both an amino group and a carboxyl group. The following are two structural formulae for the amino acid valine. **LO 1**

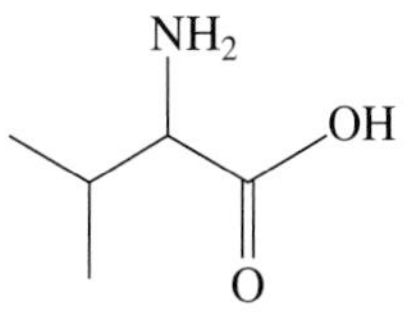

NH_2 OH O $\overset{+}{N}H_3$ O^- O

A 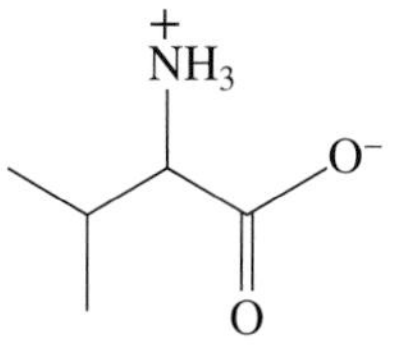   B

Is valine better represented by structural formula A or B? Explain.

23.63 The following is the structural formula for the amino acid phenylalanine in the form of a zwitterion. **LO 3**

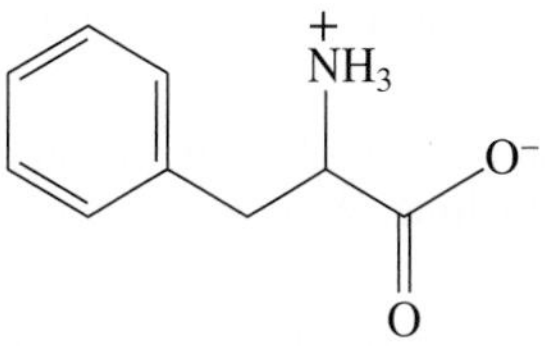

$\overset{+}{N}H_3$ O^- O   phenylalanine

What would you expect to be the major form of phenylalanine present in aqueous solution at: (a) pH = 2.0, (b) pH = 5–6 and (c) pH = 11.0? Explain.

23.64 Explain why α-amino acids, the building blocks of proteins (chapter 24), have pK_a values nearly a thousand times higher than aliphatic carboxylic acids. **LO 3**

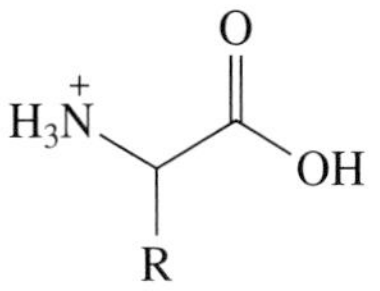

$H_3\overset{+}{N}$ O OH R   R O OH R

a protonated α-amino acid $pK_a \approx 2$ an aliphatic acid $pK_a \approx 5$

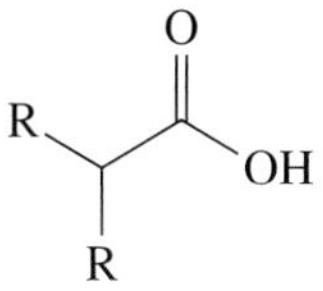

23.65 Formic acid is one of the components responsible for the sting of biting ants and is injected under the skin by bees and wasps. A way to relieve the pain is to rub the area of the sting with a paste of baking soda (sodium hydrogen carbonate) and water, which neutralises the acid. Write an equation for this reaction. **LO 5**

23.66 Show how to convert *cis*-pent-3-enoic acid to each of the following compounds. **LO 4**

(a)  COOH   (b) 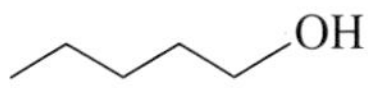 OH   (c) OH

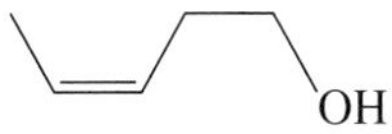

23.67 Show how to convert 3-oxobutanoic acid (acetoacetic acid) to each of the following compounds. **LO 4**

(a) $CH_3CH(OH)CH_2COOH$

(b) $CH_3CH(OH)CH_2CH_2OH$

(c) $CH_3CH{=}CHCOOH$

23.68 Methyl 2-hydroxybenzoate (methyl salicylate) has the odour of oil of wintergreen (Dencorub®). This ester is prepared by Fischer esterification of 2-hydroxybenzoic acid (salicylic acid) with methanol. Draw the structural formula for methyl 2-hydroxybenzoate. **LO1,4**

23.69 Benzocaine, a topical anaesthetic, is prepared by treating 4-aminobenzoic acid with ethanol in the presence of an acid catalyst, followed by neutralisation. Draw the structural formula for benzocaine. **LO1,4**

23.70 From which carboxylic acid and alcohol is each of the following esters derived? **LO5**

(a) 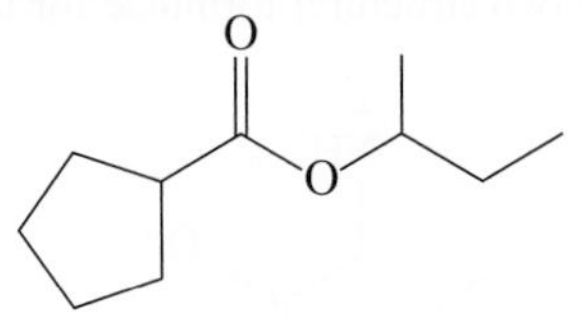 (b)

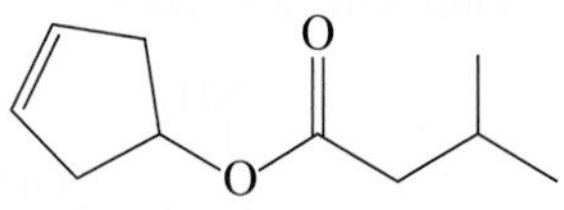

23.71 From which carboxylic acid and alcohol is each of the following esters derived? **LO5**

(a) $CH_3OCCH_2CH_2COCH_3$ (both C=O) (b) $CH_3CH_2CH{=}CHCOCH(CH_3)_2$ (C=O)

23.72 When treated with an acid catalyst, 5-hydroxypentanoic acid forms a cyclic ester (a lactone). Draw the structural formula of this lactone. **LO5**

23.73 The analgesic phenacetin is synthesised by treating 4-ethoxyaniline with acetic anhydride. Write an equation for the formation of phenacetin. **LO5**

23.74 A carboxylic acid can be converted to an ester by Fischer esterification. Show how to synthesise each of the following esters from a carboxylic acid and an alcohol by Fischer esterification. **LO5**

(a)

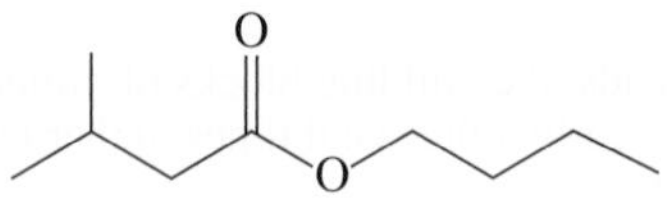

(b)

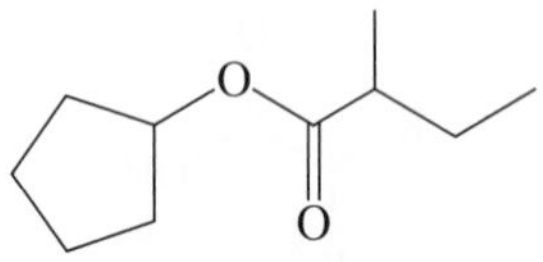

23.75 A carboxylic acid can be converted to an ester in two reactions by first converting the carboxylic acid to its acid chloride and then treating the acid chloride with an alcohol. Show how to prepare each ester in question 23.74 from a carboxylic acid and an alcohol by this two-step scheme. **LO5**

23.76 Write a mechanism for the reaction of propanoyl chloride and methylamine to give *N*-methylpropanamide and methylammonium chloride. **LO5**

23.77 Show how to prepare each of the following amides by reaction of an acid chloride with ammonia or an amine. **LO4**

(a) 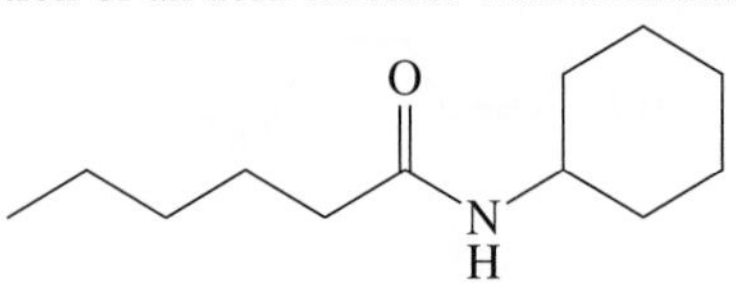 N H

(b) 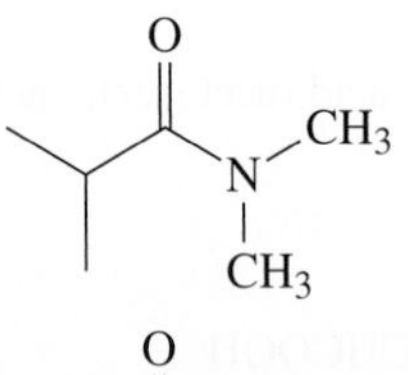 CH_3 N CH_3

(c) 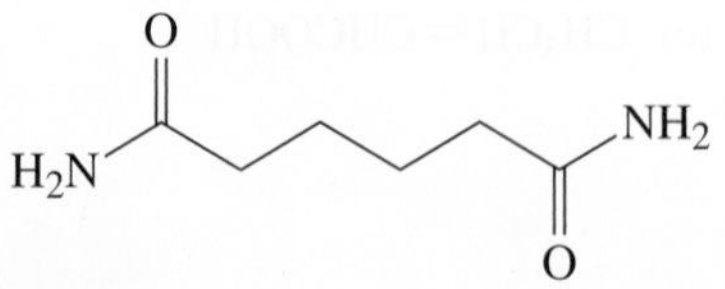 H_2N NH_2

23.78 What product is formed when pentanoyl chloride is treated with each of the following reagents? **LO4**

(a) C_6H_6, $AlCl_3$
(b) $CH_3CH_2CH_2CH_2OH$
(c) $CH_3CH_2CH_2CH_2SH$
(d) $CH_3CH_2CH_2CH_2NH_2$ (2 equivalents)
(e) H_2O
(f) 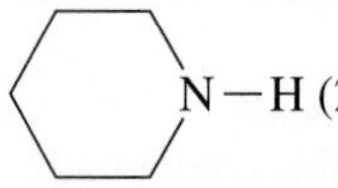 N—H (2 equivalents)

23.79 Nicotinic acid, more commonly named niacin, is one of the B vitamins. Show how nicotinic acid can be converted to ethyl nicotinate and then to nicotinamide. **LO5**

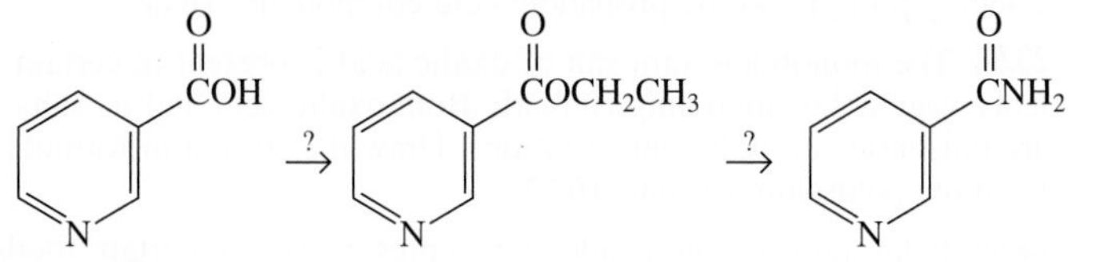

COH → ? → $COCH_2CH_3$ → ? → CNH_2

nicotinic acid (niacin) — ethyl nicotinate — nicotinamide

23.80 Complete the following reactions. **LO5**

(a) 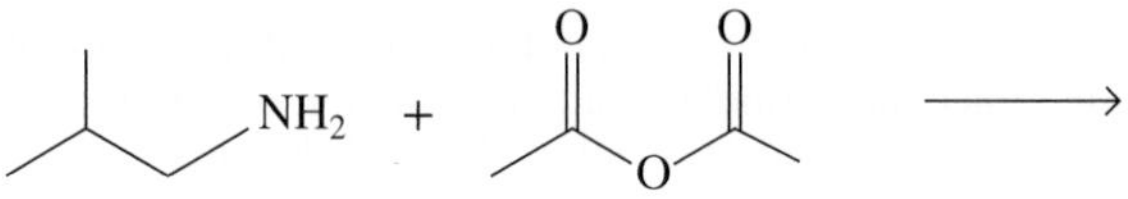 NH_2 + →

(b) 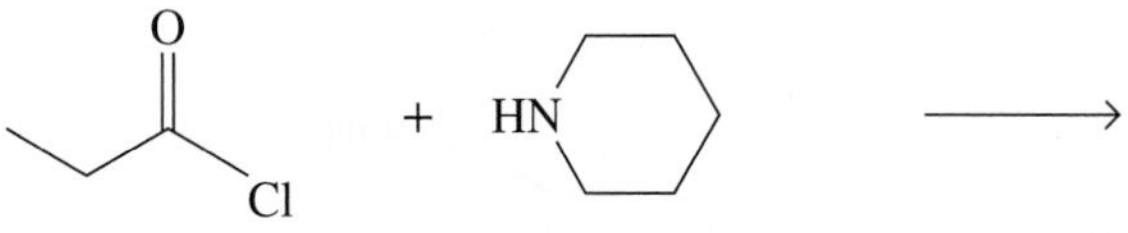 Cl + HN →

(c) 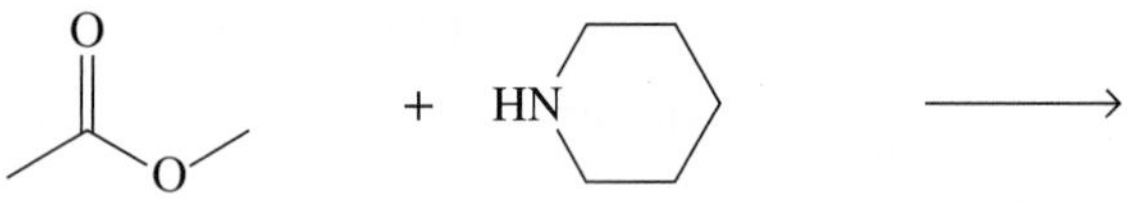 + HN →

(d) 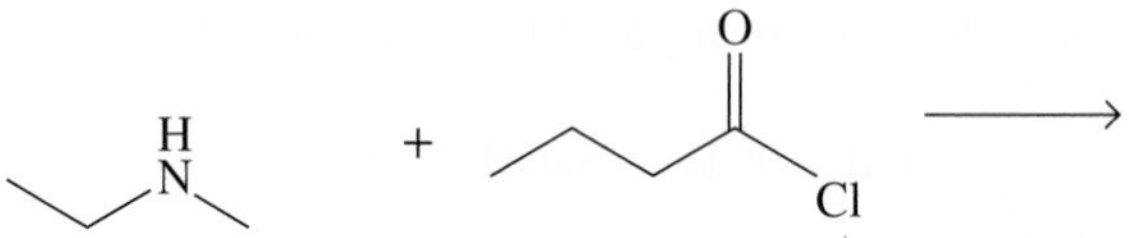 H N + Cl →

23.81 Show how to convert 2-hydroxybenzoic acid (salicylic acid) to each of the following compounds. **LO5**

(a) 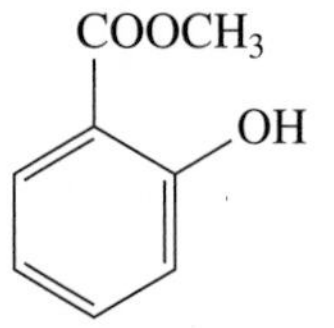 $COOCH_3$, OH — methyl salicylate (oil of wintergreen)

(b) 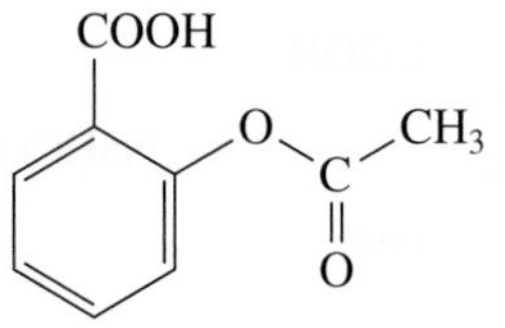 COOH, O, C, CH_3, O — acetyl salicylic acid (aspirin)

23.82 Show the product of treating 5-pentanolactone with each of the following reagents. **LO5**

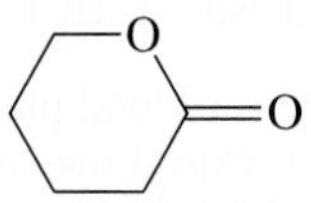

5-pentanolactone

(a) NH_3
(b) $LiAlH_4$ then H_2O
(c) NaOH, H_2O, heat

23.83 Show the product of treating *N*-methyl-γ-butyrolactam with each of the following reagents. **LO5**

(a) HCl, heat
(b) NaOH, heat
(c) $LiAlH_4$ then H_2O

23.84 Treating 5-pentanolactone with 2 equivalents of ethylmagnesium bromide, followed by hydrolysis in aqueous acid, gives a compound with the molecular formula $C_9H_{20}O_2$. **LO5**

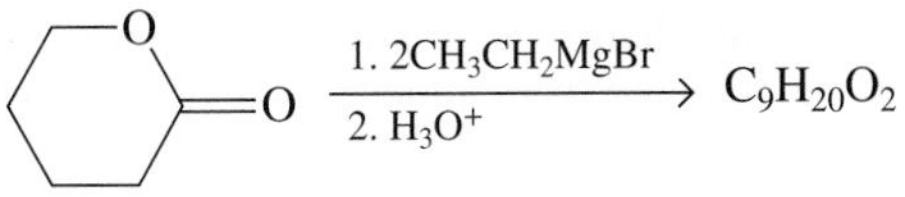

1. $2CH_3CH_2MgBr$ 2. H_3O^+ → $C_9H_{20}O_2$

Propose a structural formula for this compound.

23.85 Complete the following reactions. **LO5**

(a) 2 (allyl)MgBr, then H_2O/HCl →

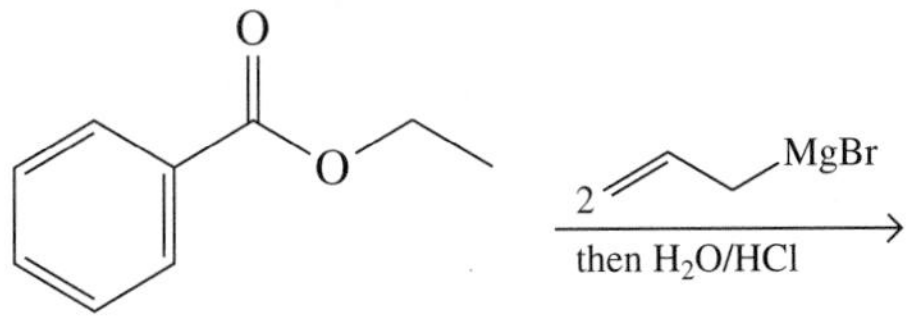

(b) $2CH_3MgBr$, then H_2O/HCl →

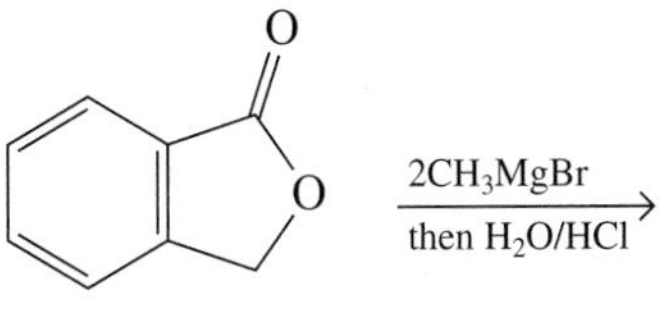

(c) $2CH_3MgBr$, then H_2O/HCl →

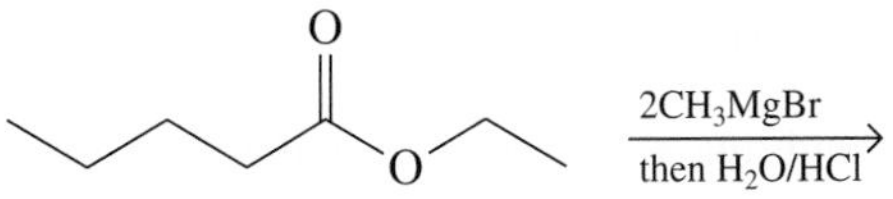

23.86 Procaine (its hydrochloride is marketed as Novocain®) was one of the first local anaesthetics developed for infiltration and regional anaesthesia. It is synthesised by the following Fischer esterification. **LO5**

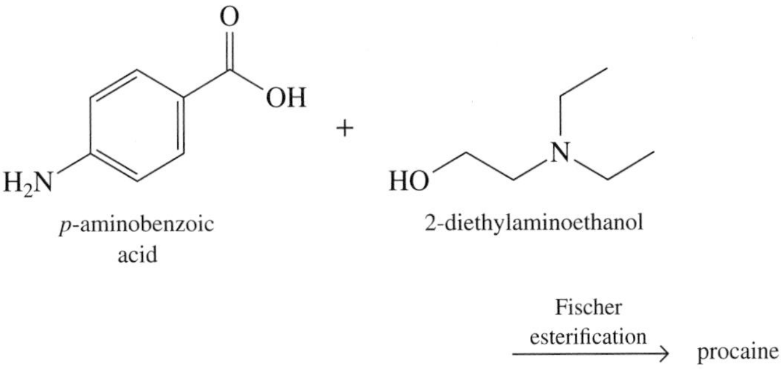

p-aminobenzoic acid + 2-diethylaminoethanol —Fischer esterification→ procaine

Draw the structural formula of procaine.

23.87 The active ingredient *N,N*-diethyl-*m*-toluamide (DEET) in several common insect repellents is synthesised from 3-methylbenzoic acid (*m*-toluic acid) and diethylamine.

Show how this synthesis can be accomplished. **LO5**

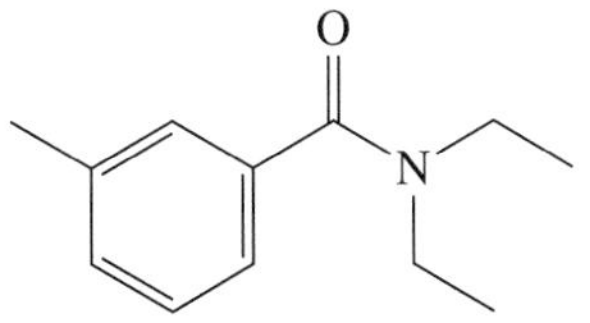

N,N-diethyl-*m*-toluamide (DEET)

23.88 Provide a mechanism for the acid catalysed hydrolysis of esters to give carboxylic acids and alcohols. **LO4**

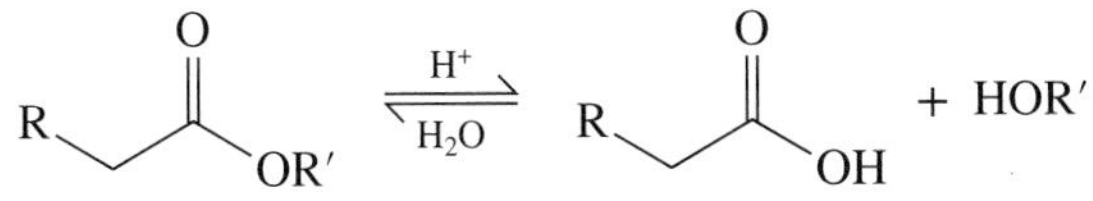

RCH_2COOR' ⇌ (H^+, H_2O) RCH_2COOH + HOR′

(*Hint:* This is the reverse of Fischer esterification.)

23.89 In section 23.5, it was suggested that the mechanism for the Fischer esterification of carboxylic acids would be a model for many of the reactions of the functional derivatives of carboxylic acids. The reaction of an acid halide with water, is one such reaction. **LO5**

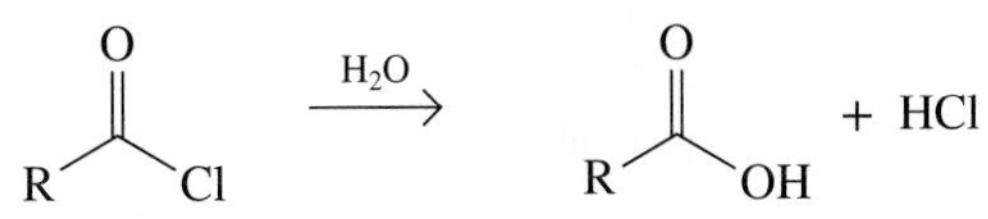

RCOCl —H_2O→ RCOOH + HCl

Suggest a mechanism for this reaction.

23.90 What combination of ester and Grignard reagent can be used to prepare each of the following alcohols? **LO5**

(a) 3-ethylpentan-3-ol
(b) 4-phenylheptan-4-ol
(c) 1,1-diphenylpropanol

23.91 Show how to convert ethyl pent-2-enoate into each of the following compounds. **LO4**

(a) (b) OH

23.92 Phosgene is highly toxic and was used as a chemical weapon in World War I. It is also a synthetic precursor used in the production of many plastics. **LO5**

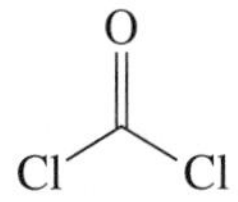

phosgene

(a) When vapours of phosgene are inhaled, the compound rapidly reacts with any nucleophilic sites present (OH groups, NH_2 groups etc.), producing HCl gas. Draw a mechanism for this process.
(b) When phosgene is treated with ethylene glycol ($HOCH_2CH_2OH$), a compound with molecular formula $C_3H_4O_3$ is obtained. Draw the structure of this product.

23.93 The analgesic acetaminophen (paracetamol) is synthesised by treating 4-aminophenol with 1 equivalent of acetic anhydride. Write an equation for the formation of acetaminophen. (*Hint:* Remember that an —NH_2 group is a better nucleophile than an —OH group.) **LO4**

23.94 The reactions of acid halides, acid anhydrides and esters with ammonia and amines are examples of the nucleophilic acyl substitution reaction. Write an equation for the reaction of each of the following. **LO5**

(a) butanoyl chloride with ammonia
(b) acetic anhydride with a methylamine
(c) ethyl benzoate with a diethylamine

In each case, identify the nucleophile, the leaving group and draw the structure of the tetrahedral carbonyl addition intermediate.

23.95 When compound *X*, $C_7H_7O_2N$, was boiled with concentrated NaOH, it dissolved readily, evolving ammonia. The resulting solution on acidification gave a white precipitate that, on refluxing with methanol and a trace of sulfuric acid, gave the characteristic smell of methyl salicylate. **LO5**

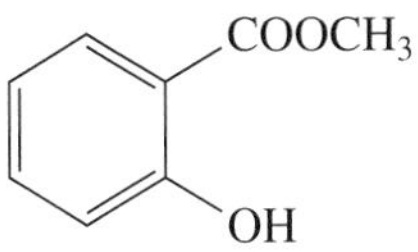

methyl salicylate

Identify *X* and trace the reaction sequence involved.

23.96 Devise chemical tests to distinguish between the following compounds, all of which are white solids. **LO5**

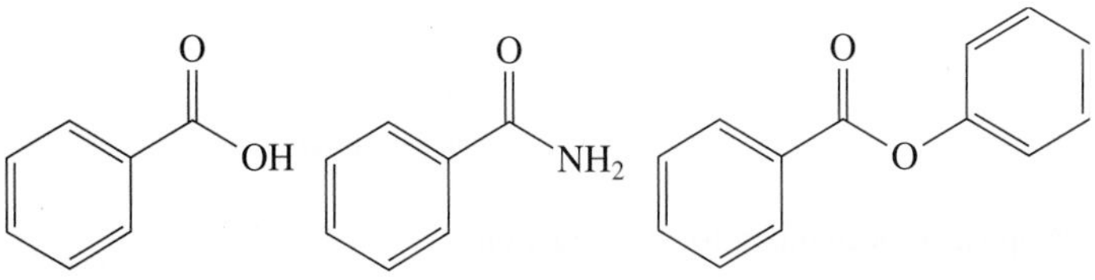

A B C

ADDITIONAL EXERCISES

23.97 Fluphenazine is an antipsychotic drug that is administered as an ester prodrug via intramuscular injection. **LO1,2**

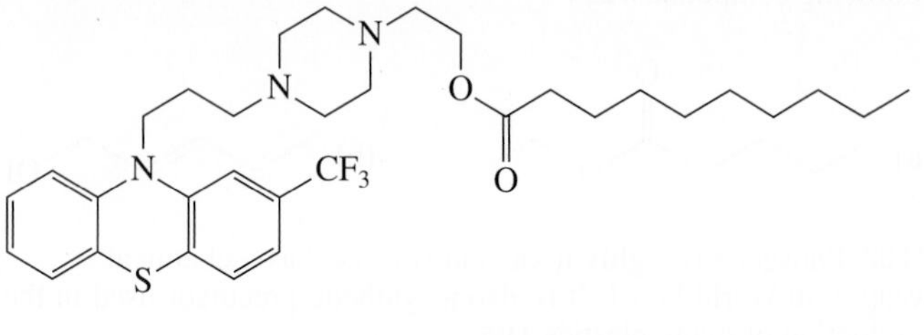

fluphenazine decanoate

The hydrophobic tail of the ester is deliberately designed to enable a slow release of the prodrug into the bloodstream, where the prodrug is rapidly hydrolysed to produce the active drug.

(a) Draw the structure of the active drug.

(b) Draw the structure of and assign a systematic name for the carboxylic acid that is produced as a by-product of the hydrolysis step.

23.98 Benzyl acetate is a pleasant-smelling ester found in the essential oil of jasmine flowers and is used in many perfume formulations. Starting with benzene and using any other reagents of your choice, design an efficient synthesis for benzyl acetate. **LO5**

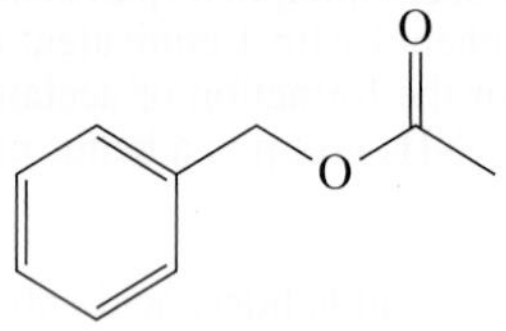

benzyl acetate

23.99 Does a nucleophilic acyl substitution occur between the ester and the nucleophile shown? **LO5**

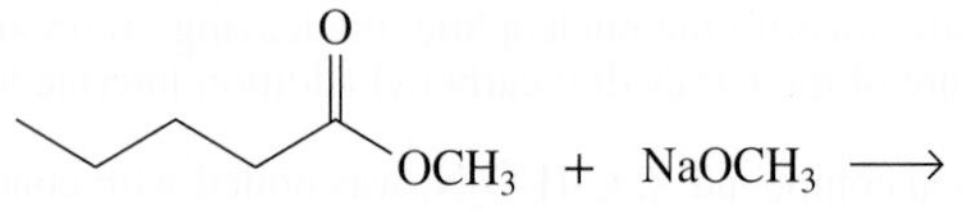

OCH_3 + $NaOCH_3$ ⟶

Propose an experiment that would verify your answer.

23.100 Reaction of a 1° or 2° amine with diethyl carbonate under controlled conditions gives a carbamic ester. **LO5**

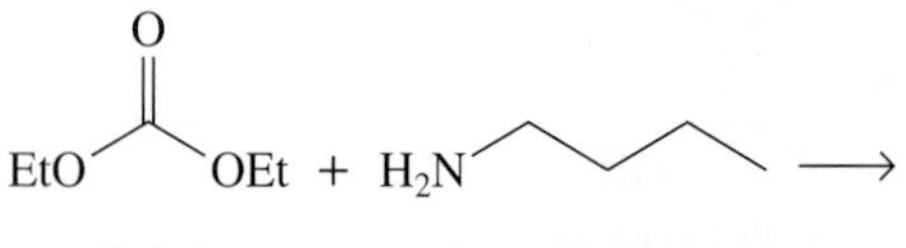

EtO OEt + H_2N ⟶

diethyl carbonate

butan-1-amine (butylamine)

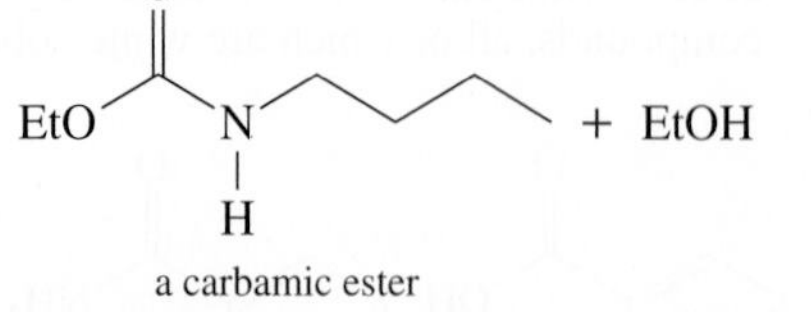

EtO N H + EtOH

a carbamic ester

Propose a mechanism for this reaction.

23.101 Methyl 2-aminobenzoate, a flavouring agent with the taste of grapes, can be prepared from toluene by the following series of steps. **LO5**

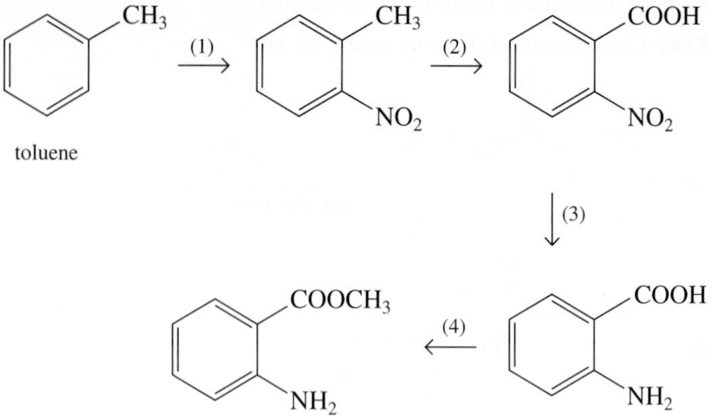

toluene CH_3 (1)⟶ CH_3, NO_2 (2)⟶ COOH, NO_2 (3)↓ COOH, NH_2 (4)⟵ $COOCH_3$, NH_2

methyl 2-aminobenzoate

Show how you might bring about each step in this synthesis.

23.102 Methyl 4-aminobenzoate and propyl 4-aminobenzoate are used as preservatives in foods, beverages and cosmetics. **LO5**

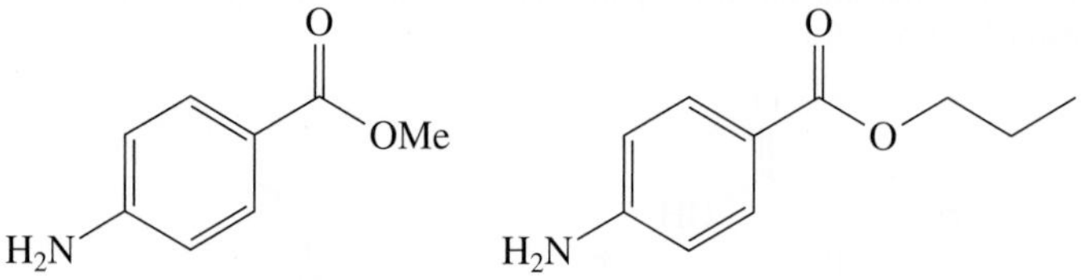

H_2N O OMe

methyl 4-aminobenzoate

H_2N O O

propyl 4-aminobenzoate

Show how the synthetic scheme in question 23.101 can be modified to give each of these compounds. (*Hint:* Nitration of toluene gives a mixture of 2- and 4-nitrotoluene.)

23.103 Starting materials for the synthesis of the herbicide propanil, used to kill weeds in rice paddies, are benzene and propanoic acid. Show the reagents (1) to (5) required for this synthesis. **LO5**

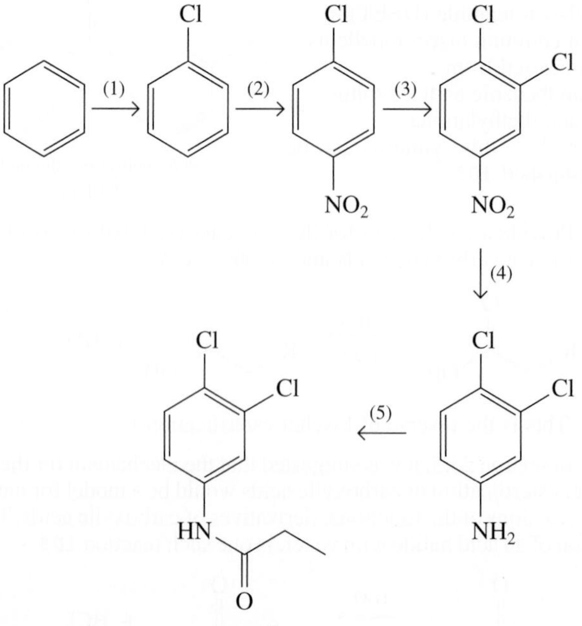

(1) Cl (2) Cl, NO_2 (3) Cl, Cl, NO_2 (4) Cl, Cl, NH_2 (5) Cl, Cl, HN, O

propanil

23.104 When chemists follow a reaction scheme, they need to verify that the product obtained in each step is actually what they were expecting. For the reaction scheme below, what evidence could be used to confirm that 4-nitrobenzoic acid has successfully been converted into methyl 4-nitrobenzoate, and then that methyl 4-nitrobenzoate has been converted into 4-nitrobenzamide? You should provide both spectroscopic evidence (chapter 20) and simple chemical tests. **LO 5**

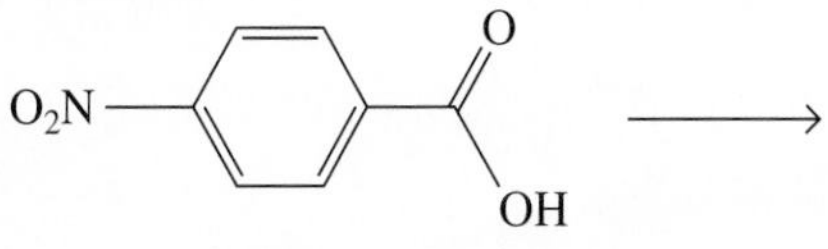

O_2N O OH

4-nitrobenzoic acid

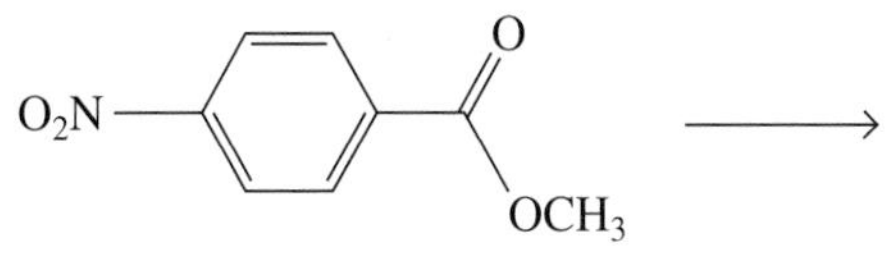

O_2N O OCH_3

methyl 4-nitrobenzoate

O_2N O NH_2

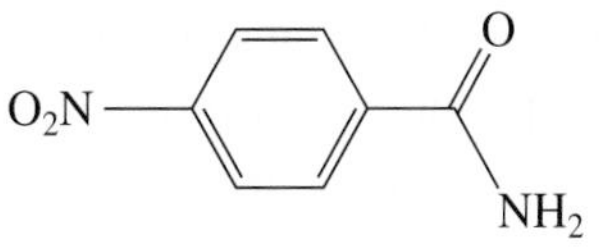

4-nitrobenzamide

23.105 The following are structural formulae of two local anaesthetics. **LO 5**

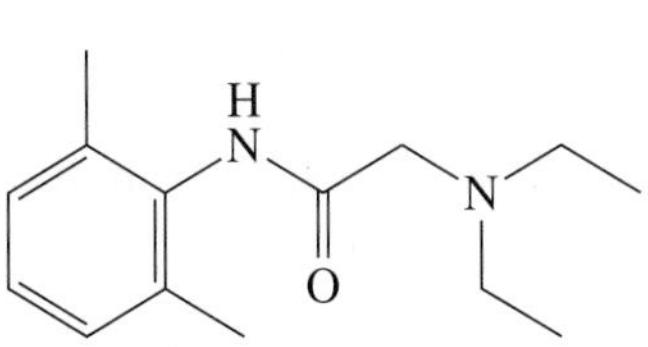

H N N O

lidocaine
(Xylocaine®)

H N N O CH_3

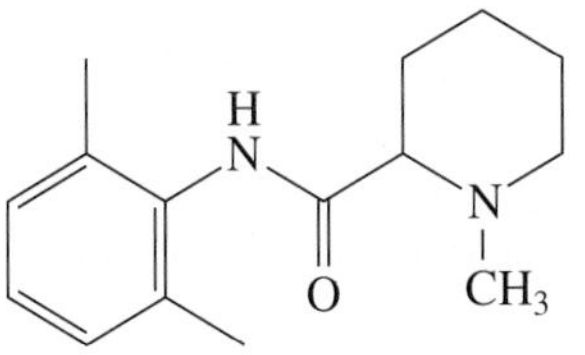

mepivacaine
(Carbocaine®)

Lidocaine was introduced in 1948 and is now the most widely used local anaesthetic for infiltration and regional anaesthesia. Its hydrochloride is marketed under the name Xylocaine. Mepivacaine (its hydrochloride is marketed as Carbocaine) acts faster than lidocaine and its effects last somewhat longer.

(a) Propose a synthesis of lidocaine from 2,6-dimethylaniline, chloroacetyl chloride ($ClCH_2COCl$) and diethylamine.
(b) Which amine and acid chloride can be reacted to give mepivacaine?

23.106 Chemists have developed several syntheses for the antiasthmatic drug salbutamol (also known as albuterol or, more commonly, Ventolin®). One of these syntheses starts with salicylic acid, the same acid that is the starting material for the synthesis of aspirin. **LO 5**

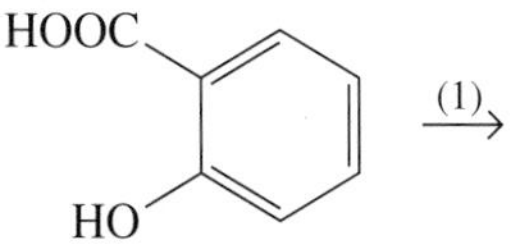

HOOC HO (1)

salicylic acid

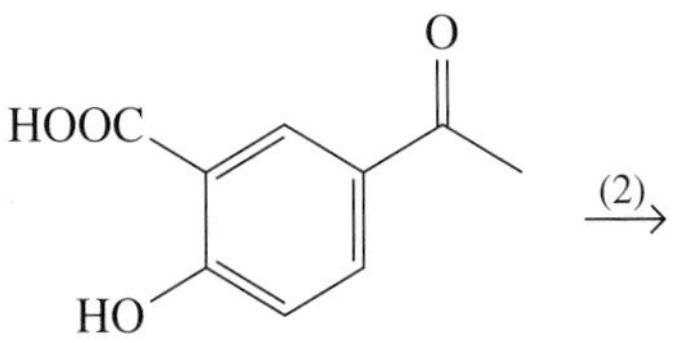

HOOC HO O (2)

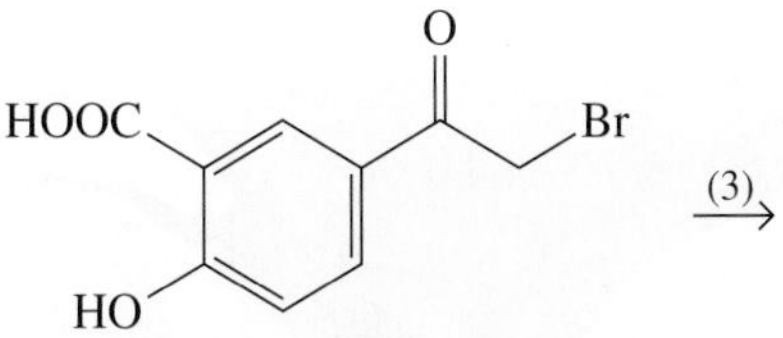

HOOC HO O Br (3)

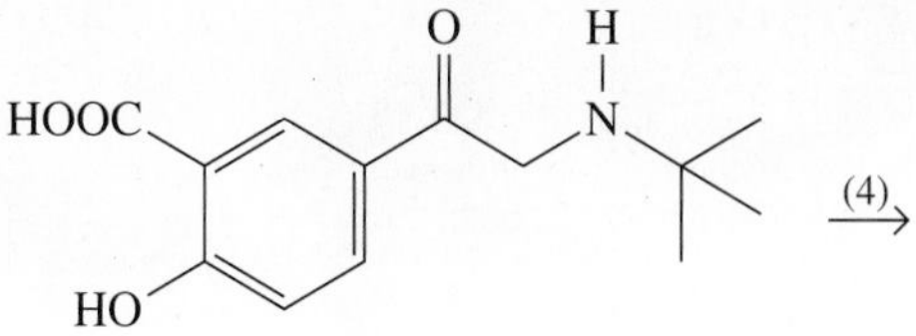

HOOC HO O H N (4)

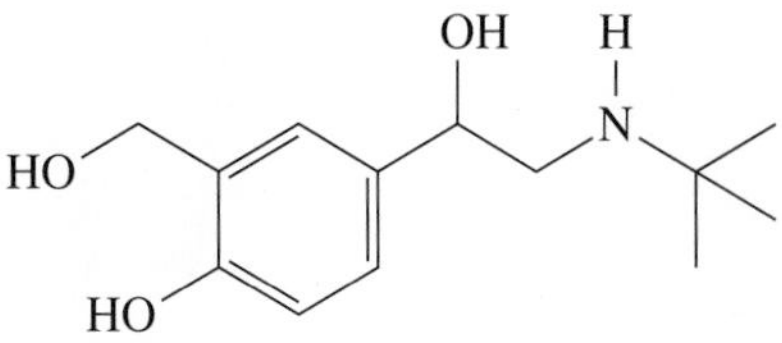

OH H N HO HO

salbutamol

(a) Propose a reagent and a catalyst for step 1. What name is given to this type of reaction?
(b) Propose a reagent for step 2.
(c) Name the amine used to bring about step 3.
(d) Step 4 is a reduction of two functional groups. Name the functional groups reduced and name the reagent that will accomplish the reduction.

23.107 By considering the relative acidity of the following compounds, briefly describe how you would separate them. Use equations to explain the chemistry involved. **LO 5**

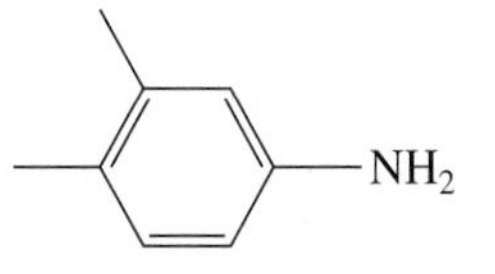

NH_2

A

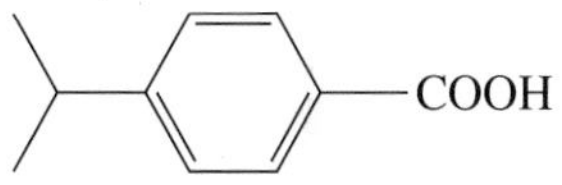

COOH

C

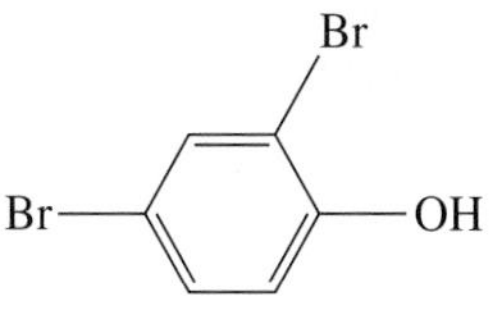

Br Br OH

B

23.108 Consider the following hydrolysis reaction. Neither of the products is appreciably soluble in water. By considering the relative acidity of the products, briefly describe how you would separate the two. **LO 5**

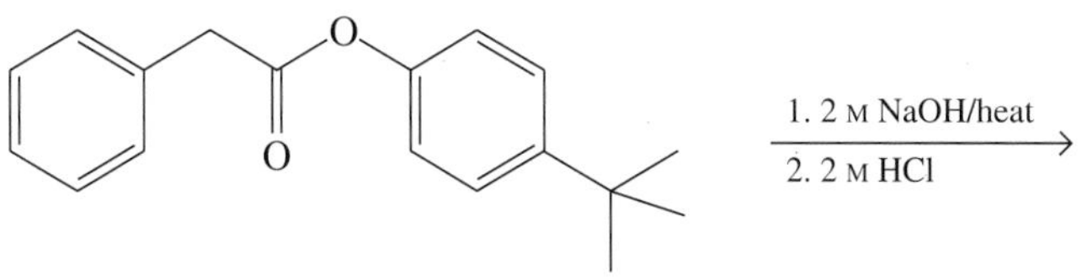

O O

4-*tert*-butylphenyl phenylacetate

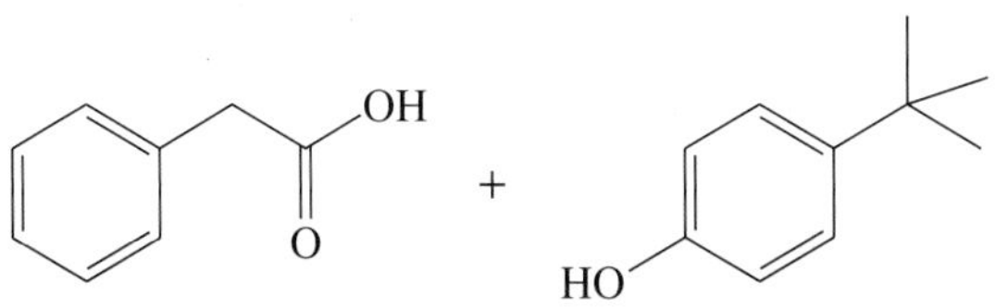

1. 2 M NaOH/heat
2. 2 M HCl

OH O + HO

phenylacetic acid

4-*tert*-butylphenol

24 Amino acids, peptides and proteins

Amino acids are molecules that combine both the amine functional group (chapter 19) and the carboxylic acid functional group (chapter 23) within the same structure. Amino acids are important in their own right, but they are also critically important for life as they provide the building blocks of peptides and proteins. In this chapter we will look at the acid–base properties of amino acids, as these control much of the nature of peptides and proteins. Understanding the chemistry of amino acids allows us in turn to understand the more complex structures and properties of peptides and proteins. Proteins are among the most important of all biological compounds as they are integrally important in the vital functions of:

- *movement:* muscle fibres are made of proteins called myosin and actin
- *catalysis:* virtually all reactions that take place in living systems are catalysed by a special group of proteins called enzymes
- *structure:* structural proteins such as collagen and keratin are the chief constituents of skin, bones, hair and fingernails
- *transport:* the protein haemoglobin is responsible for the transport of oxygen from the lungs to tissues, while other proteins transport molecules across cell membranes
- *protection:* a group of proteins called antibodies represent one of the body's major defences against disease.

Peptides are a major component of the supplements and health food sector. In the last few years certain peptides have been illegally used by athletes to boost performance.

LEARNING OBJECTIVES

After studying this chapter, you should be able to:

24.1 define the structure of amino acids

24.2 control the charge present on amino acids using pH

24.3 describe the linking of amino acids together to make peptides and proteins

24.4 categorise protein structures as 1°, 2°, 3° or 4°

24.5 explain the functional importance of the three-dimensional shape of proteins

24.6 exploit chemical and physical processes to change protein properties.

24.1 Amino acids

An **amino acid** is a compound that contains both a carboxyl group and an amino group. Although many types of amino acids are known, α-amino acids (where the amino group and the carboxyl group are attached to the same carbon atom) are by far the most significant in the biological world. This is because α-amino acids make up the building blocks from which key bioactive peptides, as well as the much larger protein molecules, are constructed. Linking just two specific amino acids together makes the peptide carnosine that is found in high levels in the brain and in muscle tissue. Three particular amino acids joined together make glutathione, one of the most important small molecules found in the body. Both glutathione and carnosine play a number of important roles, but are known to be crucial in protecting cells and tissues against harmful oxidation reactions. When many amino acids are linked together, the very large molecules thus formed are called proteins. Proteins are examples of natural polymers. Polymers are very large molecules composed of many smaller molecules repeatedly linked together. The sequence of connectivity governs the properties of the polymer. Millions of different types of proteins can be formed from combinations of only 20 types of α-amino acid. The structures of some amino acids are shown in figure 24.1, as well as how they can be combined to make small peptides.

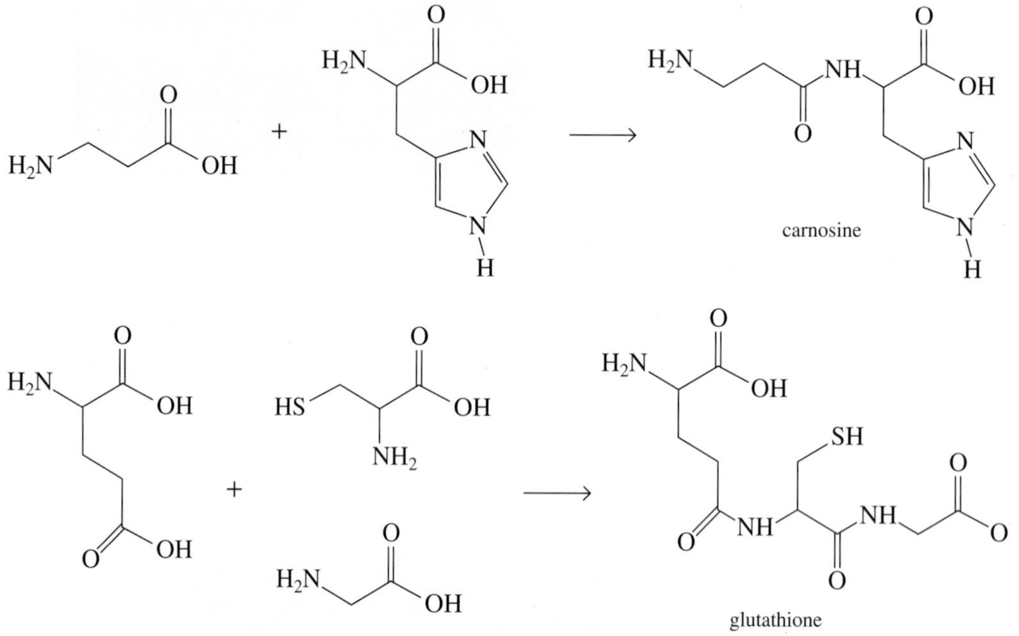

FIGURE 24.1 Carnosine and glutathione can be formed by combinations of smaller amino acids. Only one of the amino acids shown is not an α-amino acid.

Although figure 24.1 seems to depict an acceptable way to represent the structural formulae for amino acids, it is not accurate because it shows an acid functional, —COOH, and a base functional group, —NH_2, within the same molecule. These acidic and basic groups react with each other to form an internal salt (a dipolar ion). Figure 24.2 shows the more accurate general structural formula of an α-amino acid is its actual salt form. This salt is called a **zwitterion** and it is important to realise that because it possesses one positive charge and one negative charge, this makes the overall charge zero.

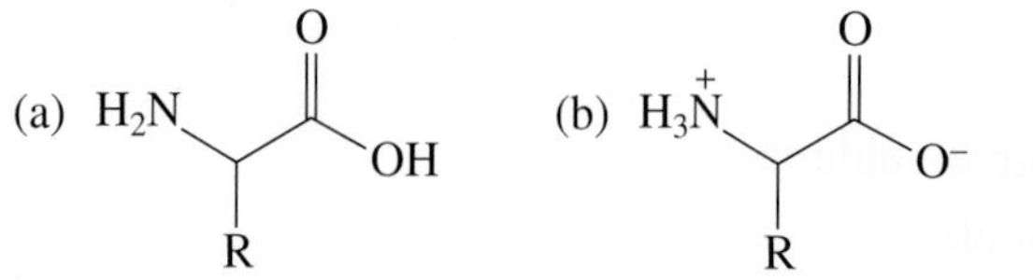

FIGURE 24.2 Two representations for an α-amino acid: (a) the (unrealistic) noncharged form and (b) the actual internal salt (zwitterion) form.

Because they exist as zwitterions, amino acids have many of the properties associated with salts. They are crystalline solids with high melting points; they are fairly soluble in water but insoluble in nonpolar organic solvents such as ether and hydrocarbon solvents.

Chirality

With the exception of glycine, $H_3\overset{+}{N}CH_2COO^-$, all protein-derived amino acids have at least one stereocentre and are, therefore, chiral (see chapter 17). Figure 24.3 shows Fischer projection formulae for the enantiomers of alanine. The vast majority of carbohydrates in the biological world are of the D-series (section 22.2), whereas the vast majority of α-amino acids in the biological world are of the L-series. From the rules we developed in chapter 17, L is the *S* stereoisomer. (Note, however, that L-cysteine is the *R* stereoisomer because of the higher priority of the sulfur atom.)

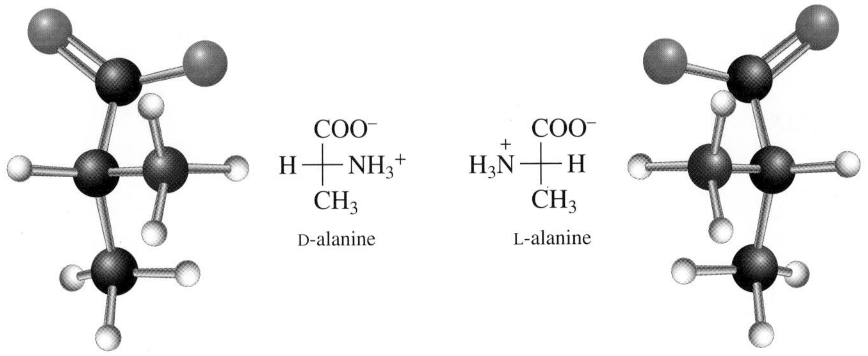

COO^- / $H-C-NH_3^+$ / CH_3 — D-alanine

COO^- / $H_3\overset{+}{N}-C-H$ / CH_3 — L-alanine

FIGURE 24.3 The enantiomers of alanine. The vast majority of α-amino acids in the biological world have the L-configuration at the α-carbon atom.

Protein-derived amino acids

Table 24.1 (overleaf) gives common names, structural formulae and standard three-letter and one-letter abbreviations for the 20 common L-amino acids found in proteins. The amino acids shown are divided into four categories: those with nonpolar side chains; those with polar, but un-ionised, side chains; those with acidic side chains; and those with basic side chains. As you study the information in this table, note the following points.

1. All 20 of these protein-derived amino acids are **α-amino acids**, meaning that the amino group is located on the carbon atom adjacent to the carboxyl group.
2. For 19 of the 20 amino acids, the α-amino atom group is primary. Proline is different; its α-amino group is secondary.
3. With the exception of glycine, the α-carbon atom of each amino acid is a stereocentre. Although not shown in the table, all 19 chiral amino acids have the same relative configuration at the α-carbon atom. In the D,L convention, all are L-amino acids.
4. Isoleucine and threonine contain a second stereocentre. Four stereoisomers are possible for each of these amino acids, but only one of the four is found in proteins.
5. The sulfhydryl group of cysteine, the imidazole group of histidine and the phenolic hydroxyl of tyrosine are partially ionised at pH = 7.0, but the ionic form is not the major form present at that pH.

WORKED EXAMPLE 24.1

Assessing the structural variations of the common amino acids

Of the 20 protein-derived amino acids shown in table 24.1, how many contain: (a) aromatic rings, (b) side-chain —OH groups, (c) phenolic —OH groups, (d) sulfur and (e) basic groups?

Analysis

Amino acids can look structurally complex, but it is important to understand their similarities as well as their differences. The type of side chain present has a substantial impact on the properties of the amino acid, and so you will need to recognise the functional groups present in the side chain. Look for the amine and the carboxylic acid linked by a single carbon atom as the common structural unit, and then assess the nature of the functional groups present in the side chain.

Solution

(a) Phenylalanine, tryptophan, tyrosine and histidine contain aromatic rings.
(b) Serine and threonine contain side-chain hydroxyl groups.
(c) Tyrosine contains a phenolic —OH group.
(d) Methionine and cysteine contain sulfur.
(e) Arginine, histidine and lysine contain basic groups.

Is our answer reasonable?

Table 24.1 shows the structures of the groups present on the amino acids. Aromatic rings require a continuous cycle of sp^2 hybridised atoms such that the total numbers of electrons in the system equals 6 (or 10, 14, 18 etc.). Consequently, histidine contains an aromatic ring as do tryptophan and tyrosine. Note that an —OH group bound to an aromatic ring imparts considerably different properties from one bound to an alkyl chain, which is why we define alcohol hydroxyl groups (serine and threonine) as different from phenolic hydroxyl groups (tyrosine).

Asparagine and glutamine possess amide side chains, so these groups are not basic since the carbonyl orbitals interact with the lone electron pair on the nitrogen atom, making it unavailable to bond to hydrogen ions, H^+.

TABLE 24.1 The 20 common amino acids found in proteins.

Nonpolar side chains

alanine (Ala, A)		phenylalanine (Phe, F)	
glycine (Gly, G)		proline (Pro, P)	
isoleucine (Ile, I)		tryptophan (Trp, W)	
leucine (Leu, L)			
methionine (Met, M)		valine (Val,V)	

Polar side chains

asparagine (Asn, N)		serine (Ser, S)	
glutamine (Gln, Q)		threonine (Thr, T)	

Acidic side chains

aspartic acid (Asp, D)
glutamic acid (Glu, E)
cysteine (Cys, C)
tyrosine (Tyr, Y)

Basic side chains

arginine (Arg, R)
histidine (His, H)
lysine (Lys, K)

Note: Each ionisable group is shown in the form present in highest concentration in aqueous solution at pH = 7.0.

Of the 20 protein-derived amino acids shown in table 24.1, which contain: (a) no stereocentre and (b) two stereocentres?

PRACTICE EXERCISE 24.1

Some other common amino acids

Although the vast majority of plant and animal proteins are constructed from just these 20 α-amino acids, many examples of other kinds of amino acid can also be found in nature. Ornithine and citrulline, for example, are found predominantly in the liver and are an integral part of the urea cycle, the metabolic pathway that converts ammonia to urea.

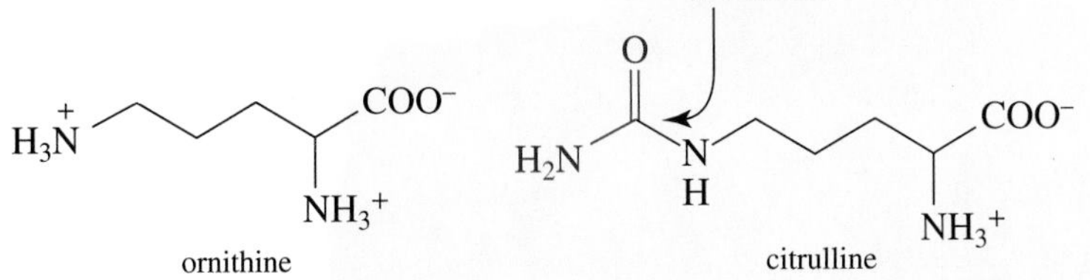

carboxamide derivative of ornithine

ornithine

citrulline

Thyroxine and triiodothyronine, two of several hormones derived from the amino acid tyrosine, are found in thyroid tissue.

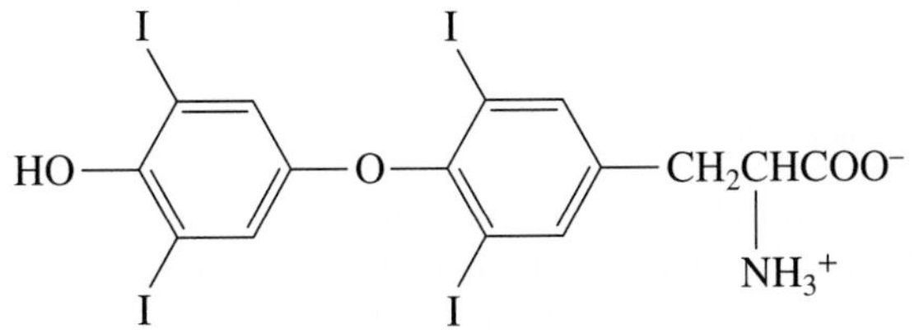

thyroxine, T_4

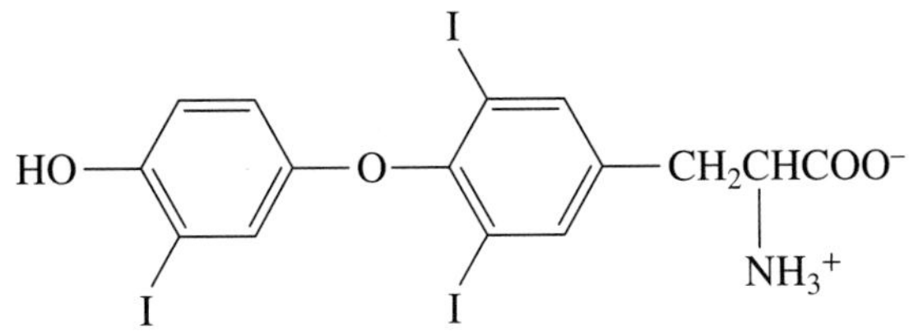

triiodothyronine, T_3

The principal function of these two hormones is to stimulate metabolism in other cells and tissues.

The amino acid 4-aminobutanoic acid (γ-aminobutyric acid, or GABA) is found in high concentration (0.8×10^{-3} M) in the brain, but in no significant amounts in any other mammalian tissue. GABA is synthesised in neural tissue by decarboxylation of the α-carboxyl group of glutamic acid and is a neurotransmitter in the central nervous system of invertebrates and possibly in humans as well.

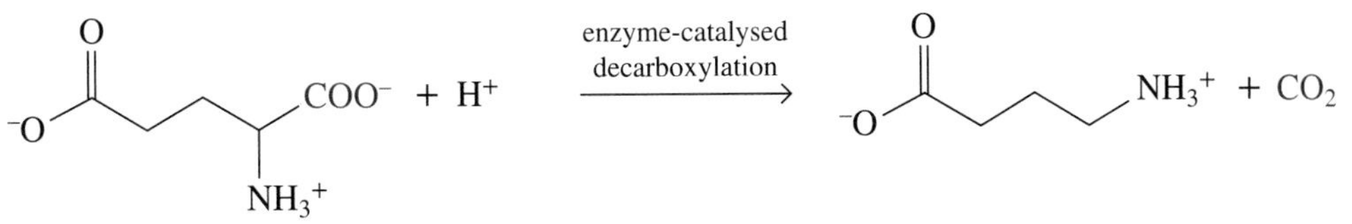

enzyme-catalysed decarboxylation

glutamic acid

4-aminobutanoic acid (γ-aminobutyric acid, GABA)

Only L-amino acids are found in proteins, and only rarely are D-amino acids a part of the metabolism of higher organisms. Several D-amino acids, however, along with their L-enantiomers, are found in lower forms of life. For example, D-alanine and D-glutamic acid are structural components of the cell walls of certain bacteria. Several D-amino acids are also present in synthetic antibiotics made from amino acids.

Chemical Connections

Amino acids from space!

Nearly 50 years ago a meteorite fell to Earth near the small Victorian town of Murchison. Although fragments of the Murchison Meteorite, as it came to be known, were scattered over a wide area, a lot of people saw it fall and raced out to collect samples (figure 24.4). In all, more than 100 kg of fragments, the largest weighing close to 7 kg, were collected by the locals. Today many of those fragments are in museums across the world. Some are for sale, occasionally popping up on eBay.

FIGURE 24.4 Part of the Murchison Meteorite.

However, what makes the Murchison Meteorite extremely important for science, even today, is that chemists were able to quickly obtain samples before they were contaminated by prolonged exposure to Earth's environment. Much of the chemical make up of the Murchison Meteorite reflects the nature of the solar system at the time the meteor formed — over 4 500 000 000 years ago. Chemists from NASA, using the best analytical chemistry tools available at the time, analysed the meteorite and shocked the world when they announced (in the scientific journal *Nature* in 1970) that in the meteorite they had found a number of amino acids — several of which were the same as the amino acids found on Earth (and some the same as found in us!). Common amino acids such as glycine, alanine and glutamic acid were detected, as well as unusual ones like isovaline and pseudoleucine.

Since these early studies and as the power of analytical scientific equipment has improved, chemists have found traces of more and more amino acids. Today more than 90 amino acids have been detected in the meteorite, as well as thousands of other organic molecules. Even more surprising is that there is growing scientific evidence that the amino acids found in the meteorite are not complete racemates (chapter 17) and actually reflect somewhat an excess of the D-enantiomers found in all life on Earth. What factors in the primordial solar system favoured the formation of specific amino acid enantiomers remains a mystery. However, these discoveries offer us tantalising glimpses of the possible roles organic molecules had in forming the Earth as we know it today.

24.2 Acid–base properties of amino acids

Amino acids are unusual molecules in that they possess both an acidic functional group and a basic functional group in the same molecule. This means that in the biological environment they can play the role of proton (H^+) donors or proton (H^+) acceptors.

Acidic and basic groups of amino acids

Among the most important chemical properties of amino acids are their acid–base properties; all can be weak polyprotic acids because of their —COOH and $—NH_3^+$ groups. The exact ability of these groups to accept or donate H^+ is indicated by their pK_a values. (Recall from chapter 11 that the smaller the pK_a the more acidic is the group. At lower pH, carboxylic acids are found in the RCOOH form and amines are found in the RNH_3^+ form. At higher pH, the opposite is true; carboxylic acids are present as the salt $RCOO^-$ and amines are present as uncharged RNH_2. Figure 24.5 on page 1069 shows how this looks at different pH). Table 24.2 gives pK_a values for each ionisable group of the 20 protein-derived amino acids.

TABLE 24.2 Values of pK_a for ionisable groups of amino acids.

Amino acid	pK_a of α-COOH	pK_a of α-NH_3^+	pK_a of side chain	Isoelectric point (pI)[a]
alanine	2.35	9.87	–[b]	6.11
arginine	2.01	9.04	12.48	10.76
asparagine	2.02	8.80	–	5.41
aspartic acid	2.10	9.82	3.86	2.98
cysteine	2.05	10.25	8.00	5.02
glutamic acid	2.10	9.47	4.07	3.08
glutamine	2.17	9.13	–	5.65
glycine	2.35	9.78	–	6.06
histidine	1.77	9.18	6.10	7.64
isoleucine	2.32	9.76	–	6.04
leucine	2.33	9.74	–	6.04
lysine	2.18	8.95	10.53	9.74
methionine	2.28	9.21	–	5.74
phenylalanine	2.58	9.24	–	5.91
proline	2.00	10.60	–	6.30
serine	2.21	9.15	–	5.68
threonine	2.09	9.10	–	5.60
tryptophan	2.38	9.39	–	5.88
tyrosine	2.20	9.11	10.07	5.63
valine	2.29	9.72	–	6.00

(a) See p. 1070.
(b) No ionisable side chain.

Acidity of α-carboxyl groups

The average value of pK_a for an α-carboxyl group of a protonated amino acid is 2.19. Thus, the α-carboxyl group is a considerably stronger acid than acetic acid (pK_a = 4.74) and other low-molar-mass aliphatic carboxylic acids. This greater acidity is accounted for by the electron-withdrawing inductive effect of the adjacent —NH_3^+ group. (Recall that we used similar reasoning in section 23.5 to account for the relative acidities of acetic acid and its mono-, di- and trichloroderivatives.)

The ammonium group has an electron-withdrawing inductive effect.

$$H_3\overset{+}{N}-\underset{\displaystyle R}{\underset{|}{C}}HCOOH + H_2O \rightleftharpoons H_3\overset{+}{N}-\underset{\displaystyle R}{\underset{|}{C}}HCOO^- + H_3O^+ \quad pK_a = 2.19$$

Acidity of side-chain carboxyl groups

Due to the electron-withdrawing inductive effect of the α-NH_3^+ group, the side-chain carboxyl groups of protonated aspartic acid and glutamic acid are also stronger acids than acetic acid (pK_a = 4.74). Notice that this acid-strengthening inductive effect decreases with increasing distance of the —COOH group from the α-NH_3^+ group. Compare the acidities of the α-COOH group of alanine (pK_a = 2.35), the γ-COOH group of aspartic acid (pK_a = 3.86) and the δ-COOH group of glutamic acid (pK_a = 4.07).

Acidity of α-ammonium groups

The average value of pK_a for an α-ammonium group, α-NH_3^+, is 9.47, compared with an average value of 10.76 for primary aliphatic ammonium ions (section 19.7).

$$H_3\overset{+}{N}-\underset{\displaystyle R}{\underset{|}{C}}HCOO^- + H_2O \rightleftharpoons H_2N-\underset{\displaystyle R}{\underset{|}{C}}HCOO^- + H_3O^+ \quad pK_a = 9.47$$

$$H_3\overset{+}{N}-\underset{\displaystyle CH_3}{\underset{|}{C}}HCH_3 + H_2O \rightleftharpoons H_2N-\underset{\displaystyle CH_3}{\underset{|}{C}}HCH_3 + H_3O^+ \quad pK_a = 10.60$$

Thus, the α-ammonium group of an amino acid is a slightly stronger acid than a primary aliphatic ammonium ion and, conversely, an α-amino group is a slightly weaker base than a primary aliphatic amine.

Basicity of the guanidine group of arginine

The side-chain guanidine group of arginine (—N=C(NH_2)$_2$) is a considerably stronger base than an aliphatic amine. Guanidine is the strongest base of any neutral compound. The remarkable basicity of the guanidine group of arginine is attributed to the large resonance stabilisation of the protonated form relative to the neutral form.

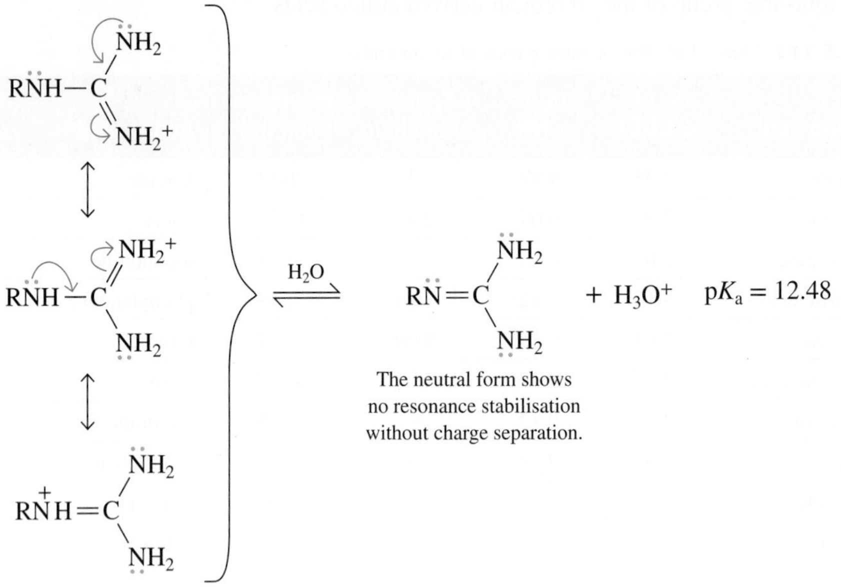

$$\left\{ RNH{-}C(NH_2){=}NH_2^+ \longleftrightarrow RNH{-}C(=NH_2^+)NH_2 \longleftrightarrow R\overset{+}{N}H{=}C(NH_2)_2 \right\} \underset{}{\overset{H_2O}{\rightleftharpoons}} RN{=}C(NH_2)_2 + H_3O^+ \quad pK_a = 12.48$$

The protonated form of the guanidinium ion side chain of arginine is a hybrid of three contributing structures.

The neutral form shows no resonance stabilisation without charge separation.

Basicity of the imidazole group of histidine

Because the imidazole group on the side chain of histidine contains six π electrons in a planar, fully conjugated ring, imidazole is classified as a heterocyclic aromatic amine (section 19.6). The unshared pair of electrons on one nitrogen atom is a part of the aromatic sextet, whereas that on the other nitrogen atom is not. It is the pair of electrons that is not part of the aromatic sextet that is responsible for the basic properties of the imidazole ring. Protonation of this nitrogen atom produces a resonance-stabilised cation.

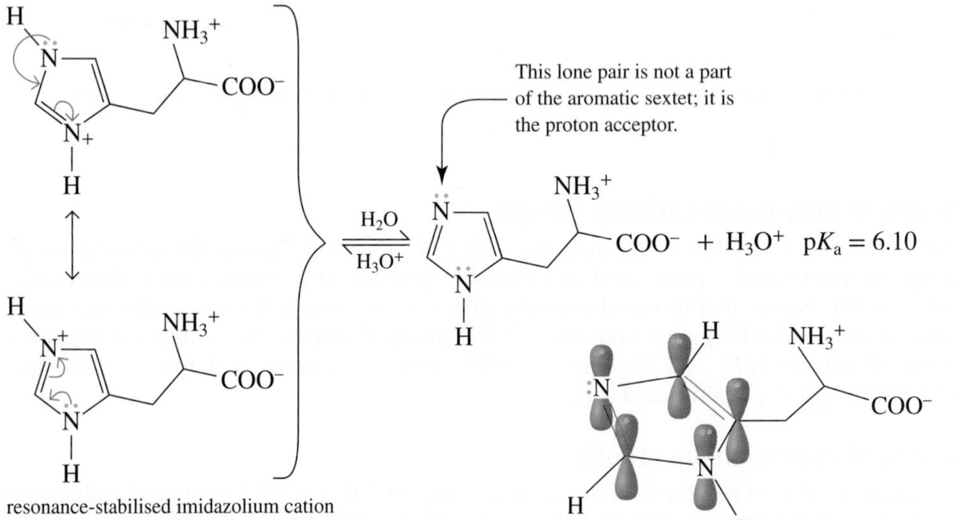

$$\text{imidazolium cation (two resonance structures)} \underset{H_3O^+}{\overset{H_2O}{\rightleftharpoons}} \text{histidine (neutral imidazole)} + H_3O^+ \quad pK_a = 6.10$$

resonance-stabilised imidazolium cation

This lone pair is not a part of the aromatic sextet; it is the proton acceptor.

Titration of amino acids

Values of pK_a for the ionisable groups of amino acids are most commonly obtained by acid–base titration (section 11.7) and by measuring the pH of the solution as a function of added base (or added acid, depending on how the titration is done). To illustrate this experimental procedure, consider a solution containing 1.00 mole of glycine to which has been added enough strong acid so that both the amino and carboxyl groups are fully protonated. Next, the solution is titrated with 1.00 M NaOH; the volume of base added and the pH of the resulting solution are recorded and then plotted as shown in figure 24.5.

The more acidic group, and the one to react first with added sodium hydroxide, is the carboxyl group. When exactly 0.50 mole of NaOH has been added, the carboxyl group is half neutralised.

At this point, the concentration of the zwitterion equals that of the positively charged ion, and the pH of 2.35 equals the pK_a value of the carboxyl group (pK_{a1}).

$$\text{At pH} = \text{p}K_{a1} \qquad \underset{\text{positive ion}}{[H_3\overset{+}{N}CH_2COOH]} = \underset{\text{zwitterion}}{[H_3\overset{+}{N}CH_2COO^-]}$$

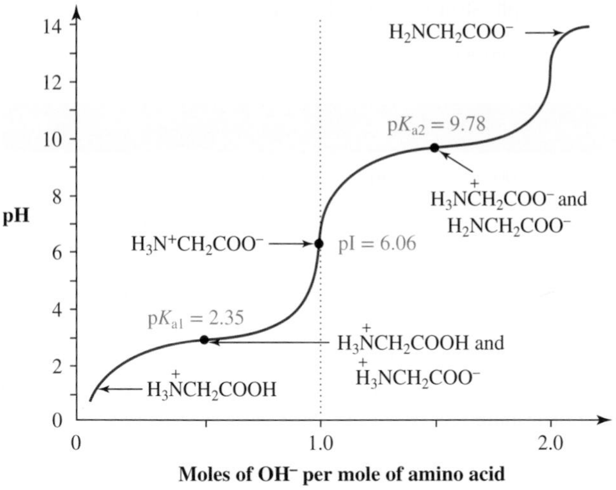

FIGURE 24.5 Titration of glycine with sodium hydroxide.

The end point of the first part of the titration is reached when 1.00 mole of NaOH has been added. At this point, the predominant species present is the zwitterion, and the observed pH of the solution is 6.06. The next section of the curve represents titration of the $—NH_3^+$ group. When another 0.50 mole of NaOH has been added (bringing the total to 1.50 moles), half of the $—NH_3^+$ groups are neutralised and converted to $—NH_2$. At this point, the concentrations of the zwitterion and negatively charged ion are equal, and the observed pH is 9.78, which equals the pK_a value of the ammonium group of glycine (pK_{a2}).

$$\text{At pH} = \text{p}K_{a2} \qquad \underset{\text{zwitterion}}{[H_3\overset{+}{N}CH_2COO^-]} = \underset{\text{negative ion}}{[H_2NCH_2COO^-]}$$

The second end point of the titration is reached when a total of 2.00 moles of NaOH have been added and glycine is converted entirely to an anion.

Amino acid charge at physiological pH

Knowing the pK_a of the functional groups allows us to understand the nature of the amino acid in a physiological context. Recall from chapter 11 that pK_a refers to the relationship between the equilibrium concentrations of the undissociated acid (HA) and the conjugate base (A^-). It is important to understand the implications of this for amino acids at the pH conditions found in biological environments. For instance, if an amino acid has a pK_a of 2, then $K_a = 10^{-2}$. The ratio of HA to A^- is derived from the Henderson–Hasselbalch equation:

$$K_a = \frac{[H_3O^+][RCOO^-]}{[RCOOH]}$$

Physiological systems are buffered and remain at a controlled pH close to 7, so $[H_3O^+] = 10^{-7}$ in these environments. Substituting into this equation gives:

$$K_a = 10^{-2} = \frac{10^{-7}[RCOO^-]}{[RCOOH]}$$

and rearranging gives:

$$\frac{10^{-2}}{10^{-7}} = 10^5 = \frac{[RCOO^-]}{[RCOOH]}$$

Consequently, the ratio of dissociated $RCOO^-$ to undissociated RCOOH is 100 000 : 1 (i.e. there is very little RCOOH present), and we can say that all of the amino acid is effectively present in the charged $RCOO^-$ form. If the pK_a value of the amino acid is equal to the pH, there is a 50 : 50 mixture of the two components:

$$K_a = 10^{-7} = \frac{10^{-7}[RCOO^-]}{[RCOOH]}$$

and rearranging gives:

$$\frac{10^{-7}}{10^{-7}} = 1 = \frac{[RCOO^-]}{[RCOOH]}$$

When the pK_a value is 1 log unit below physiological pH (pH = 7), the ratio is 90 : 10; if it is 2 log units below the pH, the ratio is 99 : 1. We can use this relationship to control the charge of the amino acid by controlling the pH. If we make the pH of the solution containing the amino acid 2 log units above its pK_a value, we know that the amino acid is essentially present (99%) in its dissociated form. To aid your understanding, table 24.3 shows the ratios of undissociated acids, HA, compared with the conjugate base, A^-, at physiological pH for various pK_a values.

TABLE 24.3 Ratios of undissociated amino acids, HA, compared with the conjugate base, A^-, at physiological pH for various pK_a values.

pK_a	4	5	6	7	8	9	10
A^-	99.9	99	90	50	10	1	0.1
HA	0.1	1	10	50	90	99	99.1

Using this analysis, it is clear why, at physiological pH, all amino-acid and side-chain carboxylic acid groups are ionised and found in the $RCOO^-$ form. Furthermore, at physiological pH, all amino, side-chain amino and guanidino groups are protonated. However, tyrosine's phenol group is not ionised, cysteine's thiol group is largely (but not completely) un-ionised, and histidine's side chain is largely (but not completely) protonated.

Isoelectric point

Chemists can control the degree of charge present on amino acids by controlling the pH of the solution in which they are dissolved. For example under very acidic conditions (pH < 1) glutamic acid has an overall charge of +1, whereas under very basic conditions (pH > 12) the charge on glutamic acid is −2. This concept is outlined in the table below using an acidic and a basic side chain amino acid.

Solution pH	pH < 1	pH ≈ 7	pH > 12
Glutamic acid	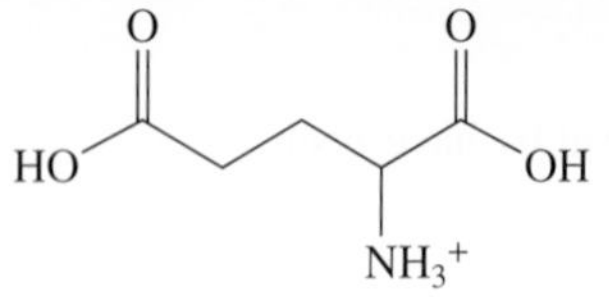 O, O, HO, OH, NH_3^+	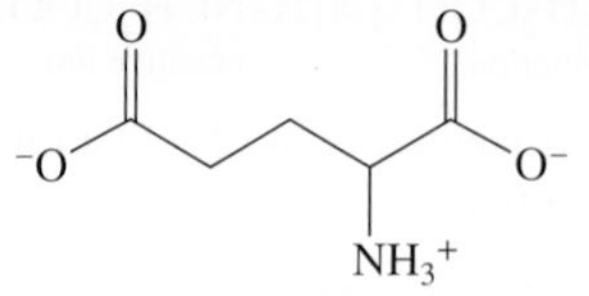 O, O, ^-O, O^-, NH_3^+	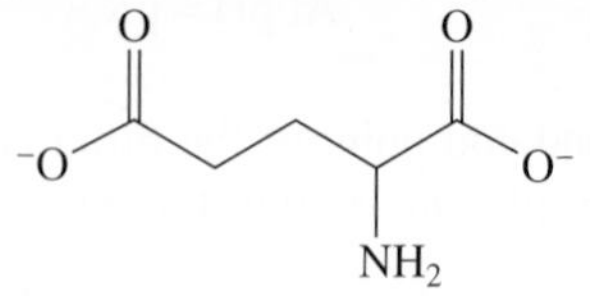 O, O, ^-O, O^-, NH_2
Overall charge	+1	−1	−2
Lysine	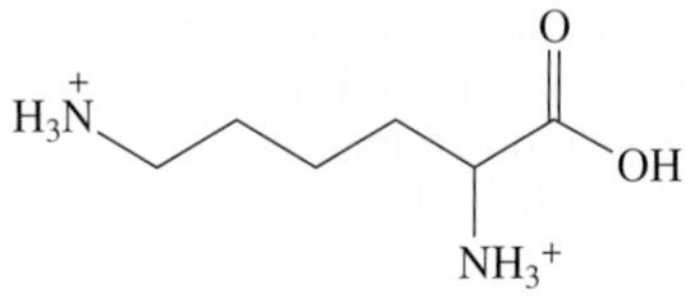 $H_3\overset{+}{N}$, O, OH, NH_3^+	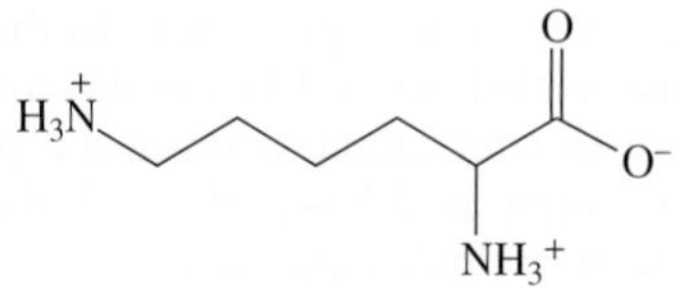 $H_3\overset{+}{N}$, O, O^-, NH_3^+	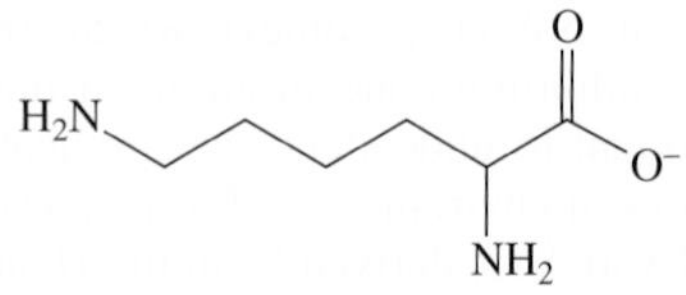 H_2N, O, O^-, NH_2
Overall charge	+2	+1	−1

By changing the pH of the solution, chemists can control the charge on the amino acids. You will note that at physiological pH, which is close to 7, both of the amino acids glutamic acid and lysine still possess an overall net charge. As the total charge possible for glutamic acid ranges from +1 to −2 (for pH between <1 and >12) clearly there must be some value of pH in between 1 and 12 where a solution of glutamic acid molecules must have (on average) no overall charge. Similarly for lysine which ranges from +2 to −1 (for pH <1 and >12) there must be some value of pH in between where the overall charge = 0. The pH at which the molecules of the amino acid in solution have on average a net charge of 0 is termed the **isoelectric point (pI)**.

Titration curves, such as that for glycine (see figure 24.5), enable us to accurately determine pK_a values for the ionisable groups of an amino acid. The pK_a values in turn allow us to determine the ratio of charged to uncharged groups in a molecule. By examining the titration curve for glycine (see figure 24.5 on the previous page), you can see that the isoelectric point for this amino acid falls halfway between the pK_a values for the carboxyl groups and the ammonium ion.

$$\begin{aligned} \text{pI} &= \frac{1}{2}(\text{p}K_a \text{ of } \alpha\text{-COOH} + \text{p}K_a \text{ of } NH_3^+) \\ &= \frac{1}{2}(2.35 + 9.78) \\ &= 6.06 \end{aligned}$$

At pH = 6.06, the predominant form of glycine molecules is the zwitterion; furthermore, at this pH, the concentration of positively charged glycine molecules equals the concentration of negatively charged glycine molecules.

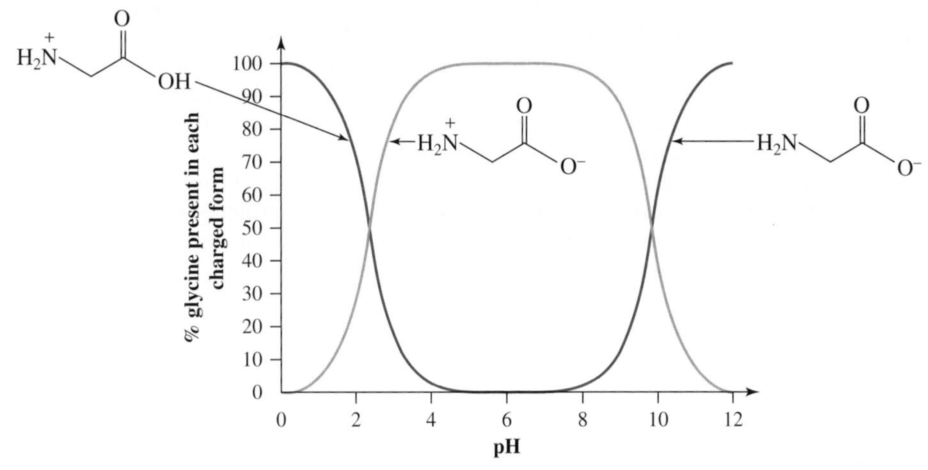

The isoelectric points for all of the common amino acids are given in table 24.2 (p. 1067). Given a value for the isoelectric point of an amino acid, it is then possible to estimate the charge on that amino acid at any pH. For example, the charge on tyrosine at pH = 5.63 (the isoelectric point of tyrosine) is 0. A small fraction of tyrosine molecules are positively charged at pH = 5.00 (0.63 of a log unit less than its pI), and virtually all are positively charged at pH = 3.63 (2.00 log units less than its pI). Looking at lysine again, the net charge on lysine is 0 at pH = 9.74. At pH values smaller than 9.74 then, an increasing fraction of lysine molecules are positively charged until they are essentially all carrying a +1 charge at pH 7. As the pH is lowered further more and more lysine molecules carry more than one positive charge. When the solution reaches pH values lower than the pK_a of 2.18, the majority of the molecules begin to carry a +2 charge.

Electrophoresis

Electrophoresis, the process of separating compounds on the basis of their electric charges, is used to isolate and identify the components present in mixtures of amino acids and proteins. Electrophoretic separations most commonly use polyacrylamide gels and the technique is referred to as PAGE, for *p*oly*a*crylamide *g*el *e*lectrophoresis. A typical electrophoresis separation set-up is shown in figure 24.6a and b.

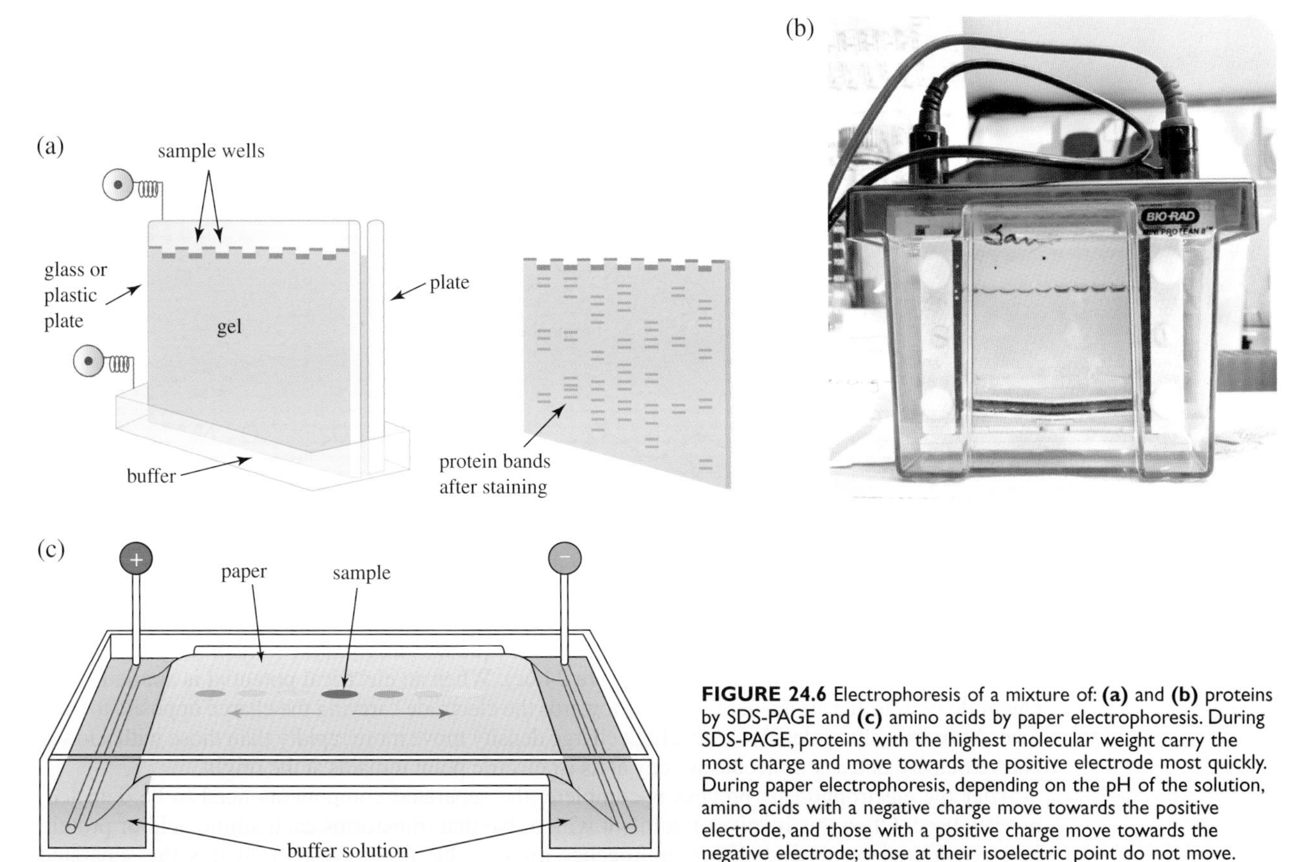

FIGURE 24.6 Electrophoresis of a mixture of: **(a)** and **(b)** proteins by SDS-PAGE and **(c)** amino acids by paper electrophoresis. During SDS-PAGE, proteins with the highest molecular weight carry the most charge and move towards the positive electrode most quickly. During paper electrophoresis, depending on the pH of the solution, amino acids with a negative charge move towards the positive electrode, and those with a positive charge move towards the negative electrode; those at their isoelectric point do not move.

The amino acids are colourless, but a blue dye, bromophenol blue, is added to aid in loading the mixture and to indicate when the process has been completed. When an electrical potential is applied to the gel, amino acids migrate towards the electrode; molecules with a high charge density move more rapidly than those with a lower charge density. Any molecule already at its isoelectric point remains at the origin. PAGE can be done on unmodified mixtures (native-PAGE), where separation is based on charge and molecular mass, or the proteins can be denatured (section 24.6) using *s*odium *d*odecyl *s*ulphate, a strong detergent (SDS-PAGE). The detergent leaves the protein with a similar linear structure and an overall negative charge. Separation then depends only on molecular weight. After the separation is complete, the gel may be treated with a dye such as Coomassie Brilliant Blue that transforms each component into a coloured band, making the separation visible. One advantage of native-PAGE over SDS-PAGE is that proteins are not altered in the separation process and may later be detected on the basis of specific enzyme interactions or subjected to further analytical techniques such as mass spectrometry (section 20.2).

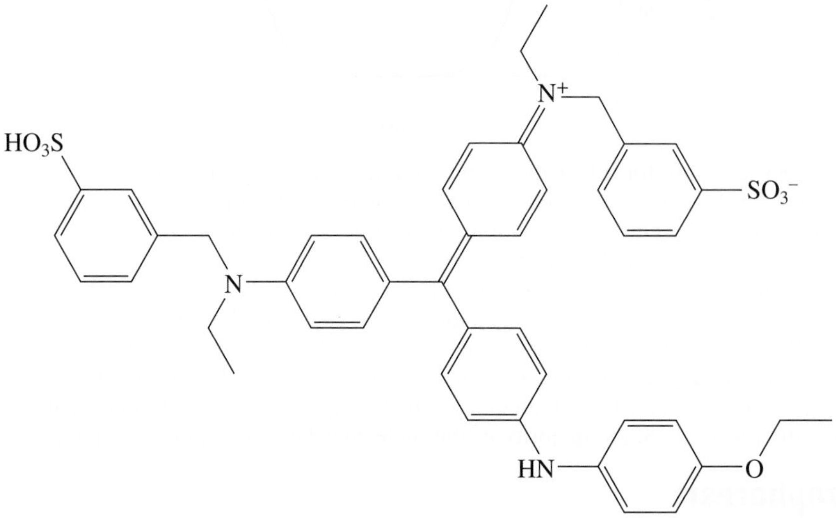

Coomassie Brilliant Blue

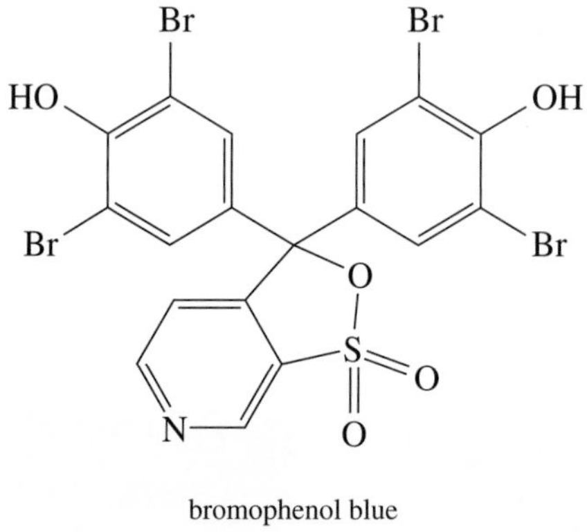

bromophenol blue

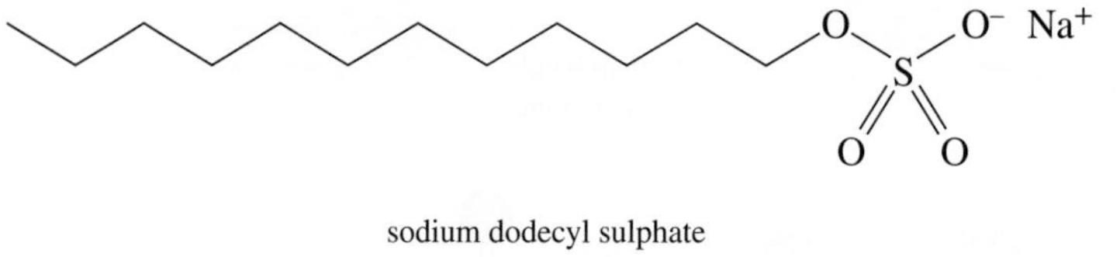

sodium dodecyl sulphate

Electrophoresis can also be performed using other polymers, starch, agar and even paper. With paper electrophoresis a paper strip saturated with an aqueous buffer of predetermined pH serves as a bridge between two electrode vessels (figure 24.6c). When an electrical potential is applied to the electrode vessels, the amino acids migrate towards the electrode carrying the charge opposite to their own. As previously, molecules with a high charge density move more rapidly than those with a lower charge density, and any molecule already at its isoelectric point remains at the origin.

After the electrophoresis process is complete, the separated components need to be detected, or visualised. Often this involves treatment with a dye that transforms each amino acid or protein into a coloured compound. Other approaches involve ‘blotting’, which transfers the separated compounds to another surface which may contain reactive dyes or enzyme-linked coloured responses. For mixtures of amino acids, treatment with ninhydrin (1,2,3-indanetrione monohydrate) is often used. Ninhydrin reacts with α-amino acids to produce an aldehyde, carbon dioxide and a purple-coloured anion.

$$RCH(NH_3^+)CO^- + 2\ \text{ninhydrin} \longrightarrow \text{purple-coloured anion} + RCHO + CO_2 + H_3O^+$$

an α-amino acid

ninhydrin

purple-coloured anion

This reaction is commonly used in both qualitative and quantitative analyses of amino acids. Of the 20 protein-derived α-amino acids, 19 have primary amino groups and give the same purple-coloured ninhydrin-derived anion. Proline, a secondary amine, gives a different, orange-coloured, compound.

WORKED EXAMPLE 24.2

Analysing amino acids based on their charge

The isoelectric point of tyrosine is 5.63. Towards which electrode does tyrosine migrate during paper electrophoresis at pH = 7.0?

Analysis

Negatively charged materials move towards the positive electrode and positively charged species move towards the negative electrode. You can determine the direction of movement during electrophoresis if you know the charge on the amino acid.

Solution

During paper electrophoresis at pH = 7.0 (more basic than its isoelectric point), tyrosine has a net negative charge so migrates towards the positive electrode.

Is our answer reasonable?

If the pH of a solution is equal to the pK_a value of an ionisable group, the group is present as a 50 : 50 mixture of charged and uncharged species. The situation becomes more complicated when there are several ionisable groups present, but the isoelectric point gives a measure of the pH required for most of the molecules to be uncharged. Here, the pH of the solution at 7.0 is more basic than the isoelectric point, so the molecule should be negatively charged.

PRACTICE EXERCISE 24.2

The isoelectric point of histidine is 7.64. Towards which electrode does histidine migrate during paper electrophoresis at pH = 7.0?

WORKED EXAMPLE 24.3

Separation of a mixture of amino acids based on their charge

The electrophoresis of a mixture of lysine, histidine and cysteine is carried out at pH = 7.64. Describe the behaviour of each amino acid under these conditions.

Analysis

Negatively charged materials move towards the positive electrode and positively charged species move towards the negative electrode. If you know the pH of the solution and the values of the isoelectric point for the amino acids involved, you can determine the charges present and the direction in which the species will move.

Solution

The isoelectric point of histidine is 7.64. At this pH, histidine has a net charge of 0 and so it does not move from the origin. The pI of cysteine is 5.02; at pH = 7.64 (more basic than its isoelectric point), cysteine has a net negative charge and so it moves towards the positive electrode. The pI of lysine is 9.74; at pH = 7.64 (more acidic than its isoelectric point), lysine has a net positive charge and so it moves towards the negative electrode.

Is our answer reasonable?

This separation can work only if the molecules present have different isoelectric points. Here the isoelectric points are 5.02, 7.64 and 9.74 so, at pH = 7.64, we have an uncharged molecule, a negatively charged molecule and a positively charged molecule.

PRACTICE EXERCISE 24.3 Describe the behaviour of a mixture of glutamic acid, arginine and valine during paper electrophoresis at pH = 6.0.

24.3 Peptides, polypeptides and proteins

In 1902, Emil Fischer proposed that proteins were long chains of amino acids joined together by amide bonds between the α-carboxyl group of one amino acid and the α-amino group of another. For these amide bonds, Fischer proposed the special name **peptide bond**. Figure 24.7 shows the peptide bond formed between serine and alanine in the molecule serylalanine.

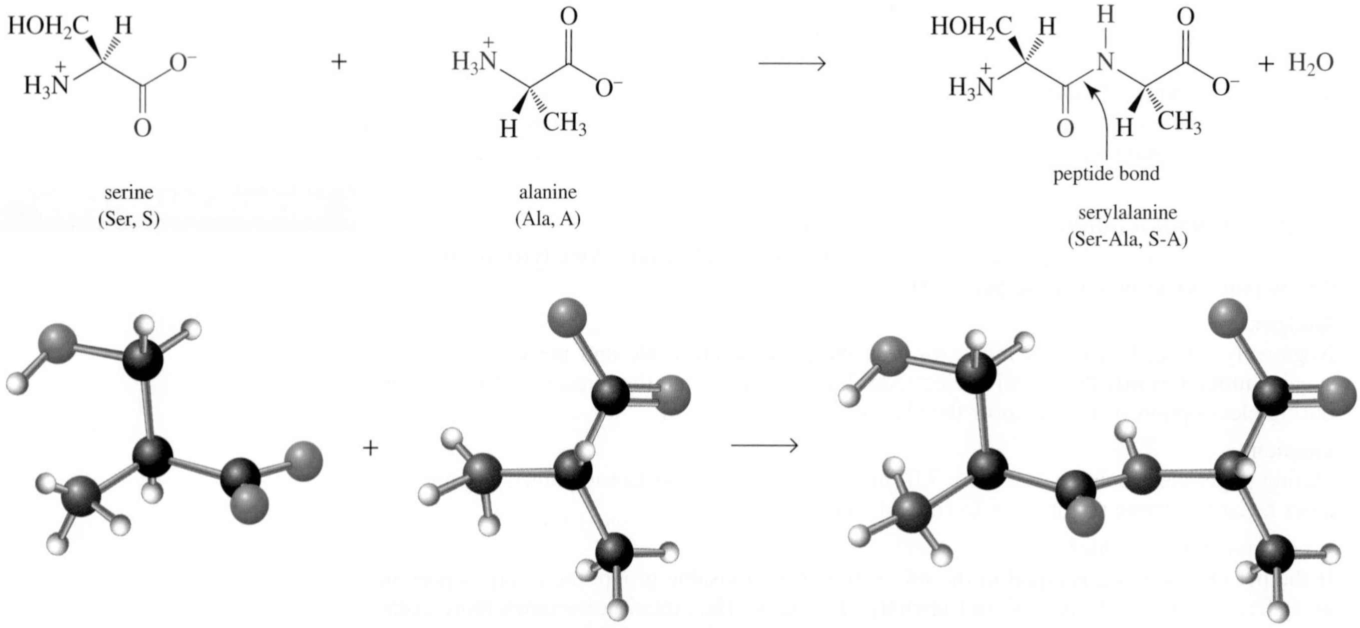

serine (Ser, S) + alanine (Ala, A) ⟶ serylalanine (Ser-Ala, S-A) + H_2O

peptide bond

FIGURE 24.7 The peptide bond in serylalanine.

Peptide is the name given to a short polymer of amino acids. We classify peptides by the number of amino acid units in their chains. A molecule containing two amino acids joined by an amide bond is called a **dipeptide**. Those containing three to ten amino acids are called **tripeptides, tetrapeptides, pentapeptides** and so on. Molecules containing more than ten but fewer than 20 amino acids are called **oligopeptides**. Those containing 20 or more amino acids are called **polypeptides**. **Proteins** are biological macromolecules with a molar mass of 5000 u (more than about 40 amino acids) or greater and consisting of one or more polypeptide chains. The distinctions in this terminology are not at all precise.

By convention, polypeptides are written from left to right, beginning with the amino acid with the free —NH_3^+ group and proceeding towards the amino acid with the free —COO^- group. The amino acid with the free —NH_3^+ group is called the ***N*-terminal amino acid**, and that with the free —COO^- group is called the ***C*-terminal amino acid**.

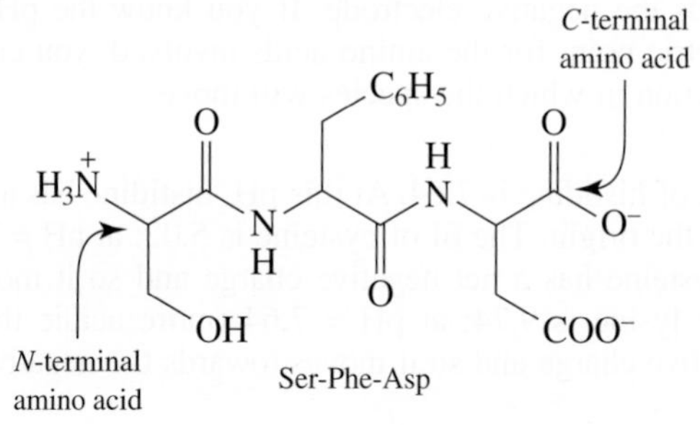

C-terminal amino acid

N-terminal amino acid

Ser-Phe-Asp

WORKED EXAMPLE 24.4

Representing the structures of peptides

Draw the structural formula for Cys-Arg-Met-Asn. Label the *N*-terminal amino acid and the *C*-terminal amino acid. What is the net charge on this tetrapeptide at pH = 6.0?

Analysis

As peptide structures can be quite complicated, it is important to represent them in as simple and common a form as possible. Start with Cys at the left-hand end — this will be the *N*-terminal amino acid. Form a peptide bond between the carboxyl carbon atom of this and the amino group of the next amino acid, Arg. Continue along the chain in this fashion until you come to the end amino acid, Asn. This will be the *C*-terminal amino acid. The net charge can be calculated by analysis of the pK_a data for all of the acidic and basic groups present in the peptide.

Solution

The backbone of Cys-Arg-Met-Asn, a tetrapeptide, is a repeating sequence of nitrogen–α-carbon–carbonyl. The following is its structural formula.

N-terminal amino acid
pK_a = 8.0
SH
C-terminal amino acid
$H_3\overset{+}{N}$
pK_a = 10.2
O^-
NH_2
pK_a of conjugate acid = 4.1
NH
pK_a = 12.5
H_2N
NH_2^+

At pH = 6.0, the basic groups present are strong enough bases (pK_a of their conjugate acids = 10.2 and 12.5) to be protonated, and the acidic group is a strong enough acid (pK_a = 4.1) to dissociate to give the conjugate base. The thiol is not acidic enough (pK_a = 8.0) to dissociate. The net charge on this tetrapeptide at pH = 6.0 is therefore +1.

Is our answer reasonable?

Linking amino acids to make a peptide removes the effect of many of the carboxylic acid and amine groups on the charge at a particular pH. You should note that a peptide (or amide) linkage is not basic, so the nitrogen atom in this part of the molecule will not be able to be protonated and become charged. Only the terminal amine and carboxylic acid, as well as their side-chain groups, govern the overall charge on a peptide.

PRACTICE EXERCISE 24.4

Draw the structural formula for Lys-Phe-Ala. Label the *N*-terminal amino acid and the *C*-terminal amino acid. What is the net charge on this tripeptide at pH = 6.0?

24.4 Primary structure of polypeptides and proteins

The **primary (1°) structure** of a polypeptide or protein is the sequence of amino acids in its polypeptide chain. In this sense, the primary structure is a complete description of all covalent bonding in a polypeptide or protein.

In 1953, Frederick Sanger (figure 24.8) of Cambridge University, England, reported the primary structure of the two polypeptide chains of the hormone insulin. This was a remarkable achievement in analytical chemistry; it also clearly established that all the molecules of a given protein have the same amino acid composition and the same amino acid sequence. Today, the amino acid sequences of over 20 000 different proteins are known, and the number is growing rapidly.

FIGURE 24.8 Frederick Sanger was awarded the Nobel Prize in chemistry in 1958 for his work on insulin and in 1980 for his work on determining the base sequences in DNA.

Amino acid analysis

The first step in determining the primary structure of a polypeptide is hydrolysis and quantitative analysis of its amino acid composition. Recall from section 23.5 (p. 1034) that amide bonds are resistant to hydrolysis. Typically, samples of protein are hydrolysed in 6 M HCl in sealed glass vials at 110 °C for 24 to 72 hours. (This hydrolysis can be done in a microwave oven in a shorter time.)

After the polypeptide is hydrolysed, the resulting mixture of amino acids is analysed by ion-exchange chromatography. In this process, the mixture of amino acids is passed through a specially packed column. Each of the 20 amino acids requires a different time to pass through the column. Amino acids are detected by reaction with ninhydrin as they emerge from the column (section 24.2), followed by absorption spectroscopy. Current procedures for the hydrolysis of polypeptides and the analysis of amino acid mixtures have been refined to the point where it is possible to determine the amino acid composition from as little as 50 nanomoles (50×10^{-9} mole) of a polypeptide. Figure 24.9 shows the analysis of a polypeptide hydrolysate by ion-exchange chromatography. Note that, during hydrolysis, the side-chain amide groups of asparagine and glutamine are hydrolysed, and these amino acids are detected as aspartic acid and glutamic acid. For each glutamine or asparagine hydrolysed, an equivalent amount of ammonium chloride is formed.

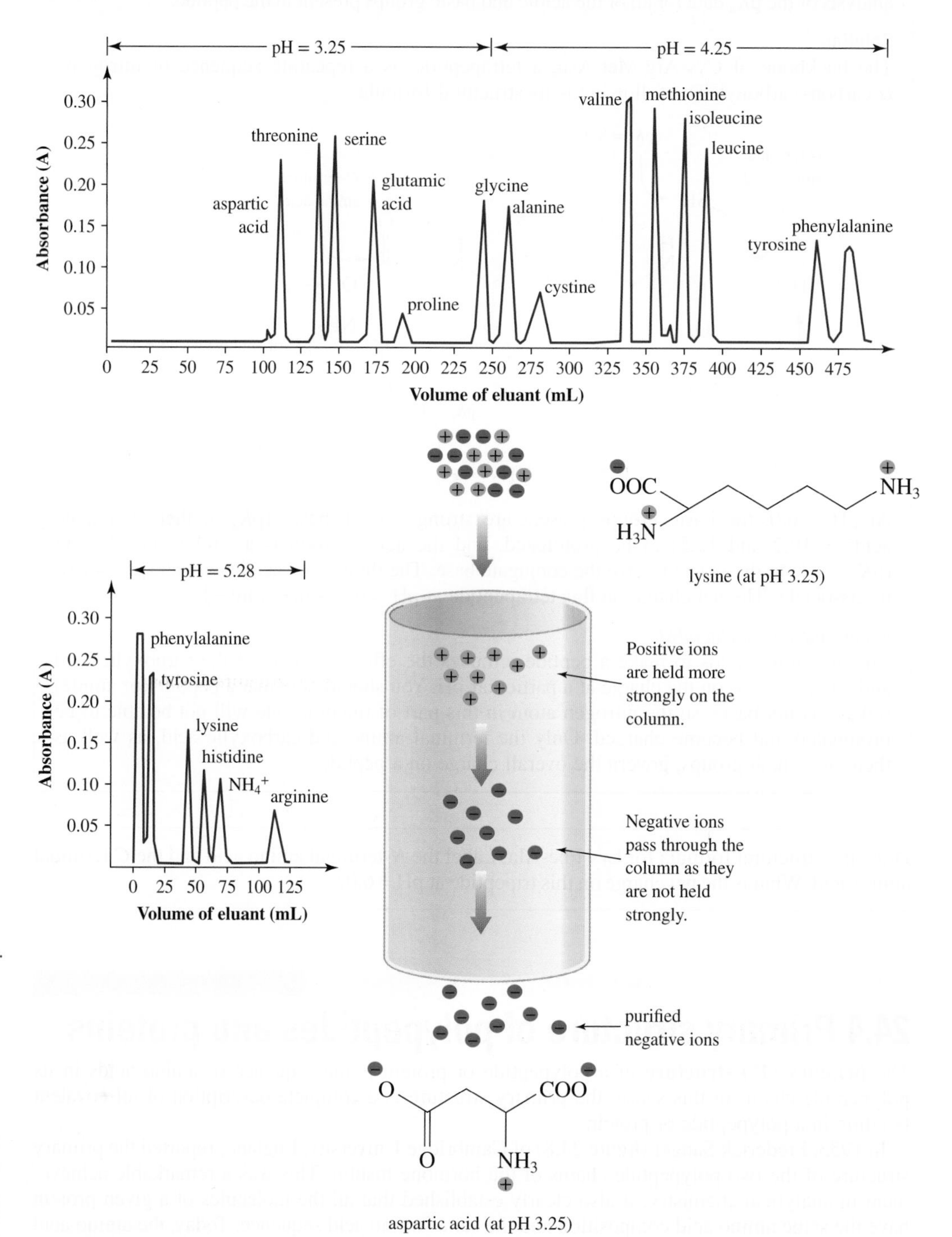

FIGURE 24.9 Analysis of a mixture of amino acids by ion-exchange chromatography using Amberlite IR-120, a sulfonated polystyrene resin. The resin contains phenyl-$SO_3^-Na^+$ groups. The amino acid mixture is applied to the column at low pH (3.25), conditions under which the acidic amino acids (Asp, Glu) are weakly bound to the resin and the basic amino acids (Lys, His, Arg) are tightly bound. Sodium citrate buffers of two different concentrations and three different values of pH are used to elute the amino acids from the column. Cysteine is determined as cystine, Cys-S-S-Cys, the disulfide of cysteine.

Sequence analysis

Once the amino acid composition of a polypeptide has been determined, the next step is to determine the order in which the amino acids are joined in the polypeptide chain. The most common sequencing strategy is to cleave the polypeptide at specific peptide bonds, determine the sequence of each fragment, and then match overlapping fragments to arrive at the sequence of the polypeptide.

24.5 Three-dimensional shapes of polypeptides and proteins

Many of the properties of polypeptides and proteins are governed by the precise three-dimensional shape of these complex molecules. The complexity of the shape arises from the nature of the peptide bond.

Geometry of a peptide bond

In the late 1930s, Linus Pauling (Nobel Prize in chemistry, 1954) began a series of studies aimed at determining the geometry of a peptide bond. One of his first discoveries was that a peptide bond is planar. As shown in figure 24.10, the four atoms of a peptide bond and the two α-carbon atoms joined to it all lie in the same plane.

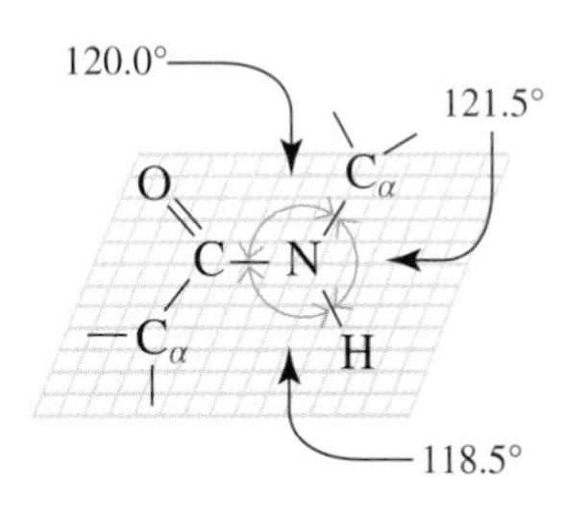

FIGURE 24.10 Planarity of a peptide bond. Bond angles around the carbonyl carbon atom and the amide nitrogen atom are approximately 120°.

Had you been asked in chapter 5 to describe the geometry of a peptide bond, you probably would have predicted bond angles of 120° around the carbonyl carbon atom and 109.5° around the amide nitrogen atom. This prediction agrees with the observed bond angles of approximately 120° around the carbonyl carbon atom. It does not agree, however, with the observed bond angles of 120° around the amide nitrogen. To account for this geometry, Pauling proposed that a peptide bond is more accurately represented as a resonance hybrid of these two contributing structures.

structure 1 $\longleftrightarrow$ structure 2 $\longleftrightarrow$ hybrid

Contributing structure 1 shows a carbon–oxygen double bond, and structure 2 shows a carbon–nitrogen double bond. The hybrid, of course, is neither of these; in the real structure, the carbon–nitrogen bond has considerable double-bond character. Accordingly, in the hybrid, the six-atom group shown in the two resonance structures is planar.

Because of the partial double-bond character, two configurations are possible for the atoms of a planar peptide bond. In one, the two α-carbon atoms are *cis* to each other; in the other, they are *trans* to each other. The *trans* configuration is more favourable because, in the *trans* configuration, the α-carbon atoms with the bulky groups bonded to them are further from each other than they are in the *cis* configuration. Virtually all peptide bonds in naturally occurring proteins studied to date have the *trans* configuration.

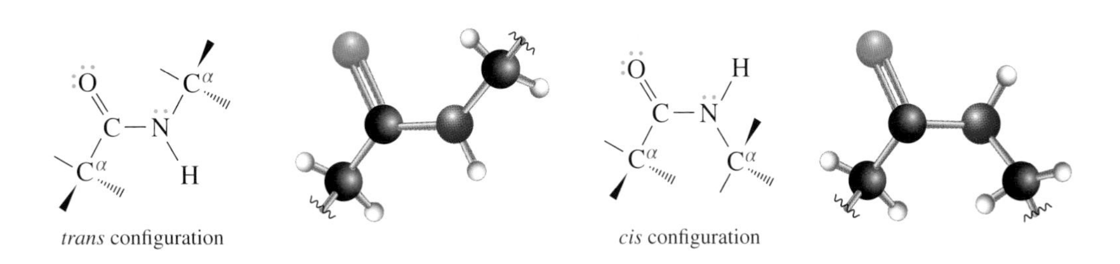

Secondary structure

Secondary (2°) structure is the ordered arrangement (conformation) of amino acids in localised regions of a polypeptide or protein molecule. The first studies of polypeptide conformations were carried out by Linus Pauling and Robert Corey, beginning in 1939. They assumed that, in conformations of greatest stability, all atoms in a peptide bond lie in the same plane and there is hydrogen bonding between the N—H of one peptide bond and the C═O of another, as shown in figure 24.11 (overleaf).

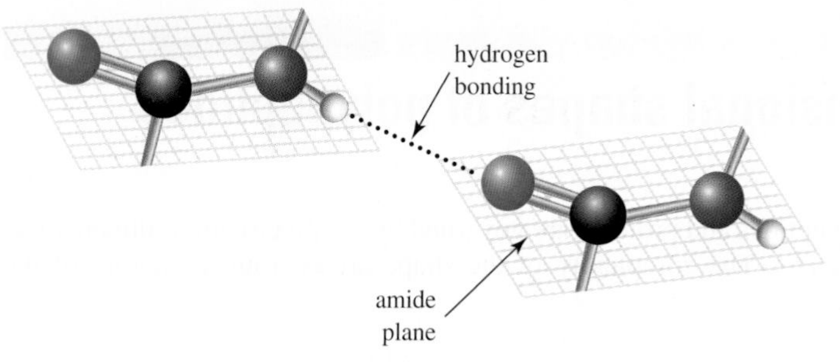

FIGURE 24.11 Hydrogen bonding between amide groups.

On the basis of model building, Pauling proposed that two types of secondary structure should be particularly stable: the α-helix and the antiparallel β-pleated sheet.

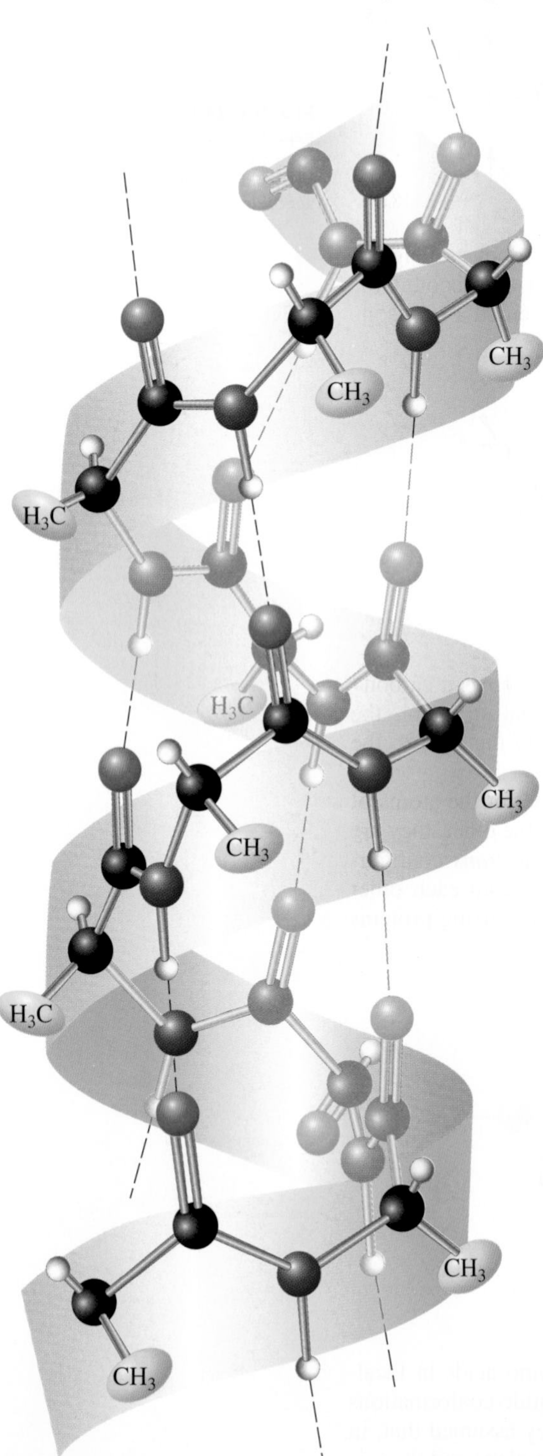

FIGURE 24.12 An α-helix. The polypeptide chain is repeating units of L-alanine. (Dashed lines indicate hydrogen bonding.)

The α-helix

In an **α-helix** pattern, shown in figure 24.12, a polypeptide chain is coiled in a spiral. As you study this section of the α-helix, note the following.

1. The helix is coiled in a clockwise, or right-handed, manner. *Right-handed* means that if you turn the helix clockwise, it twists away from you. In this sense, a right-handed helix is analogous to the right-handed thread of a common wood or machine screw.
2. There are 3.6 amino acids per turn of the helix.
3. Each peptide bond is *trans* and planar.
4. The N—H group of each peptide bond points roughly downwards, parallel to the axis of the helix, and the C═O of each peptide bond points roughly upwards, also parallel to the axis of the helix.
5. The carbonyl group of each peptide bond is hydrogen bonded to the N—H group of the peptide bond four amino acid units away from it. Hydrogen bonds are shown as dashed lines.
6. All R groups point outwards from the helix.

Almost immediately after Pauling proposed the α-helix conformation, other researchers proved the presence of α-helix conformations in keratin, the protein of hair and wool. It soon became obvious that the α-helix is one of the fundamental folding patterns of polypeptide chains.

The β-pleated sheet

An antiparallel **β-pleated sheet** consists of an extended polypeptide chain with neighbouring sections of the chain running in opposite (antiparallel) directions. In a parallel β-pleated sheet, the neighbouring sections run in the same direction. In contrast to the α-helix arrangement, N—H and C═O groups lie in the plane of the sheet and are roughly perpendicular to the long axis of the sheet. The C═O group of each peptide bond is hydrogen bonded to the N—H group of a peptide bond of a neighbouring section of the chain (figure 24.13).

As you study this section of β-pleated sheet, note the following.

1. The three sections of the polypeptide chain lie adjacent to each other and run in opposite (antiparallel) directions.
2. Each peptide bond is planar, and the α-carbon atoms are *trans* to each other.
3. The C═O and N—H groups of peptide bonds from adjacent sections point at each other and are in the same plane, so that hydrogen bonding is possible between adjacent sections.
4. The R groups on any one chain alternate, first above, then below, the plane of the sheet, and so on.

The β-pleated sheet conformation is stabilised by hydrogen bonding between N—H groups of one section of the chain and C═O groups of an adjacent section. By comparison, the α-helix is stabilised by hydrogen bonding between N—H and C═O groups within the same polypeptide chain.

Tertiary structure

Tertiary (3°) structure is the overall folding pattern and arrangement in space of all atoms in a single polypeptide chain. No sharp dividing line exists between secondary and tertiary structures. Secondary structure refers to the spatial arrangement of amino acids close to one another on a polypeptide chain, whereas tertiary structure refers to the three-dimensional arrangement of all atoms in a polypeptide chain. Among the most important factors in maintaining 3° structure are disulfide bonds, hydrophobic interactions, hydrogen bonding and salt linkages.

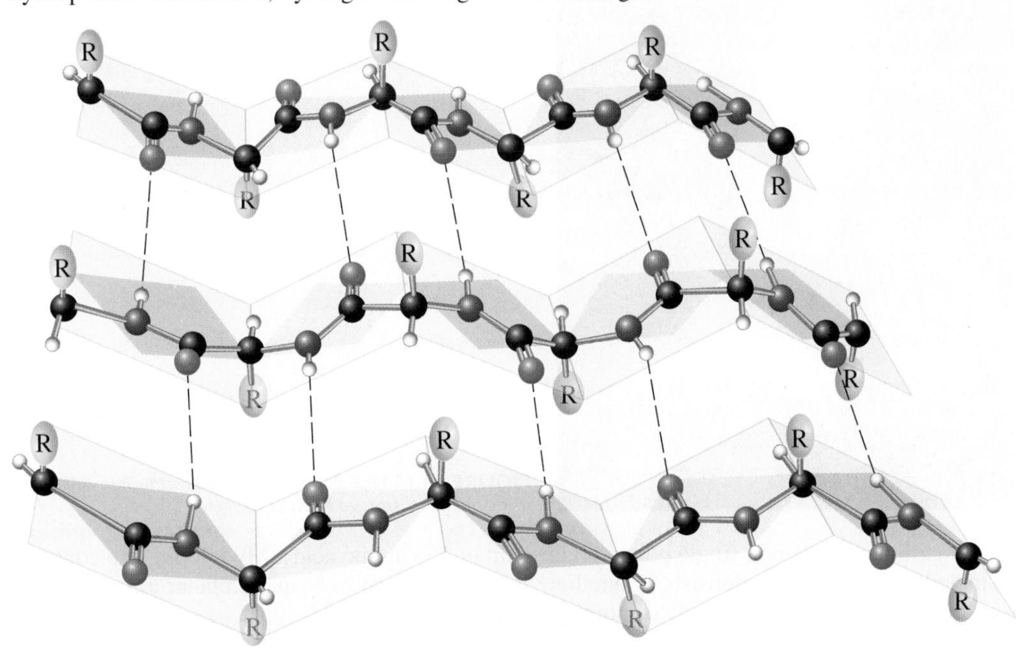

FIGURE 24.13 Three polypeptide chains running in opposite (antiparallel) directions in a β-pleated sheet conformation (dashed lines indicate hydrogen bonding; yellow shapes represent side chains).

Disulfide bonds play an important role in maintaining tertiary structure. Disulfide bonds are formed between side chains of two cysteine units by oxidation of their thiol groups (—SH) to form a disulfide bond (section 19.5). Treatment of a disulfide bond with a reducing agent regenerates the thiol groups.

side chains of cysteine

SH

SH

$\underset{\text{reduction}}{\overset{\text{oxidation}}{\rightleftharpoons}}$

S—S

a disulfide bond

Figure 24.14 shows the amino acid sequence of human insulin. This protein consists of two polypeptide chains: an A chain of 21 amino acids and a B chain of 30 amino acids. The A chain is bonded to the B chain by two interchain disulfide bonds. An intrachain disulfide bond also connects the cysteine units at positions 6 and 11 of the A chain.

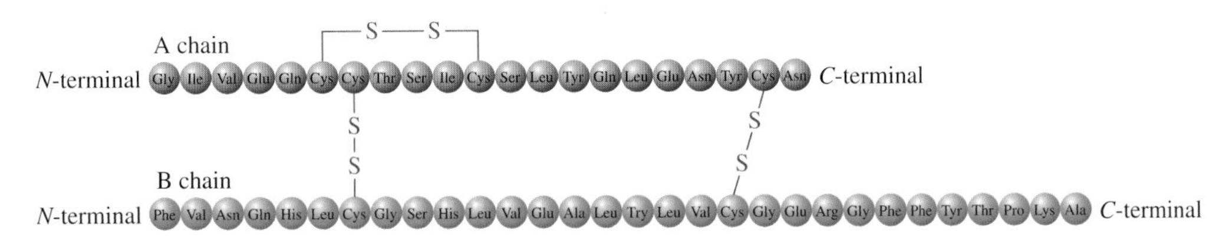

FIGURE 24.14 Human insulin. The A chain of 21 amino acids and B chain of 30 amino acids are connected by interchain disulfide bonds between A7 and B7 and between A20 and B19. In addition, a single intrachain disulfide bond occurs between A6 and A11.

As an example of 2° and 3° structure, let us look at the three-dimensional structure of myoglobin, a protein found in skeletal muscle and particularly abundant in diving mammals, such as whales (figure 24.15, overleaf), dolphins and seals. Myoglobin and its structural relative, haemoglobin, are the oxygen transport and storage molecules of vertebrates. Haemoglobin binds molecular oxygen in the lungs and transports it to myoglobin in muscles. Myoglobin stores molecular oxygen until it is required for metabolic oxidation.

FIGURE 24.15 Like all humpback whales, Migaloo relies on myoglobin as a storage form of oxygen.

Myoglobin consists of a single polypeptide chain of 153 amino acids. Myoglobin also contains a single haem unit. Haem consists of one Fe^{2+} ion, coordinated in a square planar array with the four nitrogen atoms of a molecule of porphyrin (figure 24.16).

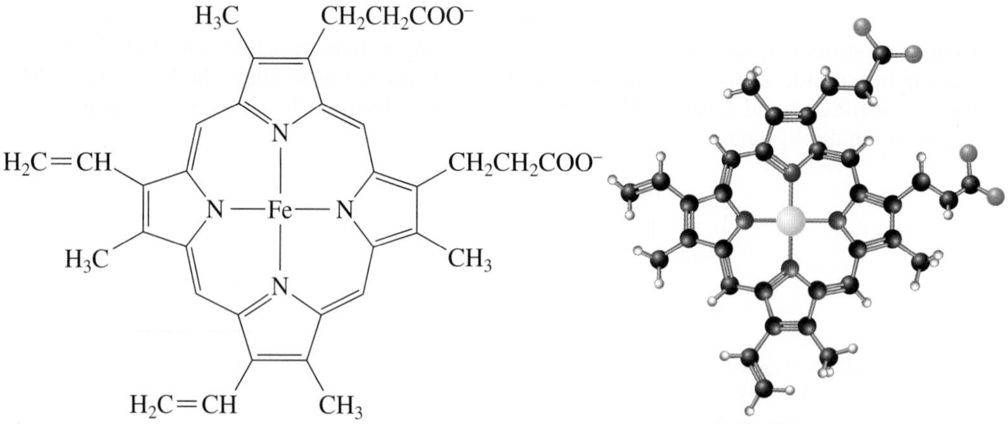

H_3C $CH_2CH_2COO^-$

$H_2C{=}CH$ N—Fe—N $CH_2CH_2COO^-$

H_3C CH_3

$H_2C{=}CH$ CH_3

FIGURE 24.16 Representations of the structure of haem found in myoglobin and haemoglobin.

Determination of the three-dimensional structure of myoglobin represented a milestone in the study of molecular architecture. For their contribution to this research, John C Kendrew and Max F Perutz, both of Britain, shared the 1962 Nobel Prize in chemistry. The secondary and tertiary structures of myoglobin are shown in figure 24.17. The single polypeptide chain is folded into a complex, almost boxlike shape.

There are several important structural features of the three-dimensional shape of myoglobin.

1. The backbone consists of eight relatively straight sections of α-helix, each separated by a bend in the polypeptide chain. The longest section of α-helix has 24 amino acids, the shortest has seven. Some 75% of the amino acids are found in these eight regions of α-helix.
2. Hydrophobic side chains of phenylalanine, alanine, valine, leucine, isoleucine and methionine are clustered in the interior of the molecule, where they are shielded from contact with water. **Hydrophobic interactions** are a major factor in directing the folding of the polypeptide chain of myoglobin into this compact, three-dimensional shape.
3. The outer surface of myoglobin is coated with hydrophilic side chains, such as those of lysine, arginine, serine, glutamic acid, histidine and glutamine, which interact with the aqueous environment by **hydrogen bonding**. The only polar side chains that point to the interior of the myoglobin molecule are those of two histidine units, which point inwards towards the haem group.

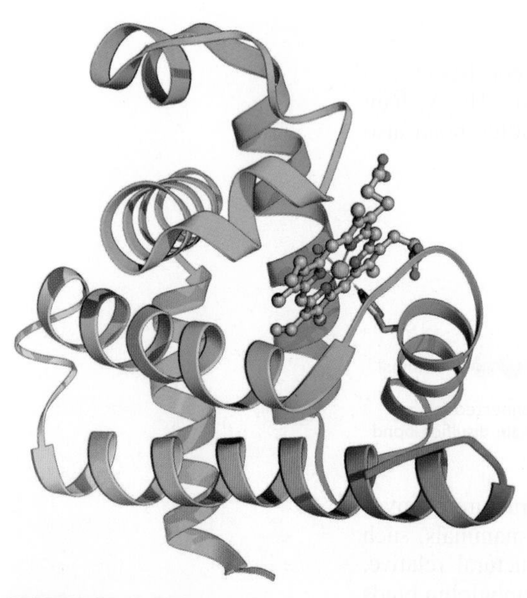

FIGURE 24.17 Ribbon model of myoglobin. The polypeptide chain is shown as a ribbon, the haem ligand as a ball-and-stick model and the Fe atom as an orange sphere.

4. Oppositely charged amino acid side chains close to each other in the three-dimensional structure interact by electrostatic attractions called **salt linkages**. An example of a salt linkage is the attraction of the side chains of lysine (—NH_3^+) and glutamic acid (—COO^-).

The tertiary structures of hundreds of proteins have also been determined. It is clear that proteins contain α-helix and β-pleated sheet structures, but that wide variations exist in the relative amounts of each. Lysozyme, with 129 amino acids in a single polypeptide chain, has only 25% of its amino acids in α-helix regions. Cytochrome, with 104 amino acids in a single polypeptide chain, has no α-helix structure, but does contain several regions of β-pleated sheet. Yet, whatever the proportions of α-helix, β-pleated sheet or other periodic structure, virtually all nonpolar side chains of water-soluble proteins are directed towards the interior of the molecule, whereas polar side chains are on the surface of the molecule, in contact with the aqueous environment.

WORKED EXAMPLE 24.5

Recognising peptide side chains that influence tertiary structure

The side chains of which of the following amino acids can form hydrogen bonds with the side chain of threonine?

(a) valine
(b) asparagine
(c) phenylalanine
(d) histidine
(e) tyrosine
(f) alanine

Analysis

Hydrogen bonds occur between heteroatoms that possess lone pairs of nonbonding electrons and polarised N—H or O—H bonds. Threonine has an —OH group on the side chain, which is a strong hydrogen bonding group. It should be able to form hydrogen bonds with any group containing lone pairs of N—H or O—H bonds.

Solution

The side chain of threonine contains a hydroxyl group that can participate in hydrogen bonding in two ways.

1. Its oxygen atom has a partial negative charge and can function as a hydrogen bond acceptor.
2. Its hydrogen atom has a partial positive charge and can function as a hydrogen bond donor.

Therefore, the side chain of threonine can form hydrogen bonds with the side chains of tyrosine, asparagine and histidine.

PRACTICE EXERCISE 24.5

The side chains of which amino acids can form salt linkages with the side chain of lysine at pH = 7.4?

Quaternary structure

Most proteins with molar mass greater than 50 000 u consist of two or more noncovalently linked polypeptide chains. The arrangement of protein monomers into an aggregation is known as **quaternary (4°) structure**. A good example is haemoglobin (figure 24.18), a protein that consists of four separate polypeptide chains: two α-chains of 141 amino acids each and two β-chains of 146 amino acids each.

The separate polypeptides pack together and are held in this arrangement by the same type of interactions seen within the tertiary structures: hydrogen bonds, salt bridges and hydrophobic interactions. For proteins in aqueous environments, aggregations in quaternary structures are stabilised mainly by hydrophobic interactions. Thus, separate polypeptide chains fold into compact three-dimensional shapes to expose polar side chains to the aqueous environment; in most cases, however, this still leaves some hydrophobic sections of the proteins in contact with water. By aggregating to form quaternary structures, they are protected from exposure.

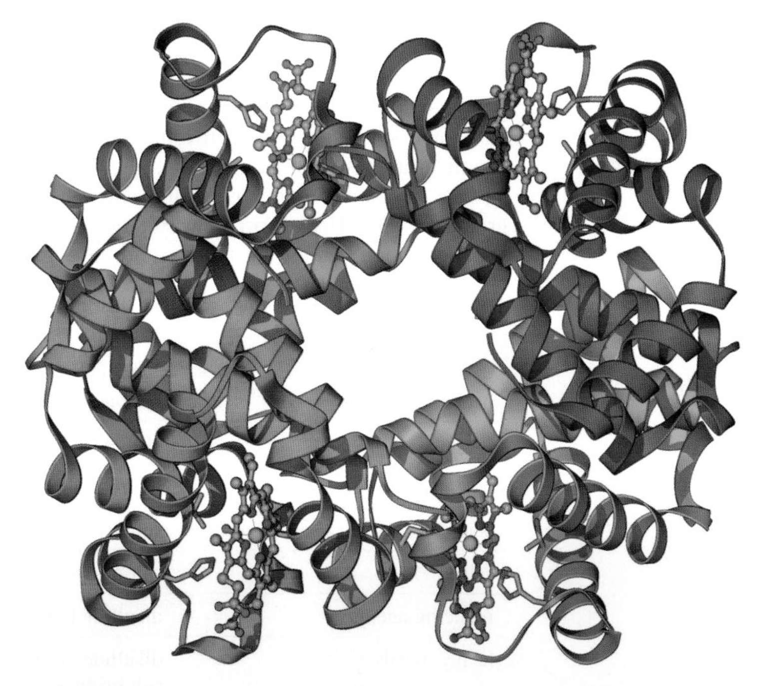

FIGURE 24.18 Ribbon model of haemoglobin. The α-chains and β-chains are shown as ribbons, the haem ligands as ball-and-stick models with the Fe atoms at their centres.

If two or more proteins assemble so that the hydrophobic sections match up, these components are then shielded from the aqueous environment. The numbers of subunits for several proteins that are known to form quaternary structures are shown in table 24.4.

TABLE 24.4 Quaternary structure of selected proteins.

Protein	Number of subunits
alcohol dehydrogenase	2
aldolase	4
haemoglobin	4
lactate dehydrogenase	4
insulin	6
glutamine synthetase	12
tobacco mosaic virus protein disc	17

On the other hand, about one-third of all proteins are actually present in the nonaqueous environment of cell membranes. Proteins that are incorporated in cell membranes are called **integral membrane proteins** because they traverse a membrane bilayer partially or completely. To keep the protein stable in the nonpolar environment of a lipid bilayer, these proteins form quaternary structures in which the outer surface is largely nonpolar and can interact favourably with the lipid environment. The integral membrane proteins aggregate because most of the polar groups turn inwards away from the nonpolar environment, thus forming the quaternary structures and shielding these groups from the lipids. Figure 24.19 represents the four levels of organisational structure found in proteins.

24.6 Denaturing proteins

The functions and properties of proteins arise from a combination of the secondary, tertiary and quaternary structures that give a protein its particular shape and conformation. Any physical or chemical agent that interferes with these stabilising structures changes the conformation of the protein and often removes the protein's functionality. We call this process **denaturation**.

Heat, for example, breaks apart hydrogen bonds, so boiling a protein destroys its α-helical and β-pleated sheet structures. The polypeptide chains of globular proteins unfold when heated; the unravelled proteins can then bind strongly to each other and precipitate or coagulate. This is what happens when an egg is boiled and the 'liquid' white is turned into a 'solid'.

Similar transformations can be achieved by the addition of denaturing chemicals. For example, aqueous urea, $H_2N—CO—NH_2$, forms strong hydrogen bonds, so it can disrupt other hydrogen bonds and cause globular proteins to unfold. Ethanol denatures proteins by coagulation. Detergents change protein conformation by disturbing the hydrophobic interactions, whereas acids, bases and salts interfere with ionic salt bridges in the tertiary and quaternary structures. Other chemical agents such as reducing agents can break the key disulfide bonds (—S—S—) holding tertiary structures together. Additionally, heavy metal ions (Pb^{2+}, Hg^{2+}, Cd^{2+} etc.) attack thiol groups (—S—H) to form new salt bridges and disrupt other salt bridges. Table 24.5 outlines some common denaturing factors and the regions of the protein that are affected.

TABLE 24.5 Protein denaturing agents and their modes of action.

Denaturing agent	Affected region
heat	H bonds
6 M urea	H bonds
detergents	hydrophobic regions
acids/bases	salt bridges and H bonds
salts	salt bridges
reducing agents	disulfide bonds
heavy metals	disulfide bonds/thiols/ salt bridges
alcohol	hydration layers

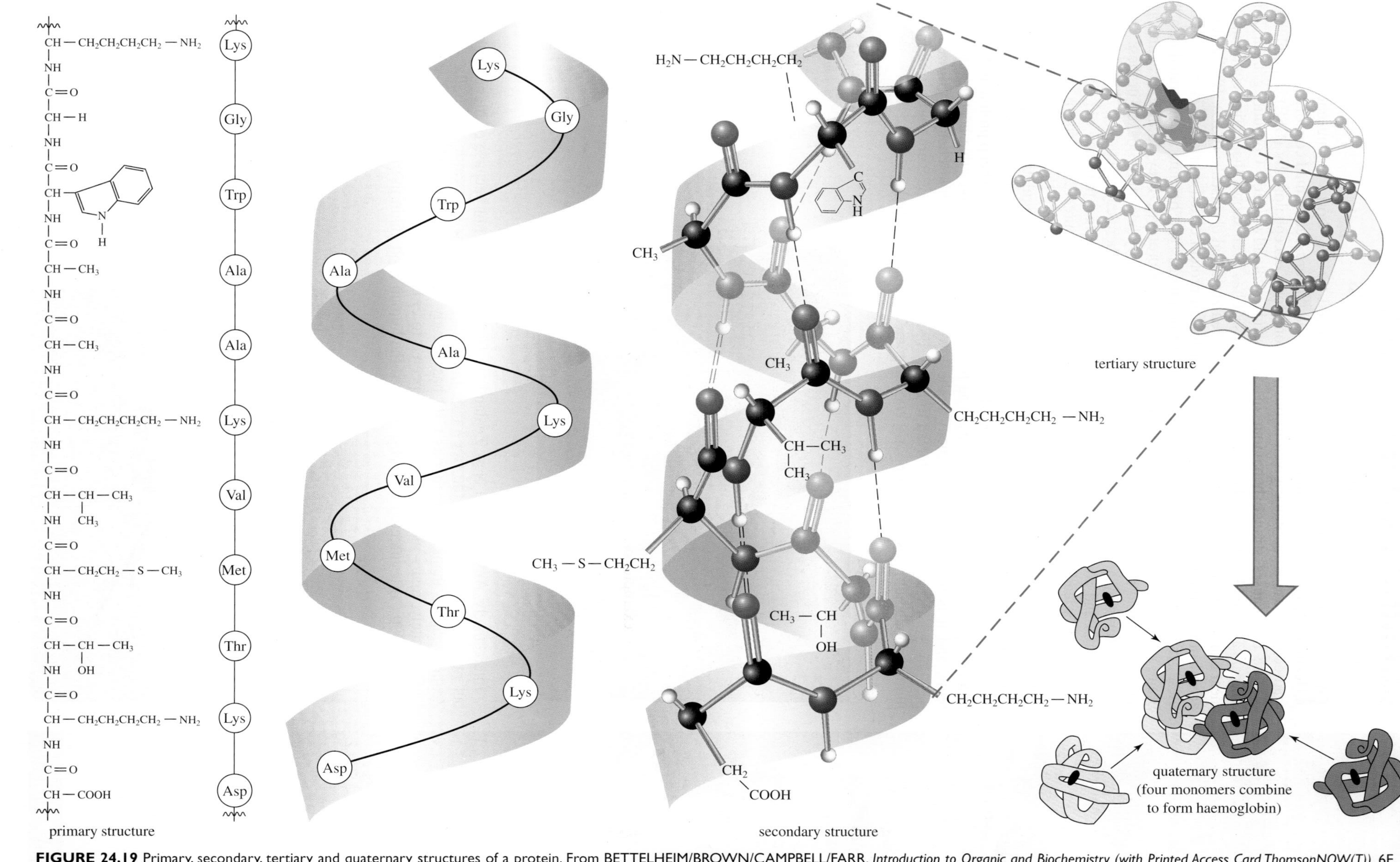

FIGURE 24.19 Primary, secondary, tertiary and quaternary structures of a protein. From BETTELHEIM/BROWN/CAMPBELL/FARR. *Introduction to Organic and Biochemistry (with Printed Access Card ThomsonNOW(T))*, 6E.

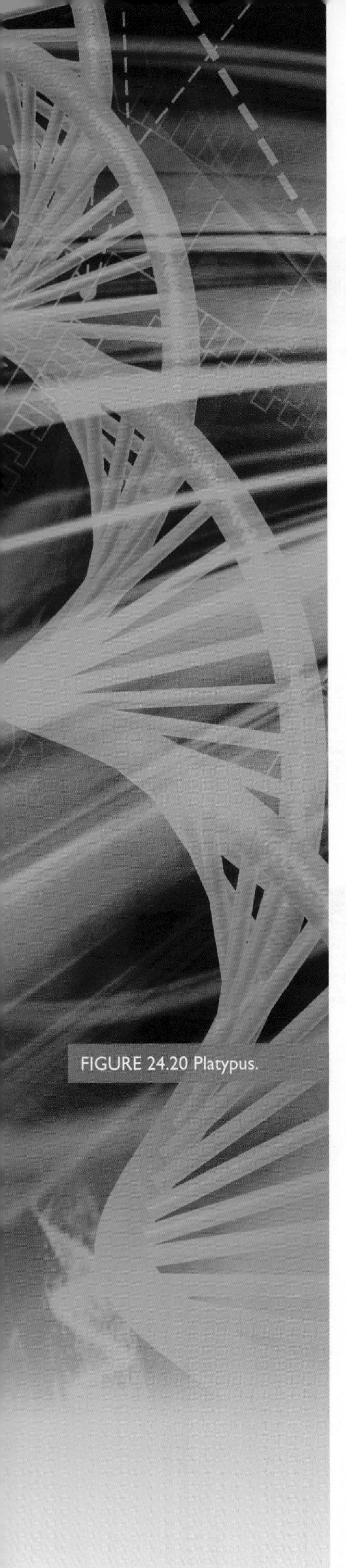

Chemical Connections

Putting the heat on poisonous proteins

In 2006, Australia and New Zealand were declared by the World Organisation for Animal Health to be two of only four countries in the world that are free of mad cow disease. So-called mad cow disease is the name given to bovine spongiform encephalopathy (BSE), a debilitating disease that causes strange behaviour in cattle and eventually death. Consuming meat from animals affected by BSE is also harmful to humans. It has been shown that eating contaminated meat can lead to a variant of the disease, called Creutzfeldt Jakob disease (CJD). BSE and CJD arise from the presence of a specific type of misfolded protein called a prion. It is believed that these small, robust proteins can survive the digestion process and become incorporated into the brain, where they lead to similar misfolding of the brain's proteins. They do this by acting as a template around which the larger brain proteins wrap, possibly becoming locked in these incorrect structures by a free radical or some other nonreversible chemical reaction.

Australia and New Zealand owe their BSE-free status to the fact that they did not follow the US and UK's lead in feeding cattle with protein supplements sourced from sheep and chickens. A similar disease called scrapie has been known in sheep for over 200 years. However, BSE and CJD have only recently become a problem due to a change in the regulations controlling the way protein supplements from sheep were treated. In a terrible example of the dangers that arise when economics overrules science, laws were changed in the 1980s to allow sheep protein to be treated at a lower temperature than previously. You have learned in this chapter that heat denatures proteins, and it is thought that treatment at the lower temperature, while cheaper, allowed prion proteins from scrapie to be transferred to cattle. From there, the prions were transferred into the human population, possibly via poorly cooked mince containing offal.

FIGURE 24.20 Platypus.

Interestingly, Australia and New Zealand are free of scrapie, not because of higher standards, but more likely because of the free-range practices used for raising cattle. Australia and New Zealand can be thankful for this as CJD has already killed more than 150 people worldwide. Unfortunately, many more sufferers are expected to emerge, as it has been estimated that more than 400 000 cattle infected with BSE entered the human food chain in the 1980s.

FIGURE 24.21 Redback spider.

Australia has a reputation of harbouring the deadliest creatures, including sharks, snakes and crocodiles. Even the timid platypus (figure 24.20) can deliver a powerful sting through a claw on its hind foot. New Zealand does not have as many dangerous creatures but it does have the katipo or red katipo (*Latrodectus katipo*), which is a venomous spider found in many parts of both islands. It is a widow spider and is related to Australia's redback spider (figure 24.21) and the USA's black widow spiders.

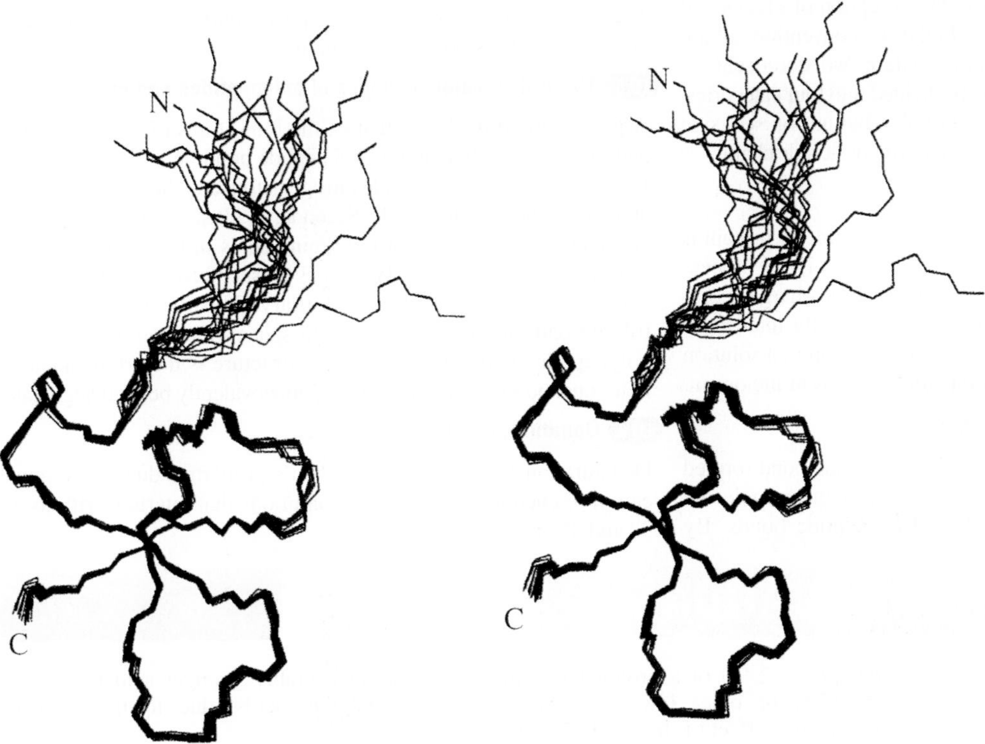

Reproduced with permission from Allan M Torres et al., *Journal of Biochemistry*, vol. 341, 1999, 785–94.
© The Biochemistry Society

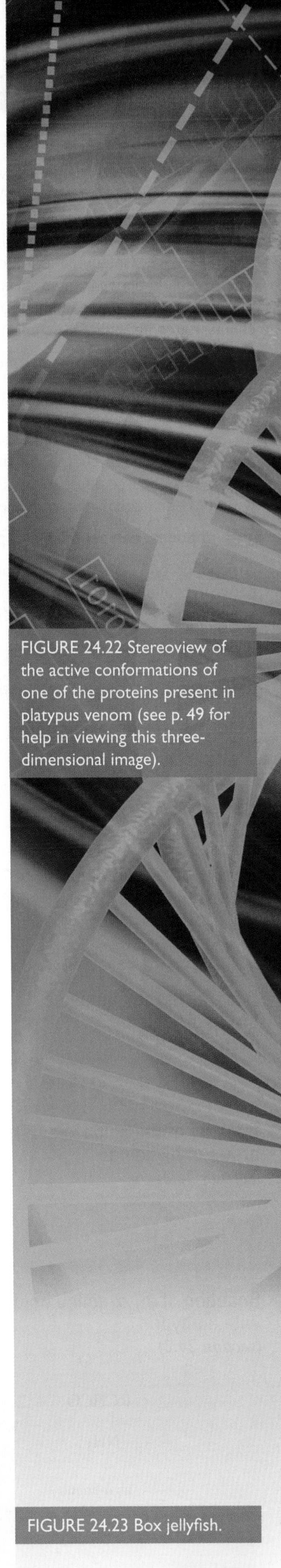

FIGURE 24.22 Stereoview of the active conformations of one of the proteins present in platypus venom (see p. 49 for help in viewing this three-dimensional image).

Interestingly, many of the venoms injected by spiders, such as the redbacks and katipos, are a cocktail of proteins that interfere with nerve transmission and cellular function. Platypus venoms (figure 24.22), snake venoms and even the stinging agents from the box jellyfish (figure 24.23) found in northern Australian waters are all similar in that they contain potent mixes of poisonous proteins, enzymes and polypeptides. While antivenoms exist for katipo, redback spider and many snake venoms, no such treatment is available for jellyfish stings. These stings cause tremendous pain, leave significant scars and have led to a number of deaths. Over the years, there have been several recommended initial treatments for such stings, including salt-bridge disrupting solutions of inorganic ions, including Stingose® and mild acid (vinegar). Currently, the recommended first aid treatment for jellyfish stings is hot water — as hot as the victim can stand. Proteins, enzymes and polypeptides must retain specific structures to maintain activity, so any action that disrupts their structure will lower or prevent their effects. Like the heat treatment needed to denature scrapie prions, hot water denatures jellyfish toxins to some extent and the heat may also penetrate into the affected area, helping to alleviate the effects of this painful sting.

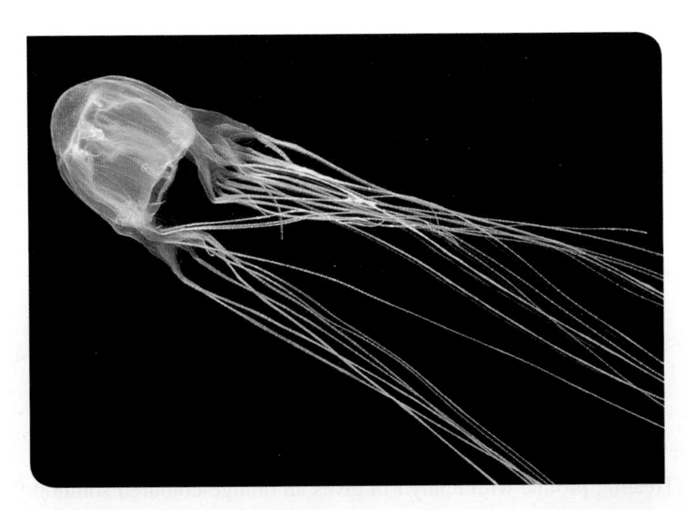

FIGURE 24.23 Box jellyfish.

SUMMARY

LO 1 Amino acids

An α-amino acid is a compound with an amino group and a carboxyl group bound to the same carbon. Amino acids exist as zwitterions, or internal salts, at physiological pH. With the exception of glycine, all protein-derived amino acids are chiral. In the D,L convention, all are L-amino acids. Isoleucine and threonine contain two stereocentres. The 20 protein-derived amino acids can be divided into four categories: nine with nonpolar side chains; four with polar, but un-ionised, side chains; four with acidic side chains; and three with basic side chains.

LO 2 Acid–base properties of amino acids

The isoelectric point, pI, of an amino acid, polypeptide or protein is the pH at which it has no net charge. Electrophoresis is the process of separating compounds on the basis of their electric charge. Compounds with a high charge density move more rapidly than those with a lower charge density. Any amino acid or protein in a solution with a pH that equals the pI of the compound remains at the origin.

LO 3 Peptides, polypeptides and proteins

A peptide bond is the special name given to the amide bond formed between α-amino acids. A polypeptide is a biological macromolecule containing 20 or more amino acids joined by peptide bonds. By convention, the sequence of amino acids in a polypeptide is written from the *N*-terminal amino acid to the *C*-terminal amino acid.

LO 4 Primary structure of polypeptides and proteins

The primary (1°) structure of a polypeptide is the sequence of amino acids in its polypeptide chain.

LO 5 Three-dimensional shapes of polypeptides and proteins

A peptide bond is planar; that is, the four atoms of the amide bond and the two α-carbon atoms of a peptide bond lie in the same plane. Bond angles around the amide nitrogen atom and the carbonyl carbon atom are approximately 120°. Secondary (2°) structure is the ordered arrangement (conformation) of amino acids in localised regions of a polypeptide or protein. Two types of secondary structure are the α-helix and the β-pleated sheet. Tertiary (3°) structure is the overall folding pattern and arrangement in space of all atoms in a single polypeptide chain. Quaternary (4°) structure is the arrangement of individual polypeptide chains into a noncovalently bonded aggregate.

LO 6 Denaturing proteins

Denaturation is the loss of a protein's properties due to the application of chemicals or physical conditions that interfere with its 2°, 3° and 4° structures.

KEY CONCEPTS AND EQUATIONS

Acidity of α-carboxyl groups *(section 24.2)*

An α-COOH (pK_a ≈ 2.19) of a protonated amino acid is a considerably stronger acid than acetic acid (pK_a = 4.74) or other low-molar-mass aliphatic carboxylic acids, due to the electron-withdrawing inductive effect of the α-NH_3^+ group.

$$H_3\overset{+}{N}-\underset{R}{CH}COOH + H_2O \rightleftharpoons H_3\overset{+}{N}-\underset{R}{CH}COO^- + H_3O^+ \quad pK_a = 2.19$$

Acidity of α-ammonium groups *(section 24.2)*

An α-NH_3^+ group (pK_a ≈ 9.47) is a slightly stronger acid than a primary aliphatic ammonium ion (pK_a ≈ 10.76).

$$H_3\overset{+}{N}-\underset{R}{CH}COO^- + H_2O \rightleftharpoons H_2N-\underset{R}{CH}COO^- + H_3O^+ \quad pK_a = 9.47$$

Basicity of the guanidine group of arginine *(section 24.2)*

The protonated form of the arginine side chain forms readily because of substantial resonance stabilisation.

$$R\overset{+}{N}H{=}C(NH_2)_2 \overset{H_2O}{\rightleftharpoons} RN{=}C(NH_2)_2 + H_3O^+ \quad pK_a = 12.48$$

Basicity of the imidazole group of histidine *(section 24.2)*

Protonation of the nitrogen atom of imidazole produces a resonance-stabilised cation.

$$\text{(imidazolium)}-CH_2CH(NH_3^+)COO^- \underset{H_3O^+}{\overset{H_2O}{\rightleftharpoons}} \text{(imidazole)}-CH_2CH(NH_3^+)COO^- + H_3O^+ \quad pK_a = 6.10$$

Reaction of an α-amino acid with ninhydrin *(section 24.2)*

Treating an α-amino acid with ninhydrin gives a purple-coloured solution.

$$RCH(NH_3^+)CO^- (=O) + 2\ \text{ninhydrin} \longrightarrow \text{purple-coloured anion} + RCH(=O) + CO_2 + H_3O^+$$

an α-amino acid — ninhydrin — purple-coloured anion

Treating proline with ninhydrin gives an orange-coloured solution.

KEY TERMS

α-amino acid *p. 1063*
α-helix *p. 1078*
amino acid *p. 1062*
β-pleated sheet *p. 1078*
C-terminal amino acid *p. 1074*
denaturation *p. 1082*
dipeptide *p. 1074*
disulfide bond *p. 1079*
electrophoresis *p. 1071*
hydrogen bonding *p. 1080*
hydrophobic interaction *p. 1080*
integral membrane protein *p. 1082*
isoelectric point (pI) *p. 1070*
N-terminal amino acid *p. 1074*
oligopeptide *p. 1074*
pentapeptide *p. 1074*
peptide *p. 1074*
peptide bond *p. 1074*
polypeptide *p. 1074*
primary (1°) structure *p. 1075*
protein *p. 1074*
quaternary (4°) structure *p. 1081*
salt linkage *p. 1081*
secondary (2°) structure *p. 1077*
tertiary (3°) structure *p. 1078*
tetrapeptide *p. 1074*
tripeptide *p. 1074*
zwitterion *p. 1062*

REVIEW QUESTIONS

LO 1 Amino acids

24.1 Which amino acid does each of the following abbreviations stand for?
(a) Phe (b) Ser (c) Asp (d) Gln (e) His (f) Gly (g) Tyr

24.2 Define the term 'zwitterion'.

24.3 Draw zwitterion forms of the following amino acids.
(a) valine (b) phenylalanine (c) glutamine

24.4 Why are Glu and Asp often called acidic amino acids?

24.5 Why is Arg often called a basic amino acid? Which two other amino acids are also basic amino acids?

24.6 What is the meaning of the alpha as it is used in α-amino acid?

24.7 Several β-amino acids exist. A unit of β-alanine, for example, is contained within the structure of coenzyme A. Write the structural formula of β-alanine.

24.8 Classify as α, β or δ the amino acids found in glutathione, carnosine and GABA.

24.9 What is the difference in structure between tyrosine and phenylalanine?

24.10 For each pair of amino acids listed below, indicate the one with the more nonpolar side chain.
(a) Ala, Ser (b) Ala, Leu (c) Val, Thr (d) Phe, Tyr (e) His, Trp

24.11 Classify the following amino acids as nonpolar, polar, neutral, acidic or basic.
(a) arginine (b) leucine (c) glutamic acid (d) asparagine (e) tyrosine (f) phenylalanine (g) glycine

LO 2 Acid–base properties of amino acids

24.12 Why does glycine have no D or L form?

24.13 Draw all possible isomers of threonine and assign *R*,*S* configurations to each chiral centre.

24.14 The common amino acids are all *S*-enantiomers except for glycine and cysteine. Cysteine is found as the *R*-stereoisomer. Does cysteine have a different three-dimensional structure to the other common α-amino acids.

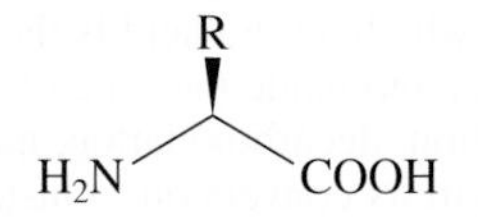

24.15 Draw the structural formula for the form of each of the following amino acids that is most prevalent at pH = 1.0.
(a) threonine (b) arginine (c) methionine (d) tyrosine

24.16 Draw the structural formula for the form of each of the following amino acids that is most prevalent at pH = 10.0.
(a) leucine (b) valine (c) proline (d) aspartic acid

24.17 Write the zwitterion form of alanine and show the reaction of 1 mole of alanine with each of the following.
(a) 1.0 mol NaOH (b) 1.0 mol HCl

24.18 Write the form of lysine most prevalent at pH = 1.0, and then show the product of the reaction of 1 mole of lysine with each of the following. (See table 24.2 for pK_a values of the ionisable groups in lysine.)
(a) 1.0 mol NaOH (b) 2.0 mol NaOH (c) 3.0 mol NaOH

24.19 Write the form of aspartic acid most prevalent at pH = 1.0, and then show the reaction of 1 mole of aspartic acid with the following. (See table 24.2 for pK_a values of the ionisable groups in aspartic acid.)
(a) 1.0 mol NaOH (b) 2.0 mol NaOH (c) 3.0 mol NaOH

24.20 Given pK_a values for ionisable groups from table 24.2, sketch curves for the titration of: (a) glutamic acid with NaOH and (b) histidine with NaOH.

24.21 Draw a structural formula for the product formed when alanine is treated with each of the following reagents.
(a) aqueous NaOH (b) aqueous HCl (c) CH_3CH_2OH, H_2SO_4 (d) $(CH_3CO)_2O$, $NaOOCCH_3$

24.22 Do the following compounds migrate to the cathode or the anode on electrophoresis at the specified pH?
(a) histidine at pH = 6.8 (b) lysine at pH = 6.8 (c) glutamic acid at pH = 4.0 (d) glutamine at pH = 4.0 (e) Glu-Ile-Val at pH = 6.0 (f) Lys-Gln-Tyr at pH = 6.0

24.23 At what pH would you carry out an electrophoresis to separate the amino acids in each of the following mixtures?
(a) Ala, His, Lys (b) Glu, Gln, Asp (c) Lys, Leu, Tyr

LO 3 Peptides, polypeptides and proteins

24.24 What is the name of the bond that links amino acids in a polypeptide or protein?

24.25 How many carbon atoms always separate the nitrogen atoms in a protein or polypeptide?

24.26 An unusual dipeptide has been isolated from an industrial bio-fermentation process. Hydrolysis reveals that the dipeptide is formed only from the amino acid glycine. Surprisingly, however, the dipeptide cannot be separated using electrophoresis. The molecular formula for the dipeptide is $C_4H_6N_2O_2$. Can you determine a structure for this dipeptide that matches this information? (*Hint:* The dipeptide does not move by electrophoresis no matter what the pH of the solution, indicating it does not carry a charge at any pH.)

24.27 An unusual dipeptide derived from leucine and arginine has been found to be present in some species of bacteria. The dipeptide has the molecular formula $C_{12}H_{23}N_5O_2$ at pH 14 and is not a zwitterion. At pH < 12, this dipeptide carries only one positive charge. Draw a structure that fits information for this novel dipeptide.

24.28 Draw all of the structures possible for a tripeptide made from threonine, arginine and methionine.

24.29 Consider the following tripeptide.

$$H_3\overset{+}{N}-\underset{\substack{|\\CH_2\\|\\CH_2\\|\\S\\|\\CH_3}}{CH}-\overset{\overset{O}{\|}}{C}-NH-\underset{\substack{|\\CH_2\\|\\CH-CH_3\\|\\CH_3}}{CH}-\overset{\overset{O}{\|}}{C}-NH-\underset{\substack{|\\CH_2\\|\\COOH}}{CH}-\overset{\overset{O}{\|}}{C}-O^-$$

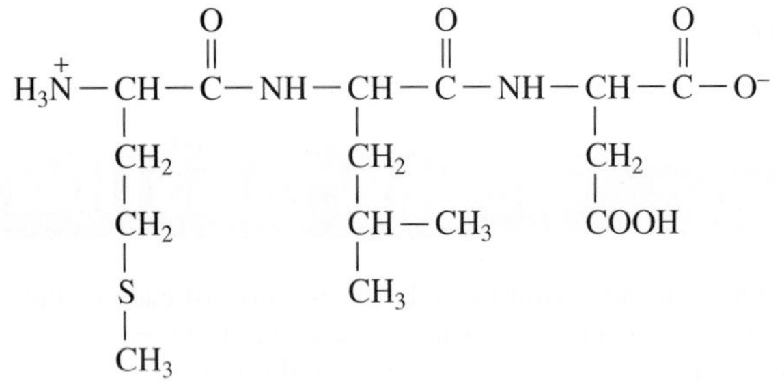

(a) Use the three-letter abbreviations for amino acids to represent the tripeptide.
(b) Which amino acid is the *C*-terminal end and which is the *N*-terminal end?

LO 4 Primary structure of polypeptides and proteins

24.30 Based on your knowledge of the chemical properties of amino acid side chains, suggest a substitution for leucine in the primary structure of a protein that would probably not change the character of the protein very much.

24.31 If a protein contains four SH groups, how many different disulfides are possible if only a single disulfide bond is formed? How many arrangements of different disulfide bonds are possible if two disulfide bonds are formed?

24.32 How many different tetrapeptides can be made in each of the following cases.
(a) The tetrapeptide contains one unit each of Asp, Glu, Pro and Phe.
(b) All 20 amino acids can be used, but each only once.

24.33 Draw a structural formula of each of the following tripeptides. Mark each peptide bond, the *N*-terminal amino acid and the *C*-terminal amino acid.
(a) Phe-Val-Asn (b) Leu-Val-Gln

24.34 Estimate the pI of each tripeptide in question 24.33.

24.35 Consider the following tripeptide.

$$H_2\overset{+}{N}-CH_2-CONH-\underset{\substack{|\\CH_2-SH}}{CH}-CONH-\overset{\overset{CH(CH_3)_2}{|}}{CH}-COO^-$$

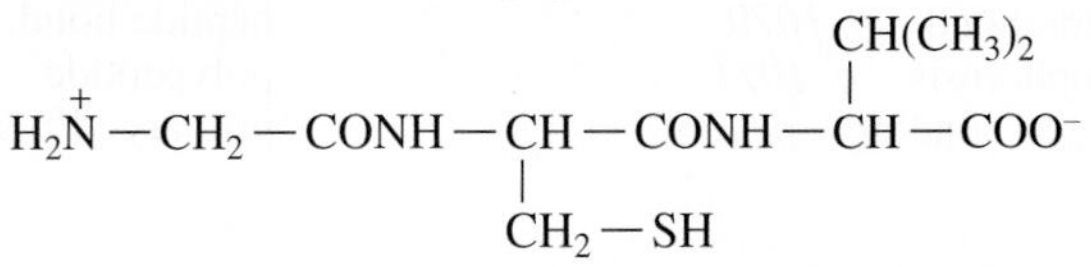

Which amino acids would be formed upon hydrolysis (digestion)?

LO 5 Three-dimensional shapes of polypeptides and proteins

24.36 Examine the α-helix conformation (p. 1078). Are amino acid side chains arranged all inside the helix, all outside the helix or randomly?

24.37 Distinguish between intermolecular and intramolecular hydrogen bonding between the backbone groups on polypeptide chains. In which type of secondary structure do you find intermolecular hydrogen bonds? In which type do you find intramolecular hydrogen bonding?

LO 6 Denaturing proteins

24.38 Enzymes are examples of proteins. Why do enzymes lose some of their activity at higher than physiological temperatures?

24.39 When an egg is boiled, the yolk changes colour and consistency but, when it is cooled to its original temperature, it does not regain its original nature. Explain.

24.40 It is essential to wear safety glasses in the laboratory. Explain why the eye must be protected from even minor amounts of acid or base.

REVIEW PROBLEMS

24.41 At what pH would it be possible to separate a mixture of aspartic acid and asparagine using paper electrophoresis? **LO2**

24.42 Methionine enkephalin is a pentapeptide that is produced by the body to control pain. From the sequence of its amino acid residues, draw a line structure of methionine enkephalin. **LO3**

N terminus ↘ Tyr-Gly-Gly-Phe-Met ↙ C terminus

methionin enkephalin

24.43 The configuration of the stereocentre in α-amino acids is most commonly specified using the D,L convention. The configuration can also be identified using the *R*,*S* convention. Does the stereocentre in L-serine have the *R* or the *S* configuration? **LO5**

24.44 Assign an *R* or *S* configuration to the stereocentre in each of the following amino acids. **LO5**
(a) L-phenylalanine
(b) L-glutamic acid
(c) L-methionine

24.45 The amino acid threonine has two stereocentres. The stereoisomer found in proteins has the configuration (2*S*,3*R*) around the two stereocentres. Draw a Fischer projection of this stereoisomer and a three-dimensional representation. **LO5**

24.46 Apart from threonine, which amino acid (or acids) contain more than one stereocentre? **LO1**

24.47 Although only L-amino acids occur in proteins, D-amino acids are often a part of the metabolism of lower organisms. The antibiotic actinomycin D, for example, contains a unit of D-valine, and the antibiotic bacitracin A contains units of D-asparagine and D-glutamic acid. Draw Fischer projections and three-dimensional representations for these three D-amino acids. **LO5**

24.48 Both norepinephrine and epinephrine are synthesised from the same protein-derived amino acid. **LO1**

(a) HO, HO, H, OH, NH_2

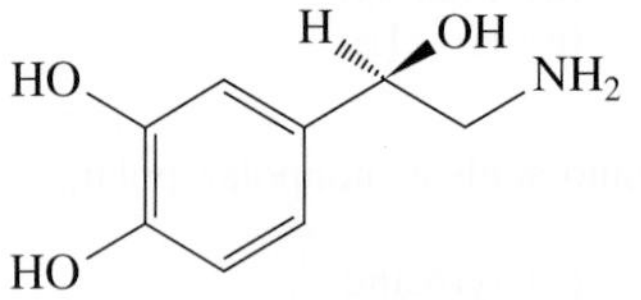

norepinephrine
(noradrenaline)

(b) HO, HO, H, OH, H, N, CH_3

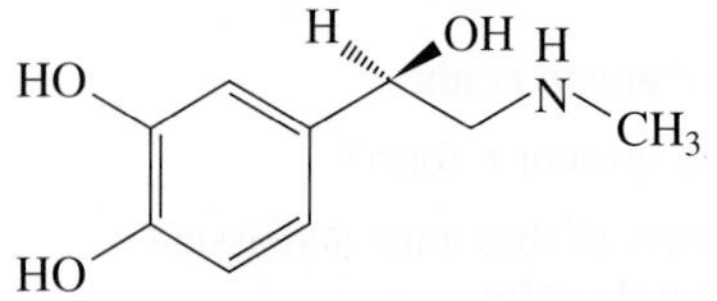

epinephrine
(adrenaline)

From which amino acid are the two compounds synthesised, and what types of reactions are involved in their biosynthesis?

24.49 Histamine is synthesised from one of the 20 protein-derived amino acids. Suggest which amino acid is the biochemical precursor of histamine, and name the type of organic reaction(s) (e.g. oxidation, reduction, decarboxylation, nucleophilic substitution) involved in its conversion to histamine. **LO1**

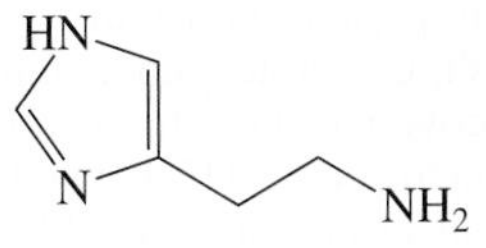

histamine

24.50 From which amino acid are serotonin and melatonin synthesised and what types of reactions are involved in their biosynthesis? **LO1**

(a) serotonin

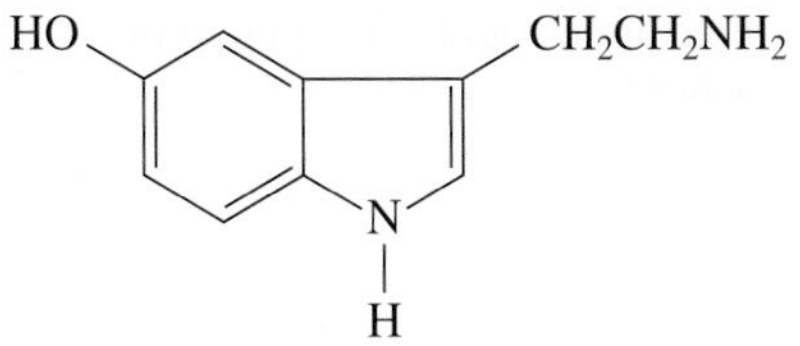

(b) melatonin

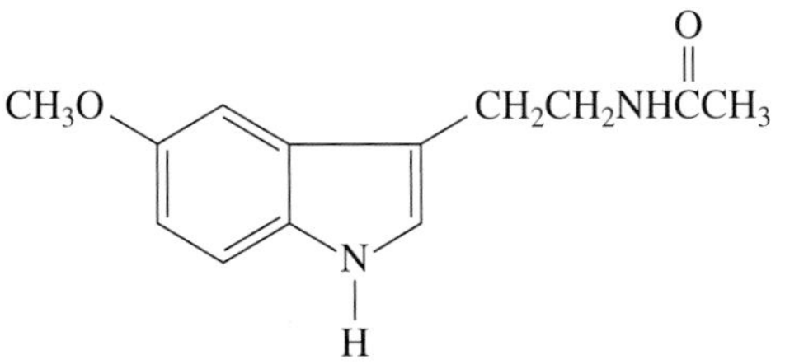

24.51 Enzymes are usually proteins and catalyse common organic reactions. Why are amino acids such as histidine, aspartic acid and serine more commonly found near the catalytic site of the enzyme but amino acids such as leucine and valine appear less often in these locations? **LO5**

24.52 Why is histidine considered a basic amino acid when the pK_a value of the protonated form of its side chain is 6.1? **LO2**

24.53 Account for the fact that the isoelectric point of glutamine (pI = 5.65) is higher than the isoelectric point of glutamic acid (pI = 3.08). **LO2**

24.54 Enzyme-catalysed decarboxylation of glutamic acid gives 4-aminobutanoic acid. Estimate the pI of 4-aminobutanoic acid. **LO2**

24.55 Guanidine and the guanidino group present in arginine are two of the strongest biochemical bases known. Account for their basicity. **LO1**

24.56 At pH = 7.4, the pH of blood plasma, do the majority of protein-derived amino acids bear a net negative charge or a net positive charge? **LO2**

24.57 Examine the amino acid sequence of human insulin (figure 24.14) and note how many Asp, Glu, His, Lys and Arg amino acids occur in this molecule. Do you expect human insulin to have an isoelectric point nearer that of the acidic amino acids (pI = 2.0–3.0), the neutral amino acids (pI = 5.5–6.5) or the basic amino acids (pI = 9.5–11.0)? **LO2**

24.58 Glutathione (GSH), one of the most common tripeptides in animals, plants and bacteria, is a scavenger of oxidising agents. **LO3**

glutathione $H_3\overset{+}{N}CH(COO^-)CH_2CH_2C(=O)NHCH(CH_2SH)C(=O)NHCH_2COO^-$

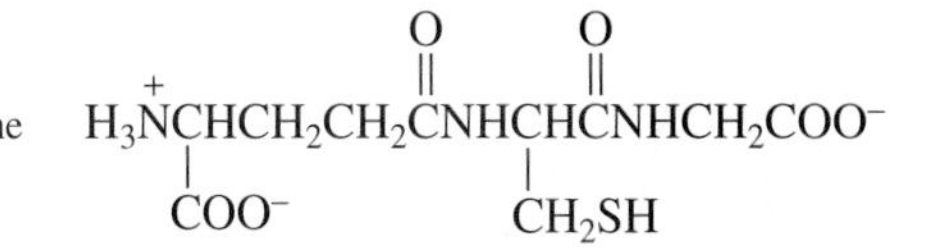

In reacting with oxidising agents, glutathione is converted to GSSG.

(a) Name the amino acids in this tripeptide.
(b) What is unusual about the peptide bond formed by the *N*-terminal amino acid?
(c) Is glutathione a biological oxidising agent or a biological reducing agent?
(d) Write a balanced equation for the reaction of glutathione with molecular oxygen, O_2, to form GSSG and H_2O. Is molecular oxygen oxidised or reduced in this reaction?

24.59 The following is the structural formula for the artificial sweetener aspartame. **LO3**

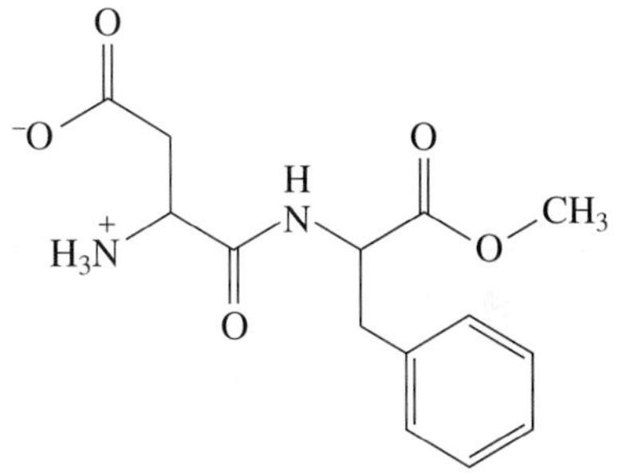

aspartame

(a) Name the two amino acids in this molecule.
(b) Estimate the isoelectric point of aspartame.
(c) Draw structural formulae for the products of the hydrolysis of aspartame in 1 M HCl.

24.60 Proline is often described as an α-helix terminator; that is, it is usually in the random-coil secondary structure following an α-helical portion of a protein chain. Why does proline not fit easily into an α-helix structure? **LO5**

24.61 Many plasma proteins found in an aqueous environment are globular in shape. Which of the following amino acid side chains would you expect to find on the surface of a globular protein, in contact with the aqueous environment, and which would you expect to find inside, shielded from the aqueous environment? Explain your answers. **LO5**

(a) Leu (b) Arg (c) Ser (d) Lys (e) Phe

ADDITIONAL EXERCISES

24.62 At pH 12, a plasma protein has a globular shape and has a surface that has a large number of Asp and Tyr amino acids. What would be the consequences of lowering the pH of plasmas to <3? **LO6**

24.63 Heating can disrupt the 3° structure of a protein. Explain the chemical processes that occur upon heating a protein. **LO6**

24.64 Some amino acids cannot be incorporated into proteins because they are self-destructive. As shown on the right, homoserine, for example, can use its side-chain —OH group in an intramolecular, nucleophilic acyl substitution to cleave the peptide bond and form a cyclic structure on one end of the chain.

Draw the cyclic structure formed and explain why serine does not suffer the same fate, as shown on the right. **LO1**

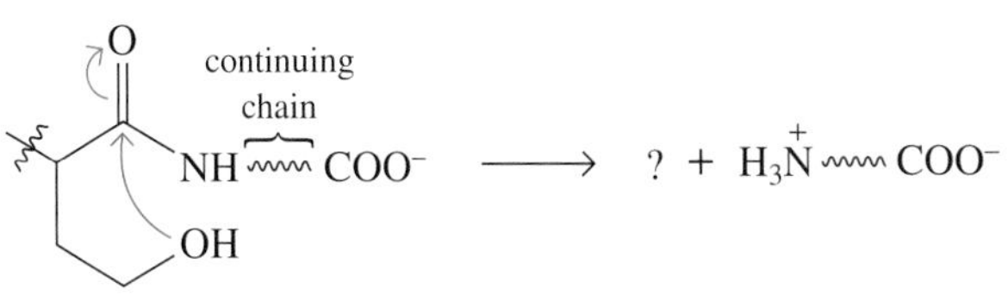

continuing chain

$\longrightarrow$? + $H_3\overset{+}{N}$~~~COO^-

homoserine residue

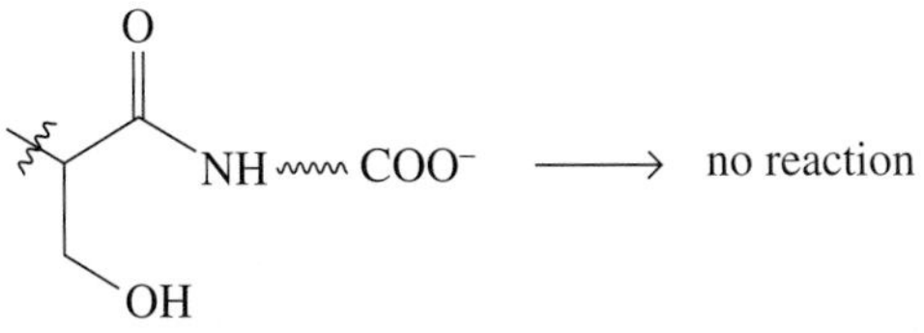

$\longrightarrow$ no reaction

serine residue

24.65 Would you expect a decapeptide of only isoleucine residues to form an α-helix? Explain. **LO5**

24.66 Which type of protein would you expect to have the following effect?

Reaction A

Reaction A + protein

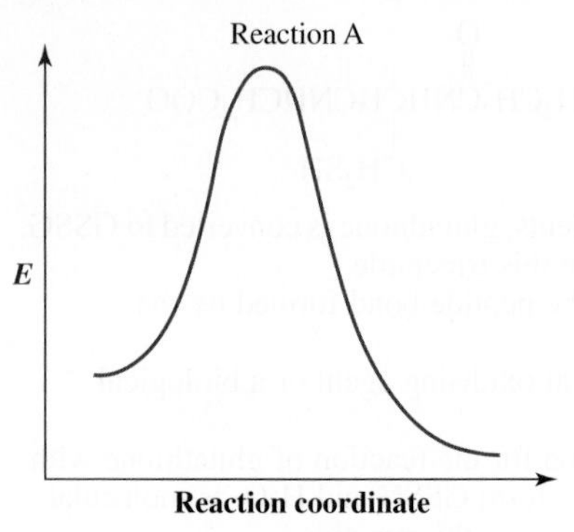

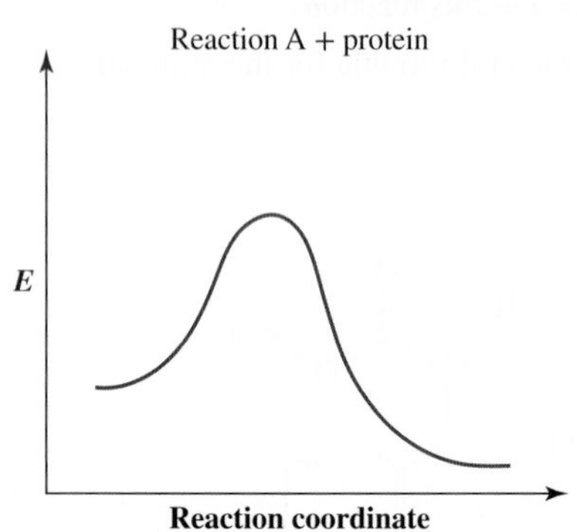

24.67 Green fluorescent protein (GFP), first isolated from bioluminescent jellyfish, is a protein containing 238 amino acid residues. The discovery of GFP has revolutionised the field of fluorescence microscopy, which enables biochemists to monitor the biosynthesis of proteins. The 2008 Nobel Prize in chemistry was awarded to Martin Chalfie, Osamu Shimomura and Roger Tsien for the discovery and development of GFP. The structural subunit of GFP responsible for fluorescence, called the fluorophore, results when three amino acid residues undergo cyclisation. Identify the three amino acids that go into the biosynthesis of this fluorophore.

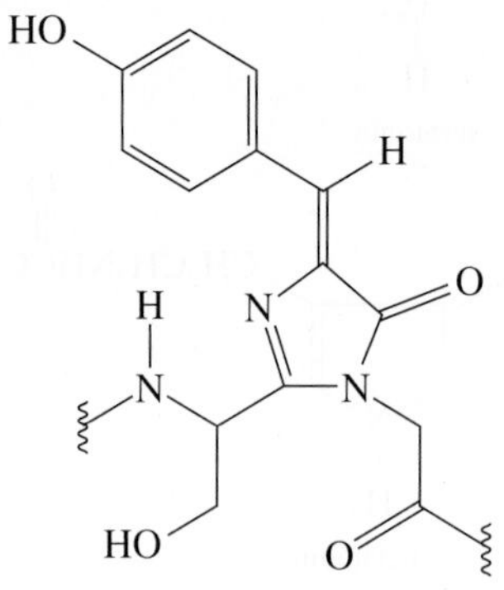

fluorophore

Chapter 1

Practice exercise 1.1
3.74 g Ti

Practice exercise 1.2
(a) number of protons = 71; number of neutrons = 106
(b) number of protons = 54; number of neutrons = 79
(c) number of protons = 77; number of neutrons = 115

Practice exercise 1.3
20.18 u

Chapter 2

Practice exercise 2.1
$kg\,m\,s^{-1}$

Practice exercise 2.2
1. (a) μg
   (b) μm
   (c) ns
2. (a) 1×10^{-9}
   (b) 1×10^{-2}
   (c) 1×10^{-12}
3. (a) cg
   (b) Mm
   (c) μs

Practice exercise 2.3
1. 1×10^{4} Pa
2. 8.56×10^{4} Pa
3. 1.43×10^{-1} atmosphere

Practice exercise 2.4
(a) $8.206 \times 10^{-5}\,m^3\,atm\,K^{-1}\,mol^{-1}$
(b) $8.314 \times 10^{-5}\,m^3\,bar\,K^{-1}\,mol^{-1}$
(c) $8.236 \times 10^{-2}\,m^3\,mmHg\,K^{-1}\,mol^{-1}$

Practice exercise 2.5
(a) $m = Mn$
(b) $q = cm\Delta T$
(c) $c = \dfrac{E\lambda}{h}$

Practice exercise 2.6
Worker C has the best precision.
Worker A has the best accuracy.

Practice exercise 2.7
(a) 7
(b) 4
(c) 8
(d) 1

Practice exercise 2.8
(a) 27.28
(b) 6.75
(c) 5.0×10^{1}
(d) 75.00 g
(e) 1.078
(f) 127

Practice exercise 2.9
(a) 56.2 g
(b) 10.5 mL
(c) 77.326 g
(d) 17.0 g
(e) $0.734\,g\,mL^{-1}$
(f) $2.5\,m^3$

Practice exercise 2.10
The student was not successful. The final concentration is $(1.00 \pm 0.03)\,g\,L^{-1}$.

Practice exercise 2.11

NH_4Br	$CaBr_2$
NH_4ClO_4	$Ca(ClO_4)_2$
$(NH_4)_2CO_3$	$CaCO_3$
$(NH_4)_3PO_4$	$Ca_3(PO_4)_2$
KBr	$AlBr_3$
$KClO_4$	$Al(ClO_4)_3$
K_2CO_3	$Al_2(CO_3)_3$
K_3PO_4	$AlPO_4$

Practice exercise 2.12

(a)

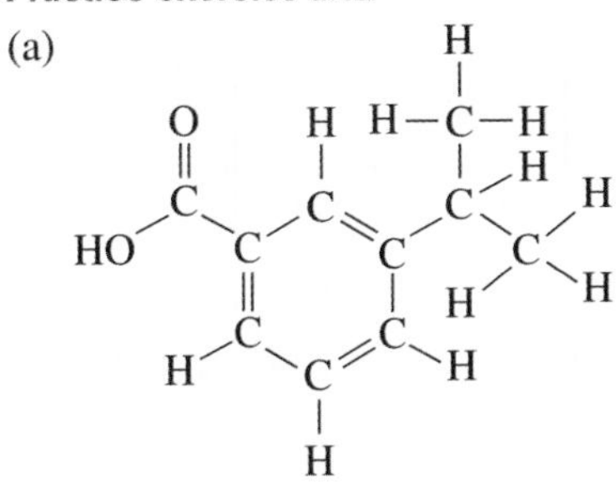

chemical formula = $C_{10}H_{12}O_2$

(b)

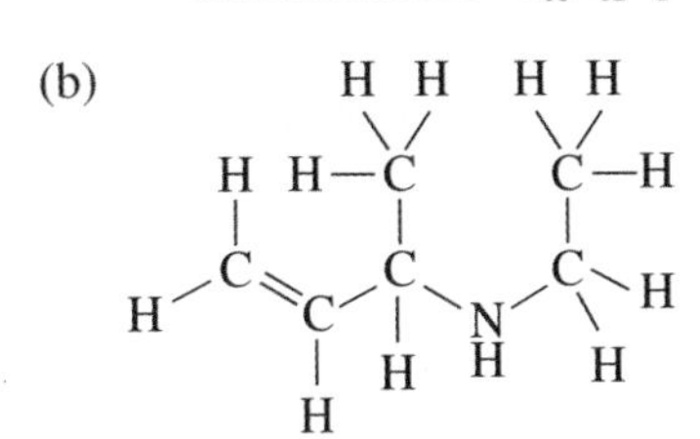

chemical formula = $C_6H_{13}N$

(c)

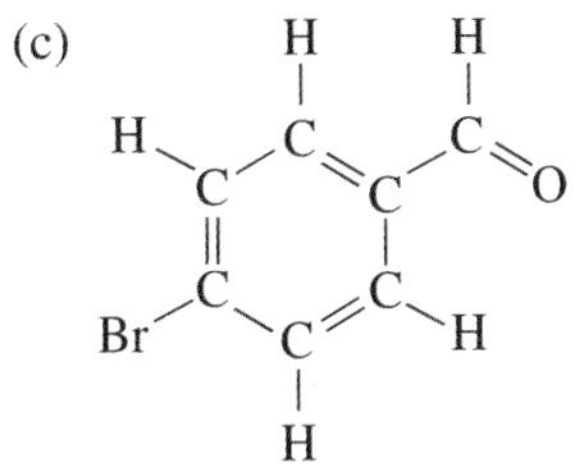

chemical formula = C_7H_5BrO

Practice exercise 2.13
(a) silicon tetrabromide
(b) sulfur trioxide
(c) chlorine trifluoride

Practice exercise 2.14

$CH_3CH_2CH_2CH_2CH_2OH$	primary
$CH_3CH_2CH_2CH(OH)CH_3$	secondary
$CH_3CH_2CH(OH)CH_2CH_3$	secondary
$(CH_3)_2CHCH_2CH_2OH$	primary
$(CH_3)_2CHCH(OH)CH_3$	secondary
$(CH_3)_2C(OH)CH_2CH_3$	tertiary
$(CH_3)(CH_2OH)CHCH_2CH_3$	primary
$CH_3C(CH_3)_2CH_2OH$	primary

Practice exercise 2.15
$CH_3CH_2COCH_2CH_3$
$CH_3COCH_2CH_2CH_3$
$(CH_3)_2CHCOCH_3$

Practice exercise 2.16
$CH_3CH_2CH_2CH_2COOH$
$CH_3CH_2CH(COOH)CH_3$
$CH_3C(COOH)(CH_3)_2$
$(CH_3)_2CHCH_2COOH$

Practice exercise 2.17
(a) 5-isopropyl-2-methyloctane
(b) 4-isopropyl-4-propyloctane

Practice exercise 2.18
(a) same compounds
(b) constitutional isomers

Practice exercise 2.19

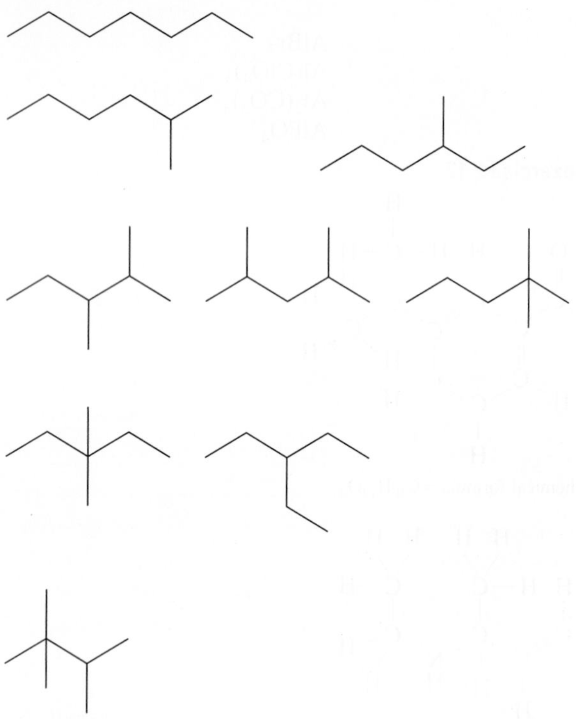

Practice exercise 2.20
(a) butanoic acid
(b) propyne

Chapter 3

Practice exercise 3.1
1 Mg, 2 O, 4 H and 2 Cl (on each side)

Practice exercise 3.2
$Mg(OH)_2(s) + 2HCl(aq) \rightarrow MgCl_2(aq) + 2H_2O(l)$

Practice exercise 3.3
$3\,BaCl_2(aq) + Al_2(SO_4)_3(aq) \rightarrow 3BaSO_4(s) + 2AlCl_3(aq)$

Practice exercise 3.4
0.139 mol

Practice exercise 3.5
13.3 g

Practice exercise 3.6
0.467 mol

Practice exercise 3.7
10.49 g

Practice exercise 3.8
59.5 g

Practice exercise 3.9
mass % (N) = 25.94%
mass % (O) = 74.06%
No other elements are present.

Practice exercise 3.10
mass % (N) = 30.45%
mass % (O) = 69.55%

Practice exercise 3.11
Na_2SO_4

Practice exercise 3.12
0.183 mol

Practice exercise 3.13
0.958 mol

Practice exercise 3.14
78.5 g

Practice exercise 3.15
30.01 g

Practice exercise 3.16
86.1%

Practice exercise 3.17
0.2498 mol L^{-1}

Practice exercise 3.18
0.53 g

Practice exercise 3.19
Mix 25.0 mL of 0.500 M H_2SO_4 with water to make 100 mL of total solution.

Practice exercise 3.20
26.8 mL

Practice exercise 3.21
0.40 M Fe^{3+}
1.2 M Cl^-

Practice exercise 3.22
0.750 M Na^+

Practice exercise 3.23
60.0 mL KOH

Practice exercise 3.24
1.20×10^{-2} mol of $BaSO_4$ will be formed.
$c_{Cl^-} = 0.480$ M
$c_{Mg^{2+}} = 0.300$ M

Practice exercise 3.25
5.41×10^{-3} mol

Chapter 4

Practice exercise 4.1
$2.14 \times 10^{10}\ s^{-1}$ or 2.14×10^{10} Hz

Practice exercise 4.2
7.82×10^{-19} J

Practice exercise 4.3
-4.56×10^{-19} J atom^{-1}
-274 kJ mol^{-1}

Practice exercise 4.4
$E_{photon} = 2.18 \times 10^{-8}$ J
$\lambda = 9.10 \times 10^{-18}$ m

Practice exercise 4.5
746 kJ mol^{-1} Hg

Practice exercise 4.6
$\lambda_{photon} = 159$ nm
$\lambda_{electron} = 0.982$ nm

Practice exercise 4.7
1.39×10^{12} m

Practice exercise 4.8

n	l	m_l	m_s
5	1	+1	+½
5	1	+1	–½
5	1	0	+½
5	1	0	–½
5	1	–1	+½
5	1	–1	–½

Practice exercise 4.9

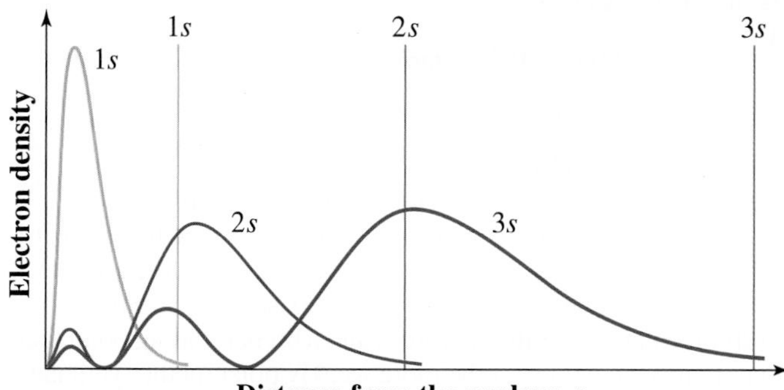

Practice exercise 4.10

Filled orbitals: 1*s*, 2*s*, 2*p*, 3*s*, 3*p*, 4*s*, 3*d*, 4*p*, 5*s*
Partially filled: 4*d* contains only 2 electrons

Practice exercise 4.11

Energy level diagram:

E
2*p* ↑↓ ↑↓ ↑
2*s* ↑↓
1*s* ↑↓

Compact notation: $1s^2 2s^2 2p^5$

Practice exercise 4.12

$[Kr]5s^2 4d^{10}$

Practice exercise 4.13

$[He]2s^2 2p^3$

$2s^2$ $n = 2, l = 0, m_l = 0, m_s = +½$
$n = 2, l = 0, m_l = 0, m_s = –½$

$2p^3$ $n = 2, l = 1, m_l = +1, m_s = +½$
$n = 2, l = 1, m_l = 0, m_s = +½$
$n = 2, l = 1, m_l = –1, m_s = +½$

(Alternatively, m_s could be –½ in each of the three electrons in the $2p^3$ set of orbitals.)

Practice exercise 4.14

$[Kr]4d^5$

Practice exercise 4.15

Sc^{3+}, Cu^+, Zn^{2+}

Practice exercise 4.16

smaller:	P and Se
larger:	Ge and Sb

Practice exercise 4.17

metals:	Ga
metalloids:	Ge and As
non-metals:	Se, Br and Kr

Chapter 5

Practice exercise 5.1

F
|
F—S—F
|
F

Practice exercise 5.2

H–C(H)(H)–C(=O)–O⁻ ⟷ H–C(H)(H)–C(–O⁻)=O

Practice exercise 5.3

O–O=O ⟷ O=O–O

Practice exercise 5.4

Cl
|
C
H H H

Practice exercise 5.5

Ethane does not have a dipole moment.
Ethanol a dipole moment.

Practice exercise 5.6

(a) Bond lengths are inversely proportional to the number of electron pairs in the bond. Since C≡C has three electron pairs in the bond and C=C has only two, C≡C is the shorter bond.
(b) Bond lengths increase with the principal quantum number (n) of the valence electrons. The valence electrons in C have $n = 2$ and valence electrons in Si have $n = 3$.
(c) Both atoms O and C are in the same row and each is bonded to an identical atom. Thus, the only difference is the orbital size of each atom. Orbital size decreases across a row, so the O—O bond is shorter than the C—C bond.

Practice exercise 5.7

Oxygen has the valence electron configuration $2s^2 2p^4$, so it uses four sp^3 hybridised orbitals. Two of these overlap with 1*s* orbitals from the hydrogen atoms to form σ bonds and the other two sp^3 hybrid orbitals contain lone pairs.

Practice exercise 5.8

The two inner C atoms have *sp* hybridisation and a triple bond, whereas the two outer C atoms have sp^3 hybrid orbitals. The two outer C atoms have three sp^3 hybrid orbitals that overlap with 1*s* orbitals from H atoms and the remaining sp^3 orbital overlaps with an *sp* orbital from the inner C.

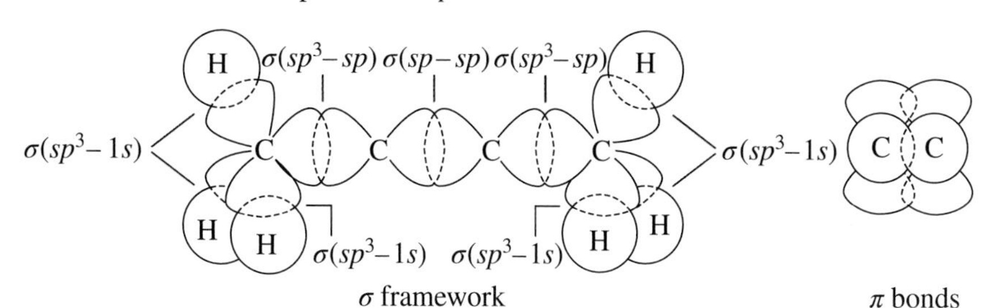

Practice exercise 5.9

H_2 has the stronger bond.

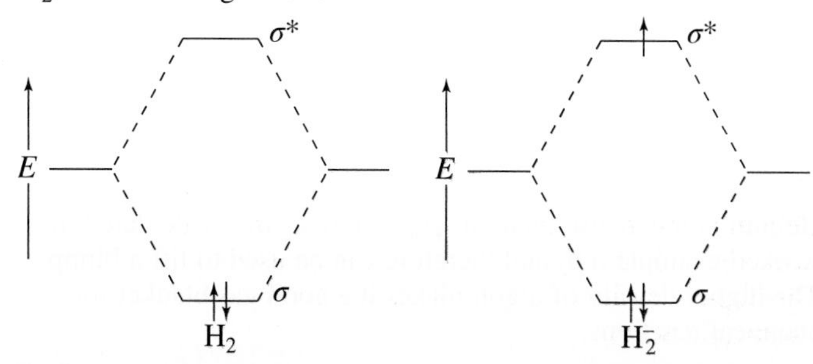

H_2 has two electrons which occupy a bonding orbital, but H_2^- has a third electron which must occupy an antibonding orbital.

Practice exercise 5.10

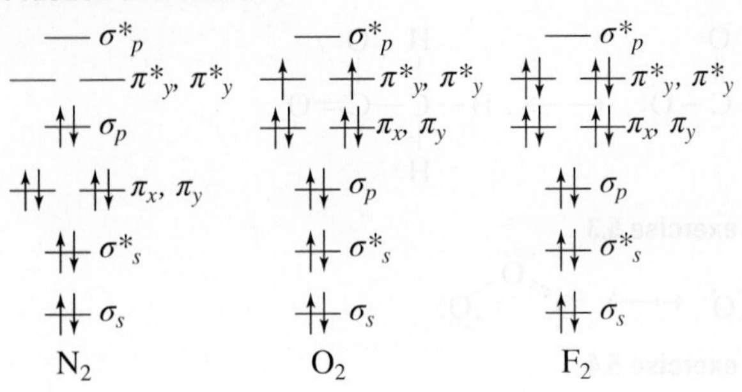

N_2 has bond order 3
O_2 has bond order 2
F_2 has bond order 1

After N_2, the additional electrons occupy antibonding orbitals, which cause the bond energy to decrease due to the destabilising effect of these electrons. Hence N_2 has the highest bond energy. The extra 2 electrons that F_2 has in π^*_y, π^*_y further decreases the bond energy in F_2 compared to O_2.

Practice exercise 5.11
CN^- is the most stable.

Chapter 6

Practice exercise 6.1
1.446×10^7 Pa

Practice exercise 6.2
4.71×10^4 Pa

Practice exercise 6.3

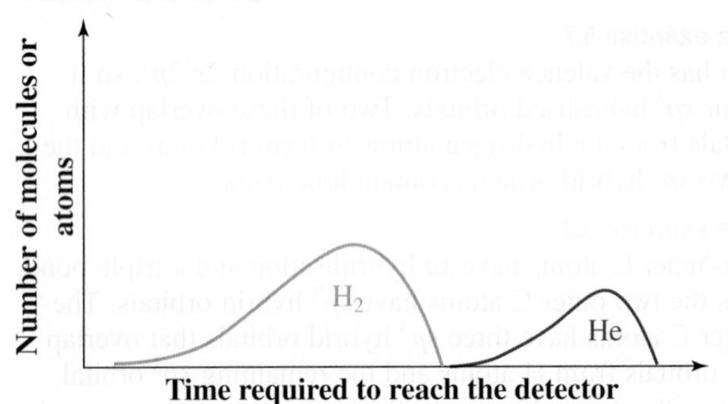

Practice exercise 6.4
$\overline{E}_{\text{kinetic}} = 4.93 \times 10^{-21}$ J
$\overline{E}_{\text{kinetic, molar}} = 2.97 \times 10^3$ J mol^{-1}

Practice exercise 6.5
$X_{O_2} = 0.273$
$X_{He} = 0.727$
$p_{O_2} = 1.16 \times 10^5$ Pa
$p_{He} = 3.09 \times 10^5$ Pa

Practice exercise 6.6
8.52×10^{-4} g or 0.852 mg

Practice exercise 6.7
28.1 g mol^{-1}

Practice exercise 6.8
$\rho_{He} = 0.163$ kg m^{-3}
$\rho_{Ar} = 1.62$ kg m^{-3}

Helium is less dense than air ($\rho_{air} = 1.18$ kg m^{-3}, calculated in worked example 6.6) and therefore can be used to lift a blimp. The higher density of argon makes it a good gas blanket for chemical reactions.

Practice exercise 6.9
1.54×10^{-3} m^3 or 1.54 L

Practice exercise 6.10
$p_{NO} = 0$ Pa (limiting reagent)
$p_{O_2} = 2.50 \times 10^5$ Pa
$p_{NO_2} = 5.00 \times 10^5$ Pa

Practice exercise 6.11
$[Mg^{2+}] = 3$ mol L^{-1}
$p_{H_2} = 2.30 \times 10^5$ Pa
$p_{total} = 3.30 \times 10^5$ Pa

Practice exercise 6.12
$$\left(\frac{V}{n}\right)_{\text{methane}} = 9.06 \times 10^{-3} \text{ m}^3 \text{ mol}^{-1}$$
$p_{methane} = 9.77 \times 10^4$ Pa

Practice exercise 6.13
Acetaldehyde is more compact than acetone and has fewer electrons, so its intermolecular forces are weaker and therefore it boils at a lower temperature than acetone. However, like acetone, acetaldehyde contains a polar C=O group, so it has stronger intermolecular dipole-dipole forces and a higher boiling point than either non-polar butane or the more weakly polar methoxyethane.

Practice exercise 6.14

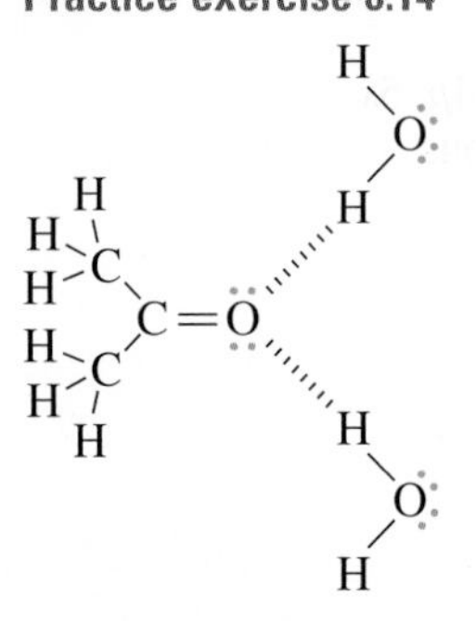

Chapter 7

Practice exercise 7.1
Heat = 41.5 kJ. Heat flows from the tray to the freezer compartment.

Practice exercise 7.2

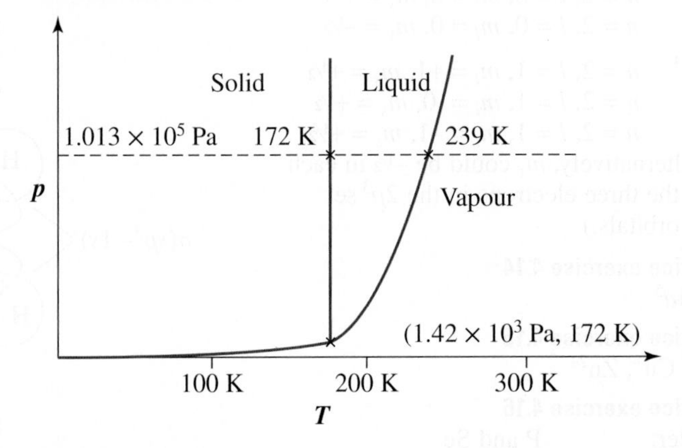

Practice exercise 7.3
The gas liquefies at $p = 1.26 \times 10^4$ Pa and the liquid freezes when the pressure is approximately 0.5×10^5 Pa. At 1.013×10^5 Pa, 63.1 K, N_2 is a solid.

Chapter 8

Practice exercise 8.1
27.9 J K^{-1}

Practice exercise 8.2
5.2 kJ

Practice exercise 8.3
-3.94×10^2 kJ mol^{-1}

Practice exercise 8.4
74 kJ mol^{-1}

Practice exercise 8.5
$3.2H_2(g) + 1.6O_2(g) \rightarrow 3.2H_2O(g) \quad \Delta_r H^\ominus = -773.8$ kJ mol^{-1}

Practice exercise 8.6
44.0 kJ mol^{-1}

Practice exercise 8.7
(a) $Na(s) + 1/2O_2(g) + 1/2H_2(g) \rightarrow NaOH(s)$
$\Delta_f H^\ominus = -426.8$ kJ mol^{-1}
(b) $2C(s, graphite) + 3H_2(g) + 1/2O_2(g) \rightarrow C_2H_5OH(l)$
$\Delta_f H^\ominus = -277.63$ kJ mol^{-1}
(c) $N_2(g) + 5/2O_2(g) \rightarrow N_2O_5(g) \quad \Delta_f H^\ominus = 11$ kJ mol^{-1}

Practice exercise 8.8
(a) −113.1 kJ mol^{-1}
(b) −177.8 kJ mol^{-1}

Practice exercise 8.9
-4.53×10^5 kJ

Practice exercise 8.10
(a) negative (b) positive

Practice exercise 8.11
(a) 161 J mol^{-1} K^{-1} (b) −232.5 J mol^{-1} K^{-1}

Practice exercise 8.12
−742 kJ mol^{-1}

Practice exercise 8.13
(a) −818.0 kJ mol^{-1} (b) −91.9 kJ mol^{-1}

Practice exercise 8.14
−787.2 kJ mol^{-1}

Practice exercise 8.15
311 K

Chapter 9

Practice exercise 9.1
(a) $K_c = \frac{[NO]^2}{[N_2][O_2]}$

(b) $K_c = \frac{[NO_2][SO_2]}{[SO_3][NO]}$

Practice exercise 9.2
$$K_p = \frac{\left(\frac{p_{HI}}{p^o}\right)^2}{\left(\frac{p_{H_2}}{p^o}\right)\left(\frac{p_{I_2}}{p^o}\right)}$$

Practice exercise 9.3
5.5×10^{-9}

Practice exercise 9.4
5.9×10^1

Practice exercise 9.5
300 K

Practice exercise 9.6
(a) $K_c = \frac{1}{[Cl_2]}$

(b) $K_c = \frac{1}{[NH_3][HCl]}$

(c) $K_c = [Na^+][OH^-][H_2]$
(d) $K_c = [Ag^+]^2[CrO_4^{2-}]$
(e) $K_c = \frac{[Ca^{+2}][HCO_3^-]^2}{[CO_2]}$

Practice exercise 9.7
$\Delta G^\ominus = -1097.4$ kJ mol^{-1}
Since $\Delta G^\ominus$ is large and negative, we would expect significant amounts of products to be formed at equilibrium. However, the reaction is very slow, thus explaining the lack of red-brown colour.

Practice exercise 9.8
$Q_p > K_p$ and thus a spontaneous change to the left will occur in order to form more reactants and decrease the value of Q_p.

Practice exercise 9.9
−33 kJ mol^{-1}

Practice exercise 9.10
1.47×10^{45}

Practice exercise 9.11
3.63×10^4

Practice exercise 9.12
(a) The amount of SO_2 will increase. The value of K_p is unchanged.
(b) The amount of SO_2 will remain constant. The value of K_p is unchanged.
(c) The amount of SO_2 will decrease. The value of K_p is decreased.
(d) The amount of SO_2 will remain constant. The value of K_p is unchanged.

Practice exercise 9.13
4.06

Practice exercise 9.14
$[O_2]$ decreases by 0.00025 mol L^{-1} and $[NO_2]$ increases by 0.00050 mol L^{-1}.

Practice exercise 9.15
(a) $[PCl_3] = 0.200$ M
$[Cl_2] = 0.100$ M
$[PCl_5] = 0.000$ M
(b) $[PCl_3]$ decreased by 0.080 M
$[Cl_2]$ decreased by 0.080 M
$[PCl_5]$ increased by 0.080 M
(c) $[PCl_3] = 0.120$ M
$[PCl_5] = 0.080$ M
$[Cl_2] = 0.020$ M
(d) 33

Practice exercise 9.16
$[H_2] = 0.022$ M
$[I_2] = 0.022$ M
$[HI] = 0.156$ M

Practice exercise 9.17
1.1×10^{-17} M

Chapter 10

Practice exercise 10.1
4.35×10^{-2} g

Practice exercise 10.2
3.2×10^{-10}

Practice exercise 10.3
(a) 1.3×10^{-3} M
(b) 8.8×10^{-11} M

Practice exercise 10.4
molar solubility of AgI in a 0.20 M NaI solution = 4.2×10^{-16} M
molar solubility of AgI in pure water = 9.1×10^{-9} M
AgI is much more soluble in pure water than in 0.20 M NaI solution.

Practice exercise 10.5
A precipitate of $NiCO_3$ will form.

Practice exercise 10.6
16.0 g

Practice exercise 10.7
1.32×10^3 Pa

Practice exercise 10.8
4.59×10^3 Pa

Practice exercise 10.9
100.29 °C

Practice exercise 10.10
157 g mol^{-1}

Practice exercise 10.11
4.02×10^4 Pa

Practice exercise 10.12
5.44×10^2 g mol^{-1}

Practice exercise 10.13
If the solute is 100 % dissociated, freezing point = −1.60 °C.
If the solute is 0 % dissociated, freezing point = −0.320 °C.

Chapter 11

Practice exercise 11.1

1. (a) OH^-
   (b) $H_2PO_4^-$
   (c) ClO_4^-
   (d) F^-
   (e) HSO_3^-
   (f) IO_3^-
2. (a) OH^-
   (b) NH_4^+
   (c) HF
   (d) H_2O
   (e) HCO_3^-
   (f) HPO_4^{2-}

Practice exercise 11.2
The Brønsted acids are $H_2PO_4^-$(aq) and H_2CO_3(aq). The Brønsted bases are HCO_3^-(aq) and HPO_4^{2-}(aq).

Practice exercise 11.3
$[H_3O^+] = 3.8 \times 10^{-6}$ M
The solution is acidic.

Practice exercise 11.4
pH = 3.44
pOH = 10.56
The solution is acidic.

Practice exercise 11.5
11.30

Practice exercise 11.6
(a) $[H_3O^+] = 5.0 \times 10^{-3}$ M
$[OH^-] = 2.0 \times 10^{-12}$ M
The solution is acidic.
(b) $[H_3O^+] = 1.4 \times 10^{-4}$ M
$[OH^-] = 7.1 \times 10^{-11}$ M
The solution is acidic.
(c) $[H_3O^+] = 1.5 \times 10^{-11}$ M
$[OH^-] = 6.7 \times 10^{-4}$ M
The solution is basic.
(d) $[H_3O^+] = 3.2 \times 10^{-4}$ M
$[OH^-] = 3.1 \times 10^{-11}$ M
The solution is acidic.
(e) $[H_3O^+] = 2.5 \times 10^{-12}$ M
$[OH^-] = 4.0 \times 10^{-3}$
The solution is basic.

Practice exercise 11.7
$[H_3O^+] = 1.3 \times 10^{-13}$ M
pH = 12.87

Practice exercise 11.8
pH = 7.00

Practice exercise 11.9
HY is the strongest acid, followed by HZ and HX.
For HX: $K_a = 7.4 \times 10^{-5}$
For HY: $K_a = 1.2 \times 10^{-4}$
For HZ: $K_a = 9.5 \times 10^{-5}$

Practice exercise 11.10
5.6×10^{-11}

Practice exercise 11.11
$[H_3O^+] = 5.9 \times 10^{-4}$ M
pH = 3.23

Practice exercise 11.12
pH = 8.62

Practice exercise 11.13
$K_b = 3.3 \times 10^{-5}$
$pK_b = 4.48$

Practice exercise 11.14
pH = 8.08

Practice exercise 11.15
pH = 5.19

Practice exercise 11.16
basic

Practice exercise 11.17
pH = 10.34

Practice exercise 11.18
(a) HBr
(b) H_2Te
(c) CH_3SH

Practice exercise 11.19
H_3PO_3

Practice exercise 11.20
(a) HIO_4
(b) H_3AsO_4

Practice exercise 11.21
4.83 (The slightly different answer from that given in worked example 11.17 is due to rounding.)

Practice exercise 11.22
8.99

Practice exercise 11.23
Ammonia and ammonium chloride make a good buffer solution since $pK_a = 9.26$ and the desired pH is within one pH unit of this value.
The mole ratio of ammonia to ammonium chloride is 1.38.
3.9 g of NH_4Cl is needed.

Practice exercise 11.24
0.09 pH units

Practice exercise 11.25

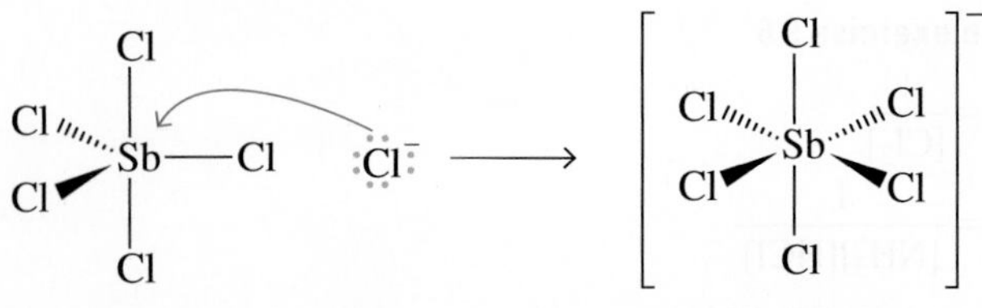

Practice exercise 11.26
H_2O, NH_3 and PH_3 are Lewis bases.
In order of increasing hardness: $PH_3 < NH_3 < H_2O$.

Chapter 12

Practice exercise 12.1

Copper is oxidised and is therefore the reducing agent.
Chlorine is reduced and is therefore the oxidising agent.

Practice exercise 12.2

(a) Ni +2, Cl −1
(b) Mg +2, Ti +4, O −2
(c) K +1, Cr +6, O −2
(d) H +1, P +5, O −2
(e) N −3 (in NH_4) H +1; Ce +4; N +5 (in NO_3), O −2
(f) K +1, Fe +2, C +2, N −3
(g) Na +1, Fe +3, C +2, N −3

Practice exercise 12.3

Sn in $SnCl_2$ is oxidised and therefore the reducing agent is $SnCl_2$. As in H_3AsO_4 is reduced and therefore the oxidising agent is H_3AsO_4.

Practice exercise 12.4

$2Al + 3Cu^{2+} \rightarrow 2Al^{3+} + 3Cu$

Practice exercise 12.5

$3Sn^{2+} + 16H^+ + 2TcO_4^- \rightarrow 2Tc^{4+} + 8H_2O + 3Sn^{4+}$

Practice exercise 12.6

$3CuS + 8NO_3^- + 11H^+ \rightarrow 3Cu^{2+} + 3HSO_4^- + 8NO + 4H_2O$

Practice exercise 12.7

$2MnO_4^- + 3C_2O_4^{2-} + 4OH^- \rightarrow 2MnO_2 + 6CO_3^{2-} + 2H_2O$

Practice exercise 12.8

anode: $Fe(s) \rightarrow Fe^{2+}(aq) + 2e^-$
cathode: $Sn^{2+}(aq) + 2e^- \rightarrow Sn(s)$
cell diagram: $Fe(s) \mid Fe^{2+}(aq) \parallel Sn^{2+}(aq) \mid Sn(s)$

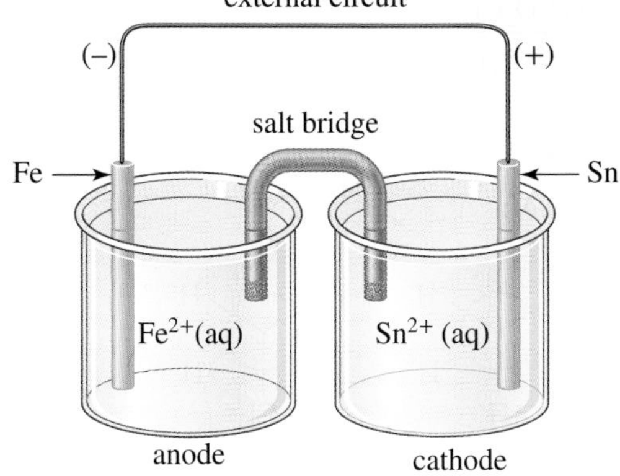

Practice exercise 12.9

−0.74 V

Practice exercise 12.10

$E^{\ominus}_{cell} = -1.93$ V. The reaction is not spontaneous.

Practice exercise 12.11

$2Cl^- + Br_2 \rightarrow Cl_2 + 2\,Br^-$

Practice exercise 12.12

Cell reaction: $3Cu^{2+} + 2\,Al \rightarrow 3Cu + 2Al^{3+}$
$E^{\ominus}_{cell} = +2.00$ V
The Al electrode is the anode.

Practice exercise 12.13

−115.8 kJ

Practice exercise 12.14

2.8×10^{-16}

Practice exercise 12.15

1.04 V

Practice exercise 12.16

6.6×10^{-4} M

Practice exercise 12.17

Cd(s) at the cathode. $O_2(g)$ at the anode.

Practice exercise 12.18

9.1 min

Chapter 13

Practice exercise 13.1

V^2: d^3
Fe^{2+}: d^6
Zn^{2+}: d^{10}
Au^{3+}: d^8
W^{3+}: d^3

Practice exercise 13.2

Only CH_3^- can function as a ligand by donating a lone pair of electrons.

Practice exercise 13.3

$[CoCl(NH_3)_5]Cl_2$: oxidation state = 3+; thus Co(III)
$[W(CO)_3(P(CH_3)_3)_3]$: oxidation state = 0; thus W(0)
$K_3[Fe(ox)_3]$: oxidation state = 3+; thus Fe(III)
$[Cr(en)_3]Cl_3$: oxidation state = 3+. thus Cr(III)
$K_3[Ni(CN)_5]$: oxidation state = 2+; thus Ni(II)

Practice exercise 13.4

(a) $[PtBrCl_3]^-$ has no stereoisomers.

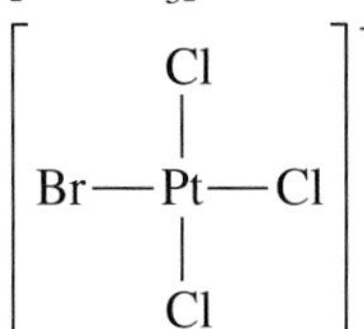

(b) $[CoCl_2(PPh_3)_2]$ has no stereoisomers.

(c) $[CoCl(NH_3)_3(OH_2)_2]^{2+}$ has 3 possible stereoisomers.

fac

mer-cis

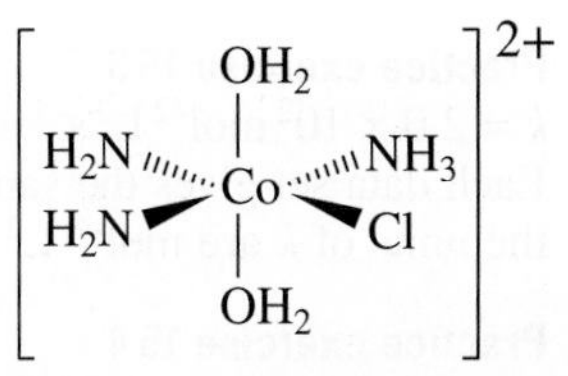

mer-trans

Practice exercise 13.5
(a) pentaamminechloridocobalt(III) chloride
(b) tricarbonylbis(trimethylphosphane)tungsten(0)
(c) potassiumtrioxalatoferrate(III)
(d) tris(ethylenediamine)chromium(III) chloride
(e) potassium pentacyanidonickelate(II)

Practice exercise 13.6
(a) $[Ru(en)_3]Cl_3$
(b) $K_2[PtCl_6]$
(c) $(NH_4)_2[Fe(CN)_4(OH_2)_2]$
(d) *trans*-$[Co(en)_2Cl_2]Cl$

Practice exercise 13.7
$[Cu^{2+}] = 4.60 \times 10^{-22}$ mol L^{-1}
$[Cu(trien)^{2+}] = 1.03 \times 10^{-2}$ mol L^{-1}

Practice exercise 13.8
The crystal field splitting energy for tetrahedral complexes (Δ_t) is related to that for octahedral complexes (Δ_o) by the equation $\Delta_t = 4/9\Delta_o$. Hence, Δ_o for $[CoCl_6]^{4-}$ is greater than Δ_t for $[CoCl_4]^{2-}$.

Practice exercise 13.9
$[Ni(NH_3)_6]^{2-}$: 2 unpaired electrons. $\mu = 2.83\ \mu_B$
$[Cu(en)_3]^{2+}$: 1 unpaired electron. $\mu = 1.73\ \mu_B$
$[ZnCl_6]^{4-}$: 0 unpaired electrons. $\mu = 0\ \mu_B$

Chapter 14

Practice exercise 14.1
Black phosphorus has the highest melting point due to its extensive cross-linking of chains of P_4 groups. Red, which consists of linear chains of P_4 molecules is next, and then white, which consists of individual P_4 molecules with relatively low intermolecular forces.
The order of increasing density is : white P < red P < black P. Black will have the highest density owing to the extensive crosslinking.

Practice exercise 14.2

O=Xe=O with =O above and =O below (Lewis structure of XeO_4)

regular tetrahedral geometry

:F–Kr–F: (Lewis structure of KrF_2)

linear geometry

Chapter 15

Practice exercise 15.1
rate of consumption of $H_2S = 0.30$ mol L^{-1} s^{-1}
rate of consumption of $O_2 = 0.45$ mol L^{-1} s^{-1}

Practice exercise 15.2
order of the reaction with respect to $[BrO_3^-] = 1$
order of the reaction with respect to $[SO_3^{2-}] = 1$
overall order of the reaction is 2

Practice exercise 15.3
$k = 2.0 \times 10^2$ mol^{-2} L^2 s^{-1}
Each data set gives the same value, i.e. $k = \text{rate}/[A][B]^2$ and the units of k are mol^{-2} L^2 s^{-1}.

Practice exercise 15.4
rate of reaction = $k\ [NO]^2\ [H_2]$

Practice exercise 15.5
rate of reaction = 0.03 mol L^{-1} min^{-1}

Practice exercise 15.6
5.33×10^{-2} M

Practice exercise 15.7
7.07×10^{-2} M

Practice exercise 15.8
3750 s or 62.5 min

Practice exercise 15.9
The reaction is first-order.

Practice exercise 15.10
(a) 1.42×10^{-4} mol L^{-1} s^{-1}
(b) 26.4 s

Practice exercise 15.11
18.2 mol^{-1} L s^{-1}

Practice exercise 15.12
The energy of the products would be higher than that of the reactants, giving a positive reaction enthalpy.

Practice exercise 15.13
rate of reaction = $k[NO][O_3]$.

Practice exercise 15.14
(a) yes
(b) $Mo(CO)_5$
(c) rate of reaction = $k_1[Mo(CO)_6]$

Practice exercise 15.15

$$\text{rate of reaction} = \frac{k[NO_2Cl]^2}{[NO_2]}$$

Chapter 16

Practice exercise 16.1
staggered — any two of:

Cl / H, Cl / H, H / H — Cl / Cl, H / H, H / H — Cl / H, H / H, H / Cl

eclipsed — any two of:

H Cl / Cl / H, H / H — Cl H / Cl, H / H, H / H — Cl Cl / H, H / H, H / H

Practice exercise 16.2
(a) $C_{10}H_{20}$: 1,2,3,4,5-pentamethylcyclopentane
(b) $C_{10}H_{20}$: 1,1-diethylcyclohexane
(c) $C_{10}H_{20}$: 1,2-diethyl-3,4-dimethylcyclobutane
(d) C_9H_{18}: isobutylcyclopentane
(e) $C_{11}H_{22}$: *sec*-butylcycloheptane
(f) C_6H_2: 1-ethyl-1-methylcyclopropane

Practice exercise 16.3
(a) (ii)
(b) (ii)
(c) (ii)
In general, the more stable conformation has alkyl group substituents in the equatorial position such that di-axial interactions are minimised.

Practice exercise 16.4

(a) *cis* and *trans* isomers are possible:

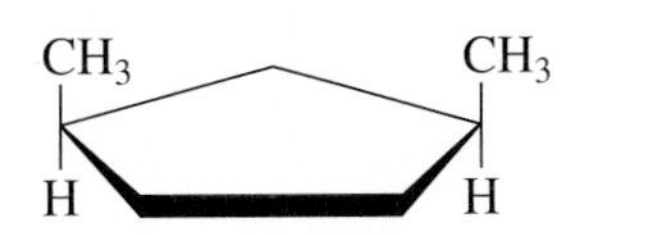

cis-1,3-dimethylcyclopentane

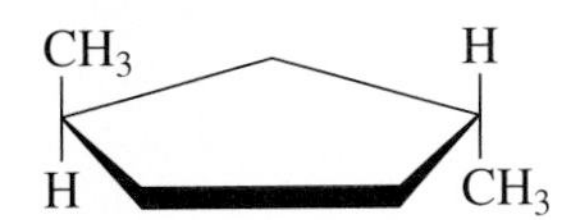

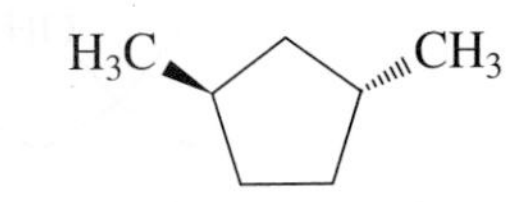

trans-1,3-dimethylcyclopentane

(b) Cannot exhibit *cis*–*trans* isomerism.

(c) *cis* and *trans* isomers are possible:

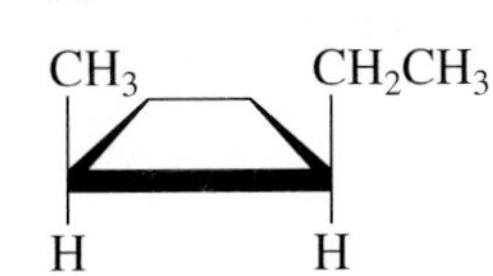

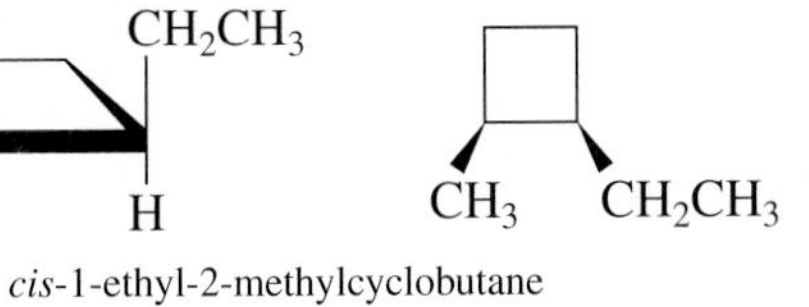

cis-1-ethyl-2-methylcyclobutane

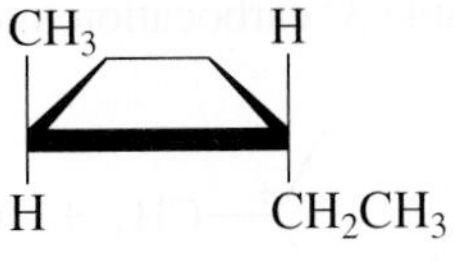

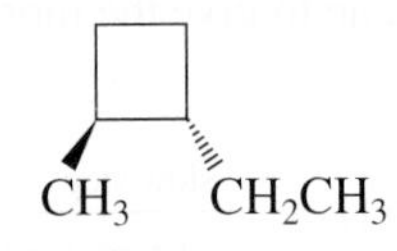

trans-1-ethyl-2-methylcyclobutane

Practice exercise 16.5

(a) 2-2-dimethylpropane < 2-methylbutane < pentane

(b) 2,2,4-trimethylhexane < 3,3-dimethylheptane < nonane

Practice exercise 16.6

(a) 3,3-dimethyl-1-pentene (b) 2,3-dimethyl-2-butene (c) 3,3-dimethyl-1-butyne

Practice exercise 16.7

(a) *cis*-4-methyl-2-pentene (b) *trans*-2,2-dimethyl-3-hexane

Practice exercise 16.8

(a) (*E*)-1-chloro-2,3-dimethyl-2-pentene (b) (*Z*)-1-bromo-1-chloropropene (c) (*E*)-2,3,4-trimethyl-3-heptene

Practice exercise 16.9

(a) 1-isopropyl-4-methyl-cyclohexene (b) 4-methylcyclooctene (c) 4-*tert*-butylcyclohexene

Practice exercise 16.10

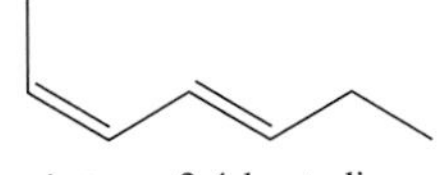

cis,trans-2,4-heptadiene

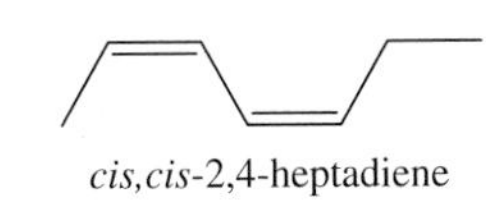

cis,cis-2,4-heptadiene

Practice exercise 16.11

4

Practice exercise 16.12

(a) 2-iodopropane

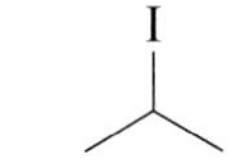

(b) 1-iodo-1-methylcyclohexane

Practice exercise 16.13

(c) < (b) < (a)

Practice exercise 16.14

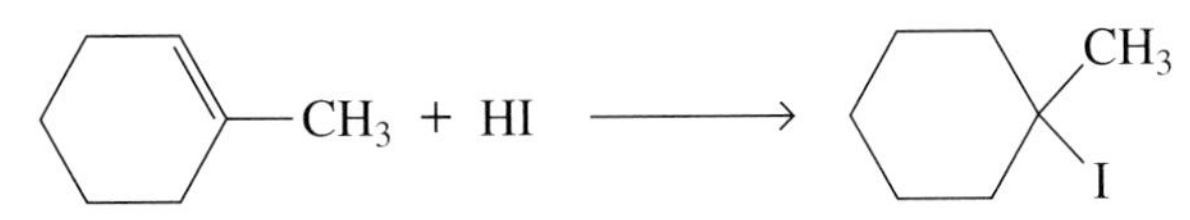

Step 1: Protonation of the alkene to give the most stable 3° carbocation intermediate.

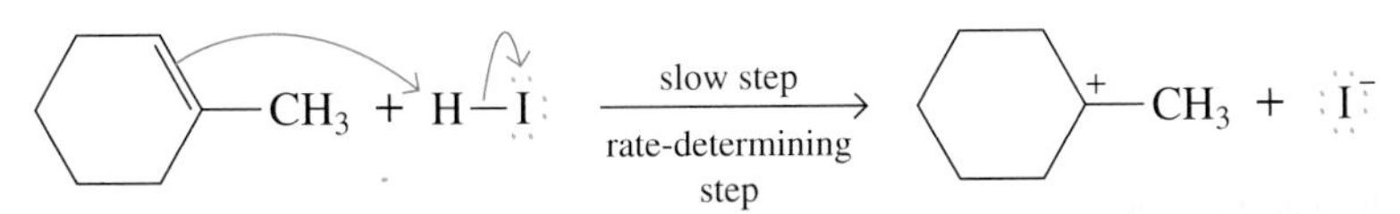

Step 2: Nucleophilic attack of the iodide anion on the 3° carbocation intermediate to give the product.

Practice exercise 16.15

(a) 2-methyl-butanol (b) 2-methyl-2-butanol

Practice exercise 16.16

Step 1: Protonation of the alkene to give the most stable 3° carbocation intermediate.

Step 2: Nucleophilic attack of the water on the 3° carbocation intermediate to give the protonated alcohol.

Step 3: The protonated alcohol loses a proton to form the product.

Practice exercise 16.17

(a) (b)

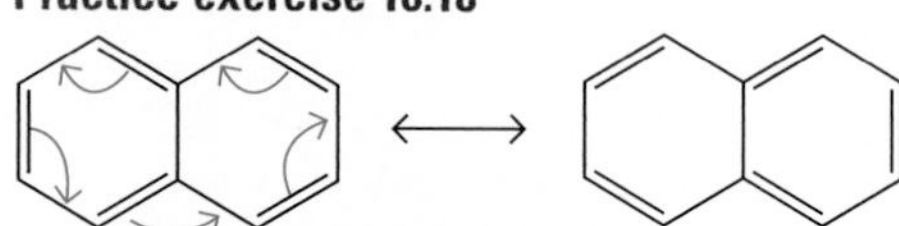

Practice exercise 16.18

Practice exercise 16.19

(a) 2-phenyl-2-propanol (b) *(E)*-3,4-diphenyl-3-hexene (c) 3-methylbenzoic acid or *m*-methylbenzoic acid

Practice exercise 16.20

Step 1: Generation of the HSO_3^+ electrophile.

Step 2: Nucleophilic attack of benzene on HSO_3^+ electrophile.

resonance-stabilised intermediate

Step 3: Loss of proton to regenerate aromatic ring (benzenesulfonic acid) and sulfuric acid.

Chapter 17

Practice exercise 17.1

(a)

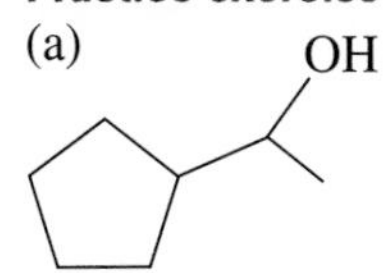

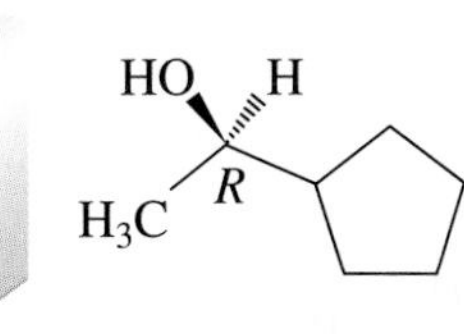

HO H
H_3C *R*

(b) OH

H OH
S CH_3

HO H
H_3C *R*

Practice exercise 17.2

(a) S (b) S (c) R

Practice exercise 17.3

(a) Compounds (i) and (iii) are enantiomers.
Compounds (ii) and (iv) are enantiomers.
(b) Compounds (i) and (ii) are diastereomers.
Compounds (i) and (iv) are diastereomers.
Compounds (ii) and (iii) are diastereomers.
Compounds (iii) and (iv) are diastereomers.

Practice exercise 17.4

(a) Compounds (ii) and (iii).
(b) Compounds (i) and (iv).
(c) Compounds (ii) and (iii).

Practice exercise 17.5

3

Practice exercise 17.6

2

Practice exercise 17.7

+6.9°

Chapter 18

Practice exercise 18.1

(a) 1-chloro-3-methylbut-2-ene
(b) 1-bromo-1-methylcyclohexane
(c) 1,2-dichloropropane
(d) 2-chlorobuta-1,3-diene

Practice exercise 18.2

(a)

$H_3C-CH(CH_3)-CH(CH_3)-CH_2-Br$

1-bromo-2,3-dimethylbutane

$H_3C-CH(CH_3)-C(CH_3)(Br)-CH_3$

2-bromo-2,3-dimethylbutane

(b)

Cl, HC, $CH-CH_3$, H_2C

1-chloro-2-methylcyclopropane

H_2C, $C-CH_3$, H_2C, Cl

1-chloro-1-methylcyclopropane

H_2C, $CH-CH_2$, H_2C, Cl

chloromethylcyclopropane

(c)

$H_2C(Cl)-CH_2-CH_2-CH_2-CH_3$

1-chloropentane

$H_3C-CH(Cl)-CH_2-CH_2-CH_3$

2-chloropentane

$H_3C-CH_2-CH(Cl)-CH_2-CH_3$

3-chloropentane

(d) H_3C, CH_3, C, CH_2, H_3C, C, Br, CH_3, CH_3

3-bromo-3,4,4-trimethylpentane

H_3C, CH_3, Br, C, CH, H_3C, CH, CH_3, CH_3

2-bromo-3,4,4-trimethylpentane

H_3C, CH_3, C, CH_2, H_3C, CH, CH_2, CH_3, Br

1-bromo-3,4,4-trimethylpentane

H_3C, CH_3, C, CH_2, H_2C, CH, CH_3, Br, CH_3

2-bromomethyl-2,3-dimethylpentane

H_3C, CH_3, C, CH_2, H_3C, CH, CH_3, $Br-CH_2$

3-bromomethyl-4,4-dimethylpentane

Practice exercise 18.3

(a) $S-CH_2CH_3 + Na^+Br^-$

(b) $O\overset{O}{\overset{\|}{C}}CH_3 + Na^+Br^-$

Practice exercise 18.4

(a) 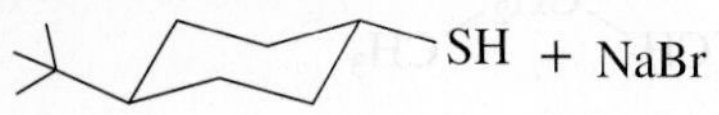 $(CH_3)_3C$–cyclohexyl–SH + NaBr

The reaction follows an S_N2 mechanism.

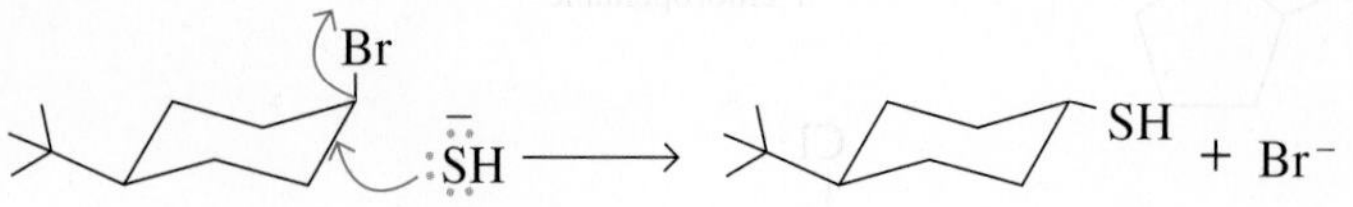

Br

:SH⁻ → SH + Br^-

(b)

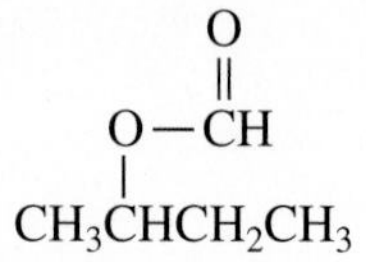

O
‖
O–CH
|
$CH_3CHCH_2CH_3$

R and *S* enantiomers

The reaction follows an S_N1 mechanism.

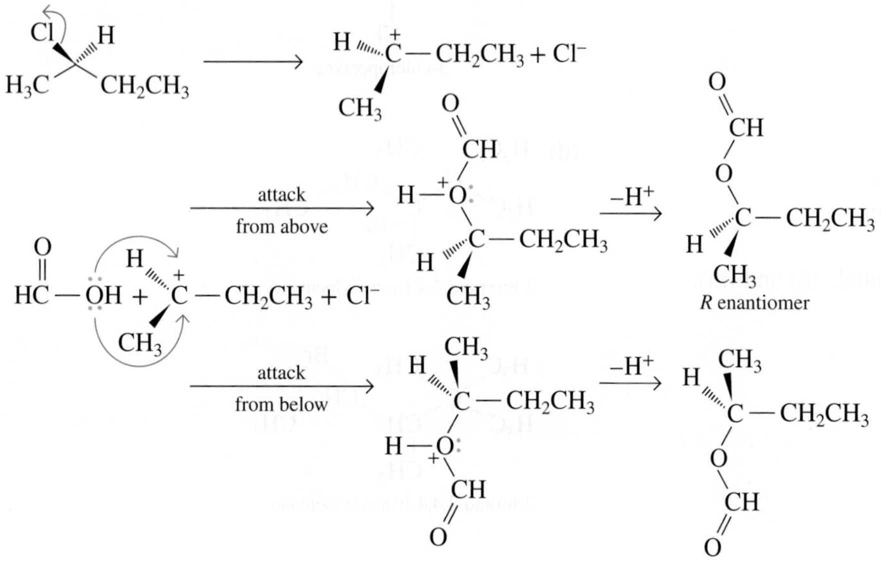

Cl, H, H_3C, CH_2CH_3 → H–$\overset{+}{C}$–CH_2CH_3 + Cl^-, CH_3

HC(=O)–OH + H–$\overset{+}{C}$(CH₃)–CH_2CH_3 + Cl^-

attack from above → H–$\overset{+}{O}$(CH=O)–C(H)(CH₃)–CH_2CH_3 $\xrightarrow{-H^+}$ HC(=O)–O–C(H)(CH₃)–CH_2CH_3

R enantiomer

attack from below → H–$\overset{+}{O}$(CH=O)–C(H)(CH₃)–CH_2CH_3 $\xrightarrow{-H^+}$ HC(=O)–O–C(H)(CH₃)–CH_2CH_3

S enantiomer

Practice exercise 18.5

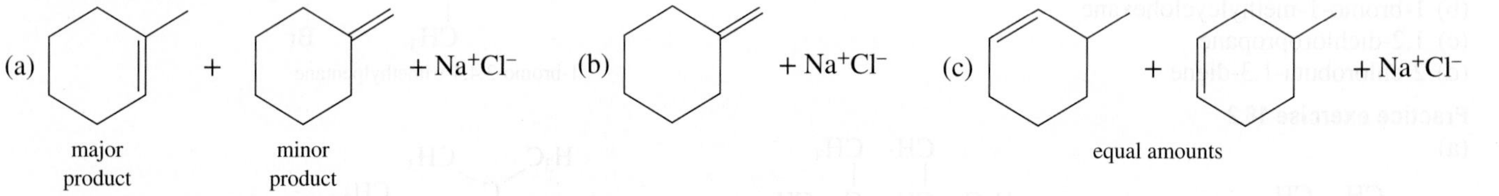

(a) [1-methylcyclohexene] + [methylenecyclohexane] + Na^+Cl^-

major product; minor product

(b) [methylenecyclohexane] + Na^+Cl^-

(c) [3-methylcyclohexene] + [4-methylcyclohexene] + Na^+Cl^-

equal amounts

Practice exercise 18.6

(a) E2

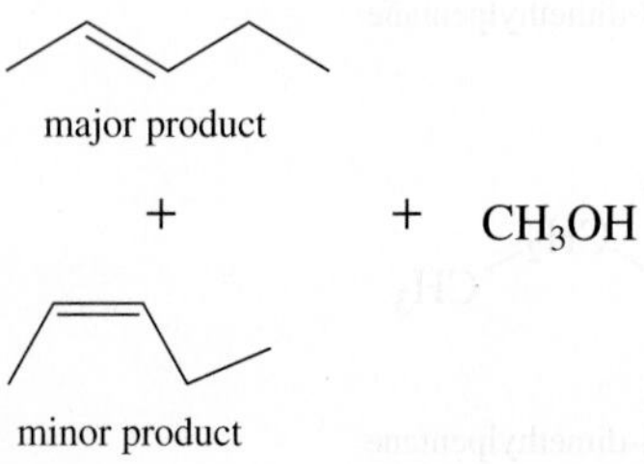

[trans-2-pentene] major product

+ [cis-2-pentene] minor product + CH_3OH

(b) E2

[4-methylcyclohexene] + CH_3CH_2OH

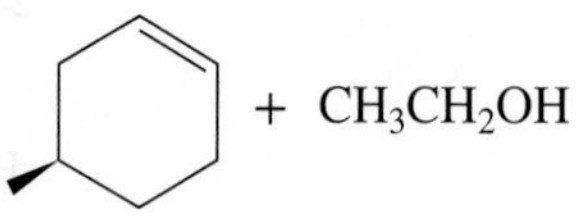

Practice exercise 18.7

(a) S_N2 and E2 are competing.

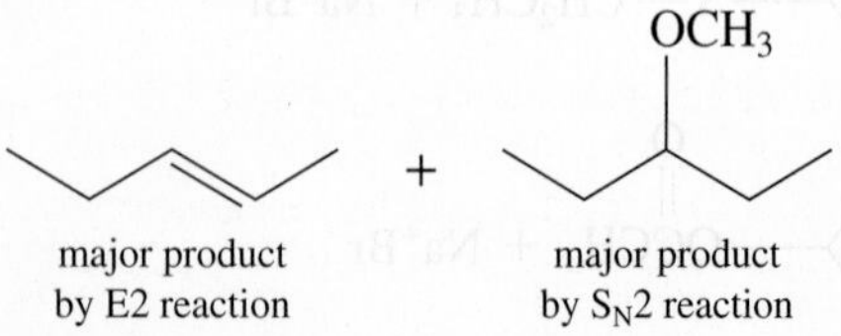

[alkene] major product by E2 reaction + [OCH₃ ether] major product by S_N2 reaction

(b) S_N2

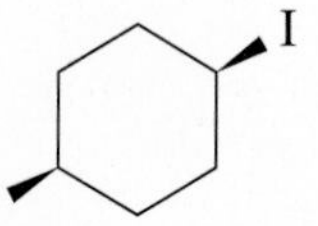

[iodo-methylcyclohexane] I

Chapter 19

Practice exercise 19.1
(a) hexan-3-ol (b) 2-methylpentan-2-ol (c) 3,3-dimethylcyclopentanol

Practice exercise 19.2
(a) primary (b) secondary (c) primary (d) tertiary

Practice exercise 19.3

(a) 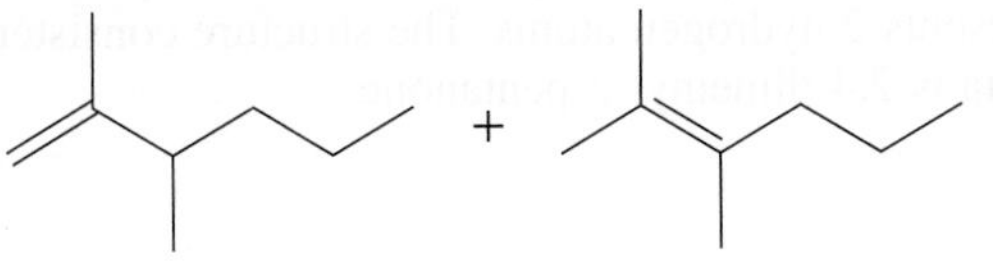 + (b) +

minor product major product minor product major product

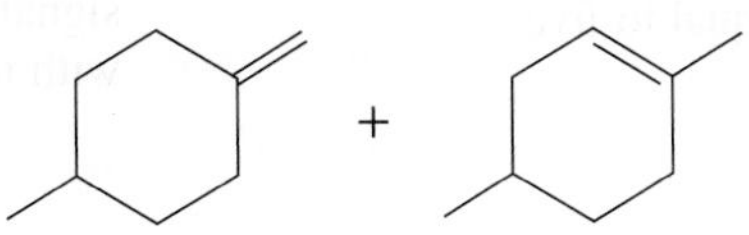

Practice exercise 19.4

(a) 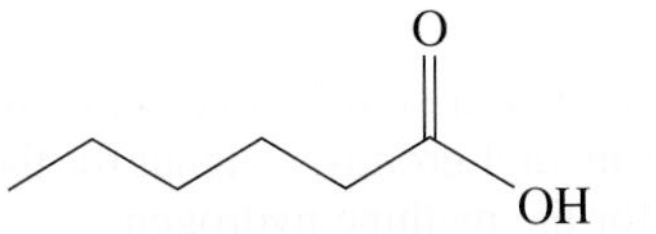 O OH (b) 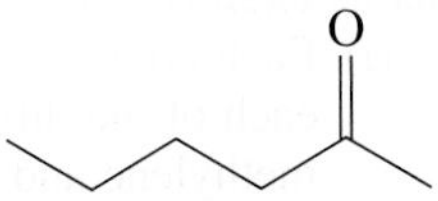 O (c) 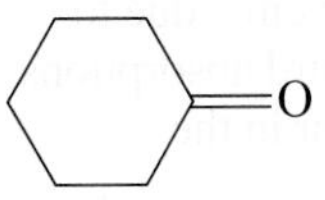 =O

Practice exercise 19.5
2,4-dimethoxyphenol < 4-chlorophenol < 2,4-dichlorophenol

Practice exercise 19.6
(a) 2-isopropoxypropane and diisopropyl ether
(b) ethoxycyclohexane and cyclohexyl ethyl ether

Practice exercise 19.7

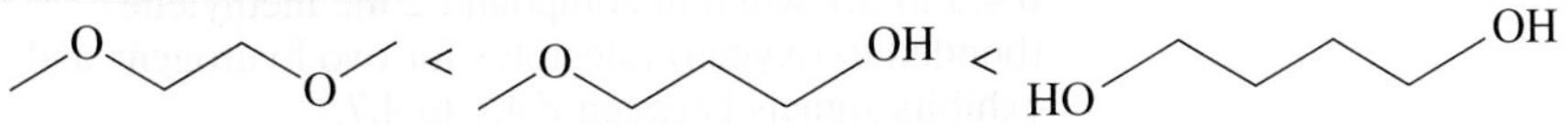

O O < O OH < HO OH

Practice exercise 19.8
(a) a tertiary heterocyclic aliphatic amine
(b) a secondary aliphatic amine
(c) one heterocyclic aromatic amine and one secondary heterocyclic aliphatic amine

Practice exercise 19.9

(a) 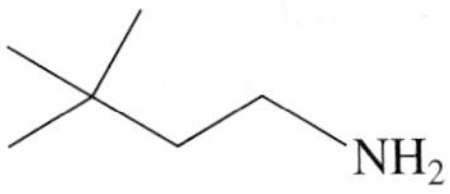 NH_2 (b) 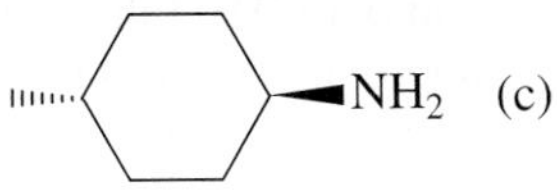 NH_2 (c) 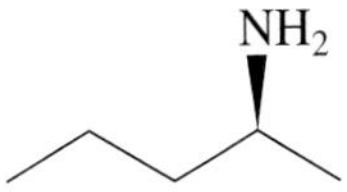 NH_2

Practice exercise 19.10
Equilibrium favours formation of the weaker acid, so equilibrium lies to the left.

Practice exercise 19.11
(a) A (b) C

Practice exercise 19.12
(a) The basic nitrogen atom and the carboxylic acid are protonated and the amino acid has a positive charge overall.
(b) Both of the acidic protons are removed and the amino acid has a negative charge overall.

Practice exercise 19.13

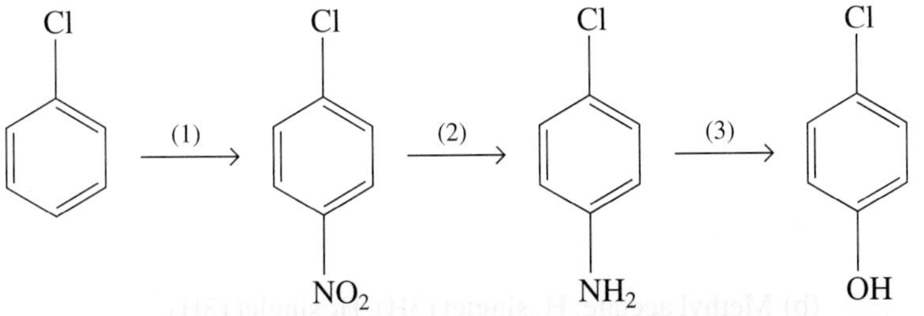

Cl —(1)→ Cl, NO_2 —(2)→ Cl, NH_2 —(3)→ Cl, OH

Step 1: Electrophilic aromatic substitution at the para position by treatment with nitric acid/sulfuric acid, followed by separation of the *para* isomer from the product mixture.
Step 2: Reduction of the *para*-nitro group by treatment with H_2 in the presence of a transition metal catalyst or using Fe, Sn, or Zn in the presence of aqueous HCl.
Step 3: Treatment with $NaNO_2$/HCl to form the diazonium ion salt and then warming the solution.

Chapter 20

Practice exercise 20.1

The index of hydrogen deficiency is four. There are three double bonds and a ring.

Practice exercise 20.2

Paracetamol has four 'double bonds' (each worth one hydrogen deficiency) and a ring (worth one hydrogen deficiency) for a total index of hydrogen deficiency equal to five.

Practice exercise 20.3

(a) $C_{13}H_{28}$
(b) $C_{11}H_{22}$

Practice exercise 20.4

$C_6H_{14}N_2$

Practice exercise 20.5

The first absorption, is a strong peak around 1715 cm^{-1} due to the C=O stretching vibration. The second and third absorptions are due to the C—O stretching vibration and occur in the region 1250–1050 cm^{-1}.

Practice exercise 20.6

Ketones have a strong absorption in the region 1705–1780 cm^{-1} due to the C=O stretching vibration. Alcohols have a broad band between 3200 and 3500 cm^{-1} due to the H—O stretching vibration.

Practice exercise 20.7

As the wavenumber increases, so does the bond strength.

Practice exercise 20.8

(a) and (b)

Practice exercise 20.9

Letter superscripts are used to distinguish non-equivalent hydrogen atoms. (Use the 'test atom' approach if you have difficulty understanding the answers given.)

(a) Methylcyclopentane has four sets of equivalent hydrogens: a methyl group (labelled a), a methine hydrogen (labelled b) and two sets of methylene hydrogen atoms (labelled c and d).

methylcyclopentane

(b) 2,2,4-trimethylpentane has four sets of equivalent hydrogens: three equivalent methyl groups (labelled a), two methylene hydrogen atoms (labelled b), one methine hydrogen (labelled c), and two equivalent methyl groups (labelled d).

2,2,4-trimethylpentane

Practice exercise 20.10

(a)

(b)

(c) Br

(d) Cl Cl Cl Cl

Practice exercise 20.11

The larger signal represents 12 hydrogen atoms. The smaller signal represents 2 hydrogen atoms. The structure consistent with the data is 2,4-dimethyl-3-pentanone.

Practice exercise 20.12

(a) (i) Each compound exhibits four signals: one signal for each of the different methyl groups, a signal for the methylene and one for the methine hydrogen.
(ii) The ratio of signals is 6:3:2:1.
(iii) In both compounds 1 and 2, there are methyl hydrogens far upfield with a signal between δ 0.8–1.0. The difference occurs with the chemical shift of the downfield methylene and methine groups, depending on which of the two is attached to the oxygen atom. In compound 1, the methine (bonded to oxygen) integrates for 1 hydrogen and exhibits signals between δ 4.5 to 5.1 while in compound 2 the methylene (bonded to oxygen) integrates for two hydrogens and exhibits signals between δ 4.1 to 4.7.

(b) (i) Methyl acetate: 2 signals, each integrating for 3H. Ethyl formate: 3 signals, with integration values of 1H, 2H and 3H for hydrogens (a), (b) and (c) respectively. Methoxyacetaldehyde: 3 signals, with integration values of 1H, 2H and 3H for hydrogens (a), (b) and (c) respectively.
(ii) Methyl acetate: 2 signals at approximately δ 2 (3H) and δ 4 (3H). Ethyl formate: 3 signals at approximately δ 8 (1H), δ 4 (2H) and δ 1 (3H). Methoxyacetaldehyde: 3 signals, at approximately δ 10 (1H), δ 6 (2H) and δ 4 (3H). The signal due to hydrogen (b) [δ 6 (2H)] is further downfield because the deshielding effect of both the carbonyl and the methoxy groups are additive.

Practice exercise 20.13

(a) The first compound has three signals, while the second has only one.

(i) a = triplet, b = quartet, c = singlet and a = singlet

(ii) a = singlet, b = singlet, c = doublet, d = triplet and a = singlet, b = singlet

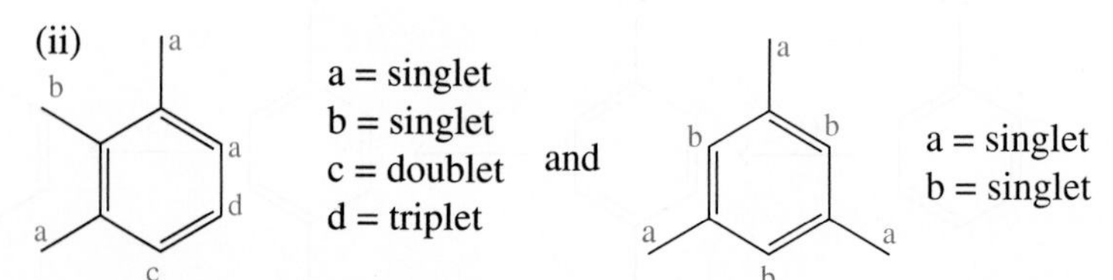

(b) Methyl acetate: H_a singlet (3H), H_b singlet (3H).
Ethyl formate: H_a singlet (1H), H_b quartet (2H), H_c triplet (3H).
Methoxyacetaldehyde: H_a triplet (1H), H_b doublet (2H), H_c singlet (3H). Note the splitting between H_a and H_b in methoxyacetaldehyde is very small and these signals may appear as singlets.

Practice exercise 20.14

(a) These molecules can be distinguished by comparing the number of ^{13}C signals. The molecule on the left has only five nonequivalent carbons (two sets of equivalent carbons due to a mirror plane of symmetry) displaying five ^{13}C signals. The molecule on the right has seven nonequivalent carbons giving seven ^{13}C signals.

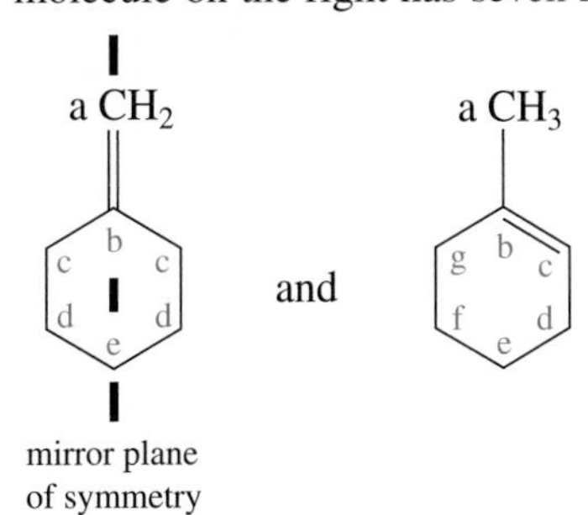

(b) The molecule on the left has lower symmetry and has six different ^{13}C signals, whereas the molecule on the right has greater symmetry (a perpendicular 180° rotation axis) and has three different ^{13}C signals.

Practice exercise 20.15

1.45×10^{-4} M

Chapter 21

Practice exercise 21.1

(a) 2,6-dimethylheptan-4-one
(b) 3,4-dibromocyclopentanecarbaldehyde
(c) (*S*)-2-hydroxybutanal

Practice exercise 21.2

The chiral aldehydes are circled and the stereocentre is indicated with an asterisk.

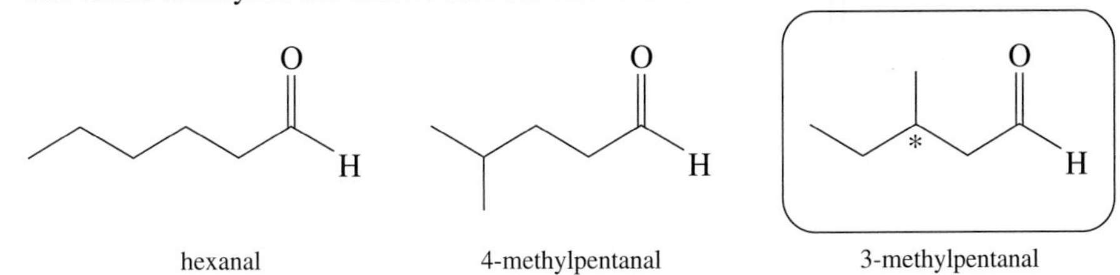

hexanal 4-methylpentanal 3-methylpentanal

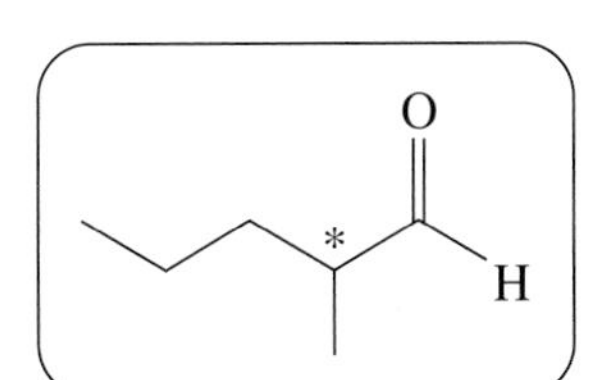

2-methylpentanal

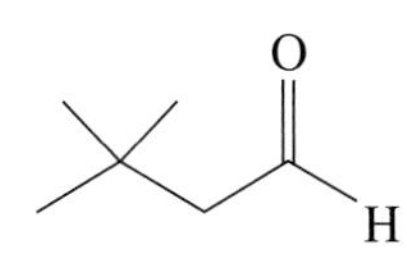

3,3-dimethylbutanal

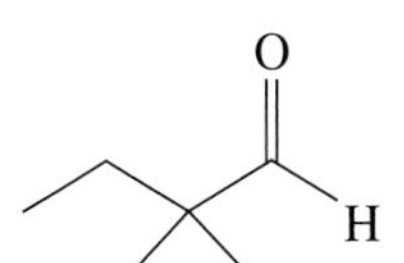

2,2-dimethylbutanal

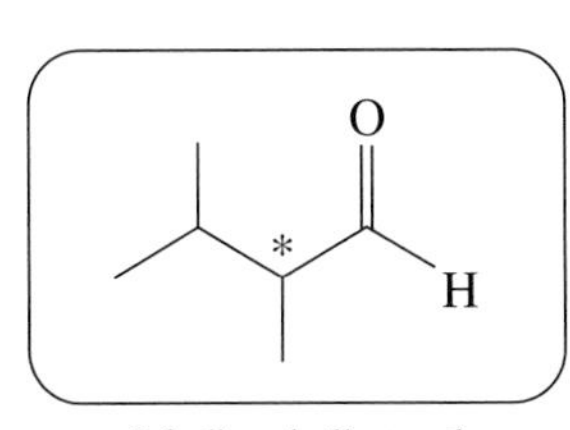

2,3-dimethylbutanal

2-ethylbutanal

Practice exercise 21.3

(a) 3-methoxy-4-hydroxybenzaldehyde
(b) 3-oxobutanoic acid
(c) (*S*)-2-aminopropanoic acid (alanine)

Practice exercise 21.4

These Grignard reagents cannot be prepared because each has an acidic functional group that would be deprotonated by the highly basic Grignard reagent portion of the molecule. In (a), the carboxylic acid would be deprotonated to give a carboxylate. In (b), the phenol would be deprotonated to give a phenoxide.

Practice exercise 21.5

(a) [cyclopentenyl]–MgBr $\xrightarrow[\text{2. dil acid}]{\text{1. } H_2C{=}O}$ [cyclopentenyl]–CH_2OH

(b) [cyclopentenyl]–MgBr $\xrightarrow[\text{2. dil acid}]{\text{1. } CH_3CH_2CH{=}O}$ [cyclopentenyl]–CH(OH)CH_2CH_3

(c) [cyclopentenyl]–MgBr $\xrightarrow[\text{2. dil acid}]{\text{1. cyclopentanone}}$ [cyclopentenyl]–C(OH) (cyclopentanol) HO

Practice exercise 21.6

(a) 1,1-dimethoxycyclohexane (OCH_3, OCH_3) $\xrightarrow{H_3O^+/H_2O}$ cyclohexanone (=O) $+ 2CH_3OH$

(b) cyclic acetal (O, O) $\xrightarrow{H_3O^+/H_2O}$ pentan-3-one (O) + HO~~~OH

(c) 2-ethoxytetrahydropyran (OCH_2CH_3) $\xrightarrow{H_3O^+/H_2O}$ HO~~~CHO (O, H) $+ CH_3CH_2OH$

Practice exercise 21.7

(a) [cyclopentyl]–$CH_2N{=}C(CH_3)_2 + H_2O \xrightarrow{HCl}$ [cyclopentyl]–CH_2NH_2 + O=C(CH_3)$_2$

(b) [phenyl]–CH=N–[cyclopentyl] $+ H_2O \xrightarrow{HCl}$ [phenyl]–C(=O)H + H_2N–[cyclopentyl]

Practice exercise 21.8

Two combinations are possible for each amine.

(a) (i) cyclopentanone (=O) + H_2N–CH(CH_3)$_2$ $\xrightarrow[\text{2. } H_2/Ni]{\text{1. } H^+ (-H_2O)}$ cyclopentyl–NH–CH(CH_3)$_2$

(ii) cyclopentylidene=NH_2 + O=C(CH_3)$_2$ $\xrightarrow[\text{2. } H_2/Ni]{\text{1. } H^+ (-H_2O)}$ cyclopentyl–NH–CH(CH_3)$_2$

(b) (i) heptan-4-one (O) $+ CH_3NH_2 \xrightarrow[\text{2. } H_2/Ni]{\text{1. } H^+ (-H_2O)}$ N-methylheptan-4-amine (HN–CH_3)

(ii) heptan-4-amine (NH_2) $+ H_2C{=}O \xrightarrow[\text{2. } H_2/Ni]{\text{1. } H^+ (-H_2O)}$ N-methylheptan-4-amine (HN–CH_3)

Practice exercise 21.9

(a) 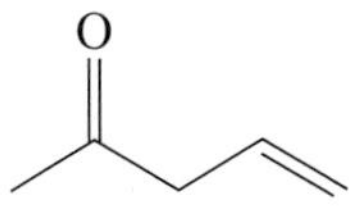 (b) 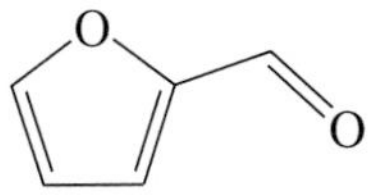 (c) 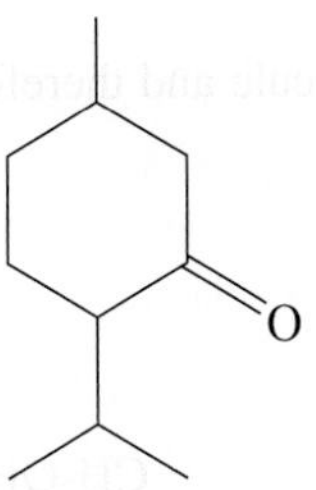

Practice exercise 21.10

(a) 2-hydroxybenzaldehyde + O_2 $\rightarrow$ 2-hydroxybenzoic acid
(b) phenylacetaldehyde + Tollen's reagent $\rightarrow$ phenylacetic acid (obtained as a carboxylate salt)

Practice exercise 21.11

(a) 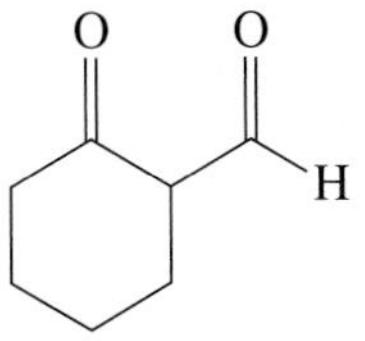 (b) 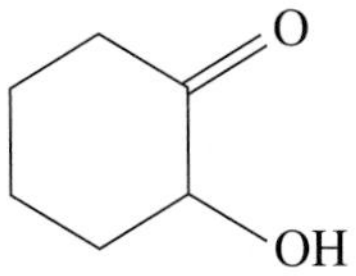 (c) 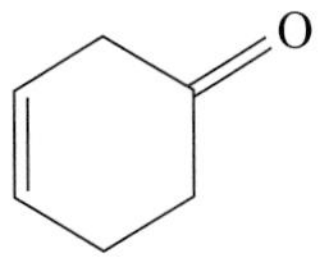

Chapter 22

Practice exercise 22.1

(a) The structures are as shown below.
(b) The labels indicate the D and L ketopentoses and the pairs of enantiomers.

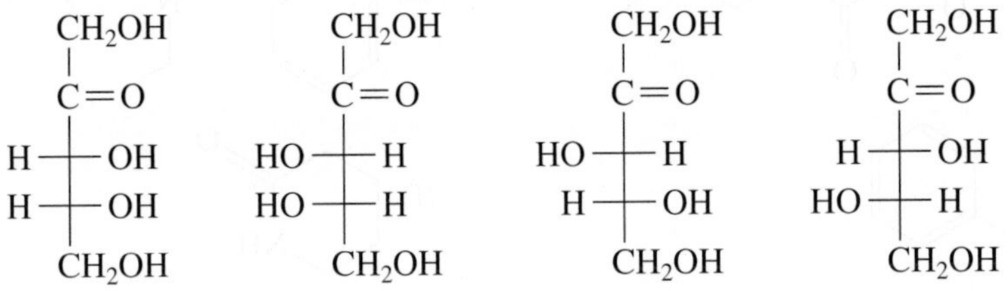

D-ribulose L-ribulose D-xylulose L-xylulose

(one pair of enantiomers) (a second pair of enantiomers)

Practice exercise 22.2

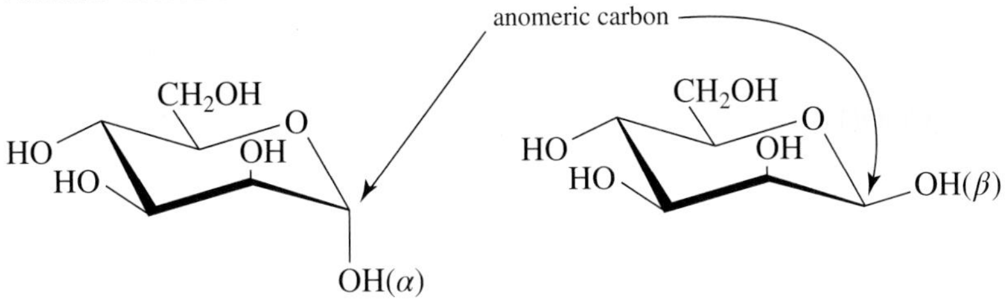

α-D-mannopyranose *β*-D-mannopyranose

Practice exercise 22.3

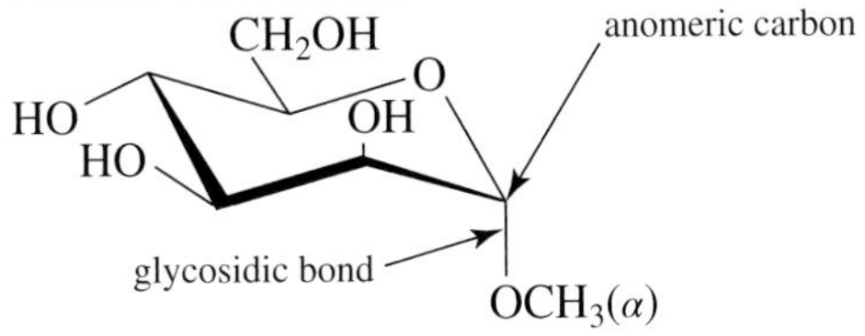

methyl *α*-D-mannopyranoside

Practice exercise 22.4

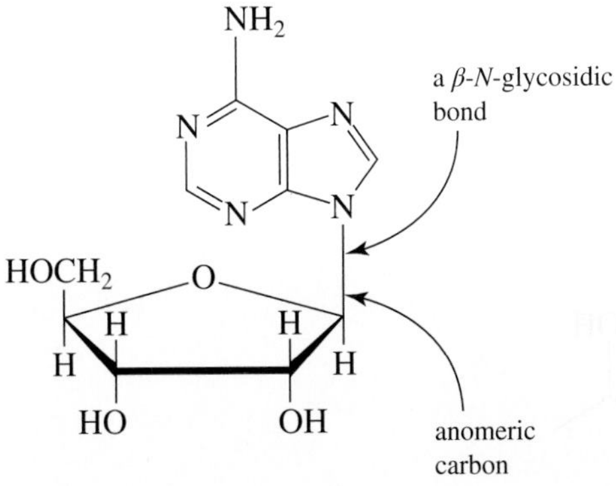

Practice exercise 22.5

Erythritol is achiral because of a mirror plane in the molecule and therefore, the product is optically inactive.

Practice exercise 22.6

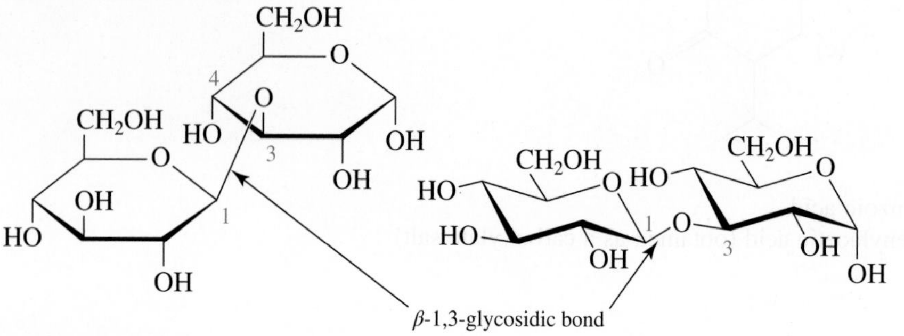

CH_2OH O 4 O CH_2OH HO 3 OH O OH OH 1 HO OH

CH_2OH O HO CH_2OH O HO 1 O HO OH 3 OH OH

β-1,3-glycosidic bond

Chapter 23

Practice exercise 23.1

(a) (*R*)-2,3-dihydroxypropanoic acid
(b) *cis*-2-butenedioic acid *or* (*Z*)-butenedioic acid
(c) (*R*)-3,5-dihydroxy-3-methylpentanoic acid

Practice exercise 23.2

(a) 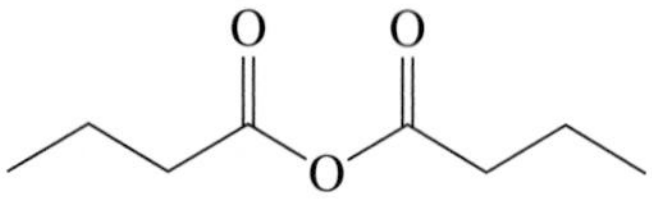 O O O

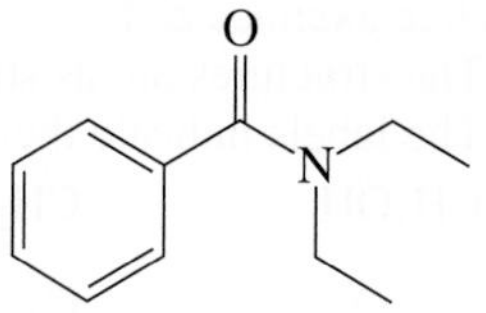

(b) 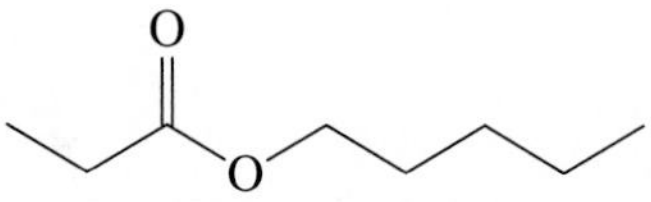 O O

(c) O O O O

(d) 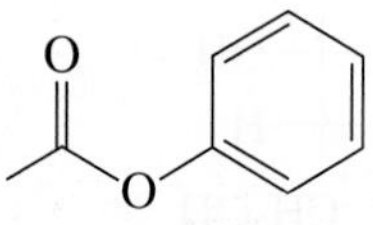 O O

(e) O N

(f) 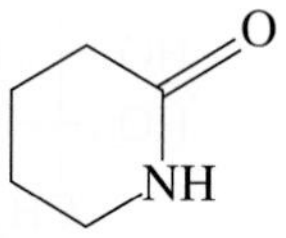 O NH

Practice exercise 23.3

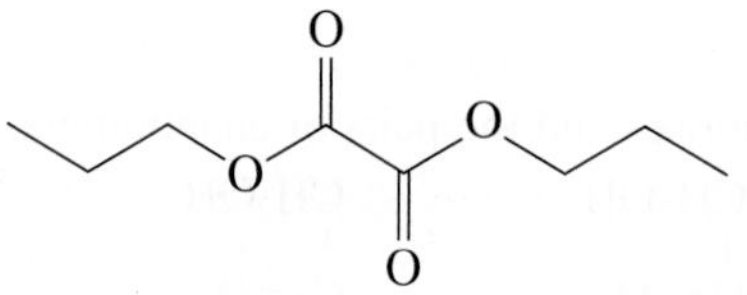

(a) 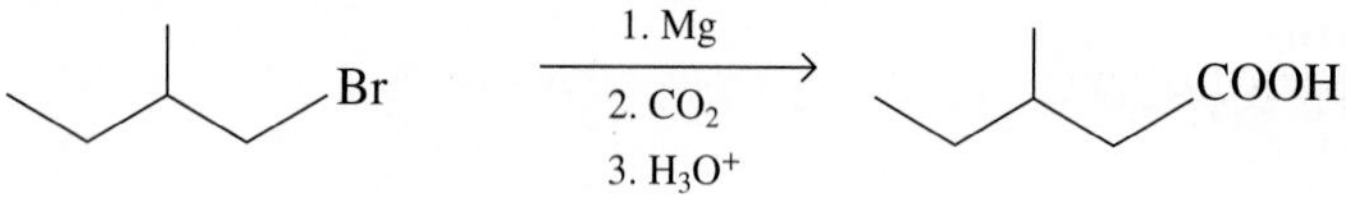 Br → 1. Mg 2. CO_2 3. H_3O^+ → COOH

(b) 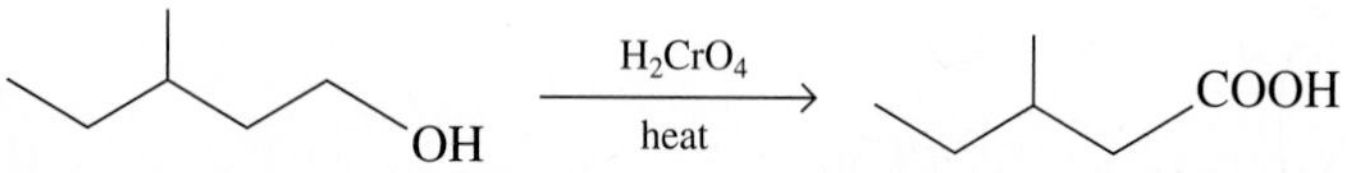 OH → H_2CrO_4 heat → COOH

Practice exercise 23.4

(a) 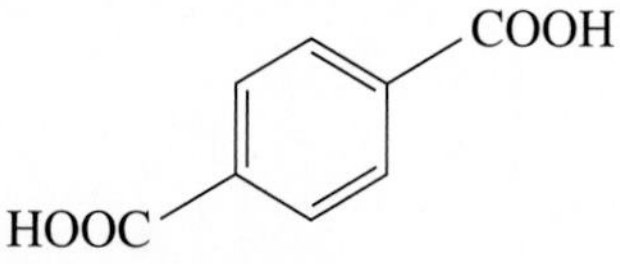 COOH HOOC

(b) 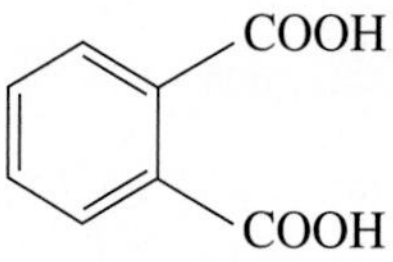 COOH COOH

Practice exercise 23.5

(a) 5.03 (b) 0.22 (c) 2.86

Practice exercise 23.6

(a) 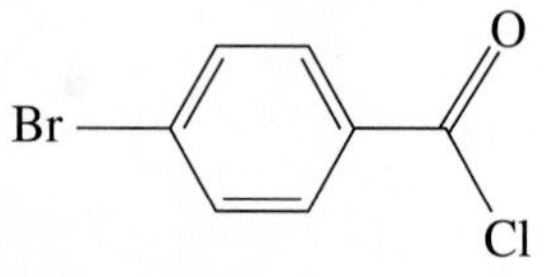 Br O Cl + SO_2 + HCl

(b) 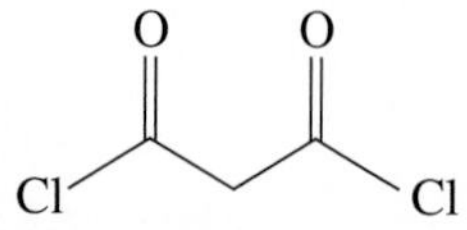 O O Cl Cl + $2POCl_3$

Practice exercise 23.7

(a) 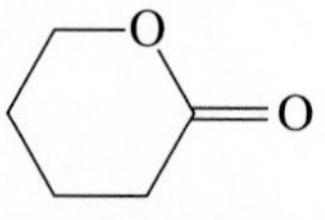 O O
a cyclic ester

(b) 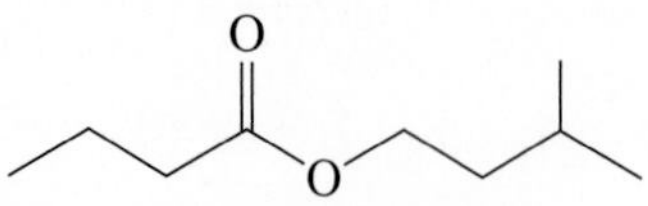 O O
an ester found in strawberries

Practice exercise 23.8

(a) 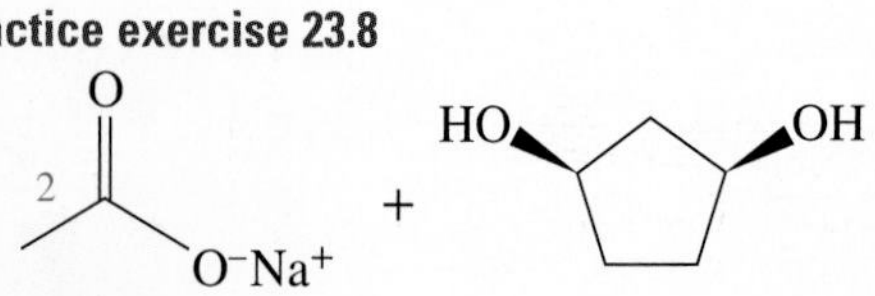 2 O O^-Na^+ + HO OH

(b) 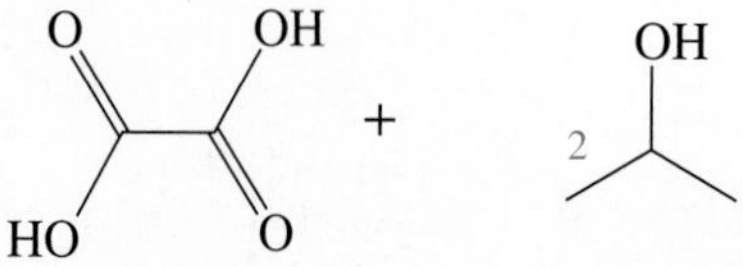 O OH HO O + 2 OH

Practice exercise 23.9

(a) 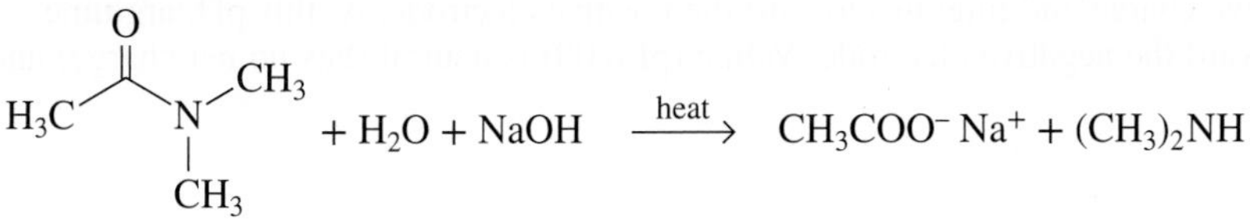 H_3C–C(=O)–N(CH_3)–CH_3 + H_2O + NaOH $\xrightarrow{\text{heat}}$ CH_3COO^- Na^+ + $(CH_3)_2NH$

(b) 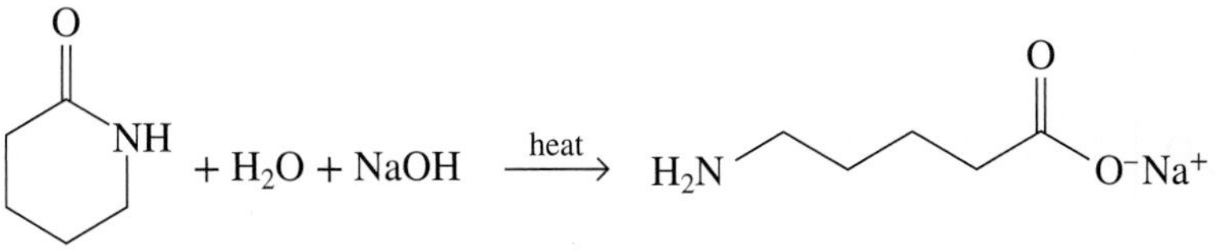 O, NH + H_2O + NaOH $\xrightarrow{\text{heat}}$ H_2N O O^-Na^+

Practice exercise 23.10

(a) 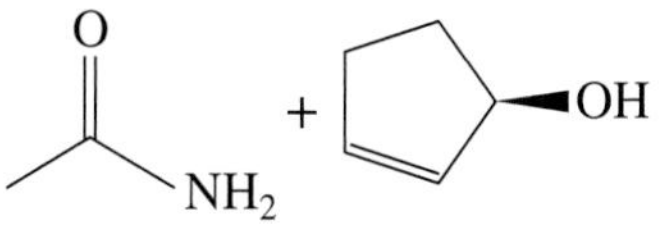 O NH_2 + OH (b) 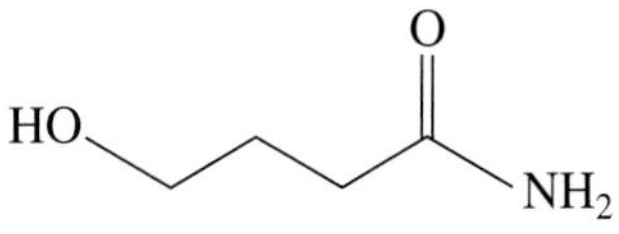 HO O NH_2

Practice exercise 23.11

(a) 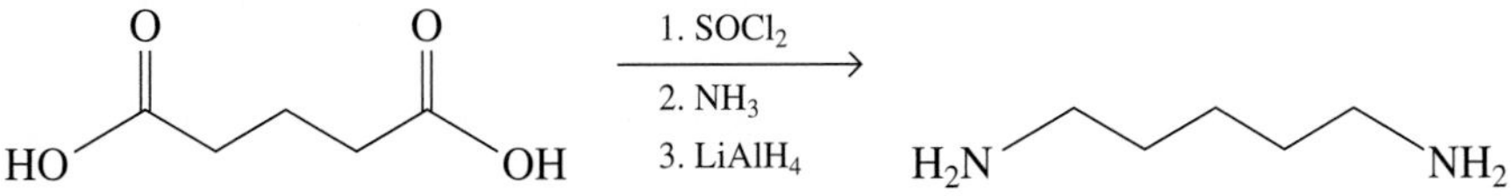 HO O O OH $\xrightarrow[\text{2. } NH_3 \text{ 3. } LiAlH_4]{\text{1. } SOCl_2}$ H_2N NH_2

(b) 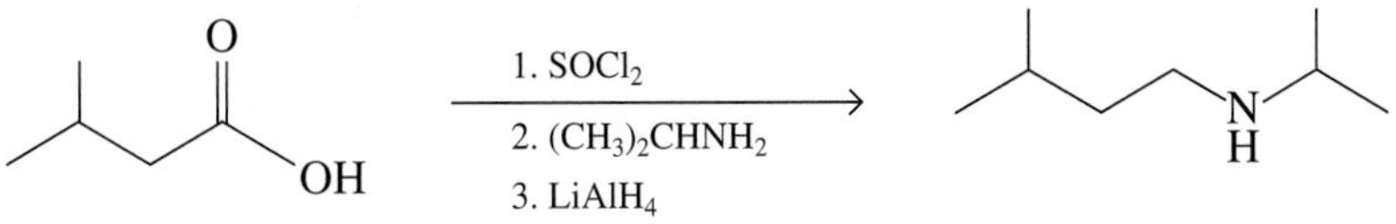 O OH $\xrightarrow[\text{2. } (CH_3)_2CHNH_2 \text{ 3. } LiAlH_4]{\text{1. } SOCl_2}$ N H

Practice exercise 23.12

(a) 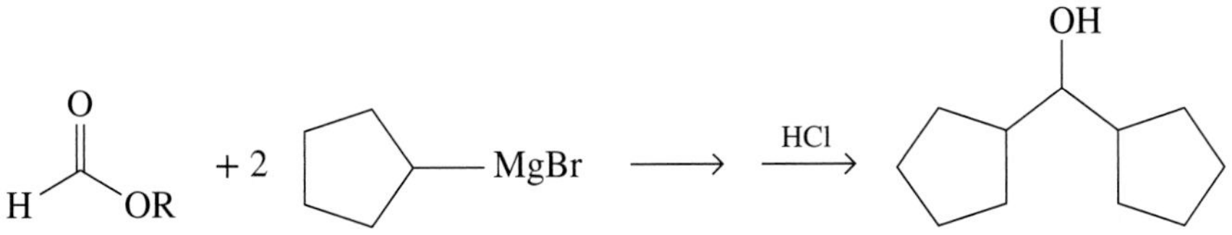 O H OR + 2 MgBr $\longrightarrow$ $\xrightarrow{\text{HCl}}$ OH

(b) 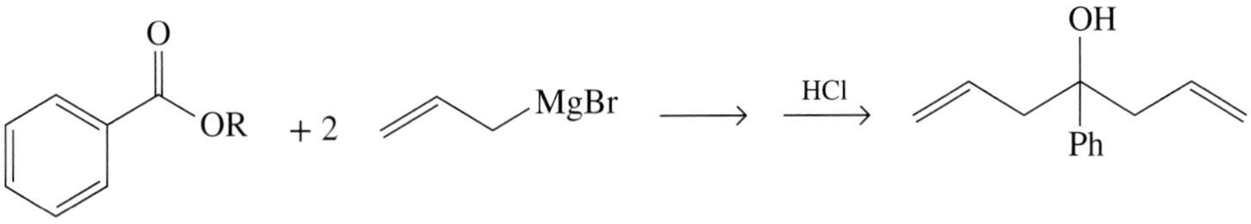 O OR + 2 MgBr $\longrightarrow$ $\xrightarrow{\text{HCl}}$ OH Ph

Practice exercise 23.13

(a) 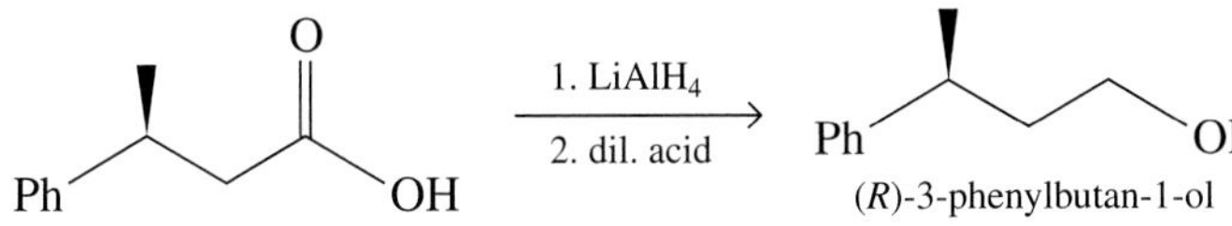 O Ph OH $\xrightarrow[\text{2. dil. acid}]{\text{1. } LiAlH_4}$ Ph OH

(*R*)-3-phenylbutan-1-ol

(b) 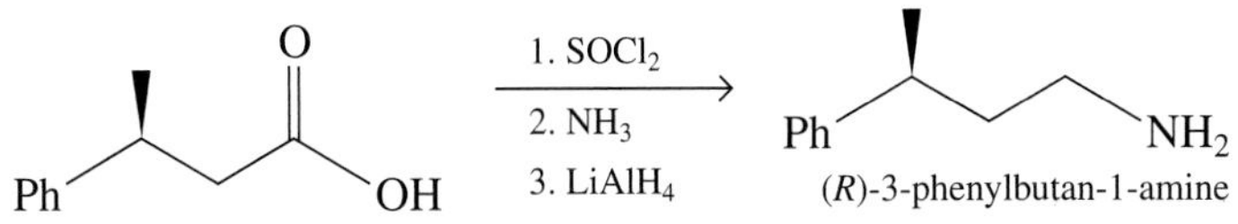 O Ph OH $\xrightarrow[\text{2. } NH_3 \text{ 3. } LiAlH_4]{\text{1. } SOCl_2}$ Ph NH_2

(*R*)-3-phenylbutan-1-amine

Practice exercise 23.14

(a) 3

(b) All 3 are chiral.

Chapter 24

Practice exercise 24.1

(a) glycine (b) isoleucine and threonine

Practice exercise 24.2

The negative electrode.

Practice exercise 24.3

At pH 6.0, glutamic acid (pI 3.08) has a net negative charge and migrates toward the positive electrode. At this pH, arginine (pI 10.76) has a net positive charge and moves toward the negative electrode. Valine (pI 6.00) is neutral (has no net charge) and therefore, remains at the origin.

Practice exercise 24.4

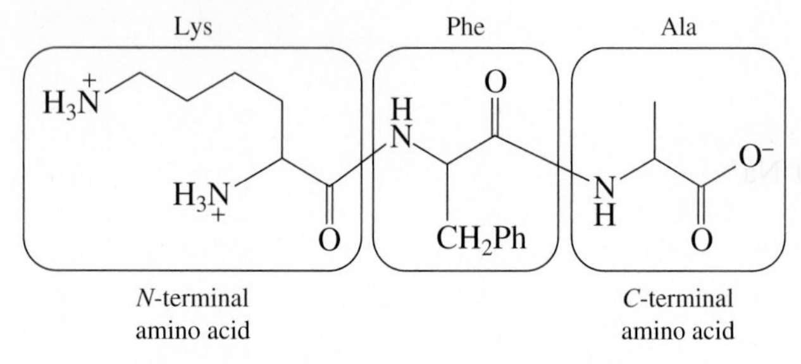

The tripeptide has a net positive charge at pH 6.0.

Practice exercise 24.5

glutamic acid and aspartic acid

Chapter 25 E-CHAPTER

Practice exercise 25.1

O
NH
N
O
HO
H
O
H
H
H
O
H
$O{=}P{-}O^-$
O^-

Practice exercise 25.2

NH_2
N
OH
O
N
5′
CH_2
O
O
CH_3
H
H
1′
HN
H
H
O
H
N
O
5′
$O{=}P{-}O{-}CH_2$
O
O^-
H
H
1′
H
H
2′
HN
N
H
O
H
H_2N
N
N
$O{=}P$
O
5′
CH_2
O
O^-
H
H
1′
H
H
3′
2′
phosphorylated 3' end
O
H
$O{=}P{-}O^-$
O^-

Practice exercise 25.3
5′-TCGTACGG-3′

Practice exercise 25.4
5′-TGGTGGACGAGTCCGGAA-3′

Practice exercise 25.5
5′-CUAACUAGCGGGUCGCCG-3′

Practice exercise 25.6
(a) 5′-UGC-UAU-AUU-CAA-AAU-UGC-CCU-CUU-GGU-UGA-3′
(b) Amino terminus-cys-tyr-ile-gln-asn-cys-pro-leu-gly-carboxyl terminus

Chapter 26 E-CHAPTER

Practice exercise 26.1
The repeating unit is –(CH_2CHCl)–.
The polymer is represented as –(CH_2CHCl)$_n$–.
The polymer is named poly(1-chloroethylene).

Practice exercise 26.2

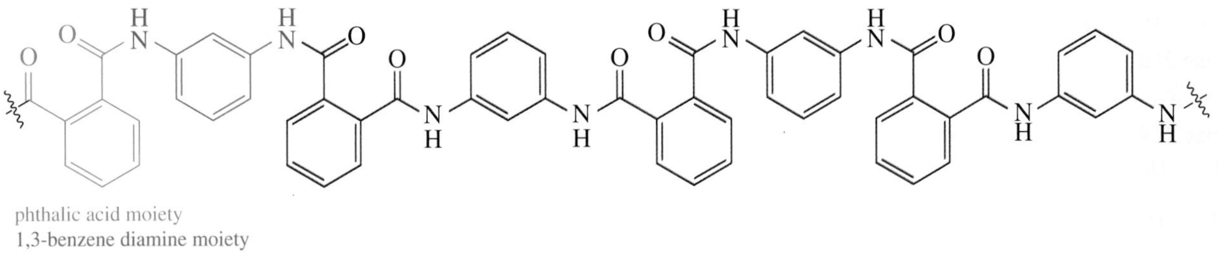

phthalic acid moiety
1,3-benzene diamine moiety

Practice exercise 26.3

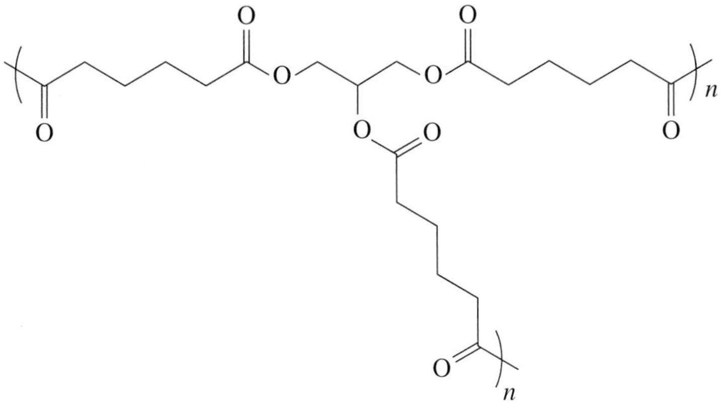

Practice exercise 26.4

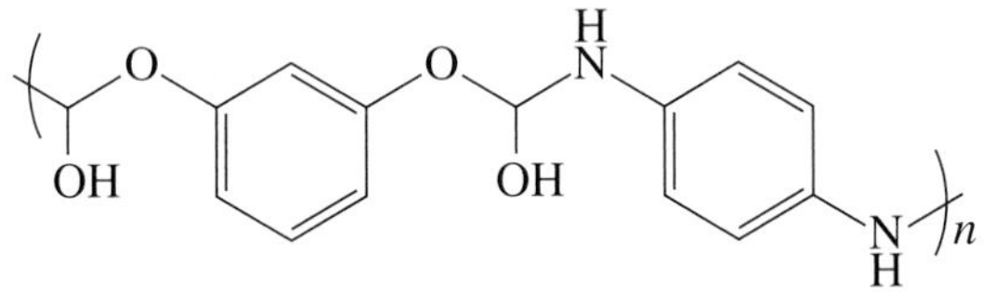

Practice exercise 26.5

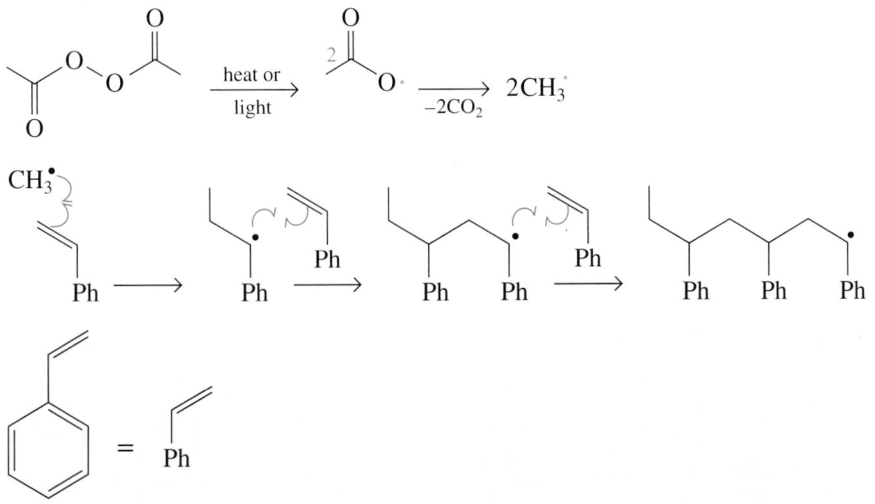

heat or light; $-2CO_2$; $2CH_3^{\bullet}$

Practice exercise 26.6

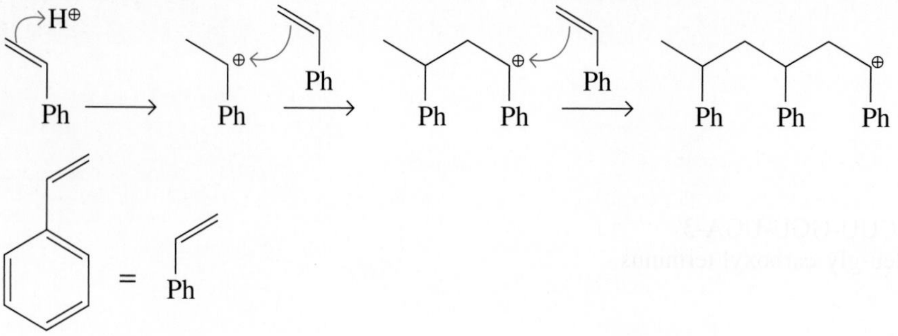

$\rightarrow H^{\oplus}$

Ph ⟶ Ph ⟶ Ph Ph ⟶ Ph Ph Ph

Ph Ph

= Ph

Chapter 27 E-CHAPTER

Practice exercise 27.1

E (1 mole) = 2.939×10^{12} J mol^{-1}

Practice exercise 27.2

$^{226}_{88}Ra \rightarrow ^{222}_{86}Rn + ^{4}_{2}He + ^{0}_{0}\gamma$

Practice exercise 27.3

$^{131}_{53}I \rightarrow ^{131}_{54}Xe + ^{0}_{-1}e$

Practice exercise 27.4

$^{214}_{84}Po \rightarrow ^{210}_{82}Pb + ^{4}_{2}He$

$^{210}_{82}Pb \rightarrow ^{210}_{83}Bi + ^{0}_{-1}\beta$

$^{210}_{83}Bi \rightarrow ^{210}_{84}Po + ^{0}_{-1}\beta$

$^{210}_{84}Po \rightarrow ^{206}_{82}Pb + ^{4}_{2}He$

Practice exercise 27.5

7.13×10^{2} yr

Appendix A

Thermodynamic data for selected elements, compounds and ions (25 °C)

Substance	$\Delta_f H^{\ominus}$(kJ mol^{-1})	$S^{\ominus}$(J mol^{-1} K^{-1})	$\Delta_f G^{\ominus}$(kJ mol^{-1})
Aluminium			
Al(s)	0	28.3	0
Al^{3+}(aq)	−524.7		−481.2
$AlCl_3$(s)	−704	110.7	−629
Al_2O_3(s)	−1669.8	51.0	−1576.4
$Al_2(SO_4)_3$(s)	−3441	239	−3100
Arsenic			
As(s)	0	35.1	0
AsH_3(g)	+66.4	223	+68.9
As_4O_6(s)	−1314	214	−1153
As_2O_5(s)	−925	105	−782
H_3AsO_3(aq)	−742.2		
H_3AsO_4(aq)	−902.5		
Barium			
Ba(s)	0	66.9	0
Ba^{2+}(aq)	−537.6	9.6	−560.8
$BaCO_3$(s)	−1219	112	−1139
$BaCrO_4$(s)	−1428.0	159	−1345
$BaCl_2$(s)	−860.2	125	−810.8
BaO(s)	−553.5	70.4	−525.1
$Ba(OH)_2$(s)	−998.22	−8	−875.3
$Ba(NO_3)_2$(s)	−992	214	−795
$BaSO_4$(s)	−1465	132	−1353
Beryllium			
Be(s)	0	9.50	0
$BeCl_2$(s)	−468.6	89.9	−426.3
BeO(s)	−611	14	−582
Bismuth			
Bi(s)	0	56.9	0
$BiCl_3$(s)	−379	177	−315
Bi_2O_3(s)	−576	151	−497
Boron			
B(s)	0	5.87	0
BCl_3(g)	−404	290	−389
B_2H_6(g)	+36	232	+87
B_2O_3(s)	−1273	53.8	−1194
$B(OH)_3$(s)	−1094	88.8	−969
Bromine			
Br_2(l)	0	152.2	0
Br_2(g)	+30.9	245.4	3.11
HBr(g)	−36	198.5	53.1
Br^-(aq)	−121.55	82.4	−103.96
Cadmium			
Cd(s)	0	51.8	0
Cd^{2+}(aq)	−75.90	−73.2	−77.61
$CdCl_2$(s)	−392	115	−344
CdO(s)	−258.2	54.8	−228.4
CdS(s)	−162	64.9	−156
$CdSO_4$(s)	−933.5	123	−822.6
Calcium			
Ca(s)	0	41.4	0
Ca^{2+}(aq)	−542.83	−53.1	−553.58
$CaCO_3$(s)	−1207	92.9	−1128.8
CaF_2(s)	−741	80.3	−1166
$CaCl_2$(s)	−795.0	114	−750.2
$CaBr_2$(s)	−682.8	130	−663.6
CaI_2(s)	−535.9	143	−529
CaO(s)	−635.5	40	−604.2
$Ca(OH)_2$(s)	−986.59	76.1	−896.76
$Ca_3(PO_4)_2$(s)	−4119	241	−3852

Substance	$\Delta_f H^{\ominus}$(kJ mol^{-1})	$S^{\ominus}$(J mol^{-1} K^{-1})	$\Delta_f G^{\ominus}$(kJ mol^{-1})
Calcium *(continued)*			
$CaSO_3$(s)	−1156		
$CaSO_4$(s)	−1432.7	107	−1320.3
$CaSO_4 \cdot \frac{1}{2}H_2O$(s)	−1575.2	131	−1435.2
$CaSO_4 \cdot 2H_2O$(s)	−2021.1	194.0	−1795.7
Carbon			
C(*s*, graphite)	0	5.69	0
C(*s*, diamond)	+1.88	2.4	+2.9
CCl_4(l)	−134	214.4	−65.3
$CHCl_3$(g)	−82.0	234.2	−58.6
CO(g)	−110.5	197.9	−137.3
CO_2(g)	−394	213.6	−394.4
CO_2(aq)	−413.8	117.6	−385.98
H_2CO_3(aq)	−699.65	187.4	−623.08
HCO_3^-(aq)	−691.99	91.2	−586.77
CO_3^{2-}(aq)	−677.14	−56.9	−527.81
CS_2(l)	+89.5	151.3	+65.3
CS_2(g)	+115.3	237.7	+67.2
HCN(g)	+135.1	201.7	+124.7
CN^-(aq)	+150.6	94.1	+172.4
CH_4(g)	−74.848	186.2	−50.79
C_2H_2(g)	+226.75	200.8	+209
C_2H_4(g)	+51.9	219.8	+68.12
C_2H_6(g)	−84.667	229.5	−32.9
C_3H_8(g)	−104	269.9	−23
C_4H_{10}(g)	−126	310.2	−17.0
C_6H_6(l)	+49.0	173.3	+124.3
CH_3OH(l)	−238.6	126.8	−166.2
C_2H_5OH(l)	−277.63	161	−174.8
HCOOH(g)	−363	251	+335
CH_3COOH(l)	−487.0	160	−392.5
HCHO(g)	−108.6	218.8	−102.5
CH_3CHO(g)	−167	250	−129
$(CH_3)_2CO$(l)	−248.1	200.4	−155.4
$C_6H_5CO_2H$(s)	−385.1	167.6	−245.3
$CO(NH_2)_2$(s)	−333.19	104.6	−197.2
$CO(NH_2)_2$(aq)	−319.2	173.8	−203.8
$CH_2(NH_2)CO_2H$(s)	−532.9	103.5	−373.4
Chlorine			
Cl_2(g)	0	223.0	0
Cl^-(aq)	−167.2	56.5	−131.2
HCl(g)	−92.30	186.7	−95.27
HCl(aq)	−167.2	56.5	−131.2
HClO(aq)	−131.3	106.8	−80.21
Chromium			
Cr(s)	0	23.8	0
Cr^{3+}(aq)	−232		
$CrCl_2$(s)	−326	115	−282
$CrCl_3$(s)	−563.2	126	−493.7
Cr_2O_3(s)	−1141	81.2	−1059
CrO_3(s)	−585.8	72.0	−506.2
$(NH_4)_2Cr_2O_7$(s)	−1807		
$K_2Cr_2O_7$(s)	−2033.01	291	−1882
Cobalt			
Co(s)	0	30.0	0
Co^{2+}(aq)	−59.4	−110	−53.6
$CoCl_2$(s)	−325.5	106	−282.4
$Co(NO_3)_2$(s)	−422.2	192	−230.5
CoO(s)	−237.9	53.0	−214.2
CoS(s)	−80.8	67.4	−82.8

(continued)

Substance	$\Delta_f H^\ominus$(kJ mol^{-1})	$S^\ominus$(J mol^{-1} K^{-1})	$\Delta_f G^\ominus$(kJ mol^{-1})
Copper			
Cu(s)	0	33.15	0
Cu^{2+}(aq)	+64.77	−99.6	+65.49
CuCl(s)	−137.2	86.2	−119.87
$CuCl_2$(s)	−172	119	−131
Cu_2O(s)	−168.6	93.1	−146.0
CuO(s)	−155.25	42.6	−127
Cu_2S(s)	−79.5	121	−86.2
CuS(s)	−53.1	66.5	−53.6
$CuSO_4$(s)	−771.4	109	−661.8
$CuSO_4 \cdot 5H_2O$(s)	−2279.7	300.4	−1879.7
Fluorine			
F_2(g)	0	202.7	0
F^-(aq)	−332.6	−13.8	−278.8
HF(g)	−271	173.5	−273
Gold			
Au(s)	0	47.7	0
Au_2O_3(s)	+80.8	125	+163
$AuCl_3$(s)	−118	148	−48.5
Hydrogen			
H_2(g)	0	130.6	0
H_2O(l)	−285.9	69.96	−237.2
H_2O(g)	−241.8	188.7	−228.6
H_2O_2(l)	−187.6	109.6	−120.3
H_2Se(g)	+76	219	+62.3
H_2Te(g)	+154	234	+138
Iodine			
I_2(s)	0	116.1	0
I_2(g)	+62.4	260.7	+19.3
HI(g)	+26.6	206	+1.30
Iron			
Fe(s)	0	27	0
Fe^{2+}(aq)	−89.1	−137.7	−78.9
Fe^{3+}(aq)	−48.5	−315.9	−4.7
Fe_2O_3(s)	−822.3	90.0	−741.0
Fe_3O_4(s)	−1118.4	146.4	−1015.4
FeS(s)	−100.0	60.3	−100.4
FeS_2(s)	−178.2	52.9	−166.9
Lead			
Pb(s)	0	64.8	0
Pb^{2+}(aq)	−1.7	10.5	−24.4
$PbCl_2$(s)	−359.4	136	−314.1
PbO(s)	−219.2	67.8	−189.3
PbO_2(s)	−277	68.6	−219
$Pb(OH)_2$(s)	−515.9	88	−420.9
PbS(s)	−100	91.2	−98.7
$PbSO_4$(s)	−920.1	149	−811.3
Lithium			
Li(s)	0	28.4	0
Li^+(aq)	−278.6	10.3	
LiF(s)	−611.7	35.7	−583.3
LiCl(s)	−407.5	59.29	−383.7
LiBr(s)	−350.3	66.9	−338.87
Li_2O(s)	−596.5	37.9	−560.5
Li_3N(s)	−199	37.7	−155.4
Magnesium			
Mg(s)	0	32.5	0
Mg^{2+}(aq)	−466.9	−138.1	−454.8
$MgCO_3$(s)	−1113	65.7	−1029
MgF_2(s)	−1124	79.9	−1056
$MgCl_2$(s)	−641.8	89.5	−592.5
$MgCl_2 \cdot 2H_2O$(s)	−1280	180	−1118
Mg_3N_2(s)	−463.2	87.9	−411
MgO(s)	−601.7	26.9	−569.4
$Mg(OH)_2$(s)	−924.7	63.1	−833.9

Substance	$\Delta_f H^\ominus$(kJ mol^{-1})	$S^\ominus$(J mol^{-1} K^{-1})	$\Delta_f G^\ominus$(kJ mol^{-1})
Manganese			
Mn(s)	0	32.0	0
Mn^{2+}(aq)	−223	−74.9	−228
MnO_4^-(aq)	−542.7	191	−449.4
$KMnO_4$(s)	−813.4	171.71	−713.8
MnO(s)	−385	60.2	−363
Mn_2O_3(s)	−959.8	110	−882.0
MnO_2(s)	−520.9	53.1	−466.1
Mn_3O_4(s)	−1387	149	−1280
$MnSO_4$(s)	−1064	112	−956
Mercury			
Hg(l)	0	76.1	0
Hg(g)	+60.84	175	+31.8
Hg_2Cl_2(s)	−265.2	192.5	−210.8
$HgCl_2$(s)	−224.3	146.0	−178.6
HgO(s)	−90.83	70.3	−58.54
HgS(*s*, red)	−58.2	82.4	−50.6
Nickel			
Ni(s)	0	30	0
$NiCl_2$(s)	−305	97.5	−259
NiO(s)	−244	38	−216
NiO_2(s)			−199
$NiSO_4$(s)	−891.2	77.8	−773.6
$NiCO_3$(s)	−664.0	91.6	−615.0
$Ni(CO)_4$(g)	−220	399	−567.4
Nitrogen			
N_2(g)	0	191.5	0
NH_3(g)	−46.0	192.5	−16.7
NH_4^+(aq)	−132.5	113	−79.37
N_2H_4(g)	+95.40	238.4	+159.3
N_2H_4(l)	+50.6	121.2	+149.4
NH_4Cl(s)	−315.4	94.6	−203.9
NO(g)	+90.37	210.6	+86.69
NO_2(g)	+33.8	240.5	+51.84
N_2O(g)	+81.57	220.0	+103.6
N_2O_4(g)	+9.67	304	+98.28
N_2O_5(g)	+11	356	+115
HNO_3(l)	−173.2	155.6	−79.91
NO_3^-(aq)	−205.0	146.4	−108.74
Oxygen			
O_2(g)	0	205.0	0
O_3(g)	+143	238.8	+163
OH^-(aq)	−230.0	−10.75	−157.24
Phosphorus			
P(*s*, white)	0	41.09	0
P_4(g)	+314.6	163.2	+278.3
PCl_3(g)	−287.0	311.8	−267.8
PCl_5(g)	−374.9	364.6	−305.0
PH_3(g)	+5.4	210.2	+12.9
P_4O_6(s)	−1640		
$POCl_3$(g)	−1109.7	646.5	−1019
$POCl_3$(l)	−1186	26.36	−1035
P_4O_{10}(s)	−3062	228.9	−2698
H_3PO_4(s)	−1279	110.5	−1119
Potassium			
K(s)	0	64.18	0
K^+(aq)	−252.4	102.5	−283.3
KF(s)	−567.3	66.6	−537.8
KCl(s)	−435.89	82.59	−408.3
KBr(s)	−393.8	95.9	−380.7
KI(s)	−327.9	106.3	−324.9
KOH(s)	−424.8	78.9	−379.1
K_2O(s)	−361	98.3	−322
K_2SO_4(s)	−1433.7	176	−1316.4
Silicon			
Si(s)	0	19	0
SiH_4(g)	+33	205	+52.3
SiO_2(s, alpha)	−910.0	41.8	−856

Substance	$\Delta_f H^\ominus$ (kJ mol^{-1})	$S^\ominus$ (J mol^{-1} K^{-1})	$\Delta_f G^\ominus$ (kJ mol^{-1})
Silver			
Ag(s)	0	42.55	0
Ag^+(aq)	+105.58	72.68	+77.11
AgCl(s)	−127.1	96.2	−109.7
AgBr(s)	−100.4	107.1	−96.9
$AgNO_3$(s)	−124	141	−32
Ag_2O(s)	−31.1	121.3	−11.2
Sodium			
Na(s)	0	51.0	0
Na^+(aq)	−240.12	59.0	−261.91
NaF(s)	−571	51.5	−545
NaCl(s)	−411.0	72.38	−384.0
NaBr(s)	−360	83.7	−349
NaI(s)	−288	91.2	−286
$NaHCO_3$(s)	−947.7	102	−851.9
Na_2CO_3(s)	−1131	136	−1048
Na_2O_2(s)	−510.9	94.6	−447.7
Na_2O(s)	−510	72.8	−376
NaOH(s)	−426.8	64.18	−382
Na_2SO_4(s)	−1384.5	149.4	−1266.83

Substance	$\Delta_f H^\ominus$ (kJ mol^{-1})	$S^\ominus$ (J mol^{-1} K^{-1})	$\Delta_f G^\ominus$ (kJ mol^{-1})
Sulfur			
S(s, rhombic)	0	31.9	0
SO_2(g)	−296.9	248.5	−300.4
SO_3(g)	−395.2	256.2	−370.4
H_2S(g)	−20.15	206	−33.6
H_2SO_4(l)	−811.32	157	−689.9
H_2SO_4(aq)	−909.3	20.1	−744.5
SF_6(g)	−1209	292	−1105
Tin			
Sn(s, white)	0	51.6	0
Sn^{2+}(aq)	−8.8	−17	−27.2
$SnCl_4$(l)	−511.3	258.6	−440.2
SnO(s)	−285.8	56.5	−256.9
SnO_2(s)	−580.7	52.3	−519.6
Zinc			
Zn(s)	0	41.6	0
Zn^{2+}(aq)	−153.9	−112.1	−147.06
$ZnCl_2$(s)	−415.1	111	−369.4
ZnO(s)	−348.3	43.6	−318.3
ZnS(s)	−205.6	57.7	−201.3
$ZnSO_4$(s)	−982.8	120	−874.5

Appendix B

Average bond enthalpies (25 °C)

Bond	Bond enthalpy (kJ mol^{-1})	Bond	Bond enthalpy (kJ mol^{-1})
C—C	348	C—Br	276
C═C	612	C—I	238
C≡C	960	H—H	436
C—H	412	H—F	565
C—N	305	H—Cl	431
C═N	613	H—Br	366
C≡N	890	H—I	299
C—O	360	H—N	388
C═O	743	H—O	463
C—F	484	H—S	338
C—Cl	338	H—Si	376

Appendix C

Solubility products (25 °C)

Salt	Solubility equilibrium	K_{sp}
Fluorides		
MgF_2	$MgF_2(s) \rightleftharpoons Mg^{2+}(aq) + 2F^-(aq)$	6.6×10^{-9}
CaF_2	$CaF_2(s) \rightleftharpoons Ca^{2+}(aq) + 2F^-(aq)$	3.9×10^{-11}
SrF_2	$SrF_2(s) \rightleftharpoons Sr^{2+}(aq) + 2F^-(aq)$	2.9×10^{-9}
BaF_2	$BaF_2(s) \rightleftharpoons Ba^{2+}(aq) + 2F^-(aq)$	1.7×10^{-6}
LiF	$LiF(s) \rightleftharpoons Li^+(aq) + F^-(aq)$	1.7×10^{-3}
PbF_2	$PbF_2(s) \rightleftharpoons Pb^{2+}(aq) + 2F^-(aq)$	3.6×10^{-8}
Chlorides		
CuCl	$CuCl(s) \rightleftharpoons Cu^+(aq) + Cl^-(aq)$	1.9×10^{-7}
AgCl	$AgCl(s) \rightleftharpoons Ag^+(aq) + Cl^-(aq)$	1.8×10^{-10}
Hg_2Cl_2	$Hg_2Cl_2(s) \rightleftharpoons Hg_2^{2+}(aq) + 2Cl^-(aq)$	1.2×10^{-18}
TlCl	$TlCl(s) \rightleftharpoons Tl^+(aq) + Cl^-(aq)$	1.8×10^{-4}
$PbCl_2$	$PbCl_2(s) \rightleftharpoons Pb^{2+}(aq) + 2Cl^-(aq)$	1.7×10^{-5}
$AuCl_3$	$AuCl_3(s) \rightleftharpoons Au^{3+}(aq) + 3Cl^-(aq)$	3.2×10^{-25}
Bromides		
CuBr	$CuBr(s) \rightleftharpoons Cu^+(aq) + Br^-(aq)$	5×10^{-9}
AgBr	$AgBr(s) \rightleftharpoons Ag^+(aq) + Br^-(aq)$	5.0×10^{-13}

(continued)

Salt	Solubility equilibrium	K_{sp}
Bromides *(continued)*		
Hg_2Br_2	$Hg_2Br_2(s) \rightleftharpoons Hg_2^{2+}(aq) + 2Br^-(aq)$	5.6×10^{-23}
$HgBr_2$	$HgBr_2(s) \rightleftharpoons Hg^{2+}(aq) + 2Br^-(aq)$	1.3×10^{-19}
$PbBr_2$	$PbBr_2(s) \rightleftharpoons Pb^{2+}(aq) + 2Br^-(aq)$	2.1×10^{-6}
Iodides		
CuI	$CuI(s) \rightleftharpoons Cu^+(aq) + I^-(aq)$	1×10^{-12}
AgI	$AgI(s) \rightleftharpoons Ag^+(aq) + I^-(aq)$	8.3×10^{-17}
Hg_2I_2	$Hg_2I_2(s) \rightleftharpoons Hg_2^{2+}(aq) + 2I^-(aq)$	4.7×10^{-29}
HgI_2	$HgI_2(s) \rightleftharpoons Hg^{2+}(aq) + 2I^-(aq)$	1.1×10^{-28}
PbI_2	$PbI_2(s) \rightleftharpoons Pb^{2+}(aq) + 2I^-(aq)$	7.9×10^{-9}
Hydroxides		
$Mg(OH)_2$	$Mg(OH)_2(s) \rightleftharpoons Mg^{2+}(aq) + 2OH^-(aq)$	7.1×10^{-12}
$Ca(OH)_2$	$Ca(OH)_2(s) \rightleftharpoons Ca^{2+}(aq) + 2OH^-(aq)$	6.5×10^{-6}
$Mn(OH)_2$	$Mn(OH)_2(s) \rightleftharpoons Mn^{2+}(aq) + 2OH^-(aq)$	1.6×10^{-13}
$Fe(OH)_2$	$Fe(OH)_2(s) \rightleftharpoons Fe^{2+}(aq) + 2OH^-(aq)$	7.9×10^{-16}
$Fe(OH)_3$	$Fe(OH)_3(s) \rightleftharpoons Fe^{3+}(aq) + 3OH^-(aq)$	1.6×10^{-39}
$Co(OH)_2$	$Co(OH)_2(s) \rightleftharpoons Co^{2+}(aq) + 2OH^-(aq)$	1×10^{-15}
$Co(OH)_3$	$Co(OH)_3(s) \rightleftharpoons Co^{3+}(aq) + 3OH^-(aq)$	3×10^{-45}
$Ni(OH)_2$	$Ni(OH)_2(s) \rightleftharpoons Ni^{2+}(aq) + 2OH^-(aq)$	6×10^{-16}
$Cu(OH)_2$	$Cu(OH)_2(s) \rightleftharpoons Cu^{2+}(aq) + 2OH^-(aq)$	4.8×10^{-20}
$V(OH)_3$	$V(OH)_3(s) \rightleftharpoons V^{3+}(aq) + 3OH^-(aq)$	4×10^{-35}
$Cr(OH)_3$	$Cr(OH)_3(s) \rightleftharpoons Cr^{3+}(aq) + 3OH^-(aq)$	2×10^{-30}
$Zn(OH)_2$	$Zn(OH)_2(s) \rightleftharpoons Zn^{2+}(aq) + 2OH^-(aq)$	3.0×10^{-16}
$Cd(OH)_2$	$Cd(OH)_2(s) \rightleftharpoons Cd^{2+}(aq) + 2OH^-(aq)$	5.0×10^{-15}
$Al(OH)_3$ (alpha form)	$Al(OH)_3(s) \rightleftharpoons Al^{3+}(aq) + 3OH^-(aq)$	3×10^{-34}
Cyanides		
$AgCN$	$AgCN(s) \rightleftharpoons Ag^+(aq) + CN^-(aq)$	2.2×10^{-16}
$Zn(CN)_2$	$Zn(CN)_2(s) \rightleftharpoons Zn^{2+}(aq) + 2CN^-(aq)$	3×10^{-16}
Sulfites		
$CaSO_3$	$CaSO_3(s) \rightleftharpoons Ca^{2+}(aq) + SO_3^{2-}(aq)$	3×10^{-7}
Ag_2SO_3	$Ag_2SO_3(s) \rightleftharpoons 2Ag^+(aq) + SO_3^{2-}(aq)$	1.5×10^{-14}
$BaSO_3$	$BaSO_3(s) \rightleftharpoons Ba^{2+}(aq) + SO_3^{2-}(aq)$	8×10^{-7}
Sulfates		
$CaSO_4$	$CaSO_4(s) \rightleftharpoons Ca^{2+}(aq) + SO_4^{2-}(aq)$	2.4×10^{-5}
$SrSO_4$	$SrSO_4(s) \rightleftharpoons Sr^{2+}(aq) + SO_4^{2-}(aq)$	3.2×10^{-7}
$BaSO_4$	$BaSO_4(s) \rightleftharpoons Ba^{2+}(aq) + SO_4^{2-}(aq)$	1.1×10^{-10}
$RaSO_4$	$RaSO_4(s) \rightleftharpoons Ra^{2+}(aq) + SO_4^{2-}(aq)$	4.3×10^{-11}
Ag_2SO_4	$Ag_2SO_4(s) \rightleftharpoons 2Ag^+(aq) + SO_4^{2-}(aq)$	1.5×10^{-5}
Hg_2SO_4	$Hg_2SO_4(s) \rightleftharpoons Hg_2^{2+}(aq) + SO_4^{2-}(aq)$	7.4×10^{-7}
$PbSO_4$	$PbSO_4(s) \rightleftharpoons Pb^{2+}(aq) + SO_4^{2-}(aq)$	6.3×10^{-7}
Chromates		
$BaCrO_4$	$BaCrO_4(s) \rightleftharpoons Ba^{2+}(aq) + CrO_4^{2-}(aq)$	2.1×10^{-10}
$CuCrO_4$	$CuCrO_4(s) \rightleftharpoons Cu^{2+}(aq) + CrO_4^{2-}(aq)$	3.6×10^{-6}
Ag_2CrO_4	$Ag_2CrO_4(s) \rightleftharpoons 2Ag^+(aq) + CrO_4^{2-}(aq)$	1.2×10^{-12}
Hg_2CrO_4	$Hg_2CrO_4(s) \rightleftharpoons Hg_2^{2+}(aq) + CrO_4^{2-}(aq)$	2.0×10^{-9}
$CaCrO_4$	$CaCrO_4(s) \rightleftharpoons Ca^{2+}(aq) + CrO_4^{2-}(aq)$	7.1×10^{-4}
$PbCrO_4$	$PbCrO_4(s) \rightleftharpoons Pb^{2+}(aq) + CrO_4^{2-}(aq)$	1.8×10^{-14}
Carbonates		
$MgCO_3$	$MgCO_3(s) \rightleftharpoons Mg^{2+}(aq) + CO_3^{2-}(aq)$	3.5×10^{-8}
$CaCO_3$	$CaCO_3(s) \rightleftharpoons Ca^{2+}(aq) + CO_3^{2-}(aq)$	4.5×10^{-9}
$SrCO_3$	$SrCO_3(s) \rightleftharpoons Sr^{2+}(aq) + CO_3^{2-}(aq)$	9.3×10^{-10}
$BaCO_3$	$BaCO_3(s) \rightleftharpoons Ba^{2+}(aq) + CO_3^{2-}(aq)$	5.0×10^{-9}
$MnCO_3$	$MnCO_3(s) \rightleftharpoons Mn^{2+}(aq) + CO_3^{2-}(aq)$	5.0×10^{-10}
$FeCO_3$	$FeCO_3(s) \rightleftharpoons Fe^{2+}(aq) + CO_3^{2-}(aq)$	2.1×10^{-11}
$CoCO_3$	$CoCO_3(s) \rightleftharpoons Co^{2+}(aq) + CO_3^{2-}(aq)$	1.0×10^{-10}
$NiCO_3$	$NiCO_3(s) \rightleftharpoons Ni^{2+}(aq) + CO_3^{2-}(aq)$	1.3×10^{-7}

Salt	Solubility equilibrium	K_{sp}
Carbonates *(continued)*		
$CuCO_3$	$CuCO_3(s) \rightleftharpoons Cu^{2+}(aq) + CO_3^{2-}(aq)$	2.3×10^{-10}
Ag_2CO_3	$Ag_2CO_3(s) \rightleftharpoons 2Ag^{+}(aq) + CO_3^{2-}(aq)$	8.1×10^{-12}
Hg_2CO_3	$Hg_2CO_3(s) \rightleftharpoons Hg_2^{2+}(aq) + CO_3^{2-}(aq)$	8.9×10^{-17}
$ZnCO_3$	$ZnCO_3(s) \rightleftharpoons Zn^{2+}(aq) + CO_3^{2-}(aq)$	1.0×10^{-10}
$CdCO_3$	$CdCO_3(s) \rightleftharpoons Cd^{2+}(aq) + CO_3^{2-}(aq)$	1.8×10^{-14}
$PbCO_3$	$PbCO_3(s) \rightleftharpoons Pb^{2+}(aq) + CO_3^{2-}(aq)$	7.4×10^{-14}
Phosphates		
$Ca_3(PO_4)_2$	$Ca_3(PO_4)_2(s) \rightleftharpoons 3Ca^{2+}(aq) + 2PO_4^{3-}(aq)$	2.0×10^{-29}
$Mg_3(PO_4)_2$	$Mg_3(PO_4)_2(s) \rightleftharpoons 3Mg^{2+}(aq) + 2PO_4^{3-}(aq)$	6.3×10^{-26}
$SrHPO_4$	$SrHPO_4(s) \rightleftharpoons Sr^{2+}(aq) + HPO_4^{2-}(aq)$	1.2×10^{-7}
$BaHPO_4$	$BaHPO_4(s) \rightleftharpoons Ba^{2+}(aq) + HPO_4^{2-}(aq)$	4.0×10^{-8}
$LaPO_4$	$LaPO_4(s) \rightleftharpoons La^{3+}(aq) + PO_4^{3-}(aq)$	3.7×10^{-23}
$Fe_3(PO_4)_2$	$Fe_3(PO_4)_2(s) \rightleftharpoons 3Fe^{2+}(aq) + 2PO_4^{3-}(aq)$	1×10^{-36}
Ag_3PO_4	$Ag_3PO_4(s) \rightleftharpoons 3Ag^{+}(aq) + PO_4^{3-}(aq)$	2.8×10^{-18}
$FePO_4$	$FePO_4(s) \rightleftharpoons Fe^{3+}(aq) + PO_4^{3-}(aq)$	4.0×10^{-27}
$Zn_3(PO_4)_2$	$Zn_3(PO_4)_2(s) \rightleftharpoons 3Zn^{2+}(aq) + 2PO_4^{3-}(aq)$	5×10^{-36}
$Pb_3(PO_4)_2$	$Pb_3(PO_4)_2(s) \rightleftharpoons 3Pb^{2+}(aq) + 2PO_4^{3-}(aq)$	3.0×10^{-44}
$Ba_3(PO_4)_2$	$Ba_3(PO_4)_2(s) \rightleftharpoons 3Ba^{2+}(aq) + 2PO_4^{3-}(aq)$	5.8×10^{-38}
Ferrocyanides		
$Zn_2[Fe(CN)_6]$	$Zn_2[Fe(CN)_6](s) \rightleftharpoons 2Zn^{2+}(aq) + [Fe(CN)_6]^{4-}(aq)$	2.1×10^{-16}
$Cd_2[Fe(CN)_6]$	$Cd_2[Fe(CN)_6](s) \rightleftharpoons 2Cd^{2+}(aq) + [Fe(CN)_6]^{4-}(aq)$	4.2×10^{-18}
$Pb_2[Fe(CN)_6]$	$Pb_2[Fe(CN)_6](s) \rightleftharpoons 2Pb^{2+}(aq) + [Fe(CN)_6]^{4-}(aq)$	9.5×10^{-19}

Appendix D

Cumulative formation constants of complexes (25 °C)

Equilibrium	β_n	n
Halide complexes		
$Al^{3+} + 6F^{-} \rightleftharpoons [AlF_6]^{3-}$	2.5×10^{4}	6
$Al^{3+} + 4F^{-} \rightleftharpoons [AlF_4]^{-}$	2.0×10^{8}	4
$Be^{2+} + 4F^{-} \rightleftharpoons [BeF_4]^{2-}$	1.3×10^{13}	4
$Sn^{4+} + 6F^{-} \rightleftharpoons [SnF_6]^{2-}$	1×10^{25}	6
$Cu^{+} + 2Cl^{-} \rightleftharpoons [CuCl_2]^{-}$	3×10^{5}	2
$Ag^{+} + 2Cl^{-} \rightleftharpoons [AgCl_2]^{-}$	1.8×10^{5}	2
$Pb^{2+} + 4Cl^{-} \rightleftharpoons [PbCl_4]^{2-}$	2.5×10^{15}	4
$Zn^{2+} + 4Cl^{-} \rightleftharpoons [ZnCl_4]^{2-}$	1.6	4
$Hg^{2+} + 4Cl^{-} \rightleftharpoons [HgCl_4]^{2-}$	5.0×10^{15}	4
$Cu^{+} + 2Br^{-} \rightleftharpoons [CuBr_2]^{-}$	8×10^{5}	2
$Ag^{+} + 2Br^{-} \rightleftharpoons [AgBr_2]^{-}$	1.7×10^{7}	2
$Hg^{2+} + 4Br^{-} \rightleftharpoons [HgBr_4]^{2-}$	1×10^{21}	4
$Cu^{+} + 2I^{-} \rightleftharpoons [CuI_2]^{-}$	8×10^{8}	2
$Ag^{+} + 2I^{-} \rightleftharpoons [AgI_2]^{-}$	1×10^{11}	2
$Pb^{2+} + 4I^{-} \rightleftharpoons [PbI_4]^{2-}$	3×10^{4}	4
$Hg^{2+} + 4I^{-} \rightleftharpoons [HgI_4]^{2-}$	1.9×10^{30}	4
Ammonia complexes		
$Ag^{+} + 2NH_3 \rightleftharpoons [Ag(NH_3)_2]^{+}$	1.6×10^{7}	2
$Zn^{2+} + 4NH_3 \rightleftharpoons [Zn(NH_3)_4]^{2+}$	7.8×10^{8}	4
$Cu^{2+} + 4NH_3 \rightleftharpoons [Cu(NH_3)_4]^{2+}$	1.1×10^{13}	4
$Hg^{2+} + 4NH_3 \rightleftharpoons [Hg(NH_3)_4]^{2+}$	1.8×10^{19}	4
$Co^{2+} + 6NH_3 \rightleftharpoons [Co(NH_3)_6]^{2+}$	5.0×10^{4}	6
$Co^{3+} + 6NH_3 \rightleftharpoons [Co(NH_3)_6]^{3+}$	4.6×10^{33}	6
$Cd^{2+} + 6NH_3 \rightleftharpoons [Cd(NH_3)_6]^{2+}$	2.6×10^{5}	6
$Ni^{2+} + 6NH_3 \rightleftharpoons [Ni(NH_3)_6]^{2+}$	2.0×10^{8}	6
Cyanide complexes		
$Fe^{2+} + 6CN^{-} \rightleftharpoons [Fe(CN)_6]^{4-}$	1.0×10^{24}	6
Cyanide complexes *(continued)*		
$Fe^{3+} + 6CN^{-} \rightleftharpoons [Fe(CN)_6]^{3-}$	1.0×10^{31}	6
$Ag^{+} + 2CN^{-} \rightleftharpoons [Ag(CN)_2]^{-}$	5.3×10^{18}	2
$Cu^{+} + 2CN^{-} \rightleftharpoons [Cu(CN)_2]^{-}$	1.0×10^{16}	2
$Cd^{2+} + 4CN^{-} \rightleftharpoons [Cd(CN)_4]^{2-}$	7.7×10^{16}	4
$Au^{+} + 2CN^{-} \rightleftharpoons [Au(CN)_2]^{-}$	2×10^{38}	2
Complexes with other monodentate ligands		
methylamine (CH_3NH_2)		
$Ag^{+} + 2CH_3NH_2 \rightleftharpoons [Ag(CH_3NH_2)_2]^{+}$	7.8×10^{6}	2
thiocyanate ion (SCN^{-})		
$Cd^{2+} + 4SCN^{-} \rightleftharpoons [Cd(SCN)_4]^{2-}$	1×10^{3}	4
$Cu^{2+} + 2SCN^{-} \rightleftharpoons [Cu(SCN)_2]$	5.6×10^{3}	2
$Fe^{3+} + 3SCN^{-} \rightleftharpoons [Fe(SCN)_3]$	2×10^{6}	3
$Hg^{2+} + 4SCN^{-} \rightleftharpoons [Hg(SCN)_4]^{2-}$	5.0×10^{21}	4
hydroxide ion (OH^{-})		
$Cu^{2+} + 4OH^{-} \rightleftharpoons [Cu(OH)_4]^{2-}$	1.3×10^{16}	4
$Zn^{2+} + 4OH^{-} \rightleftharpoons [Zn(OH)_4]^{2-}$	2×10^{20}	4
Complexes with bidentate ligands[a]		
$Mn^{2+} + 3\text{ en} \rightleftharpoons [Mn(en)_3]^{2+}$	6.5×10^{5}	3
$Fe^{2+} + 3\text{ en} \rightleftharpoons [Fe(en)_3]^{2+}$	5.2×10^{9}	3
$Co^{2+} + 3\text{ en} \rightleftharpoons [Co(en)_3]^{2+}$	1.0×10^{14}	3
$Co^{3+} + 3\text{ en} \rightleftharpoons [Co(en)_3]^{3+}$	5.0×10^{48}	3
$Ni^{2+} + 3\text{ en} \rightleftharpoons [Ni(en)_3]^{2+}$	4.1×10^{17}	3
$Cu^{2+} + 2\text{ en} \rightleftharpoons [Cu(en)_2]^{2+}$	4.0×10^{19}	2
$Mn^{2+} + 3\text{ bipy} \rightleftharpoons [Mn(bipy)_3]^{2+}$	1×10^{6}	3
$Fe^{2+} + 3\text{ bipy} \rightleftharpoons [Fe(bipy)_3]^{2+}$	1.6×10^{17}	3
$Ni^{2+} + 3\text{ bipy} \rightleftharpoons [Ni(bipy)_3]^{2+}$	3.0×10^{20}	3
$Co^{2+} + 3\text{ bipy} \rightleftharpoons [Co(bipy)_3]^{2+}$	8×10^{15}	3
$Mn^{2+} + 3\text{ phen} \rightleftharpoons [Mn(phen)_3]^{2+}$	2×10^{10}	3

(continued)

Equilibrium	β_n	n
Complexes with bidentate ligands[(a)] *(continued)*		
$Fe^{2+} + 3\ phen \rightleftharpoons [Fe(phen)_3]^{2+}$	1×10^{21}	3
$Co^{2+} + 3\ phen \rightleftharpoons [Co(phen)_3]^{2+}$	6×10^{19}	3
$Ni^{2+} + 3\ phen \rightleftharpoons [Ni(phen)_3]^{2+}$	2×10^{24}	3
$Co^{2+} + 3C_2O_4{}^{2-} \rightleftharpoons [Co(C_2O_4)_3]^{4-}$	4.5×10^{6}	3
$Fe^{3+} + 3C_2O_4{}^{2-} \rightleftharpoons [Fe(C_2O_4)_3]^{3-}$	3.3×10^{20}	3

Equilibrium	β_n	n
Complexes of other ligands		
$Zn^{2+} + EDTA^{4-} \rightleftharpoons [Zn(EDTA)]^{2-}$	3.8×10^{16}	1
$Mg^{2+} + 2NTA^{3-} \rightleftharpoons [Mg(NTA)_2]^{4-}$	1.6×10^{10}	2
$Ca^{2+} + 2NTA^{3-} \rightleftharpoons [Ca(NTA)_2]^{4-}$	3.2×10^{11}	2

(a) en = ethylenediamine
bipy = bipyridine

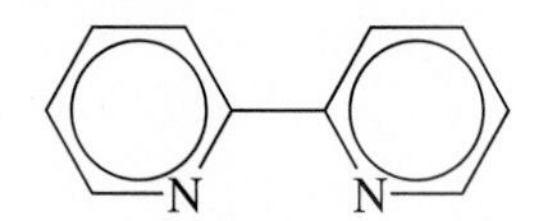

bipyridine

phen = 1,10-phenanthroline

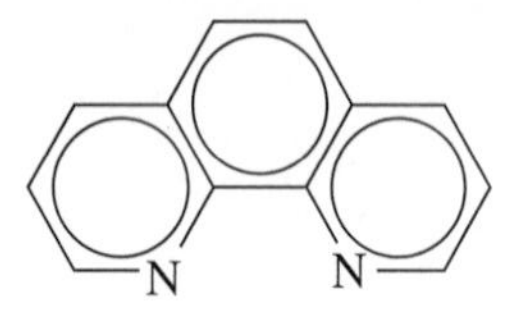

1,10-phenanthroline

$EDTA^{4-}$ = ethylenediaminetetraacetate ion
NTA^{3-} = nitrilotriacetate ion

Appendix E

Acidity and basicity constants for weak acids and bases (25 °C)

Monoprotic acids	Name	K_a
Cl_3CCOOH	trichloroacetic acid	2.2×10^{-1}
HIO_3	iodic acid	1.7×10^{-1}
$Cl_2CHCOOH$	dichloroacetic acid	5.0×10^{-2}
$ClCH_2COOH$	chloroacetic acid	1.4×10^{-3}
HNO_2	nitrous acid	7.1×10^{-4}
HF	hydrofluoric acid	6.8×10^{-4}
HOCN	cyanic acid	3.5×10^{-4}
HCOOH	formic acid	1.8×10^{-4}
$CH_3CH(OH)COOH$	lactic acid	1.4×10^{-4}
$C_4H_4N_2O_3$	barbituric acid	9.8×10^{-5}
C_6H_5COOH	benzoic acid	6.3×10^{-5}
$CH_3CH_2CH_2COOH$	butanoic acid	1.5×10^{-5}
HN_3	hydrazoic acid	1.8×10^{-5}
CH_3COOH	acetic acid	1.8×10^{-5}
CH_3CH_2COOH	propanoic acid	1.4×10^{-5}
HOCl	hypochlorous acid	3.0×10^{-8}
HOBr	hypobromous acid	2.1×10^{-9}
HCN	hydrocyanic acid	6.2×10^{-10}
C_6H_5OH	phenol	1.3×10^{-10}
HOI	hypoiodous acid	2.3×10^{-11}
H_2O_2	hydrogen peroxide	1.8×10^{-12}

Polyprotic acids	Name	K_{a1}	K_{a2}	K_{a3}
H_2SO_4	sulfuric acid	large	1.0×10^{-2}	
H_2CrO_4	chromic acid	5.0	1.5×10^{-6}	
$H_2C_2O_4$	oxalic acid	5.6×10^{-2}	5.4×10^{-5}	
H_3PO_3	phosphorous acid	3×10^{-2}	1.6×10^{-7}	
H_2SO_3 [$SO_2(aq)$]	sulfurous acid	1.2×10^{-2}	6.6×10^{-8}	
H_2SeO_3	selenous acid	4.5×10^{-3}	1.1×10^{-8}	
H_2TeO_3	tellurous acid	3.3×10^{-3}	2.0×10^{-8}	
$HOOCCH_2COOH$	malonic acid	1.4×10^{-3}	2.0×10^{-6}	
$C_6H_4(COOH)_2$	phthalic acid	1.1×10^{-3}	3.9×10^{-6}	
$HOOCCH(OH)CH(OH)COOH$	tartaric acid	9.2×10^{-4}	4.3×10^{-5}	
$C_6H_8O_6$	ascorbic acid	6.8×10^{-5}	2.7×10^{-12}	
H_2CO_3	carbonic acid	4.5×10^{-7}	4.7×10^{-11}	
H_3PO_4	phosphoric acid	7.1×10^{-3}	6.3×10^{-8}	4.5×10^{-13}
H_3AsO_4	arsenic acid	5.6×10^{-3}	1.7×10^{-7}	4.0×10^{-12}
$C_6H_8O_7$	citric acid	7.1×10^{-4}	1.7×10^{-5}	6.3×10^{-6}

Weak bases	Name	K_b
$(CH_3)_2NH$	dimethylamine	9.6×10^{-4}
CH_3NH_2	methylamine	4.4×10^{-4}
$CH_3CH_2NH_2$	ethylamine	4.3×10^{-4}
$(CH_3)_3N$	trimethylamine	7.4×10^{-5}
NH_3	ammonia	1.8×10^{-5}
N_2H_4	hydrazine	1.7×10^{-6}
NH_2OH	hydroxylamine	6.6×10^{-9}
C_5H_5N	pyridine	1.7×10^{-9}
$C_6H_5NH_2$	aniline	4.1×10^{-10}

Appendix F

Standard reduction potentials (25 °C)

Half-reaction	$E^\ominus$(volts)
$F_2(g) + 2e^- \rightleftharpoons 2F^-(aq)$	+2.87
$O_3(g) + 2H^+(aq) + 2e^- \rightleftharpoons O_2(g) + H_2O(l)$	+2.08
$S_2O_8^{2-}(aq) + 2e^- \rightleftharpoons 2SO_4^{2-}(aq)$	+2.01
$Co^{3+}(aq) + e^- \rightleftharpoons Co^{2+}(aq)$	+1.82
$H_2O_2(aq) + 2H^+(aq) + 2e^- \rightleftharpoons 2H_2O(l)$	+1.77
$PbO_2(s) + HSO_4^-(aq) + 3H^+(aq) + 2e^- \rightleftharpoons PbSO_4(s) + 2H_2O(l)$	+1.69
$2HOCl(aq) + 2H^+(aq) + 2e^- \rightleftharpoons Cl_2(g) + 2H_2O(l)$	+1.63
$Mn^{3+}(aq) + e^- \rightleftharpoons Mn^{2+}(aq)$	+1.51
$MnO_4^-(aq) + 4H^+(aq) + 3e^- \rightleftharpoons MnO_2(s) + 2H_2O(l)$	+1.51
$MnO_4^-(aq) + 8H^+(aq) + 5e^- \rightleftharpoons Mn^{2+}(aq) + 4H_2O(l)$	+1.49
$PbO_2(s) + 4H^+(aq) + 2e^- \rightleftharpoons Pb^{2+}(aq) + 2H_2O(l)$	+1.47
$BrO_3^-(aq) + 6H^+(aq) + 6e^- \rightleftharpoons Br^-(aq) + 3H_2O(l)$	+1.46
$Au^{3+}(aq) + 3e^- \rightleftharpoons Au(s)$	+1.42
$Cl_2(g) + 2e^- \rightleftharpoons 2Cl^-(aq)$	+1.36
$Cr_2O_7^{2-}(aq) + 14H^+(aq) + 6e^- \rightleftharpoons 2Cr^{3+}(aq) + 7H_2O(l)$	+1.33
$O_3(g) + H_2O(l) + 2e^- \rightleftharpoons O_2(g) + 2OH^-(aq)$	+1.24
$MnO_2(s) + 4H^+(aq) + 2e^- \rightleftharpoons Mn^{2+}(aq) + 2H_2O(l)$	+1.23
$O_2(g) + 4H^+(aq) + 4e^- \rightleftharpoons 2H_2O(l)$	+1.23
$Pt^{2+}(aq) + 2e^- \rightleftharpoons Pt(s)$	+1.20
$Br_2(aq) + 2e^- \rightleftharpoons 2Br^-(aq)$	+1.07
$NO_3^-(aq) + 4H^+(aq) + 3e^- \rightleftharpoons NO(g) + 2H_2O(l)$	+0.96
$NO_3^-(aq) + 3H^+(aq) + 2e^- \rightleftharpoons HNO_2(aq) + H_2O(l)$	+0.94
$2Hg^{2+}(aq) + 2e^- \rightleftharpoons Hg_2^{2+}(aq)$	+0.91
$HO_2^-(aq) + H_2O(l) + 2e^- \rightleftharpoons 3OH^-(aq)$	+0.87
$NO_3^-(aq) + 4H^+(aq) + 2e^- \rightleftharpoons 2NO_2(g) + 2H_2O(l)$	+0.80
$Ag^+(aq) + e^- \rightleftharpoons Ag(s)$	+0.80
$Fe^{3+}(aq) + e^- \rightleftharpoons Fe^{2+}(aq)$	+0.77
$O_2(g) + 2H^+(aq) + 2e^- \rightleftharpoons H_2O_2(aq)$	+0.69
$I_2(s) + 2e^- \rightleftharpoons 2I^-(aq)$	+0.54
$NiO_2(s) + 2H_2O(l) + 2e^- \rightleftharpoons Ni(OH)_2(s) + 2OH^-(aq)$	+0.49
$SO_2(aq) + 4H^+(aq) + 4e^- \rightleftharpoons S(s) + 2H_2O(l)$	+0.45
$O_2(g) + 2H_2O(l) + 4e^- \rightleftharpoons 4OH^-(aq)$	+0.401
$Cu^{2+}(aq) + 2e^- \rightleftharpoons Cu(s)$	+0.34
$Hg_2Cl_2(s) + 2e^- \rightleftharpoons 2Hg(l) + 2Cl^-(aq)$	+0.27

(continued)

Half-reaction	E°(volts)
$PbO_2(s) + H_2O(l) + 2e^- \rightleftharpoons PbO(s) + 2OH^-(aq)$	+0.25
$AgCl(s) + e^- \rightleftharpoons Ag(s) + Cl^-(aq)$	+0.23
$SO_4^{2-}(aq) + 4H^+(aq) + 2e^- \rightleftharpoons H_2SO_3(aq) + H_2O(l)$	+0.17
$S_4O_6^{2-}(aq) + 2e^- \rightleftharpoons 2S_2O_3^{2-}(aq)$	+0.169
$Cu^{2+}(aq) + e^- \rightleftharpoons Cu^+(aq)$	+0.16
$Sn^{4+}(aq) + 2e^- \rightleftharpoons Sn^{2+}(aq)$	+0.15
$S(s) + 2H^+(aq) + 2e^- \rightleftharpoons H_2S(g)$	+0.14
$AgBr(s) + e^- \rightleftharpoons Ag(s) + Br^-(aq)$	+0.07
$2H^+(aq) + 2e^- \rightleftharpoons H_2(g)$	0.00
$Pb^{2+}(aq) + 2e^- \rightleftharpoons Pb(s)$	−0.13
$Sn^{2+}(aq) + 2e^- \rightleftharpoons Sn(s)$	−0.14
$AgI(s) + e^- \rightleftharpoons Ag(s) + I^-(aq)$	−0.15
$Ni^{2+}(aq) + 2e^- \rightleftharpoons Ni(s)$	−0.25
$Co^{2+}(aq) + 2e^- \rightleftharpoons Co(s)$	−0.28
$In^{3+}(aq) + 3e^- \rightleftharpoons In(s)$	−0.34
$Tl^+(aq) + e^- \rightleftharpoons Tl(s)$	−0.34
$PbSO_4(s) + H^+(aq) + 2e^- \rightleftharpoons Pb(s) + HSO_4^-(aq)$	−0.36
$Cd^{2+}(aq) + 2e^- \rightleftharpoons Cd(s)$	−0.40
$Fe^{2+}(aq) + 2e^- \rightleftharpoons Fe(s)$	−0.44
$Ga^{3+}(aq) + 3e^- \rightleftharpoons Ga(s)$	−0.56
$PbO(s) + H_2O(l) + 2e^- \rightleftharpoons Pb(s) + 2OH^-(aq)$	−0.58
$Cr^{3+}(aq) + 3e^- \rightleftharpoons Cr(s)$	−0.74
$Zn^{2+}(aq) + 2e^- \rightleftharpoons Zn(s)$	−0.76
$Cd(OH)_2(s) + 2e^- \rightleftharpoons Cd(s) + 2OH^-(aq)$	−0.81
$2H_2O(l) + 2e^- \rightleftharpoons H_2(g) + 2OH^-(aq)$	−0.83
$Fe(OH)_2(s) + 2e^- \rightleftharpoons Fe(s) + 2OH^-(aq)$	−0.88
$Cr^{2+}(aq) + 2e^- \rightleftharpoons Cr(s)$	−0.91
$N_2(g) + 4H_2O(l) + 4e^- \rightleftharpoons N_2O_4(aq) + 4OH^-(aq)$	−1.16
$V^{2+}(aq) + 2e^- \rightleftharpoons V(s)$	−1.18
$ZnO_2^{2-}(aq) + 2H_2O(l) + 2e^- \rightleftharpoons Zn(s) + 4OH^-(aq)$	−1.216
$Ti^{2+}(aq) + 2e^- \rightleftharpoons Ti(s)$	−1.63
$Al^{3+}(aq) + 3e^- \rightleftharpoons Al(s)$	−1.66
$U^{3+}(aq) + 3e^- \rightleftharpoons U(s)$	−1.79
$Sc^{3+}(aq) + 3e^- \rightleftharpoons Sc(s)$	−2.02
$La^{3+}(aq) + 3e^- \rightleftharpoons La(s)$	−2.36
$Y^{3+}(aq) + 3e^- \rightleftharpoons Y(s)$	−2.37
$Mg^{2+}(aq) + 2e^- \rightleftharpoons Mg(s)$	−2.37
$Na^+(aq) + e^- \rightleftharpoons Na(s)$	−2.71
$Ca^{2+}(aq) + 2e^- \rightleftharpoons Ca(s)$	−2.76
$Sr^{2+}(aq) + 2e^- \rightleftharpoons Sr(s)$	−2.89
$Ba^{2+}(aq) + 2e^- \rightleftharpoons Ba(s)$	−2.90
$Cs^+(aq) + e^- \rightleftharpoons Cs(s)$	−2.92
$K^+(aq) + e^- \rightleftharpoons K(s)$	−2.92
$Rb^+(aq) + e^- \rightleftharpoons Rb(s)$	−2.93
$Li^+(aq) + e^- \rightleftharpoons Li(s)$	−3.05

Appendix G

Ionisation energies and electron affinities of the first 36 elements at 25 °C

Z	Symbol	EA	IE_1	IE_2	IE_3
1	H	–72.8	1312	–	–
2	He	>0	2372	5250	–
3	Li	–59.7	520.2	7298	11 815
4	Be	>0	899.4	1757	14 848
5	B	–26.8	800.6	2427	3660
6	C	–121.9	1086	2353	4620
7	N	>0	1402	2856	4578
8	O	–141.1	1314	3388	5300
9	F	–322.0	1681	3374	6050
10	Ne	>0	2080	3952	6122
11	Na	–52.9	495.6	4560	6912
12	Mg	>0	737.7	1450	7730
13	Al	–42.7	577	1817	2745
14	Si	–133.6	786.4	1577	3232
15	P	–72.0	1012	1908	2912
16	S	–200.4	999.6	2251	3357
17	Cl	–348.8	1256	2297	3822
18	Ar	>0	1520	2666	3931
19	K	–48.4	418.8	3051	4420
20	Ca	–2.4	589.8	1145	4912
21	Sc	–18.2	633	1235	2389
22	Ti	–7.7	658	1310	2653
23	V	–50.8	650	1414	2828
24	Cr	–64.4	652.8	1591	2987
25	Mn	>0	717.4	1509	3248
26	Fe	–14.6	763	1561	2957
27	Co	–63.9	758	1646	3232
28	Ni	–111.6	736.7	1753	3396
29	Cu	–119.2	745.4	1958	3554
30	Zn	>0	906.4	1733	3833
31	Ga	–28.9	578.8	1979	2963
32	Ge	–119	761	1537	3302
33	As	–78	947	1798	2736
34	Se	–195.0	940.9	2045	2974
35	Br	–324.6	1143	2103	3500
36	Kr	>0	1351	2350	3565

All values are in kJ mol^{-1}. A value of >0 means that the anion is unstable, so its electron affinity cannot be determined experimentally. Colour shading indicates the principal quantum number, n, of the electron whose ionisation energy is listed: blue, $n = 1$; green, $n = 2$; orange, $n = 3$; purple, $n = 4$; yellow, $n = 5$.

Appendix H

Characteristic infrared absorption frequencies

Bonding		Frequency (cm^{-1})	Intensity[a]
C—H	alkane	2850–3000	s
	—CH_3	1375 and 1450	w–m
	—CH_2—	1450	m
	alkene	3000–3100	w–m
		650–1000	s
	alkyne	3300	m–s
		1600–1680	w–m
	aromatic	3000–3100	s
		690–900	s
	aldehyde	2700–2800	w
		2800–2900	w
C=C	alkene	1600–1680	w–m
	aromatic	1450 and 1600	w–m
C—O	alcohol, ether, ester, carboxylic acid, anhydride	1050–1100 (sp^3 C—O)	s
		1200–1250 (sp^2 C—O)	s
C=O	amide	1630–1680	s
	carboxylic acid	1700–1725	s
	ketone	1705–1780	s
	aldehyde	1705–1740	s
	ester	1735–1750	s
	anhydride	1760 and 1800	s
O—H	alcohol, phenol		
	free	3600–3650	m
	H-bonded	3200–3500	m
	carboxylic acid	2400–3400	m
N—H	amine and amide	3100–3500	m–s

(a) m = medium, s = strong, w = weak

acidity constant for the acid–base reaction:
$HA(aq) + H_2O(l) \rightleftharpoons H_3O^+(aq) + A^-(aq)$

$$K_a = \frac{[H_3O^+][A^-]}{[HA]} \quad \text{(p. 447)}$$
$$pK_a = -\log K_a \quad \text{(p. 448)}$$

Arrhenius equation

$$k = Ae^{-E_a/RT} \quad \text{(p. 668)}$$
$$\ln\frac{k_2}{k_1} = -\frac{E_a}{R}\left(\frac{1}{T_2} - \frac{1}{T_1}\right) \quad \text{(p. 670)}$$

Avogadro's Law

$V_{gas} = k''n_{gas}$ (fixed pressure and temperature) (p. 222)

basicity constant for the acid–base reaction:
$B(aq) + H_2O(l) \rightleftharpoons BH^+(aq) + OH^-(aq)$

$$K_b = \frac{[BH^+][OH^-]}{[B]} \quad \text{(p. 447)}$$
$$pK_b = -\log K_b \quad \text{(p. 448)}$$

boiling point elevation

$\Delta T_b = K_b b$ (p. 414)

bond order

bond order = $\frac{1}{2}$(number of electrons in bonding molecular orbitals − number of electrons in antibonding molecular orbitals) (p. 206)

Boyle's Law

$p_{gas}V_{gas} = k$ (fixed temperature and amount) (p. 221)

Bragg equation

$n\lambda = 2d \sin\theta$ (p. 284)

cell potential

$$E_{cell} = E_R - E_L \quad \text{(p. 511)}$$
$$E^{\ominus}_{cell} = E^{\ominus}_R - E^{\ominus}_L \quad \text{(p. 511)}$$
$$E^{\ominus}_{cell} = \frac{RT}{zF}\ln K_c \quad \text{(p. 521)}$$

Charles' Law

$V_{gas} = k'\,T_{gas}$ (fixed pressure and amount) (p. 221)

concentration molality

$$b = \frac{\text{amount of solute (mol)}}{\text{mass of solvent (kg)}} \quad \text{(p. 410)}$$

molarity

$$c = \frac{n}{V} \quad \text{(p. 100)}$$

cumulative formation constant for the formation reaction:
$M^{x+}(aq) + nL(aq) \rightleftharpoons [ML_n]^{x+}(aq)$

$$\beta_n = \frac{[ML_n{}^{x+}(aq)]}{[M^{x+}(aq)][L(aq)]^n} \quad \text{(p. 565)}$$

Dalton's law of partial pressures

$p_{total} = p_1 + p_2 + p_3 + \ldots + p_i$ (p. 232)

de Broglie equation

$$\lambda_{particle} = \frac{h}{mu} \quad \text{(p. 133)}$$

electrostatic potential energy

$$E_{electrostatic} = k\frac{q_1q_2}{r} \quad \text{(p. 172)}$$

energy of a photon

$$E = h\nu$$
$$= \frac{hc}{\lambda} \quad \text{(p. 123)}$$

enthalpy

$H = U + pV$ (p. 308)

entropy of reaction, standard for the general reaction: $aA + bB \rightarrow cC + dD$

$$\Delta_r S^{\ominus} = cS^{\ominus}_C + dS^{\ominus}_D - (aS^{\ominus}_A + bS^{\ominus}_B) \quad \text{(p. 329)}$$

entropy and enthalpy, relationship

$$\Delta S_{surroundings} = \frac{-\Delta H_{system}}{T} \quad \text{(p. 328)}$$

equilibrium constant expression for the solution-phase reaction:
$aA + bB \rightleftharpoons cC + dD$

$$K_c = \frac{\left(\frac{[C]_e}{c^{\ominus}}\right)^c\left(\frac{[D]_e}{c^{\ominus}}\right)^d}{\left(\frac{[A]_e}{c^{\ominus}}\right)^a\left(\frac{[B]_e}{c^{\ominus}}\right)^b}$$

$$K_c = \frac{[C]^c[D]^d}{[A]^a[B]^b}, \quad \text{when } c^{\ominus} = 1 \text{ mol L}^{-1} \quad \text{(pp. 351–2)}$$

equilibrium constant for the gas-phase reaction:
$aA + bB \rightleftharpoons cC + dD$

$$K_p = \frac{\left(\frac{p_C}{p^{\ominus}}\right)^c\left(\frac{p_D}{p^{\ominus}}\right)^d}{\left(\frac{p_A}{p^{\ominus}}\right)^a\left(\frac{p_B}{p^{\ominus}}\right)^b} \quad \text{(p. 353)}$$

Faraday's law

$$\text{amount of product} = \frac{It}{zF} \quad \text{(p. 531)}$$

first law of thermodynamics

$\Delta U = q + w$ (p. 303)

freezing point depression

$\Delta T_f = K_f b$ (p. 414)

gas density

$\rho_{gas} = \frac{m}{V} = \frac{pM}{RT}$ (p. 235)

Gibbs energy

$G = H - TS$ (p. 328)
$\Delta G = \Delta H - T\Delta S$ (p. 328)
$\Delta G = -zFE_{cell}$ (p. 520)
$\Delta G^{\ominus} = -zFE^{\ominus}_{cell}$ (p. 521)
$\Delta_r G^{\ominus} = -RT\ln K$ (p. 364)
$\Delta_r G = \Delta_r G^{\ominus} + RT\ln Q$ (p. 362)

change in standard Gibbs energy

$\Delta G^{\ominus} = \Delta H^{\ominus} - T\Delta S^{\ominus}$ (p. 332)
for: $aA + bB \rightarrow cC + dD$
$\Delta_r G^{\ominus} = c\Delta_f G^{\ominus}_C + d\Delta_f G^{\ominus}_D - (a\Delta_f G^{\ominus}_A + b\Delta_f G^{\ominus}_B)$ (p. 334)

half-life for a first-order reaction

$t_{\frac{1}{2}} = \frac{\ln 2}{k} = \frac{0.693}{k}$ (p. 658)

half-life for a second-order reaction

$t_{\frac{1}{2}} = \frac{1}{k[A]_0}$ (p. 661)

half-life for a zero-order reaction

$t_{\frac{1}{2}} = \frac{[A]_0}{2k}$ (p. 663)

heat

$q = C\Delta T$ (p. 304)
$q = cm\Delta T$ (p. 305)

heat of reaction

$\Delta H = q_p$ (constant pressure) (p. 308)
$\Delta_r U = q_v$ (constant volume) (p. 306)

Henderson–Hasselbalch equation

$\text{pH} = \text{p}K_a + \log\frac{[A^-]}{[HA]}$ (p. 468)

Henry's law

$c_{gas} = k_H p_{gas}$ (constant T) (p. 395)
$\frac{c_1}{p_1} = \frac{c_2}{p_2}$ (p. 395)

Hess's law equation for the general reaction:
$aA + bB \longrightarrow cC + dD$
$\Delta_r H^{\ominus} = c\Delta_f H^{\ominus}_C + d\Delta_f H^{\ominus}_D - (a\Delta_f H^{\ominus}_A + b\Delta_f H^{\ominus}_B)$ (p. 316)

hydrogen atom emission energies

$E_n = -\frac{2.18\times10^{-18}\text{ J}}{n^2}$ (p. 129)

hydrogen atom emission frequencies

$\nu_{emission} = (3.29\times10^{15}\text{ s}^{-1})\left(\frac{1}{n_1^2} - \frac{1}{n_2^2}\right)$ (p. 128)

ideal gas equation

$pV = nRT$ (p. 222)
$p_i V_i = p_f V_f$ (constant n and T) (p. 224)

index of hydrogen deficiency (IHD)

$\text{IHD} = \frac{(H_{reference} - H_{molecule})}{2}$ (p. 876)

integrated rate law for a first-order reaction

$\ln[A]_t = -kt + \ln[A]_0$ (p. 655)
$[A]_t = [A]_0 e^{-kt}$ (p. 656)

integrated rate law for a second-order reaction

$\frac{1}{[A]_t} = kt + \frac{1}{[A]_0}$ (p. 660)

integrated rate law for a zero-order reaction

$[A]_t = -kt + [A]_0$ (p. 664)

isoelectric point of an amino acid

$\text{pI} = \frac{1}{2}(\text{p}K_a \text{ of } \alpha\text{-COOH} + \text{p}K_a \text{ of } \alpha\text{-NH}_3^+)$ (p. 1070)

kinetic energy

$E_{kinetic} = \frac{1}{2}mu^2$ (p. 134)
$E_{kinetic, molar} = \frac{3}{2}RT$, for 1 mole of gas molecules (p. 227)
$\bar{E}_{kinetic} = \frac{3RT}{2N_A}$, average of gas molecules (p. 227)

K_p and K_c, relationship

$K_p = K_c\left(\frac{RT}{p^{\ominus}}\right)^{\Delta n_g}$ (p. 355)

law of radioactive decay

$\text{activity} = -\frac{\Delta N}{\Delta t} = kN$ (e-chapter 27)

magnetic moment of a transition metal complex

$\mu = \sqrt{n(n+2)}\ \mu_B$ (p. 578)

Michaelis–Menten equation

$\frac{d[P]}{dt} = \frac{k_2[E]_0[S]}{K_M + [S]}$, where $K_M = \frac{k_2 + k_{-1}}{k_1}$ (p. 683)

molar mass

$M = \frac{m}{n}$ (p. 85)

mole fraction

$$\text{mole fraction of A} = x_A = \frac{n_A}{n_{total}} \quad \text{(p. 232)}$$

momentum of a particle

$$p = \frac{h}{\lambda} \quad \text{(p. 133)}$$

Nernst equation

$$E_{cell} = E^{\ominus}_{cell} - \frac{RT}{zF}\ln Q \quad \text{(p. 522)}$$

NMR chemical shift

$$\delta = \frac{\text{shift in frequency of a signal from TMS (Hz)}}{\text{operating frequency of the spectrometer (MHz)}} \quad \text{(p. 896)}$$

partial pressure of a component of a gas mixture

$$p_A = x_A p_{total} \quad \text{(p. 232)}$$

percentage by mass

$$\text{\% by mass of element} = \frac{\text{mass of element present in the sample}}{\text{mass of whole sample}} \times 100\% \quad \text{(p. 90)}$$

percentage yield

$$\text{percentage yield} = \frac{\text{actual yield}}{\text{theoretical yield}} \times 100\% \quad \text{(p. 98)}$$

pH

$$\text{pH} = -\log[H_3O^+] \quad \text{(p. 438)}$$
$$\text{pOH} = -\log[OH^-] \quad \text{(p. 438)}$$
$$\text{pH} + \text{pOH} = \text{p}K_w = 14.00 \text{ (at 25.0 °C)} \quad \text{(p. 439)}$$

pK_a and pK_b, relationship

$$\text{p}K_a + \text{p}K_b = \text{p}K_w = 14.00 \text{ (at 25.0 °C)} \quad \text{(p. 450)}$$

radiocarbon dating

$$\ln\frac{r_0}{r_t} = (1.21 \times 10^{-4}\ \text{yr}^{-1})t \quad \text{(e-chapter 27)}$$

Raoult's law

$$p_{solution} = X_{solvent}\, p^*_{solvent} \quad \text{(p. 411)}$$
$$p_{solution} = (1 - X_{solute})p^*_{solvent}$$
$$= p^*_{solvent} - X_{solute}\, p^*_{solvent} \quad \text{(p. 411)}$$

Raoult's law for a two-component system:

$$p_{total} = X_A p^*_A + X_B p^*_B \quad \text{(p. 413)}$$

rate laws for organic reactions

S_N1 reaction
rate of reaction = k[haloalkane] (p. 801)
S_N2 reaction
rate of reaction = k[haloalkane][nucleophile] (p. 800)
E1 reaction
rate of reaction = k[haloalkane] (p. 808)
E2 reaction
rate of reaction = k[haloalkane][base] (p. 809)

rate of reaction for the general reaction:
$aA + bB \rightarrow cC + dD$

$$\text{rate of reaction} = -\frac{1}{a}\frac{d[A]}{dt} = -\frac{1}{b}\frac{d[B]}{dt} = \frac{1}{c}\frac{d[C]}{dt} = \frac{1}{d}\frac{d[D]}{dt} \quad \text{(p. 644)}$$

reaction quotient for the solution-phase reaction:
$aA + bB \rightleftharpoons cC + dD$

$$Q_c = \frac{[C]^c[D]^d}{[A]^a[B]^b} \quad \text{(p. 353)}$$

root-mean-square speed of gas atoms/molecules

$$\bar{u}_{RMS} = \left(\frac{3RT}{M}\right)^{\frac{1}{2}} \quad \text{(p. 228)}$$

Schrödinger equation

$$\hat{H}\psi = E\psi \quad \text{(p. 136)}$$

solubility product of a salt for:
$M_aX_b(s) \rightleftharpoons aM^{b+}(aq) + bX^{a-}(aq)$

$$K_{sp} = [M^{b+}]^a[X^{a-}]^b \quad \text{(p. 403)}$$

specific heat

$$c = \frac{C}{m} \quad \text{(p. 305)}$$

specific rotation

$$\text{specific rotation} = [\alpha]^T_\lambda = \frac{\text{observed rotation (degrees)}}{\text{length (dm)} \times \text{concentration (g/mL)}} \quad \text{(p. 777)}$$

speed of a light wave

$$c = \nu\lambda \quad \text{(p. 121)}$$

van der Waals equation

$$\left(p + \frac{n^2a}{V^2}\right)(V - nb) = nRT \quad \text{(p. 243)}$$

van't Hoff equation

$$\frac{d\ln K}{dT} = \frac{\Delta H^{\ominus}}{RT^2} \quad \text{(p. 370)}$$

$$\ln K_{T_2} - \ln K_{T_1} = \frac{-\Delta_r H^{\ominus}}{R}\left(\frac{1}{T_2} - \frac{1}{T_1}\right) \quad \text{(p. 370)}$$

van't Hoff equation for osmotic pressure

$$\Pi V = nRT \quad \text{(p. 417)}$$

van't Hoff factor

$$i = \frac{(\Delta T_f)_{measured}}{(\Delta T_f)_{\text{calculated as a nonelectrolyte}}} \quad \text{(p. 422)}$$

work done in the expansion/compression of a gas

$$w = -p\Delta V \quad \text{(p. 303)}$$

GLOSSARY

• A •

absolute uncertainty An uncertainty that has the same units as the quantity being measured. (p. 35)

absorbance (*A*) A measure of the light absorbed by a solution of a compound; defined by the equation $A = \log\frac{I_0}{I}$. (p. 575)

absorption spectrum The distribution of wavelengths of light absorbed by a species. (p. 127)

accuracy A measurement is of high accuracy if it is close to the correct value. (p. 32)

acetal A molecule containing two —OR or —OAr groups bonded to the same carbon atom. (p. 955)

achiral Describes an object that lacks chirality, that is, it is superimposable on its mirror image. (p. 761)

acid Arrhenius theory — a substance that produces H^+ ions in water; Brønsted–Lowry theory — a proton donor; Lewis theory — an electron-pair acceptor. (p. 432)

acid anhydride A compound in which two acyl groups are bonded to the same oxygen atom. (p. 1012)

acid–base indicator A dye having one colour in acid and another colour in base. (p. 472)

acid–base reaction A reaction that involves the transfer of a single proton from one species to another. (Note that this is the Brønsted–Lowry definition.) (p. 432)

acid–base titration An analytical procedure involving the gradual addition of a solution of an acid or base of known concentration to a specified volume of a solution of a base or acid of unknown concentration. (p. 472)

acid halide A derivative of a carboxylic acid in which the —OH of the carboxyl group is replaced by a halogen, most commonly chlorine. (p. 1012)

acidic Describes an aqueous solution in which $[H_3O^+] > [OH^-]$. (p. 437)

acidity constant (K_a) The equilibrium constant for the dissociation of an acid in aqueous solution. For the general equilibrium:

$$HA(aq) + H_2O(l) \rightleftharpoons H_3O^+(aq) + A^-(aq)$$

K_a has the form $K_a = \frac{[H_3O^+][A^-]}{[HA]}$. (p. 447)

actinoids Elements 89 to 103 of the periodic table. (p. 16)

activation energy (E_a) The minimum kinetic energy that must be possessed by the reactants in order to give an effective collision leading to products. (p. 665)

activities Effective concentrations that give rise to thermodynamic equilibrium constants when substituted into an equilibrium constant expression. (p. 355)

actual yield The mass of a product obtained from a reaction. (p. 98)

acyclic alkanes Alkane molecules that consist solely of chains of carbon atoms. (p. 58)

acylation The process of introducing an acyl group into a compound. (p. 746)

acyl group A carbonyl group bonded to an alkyl or aryl group. (p. 1012)

addition to the carbon–carbon double bond A characteristic reaction of alkenes in which the π bond is broken and, in its place, σ bonds are formed to two new atoms or groups of atoms. (p. 723)

adduct The product of a reaction between a Lewis acid and a Lewis base. (p. 481)

alcohol A compound containing an —OH (hydroxyl) group bonded to an sp^3 hybridised carbon atom. (p. 822)

aldehyde A compound in which a carbonyl group is bonded to a hydrogen atom plus a hydrocarbon group (or a second hydrogen atom). (p. 944)

alditol The product formed when the C=O group of a monosaccharide is reduced to a CHOH group. (p. 995)

aldoses Monosaccharides containing an aldehyde group. (p. 986)

aliphatic amine An amine in which the nitrogen atom is bonded only to alkyl groups. (p. 847)

aliphatic hydrocarbon See *alkane.* (p. 698)

alkali metals The elements in group 1 (except hydrogen) of the periodic table. (p. 17)

alkaline An alternative word for basic. (p. 437)

alkaline battery A type of power cell dependent on the reaction between zinc and manganese dioxide (Zn/MnO_2). (p. 534)

alkaline dry cell See *alkaline battery.* (p. 534)

alkaline earth metals The elements in group 2 of the periodic table. (p. 17)

alkane A molecule composed only of carbon and hydrogen atoms in which all carbon–carbon bonds are single. (pp. 58, 698)

alkene An unsaturated hydrocarbon with a carbon–carbon double bond. (p. 698)

alkoxy An —OR group, where R is an alkyl group. (p. 842)

alkylamines Amines in which the nitrogen atom is bonded only to alkyl groups. (p. 849)

alkylation The transfer of an alkyl group from one compound to another. (p. 745)

alkyl group A group derived from an alkane by the removal of a hydrogen atom, commonly represented by the symbol R. (p. 59)

alkyne An unsaturated hydrocarbon with a carbon–carbon triple bond. (p. 698)

α-amino acid An amino acid in which the amino group is attached to the carbon atom adjacent to the carboxyl group. (p. 1063)

α-carbon A carbon atom adjacent to a functional group. (p. 968)

α-helix A type of secondary structure in which a section of polypeptide chain coils into a spiral, most commonly a right-handed spiral (p. 1078)

α-hydrogen A hydrogen atom bonded to an α-carbon atom. (p. 968)

alpha particle The nucleus of a helium atom (^4_2He). (p. 9)

ambidentate Describes ligands with two or more different potential donor atoms. (p. 559)

amide An acyl group bonded to a trivalent nitrogen atom. (p. 1013)

amino acid A compound that contains both a carboxyl group and an amino group. (p. 1062)

amorphous Without any organised regular repeating pattern. (p. 286)

amount of substance A base quantity in the SI system, with the unit of mole. (p. 85)

ampere (A) The SI unit of electric current; $1\ A = 1\ C\ s^{-1}$. (pp. 27, 531)

amphiprotic A solvent that can act either as a proton donor or a proton acceptor. (p. 434)

amplitude The maximum displacement of a wave from its centre. (p. 120)

angle strain The strain induced in a molecule as a result of bond angles that are distorted from the ideal values. (p. 703)

anhydrous Describes a chemical substance that contains no water. (p. 41)

aniline The common name for the monosubstituted benzene $C_6H_5NH_2$. (pp. 740, 849)

anion A negatively charged ion. (p. 4)

anisole The common name for the monosubstituted benzene $C_6H_5OCH_3$. (p. 740)

anode The positive electrode in a gas discharge tube; the electrode at which oxidation occurs during an electrochemical change. (p. 508)

anomeric carbon The hemiacetal carbon atom of the cyclic form of a monosaccharide. (p. 990)

anomers Monosaccharides that differ in configuration only at their anomeric carbon atoms. (p. 990)

***anti* addition** The addition of atoms or groups of atoms from opposite sides or faces of a carbon–carbon double bond. (p. 731)

antibonding molecular orbital A molecular orbital that has electron density concentrated outside the bonding region, making it less stable than the atomic orbitals from which it forms. (p. 206)

***anti* selectivity** The characteristic that bromine and chlorine always add *trans* to each other in cycloalkenes. (p. 731)

aprotic solvents Solvents that do not contain —OH groups and, therefore, cannot function as hydrogen-bond donors. (p. 805)

arene A compound containing one or more benzene rings. (p. 698)

aromatic Describes a compound containing one or more benzene rings. (p. 734)

aromatic amine An amine in which the nitrogen atom is bonded to at least one aromatic ring. (p. 847)

Arrhenius equation An equation that relates the rate constant of a reaction to the reaction's activation energy. (p. 668)

aryl group A group derived from an aromatic compound by the removal of a hydrogen atom. (p. 734)

association The clustering of molecular species in solution. (p. 423)

asymmetric synthesis A synthetic procedure that gives rise to predominantly one enantiomer. (p. 781)

atom A neutral particle having one nucleus; the smallest representative sample of an element. (p. 4)

atom efficiency A measure of the number of atoms in the product compared to those in the starting materials. The more atoms are transferred to the product, the higher the atom efficiency and the lower the waste or side products for the reaction. (p. 99)

atomic mass The average mass (in u) of the atoms of the isotopes of a given element as they occur naturally. (p. 12)

atomic mass unit (u) The mass ($1.660\ 54 \times 10^{-27}$ kg) equal to $\frac{1}{12}$ of the mass of one atom of ^{12}C. (p. 12)

atomic number (Z) The number of protons in a nucleus. (p. 11)

atomic radius The distance from the nucleus of an atom at which electron–electron repulsion prevents closer approach of another atom. (p. 155)

atomisation enthalpy ($\Delta_{at}H$) The enthalpy change on rupturing all of the bonds in 1 mole of a substance in the gas phase to produce its atoms, also in the gas phase. (p. 319)

Aufbau principle The statement that the most stable arrangement of electrons in an atom results from placing each successive electron in the most stable (lowest energy) available atomic orbital. (p. 145)

autoprotolysis A process involving the transfer of a proton between two identical molecules. (p. 436)

autoprotolysis constant of water (K_w) The equilibrium constant for the reaction:

$$H_2O(l) + H_2O(l) \rightleftharpoons H_3O^+(aq) + OH^-(aq)$$

It has the form $K_w = [H_3O^+][OH^-]$, where $K_w = 1.0 \times 10^{-14}$ at 25.0 °C. (p. 437)

average rate The rate of a reaction over a particular time period. (p. 643)

Avogadro constant (N_A) 6.022×10^{23}; the number of specified entities in 1 mole. (p. 85)

axial Describes the two ligands in an octahedral or trigonal bipyramidal complex that, by convention, point above and below the plane defined by the equatorial ligands; also describes the orientation of the groups bound to a cycloalkane that are above or below the plane of the ring (p. 557)

axial bonds The bonds connecting axial groups. For instance, in an octahedral or trigonal bipyramidal complex, axial bonds point above and below the plane defined by the equatorial ligands. For organic conformers, the axial bonds are those that are oriented perpendicular to the plane of the chair, boat or envelope. (p. 704)

azimuthal quantum number (l) The quantum number, restricted to integers from 0 to (n – 1), that indexes the electron distribution of an atomic orbital. (p. 136)

azo compounds Compounds of the form Ar—N=N—Ar, commonly used as dyes. (p. 860)

• B •

balanced chemical equation A chemical equation that has the same number of each type of atom and the same net charge on opposite sides of the arrow. (p. 82)

ball-and-stick model A three-dimensional representation of a molecule in which atoms are shown as balls and bonds between atoms as sticks. (p. 47)

barometer An instrument for measuring atmospheric pressure. (p. 220)

base Arrhenius theory — a substance that releases OH^- ions in water; Brønsted–Lowry theory — a proton acceptor; Lewis theory — an electron-pair donor. (p. 432)

basic Describes an aqueous solution in which $[H_3O^+] < [OH^-]$. (p. 437)

basicity constant (K_b) The equilibrium constant for the reaction of a base with water. For the general equilibrium:

$$B(aq) + H_2O(l) \rightleftharpoons BH^+(aq) + OH^-(aq)$$

$$K_b = \frac{[BH^+][OH^-]}{[B]}$$ (p. 447)

battery A group of galvanic cells usually connected in series. (p. 532)

Beer–Lambert law This relates the absorbance (A) of a solution to the molar absorption coefficient (ε) of the absorbing species, the concentration (c) of the absorbing species, and the pathlength of the cell (l), via the equation $A = \varepsilon cl$. (pp. 578, 916)

Beer's law See *Beer–Lambert law*. (p. 578)

β-elimination An elimination reaction where atoms or groups are removed from two adjacent carbon atoms. (p. 797)

bending A type of vibration that changes bond angles. (p. 883)

benzaldehyde The common name for the monosubstituted benzene C_6H_5CHO. (p. 740)

benzoic acid The common name for the monosubstituted benzene C_6H_5COOH. (p. 740)

benzyl group The $C_6H_5CH_2$— group. (p. 740)

benzylic carbon An sp^3 carbon atom bonded to a benzene ring. (p. 1020)

benzylic hydrogen A hydrogen atom bonded to a benzylic carbon atom. (p. 1020)

β-pleated sheet A type of secondary structure in which two sections of polypeptide chain are aligned parallel or antiparallel to one another. (p. 1078)

bidentate Describes ligands with two donor atoms that can be simultaneously attached to the same metal ion. (p. 551)

bimolecular reaction A reaction in which two species are involved in the transition state of the rate-determining step. (p. 800)

binary compound A compound that contains only two elements. (p. 40)

boat conformation The conformation of cyclohexane that roughly resembles a 'boat'. (p. 704)

body-centred cubic lattice (bcc) The crystal structure whose unit cell is a cube with lattice points at its corners and its centre. (p. 279)

boiling point elevation (ΔT_b) A colligative property of a solution by which the solution's boiling point is higher than that of the pure solvent. (p. 414)

bomb calorimeter Apparatus used in the determination of the heat of a reaction under constant volume conditions. (p. 306)

bond energy The energy difference between two atoms at their bond distance and the infinitely separated atoms. (p. 173)

bond enthalpy The enthalpy change on breaking 1 mole of a particular bond to give electrically neutral fragments. (p. 318)

bonding molecular orbital An orbital formed from atomic orbitals in which electron density is maximised between the bonded atoms. (p. 205)

bonding orbital An orbital with high electron density between atoms, formed by constructive interaction between atomic orbitals. (p. 196)

bond length The separation distance between two atoms that gives the maximum possible energetic advantage over infinitely separated atoms. (p. 173)

bond order $\frac{1}{2}$ (number of electrons in bonding molecular orbitals – number of electrons in antibonding molecular orbitals). (p. 206)

boundary The interface between a system and its surroundings across which energy or matter might pass. (p. 300)

boundary surface diagram A diagram that shows a solid surface that encloses most (often 90%) of an orbital's electron density. The electron density is high inside the volume enclosed by the surface but very low outside. (p. 140)

Bragg equation $n\lambda = 2d \sin \theta$. (p. 284)

bridged halonium ion A cyclic structure containing a halonium ion. (p. 731)

Brønsted–Lowry A definition of acids and bases in which an acid is a proton donor and a base is a proton acceptor. (p. 432)

buffer capacity A measure of the amount of H_3O^+ or OH^- that can be added to a buffer solution without significant change in its pH. (p. 472)

buffer solution A solution containing appreciable amounts of either a weak acid and its conjugate base, or a weak base and its conjugate acid. (p. 466)

burette A device used for the addition of accurately measured volumes of solution in a titration. (p. 472)

by-product Substances formed by side reactions. (p. 98)

• C •

^{13}C-NMR spectroscopy Nuclear magnetic resonance spectroscopy of carbon atoms within molecules. It provides information about the carbon–hydrogen framework of a molecule. (p. 894)

***C*-terminal amino acid** The amino acid of a polypeptide with the free —COO$^-$ group. (p. 1074)

calorimeter A device for measuring the heat evolved or absorbed in a chemical reaction. (p. 306)

candela The SI unit of luminous intensity. (p. 27)

capillary action The upward movement of a liquid in a narrow tube against the force of gravity. (p. 260)

carbanion An anion in which carbon has an unshared pair of electrons and bears a negative charge. (p. 952)

carbocation A species containing a positively charged carbon atom with only three bonds to it. (p. 725)

carbohydrates Polyhydroxyaldehydes, polyhydroxyketones or substances that give either of these compounds after hydrolysis. (p. 986)

carboxylic acid group See *carboxyl group*. (p. 1012)

carboxyl group A —COOH group. (p. 1012)

catalysis Rate enhancement caused by a catalyst. (p. 680)

catalyst A substance that in relatively small proportions accelerates the rate of a reaction without being chemically changed. (p. 680)

catalytic hydrogenation Reduction by hydrogen in the presence of a catalyst. (p. 731)

catalytic reduction Reduction in the presence of a catalyst. (p. 731)

catenation The linking of atoms of the same element (particularly carbon) to form chains. (p. 42)

cathode The negative electrode in a gas discharge tube; the electrode at which reduction occurs during an electrochemical change. (p. 508)

cathodic protection The process in which a structural metal, such as iron, is protected from corrosion by connecting it to a metal that has a more negative reduction half-cell potential. (p. 527)

cation A positively charged ion. (p. 4)

cell diagram A shorthand way of describing the make-up of a galvanic cell. (p. 509)

cell potential (E_{cell}) The maximum potential that a given cell can generate. (p. 511)

cell reaction The overall chemical change that takes place in an electrolytic cell or a galvanic cell. (p. 506)

ceramics Materials composed of inorganic components that have been heat treated. (p. 287)

chair conformation The most stable puckered conformation of a cyclohexane ring. (pp. 703, 991)

chalcogens The elements in group 16 of the periodic table. (p. 17)

chelate complex A complex ion containing one or more chelate rings. (p. 551)

chelate effect The larger value of the cumulative formation constant (β_n) found for complexes containing chelate rings when compared with analogous complexes containing monodentate ligands. (p. 566)

chelate ring The ring formed when a polydentate ligand coordinates to a transition metal ion. (p. 551)

chemical equation A form of notation used to describe chemical reactions, in which the reactants and products of the reaction are separated by a directional arrow, with the reactants appearing on the left-hand side. (pp. 6, 82)

chemical equilibrium A situation in which the chemical composition of a system does not change with time. (p. 350)

chemical formula A formula written using chemical symbols and subscripts that describes the composition of a chemical compound or element. (p. 40)

chemical kinetics The study of the rates of chemical reactions. (p. 642)

chemical reaction A process involving transformation of chemical species into different chemical species, usually involving the making and/or breaking of chemical bonds. (pp. 4, 82)

chemical shift (δ) The quantity used in NMR spectroscopy to identify the positions of signals produced by the nuclei of a sample. The unit of chemical shift (δ) is parts per million (ppm). (p. 896)

chemical symbol The formula of an element. (p. 11)

chemical thermodynamics The study of the role of energy in chemical change and in determining the behaviour of materials. (p. 299)

chiral Describes an object that is not superimposable on its mirror image. (p. 760)

***cis* isomer** A stereoisomer in which two groups lie on the same side of a reference plane. (p. 559)

***cis–trans* isomers** Stereoisomers that differ in the positioning of two groups with respect to a reference plane. (pp. 560, 707)

close-packed structures The most efficient arrangement for packing atoms, molecules or ions in a regular crystal. (p. 277)

colligative property A property, such as vapour pressure lowering, boiling point elevation, freezing point depression and osmotic pressure, that depends only on the ratio of the numbers of moles of solute and solvent particles and not on their chemical identities. (p. 410)

collision theory A theory that assumes that for a reaction to occur, reactant molecules must collide with proper orientation and with an energy larger than a threshold value. (p. 664)

combustion A rapid reaction with oxygen accompanied by a flame and the evolution of heat and light. (p. 92)

combustion reaction The reaction of a chemical substance with oxygen. (p. 306)

common ion An ion in a mixture of ionic substances that is common to the formulae of at least two of those substances. (p. 407)

common ion effect The effect by which the solubility of one salt is reduced by the presence of another having a common ion. (p. 407)

complementary colour The colour of the reflected or transmitted light when one component of white light is removed by absorption. (p. 575)

complex The combination of one or more anions or neutral molecules (ligands) with a transition metal ion. (p. 549)

complex splitting pattern A splitting pattern resulting from more than one set of equivalent hydrogen atoms. Signals of this type are called multiplets. (p. 903)

compound A chemical substance containing two or more elements in a definite and unchanging proportion. (p. 4)

concentration The ratio of the quantity of solute to the quantity of solution (or the quantity of solvent); see also *molal concentration; molar concentration; mole fraction*. (p. 100)

concentration cell An electrolytic cell in which the potential difference is caused by a difference in concentration of some component in the electrolyte. (p. 525)

concentration table A table that outlines the changes in concentration that occur as a chemical system comes to equilibrium. (p. 373)

condensation Conversion of a vapour into its liquid. (p. 245)

condensed structural formula A shorthand method of representing molecules in which bonds between atoms are not drawn explicitly. (p. 43)

configurational isomers Isomers that cannot be interconverted by rotation around a single bond. (p. 713)

conformation The representation of the position of atoms in a conformer. (p. 700)

conformer Any three-dimensional arrangement of atoms that results from rotation around a single bond. (p. 700)

conjugate acid–base pair Two species that differ only by a proton. (p. 434)

conjugate base The species in a conjugate acid–base pair that has the fewer number of protons. (p. 434)

constitutional isomers Isomers with different sequences of atom connectivity. (pp. 62, 760)

coordinate bond A covalent bond between a Lewis acidic metal ion and a Lewis basic ligand in which both electrons originate from the ligand. (p. 556)

coordination chemistry The study of transition metal complexes. (p. 556)

coordination compound A transition metal complex. (p. 556)

coordination isomers Isomers that result when ligands are exchanged between a complex cation and a complex anion of the same coordination compound, e.g. $[Co(NH_3)_6][Cr(CN)_6]$ and $[Cr(NH_3)_6][Co(CN)_6]$. (p. 559)

coordination number The number of donor atoms coordinated to a metal ion. (p. 557)

core electrons The inner atomic electrons with principal quantum number less than that of the valence electrons. (p. 148)

correlation table A table of data on spectroscopic absorption patterns of selected functional groups. (p. 883)

coulomb (C) The SI unit of electric charge. (p. 506)

counterion An ion that balances the charge on a cationic or anionic complex ion. (p. 556)

covalent bond A chemical bond in which two atoms share one or more pairs of electrons. (pp. 4, 172)

critical point The temperature and pressure above which the distinction between the liquid and vapour phases disappears. (p. 269)

critical pressure The pressure associated with the critical point. (p. 269)

critical temperature (T_c) The temperature associated with the critical point. (p. 269)

crystal field splitting energy (Δ_o) The difference in energy between the two sets of *d* orbitals in an octahedral or tetrahedral complex ion. (p. 571)

crystal field theory A theory that considers the effects of the polarities or the charges of the ligands in a complex ion on the energies of the *d* orbitals of the central metal ion. (p. 570)

crystal lattice The lattice of a crystalline solid. (p. 279)

crystalline defects Imperfections in a regular solid. (p. 287)

crystalline solid A solid that has a regular arrangement of unit cells and diffracts X-rays. (p. 279)

cubic close-packed (ccp) The arrangement in which close-packed layers are stacked in an ABCABC . . . pattern. (p. 276)

cumulative formation constant (β_n) The equilibrium constant for the process:

$$M^{x+}(aq) + nL(aq) \rightleftharpoons [ML_n]^{x+}(aq)$$

given by the expression $\beta_n = \dfrac{[ML_n]^{x+}}{[M^{x+}][L]^n}$ (p. 565)

cyanohydrins A compound in which both an —OH and a —CN group are bonded to the same carbon atom. (p. 955)

cyclic ether An ether in which the oxygen atom is one of the atoms of a ring. (p. 842)

cycloalkane A saturated hydrocarbon with carbon atoms joined together to form a ring. (pp. 58, 702)

cycloalkene An unsaturated hydrocarbon with carbon atoms joined together to form a ring that contains a carbon–carbon double bond. (p. 719)

• D •

***d*-block elements** Collective name for the elements in groups 3 to 12 of the periodic table. (p. 16)

***d–d* transition** An electronic transition between the t_{2g} and e_g sets of orbitals in an octahedral complex. (p. 574)

Dalton's atomic theory Matter consists of tiny, indestructible particles called atoms. All atoms of one element are identical. The atoms of different elements have different masses. Atoms combine in definite ratios of atoms when they form compounds. (p. 6)

Dalton's law of partial pressures The law stating that in a gas mixture, each gas exerts a pressure equal to the pressure that it would exert if present by itself under otherwise identical conditions. (p. 232)

dative bond A coordinate bond. (p. 556)

degenerate Describes orbitals with the same energy. (pp. 143, 570)

dehydration Elimination of a molecule of water from a compound. (p. 831)

dehydrohalogenation A process in which a halogen is removed from one carbon atom of a haloalkane and a hydrogen atom is removed from the adjacent carbon atom to form a carbon–carbon double bond. (p. 807)

delocalised bond A bond in which electron density is distributed over more than two atoms. (p. 196)

denaturation Interference with the secondary, tertiary and quaternary structures of a protein, changing its conformation and often removing its functionality. (p. 1082)

deposition The phase change in which a vapour converts directly to a solid without passing through the liquid phase. (p. 267)

deshielded Describes the situation in NMR spectroscopy in which resonance or inductive effects reduce the electron density around a nucleus, thus increasing the ability of an applied magnetic field to bring the nucleus into resonance. (p. 895)

dextrorotatory Describes the clockwise rotation of a plane of polarised light in a polarimeter. (p. 776)

dialysis The passage of solvent and small molecules and ions through a semipermeable membrane. (p. 416)

diamagnetic The quality of not being attracted to a magnetic field. Substances with no unpaired electrons are diamagnetic. (pp. 154, 548)

diastereomers A pair of stereoisomers that are not mirror images of each other. (p. 761)

differential rate law A rate law that expresses the rate of a reaction as a function of concentration. (p. 649)

diffraction pattern The pattern formed from X-rays diffracted by a crystal. (p. 284)

diffusion The movement of one type of gas molecule through molecules of another type. (p. 228)

dimensional analysis The use of units of a physical quantity to derive the equation used to determine its value. (p. 31)

dimer A close association of two similar molecules. (p. 423)

diol A compound with two alcohol groups. (p. 823)

dipeptide A molecule containing two amino acid units joined by a peptide bond. (p. 1074)

dipole–dipole force The attractive force between polar molecules that results from the negative end of one molecule aligning with the positive end of its neighbour. (p. 246)

dipole-induced dipole force The force resulting from the induction of a dipole in a molecule by a neighbouring molecule having a permanent dipole. (p. 246)

dipole moment The net electrical character arising from an asymmetric charge distribution. (p. 189)

diprotic acid An acid that can potentially donate two equivalents of protons to water. (p. 434)

diprotic base A base which is able to accept two protons. (p. 434)

disaccharides Carbohydrates containing two monosaccharide units joined by a glycosidic bond. (p. 999)

dispersion force The attraction between the negatively charged electron cloud of one molecule and the positively charged nuclei of neighbouring molecules. (p. 246)

disproportionation An electrochemical process in which a species is simultaneously oxidised and reduced to form two different products. (pp. 533, 569)

dissociate A general process in which molecules separate into smaller particles. (p. 106)

dissociation The breaking apart of a molecule or ionic solid into ions when dissolved in a solvent (usually water). (p. 420)

dissolution The process of dissolving a solute in a solvent. (p. 392)

disulfide A compound of the form RS—SR, where the R groups are either alkyl or aryl groups. (p. 847)

disulfide bond A covalent bond between two sulfur atoms, i.e. an —S—S— bond. (p. 1079)

D-monosaccharide A monosaccharide that, when written as a Fischer projection, has the —OH on its penultimate carbon atom to the right. (p. 987)

donor atom An atom of a ligand in a complex that is directly coordinated to the transition metal ion. (p. 551)

donor covalent bond A coordinate bond. (p. 556)

double bond A chemical bond containing two pairs of bonding electrons. (p. 177)

double bond equivalent See *index of hydrogen deficiency.* (p. 876)

doublet A signal in ^{1}H-NMR spectroscopy that has been split into two peaks in a ratio of 1 : 1. (p. 903)

downfield Describes a signal in NMR spectroscopy that is towards the left of the spectrum or of another signal. (p. 897)

dry cell battery A cell with an immobilised electrolyte, such as the Leclanché cell. (p. 533)

dynamic equilibrium An equilibrium in which the rates of the forward and reverse reactions are equal. (p. 350)

• E •

E1 A process in which the bond to the leaving group is completely broken before the hydrogen atom is removed and the carbon–carbon double bond is formed. (p. 808)

E2 A process in which the β-hydrogen is lost at the same time as the bond to the leaving group is broken and the carbon–carbon double bond is formed. (p. 809)

effective nuclear charge (Z_{eff}) The net positive charge, equal to the nuclear charge minus the effects of screening, that an electron in an atomic orbital experiences. (p. 144)

effusion The escape of a gas through a pinhole from a container into a vacuum. (p. 228)

electrochemical changes A chemical change that is caused by, or that produces, electricity. (p. 507)

electrochemical potential The potential difference between the electrodes in a galvanic cell. (p. 506)

electrochemistry The study of electrochemical changes. (p. 507)

electrolysis The production of a chemical change by the passage of electricity through a solution that contains ions or through a molten ionic compound. (p. 528)

electrolysis cell An apparatus for electrolysis. (p. 528)

electrolytic cell See *electrolysis cell.* (p. 528)

electrolytic conduction The transport of electric charge by ions. (p. 508)

electromagnetic radiation Energy propagated through space in the form of oscillating electric and magnetic fields. (pp. 120, 881)

electron A subatomic particle (e, ${}_{-1}^{0}e$), with a charge of −1 and mass of 5.4858×10^{-4} u (9.1094×10^{-31} kg), that is outside an atomic nucleus; the particle that moves when an electric current flows. (p. 4)

electron affinity (E_{ea}) The energy change accompanying the attachment of an electron to an atom or anion. (p. 159)

electron configuration The distribution of electrons among the various orbitals of an atom, molecule or ion. (p. 149)

electron density picture A two-dimensional dot drawing representing the distribution of electron density in an orbital. (p. 139)

electron density plot A plot of the distribution of electron probability in space around an atom or molecule. (p. 139)

electronegativity A measure of the ability of an atom in a molecule to attract the shared electrons in a chemical bond. (p. 174)

electronic conduction The conduction of electric charge by the movement of electrons. (p. 508)

electronic transition The movement of electrons between states of different energies. (pp. 19, 574)

electron paramagnetic resonance (EPR) See *electron spin resonance.* (p. 917)

electron spin resonance (ESR) A method of locating electrons within the molecules of a paramagnetic substance. (p. 917)

electrophile Any molecule or ion that can accept a pair of electrons to form a new covalent bond; a Lewis acid. (pp. 482, 723)

electrophilic aromatic substitution A reaction in which an electrophile, E^{+}, substitutes for a hydrogen on an aromatic ring. (p. 742)

electrophoresis The process of separating compounds on the basis of their electric charge. (p. 1071)

element A chemical species consisting of atoms of a single type. (p. 4)

elemental analysis A method of chemical analysis by which a given amount of a compound is decomposed chemically to find the masses of elements within it. (p. 90)

elementary reaction One of the individual steps in the mechanism of a reaction. (p. 671)

emission spectrum The distribution of wavelengths of light given off by a species in an excited state. (p. 127)

empirical formula A chemical formula that uses the smallest whole-number subscripts to give the proportions of atoms of the different elements present. (p. 87)

enantiomers A pair of stereoisomers that are nonsuperimposable mirror images of each other. (p. 760)

endergonic A process for which ΔG is positive. (p. 333)

endothermic Describes a change in which energy enters a system from the surroundings. (p. 308)

endpoint The point in a titration when the indicator changes colour. (p. 472)

energy level diagram A diagram of the specific energy levels of a species, showing increasing energy on the *y*-axis. (p. 130)

enol A molecule containing an —OH group bonded to a carbon atom of a carbon–carbon double bond. (p. 968)

enthalpy (H) The heat content of a system measured under constant pressure conditions. (p. 299)

enthalpy of hydrogenation The enthalpy of reaction, $\Delta H^{\ominus}$, with hydrogen. (p. 732)

enthalpy of solution ($\Delta_{sol}H$) The enthalpy change that accompanies the formation of a solution from a solute and a solvent. (p. 399)

entropy (S) A thermodynamic quantity related to the number of equivalent ways the energy of a system can be distributed. The greater this number, the more probable is the state and the higher is the entropy. (p. 299)

equatorial Describes a covalent bond located in the plane perpendicular to the axial ligands of an octahedral or trigonal bipyramidal molecule. (p. 557)

equatorial bonds Bonds oriented in a chair conformation of a cyclic structure in the general plane of the seat of the chair. (p. 704)

equilibrium constant (K) The value of the equilibrium constant expression when the system is at equilibrium. (p. 351)

equilibrium constant expression A fraction in which the numerator is the product of the equilibrium molar concentrations of the products, each raised to a power equal to its coefficient in the equilibrium equation, and the denominator is the product of the equilibrium molar concentrations of the reactants, each raised to a power equal to its coefficient in the equation. (For gaseous reactions, partial pressures can be used in place of molar concentrations.) (p. 351)

equivalence point The point in a titration when the stoichiometry of the reaction is satisfied. (p. 472)

equivalent hydrogen atoms Hydrogen atoms that are chemically equivalent. (p. 897)

ester A functional group generated by combining a carboxylic acid and an alcohol with the elimination of water to give a carbonyl carbon atom that is also further bound to a different saturated carbon atom via an oxygen atom linkage. (p. 1013)

ether A compound of the form ROR, where the R groups are either alkyl or aryl groups, such that an oxygen atom is bonded to two carbon atoms that are not also bonded to a heteroatom. (p. 842)

excess reagent The reagent left over when the limiting reagent has been used up. (p. 96)

excited state Any state in which a chemical system is not in its lowest possible energy state. (pp. 19, 126)

exergonic A process in which ΔG is negative. (p. 333)

exothermic Describes a change in which energy leaves a system and enters the surroundings. (p. 308)

extensive property A property of an object that is described by a physical quantity (such as mass and volume) that is proportional to the size or amount of the object. (p. 305)

***E*,*Z* system** A system used to specify the configuration of groups about a carbon–carbon double bond. (p. 718)

• F •

***f*-block elements** A collective name for the lanthanoid and actinoid elements. (p. 16)

face-centred cubic structure (fcc) A crystalline structure whose unit cell has lattice points at each corner of a cube plus one additional lattice point in the centre of each face of the cube. (p. 280)

facial (*fac*) An isomer of an octahedral complex $[ML_3X_3]^{n+}$ in which the two sets of ligands occupy the face of an octahedron. (p. 561)

Faraday constant (*F*) $1\,F = 96\,485\ \text{C mol}^{-1}$. (p. 520)

Faraday's law Allows the calculation of the amount of product formed during electrolysis; amount of product $= \frac{It}{nF}$. (p. 531)

fat A triglyceride that is semisolid or solid at room temperature. (p. 1045)

fatty acid A long, unbranched carboxylic acid, most commonly of 12 to 20 carbon atoms, derived from the hydrolysis of animal fats, vegetable oils or the phospholipids of biological membranes. (p. 1044)

ferromagnetism A property of species containing unpaired electrons where all of the electron spins are oriented in the same direction in the absence of a magnetic field. (p. 580)

fingerprint region In infrared spectroscopy, the region of the spectrum from 1000 to 400 cm^{-1}. (p. 883)

first law of thermodynamics A formal statement of the law of conservation of energy; $U = q + w$. (p. 303)

Fischer esterification The process of forming an ester by refluxing a carboxylic acid and an alcohol in the presence of an acid catalyst, commonly sulfuric acid. (p. 1028)

Fischer projections Two-dimensional representations that show the configuration of a stereocentre; horizontal lines represent bonds projecting forward from the stereocentre, vertical lines represent bonds projecting to the rear. (p. 987)

formal charge The charge that an atom in a molecule would have if each of its bonding electrons were equally shared with its bonding partner(s). (p. 179)

freezing point depression (ΔT_f) A colligative property of a liquid solution by which the freezing point of the solution is lower than that of the pure solvent. (p. 414)

frequency (ν) The number of full cycles of a wave that pass a given point in a second. (pp. 120, 881)

frequency factor The proportionality constant *A* in the Arrhenius equation. (p. 668)

Friedel–Crafts acylation The acylation of aromatic rings with an acyl chloride using anhydrous aluminium chloride as a catalyst. (p. 746)

Friedel–Crafts alkylation The alkylation of an aromatic ring and an alkyl halide using anhydrous aluminium chloride as a catalyst. (p. 745)

fuel cell An electrochemical cell that converts the chemical energy of a reaction between fuels, such as liquid hydrogen and liquid oxygen, directly and continuously into electrical energy. (p. 537)

functional group A group of atoms within an organic molecule that determines the molecule's chemical reactivity. (p. 54)

furanose a five-membered cyclic hemiacetal form of a monosaccharide. (p. 990)

• G •

galvanic cell An electrochemical cell in which a spontaneous redox reaction produces electricity. (p. 506)

galvanisation The coating of a metal, such as iron, with another, such as zinc, to protect against corrosion. (p. 527)

gas constant (*R*) The proportionality constant used in the ideal gas equation. (p. 222)

Gibbs energy (*G*) A thermodynamic quantity that relates enthalpy (*H*), entropy (*S*) and temperature (*T*) by the equation $G = H - TS$. (p. 299)

Gibbs energy diagram A plot of the changes in Gibbs energy for a system versus its composition. (p. 359)

glass An amorphous solid that has cooled to a solid state without crystallising. (p. 286)

glycol A diol in which the —OH groups are on adjacent carbon atoms. (p. 823)

glycoside A carbohydrate in which the —OH on its anomeric carbon atom is replaced by —OR. (p. 993)

glycosidic bond The bond from the anomeric carbon atom of a glycoside to an —OR group. (p. 993)

Grignard reagent An organomagnesium compound of the type RMgX or ArMgX. (p. 951)

ground state The lowest possible energy state of a chemical system. (pp. 19, 126)

group A vertical column of elements in the periodic table. (p. 16)

• H •

^{1}H-NMR spectroscopy Nuclear magnetic resonance spectroscopy of hydrogen atoms within molecules. It provides information about the carbon–hydrogen framework of a molecule. (p. 894)

half-cell That part of a galvanic cell in which either oxidation or reduction takes place. (p. 506)

half-equation A chemical equation describing either the oxidation or reduction process in a redox reaction. (p. 500)

half-life ($t_{\frac{1}{2}}$) The time required for a reactant concentration or the mass of a radionuclide to be reduced by half. (p. 646)

haloalkanes Compounds containing a halogen atom covalently bonded to an sp^3 hybridised carbon atom. (p. 792)

haloform A substituted methane molecule containing three halogen atoms of the same type. (p. 793)

halogenation A reaction in which some of the hydrogen atoms are substituted by halogen atoms. (p. 794)

halogens The elements in group 17 of the periodic table. (p. 17)

halonium ion An ion in which a halogen atom bears a positive charge. (p. 731)

hardening The process of converting oils to fats by catalytic hydrogenation of the double bonds. (p. 1046)

hard Lewis acid A Lewis acid containing an acceptor atom of low polarisibility. (p. 484)

hard Lewis base A Lewis base containing one or more electron pairs of low polarisibility. (p. 484)

hashed wedge A hashed wedge denotes a bond going into the plane of the page away from the observer. (p. 47)

Haworth projection A way of viewing furanose and pyranose forms of monosaccharides. The ring is drawn flat and viewed through its edge, with the anomeric carbon atom on the right and the oxygen atom of the ring at the rear to the right. (p. 990)

heat (*q*) A transfer of energy due to a temperature difference. (p. 299)

heat capacity (*C*) The quantity of heat needed to raise the temperature of an object by 1 K. (p. 304)

heat of reaction at constant pressure (q_p) The heat of a reaction in an open system. (p. 308)

heat of reaction at constant volume (q_v) The heat of a reaction in a sealed vessel, such as a bomb calorimeter. (p. 306)

hemiacetal A molecule containing an —OH and an —OR or —OAr group bonded to the same carbon atom. (pp. 955, 989)

Henderson–Hasselbalch equation An equation that relates the pH of a buffer solution to the pK_a of the weak acid and the ratio of the conjugate base to the acid; $pH = pK_a + \log\frac{[A^-]}{[HA]}$. (p. 468)

Henry's law The concentration of a gas dissolved in a liquid at any given temperature is directly proportional to the partial pressure of the gas above the solution. (p. 395)

Hertz (Hz) The SI unit of frequency. One hertz equals one wave cycle per second. (p. 881)

Hess's law For any reaction that can be written in steps, the standard enthalpy of reaction is the same as the sum of the standard enthalpies of reaction for the steps. (p. 313)

heteroatom Any atom in an organic molecule other than carbon or hydrogen. (p. 842)

heterocyclic amine An amine in which nitrogen is one of the atoms of a ring. (p. 847)

heterocyclic aromatic amine An amine in which nitrogen is one of the atoms of an aromatic ring. (p. 847)

heterocyclic compound An organic compound with one or more atoms other than carbon in its ring. (p. 738)

heterogeneous catalyst A catalyst that is in a different phase from the reactants. (p. 680)

heterogeneous equilibrium An equilibrium system involving more than one phase. (p. 358)

heterogeneous reaction A reaction in which not all of the chemical species are in the same phase. (pp. 358, 647)

heterolytic Describes a bond-breaking process in which both electrons of a single bond end up on one of the atoms of the bond. (p. 49)

heterolytic cleavage See *heterolytic*. (p. 461)

hexagonal close-packed (hcp) The close packing arrangement in which close-packed layers are stacked in an ABAB pattern. (p. 276)

high resolution mass spectrometry A technique that allows the precise measurement of the mass to charge ratio of ions. (p. 879)

high-spin Describes a complex ion or coordination compound with the maximum number of unpaired electrons. (p. 572)

homogeneous catalyst A catalyst that is in the same phase as the reactants. (p. 680)

homogeneous equilibrium An equilibrium system in which all components are in the same phase. (p. 358)

homogeneous reaction A chemical reaction in which all participating species are in the same phase. (pp. 358, 647)

homolytic Describes a bond-breaking process in which each of the bonded atoms receives one electron. (p. 50)

homolytic cleavage See *homolytic*. (p. 461)

Hund's rule The observation that the lowest energy arrangement of electrons among orbitals of equal energy is the one that maximises the number of unpaired electron spins. (p. 152)

hybridisation of atomic orbitals The formation of a set of hybrid orbitals with favourable directional characteristics by mixing together two or more valence orbitals of the same atom. (p. 735)

hybrid orbital An atomic orbital obtained by combining two or more valence orbitals on the same atom. (p. 197)

hydrate A solid that contains a definite number of waters of crystallisation. (p. 41)

hydration The addition of water; the development in an aqueous solution of a 'cage' of water molecules around ions or polar molecules of the solute. (pp. 398, 728)

hydration enthalpy The enthalpy change when gaseous solute particles (obtained from 1 mole of solute) are dissolved in water. (p. 399)

hydrate isomers Isomers of complex ions in which water is either coordinated to the transition metal ion or acts as water of crystallisation. (p. 559)

hydride ion A hydrogen atom with two electrons in its valence shell; $H:^-$. (p. 964)

hydrocarbon A molecule composed only of carbon and hydrogen atoms. (pp. 58, 698)

hydrogen bonding A moderately strong intermolecular attraction caused by the partial sharing of electrons between a highly electronegative atom of F, O or N and the polar hydrogen atom in a F—H, O—H or N—H bond. (pp. 825, 1080)

hydrometer An instrument used for measuring the density of a liquid. (p. 533)

hydronium ion (H_3O^+) The product of the reaction of a proton with water. It can also be thought of as a hydrated proton. (p. 432)

hydrophilic Describes a polar compound that dissolves readily in water. (p. 1018)

hydrophobic Describes a nonpolar compound that tends to be insoluble in water. (p. 1018)

hydrophobic interaction The force of attraction between nonpolar parts of a molecule or molecules. (p. 1080)

hydroxide ion (OH^-) The product of deprotonation of water. (p. 432)

hydroxyl group An —OH group. (p. 822)

hypertonic Describes a solution that has a higher osmotic pressure than cellular fluids. (p. 419)

hypotonic Describes a solution that has a lower osmotic pressure than cellular fluids. (p. 419)

• I •

ideal gas A gas in which molecular volumes and intermolecular forces are both negligible. (p. 222)

ideal gas equation The equation describing the behaviour of an ideal gas, $pV = nRT$. (p. 222)

ideal solution A hypothetical solution that obeys Raoult's law exactly. (p. 412)

immiscible Describes liquids that do not mix; see also *insoluble*. (p. 396)

index of hydrogen deficiency (IHD) The sum of the number of rings and π bonds in a molecule. (p. 876)

inductive effect The polarisation of electron density transmitted through covalent bonds caused by a nearby atom of higher electronegativity. (p. 727)

inert Used to describe a substance that is not chemically reactive. (p. 569)

inert pair effect The tendency of the electrons in the outermost *s* orbital to remain unionised or unshared in compounds of post-transition metals. (p. 620)

infrared active Any vibration that results in the absorption of infrared radiation. For a molecule to absorb infrared radiation, the bond undergoing vibration must be polar and its vibration must cause a periodic change in the bond dipole; the greater the polarity of the bond, the more intense the absorption. (p. 883)

infrared radiation The portion of the electromagnetic spectrum with wavelengths in the range 7.8×10^{-7} m to 2.0×10^{-3} m. (p. 881)

initial rate The rate of a chemical reaction immediately after mixing the reactants. (p. 648)

insoluble Describes a solute that does not dissolve in a solvent. (p. 399)

instantaneous rate of change of concentration The rate of change of concentration of a reactant or product in a chemical reaction at a particular time. (p. 644)

integral membrane protein A protein incorporated into a cell membrane. (p. 1082)

integrated rate law A rate law that expresses the rate of a reaction as a function of time. (p. 649)

integration A mathematical process used for determining the area under a signal in an NMR spectrum. (p. 899)

intensity The brightness (number of photons) of light. (p. 120)

intensive property A property of an object that is described by a physical quantity (such as density and temperature) that is independent of the size of the sample. (p. 305)

intercalation The reversible inclusion of a molecule or atom into the crystal lattice of another compound. (p. 536)

intermolecular forces Forces that exist between molecules. (p. 241)

internal energy (U) The sum of all of the kinetic energies and potential energies of the particles within a system. (p. 299)

interstitial hole The space between spheres/atoms. (p. 282)

ion A charged chemical species. (p. 4)

ionic compound A compound formed between two elements with different electronegativities. (p. 174)

ionic product An alternative name for the reaction quotient (Q_{sp}). (p. 407)

ionic solid A solid containing cations and anions that are attracted to each other by electrical interactions rather than covalent bonds. (p. 263)

ionisation energy (E_i) The energy required to remove an electron from an isolated species. (p. 143)

ionisation isomers Isomers in which a coordinated ligand is exchanged with a counterion, e.g. $[Co(NH_3)_5Br]SO_4$ and $[Co(NH_3)_5SO_4]Br$. (p. 559)

ion pair A more or less loosely associated pair of ions in a solution. (p. 421)

isoelectric point (pI) The pH at which an amino acid, a polypeptide or a protein has no net charge. (p. 1070)

isoelectronic Having the same number of electrons. (p. 153)

isomers Molecules with the same molecular formula but different structures. (pp. 43, 760)

isotonic Describes a solution that has the same osmotic pressure as cellular fluids. (p. 419)

isotopes Atoms of the same element having different numbers of neutrons in their nuclei. (p. 11)

• J •

joule (J) The SI unit of energy; $1\ J = 1\ kg\ m^2\ s^{-2}$. (pp. 300, 506)

• K •

Kekulé structure One of the two contributing structures of benzene proposed by August Kekulé. (p. 734)

kelvin The SI unit of temperature. (p. 27)

ketal A version of an acetal formed from a ketone rather than an aldehyde. (p. 956)

ketone A compound in which the carbon atom of the C=O group is bonded to two hydrocarbon groups. (p. 944)

ketoses Monosaccharides containing a ketone group. (p. 986)

kilogram The SI unit of mass. (p. 27)

kilojoule (kJ) 1000 J. (p. 300)

• L •

labile Describes a transition metal complex in which ligand exchange is rapid. (p. 568)

lactam A cyclic amide. (p. 1013)

lactone A cyclic ester. (p. 1013)

lanthanoids Elements 57 to 71 of the periodic table. (p. 16)

lattice A set of points with identical environments that describes a pattern. (p. 277)

lattice energy The energy change on converting 1 mole of an ionic solid into its constituent gaseous ions. (p. 176)

lattice enthalpy The enthalpy required to separate 1 mole of a crystalline compound into its gaseous constituent particles. (p. 399)

lattice point A point within a lattice. (p. 277)

law of conservation of mass No detectable gain or loss in mass occurs in chemical reactions. Mass is conserved. (p. 4)

law of definite proportions In a given chemical compound, the constituent elements are always combined in the same proportion by mass. (p. 4)

law of multiple proportions Whenever two elements form more than one compound, the different masses of one element that combine with the same mass of the other are in a ratio of small whole numbers. (p. 7)

Le Châtelier's principle If an outside influence upsets an equilibrium, the system undergoes a change in a direction that counteracts the disturbing influence and, if possible, returns the system to equilibrium. (p. 366)

lead storage battery A lead – sulfuric acid storage battery in a motor vehicle; usually a 12-volt battery of six cells. (p. 532)

Leclanché cell A primary battery dependent on the reaction between zinc and manganese dioxide (Zn/MnO_2); see also *dry cell battery*. (p. 534)

levorotatory Describes the anticlockwise rotation of a plane of polarised light in a polarimeter. (p. 776)

Lewis acid An electron-pair acceptor. (p. 481)

Lewis base An electron-pair donor. (p. 481)

Lewis structure A representation of covalent bonding that uses symbols for the elements, dots for nonbonding valence electrons, and lines for pairs of bonding valence electrons. (p. 177)

ligand A molecule or an anion that can coordinate to a transition metal ion to form a complex. (p. 549)

ligand-to-metal charge transfer (LMCT) transitions Electronic transitions that involve formal transfer of an electron from a ligand to a transition metal ion. (p. 577)

light Electromagnetic radiation in the visible portion of the spectrum, between $\lambda = 380$ nm and $\lambda = 780$ nm. (p. 120)

like-dissolves-like rule Strongly polar and ionic solutes tend to dissolve in polar solvents, and nonpolar solutes tend to dissolve in nonpolar solvents. (p. 396)

limiting reagent The reagent that determines how much product can form when non-stoichiometric amounts of reagents are used. (p. 96)

linear Lying along a straight line. (p. 183)

line of integration A curved line at a signal on a ^{1}H-NMR spectrum that allows the determination of the relative number of hydrogen atoms giving rise to that signal. (p. 899)

line structure A shorthand method of drawing structural formulae in which each vertex and line ending represents a carbon atom. (p. 43)

linkage isomers Isomers that result from the different ways in which an ambidentate ligand can bind to a metal ion. (p. 559)

lipid A biomolecule isolated from plant or animal sources by extraction with nonpolar organic solvents, such as diethyl ether and acetone. (p. 1043)

L-monosaccharide A monosaccharide that, when written as a Fischer projection, has the —OH on its penultimate carbon atom to the left. (p. 987)

lithium ion cell Rechargeable battery with twice the energy capacity of a nickel–cadmium (nicad) battery. (p. 536)

local magnetic field The magnetic field generated by electrons surrounding a nucleus. (p. 895)

localised bond A chemical bond that involves only two atoms. (p. 196)

lone pair A pair of valence electrons that is localised on an atom rather than involved in bonding. (p. 178)

low-spin Describes a complex ion or coordination compound with the minimum number of unpaired electrons. (p. 572)

• M •

magnetic moment (μ) A property related to the number of unpaired electrons in a transition metal complex through the equation $\mu = \sqrt{n(n+2)}\mu_B$. (p. 578)

magnetic quantum number (m_l) The quantum number, restricted to integers between $+l$ and $-l$, that indexes the orientation in space of an atomic orbital. (p. 137)

main-group metals A collective term for the metals in groups 1, 2 and 13–16 of the periodic table. (p. 548)

manometer A bent tube, containing a liquid and open at both ends, used to measure differences in pressure. (p. 220)

Markovnikov's rule In the addition of HX or H_2O to an alkene, hydrogen adds to the carbon atom of the double bond having the greater number of hydrogen atoms. (p. 724)

mass number (A) The numerical sum of the protons and neutrons in an atom of a given isotope. (p. 11)

mass percentage composition A list of the percentages by mass of the elements in a compound. (p. 90)

mass spectrometry A technique used to determine the mass of molecules and fragments of molecules. (p. 877)

matter Anything that has mass and occupies space. (p. 4)

mechanistic arrows Arrows which show the movement of electrons in bond-breaking and bond-making processes. (p. 49)

meniscus The curved surface of a liquid contained in a narrow tube. (p. 260)

mercaptan An alternative name for a thiol. (p. 845)

meridional (*mer*) An isomer of an octahedral complex $[ML_3X_3]^{n+}$ in which the donor atoms of each set of identical ligands are coplanar. (p. 561)

***meso* compound** An achiral compound with two or more stereocentres. (p. 772)

meta Describes groups occupying positions 1 and 3 on a benzene ring. (p. 740)

metallic solid An elemental solid whose atoms are held together by valence electrons occupying delocalised orbitals spread across the entire solid. (p. 263)

metalloids Elements with properties that lie between those of metals and nonmetals, and that are found in the periodic table around the diagonal line running from boron, B, to astatine, At. (p. 17)

metallurgy The science and technology of metals, the procedures and reactions that separate metals from their ores, and the operations that create practical uses for metals. (p. 584)

metals Elements that are good conductors of heat and electricity, are malleable (can be beaten into a thin sheet) and ductile (can be drawn out into a wire), and have the usual metallic lustre. (p. 17)

metal-to-ligand charge transfer (MLCT) transition An electronic transition in a transition metal complex in which an electron is transferred from the metal to the ligand. (p. 577)

metre The SI unit of length. (p. 27)

micelle A spherical arrangement of organic molecules in aqueous solution, clustered so that their hydrophobic parts are buried inside the sphere and their hydrophilic parts are on the surface of the sphere and in contact with water. (p. 1047)

Michaelis constant In enzyme reactions, the substrate concentration at half of the maximum reaction rate. (p. 683)

Michaelis–Menten equation An equation that describes the reaction rate of many enzyme-catalysed reactions. (p. 683)

Michaelis–Menten mechanism The reaction mechanism, by which many enzyme-catalysed reactions proceed. (p. 682)

miscible Describes liquids that mix completely; see also *soluble.* (p. 396)

mixed valence A chemical compound containing an element in more than one oxidation state. (p. 622)

molal boiling point elevation constant (K_b) The number of degrees (°C) per unit of the molal concentration boiling point elevation of a solution relative to the pure solvent. (p. 414)

molal concentration (b) The number of moles of solute in 1000 g of solvent; also called molality. (p. 410)

molal freezing point depression constant (K_f) The number of degrees (°C) per unit of the molal concentration freezing point depression of a solution relative to the pure solvent. (p. 414)

molality (b) Molal concentration. (p. 410)

molar concentration (c) The concentration of a solution in units of mol L^{-1}; also called molarity. (p. 410)

molar enthalpy of fusion ($\Delta_{fus}H$) The enthalpy change on melting 1 mole of a substance at its normal melting point. (p. 267)

molar enthalpy of sublimation ($\Delta_{sub}H$) The enthalpy change on sublimation of 1 mole of a substance at its normal sublimation point. (p. 267)

molar enthalpy of vaporisation ($\Delta_{vap}H$) The enthalpy change on vaporisation of 1 mole of a substance at its normal boiling point. (p. 267)

molar heat capacity The heat capacity of 1 mole of a substance. (p. 305)

molarity (c) Molar concentration. (pp. 100, 410)

molar mass (M) The mass of 1 mole of the specified substance. (p. 85)

molar solubility (s) The number of moles of solute required to prepare 1 L of a saturated solution of the solute. (p. 403)

mole SI unit of amount of substance. One mole is the amount of substance that contains the same number of elementary entities (6.022×10^{23}) as there are atoms in exactly 12 g of ^{12}C. (pp. 27, 85)

molecular formula A formula that shows the actual numbers of different types of atoms present in a molecule. (pp. 40, 91)

molecular ion An ionised molecule that produces a peak in a mass spectrum. This is usually the peak with the highest m/z value. (p. 878)

molecular orbital (MO) A three-dimensional wave that encompasses an entire molecule and describes a bound electron. (p. 205)

molecular orbital diagram An energy level diagram displaying the molecular orbitals of a particular molecule. (p. 206)

molecular orbital theory A bonding theory that considers all possible overlaps between atomic orbitals in a molecule. The overlap of two atomic orbitals leads to the formation of one bonding molecular orbital and one antibonding molecular orbital. Molecular orbitals are generally delocalised and can be spread over the entire molecule. (p. 205)

molecular solid A solid containing discrete molecules that do not have chemical bonds between them. (p. 263)

molecularity The number of colliding molecules involved in an elementary step. (p. 671)

molecule An uncharged collection of atoms bonded together in a definite structure. (p. 4)

mole fraction (x) The number of moles of a component in a solution divided by the total number of moles of material in the solution. (pp. 232, 411)

monodentate Describes a ligand that can coordinate to a metal ion using only one donor atom. (p. 551)

monoprotic acid An acid that can donate only a single proton to water. (p. 433)

monoprotic base A base that can accept only one proton. (p. 434)

monosaccharides Carbohydrates that cannot be hydrolysed to simpler compounds. (p. 986)

multiplet The splitting pattern of a complex signal in ^{1}H-NMR spectroscopy. (p. 903)

multiplicity The splitting pattern of a signal in ^{1}H-NMR spectroscopy, which can be described as, singlet, doublet, triplet, quartet, multiplet etc. (p. 904)

mutarotation The change in optical activity that occurs when an α or β form of a carbohydrate is converted to an equilibrium mixture of the two forms. (p. 992)

• N •

($n + 1$) rule A rule for determining splitting patterns in NMR spectroscopy. If a hydrogen atom has n other hydrogen atoms that are not equivalent to it, but are equivalent to each other, on the same or adjacent atom(s), its ^{1}H-NMR signal is split into ($n + 1$) peaks. (p. 904)

***N*-terminal amino acid** The amino acid of a polypeptide with the free $—NH_3^+$ group. (p. 1074)

Nernst equation $E_{cell} = E^{\ominus}_{cell} - \frac{RT}{zF} \ln Q$ (p. 523)

net ionic equation An ionic equation from which spectator ions have been omitted. It is balanced when both atoms and electrical charge balance. (p. 107)

network solid A solid containing an array of covalent bonds linking every atom to its neighbours. (p. 263)

neutral Describes a solution in which $[H_3O^+] = [OH^-]$. (p. 437)

neutron A subatomic particle (n, ${}^{1}_{0}n$), with a charge of 0 and a mass of 1.0086 u (1.6749×10^{-27} kg), that exists in all atomic nuclei except those of the ${}^{1}H$ isotope. (p. 10)

Newman projection A way to view a molecule by looking along a carbon–carbon bond. (p. 700)

nicad battery A nickel–cadmium battery, commonly used in laptop computers. (p. 535)

nickel–cadmium storage cell See *nicad battery*. (p. 535)

nitrile a —C≡N group. (p. 1021)

nitronium ion NO_2^+. (p. 744)

noble gas configuration The portion of the distribution of electrons among the orbitals of an atom that matches that of one of the noble gases. (p. 151)

noble gases The elements in group 18 of the periodic table. (p. 17)

node A point, line or surface where the electron density of an orbital is zero. (pp. 140, 205)

nomenclature A system of naming. (p. 50)

nonmetals Nonductile, nonmalleable, nonconducting elements. (p. 17)

nonoxidising acid An acid in which the anion is a poorer oxidising agent than the hydrogen ion (e.g. HCl, H_3PO_4). (p. 518)

nonvolatile Describes a substance with a high boiling point and a low vapour pressure that, consequently, does not evaporate. (p. 410)

normal boiling point The boiling point of a substance under one atmosphere ($1.013\,25 \times 10^5$ Pa) of pressure. (p. 245)

normal freezing point The freezing point of a substance under $1.013\,25 \times 10^5$ Pa pressure. (p. 245)

nuclear magnetic resonance (NMR) spectroscopy A spectroscopic technique that measures the absorption of energy by nuclei in the presence of a magnetic field. (p. 894)

nucleon A proton or a neutron. (p. 10)

nucleophile Any reagent with an unshared pair of electrons that can be donated to another atom or ion to form a new covalent bond; a Lewis base. (pp. 482, 797)

nucleophilic acyl substitution A reaction in which a nucleophile bonded to a carbonyl carbon atom is replaced by another nucleophile. (p. 1026)

nucleophilicity A term for the relative rate at which a reagent undertakes nucleophilic substitution. (p. 802)

nucleophilic substitution Any reaction in which one nucleophile is substituted for another. (p. 797)

nucleus The dense core of an atom that comprises protons and neutrons. (p. 10)

nuclide A particular atom of specified atomic number and mass number. (p. 11)

• O •

observed rotation The angle through which a compound rotates a plane of polarised light. (p. 776)

octahedral Describes a molecule in which a central atom is surrounded by six atoms located at the vertices of an imaginary octahedron. (pp. 187, 557)

oil A triglyceride that is liquid at room temperature. (p. 1045)

oligopeptide A peptide containing more than 10 but fewer than 20 amino acids. (p. 1074)

oligosaccharides Carbohydrates containing from 4 to 10 monosaccharide units joined by glycosidic bonds. (p. 999)

optical activity The ability to rotate a plane of polarised light. (p. 775)

orbital A three-dimensional wave describing a bound electron. (pp. 19, 136)

orbital mixing Interactions among *s* and *p* atomic orbitals that change the relative energies of the molecular orbitals resulting from these atomic orbitals. (p. 210)

orbital overlap The extent to which two orbitals on different atoms interact. (p. 196)

order In a reaction, the sum of the exponents in the rate law is the *overall* order. Each exponent gives the order with respect to the individual reactant. (p. 648)

order of precedence of functional groups A system for ranking functional groups in order of priority for the purposes of IUPAC nomenclature. (p. 947)

organometallic compound A compound containing a carbon–metal bond. (pp. 555, 951)

ortho Describes groups occupying positions 1 and 2 on a benzene ring. (p. 740)

osmosis The passage of solvent molecules, but not those of solutes, through a semipermeable membrane (more limiting than dialysis). (p. 416)

osmotic membrane A membrane that allows passage of solvent, but not solute particles. (p. 416)

osmotic pressure (Π) The back pressure that has to be applied to prevent osmosis; one of the colligative properties. (p. 417)

oxidation A change in which an oxidation number increases (becomes more positive or less negative); a loss of electrons. (p. 496)

oxidation number The charge that an atom in a molecule or ion would have if all of the electrons in its bonds belonged entirely to the more electronegative atoms; the oxidation state of an atom. (p. 497)

oxidation state See *oxidation number*. (p. 497)

oxidising acid An acid in which the anion is a stronger oxidising agent than H^+ (e.g. $HClO_4$, HNO_3). (p. 519)

oxidising agent The substance that causes oxidation and that is itself reduced. (p. 496)

oximes Compounds containing a C=NOH group. (p. 962)

oxoanion A negatively charged ion containing two or more elements, one of which is oxygen. (p. 54)

oxonium ion An ion in which oxygen is bonded to three other atoms and bears a positive charge. (p. 729)

• P •

***p*-block elements** A collective name for the elements in groups 13 to 18 of the periodic table. (pp. 16, 602)

pairing energy (*P*) The energy required to force two electrons to become paired and occupy the same orbital. (p. 572)

para Describes groups occupying positions 1 and 4 on a benzene ring. (p. 740)

paramagnetic Describes substances that are attracted into a magnetic field as a result of the presence of one or more unpaired electrons. (pp. 154, 548)

parent chain The longest straight carbon chain in an organic molecule. (p. 58)

partial pressure The pressure exerted by one component of a gaseous mixture. (p. 231)

parts per billion (ppb) The number of entities of one particular component present in one billion objects. (p. 233)

parts per million (ppm) The number of entities of one particular component present in one million objects. (p. 233)

pascal (Pa) The SI unit of pressure, one newton per square metre. (p. 221)

passivation The stabilisation of a substance (usually a metal) by the formation of a thin oxide layer on its surface (p. 527)

Pauli exclusion principle The requirement that no two electrons in a chemical species can be described by the same set of four quantum numbers. (p. 138)

pentapeptide A molecule containing five amino acid units joined by peptide bonds. (p. 1074)

penultimate carbon The stereocentre of a monosaccharide furthest from the carbonyl group; for example, C(5) of glucose. (p. 987)

peptide A short polymer of amino acids. (p. 1074)

peptide bond The name given to the amide bond formed between the α-amino group of one amino acid and the α-carboxyl group of another amino acid. (p. 1074)

percentage by mass The number of grams of an element present in 100 g of a compound. (p. 90)

percentage uncertainty An uncertainty expressed as a percentage. (p. 35)

percentage yield The ratio, given as a percentage, between the quantity of product actually obtained in a reaction and the theoretical yield. (p. 98)

period A horizontal row of elements in the periodic table. (p. 16)

periodic table of the elements A table in which symbols for the elements are displayed in order of increasing atomic number and arranged so that elements with similar properties lie in the same column. (p. 15)

pH $-\log[H_3O^+]$. (p. 438)

phase The starting position of a wave with respect to one wavelength. (p. 121)

phase change The transition of a substance from one phase to another. (p. 266)

phase diagram A pressure–temperature graph showing the conditions under which a substance exists as solid, liquid and gas. (p. 269)

phenol A compound that contains an —OH group bonded to a benzene ring. (pp. 740, 835)

phenyl group The C_6H_5 — group. (p. 740)

photoelectric effect The ejection of electrons from a metal surface by light. (p. 122)

photons Particles of light, characterised by energy $E = h\nu$. (p. 123)

pi (π) bond A chemical bond formed by side-by-side orbital overlap so that electron density is concentrated above and below the bond axis. (p. 203)

pK_a $-\log K_a$. (p. 448)

pK_b $-\log K_b$. (p. 448)

pK_w $-\log K_w$. (p. 439)

Planck's constant (h) The physical constant, 6.626×10^{-34} J s, that relates the energy of a photon to its frequency. (p. 123)

plane of symmetry An imaginary plane passing through an object and dividing it so that one half is the mirror image of the other half. (p. 765)

plane polarised Vibrating only in parallel planes. (p. 775)

pnictogens The elements in group 15 of the periodic table. (p. 17)

pOH $-\log[OH^-]$. (p. 438)

polar covalent bond A bond that possesses an asymmetric distribution of electrons. (p. 173)

polarimeter An instrument for measuring the ability of a compound to rotate a plane of polarised light. (p. 776)

polarisability The ease with which the electron density about an atom or molecule can be distorted. (p. 246)

polycyclic aromatic hydrocarbon (PAH) A hydrocarbon with two or more fused aromatic rings. (p. 741)

polydentate Describes a ligand containing two or more atoms that can coordinate simultaneously to a metal ion. (p. 551)

polypeptide A macromolecule containing more than 20 amino acid units, each joined to the next by a peptide bond. (p. 1074)

polyprotic acid An acid which can donate more than one proton. (p. 433)

polyprotic base A base which can accept more than one proton. (p. 434)

polysaccharides Carbohydrates containing a large number of monosaccharide units joined together by one or more glycosidic bonds. (p. 999)

potential difference Voltage, which is a measure of the amount of energy that can be delivered as a current moves through a circuit. (p. 506)

precision A group of measurements is of high precision if all the values lie close together. (p. 32)

pre-exponential factor See *frequency factor*. (p. 668)

pressure Force per unit area. (p. 220)

primary carbon atom A carbon atom attached to only one other carbon atom. (p. 55)

primary cell A nonrechargeable battery. (p. 532)

primary (1°) structure The sequence of smaller molecular components that are assembled to make a biopolymer; thus for proteins and peptides it is the specific amino acids in a polypeptide chain read from the *N*-terminal amino acid to the *C*-terminal amino acid, and for DNA and RNA it is the sequence of bases along the 2′-deoxyribose/ribose–phosphodiester backbone of a DNA or RNA molecule read from the 5′ end to the 3′ end. (p. 1075)

primitive cubic structure The crystal form with a unit cell containing one lattice point at each corner of a cube only. (p. 279)

principal quantum number (n) The quantum number, restricted to positive integers, that indexes the energy and size of an atomic orbital. (p. 136)

product The chemical species obtained as the result of a chemical reaction. (pp. 4, 82)

protein A biological macromolecule with a molar mass of 5000 u or more and consisting of one or more polypeptide chains. (p. 1074)

protic solvents Solvents that contain —OH groups and are hydrogen-bond donors. (p. 804)

proton A subatomic particle ($^1_1$p), with a charge of +1 and a mass of 1.0073 u (1.6726×10^{-27} kg), that is found in atomic nuclei. (p. 10)

pyranose A six-membered cyclic hemiacetal form of a monosaccharide. (p. 990)

pyridinium chlorochromate (PCC) A mild oxidising agent, commonly used to prepare aldehydes from primary alcohols. (p. 834)

• Q •

qualitative analysis The use of experimental procedures to determine what elements are present in a substance. (p. 82)

quantisation A phenomenon whereby the energy of a chemical system is not continuous but is restricted to certain definite values. (p. 19)

quantised Having discrete allowed values. (p. 129)

quantitative analysis The use of experimental procedures to determine the percentage composition of a compound or the percentage of a component of a mixture. (p. 82)

quantum number An integer or half-integer describing the allowed values of some quantised property. (p. 136)

quartet A signal in ^{1}H-NMR spectroscopy that has been split into four peaks in a ratio of 1 : 3 : 3 : 1. (p. 903)

quaternary (4°) structure The arrangement of polypeptide monomers into a noncovalently bonded aggregation. (p. 1081)

• R •

R Used in the *R,S* system to show that the order of priority of groups on a stereocentre is clockwise; from the Latin word *rectus* meaning 'right'. (p. 769)

racemic mixture A mixture of equal amounts of two enantiomers. (p. 778)

radical A nonmetal species containing an odd number of electrons in its outer electron shell. (p. 794)

radical cation A species with a positive charge and an unpaired electron. (p. 878)

radical substitution A substitution reaction that involves a radical in the mechanism of substitution. (p. 794)

radioactive Able to emit various atomic radiations or gamma rays. (p. 11)

radio-frequency radiation The portion of the electromagnetic spectrum with wavelengths greater than a metre. (p. 895)

radionuclide A radioactive isotope. (p. 11)

Raoult's law The vapour pressure of one component above a mixture of molecular compounds equals the product of its mole fraction and its vapour pressure when pure. (p. 411)

rare earth elements An alternative name for the lanthanoid elements. (p. 16)

rate coefficient An alternative name for the rate constant, the proportionality constant in the rate law. (p. 648)

rate constant The proportionality constant in the rate law. (p. 648)

rate-determining step The slowest step in a reaction mechanism. (p. 672)

rate law An equation that relates the rate of a reaction to the molar concentration of the reactants raised to powers. (p. 648)

rate of reaction The amount (in moles or mass units) per volume per unit time of products formed or reactants consumed in a particular reaction at a particular instant in time. (p. 644)

reactant A chemical species that is transformed in a chemical reaction. (pp. 4, 82)

reaction coordinate The horizontal axis of a potential energy diagram for a reaction. (p. 665)

reaction intermediate A species produced during a reaction that does not appear in the reaction equation because it is consumed in a subsequent step in the mechanism. (p. 666)

reaction mechanism The series of individual steps (elementary processes) in a chemical reaction that gives the net, overall change. (p. 650)

reaction quotient (*Q*) The numerical value of the equilibrium constant expression under any conditions; see also *equilibrium constant expression*. (p. 353)

reaction quotient expression A fraction in which the numerator is the product of the molar concentrations of the products, each raised to a power equal to its coefficient in the equilibrium equation, and the denominator is the product of the molar concentrations of the reactants, each raised to a power equal to its coefficient in the equation. (For gaseous reactions, partial pressures can be used in place of molar concentrations.) (p. 353)

redox reaction A reaction involving the transfer of one or more electrons between chemical species. (pp. 19, 496)

reducing agent A substance that causes reduction and is itself oxidised. (p. 496)

reducing sugar A carbohydrate that reacts with an oxidising agent to form an aldonic acid. (p. 996)

reduction A change in which an oxidation number decreases (becomes less positive or more negative); a gain of electrons. (p. 496)

reduction–oxidation reaction See *redox reaction*. (p. 496)

reduction potential (E_{red}) A measure of the tendency of a given half-reaction to occur as a reduction. (p. 512)

reductive amination The formation of an imine from an aldehyde or ketone, followed by its reduction to an amine. (p. 962)

reference compound A compound added to a sample to be studied by NMR spectroscopy, usually tetramethylsilane (TMS). The positions of the signals in the spectrum are then compared with those of the reference compound. (p. 876)

refractory A heat-resistant ceramic material. (p. 288)

regioselective reaction A reaction in which one direction of bond-breaking or bond-forming occurs in preference to all other directions. (p. 724)

resolution The separation of a racemic mixture into its enantiomers. (p. 780)

resonance The absorption of electromagnetic radiation by a spinning nucleus and the resulting flip of its nuclear spin state. (p. 895)

resonance energy The difference in energy between a resonance hybrid and the most stable of its hypothetical contributing structures. (p. 736)

resonance stabilisation The energetic stabilisation that occurs on delocalisation of negative charge over two or more atoms. (p. 464)

resonance structure One of two or more Lewis structures that are equivalent to one another. (p. 180)

reversible reaction A reaction capable of proceeding in either the forward or reverse direction. (p. 82)

root-mean-square speed Average speed obtained by taking the square root of the mean value of the squares of the individual speeds. (p. 228)

***R,S* system** A set of rules for specifying the configuration of groups around a stereocentre. (p. 768)

• S •

S Used in the *R,S* system to show that the order of priority of groups on a stereocentre is anticlockwise; from the Latin word *sinister* meaning 'left'. (p. 769)

***s*-block elements** A collective name for the elements in groups 1 and 2 of the periodic table. (p. 16)

salt bridge A tube containing an electrolyte, that connects the two half-cells of a galvanic cell. (p. 508)

salt linkage An ionic bonding interaction between charged functional groups in a peptide. (p. 1081)

saponification The hydrolysis of an ester in aqueous NaOH or KOH to an alcohol and the sodium or potassium salt of a carboxylic acid. (p. 1033)

saturated calomel electrode A reference electrode that consists of elemental mercury, mercury(I) chloride and Hg_2Cl_2 (calomel) in saturated KCl solution. (p. 516)

saturated hydrocarbon A compound composed solely of carbon and hydrogen atoms in which all C—C bonds are single; see also *alkane*. (pp. 58, 698)

saturated solution A solution in which no more solute will dissolve at a specified temperature. (p. 392)

scientific notation A method of expressing numbers in terms of powers of 10. For example, the number 3613 becomes 3.613×10^3 in scientific notation. (p. 34)

second The SI unit of time. (p. 27)

secondary carbon atom A carbon atom attached to two other carbon atoms. (p. 55)

secondary cell A battery that can be recharged. (p. 532)

secondary (2°) structure The ordered arrangement (conformation) of a biopolymer; thus the organised shape of the chain of amino acids in localised regions of a polypeptide or protein, seen as sheets or coils, and the ordered arrangement of DNA and RNA strands, to form a defined three-dimensional shape. (p. 1077)

second law of thermodynamics Whenever a spontaneous event takes place in our universe, the total entropy of the universe increases ($\Delta S_{total} > 0$). (p. 327)

seesaw shape The molecular shape that resembles a seesaw. (p. 186)

shielded Describes the situation in NMR spectroscopy in which local magnetic fields from electrons surrounding a nucleus decrease the ability of an applied magnetic field to bring the nucleus into resonance. (p. 895)

shielding The partial cancellation of the electrostatic attraction between the nucleus and an electron within an atom, caused by one or more electrons of lower principal quantum number. (p. 144)

sigma (σ) bond A bond that is totally symmetric with respect to rotation about the internuclear axis. (p. 173)

signal splitting A phenomenon in NMR spectroscopy in which the ^{1}H-NMR signal from one set of hydrogen atoms is split by the influence of neighbouring nonequivalent hydrogen atoms. (p. 904)

significant figures The figures in a physical measurement that are known to be certain plus the first figure that contains uncertainty. (p. 33)

silver-mirror test A qualitative test for an aldehyde using Tollens' reagent; the aldehyde is oxidised to a carboxylate anion, and Ag^+ is reduced to metallic silver. (p. 967)

silver/silver chloride electrode A reference electrode where an Ag rod is dipped into a mixture of AgCl and KCl of different concentrations. (p. 516)

single bond A chemical bond formed by one pair of electrons shared between two atoms. (p. 177)

singlet A signal in ^{1}H-NMR spectroscopy that has not been split. (p. 903)

SI (Système International) units A system of units based on seven base units from which all others can be derived. (p. 26)

S_N1 A type of nucleophilic substitution reaction in which only one molecule is involved in the rate-determining step. (p. 801)

S_N2 A type of nucleophilic substitution reaction in which two molecules or ions are involved in the rate-determining step. (p. 800)

soap The sodium or potassium salt of a fatty acid. (p. 1047)

soft Lewis acid A Lewis acid having an acceptor atom of high polarisibility. (p. 484)

soft Lewis base A Lewis base having a large donor atom of high polarisability and low electronegativity. (p. 484)

solid wedge A solid wedge denotes a bond coming out of the page towards the observer. (p. 47)

solubility The maximum amount of a solute that dissolves completely in a given mass or volume of solvent at a particular temperature. (p. 392)

solubility product (K_{sp}) The equilibrium constant for the dissolution of an ionic salt. (p. 403)

soluble Describes a solid or a gas that dissolves in a solvent. (p. 392)

solute The dissolved substance contained in a solution. (pp. 100, 392)

solution A homogeneous mixture in which all particles are of the size of atoms, small molecules or small ions. (pp. 100, 392)

solvation The development of a cage-like network of a solution's solvent molecules around a molecule or ion of the solute. (p. 398)

solvation enthalpy The enthalpy change due to the interaction of gaseous molecules or ions of solute with solvent molecules during the formation of a solution. (p. 399)

solvent The liquid component of a solution. (pp. 100, 392)

solvolysis A reaction in which the solvent plays the role of the nucleophile in the substitution reaction. (p. 801)

space-filling model A three-dimensional representation of a molecule that attempts to show the actual relative sizes of atoms within a molecule. (p. 47)

specific heat (c) See *specific heat capacity*. (p. 305)

specific heat capacity (c) The quantity of heat that will raise the temperature of 1 g of a substance by 1 K, usually in units of $J\,g^{-1}\,K^{-1}$; also called specific heat. (p. 305)

specific rotation [α] The observed rotation of a plane of polarised light when a sample is placed in a tube 1.0 dm long and at a concentration of 1.0 g/100 mL; if a pure sample is used, its concentration is given in g/mL (i.e. its density). (p. 777)

spectator ion An ion whose formula is identical on both sides of an ionic equation, that does not participate in the reaction and that is excluded from the net ionic equation. (p. 107)

spectrochemical series A list of ligands ordered in terms of their ability to produce a crystal field splitting. (p. 576)

***sp* hybrid orbitals** Two atomic orbitals constructed by the interaction between an *s* orbital and a *p* orbital on the same atom. (p. 201)

sp^2 hybrid orbitals Three atomic orbitals constructed by the interactions between an *s* orbital and two *p* orbitals on the same atom. (p. 200)

sp^3 hybrid orbitals Four atomic orbitals constructed by the interactions between an *s* orbital and three *p* orbitals on the same atom. (p. 197)

spin The intrinsic angular momentum of electrons and protons that gives them magnetism. (pp. 19, 137)

spin quantum number (m_s) The quantum number, restricted to either $+\frac{1}{2}$ or $-\frac{1}{2}$, which indexes the orientation of electron spin. (p. 138)

splitting pattern The pattern obtained when a signal in a ^{1}H-NMR spectrum is split by the influence of neighbouring nonequivalent hydrogen atoms. These can be described as multiplets, singlets, doublets, triplets, quartets etc. (p. 903)

spontaneous Describes a change that occurs by itself without outside assistance. (p. 298)

square planar A geometry adopted by 4-coordinate complexes in which the four M—L bonds point to the corners of a square (p. 558)

square pyramidal A geometry adopted by 5-coordinate complexes, consisting of a pyramid with four triangular sides and a square base. (p. 558)

standard cell notation A way of describing the anode and cathode half-cells in a galvanic cell. The anode half-cell is specified on the left, with the electrode material of the anode given first and a vertical bar representing the phase boundary between the electrode and the solution. Dashed double bars represent the salt bridge between the half-cells. The cathode half-cell is specified on the right, with the material of the cathode given last. Once again, a single vertical bar represents the phase boundary between the solution and the electrode. (p. 509)

standard cell potential ($E^{\ominus}_{cell}$) The potential of a galvanic cell at 25 °C, when all ionic concentrations are exactly 1 M and the partial pressures of all gases are 10^5 Pa. (p. 511)

standard enthalpy of combustion ($\Delta_c H^{\ominus}$) The enthalpy change for the combustion of 1 mole of a compound under standard conditions. (p. 318)

standard enthalpy of formation ($\Delta_f H^{\ominus}$) The enthalpy change when 1 mole of a compound is formed from its elements in their standard states. (p. 314)

standard enthalpy of reaction ($\Delta_r H^{\ominus}$) The enthalpy change of a reaction when determined with reactants and products at 10^5 Pa and on the scale of the mole quantities given by the coefficients of the balanced equation. (p. 310)

standard entropy ($S^{\ominus}$) The entropy of a substance measured under standard conditions. (p. 329)

standard entropy of formation ($\Delta_f S^{\ominus}$) The entropy change when 1 mole of a substance is formed from its elements in their standard states. (p. 329)

standard entropy of reaction ($\Delta_r S^{\ominus}$) The entropy change of a reaction when determined with reactants and products at 10^5 Pa and on the scale of the mole quantities given by the coefficients of the balanced equation. (p. 329)

standard Gibbs energy change ($\Delta G^{\ominus}$) $\Delta G^{\ominus} = \Delta H^{\ominus} - T\Delta S^{\ominus}$. (p. 333)

standard hydrogen electrode The standard of comparison for reduction potentials and for which $E^{\ominus}_{H^+/H_2}$ has a value of 0 V (25 °C, 10^5 Pa) when $[H^+] = 1$ M in the reversible half-cell reaction $2H^+(aq) + 2e^- \rightleftharpoons H_2(g)$. (p. 513)

standard reduction potential ($E^{\ominus}_{red}$) The reduction potential of a half-reaction at 25 °C when all ion concentrations are 1 M and the partial pressures of all gases are 10^5 Pa. (p. 512)

state function A quantity that depends only on the initial and final states of the system and not on the path taken by the system to get from the initial to the final state; *p*, *V*, *T*, *H*, *S* and *G* are all state functions. (p. 301)

steady-state approximation A method to derive the rate law for a reaction, using the assumption that the concentration of any intermediate remains constant as the reaction proceeds. (p. 674)

stereocentre For carbon, a tetrahedral carbon atom with four different groups bonded to it; in general terms, any atom with sufficient atoms or groups arranged around it, so that stereoisomers may be formed on exchange of any two atoms or groups. (p. 765)

stereoisomers Isomers with the same molecular formula and the same connectivity but different orientations of their atoms in space. (p. 559, 760)

stereospecific reaction A reaction in which one stereoisomer is formed or destroyed in preference to all others that might be formed or destroyed. (p. 730)

steric factor A factor that reflects the fraction of collisions with effective orientations. (p. 668)

steric hindrance The ability of groups, because of their size and shape, to hinder access to a reaction site within a molecule. (p. 803)

steric strain The strain induced in a molecule arising from the repulsion between the electrons of the bonds in the molecule. (p. 704)

stoichiometric coefficients Numbers in front of formulae in chemical equations. (p. 82)

stoichiometry A description of the relative quantities by moles of the reactants and products in a reaction as given by the stoichiometric coefficients in the balanced equation. (p. 82)

stretching A type of vibration that changes bond lengths. (p. 883)

strong acid An acid that reacts completely with water to give quantitative formation of H_3O^+. (p. 442)

strong base A base that reacts completely with water to give quantitative formation of OH^-. (p. 442)

strong-field Describes ligands that produce large crystal field splittings. (p. 576)

strong nuclear force The attractive force between protons and neutrons that holds the nucleus together. (p. 1161)

structural formula A depiction of a molecule or polyatomic ion that shows how the constituent atoms are arranged, to which other atoms they are bonded, and the kinds of bonds (single, double or triple) present. (p. 41)

structural isomers Isomers with the same molecular formula, but different orders of attachments of the constituent atoms. (p. 559)

styrene The common name for phenylethylene. (p. 740)

subatomic particles Electrons, protons and neutrons. (p. 10)

sublimation The phase change between solid and vapour. (p. 267)

substituent A group attached to the longest carbon chain of an organic molecule. (p. 58)

superconductor A material that offers no resistance to the flow of electricity. (p. 289)

supercritical fluid A phase in which the liquid–vapour transition is no longer possible. A supercritical fluid is able to expand to fill the available space (like a gas), but is resistant to further compression (like a liquid). (p. 269)

surface tension The resistance of a liquid to an increase in its surface area. (p. 260)

surroundings That part of the universe other than the system being studied and separated from the system by a real or an imaginary boundary. (p. 300)

symproportionation An electrochemical process in which two reactants that contain the same element in different oxidation states react to give a single product having an oxidation number intermediate between the two reactants. (p. 533)

***syn* addition** The addition of atoms or groups of atoms from the same side or face of a carbon–carbon double bond. (p. 732)

system That part of the universe under study and separated from the surroundings by a real or an imaginary boundary. A system may be open, closed or isolated. (p. 300)

• T •

tautomerism A form of isomerism in which constitutional isomers are in equilibrium with each other. The isomers differ in the location of a hydrogen atom and a double bond relative to a heteroatom, most commonly O, S or N. (p. 968)

tautomers Constitutional isomers that differ in the location of hydrogen and a double bond relative to O, N or S. (p. 968)

terminal carbon A carbon atom at the end of a chain of carbon atoms. (p. 42)

tertiary carbon atom A carbon atom attached to three other carbon atoms. (p. 55)

tertiary (3°) structure The three-dimensional arrangement in space of all atoms in a single polypeptide chain; the three-dimensional arrangement of all atoms of a DNA or RNA; see also *supercoiling*. (p. 1078)

tetrahedral A geometry adopted by 4-coordinate complexes in which the central metal ion is coordinated to four ligand donor atoms located at the corners of an imaginary tetrahedron. (pp. 184, 558)

tetrahedral carbonyl addition intermediate An intermediate formed from the addition of a nucleophile to the carbonyl group. (p. 951)

tetramethylsilane (TMS) The reference standard used in NMR spectroscopy. Its signal is set at $\delta = 0$. (p. 896)

tetrapeptide A molecule containing four amino acids units joined by peptide bonds. (p. 1074)

theoretical yield The amount of a product determined by the stoichiometry of the reaction. (p. 98)

theory of resonance The theory that some molecules have structures that cannot be represented by any single contributing structure so must be represented as a hybrid of two or more equivalent contributing structures. (p. 735)

thermochemical equation A balanced chemical equation accompanied by the value of $\Delta H^{\ominus}$ that corresponds to the mole quantities specified by the coefficients. (p. 310)

thermodynamic equilibrium constant An equilibrium constant defined in terms of activities. (p. 355)

thermodynamic temperature A temperature scale in which temperature is measured in kelvin. (p. 299)

thiol A compound containing an —SH (sulfhydryl) group. (p. 845)

third law of thermodynamics At absolute zero, the entropy of a perfectly ordered pure crystalline substance is 0. (p. 328)

titrant An acid or base of known concentration used in an acid–base titration. (p. 472)

titration curve For an acid–base titration, a graph of pH versus the volume of titrant added. (p. 472)

Tollens' reagent A solution of silver nitrate in aqueous ammonia. (p. 967)

toluene The common name for methylbenzene. (p. 740)

torsional strain A type of steric strain where repulsion from the electrons in bonds induces a twisting force away from an eclipsing interaction. (p. 703)

transesterification The reaction of an ester with an alcohol to produce a different ester. (p. 1031)

***trans* isomer** A stereoisomer that contains two groups that project on opposite sides of a reference plane. (p. 559)

transition metals The elements in groups 3 to 12 of the periodic table. (pp. 16, 548)

transition state The energy maximum on the potential energy surface of a reaction. It is the brief moment during an elementary process in a reaction mechanism when the species involved have acquired the minimum amount of potential energy needed for a successful reaction. (p. 665)

triacylglycerol See *triglyceride*. (p. 1043)

triglyceride An ester of glycerol with three fatty acids. (p. 1043)

trigonal bipyramidal A geometry adopted by 5-coordinate complexes in which a central metal ion is coordinated to five ligand donor atoms located at the corners of a trigonal bipyramid. (pp. 185, 558)

trigonal planar The molecular shape in which a central atom is bonded to three other atoms lying in a plane at 120° angles to one another. (p. 183)

triol A compound with three alcohol groups. (p. 823)

tripeptide A molecule containing three amino acid units joined by peptide bonds. (p. 1074)

triple bond A bond between two atoms consisting of three pairs of bonding electrons. (p. 177)

triple point The temperature and pressure at which solid, liquid and vapour can coexist at equilibrium. (p. 270)

triplet A signal in ^{1}H-NMR spectroscopy that has been split into three peaks in a ratio of 1 : 2 : 1. (p. 903)

triprotic acid An acid that can donate three protons. (p. 434)

trisaccharides Carbohydrates containing three monosaccharide units joined by glycosidic bonds. (p. 999)

tropical oil A plant oil, such as coconut oil or palm oil, that contains a relatively high proportion of low-molar-mass saturated fatty acids. (p. 1045)

• U •

uncertainty principle The assertion that position and momentum cannot both be exactly known. (p. 135)

unimolecular reaction A reaction in which only one species is involved in the transition state of the rate-determining step. (p. 801)

unit A specific standard quantity of a particular property, against which all other quantities of that property can be measured. (p. 26)

unit cell The simplest repeating unit of a regular pattern, usually within a crystal. (p. 277)

universe A system and its surroundings taken together. (p. 300)

unsaturated hydrocarbon A hydrocarbon in which at least one carbon atom does not have the maximum possible number of atoms bonded to it. (p. 698)

upfield Describes a signal in NMR spectroscopy that is towards the right of the spectrum or of another signal. (p. 897)

UV/visible spectroscopy A technique used to study compounds that absorb light in the ultraviolet–visible region. (p. 916)

• V •

valence electrons The electrons of an atom that occupy orbitals of highest principal quantum number and incompletely filled orbitals. (p. 148)

valence-shell-electron-pair-repulsion (VSEPR) The principle of minimising electron–electron repulsion by placing electron pairs as far apart as possible. (p. 182)

van der Waals equation An equation that corrects the ideal gas equation for the effects of molecular size and intermolecular forces. (p. 243)

van't Hoff equation $\frac{d\ln K}{dT} = \frac{\Delta H^{\ominus}}{RT^2}$. This shows how the value of an equilibrium constant varies with temperature. (p. 370)

van't Hoff factor (i) The ratio of the observed freezing point depression to the value calculated assuming that the solute dissolves as un-ionised molecules. (p. 422)

vaporisation Conversion of a liquid into its vapour. (p. 245)

vapour pressure The partial pressure of a vapour in equilibrium with a condensed phase. (p. 261)

vibrational infrared The portion of the infrared region with a frequency range of 400–4000 cm^{-1}. (p. 882)

viscosity The resistance to flow of a fluid. (p. 261)

volt (V) The SI unit of electric potential in joules per coulomb; $1 V = 1 J C^{-1}$. (p. 506)

voltmeter An instrument for measuring potential difference. (p. 506)

• W •

wavelength (λ) The distance between any two consecutive identical points on a wave. (pp. 120, 881)

wavenumber ($\overline{\nu}$) The number of wavelengths per centimetre. The unit of wavenumber is the reciprocal centimetre (cm^{-1}) and is commonly used in infrared spectroscopy. (p. 882)

weak acid An acid that reacts incompletely with water to form less than stoichiometric amounts of H_3O^+. (p. 442)

weak base A base that reacts incompletely with water to form less than stoichiometric amounts of OH^-. (p. 442)

weak-field Describes ligands that produce small crystal field splittings. (p. 576)

work (w) The energy expended in moving an opposing force through a particular distance. Work has units of force × distance. (p. 303)

• X •

X-ray crystallography The study of crystal structures by X-ray diffraction techniques. (p. 917)

xylene The common name for the three isomeric dimethylbenzenes. (p. 740)

• Z •

Zaitsev's rule A rule stating that the major product of a β-elimination reaction is the most stable alkene — that is, the alkene with the most substituents on the carbon–carbon double bond. (p. 807)

zinc – manganese dioxide cell See *dry cell battery*; *Leclanché cell.* (p. 534)

zwitterion An ionic salt in which both cation and anion are part of the same molecule. (pp. 449, 858, 1062)

A

B

C

D

E

F

G

H

J

K

L

M

N

O

P

Q

R

S

T

U

V

W

X

Y

Z

A selection of interconversions possible for some of the simpler functional groups

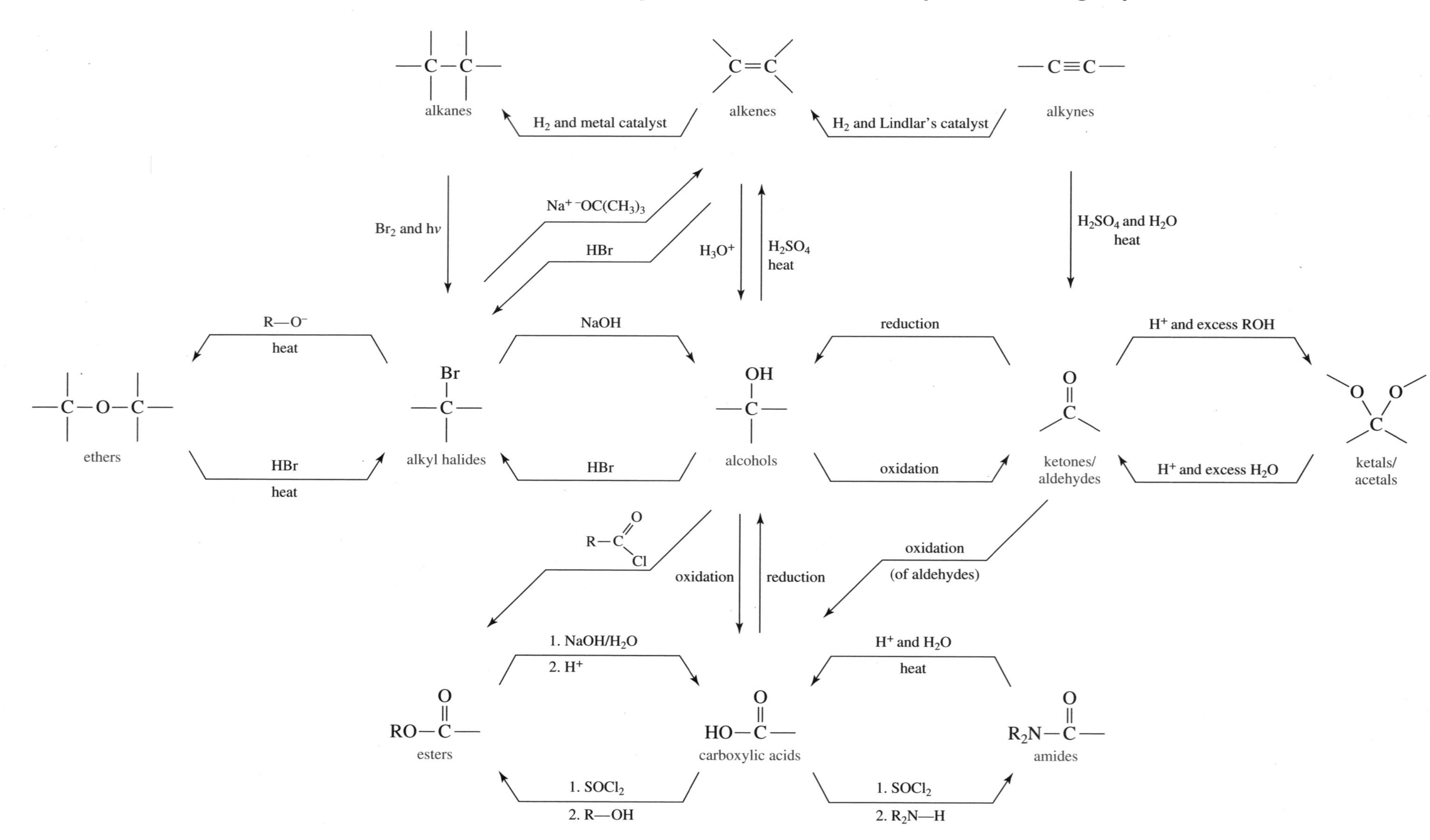